MW00445157

MARINE ENGINEERING

Written by

a Group of Authorities

EDITOR

ROY L. HARRINGTON

Newport News Shipbuilding

Published by

THE SOCIETY OF NAVAL ARCHITECTS AND MARINE ENGINEERS

601 Pavonia Avenue, Jersey City, N.J. 07306

ISBN 978-0-939773-10-7

Foreword

Beginning with the 1939 edition of *Principles of Naval Architecture*, the Society of Naval Architects and Marine Engineers has maintained a program of publishing significant treatises on the subjects of naval architecture, marine engineering, and shipbuilding. Three companion texts are dedicated to these subjects, namely, *Principles of Naval Architecture*, *Marine Engineering*, and *Ship Design and Construction*. As the base of technology has evolved with the passage of time, each of these texts has been periodically revised and updated. MARINE ENGINEERING was originally published in two volumes in 1942 and 1944, and was subsequently revised in 1971.

In accordance with Society policy, a Control Committee was formed to establish the overall guidelines and direction for this text. The members of the Control Committee were:

William G. Bullock	John C. Maxham
Allen Chin	Perry W. Nelson
Steven H. Cohen	Michael G. Parsons
Allen E. Crout	Michael Petz
John M. Dabbar	Alan L. Rowen
José Femenia, Jr.	John K. Stuart
Robert M. Freeman	Joseph Tiratto
Richard Frey	John S. Tucker
Paul R. Geiger	William Watson
John P. Hackett	John A. Youngquist
Joseph D. Hamilton	

Several factors make this edition particularly noteworthy. The technical editor, Roy L. Harrington, is the same very qualified marine engineer who, more than 20 years ago, edited the 1971 revision. Two chapter authors, Everett A. Catlin and Chester L. Long, authored corresponding chapters in the earlier edition. The Society's T&R Program, through the Ship's Machinery Committee, played a major role in the development of the book and in the chapter reviews. And, finally, less than four years elapsed from the formation of the Control Committee until publication, which was completed as the Society began its Centennial Celebration. The completion of this work within the original time schedule is a remarkable achievement to the credit of the authors and the leadership and persistence of the technical editor.

THOMAS P. MACKEY
Chairman, Marine Engineering Control Committee

Preface

More than twenty years have passed since the previous edition of MARINE ENGINEERING was published. During that period, many changes have occurred in the body of technology that is collectively known as "marine engineering." Most of the changes have been of an evolutionary nature; as examples, diesels have continued to supplant steam turbines for most ship propulsion applications in the commercial arena, and gas turbines have become more firmly established as propulsion engines, particularly for warships of the frigate class.

Because of the evolutionary nature of the technology changes, the coverage of some subjects has been reduced in comparison with that in the previous edition (e.g., Boilers and Combustion), but the coverage of other subjects has been expanded substantially. Noise Control, for example, has been given a more thorough treatment; Chapter 13 is dedicated to that subject. In addition, the introductory chapter has been broadened to include several topics, such as Design for Production, that have been given an increased emphasis since publication of the previous edition.

The purpose of this second revision remains the same as that for earlier editions, that is, to provide a basic understanding of marine engineering principles to persons who are nominally second-year engineering students.

During the process of allocating the various topics to the most appropriate chapters for coverage, the introductory chapter became the repository for subjects which, for a variety of reasons, could not suitably be published as stand-alone chapters. Notwithstanding the diverse subjects covered, however, the chapter has been shaped into a remarkably coherent theme.

The selection of the units of measure to be used in the text was the result of much deliberation. The advocates of both English and SI units held strong positions; therefore, the chapter authors were permitted to select the units used in each respective chapter, with SI units recommended where there was no strong preference. The use of dual units was considered; however, to avoid numerous duplications throughout the text and to improve the clarity of the material presented, the use of only one type of units, along with a conversion table at the end of the text, was considered preferable.

To ensure that the text was factual and accurately represented the consensus of the marine industry, each chapter was subjected to a series of reviews. The manuscripts prepared by the authors were generally reviewed within their respective organizations and then were reviewed by the technical editor, the Control Committee for the preparation of the text, and selected members of the marine industry who were recognized authorities in each particular discipline. The Society is deeply indebted to this last group, as they often transformed good manuscripts into excellent manuscripts.

Several members of the Control Committee made outstanding contributions to the development of the book. The assistance provided by Perry W. Nelson, Professor Alan L. Rowen, John Youngquist, and Richard A. Frey (and Michael Petz, who succeeded him as deputy director of the NAVSEA Machinery Group) was of particular value. John S. Tucker not only served as an active member of the Control Committee, but he and his colleagues at NASSCO also insisted upon completing the Electrical Systems chapter for Walter Schmid, the principal author, who became terminally ill.

With few exceptions it would be improper to suggest that the chapters were prepared by only the authors indicated. In several cases, the contributions of those who assisted were as large as that of the author. Professor Alan Rowen is particularly appreciative of the figures in Chapter 3 that were provided by Seaworthy Systems, Inc.; ASEA Brown-Boveri, Ltd.; Colt Industries, Fairbanks Morse Engine Division; Krupp Mak Maschinenbau GmbH; MAN-B&W Diesel A/S; and New Sulzer Diesel, Ltd. Thomas Mackey notes that Ivar Krogstad, Samuel E. Bevins, and Doug Brooke made extraordinary contributions in the preparation of Chapter 18. Henry J. Cassee acknowledges the contributions made by Edward Marino, Harold McAllister, Douglas Whittaker, and Scott Jackson in preparing the figures for Chapter 20.

An accurate listing of all who assisted in the preparation of this text would include many names. Hundreds of people made direct contributions by assisting with the preparation of manuscripts, supplying reference material, reviewing manuscripts, and providing illustrations; and hundreds more made indirect contributions. For fear of omissions, no attempt has been made to enumerate all who contributed. This text is, indeed, the result of an industry effort.

ROY L. HARRINGTON
Technical Editor

Table of Contents

CHAPTER I

J. A. Palmer | # Basic Concepts

Section 1
Introduction

1.1 Historical Perspective. In or about 1712, Thomas Newcomen, an enterprising blacksmith from Dartmoor, England, successfully developed a rudimentary steam engine for pumping water out of mines. This engine consisted essentially of a single-acting piston working in a vertical open-topped cylinder. The piston was packed with hemp since the state of the metal-working art was very primitive, and a tolerance of about one-sixteenth inch out of round or "the thickness of a thin sixpence" was about the best that could be expected. The piston was connected to one end of a rocker arm by a chain without a piston rod or guide. The differential working pressure was derived primarily from the vacuum that was created below the piston by water spray into the steam space at the end of the upstroke. The steam and water valves were worked by hand. Some 60 years later, radical improvements were made by James Watt, whose name is more frequently associated with the early development of the steam engine. In the course of time, numerous other improvements followed, of which the most important was probably the double-acting inverted vertical engine, which proved to have so many advantages that it continues to have applications.

Accounts of the work of men such as Newcomen, Watt, and others in connection with the invention and development of steam engines are truly exciting [1,2,3].[1] Despite the earlier development of steam engines, their application to the propulsion of ships was not seriously undertaken until about 1784. Attempts to adapt steam engines to ship propulsion were carried out almost simultaneously in America, Scotland, and France. At least seven reasonably practical steamboats were developed by 1807. In that year Robert Fulton inaugurated the first commercially successful use of steam marine propulsion in the small, paddle-wheel vessel *Clermont* [1]. The wooden-hulled *Clermont* operated up the Hudson River from New York to Albany, a distance of 150 miles, in about 32 hours.

Although paddle-wheel vessels were promptly adopted for river service, twelve years elapsed after the launching of the *Clermont* before the steamer *Savannah* made the first ocean voyage from America to Europe. Notable, however, is that even in this instance the machinery was not

operated continuously during the outbound leg of the trip and the inbound leg was made under sail.

The introduction of the screw propeller in 1837 was a revolutionary development, but this development also did not immediately lead to the demise of sails or paddle wheels. As late as 1860 the speed of the best clippers still exceeded that of any steamship, and the greater part of the work at sea continued to be accomplished under sail. The *Great Eastern* was an exception. It was an iron-hulled vessel, almost 700 ft long and of 22,000 tons burden, with both a screw propeller at the stern and two side paddle wheels, as well as sails arrayed on six masts. The sails, however, were little used.

By 1893, when the Society of Naval Architects and Marine Engineers was founded, a screw propeller, driven by a triple-expansion reciprocating steam engine, had become the predominant means of propulsion for seagoing ships. Paddle wheels were still used on river and excursion steamers. Steam was almost universally produced by Scotch (fire-tube) boilers, and coal was the common fuel. The steam turbine and diesel engine were yet to debut.

The decade from 1893 to 1903 was a period rich in marine engineering development. The early reciprocating steam engine reached the point of development of the six-cylinder quadruple-expansion engines of 10,000 indicated horsepower supplied with steam at 200 psi by Scotch boilers. The use of electric power generated by steam engine driven "dynamos" at 100 to 112 volts was increasing rapidly. Watertube boilers, which would eventually replace the Scotch boiler on the seas, had become established in England, France, and the United States.

By the end of the 19th century, ship's propulsion plants had become sufficiently reliable to assure the delivery of cargo to any port in the world, but often at great cost to their crews. Mechanical ventilation had not been developed, and those who shoveled coal commonly developed black lung. Steam leaks were chronic, and in warmer climates the risk of heat exhaustion was accepted by those who worked in machinery spaces. Large, exposed reciprocating mechanisms endangered the life and limb of careless crew members; and, in time, members of the machinery space crew became progressively deaf because of the exposure to excessive noise.

---
[1]Numbers in brackets refer to references at the end of the text.

The development by Sir Charles A. Parsons of the first successful application of the steam turbine to marine propulsion was another important milestone in marine engineering. This was accomplished aboard the *Turbinia*, a small vessel about the size of a torpedo boat. The rotative speed of the *Turbinia*'s three series turbines was about 2000 rpm, and they were coupled directly to relatively primitive screw propellers in a triple-shaft arrangement. Parsons had been dismayed on his earliest trials to discover that his first propellers "bored a hole in the water," developing disappointingly low driving thrust, a phenomenon now recognized as cavitation. After further developmental work, however, this new prime mover was successfully adapted to the requirements of marine propulsion.

In what must certainly be considered one of the earliest efforts at model tank testing of propellers, Parsons investigated the subject of cavitation and succeeded in redesigning his propellers. Three per shaft were ultimately employed and in 1897, at a naval review of the British fleet at Spithead, England, the *Turbinia* astounded the British admirals by steaming past smoothly at a speed of 34 knots, belching smoke like an angry bull tossing dust. Lord Kelvin described this development as "the greatest advance made in steam engine practice since the time of James Watt" [4].

Before 1893, there were a number of attempts to develop internal-combustion engines that involved fuels ranging from gunpowder to gas. One of these was of a radically different type, in which the combustion air charge was compressed to a pressure and temperature above the ignition point of the fuel. This engine was patented in 1892 by Dr. Rudolf Diesel, a German engineer. Serious difficulties had to be overcome with the diesel engine, and development proceeded slowly. Not until fifteen to sixteen years later was a successful commercial diesel engine of 25 hp produced. Once this had been achieved, however, progress became rapid. In a few years many firms in Continental Europe were actively building diesel engines with as much as 500 hp per cylinder, and experimental engines developing 2000 hp per cylinder were under test.

The challenge to the coal-fired, low-pressure reciprocating steam engine came from the steam turbine and the diesel engine at about the same time, at the turn of the century. World War I caused a greater emphasis to be placed on marine engineering developments for military applications, while in the merchant marine arena the mass production of proven designs was emphasized. However, advances in the design of steam turbines and diesel engines were made. These developments continued after the war, and were spurred on by the constraints of the Washington Treaty, by the rapid development of diesels for railroad use and for submarine propulsion, and by the realization that there would be a growing requirement for merchant ships of standard design. The advantages of oil, as fuel, became increasingly apparent. The use of oil reduced crew requirements and made fuel storage and handling an easier task. During the same period, the superior economy and performance of steam turbines and diesel engines, when compared with reciprocating steam engines, became recognized. These trends in ship propulsion continued during World War II, and by the end of the war new applications of reciprocating steam engines for propulsion applications were rare.

The next major development in marine engineering history began a few years after World War II. Under the direction of Admiral Hyman G. Rickover, the submarine *Nautilus*, the world's first nuclear-powered ship, put to sea on January 17, 1955 and transmitted the historic message "underway on nuclear power." Nuclear propulsion enabled the *Nautilus* to establish many records, including distance, speed, time traveled submerged, and passage under the geographic North Pole, which she achieved in 1958. The *Nautilus*'s pressurized-water reactor developed about 15,000 hp and was highly successful.

Nuclear propulsion revolutionized the design of submarines by allowing them to remain underwater for extended periods of time. For surface ships, however, nuclear propulsion has received mixed reviews. The U.S. Navy has successfully used nuclear propulsion for aircraft carriers, starting with the *Enterprise* in 1962. Nuclear propulsion provides aircraft carriers with an unlimited range and the ability to carry more aviation fuel and weapons than conventionally powered carriers. Nuclear propulsion has also been used for cruisers; however, when compared with gas-turbine propulsion, the advantages provided—with an increased range being the most significant—are not offset by the higher acquisition cost.

Exploratory applications of nuclear propulsion for commercial ships by the United States in 1959 (*Savannah*), Germany in 1968 (*Otto Hahn*), and Japan in 1974 (*Mutsu*) failed to establish commercial opportunities for nuclear propulsion. Nuclear propulsion was, however, proven to be advantageous by the Russian icebreaker *Lenin* in 1960, and other nuclear-powered icebreakers have subsequently been built.

The historical developments noted in the foregoing, as well as many others, were magnificent concepts and achievements, especially when viewed against the technologies and materials available at the time. No effort has been made here to include the full roster of great names and pioneering events in marine engineering. However, additional material concerning the early days in marine engineering, which conveys an indication of the ingenuity of the early practitioners, can be found in references 1–5.

1.2 Marine Engineering Defined. The concept that motivated the majority of the early advances in marine engineering was quite simple, namely, to develop a system to overcome the vagaries of the wind and the inadequacy of muscle power in the propulsion of ships. In the early history of marine engineering, the concepts formed and the decisions made, although frequently ingenious, were of sufficiently narrow scope that a single individual could become intimately familiar with all facets of the undertaking. The success of the early developments depended to a large extent upon intuitive perception and upon chance.

However, the accumulation of advances that have been made subsequently have caused the field of marine engineering to become highly sophisticated.

From a functional point of view, a ship is a most complex vehicle which must be reliably self-sustaining in its element for extended periods of time. A ship is perhaps the most multipurpose of vehicles, having more built-in functions than does any other vehicle type. As a part of a transportation or military system, the ship contains a greater variety of components than any other vehicle.

The design of a ship's mechanical, electrical, and structural systems is further complicated by the fact that they must be compatible with the marine environment. Shipboard design constraints include space and weight limitations, which can be particularly severe in the case of weight- and space-sensitive vessels such as submarines, and high-impact shock, which is a major design consideration for naval combatants. In addition, other constraints imposed by the marine environment range from hostile pitch and roll conditions resulting from violent seas, to the corrosive nature of seawater and its atmosphere. A comprehensive description of the marine environment, as it influences the design of shipboard systems, is included in reference 6.

Marine engineering is, therefore, not as simply categorized as, for example, civil, mechanical, electrical, or chemical engineering. It is an integrated engineering effort comprising parts of many engineering disciplines directed to the development and design of systems of transport, warfare, exploration, and natural resource extraction that have only one thing in common, namely, that they operate in, or upon the surface of, a body of water.

The division of responsibilities between naval architects and marine engineers is seldom sharp, and it differs from one activity to another. In any event, each needs some familiarity with the disciplines typically used by the other. Marine engineers are, in general, principally responsible for the engineering systems, including the main propulsion plant, the powering and mechanical aspects of ship functions such as steering, anchoring, cargo handling, weapon systems, heating, ventilation, air conditioning, electrical power generation and distribution, and interior and exterior communications.

Naval architects are, in general, primarily concerned with the hydrodynamic and hull form characteristics of the ship, the structural design of the hull, the maneuverability characteristics of the vehicle, and its ability to survive and endure in the marine environment. Naval architects, assisted in appropriate areas by marine engineers, are responsible for the overall exterior and interior arrangement of the ship. In addition, naval architects generally are charged with the responsibility for the overall aesthetics of the design, including the interior decorations and the pleasing quality of the architecture.

Some aspects of the design of marine vehicles are difficult to assign as the exclusive province of a naval architect or a marine engineer. The design of propellers or propulsors is one of these, being in the minds of some a hydrodynamic device in the domain of a naval architect, and in the minds of others an energy conversion device, similar to pumps, and thus in the domain of a marine engineer. Hull vibration, excited by the propeller or by the main propulsion plant, is another such area. The development of the most effective means of achieving a desired ship's speed also requires trade-offs between naval architects and marine engineers, with due regard for the vessel's intended use. Noise reduction, shock hardening, and the dynamic response of structures or machinery in general are usually the joint responsibility of both a naval architect and a marine engineer. Cargo handling, cargo pumping systems, environmental control, habitability, hotel services, and numerous other such aspects of ship design involve joint responsibilities and require close interaction between naval architects and marine engineers.

The traditional distinctions between naval architecture and marine engineering have been replaced by broader concepts of systems engineering and analysis. The multidisciplined nature of marine engineering and naval architecture has caused their definitions to evolve continuously and to assume new dimensions.

1.3 Marine Vehicle Applications and Limitations. The ranges of feasible characteristics for marine vehicles depend upon their intended use. Some of the characteristics are based upon economic comparisons with alternative modes of transportation, whereas others are derived from the laws of physics. Marine vehicles are primarily used in the following ways:

(a) *As a link in a transportation system.* In this application, the payload, speed, turnaround time, and number of vessels involved in the trade are the primary variables. These factors must be considered principally in relation to initial and operating costs.

(b) *As a mobile naval base.* Seaborne bases for warfare systems are included in this group. In this instance, the design of the ship is subordinated to the military system and weapon requirements, except for inescapable essentials such as seaworthiness and habitability. Payload and speed in this case are generally defined in terms relating to military effectiveness and the successful accomplishment of the mission.

(c) *As a special-purpose vehicle or platform.* There remain many diversified craft which have little in common beyond the fundamentals of naval architecture and marine engineering, and are, therefore, difficult to categorize. Oceangoing tugs, salvage vessels, oceanographic research ships, submersibles, offshore vessels, dredging vessels, yachts, ferryboats, towboats, pushers, barges, hydrofoil craft, and surface-effect ships are examples of such craft.

Often, depending upon the application, payload and speed may be the predominant considerations in the selection of the type of vehicle employed and may either favor or rule out some types of marine vehicles. Figure 1 is a comparison of alternative means of transportation and shows the feasible ranges of speed for various types of vehicles.

In general, size restrictions are less stringent for ships than with the alternative modes of transportation. Geometrically similar ships of different scales float at the same proportionate draft since both the amount of water

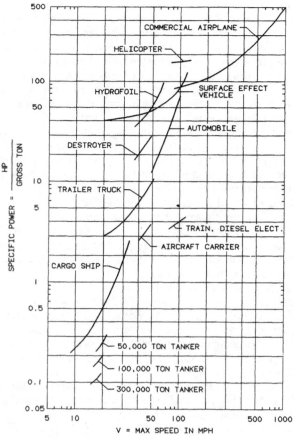

Fig. 1 Specific power versus speed for various vehicles

displaced (the buoyancy or displacement) and the weight of the ship tend to vary as the cube of the scale.

In general, displacement ships are less weight sensitive than the alternative modes of transportation. Fixed-wing aircraft, hydrofoil craft, planing boats, and surface-effect vehicles in general are weight sensitive and size limited. Such craft derive their support while in motion from lifting surfaces of various types; consequently, when geometrically similar but larger versions of a prototype are considered, the weight of the craft, including its payload, increases approximately as the cube of the scale ratio while the area of the lifting surface increases only as the square of the scale. As a result, the pressure loading on the lifting surface increases directly with the scale.

As may be evident from Fig. 1, the displacement type of vessel, while generally unlimited in size, has very definite limitations with regard to the speed at which it can be efficiently driven. The speed limitations for ships are most appropriately expressed in terms of the so-called speed-length ratio. Reference 7 contains a comprehensive treatment of this subject.

The most spectacular growth in the size of ships has been in tankers. Many tankers built in the early 1950's were in the cargo deadweight range of 20,000 to 30,000 tons, and they were appropriately called supertankers because they were twice the size of the T-2 tankers of World War II vintage. However, these "supertankers" have been eclipsed by later generations of tankers that are yet another order of magnitude larger in size. Some of the larger tankers became feasible when the natural limitations imposed by harbor depths were avoided by the use of pipelines to offshore unloading terminals.

Section 2
The Ship Design Process

2.1 Definition of Ship Design Requirements. The design of a ship is predicated upon its mission, as indicated above. But mission considerations can lead the details of the processes used to design commercial ships, naval surface combatants, and submarines, as examples, in strikingly different ways. Consequently, a comprehensive review of the design processes used for the various types of ships would be extensive in scope. To simplify this presentation and maintain a focus upon the underlying principles, the examples used in the following discussion are commonly oriented to commercial applications.

The theoretical problem of optimizing a transport system would appear to be simply that of maximizing the mathematical product of payload times mean effective speed from point to point, while at the same time minimizing initial costs and yearly operating costs. However, analyses of typical transport missions usually include other factors that put a great premium on higher speed. These include flexibility, or the ability to be in the right place at the right time with the right payload. The great increase

in the speed of communications in recent decades has placed a higher premium on speed; however, higher speeds entail higher transportation costs. Higher transportation costs are in part the result of increased power requirements, which entail both additional capital expenses (e.g., propulsion plant rating) and operating expenses (e.g., fuel consumption). Other contributions to the increased costs of higher speeds are the increased first costs and reduced payload capacity of higher-speed vehicles. To obtain feasible resistance characteristics for higher-speed ships, their lines must be finer and have more curvature, which tends to increase the cost of building the ships and to decrease their cargo capacity [8]. As a result of these factors, aircraft usually transport most of the people, special equipment, and precious commodities for which speed is of great importance, while ships continue to carry the larger portion of the heavy commodities and bulk cargoes in both military and nonmilitary services.

Ship design requirements are defined in terms that relate the ship to the overall mission objectives. Ship design requirements for both commercial and naval applications are established in a similar manner. Cargo ships, for example, must form an effective link in the cargo transportation system, but there is generally no uniquely superior or one best way in which this might be achieved. For most applications, many different combinations of ship speed, ship size, and numbers of ships may be used to transport a given amount of cargo at a given rate from one point to another.

In addition to a time schedule for the movement of cargo, the ship design requirements must also include a statement of other aspects of the mission that influence the principal features of the ship. An example is the interfaces that must be established for integrated general cargo transportation systems in which cargo is packed in containers that may then be transported by trucks, ships, barges, and trains, in any combination, before being delivered and unpacked. Effective provisions for the loading, storage, and discharge of cargo are important ship design considerations.

The constraints imposed by the intended application, and by the limitations relating to payload and speed, largely determine a feasible range of the basic ship design requirements. The basic ship design requirements specified for a merchant ship are broadly stated design criteria for the following typical characteristics:

- Payload (cargo or passenger capacity and description)
- Sustained sea speed and endurance
- Limits to overall ship dimensions (length, beam, and draft) for operation in the intended service
- Loading-discharging methods and capacities
- Number of holds, tanks, or other cargo spaces for balanced service
- Crew or manning requirements/level of automation
- Hotel requirements such as heating, ventilation, air conditioning, galley, public spaces, power, and lighting
- Special requirements for navigation and communications
- Maneuverability requirements (steering, handling, and mooring)
- Reliability and logistics support objectives

A more detailed discussion concerning the factors to be considered when establishing the basic ship design requirements is presented in reference 9.

2.2 The Ship Design Procedure. The broad ship design requirements, which are necessary to ensure that the ship can support the overall mission objectives, must be translated into specific ship performance capabilities by marine engineers and naval architects. Inputs from other perspectives, such as owner preferences, must also be considered. In-depth analyses that may involve many disciplines are required to provide confidence that the ship design requirements will support the mission objectives. The analysis of these criteria, which leads to a synthesis of mission requirements and constraints, takes the form of

an iterative process entailing successive steps of reassessment and adjustment. Most ship design requirements are interdependent and cannot be analyzed without considering the others. Investigations may demonstrate that some of the basic ship requirements are not feasible, and a change in the overall strategy is necessary. In the event of such a change, all of the ship design requirements must be reevaluated to ensure that a balanced set of ship design requirements is maintained.

The procedure used by marine engineers and naval architects to translate the ship design requirements into specific ship design criteria is often described as a design spiral [10]. The traditional method of illustrating the ship design procedure represents the process as a reiterative series of engineering and design activities that proceed through progressively more definitive phases and spiral inward to a solution. Figure 2 is a variation on this theme in which a set of concentric circular bands is used to represent the successive stages of engineering activity. As the ship design develops and progresses through the inner stages, the feasible range of each decision becomes more restricted and the engineering analyses become progressively more detailed.

The ship design process is generally considered to progress through four stages, namely, *conceptual design, preliminary design, contract design,* and *detail design.* Each of these design stages is represented in Fig. 2 by a band of engineering activity that is separated from other bands of activity by an approval of the preceding stage or by some other authorizing action. The design stages are shown separated because there is no assurance that design conditions can be developed that will satisfy all of the requirements established for any stage. Instead, in some cases, no amount of analysis and reassessment will result in a ship design that satisfies the established criteria and is acceptable to those with the authority to initiate the next stage. The authorization to proceed to a subsequent stage is normally in the form of a formal approval or contract award. These authorizations are significant events because not only do they constitute the acceptance and conclusion of that stage, but they also initiate the authorization to expend the resources required to execute the next stage. While the conceptual design stage may be performed by a small group of people for a few hundred man-hours, each subsequent stage may entail an order-of-magnitude increase in engineering effort. The award of a ship construction contract is normally the authorization to proceed with the detail design stage.

As indicated by Fig. 2, the ship design process begins with a definition of the ship design requirements. Based upon the required payload delivery schedule, limiting hull dimensions (e.g., draft restrictions), and other stipulations stated as the ship design requirements for commercial ships, the hull form and dimensions of candidate ship designs can be determined. From these, with further modifications and compromises, the most economical ship design may be established by using procedures such as those outlined in reference 9. However, before the dimensions,

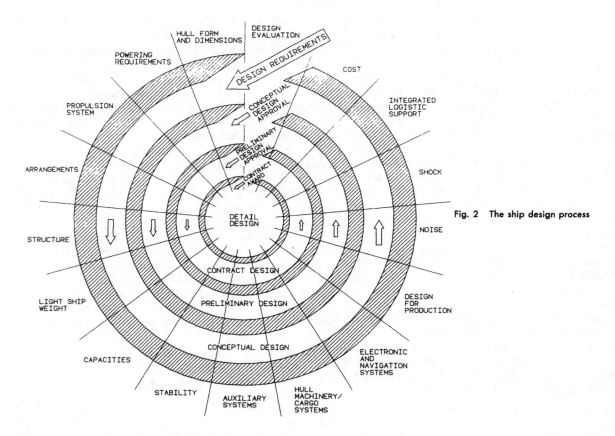

Fig. 2 The ship design process

form, and characteristics of a candidate ship can be established, as-yet unknown secondary data concerning the machinery space, fuel consumption, and other engineering features must be available. To overcome this seemingly impossible situation, and to permit the analysis to proceed, the unknown characteristics must be initially approximated and later refined as the analysis progresses. Estimates that are based on sophisticated and time-consuming analytical procedures may not be warranted during the formulative phases of a design analysis because the supporting data may change and nullify the results obtained. Parametric studies of similar ship designs, and other overall methods of comparison, are generally adequate, and are more appropriate for early approximations. Of course, as the design analysis progresses and the supporting data become more firm, increasingly rigorous analytical procedures become appropriate.

During the *conceptual design* stage, feasibility studies are conducted, usually by a small group of people, to identify alternative ship configurations that might satisfy the ship design requirements. This study effort may require the development of an array of alternative design configurations that involve various combinations of the number and size of ships and ship speed, specialized hull forms, and optional propulsion and machinery concepts. All of the ship features indicated on Fig. 2 may not be analyzed to the same depth for each candidate ship configuration, since reasonable approximations for some features may be adequate for the intended purpose. However, during the conceptual design stage, the number of

alternative configurations is narrowed to those with combinations of principal characteristics that have the greatest potential for satisfying the cost and performance criteria.

The products of the conceptual design stage generally include: hull dimensions and propulsion power requirements; general arrangement drawings and an interior allocation of space by function (e.g., machinery rooms, cargo or combat systems volume, crew accommodations); an estimate of total manning requirements; a high-level, light-ship weight estimate (i.e., the weight of the steel, outfitting, and machinery); an estimate of construction costs; and similar high-level characteristics. The conceptual design serves as a basis to determine the viability of the overall ship acquisition program and to identify candidate alternative concepts, such as optional types of propulsion plants to be evaluated in the preliminary design stage, which follows.

The conceptual design stage may continue for any number of iterations around the band indicated on Fig. 2. In fact, the ship design requirements may prove to be impracticable as initially specified, in which event the analysis would be curtailed during this stage. If, instead, an acceptable, balanced conceptual ship design is developed and approved, then authorization to proceed with the preliminary design stage may be given.

The *preliminary design* stage entails more detailed analyses and, therefore, requires a much greater engineering and design effort than does the conceptual design. Trade-off studies are made during the preliminary design stage to evaluate the relative merits of the various alternative ship systems. Once a reasonably definitive preliminary design has been developed, a tentative listing of the machinery components may be developed. Tables 1 and 2 list types that were prepared for a 42,000-dwt products tanker. In addition, preliminary sketches that depict the arrangement of major components and the larger and more important auxiliary units may be prepared. Many aspects of the ship design remain tentative during this stage, and it is essential that each member of the design team maintain close communications with all others. For example, changes in the hull form or dimensions may result in a required change in the rating of the propulsion plant or plant configuration. During the preliminary design stage, the ship design is developed to the extent necessary to ensure that it is technically sound and can meet all of the ship design requirements. When an acceptable preliminary ship design has been confirmed and approved, the development of a contract design may be authorized.

While the basic design characteristics of the ship are generally defined at the completion of the preliminary design stage, the design and performance parameters are not developed in sufficient detail to serve as a basis for a construction contract. During the *contract design* stage, the objective is to develop the ship design in sufficient detail to permit a shipbuilder to have a clear understanding of the significant ship features and to estimate accurately the cost of building the ship. This requires a full definition of the ship's features and is a major engineering effort. For merchant ships, the invoked requirements of the classification societies, such as the American Bureau of Shipping or Lloyd's Register of Shipping, as well as the requirements of the regulatory bodies, such as the United States Coast Guard and the Public Health Service, establish minimum requirements. The General Specifications for Building Ships for the U.S. Navy [11] provide a corresponding base for naval ships. Contract designs are usually performed by shipyards that have engineering departments sufficiently large and experienced, or by independent design agents.

Some details of a ship design are not readily described in words; examples are hull lines, space arrangements, and piping schemes. In such cases, contract drawings or contract guidance drawings may be prepared. Contract drawings describe precisely the design desired by the owners, and the shipbuilder must deliver a design that conforms to that described; hull lines are often specified in this way. Contract guidance drawings, on the other hand, are usually not contractually obligatory but only illustrate a design that is acceptable to the owner. The shipbuilder is expected to closely analyze contract guidance drawings and consider design alternatives to ensure that the design is proper. The contract design products typically include a set of contract specifications and drawings, the results of preliminary model tests, a list of machinery and equipment (including estimated ratings), and a contract weight estimate.

Ship specifications describe the work to be done and the performance expected; and they define standards of design, construction, materials, and workmanship. The preparation of a contract design package requires a thorough knowledge of the ship, the ability to discriminate between the essential and the trivial, a habit of thoroughness and accuracy, and a talent for clear, concise presentation. A contract design package is developed in sufficient detail to fully define the ship; however, unessential restrictions must be avoided, so as to accommodate acceptable alternative approaches that are compatible with construction procedures found to be most effective by the shipbuilder.

Design standards, which may be either a fully specified design for components that are used repeatedly, or a parametric definition of elements that with variations have multiple applications, are effective supplements to ship specifications. Standards are well known and accepted by the industry, and they avoid the necessity for redundant specification words. Descriptions of systems must state the essential system characteristics and may include specific descriptions of equipment where essential. Citing a specific manufacturer's product could avoid the necessity for a long, detailed description; but doing so has the undesired effect of limiting competition. A preferred practice is to add the words "or equal." Thus: "The distilling plant shall be an 8000 gpd, double-effect, XYZ Co. unit, or equal." With this specification wording, the owner would be obligated to accept the XYZ Co. unit, if offered, but other, comparable designs would also be considered.

To complete the contract design phase, several iterations around the contract design band indicated on Fig. 2 are required, until all aspects of the ship's characteristics are proven to be mutually compatible. The significant ship characteristics desired must be accurately reflected in the contract design package because subsequent corrections might disrupt other design work previously completed. This could necessitate a contract change with the shipbuilder, usually incurring additional expenses.

The *detail design* stage is initiated after the completion and approval of the contract design package, and usually begins immediately after the award of a construction contract. The detail design is usually conducted by the shipbuilder, but it may be performed by an independent design agent. During this stage, working drawings and other data required to construct the ship are developed. As indicated by Fig. 2, this is the final design stage, and all of the major design decisions (such as the selection of the types and ratings of machinery) have already been made and confirmed as satisfactory. However, a large volume of engineering and design work remains to be done to ensure that distributive systems can accomplish their intended functions, that individual components are technically adequate, that stress levels are within acceptable limits, that acoustic objectives are met, that regulatory body and classification society rules are satisfied, and that unambiguous instructions are available to the construction tradesmen (usually in the form of working drawings and supporting data). To the extent that the preceding stages of the design process have been accomplished properly, the detail design phase is a relatively straightforward, but nonetheless challenging, process.

Table 1   List of machinery for a 42,000-dwt products tanker

CENTRALIZED CONTROL AND AUTOMATION SYSTEM

Machinery control room — Totally enclosed, acoustically insulated, air conditioned, housing consoles, switchboards, distribution panels, and other control and monitoring equipment as required

Machinery control console — Contains instrumentation, alarm and principal control systems for main engine, ancillary, and auxiliary plant

Bridge console — Contains main engine starting, reversing, and speed controls and selected indication and alarms for main, ancillary, and auxiliary plant

Diesel generator auto-start system — Arranged to start idle diesel generators and automatically parallel them with bus in the event of overload or imminent failure of electrical supply

Preferential trip system — Arranged to automatically shed nonessential electrical load in stages, if sustained load reaches 90% of connected generating capacity

Automatic standby system — Arranged for essential motor-driven pumps, to automatically start idle units when imminent failure of running unit is indicated

Fire detection and alarm systems — Fitted in compliance with regulatory-body requirements

PROPULSION UNIT

Diesel engine one (1) — Single-acting, two-stroke, direct-reversible, turbocharged, low-speed diesel engine

Contract maximum continuous rating (CMCR), metric horsepower, rpm	15,300 / 102
Normal continuous rating (NCR), metric horsepower, rpm	12,240 / 99

Thrust bearing — Integral with main engine, pivoted shoe type

SHAFTING

Line shaft one (1) — Solid forged steel, ABS Grade 2

Diameter, in.	18¾
Length, ft	19.0

Propeller shaft one (1) — Solid forged steel, ABS Grade 2

Diameter, in.	22⅞
Length, ft	22.25

Line shaft bearing one (1) — Cast iron, tin-based babbit lined, disk lubricated

Shaft diameter, in.	18¾
Length, in.	18¾
Diametrical clearance, in.	0.019 − 0.029

Stern tube bearing one (1) — Cast ductile iron shell w/centrifically cast tin-based babbit, oil lubricated

Shaft diameter, in.	22⅞
Length, in.	35
Diametrical clearance, in.	0.023 − 0.033

Propeller one (1) — Fixed pitch, nickel-aluminum-bronze, keyless

Diameter, ft	22.25
Number of blades	4
Pitch ratio	0.761
Developed area ratio	0.54
Rated power, metric hp	15,300
Rated speed, rpm	102

DISTILLING PLANTS

Jacket-water heated evaporator one (1) — Jacket-water heated, plate or flash type, package complete with ejectors, salinity cells, and controls

Capacity, gpd	3000
Sea salt distillate max grains/gal	0.25
Feed temperature, °F	85
Jacket-water heat available, Btu/hr	2.7×10^{6}
Quantity jacket-water available, gpm	460
Jacket-water inlet temperature, °F	175

Reverse osmosis desalination plant one (1) — Membrane type, high-pressure system, complete with prefilter, controls, pressure pump, water quality indicator, etc.

Rated capacity, gpd	3000
Water quality, max grains/gal	0.25
Feed temperature, °F	85

FUEL-OIL SYSTEMS

Fuel-oil purifier module one (1) — Purifier module to consist of three (3) centrifugal, self-cleaning type purifiers complete with strainers, preheaters, built-in discharge pumps, motors, controls, etc. For use with heavy fuel oil and arranged for parallel or series operation

Capacity, gph (each purifier)	800
F. O. viscosity, cSt @ 122°F	700
F. O. temp. to purifier bowl, °F	208
F. O. max specific gravity @ 60°F	1.010
Motor rating, hp (each purifier)	20

Fuel-oil supply unit one (1) — Fuel-oil supply module complete with all necessary pumps, motors, filters, valves, controls, piping, etc.

Max F. O. viscosity, cSt @ 122°F	700
F. O. supply pumps (2): capacity, gpm	20
motor rating, hp	3
F. O. circulating pumps (2): capacity, gpm	35
motor rating, hp	5
F. O. heaters (2): F. O. inlet temp., °F	212
F. O. outlet temp., °F	302

Aux diesel engine fuel-oil heater one (1) — Plate or shell-and-tube type

Fuel oil flow, gpm (HFO)	7.5
Fuel oil inlet temp., °F	125
Fuel oil outlet temp., °F	320
Steam pressure, psig	100

In-port aux. diesel engine fuel-oil heater one (1) — Electric type, kW — 25

Inert-gas plant fuel-oil heater one (1) — Plate or shell-and-tube type

Fuel-oil flow, gpm (HFO)	5
Fuel-oil inlet temp., °F	125
Fuel-oil outlet temp., °F	200
Steam pressure, psig	25

LUBRICATION-OIL SYSTEMS

Main lube-oil cooler one (1) — Plate type, lube-oil/fresh-water

Heat dissipated, Btu/hr	2,500,000
Lube oil flow, gpm	1100
Oil viscosity, SSU @ 100°F	600
Oil outlet temp., °F	113
Freshwater inlet temp., °F	110
Plate material	stainless steel

Camshaft lube-oil unit one (1) — Camshaft L. O. module complete with all necessary pumps, motors, filters, alarms, valves, piping, etc.

Oil viscosity, SSU @ 100°F	600
Lube-oil circulating pumps (2): capacity, gpm	35
motor rating, hp	5
Lube-oil cooler (1): L. O. inlet temp., °F	117
L. O. outlet temp., °F	113

Main engine lube-oil purification module one (1) — Purifier module to be complete with two centrifugal, self-cleaning type purifiers complete with strainers, preheaters, built-in discharge pumps, motors, controls, etc.

Capacity, gpd (each purifier)	500
L. O. viscosity, SSU	600
Inlet L. O. temperature, °F	120
Motor rating, hp (each purifier)	15

Aux diesel engine lube-oil purification module one (1) — Purifier module to be complete with two (2) centrifugal, self-cleaning type purifiers complete with strainers, preheaters, built-in discharge pumps, motors, controls, etc.

Capacity, gpd (each purifier)	250
L. O. viscosity, SSU	500
Inlet L. O. temperature, °F	120
Motor rating, hp (each purifier)	7½

Piston-rod stuffing box drn. oil filtration unit one (1) — Modular unit complete with pump, motor, tanks, filter, controls, valves, piping, etc.

Circulating pump (1): capacity, gpm	1
motor rating, hp	¼
Circulating tank (1): capacity, gal	185
Drain tank (1): capacity, gal	160

FRESHWATER SYSTEMS

Jacket-water preheater one (1) — Plate type

Heat dissipated, Btu/hr	384,000
Jacket-water flow, gpm	40
Jacket-water temp. rise, °F	16
Plate material	stainless steel
Steam pressure, psig	50

Freshwater expansion tank one (1) — Welded steel, self-standing, fitted with level indicator, high- and low-level alarms

Capacity, gal	500

Jacket-water air separator one (1) — Welded, full-flow cyclone type

Capacity, gal	375

Potable-water pressure tank one (1) — Steel construction, complete with relief valve, pressure switch, air supply, vertically mounted

Tank capacity, gal	200
Tank operating pressure, psig	100
Tank design pressure, psig	150

Potable-water purification unit one (1) — Automatic bromination proportioning type

Capacity, gpm	2

Hot-water heater one (1)

Storage tank capacity, gal	200
Heating capacity, gph	540
Inlet temperature, °F	50
Outlet temperature, °F	140
Steam pressure, psig	25

SEAWATER SYSTEMS

Central FW/SW heat exchanger one (1) — Plate type, freshwater/seawater

Heat dissipated Btu/hr	29,000,000
Freshwater flow, gpm	2800
Freshwater inlet temperature, °F	130
Freshwater outlet temperature, °F	110
Seawater inlet temperature, °F	85
Plate material	titanium

AUXILIARY STEAM SYSTEM

Oil-fired auxiliary boiler one (1) — Auxiliary boiler to be complete package with all controls, etc., required for automatic operation with heavy fuel oil (HFO)

Capacity, lb/hr (saturated steam)	65,000
Steam pressure, psig	250
Feedwater temperature, °F	265

Heat-recovery boiler one (1) — Mounted in exhaust pipe of main propulsion diesel engine. Capable of automatic control.

Capacity, lb/hr (saturated steam)	6000
Steam pressure, psig	250
Feedwater temperature, °F	265
Diesel exhaust-gas temperature, °F	525

Deaerating feed tank one (1) — Vertical, spray type

Storage capacity, gal	1500
Feedwater inlet temperature, °F	200
Feedwater flow, lb/hr	71,000
Pressure, psig	25

Atmospheric drain collecting tank — Fabricated steel construction, fitted with cooling coil

Storage capacity, gal	600
Drains, lb/hr	71,000
Drains inlet temperature, °F	330
Drains outlet temperature, °F	200
S.W. cooling temperature, °F	85

Table 1 *(cont.)*

	COMPRESSED-AIR SYSTEMS	
Diesel starting compressors two (2)	Multistage, positive-displacement, motor-driven, water-cooled, direct or belt-driven	
	Capacity, cfm (free air @ discharge pressure)	115
	Discharge pressure, psig	435
	Motor rating, hp	50
	Cooling water @ 100°F, gpm	45
Main starting air receiver two (2)	Welded steel, vertical or horizontal	
	Capacity, ft³	225
	Operating pressure, psig	435
	Design pressure, psig	600
Auxiliary diesel air start receiver one (1)	Welded steel, vertical or horizontal	
	Capacity, ft³	30
	Operating pressure, psig	435
Ship-service air receiver one (1)	Welded steel, vertical or horizontal	
	Capacity, ft³	80
	Operating pressure, psig	100
	Design pressure, psig	150
Control air compressor one (1)	Liquid-ring, oil-free type	100
	Capacity, scfm	80
	Discharge pressure, psig	
	Motor rating, hp	50
Control air receiver one (1)	Welded steel, vertical or horizontal	
	Capacity, ft³	80
	Operating pressure, psig	80
Control air dehydrator one (1)	Refrigeration type	
	Capacity, scfm	100
	Dewpoint @ atmospheric pressure, °F	−25
Nitrogen generator one (1)	Membrane separator type	175
	Nitrogen capacity, scfh	270
	Inlet air pressure, psig	
	Nitrogen pressure, psig	260
	Nitrogen purity, %	95
Whistle one (1)	Air Tyfon type	
	AUXILIARY DIESEL GENERATORS	
Auxiliary diesel generator three (3)	Air-starting, nonreversing, turbocharged, four-cycle with attached heat exchangers, pumps, and controls. Capable of burning heavy fuel oil	
	Continuous rating, kW	1000
	kVA	1250
	Volts	480
	Hertz	60
	Phase	3
	Speed, rpm	900
	Cooling-water flow, gpm	390
Auxiliary generator transformer bank one (1)	One bank of three (3) single-phase 480/120-volt transformers	
Emergency diesel generator one (1)	Automatic hydraulic start, radiator cooled, nonreversing, capable of burning marine diesel oil; located outside of machinery space	
	Continuous rating, kW	250
	kVA	313
	Volts	480
	Hertz	60
	Phase	3
Emergency generator transformer bank one (1)	One bank of three (3) single-phase 480/120-volt transformers	
	HYDRAULIC POWER PACKS	
Hydraulic power packs five (5)	Electric-motor-driven	
	Rated capacity (each), kW	300
	Hydraulic discharge pressure, psig	3263
	Relief valve setting, psig	3550
Valve operation pack one (1)	Electric-motor driven for cargo-handling system valve operation	
	Working pressure, psi	1800
	Motor size, hp	5
	VENTILATION FANS	
Engine room supply fans four (4)	Axial-flow, 2-speed, motor driven, MARAD type	
	Capacity, cfm	47,500
	Static pressure, in. WG	3
	Motor rating, hp	40
Engine room exhaust fans four (4)	Axial-flow, motor-driven, MARAD type	
	Capacity, cfm	30,400
	Static pressure, in. WG	3
	Motor rating, hp	20
Auxiliary diesel room exhaust fan one (1)	Axial-flow, motor-driven, MARAD type	
	Capacity, cfm	12,080
	Static pressure, in. WG	2
	Motor rating, hp	7½
LO/FO handling room exhaust fan one (1)	Axial-flow, motor-driven, MARAD type	
	Capacity, cfm	4300
	Static pressure, in. WG	3
	Motor rating, hp	5
	AIR-CONDITIONING AND REFRIGERATION MACHINERY	
Air-conditioning plant two (2)	Direct-expansion, motor-driven, freshwater cooled; each unit rated at one-half maximum system load	
	Capacity (each) @ 33°F suction temperature, tons	25
	Refrigerant type	R22
	Cooling-water flow, gpm	60
	Cooling-water temp., °F	110
	Motor rating, hp	40
Ship-stores refrigeration plant two (2)	Condensing type, motor-driven, freshwater cooled	
	Capacity (each) @ −20°F suction temp., tons	1.0
	Refrigerant	R22
	Cooling-water flow, gpm	25
	Cooling-water temp., °F	110
	Motor rating, hp	5
	FIRE EXTINGUISHING	
Carbon-dioxide fire-fighting system	High-pressure, manifolded cylinders, mounted vertically, located external to machinery space	
	Cylinder size, lb	100
	Number of cylinders	188
Aqueous film forming foam (AFFF) system one (1)	Unit mounted complete with storage tank, proportioning pump and motor, preparation, and associated controls	
	Storage tank capacity, gal	900
	AFFF solution, %	3
	Proportioner flow rate, gpm	33
	Motor rating, hp	10
	AFFF application rate, gpm	1450
	WORKSHOP EQUIPMENT	
Workshop tools	Workshop to be outfitted with the following tools:	
	Lathe	1
	Drill press	1
	Grinder	1
	Power hacksaw	1
	Welding machine	1
	CARGO SYSTEM	
Inert-gas plant one (1)	Multi-inert type, capable of utilizing auxiliary diesel exhaust or burning heavy fuel oil to produce inert gas, complete with gas generator blowers, water seal, deck seal, oxygen analyzer, alarms, controls necessary to comprise a complete unit; skid mounted	
	Inert-gas capacity, cfm	2250
	Gas pressure, psi	1.7
	Gas temperature, °F	100
	Maximum oxygen content, %	4
	Number of blowers	2
	Capacity, each blower, cfm	2250
	Blower motor rating, hp	150
	POLLUTION-ABATEMENT EQUIPMENT	
Incinerator one (1)	Capable of burning waste oil and solids; diesel-oil fired	
	Waste-oil capacity, lb/day	1200
	Solid-waste capacity, lb/day	200
Sewage-treatment system one (1)	Marine sanitation device, capable of treating black water	
	Crew size, persons	30
Engine room bilge oil/water separator and oil content monitor one (1)	Package type unit, complete with pumps, motors, controls, etc.	
	Max oil content in effluent, ppm	15
	Pump (1):	10
	capacity, gpm	1½
	motor rating, hp	
	DECK MACHINERY	
Steering gear one (1)	Rapson-slide, 4-cyl, dual electro-hydraulic	
	Rudder torque (ahead) in.-lb	6.8×10^6
	Rudder torque (astern) in.-lb	12.3×10^6
	Electric motor rating, hp	75
Anchor windlass/mooring winch two (2)	Hydraulic operated from central system, combined anchor windlass and mooring winch; each unit to have one wire rope drum and one gypsy head; winch to be constant tensioning.	
	Hydraulic pressure (nominal), psi	3263
	Anchor weight, lb (high hold power)	16,350
	Chain size, in. (extra high strength)	3⁵⁄₁₆
	Chain speed @ 30 fathoms, fpm	30
	Total chain length, fathoms	360
	Drum line pull, long tons	15
	Drum line speed, fpm	40
	Wire rope size, in IWRC	1⅜
	Hawser break strength, lb	124,600
	Gypsy line pull, lb	20,000
Mooring winches six (6)	Hydraulic operated from central system, with single wire rope drum and single gypsy head; constant tensioning	
	Hydraulic pressure (nominal, psi)	3263
	Drum line pull, long tons	15
	Drum line speed, fpm	40
	Gypsy line pull, lb	20,400
	Wire rope size, in IWRC	1⅜
	Hawser breaking strength, lb (synthetic, double braided)	124,600
Hose handling cranes one (1)	Hydraulic operated from central system	
	Hydraulic pressure (nominal), psi	3263
	Capacity, metric tons	10
	Boom length, ft	80
Stores handling crane two (2)	Hydraulic operated from central system	
	Hydraulic pressure (nominal), psi	3263
	Capacity, lb	1000
	Boom length, ft	34
Engine room traveling jib crane one (1)	Electric, on rails in overhead of engine casing, capable of traversing entire length and width of main engine	
	Capacity, metric tons	4
	Electrical consumption, kW	10
Accommodation ladders two (2)	Compressed air operated	49
	Length, ft	100
	Air pressure, psig	
Traction winch one (1)	Hydraulic operated from central system	
	Hydraulic pressure, psig	3263
	Line pull, lb	50,000
	Line speed, fpm	20
	Line size, in.	3¾
Engine room elevator one (1)	Electric type, geared traction drive	
	Capacity, lb	500

Table 2   Pump list for a 42,000-dwt products tanker

No. Units	Service	Type	Drive	Fluid	Capacity, gpm	Head, psi	Driver hp	Suction Condition
2	main seawater	cent.ᵃ	elec. motor	SW	3 000	25	75	flooded
2	central FW cooling	cent.ᵇ	elec. motor	FW	3 200	50	150	flooded
2	jacket-water cooling	cent.ᵇ	elec. motor	FW	375	40	15	flooded
1	fire and foam	cent.ᶜ	elec. motor	SW	1 300	135	150	flooded
1	emergency fire	cent.ᶜ	elec. motor	SW	550	160	100	flooded
2	inert-gas plant scrubber	cent.ᶜ	elec. motor	SW	1 150	65	75	flooded
14	cargo	cent.ᵈ	hyd. motor	CO	700	400 ft	...	flooded
8	cargo	cent.ᵈ	hyd. motor	CO	270	400 ft	...	flooded
2	cargo/slops	cent.ᵈ	hyd. motor	CO	270	400 ft	...	flooded
8	segregated ballast	cent.ᵃ	hyd. motor	SW	2 000	40 ft	...	flooded
2	engine room bilge and ballast	cent.ᵃ	elec. motor	SW	400	45	20	5-ft lift (self-prim.)
1	engine room bilge sump	cent.ᵃ	elec. motor	SW	50	15	¾	flooded
2	evaporator feed	cent.ᵃ	elec. motor	SW	100	50	5	flooded
2	auxiliary blr. feed	cent.ᵉ	elec. motor	FW	175	350	50	20-ft NPSH
2	auxiliary drain	cent.ᵇ	elec. motor	FW	175	40	7½	flooded
1	inert-gas deck seal supply	cent.ᵃ	elec. motor	SW	15	25	½	flooded
1	sewage	cent.ᵃ	elec. motor	SW	25	15	½	flooded
2	potable water	cent.ᵇ	elec. motor	FW	40	90	5	flooded
1	hot-water circ.	cent.ᵇ	elec. motor	FW	5	15 ft	⅓	flooded
1	distillate	cent.ᵇ	elec. motor	FW	50	40	¼	2-ft NPSH
2	main lube oil	rotaryᶠ	elec. motor	LO	1 100	70	100	10-in. Hg lift
2	camshaft lube-oil supply	rotaryᶠ	elec. motor	LO	35	60	3	10-in. Hg lift
1	cylinder lube oil	rotaryᶠ	elec. motor	LO	5	30	¼	10-in. Hg lift
2	lube-oil transfer	rotaryᶠ	elec. motor	LO	30	40	1½	flooded
2	fuel-oil transfer	rotaryᶠ	elec. motor	HFO	100	50	10	5-ft lift
2	fuel-oil circulating	rotaryᶠ	elec. motor	HFO	30	145	5	flooded
2	fuel-oil supply	rotaryᶠ	elec. motor	HFO	15	60	2	10-in. lift
1	sludge	rotaryᶠ	elec. motor	HFO	25	60	5	5-ft Hg lift
2	inert-gas plant fuel-oil supply	rotaryᶠ	elec. motor	HFO	5	30	¼	10-in. Hg lift
2	auxiliary diesel fuel-oil supply	rotaryᶠ	elec. motor	HFO	5	30	¼	10-in. Hg lift
3	fuel-oil purifier supply	rotaryᶠ	elec. motor	HFO	13.3	45	3	flooded
2	main eng. L.O. purifier supply	rotaryᶠ	elec. motor	LO	8.3	45	1	5 lift
2	aux. diesel L.O. purifier supply	rotaryᶠ	elec. motor	LO	5	45	7½	flooded
2	aux. blr F.O. supply	rotaryᶠ	elec. motor	HFO	8	250	3	flooded

MATERIAL KEY

ᵃ Casing: NiAlBrz, Brz, CuNi
 Impeller: Brz, NiAlBrz
 Shaft: Monel, K-Monel

ᵇ Casing: Brz, NiAlBrz, CuNi
 Impeller: Brz, NiAlBrz
 Shaft: 304,316 S.S.

ᶜ Casing: NiAlBrz Alloy 20, CuNi
 Impeller: NiAlBrz Alloy 20, E-Monel
 Shaft: Monel, K-Monel

ᵈ Stainless steel, AISI 316

ᵉ Casing: 12% Cr steel
 Impeller: 12% Cr steel
 Shaft: 410,416 S.S.

ᶠ Casing: ductile iron, steel
 Rotor: ductile iron, steel
 Shaft: steel

2.3 Specific Marine Engineering Requirements. Returning to the ship design process illustrated by Fig. 2, marine engineers are usually assigned the responsibility of translating the broadly defined ship design requirements and cost objectives into specific requirements for the various shipboard systems. Such requirements typically fit the following scheme of classification:

Main Propulsion System
    Type and number of propulsors
    Shaft power
    Propeller diameter and speed
    Type of propulsion plant
    Automation objectives
    Fuel consumption objectives
    Noise objectives
    Space and weight objectives

Auxiliary Ship Systems
    Electric plant
    Steam
    Fuel handling and treatment
    Lubricating-oil
    Heating, ventilation, and air conditioning
    Firefighting, bilge, and ballasting
    Pollution control
    Seawater
    Fresh water

Hull Machinery Systems
    Steering gear
    Bow or stern thrusters
    Anchor handling
    Mooring
    Ship stabilization

Cargo Systems
    Tanker systems, such as cargo and ballast piping and pumps, tank-cleaning provisions, and inert-gas provisions
    Cargo handling

Naval-Ship Combat Support Systems
    Weapon elevators
    Aircraft elevators
    Aircraft catapults
    Torpedo handling
    Underway replenishment

Military Electronics
    Command and control
    Weapon directors
    Tactical data
    Electronic countermeasures

Communications and Navigation Systems
    Interior communications
    Exterior communications
    Navigation

The requirements specified for these representative systems must ensure a compatibility with the marine environment. Among the factors to be considered are the ship's motion (pitch and roll) as well as trim and list, possible exposure to seawater and a salt atmosphere, and a service life that often exceeds 30 years.

2.4 Equipment Procurement. Shipbuilders customarily purchase from other manufacturers much of the machinery and equipment required to construct a ship; however, the extent of the equipment purchased varies with the policy, facilities, and staff of the shipbuilder. During the 1930s, the larger shipyards manufactured a very high percentage of the equipment used to construct their ships. Two factors especially helped make this policy feasible: ships were generally of a lower power level and much less sophisticated in their design, and there was a preference by shipyard management to maintain control over their work. By the 1940s, however, the continual escalation in the complexity of equipment, and the economies derived from specialization and standardization, resulted in a continuing decline in the percentage of the equipment manufactured by shipbuilders. As a consequence, many shipyards purchase substantially all machinery components except for those of high volume (e.g., piping system elements) that some shipyards have elected to develop a specialized capability to produce.

Some shipyards have maintained the technical and production capabilities necessary to manufacture a great majority of ship components; however, the components can often be purchased from specialty manufacturers at a lower cost. Nevertheless, a much broader manufacturing capability must still be maintained by shipyards, because in critical production situations the shipbuilder often becomes the manufacturer of last resort. This may occur if, for example, labor disputes, business failures, or design deficiencies prevent a manufacturer from meeting an equipment delivery date. In such an event, the cost of the component, itself, may become of secondary importance as it is overshadowed by the costs that would be incurred in delaying the ship.

With the exception of standard components, purchased equipment is usually built in compliance with a special purchase order that satisfies the ship requirements. Marine engineers play important roles in the procurement process. They provide technical direction for the development of detailed procurement specifications that define the important equipment characteristics, including the features, duty, construction, materials, manufacturing standards, and test requirements. This responsibility requires a comprehensive analysis of total systems (e.g., propulsion plant, piping, ventilation, and electrical) to establish the performance requirements for specific components.

Ensuring the functional adequacy of a design is a major responsibility of marine engineers. The development of piping diagrams is an important example of a marine engineer's work. Piping diagrams present, in a convenient, usable form, the pertinent requirements of the specifications and regulatory bodies; they usually also show, in elementary form, the piping connections between the various units, the types of valves and regulating devices, and the equipment ratings as well as their sizes, materials, and pressure standards. Many major design decisions must be made to support the development of diagrams; however, the preparation of diagrams facilitates an early identification and resolution of design problems that would become increasingly disruptive to correct as the design progresses. Piping and other system diagrams, which are illustrated in Chapter 20, reflect functional characteristics only and do not explicitly describe the arrangement of a system.

Production schedules usually require equipment purchase orders to be placed before the design of systems has been completed; therefore, the equipment ratings as specified often include a prudent margin that is based mainly on the marine engineer's experience.

The purchasing agent uses the procurement specifications to obtain price quotations from prospective manufacturers, whose products are evaluated for their technical acceptability by marine engineers. After a purchase order has been placed, the detailed plans developed by the manufacturer are reviewed to confirm that the design complies with the system and purchase order requirements. When this compliance has been confirmed, the release for manufacture is authorized. The shipbuilder's marine engineers follow the progress of the equipment through the production process and may work with the manufacturer in the resolution of any production problems. A marine engineer usually will travel to the manufacturer's plant to inspect the completed equipment and observe the completion tests before it is transported to the shipbuilder for installation.

Section 3
Development of Main Propulsion System

3.1 Overall Considerations. The basic operating requirement for a main propulsion system is to propel the vessel at the required sustained sea speed for the range (or endurance) required and to provide stopping, backing, and maneuvering capabilities. All of these operations must be accomplished reliably, and the operation and onboard maintenance requirements of the propulsion system must be within the capabilities of the crew. Finally, the main propulsion system must fulfill all of the basic operating requirements at a cost within that allocated during the preliminary studies of the ship system; otherwise the preliminary studies must be reevaluated.

Before a main propulsion plant can be designed, the power required for sustained operation and endurance must be tentatively determined. However, there is an interdependent relationship involved: the space and weight requirements for the propulsion plant vary with the power rating and can have a significant effect on the ship configuration, while the dimensional and form characteristics of the hull and its approximate displacement are required to establish an estimate of the propulsive power required. It is apparent that marine engineers must coordinate their activities with those of naval architects from the earliest conceptual design stage.

3.2 Propulsion Power Profile. Except for short periods when approaching or departing port, most merchant ships operate at a high percentage of rated power. Occasionally, the operating schedule may include periods of operation at reduced speed, but seldom at speeds below that corresponding to about one-half power. Consequently, for merchant ships economical operation at the sustained sea speed corresponding to the intended trade route is of primary importance.

The situation with naval ships is entirely different. The power plants for naval ships must be designed to satisfy the highest speed requirements projected for them, but the maximum power capability provided is seldom used. During the majority of the life of a naval ship, the ship is operated at cruising speeds, which correspond to approximately 60% of maximum speed or about 20% of the propulsion plant rating. Consequently, good economy at the cruising speed is of major importance because the fuel consumption at this speed may control the range of the ship during many operations.

The power requirements for special-purpose vessels, such as towboats and icebreakers, are clearly established by extremes in the towing and icebreaking modes of operation, but free-route operations must also be given consideration.

The design of a power plant must fully reflect the operating profile of a ship. This means that for ships which must have a very high power capability, but only for rare and brief periods, the acceptable component service lives for full-power operation may be relatively low, and the acceptable rate of fuel consumption at full power may be high. Economy measures, such as waste-heat recovery, would generally not be considered for the maximum power condition. On the other hand, economy would be given major emphasis for the more extensively used, lower-power, cruising mode of operation.

3.3 Determination of Ship Resistance. The general subject of ship resistance falls within the domain of naval architecture as opposed to marine engineering. For this reason, a detailed treatment of the subject is presented in reference 7; but for completeness purposes, some of the considerations involved warrant a brief review.

The most reliable means of determining the resistance of a ship is to construct a scale model of the underwater portions of the ship and conduct model resistance tests in a towing tank. However, because of the time and cost involved, and because the ship dimensions and hull form are subject to frequent adjustment during the preliminary design stage, model testing is generally postponed to a later design stage. For preliminary purposes, sufficient accuracy can generally be obtained from an approximate assessment of the ship's resistance that is based on a published series of test results derived from systematically varied hull forms, using tentative values that have been established for the ship payload, sustained sea speed, and principal dimensions. Two such test series that can be used to assess the resistance of fine to moderately full-form displacement hulls are known as the Taylor's Standard Series [12], and Series 60 [13]. These systematic series are well known and widely used; however, other series data have been developed that are applicable to specific types of hull forms. Reference 14 is one such series that relates to full hull forms.

Although series test data can be used to estimate the resistance of ships in a straightforward manner, the process requires estimates to be developed for the major hull characteristics. If the accuracy of an estimate is less important than the rapidity with which it can be made, a statistical method may be used. Reference 15 is an illustration of a statistical method. By averaging plots of a substantial number of actual designs, an approximate statistical technique was developed to estimate the weight, displacement, speed, power, and other principal characteristics of a wide variety of cargo ships and tankers. Marine design activities commonly summarize their data on existing designs in a similar form to permit approximate parametric investigations to be made rapidly.

3.4 Selection of the Propulsor. Once the ship speed requirements and resistance have been tentatively established, the next step is to select the type of propulsor. As indicated by Fig. 3, some types of propulsors are inherently more efficient than others for particular applications [16]. Figure 3 shows that relatively slow but high-powered vessels such as trawlers and tugs have inherently lower propeller efficiencies, but improved efficiencies are achievable by using propellers in nozzles. For higher-speed ships, however, contrarotating propellers are seen to be more efficient.

The selection of the propulsor may not be a simple process, because in order to establish the type of propulsor, it may be necessary to at least tentatively select the type of main propulsion machinery. For example, the gain in efficiency offered by selecting contrarotating propellers or other complex propulsors must be traded off against the advantage of simplicity provided by conventional main propulsion machinery and shafting arrangements.

Similarly, the selection of the number of propulsors may also involve a trade-off. In general, vessels may be single, twin, triple, or quadruple screw. That is to say, the total power required to propel a vessel may be distributed (usually equally) among as many as four propulsors. From the point of view of initial and operating costs, fewer propellers are preferred; however, the magnitude of the power requirement, or the constraints on the propeller diameter, may force the selection of a multiple-screw arrangement to avoid an excessive propeller loading and the attendant cavitation that could otherwise result. In addition, there may be other factors in a given case, such as reduced vulnerability or improved maneuverability, that may favor the use of multiple propellers.

Several systematic series of fixed-pitch propellers have been model tested, and the resulting data have been presented in a form that is convenient for design selection purposes. Of these, probably the most suitable for design approximations is the Troost B Series of three, four, five, six, and seven-bladed propellers, although there are others that may be used [17]. In the usual case, the maximum propeller diameter that will allow adequate propeller submergence for the operating draft of the vessel, allow ample tip clearances, and also adapt to the stern configuration of the vessel so as to minimize propeller-blade-frequency excitation forces, would be the propeller selected.

A study must be made of the trade-off between the propeller rotative speed required to achieve a maximum propulsive efficiency and the constraints on propeller rotative speed that are imposed by size, weight, and cost considerations of the prime mover or transmission. The propeller rotative speed that is necessary to achieve a maximum propulsive efficiency (usually in conjunction with the largest propeller that can be accommodated by the hull afterbody) is frequently considerably lower than that which is feasible from the viewpoint of the prime mover or transmission (because of the larger torque and hence machinery size associated with lower propeller speeds). Furthermore, attaining the maximum propulsive efficiency does not necessarily constitute the most cost-effective system. In general, propeller characteristics are such that the propeller can be designed to operate at a rotative speed somewhat higher than that corresponding to the maximum propulsive efficiency without incurring a serious penalty in the overall efficiency. That is, the propeller efficiency generally peaks over a rather broad range of propeller speed. While no significant penalty in efficiency is incurred when the propeller design speed is somewhat higher than that for peak efficiency, significant savings can be realized in the first cost, size, and weight of the prime mover or transmission because of the lower torque rating (with the power remaining the same). The most cost-effective propeller speed is selected by conducting a trade-off study that balances the propulsive efficiency against the size, weight, and cost of the prime mover or transmission.

3.5 Establishment of Propulsion Plant Power Rating. Good practice dictates that a ship's propulsion plant be rated such that the desired ship speed can be attained with additional power capabilities held in reserve to allow for a degradation of performance with time. Factors to be considered in establishing the reserve capability include fouling and roughening of the hull, roughening of the working sections of the propeller caused by cavitation or erosion, and reductions in the performance capability of the prime mover, which taken together can result in a significant performance degradation in time. It is also

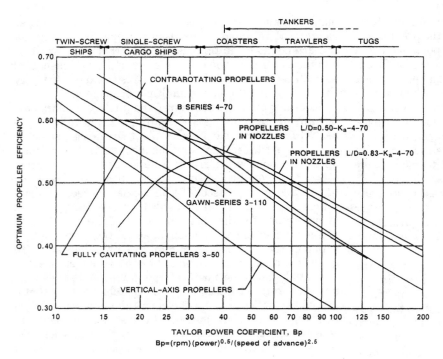

Fig. 3 Comparison of optimum efficiency values for different types of propulsors

important that the vessel have a reasonable ability to maintain speed in moderately rough seas and under adverse weather conditions. The usual practice for providing such a margin is to use the parameter *sustained sea speed*, which is defined as that speed which is obtained at some percentage of the normal continuous or service shaft power rating, during trials, at design load draft, under favorable weather conditions, when the vessel and propulsion plant are new and the hull and propeller are clean and smooth. The relationship between the trial speed at rated power and the sustained sea speed is indicated by Fig. 4. The percentage (or so-called service factor) of the normal continuous or service shaft power rating used to establish the sustained sea speed is often taken to be about 80% (0.80 service factor) for container and breakbulk cargo ships, which may be loaded to the design draft during the various legs of a voyage, and about 90% for tankers and bulk carriers, which in general operate at design draft during the loaded leg of a voyage but in ballast during the return leg. A value of 80% is generally used in the design of naval ships. However, depending on the itinerary intended, type of maintenance plan envisioned, and mean time between dry dockings and overhauls contemplated, the service factor used in a particular case may be somewhat different.

Gas turbines and steam turbines are generally expected to be operated at power levels near their maximum ratings, but such is not the case with diesel engines. A diesel engine is generally selected so that it is operated at no more than about 90% of its maximum continuous power rating, and somewhat below its rated speed. Therefore, the installed maximum continuous power of a diesel-driven ship exceeds that of an otherwise similar turbine-driven ship. This operating region usually coincides with

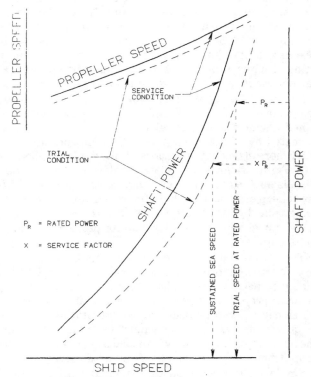

Fig. 4 Ship speed-power characteristics

the engine's range of the lowest specific fuel consumption. In addition, the engine component service-life projections and service recommendations for inspection, maintenance, and overhaul intervals are normally based upon operation in this region. The power level during normal operations is often termed the continuous service power, and the difference between the continuous service power and the maximum continuous rating is called the engine margin.

Gas turbines and steam turbines tend to be constant-power machines. That is, within limits, if rated torque is reached at a turbine speed that is somewhat less than rated, additional torque can be developed until rated power, corresponding to the flow through the power turbine, is reached. For this reason, turbine-driven propellers are usually designed so that under sea-trial conditions, when they are new, they absorb rated power at a rotative speed that is a small percentage greater than rated. During service, as the ship slows and the propeller roughens, to a limited extent rated power can continue to be developed at progressively lower engine speeds, but still remain within the continuous service power rating (torque) of the main engine.

Recognition of the fact that, while in service, a propeller torque of a given magnitude is developed at progressively lower propeller speeds is important in the design of all types of propulsion plants, but of particular importance with diesel-driven ships. Diesel engines have a limited overtorque capability and are adversely affected by continuous torque loads that exceed their torque rating. Therefore, the power that can be developed and, consequently, the speed of diesel-driven ships can in time be limited by engine torque restrictions.

3.6 Selection of Main Propulsion Plant. Like many other general design projects, the design of a machinery plant largely consists of the integration of a number of units and elements into a functioning system. The process entails selecting components, adjusting each to the constraints imposed by all others, and arranging them to achieve the required system performance, a satisfactory configuration, and an acceptable life cycle cost.

A number of design decisions must be made when formulating a main propulsion plant design. For example, the prime mover must be selected, with the common alternatives being a diesel engine, an oil-fired steam plant, a nuclear-fueled steam plant, a gas turbine, or a plant combining these alternatives. Less-conventional designs may also be considered. Once the generic type of plant has been established, then the major characteristics of that plant must be selected. For example, should a diesel plant be high speed, medium speed, or low speed? Or, in connection with a fossil-fueled steam turbine plant, should the plant have a reheat or non-reheat cycle? With gas turbine installations, there is the choice between aircraft-derivative gas turbines and industrial gas turbines, and simple or regenerative cycles.

Of the factors that must be considered when selecting the most suitable type of machinery, reliability in service is one of the most important, and must be given appropriate emphasis. A failure in the propulsion machinery plant may mean the loss of ship availability, and may even jeopardize the safety of the vessel. Other considerations, such as fuel economy, weight, space, and first cost, usually become of secondary importance when compared with the consequences of service interruptions that can result from inadequate reliability. Accordingly, only proven mechanical concepts are selected for ships; developmental features are properly proven in prototype plants ashore, where failures are of little consequence as compared with failures at sea.

The selection of a ship's main propulsion plant may be influenced by previous practice, as is the case with most other complicated engineering systems. Ordinarily, pertinent plans and essential data relating to the machinery of other ships, some perhaps rather similar to the one in question, will be available. If this information is available and in a usable form, first approximations can often be made without detailed study, thereby reducing the range and number of variables that must be given detailed consideration in the preliminary design stage.

Space and arrangement requirements. In general, the space required for the propulsion plant is considered to be space forfeited from what could otherwise be used for more productive purposes (e.g., cargo or weapon systems), and a maximum effort is accordingly made to restrain the dimensions of the machinery space.

Minimum space requirements are almost impossible to generalize satisfactorily for different types of power plants. It is usually necessary to make a preliminary ship arrangement layout to determine the effect of the power plant selection on the overall machinery space configuration. To illustrate the general differences in this respect among principal propulsion plant types, machinery arrangements from typical merchant vessels of comparable power are shown in Figs. 5, 6, 7, 8, and 9 for gas turbine, oil-fired steam turbine, low-speed diesel, medium-speed diesel, and nuclear-fueled steam turbine plants, respectively. There is a wide range of flexibility in the design of the propulsion plants illustrated, and for this reason the configurations and dimensions shown should be considered representative only of plants with comparable power.

Propulsion machinery configuration. Fundamental to the design of a main propulsion plant is the coordination of the prime mover with a transmission system and a propulsor. A number of possible machinery combinations may be considered by the marine engineer in making the selection. As indicated in Fig. 10, even with the range of considerations confined to those most commonly considered, a large number of alternatives is feasible.

As also indicated by Fig. 10, slow-speed diesel engines are the only commonly used prime movers that are directly connected to the propeller shaft. For other types of prime movers, transmission systems such as mechanical speed-reducing gears or electrical generator/motor transmissions are required to make compatible the relatively high rotative speed necessary for an economical, efficient, and small prime mover and the relatively low propeller speed necessary to achieve a high propeller efficiency. Overall speed-reduction ratios vary from relatively low values for medium-speed diesels, which generally need only a single speed reduction, to approximately 80 to 1 for turbines, which require a double reduction.

Medium- and high-speed diesel engines are mass produced and have many diverse applications. Medium-speed diesels operate in the approximate range of 300 to 1200

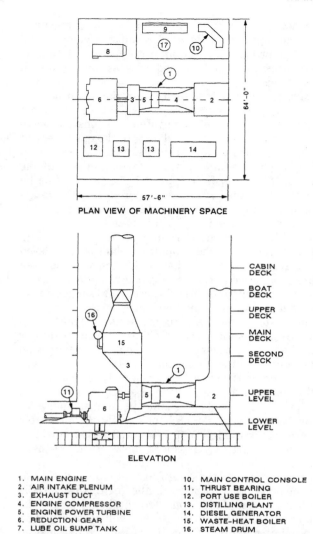

PLAN VIEW OF MACHINERY SPACE

ELEVATION

1. MAIN ENGINE
2. AIR INTAKE PLENUM
3. EXHAUST DUCT
4. ENGINE COMPRESSOR
5. ENGINE POWER TURBINE
6. REDUCTION GEAR
7. LUBE OIL SUMP TANK
8. STEAM TURBO-GENERATOR
9. MAIN SWITCHBOARD

10. MAIN CONTROL CONSOLE
11. THRUST BEARING
12. PORT USE BOILER
13. DISTILLING PLANT
14. DIESEL GENERATOR
15. WASTE-HEAT BOILER
16. STEAM DRUM
17. ENCLOSED OPERATING STATION

Fig. 5   Gas turbine power plant

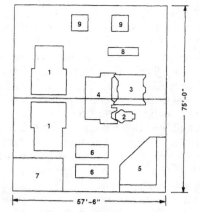

PLAN VIEW OF MACHINERY SPACE AT UPPER LEVEL

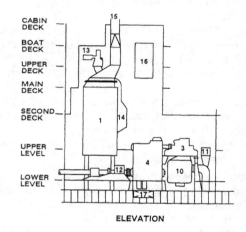

ELEVATION

1. BOILER
2. H.P. TURBINE
3. L.P. TURBINE
4. REDUCTION GEAR
5. CENTRAL CONTROL ROOM
6. STEAM TURBO-GENERATOR
7. WORKSHOP
8. CONTAMINATED STEAM GENERATOR
9. DISTILLING PLANT

10. MAIN CONDENSER
11. MAIN CIRCULATING PUMP
12. THRUST BEARING
13. FORCED DRAFT FAN
14. STEAM AIR HEATER
15. UPTAKE
16. DEAERATING FEED HEATER
17. LUBE OIL SUMP TANK

Fig. 6   Steam turbine power plant

rpm, with engines turning faster than about 1200 rpm usually classified as high-speed engines (the precise distinction in speed category is arbitrary and of no particular significance). Standardized parts and spares are readily available for diesel engines, which is highly advantageous. Engines that operate at progressively higher speeds generally have a higher wear rate, higher fuel consumption, and a lower tolerance for low-quality fuels because of their closer clearances and the shorter time period available for complete fuel combustion.

Low-speed diesel engines were specifically developed for marine propulsion purposes, and are generally more tolerant of low-quality fuel than higher-speed diesels. Low-speed diesels are designed to be directly connected to the propeller shaft, and they turn at speeds sufficiently low to provide a high propeller efficiency, thus avoiding the necessity for reduction gears. In addition, their large piston diameters and long strokes are compatible with the reliable combustion of heavy fuels. Their inherent fuel

economy is highly competitive, and their simplicity facilitates automation. In may propulsion applications, these advantages are sufficient to offset their disadvantages of increased size, weight, and first cost when compared with higher-speed engines.

An electrical transmission has features that may be attractive in some applications. In this case, a prime mover drives a generator or alternator, which in turn drives a propulsion motor that is either coupled directly to the propeller or drives the propeller through a low-ratio reduction gear. The use of an electric drive provides additional flexibility in the arrangement of the machinery, since the prime mover is not mechanically linked to the propulsion shafting. This flexibility can be of major significance in the arrangement of some types of ships, such as SWATH ships, which have twin, narrow underwater hull structures where electric motors can be installed to drive the propellers, with the generators and their drives more conveniently arranged higher in the ship.

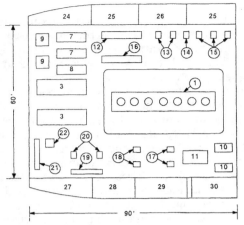

PLAN VIEW OF MACHINERY SPACE AT UPPER LEVEL

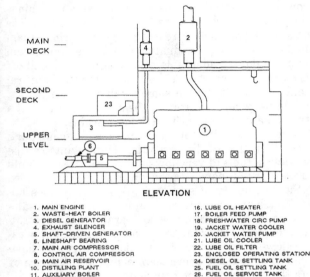

ELEVATION

1. MAIN ENGINE
2. WASTE-HEAT BOILER
3. DIESEL GENERATOR
4. EXHAUST SILENCER
5. SHAFT-DRIVEN GENERATOR
6. LINESHAFT BEARING
7. MAIN AIR COMPRESSOR
8. CONTROL AIR COMPRESSOR
9. MAIN AIR RESERVOIR
10. DISTILLING PLANT
11. AUXILIARY BOILER
12. FUEL OIL HEATER
13. LUBE OIL PURIFIER
14. DIESEL OIL PURIFIER
15. FUEL OIL PURIFIER
16. LUBE OIL HEATER
17. BOILER FEED PUMP
18. FRESHWATER CIRC PUMP
19. JACKET WATER COOLER
20. JACKET WATER PUMP
21. LUBE OIL COOLER
22. LUBE OIL FILTER
23. ENCLOSED OPERATING STATION
24. DIESEL OIL SETTLING TANK
25. FUEL OIL SETTLING TANK
26. FUEL OIL SERVICE TANK
27. AUX ENGINE LUBE OIL TANK
28. CYLINDER OIL TANK
29. MAIN ENGINE LUBE OIL TANK
30. BOILER FEED WATER TANK

Fig. 7 Low-speed diesel power plant

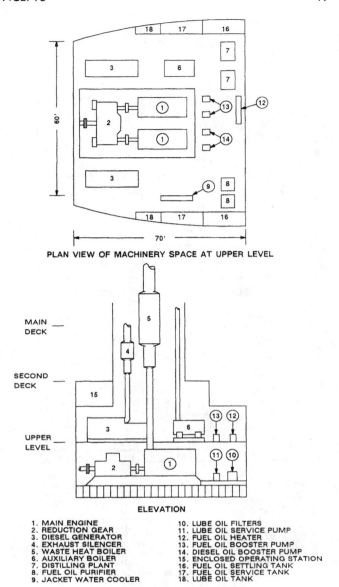

ELEVATION

1. MAIN ENGINE
2. REDUCTION GEAR
3. DIESEL GENERATOR
4. EXHAUST SILENCER
5. WASTE HEAT BOILER
6. AUXILIARY BOILER
7. DISTILLING PLANT
8. FUEL OIL PURIFIER
9. JACKET WATER COOLER
10. LUBE OIL FILTERS
11. LUBE OIL SERVICE PUMP
12. FUEL OIL HEATER
13. FUEL OIL BOOSTER PUMP
14. DIESEL OIL BOOSTER PUMP
15. ENCLOSED OPERATING STATION
16. FUEL OIL SETTLING TANK
17. FUEL OIL SERVICE TANK
18. LUBE OIL TANK

Fig. 8 Medium-speed diesel power plant

Also, with electrical transmissions, the peak power requirement can be satisfied by multiple prime-mover/power-generating sets. This permits some of the prime movers to be taken off line during partial-power operation, providing the additional advantage of enabling those remaining on line to be operated in the range of their best efficiency. As an example of electric-drive installations with multiple prime movers, submarine tenders, which are seldom moved once positioned on station, but must meet high electric-load requirements when on station, have been built with six diesel-generator sets driving a common electric propulsion motor. The six diesel-generators are identical, with all six being required to develop rated shaft power; but when on station, the six engines provide much flexibility in satisfying the varying ship-service electric-load requirements.

Electric drives may be either a-c or d-c. An a-c transmission is generally lighter and more efficient, but a d-c drive can be advantageous for applications that require a high

maneuvering torque. Examples of d-c drive applications are ferries and various types of tenders and service craft.

Several practical considerations constrain the ratings of the various alternative power plant selections. The simplicity and fuel economy of diesel engines make them well suited for many applications, particularly so for medium- and high-speed engines at the lower power ratings, where their size and weight requirements are competitive with other types of plants. However, in the high-power range, the weight and space requirements of diesel plants exceed those of comparable fossil-fueled steam and gas turbine plants.

Gas turbines offer the advantages of low weight and space requirements, but they have little tolerance for low-quality fuels. An additional constraint with gas turbines results from the fact that these engines are not developed to satisfy the requirements of a specific ship application.

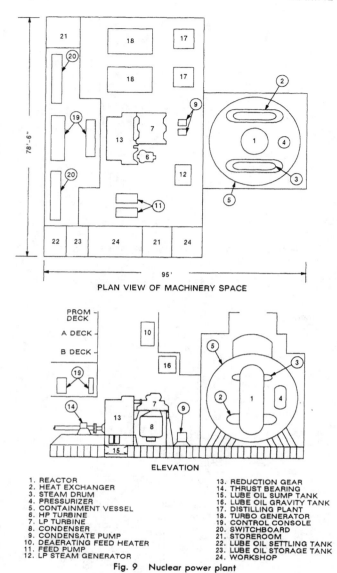

PLAN VIEW OF MACHINERY SPACE

ELEVATION

1. REACTOR	13. REDUCTION GEAR
2. HEAT EXCHANGER	14. THRUST BEARING
3. STEAM DRUM	15. LUBE OIL SUMP TANK
4. PRESSURIZER	16. LUBE OIL GRAVITY TANK
5. CONTAINMENT VESSEL	17. DISTILLING PLANT
6. HP TURBINE	18. TURBO GENERATOR
7. LP TURBINE	19. CONTROL CONSOLE
8. CONDENSER	20. SWITCHBOARD
9. CONDENSATE PUMP	21. STOREROOM
10. DEARATING FEED HEATER	22. LUBE OIL SETTLING TANK
11. FEED PUMP	23. LUBE OIL STORAGE TANK
12. LP STEAM GENERATOR	24. WORKSHOP

Fig. 9   Nuclear power plant

Gas turbines are sophisticated machines that require many years of evolutionary development at an enormous expense. No specific ship project could support the expense required to develop a unique gas turbine for a particular application. Most of the gas turbines used for marine propulsion were derived from gas turbines that were developed for aircraft applications. The incorporation of a power turbine and material changes were the principal modifications made to "aircraft" gas turbines to form the "marinized" or "aircraft-derivative" gas turbines which are prevalently used for naval combatants. Some "heavy duty" or industrial gas turbines have also been marinized and used for ship propulsion. Only those engine designs that were forecasted to be in continuing demand, with relative minor adaptations and variations to suit numerous ship applications were marinized. As a result, the available power ratings of gas turbines tend to be discrete and may not closely match a particular requirement. For example, if an analysis of the powering requirements for

a ship indicates that the maximum power requirement is 20,000 hp, the most appropriate gas turbine selection may have a manufacturer's rating of 23,000 hp. The higher-rated gas turbine would be less expensive than one specifically designed for the lower power requirement, and the higher engine capacity would provide increased service margins, which would reduce stress levels and facilitate a longer engine life; however, the specific fuel consumption of the larger gas turbine could be higher than for one designed for the particular requirements.

Steam turbine plants are generally optimized for a specific power output, and, as a result, the power ratings of steam turbine plants are generally close to their design capacity. There is no practical upper limit to the maximum power rating of marine steam turbine plants other than the source of steam and the ability to effectively transfer the power to a propulsor. The steam for steam turbines may be generated from boilers fired with any of a variety of fuels, or from nuclear reactors. Most marine steam plants use low-quality residual oil as the fuel, but steam plants are uniquely adaptable to coal firing, and boil-off gas is commonly used to fuel boilers on liquefied natural gas tankers. In addition, in some combined cycle plants, steam can be generated from the heat in the exhaust of a gas turbine and used to drive a steam turbine.

Although the specific fuel consumption of steam plants is higher than that of diesel plants, steam plants can be operated with fuels of very low quality, and steam plants require less maintenance. In the lower-power range, however, the complexity of steam plants generally does not offset the advantages provided. Many successful steamships were built, and continue to operate, that have ratings in the lower-power range; however, for most applications, slow-speed, direct-drive diesel engines have established dominance over steam plants in the range of powers that are within the capabilities of diesels.

When establishing ratings of the various power plant components, the components should be analyzed for service and design margins to ensure the high degree of reliability required for the safety of the vessel. Reasonably conservative ratings should be used for design purposes since in some cases there is a tendency for manufacturers' ratings to be derived from tests under laboratory conditions, as opposed to the less-than-ideal conditions encountered in marine service. An additional factor to be considered for gas turbines is the sensitivity of these engines to the ambient temperature; increasing ambient temperatures limit their power output capability and adversely affect their specific fuel consumption.

The cost of fuel and materials required for nuclear reactors as well as their stringent quality and safety standards make that source of steam relatively expensive. However, nuclear power provides features that are advantageous for certain applications. Nuclear reactors require no supply of oxygen for combustion, and they can operate for many years between refuelings. Nuclear power has revolutionized the military effectiveness of submarines by providing them with the additional dimension of virtually unlimited submerged endurance. Nuclear power is also beneficial with respect to high-powered surface ships that

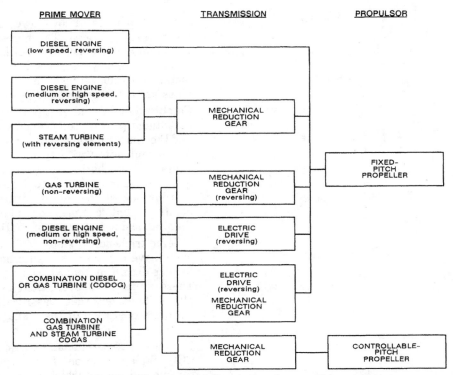

Fig. 10 Principal alternatives in the selection of a propulsion arrangement

require extended endurance, such as icebreakers and air-craft carriers, where the refueling requirements of conventional power plants could prevent them from accomplishing their missions.

Significant advances have been made in nuclear propulsion technology. USS *Enterprise*, the first nuclear-powered aircraft carrier, used eight reactors. The USS *Nimitz* class aircraft carriers were subsequently designed to have only two, but produce the same propulsion power. The use of a single reactor for aircraft carriers would not be considered, because of operational flexibility and damaged-ship survivability requirements. The first *Enterprise* reactor cores had service lives of only a few years, but the later *Enterprise* cores and those of the *Nimitz* can operate for two decades between refuelings. In the lower-power range, nuclear power is generally not competitive; however, for large, high-powered ships, such as aircraft carriers, the total life-cycle costs of nuclear propulsion are less than for fossil-fuel plants. The trade-off between economic considerations and the operational superiority provided by nuclear propulsion must be made for each particular ship application. However, the relatively higher initial and operating costs, as well as the social and environmental issues involved with nuclear plants, generally confine their application to warships.

Combined plants. The basic choices when selecting a ship's prime mover are a diesel engine, gas turbine, or steam turbine; however, for some service requirements it may be a practical impossibility for any single prime mover to acceptably satisfy all of the operating conditions. As examples, Coast Guard cutters and naval combatants may be required to operate for extended periods of time at low, economical cruising speeds, but upon command may be required to reach maximum power, which can be several hundred percent higher, in a matter of minutes to respond to an emergency. A single prime mover may not be suitable for such extreme service requirements, and a combined plant may be a preferred choice. In combined plants, two different prime movers (although they may be different sizes of the same type) are usually connected to the propeller shaft through a common transmission system, as illustrated by Fig. 11, to take advantage of the desirable features of each prime mover. A combined diesel or gas turbine (CODOG) plant may be a preferred choice for diverse service requirements such as those associated with Coast Guard cutters and small naval combatants. In these cases a relatively small diesel engine (the "cruise" engine) would be used during the low-speed, high-endurance cruising mode, to take advantage of the economy of a diesel engine; or, for maximum-power requirements, the higher-power gas turbine (the "boost" engine) would be brought on line. A mechanism such as an overriding clutch would be used to ensure that either the diesel or the gas turbine, but not both, drive the propeller. While the fuel economy and endurance of the gas turbine would be less than those of the diesel, these would be secondary considerations because of the gas turbine's infrequent use. CODOG plants make it possible to minimize the operating hours on the large gas turbine, which is required only for high power requirements, and the presence of two prime movers provides a degree of redundancy. These features are advantageous in many practical situations, and CODOG plants are commonly used.

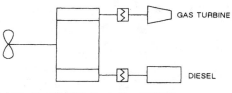

(a) COmbined Diesel Or Gas turbine (CODOG)

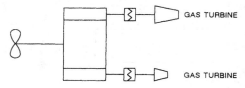

(b) COmbined Gas turbine Or Gas turbine (COGOG)

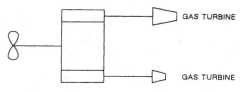

(c) COmbined Gas turbine And Gas turbine (COGAG)

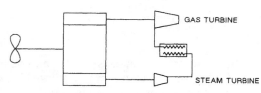

(d) COmbined Gas turbine And Steam turbine (COGAS)

Fig. 11   Typical combined propulsion plant alternatives

In addition to CODOG plants, other typical combined plants are illustrated by Fig. 11; noteworthy features of several combined plants are as follows:

- Combined gas turbine or gas turbine (COGOG). Applications for COGOG plants are similar to those for CODOG plants, with the difference being that a small gas turbine is used, instead of a diesel engine, as the cruise engine, with one or more larger gas turbines providing the boost power. At cruising speeds, the fuel economy of the small gas turbine, which operates near its rating, is better than could be obtained with a larger engine operating at a small fraction of its rating.
- Combined gas turbine and gas turbine (COGAG). COGAG plants contain one or more small gas turbines which are used for economical operations at very low cruising power levels, and one or more large gas turbines which are used occasionally at boost power levels. All of the gas turbines drive through a common reduction gear. COGAG differs from COGOG in that a COGAG plant is designed to require all gas turbines to be on line to develop maximum power. The principal advantages of CO-GAG plants are the economy realized during extended small-engine operations, the ability to de-clutch and secure unneeded engines during

extended periods of low-power operation, and the redundancy provided. The necessity to coordinate the simultaneous operations of large and small gas turbines can, however, be a disadvantage. Note that installations consisting of multiple, identical gas turbines are not considered to be combined plants.

- Combined diesel and gas (CODAG). CODAG plants potentially provide the economy of a diesel engine and the almost instant boost power of a gas turbine on demand; however, they entail a significant disadvantage. With both the diesel and gas turbine required to be in operation to develop full power, the coordination of the two dissimilar types of power plants presents complex control problems that limit the practical significance of CODAG plants.

Some triple-screw ships deployed by the Russian Navy have diesel-driven centerline shafts and gas-turbine-driven outboard shafts, and conversely. These ships have been cited as having CODAG plants; however, that classification is questioned since the operation of the plants is not directly related.

- Combined gas turbine and steam turbine (COGAS). COGAS plants can be of two types: Either the gas turbine and steam turbine can be combined mechanically but be kept thermodynamically independent, or they can be combined both mechanically and thermodynamically. In a thermodynamically independent plant, the gas turbine and steam turbine drive a common reduction gear, with both being required to develop full power; plants of this type were used in some foreign navies, but their use has been discontinued. A combined nuclear and gas turbine (CONAG) plant is a variation of this scheme in which the steam is generated by a nuclear reactor; however, no such plants have been built.

A thermodynamically combined cycle, which has also been called a STAG (steam and gas turbine) and a RACER (RAnkine Cycle Energy Recovery) cycle, uses both a gas turbine and a steam turbine to develop full power [18,19]. As indicated by Fig. 12, the exhaust gas from the gas turbine is used to generate the steam that drives the steam turbine. The power that can be developed by the steam turbine is in the range of 20% to 35% of that produced by the gas turbine. The principal advantage with this type of plant is the fuel economy that is achievable. For redundancy to be provided, which is usually desirable, the waste-heat boiler must be designed with the ability to be independently fired, because there would otherwise be no source of steam when the gas turbine is inoperable.

There are numerous variations of combined plants, any of which might be advantageous in particular circumstances.

Reversing capability. The means selected to stop and reverse a ship are closely related to the choice of prime

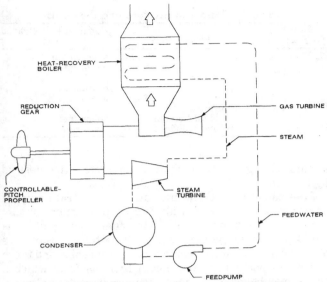

Fig. 12 Combined gas turbine and steam turbine (COGAS) cycle

mover. Most direct-drive diesel engines are readily reversible, as are electric drives. Steam turbines and gas turbines, on the other hand, cannot be directly reversed and require special reversing provisions. The usual solution with steam turbines is to locate special rows of astern blading in the region of the turbine casing that, during ahead operations, are at the exhaust end of the turbine and, therefore, in a vacuum atmosphere that is only slightly above condenser pressure. To reverse, steam is admitted to the astern blading instead of the ahead blading. With this arrangement, the windage losses during ahead operations caused by the astern blading rotating backwards in the low-pressure (and hence low-density) steam are acceptable. When operated in the astern mode, the windage losses caused by the ahead blading turning backwards are relatively high, but of little consequence since astern operations are infrequent and of short duration.

Several attempts have been made to design gas turbines with a reversing capability, but no practicable design has been developed. One such attempt was a project sponsored by the Maritime Administration in which the shanks of some of the ahead blading were shaped to form astern elements. In that arrangement the turbine efficiency proved to be low, and the astern elements unduly complicated the design. In propulsion applications, therefore, gas turbines require a supplementary reversing capability, which can be in the form of a reversing reduction gear, a reversible hydraulic coupling, a reversing electric drive, or a controllable-pitch propeller.

Reversing reduction gears are designed to reverse the direction of shaft rotation by following a series of discrete steps [20]. First, the engine throttle is reduced to idle in preparation for reversing. Next, when the shaft speed has reduced such that the torque in the main engine output shaft is minimal, the ahead drive train is decoupled and the reverse drive train is coupled. And finally, the throttle is increased to drive the reverse train. In many reversing

reduction gear arrangements, friction clutches or brakes, which permit relative motion between the connected shafts, or slip, are used to control and compensate for any difference between the speed of the driver shaft and that of the engine input shaft during the transition period as the drive trains are reversed. This process requires the absorption and dissipation of a large amount of kinetic energy in a brief period of time. Reversing reduction gears have also been designed with hydraulic couplings that have a reversing capability. Hydraulic couplings inherently slip; therefore, in applications that require extended periods of operation in a maneuvering mode, the efficiency loss may be a serious concern. When operating in a cruising mode, as at sea, the hydraulic coupling is bypassed and the main engine is directly connected to the propeller [21]. Reversing reduction gears have been proved feasible up to substantial power levels.

Electric drives are reversed by dynamically braking the propulsion motor and energizing the electric motor in the reverse direction.

Controllable-pitch propellers provide a more responsive reversing capability than do the alternatives. Controllable-pitch propellers are more complex and understandably more expensive than fixed-pitch propellers, but their design has evolved such that they are reliable. Notwithstanding the additional expense, controllable-pitch propellers are sometimes used in conjunction with prime movers that are cumbersome to reverse and in applications where maneuverability requirements are severe (e.g., some tugboats, vessels that frequently pass through locks, or vessels that can avoid the necessity for tugboats when provided the additional maneuverability). Controllable-pitch propellers are discussed in further detail in Chapter 10.

Auxiliary equipment. Considerable auxiliary equipment is needed to support the propulsion system and to provide other ship-support services. Since, in many instances, there is a choice when selecting the type of prime mover for the auxiliary equipment, interrelations between the auxiliary equipment and the main propulsion plant must be considered to ensure that the overall ship is designed in the most effective manner.

Many powered auxiliaries can be driven by either steam or electric power. If the main engines are driven by steam, or if steam is generated by waste heat, or is needed for cargo services, it may be desirable to also drive some auxiliaries by steam. Examples of such steam-driven auxiliaries include main feed pumps and forced-draft blowers on naval ships and cargo and ballast pumps on tankers. However, for many diesel plants and most gas turbine plants, where steam is not readily available, electrically driven auxiliaries are usually more appropriate. Also, some auxiliaries, including electric power generators, may be driven by the main propulsion engine through power take-off drives.

During the preliminary design stage, it is a marine engineer's responsibility to select the type of drives for the powered auxiliaries and to establish the arrangements and tentative ratings for the auxiliary equipment. To ensure an acceptable degree of reliability, redundancy is provided for some vital auxiliary services. For some functions, the redundancy may be in the form of a standby duplicate of the operating unit. This may be an appropriate provision for services such as lubricating-oil pumps, coolers, strainers, and fuel-oil service pumps. For some items, redundancy may be provided by having two or more units connected in parallel for normal service so that, upon the failure of one, the ship may continue to be operated, but at reduced power. Additional means of providing redundancy include arrangements where standby service is provided by emergency connections to units that are primarily installed for other purposes. Using the ship's firemain for some emergency cooling-water applications is an example.

When specifying the ratings of auxiliary equipment, marine engineers must use prudent judgment. The specified rating must include a margin that is added to the net calculated performance requirements to allow for subsequent minor adjustments and refinements in the plant design, as well as for performance degradation in service, but reasonable precautions should be taken to avoid requirements for unusable excess capacity. The efficiency of many mechanical components significantly declines when operated away from the design rating. Also, while it may appear to be good practice to maintain large equipment margins until the design is firm and it is known with certainty that the design rating can be reduced, such is not the case. By the time the ship design has progressed to the point that the required ratings of some components are accurately known, interfacing elements aboard ship may have been designed to accommodate the oversized units.

To support the development of a machinery layout, approximate outlines of all major auxiliaries are required, and some insight concerning the dimensions of even the minor items is needed. The degree of accuracy required in the preliminary machinery layouts depends on the type of ship and the compactness of the machinery space. For most merchant ships, outline drawings of similar equipment used on earlier ships are usually of sufficient accuracy for the purpose. At the other end of the spectrum, to support the design of the compact machinery spaces in submarines, even during the preliminary design stage, the configuration of auxiliaries must often be known within inches to support the development of a meaningful space arrangement. To do so often requires inputs from prospective machinery manufacturers. When that is impracticable, the marine engineer may be required to knowledgeably estimate the shipboard interfaces and the space envelope that the equipment manufacturer must subsequently match.

As a ship design progresses through increasingly detailed phases, the characteristics of the auxiliary equipment become better defined. At the end of the contract design stage, a complete list of the auxiliary equipment required, including the number of units, types, and ratings, must be specified. This information is needed by prospective shipbuilders to obtain estimates from equipment manufacturers and to ensure that there is a clear

Table 3   Petroleum fractions and their uses

Petroleum Fraction	Classification	Common Unrestricted Uses
Light distillate	intermediate naphthas	aviation gasoline
		motor gasoline
	kerosene	gas turbine fuel
	gas oil	heating fuel
		diesel fuel
Heavy distillate	lubricating oils	not used as fuel
	residual fuel oils	boiler fuel
Residues	refinery sludges	refinery fuel

understanding of the equipment characteristics desired by the owner.

Fuel selection.   Coal, uranium, and natural gas play important roles in worldwide energy production, but by far the greatest proportion of the fuel used by ships is liquid petroleum fuel. Virtually all petroleum fuels are obtained by fractionating or cracking crude oils, and there is a wide spectrum of petroleum fuel oils from which a choice may be made. Some of the more important alternatives are given in Table 3.

In general, fuel oils of higher viscosities are less expensive; however, another major consideration is that higher-viscosity fuels tend to have greater concentrations of impurities and harmful constituents. The fuel oil selected should be determined on the basis of the lowest overall cost, with consideration given to factors such as initial costs, handling costs, equipment maintenance costs, and additional operational complexity that can be attributed to the fuel. The factors to be considered relative to handling and equipment costs include fuel constituents, types of metals that will be in contact with the fuel and with the products of combustion, provisions available for fuel heating and treating (removal or neutralization of objectionable constituents), and provisions contemplated for corrosion protection and slag removal. The fuel quality, therefore, can impose additional requirements and constraints on the plant design.

The ash content, which represents certain incombustibles present in the fuel, may vary widely depending upon the geographical source of the fuel. Both the quantity and the chemical composition of the ash can have a marked influence on equipment life and performance. Sodium and vanadium compounds, which form in the products of combustion, tend to have relatively low fusion temperatures, and consequently can play important roles in the initiation and subsequent accumulation of slag formations. Furthermore, vanadium compounds can be highly corrosive at temperatures above their melting points. Therefore, sodium and vanadium are objectionable fuel constituents in the higher-temperature range. During the combustion process the fuel sulfur content is oxidized; and, if the temperature of the products of combustion should subsequently become lower than the acid dew point, the sulfur oxides may be hydrolyzed to form sulfuric acid, which is highly corrosive. Consequently, the fuel sulfur content is of great importance in the low-temperature range. Also, low-quality fuel can, in time, cause a degradation of the plant thermal efficiency, with the severity of the degradation being dependent upon the type of prime mover and its design parameters. Further details concerning petroleum fuels are included in Chapter 12.

Nuclear-fueled propulsion plants are well suited for many naval ships, but have not proved to be economically justified for merchant ships. Additional applications of nuclear power in merchant ships must await further advances in nuclear plant technology and applications.

The selection of a fuel is a multifaceted process that could greatly influence the success of the ship. Analyses of life-cycle costs must consider operational complexities and maintenance factors associated with a fuel, as well as the various aspects of the fuel itself.

Fuel Consumption. Different types of propulsion plants have inherently different thermal efficiencies and specific fuel-consumption rates, but specific design features may result in overlaps in the ranges of practical plant efficiencies of the different types. A heat balance is the fundamental analytical procedure used to determine the fuel-consumption of steam plants, and this subject is given a detailed treatment in Chapter 2. Fuel-consumption data for gas turbines and diesel engines are derived from information provided by their manufacturers.

Generalized fuel-consumption characteristics of various types of propulsion plants are shown by Fig. 13, which illustrates the relationship between fuel consumption and rating for the more common plant alternatives, with reciprocating steam engines included as an historical point of reference. The fuel consumption indicated in Fig. 13 corresponds to the fuel type most suitable for each respective plant, and includes that required for the main propulsion plant, auxiliaries, and normal hotel loads; no allowance has been made for extraordinary service, such as the hotel load on passenger ships, cargo heating and tank cleaning on tankers, and cargo refrigeration. Also not reflected in Fig. 13 is the consumption of lubricating oil, which is of minor significance for all prime movers shown except diesel engines; for diesels, however, lubricating oil is consumed at the rate of from ½% to 1% of the fuel-oil consumption. This rate may appear small, but lubricating oil costs much more than fuel oil and thus amounts to a significant percentage of the fuel-oil costs for diesel engines. Figure 13 illustrates the general fuel-consumption characteristics of the various propulsion plant alternatives; however, because of the different types of fuel used by the different power plants, the data cannot be used directly for cost comparisons. It is further noted that many factors can affect the fuel-consumption data for any specific application.

Once the general type of propulsion plant has been tentatively chosen, several design features remain to be selected that have the potential of improving the fuel-consumption characteristics of the plant. For example, in the case of a steam turbine propulsion plant, regenerative feedwater heating using extraction steam or reheating of the steam in the boiler after a portion of expansion work has been extracted in the turbines are among the methods by which the thermal efficiency of a steam cycle can be improved. In general, trade-off studies are needed to determine the most appropriate steam cycle. These studies

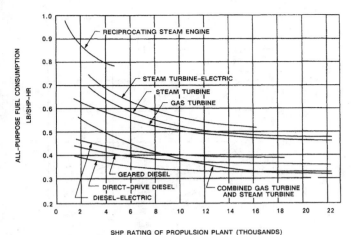

Fig. 13 All-purpose fuel consumption

could consider such parameters as boiler superheater outlet pressure and temperature, condenser vacuum, main turbine efficiency, number of stages of regenerative feed heating, and selection of extraction points.

Trade-off studies for the purpose of improving fuel economy should likewise be conducted for gas turbine or diesel propulsion plants. Cycles employing diesel prime movers tend to have higher thermal efficiencies than those using steam turbines, but the overall efficiency of a total diesel power plant can nevertheless be improved by the use of waste-heat boilers, exhaust-gas turbines, and by directly driving major auxiliaries from main-engine power take-offs. The principal means of improving the efficiency of a gas turbine plant is to recover heat from the exhaust gas. The recovery of waste heat is discussed further in Chapters 2, 3, and 4.

Weight requirements. The importance of the weight of a main propulsion plant depends upon the particular application. In the case of tankers, whose cargo capacity is limited by draft restrictions, the weight of the main propulsion machinery generally represents that of cargo foregone. Some cargo vessels, on the other hand, seldom operate at their full-load draft, and the weight of the main propulsion machinery, as such, may be mildly advantageous in that its low location improves the stability of the ship.

Naval vessels tend to have weight and stability problems, particularly since the advent of the major emphasis on shock resistance and noise control, both of which entail significant weight requirements, and the substantial growth in the weight of electronic gear. Consequently, weight control and reduction are major considerations during the design of naval ships, and are of critical importance in the design of submarines.

Representative propulsion plant weights (without fuel) are shown in Fig. 14, where the specific weight (the weight of the complete propulsion plant per unit of rated shaft horsepower) is plotted against shaft horsepower rating. Representative propulsion plant weights, with the fuel weight included, versus the plant shaft horsepower rating

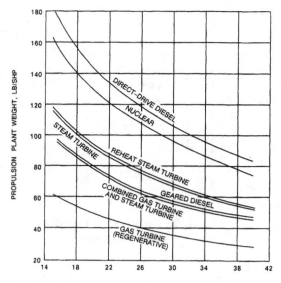

Fig. 14   Specific weight of propulsion plants

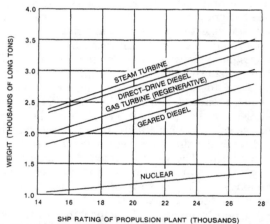

Fig. 15   Weight of cargo ship propulsion machinery plus fuel for a 10,000-mile voyage

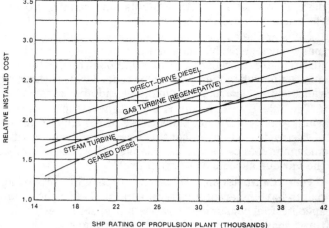

Fig. 16   Relative installed costs of propulsion plants

are shown in Fig. 15. This plot permits a proper comparison to be made between petroleum-fueled plants and nuclear plants. For the latter, the weight of fuel is not significant.

Costs. The installed cost, which is one of the most important considerations when making trade-off studies, is also the most volatile. Propulsion plant price levels are strongly influenced by factors such as material and labor costs, the similarity of a plant to those previously produced, and the manufacturer's existing work backlog. Consequently, the installed costs are subject to fluctuations that depend on the current status of the industry. Nevertheless, the approximate relative costs of the various types of plants as they vary with size are illustrated by Fig. 16.

For merchant ships in general, and in some cases for naval ships, economics is the underlying consideration when evaluating candidate types of propulsion plants. Three types of costs must be evaluated: initial costs (e.g., installed costs), recurring costs (e.g., fuel, maintenance, and crew expenses), and contingency costs (e.g., most aspects of reliability). By applying the present-value concept, the costs to be incurred in the future can be discounted and related to their present value so that the costs associated with the various design alternatives can be totaled and compared on a common baseline to establish the most advantageous alternative. There are cases, however, where the costs associated with a type of propulsion plant are of secondary importance because the ship's mission makes a particular type of propulsion plant mandatory. Such is the case with strategic-missile submarines, which must have nuclear propulsion to permit extended periods of underwater operations.

Section 4
Other Design Considerations of Noteworthy Significance

4.1   General. The main propulsion system may be the most complicated of the shipboard systems; however, the importance of other ship systems must also be recognized. Their interrelations with the propulsion system and the auxiliary systems are essential considerations in the design of the propulsion plant. In addition, some design disciplines, such as noise and shock, have pervasive effects

that influence the design of many shipboard systems. Several of the more important design considerations of this type, from a marine engineering perspective, are reviewed in the following paragraphs.

4.2   Electric Plant. The analytical procedure used to ensure that the electric plant has sufficient capacity to accommodate all operational requirements is known as an *electric load analysis.* An electric load analysis consists of an estimation of the ship's electric power requirements under a number of operational conditions. As an example, for a cargo ship, electric power estimates are typically made for the at-sea, in-port, and cargo-handling (transferring) operations as well as other conditions that impose significant electric loads. An electric load analysis is prepared by listing all of the electric power-consuming components, along with their respective loads, and applying appropriate use or weighting factors to the component loads that correspond to a particular operational condition. The weighted connected component loads are then summed as estimates of the electric power requirements for the different operational conditions.

During the early ship design stages, the scant design details available are insufficient to conduct a comprehensive electric load analysis. Therefore, early estimates must be made on the basis of former successful practice. SNAME Technical and Research Bulletins 3-11 [22] and 3-49 [23] provide guidance for the development of electric load estimates.

While it is often technically possible to install a single generator of sufficient capacity to satisfy all requirements, multiple generators are normally installed. One reason for multiple units is that a single generator failure would otherwise incapacitate the entire electric plant; another is that the part-load (as might occur in port) operation of a single generator would be at an excessively low power level. Based on the power requirements for the various operational conditions, the number and rating of the ship-service generators can be selected not only to satisfy the peak power requirements, but also to allow suitable redundancy of supply and economical operation under normal loads.

Multiple levels of redundancy are provided to ensure a continued supply of electrical power. The ship-service generators are often sized such that a minimum of two equally rated units can satisfy peak load requirements. It is considered good practice, and a requirement of some regulatory bodies, that one complete ship-service generator unit be provided in reserve, of sufficient capacity to meet the essential power requirements, including propulsion, under emergency-power conditions. In addition, an emergency generator of sufficient capacity to carry emergency and essential loads is required for the sole purpose of maintaining communications, lighting, and control of the ship under emergency conditions. The ship-service generators and the emergency generator should not be located in the same space.

The conceptual design of an electric plant is necessarily linked to that of the propulsion plant. If steam turbines are the propulsion plant prime movers, then steam-turbine-driven or attached (either directly driven by the propulsion shafting or by a higher-speed shaft from the reduction gear) ship-service generators would be the logical choices. Ships driven by diesels or gas turbines often have diesel-driven generators, but attached generators and

steam-turbine-driven generators, with the steam produced in an exhaust-gas boiler, are alternatives. Also, some ships have gas-turbine-driven generators.

For ships that have an electric propulsion drive, integration of the propulsion power system and the ship-service electric power system may be considered. This is particularly attractive when the peak propulsion power electric load does not coincide with the peak ship-service load, or when the ship-service load is large in comparison to the propulsion load.

When an attached electric generator is installed, the generator may be sized to satisfy the maximum at-sea loads with the ship operating at cruising speed. This arrangement has the advantage of not requiring additional generators to be on line when at sea, and it improves overall fuel economy by using the inherently more efficient main engine to drive the generator, instead of using a smaller engine. When there is a large difference between the maximum at-sea electrical loads for different operating conditions, as with ships that carry refrigerated cargo only on some legs of a voyage, an attached generator may be sized for the continuous base load, with an independently driven generator used in parallel to provide the peak-demand load. Several alternative arrangements can be used to drive a generator by the main propulsion engine. A generator can be attached directly to the engine, driven by a power take-off from the reduction gearing (if used), or driven by the propulsion shafting. Higher-speed generators are smaller and usually less expensive than lower-speed generators.

4.3 Steam System. On most vessels, a supply of steam is required to heat spaces and fuel; on some tankers, steam may also be needed for cargo heating. The quantity of steam required depends on the type of vessel and the service for which it is intended. If the vessel is driven by steam turbines, the steam supply is easily taken from the main steam system. For diesel or gas turbine-driven ships, boilers are required to provide the steam. Often, in diesel or gas turbine plants, the heat in the hot exhaust gas from the main engines is recovered by passing the exhaust gases through an exhaust-gas boiler designed for this purpose. Such a boiler may also be provided with an oil burner to make up for any heat deficiency, and to operate in port when the main engines are secured.

For some tankers a large steam capacity is required to heat the cargo, and rather large quantities of hot water are sometimes required for cleaning the cargo tanks. The ratings of boilers for steam-driven tankers may need to be increased well above that required for propulsion purposes for this additional load. If the tanker is driven by a diesel engine or a gas turbine, a large boiler may be required especially for this auxiliary steam purpose.

4.4 Hull Machinery and Cargo Systems. The steering gear and the bridge control of the steering gear are essential for the safe operation of a ship, and the regulatory bodies consequently require that redundant power capabilities and controls be provided for these functions. Because of the complexity of the flow of the water over the rudder during maneuvers, the required rudder torque cannot be reliably predicted with accuracy; in some cases, the margin of error can be rather large. The torque required of the steering gear to operate the rudder is, therefore, estimated by using semi-empirical methods in conjunction with experience factors derived from full-scale torque measurements [24,25]. Rudder-torque tests are sometimes conducted on a ship model, particularly for higher-speed naval ships; but because of the low Reynolds numbers inherent with such model tests, the results are often suspect. Consequently, good judgment requires that a prudent margin be incorporated when selecting the size and rating of the steering equipment.

After the integrity of the hull, the assured availability of the main engines, and the reliability of steering, many mariners would next value the effectiveness of the anchoring equipment. Anchoring equipment is normally used to secure a ship in a protected roadstead while waiting berth clearance and in other similar circumstances; however, in many emergency situations, anchoring becomes of vital importance. If the main engines should fail while the ship is in restricted waters, the ability of the anchor to hold could prevent the ship from being driven aground or into traffic channels.

Cargo-handling costs are one of the most significant operating expenses for a merchant ship; therefore, the cargo-handling operations must be analyzed in conjunction with the total ship system to establish the most cost-effective overall design. Cargo handling system designs vary widely depending upon the type of cargo handled, but some of the more common systems are as follows:

- Tankers generally are fitted with piping systems and pumps to load and discharge the cargo, with small cranes to handle hoses from shore, and with tank-cleaning machines and inert-gas generating systems. Many tankers are fitted with cargo-heating systems.
- Break-bulk cargo ships are fitted with cranes (or booms) for loading and discharging cargo.
- Containerships may be fitted with cranes to handle cargo containers, although most containerships rely exclusively on shoreside cranes.
- Many containerships and break-bulk cargo ships have "roll-on/roll-off" (RO/RO) features, which are systems of ramps and, sometimes, elevators that enable tractor-trailer vehicles to be driven into place aboard the ship and driven off.
- Most dry-bulk carriers depend on shoreside facilities for cargo loading and discharge, but some bulk carriers have self-unloading features with conveyors below the cargo holds, or with cranes on deck, to unload the cargo.
- Barge carriers are fitted with large cranes or elevators to lift the barges aboard ship.
- Refrigerated-cargo ships may carry independently refrigerated containers that are connected to the ship's electrical supply once loaded on board, or the ships may be designed with refrigerated cargo holds

that are fitted with large cargo-refrigeration systems.

A more detailed discussion concerning cargo-handling operations is presented in reference 9.

The relative merits of the alternatives in the design of cargo-handling gear, such as hydraulically operated hatch covers and special types of cranes, elevators, conveyors, and cargo pumping systems, should be rigorously analyzed during the preliminary design stage. The space, weight, and power requirements of cargo-handling gear must be estimated very early in the design of a ship because they may have important effects on the deck arrangement, the size of the electrical generating plant, and indeed the configuration of the vessel itself. In addition to the equipment directly associated with cargo handling, the requirements for associated auxiliary systems such as ballasting and deballasting, inert-gas generation and distribution, and hose-handling (which may require cranes) must be analyzed.

4.5 Automation. Automation can be defined as an apparatus, process, or system that is self-acting or self-regulating generally through the employment of mechanical or electrical devices that take the place of human observation, effort, and decision making. This is done through the application of various control schemes generally referred to as "closed-loop" or "open-loop" controls. A closed-loop control system maintains the desired operation by comparing the actual output (the controlled variable) to the desired output (the set point). The difference between the actual and desired outputs creates an error signal. That error signal is fed back to the control elements, which act upon it to reduce the error signal within an acceptable tolerance band. Such a system is self-regulating. Open-loop control systems do not employ the feedback principle. A simple switch is an example of an "open-loop" control. Open-loop controls are used in automated systems where system operation can be predicted with a high degree of certainty. For example, engine room ventilation fans are often driven by two-speed motors which, if started across the line on high speed, might impose an unacceptably large voltage dip on the electrical distribution system. An open-loop control scheme could be devised which, upon depression of the "fast-speed" pushbutton, would cause the motor to start on low speed but, after sufficient time has elapsed for the fan to accelerate, to transfer to high-speed automatically. In this example there is no automatic comparison of motor speed with a set point to ensure that the fan has, in fact, reached the desired low speed before the transfer to high speed is initiated; this is not necessary because the time required for the motor to accelerate can be predicted with a high degree of accuracy. Open-loop control systems are self-acting but not self-regulating.

When designing automated systems, in addition to the purely engineering considerations involved, marine engineers must have an appreciation for the skill level and training of the crew, the interests of labor unions, and the requirements of the regulatory bodies. American-flag ships and foreign ships trading in American waters must meet the regulations of the United States Coast Guard [26,27,28]. In addition, merchant ships must meet the requirements of the applicable classification society; American-flag ships are commonly classed by the American Bureau of Shipping [29].

Automation is an essential and integral aspect of the design of virtually all shipboard equipment and systems. Automated systems that are self-regulating can hold process variables within very close tolerances—tolerances that manual control could not maintain indefinitely. An example of such control is the regulation of the voltage and frequency of the ship's electrical generators under varying load conditions.

Automation can enhance safety by constantly and simultaneously monitoring hundreds of operational parameters, such as the operational conditions of a propulsion plant. Rather than simply displaying raw data, automated systems are generally designed to provide information that is needed to make a decision during an emergency. For example, a collision avoidance radar system warns the watch officer of other ships operating in the vicinity, automatically computes the course and speed of other ships in the area, and displays information giving the closest point and time of approach for each target.

In addition to enhancing ship safety, automated systems serve numerous other functions. Those that monitor the stack emissions and the oil content of ballast water discharged overboard guard against potential environmental hazards. Other types of automated systems are used to ensure that allowable hull stresses are not exceeded because of the improper loading of cargo. More mundane yet essential tasks such as spare-parts inventory and crew timekeeping are also reliant upon automation for effective execution.

Ship operations can be automated such that the onboard crew required to perform routine functions is minimal. Automation is an effective means of reducing crew requirements and hence operating costs for merchant ships; however, such is not the case for many types of naval ships because trained operating personnel must be available in the event that battle damage incapacitates the automation system.

Obviously, as the size of the crew is reduced, there is a corresponding reduction in the ability to conduct shipboard maintenance. The reliability of ship systems and components, therefore, becomes of greater concern, and redundancy for vital components is essential. The automation concept requires that shipboard equipment be designed to be highly reliable, and where a sufficient level of reliability is not feasible, the equipment must be designed to be easily replaceable. The routine onboard repair of failed equipment is not compatible with a high degree of automation.

Trade-off studies are required to establish the appropriate level of automation for any particular application. Factors to be considered include:

- Reduced costs derived from smaller crew
- Higher costs associated with more highly trained crew

Fig. 17 Propulsion plant control console in an Enclosed Operating Station of early design

- Higher costs of more sophisticated shipboard systems
- Reduced costs derived from eliminating shipboard maintenance
- Increased costs of shore-based maintenance program
- Reduced costs resulting from more effective operations

A detailed evaluation of these considerations can provide guidance concerning the desirable level of automation for a specific instance.

Automated shipboard systems have generally been developed within distinct disciplines: propulsion system, navigation, vessel management and cargo control. These have remained separate areas because they have minimal functional overlaps and, for the purposes of this discussion, each will likewise be treated separately.

a. Propulsion systems. One of the first approaches used to reduce the size of the engine-room crew was the centralization of the indicating devices and manual controls, such as meters, gauges, and switches, which had been located adjacent to the equipment affected. At first, the resulting console was relatively small and was placed at a convenient location in the engine room, generally near the main engine. But, as the size of the console grew to include more devices, the console and usually the main electrical switchboard were housed in a separate compartment. That compartment has been designated by various names, most commonly the "Enclosed Operating Station" (EOS).

Figure 17 is a photograph of the propulsion plant control console in an Enclosed Operating Station typical of

Fig. 18 Microprocessor-controlled propulsion plant console

The failed controller can be removed and replaced under power. When a controller is replaced, the primary controller downloads its configuration (programming) to the new controller, and the new controller becomes the secondary controller on "hot standby."

In addition to the direct monitoring and control of engine room systems, a microprocessor also provides a trending capability. The data collected and stored with this capability can be averages, minimums, maximums, or samples. Trending can give advance warnings of impending component failure or other system problems. For automated ships that have small crews with limited repair capabilities, the replacement of a suspect component at a convenient time is much preferable to an unplanned outage.

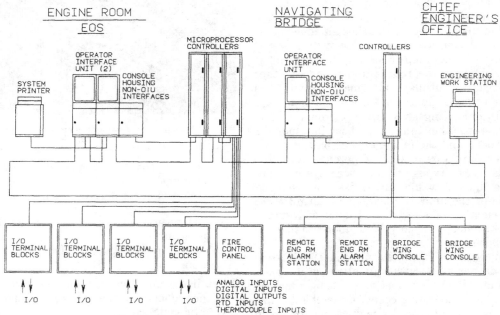

Fig. 19 Typical propulsion plant control system schematic

the early designs of this type. The centralization of controls onto a single console caused the length of some consoles to exceed 25 ft. However, on ships of more recent design, the control devices shown in Fig. 17 are generally replaced by a microprocessor-controlled CRT display and keyboard, which results in a more compact console, such as illustrated by Fig. 18.

The fundamental principles of shipboard automation are based on computer technology, and this technology enables engineers to design truly integrated and automated engine room control systems using off-the-shelf devices. Figure 19 is a schematic of a typical system layout. The microprocessor controllers are the principal devices used to control various processes. Each controller is able to receive analog and digital inputs from the field, perform processing and alarm checking, perform control algorithms, and output to motor controllers, valves, and other actuators. Each controller has a one-for-one redundant controller, and when either controller fails, the other controller automatically becomes the primary controller.

The "operator interface unit" is composed of the CRT display, disk drives, and keyboard. Typically two such units are located in the engine room's Enclosed Operating Station and another on the navigating bridge. Full control is possible from either location, and limited control functions can be provided from the bridge wings, if desired. The operator is able to monitor inputs into the system, interact with alarms, initiate or terminate control processes, trend data, and log reports at any operator interface unit.

The engineering work station, which is located in the chief engineer's office, provides restricted access to change control modes, set points, and parameters, and to reconfigure the graphic displays.

The microprocessor allows the user to custom design graphic displays of the various shipboard systems. These displays are dynamic; that is, the values, symbols, and colors change to represent the true state of the system process. There are two operator interface units in the engine room to provide full redundancy at the primary control location. To transfer control from the Enclosed Operating Station to the bridge, the EOS must first enable transfer and then the bridge may take control. The EOS can retake control at any time.

Steam propulsion plant. Control systems that are automated in steam-turbine-driven ships include combustion control, feedwater regulation, steam temperature, power-actuated sootblowers, and safety shutdowns and alarms. Specific guidance for the automation and control of steam propulsion plants is given in reference 30.

The boiler feed pump is a vital auxiliary and in most cases is located remote from the control station. Plants equipped for a one-man watch may be provided with means for remotely starting the pumps. Most plants have steam-turbine-driven feed pumps and their starting requires a control system that provides the sequential steps necessary. Such systems are usually referred to as logic systems since they contain the programming required to initiate each step in proper order and to check if each step is accomplished satisfactorily. Figure 20 is a logic diagram showing the steps that might be required to remotely start a boiler feed pump; obviously the specific steps in such a system vary, depending upon the particular feed pump and driver installed.

In the example provided, four inputs to the system indicate that the power and piping are set up in a ready state. With the "permissives" in proper array, the start switch may be manually actuated to initiate the starting cycle, which begins with the running of an auxiliary lube-oil pump. The establishment of normal lube-oil pressure permits the sequence to continue and energize a gland-seal control valve motor operator. If a proper lube-oil pressure is not obtained within a specified time (time delay unit) the procedure stops, an alarm is sounded, and the cause of the failure is usually indicated.

The described system is a relatively simple one; however, if the desire for automatic operation is extended, the system required can become highly complex. In the example given, the system was kept reasonably simple by requiring manual setup for certain valves and power supplies, and manually initiating the starting sequence. Where an unattended watch is desired, the automatic start-up of a standby unit in the event of a failure of an operating unit would also be necessary. Such a feature would have to include the ability to designate which of the installed units will operate and which will serve in a standby status.

For an unattended boiler, additional protective features are also required to monitor the critical conditions. To substitute for the fireman's eye, each burner is equipped with a flame detector. The detectors are sensitive to the ultraviolet emission from the flame, and automatically shut off the fuel and trigger an alarm in event of a flameout.

A more detailed description of boilers and their operation, including automation features, is included in Chapter 5.

The main propulsion turbine and gear unit on an automated vessel has the following features not provided on a nonautomated vessel:

1. Remote operation of the ahead and astern throttle valves.
2. Automatic operation of astern guardian and bleeder valves.
3. Extensive monitoring systems.
4. Remote emergency shutdown.

The throttle remote-operating system is the major addition. Usually there are two remote-control stations, one at the control center and one in the wheelhouse.

The throttle control system may be arranged to also accomplish allied functions, such as:

1. Automatic closing of extraction and opening of astern guardian valves upon reduction in power.
2. Automatic intermittent rotation of the turbine when under a stopped-shaft condition.
3. Throttle closure or modulation upon loss of boiler fires, low steam pressure, overspeed, etc.

Depending upon the plant design and the desired manning level, auxiliary automatic controls may be provided for start-up of standby condensate and lubricating-oil pumps, gland-seal pressure control, condensate recirculation for air-ejector condenser, lubricating-oil temperature control, etc.

The control and monitoring provided for other auxiliary systems depend upon the manning level, type of machinery, and machinery arrangement. The extent to which automatic or remote operation is applied is limited only by economics. Some of the factors to be considered when selecting the desired level of automation for the distilling plant, for example, are outlined in Chapter 17.

Diesel propulsion plant. The basic concepts described for a steam propulsion plant are equally applicable to a diesel installation; however, the specific instrumentation is considerably different. As in a steam plant, there are many varied combinations of engines, drives, and associated auxiliaries, and each entails requirements for control and information systems.

A diesel engine may be considered a self-contained power generator, with the only requirement being a supply of fuel, air, and cooling medium. A diesel plant is, therefore, less complicated than a steam plant from the standpoint of controls. Consequently, it is not surprising that diesel-propelled ships operated with two watchstanding personnel (one less than steam) before the general acceptance of the term "automation." Thus in the case of diesel propulsion plants, "automation" relates to two levels of watchstanding personnel: one-man and unattended.

The throttle control is the prime element in a diesel control system. Pilothouse control has been commonplace for some time for smaller engines. Such arrangements were simply remotely controlled means to manipulate the usual fuel-setting and reversing levers on the engine. Automated systems, however, are more sophisticated and operate through sequencing devices and interlocks for maximum engine protection while permitting bridge personnel to maneuver the vessel without concern for engine operation.

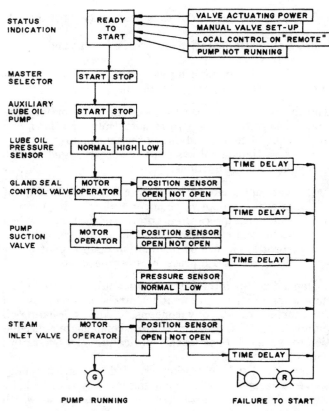

STATUS INDICATION — READY TO START — VALVE ACTUATING POWER / MANUAL VALVE SET-UP / LOCAL CONTROL ON "REMOTE" / PUMP NOT RUNNING

MASTER SELECTOR — START | STOP

AUXILIARY LUBE OIL PUMP — START | STOP

LUBE OIL PRESSURE SENSOR — NORMAL | HIGH | LOW

GLAND SEAL CONTROL VALVE — MOTOR OPERATOR — POSITION SENSOR OPEN | NOT OPEN — TIME DELAY

PUMP SUCTION VALVE — MOTOR OPERATOR — POSITION SENSOR OPEN | NOT OPEN — TIME DELAY

PRESSURE SENSOR NORMAL | LOW — TIME DELAY

STEAM INLET VALVE — MOTOR OPERATOR — POSITION SENSOR OPEN | NOT OPEN — TIME DELAY

G — PUMP RUNNING   R — FAILURE TO START

Fig. 20   Logic diagram for boiler feed pump remote start-up

For example, a directly reversible engine control system may accomplish the following sequential steps when the control mechanism is given an astern command from an ahead position:

1. Move the fuel control to stop, shutting off fuel and removing a blocking device from the reversing mechanism.
2. Move the reversing mechanism to astern, blocking the fuel control from the fuel position and positioning the starting air distributor and camshaft brake pilot.
3. Move the fuel control to air start, applying starting air, verifying rotation and direction of rotation, and removing the fuel block.
4. Move the fuel control to fuel position, setting the governor, and then controlling acceleration in accordance with a torque limitation program.

Automatic controls are provided for various auxiliary systems, including fuel oil, lubricating oil, steam generating plant (waste-heat and oil-fired boilers), and the electric generating plant, as required to support the desired manning level.

Specific guidance for the automation of diesel-propelled ships is given in reference 31.

Gas turbine propulsion plant. A gas turbine, like a diesel engine, is a self-contained prime mover. Gas turbines require close-tolerance automatic controls to function properly; these control systems are an integral part of the engine and are normally provided as part of the

engine package. Adaptation of these controls to suit marine automation depends on the propulsion arrangement selected to transmit the power to the propeller. However, as with a diesel installation, automation principally involves the provision of remote (engine room and bridge) control systems to transmit the command input to the engine control system.

Starting cycles are automatically sequenced. Remote start, therefore, involves the selection of the location of the remote controls and the information systems required to monitor the engine performance. This selection may depend to some extent on the degree of automation provided for the auxiliary systems, such as the fuel handling, lubrication, cooling, and compressed-air systems. The automation of gas turbine propulsion plants is further discussed in reference 32.

b. Navigation. The term "automation" is seldom used when discussing navigation systems but, in fact, advances in electronic aids to navigation have taken the place of human observation, effort, and decision making in most if not all aspects of ship navigation.

Radar extends man's vision not only in range but into darkness and fog; echo depth sounders provide the depth of water under the keel, and sonar maps the contour of the bottom. Satellite navigators and other systems such as Loran and Decca™ give the ship position and render the sextant and chronometer obsolete. The gyro-pilot keeps the vessel on the correct heading without the helmsman's intervention. Collision-avoidance radar systems act as "lookouts" able to automatically acquire and constantly monitor a number of targets, plot their speeds and courses, present these as vectors on the display screen, and calculate their closest points of approach to own ship and the time before that will occur. An audible alarm sounds when any target enters a "guard ring" preestablished at a selectable radius surrounding the vessel. Such systems are often referred to by the acronym "ARPA," which stands for Automatic Radar Plotting Aid.

On an integrated bridge all electronic inputs—from radar, gyro-compass, speed-log, depth sounder, satellite navigator, Loran-C, wind-speed/direction, outside temperature, and barometric reading—are combined in a common control/display. Figure 21 is a schematic representation of a typical integrated bridge system. All of the navigation sensors are connected to a common database, and the status of the engine room is conveniently displayed. Figure 22 is a photograph of a typical integrated bridge arrangement. This system permits voyages to be preplanned from berth to berth and, barring any unforeseen circumstance, the vessel will follow the predetermined course from way point to way point, making course changes without human intervention. In the event of a potential collision, the operator is alerted and an alternative course to a new way point is computed.

c. Vessel management. The same computer used in the navigation or engine-room automation packages can also be used, with appropriate software, to enhance vessel management. A typical application is a spare-parts inventory system, which can record and control spare-parts

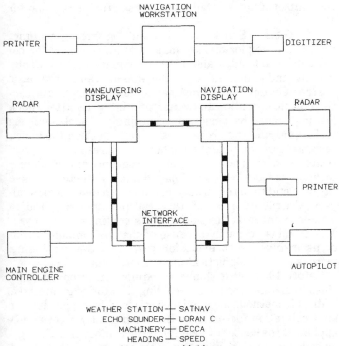

Fig. 21 Integrated bridge system

Fig. 22 Typical integrated bridge arrangement

availability not only on an own-ship basis, but on a fleet-wide basis as well. Duplicate databases often are maintained in the shore office and on the ship to provide an accurate control of supplies. Another example is a planned maintenance program, which is also designed to control maintenance activity on a fleetwide basis. With a reduced crew, maintenance is often accomplished while in port by shore-based personnel or by trained riding crews that are brought aboard and ride the ship as necessary to carry out planned maintenance. The coordination of the availability of the riding crews, the ship's schedule, the required parts, etc. is accomplished using the onboard computer. Specific guidance concerning the planning and scheduling of maintenance operations, procedures for the maintenance and testing of machinery components and

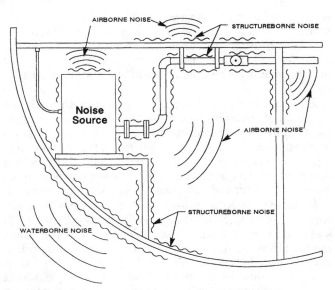

Fig. 23 Typical modes of noise transmission

control systems, and other maintenance-related activities for propulsion plants can be found in reference 33.

d. Cargo control. Just as with engine-room automation systems, cargo automation systems are microprocessor based. On tank vessels mimic displays of pumps and valves are enhanced with real-time status information of that equipment, which appears on the same screen with tank ullage data.

Tank-level information can be automatically transmitted to a loading computer that calculates the ship trim and stability and hull stresses during the cargo loading and unloading operations. On tankers and on other types of cargo ships, such as containerships and RO/ROs, loading sequence data can be entered manually, to explore any potential adverse effects on a planned loading scheme, before actually taking on cargo.

4.6 Noise. The control of noise is an important consideration during the design of most types of ships. Noise-control initiatives must often begin during the conceptual design stage and continue through the detail design process. Naval combatant ships, particularly submarines, require the most rigorous noise control design criteria; however, the control of noise must be considered to some degree in all ships, including small craft. Design decisions that range from ship arrangements to detailed material selections can have an effect on a ship's noise characteristics.

Noise is conventionally categorized by the way it is transmitted from a source to a receiver, as illustrated by Fig. 23. In the most common form, noise is transmitted through the air and is called airborne noise. Where voice communications are essential, airborne noise is undesirable, and provisions must be made to limit such noise to a tolerable level. On large ships as well as small craft, audible navigation signals such as bells, whistles, and foghorns must be heard above the normal shipboard airborne noise levels. In machinery spaces and other spaces that characteristically have high noise levels, airborne noise must be limited to an acceptable level to avoid watchstander hearing damage; otherwise, the watchstanders must use hearing protection. On cruise ships, the control of undesired airborne noise in passenger quarters and lounge areas is a strict requirement to maintain passenger comfort.

A second mode of noise transmission is through the vibration of structure, and noise transmitted in this manner is known as structureborne noise. This type of noise is transmitted through bulkheads, decks, foundations, piping, or machinery components. Structureborne noise that occurs at low frequencies may not be heard by personnel; however, this type of noise can affect passenger comfort.

Noise in a third mode—through the water—is called waterborne noise. Since this type of noise is radiated from a ship, it is also known as radiated noise. Radiated noise results primarily from structureborne noise that is transmitted to the hull, but it can also be generated from flow over discontinuities on the hull or from other external sources such as propeller cavitation. Radiated noise is a significant concern for combatant ships, particularly submarine and antisubmarine warfare, because it can reveal a ship's location as well as impair the ability of the ship's own sonar system to detect noise from other sources.

The level of attention given to noise control during the design stage depends upon the type of ship under consideration. A ship that is intended to transport cargo, for instance, would be expected to incorporate the degree of airborne noise control necessary to prevent hearing impairment of the crew, to provide an acceptable environment for crew rest, and to ensure effective speech communications where required for the safe operation of the vessel. Consideration of structureborne noise would be directed only at vibrations that affect the function of equipment and machinery or the health of personnel or that which might cause structural fatigue. Waterborne noise would likely not be considered. A cruise liner would be expected to require provisions for a pleasant environment in all passenger spaces, and would, therefore, require considerable attention to airborne and structureborne noise design details. The noise control requirements associated with submarines go well beyond those necessary for any other type of ship. In their case, state-of-the-art equipment and design techniques are necessary to meet the rigorous standards for airborne, structureborne, and waterborne noise.

When designing a ship to meet particular noise requirements, a number of options are typically considered. These range from the selection of individual components to structural design and space arrangements. For instance, when considering the trade-offs associated with particularly noisy equipment, it may be advantageous to arrange spaces that require low noise levels away from spaces containing the noisy equipment. Alternatively, applications of noise treatments such as vibration isolation or the use of acoustic enclosures may be considered. Finally, the procurement of specially designed quiet components may be possible. The various considerations that

the control of noise entail are discussed further in Chapter 13.

4.7 Shock. Shock loadings resulting from noncontact underwater explosions are major design concerns for naval combatant ships and affect the space requirements, weight, and arrangement of the ship as well as the basic design of equipment. Components that are critical for continued ship operation and watertight integrity after shock loadings must be designed accordingly. Shock analysis techniques and the procedures used to confirm the shock resistance of equipment have become increasingly sophisticated and accurate. Computer programs and modeling techniques have been developed that continue to decrease the cost and time required to conduct dynamic shock analyses. On the other hand, the cost and time required to conduct noncontact explosion tests continue to increase. This trend is primarily because of the environmental concerns involved with explosion tests and the continuing increase in the sophistication of such tests.

Before 1960, most shipboard equipment intended to be shock resistant was designed using the static "g" method. With this method, the applied static load is the product of an acceleration factor times the component weight and is applied at the center of gravity of the component. The "g" level depends on the direction of the shock loading relative to the ship and the component's weight, but not on the component's natural frequency. No consideration is given to the dynamic interactions between equipment elements. When used, the design "g" levels are stated in the ship specifications.

The inadequacies of the static "g" method led to the development of the Dynamic Design Analysis Method (DDAM) [34,35,36,37]. DDAM was formulated to represent more closely the dynamic interaction of the equipment and the ship structure. DDAM is basically a simplified modal analysis method with shock inputs empirically derived from underwater explosion tests. It is assumed that the equipment and its foundation together make up a system that responds as a linear elastic structure to the known motion of a "fixed base," with the "fixed base" being the deck, hull frame, or shell of the ship. The motion of the ship's deck, frame, or shell can be characterized in terms of a design shock spectrum. This spectrum can distinguish between shock direction (fore and aft, athwartship, or vertical), ship type (surface or submarine), and location within the ship (deck, hull frame, shell). Thus, the steps followed when conducting a DDAM are to (1) consider the equipment as an elastic system, (2) determine the normal modes and natural frequencies of this system, (3) compute the response of each normal mode to the design spectrum, and (4) superpose the resultant responses.

The DDAM procedure takes into account the shock loads applied to shipboard equipment by the motion of the ship structure itself. The design spectrum was developed from experimentally measured ship motions resulting from underwater explosions, but the spectrum is applicable to interior equipment only. Exterior, wetted equipment that is outside the hull is subjected to a different environment. Sonar transducers, rudders, control planes, auxiliary propulsion systems, and hull penetrations are typical equipment that must also be designed to withstand the direct pressure loading of the shock wave from the underwater explosion.

Underwater shock analyses of externally wetted components can be conducted using a method that is based on one-dimensional fluid/structure theory of the motion of an air-backed or water-backed plate loaded by an exponential plane wave. From the component response the average acceleration can be determined and applied to the component's center of gravity. This approach assumes that the mounting of the component is very stiff and results in high accelerations. This method is conservative, but it may be used to design hull penetrations because the consequences of a leak in a ship hull are considered to justify a conservative approach. For wetted components that are flexible, however, the method is excessively conservative and indicates significantly higher loads than those experimentally measured. As a consequence, a static "g" approach is generally used for the design of flexible external components, with the "g" levels determined from experience.

The need for rigorous procedures that can be used to analyze and design wetted components has been recognized, and research sponsored by the Defense Nuclear Agency has resulted in the development of several computer codes for underwater shock analysis. One of the most useful of these computer codes is the Underwater Shock Analysis (USA) Code [38]. This code is based on the Doubly Asymptotic Approximation [39], which is efficient and cost-effective in the solution of the shock problem. The USA code can be used to model a shock wave that has any arbitrary pressure-time history, and can be used to determine the response of a total ship or a component of a ship. The component can be completely or partly submerged, air-backed (such as the hull of a surface ship or the pressure hull of a submarine), or water-backed (e.g., the non-pressure hull of a submarine).

During the conceptual design stage, when the effects of shock loadings must be acknowledged and preliminary shock design criteria developed, the information available is insufficient to permit detailed shock analyses to be made; for this reason, the sizes of equipments and their foundations are generally approximated based on previous experience and engineering judgment.

At the beginning of contract design, a Doubly Asymptotic Approximation analysis of the entire ship may be performed to determine the shock spectra for input to DDAM for external equipment. This analysis may also be used to determine the shock spectra for internal equipment, or a predetermined spectrum may be used. Throughout the contract design stage, static "g" calculations are used to determine the approximate size and weight of equipment. Only during detail design, when specific information becomes available, can the majority of the shock analyses be performed. DDAM is used to design the foundations of both internal and external equipment. Hull penetrations are designed using exponential plane wave theory, because the "blast wave" pressure is the dominant design criterion.

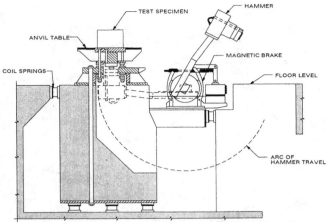

Fig. 24 High-impact shock machine for medium-weight equipment

Several methods of qualification are used to demonstrate that shipboard equipment has been designed to incorporate the desired level of shock resistance. Ordinarily, equipment may be qualified either by dynamic shock analysis or by shock testing; however, in some cases a component can be shock qualified by extending the applicability of the qualification of a similar item that was previously shock tested.

Several kinds of shock tests are performed. Military specification MIL-S-901 [40] describes detailed shock testing requirements for all equipment on naval surface ships and for equipment installed inside the pressure hull of submarines. Equipment subjected to the tests of MIL-S-901 are categorized by weight as lightweight, medium-weight, or heavyweight items. Lightweight and medium-weight items are tested on shock testing machines; a medium-weight shock test machine is shown in Fig. 24. Heavyweight items are subjected to noncontact underwater explosion tests while installed on the deck of a floating barge. The test arrangement of a floating shock platform is indicated by Fig. 25.

Critical components that penetrate the pressure hull of submarines are subjected to higher levels of underwater-explosion shock-qualification tests. These hull-integrity items are tested on submerged test vehicles that are representative of the ships on which the components will be installed. The detailed test requirements for hull-integrity components are defined in the specifications for particular ships.

Equipment installed external to the pressure hull of submarines may also become shock qualified by passing underwater explosion tests while installed on the underside of a floating shock platform or while installed outboard of submerged test vehicles.

4.8 Pollution Control. Marine engineers are responsible for the design of many ship systems that have the potential of causing a degrading effect upon the environment; therefore, for such systems the control of harmful pollutants, which could be discharged either into the air or water from marine vehicles, is a principal design consideration. Most notable among the potential sources of pollution are the cargo-oil piping and ballast systems on

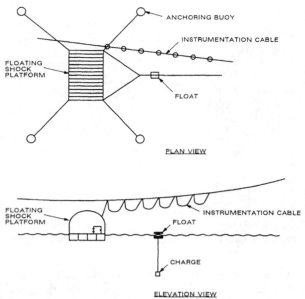

PLAN VIEW

ELEVATION VIEW

Fig. 25 Floating shock platform

crude-oil tankers, but other potential sources such as bilge systems, sewage systems, and the stack emissions from engines and boilers also require a rigorous analysis from this perspective.

Local governments commonly impose highly restrictive limits on the permissible discharges from ships. These regulations generally preclude discharges into the water and require that special clean fuels be burned while in the restricted areas to avoid affecting the air quality. The absolute prohibition of discharges into the water requires that the substances be either retained aboard ship or transmitted to shore facilities.

International organizations have taken an active interest in ship discharges that contaminate the ocean environment, particularly the discharge of oil. The International Convention for the Prevention of Pollution of the Sea by Oil, 1954, was one of the most significant early efforts. However, this effort was superseded by the International Convention for the Prevention of Pollution from Ships, as modified by the Protocol of 1978, which is known as the MARPOL 73/78 protocol, and went into force in 1983. The MARPOL 1973/1978 protocol prohibits the discharge of various pollutants, from marine vehicles, except under specified conditions. In the case of oil tankers, the discharge of oil or oily mixtures is prohibited except when all of the following conditions are satisfied [41,42]:

(i)   The tanker is not within a special area. "Special areas" are specifically designated, somewhat contained bodies of water such as regions of the Mediterranean, Baltic, Black, Red and Caribbean Seas. While in these areas, oil and oily mixtures must be retained on board.

(ii) The tanker is more than 50 nautical miles from the nearest land.

(iii) The tanker is proceeding en route.

(iv) The instantaneous rate of discharge of oil content does not exceed 60 litres per nautical mile.

(v)   The total quantity of oil discharged to the sea from new tankers does not exceed 1/30,000 of the total quantity of the particular cargo of which the residue formed a part. This requirement prohibits the discharge of a large total quantity of oil, even when dispersed over a large area.

(vi) The tanker has in operation an oil discharge monitoring and control system and a slop-tank arrangement. The monitoring and control system is used to measure and continuously record the oil content and rate of discharge. If the oil content exceeds the permissible level, the discharge must cease.

To limit the amount of oily-water mixtures that are created aboard ship, the following are among the restrictions placed upon the design of oil tankers.

• Segregated ballast tanks, that is, tanks which are designed to carry ballast water only, are required in appropriate size and location to require no ballast water to be carried in cargo tanks except possibly under extreme weather conditions.

• Slop tanks must be provided for the retention of oily-water mixtures.

• Crude-oil washing provisions must be provided. The amount of sludge that accumulates on the structure and bottom of the crude-oil tanks can be reduced by dislodging the sludge residue with a spray of the cargo oil itself. By using cargo oil, instead of seawater, as the spray medium, a higher percentage of the cargo oil is delivered. Also, a smaller amount of oily-water mixture and sludge must be handled aboard ship.

MARPOL 73/78 prescribes requirements, which are similar to those for oil tankers, that are applicable to tankers that carry noxious liquid substances in bulk.

For all ships, the overboard discharge of liquids from machinery-space bilges is prohibited except when all of the following conditions are satisfied:

(i)   The ship is not within a special area.

(ii) The ship is more than 12 nautical miles from the nearest land.

(iii) The ship is proceeding en route.

(iv) The oil content of the effluent is less than 100 parts per million (PPM).

(v)   The ship has in operation an oil-discharge monitoring and control system, oily-water separating equipment, oil filtering system, or other installation.

Separators and filters are used to process oil-contaminated water to reduce the oil content. Separators are commonly of the centrifugal type and, to be effective, rely upon the difference in density between oil and water. Separators are usually designed to provide an effluent with a maximum oil content of 100 ppm. Filters, which may be operated in concert with separators, are usually designed to produce an effluent with a maximum oil content of 15 ppm.

The MARPOL 73/78 protocol prohibits the discharge of sewage into the sea except for comminuted and disinfected sewage, which must be discharged at a distance more than 4 nautical miles from the nearest shore; sewage that is not comminuted and disinfected must be discharged at a distance greater than 12 nautical miles from shore. In either case, sewage discharges must be made at a moderate rate, while the ship is en route at a speed of at least 4 knots, and the discharges must leave no visible evidence in the surrounding waters. In addition to sewage handling and treatment provisions, this regulation, in effect, requires that a sewage collection and holding tank be provided to hold sewage while the ship is near shore or in other regions where discharges are not made; the holding tank is subsequently discharged either to shore facilities while in port, or at sea in an appropriate manner.

Among the most significant pollutants discharged to the atmosphere are the following:

- Chlorofluorocarbons, such as used in some air-conditioning and refrigeration systems, which tend to deplete ozone when released to the atmosphere.
- Volatile organic compounds that are released from crude oil.
- Products of combustion resulting from the incineration of garbage and other shipboard wastes.
- Sulfur oxides (SO_2) and nitrogen oxides (NO_x), which are in the exhaust emissions from engines and boilers.

The treatment of exhaust gases to remove oxides of sulfur and nitrogen after they have formed requires the use of equipment that is both bulky and expensive. In many cases the preferred means of reducing the sulfur oxides in the exhaust gases, for example, is to use low-sulfur fuel, even though it is generally more expensive.

Details concerning the measures taken to control the discharge of pollutants from naval ships are contained in reference 43.

4.9 Design for Production. Because of pressures to reduce ship construction costs without adversely affecting product quality, there is an increased emphasis on the adoption of design practices that facilitate ship production procedures. Both the technical content and the manner in which design data are presented can affect production costs. For example, the use of standard construction elements and components clearly simplifies construction procedures. Also, the use of computer-aided manufacturing applications that are effectively linked to computer-aided design systems can be used to reduce manufacturing costs.

As indicated by Fig. 26, when a central computer-based product model of a ship is developed, the geometry of the ship elements and the material identifications are established during the detail design stage. This geometry can be accessed and obtained in a form that is directly usable to drive manufacturing tools, such as fabricating equipment for structural elements, machine tools, and pipe-bending machines, either directly from the computer database or with numerically controlled tapes. By directly using design data in this way, the expense of reformatting

design data is avoided, opportunities for the introduction of errors are minimized, and there is increased assurance that the data are accurate.

In addition to the means of presenting the design data, the selection of the design details incorporated in the configuration of ship systems also can significantly affect construction costs. For example, by becoming familiar with the ship-production strategy during the design stage, a design engineer can often locate the required piping joints (preferably of the mechanical type, where permitted) to align with the location of planned construction breaks. This practice is an effective means of avoiding the expense of unnecessary piping joints while enhancing the efficiency of the overall construction plan.

Marine engineers must have an understanding of the ship production strategy to facilitate the development of design details that are consistent with the construction plan. Key elements of a production strategy are:

a. Modular construction. To develop a ship design that can be constructed at minimum cost, many aspects of the construction methods to be used must be reflected in the design details [44]. As soon as sufficient design information is available to lay out the arrangement of a ship, a basic erection plan is developed that defines the largest sections of the ship, or building blocks, that will be used during construction. The erection plan establishes the major construction breaks, or joints, between the ship sections. The major construction breaks should not cross large components, or complex piping assemblies, because their installation would otherwise be delayed until the ship sections are joined. For this reason the ship designers and the production planners must resolve the locations of construction breaks early in the design process.

With the major construction breaks established, a layout can be made of the larger components and piping runs. Inputs of production planners are again required to ensure that the layouts permit the equipment to be constructed as modules. Modules are major groupings of all elements, including the equipment, distributed systems, and often some structure, in a portion of the ship. The engine room module illustrated in Fig. 27, which consists of the product breakdown structure indicated in Fig. 28, is typical of those used during construction. By designing the ship so that it can be assembled using such modules, more of the work can be performed in a shop, or other appropriately equipped and supported staging area, which requires less work to be done aboard ship. This is an effective means of reducing shipbuilding costs, since outfitting work performed under such conditions can require as little as one-fourth the man-hours required to perform the same work aboard ship.

b. Computer model. When establishing a production plan for a complex ship, a three-dimensional computer product model of the ship can conveniently be used as the means of communication between ship designers, production planners, and others, with a flow of technical data as indicated by Fig. 26. As the ship design is formulated, the computer model is begun by laying in the hull structure and the arrangement of the major spaces. Thereafter, increasingly detailed design data are added to the central

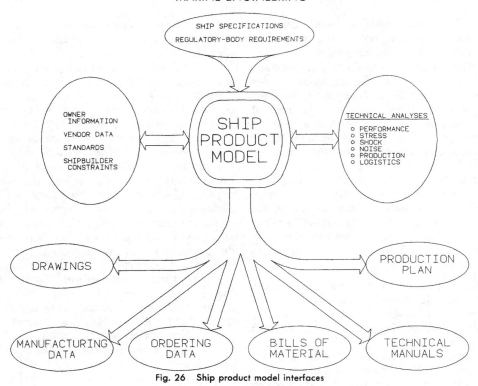

Fig. 26   Ship product model interfaces

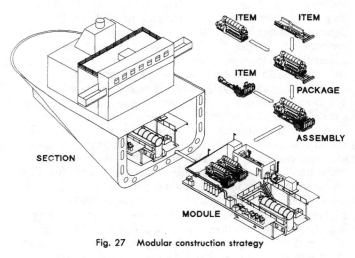

Fig. 27   Modular construction strategy

computer model, beginning with the larger components, which are the more difficult to arrange, and those that have restrictions on their shipboard position. A comprehensive discussion of the computer system architecture appropriate for naval ships is presented in reference 45.

During the course of developing a ship arrangement, marine engineers translate piping diagrams, and other design data that concern functional characteristics, into detailed construction drawings. Since piping diagrams define system features, but not shipboard arrangements, marine engineers must ensure that the construction drawings not only reflect the characteristics indicated on the diagrams, but also present an arrangement that can be

constructed economically and operated and maintained effectively.

Continuing communications are required between the ship designers and construction planners as the computer model of the ship evolves. Of course, continuing communications are also required between the hull, mechanical, electrical, and piping designers to ensure that their designs reflect reasonable compromises. That is, the routing of the major systems such as piping, ventilation, and wireway systems must be given priority over the routing of smaller secondary runs. This prioritization may require that smaller piping runs be rerouted to avoid the necessity for bends in larger piping. The continual rerouting of piping, ventilation, and wireway systems, which is extremely disruptive to the detail design process, can be minimized by laying in the larger elements first.

To maintain the integrity, and thus the validity, of the computer model, rigorous procedures must be established and enforced to control the means of altering the data in the model as the design evolves. Normally, only the person with the primary responsibility for each system's arrangement is authorized to modify the computer model. However, to facilitate communications, read-only privileges may be provided to those who have secondary responsibilities concerning the design. This practice enables the ship computer model to be accessed and evaluated by all interested parties at their convenience so that suggested changes can be accommodated with minimal disruption. The read-only privileges also permit the data in the three-dimensional computer model to be down-loaded to applications computer programs to conduct pipe-stress

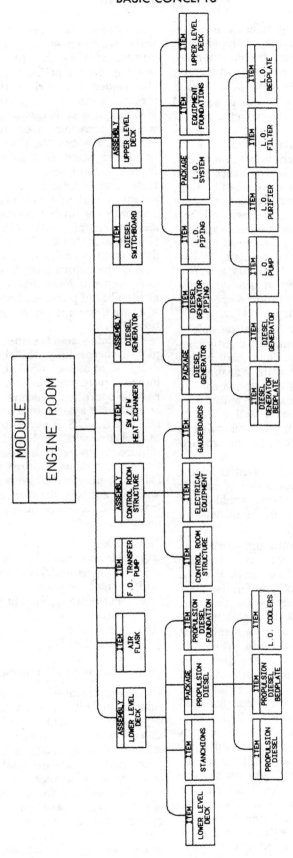

Fig. 28   Product breakdown structure of engine-room module

analyses, perform pipe pressure-drop calculations, prepare pipe-bending instructions, conduct finite-element analyses, perform interference checks, prepare numerically controlled cutting tapes, ensure material availability, prepare working drawings, and serve other purposes.

c. Standardization. The standardization of material selections is of great importance from the perspectives of material procurement, ship construction, ship maintenance, and even the design process itself. While most of the advantages of standardized materials in the material procurement, ship construction, and maintenance operations are obvious, some may not be. In the case of bolting, for example, rather than specifying the use of an array of steel bolts having the same diameter but with slightly different yield strengths and temperature limits, it may be more economical to standardize on a reduced number of alternatives having only the higher-strength, higher-temperature properties. The increased unit material cost can be offset by the reduced variety of stock requirements, larger manufacturing lots, and reduced opportunities for material misapplications during installation.

The standardization of component foundation bolting patterns also requires special attention. Apparently identical components, such as ventilation fans, may not be interchangeable unless provisions are made to ensure that the pattern of the foundation bolting and bolted connections are within tolerance limits. Standardized bolting patterns can be ensured conveniently by including the required bolting pattern, which may be as recommended by the equipment manufacturer, in the equipment procurement specifications.

The repeated use of standard parts is essential for the efficient use of computers in the development of ship arrangements, because an excessive effort would otherwise be required to create unique computer models for numerous similar parts.

d. Scheduling. Production planners' inputs are also required to establish the format and content of the construction drawings as well as a schedule for their issue. When planning the construction of a ship, production planners develop a work breakdown structure for the ship. The work breakdown structure indicates the plan for constructing the ship, usually beginning with elements that are produced by a single construction trade, which are often known as Items. Commonly, Items are then combined to progressively form larger interim products called Packages, Assemblies, Modules, and Sections. Finally, the Sections are joined together to form the ship. (Some of the intermediate stages may be omitted in a logical construction plan; e.g., some Items are installed directly in the ship.) Figures 27 and 28 illustrate the principles involved in this process. To maintain control of such a complex task, a Critical-Path Methodology network is commonly used to relate the interdependencies of the various activities. The use of such a network ensures that each step in the work sequence is completed in time to support the next. The network is also used to establish the need dates for all drawings and materials, including those purchased and those manufactured by the shipbuilder [46].

e. Scope of working drawings. Individual work tasks, which collectively constitute the work breakdown structure for the ship, are used to define the scope of the working drawings. That is, to facilitate their use by the shipbuilder, the working drawings present complete instructions required to perform specific tasks. When constructing a module that is a complete region of a ship, the working drawings used to assemble the module would include the segments of any piping system (or other distributed system) that passes through that region. This means that the scope of the working drawings must be in agreement with the production plan and requires the establishment of a production plan before the working drawings are started. As discussed, a central three-dimensional computer model of the ship arrangement is a convenient means for production planners to break the ship down into the preferred work-package groupings. By conveying this information to the ship designers at an early date, any problems that ensue (perhaps concerning the availability of purchased equipment) can be resolved, and working drawings can be issued that support the production plan.

4.10 Integrated Logistics Support. For a ship to be deployed successfully, the ship systems must not only be designed to be technically capable of accomplishing their intended purposes, but they must also be designed so that reliable operation is assured. In addition, provisions must be made for system operation and maintenance. For the ships of the U.S. Navy, Integrated Logistics Support activities deal with the investigations and supporting data required to ensure that the ship systems are always ready to be operated as required to support the ship's mission [47]. The principal objective is to provide cost-effective logistics support throughout the life cycle of the ship. Integrated Logistics Support programs may include elements such as:

(a) Design Interface
(b) Maintenance Planning
(c) Support and Test Equipment
(d) Supply Support
(e) Packaging, Handling, Stowage, and Transportation
(f) Computer Resources Support
(g) Technical Data
(h) Facilities
(i) Manpower, Personnel, and Training

The major features of these elements, as prescribed for ships in the U.S. Navy, are briefly described in the following paragraphs. Programs having similar objectives are also necessary for commercial ships; however, since maintaining a defense readiness is not a requirement, the program elements are generally handled in a more limited, less expensive manner.

a. Design interface. Throughout the iterative steps of the design process, continuing communications are required between the personnel specifying and developing the ship's design characteristics and those working in the Integrated Logistics Support program. These communications are necessary to ensure that the ship is designed so it can be operated and supported in the most effective, timely, and economical manner. The major components of the Design Interface element include: Logistics Support Analysis, Reliability and Maintainability, System Safety, and Human Engineering (also known as Human Factors).

Logistic Support Analysis. Logistic Support Analysis includes the actions taken to define, analyze, and quantify logistic support requirements throughout the design process. The primary purpose of the Logistic Support Analysis effort is to ensure that support considerations are given appropriate emphasis during the design of the ship and ship systems. Problem areas are identified as the design evolves and may require trade-offs with other design features. As the design of the ship systems progresses and becomes firm, the emphasis is shifted to the identification of specific ship-support resource requirements. This information is used to plan for, acquire, and position the support resources (e.g., personnel and material) required to ensure that readiness requirements can be met. Further details concerning Logistic Support Analysis Activities are given in MIL-STD-1388 [48].

Reliability and maintainability. Many aspects of reliability and maintainability considerations consist of design analyses that result in recommended design modifications. Therefore, to be of value, these analyses must be conducted early. After a system design is complete, the implementation of late-developing design changes is frequently impracticable.

The analytical procedures used to assess the reliability and maintainability characteristics of ship systems are based on a mathematical model that incorporates equipment failure rates (which are expressed in terms of the mean-time-between-failures or MTBF), equipment repair times (mean-time-to-repair or MTTR), and system operating scenarios. The reliability assessment process is illustrated by Fig. 29. Availability is defined as the probability, in random time, that a system or equipment will be in an operating condition. The long-term value for availability is usually expressed as:

$$\text{Availability} = \frac{\text{Uptime}}{\text{Uptime plus Downtime}}$$
$$= \frac{\text{MTBF}}{\text{MTBF plus MTTR}}$$

where all time values are usually in hours. An estimate of availability may be more relevant than a reliability prediction, especially for systems that can be repaired quickly and for which an interruption of operation can be tolerated.

To be effective, reliability and maintainability considerations must be integrated into the ship design process. As the ship design is developed, design features can be incorporated that reduce the requirements for maintenance and logistic support, and equipment can be identified that degrades the overall system performance. In addition to prediction techniques, Failure Modes, Effects and Criticality Analyses can also be performed to identify potential failure modes and evaluate their degree of severity. Formalized procedures for conducting such analyses are prescribed in MIL-STD 1629 [49].

System safety. System safety analyses are systematic procedures that are conducted during the ship design process to identify and develop resolutions for potential safety hazards. The safety engineering effort is focused on the ship systems, such as weapon systems for a naval combatant, that are of critical importance from a safety perspective. Guidelines for the implementation of a system safety program are provided by MIL-STD 882 [50]. Formalized hazard analysis techniques can be used to evaluate the safety aspects of a system design, to ensure that equipment will operate safely under all envisioned operating conditions. A fault-tree analysis is one such technique, which involves the use of a logic diagram as shown in Fig. 30 to analyze the possible causes of a specific system failure or other undesired event.

Human engineering. Human engineering (or human factors) assessments require an evaluation of man-machine interfaces and the optimization of personnel capabilities and limitations with respect to system design and equipment operation. The objective is to ensure that man-machine interfaces have been designed in accordance with principles that allow systems and equipment to be effectively and safely operated and maintained. Human engineering covers a wide range of design and operational issues, including ship bridge arrangements, maintenance access, environmental issues, automation, manning considerations, anthropometry, human error reduction, and improved training and simulation.

Guidelines, such as those illustrated by Fig. 31, are used to design an operator's work station. While most man-machine design issues can be resolved by analyzing conventional drawings or three-dimensional computer models, some complex circumstances require a full-scale mock-up. Further details concerning anthropometric guidelines for marine applications are provided in ASTM F 1166 [51] and MIL-STD 1472 [52].

b. Maintenance planning. Maintenance Planning is the process used to develop a maintenance program for the life-cycle support of ship systems and equipment, and may include the elements outlined by Fig. 32. The purpose of a maintenance program is to increase operational availability, extend the time between major overhauls, and reduce operating and support costs. The elements of a maintenance program include the identification of maintenance-worthy equipment items, a consideration of the maintenance history of similar equipment, an analysis of equipment design and performance parameters, and an evaluation of maintenance requirements of systems and equipment. Organizational level (on-ship), intermediate level (shore or ship-tender based), and depot level (shipyard) maintenance requirements are identified. The resulting program provides planned maintenance requirements, which enable the ship operator to efficiently schedule repair and overhaul activities over the entire service life of the ship.

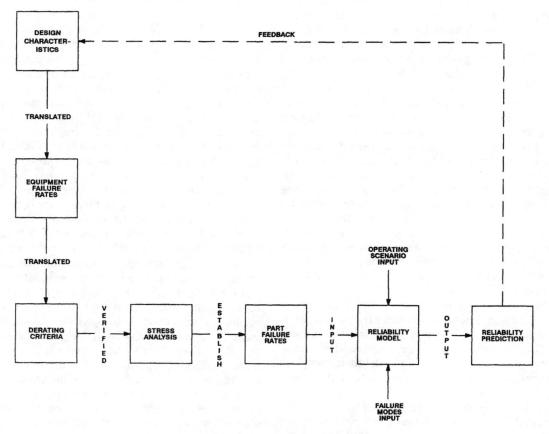

Fig. 29   Reliability assessment process

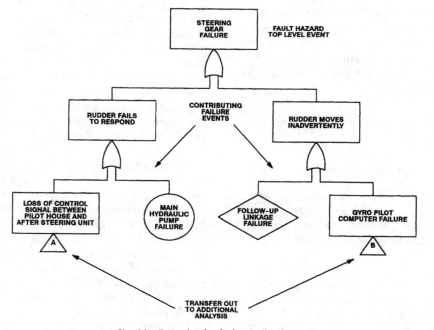

Fig. 30   Example of a fault-tree development

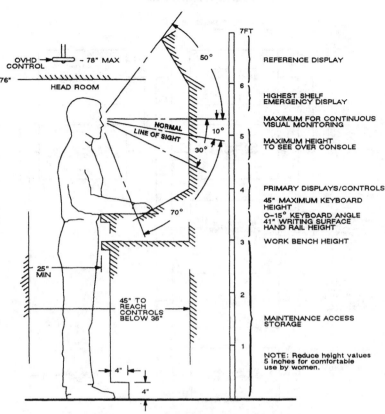

Fig. 31 Anthropometric guidelines

The provisions made for equipment access and removal are an integral part of Maintenance Planning. Maintenance access control is monitored during the ship design process to ensure that access and removal flow paths are provided, adequate hatches and access space for equipment removal and installation are available, and required maintenance access is incorporated in the design. Equipment removal is facilitated by component designs that permit rapid connection and disconnection, and by design provisions for the removal of obstructing equipment. Planned equipment access and removal significantly enhance shipboard maintainability and avoid the necessity for making hull cuts to replace equipment.

Maintainability is enhanced by the use of standard equipment and components, which also results in improved logistics support, reduction of on-board spare parts, ready interchangeability, rapid equipment change-out during overhauls, and reduced life-cycle costs.

c. Support and test equipment. The objective of this activity is to identify the test, measurement, and diagnostic equipment and programs required to properly maintain equipment and systems. The use of general-purpose, standard support equipment is preferred but is sometimes not feasible. Highly technical systems, such as weapon systems on naval combatants, often require unique support equipment. When required, the characteristics of support equipment must be identified during the design stage to ensure that suitable laboratory and storage space is allocated for this purpose.

d. Supply support. Supply support requirements are developed from the Logistic Support Analysis effort and include the identification of the supplies necessary for periodic maintenance and repair. Of particular concern is the need for support items that are not available as standard inventory, and which must be identified early for them to be acquired and stocked.

e. Packaging, handling, storage, and transportation. The purpose of this element is to provide technical instructions that will ensure the safe packaging, handling, storage, and transportation of equipment and repair parts. Of particular concern is the protection of equipment from deleterious storage environments. Early planning is essential for the allocation of space necessary for proper shipboard storage.

f. Computer resources support. The principal objective of this activity is to facilitate the standardization of the computer hardware and software used aboard ship. Standardization improves the maintainability of computer resources and reduces their support costs.

g. Technical data. The design, construction, and operation of ships require the development of an array of technical data. Ship construction drawings, system diagrams, operation manuals, maintenance and repair manuals are examples of such data. The technical data needed must be identified during the design stage to ensure that it is developed and acquired and to provide suitable storage space aboard ship. Automated on-board logistic data systems provide a means for the electronic storage of

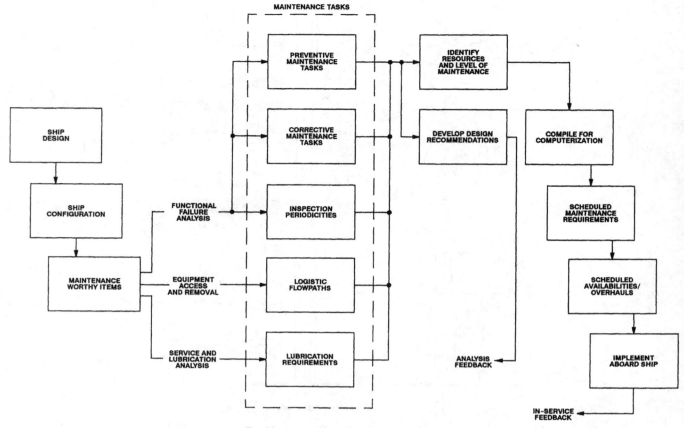

Fig. 32   Maintenance program development

technical data, thereby reducing the stowage volume required for hard-copy data.

h. Facilities. Insofar as practicable, existing facilities are used to support ships and their systems and components. However, in cases where the existing facilities are unsuitable, the requirements for new home-port facilities or modifications to those existing must be identified sufficiently early for them to be available when needed.

i. Manpower, personnel, and training. The two principal objectives of this activity are (1) to establish the crew size and skill specialities needed to effectively operate and maintain the ship, and (2) to conduct planning and crew training sufficiently in advance of ship delivery to ensure that these resources are available to support initial ship operation. These considerations are significantly influenced by the ship's complexity, operational requirements, degree of automation, and maintenance requirements.

Crew size is generally determined during the ship design process based on the ship's features, manning requirements for the engineering plant and deck operations, equipment maintenance workload and, for naval vessels, watchstation requirements for its mission. Social traditions can also have an influence, and, in the case of merchant ships, union agreements must be considered. Centralized control of shipboard machinery, automatic data recording, electronic performance monitoring, and shore-based maintenance support have allowed reduced manning in merchant ships. However, naval combatants require more manpower because of the at-sea battle-station manning condition, on-board weapon systems, and the greater complexity of naval ship systems. Manning requirements for naval combatant damage control and repair functions are not a consideration for merchant ships.

The necessary skill levels and training requirements for ship operating personnel are established by analyzing the shipboard systems and establishing the necessity for shore-based and on-board training for new equipment. Initial training for equipment operation and maintenance may be provided by the equipment manufacturer. Permanent training sites and course materials and training aids may also be established to ensure that qualified replacement personnel will be available to support continuing ship operations.

Section 5
Tests and Trials

The construction of a ship is concluded by a broad array of tests to demonstrate that the ship meets contract requirements. Some of the more simple tests are those of a quality-control nature, which are conducted to ensure the conformance of material properties to specified requirements, the soundness of castings, dimensional accuracy, and other such characteristics. Conducting quality-control tests is not peculiar to marine equipment. The more complicated program of tests and trials that is subsequently conducted does, however, reflect the unique requirements associated with ship construction. This program includes shop tests, installation tests, dock trials and, finally, sea trials.

Tests are preferably scheduled as early as feasible during the ship construction process, because early testing allows more time to evaluate and develop resolutions for design or material problems with a minimal disruption to ship construction. Shop tests for purchased equipment are advantageously conducted at the manufacturer's facility, where any corrections or adjustments can be expeditiously handled. A major advantage associated with the use of modular-construction methods in shipbuilding is the ability to conduct off-ship tests on construction modules that may include a number of components and their piping connections.

Shop tests are conducted for purposes such as confirming that assemblies are correctly built, verifying strength and tightness requirements, and demonstrating that controls and safety devices are functional and properly adjusted. Components can often be tested more economically under shop conditions, and the shop test environment is usually cleaner and less congested than that of a ship under construction. Components that have restrictive noise limits are commonly operated in a noise-test shop to confirm that they have acceptable noise characteristics before they are installed. After shop testing, components are given a protective covering that is not removed until installation aboard ship.

When warranted by technical complexity and risk, land-based test sites may be constructed to test propulsion plants or combat systems for new classes of naval ships. For example, one complete propulsion unit of equipment, including the main engines, reduction gear, associated couplings, clutches, and the associated auxiliaries and propulsion control system, may be tested with water brakes used to absorb the power produced. Extensive data are recorded to ensure that all system components are properly designed and integrated for the lead ship of the class.

Tests may also be conducted to demonstrate reliability and maintainability. Reliability testing is sometimes used for developmental items, which have no service history, to provide a means of identifying and resolving equipment reliability problems early in the development period. Prototype components are commonly subjected to reliability demonstration tests of extended duration in which service conditions are simulated. Maintainability demonstrations are sometimes conducted to confirm compliance with specified maintainability requirements and equipment accessibility, particularly for complex equipment.

Installation tests are conducted to confirm that the installation is in conformance with specified requirements, to verify system strength and tightness, to confirm the cleanliness of piping systems, to demonstrate adequate access for equipment operation, maintenance, and removal, and to confirm that components and systems function as specified. To the extent practicable, all components and systems are subjected to operational tests under conditions typical of those expected in service, to provide confidence that the design and construction are sound.

After the ship is launched and construction is essentially complete, and after the installation tests have been conducted, dock trials are scheduled to establish that the propulsion plant and its auxiliaries are ready for sea trials. During dock trials, the ship is secured to the pier with mooring lines while the main engines are used to drive the propeller. The propulsion system is loaded, until a limiting condition is reached. With the ship secured to the pier, the apparent slip of the propeller is 100 percent, instead of the 20 to 40 percent in service, with a resultant effect on the power-rpm relationship of the propeller. Because of the high-slip conditions, for a given rpm the propeller develops a substantially higher thrust and torque than when in normal service. Consequently, the allowable thrust or torque may limit the power that can be developed during dock trials. Mooring arrangements and environmental concerns (e.g., sediment disturbance) may also impose limitations during these tests.

Most shipboard components and systems can be subjected to tests in the shipyard that closely approximate service conditions, but there are exceptions. The propulsion plant, anchor windlass, and steering gear are notable examples. To the extent practicable, the propulsion plant is tested during dock trials. The anchor windlass is also tested with the ship moored, but there is not enough water depth at pierside to fully test the anchor windlass. Therefore, pierside tests are conducted with additional weights attached to the anchor to represent the weight of the anchor chain payed out during retrievals from greater depths. There is no practicable way of demonstrating that the steering gear will function properly without taking the ship to sea. The most significant unknown element of the steering gear performance is the magnitude of the torque required to move the rudder while the ship is underway at full power. Some assurance can be derived from model tests, but analytical procedures that have been correlated with full-scale test results are the primary means used in the design stage to ensure that a steering gear has an appropriate rating.

Sea trials are conducted to demonstrate the performance and adequacy of those aspects of a ship that cannot be realistically tested at pierside. The tests typically conducted during sea trials are as follows:

Speed-power standardization tests
Economy power tests
Full-power endurance tests
Ahead steering and maneuverability tests
Quick reversal astern and head reach
Astern endurance tests
Astern steering tests
Quick reversal ahead and stern reach
Anchor windlass tests
Distilling plant tests
Calibration of navigation equipment

Since sea trials are the first opportunity to observe operation of the propulsion plant and the ship's maneuvering characteristics, the sea trials are generally conducted in phases. The first phase, the Builder's Trials, is conducted to observe the ship's performance and to identify and develop resolutions for any deficiencies noted. The shipowner's representatives usually ride the ship during the Builder's Trials as observers. There may also be intermediate phases of sea trials, such as Noise Trials for naval ships that have stringent noise requirements.

Sea trials are concluded by an Acceptance Trial. The Acceptance Trial is largely a repeat of the tests conducted earlier during the Builder's Trial; therefore, few if any serious problems are likely to emerge during an Acceptance Trial. The purpose of Acceptance Trials is to demonstrate to the owner that the ship meets the contractual requirements. The Acceptance Trials for naval ships are conducted under the direction of an Inspection and Survey (INSURV) Board, which consists of experienced personnel from various Navy organizations. The INSURV Board has the responsibility to verify the proper correction of any deficiency observed during earlier trials and to confirm the acceptability of the ship. A similar approach is used for the conduct of Acceptance Trials for merchant ships.

Much of the responsibility for the planning and execution of ship tests and trials falls to marine engineers, who analyze the ship's specifications, invoked standards, technical manuals, purchase specifications, system diagrams, and design drawings and develop a test program that can be used to demonstrate compliance with all requirements in an effective manner. To do this, marine engineers often prepare test procedures describing the tests that will be conducted. Test procedures fully define the tests to be conducted, the purpose of the tests, preliminary work that must be accomplished before each test, safety precautions to be observed, instrumentation to be used, data to be recorded, and all other significant details of the tests. A test schedule, which coordinates the sequence of tests with the construction progress of the ship, is an important aspect of the test program. References 53 through 58 contain further information concerning tests and trials.

References

1 John F. Nichols, "The Development of Marine Engineering," SNAME *Historical Transactions*, 1893–1943.

2 Edgar C. Smith, "The Bicentenary of Thomas Newcomen," *Engineering*, July 19, 1929.

3 A. M. Green, Jr., *Pumping Machinery*, John Wiley and Sons, Inc., London, 1919.

4 Alexander Richardson, *The Evaluation of the Parsons Steam Turbine*, London, 1911.

5 C. W. Dyson, "Fifty-Year Retrospect of Naval Marine Engineering," ASNE *Journal*, 1918.

6 J. D. Irish and W. S. Brown, "The Environment" in *Submersible Vehicle Systems Design*, E. Eugene Allmendinger, Ed., SNAME, 1990.

7 J. D. van Manen and P. van Oossanen, "Resistance" in *Principles of Naval Architecture*, Vol. 2, E. V. Lewis, Ed., SNAME, 1988.

8 Alexander C. Landsburg, Eric Gabler, George Levine, Richard Sonnenschein, and Earl Simmons, "U.S. Commercial Ships for Tomorrow," *Marine Technology*, May 1990.

9 Ronald K. Kiss, "Mission Analysis and Basic Design," in *Ship Design and Construction*, Robert Taggart, Ed., SNAME, 1980.

10 J. Harvey Evans, "Basic Design Concepts," ASNE *Journal*, Nov. 1959.

11 General Specifications for Ships of the U.S. Navy, NAVSEA S9AAO-AA-SPN-010/GEN-SPEC.

12 Morton Gertler, "A Re-analysis of the Original Test Data for the Taylor Standard Series," DTMB Report 806, March 1954.

13 F. H. Todd, G. R. Stuntz, and P. C. Pien, "Series 60-The Effect upon Resistance and Power of Variation in Ship Proportions," *Trans. SNAME*, Vol. 65, 1957.

14 D. P. Roseman, *The MarAd Systematic Series of Full-Form Ship Models*, SNAME, 1987.

15 Roger P. Johnson and Henry P. Rumble, "Weight, Cost and Design Characteristics of Tankers and Dry-Cargo Ships," *Marine Technology*, Vol. 2, No. 2, April 1965.

16 J. D. van Manen, "The Choice of the Propeller," *Marine Technology*, Vol. 3, No. 2, April 1966.

17 J. D. van Manen and P. van Oossanen, "Propulsion" in *Principles of Naval Architecture*, Vol. 2, E. V. Lewis, Ed., SNAME, 1988.

18 Robert P. Giblon and I. Hilary Rolih, "COGAS: Marine Power Plant for Energy Savings," *Marine Technology*, July 1979.

19 Herman C. Schlappi, "An Innovative Energy Saving Propulsion System for Naval Ships," ASNE *Journal*, April 1982.

20 C. L. Stahly, "The Reversing Propulsion Gear," *Trans. SNAME*, Vol. 77, 1969.

21 Robert F. Nufrio, "Test and Evaluation of the Reversible Converter Coupling Reverse Reduction Gear," ASNE *Journal*, May 1988.

22 "Marine Steam Power Plant Heat Balance Practices," Technical and Research Bulletin 3-11, SNAME.

23 "Marine Diesel Power Plant Practices," Technical and Research Bulletin 3-49, SNAME.

24 R. L. Harrington, "Rudder Torque Prediction," *Trans.* SNAME, Vol. 89, 1981.

25 C. L. Crane, H. Eda, and A. Landsburg, "Controllability" in *Principles of Naval Architecture*, Vol. 3, E. V. Lewis, Ed., SNAME, 1988.

26 "Shipping," Code of Federal Regulations Title 46.

27 "Navigation and Navigable Waters," Code of Federal Regulations Title 33.

28 "Automated Main and Auxiliary Machinery," Navigation and Vessel Inspection Circular (NVIC) Nos. 1-69 and 6-84, U.S. Coast Guard.

29 "Guide for Automatic and Remote Control Systems for Integrated Propulsion Systems," Appendix D to: *Rules for Building and Classing Steel Vessels*, American Bureau of Shipping.

30 "Guide for Centralized Control and Automation of Ship's Steam Propulsion Plants," SNAME T&R Bulletin 3-41.

31 "Diesel Ship Automation Guide," SNAME T&R Bulletin 3-46.

32 "Guide for Centralized Control and Automation of Ship's Gas Turbine Propulsion Plants," SNAME T&R Bulletin 3-29.

33 "Maintenance Management for Automated Power Plants," SNAME T&R Bulletin 3-40.

34 R. L. Bort, "Assessment of Shock-Design Methods and Shock Specifications," *Trans.* SNAME, Vol. 70, 1962.

35 "Shock Design of Shipboard Equipment, Dynamic Design-Analysis Method," Bureau of Ships, NAVSHIPS, 250-423-30, May 1961.

36 R. L. Harrington and W. S. Vorus, "Dynamic Shock Analysis of Shipboard Equipment," *Marine Technology*, Vol. 4, No. 4, Oct. 1967.

37 "Shock Design Criteria for Trident Submarine," SUPSHIP Three, 280-8, May 1973.

38 J. A. DeRuntz, T. L. Geers, and C. A. Felippa, "The Underwater Shock Analysis (USA-Version 3) Code: A Reference Manual," DNA 5615F, Defense Nuclear Agency, Sept. 1980.

39 T. L. Geers, "Residual Potential and Approximate Methods for Three-Dimensional Fluid-Structural Interaction Problems," *Journal. Acoustical Society of America*, May 1971.

40 "Shock Tests, H. I. (High-Impact); Requirements for Shipboard Machinery, Equipment and Systems," Military Specification MIL-S-901 (Navy).

41 "International Conference on Marine Pollution, 1973," Intergovernmental Maritime Consultative Organization, 1977.

42 "Clean Seas Guide for Oil Tankers," International Chamber of Shipping/Oil Companies International Marine Forum, State Mutual Books, 1989.

43 "Pollution Control," Naval Ship's Technical Manual S9086-T8-STM-010, Chapter 593.

44 Richard Lee Storch, Colin P. Hammon, and Howard Bunch, *Ship Production*, Cornell Maritime Press, 1988.

45 Daniel W. Billingsley and J. Christopher Ryan, "A Computer System Architecture for Naval Ship Design, Construction, and Service Life Support," *Trans.* SNAME, Vol. 94, 1986.

46 James P. Chappell, "The Application of Critical Path Methodology to the Management of Ship Design Programs," *Marine Technology*, Vol. 28, No. 3, May 1991.

47 "Integrated Logistic Support Programs for Equipment, Subsystems, and Systems," Military Standard, DOD-STD-1702(NS).

48 "DOD Requirement for a Logistic Support Analysis Record," Military Standard, MIL-STD-1388.

49 "Procedures for Performing a Failure Mode, Effects and Criticality Analysis," Military Standard, MIL-STD 1629 (Navy).

50 "System Safety Program Requirements," Military Standard, MIL-STD 882 (Navy).

51 "Standard Practice for Human Engineering Design for Marine Systems, Equipment and Facilities," ASTM Standard, ASTM F 1166-88, American Society for Testing and Materials, Philadelphia.

52 "Human Engineering Design Criteria for Military Systems, Equipment and Facilities," Military Standard, MIL-STD 1472.

53 "Guide for Shop and Installation Tests," Technical and Research Bulletin 3-39, SNAME.

54 "Guide for Sea Trials," Technical and Research Bulletin 3-47, SNAME.

55 "Development of Shipboard Industrial Test Procedures," DOD-STD-21-6 (Navy), 31 July 1986.

56 "Total Ship Test Program for Ship Production, Ship Construction Test and Trials Manual," NAVSEA 0900-LP-095-2010, April 1977.

57 "Total Ship Test Program, Ship Test Evaluation Planning Guide," NAVSEA 0900-LP-095-4010, June 1985.

58 "Maritime Administration, Supplementary Procedure for Testing Machinery," Division of Engineering Office of Ship Construction, March 1989.

CHAPTER II

José Femenia

Thermodynamics and Heat Engineering

Section 1
Review of Fundamentals

1.1 Basic Equations. The applied thermodynamics problems of marine engineering depend on the conservation of mass and the conservation of energy. The first of these is conveniently expressed by the one-dimensional steady-flow continuity equation

$$W = AV/v \tag{1}$$

where

A = flow area, ft²
V = flow velocity, fps
v = specific volume of the fluid, ft³/lb
W = flow rate, lb/sec

The second is conveniently expressed for the usual steady one-dimensional situation by the general energy equation

$$\frac{V_1^2}{2gJ} + h_1 + \frac{z_1}{J} + Q_{1,2} = \frac{V_2^2}{2gJ} + h_2 + \frac{z_2}{J} + W_{k_{1,2}} \tag{2}$$

where

h = enthalpy of the fluid, Btu/lb
 = $u + Pv/J$
u = internal energy, Btu/lb
P = pressure, psfa
J = mechanical equivalent of heat
 = 778 ft-lb/Btu
g = gravitational constant
 = 32.17 ft/sec²
z = height above an arbitrary datum, ft
Q = heat transferred, Btu/lb
W_k = external work done, Btu/lb

The subscripts 1 and 2 refer to two arbitrarily chosen points along a flow path; $Q_{1,2}$ and $W_{k_{1,2}}$ therefore designate the heat transferred and work done between these points.

A condensation frequently used is the stagnation enthalpy h^0, defined by

$$h^0 = h + \frac{V^2}{2gJ} \tag{3}$$

giving

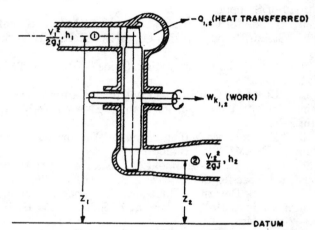

Fig. 1 Energy equation as applied to a single-stage turbine

$$h_1^0 + \frac{z_1}{J} + Q_{1,2} = h_2^0 + \frac{z_2}{J} + W_{k_{1,2}} \tag{4}$$

Typical applications of the general energy equation occur where the working fluid is being heated without work being done (a heat exchanger), where work is being done under adiabatic conditions (turbine wheel), or where mechanical energy is being degraded under adiabatic conditions and without work being done (flow against friction). The equations that apply in these situations are easily found by eliminating the inappropriate terms from equation (4). An application is illustrated by Fig. 1. Note the sign of the heat transfer term.

Evaluation of the general energy equation usually requires assistance from other equations. The continuity equation is one. Equations of state for the fluid involved are also frequently needed. The simplest form is the familiar perfect gas equation

$$Pv = RT \tag{5}$$

where

T = absolute temperature, deg R
R = a constant characteristic of a particular gas, ft-lbf/lb$_m$°R

Typical values of R are 53.34 for dry air, 53.5 for wet air

Table 1 Example calculation of an exhaust-gas mixture constant

Constituent	① Percent by Volume	② Mole Fraction	③ Molecular Weight	④ Mass lb_m per Mole of Mixture	⑤ Mass Fraction
		① ÷ 100		② × ③	④ ÷ Σ ④
N_2	72.8	0.728	28	20.38	0.730
O_2	12.0	0.120	32	3.84	0.138
CO_2	3.7	0.037	44	1.64	0.059
H_2O	11.5	0.115	18	2.07	0.074
				Σ = 27.93	

From equation (6), the exhaust-gas mixture

$$\text{constant} = \frac{1545}{27.93} = 55.32 \; \frac{\text{ft-lb}_f}{\text{lb}_m \, °R}$$

(40 percent humidity, 100 F); and 50.3 for flue gas (15 percent excess wet air and standard fuel oil). For other gas mixtures, such as engine exhaust mixtures, the gas constant R, must be derived from the universal gas constant $\bar{R}$ which is

$$\bar{R} = 1545 \; \frac{\text{ft-lbf}}{\text{lb}_m \, \text{mol} \, °R}$$

The specific gas constant is

$$R = \bar{R}/M \qquad (6)$$

where M is the molecular weight of the mixture. For a gas mixture it is convenient to note that in a mixture of ideal gases the mole fraction of a component is exactly equal to the volume fraction. An exhaust-gas analysis can be presented as shown in Table 1.

For an ideal gas undergoing a polytropic process, the functional relationship is

$$PV^n = \text{constant}$$

where $\qquad (7)$

$$V = \text{volume, ft}^3$$

For a reversible adiabatic process the exponent, n, is the ratio of the constant-pressure to the constant-volume specific heats at zero pressure, k

$$k = \frac{C_{po}}{C_{vo}} \qquad (8)$$

For air k is 1.4. For other familiar processes the values of n are:

Isobaric process	$n = 0$
Isothermal process	$n = 1.0$
Cold engine, starting	$n = 1.2$
Air-cooled compressor	$n = 1.25$
Hot engine, compression and expansion; and water-cooled compressor	$n = 1.30$ to 1.35

From equations (5) and (7), the following expressions relating the initial and final states of a polytropic process can be derived:

$$\frac{P_2}{P_1} = \left(\frac{V_1}{V_2}\right)^n$$

$$\frac{T_2}{T_1} = \left(\frac{P_2}{P_1}\right)^{(n-1)/n} = \left(\frac{V_1}{V_2}\right)^{n-1} \qquad (9)$$

For perfect gases, the following state relations also hold:

$$h = C_p T \qquad (10)$$

$$u = C_v T \qquad (11)$$

where C_p and C_v are the specific heats for constant-pressure processes and constant-volume processes, respectively. For mixtures of gases the respective specific heat may be calculated by

$$C = \sum_i f_i C_i \qquad (12)$$

where

f_i = mass fraction of each constituent
C_i = specific heat of each constituent, Btu/lb_m°R

The specific heats are functions of temperature alone but are often treated as constants when the temperature range is small. All gases obey this equation reasonably well at low pressures, and at higher pressures when the state is far from the critical point. Otherwise the tabulated properties of fluids, such as those found in the *Steam Tables* [1], *Gas Tables* [2], and *Thermodynamic Charts* [3], must be used. An alternative, particularly adaptable to turbine design work when calculations are made by computer, is to use the equations from which these tabulations are made.

Special relations for steam that are useful in nozzle design are the equation of state

$$v = 1.222 \, (h - 823) \qquad (13)$$

and the equation for isentropic expansion

$$v^{1.3} = \text{constant} \qquad (14)$$

The following two are the corresponding relations for the wet region

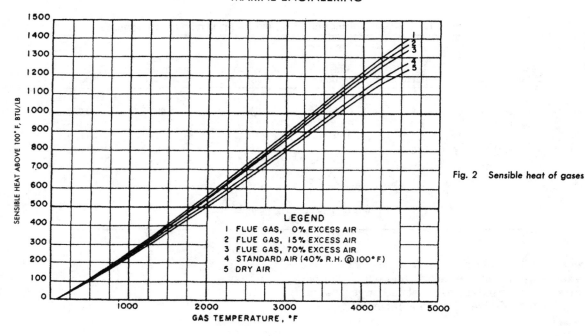

Fig. 2   Sensible heat of gases

LEGEND
1 FLUE GAS, 0% EXCESS AIR
2 FLUE GAS, 15% EXCESS AIR
3 FLUE GAS, 70% EXCESS AIR
4 STANDARD AIR (40% R.H. @ 100° F)
5 DRY AIR

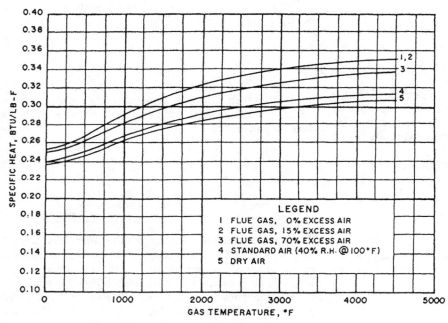

Fig. 3   Instantaneous specific heat of gases

LEGEND
1 FLUE GAS, 0% EXCESS AIR
2 FLUE GAS, 15% EXCESS AIR
3 FLUE GAS, 70% EXCESS AIR
4 STANDARD AIR (40% R.H. @ 100° F)
5 DRY AIR

$$p^{0.97}v = 0.467\,(h - 366) \qquad (15)$$

$$pv^{1.13} = \text{constant} \qquad (16)$$

Units are psi for p, ft³/lb for v, and Btu/lb for h. These equations are for use only in the vicinity of normal turbine state lines, and not for use at high superheat with low pressure, with very wet steam, or in the reheat region.

In boiler design work, the sensible heat and specific heat of the flue gas must be known. These are presented in Figs. 2 and 3 for a standard grade 6 or residual fuel oil of the composition (by weight) tabulated as follows when burned in air with a 40 percent relative humidity at a temperature of 100 F.

Carbon	0.8775
Hydrogen	0.1050
Sulfur	0.0120
Oxygen	0.0040
Nitrogen	0.0015
Free moisture	—
	1.0000

Other properties of flue gas, such as its viscosity and thermal conductivity, are also needed, and are given in Fig. 4. Values for steam and air can be found in the *Steam Tables* [1] and *Gas Tables* [2], respectively.

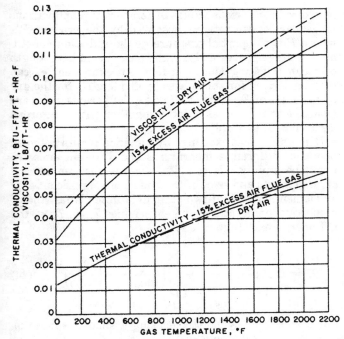

Fig. 4 — Viscosity and thermal conductivity of flue gas and air

1.2 Heat Transfer. An investigation of the $Q_{1,2}$ term in equations (2) or (4) entails a consideration of the principles of heat transfer. The transfer takes place by molecular diffusion between bodies in contact, or by electromagnetic radiation between separated bodies. Diffusion between solids is called *conduction*. When one or both of the bodies are fluids, conduction is nearly always grossly modified by the transport of heat by fluid in motion; this phenomenon is called *convection*.

a. Conduction. Conduction follows Fourier's Law, which states that heat is diffused at a rate proportional to the temperature gradient; the factor of proportionality is known as the *thermal conductivity*, and is a property of the material conducting the heat. It is generally a function of temperature, particularly for liquids and gases, but the effect of temperature is sufficiently weak that conductivity can be treated as a constant in most problems. Fourier's Law can be expressed for one-dimensional problems as

$$Q = -kS \frac{dT}{dx} \tag{17}$$

where

$$k = \text{conductivity, Btu-ft/hr-ft}^2\text{-deg F}$$
$$S = \text{conducting area, ft}^2$$
$$dT/dx = \text{temperature gradient, deg F/ft}$$

If k is constant, this equation can be integrated for a slab of thickness x, having a temperature difference between faces of $T_1 - T_2$, to obtain

$$Q = \frac{k}{x} S (T_1 - T_2) \tag{18}$$

If the conducting body is circular, as when heat is transferred through tube walls, equation (17) is modified to

$$Q = -k2\pi rL \frac{dT}{dr} \tag{19}$$

where r is the radius dimension, and L is the length of the tube. Integration of equation (19) gives

$$Q = k \frac{2\pi L}{\log_e \frac{r_o}{r_i}} (T_i - T_o) \tag{20}$$

where the subscripts o and i designate the outside and inside surfaces of the tube.

Heat transfer problems frequently involve conduction through successive layers of distinctly different conductivity. Formulas for this type of problem are readily derived, as are formulas for the transfer of heat through cylindric composite walls.

b. Convection. The convective heat transfer between a fluid at a largely constant *bulk temperature* T_B and a surface at temperature T_S is expressed by

$$Q = h_f S(T_B - T_S) \tag{21}$$

where h_f is the *film coefficient* of convective heat transfer. The major practical problem in applying equation (21) is the evaluation of the film coefficient for the several distinct mechanisms of flow and thermal behavior possible in the fluid.

Single-phase convection occurs when the fluid involved neither boils nor condenses at the solid surface. Familiar examples abound aboard ship; for instance, the water side of condenser tubes, both sides of the tubes or plates in liquid-to-liquid heat exchangers such as lube-oil coolers, and the gas side of convective heating surface in boilers are typical locations where this mechanism is prominent. The value of h_f is generally a function of fluid properties, of the fluid velocity, and of its degree of turbulence, as noted in Table 2 of the Heat Exchangers chapter. Under conditions existing in a typical condenser tube, for example, the value of h_f is likely to be in the neighborhood of 1000 Btu/hr-ft²-deg F, while on the gas side of a boiler tube, the value of h_f can be 10 Btu/hr-ft²-deg F.

When the fluid is moved over the heat transfer surface by an external force, such as that provided by a pump, the situation is described as *forced convection*. When density differences caused by expansion or contraction of the fluid near the surface are the principal source of the driving force, the situation is described as *natural convection*.

Two-phase convection occurs under conditions of surface temperature and heat flow that produce a change of phase, either boiling of a liquid or condensation of a vapor. *Nucleate boiling* describes the phenomenon in which bubbles of vapor form on a hot surface, grow in size, then break away to rise through the adjacent liquid. The turbulent mixing of fluid near the surface results in rates of heat transfer distinctly higher than in single-phase convection.

At high rates of heat transfer, or *heat flux* (Q/S), the bubbles can form so rapidly and in such great number that they effectively blanket the surface with a vapor film, and the phenomenon of *film boiling* ensues. Since liquid can not impinge on the surface, h_f is significantly less than in nucleate boiling, and in consequence, film boiling is usually avoided in practical heat transfer apparatus.

Condensation of vapors on a cool surface occurs as either *dropwise* or *film* condensation. The names are quite descriptive of the processes. The rate of heat transfer is much higher for dropwise condensation, and is comparable to that for nucleate boiling, since the drops quickly fall off as they form and thereby expose the surface to more vapor. In film condensation, the condensed film tends to cling evenly to the surface, and so forms a barrier between the surface and the vapor.

c. Radiation. All matter emits radiation of one or more kinds. The thermal radiation of practical concern requires only that the matter be at a temperature above absolute zero, and so is characteristic of all bodies. The radiation is electromagnetic, and at industrial temperatures lies within the infrared part of the electromagnetic spectrum. The wavelength is a function of temperature, and at higher temperatures it falls within the range of visible light.

The radiation is not, however, monochromatic. A curve of its intensity, I_λ, against wavelength, λ, shows a considerable spread with a peak intensity at a wavelength that is a function of temperature. The total energy emitted is thus the integral of I_λ over all wavelengths. For a black-body radiator, i.e., one that emits at the maximum intensity at all wavelengths, the integration produces the Stefan-Boltzmann relation

$$E_b = 1713 \left[\frac{T}{1000} \right]^4 \qquad (22)$$

for T in degrees Rankine and E_b in Btu/ft²-hr.

But actual bodies are not black-body radiators, and their degree of imperfection must be accounted for by their surface emissivity ϵ_λ. If the emissivity is independent of the wavelength λ, or is averaged over the spectrum, equation (22) becomes

$$E = 1713 \, \epsilon \left[\frac{T}{1000} \right]^4 \qquad (23)$$

Bodies for which this equation holds are said to be *grey radiators*.

Surfaces of solids and liquids usually have high values of emissivity and are grey. In the case of gases, ϵ_λ is not a constant but a function of molecular concentration, temperature, and wavelengths. In the industrial range of temperatures, only those gases composed of more than one type of atom, such as carbon dioxide (CO_2) or water (H_2O), radiate with an appreciable strength and only in certain wavelengths.

For gases, E_G can be measured directly, and is found to vary with temperature and the concentration of molecules in a given radiation path. Measured values of E_G can

be plotted for various values of temperature and various molecular concentrations measured by the product PL, where P is the partial pressure of the radiating constituent in atmospheres and L is the mean beam path length or mean radiating length in feet. An equivalent ϵ_G can then be derived for such gases by dividing E_G for the gas by the value of E_b. Plots of E_G and ϵ_G for various gases are available in references 4 and 5 together with a rigorous discussion of gas radiation.

In practice, heat exchange by radiation occurs between bodies of different temperatures and different emissivities. The situation is complex because the geometrical arrangements and sizes of the bodies are significant. For an elementary case of two parallel infinite planes, and of respective temperatures and emissivities T_1, ϵ_1, T_2, ϵ_2, the net energy exchange rate is

$$E = 1713 \left[\frac{1}{\dfrac{1}{\epsilon_1} + \dfrac{1}{\epsilon_2} - 1} \right] \left[\left(\frac{T_1}{1000} \right)^4 - \left(\frac{T_2}{1000} \right)^4 \right] \qquad (24)$$

For a sphere or cylinder, enclosing a smaller sphere or cylinder, the equation is

$$E = 1713 \left[\frac{1}{\dfrac{1}{\epsilon_1} + \dfrac{S_1}{S_2} \left(\dfrac{1}{\epsilon_2} - 1 \right)} \right] \left[\left(\frac{T_1}{1000} \right)^4 - \left(\frac{T_2}{1000} \right)^4 \right] \qquad (25)$$

where S_1 and S_2 are the areas of the respective spheres or cylinders.

In boiler tube banks where the heat transfer fluid is a radiating gas, heat transfer simultaneously occurs by both radiation and convection. Under these conditions (in order for the two heat transfer coefficients to be directly additive) it is often convenient to express the radiation heat transfer in the form of the artificial heat transfer coefficient

$$h_r = \left(\frac{E_G}{T_G - T_S} \right) \epsilon_S = \frac{1713 \epsilon_G \epsilon_S}{T_G - T_S} \left[\left(\frac{T_G}{1000} \right)^4 - \left(\frac{T_S}{1000} \right)^4 \right] \qquad (26)$$

where the subscript G refers to the radiating gas and S refers to the tube surface (see Subsection 2.3 for further discussion on this subject).

d. Overall heat transfer coefficient. The typical occurrence of heat transfer in power plant apparatus is between a hot fluid and a cold fluid through an intervening tube wall. Convection and radiation are involved at the inner and outer surfaces, and conduction is involved within the tube metal. The rate of heat flow is summarized succinctly by

$$Q = US(T - t) \qquad (27)$$

where T and t are the bulk temperatures of the two fluids, S is the surface area, and U is the *overall heat transfer coefficient*. U is the net effect of the conduction, convection, and radiation contributions. To illustrate the makeup of U, consider the transfer of heat from a hot gas outside a clean tube to a second fluid inside the tube. As a preliminary, note that the artificial radiation coefficient h_r, of the same dimensions as the convection coefficient h_f, is used so that the radiation and convection contributions will be additive. Thus, starting at the outer surface of the tube, the three heat transfer paths are

$$Q_o = (h_{fo} + h_r)S(T - T_o)$$

$$Q_w = \frac{k}{X_e} S(T_o - t_i)$$

$$Q_i = h_{fi}S(t_i - t)\frac{D_i}{D_o}$$

where

h_{fo}, h_{fi} = convective surface coefficients at tube outside and inside, respectively

T, t = temperatures of hot gas and cold fluid, respectively

T_o, t_i = metal temperatures at tube outside and inside surfaces respectively

k = conductivity of the tube wall

X_e = equivalent thickness for the circular tube
$$= \frac{1}{2} D_o \log_e \frac{D_o}{D_i}$$

Addition eliminates intermediate temperatures, allowing assessment of U as

$$U = \frac{1}{\dfrac{1}{h_{fo} + h_r} + \dfrac{X_e}{k} + \dfrac{1}{h_{fi}} \cdot \dfrac{D_o}{D_i}} \qquad (28)$$

By a similar process, U can be written for any number of layers, including surface fouling.

The practical process of heat exchanger design is often aided by empirical formulas for U, which give results of sufficient accuracy for industrial purposes. They are usually promulgated by manufacturer's associations to standardize methods of calculation, and are found in publications such as references 6 and 7.

e. Log mean temperature difference. The fluid flowing through a heat exchanger undergoes either a change in temperature or a change in state in response to the heat that it receives or rejects. If the fluids on both the hot and cold sides of the heat exchanger undergo a change of state, their respective temperatures are constant, and equation (27) applies without change if S is understood to mean the total heat transfer area. On the other hand, if there are temperature changes, the temperature difference in equation (27) is not constant throughout the heat exchanger, and in consequence this equation must be integrated for application to the entire apparatus.

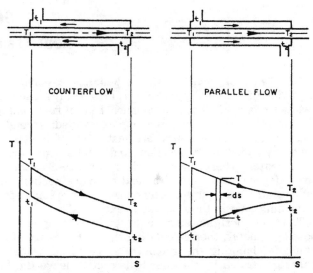

Fig. 5   Simple counterflow and parallel-flow heat exchangers

The case where there is no change of state is illustrated by a simple concentric-pipe heat exchanger, Fig. 5, in which the two fluids flow either in the same direction (parallel flow) or in opposite directions (counterflow). The temperatures of the two fluids are plotted as a function of position for both exchangers. Such a temperature difference integrated over the length of the heat exchanger produces a mean temperature difference; because of its logarithmic term it is familiarly known as the *log mean temperature difference*. In the general case, the log mean temperature difference can be written as

$$\Delta T_m = \frac{\Delta T_{\max} - \Delta T_{\min}}{\log_e \dfrac{\Delta T_{\max}}{\Delta T_{\min}}} \qquad (29)$$

Equation (29) is the general expression for ΔT_m for both simple counterflow and parallel-flow exchangers. In condensers, boilers, and feed heaters, to list several prominent examples, where a change of state rather than a temperature change occurs on one side of the tube wall, a derivation of the log mean temperature difference again produces equation (29). If the heat exchanger is multipass, equation (29) must be modified (see Section 3.2 of the Heat Exchangers chapter.)

In any case, equation (27), when applied to the heat exchanger as a whole, is written as

$$Q = US\Delta T_m \qquad (30)$$

Section 2
Heat Transfer in Boilers

2.1 Types of Heat Transfer in Boilers. A boiler may be divided functionally into four parts: first, a chemical reaction chamber where the chemical heat of fuel combustion is released and the reaction controlled; second, a steam generating section where heat is transferred to the tubes by radiation, convection, and conduction; third, a superheater, where the steam is superheated to the desired degree; and fourth, a heat-recovery section, employing air heaters and/or economizers where some of the remaining heat in the flue gas is extracted.

The boiler designer starts a design by setting up a proposed geometry, then calculating the performance of the unit and changing the geometry as required. Once the gases generated by the chemical reaction leave the furnace, the problem becomes one of convection where an overall U is estimated and an exit temperature from each bank of tubes is calculated. The designer must first estimate the performance of the furnace and calculate the gas temperature entering the first bank of tubes.

2.2 Heat Transfer in Boiler Furnaces. Furnace heat transfer is principally radiation, and it is possible to adapt the basic methods of Hottel in reference 5 to evaluate a total emissivity in terms of furnace conditions. The problem consists of equating the heat given up by the combustion gases to the heat transferred by radiation and convection to the furnace surfaces. The equation is

$$1713 \, S_C F_E F_A \left[\left(\frac{T_F}{1000} \right)^4 - \left(\frac{T_C}{1000} \right)^4 \right] + U S_W (T_E - T_C)$$

$$= W_F(R + 1) \left[\frac{LHV + q_F + (t_a - t_o)C_p R}{R + 1} - q_{T_E} \right] \quad (31)$$

where

U = convection heat transfer coefficient
T_C = furnace surface temperature
T_E = furnace exit temperature
T_F = effective flame radiating temperature
S_W = convection surface area
S_C = radiant heat absorbing surface (RHAS)
W_F = weight of fuel burned per hour
LHV = lower heating value of fuel
t_o = base temperature
t_a = temperature of air entering furnace
q_F = sensible heat of fuel above t_o
q_{T_E} = sensible heat of gas above t_o
C_p = average specific heat of combustion air
R = air-fuel ratio
F_A = arrangement factor
F_E = emissivity factor

The heat given up by the combustion gas is evaluated by ordinary stoichiometric means and the use of a set of sensible heat curves (Fig. 2).

The shape emissivity factor, $F_E F_A$, has been treated by Hottel, and if the flame fills the furnace, it has been demonstrated that

$$F_E F_A = \frac{1}{\dfrac{1}{F_{CF}} + \dfrac{1}{\epsilon_C} - 1} \quad (32)$$

$$F_{CF} = \left[1 + \frac{S_R/S_C}{1 + \left(\dfrac{\epsilon_F}{1 - \epsilon_F} \right) \dfrac{1}{F_{RC}}} \right] \epsilon_F \quad (33)$$

where

ϵ_C = emissivity of the heat-absorbing surface
ϵ_F = flame emissivity
S_R = refractory surface area
S_C = cooled surface area
F_{RC} = a geometric factor, dependent on the extent of cooled surface

An approximation of F_{RC} to a reasonable degree of accuracy is

$$F_{RC} = \frac{S_C}{S_T} \qquad \text{when } 0 < \frac{S_R}{S_C} < 0.5 \quad (34)$$

$$F_{RC} = \frac{S_C}{S_R} \qquad \text{when } 4 < \frac{S_R}{S_C} < 7 \quad (35)$$

where $S_T = S_R + S_C$.

Faired intermediate values may be taken between the two sets of limits quoted, as illustrated by Fig. 6.

The radiating temperature T_F may be approximated by

$$T_F = (T'_A T_E)^{1/2} \quad (36)$$

where T'_A = adiabatic flame temperature with 100 percent theoretical air.

Evaluation of the flame total emissivity presents a complex problem. The flame cloud consists of droplets of fuel from the burner nozzle which in turn are reduced to smaller fragments by various air and gas currents and by the chemical breakdown of the fuel to basic constituents. The flame mass then consists of a cloud of flaming fuel, carbon, some ash particles, and molecules of carbon dioxide, water vapor, sulfur dioxide, oxygen, and nitrogen. Of these constituents, the fuel, carbon, and ash particles and the carbon dioxide, water vapor, and sulfur dioxide molecules radiate. The gas molecules radiate only in certain wavelengths, that is, they are not grey. The solid particles radiate in all wavelengths. These radiations are superimposed upon each other, resulting in an overall radiation which is essentially grey in character, and the resulting emissivity is independent of temperature.

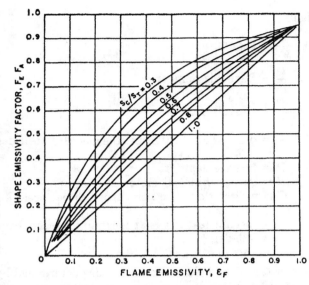

Fig. 6 Shape emissivity factor versus flame emissivity for various values of cooled surface to cooled surface plus refractory surface ratio (S_C/S_T)

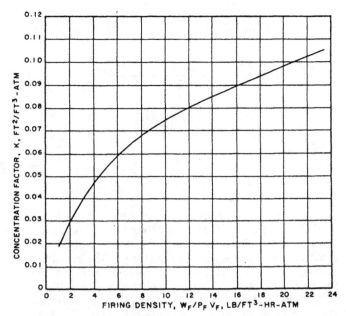

Fig. 7 Furnace concentration factor

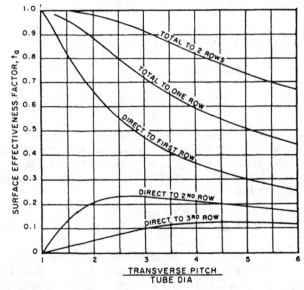

Fig. 8 Effectiveness factors for water walls based on total projected area

Combustion of oil is not instantaneous, especially when residual oils are fired. The oil droplet first ignites, then burns and breaks down into carbon and hydrogen. The carbon appears as minute flecks. These small particles make up most of the radiation. Their concentration is a function of burning time, and of the rate of flow of the gases through the furnace.

An expression derived for ϵ_F by applying probability theory is

$$\epsilon_F = \epsilon_\infty (1 - e^{-KP_F L}) \qquad (37)$$

where

ϵ_∞ = emissivity of a cloud of infinite thickness, assumed to be 0.95

P_F = furnace pressure, atmospheres

L = mean radiating path length; for ordinary marine furnaces, $L = 0.6\sqrt[3]{V_F}$

V_F = furnace volume, ft³

K = an empirical concentration factor, a function of a time parameter $W_F/P_F V_F$ with W_F representing the pounds of fuel burned per hour. $W_F/P_F V_F$ is a crude measure of particle life, but better data on the flame path are lacking.

The concentration factor, K, is evaluated from test results on various boilers and plotted against the firing density $W_F/P_F V_F$ as on Fig. 7. This plot was calculated from the test results on five different boilers, all burning residual fuels. The curve shown represents an average of the test results with 10 to 20% excess air.

It is necessary also to consider the question of effective cooled surface. A water wall consisting of tangent tubes may be treated as a surface having an area equal to the projected area of the surface. If the tubes are widely spaced, exposing the refractory surface behind the tubes, the simple projected area of the tubes is not sufficient since the refractory receives some of the direct radiation from the surface and returns only a portion of this heat

to the furnace; the remainder goes to the tubes. The effective radiant heat absorbing surface (RHAS) may be calculated by multiplying the projected area of the walls, including backing refractory, by an arrangement factor from Fig. 8, for each area making up the furnace envelope.

Solution of equation (31) is best accomplished by trial-and-error methods by breaking up the equation into three simultaneous equations, as follows:

$$\left(\frac{Q_C}{S_C}\right)_1 = 1713 F_E F_A \left[\left(\frac{T_F}{1000}\right)^4 - \left(\frac{T_C}{1000}\right)^4 \right]$$
$$+ U \frac{S_W}{S_C}(T_E - T_C) \quad (38)$$

$$\left(\frac{Q_C}{S_C}\right)_2 = \frac{W_F(R+1)}{S_C}$$

$$\times \left[\frac{LHV + q_F + (t_a - t_o)C_p R}{R+1} - q_{T_E}\right] \quad (39)$$

$$\left(\frac{Q_C}{S_C}\right)_1 = \left(\frac{Q_C}{S_C}\right)_2 \quad (40)$$

The term

$$U\frac{S_W}{S_C}(T_E - T_C)$$

in equation (38) is generally negligible except for rear-wall impingement effects. It is convenient to drop the term at this point and correct for the effect later. For most marine boilers the temperature of the radiant heat absorbing surface (RHAS) is close to 1000 R (540 F); so the term $T_C/1000$ is approximately unity. Since the value of $T_F/1000$ is between 3 and 4, the relative value of $(T_F/1000)^4$ is so much higher than 1 that the term $T_C/1000$ can be taken as equal to 1 with little error. Further noting that $T_F = (T_A T_E)^{1/2}$, equation (38) becomes

$$(Q_C/S_C)_1 = 1713 F_E F_A\left[\left(\frac{T_A}{1000}\right)^2 \left(\frac{T_E}{1000}\right)^2 - 1\right] \quad (41)$$

In equation (39), the term

$$\frac{LHV + q_F + (t_a - t_o)C_p R}{R+1}$$

is the total sensible heat released to the furnace per pound of combustion products and may be replaced by q_{T_A}, the adiabatic sensible heat. q_{T_E} may be read from Fig. 2 at any assumed value of T_E. With these simplifications, equation (39) reduces to

$$(Q_C/S_C)_2 = \frac{W_F(R+1)}{S_C}[q_{T_A} - q_{T_E}] \quad (42)$$

The solution may then be achieved by assuming values of T_E and plotting solutions for equations (41) and (42). The point of intersection of the two equations is the solution. $T_{A'}$ may be evaluated by calculating the adiabatic sensible heat

$$q_{T_{A'}} = \frac{LHV + q_F + (t_a - t_o)C_p R}{R+1} \quad (43)$$

then $T_{A'}$ may be read from Fig. 2.

Usually, the convective term in equation (31) is negligible; but when a rear wall is fitted, especially in a shallow furnace, the convection effect of the flame blasting against the rear wall may be significant. An equation for the surface heat transfer coefficient h_{RW}, based on the actual surface exposed to the gas, is

$$\frac{h_{RW}D}{k} = C\left(\frac{GD}{\mu}\right)^a \left(\frac{C_p \mu}{k}\right)^b \left(\frac{D}{d}\right)^c \quad (44)$$

where

$$\frac{C_p \mu}{k} = \text{Prandtl number}$$

$$\frac{GD}{\mu} = \text{Reynolds number}$$

$$\frac{h_{RW}D}{k} = \text{Nusselt number}$$

k = thermal conductivity of the gas, Btu/ft-hr-deg F
D = furnace depth, ft
d = burner throat diameter, ft
G = weight flow rate per burner throat area, lb/ft²-hr
C_p = specific heat of gas at constant pressure, Btu/lb-deg F
μ = viscosity of gas, lb/ft-hr

The constant C and exponents a, b, c are determined by test.

An adequate simplified form of this expression is

$$h_{RW} = \frac{G^{0.53}}{D^{0.47}} f_T \quad (45)$$

where f_T is a temperature coefficient given by

$$f_T = 0.00003875 T_{f'} + 0.1035 \quad (46)$$

The film temperature $T_{f'}$, is estimated by means of a conduction calculation through the tube wall, starting with the fluid saturation temperature in the tube, and assuming that the gas temperature is the same as the furnace exit temperature. This is not rigorous, but serves to keep the convection calculation simple. Thus, $T_{f'} = \frac{1}{2}(T_E + T_S)$, where T_S is the surface temperature of the wall.

2.3 Tube Banks. The transfer of heat to a bank of tubes from a hot fluid is a combination of a number of heat transfer phenomena. Included in this combination is the convection film coefficient, h_f, intertube radiation conductance h_r, and conduction, all combined in accordance with equation (28).

In evaluating heat transfer in tube banks, the geometry of the bank is generally fixed first, and the value of the overall transfer coefficient is deduced for this geometry. Thus, for a given geometry, it is necessary to deduce the convection film coefficient for the outside of the tube, the radiation conductance (if present), conductance through the scale, tube wall, and inside scale, and the inside convection film coefficient.

For boiler and superheater tube banks, this entire set of evaluations must be made. Normally, for boiler generating tubes, the outside convection film coefficient and the radiation conductance are so much smaller than the metal conductance and inside boiling water coefficient that in calculating the overall U, the metal conductance and inside film may be neglected. In the case of superheaters, however, the metal conductance and inside film coefficient have a marked effect and must be included.

Section 1 defines, by equation (21), the convection heat transfer coefficient, h_f (or the film coefficient). This coefficient is evaluated by the Nusselt expression

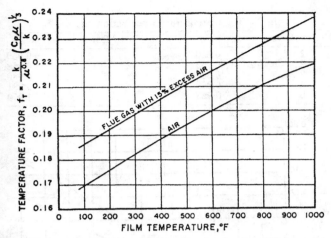

Fig. 9 Longitudinal flow convection inside tubes for air and flue gas with 15% excess air

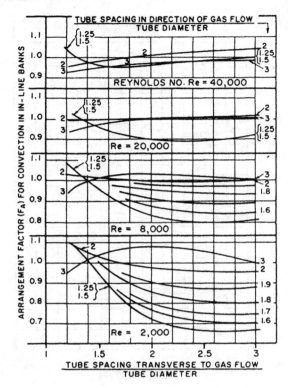

Fig. 10 Arrangement factor F_A for convection heat transfer for in-line tube banks

$$\frac{h_f D}{k} = C \left(\frac{GD}{\mu}\right)^a \left(\frac{C_p \mu}{k}\right)^b \left(\frac{D}{L}\right)^c \qquad (47)$$

where

h_f = outside film coefficient, Btu/ft²-hr-deg F
k = thermal conductivity of gas, Btu/ft-hr-deg F
G = weight flow of gas, lb/ft²-hr
μ = viscosity of gas, lb/ft-hr
C_p = specific heat of gas at constant pressure, Btu/lb-deg F
D = tube diameter, ft
L = tube length, ft

k, μ, and C_p should be evaluated at the so-called "film temperature," the arithmetic average between the average surface temperature and the average fluid bulk temperature. The constant C and the exponents a, b, and c may be determined from model tests. Note that this basic equation was used previously in the discussion of furnace convection, and a special form [equation (44)] was given for that situation. Another form of this equation that is generally useful in connection with the analysis of turbulent fluids inside tubes is the Colburn equation

$$\frac{h_f D}{k} = 0.023 \left(\frac{GD}{\mu}\right)^{0.8} \left(\frac{C_p \mu}{k}\right)^{1/3} \qquad (48)$$

where h_f refers to the inside film coefficient of the tube and D is the tube inside diameter. Equation (48) may be rearranged as follows

$$h_f = 0.023 \frac{G^{0.8}}{D^{0.2}} \frac{k}{\mu^{0.8}} \left(\frac{C_p \mu}{k}\right)^{1/3} \qquad (49)$$

or

$$h_f = 0.023 \frac{G^{0.8}}{D^{0.2}} f_T \qquad (50)$$

where f_T is a temperature factor that can be read from Fig. 9 as a function of the film temperature; the film temperature may be taken as the average of the average gas temperature and the average surface temperature.

For the flow of fluids across a bank of tubes placed perpendicularly to the direction of flow, the modified Grimison equation has been found useful, especially for gases. It is of the form

$$\frac{h_f D}{k} = 0.292 \, F_A F_D \left(\frac{GD}{\mu}\right)^{0.6} \left(\frac{C_p \mu}{k}\right)^{1/3} \qquad (51)$$

The weight flow rate per unit area, G, is based on the minimum free flow area between tubes. In practice, this equation is put in the form

$$h_f = 0.292 \, F_A F_D f_E \frac{G^{0.6}}{D^{0.4}} \qquad (52)$$

The arrangement factor, F_A, for various tube configurations can be read from Figs. 10, 11, and 12. Values for the tube depth correction factor, F_D, are given by Table 2; and the temperature factor, f_E, is plotted on Fig. 13 versus the gas film temperature (average of the average gas and surface temperatures).

Radiation conductance was defined earlier. In the passage of gases through a tube bank, the gas itself is a radiation source. For such tube banks the mean beam path length or mean radiating length can be taken as 0.85 times the hydraulic diameter. For in-line or staggered tubes, the mean radiating length in feet is

$$L = 0.071 \left[\frac{4 S_L S_T - \pi d^2}{\pi d}\right] \qquad (53)$$

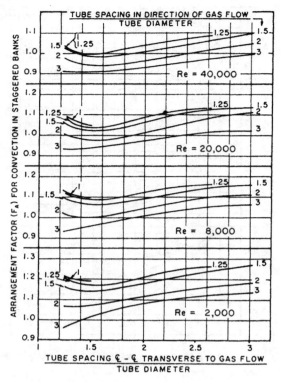

Fig. 11 Arrangement factor F_A for convection heat transfer for staggered tube banks with minimum gas flow area in transverse openings

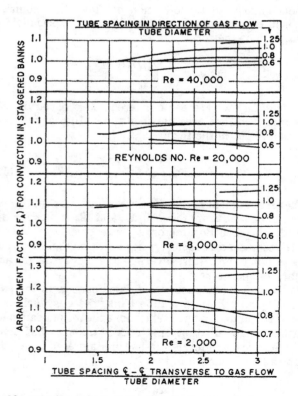

Fig. 12 Arrangement factor F_A for convection heat transfer for staggered tube banks with minimum gas flow area in diagonal openings

Table 2 Tube bank depth correction factor, F_D

No. of Rows	F_D	No. of Rows	F_D
1	0.650	6	0.950
2	0.765	7	0.968
3	0.840	8	0.980
4	0.886	9	0.991
5	0.923	10	1.000

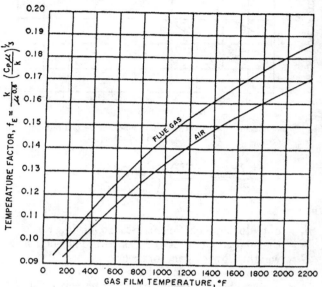

Fig. 13 Temperature factor

where

S_T = transverse pitch, in.
S_L = longitudinal pitch, in.
d = tube diameter, in.

The emissivity ϵ_G of flue gas is a function of its temperature, the mean radiating length L, and the partial pressure P_R of its radiating constituents (primarily water vapor and carbon dioxide). The flue gas produced when burning standard fuel oil in 15 percent excess air has a water vapor partial pressure of 0.114 atm/atm and a carbon dioxide partial pressure of 0.125 atm/atm, a total of 0.239 atm/atm. Values of ϵ_G for this mixture are plotted in Fig. 14 for a range of gas bulk temperatures and a range of $P_R L$ values, where P_R is in atm/atm and L is in feet. The curves are usable from 10 to 20 percent excess air without appreciable error.

Plotted on the same figure is the value

$$\frac{h_r}{\epsilon_G} = \frac{E_G \epsilon_S}{(T_G - T_S)\epsilon_G} = \frac{1713\epsilon_S}{T_G - T_S}\left[\left(\frac{T_G}{1000}\right)^4 - \left(\frac{T_S}{1000}\right)^4\right]$$

which is another form of equation (26). To determine h_r, calculate L from equation (53) and multiply L by P_R, the total partial pressure of CO_2 and H_2O. Enter Fig. 14 at the average gas bulk temperature and, at the proper $P_R L$ value, read ϵ_G on the left scale. Then reenter at the gas bulk temperature and read h_r/ϵ_G on the right scale at the appropriate tube surface temperature. h_r is equal to the product $\epsilon_G \times h_r/\epsilon_G$; included in the result is a tube surface emissivity of 0.95.

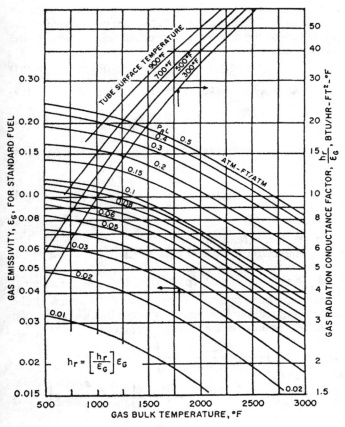

Fig. 14   Gas radiation conductance

$$h_r = \left[\frac{h_r}{\varepsilon_G}\right]\varepsilon_G$$

2.4 Heat-Recovery Equipment. That portion of the heat transfer equipment that absorbs heat at temperatures below the saturation temperature of the generated steam is considered to be heat-recovery equipment. Generally, such equipment absorbs the heat into the combustion air (gas air heater), into the incoming feedwater (economizer), or into a combination of the two.

Combustion air heaters generally are of either the rotary regenerative or tubular type, with air flowing through tubes heated by combustion gas passing around the outside of the tubes in cross flow. The rotary regenerative type is proprietary and performance data are best acquired from the proposed manufacturer. Tubular air heaters can be readily evaluated by the methods reviewed in the foregoing for tube banks. Most steam vessels fitted with air-heating heat-recovery equipment use the rotary regenerative type.

Extended-surface economizers are employed almost to the exclusion of bare-tube units. The feedwater is invariably inside the tubes, with some type of extended surface outside. There is a large variety of extended-surface types, ranging from cast-iron fins shrunk on steel tubes, to stud fins, aluminum fins, and spiral-welded steel fins. Such elements are proprietary in nature and performance data must be obtained from their manufacturers. Extended-surface performance data are usually acquired by full-scale tests of the particular geometric design. Of principal concern in fitting heat-recovery equipment on boilers is the dew point of stack gas acids. For further discussion of preventing acid attack on economizers, see Section 5.2.

For an example calculation, which illustrates the considerations involved in designing boiler heating surfaces, see Chapter 5.

Section 3
Internal Thermodynamics of the Steam Turbine

3.1 Nozzle Flow. A nozzle is a short flow passage of converging or converging-diverging flow area whose function is to convert thermal or pressure energy into kinetic energy. It thus forms an essential feature of both steam turbines and gas turbines.

As the fluid passes through a nozzle, no external work is performed, and no heat is transferred, so that the general energy equation (2) reduces to

$$\frac{V_0^2}{2gJ} + h_0 = \frac{V_1^2}{2gJ} + h_1 \tag{54}$$

or, using the concept of stagnation enthalpy, simply

$$h_0^0 = h_1^0 \tag{55}$$

In equation (54), V_0 is the approach velocity, which is the velocity of the fluid immediately upstream of the nozzle. If the approach velocity is taken as zero, equation (54) can be written as

$$V_1 = 223.7\sqrt{\Delta h} \tag{56}$$

If the fluid is turned through an angle, or is decelerated in a nozzle chest, before entering the nozzle, there is some degradation of the energy possessed by the fluid upstream. This degradation is evident as a loss of stagnation pressure and, hence, of the pressure difference available to cause flow through the nozzle. It is indicated on the enthalpy-entropy plot of the nozzle process shown by Fig. 15.

There is also degradation of energy within the nozzle itself, so that the exit velocity is not as high as ideally possible. The total degree of degradation is expressed by the nozzle efficiency, which is thus the ratio of the energy actually converted to kinetic energy to that theoretically possible. In equation form, the definition of the nozzle efficiency, η_N is

$$\eta_N = \frac{V_1^2/(2gJ)}{h_0^0 - h_1'} \tag{57}$$

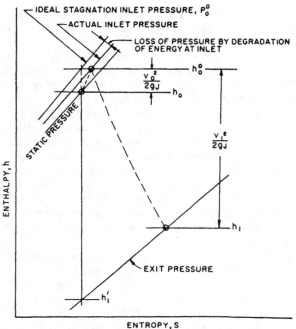

Fig. 15   The nozzle flow process

The meaning of h_1' and the derivation of equation (57) are evident in Fig. 15.

An alternative designation of the degree of energy conversion is given by the velocity coefficient k_N, which is the ratio of nozzle exit velocity to that ideally attainable, i.e.

$$k_N = \frac{V_1}{[2gJ(h_0^0 - h_1')]^{1/2}} \tag{58}$$

It may be seen that k_N is simply the square root of η_N.

The nozzle velocity coefficient is a function of factors such as the angle through which the fluid is turned, nozzle dimensions, and the ratio of approach kinetic energy to the total kinetic energy developed. Empirical curves, such as Fig. 16, give nozzle efficiencies for blade-type turbine nozzles for dry or superheated steam. To determine the efficiency of a nozzle (either fixed or moving), the basic nozzle efficiency k^2 and height correction factor f_L are read from Fig. 16, and the nozzle efficiency is computed as

$$k_N^2 = f_L k^2 \tag{59}$$

For wet steam, a correction is necessary to account for the impingement of the slower-moving droplets of water on the back of the blades. This correction is taken by some authorities to be

$$k_{NW}^2 = k_N^2[x + f(1 - x)] \tag{60}$$

where

k_{NW}^2 = nozzle efficiency for wet steam
x = steam quality
f = ratio of velocity of water to velocity of steam. A value of 0.15 may be used for design purposes.

The usual nozzle analysis procedure in turbine design is to start with a known geometry and initial conditions, then calculate the exit velocity and flow rate. Equations

of state suitable for nozzle calculations with superheated steam were given in Section 1.1. These, together with the expansion ratio $r = P_1/P_0^0$, where P_1 is the discharge pressure and P_0^0 is the ideal stagnation pressure, and the definition of the nozzle velocity coefficient k_N, lead to the approximations

$$V_1 = [2gJk_M^2(1 - r^{0.231})(h_0^0 - 823)]^{1/2} \tag{61}$$

$$h_1 = h_0^0 - k_M^2(1 - r^{0.231})(h_0^0 - 823) \tag{62}$$

$$v_1 = \frac{1.222}{rP_0^0}(h_0^0 - 823)[1 - k_M^2(1 - r^{0.231})] \tag{63}$$

for conditions at the nozzle exit. The continuity equation is

$$W_s = \frac{25A V_1}{v_1} \tag{64}$$

where

W_s = flow rate, lb/hr
v_1 = specific volume at exit, ft³/lb
A = nozzle exit area, in.²
V_1 = velocity at exit, fps

These equations determine the state of the steam at the nozzle exit and the flow rate. Combined into a single equation, W_s for superheated steam becomes

$$W_s = P_0^0 A \left[\frac{25(2gJ)^{1/2}}{1.222(h_0^0 - 823)^{1/2}}\right]$$
$$\times \left[\frac{r[k_N^2(1 - r^{0.231})]^{1/2}}{1 - k_N^2(1 - r^{0.231})}\right] \tag{65}$$

or

$$W_s = P_0^0 AQR \tag{66}$$

where

Q = first bracketed term, a function of h_0^0 alone
R = second bracketed term, a function of r alone

Note that P_0^0 is in units of psia and h_0^0 is in Btu/lb.

Employing a similar procedure, the flow of wet steam through a nozzle can be expressed as

$$W_s = P_0^0 A \left[\frac{12,000(P_1^0)^{-0.03}}{(h_1^0 - 366)^{1/2}}\right]$$
$$\times \left[\frac{r^{0.97}(k_{NW}^2(1 - r^{0.085}))^{1/2}}{1 - k_{NW}^2(1 - r^{0.085})}\right] \tag{67}$$

The phenomenon of choking limits the maximum flow rate to that obtained at the critical pressure ratio, or that value of r at which the fluid reaches the speed of sound in the smallest flow area of the nozzle. This value of r can be found by evaluating the maximum of dW_s/dr from equations (65) or (67). The critical values of r thus found are a function of the nozzle efficiency k_N^2.

3.2   Vector Diagrams. An analysis of the energy process through one stage of a turbine (fixed nozzles plus the rotating wheel nozzles or blades) illustrates the application of nozzle flow theory. A stage is shown, in the form of the cross sections of one fixed and one moving nozzle, in Fig. 17. The velocities and enthalpies of the stage may be plotted on an enthalpy-entropy diagram as in Fig. 18.

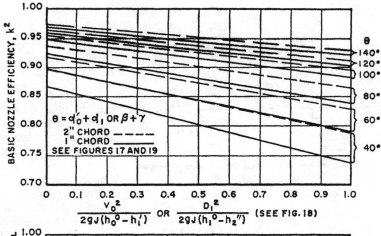

Fig. 16 Nozzle efficiency

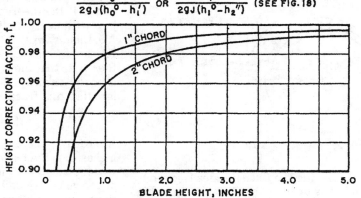

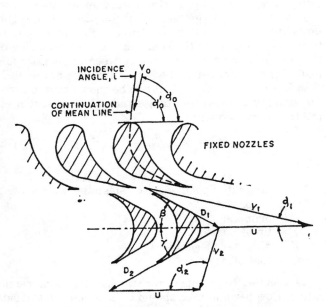

Fig. 17 Turbine blading diagram for one stage

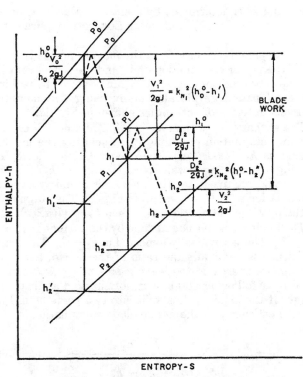

Fig. 18 Enthalpy-entropy diagram for flow through one turbine stage

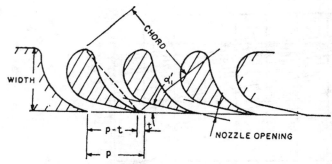

Fig. 19 Blade and nozzle partition nomenclature for a typical converging nozzle, section taken at mean diameter

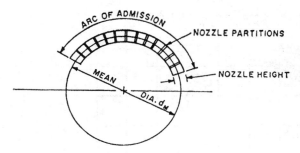

Fig. 20 Nozzle nomenclature

The steam velocity leaving the fixed nozzles is given by equation (54) or, in terms of the quantities shown in Fig. 18, by

$$V_1 = k_{N_1} [2gJ(h_0^0 - h_1')]^{1/2} \qquad (68)$$

assuming h_1' is determined by a known value of r, based on the nozzle area, k_N^2, and steam flow rate requirements or on P_1.

The approach velocity D_1 to the moving nozzle is, however, the vector combination of V_1 and the wheel velocity U. The magnitude of this velocity can be found by making the combination graphically, or by using trigonometric analysis in terms of the angles shown in Fig. 17.

The enthalpy h_1 is then $h_0^0 - (V_1^2/2gJ)$ and the corresponding stagnation enthalpy h_1^0, as seen by the moving nozzle, is $h_1 + (D_1^2/2gJ)$. Then the relative velocity D_2 leaving the moving nozzle is $k_{N_2}[2gJ(h_1^0 - h_2'')]^{1/2}$. The velocity V_2 relative to the fixed nozzle is found by a vector combination with U, as evident in Fig. 17. The stagnation enthalpy h_2^0 at the nozzle exit is then $h_2 + (V_2^2/2gJ)$.

The blade work per pound done by the turbine is simply $h_0^0 - h_2^0$. The work ideally expected is $h_0^0 - h_2'$. The stage efficiency is nominally the ratio of these two, but since the kinetic energy in the leaving steam, $V_0^2/2gJ$, can be used by a following stage, it is not charged against this stage. If the ideal work is split into two parts as in Fig. 18, the efficiency of the nozzle-blade combination is

$$\eta_{NB} = \frac{h_0^0 - h_2^0}{h_0^0 - h_1' + h_1 - h_2'' - \dfrac{V_2^2}{2gJ}} \qquad (69)$$

or very nearly

$$\eta_{NB} = \frac{h_0^0 - h_2^0}{h_0^0 - h_2' - \dfrac{V_2^2}{2gJ}} \qquad (70)$$

3.3 Nozzles and Blading. As has been pointed out in the previous sections, the difference between nozzles and blades is a relative one, nozzles being considered as fixed and blades (or buckets) as moving.

For nozzles or blades below the critical pressure ratio, the area at exit is of primary importance. From Figs. 19 and 20, the nozzle exit area is $A = [N(p - t) \sin \alpha_1']L$, where N is the number of nozzles, p is the pitch of the nozzles, t is the thickness of the nozzle partition at exit, and α_1' is the theoretical steam outlet angle, all measured in the plane of the turbine wheel. Let $m = [N(p - t)]/Np$; then $Npm = N(p - t)$, where m is a factor to correct for the thickness of nozzle (or blade) edge thickness. Further, note that the arc of admission of steam to the nozzles is not necessarily 360 deg, but some fraction thereof. Designating this fraction as E

$$A = Ld_M \pi m E \sin \alpha_1 \qquad (71)$$

where A is the area at exit from the nozzle, in square inches, L is the blade height in inches, and d_M is the mean wheel diameter. α_1 is the angle of the steam to the plane of the wheel. Usually there is a small difference between the actual steam angle α_1 and the geometric angle α_1'. This angle $\alpha_1 - \alpha_1' = \delta$ is known as the deviation angle and is a function of both the angle through which the steam is turned and the Mach number, and approaches zero as the Mach number approaches 1.0. In equation (71), α_1 should be used when its value is known.

For short blades and nozzles, there is little difference in the blade velocity between the root and tip of the blades; however, with long blades, the blade velocity varies appreciably from root to tip. Further, the flow of steam from the nozzles has a large tangential component (V_{1u}). This tangential component is constrained by the casing to travel in a circular path. Thus, on the discharge side of the nozzle, in the annular space between the nozzles and blades, the pressure must be higher at the tip of the blade than it is at the blade root. It can be shown that in order for stability to exist, the tangential velocity must meet the free vortex condition.

$$V_{1uT} d_T = V_{1uM} d_M = V_{1uR} d_R \qquad (72)$$

where

V_{1u} = tangential component of steam velocity leaving nozzles
d = diameter
T = tip of nozzle
M = diameter of nozzle
R = root of nozzle

As a further requirement for stability, the axial weight rate of flow per unit area should be constant. This avoids radial components. Thus

$$\frac{W_s}{A_x} = \frac{V_{1xT}}{v_{1nT}} = \frac{V_{1xM}}{v_{1nM}} = \frac{V_{1xR}}{v_{1nR}} \qquad (73)$$

where the subscript designations are as before and

V_{1x} = axial component of steam velocity leaving nozzles

v_{1n} = specific volume of steam leaving nozzles

A_x = axial flow area in plane at wheel

W_s = total weight of steam flow

The result of these requirements is a warped blade, with generally pure impulse at the root and with a large degree of reaction at the tip. Normally, the last few stages of the LP turbine are based on the free vortex condition, with the other stages having reasonable approximations of this flow. Obviously, as the turbine size increases, the blade lengths increase and the free vortex design may be extended into higher-pressure stages.

3.4 Windage, Friction, and Leakage. Considerable power is required to rotate even a smooth disk at high speed in an atmosphere of steam or air. This power is necessary to overcome the fluid friction of the disk and to overcome the centrifugal pumping action of the disk that tends to make the steam move radially outward. Reasonably close clearances reduce this latter action.

In addition to this friction and pumping action, a second large rotation loss arises from the fanning of idle blades outside the arc of admission. This "windage loss" is considerable and increases with blade speed and blade length. With full peripheral steam admission, the windage loss is negligible, but it increases rapidly with a larger inactive arc. If the inactive arc is well shielded, the loss is much reduced.

The combined windage and friction losses may be estimated by Kerr's equation:

$$hp_f = [k_1 d_R^2 + n k_2 (1 - E) L^{1.5} d_M] \left(\frac{U}{100}\right)^3 \frac{1}{v} \qquad (74)$$

where

hp_f = horsepower loss

d_M = mean blade ring diameter, in.

d_R = disk diameter to root of blades, in.

E = peripheral admission fraction

L = blade height, in.

U = blade speed, fps

v = specific volume of surrounding atmosphere

k_1 = a constant: 0.000433 for steam, 0.000422 for air

k_2 = a constant: 0.0388 for steam and air, unshielded; 0.0194 for shielded wheels

n = a constant dependent on the number of rows of moving blades; it is 1.0 for one row, 1.23 for two rows, and 1.8 for three rows in velocity-compounded stages

Theoretically, a small correction should be included for moisture in the steam, but in normal marine turbines the correction is negligible.

Leakage between stages and through shaft glands constitutes a large energy loss in steam turbines. Where leakage is out of the shaft glands, the end result is a reduction in steam flow through the entire following stages. Stage leakage through diaphragm interstage packing only affects the efficiency of each particular stage since the leakage steam stays in the turbine.

To control leakage, diaphragm packings of the stepped labyrinth type (Fig. 21) are commonly used. These packings consist of a series of thin strips or disks fixed alternately to the diaphragm and rotor to maintain the smallest possible leakage area without contact between parts. The small constrictions accelerate the leaking fluid; its resulting high velocity is then dissipated in the spaces between the strips. The result is a controlled leakage rate kept within tolerable limits. Egli [8] evolved a very satisfactory method of estimating leakage in stepped and unstepped labyrinths. The pertinent data for this method are given in Fig. 21 and are self explanatory.

3.5 State Lines. Since the blade work is $h_0^0 - h_2^0$, the power developed by a turbine stage, or its wheel horsepower, WHP, is

$$WHP = \frac{(W_s - W_L)(h_0^0 - h_2^0)}{2544} - hp_f \qquad (75)$$

where

W_s = total steam flow entering stage, lb/hr

W_L = leakage flow, i.e., flow that bypasses the nozzles or moving blades, lb/hr

hp_f = power absorbed by windage

The wheel work per pound, based on total flow, is

$$\Delta h_W = \frac{(WHP)(2544)}{W_s} \qquad (76)$$

and the stagnation enthalpy entering the following stage is $h_0^0 - \Delta h_W$. Included in this stagnation enthalpy is the velocity contribution from the preceding stage, and this should be corrected for the leakage flow introduced in equation (75) by

$$\frac{V_0^2}{2gJ} = \frac{W_s - W_L}{W_s} \frac{V_2^2}{2gJ} \qquad (77)$$

In this relation, the zero subscript now refers to the entrance to the following stage. Usually the fraction $(W_s - W_L)/W_s$ is near enough to unity that it can be omitted. In any case, the change in stagnation enthalpies can be found from stage to stage by noting the work done per stage, and the static enthalpies can be found by subtracting the velocity components from the stagnation enthalpies.

The overall stage efficiency is then the ratio of actual work delivered to the shaft to the available energy per stage, or

$$\eta_{st} = \frac{\Delta h_w}{h_0^0 - h_1' + h_1 - h_2'' - \frac{V_2^2}{2gJ}\left(\frac{W_s - W_L}{W_s}\right)} \qquad (78)$$

$W_L = A \alpha \phi \gamma \sqrt{g \frac{P_i}{V_i}}$

W_L = LEAKAGE, LBS/SEC.

A = LEAKAGE AREA, FT.^{2} = $\frac{\pi D \delta}{144}$

   δ = RADIAL CLEARANCE, IN.
   D = CLEARANCE DIAMETER, IN.

α = CONTRACTION FACTOR DUE TO THROTTLING

ϕ = OVERALL LABYRINTH PRESSURE RATIO FACTOR

γ = CARRY-OVER CORRECTION FACTOR FOR STRAIGHT-
   THROUGH LABYRINTH; UNITY FOR STAGGERED TYPE

g = ACCELERATION DUE TO GRAVITY = 32.2 FT./SEC.^{2}

P_i = ABSOLUTE PRESSURE BEFORE LABYRINTH, LBS/FT^2

P_f = ABSOLUTE PRESSURE AFTER LABYRINTH, LBS/FT^2

Δ = STRIP THICKNESS AT TIP, IN.

S = PITCH OF STRIPS, IN.

V_i = SPECIFIC VOLUME BEFORE LABYRINTH, FT.^{3}/LB.

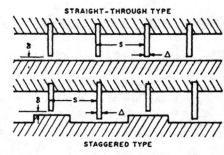

TYPES OF LABYRINTHS

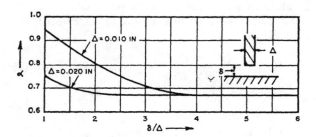

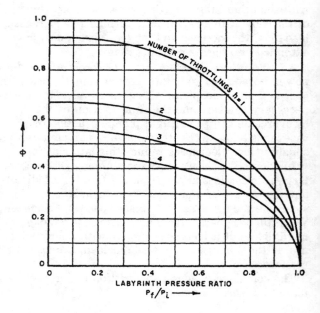

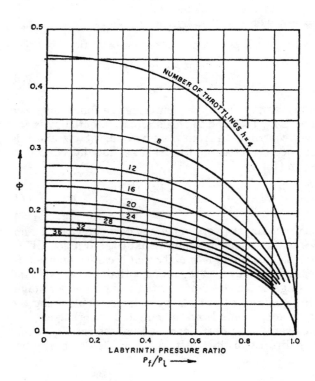

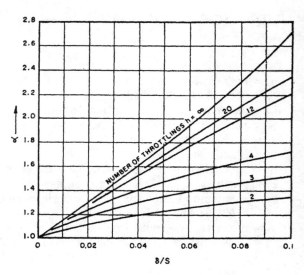

Fig. 21   Labyrinth leakage losses

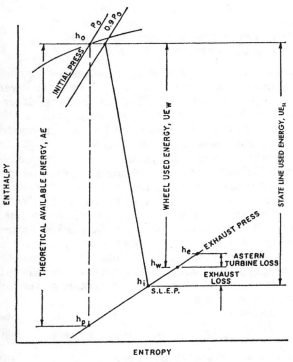

Fig. 22   Propulsion turbine state line

or, as a close approximation

$$\eta_{st} = \frac{\Delta h_w}{h_0^0 - h_2' - \dfrac{V_2^2}{2gJ}\left(\dfrac{W_s - W_L}{W_s}\right)} \qquad (79)$$

If the pressures P_0 for each stage are known, as they must be in the design process, then a stage-by-stage plot of the conditions for each stage can be made on the Mollier chart (h–s plane; see Fig. 18). When completed, the plot is known as the state line, or condition line, for the turbine, as shown by Fig. 22. It is particularly useful in the analysis of extraction point conditions, since the usual requirement is knowledge of enthalpies as functions of pressures. However, the end of this line, known as the state line end point (SLEP), represents the static enthalpy only, whereas the stagnation enthalpy of the exhausting steam must be known for such uses as condenser design. Thus the $V_2^2/2gJ$ component is usually added, and the resulting stagnation enthalpy at exhaust is also plotted on the state line diagram at exhaust pressure. This component is that which represents the approach-velocity input to a following stage, as in equation (69), but which, in the last stage, must be wasted. It thus forms the major part of the turbine leaving or exhaust loss.

Section 4
External Thermodynamics of the Steam Turbine

4.1 State Line for the Entire Unit. The wheel horsepower of a turbine stage is given by equation (75). For each stage, the values of W_S, W_L, h_0^0, h_2^0, and hp_f are different. The total power delivered into the turbine shafts thus must be expressed as a summation of the stage wheel horsepowers by

$$\Sigma WHP = \sum_{i=1}^{N} \frac{(W_{S_i} - W_{L_i})(h_{0_i}^0 - h_{2_i}^0)}{2544} - hp_{f_i} \qquad (80)$$

where i designates the individual stages. The delivered horsepower of the unit is then $(\Sigma WHP - RLP)\eta_M$ where RLP is the reversing turbine windage loss in horsepower, and η_M is the mechanical efficiency.

The astern turbine loss is the energy required to spin this turbine backwards in the atmosphere of exhaust steam. The energy is absorbed by the steam, and subsequently by the condenser.

All losses of mechanical power that occur between the turbine wheel and the point of measurement are included in the determination of η_M. The output of a geared propulsion turbine is measured aft of the low-speed gear and thrust bearing. The mechanical efficiency used in determining this output thus includes reduction gear losses, as well as bearing friction and governor-drive power of

the turbine itself. In the case of a generator turbine, output is measured at the generator terminals, so that η_M must also include the generator efficiency.

The ship designer usually cannot use equation (80) because the terms cannot be directly evaluated. The quantities that can be measured are the turbine shell pressures, throttle flow (perhaps found by measuring the condensate flow), power output, exhaust pressure, throttle pressure and temperature, and crossover pressure and temperature. Although the details of the steam conditions implicit in equation (80) cannot be constructed from these measurements, the turbine state line can, and it represents these conditions sufficiently well for design purposes. Indeed, knowledge of steam flow, power output, throttle pressure and temperature, and exhaust pressure are sufficient for this task.

A preliminary step is to express steam flow in unit form as a steam rate (or water rate) thusly:

$$SR = W/SHP \qquad (81)$$

where

$\quad SR$ = steam rate, lb/shp-hr
$\quad W$ = throttle flow, lb/hr
SHP = shaft horsepower

If RLP and shaft gland leakage are neglected, the wheel used energy, or the energy extracted from each pound of steam, is

$$UE'_W = \frac{2544}{SR\eta_M} \tag{82}$$

Adding RLP, and expressing gland leakage by an efficiency, alters this to

$$UE_W = \frac{2544}{SR\eta_M\eta_L} + \frac{2544\,RLP}{W\eta_M\eta_L} \tag{83}$$

where

UE_W = wheel used energy (see Fig. 22), Btu/lb
η_L = leakage efficiency

UE_W corresponds to

$$\sum_{i=1}^{N}(h^0_{0_i} - h^0_{2_i})$$

in equation (80). Subtraction of UE_W from the throttle enthalpy gives the stagnation enthalpy of the exhaust steam. Subtraction of the exhaust loss, which consists mainly of the leaving kinetic energy, gives the static enthalpy, and thus the state line end point.

The engine efficiency is the ratio of the power output at the point of measurement to the power theoretically developed by an isentropic expansion of steam between the initial conditions and exhaust pressure. The concept is illustrated by Fig. 22. The engine efficiency E is

$$E = \frac{2544\,SHP}{W(h_0 - h_p)} = \frac{2544}{(SR)(AE)} \tag{84}$$

where

$AE = h_0 - h_p$ = available energy, Btu/lb
W = throttle flow, lb/hr
SR = steam rate, lb/shp-hr

If the steam rate or engine efficiency is given or can be estimated, the condition line can be constructed from knowledge of the throttle steam conditions and the exhaust pressure.

4.2 State Line Estimates. In designing a steam propulsion plant, it is usually necessary to proceed with accurate heat balances before the propulsion turbine is designed. The actual state line is thus not available when first needed, but since all high-quality turbines have essentially the same efficiency for the same steam conditions, it is possible to construct a state line from standard efficiency relationships. These relationships are: basic efficiency as a function of power rating, with throttle pressure as a parameter (Fig. 23); corrections for throttle temperature (Fig. 24); exhaust loss as a function of an exhaust flow parameter (Fig. 25); and astern turbine loss as a function of exhaust pressure. The astern turbine loss can be estimated from the expression

$$RL = 0.33P \tag{85}$$

where

RL = astern turbine windage loss, percent
P = exhaust pressure, in. Hg, abs

The basic efficiency of turbines increases with higher power ratings because of the longer blades and the lower relative importance of internal steam leakage. It decreases with higher throttle pressures because of the smaller volume flow, increased leakage, and increased moisture of the steam in the later stages.

Corrections for throttle temperature show improved efficiency at higher temperatures because of increased volume flow, decreased leakage, and decreased moisture.

The exhaust loss is a function of the parameter Wv/A where W = exhaust flow, v = exhaust specific volume, and A = last stage annulus area. The data in Fig. 25 are plotted against W/pA since p, the exhaust pressure, is more readily known than v, and is closely proportional to its inverse. Typical values of the parameter give exhaust losses in the range of 10 to 17 Btu/lb at rated power for nonextraction operation.

The state line end point is given by

$$SLEP = h_0 - AE \cdot E_b \cdot f_t = h_0 - UE_{SL} = h_i \tag{86}$$

and the wheel used energy becomes

$$UE_W = h_0 - (SLEP + EL) \tag{87}$$

where

E_b = basic efficiency, from Fig. 23
f_t = temperature correction, from Fig. 24
UE_{SL} = state line used energy
EL = exhaust loss, from Fig. 25, in Btu/lb

The apparent wheel horsepower is $UE_W \cdot W/2544$, and it is necessary to subtract leakage losses and astern turbine losses to get the shaft horsepower. These losses, L & W, are given by

$$L\ \&\ W = \frac{(3.5 + RL)SHP}{100 - (3.5 + RL)} \tag{88}$$

where RL is the astern turbine loss as a percentage of the shaft horsepower from equation (85), and the 3.5 is a percentage allowance for combined leakage and mechanical losses. Thus the wheel used energy is related to the shaft horsepower by

$$SHP = \frac{UE_W \cdot W}{2544} - \frac{(3.5 + RL)SHP}{100 - (3.5 + RL)} \tag{89}$$

and

$$SR = \frac{W}{SHP} = \frac{\left[1 + \dfrac{(3.5 + RL)}{100 - (3.5 + RL)}\right]}{\dfrac{2544}{(AE \cdot E_b \cdot f_t - EL)}} \tag{90}$$

To evaluate EL for a given turbine design, it is necessary to estimate the value of SR or W. By assuming that EL is 2 percent of the state line used energy, the weight flow, W, can be closely estimated by

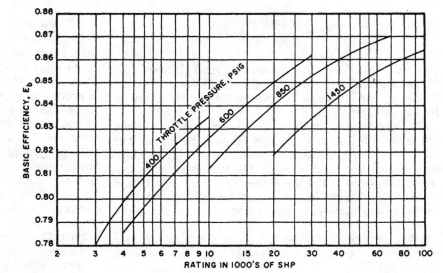

Fig. 23   Basic state line efficiency at rated load

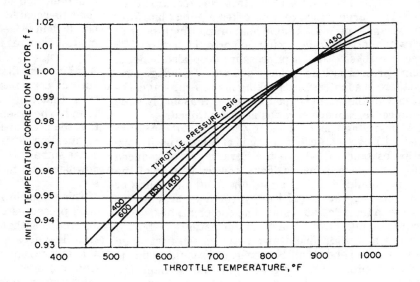

Fig. 24   Initial temperature correction factor

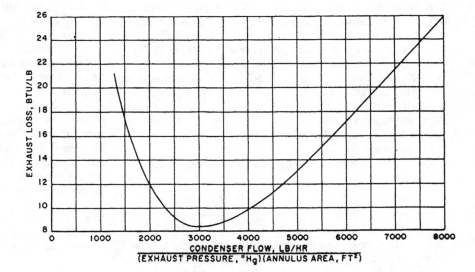

Fig. 25   Exhaust loss at rated load

$$W \approx \frac{2544 \, WHP}{0.98 \, UE_{SL}} \qquad (91)$$

This value of W may then be used to read EL from Fig. 25, assuming a value of exhaust annulus area, A, based on the expected frame size. Frame sizes are commonly 13, 18, 25, or 35 ft².

The foregoing methods can be used to locate the state line end point. The initial point is taken to be the throttle conditions, less a ten percent pressure drop at constant enthalpy. This drop represents the loss in pressure between the throttle and the first-stage nozzle block, and also serves to adjust the subsequent straight-line approximation to a more accurate position at the lower end of the line.

After the upper and lower end points are found, further consideration is required to establish the line between these points. The first stage, or control stage, exhaust is usually at about two thirds of the throttle pressure at full power. This large pressure drop takes a good "bite" out of the steam to permit a usable heat drop at high rates of flow without the stage going supersonic before 75% power is reached. The large pressure drop also benefits the high-pressure gland leakage and wheel windage loss. Since there is no interstage packing in the first stage, the efficiency of the stage is good, provided reasonably good blade-speed/steam-speed ratios are maintained.

The first stage is followed by a series of pressure stages, with each stage increasing in efficiency until moisture losses tend to balance the advantages of reduced leakages and longer blades. Thus, the true shape of the state line should be a first stage for about two thirds of throttle pressure at 70% efficiency, followed by a sweeping curve to the calculated state line end point; the sweeping curve should have a maximum deviation of about 6 Btu below a straight line drawn between the first-stage exhaust and the end point. However, the resulting line is so nearly approximated by a straight line between the nozzle inlet point and the state line end point that most designers and turbine builders prefer to use the straight line. Since the state line is used primarily for heat balance work, an error of 6 Btu in estimating the enthalpy at a given point results in a negligible error in extraction flow.

To arrive at a satisfactory state line for full power, it is then only necessary to connect, on a Mollier diagram, the point of initial pressure and enthalpy, h_0, with the point of throttle enthalpy at 90 percent of throttle pressure, and the state line end point (see Fig. 22). Extraction enthalpies can then be read at the appropriate shell or stage pressure.

The exhaust steam leaves the turbine at a total enthalpy of

$$h_c = h_i + EL + \frac{2544 \, SHP}{W} \left[\frac{RL}{100 - (3.5 + RL)} \right] \qquad (92)$$

This is the total exhaust enthalpy that should be used when establishing condenser duty. The last term represents the reheating of the exhaust steam due to the windage losses in the astern turbine.

4.3 Partial Flow and Extraction Pressures. A change in flow through a turbine, brought about by extraction of steam for feed heating, causes a change in pressure at every point (provided that flow is subsonic), even though the position of the state line is not significantly affected. A prediction of the pressure change as a function of flow can be made by considering the turbine as a nozzle. It is, of course, a collection of many nozzles, but the general behavior of its flow can be described by considering them to be lumped into a single overall nozzle. Also, the part of the turbine from any point, such as an extraction opening, to the condenser can be similarly considered as a single nozzle for the purpose being discussed.

Since the pressure at the condenser is very low, the pressure ratio from the point of interest to the condenser is typically supercritical, and maximum flow exists for the pressure at that point. Under such conditions, nozzle flow theory predicts that the flow parameter $(W/A)(\sqrt{T_0}/P_0)$ is constant, and in fact has a value of approximately 0.40 for superheated steam when T_0 is in degrees R, P_0 is in psi, W is in lb/sec, and A is in square inches. Further, for modest changes in conditions at a point in the turbine, the change in $\sqrt{T_0}$ is small compared to the change in P_0, and A is fixed, so that the relation $W \propto P_0$ follows. The pressure at a point should thus be proportional to the flow from that point to the condenser, and this prediction is found to be essentially true in practice; it is further found that the pressure is proportional to the flow past the point in question. This additional distinction is necessary because some of the steam passing a point may be extracted downstream, and therefore does not reach the condenser.

The principle stated in the foregoing is used to find shell pressures at extraction points, and from them the extraction enthalpies following small changes in flow. The state line does not shift significantly because of reasonable extraction flows, so that the enthalpies can be read from the intersections of the shifted pressure lines with the nonextraction state line.

The approximation outlined here is not quite adequate for reduced-power conditions, or at unusually large extraction flows. Other techniques, such as discussed next, must be used.

4.4 Lambda Ratio. For large variations in flow such as occur when reducing to 80% power or less, it is necessary to account for the change in efficiency because of the change in the ratio of blade speed to steam speed in successive stages. This can, of course, be done by returning to the original design and applying the theory discussed in Section 3 again. However, this normally cannot be accomplished in a timely manner, especially for preliminary work. The designer must therefore resort to other techniques based on external turbine characteristics. For this purpose, the stage group efficiency is related to a parameter known as the Lambda ratio.

In any given stage of a turbine, the efficiency is a function of the blade-speed/steam-speed ratio. This ratio is

$$\frac{U}{V} = \frac{\pi d_M N}{720(2gJ\Delta h')^{1/2}} \qquad (93)$$

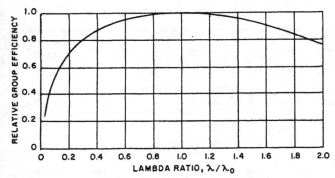

Fig. 26 Variation of group efficiency with Lambda ratio

where

d_M = mean blade diameter, in.
N = turbine rpm
$\Delta h'$ = isentropic expansion enthalpy difference, Btu/lb
U = blade speed, fps
V' = theoretical steam speed, from isentropic expansion, fps

Every turbine stage has a value of U/V' for which its efficiency is a maximum (this is shown by Fig. 18 of Chapter VI). For example, it is 0.5 for an ideal impulse stage. And although a propulsion turbine consists of a number of stages for each of which the ratio may be different, as when impulse and reaction stages are used in the same machine, there is always some value of

$$\sum \frac{d_M N}{(\Delta h')^{1/2}}$$

for which the efficiency of the entire turbine is a maximum (Σ implies summation over all stages). This parameter is known as *Lambda*, and is conventionally expressed as

$$\lambda_0 = \frac{1}{658 \cdot 10^6} \sum \frac{d_M^2 N^2}{\Delta h_0'} \qquad (94)$$

wherein the constants, including 0.5 for U/V', are included in the numerical coefficient. The efficiency at off-design points is a unique function of the Lambda ratio λ/λ_0, i.e.

$$\frac{\lambda}{\lambda_0} = \left(\frac{N}{N_0}\right)^2 \frac{\sum \Delta h_0'}{\sum \Delta h'} \qquad (95)$$

Figure 26 is a plot of the relative efficiency of a turbine against λ/λ_0. This plot is representative of impulse stage groups, but can be used to estimate the efficiency variation of any group with good accuracy.

Single stages, such as the first stage or control stage, similarly undergo variations in efficiency as a function of the Lambda ratio (or the velocity ratio U/V'). For further discussion concerning turbine control, see Chapter VI.

4.5 Nozzle Control and Reduced-Power Operation. Under rated conditions, the pressure drop from the throttle to the condenser is distributed among the stages with the first stage usually given a large share of the drop. At reduced power, which is obtained by reducing the first-stage nozzle area, the pressure distribution is altered, with an increased share of the pressure drop appearing across the first stage. If the reduction in nozzle area is known, the pressure at the first-stage exit can be calculated from principles previously stated, and from this the new pressure distribution can be found. The method of calculation is trial-and-error, and consists of choosing a first-stage exit pressure, then checking the choice by the following principle: the flow that this predicts for the first stage must equal the flow that it predicts for the remaining stages.

The flow through the first stage obeys the relation $W \propto AR$, which is a condensation of equation (66). A is the nozzle area, whose variation is assumed to be given; R is the last bracketed term in equation (65), and is seen to be a function of the pressure ratio. A choice of pressure ratio therefore determines a first-stage flow.

Critical flow can be assumed to occur through the remaining stages, and thus, by the discussion of Section 4.3, and flow conforms to the relationship $W \propto P_0/\sqrt{T_0}$, where P_0 and T_0 are the stagnation conditions at the entrance to the second stage. Knowledge of the first-stage efficiency is required in order to find values of P_0 and T_0. The efficiency under rated conditions is assumed to be known, and the efficiency under the reduced-flow conditions may be determined by means of the relationship between stage efficiency and velocity ratio as illustrated by Fig. 18 of Chapter VI. A prior step is to make an estimate of the turbine rpm, since reducing the nozzle area reduces it. The turbine rpm may be related to other parameters by the following approximations: (1) steam flow $\propto A$, (2) power $\propto$ flow, and (3) rpm $\propto \sqrt[3]{\text{power}}$. The stage efficiency and pressure ratio then determine the first-stage exit conditions, i.e., P_0 and T_0. It is better to take these to be the static conditions, since the velocity component is not likely to be recovered in the transition to the second-stage inlet.

Once P_0 is determined for the reduced nozzle area, other stage pressures can be estimated by assuming a linear pressure distribution between the first-stage exit and the condenser.

4.6 Auxiliary Turbines. Auxiliary turbines fall into two classifications; multistage machines where a significant effort is made to achieve good efficiency because the steam consumption of the unit is a fairly large proportion of the total, and single-stage units (usually velocity compounded) used to drive smaller auxiliaries or those infrequently used. Single-stage machines generally exhaust to an auxiliary exhaust system at fairly high back pressures with the exhaust steam being used for feedwater heating. In some cases single-stage machines exhaust into atmospheric condensers.

The principal application of multistage machines is for driving ship-service generators. They usually exhaust to an auxiliary condenser. In some applications, however, the turbine exhausts at pressures around 10 psia, and the exhaust is used in the low-pressure feed heater. To achieve an acceptable efficiency at these powers (generally 300 to 3000 kW), it is necessary to use small-diameter

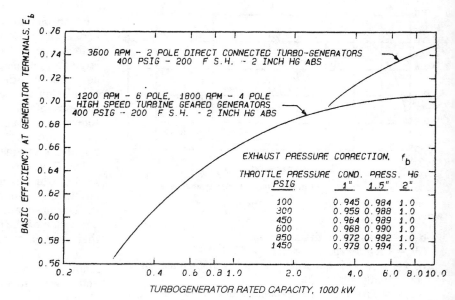

3600 RPM – 2 POLE DIRECT CONNECTED TURBO-GENERATORS
400 PSIG – 200 F S.H. – 2 INCH HG ABS

1200 RPM – 6 POLE, 1800 RPM – 4 POLE
HIGH SPEED TURBINE GEARED GENERATORS
400 PSIG – 200 F S.H. – 2 INCH HG ABS

EXHAUST PRESSURE CORRECTION, f_b

THROTTLE PRESSURE PSIG	COND. PRESS. HG		
	1"	1.5"	2"
100	0.945	0.984	1.0
300	0.959	0.988	1.0
450	0.964	0.989	1.0
600	0.968	0.990	1.0
850	0.972	0.992	1.0
1450	0.979	0.994	1.0

TURBOGENERATOR RATED CAPACITY, 1000 kW

Fig. 27 Basic efficiency of multistage condensing turbogenerators at rated load

rotors at high rotative speeds, with reduction gears. Occasionally, such a machine will also be used, direct coupled, for driving feed pumps. However, single-stage machines generally are used to drive feed pumps, as well as forced-draft fans and miscellaneous auxiliaries.

Steam rates are estimated by the same methods as outlined for propulsion turbines, although with less refinement. For multistage generator turbines, the efficiency is given for the output in kilowatts at the generator terminals, and the steam rate is expressed as

$$SR = \frac{3412}{AE \cdot E} \tag{96}$$

where

AE = available energy
E = turbogenerator efficiency

Following the procedure in reference 9, Fig. 27 gives the basic efficiency of a turbogenerator, E_b, as a function of rated capacity in kilowatts.

Correction factors for other than standard back pressures, f_b, are also given in Fig. 27. Figure 28 shows a correction factor, f_t, for the initial steam temperature, and Fig. 29 gives a correction factor, f_p, for the initial throttle pressure. In addition Fig. 30 gives a correction factor, f_L, for other than rated capacity. Thus

$$E = E_b f_b f_t f_p f_L \tag{97}$$

A calculation of the exhaust enthalpy from the turbogenerator is necessary to establish the auxiliary condenser requirements. The turbogenerator exhaust enthalpy is given by

$$h_w = h_0 - \frac{AE \cdot E}{\eta_M \eta_g} \tag{98}$$

where

h_w = exhaust enthalpy

h_0 = throttle enthalpy
η_M = mechanical efficiency
η_g = generator efficiency
= $e_n f_g$; see Fig. 31

The efficiency of a single-stage velocity-compounded machine is a function of the blade-speed/steam-speed ratio and the windage losses. The blade-speed/steam-speed ratio may be represented by the parameter

$$y = \frac{RPM \cdot DIA}{(\Delta h')^{1/2}} \tag{99}$$

where

RPM = turbine speed, rpm
DIA = mean wheel diameter, in.
$\Delta h' = h_0 - h_p$ = theoretical energy available, Btu/lb, based on an isentropic expansion from nozzle pressure to back pressure. Customarily, an allowance of 10 percent in pressure drop is made between the throttle valve inlet and the nozzles

Figure 32 is a plot of basic efficiency, E_b, against the parameter, $RPM \cdot DIA/(\Delta h')^{1/2}$ for conventional velocity-compounded two-moving-row impulse turbines with a throttle valve control. This basic efficiency is corrected for superheat by the factor, f_s, from Fig. 33 and then corrected for the operating load by the data, f_L, given in Fig. 34. The windage loss, LHP, is based on a standard 30 psia exhaust pressure; these losses are given by Fig. 35. Since the peripheral admission is usually small, windage losses are large. For other than standard 30 psia exhaust pressures, a correction must also be made for the pressure difference.

The turbine efficiency at operating load is then

$$E = E_b f_s f_L \frac{BHP}{BHP + LHP\left(\dfrac{13.7}{v}\right)} \tag{100}$$

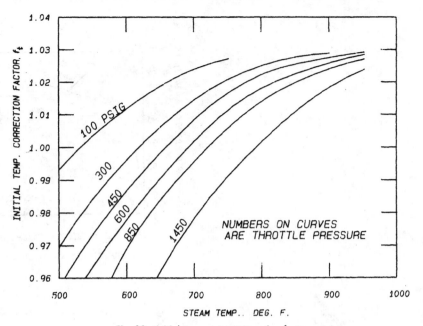

Fig. 28   Initial temperature correction factor

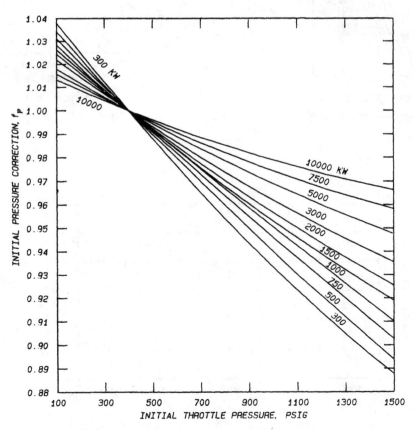

Fig. 29   Multistage condensing turbogenerator initial pressure correction factor

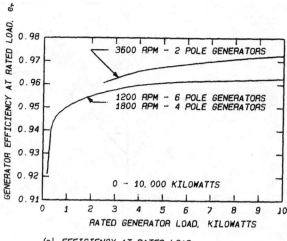

(a) EFFICIENCY AT RATED LOAD

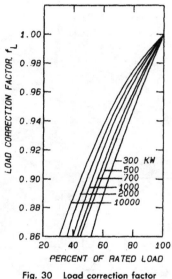

Fig. 30   Load correction factor

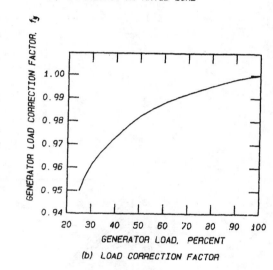

(b) LOAD CORRECTION FACTOR

Fig. 31   Efficiency of a-c generators

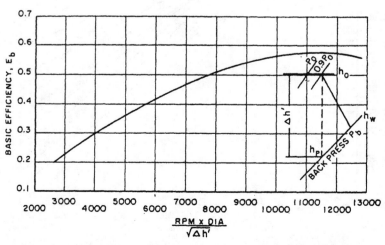

Fig. 32   Efficiency of single-stage (2-moving-row velocity compounded) turbines

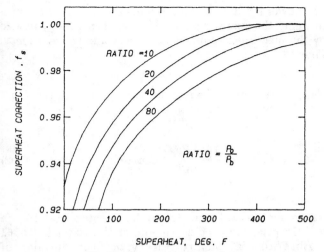

Fig. 33 Superheat correction factor

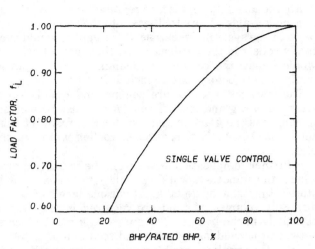

Fig. 34 Fractional load correction factor

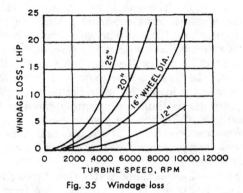

Fig. 35 Windage loss

where v is the specific volume, cubic feet per pound, of steam. For a first approximation, assume saturated vapor at the turbine back pressure.

The exhaust enthalpy is given by

$$h_w = h_0 - \Delta h'(E) \qquad (101)$$

where

h_w = exhaust enthalpy
h_0 = initial enthalpy at throttle

The methods given in this section for estimating the performance of main and auxiliary turbines are of acceptable accuracy for preliminary designs. Where specific data from a chosen vendor are available, they should be used instead.

Section 5
Thermodynamics of Steam Cycles

5.1 The Simple Steam Cycle. The most simple steam cycle possible consists of a steam turbine, condenser, condensate pump, surge tank, feed pump, and a boiler. However, most of these require auxiliary components in order to function. For example, the turbine requires a lubricating-oil system, with motor- or turbine-driven lubricating-oil pumps; the condenser requires cooling water (circulating water), which must be either pumped or scooped up by the forward motion of the ship; the condensate pump and feed pump require a source of power; and the boiler must have fuel-oil pumps and forced-draft fans. The pumps and blowers can be motor driven, with electric power being furnished by a generator driven off the shaft of the turbine, or they may be driven by small steam turbines or engines. It is also possible to include a turbine-driven generator set with its own auxiliaries.

Additional equipment is also required to remove air and noncondensable vapors that leak into the subatmospheric parts of the system. Otherwise, these vapors would eventually collect in the condenser and their partial pressure would impair the condenser vacuum. Therefore, an air pump or air ejector must be included.

The turbines also require sealing and gland leakoff steam, and a gland leakoff condenser must be included to condense the gland steam. Condensate from the main condenser is usually the source of cooling water for this unit.

A further complication arises from the need for a place where the unneeded working fluid (feedwater) can be stored, that is, a surge tank. At various levels of power, different volumes of fluid are required in the system, especially in the boilers. This surge tank can be located between the condensate pump and feed pump.

Thus, the simplest possible power plant is already quite complex. The thermal arrangement of the plant is illustrated in Fig. 36. The unit is assumed to develop 30,000 shp. The steam rate of the main propulsion turbines is calculated to be 5.46 lb/shp-hr with throttle conditions of 850 psia and 950 F and with the turbine exhausting to the condenser at 1.5 in. Hg absolute (see Fig. 37). To develop 30,000 shp, the throttle flow must be 163,800 lb/hr.

The generator load is estimated to be 1200 kW and the turbogenerator is thus appropriately rated at 1500 kW. The methods of Section 4.4 give a steam flow of 10,920 lb/hr for steam conditions of 850 psig and 950 F and a 1.5 in. Hg abs back pressure. The total steam flow is therefore 174,720 lb/hr.

Now trace the steam and water flow through the cycle illustrated by Fig. 36. The flow exhausting from the main turbine is 163,800 − 250 = 163,550 lb/hr (250 lb/hr to the gland condenser) and from the auxiliary turbine is 10,870 lb/hr (50 lb/hr to the gland condenser). The two gland leakoff flows return from the gland leakoff condenser to the main condenser. The condensate flow leaving the main condenser totals 174,720 lb/hr.

It is customary to assume that the main condenser drains are at saturation temperature and the auxiliary condensers have a 1 deg F drainwell depression in condensate temperature. At 1.5 in. Hg abs the saturation temperature is 91.7 F. Note that in the simple cycle illustrated in Fig. 36 the turbogenerator exhausts into the main condenser.

Entering the gland condenser there is a total enthalpy of 174,720 × 59.7 = 10,430,784 Btu/hr. The gland condenser receives gland steam at 1281 Btu/lb (assumed 200 Btu/lb less than throttle steam) and drains at a 10 deg F terminal difference or 101.7 F. This adds a total of 300 × (1281 − 69.7) or 363,390 Btu/lb to the condensate, making a total of 10,794,174 Btu/hr entering the surge tank. The feed leaves the surge tank at the same enthalpy with which it enters.

The feed pump puts an amount of heat into the feedwater equal to the total power of the pump, less any friction in the drive system. This friction work can be neglected but the heat from the power input is a significant quantity.

The power input is the total pump head in feet of feedwater times the quantity pumped in pounds per hour, divided by the mechanical equivalent of heat and the efficiency. Thus

Heat equivalent of feed pump work

$$= \frac{144 \Delta P v_f Q}{778 E} \text{ Btu/hr} \qquad (102)$$

Assuming the feed pump raises the pressure from 15 psia to 1015 psia, the specific volume of the water is 0.0161 ft³/lb. With a 50% pump efficiency, the heat equivalent of the feed pump work is 1,041,313 Btu/hr. This addition of heat gives a total of 11,835,487 Btu/hr entering the boiler. Assuming no leakage of steam, the steam leaves the boiler at 1481.2 Btu/lb, with a total thermal energy flow of 1481.2 × 174,720 = 258,795,264 Btu/hr. The difference between this total heat and that entering (258,795,264 − 11,835,487 = 246,959,777 Btu/hr) is the net heat added in the boiler by the fuel. With a boiler efficiency of 88 percent and a fuel having a higher heating value of 18,500 Btu/lb, the quantity of fuel burned is:

$$\text{Weight of fuel/hr} = \frac{246,959,777}{(18,500)(0.88)} = 15,170 \text{ lb/hr}$$

The specific fuel rate is the weight of fuel per hour divided by the net shaft horsepower, or 15,170/30,000 = 0.506 lb/shp-hr. The heat rate is the quantity of heat expended to produce one horsepower per hour and is calculated by dividing the net heat added to the plant, per hour, by the horsepower produced.

$$\text{Heat rate} = \frac{(15,170 \text{ lb/hr})(18,500 \text{ Btu/lb})}{30,000 \text{ shp}}$$

$$= 9,355 \text{ Btu/shp-hr}$$

This simple cycle omits many details that must necessarily be included in an actual steam plant. System losses and soot-blowing losses have been omitted, as well as steam for fuel-oil heating Steam and power must also be expended for ship services. Thus, the plant efficiency does not include all the energy that must be expended to produce the power.

These parasitic load requirements can be visualized if they are set down in a flow diagram such as Fig. 38.

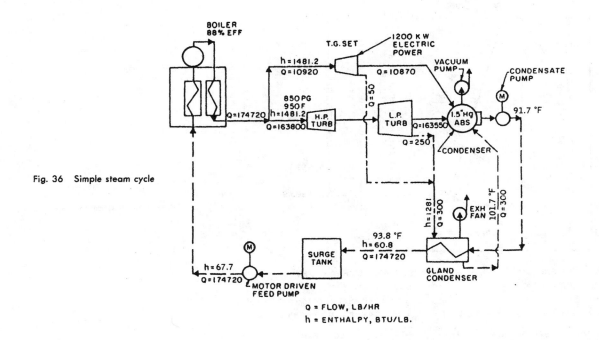

Fig. 36   Simple steam cycle

Q = FLOW, LB/HR
h = ENTHALPY, BTU/LB.

Figure 38 includes fuel-oil heating, steam leakage losses, and soot-blowing losses and has introduced a steam air ejector instead of an air pump. Provision has been made for makeup feed. Each stage of the air ejector uses 245 lb/hr of desuperheated steam at 1250 Btu/lb enthalpy. The first stage drains at 125 F; the second stage at 200 F. Fuel oil is heated by desuperheated steam, 1050 lb/hr being required, draining to a surge tank. Soot-blowing steam and system leakage are based on reference 9. Soot blowing requires 760 lb/hr, and system leakage losses are taken as ½% of the total steam flow, or about 900 lb/hr. Both of these items are lost and must be replaced by makeup feedwater introduced to the condenser.

The thermal energy added to each pound of water by the feed pump is the same as in the previous example, so that the thermal energy added to 177,920 lb/hr is 1,060,384 Btu/hr.

Allowance must be made for pressure and temperature drops in the main steam line. It is customary to allow about 2.5% on pressure, rounded up to the nearest 5 psi, and 5 deg F for temperature. Thus, the superheater outlet conditions are taken to be 875 psig and 955 F. Desuperheated steam system allowances are usually taken as 2.5% on pressure, rounded up to the nearest 5 psi with constant enthalpy, so the desuperheater outlet enthalpy is 1250 Btu/lb. The total superheated steam flow is 175,620 lb/hr and the desuperheated flow is 2300 lb/hr.

The flow of heat in the system illustrated by Fig. 38 may be tabulated as shown in Table 3.

The quantity of fuel required is determined by dividing the fuel heat output by the boiler efficiency of 0.88 and the higher heating value of standard fuel of 18,500 Btu/lb, to which is added 46 Btu/lb to account for the sensible heat added by the fuel-oil heaters (100 deg F rise at 0.46 specific heat):

$$\text{Fuel required} = \frac{250,668,237}{(0.88)(18,546)} = 15,359 \text{ lb/hr}$$

Dividing by the 30,000 shp output, the specific fuel consumption is found to be 0.512 lb/shp-hr.

5.2 The Regenerative Cycle. The power cycle shown in Fig. 38 is complete, but certain problems would arise if such a cycle were used. The feed temperature is extremely low; as a result, the economizer in the boiler would condense sulfur products from the flue gas, which would cause corrosion.

The products of combustion of marine fuels include sulfur dioxide (SO_2), which, when combined with water vapor, forms acids such as sulfurous acid (H_2SO_3) and sulfuric acid (H_2SO_4). These acids cause cold-end corrosion if allowed to condense on boiler components or flue gas ducting. Figure 39 gives the flue-gas acid dew point as a function of fuel-oil sulfur content and percent excess air. Boiler heat-recovery devices, such as economizers and gas air heaters, are designed to avoid lowering the flue gas temperature below the expected dew-point temperature of the flue-gas acids.

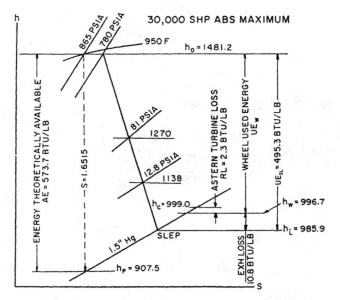

Fig. 37   State line and steam rate for cycle calculations

$E_b = 0.8527$ (SEE FIG. 23)

$f_T = 1.0125$ (SEE FIG. 24)

$RL = 0.35 \times 1.5 = 0.5$

LEAKAGE & MECHANICAL LOSSES = 3.5%

EXHAUST ANNULUS AREA = 25 FT²

$AE = 1481.2 - 907.5 = 573.7$

$UE_{SL} = E_b \times f_T \times AE = 0.8527 \times 1.0125 \times 573.7 = 495.3$

$h_i = h_o - UE_{SL} = 1481.2 - 495.3 = 985.9$

FIRST ITERATION, STEAM FLOW $= (1.04)\left(\dfrac{2544 \times 30,000}{495.3 \times 0.98}\right) = 163,600$ LB/HR

$\dfrac{W}{P_b \times A} = \dfrac{163,600}{1.5 \times 25.0} = 4360$ $\therefore$ EL = 10.8 (SEE FIG. 25)

SECOND ITERATION, STEAM FLOW $= (1.04)\left(\dfrac{2544 \times 30,000}{495.3 - 10.8}\right) = 163,800$ LB/HR

STEAM RATE $= \dfrac{163,800}{30,000} = 5.46$ LB/SHP-HR $h_c = h_i + EL + \dfrac{RL \times 2544}{\text{STEAM RATE}} = 999.0$

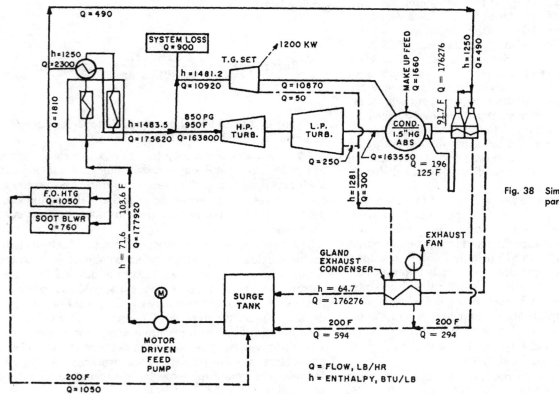

Fig. 38   Simple steam cycle with parasitic loads

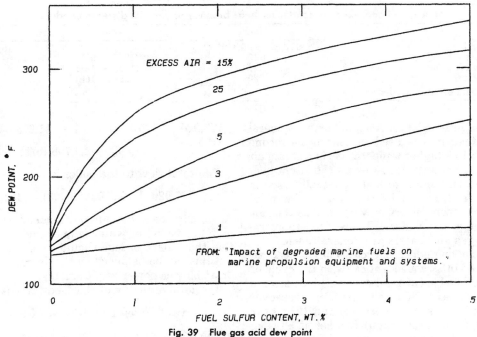

Fig. 39 Flue gas acid dew point

Further, the feedwater would have a high oxygen content (no deaeration is provided), which would cause corrosion and pitting in the boiler water sides. Entrained oxygen and air can be released by bringing the feedwater to a boil. By using steam, bled from the turbines, the feedwater can be raised to the boiling temperature and held there in a deaerating feedwater heater in an efficient manner as the bled steam has already done useful work in the high-pressure turbine before being used for feed heating.

Bleed feed heating may also be done after the deaerating feed heater and feed pump. For highest efficiency,

there should be $N-1$ feed heaters, where N is the number of turbine stages, since this leads to maximum regeneration, but such an array of heaters and bleed points is not justified in marine service. Stationary practice employs an extensive number of heaters, but such plants are not restricted by the space limitations of a ship's engine spaces, and they develop much higher power, so that additional complexity is justified.

The general practice of bleeding can also be used for miscellaneous heating services, such as fuel heating, hotel services, and distilling plants; therefore, these services

Table 3   Heat flow tabulation for simple steam cycle shown in Fig. 38

| | Condensate Flow | | |
	lb/hr	Btu/lb	Btu/hr
Leaving main condenser	176,276	59.7	10,523,677
Air ejector intercondenser	196	(1250 − 93)	226,772
Air ejector after condenser	294	(1250 − 168)	318,108
Gland exhaust condenser	300	(1281 − 168)	333,900
Condensate to surge tank	176,276	64.7	11,402,457
Fuel oil heating drain at 200 F	1,050	168	176,400
Gland condenser drain at 200 F	300	168	50,400
Air ejector after condenser drain at 200 F	294	168	49,392
Total leaving surge tank and entering feed pumps	177,920	65.6	11,578,649
Feed pump regain			1,060,384
Total to boiler	177,920	71.6	12,739,033

The boiler output is then: superheated steam $\qquad 175,620 \times 1483.5 = 260,532,270$ Btu/hr
desuperheated steam $\qquad \underline{2,300} \times 1250 \quad = \quad \underline{2,875,000}$

Total $\qquad 177,920$ lb/hr $\qquad 263,407,270$ Btu/hr

Feed input $\qquad \underline{12,739,033}$
Net heat output from fuel $\qquad 250,668,237$ Btu/hr

Table 4   Gland exhaust condenser heat balance under bleed-steam conditions

	lb/hr	Btu/lb	Btu/hr
Leaving the main condenser	$176{,}276 - 0.4297\ Q_I$	59.7	$10{,}523{,}677 - 25.65\ Q_I$
Air ejector intercondenser	196	$(1250 - 93)$	226,772
Air ejector after condenser	294	$(1250 - 168)$	318,108
Gland exhaust condenser	300	$(1281 - 168)$	333,900
Total, leaving gland exhaust condenser	$176{,}276 - 0.4297\ Q_I$		$11{,}402{,}457 - 25.65\ Q_I$

are included in the heat balance. Also, in some regenerative cycles bleed steam may be used for heating the incoming combustion air to a higher temperature in a steam air heater. This type of heating also recovers the latent heat of vaporization. Steam air heaters are generally used in cycles where the feedwater is heated to a maximum of about 285 F, so that economizers may be used as the sole heat-recovery unit in the boilers. Where feed temperatures are higher, gas air heaters are generally used.

Consider a simple single-heater cycle, using a deaerating feed heater, and otherwise identical with the cycle of Fig. 38. Steam could be bled at the crossover pipe between the HP and LP turbines at 86.5 psia. (Assume crossover pressure = 0.1 initial absolute pressure) and 1273 Btu/lb, and led through a pressure regulator, set at 50 psia, to a deaerating heater. The bled steam and the incoming feedwater could be sprayed together, resulting in a saturation temperature of 281 F, 250.1 Btu/lb enthalpy. Bleeding steam from the crossover would reduce the horsepower developed since less steam passes through the low-pressure unit. To compensate for this, the throttle flow must be increased.

Let Q_I be the quantity of steam bled in lb/hr and ΔQ_t the increase in main throttle flow required. Then, using the figures developed in Fig. 37, the reduction in heat available to the turbine is

$$\text{Loss heat} = Q_I(1273 - 996.7)$$

and the flow needed to replace this heat is

$$\Delta Q_t(1481.2 - 996.7)$$

Equating these gives

$$\Delta Q_t = 0.5703\ Q_I$$

Then, leaving the exhaust of the LP turbine, the steam flow would be

$$163{,}550 + \Delta Q_t - Q_I = 163{,}550 - 0.4297 Q_I$$

Employing the procedure used in Section 5.1, the condensate heat entering the gland exhaust condenser is as given in Table 4.

Since the after condenser drains and gland exhaust condenser drains could no longer flow to a surge tank (now the deaerating feed heater), they would be collected in a freshwater drain collecting tank (FWDCT) and pumped into the deaerating feed heater. The deaerating feed heater would also be enlarged to allow it to act as a surge tank, and the combined unit would be called a deaerating feed tank (DFT). A balance around the DFT gives the values shown in Table 5.

Since the DFT enthalpy is 250.1 Btu/lb:

$$(177{,}920 + 0.5703\ Q_I)250.1 = 11{,}678{,}649 + 1247.35\ Q_I$$

$$Q_I = 29{,}708\ \text{lb/hr}$$

The total flow to the boiler is

$$177{,}920 + (0.5703)(29{,}708) = 194{,}862\ \text{lb/hr}$$

Assuming the same pump efficiency, the feed pump regain per pound of water is unchanged (i.e., 6 Btu/lb), so that the generator load should have increased somewhat, but this can be balanced by the decrease in boiler forced-draft blower power requirements.

The total enthalpy of the feedwater is

$$194{,}862(250.1 + 6) = 49{,}904{,}158\ \text{Btu/hr}$$

to the boiler, and the boiler output is

$$
\begin{aligned}
(194{,}862 - 2300)(1483.5) &= 285{,}665{,}727\ \text{Btu/hr} \\
(2300)(1250) &= \underline{2{,}875{,}000} \\
\text{Total boiler output} &= \overline{288{,}540{,}727} \\
\text{less feed input} &= \underline{49{,}904{,}158} \\
\text{Net heat input to boiler} &= \overline{238{,}636{,}569}\ \text{Btu/hr}
\end{aligned}
$$

Dividing the net boiler heat input by a fuel heating value of 18,546 Btu/lb and a boiler efficiency of 88% gives fuel oil requirements of 14,622 lb/hr. Dividing by the 30,000 shp output results in a specific fuel rate of 0.4874 lb/shp-hr. This is a saving of 4.8% over the simple cycle, and in addition the boiler is protected from corrosion.

A further gain in efficiency accrues in this cycle. In the high-pressure turbine, more steam passes than in the nonbleed condition, and less passes through the low-pressure turbine. Since the exhaust annulus is the same as before, the volume flow is reduced, and the leaving velocity must be less also. The apparent exhaust flow is 163,800 − 0.4297 × 29,708 = 151,034 lb/hr (apparent flow is throttle flow less any bleed but including the gland leak-off steam). The exhaust annulus is 25 ft^2, so 151,034/(1.5)(25) = 4028, and from Fig. 25 the exhaust loss = 9.9 Btu/lb and h_w = 985.9 + 9.9 = 995.8 Btu/lb, vice the 996.7 Btu/lb shown by Fig. 37.

Then, the horsepower developed by the LP turbine is

$$\frac{1}{1.04}(1273 - 995.8)(151{,}034)/2544 = 15{,}824$$

and the horsepower developed by the HP turbine is

$$\frac{1}{1.04}(1481.2 - 1273)(151{,}034 + 29{,}708)/2544 = 14{,}223$$

and the total is 30,047 shp.

Thus, the reduced steam flow in the LP turbine slightly increases the turbine efficiency, and the calculation must

Table 5 Surge tank (deaerating feed tank) heat balance under bleed-steam conditions

	lb/hr	Btu/lb	Btu/hr
Leaving gland exhaust condenser	$176{,}276 - 0.4297\,Q_I$		$11{,}402{,}457 - 25.65\,Q_I$
After condenser drain	294	168	49,392
Gland condenser drain	300	168	50,400
Bleed flow	$+\,Q_I$	1273	$+\,1273\,Q_I$
F.O. heater drain	1050	168	176,400
Total leaving surge tank	$177{,}920 + 0.5703\,Q_I$		$11{,}678{,}649 + 1247.35\,Q_I$

Table 6 Revised surge tank (deaerating feed tank) heat balance with corrected flow rates

	lb/hr	Btu/lb	Btu/hr
Leaving main condenser	$176{,}006 - 0.4289\,Q_I$	59.7	$10{,}507{,}558 - 25.61\,Q_I$
Air ejector intercondenser	196	(1250 − 93)	226,772
Air ejector after condenser	294	(1250 − 168)	318,108
Gland exhaust condenser	300	(1281 − 168)	333,900
Total to surge tank	$176{,}006 - 0.4289\,Q_I$		$11{,}386{,}338 - 25.61\,Q_I$
After condenser	294	168	49,392
Gland condenser drain	300	168	50,400
Bleed flow	$+\,Q_I$	1273	$+\,1273\,Q_I$
Fuel oil heater drain	1050	168	176,400
Total leaving surge tank	$177{,}650 + 0.5711\,Q_I$		$11{,}662{,}530 + 1247.39\,Q_I$

be repeated with a new ratio of $\Delta Q_t/Q_I$ and a new non-bleed flow. Since the exhaust enthalpy has been changed, the equivalent nonbleed water rate is

$$WR = \frac{1.04 \times 2544}{1481.2 - 995.8} = 5.451 \text{ lb/shp-hr}$$

Then, the throttle flow is

$$5.451 \times 30{,}000 + \Delta Q_t = 163{,}530 + \Delta Q_t$$

and

$$\Delta Q_t = (277.2/485.4)\,Q_I = 0.5711\,Q_I$$

and the exhaust flow is

$$163{,}530 - 250 + \Delta Q_t - Q_I = 163{,}280 - 0.4289\,Q_I$$

where the 250 lb/hr is the gland leakoff steam.

The flow leaving the condenser is:

Main turbine exhaust.............. $163{,}280 - 0.4289\,Q_I$
Turbogenerator exhaust........... 10,870
Makeup feed................................ 1,660
Air ejector drain............................ 196
Total.................................. $176{,}006 - 0.4289\,Q_I$ lb/hr

A revised surge tank heat balance that incorporates the flow corrections is shown by Table 6.

The DFT outlet enthalpy is 250.1 Btu/lb; therefore

$$(177{,}650 + 0.5711\,Q_I)250.1 = 11{,}662{,}530 + 1247.39\,Q_I$$

Solving this equation for the extraction flow Q_I gives a value of 29,666 lb/hr; consequently, with the 6 Btu/lb feed pump regain, the total energy flow to the boiler is $[177{,}650 + 0.5711(29{,}666)][(250.1 + 6) = (194{,}592)(256.1) = 49{,}835{,}011$ Btu/hr. The boiler output is

$$(194{,}592 - 2300)1483.5 = 285{,}265{,}182 \text{ Btu/hr}$$
$$(2300)(1250) = 2{,}875{,}000$$
Total output = 288,140,182 Btu/hr
input = 49,835,011
Net heat from boiler = 238,305,171 Btu/hr

The fuel consumption is now 14,602 lb/hr, and the specific fuel consumption is 0.4867 lb/shp-hr.

A second iteration need not be carried out since the further change in exhaust loss is now very small.

5.3 Heat Balance Calculations. The fuel rate calculations of Section 5.2, with the accompanying flow diagram, are properly called a heat balance. Most heat balances are completed by placing the estimated and derived values on the flow diagrams. Such a diagram is of major importance in the design of the power plant since the operating characteristics of the plant and its components are determined from heat balance values.

The preparation of a heat balance is usually the first step in initiating the design of a steam propulsion plant. The results of the preliminary heat balance are the fundamental input to purchase inquiries, and also for such plant design tasks as sizing of piping. The first heat balance may be done from the approximate data in reference 9, but, subsequently, data supplied by the component vendors are used to update the calculation.

In the last example in 5.2 a direct solution of the heat balance problem was presented. Obviously this problem would be more difficult if several bleed points are needed and more heaters employed, especially if the bleed pressures vary with flow. The problem becomes even more complex if ships' service steam is added to the balance.

Some marine engineers prefer to use an indirect trial-and-error solution by assuming a condensate flow leaving the main condenser and then deriving the bleed requirements by applying the first law of thermodynamics to each heater. The calculation is completed by calculating the shp that would result from this trial flow. A second or third calculation is then carried out until the calculated shp is that desired. Other marine engineers prefer to use a direct solution by developing a set of simultaneous equations in terms of constants and an unknown boiler evaporation rate, E. The equations are then solved to obtain the

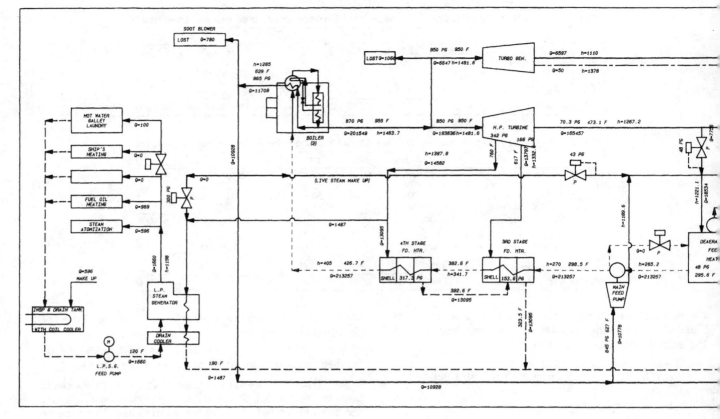

Fig. 40   Heat balance

total boiler evaporation, E, and the various other flows. For a description of this method, see reference 10.

The direct solution method given in reference 10 can be further simplified by expanding the equation for apparent wheel horsepower

$$WHP = \frac{\sum\limits^{i} Q_i h_i}{2544} \qquad (103)$$

where Q_i is the flow through a given stage and h_i the enthalpy drop through a given stage. The resulting equation is

$$WHP = \frac{\begin{aligned}(E - G)(h_0 - h_1) + (E - G - B_1)(h_1 - h_2) \\ + (E - G - B_1 - B_2)(h_2 - h_3) + \ldots \\ + (E - G - B_1 - B_2 - \ldots - B_n)(h_n - h_w)\end{aligned}}{2544} \qquad (103a)$$

where

E = total boiler evaporation
G = steam from boiler to auxiliaries
B_n = extraction steam flows at various bleed points
n = number of extractions
h_0 = throttle steam enthalpy
h_n = enthalpy at extractions
h_w = $h_0 - UE_w$

In application the required wheel horsepower is first calculated by adding the mechanical and external losses to the desired shaft horsepower. For marine turbines the mechanical losses (turbine bearings, gear bearings, tooth mesh, and windage) and part of the external losses (turbine radiation and net end packing leakage) are assumed to be a constant 3.5%. The remaining external loss, astern turbine windage (RL), in percent is a function of exhaust pressure and is determined from equation (85). The wheel horsepower then becomes

$$WHP = SHP \left[1 + \frac{3.5 + RL}{100 - (3.5 + RL)} \right] \qquad (104)$$

The enthalpies at respective extraction points are determined from the Mollier diagram by noting the intersection of the turbine state line and the extraction pressure. The steam flows to auxiliaries, direct from the boilers, are estimated from their respective duties. The extraction steam flow from each bleed point is calculated, in terms of E, from the duty of the feed heater or air heater served by the extraction and the consumption of other auxiliaries being served by the extraction. It is possible to get all extraction flows in terms of E and one or more constants.

Figure 40 is a completed flow diagram for a marine steam turbine power plant. The example plant produces 30,000 shp in a cycle with four stages of feedwater heating, which yields a final feedwater temperature of 426.7 F. The plant is fitted with a rotary regenerative air heater and operates with 850 psig and 950 F throttle conditions

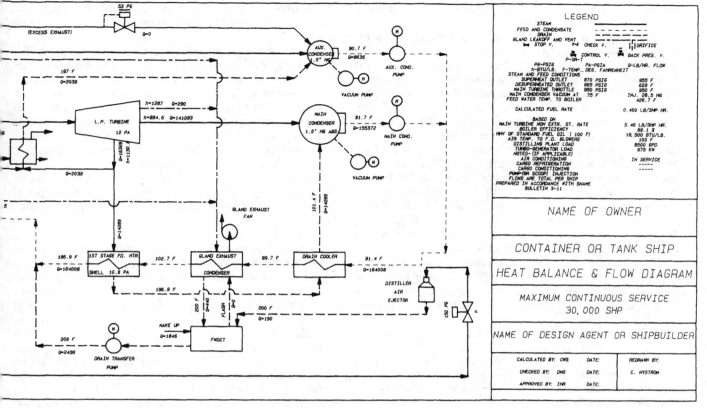

flow diagram

and 1½ in. Hg condenser pressure. An overview of the heat balance calculation follows.

The first step is to estimate optimum bleed points by starting with an assumed crossover pressure of 10% of the main throttle pressure. Since the chosen plant uses a rotary regenerative air heater, the feed temperature leaving the deaerating feed tank is independent of the condensation temperature of the sulfur products in the flue gas. Ideally it would be desirable to have the deaerator operating at the main turbine crossover pressure minus system pressure losses. To prevent deaerating pressure fluctuation, as power levels change, and consequential boiler feed pump suction pressure problems, the deaerating pressure is set at 62.7 psia, about 75% of the crossover pressure. This pressure setting will allow for continued extraction to an approximately 75% flow through the turbine, which corresponds to about 75% of rated power. To provide heating steam when no bleed steam is available (as for example, when going astern), makeup steam is supplied from the desuperheated system through a pressure regulator set at 57.7 psia. At certain times, there may be too much auxiliary exhaust steam, so a back-pressure regulator, set at 67.7 psia, will discharge excess steam to the auxiliary condenser. Thus the auxiliary exhaust system can fluctuate only between 57.7 and 67.7 psia. This limit should be sufficient to prevent the feed suction water from the DFT from flashing during maneuvering, especially if the DFT is placed well above

the pump (40 to 75 ft). Note that since this is a direct-contact heater, there is no terminal temperature difference and the deaerating feed tank outlet temperature is 295.6 F.

When preliminary estimates are made, and a specific turbine has not been selected, the design of the turbine stages and their corresponding pressures are not known. It should be assumed that the pressures most desirable for the overall cycle performance are available. These are obtained in an optimized cycle by dividing the state line in equal enthalpy rises above and below the crossover point. Once the optimized stage enthalpies are known, they are marked off on the state line and the corresponding stage pressure noted. For bleeds over 30 psia assume a 7% pressure drop at constant enthalpy to the feedwater heater being served. For other bleeds assume a 10% pressure drop at constant enthalpy.

The optimum bleed point for the low-pressure stage is selected by locating the 1130.8 Btu point (half way between crossover bleed and state line end point) on the state line and noting this point corresponds to the 12 psia pressure line. Using a 10% pressure drop the first-stage (LP) heater shell pressure is calculated to be 10.8 psia. The corresponding saturation temperature is 196.9 F. Since a 10 deg F terminal difference is usually needed between the heating steam and heated water in a shell-and-tube type heat exchanger, the condensate outlet temperature is 186.9 F. The heater drain is assumed to have no subcooling and is, therefore, at 196.9 F.

As with the LP heaters, the optimum bleed points for the HP heaters are selected so that the turbine state line is divided into equal enthalpy drops. Other considerations may result in the first extraction (final heater) enthalpy to be rounded lower than that necessary for equal enthalpy drops. The first-extraction stage pressure of 356.7 psia and a second-extraction stage pressure of 180.7 psia are obtained from the Mollier diagram. Since the third- and fourth-stage heater supply steam have over 200 F of superheat (760 F actual temperature and 426.7 F saturation temperature; 617 F actual temperature and 367.6 F saturation temperature, respectively), the heat exchangers have a desuperheater section and a 0 deg F terminal temperature difference is used. Using a 7% system pressure drop yields shell pressures of 332 psia for the fourth-stage and 168.3 psia for the third-stage heaters.

The amount of bleed steam necessary for the heaters is obtained by performing a weight flow and heat flow balance for each heater. In all cases simultaneous equations in terms of E can be written, inserted into equation (103a), and solved.

As noted on the heat balance flow diagram, Fig. 40, the calculated overall fuel rate for this 30,000 shp four-heater cycle plant is 0.460 lb/shp-hr. This result is based on fixed bleed points, which are assumed to be available, and this preliminary heat balance can be considered suitable for preliminary specifications and contract negotiations. Once the contract is settled and the steam turbine manufacturer is chosen, the balance must be reworked. In this instance, the bleed pressure is considered to be a variable that is linearly proportional to the flow beyond the bleed point. Thus, when the apparent exhaust steam flow is estimated, the pressure can be read directly from shell pressure curves such as those given in Fig. 12 of Chapter 6. Then the enthalpy of the bleed can be read from a state line. This permits a calculation of the low-pressure bleed flow; therefore, the flow past the intermediate bleed and high-pressure points can be derived and the bleed pressures and enthalpies can be read. The remainder of the calculation proceeds as before.

5.4 High-Performance Cycles. The choice of the steam propulsion plant cycle arrangement requires a careful economic balance between the relatively simple cycles having two stages of feedwater heating and the more complex four- and five-heater cycles. The cycles having larger numbers of feedwater heaters are higher in first cost and in maintenance costs, but lower in fuel costs. They are usually favored for ships of high power, which spend a large fraction of their time at sea.

Once the gains in fuel economy from regenerative feed heating have been fully exploited, additional gains must come from raising the initial pressure and temperature of the steam supplied to the turbine. However, vanadium attack on superheater tubes effectively limits the maximum practical steam temperature for marine plants to about 950 F, so that raising the temperature above this value must await significant improvements in metals and fuels technology. On the other hand, the pressure can be raised and, if higher pressures are used in conjunction with reheating, worthwhile improvements in the fuel rate can be achieved.

Reheating involves returning the steam to the boiler following partial expansion through the turbine. The partially expanded steam is heated to a temperature approximately equal to the initial temperature in a separate superheater (the *reheater*), then returned to complete its expansion through the turbine. Typical reheat cycles use steam initially at 1435 psig, 950 F, and four or five stages of feedwater heating. Fuel rates of about 0.39 lb/shp-hr are possible under such conditions.

Reheat plants are common in stationary practice, many using two stages of reheat and twelve or more stages of feedwater heating. Such plants are economically justified because the total horsepower ratings are much higher than in marine practice.

Because it is impractical to pass steam back to the reheater during astern operation, the reheater must be protected by means such as bypassing the hot gas away from it when the ahead throttle is closed. Other solutions involve controllable-pitch propellers, reversing gears, and electric transmission; all of these methods of astern operation allow the turbine to run in a single direction, therefore steam always passes through the reheater.

With the usual cross-compound articulated reduction gear, a convenient arrangement for a reheat plant would be one with the reheater receiving the high-pressure turbine exhaust and returning it to the low-pressure turbine. This arrangement is not used, however, because it would require a reheat pressure of about 50 psia, necessitating large connecting pipes, a large reheater, and exhaust steam that is slightly superheated. A condenser with a desuperheating section would be much larger than the conventional unit. The turbine arrangement normally used with 1450 psia, 950 F steam consists of a high-pressure turbine that exhausts to the reheater, an intermediate-pressure turbine that receives the reheated steam, and a low-pressure turbine. The HP and IP units can be mounted on the same shaft in the same casing, with their high-temperature ends back to back. Thus there are only two input pinions, as with the conventional nonreheat turbine.

Predictions of a reheat turbine state line during preliminary design studies are more difficult than for nonreheat turbines, since there is a wider range of characteristics that affect the line. Nonetheless, as a first estimate of a suitable state line, the high-pressure turbine exhaust pressure can be selected at about 20% of the throttle pressure, and a state line can be constructed with an efficiency of 70% (excluding bearing and gear losses). A 10% loss of pressure in the reheater may be assumed. The balance of the state line can be constructed using the method outlined in Section 4.2.

Figure 41 illustrates the state line of a propulsion turbine with reheat. Figure 41 is for a total shaft horsepower of 30,000 hp, throttle conditions of 1450 psia and 950 F, and an exhaust pressure of 1.5 in. Hg. The turbine exhaust area is taken as 18 ft². With a high-pressure turbine exhaust pressure equal to about 20% of the initial pressure and a reheater pressure loss of about 10%, the pressures to and from the reheater are determined to be 290 psia and 260 psia, respectively.

Employing the data shown on Fig. 41, the reheat turbine basic efficiency may be computed from the expression

$$E_b = \left[0.742 + 0.032\left(\frac{SHP \times AE)}{1000(AE + \Delta h')}\right)^{0.4}\right]\left[1 + \frac{4}{P_2}\right] \quad (105)$$

$$E_b = 0.865$$

and the temperature correction may be computed as

$$f_t = \frac{T + 4100}{5000} \quad (106)$$

$$f_t = 1.01$$

The state line used energy for the low-pressure turbine is

$$UE_{SL} = E_b f_t (AE) \quad (107)$$

$$UE_{SL} = 449 \text{ Btu/lb}$$

With an astern turbine loss of 0.5% [see equation (85)], a first estimate of the steam flow is determined to be 138,400 lb/hr from equation (91); therefore

$$\frac{W}{P_b A} = \frac{138,400}{(1.5)(18)} = 5130$$

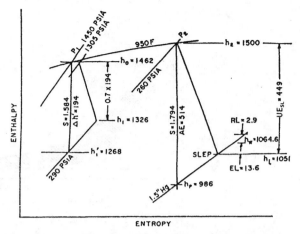

Fig. 41   Estimate of state line for propulsion turbine with reheat

and the exhaust loss, EL, is 13.6 Btu/lb from Fig. 25. Finally, the steam rate for the reheat turbine can be computed to be 4.63 lb/shp-hr.

The power consumed by the feed pumps is high because of the high boiler pressure; consequently, pump drives having a high efficiency are desirable. To accomplish this, the feed pumps may be driven by a pinion from the main reduction gear. In other designs the feed pumps are driven by a shaft extension from the turbogenerator.

Guidelines for performing reheat power plant heat balances are given in reference 11.

Section 6
Diesel Engine Thermodynamics

6.1   Diesel Cycle. The thermodynamic cycle on which diesel engines operate is based on the work of Rudolph Diesel (1858–1913). Whereas Dr. Diesel used a four-stroke mechanical cycle engine in his work, the thermodynamic diesel cycle is also applicable to the two-stroke mechanical cycle.

The thermodynamic diesel cycle is graphically shown in Fig. 42. A cycle that more closely approximates the actual processes in diesel engines is the dual-combustion cycle, which is analyzed in Chapter 3. The diesel cycle consists of an isentropic compression process (1-2), a constant-pressure combustion process (2-3) followed by an isentropic expansion process (3-4), and finally a constant-volume exhaust process (4-1). The application of the four-stroke mechanical cycle to a diesel thermodynamic cycle engine can be related to Fig. 42(a) as intake stroke (0-1), compression stroke (1-2), power stroke (2-3-4), exhaust and exhaust stroke (4-1-0). The application of the two-stroke mechanical cycle to the diesel thermodynamic cycle can be related to Fig. 42(a) as compression stroke (1-2) and power stroke (2-3-4). Exhaust and scavenging take place in the constant-volume process (4-1), theoretically, when the piston is at the bottom-dead-center (BDC) position. Figure 42(b)

shows the thermodynamic cycle on the temperature-entropy, T-S, plane.

The characteristic feature of the diesel cycle is injection of the fuel, theoretically starting at the end of the compression stroke [top dead center (TDC)], and continuing at such a rate that the burning process proceeds at constant pressure (2-3). Normally only air is in the cylinder during the compression stroke. To initiate combustion the temperature of compression, T_2, must be above the auto-ignition temperature of the fuel.

6.2   Diesel Cycle Thermal Efficiency. For a first approximation it is common practice to analyze the diesel cycle using the concepts of the air-standard cycle. In the air-standard cycle the mass and properties of the fuel as injected and the working fluid are assumed to have the properties of air throughout. Furthermore, the heat added is assumed to be the energy released by the reaction of the fuel with the air. Although gas (air) tables should be used, it is customary to assume constant specific heats for the air.

For isentropic compression

$$T_2 = T_1\left(\frac{V_1}{V_2}\right)^{k-1} = T_1 r_c^{k-1} \quad (108)$$

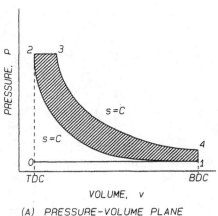

(A) PRESSURE-VOLUME PLANE

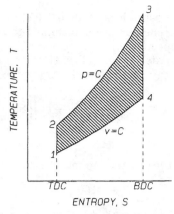

(B) TEMPERATURE-ENTROPY PLANE

Fig. 42   Diesel cycle

where k is the ratio of specific heat at constant pressure (c_p) of air to the specific heat at constant volume (c_v) of air. For air the value of k is approximately 1.4. The term r_c is called the compression ratio. Note that the temperature terms in equation (108) are in degrees absolute.

The thermal efficiency of the air-standard diesel cycle is:

$$\eta = \frac{W}{Q_A} = \frac{Q_A - Q_R}{Q_A} \qquad (109)$$

where the work, W, is the difference between the heat added, Q_A, and the heat rejected, Q_R. The heat rejection process is a constant-pressure process and consequently $Q_R = U_4 - U_1$. Accordingly the cycle efficiency can be expressed as:

$$\eta = 1 - \frac{Q_R}{Q_A} = 1 - \frac{u_4 - u_1}{h_3 - h_2} = 1 - \frac{c_v(T_4 - T_1)}{c_p(T_3 - T_2)}$$

$$= 1 - \frac{(T_4 - T_1)}{k(T_3 - T_1)} \qquad (110)$$

where u is the internal energy and h is enthalpy. Application of the isentropic compression equation (108) and Charles' Law, $T_3/T_2 = V_3/V_2$, allows T_3 to be written in terms of T_1:

$$T_3 = T_2\left(\frac{V_3}{V_2}\right) = T_1\, r_c^{k-1}\left(\frac{V_3}{V_2}\right) = T_1\, r_c^{k-1}\, r_k \qquad (111)$$

The ratio $r_k = V_3/V_2$ is called the fuel cutoff ratio. By using the converse of equation (108) for isentropic expansion and noting $V_4 = V_1$, T_4 can be expressed as T_1:

$$T_4 = T_3\left(\frac{V_3}{V_4}\right)^{k-1} = T_3\left(\frac{r_k V_2}{V_1}\right)^{k-1} = T_1\, r_c^{k-1}\, r_k\left(\frac{r_k V_2}{V_1}\right)^{k-1}$$

$$= T_1 r_k^k \qquad (112)$$

Substituting equations (108), (111), and (112) into equation (110) yields:

$$\eta = 1 - \frac{T_1 r_k^k - T_1}{k(T_1 r_c^{k-1}\, r_k - T_1\, r_c^{k-1})}$$

$$= 1 - \frac{1}{r_c^{k-1}}\left[\frac{r_k^k - 1}{k(r_k - 1)}\right] \qquad (113)$$

The efficiency of the Otto cycle, a constant-volume, heat-addition, spark-ignition cycle, is the same as equation (113) without the bracketed factor

$$\eta = 1 - \frac{1}{r_c^k - 1} \qquad (114)$$

Since the bracketed factor is always greater than 1, it follows that for a constant compression ratio, r_c, the Otto cycle has a higher efficiency than the diesel cycle. Actual diesel engines must reach temperatures above the auto-ignition temperature of the fuel at the end of the compression stroke, T_2, and since only air is being compressed, precluding detonation of a fuel-air mixture, they operate at significantly higher compression ratios. Diesel engines generally have compression ratios above 13:1; however, some large turbocharged engines have compression ratios that are somewhat lower. Most naturally aspirated, high-speed diesel engines operate with compression ratios of 19-20:1. Although the power of a diesel cycle increases with increased cutoff ratio, the thermal efficiency of the diesel cycle decreases. For actual engines the maximum practical cutoff ratio is on the order of 10% of the stroke. Above 10%, excessive fuel will be injected in the cylinder and smoking will occur.

As an example of the application of equation (113), assume an engine operating on the air-standard diesel cycle has a compression ratio, r_c, of 20:1, a cutoff at 10% of stroke, the engine is at operating temperature, and $k = 1.35$. (See Section 6.3 for an explanation of the reason for assuming that k has a value of 1.35, instead of 1.4.) Note that with a 20:1 compression ratio, the V_2 is at 5% of the stroke from TDC; therefore, $r_k = 0.10/0.05 = 2$:1:

$$\eta = 1 - \frac{1}{20^{1.35-1}} \left[\frac{2^{1.35} - 1}{1.35(2-1)} \right] = 1 - \frac{1}{2.85} \left[\frac{2.54 - 1}{1.35} \right]$$

$$= 1 - 0.40 = 0.60$$

It must be noted that the 60% thermal efficiency calculated above does not account for the combustion efficiency or the mechanical efficiency of the engine. The most efficient diesel engines can be expected to have overall efficiencies on the order of 50% and corresponding specific fuel rates of 0.3 lb/bhp-hr or somewhat less.

6.3 Diesel Cycle Ignition Requirements.

The ideal air-standard diesel cycle incorporates two isentropic processes. The equations governing isentropic compression and expansion are written in terms of dimensionless ratios raised to powers, which are a function of the ratio, k, of the specific heats. Since diesel engine cylinders are cooled to temperatures below the average temperature of the cylinder gas during compression and expansion, it is more accurate to consider the processes polytropic and replace k by n in the appropriate equations noted above. For most diesel engines at operating temperature the polytropic constant, n, is between 1.3 and 1.35. For cold engines, as during starting, the polytropic constant is approximately 1.2.

To further illustrate the use of the above equations, one can look at a hypothetical diesel engine having a compression ratio of 20:1. If one assumes that the engine is "cold" and that the air in the cylinder at the start of compression is at 14.7 psia and at 60 F, equation (108) can be used to calculate the temperature of the air charge at the end of the compression stroke.

$$T_2 = T_1 \left(\frac{V_1}{V_2} \right)^{n-1} = (60 + 460) \left(\frac{20}{1} \right)^{1.2-1} = 947\ R = 487\ F$$

For ignition to take place, upon the injection of the fuel into the cylinder, the auto-ignition temperature of the fuel must be below 487 F. Some fuel oils have auto-ignition temperatures below 487 F, but to insure auto-ignition, it will be necessary to ensure that the fuel being used has an ignition quality index (formerly stated as cetane number) high enough to yield combustion with this relatively low compression temperature. The ignition quality of a fuel influences the starting and combustion roughness of an engine.

To ensure starting under cold-starting conditions, the designer of a diesel engine provides some method of adding heat to the compression gas. These methods include the use of jacket-water heaters, air heaters, glow plugs, or auxiliary fuel and are most important when it is desired to use fuels with a low ignition quality, or when the engines are to be started in a cold environment. An alternative to using one of the above methods of assisting the starting process is to raise the compression ratio of the engine. It must be noted that unnecessary raising of engine compression ratio results in undue engine loads and increased engine weight. One need only to examine the pressure at the end of the compression stroke of the above

hypothetical engine in the hot running condition to appreciate the forces involved. The pressure at the end of compression is

$$P_2 = P_1 \left(\frac{V_1}{V_2} \right)^n = (14.7) \left(\frac{20}{1} \right)^{1.35} = 839.4\ \text{psia}$$

6.4 Mean Effective Pressure.

As noted above, the efficiency of the diesel cycle depends upon the compression ratio, r_c, and the cutoff ratio, r_k. A quantity of special interest to designers and operators of diesel engines is the work, W, done on the piston divided by the displacement volume, $V_1 - V_2$. The dimensions of this quotient are those of pressure. The essence of this quantity is an effective constant pressure which, if exerted on the piston for the entire power stroke, would yield work equal to the work of the cycle. This quantity is called the mean effective pressure, MEP:

$$\text{MEP} = \frac{W}{V_1 - V_2} \tag{115}$$

The MEP represents the height of a rectangle whose enclosed area is equal to the enclosed area of the P-V diagram (1-2-3-4-1) for the diesel cycle illustrated by Fig. 42(a). Since it represents the work the gas does on the piston acting through one stroke, it can be related to the power produced by an engine by the expression

$$hp = \frac{PLAN}{33,000} \tag{116}$$

where

hp = horsepower
p = mean effective pressure (MEP), psi
L = length of stroke, ft
A = area of piston, in.^{2}
N = number of power strokes per minute

In slow-speed diesels and larger medium-speed engines, the mean effective pressure can be obtained by attaching an indicator or an IMEP meter to each cylinder of the engine. In higher-speed engines, indicator cards (P-V diagrams) may be obtained electronically. The pressure so determined is called the indicated mean effective pressure, IMEP. The IMEP is a measure of the energy applied to the engine by the working fluid. A correlative to the IMEP is the brake mean effective pressure, BMEP. The BMEP is found by measuring the engine power with a dynamometer and solving equation (116) for the mean effective pressure. If the BMEP is divided by the IMEP, the result is the mechanical efficiency of the engine.

Examining the P-V diagram for a diesel cycle shows that varying the cutoff ratio, r_k, has a profound effect on the area under the curve and, therefore, the mean effective pressure and power produced by the engine.

As an example consider a six-cylinder, two-stroke engine with a 25-in. bore, and a 95-in. stroke operating with an IMEP of 200 psi at 120 rpm. To calculate the indicated power produced, the number of power strokes per minute must be determined and substituted in equation (116).

Since each cylinder makes one power stroke per revolution in a two-stroke engine, the number of power strokes per minute is

$$N = (6 \text{ power strokes/rev}) (120 \text{ rpm})$$
$$= 720 \text{ power strokes/min}$$

The indicated power developed by the engine is

$$ihp = (200 \text{ lb/in.}^2) \left(\frac{95}{12} \text{ ft}\right) \left(\frac{\pi}{4}\right) \frac{(25 \text{ in.})^2 (720/\text{min})}{(33,000 \text{ ft-lb/min-hp})}$$

$$= 16,957 \text{ hp}$$

Section 7
Gas Turbine Thermodynamics

7.1 Brayton (Gas Turbine) Cycle. Gas turbines operate on a thermodynamic cycle that can be idealized as the Brayton cycle. As indicated by Fig. 43(a), the Brayton cycle consists of an isentropic compression process (1-2), a constant-pressure heat-addition process (2-3), and an isentropic expansion process (3-4) followed by a constant-pressure heat-rejection process (4-1). Figure 43(b) shows the cycle on a temperature-entropy, T-S, plane. To achieve the processes required of the Brayton cycle, four basic components must be included, namely, a compressor (1-2), a combustor or heat source (2-3), a turbine (3-4), and a

heat sink (4-1). The net work, W_{net}, is the difference between the expansion work, W_t, and the compression work, W_c. The pumping power, W_c, required in a Brayton cycle is high, consuming between 40% and 70% of the expansion work, W_t, depending on the pressure used in the cycle. If the cycle is a closed cycle, as in Fig. 44(a), the heat sink is a simple heat exchanger. If the cycle is an open cycle, as in Fig. 44(b), the heat sink is the atmosphere, which receives the turbine discharge at a high temperature and supplies the compressor with cool air. For the Brayton cycle to be practical, two criteria must be met: the turbine

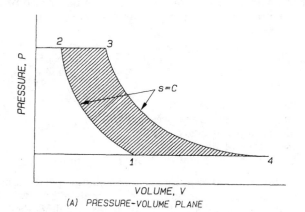

(A) PRESSURE-VOLUME PLANE

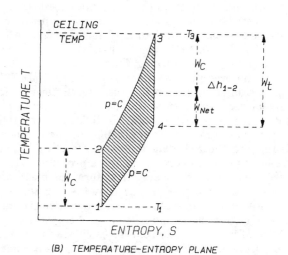

(B) TEMPERATURE-ENTROPY PLANE

Fig. 43 Brayton cycle

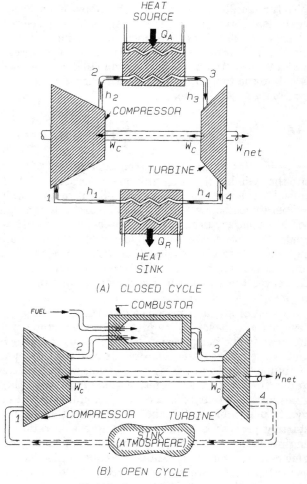

(A) CLOSED CYCLE

(B) OPEN CYCLE

Fig. 44 Gas turbine cycles

inlet temperature, T_3, must be high and the efficiencies of the compressor and turbine must be high.

7.2 Ideal Brayton Cycle Efficiency. The Brayton cycle is commonly analyzed as a closed air-standard cycle. In an air-standard cycle the working fluid is assumed to be air throughout and often the specific heat of the air is assumed to be constant. In an open air-standard cycle the mass and properties of the fuel are neglected and the heat added is assumed to be the energy released by the combustion of the fuel.

The thermal efficiency of the Brayton cycle is:

$$\eta = \frac{W_{\text{net}}}{Q_A} = \frac{Q_A - Q_R}{Q_A} = \frac{(h_3 - h_2) - (h_4 - h_1)}{(h_3 - h_2)}$$

$$= 1 - \frac{h_4 - h_1}{h_3 - h_2} \tag{117}$$

where

Q_A = heat added to cycle
Q_R = heat rejected from cycle
h = enthalpy at select cycle state points

Assuming a constant specific heat, equation (117) can be written as:

$$\eta = 1 - \frac{T_4 - T_1}{T_3 - T_2} \tag{118}$$

Noting that the processes (1-2) and (3-4) are isentropic processes and that the pressure ratio, r_p, is $p_2/p_1 = p_3/p_4$, the relationships for the isentropic process temperatures are:

for compression: $\dfrac{T_2}{T_1} = \left(\dfrac{p_2}{p_1}\right)^{(k-1)/k} = r_p^{(k-1)/k}$ (119)

for expansion: $\dfrac{T_3}{T_4} = \left(\dfrac{p_3}{p_4}\right)^{(k-1)/k} = r_p^{(k-1)/k}$ (120)

From equations (119) and (120) it can be seen that

$$\frac{T_2}{T_1} = \frac{T_3}{T_4} \quad \text{or} \quad \frac{T_4}{T_1} = \frac{T_3}{T_2} \tag{121}$$

The use of some algebra yields

$$\frac{T_4 - T_1}{T_3 - T_2} = \frac{T_1}{T_2} = \frac{T_4}{T_3} \tag{122}$$

from which equation (118) can be rewritten in the following form:

$$\eta = 1 - \frac{T_1}{T_2} = 1 - \frac{T_4}{T_3} \tag{123}$$

Writing the temperature ratio of equation (123) in terms of the pressure ratio, as noted by equation (120) yields

$$\eta = 1 - \frac{1}{r_p^{(k-1)/k}} \tag{124}$$

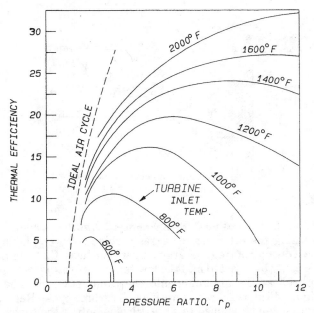

Fig. 45   Gas turbine efficiency versus pressure ratio

7.3 Efficiency of a Simple Cycle Gas Turbine. Examination of equation (124) suggests that increasing the compression ratio of a gas turbine will increase the thermal efficiency. Although true for the ideal Brayton cycle, one must note that the improvement in efficiency is followed by a decrease in work per unit mass of the working fluid, air. In addition one must note that for an actual turbine the Brayton cycle efficiency is modified to reflect the effect of compressor efficiency, η_c, turbine efficiency, η_t, and pressure ratio, r_p, as well as the effect of the compressor inlet temperature, T_1, and the turbine inlet temperature T_3. The gas turbine simple cycle efficiency can be written as:

$$\eta = \left[\frac{\eta_t T_3 - T_1 r_p^{(k-1)/k}/\eta_c}{T_3 - T_1 - T_1(r_p^{(k-1)/k} - 1)\eta_c}\right]\left[1 - \frac{1}{r_p^{(k-1)/k}}\right] \tag{125}$$

Examining the Brayton cycle efficiency modifying term, the left side of equation (125), one notes that in addition to a high compressor efficiency, turbine efficiency, and pressure ratio, the higher the turbine inlet temperature, T_3, and the lower the compressor inlet temperature, T_1, the more efficient the cycle will be. Since the compressor inlet temperature is set by the local ambient temperature, the variable temperature is the turbine inlet temperature, T_3. In a gas turbine the turbine inlet temperature is limited by material limitations and component cooling techniques. Gas turbine designers can control the turbine inlet temperature by adjusting the amount of excess air supplied to the combustor. The more excess air supplied, the lower the turbine inlet temperature.

The result of equation (125), assuming 85% for the compressor and turbine efficiencies, is shown graphically in Fig. 45. Examination of Fig. 45 shows that for each turbine inlet temperature there is pressure ratio that yields a maximum thermal efficiency.

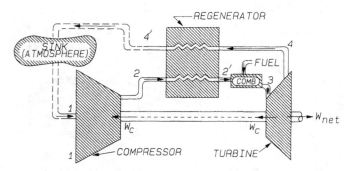

Fig. 46   Regenerative gas turbine cycle

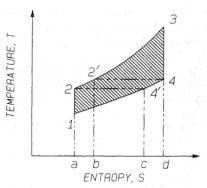

Fig. 47   Ideal regenerative gas turbine cycle

7.4   Regenerative Cycle Efficiency. Another way of increasing the efficiency of a gas turbine is by employing the concept of regeneration, in which heat that would normally be rejected into the low-temperature sink is used to preheat the air going into the combustion chamber. Figure 46 shows a regenerative gas turbine cycle. Regeneration can be applied to a gas turbine only if the turbine exhaust temperature T_4 is higher than the compressor outlet temperature, T_2. From the equation for isentropic compression, equation (119), T_1 and T_2 can be seen to be related by the pressure ratio, r_p.

The thermal efficiency of a regenerative gas turbine cycle is

$$\eta = \frac{W}{Q_A} = \frac{Q_A - Q_R}{Q_A} = \frac{(h_3 - h_2') - (h_4' - h_1)}{(h_3 - h_2')} \quad (126)$$

Assuming a constant specific heat, equation (126) can be written as

$$\eta = \frac{(T_3 - T_2') - (T_4' - T_1)}{(T_3 - T_2')} \quad (127)$$

On a temperature-entropy plane the ideal regenerative cycle appears as shown in Fig. 47. The area under curve 4-4' represents the heat absorbed from the exhaust and added to the combustion air. The area under curve 2-2' represents the heat absorbed by the combustion air. Note the two areas are equal. Consequently, equation (127) can be written as

$$\eta = \frac{(T_3 - T_4) - (T_2 - T_1)}{(T_3 - T_4)} = 1 - \left(\frac{T_2 - T_1}{T_3 - T_4}\right) \quad (128)$$

Equation (128) can be rewritten as

$$\eta = 1 - \frac{T_1}{T_3}\left[\frac{(T_2/T_1 - 1)}{1 - T_4/T_3}\right] \quad (129)$$

Using equations (119) and (120) it can be shown that

$$\eta = 1 - \frac{T_1}{T_3} r_p^{(k-1)/k} \quad (130)$$

It is interesting to note that as with a regenerative cycle, the efficiency increases with increased turbine inlet temperature but decreases with pressure ratio. With a particular turbine inlet temperature, the thermal efficiency of a regenerative gas turbine cycle peaks as a pressure ratio much lower than with a simple Brayton cycle. Since most aircraft-derivative gas turbines have high pressure ratios, they cannot be operated regeneratively.

Section 8
Waste Heat from Diesel and Gas Turbine Engines

8.1   Internal-Combustion Engine Waste Heat. A large fraction of the heat input to an internal-combustion engine, either diesel engine or gas turbine, is wasted as sensible and latent heat in the exhaust gases. Smaller but significant fractions are lost via cooling of the jacket water, lube oil, and inlet air (turbocharged engines with aftercoolers only). For example, the fractions of heat input going into the waste heat streams of an internal combustion engine are approximately as follows:

	HIGH-SPEED ENGINE	LOW-SPEED ENGINE	GAS TURBINE
to exhaust	0.35	0.25	0.65
to jacket water	0.15	0.07	...
to lube oil	0.05	0.03	0.005
to aftercooler	0.05	0.13	...

As the internal-combustion engine becomes more efficient, the amount of waste heat decreases and the availability of the waste heat also might decrease. Of all the waste heat available from internal-combustion engines, the exhaust heat has the greatest availability and can easily be used to produce usable steam. Whereas high-speed, four-stroke diesel engines have full-load exhaust temperatures of approximately 1000 F, low-speed, two-stroke, long-stroke diesel engines have full-load exhaust temperatures of approximately 500 F. Gas turbines have full-load exhaust temperatures of approximately 900 to 1000 F. In evaluating the amount of exhaust waste heat available, one needs to also examine the amount of exhaust gas flow. Four-stroke diesel engines have exhaust-gas flows of about 10 to 15 lb/bhp-hr, two-stroke engines about 15 to 18 lb/bhp-hr, and gas turbines about 20 to 35 lb/bhp-hr.

The heat rejected by internal-combustion engines to jacket water (diesels), to lube oil (diesels and gas turbines), and aftercoolers (diesels) is low-availability heat (90 to 300 F) and, consequently, has only limited uses aboard ship.

8.2 Steam from Waste Heat. Steam can be produced in a heat exchanger (waste-heat boiler) located in the exhaust ducts of diesels and gas turbines. The maximum steam pressure obtainable in a simple waste-heat boiler is limited by the exhaust-gas temperature, but otherwise the pressure is set by considerations regarding the use of the steam and the quantity needed. Supplemental firing may be used to increase the amount of steam generated, but when supplemental firing is used, the fuel so consumed must be added to the fuel consumed by the prime mover when calculating the overall plant efficiency and fuel rate.

If steam is to be used solely for low-temperature heating purposes, a relatively low pressure, say 15 psig, may be adequate, but usually the heat available is far in excess of low-pressure heating needs. On diesel ships that burn high-viscosity fuels, a minimum steam pressure might be set by the need to heat the fuel to the range of 250 to 300 F (100 psi). Often, in ships with a high propulsion power, the ship-service electrical needs at sea can be met by waste-heat steam supplied to a turbine generator. The higher the steam pressure, the lower will be the turbine steam consumption, but also the lower will be the quantity of steam that can be produced. Figure 48 illustrates alternative steam production at 50, 100, and 150 psig, showing that with the same inlet temperatures, and same "pinch point," more steam is produced and more energy is absorbed at progressively lower pressures. The minimum temperature difference, or "pinch point," as indicated, is usually the governing consideration in the steam quantity that can be produced. Pinch points from 20 to 90 deg F are reasonable with typical values being 40 to 50 deg F. Large pinch point values result in smaller (less expensive) waste-heat boilers, but less heat absorbed or lower steam pressures. If high-pressure steam is required of a waste-heat recovery unit, additional steam is sometimes obtained at a lower pressure in a second boiler downstream of the first. Multi-pressure waste-heat boilers are commonly used in shoreside cogeneration power plants to optimize the thermal efficiency of the plant and occasionally have been used in ships.

The minimum temperature to which the exhaust gas is cooled is also a limitation since, in order to avoid corrosion

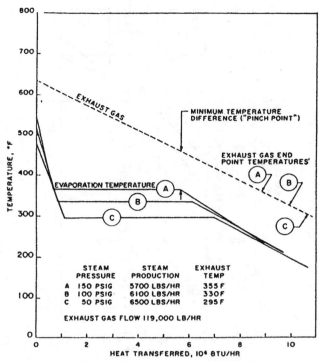

Fig. 48 Steam and engine exhaust-gas temperatures and heat transferred for different steam pressures for a 7500-bhp diesel engine

in the cold end of the boiler and stack, the temperature should not be allowed to drop below the acid dew point.

Waste-heat steam systems are designed in a variety of forms, but generally contain the components expected in a self-contained system. The designer, in making a heat balance, applies the same techniques outlined earlier in this chapter. Provisions must be made to control the steam pressure and temperature, which would otherwise vary with the steam demand, and with the quantity and temperature of the exhaust gas that the engine supplies in meeting the propulsion load. If a reasonably constant steam pressure is desired, and waste-heat energy is available, an automatic steam dump (to condenser) system can be provided. In some fire-tube arrangements, the boiler water level can be varied to control the steam pressure and quantity. An alternative is to provide for supplemental firing prior to the waste-heat boiler or to fire the standby oil-fired boiler when the pressure drops below a preset value.

Design calculations require knowledge of the exhaust-gas quantity, temperature, and specific heat. These will differ among different engines, but Figs. 49, 50, 51, and 52 (from reference 12) give fairly typical values for large two-stroke turbocharged marine engines. The exhaust gas from two-stroke diesels generally has properties that are about the same as boiler flue gas with 100% excess air.

The design of the boiler heat transfer surface is largely a problem of convective heat transfer, and in principle is the same as for the oil-fired boiler designs discussed earlier, except that the intertube radiative conductance will be so low that it can be neglected.

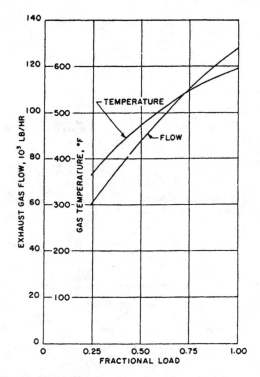

Fig. 49   Typical diesel engine exhaust flow and temperature

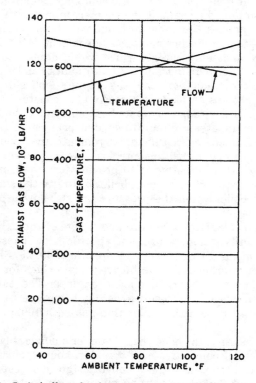

Fig. 50   Typical effect of ambient temperature on diesel engine exhaust
flow and temperature

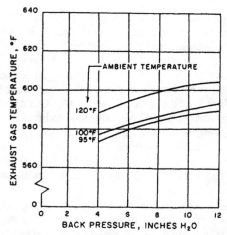

Fig. 51   Typical effect of back pressure and aftercooler outlet temperature
on diesel engine exhaust temperature

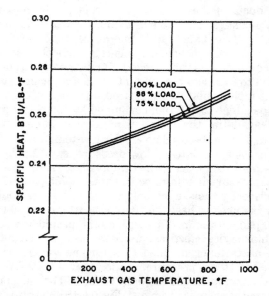

Fig. 52   Typical diesel engine exhaust-gas specific heats for a two-stroke
turbocharged engine

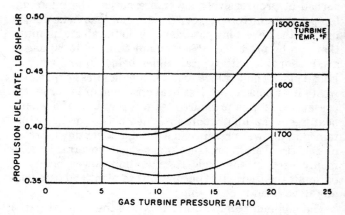

Fig. 53   Influence of gas turbine pressure ratio on COGAS-cycle fuel rate

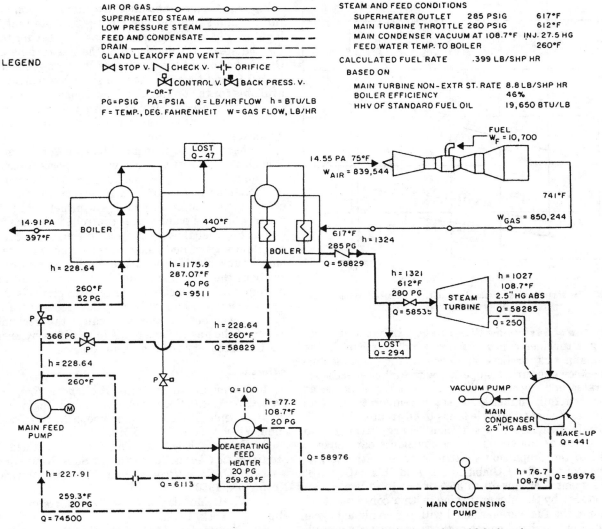

LEGEND

AIR OR GAS
SUPERHEATED STEAM
LOW PRESSURE STEAM
FEED AND CONDENSATE
DRAIN
GLAND LEAKOFF AND VENT
◁▷ STOP V. ◁◁ CHECK V. ⊣├ ORIFICE
◁▷ CONTROL V. ◁▷ BACK PRESS. V.
P-OR-T
PG = PSIG   PA = PSIA   Q = LB/HR FLOW   h = BTU/LB
F = TEMP., DEG. FAHRENHEIT   W = GAS FLOW, LB/HR

STEAM AND FEED CONDITIONS
SUPERHEATER OUTLET	285 PSIG	617°F
MAIN TURBINE THROTTLE	280 PSIG	612°F
MAIN CONDENSER VACUUM AT 108.7°F	INJ. 27.5 HG	
FEED WATER TEMP. TO BOILER		260°F

CALCULATED FUEL RATE   .399 LB/SHP HR
BASED ON

MAIN TURBINE NON-EXTR ST. RATE 8.8 LB/SHP HR
BOILER EFFICIENCY   46%
HHV OF STANDARD FUEL OIL   19,650 BTU/LB

Fig. 54   Design-point heat-balance diagram for a combined gas turbine and steam turbine (COGAS) cycle

8.3   Use of Diesel Engine Cooling Water Heat.

Approximately 20 to 25% of the total energy supplied to some diesel engines is lost through the cooling water. Depending on the engine design, there may be various cooling-water circuits servicing the engines, each operating at a different temperature and, consequently, different waste-heat availability. The most typical cooling-water circuits are the jacket-water circuit, the lubricating-oil circuit, and the turbocharger aftercooler circuit.

Diesel engines operate with jacket-water temperatures of 150 to 250 F. At the low-temperature range the availability of the coolant is so low that the only practical use of the heat is as a source of heat for the ship's vacuum freshwater distilling plant. Engines using high-temperature jacket cooling systems can supply jacket-water heat to absorption air-conditioning units in addition to distilling plants.

As with jacket-water heat rejection, the lubricating-oil heat rejection can vary in temperature (110 to 190 F) and,

consequently, the availability and quantity of the energy vary. For practical purposes the availability of this energy is too low to be effectively used aboard ship.

Cooling water coming from turbocharger aftercoolers of highly turbocharged engines, while low in quantity, is often of sufficiently high temperature (250 to 300 F) to be used for preheating waste-heat boiler feedwater, for an absorption air-conditioning system, or for other relatively high-temperature, low-demand services.

It must be noted that when using cooling-water waste heat, the device extracting the heat (distilling plant or absorption air-conditioning system) acts as a variable heat sink. The system must be designed so that under all operating conditions, the appropriate amount of heat is dissipated from the system, and the addition of a heat-reclaiming device in a cooling system cannot endanger the prime mover. Fully equipped, full-sized cooling systems are, therefore, installed in addition to any heat-recovery provisions.

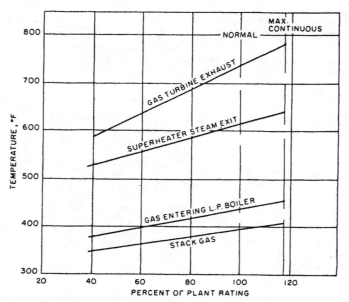

Fig. 55 Performance of a combined gas turbine and steam turbine cycle
at fractional power

8.4 Use of Gas Turbine Waste Heat. Exhaust-gas heat from gas turbines can be used to produce steam in the same manner as for diesel engines, and for the same purposes. Since gas turbines operate with higher specific air flow rates and, therefore, higher exhaust-gas flows, as well as generally higher exhaust-gas temperatures, than do diesels, the heat available tends to be greater than with diesels. In fact, there is sufficient energy available to suggest use of the steam in a propulsion steam turbine geared to the propulsion shaft in parallel with the gas turbine in a so-called COmbined Gas And Steam (COGAS) arrangement. Perhaps 20 to 35% of the total power can be produced by the steam turbine, with a consequent reduction in the fuel rate obtained with the combined cycle system. For a high utilization of the waste heat in a propulsion turbine, a multi-pressure waste-heat boiler and a mixed-pressure steam turbine (intermediate-pressure steam inducted at an appropriate stage) must be used.

The design objective in a combined gas turbine and steam turbine plant is usually a minimum propulsion fuel rate, and its attainment requires balancing between gas turbine and steam turbine requirements. For instance, the thermal energy in the exhaust is an inverse function of the gas turbine efficiency, so that the less efficient the gas turbine, the greater the energy available to the steam turbine. Figure 53, which was taken from an analysis in reference 13, illustrates this point. The figure shows the combined fuel rate as a function of gas turbine pressure ratio. For a simple gas turbine plant with typical component efficiencies, these curves would show a sharp decrease in fuel rate as the pressure ratio is increased from low values, until a minimum is reached at a pressure ratio in the neighborhood of 12 to 14. Instead, the curves are essentially flat in this range. The explanation is that an increasing efficiency for the gas turbine is accompanied by less energy in the exhaust and, consequently, less contribution to shaft horsepower from the steam turbine.

Figure 54 is a heat balance diagram for a combined gas turbine and steam turbine plant [13]. Note that the fuel input is only to the gas turbine, as the heat for the boiler for the steam turbine is from the gas-turbine exhaust. Observe also that the gas leaves the high-pressure boiler at 440 F, and thus still has considerable thermal energy available for the production of additional steam at a lower pressure. A second, low-pressure boiler is provided to make steam for the deaerating feed heater or for use in a mixed-pressure steam turbine.

The heat balance shown in Fig. 54 is for design power. It is also of interest to see how the important parameters change as the load is reduced. Figure 55, also from reference 13, illustrates part-load performance. Actually, the effect on the system parameters is influenced by the manner in which the plant load is controlled. For the example given, the fuel flow to the gas turbine is controlled to maintain a governed gas-generator rpm. The steam turbine is uncontrolled, with the output being determined solely by the energy available in the gas turbine exhaust. An important characteristic of COGAS plants is their ability to maintain a high efficiency down to low fractions of rated power output; this is because the declining efficiency of the gas turbine tends to be offset by an increasing heat recovery for the steam turbine.

References

1 *Thermodynamic and Transport Properties of Steam*, American Society of Mechanical Engineers, 1967.

2 Joseph H. Keenan and Joseph Kaye, *Gas Tables*, John Wiley & Sons, New York.

3 Frank O. Ellenwood and Charles O. Mackey, *Thermodynamic Charts*, John Wiley & Sons, New York, 1944.

4 D. Q. Kern, *Process Heat Transfer*, McGraw-Hill, New York, 1950.

5 W. H. McAdams, *Heat Transmission*, McGraw-Hill, New York, 1942.

6 "Standards for Steam Surface Condensers," Heat Exchange Institute.

7 "Standards of the Tubular Exchanger Manufacturers Association," Tubular Manufacturers Association.

8 A. Egli, "The Leakage of Steam through Labyrinth Seals," *Trans.* American Society of Mechanical Engineers, 1935.

9 "Recommended Practices for Preparing Marine Steam Power Plant Heat Balances," SNAME Technical and Research Publication No. 3-11.

10 A. S. Thaeler and D. C. MacMillan, "Heat Balance Calculations for Marine Steam Plants," *Trans.* SNAME, Vol. 46, 1938.

11 "Reheat Supplement to 3-11 (Marine Balance Practice)," SNAME Technical and Research Publication No. 3-36.

12 A. Norris, "Developments in Waste Heat Systems for Motor Tankers," *Trans.* Institute of Marine Engineers, 1964.

13 R. M. Marwood and C. A. Bassilakis, "The Thermodynamic Design of a Combined Steam and Gas Turbine Marine Propulsion System," ASME Paper 67-GT-16, American Society of Mechanical Engineers, 1967.

CHAPTER III

Alan L. Rowen | # Diesel Engines

Section 1
Introduction

1.1 Scope. This chapter is intended to provide the marine engineer with information relevant to the application of diesel engines of conventional and common design to ships. Engines of conventional and common design are taken as reciprocating, compression-ignition engines with stationary cylinders arrayed in-line or in a V-configuration, with the pistons driving a horizontal, rotating crankshaft located below the cylinders. Only generalized performance data and details are included. For specific data, reference should be made to the publications of the engine manufacturers.

1.2 Basic Principles and Description. Diesel engines of conventional design are shown in Figs. 1, 2, 3, and, schematically, in Fig. 4. In these engines, gas pressure in the cylinder acts on the piston, forcing it down during the power stroke to drive the crankshaft through connecting rods. The extreme positions reached by the piston correspond to the top and bottom dead center positions (TDC and BDC) of the crank, and are so designated. The inside diameter of the cylinder is the bore. The distance traveled between dead centers is the stroke, and the corresponding volume is the swept volume, or displacement, of the cylinder. The volumes above the piston at the extremities of its travel are the clearance volume or compression volume at TDC, and the cylinder volume at BDC. The ratio of cylinder volume to clearance volume is the nominal compression ratio.

The cycle on which diesel engines operate is illustrated in Fig. 5, in which the pressure in the cylinder is plotted against the volume of the combustion space. The solid line is an idealization of the cycle, discussed in detail in Section 2; the actual cycle is more realistically represented by the dashed trace. A charge of air, trapped in the cylinder when the air valves or ports are closed, is compressed on an upward stroke of the piston (1–2 in Fig. 5), until its temperature is sufficiently high to ignite fuel, which is injected as the piston approaches TDC. Initially, the fuel burns rapidly, leading to a corresponding rapid rise in pressure as the piston passes through TDC (2–3). Fuel injection is continued for a period determined by the power required from the engine, so that the cylinder pressure remains high even as the piston begins to descend, until all of the fuel has been burned at (4). The gases then expand, continuing to force the piston down to complete the power stroke (4–5). At the end of the power stroke, the exhaust ports or valves are opened at (5), and the gas pressure is released (5–1).

The term compression-ignition refers to the fact that the fuel is ignited only by the high temperature of the compressed charge, and is one characteristic of the cycle. Another characteristic is the extension of the fuel injec-

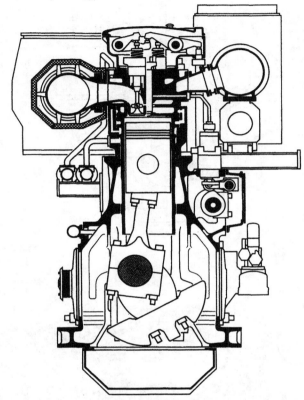

Fig. 1 Medium-speed engine with cylinders in-line

tion, and therefore of the combustion period, into the power stroke. In fact, as originally conceived by Rudolf Diesel, there was to be no initial rapid combustion, and all of the fuel was to burn as the piston descended. While the solid trace of Fig. 5, with combustion divided into the initial, constant-volume phase followed by the constant-pressure phase, is more representative of modern diesel engines, the term *diesel cycle* is reserved for cycles in which all combustion occurs at constant pressure.

All the processes of the cycle can be made to occur in one cylinder during a single revolution of the crankshaft, corresponding to two strokes of the piston, or they may occur over two revolutions, that is, in four strokes. Both versions of the cycle are illustrated in Fig. 4. In the four-stroke cycle an initial downward stroke of the piston draws in the air to charge the cylinder; this is followed by the compression stroke, with the fuel being injected as TDC is approached, then a downward power stroke, and an upward exhaust stroke. In the two-stroke cycle the charging and exhaust processes must occur without piston assistance, while the piston passes through BDC. The

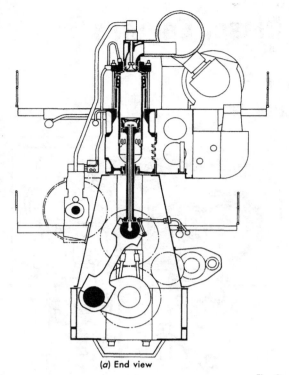

(a) End view

(b) Side view

Fig. 2   Low-speed engine

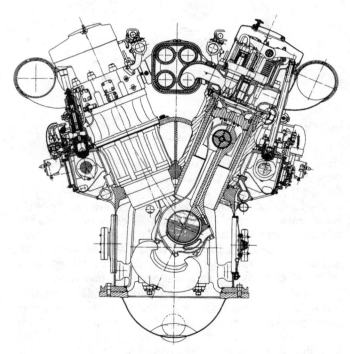

Fig. 3   Medium-speed engine with cylinders in a V-configuration

air supply for a two-stroke engine must, therefore, be under pressure to scavenge the cylinder of exhaust gas, and then fill it with a charge of air. It would seem that, with each cylinder of a two-stroke engine undergoing a power stroke in every revolution, a two-stroke engine

should develop twice the power of an otherwise similar four-stroke engine; however, the intervening strokes of the latter permit higher component stresses and, therefore, more intensive power strokes, which provide a compensating influence. Both versions of the cycle have their places in the spectrum of engines produced.

While the air for a two-stroke engine always enters through ports near BDC, the exhaust gas may exit through valve-controlled ports in the cylinder head as in Fig. 4, or through a second row of ports near BDC, as in Fig. 6. The first arrangement is referred to as uniflow scavenging, and the second, usually, as loop scavenging. While loop scavenging offers the advantage of simplicity, it generally results in cycles that are less efficient than those with uniflow scavenging.

In general, a cylinder consists of a cylinder liner that is inserted in a cylinder block and enclosed at the top by a cylinder head to form the combustion space above the piston. One exception to this description is the opposed-piston engine, represented by Fig. 7, in which each cylinder houses two pistons that move in opposite directions to form the combustion space between them.

Most engines are single-acting, which means that the combustion space exists only at one end of the cylinder. Few, if any, double-acting engines remain in marine service.

The reciprocating motion of the piston is converted to the rotary motion that drives the crankshaft by the wrist pin (also called the crosshead pin, piston pin, or gudgeon pin), the connecting rod, and the crank, which consists of the crankpin and webs on the crankshaft. If the wrist pin

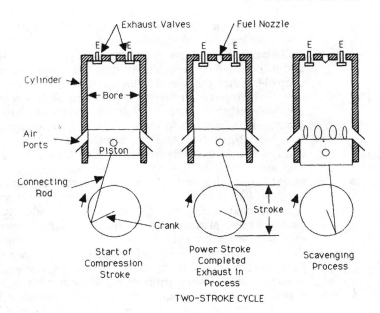

Fig. 4   Two-stroke and four-stroke cycles

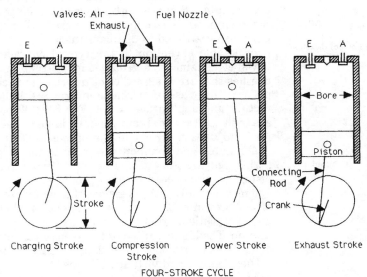

is carried by the piston itself, as in Figs. 1 and 3, the engine is called a trunk-piston engine, and the alternating transverse force, which is called the normal force because it is normal to the direction of piston motion, and which is developed at the wrist pin as the connecting rod turns the crankshaft, must be transmitted through the lower portion of the piston (the piston skirt) and cylinder liner to the engine frame and foundation.

Figure 2 shows a crosshead engine, in which a piston is rigidly connected to a crosshead by a piston rod, and the wrist pin is carried by the crosshead. The crosshead is constrained to reciprocate by stationary crosshead guides, on which its shoes or slippers slide, transmitting the normal force to the engine frame and relieving the cylinder liner and piston skirt of this load. The piston, piston rod, crosshead, connecting rod, and crankshaft are known collectively as the running gear.

In the largest engines, as illustrated in Fig. 2, the crankshaft is supported by main bearings that are carried in the engine bedplate. Vertical frames or columns are bolted to the bedplate and are connected longitudinally to provide a stiff crankcase, which constitutes the principal support structure for the cylinder blocks, and which carries the crosshead guides. In smaller engines a single cast or fabricated frame may serve as the support structure, as shown in Figs. 1 and 3.

Cycle events are made to occur in correctly timed sequence by the camshaft, which is carried in bearings that are supported in a camshaft housing. The camshaft is driven from the crankshaft, usually by timing gears, but on some engines by a chain and sprocket drive. V-engines usually have a separate camshaft for each bank of cylinders, as in Fig. 3.

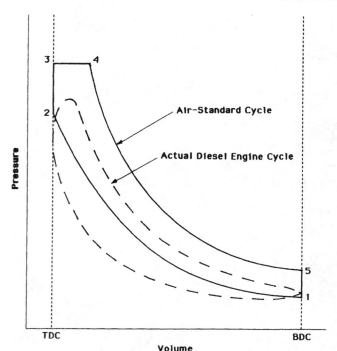

Fig. 5   Pressure-volume diagrams of air-standard cycle and actual diesel engine cycle

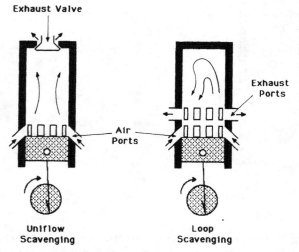

Fig. 6   Two-stroke cycle: loop versus uniflow scavenging

1.3   Engine Types. Broadly speaking, engines of conventional design fall into either a low-speed category or the medium- and high-speed category. Low-speed engines, as in Fig. 2, are generally intended for the direct drive of ships' propellers without any speed-changing device and, therefore, are restricted to an rpm range for which efficient propellers can be designed, generally below 300 rpm and possibly as low as 55 rpm at rated power. Low-speed engines are two-stroke, crosshead engines, with four to twelve cylinders, which are always in-line, always turbocharged, and always aftercooled. The stroke-to-bore ratio of low-speed engines in current production is about 3:1. These engines are heavy (see Fig. 8) and very large, but

they are well suited to operation on low-quality fuels and generally require only modest levels of maintenance.

Medium- and high-speed engines are typified by Figs. 1 and 3. Because of their higher rpm, these engines drive propellers through speed-reduction gears, but they are directly connected for driving generators. With a few but notable exceptions, these are four-stroke, trunk-piston engines, which have up to ten cylinders in-line or up to 24 in a V-configuration, and are mostly turbocharged and aftercooled. Stroke-to-bore ratios range from 1:1 to 1.5:1. The upper limit of the medium-speed category, and the start of the high-speed category, is generally placed in the range of 900 to 1200 rpm, but there are no clear physical features that enable the distinction to be made. At one time, the mean piston speed was used as a defining parameter, but the traditional boundaries fail for many engines. These engines tend to be lighter (see Fig. 8), more compact, and lower in acquisition cost than low-speed engines of comparable power output. Many of these engines have a proven heavy-fuel capability, but most evidence indicates that maintenance costs are higher than those of low-speed engines that are run on fuels of similar poor quality. Some engines, especially those in the higher-speed category, are restricted to distillate fuels.

A notable exception to the typical characteristics of low-speed engines is a group of four-stroke, trunk-piston engines that are designed to be directly connected to propellers. This class of engine is indigenous, in both manufacture and application, to the Far East. These engines are built with six or eight cylinders, in-line; they are rated for 70 to 700 kW per cylinder at speeds of 200 to 500 rpm.

A notable and common exception in the medium- and high-speed category is the series of two-stroke, trunk-piston, medium-speed engines produced in turbocharged and mechanically blown versions that dominate their field of application in American waters, despite their requirement for distillate fuels; these are built only in a V-configuration. The highest rated of these engines, with 20 cylinders, has an output of more than 3000 kW at 900 rpm.

Diesel engines of virtually all types have been used for driving generators, whether for diesel-electric propulsion applications (see Chapter 8) or for ships' services.

1.4   Turbocharging. Although some applications may be best served by normally aspirated or mechanically blown engines, the vast majority of main propulsion engines and generator-drive engines are turbocharged and aftercooled. An engine and its turbocharger(s) are interdependent in their performance; a defective or mismatched turbocharger will preclude proper engine performance, while, at the design level, the availability of turbochargers with improved performance characteristics may enable an engine design to be upgraded to a higher performance level. Figure 9 shows a typical turbocharger. The rotor is driven solely by the engine exhaust gases.

The principal reason for turbocharging, or for supercharging in any form, is to increase the power output of an engine of given size and speed, by enabling the cylinders to be charged with air at an elevated pressure (the boost pressure) and, therefore, at a higher density than air at atmospheric pressure. (The ratio of boost pressure

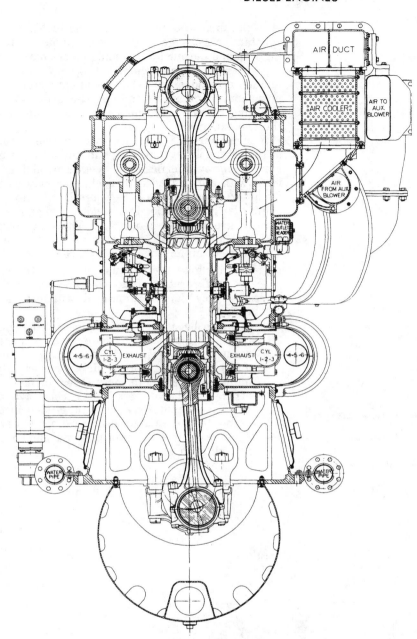

Fig. 7   Opposed-piston engine

to atmospheric pressure is the boost ratio.) Since the cylinders then contain a greater mass of air, a correspondingly greater mass of fuel can be burned, and the engine output can be higher. Some of this effect will be lost, however, if the air leaving the compressor is not cooled, because the temperature rise of the air during the compression process has the unwanted effect of decreasing the air density. In most applications, aftercoolers are fitted at the compressor discharge to remove the heat of compression and thereby increase the density of the delivered air.

Turbocharging tends to reduce fuel consumption because the friction losses of the turbocharged engine do not increase as fast as the increase in power output, and because the improved charging results in better combustion conditions.

A turbocharger may be operated on the constant-pressure principle or the pulse principle. When the exhaust valve or port of a cylinder first opens, the gas enters the exhaust system at relatively high pressure and velocity, which then diminish quickly as the exhaust process continues, leaving a brief energy pulse. In constant-pressure turbocharging, the exhaust gas from all of the cylinders enters a common manifold, from which the turbocharger is supplied. The pulses dissipate into the manifold with only limited useful recovery of the energy in the pulse, although even this recovery can be enhanced if the entry is via a diffusing section, as in Fig. 1. The energy can be recovered more effectively if carried right through to the turbine in small-bore exhaust pipes, as shown in Figs. 3 and 7. The principal tradeoff between the two systems

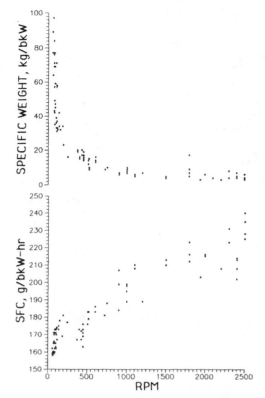

Notes: (1) The specific weight plotted is for the bare engine, without auxiliaries, accessories, foundations, or liquids.

(2) The specific fuel consumption is that published by the manufacturers,
based on shop tests with distillate fuel.

Fig. 8   Specific weight and fuel consumption of diesel engines

is in the turbine operation, with the turbine supplied at constant pressure generally operating more efficiently at high engine output levels, while pulse-charged turbines are generally more efficient at low engine output levels, and more responsive to changes in engine output. When pulse converters are fitted in pulse-charged systems, the resulting turbine performance is a compromise between the two extremes.

A turbocharger is matched to an engine as the design of the engine is developed. If, subsequently, a turbocharger of improved efficiency becomes available, replacement with the upgraded turbocharger may be feasible. Such a replacement has the potential of improving the engine design and rating by providing a higher boost pressure or greater rate of air flow. There are three ways that the operation of an engine can be altered to take advantage of an improved turbocharger:

- By changing the exhaust timing of the engine or the configuration of its exhaust system, an engine's efficiency can be improved without altering its rating.

- A higher boost pressure may permit the engine output to be increased; if the engine components are redesigned appropriately, a new generation of engines can result, with a higher power output and higher efficiency. In most cases, in fact, as more efficient turbochargers have become available, engine components have been upgraded in order to permit the potential for higher ratings to be realized.

- An excess turbine output potential can be exploited by using an exhaust gas turbine as described in Section 9.7.

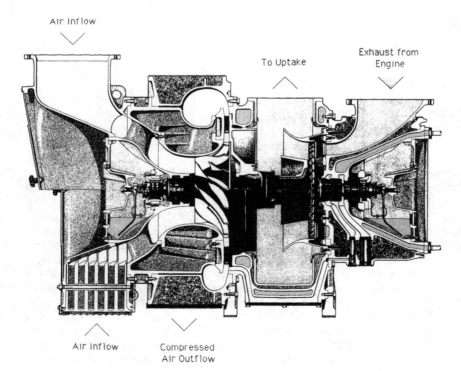

Fig. 9   Turbocharger with "uncooled" casing that is
cooled only in way of the turbine bearing. In
way of the compressor intake, the upper half
illustrates the connection for external supply;
the lower half, a direct air intake

Two-stage turbocharging arrangements, in which two turbochargers are connected in series, usually with charge air coolers at each compressor discharge, have been used to attain higher boost pressures than even the most efficient turbochargers could achieve in a single stage. In most applications the purpose of two-stage turbocharging is to obtain very high power output, which is used infrequently, from engines that are constrained in space and weight. However, the high power output is generally achieved at a reduced efficiency.

Section 2
Thermodynamic Analysis

2.1 The Air-Standard Cycle. The air-standard, dual-combustion cycle, which appears as the solid trace in Fig. 5, is a widely used theoretical representation of the events that occur in a diesel cylinder. Although air-standard cycles are completely idealized, they are useful for analysis and for demonstrating principles. More rigorous theoretical treatments can be found in references 1 and 2.

a. Cycle description. The initial pressure in the cylinder at the BDC position, P_1, is the charge pressure. Ideally, it would be equal to the ambient pressure in normally aspirated engines, and to the boost pressure in supercharged engines. In the air-standard analysis, the compression stroke is assumed to commence immediately, as soon as the piston moves away from BDC. The compression process, 1–2, is then assumed to occur adiabatically and isentropically, that is, without friction, turbulence, or heat transfer. The pressure reached at TDC, P_2, is the compression pressure, and the corresponding temperature is the terminal compression temperature. The constant-volume process, 2–3, and the constant-pressure process, 3–4, are heat-addition processes representing the heat released in combustion during the initial process of rapid combustion, and the subsequent process in which fuel burns as it is injected, while the piston begins its descent. The constant pressure, which is maintained from 3 to 4 in the air-standard cycle, is called the maximum pressure. At 4, the addition of heat ceases and the air is assumed to expand isentropically, forcing the piston down to BDC at 5. A constant-volume heat-rejection process, 5–1, representing the exhaust of the hot gases and the charging of the cylinder with fresh air, completes the cycle.

The dashed trace in Fig. 5 shows the pressure-volume relationship actually achievable in an actual cycle with the same heat input, and in a cylinder of equal dimensions. The largest discrepancy between the two traces results from the fact that the compression stroke in the actual cycle cannot begin until the air ports or valves are closed, by which time the piston will have traveled 10% to 20% of its stroke.

b. Cycle parameters. The dual-combustion cycle can be described by several ratios of pressures and volumes:

$$\text{The compression ratio, } r_c = V_1/V_2 \qquad (1)$$
$$= V_{\text{BDC}}/V_{\text{TDC}}$$

Representative values of compression ratios are in the range of 20 for normally aspirated engines, but are lower for turbocharged engines, to limit the compression and maximum pressures. A minimum value of about 8 is necessary to ensure that the terminal compression temperature is sufficient to ignite the fuel.

$$\text{The pressure ratio or explosion ratio, } r_x = P_3/P_2 \quad (2)$$

The pressure ratio characterizes the amount of heat addition at TDC and, therefore, represents the extent to which fuel burns rapidly. Representative values of the pressure ratio range from about 1.2 to 1.8.

$$\text{The cutoff ratio, } r_k = V_4/V_3 \qquad (3)$$

The cutoff ratio indicates the extent to which heat addition continues during the power stroke. A ratio of 2 would represent an extended period of heat addition, analogous to the injection of a large quantity of fuel and, therefore, to a cylinder that is very heavily loaded. There is an inverse relation between the pressure ratio and the cutoff ratio, which is reasonable in that, for a given amount of fuel injected in a cylinder, if more of it is burned rapidly (a high r_x), less of it remains to burn steadily (a low r_k).

$$\text{The boost ratio, } r_b = P_1/P_0 \qquad (4)$$

The boost ratio, which is the ratio of the charge pressure to the ambient air pressure, P_0, would ideally be unity for normally aspirated engines. Single-stage turbochargers can develop boost ratios higher than 4, but values from 2 to 3.5 are typical in normal operation.

The displacement, or volume displaced by the piston in its travel from one dead center position to the other, can be expressed in terms of the compression ratio, by substitution from equation (1):

$$V_1 - V_2 = V_1(r_c - 1)/r_c \qquad (5)$$

The compression pressure, P_2, can also be expressed in terms of the compression ratio, with k used for the ratio of specific heats, c_p/c_v. For the isentropic compression from 1–2:

$$P_2/P_1 = (V_1/V_2)^k = r_c^k$$

and so the compression pressure can be expressed as

$$P_2 = P_1 r_c^k \qquad (6)$$

The maximum pressure, $P_3 = P_4$, can be expressed in terms of the ratios, using equations (2), (6), and (4):

$$P_3 = r_x P_2 = P_1 r_x r_c^k = r_b P_0 r_x r_c^k \qquad (7)$$

The maximum pressure is a design condition for the mechanical strength of the engine components. Equations (6) and (7) show how the aforementioned reduction in compression ratio, which is here raised to the k power ($k =$

1.4 for air under standard conditions), will limit both the compression pressure, P_2, and the maximum pressure, P_3, in a supercharged engine.

c. Temperatures. The temperatures can be expressed in terms of the ratios of equations (1) through (4) as follows:

for the isentropic compression, 1–2:

$$T_2/T_1 = (V_1/V_2)^{k-1} = r_c^{k-1}$$
$$T_2 = T_1 r_c^{k-1} \tag{8}$$

for the constant-volume heat addition, 2–3:

$$T_3/T_2 = P_3/P_2 = r_x$$
$$T_3 = r_x T_2 = T_1 r_x r_c^{k-1} \tag{9}$$

for the constant-pressure heat addition, 3–4:

$$T_4/T_3 = V_4/V_3 = r_k$$
$$T_4 = T_3 r_k = T_1 r_x r_k r_c^{k-1} \tag{10}$$

for the isentropic expansion, 4–1:

$$T_5/T_4 = (V_4/V_5)^{k-1} = (V_4/V_1)^{k-1} = [(V_4/V_3)(V_3/V_1)]^{k-1}$$
$$= [(V_4/V_3)(V_2/V_1)]^{k-1} = (r_k/r_c)^{k-1}$$
$$T_5 = T_4(r_k/r_c)^{k-1} = T_1 r_x r_k^k \tag{11}$$

d. Net work and heat addition. The work done and heat transferred in each process in the idealized cycle can be calculated using basic thermodynamic principles. As a simplification, the air is treated as though its specific heats, c_p and c_v, each held constant throughout. Taking m_{air} as the mass of air in the cylinder, the following relation for the net work of the cycle, W_{net}, which is the sum of the work output during the power stroke, 3–4–5, less the work input during the compression stroke, 1–2, results:

$$W_{net} = m_{air}[c_v(T_3 - T_2) + c_p(T_4 - T_3)$$
$$- c_v(T_5 - T_1)] \tag{12}$$

An analogous relation for Q_A, the total heat added during the cycle at TDC and during the constant-pressure portion of the power stroke, 2–3–4, but not including the heat rejected at BDC, is

$$Q_A = m_{air}[c_v(T_3 - T_2) + c_p(T_4 - T_3)] \tag{13}$$

The difference between the two relations is the heat rejected, Q_R, which simply illustrates the First Law of Thermodynamics as applied to cycles:

$$W_{net} = Q_A - Q_R \tag{14}$$

From equation (14), the net work increases as the heat added is increased relative to the heat rejected. However, there is an upper limit on the heat that can be added in the cycle, which derives from the amount of air needed for the combustion of the fuel, that is, the air-fuel ratio. This ratio can be calculated from the chemical composition of the fuel and, under ideal circumstances where there is just enough oxygen present with no excess, is called the stoichiometric air-fuel ratio. For petroleum fuels, the stoichiometric air-fuel ratio is 14 to 15, but because conditions for combustion in diesel cylinders are far from ideal, the actual mass ratio of air trapped at the start of compression

to fuel required for maximum output is generally in the range of 25 to 30.

The maximum mass of fuel that could be burned in a diesel cylinder during one cycle is equal to the mass of air trapped divided by the trapped air-fuel ratio, AF, or

$$m_{fuel} = m_{air}/AF \tag{15}$$

The mass of air in the cylinder of the air-standard engine can be calculated from the specific volume, v_1, of the air at the start of the compression stroke (the specific volume is the reciprocal of the density), and can be obtained by applying the ideal gas law:

$$m_{air} = V_1/v_1 = \frac{V_1 P_1}{RT_1} \tag{16}$$

where R is the gas constant for air.

The upper limit of heat added to the cycle is equal to that released by the mass of fuel, based on the lower heating value, LHV, of the fuel:

$$Q_A = m_{fuel}LHV \tag{17}$$

or, using equations (15) and (16):

$$Q_A = \frac{LHV}{R(AF)} \frac{V_1 P_1}{T_1} \tag{18}$$

The limit on heat (fuel) that may be added per cycle is further discussed in Section 2.2, but manifests itself in engine operation by the tendency of engines to exhaust smoke when loaded to the limits of their output at any given rpm. The smoke indicates that more fuel is being injected per cycle than can be completely burned in the air in the cylinder. Accordingly, a theoretical limit exists to the output of an engine at any rpm, based on the air available in the cylinders, and is independent of the mechanical strength of the engine components.

e. Mean indicated pressure, mean effective pressure, and brake work. The pressure in the cylinder, when averaged over the entire cycle, including the compression stroke (and, in four-cycle engines, the charging and exhaust strokes) as well as the power stroke, is the mean indicated pressure (MIP). Graphically, with reference to Fig. 5, the MIP can be calculated as proportional to the area enclosed within the pressure-volume trace, divided by the length of the trace, which represents the displacement volume. Because work in an ideal cycle is equal to the instantaneous pressure integrated over the volume, the area enclosed within the pressure-volume trace in the figure represents the net work of the cycle. The MIP is, therefore:

$$MIP = W_{net}/(V_1 - V_2) = W_{net}/(V_{disp}) \tag{19}$$

and

$$W_{net} = MIP(V_1 - V_2) = MIP(V_{disp}) \tag{20}$$

The net work in all of the foregoing expressions is that developed by the gas pressure in the cylinder. When this net work is multiplied by the mechanical efficiency of the engine, the result is the brake work, W_{brake}. It follows that, when the MIP is multiplied by the mechanical efficiency, the result is the mean effective pressure (MEP).

In the idealized representation, with a mechanical efficiency of 100% by definition, the net work and the brake work are equal, as are the MIP and MEP. In general, however:

$$W_{\text{brake}} = \text{MEP}(V_{\text{disp}}) \qquad (21)$$

f. Efficiency. The thermal efficiency of a cycle, η, is defined as the net work divided by the heat added. Using the First Law relation of equation (14), the following expression can be derived:

$$\eta = \frac{W_{\text{net}}}{Q_A} = \frac{Q_A - Q_R}{Q_A} = 1 - \frac{Q_R}{Q_A} \qquad (22)$$

g. Power. The power developed by an engine is the product of the work per cycle times the number of cycles completed per unit time. The brake power output of an engine with n identical, single-acting cylinders is

$$P_{\text{brake}} = n\text{MEP}(V_{\text{disp}})N \qquad (23)$$

where N is the number of cycles completed per unit time, which is equal to the rpm of a two-stroke engine, or half the rpm of a four-stroke engine.

Because the aforementioned limit on heat addition also limits the MIP and, therefore, the MEP, and since the torque is directly proportional to the MEP, diesel engines are torque-limited.

Although a two-stroke engine would appear to have the potential to develop twice the power of an otherwise similar four-stroke engine, in actuality two-stroke engines are unable to achieve levels of MEP as high as those of four-stroke engines. The reasons for this shortfall include, for four-stroke engines, more effective scavenging and charging, better cooling afforded between compression and power strokes by the charging and exhaust strokes, an earlier start to the compression stroke, and a smooth cylinder liner that is uninterrupted by ports. Among the two-stroke engines, somewhat higher levels of MEP can be achieved in the low-speed, crosshead engines, and, indeed, all of these are two-stroke, while most medium- and high-speed engines are four-stroke. The few two-stroke higher-speed designs are roughly comparable in weight and size to their four-stroke counterparts.

The rpm is limited by mechanical and practical considerations. A high rpm imposes high stresses on running gear and bearings, related to the inertia of the reciprocating components. As a result, there is a generally inverse relation between engine size and rpm, as illustrated by Fig. 8. In some applications, the engine rpm is limited by the intended application, as for low-speed engines directly coupled to propellers. Free of such constraints, however, most medium- and high-speed engine designs have evolved in the direction of increasing rpm.

2.2 Cycle Performance. Cycle performance can be considered optimum if the net work is developed at the lowest unit energy cost, with appropriate consideration for the engine size, weight, and, therefore, the engine cost. The cycle thermal efficiency serves as a measure of the energy costs, while the MIP, the ratio of net work to displacement volume, when multiplied by the rpm, provides a measure for the cost of the engine. To understand how the efficiency and MIP are linked to each other, and to the aforementioned limits on maximum pressure and heat addition, the temperature expressions, equations (8) through (11), can be substituted for the temperatures in equations (12) and (13), for net work and heat added:

$$W_{\text{net}} = m_{\text{air}}c_v T_1\{r_c^{k-1}[(r_x - 1) + r_x k(r_k - 1)] - (r_x r_k^k - 1)\} \qquad (24)$$

$$Q_A = m_{\text{air}}c_v T_1\{r_c^{k-1}[(r_x - 1) + r_x k(r_k - 1)]\} \qquad (25)$$

The efficiency can then be expressed as follows:

$$\eta = \frac{W_{\text{net}}}{Q_A} = 1 - \frac{r_x r_k^k - 1}{r_c^{k-1}[(r_x - 1) + r_x k(r_k - 1)]} \qquad (26)$$

An expression for the MIP can be obtained, by combining equations (24), (19), (5), and (16), and recalling that, for an ideal gas, $c_p - c_v = R$:

$$\text{MIP} = \frac{W_{\text{net}}}{V_1 - V_2} = \frac{r_b P_0}{k-1}\frac{r_c}{r_c - 1}\{r_c^{k-1}[(r_x - 1) + r_x k(r_k - 1)] - (r_x r_k^k - 1)\} \qquad (27)$$

The usefulness of equations (26) and (27) is that they express the performance of the air-standard cycle in terms of the descriptive ratios. An examination of the effects that variations in these ratios have on the performance parameters can then be made, with the following comments and results:

• The limit on heat addition arising from the available air must be considered, as it links the pressure ratio, r_x, to the cutoff ratio, r_k, so that an increase in one forces a decrease in the other. If all other parameters are held constant, however, it can be shown that efficiency rises as r_k diminishes and r_x rises. A further limit is the maximum pressure, expressed by equation (7):

$$P_3 = r_x P_2 = P_1 r_x r_c^k = r_b P_0 r_x r_c^k$$

These theoretical effects are borne out in actual practice: within the constraints permitted by fuel quality and combustion conditions, the trend in engine design has been towards short injection and combustion periods, with high maximum pressures, so that the expansion process begins early in the power stroke. In many engines, load control is designed to alter the injection timing to maintain maximum pressure even as the load is reduced.

• The boost ratio, r_b, is not a factor in the relation for theoretical cycle thermal efficiency. This effect is not borne out in actuality. In fact, turbocharging tends to reduce the specific fuel consumption.

• Increases in the compression ratio, r_c, raise both the theoretical efficiency and the MIP. In practice, however, the effect on efficiency is partly offset by an increase in the friction between piston rings and cylinder liners, which decreases the mechanical efficiency. In addition, with reference to equation (7), because of maximum pressure limits, any increase in r_c must be traded off against a decrease in the boost ratio, r_b. However, an analysis of equations (27) and (7) shows that increases in r_b result in large increases in the MIP with smaller increases in the maximum pressure, while increases in r_c have the opposite effect. Consequently, evolutionary trends in diesel engines have mainly been in the direction of increasing the boost ratio and decreasing the compression ratio.

Section 3
Performance Characteristics of Diesel Engines

3.1 Engine Ratings. The rating stated for an engine by its manufacturer and approved by the regulatory bodies reflects their confidence that the engine will perform reliably at that level, under specified conditions. The rating is usually expressed as a continuous power output at a specified engine speed, and is usually called the maximum continuous rating (MCR). The rating of any particular model may later be increased to reflect component improvements or service experience. An engine may be given different ratings for different applications; for example, a high-performance rating may be given to an engine that is intended for a warship application in which the engine may be operated under conditions of sustained overload for limited periods of time, but with reduced intervals between overhauls. This might be in contrast to a lower rating assigned to the same engine for a merchant ship propulsion application, where the engine will be operated for long periods at a more modest power output, and with component lives and service intervals that are acceptable in a commercial operation. Such considerations sometimes lead to the definition of a continuous service rating that is lower than the MCR by a percentage called the engine margin.

The pressure in the cylinder of a diesel engine, when averaged over the entire cycle, is the mean indicated pressure (MIP). When the MIP is multiplied by the mechanical efficiency of the engine, the result is the mean effective pressure (MEP). The MEP is directly proportional to the torque applied to the drive shaft, so that the product of MEP and rpm is directly proportional to brake power output. Stated another way, the MEP is directly proportional to the brake power output divided by the rpm. This relation is shown in Fig. 10. When an engine is run at rated rpm and rated torque (rated MEP), 100 percent brake power (MCR) is developed. The figure shows that, within the limit of rated MEP, an engine can achieve its rated power output only at, or above, its rated rpm; at a lower rpm the power that an engine can develop is limited by the MEP. For this reason, diesel engines are often said to be torque limited [see discussion following equation (23)].

The maximum torque that a cylinder can develop, and therefore its limiting MEP, is a direct result of the amount of fuel that can be burned in the cylinder during each cycle. This amount is limited by the amount of air trapped in the cylinder (and by the ability of the fuel to combine with sufficient oxygen in the time available), and also by the ability of the cylinder components to withstand the higher temperature and extended duration of high cylinder pressure that results. Sustained operation above the fuel-rack setting corresponding to rated MEP will result in poor combustion (reflected by carbon deposits and perhaps smoke); higher exhaust gas temperatures; higher metal temperatures; shorter component lives of pistons, cylinder heads, cylinder liners, and exhaust valves and

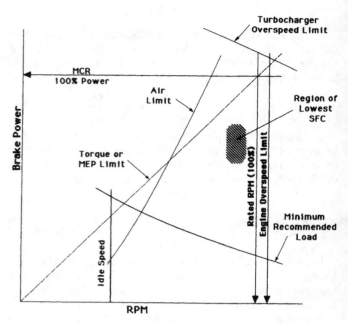

Fig. 10   Engine performance limits

seats; and greater-than-anticipated wear rates of piston rings and their grooves, cylinder liners, and bearings.

An engine may be "derated," that is, assigned a rating lower than that normally assigned, to optimize it for a particular application. Derating may or may not result in components that are different from the standard design. In general, derating follows one of the following patterns:

- An engine may be derated to a lower power output at rated rpm to reduce the in-service MEP. A lower specific fuel consumption, SFC (fuel consumed per unit power output), is likely, and lower maintenance costs may be expected.
- An engine may be derated to a lower rpm, but at the rated MEP. In the case of engines directly connected to the propeller, the lower rpm may permit the use of a propeller of higher efficiency. The SFC will most likely be reduced, but the maintenance expenses will most likely be unaffected.
- An engine may also be derated to develop a reduced MEP at a reduced rpm, thereby yielding all of the above advantages, that is, reduced SFC, lower maintenance costs, and the potential for higher propeller efficiency.

While derating provides some advantages, it may also impose disadvantages. When a given amount of power is to be obtained from a derated engine, that engine will generally be larger and heavier and have a higher acquisition cost per unit power output than an alternative engine that provides the required power at its normal rating.

The size of the engine auxiliary equipment and the design of the shafting are based on the engine rating. To reserve the option of restoring a derated engine to its normal rating to meet a future requirement, it may be a prudent design decision to use the normal MCR for these purposes.

3.2 Limits of Engine Performance. Limits defining the operating envelope for an engine are identified in Fig. 10. Engine rpm is limited by the mechanical stress on running gear, by bearing loads, and by wear rates of piston rings and cylinder liners. The turbocharger overspeed limit is usually determined by the turbine blade root strength. The setting of the engine overspeed governor or trip is usually 115% to 120% of rated rpm. Except when an engine that drives a fixed-pitch propeller is run at a modest overspeed on trials in order to maximize the load, operation beyond rated rpm is unusual.

Operation beyond the limits of rated power (MCR) and rated MEP constitutes an overload. Overload operation results in increased stresses and higher temperatures for the cylinder heads and liners and for the piston crowns, and is a condition often referred to as high thermal load. However, overload operation causes increased mechanical loadings on other engine components as well. Cylinder liner and piston ring wear can be expected to increase, and the lives of bearings and exhaust valves to be reduced, as a result of engine overload. The most convenient indication of high thermal load is the cylinder exhaust temperature since it correlates well with MEP and is readily measured.

The air limit in turbocharged engines is set by applying a margin to the surge limit of the turbocharger (defined in Section 3.6). Operation to the left of the air limit curve in Fig. 10 incurs an increased likelihood of combustion chamber and turbine fouling and smoke emission.

The engine idle speed (or minimum engine speed for engines that are directly connected to propellers) is typically 25% to 40% of rated rpm. At lower engine speeds, the leakage of air past the piston rings during the compression stroke can result in low temperatures at the time of injection, erratic combustion, and an accumulation of unburned fuel and carbon in the exhaust system.

The bottom of the operating envelope is a light-load limit. Sustained light-load operation, particularly with heavy or sulfur-bearing fuels, is not recommended because combustion chamber temperatures at low loads are likely to be too low to effectively burn some fuel constituents. Fouling of the combustion areas, exhaust path, and turbine can result. In addition, the low temperatures toward the bottom of the cylinder liners can cause condensation of sulfur compounds in the exhaust, leading to sulfuric acid attack.

Limited operation outside the operating envelope will generally result in decreased component durability, which is reflected in increased requirements for inspection and maintenance. A catastrophic failure of a properly maintained engine under these conditions is unlikely because of the design margins, and because periodic scheduled inspections reveal such effects as burning, cracking, or distortion in time for component renewal. Most manufacturers consider a certain amount of operation beyond the envelope in the directions of high MEP, high power, and with marginal combustion air inevitable, and take this into account in their service recommendations. Statements permitting limited operation in these regions, limited to perhaps one hour in ten or twelve, or a cumulative total of perhaps 500 to 2000 hours per year, are typical.

3.3 Fuel Consumption. The specific fuel consumption (SFC) is the amount of fuel consumed over a period of time, divided by the power output of the engine. It is most often quoted as grams of fuel per brake kilowatt-hour or per brake horsepower-hour. A quoted value will usually have been derived from test-bed measurements, on distillate fuel, and under controlled conditions that are in accordance with an established power test code. Quoted values of SFC for a number of engines are plotted in Fig. 8; these values must be used with care because of variations, even among similar engines, and because of continual improvements in SFC which result from more advanced technology. When interpreting the significance of the quoted SFC values, a number of factors must be given serious consideration.

• The SFC units must be clearly stated; in particular, an SFC that is quoted on the basis of metric horsepower will be numerically lower than if quoted on the basis of British horsepower.

• The SFC will vary with engine output and rpm, generally being the lowest at 75% to 85% MEP and about 90% rpm, as indicated in Fig. 10. Most manufacturers quote the SFC at rated output, but some furnish data that cover the operating range. The SFC at rating may be up to 5% higher than the lowest value, with all other factors equal.

• The SFC of a bare engine, without any attached pumps, is lower than that of the engine with pumps attached. Most manufacturers quote the SFC for their standard configuration, or for the bare engine, and provide estimates for other configurations.

• Low-speed engines are usually tested with the thrust bearing in place, but unloaded. The brake output of the engine in service, when it is directly driving a propeller and with the bearing loaded, will be slightly lower than during test conditions, and the SFC will be correspondingly higher.

• Differences from one power test code to another, such as different ambient conditions, should be normalized using correction methods that are recommended for the particular engine by the manufacturer. Similar correction procedures are used to permit adjustment of the SFC to expected service conditions.

• To normalize a difference in SFC arising from a variation in the heating value of the fuel, the SFC should be multiplied by the ratio of the heating value of the subject fuel to that of the specified fuel. Usually, the lower heating value of the fuel is used.

• To determine the SFC of an engine that is capable of operation with blended or heavy fuels, the SFC should be divided by the ratio of the heating value of the heavier fuel to that of the specified fuel. It is not usually necessary to apply any further correction for the poorer combustion conditions for the heavier fuel.

- In cases where the fuel consumption is quoted on a volumetric basis, the quoted value must be adjusted by the ratio of fuel densities.

- In most cases the SFC will be quoted or guaranteed with a tolerance of 3 to 5%. Because manufacturers can be confident of achieving a test-bed SFC within a fraction of 1%, some manufacturers add very little margin to their measured data when establishing their guaranteed value.

- In general, diesel engines must be kept in reasonable repair if they are to operate at all. Therefore, the addition of a further margin to the SFC, to account for performance deterioration in service, is usually unnecessary once the above adjustments are made.

- In using the SFC to determine fuel consumption, the relationship between brake output power delivered at the crankshaft flange of the engine and the power absorbed by the load (the shaft power, for propulsion engines) must be considered. Specifically, the loads of attached auxiliaries, and the losses of gearing or other speed-changing devices, must be taken into account. Guidance in estimating these loads can be found in reference 3.

- Most fuels, including clean distillates, contain some water and solids when delivered. Most of the water and sediment, together with any additional impurities introduced during storage aboard the ship, will be removed during settling, purification, and filtration. This removal of impurities from the fuel, which leads to an apparent increase in fuel consumption, should not be charged against the SFC of the engine, but should be allowed for in determining the overall fuel consumption and storage capacity.

3.4 Lubricating-Oil Consumption. The rate at which lubricating oil is consumed by a diesel engine will generally be ½% to 1% of the rate of fuel consumption. Because of the higher unit cost of the lubricating oil, however, the cost of the lubricating oil consumed will be a much more significant percentage of the cost of the fuel consumed. Lubricating oil consumption rates are generally quoted by manufacturers for guidance only. Consumption rates are usually derived from service experience or, for new engine models, extrapolations of experience with similar engines. In some cases consumption rates substantially greater than those published have been found to be necessary. The rate of lubricating oil consumption in service is under the control of the operator, and can be varied over a wide range, with effects that are evident only in the long term. It has been demonstrated that increased lubricating oil consumption frequently correlates with lower wear rates for the lubricated components.

Lubricating oil systems for diesel engines are the subject of Section 10.3. Lubricating oil is consumed in diesel engines in several ways:

- Crosshead engines, and some of the larger trunk-piston engines, are fitted with separate cylinder oil systems that lubricate the piston rings. In engines

that are in good condition, the cylinder oil consumption typically ranges from 0.7 to 1.4 g/bkW-h.

- In trunk-piston engines that have no separate cylinder oil systems, lubrication of the cylinders consumes about 1.5 g/bkW-h of circulating oil.

- In crosshead engines replacement of the circulating oil is rarely required during normal operations.

- In trunk piston engines, the contamination of the oil by combustion products is usually inevitable, and the oil must be renewed periodically. The oil renewal interval depends on factors that include the service that the engine performs, the type of fuel used, the amount of oil in circulation, the rate at which fresh makeup oil is added to compensate for oil consumed in cylinder lubrication, the effectiveness of the lubricating oil filtration and purification equipment, the condition of the engine, and the practices of the operators.

3.5 Intake Air Requirements. When operating near their rating, engines take in 30 to 50 kg of air per kg of fuel under standard conditions, of which 25 to 30 kg will be trapped for combustion, with the remainder serving to scavenge the cylinders and to help cool the exhaust valves. In some engine designs the ratio of air to fuel is nearly independent of load, while in others it tends to rise as load is reduced. Engines that have been derated to lower values of MEP tend to have higher air-fuel ratios throughout their load range than higher-rated models of the same engines.

Intake air flow rates are specified by the manufacturer and are usually derived from test-bed measurements of an engine run on distillate fuel, under controlled conditions, and in accordance with an established power test code. The value must be adjusted for ambient conditions and for any pressure drops resulting from intake ducting and filters, as discussed in Section 3.8.

Where the intake air is to be taken from the engine room (the typical case), it is common practice in the design of the machinery space ventilation system to duct fresh air to the vicinity of the engine intakes. In a well-designed engine room, the engine intake air temperature in service will, therefore, be between the outside air and general engine room temperatures. Reasonable values might be 35°C and 25°C, in summer and winter, respectively. If there are constraints on the fresh air supplied to the engine, intake air temperatures can exceed 45°C, which is the rated value for many engines. Where the intake air is drawn directly from the weather, the pressure drop in the intake ducting may be significant, and the variation in air temperature will be much higher than when air is taken from the engine room. In either case, the intake air must be filtered.

3.6 Turbocharger Performance. Most compressors that are used for turbocharging are of the centrifugal type, with characteristic curves as shown in Fig. 11. At a constant speed the discharge pressure first rises as the volumetric flow increases, then drops sharply. Operation is practical only to the right of the peak in the pressure curve, and only that portion of the characteristic curves is shown in Fig. 11, as is the customary practice. A dashed

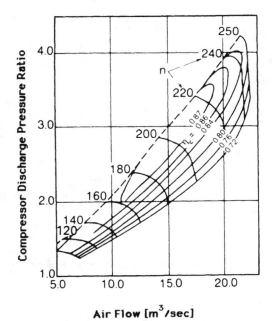

Fig. 11   Typical turbocharger compressor performance map; shaft speed, n, is in revolutions per second

curve drawn through the pressure characteristics slightly to the right of the peak, is called the surge limit. Proper selection of a turbocharger requires that there be adequate margin between the surge limit and the compressor operating line. Inadequate margin invites turbocharger surge under service conditions, while excessive margin places the compressor in a region of low efficiency. The compressor efficiency is plotted as contours in the figure, and is highest in the region slightly to the right of the surge line.

The phenomenon of compressor surge can be explained as follows: the power consumed by the compressor is related to the product of the discharge pressure and the flow rate. In the region to the right of the peak in the pressure curve, operation will be stable. In this region a momentary drop in volumetric flow rate, perhaps brought on by a momentary reduction in engine speed, will be balanced by a rise in pressure, with little or no effect on the turbine. In the region to the left of the pressure peak, however, a momentary drop in volumetric flow rate will be accompanied by a drop in discharge pressure, and if the pressure at the compressor discharge falls below that downstream, the flow will reverse. The result can be a pulsation in compressor discharge pressure if the situation is not severe or of long duration, or, if the flow reverses, a pulse can continue back through the compressor to the air intake and become audible, as a sneezing or backfiring sound.

The performance of a particular compressor, which is characterized by a rotor of given design and diameter and a diffuser with vanes of given height, is described by Fig. 11. For each compressor rotor, a narrow range of performance variations is possible by exchanging the diffuser for one with a different vane height. In general, the selection of a compressor rotor diameter predetermines

the basic turbine dimensions, which are then matched by appropriate selections for the nozzle and blade characteristics.

A turbocharger is necessarily matched to an engine for a particular set of conditions. Engine operation under different conditions will result in less-than-optimum turbocharger performance and may, in extreme cases, be so unsatisfactory as to justify the retrofit of a new turbocharger that is matched to the new conditions. Off-design point engine operation may be permanent or temporary, intentional or inadvertent. Some examples that can lead to off-design point engine operation are as follows:

- An overpitched propeller, a heavily fouled hull, single-screw operation of a twin-screw ship, or single-engine operation of a pair of engines geared together are conditions that require an engine to deliver higher than anticipated power at reduced rpm. Turbocharger surge under such circumstances is common.
- Long or tortuous runs of intake ducting, elevated intake air temperatures, fouled intake air filters, or a fouled or damaged compressor can result in air manifold pressure that is lower than expected.
- A fouled air cooler can force the compressor to operate at a higher rpm in order to supply the required flow at a higher pressure ratio. The operating point on the higher rpm curve will then approach the surge line.
- Fouled turbine nozzles can cause excessive exhaust pressure in a four-stroke engine, forcing the turbocharger into overspeed and surge; in a two-stroke engine the result will be poor turbine performance reflected in reduced air manifold pressure.
- Fouled turbine blades, a heavily fouled waste-heat boiler, or a constricted exhaust-gas uptake can prevent the turbine from reaching projected performance, and might first be reflected in a low air-manifold pressure.

The interdependent relationship of the engine and turbocharger, and of the turbine and compressor, renders the system prone to chain reactions. As an example, a fouled turbocharger turbine may be the cause of reduced air flow at a given engine output. The resulting air starvation can lead to poor combustion, resulting in further fouling of the turbine with carbon deposits.

As the engine output is reduced, the reduced exhaust-gas flow and enthalpy at the turbine result in a lower boost pressure. While four-stroke engines, by virtue of piston movement on charging and exhaust strokes, will continue to draw in their own charge air and expel most of the exhaust gases, two-stroke engines rely on an elevated charge air pressure to scavenge and charge the cylinder. Below about half power, therefore, two-stroke engines must be provided with an auxiliary means of pressurizing the air manifold. In most engines this takes the form of an electric motor-driven boost blower, which automatically starts in response to a low air-manifold pressure. In some smaller engines the boost pressure is provided at low

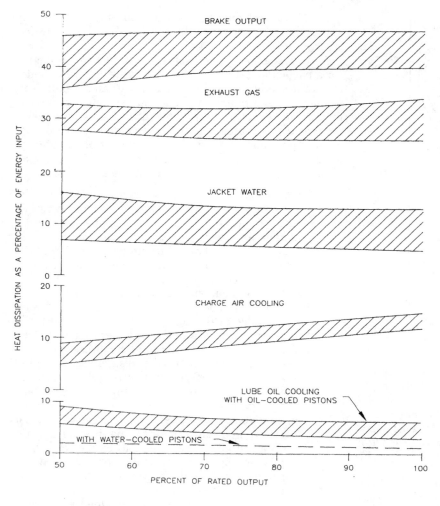

Fig. 12   Distribution of energy for diesel engines

alternative methods formerly practiced, such as the use of the piston undersides as reciprocating pumps, and the provision of reciprocating pumps that are driven by links from the crossheads, are now considered obsolete, at least for the larger engines.

In order to avoid the decline in performance that would be caused by fouled turbines and compressors, many engines, including most that are intended for operation on heavier fuels, are fitted with water-washing systems. These water-washing systems commonly take the form of small tanks that are piped to the compressor inlet and the turbine inlet, and are fitted with water and compressed air connections. In use, the engine load is reduced, and the charge of water, which is determined by the size of the tank, is injected over a period of about one minute, using compressed air. The frequency of use depends on the rate of fouling, which is determined from experience. In the case of engines run on the heaviest fuels, the turbine is often washed daily. Water-washing of the compressor is usually required infrequently.

3.7 Heat Balance. Figure 12 provides generalized data for the heat balance of a wide range of engines. Heat balance data are necessary in sizing cooling systems and establishing the feasibility of waste-heat recovery systems. Heat balance data are highly engine-specific and very sensitive to ambient conditions, fuel quality, and load pattern. For these reasons, only the data provided by the engine manufacturer, and corrected to service conditions, should be used for design purposes.

For turbocharged engines the exhaust-gas temperature leaving the turbocharger typically falls in the range of 250 to 500°C. Generally, two-stroke engines, with high air flows to assist scavenging and exhaust port and valve cooling, tend to have lower exhaust temperatures than four-stroke engines, and slower-running engines tend to have lower exhaust temperatures than higher-speed engines, but there are many exceptions to both generalizations. For any particular engine, when heavy fuel is used, the exhaust-gas temperature will typically be 10 to 20°C higher than when distillate fuel is used, at the same output. Engines that are derated to lower values of MEP tend to have lower final exhaust temperatures throughout their load range than do the normally rated models of the same engine.

Jacket cooling water generally leaves an engine at 80 to 90°C. The exit temperature is usually maintained in this range, independently of engine output, by temperature-regulating valves that control the flow of raw water at the cooler. With the rate of jacket cooling water flow held constant, the inlet temperature is, consequently, higher at lower levels of engine output, since the heat to be dissipated is also lower. This arrangement provides inherent load-dependent cooling.

At rated engine output, the lubricating oil may leave a low-speed engine at a temperature of 50 to 60°C, and a medium-speed engine at 70 to 80°C. For the lubricating oil, it is usually the engine inlet temperature that is maintained, generally 10 to 15°C below the maximum engine outlet temperature. As a result, the lubricating oil outlet temperature declines as the engine output is reduced.

The air leaving the turbocharger compressor and entering the air cooler will be at a temperature that is affected by ambient conditions but is mostly dependent on the boost ratio of the compressor. The boost ratio varies, roughly in proportion to the percentage of rated engine output. When the boost ratio is 2.5 to 3.5, the air entering the cooler will generally be in the range of 130 to 200°C.

Heat rejected from the hot engine surfaces to the surroundings is not separately represented in Fig. 12. Some manufacturers allow up to 5% of fuel heat input for these losses, but more commonly, values of 1% to 2% are allowed. A high value is appropriate if the exhaust manifold is not water-jacketed or insulated.

3.8 Noise. Noise attributed to diesel engines may be divided between structureborne and airborne noise produced by the engine itself, and the noise caused by air intake and exhaust systems. Generally, the higher-speed engines are more troublesome in this connection than the lower-speed engines.

a. Engine noise. The noise produced by a diesel engine originates from the rapid changes in gas pressure in the cylinders during the cycle, primarily those that accompany combustion and exhaust, and from the mechanical action of the engine components. Noise transmitted through an engine foundation and connected equipment and piping, which is referred to as structureborne noise, must be considered together with engine vibration. While some reduction in engine noise can be achieved through component design, greater reductions in structureborne noise are possible with an engine and its connected equipment resiliently mounted. This matter is further addressed in Sections 6.2 and 11.4.

Airborne noise originating within an engine is often magnified by the resonance of stationary engine components of relatively low mass, such as crankcase doors, oil pans, rocker covers, and camshaft covers. Some reduction in airborne noise is, therefore, possible by stiffening, lining, or resiliently mounting these components, or adding to their mass. Further airborne-noise reduction can be achieved by acoustic shielding of an engine. When an enclosure takes the form of a casing fitted around an engine, measures must be considered to ensure adequate access for maintenance and inspection. Other forms of shielding include the siting of acoustic baffles in the vicinity of an engine, and the lining of the bulkheads and overhead of the space in which an engine is located. A frequently used method of airborne-noise reduction—most useful with auxiliary engines—is to site them on lower engine-room flats, away from shops, control stations, and accommodations.

b. Air intake and exhaust noise. The dry-media, air-intake filters fitted to turbochargers with direct intakes from machinery spaces are designed to also achieve a measure of silencing. Where greater silencing is required, air-intake plenums lined with sound-absorbing media or internal baffles may be provided. Where an air intake is ducted to an engine, the intake and ducting must be designed for limited turbulence and velocities, and with adequate rigidity and support, to achieve required levels of silencing.

Exhaust noise is largely the result of pulsations in the gas stream, and is, therefore, less difficult to silence in turbocharged engines, where the pulsations have been smoothed or eliminated in the exhaust manifold and turbine. Further attenuation of exhaust stream pulsation, and reduction of outlet noise, is achieved by a muffler, which most often consists of a tubular steel chamber with internal baffles. Where a waste-heat boiler is fitted in an engine uptake, similar results may be achieved, and a muffler may be unnecessary. Wet mufflers, in which seawater is injected to cool the gas, are often convenient in small craft, although they require a horizontal mounting, and are especially prone to corrosion. In any event, exhaust ducts, like intake ducts, must be designed for limited turbulence and velocities, and with adequate rigidity and support, to achieve required levels of silencing.

3.9 Ambient Conditions. Most engines will be subjected to service ambient conditions that are not the same as those for which the published engine characteristics were established. Procedures that can be used to assess the effect of off-design ambient conditions may be recommended by the manufacturer. In the absence of specifically applicable guidance, a rise in intake air temperature at constant power output can be expected to:

—increase the fuel consumption by 0.2% to 1.0% for every 10°C rise in air temperature
—decrease the air and exhaust-gas mass flow rate, which will vary approximately inversely with the ratio of absolute air temperatures
—increase the exhaust-gas temperature by 1.5 to 2.0°C for every 1°C increase in the air temperature.

A rise in the cooling water temperature to the charge air cooler at constant power output can be expected to:

—increase the fuel consumption by 0.2% to 0.6% for every 10°C rise in water temperature
—decrease the air and exhaust-gas mass flow rate by 0.5% to 2% for every 10°C rise in water temperature
—increase the exhaust-gas temperature by about 3 to 6°C for every 10°C rise in water temperature.

It is usually not necessary to make allowances in engine output for off-design service conditions as most engines are designed to achieve their rated output at elevated ambient conditions, which are most often cited as an intake air temperature of 45°C and a charge air cooling water temperature of 32°C.

A depressed intake air pressure (below normal barometric pressure at sea level) or an elevated exhaust back pressure will increase the fuel consumption and increase the exhaust temperature, but as long as the deviations are not severe the effects are less significant than those caused by air and water temperature variations. If the intake depression exceeds about 0.01 bar, or the exhaust back pressure exceeds about 0.03 bar, the performance characteristics, including the engine output, may be significantly affected.

The effect of humidity on engine performance is usually insignificant, especially for turbocharged engines.

Section 4
Fuel Characteristics, Injection, and Combustion

4.1 Diesel Engine Fuels. Fuels are covered extensively in Chapter 12, where the terminology is defined; the coverage here focuses on the fuels that are commonly used in marine diesel engines, and on those of their characteristics most relevant to diesel engines. The basic coverage is not repeated here.

Although the great majority of low-speed engines and many medium-speed engines are operated on heavy fuels, and although an increasing number of high-speed engines have been proven capable of operation on at least the lighter blends, there is still a balance to be struck between the lower cost of the heavier fuels and the inconvenience and greater cost of fuel treatment and increased engine maintenance that inevitably result from their use.

Distillate fuels are generally suitable for use in diesel engines without preheating (except in the coldest climates), so that treatment can be limited to settling and filtering operations. Even so, it is considered good practice to centrifuge even the distillate fuels. Residual fuel is rarely used in diesel engines without being blended with distillate fuel to form an intermediate fuel, which, depending on the proportions used, can itself be described as light or heavy. A residual fuel may contain significant amounts of the undesirable constituents of the crude, and may be contaminated with catalytic fines. Since even the lighter intermediate blends require preheating before pumping, settling, centrifuging, and burning, it is reasonable to treat any intermediate fuel as a heavy fuel.

a. Viscosity. The viscosity of a fuel may, by itself, present no difficulty as long as the fuel can be heated sufficiently at each point in the system to permit pumping, settling, filtering, centrifuging, and atomizing. At the very high end of the viscosity spectrum, troubles that may arise include heater fouling, gassing of the fuel, and thermal expansion of injection pump components.

When burning heavy fuel in a diesel engine, it is essential that the viscosity at the injection pumps and injectors be within design limits at all times. The volume of fuel burned in an engine is small in relation to the volume that is available in the piping, and in installations intended for operation on heavy fuels, the transit time between the heaters and the injectors could be sufficiently long, especially at low loads, for the fuel to cool. To prevent this undesired cooling, a flow rate is maintained that is much higher than that needed for injection, that is, about two or three times engine consumption at MCR, with the unconsumed excess that leaves the spill valves of the injection pumps recirculated back to the booster pump suction.

b. Density. The ability to separate water and solids from a fuel by settling and centrifuging is dependent on the differences between the densities of the impurities and that of the fuel. The differences are magnified by heating because the density of the fuel decreases more rapidly than those of the impurities, most notably water. Heating enables conventional centrifuges to achieve adequate separation of water from heated fuel that has a specific gravity at ambient temperature as high as 0.995. More sophisticated centrifuges having water-sensing controls can separate even heavier fuels.

Because injection pumps are volume-metering devices, rack settings vary for a constant engine output, depending on both the density and the heating value of the fuel.

c. Heating value. The heating value, which is usually assessed as energy per unit mass, is typically 5% to 7% lower for residual fuels than for distillates, a difference that carries over proportionately to the blended fuels. Most published data for specific fuel consumption are based on a distillate fuel of standard heating value. The specific fuel consumption that is determined for an engine in service must be corrected accordingly in order to be comparable to published data. Customarily, the lower heating value of the fuel is used.

d. Ignition quality. The ignition delay associated with fuels of low ignition quality results in a late and, therefore, more explosive start to the combustion period. This pattern of combustion results in higher peak pressures and consequent rough and noisy operation that, if sustained, can cause damage to the cylinder heads, liners, pistons, rings, and bearings. Poor ignition quality can also cause the combustion period to be extended, which would result in rough and incomplete combustion, high fuel consumption, and fouling of the combustion chamber. Because the effects of ignition quality are related to time, slower-turning engines are more tolerant of fuels that have a low ignition quality; to some extent the injection timing in slower engines can be advanced to compensate for the ignition delay. Conversely, most higher-speed engines require fuels of higher ignition quality.

The ignition delay associated with a fuel can be reduced if the fuel is heated, and some manufacturers recommend that, for operation on low ignition-quality fuel at low loads, the temperatures of the jacket and piston coolants be maintained at higher levels, and that the temperature of the charge air leaving the charge air cooler be increased.

e. Carbon residue. Fuel with a high carbon-residue index would be expected to leave more deposits after combustion, which result in fouling and wearing of cylinder liners, rings, ring grooves, exhaust valves, and turbocharger turbine nozzles. The adverse effects on cylinder components can be reduced by the use of detergent cylinder oils. Turbocharger fouling is countered by frequent water-washing. The deposits that accumulate on fuel nozzles are removed when the injectors are withdrawn for cleaning.

f. Catalytic fines, solids, and ash. Solid particles carried into the engine with the fuel can cause abrasive wear of fuel injection pumps, injectors, cylinder liners and rings, exhaust valve seats, and turbochargers. The larger solid particles can be removed by settling, filtration, and centrifugal purification. For smaller solid particles, including catalytic fines, the most effective procedure for reducing their presence involves the continuous use of multiple centrifuges that are arranged to process the fuel in series. The first centrifuge is set up as a purifier and subsequent units as clarifiers, which are followed by fine filters.

Fine filters, used alone, provide adequate protection only for engines that burn the cleanest fuels. When fitted as the sole means of protection for engines that use lower-quality fuels, fine filters may clog at inconveniently short intervals.

g. Sulfur. Sulfur is troublesome because of its potential for cold-end corrosion and acid contamination of the lubricating oil. These effects can occur when the sulfur-bearing products of combustion are cooled below their dew point and form sulfuric acid. Engine components that are most vulnerable to cold-end corrosion include the lower ends of the cylinder liners and pistons. The potential for this trouble is highest for engines operated at low power levels for sustained periods, but is countered by maintaining the temperature of the jacket and piston coolants at high levels even at low output, and by increasing the temperature of the air leaving the charge air cooler. In addition, the oil used for cylinder lubrication should have a high alkaline content (high total base number, TBN) in order to neutralize the acids that form.

A converse difficulty can develop from using lubricating oils that have a high TBN: if there is insufficient sulfur present in the fuel to neutralize the alkaline ingredients of the lubricating oil, the resulting deposits of the alkaline ingredients can cause scoring of the liner and wear on the rings. This trouble arises when an engine that is normally operated on high-sulfur fuel is supplied with low-sulfur fuel for an extended period without changing the oil used to lubricate the cylinders.

In crosshead engines the purely vertical movement of the piston rod permits a packing gland to be fitted to separate the combustion space from the crankcase, thereby preventing combustion blow-by and excess cylinder oil from reaching the crankcase. Crosshead engines do not, therefore, require a crankcase oil of high TBN; instead, a high-TBN cylinder oil may be used in a separate cylinder oil system. In contrast, sulfur contamination of the crankcase oil can be troublesome with trunk piston engines; these, therefore, are usually supplied with crankcase oils that have a high TBN.

h. Vanadium. During the combustion process, and especially in the presence of sodium, gaseous oxides of vanadium form, some of which change phase and form solid deposits that adhere to surfaces at temperatures as low as 500°C or even lower. These surfaces include piston crowns, exhaust valves, and turbocharger turbine nozzles and blades. When the deposits occur on the seating surfaces of the exhaust valves, the adverse effects may become immediately apparent as the valves overheat and burn through.

Vanadium troubles will be minimized where temperatures of the surfaces in question are kept below about 500°C. Valve cooling is usually not difficult in large low-speed engines, but is difficult in high-speed engines and some medium-speed engines.

i. Sodium. Most of the sodium in fuels is introduced through seawater contamination, and most of it can be removed with the seawater if settling and centrifuging procedures are adequate. The principal trouble with sodium is its combination with vanadium, as the eutectic compounds then formed will adhere to surfaces at temperatures well below those of the simple vanadium oxides. A rule-of-thumb limits sodium content to less than a third of the vanadium content.

j. Flash point. The temperature of the fuel in fuel tanks is usually kept below the flash point to limit the hazard of fire. A hazard can arise in a diesel plant burning heavy fuel if the fuel leaving the centrifuge, where it may be heated to 98°C, in turn raises the day-tank temperature above the flash point. The hazard can be eliminated by fitting a cooler in the line to the day tank after the purifier.

In some installations, particularly where plants have been converted from distillate to heavy oil, the heated returns from the engine are returned to the day tank, heating it above the flash point. A preferred arrangement includes a mixing tank of limited capacity, so that its high temperature is less hazardous.

k. Pour point. To keep the temperature of fuel above the pour point, heavy fuel tanks are heated, and heavy fuel lines are steam-traced or electrically heated beneath the insulation.

l. Compatibility. An incompatibility among the constituents of a fuel is revealed by sludge accumulations in tanks, filters, and centrifuges, by fluctuating pump discharge pressures, and by frequent viscosimeter excursions. The fuel supplier is expected to ensure that fuels blended ashore do not contain incompatible constituents. Aboard ship it is important to avoid mixing fuels from different deliveries, or blending fuels, without first undertaking a spot test for compatibility. Other than discharging a fuel with incompatible constituents ashore at the next opportunity, the only practical solution for the operator is to cope with the incompatibility while the fuel is consumed.

4.2 Fuel Injection. Fuel is injected into the cylinders of a diesel engine towards the end of the compression stroke, when the compressed charge has reached a temperature sufficient to ensure the ignition and combustion of the fuel. The requirements that the fuel injection system must meet are as follows:

—the fuel must be accurately metered in response to the engine output required;

—the injection must commence at precisely the correct moment during the compression stroke;

—the fuel must enter the cylinder as a finely atomized spray of minute droplets, from the beginning of the injection period through to the end;

—the droplets must penetrate far enough into the combustion space to ensure that they are evenly distributed, without penetrating so far that they impinge on the surrounding surfaces;

—the fuel must be supplied to the cylinder at a predetermined rate; and

—at the end of the injection period, the injection must terminate sharply and completely.

"Solid" injection systems, in which the fuel oil is injected under pumped pressure (as opposed to the generally obsolete air-blast injection systems), are prevalent in diesel engines. Of the solid injection systems, the jerk pump system, in which an individual high-pressure injection pump is provided for each cylinder, is the most common, and will be further described. Other solid injection systems are the common-rail system, in which a manifold is pressurized to serve all of the cylinders, and the distributor system, in which a single pump serves all of the cylinders through a rotating distributor.

In the jerk pump system, fuel is delivered at low pressure to the injection pump, which is a reciprocating, positive-displacement, plunger pump, driven directly from the camshaft by a cam follower. Figure 13 shows a typical helix-controlled injection pump, which is the most common type. Like most other types of injection pumps, it has a stroke of constant length, with metering of the quantity of fuel to be discharged governed by controlling the amount of the stroke that is effective. This control is achieved by closing, then opening, spill ports during the stroke. Figure 14 shows the operation of the helix-controlled pump, in which the helix is the recess in the periphery of the plunger. It can be seen that, as the plunger rises, the spill port will be closed when the top of the plunger passes it. This closure traps the fuel above the plunger and initiates the effective portion of the stroke. The rise in fuel pressure as the plunger continues its stroke will be very sharp, since the fuel is almost incompressible. When the edge of the recess in the plunger exposes the spill port, the effective stroke terminates with a sharp pressure drop. Most injection pumps are fitted with a spring-loaded discharge check valve, which will then close. Because of the helical shape of the recess,

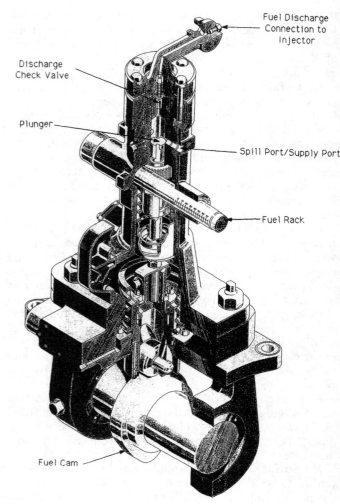

Fig. 13   Fuel injection pump

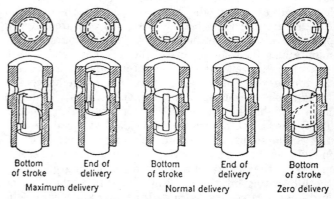

Fig. 14   Fuel pump plunger positions

rotation of the plunger will alter the length of the effective stroke and, therefore, meter the amount of fuel injected; when the vertical edge of the recess is aligned with the spill port, no fuel is injected. Rotation of the plunger is achieved by lateral movement of the fuel rack, which is in mesh with a pinion on the plunger shaft.

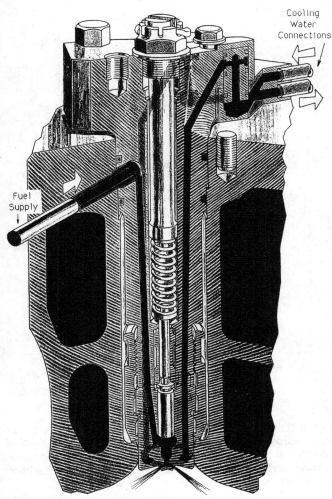

Fig. 15   Fuel injector

cylinder head, and by orienting the air inlet ports to induce a swirling motion to the air.

The discharge check valve ensures that a residual pressure is maintained in the high-pressure fuel line between injections. This residual pressure aids in ensuring a prompt beginning of each injection and also helps to avoid the cavitation that would be likely if line pressure dropped too low as the pump plunger descended, leaving the spill ports uncovered.

In the injection pump of Fig. 13, the top of the plunger closes the spill port at the same point regardless of its angular position, so that the injection always begins at the same time in the cycle regardless of engine output. It is increasingly common for the top of the injection pump plunger to be shaped to vary the beginning of the effective stroke, hence the timing of the start of injection. A common pattern advances the injection most at settings corresponding to about 80% of engine output in order to maintain maximum cylinder pressure throughout the upper end of the engine output range. This pattern of injection improves the specific fuel consumption by injecting a larger fraction of the fuel before the piston reaches its TDC position, so that the engine's design maximum pressure is achieved even at reduced output and, along with the pressure, the high temperatures conducive to complete combustion. In addition, the early end of the injection and combustion periods results in an extended expansion phase, conducive to high thermal efficiency.

Injectors may have internal circuits for water cooling. In many engines, however, sufficient cooling can be obtained by conduction to the cylinder head cooling circuit, thereby simplifying the design, manufacture, and replacement of the injector.

A high-pressure fuel line connects the injection pump to the injector. In most engines the injection pumps are fitted on the side of the engine, convenient to the camshaft, but necessitating high-pressure piping of some length. The design of this high-pressure piping requires careful evaluation from a number of perspectives:

• Even in the largest engines, the quantity of fuel injected per stroke is small relative to the volume of the fuel in the high-pressure line. The fuel discharged by the pump displaces fuel already in the injector, which opens in response to the pressure pulse that travels the length of the line. The time required for this pulse to travel the length of the pipe causes injection lag, which is the delay between spill port closure at the pump and the beginning of injection at the cylinder.

• An irregularity in the interior of the high-pressure passage can set up a reflected pressure pulse which, on reaching the injector after it closes, can cause the injector to reopen, resulting in a secondary injection.

• High-pressure fuel line leaks are fire hazards of great concern in motorships. To safeguard against these fire hazards, fuel lines are double-walled or fitted with shielding.

On some engines difficulties relating to the high-pressure piping between the injection pump and the injector are avoided or minimized by combining the injection pump

The discharge of the injection pump is led directly to the injector, which comprises a spring-loaded fuel valve mounted on the fuel nozzle, as in Fig. 15. The injector opens in response to the discharge pressure from the injection pump, and is closed sharply by the spring when the pump pressure diminishes to a set value.

The timing of the beginning and end of injection depends, first, on proper cam timing, and then on the correct internal calibration of the injection pump to ensure that spill-port operation occurs correctly relative to plunger movement. Proper fuel atomization and penetration are obtained by forcing the fuel through the fine holes of the fuel nozzle at a very high pressure, in the range of 300 to 2500 bars. Good atomization from the beginning of the injection period to the end is achieved by a sharp rise and fall of the injection pump discharge pressure, accompanied by a rapid opening and closing of the injector. The spray pattern is a function of the configuration of the nozzle, which must be designed to avoid droplet impingement on the liner or the piston crown. The distribution of the fuel droplets is assisted by air turbulence, which can be obtained by suitably shaping the piston crown and the

and injector in a single unit injector. Camshaft motion is usually then brought to the injection pump by pushrods and rocker arms.

4.3 Combustion. Fuel combustion does not take place at the tip of the injector, but rather at a distance away from it. This delay occurs because the individual fuel droplets must diffuse through the hot cylinder contents for a sufficient time to heat, vaporize, mix with air, and finally ignite.

The combustion process in a diesel cylinder is considered to occur in four phases, which begin during the compression process and end during the expansion process, as shown in Fig. 16. The four phases are the ignition-delay period (when no combustion occurs), the rapid-combustion period (which begins with ignition), the steady-combustion period, and the afterburning period.

a. The ignition-delay period. The ignition-delay period is the interval between injector opening and the start of ignition. During this phase the first droplets to enter the cylinder are heated by the surrounding charge of compressed air as they disperse and vaporize. Until ignition occurs, there is no increase in the cylinder pressure above what it would have been had injection not occurred.

The ignition-delay period is primarily a function of the ignition quality of the fuel, which is related to its chemical composition. Fuels of lower ignition quality require more preparation time, and the delay period is therefore longer. It is important to note that in a higher-speed engine, the time available for this preparation of the fuel for ignition is lower than in a lower-speed engine, a fact which helps to explain the generally lower tolerance of high-speed engines for fuel of low ignition quality.

b. The rapid-combustion period. During the rapid-combustion period, the fuel that has accumulated in the cylinder during the delay period ignites and burns rapidly. Because the accumulated fuel has already mixed with the charge air, this phase is sometimes called the premixed combustion period. The rapid combustion is accompanied by a sharp rise in cylinder temperature and pressure. If the pressure rises too sharply, the combustion becomes audible, a phenomenon known as diesel knock.

c. The steady-combustion period. Once combustion has been established in the cylinder, the ignition of further fuel droplets entering the cylinder lags the injection rate

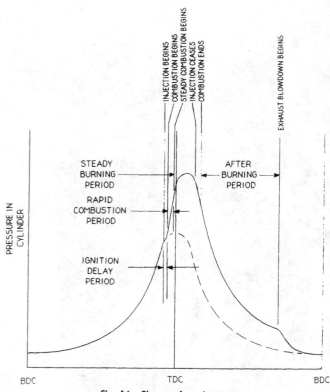

Fig. 16 Phases of combustion

by the time required for the fuel to mix, heat, and vaporize. Because the droplets burn as they diffuse into the cylinder, this phase is sometimes called the diffusion combustion period. This period ends shortly after the injector closes, when the last of the fuel has burned.

The cylinder pressure usually peaks just beyond the TDC position and near the middle of the steady-combustion period. The cylinder pressure then declines as the expansion process proceeds.

d. The afterburning period. If all of the fuel has burned cleanly and completely by the end of the steady-combustion period, the pressure profile will be smooth through the expansion stroke. Typically, however, there will be some pressure fluctuations resulting from the combustion of incompletely burned fuel or of intermediate combustion products, and some delayed chemical reactions. It is during this period that soot and other pollutants are produced.

Section 5
Engine Matching and Selection

5.1 Margins. Generally, diesel engines are so matched to their loads that normal operation in service is at some high fraction of rated output, typically in the range of 80% to 90% of the maximum continuous rating (MCR), at a speed slightly below rated rpm. This region of operation usually coincides with the best range of specific fuel consumption. Anticipated component lives and

service recommendations for inspection, maintenance, renewal, and overhaul intervals are based upon operation in this range. The difference between the power at MCR and the power level established for normal operation (which is sometimes called the continuous service power) is the engine margin.

Figure 17 shows the speed-power curves for a ship. The

curves can be projected at the design stage by methods described in Chapter 1. The power absorbed by the propeller is less than the brake power because of transmission and shafting losses and attached loads. The average service condition curve reflects the fact that more power will be required for a given ship speed to be achieved in service than on trials, as described in Chapter 1. The allowance for service conditions, which is applied to the power estimate for the trial condition, is called the service margin, and is applied in addition to the engine margin. The total margin is the difference between the installed power (or MCR) and the power required to achieve service speed on trials with the hull and propeller clean and smooth, and is equal to the sum of the engine margin and the service margin.

In practice, many operators will accept the use of the power reserve that is incorporated in the engine margin to meet required service speeds as the hull performance deteriorates. In fact, the division between engine margin and service margin is not consistently defined, since the continuous service power is arbitrarily determined. The only important consideration is that the total margin be adequate.

5.2 Relation Between Ship Speed and Engine Performance. The relation of the rpm of a fixed-pitch propeller to the ship speed at a particular draft and trim is illustrated in Fig. 17. In service, as the hull and propeller roughen and drag increases, the rpm required at any given ship speed rises slightly (the slip increases).

The average service speed is indicated in Fig. 17. If an engine were matched to run at 100% of rated rpm at the required service speed under trial conditions, it can be seen that to maintain the required service speed the engine would necessarily exceed its rated rpm increasingly thereafter, as the hull and propeller roughen. Consequently, propulsion engines must be matched to their propellers so that the ship speed achieved at 100% rpm under trial conditions (the trial speed) exceeds the required service speed.

With reference to Section 3.2, an engine is normally limited in its power output by constraints on thermal overload that are most conveniently expressed as an MEP limit. The MEP, like the torque, is proportional to the power developed divided by the rpm. Since the service speed is below the trial speed, it is normally the MEP limit that will be reached first as the hull and propeller performance deteriorate in service (see Fig. 10). Thus an engine can be forced into a condition of excessive torque and MEP without exceeding rated power. The consequences of operation beyond the MEP limit are discussed in Section 3.2 of this chapter.

5.3 Required Engine Rating. The engine rating is generally determined so that, in the trial condition of the hull and propeller, at loaded draft and trim, the power required to drive the propeller, allowing for transmission and shafting losses and any attached auxiliaries, will be between 80% and 90% of the MCR, at rated rpm. This allowance will usually result in adequate margins. (Transmission and shafting losses are discussed in Chapters 9 and 10,

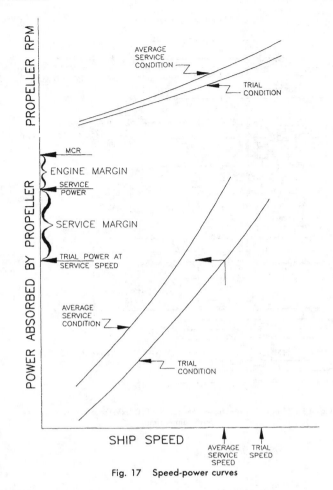

Fig. 17 Speed-power curves

and in reference 3.) Engines are often selected so that the power required under rated rpm at trial conditions is even less than 80% to 90% of the MCR, if:

— the ship must maintain rigorous schedules;
— the long-term effects of increased hull and propeller roughness are expected to be large;
— the ship is expected to be drydocked infrequently;
— a large allowance for adverse weather conditions is appropriate; or if
— the intended trade will take the ship for extended stays into warm, seawater ports or anchorages, where increased hull fouling is likely.

When the rating is determined so that the power absorbed at rated rpm under trial conditions is less than the MCR, higher average power outputs can be utilized in service as the hull and propeller roughen, without excessive torque (as reflected in high MEP and cylinder exhaust temperatures), enabling higher ship speeds to be achieved. However, more power must be installed, and so acquisition cost and plant weight will be higher. As a further consequence, in single-screw, single-engine installations, even if the ship can be ballasted down to loaded draft on trials, it may not be feasible to achieve MCR without overspeeding the engine.

REDUCTION GEAR OUTPUT 170 RPM

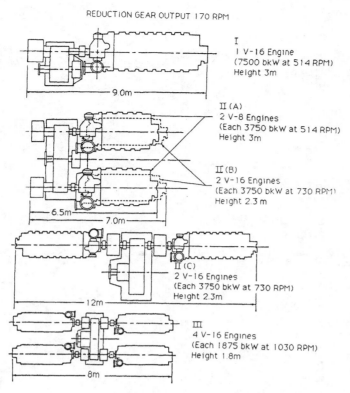

I
1 V-16 Engine
(7500 bkW at 514 RPM)
Height 3m

9.0m

II (A)
2 V-8 Engines
(Each 3750 bkW at 514 RPM)
Height 3m

II (B)
2 V-16 Engines
(Each 3750 bkW at 730 RPM)
Height 2.3 m

6.5m 7.0m

II (C)
2 V-16 Engines
(Each 3750 bkW at 730 RPM)
Height 2.3m

12m

III
4 V-16 Engines
(Each 1875 bkW at 1030 RPM)
Height 1.8m

8m

Fig. 18   Alternative arrangements for a 7500 bkW plant, with approximate dimensions

ability, maintenance requirements, and present and future parts cost and availability. Trade-offs are usually necessary. A requirement for low weight or minimum machinery volume may be achieved at the expense of high fuel consumption or high maintenance requirements. It is imperative that the engine be considered within the physical and operating context of the whole installation; selection of propulsion engines of light weight or low specific fuel consumption, for example, may not result in the lightest or most cost-effective power plant. These issues are discussed in greater detail in Chapter 1.

The operating profile of an engine assesses the time spent in various operating modes. All important modes must be considered, and periods of sustained idle or low-load operation must be included as well as those at high loads. For propulsion engines, operating modes may include conditions of deep and light draft, clean and fouled hull, calm and heavy weather, cruising and high ship speed, towing or icebreaking and running free, and operation with and without attached auxiliaries. The plant design and engine selection will be affected if the profile includes frequent or extended periods of maneuvering or astern running.

In selecting propulsion engines, consideration must be given to whether a single engine of the low-speed, direct-coupled type is most suitable, or if requirements are better met by one or more medium- or high-speed engines driving the propeller through gearing or electric drive.

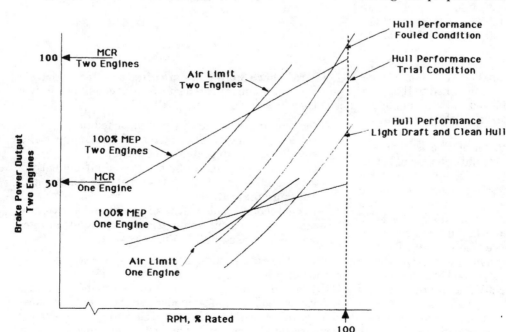

Fig. 19   Speed-power curves for two engines geared to a single fixed-pitch propeller

5.4 Engine Selection. Once the required engine rating has been established, other factors that affect the selection of engines for a particular application must be considered. Among these are the engine operating profile, the plant weight, the machinery space volume and configuration, fuel quality and consumption, acquisition cost, reli-

The decision may be dictated by the available space. Using geared engines as an example, Fig. 18 illustrates the flexibility that can be achieved. With an electric drive, there is even greater flexibility, as discussed in Chapter 8.

If low-speed operation is required for substantial periods of time, a multiple-engine installation should be considered, since latitude is provided in matching the number of engines in service to the output required at different ship speeds. Unneeded engines can be in a standby capacity and quickly started when required. With a controllable-pitch propeller installation, additional flexibility is provided in matching the number of engines in service to the ship speed, since the propeller thrust can be controlled by adjusting the propeller pitch, thereby enabling the engine rpm to be maintained within practical limits. When two identical engines are connected to drive a fixed-pitch pro-

peller, however, single-engine operation may be limited or not possible in some cases, as illustrated in Fig. 19, because one engine, run alone, would be insufficient or overloaded. Alternative arrangements for this particular application, to permit the number of engines to be matched to ship speed with greater flexibility, include two larger engines, one larger engine (which alone can provide a wider range of low-speed operation) with a smaller engine, a two-speed transmission, or a larger number of smaller engines.

Section 6
Moments, Forces, and Vibration

6.1 Introduction. The loads imposed by a diesel engine on its foundation and on connected equipment are predictable in nature, amplitude, and frequency and are usually among the data available from the manufacturer. Whether trouble will arise depends on the response of connected structure and equipment. If the frequency of a vibratory load, or one of its harmonics, is close to the natural frequency of connected structure or equipment, then even a small disturbance can excite a resonant response.

Engine-imposed disturbances may be divided between (1) external forces and moments, which can excite a response from hull structure, and (2) torsional vibration in the propulsion drive train, which usually affects only shaft-connected equipment. Generally, forces and moments internal to an engine are absorbed by the engine itself.

The frequencies of engine-produced disturbances are related to the rpm of the engine, and are defined relative to that rpm by their order, that is, their frequency as a multiple of engine rpm. Fractional orders are encountered in the case of four-stroke engines.

6.2 External Forces and Moments. External forces and moments arise from the reciprocating motion of the pistons and running gear, and would cause an unrestrained engine to pitch, roll, or yaw. With the engine installed in the ship, these disturbances can excite a response from the hull structure.

a. Forces and moments resulting from piston motion. An analysis of the motion of a piston and crank (see references 4 and 5) shows that a fluctuating and reversing force is developed, acting in the line of, but opposite to, the direction of piston motion. Since this force opposes the motion of the piston, it is generally called an inertia force, but may be more accurately termed a reciprocating force. The reciprocating force includes a first-order component with an amplitude proportional to the reciprocating mass multiplied by the square of the engine rpm, and second- and higher-order components of successively lower magnitudes.

At each cylinder of an engine, the reciprocating force is transmitted through the running gear to the engine structure as a disturbance that fluctuates in magnitude and direction at single and higher multiples of crankshaft rpm. The reciprocating force is a function of reciprocating mass, engine geometry (specifically the ratio of connecting rod length to crank throw), and rpm, so that a second, identical cylinder, running 180 degrees out of phase with the first, develops an out-of-phase but otherwise identical reciprocating force. The first-order component of this force is equal but opposite to that of the force developed by the first cylinder. However, second-order components from both cylinders would coincide to reinforce each other. This situation would be resolved if a pair of additional cylinders, 180 degrees out of phase with each other, were added 90 degrees out of phase with the initial pair of cylinders, so that the reinforced second-order components of the new pair opposed those of the first pair. This arrangement would produce a four-cylinder engine which would have balanced first- and second-order reciprocating forces.

The reciprocating forces at each cylinder act on the axis of the cylinder, and are, therefore, displaced along the length of the crankshaft. Consequently, when the reciprocating forces acting on the cylinders of a multicylinder engine are balanced as described in the previous paragraph, couples are created in the vertical, longitudinal plane that would cause an unrestrained engine to pitch. These couples are called vertical couples or inertia couples, because of their origin in the reciprocating force components, but are perhaps best described as pitching couples. If an engine has enough cylinders, they can be so arranged that first- and second-order pitching couples can be balanced, although this cannot be done with two-stroke engines having four, five, and six cylinders, or with five-cylinder, four-stroke engines.

For in-line engines it is often advantageous to convert first-order pitching couples to yawing couples, in whole or in part, since many ships sensitive to vertical disturbances at the frequencies in question may be less sensitive to athwartships disturbances. The conversion is achieved with additional counterweights on the crankshaft (that is, in addition to the counterweights fitted to balance the crankpins and webs). The additional counterweights, as they rotate, generate a first-order rotating couple, only the vertical components of which can be used to balance

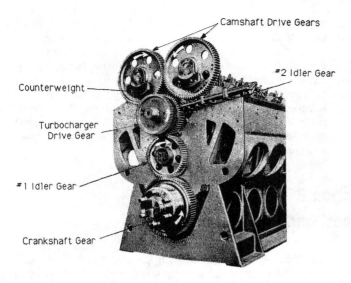

Fig. 20   Camshaft drive gears with counterweights

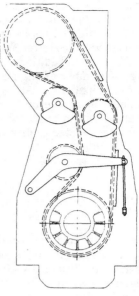

Fig. 21   Lanchester balancer

the pitching couple, leaving an unbalanced first-order couple in the horizontal plane. The manufacturers of the large two-stroke engines consider, as their standard arrangement, additional counterweights sufficient to cancel half of the first-order pitching couple, thereby imposing a first-order yawing couple of equal magnitude.

Additional counterweights are especially useful on V-engines, which, because of the inclined banks of cylinders, can develop both yaw and pitch couples that have large first-order components. Counterweights on the crankshaft are used to eliminate the yaw couple and reduce the pitch couple, or to convert the remaining pitch couple to a yaw couple.

A first-order pitching couple can be canceled by two pairs of counterweights rotating in opposite directions at crankshaft rpm. Figure 20 illustrates an arrangement on a two-stroke V-engine, where the camshafts rotate at crankshaft rpm. One weight is on each of the camshafts, which are geared to run in opposite directions. Since the weights rotate in opposite directions in the same transverse plane, the horizontal components of the forces they generate always cancel, leaving only a vertical force fluctuating at first-order frequency. By fitting two pairs of such weights, longitudinally separated along the crankshaft, the first-order pitching couple can be canceled without generating a yawing couple. The same principle can be used to balance the second-order pitching couples if the pairs of opposing counterweights are driven at twice crankshaft rpm, an arrangement that is known as a Lanchester balancer. A chain-driven Lanchester balancer is shown in Fig. 21. In principle, the higher-order pitching couples could be countered by similar means; in fact they are not often of sufficient magnitude to cause concern.

b. Transverse forces and resulting moments. Because the upper end of the connecting rod is constrained to reciprocate, a force normal to the direction of piston motion is imposed on the engine structure by the crosshead or, in a trunk-piston engine, by the piston skirt. This force, called the normal force or guide force, reflects the

gas force on the piston and the angle of the connecting rod, and reaches a peak near the middle of the power stroke, with a lower peak in the opposite direction near the middle of the compression stroke. Since the reciprocating force is superimposed on the gas force, the normal force has a major first-order component, and higher-order components of successively lower magnitude. Although the normal force is balanced at all times by an equal and opposite horizontal force at the main bearing, the vertical displacement of these two forces produces a couple acting in the vertical, transverse plane (that is, the roll plane) at each cylinder. This couple may be called a roll, capsizing, or tipping couple or, because its peak amplitude occurs during the power stroke, and is therefore proportional to the torque produced, a torque-reaction couple. Two-stroke, in-line engines have roll couples whose largest component is at an order equal to the number of cylinders; for four-stroke, in-line engines, the order of the largest component is equal to half the number of cylinders. In V-engines, where two cylinders operate each crank, the largest roll couples are at an order equal to the number of cranks for two-stroke engines, and at an order equal to half the number of cranks for four-stroke engines.

In addition to the roll couple, because of the longitudinal displacement of the cylinders in a multicylinder engine, the guide forces and the horizontal bearing reaction forces generate equal but opposite yaw moments, one moment acting at the height of the wrist pins and the other moment acting at the height of the main bearings. The resulting couple can force the top of a tall engine into a racking motion (that is, an X-pattern when viewed from above) since the base of the engine is constrained.

Roll couples and racking couples can be countered when necessary by bracing the top of a crosshead engine to adjacent hull structure with transverse struts. Because of the inherently more rigid structure of trunk-piston engines compared with the tall crosshead engines, and also because of the higher frequencies resulting from the higher rpm and greater number of cylinders, these couples are not normally troublesome with these engines.

c. Longitudinal forces. A pulsating longitudinal force is produced by the deflection of the crank webs under load, and contains first- and higher-order components, all of low magnitude and, therefore, not normally a source of trouble. Occasionally, however, a higher-order component of this longitudinal force may coincide with the natural frequency of the crankshaft itself. The usual solution in this case is to fit a damper, comprising a dummy piston under engine oil pressure, at the free end of the crankshaft.

d. Summary of external forces and moments: in-line engines. For in-line, vertical engines with evenly spaced power strokes and no balancing gear or additional counterweights, the most serious external moments and forces generally encountered are listed below. It should be noted that most of these disturbances can be corrected as described above. Higher-order disturbances may be present in all planes, including the longitudinal, but, because of their low magnitudes, these only occasionally cause concern.

- Two-stroke in-line engines:
  —four-cylinder engines usually have severe first- and second-order pitching couples and a severe fourth-order roll couple;
  —five-cylinder engines usually have a moderate first-order pitching couple, but severe second-order pitching and fifth-order roll couples;
  —six-cylinder engines usually have no first-order pitching couple, but a severe second-order pitching couple and a severe sixth-order roll couple;
  —engines with seven or more cylinders usually have moderate or negligible first- and second-order pitching couples and moderate roll couples at an order equal to the number of cylinders; and
  —eight- and twelve-cylinder engines may have racking moments sufficient to require countermeasures, typically at third, fourth, and fifth orders.

- Four-stroke in-line engines:
  —four-cylinder engines usually have a severe second-order vertical force and a severe second-order roll couple;
  —five-cylinder engines usually have a moderate first-order pitching couple, but a severe second-order pitching couple and a severe 2.5-order roll couple; and
  —engines with six or more cylinders will usually have moderate or negligible first- and second-order couples and moderate roll couples at an order equal to half the number of cylinders.

e. Summary of external forces and moments: V-engines. V-engines are arranged with opposite cylinders in each bank acting on the same crank and generally, therefore, firing in succession. For an even distribution of power strokes, the angle between the banks would have to be equal to the firing interval, but to simplify manufacturing a constant V-angle (typically 45 or 50 deg) is usually used regardless of the number of cylinders. Where an uneven distribution of power strokes results, the effect is mitigated by the large number of cylinders present.

The most serious external moments and forces generally encountered in four-stroke V-engines without balancing gear or additional counterweights beyond those necessary to reduce first-order rotating couples, and with a typical bank angle of 45 to 50 deg, are listed below. Most first- and second-order disturbances can be corrected using combinations of oppositely rotating counterweights. Higher-order components may be present in all planes but, because of their low magnitudes, only occasionally cause trouble.

- Four-stroke V-engines:
  —eight-cylinder engines can be balanced in regard to first- and second-order pitching and yawing couples, but only if the bank angle is equal to the firing interval of 90 deg; smaller angles can result in first-order disturbances sufficient to require correction;
  —ten-cylinder engines usually have moderate first- and second-order pitching and yawing couples and a moderate 2.5-order roll couple;
  —12-, 16-, and 24-cylinder engines are generally balanced in regard to low-order pitching and yawing couples, although they usually have low to moderate roll couples at an order equal to half the number of cranks (that is, at a quarter the number of cylinders); and
  —14-, 18-, and 20-cylinder engines usually have moderate or negligible first- and second-order pitching and yawing couples as well as low-to-moderate roll couples.

There are far fewer designs of two-stroke V-engines than four-stroke, but many two-stroke V-engines are found in marine service. These engines have a firing order selected to eliminate secondary forces and couples, but with a 45-deg bank angle they would have a rather large primary pitching couple even after fitting crankshaft counterweights, were it not for the use of camshaft counterweights (see Fig. 20) to provide first-order moment compensation. This is possible because the camshafts run at engine rpm and are geared to run in opposite directions. The 45-deg bank angle and the selected firing order result in evenly distributed power strokes in the V8 and V16 engines, but not in the V12 and V20 engines, where the large number of cylinders provides adequate compensation.

6.3 Torsional Vibration. Torsional vibration arises from periodically varying torque superimposed on the steady torque being transmitted by an engine to its load. The sources of this varying torque include the discrete power strokes of the engine, which generate torque pulsations once per crank throw per cycle, and higher orders of this frequency. In ships with direct-connected low-speed diesels, this is usually the dominant source of torque variation. Torsional vibration calculations are required at an early stage in the design process, as soon as the engine has been selected and the configuration of the rotating system, including shafting, couplings, clutches, gearing, bearings, and propeller, is known. Flexible couplings and spring-loaded torsional dampers are both susceptible to low-frequency excitation, such as that produced when one cylinder is taken out of service. Consequently, torsional vibration calculations are usually required for operation with a cylinder out of service, and additional barred speed ranges may then be imposed. More details concerning torsional vibration considerations are included in Chapter 10.

Section 7
Engine Components

7.1 Cylinders and Pistons. Cylinder liners are usually made of centrifugally cast iron, with a collar at the top, and are inserted downward into the cylinder block and secured by the cylinder head. Most liners are of the wet type, forming the inside of a water-circulated jacket for the cylinder, the outside of which is formed by the cylinder block. The collar of the liner, which can be quite deep as in Fig. 2, may be bore cooled, that is, drilled for cooling passages. This technique provides the high strength of a thick section while bringing cooling water close to the heated surface. Liners for two-stroke engines have ports at the BDC position for air admission; liners for loop-scavenged engines have a second set of ports, higher than the air ports, for exhaust gas.

On large engines, a separate cylinder head is provided for each cylinder. A head may be made in one piece or may consist of multiple parts of different materials. The head of a four-stroke engine carries the air and exhaust valves and the fuel injector and, in large engines, a starting air valve, an indicator cock, and a relief valve, and must be internally cooled. The resulting complex shape is usually made of high-strength cast iron, although cast steel has been used. Heads of two-stroke engines are usually simpler, particularly in loop-scavenged engines where the head carries only the injector, the starting air valve, indicator cock, relief valve, and cooling passages; and in these engines the head may be a steel forging with bored cooling passages. The head is bolted to the cylinder block through holes around its periphery, securing the liner collar against the block, and compressing the gaskets. When the head is in place, its cooling passages are continuous with those of the cylinder block and liner; the water is introduced at the bottom of the cylinder and rises to exit at the top of the head.

Cylinder blocks of large engines are usually separate for each cylinder, and are generally of cast iron. Cylinder blocks of low-speed engines are generally bolted to each other as well as to the supporting structure of the frames and bedplate, to add to the longitudinal strength of the engine. In some medium- and high-speed engines, the cylinder blocks are an integral part of the frame, into which the liners are inserted, and to which the heads are bolted. In two-stroke engines, in addition to forming the outside of the cooling passage, the cylinder block forms part of the air supply trunk and, in loop-scavenged engines, also

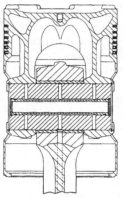

Fig. 22 Trunk piston

carries exhaust gas from the liner ports to the exhaust manifold.

The bottom of the combustion space is formed by the piston crown, which may be shaped to aid the combustion process, as in Figs. 22 and 23. One-piece pistons are usually of cast iron, although one-piece, cast-aluminum pistons are used in some light-duty engines. Most two-piece pistons have a crown of forged or cast steel and a cast iron or cast aluminum lower part, called the piston skirt.

The piston-to-liner clearance is sealed by circular iron piston rings, which are compressed to fit the bore but are otherwise free to move in the ring grooves of the piston. The resulting pressure against the liner is supplemented by gas pressure acting on the inner circumference of the rings to seal the cylinder as the piston reciprocates. The movement of the rings in the grooves results in wear, and the condition of the grooves is one factor governing the life of the piston. The grooves may be plated or locally hardened. When rings are fitted in an aluminum piston or skirt, a separate, hardened steel insert may be fitted to provide grooves for the piston rings. The piston rings themselves are wearing components, which may be plated and shaped to improve their initial running-in and their subsequent performance. Trunk pistons may be fitted with oil rings above or below the wrist pin to control the amount of oil used for cylinder lubrication.

Piston crowns must be internally cooled. In crosshead engines, the bottom of the piston is sealed, and lubricating oil or water is circulated through the cooling passages below the crown, brought in and out through telescopic

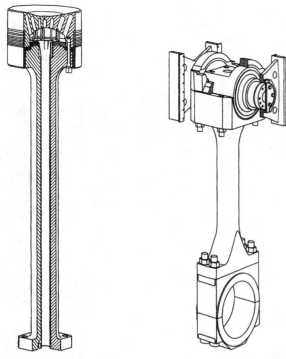

Fig. 23 Piston for crosshead engine, with piston rod

Fig. 24 Forked connecting rod with marine bottom, and crosshead

pipes or internal borings in the piston rod. Trunk pistons are oil-cooled, with the oil circulated up through the connecting rods or sprayed directly against the bottom of the crown from stationary jets. Piston cooling passages may be formed in the casting or forging process, or by boring. The internal cooling is supplemented by conduction through the rings to the cylinder liner.

Trunk pistons carry the wrist pin and transmit the normal force from the connecting rod to the cylinder liner. The pin is usually carried in bosses cast into the piston or skirt, as in Fig. 22, or, in some designs, in a carrier inserted into the piston skirt. The pin may be secured to the piston or carrier, or to the connecting rod, or may float between axial constraints in the piston. In some two-stroke engines the piston is free to rotate around the carrier; in some four-stroke engines the "pin" is in fact spherical, and the piston is mechanically rotated in small increments with each revolution. Crosshead engine pistons can be short, as in Fig. 23, since the normal force is transmitted through the crosshead and its guides.

The cylinders of crosshead engines and some trunk piston engines are lubricated by oil injected through holes in the liner, usually arranged in rows of six or eight holes, near the middle of the stroke (see Fig. 43). The cylinders of most trunk piston engines are lubricated by controlled leakage from the wrist pin bearing, or by the cylinder cooling oil, passed through holes in the skirt, or by spray or splash from the crankcase.

7.2 Piston Rods and Crossheads. Piston rods and crossheads are found only in crosshead engines. Piston rods are manufactured from steel forgings, and are bolted

to the piston and to the crosshead. With oil-cooled pistons, the rod may be bored to carry the oil.

A crosshead is shown in Fig. 24. The core of the crosshead is a steel forging. The wrist pin, more properly called a crosshead pin in this application, may be integral with the crosshead block. The crosshead shoes or slippers are cast iron or steel, and are faced with white metal (babbitt) bearing surfaces. Crosshead guides may be an integral part of the engine frame, or they may be bolted on.

Because of the purely linear motion of the piston rod, the bottoms of the cylinder blocks can be isolated from the crankcase by a diaphragm and packing gland, as shown in Fig. 2. This construction prevents contamination of the crankcase oil by combustion products blowing by the piston rings, and by excess cylinder lubricating oil running off the liner. In trunk-piston engines, this kind of contamination is all but inevitable.

7.3 Connecting Rods, Crankshafts, and Bearings. Connecting rods are manufactured from steel forgings or castings, with a bearing at the bottom for the crankpin. When the bearing housing is separate from the shank, as in Fig. 24, it is said to be of the marine type. The arrangement in Fig. 3 is frequently used in trunk-piston engines, as it permits the connecting rod to be withdrawn through the cylinder bore with the piston. The configuration at the top of the rod depends on the connection to the wrist pin: in Fig. 24, the top is forked to clear the piston-rod attachment to the crosshead. In trunk-piston engines, the connecting rod may be fixed to a pin, which is free to swing in the piston bosses, or it may be free to pivot about the pin, as in Fig. 22. In V-engines, where the connecting rods of adjacent cylinders share the same crankpin, the cylinders may be offset longitudinally to permit the connecting rod lower ends to be side-by-side, as in Fig. 3. When the cylinders are not offset, the bottom end configuration is complex.

Crankshafts are usually of steel, forged in one piece, even for very large engines, or they may be built up, as in Fig. 25, usually by shrink-fitting separately forged crankpins and main journals to cast steel crank webs. When each crank throw (crankpin and two webs) is a single piece, and these are shrunk or welded to the main journals, the term semi-built is used. Some smaller engines have ductile cast iron crankshafts. In large, low-speed engines, very long crankshafts (for more than 6 to 8 cylinders) are usually manufactured in two sections, flanged and bolted at mid-length. Counterweights are sometimes integral with the crankshaft, but for most large engines, are bolted on, as in Figs. 1 and 3. Most crankshafts are drilled for oil passages. In some cases the diameter of the internal drilling of crankpins is varied to assist in counterweighting.

Crankshafts of low-speed direct-connected engines incorporate the main thrust collar. For other engines the aftermost bearing may include shallow thrust collars intended solely to maintain the axial position of the crankshaft. Integral flanges permit attachment of flywheels, timing gears or sprockets, and torsional dampers, as well as coupling the engine to the load.

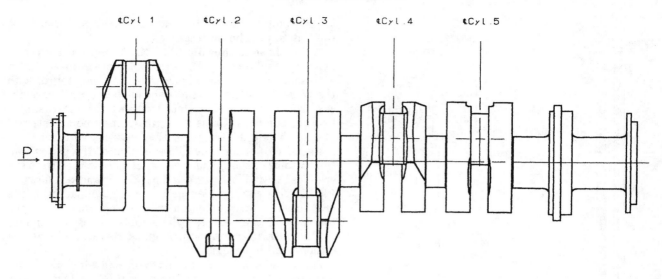

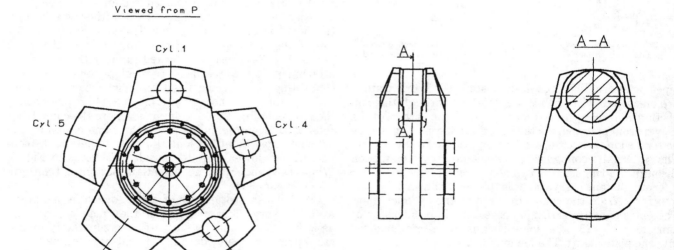

Fig. 25   Built-up crankshaft

The major bearings in most engines are of the precision type, with split, renewable-steel shells lined with alloys of tin, lead, copper, and aluminum, often with an overlay on the lining to prevent corrosion in storage and to improve running-in. Where loadings permit, as is often the case for low-speed engine main and crankpin bearings, white metal (babbitt) linings are used.

7.4 Bedplates and Frames. Bedplates and frames of the largest engines are each separate entities, as illustrated in Fig. 2, which together form the crankcase, while in smaller engines a single fabricated assembly or casting may combine the functions of bedplate and frame, as shown in Figs. 1 and 3. Fabricated assemblies may be of steel, combining plate, structural sections, and forged and cast elements, or they may include cast-iron elements.

Bedplates and crankcases of very large engines may be divided and flanged at mid-length. The engine is bolted to the foundation through a flange on the bedplate of large engines or, where there is no separate bedplate, through a flange or brackets on the frame. Cylinder blocks are bolted to the top of the frame, and the cylinder heads are bolted to the cylinder blocks. Vertical tie rods connect the tops of the cylinder blocks to the bottom of the frame or bedplate, and resist the gas forces acting upward on the cylinder head, and downward, via the running gear, on the main bearing supports.

Below each cylinder, usually on both sides of the crankcase, large doors permit access for maintenance and inspection. Pressure relief valves are fitted to the crankcase to minimize damage from an excessive pressure condition

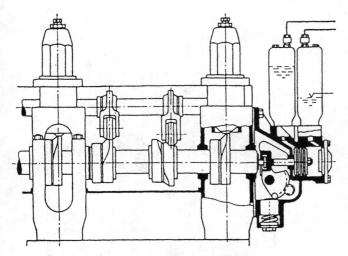

Fig. 26   Camshaft with axially displaced reversing cams

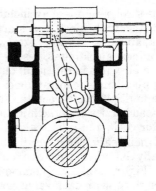

Fig. 27   Fuel pump cam, with reversing follower

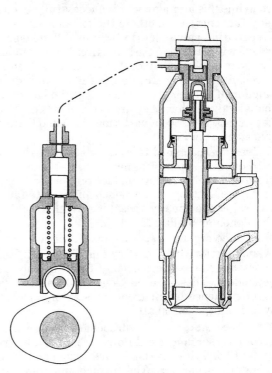

Fig. 28   Hydraulic valve gear with air spring

or explosion. This condition can arise when an overheating bearing vaporizes oil, creating an explosive mixture, which can then be ignited by further overheating of the bearing or, in trunk piston engines, by blow-by. An ignition would result in an initial rise in pressure that would be limited by the air within the crankcase. To relieve the pressure without admitting air to feed a secondary explosion, the relief valves are spring-loaded to open and then snap shut.

7.5   Camshafts and Valve Gear.   Cams may be made of hardened steel castings or forgings, shrunk onto the camshaft, or they may be integral with the camshaft. The camshaft may be divided into segments serving each cylinder, or may be continuous.

At each cylinder, separate cams are fitted for each purpose: for four-stroke engines, there are cams for the fuel injection pump, for the air valves, and for the exhaust valves; two-stroke engines with uniflow scavenging have cams for the fuel pump and exhaust valves, but two-stroke engines with loop scavenging have cams for the fuel pump only. Where each cylinder has two air valves or two exhaust valves, both are most often operated by a single cam. The camshafts may also drive the governor, a lubricating-oil pump for the valve gear, and, on air-started engines, the starting-air distributor. V-engines usually have a camshaft for each bank of cylinders.

Four-stroke reversing engines most often have a second full set of cams axially displaced from the first, on each camshaft, as in Fig. 26, with reverse running requiring that the camshaft be shifted axially. To reverse two-stroke engines, it is often sufficient to change the timing of the camshaft by an angular shift in its relation to the drive gear. On some two-stroke reversing engines, the exhaust timing is sufficiently symmetrical about BDC to remain unchanged, and the shift in fuel pump timing is achieved by displacing the follower by the required angle, as in Fig. 27. Reversing is further discussed in Section 8.

Camshafts are driven from the crankshaft, usually by timing gears, but on some engines by a chain-and-sprocket drive. To isolate the timing from the effects of torque variation in the crankshaft, a torsional damper may be fitted to the camshaft. On four-stroke engines the camshaft runs at half crankshaft rpm; on two-stroke engines the camshaft and crankshaft run at equal rpm. The camshaft drive may also be used to drive engine-mounted auxiliaries or balancing gear.

Fuel-injection pumps are usually mounted directly above their cams, with the cam followers attached to the pump plungers, as in Fig. 13. Fuel-injection equipment is further discussed in Section 4. Valves of medium- and high-speed engines are usually opened by push rods and rockers, as in Figs. 1 and 3, as are valves of older low-speed engines, but most low-speed engines use a hydraulic drive, which is illustrated in Fig. 28. Valves are generally closed by steel springs, but, in most low-speed engines, compressed air is trapped below a piston on the valve stem to serve as an air spring, also illustrated in Fig. 28.

7.6   Cylinder-Head Mountings.   Cylinder-head mount-

ings of a four-stroke engine may include the air and exhaust valves, the fuel injector, a starting-air valve, an
indicator cock, and a relief valve. For two-stroke uniflow-
scavenged engines, air admission is by ports in the liner,
and there is no air valve in the head, but if the exhaust is
then handled by a single, centrally located valve (Fig. 2),
there may be multiple fuel injectors. For loop-scavenged
engines the head carries only the injector, the starting-air
valve, indicator cock, and relief valve.

Valves and their seating surfaces can be of critical importance, especially the exhaust valves of engines that operate
on heavy fuel. Figure 29 shows a common arrangement for
four-stroke engines, with each valve carried in a cast-iron
guide inserted in the head, and seating on a renewable insert. While this arrangement is successfully used in many
heavy-fuel engines, in others the exhaust valve is carried
and seated in a valve cage (Fig. 30), enabling inspection and
renewal without lifting the head. In either arrangement the
valve seat is intensively cooled, as may be seen from the
figures. Exhaust-valve cooling is further improved by the
reduction of hot spots; this is accomplished by rotating the
exhaust valves when they are off their seats. Mechanical
rotators may be used, or vanes may be fitted to the exhaust
valve stem, as in Fig. 1.

Valves are machined from alloy steels. Exhaust valves
are frequently made from nimonic steel, and may have a
welded-in Stellite seating surface on the head. Valve seat
inserts are usually of hardened or Stellite-faced steel.
Valve cages are usually of cast iron, with a welded-in seat
of hardened or Stellite-faced steel.

Fuel injectors are discussed in Section 4.

A starting-air valve is shown in Fig. 31. In V-engines,
starting-air valves are normally fitted in only one bank
of cylinders. These are pilot-operated valves, opened in
correct sequence by air from the starting-air distributor
driven from the camshaft.

7.7 Turbochargers. A turbocharger with a centrifugal compressor and axial-flow turbine is illustrated by
Fig. 9. Radial-flow turbines are also used, usually in
smaller turbochargers. The turbine and compressor are
mounted on a short, stiff shaft to form the rotor. Rotors
are usually built up from independent elements, all of
corrosion-resisting steel. Casings are usually cast-iron
barrels, arranged for end-withdrawal of the rotor. Casings may have a water jacket for the whole turbine, or
may be cooled only in the vicinity of the turbine gland and
bearing, as shown in Fig. 9, to permit better waste-heat
utilization (see Section 9). Bearings may be of the sleeve
or antifriction types, and may be at the extreme ends of
the shaft as shown in Fig. 9, or between the turbine and
compressor. One bearing must incorporate a thrust element.

Lubricating oil and cooling water for turbochargers are
generally supplied from the engine systems, but occasionally the turbocharger is fitted with an independent oil
system. In either case means must be provided to maintain
lubrication during a failure, until the rotor coasts to a
stop.

Turbocharger performance is discussed in Section 3.6.

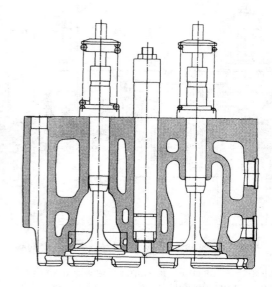

Fig. 29   Medium-speed engine cylinder head and valves. Note that the exhaust-valve seat contains a cooling passage

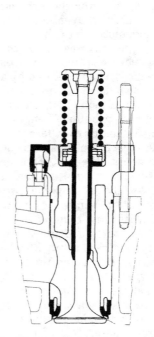

Fig. 30   Caged exhaust-valve with
cooled seat and rotator

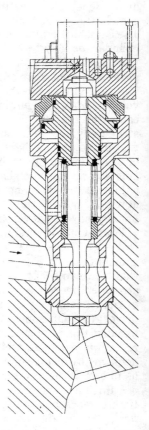

Fig. 31   Starting-air valve

Section 8
Engine Control

8.1 Starting and Reversing Procedures. To start a diesel engine it must be rotated at sufficient speed to compress the air in the cylinders (at very low speeds, leakage past the piston rings becomes very significant) so that a temperature is reached that is high enough to ignite the fuel when it is injected. Some engines, mostly smaller engines, are started by cranking motors, which may be battery, hydraulically, pneumatically, or mechanically driven. Most larger engines, however, are started by the timed introduction of compressed air directly into cylinders that are positioned at the beginning of their power strokes in the selected direction of rotation. The compressed air then drives those pistons down, in firing-order sequence, thereby compressing the air trapped in other cylinders. After one or two revolutions have been completed, fuel is introduced in the normal manner into those cylinders whose pistons are completing a compression stroke, which then fire. The starting air is cut off, and the engine accelerates to idle or operating speed under the control of the governor. Typically each cylinder of an in-line engine is fitted with a starting-air valve, but commonly only the cylinders of one bank of a V-engine are so fitted. The valves are usually opened by pilot air that is supplied in correct sequence via a camshaft-driven starting air distributor.

In direct-reversing engines the starting-air distributor timing is shifted for reverse rotation (together with the timing of the inlet and exhaust valves and the fuel injection pumps), directing pilot air to those cylinders whose pistons have stopped just short of TDC on their up-strokes, so that the engine is rotated in the opposite direction for starting.

Starting systems of direct-connected, direct-reversing propulsion engines must be able to rotate the engine against the torque developed by the propeller. Where an engine is clutched to the propeller shaft, it is usually disengaged for reversing, while the shaft is held by a brake, then clutched-in after starting in the reverse direction. See Section 11.6 for further discussion of reversing sequences.

Because the maneuverability of a ship is affected by the ability to start and reverse the main engine, regulatory bodies have set pertinent requirements. Typically, starting systems for direct-reversing main engines must be capable of twelve consecutive starts without recharging, and for nonreversing engines the capability for six consecutive starts is required.

8.2 Method of Engine Control. The control of diesel engines centers on the regulation of the amount of fuel injected into each cylinder during each cycle. This is usually accomplished by positioning the fuel racks of all of the injection pumps simultaneously, thereby regulating their effective strokes (see Section 4.2). For constant-speed engines, such as those driving generators, the racks are positioned by a governor. For variable-speed engines,

in some cases the racks may be positioned directly by the operator, but in most cases regulatory bodies require that the racks be positioned by a governor whose set point is adjusted by the operator.

8.3 Control System Safety Devices. In addition to the aforementioned speed-regulating governor, most engines are fitted with some or all of the following additional safety devices, some of which are regulatory body requirements, and some of which are required only for propulsion engines:

- Interlocks:
    to prevent starting if the turning or barring gear is engaged;
    to prevent starting of a direct-reversing propulsion engine if the camshaft position (ahead or astern) is not in compliance with the engine order telegraph;
    to prevent fuel injection to a direct-reversing propulsion engine if the engine rotation is incorrect (because of way on the ship), even if the camshafts are correctly set; and
    to prevent reversing of a direct-reversing propulsion engine if the fuel rack is not in its zero injection position.

- Devices to stop the engine or to reduce its speed to a low level in the event of:
    low lubricating oil pressure,
    overspeed (independent of the governor),
    loss of cylinder oil flow,
    low coolant pressure,
    high coolant temperature,
    high charge air temperature,
    high air manifold temperature,
    high exhaust gas temperature, and
    deviation in exhaust gas temperature of one cylinder from the average.

- A maximum fuel injection limit (which corresponds to a torque limit) that is set by a mechanical stop that blocks movement of the fuel rack to prevent extreme overloads.

8.4 Engine Control, Instrumentation, and Automation. Even in highly automated plants, all essential data should be monitored by local, independent indicators (Bourdon-tube pressure gages and mercury thermometers, for example) and all essential functions should be capable of manual control. Where direct manual control is not feasible, and a pneumatic, hydraulic, or electrical link is present, suitable local backup should be present.

In addition to local instrumentation of the support systems (see Section 10), local instrumentation on a propulsion engine typically includes the following:

- Thermometers for:
    jacket water supply (inlet to engine);

jacket water outlet of each cylinder;
cooling water outlet of each turbocharger;
injector coolant supply;
injector coolant outlet for each cylinder;
bearing lubricating oil and control oil supplies;
lubricating oil outlet of each main bearing, crankpin bearing, crosshead bearing, turbocharger
  bearing, and the thrust bearing;
piston coolant supply;
piston coolant outlet for each piston;
fuel oil supply;
air cooler air inlet and outlet;
air cooler cooling water supply;
air manifold (for fire detection);
exhaust gas at each cylinder; and
exhaust gas before and after each turbocharger
  turbine.

- Pressure gages (or manometers) for:
  bearing lubricating oil and control oil supplies,
  seawater supply,
  injector coolant supply,
  piston coolant supply,
  fuel oil supply,
  air intake filter pressure drop,
  air cooler air pressure drop,
  air manifold,
  exhaust manifold,
  control air supply, and
  starting air supply.

- Dipsticks, gage glasses, or sight-flow indicators for:
  lubricating oil sump level,
  turbocharger gravity tank level,
  discharge of each cylinder oil lubricator, and
  air manifold level (to warn of oil or water accumu-
    lation).

For geared and clutched engines, local instrumentation
typically includes the following:

- Thermometers for the reduction gear and thrust
bearing lubricating oil supply, and for the lubricating oil
outlet of each reduction gear bearing, and the thrust
bearing.
- Pressure gages for the reduction gear and thrust
bearing lubricating oil supply, and for the clutch operating
and control air supply.
- Dipsticks for gear and thrust bearing oil sumps.

Local instrumentation and sensors fitted for remote
instrumentation, for alarms, and for the initiation of automatic sequences should be independent of each other.

A local control console at an engine might provide the
following control and instrumentation functions:

- Control functions:
  selection of local or remote control,
  engine order telegraph,
  positioning of camshaft for ahead or astern,
  starting air valve opening and closing,
  governor speed set-point selection,
  direct fuel rack positioning,
  engine starting,

emergency stop, and
override automatic stop or slowdown.

- Instrumentation functions:
  bearing lubricating oil supply pressure,
  jacket water outlet temperature,
  piston coolant outlet temperature,
  starting air supply pressure, and
  engine rpm.

Where plant control is centralized (the normal case),
usually in an acoustically isolated, air-conditioned control
room, and usually with bridge control of the main engine,
the main engine control console may provide the following
control and instrumentation functions:

—selection of local, control room, or bridge control;
—selection of direction of engine rotation (direct-reversing engine);
—initiation of a programed start sequence;
—manual start;
—speed setting;
—control of clutches;
—controllable-pitch propeller pitch selection and indication;
—engine order telegraph;
—emergency stop;
—override of automatic stop or slowdown;
—continuous display of key parameters, which may include:
      engine rpm,
      turbocharger rpm,
      propeller rpm (where clutches are fitted),
      starting air supply pressure,
      control air pressure, and
      air manifold pressure;
—demand display of all of the parameters listed above
   for local instrumentation (frequently using listings
   or schematics on a visual display terminal); and
—alarm annunciator for all alarmed points, which will
   include most or all of the parameters listed above
   for local instrumentation, depending on the level of
   automation, and, in addition, for a crankcase oil-
   mist detector.

Usually the main engine control console is incorporated
in a larger control console for the entire plant, with controls and continuous display instrumentation for auxiliary
systems arrayed in rational order. Demand display and
alarm annunciation for these systems are usually integrated with those for the main engine. In the most common arrangement, the ship's service generator remote
control panels, the main switchboard, and grouped electrical distribution panels are also located in the control room,
opposite the plant control console. In addition, consoles for
the control of bilge and ballast systems, and the control of
fuel filling and transfer, may be installed in the control
room.

Where bridge control is installed, the bridge control
console may provide the following main engine control
and instrumentation functions:

—selection of engine or bridge control;

—single-lever or telegraph control of:
>   direction of engine rotation (direct-reversing engine),
>   programed start sequence,
>   speed setting,
>   clutches, and
>   controllable-pitch propeller blade position;

—engine order telegraph;

—emergency stop;

—override of automatic stop or slowdown;

—continuous display of key parameters, which may include:
>   engine rpm,
>   propeller rpm (where clutches are fitted),
>   controllable-pitch propeller blade position,
>   starting air supply pressure,
>   clutch air supply pressure;

—engine alarm condition; and

—auxiliary system alarm condition.

Section 9
Waste-Heat Recovery and Utilization; Steam Systems

9.1 Sources and Uses of Waste Heat. The points of heat rejection from diesel engines that are normally considered to have a practical potential for waste-heat recovery are the engine exhaust, the charge-air cooler, and the jacket coolant.

- Heat usually is recovered from the exhaust gas of a main propulsion engine to generate steam for fuel and lubricating oil heating and hotel services. Excess steam can contribute to cargo heating or other cargo service needs where these are appropriate.

- Where the diesel plant is large enough and the electrical needs sufficiently limited, enough steam can be generated by the exhaust gases to meet the electrical load with a waste-heat turbogenerator (WHTG). Where a WHTG can meet only part of the demand with steam from the waste-heat boiler, it may still be economically justified, with the balance of the electrical demand met by supplemental steam from oil-fired boilers, from an attached generator, from an exhaust-gas turbine-driven generator (see Section 9.7), or from SSDGs.

- Heat recovered from the charge-air cooler of a main engine is most often considered for the preheating of feedwater for an exhaust-gas boiler, for fuel tank heating, for lubricating oil purifier heating, for absorption refrigeration systems, for quarters heating, or for producing hot water.

- Heat is most often recovered from the jacket water of a main engine for use in freshwater generating plants.

The practicality of waste-heat recovery from auxiliary engines depends on their size and utilization. Passenger and naval vessels, with their large electrical plants and high demand for heat to generate fresh water, are likely candidates for auxiliary engine waste-heat recovery.

9.2 Steam Systems. Figure 32 illustrates a simple steam system used frequently on ships where steam requirements are few, and in which an oil-fired package boiler serves as the steam drum for a forced-circulation, water-tube, waste-heat boiler. This type of system is easily automated; any excess steam produced is dumped through a pressure-regulating valve to the seawater-circulated condenser, while a shortage of steam triggers a pressure switch to supplementally fire the oil-fired boiler.

The fact that the oil-fired boiler is filled with saturated steam and water at all times facilitates a rapid response. In some installations gas bypasses are fitted to the waste-heat boiler to control the production of steam. Where sufficient waste heat is available for a turbogenerator to be fitted, the system can be as simple as in Fig. 33, or more complex, as in Fig. 34.

Some ships are fitted with waste-heat boilers of the gas-tube type in which the production of steam can be controlled by varying the water level. The oil-fired boiler would not necessarily be in the circuit, but it can be kept warm by a steam heating coil. Where the steam demand is in the range of about 2000 kg/h or less (the usual case on dry-cargo vessels), the oil-fired boiler is most often of the fire-tube type.

Figure 35 illustrates a typical tanker steam plant that supplies steam for cargo heating, cargo and ballast pumping, and tank cleaning. Two oil-fired boilers supply steam at a pressure sufficient for cargo and ballast pump turbines, and also serve to supplement the output of the waste-heat boiler. Because the waste-heat boiler operates at lower pressure, it is provided with its own steam drum, and the oil-fired boilers are fitted with heating coils. The oil-fired boilers in this example are of the double-circuit type; an oil-fired, closed primary steam circuit generates steam in an attached secondary drum (also called a reboiler) to preclude contamination of the primary steam circuit from a leaking cargo heating coil. The same results can be achieved by using a contaminated evaporator (low-pressure steam generator) to supply cargo heating steam.

The design pressure of a ship's steam system may be dependent on its use for fuel heating, as discussed below. Heating coils or pipes are fitted in drain tanks to maintain the feed temperature close to 100°C to provide a measure of deaeration. Because of the danger of sulfuric acid attack on cold economizer tubes, the system of Fig. 33 incorporates an economizer recirculation valve which passes water from the saturated circuit into the feed upstream of the economizer, thereby enabling the feed temperature to be maintained even at a low engine output.

Local pressure gages are fitted on each steam drum, on each steam main, after each pressure-reducing or regulating valve, and at each pump suction and discharge. Thermometers are fitted at the superheater outlet, condenser,

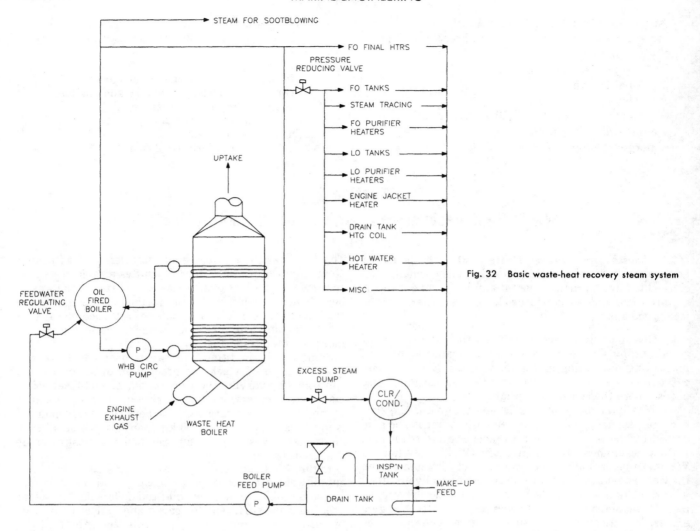

Fig. 32 Basic waste-heat recovery steam system

drain tank, feedwater heater, and at the feedwater inlet and economizer outlet. Gage glasses are provided for the condenser, for the drain tank, and for steam drums. Alarms are provided for high steam pressure, high superheater outlet temperature, low feed-pump discharge pressure, high and low levels in steam drums and in the drain tank, and opacity or oil content at the inspection tank.

9.3 Heat Recovery for Basic Ship's Services. The quantity and temperature of the exhaust heat vary with ambient conditions and with engine output; the steam demand also varies. But usually, sufficient heat can be recovered from the engine exhaust to supply the basic ship's services under most, if not all, expected operating conditions. If it appears that sufficient heat can be recovered for basic services for some conditions, but is marginal for others, then a modest sophistication of the heat-recovery cycle might be justified along the lines laid out below. If the marginal condition might only occasionally be encountered—for example, only in the coldest environment—then considering the frequency with which this might occur, the potential use of an oil-fired boiler to supplement the steam supply should be considered.

Some design constraints must be kept in mind in regard to waste-heat recovery:

• The steam temperature and, therefore, its pressure must be high enough to ensure that the highest-viscosity fuels that will be used can be adequately heated. The highest-viscosity fuels may have to be heated to about 160°C, requiring that saturated steam be generated at 8 bars, to reach the fuel-oil heaters at about 170°C. (Difficulties arising from using such high fuel temperatures are discussed in Chapter 12.)

• Since the generating tubes contain water and steam at the saturation temperature corresponding to the steam pressure, a waste-heat boiler without an economizer can recover heat only to the point at which the exit gas temperature is reduced to the saturation temperature of the steam, plus a terminal temperature difference of about 10°C. (The closest approach of the gas temperature to the steam temperature is called the pinch point, and is illustrated in Fig. 36.) Consequently, if the steam is generated at 8 bars, the lowest exit gas temperature achievable without an economizer would be about 180°C.

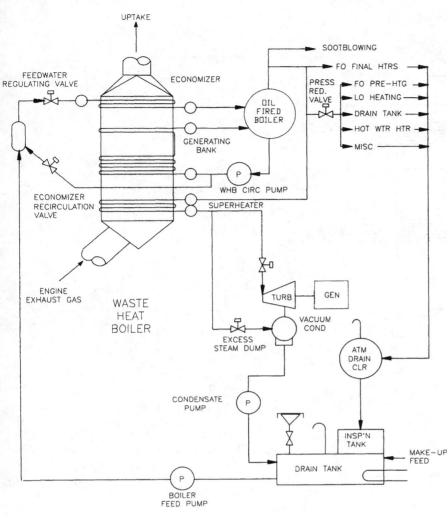

UPTAKE

SOOTBLOWING

FO FINAL HTRS

FEEDWATER
REGULATING VALVE

ECONOMIZER

PRESS
RED.
VALVE

FO PRE-HTG

LO HEATING

OIL
FIRED
BOILER

DRAIN TANK

HOT WTR HTR

GENERATING
BANK

MISC

ECONOMIZER
RECIRCULATION
VALVE

WHB CIRC PUMP

SUPERHEATER

ENGINE
EXHAUST GAS

WASTE
HEAT
BOILER

TURB

GEN

VACUUM
COND

EXCESS
STEAM DUMP

ATM
DRAIN
CLR

CONDENSATE
PUMP

INSP'N
TANK

MAKE-UP
FEED

DRAIN TANK

BOILER
FEED PUMP

Fig. 33   Waste-heat steam plant with turbine-generator and single-pressure boiler

• When lower generating pressures are used, or when waste-heat boilers are fitted with economizers, lower gas exit temperatures are possible. In any event, however, the gas exit temperature should not be permitted to fall below about 160°C because of the danger of acid attack when sulfur-bearing fuels are used. Even distillate fuels can contain sulfur.

9.4 Enhanced Waste-Heat Recovery. An increased use of waste heat, beyond meeting the basic needs for fuel and lubricating-oil heating and hotel services, requires a heat-recovery plant of increased complexity, size, weight, and cost. The extent to which such a plant is justified is a matter that must be resolved from an examination of the trade-offs involved, which may be largely economic in nature.

a. Economizers. Because the lowest gas temperature leaving the generating bank will exceed the saturation temperature of steam by the pinch point value, considerable heat may remain available in the gas. An economizer will enable further heat to be recovered from the exhaust gas until its temperature is reduced to the aforementioned limit, as illustrated in Fig. 36. However, with feedwater entering the economizer at the drain tank temperature (or even up to about 140°C), the outside surface of the economizer tubes will be below the acid dew point temperature even though the bulk temperature in the gas stream is above it, thereby inviting acid condensation and attack. One means of avoiding this situation is by economizer recirculation; in the form shown in Fig. 33, water from the discharge of the circulating pump is injected into the feedwater stream at the economizer inlet.

b. Feedwater heaters. The amount of steam produced by a waste-heat boiler can often be increased by preheating the feedwater, but if the feedwater is preheated above a temperature of about 150°C, it cannot be used to reduce the gas exit temperature leaving an economizer to the lower limit of about 160°C so that, in terms of efficiency, nothing would be gained, although the problem of acid condensation would be largely avoided.

Drain tanks of systems incorporating a turbogenerator with a vacuum condenser are at temperatures considerably lower than saturation, eliminating any possibility of deaeration in the tank. In these cycles, a deaerating feed heater may be justified.

c. Multi-pressure boilers. The quantity of steam that can be produced from exhaust gas can be enhanced

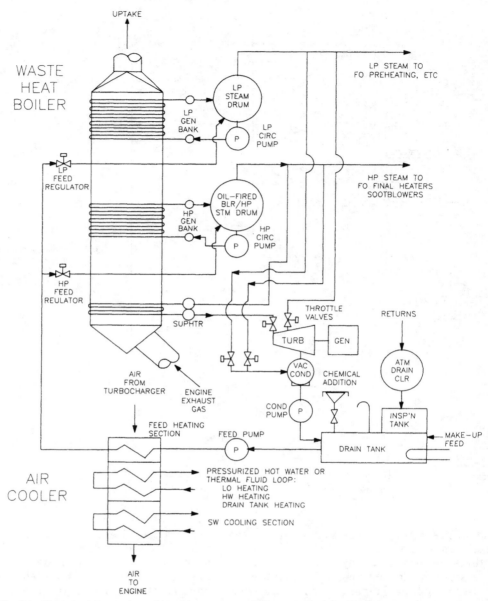

Fig. 34   Waste-heat steam plant with turbine-generator, dual-pressure boiler, and heat recovery at air cooler

if it is recognized that, while there is a need for steam at a sufficiently high pressure to ensure adequate fuel heating, other services can utilize steam generated at lower pressures. Figure 37 is a temperature diagram for a dual-pressure boiler fitted with a superheater, which is shown schematically in Fig. 34. In this example, feedwater is supplied to both steam drums directly from feed heaters, and is already at 150°C. With a careful selection of parameters and the use of charge air feed heating, a dual-pressure boiler might produce 20% to 30% more steam than a single-pressure boiler constrained to the higher pressure, for the same application.

9.5 Heat Recovery at Charge Air Coolers. The temperature at which air enters the coolers of a turbocharged engine at high output may make it an attractive source for heat recovery. Most engines are designed for the air

to be cooled to 45 to 50°C, although higher temperatures may be encountered in engines using output-dependent cooling systems, when operated at reduced output. Because only a portion of the heat can be usefully recovered, a heat-recovery cooler is built with its cooling side divided into sections, with the final section in the direction of air flow reserved for engine cooling water. The heat-recovery sections of the cooler are then arranged in series, with the highest-temperature section at the inlet. These sections may be circulated with the fluid being heated (feedwater, for a feed heater) or with an intermediate fluid, which might be pressurized water or another fluid.

9.6 Waste-Heat Turbogenerator (WHTG) Cycles. The power level at which sufficient waste heat can be recovered to satisfy an electrical load is highly plant-specific, depending on exhaust-gas quantity and temperature,

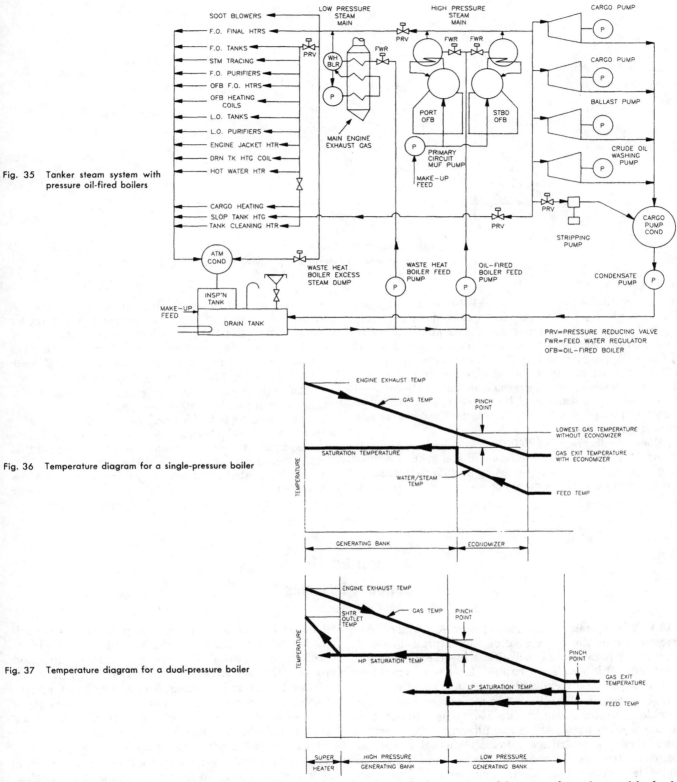

Fig. 35 Tanker steam system with pressure oil-fired boilers

PRV=PRESSURE REDUCING VALVE
FWR=FEED WATER REGULATOR
OFB=OIL-FIRED BOILER

Fig. 36 Temperature diagram for a single-pressure boiler

Fig. 37 Temperature diagram for a dual-pressure boiler

steam conditions, competing needs for heating steam, the actual configuration of the heat-recovery plant, and the efficiency of the turbogenerator, as well as on the ship's electrical requirements. Figure 38 illustrates the electrical load that can be supported by an unsupplemented WHTG for a particular range of low-speed engines with dual-pressure boilers, using heat recovery at the engine air cooler. In this example, the available turbogenerator output under the most favorable ambient conditions (summer) matches the minimum average service load with a

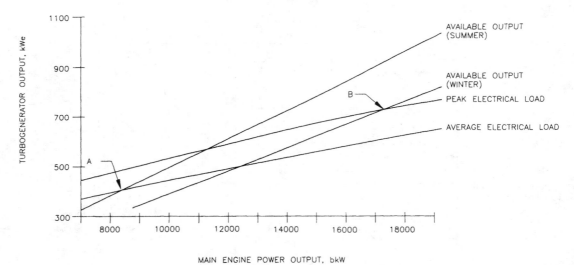

Fig. 38 Output potential of waste-heat turbogenerator

main engine output of some 8000 bkW (point A in Fig. 38). The power level at which the electrical load is completely covered under all reasonable conditions and with sufficient margin to meet normal peak load peaks is about 17,000 bkW (point B in Fig. 38). These crossover points will be at considerably higher engine power levels with simpler cycles or higher electrical load, but can be lowered by further sophistication of the waste-heat cycle or by reducing the electrical load by increasing the use of power takeoff for auxiliaries.

Schematic diagrams of WHTG plants appear as Figs. 33 and 34. Guidance in the selection of steam and exhaust conditions and methods of assessing WHTG performance and the output of these plants is presented in reference 3.

9.7 Exhaust-Gas Turbines. High-efficiency turbochargers are often able to supply sufficient combustion air without using all of the exhaust-gas flow. In these cases there is the potential for excess exhaust gas to be used in an exhaust-gas turbine that drives a mechanical load. Three of the many possible configurations are:

• An exhaust-gas turbine-driven generator may be fitted, serving in parallel with other generators.

• An exhaust-gas turbine may be geared to the engine output shaft, thereby forming, in effect, a combined cycle or turbo-compound arrangement; the power contributed by the turbine can be up to 5% of the total.

• An integrated plant may be arranged in which the exhaust-gas turbine is connected at the power takeoff gear of an engine fitted with a shaft-driven generator.

Section 10
Support Systems

10.1 Fuel Systems. The fuel system for a diesel engine must typically be capable of handling blended and other heavy fuels as well as distillate fuels. Fuel systems are discussed in detail in Chapter 12, and the coverage here is limited to principles that warrant emphasis.

While clean distillate fuels are sometimes considered suitable for combustion in diesel engines without any treatment other than settling and filtering, it is nevertheless advisable to centrifuge even distillate fuel. In normal operation fuel passes to the day tanks, from which the engine is supplied, only via the purifiers. At least two purifiers should be provided for continuous operation, either in series or in parallel. The rated capacity of each purifier should include sufficient margin above the main engine consumption at maximum continuous rating

(MCR) to allow for maintenance. The performance of purifiers is inversely related to their volume of throughput; therefore, it is desirable to install purifiers with rated capacities that are larger than required.

Heavy fuel must be heated close to 100°C to facilitate purification, and this hot oil input to the day tank could force the day-tank bulk temperature to rise above the flash point. For this reason some installations are fitted with a cooler in the heavy fuel purifier discharge line.

The fuel service system supplies fuel to the engine at the correct viscosity. To limit the cooling of heavy fuel between the heaters and the engine, and to ensure a constant pressure at the injection pumps, a flow rate equal to two or three times engine consumption at MCR is maintained. The unconsumed fuel is recirculated back to the

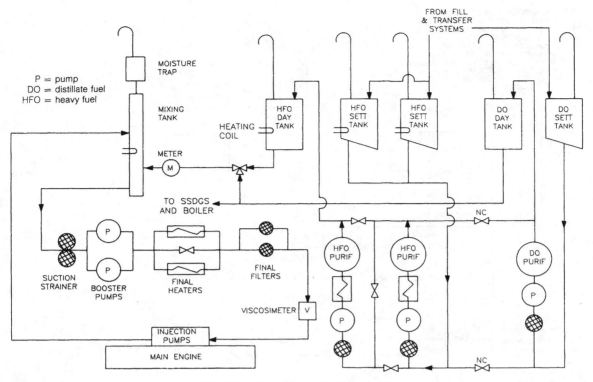

Fig. 39 Fuel treatment and service system

mixing tank. The functions of the mixing tank are to prevent overheating the day tank, to enable the changeover from hot heavy fuel to cold distillate fuel to occur gradually, and to permit vaporized lighter fuel fractions to vent. In some plants there is no mixing tank and recirculated fuel returns directly to the day tank; however, this return of hot oil can lead to an objectionably high day-tank temperature. When the mixing tank is at atmospheric pressure, it is supplied by gravity from the day tank, and its bottom is therefore lower than the bottom of the day tank, as shown in Fig. 39, while its top must be higher than the top of the day tank to permit the venting of light fractions. To facilitate a rapid changeover from one fuel to the other, the volume of the mixing tank must be kept small; consequently, the mixing tank is often configured as a pipe column. Because of the high temperatures to which the heaviest fuels must be heated and the increased amount of gas formation that would occur at atmospheric pressures, some engine manufacturers recommend that the hot circuit of the service system be designed as a closed and pressurized loop. This pressurization requires an additional set of low-pressure booster pumps to feed the mixing tank from the day tank, but the mixing tank can then be more conveniently proportioned and located. Venting of gas, which forms in the pressurized loop, must then be through a relief valve.

At least two fuel-oil heaters should be provided, each with sufficient capacity to heat the heaviest fuel likely to be encountered, and each conservatively rated in regard to fouling margins. The steam supply to the heaters is controlled by the viscosimeter, which must be in close proximity to the injection pumps.

All of the heavy fuel piping in the service system, possibly including the filters and pumps, may require steam (or electric) trace heating and insulation.

10.2 Compressed-Air Systems. A typical compressed-air system is shown in Fig. 40. The system can be divided into three segments that provide air for main and auxiliary engine starting, air for instrumentation and control, and air for miscellaneous ship's services.

Because the maneuverability of a ship is dependent on the availability of starting air, the minimum number and size of the starting air receivers must comply with regulatory body requirements. Typically, sufficient air must be available to enable at least six consecutive starts of a nonreversing engine, or twelve of a direct-reversing engine. This volume of air must be available without recharging, and be stored in at least two receivers. The main starting air receivers are normally high pressure, usually about 30 bars, in order to reduce the required size of the receivers. The main starting air receivers are normally charged by two air compressors. If normal ship's service air consumption is low, this air may be supplied from the main starting air receivers through a pressure-reducing valve. In this case an additional, lower-capacity, topping air compressor may be installed. If the ship's service usage is expected to be large, provision of a separate low-pressure air compressor may be more appropriate.

A separate, small auxiliary (or emergency) air receiver is usually provided for starting the auxiliary engines; it

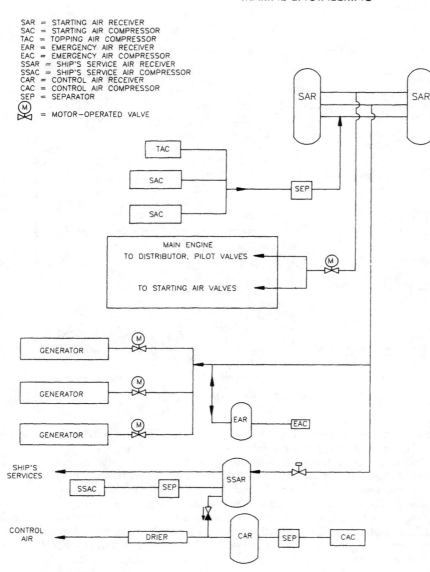

SAR = STARTING AIR RECEIVER
SAC = STARTING AIR COMPRESSOR
TAC = TOPPING AIR COMPRESSOR
EAR = EMERGENCY AIR RECEIVER
EAC = EMERGENCY AIR COMPRESSOR
SSAR = SHIP'S SERVICE AIR RECEIVER
SSAC = SHIP'S SERVICE AIR COMPRESSOR
CAR = CONTROL AIR RECEIVER
CAC = CONTROL AIR COMPRESSOR
SEP = SEPARATOR
Ⓜ = MOTOR–OPERATED VALVE

Fig. 40   Compressed-air system

is normally supplied from the main starting air system through a stop-check valve. For cold ship start-up, the auxiliary air receiver may be supplied by an emergency air compressor, which may be driven by hand, by a hand-started diesel engine, or by a motor supplied from the emergency switchboard.

The control air system is most often supplied from an independent, oil-free control air compressor and a separate receiver, via a drier. Standby service may be through a cross-connection from the ship's service system, or through a reducing valve from the starting air system.

Local instrumentation may include an air pressure gage at each compressor discharge, at each receiver, and after each reducing valve; an oil pressure gage and an oil sump dipstick for each compressor; and thermometers for cooling water at water-cooled compressors. Starting air pressure and control air pressure are indicated at central, local, and remote engine control stations. Alarms are fitted for low starting air pressure and control air pressure.

Except for the emergency compressor, compressors are usually arranged to start and stop automatically in response to receiver pressure. Starting air compressors are usually started in sequence as the receiver pressure falls. Where the starting load of large starting air compressors is a concern, they may be arranged to start automatically and then cycle on their discharge pressure unloaders. Automatic drain traps are fitted at moisture separators and at receivers.

10.3 Lubrication Systems. Motorship lubricating-oil systems are complex because of the number of grades of oil required. A geared, medium-speed diesel plant may require different grades of oil for the main and auxiliary engines, and other grades for gearing and miscellaneous uses. A lubricating-oil system for a low-speed diesel plant involves at least two grades of oil for the main engine (one for the circulating system and the other for the cylinders) and a third for the auxiliary engines.

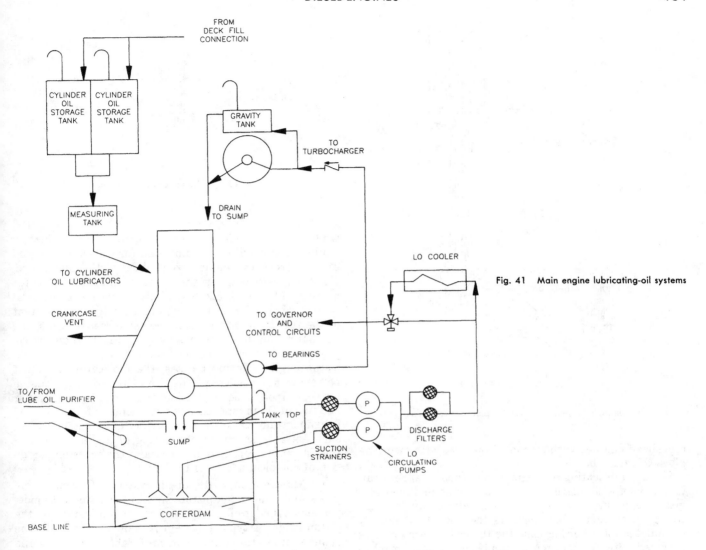

FROM
DECK FILL
CONNECTION

CYLINDER OIL STORAGE TANK | CYLINDER OIL STORAGE TANK

GRAVITY TANK

TO TURBOCHARGER

MEASURING TANK

DRAIN TO SUMP

TO CYLINDER OIL LUBRICATORS

LO COOLER

Fig. 41   Main engine lubricating-oil systems

CRANKCASE VENT

TO GOVERNOR AND CONTROL CIRCUITS

TO BEARINGS

TO/FROM LUBE OIL PURIFIER

TANK TOP

SUMP

DISCHARGE FILTERS

SUCTION STRAINERS

LO CIRCULATING PUMPS

COFFERDAM

BASE LINE

a. Main engine lubricating-oil circulating system. A typical circulating oil system for a low-speed engine is illustrated by Fig. 41. Oil draining from bearings and cooling passages to the bottom of the crankcase passes into an independent sump that is built into the double bottom below the engine, from which it is drawn by the lubricating-oil circulating pump for redistribution via a filter and a cooler. The suction bellmouths clear the bottom by about 100 mm to avoid ingesting water or sediment. In smaller engines, the additional cost and complexity of an independent sump are often avoided in favor of a wet sump, as formed by the bottom of the crankcase. The sump must be so designed that the lubricating-oil pump suction will be maintained at the most extreme conditions of list, trim, and sloshing.

For a wide range of low-speed engines and large medium-speed engines, independent sump tanks are designed to contain an amount of oil in the range of 0.5 to 1.5 kg/bkW, at the highest rating of the engine.

The lubricating-oil circulating pumps are most often positive-displacement rotary pumps, and in larger plants

are fitted in duplicate; however, as an alternative to the positive-displacement pumps, deep-well centrifugal pumps have also been used. Both pumps are motor-driven in installations with low-speed diesels, but higher-speed engines are often fitted with one engine-driven pump, relying on a second motor-driven pump for standby service. With very large engines, three motor-driven pumps may be used to ease the starting load, with two running and one on standby. Each pump has a coarse suction strainer for its own protection.

A full-flow filter is provided in the pump discharge line. It may be of the duplex, basket type, but better filtration is usually provided by a disposable-element or self-cleaning simplex unit, with a standby filter in a bypass. Filtered oil is distributed to engine bearings, for governing and control service, to valve gear, and, on trunk-piston engines as well as some crosshead engines, for piston cooling. Some of these services may require higher-pressure oil, which can be obtained by fitting booster pumps in the line, or by providing a second, higher-pressure circulating system. Figure 42 illustrates a common arrangement for

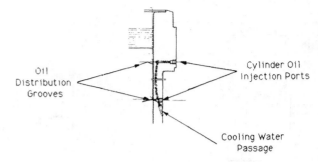

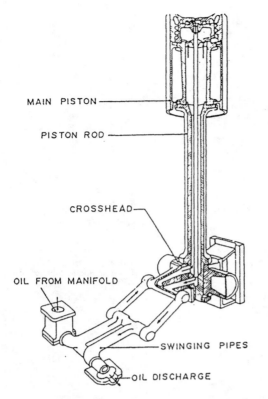

Fig. 42   Oil-cooled piston with swinging pipes for crosshead engine

Fig. 43   Lubrication of a cylinder liner

crosshead engines with oil-cooled pistons, with swinging pipes to carry the oil.

Usually lubricating oil is supplied to turbocharger bearings from the circulating-oil system, but some turbochargers are provided with their own, self-contained lubricating-oil system, which permits the use of differently formulated oil. In either case the turbochargers must be provided with an emergency supply of lubricating oil to provide lubrication as they continue to spin at high speed following a lubricating-oil failure. The system shown in Fig. 41 includes a gravity tank for this purpose.

Pressure gages are fitted at pump suctions and discharges, and across filters and strainers. Thermometers are located before and after the coolers. A sounding tube or dipstick is provided for the main engine sump. Sight flow indicators may be fitted to gravity tank overflow lines. Arrangements are made for the automatic starting of the standby pump. Alarms are fitted for low pump discharge pressure, high oil temperature, and high strainer and filter differential pressure. The oil-mist concentration in the crankcase is metered and alarmed, and centralized indication and alarms are provided for the oil supply pressure at the engine and for the outlet temperatures from the bearings as part of the engine control system (see Section 8). The oil pressure may also be indicated at local and remote engine control stations.

In crosshead engines, the crankcase and the circulating oil are protected from contamination by combustion products blown past the piston rings, by the piston rod packing and scraper rings, whereas this kind of contamination in

trunk-piston engines is almost inevitable. Consequently, while a straight mineral oil with corrosion and oxidation inhibitors is usually recommended for the circulating oil of most crosshead engines regardless of the fuel in use, for trunk-piston engines the usual recommendation is for a detergent oil with alkaline additives (measured as total base number or TBN) matched to the likely sulfur content of the fuel to be burned. For the same reason, while the circulating oil in a crosshead engine rarely requires replacement in the normal course of events, with trunk-piston engines, in most cases, the lubricating oil must be renewed periodically. The life of trunk-piston engine lubricating oil, which is routinely extended by the regular addition of fresh makeup oil to compensate for oil burned in cylinder lubrication, can be further extended by taking such extra measures as the fitting of additional, extra-fine filtration loops and, if necessary, the occasional addition of chemical additives to the oil.

b. Main engine cylinder-oil system.   Crosshead engines, and some trunk-piston engines, are fitted with independent cylinder-oil systems, for the lubrication of the piston rings. A typical system is shown in Fig. 41. The cylinder oil is stored in one or, preferably, two tanks and is transferred daily to a small-capacity measuring tank from which it passes by gravity to the cylinder lubricators on the engine. A separate tank, sized to meet the running-in requirements of one or two cylinders, may also be provided. The lubricators are mechanically driven by the engine to inject a metered quantity of oil into the cylinder as the piston ring pack passes the injection points. Figure 43 shows an arrangement with two rows of injection ports and distribution grooves. Most of the oil is burned. Because the quantities of oil injected per stroke are small, the measuring tank enables the total consumption to be determined accurately as a drop in level over an elapsed time period.

Cylinder oil is a high-viscosity mineral oil with a TBN that is matched to the anticipated sulfur content of the fuel. Two cylinder-oil storage tanks provide flexibility in this regard by enabling cylinder oils of differing TBN to be carried. Cylinder-oil storage tanks are often filled from the deck by gravity, in an arrangement that may preclude filling the measuring tank from the storage tanks by gravity as well, thus necessitating a small hand- or motor-driven transfer pump.

Cylinder-oil tanks are fitted with local level indicators, and the measuring tank is fitted with a low-level alarm. Sight flow indicators and flow failure alarms are fitted in the discharge lines of the cylinder lubricators as part of the engine monitoring system.

In most trunk-piston engines, cylinder lubrication is accomplished with circulating oil, which usually reaches the ring pack and cylinder liner walls by a controlled leakage from the wrist-pin bearing. In some trunk-piston engines, oil is injected for cylinder lubrication in the same manner as cylinder oil in low-speed engines, in most cases using oil taken from the circulating system, although some trunk-piston engines have independent cylinder-oil supplies.

c. Lubricating-oil storage, transfer, and purification system. Smaller high-speed engines may rely solely on filtration and occasional oil changes to maintain the quality of the circulating oil, but most larger engines are arranged for continuous bypass purification using centrifugal purifiers, as shown in Fig. 44. Two purifiers are shown, but one is a standby unit. In multiple-engine installations, individual purifiers may be installed for each engine. Normally, the purifier draws from a bellmouth at the aft end of the sump, returning the oil to the forward end of the sump to avoid short-circuiting. The purifier suction is generally located aft of the circulating pump suction, and lower, about 50 mm above the tank bottom, to remove water before it reaches the circulating pump suctions. The suction line velocity must be sufficient to carry water and sediment to the purifier. Sometimes a hand pump is fitted to the drain tank or sump to facilitate the removal of water that has settled during extended idle periods.

The main lubricating-oil storage tank, with a capacity at least equal to one charge for each engine it serves, plus sufficient margin to meet miscellaneous needs, is filled from the deck by gravity, while the settling tank is normally empty. Should an engine's circulating oil be massively contaminated, for example by water, it can be transferred to the settling tank by the transfer pump, and fresh oil can be transferred from the storage tank. If the oil cannot then be salvaged by a combination of settling and purification, it can be discharged ashore or to a barge for reclamation.

Lubricating-oil purifiers, with their attached pumps, must be located low in the ship to minimize the suction lift required from the main engine sump. As an alternative, a motor-driven purifier feed pump may be arranged near the sump, enabling the lubricating-oil purifiers to be located on an upper flat, often adjacent to the fuel-oil purifiers. In most installations, the oil leaves the purifiers under sufficient head to reach the settling tank, but where this is not the case, a pump must be installed at the purifier discharge. Each purifier is normally sized to circulate the main engine drain tank three to five times per day.

Purifiers are usually fully automatic in their operation, once started, and are programed to shut down and alarm when malfunctions, such as water at the oil outlet, oil at the water outlet, or excessive vibration, are sensed. Self-cleaning units cycle through sludge-ejection sequences automatically at preset intervals. Local pressure gages are fitted at the purifier inlet and discharge, and at the suction and discharge of independently driven feed and discharge pumps. Heaters are fitted with thermometers and thermostatic control.

Purifier connections can be provided for batch purification of auxiliary engine lubricating oil but are unlikely to be used, given the relatively small quantities involved, the fact that the alkaline-additive, detergent oil will require renewal at intervals in any event, and the risk potential for contamination of the main engine oil. When auxiliary engines of substantial size are involved, however, installation of a separate purifier may be justified. The fixed piping shown in Fig. 44 for filling and draining the auxiliary engine sumps may not be provided for smaller installations.

An analogous problem exists in medium-speed diesel plants, where the reduction gear contains a straight mineral oil that is subject to water contamination from condensation within the gear case, but which would otherwise last indefinitely. Again, because of the danger of contamination (in this case contamination of the gear oil by the alkaline main engine oil) even where purifier connections are fitted, they are unlikely to be used. Some operators fit water-absorbing (coalescing) filters in the gear lubrication system; others rely on occasional batch purification using a portable purifier or filter.

10.4 Freshwater Cooling Systems. Figure 45 illustrates a typical freshwater cooling system for a plant with a crosshead engine having water-cooled pistons and three diesel generators. The cooling system shown can be divided into separate subsystems for cooling the main engine cylinder jackets, the main engine pistons, the main engine injectors, and the generator engines.

Fresh water is often used to cool equipment that was formerly cooled exclusively by seawater. In some applications a freshwater cooling system is used in which fresh water is circulated from a seawater-cooled central heat exchanger to the main engine and auxiliary machinery coolers. The central system may be divided into separate high- and low-temperature subsystems. The benefit of a central freshwater cooling system is in reduced maintenance, since only the central cooler is exposed to seawater. The added acquisition cost of a central cooling system may be partly offset by the reduced use of corrosion-resistant materials.

Local instrumentation in each system includes pressure gages at pump suctions and discharges, thermometers before and after heat exchangers, and gage glasses at expansion and drain tanks. Temperature control is usually achieved automatically by three-way thermostatic valves permitting some flow to bypass the cooler. Duplicate pumps are arranged for automatic starting of the standby pump. Alarms are fitted in each system for low pump discharge pressure, high (and sometimes low) temperature, and low levels in tanks. Additional alarms and remote instrumentation are fitted as part of the engine monitoring system.

a. Jacket water cooling system. The freshwater jacket cooling system shown in Fig. 45 is independent of the lubricating-oil cooler and charge air cooler, which are,

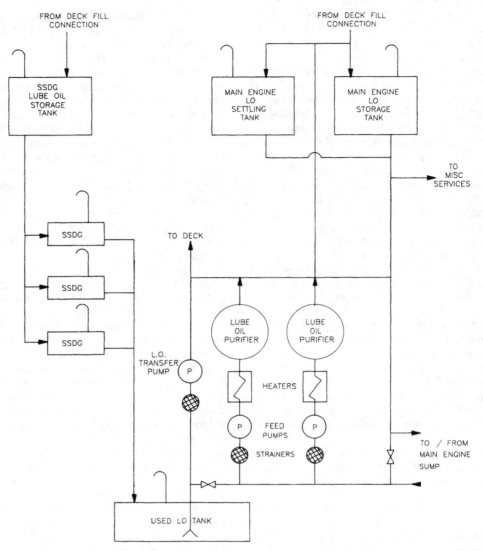

Fig. 44   Lubricating-oil storage and treatment system

in this case, directly cooled by seawater. In other arrangements the jacket water may be the coolant for the lubricating-oil cooler and charge air cooler.

An elevated expansion tank maintains a static head on the suction side of the system, and provides a convenient point for collecting vents and adding make-up feed. Chemicals for corrosion protection and scale inhibition must be added to the water and, if there is sufficient circulation through the tank, they may be added at the expansion tank. Alternatively, the chemicals may be injected into the system under pressure.

The turbocharger supply and return lines are shown in Fig. 45, since even turbochargers with uncooled casings (see Section 7.7) usually require cooling water for the turbine-end bearing.

The jacket water circulating pumps are usually centrifugal pumps, and in larger plants are fitted in duplicate.

Both pumps are motor-driven in low-speed diesel installations, but medium- and high-speed engines are often fitted with one engine-driven pump and one motor-driven pump for standby service. The pumps may precede or follow the cooler and evaporator.

Most seagoing motorships recover heat from the main engine jacket water to produce fresh water. The evaporator is located ahead of the jacket cooler, and may be fitted with a supplemental supply of steam or hot water heating coils for use when insufficient jacket water heat is available.

A jacket water heater may be fitted for use when the engine is idle. Maintaining the engine in a warm condition assists in minimizing corrosion and facilitates starting.

b. Piston cooling water system. While all trunk-piston engines, as well as some crosshead engines, use oil to cool the pistons, a number of crosshead engines use a

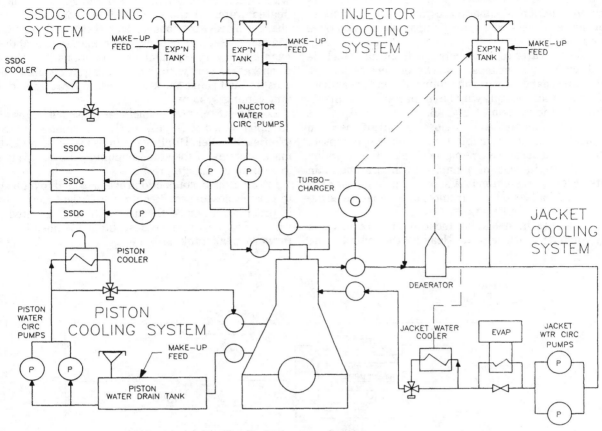

Fig. 45 Freshwater cooling system

cooling water system separate from the jacket water system. The water reaches and leaves the pistons through telescoping tubes enclosed within compartments inside the crankcase, in order to avoid contamination of the lubricating oil should a gland fail.

Because of the high temperature of the water draining from the pistons, and the resulting potential for flashing at the pump suction, some manufacturers recommend the use of deep-well pumps immersed in the tank. Two motor-driven pumps are generally provided.

c. Injector cooling system. Some engines, but not all, are fitted with fuel-injector cooling systems. In some cases injectors are cooled only through conduction to the heads, in other cases they are circulated with water as part of the jacket cooling system, and in still others they are circulated with diesel oil in a closed loop. When a separate injector water circuit is fitted, it is a scaled-down version of the jacket cooling system, often without a cooler, and with a steam coil in the expansion tank to maintain a sufficiently high temperature at low engine output when a heavy fuel is used.

d. Diesel generator cooling system. Auxiliary engines generally have self-contained cooling circuits, with the charge air cooler, cylinder jackets, and lubricating-oil cooler circulated by a single cooling pump on each engine.

The SSDG system shown in Fig. 45 combines these circuits into a common system with a central generator engine cooler and expansion tank, and has a motor-driven pump for each engine. This arrangement allows circulation of idle generator engines with warm cooling water as protection against corrosion, and enables the standby engine to start and pick up load more rapidly. Of course, there are other means of accomplishing these objectives. For reasons of reliability, and to facilitate maintenance and diagnosis, many operators prefer that the auxiliary engines have separate cooling systems.

10.5 Seawater Systems. Seawater systems include cooling services in the engine room and such ship-service systems as ballast and fire main, as illustrated by Fig. 46. Water enters the system through high and low sea chests on opposite sides of the ship. Each sea chest is fitted with sea valves, vents, valves for steam or air blow-out, and gratings mounted flush with the hull. On tankers the high sea suction should be on the opposite side of the ship from the pump room ballast discharge, since both are likely to be used in port. Where operation in freezing water is frequent, the sea chests may be fitted with piped connections to introduce hot water from the overboard discharge or from separate steam heaters, or for direct steam injection. The low sea suction is used at sea where it is more likely to remain immersed as the ship rolls and pitches. The sea chests are connected by a large-diameter cross connection. Suction strainers are installed adjacent to each of the sea chests. Seawater pumps take suction from the cross connection.

The main seawater cooling system will normally include two full-size pumps. Pumps in seawater service tend to be high-maintenance items and pumps of similar capacity should be identical, simplifying spare parts requirements in service. In the system of Fig. 46, the required ballast pump capacity was sufficiently close to that of the main engine cooling pump to enable three identical pumps to be fitted, with one ballast pump designated as standby for the main engine cooling pump. On tankers, the main seawater cooling system may also serve a condenser for the cargo and ballast pump turbines, or a separate circulating system may be fitted.

The heated seawater leaving most of the auxiliaries is usually combined with seawater leaving the main engine coolers, upstream of a thermostatically controlled three-way recirculating valve, which can return some of the heated water to the suction cross connection, and discharge the rest overboard. This recirculation enables the seawater used for cooling to be maintained above 20°C, even under very cold ambient conditions. For the system to operate in port with the main engine secured, the thermostat must sense the seawater temperature in the auxiliary cooling system.

Local instrumentation includes pressure gages at pump suctions and discharges, and thermometers before and after each cooler. Duplicate pumps are arranged for automatic starting of the standby pump. Alarms are fitted for low pump discharge pressure.

In some ships, connections are provided to ballast tanks to permit closed-circuit cooling sufficient for in-port use, in order to minimize the use of fouled or silted harbor water. In cold-water trades, similar connections to the forward peak tank may be used at sea, for main and

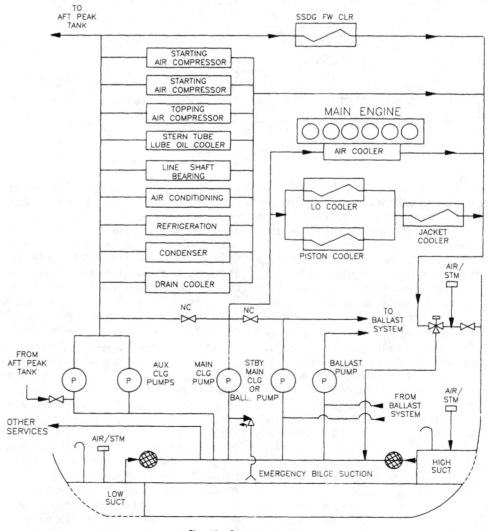

Fig. 46   Seawater system

auxiliary cooling. Connections to the aft peak tank may be provided from the auxiliary cooling water system to enable limited cooling services to be maintained while in dry dock, but, in any case, hose connections are provided in the piping to allow use of water from shore.

Cooling water for refrigeration machinery, often on a higher flat in the engine room, and for the inert gas system, may be supplied from separate pumps to avoid imposing the higher discharge head requirement on the rest of the system. Separate pumps are often fitted for evaporators.

A set of cross-connections between the various seawater pumps is fitted to allow continuous operation under emergency conditions.

Corrosion is troublesome in seawater systems. Therefore, piping is normally of copper-nickel, or of steel that is internally coated with an inert material such as polyethylene. Galvanized piping is unlikely to last long enough to be cost-effective.

Marine growth is also troublesome in seawater systems. Systems often are fitted with connections for the continuous injection of chemical poisons.

Section 11
Installation, Operation, and Maintenance Practices

11.1 Machinery Arrangement. The governing consideration in a machinery arrangement should not be simply to minimize engine room volume or length, but to do so while also achieving a rational layout of main and auxiliary machinery, one that is well suited to preoutfitting and modular construction, while providing the best possible access for operation, maintenance, repair, and removals.

If the machinery is fully aft in the hull, the aftermost location for direct-connected or geared diesels driving fixed-pitch propellers is usually determined by the need to draw the tailshaft inboard in a straight line, clearing the engine or bull gear flange at the forward end and the stern tube forward seal at its after end. The engine can be positioned close to the after bulkhead if there is a shaft alley, or if the tailshaft is drawn outboard. With geared engines in a narrow hull, the width of the gearing may govern how far aft the gearing can be placed. The location of the forward bulkhead with respect to the engine must allow for athwartships access for piping and personnel and, in single-screw ships with a pipe tunnel, for the access hatch and trunk on or adjacent to the centerline.

Most engine manufacturers permit an engine inclination of up to 5 deg, and when the propeller centerline is low, the shaft line and engine may be inclined to permit a double bottom of adequate depth beneath the engine. More often, the shaft line is horizontal and, with low-speed, direct-connected engines, the double-bottom depth in the engine room can be raised to suit.

The control room should be located with good access from the accommodations, yet close to those auxiliaries that are most essential or that may require attention most frequently. Almost invariably, these considerations lead to a location on an intermediate or upper flat. On many ships with low-speed engines, with little space available forward or aft of the engine, the control room is on the port side, in proximity to the camshaft and engine controls.

Auxiliaries and tanks are best arranged in groups by function, to simplify piping runs and facilitate operation and maintenance. Locations may be dictated by proximity to piping connections on related equipment, by pump suction or discharge conditions, by gravity runs, by static head, or by the need for a sea suction.

Large tanks and heavy machinery placed high in the ship may decrease stability. Built-in tanks are coordinated with ship's structure, spanning whole frame spaces, and tanks at the ship's side may incorporate the side shell and structure. Heavy-fuel settling and service tanks should be kept clear of the side shell and other weather-exposed surfaces, because of the disruptive thermal currents that might otherwise result. Tanks must be sufficiently clear of the overhead structure to enable piping to be run and to permit access. The engine room double bottom may be used for waste oil tanks, distillate fuel storage, reserve feed tanks, the bilge holding tank, and the main engine sump with its surrounding cofferdams. Small drain tanks, such as those for the main engine, should not be double-bottom tanks but may be placed in the bilge if a low location is needed. The purifier sludge tank is usually immediately below the purifiers, either as part of their foundation or hung below the deck supporting the purifiers.

With the main machinery fully aft, the ship's service diesel generators are often placed aft of the main engine, on an intermediate or upper flat where the hull is sufficiently wide. The space available on these flats can often be increased by stepping the after bulkhead of the machinery space farther aft into the transom. With machinery amidships, the generators tend to be placed outboard of the engine. A common practice is to separate the ship's service diesel generators from the rest of the machinery space by a bulkhead or an acoustic partition.

In some cases, machinery of like type but serving different functions might best be placed together to simplify maintenance. A case in point concerns purifiers. Suction lift requirements would locate the lubricating-oil purifiers on the floor plates; however, a common practice is to group all of the purifiers in one space, often on an upper flat. In this arrangement, separate feed pumps must consequently be provided at the floor plate level to draw from the sump.

Workshops and storerooms are located in relation to access routes for personnel and heavy components. The storeroom should be adjacent to the workshop. Consideration should be given to air-conditioning workshops and storerooms.

11.2   Access and Overhauling Gear.   During the life of a ship almost every component on board will likely need to be repaired or replaced. Adequate lifting gear, designated landing and storage areas, and sufficient room for access must be allotted in the early design stages, an effort which will be repaid by ease of repair or replacement in service.

Sufficient height must be provided over the main and auxiliary engines to remove the pistons. Main engines that are tall relative to the depth of the hull may require a trunk above the main deck level. Intermediate flats must be positioned with sufficient headroom for the disassembly of machinery, and be properly integrated with ship's structure.

Normally an overhead gantry crane is installed over the main engine for main engine maintenance. By extending the longitudinal and transverse runs of the crane, and by providing sufficient wire to allow the hook to reach down to the floor plates, the crane can be useful for other purposes as well. A second carriage on the crane facilitates maintenance by permitting simultaneous lifts. The lifting capacity of the crane must be matched to the heaviest assembly for which it might be used.

Trolleys, lifting beams, or pad eyes should be located over the diesel generators, the purifiers, and other large auxiliaries, to facilitate their maintenance. Pad eyes should be placed over each piece of equipment that cannot easily be moved by hand. Heat exchanger tube bundle removal normally requires multiple pad eyes, as well as space to perform this operation without removing any other equipment or structure. Deck plating in designated landing areas must be suitably reinforced.

At least one of the ship's stores cranes should be arranged to allow the direct transfer of parts and stores to the storeroom level of the engine room, often through a hatch on deck and a vertical trunk. A trolley in the engine room can be provided to transfer these parts and stores to the engineer's storeroom, or the main engine gantry crane may serve this purpose. To allow flexibility of use, sufficient wire should be installed on the drum of the provision crane to allow the hook to reach the lowest engine room flat that it plumbs.

11.3   Foundations.   A propulsion engine is normally seated on the upper flanges of a rigid box girder that is formed as an integral part of the double-bottom structure. Figures 47 and 48 show typical examples of engine foundations, viewed in transverse section. The seating flanges for low-speed engines are usually insert plates in the tank top; flanges for medium- and high-speed engines are usually elevated above the tank top.

The engine foundation must be sufficiently stiff to absorb forces and moments generated by the engine and connected components, while resisting the transfer of bending moments from the hull to the engine. Low-speed direct-connected engines transmit propeller thrust through the foundation, using fitted bolts, brackets, or end stops in way of the thrust bearing. Foundations for geared medium- and high-speed engines are usually integral with the foundations of the gearing and propeller thrust bearing.

Hold-down bolts must be through-bolts, either headed bolts inserted upward through the seating flange, or studs secured by nuts at both ends. To avoid imposing a bending load on the bolts, the seating surface for the bolt head or lower nut to the underside of the seating flange is machined and, in some cases, the bolt is seated upward against the flange using spherical nuts and washers; welding of the head or nut to the flange is not a recommended practice. Fitted bolts are used to secure the aft end of the engine, but other bolts have clearances to accommodate thermal expansion.

When an engine is rigidly mounted, rather than resiliently mounted, a series of cast-iron or steel foundation chocks is used to provide solid contact between the base of the engine and the top of the ship foundation. The chocks are individually machined to precisely fit each location after the engine is aligned. Either the engine or foundation seating flanges are tapered or sloped slightly towards their outer edges to facilitate the installation of the chocks. Each chock spans at least two hold-down bolts in order to positively secure it.

Solid chocks may also be formed of an epoxy resin, which is poured in place after the engine is aligned on temporary supports. Epoxy chocking materials are carefully selected, based on negligible shrinkage, limited creep, compatibility of thermal expansion with steel and cast iron, and resistance to oil. With epoxy chocking, the permissible preload in the hold-down bolts is limited, and positive locking devices on the bolts may be required.

A number of side stops are welded to the seating flange of the foundation, along each side of the engine, but clear of the engine to permit the insertion of tapered keys. When the engine has been aligned, the keys are tack-welded in place.

11.4   Resilient Mounting.   Most engines are rigidly mounted to the seating flange through solid chocks, although a common practice for medium- and high-speed engines is to mount them resiliently, through a vibration-absorbing material. Resilient mounting is used to reduce the structureborne vibration that an engine would transmit to the hull, but it is feasible only where the engine itself is sufficiently rigid in bending and torsion. Low-speed diesel engines usually require solid foundations.

For generating sets, the engine and generator are usually aligned on a common, rigid bedplate, which is usually a stiff, steel structure. The bedplate is then mounted to the hull through an intermediate flexible element. When a resiliently mounted engine is connected to drive rigidly mounted gearing or a generator, the mounts are usually connected directly to the engine seating flange. A flexible mount may consist of upper and lower steel or cast-iron plates that are separated by a resilient element. The upper plate is bolted to the bedplate or engine, and the lower one to the foundation. Rubber elements may be simple rectangular blocks or more complex shapes, and are loaded principally in compression. Extreme motions, such as those caused by ship motions, are limited in all directions by solid stops, which may be integral with the

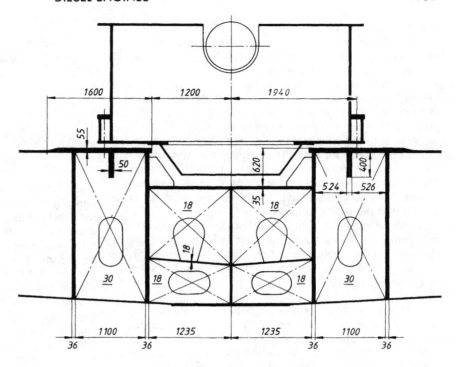

Fig. 47 Typical low-speed engine foundation. The space between the engine bedplate and the insert plates is reserved for chocks

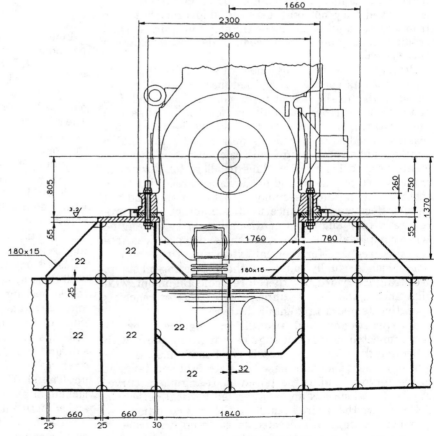

Fig. 48 Typical medium-speed engine foundation

mounts or separate. Alternatively, the resilient mounting may be in the form of a distributed isolation material (DIM), of which there are several types. In principle, DIM is an elastomer that is produced in sheets of various thicknesses, and is cut to fit the contact area between the engine and foundation with a pressure loading that is within prescribed limits.

Connections made to resiliently mounted engines from rigid piping and control systems must be through flexible pipe couplings or other provisions for flexibility that are capable of accepting the full range of engine motions without imposing intolerable loads. For engines connected to rigidly mounted gearing or generators, a coupling of similar flexibility is necessary.

Resonance is potentially troublesome with resilient mounts because the natural frequency of the mounting system, which is relatively low, may coincide with the lower orders of engine-produced vibrations. With generator sets running at constant speed and, therefore, with easily predicted vibration frequencies, the mounting system can be designed with its frequency below these values. With variable-speed propulsion engines, resonance can be more difficult to avoid, and may restrict the type of resilient mounting selected.

As described in Section 6 of this chapter, low-order engine-produced vibrations that may induce resonance in resilient mounts include all of the first- and second-order forces and couples, torque variations, fractional-order torque-reaction couples, and axial pulsation of the crankshaft.

11.5   Engine Alignment. The objective in an engine alignment procedure is to ensure that when the engine is in service and under load, its crankshaft axis is straight, uniformly supported by the bearings, and properly positioned longitudinally. The methods by which this objective is achieved vary not only between shipyards, but even from ship to ship, depending on the ship configuration and production schedules. The procedures described below are only typical.

Preferably, the final alignment of an engine should be made after its connected load has been aligned and secured, with the ship afloat in its normal load condition, and with the surrounding hull and foundation at service temperatures, but this set of conditions is usually not feasible. Production schedules may call for the engine alignment at an early stage, with calculations or measurements derived from sister ships to allow for deviations from service conditions and for deviations resulting from further welding and assembly of both the ship and the engine. Because of the uncertainties inherent in this procedure, a final alignment confirmation, after the engine is in a completed condition, is recommended practice. Analytical corrections for thermal expansion of the foundation, which is usually caused by the lubricating-oil sump, are straightforward, and therefore a cold alignment, with allowance made for the anticipated effects of reaching operating temperature, entails little risk.

a.   Direct-connected propulsion engines. For direct-connected propulsion engines, especially those with short and therefore more nearly rigid shaft lines, the engine alignment should be based on a straight line, tangent to the forward end of the faired shaft-alignment curve. Figure 49, which has an exaggerated vertical deflection scale, shows the desired result, in the cold condition.

When the engine is erected in the ship, the base of the engine is placed on the foundation in its approximate position in the ship. An initial reference line is established using optical or laser sighting, or a taut wire, perhaps through the center of the stern-tube boss to its projection on the forward engine room bulkhead. The engine base is

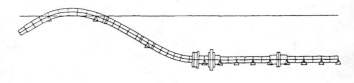

Fig. 49   Cold alignment of a low-speed diesel engine and its shaft line; vertical dimensions are exaggerated

then shifted until it is in the correct position longitudinally and athwartships, and temporary side- and end-stops are tack-welded in place. Jack screws are then used to adjust the position of the engine base vertically until the bearing centers are displaced vertically from the reference line in accordance with the alignment calculations, until it is flat and parallel to the bearing centerline, and until it otherwise reproduces shop test conditions.

When the engine is installed fully assembled, or in subassemblies with the crankshaft in place, the initial reference line may be established by optical or laser sighting from the stern-tube boss to the engine flange; the position of the engine is adjusted until the centerline of the crankshaft is coincident with the reference line for its entire length.

Final alignment confirmation checks are preferably postponed until the engine is fully assembled, the stern tube is bored, the tailshaft and propeller are in place, the line shafting is in place with line shaft bearings temporarily positioned, and the ship is afloat and essentially complete with regard to major welding and weight changes. The shafting is aligned first, usually by use of calculated drop and gap measurements at each coupling, bringing the forward-most shaft flange into position on a temporary support, or on its own bearing temporarily positioned, and all couplings aft of the engine coupling are made up. The internal alignment of the engine is verified; with adequate lubrication of the bearings assured, by temporary measures if necessary, crankshaft deflections (see 11.7) are taken and proven to be within the engine manufacturer's tolerances. The position of the engine is then adjusted longitudinally, transversely, and vertically, until, with the crankshaft pressed forward to bear on the ahead thrust bearing, measurements between the engine flange and shaft flange show the drop, gap, and transverse deviations to be within calculated tolerances. The line shaft bearings are positioned and secured, permanent engine side and end-stops are welded in place and their wedges or shims fitted, and the engine-to-shaft coupling is made up. The shafting alignment is then checked, either by jacking or by strain gage measurements (see Chapter 10), and adjustments are made as required. At this point, after a final check of shaft and engine alignment, hold-down bolt holes are bored and the bolts are fitted, chocks are fitted or poured, the bolts are torqued, and side- and end-stop wedges are tack-welded in place.

b.   Engines driving gears or generators. The alignment of assembled medium- and high-speed propulsion engines to gears or generators already in position is relatively straightforward. The alignment is preferably conducted with the ship afloat. If the engine foundation cannot be heated to its operating temperature, calculated corrections may be required for the deviation, which may have components athwartships as well as vertically. The engine, usually stripped of cantilevered shaft-mounted equipment such as flywheels and couplings that would distort the shaft, is approximately positioned on its jack screws. In multiple-engine installations, all engines should be so placed before alignment in order to avoid such effects of hull distortion as their installation might introduce. The engine position is adjusted longitudinally, transversely, and vertically until measurements between the crankshaft flange and pinion show drop, axial, and transverse deviations within tolerances, when the crankshaft and pinion are in their specified axial positions. The engine side- and end-stops are then welded in place and their wedges or shims are fitted. With the flywheel restored and the coupling installed and made up, the alignment is checked and adjusted, if necessary. With adequate lubrication of the bearings assured, by temporary measures if necessary, crankshaft deflections (see 11.7) are taken and proven to be within the engine manufacturer's tolerances. At this point, after a final alignment confirmation, hold-down bolt holes are bored and the bolts are fitted and lightly torqued, the chocks are fitted or poured, the bolts are torqued, and the side- and end-stop wedges are tack-welded in place.

11.6 Engine Operation. The procedures described here are general and abbreviated, and are intended to provide background information only.

a. Starting. Before starting an engine that has been shut down for an extended period, or which has been overhauled, it is thoroughly inspected. If the crankcase has been opened or entered, it must be inspected for overlooked tools or debris. Fluid levels and fluid quality are verified. Checks are made that gage valves are open, that sensors are connected, and that standby systems are lined up. Safety devices, interlocks, and alarms are tested.

On large engines, the lubricating oil and coolants are circulated and heated to approximately the normal running temperature. Seawater pumps are started. Where fitted, pre-lubrication pumps are operated until all bearings receive oil. Cylinder lubricators are operated manually until all cylinders receive oil.

Drains in air and exhaust manifolds, and the indicator cock on each cylinder, are opened. After a check to see that all hands are clear, the engine is rotated through several revolutions with the turning gear to confirm freedom from interference, to help establish oil films at bearings, and to ensure that cylinders and manifolds are free of water. If the engine is direct-connected, permission to rotate the engine must be obtained from the bridge.

For engines that are started on heavy fuel, a booster pump is started with the recirculating line open. Steam is lined up to heaters, to steam-traced fuel lines and, where appropriate, to the injector cooling circuit. The fuel is recirculated until hot fuel is present at the injectors, or as close to the injectors as the system permits.

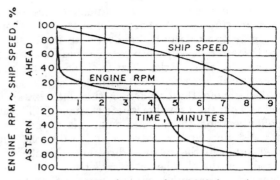

Fig. 50   Reversing performance of a 65,000 dwt tanker

On air-started engines the starting air receivers are charged and drained.

Before starting propulsion engines that are directly connected to the propeller, permission must again be obtained from the bridge. With fixed-pitch propeller arrangements, the engine speed must be limited to avoid excessive strain on mooring lines and bollards.

On reversing engines, the camshaft and starting air distributor are set in the desired direction. The fuel rack is set to a starting position, usually about 25% of maximum. Starting air is lined up to the starting valve. When the starting valve is opened, the engine will crank through one or two revolutions before firing. As the engine accelerates, the fuel rack is repositioned to the idle-speed setting.

On engines that are fitted with attached pumps but are started with independent motor-driven pumps in use, the motor-driven pumps can be stopped once the maneuvering period has ended.

b. Maneuvering. Propulsion engine control is discussed in Section 8.

Engines that directly drive fixed-pitch propellers without reduction gearing are necessarily reversing engines. Figure 50 shows the reversing performance of a tanker that is so propelled. The first step in reversing is to interrupt the flow of fuel to the engine by moving the fuel racks to a zero position. The ship begins a gradual decrease in speed, but the engine rpm abruptly drops until the propeller takes charge of the engine and causes it to continue rotating in the ahead direction at about 10% to 15% of the initial rpm. As the ship speed decreases, the hydrodynamic torque developed by the propeller decreases until the reversing speed is reached. The reversing speed is the speed at which the torque that can be developed by admitting starting air to the engine in the reverse direction is sufficient to stop the engine (and propeller) and reverse it. Once initiated, the reversing process is accomplished rapidly, as can be seen in Fig. 50. In an emergency, starting air can be admitted to the engine in the reverse direction sooner, before the engine can actually take charge of the propeller, as a means of braking the ship and decreasing the time required to reach the reversing speed.

In fixed-pitch propeller installations where reversing engines are clutched to the propeller shaft, normal maneuvering may be as described above, with the clutches engaged throughout. Alternatively, if a shaft brake of sufficient capacity to stop the shaft is fitted, a stop may be achieved by slowing the shaft with the engines engaged, then declutching the engines as the shaft brake is engaged to hold the propeller shaft stationary. To reverse propeller rotation, the engines must be stopped, the camshafts and starting air distributors shifted, the engines started again in the reverse direction, the shaft brake released, and the clutches reengaged.

Where a nonreversing engine drives a fixed-pitch propeller through reverse gearing, a shaft brake must be fitted. Reversing is achieved by reducing the engine speed setting to its minimum, declutching the ahead gear train, applying the shaft brake to stop the shaft, engaging the clutch on the astern gear train, and then raising the engine speed to the desired setting. In this arrangement the clutch movements and shaft brake application would normally be programed to operate in sequence, with appropriate time intervals, from a single lever.

The minimum engine speed is typically 25% to 40% of rated rpm and, unless a slip clutch is included in the drive train, or electric drive is used, this minimum rpm determines the minimum shaft speed. To maintain extremely low ship speeds, most ships with fixed-pitch propellers require that the engines be stopped and started repeatedly or, where clutches are fitted, that they be clutched in and out repeatedly, unless they are of the continuously slipping type. A ship with a controllable-pitch propeller may be maneuvered extensively at low power levels by pitch control, usually with the engine speed set at a constant but modest rpm throughout the maneuvering period.

At the close of a departure maneuvering period, engines should be loaded in increments extending over at least the first hour at sea. If there is a barred speed range, it is passed through quickly.

c. Running-in after an overhaul. Running-in is necessary to enable new piston rings to wear to conformity with the cylinder liner. An engine with one or more pistons fitted with new rings should be started and run on distillate fuel, with the load applied gradually over a period of at least six hours. On engines with separate cylinder-oil lubricators, cylinder oil should be fed to the re-ringed cylinders at an increased rate during the running-in period.

d. Underway. While underway, the engine and all systems are monitored. Even in fully automated plants, periodic inspections are made. Trends are noted as an aid in planning maintenance and in avoiding unexpected failures.

Fuel-oil consumption is calculated daily. In crosshead engine installations, the cylinder-oil measuring tank is refilled daily, consumption is calculated, and cylinder lubricators are adjusted as required.

On engines burning heavy fuel, the turbocharger turbines are periodically water-washed, often every day. The compressor requires cleaning less frequently.

Approximately weekly, indicator cards, or the electronic equivalent, or maximum pressure readings, are taken in order to confirm that the load is balanced among the cylinders.

Self-cleaning purifiers are opened for manual cleaning at intervals determined by experience. Other purifiers are cleaned at shorter intervals that are also determined by experience.

e. Stopping. Propulsion engines cannot be secured until the bridge has given and confirmed the order "finished with engines." Once an engine is secured, fuel pumps are stopped, the starting air stop valve is shut, and the turning gear is engaged to prevent an accidental rotation of the engine. Cylinder indicator cocks and drains at air coolers and at air and exhaust manifolds are opened and secured in the open position.

Before stopping an engine that is run on heavy fuel but started on distillate fuel, it is necessary to change over to distillate fuel. This may be advisable even for engines that normally are started on heavy fuel, to clear the lines of heavy fuel and to facilitate maintenance.

Care must be taken on stopping, to ensure that the engine is in a position to be restarted.

Because of the danger of a crankcase explosion, crankcase doors must not be opened until after the crankcase has cooled down.

When steam to fuel heaters and for fuel-line tracing is no longer needed, it is secured. Independent cooling water and lubricating-oil circulating pumps are kept in operation for about a half hour after the engine has been stopped. Lubricating-oil purifiers on propulsion engines are kept in operation for 12 to 24 hours after the engine is stopped.

f. Emergency operation. It is often necessary to put an engine into temporary service with one or more cylinders out of service, or with a turbocharger secured. Depending on the nature of the problem, it is usually better from the standpoint of engine balance if the piston and running gear of the out-of-service cylinder remain in place, with cylinder lubrication continued. If a piston must be hung up or removed, the starting air valve must be isolated.

Where machinery is normally unattended, watches should be set for the duration of the emergency. The engine output must be reduced to avoid overloading the remaining cylinders, and reduced further if limits are approached. The rpm must be changed immediately if vibration intensifies. If barred speed ranges exist for operation with a reduced number of cylinders, these ranges must be ascertained and then avoided.

11.7 Maintenance Practices. Preventive maintenance is performed according to a schedule as recommended by the manufacturer and as amended by the operator in light of continuing experience. Additional guidance in the scheduling of preventive maintenance is provided by analysis of performance data (condition monitoring or trend analysis).

a. Performance data analysis. The analysis of performance data may be limited to an observation of trends in pressures and temperatures that are read from thermometers and pressure gages, and recorded in the logbook, but more comprehensive methods are also used. As more sophisticated measuring techniques and instruments have become available and proven to be effective and reliable in service, the number of parameters monitored has increased, and the use and effectiveness of condition monitoring have grown. For example, some engines are fitted with proximity detectors to locate piston rings and measure their wear, and others to measure valve and fuel plunger positions and accelerations. Some engines are fitted with pressure sensors in cylinders and high-pressure fuel lines, together with position sensors on injectors and valve stems, that are keyed to crank angle sensors, enabling indicator cards, crank angle diagrams, fuel line pressure diagrams, and needle and valve lift diagrams to be produced accurately and instantly. Where sufficient data can be gathered, a computer can track performance, detect ominous trends, diagnose likely causes, and provide guidance concerning possible corrective action or amended practice.

Performance data analysis is useful in that it increases the ability to avoid unnecessary maintenance and renewals, to warn of early failure, and to enable scheduled maintenance intervals to be adjusted.

One of the most important parameters to provide an indication of engine condition is the exhaust-gas temperature. In the hands of an experienced operator, and taken together with other data, exhaust temperatures may indicate such diverse problems as fouled air coolers, damaged turbochargers, fouled or blocked ports, a maladjusted fuel rack, defective injection pumps or injectors, worn fuel cams, leaking or burned exhaust valves, broken or stuck piston rings, sticking intake valves, or leaking gaskets.

Indicator cards or their electronic equivalent or, less effectively, maximum pressure readings, are used to detect problems with valves and combustion, and to balance the cylinder outputs. Cylinder compression readings can be used to indicate whether rings and valves are sealing tightly.

An unexpected change in the oil sump level may reflect a leakage of oil, leakage of water from a heating coil or heat exchanger, contamination by distillate fuel oil, worn or stuck piston rings, or worn valve guides.

Lubricating-oil samples are analyzed to indicate bearing wear, liner or ring wear, piston wear, water leakage, fuel dilution, or stuck or worn piston rings.

A gradually rising pressure in the crankcase of trunk piston engines usually indicates blow-by.

Smoke observations provide a coarse indication of problems that, in reasonably well-instrumented engines, are more likely to have been detected earlier. Smoke may indicate overload, air starvation, defective or maladjusted injection equipment, low fuel temperature, worn or stuck piston rings, worn valve guides, maladjusted cylinder lubricators, low combustion temperature, advanced or retarded injection, or a leaking jacket, cylinder head, turbocharger, or exhaust-gas boiler.

Falling pressures in systems serviced by pumps may be indicative of pump wear, leakage, improperly set stop valves, or blockage at a strainer, heat exchanger, or elsewhere in the system. A falling pressure in a lubricating-oil system could also be the result of excessive bearing wear, of overheating of the oil, or of dilution of the oil by fuel. Discharge filters in lubricating-oil and fuel systems, and charge air intake filters and coolers, are normally fitted with pressure gages or manometers that indicate the differential pressure across the unit.

Vibration sensing and analysis has proven to be a very reliable monitoring technique for rotating machinery, indicating bearing deterioration or imbalances resulting from damage or deposits. The technique is widely applied to auxiliary machinery, but its usefulness on engine-driven or otherwise attached equipment is limited by the extent to which the inherent vibration of the engine's many reciprocating components can be filtered out. Nevertheless, vibration-monitoring techniques have successfully detected bearing wear and piston skirt wear in high-speed engines, and wider applications to large engines for these purposes are feasible.

b. Maintenance schedules. Engine manufacturers' maintenance schedules are usually predicated on the basis of a specified fuel, and for operation at 80% to 90% of rating. For engines using fuels of poor quality or operating under such adverse conditions as sustained overload or low load, or subjected to frequent, cyclic load changes, more frequent maintenance may be necessary. In general, lower-speed engines require less maintenance than higher-speed engines when run on fuel of poor quality. The intervals cited below are typical, but are intended to provide background information only.

- At intervals, typically of 500 to 1000 operating hours, samples of lubricating oil are taken for analysis; oil is renewed if required. At similar intervals, fresh water in each cooling circuit is sampled, tested, and treated or renewed.

- At intervals of 1000 to 2000 hours on engines burning heavy fuel, or as experience dictates, the fuel injectors must be removed, tested, and reconditioned as required.

- On medium- and higher-speed engines burning high-vanadium fuel and fitted with exhaust valves, exhaust valve performance may become suspect at intervals as short as 1000 hours or less, depending on factors such as fuel quality, valve materials, valve cooling, and operating practices. If a defective valve is confirmed, the valve cage or cylinder head must be lifted to clean, lap, grind or replace the valve or the seat.

- At intervals of 1000 to 3000 hours, air manifolds and, on crosshead engines, under-piston spaces, are inspected and cleaned. On large two-stroke engines the air ports usually permit a limited visual inspection of the cylinder liner, piston crown, rings, and skirt from the air manifold. Cam surfaces are inspected for pitting; timing gears or sprockets are inspected for tooth wear and cracks; timing chain tension is checked and adjusted; tappet clearances are checked and adjusted; on engines with valves, the valve gear is inspected for free and proper operation; valves are confirmed to move easily but not loosely in their guides; exhaust-valve rotators are checked

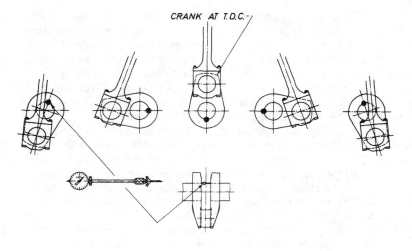

Fig. 51 Crankshaft deflection measurements

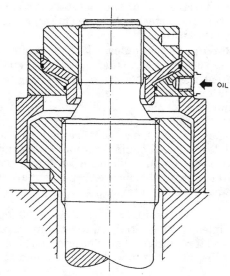

Fig. 52 Hydraulic jack for preloading a stud or bolt

for proper functioning; and the valve and injector timing is checked.

• At intervals of 3000 to 6000 hours, major nuts and bolts, including the foundation bolts and tie rods, are checked to ensure that they remain properly tightened; crankshaft deflection readings are taken and suspect main bearing clearances are measured; telescopic tubes or swinging pipes used for oil and water delivery to pistons and crossheads of crosshead engines are examined; and glands are tightened or packed.

• At intervals of 6000 to 12,000 hours in engines running on heavy fuel, or up to 25,000 hours in distillate-fueled engines, cylinder heads are removed, inspected, and rebuilt with new parts as appropriate. Pistons are pulled, inspected, and measured; the condition of the wrist pins and clearances in trunk pistons are determined; acceptable pistons are usually returned to service with new rings. Cylinder liners are inspected, measured, dressed, honed, or replaced, as required. Cylinders are reassembled with new gaskets and seals. Usually in coordination with a cylinder overhaul, the condition of connecting-rod

bearings and their clearances are checked; in crosshead engines, crosshead guides are also examined.

• At intervals of 8000 to 12,000 hours in engines running on heavy fuel, or up to 25,000 hours in engines burning distillate fuel, the turbochargers are disassembled, cleaned, inspected, and, if acceptable, balanced and reassembled with new bearings and seals.

• At intervals of 12,000 to 25,000 hours the main bearings and the thrust bearing are inspected and their clearances measured. At similar intervals attached oil and water pumps are inspected and overhauled as required; the injection pumps and the governor are also overhauled and recalibrated.

In general, major machinery maintenance tasks aboard ship are staggered to provide a manageable amount of work during each port visit. Classification societies offer continuous machinery survey provisions to suit this practice. Operators of ships in seasonal trades usually attempt to restrict all planned maintenance to the lay-up period, when complete overhauls are more conveniently undertaken.

Spares are kept in an overhauled, subassembled condition, ready for use, to expedite both emergency repairs and staggered maintenance schedules. A used component that is withdrawn from the engine is either reconditioned to become the next spare or is scrapped.

c. **Crankshaft deflection measurement.** Crankshaft deflection measurements are taken periodically as a means of checking main bearing alignment. The measurements are taken between the webs of each crank, opposite the crankpin, at a designated distance from the crankshaft centerline, as shown in Fig. 51. The designated distance is usually established by centerpunch marks. A dial indicator, with extension bars as required, is inserted within the marks, and the crankshaft is rotated through a revolution, stopping every 90 deg, at or close to each dead center and mid-stroke position. Perfect alignment would be indicated if all of the readings at each crank were equal, but an acceptable alignment is considered to exist when variations are within manufacturer's limits, which may range from about 0.03 mm for smaller engines up to about 0.5 mm for the largest engines. During the process, the crankshaft must be in contact with the bottoms of the adjacent main bearings, a requirement which is usually met by component weight.

d. Special tools and maintenance aids. The premier maintenance aid in a diesel plant is the overhead crane. While a simple overhead beam fitted with a trolley for a chain fall may suffice for smaller engines, a permanently installed gantry crane, electrically or pneumatically powered, with mobile remote control and precise longitudinal and transverse positioning, is common in larger plants. With eight or more main engine cylinders and short port stays, a second gantry carriage fitted on the same rails will enable work on two different cylinders simultaneously. Auxiliary engines should have their own lifting gear, to avoid interfering with main engine maintenance schedules.

Maintenance is simplified by special tools and access gear, which usually are supplied by the engine builder. Access gear generally is intended for use with the overhead crane and may include built-in rails, lifting points, and jacking pads within the crankcase to facilitate the removal of large items. Equipment for a large engine might include lifting beams and brackets, tensioning devices (hydraulic or electric) for all major studs and bolts, strongbacks for removing components whose withdrawal might be difficult, motor-driven grinders for exhaust valves and seats, an injector testing unit, mandrels for lapping fuel and starting-air valve seats, piston ring spreading and compression devices, and measuring devices, including trammels, bridge gages, brackets for jacks and dial indicators used in measuring bearing clearances, and go/no-go gages for valve guides and small bearings.

Smaller bolts and studs can be tensioned by torque wrenches, but larger bolts and studs are most often tensioned by hydraulic jacks, as illustrated by Fig. 52, to stretch the bolt or stud a predetermined amount, which coincides with a specific pressure developed by a hand pump. At the specified tension the nut is easily run up, the tension is released, and the bolt or stud remains prestressed; for removal, the process is reversed. An alternative arrangement, frequently fitted, is a hydraulic nut, illustrated in Chapter 10. In some applications the tensioning device is an electric heating rod that is inserted in a central boring of the bolt or stud to elongate it by thermal expansion. Another procedure, used to confirm that the desired amount of bolt preload has been achieved, entails the use of bolts that have a small, blind hole bored from the head end, which extends into the shank of the bolt. The depth of the hole is measured before and after it is installed, thus permitting the bolt strain, and therefore the preload stress, to be confirmed. This technique is further discussed in Chapter 10.

References

1 R. S. Benson and N. D. Whitehouse, *Internal Combustion Engines*, Pergamon Press, New York, 1979.

2 C. F. Taylor and E. S. Taylor, *The Internal Combustion Engine*, International Textbook, 1961.

3 "Marine Diesel Power Plant Practices," SNAME T&R Bulletin No. 3-49, 1990.

4 S. H. Henshall, *Medium and High Speed Diesel Engines for Marine Use*, Institute of Marine Engineers, 1972.

5 A. L. Rowen, *Notes on Marine Diesel Engines*, Webb Institute of Naval Architecture, 1989.

6 K. W. Stinson, *Diesel Engineering Handbook*, 12th ed., Business Journals, 1983.

CHAPTER IV

Dan A. Groghan | # Gas Turbines

Section 1
Introduction

1.1 Basic Considerations. Since the early 1970's gas turbine prime movers have been used increasingly in main propulsion and generator-drive applications for surface ships. While commercial shipping has turned to diesel engine propulsion as the preferred prime mover, the world's navies have selected gas turbines for the majority of their non-nuclear propulsion applications. The U.S. Navy first introduced gas turbine propulsion and ship-service generators on the DD 963 *Spruance* Class destroyers, and since that time, many major combatants (destroyers, frigates, and cruisers) have utilized gas turbine engines. Gas turbines are the dominant main propulsion and ship-service prime movers for destroyers and cruisers. Gas turbines are also used for main propulsion for frigates as well as the foilborne engine for hydrofoil craft. These aircraft-derivative engines are also in service as peaking power plants for electrical utilities and for gas transmission and compression in the oil and gas industry. The gas turbine is an attractive prime mover due to its high power-to-weight ratio, quick-starting capability, acceleration characteristics, and reliable operation.

All gas turbines are practical applications of the Brayton thermodynamic cycle, which is illustrated by Fig. 1. The Brayton cycle is an ideal cycle in which the working fluid is a perfect gas (air in most cases), which is compressed isentropically by a compressor (from 1 to 2), heated at constant pressure in a combustion chamber (from 2 to 3), then allowed to isentropically expand through a turbine back to compressor inlet pressure (from 3 to 4). The power produced by the turbine exceeds the requirements to drive the compressor, thus providing excess power for ship propulsion or ship-service generator drives.

Over the last 50 years progressive improvements in

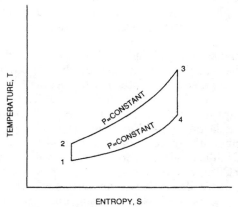

Fig. 1 Brayton thermodynamic cycle

gas turbine design have resulted in continual increases in cycle efficiency. These improvements include:

- Higher compressor pressure ratios.
- Higher turbine inlet temperatures.
- Improved compressor and turbine stage efficiencies.
- Reduced number of compressor and turbine stages through higher loading.
- Introduction of intercooling in the compression process.
- Introduction of recuperation from the exhaust gases.
- Other combined-cycle waste-heat-recovery features.

Different designs have used various combinations of the foregoing to provide improvements in the cycle efficiency, power-to-weight ratio, and specific fuel consumption. These combinations modify but do not change the basic concepts of the Brayton cycle.

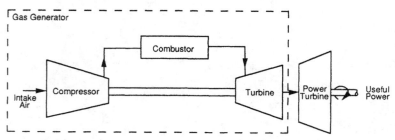

Fig. 2 Gas turbine engine schematic

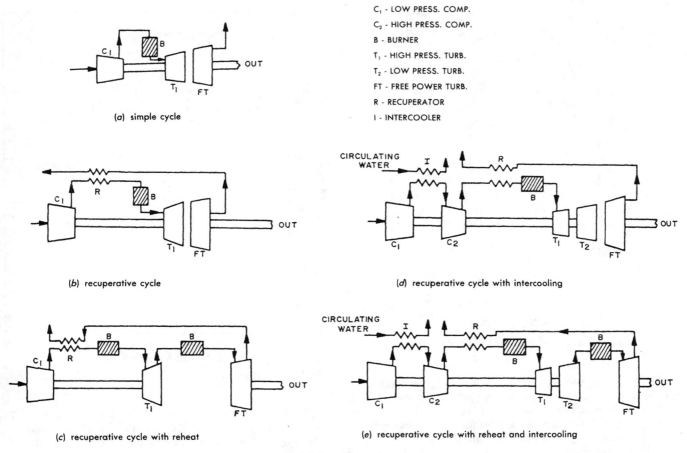

C₁ - LOW PRESS. COMP.
C₂ - HIGH PRESS. COMP.
B - BURNER
T₁ - HIGH PRESS. TURB.
T₂ - LOW PRESS. TURB.
FT - FREE POWER TURB.
R - RECUPERATOR
I - INTERCOOLER

(a) simple cycle

(b) recuperative cycle

(c) recuperative cycle with reheat

(d) recuperative cycle with intercooling

(e) recuperative cycle with reheat and intercooling

Fig. 3   Alternative gas turbine configurations

Basically, a gas turbine engine consists of two parts: (1) a gas generator to provide gas at high pressure and temperature, to be expanded through (2) a turbine to deliver useful work. In marine propulsion applications, the power-producing turbine, or power turbine, can be mounted on a separate shaft from the gas generator, as shown by Fig. 2. The gas generator and the power turbine are coupled aerodynamically but not mechanically in what is known as a "free power turbine" arrangement.

For electrical generator applications, the gas generator and the power turbine can be mounted on a common shaft in a "single-spool" gas turbine arrangement, or they can be mounted on separate shafts in a "free power turbine" arrangement.

1.2 Cycle Performance. Several potential cycles for marine gas turbines are shown in Fig. 3. The simple cycle shown in Fig. 3(a) is well suited for naval applications because of its simple design and easy operation. This engine can achieve a specific fuel consumption at full power rating conditions of approximately 0.40 lb/bhp-hr at a 16-to-1 compressor pressure ratio with liquid fuel having a lower heating value of 18,400 Btu/lb.

Figure 3(b) depicts the addition of a recuperator to the simple cycle, which is used to transfer heat from the ex-

haust gas to the compressed air entering the combustor and thereby reduce the amount of fuel required to be burned in the combustor. The addition of a recuperator can increase the thermal efficiency of a gas turbine cycle by 20 to 30% if the compression ratio is optimized for the desired cycle; however, as indicated in Fig. 4, the advantages of a recuperated cycle diminish as the compression ratio rises [1]. This occurs because the compressor discharge temperature begins to approach the exhaust temperature as the pressure ratio rises. This reduces the amount of heat that can be transferred from the exhaust gas to the compressor exit air. Most recuperated gas turbines have pressure ratios in the range of 7 to 12. The effectiveness and pressure drop of the recuperator will also have an effect on the cycle performance.

Figure 3(c) illustrates the recuperative cycle with reheat. This cycle is obtained by adding a secondary combustion chamber between the high-pressure turbine and the power turbine to reheat the gas back to a temperature nearly equal to the high-pressure turbine inlet temperature. This results in increasing the total available energy of the power turbine; however, the improvement in fuel rate is not substantial because of the added pressure losses of the reheater.

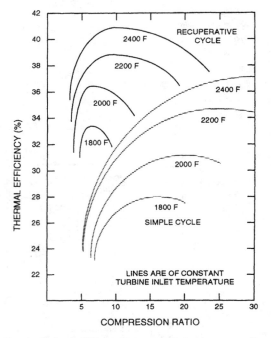

Fig. 4   Thermal efficiency improvements due to recuperation

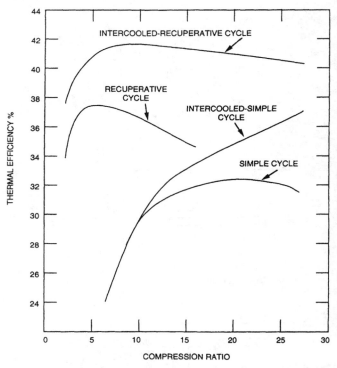

Fig. 5   Effect of intercooling on simple-cycle and regenerative-cycle thermal efficiency

To obtain higher efficiencies at higher pressure ratios, intercooling can be added to the cycle as shown in Fig. 3(d). The intercooler is usually a water-cooled heat exchanger, which reduces the air temperature between the low-pressure compressor and the high-pressure compressor. This effectively reduces the amount of work required of the compressor to deliver the same pressure ratio because the gas volume is reduced. Intercooling also reduces the discharge temperature of the high-pressure compressor, which allows the recuperator to effectively transfer heat from the exhaust gases back to the compressor discharge air. The combination of intercooling and recuperation results in significantly higher thermal efficiencies for higher pressure ratio gas turbines as shown in Fig. 5 [2].

A further improvement in off-design efficiency can be obtained in the intercooled recuperative gas turbine by providing variable-area geometry in the power turbine. This variable geometry, which consists of one or more stages of variable vanes, allows the engine to maintain higher turbine inlet temperatures at lower power settings and thus raises the thermal efficiency at off-design operation. The effect of adding variable geometry to the power turbine is shown in Fig. 6. This increased performance at low power points is very important to naval ship designers since the majority of operating time for naval ships is spent at part-load power points.

Figure 3(e) shows a complex recuperative cycle with reheat and intercooling. Although this cycle can reduce the specific fuel consumption and specific airflow significantly when compared to a simple cycle, the added equipment and increased complexity tend to offset these advantages.

More extensive discussions on gas turbine cycle performance can be found in references 3 through 6.

1.3   Single-spool and Multi-spool Gas Turbines.   The simplest gas turbine design is that of a single-spool engine. Figure 7 is an example of a single-spool engine. A single shaft connects the compressor and the turbine to a ship-service generator. The excess power developed by the turbine over the compressor power requirements is used to drive an a-c generator through a reduction gearbox. Figure 8 illustrates a single-spool gas generator and a free power turbine. A free power turbine has an independent shaft and uses the excess energy at the discharge of the gas generator turbine to produce output shaft power in its power turbine. The power turbine shaft is connected to the main reduction gear and the propeller shaft of the ship.

Many aircraft-derivative engines utilize two-spool gas generators and a free power turbine. These engines consist of a low-pressure compressor, high-pressure compressor, combustor, high-pressure turbine, low-pressure turbine, and free power turbine. The use of multi-spool compressors allows higher compression ratio designs to operate efficiently over the entire speed range by allowing the low-pressure compressor and high-pressure compressor to operate at their own optimum speed.

1.4   Effect of Ambient Conditions and Duct Losses on Performance.   The effect of ambient conditions on the performance of gas turbines can be quite significant, and ambient conditions must be considered in the selection and application of gas turbines on any ship. The power output of any gas turbine is directly proportional to the airflow rate through the machine. Therefore, changes in barometric pressure directly affect the capacity of the engine, but not the efficiency or specific fuel consumption.

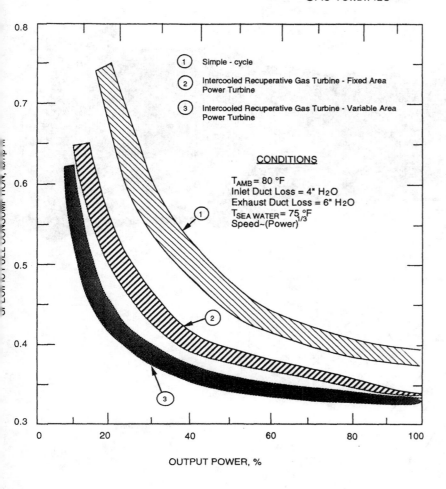

Fig. 6   Part-load fuel consumption characteristics of
cycle variations

① Simple - cycle

② Intercooled Recuperative Gas Turbine - Fixed Area
Power Turbine

③ Intercooled Recuperative Gas Turbine - Variable Area
Power Turbine

CONDITIONS

T_{AMB} = 80 °F
Inlet Duct Loss = 4" H_2O
Exhaust Duct Loss = 6" H_2O
$T_{SEA\ WATER}$ = 75 °F
Speed~(Power)$^{1/3}$

OUTPUT POWER, %

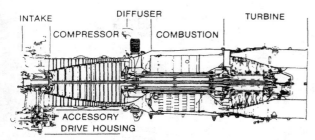

Fig. 7   Single-spool gas turbine with axial compressor

The efficiency and specific fuel consumption are not affected because the fuel-to-air mixture is maintained to hold the turbine inlet temperature constant. Therefore the fuel flow is reduced proportionally to the airflow.

The ambient temperature, or compressor inlet temperature has a pronounced effect on a gas turbine's overall performance. A change of 10 deg F in ambient temperature can change the power output of a gas turbine by as much as 5% for a constant power turbine inlet temperature [7]. Figure 9 shows the effect of inlet ambient temperature on gas turbine performance. These curves show the reductions in output power that must be made, as the ambient temperature rises, to maintain a constant power turbine inlet temperature.

Both the power output and efficiency are very sensitive to pressure drops anywhere in the cycle. The inlet and exhaust system losses must, therefore, be carefully controlled in the ship design process in order to reduce their effect on the gas turbine's performance. The inlet pressure drop is extremely critical since it not only introduces an efficiency loss into the cycle, but it also reduces the airflow through the engine. A pressure drop of 1% (4 in. of water) in the inlet reduces the power output by 1.5 to 2% and increases the heat rate (thermal energy input/ power output) by 0.5 to 1.0%. The same pressure drop at the exhaust reduces the power output by 0.5 to 0.75% and increases the heat rate by 0.5 to 0.75%. Typical correction curves for inlet and exhaust duct losses are shown in Figs. 10 and 11.

To provide a common baseline for design analyses, the U.S. Navy has defined the following standard rating conditions to establish the maximum power output of propulsion and auxiliary gas turbines [8]:

- Ambient air temperature   100 F
- Ambient air pressure   29.92 in. of Hg abs
- Inlet pressure loss   4 in. water
- Exhaust pressure loss   6 in. water

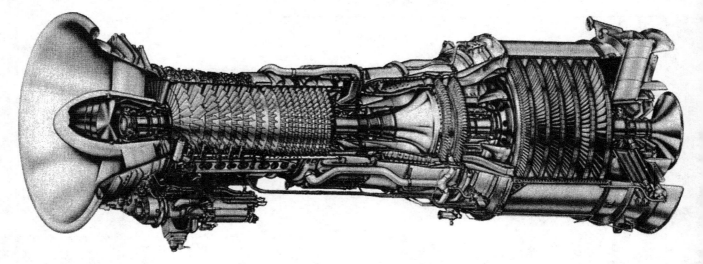

Fig. 8   Large gas turbine with axial-flow compressor

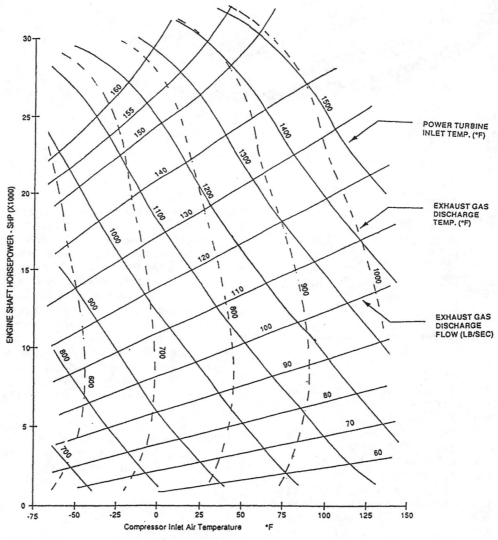

Fig. 9   Typical effect of inlet air temperature on gas turbine performance [7]

NOTE: ESTIMATES BASED ON 7LM2500PC INLET DUCT LOSS CHARACTERISTICS
CONSTANT POWER TURBINE INLET TEMPERATURE AND SPEED

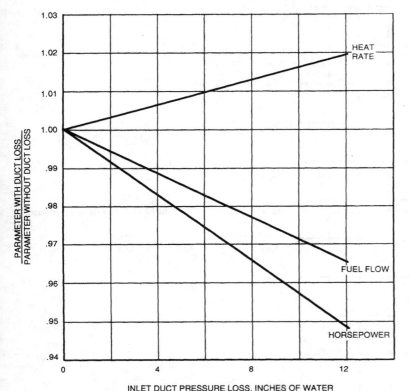

Fig. 10   Effect of inlet duct loss on gas turbine performance

These rating conditions are significantly more demanding than the ones used by the International Standards Organization (ISO), which are 59 F, 29.92 in. of Hg, with no inlet or exhaust losses specified. As a result, a specific gas turbine would have a lower rating for Navy applications than indicated by ISO standards.

1.5 Combined Cycles. A gas turbine is a flexible prime mover that can be applied not only alone but also in combination with other prime movers. Several combinations have been applied successfully in marine applications, and others have been investigated. Some of the combinations that have been considered include:

COmbined Gas turbine And Gas turbine plants (COGAG)

COmbined Diesel And Gas turbine plants (CODAG)
COmbined Steam And Gas turbine plants (COSAG)
COmbined Gas turbine And Steam plants (COGAS)
COmbined Diesel Or Gas turbine plants (CODOG)
COmbined Gas turbine Or Gas turbine plants (COGOG)
COmbined Nuclear And Gas turbine plants (CONAG)

The ratings of gas turbines tend to be discrete and limited because the cost to develop a new gas turbine for a specific power is very large. As a result, to satisfy larger power requirements, multiple gas turbine engines are frequently used. The engines may or may not be of the same rating. In a multi-engine arrangement, one engine may be used at low speeds, and additional power requirements are met by bringing additional gas turbine engines on line as required. The reason for this is that the specific fuel consumption of a gas turbine engine is poor at partial power, therefore, by using multiple engines and progressively bringing them on line as more power is required, each engine can operate near its capacity resulting in reduced overall fuel consumption. As an alternative to installing multiple engines of equal rating, a small gas turbine engine (or engines) may be installed in combination with a large gas turbine engine (or engines). When the small gas turbine(s) is used only in the cruising mode and the large gas turbine(s) is used for maximum-power situations, this is known as a COGOG arrangement. When both the small engine(s) and the large engine(s) are required to develop full power, that is known as a COGAG arrangement. Gas turbines may also be combined with other types of prime movers. CODAG plants combine a diesel engine(s) with a gas turbine(s) to obtain increased range at low speed with the diesel(s) while maintaining a high-speed capability with the gas turbine(s). Both the diesel(s) and the gas turbine(s) are required to develop full power. The CODOG plant refers to an arrangement where a separate smaller diesel(s) is used for cruise conditions and a larger gas turbine(s) is used at high power conditions. This type of plant has been successfully installed on various naval ships. COSAG plants refer to the combination of a steam plant and a gas turbine where the

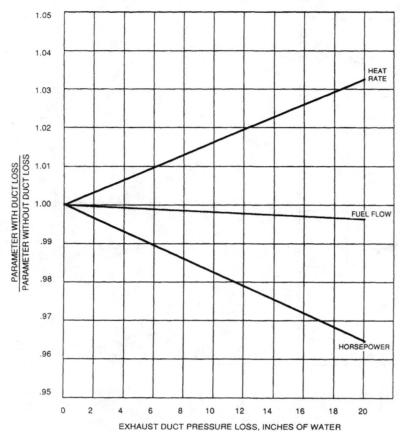

Fig. 11   Effect of exhaust duct loss on gas turbine performance

steam plant is sized for cruise conditions and the gas turbine provides extra power for high speed operations. These plants have been considered for some ship designs but are not very attractive because steam turbines are not well suited for low-power applications. COGAS plants are combined cycles in which waste-heat boilers in the exhaust ducts of the gas turbines are used to generate additional shaft power through a steam turbine. The U.S. Navy investigated this type of plant in their Rankine Cycle Energy Recovery (RACER) program in order to reduce the fuel consumption of gas turbine ships at low power conditions. CONAG plants have been studied in which small nuclear plants are used to power the ship for extended cruise performance and gas turbines are used to achieve higher top speeds. Further discussion of combined cycles is provided in Chapter 1.

1.6 Bleed Air. A gas turbine may offer an advantage to the ship designer in the form of bleed air. Most gas turbines are capable of providing up to 10% of their airflow in the form of high-pressure bleed from the compressor discharge. Some engines also offer inter-compressor bleeds at lower pressures and usually lower flow rates. The bleed air, which can range from 50 to 300 psig, can be used for a variety of purposes. On U.S. Navy ships, bleed air is used to start other gas turbines, as anti-icing air for the intake duct during cold weather operation, and

as an air blanket system called the prairie/masker system, which is used to attenuate machinery and propeller noise.

The availability of gas turbine bleed air entails a penalty in gas turbine performance. Figure 12 shows the effect of bleed air extraction from compressor discharge of a typical gas turbine. This figure shows that for a 10% bleed air extraction rate, a 14% increase in specific fuel consumption occurs.

1.7 Quick Starting. A major advantage of a gas turbine over a steam plant is the ability to start and be put on line quickly. A gas turbine has no large masses that must be heated slowly; therefore, the time required to reach full speed and accept load is limited almost entirely by the rate at which energy can be supplied to accelerate the rotating components. A gas turbine is designed for rapid transients; therefore, there is no concern for the rotor bowing or other transient effects which limit steam turbines.

The speed with which a gas turbine can be accelerated is, therefore, a function primarily of the starting power available. Small units, which have a high power rating in relation to the size of the engine, can be brought up to speed in a few seconds. But even the larger machines can be started and brought to full power within two minutes.

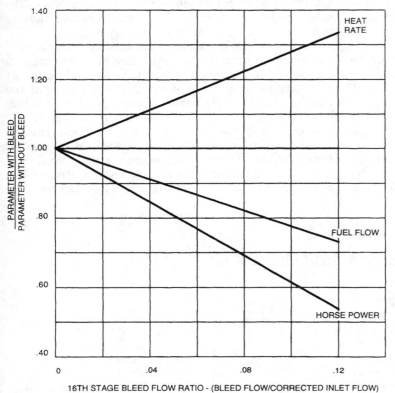

NOTE: ESTIMATES BASED ON 7LM2500PC 16TH STAGE BLEED DERIVATIVES
CONSTANT POWER TURBINE INLET TEMPERATURE AND SPEED

16TH STAGE BLEED FLOW RATIO - (BLEED FLOW/CORRECTED INLET FLOW)

Fig. 12   Effect of bleed air extraction on gas turbine performance

Of prime importance in starting a gas turbine is a close and accurate control of the fuel schedule, and the availability of adequate starting torque under all conditions. A small excess of fuel can result in elevated temperatures and damage to the machine. A low starting torque can result in a low airflow or high fuel flow at some point in the cycle, which can also over heat the engine. Temperature control of the fuel schedule, which controls the amount of fuel delivered to the engine based on the turbine exhaust temperature, offers the best assurance of quick starting without excessive temperatures.

1.8   Fuels and Fuel Treatment. A discussion will be given later of the effect of some fuels on the life and corrosion rate of hot gas-path parts. This is a particularly important consideration in marine applications since one of the worst elements in the corrosive attack of gas turbine parts is sodium, which is always present in the marine environment. Another damaging element is vanadium, which is frequently present in petroleum oils, particularly crudes and residuals.

Since gas turbines are sensitive to the quality of fuel, most marine applications utilize lighter distillate fuels. Typical fuels used are JP-4 and JP-5 aviation fuels and diesel fuel marine (DFM).

1.9   Installation. Gas turbine engines are usually installed as self-contained complete assemblies. Most arrangements are designed for the gas turbine to be installed in an enclosure for noise suppression and containment in case of fires or rotating-part failures. Figure 13

Fig. 13   Gas turbine base/enclosure assembly

is a gas turbine assembly of the type commonly installed aboard ship. This gas turbine also has its own self-contained lubricating-oil system since gas turbines utilize high-temperature synthetic lubricating oils, which are not common on other shipboard machinery systems.

One of the disadvantages of a gas turbine is the relatively high airflow requirements compared to other prime movers. This results in higher volume requirements for both the air intake and exhaust uptake, which often results in arrangement problems.

Attention must be given to the design of the inlet and exhaust ducts and the flow of air into the gas turbine module since the engine performance is very sensitive to pressure losses in airflow ducting, as shown in Figs. 10 and 11. The inlet ducting is usually designed to accommodate engine removal because, in most arrangements, this is the path usually used to remove a gas turbine from the ship.

As discussed in Chapter 1, gas turbine plants have a light weight, and facilitate compact arrangements with low space requirements since few major supporting auxiliaries are required.

1.10 Operation and Maintenance. Gas turbine propulsion plants have low manning requirements. This is due to the adaptability of gas turbines to automation and their low maintenance requirements. As discussed in Section 4, the gas turbine inherently requires built-in automatic controls to protect it during starting, stopping, and normal operation. Manual controls cannot respond fast enough to meet the requirements of gas turbines. The extension of the gas turbine controls to include the gas turbine auxiliaries is easily accomplished.

The gas turbine has attained outstanding reliability and maintainability records through years of development in commercial and military aviation. These high levels of reliability have also been demonstrated on ships of the U.S. Navy and foreign navies. The design of the gas turbine lends itself to easy removal usually through the intake ducting of the ship. Typically the gas generator is disconnected from the power turbine and each is removed as an assembly. Onboard maintenance usually consists of removal and replacement of accessories or control components and minimal preventive maintenance actions.

Section 2
Arrangement and Structural Details

2.1 General Arrangement. The physical arrangement of a gas turbine on board ship is influenced by mechanical considerations (i.e., shafts must have adequate bearings, seals, etc.), by structural considerations (i.e., proper mounting to avoid misalignment in the turbine), and by the necessity to conduct very large air and gas flows to and from the various components. Figure 14 shows a typical arrangement of a propulsion marine gas turbine on a U.S. Navy ship [9]. Sections of the gas turbine arrangement on board ship include the air intake, exhaust, mounts, reduction gears, and other components.

2.2 Air Intake (or Inlet) Section. The air-intake section must supply the gas turbine with a high volume of air from the atmosphere with a minimal pressure drop. The air-intake section should be designed to provide the gas turbine with a uniform pressure and velocity flowfield at the compressor inlet. Unequal airflow into the inlet annulus can reduce the efficiency of the compressor and cause blade vibrations that can lead to early blade failure. Ideally the compressor inlet should pull from an infinite plenum, as with an aircraft jet engine in flight. Practically, the engine must pull the air through a duct system; therefore, some form of air-inlet housing must be used. If a large plenum can be used, the compressor inlet can be inserted in one wall of the plenum so that the air flows axially into the compressor annulus. Model tests of the inlet configuration are often conducted to identify undesirable flow distortions. If tests are not conducted, experience dictates that the engine inlet should be at least two engine inlet diameters away from the far side of the plenum.

In order for the engine intake section to consistently provide the quality of air required, several features are often included in the inlet ducting. Figure 15 shows a typical intake arrangement. Included in the intake section are the filtration provisions (louvers and demisters), anti-icing manifold, enclosure cooling air ducting, and silencers

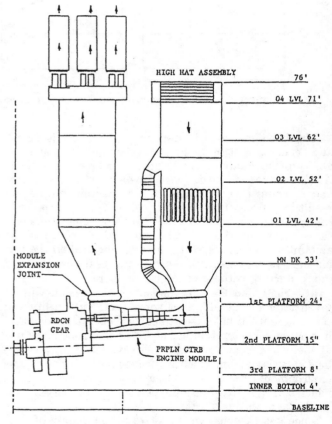

Fig. 14 Typical arrangement of a marine propulsion gas turbine

[9]. Each of these features is discussed in Section 3 in more detail.

The plenum is a chamber of sufficient volume to reduce the air velocity and suppress vortices before it enters the compressor. The geometry of the plenum is designed to transition the airflow to the front of the compressor so

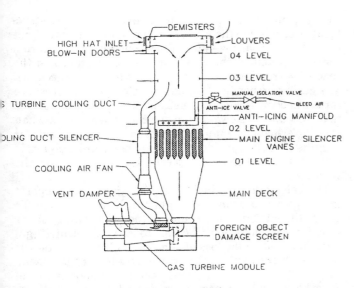

Fig. 15 Typical gas turbine intake-duct arrangement

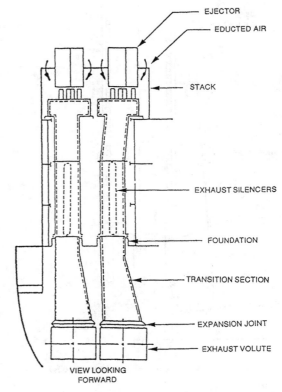

Fig. 16 Typical exhaust duct

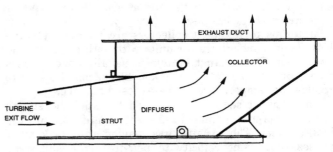

Fig. 17 Exhaust collector

that it is evenly distributed and aligned to the centerline of the gas turbine.

To reduce the axial length of the air intake system, a radial intake design may be considered. The highly three-dimensional flow characteristics of this type of arrangement usually require model testing to obtain designs with satisfactory airflow characteristics entering the compressor.

The air intake ducting is often used as a removal path for the gas turbine. This is accomplished with a rail system, which must be in place when an engine is to be pulled up through the inlet ducting. When this engine-removal technique is used, the ducting cross-sectional area requirement is typically determined by the engine extraction envelope, rather than the airflow requirements.

2.3 Exhaust Uptake. Due to the higher temperatures, the exhaust volume flow is larger than that of the inlet. Maintaining velocities that provide reasonable pressure drops is, therefore, more difficult for the exhaust uptake. The high temperature also introduces expansion problems since the movement of the ducting due to thermal growth must be accommodated without introducing high forces into the turbine structure. A typical exhaust uptake is shown in Fig. 16. The exhaust uptake consists of an exhaust collector, transition section, eductor, expansion joints, and ejector.

The exhaust collector consists of a diffuser, which recovers the velocity energy of the turbine exit flow, and a collector, which collects the turbine exit flow and turns the flow to the exhaust ducting as shown in Fig. 17. In addition to its aerodynamic function as a diffuser and collector, the exhaust collector must frequently act as a structural member and carry loads and bending moments through the gas path. Struts through the collector are often used to support the power turbine bearing housing and the inner wall of the diffuser. Such struts should be located as far down the diffuser as possible, as any

obstruction in a diffuser, even when streamlined, markedly reduces its recovery efficiency. In addition, provision must be made for the thermal expansion (especially during starting) of the struts, which are completely immersed in the hot gas stream. In small units, distortion of the casings at the attachment points may be insignificant; but in large units some radial flexibility is generally allowed. Tangential struts or a radial strut with a tangential spring member at the outer end are possible arrangements. Air-cooled struts attached to cylindrical inner and outer members, which can grow sufficiently to allow for some expansion, can also be used.

The exhaust collector is usually fabricated from relatively thin material with suitable stiffening ribs formed in the material or welded on externally. This serves to stiffen the structure to prevent drumming or resonances and also enables the structure to withstand the internal pressure due to back pressure imposed on the unit when

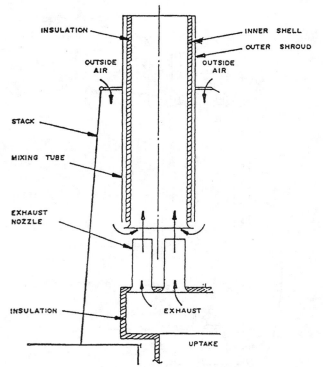

Fig. 18 Typical exhaust eductor details

heat-recovery equipment is used. Large flat surfaces should be designed to withstand at least 20 in. of water pressure without bulging.

Materials for exhaust collectors are frequently stainless steels, particularly for units with high exhaust temperatures. The exhaust system is usually covered with some form of insulation (blanket, block, plastic, etc.) to reduce the temperature of the exposed surface.

The exhaust section is designed to expel the engine exhaust gas in a manner to prevent reingestion into the intake ducts. The exhaust gas leaving the stack must possess sufficient velocity to penetrate the ship's air boundary layer to prevent downwash and subsequent reingestion by the intake system. One method used to avoid reingestion is to have the exhaust outlets at a higher level than the intake assembly. For military ships, sound suppression and temperature reduction of the gas are also part of the exhaust section. Sound suppression is typically accomplished with insulation and a vane type silencer. The sound suppression system is described in more detail in Section 3. Exhaust temperature reduction starts by mixing the enclosure cooling air into the exhaust flow at the exhaust collector. Additional reduction is accomplished with an eductor system. A typical eductor system is shown in Fig. 18. This system mixes the outside air with the exhaust gases. The eductor is a mixing tube positioned to draw outside air into the exhaust gas stream as it enters the mixing tube. This exhaust temperature reduction is needed to reduce the infrared signature of the exhaust plume.

2.4 Structural Arrangement. Structurally, the stationary gas turbine parts, compressor casings, combustion casings, turbine casings or shells, and the related structural supports for the rotor bearings must withstand not only the internal pressure forces but also the external forces imposed on the unit from its own weight and the reactions from engine torque and external connections. The casings must be designed for the internal pressure forces and must also be checked as a beam, under the reactions due to the weight of the components, plus whatever "g" loading may be imposed. Rotor and stator weights must be considered, and supports are frequently located so as to minimize the bending moments in the structure.

The calculation of the bearing housings and supports cannot be based upon the weights of the rotors alone. To insure the integrity of the unit in the event of a blade or vane failure, they should be able to carry the centrifugal loads imposed by the loss of some credible combination of vanes or blades within the tensile strength of the members. The loss of two adjacent vane sections or one complete blade and dovetail is considered a reasonable assumption.

Mounting of the gas turbine to a base can take many forms. The mounting system must support the unit and maintain it in line with the driven equipment, while allowing for the axial and radial thermal growth of the unit from cold to normal operating temperatures. Several methods of support are shown in the various illustrations, and it will be seen that small units frequently use three points of support with one centering key or gib, while larger units usually use at least four supports; more are used if the whole unit is divided into several casings as in multi-engine arrangements. An example of a gas turbine mounted to a base is shown in Fig. 19.

Mounting of the base to the hull may be either accomplished with hard mounts or with resilient mounts. Resilient mounting is used when either structureborne noise attenuation or high-intensity shock protection is a requirement. Resilient mounts typically consist of sets of spring washers attached to resilient neoprene shock mounts.

2.5 Intercoolers and Recuperators. Marine gas turbines are typically derivative designs of aircraft gas turbines. However, because the size and weight restrictions of marine gas turbines are more flexible than for aircraft gas turbines, enhancements have been made to the simple gas turbine cycle of the aircraft engine to improve performance. As illustrated in Fig. 3, two such enhancements are the additions of an intercooler and a recuperator. An intercooler is an air-to-liquid heat exchanger used during the compression process to enhance the specific power of the gas turbine. This is accomplished by cooling the gas turbine airflow midway through the compression process, thus reducing the specific volume of the compressed air and thereby reducing the energy required to complete the later stage of compression. The reduction in energy required during the compression process is then reflected in an increase in the power output.

Intercoolers divert airflow from the compression process to a heat exchanger and then back to the compressor section. This diversion causes a significant change in the thermodynamic matching of the components, making it

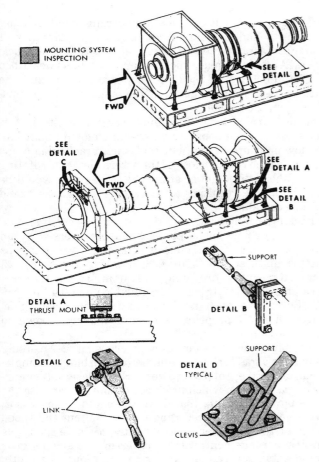

DETAIL A
THRUST MOUNT

DETAIL B

DETAIL C

DETAIL D
TYPICAL

LINK

CLEVIS

Fig. 19   Typical gas turbine mounting system

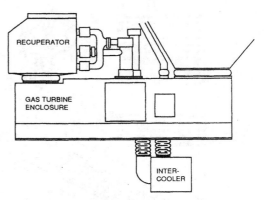

Fig. 20   Off-engine intercooler design

more practical for engines with two compressors on separate shafts. Typically, intercoolers are shell-and-tube or plate-fin heat exchangers with the air flowing on one side transferring heat to the liquid flowing through the other. The liquid in the tubes can be seawater from an open seawater loop, or fresh water or glycol from a closed loop. If a closed loop is used, a second heat-exchanger loop is required to remove the heat from the closed-loop fluid. The second loop is usually an open seawater loop.

An intercooler can be an on-engine design or an off-engine design. An on-engine design is contained within the engine module and because of size considerations is typically a closed heat exchanger loop on the engine and a secondary loop off the engine. The off-engine design diverts the airflow from the core engine, through the heat exchanger, and then returns it back into the core engine. An off-engine intercooler design is shown in Fig. 20 [10].

A recuperator is used to transfer heat from the exhaust gas to the engine airflow to reduce the fuel required during combustion. A recuperator is a fixed-surface heat exchanger in which the hot exhaust gas is on one side and the engine airflow is on the other, and the heat is transferred through the wall by conduction. Rotary heat exchangers in which a heat-storage matrix is alternately exposed to the hot exhaust gases and then to the engine airflow, transferring heat from the former to the latter, are called "regenerators." Both recuperators and regenerators have been successfully used with gas turbines although a recuperator is far more common. The choice of one or the other is determined primarily by the installation arrangement and consideration of the leakage and carryover problems experienced by regenerators in higher pressure ratio gas turbines [11].

Heat exchangers for recuperators include the plate-fin, primary-surface, and profile-tube types. A plate-fin heat exchanger consists of alternating flows of exhaust gas and engine airflow between corrugated sheets of stainless steel. A primary-surface heat exchanger is shown in Fig. 21(a) and is very similar to a plate-fin design, except that the primary-surface is a continuous plate folded to create alternating passages of exhaust gases and engine air flow [12]. Figure 21(b) shows a profile-tube design which is similar to a shell-and-tube design, except the tubes are aerodynamically shaped and stacked in a compact manner [13].

The selection of the materials used in recuperators depends upon the operating temperature range of the recuperator. Where the maximum turbine exhaust temperature is 1000 F or below, low-alloy carbon steels can be used, but the metal thicknesses chosen should provide an adequate allowance for minor corrosion. For turbine exhaust temperatures over 1000 F, or where the design has been optimized for minimum weight with resulting thin-gage materials, corrosion-resistant materials such as stainless steel or one of the Inconels are necessary.

Another concern in the design of the heat exchangers for recuperators is the fouling build up over numerous hours of operation. Experimental results from research conducted at the David Taylor Research Center show that fouling does occur and causes a performance degradation for the type of heat exchanger designs used in marine gas turbines. However, the buildup of contaminant can be cleaned (burned off) by running the gas turbine with only the hot exhaust gases allowed to flow through the heat exchanger [14].

2.6 Reduction Gearing and Reversing Considerations. A gas turbine is a high-speed machine with output shaft speeds ranging from about 3600 rpm for large machines up to 100,000 rpm for very small machines. Approximately 25,000 rpm is an upper limit for units suitable for

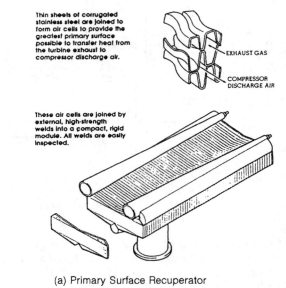

Thin sheets of corrugated stainless steel are joined to form air cells to provide the greatest primary surface possible to transfer heat from the turbine exhaust to compressor discharge air.

EXHAUST GAS

COMPRESSOR DISCHARGE AIR

These air cells are joined by external, high-strength welds into a compact, rigid module. All welds are easily inspected.

(a) Primary Surface Recuperator

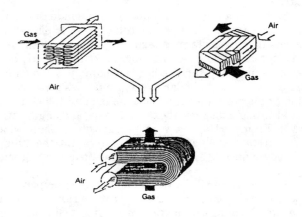

Gas

Air

Air

Gas

Air

Gas

(b) Profile Tube Recuperator

Fig. 21   Alternative recuperator designs

the propulsion of small boats. With these output speeds, a reduction gear is necessary to reduce the speed to the range suitable for a propeller. Smaller units suitable for boats or for driving auxiliary units, such as generators in larger vessels, frequently have a reduction gear built integral with the unit. Larger units normally require a separate reduction gear.

The gearing itself can be of any arrangement. Smaller units with built-in gears frequently use a planetary or star gear arrangement. Larger units use double-helical gears. Any of the gear types and arrangements described in Chapter 9 can be used with a gas turbine to suit the rating, speed ratio required, and arrangement of the machinery in the vessel.

A gas turbine is not inherently reversible. Steam turbines can provide separate reversing elements built into the low-pressure casing, but this is not practical in a gas turbine as the rotation loss of the astern elements rotating in the ahead direction at atmospheric pressure would be very high. The resulting temperatures and losses would be unacceptable.

Lacking a practical internal reversing method, propulsion gas turbines installations must be reversed by an external means. Electric drives offer ready reversing, as further discussed in Chapter 8; however, the two most common means of providing a reversing capability are a reversing gear [15,16] or a controllable-pitch (CP) propeller. Both have been used successfully in gas-turbine-driven ships. CP propellers are common in ships powered by gas turbines.

An important consideration in choosing a reversing means for a gas turbine is whether the turbine is a single-spool, or multi-spool (free power turbine) design. A single-spool gas turbine, which produces output power with the same power elements that drive the engine compressor, has a very limited speed range, thus limiting the speed range of the propeller. Consequently, a single-spool gas turbine arrangement with a reversing reduction gear would entail severe maneuvering restrictions; therefore, a CP propeller may be more appropriate when maneuvering is a concern. A gas turbine engine with a free power turbine can be applied equally well with a reversing gear or a CP propeller.

2.7   Engine Module.   In most ship applications marine gas turbines are enclosed and mounted on an independent base. A typical engine module is shown in Fig. 13. The base and enclosure are integral elements of a gas turbine module. The engine enclosure minimizes physical damage, suppresses engine noise, provides thermal isolation, and acts as a fire-containment system. The enclosure is equipped with a fire-detection system and contains cooling-air passages for the secondary inlet flow around the gas turbine and back into the exhaust. An access door is provided for on-board maintenance. Some small and special-purpose craft, which have severe space limitations, do not use enclosures.

Section 3
Accessories

3.1 Auxiliary Pumps and Drives. A gas turbine, while basically a complete, self-contained power plant, requires certain accessories for its operation. In particular:

- A fuel boost pump is used to provide fuel oil at a regulated positive pressure to the fuel injectors or nozzles. The size of the pump is usually determined by engine start-up, rather than engine full-power requirements.

- A lubricating-oil pump is used to provide oil to gears and bearings (see Section 8.6 for lubrication system description).

- A deaerator is used to remove entrained air from the lubricating oil.

- A start motor is used to bring the engine to a self-sustaining speed during the engine start sequence.

- Hydraulic pumps, governors, speed switches, tachometers, and other engine-driven devices may be required for operation of the engine's control system.

Although the auxiliary pumps could be driven by electric motors, the entire set of accessories is normally driven by a reduction gear that is connected to the high-pressure compressor shaft. The gear is designed for the duty and life required of the engine itself and is provided by the engine manufacturer as part of the engine.

3.2 Fuel System. The fuel boost pump has little or no suction lift capability; separate arrangements must be made to keep the pump suction flooded with fuel oil. This is done by the ship's fuel service system, consisting of service tanks, pumps, heaters, filter-separators, and interconnecting piping and valves as shown in Fig. 22. The heaters warm the fuel oil, if required, to the point that it can be efficiently handled by the filter-separators, which remove water and contaminants from the fuel oil. The fuel boost pump takes suction from the tanks and delivers fuel oil to the engine.

Some fuel service systems have head tanks between the filter-separator discharge and the engine module fuel-oil inlet. The tank provides fuel at the required pressure to the engine, even in the absence of ship service power to drive the fuel service pumps.

Electric heaters are often provided within the engine module. The heaters are thermostatically controlled, and set to ensure that fuel within the module does not become too cold (and viscous) when the module temperature drops (i.e., when the engine is shut down).

A "final" fuel filter is included in the module, to remove contaminants from the fuel delivered to the engine. The filter mesh used varies with fuel type; for the diesel fuel marine (DFM) used by the U.S. Navy, a 5-μm mesh is typical.

Engines that have more than one combustor must have provisions for dividing the fuel equally among the combustors. This is particularly important, since even individual nozzle arcs can only reliably meter the airflow to within 3 to 4% accuracy, so that a 5% difference in the fuel nozzle flow can result in an 8 to 9% variation in temperature rise. On a 1000 deg F rise this variation is 80 to 90 deg F, an amount large enough to affect the life of the hot gas path parts. Although various techniques for fuel division have been used, the most common solution is to use a multiple-section positive-displacement (typically gear-type) fuel boost pump, with one section dedicated to each combustor. Fuel delivery to each combustor then becomes dependent only on the speed of the pump, and not on pressure variations caused by differences in individual fuel nozzles.

3.3 Starting Devices. A gas turbine, like other internal combustion engines, is not self-starting, and external means must be utilized to bring it up to the self-sustaining speed. This is the speed from which the rotors can be accelerated by the addition of fuel alone, without external assistance, and it is usually about 30 to 50% of the gas-generator's full speed. On units with two or more compressor shafts, only one starting device is used in order to maintain the proper airflow match between the two

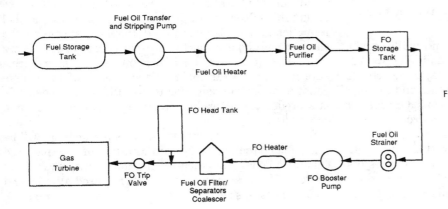

Fig. 22   Schematic of a gas turbine fuel system

compressors. The starter is used to accelerate one shaft and the induced airflow through the engine accelerates the other shaft to the proper speed.

Starting devices in common use include electric motors, air motors operating on compressed air, and small independent gas turbine engines, which must, in turn, have their own starting systems. Other starting systems that may be considered include hydraulic motors fed from high-pressure pumps or accumulator systems, and special rotary-type starting motors fed from high-pressure air supplies, some of which include the combustion of fuel to provide the starting energy. In any case, it is important that the starting device have adequate power to reliably bring the unit to the self-sustaining speed because a dependence upon energy from combustion during the critical start-up phase could result in excessive temperatures at the turbine inlet. That is, the fuel/air ratio during the starting cycle should be held close to normal limits. This requires a relatively large energy input from the starting device. Since the operating time of the starting device is of relatively short duration, the starting device can be highly loaded or peak-load rated, particularly if it is an electric motor. Starting times range from seconds on a very small gas turbine to 1 to 2 minutes on large aircraft-type engines.

Since the starting device is normally required only up to about 50% speed, it is usually connected to the turbine through a clutch, which allows it to be disconnected during normal operation. The simplest, and probably most satisfactory, form of clutch is a simple jaw clutch that is magnetically or pneumatically engaged and spring disengaged. Provisions should be made for rotating the starter slowly during engagement to make sure the jaws are fully engaged before full torque is applied; otherwise, severe damage could result.

Alternatively, an overrunning clutch could be used, but unless the clutch is a self-synchronizing form, such as illustrated in Chapter 9, it is susceptible to damage if the starting device is energized while the gas turbine is still decelerating. The shock load under such conditions can be very severe, and few overrunning clutch designs have the torque margin to withstand such a load. The clutch should also disengage completely at some speed below the minimum operating speed of the power turbine shaft but above the self-sustaining speed of the gas turbine to avoid excessive wear on the mechanism.

3.4 Inlet Air Filters. Additional items normally considered as accessories are inlet air filters and inlet and exhaust silencers. Pressure drops are of major importance in gas turbine operation; therefore, the inlet and exhaust gas paths must be designed with careful consideration for the economic balance of size and pressure drops.

Gas turbines require clean air, as otherwise the compressor will eventually become coated with foreign matter, which reduces its capacity and efficiency and results in a degradation of the entire engine or, in an extreme case, may even cause a compressor blade failure due to

stall. In marine applications the most important requirement is to keep salt particles and water, even small drops, from entering the compressor. For this reason air inlets should be placed as high in the ship as possible and must be equipped with effective baffles or eliminators to prevent the entry of water. Behind the eliminators a demister should be installed to coalesce fine (mist) droplets of water into larger drops, which are then either drained off or blown through to the next stage. The filter system can consist of an inertial type separator or, a pad arrangement (similar to filter pads) may be used that consists of metal or synthetic fibers of controlled size and spacing to effectively control the size of droplets passed. Multiple stages of inertial type separators and pads may be utilized as well. If the demister pads are kept wet, even with seawater, they are also effective in stopping the ingestion of salt and other foreign particles and thereby serve as a filter medium. The type of ship under consideration and its above-water profile and height above the waterline must be considered when selecting the type of mist eliminator to be used.

If the engine room is used as a plenum, oil vapors from other equipment can also adversely affect engine performance. The oil vapor deposits on the compressor blades and causes other particles to deposit on the blades which reduces the compressor efficiency.

3.5 Inlet and Exhaust Silencers. A gas turbine, being a high-speed machine, generates a relatively large amount of noise with a wide frequency spectrum. The sound-power level of a given gas turbine design is a function of its size or power and is approximately proportional to its rating. Most of the noise is generated aerodynamically and is related to blade passing frequency and, therefore, is in the high-frequency range [17].

The major sources of the noise radiated to the surroundings are the inlet openings, exhaust openings, and gears. However, the entire machine radiates noise; the intensity of the noise radiated from the engine is inversely related to the casing thickness or, more exactly, to the casing mass. Noise radiated from the casing can be reduced and controlled by appropriate sound treatment and is usually effectively absorbed by the enclosure.

The airborne noise, in both the inlet and exhaust, can be attenuated to almost any required level by the use of suitable silencers. In general, the greater the decibel reduction in noise level required, the more expensive the silencer and the greater the pressure drop. In specifying silencer performance, it is important that the sound-pressure level at the turbine be given in each of the octave bands and that the reduction to give the required decibels at a predetermined radius be also specified for the same octave bands.

Since the sound attenuation in the surroundings will be somewhat directional, the configuration of the inlet and exhaust openings and their orientation should be carefully chosen. Of course, the sound levels required also depend upon the vessel's service (e.g., cargo, passenger, or naval).

Section 4
Controls

4.1 Control System Functions. Every gas turbine has a control system, which performs two functions: operational control and safety control. A third function, monitoring, is often assigned to the control system. On some engines, monitoring is separate from control. "Operational control" means the adjustment of the engine component parts so that the engine starts and stops when commanded, produces the desired output power while running, and does no damage to itself or its surroundings. "Safety control" means the detection of, and initial response to, engine failures. "Monitoring" means the provision of information about the condition of the engine.

4.2 Control System Hardware. Gas turbine control systems consist of three types of components: actuators, sensors, and logic as shown in Fig. 23.

Control system actuators (which are sometimes called "final control elements") physically change engine variables, such as fuel flow or compressor bleed flow, in response to commands issued by control system logic. Most gas turbine actuators are valves of various sorts. Electrical components, such as igniters and cooling fans, are usually actuated by electrical relays.

Control system sensors measure information about key engine variables (temperatures, pressures, positions, etc.) and transmit the status data to the control system logic.

The control system logic generates commands to the actuators so that the gas turbine will operate safely and as commanded by the turbine operator.

All three component types (actuators, sensors, and logic) can be mechanical, pneumatic, hydraulic, or electrical/electronic. The control systems that were used on earlier gas turbine designs largely consisted of mechanical

components; however, mechanical components have generally been replaced by electronics, including microprocessors.

Gas turbine control systems are designed so that they can accept commands from remote control systems, such as a ship's propulsion or electric plant control system. Some turbines also have local control panels, which allow a gas turbine operator or maintenance personnel to operate the engine from the immediate vicinity of the engine.

4.3 Operational Control. Gas turbine operational control functions can be conveniently divided into those associated with starting, stopping, and running the engine. A typical gas turbine start sequence involves the following steps:

- If necessary, the engine is purged of any low-temperature, high-viscosity fuel remaining in the fuel lines (most engines have more restrictive fuel limits for starting than for running). Purging can be done by recirculating the fuel back to the fuel service tank, by draining the fuel to the oily-waste system, or by a combination of both.

- The turbine start motor is engaged and energized, causing the spool to begin rotating (on multi-spool engines, the start motor drives the high-pressure compressor/high-pressure turbine spool).

- When the spool is turning fast enough to provide an adequate airflow to the combustors, the igniters are energized, and the master fuel valve is opened. The engine fuel flow control function, operating under a separate "engine start" schedule, allows a small amount of fuel into the engine.

- The fuel ignites, and the engine continues to accelerate under the combined influences of the start motor and the burning fuel. The fuel flow control, still operating on the start schedule, meters increasing amounts of fuel into the combustion chamber.

- When the spool has reached a self-sustaining speed, the igniters are de-energized, and the start motor is de-energized and disengaged.

- The engine accelerates until it reaches idle speed. On multi-spool engines, the airflow induced by the rotating spool provides the energy to accelerate the remaining spools. The fuel flow control transitions from the "engine start" schedule to the "engine run" schedule.

All gas turbine control systems have an "engine start" fuel control schedule similar to that described above. Most control systems also include an engine start function, which automatically steps the engine through the start sequence in response to a single "start" command from the engine operator or remote control system. The automatic start function usually includes provisions for aborting the start if one or more of the steps takes an unusually long time; this is called a "hung start." Less-sophisticated

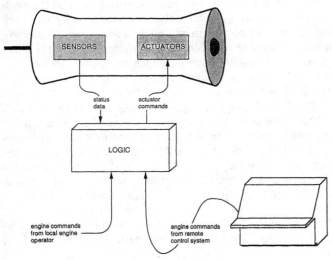

Fig. 23   Gas turbine control system schematic

control systems, which do not have the start function, require the operator to control each step.

Stopping a gas turbine is done by simply closing the engine's master fuel valve. Without fuel, the engine decelerates rapidly. Because of the high initial speeds, however, spool rotation continues for some time. Generally, all components come to a complete stop within five minutes after the fuel supply is cut.

Extra attention must be paid when shutting down engines from high power. At high power, both the turbine blades and the engine case have experienced significant thermal growth. When the engine is shut down, spool rotation will continue to draw cooling air through the engine core, and engine module cooling air will cool the engine case. Once spool rotation stops, the engine core's cooling rate drops. However, the engine module fan will continue to run, maintaining engine case cooling. On second generation and later engines (generally, those designed after 1965), which have tight blade-to-case clearances, the cooled case can shrink onto the hot turbine blades, and make hard contact. Once the blades cool and shrink sufficiently, the original clearances will be restored; however, to avoid the problem altogether, engines are idled for several minutes before being shut down. If idling is impossible, spool rotation is maintained after shutdown by "motoring" the engine on the starter. Some engine control systems have provisions for both "normal" stop (which automatically maintains the engine at idle for a fixed period before shutting the engine down) and "emergency" stop (which initiates an immediate shutdown).

While a gas turbine is running, only two basic parameters can be controlled: the flow of fuel, and the flow of air. All turbine control systems have a fuel control function; however, all do not have air control capabilities.

The fuel control function regulates the flow of fuel into the engine, ensuring that the engine's power output matches the power output requested by the engine operator or remote control system. Steady power demands, therefore, require the fuel control to provide a constant fuel flow; but power demands that are not constant require a change of fuel flow that closely conforms to the power requirement. The most complex portion of the fuel control function is involved with making appropriate responses to changes in power demands.

A suddenly increased power demand, for example, requires a corresponding increase in fuel flow. However, if the fuel flow is allowed to increase much faster than the airflow, the engine combustion zone temperatures, and thus the turbine inlet temperatures, will rise above acceptable limits, and the life of the engine will be reduced. Also, a rapid increase in the combustion zone temperature is accompanied by a corresponding combustion zone pressure increase. If the combustion zone pressure exceeds the compressor discharge pressure, the air will flow backwards through the compressor, causing a compressor stall and possible engine damage.

Similarly, a suddenly decreased power demand requires a corresponding rapid decrease in fuel flow. If the fuel flow is allowed to decrease much faster than the airflow,

the excess air can blow out the combustion flame and cause an engine flameout.

The basic purpose of the fuel control function, therefore, is to keep the engine's fuel-air ratio within prescribed limits. Restrictive fuel-air ratio limits increase engine safety, but decrease engine responsiveness; wide limits decrease safety while increasing responsiveness. Limits are set in accordance with the engine design. A large aircraft-derivative gas turbine, for example, can be commanded to go from idle to full (compressor) power in 7½ seconds. Most fuel control functions estimate airflow by sensing the compressor discharge pressure. The fuel control function operates both during normal engine operation, and during engine start-up.

There are three basic ways to adjust the gas turbine airflow: varying the compressor spool speed, varying the engine internal geometry, and bleeding air from the compressor. Engines, and their control systems, may use any combination of these techniques.

Varying the compressor spool speed is the primary means of varying the airflow. However, since the compressor has considerable inertia, spool speed control is not always adequate for transient responses; it is essentially limited to adjustment of steady-state conditions.

Varying the engine internal geometry is normally done by providing the engine with variable stator vanes, variable-area power-turbine nozzles, or both. The vanes adjust airflow into the compressor; the nozzles adjust airflow into the power turbine. Internal geometry controls can respond quickly, and are, therefore, usable for both transient and steady-state control.

Compressor air bleeding is done by providing the engine with one or more compressor interstage bleed valves. In most cases, the valves are solenoid valves, which can respond quickly but have only two positions: fully open, or fully closed. Interstage bleeding, therefore, is not usable for small adjustments to airflow; it is typically used to stabilize the compressors during off-design operation, such as during engine starts.

The air-control function regulates airflow, and the fuel-control function adjusts fuel in accordance with the fuel-air ratio; the two functions are interdependent. Many engines, therefore, combine the air- and fuel-control functions into a single function, which controls both parameters. The fuel-control function typically commands both the fuel valve and the variable stator vanes.

The U.S. Navy uses gas turbines for the generation of both electrical and propulsion power. Electrical power generators are required to produce power which conforms to DOD-STD-1399 Section 300 Type I, which mandates constant frequency parameters. Since the electrical power frequency is directly proportional to the engine output shaft speed, the output shaft's speed must also be constant. Generator control systems, therefore, employ "speed governing," which means fuel flow is adjusted to match the engine output power with the connected load while maintaining the engine output shaft speed at the set value.

Propulsion power generation is similar to electrical power generation in that the engines are expected to produce exactly the power required to keep the propeller shaft speed at the value commanded by the bridge. The power requirements vary: each time the ship encounters a wave peak, the ship's resistance to forward movement increases, and the propellers attempt to slow down; each time the ship enters a wave trough, the resistance decreases, and the propellers attempt to speed up. If speed governing were used in a propulsion application, the engine power command would cycle each time a wave is encountered. Such continuous variations in the engine power command is undesired and avoidable. The objective of keeping the average ship speed constant can be met by keeping the engine output power constant, recognizing that the short-term, periodic, wave-induced shaft speed changes have little net effect. Control systems that maintain constant engine output power are said to be "power governed." All U.S. Navy propulsion gas turbines are power governed. Some ships, however, have a speed-governed propulsion control system (which converts the bridge's shaft speed orders into engine power commands). The combination of the speed-governed propulsion control system and the power-governed engines makes the engines perform as if they were speed governed.

Operational control also involves two "miscellaneous" functions: bleed-air control, and module cooling control.

Most gas turbines are designed such that air can be "bled" from the compressor discharge, and used for purposes unrelated to the gas turbine. The gas turbine control system usually controls the position of the bleed air valve, opening and closing it as commanded by the engine operator. Some engines require the bleed air valve to be closed during engine start and, occasionally, during low-power operation. When required, the bleed control function automatically closes the bleed air valve, and releases the valve for operator or remote control when the engine power has reached a sustaining level.

Most gas turbines are mounted in modules, which are enclosures that contain the engine and some of its accessories. The modules are equipped with large electric fans, which cool the engine and module. Some turbine control systems automatically turn these fans on and off depending on the engine status, module temperature, engine power level, and other parameters.

4.4 Safety Control. Most gas turbine control systems have features that protect the engine from excessive speeds, excessive temperatures, and lack of lubrication. Some systems incorporate other features, including protection against excessive vibration and torque, and detection of engine flameout, engine icing, and module fires.

Gas turbine blades are attached to disks, which rotate with their spool shafts. If a spool overspeeds, the centrifugal force exerted by the blades can cause disks to rupture, and release the blades. The tremendous kinetic energy of the blades can drive them through bulkheads and decks, destroying equipment and killing personnel. Therefore, the prevention of overspeeding is one of the most important, if not the most important, function performed by a gas turbine control system. Usually, two forms of

overspeed protection are provided: limiting, and shutdown.

Overspeed limiting functions reduce the engine power command as required to avoid overspeed: as the danger of overspeed increases, the amount of the reduction increases. Overspeed limiting logic is normally designed to keep spool speeds below 103% of their rated value.

Overspeed shutdown functions simply shut the engine down (by closing the master fuel valve) upon detection of overspeed. Overspeed shutdown is essentially a backup feature, in that an overspeed shutdown should not occur unless both the normal engine governing *and* the overspeed limiting functions are unable to control the overspeed failure. Overspeed shutdown logic is normally designed to 110% of the rated rpm value. The overspeed shutdown function is of critical importance and is normally designed so that if the hardware used to implement the function fails, the engine automatically shuts down.

Overspeed limiting and shutdown functions are normally provided for each turbine spool. The functions protect the engine both from overspeeds caused by fuel control failures that result in excessive engine power commands, and from overspeeds caused by sudden drops in engine load (as may occur, for example, when a generator breaker trips on overload). Drop-load protection is particularly important on engines that have free turbines, since an unloaded free turbine can be driven into overspeed at low engine power. The tremendous acceleration associated with a suddenly unloaded free turbine requires the overspeed trip circuitry to have a response time, and the main fuel valve to have a closing time, that is measured in milliseconds.

Related to, but separate from, overspeed limiting is torque limiting. It is used only on engines with free turbines. The only couple between a free turbine and the remainder of the engine is the hot gas that flows from the gas generator turbine into the free turbine. This loose aerodynamic coupling allows high-power airflows to be generated at low free-turbine speeds. The combination of high airflow and low free-turbine speed results in a high free-turbine torque capability, since power increases as the airflow increases but a reduced rpm can result in a high torque, which can damage torque-sensitive components. In propulsion applications, damaging torques can occur during rapid accelerations from low ship speeds, and during high-speed turns. Some gas turbine control systems automatically reduce the engine power command as required to avoid excessive torques. The torque is not measured directly; it is computed from measurements of free-turbine speed and inlet pressure. The overtorque power command reduction is performed in the same way the overspeed power reduction is done: the greater the risk of overtorque, the greater the power command reduction.

High torques are not a problem when turbines are used to drive electrical generators, since the generators operate at relatively high rotational speeds and the power turbines are loaded only at rated speed. This avoids the excessive torques which could possibly occur during high-power, low-speed operation.

Overtemperature protection is the second most important safety function, after overspeed protection. In general, gas turbine output power is limited by materials considerations: operation at higher-than-rated power produces high internal temperatures, which induce rapid failure of key turbine components, such as the high-pressure turbine. The gas turbine control system, therefore, is required to protect the engine from excessive internal temperatures. As with overspeed, the protection can be by limiting, by shutdown, or both. As was noted earlier, temperature limiting is inherently incorporated into the fuel control function. Overtemperature shutdown is a separate safety feature which may share some hardware with the temperature-limiting function.

The combustors are the hottest components in a gas turbine. Direct measurements of combustion-zone temperatures, however, are generally not done, because combustion-zone temperatures would rapidly destroy the thermocouples used for temperature measurement. This could have highly undesired consequences in that a mechanical failure of a combustion-zone thermocouple would cause the airstream to sweep broken thermocouple parts through the remainder of the engine, possibly damaging expansion-section blades and vanes. Instead, the thermocouples are placed farther downstream in the engine, usually just ahead of the lowest-pressure turbine. That placement, while providing adequate thermocouple life, also "scales down" the temperature measurement; any overheated air must pass through most of the expansion section of the engine, cooling along the way, before it reaches the thermocouples. Also, since the airstream is twisted by spool rotation, the location of the hottest portion of the airstream, the portion coming from the combustors, will vary with spool speed. Airstream temperature, therefore, is measured by several thermocouples, spaced around the engine so that at least one of them is close to a "hot spot" at all times. Small thermocouples are used to measure post-combustion airstream temperatures so as to minimize their mass and increase the responsiveness of the post-combustion airstream temperature measurement.

Overtemperature shutdown functions only protect the turbine; it is the only component that is damaged in an overtemperature failure. Since a naval ship may be placed in a situation where it is necessary to sacrifice the engine in order to save the ship, many propulsion gas turbine control systems are equipped with a "battle override" feature, which disables the overtemperature shutdown function. The battle override is set by the engine operator during battle conditions and during maneuvering in restricted waters.

Most naval gas turbine control systems include a function that shuts the engine down when the engine lube-oil supply pressure is too low for safe engine operation. The function is also usually designed so that it can be disabled by a battle-override setting.

Most naval gas turbine control systems also include functions that protect the engine from operation with excessive engine vibration. The most complex implementation has both a power-reduction feature, which is energized in the event of high vibration level, and a shutdown feature, which is energized in the event of even higher vibration levels. However, some control systems only include a shutdown function. Vibration protection functions usually have a battle override.

Some gas turbine controls include a flameout protection function. This function automatically shuts the engine down if combustion fails during engine operation, thus preventing a damaging, explosive "hot relight." Since the presence of flame is not sensed directly, the function uses the engine temperature measurement to determine if a flameout has occurred. Flameout protection does not have a battle override.

Engine modules always include a fire-extinguishing system. All module-equipped turbine control systems include provisions for actuating the extinguishing system. Upon detection of a fire in the module, some systems automatically shut the engine down, stop the module cooling fan, close the module cooling vent, and then release the fire-extinguishing agent.

4.5 Monitoring. Marine gas turbines are generally derived from gas turbine designs that are used in aircraft applications. In aircraft applications, all control and monitoring functions are performed in the cockpit; there are no meters, indicating lights, or other displays on the engine itself. In ship applications, the same situation exists: control and monitoring are performed away from the engine, either from panels in the immediate vicinity of the engine module, from a console in the space where the engine is located, from a console in the Central Control Station or Enclosed Operating Station, or (for propulsion power control and monitoring only) from the Ship Control Console on the bridge. Although it is physically possible for an operator to be in the engine module or room, engine noise and heat make it impractical to station a watch stander there.

In order for an operator to effectively start, stop, and operate a gas turbine, monitoring information is needed that describes the current state of the engine. The required data are obtained from sensors mounted on the engine, which drive displays located at each engine control location. Some of these sensors, e.g., the engine speed sensors, are also used in the operational and safety control functions described above; others are used only for monitoring.

The set of monitoring instrumentation associated with each gas turbine design may be unique, but most gas turbine monitoring systems include at least the following:

• An indication of the speed of each spool, along with alarms to indicate spool overspeed and overspeed trip.

• An indication of the post-combustion airstream temperature, along with alarms to indicate unusually high temperatures and overtemperature trip.

• An indication of the status (opened-closed) of the engine main fuel valve.

• One or more indications of engine power (in addition to the spool speeds discussed above). The actual signals chosen vary with the engine; examples of parameters that are used include compressor discharge pressure, power-turbine inlet pressure, and engine power command.

- An indication of the temperature of the air entering the engine. Although this temperature can be used to indicate the risk of engine icing (see below), it is more important for its usefulness in calculating the available engine power. The performance of gas turbines is very sensitive to ambient conditions, particularly to the ambient (that is, engine inlet) air temperature.

- An indication of the engine lube-oil supply pressure, along with alarms to indicate unusually low-pressure and low-pressure trip.

- One or more indications of engine lube-oil temperature, along with alarms to indicate unusually high lube-oil temperatures. Temperature measurements are usually made at the engine lube-oil inlet, to detect an oil-cooler failure, and at the oil discharge from key bearings, to detect a bearing failure.

- One or more indications of engine vibration, plus alarms to indicate unusually high vibration, and high vibration trip (if provided).

- An indication of the status (opened-closed) of the bleed-air valve.

- An alarm to indicate detection of an engine module or engine room fire.

- An alarm to indicate that potential icing conditions exist at the engine inlet. The ingestion of ice or any other incompressible substance can cause severe "foreign object" damage inside an engine. The formation of ice on the air inlet could also block airflow into the engine causing the engine to be "choked." Icing can be prevented by mixing hot engine bleed air with the engine intake air. The activation of the icing alarm is an indication that the anti-icing system must be energized.

Section 5
Compressor Design

5.1 The Centrifugal Compressor. The compressor is one of the three basic components of a simple gas turbine, the other two being the turbine and the combustion system. The efficiency of the compressor is an important consideration, as is the airflow capacity and the required pressure ratio. Mechanical integrity and reliability are also very important.

A centrifugal type of compressor design is frequently used on smaller units. A centrifugal compressor is analogous to a centrifugal pump, and consists of an impeller with an axial inlet and essentially radial discharge, with passages formed by approximately radial vanes. The impeller is fitted in a casing, which forms the axial entry, with a diffuser at the exit to slow down the high radial velocity leaving the impeller and convert the kinetic energy into a pressure rise. The diffuser is followed by a collector to collect the flow from the exit of the diffuser and direct it to the combustor or, in some cases, another succeeding compressor stage. A typical centrifugal compressor is shown as the last stage of compression in the gas turbine shown in Fig. 24.

The performance of a centrifugal compressor can be presented in various ways, but the most common is a plot of the pressure ratio ($P_{\text{disch}}/P_{\text{inlet}}$) versus airflow rate at the inlet for various rotor tip speeds, such a plot is shown in Fig. 25. Lines of constant adiabatic efficiency can then be superimposed onto this plot. The compressor performance can also be presented as a dimensionless plot by expressing the flow as a percentage of design flow, and speed as a percentage of design speed.

The airflow rate at specified conditions may also be used as the flow parameter or, to render it more universal, it may be expressed as:

$$W(14.7/P_{\text{I}})\sqrt{(T_{\text{I}}/560)} \qquad (1)$$

where

W = weight flow, lb/sec
P_{I} = inlet pressure, psia
T_{I} = inlet temperature, deg R

An important characteristic of all continuous-flow compressors is the surge line. This is the line representing the relation between pressure rise and airflow above which operation is unstable. Under these conditions the airflow surges or pulsates, often with destructive effects on the unit. The limit must be determined by test and is nearly always shown on the plot of the compressor's characteristic performance curve.

While a centrifugal compressor is simple in construction, typically comprising only a single rotor in a casing, the achievement of high pressure ratios and high efficiency is not simple. Since the rotor or impeller imparts kinetic energy and some static pressure rise to the fluid, the diffuser and scroll, which are the major stationary parts, must convert the kinetic energy into potential energy in the form of a static pressure rise. Therefore, the higher the pressure ratio required, the more important is the design and corresponding performance of the diffuser.

The forward section of the impeller, usually called the inducer, may be separate from or integral with the rest of the impeller. In either case, the aerodynamic shape of the inducer is very important insofar as the overall efficiency, capacity, pressure rise, and surge performance of the machine are concerned.

a. Centrifugal impeller design. To achieve good performance and high efficiency, the detailed aerodynamic design of the impeller (rotor) and diffuser is necessarily based on experience and usually a long development period. Many approaches have been taken in designing centrifugal compressor impellers. The usual method employs a combination of velocity diagrams at the entrance

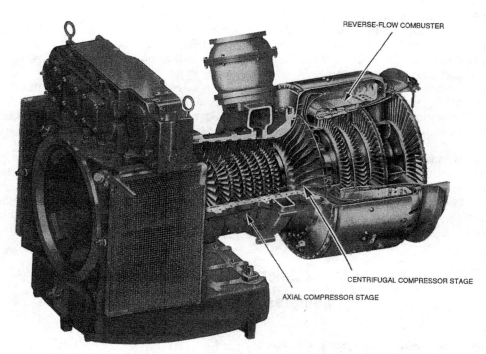

Fig. 24   Typical gas turbine with an axial/centrifugal compressor and reverse-flow combustor

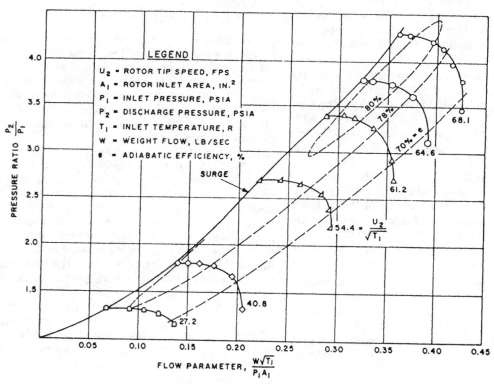

Fig. 25   Typical performance map for a centrifugal compressor

and exit which, with the principle of angular momentum, lead to an overall design. While refinements permit an evaluation of various loss factors, the method does not give detailed information regarding the nature of the flow in the impeller but does permit the evaluation of general

effects. It thereby indicates areas in which improvements can be anticipated, empirically or analytically.

The axial thrust generated by the pressure differential across the rotor disk can be reduced by balance holes to the back, or low-pressure side of the impeller, to create

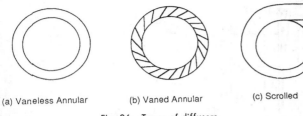

(a) Vaneless Annular   (b) Vaned Annular   (c) Scrolled

Fig. 26   Types of diffusers

an opposing pressure gradient. The thrust of a double-flow impeller, where the flow enters at the center of a compressor and symmetrically flows to each end, is inherently balanced.

The use of pre-swirl guide vanes, with rotation, reduces the inlet relative Mach number but also reduces the work input for a given tip speed. This design approach allows higher tip speeds, so higher pressure ratios can be obtained in this manner.

b. Diffuser design. The diffuser used with a centrifugal compressor plays an important part in the determination of the overall performance and efficiency of the compressor. However, the design of the diffusing section of the machine is usually dictated by space and configuration limitations as well as by aerodynamic considerations.

Diffusers can be classified as vaneless annular, vaned annular, or scroll as shown in Fig. 26. These fundamental types may be used separately or in combination. An annular diffuser may have a vaneless section preceding or following a vaned section, and a scroll may be used as a secondary diffuser after either of the annular types, or a combination.

Scroll diffusers are used, either alone or in combination with radial diffusers, where the air leaving the compressor must be collected in a duct and redirected. One arrangement is a single scroll wrapped around the compressor periphery. The air velocity may be diffused in the scroll, or after it, or the scroll may be combined with a vaned or vaneless diffuser which precedes it.

In comparing diffusers, a scroll diffuser will usually accomplish a given amount of diffusion in a smaller radius ratio than needed for a vaned or vaneless diffuser. They are apparently not as efficient as comparable vaned diffusers, but tests have shown that a 20-in. vaneless diffuser and scroll has approximately the same performance as a 34-in. vaneless diffuser and collector ring and better performance than a 17-in. vaned diffuser and collector ring [18].

One further point to emphasize is that the overall compressor performance depends not only upon the performance of the individual components (impeller and diffuser), but upon interactions between them. If they peak together, the compressor will have a higher maximum efficiency, but a sharper drop-off at off-design conditions than if they are slightly mismatched. Also, since the flow from the impeller is unsteady, with peak-to-peak variations of 10 to 30% of the mean velocity and corresponding changes in airflow direction, the performance of a vaned diffuser will be adversely affected by unsteady effects.

c. Rotor design and stress analysis. At the pressure ratios commonly used in gas turbines, the tip speed of the impeller in a centrifugal compressor must be high (1800 to 2100 fps), resulting in high rotor centrifugal stresses. Steel rotors are sometimes used, particularly for very high pressure ratios, or where a centrifugal compressor

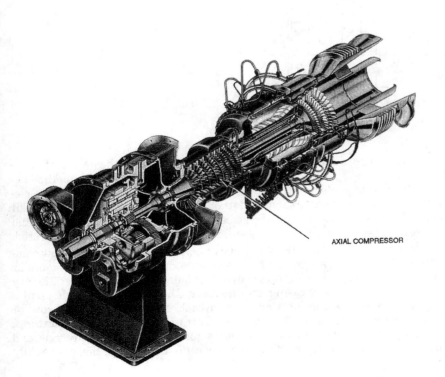

Fig. 27   Gas turbine with axial compressor

AXIAL COMPRESSOR

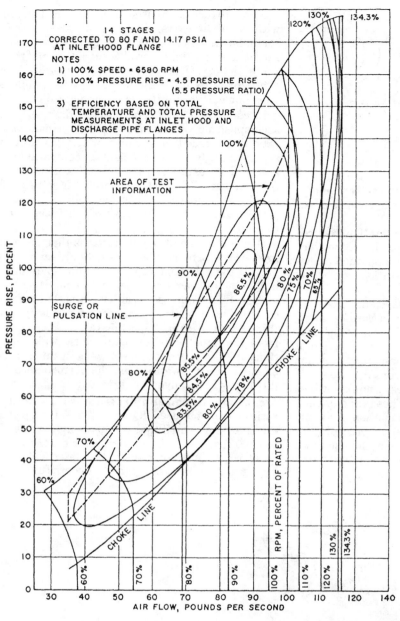

Fig. 28 Typical performance map for an axial-flow compressor

is used for the final stage of compression and discharge temperatures are high. Titanium is also an excellent material for compressor rotors due to its high strength/weight ratio. In many applications a high-strength aluminum alloy may be used, either forged and machined, cast in plaster molds, or cast by investment casting techniques.

Since stresses are high, their accurate calculation is important, and the rotor profile must be accurately determined to keep the stresses, particularly the radial stresses at the center, within limits.

One way to calculate rotor stresses is by the use of the Manson method, using a density correction factor to account for the radial load of the vanes [19,20]. The Manson method was developed for the analysis of symmetrical

disks and has commonly been used for the design of axial-flow compressors and turbines. However, this method neglects stress variations in an axial direction. Finite-element analysis methods are a more versatile means of calculating the stresses in all parts of a compressor and may be used regardless of how the stresses are generated.

5.2 The Axial-Flow Compressor. In an axial-flow compressor, the flow is in an axial direction, in the annular space occupied by the blades, between the stationary outer casing and the rotor. The flow path is made up of a series of stages, with each stage comprising a row of rotating blades and a row of stationary vanes (stators); each stage contributes to the overall pressure rise and corresponding pressure ratio. In addition, there is usually a set of stationary inlet guide vanes to direct the flow into the first rotor row, and frequently there are one or more rows of stationary exit guide vanes that serve to decelerate the discharge flow and convert the velocity head leaving the last stage to a static pressure rise. The number of stages varies from about 6 to as many as 20 or more, depending upon the overall pressure ratio and aerodynamic characteristics of the blading. Typical axial-flow compressors are shown in Figs. 7, 8 and 27. A performance map for a typical axial-flow compressor is shown by Fig. 28.

a. Blading design. The aerodynamic design of an axial-flow compressor is based fundamentally upon velocity diagrams for the rotors and stators coupled with the airfoil characteristics or, commonly, the cascade performance of the individual blade rows.

An important consideration in the design of axial-compressor blading is that of blade loading. Early data on low-speed cascades [21] were extended by later National Advisory Committee for Aeronautics (NACA) programs, including information on blade sections operating at higher Mach numbers and higher blade-setting angles [22–24]. From the early data, Lieblein and others developed a diffusion parameter that is frequently used as a blade loading limit [25].

In older gas turbines the rotor and stator airfoil sections were commonly a NACA-65 section or the English C-4 or C-7 section. The length of the blades is set to give the design axial velocity at a particular location. Later designs utilize diffusion controlled airfoil shapes to further increase the aerodynamic performance of the compressor.

A compressor is designed with a decreasing annulus area with the maximum flow path area at the inlet where the air density is lowest to a minimum flow path area at the discharge where the air density is highest. To optimize performance, compressor airfoil paths are often designed with a constant mean diameter configuration. This is accomplished by decreasing the tip diameter and increasing the root diameter of succeeding stages from the inlet end to the discharge end. However, some older designs had constant blade-tip diameter or constant hub diameter configurations.

The thickness of the airfoil sections is chosen to meet vibration criteria while maintaining good aerodynamic performance. The blade chord must then be sufficient to maintain the blade bending stresses at an acceptable level to insure freedom from fatigue failures. Sample blades are commonly constructed and their natural frequencies checked for all the lower modes and, if necessary, tuned by changing the thickness to avoid resonances at running speeds as shown by a Campbell diagram [26]. Even this cannot give positive assurance that all resonances are avoided, particularly in variable-speed units. Even constant-speed units must often pass through resonant speeds when starting up and shutting down.

The allowable blade bending stress chosen depends upon the material and experience. Generally the use of steel, titanium, or high-strength aluminum is necessary to satisfy requirements for strength, corrosion resistance, and foreign-object damage resistance.

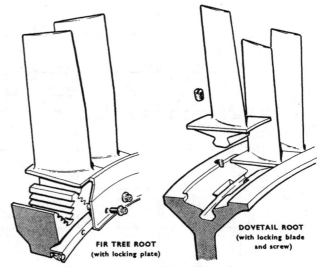

Fig. 29   Methods of securing blades to rotor

A blade is commonly attached to the rotor by an enlarged section or root in the form of a dovetail or fir tree as shown in Fig. 29. Other forms of attachment such as a bulb root, cylindrical root, and bolted attachment have also been used. The dovetail and fir-tree sections must be accurately manufactured, and good fits in the rotor slots are essential; axial rotor slots are commonly broached, whereas circumferential grooves are machined in a lathe.

The stresses in the dovetail comprise shearing stresses and crushing or compressive stresses. The total load on the dovetail is determined by adding the centrifugal force due to the blade and that of the dovetail itself. The shear and crushing stresses are then calculated by applying this total load to the geometrically determined areas. The dovetail neck shear stress at maximum speed should be less than 30% of the minimum yield stress of the material.

In the layout of the blade sections, the sections are usually radially stacked so that a line through the centroids of the sections is a straight line that is either radial or with a slight inclination to compensate for the aerodynamic bending moments.

b. Rotor design. Many compressors have a rotor that is made up of a series of individual disks. The individual disks are held together by a central bolt or series of bolts at an intermediate radius. In either case, the bolts are prestressed at assembly so that the total bolt load is sufficient to keep the disks from separating under the highest bending moment that is likely to be imposed on the rotor.

The individual disks are usually contoured to give an approximately constant radial stress in the wheel. Various methods may be used for the calculation of the wheel stresses, including the centrifugal loading of the blades, but the one most commonly used is finite-element analysis. The average tangential stress is of particular significance, since, for the ductile materials usually chosen, it is generally considered that the wheel bursting speed is reached when this stress exceeds a value in the range of 85 to 100% of the ultimate strength of material. Typical stress distribution curves for a compressor wheel are

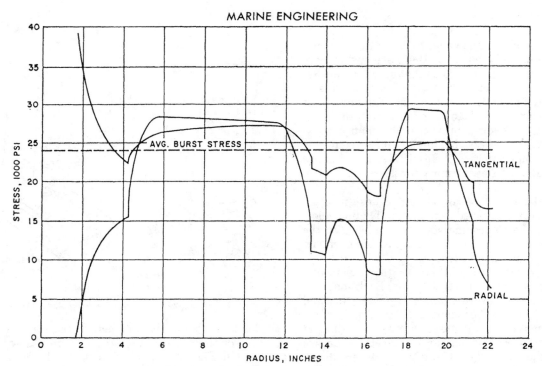

Fig. 30   Stress distribution in a typical axial-flow compressor disk

given in Fig. 30. This is a wheel with a central hole, and the tangential stress at the bore is quite high, approaching twice the stress in the center of a solid disk.

Compressors have been built with drum rotors, similar to some steam turbine rotors, and solid rotors (blisks) have been designed, although actual applications are rare. Stresses in drum rotors are usually based on unsupported ring theory, and are appreciably higher than for individual disks for the same peripheral speeds. Consequently, they are used only in compressors with low blade speeds and a large number of stages for a given overall pressure ratio. Such designs can be classed as very conservative.

c. Stator design.   Stator blades are not subject to the centrifugal stress that rotor blades experience; therefore, their mechanical design and aerodynamic design are not as limited. Nevertheless, bending stresses due to aerodynamic loading must be calculated and kept to conservative values to allow for unknown vibratory stresses. Blade vibration modes and frequencies must be estimated for a preliminary design, and checked for sample blades, so as to avoid lower-order resonances within the normal running speed range. There is, of course, no stiffening effect due to centrifugal forces as there is for rotor blades.

The method of attachment is usually more simple than that of rotor blades, frequently using a T-slot root or tip attachment. Blades may be mounted either directly in slots that are machined in the casing, mounted in separate blade rings by T-slot attachments, or brazed into the rings. The rings are made in two or more segments, which are then slid into grooves machined in the casings.

Some compressor designs use variable-angle stator blades in one or more rows so that the blade angles can be adjusted to match the operating conditions, particularly during starting. In such cases, the blades have cylindrical

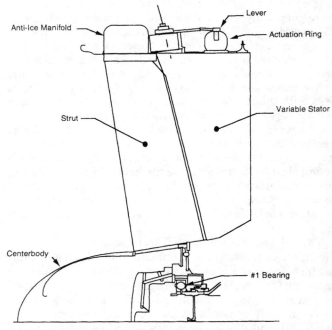

Fig. 31   Variable-angle stator blades with an external control mechanism

shanks that extend radially outward through holes in a casing. Levers are attached to the outer ends, which are linked together, so that all stators in one or more rows can be adjusted simultaneously by an external control mechanism as shown in Fig. 31.

d. Casing design.   The compressor casing must be designed to hold the stator blades or stator blade rings and contain the pressures developed in the various stages. The casing must also be designed to contain fragments of the engine or foreign objects in the event of mechanical failures, such as blades breaking off. Smaller machines are frequently made with a one-piece casing, the rotor and stator blade rings being assembled external to the casing and then assembled endwise into the casing as a unit. The compressor casing is then assembled to an air inlet casing and a discharge diffuser that locate and restrain the blade rings endwise. This design has the advantage of freedom from distortion of the casing, due to its symmetry, so that closer tip clearances over the blading can be maintained in operation. It is also somewhat lighter.

Larger units commonly have the compressor casing split on the horizontal centerline, with each half containing half the stator blades. The two halves are bolted together with a horizontal flange. The horizontal flanges must be designed to transmit the circumferential hoop stress, due to the internal pressure, across the joint without leaking. The joint itself is always made with a metal-to-metal contact, and the joint is carefully finished to avoid leakage. The design of the flange and bolting is usually based on experiments as well as stress analyses [27].

5.3 Other Compressor Types and Combinations. Other types of compressors have, from time to time, been proposed for application to a gas turbine cycle. In all cases they incorporate a combination of axial and centrifugal stages with the final stage, or stages, being of the centrifugal type.

Section 6
Turbine Design and Construction

6.1 Types of Turbines. The turbine converts the thermal energy in the hot, high-pressure gas stream into mechanical work. Part of the work output produced goes into driving the compressor (or compressors), and the remainder is available as useful shaft work. The turbine may be of only one stage, or many stages. A stage is a combination of a single stationary and a single rotating element that is arranged to absorb the energy released in a single expansion between two pressure levels. The stages all may be on a single shaft, or on two or more shafts.

Turbines are generally of the axial-flow type, although on occasions engines have used radial-flow turbines. Gas turbines generally have reaction type axial-flow stages unlike the axial-flow impulse type of the steam turbine. The axial-flow stages are designed with a varying reaction from the root to the tip to give the optimum airfoil shape.

Axial-flow turbine design practice can utilize a small number of highly loaded stages or a larger number of lightly loaded stages. Turbines with highly loaded stages generally have one stage versus two or more stages for a lightly loaded stage design. Turbines with lightly loaded stages tend to be more efficient than those with highly loaded stages; however sophisticated design techniques, such as three-dimensional flow analysis, have led to improved blading and nozzle designs that minimize shock losses due to high velocities and have led to efficiencies in highly loaded turbines approximately matching efficiencies in lightly loaded turbines. A highly loaded stage turbine is the preferred design because of its size and weight advantage over a comparable turbine with lightly loaded stages. Highly loaded stages imply high velocities (often above sonic velocity) and these designs emphasize the aerodynamic details of the nozzle and blade profiles to minimize shock and endwall losses.

6.2 Aerodynamic Design. Gas turbines can be designed with either of two types of turbines, radial flow or axial flow. Axial-flow turbines are commonly used for

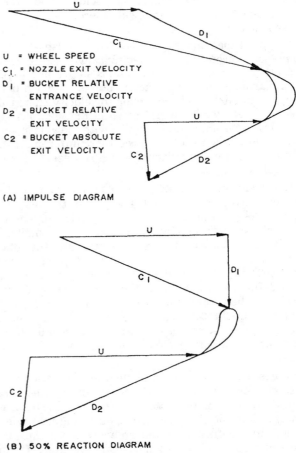

U = WHEEL SPEED
C_1 = NOZZLE EXIT VELOCITY
D_1 = BUCKET RELATIVE ENTRANCE VELOCITY
D_2 = BUCKET RELATIVE EXIT VELOCITY
C_2 = BUCKET ABSOLUTE EXIT VELOCITY

(A) IMPULSE DIAGRAM

(B) 50% REACTION DIAGRAM

Fig. 32 Axial-flow turbine velocity diagrams

larger engines specifically because of their higher adiabatic efficiency at higher specific speeds. The selection of an axial-flow turbine for larger engines is dictated also by

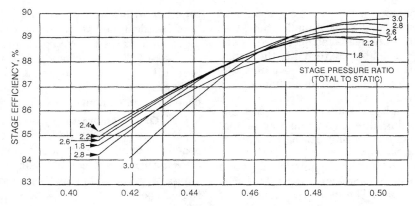

Fig. 33 Axial-flow turbine stage reversible adiabatic efficiency

influences of overall size (i.e., Reynolds number, weight, manufacturing considerations, and detail design). In larger-diameter, higher flow-rate machines, radial-flow turbines weigh too much to be considered in aviation applications. Radial-flow turbines are seldom used for the larger marine engines and will, therefore, not be discussed further.

A typical diagram for a so-called impulse stage (i.e., no pressure drop or corresponding conversion to velocity energy in the rotor blade passage) is shown in Fig. 32(a). A similar diagram for a reaction stage, in which part of the pressure drop and conversion to velocity energy does occur in the rotor blade, is shown in Fig. 32(b). A stage will vary in the amount of reaction with the velocity ratio, U/C_1, that is, the ratio of the rotor circumferential velocity to the nozzle exit velocity, which represents the total available energy to the stage.

The prediction of the stage and overall turbine performance can be considered an integral part of the design. For a given design, the flow capacity is predetermined from the design calculations within rather narrow limits. The efficiency varies with the velocity ratio and the total-to-static pressure ratio as shown by the typical curves of Fig. 33. The reversible adiabatic efficiency as used in Fig. 33 is defined as the ratio of the work developed by the turbine to the theoretical work available when expanding from the initial total pressure to the final total pressure. The total pressure equals the static pressure plus the total head corresponding to the absolute gas velocity leaving the stage.

Stages designed with a high degree of reaction have a flatter efficiency curve when plotted versus velocity ratio than those of an impulse design. The actual level of efficiency is determined by the details of a design. Nozzle and rotor blade profiles, root and tip endwall overlaps, rotor blade tip clearance, etc., each have an effect on efficiency.

There are probably as many techniques that can be used to predict stage performance as there are turbine manufacturers. Irrespective of the method used, however, the following general types of losses characteristic of turbines should be evaluated:

1. Friction losses on all gas-path surfaces. This may be assessed by using an "effective" drag or surface friction coefficient.
2. Flow separation and mixing losses that result from excessive trailing edge thickness, diffusing interstage passages, and divergence of endwalls in vanes and blades.
3. Leakage losses such as nozzle seal leakages, disk leakage into flowpath, and blade tip-clearance leakage.
4. Shock and attendant high Mach number losses that may occur from the nozzle throat up to the blade entrance.

The following possible sources of losses probably cannot be evaluated directly.

1. Non-steady-state effects that may produce a nozzle/blade "interference" loss.
2. Non-uniformity of nozzle exit stream within the nozzle pitch.
3. Significant flow separation from nozzle or blade profiles due to either a poor profile design or an excessively low solidity (wide airfoil spacing).

6.3 Nozzle Design and Construction. The turbine nozzles form the stationary elements in the flow path of the turbine. The velocity of the stream is accelerated in the nozzle by the release of available energy.

Nozzles are usually constructed using castings to form the airfoil shape with the endwall segments on either end.

The material of the nozzles can be any of the commonly used high-temperatures alloys, depending upon the design inlet temperature and, to some extent, on the fuel used. For fabricated construction various Inconels such as 706 or 718 are suitable, while for cast construction X-40, Rene' 80 and Rene' 77 or similar alloys are commonly used [28].

In an axial-flow turbine, the nozzles are formed into an annular ring and serve to turn the flow from an essentially axial direction to that required by the blade entrance conditions. The actual physical arrangement of the nozzles can take many forms. A nozzle must be supported within the turbine casing and held against the pressure forces due to the pressure drop across the nozzle; at the same time it must be free to expand with temperature changes. In large units operating at high inlet temperatures, the nozzle assembly may expand over an inch on the diameter between cold conditions and operating temperature. To allow for this expansion, and still keep the nozzles concentric with the rotor axis and with a minimum amount of leakage, requires skillful detail design. Nozzles of one-piece cast or welded construction require slots or cuts in the inner or outer wall to allow for differential expansion. For best performance these slots should be sealed against leakage.

In the case of multistage machines with rotors made up of individual wheels, the intermediate nozzles usually are built with a diaphragm, either integral with the inner wall or mechanically attached to allow for differential expansion. The diaphragm extends to the hub of the wheel and carries some form of labyrinth seal to control the leakage of gas along the hub or shaft.

In the case of two-shaft machines, which are usually constructed with the gas generator turbine and the power turbine separate but with a common axis of rotation, the nozzle for the power turbine can be of either fixed or variable area. The fixed-area nozzle is constructed the same as other intermediate nozzles, except that the diaphragm portion is usually solid as it normally does not encircle a shaft. For greater operating flexibility, the power turbine can be built with variable-angle nozzles in the first and perhaps the second stage to give a variable area; in this way the energy distribution and hence the relative speeds between the gas-generator turbine and the power turbine can be varied by the control system. The advantages of this arrangement are discussed in Section 1.3.

The selection of nozzle materials depends upon the operating temperature, and consideration must be given to the fuel used, since some elements in the fuel can cause rapid corrosion or intergranular attack of otherwise suitable materials.

For long-term operation at high turbine-inlet temperatures (i.e., 2000 to 3000 F or more) some form of nozzle cooling is required to keep metal temperatures below the peak gas temperatures so as to prevent rapid deterioration of nozzle materials. The simplest and most commonly used method is air cooling by means of internal passages that are cast into the nozzles. A typical arrangement is one in which high-pressure cooling air is passed through a fabricated core, with a cast shell forming the partitions, and is bled into the gas stream through holes in the leading edge and near the trailing edge to keep the thin edge cool, as shown in Fig. 34.

6.4 Rotor Design. In an axial-flow turbine, the blades are usually attached to the wheel or rotor mechanically by what is commonly known as a "fir tree" or "dovetail" attachment. But sometimes on smaller units the blades are attached to the rotor by welding, or even cast integrally with the wheel.

Good blade design is a difficult and complex process. Not only must the aerodynamic conditions be satisfied as to entrance and exit angles and passage shapes, but

equally important the airfoil section must have centrifugal and bending stresses below the allowable limits for the material and service involved. There must also be no vibration modes, of any significance, that are resonant to stimuli in the operating speed range. The blade design is by necessity a process of design compromises.

The final design of a blade and its attachment entails plotting the calculated or measured vibration frequencies on a Campbell diagram to determine possible resonances [26]. It is generally considered advisable to keep the three fundamental frequencies from coinciding with passing frequencies of combustion chambers, nozzles, struts, and rotors, etc. above 50% speed. First-to-third-order frequencies are also to be avoided at running speed. Even at best all stimuli cannot be avoided, therefore some manufacturers incorporate tip dampening in long blades or special root dampening devices in short ones. Interlocking, integral shrouds, which are practical with precision cast blades, can also be used to control vibration.

With the airfoil sections determined, their properties at various radii can be calculated (i.e., section areas, section moduli, and centers of gravity). The areas can then be used to calculate the centrifugal stresses along the airfoil section, and the section moduli in combination with the gas bending forces are used to determine the bending stresses. Usually the sections are "stacked" with respect to their centers of gravity so that the centrifugal forces create a moment that offsets the gas bending moment and reduces the stresses in the leading and trailing edges.

The root attachment, frequently called the dovetail or fir tree, must carry the centrifugal loading of the airfoil section plus that of the platform and the dovetail or fir tree itself into the wheel. This must be accomplished without exceeding the allowable stresses in the dovetail or fir tree or the associated wheel rim sections. Frequently, the blade material and the wheel material are not the same, so that the allowable stresses in the parts differ even though the temperatures are generally assumed the same. Stresses in both parts must, therefore, be checked. An extended or "long-shank" blade design may be used to reduce the temperature at the dovetail or fir tree by the temperature drop in the shank.

To attain the required aerodynamic shape within allowable stress limits, temperature limits, and vibration characteristics, the airfoils are usually cast with hollow internal cooling passages. These passages may be used to direct cooling flow through multi-pass cooling schemes and out through leading edge and trailing edge cooling holes as shown in Fig. 35.

There is a wide range of blade materials available, the choice depends upon the temperature level and consideration for the effect of the fuel used. Both nickel-based and cobalt-based alloys are used for both cast and forged blades. As turbine inlet temperatures and stresses increase, high-strength nickel-based alloys such as Rene' 80 (R80) are used as are equiaxed grain structures and directionally solidified materials, for example, DSR80. Diffused aluminum coatings are often used to increase the hot corrosion life of nickel-based alloys. For lower-temperature applications, as in last stages, Rene' 77,

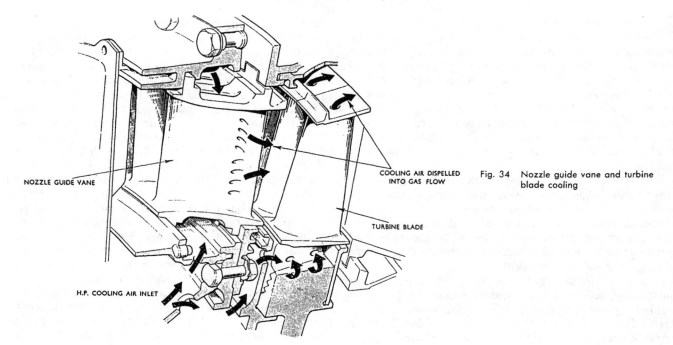

NOZZLE GUIDE VANE

COOLING AIR DISPELLED INTO GAS FLOW

TURBINE BLADE

H.P. COOLING AIR INLET

Fig. 34   Nozzle guide vane and turbine blade cooling

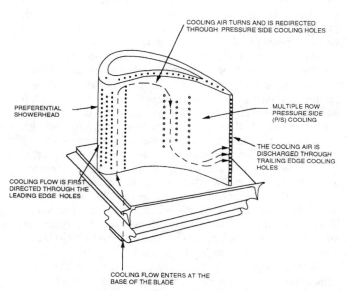

COOLING AIR TURNS AND IS REDIRECTED THROUGH PRESSURE SIDE COOLING HOLES

PREFERENTIAL SHOWERHEAD

MULTIPLE ROW PRESSURE SIDE (P/S) COOLING

THE COOLING AIR IS DISCHARGED THROUGH TRAILING EDGE COOLING HOLES

COOLING FLOW IS FIRST DIRECTED THROUGH THE LEADING EDGE HOLES

COOLING FLOW ENTERS AT THE BASE OF THE BLADE

Fig. 35   Multi-pass cooling scheme of airfoil

Rene' 41, and Waspalloy materials have been very successful within their temperature limitations.

Most gas turbine blades are unshrouded in the gas-generator section. Leakage past the tips of the blades is limited by maintaining closely controlled clearances between the tips of the blades and the stationary endwall in the turbine casing. Power turbines, which operate at lower temperatures, often use shrouded blades to reduce tip leakage and to provide vibration damping. Several radial knife-edge seals may be added on the cover of a shrouded blade to improve the efficiency of the stage.

Smaller rotors are usually held together by a single central bolt, with the axial location of the individual wheels maintained by dowels or couplings. With large rotors, a single central bolt becomes impractical; therefore, a multiplicity of smaller bolts at an intermediate radius is usually used. In either case, the bolt or bolts are stressed to a point where the resulting preload will positively prevent the wheels from separating under all normal operating conditions. Dynamically, the rotor then acts as a solid or one-piece rotor, and calculations such as that for critical speed can be made on the basis of the section inertia of the rotor at the contacting points, or lands, between the wheels. The tensile load in the bolts is usually checked at assembly by measuring the actual stretch or elongation of the bolts as torque is applied to the nuts. For very large rotors, the bolts may be stretched hydraulically, the nuts seated, and the elongation checked after the hydraulic load is released.

The centrifugal load of the complete blade (airfoil, platform, shank, and dovetail), plus the interrupted portion of the wheel rim between the dovetail slots, is carried by the wheel disk. This loading can be represented as a distributed load around the wheel circumference. The stresses in the wheel disk are usually calculated by the finite-element analysis method. However, in turbine wheels the thermal stresses are of considerably greater magnitude and must be accurately assessed. Methods of calculating heat transfer from a rotating disk are given by Kreith and Taylor [29] and a method for calculating stress in disks subjected to creep is given by Wahl [30]. It should be pointed out that for turbines used in marine service, stress levels should be based on long-time operation and creep must be considered.

The large thermal growth of the nozzles due to their high temperature and the thermal and elastic growth of the rotor wheels make it difficult to ensure the desired overlap of the nozzles and blades at the root of the gas path under all conditions. A uniform overlap around the circumference is particularly difficult to maintain.

The critical speeds of the rotors must be carefully calculated to avoid resonance at normal running speeds. These calculations are usually based on the traditional Rayleigh method, taking bearing flexibilities into account. While exact bearing flexibility is rarely known, a range of values usually can be estimated from experience so that meaningful values of actual running critical speeds can be determined. Critical speeds calculated on the assumption of rigid bearings are inaccurate and of little significance.

Section 7
Combustion Systems

7.1 Combustion Chamber Configurations. The term "gas turbine" is a contraction of the more explicit term "combustion gas turbine," and the process of combustion is an important part of the gas turbine system. A gas turbine is basically an internal-combustion engine as the combustion, which releases the energy in the cycle, takes place inside the machine. In gas turbines, combustion is a continuous process, unlike the intermittent process that takes place in a reciprocating engine. The combustion also takes place in a very confined space, in contrast to most other continuous-process combustion systems, such as the furnace of a conventional boiler. The resulting continuous high rates of heat release make good combustion and cooling of the combustion chamber major design considerations. These areas of concern have been successfully dealt with in a number of ways.

Combustion chamber design is as much an art as science and, perhaps for that reason, a wide variety of configurations is used for gas turbines. In some cases the design is dictated more by the experience and practice of the manufacturer than the inherent advantages of a particular design.

Combustion systems are generally built as an integral part of the unit and are combined structurally with the compressor and turbine. Two basic types of combustion systems are used: the annular, as illustrated by Fig. 36, and the can-annular, as illustrated by Fig. 7. The annular combustion system consists of liners in the form of an inner and outer cylinder, with a half-doughnut-shaped dome at the inlet, and a downstream end that merges directly into the first-stage turbine nozzle. Combustion takes place in the annular space between the two liners. The can-annular combustion system consists of a number of individual liners in the form of cylinders or "cans" that are located in an annular space between an inner and outer combustor casing. Each of the cans has a dome end through which air is introduced and flows along the liner, gradually passing through the mixing and cooling holes, until, at the turbine end, it is all inside the liner from which it passes out as hot gas to the turbine nozzle.

Another arrangement of the foregoing combustion systems is a "reverse flow" combustor where the direction of airflow through the combustor can vary from greater than 90 deg relative to the centerline of the engine to a combined total turn angle of 270 deg. This is accomplished by the physical design of the combustion chamber and liner. This arrangement is used to more readily accept air from a recuperator or to limit the overall length of the engine without affecting the efficient burning and mixing of the gases within the burner. An example of a reverse-flow combustor is shown by Fig. 24.

7.2 Design Parameters. While the design of a combustion system depends largely upon a combination of skill and trial-and-error, certain design parameters have been established that can form the basis for a design. However, these parameters can only be given as ranges in which successful designs of combustion systems fall.

The most basic design criteria are the heat release in terms of Btu/hr-ft²-atm or Btu/hr-ft³-atm. Sometimes comparison is made without correcting for the operating pressure in atmospheres. Usually, the pressure is taken as atmospheres to the 1.8 power, that is:

$$I = \text{Btu/hr-ft}^3\text{-atm}^{1.8} \qquad (5)$$

or

$$I' = \text{Btu/hr-ft}^2\text{-atm}^{1.8} \qquad (6)$$

Values per cubic foot-atmosphere range from 1 to 10 × 10^6 Btu/hr while, if the pressure is taken as atm$^{1.8}$, values

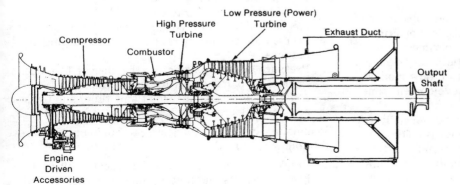

Fig. 36 Gas turbine with an annular combustor

Compressor

High Pressure Turbine

Low Pressure (Power) Turbine

Combustor

Exhaust Duct

Output Shaft

Engine Driven Accessories

of the heat release, I, range from about 0.5, or less, to 5 $\times 10^6$ Btu/hr. Typical values per square foot-atmosphere of flow area inside the liner are approximately 4×10^6 Btu/hr.

In small chambers, the radiant heat flux is in the range of 80,000 Btu/hr-ft^2, which increases with the size of the chamber due to the greater thickness of the body of radiating flame. It also increases with the luminosity of the flame. Liquid fuels, in general, give a more luminous flame than gaseous fuels; the heavy oils, particularly the residuals known as Bunker C, produce highly luminous flames. The result is that large single combustors intended to burn heavy oils are designed for rather low rates of heat release compared to the small individual chambers used in engines intended to burn lighter fuels.

Most combustion chambers are designed so that a small amount of primary air is admitted, usually with some swirl velocity, at the upstream end of the chamber in what is usually called the dome. In the dome the air is mixed with the fuel, which then burns nearly stoichiometrically at temperatures of 3,000 to 3,500 F. The quantity of air admitted in the primary zone may give up to 20 to 30% excess air at that point. Of the remaining air, part is used for cooling the liner (perhaps 15 to 30% of the total air) and the balance is injected through mixing or dilution holes to thoroughly mix with the products of combustion and reduce the average temperature to the allowable turbine inlet temperature. The overall fuel/air ratio ranges from perhaps 0.005 to 0.020 depending upon the operating cycle and the load on the unit.

Both the dilution air and the cooling air perform important functions; one is to give as near as possible the desired temperature profile at the turbine inlet. Another is to keep the metal walls of the liner below the allowable temperature limits for the material. It is the balance of these two flows, with proper mixing in the primary zone, that results in a successful combustion chamber design. Advancements that have been made in computational fluid dynamics modeling have greatly improved the design techniques for combustors.

Flow velocities, based on the combustor cross section, at the combustor inlet temperature and pressure, range from 15 to 20 fps to 100 to 120 fps in some small compact designs. Annulus velocities (around the liner) run from 30 to 250 fps or higher, with the velocities held as low as the space allows to keep the flow distribution irregularities into the liner to a minimum.

With the complicated aerodynamic conditions in the combustion system, trial-and-error methods, sometimes involving rather long and expensive experimental programs, are required to develop a fully satisfactory design. The use of flow models, particularly water flow models, to help visualize the flow reduces the development time considerably.

Combustion efficiency in a gas turbine is normally so high (97% or more) that the actual efficiency is difficult to measure. Even considerable smoke and carbon formation, otherwise very undesirable traits in a combustion chamber, do not reduce the combustion efficiency appreciably.

7.3 Design Objectives. The physical configuration of representative combustion systems and the more important parameters used in their design have been described, but specific design objectives have not been discussed. In addition to the obvious primary objective of burning the required amount of fuel, the objectives in combustion chamber design, not necessarily in order of importance, may be listed as:

1. Effective release of chemical energy by the combustion of fuel within highly confined spaces.
2. Stable operation over a wide range of fuel/air ratios.
3. Minimum pressure drop compatible with the other requirements.
4. A controlled temperature distribution at the turbine inlet.
5. Short flame length to prevent discharging flames into the turbine.
6. Clean burning of fuel, with negligible smoke and no carbon formation under all operating conditions.
7. Dependable, straightforward ignition to give reliable starting and restart capability.
8. Durable components with low maintenance.

The accomplishment of these objectives entails compromises. Smoke-free combustion is generally in conflict with lean-limit stability. Similarly, a low pressure loss and short flame length are difficult to attain simultaneously.

The aerodynamic processes in a combustion system present a most difficult set of design problems. The provisions for the interaction of fuel and air by recirculation in the primary zone to provide the correct spatial arrangement of the reactants, the necessary flame stability, the necessary mixing and dilution downstream to give the required temperature distribution at the combustor exit, the maintenance of metal parts at proper operating temperatures, and the low pressure drop requirements combine to present a complex design challenge.

7.4 Mechanical Details and Construction. Structurally, a gas turbine combustion system is rather simple, but the design is complicated by the varied and rather severe conditions it must meet. Typically the design must provide for:

1. Flame temperatures of 3,000 F or more.
2. Cyclic temperature variations.
3. Metal temperatures during operation of 1,300 to 1,800 F with peaks to 2,050 F.
4. Metal temperature gradients of 540 F/in.
5. Mechanical or aerodynamic exciting frequencies over a wide frequency range and with varying amplitudes.
6. Pressure differentials across the liner of 1 to 5 psi.
7. Thermal expansion relative to the casing.
8. Axial forces on the liner (or flame tube) due to the pressure differential.

Combustion system components can fail in one or more of the following ways:

1. Cracks, particularly at holes or discontinuities.
2. Fretting at fits and mating joints between parts.

Fig. 37 Annular combustor liner with plunged dilution holes (inner liner half)

3. Structural failure due to pressure forces.
4. Distortion and buckling due to temperature gradients.
5. Overheating resulting in local metal failure.
6. High-temperature corrosion.

Pressure loads that tend to collapse the liner and create axial forces due to the pressure difference over the dome area are particularly important in the case of annular designs. The critical pressure difference that the liner can withstand depends upon the proportions (length/diameter and thickness/diameter), the degree of support and stiffening, manufacturing eccentricities, discontinuities, and Young's modulus at the operating temperatures. Collapse is usually due to yielding at local eccentricities rather than elastic instability.

In the early days of gas turbines, very thick liner material was sometimes used in an attempt to obtain longer life and reliability, but experience has shown that thinner liner materials are more heat resistant, and except in large single combustion chambers, thicknesses of more than 0.050 in. to 0.060 in. are rarely required.

The liner must be supported in the combustor casing while being allowed to expand freely. Bearing pressures on suspension pins or fittings should normally be limited to 250–500 psi. Sometimes surface treatment of the material is used to reduce fretting at points of contact.

Thermal stresses in the liner cause cracks, which usually originate at dilution or secondary air holes. The dilution air holes reduce the local thermal stresses and increase the fatigue life of the liner. Plunged holes as shown in Fig. 37 are preferred in that they have higher flow coefficients than simple sharp-edged holes.

Liner materials can be any of a number of high-tempera-

ture, corrosion-resisting alloys. Typical materials include Hastalloy X, HS188, and INCO 718.

Occasionally, combustor parts are attacked by corrosion due to constituents of the fuel. Attack by hydrogen sulfide (H_2S) is particularly virulent, but is likely only in the primary zone, or over-rich pockets. H_2S corrosion is negligible with metal temperatures below 1100 F or above 2000 F. Attack may also occur from vanadium pentoxide, which may be formed from the vanadium that is frequently present in residual fuels.

7.5 Fuel Nozzles. The fuel nozzles, or fuel injectors, must introduce the fuel into the combustion chamber over the entire range from light-off to maximum load in a way that is compatible with the basic objectives previously listed. The fuel nozzles can have a marked effect on the performance of combustion systems.

Gas turbines are well suited for the propulsion of liquefied natural gas carriers as they can readily burn the gas boil-off [31]; however, aside from this application, gaseous fuels are not available for most marine applications. Liquid fuels commonly used in marine installations are DFM (Diesel Fuel Marine), JP-4, JP-5, distillate fuels, and residual fuels.

The nozzle must atomize the liquid fuel into droplets small enough to insure complete combustion of the fuel in the residence time available, provide a spray pattern that will insure adequate mixing of the fuel in the residence time available, and provide an intimate mixing of fuel with the available air under the full- and part-load aerodynamic conditions existing in the primary zone. The nozzle spray must not strike the walls of the combustion chamber, and must not cause excessive temperature variations.

Five types of liquid-fuel nozzles have been applied in gas turbine combustors. Each has advantages and disadvantages as follows:

1. Pressure-Atomizing Nozzles. There are two types of pressure-atomizing nozzles:
   (a) Simplex (or single-orifice) nozzles. The simplest type but not suitable for the wide flow range usually required.
   (b) Duplex (or dual-orifice) nozzles. With either internal or external flow dividers, these give a much wider operating range and are usually satisfactory for the lower-viscosity fuels.

2. Air-Blast (Air-Atomizing) Nozzles. These nozzles utilize the combustion-chamber pressure drop to create an airstream that is used to assist in atomizing the fuel. The airflow around the nozzle body breaks up the fuel spray, so the fuel distribution is determined mainly by the airflow pattern. Combustion is characterized by a blue flame of low luminosity, cool walls, and minimum smoke. It has, however, poor "lean blowout" and poor atomization at starting.

3. Air-Assist (Air-Atomizing) Nozzles. These are single-orifice nozzles, with small quantities of air introduced internally via separate passages from a separate external compressor. At low fuel flows and

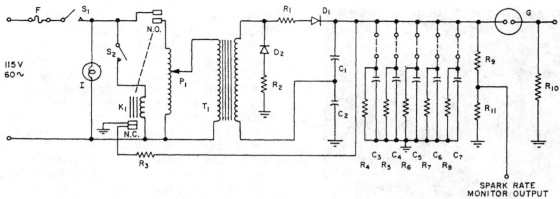

INPUT VOLTAGE:	110–120 VOLTS–50–60 CYCLES
INPUT CURRENT:	7 AMPS, MAX.
STORED ENERGY RANGE:	1–20 JOULES, INCREMENTS OF 1 JOULE
SPARK RATE RANGE:	0–100 SPARKS/SEC AT 1 JOULE
	0–5 SPARKS/SEC AT 20 JOULES
GAP VOLTAGE:	3000–3100 VOLTS
SPARK RATE MONITOR OUTPUT	30 VOLTS PEAK

Fig. 38   Typical electrical circuit for "high-energy" ignition system

pressures, the air atomizes the fuel, while at high flows the air can be cut off and the fuel pressure relied upon for atomization. For heavy residual fuels, air-assist nozzles can be designed with a separate air compressor that takes air from the main compressor discharge, and provides a pressure ratio of about 2:1 over the entire operating range; this gives efficient atomization (a Sauter Mean Diameter of the fuel droplets of 60 to 80 μm) over the entire firing range. The complication and weight of the additional atomizing air compressor are disadvantages with this arrangement, but it is the only arrangement suitable for burning Bunker C in high-intensity combustion systems. Air-assist nozzles may also be used in machines that burn distillates to assure smokeless combustion throughout the load range.

4. Vaporizing Nozzles. In vaporizing nozzles the fuel is introduced in a tube or passage and vaporized by heat from the combustion chamber. The advantages are: (a) the vaporizing tubes contain both fuel and air and very rich mixtures are avoided, (b) a blue flame is produced and problems of smoke and radiation are diminished, and (c) dispersal of the fuel is independent of fuel flow. The outlet temperature profile is, therefore, not sensitive to fuel flow. Vaporizing nozzles have the disadvantages of not being able to operate at weak mixture ratios and a given design being able to operate satisfactory only with minimal variation in fuel properties.

5. Centrifugal Atomizers. These employ the centrifugal force of a rotating cup to atomize the fuel. This arrangement gives a very uniform temperature distribution (25 F in a radial direction and 75 F circumferentially). Centrifugal atomizers are adaptable to a wide range of fuels, from liquid propane to SAE

30 oil, with no apparent trouble from combustion-chamber deposits. They are, however, only suitable for annular combustors with a rather restricted configuration.

All of these systems have been used in gas turbines, and the choice depends upon the fuel to be used, the arrangement of the combustion system chosen, the range of fuel flows over which satisfactory operation must be obtained, and the ambient conditions (particularly the temperature at which reliable light-off must be obtained).

7.6 Ignition System. Normally the combustion process in a gas turbine is self-sustaining and continuous, but it must be initiated by an external means, which is the function performed by the ignition system. Electric ignition is almost universally used, with the ignition system consisting of an ignitor or ignitors in the combustion chamber and a source of high voltage to create the spark at the electrodes. The voltage used ranges from 3,500 V for a low-voltage system to 18,000 V in a high-voltage system. The energy required varies from 0.2 to 4.0 joules (J) per spark and up to 12 J can be released in about 100 μsec, that is, 100 kW at peak spark discharge in a high-energy system. However, only about 7% of the total heat energy appears at the ignitor face.

The energy can be furnished by a high-voltage transformer, sometimes with a capacitor to store the energy, or in some cases a magneto driven by the engine or starting device is used. The electrical circuit for a typical high-energy system is shown in Fig. 38.

The ignitor can be of several types. An ignitor with a central electrode in an insulator, discharging to another electrode on the body (similar to an automobile distributor and spark plug), can be used. For large chambers, where it is difficult to insure the presence of a combustible mixture at the ignitor mounted in the combustor case, a retractable ignitor is used. The ignitor can be spring-injected and retracted by compressor discharge pressure acting on a piston, or it can be pressure-injected and spring returned. In either case, the electrodes must be inserted far enough into the chamber to insure ignition without danger of them being burned during operation.

For high-energy systems a surface-discharge ignitor can be used. This type is less susceptible to fouling or burning. It furnishes a high-energy source to ignite the fuel but can be used only with a high-energy, capacitor-discharge electrical system.

Section 8
Bearings, Seals, and Lubrication

8.1 Bearing Types. The bearings used in gas turbines fall into two classes: (i) antifriction or rolling-contact bearings and (ii) sleeve or journal bearings and the corresponding oil-film thrust bearing. Gas turbines adapted from aircraft designs generally use antifriction bearings because of their high capacity, low space requirement, and lighter overall engine weight, although in small engines this difference is negligible.

8.2 Antifriction Bearings. Antifriction bearings, whether ball or roller, can and do provide long, reliable service, but by their very nature they have a defined life. For a given load and speed, the bearing manufacturers specify a B-10 life, which 90% of bearings operating with proper lubrication and temperatures will meet or exceed.

For antifriction bearing applications, each rotor is supported by two or more bearings. One bearing is frequently a cylindrical roller bearing to provide for axial movement of the rotor, and the other bearing, which locates the rotor and takes the axial thrust, is some form of ball bearing. Roller bearings are frequently used adjacent to the turbine wheels, since they are usually considered better able to withstand high operating temperatures and higher soak-back temperatures after shutting down. In addition, they have a somewhat greater overload capacity, which may prevent complete bearing failure in the event of damage and loss of balance.

While antifriction bearings require only small amounts of oil for lubrication, a means must be provided to keep them cool in locations where heat from the turbine can flow into them through the shaft or housing. A carefully controlled amount of cooling oil is required to maintain their internal clearances and, at the same time, avoid an excess that can be trapped between the rolling elements (particularly rollers) and the raceways with resulting hydraulic lock and noise. The oil jets should be directed at the shaft adjacent to the inner race and around the housing, to keep the bearing cool without excess oil being forced into the bearing.

8.3 Sleeve Bearings. Properly designed oil-film bearings, appropriately applied and supplied with an adequate quantity of clean oil of suitable viscosity, will run almost forever. Gas turbine sleeve bearings usually follow normal bearing practice. In large machines, the journal bearings consist of babbitted shells (frequently with spherical, self-aligning seats) held in bearing housings of cast or fabricated construction. Bearing loadings are usually below 200 psi of projected bearing area. The bearing bore is frequently elliptical, and where very lightly loaded it may incorporate special pressure pockets or other means to ensure stability. Since the rotor weight goes down as the cube of the scale while the bearing area reduces only as the square, bearing loadings are very low in small units even though L/d ratios less than 0.4 are frequently adopted. Lightly loaded bearings are subject to oil whip and other instabilities, so special measures must sometimes be taken to provide stable, satisfactory operation. Three-lobed bearings, floating bushes, and pivoted-shoe journal bearings have been employed in various units.

8.4 Thrust Bearings. With antifriction bearing arrangements, one of the ball-bearing assemblies is generally designed to take the net thrust. The net thrust is the difference between the compressor thrust and the turbine thrust in the case of gas generators, or the turbine thrust plus or minus any gear or coupling thrust in case of power turbines. With oil-film bearings, a separate thrust bearing is usually provided for each rotor. This bearing normally comprises a thrust collar firmly attached to the shaft with thrust washers or thrust plates on either side. For lightly loaded thrust bearings, these can be simple babbitted flat plates with radial grooves for oil distribution and flow. However, it is preferred practice to use tapered-land thrust shoes in which each land has a slight circumferential taper. While tapered-land thrust bearings can carry very high loadings (above 750 psi), they are not inherently self-aligning and require careful manufacture as the taper required is very slight. Consequently, for highly loaded thrust bearings, a multiple, pivoted-shoe thrust bearing is often adopted. Since the thrust is usually greater in one direction than the other, the more lightly loaded side is sometimes made with fewer pads or shoes than the loaded side.

Sleeve bearings and thrust bearings require an ample supply of clean cool oil for both lubrication and cooling. These bearings operate with a hydrodynamic film; and various factors, such as an interruption in the oil supply, high oil temperatures, or low oil viscosity, can result in bearing wear and damage or even failure (by unduly reducing the oil-film thickness).

Small gas turbines frequently run with light oils (SAE 10 or even lighter) and rather high temperatures. Aircraft-derivative gas turbines are designed to operate with a synthetic oil.

8.5 Shaft Seals. Shaft seals serve the purpose of preventing or controlling fluid leakage along a shaft where it passes through a wall or diaphragm that separates regions, which are at different pressure levels or contain different fluids. Shaft seals are used where shafts enter a bearing housing, where they enter a compressor or turbine casing, and between individual stages of a compressor or turbine.

Shaft seals can be divided into two general classes: contact seals and labyrinth seals. Contact seals usually consist of a carbon or graphite ring with a flat face that is held by a spring in contact with a metal face or sealing ring on the shaft, which has been lapped almost absolutely flat. Contact is maintained between the two faces to prevent leakage. The metal face is hardened and the carbon-ring material is selected to give minimum friction and wear. These seals will operate with essentially no lubrication, although they are most frequently applied where a liquid is present on at least one side of the seal.

A labyrinth seal works on the principle of a series of throttlings, which are produced by a series of teeth on the stationary member, the shaft, or sometimes both. The teeth break down the total pressure difference between the fluid on the two sides of the seal into a series of steps to control the flow through the clearance space between the tip of the tooth and the mating surface. The velocity created in the clearance by the pressure drop is at least partially dissipated in turbulence in the volume between adjacent teeth, thereby minimizing the flow. A more detailed discussion concerning the technical aspects of labyrinth seals is included in Chapter 2.

The clearance that can be maintained between the shaft and the stationary member depends upon the specific machine configuration used, particularly the location of the seal with respect to the bearings and the clearance in the bearings. For small machines with small shafts and seals located immediately adjacent to the bearings, a total clearance of 1.3 to 1.5 mils per inch of shaft diameter is usually satisfactory. For large machines with a considerable distance between the seal and a bearing, a radial clearance of 2 mils per foot of shaft span may be necessary.

A variety of materials can be used for labyrinth seals. Simple bearing housing seals, as shown in Fig. 39, that are intended primarily to prevent the leakage of oil out of the bearing housing, can have the stationary member made of a solid piece of brass or aluminum. For seals located remote from bearings, such as in turbine diaphragms, the stationary member is frequently segmented and spring-supported so as to limit the contact pressure in the event of contact with the shaft; such a seal construction is shown in Fig. 40. At low temperatures, however, these can also be made of brass, leaded bronze, or even plastic, and at high temperatures, ferritic or even austenitic materials with a chrome-moly steel shaft are used. It is also common practice to put the teeth on the shaft, as shown in Fig. 41, to minimize the heating and resulting bowing of the shaft in the event of contact.

A third type, which involves a combination of the two principles, is the carbon-ring seal. In this case a carbon or graphite ring, which can be either solid or segmented and held together by a garter spring, is bored to have a close clearance to the shaft, so as to control the leakage by laminar flow through the clearance space. The ring floats on the shaft and is free to turn in a groove in the housing, but the pressure difference holds it against one side of the groove and seals off leakage by that path. This type of seal is compact and is sometimes used in smaller machines where length is important. This type of seal has also frequently been used in steam turbine practice.

8.6 Lubrication Systems. Proper lubrication is vital to the operation of gas turbines, whether equipped with rolling-contact or fluid-film bearings. A continuous supply of the proper grade of lubricant at the proper pressure and temperature is so important that most gas turbines are equipped with their own integral lubrication system. Where aircraft gas turbines have been adapted to industrial or marine use, the gas turbine used as a gas generator frequently has different lubrication requirements from the power transmission system and, therefore, usually retains an independent lubrication system.

Engines equipped with rolling-contact bearings require less oil, and usually a different grade, than those with fluid-film bearings. Many antifriction-bearing engines are designed to operate with synthetic lubricants. It is important that synthetic lubricants be used only in engines equipped with suitable gaskets, O-rings, seals, etc., as synthetic lubricants can attack and cause rapid deterioration and failure of many common gasket and O-ring materials.

Basically, the lubrication system consists of an oil reservoir, a pump or pumps (for pumping oil from the reservoir to the bearings, gears, and control systems), pressure regulators (to control the supply pressure to various components), an oil cooler or coolers (to control the oil temperature), and a filter or filters (to assure clean oil).

The lube-oil flow is determined by the quantity required to absorb the losses of the bearings, plus heat pickup from the surroundings, within the allowable temperature rises in the bearings. The oil required by all control devices such as governors, hydraulic actuating cylinders, etc., must also be provided. With large units, the pump size may be determined by the oil flow requirements of hydraulic cylinders used to actuate variable-angle nozzles and similar devices. The oil pump is always sized to deliver more than the calculated requirements, the excess being returned to the oil tank (reservoir) by a pressure-regulating valve.

The main lube-oil pumps are generally of the gear type and at least one is almost always driven directly by the main gas turbine shaft to ensure that the pump is driven as long as the turbine shaft rotates. Shaft-driven centrifugal pumps have also been used for the main lube-oil pump, and centrifugal pumps, usually electrically driven, are frequently used for auxiliary pumps where the capacity required warrants it. In smaller sizes, motor-driven gear

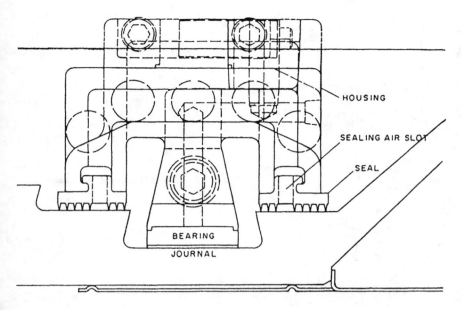

Fig. 39 Labyrinth bearing sealing arrangement

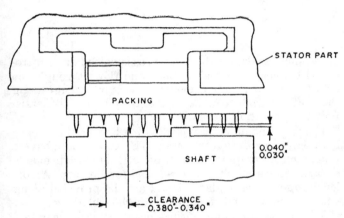

Fig. 40 Typical high-low labyrinth seal with stationary teeth

or vane-type pumps are used for the auxiliary or emergency supply.

Sleeve bearing units should always have a supply of oil to the bearings before the unit is started, and some designs also require control oil pressure before starting. This is the purpose of the auxiliary pump, which is usually under control of a pressure switch so that it runs whenever the control system is energized and no main pump pressure is available. The auxiliary pump will also start on loss of oil pressure while the unit is operating and thereby furnish lubrication until the unit can be shut down. Sometimes the auxiliary oil pump, or a supplementary small "cooldown" pump, is operated for a period of time to circulate oil to the bearings in order to remove the heat that flows into them from the hot parts, particularly the wheels of the turbine. This keeps the babbitt in the bearings below 250 F and prevents the damage that will otherwise result from exposure to higher temperatures while the machine cools down.

Antifriction-bearing machines generally do not require pre- or post-lubrication for the protection of the bearings,

since rolling-contact bearings adjacent to hot parts of the machine are usually stabilized at temperatures of 350 to 450 F. However, lube oil deteriorates at the temperatures that are frequently reached after shutdown. Therefore, in some cases, post-lubrication after shutdown is used to remove the heat and keep the bearing area cool to prevent varnishing and carbonizing of the oil.

The heat generated in the bearings and gear meshes, plus the heat flowing in from the hot parts of the machine are absorbed by the oil cooler. The latter source can amount to $\frac{1}{3}$ of the total heat absorption. For certain installations oil coolers are direct oil-to-air radiators, but for marine installations water-cooled or oil-to-oil heat exchangers are the logical choice. The heat exchanger must be capable of rejecting all the heat absorbed by the oil and provide a proper oil-cooler discharge temperature (usually 135–220 F). The heat exchangers are generally of the shell-and-tube type with the tubes readily accessible for cleaning. Sometimes, the oil coolers are in duplicate, with quick-changeover valves, so that the machine can operate with either cooler while the other is being cleaned.

Oil filters are always installed to insure clean, particle-free lube-oil. The filter must have adequate capacity for the full oil flow within the manufacturer's pressure-drop limitations. Frequently, dual filters are installed with quick transfer (four-way) valves so that one filter can be cleaned or changed while the other is in service. The filtration system should be chosen with consideration for the minimum clearances in the machine. On large machines, which may not require very fine (below 10 μm) filtration for the bearing oil, an additional finer filter (down to 2 μm) may be added in the circuit to other hydraulic devices to protect their very close clearances and fine finishes.

Filter bypasses, particularly internal bypasses, are not recommended even on full-flow filters. When the filter is plugged with dirt and the bypass opens, large quantities

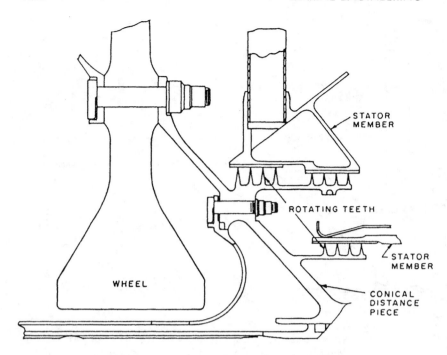

Fig. 41   Rotating-tooth labyrinth-type seal

STATOR
MEMBER

ROTATING TEETH

STATOR
MEMBER

CONICAL
DISTANCE
PIECE

WHEEL

of contaminants may go through the bypass and into the bearing system. It is preferable to monitor the pressure drop across the filter and provide dual filters if it is necessary to assure continuous operation under all conditions.

Pressure regulators are usually simple spring-loaded relief valves as close regulation of the pressure level is not important. Some systems use two pressure levels, one for the control functions and the other at a lower pressure for lubrication. The control circuit is usually arranged to have top priority on the oil supply, since operation of the controls is vital to the operation of the unit, and the lubrication supply can be reduced or even cut off for the fraction of a second it takes the controls to operate.

The oil reservoir, or lube-oil tank, is usually located below the unit although with positively scavenged (drained) systems it can be located anywhere. With jet engines it is frequently fastened to the side of the compressor casing. With gravity drain systems, it must be located a sufficient distance below the bearings to allow a positive slope of at least ¼ in. per foot to the drain lines under all conditions of pitch and roll.

The capacity of the tank is usually based on the main lube-oil pump flow. Where possible, a capacity of four times the oil pump capacity in gpm should be used. This gives what is known as a four-minute supply; i.e., in the event of failure of the drain system, the tank will provide a four-minute supply of lubricant. Smaller units, particularly those mounted in a package, such as shipboard generating sets, may have smaller tanks. Due to space limitations, these may be as small as a two-minute supply or less. Adequate deaeration of the oil is difficult in tanks this small, although the carry-over of mist out the vent can be minimized by locating baffling properly, by locating the oil drains and the tank vent connection as far apart as possible, and by providing a deaeration tray. Pressurized

outer shaft seals, in which compressor bleed air is introduced between two seal sections and flows through one side into the bearing housing (to prevent oil leakage along the shaft), introduce extra air into the oil and make satisfactory deaeration more difficult.

The tank should be provided with a bottom that slopes both ways to a drain connection. The tank should have a removable cover or access door of sufficient size to enable every part of the tank to be reached for cleaning. An oil-level gage or sight glass should also be provided along with a low-level and sometimes a high-level alarm.

Supply and drain piping is preferably made of seamless tubing. Stainless-steel tubing is frequently used in the smaller sizes. AN-type flexible-hose connections are also suitable in the smaller sizes (below about 1.5 in.). In large units, it is good practice to run the pressure feed lines inside the drain lines to the extent possible. The drain line then acts as a guard line in the event of a leak or failure of the feed line, which otherwise could spray oil onto hot parts of the machine and cause a fire.

In order to avoid fire hazards, the number of pipe joints should be minimized. Where joints are necessary, welded flanged connections are preferred, with an SAE four-bolt split-flange connection being the second choice. Compression-type fittings are satisfactory in the smaller stainless steel lines. Flexible lines usually are provided with standard AN-type fittings. Threaded pipe joints, and particularly pipe nipples, should be used only where unavoidable; and then extra-heavy or double-extra-heavy pipe should be used to ensure adequate wall thickness under the threads to avoid fatigue failures originating in the threads.

Pipe sizes, both feed and drain, should be sized for low velocities at full flow. A velocity of not more than 6 fps in feed lines and 2 fps in drain lines will keep system pressure drops to reasonable values and provide free and complete drainage from bearings and gear housings. Poor drainage can cause oil leakage along shafts and additional losses and heating in gear systems.

References

1 T. L. Bowen and J. C. Ness, "Regenerated Marine Gas Turbines, Part I: Cycle Selection and Performance Estimation," ASME 82-GT-306, American Society of Mechanical Engineers, April 1982.

2 T. L. Bowen and D. A. Groghan, "Advanced Cycle Gas Turbines for Naval Ship Propulsion," ASNE *Journal*, May 1984.

3 H. Cohen, et al, *Gas Turbine Theory*, Longman Scientific & Technical Publisher, 1987.

4 R. T. C. Harman, *Gas Turbine Engineering*, John Wiley & Sons, New York, 1981.

5 M. P. Boyce, *Gas Turbine Engineering Handbook*, Gulf Publishing Company, 1982.

6 D. G. Wilson, *The Design of High Efficiency Turbomachinery and Gas Turbines*, MIT Press, Cambridge, Mass., 1984.

7 "7LM2500PC101/102/103 and 104 Gas Turbine Performance Data," General Electric Report MID-TD-2500-5, General Electric Co.

8 "Engines, Gas Turbine, Propulsion and Auxiliary Naval Shipboard," Military Specification, Mil-E-17341.

9 "LM2500 Marine Gas Turbine Installation Design Manual," General Electric, MID-IDM-2500-2, General Electric Co.

10 "Marine Gas Turbine: the 1990's and beyond," *Automotive Engineering*, Vol. 95, No. 9, Sept. 1987.

11 J. W. Watts, and T. L. Bowen, "Regenerating Marine Gas Turbines, Part II: Regenerator Technology and Heat Exchanger Sizing," ASME Paper No. 82-GT-314, American Society of Mechanical Engineers, 1982.

12 R. G. Mills and K. W. Karstensen, "Intercooled/Recuperated Shipboard Generator Drive Engine," ASME Paper No. 86-GT-203, presented at the International Gas Turbine Conference and Exhibit, Dusseldorf, West Germany, American Society of Mechanical Engineers, June 1986.

13 R. Eggebrecht and W. Schlosser, "Compact High Temperature Heat Exchanger for Turboshaft Engines," *MTZ Motortechnische Zeitschrift*, Vol. 47, No. 6, 1986.

14 T. L. Bowen, D. P. Guimond, and R. K. Muench, "Experimental Investigation of Gas Turbine Recuperator Fouling," ASME Paper No. 87-GT-97, presented at the Gas Turbine Conference and Exhibit, Anaheim, Calif., American Society of Mechanical Engineers, June 1987.

15 C. Zeien, H. F. Smith, and F. W. Hirst, "The Gas Turbine Ship *Callaghan's* First Two Years of Operation," *Trans.* SNAME, Vol. 77, 1969.

16 W. S. Richardson, "The Friction Clutch Reverse-Reduction Gears In the GTS," ASME Paper No. 69-GT-5, American Society of Mechanical Engineers, 1969.

17 M. J. T. Smith and M. E. House, "Internally Generated Noise from Gas Turbine Engine," ASME Paper 66 GT/N-43, American Society of Mechanical Engineers, 1966.

18 W. B. Brown and G. R. Bradshaw, "Design and Performance of a Family of Diffusing Scrolls with Mixed Flow Impeller and Vaneless Diffuser," NACA Report 936, National Advisory Committee on Aeronautics, 1949.

19 S. S. Manson, "Determination of Elastic Stresses in Gas Turbine Disks," NACA TN 1279, National Advisory Committee on Aeronautics, 1947.

20 M. B. Millenson and S. S. Manson, "Determination of Stresses in Gas Turbine Disks Subject to Plastic Flow and Creep," NACA TN 1636, National Advisory Committee on Aeronautics, 1948.

21 L. J. Herric, J. C. Emery, and J. R. Erwin, "Systematic Two-Dimensional Cascade Test of NACA 65-Series Compressor Blades at Low Speeds," NACA TN 3916, National Advisory Committee on Aeronautics, 1957.

22 J. C. Emery, "Low Cascade Investigation of Thin Low Camber NACA 65-Series Blade Sections At High Inlet Angles," NACA RM L57E03, National Advisory Committee on Aeronautics, 1957.

23 J. C. Dunavant, J. C. Emery, H. C. Walch, and W. R. Westphal, "High Speed Cascade Tests of the NACA 65-$(12A_{10})$ 10 and NACA 65-$(12A_2I_{8b})$ 10 Compressor Blade Sections," NACA RM L55108, National Advisory Committee on Aeronautics, 1955.

24 J. C. Emery and J. C. Dunavant, "Two Dimensional Cascade Tests of NACA 65-$C_{10}A_{10}$) 10 Blade Sections at Typical Compressor Hub Conditions for Speeds up to Choking," NACA RM L57H05, National Advisory Committee on Aeronautics, 1957.

25 S. Lieblein, F. C. Schwenk, and R. L. Broderick, "Diffusion Factor or Estimating Losses and Limiting Blade Loadings in Axial-Flow Compressor Blade Elements," NACA RM E53D01, National Advisory Committee on Aeronautics, June 1953.

26 R. W. Nolan, "Vibration of Marine Turbine Blading," *Trans.* SNAME, Vol. 57, 1949.

27 P. N. Bright, "Structural Design Problems in Gas Turbine Engines," ASME Paper 54-A-152, American Society of Mechanical Engineers, 1954.

28 O. Y. Chen and D. N. Duhl, "Single Crystal Superalloy Turbine Airfoils for Marine Gas Turbine Engines," Pratt & Whitney Report FR-17370, Oct. 1983.

29 F. Kreith and J. H. Taylor, Jr., "Heat Transfer from a Rotating Disk in Turbulent Flow," ASME Paper 65-A-146, American Society of Mechanical Engineers, 1965.

30 A. M. Wahl, "Stress Distributions in Rotating Disks Subjected to Creep Including Effects of Variable Thickness and Temperature," ASME Paper 56-A-162, American Society of Mechanical Engineers, 1956.

31 H. F. Smith, "Gas Turbine Propulsion of LNG Tankers," ASME Paper No. 69-GT-47, American Society of Mechanical Engineers, 1969.

CHAPTER V

E. A. Catlin | # Boilers and Combustion

Section 1
Classification of Marine Steam Generators

1.1 Boiler History. From the dawn of the Industrial Revolution the attention of many individuals focused on the advantages of powering ships by steam. A study of the history of early marine boilers reveals that early designers and engineers did not lack novel and ingenious ideas for steam propulsion equipment. However, they did lack the materials and machine tools with which to implement these ideas. Dependable boilers were sorely needed in the development of the steamboat, which began in the United States with the work of James Rumsey (1743–1792) on the Potomac and of John Fitch (1743–1798) on the Delaware.

Robert Fulton inaugurated steam navigation on a regular basis in 1807 with the *Clermont*, popularly called "Fulton's Folly," whose engine and boiler were imported from England. The success of this vessel prompted others to follow Fulton's lead and soon steamboats were navigating all the great waterways of the North American continent, and a new industry was born.

a. Flue boilers. By 1835 there were about 700 steam-powered vessels in use in the United States. Steam pressures were low—a few pounds above atmospheric—and the boilers were simply rectangular boxes with square furnaces and low winding flues large enough for a man to pass through to permit cleaning. They were made of riveted copper or iron plates and were nearly always fed with salt water or raw water. Coal and wood were the usual fuels.

The famous vessel *Monitor* was equipped in 1861 with two firetube boilers which were typical of that era, and the world awoke to the value of steam-powered iron ships for naval service following her encounter with the CS *Virginia* (ex-*Merrimac*). The demands for higher power and speeds resulted in many ingenious boiler designs. In 1861 the U.S. Navy cruiser *Wampanoag* represented the zenith of marine engineering achievement in the Civil War era. This fine vessel was the fastest steamer of her day and during trials reached a maximum speed of 19.5 knots, her geared engines supplied with superheated steam from four superheating boilers and eight vertical watertube boilers. In reality these later units were a combination of watertubes with cylindrical furnaces similar to those found in firetube and return-flue-type boilers. Boiler pressure was maintained during the record-breaking runs at

about 30 psig, which was about the upper limit of pressure during the Civil War era.

b. Scotch boilers. Following the Civil War, continued advances in metallurgy and engineering resulted in the cylindrical firetube boiler or "Scotch" boiler, as illustrated by Fig. 1, becoming most popular. In the late 1800's and early 1900's coal-fired Scotch boilers were used for pressures of up to 250 psig and steam temperatures of up to 650 F. Their tolerance for poor water, lubricating oil, and general abuse made them ideal for use with the steam engines which also reached their peak of popularity in the same period. When equipped with a superheater and air heater, and firing oil, the Scotch boiler had an efficiency of about 80%.

c. Sectional header boilers. The development of the steam turbine created a need for higher pressures and steam temperatures, and boilers of the Scotch type were superseded by watertube boilers. Many of the early attempts at building watertube boilers were doomed to failure due to unrecognized circulation deficiencies, inadequate water treatment, and poor tube arrangements which made repairs difficult. The first successful marine watertube boiler was of the cross-drum straight-tube sinuous-header type and was developed to avoid these difficulties. It provided, by means of its straight tubes, easy inspection of the watersides and permitted renewal of individual boiler tubes.

First tried in the steam yacht *Reverie* in 1889, this concept rapidly developed for naval service in both the United States and Great Britain as well as for merchant ships. RADM George W. Melville, the famous Arctic explorer and Engineer-in-Chief of the U.S. Navy from 1887 to 1903, summarized the basic requirements for a marine boiler of this type, which time has not changed greatly, in the following way [1]:

"From the study of the subject, I concluded that the thoroughly satisfactory watertube boiler should possess, among others, the following characteristics:

"Reasonable lightness, with scantlings sufficient to promise longevity;

"An adequate amount of water, so that failure of the feed supply or any inattention thereto would not immediately cause trouble;

"Accessibility for cleaning and repairs on both water and firesides;

184

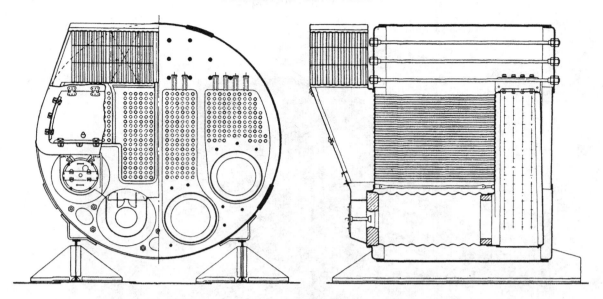

Fig. 1   Front view and longitudinal section of an oil-burning Scotch boiler fitted with an air heater

"Straight tubes, with no screw joints in the fire but the simple expanded joints so well tested for years;

"No cast metal, either iron or steel, subjected to pressure;

"Ability to raise steam quickly;

"High economy of space;

"Interchangeability of parts, and, as far as possible the use of regular commercial sizes, so that repair materials could be procured anywhere;

"The ability to stand severe forcing without injury;

"The ability to stand abuse, that is, to be of rugged construction and not so delicate as to require skilled mechanics to run it;

"Safety against disastrous explosion, meaning that only the part of the boiler which gave way would be damaged."

The performance of the header-type design was improved through the years by the use of 2-in. tubes with three gas passes, and later, 1¼-in. or 1-in. tubes in a single gas pass, as in Fig. 2, to improve efficiency and increase capacity. The basic soundness of this design is indicated by the fact that many boilers of this type are still in service for propulsion purposes on older vessels.

d. Bent-tube boilers. The drum-type watertube boiler, which had been experimented with from the earliest days, finally came into practical being in the 1890's. The chief impetus for this was the development of the high-speed torpedo boat. High steam outputs at about 250 psig, which for that time was high pressure, resulted in the application of both natural-circulation watertube boilers and forced-circulation once-through boilers to some of these craft.

With once-through forced-circulation boilers the problems of control, feedwater treatment, start-up, maneuvering, and the lack of dependable pumps presented formidable obstacles. Accordingly, the major efforts of boiler

Fig. 2   Single-pass sectional header boiler with water-cooled furnace

designers were directed to the development of natural-circulation watertube boilers which were simpler to operate and maintain. The watertube natural-circulation boiler has taken many forms through the years. The early designs were generally of the single-furnace three-drum double-uptake type with the superheater in one side. As steam temperatures increased, it became necessary to develop an economical method for controlling the steam temperature while maneuvering and particularly during astern operations. This was accomplished initially by the use of separate superheater boilers in which the firing rate was reduced when going astern. Later, a two-furnace

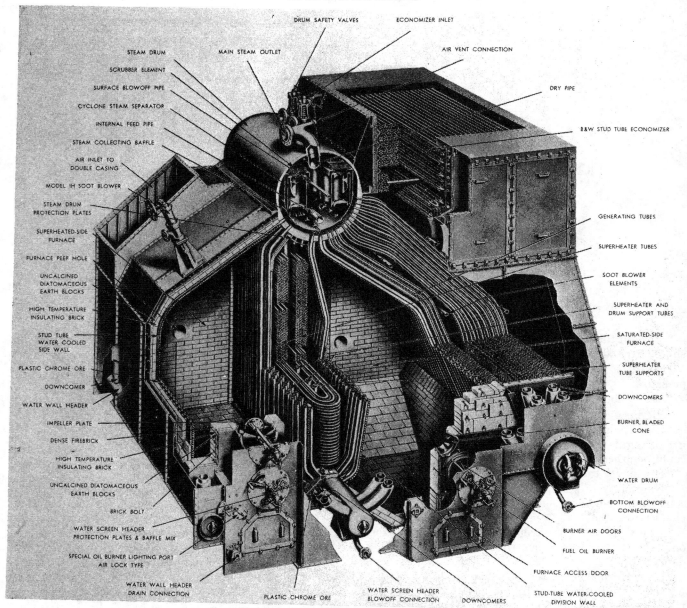

DRUM SAFETY VALVES   ECONOMIZER INLET

STEAM DRUM   MAIN STEAM OUTLET   AIR VENT CONNECTION

SCRUBBER ELEMENT

SURFACE BLOWOFF PIPE   DRY PIPE

CYCLONE STEAM SEPARATOR

INTERNAL FEED PIPE   B&W STUD TUBE ECONOMIZER

STEAM COLLECTING BAFFLE

AIR INLET TO
DOUBLE CASING   GENERATING TUBES

MODEL 1H SOOT BLOWER   SUPERHEATER TUBES

STEAM DRUM
PROTECTION PLATES   SOOT BLOWER
                                                                      ELEMENTS
SUPERHEATED-SIDE
FURNACE   SUPERHEATER AND
                                                                DRUM SUPPORT TUBES
FURNACE PEEP HOLE
                                                                 SATURATED-SIDE
UNCALCINED   FURNACE
DIATOMACEOUS
EARTH BLOCKS   SUPERHEATER
                                                                 TUBE SUPPORTS
HIGH TEMPERATURE
INSULATING BRICK   DOWNCOMERS

STUD TUBE   BURNER, BLADED
WATER COOLED   CONE
SIDE WALL

PLASTIC CHROME ORE   WATER DRUM

DOWNCOMER   BOTTOM BLOWOFF
                                                                  CONNECTION
WATER WALL HEADER

IMPELLER PLATE   BURNER AIR DOORS

DENSE FIREBRICK   FUEL OIL BURNER

HIGH TEMPERATURE
INSULATING BRICK   FURNACE ACCESS DOOR

UNCALCINED DIATOMACEOUS
EARTH BLOCKS   STUD-TUBE WATER-COOLED
                                                                  DIVISION WALL
BRICK BOLT

WATER SCREEN HEADER
PROTECTION PLATES & BAFFLE MIX

SPECIAL OIL BURNER LIGHTING PORT
AIR LOCK TYPE

WATER WALL HEADER   WATER SCREEN HEADER
DRAIN CONNECTION   PLASTIC CHROME ORE   BLOWOFF CONNECTION   DOWNCOMERS

Fig. 3   Two-furnace, single-uptake controlled superheat boiler

boiler, in which the steam temperature was controlled by varying the firing rate in its two furnaces, was evolved in a continuing effort to save space and weight. These early arrangements used double uptakes with the superheater section located in one of the gas paths. Subsequently, the two-furnace single-uptake boilers shown in Fig. 3 was developed. Boilers of this general design were installed in almost all of the combatant ships built for the U.S. Navy in World War II [2].

With the development of steam turbines capable of using full steam temperature during astern operations, it became possible to further reduce the weight and size of the drum type of boiler. The two-drum single-furnace boiler with an integral superheater, as in Fig. 4, was the result. The location of the superheater near the furnace provided a relatively constant steam temperature over a

wide range of operation due to the combination of its convection and radiant heat-transfer characteristics. Most current merchant marine and naval boilers form solid links in the chain of evolution of this type of boiler.

While many variations of the foregoing boiler types have been employed throughout the world, the types discussed are fairly representative and provide an adequate background for an understanding of steam generator types and characteristics.

1.2 Current Types of Merchant and Naval Boilers.   In the past 100 years steam pressures and temperatures have increased from 30 psig saturated to 870 psig–950 F in most merchant vessels, and to 1200 psig–1000 F maximum (950 F nominal) in most post World War II naval combatant vessels. In large, high-power installations steam at

Fig. 4   Two-drum, single-furnace boiler with horizontal superheater

1500 psig–950 F, and in some instances reheat to 950 F, appears feasible.

For the most part, widespread use of water cooling in the furnaces is employed to reduce refractory maintenance. Economizers and air heaters singly, or in combination, are used to obtain the desired overall steam generator efficiency. Attemperators are employed to control the steam temperature over a wide operating range and thereby improve turbine performance. Desuperheaters are installed to provide low-temperature steam for auxiliary purposes throughout the ship.

Bunker C residual oil is the most widely used fuel in commercial applications, while marine diesel and other light fuel oils are widely used in naval applications. Steam-atomizing oil burners, first used aboard ships in the late 1800's, have returned to favor with the advent of high-capacity low-cost evaporators to supply the necessary water. This type of atomizer, while providing an extremely wide range of operation, results in a loss of distilled water which was, until recently, too big a penalty to pay for its advantages. However, improvements to reduce the consumption of steam, coupled with abundant distilled water, have led to its widespread use, particularly in automated boilers.

a. Two-drum boiler. Two-drum integral-furnace boilers, or D-type boilers as they are often called, usually consist of a steam drum and water drum connected by water screen and boiler bank tubes. Superheaters are installed between the water screen and the boiler bank and may have tubes arranged either vertically or horizontally, depending in part on which arrangement best fits the machinery arrangement. Where required, the steam temperature may be controlled by means of a control desuperheater or attemperator located in either the steam or water drum. The firing front or location of the oil burners is frequently dependent on the machinery arrangement and may be in the front wall (most conventional), roof, or sidewall. Figures 4 and 5 indicate some of these variations.

In most installations some form of air heater is used with an economizer. The type and proportions of these auxiliary heat exchangers depend on the cycle arrangement. If two stages of feed heating are selected, a steam air heater and an economizer are often used. In the steam air heater, condensing low-pressure steam (40 to 65 psia) heats the incoming air. Where three or four stages of feed heating are employed, it is generally advantageous to use an air heater of the recuperative or regenerative type. A small economizer may be used in such instances to hold

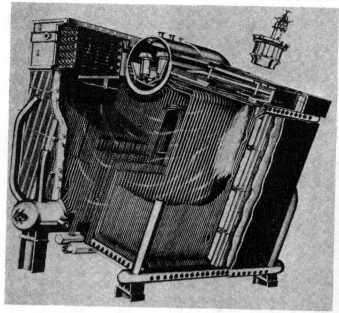

Fig. 5 Two-drum, top-fired boiler with vertical superheater and economizer

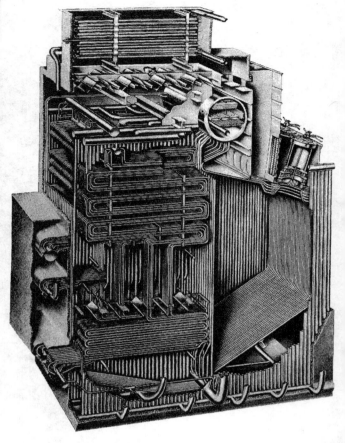

Fig. 6   Single-furnace, gas-bypass reheat boiler

the air heater down to a practical size. Air heaters are gas-to-gas heat exchangers and tend to be quite large due to their relatively low heat-transfer rates.

Natural circulation is employed exclusively in two-drum boilers and is dependent on downcomers, which may be of the internal (heated) variety in units of low rating or of the external (unheated) variety in highly rated merchant and naval units. Economizers are force fed by the boiler feed pump in all cases.

Cleaning of external surfaces exposed to fuel-oil ash and soot is accomplished by means of steam sootblowers. Those in the superheater zone are of the retractable mass-action type to provide sufficient cleaning power to remove slag. Protection from the high gas temperatures in this area is provided by withdrawing them when out of service.

An auxiliary desuperheater is installed in the steam or water drum in most vessels to supply low-temperature steam for purposes other than main propulsion.

Casings surround the pressure parts and form a gas-tight envelope to contain the products of combustion. They can be of varied construction; however, a double casing is generally used to avoid any possibility of flue gas leakage into the engine room. Where a single casing is used, expansion joints, access openings, etc., are supplied with pressurized-air seals to prevent leakage.

b.   Reheat boilers.   In the reheat cycle, steam is generated at a high pressure, superheated in conventional boiler and superheater elements, and expanded through the high-pressure elements of the turbine to the reheater inlet pressure. It is then reheated at this reduced pressure and expanded through the low-pressure elements of the turbine (see Chapter 6 for a discussion of reheat turbines). For satisfactory and reliable operation, means must be provided to protect the reheater from overheating during maneuvering and astern operation when the reheater steam flow is reduced or nonexistent. The design of a

suitable reheat boiler, therefore, is somewhat more complicated than that of a superheater for a conventional nonreheat cycle.

Reheating of steam becomes increasingly more attractive as the installed horsepower increases. The fuel savings to be gained by reheat are then sufficient to justify the more complicated machinery. For the most part, steam generators to supply reheat plants have been adaptations of the two-drum integral-furnace boiler. One or two furnaces can be used. The single-furnace concept utilizes a divided gas flow path beyond the furnace with the superheater and reheater placed in separate sections. The flow of gas in these sections is regulated by dampers, thereby controlling the superheater and reheat steam temperatures. It then combines to flow to the auxiliary heat exchangers. By burning all the fuel in a single furnace, the control of the oil burners and the forced-draft air supply system is simplified. Figure 6 illustrates a boiler of this type.

In the divided-furnace boiler, a form of two-furnace boiler, one furnace supplies heat to the reheater and the other furnace supplies heat to the superheater. Some designs incorporate a part of the superheater (called a primary superheater) in the reheater zone to provide additional protection for the reheater and to obtain the desired steam temperature characteristics [3]. The gas flowing from both the reheater and superheater combine in the

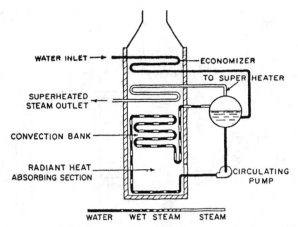

Fig. 7   Schematic of LaMont forced-circulation boiler with economizer and superheater

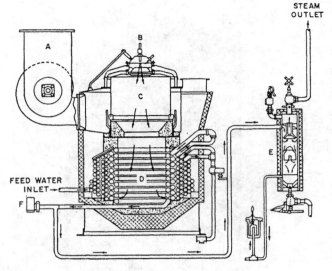

A — FAN
B — OIL BURNER WITH IGNITOR AND FLAME SCANNER
C — FURNACE
D — GENERATING COILS
E — STEAM SEPARATOR
F — STEAM TEMPERATURE LIMIT CONTROL

Fig. 8   Once-through boiler

main generating tube bank, and a single gas flow path is maintained through the auxiliary heat exchangers as in the single-furnace design.

c. Forced-circulation boilers. Ever since the first boiler was used aboard ship, marine boiler designers have investigated and experimented with various means to reduce the size and weight of boilers. A boiler arranged for natural circulation of the water and steam requires low waterside pressure drops, which can only be obtained by installing sufficient downcomers and risers. This adversely affects size and weight. By supplying a pump to either augment or supplant natural circulation, a smaller and lighter boiler can be designed for a given steam output [4]. The circulation in such a boiler is said to be controlled or forced. The chief advantages of this are that very small-diameter tubes with a high resistance to flow can be used in arrangements of heating surfaces and steam drum locations, which would be incompatible with natural circulation. The greatest disadvantage is the circulating pump itself, which is a potential source of trouble and maintenance.

The LaMont boiler, shown schematically in Fig. 7, is a typical example of the forced-circulation type. While used abroad, it has not found wide application in the marine field in the U.S. The LaMont boiler uses a single drum into which the heating surface discharges a mixture of steam and water. The circulating pump suction is supplied by gravity from this drum and forces water through the generating tube surface, which is composed of a number of tube circuits arranged between a distributing header and the steam drum. The inlet of each tube is fitted with an orifice to balance the flow resistance within the various circuits. This is necessary to obtain an adequate flow of water in each tube depending on its expected heat input. The furnace, oil burners, superheater, and economizer are similar to those of natural-circulation boilers.

d. Once-through boilers. The boiler in Fig. 8 is an example of once-through boilers used for auxiliary steam. Water is passed through the heating surface in one continuous circuit by the feed pump. The boiler is basically one long spiral tube arrangement composed of a steaming

economizer and a transition zone, where evaporation is completed, which surrounds the furnace. The feed pump pressure determines the outlet steam pressure, which may be 1200 to 1800 psig, although for the usual marine installation the pressure is in the range of 150 to 300 psig. Boilers of this type are usually built only in small sizes and supply up to 7500 lb of saturated steam per hour.

Because of the difficulties in maintaining feedwater chemistry, adequate water flow through parallel tube circuits, which would be required for higher capacity boilers of this type, and the control of superheated steam temperatures, the once-through boiler is not well suited for marine propulsion purposes.

e. Supercharged boilers. The supercharged boiler has the characteristic of using combustion pressures higher than one atmosphere in the furnace to take advantage of higher gas densities and higher gas velocities than are available in the usual marine boiler. Figure 9 is a typical supercharged boiler. This unit is an outgrowth of the Velox boiler which has been used in a few stationary power plants for a number of years. In this type of boiler, the exhaust gases are at a pressure high enough to drive a gas turbine, which in turn drives an axial-flow compressor supplying the combustion air. Combustion pressures of up to five atmospheres are readily obtained and permit the size of the boiler to be reduced considerably when compared to a conventional unit [5]. The generating tube spacing can be reduced since draft losses become a secondary consideration. The higher gas velocities, which result from a tight tube spacing, result in a much higher rate of convection heat transfer so that the installed surface for a given performance can be reduced to about one third to one fourth of that in the usual boiler.

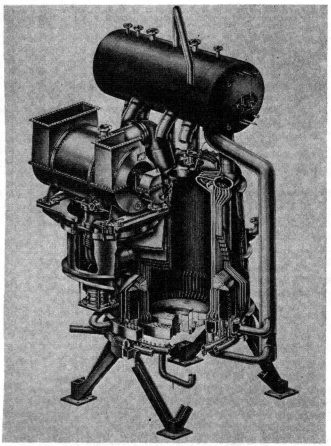

Fig. 9   Supercharged boiler with gas-turbine-driven air compressor

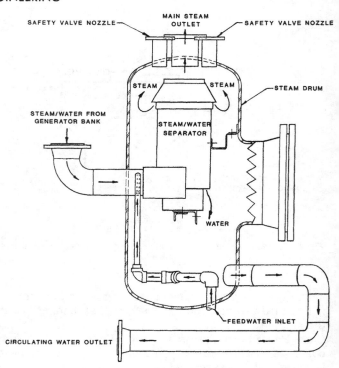

Fig. 10   Steam drum module for waste-heat boiler

The work of compression shows up, in good measure, as an increased temperature of the combustion air. As a consequence, the compressor acts as a regenerative air heater. It is, therefore, possible to obtain a high boiler efficiency without using an air heater or economizer. Boiler overall weight is markedly reduced by the use of this cycle, and it has found application in weight-sensitive naval vessels.

The original Velox boiler, from which supercharged boilers evolved, was a forced-circulation boiler. However, subsequent supercharged units have employed natural circulation to avoid the extra complication of the circulation pump.

f. Waste-heat and auxiliary boilers. Vessels with diesel, gas turbine, or nuclear propulsion usually require steam for auxiliary purposes such as hotel services, cargo or bunker oil heating, cargo pumps, evaporators, and deck machinery. The exhaust gases from diesel or gas turbine engines contain considerable available latent heat. Boilers placed in the stack to collect this otherwise lost heat are called waste-heat boilers. Where the required capacity exceeds that available from the waste heat, or where waste heat is not used, an auxiliary oil-fired boiler may be installed. This unit provides steam when the main engines are shut down and may supplement that available from the waste-heat units when at low power. Nuclear

vessels may use an auxiliary oil-fired boiler when the reactors are secured.

Waste-heat boilers usually consist of banks of tubes, similar to those in an economizer, connected to a suitable steam drum. Some units may be designed to burn oil to supplement the available waste heat or replace it when the main unit is shut down. Either natural or forced circulation may be used. Typically, waste-heat boilers are furnished complete with controls, feed pumps, safety valves, etc. and are skid-mounted for easy installation [6]. The steam drum and steam-generating modules installed in the U.S. Navy's CG-47 class gas turbine-powered vessels are shown schematically in Figs. 10 and 11. Forced circulation is employed in these units [7].

Auxiliary oil-fired boilers are usually of the two-drum type and generate saturated steam. They may be supplied built-up and skid-mounted with burner, fan, controls, pumps, etc. ready to fire in capacities up to about 135,000 lb/hr. For higher steam capacities field assembly is usually required. Natural circulation is generally employed for all capacities. Forced circulation is generally limited to small boilers with low steam outputs.

1.3 Auxiliary Heat Exchangers. In addition to the steam generator, several forms of auxiliary heat exchangers are incorporated in boilers to improve the efficiency and the overall operation of the plant [8]. Economizers of either the bare-tube or extended-surface type are used to increase the temperature of the incoming feedwater by cooling the flue gases leaving the boiler. Air heaters are used to increase the temperature of the combustion air so as to promote better combustion of the fuel. In the case of gas-to-air heat exchangers, air heaters also improve the boiler efficiency by reducing the temperature

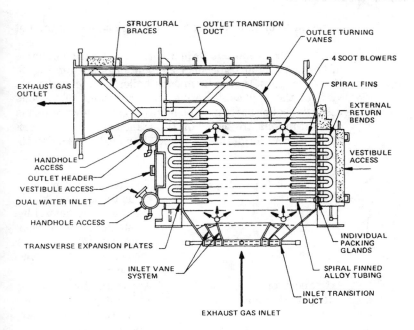

STRUCTURAL BRACES

OUTLET TRANSITION DUCT

OUTLET TURNING VANES

4 SOOT BLOWERS

SPIRAL FINS

EXTERNAL RETURN BENDS

EXHAUST GAS OUTLET

VESTIBULE ACCESS

HANDHOLE ACCESS

OUTLET HEADER

VESTIBULE ACCESS

DUAL WATER INLET

HANDHOLE ACCESS

TRANSVERSE EXPANSION PLATES

INDIVIDUAL PACKING GLANDS

INLET VANE SYSTEM

SPIRAL FINNED ALLOY TUBING

INLET TRANSITION DUCT

EXHAUST GAS INLET

Fig. 11   Steam generating module for waste-heat boiler

of the flue gases. By using low-pressure, low-temperature exhaust or turbine bleed steam to heat combustion air, as in the case of the steam air heater, the overall cycle efficiency is improved. These various types of heat exchangers may be used singly or in combination with each other.

a. Economizers. An economizer is a simple heat exchanger consisting of a bank of tubes connecting an inlet and outlet header located in a relatively cool gas temperature zone beyond the boiler main generating bank. Supplied with water at a temperature near that leaving the last feedwater heater, the economizer supplies additional heat to the feedwater by cooling the flue gas. In many installations the economizer is the final heat exchanger in the exhaust gas path. It may, however, be followed by an air heater where a higher efficiency is desired.

Economizers can be divided into two general categories: the bare-tube and the extended-surface types. Both types are forced circulated by the main feed pump. In general, they are designed to heat the incoming feedwater to within about 35 deg F of saturation temperature. They are arranged for counterflow of the water and the products of combustion since larger temperature differentials and a greater heat absorption can be obtained thereby. This serves to provide the highest boiler efficiency for a given economizer size since the exit gas temperature tends to approach that of the incoming feedwater.

The simplest economizer arrangement is the bare-tube type and this was the form the first economizers took. However, it was recognized that the use of extended surface to increase the total heat-transfer surface for a given length of tube would provide significant increases in performance without penalizing weight and space considerations adversely. Figure 12(a) shows an efficient form of extended surface in which flat studs are spaced at 45-deg angles around the circumference and at ½-in. intervals along the tube.

Extended surface can also take the form of spiral fins welded on the tubes or of cast iron or aluminum gill rings bonded or shrunk onto steel tubes as shown by Fig. 12(b).

b. Air heaters. The cooling of hot flue gases by the incoming combustion air is one of the oldest of concepts to improve boiler efficiency. In addition, heated air provides an additional beneficial effect by promoting rapid and complete combustion of the fuel. This can be of importance in the relatively small furnaces used in marine boilers.

Air heaters fall into two broad classifications: the recuperative and the regenerative. In the recuperative type, heat from the products of combustion passes through a partition which separates the products from the air. Tubular and plate-type air heaters are examples of recuperative air heaters. In the tubular heater (Fig. 13) the walls of the tubes transfer the heat from the gas to the air. In heaters of the plate type, the air and gas are separated by plates through which the heat flows. However, recuperative air heaters have been largely superseded by the regenerative type.

In the regenerative air heater, heat is first stored in the structure of the heater itself as it passes through the hot gas stream. The heat is then given up to the air as the structure turns through the airstream. The air preheater shown in Fig. 14 is an example of this type [9]. It consists of closely spaced heating elements packed into a revolving frame. The frame speed is constant and is controlled by a small electric motor. The frame speed is selected such that the elements will absorb heat from the gas with a good temperature differential and, at the same time, the elements will heat the incoming combustion air to the highest possible extent. The upper section of the air heater is in the cold-air zone and also "sees" the coolest gas. It is usually arranged so that the heat-transfer surface can be conveniently removed in easily handled sections—called "baskets"—since corrosion and fouling may occur there.

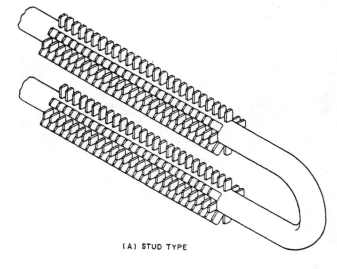

(A) STUD TYPE

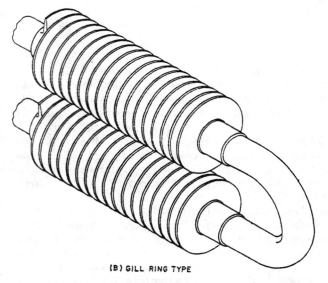

(B) GILL RING TYPE

Fig. 12   Typical extended-surface economizer elements

These baskets may be provided, in addition, with a ceramic coating similar to porcelain enamel for protection against the corrosive effects of the sulfuric acid, which can form from the products of combustion. Air bypasses may be provided to avoid excessive cooling and condensation of the exhaust gases during low loads and port or cold-weather operations.

The steam air heater (Fig. 15) is used where an adequate quantity of low-pressure, low-temperature steam is available, and it is desired to heat the combustion air. If exhaust or turbine-bleed steam is used, there is a slight improvement in the overall cycle efficiency as well. The heater consists of coils of tubes fitted with extended surface, usually comprising spirally wrapped strips of material, welded or brazed to copper-nickel tubes. The tubes are U-shaped and arranged between inlet and outlet headers. The steam supplied is condensed by the incoming cold air and the condensate is removed by steam traps. The latent heat of this steam, which would otherwise be re-

jected in a condenser, is returned to the boiler via the hot air.

1.4   Boiler Terms and Definitions.   The location of some of the more important boiler elements is shown in Fig. 3. For an understanding of marine boiler technology, a review of the applicable terms and definitions of various essential boiler parts may be helpful. The following terms and definitions are based on the standards of the American Boiler Manufacturers Association [10] and on day-to-day usage:

Air (pre) heater	The heat-transfer apparatus through which air is passed and heated by a medium of higher temperature, such as the products of combustion or steam.
Attemperator (desuperheater)	Apparatus for reducing and controlling the temperature of a superheated vapor.
Brickpan	The plate and structural steel work which supports the furnace floor.
Brickwork	The refractory linings of the furnace.
Casing	The covering of metal plates and structure used to enclose all or a portion of a steam generator unit.
Chemical feed pipe	A pipe inside a boiler drum through which chemicals for treating the boiler water are introduced.
Circulation ratio	The ratio of water entering a circuit to the steam generated within that circuit.
Downcomer	A tube in a boiler or waterwall system through which fluid flows downward.
Dry pipe	A perforated or slotted pipe or box inside the steam drum which is connected to the steam outlet.
Economizer	A heat recovery device designed to transfer heat from the products of combustion to a fluid, usually feedwater.
Feed pipe	The pipe used to distribute the feedwater inside the boiler steam drum.

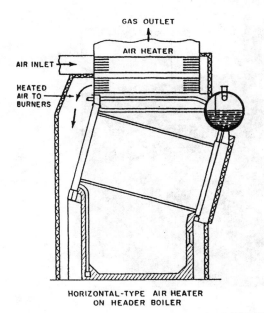

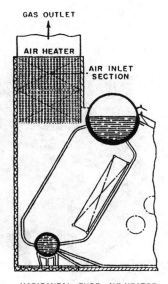

Fig. 13 Tubular air heaters

Firetube . . . . A tube in a boiler having water on the outside and carrying the products of combustion on the inside.

Floor tubes . . . . Those tubes in the furnace floor which if exposed to the products of combustion are generating tubes but if arranged beneath refractory are used as supply tubes to supply water to a drum or header.

Forced circulation . . . . Circulation in a boiler by mechanical means external to the boiler.

Furnace screen . . . . One or more rows of tubes arranged across the furnace gas outlet.

Furnace volume . . . . The volume contents of the furnace or combustion chamber.

Generating tube . . . . A tube in which steam is generated.

Header . . . . A drum too small to permit entry through a manhole.

Heat release . . . . The total quantity of thermal energy above a fixed datum introduced into a furnace by the fuel. It is considered to be the product of the fuel delivered per hour and the fuel higher heating value, expressed in Btu per hour per cubic foot of furnace volume.

Heated downcomer . . . . Any tube in a boiler generating bank in which water may flow from the steam drum to the water drum or header.

Heating surface . . . . That surface which is exposed to the heating medium for absorption and transfer of heat to the heated medium, including any fins, gills, studs, etc. attached to the outside of the tube for the purpose of increasing the heating surface per unit length of tube.

Ligament (tube) . . . . The minimum distance between two adjacent tubes.

Moisture-in-steam . . . . Particles of water carried in steam, usually expressed as the percentage by weight.

Mud, lower, or water drum . A pressure chamber of a drum or header type located at the lower extremity of a watertube boiler convection bank which is normally provided with a blowoff

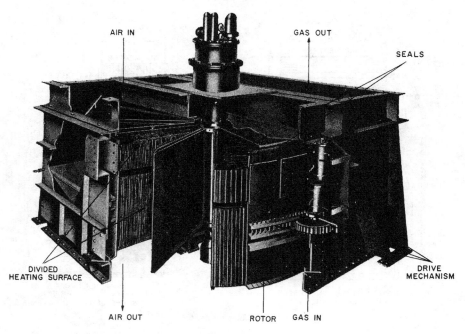

AIR IN

GAS OUT

SEALS

DIVIDED
HEATING SURFACE

DRIVE
MECHANISM

AIR OUT   ROTOR   GAS IN

(a) Assembly

(b) Replaceable baskets

Fig. 14   Rotary regenerative air heater with replaceable cold-end baskets

valve for periodic removal of sediment collecting in the bottom of the drum.

Natural circulation........ Circulation of water in a boiler caused by the difference in density between the water in the downcomers and the water-steam mixture in the generating tubes.

Radiant heat absorbing .... surface (RHAS) The projected area of tubes and extended metallic surfaces as viewed from the furnace. Included are the walls, floor, roof, and partition walls in the plane of the furnace exit screen.

Reheater............... Heat-transfer apparatus for heating steam after it has given up some of its original heat in doing work.

Riser................. A tube through which steam and water passes from an upper waterwall header to the steam drum.

Steam baffling.......... The plates, centrifugal separators, or baffles arranged to remove entrained water from the steam.

(a) Assembly of typical section

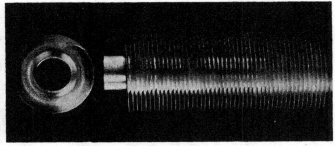

(b) Smooth spiral fin

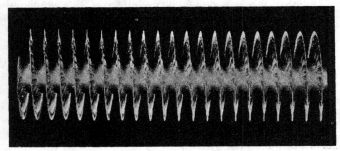

(c) Crimped spiral fin

Fig. 15   Steam air heater

Steam or steam-and-water drum . . . . A pressure chamber located at the upper extremity of a boiler circulatory system in which the steam generated in the boiler is separated from the water and from which steam is discharged at a position above a water level maintained therein.

Superheater . . . . A group of tubes which absorbs heat from the products of combustion to raise the temperature of the vapor passing through the tubes above the saturation temperature corresponding to its pressure.

Tangent-tube wall . . . . A waterwall in which the tubes are substantially tangent to each other with practically no space between the tubes.

Tube bank . . . . A group of two or more rows of tubes forming part of a watertube boiler circulatory system and to which heat is transferred from the products of combustion mainly by convection.

Tube sheet . . . . The part of the drum or header which the ends of the tubes penetrate.

Unheated downcomer . . . . . . A tube not exposed to the products of combustion in which water may flow from the steam drum to the water drum or header.

Watertube . . . . A tube in a boiler having the water and steam on the inside and the products of combustion on the outside.

Water-cooled furnace . . . . . . A furnace wall containing watertubes arranged to form a waterwall.

Welded, mono-wall, or membrane wall . . . . . A waterwall in which the tubes are welded together (or to filler bars between them) to form a continuous furnace wall.

Section 2
Considerations in the Selection of a Boiler

2.1 General. Many factors influence the design and selection of steam generating equipment to produce the required quantities of steam at the design pressure and temperatures for a particular installation. Efficient operation when burning the various fuels available throughout the world is a requirement. The boiler also must fit easily and conveniently within a minimum of engine room space, yet be accessible for operation, inspection, and maintenance. Although light in weight, it has to be sufficiently rugged to operate dependably under adverse sea conditions. Operation over a wide load range, with a minimum of attention, and operating characteristics compatible with a high degree of automation are also required. The factors used in both the thermal and structural design must be conservative to provide assurance that continuous operation over extended periods of time will be provided with minimum maintenance. Finally, the boiler must meet the rules and regulations of the regulatory bodies.

2.2 Cycle Requirements. The design of a marine boiler is directly affected by the heat cycle selected by the ship's designer. Higher steam pressures and temperatures may make reductions in the size and weight of a given propulsion plant possible, or permit a higher horsepower installation in the same space. During World War II, most combat naval vessels operated at 600 psig–850 F while steam to 450 psig–750 F was widely used in merchant ships. In the postwar era the Navy advanced to 1200 psig–950 F (nominal) for its combat vessel construction. In the late 1940's and 1950's a significant number of merchant vessels appeared using steam at 600 psig–850 F and 850 psig–850 F. By the 1960's almost all new construction used 850 psig–950 F steam; several large vessels used steam (in some cases with reheat) at 1500 psig–950 F. Machinery plants utilizing steam at pressures of 850 to 1500 psig and temperatures from 950 to 1000 F are characteristic of most commercial steamships built in recent years.

The quantity of steam produced by a marine boiler can range from as little as 1500 lb/hr in small auxiliary boilers to over 400,000 lb/hr in large main propulsion boilers. Steam outputs of 750,000 lb/hr or more per boiler are practical for high-power installations.

2.3 Heat Balances. The fuel cost per shaft horsepower is one of the deciding factors in establishing the characteristics of the boiler installation and whether or not the installation is economically sound. The fuel rate can be decreased by the use of higher steam pressures and temperatures or a more sophisticated cycle can be employed by the use of reheating, economizers, air heaters, more stages of feed heating, etc. The designer must analyze these factors in light of initial cost, maintenance, weight, and space requirements versus the savings resulting from increased thermal efficiency.

As steam pressures increase, it is essential to use additional heat-reclaiming equipment in the boiler unit. This is because of the corresponding increase in saturated steam temperature, which results in a higher gas temperature leaving the boiler bank and thereby reduces the boiler efficiency at a given firing rate.

Reheating the steam improves the thermal efficiency but requires larger boilers and special provisions to protect the reheater during astern operation. High steam pressures and temperatures, along with reheating, are more likely to be used in installations of 30,000 shp and up, where the value of the fuel saved may well justify greater initial cost and cycle complication. In addition, the load factor in such vessels is apt to be much higher, giving added impetus to the establishment of more efficient designs [11].

It is from the detailed heat balances prepared by the marine engineer that the quantities of steam and feedwater flow are determined. In the usual plant from two to four stages of feedwater heating are used to supply water to the boiler at temperatures from 270 to 400 F.

Boiler efficiencies of over 90% are possible. However, to minimize corrosion and maintenance in the cold-end heat exchangers and uptakes, it may prove advantageous to limit the boiler efficiency to 88.5–90% with some fuels. Fuel oils vary widely in quality and often contain significant amounts of sulfur, which can form sulfuric acid if there is condensation in the exhaust gas path. Corrosion and maintenance costs should be balanced against the possible savings in fuel costs derived from a higher boiler efficiency.

2.4 Fuels and Methods of Firing. The characteristics of the fuels that will be available to the ship in its usual trade should be established early in the design process. This will permit the optimum selection of equipment for burning the fuel and cleaning the boiler. In addition, a suitable selection of uptake temperatures and materials can be made for the entire boiler plant so as to reduce corrosion and maintenance problems.

Most marine boilers are oil fired, with gas- and coal-fired boilers less common. At sea, tankers designed to carry liquefied natural gas (LNG) may use the natural boil-off from their cargo gas tanks as a supplemental fuel. This cargo gas boil-off is collected and pumped to the boilers where it is burned either by itself or in combination with oil. The quantity of boil-off available from the liquefied natural gas is a function of ambient sea and air temperatures, the ship's motion, and the cargo loading, among other things, and may vary from day to day.

For years boilers using coal for fuel were hand or stoker fired. As oil became readily available and relatively inexpensive, and as labor and air-pollution control costs rose, the use of coal declined. Various attempts were made to use pulverized-coal firing, which was widely used ashore, but the results were poor since the furnace volume necessary for flame travel, low heat release rates, and satisfactory combustion when firing with pulverized coal required a considerably larger furnace than could be installed in the space usually allocated to a marine boiler.

Drawing from shoreside practice, radiant boilers with very large furnaces and relatively small banks of generating surface have been provided for recently built coal-fired ships. These units were built with updated stokers, controls, and coal-handling equipment, which demonstrated the capability of operating in a periodically unmanned engine room environment.

Pulverized-coal firing results in a high fly-ash loading of the flue gas, which aggravates tube erosion, slagging, and stack-emission problems. These are important concerns in marine boiler applications. However, a fluidized-bed combustion technique may permit increased furnace loadings with significant reductions in slag and ash carryover to the generating banks, as well as reduced stack emissions. The combustion characteristics of other low-grade fuels may also be enhanced by the use of a fluidized-bed fired boiler.

Oils were used as boiler fuels as early as the 1870's but did not achieve widespread use until the automobile age required a worldwide petroleum industry. Compared with other fuels, oil is easily loaded aboard ship, stored, and introduced into the furnace; and the firing equipment requires little costly maintenance. The small amount of ash and contaminants it contains does not require the extensive ash-handling facilities required for coal firing.

It should be recognized that fuel oils from different sources, while similar in heating value, have varying amounts of contaminants, which may be harmful in various ways. The major contaminants consist of salts of vanadium and sodium. As a class, they are called "ash" and their presence must be fully taken into account by the designer. Likewise, the sulfur content may vary over a range from almost none to as much as 6–7% in "sour" crudes; sulfur has a decided effect on the cycle efficiency which can be obtained without serious corrosion in the economizer, air heater, and uptakes.

The compounds of vanadium and sodium affect the design of the superheater. If oils to be burned in a particular trade are especially rich in these constituents, the superheater can be designed with tube metal temperatures lower than normal to avoid the possibility of severe slagging and tube metal corrosion problems. Cold-end heat exchangers designed with full recognition of the sulfur content present in the fuel will experience a minimum of corrosion and expensive maintenance.

A boiler designed to take advantage of low-cost residual fuel oils can always burn lighter fuels if the situation justifies it. However, a boiler with tightly packed heating surfaces designed for light oils such as diesel or aviation turbine fuels would not perform satisfactorily on residual fuels for very long. Gas-side fouling and oil burner and combustion problems in the furnace could be anticipated.

2.5 Effect of Ship Design and Other Machinery on Boiler Design.   Factors such as space, weight, and the requirements of the regulatory bodies are major considerations in the design of a boiler. In addition, however, the prospective vessel owner or his naval architect may have preferences regarding the boiler design and specific design requirements. These preferences may include the number of boilers, types of boilers and their arrangement, locations of major connections, the use of economizers or air heaters or both, firing, and evaporative ratings, and the type and method of firing. Life-cycle costs can have a bearing on the preference likewise, since the total cost and labor involved in maintaining a previous design or construction may be reflected in the owner's specifications and result in the selection of an improved design and construction.

a. Space.   The space provided for the machinery is held to a minimum by the naval architect because the space occupied by the machinery produces no revenue. The boiler designer is usually required to adapt the boiler design to the available space. The boiler height may be limited by deck or machinery casing locations. The fore-and-aft or depth dimension of the boilers may be controlled by bulkhead locations, access, or tube renewal space requirements as well as the location of control consoles, main engines, etc.

To a large extent the available space determines the economy of the design. A height restriction is particularly serious, since it usually necessitates increased boiler width or length to obtain the required heating surface. This generally results in a marked increase in boiler cost, weight, and the base area occupied.

b. Weight.   With drum-type boilers, the minimum weight for maximum efficiency is obtained with minimum furnace depth, maximum tube length, and the maximum number of tube rows. Limiting the height may restrict capacity because of reduced circulation. It may also result in tube slopes and in burner clearances less than the minimum necessary for a good design.

The minimum weight of any type of boiler will vary considerably with design conditions; increases in evaporative rating, burner capacity, or air pressure decrease the weight of a boiler designed for a specified steam output.

With a fixed evaporative rate per square foot of heat-absorbing surface, the weight of a boiler per pound of steam generated will be less for boilers with greater steam output, since certain boiler parts remain fixed in size and weight over a reasonable range in capacity.

Weight is greatly dependent on space also. Generally the larger the physical dimensions of a boiler for a given output, the greater its weight.

c. Regulations.   The ocean environment is no place to test unproven principles. This became evident in the early days of steamship construction when it was recognized that some rules and regulations were necessary to protect life and property. These rules were not intended to inhibit the designer or innovator but rather to provide a sound basis and yardstick for comparison of new designs with older successful designs.

Disastrous boiler explosions, common to both marine and stationary boilers, resulted in the establishment of a steam boiler inspection service and strict regulations governing the construction, care, and operation of steam boilers. In the design of marine boilers the applicable regulatory rules and standards must be rigorously followed. Most units built for American-flag ships meet the requirements of the United States Coast Guard and the American Bureau of Shipping.

Boilers for naval combatant ships are built in accordance with Navy specifications, although for auxiliary naval vessels the use of the United States Coast Guard or the American Society of Mechanical Engineers codes often is permissible. For foreign-flag ships, the rules and regulations of other regulatory bodies would apply. In addition, many shipyards and operators of large fleets have established their own supplementary requirements.

Most rules pertain to construction and the inspection and approval of materials, and establish very few performance limitations. Although Navy specifications limit the heat release rates per cubic foot of furnace volume, per square foot of radiant heat absorbing surface, and per square foot of total heating surface, these limits may be modified in the special specifications issued for a particular class of vessel. The Maritime Administration follows a somewhat similar procedure and usually establishes evaporative and furnace heat release rates for each design.

2.6 Boiler Design Criteria. Theoretical and practical considerations have led to the establishment of boiler design criteria in a number of areas not directly associated with the regulatory bodies' rules, which concern mainly pressure-part scantlings and construction techniques. The design criteria are most important in the areas of combustion, heat absorption rates, circulation, and pressure drops through the boiler system. They provide the yardstick by which various boiler designs can be compared for their suitability for specific applications.

a. Combustion. At the heart of a successful boiler is a properly designed furnace and fuel burning system. If the fuel supplied to the furnace is not burned cleanly and completely within the furnace throughout the range of operation, it will not be possible to accurately predict the performance of the evaporator-superheater combination. For example, the total steam generated may be insufficient, the steam temperature may be incorrect, or the efficiency may be lowered by incomplete combustion or improper excess air.

A number of criteria by which combustion in furnaces can be gaged and by which different furnaces can be compared have been developed [12]. In general, with the exception of the furnace heat absorption rate which is derived from the actual heat-transfer calculations developed for the furnace, they are empirical relationships with little theoretical value; however, they can be used to compare similar boiler designs provided their limitations are recognized.

The criteria most frequently used for these comparisons are:

- Heat release rate per cubic foot of furnace volume.
- Firing rate per square foot of radiant heat absorbing surface.
- Heat absorption rate per square foot of radiant heat absorbing surface.

A brief review of these factors will serve to indicate their importance and usefulness.

The heat release rate per cubic foot of furnace volume is useful in comparing geometrically similar furnaces, but while widely used because of its simplicity, it is not an important criterion. The heat released is the product of the hourly fuel rate and its higher heating value, ignoring any heat above 100 F in the combustion air.

If radiant heat absorption rates, furnace gas temperatures, and furnace tube metal temperatures are satisfactory, the only limitation on the heat release rate per cubic foot of furnace volume should be that imposed by the ability of the firing equipment to maintain good combustion conditions. The use of a high, yet satisfactory, furnace volume heat release rate greatly facilitates the installation of high-capacity, lightweight boilers in a minimum of space.

The temperature within a boiler furnace can be controlled to a large extent by the effective radiant heat absorbing surface (RHAS) present in the furnace [12]. Heat is radiated from the flame envelope to the heat-absorbing surfaces with the uncooled refractory surfaces acting as an intermediary, receiving heat from the flame and then re-radiating most of the received heat back to the flame and cold surfaces. For a given heat input or firing rate, the heat absorbed per unit area decreases with an increase in total RHAS. The greater the RHAS the greater will be the total amount of heat absorbed by the furnace. Therefore, the temperature of the gases leaving the furnace will be lower.

As marine boilers are normally fired with residual fuel oils of nearly uniform heating values, the Btu input can be reduced to pounds of fuel oil per hour and the criterion "firing rate per square foot of radiant heat absorbing surface" is obtained. This popular criterion has little basis in theoretical heat-transfer analysis but is widely used.

The heat absorption rate per square foot of radiant heat absorbing surface received very little attention for many years because furnace designs were conservative and steam pressures were low. However, it is mandatory to determine the furnace heat absorption rates for high-pressure and highly rated boiler units. These rates offer a good basis for comparing furnace performance since the relationships between heat-absorbing and radiant surfaces, as well as the effect of heated combustion air, are taken into account.

The designer is concerned primarily with furnace heat absorption by radiation to the furnace wall tubes and by radiant and convection heat transfer to the furnace rows of the boiler tube bank—because these are the base from which the furnace exit gas temperature, tube metal temperatures, thermal stresses, and allowable thickness of internal scales are calculated.

The higher the absorption rates, the better the water chemistry has to be to avoid scale and sludge deposits which retard heat transfer through the furnace tube walls. If the resulting tube metal temperatures exceed the allowable limits, early and sudden tube failures will result.

b. Circulation. Natural-circulation boilers depend on the heat input to establish differential densities between the steam-and-water mixture flowing in the riser or generating circuits and the water flowing downward in the downcomer circuits. The geometry of the boiler plays a part in this since the boiler must be designed to circulate regardless of the angle of list or trim of the vessel and the roll or pitch it may experience in a seaway. In this regard the circulation of a marine boiler has more demands placed on it than a corresponding shoreside boiler. In addition, the heat input and the steam output of the marine boiler are probably higher than for a comparable application ashore.

It is customary to consider a momentary roll of 30 deg from the horizontal and a momentary pitch of ±5 deg when computing static and dynamic loads. In establishing circulation, boilers are usually designed for a permanent list of 15 deg and a permanent trim by the bow or stern of 5 deg. The latter, when coupled with the momentary pitch of 5 deg, means that in the fore-and-aft direction the boiler may be as much as 10 deg from the horizontal. The arrangement of the tubes and steam-and-water drum must take this into account. In addition, drains must be located so that under normal conditions the boiler and its component parts can be drained when necessary for cleaning and inspection.

The size of the steam drum and the location of the design water level must be compatible with the list and trim requirements of the vessel and provide a safe working level of water adequate to cover the downcomers at all times. The location of the gage glasses showing this working water level should be such as to assure that a safe level is continuously maintained within the steam drum [13].

c. Pressure drops. The flow of water, steam, air, and flue gas within a boiler entails losses in pressure. A good design limits these losses to values consistent with the benefits obtained by incurring them. On the steam side, the pressure drop of water flowing through the economizer and of steam flowing through the superheater provides even distribution between parallel tube circuits and thereby assures that the tubes will be uniformly cooled by the particular fluid. Excessive pressure drops on the steam side result in an unwarranted increase in the design pressure of the boiler and the economizer. This increases the first cost and results in a heavier boiler and thicker piping, as well as additional feed pump power requirements.

Conversely, high combustion air and flue gas pressure losses can result in a smaller boiler since the tube banks can be arranged with less free space between adjacent tubes and the high velocities resulting from the higher pressure drops increase the convection heat-transfer rates. These higher heat-transfer rates permit the use of

less surface and result in a smaller, lighter boiler for any given steam output. The increased life-cycle cost associated with this approach is reflected in the additional horsepower required to drive the forced-draft blower over the life of the vessel.

d. Duty cycle. The utilization or duty cycle of a ship's propulsion system is an important factor in establishing the type of boiler to be used, its firing equipment, and the extent of the automatic controls to be applied. The boiler must be capable of rapid response to changes in demand for steam output. This is necessary because of the rapid and wide changes in maneuvering power requirements of large, high-speed ships. The boiler must likewise be capable of prolonged periods of steady operation at its design rating. Also, in port it may be subjected to long periods of operation at low or minimum outputs.

Cleaning, with the exception of the daily use of the soot blowers or occasional attention to the atomizers in the oil burners, is normally deferred to the annual or biannual period when the vessel is in a shipyard for other maintenance. This must be fully taken into account by properly locating soot blowers so they are effective; by using the optimum burner combinations for the range of fuel-oil types anticipated to be bunkered; and by using the best possible arrangements of economizer, air heater, boiler furnace, and generating surfaces to minimize fouling.

The duty cycle may also have a pronounced effect on the number of boilers selected. A single boiler may be employed in ships of up to about 90,000 shp. Two or more boilers may be selected for higher power levels or where redundancy is desired or required. Single-boiler vessels have proven reliable in service and should continue to do so. This is in part due to the fact that a boiler kept continuously in service reaches thermal equilibrium and can have the waterside chemistry optimized. In general, from a boiler performance point of view, the least number of boilers that can deliver the required steam will prove to be the best selection for any particular vessel. However, most operators prefer more than one boiler to permit maintenance work to be accomplished without shutting down the entire power plant.

e. Automation. The widespread use of automatic controls and monitoring equipment has made bridge control of the power plant a reality and has permitted a reduction in the number of watch standers in the machinery space. These desirable improvements have added additional considerations to the problem of designing a suitable boiler.

Of prime importance is a fuel-burning system that can respond rapidly throughout the range of operation from standby to maximum power without a fireman's attention. It must do so to prevent excursions in steam pressure and reduce water level fluctuations (shrink and swell due to changes of the volume of steam present in the boiler), which might result in water carry-over into the superheater [14].

Burners can be designed to operate over the full boiler range with all burners in service, or other burner types with less range can be sequenced, that is, placed in or out of service on command by the control system. Suitable flame-monitoring safeguards and purge interlocks are necessary in varying degrees of complexity depending on the extent of manual supervision desired.

Feedwater regulators, steam temperature controls, data logging equipment for flows, pressures, tempera-tures, levels, etc. are all available from the simple to the ultrasophisticated. The owner and his naval architect usually select the scope of equipment and advise the boiler designer so that the boiler and burner combination can be made compatible with it.

Section 3
Boiler Design

3.1 General. The fundamental boiler design problem is to determine the proper proportions of the various heat-absorbing surfaces to use the maximum heat available in the products of combustion. A proper design will accomplish this at the lowest cost on a life-cycle basis. Each component must be integrated with the other elements of the unit to provide a balanced design in which the first costs and fuel, maintenance, and operational costs will be a minimum over the useful life of the ship. In no way must safety or reliability be compromised by these cost considerations.

For the steam generator system, the following must be considered:

1. Fuel burning equipment
2. Furnace
3. Boiler generating surface
4. Superheater (and reheater if used)
5. Economizer and air heater
6. Attemperator (or control) and auxiliary desuper-heaters
7. Circulatory and steam separator system
8. Casing and setting
9. Cleaning equipment
10. Safety valves and other mountings
11. Feedwater and treatment
12. Foundations and supports
13. Combustion air supply system
14. Uptake gas duct system and stack

These considerations require many interrelated steps. In most cases, a number of assumptions must be made in order to initiate the design. As the design calculations proceed, the assumptions are refined to achieve the desired accuracy in the final analysis.

The first step is the selection of the basic type of boiler, superheater, and economizer or air heater (or both) to be used. This selection is based in part on preference and in part on the space available for the installation and its operating requirements.

The quantity of fuel required is determined from the desired steam generator efficiency, the given steam pressure, temperature, and flow, the feedwater temperature, and the heating value of the fuel.

The fuel characteristics and quantities establish the fuel burning equipment to be employed. This in turn sets the excess air requirements. Combustion calculations are next made to determine the hourly quantities of flue gas flowing through the unit. The exit or stack gas tempera-

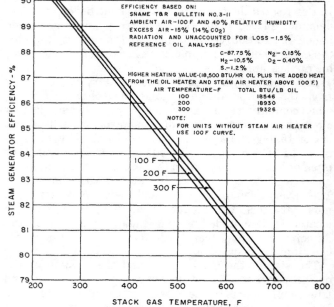

Fig. 16 Efficiency versus stack-gas temperature

ture to which the flue gas must be cooled to achieve the desired efficiency is determined (Fig. 16); and if experience indicates that it is attainable or otherwise satisfactory, the design can proceed. If not, another selection of efficiency must be made and the calculations repeated.

The furnace exit gas temperature is next calculated. Its value is dependent on the radiant and convection heat-transfer surface installed in the waterwalls, floor, roof, and screen (radiant only) as well as the extent of refractory present. Next, the gas temperature drops and the heat absorbed by the screen and superheater are determined. The size and spacing of tubes and the amount of surface are assumed initially. These are then modified to provide the desired steam temperature and conservative tube metal temperatures as necessary. Usually several screen and superheater combinations are investigated to determine the most economical solution.

The boiler bank, economizer, and air heater surfaces are then sized to provide the final uptake gas temperature required. In each of the steps just outlined, initial choices of thicknesses and types of materials for tubes, headers, and drums are made.

With the heating surfaces established, the draft loss through all components is calculated. If the draft loss exceeds the capability of the fan desired, the heat transfer and drafts previously calculated are adjusted by changing the tube spacing, number of rows crossed or the depth or height of the boiler components. A different number or size of oil burners may be necessary to assist in achieving a final balance of draft requirements and fan capabilities.

Pressure drops of water and steam through all components from the economizer feedwater inlet to the superheater outlet are next computed. They, in turn, establish the required boiler and economizer design pressures and the safety valve settings. A circulation analysis is then prepared using the heat absorptions determined from the heat-transfer calculations. From this, the size and number of supply and riser tubes are adjusted as required.

The foregoing steps are followed for each design. However, with experience the designer can make very close first approximations and substantially reduce the time required to prepare a design.

3.2 Fuel Combustion. The basic function of a boiler furnace is to generate the maximum amount of heat from a given quantity of a specific fuel. A useful secondary function is to generate steam in the furnace wall tube circuits. The theoretical aspects of combustion have been well known for many years. However, the achievement of good combustion within the furnace of a relatively small marine boiler requires practical knowledge and experience. Complete combustion can be obtained provided there is sufficient time (a function of furnace volume), turbulence (provided by the geometry of the burner assembly), and a temperature high enough to provide ignition.

Combustion may be defined as the chemical combination of oxygen with the combustible elements in the fuel. The common fuels have only three elemental constituents which unite with oxygen to produce heat. The elements and their compounds, as well as their molecular weights and combustion constants, including heating values, are given in Table 1.

Oxygen combines with the combustible elements and their compounds in accordance with the laws of chemistry. Typical reactions for the combustible constituents of fuel oil, based on the assumption that the reaction is completed with the exact amount of oxygen required, are:

for carbon (to CO_2) $C + O_2 = CO_2 + \Delta Q$
for hydrogen (to H_2O) $2H_2 + O_2 = 2H_2O + \Delta Q$
for sulfur (to SO_3) $2S + 3O_2 = 2SO_3 + \Delta Q$

where ΔQ is the heat evolved by the reaction.

The heat evolved or heat of combustion is commonly called the "fuel heating value" and is the sum of the heats of reaction of the various constituents for one pound of the fuel considered. The heating value of a fuel may be calculated from theoretical considerations or may be determined, for an actual oil, by burning a sample in a bomb calorimeter (see Chapter 12 for additional discussion in this regard).

In testing fuels by a bomb calorimeter to determine the heat given up, two values may be reported: the higher (or gross or upper) heating value and the lower or net heating value. For the higher heating value, it is assumed that any water vapor formed by burning the hydrogen constituent is all condensed and cooled to the initial temperature in the calorimeter at the end of the test. The heat of vaporization, about 970 Btu/lb oil, is included in the reported heating value. For the lower heating value, it is assumed that none of the water vapor condenses and that all the products of combustion remain in a gaseous state. In the United States higher heating values are used as they are available directly from the calorimeter determinations and because of the established practice of buying fuel on a higher heating value basis. The lower heating values are generally used in European practice.

a. Fuel analysis. For design and comparative purposes, the standard reference fuel oil is #6 fuel oil (Bunker C) having the following characteristics [15]:

CHEMICAL COMPOSITION
(percent by weight)

Carbon	87.75
Hydrogen	10.50
Sulfur	1.20
Oxygen	0.40
Nitrogen	0.15
Moisture	—
Total	100.00

The higher heating value of the standard reference fuel as determined by a bomb calorimeter and corrected for specific heat at constant pressure is 18,500 Btu per lb. The base temperature for heat content is established as 100 F. For design and heat balance calculations the heating value of the oil is corrected for the additional heat added (in Btu/lb) in heating the oil to the temperature (assumed to be 200 F) necessary for proper atomization by the following expression:

Added heat = 0.46 (atomizing temperature − 100 F)

The total heating value of the reference oil is, therefore, 18,546 Btu per lb and is used for all types of atomization including steam atomization.

b. Combustion air. The oxygen required for combustion is supplied by the combustion air. The other constituents of the air act as diluents. Atmospheric air is a mixture—as distinguished from a chemical compound—of oxygen, nitrogen, and small quantities of carbon dioxide, water vapor, argon, and other inert gases. The basic composition of dry air for combustion purposes is considered to be:

	% OXYGEN	% NITROGEN
By weight	23.15	76.85
By volume	21.00	79.00

The rare gases are included as part of the nitrogen constituent.

Air is assumed to be supplied to the forced-draft fan at a temperature of 100 F, a relative humidity of 40 percent, and a barometric pressure of 29.92 in. Hg. Under such conditions air has the following physical properties:

Table 1   Combustion constants

No.	Substance	Formula	Molecular Weight	Lb per Cu Ft	Cu Ft per Lb	Sp Gr Air = 1.0000	Heat of Combustion Btu per Cu Ft Gross (High)	Btu per Cu Ft Net (Low)	Btu per Lb Gross (High)	Btu per Lb Net (Low)	For 100% Total Air — Moles per mole of Combustible or Cu Ft per Cu Ft of Combustible — Required for Combustion O₂	N₂	Air	Flue Products CO₂	H₂O	N₂	For 100% Total Air — Lb per Lb of Combustible — Required for Combustion O₂	N₂	Air	Flue Products CO₂	H₂O	N₂
1	Carbon*	C	12.01	...	...	...	...	...	14,093	14,093	1.0	3.76	4.76	1.0	...	3.76	2.66	8.86	11.53	3.66	...	8.86
2	Hydrogen	H_2	2.016	0.0053	187.723	0.0696	325	275	61,095	51,623	0.5	1.88	2.38	...	1.0	1.88	7.94	26.41	34.34	...	8.94	26.41
3	Oxygen	O_2	32.00	0.0846	11.819	1.1053	...	...	...	...	...	...	...	...	...	...	...	...	...	...	...	...
4	Nitrogen (atm)	N_2	28.01	0.0744	13.443	0.9718	...	...	...	...	...	...	...	...	...	...	...	...	...	...	...	...
5	Carbon monoxide	CO	28.01	0.0740	13.506	0.9672	321	321	4,347	4,347	0.5	1.88	2.38	1.0	...	1.88	0.57	1.90	2.47	1.57	...	1.90
6	Carbon dioxide	CO_2	44.01	0.1170	8.548	1.5282	...	...	...	...	...	...	...	...	...	...	...	...	...	...	...	...
PARAFFIN SERIES																						
7	Methane	CH_4	16.04	0.0425	23.552	0.5543	1012	911	23,875	21,495	2.0	7.53	9.53	1.0	2.0	7.53	3.99	13.28	17.27	2.74	2.25	13.28
8	Ethane	C_2H_6	30.07	0.0803	12.455	1.0488	1773	1622	22,323	20,418	3.5	13.18	16.68	2.0	3.0	13.18	3.73	12.39	16.12	2.93	1.80	12.39
9	Propane	C_3H_8	44.09	0.1196	8.365	1.5617	2524	2322	21,669	19,937	5.0	18.82	23.82	3.0	4.0	18.82	3.63	12.07	15.70	2.99	1.63	12.07
10	n-Butane	C_4H_{10}	58.12	0.1582	6.321	2.0665	3271	3018	21,321	19,678	6.5	24.47	30.97	4.0	5.0	24.47	3.58	11.91	15.49	3.03	1.55	11.91
11	Isobutane	C_4H_{10}	58.12	0.1582	6.321	2.0665	3261	3009	21,271	19,628	6.5	24.47	30.97	4.0	5.0	24.47	3.58	11.91	15.49	3.03	1.55	11.91
12	n-Pentane	C_5H_{12}	72.15	0.1904	5.252	2.4872	4020	3717	21,095	19,507	8.0	30.11	38.11	5.0	6.0	30.11	3.55	11.81	15.35	3.03	1.50	11.81
13	Isopentane	C_5H_{12}	72.15	0.1904	5.252	2.4872	4011	3708	21,047	19,459	8.0	30.11	38.11	5.0	6.0	30.11	3.55	11.81	15.35	3.05	1.50	11.81
14	Neopentane	C_5H_{12}	72.15	0.1904	5.252	2.4872	3994	3692	20,978	19,390	8.0	30.11	38.11	5.0	6.0	30.11	3.55	11.81	15.35	3.05	1.50	11.81
15	n-Hexane	C_6H_{14}	86.17	0.2274	4.398	2.9704	4768	4415	20,966	19,415	9.5	35.76	45.26	6.0	7.0	35.76	3.53	11.74	15.27	3.06	1.46	11.74
OLEFIN SERIES																						
16	Ethylene	C_2H_4	28.05	0.0742	13.475	0.9740	1604	1503	21,636	20,275	3.0	11.29	14.29	2.0	2.0	11.29	3.42	11.39	14.81	3.14	1.29	11.39
17	Propylene	C_3H_6	42.08	0.1110	9.007	1.4504	2340	2188	21,048	19,687	4.5	16.94	21.44	3.0	3.0	16.94	3.42	11.39	14.81	3.14	1.29	11.39
18	n-Butene	C_4H_8	56.10	0.1480	6.756	1.9336	3084	2885	20,854	19,493	6.0	22.59	28.59	4.0	4.0	22.59	3.42	11.39	14.81	3.14	1.29	11.39
19	Isobutene	C_4H_8	56.10	0.1480	6.756	1.9336	3069	2868	20,737	19,376	6.0	22.59	28.59	4.0	4.0	22.59	3.42	11.39	14.81	3.14	1.29	11.39
20	n-Pentene	C_5H_{10}	70.13	0.1852	5.400	2.4190	3837	3585	20,720	19,359	7.5	28.23	35.73	5.0	5.0	28.23	3.42	11.39	14.81	3.14	1.29	11.39
AROMATIC SERIES																						
21	Benzene	C_6H_6	78.11	0.2060	4.852	2.6920	3752	3601	18,184	17,451	7.5	28.23	35.73	6.0	3.0	28.23	3.07	10.22	13.30	3.38	0.69	10.22
22	Toluene	C_7H_8	92.13	0.2431	4.113	3.1760	4486	4285	18,501	17,672	9.0	33.88	42.88	7.0	4.0	33.88	3.13	10.40	13.53	3.34	0.78	10.40
23	Xylene	C_8H_{10}	106.16	0.2803	3.567	3.6618	5230	4980	18,650	17,760	10.5	39.52	50.02	8.0	5.0	39.52	3.17	10.53	13.70	3.32	0.85	10.53
MISCELLANEOUS GASES																						
24	Acetylene	C_2H_2	26.04	0.0697	14.344	0.9107	1477	1426	21,502	20,769	2.5	9.41	11.91	2.0	1.0	9.41	3.07	10.22	13.30	3.38	0.69	10.22
25	Naphthalene	$C_{10}H_8$	128.16	0.3384	2.955	4.4208	5854	5654	17,303	16,708	12.0	45.17	57.17	10.0	4.0	45.17	3.00	9.97	12.96	3.43	0.56	9.97
26	Methyl alcohol	CH_3OH	32.04	0.0846	11.820	1.1052	868	767	10,258	9,066	1.5	5.65	7.15	1.0	2.0	5.65	1.50	4.98	6.48	1.37	1.13	4.98
27	Ethyl alcohol	C_2H_5OH	46.07	0.1216	8.221	1.5890	1600	1449	13,161	11,917	3.0	11.29	14.29	2.0	3.0	11.29	2.08	6.93	9.02	1.92	1.17	6.93
28	Ammonia	NH_3	17.03	0.0456	21.914	0.5961	441	364	9,667	7,985	0.75	2.82	3.57	...	1.5	3.32	1.41	4.69	6.10	...	1.59	5.51
29	Sulfur*	S	32.06	...	...	...	...	...	3,980	3,980	1.0	3.76	4.76	SO₂ 1.0	...	3.76	1.00	3.29	4.29	SO₂ 2.00	...	3.29
30	Hydrogen sulfide	H_2S	34.08	0.0911	10.979	1.1898	646	595	7,097	6,537	1.5	5.65	7.15	SO₂ 1.0	1.0	5.65	1.41	4.69	6.10	SO₂ 1.88	0.53	4.69
31	Sulfur dioxide	SO_2	64.06	0.1733	5.770	2.2640	...	...	...	...	...	...	...	...	...	...	...	...	...	...	...	...
32	Water vapor	H_2O	18.02	0.0476	21.017	0.6215	...	...	...	...	...	...	...	...	...	...	...	...	...	...	...	...
33	Air	...	...	0.0766	13.063	1.0000	...	...	...	...	...	...	...	...	...	...	...	...	...	...	...	...

* Carbon and sulfur are considered as gases for molal calculations only.

NOTES: (1) All gas volumes corrected to 60 F and 30 in. Hg dry.
(2) This table is included by courtesy of the American Gas Association and the Industrial Press. The format and data are taken principally from "Fuel Flue Gases," 1941 Edition, American Gas Association, with modifications, especially in the four columns labeled "Heat of Combustion," using data from *Gas Engineers Handbook*, The Industrial Press, 1965.

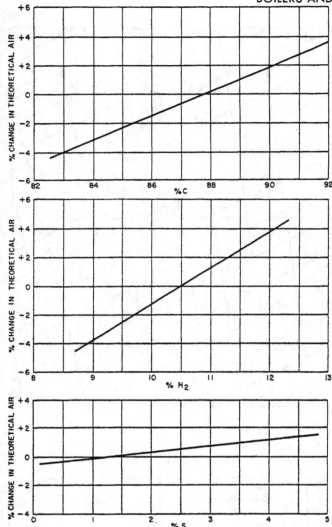

Fig. 17 Percent change in theoretical air for combustion versus various C, H_2, and S contents of fuel oil

Dry-air density, lb/ft³	0.0709
Moisture, lb/lb of dry air	0.0165
Mixture density, lb/ft³	0.0701
Specific heat	See Fig. 3 of Chapter 2

Based on the foregoing fuel and air standards, analysis will show that the stoichiometrical or theoretical quantity of dry air to burn one pound of fuel is 13.75 lb. From this, the following quantities of air for various excess percentages are determined:

Excess air, percent	0	5	10	15	20
Dry air, lb	13.75	14.44	15.13	15.81	16.50
Moisture, lb	0.23	0.24	0.25	0.26	0.27
Moist air, lb	13.98	14.68	15.38	16.07	16.77
Volume, ft³ (at 100 F, 29.92 in. Hg)					
dry air	194	204	213	223	233
moist air (40% RH)	200	210	220	230	240

The ultimate analysis of the fuels actually encountered in service varies from that of the standard reference fuel.

Figure 17 shows the effect of these variations on the theoretical air required for combustion. As an example, a fuel consisting of 87.25 C, 12.0 H_2, 0.2 S, 0.4 O_2, and 0.15 N_2 would require 3.0 percent more air for stoichiometric combustion (+3.8% for H_2, −0.4% for C, −0.4% for S) [16].

To reduce the dry gas loss of heat up the stack, the weight of flue gases should be held to a minimum consistent with supplying enough air to completely burn the fuel. Recognizing the foregoing, an operator should observe the results with the particular fuel oil bunkered and adjust the excess air to achieve complete combustion. However, in most cases, the design of a boiler is based on an air-fuel ratio sufficient to provide 15% excess air. While many oil burners and combustion control systems can operate successfully with less excess air, the use of 15% for design purposes assures sufficient heat-transfer surface and forced-draft fans with adequate capacity. For additional margin, when no air heater is installed, 20% excess air is frequently used.

Often the excess air or air-fuel ratio is discussed in terms of CO_2, which is readily obtained from an operating boiler by means of an Orsat analysis. An Orsat reading of 14% CO_2 corresponds to approximately 15% excess air. Figure 18 shows the relationship between CO_2, O_2, and excess air. As the heat-transfer and draft calculations are based on the weight of air and flue gas, the use of the terminology "percent CO_2," which is a volumetric measure, is of importance only in comparing oil burner performance. It is most useful where the oils used are of widely varying analysis from that of the standard reference fuel. The excess air, or air-fuel ratio, can also be determined conveniently by using an oxygen analyzer; a reading of 3% oxygen corresponds to approximately 15% excess air.

c. Efficiency. Boiler efficiency is defined as the ratio of the heat output to the heat input. The heat output is also equivalent to the heat input minus the losses. Stated in equation form:

$$\text{Efficiency} = \frac{\text{Heat output}}{\text{Heat input}} \qquad (1)$$

$$= \frac{\text{Heat input} - \text{Heat losses}}{\text{Heat input}}$$

For boilers not equipped with steam air heaters, this basic expression becomes:

$$\text{Efficiency} = \frac{H_i - H_L}{H_i} = \frac{H_i - (H_g + H_u)}{H_i} \qquad (2)$$

where

H_i = heat input, higher heating value of fuel burned corrected for specific heat at constant pressure, plus heat above 100 F in the fuel injected

H_L = heat loss

H_g = stack loss

H_u = radiation losses unaccounted for and manufacturer's margin

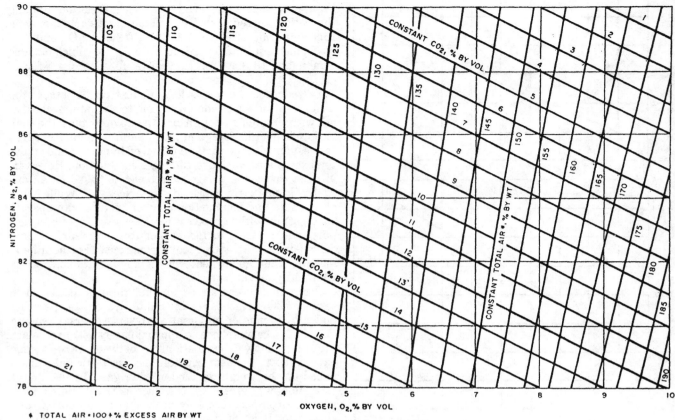

* TOTAL AIR = 100 + % EXCESS AIR BY WT

Fig. 18 CO_2, O_2, and excess air

The heat output may also be defined as the difference in enthalpy between the incoming feedwater to the boiler or economizer, if fitted, and the steam leaving the boiler (both superheated and desuperheated).

When steam air heaters are installed, the heat input from the steam is charged to the boiler total heat input and the efficiency becomes:

$$\text{Efficiency} = \frac{(H_i + H_a) - H_L}{H_i + H_a}$$

$$= \frac{(H_i + H_a) - (H_g + H_u)}{H_i + H_a} \quad (3)$$

where H_a is the heat added above 100 F to the combustion air by the steam air heater.

In the preliminary design process, one of these expressions is solved for the heat input H_i, from which the weight of oil fired is readily determined by dividing by the heating value of the design fuel, usually 18,546 Btu/lb. All quantities determined are based on hourly flow rates.

The required boiler efficiency is usually established by the specifications or heat balance. Along with the design steam pressure and temperature, it establishes the amount and arrangement of the heating surface installed in the boiler and economizer. The design steam pressure and the corresponding saturation temperature set the effective "sink" temperature of the boiler generating bank; and the feedwater sets that of the economizer. In the case

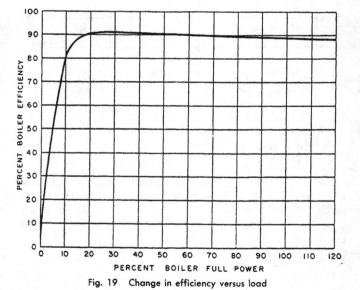

Fig. 19   Change in efficiency versus load

of an air heater installation, the sink is the temperature of the inlet air to it, usually 100 F. A typical curve of efficiency versus load for a steam generator is shown in Fig. 19. Note that the efficiency decreases with increased steam output. The quantity of hot flue gas to be cooled increases as more fuel is consumed to increase the steam output. As this occurs, the effectiveness of the fixed amount of heating surface installed decreases and the efficiency drops. It is usual to size the surface of boilers for merchant ships to provide the desired efficiency at the rating corresponding to "ABS power." The efficiency at the maximum or minimum rates is then a function of this design point and must lie on the characteristic efficiency curve.

A practical ceiling on the boiler efficiency is imposed by the requirement to maintain the temperature of the uptake gas above the dew point of the flue gas. This minimizes sulfur-bearing deposits and corrosion of the cold end of the heat exchanger and ductwork. In economizers, corrosion leads to leakage and forced boiler outages; therefore, it is common practice to maintain a minimum feedwater inlet temperature of about 280 F, which results in a flue-gas temperature of about 315 to 320 F and limits the risk of corrosion.

In a rotary regenerative air heater, a corrosion failure is non-catastrophic; therefore, a lower stack temperature (280 F or less) is practical and a correspondingly higher boiler efficiency is obtained. The cycle efficiency can be further improved through the use of high-pressure feedwater heaters to supply feedwater at temperatures which would be impractically high in an economizer cycle.

d. Oil burner selection. The choice of the type and number of oil burners to be used is dependent on the available draft loss, furnace dimensions, and the boiler firing rate. High-capacity, wide-range burners are usually selected for most installations to reduce the number of burners required and simplify maintenance and operation. The cost of control and flame safety equipment, as well as its maintenance, is thereby kept to a minimum.

The size and arrangement of the engine room often affects the burner location. It is desirable to locate the burners adjacent to the control console for ease of visual monitoring and accessibility. In the two-drum boiler, burners can be installed in the furnace front wall, roof, or sidewall.

In the front-fired boiler, the gases are fired parallel to the boiler bank. They then make a 90-deg turn to enter the screen rows, and as the furnace depth is usually the shortest dimension the gas tends to pile up on the rear wall. This heavy concentration of gas in the rear unbalances the gas temperatures, and makes the prediction of the steam temperature and superheater tube metal temperatures more difficult. On the other hand, with roof firing the gases are uniformly distributed over the depth of the boiler. Since furnace height is usually the longest dimension, there is less tendency to concentrate the gases before they turn into the superheater screen.

Side firing, with the burners in the sidewall, requires that careful attention be given to design details. Since the gases make no turns before entering the screen, flame lengths tend to be extremely long. This can result in flame penetration of the screen and superheater banks with an adverse effect on the superheater tube temperatures and steam temperature [3].

Usually, at least two oil burners are used so that one burner can be fired when cleaning or changing sprayer plates in the other. Ideally, a single burner per boiler

Table 2 Oil burner clearances

Throat Diameter, in.	Centerline Spacing, in.	Wall Clearance, in.
16	30	30
20	36	34
24	42	38
30	48	42
36	54	46
42	60	50

would greatly simplify the installation of controls and permit the furnace arrangement to be optimized. In relatively small boilers, it is possible to obtain the required oil rates with a single burner. In high-capacity boilers the large amount of fuel and air to be introduced into the furnace necessitates a multiple burner installation.

Each size burner has a minimum rate of operation below which the flame becomes unstable and there is risk of losing ignition. In part this is a characteristic of the burner, but the forced-draft, fuel, and control systems also have an influence. The minimum rate is of great importance since a much simpler plant results when all burners can be left in service at all times. When in port or during maneuvering conditions, the minimum oil flow capability must be less than that required by the plant demand, if frequent safety valve popping or steam dumping is to be avoided. Both of these actions waste steam and lead to increased maintenance.

Burner sequencing can be used effectively to follow the load demand where burners with limited range or higher-than-desired minimum flows are used. Solid-state, computer-controlled logic systems are often used to sequence burners; however, this equipment can increase costs considerably [17].

Care must be taken in arranging the burners to provide for even air distribution to each burner within the windbox to optimize combustion with a minimum of excess air. The clearances between the burners and the furnace walls must be sufficient to prevent interference and impingement. The furnace volume must be large enough to provide the time necessary for complete combustion to take place before the gases enter the superheater screen. Satisfactory combustion has been obtained at furnace release rates of up to 1,500,000 Btu/ft^3 in marine boilers.

Each burner manufacturer has his own recommended clearances and the shape of the flame can be adjusted to some extent to modify them when necessary. This is done by changing the spray angle of the atomizer. A wider angle is employed to shorten the flame length and produce a wide bushy flame while a narrower angle increases flame length and decreases width. The burner manufacturer should always be given the opportunity to review the projected furnace design so the best possible installation can be obtained. Generally suitable burner clearances are shown in Table 2. When firing Bunker C oil, it is customary to use the minimum clearances established by experience. These may be decreased, perhaps by six inches, if distillate oils are fired. Furnace depths of watertube boilers, which are front-fired, are usually limited to a minimum of six feet although there are highly rated boilers in service with furnace depths of only five feet.

The selection of the oil burner must also be compatible with the type of atomizer to be used. The atomizer alternatives include: steam atomization (internal mix), steam mechanical (external mix), mechanical wide-range vented-plunger, rotary cup, and others. Of these types, the internal-mix steam atomizer and the vented-plunger atomizer have the highest turndown (about 12 to 1) and provide the smallest and most uniform particle size over their range of operation. Finely atomized fuel droplets provide more surface area for combustion and permit less excess air to be used, thereby reducing the draft loss, fan power requirement, and stack dry-gas loss.

The number of burners selected usually results in a burner draft loss equivalent to about 35 to 50% of the total draft loss of the boiler unit. The burner draft loss varies with the volumetric flow of air through it. At any given airflow, a change in the temperature of the air will increase or decrease the draft loss in the ratio of the change of absolute temperatures. In designing a boiler with an air heater, it is standard practice to limit the air temperature leaving the air heater and entering the burners to no more than 600 F and preferably less to assure long life and prevent overheating of the burner parts. If the preliminary design yields an excessive air temperature, the designer must reapportion the surfaces, possibly adding a small economizer, to reduce the air heater air outlet temperature to an acceptable value.

3.3 Furnace Design. After the firing rate and the number and type of oil burners are established, the design of the furnace is undertaken. As the amount of radiant heat absorbing surface provided determines the furnace exit gas temperature to a large extent, a selection from possible furnace construction alternatives must be made. There are two general types of boiler furnace wall construction: the refractory wall and the water-cooled wall. Originally all furnaces were simple brick-lined chambers. As firing rates were pushed higher and the output of a given size boiler was increased, the life of refractory walls and maintenance costs became totally unsatisfactory.

To improve refractory life, water-cooling in the form of spaced tubes was arranged to absorb heat by radiation directly from the flame cloud. By this means the furnace temperatures were lowered and it was possible to increase the firing rate within a given furnace envelope [2,3]. Outages for repairs were minimized and the use of lower grade fuels became possible.

With ever-increasing firing rates, the spaced tubes gave way to tangent-tube construction. Tangent tubes are defined as tubes so arranged that the gaps between them do not exceed $\frac{1}{4}$ in. [12]. Refractory and insulating materials are used behind the tubes and, in this protected position, they have an almost unlimited life. Another form of water-cooled wall is the welded wall, as illustrated by Fig. 20, which is prevalent in boilers of larger size. In this type of construction, the tubes are assembled into panels by spacing the tubes and shop-welding filler bars between them on automatic welding machines. The tubes of the assembled panels are then field-welded to stub nozzles on the waterwall headers and drums when the unit is erected in the shipyard. From a performance standpoint, this type of construction is equivalent to the tangent-tube wall.

For a number of years furnace floors have been water-cooled by means of horizontal tubes installed below a refractory covering. The water cooling, although slight, provided increased life for the floor refractory. The refractory was absolutely necessary to ensure long life of the floor tubes since a horizontal tube cannot tolerate high heat inputs to its upper surface. Any steam formed there tends to blanket the surface, driving up the tube metal temperature and leading to early failure [8].

After the selection of the type of water-cooled surface to be used and a preliminary furnace size and contour have been estimated on the basis of the number and arrangement of oil burners, it is possible to estimate the furnace exit gas temperature and heat release and heat absorption rates. Based upon the furnace and burner designs, the furnace heat release rates should be limited to those which will result in good combustion conditions and a boiler of minimum size.

a. Exit gas temperature. For many years, accurate estimates of furnace exit gas temperatures were not necessary because of conservative firing rates and the use of saturated steam. Those units which generated superheated steam usually had several rows of boiler tubes between the superheater and the furnace. Consequently, a large error in the calculated furnace exit gas temperature had very little effect upon superheater performance. In units with superheaters located close to the furnace, however, the furnace exit gas temperature must be determined accurately to assure a satisfactory superheater design. In addition, an accurate determination of the heat absorption in the various furnace waterwall areas is necessary to provide adequate water circulation with a practical number of supply and riser tubes.

When estimating the furnace gas temperature, most designers use formulas based upon the Stefan-Boltzmann law, which states that the heat absorbed by radiation is proportional to the difference between the fourth powers of the absolute temperatures of the radiating bodies and receiving surfaces (see Chapter 2). However, in a boiler furnace the exact determination of radiant heat transfer, or heat absorption, is extremely complex and depends upon: the furnace size and shape; the radiant beam (mean distance from the radiating gas mass to the absorbing and the re-radiating surfaces); the partial pressure of the products of combustion; the amount, type, and effectiveness of the heat absorbing surfaces; the ratio of the heat absorbing to the refractory surfaces; the type, quantity, and heat content of the fuel; the amount of excess air; the temperature of the combustion air; the latent heat losses; the emissivity of the various surfaces and the radiating mass of gas; and the flame luminosity. Designers usually calculate furnace exit gas temperatures and heat absorptions by rational methods and then, as a check, plot the calculated values against empirical data derived from boiler tests [2].

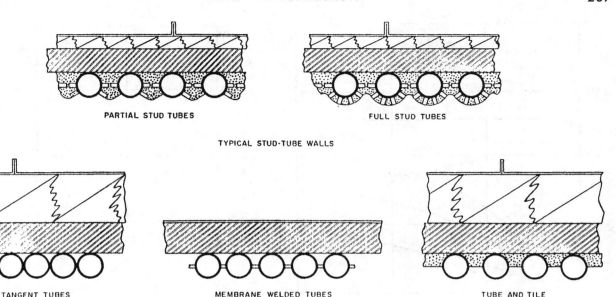

PARTIAL STUD TUBES   FULL STUD TUBES

TYPICAL STUD-TUBE WALLS

TANGENT TUBES   MEMBRANE WELDED TUBES   TUBE AND TILE

TYPICAL BARE-TUBE WALLS

Fig. 20 Furnace wall construction

b. Radiant heat-absorbing surface. In evaluating the radiant heat-absorbing surface, the flat projected areas of the walls and tube banks are used [12]. The spacing of the tubes in the boiler bank adjacent to the furnace has no effect upon the furnace temperature, but with widely pitched boiler tubes a large percentage of the radiant heat is absorbed in the tube rows behind the furnace row. Furnace waterwalls and roofs usually consist of bare or covered tubes (Fig. 20) and, with the exception of bare tangent tubes or welded walls, the effectiveness of the absorbing surfaces is less than the black-body coefficient of 1.0 considered for the furnace rows of boiler tubes.

The furnace gas temperatures usually are not accurately estimated in preliminary analyses since the general design characteristics are of primary interest, and an approximate estimate of furnace gas temperatures and heat absorption rates can be made with knowledge of the boiler and the firing conditions. Thus, with the assumed excess air, the heat content of the products of combustion and the adiabatic temperature can be determined. Further, the approximate furnace size provides an indication of the water-cooled surfaces and estimates can be made of the surface absorption effectiveness and the expected furnace gas temperature. In approximations of this nature it is usually desirable to estimate both the furnace temperature and the heat-absorbing surface on the low side when firing oil. This increases the estimated furnace heat absorption and assures a margin of reserve in the final design. However, with coal firing it is more important to estimate the furnace gas temperature on the high side to preclude the possibility of operating with furnace temperatures above the initial ash deformation temperature.

In a boiler furnace, both the furnace exit gas temperature and the heat absorption can be changed appreciably, for a given firing rate, by varying the amount of radiant

heat absorbing surface. The furnace gas temperature and heat absorption also can be lowered, at any firing rate, by increasing the excess air (Fig. 21), except when operating with a deficiency of air. The additional air increases the weight of the products of combustion per pound of fuel fired. This decreases the adiabatic temperature since there is less heat available per pound of products of combustion; and, as indicated by the Stefan-Boltzmann law, lowering the radiating temperature reduces the heat absorption rate. Generally, the radiating temperature is assumed equal to one third of the adiabatic temperature plus two thirds of the furnace exit gas temperature.

c. Heat absorption rates. The furnace heat absorption rate per square foot of radiant heat absorbing surface increases with larger heat release rates. However, the percentage of the total heat released, which is absorbed in the boiler by radiation, decreases with an increase in firing rate, and varies from as much as 50%, or more, at the lower firing rates to about 15% at the higher firing rates; see Fig. 22. This results from the fact that the adiabatic temperature remains practically constant, except for changes due to variations in excess air and combustion air temperatures, over the entire range of boiler operation, while the temperature of the gases leaving the furnace and entering the tube bank increases with the firing rate.

Even though the furnace heat absorption rates may be conservative, the furnace exit gas temperatures may be excessive with respect to ash fusion temperatures and slagging. This is true particularly in coal-fired boilers where the gas temperatures entering the tube bank should be less than the initial ash deformation temperature. Because of the lower ash fusion temperatures of oil slags, they pass out of the furnace in a gaseous or molten state and are not amenable to control by reducing the

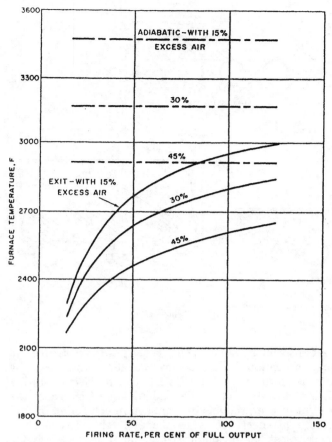

Fig. 21   Effect of excess air on adiabatic and furnace gas temperature

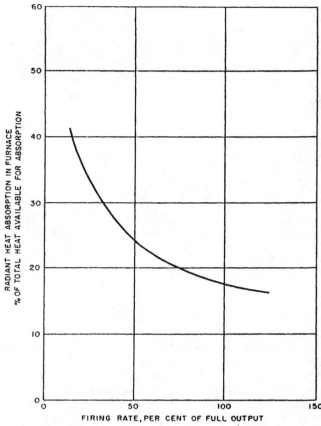

Fig. 22   Relationship of radiant heat absorption and firing rate

furnace exit gas temperature. They must be considered in the design of the superheater.

d. Tube metal temperatures. In boilers, the heat-transfer rate across the boiling water film on the inside of the tubes may be as high as 20,000 Btu/ft²-hr-F; however, when estimating tube metal temperatures, a transfer rate of only 2000 Btu/ft²-hr-F is usually assumed in order to provide a margin against the resistance of possible internal deposits to heat transfer. Thus, with a heat absorption rate of 120,000 Btu/ft²-hr, the temperature drop across the water film is about 60 deg F and, unless exceptionally heavy tube walls are used, the temperature gradient across the tube metal is about the same. Consequently, with a steam pressure of 600 psig (489 F saturated steam temperature) and a heat input of 120,000 Btu/ft²-hr, the outside surface metal temperature of the furnace tubes usually does not exceed 620 F. With a tube metal temperature of 620 F, there is a margin of about 330 deg F between the metal temperature and the allowable oxidation temperature of carbon steel. However, the resistance to heat flow across internal tube scales is appreciable and it is good design practice to provide a tolerance for variations in the quality of the feedwater [8].

With adequate boiler circulation and proper feedwater quality, the heat input to the furnace tubes is limited only by the tube metal temperatures, thermal creep, and pressure stresses.

e. Design limitations. While there are no specific furnace exit gas temperatures which can be used as design criteria for all types of boilers, they should be sufficiently high to maintain good combustion at all ratings, including port loadings. However, at the same time they should not be so high as to cause high casing temperatures or excessive furnace maintenance.

Because of the requirements for exceedingly light-weight and compact units for naval installations, evaporative ratings in naval boilers are three to four times greater than those common to most merchant installations. Consequently, the furnace exit gas temperatures in the full-power to overload range are about 2800 to 3050 F when firing oil with approximately 15% excess air. Adiabatic, or theoretical, flame temperatures are about 3450 to 3500 F with oil firing, 15% excess air, and 100 F combustion air. With combustion air temperatures of 300 to 350 F, the adiabatic temperatures increase to approximately 3650 to 3700 F.

Although furnace heat release rates vary considerably, practically all oil-fired merchant boilers are designed for heat release rates of 65,000 to 125,000 Btu per cubic foot of furnace volume per hour at normal rating—approximately 15 to 20% of the corresponding full-power heat release rates on naval boilers [12].

The heat release rate per square foot of radiant heat absorbing surface is generally in the range of 200,000 to

250,000 Btu per hour on merchant boiler designs. Naval boilers are designed for ratings four to five times greater than those used for merchant marine boilers.

Radiant heat absorption rates vary greatly, depending upon the firing rate and the amount of cold (water-cooled) surface in the furnace. Generally, a radiant heat absorption of 120,000 Btu per square foot of cold surface per hour is considered satisfactory for continuous overload operation of merchant boilers with treated evaporated feedwater. This results in an absorption of about 100,000 Btu per square foot of cold surface per hour at the full-load rating.

There are merchant boilers in continuous service with radiant heat absorptions of approximately 150,000 Btu per square foot of cold surface per hour; and most naval boilers have been designed for radiant heat absorption rates of 150,000 to 200,000 Btu per square foot of cold surface per hour at overload rating, but operation at this rating is infrequent.

3.4 Boiler Tube Bank. The arrangement of the boiler tube banks is established after development of the preliminary furnace size. The simplest type of tube bank is that of a boiler delivering saturated steam. Usually two sizes of tubes are used in such banks. The tubes in the rows adjacent to the furnace absorb considerably more heat than those in the other rows and, therefore, should be of larger diameter to increase the water flow. The total heat input to the furnace row tubes is the sum of the radiant and convection heat transfers; in general, the convection heat transfer is approximately 5 to 20% of the radiant heat transfer. This relatively wide range in convection heat transfer results from variations in tube diameter, tube pitch, gas mass flow rate, and the temperature difference between the products of combustion and the tube surface.

The number of tube rows installed is primarily dependent upon the circulatory system and the desired gas temperature leaving the tube bank. The gas temperature leaving the boiler tube bank varies with changes in steam pressure, firing rate, and tube size and arrangement (the tube arrangement may be either staggered or in-line). However, sufficient boiler heating surface must be installed to obtain exit gas temperatures which result in economical operating efficiencies and do not require excessive stack and breeching insulation. Generally, the exit gas temperatures should not exceed 750 F unless economizers or air heaters are used.

The resistance to gas flow can be varied appreciably by changing the pitch of the tubes in a direction perpendicular to the gas flow or from changes in boiler width, tube length, and the number of tube rows.

Moderately rated drum-type boilers usually have 1½-in. tubes in the furnace rows, but these are increased to 2 in. in boilers of higher rating. One-inch and 1¼-in. tubes are common in the main tube banks. There is no standard pitch for tubes. However, it is customary to use the minimum longitudinal tube pitch (direction parallel to the drum and perpendicular to the gas flow) consistent with good manufacturing practice and acceptable drum design, unless the draft requirement or the type of fuel fired dictates the use of a greater pitch. Manufacturing and fabricating practices permit the use of ½-in. metal ligaments between 1-in. or 1¼-in.-OD tubes.

The circumferential, or back, pitch (direction parallel to the gas flow) of the tube usually is set to maintain circumferential or diagonal ligament efficiencies[1] equal to, or better than, the longitudinal ligament efficiency in the drums. Tube arrangements utilizing a minimum back pitch reduce the drum periphery required for a given number of tube rows and allow the use of smaller-diameter steam drums provided the steam drum release rates are satisfactory. With such arrangements, the size and weight of the boiler can be reduced.

When designing for high steam pressures, it is often necessary to increase the tube spacing in order to improve the ligament efficiency and reduce the thickness of the drum tube sheet [18]. If this is not done, large thermal stresses may be set up in the tube sheet. It also is possible to maintain close tube spacing and yet reduce the drum tube sheet thickness by using tubes with the ends swaged to a smaller diameter.

The number of tube rows installed should be limited so that an impractically large steam drum diameter is not required and so that heat absorption in the last tube rows is adequate to maintain good circulation. The tube length should be such that the total absorption per tube does not result in too high a proportion of steam in the water-steam mixture leaving the upper end of the tubes.

Most marine boilers deliver superheated steam from convection-type superheaters. In these boilers, the generating tube bank is arranged in two sections. The section between the furnace and the superheater is known as the "waterscreen" and the other section, installed beyond the superheater, is called the "boiler bank" or "generating bank."

The size and arrangement of the waterscreen greatly affects the design of the superheater. A superheater located closer to the furnace behind a few rows of widely pitched tubes in the waterscreen provides a relatively flat steam temperature characteristic over a wide range of rating since the radiant and convection heat-transfer rates tend to complement each other. However, a superheater located farther away from the furnace radiation behind a deeper waterscreen has a steam temperature characteristic which rises steeply with increased rating, due to the greater effect of convection and the reduction in radiation heat-transfer rates.

Naval boilers usually have waterscreens consisting of three or four rows of tubes and merchant marine boilers generally have two- or three-row waterscreens in front of the superheater. Most of the heat transfer in the superheater is due to convection and inter-tube radiation; however, a spacing of the screen tubes usually can be selected to permit sufficient furnace radiation to the superheater to provide a relatively constant steam temperature over a wide range of rating.

---

[1] Ligament efficiency is the relative strength of the ligaments between adjacent tube holes in a drum or header as compared with a drum or header having no holes.

3.5 Superheaters. The superheater must deliver the specified steam temperature during the operating life of the boiler—not just during the initial trials or test operations—and predicted performance must be maintained continuously with minimum variations in firing rate, air pressure, burner settings, and excess air. The design should avoid the necessity of unscheduled outages for cleaning, etc., in order to maintain performance. Of course, scheduled periodic boiler outages are required in any well-organized operating schedule.

The design of the superheater is the most difficult and complicated of any of the boiler components as it affects many of the boiler's functional and mechanical features. An adequate steam pressure drop through the superheater is required for good steam distribution and satisfactory tube metal temperatures and is an important factor in establishing the boiler's design pressure. The design pressure dictates the thickness of the superheater tubes, which in turn is an important factor in the determination of superheater pressure drop and tube metal temperatures.

The superheater's location affects its size and tube metal temperatures [2,3]. It also affects the design of the waterscreen and furnace, especially in high-temperature units. The location and tube arrangement have an important bearing upon possible slagging and this directly affects maintenance and outages.

There are fundamental considerations common to all types of superheaters, which are designed to have a minimum of heat absorbing surface so as to reduce cost, size, and weight. Minimum surface can be obtained by increasing the heat-transfer coefficient and the temperature differential between the products of combustion and the steam since the total heat absorbed is the product of these two factors and the surface. Increasing the temperature differential takes advantage of the available temperature potential, while an increase of the heat-transfer coefficient necessitates a larger resistance to gas flow. Full advantage should be taken of a high temperature difference, but the entering gas temperature should not be so high as to result in excessive tube metal temperatures or high-temperature fuel ash corrosion (these are primarily a matter of location). The change in steam temperature with firing rate should be a minimum in order to prevent excessive temperatures during maneuvering and, again, this depends upon location. Steam velocities should provide for good distribution of steam, minimum tube metal temperatures, and acceptable steam pressure drops, all of which require correlating the effects of size, location, and the arrangement of the steam passes.

a. Types and characteristics. The radiant and convection-type superheaters are the two basic types. They are, as their names imply, superheaters which receive heat by radiant or convection heat transfer and they may be arranged horizontally or vertically.

In the radiant type the steam temperature decreases with increased rating since the quantity of heat absorbed by radiation does not increase proportionally with steam

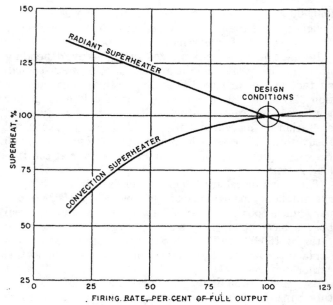

Fig. 23   Temperature characteristics of radiant and convection superheaters

flow; see Fig. 23. In the convection type, the steam temperature generally increases with increased rating because the heat absorption, due to greater heat-transfer coefficients and higher inlet gas temperatures, increases at a faster rate than the steam output.

Most superheaters are a combination of the two basic types in which the designer builds in a radiant component to achieve a flatter temperature characteristic. Generally, "hairpin" tube arrangements, as illustrated by Fig. 24, in which the headers are connected to each other by U-shaped tubes, are used. Multiple-bend, or continuous-loop, designs are frequently used for high-pressure, high-temperature installations to reduce the number of tube joints and the thickness of the headers.

b. Arrangement of steam passes. The steam pass arrangement must result in acceptable pressure drops and satisfactory tube metal temperatures. A change in the number of steam passes greatly affects the resistance to steam flow and tube metal temperatures but, in general, has only a small effect on the heat-transfer rate.

The arrangement of the steam passes provides a good steam distribution if the resistance to flow in the tubes is high compared to that in the headers. The location of the inlet and outlet header connections also affects steam distribution. The aggregate tube flow area per pass should be less than the flow area in the inlet header in order to minimize the tendency for steam to bypass certain tubes.

The heat absorbed in each steam pass can be assumed proportional to the heat absorbing surface, unless there is maldistribution of the gas flow or the temperature differences between the products of combustion and the steam vary appreciably between steam passes. However, the increase in steam temperature per pass is not proportional to the heating surface, since the specific heat of superheated steam decreases with increased temperature.

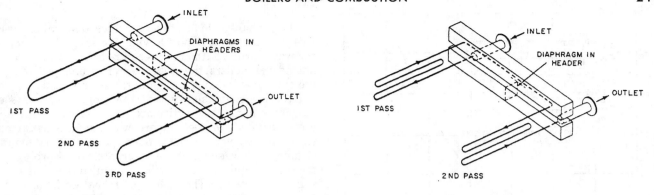

(a) Three-pass hairpin loop type   (b) Two-pass continuous loop type

Fig. 24 Schematic arrangement of hairpin and continuous-loop superheaters

Accurate estimates of the steam temperature entering and leaving each pass are necessary to determine the resistance to flow and to design and arrange the diaphragms in the superheater headers. If large quantities of heat are absorbed in each pass, the temperature differentials across the diaphragms are appreciable and high thermal stresses are set up in the headers. These stresses may be satisfactory insofar as strength is concerned, but they can cause leakage in superheater tube seats. When the temperature differential across the diaphragms is in excess of 175 deg F, and additional steam passes cannot be used, it is customary to use separate headers. Although this arrangement reduces thermal stresses and eliminates tube seat leakage, wide gas lanes are formed at the junction of the headers. Therefore, the tubes bordering the lanes will have higher metal temperatures than the average tubes because of the greater gas flow past the lane tubes.

c. Tube temperatures, materials, and attachments to headers. The tube metal temperatures depend upon the adjacent gas and steam temperatures; the size, thickness, and material of the tubes; the thermal conductivity of the metal; the steam film heat-transfer rate; and the overall heat input. As the resistance to heat flow through the metal is usually much less than that across the steam film, elevated tube metal temperatures are due primarily to the high temperature drop across the steam film. An increase in the steam mass flow, obtained at the expense of a higher pressure drop, will increase the steam film transfer rate and decrease both the temperature gradient across the steam film and the outside metal surface temperature. Therefore, good design practice dictates the use of the highest practicable steam pressure drop in order to minimize tube metal temperatures and the need for high-grade alloy materials.

Allowance must be made for the possible maldistribution of both the steam and gas flows in calculating tube metal temperatures. Unless there are exceptional conditions, it is customary to consider a total maldistribution of 20% on the gas side. On the steam side, a calculated unbalance, which is dependent upon the tube and header arrangement, is used. The highest tube metal temperature is usually encountered in a tube having the minimum

percentage of the steam flow and which receives about 110% of the average gas flow.

From the viewpoint of heat transfer, it is desirable to use a counterflow of the steam and the products of combustion in order to increase the temperature differential between the steam and the gas and thus reduce the required amount of heat absorbing surface. However, with high steam temperatures this may result in excessive tube metal temperatures since the greatest amount of heat will be transferred to those tubes carrying the highest-temperature steam. Therefore, parallel flow is often used in the last steam pass. A small amount of additional heating surface will be required to compensate for the use of a parallel flow; however, a lower-grade alloy tube can be used.

Generally, tubes are attached to the headers by rolling when steam temperatures are below 850 F, and are welded above this temperature. If, because of the temperature or the service intended, it is necessary to weld the tubes to the headers, special consideration must be given to the materials and the methods of welding. Similar materials present few problems; however, dissimilar materials are used in the tubes and headers. The difficulties encountered in welding tubes to headers are minimized by making the tubes "safe-ended." A safe-end is a short section of tubing of a material compatible with that of the header, which facilitates field welding. The more difficult dissimilar weld between the tube and the safe-end is made under controlled conditions in the shop.

d. Supports. Most superheaters have water-cooled supports consisting of tubes fitted with a number of alloy brackets. Often, the tip temperatures of these alloy brackets are in excess of 1700 F and the alloys used must be capable of withstanding considerably higher temperatures in order to provide an operating margin against maldistribution of gas flow and temperature.

e. Location of headers. In boilers with horizontal superheaters, the headers are generally located at the rear, and tube renewal is provided at the front of the boiler. Where vertical superheaters are used, the screen tubes may be arranged to permit the superheater tubes to be renewed through the furnace at the front or the rear of the boiler. In other arrangements, superheater tubes

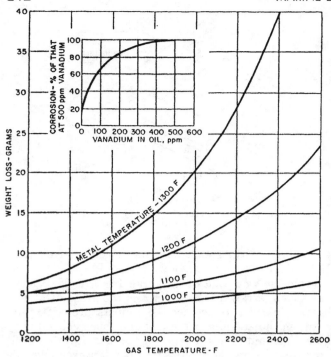

Fig. 25   Effect of gas temperature, metal temperature, and vanadium in oil on corrosion

may be renewed through the superheater cavity. Use of the firing aisle for superheater tube renewal reduces the space required for the boiler installation.

f. Slagging and high-temperature corrosion. Slagging and high-temperature corrosion of the tubes and supports vary appreciably with the type of fuel oil used and the amount of fuel-oil contamination. High ash contents and contamination, particularly by sodium chloride and vanadium salts, usually result in heavy slagging and corrosion.

Laboratory investigations have shown that for a given tube material the rate of corrosion increases with an increase in gas temperature or metal temperature. Further, the rate of corrosion increases greatly with an increase of vanadium in the fuel oil as shown by Fig. 25.

In oil-fired units, the boiler and superheater tubes are usually pitched on the minimum practical centers to reduce boiler size and weight. Consequently, as firing and evaporative ratings increase and the quality of fuel oil deteriorates, slag accumulations in superheaters become a major design and maintenance problem. To overcome this, superheater designs have been developed with an "in-line" instead of a "staggered" tube arrangement, increased tube pitch, and with the superheater located in the lowest practical gas temperature zone so as to provide the most favorable gas-metal temperature relationship. For the usual materials employed, mean metal temperatures are limited to 1050 F. The cavity or "walk-in" type superheater (Figs. 4 and 5) incorporates these features and greatly improves the accessibility for cleaning. A further improvement for cleaning and maintenance is the provision of a walk-in cavity between the superheater and

the boiler bank. This cavity facilitates removal of corrosive slag and soot deposits which accumulate on top of the water drum. Although these features result in larger and heavier boilers, the units are more economical to operate since maintenance and outages are reduced.

Experience has shown that the diligent use of soot-blowing equipment (particularly mass-action retractable units) usually can keep superheater surfaces satisfactorily clean for a year, or more, of operation and that manual cleaning and washing of the external heat absorbing surfaces are required only during scheduled overhauls.

g. Reheaters. The design of reheaters involves the same procedures and considerations that are pertinent to superheater design. However, the steam distribution and tube metal temperature problems are more critical since reheaters must be designed for exceptionally low steam pressure losses if a high cycle efficiency is to be obtained.

Steam or combustion gas can be used as the heating medium in reheaters. When steam heating is used, the temperature of the reheated steam usually is limited to 550 to 600 F, since it is customary to use condensing rather than superheated steam as the heating medium because of the much higher rate of heat transfer.

The use of gas reheaters is necessary if high reheat steam temperatures and cycle efficiencies are required. Such reheaters may be fired separately or installed in the boiler proper. Separately fired reheaters are not common because they require an individual firing aisle and renewal clearances, as well as additional piping, controls, breechings, firing equipment, fans, etc.

3.6 Air Heaters and Economizers. Air heaters or economizers or both are necessary to obtain high boiler efficiencies.

The temperature of saturated steam at a pressure of 850 psig is 528 F and the temperature of the products of combustion leaving the boiler tube bank would be, for a conservative boiler design, approximately 150 deg F above this value, or about 675 F. When firing oil, and operating with 14.0 percent CO_2 in the products of combustion (approximately 15 percent excess air), this uptake gas temperature would result in an operating efficiency of only about 80 percent as can be seen from Fig. 16.

If the uptake gas could be cooled to a temperature equal to the steam's saturation temperature by the use of an infinite amount of heat absorbing surface, the improved efficiency would only be 83.75%. Therefore, air heaters or economizers must be installed to increase full-load efficiencies to the 88–90% range usually desired. Further, the use of high evaporative ratings at any given steam pressure increases the need for additional heat-reclaiming equipment.

When air heaters or economizers are installed, the proportions of the boiler, air heater, and economizer surfaces must be balanced. Usually, the temperature differential between the products of combustion and the heat-absorbing fluids in the economizer and air heater is greater than that in the last section of the boiler tube bank. This is advantageous in reducing the heat-absorbing surface required for a given heat recovery. In air heaters, part of the advantage resulting from the improved temperature difference is offset by the high resistance to heat flow across the air film [19]. Therefore, the proportions of component surfaces must be studied carefully to obtain the most economical overall arrangement.

The minimum temperature of the feedwater to most merchant marine economizers varies between 270 and 280 F. The standard feedwater temperature for most naval installations is 246 F. This lower temperature is satisfactory because a premium fuel with a low sulfur content is used.

Since the gas temperature leaving the economizer cannot be less than the inlet water temperature, it follows that high feedwater temperatures limit the obtainable efficiency. Consequently, with high feedwater temperatures, economizers are not often used unless they are installed in conjunction with air heaters.

In an air heater, the minimum uptake gas temperature is dependent on the entering air temperature. Therefore, the attractiveness of air heater installations is due to the possibility of operating with a high boiler efficiency when using feedwater temperatures in the range of 300 to 450 F.

When steam turbines are bled for regenerative feed heating, the plant efficiency is increased about 1% for each 100 deg F rise in feed temperature due to the reduced heat loss in the condensers. Whether this improvement in efficiency warrants the expenditure required for additional feed heating and other equipment should be carefully weighed for each application.

The use of an air heater necessitates an increased air pressure to the boiler unit because of the additional resistance to air flow through the air heater. Air pressures also must be increased when using economizers because of the relatively high resistance to the gas flow across the economizer, but, for boilers of the same size operating at comparable firing rates, an air heater installation will usually require a higher total air pressure than will a unit fitted with an economizer.

Air heaters are not pressure vessels, so the tubes can be fabricated from mechanical tubing (less expensive than pressure tubing) that is lightly expanded into the tube sheets. However, economizers are part of the pressure system and must be designed to withstand the main feed pump discharge pressure, to operate without leakage, and to withstand thermal shock.

a. Air heaters. Preheated air can improve combustion, reduce boiler sooting, and reduce the possibility of ignition loss particularly at the extreme low end of the firing range.

Practically all of the older marine air heaters were of the tubular type; however, the rotary regenerative air preheater has found wide application. A typical example of a regenerative air heater is shown in Fig. 14. Its gastight casing forms part of the boiler forced-draft air and uptake gas ducts. The heater is separately mounted above the boiler and suitable expansion joints are used in the ducts joining the two [9].

The essential component of the heater is the rotor in which the heat-transfer plate elements are packed. The air for combustion is passed axially through one side of the rotor while the flue gas is passed through the other side in the opposite direction. As the rotor turns, heat is continuously transferred from the gas to the heating surface; heat is also continuously given up to the air as the heated plates traverse the air side. Counterflow of the gas and air insures efficient heat transfer.

The heat-transfer elements are made of corrugated and flat sheets, which are alternately packed in the main section of the heater and in the cold-end baskets. The cold-end basket is designed to be readily removable for cleaning or replacement when conditions warrant. For daily cleaning, a cleaning device consisting of a mass-action soot blower is installed. Air and gas bypass dampers are an integral part of the preheater and are useful in maintaining the heat-transfer metal surface temperature above the gas dew point. This minimizes corrosion at low rates of operation and also helps minimize soot buildup. Operation of these dampers can be made fully automatic.

Most tubular air heaters are of the horizontal type (Fig. 13). The vertical type is not often used since it is necessary to install considerably more surface for a required heat absorption than would be needed for the horizontal type. In the horizontal type, it is customary to pass the air through the tubes and the gas across the tubes. In the vertical type the gas usually passes through the tubes and the air crosses the tubes.

Horizontal tubular air heaters generally utilize in-line tube arrangements. These facilitate cleaning of the external heating surfaces, a feature considered far more advantageous than the slightly higher heat transfer obtained with staggered-tube arrangements.

Air heater tubes range in size from 1½-in. to 2½-in. outside diameter, with most installations using 1½-in. tubes. If the resistance to internal flow must be reduced, large tubes are preferable since, for a given gas mass flow, the resistance varies inversely as the inside diameter. However, compactness is a prime requisite of marine boilers and, therefore, the use of small tubes may be preferable to permit the installation of maximum surface in the space available. Further, the heat-transfer coefficients of the gas and air flowing across and through the tubes, respectively, vary inversely as about the 0.34 and 0.22 powers of the tube diameter and, thus, the heat transfer at any mass flow rate increases with a reduction in tube diameter [2,19].

In preliminary air heater designs, both the tube size and tube pitch are assumed. It is satisfactory, in most cases, to consider 1½-in. to 2-in. tubes with ½-in. tube ligaments. An estimate is then made of the length of the tubes, the number of tubes per row, the number of tube rows, and the number of gas and air passes. This facilitates determination of the heat-transfer rates and the heating surface. The preliminary assumptions are then adjusted, if necessary, so that the surface arrangement and heat transfer provide the required heat absorption.

Gas and air flow patterns also must be analyzed since maldistribution could reduce heat absorption, increase fan power, reduce or elevate tube metal temperatures, or restrict the capacity of the boiler unit.

Air heater designs are usually predicated upon inlet air temperatures of 100 F, and exit air temperatures ranging from 300 to 450 F at the normal full-load operating rate. Design exit gas temperatures of 290 to 320 F are common for tubular air heaters and result in boiler efficiencies of 88.5 to 88%. Regenerative air heaters can be designed for lower uptake gas temperatures for a given risk of corrosion since for the same air and gas temperatures the heating surface metal temperature is somewhat higher than that of the tubular heater. Gas temperatures from 240 to 260 F are common for regenerative air heaters with boiler efficiencies of 90 to 89.5%, respectively.

Both the weight of the gas produced and the specific heat of the flue gas are greater than that of combustion air. Therefore, when firing oil with about 15% excess air, the reduction in the temperature of the products of combustion passing through the air heater is about 13% less than the rise in air temperature.

In air heaters the heat-transfer coefficients across the gas and air films are of about the same magnitude, and high resistance to heat flow is encountered in the gas film on both sides of the tube.

b. Economizers. Marine economizers can be grouped into two general classifications, the "bare-tube" and the "extended-surface" types. They are generally nonsteaming and are usually arranged for counterflow of the water and the products of combustion. This results in larger temperature differentials, and greater heat absorption can be obtained. The counterflow arrangement permits a higher boiler efficiency because the exit gas temperature can approach that of the inlet water.

Economizers use tubes ranging in size from 1½ to 2 in. that are arranged in the form of either hairpins or continuous loops. The hairpin type consists of U-bend tubes that are welded, or expanded, into headers. Single or multiple rows of loops can be used as well as two or more headers. In the continuous-loop type, each tube element consists of a length of tubing bent back and forth to form the desired number of rows; the ends of the tube are attached to the inlet and outlet headers, usually by welding. Since only two headers are required, the number of tube joints is greatly reduced as may be noted from Fig. 26.

There are many types of extended-surface economizers. The most prominent are those having steel studs or helically wound spiral steel fins welded to the tubes (see Fig. 12). Features common to all extended-surface economizers include the concentration of heating surface, the minimizing of tube joints, and the use of only two main headers.

Most marine economizers use counterflow arrangements with up-flow gas and down-flow water. The water pressure drop at about 25% of the normal full-load operating rate should be equal to, or greater than, the static water head in order to prevent recirculation. This minimum pressure drop requirement is not necessary if parallel-flow, up-flow gas and water, nonsteaming economizers are used, since the water pressure in the outlet header always will be less than that in the inlet header.

Fig. 26   Continuous-loop, stud-tube economizer

Multiple water passes are often used in hairpin-type economizers to obtain satisfactory water velocities and pressure drops. These arrangements have both counter- and parallel-flow relations between the water and the products of combustion, and the calculated heat transfer should be based on the average of the flow arrangements. Most continuous-loop and extended-surface type economizers have a single water pass arranged for flow counter to that of the gas.

Although economizers operate with feedwater temperatures as low as 180 F, it is customary to restrict the feed temperature to a minimum of 220 F. This temperature allows deaeration and, therefore, is sufficiently high to prevent, or minimize, internal oxygen corrosion. However, with residual fuel oils it is necessary to use feedwater temperatures in the range of 270 to 280 F to maintain the tube metal temperatures above the dew point of the gaseous products of combustion. With this range of feedwater temperature, economical counterflow economizers can be designed to reduce the exit gas temperature to approximately 320 F (which corresponds to a boiler efficiency of about 88%) at the full-load rating.

When both economizers and air heaters are installed, feedwater temperatures of 350 to 400 F generally are used to take advantage of regenerative feed heating.

In counterflow economizers (down-flow water, up-flow gas) a differential temperature of at least 50 deg F should be maintained between the exit water and the saturated steam temperatures to prevent steaming, water hammer, and thermal shock, particularly during overloads and rapid maneuvers. If it is necessary for the exit water temperature to approach closely the saturated steam temperature, parallel-flow (up-flow water and gas) or a combination of counter- and parallel-flow economizers must be used.

When firing oil, the rise in economizer water temperature is about one third the gas temperature drop across the economizer. Since economizers are generally designed to effect gas temperature reductions ranging between

200 and 300 deg F, the water temperature rise varies between 70 to 100 F. Thus, with an inlet water temperature of about 280 F, the exit water temperature is approximately 350 F. This is more than 100 F below the saturated steam temperatures corresponding to pressures of 600 psig and above.

In bare-tube economizers the temperature drop across the tube wall is small and, for all practical purposes, the tube metal temperature can be considered the same as that of the water it carries. Tube metal temperatures of extended-surface elements also are about the same as the adjacent water, although the tip temperatures of the extended surface are considerably higher.

A bypass line around the economizer will allow operation of the boiler with an economizer outage. However, few economizers are fitted with bypass lines because of their cost and the piping complications involved.

If the economizer is bypassed, resulting in a loss of efficiency, approximately 5 to 7% additional fuel must be fired to maintain the required evaporative rating and the fans must supply proportionally more air against an increased static pressure. The temperature of the gas entering the economizer is usually less than 950 F and, therefore, tube metal oxidation should not be experienced during bypass operations.

Of paramount importance, when operating with the economizer bypassed, is the increase in the superheater outlet temperature. This results from the greater gas flow which, in turn, causes an increase in the ratio of the gas to the steam weight and, thus, the transfer of more heat to the steam.

While maneuvering, economizers may be subjected to thermal shock as a result of rapid fluctuations in the water temperature. For example, with a momentary stoppage of the water flow, the economizer may start to steam; then, as the load increases, the sudden introduction of feedwater into the economizer may cause water hammer and high thermal stresses. Because of such possibilities, means must be provided to prevent joint and fitting leakage; therefore the tubes are welded to the headers, and soft sheet gaskets generally are used in all fittings, or weldable fittings may be employed.

If the feedwater to the economizer contains oxygen, the oxygen is released when heat is applied and may result in tube and header corrosion. Therefore, when economizers are used, it is always necessary to make provisions for deaerating the feedwater.

3.7 Desuperheaters and Attemperators. Desuperheaters and attemperators are heat exchangers which reduce and control the temperature of superheated steam. Two major types are employed: the drum (or internal) type, which is installed in either the steam or water drum; and the external type, which is located in the piping system external to the boiler.

Internal desuperheaters may consist of a single pipe or may comprise a number of small-diameter tubes rolled or welded to manifolds and installed below the water level in the steam drum or water drum as in Fig. 27. They may be arranged with bends to provide the required heating surface within the available straight length of the drum.

Fig. 27   Drum-type desuperheater

The auxiliary desuperheater's function is to reduce the temperature of a portion of the superheated steam output of the boiler for use in auxiliary machinery, general heating, etc. It is usually designed to provide no more than 50 to 75 F residual superheat at its designed maximum flow with a pressure drop of 75 to 100 psi below the superheater outlet steam pressure.

Pipe-type desuperheaters are satisfactory for relatively low flows of 5000 to 20,000 lb/hr. For greater flows and to limit the pressure drop to reasonable values, the multitube bundle-type desuperheater is used. Maximum flows in this type may reach 150,000 lb/hr in large tankers where cargo heating and auxiliary pumping loads may be very high. No controls are required as an auxiliary desuperheater cannot cool the steam below the drum saturation temperature. Owing to the pressure drop through the superheater, connecting piping, and desuperheater, the desuperheated steam is slightly superheated at all times. This helps insure dry steam at the auxiliaries.

External desuperheaters are normally of the spray type and are used to provide large quantities of desuperheated steam where the steam quality is relatively unimportant. An external desuperheater is shown in Fig. 28.

An attemperator (control desuperheater) is a desuperheater used for close control of the final steam temperature to the design value. The temperature of superheated steam is a function of rating and for the usual marine boiler rises as shown by the "uncontrolled curve" in Fig. 29. To make the most effective use of the materials in the superheater and main steam piping, the final steam temperature can be controlled so as not to exceed the design value. This can be accomplished by passing a portion of the superheated steam through a desuperheater in the drum. The location of the outlet and inlet connections is usually "interpass"; a typical arrangement is illustrated in Fig. 30. The desuperheated steam is returned to

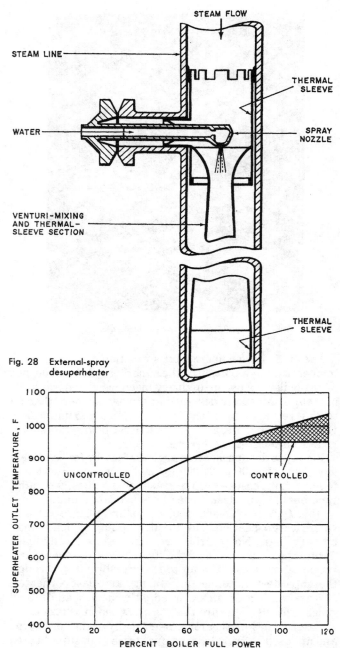

Fig. 28 External-spray
desuperheater

Fig. 29 Uncontrolled and controlled steam temperature

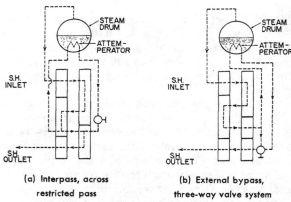

(a) Interpass, across
restricted pass

(b) External bypass,
three-way valve system

Fig. 30 Interpass control desuperheater arrangements

the circulatory system, simultaneous analyses are made. Circulation calculation procedures are in part empirical and in part theoretical. The purpose of the analysis is to establish a system of downcomers, risers, and generating tubes, which will insure that each tube receives an adequate supply of water in relation to the maximum heat absorbed.

a. Circulation: boiler tube banks and furnace water-walls. The circulation characteristics of furnace waterwalls and boiler tube banks are determined by the same procedure and, since the water-steam ratio decreases with increased rating, the characteristics must be established for the maximum contemplated rating.

In analyzing boiler circulation, it can be assumed that each circulating system is, in effect, a U-tube [8,20]. The riser section of the U-tube is that portion of the tube bank in which the flow of steam and water is upward as heat is applied. The downcomer section consists of unheated tubes or those portions of the tube banks in which the heat absorption is considerably lower than in the riser section. Because of the difference in fluid densities, heated tubes can act as downcomers for the riser sections and there is a definite transition zone between the heated downcomers and the riser tubes, the location of which varies considerably with changes in the boiler firing rate.

In the U-tube analogy, there is initially a vertical pressure plane at the bottom on which the pressures exerted by the hot and cold water legs are equal. As heat is applied and the water begins to circulate, resistance to flow is encountered. Thus, at the hypothetical equal pressure plane in the lower water drums, or headers, the pressure corresponding to the flow of water through the downcomers is equal to the product of the head of water and its density minus the resistance to flow. This pressure must balance the product of the head of water in the risers and its density plus the resistance to flow. By equating these two quantities and solving for the friction loss in the downcomers, it is evident that the downcomer friction loss is equal to the product of the water head and the difference in the downcomer and riser densities, minus the riser friction loss—a quantity known as the net available circulation head [3].

In most circulation analyses the steam generated in the riser tubes is calculated and the water-steam flow, as well

the last passes of the superheater where it mixes with the main flow to deliver the design temperature [2,3,8].

A manually operated valve or an automatically controlled valve is used to regulate the temperature at all rates above the "control point" (that point on the uncontrolled steam temperature characteristic curve which crosses the desired controlled temperature line).

3.8 Circulation and Steam Baffles. The natural circulation characteristics of the boiler and the type of steam drum baffling are determined after the arrangement of the heat-absorbing surfaces has been established. Generally, because of the effect of the steam drum baffles upon

NOTE
IN A TYPICAL BOILER (SEE FIG 44) THE SIDEWALL AND ROOF CIRCUIT A, AND
THE SCREEN AND FLOOR CIRCUIT B ARE SUPPLIED BY TWO
DOWNCOMERS C. FURNACE FRONT AND REAR WALLS D AND GENERATING
BANK E ARE SUPPLIED BY HEATED BANK DOWNCOMERS F. A HEAD TO
WATER-STEAM MIXTURE FLOW CURVE IS REQUIRED FOR EACH
INDEPENDENT CIRCULATING SYSTEM, AND WOULD BE SIMILAR TO
THAT BELOW.

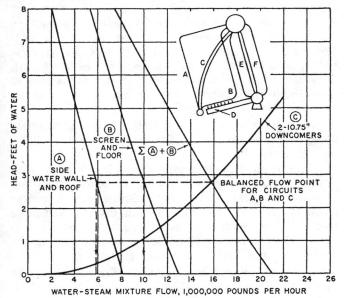

Fig. 31 Characteristic head versus water-steam mixture flow for circulation
calculations

as the net available head, is then determined for various water-steam ratios. In analyzing circulatory characteristics, it is customary to graphically plot both the downcomer friction losses and the net available circulation heads for the assumed water-steam mixture flows. As shown by Fig. 31, the flow at which the available head minus the resistance to flow through the steam baffles equals the resistance to downcomer flow is that required to balance the circulatory system. From the flows at the balance point the percentages of steam by volume at the top of the riser tubes can be calculated.

The percentage of steam by volume at the top of the riser tubes must be such as to preclude overheating of the tubes. If the quantity is excessive, the circulatory system must be redesigned to provide additional downcomers, or the size and contour of the downcomers must be changed to reduce the resistance to flow. It also may be necessary to change the location, size, and contour of the boiler tubes to redistribute the heat absorption and reduce flow resistance.

In a satisfactory circulatory system, an adequate amount of water must be supplied for each pound of steam generated. Therefore, if the percentage of steam by volume at the exit of the riser tubes is used as a design criterion, it is necessary to vary the allowable percentage as the pressure changes since the percentage of steam by volume will increase as the pressure is reduced because of the increased specific volume of the steam. Naval boilers are usually designed for water-steam ratios (i.e., weight of water/weight of steam passing through the

generating tubes) ranging between 5.0 and 10.0, and merchant units usually fall in the range of 15.0 to 20.0 at the overload rates of operation. Lower water-steam ratios are used on naval boilers in order to reduce the boiler size and weight by minimizing downcomer requirements.

b. Heated downcomers. If evaporative ratings are conservative and the gas temperatures leaving the boiler do not exceed about 750 F, the first several rows of tubes will function as risers with the remainder serving as heated downcomers. As the firing rate increases, the high-temperature gas zone moves farther back into the tube bank and additional tubes become risers while a corresponding lesser number act as downcomers. If the firing rate is further increased, the number of downcomers becomes inadequate, circulation is impeded, and tube casualties may occur; when design analyses indicate such circumstances, external or unheated internal downcomers must be installed.

c. External and unheated internal downcomers. With conservative evaporative ratings, external downcomers are required for only those portions of the boiler in which the tubes cannot act as downcomers (i.e., a single tube row forming a furnace boundary, a shallow tube bank installed between two furnaces, or tube banks shielding a superheater from two furnaces).

If downcomers are required for the main tube bank, they usually are located external to the tube bank even though the arrangement requires longer boiler drums. The use of unheated internal downcomers minimizes the drum length and eliminates tubes in the main boiler bank; however, unheated internal downcomers usually enter the steam drum at high water levels and they may lose water during heavy rolls or inadvertent reductions in water level. Further, the use of unheated internal downcomers complicates tube bank arrangements, increases resistance to gas flow, and reduces the boiler heat-absorbing surface. Heat transfer to internal downcomers can be minimized by using plate, stud-tube, or finned-tube protection baffles.

d. Steam drum baffles. The steam drum baffling used in most marine boilers is simple in construction and arrangement. A common type used in header-type boilers is the "vertical baffle," which is located between the dry pipe and the discharge of the circulator tubes. The only design check required when using a vertical baffle is the determination of steam velocities behind and around the ends of the baffle. These velocities, based on the maximum steam output of the boiler, should be less than the critical velocity at which the steam picks up water; the steam velocity can be reduced by increasing the size of the steam drum or by sloping the baffle.

Single and multiple perforated-plate baffles, as in Fig. 32(a), are used in most drum-type boilers operating at conservative steam ratings; these baffles are dependent upon the natural separation of steam and water. For higher boiler ratings a positive means of steam separation is required and compartment-type baffles, Fig. 32(b), are frequently used.

Centrifugal steam separators are used primarily in highly rated merchant and Navy drum-type boilers; they

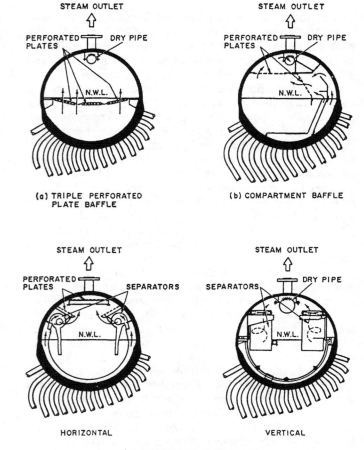

(a) TRIPLE PERFORATED PLATE BAFFLE

(b) COMPARTMENT BAFFLE

HORIZONTAL

VERTICAL

(c) CENTRIFUGAL SEPARATORS

Fig. 32 Typical steam separation equipment

are particularly desirable for boilers subjected to rapid maneuvering, fluctuating water levels, or high solids concentrations in the boiler water. Centrifugal steam separators may be arranged either horizontally or vertically in the steam drum as in Fig. 32(c).

The resistance to flow through centrifugal separators is greater than that through plate- or compartment-type baffles. This would tend to increase downcomer requirements, or impede circulation; but the bottom discharge from centrifugal separators is practically steam-free water, and thus the available head for circulation is increased because the density of the water supplied to the downcomers is greater than that of the "frothy" water-steam mixture discharged from perforated-plate and compartment-type baffles.

The steam-water flow through the steam drum baffles is in series with all of the flow circuits in the circulatory system. Thus, if the flow through one of the circuits is increased, for example, by the installation of additional downcomers, the flow through the steam baffles also is increased. This imposes an additional resistance in the overall circulatory system with the result that the flow in downcomers will not increase in direct proportion to the additions made.

3.9 Construction and Physical Requirements. The structural design of drums, headers and tubes must be in accordance with the rules of the regulatory bodies governing the construction of the vessel (e.g., USCG, ABS, USN, Lloyd's, etc.)

a. Drums. The construction of steam and water drums is basically similar. A drum is a cylinder with the ends closed by heads of either semi-elliptical or hemispherical form. The drum shell is usually made of plates called the wrapper and tube sheets. The tube sheet thickness is greater than the wrapper sheet to provide the requisite strength in way of the tube holes. For the most part, drums are of welded construction although in the smaller sizes single-piece hollow forgings may be used. Drum heads are usually forged.

For commercial work 70,000 tensile steel is used extensively in drum construction; for weight reduction, 80,000 tensile steel is used.

Steam drum diameters range from 36 to 72 in. with most merchant units having 48-in. to 54-in. drums and most naval boilers using 46-in to 60-in. drums. As power levels increase, drums of 60 to 72 in. dia are used more frequently to provide the necessary room for steam baffles and to provide the capability of accommodating the shrink and swell that occurs when maneuvering.

b. Headers and tubes. Headers for waterwalls or economizers are usually fabricated from pipe stock. Hollow forgings may also be used especially for superheaters. They may be round or forged to a rectangular or other cross section to facilitate tube installation. Tubes are installed by expanding or by welding.

Standard boiler and economizer tubes are fabricated from either electric resistance welded or seamless stock. Electric resistance welded tubes are less expensive and have been proven to be as dependable as seamless tubes in boilers and economizers. Superheater tubes are made from seamless steel or alloy tube stock, as required by the metal temperatures involved.

Gas air heater tubes are normally made from welded mechanical tubing since the differential pressures between the air and gas are slight, and do not warrant the expense of pressure tubing.

c. Casing design. The envelope that contains the combustion air and flue gases is known as the casing. Its chief function is to contain and channel the combustion air and flue gases through the pressure parts. An important secondary function is to reduce heat losses to the machinery space, thereby improving both the boiler efficiency and the habitability of the engine room. Good design practice limits the average casing outer surface temperature to 130 F or less. Local areas, for instance where superheater inlet or outlet nozzles penetrate the casing, may be hotter due to "through steel" which does not permit insulation to be effectively applied.

Most boilers are of double-casing construction. An inner and outer casing is used to form an air space surrounding the boiler. This space is pressurized with combustion air, which is at a higher pressure than the flue gases (by no less than the oil burner draft loss). Any tendency for leakage at seals, access doors, etc. will entail the leakage

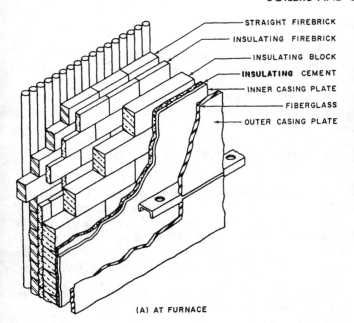

STRAIGHT FIREBRICK
INSULATING FIREBRICK
INSULATING BLOCK
INSULATING CEMENT
INNER CASING PLATE
FIBERGLASS
OUTER CASING PLATE

(A) AT FURNACE

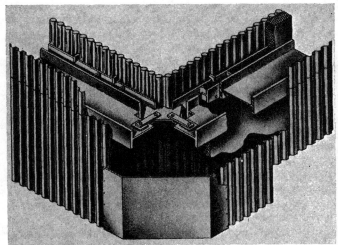

Fig. 34 Membrane welded-wall construction showing insulation and metal lagging

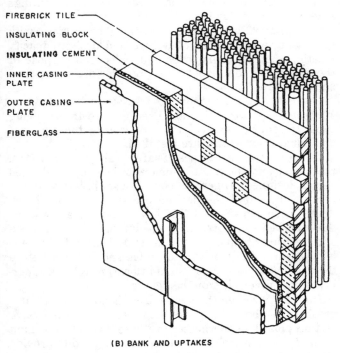

FIREBRICK TILE
INSULATING BLOCK
INSULATING CEMENT
INNER CASING PLATE
OUTER CASING PLATE
FIBERGLASS

(B) BANK AND UPTAKES

Fig. 33 Typical sections of boiler casings

of combustion air into the machinery space or to the firesides. On the other hand, with single-cased boilers, combustion gases are discharged into the machinery space in the event of a leak.

Figure 33 shows sections of a typical double-cased boiler construction. The thicknesses of insulation and refractory materials vary to suit the application; those necessary for a particular unit can be readily determined.

Structural or strength members of the casing are used to support some of the loads of the pressure parts. The generating bank and screen and furnace walls are mainly self-supporting; however, the casing may lend support to these parts during rolling and pitching of the ship or whenever it is not on an even keel. It is usual practice to support the superheater headers and the superheater tubes (wholly or in part), as well as the economizer or air heater (steam or tubular), on the casing structure.

Suitable access and inspection doors are required and their location is an important practical aspect of casing design. Provisions must also be made for differential expansion between the pressure parts and the casing and between the casing and the boiler foundation and surrounding decks, platforms, piping, etc.

In large boilers where welded walls are used, another type of casing construction can be employed; Fig. 34 indicates this construction. Insulation is used outside the welded tubes, which form the gastight envelope. Structural members called "buckstays" give the necessary rigidity to the wall panels to overcome the pressure of the combustion gases. It is necessary to provide air seals at all junctures where leakage is possible or to provide an outer air casing supported by the buckstays.

Normally, mild steel plate is used in casing construction. Thicknesses may vary from one design to another since structural strength requirements may vary. However, all casings should be designed to withstand the shutoff head of the forced-draft blower.

3.10 Oil Burners. Oil burners consist of two principal parts, the atomizer and the air register assembly. The atomizer serves to break up the fuel into fine particles and inject them into the furnace. The air register is an assembly of the throat, air doors, vanes, impeller or air diffusers, and cover plate which directs the air stream to facilitate intimate mixing with the atomized fuel droplets.

In addition to the atomizer and air register, various valves, fittings, and safety couplings are necessary. The safety coupling is a device designed to prevent leakage of steam or oil if the atomizer is removed for cleaning with the supply valves accidentally left open. Such leakage would be a serious fire threat and a hazard to personnel.

Several types of fuel atomizers are used in marine boilers. Most common in older boilers is the mechanical type, which operates over a range of oil pressure from 100 to 300 psi in merchant vessels and up to 600 psi in naval ships. The oil is atomized in the sprayer plate by channeling it through tangential slots into a central whirl chamber and discharging it through a suitable orifice. Capacity is a function of the orifice diameter and the number of slots used (usually four or six). The pressure range provides only limited changes in the flow rate (usually less than 2 : 1); therefore, the boiler firing rate is controlled by changing the number of burners and the size of the burner sprayer plates in use.

A wide-range mechanical atomizer is used on some U.S. Navy vessels; it provides a turndown ratio of 12 : 1. In this atomizer the oil supply pressure varies from 100 to 400 psi in response to steam demand and acts to position a spring-loaded plunger. As the plunger moves in the whirl chamber, it covers and uncovers tangentially drilled whirl holes and varies the slot area of the atomizer. The turndown range obtainable is sufficient to satisfy all underway steaming conditions without cutting burners in or out or changing sprayer-plate sizes.

Steam-assist or steam-mechanical atomizers are basically straight mechanical atomizers to which steam jets have been added. The oil and steam mix externally where the steam increases the turbulence of the air-oil mixture. The range is somewhat greater than that of the mechanical atomizer but not as great as that of the internal mixing steam atomizer.

In a typical internal-mixing steam-atomizing burner, the steam entrains the oil and the mixture expands through the outlet nozzles at a high velocity such that the oil spray is finely divided. Oil can be supplied at up to 300 psi and the steam pressure required is normally kept constant at 125 psi.

The rotary-cup type of atomizer is another form of mechanical atomizer. The oil is atomized by slinging it off the edge of a rapidly rotating cup. Sometimes jets of high-pressure air are used to increase the effectiveness of the rotary cup. This type of atomizer finds its greatest application in small auxiliary boilers and is not an important type for main propulsion boilers.

The two basic types of air register assemblies are shown in Fig. 35. One uses radial air doors and the other uses cylindrical air doors. In normal operation the air doors are kept in the wide-open position. Their principal purpose is to shut off the air supply when a burner is out of service. In automated boilers, air cylinder actuators are applied to the doors and the doors are remotely controlled by the burner sequencing controls. In other installations the doors are manually operated.

3.11 Boiler Mountings. Boiler mountings are the valves, fittings, and trim items which are mounted on the pressure parts or close to the pressure parts. Boiler mountings include:

- stop valves
- feed check valves
- feedwater regulator

- safety valves
- sentinel valves
- high- and low-water-level alarms
- pressure gages
- vent and drain valves
- blowdown valves
- water-level indicators
- water-sampling connections
- soot blowers
- burner flame scanners and ignitors
- instrumentation

In general the boiler mountings are "accessories," which are important to the control and operation of the boiler as well as to the safety of operation. The term "mounting" is synonymous with "fitting." Reliable boiler operation can be obtained only if the mountings are selected with regard for their quality and functional suitability.

When installing mountings, consideration should be given to their function and location for ease of operation and maintenance.

A discussion of some of the more important mountings is given in the following.

a. Safety valves. Every boiler must be equipped with sufficient safety valves of a type and design capacity meeting the applicable codes and rules. The purpose of these valves is to prevent a rise in boiler pressure above specified safe limits.

Normally, two drum-mounted safety valves and one superheater outlet valve are used. Where the relieving capacity required exceeds that provided by these valves, additional drum valves are installed.

There are two basic types of safety valve systems. One, the oldest and the most common, utilizes spring-loaded valves designed to open or "pop" at a set pressure and to remain open until the desired pressure drop or "blowdown" has been reached. The lowest set valve is always the superheater safety valve, which is usually set to open at a pressure about 4% above the superheater design outlet pressure. The superheater valve should open before the drum valves to assure steam flow through the superheater to prevent it from being overheated.

The first valve on the steam drum is normally set at a pressure 2% above the sum of the superheater outlet valve setting plus the superheater pressure drop at maximum rate. If one or more drum valves are required to provide the relieving capacity, they are set at pressures 5 to 10 psi apart to assure crisp operation on opening and closing.

The spring-loaded superheater valve setting may be reached frequently in a rapidly maneuvering situation due to control (or operation) response time. Frequent operation of the valves causes leakage and maintenance problems as the valves are primarily "safety" valves and are not designed or built to be "pressure control" valves.

To increase the spread between the superheater outlet pressure and the set pressure of the superheater safety valve (so as to provide more margin for transient pressure excursions), a pilot-operated safety valve system should be used. A pilot-operated valve consists of a small safety

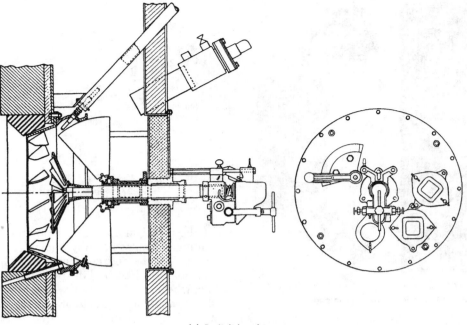

(a) Radial door burner

(b) Cylindrical door burner

Fig. 35   Radial and cylindrical door burners

valve on the steam drum (called a drum pilot valve) and an unloading valve at the superheater outlet. The drum pilot is set at a pressure equal to the sum of the superheater outlet steam pressure, the maximum superheater pressure drop at overload, and a 4% to 5% margin on the outlet steam pressure.

In operation, the pilot pops when its set pressure is reached and almost instantly triggers the opening of the superheater unloading valve. If the pressure continues to rise, the drum safety valves pop in their set sequence.

b.   Water level indicators.   It is essential that sufficient water be in the boiler at all times for safe operation. The normal working range is indicated visually by means of direct-reading gages mounted on the steam drum and by indirect remote indicators mounted for example at the main control console (or, in the case of bridge control of the main engine, on the bridge). Typical direct-reading and remote-reading gages are shown in Fig. 36.

Too low a water level could result in loss of circulating head and tube failures if the downcomers are uncovered. Too high a level could cause carry-over with a subsequent thermal shock to the superheater or, if severe enough, to the main engine. Maintaining close control of the water level is mandatory and the water level indicators must be kept clean and in good condition at all times.

Every boiler must have at least two independent means of showing the water level. The usual practice is to use two separate direct-reading gage glasses and as many remote gages as the particular operating conditions warrant.

c.   Smoke indicators.   The ability to view the stack gas discharge is a distinct aid to boiler operation. Sudden

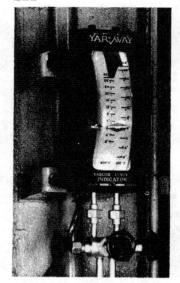

(a) Indirect or remote reading gage   (b) Direct reading gage

Fig. 36   Direct and remote-reading gages

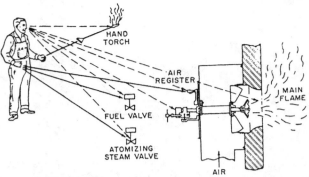

Fig. 37   Burner operation-local manual control

changes from clean to dark stack gases may evidence maloperations such as dirty oil burners or forced-draft or control difficulty. To permit viewing the stack without the necessity of leaving the operating station, smoke indicators are installed. The direct-reading indicator is basically a periscope arranged to give the operator a direct view of a light source which shines through the boiler uptake and the combustion gases. Another type employs a photoelectric cell and provides a readout on a meter scale calibrated in smoke density units; it may also be fitted to sound an alarm when a certain smoke density is reached.

d. Instrumentation and controls. The need for operating instruments and manual or automatic controls varies with the size and type of equipment, the method of firing, the proficiency of the operating personnel, and the desired degree of automation.

For safe operation and efficient performance, information is required relative to the water level in the boiler drum; burner performance; pressures of the steam and the feedwater; temperature of the superheated (and reheated) steam; pressures of the gas and air entering and leaving the principal components; feedwater and boiler water chemical conditions and particle carry-over; operation of feed pumps, fans, fuel burning, and fuel preparation equipment; relationship of the actual combustion air passing through the furnace to that theoretically required for the fuel fired; temperatures of the water, gas, fuel, and air entering and leaving the principal component parts of the unit; and feedwater, steam, fuel, and air flows.

For many years, marine boilers have been equipped with control equipment permitting steady operation at sea with little operator participation except while maneuvering. However, complete automation of the boilers may be desired so that, with the exception of starting up, they can be operated throughout the full range from standby to full load without manual adjustments.

To attain fully automatic operation, adequate control components are essential. The operating characteristics

of the principal and auxiliary items of steam-generating equipment must be fully known since these characteristics affect the degree of controllability, the scope of the controls required, and the response obtained. As an example, where the burners have a range of operation or turndown capability equal to or greater than that required by the boiler, the necessity to sequence burners (or take them out of service) is eliminated. This, in turn, eliminates many decisions and functions that would otherwise be required of an automatic burner management system, and a simpler system may be selected.

The degrees of control which can be achieved, in ascending order of sophistication, are manual, local supervised manual, remote supervised manual, automatic (nonrecycling), and automatic (recycling). These various types of control can best be delineated by relating their functions to burner operation.

With the manual type of control, Fig. 37, a burner is manually purged and ignited. It may be automatically modulated but it is stopped manually. Although no operator function is performed automatically, wide-range burners can be used with automatic combustion controls to facilitate dock-to-dock operation without manual participation. However, without boiler and burner monitoring devices, the operator must remain in close proximity to the boiler to provide the necessary surveillance.

In the local supervised manual system, Fig. 38, a burner is purged and manually ignited, but certain procedures and conditions are supervised by safety interlocks. All manual functions are performed and checked by the operator at the burner station during normal operation, and if the demand for steam is within the capability of the burners, unattended boiler operation is attained. Monitoring and safety interlocks are provided to alter the operation if an unsafe condition develops, and to trip the burner or the boiler, if necessary. After a trip-out, the operator must take the necessary corrective action to clear the interlocks and recycle the burner or the boiler.

The remote supervised manual system, Fig. 39, allows a burner to be purged and ignited by a pushbutton or selector switch, modulated automatically, and secured by a remote manually actuated pushbutton or selector switch. It also provides supervision of procedures by safety interlocks. The burner is mechanized and all operating functions are performed by mechanical devices

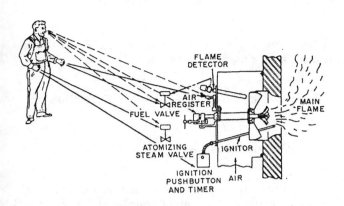

Fig. 38 Burner operation-local supervised manual control

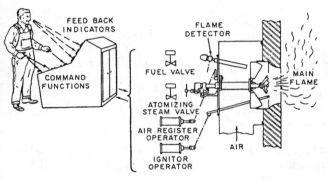

Fig. 39 Burner operation-remote supervised manual control

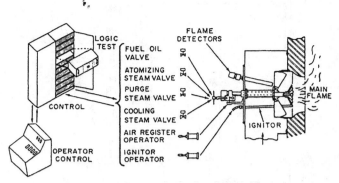

Fig. 40 Burner operation-automatic (nonrecycling) (recycling)

initiated from a remote control station, which indicates whether or not each function has been performed correctly. This system of control does not relieve the operator of burner manipulation. He must devote his undivided attention to the step-by-step procedures for starting and securing burners, which is a time-consuming process. This control system can be justified only in installations where the turndown capabilities of the burners do not match the turndown requirements of the boiler and where the burners must be manipulated to cover the operating range. Its application will not meet the USCG requirements for an automatic boiler.

The automatic (nonrecycling) control system, Fig. 40, involves a burner which, when actuated manually by a pushbutton, is purged, ignited, and modulated automatically; and although secured either automatically or remote-manually, the burner does not recycle automatically. When start and stop sequences are manually initiated from a remote control station, each function in the start-up and stop sequence is performed and checked automatically and all procedures and conditions are supervised by safety interlocks. Since the operator may be required to initiate the start-up and securing of a burner to meet load requirements, this control system does not meet the USCG definition of an automatic boiler.

With an automatic (recycling) type of control system, a burner is purged, ignited, modulated, and stopped automatically, and the burner recycles within a prescribed load range.

3.12 Sample Design Problem. The steps followed in developing the heating surfaces of a steam generator designed to meet specified cycle performance requirements are best illustrated by an example. The basic performance requirements of a boiler are normally provided by the ship's specifications; however, as an example the requirements of a 30,000 shp plant (see the heat balance example in Chapter 2) will be used.

It is assumed that a single boiler is required to furnish the steam necessary for main propulsion and other ship's services. The unit is to be a two-drum integral-furnace boiler equipped with a steam air heater and an economizer. The furnace is to be completely water cooled and two wide-range burners are to be used. The superheater will be installed vertically and one access cavity will be provided.

From the preliminary heat balance for an ABS rating of 30,000 shp, the following operating requirements are to be met:

Steam pressure, drum, approx........	960 psig
Steam pressure, superheater outlet....	875 psig
Steam temperature, superheater outlet.	955 F
Steam flows	
superheated 185,520 lb/hr	
desuperheated 16,870 lb/hr	
Total 202,390 lb/hr	
Feedwater temperature	284 F
Efficiency (based on 1½% radiation and unaccounted for losses and 15% excess air)	88.5%
Fuel total heating value (standard Bunker C + added heat in air)	19,264 Btu/lb
Fuel required	14,349 lb/hr
Air temperature, leaving steam air heater	278 F
Air flow (16.07 lb/lb oil at 15% excess air)	230,600
Flue gas flow = 244,937, say........	245,000 lb/hr

For the example, only one rate of operation will be calculated although for an actual boiler design it is not

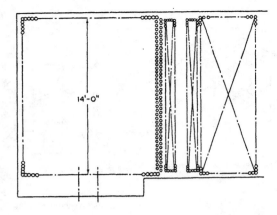

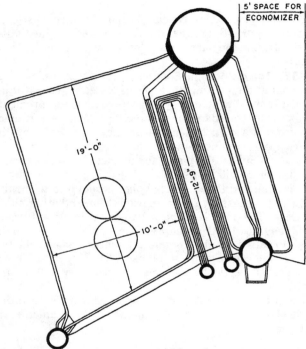

Fig. 41 Boiler layout for sample problem

unusual to calculate three or more rates to establish characteristic curves of performance. Rated power will be calculated since this establishes the design meeting the specified efficiency and steam temperature. The heat-transfer data are derived from the curves and procedures of Chapter 2.

a. Boiler layout. Two oil burners will be used to supply the total oil flow of 14,349 lb/hr at rated power and about 8000 lb/hr each at overload. The necessary clearances for burners of this capacity are obtained from the selected burner manufacturer. Based on this information and experience, an approximate furnace and boiler layout is prepared (see Fig. 41) from which the furnace volume and heating surfaces can be estimated.

b. Furnace calculations. The furnace volume, cold surface, and radiant heat absorbing surface (RHAS) are determined by the methods of Section 2 of Chapter 2 or T&R Bulletin 3–14 [12] to be:

$$\text{Furnace volume} = 2655 \text{ ft}^3$$
$$\text{Projected surface} = 1200 \text{ ft}^2$$
$$\text{RHAS} = 1175 \text{ ft}^2$$

With a fuel higher rating value of 18,500 Btu/lb, the furnace ratings at rated power are:

$$\text{Release rate} = \frac{14,349 \times 18,500}{2655} = 99,985 \text{ Btu/ft}^3$$

$$\text{Oil rate/RHAS} = \frac{14,349}{1175} = 12.2 \text{ lb/ft}^2$$

As these are satisfactory, the furnace exit gas temperature and heat absorption can be calculated (See Section 2 of Chapter 2), based on the following furnace surface areas:

	PROJECTED AREA	SURFACE EFFECTIVENESS FACTOR (see Fig. 8 of Chapter 2)	RHAS
Rear waterwall	190	1.00	190
Front waterwall (2-in. tubes with 2⅞-in. avg. spacing)	175	0.856	150
Screen and floor	435	1.00	435
Side waterwall and roof	400	1.00	400
	1200 ft²		1175 ft²

The adiabatic sensible heat in the combustion gases can be computed from equation (43) of Chapter 2. With a fuel lower heating value of 17,500 Btu/lb and a fuel sensible heat of 46 Btu/lb (100 deg F rise at 0.46 specific heat), for perfect combustion the sensible heat becomes

$$q_{T_{A'}} = \frac{\text{LHV} + q_F + (t_a - t_0)C_pR}{R + 1}$$

$$q_{T_{A'}} = \frac{17,500 + 46 + (278 - 80)(0.2445)(13.98)}{13.98 + 1}$$

$$= 1216 \text{ Btu/lb}$$

From Fig. 2 of Chapter 2 the adiabatic flame temperature, $T_{A'}$, is found to be 3990 F or 4450 R.

With 15% excess air

$$q_{T_A} = \frac{17,500 + 46 + (278 - 80)(0.2445)(16.07)}{16.07 + 1}$$

$$= 1073 \text{ Btu/lb}$$

To determine the shape emissivity factor, $F_E F_A$, the following data are required:

$$V_F = 2655 \text{ ft}^3$$
$$S_T = 1200 \text{ ft}^2$$

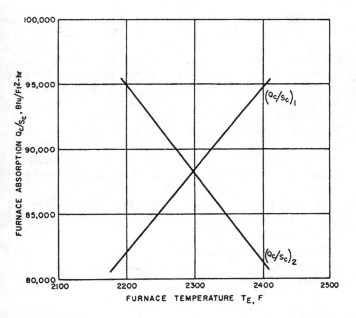

Fig. 42 Determination of furnace exit temperature and furnace absorption

$$S_c = 1175 \text{ ft}^2$$
$$P_F = 1 \text{ atm}$$

Therefore the firing density is [see equation (37) of Chapter 2]

$$\frac{W_F}{P_F V_F} = \frac{14{,}349}{1 \times 2655} = 5.40$$

and from Fig. 7 of Chapter 2 the concentration factor K is 0.056. The mean radiating length is $L = 0.6 \sqrt[3]{2655} = 8.3$ ft. Equation (37) of Chapter 2 can now be evaluated to determine the flame emissivity

$$\epsilon_F = 0.95(1 - e^{-(0.056)(1)(8.3)}) = 0.353$$

and for an S_c/S_T value of 0.98, $F_E F_A$ is determined to be 0.34 from Fig. 6 of Chapter 2.

In order for the calculation to proceed, it is necessary to assume several values of the furnace exit temperature. Making this assumption

Assumed T_E, F	2,200	2,300	2,400
Assumed T_E, R	2,660	2,760	2,860
$T_{A'}$, R	4,450	4,450	4,450
$(Q_c/S_c)_1$; equation (41) of Chapter 2; Btu/ft²-hr	81,800	88,100	94,700
q_{T_A}; see above; Btu/lb	1,073	1,073	1,073
$q_{T_{E'}}$; Fig. 2 of Chapter 2; Btu)/lb	617	651	683
$(Q_c/S_c)_2$; equation (42) of Chapter 2; Btu/ft²-hr	95,100	88,000	81,300

By plotting the values of $(Q_c/S_c)_1$ and $(Q_c/S_c)_2$ for the assumed furnace exit temperatures, as illustrated by Fig. 42, the uncorrected furnace exit temperature is determined to be $T_E = 2298$ F and the furnace absorption is found to be 88,100 Btu/ft²-hr.

The correction to be applied to the furnace exit temperature due to the effect of rear-wall convection is determined next (see Chapter 2, Subsection 2.2). The furnace consists of 2-in.-OD, 0.165-in.-thick, ASME SA 178 A, electric resistance welded tubes. The projected area of the rear wall is 190 ft², and the front half of the wall surface, $S_W = 284$ ft².

From Subsection 1.2(d) of Chapter 2, the tube surface temperature in the furnace is

$$T_S = t_S + \left(\frac{Q_c}{S_c}\right)\frac{X_e}{k} = 592 \text{ F}$$

where

$t_S = 541$ F at drum saturation pressure of 975 psia
$Q_c/S_c = 88{,}100$ Btu/ft²-hr

$$X_e = \text{tube equivalent thickness} = \frac{1}{2}D_o \log_e \frac{D_o}{D_i}$$
$$= \frac{2}{2}\log_e \frac{2}{1.67} = 0.180$$

$k = $ tube conductivity $= 310$ Btu/hr-ft-F

Next, by estimating the corrected furnace exit temperature, $T_{E'}$, to be 2200 F the tube film temperature can be approximated as

$$T_f = \frac{\dfrac{T_E + T_{E'}}{2} + T_S}{2} = \frac{\dfrac{2298 + 2200}{2} + 592}{2}$$
$$= 1412 \text{ F}$$

The temperature coefficient f_T [equation (46) of Chapter 2] then becomes

$$f_T = 0.00003875 T_f + 0.1035 = 0.1586$$

With a flue gas flow W_g of 245,000 lb/hr and two burners having 2-ft throat diameters, the flue gas weight flow rate G is 39,000 lb/ft²-hr. Since the furnace depth D is 14 ft, the surface heat-transfer coefficient h_{RW} can be computed from equation (45) of Chapter 2 as

$$h_{RW} = \frac{G^{0.53}}{D^{0.47}} f_T = 12.4 \text{ Btu/hr-ft}^2\text{-F}$$

From Fig. 3 of Chapter 2 the specific heat C_p of the flue gas at T_f is 0.314 Btu/lb-F; therefore, the corrected furnace exit temperature $T_{E'}$ can be determined

$$T_{E'} = \frac{T_E - T_S}{e^{\frac{h_{RW} S_W}{W_g C_p}}} + T_S = \frac{2298 - 592}{e^{\frac{12.4 \times 284}{245{,}000 \times 0.314}}} + 592$$

$$T_{E'} = 2222 \text{ F}$$

The estimated $T_{E'}$ value of 2200 F was sufficiently accurate to result in a negligible error.

c. Heating surfaces. After determining the furnace exit temperature, the performance of the boiler in the screen, superheater, and generating bank is evaluated in that sequence. From the approximate boiler layout, Fig. 41, the following data are determined (all tube banks are arranged in-line rather than staggered for ease of cleaning):

	TUBE DIAMETER, in.	TRANSVERSE PITCH, in.	BACK PITCH, in.	GAS FLOW AREA, ft²	HEATING SURFACE ft²
Screen (2 rows)	2	3.5	2.5	100	765
Superheater (10 rows)	1.25	1.875	1.75	70	4,065
Generating bank (24 rows)	1.25	1.875	2.37	75	11,460

The screen, superheater, and generating bank performance calculations may be conducted as shown in Table 3.

The estimated screen, superheater, and generating bank outlet temperatures are sufficiently close to the calculated values to result in no significant errors.

The final steam temperature is next determined. The heat absorbed by the steam in the superheater by direct radiation from the furnace is

$$Q_{SH_R} = AF_L Q_c/S_c$$

where

A = screen projected area = $12.9 \times 14 = 180$ ft²
F_L = screen leakage factor for a tube pitch/diameter ratio of 1.75 (Fig. 8—Chap. 2)

$$F_L = 1 - F_{A_{\text{Direct to first row}}} - F_{A_{\text{Direct to second row}}}$$

$$F_L = 1 - 0.73 - 0.18 = 0.09$$

$$Q_c/S_c = 88,100 \text{ Btu/hr-ft}^2$$

Therefore

$$Q_{SH_R} = (180)(0.09)(88,100) = 1,427,000 \text{ Btu/hr}$$

The heat absorbed in the superheater by convection from the flue gas as it flows through is

$$Q_{SH_c} = W_g C_p (T_1 - T_2)$$

$$Q_{SH_c} = (245,000)(0.316)(2084 - 1339)$$

$$= 57,680,000 \text{ Btu/hr}$$

Consequently, the total heat absorbed in the superheater is $Q_{SH_R} + Q_{SH_c} = 59,107,000$ Btu/hr. With a superheater steam flow of 202,390 lb/hr, the heat rise of the steam passing through the superheater is 292 Btu/lb; therefore, since the saturated steam entering the superheater has an enthalpy of 1193 Btu/lb, the final enthalpy of the superheated steam is 1485 Btu/lb, which corresponds to a final steam temperature of 958 F at a pressure of 890 psia.

d. Economizer. An extended-surface economizer is to be used for the final heat recovery. The economizer will be designed to reduce the flue gas temperature to the 316 F required to obtain the 88.5-percent boiler efficiency desired (see Fig. 16). The 1.5-in.-OD economizer tubing selected has 1.75 ft² of surface per lineal foot of tubing. In the space indicated for the economizer in Fig. 41, a unit 18 rows wide having a heated length of 14 ft can be installed. Under the conditions which the economizer will experience in service, the unit will have an overall heat-transfer rate of 15 Btu/hr-ft²-F. The total economizer heating surface required, economizer height, and the final water temperature are to be determined.

The economizer calculations may proceed as follows:

Gas temp. entering, T_1	631 F
Gas temp. leaving, T_2	316 F
Change in gas temp., ΔT_g	315 F
Gas sensible heat entering (Fig. 2—Chap. 2), Q_{g1}	149 Btu/lb
Gas sensible heat leaving, Q_{g2}	65 Btu/lb
Change in gas sensible heat, ΔQ_g	84 Btu/lb
Gas flow, W_g	245,000 lb/hr
Feed temp. entering, t_1	284 F
Feed enthalpy entering, h_1	255.3 Btu/lb
Feedwater flow, W_w	202,390 lb/hr
Feedwater enthalpy rise ($\Delta Q_g \times W_g/W_w$), Δh	101.7 Btu/lb
Feed enthalpy leaving ($h_1 + \Delta h$), h_2	357.0 Btu/lb
Feed pump discharge pressure, p	1200 psig
Feedwater temp. leaving (at h_2 and p), t_2	382 F

$$\text{LMTD} = \frac{(T_1 - t_2) - (T_2 - t_1)}{\log_e\left(\dfrac{T_1 - t_2}{T_2 - t_1}\right)} \qquad 105.6 \text{ F}$$

Economizer heat-transfer coeff., U	15 Btu/hr-ft²-F
Economizer surface required ($\Delta Q_g \times W_g/U \times$ LMTD), S	12,990 ft²
Economizer heated length, L	14 ft
No. tube rows wide, N_w	18
Area per foot of tube, A	1.75 ft²/ft
Required no. tube rows high (S/LN_wA)	29.5
No. tube rows high	30

(an even number must be used as the inlet and outlet are on the same end)

The preliminary design of the boiler is now complete. Additional boiler ratings, i.e., part loads or overloads, would be calculated next. Following this, the draft losses, circulation characteristics, tube metal temperatures, control and auxiliary desuperheater sizes, safety valve settings, miscellaneous pressure drops, and similar considerations would be investigated.

Table 3 Screen, superheater, and generating bank performance calculations

TEMPERATURE DATA			SCREEN	SUPERHEATER	GENERATING BANK
Inlet gas temp. ($T_{E'}$ for screen)	T_1	F	2222	2084	1339
Estimated outlet temp.	T_2	F	2090	1340	635
Mean temp. $= \frac{1}{2}(T_1 + T_2)$	T_{AV}	F	2156	1712	987
Specific heat at T_{AV} (Fig. 3—Chap. 2)	C_p	Btu/lb-F	0.327	0.316	0.290
Surface temp. (sat. temp.)	T_S	F	541	—	541
Surface temp. $\sim$ superheater					
$T_S = \dfrac{t_1 + t_2}{2} + \Delta t_w$	T_S	F	—	848	—
where					
saturation temp.	t_1	F	—	541	—
desired steam temp.	t_2	F	—	955	—
assumed wall drop	Δt_w	F	—	100	—
Avg. steam temp. $= \frac{1}{2}(t_1 + t_2)$	t_s	F	—	748	—
Film temp. $= \frac{1}{2}(T_{AV} + t_s + \Delta t_w)$	T_f	F	—	1280	—
$\quad = \frac{1}{2}(T_{AV} + T_S)$	T_f	F	1349	—	764

HEAT TRANSFER DATA
Convection:

			SCREEN	SUPERHEATER	GENERATING BANK
Flue gas flow	W_g	lb/hr	245,000	245,000	245,000
Gas flow area	A_g	ft^2	100	70	75
Gas flow rate $= W_g/A_g$	G	lb/ft^2-hr	2450	3500	3270
Tube OD	D	ft	0.167	0.104	0.104
Gas viscosity at T_f (Fig. 4—Chap. 2)	μ	lb/ft-hr	0.092	0.087	0.070
Reynolds no. $= GD/\mu$	Re	—	4450	4180	4860
Tube transverse spacing/dia ratio	a_1	—	1.75	1.5	1.5
Tube back spacing/dia ratio	a_2	—	1.25	1.4	1.9
Tube arrangement factor (Fig. 10—Chap. 2)	F_A	—	0.83	0.96	1.01
Tube bank depth factor (Table 2—Chap. 2)	F_D	—	0.765	1.0	1.0
Temp. factor (Fig. 13—Chap. 2)	f_E	—	0.159	0.1565	0.133
Convection heat-transfer coeff.					
$0.292 F_A F_D f_E \dfrac{G^{0.6}}{D^{0.4}}$ [equation (52)—Chap. 2]	h_f	$\dfrac{\text{Btu}}{\text{hr-ft}^2\text{-F}}$	6.52	14.56	12.5

Radiation:

			SCREEN	SUPERHEATER	GENERATING BANK
Mean radiating length [eqn. (53)—Chap. 2]	L	ft	0.253	0.148	0.233
Partial pressure of radiating constituents (Subsection 2.3—Chap. 2)	P_R	$\dfrac{\text{atm}}{\text{atm}}$	0.239	0.239	0.239
$P_R L$	—	$\dfrac{\text{atm-ft}}{\text{atm}}$	0.0605	0.0354	0.0557
Gas emissivity (Fig. 14—Chap. 2 at T_{AV})	ϵ_G	—	0.051	0.048	0.081
Radiation conductance factor (Fig. 14—Chap. 2)	h_r/ϵ_G	$\dfrac{\text{Btu}}{\text{hr-ft}^2\text{-F}}$	48	37	13
Radiation heat-transfer coeff. $h_r = \epsilon_G(h_r/\epsilon_G)$	h_r	$\dfrac{\text{Btu}}{\text{hr-ft}^2\text{-F}}$	2.45	1.78	1.05

Total Heat Transfer:

			SCREEN	SUPERHEATER	GENERATING BANK
Steam film conductance (ref. [8])	h_s	$\dfrac{\text{Btu}}{\text{hr-ft}^2\text{-F}}$	—	570	—
Metal conductance (1¼ Cr-Mo tubes)	h_m	$\dfrac{\text{Btu}}{\text{hr-ft}^2\text{-F}}$	—	1260	—
Total heat-transfer coeff.:					
$\dfrac{1}{U} = \dfrac{1}{h_f + h_r} + \dfrac{1}{h_s} + \dfrac{1}{h_m}$	U	$\dfrac{\text{Btu}}{\text{hr-ft}^2\text{-F}}$	8.97	15.69	13.55

EVALUATION OF OUTLET TEMPERATURE

			SCREEN	SUPERHEATER	GENERATING BANK
Heating surface	S	ft^2	765	4065	11,460
$\dfrac{T_1 - T_S}{T_2 - T_S} = e^{\frac{US}{W_g C_p}}$	r	—	1.089	—	8.9
$T_2 - T_S = \dfrac{1}{r}(T_1 - T_S)$	—	—	1543	—	90
Outlet temp. $= (T_2 - T_S) + T_S$	T_2	F	2084	—	631

Table 3—(continued)

TEMPERATURE DATA			SCREEN	SUPERHEATER	GENERATING BANK
The superheater is considered to be an average of a parallel-flow and a counterflow condition.					
Steam flow	W_s	lb/hr	—	202,390	—
Steam average specific heat	C_{ps}	Btu/lb-F	—	0.702	—
$v = W_g C_p / W_s C_{ps}$	v	—	—	0.545	—
For parallel flow; $$\frac{T_1 - T_{2p}}{T_1 - t_1} = \frac{1 - e^{-\frac{US}{W_s C_p}(1+v)}}{1+v}$$	T_{2p}	F	—	1365	—
For counterflow; $$\frac{T_1 - T_{2c}}{T_1 - t_1} = \frac{1 - e^{-\frac{US}{W_s C_p}(1-v)}}{1 - ve^{-\frac{US}{W_s C_p}(1-v)}}$$	T_{2c}	F	—	1313	—
Superheater outlet temp. $= \frac{1}{2}(T_{2p} + T_{2c})$	T_2	F	—	1339	—

Section 4
Boiler Operation

4.1 Water Treatment. In boilers, water is converted into steam, which leaves the drum in a relatively pure state. Impurities, other than the gases which enter with the feedwater, are retained and concentrated in the boiler water. High concentrations of foam-producing solids in the boiler water aggravate water carry-over and contaminate the steam. Chemical and solubility changes also take place as the temperature is increased.

Natural waters contain impurities, which may be harmful in boiler operation. These impurities originate from the earth and the atmosphere (or from municipal and industrial wastes), and are broadly classified as suspended or dissolved organic and inorganic matter, and dissolved gases.

With few exceptions, the waters found in nature are not suitable for use as boiler feedwater but they can be used after proper treatment [21,22]. In essence, this entails: the removal from the raw water of those constituents which are known to be harmful; supplementary treatment (within the boiler or connected system) of residual impurities to convert them into harmless forms; and systematic removal, by blowdown of boiler water concentrates, to prevent excessive accumulation of solids within the unit.

The ultimate purpose of feedwater and boiler water treatment is to keep the internal surfaces free from deposits of scale or sludge and to prevent the corrosion of these surfaces. Hard-scale formations, formed by certain constituents in zones of high heat input, retard the flow of heat and raise the metal to higher-than-normal temperatures. This can cause overheating and the failure of pressure parts. Sludge, or solid particles normally carried in suspension, may settle locally and restrict the flow of cooling water or, in some cases, may deposit in the form of insulating layers with an effect similar to that of hard

scale. Oil and grease prevent adequate wetting of the internal surfaces and, in areas of high input, cause overheating; they also may carbonize and form a tightly adherent insulating coating. Corrosion due to acidic conditions, or to dissolved gases, can weaken the boiler by the removal of metal. This usually occurs in localized areas in the form of cavities and pits which if unchecked may result in complete penetration and leakage. Certain chemical reactions produce an intergranular attack on the metal, leading to embrittlement and fracture.

4.2 Feedwater. Virtually all oceangoing vessels use feedwater evaporated from seawater for the boilers, and thus, feedwater treatment is minimized. Some contamination may be encountered in the distillate due to the carryover of water particles with the vapor and the reabsorption of noncondensable gases but additional solids removal is not required. However, dissolved gases must be removed to prevent corrosion.

Dissolved oxygen is usually the greatest factor in the corrosion of boiler surfaces in contact with water. It may be in the makeup water or in the feedwater, as a result of previous contacts with atmospheric air, or it may be added to the water by leakage into the system through low-pressure pump seals, storage tanks, etc. Fortunately, most of the oxygen can be readily removed from the water by the use of deaerating-type feedwater heaters.

Corrosion may be experienced in the condensate piping and the preboiler system due to dissolved gases, such as carbon dioxide, sulfur dioxide, or hydrogen sulfide, in the water. These gases originate from the atmosphere or from constituents in the boiler water. They are released in the steam generators, intimately mixed with the outgoing steam, and finally exhausted to the condenser.

4.3 Boiler Water. Boiler water is treated within the boiler to prevent corrosion, the fouling of heat-absorbing surfaces, and the contamination of steam. This requires the injection of chemicals into the steam drum where they react with the residual impurities in the feedwater. Properly controlled, internal treatment can maintain boiler water conditions within satisfactory limits [17,20].

Corrosion is minimized by maintaining an alkaline boiler water and this condition is usually expressed in terms of "pH" or "total alkalinity." The pH of boiler water usually is maintained within the range of 10.2 to 11.5.

The removal of dissolved oxygen is desirable in all boilers but it is mandatory for high-pressure units. It is customary in removing oxygen to supplement feedwater deaeration by internal chemical treatment of the water, using a scavenging agent such as sodium sulfite, which combines with the oxygen to form a stable sodium sulfate. Hydrazine also may be used for the purpose yielding end products of water and inert nitrogen. These chemicals prevent the entrance or the retention of dissolved oxygen and are maintained in the boiler water with a small marginal excess.

The elimination of hardness in the boiler water is necessary to prevent scale and it can be removed by injecting one of the combinations of sodium or potassium phosphate and thoroughly mixing the compound with the boiler water. If the alkalinity is maintained at a pH of 10 or higher, the residual calcium ions entering with the feedwater are precipitated as an insoluble phosphate sludge and the magnesium is precipitated as a nonadherent magnesium hydroxide. Routine control requires the adjustment of the pH by the addition of sodium hydroxide, or its equivalent, and the maintenance of a moderate excess of phosphate ions in the boiler water.

The treatment best suited to any boiler plant depends upon many factors, and a feedwater specialist should be consulted to establish specific procedures. However, the results obtained will depend upon the diligence and integrity of the routine sampling and the control measures conducted by the operating personnel.

4.4 Boiler Operation and Care. The boiler design should reflect careful consideration of the factors affecting the operation and care of boilers. These factors include: initial preparation of new equipment for service; normal operation, including routine start-up and shutdown; emergency operation; inspection and maintenance; and idle storage. In all phases the handling of the equipment is the responsibility of the operator, but the overall boiler design reflects the recommendations and operation instructions supplied by the manufacturer. These should be thoroughly understood and carefully followed.

a. Initial preparation. The initial preparation of new units for service, or that of older equipment after major alterations or repairs, entails the removal of foreign material from both the casing and the interior of pressure parts; hydrostatic testing and inspection for leaks; and the boiling out of the unit with a caustic solution to remove grease and other deposits, which may be present in the economizer and steam-generating pressure parts. During boil-out the unit is fired at a low rate to maintain about 50% of the normal operating pressure. This procedure facilitates the desired slow drying of any refractories used in the boiler setting. During the boil-out period, which is usually from 12 to 36 hr duration, the boiler is blown down periodically through all of the blowdown connections in rotation to eliminate the sediment removed from the surfaces. If necessary, the boil-out may be supplemented by an inhibited acid cleaning to remove mill scale.

Following the boil-out, it is general practice to reduce the concentration of boil-out chemicals to a satisfactory operating level by blowing down and replenishing with fresh water. The boiler pressure is then increased to test and set the safety valves. Next, the superheater and the steam piping are blown out to remove foreign material, and the boiler is placed on the line for a period of low-load operation during which the auxiliary equipment, controls, and interlocks are test operated. After this operation and testing, it is customary to shut down, cool, and drain the unit and then thoroughly inspect the internal and external surfaces before the start of normal operation.

b. Normal operation. Normal operation includes the orderly start-up and shutdown of equipment and its operation, under controlled conditions, to meet load requirements. Since about 80% of all furnace explosions occur during start-up and low-load operation, particular care must be taken during these periods.

The rate of firing during start-up is restricted to prevent overheating of the superheater (and reheater) metals when there may be little or no cooling steam flow. To prevent excessive temperature differentials and possible high thermal stresses in the pressure parts, excessive firing rates are to be avoided. These considerations govern the time required for start-up and also, to some extent, for cooling after shutdown.

c. Boiler cleaning. For satisfactory and efficient operation, a boiler must be kept clean on both the waterside and fireside. With adequate attention to the preboiler feed system and by maintaining the boiler waterside chemistry within prescribed limits, there should be little need to clean the waterside. The fireside, on the other hand, requires daily attention if the steam temperature and boiler efficiency are to be maintained at their optimum values.

Only distilled and deaerated water should be used for feeding the boiler and for feed makeup. Total solids in the boiler water should not exceed a maximum of 500 ppm during normal operation. Suspended solids should be zero but not greater than 5% of the total solids. Chlorides should be lower than 2 ppm and phosphate should be in the range of 10 to 25 ppm. The pH of the boiler water should be in range of 10.2 to 11.5. An oxygen scavenger (sodium sulfite) should range from 30 to 50 ppm measured as SO_3. The boiler water, maintained within these limits, will not form scale or baked-on sludge deposits on the tube surfaces.

To assist in maintaining clean waterside conditions, a steaming boiler should be given a surface blow each day. A test for total dissolved solids made before and after the blow will indicate if additional attention is required. The water drum bottom blowoff connection and waterwall blowdowns should be used as necessary in controlling suspended or total solids in the boiler water.

If conditions are allowed to deteriorate to the point that scale or baked-on sludges are found during waterside inspections, chemical analysis of the deposits will indicate the cleaning method best suited for their removal. Tubes may be cleaned by passing air-turbine-driven brushes and scale cutters through each tube, and flushing with a high-pressure water hose. The entire boiler can be cleaned as a unit more quickly and efficiently by acid cleaning. A specialist should be consulted to accomplish this procedure, which entails the use of acid and neutralizing rinsing agents. The acid strength, neutralizers, and the temperature at which they are used are of vital importance if the cleaning process is to be kept within safe limits. Excessive acid strength or unneutralized acid remaining after cleaning will pit and attack the metal possibly to the point that replacement of parts is required.

A boiler should be designed to facilitate the cleaning of the fireside. The heating surfaces of the superheater boiler generating bank as well as the economizer and air heater should be arranged in in-line patterns, which provide clear lanes through which inspection and cleaning can be accomplished. Staggered patterns are slightly more efficient from a heat-transfer standpoint but are more difficult to inspect and clean.

In extreme conditions, hand lancing or water-washing with high-pressure water may be required; however, improvements in burners and in the arrangement of the boiler and its cleaning equipment have minimized the need for hand cleaning. Soot blowers are used to clean the fireside at regular intervals. The frequency depends on the fuel ash characteristics, combustion efficiency, and the rates of operation.

Air or steam can be used as the blowing medium; however, oil-fired boilers almost universally use steam. Steam is available in large quantities and at a low cost. Air, often used in coal-fired units, is "puffed" intermittently to permit repressurization of the air receivers by the air compressor.

Superheated or desuperheated steam can be used with good effect. The steam should be supplied in a dry state, and the supply system must have adequate traps or be fitted with orificed drains to remove condensate so as to prevent it from reaching the blower elements.

Three basic types of steam soot blowers are used. The long, retracting, mass-action type is used in superheaters; the rotary, valve-in-head, line blower is used in boiler banks, economizers, and tubular air heaters; and the stationary-type unit is used in hoppers and where fixed directions of blowing are desired to remove localized deposits, such as those forming on top of the water drum.

The soot blowing system can be manually operated or sequential pushbutton controls can be employed to automatically program the cleaning process. Once initiated, the automatic sequencing control opens the steam supply valve, warms the lines, blows the soot blowers in sequence, and then shuts down the steam supply.

4.5 Boiler Storage

a. Dry storage. When a boiler will be idle for a considerable length of time and there will be ample time

available to prepare for its return to service, the dry-storage method is recommended. To accomplish this, the unit is emptied, thoroughly cleaned internally and externally, dried, and then closed tightly to exclude both moisture and air.

Trays of lime, silica gel, or other moisture absorbents are placed in the drums to collect the moisture trapped in the air when closing the boiler. To insure against a possible overflow of corrosive liquid after the moisture has been absorbed, not more than 75% of the tray capacity should be filled with the dry absorbent. Care must be taken to prevent water, steam, or air leakage into the unit, and periodic inspections should be made to make sure that there is no corrosive action. The absorbent should be replenished as required.

b. Wet storage. If boilers are to be placed in standby service but must be available for immediate operation, before shutting down they should be steamed to stabilize the boiler water conditions and to remove oxygen bubbles from the internal surfaces. The boiler firing rate should then be decreased slowly and the steam drum water level should be raised as high in the gage glass as is consistent with safe operation while still passing steam to the line. The hydrate alkalinity in the boiler water should be increased to a minimum of 400 ppm, and, with the addition of sodium sulfite in the amount of 100 ppm, oxygen corrosion can be prevented.

During storage, boiler connections should be checked for leakage and frequent samples of boiler water should be taken and analyzed. If analyses indicate that the hydrate alkalinity is less than 250 ppm, the water in the steam drum should be lowered to the normal operating level, and chemicals should be injected to bring the hydrate alkalinity back to 400 ppm. The boiler should then be steamed sufficiently to circulate the added chemicals, following which the process of wet storage should be completed in the usual manner.

c. Steam blanket. The steam blanket method provides excellent protection for short-time idle storage, but requires a continuous source of low-pressure steam (in order of 150 psig) and connections for maintaining this steam pressure in the stored boiler. All vents and drains should normally be closed to allow the boiler and superheater to fill with condensate but the boiler can be drained periodically if desired.

d. Nitrogen blanket. The oxygen-free nitrogen storage method is one in which nitrogen gas at a pressure of 10 to 15 psig is maintained in the unit at all times during its idle status. It can be used with very satisfactory results if the boiler, terminal valves, and fittings are tight under normal hydrostatic pressure.

The boiler can either be emptied or a normal water level maintained in the steam drum. The nitrogen is admitted when the boiler pressure has dropped below the gas pressure which will be maintained in the unit. Satisfactory protection against corrosion depends upon system checks and the renewal of nitrogen, as necessary.

To ready a boiler for service after storage, the nitrogen supply is secured and the water level in the steam drum is raised to that required for lighting-off. Any nitrogen in the steam drum and superheater will be displaced by the steam generated during the customary venting of the steam drum and the superheater as steam pressure is increased [23].

References

1 George W. Melville, "Development of the Marine Boiler in the Last Quarter Century," *The Engineer*, 1911.

2 G. W. Kessler, "Procedures and Influencing Factors in the Design of Marine Boilers," *Trans. SNAME*, Vol. 56, 1948.

3 W. I. Signell, "Marine Boiler Design Today," *Trans. SNAME*, Vol. 76, 1968.

4 Carl D. Shields, "Boilers, Types, Characteristics and Functions," F. W. Dodge Corp., 1961.

5 W. A. Fritz, Jr., and L. Cohen, "Development and Evaluation of a Supercharged Steam Generating System," ASME Paper 62-WA-279, American Society of Mechanical Engineers, Nov. 1962.

6 "Guidelines for the Use and Applications of Marine Waste Heat Boilers and Waste Heat/Auxiliary Boiler Arrangements", SNAME T&R Bulletin No. 3–38, 1985.

7 Thomas P. Mastronarde, "Energy Conservation Using Waste Heat Boilers: The Challenges, Problems and Solutions", *ASNE Journal*, April 1982.

8 *Steam, Its Generation and Use*, Babcock & Wilcox Co., Barberton, Ohio.

9 F. P. Bergonzi, G. Cooper, and J. F. Mooray, "Heat Recovery Equipment for Modern Marine Cycles," SNAME Southern California Section, Nov. 1966.

10 "Lexicon—Steam Generating Equipment," American Boiler Manufacturers Association, Newark, N.J.

11 "Marine Steam Power Plant State of the Art Seminar," General Electric and Babcock & Wilcox, 1969.

12 "Furnace Performance Criteria for Gas, Oil, and Coal Fired Boilers", SNAME T&R Bulletin No. 3–32, July 1981.

13 *Code of Federal Regulations, Title 46—Shipping*, published by the Office of the Federal Register.

14 W. O. Nichols, M. L. Rubin, and R. V. Danielson, "Some Aspects of Large Tanker Design," *Trans. SNAME*, Vol. 68, 1960.

15 "Recommended Practices for Preparing Marine Steam Power Plant Heat Balances," SNAME T&R Bulletin No. 3–11.

16 R. P. Giblon, K. M. Shauer, and I. H. Rolih, "Design Considerations for Boiler Forced-Draft Systems," *Marine Technology*, Vol. 6, No. 4, Oct. 1969, p. 406.

17 Thomas C. Elliot, *Standard Handbook of Powerplant Engineering*, McGraw-Hill Publishing Co., New York, 1989.

18 "Section I ASME Boiler and Pressure Vessel Code—Power Boilers," American Society of Mechanical Engineers.

19 William H. McAdams, *Heat Transmission*, McGraw-Hill Book Co., Inc., New York, 1954.

20 G. R. Fryling, *Combustion Engineering*, Combustion Engineering Inc., New York.

21 *Handbook: Impurities in Water and Means for Their Removal*, Drew Chemical Corp., Boonton, N.J., 1972.

22 *Handbook of Industrial Water Conditioning*, Betz Laboratories, Inc., Trevose, Pa., 1980.

23 "Guidelines for the Preservation of Marine Boilers and Boiler Components," SNAME T&R Bulletin No. 3–30, April 1980.

CHAPTER VI

W. G. Catlow | # Steam Turbines

Section 1
Arrangements and Cycles

1.1 Introduction. Although the early development of steam power was based on the use of a reciprocating steam engine as a prime mover, the inherent advantages of a steam turbine soon became apparent and have made it the choice for steam propulsion plants. Turbines are not size limited and can be provided for any power rating up to the maximum likely to be encountered in marine service. High steam pressures and temperatures can be accommodated safely and are limited only by boiler problems. Rotary motion is simpler than reciprocating motion; and the unbalanced forces (that produce vibration), which are present in many reciprocating machines, can be eliminated in the turbine. In addition, a turbine can efficiently use a low exhaust pressure and is characterized by light weight, minimum space, and low maintenance.

A marine steam turbine operates in accordance with the same basic fundamentals as its land-based counterparts in central station and industrial applications but differs in many important respects. This chapter emphasizes those features and characteristics peculiar to marine applications, which are derived from the special requirements of marine propulsion or auxiliary drives.

The science, and often art, of turbine design is a highly specialized field, and a number of textbooks have been devoted almost entirely to this subject [1–3]. This chapter is intended to provide information useful to marine engineers, owners, and operators in connection with the application and operation of marine steam turbines.

Turbines are used to propel many types of vessels having widely varying requirements with respect to power, economy, weight, and arrangement. In the selection and development of a turbine design for a specific application, the following factors must be considered:

(1) The maximum ahead power needed to provide the desired ship's speed.
(2) The relative amounts of time spent at maximum power and reduced cruising powers.
(3) The turbine throttle steam pressure and temperature.
(4) The steam cycle arrangement, together with the number and location of extraction points and corresponding steam flows.
(5) The turbine exhaust vacuum for design purposes.
(6) The type of power transmission to the propeller.
(7) The astern operating requirements.
(8) Space limitations of the engine room arrangement.

(9) The importance of machinery weight and size.

This chapter deals with the effects of these and other factors upon the design of turbines.

1.2 Machinery Arrangements. A steam turbine is essentially a high-speed machine, whereas a propeller is more efficient at lower speeds. A speed reducer is, therefore, required to optimize the propulsion system. Reduction gearing is the common choice for most marine applications; however, other forms of power transmission are possible. Many applications have been built with an electric drive, for example. With a conventional electric drive, the reversal of shaft rotation is obtained electrically, and the turbine may be undirectional. This, coupled with more arrangement flexibility and improved technology, has made an electric drive attractive for certain applications, as discussed further in Chapter 8.

The most common steam turbine arrangement is a cross-compound unit consisting of a high-pressure, high-speed turbine, as shown in Fig. 1, and a low-pressure, low-speed turbine, as shown in Fig. 2, which drives a single fixed-pitch propeller through reduction gears and shafting. A complete astern turbine is generally provided in the low-pressure turbine casing.

The efficiency of a cross-compound unit is maximized by providing sufficient stages to achieve an optimum blade speed. In addition, the high-pressure portion of the turbine can be made smaller, lighter, and more efficient by designing it to operate at a higher speed than the low-pressure turbine.

As the rated power and steam flow of a cross-compound turbine are increased, the required diameter at the exhaust increases correspondingly; and a point is reached where the size becomes objectionable from both an arrangement and a manufacturing viewpoint. To provide additional exhaust area, the low-pressure turbines of high-powered cross-compound sets may be designed to have a double-flow exhaust as shown in Fig. 3. With this arrangement, the steam flow is divided and flows through two equal-capacity low-pressure elements to the condenser. Since the same total exhaust area is provided at a reduced diameter, this construction permits the low-pressure turbine to operate at a higher rpm, which, in turn makes possible a reduction in weight.

It is customary to provide an astern turbine at each end of a double-flow rotor. The astern steam flow is controlled by a single throttle and the flow is divided equally between

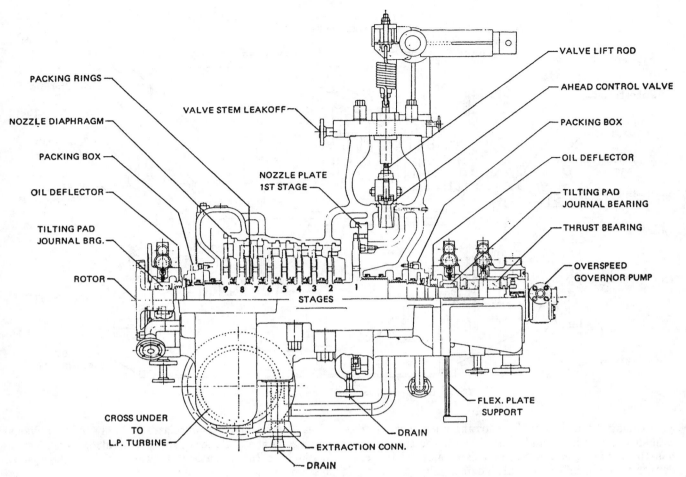

Fig. 1 Typical nonreheat high-pressure turbine—impulse type

the two turbines. The symmetrical arrangement and equal division of flow results in the same pressure at each ahead exhaust and, therefore, no pressure differential across the ahead blading.

In lieu of a cross-compound arrangement, all of the ahead and astern blading can be provided in a single casing. While such an arrangement could be built for any power output, single-casing turbines generally are not considered for powers above 20,000 to 25,000 shp. At powers below this range, a single-casing turbine has some definite advantages, such as reduced initial installation and maintenance costs and more simple gland sealing, gland exhaust, lubricating-oil supply and drain systems, overspeed protection system, and machinery foundations.

A cross section of a typical single-casing turbine is shown in Fig. 4. Thirteen impulse stages are provided in the ahead turbine; these consist of a two-row wheel followed by twelve single-row wheels. The astern turbine consists of two impulse stages: a two-row wheel, followed by a single-row wheel.

The steam rate for a single-casing turbine is approximately 1% lower than a comparable two-casing or cross-compound turbine. This lower steam rate is due to several factors: (1) the total blade speed is limited by the number of stages that can be accommodated on a single rotor of

practical length, and also by the maximum rotor speed for which the ahead exhaust stages can be designed; (2) the increased rotor length requires a larger-diameter shaft and, consequently, interstage leakage losses are greater; and (3) some compromise is necessary with respect to the blade height-to-diameter ratio selection.

When developing the best overall turbine design for a given set of conditions, the designer must select a great many variables including the number of casings, revolutions per minute for each rotor, number of stages, and the nozzle and blade height for each stage. An optimum design could be formulated by an iterative process consisting of a comparison of a series of turbine designs in which each of the principal variables is individually tested through an appropriate range. In the evaluation of the results, proper consideration is given to weight, size, and cost as well as efficiency. However, a complex study of this type is seldom necessary because experience and comparisons with similar designs, aided by the judgment of the turbine designer, make short-cut procedures possible.

At least three arrangements of the low-pressure turbine and condenser are in use. In one, the low-pressure turbine is supported by longitudinal girders forming an integral part of its lower casing; the girders are supported by a foundation structure at the forward end and by the

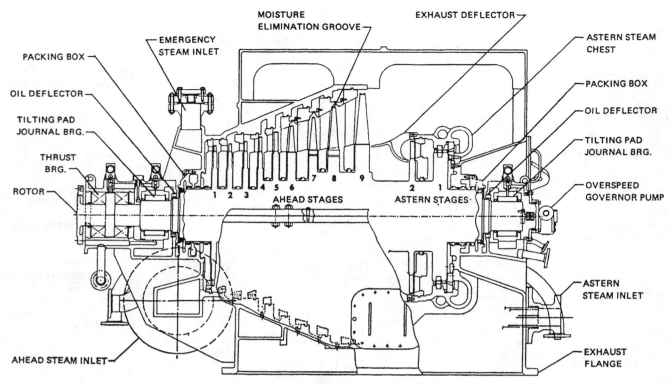

Fig. 2   Typical nonreheat low-pressure turbine—impulse type

gear casing at the aft end. This arrangement permits the condenser to be hung and located below the turbine and has the advantage that thermal expansion of the condenser does not affect the turbine-gear alignment.

As an alternative, the turbine may be supported by the condenser. In this case, thermal expansion of the condenser will raise the turbine centerline with respect to the pinion, and this must be considered in the design of the flexible couplings between the turbine and pinion.

In a third arrangement, the condenser is located forward of the low-pressure turbine such that the turbine exhausts axially into the condenser. This arrangement has the advantage of reduced overall height, but the disadvantage of increased machinery length.

The detail design of foundations for machinery is the responsibility of the ship designer; however, it is in order for the machinery supplier to review and comment upon the machinery foundation drawings to ensure that proper support is afforded the equipment and that no undesirable restraints are imposed. Emphasis must be given to the provision of adequate foundation rigidity to avoid vibration conditions. This is particularly important with respect to periodic variations in propeller thrust, which may excite longitudinal vibrations in the propulsion system. A more complete discussion of the principles and problems associated with the design of equipment foundations is presented in reference 4.

Steam turbines, reduction gears, shafting, and propellers form a closely related system; and to develop an optimum overall system design, it is important that the total system requirements be considered in the development

of each component. This relates not only to the physical arrangement and choice of speeds, but also to such considerations as the vibratory characteristics of the overall system.

1.3   Reheat Turbines.   Reheat is usually accomplished by passing the steam through a section of the turbine, returning it to a special section of the boiler (called the reheater) where its temperature is increased by flue gases, and passing the reheated steam through the remainder of the turbine. Reheat may also be accomplished by increasing the temperature of the returned steam in a separate heat exchanger that is heated by steam taken directly from the boiler. With a gas reheater, the steam can be reheated to the initial temperature; but with a steam reheater, the returned steam can only be reheated up to the saturation temperature of the steam taken from the boiler.

A reheat cycle is obviously more complex than a nonreheat cycle; however, a reheat cycle can provide a 10 to 12% gain in plant efficiency. The increase in cycle efficiency is due, primarily, to an increase in the mean effective temperature at which the heat is added since the reheat part of the cycle increases the quantity of heat added at higher temperatures. A secondary, but important, effect is the reduced formation of moisture in the last stages of the turbine, which improves the efficiency of these stages.

A reheat cycle is primarily used in higher-powered units (i.e., 25,000 shp and above). As vessels increase in size and speed, their power requirements, and therefore, fuel costs, often increase faster than other operating costs;

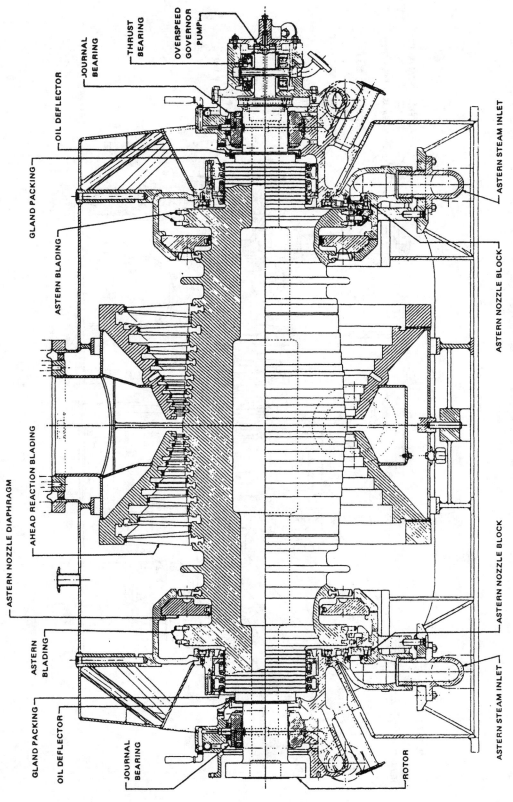

Fig. 3 Typical double-flow low-pressure turbine—reaction type

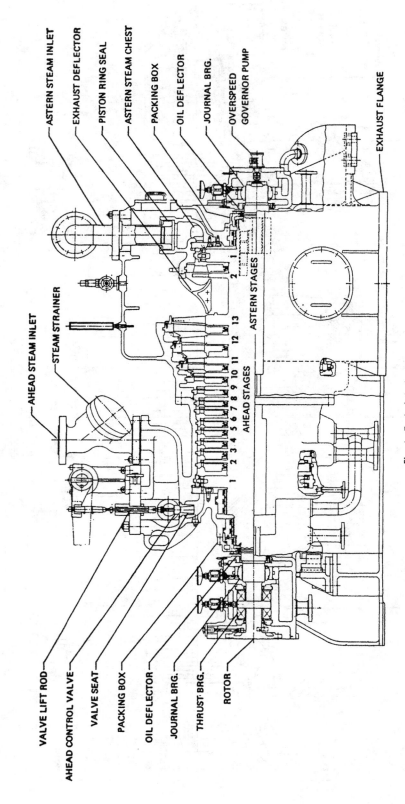

Fig. 4 Typical single-casing turbine

consequently, fuel savings can become relatively more important. A reheat cycle offers the maximum economic thermal efficiency that can be provided by a steam plant. Reheat cycles are not used in naval designs because the improvement in efficiency is not considered to warrant the additional complexity.

The considerations associated with the selection of initial steam conditions and vacuum for reheat plants are similar to those for nonreheat units. Marine reheat plants probably will continue to have more modest steam conditions than land-based applications because of lower power ratings and the fact that the safety of a ship is dependent upon the reliability of its power plant.

a. Gas reheat. The improvement in plant performance offered by a gas reheat cycle is equivalent to an increase of about 125 deg F in the initial steam temperature of a nonreheat plant. In other words, a reheat plant with an initial temperature of 950 F and a single reheat to 950 F will have about the same fuel consumption as a nonreheat plant with an initial temperature of 1075 F. It is unlikely that marine plants will include more than a single stage of reheat because of piping complications and the fact that while the addition of the first reheat reduces fuel consumption by about 5%, the improvement resulting from the use of a second stage would be less than half this value.

The selection of the appropriate reheat pressure involves several considerations. Reheating at high pressures reduces the volume of steam and, therefore, smaller pipe diameters may be used, while the reverse is true when lower reheat pressures are selected. The improved flexibility of smaller-diameter piping is advantageous since thermal expansion may be accommodated more easily; however, a higher design pressure may require an increase in the pressure standard for piping, valves, and fittings. Generally, the selected reheat pressure falls between 15 and 20% of the throttle pressure, and within these limits plant efficiency is not sensitive to changes of as much as plus or minus 50 psi.

When marine boilers and turbines are located adjacent to each other, which is desirable to minimize the amount of piping involved, it is generally possible to avoid a problem sometimes encountered with land-based plants. The large amount of steam contained in long piping runs to and from the reheater constitutes a hazard if the load is accidentally lost. The work, which can be done by the expansion of this steam through the remainder of the turbine, may be sufficient to overspeed the unloaded turbine. Automatically controlled intercept valves provided at reentry to the turbine in land practice have not been required in marine installations because of shorter piping runs.

A gas-reheat cycle subjects a larger portion of the plant to high temperatures and proper attention must be given to thermal expansion. There is also a wider use of high-temperature materials. A reheat cycle entails a large change in enthalpy within the turbine, and this requires either more stages or more blade speed to efficiently convert this additional heat to useful work.

A three-casing arrangement was selected for one of the first marine turbines designed for a gas-reheat cycle [5],

which allowed reheating at approximately optimum pressure between the high-pressure and intermediate-pressure turbines. The relatively short turbine rotors were considered to be more rugged; however, the increased number of casings, bearings, couplings, and rotating parts increased the initial cost.

Subsequent marine reheat installations have used a two-casing turbine arrangement [6], which was based upon the successful naval experience with the series-parallel type of high-pressure turbine. A typical HP-IP turbine for reheat applications is shown in Fig. 5. In this arrangement, steam enters at the center of the unit and exhausts from one end to the boiler, where it is reheated. It is then returned to the turbine at the center. After expansion through the remaining stages, the steam flow leaves from the other end and is led to the low-pressure turbine inlet by a cross-under connection. High temperatures and pressures are confined to the central portion of the turbine and do not influence the glands or bearings.

No significant turbine problems are associated with maneuvering and astern operations that are peculiar to gas-reheat plants, except that adequate axial clearances must be provided to assure that rubs will not occur due to differential expansion of the casing and rotor should rapid changes in the reheat temperature occur during transient conditions.

The boiler designer, however, must deal with the fact that there will be no flow through the reheater section of the boiler during astern operations, and suitable protection must be provided to prevent overheating and damage to the reheater tubes. Steam flowing through the astern turbine is not reheated for several reasons: (a) a nonreheat astern turbine is simple, compact, easy to control, and entirely adequate; (b) switching a single reheater from ahead to astern flow circuits would be impractical; (c) astern operation is infrequent and of short duration, therefore the expense incurred to improve its performance cannot be justified; and furthermore, (d) the gains due to reheat would be insignificant since the conditions curve for an astern turbine usually lies entirely in the superheated region and the addition of reheat would only result in undesirably high exhaust temperatures.

b. Steam reheat. Steam leaving the high-pressure turbine may be reheated by saturated steam at boiler pressure in a separate heat exchanger in lieu of being passed through a reheater located in the boiler gas path. With a gas reheater the steam can be reheated to the initial temperature, but with a steam reheater the temperature of the steam returned to the turbine is limited to the saturation temperature corresponding to the boiler drum pressure.

Steam reheat has advantages when maneuvering in that the reheater elements cannot become any hotter than the saturation temperature of boiler pressure steam, and no special precautions are necessary to protect the reheaters. The reheater size is reasonable and it may be placed adjacent to the turbines to form a portion of the flow path between casings. A conventional boiler with a normal superheater may be used, and operation of the plant is

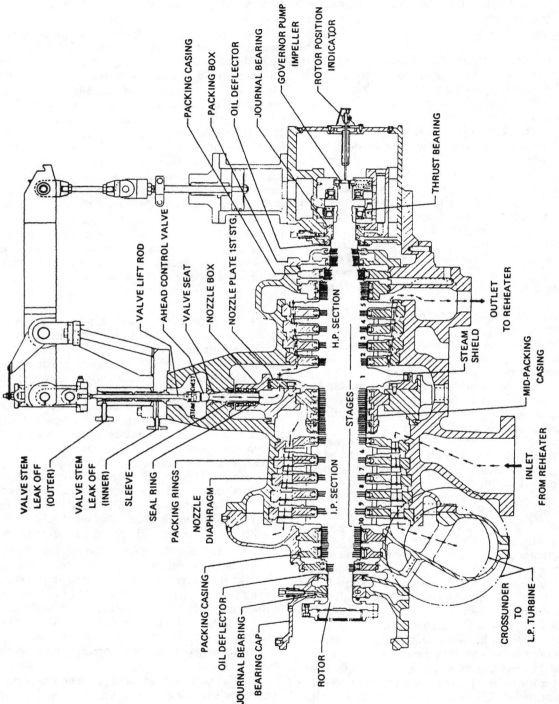

VALVE STEM LEAK OFF (OUTER)

VALVE STEM LEAK OFF (INNER)

SLEEVE

SEAL RING

PACKING RINGS

NOZZLE DIAPHRAGM

PACKING CASING

OIL DEFLECTOR

JOURNAL BEARING

BEARING CAP

ROTOR

VALVE LIFT ROD

AHEAD CONTROL VALVE

VALVE SEAT

NOZZLE BOX

NOZZLE PLATE 1ST STG.

PACKING CASING

PACKING BOX

OIL DEFLECTOR

JOURNAL BEARING

GOVERNOR PUMP IMPELLER

ROTOR POSITION INDICATOR

THRUST BEARING

H.P. SECTION

I.P. SECTION

STAGES

STEAM SHIELD

MID-PACKING CASING

OUTLET TO REHEATER

INLET FROM REHEATER

CROSSUNDER TO L.P. TURBINE

Fig. 5   Typical HP-IP reheat turbine

no more sensitive or complicated than for a conventional nonreheat plant.

There are several disadvantages associated with steam reheat. One is due to the need for a reheater drain pump, which must operate at a high discharge pressure to return the live steam drains to the boiler drum to avoid the losses that would occur if the drains were allowed to cascade through the feed heaters. In addition, and perhaps more important, the steam-reheat cycle with one stage of reheat is about 4% less efficient than gas reheat at the same initial temperature, although this difference may be cut in half by the use of two steam-reheat stages [7,8]. Since cycle studies indicate that this discrepancy increases as the initial pressure and temperature increase, steam reheat becomes less attractive at higher initial conditions.

1.4 Nuclear Cycles. The introduction of pressurized-water nuclear reactors for submarines and for naval and merchant surface vessels involves complications for a turbine designer. These problems are not new; indeed, since the turbine inlet steam is saturated, the high-temperature problems associated with oil-fired steam plants are replaced by the problems of excessive moisture that were common during the early days of turbine development.

The steam pressure developed by a pressurized-water reactor plant varies with output; the maximum pressure occurs at no load and decreases approximately linearly with load to a minimum at full power. The merchant vessel NS *Savannah* was designed for 715 psig and 445 psig at no load and full power, respectively.

Variations in inlet pressure require special attention during the design of the control stage. Blade stresses in the control stage of a nuclear turbine increase more rapidly with a decrease in power than a conventional turbine because not only does the outlet pressure of the stage decrease but also the inlet pressure increases, thus accentuating the increase in heat drop and steam velocity with corresponding increases in blade stresses at lower powers.

a. Steam moisture considerations. The steam produced by a pressurized-water reactor plant is saturated, and with careful design its moisture content can be kept to less than 1 percent at the turbine inlet. As the steam expands through the turbine, the moisture content increases, reaching as much as 18–20% in the exhaust to the condenser if no steps are taken to reduce the moisture, as may be seen from Fig. 6. The high moisture content in the turbine steam path is a disadvantage of the pressurized-water reactor cycle and requires careful attention in the design of the turbine.

Moisture causes a loss in the turbine stage efficiency, principally due to the loss of a portion of the steam, or working fluid, but also because moisture travels through the turbine at a lower speed than the steam and tends to hit the backs of the inlet edges of the moving blades. The overall loss amounts to a reduction in stage efficiency of approximately 1.2% for every average percent of moisture in the stage.

The action of the blades striking the slower-moving water droplets at a high relative speed causes erosion of the leading edges of the blades, particularly near the tips

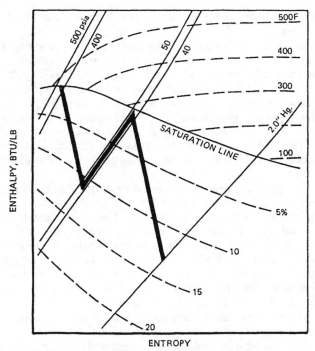

Fig. 6   Turbine condition curve—nuclear cycle—pressurized-water reactor

where speeds are greatest and where a concentration of moisture develops due to centrifugal action in the moving blades. When the acceptable limits of speed and moisture for an unprotected blade must be exceeded, local hardening or shields of hard material may be secured to the blade to extend these limits.

Moisture creates still other erosion problems. Some internal steam leakage may occur at the horizontal joints of diaphragms, gland packing boxes, and other internal joints subject to a pressure differential. This leakage is normally so small as to be negligible in a properly constructed turbine and will not increase significantly if the leakage steam flow is dry; however, the presence of moisture will create a corrosion-erosion problem if the surfaces subject to leakage are of plain carbon steel, and the leakage area and flow will increase continuously until the loss of efficiency requires corrective action.

Moisture in the leakage flow through the interstage packing has also been found to cause unacceptable corrosion-erosion of the sides of the turbine wheels as the moisture is thrown outward.

Experience has shown that it is desirable to use a corrosion-resistant material for all surfaces subject to moisture impingement. Rotors and casings may be machined from stainless steel, or carbon steel casings may be faced in critical areas with a welded inlay of stainless steel.

Although a proper choice of materials alleviates the corrosion-erosion problem, it does nothing for the loss in efficiency, and therefore steps must be taken to remove as much moisture as possible from the steam path. In general, two approaches are possible: (1) internal separation, or moisture separators, which are used for both nuclear and nonnuclear cycles, as described in Section 5.4;

and (2) external separation, which is treated in the following.

b. External moisture separation. Moisture may be extracted by removing the steam from the turbine and treating it in an external separator. This method is applicable to the cross-compound type of unit where the pressure in the high-pressure turbine exhaust is about 45 psia and the moisture may reach 12% at full power as illustrated by Fig. 6. If this moisture were not removed, it would increase still further when expanding through the low-pressure turbine and cause serious erosion problems and a loss in turbine efficiency.

The ideal separator should be designed for maximum moisture removal coupled with minimum pressure drop and minimum space requirements. The power loss to the overall plant is about 1% for every 1-psi pressure drop, compared with a power loss of 0.6% for every 1% of moisture entering the low-pressure turbine. Thus, if a net gain in efficiency is to be realized, the sum of moisture and pressure-drop losses with a separator must be less than the moisture loss with no separator.

Many types and arrangements have been developed, but in general external separators consist of a pressure vessel having inlet and outlet connections to the crossover piping, internals arranged to remove moisture, and a drainage system. The separator may be located forward of the high-pressure turbine and supported on the same longitudinal girders, or arranged at the side of the unit and supported by ship's structure.

The separator internals may include centrifugal, baffle, or wire-mesh devices. Centrifugal devices rotate the steam flow, which tends to drive the heavier water droplets to the outer diameter where they are drained off. The baffle types function by collecting moisture on their surfaces and then allowing it to drain off. Baffles vary from simple chevron styles to more elaborate arrangements having hooks and partially sheltered passages for drainage as shown in Fig. 7. Wire-mesh types function similarly by collecting moisture through surface contact.

The design of moisture separators is not an exact science but rather the result of a great deal of testing and experience. It has been found that the steam velocity configuration in the crossover pipe is complex and includes secondary flows induced by the exhaust elbow and by the turbine-exhaust hood. This uneven velocity distribution

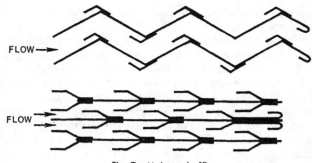

Fig. 7   Moisture baffles

may overload some sections of the separator. A large portion of the moisture in the high-pressure turbine exhaust, up to 80%, has been found to consist of water running along the surface of the pipe wall. This suggests the use of a skimmer at the inlet of the separator to avoid overloading the inlet portions of the separator.

Reentrainment of separated moisture may occur when water on an internal surface is swept back into the vapor stream before it can be drained, and must be avoided because droplets may be produced which are too small to respond to subsequent separating efforts, or they may be formed in a late stage and thus escape.

Large droplets may split into smaller ones, a process called "droplet fracture" that makes separation more difficult. Such fracture occurs when there is a large relative difference in velocity between the vapor and the droplet, such as may occur when large drops are swept from a trailing edge. In addition, mechanical fracture, or split-up, may occur by collision of a large drop with an obstacle.

Testing and performance verification of moisture separators involve practical problems. The performance of separators is highly sensitive to the moisture particle size. Since both the measurement of drop size and the artificial creation of moisture truly representative of that found in actual turbine exhausts are difficult, if not impossible, it has been found necessary to supply saturated steam to a high-pressure turbine coupled to a power absorption device, then lead the exhaust containing the required moisture to the separator on test. After passing through the separator, the steam is throttled through a valve to a low back pressure at which the steam becomes superheated, and its temperature may be used to calculate the residual moisture at the separator outlet.

Section 2
Turbine Performance

2.1 Steam Conditions. Ever since the early days of the steam turbine, efforts have been made to improve the steam conditions in order to increase the economy of the power plant. Steam conditions of marine steam power plants have tended to advance by steps, with each advance followed by a period during which experience has been obtained and the knowledge gathered for another forward step.

Increasing the steam pressure will increase turbine efficiency until a pressure of about 2500 psig is reached. A rough rule states that "Doubling the pressure will reduce the heat rate by 4–6 percent." More accurately, a 100-psi increase in initial pressure will reduce the steam rate by the percentages given in Fig. 8.

The gains decrease as the pressure increases because the turbine efficiency declines at higher pressures. As

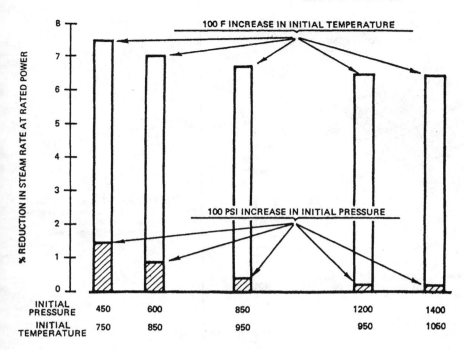

Fig. 8 Reduction in steam rate for increased initial pressure or temperature

Table 1 Recommended initial steam pressures

RATED SHP	INITIAL PRESSURE, PSIG
2500–4000	400
4000–10,000	600
10,000–20,000	900
20,000 and up	1500

steam pressures increase, the specific volume of the steam decreases; therefore, the nozzles and blades become smaller and less efficient. A limiting pressure is reached for every capacity of turbine at which the gain due to the improvement in pressure is offset by the decrease in internal efficiency. For this reason, higher initial pressures may be used more effectively on larger turbines.

The initial pressures given in Table 1 are recommended as practical upper limits for various sizes of propulsion units. It should be noted, however, that somewhat lower pressures are determined to be optimum when all of the economic factors are considered. The pressures given in the table have been chosen because they permit reasonable utilization of the pipe flange and valve dimensional standards of the American National Standards Institute as shown in Fig. 9. The regulatory bodies require adherence to these standards for merchant marine propulsion units.

Increasing the initial temperature will also reduce the heat rate and steam rate. As an approximation, a 15 deg F increase in temperature at full power or a 25 deg F increase at lower powers will result in a 1% decrease in steam rate. More accurately, a 100 deg F increase in initial temperature will reduce the steam rate by the percentages given in Fig. 8.

Unlike a change in initial pressure, a change in temperature affects the specific volume only a relatively small amount. As a result, the physical dimensions of the parts are changed only slightly, and an increase in initial temperature will improve the economy of both large and small turbines about equally. Part of the reduction in heat rate resulting from an increase in initial temperature (at constant initial pressure) is caused by an increase in turbine efficiency resulting from a reduction in the moisture present in the lower pressure stages. The remainder is due to the increase in the available energy.

For a given initial pressure, there is a minimum initial temperature below which the moisture content in the low-pressure end of the turbine is sufficient to cause undesirable erosion of turbine blades and loss of stage efficiency. A moisture content of 12% in the exhaust is often accepted as limiting, and the corresponding minimum initial temperatures may be noted from Fig. 9. Certain combinations of pressure and temperature have become widely accepted; these standardized conditions are indicated in Fig. 9. Although marine steam turbines can be designed to operate at temperatures of 1040 F or higher, practical problems in the superheaters of marine boilers tend to impose an upper limit of 950 F.

Concern has been expressed regarding the sudden imposition of high temperature upon a normally cool astern turbine and exhaust casing; however, it is possible to accommodate the thermal shock and rapid expansion without either distress or distortion.

2.2 Exhaust Vacuum. A moderate vacuum of 28½ in. Hg has become generally accepted as a design basis for merchant propulsion turbines. This selection is considered a reasonable economic compromise considering the worldwide variation in seawater temperature and the size, weight, and cost of turbines and condensing equipment.

Low seawater temperatures permit high vacuum and, conversely, high seawater temperatures may limit the attainable vacuum. It is often difficult at the construction

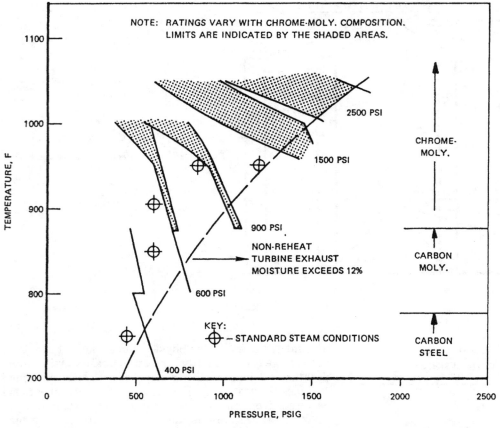

NOTE: RATINGS VARY WITH CHROME-MOLY. COMPOSITION.
LIMITS ARE INDICATED BY THE SHADED AREAS.

2500 PSI

CHROME-
MOLY.

1500 PSI

900 PSI

NON-REHEAT
TURBINE EXHAUST
MOISTURE EXCEEDS 12%

CARBON
MOLY.

600 PSI

KEY:
⊕ – STANDARD STEAM CONDITIONS

CARBON
STEEL

400 PSI

TEMPERATURE, F

PRESSURE, PSIG

Fig. 9 Standard steam conditions in relation to pressure-temperature service ratings for ANSI Standard flanges, valves, and fittings

stage to assure that the vessel will remain on a specific trade route throughout its useful life; therefore, it is generally considered desirable to design for a moderate vacuum. In special cases where service is limited to areas of low seawater temperature, the increased cost and weight of turbines designed for higher than standard vacuum should be subjected to an economic evaluation.

The specific volume of steam increases rapidly as the vacuum is improved. For example, an increase of from 28 to 29 in. of mercury (referred to a 30-in. barometer) practically doubles the specific volume. To handle this increased volume efficiently at full load, it is necessary to increase proportionally the flow areas of the turbine stages at the exhaust end. A more detailed discussion of exhaust losses is given in Section 2.6.

It is customary for high-speed, lightweight naval vessels to make some sacrifice in economy by accepting a higher exhaust pressure (generally 2.5 psia) at high power in order to reduce the weight and size of turbines and condensers. At cruising powers where economy is more important, the vacuum approaches merchant levels due to the reduced condenser loading.

Good vacuum is important for reasons other than economy. When the astern element is developing power, the ahead blading is being driven backward in steam at essentially exhaust pressure. If the vacuum is poor, the windage losses of the ahead stages will cause more rapid heating and may limit the allowable speed or period of astern operation.

2.3 Nonextraction Steam Rate. When the rated full power, the initial steam conditions, and the exhaust vacuum have been selected, it is possible to establish the steam rate that may be expected from well-designed equipment. Figure 10 indicates typical nonextraction steam rates for merchant type, geared-turbine units designed for optimum performance at full power with a proper balance between efficiency, size, weight, and cost. The reduction in steam rate with improved steam conditions is easily seen as well as the decrease in steam rate, which is possible with higher powered units. Methods for estimating steam rates at other conditions are given in reference 9 and Chapter 2.

Merchant vessels generally operate at or near full power most of their service life; therefore, performance at partial loads tends to be less important. A typical variation in steam rate at fractional powers that is representative of turbine designs, which incorporate no special features to enhance partial-power performance other than first-stage nozzle control, is shown by the curve marked "straight-through" in Fig. 11. When partial-power performance is important, as in the case of naval combatant vessels, several means may be employed to move the point of optimum steam rate to a partial power and to reduce the low-power steam rates. Possibilities include an excess in total wheel speed at full power, an interstage bypass, a series-parallel turbine, and a two-row/one-row control stage. Discussions of these arrangements are given in the

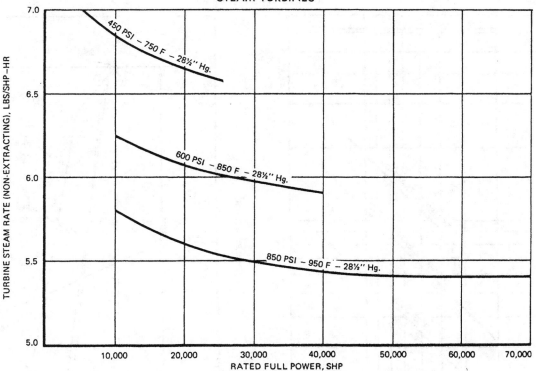

Fig. 10   Turbine steam rate, nonextracting, for merchant-type geared steam turbines

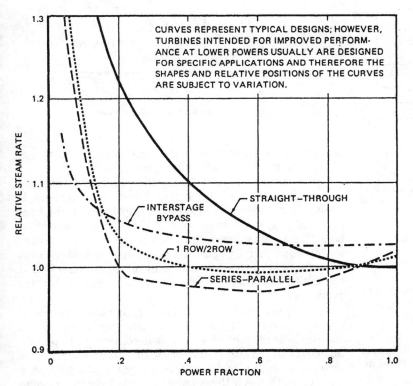

CURVES REPRESENT TYPICAL DESIGNS; HOWEVER, TURBINES INTENDED FOR IMPROVED PERFORMANCE AT LOWER POWERS USUALLY ARE DESIGNED FOR SPECIFIC APPLICATIONS AND THEREFORE THE SHAPES AND RELATIVE POSITIONS OF THE CURVES ARE SUBJECT TO VARIATION.

Fig. 11   Relative steam rates versus power fraction of various types of turbines

following sections, and typical relative steam rates are shown in Fig. 11.

2.4 Extraction of Steam. Steam may be extracted from the turbine for feed heating and for ship's services. The number of extraction points and the required pressure for each system are dependent upon the selected steam plant cycle as discussed in Chapter 2. Connections for this purpose are provided at turbine stages where the pressure is appropriate to the intended use. Often, it will be found that the number of stages and the corresponding stage pressures selected by the turbine designer to give optimum turbine performance will result in the desired

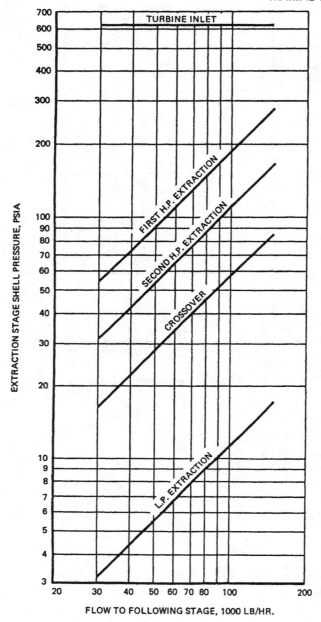

Fig. 12   Typical extraction stage pressure curve

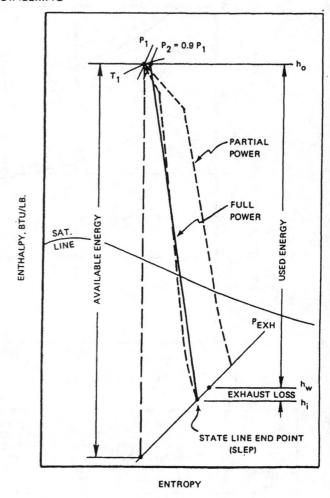

Fig. 13   Expansion line—nonreheat cycle

extraction pressure falling between two stages. In this case, it is usually best to select either the higher or lower stage pressure rather than disturb the turbine design.

In general, stage pressures vary almost linearly with the apparent flow beyond the extraction point, and a curve such as Fig. 12 may be obtained from the turbine designer. In choosing the proper turbine stage for each extraction point, it is customary to select a stage having a full-power pressure somewhat higher than that required so that extraction may be possible in a reasonable operating range below full power.

Extraction connections often are omitted for simplicity on naval turbines, because the broad power range in which extraction is desired would necessitate a complex valve and piping arrangement to shift each system from one turbine stage to another as the power varies.

The converse of extraction, called induction, wherein excess steam in the plant is introduced to the turbine, is generally discouraged. While this may improve the heat balance, it (1) tends to congest the last stages of the turbine, (2) may introduce large slugs of water from auxiliary systems and thereby cause blade damage, and (3) requires an automatic control valve to prevent induction when at standstill, operating astern, or in the event of turbine overspeed.

2.5 Expansion Line. The Mollier chart is a graphical representation of pressures, temperatures, superheat, and the heat content of steam. It is convenient to plot on this chart the "state point" of steam at any stage in the turbine; when this is done for all stages of the turbine, a line drawn through these points is called the "expansion line" or "state line."

The "expansion lines" for rated full power and for partial powers are usually obtained from the turbine designer. They are useful in defining the characteristics of steam throughout the turbine, particularly at extraction points and at the turbine exhaust.

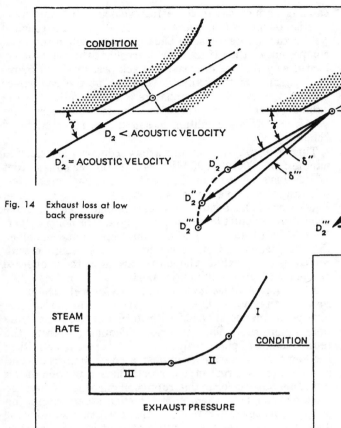

Fig. 14 Exhaust loss at low back pressure

In the event that an expansion curve is not available, an approximation of the full-power expansion curve may be made for preliminary purposes as indicated in Fig. 13. The initial point is established at the intersection of the initial pressure (P_1) and initial temperature (T_1). A state line "top point" is then plotted at initial enthalpy and 90% of the initial pressure.

The "used energy" per pound of steam may be found from

$$\frac{2544}{\text{steam rate} \times \text{external efficiency}} \qquad (1)$$

The external efficiency accounts for turbine bearing losses as well as mechanical losses in reduction gears, and for electric drive installations includes the motor and generator losses. If these external losses are not known, the following assumptions may be made:

TYPE OF DRIVE	EXTERNAL EFFICIENCY
Gear drive, single reduction	0.97
Gear drive, double reduction	0.96
Electric drive	0.92

The external efficiency at partial powers may be approximated by varying the external loss (one minus external efficiency) at full power as the 1.7 power of the propeller speed.

The enthalpy of steam exhausted to the main condenser is

$$h_w = h_o - \text{used energy} \qquad (2)$$

The exhaust loss is caused by the velocity energy in the steam leaving the last row of blades, which cannot be recovered as useful work, but instead is converted to heat by dissipation in eddies and friction. The exhaust loss also includes any pressure-drop losses between the last wheel exit and the exhaust flange. The magnitude of the overall loss depends upon the particular turbine design, steam flow, and exhaust vacuum.

For a typical merchant propulsion turbine design, the total exhaust loss (EL) at full power with rated exhaust vacuum is about 12 Btu/lb and at partial powers varies approximately as the square of the ratio of steam flow to the absolute exhaust pressure.

The enthalpy of steam at the "state line end point" is

$$h_i = h_w - \text{EL} \qquad (3)$$

A straight line joining the "top point" and the "state line end point" gives a reasonable approximation of the expansion curve. The nature of the deviation from an accurate curve is indicated by the dashed line in Fig. 13 and is caused by the inability to obtain an average stage efficiency in the first and last stages of the turbine.

Figure 13 also indicates the trend of the condition curve at partial powers. Note that at very low powers, if the initial temperature remains constant, the exhaust may be superheated.

2.6 Exhaust Loss. Among the factors that determine the efficiency and size of a turbine, the exhaust loss is one of the more important. Machines designed to operate economically at high vacuums are inherently large, but more efficient. For this reason the designer must strike a balance between the required economy and the weight and size of the unit. In general, units designed to operate during a large portion of their life at higher powers should have ample exhaust areas, whereas units that generally operate at reduced power may be designed with smaller

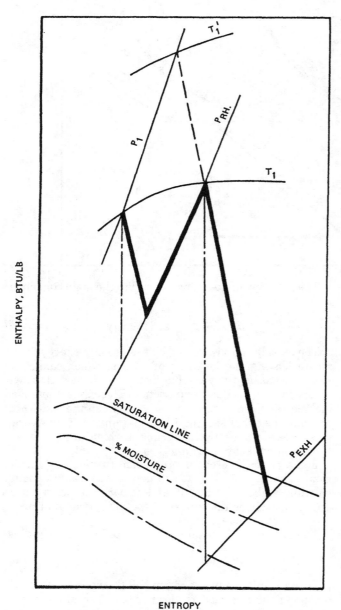

Fig. 15   Turbine expansion line—reheat cycle

causing an increase in the efflux velocity (D_2'') and a deflection of the jet to angle $\gamma + \delta''$. Condition III is reached with a further decrease in back pressure, when the axial component of the efflux velocity reaches the acoustic velocity. Any further decrease in back pressure will result in expansion in the exhaust trunk or at the condenser inlet, but there will be no change in the conditions at the exhaust annulus; and, therefore, the steam rate will not be affected. This condition is sometimes referred to as "choking."

Turbine last-stage annulus areas ordinarily are sized to handle from 6000 to 8000 lb/hr of steam per square foot of annulus area at 1.5 in. Hg abs; for other back pressures, this range would vary inversely as the absolute pressure. The lower of these values represents machines that have a very low exhaust loss and are designed for normal operation at full load. The higher value represents machines that, for economy in weight and size, are designed with relatively small exhaust hoods and are expected to operate at reduced load for the major portion of their life.

2.7   Equivalent Nonextraction Steam Rate of Reheat Cycles. In the case of nonreheat marine power plants, turbine performance is specified in terms of the nonextraction steam rate. This criterion adequately meets the requirements of plant designers and purchasing activities; furthermore, it is easily demonstrated on ship trials, and its variations are reflected correctly in that portion of the total fuel burned for main-propulsion purposes. When the reheat principle is used, nonextraction steam rate is no longer a proper criterion of turbine performance, since it does not recognize the addition of heat in the reheater. The steam rate varies with the reheat pressure that is selected, decreasing as the pressure is dropped; and even if, in addition to defined steam conditions and vacuum, a specific reheat pressure is associated with each steam rate, the result could not be used for a comparison of competitive turbine designs. This may be seen by consideration of the reheat turbine condition curve, Fig. 15. For example, two turbine designs could be developed such that each has the same initial and reheat steam conditions, the same condenser vacuum, and the same steam rate, but the corresponding plant efficiencies and fuel rates could differ. The turbine having a higher efficiency in the stages prior to the reheater and a lower efficiency following this point would require a greater amount of heat to be added in the reheater at the expense of increased fuel to the boiler.

It becomes necessary, therefore, to use some form of heat rate as the criterion of reheat turbine performance. Heat rate may be defined in several ways, but a standardized method for marine units has been suggested [10]. In general, the turbine heat rate in Btu/shp-hr is expressed as

Turbine heat rate

$$= \frac{\text{Heat added to turbine cycle by boiler (Btu/hr)}}{\text{Power output (shp)}} \quad (4)$$

The heat added is defined as

$$Q_T(H_T - h_{FW}) + Q_{Rhtr}(H_{HR} - H_{CR}) \quad (5)$$

where

Q_T = throttle flow, lb/hr

exhausts, because the exhaust loss reduces rapidly as the load decreases.

An understanding of the effects of high volumetric loading per unit of exhaust annulus area is useful. To illustrate, assume that the exhaust pressure of a typical turbine is reduced in a series of steps. Three conditions are encountered, as illustrated in Fig. 14. In condition I, as the back pressure decreases, the steam velocity (D_2) at the throat of the last row increases until it equals the acoustic velocity corresponding to the steam conditions at this point. The steam jet leaves at the blade exit angle (γ). With a further decrease in back pressure, represented by condition II, the throat pressure and velocity remain constant, but expansion now takes place beyond the throat

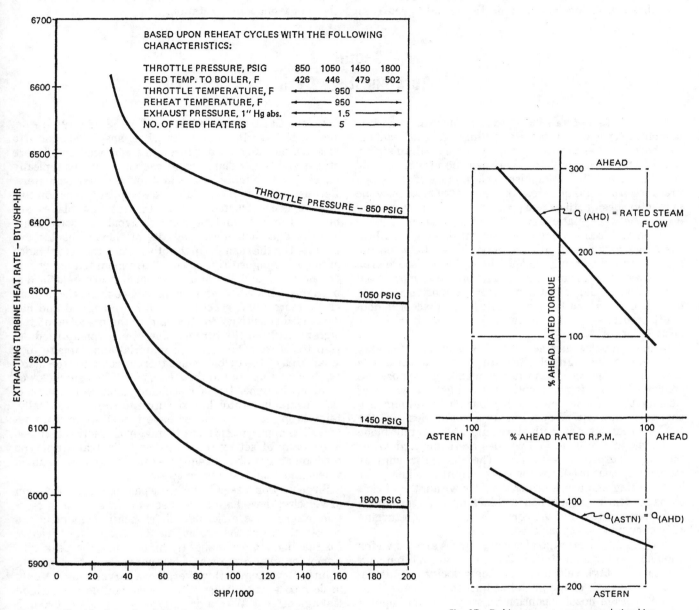

Fig. 16 Extracting turbine heat rate versus shp for reheat cycles

Fig. 17 Turbine torque-versus-rpm relationships

Q_{Rhtr} = reheater flow, lb/hr
H_T = throttle enthalpy, Btu/lb
h_{FW} = final feedwater enthalpy, Btu/lb
H_{HR} = enthalpy leaving reheater, Btu/lb
H_{CR} = enthalpy entering reheater, Btu/lb

The turbine heat rate may be defined on a nonextraction basis, which is relatively simple in that other components of the cycle are not involved. The demonstration of the nonextraction heat rate requires a special test with the bleeders closed. Alternatively, an extracting turbine heat rate may be defined in a manner similar to land practice

wherein the feed heating arrangement must be completely specified [10]. This type of heat rate can be demonstrated during shipboard trials in the course of regular economy runs but requires additional measurements, more complex calculations, and more involved corrective procedures. Typical extracting turbine heat rates are given in Fig. 16 for a five-feed-heater cycle.

2.8 Torque and Speed Characteristics. The inherent ability of steam turbines to maneuver rapidly is due to their speed-torque characteristics. Curves of turbine output torque for typical ahead and astern turbines are plotted against speed in Fig. 17 at rated ahead steam flow.

The curves for the astern turbine show that the torque available to decelerate the unit while it is still rotating in the ahead direction increases as the ahead speed from which the maneuver starts is increased. A similar relationship for the ahead turbine torque assists in decelerating the unit from astern rotation.

Section 3
Turbine Stage Design

3.1 Impulse and Reaction Stages. There are two fundamental types of turbine stages: impulse and reaction. The basic difference between the two types is the distribution of stage pressure drop between the stationary and rotating blades. In an impulse design, most of the stage pressure drop is taken across the stationary nozzles, while in reaction design, the stage pressure drop is more equally divided between stationary and rotating blades. Typical reaction turbines employ 50% reaction blading, in which the stage pressure drop is equally divided between the stationary and rotating blades. The higher pressure drop across reaction turbine rotor blades produces more axial thrust, and a balance piston on the rotor is often required to reduce the thrust to manageable levels (except in double-flow turbines, such as shown in Fig. 3, which are inherently balanced).

In pure impulse turbines, the flow is accelerated only in the nozzle, and high velocities are achieved at the nozzle exit. The kinetic energy in the steam exiting the nozzle is converted to shaft work entirely through the action of turning the flow in the rotating bucket, resulting in a change in angular momentum. In reaction designs, the flow is accelerated in both stationary and rotating blade passages, and in general velocities are somewhat lower than in a pure impulse design. The change in angular momentum that produces the work output is achieved by both turning and accelerating the flow within the bucket passage. Marine propulsion turbines operate through a wide speed range, and the consequent effects on turbine efficiency must be considered.

In Fig. 18, typical stage efficiency curves are shown for impulse and reaction stages. These data are plotted as a function of stage velocity ratio. The velocity ratio is defined as the ratio of the pitch-line bucket wheel speed, u, to the average nozzle spouting velocity. For an impulse stage, the peak efficiency occurs at a velocity ratio of 0.5, while for a reaction stage, the peak efficiency occurs at a velocity ratio of 0.707. Due to theoretical considerations, a 50% reaction stage with the same pitch diameter as an impulse stage must be run at a speed the square root of two (1.414) times as fast to produce the same output, and the peak efficiency occurs at the higher velocity ratio. A 50% reaction stage with the same pitch diameter as an impulse stage and running at the same speed will only produce half as much output as the impulse stage, so twice as many reaction stages are needed to generate the same total output. There are pronounced differences in the number of stages and the constructional features of impulse and reaction turbines, as may be noted from Figs. 1, 2, 3, and 19.

In practice, reaction turbines (Figs. 3 and 19) are designed to have the additional wheel speed necessary to produce the same output through a combination of more stages and larger diameters. Reaction machines typically have 75 to 85% more stages at 4 to 7% larger diameters. Because so many more stages are required, reaction turbines are constructed so that the rotating blades are mounted on a drum-type rotor to avoid excessive shaft lengths and diameters, and the stationary blades are mounted on the casing. On the other hand, the rotor blades in impulse stages (Figs. 1 and 2) are mounted on wheels on the rotor and the stationary blades are held in diaphragms. This wheel-and-diaphragm construction is generally more reliable because it is more rugged and has decreased sensitivity to rubs and to the presence of foreign material in the turbine. The compartmentalized design tends to isolate foreign material and damage to a small area of the turbine since there is room for the material to centrifuge out of the steam path. The usually lower rotor critical speed of a wheel-and-diaphragm construction results in a rotor less sensitive to rubs. Historically, failures in reaction turbines have tended to be more severe. The apparent size disadvantage of reaction turbines is partially offset by the fact that individual drum-type reaction stages do not require as much axial space as do wheel-and-diaphragm impulse stages.

Since friction losses in the blade passages are a function of velocity, these losses are theoretically smaller in reaction stages. This efficiency difference is evident in the curves shown at the top of Fig. 18. The peak efficiency for reaction stages is slightly higher than the peak efficiency for impulse stages. This is particularly true in high-volume-flow stages. However, these curves consider only basic nozzle and bucket efficiencies, and do not include leakage effects. With a drum-type construction, leakage areas are larger, and the larger pressure drops across the bucket tips increase the tip leakage flow. When leakage effects are taken into account, reaction turbines have no efficiency advantage over impulse turbines. For high-volume-flow low-pressure stages, both impulse and reaction turbine manufacturers use very similar staging, with low reaction at the bucket root and high reaction at the bucket tip. All things considered, both types of turbines can be designed with equally high levels of efficiency. Impulse turbines, however, generally have higher sustained efficiencies over their lifetimes due to their more rugged construction and reduced sensitivity to rubs.

3.2 Number of Stages. With the high initial pressure and low exhaust pressure used in marine power plants, large heat drops are available to the turbine. If a single

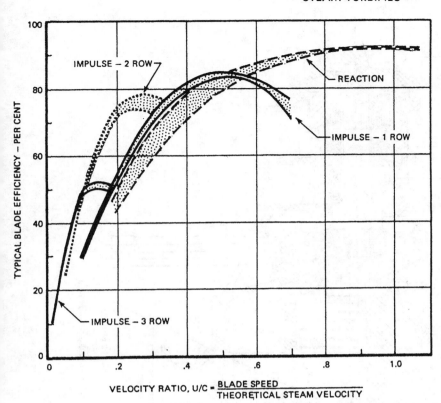

Fig. 18 Velocity ratio versus blade efficiency

stage were used, the theoretical steam velocity would be extremely high for good efficiency. Blade speed is limited by mechanical difficulties; if blade speeds are too high, practical limitations are reached. Thus, with a single-stage turbine the velocity ratio would be low and the efficiency would be poor. To avoid low speed ratios, it is common practice to split the total heat drop among a number of stages placed in series.

An important design consideration is the total or cumulative blade speed of a number of stages required to give an average overall velocity ratio that will provide a satisfactory efficiency. The total blade speed needed in a turbine may be obtained either by having large-diameter wheels operating at low rotational speeds or by using small wheels that operate at high rotational speeds. The latter approach can be taken to reduce weight, size, and cost. The use of double-reduction gears allows a wide freedom of choice in selecting turbine speed. A higher turbine speed increases the overall gear ratio and the size of the gears, but this is usually not a determining factor; instead, stress considerations in the turbine tend to limit the rotational speed. If permitted by stress considerations, increasing the rpm is usually the most effective way to obtain an increased wheel speed with a minimum increase in weight.

3.3 Velocity-Compounded Stages. An inspection of the velocity diagram for a single-row impulse stage indicates that the steam leaving the stage retains considerable absolute velocity. If the arrangement is such that this velocity is carried over into the nozzles of the next stage, it can be utilized effectively; otherwise, it results

in a loss. If a stationary row of blades is placed after the first moving row, the steam flow may be redirected and passed through a second row of moving blades. Such a two-row impulse stage (sometimes called a Curtis stage) has a lower peak efficiency, as may be noted from Fig. 18.

Frequently, a two-row impulse wheel is used for the first or "control" stage. Theoretically it has the energy-absorbing capacity of three single-row wheels and requires less space. In addition, it is useful for control because it permits the use of a lower first-stage exhaust pressure and temperature, which reduces leakage and rotation losses. Because of this reduction in losses, there is very little difference in the overall stage efficiency of a Curtis control stage and the equivalent Rateau stages (single-row stages) at the design point for units under 30,000 shp. The overall efficiency of a Curtis control stage at part load exceeds that of an equivalent Rateau stage.

In some astern turbines, three-row velocity-compounded wheels are used. For this type, the maximum efficiency is reached theoretically when the velocity ratio equals approximately 0.16, as may be seen from Fig. 18. The peak efficiency is less than that for a two-row wheel, but it has the energy-absorbing capacity of nine single-row wheels. Experience indicates that two- and three-row wheels reach their peak efficiencies when the velocity ratios are somewhat higher than the theoretical values.

3.4 Stage Efficiency Features. Once the velocity ratios are optimized in the individual stages, the thermodynamic efficiency of the turbine is to a great extent determined by the design of the steam path parts, which include the

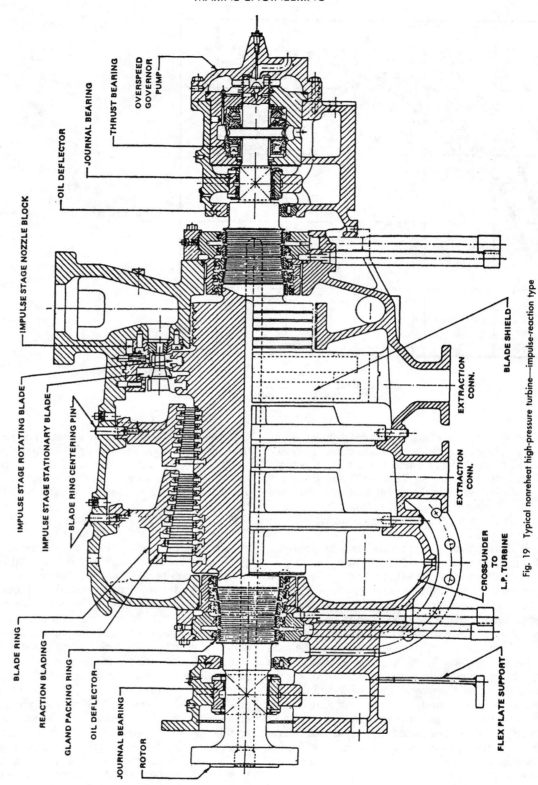

IMPULSE STAGE NOZZLE BLOCK

OIL DEFLECTOR

JOURNAL BEARING

THRUST BEARING

OVERSPEED GOVERNOR PUMP

IMPULSE STAGE ROTATING BLADE

IMPULSE STAGE STATIONARY BLADE

BLADE RING CENTERING PIN

BLADE SHIELD

EXTRACTION CONN.

EXTRACTION CONN.

CROSS-UNDER TO L.P. TURBINE

BLADE RING

REACTION BLADING

GLAND PACKING RING

OIL DEFLECTOR

JOURNAL BEARING

ROTOR

FLEX PLATE SUPPORT

Fig. 19 Typical nonreheat high-pressure turbine—impulse-reaction type

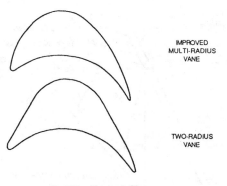

Fig. 20   Vane profiles

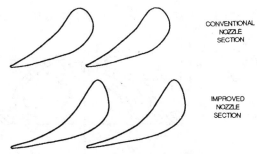

Fig. 21   Diaphragm nozzle sections

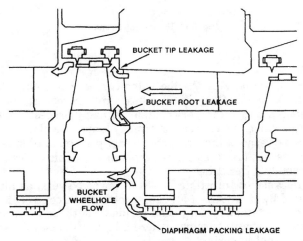

Fig. 22   Stage leakage flows

nozzle and bucket shapes and steam leakage control devices. Nozzle and bucket aerodynamic profile losses and leakage losses account for the majority of the total stage loss. Hence, to ensure high-efficiency turbine designs, it is necessary to use the most efficient nozzle and bucket profiles available, and to minimize leakage flows without compromising turbine reliability.

Analytical techniques have been developed to optimize nozzle and bucket vane profiles. Figure 20 shows a comparison between a traditional "two-radius" bucket vane shape and a more effective multi-radius vane section. The shapes of the cross sections of turbine blades are partially influenced by the required inlet and outlet angles determined from velocity diagrams as described in Chapter 2. The improved steam guidance and reduced boundary-layer flow separation in the multi-radius design can typically result in a 1% gain in stage efficiency. The twisted and tapered airfoils that are applied to the longer, latter stages are also optimized to have a radial pressure distribution that offsets the centrifugal force on the steam flow through the bucket passage. This eliminates losses due to the radial flow of steam in the bucket passage, which also contributes to turbine blade inefficiencies.

Concurrent with the development of improved bucket blade airfoils, improved nozzle shapes were also developed. The differences in contour between a typical standard and an improved section are shown in Fig. 21. The improved section has an extended and contoured thin edge for increased steam guidance and reduced turbulence in

the vicinity of the nozzle wake. Improvements in this area have also resulted in stage efficiency gains, which typically are in the 1% range.

Steam leakage through the seals between stationary and rotating components of the turbine constitutes an efficiency loss because the leakage flow does not contribute to the work output of the stage (see Fig. 22). Several measures can be taken to minimize this leakage. Double radial spillband tip leakage control, as shown in Fig. 22, has been used in some applications. This feature, coupled with more effective bucket root leakage control, has resulted in stage efficiency gains from 1% to 3%, depending on the application.

Section 4
Turbine Control

4.1   Power and Speed.   Means must be provided to vary the flow of steam through the turbine so that its power output and speed can be controlled. Steam flow may be varied by:

(a)   A separate throttle valve

(b)   A separate throttle valve plus hand control valves

(c)   Integral bar-lift valves and cam-lift valves

(d)   Bypass valves

(e)   Variable boiler pressure

   a.   Throttle valve control.   The most simple method of regulating steam flow is by a separate throttle valve in the steam supply to the turbine, as illustrated by Fig. 23. If properly sized, the valve will have little pressure drop when wide open; therefore, at maximum power, practically full boiler pressure will exist at the inlet to the

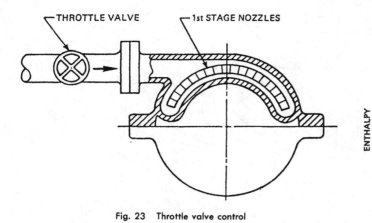

Fig. 23   Throttle valve control

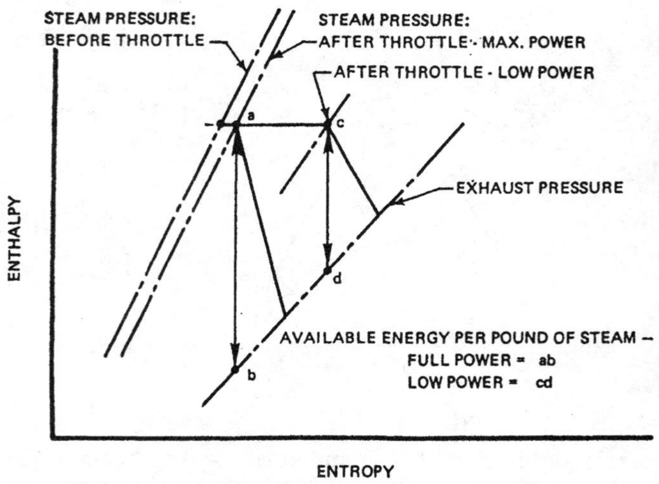

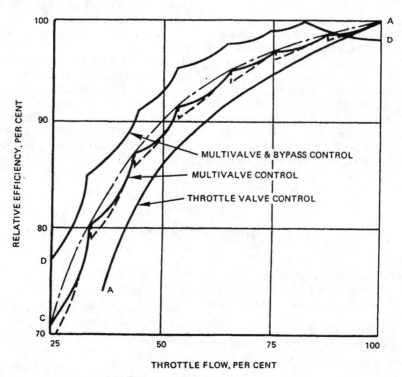

Fig. 24   Effect of throttle governing

first-stage nozzles. As the valve is closed to reduce the rate of steam flow, its pressure drop increases; consequently, a throttling or constant enthalpy process occurs at the valve and causes a thermodynamic loss since there is a decrease in the available energy per pound of steam. Figure 24 illustrates the reduction in available energy as a result of throttling.

Because of throttling losses at lower powers, as illustrated by curve *AA* of Fig. 25, throttle valve control alone is not satisfactory for ahead turbines, but it is generally used for astern turbines where a high efficiency at part load is not important.

b.   Throttle valve plus hand-control valves.   Throttling losses at reduced powers can be minimized if the first stage is of the impulse type and its total nozzle area is divided into arcs as illustrated by Fig. 26. One nozzle group normally has about one half of the total nozzle area and is controlled only by the throttle valve, while each of the remaining groups is controlled by the throttle valve and a hand-control valve. Thus, if the throttle is wide open, the nozzle area and, therefore, the steam flow may be varied in a series of steps by opening each hand-control

Fig. 25   Typical efficiency curves for various types of control

HAND CONTROL VALVES

THROTTLE VALVE

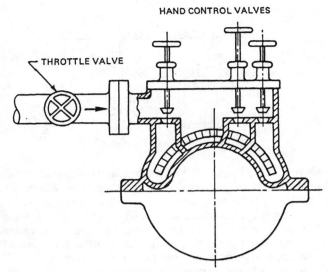

Fig. 26 Throttle valve plus hand-control valves

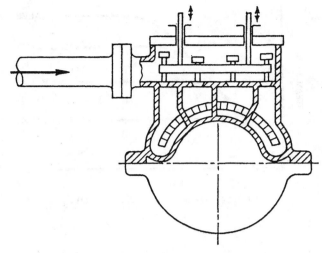

Fig. 27 Bar-lift control valves

valve in proper sequence. When the throttle valve and all hand-control valves in service at a particular point are wide open, throttling losses will be at a minimum. The smooth curve AC in Fig. 25 would result if it were possible to have an infinite number of hand-control valves.

With the limited number of valves that it is practical to use, there are power levels that cannot be obtained by having combinations of valves wide open or shut. Two modes of operation are possible for these intermediate powers. One procedure is to fully open as many control valves as can be utilized and then partially open one additional control valve to get the exact power desired. The throttling loss of the partially opened valve produces the scalloped effect (or valve loops, as they are commonly called) in the curve AC of Fig. 25. Alternatively, the next largest hand-control valve combination may be opened wide and the power reduced to the desired value by closing the throttle valve as necessary. Although this results in a slightly greater throttling loss over a portion of the power range and produces the sawtooth effect shown in Fig. 25, it is sometimes preferred since all speed adjustments are made at the throttle valve. With this method the nozzle area is known; therefore, the observed chest pressure can be used to compute the steam flow and estimate the power output of the turbine.

The number of hand-control valves that should be provided depends upon the importance of efficient operation at lower powers. Three hand-control valves have been provided in many merchant marine applications; however, where the turbine operates at full power most of the time, two hand-control valves have proven satisfactory. Besides being simpler, a smaller number of hand-control valves reduces the energy input to blade vibration from partial-arc loading.

The ahead and astern throttles, governor valve, and steam strainer may be grouped in a manifold arrangement and mounted on a convenient bulkhead adjacent to the main engine control point. This permits a preferred arrangement in that the design of the various components

can be coordinated, flanges eliminated, piping design simplified, and the controls concentrated.

c. Bar- and cam-lift control valves. The requirement for the proper selection and manual operation of individual hand-control valves can be eliminated by the use of bar- or cam-lift control valves. Bar- and cam-lift control valves, as illustrated by Figs. 27 and 28, commonly are provided for naval turbines, and this type of control is widely preferred for merchant turbines since it adapts readily to remote operation.

The individual nozzle control valve stems in a bar-lift arrangement pass through openings in the bar and have different lengths such that when the bar is lifted by two rods, which pass through seals to overhead valve gear, the nozzle valves are lifted in a predetermined sequence. The total nozzle area of the first stage is controlled in this way, and there is no uncontrollable nozzle area. The bar-lift control valves thus serve to regulate the steam flow throughout the entire power range, and a separate throttle valve preceding the turbine is not required.

The valve opening sequence is determined by the gaps between the top of the bar and the valve stem backseats. These backseats are either integral or adjustable and many designs have a spherical contact to increase valve stability. Valve contact areas are generally hardened to

LIFT LEVER ROLLER CAM

SPRING BIAS

TORQUE TUBE

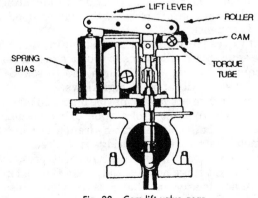

Fig. 28 Cam-lift valve gear

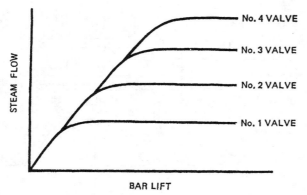

Fig. 29   Bar-lift control—flow versus lift

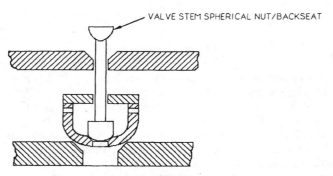

Fig. 30   Nozzle valve with pilot

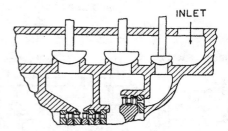

Fig. 31   Bypass valve control

minimize wear and steam erosion, often by using a stellite inlay or by nitriding. Valves are timed to open so as to secure a relatively smooth and linear relationship between steam flow and bar lift as illustrated by Fig. 29. When fine control is required at very low speeds, a small pilot valve, which opens first, may be built into the initial nozzle valve; a valve of this type is shown in Fig. 30. This valve is sometimes called a "double-lift" valve. This arrangement is also used to reduce the force required to unseat the valve and to limit the size of the hydraulic operator.

Cam-lift valves can also be used to provide a sequenced lifting of the valves. In cam-lift machines the cams are attached to a torque tube, as shown by Fig. 28. Lift levers for each individual valve have rollers on one end that ride on individual cams. Torque from the hydraulic operating cylinder (not shown) rotates the torque tube and attached cams. As the rotation progresses, the valves open in a prescribed sequence.

Cam-lift valves are used most extensively in higher-flow and higher-pressure applications to produce the larger lift forces required for higher pressures and to reduce the potential for valve vibration at high flows.

Both bar- and cam-lift valves may be operated manually from the engine room control station using mechanical operating gear or by an electro-hydraulic servomotor, when remote control from the bridge is desired.

d.   Bypass valves.   When the full chest pressure is applied to all first-stage nozzles, the steam flow is limited by the total nozzle area and the pressure drop across the nozzles. In order to further increase the steam flow, valves may be installed, which allow most of the steam to bypass the first stage and enter a later stage where the nozzle area is large enough to pass the desired flow.

Bypass valves, such as in Fig. 31, may be added when efficient operation is required throughout a greater range of power. Bypass provisions generally move the point of optimum efficiency to a lower power, as shown in Fig. 25. On some turbines, several bypasses may be used to further extend the economical range. The use of bypass valves is common for naval turbines, and they are sometimes provided for merchant units where significant overload capability is required, but is seldom used.

The use of bypass provisions is compatible with velocity ratio considerations since the bypassed stage is designed to operate with an approximately optimum velocity ratio just before the bypass valve is opened. If, for example, instead of bypassing a first stage, additional flow were introduced by increasing the first-stage nozzle area, the second-stage inlet pressure would increase proportionately; therefore, the pressure drop, heat drop, and steam velocity of the first-stage nozzles would be reduced at the same time that the blade speed was being increased. The consequent increase in velocity ratio would result in a decrease in efficiency, as may be observed from Fig. 18.

When stages are bypassed, arrangements must be made to provide sufficient steam flow through the idle bypassed stages to keep them from overheating. This may be accomplished by limiting the size of the bypass valve, thus maintaining a pressure drop across the bypassed stages and causing flow through them.

e.   Variable boiler pressure.   Consideration has been given to the control of power output by varying the boiler pressure in proportion to the load requirements. This method has not been used because the response of the system is too slow, it is less efficient at low loads, and the resulting complications in subsystem controls and operating procedures do not justify its use.

4.2   Overspeed and Low Oil Pressure Protection. Overspeed protection is desirable for every turbine that can reach a dangerous speed upon a loss of load. In the case of geared propulsion turbines, this can occur if a propeller is lost or a shaft is broken, and periodic racing can occur in heavy weather when the propeller intermittently emerges from the water due to the ship pitching. A speed-limiting governor is best suited for this purpose since it will prevent an excessive speed while still allowing continuous operation at the maximum governor setting (usually 10 to 15% above the maximum continuous rating). Trip devices, which shut off steam flow completely, are

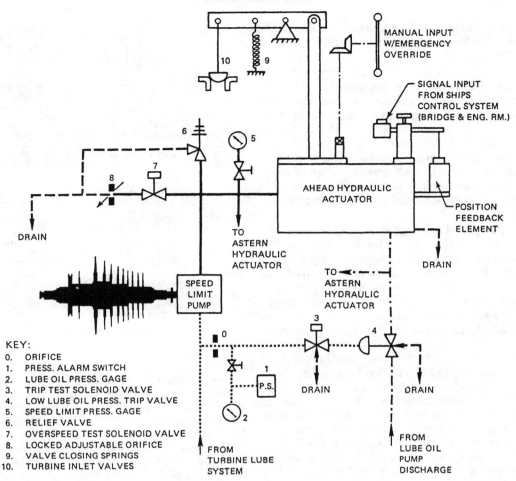

MANUAL INPUT
W/EMERGENCY
OVERRIDE

SIGNAL INPUT
FROM SHIPS
CONTROL SYSTEM
(BRIDGE & ENG. RM.)

AHEAD HYDRAULIC
ACTUATOR

POSITION
FEEDBACK
ELEMENT

DRAIN

DRAIN

TO
ASTERN
HYDRAULIC
ACTUATOR

TO
ASTERN
HYDRAULIC
ACTUATOR

SPEED
LIMIT
PUMP

P.S.

DRAIN

DRAIN

FROM
TURBINE LUBE
SYSTEM

FROM
LUBE OIL
PUMP
DISCHARGE

KEY:
0. ORIFICE
1. PRESS. ALARM SWITCH
2. LUBE OIL PRESS. GAGE
3. TRIP TEST SOLENOID VALVE
4. LOW LUBE OIL PRESS. TRIP VALVE
5. SPEED LIMIT PRESS. GAGE
6. RELIEF VALVE
7. OVERSPEED TEST SOLENOID VALVE
8. LOCKED ADJUSTABLE ORIFICE
9. VALVE CLOSING SPRINGS
10. TURBINE INLET VALVES

Fig. 32 Control and protection system

not satisfactory for the primary protection of propulsion machinery. Speed limiters are designed such that if there is a catastrophic failure in the turbine coupling, the turbine will not overspeed to destruction. Overspeed protection is an essential requirement when an electric drive is used, since there is the additional possibility that the generator may lose its electrical load.

A typical overspeed protection system is shown in Fig. 32. In this example, the speed sensors are small positive-displacement pumps each driven by its corresponding turbine rotor and supplied from the main lubricating-oil system. The pump discharge pressures at any given speed may be varied by adjustment of the variable orifices. The discharge pressures operate a pilot valve by acting against a spring-loaded piston. The pilot valve in turn directs power oil to the proper side of a hydraulic servomotor piston, which opens or closes the throttle as required. The governor may be checked and adjusted in service by closing the test valve until the throttle begins to close and noting the corresponding discharge pressure. The test valve is then opened; and the speed and corresponding pressure at various points within the operating speed range are noted. The variable orifices are adjusted until a plot of speed versus pressure indicates by extrapolation that the throttle will close at the limiting speed.

Alternatively, a centrifugal pump may be used, in which case the discharge pressure acts against a piston and an adjustable spring. The overspeed setting is generally made during the turbine spin test. A centrifugal pump tends to act as a centrifuge, and the small air bubbles present in the lubricating oil tend to collect at the center of the impeller. Unless this air is vented, the pressure-speed relationship will be affected.

It has generally not been considered necessary to provide overspeed protection while operating astern; however, when the astern throttle is power operated for remote-control purposes, it is a relatively simple matter to include astern overspeed protection.

A continuous supply of lubricating oil is essential for the safe operation of a turbine installation. Without an adequate supply of lubricating oil, turbine bearings may fail in a matter of minutes. If the lubricating-oil system pressure drops below a safe value, a spring-loaded piston will actuate the pilot valve distributing oil to the servomotor, and the ahead throttle valve will close. Oil failure may occur with the vessel operating at a high ahead speed, and it is important that astern steam be available to stop rotation of the shaft until the vessel is dead in the water. If this is not done, the coasting period of most large vessels is so long that the supply of emergency lubricating oil will be used up as the propeller is dragged through the water, and the resulting rotation may damage the gear and turbine bearings.

4.3 Governors. A governor generally is provided for the control of a turbine-electric propulsion unit. The governor is adjustable so that it will operate through a wide range of speed and is designed so that it will hold approximately constant turbine rpm irrespective of the load requirements at the selected speed setting. In principle, such a governor is similar to the speed controls of turbine-generator sets, but it has a much greater range of speed adjustment. Governors seldom are furnished for geared propulsion turbine installations, except for overspeed-limiting protection.

4.4 Valve Design. Three types of valves are commonly used to control the steam flow to a turbine. The single-seated valve shown in Fig. 33(a) is simple, but it requires a large lifting force because of the unbalanced pressure across the valve when in the closed position.

To reduce the force required to open the valve, single-seated, balanced valves having an internal pilot valve are used. Figure 33(b) shows this design in diagrammatic form. When the pilot valve B is closed, the pressure in the dotted area equals the inlet pressure, and the pilot and main valves are held tightly down upon their seats. When the valve stem is lifted, the pilot valve opens first, and the pressure within the dotted area drops. It is necessary to control the pressure in the balance chamber by limiting the lift of the pilot valve so that a sufficient steam force is exerted downward on the disk to prevent chattering, which can be caused by instability of the steam flow as the main valve starts to open. This type of valve is widely used with marine turbines.

It is possible to substitute an external hand-operated bypass valve for the internal pilot, as shown in Fig. 33(c).

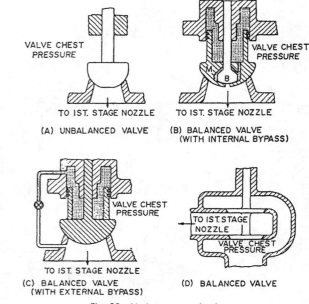

Fig. 33   Various types of valves

Although it can be more nearly balanced because the main valve disk is fixed to the valve stem, and therefore is not susceptible to chattering, a hand-operated bypass requires a separate control, and it is possible that the operator may forget to close it.

The double-seated balanced valve can be used to minimize the force required to open the valve, but it is seldom used with high-temperature, high-pressure marine turbines at locations where both seats must be tight under all conditions of operation. A valve of this design is shown in diagrammatic form in Fig. 33(d).

Section 5
Rotors and Blades

5.1 Blade Design. The length of each row of turbine blades is governed by the volume flow, the mean diameter of the flow path, the velocity of the steam, and the active arc through which flow takes place. Blade lengths generally increase from the high-pressure to the low-pressure end of the turbine, and the length of the last-stage blades is determined by the selected level of exhaust leaving loss. Since each blade extends radially from the wheel, the blade pitch is greater at the tip than at the root. To keep this disparity from causing too great a loss in efficiency, the length of blades having a uniform cross section throughout their length is generally limited to cases where the ratio of blade-tip radius to blade root radius is less than 1.2. This is called the blade-radius ratio, and for greater radius ratios the blades are designed to be tapered and twisted, having angles that vary from the root to the tip as necessary to suit the velocity of steam relative to

the blade at each point along the length. A tapered blade has reduced cross sections from the root to the tip, which decrease the centrifugal stress at the root and permit a longer blade for the same limiting stress. The lengths of tapered and twisted blades have been as tall as 33% of the mean diameter of the flow path.

The shapes of the cross sections of turbine blades are partially influenced by the required inlet and outlet angles determined from velocity diagrams. In addition, as described in Section 3.4, the blading is also optimized to minimize losses due to flow separation and radial flow in the blade passage.

The final optimization of a turbine blade is a combination of mechanical as well as thermodynamic considerations.

The mechanical integrity of blading is evaluated by considering the steady stress due to steam and centrifugal force loadings and the alternating stresses due to stimuli in the steam path. Steady bending stresses on the turbine blades are caused by axial and tangential forces exerted by the steam. These forces are normally highest at full power except in partial-arc control stages, where the highest blade loadings are experienced at reduced power with small partial arcs and high pressure drops. When the centers of gravity of the vane sections at various radii do not fall on a radial line, bending stresses in the vane will also be introduced by centrifugal forces.

The centrifugal force due to the mass of the blade and its rotation causes a tensile stress in the blade that is particularly significant at the blade root and dovetail. This stress varies with the blade length, the wheel diameter, and the square of the blade speed. Centrifugal stresses are relatively steady in nature and do not cause vibration or fatigue failures. They are generally limited to one half of the yield strength of the material.

5.2 Blade Vibration. Steam forces acting on a blade may vary widely throughout a revolution, as in the case of partial admission, or to a much lesser extent as the blade passes through the disturbed flow in the wake of each nozzle partition. Irregularities in the flow path also may be caused by obstructions such as struts and by the extraction of large steam quantities from only a portion of the circumference of the flow path. Periodic variation in steam forces may cause potentially damaging vibration of the blades, or blades and disk, if resonance occurs.

In the case of partial admission, the steam force on each blade rises from zero to a maximum as it enters the active steam arc, and drops to zero upon leaving. This very sudden application and removal of force causes the blade to vibrate at its natural frequency, and if these changes in force are in phase with the motion of the blade, energy will be added to the vibrating system [11].

The major type of vibration stimulus for shorter blades arises from the disturbance in the flow path following stationary blades or nozzles [12–15]. The periodic variation in steam force imposed upon each blade has a fundamental frequency equal to the number of nozzles, which any blade passes per unit of time; this is generally called the "nozzle-passing frequency." If the nozzle-passing frequency stimulus is coincident with a turbine blade natural frequency, resonance will occur. In some cases, even the second and third harmonics of the nozzle-passing stimulus are considered.

Individual blades have many natural frequencies. Vibration may be of either the axial, tangential, or torsional type; typical modes of vibration for banded groups of blades are shown in Fig. 34. When the effects of shrouds or tie wires are considered and the blades are treated in groups, a large number of vibrational modes must be evaluated, and the evaluation of blade strength and reliability becomes complex. In addition, coupling of blade and wheel axial vibration sometimes may occur, such that the blades and wheel vibrate as a system.

Since blade heights increase from stage to stage and natural frequencies decrease with height, and because each blade has a large number of natural frequencies, it follows that at any given speed, one or more stages will

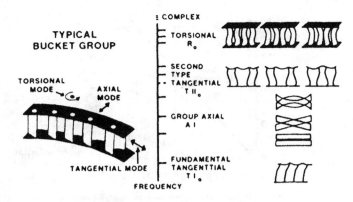

Fig. 34 Vibration modes of banded blades

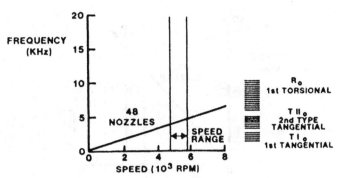

Fig. 35 Campbell diagram with second type tangential resonance of short buckets

run with a natural frequency close to the nozzle-passing frequency. This is particularly true of marine turbines, which must be capable of continuous operation at any speed up to the maximum speed. Marine turbine manufacturers have developed analytical tools to predict the vibratory stress for all potential resonances for comparison to the fatigue strength of the bucket material. This is a practical necessity because it is impossible to avoid all resonances.

A graphical representation often called a "Campbell diagram" is useful to display the vibrational situation for a specific stage. A typical Campbell diagram for relatively short blades is shown in Fig. 35. In general, it is desirable to avoid resonance with the nozzle-passing frequency for the higher responding modes of vibration. The vibrational stress levels must be analyzed carefully for all potential resonances to ascertain that the design is satisfactory. Problem vibration modes can be avoided by modifying the number of nozzles in a stage to change the nozzle-passing frequency. In the example of Fig. 35, there was a resonance with the second type tangential mode of vibration. By changing the nozzle pitch, this resonance could be avoided as shown in Fig. 36.

Exhaust-end buckets in a condensing turbine are typically four to six times longer than the buckets in the high-pressure end of the machine. These longer blades have natural frequencies, which are much lower than the high-pressure stages. A typical Campbell diagram for a long, last-stage blade with 82 nozzles is shown in Fig. 37. Since

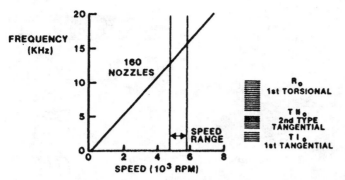

Fig. 36 Campbell diagram with second type tangential resonance of short buckets moved out of operating range

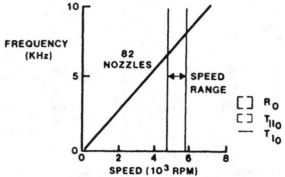

Fig. 37 Campbell diagram for nozzle-passing frequency stimulus on long blades

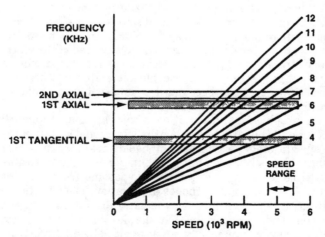

Fig. 38 Campbell diagram for low per-rev stimulus on long blades

these longer blades have lower resonant frequencies, the coincidence with nozzle-passing frequency does not occur within the operating speed range. As shown in Fig. 38, for this example, the bucket frequencies are most closely resonant with four through eight multiples of rotor speed, which are commonly referred to as 4/rev to 8/rev resonances.

As pointed out above, with variable-speed marine turbines, for either long or short blades it is impossible to

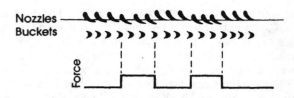

Fig. 39 Example of a 2/rev nozzle stimulus

avoid all resonances. At a resonant condition, the amplitude of vibration is related primarily to the magnitude of the stimulus, the damping present in the system, and the resonant response factor or phase relationship between the excitation force or stimulus and the mode of vibration. The response is directly proportional to the stimulus and resonant response factor while it is inversely proportional to the damping. The primary source of damping is friction at the banded tip connections and blade dovetail. The proper grouping of buckets minimizes the resonance response factor.

For long blades, the low per-rev stimulus can be due to a variety of causes such as:

• Struts in the exhaust casing
• Nonuniform upstream flow due to extractions
• Discontinuities at the horizontal joint
• Coupling excitation at the shaft end
• Rotor torsional resonances
• Upstream partial-arc stages
• Stationary nozzle blades
• Nonuniform exhaust flow

While most of these sources can be minimized in the design phase, the stimulus from the stationary nozzle blades is the major source of low per-rev stimulus and must be considered during the manufacturing cycle of the diaphragms. The basic simplified flow diagram shown in Fig. 39 can be used to illustrate the phenomenon. If the exaggerated circumferential nozzle spacing variation is as shown, the bucket will be subjected to two force pulsations per revolution and thus a 2/rev stimulus. This is analogous to the nozzle-passing frequency where the 2/rev is equivalent to two nozzles per 360 deg. In an actual application, small, more random variations in nozzle blade positions must also be considered. While these variances are small and within tolerances, the pattern of the variations is important and can cause a low-frequency harmonic stimulus. These variations result in flow, flow angle, and bucket reaction variations around the stage, thereby causing potential tangential and axial force variations that must be controlled. Turbine manufacturers perform a harmonic analysis of their production diaphragms to insure that they do not have high low-per-rev stimuli at potential resonances. A simple example is shown in Fig. 40. In this case, the nozzle spacing is such that the force is different between the upper and lower halves of the nozzle blade assembly. A harmonic analysis of the force wave shows the relative magnitude of the harmonics. If one of these harmonics is coincident with a turbine blade natural frequency, resonance will occur. Actual nozzle dimensions vary in a much more complex fashion, but the

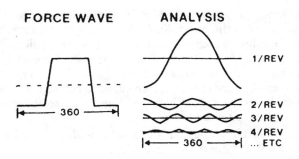

Fig. 40 Example of harmonic stimuli

stimulus force of low-order harmonics does cause bucket excitation.

5.3 Blade Material. The ideal blading material has: (1) a high yield strength at all operating temperatures, (2) good ductility, (3) high fatigue strength, (4) good resistance to corrosion, (5) resistance to erosion due to wet steam, (6) machineability, and (7) high internal damping capacity to absorb vibrational energy. The several types of corrosion-resisting steels, having approximately 12 to 13% chromium content, which are in common use for turbine blading, satisfy most of these requirements. Molybdenum is added to achieve higher creep strength at elevated temperatures. The corrosion resistance of steels having 12 to 13% chromium is good in normal clean steam; however, steels of this type are susceptible to attack by chlorides carried over from the boiler and deposited on the blading. These deposits are hygroscopic, and if turbines are laid up or stored with chloride deposits, concentrated pitting action may cause severe damage. Furthermore, a high concentration of chlorides in the steam can cause stress corrosion cracking, particularly in the highly stressed dovetail region on stages where the steam is at the transition zone between wet and dry. This can cause catastrophic turbine blade failures and is characterized by evidence of intergranular cracking. Periodic inspection of turbines for corrosion, erosion, solid particle erosion, and fatigue is critical [16–18].

Other, more advanced materials have also been used. These include Jetheat M152 and titanium. M152 offers the advantage of higher strength and resulting greater speed capability. Titanium (6Al/4V) has also been used in steam turbine applications to gain an even higher speed capability. While titanium cannot be used in applications above 600 F, due to susceptibility to hydrogen embrittlement, it is an excellent material for lower-temperature applications. Titanium is immune to chlorides and is, therefore, an option when pitting or stress corrosion cracking is a concern. It also offers the advantage of higher fatigue strength, particularly in wet stages where its fatigue strength does not degrade in a moisture environment— unlike 12% chrome steels, which typically degrade by 33%. The primary disadvantages of titanium are its cost, machinability, and higher notch sensitivity.

5.4 Blade Erosion. After passing the intersection of the state line and the saturation line, steam expanding in a turbine transforms into moisture, as shown by a Mollier chart. Some of the finely divided moisture particles strike the metallic surfaces and collect to form small drops that are subsequently swept along with the steam flow. Water traveling in the form of drops moves at only a small fraction of the steam velocity. If separate velocity diagrams are drawn for the steam and water, it is seen that while steam encounters a moving blade with little or no impact under normal conditions, water drops tend to hit the back of the blade perpendicular to the surface and with a speed only slightly less than blade velocity. The leading edge, particularly at the tip where the blade speed is highest, is subject to a barrage of water droplets striking the surface at a high velocity. When a water droplet hits the entrance to a microscopic intergranular crack in the blade material, it tends to penetrate the crack. The resulting stress in the metal is magnified at the sharp end of the crack, and the crack, therefore, propagates inward until several cracks meet, with the result that a small particle of metal breaks away. Repeated again and again, this leads to erosion, pitting, and deterioration of the blade tip.

Blade erosion is accelerated by a high blade-tip speed, high moisture content of the steam, or soft blade material. Other parameters also affecting blade erosion include stage reaction, clearance between the nozzle trailing edge and bucket leading edge, and the geometry of the steam path at the outer diameter. Since moisture is thrown from the moving blades toward the casing by centrifugal force, annular pockets or grooves are usually machined in the diaphragms where there is significant steam moisture and are arranged to collect and discharge as much water as possible to the condenser. This reduces the moisture content in the latter stages with higher tip speeds and thus reduces the tendency for blade-tip erosion in those stages. It also improves the thermodynamic performance. Several shapes and arrangements of moisture grooves or moisture separators may be seen by inspection of Figs. 2, 3, 4, and 50. When required, the leading edge of the blade at the tip is hardened to increase erosion resistance. This includes either locally flame hardening the blade material or using specially shaped pieces of hard material, called moisture shields, which are brazed to the leading edge of the blade. Stellite, a material used for tipping machine tools and for other applications where an extremely hard material is needed, is used for this purpose. The use of flame hardening or moisture shields is not necessary when a harder 12% chrome alloy is used as the blade material. Figure 41 shows a typical curve that defines the limits for blade erosion protection as a function of blade-tip speed and moisture at the inlet to the stage for a straight 403 stainless steel. This curve will change depending on the blade material and the other variables that influence blade erosion as noted above.

5.5 Blade Fastenings. The many types of blade fastenings that have been developed may be classified in two general groups, "circumferential entry" and "axial entry." Blades held by a circumferential-entry fastening system are inserted radially in a circumferential groove and then moved around the rotor into final position, while axial-entry system blades are held in axial slots spaced uniformly around the periphery of the rotor. The most

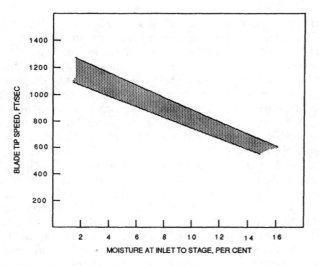

Fig. 41   Acceptable tip speed versus moisture at stage inlet for straight
403 stainless steel

the blade to as nearly a correct shape as possible, then machine it to the exact dimensions where necessary. The third method, which is used occasionally for auxiliary turbines, is to cut the blades from long bars of rolled stock having a suitable cross section. These pieces are then dovetailed at one end and inserted in the rotor groove with suitable spacer pieces between them to give the blades the correct pitch and angle.

Impulse blading generally is fitted with a shroud band at the outer diameter to stiffen the blades, suppress potential blade vibration, and to provide an enclosed flow path. Usually, holes are drilled or punched in the shroud bands to match the position of tenons on the blades. When the bands are in position, the tenons are peened over to secure the bands. In early practices the bands were sometimes welded to the blades, but this requires special techniques to avoid air hardening due to the welding heat. Shroud bands typically span from 4 blades up to as many as 25. Long exhaust-end blades typically use the longer banded segments to suppress low per-rev blade resonances. Special attention must be given to these longer segments during assembly to insure that the assembly stresses are minimized. The maximum shroud length may be limited by the differential expansion between the shrouds, blades, and rotor that result from rapid changes in temperature. Shrouds are made of corrosion-resisting steel similar in composition to the blades.

The efficiency of impulse blading may be improved in certain cases when the blading is designed to include some reaction and there is a pressure drop across the moving blades. In this case, either axial or radial seal strips are used to minimize steam leakage and interference with the boundary layers of the flow path. Such seals have thinned

common forms of blade fastenings are shown in Fig. 42. The dovetail, T-slot, and straddle are of the circumferential-entry type while the fir-tree and bulb-and-shank are the most common types of axial-entry fastenings.

Dovetail designs are optimized to have minimal stress concentrations in fillets through use of sophisticated techniques such as finite-element analysis and photoelastic evaluation. The design guidance provided by these analysis techniques can substantially increase the fatigue life of the dovetail and improve its reliability.

5.6   Impulse Blading.   Impulse blades are manufactured in a variety of ways. The most common method is to machine the blades from rectangular bars of stainless steel. Another method frequently used is to drop forge

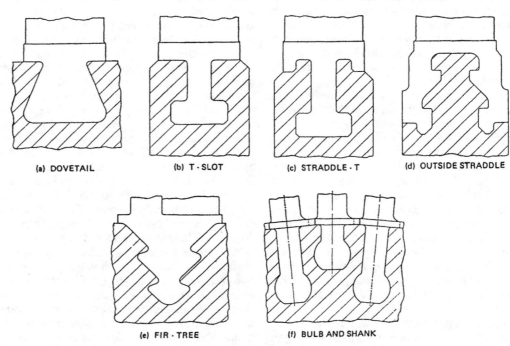

(a) DOVETAIL   (b) T-SLOT   (c) STRADDLE-T   (d) OUTSIDE STRADDLE

(e) FIR-TREE   (f) BULB AND SHANK

Fig. 42   Typical blade fastenings

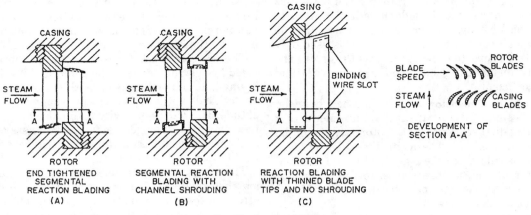

Fig. 43   Reaction blading

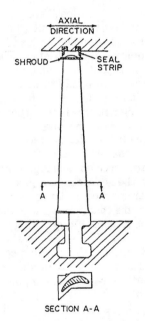

Fig. 44   Reaction blade—T-type fastening

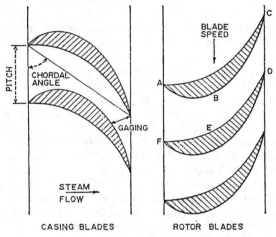

Fig. 45   Reaction nozzle and blade section

edges to reduce heating effects in the event of accidental contact between moving and stationary parts.

5.7 Reaction Blading. Purely reaction blading is generally no longer manufactured; however, hundreds of units with reaction blading in the high-pressure stages are in operation and will continue to operate for many years. Accordingly, it is necessary to describe the more common forms of reaction blading.

Many methods of securing reaction blading to the rotor are in use, but Figs. 43 and 44 represent the most common arrangements. Figures 43(a) and 43(b) show the segmental type of reaction blading. End-tightened blading is no longer used because of the difficulty in maintaining the axial clearance with higher steam temperatures. The individual blades are cut to length from long bars of rolled stock having a suitable cross section. A typical Parsons section is shown in Fig. 45. The blades are spaced and positioned by packing pieces located at the root. The packing pieces are cut from rolled bar stock of a shape such that the correct pitch and angle is obtained when the blades and packing pieces are assembled in a form. The shrouding is brazed by dropping the assembly in a pot of molten brass. The solid root portion is then machined to fit the rotor.

Figure 43(c) illustrates a somewhat similar construction for blades having moderate speeds and stresses. Each packing piece is brazed to a single blade and then machined to fit the groove. The blades and packing piece units are then driven up individually in the groove and the binding wire or shroud secured after the entire row is installed.

Figure 44 shows a reaction blade having the spacer piece at the root formed integrally with a T-slot root fastening. A blade of this type may be machined from a solid rectangular bar, or forged and then finish machined. The fastening is suitable for long blades at high blade speeds.

The pressure drop in either fixed or moving reaction blade rows causes leakage by the tips of the blades, which constitutes a loss in efficiency. Close clearances are necessary to reduce this loss to an acceptable level. One method, illustrated by Fig. 43(c), involves thinning the blade tips to a fine edge to prevent serious damage in the event of a rub. Since this destroys the shape of the blade and the leakage flow disturbs the boundary layer, it is used only on long blades where the clearance loss is small. The shrouded reaction blades as shown in Fig. 43(a), 43(b), and 44 are more efficient.

5.8 Rotor Axial Dimensions. The distance between bearings is the sum of the axial space requirements of a number of individual sections of the rotor and should be kept as short as possible. The axial dimension of each impulse wheel is related to the blade width, the size and type of blade root and, in some cases, to the hub width necessary to keep stresses in the rotor due to centrifugal forces at a reasonable value. The thickness of each diaphragm is determined by limiting the deflection caused by steam forces due to the pressure drop across the diaphragm and by stress considerations. The axial lengths required for high-pressure packing, interstage packing, and glands of the labyrinth type are chiefly dependent upon the number of sealing teeth. Finally, the length between bearings is affected by the axial clearances provided between stationary and moving parts. Since there are a large number of such clearances, it is important to keep the clearances as small as possible; however, they must be sufficient to avoid rubbing contact under all operating conditions. When relatively rapid variations in steam temperatures occur, during either maneuvering or sudden large changes in power output, the rotor and casing usually will adjust to the new temperatures at different rates such that the resulting differential expansion may seriously reduce some of the running clearances. The selection of design, or cold, clearances is dependent upon calculations for the rates of change in temperature of all parts involved.

5.9 Rotor Critical Speed. The selection of the shaft diameter for a turbine rotor involves consideration of the critical speeds of the rotor. The simplest method of analyzing the critical speed of a rotor is to consider it as a beam of variable cross section, freely supported by a rigid bearing at each end and subject to transverse vibratory forces of variable frequency, whereupon it will be found that the amplitude near the center of the bearing span will reach a maximum at a resonant frequency called the "first critical speed." At a higher frequency, called the "second critical speed," the rotor will vibrate in resonance with nodes at each bearing and the midpoint of the span. Still higher critical speeds are possible, but are seldom encountered in marine turbines. In actual practice the bearings and their supports are not rigid and the flexibility of the fluid film in the bearing alone may lower the rigid-bearing critical speeds by as much as 50%.

When a rotor is brought up to speed, the centrifugal forces due to unbalance in any section of the rotor cause it to deflect, and the bent rotor then whirls around its neutral axis at a rotational speed in what is called "synchronous whirl." The theoretically infinite amplitude at a critical speed is limited in practice by the presence of damping in the system. It should be noted that damping

inherent in the rotor material does not contribute to limiting the amplitude since the rotor shape does not change, but merely whirls. Instead, damping is contributed principally by the bearings and bearing supports [19].

The exciting force for rotor whirling vibration is provided by the unbalance in the rotor; and if this unbalance is small, as is usually the case, the rotor may go through its "critical" without a noticeable change in vibration amplitude. The rotor sensitivity to unbalance is dependent on the particular rotor geometry, the stiffness and damping of the bearings and their support structure, and the operational loads on the bearings.

The dynamic characteristics of the bearings are matched to specific rotor characteristics to evaluate the effect on the vibration response of the rotor-bearing system. For example, tilting-pad bearings (see Section 8.2) were introduced to avoid half-frequency, oil-whip problems with sleeve bearings [20]. Their use was originally limited to high-speed, light-load applications; however, tilting-pad bearings have subsequently been used in wider applications regardless of speed or load because of their stiffness and damping characteristics. They have an effect on the rotor sensitivity to unbalance, and they also have a wider tolerance to misalignment.

The operational loads on a bearing have also been recognized as being a significant factor that influences the bearing stiffness and damping and, thus, the rotor sensitivity to unbalance [19]. Under part-load operation, a turbine operates with one or more of its valves closed; this creates a partial-arc force on bearings that can be significant, especially if these forces are large relative to the rotor weight. Other sources of partial-arc forces are fixed partial-arc diaphragms that may be used in the steam path for thermodynamic reasons along with any partial-arc extractions or admissions. The effect of these forces must be considered; not only for their influence on the bearing dynamic characteristics used to predict rotor response, but also for their influence on bearing operational temperatures and reliability.

A valve opening sequence is illustrated by Fig. 46 for a machine with a 4-pad, tilt-pad bearing, as shown in Fig. 47. Figure 48 shows the force vector change for the case with the first two valves open. As the second valve opens in the lower half, the rotor gravity load vector and torque force vector combine to have a resultant force vector, which rotates toward the pivot point of the adjacent tilt pad. For this case the bearing characteristics have then changed from a 4-pad load-between-pad bearing to a 4-pad load-on-pad configuration.

When the rotor force vector is directed to be a load-between-pad bearing, the oil-film stiffness and damping coefficients are nearly symmetrical, and the shaft orbits are reasonably circular. When the force vector is directed toward the pad, to be a load-on-pad configuration, the coefficients in line with the pad are vastly different and produce a narrow, elliptical orbit. The two extremes bound the potential variation of stiffness and damping that can occur in the bearing due to partial-arc effects. The direct results of these variations are changes in the rotor critical speed and response to unbalance. This condition will not

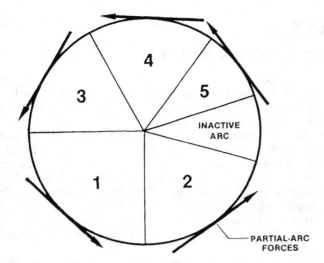

Fig. 46 Nozzle box valve arcs showing opening sequence and direction of tangential forces

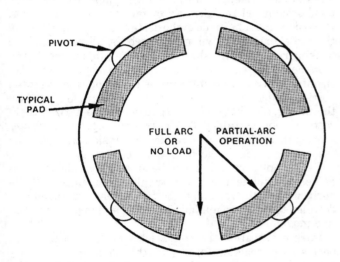

Fig. 47 Rotation of bearing force vector due to effect of partial-arc steam forces

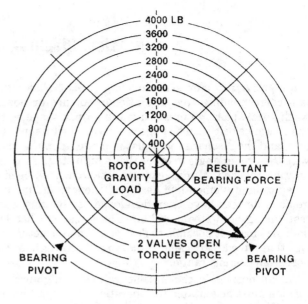

Fig. 48 Resultant bearing load from combined gravity and steam torque forces with two valves open at rated speed and steam conditions. Lower half bearing pivot locations shown for 4-pad, load-between-pads bearing

occur during a factory no-load test, but only during operation with load.

5.10 Rotor Balance. Finished turbine rotors are balanced in special machines developed for this purpose before being run for the first time. Relatively stiff rotors may have correction weights located only in two planes, one at each end of the rotor. However, when the rotor is flexible and very low vibration levels are required, it is often desirable to balance the unbladed rotor first, then rebalance after each stage of blading is installed, making the weight corrections in the plane of each respective row. The final check of the rotor balance is made when the turbine is tested at design speed in its own casing and bearings or in a high-speed balance facility.

Rotors in service may be balanced in their own casings by using special instrumentation designed for this purpose. This equipment includes a speed pickup and vibration probes at each end of the rotor. Generally two test runs are required with known test weights applied to the rotor in different positions for each test, then a determination of the final correction weights may be made by calculation. Some turbines have provisions to install correction weights without the necessity of lifting the upper turbine casing, which greatly expedites the balancing process.

Some machines that have been well balanced initially tend to become out of balance with time. The unbalance may result from creep, unsymmetrical thermal or pressure distortion, erosion, corrosion, wear, material deposits, or other causes.

5.11 Rotor Material. Most marine turbines are machined from solid forgings. But, in smaller turbines or older machines the disks may be shrunk and keyed to the shaft. Reaction turbines usually are made with forged one-piece rotors, but in some cases a hollow forging has been used with the shaft end (gland and bearing) fitted into it at one or both ends. The rotor material must be suitable for the range of temperatures at which the turbine will operate. Fortunately, the blades at the hot end of the turbine are short and the wheel stresses may be kept reasonably low and commensurate with the reduced physical properties of the forging at higher temperatures, while the reverse is true at the cold, or exhaust, end. Forgings for use above 700 F generally contain 0.5% molybdenum, and for higher temperatures and greater strength may contain various amounts of vanadium, chromium, and nickel.

Rotors operating at speed have tangential and radial stresses due to centrifugal forces. These stresses are evaluated at each wheel to insure that they do not exceed established allowable limits. In addition, for high-temperature rotors, care must be taken to select rotor wheel fillets that have low stress concentration factors to minimize the potential for thermal cracking of the rotor due to low-cycle fatigue.

Section 6
Nozzles, Diaphragms, and Stationary Blading

6.1 Nozzles. For every moving row of blades there is a corresponding row of stationary blades or nozzles. Each nozzle passage must have the proper shape so that the steam velocity can be generated with the least possible loss, and the discharge jet must impinge upon the blades with the proper angle in order that a high stage efficiency may be realized. Many years of research have been devoted to improvements in nozzle efficiency, and many ingenious devices have been developed for making nozzle tests using both steam and air. The work that has been done has made it possible to design and manufacture nozzles that have an efficiency of 96 to 97% throughout their normal operating range.

Ahead nozzles in the first stage of the high-pressure turbine must be suitable for operation at a high temperature. The nozzle arc is usually less than 180 deg and is divided into a number of segments such that steam flow to each segment is controlled by a separate valve. When the turbine is carrying a light load and only one of the valves is open, the pressure drop is large across the few nozzle partitions that are operating. These partitions must be strong enough to withstand the pressure drop and stiff enough to prevent the discharge edges from vibrating and breaking. The cross section of the flow passage generally is rectangular, and the nozzles are assembled from separately machined partitions or vanes welded together to form a nozzle plate, which is secured to the nozzle chest. The nozzle material generally is 12 to 13% chrome stainless steel.

The nozzles at the inlet of the astern turbine must withstand large pressure drops, often the full pressure drop from main steam pressure to condenser pressure. As a result, at full astern power the discharge velocity of the steam as it leaves the nozzles is well above the velocity of sound. To generate this high velocity efficiently, convergent-divergent nozzles (nozzles having a discharge area larger than the throat area) are used. The cross section of the flow passage may be rectangular or may be round with a conical enlargement in the divergent section. The round type, generally called a "reamed nozzle" because of the method used in manufacture, has a somewhat lower efficiency, greater length, and cannot utilize the area of a given flow annulus as effectively. These disadvantages are not significant in astern turbine applications and are offset by lower cost and ease of manufacture. Astern nozzles normally are surrounded by cold, wet steam during ahead operation; and to avoid excessive corrosion and loss of efficiency, the nozzles are made from corrosion-resisting steel. To prevent steam leakage during astern operation, the astern nozzle block must be held securely to its supporting member. In addition, the combination of the two members must be able to expand freely without changing the relationship between the nozzles and blades when the astern turbine is in use and the temperature at the inlet changes rapidly from condenser to main steam temperature.

6.2 Diaphragms. A typical turbine diaphragm is illustrated by Fig. 49. The packing diameters between impulse turbine wheels are made as small as the rotor critical speed calculations will permit to minimize the packing leakage from stage to stage. The pressure drop across the diaphragm causes it to deflect axially, and the thickness of the web and the strength of the nozzle partitions must be sufficient to limit stresses and deflections to acceptable values.

The diaphragm has more surface exposed to the steam than that portion of the casing in which it is supported. As a result each diaphragm will change temperature and expand or contract radially at a faster rate than the casing. Provisions must be made in the design so that this differential radial expansion will not disturb the alignment of the packing relative to the shaft and cause a rub. The most common method is to support the diaphragm just below the horizontal joint at two points, one on each side, such that the vertical and axial positions of the diaphragm are fixed by the casing, and differential expansion in the horizontal direction is not restrained. A third point of support at the bottom centerline allows free vertical expansion and positions the diaphragm in the horizontal direction.

Each diaphragm must also be arranged so that its two halves will not separate and permit steam leakage at the horizontal joint. Normally, the two halves are keyed and doweled together to minimize leakage and to prevent relative movement.

During operation the steam jets from the nozzles impose a large reaction force upon the diaphragm. The tangential component of this force would tend to turn the diaphragm if some means were not provided to prevent its rotation. When the diaphragm is center supported, the arrangement inherently prevents rotation; but if the diaphragm simply rests in a groove in the casing, it is necessary to rely upon locking devices, which are generally used to hold the upper-half diaphragms in the upper casing when the casing is lifted.

6.3 Intermediate Blades. Stationary blades, often called "intermediates," are located between the moving rows of blades on two- and three-row velocity-compounded wheels, as may be noted from Fig. 50. Their function is to redirect the steam from the first row of moving blades so that it will impinge upon the second moving row at the proper angle. If the first-stage ahead or astern nozzles cover only a portion of the arc, the intermediate blades will cover a slightly greater arc, which is oriented to allow for the movement of steam around the circumference as it passes through the first row of blades. Since stresses are low, it is common practice to use dovetail-type fastenings. The blades may be inserted directly into a dovetail groove in the casing or fitted in a separate holder.

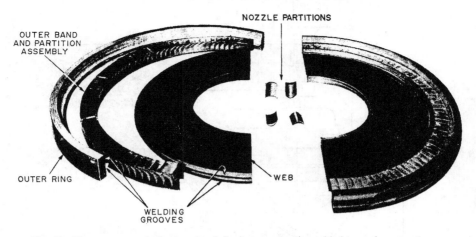

Fig. 49 Necessary parts and a completed diaphragm using the welded type of construction

Section 7
Casings and Packings

7.1 Casings. The turbine casing serves to position the blades, diaphragms, gland packing, bearings, and other stationary parts as well as to contain and direct the flow of steam. At the inlet end the casing is subject to high pressures and temperatures and must be designed to contain the steam safely and without leakage. At the exhaust end of the turbine, the steam pressure is subatmospheric and the casing must be designed so that it will not collapse or allow air to leak into the condenser and impair the vacuum and, at the same time, support the low-pressure end of the rotor.

High-pressure turbine casings and many low-pressure turbine casings are made from cast steel. Cast iron is no longer used for the casings of main propulsion units since it is brittle and cracks under the impact of shock forces, and because above 450 F it grows in size, distorts, and weakens. Cast carbon steel is used when steam temperatures are below 775 F, but it is not satisfactory above this temperature because the material yields or creeps. For low-temperature nuclear applications, where the steam is saturated and moisture erosion is of concern, 12-chrome steel is often used. Carbon molybdenum steel containing not less than 0.5 percent molybdenum may be used for temperatures up to 875 F; and for still higher temperatures, various alloys of chrome molybdenum are suitable. The creep characteristics of the casing material should be considered in the range of 800–950 F and become highly important above this range.

For assembly and machining purposes, the turbine casings are made in two halves. The flanges and bolts that hold these two halves together must be designed so that the joints will not leak under any operating condition. The joint finish is important and several procedures may be used to ensure a proper fit. In the past, joints have been planed, then hand scraped to achieve a good contact between upper and lower halves. As units have increased in size it has been found more practical to grind the joints on a large surface grinder.

Calculations and experience influence the flange design. It must be stiff and the joint faces must be flat and mate properly with those on the other half. The joint bolts must be substantial and placed as near the inside wall of the flange as possible. A high grade of creep-resisting bolt material is necessary, since the temperature of the bolts is only slightly less than that of the steam inside the casing. Experience has indicated that when the nuts are tightened by heavy sledging on a wrench, the thread surfaces are damaged. Consequently, bolt tightening procedures generally involve the use of hydraulic wrenches or, in the case of bolts 2 in. in diameter and larger, bolt heating. This involves tightening snugly but not heavily, then marking the angular position of the nut. An electric or gas heating element is inserted in an axial hole in the bolt. When the bolt has become heated and expanded, the nut is turned a predetermined amount that is calculated to give the desired stress in the bolt and compression in the joint. The bolt preload stress can be checked by measuring the elongation of the bolt. The maximum allowable stress in the bolt upon cooling is related to the creep characteristics of the material but is typically in the range of 30,000–60,000 psi depending on the application. As noted above, the use of hydraulic wrenches has also become a common method for bolt tightening. This approach can be very successful where there is adequate space to accommodate the wrench. It is also sensitive to having proper lubrication of the threads and nut faces to insure that the torque is effective in prestressing the bolt to the desired level of stress. Bolts have a tendency to creep rather rapidly at

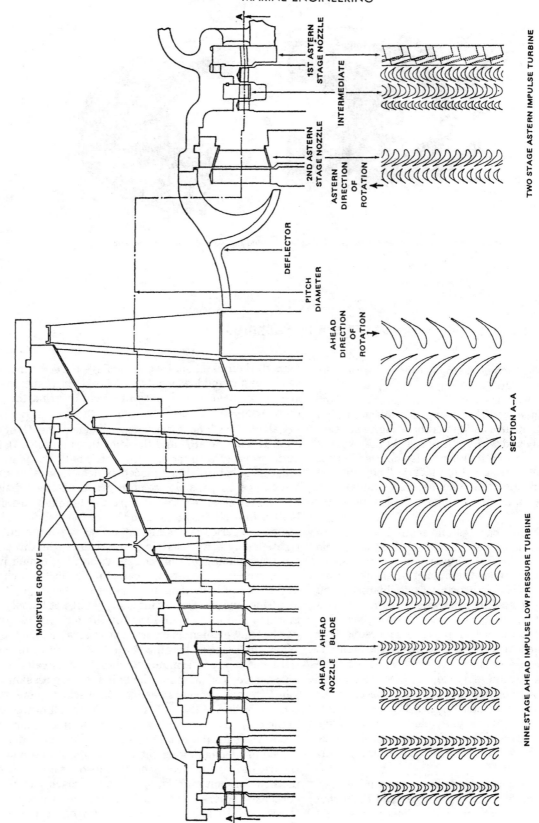

Fig. 50   Steam path through impulse turbine

first and should be checked and tightened if necessary after several months of operation.

The low-pressure turbine casings are cast or fabricated, and the associated exhaust casing is generally fabricated. In order to withstand the difference between atmospheric and condenser pressure, the exhaust casing must be adequately braced. These braces must not interfere with the steam flow from the last stage to avoid stimulating vibration in the last-stage blades. If the exhaust casing must support the condenser weight, this factor must be taken into account in its design.

As the temperature of a turbine changes, the casings must be free to expand and contract freely. Between ambient and operating temperatures, a single-casing turbine will expand lengthwise from $\frac{1}{4}$ to $\frac{3}{8}$ in. and proportional amounts in other directions. Thermal expansion cannot be restrained because the forces involved would be enormous.

To provide for thermal expansion, turbines are anchored firmly to their foundations at one point, usually at the end next to the driven unit. With this point as a base the casings are guided so that their expansion is in the desired direction. A grooved sliding seat at the forward end is sometimes used, but more often the forward end of the casing is supported by a deep flexible I-beam, or equivalent, installed with its longitudinal axis athwartships, an arrangement that allows for free fore-and-aft movement while positioning the unit vertically and athwartships.

There are several other miscellaneous requirements for a good casing design. Adequate provisions must be made for draining the casing during shutdown, starting, and while in operation. It must be free from pockets that would collect water while the ship is listing and then empty suddenly when the ship returns to an even keel.

Suitable lifting lugs and guide pins must be provided so that the upper half shells can be lifted with the limited headroom and lifting facilities in the engine room of the ship.

If possible, openings should be provided in the casings so that balance weights can be changed without lifting the upper casing. This is not difficult in the exhaust casing, where manholes are provided for access to the condenser.

At the high-pressure end of the turbine, small borescope openings are sometimes provided to check the axial clearances between the rotating and stationary parts of the control stage and for general internal inspection. These openings are particularly useful when adjustments are made to the turbine thrust bearing.

The amount of heat that is radiated from a turbine is not enough to affect its efficiency to any measurable extent. However, turbines are thermally insulated to reduce the heat load imposed upon the engine-room ventilation system, to protect personnel from hot surfaces, and to prevent abrupt temperature gradients in the turbine casings, which could cause distortions and joint leaks. Various types of thermal insulation materials are used to cover hot surfaces. Plastic insulation made from high-grade cement with a mineral wool base may be used at any operating

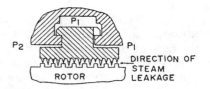

Fig. 51 High-low tooth packing

steam temperature. Spun glass or glass fiber in glass-woven fabric bags makes a good installation, even next to the hottest surfaces, although they have shown a tendency to disintegrate if bent around too sharp a corner. Eighty-five percent magnesia is suitable for use where the steam temperatures do not exceed 450–500 F. In some cases a high-temperature-resistant insulation is used for the first or inner coat and a lower quality is used on the outside. Metal lagging is sometimes applied, chiefly for appearance purposes.

7.2 Packings. All of the steam that passes through a multistage turbine should perform useful work on the blades, but unfortunately this ideal is never reached. There are a number of points where it is necessary to prevent the bypassing of steam through leakage between the stationary and rotating parts. It is not practical to reduce this leakage to zero by the use of seals or packings that require contact between the moving surfaces, because the high relative rubbing speeds would generate excessive heat and the sealing surfaces would deteriorate in a very short time. Hence, clearances must be provided between the rotating and stationary parts and suitable arrangements are needed to minimize the leakage flow.

The most commonly used form of packing for marine turbines is the labyrinth seal. In principle, the labyrinth seal consists of a series of small clearances that serve to constrict the leakage flow. At each clearance the velocity of the leakage fluid is increased by the corresponding pressure drop, only to have this velocity energy converted, or partially converted, to heat by turbulence and eddying in the space that follows each point, thus effectively throttling the fluid. The leakage flow is dependent upon a number of factors and may be estimated as described in Chapter 2. Two arrangements of the seal points are possible. The most common arrangement, called a "stepped labyrinth" packing, is shown in Fig. 51 and consists of a stationary ring having a series of fins tapered to thin edges so that in the event of an accidental rub there will be a minimum of heat generated. High and low points alternate and corresponding lands and grooves are machined in the shaft to suit. With this arrangement, the flow from each point does not impinge upon the following clearance. Sufficient axial clearance must be provided between the long points and the lands to avoid contact when differential expansion of the casing and rotor takes place due to the temperature differences in these parts. In addition, the lands must be wide enough so that the fins will not move beyond the lands and become ineffective. When there is a considerable differential axial movement, a "straight-through" labyrinth-type packing is sometimes used with a smooth shaft, and all seal points are of the same radial length. In addition to the factors governing flow through a stepped labyrinth, the leakage flow through a straight-through labyrinth packing is dependent upon the ratio of the clearance to the axial pitch of the points because of the velocity-carryover effect.

Each labyrinth packing ring is cut into four or six segments, which are held in position in the casing groove by either radial or flat leaf-type springs. If rubbing should occur, the segments are pushed out to a larger diameter and return to their normal position when contact with the shaft ceases. This flexibility minimizes the wear down of the points and the consequent increase in clearance. If the coefficient of expansion of the packing ring material is higher than that of its supporting member, clearance must be provided at the circumferential ends of the segments. This is sometimes accomplished by "archbound" packing where the circumferential contact radius of the packing is less than the radius of the supporting lip. This feature permits thermal growth and radius straightening during heating. In operation, the segments are surrounded by steam at various pressures (see Fig. 51) and care must be used in the design to assure that the resultant force will always push the segments toward the shaft.

It is important that the material used for the stationary replaceable points have good rubbing characteristics, i.e., minimum tendency to gall or tear in the event of accidental contact with the shaft. It is customary to use a leaded nickel-brass casting (for example, 6Pb-13Ni-65Cu) when steam temperatures do not exceed 750 F. In some cases leaded nickel brass is used up to 800 F, but this should be done with care since the strength of this material declines at higher temperatures. A 22% nickel ductile iron casting is used for higher temperatures.

The loss due to steam leakage through the diaphragm packings of an impulse turbine averages between 1 and 3%. The loss in individual stages varies through much wider limits. In the low-pressure stages, where the specific volume of the steam is high, the percentage of steam that passes through the packings is small. In the high-pressure stages, the percentage is greater, and it is desirable to use more points and smaller clearances.

Labyrinth packings are also used for the high-pressure packing of impulse turbines and the dummy packing of reaction turbines. In each case the steam that passes through these packings is discharged to a low-pressure stage, and therefore part of the available energy in the steam is recovered as it flows through the remaining stages to the condenser.

Labyrinth seals are used also in the shaft glands where the rotor passes through the ends of the casing to prevent steam leaking out when the internal pressure is greater than atmospheric, and to prevent air from leaking into the turbine when the casing pressure is subatmospheric. The internal arrangement of a typical gland is shown in Fig. 52, and a diagrammatic arrangement of a typical gland seal system for a cross-compound turbine installation is shown in Fig. 53. The directions of flows at various powers are indicated, and it can be seen that steam must be supplied from an external source for operation at low ahead powers, standby, and astern powers. Excess steam must

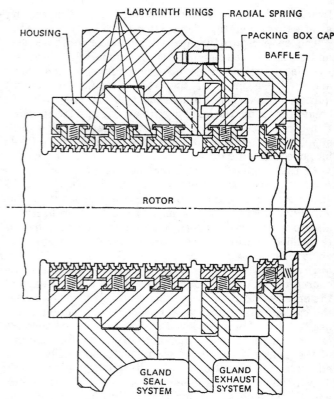

Fig. 52   Typical turbine gland—labyrinth packing

be dumped at high ahead powers. Adjustment of the seal steam pressure to 0.5–1.0 psig may be manual or automatic, but in either case it is desirable to interlock the steam supply and steam dump valves to prevent misoperation, which could result in both valves being open with a consequent loss of live steam to the condenser. The gland exhaust or leakoff system prevents the loss of vapors to the engine room; it discharges to a small condensate-cooled condenser, where the air is pumped out by a motor-driven fan.

Thus far, the discussion of labyrinth packings has been confined to multitooth packings used to reduce steam leakage along the shaft. Single- or double-tooth labyrinth packings are commonly used to reduce steam leakage around the rotating blades in an impulse turbine or the stationary and rotating blades in a reaction turbine. These teeth, or spill strips as they are sometimes called, may be fastened radially into the casing or diaphragm, or they may be attached to or form a part of the blade or its cover, as illustrated by Figs. 43 and 44.

Another type of packing, which is sometimes used for glands, consists of rings of carbon carefully fitted with close clearances about the shaft and held in a suitable housing. Each carbon ring is located in a separate compartment and is made up of segments held together by a garter spring. Each ring is free to move in a radial direction such that minimum contact with the shaft takes place and at the same time steam pressure forces the ring against the downstream side of its compartment to hold the ring in this position and to effect a seal. The quality

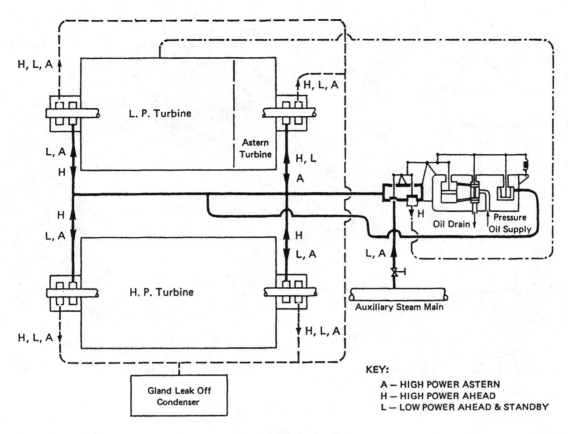

KEY:
A — HIGH POWER ASTERN
H — HIGH POWER AHEAD
L — LOW POWER AHEAD & STANDBY

Fig. 53   Typical gland seal system

of the carbon used in a carbon packing is important. Carbon that is too hard will generate too much heat and may score the shaft when rubbing occurs, whereas excessively soft carbon will produce a packing that is easily damaged either by handling or in service. Carbon glands are sensitive to any dirt particles that pass through the clearance space and to roughness of the shaft surface. Because of maintenance problems, carbon glands are not widely used for main propulsion units or ship service turbine-generator sets.

Section 8
Lubrication and Bearings

8.1 Lubrication System. Oil is used to lubricate the turbine journal and thrust bearings and the main reduction gear, and to furnish power to valve-operating mechanisms and protective devices. There is no direct contact between the lubricating oil and the steam and, therefore, with proper design, there is no danger of contaminating the boiler feed water.

The gravity-tank system frequently used in merchant marine applications uses an overhead tank from which the bearings, oil sprays, etc. are fed by gravity. This tank is located high enough to ensure at least 10 psig pressure at the highest bearing, and is large enough to supply oil for three to five minutes when the propulsion unit is operating at maximum power. With a gravity tank, a failure of the independently driven lubricating-oil service pumps does not immediately interrupt the flow of oil to the bearings, and time is available to take corrective steps before major damage occurs.

An alternative system, sometimes referred to as a pressure system, has no overhead tank, but pumps the oil to a constant-pressure head. This system is frequently used for naval vessels where the overhead tank would constitute a fire hazard in the event of battle damage and where the necessary space and elevation are not generally available. When operating at higher speeds, some improvement in reliability may be obtained by driving the lubricating-oil service pump from the reduction gear or the main shaft, with independently driven standby pumps for lower speeds, astern, and emergency operation.

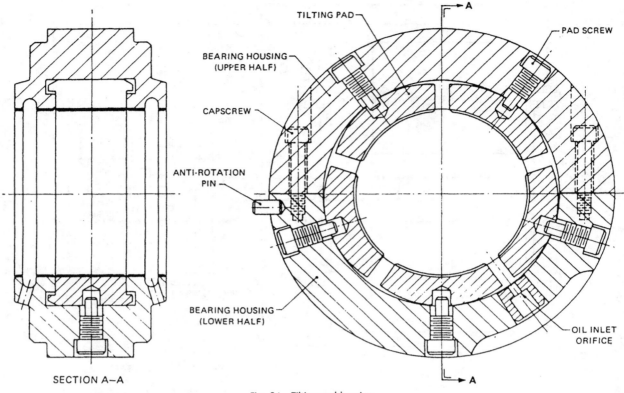

TILTING PAD

PAD SCREW

BEARING HOUSING
(UPPER HALF)

CAPSCREW

ANTI-ROTATION
PIN

BEARING HOUSING
(LOWER HALF)

OIL INLET
ORIFICE

SECTION A–A

Fig. 54   Tilting-pad bearing

Still another version of the pressure system, sometimes used for merchant vessels, compensates for the elimination of the gravity tank and its emergency supply by fitting a reduced-capacity, direct-current, motor-driven emergency pump capable of limited operation by storage battery.

Turbine manufacturers generally do not specify the characteristics of lubricating oil in great detail, but instead recommend only viscosity and the use of a reliable, recognized type of marine oil. For turbines a viscosity range of 150–250 SSU at 100 F would be satisfactory; however, if mechanical reduction gears are lubricated by the same system, a heavier oil having a viscosity of 350–400 SSU at 100 F is desirable. The oil should be free of water, sediment, acids, soap, resins, and other injurious materials. There should be no tendency toward permanent emulsification, or rapid oxidation with the formation of sludge.

The oil pressure typically is 30 to 50 psig at a point ahead of the oil coolers and 10 to 15 psig at turbine bearings. The required oil quantity is dependent upon many factors such as bearing size, speed, clearance, and internal arrangement, as well as the gear design.

A vacuum usually exists in the unloaded half of a high-speed sleeve-type journal bearing. The vacuum draws into the oil film a significant quantity of air that often appears as very small bubbles in the drain oil. This entrained air can disturb pump and hydraulic governor operation, and

become a semipermanent froth with undesirable types of oil. A large sump capacity, equivalent to three to five minutes' supply for the lubricating-oil service pumps, and a sump arrangement, which gives as much opportunity as possible for the release of entrained air, are necessary to ensure reliable operation.

The outlet temperature of the oil from the cooler should not exceed 130 F. Incoming oil to the cooler is generally in the range of 140–160 F.

8.2   Journal Bearings.   A turbine rotor is supported in bearings, which carry the weight of the rotor and maintain the correct radial clearances between the rotor and casing. Normally there is a bearing at each end of the rotor, except in the case of some auxiliary units that use overhung turbine wheels.

Two types of sleeve bearings are in common use, namely, cylindrically and spherically seated. The length-to-diameter ratios of turbine bearings lie in the range of 0.8 to 1.75; more efficient performance is obtained with the lower ratios. The load per square inch of projected bearing area is in the range of 100 to 200 psi. The clearance between the journal and the bearing is kept as small as possible to allow accurate radial positioning of the rotor, and is typically 1.5 to 2 mils per inch of journal diameter.

Figure 19 illustrates a typical fixed-sleeve bearing. Spherically seated bearings are shown in Figs. 3 and 4. The latter type, sometimes called a self-aligning bearing, permits a small amount of shell movement to compensate for minor misalignment with the shaft. This feature is generally effective only for adjustments during assembly, because for many designs the bearing cap has such a tight fit that movement in service is unlikely.

Bearings normally wear at a very slow rate; indeed, the initial surface is sometimes visible after years of service. Wear occurs chiefly during starting, before an oil film is established, but is also caused by foreign particles in the oil. The amount of wear may be determined by the use of special bridge gages placed at each journal to confirm the proper rotor position, by measurements of the bearing shell thickness at prescribed points on the circumference, or by a depth micrometer mounted on the bearing cap.

Multishoe, tilting-pad radial bearings have become increasingly common in steam turbine applications [21]. This type of bearing resembles the well-known multishoe tilting-pad thrust bearing, except that it is wrapped around a circular shaft. Several identical shoes are enclosed in a circumferential space filled with oil. As illustrated by Fig. 54, each pad is pivoted either by line or point contact to tilt and form a wedge-shaped film of oil to support the rotating shaft. This type of bearing has high internal damping and, because of its inherent stability and freedom from oil-film whirl and other vibrational phenomena, it has been used to replace sleeve bearings when such difficulties have become apparent. This is particularly true in the case of lightly loaded high-speed bearings.

8.3 Turbine Thrust Bearings.

To hold the turbine rotor in the correct axial position relative to the stationary parts and to absorb the unbalanced axial forces imposed upon the rotor, some form of thrust bearing is necessary.

In a reaction type of turbine, a large portion of the total pressure drop occurs in the moving blades and produces a sizable axial force, which is counteracted by a dummy piston that is designed to produce an axial force in the opposite direction. The diameter of the dummy piston is selected to give a net axial thrust such that the calculated loading of a thrust bearing of reasonable size will be conservative. It must be recognized that the net thrust is the difference of two large quantities; therefore, a small-percentage variation in either quantity will produce a large-percentage change in the thrust loading.

In the impulse type of turbine, the pressure drop across the blades is much less, but due to the large total area of all the wheels, even a small pressure drop across each will produce a fairly large total force. In addition, the diaphragm packings sometimes increase in diameter from the high-pressure to the low-pressure stages and produce additional axial forces. The diameter of the packing at the high-pressure end may be adjusted to produce a net thrust, which can be carried by the thrust bearing. To reduce the pressure drop across the wheels, it is common practice to drill several steam pressure equalizing holes in each wheel. These holes are reamed and the edges carefully polished to avoid stress concentrations and the development of cracks. Pressure equalizing holes are not necessary in double-flow turbines; the thrust from one end of a double-flow turbine counterbalances the thrust

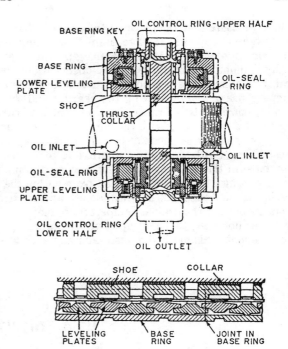

Fig. 55 Multishoe-type thrust bearing

from the other end, and the thrust bearing need only hold the rotor in position.

The most commonly used type of turbine thrust bearing is the pivoted segmental shoe or Kingsbury type. Figure 55 illustrates such a bearing and shows the basic elements, which consist of a collar, which may be either separate from, or integral with, the shaft, and stationary segments or shoes, which bear against the collar and absorb the thrust. The shoes are free to tilt and form a wedge-shaped hydrodynamic oil film. The leveling plates distribute the load equally among the shoes. The load that can be carried safely by a thrust bearing depends upon the bearing size, shaft speed, and oil viscosity. The load per square inch of bearing surface is often used as a design criterion, although it does not account for the last two variables. At turbine speeds, pressures of up to 1000 psi have been demonstrated successfully in tests; however, turbine thrust bearings are seldom continuously loaded above 250–400 psi during normal steady-state operation.

Some multishoe marine thrust bearings operate with complete submergence, the oil being admitted in the bottom half and arranged to discharge through a controlled area in the top half. This type of arrangement is called "flooded." An "unflooded" thrust bearing has the discharge in the bottom half radially outward from the thrust collar; this design has been shown to result in lower power loss and a cooler-running bearing.

8.4 Pressure and Temperature Indication. To confirm oil flow, an inlet pressure gage, sight-flow fitting, or combination thereof may be used. Bearing performance generally is monitored by temperature observations. In general, two approaches are possible; one is to measure the temperature of the oil leaving the bearing and the other is to measure the temperature in way of the load-carrying region of the bearing. With the former, the temperature of an oil mixture is measured, only part of which has passed through the load area where, due to shearing action, most of the heat is added. This may be accomplished by extracting a small amount of oil from within the bearing by special passages in the babbitt and shell or by measurement of the drains. Temperatures may be sensed by a thermometer, thermocouple, or resistance temperature element. This method gives readings which are relatively stable and can be arranged so that temperatures do not normally exceed 180 F. The absolute reading, however, is not as important as a sudden rise in temperature with no corresponding change in operating conditions, since this would normally signify abnormal operation such as an internal failure or a wiped bearing.

Alternatively, the temperature of the oil film at the load point, or the temperature of the bearing metal at this point, may be taken by a thermocouple or resistance temperature element embedded in the babbitt [22,23]. Temperatures measured in this manner are generally in the range of 180 to 250 F for a loaded thrust bearing. If the temperature is higher than 250 F, or if it is abnormally high, there is cause for concern. For journal bearings, the typical temperature range is 180 to 230 F.

Section 9
Main Propulsion Turbine Operation

9.1 Preliminary Procedures. The procedures for starting, operating, and shutting down a main propulsion turbine unit vary somewhat depending upon the particular installation; however, the following procedures apply generally to any conventional, geared-turbine marine set.

Before starting a turbine, the following checks are made:

(1) Ensure that there is an adequate amount of lubricating oil in the system.
(2) Ensure that the temperature of the oil supplied to the unit is at least 90 F before the unit is turned.
(3) Remove any water accumulated at the bottom of the supply, gravity, and sump tanks as applicable.
(4) Check the availability of circulating water to the oil cooler. If the oil temperature from the cooler is manually controlled, do not turn on the circulating water until necessary.
(5) Start the lubricating-oil pump and make sure that oil is flowing freely to all bearings.
(6) Use the rotor position indicators to establish that the turbine rotors are in their proper positions.
(7) Check with the bridge to ensure that there are no obstructions in the way of the propeller.
(8) Engage and start the motor-driven turning gear. Check to see that turbines and gears are rotating freely. Listen for any unusual sounds of rubbing or other indications of trouble. Continue rolling the turbines with the turning gear.
(9) Make certain that the main steam line stop valves to the propulsion unit are closed and tight. Check to be sure that there is no steam pressure in the steam line to the turbines, then open wide and close the ahead throttle, astern throttle, and astern guarding valves to make sure that they operate properly before steam is admitted to the turbines. After checking, close the valves.

9.2 Warming the Turbines. Although it is probable that a turbine, if its rotor is straight, can be started from a cold condition without warming up, such operation does not contribute to continued successful operation and should be done only in an emergency. The following procedure should be observed after the preceding preliminary steps:

(1) Open all drains on main steam lines, turbine casings, and gland steam seal lines to remove condensation.
(2) Start the gland exhauster fan and place the steam sealing system in operation to maintain a seal pressure of ½ to 1½ psig.
(3) Put the condensing equipment in service to provide a vacuum of about 10 in. Hg.; higher vacuum during the warming period would lengthen the time required to heat the turbine, and might cause inequalities in temperature that would result in distortion.
(4) Continue the use of the turning gear for about 15 minutes.
(5) Obtain clearance from the bridge to run the main engine.
(6) Stop the turning gear, disengage, and secure.
(7) Open the main steam stop valve and admit steam to the ahead and astern throttle valves.
(8) Alternately open the ahead and astern throttle valves sufficiently to turn the unit slowly for a few seconds, first in one direction and then in the other. This will tend to avoid putting way on the ship, straining dock lines, or riding up over the anchor if moored. Repeat this spinning procedure every five minutes for at least 20 to 30 minutes if possible, or until the ship is ready to get underway.
(9) Before getting underway bring the vacuum to normal.

9.3 Standing By. Keep the turbine drains open and continue turning the unit ahead and astern at intervals not exceeding five minutes. With steam supplied to the gland seal system, the turbine rotors should not be allowed to remain stationary for more than five minutes.

The continued operating efficiency of steam turbines depends upon the maintenance of the relatively small radial packing clearances. These clearances can be maintained at the designed values only by keeping the turbine rotors straight, and to preserve this condition it is necessary to maintain a uniform temperature around the circumference to the rotor. When a rotor is stationary, the heat tends to be concentrated on the top. This uneven heating results in temporary bowing or distortion of the rotor with consequent rubbing and wear of the packing when rotation is resumed.

9.4 Underway at Sea. When maneuvering has been completed and the ship is proceeding to sea, close all turbine drains and make sure that the astern throttle and the astern guarding valve are tightly closed; then the unit may be brought up to about half speed. Higher speeds should be reached gradually, and at least 15 minutes should be taken to reach full speed, except in emergencies.

Check the axial location of the turbine rotors with the rotor position indicators, and repeat at least once each day with measurements taken at approximately the same rpm.

Check pressures, temperatures, oil levels, and flow indicators periodically to see that they remain normal. If any bearing shows signs of excessive heating or unexplained temperature changes, slow down immediately and ascertain the cause.

Inspect and clean oil strainers regularly. Foreign matter found in the strainer should be examined and traced to its source. Traces of water should be removed by operation of the purifier. At regular intervals oil samples should be analyzed to check Ph, viscosity, additives, water content, and other properties.

In the event that the lubricating-oil pressure is lost for any reason, the low oil pressure trip will shut off the ahead steam. If the vessel is underway ahead, it will continue to coast for some time. Due to the hydrodynamic action of the water on the propeller, the propeller will continue to turn in the ahead direction and will rotate the engine. To avoid bearing failures, it is extremely important that shaft rotation be stopped by the use of astern steam until the vessel stops or oil pressure is restored.

The inlet steam conditions should be periodically monitored. If an abnormally high inlet steam temperature is permitted over an extended period, damage may result. If the inlet temperature is too low, then moisture erosion will increase in the last stages of the low-pressure turbine.

The operator should be constantly alert for any abnormal change in noise level, for unusual sounds, and for indications of increased vibration, particularly during maneuvering. If such are noted, slow down until the noise or vibration disappears. Operate for 10 to 15 minutes at this reduced speed, then slowly increase speed, taking at least another 15 minutes to reach operating power.

If the rotor becomes temporarily bent due to thermal conditions and rubs on the packing strips, heat will be generated at the shaft surface on a small segment of its circumference. This will increase the shaft distortion and cause a harder rub, which will generate additional heat such that the rub becomes progressively worse, possibly resulting in a severe casualty. Hence, it is necessary to slow down, allow time for temperatures in the shaft to equalize, and thus permit the shaft to straighten.

9.5 Prolonged Astern Operation. Main propulsion steam turbines designed for merchant ships generally are capable of continuous astern operation at 70% of the ahead speed for one hour without danger of rotation losses causing overheating of the idle ahead blading. This performance is contingent upon the exhaust vacuum being at or near the design value. In addition, there must be no steam leakage into the ahead turbine through the ahead throttle or extraction valves. If temperatures in the crossover pipe and high-pressure turbine exceed allowable values, the speed should be reduced.

It should be noted that if the inlet steam temperature is constant, the astern exhaust temperature will rise with a drop in speed since the exhaust is superheated and the turbine efficiency decreases.

9.6 Securing the Turbine. The following procedure should be followed when securing a turbine:

(1) Close all turbine control valves and valves in the main steam line to the turbine.

(2) Open all turbine drains.

(3) Engage and start the turning gear. This allows the turbine rotors to cool uniformly while the oil circulation enables the heat transmitted through the shafts to be carried away from the bearings and thus avoid possible damage to the babbitt lining.

(4) Secure the gland sealing and exhaust systems.

(5) Keep the condenser circulating and condensate pumps in operation at minimum speeds until the turbines are drained, then secure.

(6) Secure the first-stage air ejector jets but leave the second-stage jets in service for a few hours to draw air through the turbines. This should be repeated every two or three days to keep the turbines dry.

(7) When the turbines have cooled sufficiently to avoid bowing, secure all associated equipment.

(8) Circulate oil and operate the turning gear every two or three days in port, covering all applicable parts with lubricant to prevent rusting.

9.7 Emergency Operation. If either turbine of a cross-compound unit is damaged to the extent that it cannot be operated, the other turbine can be run on high-pressure steam by rearranging the steam and exhaust connections as necessary. The damaged turbine is disconnected from the reduction gear and remains idle.

When operating with the high-pressure turbine alone, a special pipe is provided to exhaust directly to the condenser. If the astern turbine is located in the low-pressure turbine, and this is usually the case, no astern operation is possible and the astern throttle should be wired shut to prevent its being opened by mistake.

When the high-pressure turbine is out of service, high-pressure steam is admitted directly to the inlet of the low-pressure turbine and controlled by a valve in the supply line, which serves temporarily as a throttle. An orifice is generally fitted after the valve to limit the pressure to an allowable value.

The power output is reduced not only by the decreased turbine efficiency but also by consideration of the gear loadings when operating with a single turbine. Gears driven by a single turbine are loaded to design torque values when the propeller speed is about 70% and the corresponding power about 35% of the normal ahead rating. It is generally recommended that the emergency speed should not exceed 70–75% of the normal ahead rating.

Section 10
Auxiliary Turbines

10.1 Introduction. For steam power plants, steam turbines are also commonly selected as prime movers for auxiliaries such as electric generators, feed pumps, and the cargo-oil pumps of tankers. Many of the basic principles of steam turbine design and construction outlined in previous sections apply generally to the smaller units, but the design criteria may be modified because of the reduced power output and because of economic considerations. Some of the more important considerations in this regard are discussed in the following.

10.2 Ship-Service Turbine Generators. The ship-service turbine generator (SSTG) provides electric power for the operation of motors, lighting, communications, and hotel services. The electrical generating capacity required for a particular vessel depends upon its type, its size, and its propulsion power, but in most cases the electric power requirements can be met by the selection of a unit from a series of standard ratings, which is as follows: 500, 600, 750, 1000, 1250, 1500, 2000, and 2500 kW. These ratings have been chosen for standardization purposes, as it minimizes the number of frame sizes required to be offered by manufacturers and thus reduces development costs. It is possible, of course, to design and build nonstandard units of any size that may be required, both above and below this standard range.

A typical SSTG consists of a high-speed, multistage condensing turbine driving a generator through a single-reduction gear. These components are mounted on a bedplate together with the turbine drain, lubricating oil, gland seal, and gland exhaust systems to form an integral unit. The turbine may exhaust to the main condenser or to an auxiliary condenser. When the auxiliary condenser is supported by the bedplate, the turbine-gear-generator-condenser assembly is called a "packaged unit."

Savings in initial costs may be realized by the installation of a multistage turbine for normal service and a single-stage, back-pressure-type turbine for standby service in lieu of two multistage units.

Steam and vacuum conditions for the SSTG normally are the same as for the propulsion plant. A single casing, multistage condensing turbine generator has a lower efficiency than the main unit primarily because its rated output is much smaller. An SSTG turbine generally consists of five to eight impulse stages and operates at 8000 to 12,000 rpm. A cross section of a typical multistage condensing turbine is shown in Fig. 56. The performance that may be expected from properly designed multistage turbines is indicated in Fig. 57.

Accurate control of speed is essential to maintain constant and correct frequency in an a-c electrical system. Speed control is accomplished by regulating steam flow to the unit as directed by a control system utilizing the input from a speed sensor. Although there are several types of sensors and systems, in each case the flow regulation is achieved by the operation of nozzle control valves supplying steam to the first stage of the turbine.

A simple mechanical system is shown in Fig. 58(a), in which a flyweight assembly senses shaft speed. Two flyweights are mounted on a plate, which turns about a vertical axis driven by the turbine shaft through a worm and gear. The centrifugal force throws the weights outward and then compresses the stationary spring, thus lifting the vertical rod and moving the linkage and control valve until an equilibrium position is reached that corresponds to the speed. The speed setting may be changed by adjustment of the speed changer. This simple form of governor is used for small mechanical-drive turbines but does not have sufficient force to operate large steam-control valves. To overcome this difficulty, a pilot valve and servomotor may be added as shown in Fig. 58(b) to form a mechanical-hydraulic system. The vertical rod operates the pilot valve to admit (or drain) high-pressure oil to (or from) the spring-loaded servomotor cylinder. As the servomotor piston responds, it tends to restore the pilot valve to the neutral position. In some cases the pilot valve is double ported, and high-pressure oil is directed to either the top or bottom of the servomotor piston as required.

With any of the preceding arrangements, the speed will vary slightly with load. The difference in speed between rated load and no load divided by the rated speed is called the "regulation" or "speed droop" and is usually about 3 to 4%. The amount of friction in the mechanism is important. The speed change above and below a mean required to produce corrective action is termed the "dead band" and is a measure of the "sensitivity." When a sudden change in load occurs and the governor overcorrects followed by undercorrection, perhaps continuing for several oscillations, the action is called "hunting." A certain amount of regulation is essential to minimize hunting. As

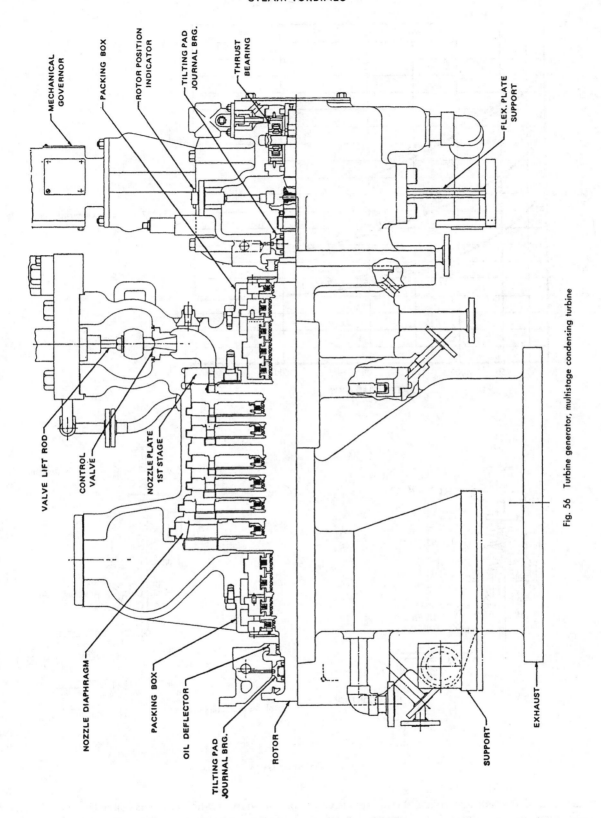

MECHANICAL GOVERNOR

PACKING BOX

ROTOR POSITION INDICATOR

TILTING PAD JOURNAL BRG.

THRUST BEARING

FLEX. PLATE SUPPORT

VALVE LIFT ROD

CONTROL VALVE

NOZZLE PLATE 1ST STAGE

NOZZLE DIAPHRAGM

PACKING BOX

OIL DEFLECTOR

TILTING PAD JOURNAL BRG.

ROTOR

SUPPORT

EXHAUST

Fig. 56 Turbine generator, multistage condensing turbine

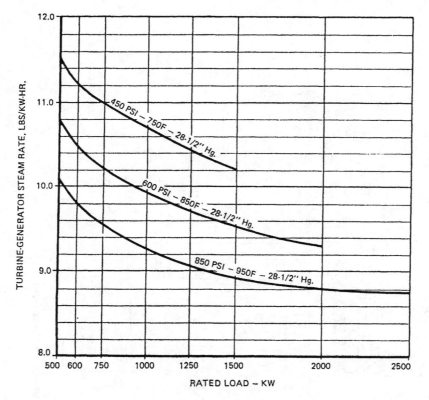

Fig. 57   Turbine-generator steam rate

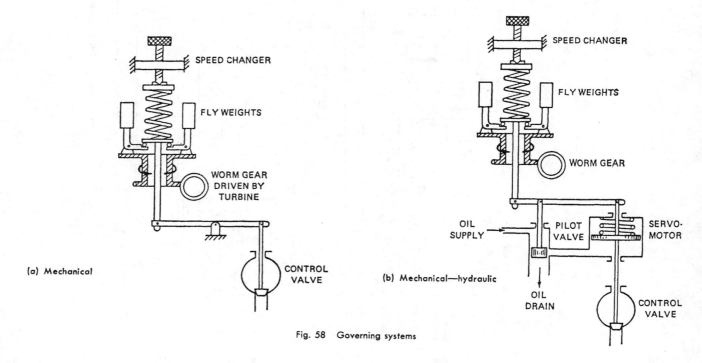

Fig. 58   Governing systems

both regulation and friction are reduced, the sensitivity is increased; however, the stability is decreased, thus a compromise is usually necessary, with the regulation being kept as small as stability or freedom-from-hunting considerations will allow. When a-c generator sets are operated in parallel, it is necessary that each speed governor be adjusted for the same speed regulation if each set is to take an equal share of the load regardless of the load variation.

In addition to the flyweight or mechanical type of speed

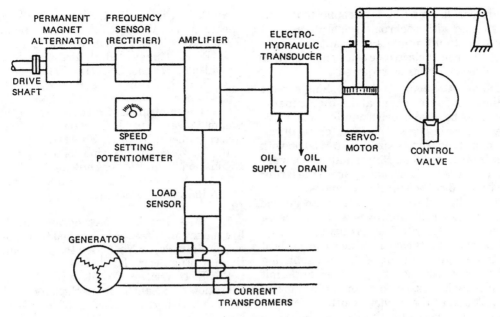

Fig. 59 Electric governor

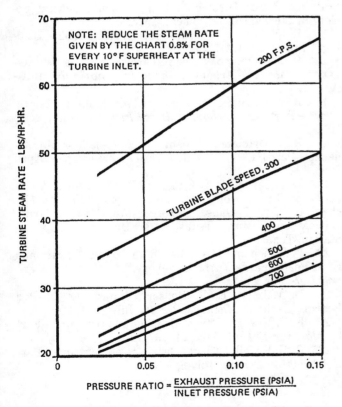

NOTE: REDUCE THE STEAM RATE GIVEN BY THE CHART 0.8% FOR EVERY 10°F SUPERHEAT AT THE TURBINE INLET.

200 F.P.S.

TURBINE BLADE SPEED, 300

400
500
600
700

TURBINE STEAM RATE – LBS/HP-HR.

PRESSURE RATIO = $\dfrac{\text{EXHAUST PRESSURE (PSIA)}}{\text{INLET PRESSURE (PSIA)}}$

Fig. 60 Turbine steam rate for single-stage (2-row) auxiliary units

sensor, a hydraulic pump driven from the turbine shaft may be used in a hydraulic system. This pump may be either a positive-displacement type or a centrifugal type and the system may be similar to the speed-limiting governor for main propulsion units described in Section 4.2

With a third basic type of governor, speed or frequency control is accomplished by a combination of electric and hydraulic components (see Fig. 59). The speed signal is obtained from the frequency of a small permanent-magnet alternator driven by the turbine rotor or speed-sensing magnetic pickup; its a-c voltage impulses are converted into a d-c voltage, which is proportional to speed. A reference d-c voltage of opposite polarity, which is representative of the desired operating speed, is established by manual adjustment of a speed-setting potentiometer. These two voltages are connected to the input of an electronic amplifier. If the two voltages are equal and opposite, as occurs during steady-state operation, they cancel and there is no voltage input to the amplifier and, therefore, no change in its output voltage. The amplifier output voltage drives an electro-hydraulic transducer, which directs the flow of oil to a servomotor that adjusts the governor steam valves to maintain the turbines at the speed corresponding to the position of the speed-setting potentiometer. If the turbine speed changes, the speed signal frequency and, therefore, the voltage supplied to the amplifier change. The difference between this voltage and the reference voltage is supplied to the amplifier. The amplifier then supplies an output voltage to the electro-hydraulic transducer, which causes the steam valves to increase or decrease the steam flow to return the turbine speed to the set value. Stability is achieved by a time delay in the negative feedback around the amplifier. Since there is only one speed at which the speed signal and reference voltages are equal and opposite, this type of control is "isochronous"; that is, it maintains the same turbine speed regardless of load variation. A load sensor measuring current in each lead of the generator is used to anticipate speed changes and thus improve the dynamic response of the control system. The load sensors of several similar units operating in parallel may be interconnected to ensure equal load sharing with isochronous operation. Operation in parallel with an infinite bus or dissimilar governors is possible by the use of electronic components, which introduce droop characteristics as required.

10.3 Single-Stage Auxiliary Turbines. Single-stage turbines, sometimes called mechanical-drive or general-purpose turbines, may be used to drive pumps, fans, blowers, and standby generating sets. The need for small turbines has resulted in standardized sizes up to 1500 hp with wheel diameters from 12 to 36 in. Rotational speeds vary from 600 to 7200 rpm; the lower speeds apply to the larger wheel sizes used with direct-connected turbines and the higher speeds to smaller wheels associated with geared units. The efficiency generally improves with increasing blade speed, as shown in Fig. 60. There is usually a large energy drop with auxiliary turbines, which can be best handled in a two-row stage using nozzles of the expanding type. Individual hand valves may be provided to permit the opening and closing of nozzles to accommodate major changes in load. The speed governor is often mounted on the turbine shaft and acts directly through levers to actuate the inlet valve, usually a balanced single-seated throttle valve. Speed regulation is generally in the range of 5 to 6%. Mechanical-drive turbines are designed as complete units arranged for coupling to the driven unit.

A close-coupled, integral type of turbine-driven pump, consisting of a single-stage steam turbine and a single- or two-stage centrifugal pump mounted on the same shaft, has been widely applied for boiler feed, fire, and tank cleaning services. Packaged units of this type are supplied with a forced-feed lubricating system, speed controls for either constant or differential pressure regulation, speed limiting governor, and back-pressure trip.

References

1   John F. Lee, *Theory and Design of Steam and Gas Turbines*, McGraw-Hill Book Company, Inc., New York, 1954.

2   E. F. Church, Jr., *Steam Turbines*, McGraw-Hill Book Company, Inc., New York, 1950.

3   J. K. Salisbury, *Steam Turbines and Their Cycles*, John Wiley & Sons, Inc., New York, 1950.

4   R. M. Cashman, "Design of Marine Machinery Foundations," *Trans.* SNAME, 1962.

5   B. Fox and R. H. Tingey, "A 1200 Pound Reheat Marine Installation," *Trans.* SNAME, 1941.

6   R. P. Giblon, W. I. Signell, N. A. Smith, H. C. K. Spears, and C. W. Stott, "A Modern Steam Reheat Power Plant," New York Section, SNAME, Oct. 1965.

7   R. H. Tingey, "High Pressure Steam for Marine Propulsion," *Trans.* SNAME, 1943.

8   H. F. Robinson and E. P. Worthen, "The Ore Carrier S.S. *Venore*," *Trans.* SNAME, 1945.

9   "Recommended Practices for Preparing Marine Steam Power Plant Heat Balances," Technical and Research Bulletin No. 3-11, SNAME.

10   C. W. Stott, "Marine Reheat Cycles and Systems Evaluation," *Marine Technology*, Vol. 7, No. 3, July 1970.

11   R. P. Kroon, "Turbine-Blade Vibration Due to Partial Admission," *Journal of Applied Mechanics*, Dec. 1940.

12   M. A. Prohl, "A Method for Calculating Vibration Frequency and Stress of a Banded Group of Turbine Buckets," ASME Paper No. 56-A-116, American Society of Mechanical Engineers, Nov. 1956.

13   F. L. Weaver and M. A. Prohl, "High Frequency Vibration of Steam-Turbine Buckets," ASME Paper No. 56-A-119, American Society of Mechanical Engineers, Nov. 1956.

14   W. Campbell and W. C. Heckman, "Tangential Vibration of Steam Turbine Buckets," *Trans.* American Society of Mechanical Engineers, 1925.

15   R. W. Nolan, "Vibration of Marine-Turbine Blading," *Trans.* SNAME, 1949.

16   R. J. Ortolano, "Steam Turbine Blading Maintenance—Part I," *Turbomachinery International*, July-Aug. 1979.

17   R. J. Ortolano, "Steam Turbine Blading Maintenance—Part II," *Turbomachinery International*, Sept. 1979.

18   R. J. Ortolano, "Steam Turbine Blading Maintenance—Part III," *Turbomachinery International*, April 1983.

19   W. J. Caruso, B. E. Gans, and W. G. Catlow, "Application of Recent Rotor Dynamics Developments to Mechanical Drive Turbines," *Proceedings*, 11th Turbomachinery Symposium, Texas A&M University, College Station, Texas.

20   F. Ehrich and D. Childs, "Self-Excited Vibration in High Performance Turbomachinery," *Mechanical Engineering*, May 1984.

21   D. Wilcock, E. R. Booser, "Improving Tilt-Pad Bearing Design," *Machine Design*, July 12, 1990.

22   R. C. Elwell, "Thrust-Bearing Temperature—Part I," *Machine Design*, June 24, 1971.

23   R. C. Elwell, "Thrust-Bearing Temperature—Part II," *Machine Design*, July 8, 1971.

CHAPTER VII

K. L. Harrington

Nuclear Marine Propulsion

Section 1
Fundamental Principles

1.1 Scope. The purpose of this chapter is to present a survey of the application of nuclear energy to marine propulsion. The chapter is directed towards persons having an engineering background but no experience regarding nuclear reactors. Fundamental atomic theory and nuclear reactor theory are described, and an overview of several nuclear reactor designs is included. For a more detailed treatment of highly specialized considerations of reactor engineering, including fuel design, reactor design, coolant chemistry, and nuclear instrumentation, reference is made to comprehensive textbooks that are available on the subject [1–7].

In order to avoid security classification problems and problems with 15 CFR Part 385, U.S. Export Regulations, primary emphasis is placed on nuclear applications to commercial merchant ships, in contrast to military applications. However, a discussion of the differences in these requirements is included in a paper presented to the Society by ADM H. G. Rickover et al [8].

1.2 Introduction. Only one merchant vessel, the NS *Savannah*, has been built in the United States using nuclear power for propulsion. The *Savannah* was built as a part of the Atoms for Peace Program to demonstrate the feasibility of nuclear propulsion in commercial applications, and traveled over 450,000 miles visiting many countries before being retired from active service in 1971.

Although nuclear propulsion is more attractive as a general concept for merchant ships having long trade routes, even in this case nuclear propulsion is not economically competitive with fossil-fuel propulsion plants under the prevailing technical, economic, and political conditions. The initial construction cost of a nuclear-powered ship can exceed the construction cost of a conventional ship by 50% or more. In addition, the extensive and prolonged nuclear plant licensing process can intolerably extend a construction schedule. With the large difference in initial capital requirements, the potential savings associated with long trade routes and high speeds may not be of sufficient magnitude to make nuclear propulsion economically viable. In addition, nuclear ships entail a final major expense that is associated with decommissioning the vessel and disposing of the radioactive materials.

One of the most significant characteristics of nuclear power for maritime applications is the compact nature of the energy source. The power produced by the fission of one gram of uranium-235 per day is equivalent to about one megawatt. In other terms, the fission of one pound of uranium-235 is equivalent to the combustion of 900 tons of 18,500 Btu/lb fuel oil or 1,100 tons of 15,000 Btu/lb coal. Refueling for a nuclear ship occurs during a scheduled overhaul period. The NS *Savannah* traveled over 350,000 miles before having a partial refueling overhaul and another 100,000 miles after the partial refueling was accomplished.

Safety is a major consideration with nuclear reactors because of the emission of radiation, consisting primarily of neutrons and beta and gamma radiations, from the fissions and fission products. Furthermore, the fission-product radiation must be considered for a long time after the reactor is shut down and the spent fuel elements are removed. Operating personnel must be protected from the radiation by suitable shielding; the shielding may consist of lead, water, steel, concrete, and other radiation-absorbing materials. This shield adds considerably to the size and weight of the reactor, and represents one of the significant engineering problems in the design of a marine nuclear power plant.

In addition to plant safety, accessibility for maintenance and the provision of a sufficiently high level of reliability to ensure a long service life must be considered not only in the formulation of the basic concept of the plant, but also in the procurement of equipment and components. Strict adherence to codes and standards and compliance with rigorous quality assurance programs during construction are the means used to ensure plants of high quality.

A major engineering consideration for nuclear reactors is to provide under all circumstances for the removal of heat from the nuclear fuel. One of the significant differences between nuclear and fossil-fuel propulsion is the problem of decay-heat removal after reactor shutdown. Unlike an oil-fired boiler or combustion turbine, the nuclear reactor cooling circuits must be operated and monitored for an extended period after shutdown for the safe removal of decay heat.

The high power density potential of the fuel and its theoretical potential to release its contained energy in a short time result in the necessity to provide efficient, highly reliable, and sometimes unusual heat-transfer systems not only for the steady-state power operation and normal heat removal after shutdown, but also for all emergency and credible accident conditions.

There is also an inherently larger annual operating cost associated with nuclear-powered ships as well as an increased cost of insurance (in recognition of the higher value of the ship), additional insurance to cover third-party liability, additional expenses required for special nuclear material handling facilities, and the additional expense required to maintain a nuclear-qualified crew. The large crew expenses are due both to the increased compensation required to attract highly capable personnel and the time involved in training the operators to comply with the strict regulations necessary to operate a nuclear reactor safely.

Another problem associated with nuclear merchant-ship trade routes is that political climates and social attitudes in some countries may restrict port entry. The complex legal, political, and social situations concerning the operation of nuclear ships will have to be clarified and defined before commercial operators will have any incentive to expose themselves to the liabilities that nuclear energy currently entails. For a more detailed review of the non-technical factors that impact the commercial application of nuclear propulsion, see references 9 and 10.

Economic studies indicate that the cost penalties associated with nuclear propulsion are of such a magnitude that further innovations will be required before nuclear power will be able to economically compete with fossil-fueled power systems for ship propulsion; therefore, nuclear power is attractive only where the advantages of high power, endurance, or submerged operations capabilities override purely economic considerations. This is often the case for military purposes; and in some commercial situations, these advantages can be of substantial importance. Reference 11 is a report of feasibility studies to transport Arctic crude oil by icebreaking tankers, which require high propulsion power, and by submarine tankers, which require a substantially lower power but also require an extended under-ice operation capability. The feasibility of transporting liquefied natural gas (LNG) from the Arctic by nuclear submarine tankers was investigated and reported in reference 12. The conclusions reached were that an under-ice transport system would be safe and reliable and would permit a scheduled movement of natural gas that is unimpeded by surface weather conditions.

In the event that technological changes in shipping procedures occur which place an increased importance upon higher ship speeds through total system integration, it is possible that an increased utilization and higher propulsion power requirements could tend to make nuclear propulsion more attractive for commercial applications on a broad scale.

Aside from the wealth of operating experience accumulated in naval nuclear-powered ships, commercial ships such as the *Savannah* have provided a baseline against which safety issues can be judged. Although no nuclear commercial ship ventures are expected in the near term, the long-term scarcity of petroleum fuels could force the need to develop alternative means of propulsion that include nuclear power. Safety considerations relative to nuclear propulsion have become more stringent but better understood, and naval nuclear propulsion technology can be rapidly converted to meet commercial needs. Long-life reactor cores and improved system designs can facilitate the optimization of operation and maintenance costs; and future systems can be designed to be simple, reliable, and easily tested at a minimum cost while assuring both operational continuity and public safety.

1.3 Glossary. Many of the terms used in connection with nuclear science and its applications require definition, as their meanings are not self-evident. The following glossary, which was taken largely from reference 13, consists of definitions that have special meanings in nuclear applications.

Absorber. Any material that absorbs or diminishes the intensity of ionizing radiation. Neutron absorbers, such as boron, hafnium, and cadmium, are used in control rods and for the absorption of neutrons in reactor shields. A sheet of paper will absorb or attenuate alpha particles and a thin sheet of metal will stop all except the most energetic beta particles. Gamma radiations must be attenuated by high-mass materials.

Absorption. The process by which the number of particles or photons entering a body of matter is reduced by interaction of the particles with the matter; similarly, the reduction of the energy of a particle while traversing a body of matter.

Alpha decay. A radioactive process that produces a positively charged alpha particle.

Alpha particle. A positively charged particle emitted by certain radioactive materials. It is made up of two neutrons and two protons bound together, hence it is identical with the nucleus of a helium atom. It is the least penetrating of the three common types of radiation (alpha, beta, gamma) emitted by radioactive material.

Background radiation. The radiation in man's natural environment, including cosmic rays and radiation from the naturally radioactive elements, both outside and inside the bodies of men and animals. It is also called natural radiation. The nominal level of natural radiation is 100 millirem per year to each person. Background radiation may also refer to radiation that is unrelated to a specific experiment.

Beta decay. A radioactive process that produces a beta particle with a single unit charge and the mass of an electron.

Beta particle. An elementary particle emitted from a nucleus during radioactive decay, with a single electrical charge and a mass equal to 1/1,837 that of a proton. A negatively charged beta particle is identical to an electron. A positively charged beta particle is called a positron. All except the most energetic beta particles are easily stopped by a thin sheet of metal.

Binding energy. The binding energy of a nucleus is the minimum energy required to dissociate it into its component neutrons and protons. Neutron or proton binding energies are those required to remove a neutron or a proton, respectively, from a nucleus. Electron binding energy is that required to remove an electron from an atom or a molecule.

Biological dose. The radiation dose absorbed in biological material; it is measured in rems.

Biological shield. A mass of absorbing material placed around a reactor or radioactive source to reduce the radiation to a level that is safe for humans.

Boiling-water reactor. A reactor in which water, used as both coolant and moderator, is allowed to boil to remove heat from the core.

Breeder reactor. A reactor that produces fissionable fuel as well as consuming it, especially one that creates more than it consumes. The new fissionable material is created by capture in fertile materials of neutrons from fission. The process by which this occurs is known as breeding.

Breeding ratio. The ratio of the number of fissionable atoms produced in a breeder reactor to the number of fissionable atoms consumed in the reactor. Breeding gain is the breeding ratio minus one.

Burnable poison. A neutron absorber (or poison), such as boron, which, when purposely incorporated in the fuel or fuel cladding of a nuclear reactor, "burns up" (is changed into nonabsorbing material) gradually under neutron irradiation. This process compensates for the loss of reactivity that occurs as fuel is consumed and fission-product poisons accumulate, and keeps the overall reactivity characteristics of the reactor nearly constant during its life.

Capture. A process in which an atomic or nuclear system acquires an additional particle; for example, the capture of electrons by positive ions, or the capture of neutrons by nuclei.

Chain reaction. A reaction that stimulates its own repetition. In a fission chain reaction a fissionable nucleus absorbs a neutron and fissions, releasing additional neutrons. These in turn can be absorbed by other fissionable nuclei, releasing still more neutrons. A fission chain reaction is self-sustaining when the number of neutrons released in a given time equals or exceeds the number of neutrons lost by absorption in fissioning and nonfissioning material and by escape from the system.

Chemical shim. Chemicals, such as boric acid, that are introduced into a core via the coolant to control the reactor. Chemical shims are used in conjunction with mechanical control rods to reduce the number of rods needed and to create a more uniform power distribution in the core.

Cladding. Outer protective jacket of nuclear fuel elements. Cladding prevents the corrosion of the fuel, gives structural strength to the fuel, and prevents the release of fission products. Common cladding materials are aluminum alloys, stainless steel, and zirconium alloys.

Collision. A close approach of two or more particles, photons, atoms, or nuclei, during which such quantities as energy, momentum, and charge may be exchanged.

Containment. A gastight shell or other enclosure around a reactor to confine fission products and other radioactive materials that otherwise might possibly be released in the event of an accident.

Control rod. A rod, plate, or tube containing a material (hafnium, boron, etc.) that has a large neutron absorption cross section and hence readily absorbs neutrons. Control rods are used to control the power of a nuclear reactor. By absorbing neutrons, a control rod prevents the neutrons from causing further fission.

Core. The central portion of a nuclear reactor that contains the fuel elements, control rods, and passages for the moderator and coolant to flow through.

Critical energy. The minimum neutron energy that will overcome the binding energy of the nucleus and cause fission to occur.

Critical mass. The smallest mass of fissionable material that will support a self-sustaining chain reaction (multiplication factor equal to one) under stated conditions.

Cross section. The probability that a nuclear reaction will occur as an oncoming particle travels through an atom. It is a measure of the effective area presented by a target nucleus and is measured in barns. The cross-section value for a given atomic isotope will depend on the type of oncoming particle and the energy of the particle. Cross sections are calculated for fission, absorption, scatter, and capture.

Curie. The basic unit used to describe the intensity of radioactivity in a sample of material. The curie is equal to 37 billion disintegrations per second, which is approximately the rate of decay of 1 gram of radium. This unit was named for Marie and Pierre Curie, who discovered radium in 1898.

Decay heat. The heat produced in the core by the continuing decay of accumulated fission products after the reactor has been shut down. The energy from decay heat can equate to 7% of the reactor output under some conditions. Therefore, core cooling must be provided after the reactor is shut down; otherwise, the temperature rise could cause severe fuel plate damage.

Delayed neutrons. Neutrons emitted by radioactive fission products in a reactor over a period of seconds or minutes after a fission takes place. Fewer than 1% of the neutrons are delayed, the majority being prompt neutrons. Delayed neutrons are important considerations in reactor design and control.

Depleted uranium. Uranium having a smaller percentage of uranium-235 than the 0.7% found in natural uranium. It is obtained from the spent (used) fuel elements or as a residue from uranium isotope separation.

Deuterium. An isotope of hydrogen whose nucleus contains one neutron and one proton and is therefore about twice as heavy as the nucleus of normal hydrogen, which is only a single proton. Deuterium is often referred to as heavy hydrogen; it occurs naturally as 1 atom to 6,500 atoms of normal hydrogen. It is nonradioactive.

Direct-cycle reactor system. A nuclear power plant system in which the coolant or heat-transfer fluid circulates first through the reactor and then directly to an energy extraction device such as a turbine.

Doppler effect. The shift with temperature of the interaction rate between neutrons and reactor materials, including fuel rods, structural materials, and fertile materials. This shift can affect appreciably the reactivity of reactors.

Dose rate. The radiation dose delivered per unit time, which is typically measured in rems per hour.

Electron volt. The amount of kinetic energy gained by an electron when it is accelerated through an electrical potential difference of 1 volt. It is equivalent to 1.603×10^{-12} erg. It is a unit of energy, or work, not of voltage.

Enriched material. Material in which the percentage of a given isotope present has been artificially increased so that it is higher than the percentage of that isotope naturally found in the material. Enriched uranium contains more of the fissionable isotope uranium-235 than the naturally occurring percentage (0.7%).

Excess reactivity. The amount of reactivity that exceeds the amount needed to achieve criticality. Excess reactivity is incorporated into the reactor by using extra fuel to compensate for fuel burnup and the accumulation of fission-product poisons throughout core life.

Exclusion area. An area immediately surrounding a nuclear reactor where human habitation is prohibited to assure safety in the event of an accident or to assure safety during normal operations.

Excursion. A sudden, very rapid rise in the power level of a reactor caused by supercriticality. Excursions are usually quickly suppressed by the negative temperature coefficient of the reactor or by automatic control rods or by both.

Fast neutron. A neutron with kinetic energy greater than 100,000 electron-volts. Most neutrons released from fission are fast neutrons.

Fast reactor. A reactor in which the fission chain reaction is sustained primarily by fast neutrons rather than by thermal or intermediate neutrons. Fast reactors contain little or no moderator to slow down the neutrons from the speeds at which they are ejected from fissioning nuclei. Fast reactors are usually linked to breeder reactors.

Fissile material. While sometimes used as a synonym for fissionable material, this term has also acquired a more restricted meaning; namely, any material fissionable by neutrons of all energies, including (and especially) thermal (slow) neutrons as well as fast neutrons—for example, uranium-235 and plutonium-239.

Fission. The splitting of a heavy nucleus into two or more parts (which are nuclei of lighter elements), accompanied by the release of a relatively large amount of energy and generally one or more neutrons. Fission products are frequently radioactive. Fission can occur spontaneously, but usually is caused by incoming neutrons, or other particles.

Fission products. The nuclei (fission fragments) formed by the fission of heavy elements and the nuclides formed by the radioactive decay of fission fragments.

Fuel cycle. The series of steps involved in supplying fuel for nuclear power reactors. It includes mining, refining, enriching, the original fabrication of fuel elements, their use in a reactor, chemical processing to recover the fissionable material remaining in the spent fuel, re-enrichment of the fuel material, and refabrication into new fuel elements.

Fuel element. A rod, tube, plate, or other mechanical shape or form into which nuclear fuel is fabricated for use in a reactor.

Fusion. The formation of a heavier nucleus from two lighter nuclei (such as hydrogen isotopes), with the attendant release of energy.

Gamma rays. High-energy, short-wavelength electromagnetic radiation. Gamma radiation frequently accompanies alpha and beta emissions and always accompanies fission. Gamma rays are very penetrating and are best stopped or shielded against by dense materials, such as lead or depleted uranium. Gamma rays are essentially similar to X-rays, but are usually more energetic and are nuclear in origin.

Half-life. The time in which half the atoms of a particular radioactive substance disintegrate to another nuclear form. Measured half-lives vary from millionths of a second to billions of years.

Heavy water. Water containing significantly more than the natural proportion (one in 6,500) of heavy hydrogen (deuterium) atoms to ordinary hydrogen atoms. Heavy water is used as a moderator in some reactors because it slows down neutrons effectively and also has a low cross section for the absorption of neutrons.

Intermediate (epithermal) neutron. A neutron having energy greater than that of a thermal neutron but less than that of a fast neutron. The range is generally considered to be between about 0.5 and 100,000 electron volts. Some neutrons are born with this energy, but most are in transition to thermal neutrons.

Ion. An atom or molecule that has lost or gained one or more electrons. By this ionization it becomes electrically charged.

Examples: an alpha particle, which is a helium atom minus two electrons; a proton, which is a hydrogen atom minus its electron.

Ionization chamber. An instrument that detects and measures ionizing radiation by measuring the electrical current that flows when radiation ionizes gas in a chamber, making the gas a conductor of electricity.

Isotope. One or two or more atoms with the same atomic number (the same chemical element) but with different atomic weights. Isotopes usually have very nearly the same chemical properties, but somewhat different physical and nuclear properties.

Leakage. The escape of neutrons from a reactor core. Reducing leakage improves the reactivity of a reactor.

Man-made radiation. Radiation exposure that includes radiation produced for medical and other purposes.

Maximum credible accident. The most serious reactor accident that can reasonably be imagined from any adverse combination of equipment malfunctions, operating errors, and other foreseeable causes. The term is used to analyze the safety characteristics of a reactor. Reactors are designed to be safe even if a maximum credible accident should occur.

Moderator. Material used in a nuclear reactor to moderate, that is, slow down, neutrons from the high energies at which they are released. Neutrons lose energy by scattering collisions with nuclei of the moderator. A good moderator has a high scattering cross section and low atomic weight. In each collision there is a chance of absorption. To reduce this loss of neutrons during the slowing-down process, the moderator atoms also should have a low neutron-absorption cross section. A high scattering cross section implies frequent collisions; these give the neutron a better chance of being slowed down before it is captured and also reduce the average net distance traveled in slowing down so that leakage is reduced. Some practical materials are carbon (used in the form of graphite), beryllium and its compounds, and water.

Molecule. A group of atoms held together by chemical forces, and the smallest unit of matter which can exist by itself and retain all of its chemical properties.

Multiplication factor. The ratio of the number of fissions in one generation divided by the number of fissions in the preceding generation. When the multiplication factor is 1, the number of fissions and the energy released remain constant with time, and the reactor is defined as being critical.

Natural circulation reactor. A reactor in which the coolant (usually water) is made to circulate without pumping, that is, by natural convection.

Neutron. An uncharged elementary particle that has a mass slightly greater than that of the proton and is found in the nucleus of every atom heavier than hydrogen. A neutron is stable as long as it is bound to an atomic nucleus. A free neutron is unstable and decays with a half-life of about 13 minutes into an electron, proton, and neutrino. Neutrons sustain the fission chain reaction in a nuclear reactor.

Neutron capture. The process in which an atomic nucleus absorbs or captures a neutron. The probability that a given material will capture neutrons is measured by its neutron capture cross section, which depends on the energy of the neutrons and on the nature of the material.

Neutron economy. The degree to which neutrons in a reactor are used for desired ends instead of being lost by leakage or nonproductive absorption. The desired ends may include propagation of the chain reaction, converting fertile to fissionable material, or producing isotopes.

Neutron flux. A measure of the intensity of neutron radiation. It is the number of neutrons passing through one square centimeter of a given target in one second. Expressed as nv, where n is the number of neutrons per cubic centimeter and v their velocity in centimeters per second.

Nuclear reactor. A device designed to allow a fission chain reaction to be maintained in a controlled manner. The basic components of a nuclear reactor are the fuel, moderator, reflector, coolant, shielding, and control medium contained inside a pressure vessel.

Plutonium (Pu). A heavy, radioactive, man-made metallic element with atomic number 94. Its most important isotope is fissionable plutonium-239, produced by neutron irradiation of uranium-238. It is used for reactor fuel and in weapons.

Poison. Any material of high absorption cross section that absorbs neutrons unproductively and hence removes them from the fission chain reaction in a reactor, thereby decreasing its reactivity.

Power density. The rate that energy is generated in the reactor core per unit volume. Power density is usually measured in watts per cubic meter.

Pressure vessel. A sturdy container housing the core of most types of power reactors; it usually also contains the moderator, reflector, thermal shield, and control rods.

Pressurized-water reactor (PWR). A power reactor in which heat is transferred from the core to a heat exchanger by water kept under high pressure to achieve a high temperature without boiling in the primary system. Steam is generated in a secondary circuit. Many reactors producing electric power are pressurized-water reactors.

Prompt criticality. A reactor state when criticality is sustained solely by prompt neutrons, that is, without the presence of delayed neutrons. In this state, reactor control is compromised as the reactor period is reduced to microseconds.

Prompt neutrons. Neutrons that are emitted essentially at the instant of fission and account for more than 99% of all fission neutrons. The lifetime of a prompt neutron is several microseconds.

Rad. The basic unit of absorbed dose of ionizing radiation. A dose of one rad means the absorption of 100 ergs of radiation energy per gram of absorbing material.

Radioactive decay. The lowering of the energy state of an unstable nucleus to one that is more stable with an accompanying release of particles or energy. Common forms are alpha, beta, and gamma. A decay chain describes several sequential decays that result in a stable nucleus.

Radioisotope. A radioactive isotope. An unstable isotope of an element that decays or disintegrates spontaneously, emitting radiation. More than 1,300 natural and artificial radioisotopes have been identified.

Reactor period. The time required for the neutron population to increase by a specified factor.

Reactivity. A measure of the departure of a nuclear reactor from criticality. It is about equal to the effective multiplication factor minus one and is thus precisely zero at criticality. If there is excess reactivity (positive reactivity), the reactor is supercritical and its power will rise. Negative reactivity (subcriticality) will result in a decreasing power level.

Reflector. A layer or structure of material surrounding the core of a reactor to reduce the leakage of neutrons. The reflector is located between the core and the shield. Neutrons entering the reflector are scattered randomly, some of them many times; and a large fraction of them ultimately may return to the core. It is possible to design a reflector such that more than 90% of the neutrons that would be lost may be returned.

The returned neutrons can then cause more fissions and improve the neutron economy of the reactor. Common reflector materials are graphite, beryllium, and natural uranium.

Rem. The unit of dose of any ionizing radiation which produces the same biological effect as one roentgen of absorbed dose of ordinary X-rays. Radiation tolerance is usually expressed in millirems per week or year.

Resonance. The phenomenon whereby particles such as neutrons exhibit a very high interaction probability with nuclei at specific kinetic energies of the particles. Cross sections for neutron capture and scattering, for example, exhibit peaks at these so-called resonance energies and have relatively low values between the peaks.

Roentgen. A unit of exposure to ionizing radiation. It is that amount of gamma or X-rays required to produce ions carrying one electrostatic unit of electrical charge (either positive or negative) in one cubic centimeter of dry air under standard conditions. Named after Wilhelm Roentgen, the German scientist who discovered X-rays in 1895.

Safety rod. A standby control rod that can be used to shut down a nuclear reactor rapidly in an emergency.

Scram. The sudden shutdown of a nuclear reactor, usually by the rapid insertion of the safety rods. Emergencies or deviations from normal reactor operations cause the reactor operator or automatic control equipment to scram the reactor.

Seed (and blanket) core. A reactor core which includes a relatively small volume of highly enriched uranium (the seed) surrounded by a much larger volume of natural uranium or thorium (the blanket). As a result of fissions in the seed, neutrons are supplied to the blanket where more fissions take place. In this way, the blanket is made to furnish a substantial fraction of the total power of the reactor. It is also called a spiked core.

Shield. Any material used to reduce the amount of radiation reaching one region of space from another region of space.

Special nuclear material. In atomic energy law, this term refers to plutonium-239, uranium-233, uranium containing more than the natural abundance of uranium-235, or any material artificially enriched in any of these substances.

Temperature coefficient of reactivity. The change in a reactor reactivity (per degree of temperature) occurring when the operating temperature changes. The coefficient is positive when an increase in temperature increases the reactivity, and negative when an increase in temperature decreases reactivity. Negative temperature coefficients are desirable because they help to prevent power excursions.

Thermal neutron. A neutron in thermal equilibrium with the substance in which it exists; most commonly, a neutron with a kinetic energy that is less than 0.5 electron volts.

Thermal reactor. A reactor in which the fission chain reaction is sustained primarily by thermal neutrons. Most reactors are thermal reactors; this is because the fission cross section of uranium-235 is about one hundred times greater for thermal neutrons than for fast neutrons.

Thorium. A naturally radioactive element with atomic number 90 and, as found in nature, an atomic weight of approximately 232. The fertile thorium-232 isotope is abundant and can be transmuted to fissionable uranium-233 by neutron irradiation.

Uranium. A metal symbol U, ninety-second element of the atomic series. Natural uranium is a mixture principally of the isotopes U-235 and U-238, the former being about 1/140 of the total. The nucleus of U-235 is capable of absorbing a neutron of thermal energy and thereupon undergoing fission into two fragments, which fly apart with great energy. The fragments are highly radioactive. Some neutrons are released almost immediately in each fission (the prompt neutrons). A small fraction (delayed neutrons) is released later in the radioactive decay of some of the fission products. The fact that fission is induced by one neutron but releases more than one makes possible a chain reaction. The fact that some are delayed simplifies control problems. The proportion of U-235 in natural uranium may be increased artificially by any of several processes, of which gaseous diffusion is the most important. The resulting material is called enriched uranium. U-238, which occasionally undergoes fission with fast neutrons, may absorb slower neutrons and, decaying through the intermediate element neptunium, yield plutonium, which is fissionable and can be separated from uranium by chemical methods.

Xenon. A naturally occurring element with atomic number 54. One isotope, xenon-135, is produced by fission or the decay of the fission fragment iodine-135 and has a neutron absorption cross section larger than the fission cross section of uranium-235. It, therefore, acts as a reactor poison by reducing the neutron flux.

Xenon effect. In a light-water reactor, xenon-135 is continuously produced and burned out. During steady-state operations, the xenon level reaches an equilibrium level after about 40 hours of operation. However, after a reactor shutdown, xenon-135 continues to be produced from iodine-135 and reaches a peak level about eleven hours after shutdown. Late in core life this xenon effect may preclude reactor criticality until the xenon-135 has decayed and there is sufficient excess reactivity to overcome the poison buildup.

1.4 The Fission Process. In the fission reaction a fuel nucleus absorbs a neutron that has sufficient energy to break apart the absorbing nucleus; this energy level is called the binding energy. The resulting compound nucleus breaks up into two or more parts with the liberation of a considerable amount of energy. At the same time, two or three neutrons are usually emitted, mostly (although not entirely) at the instant of fission. Instantaneous gamma radiation also accompanies the fission process, and beta particles and gamma rays are produced over a period of time as the radioactive fission products decay.

The "cross section" of an atom is an important consideration in the design of a nuclear reactor. The cross section of an atom is a measure of its effective area that is presented to an incoming particle. This effective area is related to the probability of interaction and depends on such parameters as the type of particle, particle speed, particle energy, and the type of interaction. Cross sections are known for most isotopes of atoms, and components of the reactor core and structure are chosen with due regard for the cross-section information. For example, U-235 is chosen for the fuel because it has a much higher fission cross section than other isotopes, while the structural materials are chosen that have small absorption cross sections to minimize the number of neutrons removed from the fission process in the core.

If at least one of the neutrons emitted in each fission is captured by another fissionable nucleus, a fission chain reaction becomes possible. Because of the loss of neutrons from the system, a chain reaction can be maintained only if the system exceeds a certain "critical" size. This size has a definite value for a particular system depending on the nature and amounts of the materials present and their geometrical configuration. The amount of fuel to be consumed for energy production must be present in addition to the "critical" quantity since this is the essential minimum. Although a reactor consisting largely of pure fissionable material (uranium-235 or plutonium-239) can have a critical size that is smaller than a football, the nonfissionable content of practical power-production reactors may result in cores containing from several hundred pounds to several tons of material. If the amount is less than the critical, the fission chain cannot be maintained and there can be no continuous production of energy.

The kinetic energy of neutrons emitted in fission is very high; however, the fission cross sections of fissionable materials are very much lower at high neutron energy levels than at lower levels. In their passage through matter, neutrons collide with atomic nuclei and scattering occurs in which there is a transfer of energy from the fast neutrons to other relatively slow-moving nuclei. Therefore, in most reactors the "fast" neutrons are slowed down to "thermal" energy levels to increase the probability of a fission capture. The material used to slow down the fast neutrons is called a moderator. The process of slowing down as a result of scattering collisions is referred to as moderation. The neutrons must be slowed quickly to increase the probability that they will produce further fissions and not be absorbed by nonfissioning atoms or lost by leakage. A good moderator is a material that reduces the speed of fast neutrons in a small number of collisions so as to increase the probability of fissile captures. Consequently, materials consisting of atoms of low mass number and a high scattering cross section are the best moderators. Ordinary water, heavy water, beryllium oxide, and carbon (i.e., graphite) have been employed as moderators. Without a moderator the inventory of fissionable material required to maintain a chain reaction would be significantly increased.

An important element of core design is the use of reflectors to return leaking neutrons to the core, to improve neutron economy, and therefore to reduce the amount of U-235 required to sustain a chain reaction. Reflectors, like moderators, are made of materials with high scattering cross sections but are configured to redirect neutrons that are leaving the core, back towards the active core region.

1.5 Types of Fissionable Material. A nuclear reactor consists of fuel containing fissionable material, a moderator to slow down neutrons (except in the case of fast reactors), a coolant to remove the heat generated by fission, a neutron absorber or neutron leakage control device, and the necessary structural materials. Although the fuel form is different for various types of reactors and is closely associated with the characteristics of the reactor coolant, practical considerations limit the choice of the fuel material subject to fission to three possibilities; these are uranium-235, uranium-233 (an artificial material produced in a reactor by irradiation of the fertile material thorium), and plutonium-239 (an artificial material resulting from the conversion of the essentially nonfissionable U-238 by neutron absorption).

1.6 Conversion and Breeding. Uranium-235 is the only fissionable material occurring in nature to any appreciable extent. Since natural uranium contains only 0.7% of uranium-235, it is expected that the use of nuclear reactors for power production will eventually consume the available uranium-235. Since nearly 140 times more nonfissionable uranium-238 is available than is the fissionable uranium-235, it is of major significance to consider ways in which uranium-238 could be utilized for power production. In plutonium production reactors, such as those operated by the U.S. Department of Energy at Savannah River, excess fission neutrons are captured by fertile uranium-238, and, via a beta decay chain, uranium-238 is converted into plutonium-239, which has a high fission cross section. Using this procedure, high-concentration uranium-238 cores actually create fuel as fuel is consumed.

Although not feasible for marine propulsion applications, a prospect of great interest is that of a regenerative reactor with a conversion efficiency exceeding 100% where more fissionable material is produced than is consumed in maintaining the fission chain; this process is known as "breeding." Through the operation of breeder reactors, the stockpile of fissionable material can be increased until essentially no more uranium-238 remains.

1.7 Radioactive Decay. In a nuclear reactor the fission process results in the liberation of energy and also the emission of nuclear radiation of different kinds. In general, the remarkably large amount of energy released in fission (about 200 MeV per fission-nucleus) manifests itself in the form of heat that results from the kinetic energy of fission fragments. The radioactive decay, neutron reactions, and radioactive emissions are not productive in a reactor that is primarily intended to produce power and must be dealt with by the shielding, the reactor operating characteristics, or otherwise in the design of the reactor.

The majority of the elements that occur naturally are stable except for several elements of high atomic weight, such as radium. An unstable element undergoes spontaneous radioactive disintegration at a definite rate with the emission from the nucleus of an electrically charged particle (either an alpha particle, i.e., a helium nucleus, or a beta particle, i.e., an electron). Often, the products of decay are themselves radioactive, expelling either an alpha or a beta particle. After a number of stages of disintegration, an atomic species with a stable nucleus is formed.

In a given specimen, the rate of decay at any instant is always directly proportional to the number of parent radioactive atoms of the isotope under consideration present at that instant. For a given radioactive species, every nucleus has a definite probability of decaying in a given time; this decay probability has a constant value that is characteristic of the particular radioisotope. The most widely used method for representing the rate of radioactive decay is by means of the half-life, which is the time required for the number of radioactive nuclei of a given kind to decay to half its initial value. Because of the exponential nature of the decay, this time is independent of the amount of the radioisotope present.

Gamma rays are the electromagnetic radiation released when an excited nucleus emits its excess energy. Gamma rays are similar in character to X-rays; they are highly penetrating and have short wavelengths. Although the term "gamma ray" was originally used to describe the electromagnetic radiation that frequently accompanies radioactive decay, the definition has now been extended to include all electromagnetic radiations of nuclear origin. The rays are emitted when a nucleus undergoes transition from a higher-energy to a lower-energy state.

Gamma radiation is described in terms of its photon energy; for example, "1-MeV gamma rays." Apart from the fact that X-rays frequently have lower energies, the essential difference between gamma rays and X-rays is that X-rays are produced outside the atomic nucleus. The characteristic X-rays, which as their name implies have definite energies (and wavelengths) that are characteristic of the particular element concerned, result from transitions between electron energy levels of the atoms.

Section 2
Reactor Design Considerations

2.1 Basic Reactor Arrangement. Although there are a number of special-purpose reactors, all power reactors have certain characteristic features. Figure 1 is a simplified schematic diagram that provides a framework for the identification of the basic reactor components.

The fuel (1) is arranged in intimate contact with both the coolant and moderator. The fuel is composed of a fissionable material, which is protected as necessary (by fuel cladding) from corrosion by the coolant and from the loss of fission products to the coolant. The moderator (2) is a low-mass-density material that slows down the fast neutrons formed at fission to an energy range that increases the probability of a fissile capture by U-235. The coolant (3) provides for heat transfer and the removal of

heat generated in the fuel as necessary to maintain a desirable fuel temperature.

The active core region (4) consists of an assembly of fuel, control rods, reflector, and flow orifices in an arrangement that permits intimate contact of the fuel, moderator, and coolant. The control rods (5), which are actuated by external control rod drives (6), are mechanically withdrawn from the fuel region to start up or maneuver the reactor and are inserted into the fuel region to shut down the reactor. The assembly of control rods, fuel, moderator, and coolant is surrounded by a reflector region, which is composed of light and heavy materials that minimize the loss of neutrons from the system by reflecting neutrons back into the active core region.

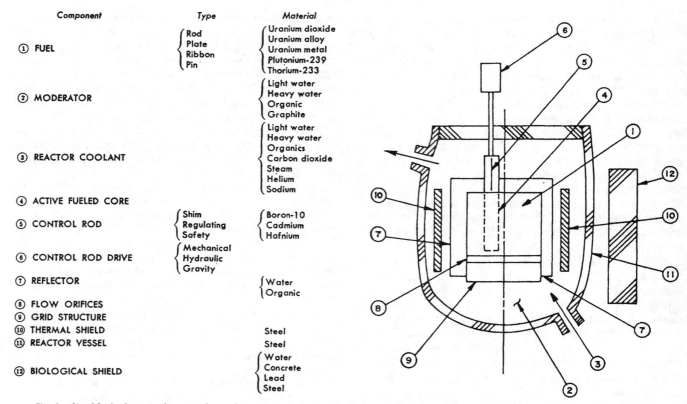

Component	Type	Material
① FUEL	Rod Plate Ribbon Pin	Uranium dioxide Uranium alloy Uranium metal Plutonium-239 Thorium-233
② MODERATOR		Light water Heavy water Organic Graphite
③ REACTOR COOLANT		Light water Heavy water Organics Carbon dioxide Steam Helium Sodium
④ ACTIVE FUELED CORE		
⑤ CONTROL ROD	Shim Regulating Safety	Boron-10 Cadmium Hafnium
⑥ CONTROL ROD DRIVE	Mechanical Hydraulic Gravity	
⑦ REFLECTOR		Water Organic
⑧ FLOW ORIFICES		
⑨ GRID STRUCTURE		
⑩ THERMAL SHIELD		Steel
⑪ REACTOR VESSEL		Steel
⑫ BIOLOGICAL SHIELD		Water Concrete Lead Steel

Fig. 1   Simplified schematic diagram of a nuclear reactor (boiling water, pressurized water, organic moderated, gas cooled, or sodium cooled)

Flow orifices (8) are sometimes used to provide a higher coolant velocity through the central regions of the core than in the outer regions. This may be desirable because there is a lower neutron density at the outside of the active core region because of neutron leakage. The fuel elements in the center region are, therefore, more intensively cooled than those in the outer fuel region, because maximum reactor power is based on the limiting operating conditions of the hottest fuel element.

The active core region (4), its support structure (9), and the thermal shield (10) are supported by the reactor pressure vessel (11). The support structure is minimized in the active core region to reduce parasitic neutron capture, but the structure must provide sufficient strength and rigidity to maintain fuel alignment for control rod entry and also to eliminate objectional vibrations of the relatively flexible fuel elements in the high-velocity coolant. The pressure vessel must be strong enough to not only maintain the high system pressures (2000 psi for pressurized-water reactors and 1000 psi for boiling-water reactors) encountered during normal operation, but also to withstand abnormal conditions such as overpower, thermal shock, mechanical shock, and vibration while maintaining integrity against the loss of reactor coolant and the loss of control rod alignment.

The biological shield (12) completely surrounds the reactor system. The biological shield is provided to reduce the radiation levels from the active core region and the primary coolant system to acceptable levels as necessary to conduct ship operations.

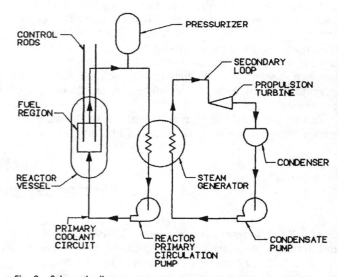

Fig. 2   Schematic diagram of a pressurized-water reactor power system

A simplified schematic diagram of a pressurized-water power cycle is illustrated by Fig. 2. The reactor vessel contains the fuel region and is designed to contain the moderator and coolant at a pressure that is sufficiently high to suppress boiling. The reactor primary circulation pump is used to circulate the primary coolant and overcome the system pressure-drop losses. The primary coolant, which is subjected to radioactive exposure, removes heat from the reactor and conveys the heat to the steam generator, where it is used to produce steam in the secondary power cycle. Steam from the steam generator is expanded in the propulsion turbine to produce power and is exhausted to the condenser, where it is condensed and then pumped back through the steam generator. The pressurizer maintains the pressure of the primary systems within allowable limits. Many additional components are required to make a pressurized-water reactor system functional; however, Fig. 2 illustrates the basic principles involved.

A major advantage of the primary/secondary loop concept is that the steam generator forms a boundary that isolates the radioactive coolant in the primary loop from the steam in the nonradioactive secondary loop, which is used to drive the propulsion turbine and is subject to small amounts of leakage through packing glands, etc.

2.2 Nuclear Fuels. For some applications mixed plutonium-uranium or thorium fuel forms may have important advantages; however, because of its widespread commercial applications, only uranium will be considered in the following. The oxide fuel form is usually selected for low enrichments, because it offers good fuel economy, including a low fabrication cost, and has demonstrated a high burnup capability. Of major consequence is the very extensive background of design and operating experience of low-enrichment oxide fuel from central-station reactors of the pressurized-water and boiling-water types.

Uranium dioxide (UO_2) is the most common fuel for pressurized-water and boiling-water reactors and is usually fabricated in a ceramic fuel form by sintering pellets at a high temperature. The UO_2 pellets have the usual ceramic characteristics of being very hard with a typically low thermal conductivity and high melting point; they also possess a good chemical stability and have good dimensional stability when irradiated in service. In addition, UO_2 fuels generally do not react with water or fuel cladding material. They exhibit an ability to retain fission gas and have demonstrated satisfactory operation to burnups of economic interest.

Low-enrichment uranium fuel has the advantage of a negative fuel temperature coefficient, which means that a reaction will slow as the temperature increases; this facilitates the control of a reaction. It occurs in part because of the "Doppler broadening" of U-238 neutron capture resonance with increased temperature, which is a result of the increased random motion of the U-238 nuclei. Normally the cross section of an element decreases with increasing temperature, but in the case of U-238 there is a neutron absorption "resonance" above the thermal energy range so that, as the fuel temperature is increased, the probability of a nonfission capture of a neutron in U-238 is increased.

Low-enrichment UO_2 fuel is usually fabricated in the form of short, small-diameter (0.3 to 0.6 in.) pellets that are stacked in either stainless steel or zirconium fuel cladding. The fuel containers are sealed to prevent a loss of fuel or fission products to the reactor coolant, and leakage of coolant to the fuel, under all normal operating conditions. The fuel elements are assembled in fuel bundles that consist of a number of fuel rods that are usually assembled in basically a square array. Zirconium has some advantages as compared to stainless steel because of its lower parasitic neutron capture characteristics.

There is a significant amount of experience with metallic fuel types from the Naval Reactor Program. These fuels are characterized by a high enrichment of U-235. Characteristics generally attributed to metal fuels are (1) a high heavy-atom density, (2) a significant and reliable thermal-expansion coefficient, (3) an amenability to potentially inexpensive fabrication methods, and (4) a high thermal conductivity. General disadvantages of metal fuels are (1) a low melting temperature, (2) a high rate of radiation-induced swelling, and (3) a poor high-temperature compatibility with austenitic stainless steels. A high thermal conductivity and a low melting temperature tend to offset each other in terms of the specific power attainable, but metal fuels have the potential for somewhat higher specific powers than do oxides.

There are other types of fuels that may be considered. Uranium carbide fuels do not have as extensive a background of experience as either oxide or metallic fuels, and for marine propulsion reactors they should be considered developmental. However, carbide fuels do offer potential advantages because of their good nuclear characteristics, good thermal conductivity, high heavy-atom density, high melting point, and potential for significant increases in fuel specific power. Problems that have been identified but not completely solved for carbide fuels are the accommodation of irradiation-induced swelling, the carburization of cladding material, and the development of a higher conductivity gap between the fuel and cladding as necessary to take advantage of the improved thermal conductivity of the carbide fuel.

Nitride fuels also have good nuclear characteristics, good thermal conductivity, and a high melting point. It is expected that nitrides may be more compatible with cladding material than carbides because of the lower strength of nitrides. However, because of the lack of related engineering data, uranium nitride fuels should be considered even more developmental than carbide fuels.

2.3 Reactor Coolants. One of the unique engineering problems of a nuclear reactor is that, compared with a conventional plant, where the maximum temperature is limited by the chemical reaction of fuel oxidation and the rate of energy release is a direct function of the rate of fuel injection, a nuclear reactor has no such limitation. A nuclear reactor has a large quantity of energy stored in the fuel contained within the reactor, and the maximum temperature of the reaction is limited only by the ability to remove heat or, more properly, by the correspondence between the heat removed from the fuel by the coolant and the power level of the fuel as a function of the excess reactivity neutronics of the system. This should be recognized as being true only on a theoretical basis, since for power reactors of interest for marine propulsion the neutronics of the system are such that operation at power levels above full power results in the removal of the moderator, which has a negative or "shutdown" effect on the reactor, and all major systems are designed to fail safe or shut the reactor down. Nevertheless, the point is still valid that generally the most important aspect of reactor design is heat removal and the most important single aspect of heat removal is the coolant selection.

The selection of light water as the prime candidate for the reactor coolant for a marine reactor is based on a large number of considerations, not the least of which is that light water has a background of operating and design experience obtained in the U.S. Naval Reactor Program, all of the merchant nuclear-powered ships that have been built, and also a large background of experience from pressurized-water and boiling-water reactors in central-station nuclear power plants. The technical bases for the selection of water are:

- Water is an excellent moderator, which permits the use of nonmoderating reactor internal structure materials.
- Water has good heat-transfer properties and a high volumetric heat capacity.
- Water is readily available at a low cost.
- Water-moderated reactors can have large negative temperature coefficients.
- Water is noncombustible and relatively insensitive to radiation damage.
- Water provides some lubricating qualities and can be pumped efficiently.
- Water does not freeze in the range of temperatures of interest.
- Water retains many soluble and solid fission products and generally releases only noble gas fission products in the event of a loss of fuel integrity.
- A water coolant provides the capability of direct steam generation in a boiling-water reactor.
- Water technology is well known and system components are available, reliable, and relatively inexpensive.

There are, however, a number of disadvantages of water as a reactor coolant. As more advanced technology is developed, it is probable that water will be replaced by a reactor coolant that will permit more compact reactors. The general limitations associated with the use of water as a reactor coolant are:

- Water reacts with metallic uranium and may have chemical reactions with other fuel forms under some conditions.
- Water may react with fuel cladding or other reactor construction materials at high temperatures.
- Water must be pressurized to be maintained in the liquid phase at temperatures above 212 F; this requires high system pressures at temperatures of interest.
- The primary system must be shielded against radioactivity from short-lived neutron activation of corrosion products carried with primary coolant water into the active core region.

A number of gases are also candidate reactor system coolants; however, most of the possibilities can be eliminated, either by chemical or metallurgical evaluations (air,

hydrogen, carbon monoxide) or by heat-transfer considerations (neon, argon). Properties of gases that are suitable for reactor cooling are given in Table 1 [14].

In addition to relatively poor heat transfer, argon also has problems of neutron activation, and neon is very expensive. Nitrogen has a high neutron-absorption cross section and might cause nitriding at high temperatures. Thus, the list of gaseous coolants of interest for marine propulsion can be reduced to carbon dioxide (CO_2), helium, and steam. From purely thermal and cost standpoints, there appears to be little choice between CO_2 and helium. CO_2 cooling leads to higher core pressure drops, high fuel ratings, and greater potential for flow-induced vibrations. The biggest advantage of steam as a reactor coolant is that it provides a gaseous coolant with a system utilizing the large background of light-water reactor technology and marine propulsion experience with steam systems. The steam-cooling technology is simple in that steam-cooled reactors operate on a direct cycle where primary heat exchangers are not required. This could be very important from a weight point of view. Additional incentives include the inert chemical nature of the coolant, the possibility of flooding the reactor with a transparent medium (water) during refueling and maintenance operations, and the possibility of using water spray systems for emergency cooling.

Disadvantages of steam-cooled reactors include the high operating pressure and high velocity flow of the coolant, and safety problems arising from the possibility of loss-of-coolant or depressurization accidents.

Incentives for using helium as a reactor coolant include the fact that helium does not undergo a change in phase during the power cycle (unlike water), and since helium remains a gas for the entire power cycle, better thermal efficiencies are achieved. Advances in fuel design for gas reactors have improved the retention of fission products in the fuel. The use of helium as the reactor coolant results in reduced shielding requirements and lower exposure levels for personnel. In fact, exposure levels are lower in gas reactors by a factor of ten when compared with light-water reactors. Another incentive for using helium is the availability of system components, including helium circulators, steam generators, and helium piping, which draw heavily on technology from existing plants such as the advanced gas-cooled reactors, operational high-temperature gas-cooled reactors, and the developmental modular high-temperature gas-cooled reactors coupled with closed-cycle gas turbines [15]. Disadvantages of a helium-cooled marine propulsion reactor include the high operating pressure of the coolant and the safety problems arising from the possibility of loss-of-coolant or depressurization accidents. Considerable developmental work on gas-cooled reactor systems has been done by the U.S. Maritime Administration on the Maritime Gas Cooled Reactor [15].

Reactor coolants other than gases and water are also feasible. Liquid-metal coolants have a number of advantages, with the most significant one being excellent heat-transfer properties that could permit the achievement of highly compact reactor power systems [16]. However, for

Table 1 Properties of gases suitable for reactor cooling

GAS	H_2	He	N_2	Air	CO	CO_2	A
Molecular weight	2	4	28	29	28	44	40
Thermal conductivity, Btu/hr-ft²-F/ft							
200 F	0.125	0.097	0.018	0.018	0.017	0.013	0.012
700 F	0.199	0.135	0.028	0.028	0.027	0.028	0.018
1330 F	...	0.172	0.037	0.039	...	0.042	0.025
Viscosity, centipoises at:							
200 F	0.010	0.023	0.020	0.021	0.020	0.017	0.027
700 F	0.015	0.033	0.031	0.032	0.031	0.028	0.041
1330 F	0.020	0.044	0.041	0.042	0.044	0.041	0.054
Specific heat, Btu/lb-F, at:							
200 F	3.47	1.24	0.249	0.241	0.250	0.217	0.124
700 F	3.51	1.24	0.259	0.254	0.262	0.262	0.124
1330 F	3.60	1.24	0.279	0.272	0.283	0.295	0.124
Density at STP, lb/ft³	0.0052	0.0104	0.0727	0.0748	0.0727	0.114	0.104
Volumetric specific heat at STP, Btu/ft³-F	0.0178	0.0129	0.0180	0.0179	0.0180	0.0238	0.0129
Relative heat transfer coefficient compared to He for same gas temperature and same power output	1.19	1.00	0.73	0.73	0.72	0.79	0.68
Relative pumping power compared to He for same gas temperature and same power output	0.17	1.00	2.2	2.2	2.2	0.88	10
Relative pumping power compared to He	0.17	1.0	4.0	4.0	4.0	1.8	24
Relative cost of gas per 1000 ft³ at STP	6	22.7	10	0	60	5	40
Relative total activity	4.53×10^{-4}	18.5	9294	7225	0.51	1.0	1392
Relative gamma activity	0	0	0.0456	1284	0.5	1.0	137,065

commercial marine propulsion applications liquid-metal-cooled reactors should be considered developmental. It is expected that when liquid-metal fast-breeder reactors have been demonstrated by land-based central-station applications, liquid-metal-cooled reactors for marine propulsion will be seriously reconsidered. A considerable amount of liquid-metal technology was developed by the Naval Reactor Branch, AEC on the Submarine Intermediate Reactor Development (SIR) program. A liquid-metal-cooled reactor was used on the second nuclear-powered submarine in the U.S. Navy, the *Seawolf*, and has been used on several classes of Russian submarines.

Organic moderated and cooled reactor systems have also been considered. Based on the same arguments presented in the previous paragraph, these systems continue to be too developmental to be considered for marine propulsion.

2.4 Reactor Control. In order to be a useful method of providing energy, a nuclear reactor must be under complete control during start-up, shutdown, and steady-state operating conditions. Unlike most large commercial reactors that operate at the same power level for extended time periods, the power level of propulsion reactors must be constantly adjusted to meet changing ship speed requirements. An additional consideration, which complicates reactor control, is the fact that nuclear fuel cannot conveniently be replaced as it is used; this means that the total amount of fissionable material needed to produce the power required between refuelings must be loaded in the reactor at one time. The amount of fuel loaded must include sufficient fuel to constitute a "critical mass," fuel to produce power, and fuel to overcome poison effects. The critical mass is the minimum amount of fuel in the core necessary to keep the reactor critical throughout the

core life. The power fuel is used to provide propulsion and other shipboard power loads. Since the excess fuel at start-up provides reactivity in excess of that required to maintain criticality, an essential aspect of reactor control is to provide margin for shutdown at all conditions. This is achieved by the addition of poisons and the use of control rods and chemical shims.

Burnable poisons are added to the core at initial core loading to "burn out" under neutron irradiation at a rate that compensates for fuel use. Poisons are atoms with larger cross sections for neutron absorption than the fuel. Adding poisons such as boron, therefore, reduces the overall reactivity of the core at the beginning of core life. But since the poisons are quickly burned out, their use allows for a greater fuel loading for a given power output and thereby helps to lengthen core life. This also helps to create a more even neutron flux distribution throughout core life. Fission-product poisons are the result of the fission process and act as neutron absorbers. Xenon-135 and samarium-149 are two fission-product poisons that build to appreciable levels in the core and must be taken into consideration when designing the core. Since fission-product levels and their effect on reactivity are dependent on power level, reactor control must be established for all possible levels of burnable and fission-product poisons. Unless power levels stay constant for greater than approximately eleven hours, the concentration of xenon continuously changes; consequently, the reactivity of the core also changes continuously.

Two common methods of reactivity control involve the use of control rods and chemical shims. Control rods are rods, or plates, that contain a material, usually hafnium or boron, which has a large neutron absorption cross section and hence readily absorbs neutrons. In addition to their use in regulating reactor output, control rods are also mechanically withdrawn or inserted in the core to compensate for changes in the reactor properties throughout core life. For all reactor designs, the control rod drive system is designed such that any system failure, such as a loss of power, causes the control rods to be fully inserted into the core, thereby ensuring reactor shutdown. In order to reduce the number of moving control rods, chemical shims can also be used. A chemical shim is a neutron-absorbing chemical such as boric acid, which is used in varying concentrations in the moderator or coolant to adjust the reactivity of the core. The concentration of the boric acid is controlled by the operator; and since the poison is carried by the coolant, the poison reduces reactivity equally throughout the core.

Another important aspect of reactor control is the temperature coefficient of reactivity or the amount of reactivity change per degree of temperature change. The temperature coefficient is a design parameter and is dependent on the size and shape of the core. A positive temperature coefficient indicates that an increase in temperature results in additional reactivity. A positive coefficient is undesired and can create a dangerous situation, since as the temperature rises, there is more reactivity, which raises the temperature further and adds yet more reactivity, which could eventually lead to a catastrophic failure. A negative temperature coefficient, however, facilitates reactor stability; in this case as the temperature rises, reactivity is reduced, which lowers the reactor power, lowers temperatures, and helps to maintain the plant within operating parameters.

A negative temperature coefficient is highly desirable in plants that are designed so that reactor power follows steam demand. As the propulsion throttles are opened, more steam is removed from the steam generator, which in turn lowers the temperature of the return leg of the primary loop into the core. This adds positive reactivity, which raises the power level and heats up the temperature of the primary coolant traveling back into the steam generator. This self-stabilizing effect continues until the energy transferred across the steam generator is equal to the energy transferred from the reactor to the coolant.

Light-water reactors are designed to have a negative moderator temperature coefficient. Therefore, a cold reactor that is critical but has not reached operating temperature will be subcritical at operating temperature. Although this provides good operating characteristics, depending upon the magnitude of reactivity swing between hot and cold conditions, it does require sufficient excess reactivity control (i.e., control rods) to shut down in the cold condition.

The reactor control provisions are designed to safely operate the reactor during all stages of core life and at all power levels. As described, a system with a negative temperature coefficient inherently facilitates the stabilization of steam demand and reactor output while maintaining design core parameters. As the core ages and fuel burns out, control rods and chemical shims are used to maintain the design core parameters, to overcome the effects of fission-product concentrations and fuel burnout

on the primary plant temperatures, and to ensure that the reactor can be shut down during all operational conditions.

2.5 Shielding. For compact reactors, such as marine reactors, considerable design attention must be given to ensure the attenuation of emitted nuclear radiations by some form of shielding. Shielding is necessary for the protection of personnel and because a high radiation background will interfere with the operation of instruments that are used in various aspects of reactor operation and control. Although the radiation from a reactor system includes alpha and beta particles, gamma rays, and neutrons of various energies, only gamma rays and neutrons need be considered since these are by far the most penetrating. Any material that attenuates these radiations to a sufficient extent will automatically reduce all the others to a negligible value. Some shielding is inherent in the basic reactor construction, while other layers of shielding are provided solely for that purpose.

Shielding the reactor involves three aspects: slowing down the fast neutrons, capturing the slowed-down neutrons, and absorbing all forms of gamma radiation. Since elements of low mass numbers are the best moderators, hydrogen in the form of water can suitably be used as the shield constituent for slowing down fast neutrons. However, at high neutron energies, the scattering cross section of hydrogen is very small; therefore, a considerable thickness of hydrogeneous material would be required to slow down the fission neutrons of highest energy. The situation can be greatly improved by introducing an element of fairly high mass number. Although such substances are not good moderators, they reduce the energies of very fast neutrons as a result of inelastic-scattering collisions. Elements such as lead, barium, or iron readily decrease the neutron energy down to about 0.05 MeV where the hydrogen (elastic) scattering cross section is relatively large. Hence a combination of a moderately heavy or heavy element with hydrogen will effectively slow down even neutrons of very high energies.

In marine propulsion reactors, the weight of the shield is of major importance. In this instance, the cost of the shield may be secondary in significance. In addition, shielding that results in a concentrated loading distribution may lead to problems with the ship's structure.

Wherever possible, advantage is taken in shield design of the attenuating effect of distance, according to the inverse-square law, on the radiation intensity or flux. If the operating personnel can be kept at an appreciable distance while the reactor is in operation, a significant saving in thickness of the shielding may be feasible. For example, a shield may be made thinner at the top and bottom if access is restricted to the sides. If the circumstances are such that personnel can be kept at a good distance from a reactor when it is in operation, it is usually desirable to do so. This may be accomplished by designating exclusion areas of several maximum permissible radiation levels for passengers, ship's crew, reactor operators on watch, and shore personnel.

To protect the heavy structural components surrounding the core, including the reactor vessel itself, from possible damage from the heat liberated upon absorption of radiation, a thermal shield is usually placed between the reactor core and the reactor vessel (see Fig. 1). The thermal shield consists of a thick layer of steel, which effectively absorbs gamma radiation and inelastically scatters fast neutrons. Since these two types of radiation carry most of the energy leaking from the reactor, a large amount of the heat produced in the shielding is released in the thermal shield and transferred to the coolant.

The reflector makes an important contribution to fast-neutron shielding. The reflector, especially for a thermal reactor, is invariably a good moderator (e.g., water, heavy water, beryllium, beryllium oxide, or graphite) so that it will slow down an appreciable portion of the moderately fast neutrons escaping from the core. Because of scattering, many of these slowed-down neutrons are returned to the core, thereby easing the shielding problem.

The penetration of gamma rays is a function of their energy but they are effectively absorbed by a material of high density. The shield material, such as iron or lead, which serves as the inelastic scatterer of neutrons, will also function as the absorber of gamma radiation. Within the energy range of interest, gamma absorption is determined essentially by the mass of the shielding material. The thickness of shield required to produce a specified absorption of gamma rays is inversely proportional to the density of the shielding material. Thus, a smaller volume of lead than of iron would be required, but the masses would be approximately the same.

Because of its high density and ease of fabrication, lead is a good shield component. For gamma rays with energies in the region of 2 MeV, roughly the same mass of lead as of iron is required to absorb a specified fraction of the radiation. However, at both higher and lower energies, the mass absorption efficiency of lead is appreciably greater than that of iron. The disadvantages of lead in reactor shields are its relatively low melting point and its softness. It cannot carry any appreciable portion of the reactor system structural load, and, because of relatively low temperature limits, it may require cooling.

Polyurethane and other plastics containing boron are often used for shielding. Polyurethane is a good neutron attenuator and is relatively light in weight, which makes it well suited for marine applications.

Masonite, with a density of about 1.3 g/cm³, was used as the hydrogenous material in some of the early reactors. The number of hydrogen atoms per cubic centimeter is not much less than for water. In addition it contains both carbon and oxygen, which can act as moderators.

As a general shield material, there is much to recommend concrete since it is strong, inexpensive, and adaptable to both block and monolithic types of construction. Ordinary concrete of 2.3 g/cm³ density contains somewhat less than 10% by weight of water when cured. Although the hydrogen concentration in concrete is considerably less than the concentration in water, the larger proportion of oxygen (which acts as an additional moderator) and the calcium and silicon in concrete compensate, to

a great extent, for the difference. Nevertheless, ordinary concrete alone is not very efficient as a reactor shield material since it normally contains no element of high mass number.

Various special ("heavy") concretes incorporating heavy elements have been developed for reactor shielding. In barytes concrete, for example, the mineral barytes, consisting mainly of barium sulfate, largely replaces the sand and gravel aggregate in ordinary concrete. The density of barytes concrete is about 3.5 g/cm³. Thus a shield of barytes concrete would have to be no thicker than an iron-water shield of the same effectiveness although the total weight of the barytes concrete shield would be greater.

2.6 Safety [17]. Nuclear ships must comply with the rules and regulations of the cognizant agencies, which for U.S.-flag ships include: the United States Department of Energy [18–22]; United States Coast Guard [23]; United States Department of Commerce; National Bureau of Standards [24]; International Convention for the Safety Life at Sea [25]; the classification societies [26]; and the rules and regulations of agencies having cognizance over the ports of call [27]. Industrial safety codes may also be applicable in part; codes of this type include the American Society of Mechanical Engineers Boiler and Pressure Vessel Code [28], and the applicable ANSI and IEEE codes [29,30].

The Nuclear Regulatory Commission regulates commercial nuclear activities in the United States, and the Commission has promulgated an enormous body of regulations for the design and operation of power reactors. Any U.S.-flag commercial nuclear ship would also be required to establish a set of applicable regulating guidelines in addition to other regulatory requirements.

A nuclear ship must also provide a degree of safety for the nonnuclear portions that is sufficiently high to ensure safe operation of the entire ship. In this respect provisions such as watertight subdivision, stability, fire protection, bilge pumping, fire extinguishing, electrical installations, steering gear, astern power, and navigational aids should be evaluated in order to provide for the maximum practicable safety for the ship.

a. Containment. Containment constitutes the outer enclosure or other systems or arrangements that are provided to prevent the uncontrolled release of hazardous amounts of radioactivity to normally accessible spaces or the ship's environment in the event of an accident or malfunction of the nuclear system. It is recognized that any one of several containment methods may be most suitable for a particular application. Separate pressure-tight containment vessels or containments utilizing integral portions of the ship's structure are examples of containment systems that may be utilized. In the design of a containment system, the effects of purification of radioactive loops, pressure relief or suppression systems, and systems that effectively prevent core meltdown or its consequences, must be considered.

The containment system is designed to ensure that the basic integrity of the containment will be maintained for any credible operating or accident condition. The following factors are typical of those which should be considered:

- Maximum credible pressure buildup within the containment due to an accident to the nuclear system.
- Maximum credible internal missile.
- Location as regards collision or grounding damage.
- Rupture of piping, ducts, or similar components outside of the containment, and such components connected to and passing through the containment.
- External fires and explosions on board.
- Fires within the containment.
- Sinking of the ship.
- Forces due to ship motion.
- Removal of reactor decay heat in the event of a loss of coolant circulation and provisions for preventing the reactor core from melting through the containment.
- Leakage and measurement of leakage rate.

All nuclear systems producing useful power contain stored energy that is indicated by pressure and temperature. A sudden uncontrolled release of this energy and any additional energy that might be generated in a nuclear accident provides a potential mechanism for the dispersion of radioactive material. The containment system is designed to contain, control, and possibly suppress the release of any radioactive material that could result from any credible accident. Consideration should be given to (1) the pressure and temperature of the coolant, (2) the energy released as a result of any chemical reaction within the system, (3) the nuclear heat generation, including decay heat, and (4) the energy stored in the structure. The processes involved in the release of this energy are heavily dependent upon the type and specific design of the nuclear power plant. Each system is evaluated on an individual basis to determine the pressure buildup in relation to the containment design.

Missiles resulting from a malfunction of the system components should not result in the release of hazardous amounts of radioactive or toxic materials to occupied spaces or the ship's environment. The following components are typical of those that may be considered as potential sources of missiles:

- High-speed rotating equipment. The installation of such equipment within the containment should be kept to a minimum, but, if installed within the containment, should be designed to reduce the probability of rupture of the containment wall due to a failure of any rotating element.
- Control rods. Positive means should be provided to prevent the control rods from being ejected.
- Fittings within the pressurized system. These should be located or protected so as to minimize the probability of damage to the containment walls in case of failure.

The primary objective under these circumstances is to maintain the integrity of the containment and, insofar as practicable, to prevent impairment of the secondary shielding when materials that are particularly susceptible to fire damage are used (e.g., lead, polyethylene, or nonferrous materials).

The containment is designed to remain intact if the ship sinks in shallow water, and consideration is given to provisions for decay heat removal. Containment integrity should be maintained for a period of several years following such an incident in order to provide sufficient time for salvage operations.

b. Shielding and radiological safety. Shipboard shielding and radiological safety are intended to provide standards for protection against nuclear radiation for personnel on board ship and for persons in the vicinity of such ships in conformance with the cognizant regulatory agencies. Inasmuch as all regulatory agencies normally follow the recommendations of the Federal Radiation Council [19,20], the recommendations of the Federal Radiation Council should be considered to anticipate changes to the criteria specified by the regulatory agencies. It is the intent to provide standards for protection by means of shielding and control of personnel access so that passengers and shore personnel will not be exposed to radiation exceeding recommendations for the general population, and so that operating, maintenance, and other authorized personnel will not be exposed to radiation exceeding the recommendations for occupational exposure.

The primary function of shielding is to attenuate all nuclear radiations resulting from reactor operations to levels that will not be hazardous to personnel or equipment. Shielding may be provided by ship's cargo, ship's stores, structure, and reactor and machinery components provided that toxic, explosive, or other hazardous materials are not produced from the effects of radiation on the shield materials. Shielding may also be used to protect against missiles and to provide containment in case of an accident to reactor components.

c. Radioactive waste disposal. Radioactive wastes are defined as the end products that result from the fission process and contain radioisotopes in significant quantities. Radioactive wastes include solids, liquids, and gases. Some examples of solid wastes are contaminated dirt, chips, or other small particles, and spent ion exchange resins. Equipment that has become contaminated or radioactive may also have to be treated in the same manner as waste.

The purpose of a radioactive waste disposal system is to collect, audit, confine, and dispose of waste material in a manner that limits the contamination of any area outside of the controlled system. Systems and procedures for the transfer, collection, and storage or disposal of radioactive material must be subject to hazard evaluation. Disposal of wastes is acceptable if it can be accomplished in such a manner that no environmental hazard results.

In general, airborne radioactive wastes dispersed to the atmosphere must result in activity concentrations in the atmosphere not exceeding the applicable radioactivity concentration guides at locations where personnel exposure can occur. This criterion necessitates separate disposal guides depending upon whether the ship is in locations where the surrounding environment is inhabited, such as in harbors or in channels, or whether the ship is

Table 2 Radioactive liquid waste disposal limits for discharge to the sea as specified for the NS _Savannah_

Zone Number	Zone Defined	Limits on Discharge Concentration	
		Unidentified Isotopes[a]	Identified Isotopes[b]
1	Harbors, estuaries, and up to 2 miles from shoreline [c]	$\leq 1 \times 10^{-7}$ μCi/cc	≤ 10 CFR 20 limits
2	Two to 12 miles from shoreline[d]	$\leq 1 \times 10^{-4}$ μCi/cc	$\leq 10^{3} \times 10$ CFR 20 limits
3	Twelve miles from shoreline to 200-fathom depth contour[d]	$\leq 10^{-1}$ μCi/cc	$\leq 10^{6} \times 10$ CFR 20 limits
4	More than 12 miles from shoreline at depths greater than 200 fathoms[d]	10^{3} μCi/cc	$\leq 10^{10} \times 10$ CFR 20 limits

[a] "μCi/cc" denotes micro curies per cubic centimeter.
[b] "10 CFR 20 limits" denotes the limits specified in the Code of Federal Regulations Title 10, Part 20.
[c] Liquid waste with concentrations above the limits specified can be discharged to a properly licensed contractor in port.
[d] Liquid waste discharges are to be made while the ship is underway.

Table 3 Radioactive gaseous waste disposal limits as specified for the NS _Savannah_

Zone Number	Zone Defined	Limits on Discharge Concentration	
		Unidentified Isotopes	Identified Isotopes
1	Harbors, estuaries, and up to 2 miles from shoreline	8×10^{-6} μCi/cc	$8 \times 10^{5} \times 10$ CFR 20 limits
2	Greater than 2 miles from shoreline[a]	8×10^{-4} μCi/cc	$8 \times 10^{6} \times 10$ CFR 20 limits

[a] Gaseous waste discharges are to be made while the ship is underway.

on the high seas. While the published radioactivity concentration guides provide some assistance to the development of such standards, specific limits for the release of radioactive effluent to the atmosphere should be determined for each particular case on the basis of the isotopic content, magnitude of the anticipated release, known principles of atmospheric diffusion, and source-to-receptor geometry. Radioactive wastes dispersed to the atmosphere at sea are ultimately deposited in the seawater. Limits regarding the disposal of radioactive liquid wastes and radioactive gaseous wastes, as specified for the NS _Savannah_, are given by Tables 2 and 3, respectively.

d. Health physics. The prime function of health physics is to safeguard the health of all individuals whose work is likely to involve exposure to nuclear radiations by taking steps necessary to minimize such exposure. In addition there is the responsibility of making sure that nothing escaping from the nuclear plant, even in the event of an accident, would represent a radiological hazard.

The regulatory bodies stipulate the maximum radiation exposure limits for personnel, maximum permissible concentrations of certain radioisotopes in air and water, and maximum permissible amounts of such isotopes that may accumulate in the human body. Such recommendations are subject to regular review as increasing knowledge is gained of the effects of nuclear radiation on the human body. Dosages are set at such low levels that daily exposure over many years is unlikely to cause any injury. On the other hand, the levels cannot be so low as to make operation of a plant impossible.

Radiation dosage is measured in terms of the energy absorbed from the radiation, and the dose rate is the time rate at which such energy is absorbed. In general, the total dose (or dosage) received is the product of the dose rate and the exposure time. All persons likely to be exposed to radiation are provided with meters whereby the dose received can be measured, and any individual who has received an appreciable dose on a given day is notified so that appropriate steps may be taken to avoid overexposure. In addition to these meters, which record the total radiation dose received over a period of time, individuals leaving a contaminated area are surveyed by appropriate hand and foot counters. In special cases it may be necessary to measure the radioactivity of radioiodine, and an analysis of urine and feces may be required to determine if certain radioactive species are accumulating in the body.

All accessible areas around a reactor and its associated equipment and laboratories are surveyed at frequent intervals to determine the radioactive contamination of surfaces and the amount of activity in the air. A regular survey must also be made of the radioactivity of the solid, liquid, and gaseous materials discharged from the plant so as to make certain that they will not constitute an environmental hazard.

e. Radiation detection. The ability to measure the different types of radiation emitted from nuclear power plants is essential to protect personnel, ensure safe operation, and monitor power levels. There are three common ways to monitor the radiation exposure to personnel: film badges, pocket dosimeters, and thermoluminescent dosimeters.

Film-badge dosimeters. Photographic emulsions are widely used in radiation dosimetry. A "film badge" consists of a small piece of film that is sealed inside an opaque container, which is clipped onto personnel's clothing. A film badge is used to measure photons and beta particles, and, by adding special filters, can also be used to measure thermal as well as fast neutrons. Film badges measure the cumulative dose over the exposure period, usually a month. At the end of this period, the amount of exposure of the film corresponds to the dose received by the wearer. A major limitation of the film badge is that the exposure cannot be continuously read.

Pocket dosimeters. Pocket dosimeters are small ionization chambers that are often used to measure the gamma dose absorbed by the wearer. These pocket dosimeters are widely used in health physics work. Unlike film badges, the pocket dosimeters can be constantly monitored.

Thermoluminescent dosimeters. Thermoluminescent

dosimeters (TLDs) are made of crystalline materials that trap radiation energy in their lattice structure. When the exposed TLDs are heated, the trapped energy is released in the form of light, which can be counted and equated to the dose level received. TLDs are reusable after they have been annealed, and have largely replaced film badges in industrial applications.

Besides monitoring exposure to individuals, radiation detection is also vital for other reasons. The reactor fluids must be checked for unusually high radiation levels, which could indicate that there is leakage between the radioactive and nonradioactive systems. The atmosphere must also be checked for radiation. High airborne radiation levels not only can endanger the crew but also can indicate a leakage of reactor fluids into the environment. Sample swipes of dust are taken and measured to monitor airborne radiation.

Finally, radiation detectors are used to monitor plant operation by providing information concerning plant parameters. Inside the core, various detectors are used to count the number of fissions occurring and thus measure the power level of the reactor.

Section 3
Nuclear Propulsion Applications

3.1 Naval Applications. In naval circles, the advantages that nuclear propulsion affords, in terms of almost indefinite underwater operations for submarines and the elimination of routine refuelings for surface ships, were immediately recognized. The early research and development efforts that were undertaken led to the launching of the nuclear submarine *Nautilus* with its pressurized-water reactor in 1954, with at-sea operations commencing in 1955. Since then, nations have developed and deployed many types of nuclear-powered submarines and surface ships around the globe. Table 4 is a listing of representative nuclear-powered naval ships.

3.2 NS *Savannah*—Pressurized-Water Reactor [31–32]. The potential of nuclear energy for commercial shipping influenced the then President, Eisenhower, to recommend the construction of a nuclear-powered merchant ship in the United States. In 1956 the Department of Commerce was authorized to develop and construct such a vessel for the purposes of demonstrating the peaceful use of nuclear energy and resolving developmental problems confronting commercial nuclear marine operations.

The NS *Savannah* had a length of 595 ft, a beam of 78 ft, and drew 29.5 ft of water. She was designed to carry a crew of 110 and 9300 tons of dry cargo. Fully loaded, she displaced 20,000 tons. Like many cargo ships of her era, she had the capability of carrying passengers with cabins for 60. Her turbines developed 22,000 shp and her cruising speed was about 21 knots. The machinery space arrangement of the *Savannah* is illustrated by Fig. 9 of Chapter 1.

The *Savannah*'s nuclear power plant was simple in principle. Uranium, artificially enriched to ensure readily fissionable atoms, was contained in fuel elements within the core. When the rods were withdrawn (see Fig. 3) a chain reaction started in the fuel. Fissioning uranium quickly heated the surrounding water to a high temperature; however, a pressurizer kept the water under enough pressure to prevent boiling (hence the name pressurized-water reactor). The primary hot water was circulated through the steam generators, as illustrated by Figs. 4, 5, and 6, where it gave up part of its heat to generate steam in a secondary loop. Steam from the steam generators drove the main turbines and the turbogenerators. After passing through the turbines, the steam was condensed and fed back to the steam generators. At full power the *Savannah*'s reactor core gave off heat energy equivalent to 80 MW.

Fuel for the *Savannah* was uranium that had zones of enrichment of 4.19 and 4.59% with an average value of 4.4%. This slight enrichment simplified the design of the fuel elements and the reactor by permitting the use of structural materials that were resistant to corrosion and radiation. The uranium, in the form of uranium dioxide, was compressed into pellets. The pellets were slipped into tubes of stainless steel called fuel pins. Uranium dioxide was chosen because it does not react chemically with water, has a high melting point, and can hold its shape at the high-temperature and high-radiation levels within a reactor.

The *Savannah*'s first core contained 17,000 lb of uranium, of which 668 lb was uranium-235. During its useful life, about 130 lb of uranium-235 could be fissioned within the core. With the first core loading, she was designed to cruise 300,000 nautical miles at a speed of 21 knots, increasing to 23 knots when necessary. The NS *Savannah* actually travelled 350,000 miles before the first refueling. To travel this distance any ship of the same size would consume four to five times its own weight in fuel; however, the *Savannah* used a quantity of nuclear fuel weighing less than one of her passengers!

Hot water from the reactor primary loop was circulated through two separate steam generators where some of its heat was given up to make steam in these separate secondary loops. This design isolated the turbine loop and engine room from any radioactive materials in the reactor primary, for there was no open path between the primary loop and the secondary loop. Two heat exchangers with independent pumps were installed to ensure reliable cooling of the reactor core. The steam generators generated up to 265,850 lb of steam per hour at pressures varying from 715 to 445 psi.

Table 4   Representative naval ships with nuclear propulsion[a]

Ship	Reactor Type/Number		Power, shp	Displacement, tons	Type	Flag
Seawolf	PWR[b]	1	60,000	9,150	fleet submarine	U.S.A.
Los Angeles	PWR	1	35,000	6,927	fleet submarine	U.S.A.
Ohio	PWR	1	60,000	18,700	ballistic missile submarine	U.S.A.
Nimitz	PWR	2	260,000	96,386	aircraft carrier	U.S.A.
Virginia	PWR	2	70,000	11,300	cruiser	U.S.A.
Vanguard	PWR	1	27,500	15,000	ballistic missile submarine	U.K.
Trafalgar	PWR	1	15,000	5,200	fleet submarine	U.K.
Le Triomphant	PWR	1	41,000	14,200	ballistic missile submarine	France
Rubis	PWR	1	9,500	2,670	fleet submarine	France
Charles De Gaulle	PWR	2	82,000	36,000	aircraft carrier	France
Han	PWR	1	. . .	5,000	fleet submarine	P.R.C.
Xia	PWR	1	40,000	8,000	ballistic missile submarine	P.R.C.
Kirov	PWR	2	150,000	28,000	cruiser	Russia
Typhoon	PWR	2	80,000	26,500	ballistic missile submarine	Russia
Delta III	PWR	2	60,000	12,150	ballistic missile submarine	Russia
Sierra	LMR[c]	2	40,000	8,100	fleet submarine	Russia

[a] Data obtained from *Jane's Fighting Ships*.
[b] Pressurized-water reactor.
[c] Liquid-metal reactor.

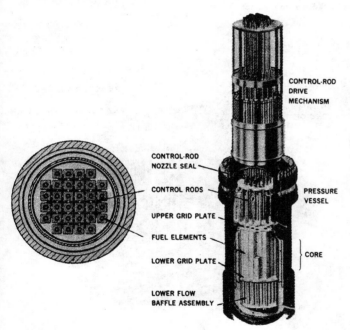

Fig. 3   Cutaway of NS *Savannah*'s complete reactor. Note cross-shaped control rods that fit between fuel elements

The *Savannah*'s reactor and other system components that may contain radioactive materials were enclosed in a containment vessel. In it were the reactor core in its pressure vessel, the water pumps, the steam generators, the steam drums, and the pressurizer. Figure 6 shows how compactly these parts were fitted into the containment vessel.

The *Savannah* had two distinct sets of shields. The first was built around the reactor vessel. It reduced the escape of neutrons and gamma radiation sufficiently to permit the crew to enter the containment vessel for short times after the reactor was shut down. The secondary shielding was outside the containment vessel. It would serve to reduce personnel exposure to radiation should a reactor accident release radioactive materials within the containment vessel. The primary shield consisted of a layer of water 33 in. thick, supplemented by a layer of lead. The secondary shielding was a combination of lead and polyethylene around the upper part of the containment vessel and concrete around the lower portion.

In 1971 the *Savannah* was retired from active service and in 1981 was put on display at Patriots Point Naval and Maritime Museum in Mount Pleasant, which is near Charleston, South Carolina.

The operating experience provided by the *Savannah* demonstrated that the pressurized-water reactor is a viable option candidate for merchant ship propulsion. The PWR offers the attractive characteristics of having a light-water moderator and coolant, high power density, and the ability to follow the load. The PWR is also characterized by a high capital cost, a high stored energy in the coolant, and the production of low-pressure saturated steam.

Since the development of the *Savannah*, considerable study work has been performed to improve the design of marine pressurized-water reactors. Many innovations have been made; however, the one having the most impact has been the inclusion of a once-through-type heat exchanger within the reactor pressure vessel.

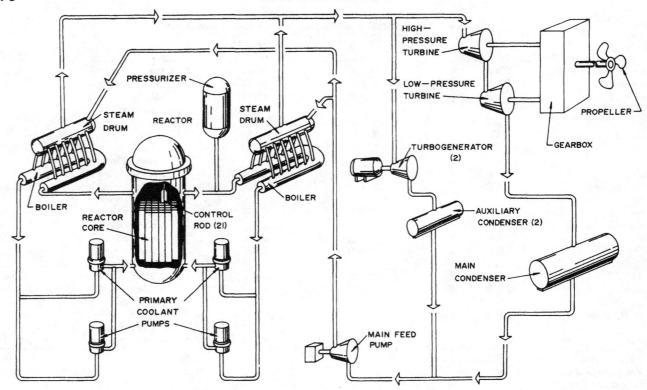

Fig. 4   Schematic diagram of NS *Savannah* reactor circuit

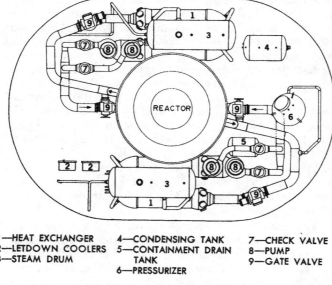

1—HEAT EXCHANGER   4—CONDENSING TANK   7—CHECK VALVE
2—LETDOWN COOLERS   5—CONTAINMENT DRAIN   8—PUMP
3—STEAM DRUM   TANK   9—GATE VALVE
                      6—PRESSURIZER

Fig. 5   NS *Savannah* system arrangement

Fig. 6   NS *Savannah* steam generating equipment

3.3   Non-Naval, Nuclear-Powered Soviet-Flag Ships
[33,34].   The first application of nuclear power to surface ships was aboard the *Lenin*, which the U.S.S.R. put into operation in 1960. The *Lenin* is a 44,000-shp icebreaker that displaces 19,200 tons. The thermal output of the three original pressurized-water reactors totaled 90 MW (thermal) and was used to generate steam in the secondary loop at 398 psi and 590 F.

The *Lenin* had a major problem in the 1960s and was not seen for many years. The *Lenin* plant was extensively modified in 1966, and subsequent operation was with two pressurized-water reactors instead of three. Also, the operating pressure in the primary loop was reduced from 2840 to 1846 psi. The reactor fuel for the *Lenin* is uranium-dioxide that is enriched to a level of 5% uranium-235 and clad by Zircaloy.

In 1974 a second nuclear icebreaker, the *Arktika*, of 23,500-tons displacement and 75,000 shp, was put into service in the U.S.S.R. and was equipped with the same type of reactor plant as the *Lenin*.

The *Taymyr*, the lead ship of a third class of Soviet-flag nuclear icebreakers, was put into service in 1989. The *Taymyr* is a 48,000-shp, triple-screw, shallow-draft polar icebreaker that was designed and built by Wärtsila Marine Industries in Helsinki with the exception of the reactor and turbines, which were installed by the Admiralty Shipyard in Leningrad. Previous generations of nuclear icebreakers had two reactors; however, there is only one pressurized-water reactor on *Taymyr*. The turbines are of a single-casing, double-flow design to take best advantage of the low-quality steam produced. The turbo-electric ac/ac propulsion machinery installed on *Taymyr* was determined to be superior to mechanical transmissions because of its ability to develop high outputs at low speeds.

The Soviet Union (now Russia) also has a 61,000-dwt nuclear-powered barge carrier in operation on their eastern seaboard. The *Sevmorput* was built at the A. Zaliv Shipyard at Kerch on the Black Sea. This ship was built to the highest level of ice class and is intended for service on the northern sea route. The *Sevmorput* is a single-screw, 40,000-shp barge carrier that uses the LASH (lighter-aboard-ship) system, which is suitable for operations in ice-laden waters. The single 135-MW pressurized-water reactor produces steam at 511 psi and 545 F in the secondary loop. Local environmental objections have encumbered port access to the *Sevmorput*.

3.4 The Babcock & Wilcox CNSG Pressurized-Water Reactor [35].

The Consolidated Nuclear Steam Generator (CNSG), which was developed by the Babcock & Wilcox Co. in conjunction with research and development support from the Maritime Administration, is a compact pressurized-water reactor that is designed for merchant marine applications. The CNSG reflects the experience gained during the design and operation of the *Savannah*, as well as central-station practices. The CNSG incorporates a once-through steam generator, which produces superheated steam at a constant pressure over the entire operating load range. The complete CNSG system consists of the reactor with its integral steam generator, pressurizer, reactor coolant pumps, control and safety systems, auxiliary systems, and instrumentation. The reactor can be shop-assembled to improve the quality control and minimize the erection time. The compact vapor suppression system provides both neutron attenuation and energy containment at greatly reduced post-accident pressures.

The CNSG discussed herein was designed to power a 105,000-shp containership. The reactor power level for this shaft horsepower requirement is 270 MW.

The unique feature of the CNSG design is that the major components of the steam system are located within the reactor vessel as shown in Fig. 7. Primary pumps and control rod drives are on the top of the pressure vessel. The reactor fuel is low-enriched uranium dioxide clad with Zircaloy. Reactivity control is accomplished by movable-cluster control rods and fixed lumped burnable poisons.

The steam generator is of the once-through, forced-circulation type and is located in the annular space between the core and pressure vessel. The once-through design enables the system to produce superheated steam, which permits an improved turbine efficiency and provides a greater margin for load changes without moisture carry-over.

The reactor vessel contains the reactor core and control rods, steam generator, pressurizer steam space, and all internal supports for the core, steam generator, and control rod drive line. The vessel is designed in accordance with the ASME Boiler and Pressure Vessel Code, Section III, Nuclear Vessels. The vessel is fabricated from carbon steel and is clad on the inside with stainless steel.

The suppression chamber of the containment is formed by the dry-well vessel and an outer concentric cylindrical pressure vessel. Venting connections from the dry well to the suppression chamber consist of pipes attached to openings in the dry-well vessel wall and extending into the suppression water. Low-pressure rupture disks normally seal these vent penetrations for humidity control in the dry well and prevent backflow of the suppression water at extreme ship roll attitudes.

The shielding design of the CNSG includes the use of lead, water, concrete, and steel. Operational shield requirements at core elevation are met by the considerable quantity of structural steel in the equipment, the suppression water in the containment, and the concrete shell wall. The suppression water effectively attenuates the neutron flux, and the concrete shell is sized to reduce secondary gamma rays and operating (primary) gamma rays to acceptable levels. The lead shield furnished is for post-accident shielding.

The 270-MW CNSG core is composed of 32 fuel assemblies and 32 cluster control rods. The core has an equivalent diameter of about 70 in. and an active length of 84 in. The total contained uranium is 12,583 kg at an average enrichment of about 4.1%. Radial power flattening is accomplished by zone loading with three enrichments.

Reactivity control is accomplished by a combination of fixed burnable poisons for lifetime reactivity loss and movable control rods for the temperature defect, void and Doppler deficits, xenon and samarium, and the subcritical shutdown margin.

Besides reducing the required space and weight for the ship's energy system, the consolidation and physical confinement of the CNSG's entire reactor coolant system within a single high-integrity vessel without external circulation loops provides such safety advantages as:

- The elimination of all large penetrations to the reactor vessel and reactor coolant system, thus significantly limiting the consequences of a credible loss-of-coolant accident.

- The elimination of piping, pump casings, and other secondary vessels as part of the reactor vessel wall.

- The elimination of field assembly and welding of the primary system with the attendant problems of quality control for the reactor coolant boundary.

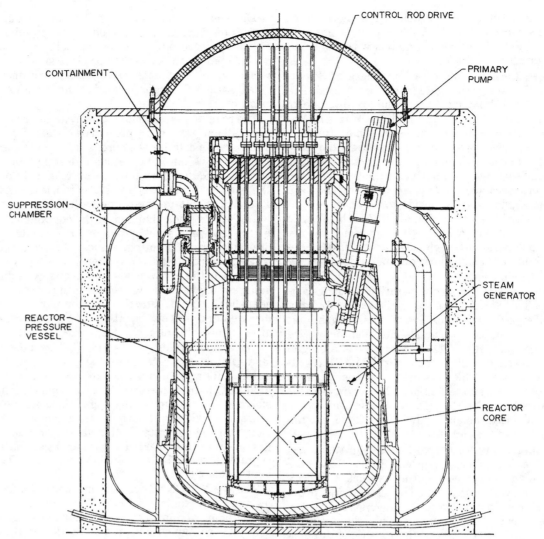

Fig. 7   Cross section of Babcock & Wilcox CNSG III reactor

- The elimination of any significant source of cold reactor coolant water that might cause a reactivity excursion.
- A reduction of the physical "target" size of the reactor coolant system as a consideration in ship collision accidents.

In addition to the foregoing features, the compact pressure-suppression arrangement for the CNSG design provides for energy containment at greatly reduced post-incident pressures and thus decreases the magnitude of the driving force for fission product leakage and of resulting fission product dispersion. The decreased size of the containment structure also enhances the ability of the surrounding ship's structure to provide collision protection.

Continual design developments have been incorporated into the CNSG III design illustrated by Fig. 7 to improve performance and to reduce weight, space, and cost for a given power output. Figure 8 is a cross section through

the CNSG IVA design. Some of the notable features of the CNSG IVA are:

1. The use of control rod assemblies and the elimination of follower rods.

2. A lowering of the core in the vessel, permitting a reduction in vessel height and water inventory.

3. The placement of pumps in the head of the vessel instead of at the side, permitting a reduction in the size and weight of the containment and shielding.

4. Electrically heated separate steam pressurizers.

5. Vapor suppression containment.

6. Steel and water around the entire reactor containment.

7. The use of lumped burnable poison in the core design to attain a high core burnup and attractive fuel economics.

8. Use of modular once-through steam generators derived from central station practice.

3.5 NS *Otto Hahn*—Pressurized-Water Reactor [33].

The nuclear-powered merchant ship, NS *Otto Hahn*, was

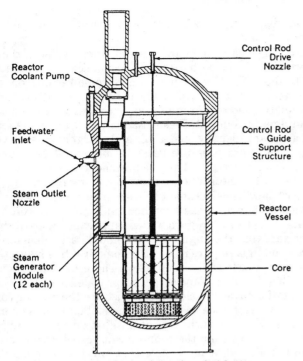

Reactor
Coolant Pump

Feedwater
Inlet

Steam Outlet
Nozzle

Steam
Generator
Module
(12 each)

Control Rod
Drive
Nozzle

Control Rod
Guide
Support
Structure

Reactor
Vessel

Core

Fig. 8   Cross section of Babcock & Wilcox CNSG IVA reactor

built in Germany, primarily as a research vessel and secondarily as a 14,000-dwt ore carrier. The Babcock & Wilcox CNSG design, with some modifications that included a dry-containment arrangement, was incorporated in the nuclear plant of the *Otto Hahn*.

The power level of the *Otto Hahn* reactor was 38 MW, which produced a normal shaft horsepower rating of 10,000 hp. The reactor coolant pressure was 903 psig at a core temperature of 533 F. To obtain flat neutron flux distribution and uniform fuel consumption, the core was divided into four radial enrichment zones having values of 2.77, 3.20, 3.89, and 4.87 percent, with an average value of 4.03% uranium-235. The initial core was designed for a full-power life of 500 days.

Several advances that had been established in the field of marine reactor design since the construction of the *Savannah* were incorporated in the *Otto Hahn*. Improvements of minor significance were effected in areas such as core design, flux shaping, burnup, lifetime, and control effectiveness; however, of major significance was the consolidation of the reactor-steam plant that was provided by the CNSG concept. Previous dispersely arranged pressurized-water reactors included a pressurizer and a steam generator, which was located outside the reactor pressure vessel. On the *Otto Hahn*, however, the pressurizer was eliminated and the steam generator was inside the reactor pressure vessel. The pressure in the primary loop was lower, and the use of a once-through steam generator permitted the use of a modest degree of superheat. The integral arrangement of the steam-generator over the reactor core facilitated primary-loop circulation such that about 25% of full power could be produced by natural circulation without using the main coolant pumps. The

consolidated arrangement of the reactor-steam plant enhanced the safety of the *Otto Hahn* by reducing the probability of an accident that would affect the primary coolant boundary.

The *Otto Hahn* was delivered in December 1968, and the first core was removed in October 1972. The second core went critical in March 1973 and logged 1,141 days of full-power operation before the *Otto Hahn* was taken out of service in 1979.

The *Otto Hahn* was not intended to be an economically competitive commercial ship. The ship was built to obtain the technical and practical knowledge required for the development of economically competitive nuclear merchant ships, and the *Otto Hahn* satisfied all expectations of both the builder and the operator. A comprehensive description of the design and early operational performance of the *Otto Hahn* is given in reference 33.

3.6   NS *Mutsu*—Pressurized-Water Reactor.   Japan's nuclear-powered ship, the NS *Mutsu*, was built to gain experience in nuclear shipbuilding, propulsion reactor design and operation, and nuclear-ship crew training and to acquire data necessary for future studies and work. The *Mutsu* was essentially a nuclear-powered demonstration ship that had special cargo transport and crew-training facilities. The *Mutsu* displaced 10,400 tons and had a propulsion plant that was rated at 10,000 shp. The *Mutsu* was equipped with the first marine power reactor to be built in Japan. The pressurized-water reactor had a rating of 36 MW and was used to generate steam at 738 psi and 590 F. The reactor fuel was stainless steel-clad uranium dioxide having enrichment zones of 3.2% and 4.4% [33].

The structure of the *Mutsu* was analyzed in great detail in terms of anticollision and stranding safeguard features.

The *Mutsu* was completed in September 1972 but was unable to put to sea until August 1974 because of protests from local fishermen. Soon after the reactor went critical during sea trials, alarms sounded, indicating an excessive radiation leakage. Subsequent investigations confirmed that the shielding, itself, performed as predicted; however, there was a leakage of high-energy neutrons (0.01 to 1 MeV) through a gap between the reactor vessel and the primary shield. Leaking neutrons were also evidently being reflected from the steel plates of the reactor containment and the ship's hull. The gamma rays, which were detected on the upper deck, were gamma rays that were produced secondarily as a result of the neutrons leaking through the gap in the primary shield being absorbed by the steel structure comprising the secondary shield. Technical corrections to the design errors in the *Mutsu* shielding were readily developed [36,37].

3.7   Application of Boiling-Water Reactors to Ship Propulsion.   Interest in the application of boiling-water reactors (BWR) to marine propulsion developed initially because of the possibility of reductions in capital cost as compared with that of a pressurized-water reactor (PWR). This reduction could result from the elimination of the heat exchanger and the associated piping inherent in changing from an indirect to a direct cycle.

The development work on the marine BWR has been relatively small and conducted principally by the European nations. The primary interest has been in the area of reactor response to the sinusoidal motion of the ship. Very little has been accomplished in reducing volume requirements, or capital cost reduction. The volume constraint on a marine reactor is considerably different than the constraint on a stationary plant, although both systems are improved if the power density is increased. Stationary boiling-water reactors have capital costs that are more sensitive to the diameter of the vessel than to its height. A marine reactor design is sensitive to both dimensions, and it is not possible to arbitrarily increase the height and reduce the vessel diameter. Also, with a stationary plant, as much importance is not placed on compactness of the entire steam generating system as is necessary for a marine plant.

The higher power levels present problems to the BWR from the physical size standpoint. The low power density of the BWR does not makes its physical size particularly compatible with ship dimensions at the high power ratings. Boiling-water reactors that have been designed for 200 MW are about 60 ft high, which is about twice the height that would be desirable for ship application. The PWR, because of its higher power density, is not pressed quite as hard to fit in the allowable space at the increased power.

There are two basically different means of obtaining coolant movement in a BWR: natural and forced circulation. In the natural-circulation system, the pumping cost is reduced, while in the forced-circulation system there is generally a reduction in size. Either system can be applied to marine propulsion. There is a reduced economic incentive for the forced-circulation plant because marine BWR plants are relatively small and the capital cost is therefore sensitive to additional components. On the other hand, natural-circulation reactors that must use free-surface steam separation and gravity-dependent driving heads have reduced performance as a result of the ship's motion. Both reactors suffer from the fact that when operating in a direct cycle they do not naturally follow load changes. This is not significant in stationary plants, but is of importance in marine plants because of the more frequent power changes. The forced-circulation plant can use recirculation flow control to overcome this disadvantage, without requiring rod movement.

Experiments and analyses have been conducted that indicate, for marine boiling-water reactors experiencing variations in the effective gravitational field due to ship motions, leaving velocities in the neighborhood of 1 fps can be considered as an upper limit. For a given reactor power and free-surface area, a vessel diameter can be calculated for this upper value of effective steam velocity. It can be seen that if the steam separation area requirements are reduced, a reduction of 20 to 30% can be made in the pressure vessel diameter depending on core shape and power density. The separation area requirements can be reduced by mechanical separation. If mechanical separation is employed to realize this diameter reduction in

a marine BWR, it will then be necessary to use forced circulation.

In either natural- or forced-circulation systems, the direct cycle must be used to be economically competitive with the PWR. The savings which would result from the elimination of recirculation lines and pumps in the natural-circulation indirect system would not overcome the disadvantages introduced. These disadvantages are in the form of a variable-density moderator and a larger pressure vessel. The larger pressure vessel would result because of the phase change and the requirement to condense the steam in the internal once-through steam generator.

Although considerable experience with central-station direct-cycle plants has been accumulated, there remains some reserve with regard to their application aboard ship. The fact that the engine room would be a limited-access area and thus require more planning of routine inspection is contrary to the usual ship policy. There also exists doubt whether isolation valves can be provided which, in the event of a nuclear accident, could isolate the reactor compartment from the machinery room. Plant components through which the primary coolant flows may be required to be located within the containment vessel. If this practice is followed, it is possible that only motor-generator drives could be considered for system components within a direct-cycle marine BWR.

3.8 Marine Gas-Cooled Reactor Concepts. Gas-cooled graphite-moderated reactors, in which the heat generated in the core is removed by carbon dioxide and transferred to the steam generators, have been used in the United Kingdom and France. Also, in the United States gas-cooled reactors have been developed using helium as the coolant.

A marine gas-cooled reactor (MGCR) program was undertaken to develop a conceptual design for a high-temperature, gas-cooled, closed-cycle gas turbine power plant [38]. The reactor was BeO moderated and was a 10.5% enriched, uranium dioxide, helium-cooled plant. The reactor exit temperature was to be 1500 F, with a cycle efficiency of 36.2%. Considerable materials development work in both the reactor and turbomachinery was necessary. Although the reactor and plant were attractive from a weight and volume standpoint, the BeO moderator and 10.5% enriched fuel entailed a fuel cycle cost and capital cost that were not competitive with other nuclear systems.

Borrowing upon technology that was developed for other applications, a conceptual design of a lightweight nuclear propulsion (LWNP) system for marine installations has been developed, as reported in reference 39. The power plant concept was based upon nuclear-rocket reactor technology and included a gas-cooled nuclear reactor with an integrally packaged gas turbine propulsion system (Fig. 9).

A compact closed Brayton-cycle power conversion system, using helium as the working medium, with recuperation and one stage of intercooling was determined to be the most advantageous. Figure 10 is a schematic diagram of the power plant cycle that was developed. Helium leaves the reactor at 1700 F and 1500 psi and enters the

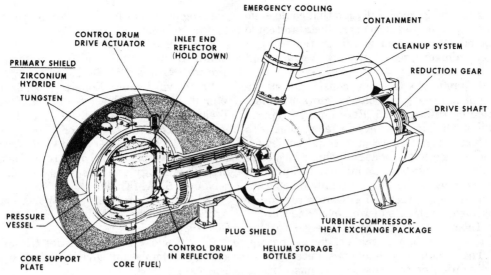

Fig. 9 Gas-cooled reactor system arrangement

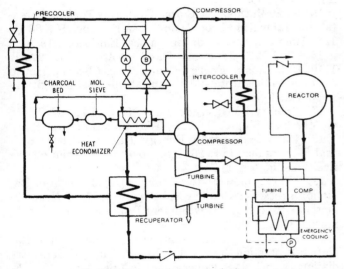

Fig. 10 Gas-cooled reactor cycle diagram

high-pressure turbine. These conditions are within the capability of state-of-the-art turbine blade materials and provide an attractive overall cycle efficiency. The high-pressure turbine drives the low-pressure and high-pressure compressors. The helium leaving the high-pressure turbine enters the low-pressure free power turbine and is then directed to the recuperator, which transfers heat to the reactor inlet gas, and then the precooler before entering the low-pressure compressor. An intercooler is located between the low- and high-pressure compressors to reduce the power requirements of the high-pressure compressor. The heat removed by the precooler and intercooler is transferred to an intermediate heat transfer circuit, which rejects the heat to a seawater-cooled heat exchanger.

The molecular sieve and cooled charcoal bed reduce the small amount of activity that could be in the primary system due to fission products. The clean helium leaving

the charcoal bed may be either returned to the cycle or directed to storage bank A or B.

The emergency cooling system provides the capability to remove reactor decay heat. This system transfers the heat to its own intermediate heat transfer system and from there to the ambient air via heat exchangers.

Figure 9 is an illustration of the lightweight nuclear power plant conceptual design that was developed to be a replacement module for conventional power plants in the range of 25,000 to 30,000 shp. The maximum dimensions of the plant are 8.75 by 14 by 26.5 ft, and the plant has a specific weight of 25 lb/hp. For clarity, some of the auxiliary components are deleted from Fig. 9.

The gas-cooled reactor shown by Fig. 9 is graphite moderated with coated fuel beads dispersed in graphite elements. The fuel beads used were developed and applied in commercial gas-cooled reactors and were designed to have a high retention of fission products within the fuel beads themselves. This forms a barrier to fission product release and is a significant safety feature of the system.

The reactor assembly consists of a fueled core that is surrounded by a reflector and internal shield. A pressure vessel encloses the reactor assembly. The reactor core consists of hexagonal fuel elements that are 30 in. long and measure 0.75 in. across the flats. The diameter of the fuel-element bundle is 35 in.

The free power turbine drives an internal planetary reduction gear, which provides rated power output at 3600 rpm and connects to external reduction gears. The use of a free low-pressure power turbine, which is not coupled to the high-pressure turbine that drives the compressors, avoids output shaft rpm restrictions that would otherwise be imposed by requirements for compressed gas.

The primary shielding function is accomplished by shielding internal to the reactor assembly and the borated zirconium hydride and borated water shields that are immediately around the reactor but outside the containment vessel. Shielding for the gamma radiation that could emanate from the small quantity of fission products, and could possibly contaminate the helium gas, is in the form of the thick-wall containment vessel and other component structures.

A survey of the numerous technical considerations entailed with the application of gas-cooled reactors in submarines is presented by reference 15. Reference 15 reports that the coupling of a helium closed-cycle gas turbine with a high-temperature gas-cooled reactor could achieve a power plant specific weight of about 20 lb/hp, which compares with a specific weight of about 100 lb/hp for a conventional pressurized-water reactor plant. When compared with a conventional pressurized-water reactor, a reduction in space requirements of well over one third is also achievable. In addition, a nuclear marine gas turbine has the capability of achieving a cycle efficiency that approaches 38%. The enabling technologies necessary to support the exploitation of the closed-cycle gas turbine potential are in varying stages of development, but their maturity is feasible.

Many details concerning a gas-cooled reactor propulsion plant conceptual design would have to be defined to permit meaningful cost estimates to be made; however, preliminary estimates indicate that the cost of a gas-cooled reactor plant would be about the same as the cost of a conventional pressurized water reactor plant. The principal reasons for the costs being the same are that the costs of the more expensive materials used in the gas-cooled reactor are offset by a reduction in material requirements, and that the predominant fuel cost is uranium, which is about the same in either case [39].

3.9 Appraisal of Other Marine Reactor Concepts. All reactor concepts other than the PWR, BWR, or gas-cooled thermal reactor have the disadvantage of not having land-based predecessors. It is highly unlikely that such systems would be able to obtain acceptance for marine applications without first being operated as central-station plants, unless they possess extremely attractive characteristics.

References

1   S. Glasstone and A. Sesonske, *Nuclear Reactor Engineering*, Van Nostrand, New York, 1967.

2   J. M. Harrer, *Nuclear Reactor Control Engineering*, Van Nostrand, New York, 1963.

3   M. A. Schultz, *Control of Nuclear Reactors and Power Plants*, McGraw-Hill, New York, 1961.

4   Frank Rahn et al, *A Guide to Nuclear Power Technology*, Electric Power Research Institute, Wiley-Interscience, New York, 1984.

5   J. G. Yevick, *Fast Reactor Technology*, MIT Press, Cambridge, Mass., 1966.

6   Raymond L. Murray, *Nuclear Energy*, Pergamon Press, New York, 1988.

7   John R. Lamarsh, *Introduction to Nuclear Engineering*, Addison-Wesley Publishing Co., Mass., 1983.

8   H. G. Rickover, J. M. Dunford, Theodore Rockwell, III, W. C. Barnes, and Milton Shaw, "Some Problems in

the Application of Nuclear Propulsion to Naval Vessels," *Trans.* SNAME, Vol. 65, 1957.

9   Alvin G. Kalmanson, "Nuclear-Powered Merchant Ships: Some Legal and Regulatory Considerations," *Trans.* SNAME, Vol. 83, 1975.

10   Zelvin Levine, Robert W. Dickinson, and Alvin O. Winall, "Nuclear Energy—A Viable Alternative," Ship Technology and Research (STAR) Symposium Proceedings, SNAME, 1977.

11   J. B. Montgomery and C. R. Jordan, "Commercial Marine Transportation of Arctic Natural Resources," Ship Technology and Research (STAR) Symposium Proceedings, SNAME, 1981.

12   L. Jacobsen et al, "Transportation of LNG from the Arctic by Commercial Submarine," *Marine Technology*, Oct. 1983.

13   "Nuclear Terms, A Brief Glossary," U.S. Atomic Energy Commission Division of Technical Information, Tennessee, 1969.

14   "The ORNL Gas Cooled Reactor," ORNL-2500, Oak Ridge National Laboratory, U.S. Government Printing Office, 1958.

15   C. F. McDonald, "Closed-Cycle Gas Turbine Potential for Submarine Propulsion," ASME, Paper No. 88-gt-126, American Society of Mechanical Engineers, 1988.

16   "LMFBR Program Plan," WASH 1101-WASH 1110, U.S. Atomic Energy Commission, Clearing House for Federal Scientific and Technical Information, Springfield, Va., 1968.

17   "Safety Considerations for Nuclear Power Plants on Merchant Ships," SNAME T&R Bulletin No. 3-18.

18   U.S. Department of Energy, Title 10—Code of Federal Regulations; Part 20, U.S. Government Printing Office, Washington, D.C.

19   "Background Material for the Development of Radiation Protection Standards," Report No. 1, Federal Radiation Council, U.S. Government Printing Office, Washington, D.C., 1960.

20   "Background Material for the Development of Radiation Protection Standards," Report No. 5, Federal Radiation Council, U.S. Government Printing Office, Washington, D.C., 1964.

21   "Report of Committee II on Permissible Dose for Internal Radiation, 1959." Publication 2, International Commission on Radiological Protection, Pergamon Press, New York, 1960.

22   "General Design Criteria for Nuclear Power Plant Construction Permits," Federal Register 32FR10213, July 1967.

23   "Part 55—Nuclear Pressure Vessels," Title 46, Code of Federal Regulations.

24   "Maximum Permissible Body Burdens and Maximum Permissible Concentrations of Radio-Nuclides in Air and in Water for Occupational Exposure," *NBS Handbook*, 69, U.S. Department of Commerce, U.S. Government Printing Office.

25   "International Conference on Safety of Life at Sea, 1960," Chapter 8 and Annex C.

26   "Guide for Nuclear Ship Classification," American Bureau of Shipping.

27 "FAST-2 NS *Savannah*—Port Operations Criteria," Contract No. MA-3890, April 1965.

28 "Nuclear Vessels," Section III, ASME Boiler and Pressure Vessel Code, American Society of Mechanical Engineers.

29 "Code for Nuclear Power Piping," ANSI B31.7, American National Standards Institute.

30 "Criteria for Nuclear Power Plant Protection Systems," No. 279, Institute of Electrical and Electronic Engineers.

31 "The Power Plant for the First Nuclear Merchant Ship (NS *Savannah*)," Nuclear Merchant Ship Symposium, Contract No. AT(30-3)-274, Aug. 1958.

32 G. E. Kulynych, "Description of the NS *Savannah* NMSR," BAW-1164, Final Safeguards Report, Atomic Energy Commission, Vol. I, III through VIII, Contract No. AT(30-3)-274, June 1960.

33 D. Ulken, "NS *Otto Hahn*," *Trans.* Institute of Marine Engineers, 1971.

34 "*Taymyr:* A New Nuclear Icebreaker for the Soviet Union," *The Naval Architect*, Sept. 1989.

35 "CNSG—Development and Status 1969–1977, Competitive Nuclear Merchant Ship Program," Babcock & Wilcox Co., Maritime Administration Office of Advance Ship Development, Contract 4-37067, March 1979.

36 "Nuclear-Powered Ships," *Bulletin of the Marine Engineering Society of Japan*, Sept. 1975.

37 W. Jager and E. Shroder, "Waiting for the Ice to Thaw," *Marine Week*, Aug. 29, 1975.

38 K. A. Trickett, "A Review of the Maritime Gas-Cooled Reactor Program," Maritime Gas-Cooled Reactor Program, General Atomic Division, GA-2603, Contract No. AT(04-3)-187, Dec. 1961.

39 Robert E. Thompson and Richards T. Miller, "Marine Applications of an Advanced Gas-Cooled Reactor Propulsion System," Ship Technology and Research (STAR) Symposium Proceedings, SNAME, 1977.

CHAPTER VIII

J. A. Beverley

Electric Propulsion Drives

Section 1
Basic Considerations

1.1 Significant Features. Electric propulsion drives offer a number of important advantages that have the potential of outweighing the inherently higher first cost, increased weight and space, and the higher transmission losses of such systems, when compared with other conventional types of propulsion drive alternatives. The electrical connection between the generator and the propulsion motor provides a freedom of arrangement not offered in a mechanical drive system. Power requirements can be met by one or more generating sets that are located in compliance with space restrictions, or to alleviate weight and stability problems. Combinations of different types of prime movers such as diesels, gas turbines, and steam turbines are easily accommodated once their mechanical output has been converted to the common denominator of electric power. For unidirectional prime movers, an electric drive has the inherent ability to provide the required reverse rotation of the propeller by relatively simple control means.

In cases where the development of the desired propeller power requires the use of multiple prime movers, a typical situation in medium- and high-speed diesel drives, an electric-drive provides a convenient means of coupling several units to the propeller without the use of mechanical clutches or couplings. An electric-drive system is normally arranged so that vessel operation at less than full power can be accomplished with a minimum number of prime movers in service, each operating near peak efficiency. This contributes to more efficient vessel operation and affords downtime for scheduled maintenance on units not required for propulsion. The higher transmission loss of the electric system may in fact be more than offset by a better match between the prime mover capacity and power demand.

The ease and convenience by which the propeller speed and direction of rotation are controllable in an electric drive are among its attractive features. This control can be effected from a number of remote locations, and thus put directly in the hands of the vessel navigator or operator as well as dynamic-positioning computers. The electric control system lends itself to any desired speed of response and to the incorporation of time constants or limits, so that the machinery is not subjected to abusive or dangerous operation.

It is possible, and oftentimes convenient, to select a low-speed motor that matches the desired propeller speed directly without the use of reduction gearing. This motor can be supplied with power from high-speed generators, thus providing the necessary speed reduction between the prime mover and the propeller. Geared motors can be considered as a means of reducing the size and cost of the electric motors. High-speed motors for use with reduction gears are lighter and less costly than their low-speed direct-drive counterparts, but any comparison must include the cost and weight of the added gearing and foundations.

Main propulsion generator sets are usable as a source of power for other functions when their full output is not required for propulsion power. Electrical power for cargo pumps, cargo refrigeration, or dredge-pump operations is illustrative of such supplemental services. Ship-service power derived from the propulsion bus is an attractive and often-used feature on many electric-drive vessels.

Electric drives have been built in sizes to 60,000 hp per shaft. However, while such high power applications are noteworthy, most electric drives are of more modest ratings.

1.2 Electric-Drive Applications. Electric-drive systems are applied when their ease of control, flexibility of arrangement, and adaptability to multiple use of the prime mover generator sets provide a distinct advantage. In general, propulsion electric drives are used in the following types of applications:

• *Vessels requiring a high degree of maneuverability*— Vessels such as ferries, icebreakers, tugs, oceanographic vessels, and cable-layers are typical of this group. In these cases, changes in speed and in direction of propeller rotation are frequent and of vital importance to the successful operation of the vessel. Most vessels of this type have used d-c machinery because of the superior speed control inherent with such machines. Even with the availability of a-c static power converters, the d-c motor, with its high transient torque capability, is often the machinery of choice.

- *Vessels requiring large amounts of special-purpose power*—Vessels such as self-unloaders, fireboats, self-propelled dredges, drill ships, oil-recovery vessels, large tankers, and tenders often have large electric power requirements that are not coincident with maximum propulsion power requirements. The nature of the operational requirements for such vessels permits the generating sets to be applied to supply large amounts of power for pumping or cargo handling when propulsion power demands are low or even nonexistent. This "either-or" requirement for power allows a capital cost saving by reducing the total amount of power capacity that is installed.

- *Vessels with large hotel loads*—The large hotel loads on cruise ships, coupled with varying propulsion power needs, present the opportunity for scheduling the on-line operation of multiple prime movers to meet the cyclic load demands with the most efficient prime mover selections. Unlike the vessels described in the preceding paragraph, the total installed horsepower of a cruise ship must meet the simultaneous demands of both propulsion and ship-service loads. The power generation plant is, in effect, a small utility, delivering power to electric propulsion motors and the hotel loads at the most economical fuel rate. Multiple diesel-generator sets are commonly used in such applications; however, simple- and combined-cycle gas turbine and heat-recovery steam turbine (COGAS) systems are among other feasible alternatives.

- *Vessels using nonreversing, high-speed, and multiple prime movers*—Gas turbines and many high-speed diesel engines are of the unidirectional type and frequently are installed as multiple units to produce the required prime-mover power. In such cases, an electric drive provides the means for reversing the propeller, converting the high speed of the prime movers to suitable propeller speeds, and also electrically coupling the multiple units to a single drive.

- *Deep-submergence vehicles*—Deep-submergence vehicles usually employ relatively small amounts of propulsion power and move at low speeds. The energy is obtained from batteries and is applied to multiple motors either directly or through power converters.

1.3 Types of Electric-Drive Systems.

Generating sets on electric propulsion systems are arranged in either dedicated or integrated configurations. A dedicated system is one in which the propulsion generators do not simultaneously also serve ship-service loads. For example, a propulsion generator directly connected to a high-speed turbine may produce power at a frequency considerably higher than the standard values of 50 to 60 Hz, making such power unsuitable for use as ship-service power. In other cases, a dedicated configuration may also be required if transient excursions of either frequency or voltage on the propulsion bus make the power unsuitable for direct use on the ship-service bus.

An integrated system uses one or more buses that are electrically connected to distribute power to both propulsion and ship-service loads. The bus frequency and voltage are maintained within tolerances that permit the bulk of the ship-service loads to be served directly or through step-down transformers. However, some sensitive loads on the vessel may require isolation from harmonics generated by large propulsion power converters. Because the sensitive loads are generally not large compared to the size of the power system, common practice is to provide motor-generator sets to achieve complete isolation for the sensitive loads.

A variation of the integrated design is made possible by designing suitable switching arrangements to allow a generator to be switched to either propulsion or ship-service switchboards, but not both simultaneously. A tie circuit breaker between propulsion and ship-service switchboards may be added to allow the operator to select between dedicated and integrated plant operation. However, this additional flexibility is gained at the expense of additional control complexity, space and weight requirements, and cost.

1.4 Power Generation and Conversion.

Nearly all power generated on ships is accomplished using a-c generators. The term "d-c propulsion system" implies that a propulsion system includes a power conditioner to convert a-c to d-c, such that a d-c motor can be used. An "a-c propulsion system" consists of a different type of power conditioner and an a-c motor. An additional descriptive term is often included in relation to a-c propulsion systems, to further describe and classify the type of a-c power conditioner. Alternating-current power conditioners include cycloconverters, pulse-width-modulated inverters, and load-commutated inverters. These three basic generic types of power conditioners, and variations thereof, are manufactured under various trade names for use in marine electric propulsion systems.

Basic power conditioners, such as a cycloconverter or a load-commutated inverter, have been used since the early part of the century. The trend to the almost exclusive use of power conditioners in the electric propulsion of ships is largely the result of remarkable developments in two branches of technology: power electronics and microelectronics. Together they have provided the capability to achieve the dense packaging and computational ability necessary to control large amounts of a-c power within the often restricted confines of a ship.

Direct-current propulsion systems are normally applied in the low- and moderate-power ranges. The usual applications of d-c motors are on vessels requiring between 600 and 10,000 hp, while a-c motors are generally used in installations above 10,000 hp. No rigid rules govern the type of propulsion motor, because both types of machines are capable of delivering excellent performance and reliable service. However, physical limits do exist on the maximum practical size of d-c machines.

Figure 1 shows an electric drive arrangement in basic block diagram form, indicating the relative placement of major system elements: prime movers, switchgear, power converters, and motors. Although the illustration shows multiple prime movers and a single motor, any combination of units may be used as long as the total developed power can satisfy the connected load. Also, there are no restrictions concerning the variety of prime movers used to drive the generators.

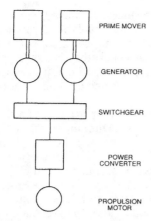

PRIME MOVER

GENERATOR

SWITCHGEAR

POWER
CONVERTER

PROPULSION
MOTOR

Fig. 1 Basic electric drive with multiple prime movers

The use of higher-speed prime movers results in a considerable reduction in the size and weight of machinery, and the use of engines with rotational speeds as high as practicable, consistent with reasonable maintenance and life, is usually desirable. The rated capacities of generator sets may be selected on the basis of an operating profile that allows some units to be secured during reduced-power operations. To facilitate load sharing among generator sets, an electric governor operating in the isochronous mode with automatic load sharing is usually provided.

1.5 System Voltage Selection. The voltage selected for an electric propulsion system must reflect the necessity for reasonable cable sizes and efficient rotating machine design and take into account national standards that define available ratings for circuit breakers. As the size of a power system increases, the voltage level required on the main bus also increases. The relationship is not linear, because the voltage classes and current ratings of switchgear are in discrete steps. For example, U.S. standards define preferred ratings of 480, 600, 4160, 6900, and 13,800 volts [1,2].

For marine power systems, the maximum current rating of available circuit breakers is often the determining factor in choosing the voltage for the main power bus. For example, on low-voltage (600 V) systems, the largest frame size of the circuit breakers listed in the preferred rating table of ANSI/IEEE C37.16 would restrict the generator to about 3300 kVA in a 50 C ambient [2]. Some manufacturers have circuit-breaker products available in current ratings, in excess of the published tables, which allow the use of higher-rated generators at the 600 V level. When multiple low-voltage machines are connected in parallel on the same bus, the resulting increase in short-circuit current may also be a deciding factor in selecting a higher voltage for the power system.

Breakers in the higher-voltage classes are available with current ratings that will accommodate generators rated well in excess of 30,000 kVA. However, even in the higher-voltage classes, careful attention must also be paid to the short-circuit capabilities of the switchgear, to insure that equipment ratings are not exceeded. The short-circuit calculation techniques and procedures recommended by IEEE [3], or other appropriate standards, can be used to confirm that suitable short-circuit capabilities have been provided at the voltage level under consideration.

Circuit breakers in the 600 V class have integral overcurrent tripping devices and do not require an external source of power to open under fault conditions. Higher-voltage circuit breakers generally use externally mounted current transformers with protective relays to sense a fault condition and must have power applied to a tripping coil to open. A reliable source of tripping power, such as a battery, is therefore required to ensure operation of the higher voltage circuit breakers under fault conditions.

1.6 Control and Regulating Systems. The successful operation of an electric propulsion system requires the interaction of a number of regulating and control systems acting on the prime movers, generators, and power converters. Some of these systems enhance the performance and stability of the power system, while others contribute to the reliability and availability of the power system by preventing unintentional overloads that could ultimately shut down the power system. The most significant of these regulating and control systems, from a system design point of view, include the following:

• *Prime-mover load sharing*—A governor is used to control the real power (kilowatts) delivered by the generator set to the power system. Prime-mover governing systems of the electronic type are commonly used to provide constant frequency on the main bus. The amount of load being carried by each prime mover is continuously monitored and the fuel supplied to each unit is adjusted to ensure that the load is proportionally shared. While operating in this manner, the governor maintains constant speed for all loads within the capability of the prime mover. This mode of operation is called "isochronous" and is preferred to a "drooping" characteristic in which the prime mover's speed is allowed to decrease as the load is increased. Often, the electronics for the governor are located in the switchgear, providing convenient access to the generator's current and voltage transformers, which are used by the governor to sense load.

• *Generator load sharing*—Automatic generator voltage regulators are required to maintain the terminal voltage of the generator at rated conditions. When generators are connected in parallel on a common bus, the voltage regulator also provides the means to proportionally share current among generators. In an arrangement which is analogous to the governor's proportional load-sharing feature, the voltage regulators for all generators have common generator current sensors that are used to adjust the field current in each generator to achieve a proportional current division. This mode of regulator operation is called "crosscurrent compensation" and is preferred to a "drooping" characteristic in which the generator's voltage is allowed to decrease as the load is increased. The voltage regulator requires an input from the generator's current and voltage transformers; therefore, it is commonly mounted in the switchgear.

• *Automatic load shedding*—Automatic load shedding is the process of disconnecting preselected loads from the system in response to an abnormal overload condition, such as may be caused by an unscheduled shutdown of a generator set, in order to maintain the load within the system's capability. A power system will operate at a progressively lower frequency when the kilowatts demanded by the load exceed the power available from the prime movers. This frequency reduction will generally lead to a shutdown of all generators by the underfrequency protective system, unless corrective action is taken to restore a balance between the power generated and the load. In an integrated power system, the removal of preselected ship-service loads may not be sufficient to balance the system. A reduction of propulsion power also may be required, because it is likely to be the single largest load. Regulators called power limiters (which are discussed in the following paragraph) are located in the propulsion power converters and are designed to take advantage of the significant drop in shaft horsepower that accompanies a modest reduction in the propeller shaft speed.

• *Power limiting*—A reduction of the propulsion motor speed by an automatic power-limiting regulator is a technique commonly used to compensate for a sudden loss of generating capacity such as might occur during an unscheduled shutdown of a diesel-generator set. The same regulators prevent overloading the system in the event that the throttle is advanced to call for more shaft power than is currently available from the power system. Two regulator functions are usually combined to monitor the status of generator sets and make adjustments in the propulsor's speed. The first function, called "kilowatt" or "real-power" limiting, provides a limit to the amount of power that can be drawn from the prime mover. The second is "kVA" or "current" limiting, which provides a limit to the amount of current (and therefore a limit on reactive power) drawn from the generator. By including both functions in the regulator, provisions are made for a variety of abnormal operating conditions. Both of these features have been widely applied. An alarm is usually sounded to inform the operator that the regulator is controlling the power or current demand.

The application of an automatic regulator requires kilowatt and kVA transducers to be installed in the switchboard so that signals proportional to load demanded can be sent to the power-converter regulator. When the regulator determines that the power demand is exceeding a preset value, such as 95% of the total on-line generator capacity, the regulator reduces the output of the power converter to maintain the generator set loads at 95% of rated capacity. The change in power is accomplished in a very brief time. The rationale for this approach is that it would be preferable to reduce the load than experience a complete loss of power. If another generator set were available, it would be placed on line by either an operator or an automatic power management system. At power levels below the set point, the regulator has no effect on operation of the propulsion plant.

• *Throttle controls*—The electronic regulators supplied with power converters are usually fitted with features to promote smooth yet responsive control of the main propulsion motors in response to movements of the throttle. Included in the features are preset timing ramps that provide a controlled rate of motor speed change that compensates for rapid throttle movements, permitting machinery such as diesel governors, turbochargers, and voltage regulators to follow the increased power demands in an orderly manner.

In addition to timing ramps, throttle controls usually include current-limiting circuits to provide continuous and nonintrusive overload protection for motors by allowing either a programmed value or full torque to be developed at any shaft speed. This is especially useful during vessel acceleration, and in heavy seas. When studies indicate that regenerative power during a full-speed shaft reversal is likely to raise the bus frequency beyond desired limits, an automatic propeller reversal limit may be programmed into the regulator to limit the bus frequency or add load-absorbing resistors to the bus.

• *Propeller blade-position controls*—For multiple-shaft vessels, the rotational speed of the shafts can be synchronized by the regulators to eliminate vibrational beats resulting from propeller blade-rate excitation that is caused by slight differences in propeller speeds. On some ships where the pressure pulsations due to blades passing by hull appendages have caused objectionable structural vibration, regulators have been programmed to control the relative blade position of one shaft versus the other. The objective in this instance is to offset the blade positions and thereby cause the excitation from the two shafts to be out of phase, eliminating the reinforcement of pressure pulsations.

1.7 Generators for Electric Drives. Generators applied to most ship-service power systems are designed to supply a normal mix of induction motors and lighting loads. The reactive power demands of these systems can usually be satisfied with industry standard designs rated at 0.8 power factor. However, the nonlinear nature of current flow in static power converters may result in increased demand for reactive power and increased generator heating in electric-drive vessels. Therefore, generators for electric-drive ships should be designed with a power factor capability to specifically meet the needs of the electric-drive system it is intended to support. Most electric-drive systems, with the exception of icebreakers, use either 0.8 or 0.7 power factor machines. Icebreakers, because of their need for high overload torques, may require machines with power factor ratings as low as 0.5.

In addition to being designed for the proper power factor, the design of a generator used with a static power converter must include allowances for the extra heating that occurs as a result of harmonic currents flowing in the windings. The degree that the converter configuration affects the design of the generator is generally an inverse function of the number of pulses produced by the converter: as the number of pulses increases, the wave form has harmonics of smaller amplitude, which have less effect on the design of the generator.

Turbine-driven, 3600-rpm generators are especially sensitive to harmonic currents. Because of the high mechanical stresses in 3600-rpm machines, the rotors are made from solid forgings, which tend to have higher losses for harmonic currents. The windings on these rotors are held in place with steel wedges, which exhibit similar high loss characteristics. Generators can readily be designed for power converter loads that have a minimum of 24 pulses per cycle; however, for loads having less than 24 pulses per cycle, 3000- or 3600-rpm generators usually have custom-design features.

Excitation for the generator is often provided by a brushless exciter mounted directly on the generator shaft. The voltage regulator, which controls the generator's output by varying current into a small winding on the brushless exciter, is usually mounted in the switchboard, since voltage and current sensing is also performed there. Generator reactances for machines in this service are usually designed to maintain the machine direct-axis subtransient reactance in the range of 12 to 20%. A low value is preferred since it directly affects the voltage distortion on the generator bus. Unfortunately, a low subtransient reactance yields high short-circuit levels, which tend to increase the cost of the switchboard and power converter. The two objectives, low voltage distortion and tolerable short-circuit levels, must be in balance to properly control this aspect of the generator design. A suitable compromise is usually established within the range of 12 to 20%.

Smaller synchronous generators can be ventilated with ambient air by a shaft-mounted fan. To remove the heat, the generator space should be supplied with 70 to 100 cfm of air per kilowatt of generator loss. In larger installations where the removal of the heat losses by ventilating air is not practical, generators are fitted with a closed ventilating system that includes a shaft-mounted fan, which forces air through a water-to-air heat exchanger. This arrangement is referred to as a totally enclosed, water-to-air-cooled (TEWAC) machine [4]. Heat exchangers for these machines are usually furnished in a double-tube construction in order to ensure that water from tube leakage cannot reach the windings, where it could cause a failure of the insulation provisions.

1.8 Transformers for Electric Drives. In many of the larger d-c and nearly all a-c electric-drive installations, a transformer is used to match the higher bus voltage to the lower voltage required by the static power converter. Transformers are also used to provide phase shifting of the a-c supply voltage for drives with more than six pulses. These transformers are, therefore, subject to the harmonic currents of the power converters, and are usually designed in accordance with the practices specified in standard ANSI/IEEE C57.110 [5]. Transformers not designed in accordance with this standard are not excluded from converter application; however, they are often derated by following specified procedures. Transformers designed for marine service must be specially braced and reinforced to withstand the inertia forces developed by the heavy coils and magnetic structure as a result of ship motions.

Dry-type transformers are generally applied in shipboard drive applications. However, liquid-filled transformers are routinely used in industrial applications particularly when the voltage level exceeds 4160 V or the size exceeds 10 MVA. If liquid-filled units are considered for marine drives, the tank must be designed to insure that the coils are covered by fluid under all ship motions. Flammability varies among transformer fluids and must be considered when locating the unit. When air-cooled transformers are used, ventilating systems should be arranged to maintain the temperature of the space well within the transformer ratings under all operating conditions.

An estimate of the transformer losses should be obtained from the manufacturer, since for large systems even a 1% loss can represent a significant load on the HVAC system.

1.9 Power System Harmonics and Filtering. All static power converters, including those applied to marine drives, appear as nonlinear loads to the power system and, therefore, have the potential of introducing harmonic currents into the power system. Harmonic currents can adversely affect the quality of ship-service power and the operation of instrumentation and protection features, and furthermore can introduce electromagnetic interference [6].

Experience indicates that the miscellaneous induction motors of the size found on ships are not adversely affected by the voltage distortion that is common on vessels with power converters, and no special precautions have been required [7]. Energy-efficient motors, because of their generally more conservative designs, have a greater tolerance for systems where harmonics are present, and are usually recommended for these applications.

On vessels where the ship-service power is derived from the propulsion bus, an investigation of the power quality requirements of the loads served must be made to ensure compatibility. Unfortunately, many equipment manufacturers have not tested or designed equipment to operate in an industrial or marine environment and have no specific knowledge concerning the power quality tolerance limits of their equipment. When power quality tolerance limits are not known, the 120 V loads composed of entertainment, navigation, and communication equipment are usually isolated by supplying them from small motor-generator sets, while the less-sensitive induction-motor loads are usually fed directly from, or via voltage-matching transformers connected to, the main bus. Uninterruptable power supplies (UPS) also isolate loads from the power bus; however, most UPS equipments use static power converters as input devices and are, therefore, themselves susceptible to, and also a source of, harmonic currents.

In some cases filters are added to the power system to reduce the amount of voltage distortion caused by harmonic current flow from d-c and a-c drives using load-commutated inverter or pulse-width modulated technology. Filters provide a low-impedance path for harmonic currents, which flow mostly through the filter, bypassing the other system components. Filters consist of capacitor, inductor, and resistor assemblies that are chosen to resonate at or near the harmonic frequency requiring suppression. To have a significant effect on the power system, three or more filters tuned to different frequencies are often required. When more than one voltage level is used on a ship, the filters should generally operate on the same voltage level as the power generation. The disadvantage of filters is that they are costly, large, and introduce losses. Also, the dynamic interactions between the filter capacitors and other system inductances must be carefully considered when designing a filter system to ensure that a filter does not exacerbate the voltage distortion.

When the voltage and current of a power system contain significant harmonics in addition to the fundamental, the peak values of the voltage and current waves may not equal the square root of two times their root-mean-square values [7,8]. As a consequence, instruments that only measure the peak or average value of a wave form may not read as accurately as those designed to be unaffected by harmonics. Attention should be paid to the selection of measuring devices for current, voltage, real power, and reactive power for switchboards as well as remote monitoring and alarm systems to be sure that their internal circuits are designed to provide accurate readings on power systems with a significant harmonic content. Protective relays and integral trip units on circuit breakers should be selected from those devices that respond to the root-mean-square value of the current flowing, so that they will not be subject to false tripping in systems with static converter loads.

The harmonic currents of the largest amplitudes associated with power converters occur at relatively low multiples of the fundamental power frequency and are within the audio frequency spectrum. A higher-frequency phenomenon, commonly referred to as "commutation notching," is also produced by the power converter because of the rapid changes in current that are associated with the physics of the converter's operation [9]. The higher frequencies are likely to affect the operation of equipment susceptible to either audio- or low-frequency radio interference. A proper system layout and physical separation of power cables from communication cables are important and usually sufficient steps in minimizing the effects of this phenomenon. To minimize the propagation of higher frequencies to other voltage levels, by capacitance coupling between transformer windings, grounded shields between the primary and secondary windings are recommended.

1.10 Grounding. Resistance grounding of the power system is recommended to limit the fault-energy release in equipment and ground fault currents through the ship structure. Grounding resistors are selected to overdampen the naturally occurring capacitance in the cables and rotating equipment [10] as well as to thermally withstand the power dissipation occurring on a bolted line-to-ground fault. Usually, the grounding scheme involves detection and alarm rather than tripping of the circuit breaker. System designs where the continuous line-to-ground fault current exceeds the nominal recommended range of 5 to 10 amperes should be avoided because of the potential of damage caused by the magnitude of the energy released at the point of the fault.

Section 2
Direct-Current Propulsion-Drive Systems

2.1 System Characteristics. Equipment for d-c power generating systems usually consists of multiple medium- or high-speed diesel generator sets paralleled on a common bus, producing a-c power which is then converted for propulsion. The sizes of the engines may be selected on the basis of an operating profile that allows some units to be shut down when a particular mode of operation demands less than the total installed power. To take full advantage of this design approach, the generator sets are sometimes selected at two different ratings.

The electric output of the generator sets is connected to the main power bus by circuit breakers located in one or more switchboards. These circuit breakers provide the means to connect and disconnect the sets and satisfy requirements for overload and short-circuit protection of the machinery. Frequently, an integrated bus system is provided to enable the ship-service power to be taken from the same bus via transformers or motor-generator sets.

Because of advantages in size, weight, and cost, 600 V systems are preferred for small- and medium-sized systems. However, when the generator rating exceeds approximately 3300 kVA, which may occur with medium-speed engines, the full-load or fault current can exceed the rating of available 600 V circuit breakers. In that case the next-higher voltage class, 4160 V, is chosen. All 4160 V systems require a transformer to reduce the voltage to a suitable drive input level, usually 600 V for drives above 500 hp.

An adjustable-voltage d-c power converter is connected to the bus by a circuit breaker or by a combination of a circuit breaker and a transformer. Omission of the transformer, where appropriate, is desirable to save cost and space. The converter rectifies the incoming a-c voltage to produce adjustable d-c for the armature of the propulsion motor.

The most common converter configuration used for motors rated up to several thousand horsepower is a six-pulse converter. Figure 2 shows the arrangement and ratings of the d-c electric-drive equipment used on a typical research vessel. Six-pulse converters are used for adjustable speed control of a bow thruster and two direct-drive, d-c main propulsion motors. In this case the generators are arranged in parallel on a common 600 V bus that also supplies 480 V ship-service power through step-down

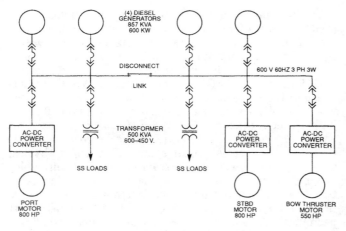

Fig. 2   Basic d-c drive for a research vessel

transformers. Each of the three six-pulse power convert-ers is directly connected to the 600 V bus to avoid the additional space and weight associated with transformers.

Higher-horsepower motors and motors with special acoustic requirements are typically supplied with d-c from converters with twelve or more pulses. The choice is often one of economics, with twelve-pulse current being favored only on larger systems, since a transformer is necessary with a twelve-pulse arrangement.

The d-c motor armature voltage selection is influenced by two factors. First, motor voltage design considerations have resulted in a somewhat arbitrary upper limit of 1000 volts. This limit is considered prudent for d-c machines that are to operate under shipboard conditions of vibra-tion, high humidity, and frequent maneuvering, and has been established by the IEEE Committee on Marine Transportation [11].

Second, for the many systems with power converters that are directly connected to the bus without going through a transformer, the highest d-c output voltage available using the U.S. standard 600 V line-to-line a-c bus voltage is generally 750 V d-c, a ratio of 1 : 1.25. An input-to-output voltage ratio of greater than one is due to the rectification process, which produces an output wave form that is closer to the peak of the supply voltage. A descrip-tion of the technical basis for these values is contained in a number of sources as well as the standards developed by the IEEE Thyristor Motor Drive Subcommittee [9].

If a transformer is placed between the converter and the power source, the transformer voltage ratio may be chosen to allow either an increase or decrease in the mo-tor's rated armature voltage. Using the flexibility af-forded by a transformer, the motor voltage may then be specified over the full design range up to 1000 V, although 750 V is the most common voltage chosen. For any given level of power transmission, a lowering of the voltage level requires a proportionate increase of the current, and higher currents require larger commutators, more brushes, and larger cables. Also, transformers contribute to the cost, space requirements, and weight of a system; therefore, most small- and medium-capacity systems are

designed with power converters directly connected to the a-c bus. The tendency, then, is to use the highest motor voltage consistent with the voltage level of the vessel's power system.

Once the power and voltage rating of a d-c motor has been established, the system designer must select cables and size a-c circuit breakers for the vessel's one-line dia-gram. The armature current to be carried by the d-c cables is obtained from manufacturer's data or estimated by calculation. If it is necessary to calculate the motor cur-rent, the shaft horsepower of the motor is usually rounded up to a convenient whole number using an increment such as 50 hp. Calculation of estimated armature current is obtained by converting the motor-rated horsepower to kilowatts and dividing by the d-c voltage and motor effi-ciency.

The efficiency of most d-c propulsion motors falls in the range of 92 to 96%, with direct-drive, low-speed machines having the lowest efficiency, and 400- to 900-rpm ma-chines, which must drive the propeller through reduction gears, having the highest. Manufacturer's data should be used as a guide when available, and ample margin should be included when they are not.

Shunt-wound motors are generally used in d-c drives. The speed of a d-c shunt-wound motor is nearly directly proportional to the voltage applied to its armature. There-fore, speed control of a d-c electric-drive system is readily accomplished by adjusting the output voltage of the power source in response to a small reference voltage developed from a potentiometer or digital position encoder located on the throttle shaft.

To reverse the direction of rotation of a d-c motor, as is usually required for fixed-pitch propellers, two ap-proaches may be used. The first and most common ar-rangement is to reverse the direction of current flow in the motor's field winding. Since the application of reversing torque to the propeller does not take place until the field polarity actually reverses, a delay of several seconds is usually experienced. In most marine drives, however, this time delay is insignificant compared to the overall time required to effect a change in vessel speed. The second method of reversing the direction of rotation of a d-c mo-tor is to reverse the direction of current flow through the armature. To accomplish this, the power converter is fitted with a duplicate set of main current-carrying de-vices arranged to conduct armature current in the oppo-site direction. This second set of cells is referred to as the reverse-bridge while the main set is referred to as the forward-bridge. Reversing current and the resulting re-verse torque are established in a fraction of a second. Economics usually favor the field reversal approach; how-ever, either approach is technically adequate.

A d-c electric-drive system, in addition to providing the usual electric-drive features, is occasionally furnished with a shunt field regulator that allows the motor to de-liver constant power over a range of propeller speeds. The mechanical analogy of this feature is an infinitely adjustable-ratio reduction gear, which allows the prime mover to operate at rated speed while the propeller varies

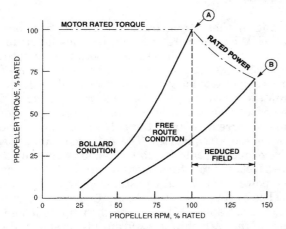

Fig. 3 Propulsion motor features adapted for propeller torque-rpm characteristics of tow vessels

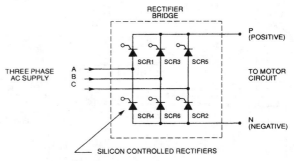

Fig. 4 Six-pulse d-c power-converter circuit

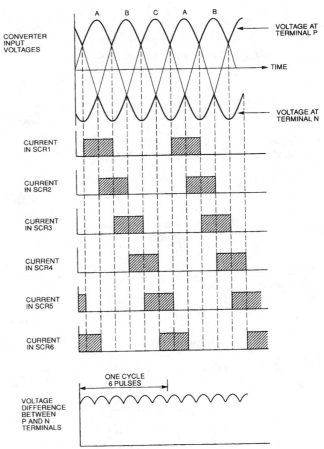

Fig. 5 Six-pulse d-c converter wave forms

in speed due to changes in loading. Icebreakers and towing vessels are among the applications that can benefit from this feature. In such cases the restraining effect of heavy tows or passage through ice causes wide variations in the power-rpm characteristics. On stalled vessels (bollard condition) the propeller may develop full-power torque at approximately 70% of the free-route propeller rpm.

The adaptability of the propulsion motor characteristics illustrated in Fig. 3 is obtained by designing the motor so that it can develop full power at the speed corresponding to the bollard condition, point "A," and then weaken the field strength of the motor to match any other full-power propeller speed up to that of the free-route condition, point "B." Thus the motor has a constant-power range from point A to point B. This permits the utilization of full rated capacity through a varying propeller characteristic speed range; however, a motor designed for this operation would be larger and more expensive than a motor designed for operation at point B only.

On vessels where the ship-service power is derived from the propulsion bus, an isolation transformer is recommended to provide ground isolation between the two buses. On a relatively small number of d-c drives, filters are added to the power system to reduce the amount of voltage distortion in the ship-service system due to the harmonic current flow from d-c drives. Power quality and filtering are discussed in Section 1.

2.2 Power Converters for Direct-Current Drives.

Direct-current systems have been used on by far the greatest number and variety of installations. The most common d-c system consists of multiple high-speed diesels driving direct-coupled a-c generators synchronized to a common bus. Connected to this bus are static power converters, which rectify the a-c to d-c and in turn feed power to one or more d-c motors driving the propeller shaft.

The six silicon-controlled rectifiers (SCRs) in the power-converter circuit illustrated by Fig. 4 are arranged to conduct in pairs to connect the terminals of the d-c propulsion motor sequentially to successive phases of the constant-frequency a-c power source. The SCRs in Fig. 4 are

numbered in the order in which they conduct [12]. The numbering sequence is continued in Fig. 5, which illustrates the sequence of current in each cell. This process produces a d-c output voltage with a characteristic ripple of six times the input frequency, as shown in Fig. 5.

Figure 6 shows a power circuit that is most commonly used on higher-power twelve-pulse marine drives [9]. As illustrated, the twelve-pulse converter circuit uses two six-pulse converters fed by separate transformer windings connected so as to produce two output voltages displaced by 30 deg. The voltage applied to one bridge by the delta-

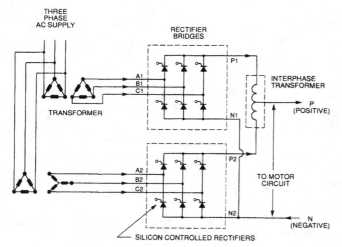

Fig. 6   Twelve-pulse d-c power-converter circuit

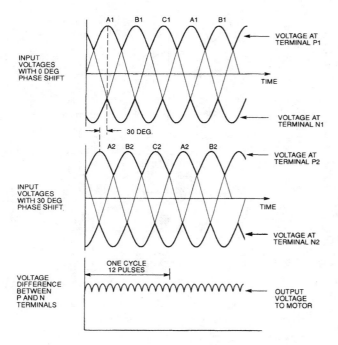

Fig. 7   Twelve-pulse wave forms

differing in phase from the other. In addition to ripple isolation, the interphase transformer adds inductance to the motor circuit, which acts to smooth the d-c output.

As a consequence of constructing the output wave form from evenly spaced segments of the supply voltage, the current is drawn in generally uniform segments from each phase of the input. Therefore, the frequencies of the harmonic currents in the a-c supply are a simple series of odd-numbered frequencies. The expected amplitudes of the harmonic currents for both six- and twelve-pulse converters (relative to a fundamental current of 1.0) flowing in the power system can be estimated by using the data published by the IEEE Static Power Converter Committee and listed in Table 1 [7]. Because the amplitude and frequency of the harmonics are known, it is possible to design tuned filters to improve the quality of the propulsion bus power.

An inspection of Table 1 shows that the amplitudes of the 5th and 7th harmonics for a twelve-pulse converter are expected to be considerably lower than those of a six-pulse circuit. This occurs when two six-pulse power converters are operated in parallel but phase displaced by 30 electrical degrees, which is ordinarily accomplished by using delta-delta and delta-wye transformer connections. A convenient way of visualizing this is to consider the 5th harmonic current as a vector rotating in the direction opposite the fundamental at 5 times the fundamental frequency. Likewise the 7th harmonic current is a vector rotating in the same direction as the fundamental at 7 times the fundamental frequency. When the fundamental is phase shifted by 30 deg by the delta-wye winding of the transformer, the 5th harmonic component appears to be rotated by $(5 \times 30) + 30$ or 180 deg when viewed from the primary of the transformer. The 5th component phase shifted by 180 deg then cancels the 5th component from the other circuit that is not phase shifted. The 7th harmonic rotates in the same direction from the fundamental and is, therefore, rotated by $(7 \times 30) - 30$ or 180 deg, providing cancellation for the 7th harmonic current.

Power converters are generally housed in cabinets, which are cooled by internal fans that force ambient air over the semiconductor devices and their mounting structures; however, in some cases, liquid-cooled converters are used. Marine equipment is usually rated for operation in a 50 C ambient. Ventilating systems should be arranged to maintain the room temperature well within the equipment ratings under all ambient conditions. Estimates of the losses from the converters can be obtained from the manufacturers or assumed to be approximately 1% of the drive rating. Converters should be located in spaces that are protected from the oil vapors that are common in engine rooms.

2.3 Physical Characteristics of Motors for D-C Drives. Motors are manufactured with one or two armatures mounted on a single shaft, which is generally supported by two bearings. From the standpoint of minimum cost and weight, a single-armature motor is preferred; however, a double-armature motor may be used where the diameter is restricted or where the added reliability of two separate electrical units is desirable. Figure 8 can

connected set of windings is in phase with the primary voltage, while the wye voltage is phase displaced by 30 deg.

The phase shift of 30 deg between secondary voltages produced by the transformer, which is indicated in Fig. 6, is the key to the converter being able to operate in a twelve-pulse mode, because it is that phase displacement which shifts the position of the d-c ripple voltages to give a characteristic fundamental ripple of twelve times the input frequency, as illustrated in Fig. 7 [8].

The interphase transformer connecting the converters to the motor in Fig. 6 is required to allow each of the two six-pulse power converters to have a d-c output ripple

Table 1 Amplitudes of harmonic currents (relative to a fundamental value of 1.0) [7]

Harmonic No.	6-Pulse	12-Pulse	Harmonic No.	6-Pulse	12-Pulse
5	0.192	0.0192	29	0.014	0.0014
7	0.132	0.0132	31	0.012	0.0012
11	0.073	0.073	35	0.011	0.011
13	0.057	0.057	37	0.01	0.01
17	0.035	0.0035	41	0.009	0.0009
19	0.027	0.0027	43	0.008	0.0008
23	0.02	0.02	47	0.008	0.008
25	0.016	0.016	49	0.007	0.007

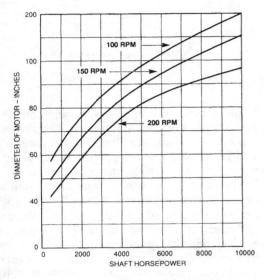

Fig. 8 Approximate diameter of d-c propulsion motors

be used to determine the approximate diameter of a single-armature propulsion motor that has Class F insulation. The data in Fig. 8 may be useful when analyzing arrangements that impose restrictions on the motor diameter.

An alternative arrangement that may be used where space and weight limitations are severe is one in which one or more high-speed motors are connected to the propeller shaft through reduction gears. Higher-speed motors usually offer reduced cost, weight, and space requirements. However, evaluations for such proposals should take into account the additional cost, weight, and space required for the reduction gear and lubrication system. The type of d-c motors used with reduction gears often include grease-lubricated antifriction bearings, which require no external lubrication system.

When small- to medium-sized directly connected propulsion motors are used, it is common practice to include the main thrust bearing in the motor assembly. The thrust bearing can be located at either end of the motor, but positioning the thrust bearing opposite the drive end of the motor permits the use of a smaller thrust collar and one that can be removed readily from the end of the motor shaft. When geared motors are used, the thrust bearing is ordinarily located in the reduction gear. In some higher-

horsepower, direct-drive arrangements, it may not be feasible to provide sufficient rigidity in the motor support structure to also serve as the thrust bearing foundation. For this reason the main thrust bearing on large d-c machines may be located aft, separate from the motor. Both the thrust bearing and motor bearings have similar lubricating-oil requirements and can be served from a common lubrication system.

Motors for use with power converters require special designs that take into account the effects of the power converter's inherent a-c ripple component superimposed on its d-c output. The ripple a-c component must pass through the brushes and commutator of the d-c machine; therefore, allowances must be made in the design and material selections so that the service life of the commutator is not consequently shortened. Some d-c machines require an external d-c reactor to reduce the ripple current, in order to be compatible with the design of the commutator. Additional heating takes place in the d-c machine because of the a-c ripple component introducing hysteresis and eddy-current losses in the magnetic structure. Because of this, extra material and attention to adequate cooling are required to ensure that copper windings operate within allowable temperature limits. A six-pulse power converter has a higher a-c ripple amplitude than a twelve-pulse converter and, therefore, imposes more restrictions on the design of a d-c motor. Motors are normally of the shunt-wound type, separately excited by power supplies located in the power-conversion equipment. Excitation equipment includes redundancy and other features to ensure continuity of power under almost all circumstances, which is consistent with the importance of maintaining propulsion power control.

Smaller d-c propulsion motors are usually fitted with separate a-c motor-driven blowers, which ventilate the machine with room air. Since the blower must be running whenever the motor is energized, a small a-c motor starter for the blower is usually located in, and controlled from, the power converter. Fans mounted on the shaft of the d-c motor are not recommended for cooling since the shunt field, with its constant losses, would not be adequately cooled at low motor speeds. In larger installations, where the heat loss is considerable and the machine is likely to be located some distance from the source of outside air, motors are fitted with a closed ventilating system where an a-c driven blower forces air through a water-to-air heat exchanger.

Section 3
Alternating-Current Propulsion-Drive Systems

3.1 System Characteristics. Alternating-current propulsion-drive systems are usually applied to vessels in the moderate- to high-power range. Alternating-current motors can be readily built for a rating of 60,000 hp, and there is no significant physical constraint concerning the maximum propulsion power that can be installed. Either integrated or dedicated power system arrangements (which are discussed in Section 1) can be used with a-c propulsion; however, integrated systems are usually selected, particularly for passenger ships. As is the practice with integrated power systems, large ship-service motor loads are connected directly at the main bus voltage level, with smaller loads served from step-down transformers and motor-generator sets provided for sensitive loads. There are no restrictions on the types of prime movers that can be used, but medium-speed diesel engines are the common generator-drive selection. The generating capacity required for most large systems necessitates the use of medium-voltage (4160 or 6900 V) equipment on the main switchboard.

Alternating-current motors have lower losses than d-c motors and, therefore, provide a higher overall transmission efficiency. The total loss in an a-c electrical transmission system, between the prime mover and the propeller shaft, is in the range of 6 to 8%. Another advantage with the selection of an a-c motor is the elimination of the maintenance associated with a d-c motor commutator.

Within the moderate- to high-power range of electric propulsion systems, the majority of motors are of the synchronous type with wound fields. The typical 96 to 98% efficiency of a direct-drive synchronous motor is usually 3 to 4% higher than for a comparable induction motor. In addition to the higher efficiency, the relatively larger air gap between the rotor and stator of a synchronous motor is more tolerant of misalignment and is, therefore, conducive to a more satisfactory installation and reduced maintenance.

In the early applications, a-c electric drives were generally associated with the use of turbine prime movers. The usual system consisted of a single turbine driving a directly connected, high-speed generator which furnished power to a single, low-speed, direct-drive motor of the synchronous type. The effective gear ratio of turbine speed to motor speed was fixed by the ratio of generator poles to motor poles. Speed control of the propeller was obtained by varying the turbine speed. Without power-conversion technology, these early systems were limited to one or possibly two fixed speed ratios. The second speed ratio was available only on induction motors equipped with a pole-changing winding.

More recent applications use power converters to change constant-frequency a-c power into adjustable-frequency a-c power, also for use by a direct-drive motor of

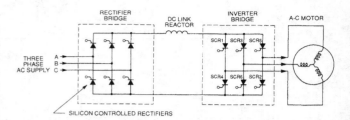

Fig. 9 Six-pulse load-commutated inverter circuit

the synchronous type. The speed of a synchronous a-c motor is directly proportional to the frequency applied to its stator windings. Therefore, the speed control of an a-c electric-drive system is accomplished by adjusting the output frequency of the power converter in response to a small reference voltage developed from a potentiometer or digital position encoder located on the throttle shaft.

To reverse the direction of a-c motor rotation, the phase sequence of voltages generated by the power converter is reversed electronically by the converter's built-in regulator. The rotor of a synchronous a-c motor simply follows the frequency and phase sequence rotation of voltage applied to the stator windings.

3.2 Power Converters for Alternating-Current Drives. Power converters change a-c power to d-c power or vice versa, or change power from one frequency to another. A typical converter used on marine propulsion applications consists of an array of controlled rectifier devices arranged as shown in Fig. 4. When this same physical arrangement of rectifier devices is used in a-c converter arrangements, such as load-commutated inverters (LCIs), as illustrated by Fig. 9, or pulse-width modulated inverters (PWMs), which are shown by Fig. 10, the single-stage power converter is referred to as either a "source converter" or "load converter," depending on its location in the circuit. Conventionally, the source converter is connected to the incoming power, and the load converter is connected to the motor. LCIs and PWMs are called "double-stage converters" because they must convert the power twice, once by the source converter and again by the load converter. A cycloconverter (CCV) is another class of power converters that is categorized as a single-stage power converter, because the power is converted only once. Synchronous motors can be used with either LCI, PWM, or CCV converters; however, only PWM and CCV converters are suitable for induction motors.

Since the speed of an a-c motor follows the electrical power frequency, the combination of power converter and a-c motor allows the propeller speed to be adjusted

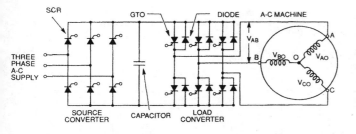

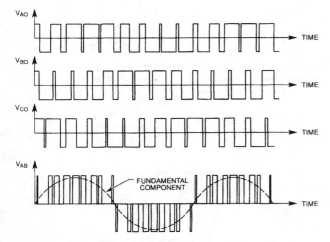

Fig. 10 Pulse-width modulated circuit and output wave forms

software that enable the operator to perform all but the major maintenance activities.

The heat generated in the cells of a power converter is removed by either air- or water-cooling provisions. Air cooling generally results in larger equipment for a given rating and imposes an additional heat load on the ship's HVAC system; however, the majority of converters are cooled by internal fans that force ambient air over the semiconductor devices and their heat sinks. Water cooling permits a more compact equipment arrangement, but requires a closed primary cooling loop to circulate deionized water through the semiconductor heat sinks and through a water-to-water heat exchanger connected to the ship's chilled-water system. The flow, temperature, and conductivity of the deionized water system must be monitored. Figure 11 is a diagram for a deionized-water cooling system as typically used in a load-commutated inverter installation. The interfaces between the deionized water and the semiconductor heat sinks must be designed to ensure leak-free service.

Figure 12 is a photograph of a silicon-controlled rectifier, with gate-control wires attached, which is typical of those used in marine propulsion power converters. Figure 13 illustrates an air-cooled heat-sink arrangement for the rectifiers, and a water-cooled arrangement is shown by Fig. 14.

Cycloconverters. Although cycloconverters (CCVs) were applied to synchronous motor drives as early as the 1930s [13], it was the more recent development of solid-state power electronic devices, namely, the silicon-controlled rectifier, which enabled the cycloconverter to become an attractive technology for use on adjustable-speed marine propulsion motors. Several icebreakers and cruise ships have been fitted with CCV drives for main propulsion.

A cycloconverter uses silicon-controlled rectifiers (SCRs) as switches to connect the terminals of the a-c propulsion motor to successive phases of the constant-frequency a-c power source, thereby fabricating a lower-frequency adjustable alternating voltage wave form for use by the propulsion motor. For purposes of illustrating the operating principle and wave forms of a CCV, a single phase of a three-phase six-pulse CCV is connected as shown in the circuit of Fig. 15. This basic configuration uses both forward and reverse conducting bridges to construct an output wave form with both positive and negative parts. During operation, the conduction period of each cell is controlled so that the output wave form is constructed as an approximation of a sinusoid. The fundamental component as well as the output wave form is shown in Fig. 16.

As a consequence of this method of constructing the output wave form, the current drawn from each phase of the input is composed of generally nonuniform and nonsynchronized segments of the output current wave. The significance of this is that the harmonic currents drawn by the CCV are no longer the simple series of harmonics of a rectifier, but are a spectrum of components

smoothly over the entire speed range. The usual method of controlling motor speed is to have the power converter respond to a small reference voltage developed from a potentiometer or digital position encoder located on the throttle shaft. On controllable-pitch propeller systems, the converter reference may also be linked to the pitch control to optimize propulsive performance. For fixed-pitch propeller installations, the direction of motor rotation is easily changed by electronically reversing the phase sequence applied to the motor.

The use of microprocessors in the control of power converters has largely replaced the use of analog circuits. In addition to allowing smooth and responsive control of the propulsion motors, the use of microprocessors has allowed software to accomplish a number of useful functions and features that previously were too difficult or expensive to achieve in hardware. For example, programs executed by the microprocessors are able to dynamically predict the motor temperature by feeding data into a mathematical thermal model of the motor that provides an advance warning when motor overheating can be expected. A similar modeling technique can be used to provide advance warnings for the power-handling devices in the power converter, enhancing availability of the equipment. Microprocessors are also used to perform self-checks and advise the operator if any abnormal conditions are found. Much emphasis is placed on the built-in diagnostics and repair

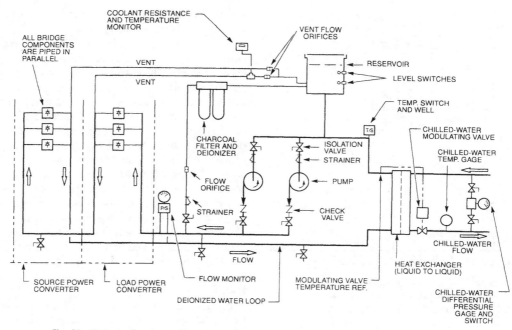

Fig. 11   Deionized-water cooling system diagram for a load-commutated inverter installation

Fig. 12   Typical silicon-controlled rectifier used in marine power converters

Fig. 13   Silicon-controlled rectifier with air-cooled heat sink

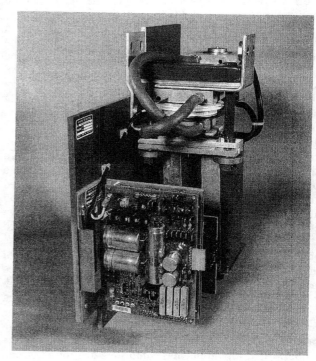

Fig. 14   Silicon-controlled rectifier with water-cooled heat sink

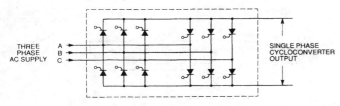

Fig. 15 Single-phase circuit of a cycloconverter

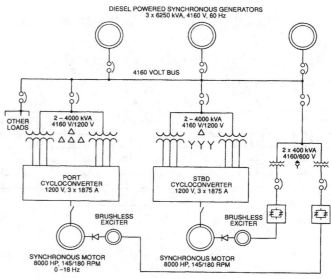

Fig. 17 Typical cycloconverter drive system for a light-duty icebreaker

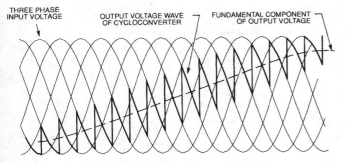

Fig. 16 Single-phase cycloconverter output wave form

approaching a continuous function when changes of amplitude as well as frequency are considered. Consequently, tuned filters, such as those used to improve power quality on d-c, LCI, and PWM drives, are not an effective means of performing the same function on CCV drives.

A twelve-pulse CCV frequency changer uses twice the number of SCR cells as a six-pulse circuit. As is the case with d-c to a-c converters, an increase in the number of pulses leads to reduced harmonic amplitudes in the output wave form. In this respect a twelve-pulse CCV will produce fewer harmonics than a six-pulse converter of the same rating. It should be noted that, depending on the control scheme used by the designer to sequence the conduction of the SCRs, the magnitude of harmonic amplitudes and input displacement power factor can vary significantly [14].

Both the output frequency and voltage of a CCV are controlled by varying the angle at which the SCRs are allowed to start conducting. By delaying the conduction, a lag between the current and input voltage is introduced in the input power system. This characteristic is most notable in icebreaker applications where ice conditions produce a need for high torques at low propeller speeds, which are reflected to the CCV as high currents at low voltages. The simultaneous demands for high current and low voltage place extra demands on the power system, which must be satisfied by larger and more costly generator designs. However, the ability to develop a high torque at zero speed is not an important feature for most ships.

When applied with an appropriate control system, a cycloconverter can mimic the dynamic characteristics of a d-c motor by providing high shaft torques from zero to full shaft rpm, and can quickly reverse the rotation of the propeller [15]. Figure 17 shows the arrangement and ratings of the electrical equipment on a typical light-duty

icebreaker with a CCV-type static frequency changer. Transformers are used between the main bus and the input to the power converter to match the 4160 V generator voltage to 1200 V required by the design of the CCV. Wye and delta connections of the transformer windings are used to phase shift the output voltage to achieve harmonic cancellations.

Three-phase to three-phase cycloconverters are commonly built in two forms, either with 18 or 36 thyristor switched paths. The two forms differ in the power-factor demands from the power system as well as in the input power and motor torque pulsation characteristics [16]. A twelve-pulse 36-leg CCV has been used on cruise ship applications and produces a satisfactory output wave form up to approximately 20 Hz.

The basic building block of a CCV frequency changer is identical to d-c power converters that have both forward and reverse conducting bridges. This is illustrated in the circuit of Fig. 15 for a single phase of a CCV, and in Fig. 18 for a complete 36-leg twelve-pulse CCV drive.

The cycloconverter typically used for ship propulsion belongs to the class of frequency converters that depends on the supply voltage to commutate the current from one cell to the next. The generator that supplies this voltage and the associated reactive power sense the CCV as a load with a power factor ranging between 0.75 to 0.85 at full motor speed, with the power factor decreasing as the speed is reduced. This tends to increase the size of the generators required. The motor, however, benefits from the effects of the CCV since it does not have to be designed to deliver any reactive power to the CCV. The power factor of the synchronous motor is usually specified as unity (1.0) when it is to be driven by a CCV. In contrast to this, a motor with the same power rating for use with a load-commutated inverter drive would be designed for a 0.9 power factor.

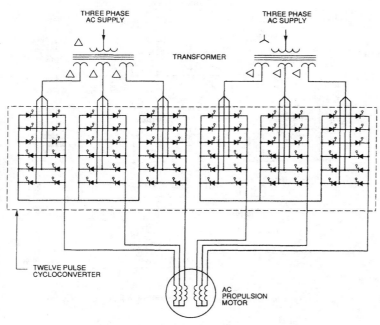

Fig. 18   Twelve-pulse cycloconverter drive circuit

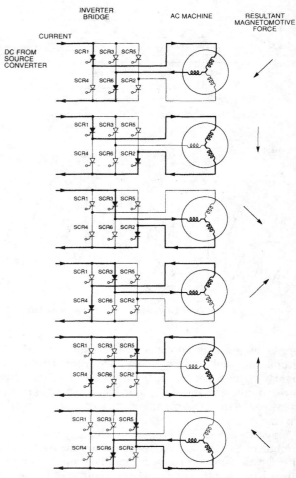

Fig. 19   Load-commutated inverter conduction sequence

Load-commutated inverters. The basic circuit of a load-commutated inverter (LCI) is constructed from two power converters—a source converter and a load converter—as shown in Fig. 9 [12]. The source converter is connected to the incoming power to rectify a-c to d-c, and the load converter is connected to invert the d-c back to a-c for use by the propulsion motor. A d-c link reactor connects the d-c output of the source converter to the d-c input of the load converter. The link reactor is an energy storage device that attenuates the d-c output ripple of the source converter and also helps isolate the power system from any torque pulsations that may originate in the propulsion motor or be transmitted by the propulsion shafting system.

The commutation of current (the transfer of current from phase to phase) in the load converter depends on the synchronous propulsion motor (the load) supplying voltage for commutation by operating at a leading power factor. When operating under these conditions, the sequence of the silicon-controlled rectifier (SCR) conduction and paths of current flow during a complete electrical cycle of the LCI circuit will follow the illustration of Fig. 19 for a six-pulse configuration. Heavier-weight lines in the figure indicate the paths carrying current in each of the six possible combinations shown. The resultant magnetomotive force represents the rotation of the magnetic field within the motor, which produces shaft rotation. The simplified motor represented in the figure implies one mechanical shaft rotation for each electrical cycle, a two-pole machine. In practice the motor will have many pairs of poles.

Below approximately 10% speed, the motor (load) cannot generate enough voltage to commutate the d-c to a-c inverter bridge. Therefore, operation of the motor below

10% speed is accomplished by synthesizing a low-frequency a-c wave by sequentially turning on the rectifier power semiconductors and letting the naturally occurring zero crossings of the a-c line do the commutating. This scheme tends to produce somewhat pulsed motor torque characteristics at low speeds. However, acceleration through this speed is usually rapid due to the low load torque of the propeller. A more sophisticated technique, which may be employed on a twelve-pulse converter, is to operate the two converter bridges and the two-winding motor as a two-phase CCV. The software in the LCI controller would be designed to automatically switch modes from LCI to CCV operation as required by low motor speeds.

Motors for LCI service are usually rated at 0.9 power factor, since they must provide commutating power for the inverter. The generators must also supply reactive power to the rectifier section of the LCI, which also appears as a 0.9 power factor load at full load.

The proper timing of commutations is under the control of an electronic regulator, which initiates the events by applying a pulse of power to the gate terminal of the

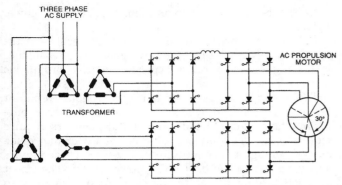

Fig. 20 Twelve-pulse load-commutated inverter circuit

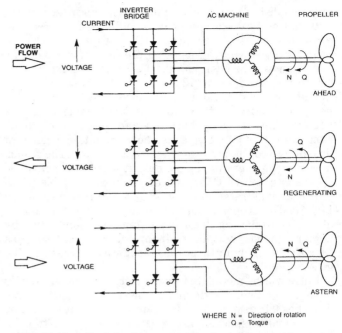

WHERE N = Direction of rotation
       Q = Torque

Fig. 21 Propeller reversal sequence with a load-commutated inverter

oncoming controlled rectifier. To aid the regulator in computing the proper time to initiate gate firing, information concerning the rotor's position is often supplied by a shaft position encoder mounted on the end of the motor shaft. The load-commutated inverter output is determined by the frequency of the motor.

Load-commutated inverters for higher-power motors are usually designed with 12-pulse converters. However, 18- or 24-pulse circuits may be constructed, using a variety of transformer and converter configurations, where the reduction in harmonic amplitudes from such arrangements is found to be advantageous. Close engineering coordination of the motor and power converter parameters is required because the motor winding will of necessity be of special design and often will require more terminal connections than one of conventional design. Several circuit configurations have been used to arrange 12-pulse power converters. Figure 20 shows one such circuit that could be applied to higher-power 12-pulse marine propulsion drives. As illustrated, the 12-pulse converter circuit makes use of two 6-pulse converters fed by separate transformer windings connected to produce two output voltages displaced by 30 deg. The voltage applied to one bridge by the delta-connected set of windings is in phase with the primary voltage, while the wye voltage is phase-displaced by 30 deg. An output transformer is not required when the motor, as shown in Fig. 20, is constructed with two sets of isolated windings, also displaced by 30 deg. Higher-order drives of 18- or 24-pulse circuits may be constructed using a variety of transformer and converter configurations [8]. As with other power converters, an increase in the number of pulses will reduce harmonic amplitudes in both the input and output wave forms but will require additional equipment.

If tuned filters are contemplated for use in the power system to lower the harmonic levels, consideration should be given to having an analysis conducted to be certain that unwanted resonances will not occur because of modulation of the main bus voltage by frequency disturbances originating in the motor/LCI/propeller system.

An LCI allows power to flow in either direction, to or from the propulsion motor. During fixed-pitch propeller reversals, the propulsion motor briefly regenerates power

into the a-c system to slow down the motor before reversing its direction of rotation. Figure 21 shows the three LCI operating modes—ahead, regenerating, and astern—for a fixed-pitch propeller application. The direction of motor rotation is reversed by changing the sequence in which gate pulses are applied to the load inverter. Thus, astern operation of the motor is determined by an electronic regulator rather than by the mechanical operation of reversing switches.

The LCI form of static power converter has been applied to a number of industrial motor-drive applications but few marine drives. The repowering of the *Queen Elizabeth 2* in the late 1980s was a notable application of LCI converters to main propulsion. On that ship, the LCI power converters are used to start the main motors and accelerate them to 72 rpm for low-power high-efficiency cruising. At 72 rpm the LCI output frequency is 30 Hz and is capable of delivering about 5.5 MW to the controllable-pitch propeller. When more power is required, the propeller pitch is reduced, to develop less torque, and the motor is brought up to full rpm by the LCI power converter, which then produces an output frequency of 60 Hz. After reaching 144 rpm the motor is synchronized to the main bus and the circuit to the LCI is opened. Once connected to the main bus, the propeller pitch may be increased to draw the full rated 44 MW from the system.

Pulse-width modulated converters. The basic circuit of a pulse-width modulated (PWM) converter is similar to an LCI circuit because it is also constructed from two power converters: a source converter and a load converter. A PWM is distinguished from an LCI by the ability of the load bridge to switch off the motor current during any part of the a-c wave. Switching of the motor current is accomplished by either forcing the current to zero using external energy storage devices or by using a rectifier element such as a gate-turnoff (GTO) thyristor, which is intrinsically capable of opposing the flow of current. Figure 10 shows the basic circuit of a PWM converter using GTO thyristor devices [12].

The source converter is connected to the incoming power to rectify a-c to d-c and the load converter is connected to invert the d-c back to a-c for use by the propulsion motor. A d-c link capacitor helps maintain a constant d-c voltage to the input of the load converter. The fundamental component of the a-c output wave is constructed by turning the GTO on and off to form variable-width pulses as illustrated in the voltage wave forms shown in Fig. 10. The proper timing of pulses is under the control of an electronic regulator, which initiates conduction by applying a positive pulse of gate current and controls turnoff by applying a pulse of negative current to the gate.

Pulse-width modulated converters characteristically have a relatively high input power factor, which approaches 0.9, as compared to systems using either LCI or CCV converters, which are in the range of 0.75 to 0.85. Higher system power factors can reduce the need for generator designs that have special power-factor ratings.

Regenerative power flowing from the propulsion motor during fixed-pitch propeller reversals will raise the d-c link capacitor voltage because energy cannot flow back into the a-c system. Either a means for dissipating this power in a resistor or a duplicate source converter is required with the ability to conduct current in the reverse direction. The direction of motor rotation is reversed by changing the sequence in which gate pulses are applied to the load inverter. Thus, astern operation of the motor is determined by an electronic regulator rather than the mechanical operation of reversing switches.

The PWM form of static power converter has been applied in a number of low- to medium-power industrial motor-drive applications but few marine propulsion drives. However, higher-speed induction motors, driving through reduction gears, have been used for propulsion power in some installations. PWM converters are applied in thruster drives where medium- to high-speed induction motors are favored because of the adverse service environment and the cost advantages. A notable early application of PWM converters was made on the dynamically positioned pipe-laying vessel *Lorelay*, which had seven adjustable-speed PWM-induction motor drives for thrusters and propulsion.

3.3 Motors for A-C Drives. The insulation systems and cooling provisions for motors that are used with either cycloconverters or load-commutated inverters require special attention to ensure operation within allowable temperature rises. Motors are normally of the wound-field type with excitation supplied by either slip rings or brushless exciters. Brushless exciters should be designed so that the required excitation power can be delivered over the full speed range of the motor, including zero speed. Excitation equipment typically includes redundancy and other features to ensure continuity of power, consistent

with the importance of maintaining propulsion power control.

Induction motors are used infrequently because of the efficiency advantage provided by synchronous machines in the higher-power range typical of marine propulsion installations.

The power factor required in a synchronous a-c motor design is determined by the type of power converter supplying the machine. Load-commutated inverters require motors that can supply a leading power factor to provide enough voltage to commutate the inverter. These motors are typically rated at 0.9 leading power factor. The effect of a 0.9 leading power factor on a motor design is to increase the current that the machine must handle, which tends to increase the motor size slightly as compared to a 1.0 power factor machine of the same frequency and rating. Cycloconverters and pulse-width modulated inverters do not depend on the motor voltage to commutate the power converters and are designed for 1.0 power factor.

The speed of the motor and the power frequency are governed by the relationship:

$$n = 120\, f/p$$

where

n = motor speed, rpm
f = power frequency, Hz
p = number of motor poles (multiples of 2)

Because the output frequency of a CCV has a practical upper limit of about 20 Hz on a 60-Hz power system, motors used with CCVs are of the direct-drive type. Motors used with LCIs are typically designed for 40 to 60 Hz and may either be direct-drive or drive through a reduction gear.

Figure 22 shows the approximate dimensions of single-armature motor designs which would be used with CCV static power converters. The motors for CCV service are designed for frequencies of 20 Hz and below while LCI motors are generally designed for frequencies above 40 Hz; therefore, the motor dimensions vary somewhat for the same power and speed. Figure 22 also shows the approximate dimensions of single-armature 100-rpm 40-Hz motors which would be used with LCI static power converters.

Higher-speed motors that drive the propeller shaft through reduction gears may also be used with LCI converters and provide some cost and weight advantages. Figure 23 shows the approximate dimensions of conventionally arranged geared-motor drives. The combined length of a motor and gear is longer than a direct-drive motor; however, the narrower equipment outline and lighter weight could be advantageous in some ship arrangements.

While it is possible to locate the main propulsion thrust bearing in one of the a-c motor bearing housings, the large diameter of higher-horsepower direct-drive motors may make it difficult to provide enough rigidity in the thrust-bearing foundation to transmit the propeller thrust into the vessel's structure. For this reason the main thrust

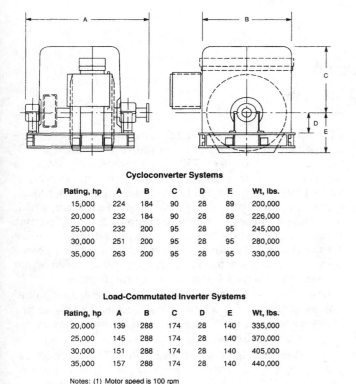

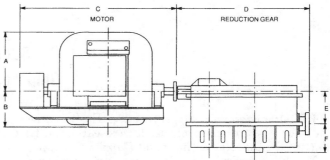

Cycloconverter Systems

Rating, hp	A	B	C	D	E	Wt, lbs.
15,000	224	184	90	28	89	200,000
20,000	232	184	90	28	89	226,000
25,000	232	200	95	28	95	245,000
30,000	251	200	95	28	95	280,000
35,000	263	200	95	28	95	330,000

Load-Commutated Inverter Systems

Rating, hp	A	B	C	D	E	Wt, lbs.
20,000	139	288	174	28	140	335,000
25,000	145	288	174	28	140	370,000
30,000	151	288	174	28	140	405,000
35,000	157	288	174	28	140	440,000

Notes: (1) Motor speed is 100 rpm
(2) Dimensions are in inches

Fig. 22   Approximate dimensions and weights of propulsion motors in cyclo-converter and load-commutated inverter systems

Output, hp	A	B	C	D	E	F	Motor Width	Gear Width	Total Wt, lbs.
30,000	50	82	197	180	52	48	106	116	222,000
35,000	55	86	190	185	52	48	116	120	240,000
45,000	55	86	205	190	54	50	116	126	287,000
50,000	55	86	210	195	58	53	116	134	306,000
60,000	55	86	224	201	62	57	116	140	353,000

Notes: (1) Gear output speed is 100 rpm
(2) Motor speed is 1200 rpm
(3) Dimensions are in inches

Fig. 23   Approximate dimensions and weights of conventionally arranged, geared propulsion motors in load-commutated inverter systems

bearing is often located separate and aft of the motor. The thrust and motor bearings have similar lubricating-oil requirements and can be served from a common lubrication system.

From a standpoint of minimum cost and weight, a single-armature motor should be used where space permits. However, a double-armature motor can be used where the diameter is restricted or where the added reliability of two separate electrical units is considered desirable.

Thyristor cell failures are rare. When one does occur, the cell nearly always fails in the shorted mode. Consequently, most a-c converters are constructed with redundant thyristor cells connected in series to allow continued operation at full rating, even with the loss of a single cell.

Because of the physical arrangement of conductors, electrical faults in the cables or motor windings tend to be single-phase in nature, producing torque pulsations at twice motor operating frequency. The duration that the shafting system is subject to excitation from fault-induced torques can approach several seconds, depending on the length of time required by the protective system to detect and clear the fault.

Because the motor can operate over a wide frequency range, faults at some operating frequencies can cause torque pulsations at frequencies that can be amplified by mechanical resonance between the motor and propeller via the line shafting [16]. This requires careful analysis of the electromechanical system to assure that the shafting system is designed properly. Mitigation can be accomplished by automatically programming the motor excitation to reduced levels when operating in the critical speed ranges, by energy-absorbing couplings between the motor and the line shaft, by avoiding continuous operation at the critical motor speed range, or by combinations of the above.

Section 4
Ship Applications

4.1 Oceanographic Ship, AGOR 23. Electric drives of the ac-dc type are commonly used for research and survey ships of the U.S. Navy. The vessel *Thomas G. Thompson*, AGOR 23, is typical of the arrangements found on these vessels. Figure 24 is a simplified one-line diagram of the electrical system on the AGOR 23. The propulsion plant contains diesel-generator sets that can be synchronized to a common 600-volt bus. Each 750 V d-c propulsion motor and bow thruster motor is powered by a direct-coupled six-pulse static power converter. The power system is arranged so that the diesel-generators can be synchronized to either the propulsion or ship-service bus section. Two step-down transformers are fitted to provide 450 V to the distribution switchboard.

4.2 Dredge, U.S. Army Corps of Engineers, *Hurley*. The dredge *Hurley* is a 300-ft self-propelled dredge designed for use on the inland waterways of the United States. In operation the dredge annually clears shipping channels of over 20 million cubic yards of material. The dredge is self-propelled by three steerable "Z"-type fixed-pitch thrusters when transiting to the work location and during maneuvering at the site. Anchors and hauling winches are used to provide local movement during dredging operations. The operating profile of the dredge, therefore, allows the power produced by the generator sets to alternatively be used for pumping or for propulsion.

Figure 25 is a simplified one-line diagram of the electrical system on the *Hurley*. The propulsion plant consists of three 900-rpm, 2450-kW, 0.8-power factor diesel-generator sets that can be synchronized to a common 600 V bus. Each of the three 750 V d-c propulsion motors is powered by a transformerless six-pulse nonreversing static power converter fed from the main bus. Other loads served by power converters include two d-c motors, which are coupled to the main dredge pump by a multiple-input reduction gear, and two smaller jetting pumps. In order for the two dredge pump motors to share load, one power converter acts as a master unit controlling speed while the second acts as a torque slave to the first.

Other uses of a-c power include a single-speed a-c induction motor coupled to a bow thruster as well as the entire ship-service load, which is connected to the main bus via step-down transformers.

4.3 Cruise Ship *Crystal Harmony*. The twin-screw cruise ship *Crystal Harmony* is powered by four 8.2-MW and one 3-MW, 0.75-power factor, 60-Hz diesel-generator sets, as indicated by Fig. 26. The electric output of the diesel-generator sets is connected to a 6.6-kV sectionalized main switchboard. Static power converters of the cyclo-converter type are transformer coupled to the main bus to provide adjustable frequency a-c power to the 12-MW direct-drive 1-kV propulsion motors. The propeller is of the controllable-pitch type. The major applications of ship-service power are connected to the main bus, while transformers and motor-generator sets serve smaller loads.

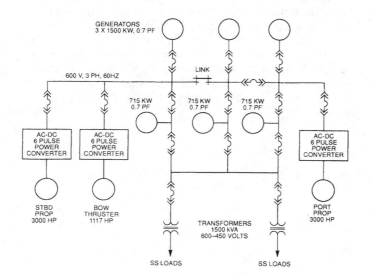

Fig. 24   One-line diagram of the electrical system of the oceanographic ship AGOR 23

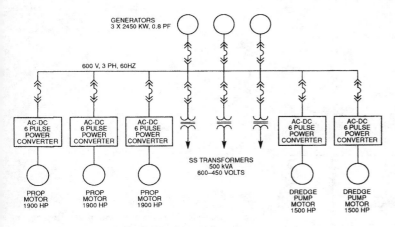

Fig. 25   One-line diagram of the electrical system of the dredge *Hurley*

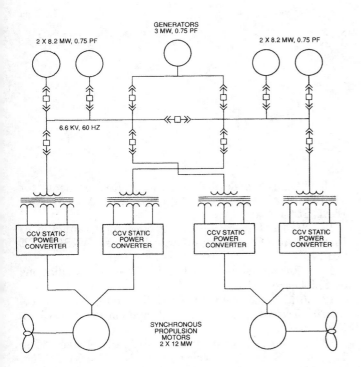

Fig. 26   One-line diagram of the propulsion system of the cruise ship *Crystal Harmony*

4.4 Cable Ships. The cable ships *Global Link* and *Sentinel* were designed to lay transoceanic submarine communication cable. Cable is payed out at relatively slow speeds, and the electric propulsion system enables the vessel to maintain proper cable tension while also placing the cable in the tract prescribed by underwater surveys of the ocean bottom. Reduced power requirements during cable-laying operations are met by securing unneeded prime movers.

Figure 27 is a simplified one-line diagram of the electrical system on the cable ship *Global Link* and shows that the propulsion plant consists of three 3750-kW main die-

sel-generator sets that can be synchronized to a common 4160 V bus. Each of the two 4700-hp 750-volt d-c propulsion motors is powered by a twelve-pulse reversing static power converter. The shafts of the two motors are coupled to the line shaft by a multiple-input reduction gear. Other loads served by power converters include two 1750-hp d-c motors for rotatable thrusters located at the stern. Additional applications of a-c power include a pair of 2000-hp single-speed a-c induction motors coupled to bow thrusters as well as the entire ship-service load, which is connected to the main bus via step-down transformers.

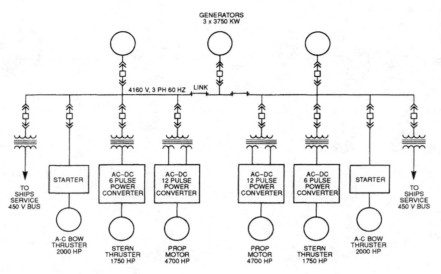

Fig. 27   One-line diagram of the electrical system of the cable ship *Global Link*

References

1 "IEEE Standard Rating Structure for AC High-Voltage Circuit Breakers Rated on a Symmetrical Current Basis," ANSI/IEEE C37.04.

2 "Preferred Ratings, Related Requirements, and Application Recommendations for Low-Voltage Power Circuit Breakers and AC Power Circuit Protectors," ANSI/IEEE C37.16.

3 "IEEE Recommended Practice for Protection and Coordination of Industrial and Commercial Power Systems," ANSI/IEEE Std 242.

4 "Motors and Generators," NEMA MG 1, National Electrical Manufacturers Association.

5 "IEEE Recommended Practice for Establishing Transformer Capability When Supplying Nonsinusoidal Load Currents," ANSI/IEEE C57.110.

6 J. Hensler, "Electric Ship Propulsion: Application and Integration," *Trans.* SNAME, Vol. 97, 1989.

7 "IEEE Guide for Harmonic Control and Reactive Compensation of Static Power Converters," IEEE Std 519.

8 *Power Converter Handbook: Theory, Design and Application*, R. W. Lye, Ed., PCH Associates, Peterborough, Ont., 1976.

9 "IEEE Standard Practices and Requirements for Thyristor Converters for Motor Drives," ANSI C34.3/IEEE Std 444.

10 "IEEE Recommended Practice for Grounding of Industrial and Commercial Power Systems," ANSI/IEEE Std 142.

11 "Recommended Practice for Electric Installations on Shipboard," IEEE Std 45.

12 J. M. D. Murphy and F. G. Turnbull, *Power Electronic Control of AC Motors*, Pergamon Press, Oxford, U.K., 1988.

13 E. F. W. Alexanderson and A. H. Mittag, "The Thyratron Motor," *Electrical Eng.* (New York) Vol. 53, Nov. 1934.

14 L. Gyugyi and B. R. Pelly, *Static Power Frequency Changers*, John Wiley & Sons, New York, 1976.

15 W. A. Hill, R. A. Turton, R. J. Dungan, and C. L. Schwalm, "A Vector-Controlled Cycloconverter Drive for an Icebreaker," *IEEE Transactions on Industry Applications*, Vol. IA-23, No. 6, Nov./Dec. 1987.

16 D. Peters and C. Vanzeyl, "Electric Propulsion Systems—The Way of the Future," 41st Annual Technical Conference, Canadian Maritime Industries Assn., Feb. 1989.

CHAPTER IX

Gary P. Mowers | # Reduction Gears

Section 1
Introduction

1.1 Early History. It is generally acknowledged that Pierre DeLaval was the first to apply a reduction gear for ship propulsion with a 15-hp experimental unit in 1892. This was followed by Sir Charles Parsons with his 10-hp experimental geared-turbine unit in 1897. These were experimental units, however, and functional marine reduction gears did not make their debut until some years later.

Just after the turn of the century, the steam turbine was being championed for ship propulsion by Parsons and others. In a study in 1904 of the probability of the steam turbine becoming a successor to the reciprocating engine, Admiral George W. Melville and Mr. John H. MacAlpine, consulting engineers, reported: "If one could devise a means of reconciling, in a practical manner, the necessary high speed of revolution of the turbine with the comparatively low rate of revolution required by an efficient propeller, the problem would be solved, and the turbine would practically wipe out the reciprocating engine for the propulsion of ships. The solution of this problem would be a stroke of great genius." Parsons in 1909 said, "The solution may be found in reverting to some description of gearing ... and if a satisfactory solution can be found, then the field of the turbine at sea will be further extended."

Parsons carried on further research and experimental work in applying the helical gear to large-scale marine installations, and in 1909–1910 he equipped the *Vespasian* with a geared-turbine plant. The gear was rated at 1095 hp and reduced the turbine speed of 1450 rpm to a propeller speed of 73 rpm.

George Westinghouse, in 1909, demonstrated in a shop test a 6000-hp gear which reduced the speed from a 1500-rpm turbine to a 300-rpm hydraulic dynamometer. This gear was the forerunner of the 6500-hp gears installed in the collier *Neptune* a short time later.

The adoption of high-speed helical reduction gears in connection with marine propulsion was rapidly accepted by engineers all over the world, and this type of equipment had a very rapid development. At the end of 1910 the total power of geared marine turbines was about 15,000 shp, whereas 30 years later marine propulsion of this type in service totaled over 100,000,000 shp. It is interesting to note that the last large ship built with direct-connected turbines was the passenger liner *Ile-de-France*, which went into service in 1927. The *Ile-de-France* had a propulsion plant of 52,000 shp divided among four screws. The turbines were designed by Parsons and were of the reaction type. The main turbines contained a total of more than 800,000 blades and weighed 1065 tons.

With the further development of the steam turbine, still higher turbine speeds could be used to advantage and the single-reduction gear no longer met the need. Engineers began development of the double-reduction gear, where practically no limits were imposed on the speed ratio that could be obtained. This permitted both the turbines and propellers to be operated at speeds suitable for their individual maximum efficiencies. Double-reduction gearing was first used about 1917. For a few years during and after the first world war, many ships were equipped with this type of gearing.

Due to many unknown factors entering into the design and use of this new type of reduction gear and also due to the unusual operating conditions during the first years of use, considerable difficulties were experienced and many casualties of reduction gears occurred, which more or less slowed the general adoption of double-reduction gearing. However, by adhering to sound design principles, it was possible to eliminate early mistakes and develop satisfactory double-reduction gears and to greatly increase the application of this type of power transmission. This is not to suggest that the development of double-reduction gears immediately made single-reduction gears obsolete. For higher-powered naval ships with propeller speeds above about 200 rpm, single-reduction gears remained in general use until the early 1930's. Then, the higher rotational speeds of the more modern steam turbines brought about the demise of single-reduction gearing for turbine drives in all categories. Single-reduction drives still remain the standard, however, for high- and intermediate-speed diesel engine service.

The development of propulsion gearing has been one of a continuous improvement and refinement in materials and in manufacturing techniques and equipment to provide greater reliability and longer life. The horsepower ratings of gears have increased to keep pace with the requirements for larger and faster ships. There are only a few step-advances that can be identified, the step from single to double reduction, the introduction of welding to the construction of gear wheels and casings, and the introduction of higher hardness pinion and gear materials with the attendant higher gear tooth loadings. The reliability, high efficiency, and long life of the modern reduction gear are well known, and its low noise level makes it completely acceptable in the engine room. These factors have been in large part responsible for the continuing popularity of the geared-turbine drive for ships.

1.2 Articulation and Gear Arrangement. The early reduction gear designs incorporated many devices to minimize the effects of bending and torsion of the pinion and of inaccuracies in machining and alignment. However, experience has demonstrated that such devices are unnecessary, and gear elements can be so proportioned and machined that uniform tooth pressures are obtained without the use of mechanical devices to compensate for pinion deflections.

Figure 1(a) represents the simplest arrangement of a marine reduction gear, i.e., one pinion meshing with a gear, as used, for instance, for connecting a propeller to a diesel engine or to an electric motor. It is not used for propelling equipment with a turbine drive but, on the other hand, it has found a wide application for turbine-driven auxiliary equipment on board ship such as generators and circulating pumps.

Figure 1(b) is a drive with two pinions as used frequently with diesel engines of comparatively large power. It is not used for direct connection to high-speed turbines, but is often used in the second reduction gear unit of a turbine drive using double-reduction gears.

Figure 1(c) represents the early type of single-reduction gear for a turbine drive, the principal difference between this reduction gear and the one shown in Fig. 1(b) being in the number of pinion bearings. The third bearing located between the two helices is necessary because of the wide tooth face in relation to the diameter of the pinion. While some ships with reduction gears built according to Fig. 1(c) are still in successful operation, this design must be considered obsolete. It was used for speed ratios up to or slightly above 20 to 1.

Figure 1(d) is the usual arrangement of a double-reduction gear for turbine-driven ships. The two input pinions are driven by the two elements (high-pressure and low-pressure turbines) of a cross-compound turbine. Power is divided between the two input pinions by the turbine characteristics and is normally split approximately equally between the two turbines. Note that the second reduction gear is common to both high-pressure and low-pressure trains but that, although it transmits the power from both turbines to the gear shaft, the tooth portion is designed to transmit the power from one turbine. The terms "tandem" and "articulated" are also applied to this arrangement; tandem because of the disposition of the first and second reductions, and articulated because a flexible coupling is generally provided between the first reduction or primary gear wheel and the second reduction or secondary pinion.

Figure 1(e) represents the "nested type" double-reduction gear, which has also been used with cross-compound turbines. The configuration shown has the second reduction helices divided to provide space for the first reduction

and is additionally referred to as a "split secondary." The nested type may also be arranged as a "split primary."

Figure 1(f) illustrates the type of gear referred to as a locked-train double-reduction gear. In it the power of the single input pinion is equally divided between the two intermediate-speed elements. Its advantage is that the gear elements are proportioned for one half of the input horsepower and are therefore smaller than would be the case with a single intermediate element. The overall size and weight are reduced, but offsetting this advantage is the added number of parts, the need to provide torsionally flexible shafts between the first and second reductions, and the need to "time" the assembly to equalize the power split between the two trains. The term "dual tandem" is also applied to this type of gear.

Figure 1(g) is a locked-train type of double-reduction gear for a cross-compound turbine or for two gas turbine prime movers. This arrangement has become standard for high-powered naval ships and has been used for higher-powered merchant ships because it minimizes the total weight and the size of the assembly.

Figure 1(h) is a planetary gear. It has a single input "sun pinion" which drives three or more "planet gears." These planet gears are mounted on a planet carrier which is solidly connected to the output coupling. The outer "ring gear" is held stationary in the gear housing. This type of gear has been applied to turbine-generator drive gears and to main turbine drive first reductions. It has also been considered for the second reduction of main reduction gears.

Many other reduction gear arrangements are possible and have been used. These can be highly varied as in cases where more than one type of prime mover is coupled to the propeller.

1.3 Methods of Manufacture. Nearly all gears produced in the United States have their teeth cut by the hobbing process. In this process the cutting tool is a hob, a rotary cutter having one or more leads, with teeth that are accurately formed to the "basic rack" tooth form selected. In the hobbing process the teeth are cut and the true involute form of the tooth flanks is generated by the continuous rotation of the hob and the gear blank.

The hob determines the dimensions of the teeth in the plane normal to the teeth. The other factors determining the tooth geometry, number of teeth, and helix angle are obtained by selecting change gear ratios for the hobbing machine; the selection of the change gear ratio provides a choice of these variables without a change of tooling. By adjusting the helix angle (which affects the tooth profile in the plane of rotation), it is possible to use a given hob (which dictates the tooth profile in the plane normal to the teeth) and produce a favorable number of teeth within rather broad limits. For this reason, manufacturers standardize with a small number of hobs.

The other cutting process which has been used in the U. S. and is still used abroad for large gears is shaping. In this process the shaping cutter is either in the form of a basic rack section or a small gear, stroking in timed relation to the rotation of the blank to generate the tooth form.

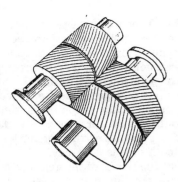

(a) Single reduction, single input

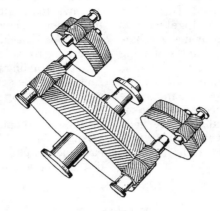

(d) Double reduction, double input, articulated

(f) Double reduction, single input, locked train

(b) Single reduction, double input

(g) Double reduction, double input, locked train

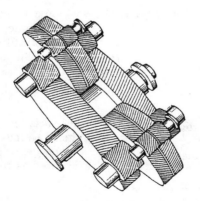

(e) Double reduction, double input, nested

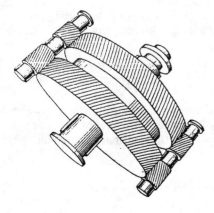

(c) Single reduction, double input, three-bearing pinions

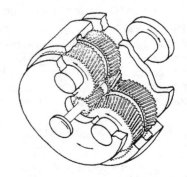

(h) Single reduction, planetary

Fig. 1 Gear arrangements

Post-cutting processes are generally applied to further refine the accuracy and surface finish of the gear teeth. In the cross-axis shaving process, a multitooth cutter in the form of a small gear is pressed tight in mesh with the gear being shaved. The surface of the tool has serrated rectangular grooves and a helix angle that is a few degrees different than the gear being shaved. This difference in angles creates a sliding motion that, together with the serrations and tight mesh condition, provides the shaving action. As the gear is rotated rapidly and the shaving cutter fed slowly across the gear face, a very light cut is taken from the tooth flanks. This results in a finer tooth surface and a more precise involute form than can be produced by hobbing. The shaving process also makes possible the correction of slight mismatch in the helix angle of the gear and pinion by selectively shaving that portion of the face width which indicates the heaviest tooth contact.

Skiving, a process which uses a carbide-tipped hob, has gained popularity as a means of removing distortions of case-hardened gear elements. However, the high accuracy requirements of marine reduction gears usually precludes the use of this process as a finishing operation.

Gear tooth rotary honing is a process often used to improve tooth surface finish. Rotary honing is similar to cross-axis shaving, but uses an abrasive tool without serrations. Special profile grinding of the honing tool is necessary to maintain tooth profile accuracy.

In the grinding process, the flanks of the gear teeth are finished by the action of a grinding machine—simultaneously generating designed tooth form and helix angle. Grinding produces very consistent and repeatable tooth profile modifications, helix angles, and end reliefs. Changes in modification magnitudes and positions are easily accomplished to achieve optimum gear tooth load distribution and lowest mesh noise excitation.

Section 2
Tooth Design Factors

2.1 Tooth Contact Pressure. The most important factor in the design of a reduction gear is the tooth contact pressure, that is, the pressure which exists between the mating tooth surfaces when force is transmitted from one to the other. This factor determines the durability of the working surfaces of the teeth.

The tangential force transmitted per unit of gear face width is determined from the expression

$$\frac{W_t}{F_e} = 126{,}050 \, \frac{HP}{RPM_p \cdot d \cdot F_e} \tag{1}$$

where

$$
\begin{aligned}
W_t &= \text{total tangential tooth load, lb} \\
F_e &= \text{effective face width (at pitch diameter), in.} \\
RPM_p &= \text{pinion revolutions per minute} \\
HP &= \text{horsepower transmitted (per mesh)} \\
d &= \text{pitch diameter of pinion, in.}
\end{aligned}
$$

The allowable tooth load per unit of face width increases with the diameter of the pinion because of the decreasing curvature of the contacting surfaces. In early gear designs, particularly in Britain, the allowable tooth pressure per unit of face width was taken as proportional to the square root of the pinion diameter, that is

$$\frac{W_t}{F_e} \text{ (allowable)} = C\sqrt{d} \tag{2}$$

where C is an experimentally determined constant.

Yet another expression related the allowable pressure to the two-thirds power of the pinion diameter. However, neither of these expressions was an accurate measure of the actual load-carrying capacity of a reduction gear because they did not take into account the contact pressure between mating teeth. The contact pressure is the proper design criterion because it is the factor that determines the satisfactory operation and durability of gears.

For many years, in the U. S., the allowable gear tooth pressure for turbine drives was related directly to the pinion diameter so that the loading was specified as "pounds per inch of face per inch of pitch diameter." This was logical since the curvature of the pinion tooth as it affects contact pressure, or more precisely the compressive stress at the contact surface, is directly proportional to the pinion diameter. Then

$$\frac{W_t}{F_e} \text{ (allowable)} = J \cdot d \tag{3}$$

where J is an experimentally determined constant, with units of pounds per inch of face per inch of diameter.

When gear dimensions are known, the J factor can be calculated as follows:

$$J = \frac{W_t}{F_e \cdot d} = \frac{126{,}050 \cdot HP}{RPM_p \cdot d^2 \cdot F_e}$$

The foregoing relationship, although an improvement, is not precise because it ignores the effect of the curvature of the mating tooth. A further refinement which takes this into account is

$$\frac{W_t}{F_e} \text{ (allowable)} = K \, \frac{R}{R+1} \, d \tag{4}$$

where R is the gear ratio and K an experimentally determined constant. This factor K, representing the allowable tooth surface stress, is the familiar "K-factor" by which gear loadings are now generally specified. Note that the K-factor is simply the loading per inch face per inch diameter, J in equation (3), multiplied by $(R + 1)/R$. Where gear

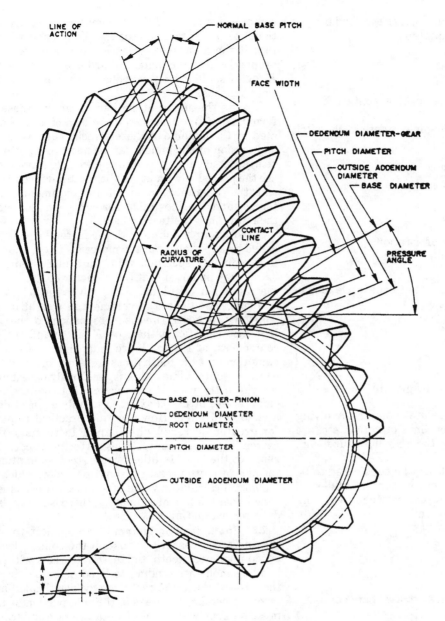

LINE OF
ACTION

NORMAL BASE PITCH

FACE WIDTH

DEDENDUM DIAMETER-GEAR
PITCH DIAMETER
OUTSIDE ADDENDUM
DIAMETER
BASE DIAMETER

CONTACT
LINE

RADIUS OF
CURVATURE

PRESSURE
ANGLE

Fig. 2 Involute geometry

BASE DIAMETER-PINION
DEDENDUM DIAMETER
ROOT DIAMETER
PITCH DIAMETER
OUTSIDE ADDENDUM DIAMETER

design details are known, the *K*-factor can be determined by the following relationships:

$$K = \frac{W_t}{F_e \cdot d} \frac{R+1}{R} = J \frac{R+1}{R}$$

$$= \frac{126{,}050 \cdot HP}{RPM_p \cdot d^2 \cdot F_e} \frac{(R+1)}{R} \qquad (5)$$

It can be shown that the *K*-factor is a good measure of tooth surface stress, i.e., the maximum compressive stress to which the tooth materials in contact are subjected.

Referring to Fig. 2 it can be seen that the total tooth loading in a helical involute gear is carried by a series of straight contact lines extending diagonally from the tip to the root of each meshing pair of teeth. The total force normal to the surfaces in contact is

$$W_n = \frac{W_t}{\cos \phi_n \cdot \cos \psi} = \frac{126{,}050 \cdot HP}{\cos \phi_n \cdot \cos \psi \cdot RPM_p \cdot d} \qquad (6)$$

where

W_n = total load normal to contact lines, lb
ϕ_n = pressure angle (plane normal to teeth)
ψ = helix angle

The average total length of all the lines of contact is

$$L_{\text{avg}} = \frac{F_e \cdot Z}{P_n} \qquad (7)$$

where

L_{avg} = average total length of all contact lines, in.
Z = length of line of action, in.
P_n = normal base pitch, in.

Dividing equation (6) by equation (7) the loading per unit of contact line length is equal to

$$W_n' = \frac{W_n}{L_{avg}} \tag{8}$$

where W_n' is the tooth load per inch of contact line in pounds per inch.

The radius of curvature of the pinion tooth at the pitch diameter is

$$\rho_p = \frac{d \sin \phi}{2 \cos \psi_b} \tag{9}$$

where

ρ_p = radius of curvature, in.
ϕ = pressure angle in plane of rotation
  $= \tan^{-1} \dfrac{\tan \phi_n}{\cos \psi}$
ψ_b = base helix angle (helix angle at base circle diameter) $= \sin^{-1} \sin \psi \cos \phi_n$

and of the gear tooth is

$$\rho_g = \frac{D \sin \phi}{2 \cos \psi_b} = R\rho_p \tag{10}$$

where D is the pitch diameter of the gear in inches and R is the gear ratio.

The relationships of helical involute geometry involving pressure angles, lines of contact, lines of action, etc. are described in numerous places in the gear literature, e.g., reference 1.

The tooth elements in contact may be considered to be elements of two tangent cylinders in contact under an applied force. The compressive stress between two cylinders is given by the Hertz equation:

$$S = \sqrt{0.175 \frac{P}{L} E \frac{r_1 + r_2}{r_1 \cdot r_2}}$$

where

S = maximum compressive stress between surfaces, psi
P/L = loading per inch of length, lb/in.
E = modulus of elasticity, psi
r_1, r_2 = radii of cylinders, in.

Substituting equations (5) to (10) in this Hertz equation, the compressive stress between the pinion and gear teeth becomes:

$$S_c = \left(4580 \sqrt{\frac{P_n}{Z \sin 2\phi}}\right) \sqrt{K} \tag{11}$$

The first term includes the modulus of elasticity and geometric factors, which are chosen by the gear designer. However, within practical limits, for steel gears with well proportioned tooth geometry, this term cannot be varied significantly.

The second term is the square root of the K-factor and shows that gears of equal K-factor will have nearly equal compressive stress. With the compressive stress proportional to the square root of the K-factor, it would follow that if the allowable stress is considered to be directly proportional to the material hardness, then the allowable K-factor should be proportional to the square of the material hardness.

Despite the apparent mathematical exactness of these formulas, many effects on tooth durability are not evaluated by them. Some of these effects, such as the bending and torsion of the pinion, can be analyzed; but others can be evaluated only by service experience. Among the latter are the precision with which the tooth surfaces are formed and the tolerance to small misalignments, vibratory forces, and the inevitable foreign particles, which find their way into the teeth mesh.

Satisfactory values for the K-factor have been established by experience for the materials in common use, and the commonly specified values are discussed in Section 3.8. It may be noted that the K-factor controls the size of the reduction gear unit. For a given set of horsepower and rpm conditions, the volume and weight of the gear will vary in nearly inverse proportion to the K-factor. It is also significant that the K-factor and the pitch, or coarseness, of the teeth are independent. Under the assumption that the tooth pressure is uniformly distributed over the contact lines, that is, uniformly distributed from the tip to the root of each contacting tooth, the contact compressive stress is affected to only a slight degree by a change in pitch. However, the practical requirements for greater tip relief with coarser teeth make the tooth extremities of coarser teeth less effective in carrying their share of the load. In other words, the assumption of uniformity of tooth pressure from tip to root, which leads to a minimal calculated value of surface stress, is less valid for coarser teeth. Unfortunately, there is no precise procedure for evaluating this effect.

2.2 Tooth Bending Strength. In addition to providing the surface necessary to sustain the contact loading imposed upon gear teeth, the teeth must also withstand the bending moments tending to bend or break the teeth at their roots. Since the teeth are cyclicly loaded at a high rate, the bending stresses in the root portion of the tooth must be kept well within the fatigue or endurance limit of the material.

To arrive at a formula for bending stress, it is necessary to make the same assumptions of uniform distribution of tooth pressure over all lines of contact. The loading per inch of contact line developed earlier is

$$W_n' = \frac{W_n}{L_{avg}} \tag{8}$$

For a spur gear tooth the highest bending stress occurs when the load is acting at the extreme tip of the tooth. The tooth form factor Y, which relates the tooth loading to bending stress at the root, is

$$Y = \frac{t^2}{6h} \tag{12}$$

where t is the tooth thickness at the root and h the tooth

Table 1 Diagonal loading factor

Helix Angle	k
0	1.000
10	0.850
20	0.770
30	0.713
40	0.670
45	0.653

height dimension, both in inches. These tooth dimensions are shown in Fig. 2. For a spur gear, the bending stress at the root is computed as

$$S_b = \frac{6h W_n}{t^2} \qquad (13)$$

This same relationship also holds for helical gears but, although the assumption that the loading is tip-applied is good for spur or low helix angle gears, it is invalid for steeper helix angles where the loading extends diagonally over a portion of the tooth. Therefore, a diagonal loading factor should be applied. This is particularly true when comparing designs with different helix angles. The stress, as given in the foregoing, should be reduced by a factor, k, which is a function of the helix angle. Equation (13) then becomes

$$S_b = \frac{k \cdot 6h W_n}{t^2} \qquad (14)$$

where k is the diagonal loading factor. Values for the diagonal loading factor are given in Table 1. Factors for intermediate values of the helix angle may be determined by interpolation.

By substituting equations (6), (7), and (8) into equation (14), the bending stress in the root of the helical gear teeth becomes

$$S_b = k \frac{W_t}{F_e} \frac{6h \cdot P_n}{Z \cdot t^2 \cdot \cos \phi_n \cdot \cos \psi} \qquad (15)$$

Equation (15) contains the important variables affecting bending stress. Further refinement, or a more precise assessment of the stress as it determines the bending fatigue strength of the teeth, can be made by including two additional factors. One is the compressive stress across the tooth root cross section due to the radial component of the tooth load, which acts to reduce the bending stress on the tension side; the other is the stress concentration created by the root radius adjacent to the critical bending cross section. Both of these additional factors are included in the bending strength derivation in the military specification for reduction gears [2].

Referring to equation (15) it can be seen that the bending stress, S_b, is directly proportional to the tangential tooth load per inch of face and inversely proportional to the first power of the tooth dimensions. Other variables are of secondary importance and change very little with well proportioned teeth in the usual range of helix angle and pressure angle. As a good approximation, the bending stress formula can thus be simplified to

$$S_b = CU \qquad (16)$$

where

$C =$ a constant depending on the tooth proportions, helix angle, pressure angle, etc.

$U =$ unit loading $= \dfrac{W_t}{F_e} NDP$

$NDP =$ normal diametral pitch of teeth

The normal diametral pitch is in inverse proportion to the linear dimensions of the tooth cross sections and is, therefore, an accurate reference for tooth size. As a result, the unit loading, which is simply the tooth loading per inch of face multiplied by the normal diametral pitch, is a convenient measure of bending stress, just as the K-factor is a measure of surface stress. The allowable unit loadings are generally in the range of 6000 to 8000. However, this range may be safely exceeded with proper standards of alignment accuracy, metallurgy, etc. High-powered naval vessels employ unit loads well above 10,000.

From the consideration of bending stress alone, it would appear quite easy to lower the bending stress simply by increasing the size of the teeth. However, this entails compromises with surface stress, scoring, and noise considerations; consequently, the tooth pitch must be selected to provide the best balance of all factors.

2.3 Tooth Scoring. The action of two involute tooth surfaces when rotating in unison is such that the contacting surfaces both roll and slide over each other. At the pitch line point of contact, the sliding component is zero and the contacting surfaces are in pure rolling contact. But the sliding component increases with the distance from the pitch line and is a maximum at the tooth extremities, tip and root. This sliding action, if sufficiently severe, can cause scoring of the tooth surfaces. This scoring or galling is an actual fusing or welding together of particles of the contacting surfaces. Under the continued motion, particles are torn from one surface and either deposited on the other surface or released.

Scoring, which results in a serious deterioration of the tooth surfaces, is not to be confused with the minor scratching of the tooth surfaces that results from the passage of minute particles between the teeth. Scratching under certain light reflection can appear to be scoring. Scoring, however, is rough to the touch. For a comprehensive discussion of the various modes of gear tooth failure and some practical experiences in this regard, see reference 3.

Scoring results from tooth pressure in conjunction with a sliding velocity. Historically, the tendency to score was assessed by means of a scoring or PVT factor, which places a numerical value on a combination of the contact pressure and sliding velocity. A definition of the terms and formulas for calculating PVT can be found in reference 4.

A number of additional factors, such as lubricant and tip relief, influence the tendency of gears to score such that considerable expertise and practical experience are required to select the tooth form, materials, surface finish, and lubricant to avoid scoring difficulties.

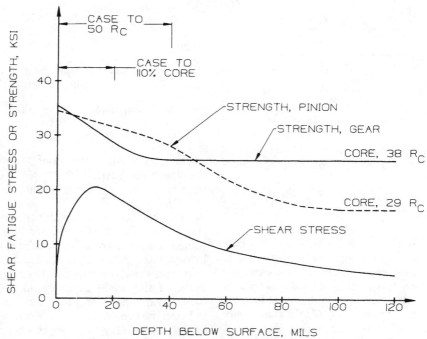

Fig. 3 Subsurface shear stress and strength versus depth

There is no universally accepted calculation procedure that can be used to assess the tendency of marine propulsion gears to score. Gear manufacturers often modify methodologies set forth by researchers and correlate the results with known scoring failures. The work of Blok [5], Dowson and Higginson [6], and Winter and Michaelis [7] are methods that have been used to evaluate scoring risks. Emerging as a measure of scoring risk is the Lambda ratio, which is as follows:

$$\text{Lambda ratio} = \frac{\text{Oil film thickness}}{\sqrt{f_1^2 + f_2^2}}$$

where f_1 is the surface finish of the pinion and f_2 the surface finish of the gear. It is generally agreed that if the Lambda ratio is greater than 1.0, the risk of scoring will be low.

The risk of scoring can be reduced with proper design attention to surface finishes, oil viscosity, and oil extreme pressure (EP) additives. As the gear tooth loading intensity is increased, these selections and tradeoff studies become increasingly important. The uniformity of load is also an important consideration. Maldistribution in load uniformity will dramatically increase scoring risk. Coatings, such as silver or copper, provide excellent break-in protection and reduce scoring risk.

2.4 Subsurface Shear Stress. The ratio of the gear tooth subsurface shear stress to the material strength is a very important design criterion, particularly with case-hardened gear elements. Marine gears are typically designed for a service life of 30 or more years and the number of fatigue cycles that a tooth is subjected to approaches 10^{11} cycles in many cases. In order to avoid subsurface fatigue, or spalling, materials and hardening processes must be evaluated to ensure that adequate margin or safety factors exist.

The subsurface shear stress is a function of surface compressive stress and varies with depth below the tooth surface. Also, the fatigue strength in shear varies with case and core hardness and, therefore, depth below the tooth surfaces. Fatigue strength is also related to cycles. A typical plot of shear stress and strength is shown in Fig. 3. The tooth strength and stress can be compared for any number of cycles. Local material strength can then be compared with local stress to determine the margin or safety factor concerning subsurface fatigue.

Section 3
Gear Design

3.1 Determination of Approximate Size of Gears. While the detail design of a reduction gear requires a high degree of skill, it is fairly easy to establish approximate dimensions of a reduction gear. As an example, consider a double-reduction gear which is to be designed to meet the following requirements:

Shaft horsepower ....25,000 hp at 108 rpm
HP turbine ....12,500 hp at 6100 rpm
LP turbine ....12,500 hp at 4100 rpm
First reduction K-factor............$K_1 = 140$
Second reduction K-factor.........$K_2 = 110$

A conventional arrangement, as illustrated by Fig. 1(d), has been selected and suitable dimensions for the pitch diameters and face widths are to be computed.

The HP and LP turbines develop equal horsepower; however, the HP turbine turns faster than the LP turbine. As a result, the HP side will require a larger gear reduction and will control the size of the second reduction elements; therefore, it will be computed first. The overall reduction ratio of the HP side is 6100 to 108. As a first approximation, the ratio of the second reduction can be taken as the square root of the overall ratio minus 1.0. (For a locked-train gear, 3.0 would be added to the square root of the overall ratio.) The second reduction ratio then becomes

$$R_2 = \sqrt{\frac{6100}{108}} - 1 = 6.52$$

and the first reduction ratio is

$$R_1 = \frac{6100}{108 \times 6.52} = 8.66$$

The loading per inch of face per inch of pitch diameter for the first reduction can now be computed:

$$J_1 = \frac{W_{t1}}{F_{e1} \cdot d_1} = K_1 \frac{R_1}{R_1 + 1} = 140 \frac{8.66}{8.66 + 1}$$
$$= 125.5 \text{ lb/in.-in.}$$

The next step is to equate two expressions for the tangential tooth load as follows,

$$W_{t1} = \frac{126,050 \cdot HP}{d_1 \cdot RPM_1} = J_1 \cdot F_{e1} \cdot d_1.$$

solving for $d_1^2 F_{e1}$:

$$d_1^2 F_{e1} = \frac{126,050 \cdot HP}{J_1 \cdot RPM_1} = \frac{(126,050)(12,500)}{(125.5)(6100)} = 2058 \text{ in.}^3$$

Generally, the most economical reduction gear is one where the pinion diameter is as small as possible with relation to its working face. However, as will be seen later, the face width-to-diameter ratio cannot be too high if excessive deflections are to be avoided. Ratios of 2.0 to 2.25 represent good practice and 2.25 is selected. With this stipulation, the computations may proceed:

$$F_{e1} = 2.25 d_1$$
$$d_1^3 = \frac{2058}{2.25} = 915 \text{ in.}^3$$

and the first reduction pinion diameter is

$$d_1 = \sqrt[3]{915} = 9.71 \text{ in.}$$

with an effective face width of

$$F_{e1} = (2.25)(9.71) = 21.8 \text{ in.}$$

The first reduction gear is next computed as

$$D_1 = R_1 d_1 = (8.66)(9.71) = 84.1 \text{ in.}$$

Similar calculations can now be made for the second reduction:

$$J_2 = K_2 \frac{R_2}{R_2 + 1} = 110 \frac{6.52}{6.52 + 1} = 95.4 \text{ lb/in.-in.}$$

$$RPM_2 = \frac{RPM_1}{R_1} = \frac{6100}{8.66} = 704 \text{ rpm}$$

$$d_2^2 F_{e2} = \frac{126,050 \cdot HP}{J_2 \cdot RPM_2} = \frac{(126,050)(12,500)}{(95.4)(704)} = 23,460 \text{ in.}^3$$

Again selecting $F_{e2} = 2.25 d_2$, the second reduction pinion diameter becomes

$$d_2^3 = \frac{23,460}{2.25} = 10,430 \text{ in.}^3$$

$$d_2 = \sqrt[3]{10,430} = 21.8 \text{ in.}$$

with an effective face width of

$$F_{e2} = (2.25)(21.8) = 49.0 \text{ in.}$$

and a second reduction gear diameter of

$$D_2 = (21.8)(6.53) = 142.1 \text{ in.}$$

The LP first reduction can be proportioned in the same manner, but it is desirable to design the arrangement such that the second reduction pinions on both the HP and LP sides are identical. Since the first reduction gear speed on the LP side must be the same as that on the HP side (704 rpm), the first LP reduction ratio will be:

$$R_1 = \frac{4100}{704} = 5.82$$

Proceeding as before

$$J_1 = K_1 \frac{R_1}{R_1 + 1} = 140 \frac{5.82}{5.82 + 1} = 119.5 \text{ lb/in.-in.}$$

$$d_1^2 F_{e1} = \frac{126,050 \cdot HP}{J_1 \cdot RPM_1} = \frac{(126,050)(12,500)}{(119.5)(4100)} = 3216 \text{ in.}^3$$

Selecting $F_{e1} = 2.25 d_1$

$$d_1^3 = \frac{3216}{2.25} = 1429 \text{ in.}^3$$

$$d_1 = \sqrt[3]{1429} = 11.26 \text{ in.}$$

$$F_{e1} = (2.25)(11.26) = 25.3 \text{ in.}$$

$$D_1 = (11.26)(5.82) = 65.5 \text{ in.}$$

It may be desirable to use the same first reduction gear on the LP side as used on the HP side. In this case,

$$d_1 = \frac{D_1}{R_1} = \frac{84.1}{5.82} = 14.45 \text{ in.}$$

$$F_{e1} = 21.8 \text{ in.}$$

$$D_1 = 84.1 \text{ in.}$$

$$W_{t1} = \frac{126{,}050 \cdot HP}{d_1 \cdot RPM_1} = \frac{(126{,}050)(12{,}500)}{(14.45)(4100)} = 26{,}600 \text{ lb}$$

$$J_1 = \frac{W_{t1}}{F_{e1} \cdot d_1} = \frac{26{,}600}{(21.8)(14.45)} = 84.7 \text{ lb/in.-in.}$$

$$K_1 = J_1 \frac{R_1 + 1}{R_1} = 84.7 \frac{5.82 + 1}{5.82} = 99.3$$

This LP first reduction is larger than it would be if it were designed to the maximum permissible K-factor, but this may be offset by the economy of using the same part for both first reduction gears.

The pitch diameters as determined in the foregoing must now be laid out to determine if centerline positions and other arrangement considerations are acceptable. The optimum gear arrangement may require adjusting the choice of ratios between the first and second reductions and the choice of face width-to-diameter ratios.

With the approximate diameters and face widths as determined in the foregoing, the designer will next check to determine that bending and torsional deflections are acceptable. Formulas for these deflections are developed in the following section. A lower L/D ratio may be selected if these deflections are too high, with the diameters and face widths adjusted accordingly.

Tooth pitch is then selected to provide the best balance between bending stress, scoring factor, and noise. The best compromise in this regard is generally the finest pitch permitted by the bending stress or unit loading limits. This will result in an acceptable bending stress, minimum scoring factor, and minimum noise level.

Tooth pitch, addendum, dedendum, pressure angle, etc., and tooth proportions, are made to suit the standards for which the manufacturer has tooling. These standards are in small enough increments that no significant compromise is involved. The numbers of teeth are chosen to provide "hunting tooth" combinations between mating pinions and gears, and diameters or helix angles are adjusted to the precise values determined by the numbers of teeth. A hunting tooth combination is one in which the numbers of pinion and gear teeth have no common prime factor. This means that each tooth will mesh with every tooth of the mating element and thus avoid any wear or tooth spacing pattern that can give rise to a sub-harmonic of the tooth meshing frequency.

As noted previously, the design of gears is based on the tooth pressure being uniformly distributed across the entire face width. Many factors adversely affect this tooth pressure distribution and must be taken into account. Among these factors are torsional and bending deflections of the pinion, accuracy of manufacture, deflections due to centrifugal force, strains due to temperature variations, and casing distortions due to temperature differences and hull deflections. Two of these factors, torsional and bending deflections of the pinion, are important in proportioning gear elements and, fortunately, are readily evaluated.

3.2 Torsional Pinion Deflection. When subjected to a uniform tooth pressure, a pinion will deflect torsionally

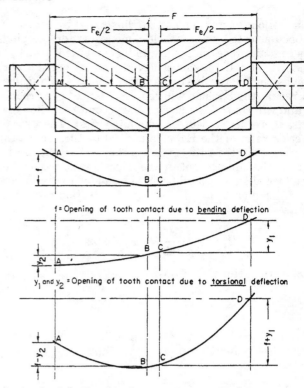

f = Opening of tooth contact due to _bending_ deflection

y_1 and y_2 = Opening of tooth contact due to _torsional_ deflection

$(f+y_1)$ and $(f-y_2)$ = Opening of tooth contact due to bending & torsion

Fig. 4   Pinion deflection

as shown in Fig. 4. The teeth will separate from the mating gear teeth by the distance y. However, since the pinion is always free to shift endwise to balance the load between the two helices, the separation after this axial shift will be y_1 on the helix next to the coupling and y_2 on the helix away from the coupling. The torsional deflection in the space between the helices has no effect on the separation. The separations will then be

$$y_1 = 7.95c \cdot J\left(\frac{F_e}{d}\right) F_e \cdot 10^{-8} \tag{17}$$

$$y_2 = 2.65c \cdot J\left(\frac{F_e}{d}\right) F_e \cdot 10^{-8} \tag{18}$$

where

y_1 = tooth separation at driving end, in.
y_2 = tooth separation opposite from driving end, in.
$c = \dfrac{d^4}{d^4 - d_0^4}$ where d_0 = diameter of pinion bore; $c = 1.0$
    for a solid pinion
J = tooth loading, lb/in.-in.
F_e = effective face width of pinion, in.
d = pitch diameter of pinion, in.

These equations are based on a uniform distribution of tooth pressure, endwise freedom to equalize the load between both helices, an effective diameter for torsion equal to the pitch diameter, and a shear modulus for steel equal to 12.0×10^6 psi.

3.3 Bending Pinion Deflection. In addition to torsional pinion deflections, the tooth loading will cause the pinion to deflect due to bending stress as shown in Fig. 4. The pinion can be assumed to be uniformly loaded, and by using the deflection equation for a simply supported, uniformly loaded beam, the tooth separation due to bending is found to be

$$f = 0.885c \cdot J \left(\frac{F}{d}\right)^3 F \cdot 10^{-8}$$

where f is the tooth separation due to deflection and F the distance between ends of bearings, both in inches. The remaining terms are as defined previously.

This expression is based on a uniform distribution of tooth pressure, the tooth pressure acting over the distance between the ends of the bearings, the effective diameter for bending equal to the pinion pitch diameter including the space between helices, the pinion simply supported at the inner ends of the bearings, and the modulus of elasticity for steel equal to 30.0×10^6 psi.

A generally accepted value for the allowable deflection due to torsion and bending is 0.001 in. However, other effects can add to these calculated values. The total effect can be observed by tooth contact patterns under full-load operation, estimated from experience on similar gears, or estimated by analysis. The sum may exceed 0.001, but the gearing can be made perfectly satisfactory by machining corrections into the helix angles so that the tooth contact will be uniform under full-load operating conditions. When this is done, the cold light-torque contact pattern will not be uniform. But since the direction and amount of the helix angle corrections are known, the light-torque contact pattern will be a good indication of the contact pattern under operating conditions.

Such a light-torque contact check will be made at the factory to confirm the correct machining and assembly of the unit, and the check will be repeated in the ship installation to confirm that the factory alignment has been duplicated. These contact checks can be made by observing the transfer of a marking compound such as Prussian Blue, or light layout lacquer, from one element to the other. Uniform transfer of compound over the full face width will indicate uniform face contact under light loads. While satisfactory contact checks can be made with very light torques, they can be made with greater reliability with higher torques. When light loads are not sufficient to bring about uniform contact, a quantitative measure of face contact can be made by gaging the opening between meshing teeth with feeler gages graduated in 0.0001-in. steps.

Despite the care which may be taken in factory and installation tests, the final quality of tooth contact must be judged after full-power operation in the ship. For this observation, the teeth of each pinion or gear may be coated in a band extending across each face with copper by the application of a weak acid copper sulfate solution, or with a thin coat of layout lacquer.

3.4 Other Deflections. There are other deflections that can act to affect the uniformity of tooth contact across the tooth faces. The gear housing structure will deflect under the forces applied to the bearings and may deflect to misalign the teeth; an example would be the case in which the support of one pinion bearing is more flexible than the support of the bearing at the opposite end of the pinion.

Gear casings are also subject to thermal strains and these can affect tooth alignment. For instance, the casing support structure for the bearings in the middle of a double-reduction gear housing may be at a higher temperature than the structure that supports the end bearings.

The rotating elements are also subject to elastic and thermal deformations. Gear rims that are attached to their hubs by a series of thin plates or cone members are deformed by the action of centrifugal forces. The design must be such that these deflections do not have a significant effect on the tooth portion.

Thermal strains can also be important, particularly with wide face widths. If a pinion is allowed to reach a temperature higher than its mating gear wheel, the uniformity of tooth contact across the faces of both helices will be affected.

3.5 Gear Alignment and Installation. An important source of misalignment in the second reduction mesh can be due to the difference in the magnitude of the forward and after slow-speed gear bearing reactions. Figure 5 is a typical bearing reaction diagram for a double-reduction gear. It may be seen that the gear bearing reactions consist of one or more components due to the torque loadings and a component due to the static weight of the pinion or gear supported. With the exception of the slow-speed gear bearing reactions, none are affected by external influences. However, such is far from the case with the slow-speed gear bearings. When the static loads imposed on the forward and after slow-speed gear bearings are different in magnitude, as opposed to being equal as shown in Fig. 5, the resultant reactions will not be in the same direction. This will cause the forward and after gear bearing journals to ride in different positions within their bearing clearances. The slow-speed pinions are not subjected to a similar influence; therefore, there results a crossed-axis condition between the slow-speed pinions and gear.

The foundations of slow-speed gear bearings and line shaft bearings are completely dissimilar. Slow-speed gear bearings are located very close to the lube oil sump tank and, therefore, their foundations become very warm when at operating temperature, causing an attendant thermal rise in the position of the slow-speed bearings. On the other hand, little heat is generated in line shaft bearings, and they operate at a temperature little above the ambient. This being the case, it is unavoidable that the line shafting have an influence on the slow-speed gear bearing reactions when the plant goes from a cold to the operating condition. When going from a cold to the operating condition, the slow-speed gear bearings will rise about 15 to 30 mils higher than the line shaft bearings.

Prior to the late 1950's, misalignments due to this source were generally disregarded and the slow-speed gear shaft was aligned concentric to the line shafting. It is easily shown that the forward slow-speed gear bearing

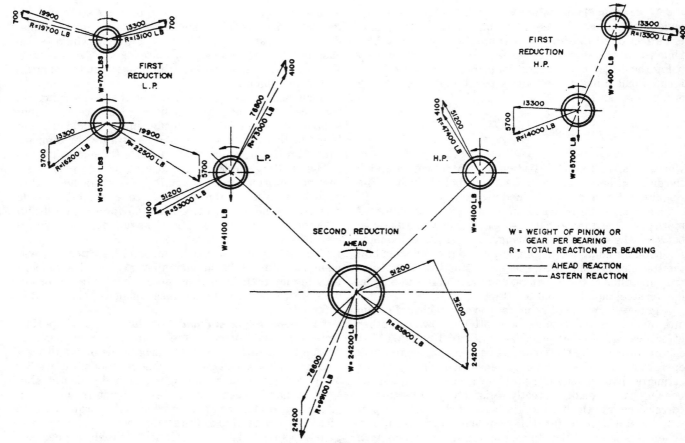

Fig. 5 Typical reduction gear bearing reaction diagram

on many of the older ships carried no static load when in the operating condition. It is speculated that the disregard of this factor led to a number of their problems.

The successful operation and reliability of main reduction gears are not only the responsibility of the gear designer, but also the naval architect and shipbuilder. Factors influencing the gear-mesh loading and distribution of load are affected by the design and manufacturing accuracy, foundation deflections, lineshaft alignment and flexibility, and installation accuracy. These factors, along with system design responsibilities, alignment techniques, installation procedures, and operational verification of uniform gear tooth loading, are reviewed in SNAME T&R Bulletin 3-43, which was prepared by the M-16 Panel [8].

The achievement of an acceptable gear-mesh load distribution, particularly in the slow-speed gear mesh, is a key factor towards reliable service, and compensation for the elastic and thermal deflections of the pinions and gears is the essential first step.

Reduction gear problems have also been encountered due to hull flexibility, excessive lineshaft stiffness, and improper alignment; therefore, as the design matures, continuous communications between the manufacturer, the architect, and the shipbuilder are important to ensure a satisfactory installation. Conducting alignment studies

and establishing proper installation and alignment procedures are of vital importance in this regard.

Finally, the verification of gear tooth contact and the uniformity of load across the face widths are essential to the achievement of high reliability and low risk of tooth wear. Brake tooth contact checking is used as an indicator of proper alignment. The verification of tooth contact and uniformity of load can be performed by using strain gages, coatings such as silver or copper, or either red or blue dykem. The "reading" of dykem and coatings requires experience and can be interpreted incorrectly by untrained personnel. Experienced personnel can judge proper tooth contacts; however, changes due to environmental conditions are difficult to detect. Strain gaging has been successfully applied to large reduction gears and is a reliable procedure for accurately determining the uniformity of load. Using frequency modulated telemetry, gear tooth strain can be monitored and quantified under all operating conditions, thus recording the influences of the external and internal factors that affect gear reliability.

3.6 Critical Speeds. Pinion and gears, designed as they are for stiffness to resist tooth forces, have lateral critical speeds that are well above any operating speed. They will run free of vibration with normal procedures for balancing. With both gas turbine or steam turbine prime movers, balance is a particularly important consideration with the first-reduction pinion. It rotates at turbine speed, and must be given the same high degree of dynamic balance as the turbine.

Coupling shafts connecting the turbine to the pinion are important elements in determining the lateral critical speeds of the turbine rotor-coupling-pinion assembly and must be considered when evaluating turbine critical speeds.

The combination of the propeller, shafting, gears, and prime mover forms a system, which can vibrate torsionally in response to the impulses from the propeller blades. With the very early gear designs, manufacturing irregularities in the gear teeth occasionally were a source of serious torsional vibration; however, the precision with which modern gears are manufactured has eliminated this as a source of torsional vibration. With gas or steam turbine-driven gears, the first three modes of torsional vibration warrant careful analysis. In the first mode of torsional vibration with a geared-turbine drive, the angular vibratory motion is greatest at the propeller, but the vibratory torque is a maximum at the reduction gear. This mode generally occurs within the operating range, being well down in the operating range with arrangements having long shafts but relatively high in the operating range and potentially dangerous with very short shafting arrangements.

The first mode of torsional vibration must be evaluated to ensure that the vibratory torque in the gear train, when added to the torque transmitted under steady power conditions, will not be deleterious to the reduction gearing.

The inertia and elastic factors of the turbines and gears have no significant effect on the first critical speed; it is controlled by the inertia of the propeller and entrained water, the stiffness of the shafting, and the number of propeller blades.

The second mode of torsional vibration is one in which the two turbine branches vibrate in opposition and it may occur in the operating range. When this is the case, vibratory torques must be evaluated as for the first critical. However, by employing a so-called "nodal drive" arrangement, it is possible to render the second mode incapable of excitation. In a nodal drive arrangement, the two turbine branches are tuned by adjusting the dimensions of the quill shafts, such that they have identical frequencies with the slow-speed gear, shafting, and propeller considered nodal points. As a result, all motion in the second mode is in the turbine branches, and propeller excitation cannot excite this mode since the propeller is on a node.

The third mode of torsional vibration, in which the slow-speed gear is an antinode, may be of concern. It is usually well above the operating range, but when an unusually large number of propeller blades are used, it may be of importance.

A description of the modes of torsional vibration and a procedure for calculating the natural frequencies and amplitudes are included in Chapter 10.

Main propulsion systems using diesel engines as prime movers require extensive torsional evaluations to ensure satisfactory operation. Diesel engines have many excitation orders. Four-stroke engines produce excitation orders of 1/2, 1, 1½, 2, 2½, etc. Two-stroke engines produce excitation orders of 1, 2, 3, etc. It is not uncommon to analyze the propulsion system for as many as twelve orders.

Nearly all diesel propulsion systems require a torsional flexible coupling with proper stiffness and damping characteristics to minimize torsional vibrations in reduction gears and power takeoff drives for generators. The selection of these couplings is very important for normal operation as well as for misfiring conditions.

The analysis of torsional vibration in diesel systems, the selection of couplings, and an analysis of misfiring are comprehensively discussed in references 9, 10, and 11.

The various considerations relating to vibration monitoring, diagnostic techniques, and predictive maintenance, of gears and other ships machinery are reviewed in reference 12.

3.7 Gear Case. The function of the gear case is to furnish adequate support for the bearings as well as to provide an oil-tight enclosure for the reduction gear. Typical gear cases may be seen in Figs. 6 and 7. All journal bearing load reactions are in planes perpendicular to the axis of the revolving shafts. In many instances, and particularly in connection with double-reduction gears, the bearing supports will have to support bearings at different elevations. It is of the utmost importance that these bearing supports including the cap have sufficient structural stiffness to prevent any measurable deflection under varying load conditions. Due to the direction of rotation of the different shafts and the location of pinions in reference to gear bearing loadings, the bearing reactions as illustrated by Fig. 5 may occur at any angle to the axis, and it is important that the bearing cap construction takes this into account. It must be borne in mind that for satisfactory operation of the gears and to minimize wear, the revolving shafts must operate continuously parallel to each other. The gear case construction is the only means provided to maintain the different shafts in their correct relation to each other.

The construction and stiffness of the gear case must be studied and compared with the structure and rigidity of the foundation below the gear case and the arrangement whereby the gear case is secured by bolting to the ship structure. The gear casing generally is rigidly bolted to the foundation to form a combined structure to prevent deflections between the gear and pinion axes, which may be caused by deflection in the ship structure when operating in a heavy sea.

Since the strains in the hull due to the loading of the ship and to the forces imposed by the seaway are imposed on the gear casing, designers are attracted to means for isolating strains in the ship's hull from the main gear casing. Two, three, or four points of support between the gear case and its foundation can effectively accomplish this isolation.

Except for small auxiliary gears, the casing has separate inspection covers for convenience in inspecting the condition of the gear teeth, and is arranged so that bearings, flexible couplings, and oil sprays may be inspected

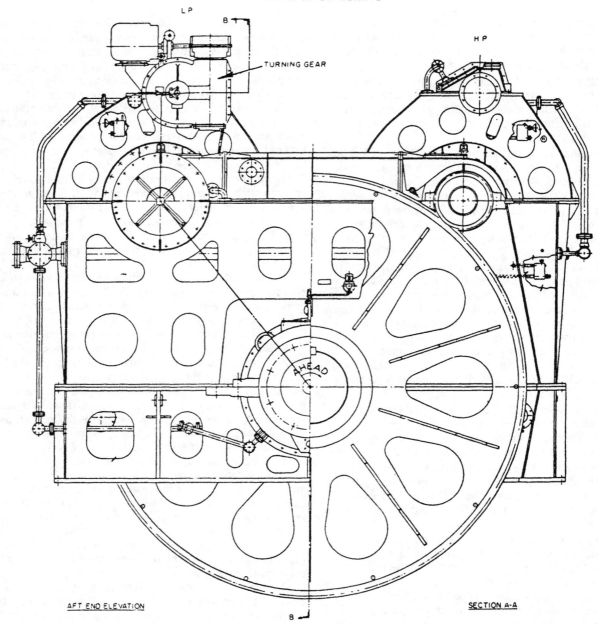

Fig. 6(a)   Articulated double-reduction gear

and replaced without having to dismantle large sections of the casing.

The casing enclosure acts to attenuate the noise that is generated by the meshing teeth. Even though the most effective means of reducing gear noise is the precision with which the teeth are machined, the gear case offers a means of further quieting by designing to minimize the transmission of sound.

Some of the smaller gear cases are made of steel castings. However, in the propulsion gear sizes, the casings are of fabricated steel construction. Some castings may be employed in the fabrication for the heavier sections, but the trend is away from castings in favor of sections burned out of heavy plate or formed from plate material.

3.8 Pinions and Gear Wheels.   Pinions are most often made of a one-piece forging and may be hollow-bored to accommodate a quill shaft. The requirements for strength and rigidity generally preclude making the tooth and journal portion of the pinion in more than one piece. The material is usually an alloy steel that is through hardened, or case hardened, to the desired hardness.

The tooth portion of the gear wheel is usually an alloy-steel forging that is either welded to a center portion, which is made up of steel plate in the larger diameters, or integral with the center portion in the smaller diameters.

For many of the early turbine-driven gears, the most common materials were pinions in the 200-240 Brinell hardness number range running with gears in the 160–

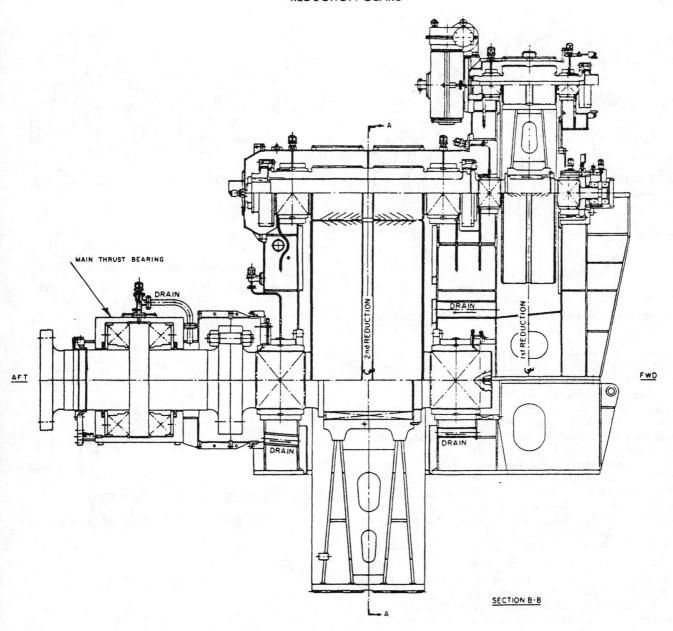

AFT   FWD

MAIN THRUST BEARING

DRAIN

2nd REDUCTION

1st REDUCTION

DRAIN

DRAIN   DRAIN

SECTION B-B

A   A

Fig. 6(*b*) Articulated double-reduction gear elevation view

190 Bhn range. With these hardnesses, *K*-factors of about 90 for the first reduction and 75 for the second reduction were generally appropriate. However, materials with higher hardness and higher *K*-factors have subsequently had wide application in naval combatant ships and for commercial service. *K*-factors of 140/110 (for the first and second reductions, respectively) have been applied using through-hardened pinions with a Bhn above 300 and through-hardened gear rims with a Bhn above 220. The higher *K*-factor gearing has the advantage of increasing the power capability of a given size of gear unit approximately in direct proportion to the *K*-factor. The higher *K*-factor gear is, therefore, more compact and lighter in weight.

As an interesting side note, the *Vespasian* had a *K*-factor of 78, and the *Neptune* had one of 125, which attests to the genius of the early inventors.

Case-hardened and ground materials are used with *K*-factors ranging from 350 to 600 in both commercial and military applications. As compared to early hobbing, shaving, and lapping techniques, the advancements that have been made in gear grinding machinery technology have made it possible to produce profile and helix modifications as required to ensure superior gear tooth load distribution and low noise levels.

The success of any marine gear is highly influenced by the selection of the correct materials and proper heat

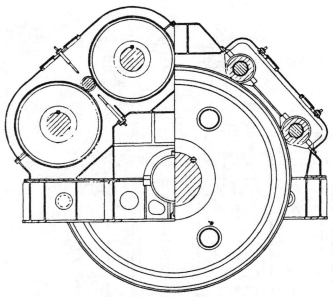

Fig. 7(a) Fore-and-aft sectional view of a locked-train reduction gear

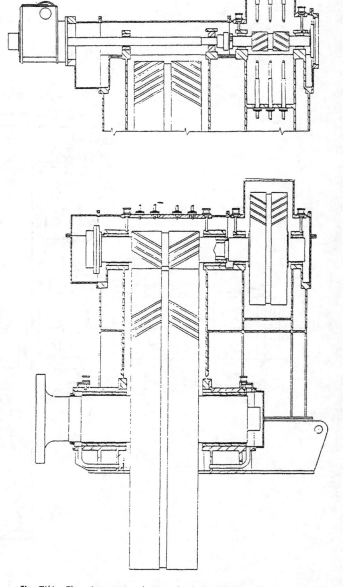

Fig. 7(b) Elevation sectional view of a locked-train reduction gear

treatment in conjunction with a Quality Assurance program, which assures that the material properties are achieved and that there are no deleterious material flaws. The unique requirements for through-hardened and case-hardened materials are delineated in reference 13. This reference standard details requirements for melting, grain size, heat treatment, mechanical properties, chemistry, workmanship, and quality control.

3.9 Journal and Thrust Bearings. Journal bearings must carry the weight of the gear elements and also transmit the large tooth meshing forces to the casing structure. Hydrodynamic bearings have been used almost exclusively in this application, and the conventional babbitt-lined, steel-shell sleeve bearing has proved to be extremely long-lived, with a high tolerance for abnormalities, such as dirt and rust, in the operating environment.

As may be noted from Fig. 5, the tooth meshing forces for ahead and astern rotation are in nearly opposite directions and generally in different directions than the weight reaction. It is necessary, therefore, to select an angular position for mounting the bearing in the housing so that the bearing areas are in the best possible relationship to the applied bearing forces.

First-reduction pinion bearings operate at high speeds. When starting and at low speeds, the bearing forces are low, consisting almost entirely of the weight components. At higher speeds, however, the bearing reactions continuously increase. These conditions are favorable for hydrodynamic bearings and permit the safe use of high unit loadings in these bearings. Unit pressures of 225–250 psi of projected area are generally acceptable for high-speed journal bearings in commercial service, and considerably higher pressures are used successfully in naval service.

The clearance ratio for these high-speed bearings should be 0.002 to 0.003 in. per inch of journal diameter. This clearance ratio is important in defining the difference in radii of curvature of the journal and bearing surfaces,

and thus the degree of convergence and divergence of the load-carrying oil film. The clearance opening, as such, in the unloaded half of the bearing affects only the quantity of oil passing through the bearing.

Second-reduction gear bearings operate at lower speeds and have relatively high static loadings due to the weight of the bull gear. These conditions are less favorable and require lower unit pressures of 150–250 psi. The clearance ratio should be about 0.001 in. per inch of journal diameter.

Intermediate-speed bearings fall between the high- and low-speed bearings, with loadings of 175–200 psi and clearances of 0.001 to 0.0015 in. per inch of journal diameter.

In case-hardened designs tooth loads and bearing loads are higher. Loadings for high- and intermediate-speed bearings are typically in the range of 350–500 psi.

In addition to carrying a load, the journal bearings must accurately position the gear and pinion journals to keep their axes precisely parallel. The replacement of bearings, therefore, must be made so as not to alter the journal position. To facilitate bearing replacement, a common practice is to stencil on the bearing shell its shell thickness at several points. Then, a replacement bearing with the same shell thickness will maintain the original journal position.

The main propeller thrust bearing is generally either integral with the gear unit or immediately adjacent to it. Its main purpose, of course, is to transmit the propeller thrust to the hull, but a secondary purpose is to hold the second reduction gear wheel in its proper axial position.

With double-helical gearing the main thrust bearing also holds the second-reduction pinions in their axial position, and further, by an axially restricted coupling the first-reduction gears can also be positioned. It is also common to couple the first-reduction gear to its second reduction pinion with a coupling, which permits endwise motion. Then, positioning thrust bearings must be provided for the first-reduction elements. This can be either a pivoted shoe or plain collar thrust bearing applied to either the first-reduction pinion or its gear wheel. This bearing must have sufficient capacity to overcome the frictional forces in the couplings, which act on the first-reduction elements.

Where adequate foundation structure can be provided, it is convenient to locate the main thrust bearing forward of the second-reduction gear, with the thrust housing an integral part of the gear casing. This location has two advantages: (a) the diameter of the thrust bearing can be smaller because the shaft portion does not have to transmit torque, and (b) the thrust collar can be a separate piece that can be readily removed over the end of the gear shaft in the event that it is necessary to replace or refinish the collar surface.

For higher powers, and where greater stiffness is required for the thrust bearing foundation, the thrust bearing is located aft of the second-reduction gear. An installation in which the main thrust bearing is located immediately aft of the slow-speed gear is shown in Fig. 6. The thrust housing structure is independent of the gear case and joins it by a flexible oil-tight connection. The bolted attachment to the foundation, which transmits the propeller thrust to the hull, is independent of the bolting attachment of the gear base to its foundation, so that the thrust bearing and its foundation can deflect as a result of the propeller thrust with no distorting effect on the gear casing.

In either location, the thrust bearing shares the lubricating system with the gear, and its oil drain discharges into the gear base. Only a single shaft oil seal is required on the output shaft.

The thrust bearing is of the pivoted-shoe type, with two sets of shoes acting on opposite sides of a thrust collar to accept thrust in either direction with a design pressure in the range of 375 to 500 psi.

All reduction gear journal and thrust bearings are force-fed from a central lubricating oil system. Each journal and thrust bearing is generally provided with a sight-flow and thermometer fitting in a visible location so as to provide an indication of performance. As a sample of oil leaving the bearing passes through the sight flow (or bubbler) it provides a visible jet of oil that can be seen at some distance, giving assurance that the bearing is being properly lubricated. This oil also passes over a thermometer well installed integral with the sight-flow fitting for sensing and indicating either locally or remotely the temperature of the oil leaving the bearing.

For highly loaded bearings the temperature is best monitored by using embedment-type resistance temperature elements (RTE). These RTE elements are positioned in the babbitt at the minimum film thickness location.

3.10 Couplings. The coupling of each gear and pinion to its connecting shaft can be of a number of types depending upon the degrees of freedom of movement that the service requires. The second-reduction gear to line shaft coupling is usually a "solid coupling" with the flanges integral with the shaft sections. This coupling provides no freedom of movement within itself either axially, angularly, or torsionally. As discussed previously, the line shaft bearings and the second-reduction gear bearings cannot be held in absolute alignment due to thermal and other distortions in the hull and foundations; however, these movements are predictable and the bearing arrangement can be designed such that, when properly aligned, the shafting can bend elastically without imposing objectionable stresses in the shaft or altering the bearing reactions in an unacceptable manner.

Thermal distortions in the gear and prime mover casings and their supporting structure create a relative movement of the rotors and high-speed pinion axes that introduces an angular offset at one or both coupling elements. In addition, the prime mover rotor is positioned axially by its thrust bearing; consequently, the thermal growth of the rotors creates a considerable end motion, which must be accommodated by endwise sliding and clearances in the coupling.

For steam turbine-driven gears the coupling to the first reduction pinion is usually a gear-tooth (dental) type flexible coupling, with two gear tooth elements separated by a length of shafting or a sleeve. Figure 8 is a typical coupling of this type. The engaging tooth elements at each end use internal and external spur gears of involute form, for convenience of manufacture, which mesh with backlash in the circumferential direction, but with closely controlled radial clearance between the tips of the external teeth and the roots of the internal teeth. Under torque the axes of the two elements are held in line by the contact on the involute tooth faces. With no torque transmitted, the axes are held in line within the limit of the radial clearance.

When running under angular misalignment, each meshing pair of teeth will slide back and forth a small amount. The angular misalignment, which this type of coupling can accept without significant wear, is limited and is dependent upon the coupling size, speed of rotation, torque,

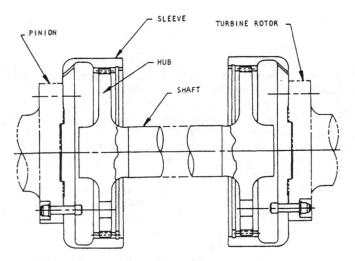

Fig. 8   Gear-tooth type of flexible coupling

Fig. 9(a)   Convolution type of diaphragm flexible coupling

and the hardness and finish of the tooth surfaces. It is obviously desirable to avoid excessively short coupling lengths, which impose high angular movements on the tooth elements. For longer coupling lengths it may be preferable to use a "single-ended" flexible coupling with a tooth-element coupling at one end and a solid coupling at the other end. In this case the long shaft can deflect elastically as a cantilever beam to accommodate the lateral offset.

Couplings between the first reduction gears and the second-reduction pinions have smaller misalignments to accommodate, but otherwise resemble the turbine-to-pinion coupling. In the single-case gears, Figs. 6 and 7, the misalignment is limited to the clearance in the journal bearings since the bearings themselves are held rigidly in line. Like the turbine-to-pinion coupling, they can be "single-ended" [as shown in Fig. 7(b)] or double-ended [as shown in Fig. 6(b)]. When the first-reduction elements have their own thrust bearing, at least one flexible element is needed for endwise freedom. When the first-reduction elements are positioned by the second-reduction and the main propeller thrust bearing, the flexible coupling elements are made with a close end clearance.

Lubrication of the coupling teeth is important even though the reciprocating sliding velocity is entirely too low to support an oil film between the surfaces in contact. Oil is held in the tooth portion by centrifugal force and an oil retaining ring keeps the sliding surfaces submerged in oil. Oil is fed to the annulus at one end of the teeth and leaves from the other end, forcing a flow endwise through the teeth for lubrication, cooling, and purging of the sludge, which tends to centrifuge and collect.

The diaphragm type of flexible coupling has been developed as an alternative to the gear-tooth type and offers significant advantages. Diaphragm couplings, such as illustrated by Figs. 9(a) and 9(b), can accommodate angular and axial misalignments. The stiffnesses of this type of coupling are relatively low, which reduces the moment transmitted to the connected components. Additional advantages include the elimination of coupling lube-oil flow

Fig. 9(b)   Contoured diaphragm type of flexible coupling

requirements and better control of concentricity and unbalance forces.

For diesel engine drives it is usually necessary to have a coupling with torsional flexibility to minimize the transmission of torque variations to the gearing. Several types of couplings, using rubber or other elastometers in compression or shear, are effective in adding both torsional resilience and damping to attenuate the torsional oscillations, which are inherent in a reciprocating engine.

Hydraulic couplings, now familiar in automotive transmissions, had one of their earliest applications in marine drives. They are effective in smoothing the torque input to the gear. However, their slip represents a direct power loss. Electric couplings have characteristics similar to the hydraulic coupling but are dependent on a source of electric power for their operation. Both hydraulic and electric couplings have the capability of providing a convenient means of disconnecting, synchronizing, and reconnecting engines in a multi-engine arrangement.

3.11 Clutches. Propeller drives that use either prime movers in combination or a prime mover that is unirotational may require a clutch to disconnect and reconnect or synchronize and reconnect the main engines to the propeller. An assortment of devices using mechanical, frictional, hydraulic, or electrical schemes is available for these purposes. Each device has its peculiar characteristics, so that the selection of the best coupling arrangement depends on the requirements in a specific case.

The hydraulic coupling transfers torque by the passage of oil between two halves of a torus. The torque transmitted can be controlled, therefore, by controlling the volume of oil in the coupling. It can be arranged to quickly discharge the oil in the coupling to disconnect the load, and to reconnect the load by readmitting oil. A hydraulic coupling will absorb energy to bring the shafts into near synchronization; however, there are limits to the hydraulic torque available for synchronization, and limits on the amount of energy that can be absorbed during the period of high slip.

An electric coupling has characteristics very similar to a hydraulic coupling. It too can act as a disconnect and as a synchronizing clutch, by controlling the current to the rotating field. Like the hydraulic coupling, there are limitations on the synchronizing torque, and the energy absorption during synchronizing.

Friction clutches use friction elements which slide under controlled pressure to bring the shafts into synchronization, and then, once synchronized, transmit torque without slip by the same friction surfaces. Figure 10 shows the application of clutches of this type in a reversing gear train. In this case the friction material is attached to the inside of an inflatable tube and is made to bear on the cylindrical drum, which it surrounds, by admitting fluid, usually air, under pressure to the tube. The inflatable tube, called the clutch gland, is made of fabric and rubber similar to an automobile tire and is bonded to a steel outer ring. Both ahead and astern clutch glands are driven by the engine. One engages with the drum driving the ahead gear train or the other engages with the drum driving the astern gear train. Reversing is accomplished by alternately admitting fluid and inflating the ahead and astern clutches. When the idle clutch is deflated, the friction surfaces are removed from contact. The operating air is admitted to the rotating shaft through a shaft seal. This is conveniently done at a shaft end as illustrated but can also be done on any available shaft portion.

With synchromesh couplings the normal torque transmission is through sets of engaging internal and external tooth elements similar to those of a "dental" flexible coupling. In addition, the coupling includes a mechanism for shifting the tooth elements axially to engage and disengage the teeth, a friction element to bring the shafts into synchronism prior to engagement, a balking mechanism to prevent tooth contact while a differential speed exists and, in the case of high-speed applications, a "transition torque control" to maintain shaft synchronism during the shift from friction to gear tooth drive. Referring to Fig.

11, the clutch is engaged by admitting air or oil to the operating cylinder. This applies pressure to the friction disks, and the torque so developed acts to synchronize shaft speeds. When synchronism is reached the accelerating torque is reduced to zero. The balking mechanism then automatically releases the coupling sleeve, and as it approaches engagement with its mating hub, the spring-loaded pins act as loose keys to maintain synchronous speed and align the hub and sleeve teeth for final engagement. After engagement the torque is transmitted through the dental coupling elements; the friction disks and balking mechanism perform no further function. The clutch is disengaged by simply shifting the sleeve endwise so that the hub and sleeve teeth at one end move out of mesh. Once engaged or disengaged, the clutch is maintained in the desired position by mechanical means and hydraulic pressure is no longer required.

The torque-transmitting capability of the synchromesh clutch is determined by the tooth elements. The torque capacity of the friction disks need only be sufficient to bring the shafts into synchronization.

A synchro-self-shifting clutch, like the synchromesh coupling, is a positive coupling in the engaged position. It is self-engaging when passing through synchronism; that is, immediately upon synchronizing the speeds of the input and output shafts, the input shaft engages the output. The clutch disengages automatically as soon as the torque reverses, that is, when the output shaft tends to drive the input shaft. In this coupling, engagement and disengagement are brought about by the relative rotation of the driving and driven ends. The primary action of the clutch, in fact, resembles that of a ratchet, which will lock up to transmit torque in one direction, but will turn freely under a torque in the other direction.

The basic clutch, Fig. 12, has a torque-transmitting sleeve, which can shift axially. One end of this sleeve is in constant engagement with the output shaft through a helical spline. The other end of the sleeve has dental coupling type teeth, which engage and disengage with mating teeth in the clutch ring, which is secured to the input shaft. The position of the sleeve is controlled by a ratchet-and-pawl arrangement, which senses the relative speeds of the input and output shafts. When the speed of the driving half overtakes the speed of the output half, the pawls engage so that further rotation of the driving half forces the sleeve to move axially on the helical spline to bring the coupling teeth into engagement. The coupling can also be made with two sets of spur dental type coupling teeth so that in the engaged position it also acts as a flexible coupling.

It can be arranged with a manual shift that will prevent engagement of the pawls and allow its driving engine to be tested without driving the propeller.

3.12 Lubrication. The main reduction gears are provided with a source of lubricating oil by a system that is either separate from, or integral with, the gear. In the case of steam turbine drives, the same source of oil also serves the turbine requirements. However, the characteristics of the oil required by the turbine is less than optimum for the gear. The distribution system for leading oil

Fig. 10 Reverse gear with friction clutches

DISENGAGED POSITION

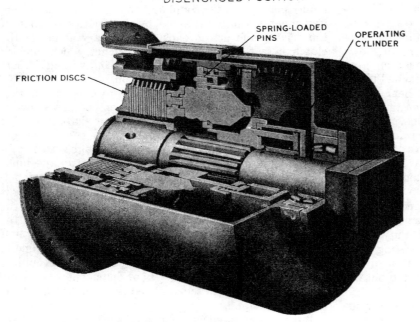

FRICTION DISCS

SPRING-LOADED
PINS

OPERATING
CYLINDER

Fig. 11 Synchromesh coupling

ENGAGED POSITION

under pressure to each bearing and to the tooth sprays, and for containing and leading the drains to the oil sump, is contained within the gear. All the gear requires is a continuous supply of clean oil at a pressure of about 10–15 psi and a temperature of about 120 F. The oil distribution system integral with the gear is designed to provide each bearing and tooth spray with the proper quantity of oil.

The temperature of the oil supply to the gear can range from 110 to 130 F. The discharge from the high-speed journal bearings may be as high as 180 F but the average temperature rise in the total flow through the gear unit is of the order of 20 to 30 deg F.

Oil in the viscosity range of 380–510 SSU at 100 F is generally suitable. It represents a compromise between

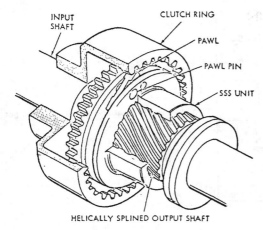

INPUT SHAFT
CLUTCH RING
PAWL
PAWL PIN
SSS UNIT
HELICALLY SPLINED OUTPUT SHAFT

Fig. 12 Synchro-self-shifting clutch

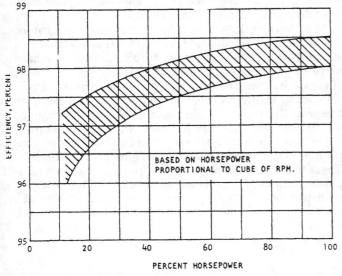

BASED ON HORSEPOWER PROPORTIONAL TO CUBE OF RPM.

Fig. 13 Efficiency of double-reduction gears

Another secondary function of the lubricating oil is the prevention of rusting of the interior surfaces of the gear. Lubricating oils have good antirust qualities even in the presence of small amounts of moisture, which cannot be avoided. They are completely effective during operation when all inside surfaces are thoroughly washed with oil. For prolonged shutdowns, however, the normal lubricating oil will drain from the steel parts and become ineffective. As a result, special precautions must be taken to prevent rust damage to the gear during prolonged shutdowns. In many cases dehumidifiers are used to control the moisture content within the gear case.

The lubrication system is essential to the performance of the reduction gear; therefore, for remote control, it is necessary to monitor the pressure and temperature of the lubricating oil supplied to the reduction gear. Consequently, most remote-control systems have provisions to monitor the temperature of the oil leaving each bearing to give an indication of the performance of each of the reduction gear bearings and the main thrust bearing.

It is general practice to provide a vent to avoid a buildup of pressure within the gear. A single vent opening from the entire gear case and sump space is sufficient. When more than one vent connection is made, all vents must be interconnected to avoid a circulation of outside air through the gear interior.

3.13 Accessories. In addition to performing its primary function of transmitting power, a number of accessory features are often provided as a part of the gear unit. For turbine applications, a motor-driven turning gear is provided with the main reduction gear. Its primary purpose is to rotate the turbine rotors slowly during warm-up and cool-down operations. It is essential that the turbine rotor be rotated when heating or cooling as the rotor will otherwise not be of a uniform temperature and will bow. The turning gear also serves the useful purpose of providing a means of turning the shaft for other purposes such as inspection of the gear teeth.

While the turning gear is a low-powered device (2.5 to 10 hp), it is mechanically capable of developing large torques in the propeller shaft by virtue of its high gear ratio. On occasions, therefore, the turning gear serves the additional purpose of being a locking device that prevents rotation of the turbine and gears. This is a desirable feature under casualty conditions when the ship is being towed or, in the case of a multiple-screw ship, when driving with other shafts.

The turning gear is generally a double-reduction gear with two worm-and-wheel reductions to connect the driving motor to the end of one of the pinions. The ratio of the turning gear is selected so that the propeller shaft turns at a rate of about one revolution in ten minutes. Figure 6 shows a turning gear, which drives the aft end of the first-reduction pinion on the low-pressure side. A disconnect clutch, usually lever operated, uncouples the turning gear for normal operation.

The turning gear requires lubrication and this is provided by the main lubricating system since the main system must be in operation even for the low turning speed.

the lighter viscosity oil, which would be optimum for high-speed journal bearings, and the somewhat heavier viscosity, which would be more favorable for the tooth meshes. Moderately loaded gear units with relatively fine pitch teeth will operate satisfactorily with a good grade of straight mineral oil; however, more heavily loaded gears, and particularly gears with teeth of coarser pitch and higher K-factor, require an oil having good "extreme pressure" or antiscuffing qualities to prevent scuffing or galling. The "EP" quality of the oil is evaluated by tests such as the "Ryder" or "Three-ball" test in which the ability to resist scuffing between two sliding steel surfaces is measured. Generally, the "EP" quality is given to the oil by the addition of chemical agents.

A secondary function of the circulating lubricant is to carry away the heat losses of the gearing and its bearings. As indicated by Fig. 13, the efficiency of reduction gears varies with the percentage of rated power delivered. In addition, the efficiency of a gear depends on the type of gear and the particular bearing and tooth pressures.

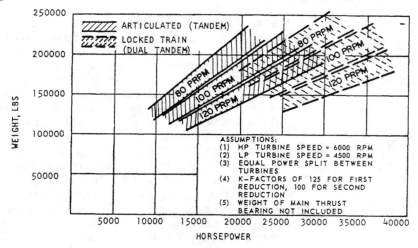

Fig. 14 Approximate weight of reduction gears

Interlocks with the prime mover or warning plates are necessary to prevent inadvertent starting of the prime mover with the turning gear engaged; otherwise, extreme damage may be done to the turning gear and motor.

The main gear also may serve as a convenient mounting for the shaft revolution counter and tachometer. They are usually mounted just forward of the slow-speed gear shaft and are driven directly or through gearing from the main shaft.

For naval applications, it is standard practice to drive the main lubricating oil pump by a train of gearing taking its power from one of the intermediate shafts. This so-called "attached pump" furnishes oil for all purposes during normal operations; however, separately driven pumps are required for low-speed, standby, and astern operation because the attached pump cannot supply an adequate oil supply under these conditions. An attached pump has the advantage of protecting against the loss of oil supply due

to an interruption of electric power or the inadvertent securing of a motor-driven pump. This feature, however, is seldom applied in merchant service.

3.14 Weight Estimates. An approximate weight of the reduction gear unit for conventional articulated and locked-train gears for steam turbine-driven merchant ships can be obtained from the curves in Fig. 14. These curves show that the gear weight varies in nearly direct proportion to the horsepower and inverse proportion to the propeller speed. Within reasonable limits, the weight of the gear will also vary inversely with the K-factor.

For more complex gear arrangements, there is no simple procedure for determining weights and dimensions. These must be determined by first establishing the principal dimensions of the rotating parts, and roughly detailing their weights. A good approximation of the total weight can then be made by doubling the weight of the rotating parts.

Section 4
Applications

4.1 Articulated Double-Reduction Gears. Most of the gear arrangements that have been used for ship propulsion are described in Section 1. The vast majority of the early turbine-driven merchant ships built in the U. S. were double-reduction with the rotating elements arranged as in Fig. 1(d). In most of these the arrangement of the rotating parts can be further categorized as "three-plane." The number of planes refers to the number of horizontal planes which contain the pinion and gear axes (with small differences in elevations ignored), and to the number of horizontal joints in the gear housing to provide for assembly.

The section drawing shown in Fig. 6 is representative of a typical double-reduction, articulated, three-plane reduction gear in a single housing structure. It was widely used with cross-compound steam turbines up to horsepowers of about 30,000 with propeller speeds conventional for merchant ships (i.e., 105 to 120 rpm for single-screw

ships). Variants of this arrangement are the single-plane and two-plane gear casing constructions, which use similar rotating parts, but with modified dispositions of the turbine and propeller shaft axes. Figure 15 illustrates the characteristics of the arrangements.

Whereas the three-plane gear is generally a single structure supported by the ship's foundation, the single and two-plane gears are more conveniently built as separate first- and second-reduction units, each independently mounted on the ship's structure.

The three-plane gear requires maximum headroom but, in return, provides good spacing between the turbines and ample space beneath the low-pressure turbine to install the condenser. At the other extreme, the single-plane gear requires a minimum of headroom and foundation structure underneath the turbines. On the other hand, additional overall length of the propulsion machinery is necessary because the condenser must be located forward of

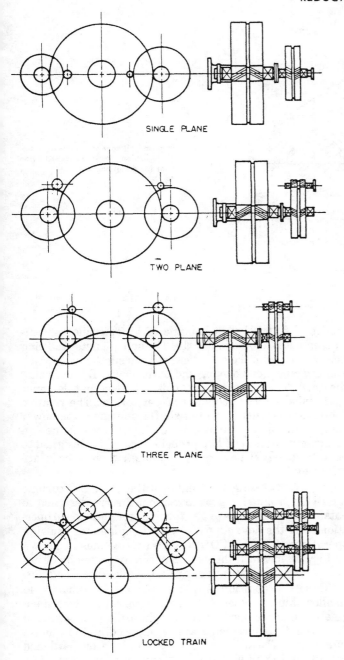

SINGLE PLANE

TWO PLANE

THREE PLANE

LOCKED TRAIN

Fig. 15 Double-reduction gear arrangements

the low-pressure turbine. The two-plane gear is intermediate in its space requirements as regards headroom, length, and width. The space available for the condenser under the LP turbine usually is less than convenient. The choice of one, two, or three planes is, therefore, largely one of arrangement.

4.2 Locked-Train Gears. For higher powers, the diameters, face widths, and tooth bending stresses of the gears in a conventional double-reduction arrangement increase to the point where a locked-train arrangement becomes the most practical. By dividing the power paths from each input pinion as illustrated by Fig. 1(g), the diameters, face

widths, and tooth loading factors become more favorable, but at the price of a greater number of parts and a more complex casing structure.

The division of power between the two intermediate shafts driven by a common high-speed pinion must be nearly equal, and to accomplish this they must be "timed." That is, the driving pinion must contact both driven gears when all backlashes are taken up. There are a number of ways in which this may be conveniently done. Once done the meshing gear and coupling teeth are match-marked so that the timed assembly can be repeated. However, if a pinion, gear, or coupling is replaced, it is necessary to retime.

The quill shafts connecting the first-reduction gears to the second-reduction pinions have considerable torsional flexibility and thus act to divide the torques equally despite slight inaccuracy in timing. Without this flexibility, timing and machining would have to be perfect to obtain equal power division.

Figure 7 is a section drawing of a typical locked-train gear for merchant marine service. It follows the construction that has been almost standard for naval combatant ships from destroyer escorts to aircraft carriers since the mid-1930's. It differs slightly in proportions since the gear ratio is higher for merchant ships than for the faster-turning propellers of naval ships.

Note from Figs. 7 and 15 that the locked-train gear is more restrictive in pinion locations (and gear arrangement) than the conventional double-reduction gear.

4.3 Gears for Diesel-Engine Drives. There are many applications of single-reduction gears for diesel engines in the moderate-power range. A typical arrangement is shown by Figs. 16(a) and 16(b). In this case, the gear combines the input of three diesel engines. The minimum center-to-center separation of the engines determines the size of the gear. The selective engagement of one or more engines is accomplished using pneumatic clutches, similar to those shown in Fig. 10. Figure 10 is a typical diesel-driven reversing reduction gear arrangement using friction clutches for alternately driving the ahead and astern gear trains.

4.4 Gears for Contrarotating Propellers. Contrarotating, coaxial propellers are often given consideration because of their improvement in propulsive efficiency. They require special gear arrangements, and many variations are possible [14].

Contrarotating gear arrangements fall into two distinct categories; one drives the two propellers at an equal or another predetermined ratio of revolutions, and the other drives the two propellers with an equal or another predetermined ratio of torque.

An example of the first category is shown in Fig. 17. Here a cross-compound steam turbine arrangement, with oppositely rotating turbines, drives the oppositely rotating propeller shafts through double-reduction gears. Note that the two first-reduction gears are in mesh to assure that the two propeller shafts will make the same number of revolutions even though the power inputs of the two turbines may not be equal.

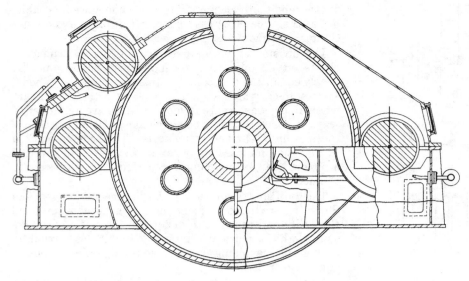

Fig. 16(a) Fore-and-aft sectional view of a single-reduction, three-input reduction gear for diesel engines

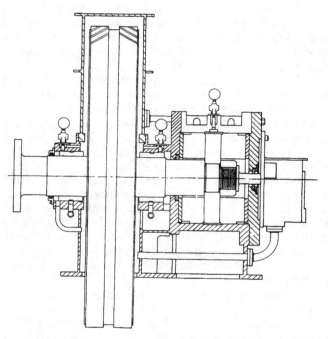

Fig. 16(b) Elevation sectional view of a reduction gear for diesel engines

As an illustration of the second category, also for a cross-compound steam turbine, each turbine drives the sun gear of a planetary first-reduction gear. Referring to Fig. 18, the planet carriers rotate in the direction of the turbines to drive one second-reduction gear. The ring gear rotates in the opposite direction to drive the other second-reduction gear. The ratios of the two second-reductions can be made to differ by the amount of the torque difference in the output shafts of each planetary gear so that equal torques are applied to each propeller shaft.

Contrarotation can also be achieved using differential epicyclic gearing, with either single or multiple prime movers [15,16,17]. A schematic of an epicyclic gear arrangement is shown in Fig. 19. The sun, planets, and ring gear rotate, with the planet carrier driving the inner shaft and the ring gear driving the outer shaft. By analyzing the gear elements and tooth loads, it is seen that the inner shaft inherently transmits a larger torque. The relative speeds of the inner and outer shafts, and hence the power split, would be determined by the hydrodynamic design of the propellers. The two propellers could be designed to operate at equal powers even though their speeds are different.

4.5 CODOG Gears. COmbined Diesel Or Gas turbine (CODOG) propulsion gear arrangements are designed to obtain the advantages of gas turbines for high-speed operations and the advantages of diesel engines for cruising operations. A typical CODOG gear arrangement is shown by Fig. 20, in which the gear is normally driven by either a single large gas turbine or by two smaller diesel engines.

CODOG propulsion systems are used primarily in smaller ships, such as frigates and corvettes, where engine room space and system weight are a premium, and the mission requirements are high speed and low noise signatures. These goals usually require case-hardened and ground gear elements. As can be seen from Figs. 20(a) and 20(b), the gas turbine and diesel engines drive through several clutches, which are arranged to provide a means of connecting and disconnecting each prime mover to the propeller system and to prevent simultaneous engagement of both the gas turbine and the diesels.

In the gas turbine mode, power is delivered through the primary gearbox to secondary gearboxes to simultaneously drive both propeller shafts, and the diesels are declutched. In the diesel mode, power is delivered through the secondary drives only, and gas turbine is declutched.

4.6 Epicyclic Gears. Epicyclic gears have been used in marine applications since the early 1960s. Epicyclics have been designed as the primary reduction in double-reduction gears and the primary and secondary reductions

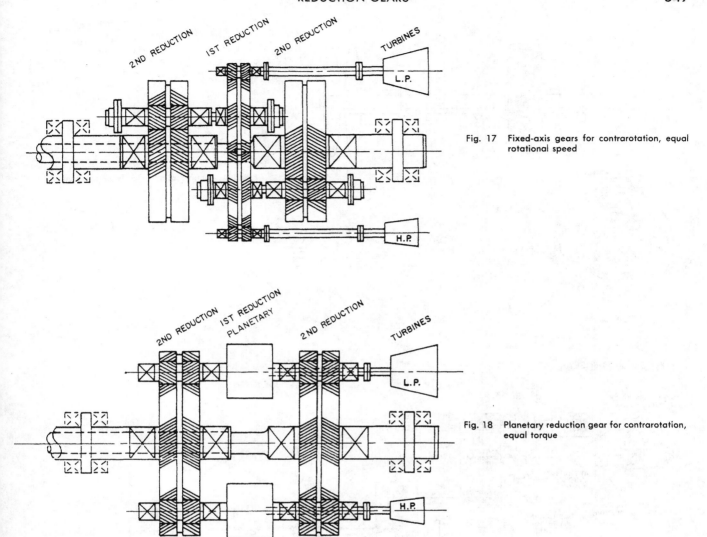

Fig. 17 Fixed-axis gears for contrarotation, equal rotational speed

Fig. 18 Planetary reduction gear for contrarotation, equal torque

in triple-reduction designs that were driven by steam turbines [18,19,20]. Epicyclics have also been used in a variety of medium-speed diesel applications in the low to medium horsepower ranges. Nearly all of the steam- and diesel-driven epicyclic arrangements have been installed in ships that were built in European shipyards. The introduction of marine epicyclic gearing into the United States has been slow and confined primarily to design studies [15].

Epicyclic gearing is 10 to 50% smaller and lighter than comparable parallel-shaft, locked-train gearing of the same horsepower rating. The torque is typically divided through three or more planets, whereas a locked-train gear is limited to two paths. Inherent in the epicyclic design is a low number of shaft support bearings. This feature is important when considering the low noise requirements for naval vessels [21].

It is interesting to note that epicyclic gears have been used in variable-ratio applications in which a generator is driven at a constant speed as the diesel engine speed varies. Such an arrangement is illustrated by Fig. 21, which is typical of those used almost exclusively with slow-speed diesel engines. The variable ratio is achieved by rotating the annulus ring. Rotation of the annulus ring is achieved through a hydrostatic motor and parallel-shaft gear system. An electronic control system monitors engine speed and adjusts the hydrostatic system to maintain a constant generator frequency.

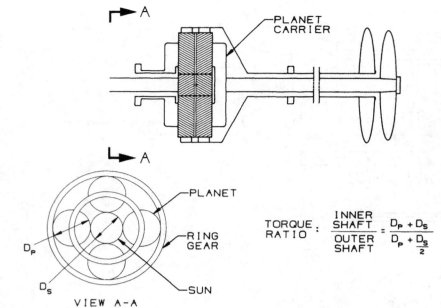

Fig. 19   Contrarotating epicyclic gear schematic

$$\text{TORQUE RATIO}: \frac{\text{INNER SHAFT}}{\text{OUTER SHAFT}} = \frac{D_P + D_S}{D_P + \dfrac{D_S}{2}}$$

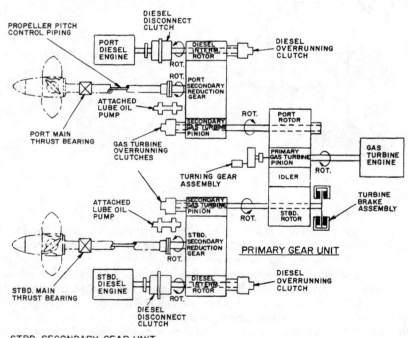

Fig. 20(a)   Schematic arrangement of a twin-screw CODOG plant

References

1   "Reference Information—Basic Gear Geometry," American Gear Manufacturers Association, Standard 115.01.

2   "Gear Assembly, Propulsion (Naval Shipboard Use)," Military Specification MIL-G-17859, Department of the Navy.

3   "Nomenclature of Gear Tooth Failure Modes," ANSI/AGMA Standard 110.04, American National Standards Institute/American Gear Manufacturers Association, Aug. 1980.

4   Darle W. Dudley, *Handbook of Practical Gear Design*, McGraw-Hill, New York, 1984.

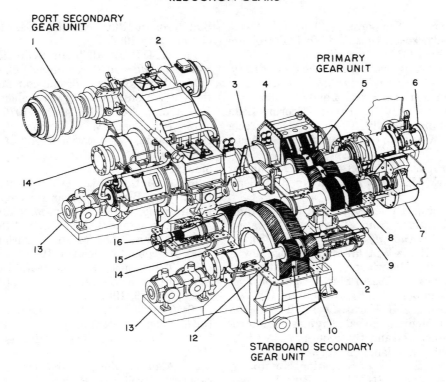

1. Diesel Disconnect Clutch and Coupling
2. Diesel Overrunning Clutch
3. Turning Device
4. Port Rotor (Inside Casing)
5. Primary Gas Turbine Pinion
6. Input Flange
7. Turbine Brake Assembly
8. Starboard Rotor
9. Idler
10. Diesel Pinion
11. Second Reduction Gear
12. Ring Gear
13. Attached Lube Oil Pump
14. Output Flange
15. Secondary Gas Turbine Pinion
16. Gas Turbine Overrunning Clutch

Fig. 20(b)   CODOG reduction gear for twin-screw propulsion

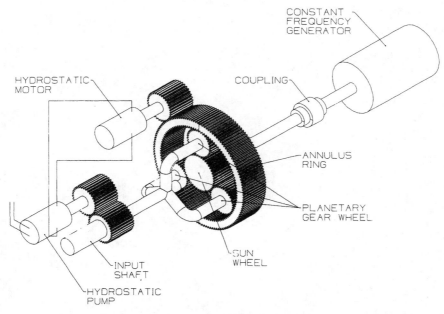

Fig. 21   Variable-ratio epicyclic gear

5   Blok, H., "Surface Temperatures Under Extreme Pressure Conditions," *Proceedings*, 2nd World Petroleum Congress, Paris, Sept. 1937.

6   Dowson, D. and Higginson, G. R., *Elastrohydrodynamic Lubrication—The Fundamentals of Roller and Gear Lubrication*, Permagon Press, London, 1966.

7   Hans Winter and Klaus Michaelis, "Scoring Load Capacity of Gears Lubricated with EP-Oils," AGMA Technical Paper P219.17, American Gear Manufacturers Association, Oct. 1983.

8   "Guide to Propulsion Reduction Gear Alignment and Installation," SNAME, T&R Bulletin No. 3-43, Sept. 1987.

9   Gary P. Mowers, "Diesel Reduction Gear Design and System Considerations from a Torsional Vibration Viewpoint," *Marine Technology*, Vol. 20, April 1983.

10   Peter Gold, K. Heinz Watzlawek, and Zbigniew J. Karaszewski, "Design, Manufacture, and Testing of Large Reduction Gears for Diesel Ship Propulsion Systems," *Marine Technology*, Vol. 27, No. 1, Jan. 1990.

11   W. F. Schafer and Z. J. Karaszewski, "Propulsion Plants with Geared Medium Speed Engines—Requirements for the Components and the System," SNAME, New York Metropolitan Section, Jan. 1989.

12   "Guidelines for the Use of Vibration Monitoring for Preventive Maintenance," SNAME, T&R Bulletin No. 3-42, May 1987.

13   "Standard for Marine Propulsion Gear Units, Part 1 Materials," ANSI/AGMA 6033-A88, American National Standards Institute/American Gear Manufacturers Association, May 1988.

14   W. I. H. Budd, "Main Reduction Gears for Contrarotation," *Marine Technology*, Vol. 6, No. 4, Oct. 1969.

15   T. W. Steele, "Propulsion Machinery Considerations for Contrarotating Propeller Systems," *Marine Technology*, Vol. 7, No. 4, Oct. 1970.

16   F. J. Dashnaw, A. W. Forrest, J. B. Hadler, and G. C. Swensson, "Application Study of Contrarotating Propeller for a U. S. Flag Merchant Ship," SNAME, Philadelphia Section, 1980.

17   C. McDowell and D. J. Folenta, "Status Report on U. S. Navy Planetary Gear Developments," Naval Ship Engineering Station Report, May 1980.

18   I. Jung and T. Lundstrom, "Recent Developments in Propulsion Gears for Steam Turbines," SNAME, New York Metropolitan Section, 1963.

19   T. P. Jones, "Fifteen Years' Development of High-Power Epicyclic Gears," *Trans.* Institute of Marine Engineers, Vol. 79, 1967.

20   T. P. Jones, "Design, Operating Experience and Development Potential of Main Propulsion Epicyclic Gears," *Trans.* Institute of Marine Engineers, Vol. 84, 1972.

21   Alan L. Marquis and Terence J. Dansdill, "Advanced Epicyclic and Parallel Offset Propulsion Gearing for Surface Ships," *Proceedings*, Destroyer, Cruiser, and Frigate Technology Symposium, American Society of Naval Engineers, Sept. 1990.

CHAPTER X

C. L. Long

Propellers, Shafting, and Shafting System Vibration Analyses

Section 1
Introduction

1.1 General. A main propulsion shafting system consists of the equipment necessary to convert the rotative power output of the main propulsion engines into thrust horsepower that is used to propel the ship. The propeller is included as an element of the shafting system, as is the means to transmit the propeller thrust to the ship structure. In the following pages, the design of a main propulsion shafting system is reviewed from the perspective of a shipbuilder undertaking the task of preparing a detailed design. It will, however, be assumed that the propeller hydrodynamic design has been developed; the hydrodynamic design of propellers and other propulsion devices is thoroughly covered by Van Manen and Oossanen in *Principles of Naval Architecture* [1] and, therefore, will not be repeated here. Although the fundamentals outlined in the following sections apply to all types of propulsors and prime movers, the discussion has been directed primarily towards an arrangement with a fixed-pitch propeller and geared, turbine-driven, propulsion machinery; however, where the selection of a diesel prime mover has an influence upon the shafting design considerations or procedures, those instances are discussed. This was necessary in order to avoid unduly confusing the discussion with details.

Due to the nonuniform wake field in which a ship's propeller operates, the propeller is a source of potentially dangerous vibratory excitations. The shafting system itself, which is inherently flexible, is extremely vulnerable to these vibratory excitations; consequently, an analysis of the dynamic characteristics of a shafting system is an integral aspect of the design process and is discussed in this chapter.

1.2 Shafting System Description. The main propulsion shafting system must accomplish a number of objectives that are vital to the ship's operation. These objectives are: (*a*) transmit the power output from the main engines to the propulsor; (*b*) support the propulsor; (*c*) transmit the thrust developed by the propulsor to the ship's hull; (*d*) safely withstand transient operating loads (e.g., high-speed maneuvers, quick reversals); (*e*) be free of deleterious modes of vibration; (*f*) provide reliable operation

throughout the operating range; and (*g*) be a low maintenance system.

Figure 1 is a shafting arrangement typical of those found on multishaft ships and single-shaft ships having transom sterns. The distinguishing characteristic of this arrangement is that the shafting must be extended outboard for a considerable distance in order to provide adequate clearance between the propeller and the hull. One or more strut bearings are required to support the outboard shafting.

A shafting arrangement typical of single-screw merchant ships is shown in Fig. 2. The arrangement illustrated corresponds to the so-called Mariner or clear-water stern design (there being no lower rudder support); however, the shafting arrangements of most merchant ships are very similar. The major difference between the shafting arrangements of various merchant ships is the location of the main engines. When the main engines are located well aft, such as on tankers, there may be as few as one or even no inboard line shaft bearings. When the main engines are located farther forward, which is predominantly the case for naval ships, a considerable length of inboard shafting may be required to accommodate the ship arrangement.

The shafting located inside the ship is termed line shafting. The outboard sections of shafting (wet shafting) are designated differently depending upon their location. The section to which the propeller is secured is the propeller shaft or tail shaft. The section passing through the stern tube is the stern tube shaft unless the propeller is supported by it (as is the case with many merchant ships), in which case it is designated as the propeller shaft or tail shaft. If there is a section of shafting between the propeller and stern tube shafts, it would be referred to as an intermediate outboard shaft.

Shafting sections are connected by means of bolted flange couplings. The coupling flanges are normally forged integrally with the shafting section; however, when required by the arrangement (e.g., stern tube shafts which require flanges on both ends and also require corrosion-resistant sleeves to be fitted to the shaft in way of

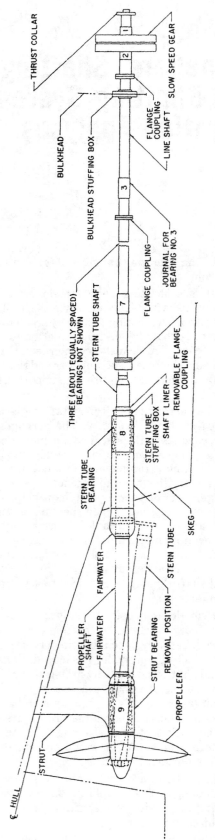

Fig. 1   Shafting arrangement with strut bearing

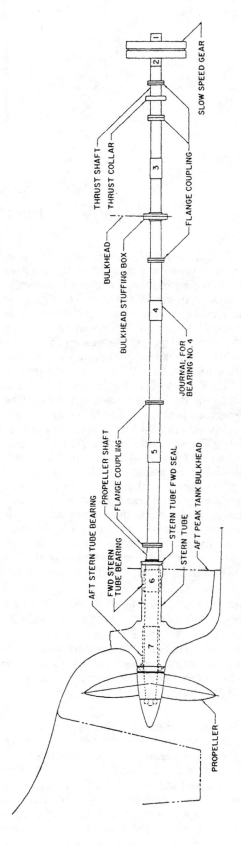

Fig. 2   Shafting arrangement without strut bearing

bearings), a removable coupling, sometimes referred to as a muff coupling, is used.

Bearings are used to support the shafting in essentially a straight line between the main propulsion engine and the desired location of the propeller. Bearings inside the ship are known by several names, with line shaft bearings, steady bearings, and spring bearings being the most popular in that order. Bearings which support outboard sections of shafting are called stern tube bearings if they are located in the stern tube and strut bearings when located in struts. Outboard bearings may be lubricated by either seawater or oil; seals having a very high reliability are required in the event the latter is used. Alternatively, the necessity for a muff coupling can be avoided by welding in place, onto the shaft, halves of a split sleeve that are made of a corrosion-resistant material; but in this case, the stern tube bore, with the bearing bushings removed, must be large enough to pass an integral flange.

In order to control flooding, in the event of a casualty, watertight bulkheads are installed within the ship. Stuffing boxes are installed where the shafting passes through these bulkheads. A seal, which is more substantial than a bulkhead stuffing box, is installed where the shafting penetrates the watertight boundary of the hull.

The propeller thrust is transmitted to the hull by means of a main thrust bearing. When the main engine drives the propeller through reduction gears, the main thrust bearing may be located either forward or aft of the slow-speed gear. If located forward, the thrust collar is detachable from the gear shaft so as to permit the installation of the slow-speed gear on the shaft and, secondarily, to permit replacement of the thrust collar if ever required. If located aft, the collar is forged integrally with either the slow-speed gear shaft or a subsequent section of shafting. Since one purpose of the main thrust bearing is to limit movement of the slow-speed gear, the main thrust bearing is usually installed close to the gear. The installation of the thrust bearing close to the gear also facilitates arrangements to lubricate the thrust bearing, as an oil line can readily be run from the gear to the thrust bearing with the oil returned by a gravity drain.

Direct-drive diesel ships have the thrust bearing aft of the engine, either attached directly to the engine crankshaft or located farther aft independent of the engine.

1.3 Design Sequence. The design of a shafting system is, by necessity, an iterative process because the various system design parameters are, to some extent, mutually dependent. The iterative design process usually followed is illustrated in Fig. 3.

As indicated by Fig. 3, the first step in the design of a shafting system is to state the performance requirements, that is, the type of propulsive system, number of shafts, type of service, and the like. Next, the design criteria to be employed must be fixed. That is, one of the various classification society rules could be followed; the type of stern tube bearings may be selected (i.e., oil lubricated or water lubricated); hollow shafting may be ruled out, and so on. In establishing the design criteria, it must be recognized that the shafting interfaces with the propulsor, the main engines, and the ship system as a whole.

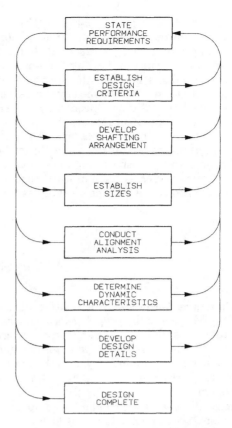

Fig. 3 Shafting system design sequence

After the design criteria are established and the general ship arrangement is available, an approximate shafting arrangement can be developed. This entails at least tentatively locating the main engine, propeller, and shaft bearings with due regard given to arrangement restrictions, clearances required, shaft rake, construction restraints, and overhaul and maintenance requirements.

Before the design can progress further, the shafting diameters, corresponding to the preliminary arrangement, must be computed along with the length of shafting sections, flange dimensions, and preliminary propeller data. Preliminary shafting system alignment calculations can be initiated, and the changes in bearing loads due to the thermal expansion of the shaft bearing foundations, particularly those in the way of the main engines, can be investigated to ensure satisfactory bearing performance under all operating conditions. In addition, when the arrangement includes water-lubricated outboard bearings that are designed to accommodate weardown, it must be ascertained that the weardown of these bearings will have no adverse effects. With these data the bearing reactions under the full range of operation conditions can be determined and the bearing dimensions and loadings can be checked. At this point, it will generally be desirable to adjust the bearing arrangement tentatively selected so as to obtain more equal bearing reactions or to alter the number of bearings.

There are three basic types of vibration that can occur in a main propulsion shafting system: torsional, longitudinal, and whirling vibration. It is essential that a preliminary vibration analysis of the shafting system be made in the early design stages because the shafting vibration characteristics are largely established by the ship parameters that are fixed at that time. Specifically, the shape of the hull afterbody, type of propeller, propeller aperture clearances, number of propeller blades, length of shafting, shaft material, position of the main thrust bearing, rigidity of main thrust bearing foundation, type and configuration of prime mover, spacing of the aftermost bearings, and type of aftermost bearings largely establish the dynamic characteristics of a shafting system. The subsequent development of design details has a relatively secondary effect as compared with these major parameters. In addition, an analysis of the system's response to shock loadings is required for naval combatant ships. An analysis of the dynamic characteristics of a shafting system can be one of the more complex aspects of the design process.

Once the arrangement, component sizes, and dynamic characteristics have been shown to comply with the design criteria, design details are developed. This entails the development of detailed designs for flange fillets, flange bolts, propeller attachment, sleeves, and the like.

Section 2
Arrangement Considerations

2.1 Main Engine Location. The location of the main engine output flange and the propeller location are essential information in establishing the shafting arrangement. The fore-and-aft position of the main engine is generally established during the preliminary design stages after studying the ship arrangement, ship light-draft trim, and shafting system.

To minimize the use of prime shipboard space for the propulsion plant, the main machinery is located as far aft as practicable. For ships that are driven through reduction gears, the limiting factor is usually the breadth of the ship in way of the reduction gears, provided that the resulting light-load trim of the ship is satisfactory. With vessels that have provisions for ballast, the light-load draft of the ship can be adjusted by taking on ballast, and the main machinery is usually confined to the stern of the ship. This arrangement requires a short run of shafting, and the number of line shaft bearings required is minimal. On the other hand, other types of vessels such as warships and dry-cargo ships have a limited ability to adjust their operating draft and trim by taking on ballast; therefore, to provide satisfactory light-load draft conditions in these cases, it is necessary to locate the main engines (and the associated weight) well forward of the stern.

Normally a main engine with a reduction gear will be set as close to the innerbottom as the configuration of the main machinery will permit to reduce shaft rake. It is possible, and it is the usual case, to have limited projections of the main machinery (e.g., the slow-speed gear lube oil sump) below the innerbottom when such projections do not excessively weaken the innerbottom (see Figs. 5, 6, and 9 of Chapter I).

Projections into the innerbottom structure are generally not required with direct-drive diesel installations, because the distance from the crankshaft to the engine base is very much less than the radius of the slow-speed gear in a reduction gear arrangement.

The main engine location in the athwartship direction is on the ship centerline of single-screw ships. On

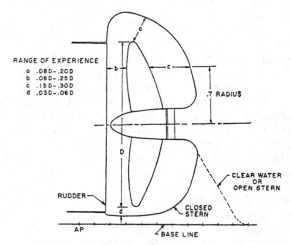

Fig. 4 Propeller aperture clearances

multiscrew ships the engines are set off the ship centerline approximately the same distance as the propellers, but the shaft centerlines usually do not parallel the centerline of the ship. The location of the engine in the athwartship direction is controlled by the propeller location, main engine details, and the machinery room arrangement requirements.

2.2 Propeller Location. The location of the propeller is determined by the propeller diameter, the acceptable clearance between the propeller and the baseline of the ship, and the acceptable clearances between the propeller and the hull in the plane of the propeller. Although the propeller diameter selected should theoretically be the one corresponding to optimum efficiency for the propeller-ship system, in practice the optimum propeller diameter is usually larger than can be accommodated. As a result, the propeller diameter selected is a compromise.

In locating the propeller in the aperture of a single-screw ship, a clearance of 6 to 12 in. is normally provided between the propeller tip and the baseline with clearwater sterns or to the rudder shoe with a closed stern (Fig. 4).

With high-speed ships, which are generally characterized by shallow draft and multiple screws, propellers are often permitted to project below the baseline in order to provide adequate clearance between the propeller and the hull. This is satisfactory provided maximum draft limitations for service routes or dry-docking are not exceeded.

One of the most effective means of ensuring a satisfactory level of vibration aboard ship is to provide adequate clearances between the propeller and the hull surface. For this reason, the subject of propeller clearances is one of overriding importance. Generally speaking, the greater the clearances, the better the performance from a vibration standpoint.

Three types of vibratory forces are generated by the propeller: (a) alternating pressure forces on the hull due to the alternating hydrodynamic pressure fields caused by the propeller blades; (b) alternating propeller-shaft bearing forces, which are primarily caused by wake irregularities; and (c) alternating forces transmitted throughout the shafting system, which are primarily caused by wake irregularities. If the frequency of the exciting force should coincide with one of the hull or shafting system natural frequencies, very objectionable vibration can occur. An analysis of the forces generated by the propeller is given by W. S. Vorus in reference 2, which includes numerous references to papers and studies that deal with ship vibratory forces and dynamic behavior.

When establishing propeller clearances, the performance experience gained during the operation of similar ships should be taken into consideration. Of course, differences between the important parameters of the ships under comparison must be assessed. Important parameters to consider are the unit thrust loading on the propeller blades, the number of propeller blades, the amount of propeller skew, the length of the ship, and the closing angle of the waterplane forward of the propeller. The importance of propeller positioning, propeller skew, etc. is given a detailed treatment in reference 2.

Figure 4, which may be used as guidance in assessing aperture clearances, shows the range of experience which has been obtained in connection with large single-screw ships. When the propeller is supported by a strut bearing, i.e., multiscrew and transom-stern vessels, two clearance dimensions warrant careful study. These dimensions and the range of experience with them are shown in Fig. 5.

2.3 Shaft Rake. In order to provide latitude when locating the propeller and the main engine, it is usually necessary to rake the shaft centerline. The shaft is generally raked downward going aft as this permits the main engines to be located higher in the ship. In multiscrew ships the shaft is generally raked in both the vertical and horizontal planes, usually downward and outboard going aft.

Large rakes should be avoided since a reduction in the propulsive efficiency is associated with rake. The introduction of rake incurs a reduction in the propulsion efficiency equal to

$$e = (1 - \cos\theta \cos\phi) \, 100 \qquad (1)$$

where θ is the shaft vertical rake angle and ϕ is the

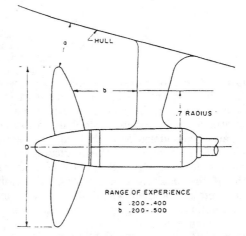

Fig. 5   Clearances of a propeller supported by a strut bearing

athwartship rake angle, both of which are measured relative to the ship centerline. It is rare for θ to exceed 3.75 deg or ϕ to exceed 2.5 deg. From rake alone the reduction in propulsion efficiency will normally not exceed 0.3%. Aside from the efficiency penalty, there is no objection to moderate amounts of rake.

2.4 Shaft Withdrawal. Occasionally shafting sections, particularly those outboard, must be withdrawn to be inspected or repaired. Consequently, provisions for removing shaft sections from the ship must be considered when developing a shafting arrangement.

On single-screw ships with shafting arrangements similar to Fig. 2, the propeller shaft is almost without exception withdrawn inboard for inspection. If repairs are necessary, the shaft is removed from the ship by cutting a hole in the side of the ship and passing the shaft through it. This technique would be used for removing line shaft sections as well.

With shafts having struts as shown in Fig. 1, a check must be made to ensure that the propeller shaft can be withdrawn from the strut after the propeller is removed. Withdrawal can be accomplished by removing the bearing bushings so that the shaft can be inclined sufficiently to allow its forward end to clear the ship's structure, the mating shaft flange, etc. This consideration can govern the length of the propeller shaft and the size of the strut barrel. Figure 1 shows the removal position of the propeller shaft.

Removal of the stern tube shaft, which must have flanges on both ends, requires a decision regarding the type of flanges to be provided on the shaft. If the shaft is manufactured with integral flanges on both ends, the stern tube barrel and bearing bushings must be sized to pass the flange diameter. Since it is desirable to pass the shaft outboard, sufficient clearance should be provided to incline the shaft such that it will clear outboard struts, etc. In order to use smaller stern tubes and bearing bushings, the stern tube shaft can be manufactured with a removable flange coupling on the forward end. Prior to unshipping the shaft, the removal coupling is detached so that it is not necessary to disturb the stern tube bearings.

Section 3
Shafting Loads

3.1 Design Considerations. In general, the dimensions of shafting are predicated on the basis of strength requirements; however, it is occasionally necessary to modify an otherwise satisfactory shafting system design due to vibration considerations. Shafting diameters usually have only a minor impact on the longitudinal vibration characteristics, due to the fact that both the stiffness and the weight of the shafting change proportionately; but the whirling and torsional modes of vibration are sensitive to shaft diameters. Shafting vibration, as such, is discussed in Sections 7, 8, and 9.

Propulsion shafting is subjected to a variety of steady and alternating loads, which induce torsional shear, axial thrust, and bending stresses in the shafting. In addition, there are radial compressive stresses between the shafting and mating elements (such as between the propeller and shaft) which, when coupled with axial strains from bending stress, are very important from a fatigue standpoint.

The steady loads represent average conditions and can be estimated with a degree of certainty as they are directly derived from the main engine torque and the propeller thrust. On the other hand, vibratory loads do not lend themselves to a precise evaluation and are difficult to treat in an absolute sense.

3.2 Propeller-Induced Loads. Aside from the alternating bending stress due to the weight of the propeller, the circumferentially nonuniform velocity of the water inflow to the propeller (wake) generates one of the most important sources of the alternating loads in the shafting system. It is, however, important to distinguish between the importance of the circumferential nonuniformity of water inflow at a particular propeller radius and the nonuniformity of the average flow at one radius as compared with another. While the former leads to vibratory propeller forces, the latter does not.

A propeller blade section working in a constant velocity field at a particular radius has a steady flow and force pattern. The average axial velocity at each radius can be different without causing alternating loads. In such a case the propeller design can be adjusted for radial variations in the inflow velocity to achieve optimum performance. However, a propeller can only be designed to satisfy average conditions at each radius.

A variation in the inflow water velocity at a particular radius results in a change in the angle of attack of the propeller blade section as the propeller makes one revolution, thereby creating alternating propeller forces. Figure 6 is an example of the axial, V_A, and tangential, V_T, inflow velocities in the plane of the propeller for a single-screw ship. The tangential velocity component is symmetric on both sides of the vertical centerline of single-screw ships and is generally upward. The symmetry of the tangential velocity component would tend to suggest that its effect is uniform, but such is not the case. For a propeller blade

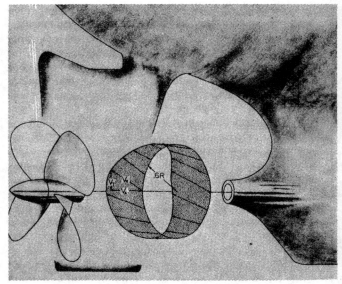

Fig. 6   Flow of water in plane of propeller

rotating clockwise looking forward, the tangential velocity component effectively reduces the angle of attack of the blade sections as they pass up the port side (reducing thrust) and increases the angle of attack of the blade sections as they pass down the starboard side (increasing thrust). Figure 7 illustrates the variable propeller loads that result from nonuniform axial and tangential wake velocities. Also, another very important fact is that the tangential velocity components shift the center of propeller thrust to the starboard side of the propeller centerline of a propeller turning clockwise on a single-screw ship. This off-center thrust gives rise to a bending moment which is imposed upon the propeller shaft.

Analyses can be made to predict the magnitude of the alternating components of torque and thrust, including the eccentricity of the resultant thrust relative to the shaft centerline. An approximation of these alternating components, which is sufficiently accurate for most applications, can be obtained by using a quasi-steady analysis approach. A quasi-steady analysis is readily comprehensible and is conducted by making an instantaneous examination of the flow velocities relative to the propeller blades at discrete angular positions of a propeller blade [3]. The inflow velocities are regarded as constant (quasi-steady) at each blade position. By using the open-water characteristics (K_T-K_Q-J diagram) of the propeller, the thrust and torque can be determined per blade, summed, and plotted as shown in Fig. 8.

Since the slowest axial inflow velocity (highest wake) of single-screw ships is generally in the region above the propeller centerline, the greatest thrust tends to be developed when the propeller blade is in the upper part of its orbit. The effect of the tangential inflow velocity is to

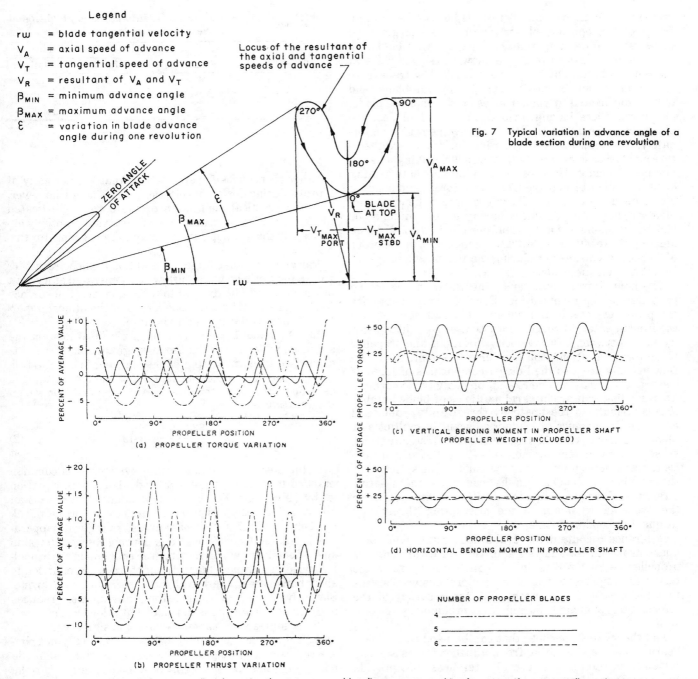

Fig. 7 Typical variation in advance angle of a blade section during one revolution

Fig. 8 Typical single-screw propeller alternating thrust, torque, and bending moments resulting from nonuniform water inflow velocities

shift the resultant thrust to the starboard side because the propeller blades develop greater thrust moving against the tangential velocity, as discussed in the foregoing. This subject is given a detailed discussion in reference 4, and it is noted that as the shape of the stern sections change from a V to a U shape, the resultant thrust center tends to move down because the inflow velocities over the bottom region of the propeller disk become more nearly equal to those in the upper region. The position of the resultant thrust is also sensitive to the ship's draft. For

instance, when a cargo ship operates lightly loaded with the propeller blades breaking the water surface, the center of thrust obviously shifts lower in the propeller disk.

Figure 8 shows that a single-screw ship with a four- or six-bladed propeller (that is, an even number of blades) has larger torque and axial thrust variations than one with a five-bladed propeller. However, the thrust eccentricity (propeller shaft bending moment) is shown to be much greater for the five-bladed propeller than for the four- or six-bladed propeller. For a single-screw ship having a propeller with an even number of blades, the fluctuating forces of two opposite blades give rise to a larger total thrust and torque amplitude because opposite blades simultaneously pass through the slow water velocities at the top and bottom of the propeller disk. The transverse force and bending moment developed by one blade tend to be compensated by similar loads on the opposite blade.

For propellers having an odd number of blades, the blades pass the upper and lower high-wake regions alternately. The total thrust and torque variations are therefore smaller as compared with a propeller having an even number of blades. However, due to the alternate loading of the propeller blades, the transverse forces and bending moments do not cancel. Therefore, larger bending moments occur with propellers having an odd number of blades. Propellers have been designed with a large amount of skew (see blade skew, Section 6.3), in the order of 70 deg, as a means of reducing the alternating propeller thrust and torque loads.

The nonuniform character of the water inflow to the propeller can be resolved into Fourier components with the propeller rotational frequency (shaft frequency) as the fundamental [5]. Since it may be assumed that linearity exists between inflow velocity variations and propeller blade force variations, the Fourier components of the inflow velocity are also the Fourier components of force of a single blade making one revolution. Only those harmonics of loading that are integral multiples of blade number (kZ) contribute to the unsteady thrust and torque, and only those harmonics of loading adjacent to multiples of blade frequency ($kZ \pm 1$) contribute to the unsteady transverse forces and bending moments [2]. All other harmonics cancel when summed over the blades. The selection of the number of blades can be based on the relative strengths of the harmonics in the inflow water velocity to the propeller to minimize the alternating thrust and torque and bending moments.

Reference 6 contains an evaluation of the value of a quasi-steady inflow analysis approach for estimating the propeller bearing forces and demonstrates by full-scale tests and comparisons with more comprehensive analyses that the method is consistent with the accuracy of the information available during the preliminary design stage.

The theory and concepts that can be used to evaluate bearing reactions resulting from propeller forces are also reviewed in reference 2; useful references that provide further information on the subject are also provided. Reference 2 describes and provides an example stripwise application of two-dimensional gust theory as developed by von Karman and Sears. This analytical procedure, which is relatively simple to apply, can be used to develop meaningful evaluations of variations in design parameters such as wake and blade skew, though it should be noted that the procedure tends to overestimate blade lift by 30 to 50% because the effects of the blade aspect ratio are not properly taken into account. However, the results obtained are useful when evaluating the characteristics of design alternatives; and, as noted by Vorus in reference 2, the two-dimensional unsteady strip method is preferred

Table 1 Ratio of shaft torque measured during high-speed maneuvers to normal torque

Ship Type	No. of Shafts	Torque Ratio	
		Inboard	Outboard
Naval	4	1.2–1.4	1.2–1.3
Naval	1	1.1–1.2	
Merchant	2	1.2–1.3	
Merchant	1	1.1–1.3	

to other procedures for applications that are normally of interest during design developments. It is noted, however, that the practical application of the more sophisticated unsteady analytical procedures has been substantially enhanced by the development of computer programs for this purpose.

Variable propeller forces, in addition to those resulting from a nonuniform water inflow, are generated as a result of the proximity of the hull to the propeller. Hull surface forces generated by the propeller are of the utmost importance when evaluating hull vibrations.

3.3 Torsional Loads. The mean torsional load on the shafting, which results in the average torsional stress, is calculated from the output of the main engine. If the full-power shaft horsepower output, H, of the main engine is developed at N rpm, then the steady torsional load, Q, on the shafting is

$$Q = \frac{63,025\,H}{N} \text{ in.-lb} \qquad (2)$$

In the design of naval shafting systems, it is common practice to increase the torque calculated with equation (2) by 20 percent [7]. The increase in design torque is an allowance in recognition of the additional torque developed during high-speed maneuvers, rough-water operations, foul-hull conditions, etc. During turns, the ship speed decreases, which results in a reduction in the propeller speed of advance; therefore, if there is no reduction in shaft horsepower, there is an increase in the shaft torque. Similarly, as the hull becomes foul, the ship speed reduces and full-power torque is developed at a lower rpm.

While high-speed maneuvers are a design criterion for naval ships, such torque increases are normally not considered in merchant practice because merchant ships do not engage in extensive high-speed maneuvers. Also, the torque increase, which is relatively small, due to hull fouling is accepted as a reduction in the factor of safety in merchant practice.

The torque increases measured during trials of single-screw and multiscrew ships in high-speed turns are given in Table 1. The torque ratio shown is the peak torque value observed during steering maneuvers divided by the torque at the start of the tests.

Although alternating torsional loads can be generated by other sources, the propeller is the only one of practical importance except in direct-drive diesel propulsion plants, where the cyclic engine torque may be significant. Alternating torsional loads generated by the propeller occur

Table 2   Propeller variable torque excitation factors

| No. of Propeller Blades | Torque Excitation Factor, r | | |
	3	4	5
Single-screw vessels	0.07–0.12	0.10–0.15	0.06–0.10
Twin-screw vessels with struts	0.02–0.05	0.02–0.05	0.02–0.04
Twin-screw vessels with bossings	0.04–0.08	0.04–0.06	0.04–0.05

NOTE: Excitation torque $= rQ$, where $Q =$ mean torque.

predominantly at blade frequency as a result of the non-uniform wake as discussed in Section 3.2. Shafting systems are carefully designed to avoid torsional resonant frequencies at full power; therefore, alternating torsional loads are not considered to be amplified by resonance. The range of the magnitude of the forced torsional alternating loads is given in Table 2. It will be noted that the variable torque can be of a significant magnitude even without magnification.

3.4   Thrust Loads. The magnitude of the mean propeller thrust load on the shafting system is equal to the towed resistance of the ship, corrected by the interaction effect between the propeller and hull as the propeller pushes the ship. This interaction effect is known as the thrust deduction [1]. The value of the design thrust can be obtained from powering calculations or from model basin tests conducted for the ship. For preliminary design purposes the propeller thrust can be estimated as

$$T = \frac{R}{1-t} = \frac{326\,E}{V(1-t)} = \frac{326\,H\,(PC)}{V(1-t)} \qquad (3)$$

where

$T =$ propeller thrust on shaft, lb
$V =$ ship speed at maximum power, knots
$R =$ hull resistance at V, lb
$E =$ hull effective horsepower at V, hp
$H =$ maximum shaft horsepower, hp
$t =$ thrust deduction fraction
$PC =$ propulsive coefficient

The value of t ranges from about 0.16 to 0.23 for single-screw ships that range from fine to full lines, respectively. Twin-screw ships have t values ranging from about 0.1 to 0.2, with the smaller value corresponding to ships with struts, and the larger value applying to ships with bossings. PC values of 0.73 for single-screw ships and 0.68 for multiscrew ships are average values and are normally found to be suitable for preliminary estimates. Reference 1, describes methods that may be used to estimate t and PC in cases where model test results or ship's performance calculations are not available.

In the design of submarines, another component of shaft load may become of sufficient magnitude to warrant explicit consideration. The shaft axial load resulting from submergence is equal to the submergence pressure times the shaft area in way of the hull-penetration shaft seal. The axial load is equal to:

$$T_s = 0.35\,SD_s^2 \qquad (4)$$

Table 3   Propeller variable thrust excitation factors

No. of Blades	Location	Thrust Excitation Factor, f
Multiple screw		
3, 4, or 5	behind struts	0.02–0.05
3	behind skegs	0.07–0.13
4 or 5	behind skegs	0.03–0.09
3	behind bossings	0.06–0.12
4	behind bossings	0.05–0.10
5	behind bossings	0.04–0.08
Single screw		
3		0.08–0.12
4 or 5[a]		0.03–0.08

[a] U-sections tend to emphasize the even-order components, and V-sections the odd-order components.

NOTE: Excitation thrust $= fT$, where $T =$ mean thrust.

where

$T_s =$ submergence-pressure shaft load, lb
$S =$ submergence design test depth, ft
$D_s =$ shaft sealing diameter, in.

The predominant alternating thrust load generated by the propeller occurs at propeller blade-rate frequency as a consequence of the nonuniform inflow water velocity to the propeller as discussed in Section 3.2. The magnitude of the variable thrust loads is dependent upon the number of propeller blades. For single-screw ships, an even number of blades will result in greater alternating thrust loads than an odd number, as noted in Section 3.2. For preliminary estimates, the magnitude of the alternating thrust as a percentage of mean thrust can be taken from Table 3.

Insofar as the strength of the shafting is concerned, neither the mean nor the alternating thrust loads are major design considerations. With merchant ships, the mean compressive stress is 1000 to 1500 psi; even in highly stressed shafts in naval ships the mean compressive stress seldom reaches 2500 psi. Torsional shear stresses are of predominant importance; and since the stresses due to thrust do not combine additively with the torsional stresses, the importance of the thrust stress is reduced even further.

3.5   Bending Loads. Loads which cause bending stresses to occur in the shafting are the result of gravity, shock, off-center thrust loads, and whirling shaft vibration. With the exception of once-per-revolution whirling vibration, all are alternating loads relative to a point on the shaft and occur at either shaft rotative frequency or once, or twice, the propeller blade frequency.

The weight of the shafting itself, which is a gravity load, is normally of minor importance with regard to inboard shafting unless there are unusual weight concentrations, such as a shaft locking device or brake drum, a pitch-control mechanism for a controllable-pitch propeller, or an exceptionally long span between bearings. When the shaft spans between bearings are essentially equal, and neglecting the weight of flanges, the maximum static bending moment occurring at the shaft bearings as a result of shaft weight can be determined approximately as

$$M = \frac{wL^2}{12} \tag{5}$$

where

M = bending moment at bearing, in.-lb
L = span between bearings, in.
w = weight per unit length of shafting, lb/in.

If the spans between bearings are not approximately equal, such a simple approach cannot be used; instead, a continually supported beam analytical technique must be used. The customary practice is to use a continuous-beam analytical procedure to calculate the bending moments at all critical shaft locations.

Gravity loads on the outboard shafting tend to be of major importance due to the large concentrated weight of the propeller. The long length of the bearings used outboard complicate an accurate definition of the bearing reaction resultant location. The assumed locations of the bearing reaction resultants are very important as they are direct determinants of the bending moments calculated. An accepted practice is to assume the reaction to be at the center of all bearings except the bearing just forward of the propeller. Because of the large weight of the propeller, the propeller shaft has a significant slope at this bearing; therefore, the resultant bearing reaction tends to be in the after region of the bearing. Water-lubricated bearings have L/D ratios of about 4 for this bearing, and the resultant reaction is usually assumed to be one shaft diameter forward of the aft bearing face. Oil-lubricated bearings have L/D ratios that range from about 1 to 2, and inspections of the shaft contact in these bearings indicate that hard contact is confined to the after region of the bearing for a length approximately equal to the diameter of the shaft. The accepted practice is to assume that the resultant bearing reaction in oil-lubricated bearings is located one-half shaft diameter from the after bearing face.

Generally the most significant shaft bending moment is due to the overhung weight of the propeller. The maximum static propeller shaft bending moment is computed as

$$M_p = W_p L_p \tag{6}$$

where

M_p = propeller overhung moment in propeller shaft, in.-lb
W_p = weight of propeller assembly in water, including shafting, abaft the aftermost bearing reaction resultant, lb
L_p = distance from CG of propeller assembly to aftermost bearing reaction resultant, in.

Equation (6) is the moment at the bearing reaction point assuming that the reaction is a point support rather than a distributed reaction over a region of the shaft. The point-support assumption is justified in that the exact load distribution on the bearing is unknown and the moment calculated in this manner is somewhat in excess of the actual value, which could be determined if the distribution of the bearing reaction was known.

There are a number of influences in addition to the gravity moment of the propeller that can have a significant impact on the propeller shaft bending stress. These are the eccentricity of thrust, water depth, sea conditions, and ship maneuvers. Under the general guidance of SNAME Panel M-8, the propeller shafts of a total of five ships have been instrumented to measure the bending stresses under actual operating conditions. Data obtained from these tests are reported in references 8 through 12. Table 4 summarizes the characteristics of the ships tested. The tests were conducted to show the significance of the ship loading, sea conditions, ship maneuvers, and thrust eccentricity.

Eccentricity of the propeller thrust produces a significant propeller shaft bending moment. With the possible exception of submarines, the propeller resultant thrust is eccentric from the propeller shaft centerline under almost all operating conditions and is usually in the upper starboard quadrant of single-screw ships. Therefore, it does not combine directly with the propeller gravity moment. Light-draft operating conditions and "U" shaped stern sections tend to bring the thrust and gravity moments closer together and make them more additive. Table 4 gives the thrust eccentricity factor, C/D, determined from full-scale test data for heavy-displacement, calm-sea conditions. The thrust eccentricity, C, shown in Table 4 is the resultant of the eccentric thrust and the gravity components.

Full-scale tests on the *Jamestown* [11] permit an evaluation of the influence of ship loading, sea conditions, and maneuvers. These factors are summarized in Table 5. The factors presented in Table 5 are the ratios of the bending stresses for the various conditions described to the bending stresses under full-load, deep-water, calm-seas, and straight-ahead operations. The extrapolated results from the *Observation Island* tests [12] generally support the factors in Table 5. It should be noted that maneuvers such as crash-backs rarely occur, and that shafting need not be designed to withstand stresses three times the normal value on a continuing basis.

The shock loadings that must be considered in the design of shafting for naval combatant ships are akin to the gravity loading and are frequently determined by multiplying the gravity force loads by a "shock" factor; however, more sophisticated methods are available for determining the shock loads through the application of dynamic analysis techniques. References 13 and 14 describe the procedures used to conduct dynamic shock analyses of shafting systems.

Misalignment in shafting systems can produce very significant bending loads and this factor is probably responsible for the majority of inboard shafting failures. The sensitivity of the shafting to misalignment should be reviewed with particular attention given to water-lubricated stern tube and strut bearings, which are subject to weardown in service. The sensitivity of the shafting to misalignment can be assessed by calculating the shafting bearing reactions and moments with the shafting in various misaligned conditions. These calculations are very

Table 4   Ships instrumented to determine tail shaft bending stresses

Ship Name	Type Ship	Δ	LBP, ft	B, ft	C_B	shp	rpm	Design Thrust, lb	Prop. Dia., D, ft	Thrust Eccent., C, ft	Thrust Eccent. Factor, C/D	Reference
Obispo	T2-SE-A2 tanker	20,300	503	68	0.74	10,000	95	184,000	19.5	0.91	0.047	[8]
Chryssi	tanker	38,100	615	84	0.77	15,000	112	246,000	22	1.31	0.060	[9]
Robinson	Victory ship	15,200	436.5	62	0.69	8,500	85	156,000	20.5	1.67	0.082	[10]
Jamestown	tanker	50,200	685	93	0.75	26,500	108.5	385,000	23	1.26	0.055	[11]
Observation Island	Mariner	16,400	528	76	0.60	22,000	110	271,000	22	1.68	0.076	[12]

Table 5   Increase in propeller shaft bending stresses due to various effects

Load	Sea Condition	Operation	Water Depth	Factor
Heavy	calm	ahead	deep	1
Heavy	calm	maneuvering	deep	2
Heavy	calm	crash back	deep	3
Light	calm	ahead	deep	$1\frac{1}{8}$
Light	calm	maneuvering	deep	$2\frac{1}{4}$
Light	calm	crash back	deep	$3\frac{3}{8}$
Heavy	stormy	ahead	deep	$2\frac{1}{3}$
Light	stormy	ahead	deep	$2\frac{2}{3}$
Heavy	calm	ahead	shoal	$1\frac{1}{2}$
Light	calm	ahead	shoal	$1\frac{3}{4}$

easily made by using the bearing reaction influence numbers, which are discussed in Section 4.4.

Lateral or whirling vibration of the shafting does not cause stress reversals, but it can result in increased bending loads in the shafting. However, since the shafting system is designed to avoid whirling criticals in the upper operating range, bending loads from shaft whirling vibration are not considered when designing the shafting.

3.6 Radial Loads. Radial loads in shafting are caused by driving the propeller onto the shaft taper, shrink-fitting sleeves on the shafting, and shrink-fitting removable-flange couplings. The radial compressive stresses resulting from these loads are normally of insignificant magnitude and are not considered in determining the shaft factor of safety. However, these radial loads can be of importance in that they can give rise to fretting corrosion when coupled with bending loads or alternating torsional loads that cause minute relative movement of the mating surfaces. Fretting corrosion can be controlled by limiting the relative motion and by cold-rolling the mating shafting surface. Cold-rolling of shafting surfaces is discussed in Section 4.7.

Another consideration is that if a radial load is applied abruptly, a stress concentration can occur. Therefore, the ends of shrunk-on hubs, liners, etc. should be designed with stress-relief grooves to minimize sudden changes in radial compression loads caused by shrink or press fits.

Section 4
Shafting Design

4.1 Shaft Materials. With the exception of naval vessels and merchant vessels of very high power, mild steel is used for both inboard and outboard shafting. In the case of high-powered ships, the inboard shafting may be made of high-strength steel; however, high-strength steel is not recommended for outboard applications. Because of the seawater environment, as well as the fretting corrosion conditions that exist at shaft sleeves and the propeller interface, the fatigue limit of high-strength steel is not reliably greater than that of mild steel, nor is the endurance limit in a fretting corrosion condition better than that of mild steel.

Considerations in the selection of shafting materials are: fatigue characteristics, weldability, the nil-ductility temperature, and the energy absorption capability. An array of chemistry and physical property standards has been established for marine shafting materials that provides a range from which shafting materials can be selected. Chapter 22 contains more specific information regarding materials.

4.2 Computation of Shaft Diameters. Shafting for merchant vessels is required to meet the minimum standards set by the classification society which classes the vessel. Classification societies use rather simple formulas to compute the minimum shaft diameters. These formulas normally contain coefficients which are changed from time to time in recognition of experience or advancements in technology. The American Bureau of Shipping (ABS) formula for propulsion shafting is of the following form [15]:

$$D = 100 K \sqrt[3]{\left(\frac{H}{R}\right)\left(\frac{3.695}{U + 23180}\right)} \qquad (7)$$

where

D = required shaft diameter, in.

K = shaft design factor, which is dependent upon design details and ranges from: 0.95 to 1.2 for line shafts, 1.15 to 1.18 for stern tube shafts, and 1.22 to 1.29 for propeller shafts

H = rated shaft horsepower

R = rated rpm

U = ultimate tensile strength of shaft material, not to exceed 60,000 psi for propeller shafts or 116,000 psi for other shafts.

It may be noted that equation (7) does not explicitly recognize bending loads and dynamic loads; nevertheless, equation (7) does provide a sound basis for the establishment of shafting diameters. This is because the predominant torsional shear stress in line shafting is properly considered and an appropriate allowance has been made for propeller bending loads and dynamic loads. The level of torsional shear stress, S_s, corresponding to equation (7) can be determined by observing that

$$S_s = \frac{160}{\pi D^3} = 321,000 \, \frac{H}{RD^3} \tag{8}$$

By substituting equation (7) into equation (8) and setting $U = 60,000$ psi for mild-steel shafting, it is seen that

$$S_s = \frac{7226}{K^3} \tag{9}$$

When $K = 1$, as for straight sections of line shafting for diesel-driven ships, the mean torsional shear stress would be 7226 psi. The minimum tensile yield stress of mild steel is 30,000 psi; consequently, it is seen that adequate margin is provided for secondary influences that may appear to be neglected.

The shaft design factor, K, specified for a propeller shaft that has a keyless propeller attachment and oil-lubricated stern tube bearings is 1.22, which means that the propeller shaft diameter is required to be 22% larger than the corresponding straight section of line shafting that is sized with $K = 1$. The 22% increase in diameter is based upon analytical and empirical data and is in recognition of influences such as the propeller shaft bending load due to the overhung propeller weight and dynamic propeller loads.

Equation (7) includes allowances for the shafting loads associated with conventional ship designs; however, no provisions are made for navigation in ice. The ABS Rules [15] acknowledge the substantial loads that ice navigation can impose upon a ship, and the design criteria for the shafting system is one of the specific aspects of a ship that is required to be strengthened for navigation in ice. There are several classes of ice strengthening, which correspond to different types of operating regions, ice conditions, and vessel service.

In no case can a designer accept classification society design criteria without question, particularly when applied to unusual designs. Classification society rules are largely established on the basis of successful experience; consequently, a designer must analyze a particular design from the perspective of conventional practice and compensate for the effects of influencing factors not appropriately considered.

Prudence requires that the propeller shaft diameter specified for a particular application be increased by a small margin (0.25 to 0.5 in.) above the minimum diameter required by the classification society rules. This margin can be used to remove a small amount of surface metal in the event that the shaft becomes superficially damaged when being handled or during service without infringing upon the classification society minimum requirements. If no margin is provided, a small damaged area could necessitate weld repairing the damaged area or perhaps scrapping an expensive shaft.

The formulas included in the ABS Rules for the establishment of minimum shafting diameters do not preclude the use of other analytical design procedures. The ABS Rules include a provision which states that, as an alternative to the use of the formulas, shafting with a minimum safety factor of 2.0, based on a detailed fatigue analysis, will be specially considered [15].

Reference 16 is a report of a service life comparison of 15 oversized propeller shafts, which have section moduli that are 74% greater than required by ABS, with 15 shafts of normal size, which have section moduli that are 11.5% greater than required. The comparison showed that the mean expected service life of the oversized shafts was less than that of the shafts of normal size. Although the statistical sample was small, the study clearly showed that propeller shaft problems are not necessarily solved by simply making the shaft larger.

The approach used to establish the size of naval shafting is considerably different from that used with merchant shafting. The procedure used to determine the size of naval shafting is delineated in reference 7. As stated in reference 7, an effort is made to assess all significant shafting loads in each particular case, although some loads are by necessity handled in an approximate manner. For example, in order to allow for the effects of off-center thrust and abnormal loadings due to rough weather and the like, the propeller shaft bending stress due to the static weight of the propeller is multiplied by a factor of 3 for single-screw ships and 2 for multiple-screw ships.

An additional difference between merchant and naval procedures is the criteria of acceptance. In naval practice, dual criteria are used. Factors of safety are specified for all shafting and, in addition, a specific bending stress limit is specified for the propeller shaft. The reason for the latter requirement is that fatigue tests conducted on models of propeller shaft assemblies and crankpins showed that alternating bending stress levels in excess of 6000 psi in surface rolled (cold-rolled) shafts result in fatigue cracks after 84 million stress reversals [10]. Therefore, normal operating bending stresses in excess of this stress level are not prudent design criteria. Furthermore, the endurance limit of a propeller shaft assembly can be essentially independent of the fatigue limit of the material in air. If seawater contacts the steel shaft, no endurance limit exists; and it is only a matter of time before cracks occur that are followed by ultimate failure.

4.3 Bearing Locations. Bearing locations have been determined by an array of criteria. It was once thought that providing two bearings per shafting span was desirable because that arrangement facilitated shafting installation and alignment. Also, bearing locations were often based upon intuitive judgment. With design criteria such as these, problems due to unloading of bearings, excessive rates of weardown, shaft whirling, and gearing misalignment were not rare. Problems were frequently related to the system having too many bearings. In order to better understand the optimum locations for bearings, designers began analyzing shafting as a continuous beam. The development and general dissemination of computer programs specifically applicable to shafting system analysis made it feasible to routinely conduct in-depth studies to optimize shafting systems as well as diagnose recurring problem areas.

Factors to be considered in determining the number and location of shaft bearings are:

(a) Ship's fixed structure and arrangement.
(b) Equality of line shaft bearing reactions.
(c) Bearing unit loads and L/D ratios.
(d) Shafting flexibility.
(e) Lateral vibration natural frequencies (shaft whirl).

The ship's fixed structure such as bulkheads and stanchions will usually require compromises in the shafting arrangement. Also, provisions for maintenance and overhaul must be considered before final bearing locations are set.

From a cost and interchangeability standpoint, all line shaft bearings should be identical. Therefore, the bearings should be spaced such that the bearing reactions are approximately equal. If this is done, the total number of bearings in the run of shafting is set by the total shaft weight, the permissible design unit load, and the acceptable L/D limits. The number, R, of line shaft bearings required to support a run of shafting can be tentatively determined as follows:

$$R = \frac{W}{pD^2L/D} \qquad (10)$$

where

W = total weight of shafting to be supported (note that gear and stern tube bearings may carry some line shaft weight)

p = design bearing pressure based on projected area (maximum permissible pressure less 5 to 10 psi to allow for variations)

D = shaft diameter in way of journal (normal practice is to increase the shaft diameter $\frac{1}{8}$ to $\frac{1}{4}$ in. in way of bearings)

L/D = bearing length/diameter ratio

After tentatively selecting the number of bearings and spacing them approximately equally, a detailed analysis of the bearing loads under all normal operating conditions is made (see Section 4.4). Particular care must be taken during the selection of the aftermost and forwardmost line shaft bearing locations to ensure that adequate shafting flexibility is provided; these bearings are subjected to a varying alignment in service. The conditions of primary importance are the cold start-up condition, the hot operating condition, and the bearing weardown and misalignment conditions. A major consideration in this analysis is the influence of the shafting on the reduction gear bearing loads or diesel engine bearing loads resulting

from the thermal change in the position of these bearings when going from the cold to the hot operating condition. Criteria for the alignment of the propulsion unit to the shafting are developed on the basis of this analysis.

Weardown of water-lubricated outboard bearings alters the loads on the aftermost line shaft bearings. The loads on the aftermost bearings must be analyzed for all operational conditions of weardown to ensure satisfactory performance. A major concern is the possibility of a bearing becoming unloaded. The effects of weardown are discussed further in the following sections.

A parametric study of minimum line shaft bearing spacing was carried out and reported in reference 17. The conclusion reached was that for shafting arrangements having one or more line shaft bearings the minimum span ratio (i.e., ratio of bearing center distance to shaft diameter) should be 14 for shafts with diameters in the range of 10 to 16 in. and 12 for shaft diameters of 16 to 30 in. The maximum span ratio could be in the range of 20 to 22 but the final determination must be based on strength, shaft slope at the bearings, and vibration characteristics.

4.4 Shafting System Calculation Output. In the course of preparing a shafting arrangement, alignment studies are conducted to determine the effects of bearing weardown and the effects of changes in the heights of the gear and steady bearings resulting from thermal expansion. The alignment studies are based upon a computer model that is developed from shafting arrangements such as illustrated by Figs. 1 and 2. The model incorporates the location of all concentrated weights, changes in shaft dimensions, and locations of bearings. The effect of buoyancy on the weight of the propeller and wet shafting is considered. Specialized computer programs that are based on continuous-beam analysis methods are generally used to calculate the data required to conduct shafting analyses.

The important output from shafting computer calculations includes the following specific data:

(a) Line-in-line reactions.
(b) Slope of shafting at discrete points.
(c) Deflection of shafting at discrete points.
(d) Moments in shafting at discrete points.
(e) Lateral natural frequency of shafting.
(f) Bearing reaction influence numbers.

The shafting line-in-line bearing reactions are the reactions with all bearings aligned concentrically, and are illustrated by the first row of Table 6 for the shafting arrangement shown in Fig. 1. The significance of the shaft slopes, shaft deflections, shaft moments, and lateral natural frequency of the shafting is apparent; however, the importance of bearing reaction influence numbers may not be as evident. Table 7 is a tabulation of the bearing reaction influence numbers for the same shafting arrangement. The numbers given in Table 7 represent the change in the magnitude of the bearing reaction of the various bearings as a result of raising any bearing one mil. Through the application of these influence numbers, which reflect the shafting system flexibility, it is possible to investigate the influence of shafting misalignment

Table 6 Tabulation of bearing reactions for shafting arrangement shown in Fig. 1

Bearing No.	Fwd Slow-Speed Gear 1	Aft Slow-Speed Gear 2	Fwd Line Shaft 3	Line Shaft 4	Line Shaft 5	Line Shaft 6	Line Shaft 7	Stern Tube 8	Strut 9
Line-in-line	26900	49500	23200	25500	25200	27200	15900	62600	89900
Cold (as aligned)	41700	29700	29600	23700	25700	27000	15900	62600	89900
Hot (as aligned)	36300	37300	26500	24900	25400	27100	15900	62600	89900
Weardown condition[a]	36300	37100	26700	24300	27200	20100	30800	51000	92200
Measured (hot)	...	38500	26700	26400	25300	30300	11200	...	...

[a] Assume a strut bearing weardown of 0.200 in. and a stern tube bearing weardown of 0.185 in. (consider weardown proportional to bearing pressure).

Table 7 Bearing reaction influence numbers for shafting arrangement shown in Fig. 1
(pounds per mil of vertical displacement)

Bearing No.	Fwd Slow-Speed Gear 1	After Slow-Speed Gear 2	Fwd Line Shaft 3	Line Shaft 4	Line Shaft 5	Line Shaft 6	Aft Line Shaft 7	Stern Tube 8	Strut 9
1	853	−1125	336	−81	21	−6	2	—	—
2	−1125	1506	−493	141	−36	10	−3	1	—
3	336	−493	259	−146	55	−15	4	−1	—
4	−81	141	−146	165	−116	48	−13	3	—
5	21	−36	55	−116	152	−115	48	−11	1
6	−6	10	−15	48	−115	158	−120	42	−4
7	2	−3	4	−13	48	−120	163	−98	16
8	—	1	−1	3	−11	42	−98	82	−18
9	—	—	—	—	1	−4	16	−18	5

NOTES:
The numbers tabulated above represent the effect of raising a given bearing one mil; e.g., if the forward line shaft bearing is raised one mil, the forward slow-speed gear bearing reaction increases 336 lb, the after slow-speed bearing reaction decreases 493 lb, the forward line shaft bearing increases 259 lb, etc.

caused by thermal expansion, weardown, and other such effects. Alignment requirements are developed on the basis of the bearing reaction influence numbers. Also, the principles employed with the hydraulic jack method of measuring bearing reactions (see Section 4.12) originate with the bearing reaction influence numbers.

There are many instances where bearing reaction influence numbers provide the basis for essential analytical capabilities; however, such data often are not available when they are needed. As an alternative to generating bearing reaction influence number data from specialized shafting analysis computer programs, it should be noted that the same data can be derived from any of the generalized finite-element analysis procedures. This is accomplished by creating a beam model that incorporates all bearing locations and all significant changes in shaft geometry. Successive calculations are then made to determine the effect upon all bearing reactions that results from each of the bearing positions being individually displaced one mil vertically. To produce a complete table of bearing reaction influence numbers, a calculation must be made for each bearing in the shafting system. To simplify the calculations, the vertical stiffnesses of the bearing housings and foundations are generally assumed to be infinitely rigid relative to the flexibility of the shafting in way of the bearings. This is generally a permissible assumption; however, unusual arrangements, such as a rubber-mounted outboard bearing, may require that the flexibility of the bearing support be considered.

4.5 Shaft Alignment. The alignment of all main propulsion shafting, including that for turbine-gear, diesel-gear, and direct-drive diesel prime movers, is carried out

following essentially the same steps. There are some unique differences in the criteria for satisfactory alignments, which primarily concern the presence of a reduction gear; however, in all ships both the outboard bearings and the line shaft bearings must be loaded within acceptable limits, the shafting must be manufactured without excessive eccentricity, and the athwartship alignment of the shafting must ensure that there will be no objectionable side loads on bearings.

a. Shaft alignment to reduction gears. The ship structure that supports propulsion reduction gears is commonly designed to serve as the main lubricating oil sump; consequently, when the lubricating oil warms to operating temperature, the containing structure expands. When the propulsion plant goes from the cold to the operating condition, the slow-speed gear bearings typically rise in the range of 15 to 30 mils relative to the line shaft bearings. This rise can significantly alter the reactions of the slow-speed gear bearings and the forward line shaft bearings. Of particular concern is the fact that the static load on the forward slow-speed gear bearing decreases while that on the after bearing increases. As can be seen from the typical reduction gear bearing reaction diagram shown in Chapter 9, this causes the slow-speed gear to assume a crossed-axis position relative to the slow-speed pinions, which are not similarly affected. As a result, the tooth load tends to be more heavy on one end of each helix. Reference 18 contains a detailed discussion of the effects of unequal gear bearing reactions on gear performance and the maximum permissible differences between the gear bearing loadings. The gear manufacturer normally states the maximum difference permitted between the loads on the slow-speed gear bearings. Reference 19 notes that no standard procedure has been established for the determination of the allowable difference between the slow-speed gear bearing reactions and outlines some of the complexities that must be taken into consideration. One possible complication is a nonparallel rise of the gear bearings due to temperature differences in way of the forward and after slow-speed gear bearings.

The flexibility factor, which has also been less appropriately called the allowable setting error, is conveniently used as an index of shafting flexibility in way of the reduction gear. The flexibility factor, *FF*, is defined as the allowable difference in the static vertical gear bearing loads divided by the difference between the bearing reaction influence number of the forward slow-speed gear bearing on itself and the after slow-speed gear bearing on itself. Therefore, the flexibility factor is determined as follows:

$$FF = \frac{\Delta R}{I_{11} - I_{22}} \qquad (11)$$

ΔR = allowable difference between the two slow-speed gear bearing static reactions, lb
I_{11} = reaction influence number of forward slow-speed gear bearing on itself, lb/in.
I_{22} = reaction influence number of aft slow-speed gear bearing on itself, lb/in.

The flexibility factor represents the total of the permissible error in estimating the thermal rise of the slow-speed gear bearings relative to the line shaft bearings and the permissible error in setting the gear to the line shafting without exceeding the maximum allowable difference in the static slow-speed gear bearing reactions. An absolute minimum acceptable value for the flexibility factor has been recognized to be 0.010 in.

If the flexibility of the shafting meets the flexibility factor criterion, the analysis proceeds to an investigation of the gear-to-shaft alignment. Beginning with the line-in-line reactions, that is, the bearing reactions with all bearings concentric, the estimated thermal rise of the gear bearings relative to the line shaft bearings when going from the cold to the operating temperature is used to establish alignment data that will provide approximately equal slow-speed gear bearing static reactions when in the operating condition. It must additionally be ascertained that the line shaft bearing reactions are satisfactory under all operating conditions.

The installation of the shafting system is usually begun by positioning the line shafting on the theoretical shafting centerline (the line-in-line position), as established by construction reference marks on the hull structure. The bearings that support the outboard shafting are also installed relative to this theoretical reference line, but not necessarily concentric with it, as required to satisfy the results of the alignment analysis. One convenient method that has been used to align the line shafting requires the shafting sections to be supported on temporary bearings such that the centers of the flanges are concentric with the shaft centerline through the support points and the flange faces are perpendicular to that centerline; while this cannot be done with absolute precision, sufficient accuracy can generally be obtained by supporting the shafting sections at approximately the ⅖ points from each end. When this is done, the flange drops and gaps are measured directly at the flanges with no corrections required. Drop is the vertical distance between the centers of adjacent flanges, and gap is the difference in opening between the top and bottom of the two flanges (nonparallelism of the flange faces). A correction usually must be made to the drop and gap measured at the stern tube shaft flange since it is impracticable to support the stern tube (or propeller) shaft such that the forward stern tube shaft flange is concentric and perpendicular to the theoretical shaft centerline. Consequently, the alignment of the stern tube shaft to the aftermost section of line shafting must include a correction for the slope and deflection of the forward stern tube flange when the stern tube shaft is supported in the as-aligned condition.

With the line shafting installed in the line-in-line condition, the positions of the slow-speed gear bearings relative to the centerline of the line shaft bearings are readily determined from the drop and gap of the slow-speed gear shaft flange relative to the line shaft flange; this is derived geometrically. Based on these geometrical data, the bearing reaction influence numbers can be used to plot the bearing loads for various alignment conditions. Such a plot is shown in Fig. 9. Figure 9 is an informative means of illustrating the effects that the thermal rise of the slow-speed gear bearings and alignment errors have upon bearing loads.

Table 6 shows the cold alignment bearing reactions with an alignment corresponding to point A on Fig. 9. The hot reactions listed in Table 6 are the estimated bearing reactions after the gear has reached operating temperature; this is point B on Fig. 9.

The alignment of the athwartship direction should be such that no significant forces are imposed on the slow-speed gear bearings in the horizontal plane.

The initial bearing reactions are often measured with the shafting coupled-up in the line-in-line condition except for the outboard bearings and the slow-speed gear bearings, which are positioned in compliance with the theoretical alignment calculations. The initially measured bearing reactions are then used to establish the bearing height changes that are required to optimize the bearing loads. After adjusting the bearing heights, the bearing reactions are measured again to confirm that the bearing reactions in the cold condition are within established criteria. As a subsequent confirmation, it may also be advantageous to measure the bearing reactions during dock trials or sea trials when the shafting system is in the hot operating condition. The methods used to measure bearing reactions are reviewed in Section 4.13.

The bearing reaction influence numbers also provide a means to study the effects of circumstances such as bearing movements resulting from hull deflections, bearing weardown, and the like. An analysis of bearing reactions

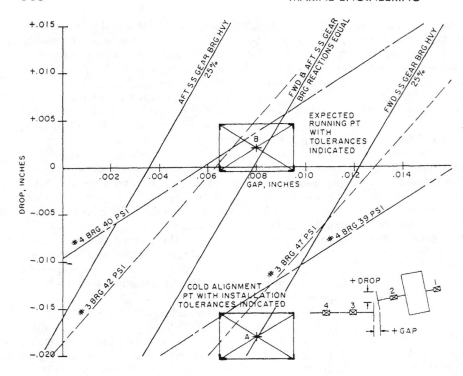

Fig. 9 Gear-to-shaft alignment analysis

with the stern tube and strut bearings worn down is given in Table 6.

The procedures used to install and align shafting arrangements that contain reduction gears are outlined in reference 20. Reference 20 does not specifically refer to direct-drive diesel shafting arrangements; however, the same principles generally apply.

b. Shaft alignment to a slow-speed diesel. The procedure used to align propulsion shafting to a direct-drive diesel engine is basically the same as that used for the alignment to reduction gears; however, there are significant differences in the procedural details. The principal considerations are to ensure that the crankshaft stresses resulting from shafting loads are not excessive and to ensure that the shafting and engine bearing loads are within acceptable limits.

The distance between the engine foundation and the crankshaft bearings is relatively small when compared with the height of the heated structure below a slow-speed gear shaft; consequently, the thermal rise of the crankshaft when going from the cold to the hot operating condition is similarly small and more accurately predictable than is the thermal rise of reduction gear bearings. The engine manufacturer normally provides an estimate of the thermal rise of the engine crankshaft centerline resulting from the engine going from the cold to the hot operating condition.

The procedure begins by aligning the outboard shafting in accordance with the theoretical alignment calculations, installing the inboard shafting in the line-and-line position, and positioning the diesel engine to obtain engine bearing loads as determined by the shafting alignment calculation output. With these initial alignment conditions and with

the engine output flange and the shafting flanges made up, the line shaft bearing reactions should be measured, as well as the aftermost engine bearing reaction, if accessible, to establish the as-installed baseline. The as-installed bearing reaction baseline is then used to establish any changes to the bearing heights that are required to ensure satisfactory bearing loads. After adjusting the bearing heights, as predicated by the measured bearing reactions and the theoretical bearing influence numbers, the bearing reactions should be measured again to confirm that the cold bearing reactions satisfy the established criteria. The bearing reactions may also be measured with the engine at the operating temperature if there is any reason for concern relative to the effects that reaching operating temperature may have on shaft alignment.

Before operating the diesel engine, the stress variation in the crankshaft, as it rotates, should be measured with the shafting made up to the engine and the engine bolted down to its foundation. This can be accomplished by measuring the variation in the axial deflections of the crankshaft throws as the crankshaft is rotated. Generally, the technical manual for the engine describes the procedure to be used to measure the throw deflections and establishes the bounds for acceptable deflection readings. Reference 21 outlines the basic process used when taking the crankshaft throw readings. If the crankshaft deflection measurements taken exceed the deflection limits established by the engine manufacturer, the underlying cause must be determined. Some of the factors that can influence the amount of crankshaft deflection are: loads imposed on the crankshaft by the shafting, engine casing distortion caused by foundation bolting loads, and improperly set or wiped crankshaft bearings.

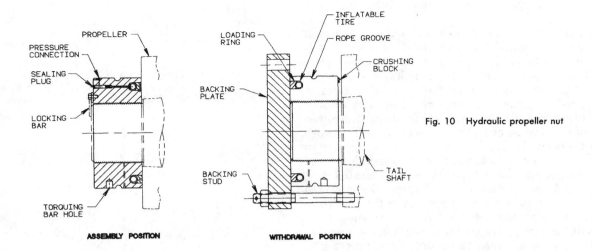

Fig. 10 Hydraulic propeller nut

When the shafting system bearing reactions are confirmed to be satisfactory and the crankshaft throw axial deflections, and hence stresses, are found to be within acceptable limits, a satisfactory shafting alignment is assured.

4.6 Propeller-to-Shaft Interface. Design details of the propeller-to-shaft interface are a critical aspect of a shafting system development. Propeller shaft failures in way of the propeller were not rare up through the 1960's and into the 1970's. However, the advances in design technology (e.g., stress relief grooves at the forward end of the propeller and the aft end of the liner, shortened and spooned keyways, slotted keys, and improved sealing methods) significantly improved the reliability of propeller shafts and increased their service lives. Also, improvements in inspection technology have provided the means to detect incipient cracks and thus have greatly reduced the loss of propellers at sea.

Details of the propeller-to-shaft interface required for naval ships are specified by references 22 and 23. The naval type of propeller-to-shaft interface is consistent with merchant practice and both have comparable service histories.

The propeller keyway introduces a stress concentration in the propeller shaft and weakens the shaft even though the keyway may have generous fillet radii and the forward end of the key may have slots to relieve the key load at the forward end. To avoid the stress concentrations associated with propeller keys and to minimize the vagaries associated with fitting the propeller on the shaft taper by brute-force hammering, propeller nuts have been developed that incorporate annular pistons, which are motivated by hydraulic oil or grease. A "hydraulic" nut provides the means to apply a large force of a known magnitude to the propeller, thereby pushing it onto the shaft taper such that the frictional force between the propeller and shaft is sufficiently large to transmit all torsional and thrust loads, and no propeller key is required. The operating principles of a "hydraulic" nut are illustrated by Fig. 10. In effect, the device is seen to be a hydraulic ram that is built into the propeller nut. The tire,

which is pressurized to about 15,000 psi, develops a large axial force under controlled conditions. A pneumatic-hydraulic pressure intensifier is used to produce the high hydraulic-oil pressure.

The first step in fitting a propeller to a shaft, with any procedure, is to ensure that there is uniform contact between the propeller and shaft over 60% of the mating surface. The propeller is then pushed firmly up on the taper, and the hydraulic nut is put in place and used to advance the propeller a specified amount that is dependent upon the ambient temperature. The temperature correction is required because of the difference in thermal expansion between the material of the propeller and that of the shaft.

Keyless propeller designs rely entirely upon the friction between the propeller and shaft to withstand the propeller torsional and thrust loads, with the torsional loads generally the larger. For a dry, greaseless, installation, the coefficient of friction between the propeller and shaft may vary from a low of 0.13 to a high of approximately 0.18, being influenced by factors such as the materials and the quality of the fit-up. However, the effective coefficient of friction has been found to be about 0.15. To provide adequate service margins, the frictional loads that can be transmitted must be 2.8 times the loads corresponding to ahead operations. Loads experienced during astern operations are generally smaller than for ahead operations, but this should be confirmed to be the case. The amount of propeller advance required to achieve the desired frictional load-carrying capability is readily calculated and is used to establish propeller installation criteria. An assessment of keyless propellers for naval applications is reported in reference 24; also, the factors of safety provided by keyed and keyless propellers are compared.

The hydraulic nut is also used to withdraw, or "jump," the propeller from the propeller shaft, as indicated by Fig. 10. The hydraulic oil pressure required to remove the propeller is 30 to 50% of that required to install the propeller, with the difference being the effect of shaft taper. The shaft taper used with keyless propellers ranges from 1 : 12 to 1 : 20, with lower tapers becoming more common.

Some earlier hydraulic nut designs entailed the use of oil pressure to expand the propeller hub while it was forced onto the taper. By expanding the propeller hub, a reduced force was required of the hydraulic nut to advance the propeller hub up the taper; and theoretically, an expanded propeller hub could be withdrawn without the use of a hydraulic nut. Some earlier designs involved the use of cast-iron sleeves, which were installed between the shaft taper and propeller hub. However, as the design of hydraulic nuts has matured, these complexities of the earlier designs have been found to be unwarranted.

4.7 Cold Rolling. Rotating shafts that are subjected to bending stresses in way of shrunk-on or tightly fitting mating components can result in fretting corrosion. Fretting corrosion occurs where there are minute amounts of relative motion between two tightly fitting materials, which causes the surface material to tear, thereby introducing a stress riser in the shaft, with the progressive ultimate consequence of a shaft failure.

Fretting corrosion was first recognized as the cause of railway car axle failures where hubs were shrunk onto the the axles. Cold-rolling the axles in the shrink-fit area was accomplished to introduce compressive stresses in the shaft surface; these compressive stresses were found to alleviate the problems experienced with railway car axles.

Conditions conducive to fretting corrosion also exist at the ends of sleeves and at the forward end of propeller hubs in marine shafting configurations; and the development of fretting fatigue cracks in the surface of the propeller shaft at the forward end of the propeller hub and under the after end of the propeller shaft sleeve, where bending stresses are normally the highest, is one of the most common modes of propeller shaft failure. The introduction of compressive stresses into a material by cold rolling the surface will not eliminate the occurrence of fatigue cracks that are caused by fretting corrosion; however, the compressive stresses will retard the formation of minute cracks by fretting corrosion and will significantly reduce the propagation rate of cracks that form. When tested and applied to marine shafting systems, cold rolling also proved to be effective in extending the life of marine shafting systems [10].

Propeller shafts are commonly cold rolled for a distance forward and aft of the shaft taper, where the propeller hub mates, and in way of the ends of shaft liners. Residual compressive stresses are introduced into the shaft to a depth of about 0.5 in. The apparatus required to cold roll shafting and the procedure to be followed are described in reference 25.

4.8 Protection from Seawater. Except in the case of designs in which all bearings are of the oil-lubricated type, outboard shafting is provided with sleeves that are shrunk on the shafting in way of bearings, stuffing boxes, and fairings. Shaft sleeves are made of bronze or other materials, which are resistant to attack by seawater.

Ships having a single short section of outboard shafting may employ a single continuous sleeve. Where continuous

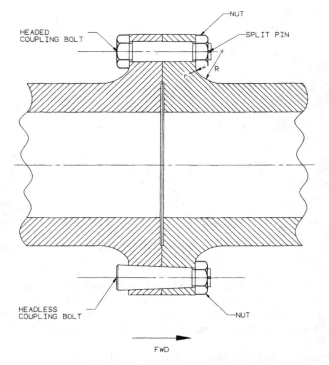

Fig. 11 Shaft bolting designs

sleeves are not used and the steel shafting would otherwise be exposed to seawater, those areas of the steel shaft are normally protected by applying a rubber [26] or plastic [27] compound directly to the shafting surface. The adequacy of both rubber and plastic protective coverings for outboard shafting has not been uniformly good, and on occasion the protection offered to outboard couplings by such coverings has been particularly unsatisfactory. In applying these shaft coverings, good quality control and proper practices are essential for good results. Shaft cleanliness (bright clean steel) at covering installation is essential for the covering to bond properly, and bonding can be further improved by grit blasting the shaft to roughen its surface. Rotating coupling covers (fairwaters), which clamp onto and rotate with the shaft, thereby eliminating the violent erosive flow of water around coupling bolts, have been used to avoid the erosive effect of the water.

A reliable static sealing arrangement at the propeller, which prevents seawater from contacting the propeller shaft, is of the utmost importance. A propeller-shaft assembly in which seawater is allowed to contact the shaft will result in the shaft not having an endurance limit, and therefore, it is only a matter of cycle accumulation before a failure occurs. Typical designs of propeller-hub sealing arrangements, which are necessary with systems utilizing water-lubricated outboard bearings, are contained in references 15 and 23.

4.9 Shaft Couplings. Except in instances where special considerations preclude their use, shafting sections are connected by means of integrally forged couplings as illustrated by Fig. 11. Although the design of virtually all integral shaft couplings is similar, the details of shaft coupling designs vary. For example, despite individual preferences, no specific number of coupling bolts has been established as optimum, and the proportions of flange dimensions and flange fillet radii may vary from one design to the next.

The headed bolt design shown by the upper half of Fig. 11 is commonly used to connect shaft sections. The diameter of the bolt is given a taper of about ⅛ in. per foot and the mated bolt holes are reamed at assembly to ensure a metal-to-metal fit. The headless bolt illustrated by the lower half of Fig. 11 is an alternate design. The taper of headless bolts is increased to about ¾ in. per foot for bolt diameters less than 3 in. and 1 in. per foot for larger bolts to provide an additional capability to transmit axial loads. The spot-faced surfaces on the coupling flanges, on which the coupling nuts land, must be perpendicular to the axes of the coupling bolts; otherwise, galling can occur at local areas of hard contact.

Guidelines for the design of flange couplings for merchant vessels are given in classification society rules such as reference 15; similar guidelines for naval vessels are given in references 7 and 22. An effort to standardize the design of shafting couplings was made in reference 28. The standardization effort never gained a significant level of support for a number of reasons, one being the strong incentive to optimize any design for that specific application; however, the data given in reference 28, are useful for preliminary design purposes.

Flange coupling bolts, such as illustrated by Fig. 11 are commonly installed by hammering custom-designed wrenches until the installing mechanic is satisfied with the "ring" of the hammer. Extensive experience confirms that satisfactory service can be obtained using this procedure; although it entails the disadvantages of having to rely upon the experience of the mechanic, subjects the bolt mating surfaces to damage, and possesses no means of confirming the amount of preload achieved.

Figure 12 is an illustration of a hydraulically preloaded bolt that may be appropriate for use in some cases to avoid bolt mating-surface damage when the connection is broken and to provide a means of confirming the amount of bolt preload achieved. With this type of bolt, a hydraulic pressure of about 25,000 psi is used to elongate the bolt, with a simultaneous known decrease in bolt diameter. A hydraulic shutoff valve maintains the bolt in the elongated condition until it is in place. After installation, the hydraulic power head is removed, and a depth gage is used to measure the length of the hole in the bolt and, consequently, the preload achieved in the installed bolt. The bolts are easily removed by reversing the process. There are numerous appropriate applications for bolts of this type.

Couplings with removable flanges are required in some instances; for example, those cases where a liner (that is not designed as halves which are welded in place) is to be installed on a shaft that requires a bolted flange on each end. Figure 13 illustrates a typical removable-flange coupling and shows the means provided to transmit both thrust and torque. Both torque and thrust are normally transmitted by friction between the shrunk-on muff and

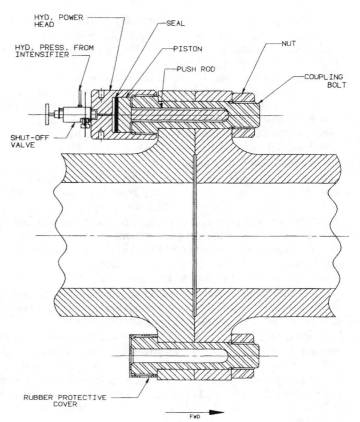

Fig. 12   Hydraulically preloaded bolt design

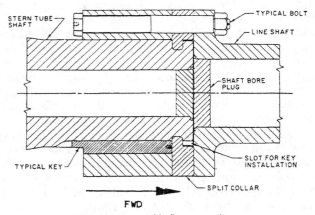

Fig. 13   Removable-flange coupling

the shaft. The keys are a backup for the transmission of torque, and the split collar is a backup for the transmission of thrust.

Removable shaft couplings have also been used that are mounted in place by hydraulic pressure to both expand the coupling hub and advance the hub up a tapered liner, which is installed on the shaft. Torque and thrust are transmitted between the coupling hub and shaft solely by friction.

Some shafting arrangements are designed such that it is necessary to remove the forward flange of a stern tube shaft in order to withdraw the shaft aft; this is undesirable when particular types of coupling designs are used because it is difficult to remove the flange without damaging the flange-shaft interface. An arrangement sometimes preferred is one in which the outside diameters of the stern tube bearing bushings are made sufficiently large so that they can be removed to permit the stern tube shaft to be withdrawn aft with the forward flange in place. Figure 24 illustrates a strut bearing with a removable bushing that permits the shaft to be withdrawn with the forward flange in place.

4.10 Shaft Axial Movements. The axial movement of the shafting relative to elements that are fixed to the hull must be considered to establish proper clearances between the propeller hub and stern frame structure and between bearing housings and rotating elements within the bearings that are secured to the shaft (oil slingers, oil disks, etc.). Several factors can contribute to the movement of the shaft relative to hull structure; these are:

1. Thrust bearing clearances. Axial clearances between the thrust collar and shoes permit a corresponding fore-and-aft movement of the entire shafting system.

2. Propeller thrust. The propeller thrust results in a small axial deflection of the shafting and thrust bearing.

3. Submergence pressure. For surface ships the submergence pressure is a minor consideration; however, for submarines the effects of the submergence pressure, both in terms of shaft axial loading and hull compression, are important.

4. Hull hogging and sagging. Hull bending loads strain the hull; however, the shafting is not similarly affected. This factor is conveniently assessed by assuming an extreme-fiber hull bending stress and the location of the neutral axis of the hull in bending; the stress, and corresponding hull strain, at the shaft centerline is then determined by interpolation.

5. Temperature differences. The shafting can be at a warmer temperature (70–80 F) relative to that of the hull structure (in the vicinity of 30 F), and the difference in thermal expansion results in relative movement.

The foregoing factors would generally not reach maximum values simultaneously, but they are prudently considered to do so. Typical axial movements of the propeller (the point at which movement is a maximum) relative to the hull range from 0.5 in., for tankers with very short shafts, to 2 in., for ships with long shafts.

4.11 Shafting Balance. Solid shafting is inherently balanced, but hollow shafting requires attention in this regard. The balance of hollow shafting is accomplished during the machining operation by shifting lathe centers prior to the final machining cuts. The amount of static unbalance in a shaft can be determined by either a static or dynamic balancing technique.

After the rough machining cuts have been made, a shafting section can be statically balanced by removing the shaft section from the lathe, placing it on rails, noting the equilibrium position of the shaft section, shifting the lathe centers to compensate for the unbalance, and then taking additional machining cuts on the shaft section to

effect a static balance. Good practice dictates that adjoining shafting sections be installed such that the residual static unbalances, as determined by a check on the rails after final machining, tend to offset.

Although shafting sections have occasionally been specified to be dynamically balanced (shaft sections rotated in a balancing machine to determine both static and dynamic unbalance), there are conflicting schools of thought regarding the necessity of a dynamic balance. The need for dynamic balancing is influenced by the maximum rotating speed of the shafting. It has been argued that the tolerances customarily imposed on the manufacture of shafting sections in conjunction with good shop practice preclude objectionable shafting dynamic unbalance.

4.12 Shafting Eccentricity. Propulsion shafting can become eccentric for a number of reasons, with accidents being the most common, particularly those while underway that involve the propeller. Shafting eccentricity, itself, is not of practical significance provided that neither the shaft balance nor the bearing reactions are adversely affected, and provided that the balance of masses, such as a propeller, attached to the shafting is not adversely affected. The eccentricity of shafting sections can be measured before the shafting is installed, and variations in bearing reactions resulting from shaft eccentricity can be determined by an alignment check.

In general, shaft eccentricity of an objectionable magnitude cannot be corrected by machining without reducing the shaft diameter below acceptable dimensions; therefore, an eccentric shaft must be straightened to be usable. Three methods have been used to straighten propulsion shafting: peening, flame heating (hot-spot method), and selective cold rolling. Peening involves hammering a local linear shaft surface area on the inside of the eccentric bend sufficiently hard to exceed the yield strength of the shaft material and thereby introduce residual compressive stresses in the peened area. Peening is more effective on smaller shafts. The mechanics of peening entails many vagaries; however, peening can often be done with the shaft installed.

With the flame heating or hot spotting method, a narrow linear shaft surface area on the outside of the eccentric bend is heated. As the local surface area heats, the metal expands and the metal yield point decreases. The objective is to heat a local area such that the restraint of the cooler surrounding metal causes the metal in the heated area to exceed the material yield strength at the elevated temperature; consequently, when cooled, residual tensile stresses are introduced in the shaft. Flame heating requires experience to accomplish satisfactorily as there is a risk of adversely affecting the metallurgy of the material; however, flame heating can often be accomplished in place.

The selective cold rolling method is the most controlled and reliable means of straightening shaft sections; however, the cold rolling procedure generally requires the shaft section to be placed in a lathe. Selective cold rolling is accomplished by cold rolling the entire shaft circumference in way of the bend, but with a much heavier imposed

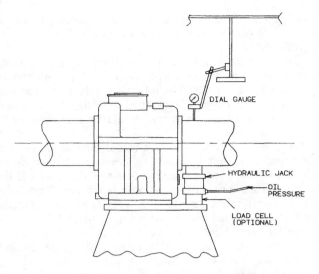

Fig. 14 Hydraulic jack method for measuring bearing reactions

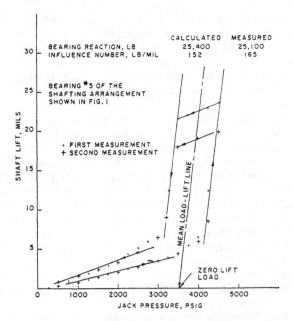

Fig. 15 Bearing reaction determined by hydraulic jack

load on the inside of the bend. Higher residual compressive stresses are, therefore, introduced on the inside of the bend, which tend to straighten the shaft.

A more detailed discussion concerning the equipment and process techniques involved with the flame heating and selective cold-rolling methods of straightening shafting is included in reference 29.

4.13 Determination of Shaft Alignment. There are basically three ways that the alignment of a completely installed shafting system can be checked. One, which is akin to the drop-and-gap method of alignment at initial installation, is to remove the bolts from select couplings and compare the relative positions of mating flanges with calculated values. Although the circumstances sometimes require this method to be used, it is the least preferred method because of the numerous opportunities for erratic readings. A second method, which is both easier to accomplish and more directly meaningful, is to measure the loads being carried by the bearings with either a calibrated hydraulic jack or a hydraulic jack in conjunction with a load cell. The hydraulic jack is used to lift the shaft off of a bearing, and the load carried by the jack is then related to the bearing reaction. A third method is to install strain gages on the shafting at strategic positions and determine the bearing reactions by utilizing the measured strain (shaft moments) to make the shafting a determinate system. A general review of the alternative shaft alignment procedures used by American shipyards was conducted by SNAME Technical and Research Panel M-16 and reported by reference 30.

a. Hydraulic jack method. When using the calibrated hydraulic jack or hydraulic jack and a load cell method, the actual bearing load is determined by placing a hydraulic jack as close to the bearing housing as possible (bearing foundations are often designed with an extension to provide a jack foundation as illustrated by Fig. 14). A dial indicator is located on the shaft immediately above the jack so as to measure vertical movement of the shaft. Where possible, the anchor point for the dial indicator

should be independent of the bearing housing because the shifting loads on the bearing foundation can cause the bearing housing to tilt, which would introduce confusing readings. Before recording any readings, the position of the shaft in the bearing clearance should be checked to ensure that the shaft is in the bottom of the bearing and is centered athwartship. The shaft should be lifted at least once to ensure that it can be lifted 20 to 30 mils without coming into contact with the upper half of the bearing; this preliminary jacking also tends to reduce hysteresis in the shaft and erratic readings. For short shaft spans, a dial indicator should also be installed on the shaft at adjacent bearings so that any rise of the shaft at these bearings can be noted, because in stiff shafting arrangements it is possible to also lift the shaft off a bearing adjacent to the one being measured.

With the dial indicators and jack in place, the shaft is raised and lowered in increments, noting the jack load corresponding to each increment of shaft rise. These data are plotted as shown in Fig. 15. The data points conform to two basic slopes. The slope of the lift-versus-load line as the load is transferred to the jack from the bearing represents the spring constant of the bearing shell, bearing housing, and the like. When the shaft lifts clear of the bearing, an abrupt change in the slope of the data points occurs. The second slope corresponds to the bearing reaction influence number for the bearing.

Due to the friction in the shafting and load-measuring system, the data points when raising the shaft do not coincide with those obtained when lowering the shaft; the characteristic pattern is the equivalent of a hysteresis loop. The deflection-versus-load plot indicates a higher jack load for a given shaft lift when raising the shaft than when lowering it. When using a calibrated hydraulic jack, the hysteresis in the load versus shaft-deflection relationship tends to be larger than when a load cell is used. This is caused by the friction in the hydraulic jack (primarily seals). Load cells are located such that they are subject to the force developed by the jack and can provide a direct readout in pounds with an accuracy of 0.5%. Calibrated hydraulic jacks are inherently less accurate than load cells; however, their accuracy is adequate for most purposes.

Experience indicates that the true relationship between the jack load and shaft lift is approximately midway between the lines determined when raising and lowering the shaft as indicated by Fig. 15. However, in cases where the increasing and decreasing load lines are significantly different, the increasing load line should be favored. With an effective mean line established that represents the true relationship between the measured load and shaft lift, the load that would be on the jack at zero shaft lift and with the bearing removed is determined by extrapolating the mean line downward to zero shaft lift. If the jack and bearing are close together, the load as determined may also be considered to be the load on the bearing if the jack were removed (or the bearing load being sought).

Under favorable jacking conditions (with no binding of the shaft in the bearing due to athwartship misalignment, interference with stuffing boxes, etc.), experience shows that the accuracy of the bearing reactions determined is usually within 10% of the actual value. However, the influence numbers obtained by jacking may not be as accurate. When the bearings being jacked are located towards the middle of the shaft and span lengths are fairly equal, jack influence numbers are generally within 30% of the calculated influence numbers. For bearings located near the ends of the shaft, the influence numbers obtained by jacking may disagree with the calculated values by 50% or more.

Both the load and influence number errors are due to inaccuracies that are inherent in the jacking procedure; e.g., the jack not being located at the bearing center, the load center in adjacent bearings shifting as the shaft is raised, and hysteresis in the shafting system and load-measuring system. Consequently, when a bearing is to be realigned, the distance that the bearing should be raised or lowered is more appropriately based on the calculated influence numbers rather than on the influence numbers determined by jacking.

When jacking bearings that are very close together or in cases where the jack must be located some distance from the bearing, a correction factor must be applied to the jack load measured to obtain an accurate assessment of the bearing reactions. The correction factor is as follows:

$$C = \frac{I_{bb}}{I_{jb}} \qquad (13)$$

where I_{bb} is the influence of bearing on bearing and I_{jb} the influence of the jack on the bearing. The influence numbers used to calculate the correction factor are determined by including both the jack and the bearing being jacked as support points in the shafting system calculations. To be theoretically accurate, this correction factor

should be used for every bearing that is jacked; however, only in the aforementioned two cases is it a factor of significance.

Table 6 contains a tabulation of the measured bearing reactions for the shafting system in Fig. 1 and illustrates typical jacking results. The oil in the reduction gear was heated to operating temperature and circulated; therefore, the measured reactions should be correlated with the hot reactions.

The hydraulic jack procedure can also be used to detect bent shafts in that the bearing reactions can be determined with the shaft rotated in 90-deg increments. If the bearing reaction changes significantly with shaft position, a bent shaft can be suspected. This technique is often appropriately used when analyzing a shaft that is suspected of being bent. However, the uniform tightness of flange coupling bolts should be confirmed before concluding that a section of shafting is bent.

b. Strain-gage method. The alignment of an installed and complete shafting system can also be determined by measuring the strain in the shaft at strategic positions. A shafting system with N bearing supports is N-2 degrees statically indeterminate, and N-2 additional data points are required to determine the shaft alignment. Referring to the shafting arrangement illustrated by Fig. 2, which has 7 bearings, an additional 5 data points must be known for the shafting alignment to be determinate. By installing strain gages at the break points indicated on the free-body diagram of Fig. 16, the moments in the shafting at these points can be measured, and the moments measured can be used to calculate the bearing reactions. The equations that can be used to relate the bearing reactions to the measured bending moments are outlined in Table 8.

In Fig. 16 there is a free section, which does not contain a shaft bearing. This section is inherently determinate and the unknown shear forces can be calculated. A free section can be used between any two bearings to act as the starting point in either direction for the determination of the shaft bearing reactions. When the free section is located adjacent to a two-bearing stern tube arrangement such as Fig. 16, a value for the unknown shear is established and only the two bearing reactions are unknown. This permits the determination of both bearing reactions, and establishes a known shear force on the forward end of the free section, which permits the remaining bearing reactions in the shafting system to be quantified.

If the propeller shaft in Fig. 16 had only one bearing support, i.e., if bearing 6 were not there, the free section would not be necessary because, with the knowledge of M_e, the aftermost free-body shaft section is determinate and the aftermost bearing reaction and the shear force at e can be determined.

Three bearings with inaccessible connecting shafting (e.g., a double stern tube and strut bearing arrangement, or the shafting arrangement shown by Fig. 1 with a second stern tube bearing added) cannot be analyzed using the strain gage method alone. For a double stern tube/strut bearing configuration, bearing reactions forward of a free section can be determined since the free section

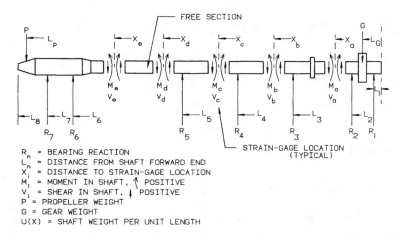

Fig. 16 Free-body diagram of shafting arrangement shown by Fig. 2 for the determination of bearing reactions using strain-gage measurements

R_n = BEARING REACTION
L_n = DISTANCE FROM SHAFT FORWARD END
X_i = DISTANCE TO STRAIN-GAGE LOCATION
M_i = MOMENT IN SHAFT, ↑ POSITIVE
V_i = SHEAR IN SHAFT, ↓ POSITIVE
P = PROPELLER WEIGHT
G = GEAR WEIGHT
$U(X)$ = SHAFT WEIGHT PER UNIT LENGTH

provides the needed shear force at its forward end. However, to determine the values of the three aftermost bearing reactions, one of the bearings, such as the forward stern tube bearing, would have to be weighed. By knowing the load on the forward stern tube bearing and by including a free section of shafting in the free-body diagram, the outboard shaft section becomes determinate since only the strut bearing and aft stern tube bearing reactions are unknowns.

Once the free-body sections are established for a given shafting arrangement, which defines the strain-gage locations, the equations for determining the bearing reactions are obtained by setting ΣM and ΣF equal to zero for each free-body section. When the specific geometry for a shaft system is incorporated into the equations, the integrals reduce to constants with the only variables being the bearing reactions, the moments to be measured, and shear forces at the locations.

Details concerning the strain-gage method of shafting alignment are contained in references 31 and 32. Reference 32 outlines a procedure that can be used to optimize the location of strain gages on a given shaft system and to estimate maximum errors in the resulting bearing reactions.

Advantages associated with the strain-gage method of determining shaft alignment are as follows:

- Both horizontal and vertical alignment can be checked simultaneously.
- Strain-gage readings can be taken at sea because of the ease with which the reading can be obtained, and there is no need for jack or dial indicator support structure.
- Once the gages are installed, a complete set of strain readings can generally be taken in less than an hour.
- Reactions of bearings that are inaccessible for jacking can be determined accurately.

As noted by reference 32, the last two items are the most important advantages associated with the strain-gage method. However, the strain-gage method requires more calculations to interpret the test data and requires a more complex measurement technique.

Section 5
Bearings

5.1 Introduction. Main propulsion shafting is supported by bearings that maintain the shafting in proper alignment. These propulsion shaft bearings are naturally divided into two groups: bearings inside the watertight boundary of the hull and bearings outside the hull watertight boundary.

The requirements imposed upon the design of propulsion shaft bearings are extremely severe. The bearings are required to operate at speeds ranging from about 0.1 rpm, when on the jacking gear, to 100 or more rpm in either direction of rotation. And, unlike some applications, the bearing loads do not vary with rpm but are essentially constant at all speeds. Reliability is heavily emphasized in the design of bearings because there is no redundancy for bearings, and a single bearing failure can incapacitate the propulsion system.

In addition to the radial bearings that support the shafting, a main thrust bearing is located inside the ship and transmits the propeller thrust from the shafting to the hull structure. Figures 1 and 2 show the two typical main thrust bearing locations. Figure 1 indicates that the thrust bearing is located such that the thrust collar is on the forward end of the reduction gear slow-speed shaft. In this arrangement, the thrust collar is necessarily removable, which may be advantageous if the thrust collar is damaged. However, the thrust bearing foundation stiffness that can be achieved with this arrangement is limited because the continuity of the supporting structure is inter-

Table 8   Equations for calculation of bearing reactions for free-body diagram shown by Fig. 16

REDUCTION GEAR SECTION

$$\sum M_{R1} = 0 = M_a + V_a(x_a - L_1) + R_2(L_2 - L_1) - G(L_G - L_1) - \int_{L_1}^{x_a} x\,u(x)\,dx + \int_0^{L_1} x\,u(x)\,dx$$

$$R_2 = \frac{-M_a - V_a(x_a - L_1) + G(L_G - L_1) + \int_{L_1}^{x_a} x\,u(x)\,dx - \int_0^{L_1} x\,u(x)\,dx}{(L_2 - L_1)}$$

$$\sum F = 0 = V_a + R_2 - G + R_1 - \int_0^{x_a} u(x)\,dx$$

$$R_1 = -V_a - R_2 + G + \int_0^{x_a} u(x)\,dx$$

LINE SHAFT SECTION WITH BEARING 3

$$\sum M_a = 0 = M_b - M_a + V_b(x_b - x_a) + R_3(L_3 - x_a) - \int_{x_a}^{x_b} x\,u(x)\,dx$$

$$R_3 = \frac{-M_b + M_a - V_b(x_b - x_a) + \int_{x_a}^{x_b} x\,u(x)\,dx}{(L_3 - x_a)}$$

$$\sum F = 0 = V_b + R_3 - V_a - \int_{x_a}^{x_b} u(x)\,dx$$

$$V_a = V_b + R_3 - \int_{x_a}^{x_b} u(x)\,dx$$

LINE SHAFT SECTION WITH BEARING 4

$$\sum M_b = 0 = M_c - M_b + V_c(x_c - x_b) + R_4(L_4 - x_b) - \int_{x_b}^{x_c} x\,u(x)\,dx$$

$$R_4 = \frac{-M_c + M_b - V_c(x_c - x_b) + \int_{x_b}^{x_c} x\,u(x)\,dx}{(L_4 - x_b)}$$

$$\sum F = 0 = V_e + R_4 - V_b - \int_{x_b}^{x_c} u(x)\,dx$$

$$V_b = V_e + R_4 - \int_{x_b}^{x_c} u(x)\,dx$$

LINE SHAFT SECTION WITH BEARING 5

$$\sum M_c = 0 = M_d - M_c + V_d(x_d - x_c) + R_5(L_5 - x_c) - \int_{x_c}^{x_d} x\,u(x)\,dx$$

$$R_5 = \frac{-M_d + M_c - V_d(x_d - x_c) + \int_{x_c}^{x_d} x\,u(x)\,dx}{(L_5 - x_c)}$$

$$\sum F = 0 = V_d + R_5 - V_c - \int_{x_c}^{x_d} u(x)\,dx$$

$$V_c = V_d + R_5 - \int_{x_c}^{x_d} u(x)\,dx$$

FREE SECTION

$$\sum M_d = 0 = M_e - M_d + V_e(x_e - x_d) - \int_{x_d}^{x_e} u(x)\,dx$$

$$V_e = \frac{-M_e + M_d + \int_{x_d}^{x_e} x\,u(x)\,dx}{(x_e - x_d)}$$

$$\sum F = 0 = V_e - V_d - \int_{x_d}^{x_e} u(x)\,dx$$

$$V_d = V_e - \int_{x_d}^{x_e} u(x)\,dx$$

STERN TUBE SECTION

$$\sum M_6 = 0 = -P(L_p - L_6) + R_7(L_7 - L_6) + V_e(L_6 - x_e) - M_e + \int_{x_e}^{L_6} x\,u(x)\,dx - \int_{L_6}^{L_8} x\,u(x)\,dx$$

$$R_7 = \frac{P(L_p - L_6) - V_e(L_6 - x_e) + M_e - \int_{x_e}^{L_6} x\,u(x)\,dx + \int_{L_6}^{L_8} x\,u(x)\,dx}{(L_7 - L_6)}$$

$$\sum F = 0 = -P + R_7 + R_6 - V_e - \int_{x_e}^{L_8} u(x)\,dx$$

$$R_6 = P - R_7 + V_e + \int_{x_e}^{L_8} u(x)\,dx$$

rupted by the reduction gear on the after side and the main condenser, which is located on the forward side of the thrust bearing.

The thrust bearing for the shafting arrangement shown in Fig. 2 is located aft of the reduction gear, and there is an independent thrust shaft. This arrangement has the potential of providing a more stiff thrust bearing foundation than when the thrust bearing is integrated with the reduction gear. Where the thrust bearing is independent of the reduction gear, with its own thrust shaft, a line shaft bearing is commonly integrated with the thrust bearing housing. If the thrust shaft is located some distance aft of the reduction gear, an independent thrust bearing oil system may be required.

5.2 Main Thrust Bearings. In general, the normal practice for propulsion main thrust bearings is to use thrust bearings with tilting pads where the individual shoes are free to pivot as the oil film dictates. There are two basic designs of tilting-pad thrust bearings, the Kingsbury and Michell types. The Michell bearing pads pivot on a radial line across the thrust pad, whereas the Kingsbury type pivots on a radiused button support. The Kingsbury type of main thrust bearing normally has the thrust pads supported on leveling links to distribute the thrust load equally to all the thrust pads; this is referred to as a self-equalizing type of thrust bearing. The Michell marine thrust bearings, which are used primarily in Europe and Japan, are normally built with the thrust pads supported directly in the thrust bearing housing without the use of leveling links. Theoretically, the self-equalizing type of bearing can carry more load than a non-self-equalizing bearing since the individual thrust-pad loads are not affected by the machining accuracy of the thrust bearing parts, or by the deflection of the thrust bearing housing and foundation under load. Figure 17 is a photograph of the base ring, leveling links, and thrust shoes of an 8-pad self-equalizing Kingsbury type main thrust bearing with two thrust pads removed. Figure 18 is a section through a main thrust bearing, and shows how the thrust-pad elements pictured in Fig. 17 are arranged in the thrust housing to carry the thrust load. Figure 19, which is a developed view of the thrust pads and leveling links, shows how the leveling links distribute the load equally among the thrust pads.

U.S. Navy practice is to apply self-equalizing main thrust bearings at unit loads up to a maximum of 500 psi, allowing higher unit loads to occur in transient conditions such as ship turns. The commercial-ship practice with self-equalizing bearings is more conservative than Navy practice, with a limit to the full-power thrust unit load of about 400 psi. Standard 6-pad self-equalizing main thrust bearings are normally designed such that the load-carrying area is equal to one half of the square of the outside diameter of the thrust pads.

Main thrust bearings can be made with any number of thrust pads. A larger number of thrust pads can, in some cases, facilitate the use of a smaller bearing outside diameter. The optimum thrust pad geometry from a hydrodynamic perspective is one with the mean circumferential pad length equal to the pad radial length. With a normal

Fig. 17   Self-equalizing main thrust bearing

thrust bearing configuration, an 8-pad thrust bearing is the preferred selection to satisfy this condition. Main thrust bearings are generally designed with 8 pads.

The thrust bearing housing and foundation deflection under maximum load must be established before the maximum design unit load is set for non-equalizing bearings; normal practice limits the average maximum loading for these bearings to about 300 psi.

For naval ships, shock loadings must be considered when evaluating the thrust bearing strength and maximum thrust-pad loads. Because of the rigid design of thrust bearing housings and foundations and the capacity of thrust pads to take large transient loads, shock loadings do not normally control the thrust bearing design.

Thrust-pad support disks can be replaced by load-measuring cells in a Kingsbury type thrust bearing to measure the load on the thrust bearing. Thrust measurements permit verification of the thrust calculations and monitoring of the propulsion system performance. The necessary number and location of the load cells depend on the application; however, because of the leveling-link concept, load cells are not required under all pads.

Thrust-pad operating temperatures can be monitored by having a thermocouple or a resistance temperature detector embedded into the babbitt of the thrust pads. This is the most common practice of monitoring bearing performance and operating limits.

Vibration reducer. Thrust bearings and the thrust bearing foundations are designed to be stiff to limit the longitudinal deflection and stress resulting from the steady and alternating thrust. A longitudinal vibration

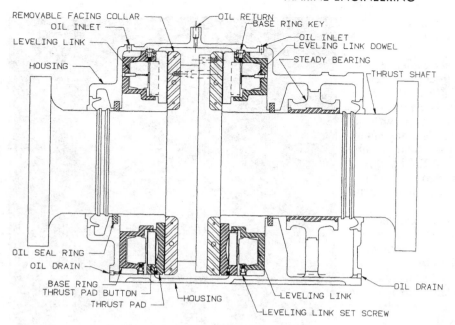

Fig. 18   Longitudinal section through thrust bearing

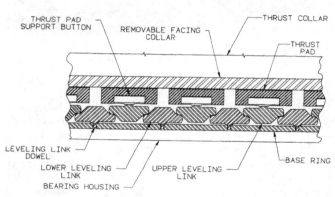

Fig. 19   Developed view of thrust pads and leveling links

Fig. 20   Vibration reducer modification of thrust bearing

resonance is intolerable in the upper propeller rpm range. One means of avoiding objectionable longitudinal vibration is to modify the main thrust bearing to incorporate a "vibration reducer," which reduces the longitudinal stiffness of the thrust bearing without increasing the thrust bearing and foundation deflections, and adds dampening to the shafting system. The increased flexibility in the shafting system shifts the resonant frequency downward, thereby reducing the alternating forces and amplitudes of vibration.

To accommodate a vibration reducer arrangement, the thrust bearing leveling links are removed, and each thrust pad is supported on a piston, as illustrated by Fig. 20. The pistons are connected to an oil manifold in the thrust bearing, which in turn is connected to oil flasks external to the thrust bearing. A hydraulic valve, which is controlled by the position of the thrust collar, adds or removes oil from the system to maintain the thrust collar in the central operating position within the thrust housing. The thrust-collar positioning system must have a source of oil at a pressure greater than the maximum operating pressure. The maximum operating pressure is equal to the maximum thrust divided by the sum of the thrust-pad piston areas. Figure 21 is a diagram of a vibration reducer system.

Reference 33 includes a discussion of the parameters and approach used to design a vibration reducer system. The reduced thrust bearing spring constant is achieved through the bulk modulus of the oil in the flasks that support the thrust pads. The effects of a vibration reducer are discussed further in Section 8.

5.3   Line Shaft Bearings.   Bearings located inside the ship's watertight boundary are called line shaft bearings, although they are sometimes referred to as steady or spring bearings. Almost without exception, these bearings are ruggedly constructed, conservatively designed, babbitt lined, and oil lubricated. Except in special cases, the bearings are self-lubricated by rings or disks arranged

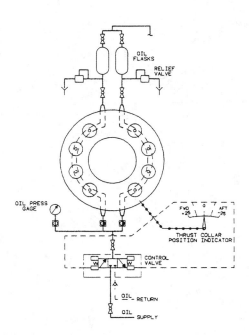

Fig. 21   Schematic of a thrust bearing vibration reducer system

in such a manner that lubrication is effected by the rotation of the shaft. Roller bearings have been used in the smaller shaft sizes, but the advantages of lighter weight and lower friction have in general not been sufficient to offset the higher reliability and lower maintenance costs of the babbitt-lined type.

Line shaft bearing housings are made of steel castings or fabricated of steel plates welded together. Completely satisfactory bearing housings are obtained by either method, and manufacturing costs govern the construction method used. Since rigidity is of more concern than strength, low-carbon steel is used as the material for bearing housings with the exception of bearings for naval combatant vessels, in which case high-impact shock requirements may necessitate the use of high-strength steel. Bearing housings are split horizontally at the shaft centerline. The bottom half of the bearing must be very ruggedly designed since it carries the vertical shaft load and any side load that occurs.

The bearing housing supports a heavy steel removable shell, which is lined with babbitt. The shaft rests on the babbitted surface. The bearing shell can be made with a self-aligning feature by providing a spherical or crowned seat at the interface between the bearing shell and housing. This allows the axis of the bearing shell to align exactly with that of the shaft. Figure 22 is a section through a bearing with a self-aligning feature, and Fig. 23 is a section through a bearing that is similar but without a self-aligning capability. The general construction of bearing housings and shells can be observed from Figs. 22 and 23. Design and maintenance criteria for bearings in naval service are provided by reference 34.

Except for the aftermost line shaft bearings in merchant applications, it is general practice to babbitt only the bottom half of the bearings since these bearings would never be expected to be loaded in the top. However, the aftermost bearing (the one closest to the stern tube) may become unloaded particularly when the stern tube and propeller bearings are water lubricated. Water-lubricated bearings are subject to a large amount of wear, which can result in severe misalignment. It is considered good practice to provide the maximum practicable amount of babbitt in the top half of the aftermost line shaft bearings when water-lubricated stern tube bearings are used. With oil-lubricated stern tube bearings, the probability of the after bearing becoming unloaded is considerably reduced. In naval practice, the top halves of line shaft bearing shells are babbitted to accommodate upward bearing loads during shock conditions.

Babbitt that is centrifugally cast onto the bearing shell is considered preferable to that which is statically poured. The former technique dependably provides a more secure bond between the babbitt and the bearing shell.

Babbitt can be of either the lead- or tin-base type. Tin-base babbitt has greater strength and is generally preferred for shaft bearings; it is specified almost exclusively for centrifugally cast bearings. Lead-base babbitt is preferred where embedding, conforming, and anti-friction are primary considerations. Lead-base babbitt has a lower yield point and a slightly better fatigue resistance.

The oil reservoir that is provided within the line shaft bearing housing must be sized to operate during extreme roll and pitch conditions without leaking oil by the shaft or disabling the bearing lubrication system. Furthermore, the oil quantity and sump surface area must be sufficient to dissipate the heat generated. Line shaft bearings are sometimes designed with cooling coils located in the sumps as shown in Fig. 23; however, experience has shown that the cooling coils are rarely, if ever, needed.

Line shaft bearings may be lubricated by means of oil rings, an oil disk, or by a supply of oil under pressure. Ring oil-lubricated bearings contain two or three metal rings with diameters of 1.25 to 1.5 times that of the shaft (the ratio decreases with larger shaft diameters). The number of rings in a bearing should be selected such that no ring is required to distribute oil for an axial distance greater than 7 in. on either side of the ring. The rings rest on top of the shaft and dip into an oil reservoir located beneath the bearing shell. Figure 23 is an example of a ring-lubricated bearing. As the shaft turns, the rings are rotated by the frictional contact with the top of the shaft. Oil that adheres to the ring in way of the oil reservoir is carried up to the top of the shaft where a part of the oil is transferred to the shaft and is subsequently carried into the contact region of the bearing. Ring-lubricated bearings have proved to be capable of accommodating large angles of list and trim and have proved to be reliable in service with design bearing unit loads of 50 psi. With regard to the possible adverse effects of trim, tests have demonstrated that ring-lubricated bearings can accommodate angles of approximately 10 deg from the horizontal with no sacrifice in performance. Reference 35 includes an evaluation of the performance of oil rings based on

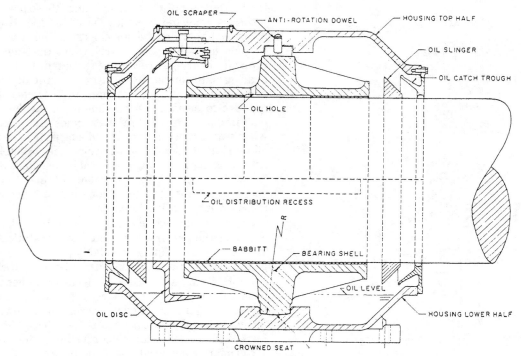

Fig. 22   Self-aligning line shaft bearing with oil-disk lubrication

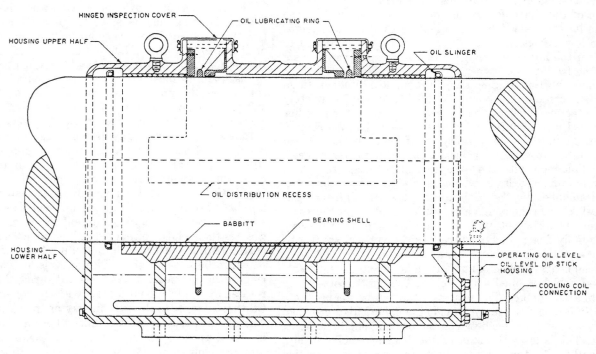

Fig. 23   Non-self-aligning line shaft bearing with oil-ring lubrication

laboratory tests and notes, among other things, the sensitivity of the quantity of oil delivered to the oil viscosity.

Disk-lubricated bearings use a metal disk that is clamped to the shaft at one end of the bearing shell. The disk may have a flange as illustrated by Fig. 22. As the shaft turns, the lower portion of the disk, which is immersed in an oil reservoir, is coated with oil. This oil is carried to the top where a metal bar scrapes the oil from the disk and guides it into passages where it is admitted to the top of the shaft and then into the contact region of the bearing. When disk-lubricated bearings were first introduced, they were designed with a unit pressure that was about the same as for a ring-lubricated bearing; however, as experience has been gained with disk-lubricated bearings, the design unit pressure specified for them has continually been increased to a value of about 100 psi.

The results of tests conducted with two sizes of disk-scraper lubrication arrangements (22 and 37 in. diameter), which are representative of those used in line shaft bearings, are presented in reference 36. During the tests an emphasis was placed on obtaining oil-flow data at low shaft speeds. In the lower rpm range (below 35 rpm for the larger disk), the oil flow varied as the 1.5 power of the disk surface velocity, the 0.5 power of the oil viscosity, and directly with the axial width of the disk. At higher shaft speeds, oil was centrifugally thrown from the disk, and the oil delivery became essentially independent of shaft speed.

In special cases, line shaft bearings may be lubricated with oil supplied by a pump. If the shafting system is very long, and the main engine lubricating oil pump is used to supply the oil, sump pumps would be required to return the oil from the bearings since a gravity drain would not be possible under all conditions of trim and pitch. Another alternative is for each line shaft bearing to have an independent closed lubricating system. While this method of lubrication assures an adequate supply of oil at all shaft speeds, and can result in smaller bearing sizes, it has the disadvantages of the extra pumps and added complexity.

The load that can be supported by a babbitted journal bearing is dependent upon the method of lubrication, the bearing configuration, the bearing length to diameter (L/D) ratio, and of course the installation workmanship. In the early designs, babbitted journal bearings had L/D ratios as large as 2, and even with such high L/D ratios, the shafting systems had very closely spaced bearings such that the bearing loads were very sensitive to alignment. The use of higher bearing pressures, in conjunction with the advent of more sophisticated techniques for the alignment of bearings, has resulted in more reliable shafting systems by affording more favorable bearing L/D ratios and more flexible shafting systems. Bearing L/D ratios are normally limited to a maximum of 1.5 in commercial ship design, but not less than 1 shaft diameter for bearings that are ring or disk lubricated to prevent end leakage of the oil from impairing adequate lubrication.

The most severe demands on the lubricating system of a line shaft bearing do not correspond to full-power, full-rpm operation, but to the condition when the shafting is rotated by the turning gear at about 0.1 rpm for extended periods of time to facilitate uniform cooling or heating of the main turbine rotors. If the lubrication system fails to deliver adequate oil to the journal under this condition, damage to the bearing surface may occur. Lubrication provisions have a strong influence on the ability of a bearing to operate satisfactorily in the critical jacking mode of operation; and, consequently, the means of lubrication strongly influences the extent to which line shaft bearings can be loaded. As a guide, it has been determined that as little as 25 drops of oil per minute on the journal surface will sustain indefinite operating in the jacking mode at bearing pressures of about 75 psi.

Tests to determine the conditions under which the transition from fluid-film lubrication to boundary lubrication (see Fig. 3 of Chapter 11) occurs in journal bearings are reported in reference 37. Both tilting-pad and sleeve bearings were tested at shaft speeds as low as those representative of turning-gear operations. The tilting-pad bearing was immersed in oil, and the sleeve bearing was disk lubricated. The bearings had 13 in. diameters and were tested at a load of 200 psi. An abrupt change in the friction coefficient occurred at 1.3 rpm for the tilting-pad bearing and 2.5 rpm for the sleeve bearing. The oil flow supplied by the disk was more than adequate, even at speeds as low as 0.1 rpm. The tilting-pad bearing was also subjected to a series of high-loading (up to 1000 psi), low-speed (0.012 to 0.2 rpm) tests, with the results showing only light polishing of the babbitt up to 300 psi and definite movement (wiping) of the babbitt above 750 psi.

With proper attention given to design details, ring-lubricated bearings, disk-lubricated bearings, and pressure-lubricated bearings can carry increasingly higher unit loads in that order. Disk-lubricated bearings can carry a higher unit load than ring-lubricated bearings based on the assumption that the oil scraper functions properly. Very close controls must be maintained in the manufacture of oil scrapers because manufacturing flaws, which are hardly perceptible, can have a large influence on their performance.

5.4 Outboard Bearings. Outboard bearings can be further classified as stern tube or strut bearings. Figures 1 and 2 show the locations of these bearings relative to the ship arrangement.

Outboard bearings can either be water lubricated or oil lubricated. Nearly all outboard bearings were water lubricated until about 1960, when a transition to oil-lubricated bearings began. This transition to oil-lubricated bearings was stimulated by the unduly short service life of many of the water-lubricated bearing assemblies during that period. It is believed that the shortened life of the water-lubricated bearings was caused by the trend to larger ship sizes, which had higher bearing loads, and more contaminated water passing through the bearings. Larger ships generally operate at deeper drafts; and with less clearance between the hull and channel bottom more contaminants, such as silt, mud, and sand, are drawn into the bearing clearance. The experience of ship operators during that time period regarding the weardown of water-lubricated outboard bearings with lignum vitae staves was generally unsatisfactory. The use of lignum vitae, which is a resinous dense hardwood, as a bearing material for water-lubricated bearings has been supplanted by the use of rubber or laminated phenolic materials. This upgrade in bearing material has substantially improved the performance of outboard bearings; however, comprehensive resolutions have not been developed for the many external factors that affect the performance of outboard bearings.

The minimization of vibration was also influential in the adoption of oil-lubricated bearings. Particularly with larger and fuller ships, variations in the water inflow velocity to the propeller generate large variable bending

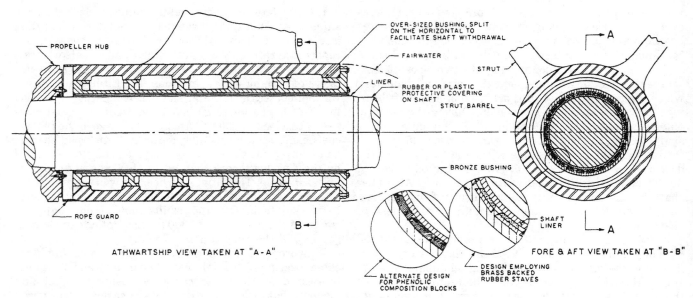

Fig. 24   Water-lubricated strut bearing

forces on the shafting. There have been many reported instances of shafting pounding in the forward stern tube bearing and the stern tube stuffing box on single-screw ships, particularly when five-bladed propellers were used. With proper initial alignment, oil-lubricated bearings, which have close bearing clearances and minimal wear-down, eliminate the pounding and associated maintenance of propeller shafts and stuffing boxes.

Oil-lubricated stern tube bearings also reduce the power losses in the shafting system. For a 22,000-shp ship an efficiency improvement of about 0.2 percent can be expected with oil-lubricated vice water-lubricated outboard bearings.

Oil-lubricated outboard bearings are favored on commercial ships, but water-lubricated bearings are prevalently used for naval ships. Figure 24 illustrates a typical water-lubricated strut bearing design. A water-lubricated stern tube bearing design is similar except that the bearing bushing is fitted inside the stern tube rather than inside the strut barrel.

Water-lubricated bearings basically consist of a nonferrous corrosion-resistant bearing bushing that retains a number of bearing contact elements, which may be a phenolic composition, or be made of rubber that is bonded to brass or nonmetallic backing strips. A sleeve is installed on the shaft to provide a corrosion-resistant contact surface.

When brass-backed rubber strip (rubber stave) bearings are used, as is common in naval practice, dove-tailed slots are accurately cut in the bushing to accommodate the bearing staves. Sufficient metal is left between each slot to hold the staves securely; the space between staves also provides a cooling water flow passage. Continual improvements have been made in the design of water-lubricated rubber bearings. Reference 38 is a report of engineering developments, such as reducing the rubber

thickness, using a more resilient compound, and using nonmetallic backing materials, that enhance the performance of rubber-stave bearings.

As indicated by Fig. 24, bearings that use phenolic materials are similar to rubber stave bearings. A "V" or "U" shaped groove is cut at the longitudinal joints of the blocks to provide lubricating and cooling water flow. Brass retaining strips are generally placed at four points around the circumference to secure the contact elements.

Phenolic bearing materials, usually installed when dry, absorb water and consequently tend to swell; therefore, swelling must be considered in the design of these bearings.

Water-lubricated bearings that had lignum vitae as the bearing material were designed with L/D ratios of approximately 4 for the bearing adjacent to the propeller and 2 for those forward of the propeller bearing. For water-lubricated bearings that have synthetic bearing materials, when substantiated by test results, an L/D ratio as low as 2 has been used for the propeller bearing, with L/D ratios of 1 for more-forward bearings. The unit loading of the propeller bearing, based on projected area (shaft diameter times bearing length), is normally under 80 psi when an L/D ratio of 2 is used; however, great care must be taken in placing importance on the absolute value of bearing contact pressures that are based on the projected area. Not only does the eccentricity of propeller thrust alter the loading, but also the load distribution is both difficult to assess and is subject to radical change. Outboard water-lubricated bearing materials may wear 0.2 to 0.5 in. before being replaced.

Water-lubricated outboard bearings can be installed to a slope that corresponds to the static slope of the shaft resulting from the weight of the propeller and shaft. The objective is to obtain a more uniform bearing load and

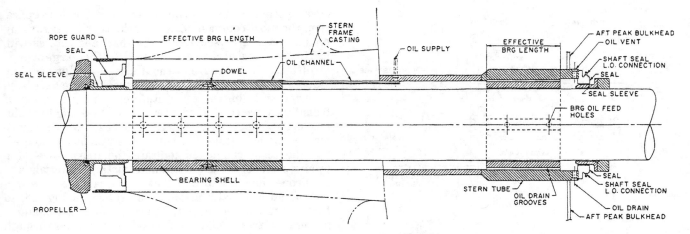

Fig. 25 Typical oil-lubricated stern tube bearing

shaft contact in the bearing when initially placed in service. Slope boring can facilitate hydrodynamic lubrication at a lower shaft rpm as a result of the lower bearing unit loads; however, this procedure has not proven entirely satisfactory from a weardown standpoint since the eccentric thrust of the propeller is not taken into account.

To provide a more uniform load distribution over the length of water-lubricated propeller shaft bearings, some naval ships have been designed with the sleeve that contains the bearing material mounted within a rubber support, which is located near the middle of the bearing. The flexible rubber support allows the bearing to conform to the slope of the shaft, and is an effective means of obtaining a more uniform bearing load distribution.

Propeller-shaft bearings of the tilting-pad type have also been used to provide a self-adjusting bearing contact feature. However, the average bearing pressure is inherently higher with tilting-pad bearings, and the bearing housing is necessarily much larger in diameter to accommodate the tilting-pad feature, which is expensive to provide and obstructs the flow of water to the propeller.

The shaft breakaway friction torque in water-lubricated bearings can indicate a coefficient of friction as high as 0.4 [39]. Depending on the shaft length, shaft torsional stiffness, bearing loads, and the boundary-lubrication coefficient of friction in the bearing, the shafting can rotate intermittently. The stick-slip shaft motion is observed most frequently when operating on the jacking gear, but it may continue up to speeds of about 10 rpm. The resulting nonuniform shaft motion is often accompanied by reduction gear backlash noises.

Water-lubricated bearing materials have upper operating temperature limits that, if exceeded, will result in material damage. Therefore, each application should be reviewed for possible cooling-water requirements. In general, the cooling-water flow requirements are no more than 1 to 10 gpm depending on the maximum design operating temperature, bearing coefficient of friction, bearing load, and shaft surface velocity. Generally, conservative design operating temperatures of 140 F or less are used. Data concerning friction coefficients for various water-lubricated bearing materials and configurations are given in reference 38. For instance, the Navy standard brass-backed rubber-stave bearing has a dynamic coefficient of friction that ranges from 0.006 to 0.013.

Oil-lubricated bearings, as illustrated by Fig. 25, have primarily been used in stern tubes and bossings, but have also been adapted for strut bearings. Oil-lubricated bearings do not require a liner to be installed on the shaft since contact with seawater does not occur, nor is there any significant shaft wear. Also, no bushing is inserted in the stern tube; the bearing shells, which have heavy wall thicknesses, are pressed directly into the stern tube. The L/D ratios of the heavily loaded after stern tube bearing have ranged widely. Early designs had ratios of 2.5 but a trend toward a value of 1.5 was subsequently established.

Although the unit bearing pressure based on the projected area normally falls in the 80 psi range for oil-lubricated bearings, the actual operating pressure is probably closer to twice this value. An inspection of the bearing contact area after operation reveals that the after bearing is loaded only on the after end for a length of about one shaft diameter; shorter bearings are often advocated for this reason. Slope boring the aftermost bearing so that it corresponds to the mean slope of the propeller shaft can provide better bearing contact and reduce the high pressures that would otherwise be applied to the after end of the bearing. Alternatively, tilting-pad propeller shaft bearings have been developed that assure a uniform contact between the bearing and shaft under all operating conditions. Tilting-pad bearings are generally designed with an L/D ratio of 1.

Figure 26 illustrates a typical lube-oil diagram for an oil-lubricated stern tube bearing. Oil-lubricated stern tube bearings are totally submerged in oil, and seals on the after and forward ends of the tube prevent the ingress of seawater and the leakage of oil. The pressure differential between the oil in the stern tube and the ambient seawater has been controlled in several ways. For ships that operate at a nearly constant draft, this has been accomplished by means of a head tank that is located about 10 ft above the full-load waterline. Ships that have large draft variations

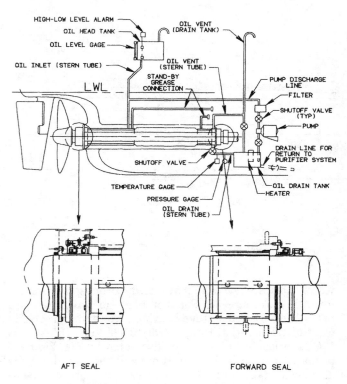

HIGH-LOW LEVEL ALARM
OIL HEAD TANK
OIL LEVEL GAGE
OIL INLET (STERN TUBE)
OIL VENT (DRAIN TANK)
OIL VENT (STERN TUBE)
STAND-BY GREASE CONNECTION
PUMP DISCHARGE LINE
FILTER
SHUTOFF VALVE (TYP)
PUMP
DRAIN LINE FOR RETURN TO PURIFIER SYSTEM
SHUTOFF VALVE
TEMPERATURE GAGE
PRESSURE GAGE
OIL DRAIN (STERN TUBE)
OIL DRAIN TANK HEATER
LWL

AFT SEAL   FORWARD SEAL

Fig. 26 Stern tube bearing lubricating oil diagram

several seal designs have been developed for this application, the most common design has lip-type seals running on a sleeve that is secured to the shaft and rotates with it. This type of seal can be subjected to rather large swings in differential pressure without damage. However, instances of excessive liner wear and seal leakage have been reported with lip-type seals. Reference 40 is a report of tests that were conducted with a special test rig to evaluate seal and liner materials. The seals used were approximately one-half scale. The test series was not completed, but the tests conducted indicate the importance of a proper selection of the contacting mating materials and the proper manufacture of these materials.

A pump is usually installed, as shown in Fig. 26, to force oil circulation through the stern tube. The piping is preferably arranged such that oil is circulated through both bearings. Many variations of this system have been used, including the deletion of the pump. Owners often specify filters, heaters, coolers, and coalescers to condition the oil as it passes through the circuit. Coolers are rarely used as the temperature leaving most stern tubes does not exceed 120 F.

A design variation that was developed to minimize the risk of leaking lubricating oil to the seawater incorporates a drained air cavity that is located in the after seal housing between the seawater and the lubricating-oil seal elements. With this arrangement any leakage past the seawater or lubricating-oil seal elements goes to the air cavity and is drained inboard. The air pressure and lubricating-oil pressure in the after seal assembly are automatically adjusted to compensate for variations of draft.

Stern-tube bearings on smaller ships have been installed by fitting the bearing bushings within the stern tube with a poured-in-place epoxy resin. Such an installation eliminates the necessity for an accurate stern tube boring operation. The stern tube bearings can be positioned to provide the best theoretical contact with the shaft, and then fixed in that position by the epoxy resin. Reference 41 relates experience with this installation technique and outlines the installation process.

may require two head tanks: one for full-draft operation and one for ballast operation. An alternative means of controlling the differential pressure between the lubricating oil and the ambient seawater at the seal is to use an automatic control system that senses the seawater pressure and adjusts the oil pressure accordingly, usually by imposing a variable air pressure on the head tank. The operating life of the shaft seals, which retain the oil in the stern tube and prevent the ingress of seawater, can be sensitive to the differential pressure between the seawater and the oil and to the shaft rubbing velocity. Although

Section 6
Propellers

6.1 Introduction. As discussed in Chapter 1, the selection of the propulsor, which converts engine torque to ship thrust, and the selection of the propulsion machinery are often closely related. Because of the interfaces between the propulsor, propulsion machinery, and hull, the design of the propeller is often a task of shared responsibility. A naval architect is usually responsible for the development of the hull lines and the propulsor hydrodynamic characteristics; but a marine engineer is generally responsible for the equipment and mechanical arrangements that properly interface with the naval architect's area of responsibility. This is the general basis upon which

the Society's two publications, *Principles of Naval Architecture* and *Marine Engineering* are presented.

Figure 3 of Chapter 1 is a comparison of the optimum efficiency values for a number of different types of propulsors. This information gives guidance concerning the relative merits of one propulsor versus another from an efficiency standpoint. However, in the preliminary design stage, more specific information is required to make the necessary trade-off studies required to support a design selection. Systematic model tests of propulsors provide the necessary information for the trade-off studies, and in many cases may be adequate for the final design [1].

6.2 Propulsor Types. As noted in Chapter 1, the type of propulsor to be used must be selected very early in the ship design process as the type of propulsor can have a strong impact on the design of the ship itself. The majority of ship propulsors are of the solid fixed-pitch propeller type. Nevertheless, there are a number of other types of propellers that may be more suitable in particular instances. A brief description of the mechanical aspects of the various types of propulsors is as follows:

Conventional fixed-pitch propellers. Most propellers are of the conventional fixed-pitch type and are made from single castings. Conventional fixed-pitch propellers usually have an efficiency, cost, and simplicity advantage over other types of propellers. The procedures used to develop the hydrodynamic design of fixed-pitch propellers and variants of fixed-pitch propellers, such as propellers in nozzles and controllable-pitch propellers, are outlined in reference 1.

Detachable-blade propellers. Detachable-blade (or built-up) propellers consist of a separately cast hub and blades. The blades are bolted to the hub to form the composite propeller. When operating conditions are such that there is a high probability of propeller blade damage, detachable-blade propellers offer the advantage that individual blades can be replaced. Also, some blade attachment designs have elongated bolt holes that offer the advantage of being able to make small modifications in blade pitch, which can be used to adjust the operating rpm. The disadvantages associated with detachable-blade propellers, as compared with propellers made from a single casting, are the greater first cost, greater complexity, and greater susceptibility to cavitation in the vicinity of the blade root because of the shorter, thicker sections necessitated by the restriction on blade bolting flange diameter.

Controllable- and reversible-pitch propellers. Controllable- and reversible-pitch propellers [or, more succinctly, controllable-pitch (CP) propellers] have a mechanism in the propeller hub that can be operated remotely to change the propeller pitch setting from a maximum design ahead pitch to a maximum design astern pitch. The pitch can be changed while the propeller rotates and develops thrust within these limits, or the pitch can be maintained at any intermediate setting for continuous operation. A CP propeller is advantageous in any of the following situations:

1. Where the operating conditions vary widely and maximum thrust is desired throughout these operating conditions, such as tug, trawler, and offshore supply boat applications.

2. Where shaft reversing capabilities are not readily provided by the main engines (e.g., gas turbines). The unidirectional rotation is also beneficial in applications that require highly skewed blades in that thicker sections from the 0.8 radius to the tip are not required on the trailing half of the blade chord.

3. Where extensive low-speed maneuvering is required for a diesel-powered vessel. The thrust can be varied continuously from ahead to astern, including zero thrust,

while operating in the minimum speed range of the diesel, thus avoiding slipping clutches or stopping and starting the main engine.

4. Where the ship will operate in ice-covered water. The unidirectional rotation of a CP propeller subjects the blades to less ice damage because the leading edges of the blades are thicker and stronger than the trailing edges and because continuous propeller rotation during thrust reversals minimizes fouling of propeller blades by blocks of ice, which sometimes impose loads on the propeller blade sections in their weakest direction.

5. Where an improved maneuverability and a minimum ship stopping distance are desired. An infinitely variable thrust capability in either direction and a more rapid response to thrust-reversal commands improve ship maneuverability and reduce the ship's headreach.

6. Where a constant shaft rpm is an advantage over a wide range of operating powers. By adjusting the propeller pitch, a constant rpm can be maintained over a range of ship speeds; this can be advantageous with shaft-driven generators. However, a propeller efficiency penalty is incurred.

At the propeller design point, the efficiency of a CP propeller approaches the efficiency of a fixed-pitch propeller. However, the larger hub normally prevents the efficiency of a CP propeller from exceeding that of a fixed-pitch propeller. Off the design point, the efficiency of a CP propeller is usually less than that of a fixed-pitch propeller designed for that operating condition. This is because all sections of a CP propeller blade are rotated through the same angle as the pitch is changed; thus, the angles of attack of the various blade sections along the propeller radii are optimum only at the design point.

The blade setting of a CP propeller is controlled by a hydraulic piston, a servo valve, an oil-distribution system, a pitch feedback indicator, and a pitch demand control loop, which is located within the ship and can be designed to respond to a variety of inputs.

Three basic arrangements of the power piston and servo valve are in general use: (1) those with the power piston and servo valve located in the hub, as in Fig. 27; (2) those with the power piston in the propeller hub and the servo valve inboard, as in Fig. 28; and (3) those with both the power piston and the servo valve located inboard, as illustrated by Figs. 29 and 30. In all cases the propeller hub houses the pitch-changing mechanism and structurally contains the torque, thrust, and centrifugal forces.

The pitch of the blades is changed by rotating the blades about their radial axis. For the configuration shown by Fig. 27, the pitch is changed with a Scotch-yoke mechanism, which is shown in more detail by Fig. 31. Link mechanisms, of the types illustrated by Figs. 28 and 29, are also commonly used. A simplified technique that can be used to analyze mechanisms of the trunnion-link type, as in Fig. 28, is presented in reference 42.

When the power piston is located in the propeller hub with the servo valve inboard, the motivating hydraulic oil is transmitted through a distribution assembly to piping that is inside the hollow propulsion shaft. The oil-distribution assembly can be located along the run of the line

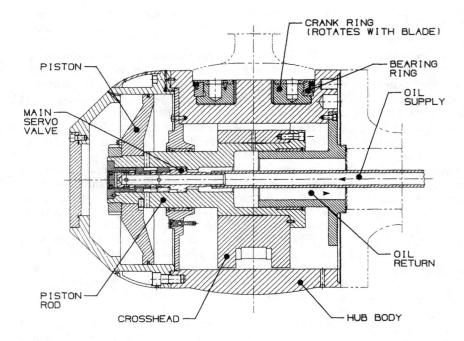

Fig. 27 CP propeller with power piston and servo valve in the hub

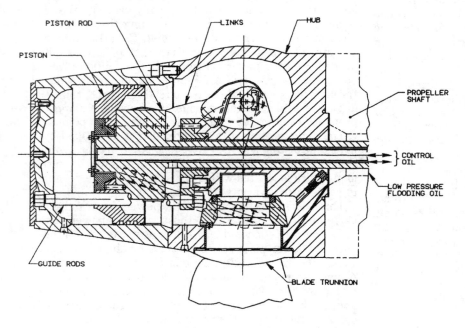

Fig. 28 CP propeller with power piston in hub and servo valve inboard

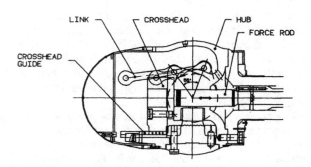

Fig. 29 CP propeller with power piston and servo valve inboard

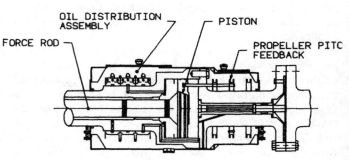

Fig. 30 Inboard pitch-control assembly

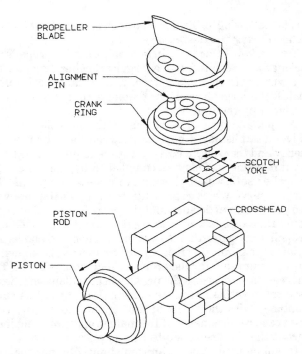

Fig. 31 Scotch-yoke blade pitch mechanism (see also Fig. 27)

shafting or it may be conveniently located at the forward end of the reduction gear slow-speed shaft, in arrangements that have geared drives.

For CP propeller arrangements that have the power pistons inboard, the force required to control the propeller pitch is developed by a pitch-control assembly, such as shown by Fig. 30, and is transmitted through a push/pull force rod that is in the hollow propulsion shafting.

The selection of the diameter, rpm, blade area, and blade thickness for a CP propeller proceeds basically the same as for a fixed-pitch propeller except that the regulatory bodies generally require that the stress in the propeller blade be calculated at the 0.35 radius, instead of the 0.25 radius, because of the inherently larger hub diameter. It is also noted that there may be an increased susceptibility to cavitation in the vicinity of the blade root of the CP propeller because of the shorter, thicker sections necessitated by the restriction of the propeller blade flange diameter.

Propellers in nozzles. Two types of arrangements fall into this category, namely, the pump jet and the Kort nozzle. In the pump jet arrangement the propeller is placed in a rather long nozzle with guide vanes either forward, aft, or in both positions relative to the propeller. A pump jet is normally considered where propeller noise is important. Because of the resistance of the nozzle and guide vanes, the overall efficiency of a pump jet arrangement is strongly dependent on particular circumstances.

Kort nozzle or ducted-propeller arrangements provide efficiency advantages in applications where the thrust loading is high; examples of such applications are tugs, trawlers, and large slow-speed ships (see Fig. 3 of Chapter I). A Kort nozzle arrangement consists of a propeller located in a nozzle of relatively short length; the length/

diameter ratio of the nozzle is in the range of 0.5 to 0.8. Kort nozzles are used extensively in connection with tugboats because the bollard pull and towing pull can be increased 30 to 40% as compared with a propeller operating alone without a nozzle. Nozzles can be of the accelerating or decelerating flow type depending on whether they accelerate or decelerate water flow through the nozzle. The accelerating nozzle also augments the forward thrust of the propeller and is used extensively in cases where the ship screw is heavily loaded [1]. Reference 43 includes a review of the design of CP propellers that operate in a duct and also outlines operational problems associated with this propulsor arrangement.

Tandem propellers. As the horsepower requirements for a ship increase, a single propeller can become inadequate because of restrictions on the propeller diameter, draft limitations, or excessive thrust loading. When this occurs, an increase in the number of propellers is required. Since a single shaft is desirable from an operating and design viewpoint, there could be justification for considering the installation of two propellers positioned in tandem on the same shaft.

Only small losses in propulsive efficiency (2.2%) were reported from model tests for a tandem arrangement as compared with a twin-screw arrangement for a large tanker [44]. The economy of a single propulsion plant, as opposed to two propulsion plants, in addition to the single-screw simplicity of the shafting arrangement are the advantages offered by tandem propellers.

Vane wheel. A vane wheel is a freely rotating propeller that is located abaft a powered propeller. The vane wheel may be supported on an extension of the powered-propeller shaft or it may be supported by the rudder support structure. The diameter of a vane wheel is larger than that of the powered propeller, and the blades of a vane wheel are designed so that the vane wheel develops torque from the powered-propeller race at the inner radii and simultaneously converts the torque into thrust by the outer radii of the vane wheel, which are outside the race of the powered propeller. Full-scale tests on a research vessel with this type of propulsor showed an improvement of 9% when compared with a conventional propeller. With optimization of the system, it was predicted that a 12% improvement could be achieved [1]. The improvement in the overall efficiency is attributed to a reduction in the rotational energy in the wake of the powered propeller. Vane wheels introduce complications in the outboard arrangement, however, and the increased clear diameter that is required in way of the propeller aperture may not be easily accommodated. The bearing-support arrangement for a vane wheel requires a comprehensive engineering analysis, as does the vibratory interaction effects introduced by a vane wheel.

Contrarotating propellers. Contrarotating propeller arrangements consist of two propellers positioned in tandem on coaxial shafts that rotate in opposite directions. Higher efficiencies can be achieved with this propeller arrangement because no rotational energy need be left in the propeller wake. Reference 44 reported a propulsion efficiency improvement of 6.7% for a 136,000-ton-displacement tanker with contrarotating propellers as compared with a conventional single-screw arrangement; similar tests for an 18,170-ton-displacement dry cargo ship indicated a 12% improvement [45].

Contrarotating propeller arrangements have not been used extensively in connection with commercial ships because of the mechanical complications involved with the coaxial propulsion system arrangement (see Chapter 9). Some naval installations have been made, but their performance has not been made public. Reference 46 outlines the design process and developments in the design theory for contrarotating propellers.

Fully cavitating propellers. The primary objection to propeller cavitation is the deleterious effect that it has on the propeller blade surfaces and overall propeller performance. Once the propeller loading conditions become such that cavitation can no longer be avoided, as may be the case with very fast ships, then rather than accept a limited amount of cavitation a more satisfactory choice is to design the propeller such that it cavitates fully. In this event, the cloud of vapor, which forms on the suction side of a blade, does not collapse until it is clear of the propeller blade, thus having no deleterious effect on the propeller blades. Operating at off-design conditions may result in severe propeller cavitation erosion, and such operations (accelerating, decelerating, etc.) cannot be entirely avoided in service. For this reason and to withstand the high stresses resulting from the large thrust load, fully cavitating propellers are frequently made of exotic, cavitation-resistant materials. Reference 1 gives the expected performance of a theoretical series of 3-bladed supercavitating propellers, which can be used to estimate their performance.

In order to achieve fully cavitating performance in a speed range too low for the usual fully cavitating propeller design, but still in the range where conventional propellers would cavitate excessively, ventilation may be considered. Ventilation is the term used to describe the introduction of air into the cavitation areas to produce a fully developed cavity. Reference 1 reports that there is little difference in efficiency between fully-cavitating and ventilated propellers once the cavity is formed. Experience with ventilated propellers is very limited, but some model testing has been carried out; reference 47 is a report of one such test.

6.3 Propeller Characteristics. The characteristics of a propeller have an important influence on the design of the shafting and bearing system. The propeller weight is carried by the propeller shaft, and propeller hydrodynamic loads are imposed on the shafting system. Standard propeller series data, such as reported in reference 1, are commonly used to develop a preliminary estimate of the basic propeller parameters and the related propeller-hull interactions. Propeller series data can be used, with a minimum of cost and time, to obtain results that are adequate for many purposes. To optimize a propeller design for a specific application, propeller calculations are often made using hydrodynamic lifting-surface theory [1].

The following propeller characteristics must be established during the design process:

Propeller diameter. In general, a higher propeller efficiency is associated with a larger propeller diameter and a lower shaft rpm. Therefore, it is usually desirable to install the largest propeller diameter that can be accommodated by the hull lines.

Propeller rpm. The choice of the propeller rpm involves establishing a balance between the propeller efficiency and the weight, cost, and space requirements of the main machinery. This is accomplished by using standard propeller series data to compute a series of points that form a curve which represents the relationship between the propulsive efficiency and the propeller speed over the rpm range of interest. These calculations are based on a propeller diameter that is selected as indicated above. The point of maximum efficiency on this curve is known as the optimum rpm for the propeller diameter selected. This curve is used to assess the penalty in propulsive efficiency associated with an increase in rpm. Data from this curve combined with the effect of the rpm on the weight, cost, and space requirements of the main propulsion machinery permit the final selection to be made.

It will be noted that at an rpm slightly higher than the optimum propeller rpm for a given propeller diameter, the propeller efficiency decreases only slightly. But on the other hand, the effect of a relatively small increase of propeller rpm (with the power remaining the same) on the weight, cost, and space requirements of the main machinery can be significant. In the case of higher-powered vessels, it is usual to select a propeller rpm that is higher than optimum and to accept some sacrifice of propeller efficiency to reduce the size of the propulsion machinery.

Number of blades. Propellers may have three, four, five, six, seven, or more blades. Over the years, the trend has been to use a larger number of blades; three blades fell into complete disuse for large ships during the 1940's, and the use of six- and seven-bladed propellers has become common. The major factor in the selection of the number of propeller blades is vibration considerations. Both the hull hydrodynamic pressure forces and the forces transmitted through the shafting system bearings are strongly influenced by the selection of the number of propeller blades. In general, the propeller exciting forces decrease rapidly with larger numbers of blades; however, there are exceptions. For more details concerning the relationship between the number of propeller blades and the vibratory forces generated, see Section 3.2. A prudent selection of the number of propeller blades is an important design criterion that can be used to avoid the excitation of natural frequencies in the propulsion system.

Propeller pitch. The selection of propeller pitch can be made when the power, ship speed, shaft rpm, propeller diameter, and general hull characteristics have been determined. The pitch ratio may be selected on the basis of standard propeller model series data. However, when a propeller is highly loaded or operates in a nonuniform wake field, it may be desirable to design a propeller with a pitch ratio and pitch distribution tailored to suit the particular operating conditions. This can be accomplished by designing a propeller in accordance with the procedures outlined in reference 1.

Blade skew. A propeller blade is termed skewed when its outline is asymmetrical with respect to a straight radial reference line in the plane of the propeller. Skew is usually introduced by successively displacing the blade sections away from the direction of rotation. Propeller blades with skew tend to enter and leave the regions of high wake more gradually. Model test results show that blade skew is an effective means of reducing the fluctuating forces and moments acting on a propeller. It is normal practice to skew propeller blades a moderate amount based on past experience, without specific knowledge regarding the benefits achieved; however, model tests may be conducted to evaluate the effects or an analytical evaluation can be made as discussed in reference 2.

Developed area. With heavily loaded propellers, which is commonly the case, the developed area must be established with care. Considerations in the selection of the propeller developed area are the penalty in efficiency associated with an excessive developed area and the effects of cavitation resulting from an inadequate developed area. The effects of an inadequate area can be of greater consequence than those of an excessive area; therefore, prudent practice dictates that a developed area be provided which is sufficiently large to incur a minimal cavitation hazard. For a more detailed discussion of propeller cavitation, and consequently developed area, see reference 1.

Propeller blade thickness. Requirements concerning the minimum allowable blade thickness are given in classification society rules, such as reference 15. A thorough discussion of the development of the classification society rules is given in reference 48, which also provides the basis for making an in-depth analysis of the propeller blade stress.

Propeller hub. The contour of the propeller hub outside diameter is shaped to maintain a smooth flow of water over the hub from the stern frame or strut barrel, and the hub length is largely controlled by the propeller blade fore-and-aft length at the interface with the hub. These parameters establish only the lower limits, and thicker hubs may be required to provide adequate strength. Excessively large propeller hubs are disadvantageous in that they increase the expense of the propeller and propeller weight (and consequently propeller shaft stress).

Propeller weight. An estimated propeller weight can be obtained in several ways. The most accurate is to calculate the weight based on detailed drawings. Unfortunately, however, the need for the propeller weight is well in advance of the time detailed drawings are available. There are a number of approaches that may be used to approximate propeller weights; one approach is given in reference 48. There are other methods such as

$$W = KD^3(MWR)(BTF) \qquad (13)$$

where

W = propeller weight (including hub), lb

K = material density factor, having a value of:
    0.26 for Mn-Brz
    0.25 for Ni-Mn-Brz
    0.235 for Ni-Al-Brz
    0.235 for Mn-Ni-Al-Brz
    0.225 for cast iron
    0.245 for stainless steel

D = propeller diameter, in.

MWR = mean width ratio
$$= \frac{\text{developed area per blade}}{D(\text{blade radius} - \text{hub radius})}$$

BTF = blade thickness fraction
$$= \frac{\text{maximum blade thickness extrapolated to shaft axis, in.}}{D}$$

Care must be exercised in the use of approximate methods because of considerations such as unusual hub dimensions and allowances for ice strengthening.

6.4 Manufacturing Tolerances. Two distinctly different methods are used to verify that propeller blades are in compliance with specified tolerances. The International Standards Organization Recommendation 484 delineates a system of tolerances that is based upon the use of a pitchometer [49]. A pitchometer is a device for determining the propeller blade pitch angle by measuring the distance from a reference plane to points on the blade pressure face. Propellers for merchant ships are almost exclusively manufactured and inspected using the pitchometer method. A description of the process used to measure propeller tolerances is outlined in reference 49, and four classes of tolerances are presented from which the one appropriate for a specific application can be selected. Table 9 is a summary of the four tolerance classes and the tolerances recommended by reference 49.

A second method to measure the accuracy of propeller blades is based upon the use of a series of sheet-metal template gages. This is the method required by the U.S. Navy [23]. Three types of sheet-metal template gages are used: (1) suction and pressure face cylindrical contour gages; (2) leading edge, trailing edge, and tip gages; and (3) fillet and hub or palm gages. The procedure used to design and manufacture the gages is outlined in reference 50. Generally, a minimum of 53 gages is required to inspect a propeller. The gage method provides considerably more insight and control over blade geometry than does the pitchometer method; however, there is also an increase in cost. The cost increase is related to: the gages, which must be manufactured for each propeller design; the increased number of measurements, with more than 2.5 times as many required; the data gathering process, which is not easily automated; and the fact that the process is subject to errors because of the number of gages and measurements required. But, for applications where propeller accuracy is of critical importance, as is the case with some naval ships, such a rigorous measurement procedure is a necessity.

A discussion and evaluation of the pitchometer and gage methods of assessing propeller blade accuracy are presented in reference 51. Included in reference 51 is a

Table 9   Summary of recommended propeller tolerances [48]

Class	Accuracy of Manufacturing	Normal Application
S	high precision	propellers of superior quality for special purposes
I and II	medium precision	for the majority of merchant vessels
III	large tolerances	for vessels without special characteristics, in general only applied to cast-iron propellers

	CLASS			
	S	I	II	III
PITCH				
Local pitch[a]	± 1.5%	± 2%	± 3%	
Mean pitch of each radius at each blade[a]	± 1%	± 1.5%	± 2%	± 5%
Mean pitch per blade[a]	± 0.75%	± 1%	± 1.5%	± 4%
Mean pitch of propeller[a]	± 0.5%	± 0.75%	± 1%	± 3%
RADIUS				
Deviation[b]	± 0.2%	± 0.3%	± 0.4%	± 0.5%
BLADE THICKNESS				
Upper deviation[c]	+2%	+2.5%	+4%	+6%
with a minimum of	+0.0787 in.	+0.0984 in.	+0.158 in.	+0.236 in.
Lower deviation[c]	−1%	−1.5%	−2%	−4%
with a minimum of	−0.0394 in.	−0.0591 in.	−0.0787 in.	−0.158 in.
BLADE (SECTION) LENGTH				
Tolerance[d]	± 1.5%	± 2%	± 3%	± 5%
with a minimum of	± 0.276 in.	± 0.394 in.	± 0.512 in.	± 0.591 in.
BLADE SKEW				
Tolerance[e]	± 3%	± 4%	± 6%	± 10%
LONGITUDINAL POSITION OF PROPELLER BLADE				
Deviation of the distance from the reference plane to the blade reference axis[f]	± 0.5%	± 1%	± 1.5%	± 3%
with a minimum of	± 0.197 in.	± 0.393 in.	± 0.591 in.	± 1.181 in.
ANGULAR DEVIATION				
Tolerance[g]	± 1 deg	± 1 deg	± 2 deg	± 2 deg
SURFACE FINISH[h]	125	250	500	1000
	hub to tip	0.3R to tip	0.4R to tip	0.5R to tip

[a] Expressed as a percentage of the design pitch at the corresponding radius. Tolerances for sections at 0.2, 0.3, and 0.4R should be increased by 50%.
[b] Expressed as a percentage of the radius of the propeller.
[c] Expressed as a percentage of the local blade thickness.
[d] Expressed as a percentage of the ratio of the propeller diameter divided by the number of propeller blades.
[e] Expressed as a percentage of the propeller diameter divided by number of propeller blades for arc length from reference line to leading edge.
[f] Expressed as a percentage of the propeller diameter. The same tolerances apply to the differences between readings on the same blade and to the differences between adjacent blades. Measurements are to be made at the 0.3, 0.6, and 0.95R.
[g] Deviation between adjacent blades.
[h] Maximum mean roughness height, microinches.

review of the adequacy of the two tolerance systems relative to their intended purpose and an assessment of the effect of a tolerance change on the manufacturing process and propeller performance. Reference 51 suggests that propeller manufacturing tolerances reflect the effort required for accomplishment and be related to performance requirements.

In addition to the tolerances governing the propeller physical dimensions, balance tolerances must also be specified. Ship's specifications usually require that propellers be balanced (with static or dynamic equipment) such that the static unbalanced force at rated rpm is no greater than 1% of the propeller weight. The following expression may be used to determine the static unbalance corresponding to an unbalanced force equal to 1% of the propeller weight:

$$U = 352W/N^2 \qquad (14)$$

where

U = static unbalance which will generate an alternating force equal to 1% of the propeller weight, in.-lb

W = propeller weight, lb
N = maximum rated propeller rpm

Limits are not generally placed on dynamic unbalance because of the large diameter-length ratio of propellers, but good practice dictates that corrections made for static unbalance be accomplished so as to improve the dynamic unbalance. Dynamic unbalance is generally not found to be a problem; nevertheless, the dynamic unbalance should be limited such that the alternating force generated at the aftermost bearing is no greater than an alternating force at the aftermost bearing corresponding to a static unbalance equal to 1% of the propeller weight. A useful expression for the maximum allowable dynamic unbalance under these conditions is

$$D = \frac{352\,W}{N^2}\left[\frac{L_1 + L_2}{1 + \dfrac{0.0000284\,WN^2 L_1 (L_1 + L_2)(0.5L_1 + L_2)}{3\,EI}}\right] \qquad (15)$$

where

D = dynamic unbalance that will produce the same

force at the aftermost bearing as a static unbalance equal to 1% of propeller weight applied at propeller center of gravity, in.-lb-in.

W = propeller weight, lb

L_1 = distance from propeller center of gravity to the aftermost bearing reaction, in.

L_2 = distance from aftermost bearing reaction to the reaction of next bearing forward, in.

N = maximum rated propeller rpm

E = shaft modulus of elasticity, psi

I = shaft rectangular moment of inertia, in.^{4}

Section 7
Torsional Vibration

7.1 General. Severe torsional vibration difficulties experienced with the early reciprocating-engine drives, particularly diesel engines, placed an emphasis on the importance of torsional vibration as a consideration in the design of shafting systems. Subsequently, the design methodology required to conduct a comprehensive torsional vibration analysis was formulated [52–61], and torsional vibration became established as a factor to be carefully considered in the design of all types of main propulsion shafting systems.

7.2 Modes of Torsional Vibration. The design of most large ships is such that one or more resonant modes of torsional vibration occur within the operating range.

With the possible exception of propulsion systems that use a power takeoff to drive a component having a large inertia through a torsionally flexible drive train, the first mode of torsional vibration in the shafting system will have the node and the highest torsional alternating stresses occur immediately aft of the propulsion engine. For vessels with fairly long runs of shafting, the first-mode frequency usually occurs at a low shaft rpm (less than 50% of rated rpm). This condition is normally not objectionable for a turbine-driven system because the propeller is the source of excitation, and the alternating torques developed are of low magnitude. In addition, the damping energy absorbed by the propeller is high enough to result in low vibration amplitudes and stresses. However, for diesel-engine drives, the engine also can excite the first mode and can cause shaft stresses that must be avoided on a continuous basis. In such cases, a barred speed range is often imposed. In the case of ships that have short runs of shafting, the first mode can occur sufficiently high in the operating range to become of significant concern. In these cases, the shafting system may require special design features to avoid deleterious torsional vibrations.

The second and third modes of torsional vibration are determined primarily by the characteristics of the prime mover. With geared-turbine drives, the turbine-gear system generally cannot be designed such that the second mode of torsional vibration is out of the operating range. Such being the case, a so-called "nodal drive" is frequently provided [52]. In a nodal drive, the turbine branches are designed to have equal frequencies, which forces the slow-speed gear to be a nodal point. The second mode of torsional vibration then consists of motion in which the two turbines vibrate so that their vibratory torques oppose

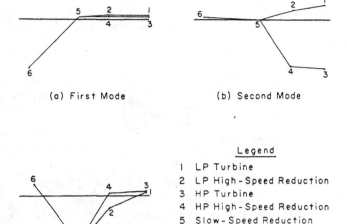

(a) First Mode   **(b) Second Mode**

Legend

1 L P Turbine
2 L P High - Speed Reduction
3 H P Turbine
4 H P High - Speed Reduction
5 Slow - Speed Reduction
6 Propeller

(c) Third Mode

Fig. 32 Mode shapes of first three modes of torsional vibration of a turbine-driven propulsion system of the nodal-drive type

each other with a nodal point at the gear. This being the case, the turbine branches cannot be excited by the propeller.

In the third mode of torsional vibration of a geared-turbine system, the vibratory torques of the propeller and turbines oppose that of the slow-speed gear. The third mode usually occurs considerably above the operating range; consequently, it is rarely of concern. However, very high propeller speeds or a large number of propeller blades can cause it to occur within the operating range. The third mode is difficult to excite because the antinode occurs at the slow-speed gear, which is not a source of excitation, and a node occurs near the propeller, which is a source of excitation. The mode shapes of the first three modes of torsional vibration for a geared-turbine propulsion system are shown by Fig. 32.

The first three modes of vibration of a slow-speed diesel propulsion system are shown by Fig. 33. The second and third modes of vibration have nodes in the engine crankshaft, but these modes can be excited by the higher orders of engine alternating torque. The strengths of the higher

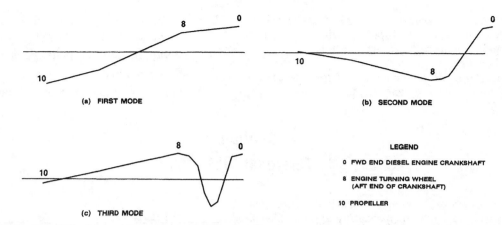

Fig. 33   Mode shapes of first three modes of torsional vibration of a direct-drive diesel propulsion system

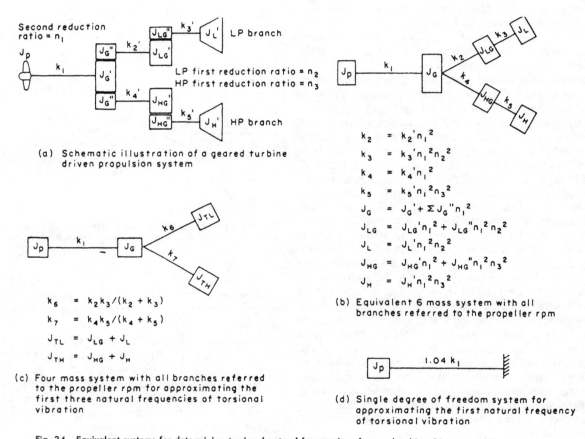

(a)   Schematic illustration of a geared turbine driven propulsion system

$k_2 = k_2' n_1^2$

$k_3 = k_3' n_1^2 n_2^2$

$k_4 = k_4' n_1^2$

$k_5 = k_5' n_1^2 n_3^2$

$J_G = J_G' + \Sigma J_G'' n_1^2$

$J_{LG} = J_{LG}' n_1^2 + J_{LG}'' n_1^2 n_2^2$

$J_L = J_L' n_1^2 n_2^2$

$J_{HG} = J_{HG}' n_1^2 + J_{HG}'' n_1^2 n_3^2$

$J_H = J_H' n_1^2 n_3^2$

(b)   Equivalent 6 mass system with all branches referred to the propeller rpm

$k_6 = k_2 k_3/(k_2 + k_3)$

$k_7 = k_4 k_5/(k_4 + k_5)$

$J_{TL} = J_{LG} + J_L$

$J_{TH} = J_{HG} + J_H$

(c)   Four mass system with all branches referred to the propeller rpm for approximating the first three natural frequencies of torsional vibration

(d)   Single degree of freedom system for approximating the first natural frequency of torsional vibration

Fig. 34   Equivalent systems for determining torsional natural frequencies of geared turbine-driven propulsion systems.

of engine alternating torque. The strengths of the higher orders of engine excitation are generally of smaller magnitude; however, some of the higher orders are potentially hazardous and must be evaluated.

7.3   Models for Torsional Vibration Analyses.

a.   **Geared-turbine drives.** A typical steam turbine propulsion system is schematically illustrated by Fig. 34(a). To simplify the analytical procedure, the system can be reduced to an equivalent model in which all elements

are referred to the same rotational speed, as illustrated by Fig. 34(b). Figure 34(b) can be used to evaluate all modes of torsional vibration that would be expected to be of practical interest. However, if only the first three modes of vibration are of interest, which would generally be the case, the model shown in Fig. 34(b) can be further simplified to that shown in Fig. 34(c) without a serious loss of accuracy because the equivalent inertias of the turbines and stiffnesses of the turbine shafts are very high compared with those of the first-reduction gear elements.

If only the first mode of torsional vibration is of interest, then it can be approximated by directly adding the equivalent inertias of the turbine branches, J_{TL} and J_{TH} in Fig. 34(c), to the slow-speed gear inertia, J_G, and making an analysis based on a two-mass system. When the inertias of the turbines and gears are not known, the nodal point in the first mode of vibration is often assumed to be aft of the slow-speed gear, a distance that is equal to 4% of the length between the slow-speed gear and the propeller. With such an assumption, the first-mode natural frequency can be simply determined by considering the system to be a one-degree-of-freedom model as shown in Fig. 34(d).

All of the system parameters needed to evaluate the torsional natural frequencies can be directly determined from the physical properties of the system except for the propeller entrained water. An assessment of the propeller entrained water can be made from the work of Burrill and Robson [58] or Parsons and Vorus [62].

To avoid the tedious labor associated with calculating the moment of inertia of the propeller in air, the propeller radius of gyration is often estimated to be between 0.40 and 0.44 of the propeller radius (the lower end of the range corresponds to larger propeller hubs and smaller numbers of blades).

b. Diesel drives. The models used to analyze the torsional vibration characteristics of diesel-driven propulsion systems are similar to those used for geared-turbine systems. If the prime mover is a medium- or high-speed diesel(s) driving through a reduction gear, the model is developed with the inertias and stiffnesses referred to the propeller rotative speed as discussed for a geared-turbine model [60].

A model representation of a typical slow-speed, direct-drive diesel propulsion system is shown by Fig. 35. The inertias of the propeller and entrained water are determined the same as for geared-turbine drives. The inertias of rotating masses and the spring constants of the shafts can be determined in a straightforward manner; however, "effective" inertias are used to represent the inertias of the pistons and connecting rods.

By inspecting Figs. 1, 2, or 3 of Chapter 3, it is evident that the reciprocating masses of the pistons transmit intermittent inertia loads into the crankpin. When a piston is at either the top- or bottom-dead-center position, there is no piston motion; however, at crank angles halfway between the top- and bottom-dead-center positions, the reciprocating parts move at approximately crank-pin velocity. When located at the crankpin, the "effective" inertia of the reciprocating mass is one-half the total reciprocating mass [59,60], that is:

$$J_r = \frac{W_r R^2}{2g} \qquad (16)$$

where

J_r = reciprocating mass effective inertia to be located at crankpin, in.-lb-sec^2

W_r = equivalent weight of reciprocating parts, lb
R = one-half of piston stroke, in.
g = gravitational constant, in./sec^2

W_r includes the total weight of purely reciprocating masses, such as the piston and crosshead; however, the weight of a connecting rod, which has one end attached to the crankpin and the other end attached to the reciprocating element, is broken down into two components. One component is placed at the crankpin and is added directly to the rotary weight of the crankpin, and the other component is placed at the reciprocating end of the attachment and is added to W_r in equation (16). The two weights representing the connecting rod are distributed so that the total connecting rod weight and its center of gravity are accurately represented [59,60].

The inertia properties of rotating masses attached to the crankshaft or shafting, such as flywheels and moment compensators, can be directly determined from equipment drawings. Also, such data are often available from the engine manufacturer.

7.4 Determination of Natural Frequencies. The Holzer method of computing the natural frequencies of lumped spring-mass systems is a convenient procedure for determining the torsional natural frequencies of propulsion systems, whether turbine or diesel powered. To illustrate the computational procedures used, a turbine-driven vessel, modeled as shown by Fig. 34(b), and a direct-drive diesel, modeled as indicated by Fig. 35, will be analyzed.

a. Geared-turbine drives. Referring to Fig. 34(b), typical values of system inertias and spring constants and calculations for the first torsional natural frequency of geared-turbine drives are given in Table 10. Since the principal source of excitation is the propeller at blade-rate frequency, following conventional practice the calculation is initiated by first assuming the resonant frequency of the system, in terms of the propeller rpm, and then calculating the corresponding vibratory torque and torsional amplitude at the slow-speed gear (inertia J_G) in terms of the amplitude at the terminal end of each branch. For convenience, the amplitudes at the terminal ends of the three branches are initially assumed to be one radian. Since the three branches (propeller, LP turbine, and HP turbine) must have the same amplitude at the slow-speed gear, the amplitudes of the three branches can be expressed as a function of the same unknown amplitude—for instance, the propeller—thereby obtaining the mode shape. The torques imposed on the slow-speed gear are then summed; if the sum is zero, a resonant condition is established. If the sum is not zero, the process is iterated until the sum is zero by assuming a different resonant frequency.

The same procedure can be repeated to determine the remaining four torsional natural frequencies but, as previously mentioned, only the first three modes would generally be of interest. The mode shapes of the first three natural torsional frequencies are shown in Fig. 32. The node in the first mode is seen to be immediately abaft the slow-speed gear. The two turbine branches are tuned in the second mode such that the slow-speed gear is a nodal

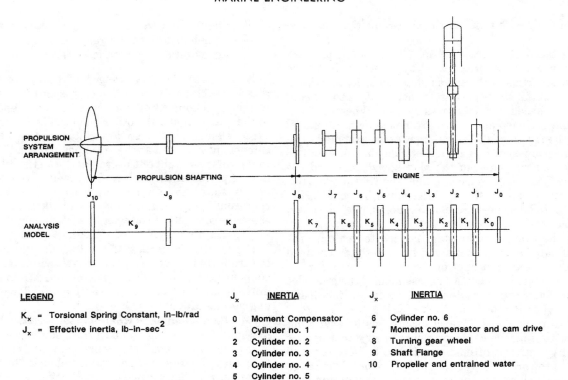

LEGEND

K_x = Torsional Spring Constant, in–lb/rad

J_x = Effective inertia, lb–in–sec^2

J_x	INERTIA	J_x	INERTIA
0	Moment Compensator	6	Cylinder no. 6
1	Cylinder no. 1	7	Moment compensator and cam drive
2	Cylinder no. 2	8	Turning gear wheel
3	Cylinder no. 3	9	Shaft Flange
4	Cylinder no. 4	10	Propeller and entrained water
5	Cylinder no. 5		

Fig. 35   Torsional vibration analysis model for a six-cylinder direct-drive diesel engine

Table 10   Determination of first natural mode of torsional vibration for a turbine-driven propulsion system modeled as shown by Fig. 34(b)

		Number of propeller blades = 6 J = in.-lb-sec^2 ÷ 10^6 k = in.-lb/rad ÷ 10^6		Assumed propeller speed at resonance = 22.38 rpm ω = 6(22.38) (2π)/60 rad/sec			
	J	$J\omega^2$	θ	$J\omega^2\theta$	$\sum J\omega^2\theta$	k	$\sum J\omega^2\theta/k$
J_L	5.640	1115	1.0000 θ_L	1115 θ_L	1115 θ_L	12,620	0.0884 θ_L
J_{LG}	0.5412	107.0	0.9116 θ_L	97.54 θ_L	1213 θ_L	4,083	0.2971 θ_L
J_G	0.4038	79.83	0.6145 θ_L				
J_H	0.7268	143.7	1.0000 θ_H	143.7 θ_H	143.7 θ_H	30,230	0.0048 θ_H
J_{HG}	0.9013	178.2	0.9952 θ_H	177.3 θ_H	321.0 θ_H	856	0.3750 θ_H
J_G	0.4038	79.83	0.6202 θ_H				
J_P	0.5159	120.0	1.0000 θ_P	102.0 θ_P	102.0 θ_P	98.1	1.0398 θ_P
J_G	0.4038	79.83	−0.0398 θ_P	−3.177 θ_P	98.8 θ_P		

NOTE THAT:

$\theta_L = (-0.0398/0.6145) \theta_P = -0.06477 \theta_P$

$\theta_H = (-0.0398/0.6202) \theta_P = -0.06417 \theta_P$

$\sum Q = 1213 \theta_L + 321.0 \theta_H + 98.8 \theta_P$

$\sum Q = -78.57 \theta_P - 20.60 \theta_P + 98.8 \theta_P$

$\sum Q \approx 0$; therefore a resonant condition

point. The third mode is the one in which the slow-speed gear is the antinode with the terminal ends of the three branches being near nodal points.

b. Diesel drives. The characteristics of a typical two-stroke, six-cylinder diesel engine are given in Table 11. A model that can be used to conduct a torsional vibration analysis of that engine is illustrated by Fig. 35. A calculation for the first mode of vibration, using the Holzer method, is given in Table 12. Unlike the case of geared-turbine drives, the calculation is usually begun by assuming the resonant frequency of the system in terms of radians per second or cpm. Next, an amplitude of one radian is assigned to the mass at the forward end of the

crankshaft, and subsequent calculations are based on the reference mass amplitude, θ_0. The $J_x\omega^2\theta_x$ column represents the inertia torque of each mass relative to the reference mass amplitude.

The $\sum J_x\omega^2\theta_x$ column represents the sum of the inertia torques of all preceding masses, and is the torque in the shaft immediately following the mass under consideration. Thus, when this torque is divided by the corresponding shaft stiffness, the incremental torsional deflection between masses, in radians, is determined. The incremental torsional deflection is subtracted from the amplitude of the mass under consideration to obtain the angular amplitude of the next mass. A system resonant frequency

Table 11   Characteristics of a typical two-cycle, six-cylinder diesel engine

Indicated horsepower	IHP	hp	18,380
Mean indicated pressure	P_m	psi	261
		bars	18
Brake horsepower	BHP	hp	17,359
Brake mean effective pressure	BMEP	psi	246.5
		bars	17
Engine rated rpm	N	rpm	74
Engine rated torque	Q_r	in.-lb	14,784,000
Piston cylinder bore	B	in.	27.56
Piston area	A	in.^{2}	596.6
Engine stroke	L	in.	105.28
Crank pin turning radius	R	in.	52.64
Connecting rod length	L_c	in.	120.7
Equivalent weight of reciprocating parts per cylinder (piston, X-head, con. rod)	W_r	lb	19,464
Effective inertia of reciprocating parts [see eq. (16)]	J_r	lb-in.-sec^2	69,791
Crankshaft rotational inertia/cyl	J_0	lb-in.-sec^2	149,575
Total rotational inertia per cylinder, $J_r + J_0$	J_c	lb-in.-sec^2	219,366
Crankshaft: outside dia.	D	in.	30.87
inside dia.	d	in.	5.91
Engine firing order			1-5-3-4-2-6

Table 12   Calculation of first mode of torsional vibration of diesel-driven propulsion system modeled as shown by Fig. 35

J_x = in.-lb-sec^2 ÷ 10^6
k_x = in.-lb/rad ÷ 10^6

Assumed system natural frequency:
ω = 25.64 rad/sec (244.8 cpm)

Mass (Spring) Number, x	J_x	$J_x\omega^2$	θ_x	$J_x\omega^2\theta_x$	$\sum J_x\omega^2\theta_x$	k_x	$\dfrac{\sum J_x\omega^2\theta_x}{k_x}$	α_x	$\frac{1}{2}\alpha_x J_x\omega^2\theta_x^2$
0	0.038	24.98	1.0000	24.98	24.98	20964	0.0012	0.02	0.250
1	0.219	143.97	0.9988	143.80	168.78	15924	0.0106	0.02	1.436
2	0.219	143.97	0.9882	142.27	311.05	15924	0.0195	0.02	1.406
3	0.219	143.97	0.9687	139.46	451.51	15924	0.0284	0.02	1.351
4	0.219	143.97	0.9403	135.37	586.88	15924	0.0369	0.02	1.273
5	0.219	143.97	0.9034	130.06	716.94	15924	0.0450	0.02	1.175
6	0.219	143.97	0.8584	123.58	840.52	20790	0.0404	0.02	1.061
7	0.088	57.85	0.8180	47.32	887.84	28090	0.0316	0.02	0.387
8	0.069	45.36	0.7864	35.67	923.51	644	1.4340	0.01	0.140
9	0.008	5.26	−0.6476	−3.41	920.10	1613	0.5704	0.01	0.011
10	1.149	755.36	−1.2180	−920.03	0.07				
							$\sum \frac{1}{2}\alpha_x J_x\omega^2\theta_x^2 =$		8.490θ_0^2

is established when the sum of the oscillating torques developed by all masses, the $\sum J_x\omega^2\theta_x$ column, is zero; that is, when no external torque is required to sustain the vibration amplitudes. The same procedure is used to establish the higher modes of system vibration [59,60].

7.5 Excitation Factors. Torsional vibrations in propulsion shafting systems are principally excited by the propeller and, in the case of diesel-driven ships, the diesel. Reduction gears are manufactured with such accuracy that gear-tooth excitation is of negligible magnitude. The strength of the propeller alternating torque is influenced by a number of factors, including propeller loading, propeller aperture clearances, appendages that influence the flow of water to the propeller, number of propeller blades, hull lines, and hull draft. The propeller torque variation is often expressed as rQ, where r is the alternating torque as a fraction of the mean torque, Q. Generalization in this area must be used with care; however, typical ranges of blade-rate torque excitation for normal ship proportions are presented in Table 2. Propeller excitation of a frequency higher than blade rate (propeller rpm times the number of propeller blades) exists and occurs at multiples of the blade-rate order, but the higher orders are generally of negligible magnitude. For additional details concerning torque excitation, see references 3 and 56.

With the propeller excitation torque expressed as rQ, the energy input, E_{ip}, from the propeller per cycle of vibration is [59]:

$$E_{ip} = \pi\, rQ\theta_p \qquad (17)$$

where rQ is the alternating torque in inch-pounds and θ_p is the propeller amplitude of vibration in radians.

For diesel-driven ships, in addition to the propeller, the diesel is a source of torsional vibration excitation. The varying piston gas pressure and the inertia loads due to the cylinder reciprocating masses cause the diesel to produce a periodically varying torque that is related to crankshaft rotation.

The periodic torque applied to the crankshaft of a diesel engine by the piston gas pressure can be analyzed in terms of the force per cylinder acting on the crankpin in a direction tangent to the crankpin turning radius. This periodic torque is commonly expressed as the tangential effort. A diagram can be drawn that traces the tangential effort acting on the crankpin, which is caused by the gas pressure (and is commonly stated per square inch of piston area), through the working cycle of one cylinder. For a two-cycle engine, the working cycle is one revolution of the crankshaft, but for a four-cycle engine, the process repeats every two revolutions.

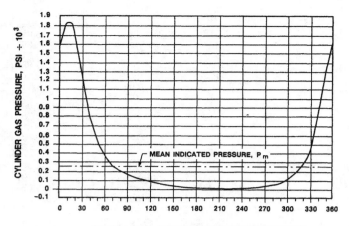

Fig. 36   Full-power indicator card for the engine described in Table 11

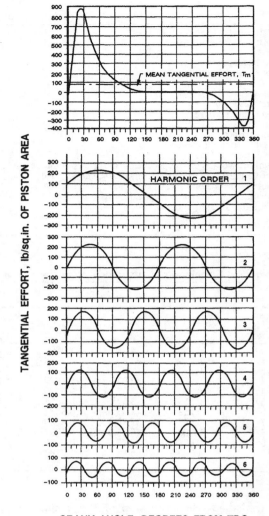

CRANK ANGLE, DEGREES FROM TDC

Fig. 37   Full-power tangential-effort diagram (with harmonic components) developed from the indicator card shown by Fig. 36

A tangential-effort diagram can be developed from a cylinder gas pressure indicator card where the cylinder pressure is related to the piston crankpin angle. Figure 36 is a cylinder indicator card for an engine with the characteristics outlined by Table 11; the mean indicated pressure is 261 psi (18 bars).

The cylinder tangential effort at any crankpin position is related to the cylinder gas pressure by a factor. The factor is a function of the ratio of the length of the connecting rod to the crank radius and the angular position of the crankpin relative to the top-dead-center position [61]. Figure 37 is a tangential-effort diagram that was developed from the indicator card, illustrated by Fig. 36. The relationship between the mean indicated pressure, P_m, and the mean tangential effort, T_m, shown in Figs. 36 and 37, respectively, is

$$IHP = \frac{P_m LAN_w}{396,000} = \frac{2\pi T_m ARN}{396,000} \qquad (18)$$

where

$$T_m = \frac{P_m}{\pi}\left(\frac{N_w}{N}\right)$$

P_m = mean indicated pressure, psi
L = stroke of piston, in.
R = crank radius = $L/2$, in.
A = area of piston, in.^{2}
N = crankshaft speed
N_w = number of working cycles per minute: $N/2$ for a 4-cycle engine; N for a 2-cycle engine

A tangential-effort diagram caused by the cylinder gas pressure, TE_g, can be represented by a Fourier series consisting of a constant term and a series of harmonically varying terms. The constant term is the mean tangential effort, T_m, and does not excite torsional vibration. The harmonically varying terms, however, are the principal source of torsional vibration and do not contribute to the useful work output of the engine.

A Fourier series representing the gas-pressure tangential-effort diagram can be written as:

$$TE_g = T_m + \Sigma (A_{gn}\sin n\theta + B_{gn}\cos n\theta) \qquad (19)$$

where A_{gn} and B_{gn} are the Fourier coefficients, θ is the crankshaft angular position from top dead center, and n is the order number of the harmonic component. The order numbers of the harmonics relate to the number of complete cycles that each respective harmonic completes per revolution of the crankshaft. Consequently, the engine speed (in rpm) for resonant harmonic excitation of any system natural frequency is determined by dividing the system natural frequency (in cpm) by the harmonic order number, n, of the engine excitation. For a two-cycle engine, where the tangential-effort curve is periodic every crankshaft revolution, there are only integer orders of the Fourier components. For a four-cycle engine, where the tangential-effort curve is periodic over two crankshaft revolutions, there are Fourier harmonic components of $\frac{1}{2}$, 1, $1\frac{1}{2}$, etc. orders.

Table 13 Diesel engine harmonic gas-pressure coefficients

Order No.	Data Item	Engine Mean Indicated Pressure, bars (1 bar = 14.5 psi)						
		6	8	10	12	14	16	18
1	TE_{g1}	78.25	97.17	119.74	142.89	167.76	193.09	223.61
	γ_{g1}	26.56	28.62	28.89	28.88	28.40	27.94	26.03
2	TE_{g2}	83.97	100.68	122.89	145.96	171.58	197.83	230.48
	γ_{g2}	2.93	3.18	3.05	2.83	2.35	1.91	−0.14
3	TE_{g3}	64.81	76.64	93.22	110.50	129.76	149.50	172.29
	γ_{g3}	−5.45	−6.20	−6.54	−6.89	−7.43	−7.91	−10.32
4	TE_{g4}	47.30	56.08	68.28	80.92	94.82	109.01	123.37
	γ_{g4}	−10.50	−11.98	−12.66	−13.26	−14.02	−14.70	−17.50
5	TE_{g5}	33.27	39.54	48.00	56.71	66.11	75.63	83.39
	γ_{g5}	−19.24	−21.82	−22.80	−23.63	−24.60	−25.46	−28.93
6	TE_{g6}	21.86	25.89	31.40	37.02	42.89	48.75	51.32
	γ_{g6}	−24.96	−28.42	−29.60	−30.62	−31.88	−33.02	−37.17

NOTES:

1. Amplitude coefficients, TE_{gn}, and phase angles, γ_{gn}, define the harmonic gas-pressure components of the tangential effort, equation (20), for the engine described by Table 11.

2. The above coefficient data items can be interpolated for intermediate pressure values. For example, at a MIP of 15 bars, the third-order harmonic component of the tangential effort caused by gas pressure, equation (20), is:

$$TE_{gn} \sin(n\theta + \gamma_{gn}) = 139.63 \sin(3\theta - 7.67)$$

Equation (19) can also be written as:

$$TE_g = T_m + \Sigma\, TE_{gn} \sin(n\theta + \gamma_{gn}) \qquad (20)$$

where

$$TE_{gn} = \sqrt{A_{gn}^2 + B_{gn}^2}$$

$$\gamma_{gn} = \tan^{-1} \frac{B_{gn}}{A_{gn}}$$

and the remaining symbols are as defined above.

The procedure used to derive the Fourier series harmonic components of the tangential-effort diagram, equation (19), is well known [59,61]. Six of the Fourier harmonic components are plotted in Fig. 37 for the tangential-effort diagram shown on the same figure. It can be seen that the six harmonic orders essentially sum to the original tangential-effort diagram with the axis being at the mean tangential effort, T_m. Normally 12 to 24 harmonic orders are used to represent the tangential-effort diagram. The engine manufacturer can often provide data for the terms TE_{gn} and γ_{gn} in equation (20) as a function of n, the harmonic order, and P_m, the mean indicated pressure. Table 13 is a portion of such data that maps the tangential-effort harmonic components for the two-cycle engine described by Table 11. The six harmonic orders shown in Fig. 37 correspond to the data given in Table 13 for a mean indicated pressure of 18 bars.

The second component of the alternating torque produced by a diesel engine is the inertia loads imposed by the reciprocating mass of each piston assembly. A tangential-effort diagram for inertia loads and their harmonic composition, can be determined analytically [54,59,61]. Normally, only the first four orders of the inertia loads have a significant magnitude. Both two- and four-cycle engines contain integer orders only, and both are represented by a sine term only; there are no half-orders and no cosine terms. The expression for the alternating inertia load is

$$TE_{in} = 0.0000284 \frac{W_r R N^2}{A} H_n \sin(n\theta) \qquad (21)$$

where

TE_{in} = tangential crankshaft force, per square inch of piston area, caused by piston assembly, acting at the crankpin turning radius, psi

H_n = harmonic order coefficient

$H_1 = 0.25/r$ $\qquad$ $H_3 = -0.75/r$

$H_2 = -0.5$ $\qquad$ $H_4 = -0.25/r^2$

r = connecting rod length/R

A, N, R, W_r, n, and θ are as defined for equations (16), (18), and (19).

To determine the total alternating torque for the first four orders, the gas-pressure and inertia tangential-effort harmonics must be combined. Above the fourth order, the gas pressure harmonics are predominant, and the inertia components can be neglected. To combine the gas and inertia tangential-effort components, the gas harmonic component must be stated in sine and cosine components, as in equation (19). The amplitudes of the sine components caused by the inertia forces, TE_{in}, and gas-pressure loads, A_{gn}, are then added to obtain the total amplitude, A_{gin}, of the sine term. The total sine component can then be combined with the cosine term of the gas-pressure component, B_{gn}, to obtain the resultant combined amplitude, TE_{gin}; and the phase angle, γ_{gin}, can be calculated from equation (20).

The procedure used to combine the harmonic tangential-efforts caused by gas pressure and inertia forces is illustrated by Table 14. It may be noted that in the example given the sine terms of the inertia components oppose those of the gas-pressure components such that the combined gas-inertia amplitude, TE_{gin}, is less than that of the gas pressure, TE_{gn}, alone.

For the first four harmonic orders, the maximum alternating torque per cylinder at order n, Q_n, is determined

Table 14   Procedure for combining gas and inertia tangential-effort harmonic components

Engine full power:						
mean indicated pressure	P_m	bars	18			
brake effective pressure	BMEP	bars	17			
friction mean effective pressure $= P_m -$ BMEP	FMEP	bars	1			
rotational speed	N	rpm	74			
first-mode resonant frequency	f_1	cpm	244.8			
Harmonic order	n	...	3^a	4	5	6
Resonant engine rpm $= f_1/n$	N_c	rpm	81.6^b	61.2	49.0	40.8
Engine mean indicated pressure at N_c	P_{mc}	bars	18^c	12.6^d	8.5	6.2
Maximum amplitude of gas-pressure tangential effort	TE_{gn}	psi	172.29^e	85.09	41.66	22.26
Phase angle of gas-pressure tangential effort harmonic	γ_{gn}	deg	-10.32^e	-13.49	-22.07	-25.31
Sine term:						
gas-pressure amplitude	A_{gn}	psi	169.50^f	82.74		
inertia amplitude	TE_{in}	psi	-106.23^g	-8.68		
total $= A_{gn} + TE_{in}$	A_{gin}	psi	63.27	74.06		
Cosine term:						
gas-pressure amplitude	B_{gn}	psi	-30.86^h	-19.85		
Combined amplitude	TE_{gin}	psi	70.39^i	76.67		
Combined amplitude phase angle	γ_{gin}	deg	26.00^j	15.00		

^{a} The first two harmonic orders are more than 40% above the engine rated speed and are of no concern.

^{b} All harmonic exciting frequencies below 140% of rated engine speed should be evaluated. The major-order critical frequency should occur at least 40% above the maximum continuous rated rpm for arrangements with "under critical" operations.

^{c} The engine mean indicated pressure is often assumed to remain constant, at the maximum rating, above rated speed.

^{d} Specific plant data should be used. When unavailable, as an approximation:

$$P_{mc} = \text{BMEP}(N_c/N)^2 + \text{FMEP}$$

^{e} From Table 13.

^{f} From equation (20):

$$A_{gn} = \sqrt{\frac{TE_{gn}^2}{1 + \tan^2 \gamma_{gn}}}$$

^{g} From equation (21), with data from Table 11.

^{h} From equation (20), $B_{gn} = A_{gn} \tan\gamma_{gn}$

^{i} $TE_{gin} = \sqrt{A_{gin}^2 + B_{gn}^2}$

^{j} $\gamma_{gin} = \tan^{-1}(B_{gin}/A_{gin})$

from the combined gas and inertia tangential-effort maximum amplitude, TE_{gin}, as follows:

$$Q_n = AR\,(TE_{gin}) \tag{22}$$

For the fifth and higher harmonic orders, only the gas-pressure forces need be considered; therefore

$$Q_n = AR\,(TE_{gn}) \tag{23}$$

Equations (22) and (23) are used to determine the alternating torques per cylinder, which can be summed to determine the combined effect of all cylinders. In a linear elastic system, the motions produced by two or more sets of periodically varying forces acting simultaneously are equal to the sum of the motions that would be produced by the separate forces acting alone, recognizing the phase relationships between the different components. Following this principle, each torque harmonic order induces in the system a forced torsional vibration of its respective frequency; consequently, the motion of the shaft is the summation of as many harmonics as are present in the applied torque. However, in general, only when the frequency of a harmonic order coincides with a natural frequency in the shafting system is the amplitude of the vibration response significant.

The work, W_{cn}, done by the cth cylinder of an engine at order n can be expressed as [59]:

$$W_{cn} = \pi\,Q_n\,\theta_c\,\sin\beta_n \tag{24}$$

where

Q_n = harmonic torque of a cylinder at order n, from equation (22) or (23)

θ_c = amplitude of vibration of cth cylinder for mode of vibration being analyzed

β_n = phase angle between cylinder harmonic torque vector, Q_n, and amplitude vector, θ_c

Q_n has the same magnitude for all cylinders since all cylinders nominally fire equally; however, the phases are different because all cylinders do not fire at the same time. At resonant vibration conditions, the amplitude vectors, θ_c, of all cylinders are in phase since all cylinders vibrate at the same frequency and go through zero amplitude at the same instant. However, the magnitude of the θ_c vector is different from cylinder to cylinder, depending on the normal elastic curve of vibration for the mode being analyzed. The frequency of the two vector rotations Q_n and θ_c is n times the crankshaft rotation and equals a natural vibration mode of the system.

To obtain the work of all engine cylinders at a given resonant mode of vibration, it is necessary to determine the resultant of all the θ_c amplitudes and the phase angle between that resultant and the Q_n vector. At resonance, the resultant phase angle is 90 deg, which makes the input work a maximum (i.e., $\sin\beta = 1$); therefore, the total

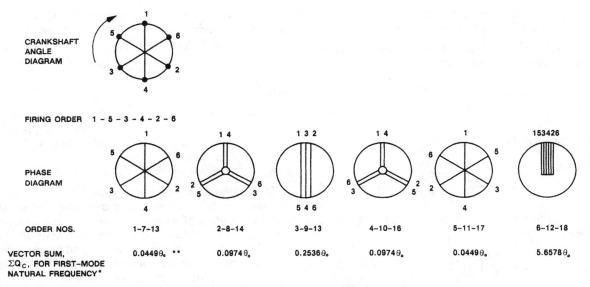

Fig. 38 Cylinder phase diagrams versus excitation order and first-mode vector sums for the propulsion system described by Tables 11 and 12

engine input work per cycle, E_{ie}, at a resonant frequency of order n is [59]:

$$E_{ie} = \pi Q_n \Sigma \theta_c \qquad (25)$$

where $\Sigma \theta_c$ is the vector sum obtained by adding vectors with the phases of the cylinder torques, Q_n, and the magnitudes of the cylinder vibration displacements, θ_c.

Figure 38 shows the crankshaft angle diagram and firing order, as well as the phase diagrams for 18 orders of excitation, for the two-cycle engine described by Table 11. The phase angles between the θ_c vectors depend upon the crank arrangement, its firing sequence, and the harmonic order number. The phase diagram is developed by assuming that the number 1 cylinder is at top dead center at its firing position; this is the zero phase angle. The phase angle of the θ_c vector corresponding to the cth cylinder is found by multiplying the number of degrees the crank must be rotated to fire that cylinder by the order number being investigated. As an example, the fifth-order location of the number three cylinder is found by observing that the number three cylinder must rotate 120 deg to the TDC firing position. The fifth-order vibration vector would rotate 5×120 or 600 deg during this period. Six hundred degrees from the TDC firing position measured counter to shaft rotation locates the number three cylinder vector at 240 deg counterclockwise from the number one cylinder firing position, as shown in Fig. 38. Note that with two-cycle engines the angle of rotation to bring a cylinder to firing position is the same as the crank angle rotation to bring that cylinder to the TDC position. With a four-cycle engine, the angle to fire may be the crank angle plus 360 deg.

As illustrated by Fig. 38, a phase diagram can be constructed that defines the phase relationships between the cylinders as a function of order. A length is assigned to each vector that is equal to the amplitude of vibration of that mass, which is taken from the mode shape of the mode of vibration being analyzed. The six cylinder amplitudes, θ_1 through θ_6, are applied to the vector lengths in the phase diagrams and summed. As an example, for the major order 6, the phase diagram shows all six vectors are in phase; therefore, the vector sum is obtained by adding the vector amplitudes. For the first-mode frequency, the sum is $5.6578 \theta_0$.

If all cylinders do not fire equally, as would be the case if the injection of fuel was not uniform for all cylinders, then the Q_n value would not be the same for all cylinders. The effects of this condition can be analyzed by determining the Fourier components of the tangential effort produced by the misfiring cylinders. When the Fourier components for all cylinders are known, the vector sum of the input work per cycle from all cylinders can be calculated from equation (25).

7.6 Damping. There are several sources of damping that control the maximum attainable amplitude of torsional vibration; one of the most important is the propeller, particularly in the first mode. For second and higher modes of vibration, damping within the system caused by elastic hysteresis, sliding fits, and friction elements can be important if these modes of vibration are excited in the operating range of the system. This is particularly true for diesel-driven ships.

Propeller damping can be determined in several ways [53,55,63]. In general, the propeller damping coefficient can be expressed as

$$b = KQ/\Omega \qquad (26)$$

where

b = propeller damping coefficient, in.-lb-sec/rad
K = a constant

Q = mean propeller torque, in.-lb
Ω = rotative speed of propeller, rad/sec

If propeller model test data are available, it may be convenient to use the relationship

$$K = \frac{(1 - s)}{Q} \left(\frac{\partial Q}{\partial s}\right) \qquad (27)$$

where s is the propeller slip.

If the propeller data are given in the form of J, K_Q curves:

$$K = 2 - \frac{J}{K_Q} \left(\frac{\partial K_Q}{\partial J}\right) \qquad (28)$$

If given in the form of a Troost diagram (B_p, δ) as in reference 1:

$$K = \frac{2\delta}{B_p} \left(\frac{\partial B_p}{\partial \delta}\right) - 3 \qquad (29)$$

As an approximation for many propellers, $K = 3.7$ to 4, which may be used in the absence of other data. A value of 4 corresponds to a damping constant that is double the slope of the torque-speed curve. In all cases the derivatives are computed at the operating point of the propeller by taking the ratio of small differences in dependent and independent variables, moving along a constant-pitch line.

The energy loss via the propeller per cycle of torsional vibration can be written as

$$E_{dp} = \pi \, b\omega\theta_p^2 \qquad (30)$$

where

ω = circular frequency of vibration, rad/sec
θ_p = amplitude of propeller vibration, radians
b = as defined for equation (26)

Energy is also dissipated as a result of elastic hysteresis in the shafting, sliding fits, etc. The elastic hysteresis energy loss per cycle is a function of the material and stress level. Damping from sliding fits is very difficult to estimate. Even for identical machinery, it will vary from one unit to another. The degree of energy loss per cycle varies with clearances, oil viscosity, and the amount of lateral motion the moving part has within the fit. However, for general guidance, the internal damping vibratory energy loss per cycle is frequently estimated to be between 1 and 5% of the total system vibratory energy. The energy dissipated because of internal damping can be expressed as [55]:

$$E_{di} = \Sigma \frac{1}{2} \alpha_x J_x \omega^2 \theta_x^2 \qquad (31)$$

where

α_x = fraction of energy dissipated at mass x
J_x = moment of inertia of mass x, lb-in.-sec^2
θ_x = amplitude of vibration of mass x, radians

The damping action of the turbines in geared-turbine drives would generally be expected to be of secondary importance especially in modes where the turbines have

small relative amplitudes; however, it may warrant assessment under some circumstances. The energy dissipated due to turbine damping can be expressed as

$$E_{dt} = \pi \, c\omega\theta_t^2 \qquad (32)$$

where

c = turbine damping constant, which can be approximated as the ratio of turbine torque to turbine rpm at the speed corresponding to the point under study, in.-lb-sec/rad
θ_t = amplitude of vibration of turbine rotor, radians

If damping is introduced into the vibration calculations, the computational procedure is modified considerably. An external source of damping, such as that at the propeller or turbines, introduces an external moment of $-jb\omega\theta$ on the respective mass concentration; and internal damping, such as shafting hysteresis, between two masses is equivalent to changing a spring constant k to a complex spring constant

$$k' = k(1 + j\alpha/2\pi) \qquad (33)$$

where α is the fraction of the elastic energy absorbed by the damper [54].

Calculations that incorporate damping as just indicated are somewhat tedious; an alternative procedure is to compute the effect of damping at resonance only by equating input energy to damping energy.

7.7 Vibratory Torque Calculations. In many cases, the torsional vibration characteristics of a shafting system can be shown to be satisfactory in the design stage with only a computation of the system natural frequencies and without predicting vibratory torques and amplitudes. Normally this is possible when a comparison is made with a similar system that has proven satisfactory in service. For designs where the system natural frequencies, vibratory excitation, or anticipated system operation may cause concern, however, an investigation of the magnitude of the vibratory torques and stresses is necessary.

In order to establish the vibratory torques and stresses in a vibrating system, the amplitudes of vibration of the system masses must be established so that the twist in the interconnecting shafts (springs) can be determined. In a system vibrating at resonance, the amplitude of vibration increases until the damping work per cycle of vibration is equal to the input work per cycle of vibration.

a. Geared-turbine drives. A geared-turbine drive, with accurately cut gears, is excited torsionally only by the propeller. Also, in the first mode, internal damping and turbine damping can be ignored without a significant loss of accuracy. Therefore, the propeller excitation energy, equation (17), can be equated to the propeller damping energy, equation (30):

$$E_{ip} = E_{dp}$$

$$\pi r Q \, \theta_p = \pi b\omega\theta_p^2$$

By substituting $4Q/\Omega$, equation (26), for the propeller damping coefficient and letting $\omega = Z\Omega$, where Z equals the number of propeller blades, the maximum propeller amplitude, θ_p, is

$$\theta_p = \frac{r}{4Z}$$

The maximum amplitude of propeller vibration can be determined from the foregoing expression. The normalized mode shape determined from the system natural frequency calculation, Table 10, can then be used to assess the vibratory torque at resonance at any element of the system. The alternating torques in the quill shafts between the high-speed gears and low-speed pinions are usually the largest from a relative viewpoint; consequently, it is customary to analyze these elements when investigating the possibility of torque reversals.

As an example, referring to the calculation in Table 10, with a propeller excitation, r, equal to 3% of the mean propeller torque, the alternating torque, q, in the low-pressure quill shaft in the first mode of torsional vibration would be:

$$q = 1213\, \theta_L \times 10^6$$
$$q = 1213\,(0.06477\,\theta_p) \times 10^6$$
$$q = 78.57\,(r/4Z) \times 10^6$$
$$q = 98,210 \text{ in.-lb}$$

This is the torque in the low-pressure quill shaft referred to line-shaft speed. With a second-reduction ratio of 7.5, the actual vibratory torque in the quill shaft will be 98,210/7.5 or 13,090 in.-lb. The mean operating torque corresponding to the resonant frequency can be approximated by determining the rated propeller torque (the torque corresponding to 22,000 shp at 115 rpm) and assuming that the propeller torque varies as the square of the propeller rpm; therefore, the mean operating propeller torque at the resonant frequency is estimated to be

$$Q = [(63,025)(22,000)/115]\,[22.38/115]^2$$
$$Q = 456,630 \text{ in.-lb}$$

In this particular case at the resonant frequency, the low-pressure turbine develops 55% of the total power delivered to the propeller, and the mean torque in the low-pressure branch at resonance is determined to be 33,490 in.-lb whereas the vibratory alternating torque is estimated to be 13,090 in.-lb; therefore, torque reversals in the low-pressure train at the first resonant mode of torsional vibration are not expected. The vibratory stress in the quill shaft can be calculated using the alternating torque across this shaft of 13,090 in.-lb.

b. Diesel drives. As with turbine drives, the propeller is the principal means of system damping in the first mode of diesel shafting systems. However, in the following discussion concerning the amplitude of vibration and the resulting stress levels in a diesel propulsion system, internal damping will be considered to evaluate its effect. The vibratory work input per cycle from a diesel engine is given by equation (25). The $\Sigma\,\theta_c$ term in equation (25) is obtained from Fig. 38. This term establishes the order

that provides the maximum energy input per cycle. In the first mode, when all engine cylinders have essentially the same amplitude, the "major orders" are the principal orders to investigate. For a two-cycle engine, the major orders are integral multiples of the number of cylinders, and for a four-cycle engine they are integral multiples of one-half the number of cylinders; however, the major orders do not always cause the most severe vibration.

Figure 38 shows the major orders to be dominant. In this case, the major orders are over 20 times larger than the vector sum of any other order. For the sixth-order, the alternating torque per cylinder, Q_n, can be determined from equation (23) and Tables 11 and 14 as

$$Q_n = AR(TE_{gn}) \qquad (23)$$
$$= (596.6)(52.64)(22.26)$$
$$= 699,100 \text{ in.-lb}$$

The vector sum $\Sigma\,\theta_c$ for the sixth order is obtained from Fig. 38, and the engine input work per cycle becomes

$$E_{ie} = \pi Q_n\,\Sigma\,\theta_c \qquad (25)$$
$$= \pi\,(699,100)(5.6578)\theta_o$$
$$= (12.43 \times 10^6)\,\theta_0 \text{ in.-lb}$$

The propeller damping work per cycle, which is the principal source of damping in the first mode, can be computed using equation (30). The propeller damping coefficient is calculated from equation (26), assuming that $K = 4$ and Q is proportional to the propeller rpm squared. With the rated torque given by Table 11:

$$b = \frac{4\left(\dfrac{40.8}{74}\right)^2 (14.784 \times 10^6)}{\dfrac{2\pi}{60}(40.8)} = 4.21 \times 10^6$$

The propeller amplitude and first mode natural frequency are given by Table 12; therefore

$$E_{dp} = \pi\,b\omega\theta_p^2 \qquad (30)$$
$$E_{dp} = \pi\,(4.21 \times 10^6)(25.64)(-1.2180\theta_0)^2$$
$$= (503.1 \times 10^6)\,\theta_0^2 \text{ in.-lb}$$

With the fraction of internal energy dissipated, α_x, as indicated in Table 12 for the various masses, the internal damping per cycle of vibration can be calculated and summed as shown in Table 12, that is:

$$E_{di} = \Sigma\,\frac{1}{2}\,\alpha_x J_x \omega^2 \theta_x^2 \qquad (31)$$
$$= (8.490 \times 10^6)\,\theta_0^2 \text{ in.-lb}$$

The absolute amplitude of vibration at the first-mode natural frequency, when excited by the sixth-order excitation, can be found by equating the engine input work per cycle to the sum of the propeller damping and the internal system damping, that is:

$$E_{ie} = E_{dp} + E_{di}$$

$$(12.43 \times 10^6) \, \theta_0 = (503.1 \times 10^6) \, \theta_0^2 + (8.490 \times 10^6) \, \theta_0^2$$

$$\theta_0 = \frac{12.43 \times 10^6}{511.6 \times 10^6} = 0.0243 \text{ radians}$$

In this mode the most highly stressed element in the shafting system is the line shaft (spring constant k_8 in Fig. 35 and Table 12). The vibratory torsional stress in the line shaft can be determined by calculating the torsional deflection of the line shaft, multiplying the line-shaft deflection by the line-shaft spring constant (which equates to the vibratory torque in the line shaft), and dividing the torque by the line-shaft section modulus. Referring to Table 12, the line-shaft deflection is

$$Q_{9\text{-}8} = |\theta_9 - \theta_8| = |-0.6476 \, \theta_0 - 0.7864 \, \theta_0|$$

$$= 1.4340 \, \theta_0$$

With a line-shaft spring constant, k_8, of 644×10^6 in.-lb/rad, and a reference amplitude, θ_0, of 0.0243 radians, the line-shaft sixth-order vibratory torque is

$$Q_6 = \theta_{9\text{-}8} \, k_8$$

$$= [(1.4340)(0.0243)](644 \times 10^6) = 22.44 \times 10^6 \text{ in.-lb}$$

(The same result can be obtained more directly by multiplying θ_0 by the $\Sigma \, J_8 \omega^2 \theta_8$ term in Table 12.) The diameter of the line shaft is 21.5 in., which provides a section modulus of 1951 in.^{3}; therefore, the vibratory torsional stress is 11,500 psi, a value sufficiently high to be of concern.

7.8 Acceptable Limits for Torsional Vibration. As a general guideline for double-reduction geared systems, untuned resonant frequencies of torsional vibration should not occur in the range of 60 to 115% of rated rpm; however, this broad guideline does not ensure satisfactory torsional vibration characteristics. Furthermore, there may be satisfactory operational systems that this limitation would exclude. Such being the case, the details of each particular shafting system must be analyzed to ensure satisfactory performance.

Specific acceptance criteria for torsional vibration in geared-turbine propulsion systems are generally governed by two considerations: (1) limiting fatigue stresses within the system to safe values, and (2) avoiding torque reversals in the reduction gear elements by ensuring that alternating torques within the reduction gear engagements do not exceed the continuous torques. Generally, these criteria can be satisfied by the appropriate selection of the number of propeller blades, propeller design details, reduction gear design details, and the shafting diameters.

For gear-driven propulsion systems, the classification societies normally require a barred speed range, independent of stress level, if the system has a resonant frequency that causes gear-tooth chatter during continuous operations.

Torsional vibration in diesel-driven propulsion systems can lead to damage or failure of propulsion elements, such as the shafting, crankshaft, gears, and couplings. Consequently, the classification societies require torsional vibration calculations to be submitted and also require verification by measurement. The classification societies prescribe criteria that govern stress limits for vibratory torsional stresses [15]; the criteria typically stipulate an upper limit for torsional stresses that occur during continuous operations, but also allow transient operations, below a higher torsional stress limit, as necessary to pass through a torsional critical frequency that occurs below 80% of rated speed. A band of propeller speeds, extending above and below the critical speed, is specified that forms the "barred speed range." Figure 39 illustrates the barred speed range for the propulsion system used for the calculations outlined in Table 12. Barred speed ranges are usually not permitted above 80% of rated speed [15].

For ships that have direct-drive diesels located aft, the torsional characteristics of the propulsion system are generally dominated by a resonance of the first mode that is excited by a major order of the engine. For two-cycle engines the major orders are multiples of the number of cylinders, and for four-cycle engines they are multiples of one-half the number of cylinders. The number of cylinders in the engine selected is, therefore, an important consideration in the design of the propulsion shafting system. Where an unacceptable critical frequency occurs in the operating range, there are basically three approaches that can be taken to achieve satisfactory propulsion system operation [64]:

1. Lower the critical frequency. This can be accomplished by decreasing the shafting diameter, as permitted by the regulatory bodies, or by adding a tuning wheel to the engine crankshaft. While this approach will normally reduce the severity of the critical frequency, a barred speed range will usually be necessary.

2. Raise the critical frequency. This can be accomplished by increasing the shafting diameter. The objective is to shift the major-order critical frequency at least 40% above the operating range. A barred speed range would be avoided, but the shafting would be very stiff. When large-diameter shafting is used, the alternating propeller thrust that is induced by torsional vibrations, and the resulting effects on the hull structure, requires a rigorous analysis.

3. Install an engine torsional vibration damper. The objective would be to reduce the torsional vibratory stresses to an acceptable level. However, satisfactory operation of the propulsion system would become dependent on the reliable operation of the torsional damper.

Of the three alternatives, the first is generally the least expensive and is preferred, if practical.

Couplings with a low torsional stiffness can be used to decouple vibrating components in a mechanical system.

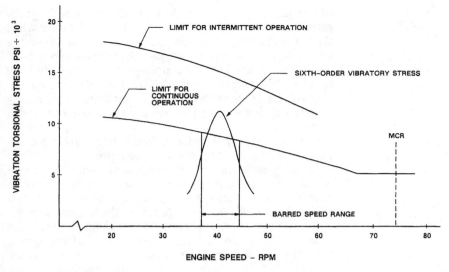

Fig. 39 Limiting torsional vibration stresses in line shaft for diesel propulsion system example

Torsionally flexible couplings are often used to shift a system natural frequency low in the operating range in arrangements where a power takeoff is driven by a slow-speed diesel, and such couplings are also used on the output shafts of medium-speed diesels that drive the propulsor through a reduction gear [60,61,64].

Section 8
Longitudinal Vibration

8.1 Introduction. Severely objectionable longitudinal vibrations in shafting systems were not encountered until the advent of several classes of large naval vessels in early 1941. Reference 65 contains a description of the difficulties experienced with these ships and also presents the most thorough treatment of longitudinal vibration that has been prepared for steam-turbine-driven ships. The works of Panagopulos [56], Rigby [66], Couchman [67], and others add to the knowledge on the subject, yet the fundamental problem areas remain as those identified by Kane and McGoldrick in reference 65.

The principles set forth by Kane and McGoldrick are primarily oriented to turbine-driven ships; however, the principles are equally applicable to diesel-driven ships. The treatment of the shafting, thrust bearing, and foundations is directly applicable; and the engine crankshaft can be modeled as masses and springs and analyzed as outlined in reference 65. An axial vibration resonance can cause high crankshaft stresses in a diesel engine and can result in large axial forces going into the thrust bearing; consequently, they are to be avoided.

The axial stresses in line and propeller shafting associated with even the most violent instance of longitudinal vibration are not sufficiently large to induce failures in the shafting itself; nevertheless, longitudinal vibration can produce effects that are destructive to engine room equipment. Shafting systems having longitudinal vibration characteristics that are resonant with diesel engine or propeller blade-rate frequency forces produce a significant magnification of the exciting forces. Such a force magnification can result in such deleterious effects as:

(*a*) Accelerated wear of gears, flexible couplings, thrust bearings, etc., and destruction of main engine clearances because of the increased relative axial movements.

(*b*) Large vibration amplitudes and stresses in attachments to the main engine, condensers, and main and auxiliary machinery, which can ultimately result in fatigue failure.

(*c*) Cracks in foundations and hull structures.

Axial vibration issues should be considered by the engine builder and ship designer during the preliminary design stage. In addition to the variations in thrust caused by irregularities in the propeller wake field, consideration must also be given to the induced propeller thrust forces resulting from propeller torsional vibration, and, for diesel-driven ships, the forces from coupled axial and torsional vibration of the diesel engine and the variable axial forces emanating from the diesel engine crankshaft.

8.2 Determination of Natural Frequencies. There are basically three approaches that may be taken to determine the natural frequencies of longitudinal vibration. The first approach would be to use a simplified method for the purpose of quickly assessing a situation. Approximate methods suitable for investigating the first mode of vibration are given in references 56 and 65.

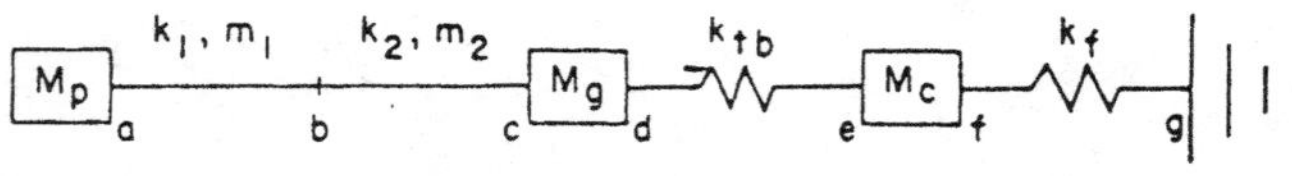

M_p = mass of propeller, 227 lb-sec^2/in

M_g = mass of gears, 147 lb-sec^2/in

M_c = mass of machinery, 767 lb-sec^2/in

m_1 = mass of larger (outboard) shafting, 301 lb-sec^2/in

m_2 = mass of smaller (inboard) shafting, 366 lb-sec^2/in

k_1 = spring constant of larger (outboard) shafting, 20.3 x 10^6 lb/in

k_2 = spring constant of smaller (inboard) shafting, 5.9 x 10^6 lb/in

k_{tb} = thrust bearing spring constant, 7 x 10^6 lb/in

k_f = thrust bearing foundation spring constant, lb/in

z = number of propeller blades, 6

Fig. 40   Representation of a geared turbine propulsion shafting system (Fig. 1) for a longitudinal vibration analysis

A second approach is to model the shafting system as discrete masses and springs and use the Holzer method to determine the system natural frequencies. The accuracy obtained with a discrete spring-mass model depends upon the technique used to establish the masses that are used to represent the shafting system.

A third approach is the mechanical impedance method, as proposed by Kane and McGoldrick [65]. This method is inherently more accurate than the Holzer method since the weight of the shafting is considered to be distributed; however, the impedance method has the slight disadvantage of being somewhat more complex and difficult to grasp. For illustrative purposes, a calculation of the natural modes of longitudinal vibration of the shafting arrangement shown in Fig. 1 has been made using the mechanical impedance method. Figure 40 is a model of the shafting arrangement, which is suitable for analysis by the mechanical impedance method. It may be noted that the difference in the diameter of the inboard and outboard shafting is taken into account; in general, especially with short spans of shafting, this additional degree of accuracy is not warranted.

The majority of the system parameters are directly calculated from the system scantlings and, therefore, there is no difficulty in establishing their values. However, the assessment of several of the system parameters can be nebulous. For example, the determination of the water entrained with the propeller does not lend itself to an accurate calculation, but as a first approximation, the entrained water weight may be assumed equal to 60% of the propeller weight. The results obtained from the experimental work of Burrill and Robson [58] are widely used in estimating propeller entrained water, and reference 65 suggests other approaches. Parsons and Vorus [62] provide an analytically derived estimate of the added mass of marine propellers when vibrating as a rigid body.

The behavior of longitudinally flexible couplings, which are affected by the friction between mating surfaces, in connection with vibratory movements similarly cannot be stated with certainty. The consequences of the response

of flexible couplings to vibratory loadings, aside from the effect on the couplings themselves, is not great when the thrust bearing is located well forward. But when the thrust bearing is located aft and there is an appreciable vibratory amplitude at the couplings, the effect of flexible coupling behavior can be significant. For a detailed discussion of the response of flexible couplings to vibratory loadings and the complications involved, see references 65 and 67.

Some machinery liquid and foundation weight participates with the shafting system when vibrating longitudinally as a consequence of being near the main thrust bearing, but assigning a magnitude to these quantities entails numerous uncertainties. An assessment of the machinery mass, M_c, to be included in the calculations requires judgment, which must be based on the specifics of each system. In instances where the shafting spring constant and the propeller mass are the major determinants of the first-mode frequency, the machinery mass has a small participation; consequently, an accurate assessment of its magnitude is not important. On the other hand, the machinery mass is expected to have a significant participation when the foundation stiffness is a major determinant of the first-mode natural frequency and in the second mode, in which cases a more accurate assessment of the machinery mass is required. In general, the first-reduction gear rotating elements, gear casing, turbines, condenser, foundation structure, or portions thereof may be included as machinery mass. Reference 65 provides guidance in the assessment of the machinery mass for turbine-driven ships. Reference 68 contains an interesting approach concerning the treatment of machinery masses in that the center of gravity of a portion of the machinery mass is displaced from the shaft centerline and given a leverage ratio relative to the centerline of the shaft.

The spring constant of the thrust bearing, k_{tb}, may be considered to consist of three components: the spring constants of the thrust-bearing housing, the thrust collar, and the thrust elements (or shoes). Aside from the tedious calculations, no difficulty is experienced in calculating the spring constants of the thrust-bearing housing and collar inasmuch as the majority of the deflections are due to shear and can be estimated on the basis of well established techniques. But the spring constant of the thrust elements can be difficult to evaluate because of the oil film and the supporting configuration of the shoes. In the absence of more specific information, the data given in reference 65 may be used for guidance in establishing the stiffness of thrust elements. However, a more accurate assessment of the thrust-shoe stiffness can be obtained by modeling the shoes using a finite-element procedure.

The determination of the thrust-bearing foundation spring constant can be a difficult and nebulous undertaking even for an experienced analyst. The foundation stiffness is determined by estimating the deflection at the thrust bearing along the shaft centerline due to:

(a)   Shear deflection of the thrust-bearing foundation structure.

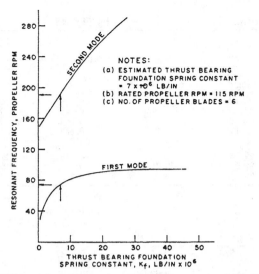

Fig. 41 Effect of thrust bearing foundation stiffness on longitudinal resonant frequencies

(b) Bending deflection of the thrust-bearing foundation structure.

(c) Hull bottom longitudinal deflection.

(d) Deflection at shaft centerline resulting from rotation (flexural bending) of the hull structure.

The model used to analyze the thrust-bearing foundation stiffness can take several forms, depending upon the analytical procedure to be used and the information available concerning the structural details. One model that has been used is based on a fore-and-aft strip through the ship bottom structure that includes the thrust-bearing foundation and the longitudinals that transmit the thrust load into the hull structure. The strip is assumed to be supported at the forward and after engine room bulkheads. The strip may extend abaft of the aft engine room bulkhead if the structure there effectively contributes to the thrust-bearing foundation stiffness or if that structure effectively restrains rotation (flexure) of the bottom structure in way of the thrust bearing. Reference 2 contains a discussion of the techniques used to calculate the main thrust-bearing foundation stiffness.

Longitudinal vibration calculations are frequently conducted such that the natural frequency is expressed in terms of the thrust-bearing foundation stiffness. This procedure permits a judgment to be made concerning the degree of accuracy required for the thrust-bearing foundation stiffness calculation. In many cases, an approximate procedure that understates the actual foundation stiffness is found to provide sufficient accuracy.

Table 15 contains a calculation for the first and second resonant modes of longitudinal vibration of the shafting system modeled as shown in Fig. 40. Table 15 is based upon the mechanical impedance method described in reference 65 and is arranged such that the resonant frequencies can be plotted in terms of the thrust-bearing foundation stiffness. Figure 41 is such a plot and indicates the accuracy required of the thrust-bearing foundation stiffness calculations. In some instances, as may be the case with a ship having a very short run of shafting, an inspection of the thrust-bearing foundation drawings may be all that is required to provide assurance that the resonant modes of longitudinal vibration will be well clear of the operating range. On the other hand, lengthy and sophisticated thrust-bearing foundation stiffness calculations may be required in order to ensure that ships with long runs of shafting have satisfactory longitudinal vibration characteristics.

The longitudinal vibration characteristics of ships driven by diesel engines are analyzed using the same analytical approach. The engine crankshaft can be modeled as a series of masses (one for each piston throw) and springs that represent the longitudinal stiffness of the crankshaft between pistons. The mass and stiffness values representing the crankshaft can be developed from engine drawings. Engine manufacturers provide this information for shafting system development.

8.3 Vibration Reducers. In cases where design constraints make it impossible to design a shafting system such that it is free of objectionable frequencies of resonant longitudinal vibration, use may be made of a "vibration reducer." Briefly, vibration reducers are thrust bearings, that are modified so that the thrust pads are supported by hydraulic pistons, as illustrated by Figs. 20 and 21. The volume of oil supporting the thrust pads and the connecting piping can be sized to alter the thrust-bearing spring constant, system effective mass, and damping to avoid objectionable resonant frequencies.

A schematic design of a vibration reducer system is illustrated by Fig. 21. The major elements establishing the characteristics of a vibration reducer are the oil flask volume and the piping size and length between the thrust bearing and the flasks. Reference 33 provides details concerning the design approach used to size these components and control the thrust vibration at the thrust bearing. A complicating factor when using vibration reducers is the necessity to add or remove oil from the oil flasks to keep the thrust pistons from bottoming out. As shown schematically in Fig. 21, a valve controlled by the thrust collar position adds or dumps oil from the active system. A supply of oil at adequate pressure must be available.

Analytical results and experience confirm that vibration reducers can be very effective in lowering the alternating forces at the thrust bearing and the vibratory amplitudes in the engine space. In general, reductions greater than 4 can be achieved.

8.4 Excitation Factors. Longitudinal vibration of propulsion shafting systems is often excited by the variable thrust developed by the propeller due to the nonuniform wake pattern in which it operates. The predominant periodic component of the thrust developed by a propeller occurs at blade-rate frequency, i.e., the number of propeller blades times the rotational frequency of the shaft. Higher harmonics of blade-rate frequency occur but, due to their relatively small magnitude, they are generally not of practical importance. A number of factors influence the magnitude of the vibratory thrust; consequently, generalizations in this area must be used with care. Nevertheless, typical ranges of thrust excitation, expressed as a

Table 15   Longitudinal vibration calculations for shafting system model shown by Fig. 40

N rpm	ω rad/sec	$\omega^2 \times 10^5$	$Z_a = \omega^2 M_p$ lb/in. $\times 10^6$	$\epsilon_1 = \omega\left(\dfrac{m_1}{k_1}\right)^{1/2}$ rad	$\epsilon_1^0 = \dfrac{180}{\pi}\epsilon_1$ deg	$k_1\epsilon_1$ lb/in. $\times 10^6$	$\tan(\alpha_1^0 - \epsilon_1^0)$ $= \dfrac{Z_a}{k_1\epsilon_1}$	$\alpha_1^0 - \epsilon_1^0 =$ $\tan^{-1}\left(\dfrac{Z_a}{k_1\epsilon_1}\right)$ deg
70	43.98	0.0193	0.4381	0.1694	9.706	3.439	0.1274	7.260
80	50.27	0.0253	0.5743	0.1936	11.09	3.930	0.1461	8.312
90	56.55	0.0320	0.7264	0.2178	12.48	4.421	0.1643	9.330
95	59.69	0.0356	0.8081	0.2298	13.17	4.665	0.1732	9.826
160	100.5	0.1010	2.293	0.3870	22.17	7.856	0.2919	16.27
200	125.7	0.1580	3.587	0.4840	27.73	9.825	0.3651	20.06
240	150.8	0.2274	5.162	0.5807	33.27	11.79	0.4378	23.64
280	175.9	0.3094	7.023	0.6773	38.81	13.75	0.5108	27.06

N rpm	$\alpha_1^0 =$ $(\alpha_1^0 - \epsilon_1^0) + \epsilon_1^0$ deg	$\tan\alpha_1^0 =$ $Z_b/k_1\epsilon_1$	$Z_b =$ $k_1\epsilon_1 \tan\alpha_1^0$ lb/in. $\times 10^6$	$\epsilon_2 = \omega\left(\dfrac{m_2}{k_2}\right)^{1/2}$ rad	$\epsilon_2^0 = \dfrac{180}{\pi}\epsilon_2$ deg	$k_2\epsilon_2$ lb/in. $\times 10^6$	$\tan(\alpha_2^0 - \epsilon_2^0)$ $= \dfrac{Z_b}{k_2\epsilon_2}$	$\alpha_2^0 - \epsilon_2^0 =$ $\tan^{-1}\left(\dfrac{Z_b}{k_2\epsilon_2}\right)$ deg
70	16.97	0.3052	1.050	0.3464	19.85	2.044	0.5137	27.19
80	19.40	0.3522	1.384	0.3959	22.68	2.336	0.5925	30.65
90	21.81	0.4002	1.769	0.4454	25.52	2.628	0.6731	33.94
95	23.00	0.4245	1.980	0.4701	26.93	2.774	0.7138	35.52
160	38.44	0.7937	6.235	0.7916	45.36	4.670	1.335	53.16
200	47.79	1.102	10.83	0.9900	56.72	5.841	1.854	61.66
240	56.91	1.535	18.10	1.188	68.07	7.009	2.582	68.83
280	65.87	2.232	30.69	1.385	79.35	8.172	3.756	75.09

N rpm	$\alpha_2^0 =$ $(\alpha_2^0 - \epsilon_2^0) + \epsilon_2^0$ deg	$\tan\alpha_2^0 =$ $Z_c/k_2\epsilon_2$	$Z_c =$ $k_2\epsilon_2 \tan\alpha_2^0$ lb/in. $\times 10^6$	$\omega^2 M_g$ lb/in. $\times 10^6$	$Z_d =$ $Z_c + \omega^2 M_g$ lb/in. $\times 10^6$	$Z_e = \dfrac{k_{tb}Z_d}{k_{tb} - Z_d}$ lb/in. $\times 10^6$	$\omega^2 M_c$ lb/in. $\times 10^6$	$k_f = Z_f =$ $Z_e + \omega^2 M_c$ lb/in. $\times 10^6$
70	47.04	1.074	2.195	0.2837	2.479	3.838	1.480	5.318
80	53.33	1.343	3.137	0.3719	3.509	7.036	1.941	8.977
90	59.46	1.695	4.454	0.4704	4.924	16.60	2.454	19.05
95	62.45	1.917	5.318	0.5233	5.841	35.28	2.731	38.01
160	98.52	-6.675	-31.17	1.485	-29.69	-5.664	7.747	2.083
200	118.4	-1.850	-10.81	2.323	-8.487	-3.836	12.12	8.284
240	136.9	-0.9358	-6.559	3.343	-3.216	-2.204	17.44	15.24
280	154.4	-0.4791	-3.915	4.548	0.633	0.6959	23.73	24.43

* Exciting frequency, $\omega = \dfrac{2\pi N}{60} \times$ number of propeller blades.

percentage of the mean thrust, are tabulated in Table 3; for additional details concerning thrust excitation, see references 69 through 72.

Slow-speed, long-stroke diesel engines can excite longitudinal vibration by:

• The combustion pressure and mass forces in the individual cylinders. When the crank throw is loaded by the gas and mass force through the connecting-rod mechanism, the arms of the crank throw deflect in the axial direction of the crankshaft, exciting axial vibrations.

• Propeller thrust variation that is induced by engine torsional vibration. This is particularly troublesome when the major-order critical is above the running speed (stiff shafting system) and the engine variable torque causes an irregular angular velocity of the propeller blades through the water.

• Torsional resonances in the engine that cause axial crankshaft deflections.

Reference 64 contains a discussion of the mechanisms by which slow-speed, two-stroke diesel engines excite longitudinal vibration.

8.5 Damping. There are several sources of damping in the longitudinal vibration system, the most important of which is the propeller. Other sources of damping such as hysteresis and sliding friction are generally relatively small, difficult to estimate, and not readily reflected in the analytical procedures; therefore, their effects are usually considered to be lumped with the propeller damping allowance.

A procedure for estimating propeller damping is described in reference 65. The procedure entails plotting the propeller thrust coefficient versus slip curve at the operating slip point. The propeller damping constant, C_p, is accordingly determined to be

$$C_p = \frac{1}{12} PnD^2 \frac{\partial C_T}{\partial s} \qquad \text{lb-sec/in.} \qquad (34)$$

where

P = propeller pitch, ft

n = propeller speed, rps

D = propeller diameter, ft

C_T = propeller thrust coefficient, $T/n^2P^2D^2$

s = propeller slip = $1 - \dfrac{V_a}{Pn}$

V_a = propeller advance velocity, fps

A somewhat different approach for determining system damping was taken by Rigby [66]. Rigby used full-scale data to calculate an equivalent propeller damping constant and concluded that for three-bladed propellers the equivalent damping constant was about 16.5 lb-sec/in. per square foot of propeller developed area; he further concluded that the propeller damping constant tends to increase with larger numbers of blades and suggested that a damping factor equal to 39 lb-sec/in. per foot of blade edge may give better results based on tests made with four- and five-bladed propellers.

Reference 62 presents an analytically developed estimate of propeller damping of longitudinal vibration based upon Wageningen Series B propeller geometry. The analytical process recognizes the actual propeller physical attributes in the estimate.

In cases where the thrust bearing is located well aft and there is a significant amplitude of vibration at the slow-speed gear, investigations made and reported in reference 65 indicate that machinery damping must also be considered; reference 65 suggests a procedure which may be used to include the effects of machinery damping.

Significant damping can also be provided by a vibration reducer, when incorporated, which must be recognized when evaluating the shafting system vibration characteristics. The absolute value of the vibration reducer damping is dependent on the specific design, and can be estimated by the procedures given in reference 33.

Most slow-speed diesel engines incorporate longitudinal vibration damping features in the crankshaft arrangement. Diesel-engine damping provisions can be effective in minimizing crankshaft axial vibration, as noted in reference 64 and the discussion provided by Guglielmotti to reference 33.

8.6 Vibratory Thrust Calculations. A meaningful preliminary indication of the importance of a resonant mode of longitudinal vibration can be obtained by assessing and comparing the alternating thrust component with the mean thrust component at the main thrust bearing and the vibratory amplitude, velocity, and acceleration at the slow-speed gear, when installed. This can readily be accomplished by assuming that the only source of system damping is the propeller (the restriction need not be quite so severe in that an "equivalent propeller damping" may be used which incorporates other system damping effects) and that the propeller is also the only source of excitation. With these assumptions, the propeller input work per cycle can be expressed as:

$$E = \pi t X_a \qquad \text{in.-lb/cycle} \qquad (35)$$

where t is the maximum amplitude of alternating thrust (pounds) and X_a is the maximum amplitude of propeller vibration (in inches). This can be equated to the damping per cycle

$$D = \pi C_p \omega X_a^2 \qquad \text{in.-lb/cycle} \qquad (36)$$

where C_p is the propeller damping constant (lb-sec/in.) and ω is the resonant frequency of vibration (rad/sec), to obtain the following expression for the amplitude of vibration at the propeller:

$$X_a = t/\omega C_p \qquad (37)$$

Once the amplitude of vibration at the propeller has been established, the alternating force at other points in the system can be determined by using the mechanical impedance method.

The procedure of determining the alternating force on the main thrust bearing may be illustrated by referring to Table 15 and Fig. 40. In addition, the following data are required:

Mean propeller thrust at 75 rpm (thrust is assumed to vary as the rpm squared) .... 110,000 lb

Propeller damping constant (based on reference 66) .... 5800 lb-sec/in.

Ratio of alternating to mean propeller thrust (based on an analysis of the propeller wake) 3%

With these data the exciting force from the propeller is established to be 3300 lb and from equation (37) the amplitude of vibration at the propeller is

$$X_a = 3300/[2\pi(75)(6)/60][5800]$$
$$X_a = 0.012 \text{ in.}$$

The procedure outlined in reference 65 can be used to establish the amplitude of vibration at the thrust bearing (point d in Fig. 40). Based on reference 65, the amplitude of vibration at point b becomes

$$X_b = X_a \cos(\tan^{-1} Z_b/k_1\epsilon_1)/\cos(\tan^{-1} Z_a/k_1\epsilon_1) \qquad (38)$$

and the amplitude of vibration at point c becomes

$$X_c = X_b \cos(\tan^{-1} Z_c/k_2\epsilon_2)/\cos(\tan^{-1} Z_b/k_2\epsilon_2) \qquad (39)$$

(see Table 15 for a definition of terms.) A numerical evaluation of the foregoing expression indicates a vibratory amplitude of ± 0.008 in. at point d (the same as c). This is the amplitude at the slow-speed gear from which the velocity and acceleration can be calculated knowing the frequency of vibration. The alternating force at point d is then determined from

$$F_d = Z_d X_d \qquad (40)$$

By interpolating the data shown in Table 15, Z_d is established to be 2.95×10^6; therefore, $F_d = 23,600$ lb and the ratio of the alternating thrust to the mean thrust is established to be 0.215, an entirely acceptable value in view of the low power level at which it occurs.

8.7 Acceptable Limits for Longitudinal Vibration. Several attempts have been made to enumerate acceptance criteria for longitudinal vibration characteristics. The most well known criteria are those specified in reference [73]. Other acceptance criteria have stated that reversals of thrust in thrust bearings are not permitted. One criterion [67] has specified that the acceleration of the slow-speed gear should not exceed 3 ft/sec²; but this recommended criterion is accompanied with the statement: "It is emphasized that each particular case warrants individual attention and that the thrust variation levels must be given equal consideration."

There are so many variables which must be considered when analyzing the vibration characteristics of a system that there appears to be no satisfactory alternative to conducting an analysis of each particular system and studying each system individually.

Section 9
Whirling Vibration

9.1 Introduction. Whirling vibration can best be visualized by considering the motion to be the resultant of two shaft vibrations each in perpendicular planes passing through the shaft neutral position. Depending upon the manner in which the vibratory motions combine in the two perpendicular planes, the resultant motion may be circular (analogous to the motion of a skip-rope), elliptical, or in a single plane (if one of the two combining vibrations is of negligible magnitude). Visualization of whirling vibration is further complicated by the fact that the whirling frequency may be either at the frequency of shaft rotation or a multiple of shaft rotation, and the whirling motion can be either in the direction of shaft rotation or opposite to the direction of shaft rotation. Shaft whirling vibration is discussed in references 56 and 74–76.

9.2 Determination of Whirling Natural Frequencies. The shafting system whirling natural frequencies are determined by analyzing a model that: considers the shafting system to be a continuous beam, represents significant concentrated masses (such as propellers), includes an allowance for the propeller entrained water, recognizes the gyroscopic effect of the propeller, and allows for flexibility in the shaft bearing supports. Computer programs are commonly used to conduct such continuous-beam calculations; however, as an alternative, a method based on the Rayleigh approximation is simple to apply and is useful as a means of assessing the potential for a whirling natural frequency. To apply this method, an assumption must be made with regard to the shape of the shaft centerline when the amplitude of vibration is at its maximum. It is not necessary that the assumed curve have exactly the same shape as the actual vibration deflection curve, but it should have the same general characteristics. In problems concerning the vibration of beams it has been found satisfactory to use the deflection curve corresponding to the static loading condition, and this assumption is considered sufficiently accurate in connection with propulsion shafting.

The total energy of vibration at any instant consists of two parts: kinetic energy due to the motion of the shaft masses and potential energy due to the bending stresses caused by the shaft deflection. At the point of maximum amplitude, the masses of the shafting are all stationary, so the kinetic energy is zero; but the potential energy stored up in the shaft is at its maximum. On the other hand, when the amplitude is zero, there is no bending in the shaft, so the potential energy is zero; but the kinetic energy is at its maximum. During vibration the total energy in the shaft system remains constant; therefore, the potential energy at the point of maximum amplitude is equal to the kinetic energy at the point of zero amplitude. The maximum kinetic and potential energies are determined by the shaft deflections, the masses carried by the shaft, and the frequency of vibration. These energies so determined may be equated, and this equation is then solved for the critical speed, N_c in cycles per minute, as follows:

$$N_c = 187.7 \left(\frac{\Sigma y \, dW}{\Sigma y^2 \, dW} \right)^{1/2} \tag{41}$$

In this formula dW is the weight of a short section of shafting whose mass may be considered concentrated or the weight of any concentrated mass carried by the shaft (such as the propeller), and y is the deflection of the center of gravity of this mass. The summations include all of the masses in the shaft system.

The static deflection curve can be calculated as described in any standard book on strength of materials. It should be noted, however, that the loads on the shafting should be reversed in direction in alternate spans; that is, the weights are assumed to act down in one span and up in the next. This reversal is necessary to produce a deflection curve that has the same general form as the vibration curve corresponding to the lowest natural frequency.

Such calculations were made for the shafting system shown in Fig. 1, and a plot of the mode shape for the whirling mode of vibration is shown in Fig. 42. It may be seen from Fig. 42 that the large amplitudes of vibration are confined to the aftermost regions of the shafting system. From an inspection of the Rayleigh equation it is evident that only the regions of the shafting system that have relatively large amplitudes have an important effect on the whirling critical frequency. This fact can be exploited to greatly simplify the calculation procedure.

The computation of whirling critical frequencies, by necessity, entails several approximations. One is the assumed location of the resultant reaction in the bearing just forward of the propeller. Assessing the load distribution in this bearing is difficult because neither the shaft alignment conditions nor the slope of the shaft in way of the bearing is known with certainty; furthermore, in the case of water-lubricated bearings, the position of the bearing reaction moves due to bearing weardown. With shafting arrangements similar to that shown in Fig. 2 (closely

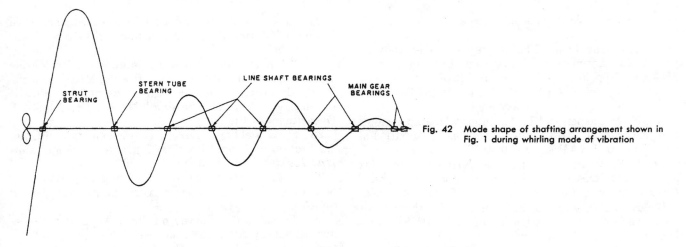

STRUT
BEARING

STERN TUBE
BEARING

LINE SHAFT BEARINGS

MAIN GEAR
BEARINGS

Fig. 42   Mode shape of shafting arrangement shown in
Fig. 1 during whirling mode of vibration

spaced stern tube bearings), the forward stern tube bearing can become unloaded or even possibly develop a downward reaction. Since the load condition of the forward stern tube bearing strongly affects the whirling natural frequency, computations are usually made with that bearing assumed to be loaded and also unloaded, particularly with water-lubricated bearings that are subject to large amounts of wear.

The effects of entrained water can be estimated [62], but are normally approximated by increasing the weight of the propeller by a percentage (usually 25%). Bearing flexibilities are normally neglected in the Rayleigh calculations as the problem would otherwise be considerably more complex. However, neglecting bearing flexibility may not be justified if the bearings are softly mounted in rubber to achieve self-aligning capabilities. Propeller gyroscopic effects, which tend to stiffen the system, are similarly neglected. Fortunately, the inaccuracies associated with considering bearings rigid and neglecting propeller gyroscopic effects tend to offset each other.

N. H. Jasper developed a calculation procedure that takes bearing flexibility and propeller gyroscopic effects into account [74,75,76]. The procedure is relatively simple due to the fact that only the aftermost region of the shafting is considered. The results obtained with the Jasper procedure are in very good agreement with the results obtained by using the Rayleigh method; this is attributed to the opposite effects of the additional factors taken into consideration.

9.3   Acceptable Limits for Whirling Vibration. Important sources of whirling vibration excitation are propeller and shafting unbalance and occur at a frequency corresponding to propeller rpm. Also, where Z is the number of propeller blades, the $kZ \pm 1$ harmonics of the propeller wake field produce exciting frequencies; however, these are generally not significant due to the small exciting force and the dampening resulting from the relatively higher frequency. Only the frequency corresponding to the propeller rpm is considered to be important by some authorities; however, this point can be debated. In any case, it must be agreed that the severity of excitation at

blade-rate frequencies must be investigated for a specific case before blade-rate vibration can be categorically dismissed.

In order for the whirling natural frequency to be coincident with the propeller rpm, conditions considerably different from those on the usual large ship must exist. For example, bearing spans would have to be abnormally long, shaft diameters would have to be abnormally small, a bearing would have to become unloaded for some reason, etc. Whirling frequencies corresponding to blade-rate frequencies can and do fall within the operating range. Fortunately, however, the exciting forces at blade rate frequencies are generally not of great severity unless coupled with other adverse conditions, such as the forward stern tube bearing becoming unloaded.

Perhaps the only generalization which can be made with respect to acceptance criteria for whirling vibration is that shafting arrangements should be designed such that, in the upper operating range, whirling resonant frequencies do not come into close proximity to the propeller rpm or the blade-rate frequencies that have strong exciting forces. Blade-rate exciting forces can be evaluated as discussed in Section 3.2.

References

1   J. D. Van Manen and P. van Oossanen, "Propulsion" in *Principles of Naval Architecture*, Vol. 2, E. V. Lewis, Ed., SNAME, 1988.

2   William S. Vorus, "Vibration" in *Principles of Naval Architecture*, Vol. 2, E. V. Lewis, Ed., SNAME 1988.

3   J. H. McCarthy, "On the Calculation of Thrust and Torque Fluctuations of Propellers in Nonuniform Wake Flow," DTMB Report 1533, David Taylor Model Basin, Bethesda, Md., Oct. 1961.

4   G. R. Stuntz, Jr., P. C. Pien, W. B. Hinterthan, and N. L. Ficken, "Series 60—The Effect of Variations in Afterbody Shape Upon Resistance, Power, Wake Distribution, and Propeller Excited Vibratory Forces," *Trans.* SNAME, Vol. 68, 1960.

5   J. B. Hadler and H. M. Cheng, "Analysis of Experimental Wake Data in Way of Propeller Plane of Single

and Twin-Screw Ship Models," *Trans.* SNAME, Vol. 73, 1965.

6 Tako Sasajima, "Usefulness of Quasi-Steady Approach for Estimation of Propeller Bearing Forces," SNAME, Propellers '78 Symposium, 1978.

7 "Propulsion Shafting," Design Data Sheet DDS 243-1, Department of the Navy.

8 N. H. Jasper and L. A. Rupp, "An Experimental and Theoretical Investigation of Propeller Shaft Failures," *Trans.* SNAME, Vol. 60, 1952.

9 E. P. Panagopulos and A. M. Nickerson, Jr., "Propeller Shaft Stresses Under Service Conditions—The SS *Chryssi* Investigation," *Trans.* SNAME, Vol. 62, 1954.

10 H. R. Neifert and J. H. Robinson, "Further Results from the Society's Investigation of Tailshaft Failures," *Trans.* SNAME, Vol. 63, 1955.

11 E. F. Noonan, "Propeller Shaft Bending Stresses on the SS *Esso Jamestown*," *ASNE Journal*, American Society of Naval Engineers, Aug. 1961.

12 R. L. Price, E. F. Noonan, and S. Feldman, "Bending and Torsional Stresses in Propeller Shaft of USS *Observation Island* (EAG 154) in Smooth and Rough Sea," DTMB Report 1596, David Taylor Model Basin, Bethesda, Md., June 1962.

13 R. L. Harrington and W. S. Vorus, "Dynamic Shock Analysis of Shipboard Equipment," *Marine Technology*, Vol. 4, No. 4, Oct. 1967.

14 "Mathematical Model and Dynamic Analysis Guide for Main Propulsion Shafting," Supervisor of Shipbuilding, U.S. Navy, Third Naval District Report SUPSHIP 280-1, July 1968.

15 *Rules for Building and Classing Steel Vessels*, American Bureau of Shipping.

16 J. W. Heck and E. Baker, "Marine Propeller Shaft Casualties," *Trans.* SNAME, Vol. 71, 1963.

17 W. E. Lehr, Jr. and E. L. Parker, "Considerations in the Design of Marine Propulsion Shaft Systems," *Trans.* SNAME, Vol. 69, 1961.

18 H. C. Andersen and J. J. Zrodowski, "Coordinated Alignment of Line Shaft, Propulsion Gear, and Turbines," *Trans.* SNAME, Vol. 67, 1959.

19 "Guide to Propulsion Reduction Gear Alignment and Installation," Technical and Research Bulletin No. 3–43, SNAME, Sept. 1987.

20 J. Cameron McKernan, "Propulsion Shafting Installation and Alignment Procedures" in *Proceedings*, Ship Technology and Research (STAR) Symposium, SNAME, 1986.

21 *Naval Ships' Technical Manual*, NAVSEA S9086-HB-STM-000, Chapter 233, Diesel Engines, Department of the Navy.

22 "Propulsion Shafting and Components," NAVSHIPS Drawing 803-2145807, Department of the Navy.

23 "Propeller," NAVSHIPS Drawing 810-4435837, Department of the Navy.

24 C. Adams, "An Assessment of Keyless Propellers for Naval Applications" in *Proceedings*, Propellers '84 Symposium, SNAME, 1984.

25 L. L. Shook, Jr. and C. L. Long, "Surface Cold Rolling of Marine Propeller Shafting," *Trans.* SNAME, Vol. 66, 1958.

26 "Rubber, Shaft Covering Materials (For Marine Propeller Shafts)," Military Specification MIL-R-15058.

27 "Instruction Manual Glass Reinforced Plastic Coatings for Propeller Shafting," NAVSHIPS 250-634-4, Department of the Navy.

28 S. A. Fielding, "Design Improvements and Standardization of Propulsion Shafting and Bearings," *Marine Technology*, Vol. 3, No. 2, April 1966.

29 J. E. Ancarrow and R. L. Harrington, "Main Propulsion Shafting Eccentricity Considerations," *Trans.* SNAME, Vol. 81, 1973.

30 "Results of a Survey on Shaft Alignment Procedures Used by American Shipyards," Technical and Research Report R-25, SNAME, March 1978.

31 Robert B. Grant, "Shaft Alignment Methods with Strain Gages and Load Cells," *Marine Technology*, Vol. 17, No. 1, Jan. 1980.

32 Albert W. Forrest, Jr. and Richard F. Labasky, "Shaft Alignment Using Strain Gages," *Marine Technology*, Vol. 18, No. 3, July 1981.

33 A. J. H. Goodwin, "The Design of a Resonance Changer to Overcome Excessive Axial Vibration of Propeller Shafting," *Trans.* Institute of Marine Engineers, Vol. 172, 1960.

34 "Shafting Bearings, and Seals," *Naval Ships' Technical Manual*, NAVSEA 0901-LP-430-0012 Chapter 9430, Department of the Navy.

35 R. Baudry and L. M. Tichvinsky, "Performance of Oil Rings," ASME paper, American Society of Mechanical Engineers, 1936.

36 W. W. Gardner, "Bearing Oil Delivery by Disk-Scraper Means" *Journal of Lubricating Technology*, April 1977.

37 W. W. Gardner, "Journal Bearing Operation at Low Sommerfeld Numbers," *Trans.* American Society of Lubrication Engineers, Vol. 19.

38 Roy L. Orndorff, Jr., "Water-Lubricated Rubber Bearings, History and New Developments," *ASNE Journal*, American Society of Naval Engineers, Nov. 1985.

39 Thomas L. Daugherty, "Friction and Wear Characteristics of Nine Water-Lubricated Bearings," DTNSRDC Report 83/114, David W. Taylor Naval Ship Research and Engineering Center, Bethesda, Md., March 1984.

40 F. J. Dashnow and L. W. Winn, "Evaluation of Selected Lip Seals/Liners for Propulsion Shafts of Merchant Ships," in *Proceedings*, Propellers '81 Symposium, SNAME, 1981.

41 J. Murray Wilson, "Precise Stern Tube and Strut Bearing Alignment Without Boring" in *Proceedings*, Propellers '84 Symposium, SNAME, 1984.

42 L. Vassilopoulos and P. K. Ghosh, "Simplified Structural Analysis Technologies for Trunnion-type Hub Mechanisms of Controllable Pitch Propellers" in *Proceedings*, Propellers '84 Symposium.

43 E. Bjarne, "Multi-Purpose Ducted CP-Propellers, Some Design and Operational Problems" in *Proceedings*, Propellers '88 Symposium, SNAME, 1988.

44 J. B. Hadler, W. B. Morgan, and K. A. Meyers, "Advanced Propeller Propulsion for High-Powered Single-Screw Ships," *Trans.* SNAME, Vol. 72, 1964.

45 H. Yeh and N. Hubble, "Propulsive Characteristics for a Cargo Ship With Contrarotating Propellers," David Taylor Model Basin Hydromechanics Laboratory Report 111-H-01, Nov. 1965.

46 B. D. Cox and A. M. Reed, "Contrarotating Propellers—Design Theory and Application" in *Proceedings*, Propellers '88 Symposium, SNAME, 1988.

47 Richard Hecker, "Powering Performance of a Ventilated Propeller," David Taylor Model Basin Report 1487, June 1961.

48 K. E. Schoenherr, "Formulation of Propeller Blade Strength," *Trans.* SNAME, Vol. 71, 1963.

49 "Shipbuilding, Ship Screw Propellers, Manufacturing Tolerances Part I: Propellers of Diameter Greater than 2.5 Meters," International Standard (ISO) R484/1981 (E).

50 "Instructions for Design of Marine Propeller Blade Gages," *Technical Manual*, NAVSEA S9245-AP-TSM-010/PROP, Department of the Navy.

51 G. V. Cole, W. S. Vorus, and R. F. Kress, "A Rational Approach to Propeller Manufacturing Tolerances" in *Proceedings*, Propellers '84 Symposium, SNAME, 1984.

52 J. H. Smith, "Nodal Arrangements of Geared Drives," *Trans.* Institute of Naval Architects, 1922.

53 F. M. Lewis, "Torsional Vibration in the Diesel Engine," *Trans.* SNAME, Vol. 33, 1925.

54 F. M. Lewis, "Dynamic Effects" in *Marine Engineering*, H. L. Seward, Ed., SNAME, 1944.

55 S. C. Powell and W. V. Bassett, "Practical Aspects of Torsional Vibration in Marine Geared-Turbine Propulsion Units," *Trans.* SNAME, Vol. 52, 1944.

56 Eugene Panagopulos, "Design-Stage Calculations of Torsional, Axial, and Lateral Vibrations of Marine Shafting," *Trans.* SNAME, Vol. 58, 1950.

57 A. D. Andriola, "Torsional Vibration in Geared-Turbine Marine Propulsion Plants—An Introduction to the Subject," *Trans.* SNAME, Vol. 58, 1950.

58 L. C. Burrill and W. Robson, "Virtual Mass and Moment of Inertia of Propellers," North East Coast Institution of Engineers and Shipbuilders, Newcastle upon Tyne, U.K., 1962.

59 J. P. Den Hartog, *Mechanical Vibrations*, McGraw-Hill, New York, 1947.

60 W. Ker Wilson, *Practical Solutions of Torsional Vibration Problems*, Vol. 1, *Frequency Calculations*, Chapman and Hall, London, 1971.

61 W. Ker Wilson, *Practical Solutions of Torsional Vibration Problems*, Vol. 2 *Amplitude Calculations*, Chapman and Hall, London, 1963.

62 Michael G. Parsons and William S. Vorus, "Added Mass and Dampening Estimates for Vibrating Propellers" in *Proceedings*, Propellers '81 Symposium, SNAME, 1981.

63 S. Archer, "Torsional Vibration Dampening Coefficients for Marine Propellers," *Engineering*, May 13, 1955.

64 Lars Bryndum and Stig Jakobsen, "Vibration Characteristics of Two-stroke Low Speed Diesel Engines," MAN B&W-Diesel A/S, Copenhagen.

65 J. R. Kane and R. T. McGoldrick, "Longitudinal Vibrations of Marine Propulsion-Shafting Systems," *Trans.* SNAME, Vol. 57, 1949.

66 C. P. Rigby, "Longitudinal Vibration of Marine Propeller Shafting," Institute of Marine Engineers, Feb. 1948.

67 A. A. J. Couchman, "Axial Shaft Vibration in Large Turbine-Powered Merchant Ships," *Trans.* Institute of Marine Engineers, March 1965.

68 G. P. Antonides, "Longitudinal Vibration of Propulsion System on USS *Simon Lake* (AS33)," David Taylor Model Basin Report 2147, Jan. 1966.

69 A. J. Johnson and W. McClimont, "Machinery Induced Vibrations," *Trans.* Institute of Marine Engineers, 1963.

70 R. Wereldsma, "Propeller Excited Shaft and Hull Vibrations of Single Screw Ships," *International Shipbuilding Progress*, Dec. 1964.

71 S. Archer, "Propeller Excited Vibration; Five Blades or Four?" *International Shipbuilding Progress*, March 1962.

72 R. T. McGoldrick, "Ship Vibration," David Taylor Model Basin Report 1451, Dec. 1960.

73 "Mechanical Vibrations of Shipboard Equipment," Military Standard, MIL-STD-167 (Ships).

74 N. H. Jasper, "A Theoretical Approach to the Problem of Critical Whirling Speeds of Shaft-Disk Systems," David Taylor Model Basin Report 827, Dec. 1954.

75 N. H. Jasper, "A Design Approach to the Problem of Critical Whirling Speeds of Shaft-Disk Systems," David Taylor Model Basin Report 890, Dec. 1954.

76 N. H. Jasper and L. A. Rupp, "An Experimental and Theoretical Investigation of Propeller Shaft Failures," *Trans.* SNAME, Vol. 60, 1952.

CHAPTER XI

James F. Dray

Bearings and Lubrication

Section 1
Review of Fundamentals

1.1 Fluid Film Formation. Fluid-film lubrication provides the basis for the operation of bearings of many types. Lubricants used aboard ship include synthetic and natural mineral oils, water, greases, gases, and solids. Fluid-film lubrication depends on the generation of sufficient fluid pressure between adjacent surfaces to transmit the load through the fluid film. Pressure for hydrostatic bearings is supplied by an external pump while hydrodynamic (self-acting) bearings generate pressure by the relative motion between the surfaces, which draws the fluid lubricant into a converging wedge-shaped space. As the flow area decreases, a positive pressure is generated, some of which is relieved by end leakage [1–5].

1.2 Film Pressure Formation. The concept of a positive pressure being generated by lubricant shear in a converging channel or wedge is basic to lubrication theory [1]. Figure 1 depicts a variety of bearings and the fluid-film shapes usually associated with their lubrication. In Fig. 1, W indicates the bearing load, N the speed of rotation, U the relative velocity of the bearing surfaces, p the pressure in the film, and h_m the minimum film thickness. The first three bearings (tapered-land thrust bearing, tilting-pad thrust bearing, and the journal bearing) are treated analytically as hydrodynamic bearings. The surfaces are considered to be smooth, rigid, and perfectly aligned. The fluid is considered to be an incompressible liquid, and in most solutions the viscosity is considered constant across the film but is allowed to vary along the film length. As a result of advances in computer technology, such as finite-element analysis, more elegant simultaneous solutions of the Reynolds and general energy equations have provided methods that allow variation in both local temperature and viscosity through the film [6].

The fourth bearing, a rolling-contact bearing, is treated as being elastohydrodynamic. In elastohydrodynamic analysis, the deformation of the surfaces and the effect of changes in lubricant viscosity due to pressure are included in the analysis by solving first the elastic equations for an approximate pressure distribution, then the hydrodynamic equations for a pressure distribution corresponding to the deflected bearing shape with the lubricant viscosity adjusted to the local pressure. The process is then repeated until an acceptably small error is obtained. It should be noted that fluid films are very thin in elastohydrodynamic lubrication, approximately the same magnitude as the surface deflections. Gear-tooth contact and the contact surface of traction drives are other examples where lubrication is elastohydrodynamic.

In inverse hydrodynamic lubrication, shown in Fig. 1, the bearing surface is easily deformed many times the film thickness, and the lubricant properties remain unchanged at the pressures involved. Blok has shown that the pressure profile attains a relatively constant value over the surface in contrast to the other lubrication regimes. Water-lubricated main propulsion shaft stern-tube and strut bearings are examples of bearings that operate with inverse hydrodynamic lubrication.

In thermal-wedge bearings, the load capacity results from changes in lubricant viscosity due to progressive heating by oil shear. Surface imperfections, such as waviness induced during finishing operations and thermal distortions induced by temperature gradients, often make significant contributions to the load capacity of this type of bearing. Flat thrust washers, such as those used in diesel engines, are examples of thermal-wedge bearings.

The pressure generated between the bearing surfaces is directly proportional to the relative velocity between the bearing surfaces and to the fluid viscosity and is a function of film shape. Figure 1 illustrates some of the pressure profiles associated with the wedge shapes shown. The maximum pressure in a hydrodynamic bearing is from 2½ to 5 times the average unit load. Since the average unit loads for oil-lubricated bearings are normally less than 600 psi, the maximum film pressures are less than 3000 psi. In dynamically loaded bearings, average unit loads often reach 4000 psi.

The maximum pressure in elastohydrodynamic bearings is less than the Hertzian contact pressure. Since the Hertzian contact pressure in a rolling-contact bearing may be as much as 300,000 psi, the film pressures are very high. The analytical treatment of rolling-contact bearings allows for both rolling and sliding in the contact zone by a simultaneous solution of the thermal Reynolds equation and the energy equation with viscosity and density functions of pressure, temperature, and the elasticity equation [7]. Inverse hydrodynamic pressures are generally low,

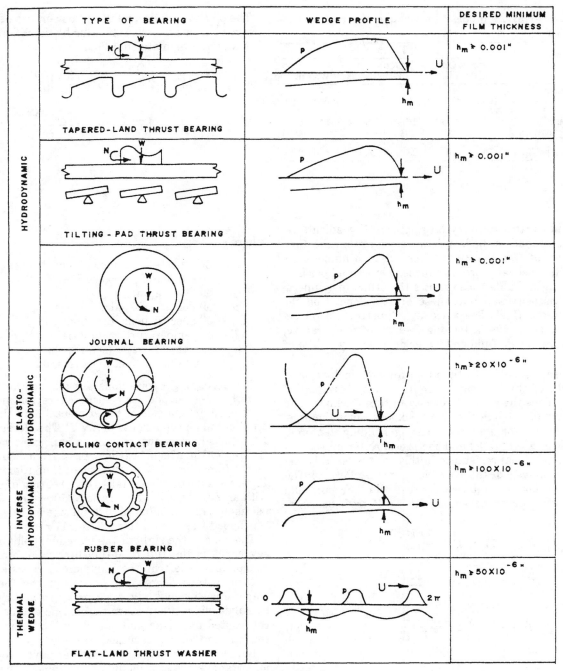

TYPE OF BEARING	WEDGE PROFILE	DESIRED MINIMUM FILM THICKNESS
HYDRODYNAMIC — TAPERED-LAND THRUST BEARING		$h_m \geqslant 0.001''$
TILTING-PAD THRUST BEARING		$h_m \geqslant 0.001''$
JOURNAL BEARING		$h_m \geqslant 0.001''$
ELASTO-HYDRODYNAMIC — ROLLING CONTACT BEARING		$h_m \geqslant 20 \times 10^{-6}''$
INVERSE HYDRODYNAMIC — RUBBER BEARING		$h_m \geqslant 100 \times 10^{-6}''$
THERMAL WEDGE — FLAT-LAND THRUST WASHER		$h_m \geqslant 50 \times 10^{-6}''$

Fig. 1 Lubrication wedges

around 100 psi. The ability of thermal-wedge thrust bearings to carry load is considerably less than that of tapered-land or tilting-pad bearings; the design loads are usually limited to about 75 psi.

1.3 Viscous Effects. The oil leakage out of the lubricant film may be determined from an equation of the form

$$u = \frac{h^2}{12\mu}\frac{dp}{dx} \qquad (1)$$

where

u = mean leakage velocity, ips
h = film thickness, in.
μ = viscosity, lb-sec/in.^{2}
dp/dx = pressure change per unit length, psi/in.

Table 1 gives the pressure gradient in psi per inch of length of flow path required to produce a mean outflow velocity of one inch per second from a unit width of film.

Table 1   Pressure gradient for leakage of 1 ips mean velocity

	FLUID					
	TURBINE OIL NS 2190 TEP		SILICONE DC 500		WATER	
Temperature, deg F	100	200	100	200	100	200
Viscosity, centipoise	98.1	8.75	16	7.3	0.69	0.31
Film thickness, in.	0.001	0.001	0.001	0.001	0.001	0.001
Pressure gradient, psi/in.	170	15	28	13	1.2	0.54

Table 1 illustrates the very large pressure gradients required to produce a modest flow velocity from the film. The values of film thickness given are also numerically equal to the discharge in cubic inches per second per inch of film length. Table 1 also shows the effect of temperature on viscosity and that these effects depend on the nature of the fluid. The viscosity of petroleum oil decreases by 11.4:1 over a 100 deg F range of temperature, whereas the silicon fluid and water decrease by a ratio of 2.2:1.

The shear stress in the fluid film for each inch of bearing length of the bearing is the sum of the product of viscosity times the ratio of velocity to film thickness and the product of one-half the film thickness times the pressure gradient. For a journal bearing running centered, the surfaces are parallel and the pressure term becomes zero. The frictional resistance in this case is entirely viscous. The analysis of this case was given by Petroff. The Petroff bearing is of interest because all other fluid-film bearings have greater friction. For the Petroff bearing

$$F = 4.77 \times 10^{-8} \frac{ZLD^2N}{C} \tag{2}$$

$$H = 3.784 \times 10^{-13} \frac{ZLD^3N^2}{C} \tag{3}$$

$$f = 4.77 \times 10^{-8} \frac{ZN}{P} \frac{D}{C} \tag{4}$$

where

F = tangential shear force on journal, lb
Z = oil viscosity, centipoise
N = journal angular velocity, rpm
L = bearing length, in.
D = journal diameter, in.
C = bearing clearance (bearing bore − journal diameter), in.
H = bearing power loss, hp
f = bearing coefficient of friction
P = bearing unit load (i.e., bearing load divided by LD), psi

The oil flow requirements and power loss of cylindrical sleeve bearings may be estimated by the following simplified and approximate equations [8]:

$$S = 2.42 \times 10^{-9} \left(\frac{ZN}{P}\right)\left(\frac{D}{C}\right)^2 \tag{5}$$

$$H = 3.77 \times 10^{-13} j \frac{ZLD^3N^2}{C} \tag{6}$$

$$Q = 42.4 \frac{H}{\rho C_p \Delta T} \tag{7}$$

$$h_{\min} = (1 - n) \frac{C}{2} \tag{8}$$

where

S = Sommerfeld number (dimensionless)
j = power loss factor (see Fig. 2)
Q = oil flow, gpm
ρ = oil density, lb/gal
C_p = oil specific heat, Btu/lb-deg F
ΔT = oil temperature rise (outlet-inlet), deg F
n = eccentricity ratio (see Fig. 2)
$h_{\min}$ = minimum oil film thickness, in.

The foregoing expressions are among the many that have been developed for bearing characteristics. Hersey developed an analysis of experimental data that showed friction to be proportional to ZN/P. The Hersey number, ZN/P, has been shown to yield an important indication of the region in which bearings pass from boundary and mixed-film lubrication into a zone of fluid-film lubrication. Figure 3 is a plot of data taken on a bearing of 60-deg arc, with an L/D of 0.5, a D/C ratio of 500, and a load of 436 psi using Navy symbol 2190 T turbine oil as a lubricant. The speed was varied from 25 to 211 rpm.

The zone to the right of the minimum point of the Hersey curve is the full fluid-film lubrication domain. The region to the left of the minimum is a zone of mixed friction with a decreasing portion of the forces being transferred by hydrodynamic pressure and the balance by solid-solid contact. Since the marine engineer is often confronted by mixed and boundary lubrication regimes, this subject will be discussed further.

Before leaving fluid films, however, several of the more important references should be mentioned. The theory of fluid-film lubrication is given its most complete analytical treatment by Pinkus and Sternlicht [2]. Purdy [9] has developed the theory in a somewhat simplified manner, but with exceptional clarity in presentation. Hersey has given the most complete record of the development of lubrication theory in all its aspects in reference 10. For bearing design and design checking, the presentations in reference 2, that by Boyd and Raimondi [3], and Rippel [5,11,12] are especially recommended. The work of Wilcox and Booser [8] is based on a large mass of experimental data and thus it offers the advantage of a verified analysis. The design of large marine and waterwheel thrust bearings is analyzed in reference 13.

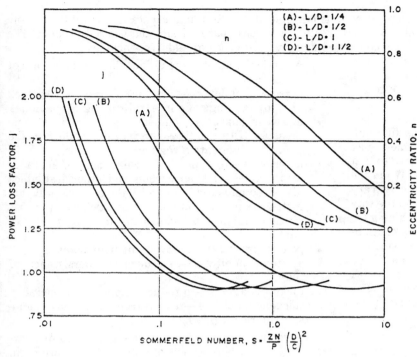

Fig. 2 Eccentricity ratio and power loss factor versus Sommerfeld number for split cylindrical bearings

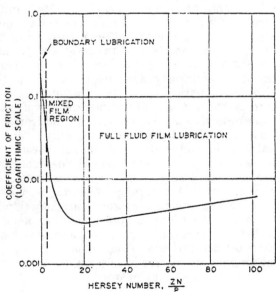

Fig. 3 Bearing friction versus Hersey number

1.4 Boundary and Mixed-Film Lubrication. Boundary lubrication processes are difficult to define with precision since the mechanisms to be included are not well understood. Generally when the film thickness becomes so small that the nature of the surface and the chemistry of the lubricant-surface combination become more important than the bulk properties of the fluid, boundary lubrication processes predominate. Friction wherein only the properties of the solid surfaces are involved is usually known as dry friction. Some dry-friction processes may be involved in boundary lubrication. Mixed film, as the name implies,

is that portion of bearing operation where both boundary and hydrodynamic lubrication processes are effective.

The requirements for a shearing velocity and an oil wedge to produce a hydrodynamic pressure were discussed in the previous section. The time required to establish pressure and the rate at which pressure decays after motion stops were not mentioned. The long retention of low starting friction of some types of bearings may be attributed to their holding a substantial portion of the film pressure trapped between the bearing and shaft surfaces. The squeeze film accounts for the operation of many bearings that experience load reversals (dynamic loads). The lubrication of diesel engine main connecting-rod bearings is assisted by squeeze films, as is piston ring flank lubrication; piston pin bushings are lubricated almost completely by squeeze films.

The significance of the high friction associated with operation in the boundary regime is that the surfaces are in intimate contact, and localized welding and then tearing of the asperity junctions occur. Material transfer and surface damage occur in proportion to the contact pressure, material couple, and sliding distance.

Just as surface roughness can penetrate oil films down to and through the boundary layer, so too can dirt particles that circulate with the oil stream. The effects produced by the passage of dirt through a film can be quite different depending on the circumstances surrounding the event. The least damaging dirt passage occurs when the dirt is smaller in size than the thickness of the film. The thinnest part of the film through which the particle passes determines the damage. If the particle is too small to contact both solid surfaces, then the force necessary for damage is not available and the dirt passes through, leaving the surfaces unchanged. When the particle size exceeds the film thickness, the least damage occurs when a ploughed track in the soft bearing material is the only result. Damage of this type can raise local temperatures, but the bearing damage is quickly smeared back to the level of the surrounding surface in most cases.

The entry of dirt into the bearing has a more serious result when the dirt particle fails to transit the film but, instead, embeds in the bearing surface. Embedment may be complete, in which case the damage is restricted to that done to the journal in the embedment process. When the embedment is incomplete, the consequences may vary from the cutting of a small groove in the shaft, all the way to a catastrophic machining-type failure. In machining (or "steel wool") type failures, the steel surface is continuously cut by particles from the steel journal surface, which are embedded in the bearing material, and those that weld together to form massive scabs. The particles are hardened by friction heating and oil quenching. The probability of a machining-type failure occurring increases with:

- dirt size, quantity, and hardness
- residual hardenability (difference in journal or thrust surface hardness and the maximum hardness attainable with the material)
- oil film temperature (which is primarily related to surface speed)
- amount of chlorine in the oil (usually an extreme-pressure additive)
- composition of rotor materials; carbide-forming elements (high-chrome steels can be especially troublesome)

When designing bearings, consideration must be given to the demands for stable, fully hydrodynamic operation in the upper operating range, the ability of the material and lubricant to provide prolonged operation in the boundary lubrication regime at slow speed, and the probability of encountering some boundary operation at high speed. The great significance of material factors in all types of boundary operation cannot be overemphasized.

1.5 Lubricants. Lubricants for marine machinery usually have a petroleum base. Crude petroleum oils are roughly classified as paraffinic, naphthenic, or mixed base; the differences being associated with the presence of paraffin wax in the paraffinic crude, asphaltic components in the naphthenic, and both in the mixed-base oils. An entire range of crudes exists, ranging from the highly paraffinic to the highly naphthenic; however, advances in refining and in additive chemistry have greatly reduced the significance of geographic source in the final application of the oil. Both the wax and the asphalts are largely removed in the manufacture of the lubricating oil. The specific property levels desired are attained by blending refined, naturally occurring oils and by the introduction of additive packages.

Additives are used to change some chemical and some physical characteristics of an oil. Several additives are usually blended into a single package suitable for converting a specific base oil into the desired product. The commonly used types of additives include:

- oxidation inhibitors—to reduce the onset and rate of oil oxidation
- corrosion inhibitors—to reduce or prevent oil attack on alloy bearings and internal rusting of machines
- antiwear improvers, oiliness agents, and extreme pressure (EP) additives—to improve sliding in the boundary lubrication regime
- detergents—to improve the cleanliness of surfaces
- dispersants—to keep carbon and other insolubles dispersed and circulating with the oil; this prevents malfunctions such as piston ring sticking
- alkaline agents—to neutralize acid from oil oxidation
- pour depressants—to lower the pour point
- antifoam additives—to reduce the persistence of foam
- tackiness agents—to reduce dripping tendencies
- thickener additives—to convert oil to a solid or semi-solid lubricant (grease)

Greases are very important in the lubrication of bearings and gears, where speeds are low and loads are high, and in rolling-contact bearings where oil can bleed rapidly enough from the grease to provide the necessary amounts of lubricant. Grease properties differ greatly from the base oil. The amount and type of thickener additive and the rate of shear of the application both affect the grease properties. Nine softness grades are recognized in the NLGI (National Lubricating Grease Institute) system ranging from 000 to 6 in order of increasing stiffness. Greases are formed as a gel of the base oil and from 4 to 25% of a soap of (in decreasing order of quantity manufactured) calcium, lithium, sodium, aluminum, barium, or other metals or through the use of finely divided inorganic or organic compounds such as modified bentonite, colloidal silica, or arylureas. A wide range of oils has been used, including petroleum, animal, vegetable, and synthetic oils. The wide variety of starting materials and the combinations that have been made have produced lubricants applicable over a temperature span from −100 to 450 F.

The factors to be considered in selecting the lubricants to be used aboard ship are as follows:

- The lubricant in each machine must have sufficient viscosity to meet the most severe anticipated operational requirements.
- The viscosity of the lubricant in each machine should be held to the minimum to minimize power losses and cooling requirements.
- The volume and variety of lubricants carried aboard ship should be minimized.

From the known loads, speeds, and temperatures of the power plant bearings and gear teeth, the physical properties of the lubricants satisfying the maintenance of a safe film thickness can be determined. The lubricant properties of primary concern, the test methods, and the significance of the results are briefly discussed in the following paragraphs.

a. Lubricant viscosity and viscosity index. In 1745, Newton measured the torque required to rotate concentric cylinders in a liquid. He determined that the force required on the rim of the cylinders was directly proportional to the speed of rotation and the area of surface in the liquid and inversely proportional to the distance between moving surfaces. The relationship from these results is $F = \mu(AV/h)$, where μ is the proportionality constant. This constant is the absolute viscosity. When A is the surface area in inches squared, V is the surface speed in inches per second, h is the oil film thickness in inches, and F is the force in pounds, the units of μ are lb-sec/in.² This is a reyn. The corresponding unit of absolute viscosity in the metric system is a poise. The units of a poise are dynes-sec/cm². The metric unit of absolute viscosity most commonly used is a centipoise, which is 1/100th of a poise.

For oils, viscosity is usually measured as the time for a fixed volume to be discharged from a standard container through a fixed orifice at a specified temperature. The viscosity of most oils is nearly independent of the rate of shear and the oils are termed Newtonian fluids. Multigrade engine oils are non-Newtonian. At high shear rates, multigrade oils show a decrease in viscosity with increasing rate of shear until the viscosity of the base oil is approached.

Grease is strongly non-Newtonian. The apparent viscosity is measured by pumping the grease through each of a set of eight capillary tubes using two pumping rates. From the observed pressures, the dimensions of the capillary tubes, and the pumping rate, the viscosity and shear stress are computed and viscosity in poises is plotted against rate of shear in reciprocal seconds on log-log paper. The viscosity of the base oil at the test temperature provides a lower bound on the grease apparent viscosity.

The viscosity index (VI) is a measure of the rate of change of viscosity with temperature. Oils from a Pennsylvania crude, which had the lowest rate of change of viscosity with temperature existing at the time the index was established (1929), were arbitrarily assigned a VI value of 100. Another series of oils from a Gulf Coast crude with the greatest change were assigned a value of 0. The viscosities of each series of oil were determined for 100 F and 210 F. The viscosity index of an oil is then computed as

$$VI = \frac{L - U}{L - H} \times 100$$

where

    U = viscosity at 100 F of the oil whose VI is to be calculated

    L = viscosity at 100 F of an oil of 0 VI having the same viscosity at 210 F as the oil whose VI is to be calculated

    H = viscosity at 100 F of an oil of 100 VI having the same viscosity at 210 F as the oil whose VI is to be calculated

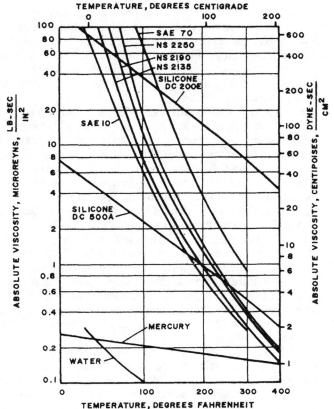

Fig. 4   Lubricating-oil viscosity-temperature chart

In many cases a sufficiently high viscosity will be ensured at the maximum temperature by specifying the minimum acceptable viscosity at each of two temperatures; however, such specifications may permit an excessive viscosity when operating at intermediate temperatures. A comparison of the economic value of lower oil costs for a lower VI oil versus the longer term value of a reduced low-temperature power loss must be made to provide a valid decision.

The effect of temperature on oil viscosity is illustrated in Fig. 4. Mercury and water have been included to illustrate low-viscosity fluids. The data given in Fig. 4 may be converted to other standards by employing the following relationships:

• To obtain reyns (lb-sec/in.²), multiply the viscosity in centipoises (dynes-sec/cm²) by 1.45×10^{-7}.
• To obtain centistokes, divide centipoises by the oil density in grams/cm³.

Above 60 centistokes (cSt), the following approximations may be used:

Saybolt Universal Seconds = 4.62 × cSt
Redwood No. 1 Seconds = 4.05 × cSt
Degrees Engler = 0.132 × cSt

b. Oxidation stability. Oxidation stability provides a measure of the lubricant's resistance to oxidation under an accelerated test. Testing is performed by exposing a small oil sample (300 mL) contaminated with water (60 mL), and containing a steel-copper wire catalyst coil, to a flow of 0.5 liters per hour of oxygen at a temperature of 203 F. Neutralization numbers are determined periodically during and at the end of the test. Limits may be set by reaching a maximum neutralization number or by time. The test provides an indication of the oxidation-limited life of a lubricant in service.

c. Neutralization number. The neutralization number is a measure of the acidity of an oil. The oil is extracted or mixed in a solvent and the number of milligrams of potassium hydroxide required to neutralize a gram of oil is reported as the neutralization number. The neutralization number can provide a useful indication of the degree of oxidation of an oil and hence to the need to replace the oil. Care should be used with additive oils since the additive may affect the neutralization number.

d. Carbon residue. Carbon residue is the percentage of carbonaceous material remaining after an oil sample has been burned. The test is best made in a small glass container (Ramsbottom method) by subjecting a weighed sample of oil to a constant temperature of 1020 F for a fixed time period. The test indicates the carbon-forming tendencies of the oil and is significant for internal combustion engine oils.

e. Demulsibility. The demulsibility of an oil is a measure of the separating tendency of oil-water mixtures or emulsions. Equal volumes (40 mL) of oil and water are stirred for five minutes at 130 F. If the volume of emulsion (unseparated oil and water) remaining is less than 3 mL when the sample is examined at 5-min intervals, the mixture is reported as breaking, and the time is noted. If the mixture fails to break after one hour, the volumes of oil, water, and emulsion remaining are reported. The maximum time for demulsification is specified for each particular oil. In used oils, the test is heavily affected by rust, oxidation products, and other less clearly defined contaminants. The test provides an indication of the effectiveness of separators in water removal and of the probable sludge-forming tendencies of the oil. Heavy sludges or emulsions can block the flow in oil return lines and reduce the effectiveness of oil coolers.

f. Foaming. The foaming tendency of a lubricating oil indicates the stability of air-oil foams. The test is performed by the formation of a foam by the passage of 94 mL per minute of air for 5 min through a 200-mL oil sample. Foam volumes are noted as formed and also after settling for 10 min. Tests are run at 75 F and at 200 F. Lubricating oils having a tendency to foam are undesired because foam can cause inadequate lubrication, oil overflow, blockage of oil return lines, loss of cooler effectiveness, air binding of pumps, etc.

g. Corrosion inhibition. The corrosion-inhibition capabilities of a lubricating oil or grease are intended to define either the tendency of the product to corrode a component of the system being lubricated or to protect the surfaces in the system from being corroded by fresh or salt water entering the system. A wide variety of tests is used on new oils or greases. Each test is closely related to the intended application. Most of the tests are for surfaces wetted by the lubricant; however, tests have been developed also to measure the vapor-space corrosion protection offered by oils that have vapor-space corrosion inhibitors included in their formulation.

All of the corrosion tests are examples of surface chemistry phenomena and, therefore, require exceptional care in the preparation of the test surfaces. The material of the specimen must be exactly to specification. The finishing method is critical in each step as to grades and types of abrasive, speeds of finishing, and scratch pattern and depth. Each of these items has an effect on the surface structure and residual stress, and hence on the corrodibility of the surface. Finally, the cleaning of the specimen must be performed exactly as specified, if all the previous care is not to be wasted.

The actual corrosion test methods are numerous. American Society for Testing and Materials (ASTM), Federal, and military specifications cover most of those tests of concern to the marine engineer. One of the most frequent tests applied to both oils and greases is the copper strip test. Copper is easily corroded and is catalytic to petroleum oils or at least to some of the constituents. The copper strip after proper finishing and cleaning is partially immersed in the lubricant and heated in a bomb pressurized with oxygen for 20 hr at 210 F. After solvent cleaning, the test strip is compared with a reference strip for evidence of discoloration, etching, and corrosion. The test will identify lubricants with sufficient activity to cause corrosion.

The corrosion-inhibition test for steam turbine oils uses a sample of oil (300 mL) contaminated with 30 mL of seawater for marine turbine oils (or with the same amount of distilled water for oils to be used with land-based turbines). The resistance to rusting of very carefully abraded and cleaned low-carbon steel rods that are suspended in the oil-water mixture for 24 hr at 140 F is used as the test criterion.

The Underwood corrosion test is intended for use with internal-combustion engine oil. The test specimens are connecting-rod shells of bearing materials known to be easily corroded. Jets spray oil on the bearings at 375 F for 30 hr.

None of the corrosion tests are expected to yield corrosion rate predictions but they do provide general guidance regarding the corrosive tendency or corrosion-inhibiting characteristics of an oil.

h. Flash, fire, and autogenous ignition points. These three points refer, respectively, to the temperature at which the vapor over a heated oil sample will flash when tested with a small flame; to the temperature at which sustained burning occurs when similarly tested; and to the temperature at which ignition occurs without external flame. These values have no meaning for lubricating processes but are significant where fire hazards exist. Marine lubricating oil distribution systems are generally considered to constitute such a hazard.

i. Pour point. The pour point defines the lowest temperature at which the oil will pour. An oil sample contained in a tube is cooled in 5-deg F increments. The tube is tilted after the temperature stabilizes and the oil behavior is observed. When the sample shows no motion after 5 sec, it is considered to have become solid. The previous 5 deg F point is reported as the pour point. The significance of the pour point in lubrication lies in the inability of pumps to move the lubricant to the bearings in time to prevent damage, when starting near the pour point. Tank heating coils are provided to avoid difficulties associated with low pour points.

j. Grease penetration. The grease penetration number is a measure of the consistency of a grease. Penetration is determined on both worked and unworked samples. Worked samples are produced by forcing the grease back and forth through a perforated disk for a specified number of strokes. Normally, only a modest number of strokes are required; however, a greater number of strokes can be specified and can reveal shear instability of some greases. The penetration is measured by allowing a standard cone to sink into the grease for a period of 5 sec. The depth of penetration in tenths of a millimeter is reported as the penetration. Grease penetration is useful in selecting greases for ball and roller bearing applications and as a quality-control check.

Greases that are too soft tend to excessively flow into the path of rollers and cause high heat through churning. Harder greases, especially those displaying channeling characteristics, run much cooler. Hardness alone will not ensure grease channeling. Channeling is a complex function of grease hardness, tack, fiber structure, and application factors including temperature, shear rate, and vibration. Since no single test suffices to predict channeling, it is often necessary to resort to experiments in critical applications.

k. Grease dropping point. The dropping point of a grease is the temperature at which the grease starts to drop out of the end of an inverted cone. The dropping point is the nearest approximation to the grease melting point. It does not provide the safe upper limit for use of a grease, but it does identify a temperature above the usable range.

l. Service simulation testing. A wide variety of service simulation tests has been included in various lubricant specifications. These tests are included to check the overall suitability of a lubricant for a particular service application, and to provide assurance that operation will be free of trouble from unspecified or unspecifiable lubricant characteristics. Many different types of engine tests and the Navy work factor test may be cited as examples of this type of test. Service simulation tests are often very arbitrary.

m. Lubricant applications. Although petroleum-based materials satisfy the great majority of marine lubrication requirements, a number of other lubricants should be mentioned. The use of synthetic lubricants has increased rapidly with advances in polymer chemistry. Synthetic lubricants in wide use include [14,15]:

- Dibasic acid esters (for high VI, low volatility, and low freezing point)
- Phosphate esters (for fire resistance and boundary lubrication)
- Silicones (for the highest VI, thermal and oxidation stability, and low freezing point)
- Polyglycol ethers (for high VI and good boundary lubrication)
- Polyphenyl ethers (for thermal and hydrolytic stability)
- Silicate esters and polysiloxanes (for excellent viscosity and low-temperature properties)
- Halogenated polyaryls (for hydrolytic stability and fire resistance)
- Polyolefins (for good VI and thermal stability)

For more than a century, stern-tube and strut bearings have commonly been water lubricated. During this time materials for this application have changed from lignum vitae (wood) to laminated phenolic (plastic) and Buna N (nitrile rubber). Buna N water-lubricated strut bearings are in wide use in dredges, tugboats, and Navy ships. Synthetic rubber has particularly good resistance to abrasive wear and hence to salt and sand carried in seawater.

Pumps are frequently lubricated by the fluid being pumped, often water. Air- and gas-lubricated bearings are widely used where rotational speeds are high. (Air has the interesting property that its viscosity increases with temperature, unlike oil and water, whose viscosities decrease with increasing temperature.) These applications include circulating fans, cryogenic turbo-expanders, gyros, computer disks, and tape drives.

Solid lubricants include organics, polymers, metals, inorganics, and glasses [16]. The most popular solid lubricants are the lamellar solids, graphite and molybdenum disulfide. The polymerized olefins, polyethylene and polytetrafluoroethylene (Teflon), are next. Among the organics are soaps, fats, and waxes; and among the inorganics are metal fluorides, chlorides, bromides, iodides, hydroxides, and sulfides. Teflon, polyethylene and metharylates are polymers while indium, tin, lead, and silver are metals. Boron oxide, phosphate, and lead borosilicate are glasses. These materials are used as solid lubricants because they form low shear strength films. These materials may be applied as bonded films, dispersed in an oil or grease, or dispersed in another bearing material, usually a polymer such as nylon. The popularity of these lubricants has increased dramatically because they allow many very severe operational requirements to be met in applications such as control linkage bearings, where speeds are too low and loads are too high for a hydrodynamic bearing.

n. Lubrication of dynamically loaded bearings. Dynamically loaded bearings, such as the bearings in a reciprocating engine or air compressor, derive their load capacity from a different effect than those having a steady load such as line shaft, turbine, and main thrust bearings. They also pose a different set of problems for the engineer. A dynamically loaded journal bearing derives its load capacity from the fact that when the shaft is pushed toward the bearing surface, the oil film takes a finite time to be squeezed out, and the oil-film thickness decays exponentially. Exploitation of this effect has resulted in engine bearings being designed for very high projected area loads and very thin oil films. Steadily loaded journal bearings, such as turbine bearings, might support a load of 150 psi, and a main propulsion thrust bearing a load of 500 psi, while diesel engine bearings designed for dynamic loads of 4000 psi are not uncommon. A turbine journal bearing typically has an operating film thickness at full power of one mil or greater while an equivalent film thickness for a diesel engine bearing would be about 0.4 mil. These operating conditions make reciprocating engine bearings more sensitive to contaminants and place special requirements on the bearing material, oil, and the journal itself. Erosion and cavitation erosion of the bearing material are not uncommon because of the dynamics of the oil film. These fluid phenomena are often controlled by the location of the oil admission hole relative to the load vector.

The bearing materials used in engine bearings must have good fatigue resistance characteristics as well as low surface layer shear strength of good embedability [20]. While both tin- and lead-base babbitt can satisfy these requirements when used in layer thicknesses less than 0.020 in., trimetal bearings are often used to improve one of these properties. A popular trimetal bearing consists of a 0.013 to 0.025 in. intermediate layer of copper-lead on a steel back with a one-mil plated overlay of lead-tin. A trimetal bearing has good fatigue life due the copper-lead intermediate and good surface shear and embedability because of the lead-tin overlay. Leaded-bronze, aluminum, and silver bearings have also been used extensively because of good fatigue properties. It should be obvious that hand working these bearings, with operating bearing metal layers as low as one mil, is highly detrimental to their life.

Engine oils must meet the severe requirements imposed by their being exposed to the high cylinder-wall temperatures. These temperatures cause more rapid oxidation of the oil and its additives. It is important to engine life that the oil be changed at the intervals recommended by the engine manufacturer. A good oil analysis program is also useful because it may detect abnormal conditions in the engine before the normal change interval is reached and thus prevent a catastrophic failure. An oil analysis that is within normal limits should not be a reason to extend the change interval, however, because the analytical equipment used in most such programs does not provide a clear distinction between concentrations of metallic elements in the additive package and those induced by wear. The qualification of an engine oil is a long and tedious process requiring laboratory bench tests, engine tests, and service experience. Therefore, oil specifications for a particular engine should not be changed without extensive supporting evidence, and this includes changes in the additives in the oil. Diesel engines have to operate with foreign particles suspended in the oil, which are introduced as products of combustion, wear debris, and any residual crankcase dirt; and to accomplish this the lubricating oil contains detergent additives. Corrosion inhibitors are also used in marine diesel engines since they operate in a seawater atmosphere, and corrosion is often a problem. The iron-oxide particles also contaminate the oil; if these particles can be carried in suspension, then a filter can be used to continuously remove them and minimize engine wear and bearing damage. Proper filtration for both diesel engines and gas turbines commonly requires 50-micron (μm) filters, or finer.

The material, hardness, and surface finish of the shaft journal in the way of bearings are critical parameters in determining the rate of bearing wear. The bearing-journal material combination determines the adhesive wear rate under boundary lubrication conditions. Because of the low minimum oil-film thicknesses, the shaft surface finish (peak-to-valley asperity height) determines how much bearing material is removed by ploughing. A surface finish of 16 RHR is often specified with 8 being needed for heavily loaded bearings. Shaft hardness is important in determining resistance to abrasives in the oil and fatigue life. Soft journals score in the presence of dirt particles that exceed the bearing minimum oil-film thickness, which in turn results in either immediate failures due to oil-film penetration or decreased fatigue life due to a decreased oil film thickness and increased oil-film pressure. Case carbonizing is a surface-hardening treatment that is commonly used on engine journals, and hardnesses in excess of 50 Rockwell "C" are achievable. Shafts operating against a hard bearing material such as 750T aluminum should have hardnesses of at least 50.

Section 2
Bearings

2.1 General. Bearings may be of either the sliding-contact type, where the bearing elements are separated by a film of oil, or of the rolling-contact type such as ball, roller, or needle bearings. Sliding-contact bearings include many varieties of sleeve or journal bearings, and they also include thrust bearings and guide bearings (e.g., bearings which guide linear motions such as crosshead guide bearings in diesel engines). The determination of the type of bearing to use is based upon a study of several characteristics relating to the mechanical requirement, the environmental conditions, and the relative cost.

Each type of bearing has its peculiar advantages with respect to each condition of application; Table 2 is a summary of the relative advantages of each [8]. Generally speaking, sliding bearings are used in propulsion steam turbines, main reduction gears, diesels, industrial-type gas turbines, turbine generator auxiliary power plants, diesel electric emergency plants, large motors or generators, boiler feed pumps, large fans and blowers, compressors and refrigeration plants, line-shaft bearings, stern-tube bearings, steering gears, and rudders. Rolling-element bearings are generally used with electric motors up

Table 2 Characteristics of sliding- and rolling-contact bearings

SERVICE FACTORS	CHARACTERISTIC		SLIDING	ROLLING
	Load	Unidirectional	Good	Excellent
		Cyclic	Good	Excellent
		Starting	Poor	Excellent
		Unbalance	Good	Excellent
		Shock	Fair	Excellent
		Emergency	Fair	Fair
	Speed limited by		Turbulence Temp. rise	Centrifugal loading Dynamic effects
	Misalignment tolerance		Fair	Poor in ball bearings except where designed for at sacrifice of load capacity. Good in spherical roller bearings. Poor in cylindrical roller bearings
	Starting friction		Poor	Good
Mechanical requirements	Space requirements (radial bearing) Radial dimension Axial dimension		Small $\frac{1}{4}$ to 2 times the shaft dia	Large $\frac{1}{5}$ to $\frac{1}{2}$ the shaft dia
	Type of failure		Often permits limited emergency operation after failure	Limited operation may continue after fatigue failure but not after lubricant failure
	Damping		Good	Poor
	Type of lubricant		Oil or other fluid, grease, dry lubricants, air, or gas	Oil or grease
	Lubrication, quantity required		Large, except in low-speed boundary-lubrication types	Very small, except where large amounts of heat must be removed
	Noise		Quiet	May be noisy, depending upon quality of bearing and resonance of mounting
	Power consumption		Varies as $\frac{N^2 D^3 L}{C}$	Varies widely depending upon type of lubrication. Varies directly as speed. Usually lower than slider bearing
Environmental conditions	Low-temp. starting		Poor	Good
	High-temp. operation		Limited by lubricant	Limited by lubricant
Economics	Life		Unlimited, except for cyclic loading	Limited by fatigue properties of bearing metal
	Maintenance		Clean lubricant required	Clean luricant required. Only occasional attention with grease
	Cost		Very small in mass-production quantities, or simple types	Intermediate, but standardized, varying little with quantity
	Ease of replacement		Function of design and installation	Function of type of installation. Usually shaft need not be replaced

to 50 hp, aircraft-type gas turbines, small gear sets, some small auxiliary steam turbines, and small mechanical equipment.

2.2 Pressure-Fed Journal Bearings. Many different types of journal bearings have been developed over the years and a rather large variety remains in use. These may be roughly classified into the three different methods of lubrication; namely, pressure-fed, nonpressurized, and externally pressurized bearings.

A number of designs of pressure-fed bearings are in common use, with the oil supply pressure normally being in the range of 15 to 30 psi. Variations in journal bearings lie in the shape and location of the oil grooves and in the geometry of the bore. Variations of grooving and bores give rise to the following common types of bearings:

a. Cylindrical bearing. This bearing, Fig. 5(a), has a cylindrical bore. It usually has two oil spreader grooves along the split line. It has a good load-carrying capacity and is used in heavily loaded gear bearings. In designs incorporating L/D ratios of 0.5 to 1.0, loadings range from

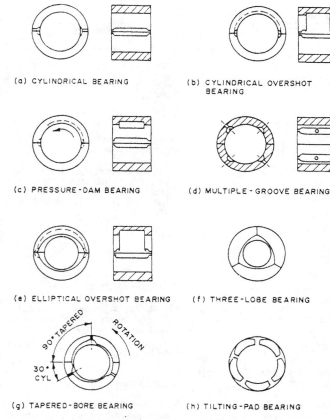

(a) CYLINDRICAL BEARING

(b) CYLINDRICAL OVERSHOT BEARING

(c) PRESSURE-DAM BEARING

(d) MULTIPLE-GROOVE BEARING

(e) ELLIPTICAL OVERSHOT BEARING

(f) THREE-LOBE BEARING

(g) TAPERED-BORE BEARING

(h) TILTING-PAD BEARING

Fig. 5   Types of pressure-fed bearings

100 to 500 psi. At light loads (under 100 psi) and higher speeds (over 3600 rpm) it is very susceptible to a bearing instability known as oil whip.[1]

b. Cylindrical overshot bearing. This bearing, Fig. 5(b), is the same as that in Fig. 5(a), except a relief groove is added in the unloaded half of the bearing. This is used to reduce the power loss and to provide extra cooling capacity. Oil is usually admitted at the trailing-edge spreader groove and allowed to flow over the shaft within the relief. The excess oil over the bearing requirement is often drained away from the leading-edge spreader groove.

c. Pressure-dam bearing. Figure 5(c) shows a pressure-dam or "Newkirk" bearing. In the pressure-dam bearing a groove over the top half of the bearing is terminated at a sharp-edged dam about 45 deg beyond the top vertical line. Stopping the oil flow in the groove at the

dam creates an oil pressure due to viscous pumping and this pressure increases the load on the bearing [17]. This design may be successful in reducing the oil-whip phenomenon; however, it is not always a cure for bearing instability, and it therefore has a limited use.

d. Multiple-groove bearing. This is another variation of the cylindrical bearing, but instead of two spreader grooves at the split line, it has four axial grooves spaced either 35 or 45 deg from the vertical centerline. It has been used successfully in steam turbine rotors at loads of 150 to 250 psi and L/D ratios of 0.4 to 0.6. It has some effectiveness in eliminating oil whip. A four-groove bearing is shown in Fig. 5(d).

e. Elliptical bearing. An elliptical bearing has a bore similar to an ellipse. It is manufactured by machining the bore with shims installed at the split line. After machining, the shims are removed so that when assembled the bearing bore has a larger clearance horizontally than vertically. Usually this clearance ratio is 2 to 1. These bearings are more stable than circular bearings and are used for loads of 100 to 300 psi in L/D ratios of 0.5 to 1.0. They are commonly used as steam turbine, industrial gas turbine, and generator bearings.

An elliptical-overshot bearing is the same as an elliptical bore bearing but with a relief groove in the upper half, similar to the cylindrical-overshot bearing. It is designed for cooler running and for a decreased power loss. An elliptical-overshot bearing is shown in Fig. 5(e).

f. Three-lobe bearing. A three-lobe bearing, shown in Fig. 5 (f), is very effective in preventing oil whip at light-load conditions. Its chief disadvantage is its manufacturing difficulty and awkwardness in handling. It is commonly made in three sections with shims in the joints.

g. Tapered-bore bearing. As a variation of the three-lobe bearing, a three-groove tapered-bore bearing has 30 deg of arc that is cylindrical and 90 deg of arc that is tapered to a depth of 0.004 to 0.005 in. Unlike the three-lobe bearing, the tapered-bore bearing can be machined in a lathe with a special bore-tapering attachment. These bearings can be made in two halves. A sketch of this bearing is shown in Fig. 5(g).

h. Tilting-pad bearing. The most effective bearing to prevent oil whip and dampen shaft vibration is the tilting-pad or multiple-shoe bearing as illustrated by Fig. 5(h). The bearing surface is divided into equal shoes or pads (3, 4, or 5 are the numbers most commonly used) and each pad is allowed to tilt freely as the load and oil-film wedge dictates. Pivot locations are normally centered, but can be shifted downstream to 0.6 the shoe length for increased load capacity. The tilting-pad bearing was designed for the light-load bearing instability problem, but its oil film operates at a lower temperature than a comparable full-sleeve journal bearing and it is capable of withstanding higher unit loads than a full-sleeve bearing.

Tilting-pad bearings are in common use on steam turbines, high-speed reduction gears, and centrifugal compressors.

The bearings as listed in items (a) through (h) are generally in the order of increased rotor damping properties. The cylindrical bearing offers the least protection against

---

[1] Any bearing instability, regardless of its nature, is commonly referred to as "oil whip" in the marine trade. Actually bearing instabilities fall into two types, half-frequency whirl and resonant whip. Half-frequency whirl is a vibration of a shaft in a fluid film journal bearing which may occur at any speed and at a frequency of one-half or near one-half journal speed. A resonant whip is a resonant vibration of a shaft in a fluid-film journal bearing which is self-starting at a shaft speed approximately twice the actual first system critical and remains at a constant frequency as the speed increases. The frequency of this vibration is equal to the first critical frequency of the shaft regardless of running speed.

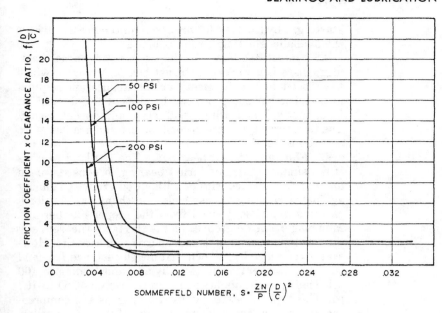

Fig. 6   Friction coefficient versus Sommerfield number for a 13-in. line-shaft bearing

oil-film whip or rotor unbalance vibration, while the tilting pad is the most satisfactory for these types of problems. Even though widely accepted, the tilting-pad bearing is often used only as a last resort, because it is more costly than other types of bearings.

2.3 Nonpressurized Journal Bearings. Nonpressurized bearings include bushings, oil-ring bearings, and disk bearings. Bushings may be oil, air, grease, or water lubricated depending upon the application or material used. Solid bronze bushings are usually grease lubricated while sintered materials are impregnated with lubricants, such as oil or Teflon. Some bushes are solid Teflon or graphite, or metals coated with molybdenum disulfide. Other nonpressurized bearings include line-shaft (also called pillow-block, tunnel, spring, or plummer block) bearings. Line-shaft bearings may be of either the oil-ring type (as in Fig. 23 of Chapter 10) or the disk type (as in Fig. 22 of Chapter 10).

The number of oil rings in a bearing should be such that no ring is required to distribute oil for an axial distance greater than 7 in. on either side of the ring. The ratio of the inside ring diameter to the diameter of the shaft should be 1.5 to 1.25, decreasing with increasing shaft diameter. The ring cross section should be such as to have sufficient weight and area to deliver the required flow at the given speeds and oil viscosities and the oil ring should be submerged a minimum of 1 in. below the oil in the sump under extreme pitch and roll conditions. Tests have shown that an effective grooving pattern for maximum oil flow is $\frac{1}{8}$-in. square grooves spaced $\frac{1}{8}$ in. apart on the ring inside diameter (ID). For low-speed operations, rings for large shafts should be 1 in. wide or more and should be grooved at the ID for increased oil delivery.

There are several disadvantages to oil rings. At low speeds, such as 0.1 rpm while on turning gear, oil delivery is uncertain. Also, rings may tend to hold up on the ring guides and thereby impair oil delivery. In addition, the oil delivery from rings is proportional to viscosity so that, as a bearing overheats and the oil temperature rises in the

sump, the oil delivered to the bearing is decreased at the very time when more oil is needed.

Disk bearings are lubricated by a plate clamped to the shaft at one end of the bearing. At the outside diameter (OD) of the plate a cylindrical section forming a part of the disk rubs against a brass scraper, which removes and directs oil into the bearing, generally at the top centerline of the bearing (as in Fig. 22 of Chapter 10). The cylindrical section of the disk is submerged below the oil sump level, and rotates with the shaft at shaft speed. Disk bearings have a greater and more positive oil flow at all speeds. Disk oil flow exceeds ring oil flow in a given bearing anywhere from 1.5 to 7 times depending on the speed.

Tests with disk-lubricated bearings show that at low speeds the oil delivery can be expressed as:

$$Q = 0.492b(DN)^{3/2}\mu^{1/2} - 0.25 \qquad (9)$$

where

Q = oil flow from disk, gal/hr
b = disk width, in.
D = disk diameter, in.
N = disk rpm
μ = oil viscosity, reyns

2.4 Externally Pressurized Bearings. Sliding bearings of the oil-film type operate on a hydrodynamic film after the surface speed is sufficient to carry the load. Under these conditions an oil-film bearing is extremely efficient and reliable. The problem arises when the speed is too low to maintain a hydrodynamic film. Because steam turbine rotors require a cooling-down period of several hours under turning-gear operation, the line-shaft bearings during this period operate on boundary films or metal-to-metal contact even with extremely low loadings (the line-shaft speeds are in the range of $\frac{1}{10}$ to $\frac{1}{6}$ rpm). Tests on a 13-in. disk-lubricated line-shaft bearing with 300 SSU oil show that mixed-film lubrication (i.e., the onset of boundary film lubrication) exists at a Sommerfeld number of about

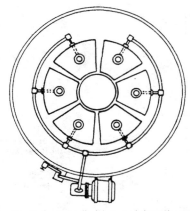

Fig. 7 Hydrostatically lubricated thrust bearing

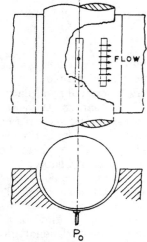

Fig. 8 Schematic diagram of a hydrostatic lift

0.008 (see Fig. 6). It is standard marine practice to limit loadings and rely on large L/D's because of the limitations resulting from continuous operation at turning-gear speed.

Some thrust bearings and journal bearings of large low-speed shafts have externally pressurized pockets in the bearings as shown in Figs. 7 and 8 for starting and stopping under hydrostatic pressure [1]. The hydrostatic lift in thrust bearings consists of a pocket at the center of each pad of approximately 2 to 4 in. dia (Fig. 7). The pads are drilled from the pad OD to the center. An orifice and check valve are mounted on the pad OD, which in turn is connected to a high-pressure manifold supplied with oil from a high-pressure pump. Figure 8 illustrates a hydrostatic lift arrangement for a journal bearing. The oil pocket at the bottom centerline has a total area of from 2 to 5% of the total projected area. Generally, 1 to 2 gpm of oil is supplied at a pressure of about 1000 psi; however, each application must be designed for the specific flow and pressure required to lift the load free of the bearing. Starting friction with hydrostatic lifts is reduced by a factor of about 10 to 1. For thrust bearings on deep-submergence vessels, which have large breakaway thrust on

starting, a hydrostatic lift may be required in cases where the available starting torque is limited.

2.5 Thrust Bearings. As in the case of journal bearings, there is a variety of thrust bearing designs, which have been in use to meet various application requirements. The various types can be classed as follows: flat land, tapered land, tilting pad, step, and pocket thrust. The first three types are in common use and are therefore covered in detail.

a. Flat-land thrust bearings. The flat-land bearing is the simplest form of thrust bearing. It consists of an annular flat surface with or without oil grooves, running against a rotating thrust collar or shoulder. The load-carrying ability is derived from the proper number and radiusing of the oil grooves on the flat face. Theoretically the flat-land bearing has no load-carrying capacity. However, experience has shown that a well designed flat-land bearing will operate successfully at a unit loading of 100 psi. Their normal application is in the range of 50 to 100 psi. This type of bearing is used on pumps and compressors on the inactive face (or idle side) of the thrust collar, where the reverse load is momentary. Where steady, heavy-thrust loads exist, a tapered-land or tilting-pad bearing is used.

The power loss for a flat-land bearing can be determined by the use of empirical data such as presented in reference 8.

b. Tapered-land thrust bearings. The tapered-land thrust bearing has the general appearance of the flat-land thrust bearing. Its surface is divided into a number of pads separated by an equal number of radial oil-feed grooves. In the tapered-land bearing, each pad is tapered in the circumferential direction so that the motion of the runner will wipe oil into the wedge area to build up load-carrying oil pressures. The taper can be either single, that is, the same amount at the ID and OD, or compound, where the taper at the ID is larger than that at the OD. For optimum load-carrying capacity, compound tapers are used. The taper usually extends for 80 to 90% of the pad with the rest of the pad remaining flat. For convenience of manufacture and to accommodate splitting in two halves, an even number of pads is always used. The ratio of pad length to pad height is kept near 1.0 for optimum load capacity. The oil groove width amounts to about 20% of the pad length. Pad tapers range from 0.005 to 0.009 in. at the ID and from 0.003 to 0.006 in. at the OD, depending on the pad size. As the pad increases in size, the amount of cooling oil required increases and thus tapers are increased to pass a sufficient amount of cooling oil. Detail calculations can be made for this type of bearing by employing the procedure outlined in reference 8 or 18.

c. Tilting-pad thrust bearings. The tilting-pad thrust bearing of the Kingsbury or Michell types (both held patents concurrently dating from 1910) differs from the tapered-land bearing in that each pad is an individual shoe that is free to pivot as the oil film dictates. The pivots can be a radial line as in the Michell type or a radiused button support as in the Kingsbury type. It has become common practice in American marine applications for tilting-pad bearings to have plates, known as leveling links, which equalize the load between pads. In Europe and Japan, most marine thrust bearings are built with thrust pads supported directly in housings without leveling links. A conventional marine main thrust bearing of the self-equalizing type is illustrated in Section 5.2 of Chapter 10.

A standard thrust bearing has 6 pads with the OD equal to twice the ID. The pads, with a 51-deg arc, have a length to radial height ratio of 1.33. Bearings of this design have an area equal to one-half the square of the bearing OD. They are capable of carrying loads of 300 to 500 psi, depending on the bearing size, speed, and the type of oil.

Larger bearings for main propulsion shafting usually have restrictions on the bearing OD; therefore, in order to minimize the bearing size, these bearings quite often have 8, 10, or 12 pads. The most common marine propeller thrust bearing for cargo ships and tankers is an 8-pad bearing ranging in OD from 41 to 61 in. There is no hydrodynamic or manufacturing limit on the size of thrust bearings.

The spherically radiused button supports of tilting-pad bearings are generally placed at the center of the pad in the circumferential direction. This is to accommodate ahead and astern rotation. A study by Raimondi and Boyd [19] showed that in theory a flat pad with a center pivot had no load-carrying capacity. In actual practice the pad crowns due to the load over the center pivot and from thermal gradients in the pad itself. Thrust bearings for unidirectional rotation can be designed with a pivot at 0.58 the shoe length for optimum load-carrying capacity.

The bearing design variables of film thickness, power loss, oil flow, and temperature rise can be determined by the hydrodynamic relations given in reference 8.

2.6 Rolling-Contact Bearings.

a. General. Rolling-contact bearings are distinguished by the use of a series of rolling elements to position the shaft with respect to the housing of the machine. The rolling elements most frequently employed include: balls, needles, and cylindrical, tapered, and convex rollers. The rolling elements of rolling-contact bearings provide much closer positioning of shafts than can be achieved with the use of self-acting sliding bearings. In addition to the close positioning, rolling-contact bearings provide a radially (or axially) stiff bearing that permits heavy loading of machine components with minimum deflection. The lubrication system is usually simpler for rolling-contact bearings, especially where the size, load, and speed are such that grease lubrication can be used. Rolling-contact bearings have much lower starting friction coefficients (0.002 to 0.006) than self-acting sliding-contact bearings (0.15 to 0.25).

The load capacity of rolling-contact bearings is fairly well-defined in terms of the cycles of operation to obtain a fatigue failure in a definite percentage of a given population at a single load level. This fatigue failure mode is the normal basis for sizing rolling-contact bearings to satisfy the requirements of a given application. Equations have also been developed to compute the static load to cause surface indentation of a size known to cause rough running. The static load capacity provides a design limit for slow-speed, high-load applications.

The size variation of rolling-contact bearing elements must be kept to very small values. Size variations within an element or between elements in a bearing must be minimized to provide a uniform distribution of load between the elements. Any lack of internal uniformity must be compensated for by compression of the rolls and deflection of the rings and supporting structure. Remarkable advances have been made in achieving uniformity of rolling-element diameters. Military specification MIL-B-17931 restricts the ball size variation within a single bearing to 10 microinches (μin.). Balls with only 3-μin. diametrical variation are available.

Unevenness in mounting surfaces also imposes an unequal load distribution on the bearing components and probably accounts for many of the premature bearing failures.

b. Mounting. The highly desirable rigidity and close-positioning capability of rolling-contact bearings have implications in the incorporation of these bearings into machinery designs. Careful analysis of the starting and operating temperature gradients of the machine design should be made to prevent internal loading of the bearings as a result of thermal expansion forces. A number of mounting designs have been developed to accommodate the most frequently encountered situations. Two basic alternative principles underlie most of the mounting arrangements, i.e., fixed-free mounting or opposed-shoulder mounting. These arrangements assume that the shaft will be supported by two bearings, one near each end of the shaft. Axial positioning will be determined by either imposing the constraint at one end, as in the fixed-free mounting, or allowing the shaft to float between the opposed shoulders of the shaft and housing. The free-end play provided in a fixed-free mounting must exceed the sum of the thermal and elastic differential motion between the shaft and the housing. In cases where the free-end bearing is to be preloaded to provide a quieter installation, the spring force should be applied so that it is reduced by differential thermal expansion within the machine.

Opposed-shoulder mountings tend to be less expensive than fixed-free mounting in manufacture; however, the axial location is not as close as may be obtained in fixed-free mounting. Some opposed-shoulder mounts are designed to allow adjustment of the free play through the use of shims between the cap shoulder and the housing to obtain the degree of axial control sought.

c. Ball bearings. Ball bearings consist of one or two rows of balls contained in grooves having a circular cross section. The grooves form raceways and are normally cut into rings that confine the balls. The radius of the raceway cross section is slightly larger than that of the ball. The largest ball possible, consistent with the other design features of the bearing, is normally used since this gives the largest load capacity. Ball bearings accept either radial or bidirectional thrust loading. Angular-contact ball bearings provide a very high axial load capacity in one direction. Duplex pairs of angular-contact ball bearings are used for very high bidirectional axial loadings.

Ball thrust bearings are designed with a row of balls running in grooved washers placed perpendicular to the axis of rotation. The bearing will accept virtually no radial load, and thrust load is limited to one direction. Two-direction thrust capacity is obtained by adding a second row of balls and a third washer.

d. Cylindrical roller bearings. Roller bearings are classed as line-contact bearings in contrast to the point-contact designation of ball bearings. Cylindrical roller bearings consist of right-circular cylindrical rollers between rings of cylindrical inside and outside diameters. The roller length is less than four times its diameter. Rolls are separated by retainers that may be positioned radially by the rolls or by either of the two rings. The rolls are restrained in an axial direction by ribs on either of the two rings. The cylindrical roller bearing has very little thrust capacity, and for this reason it is frequently used to provide longitudinal freedom in fixed-free shaft mountings. The fixed-end bearing may be any bearing providing axial location. Cylindrical roller bearings with solid rolls usually have a somewhat larger radial play than ball bearings of the same bore size. By the use of hollow rollers, cylindrical roller bearings may be given an internal radial preload. The cylindrical roller bearing has a very high radial load capacity and low friction.

e. Needle bearings. Needle bearings differ from cylindrical roller bearings in having their roll length more than four times the roll diameter. The needle bearing is normally restricted to shafts less than three inches in diameter and speeds below 3600 rpm. The needle bearing is available as a full complement of needles and as a bearing with the needles separated by a cage. The cage-type bearing is less subject to skewing of the needles than the full-complement bearing. The needle bearing occupies the least radial space of any roller bearing. Both needle and cylindrical roller bearings may be used without an inner race; in such a case the shaft must be hardened to a Rockwell C hardness of 58–65 and given a fine grind.

f. Tapered roller bearings. Tapered roller bearings use frustums of a cone as a rolling element. The races have a mating taper. The apex of the tapers on both rings and rolls meet at a single point on the axis of rotation. Cages are used to separate the rollers, and a rib is provided on the inner ring to accept the roll thrust component resulting from the small angle of divergence of the conical roller. The large end of the roll and its mating rib are shaped to provide a converging load-carrying wedge. Pairs of tapered roller bearings used as fixed-end locating bearings provide a very rigid high-load capacity unit.

g. Roller thrust bearings. Roller thrust bearings have no radial load capacity and must be used in conjunction with a radial bearing. The radial bearing must be positioned very carefully if internal loading between the radial and thrust bearing is to be avoided. Roller thrust bearings have a very high thrust capacity. Their speed limits are much lower than radial bearings and more viscous oil is usually employed to prevent smearing of the surfaces. The supporting structure must be very rigid to develop the full capacity of roller thrust bearings. Care must be taken regarding the oil circulation in large roller thrust bearings to avoid thermal distortions.

2.7 Hydrostatic Bearings. Hydrostatic thrust and journal bearings have been used for many years in special applications such as high-precision machine tool spindles. Now they are being used increasingly in marine applications because of two unique and highly useful characteristics: first, they do not depend on shaft rotation to generate a supporting pressure, and second, their vibratory characteristics are controllable within a reasonably wide range. The first of these properties results from the fact that the pressurized oil needed to support the load is supplied by an external high-pressure pump. Oil is usually provided at a pressure equal to about twice the projected bearing area unit load. The static or starting coefficient of friction is, therefore, very low. Hydrostatic lift has for many years been used in large thrust bearings to reduce the starting coefficient of friction and assist in establishing a hydrodynamic film. Bearings having high-pressure oil pockets or grooves in the bearing surface for hydrostatic lift during start-up are usually distinguished from hydrostatic bearings in that most of their running life they operate as hydrodynamic bearings. The pockets or grooves are not optimized as a hydrostatic bearing but are just large enough to provide the needed starting lift-off pressure. Gas-lubricated hydrostatic bearings are externally pressurized and have had extensive usage in gyros and guidance systems, computers and computer tape drives, and in various kinds of communication equipment. Oil-lubricated hydrostatic bearings are used in heavy-duty crane and cargo-handling equipment on ships and in ocean oil-drilling and pumping platforms. These applications have in common the advantages that drive motors can be reduced in size due to lower friction torque requirements and that very heavy loads can be carried since the bearing load capacity is limited only by the supply oil pump pressure. Some of the gains in reduced bearing size are lost because of the increased pump requirements, but in many applications the gains far outweigh the losses. A hydrostatic bearing is designed with perhaps four equispaced recesses around the circumference [5]. Each of these recesses, which is typically 10 to 20 mils deep, is connected to an oil supply manifold via a capillary or orifice restrictor (about a 1/8 in. in diameter) and thence to the pump. The stiffness and damping as well as oil flow are controlled by the design of the recess size and depth, sill width, restrictor diameter and length, and oil supply pressure. Two rows of recesses are often used around the bearing to compensate for misalignment. The bearing minimum oil-film thickness is self-adjusting with load up to the bearing ultimate capacity, since as the load increases on a recess the film thickness decreases, which increases the pressure and, therefore, the resisting force generated. Hydrostatic bearings have a good performance record, but, because of the small minimum oil-film thicknesses and restrictor sizes used, they are sensitive to dirt in the lube oil and hence have rigorous oil filtration requirements.

2.8 Magnetic Bearings. Magnetic bearings are used in applications where noise and vibration are important considerations or where lubricants are unacceptable. Like hydrostatic bearings they do not require shaft rotation to generate bearing support forces. The rotor is floated in space and levitated by magnetic forces, which are generated within the bearing by magnets or electric currents flowing through windings similar to those of an electric motor. This kind of bearing has been successfully applied to such diverse equipment as precision lathe spindles, centrifugal compressors, ventilation fans, generators, and inertial guidance systems. Magnetic bearings have the advantage of not needing a lubricant and can, therefore, operate in a vacuum and be free from contaminants. They have no mechanical losses, so energy can be saved and there are no start-stop problems such as those associated with hydrodynamic bearings [21]. Bearing stiffness and damping are controlled by electrical means, hence magnetic bearings facilitate a transition through critical speeds and may be used to stabilize a rotor or reduce or eliminate forces transmitted to a foundation and, therefore, reduce noise [21]. Magnetic bearings may be of either the active [22] or passive [23] type. A passive magnetic bearing supports the rotor in a magnetic field, and the magnets may be either attractive or repulsive. A rotor supported on passive bearings cannot be stable in all three spatial directions and, therefore, must be electrically served in the unstable directions. In a machine having active magnetic bearings, the rotor is supported on electromagnets, usually an eight-pole configuration that is driven by power amplifiers. Since feedback and active control are required, sensors and control electronics are added. The rotor is centered in the bearing by a bias current in the poles, which can be varied independently of the other quadrants. Journal displacement from the central position, either due to low-frequency drift or rotor unbalance, is sensed and corrected instantaneously and continuously by superimposing a small control current to each bias current. Passive magnetic bearing systems are adequate for many applications; they cost less and have more simple electronics than active systems. However, active magnetic bearings are necessary when the reduction of noise and vibration is a principal design requirement. Magnetic bearings are usually designed to have a projected area loading of about 30 psi.

2.9 Bearing Materials. Grade 2 babbitt is probably still the most popular machinery bearing material in use, which is amazing considering that the original British patent was granted on this material in 1839. The chemical compositions of both Grades 2 and 3 babbitt are covered by Federal Specification QQ-T-390 and also by ASTM Specification B23. Grade 3 babbitt has been used extensively by electrical manufacturers because its higher copper content makes it slightly harder than Grade 2. Grade 2, however, seems to have the ideal balance of strength or load capacity, dirt embedability, and low shear strength with a resulting lower coefficient of friction. Grade 3 has a greater tendency to score journals in the presence of dirt, and cooling rates during casting must be very carefully controlled to prevent segregation of copper cuboids at the bond interface, which often results in a brittle bond that vibration may cause to fracture.

Polymers are being used increasingly as bearing materials. Perhaps the first polymeric bearing to have extensive usage in marine applications was Buna N or nitrile rubber, which was first used as a water-lubricated stern-tube and strut bearing material by the German Navy during the Second World War. It was later adopted by the U.S. Navy for the same application and continues to be used very successfully for that purpose. It is also used extensively in commercial applications, especially in dredges and tugboats that operate in dirty water, since it has a high resistance to abrasion. Laminated phenolic materials have been used extensively as water-lubricated stern-tube and strut bearing surfaces in nested-stave configurations and as a greased rudderstock bearing. Laminated phenolic is supplied as a cotton duck fiber woven laminate in a phenolic resin matrix.

A whole group of true polymeric bearing materials such as nylon and Teflon has been adopted in a variety of marine applications, including control surfaces, winches, cranes, cargo-handling equipment, and auxiliary machinery, based on their earlier success as control surface bearings in aircraft. Included in this category are polytetrafluoroethylene (Teflon) resins, polyamides, polyimides, and polyethylenes. These materials can be strengthened by the addition of chopped or continuous fibers such as Teflon. Woven Teflon fiber itself has become an important bearing material because it has much higher strength than molded Teflon. It is immune to problems such as cold flow, which have plagued molded polymers, while retaining a low coefficient of friction and good wear resistance. The addition of Teflon fiber to a polyamide material also reduces its coefficient of friction. Teflon has a coefficient of friction that decreases with increases in load, unlike most other materials. While fiber additives are used to increase the strength of these materials, friction-reducing additives including molydisulfide and graphite enhance their bearing performance. These bearings are generally intended for high load but lower-speed applications. Many are used with design loads of 4000 psi.

Stoody and Stellite, both high cobalt transition metal alloys, are used in extremely heavily loaded applications where grease lubrication is possible. They are used in control-surface bearings at loadings of 4000 psi and characteristically have extremely long lives. These materials also have good erosion and galling resistance and may be used as-cast or as weld-deposited coatings.

Manufactured carbon-graphites are useful both as high-temperature bearings and as low-friction materials for bearings and mechanical face seals. The amount and size of the graphite crystals are controlled to obtain the required friction characteristics, and porosity is controlled by epoxy impregnation. Graphite crystals contain low shear-strength planes, which produce the desired low coefficient of friction.

2.10 Bearing Failure Modes. Among the bearing failure modes that may be expected to occur in marine applications, the most common are normal wear, bond failure, corrosion, cavitation erosion, improper lube oil, fatigue, overload, misalignment, and electrostatic discharge.

Fig. 9   Abraded thrust pad

Fig. 11   Bond failure

Fig. 10   Fretting corrosion on a thrust runner

Normal wear includes both abrasive and adhesive wear. These failures are usually defined by a bearing exceeding the allowable clearance, and the bearing and journal surface may exhibit a ploughed, abraded, or roughened surface. If the wear is dominated by particulates resulting from surface asperity adhesion and time, then the process is normal. If the abrasives (such as sand, aluminum oxide, or products of corrosion in the oil) cause the failure, then it is abnormal and may be of the type depicted in Fig. 9. Fretting corrosion, which is a form of adhesive wear, is shown in Fig. 10 and is the result of the asperities on the two surfaces adhering to each other under heat and pressure.

Bond failures may occur in all kinds of bearings, especially babbitted machinery and engine bearings (Fig. 11). The two common causes are a lack of cleanliness when preparing the steel, aluminum, or bronze back to accept the bearing material and poor deposition or cooling techniques. Since engine and other machinery bearings are manufactured using metallurgical coating techniques, the bearing back or shell must be absolutely clean, and caustic agents are commonly used to clean them. Any residual oil on the bearing back will inhibit bond formation. Also, an improper cooling rate will result in a brittle or poor bond, which may separate under vibration or dynamic loading. These failures are often easily identified because the original machining marks can be seen on the bearing back or shell.

Corrosion is a trigger mechanism in many marine bearing failures. Corrosion failures are usually initiated by the presence of iron or tin oxide in the oil or on the bearing surface. Typically, the bearing housings or shells rust, the particles of iron oxide are picked up and suspended in the oil and are carried into the bearing load zone, the iron-oxide particles score the bearing surface as they become embedded in the babbitt, and then the particles score the shaft journal area. This type of failure is normally readily identified by the presence of iron-oxide particles embedded in the bearing surface and by the circumferential score marks or grooves machined in the shaft. Tin oxide (stannous or stannic) may be formed on the babbitt surface at the elevated temperatures encountered in the bearing oil film if the oil contains small amounts of water and chlorides. These oxide surface coatings are sometimes hard enough to cut and groove the journal as illustrated by Fig. 12. They are typically black or grey in appearance, and the damaging type can be identified by erasure with an ordinary pencil eraser. If the oxide can be erased, then

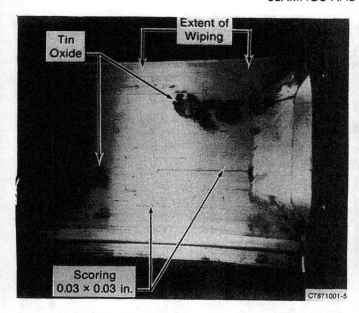

Fig. 12   Failure caused by corrosion (tin oxide)

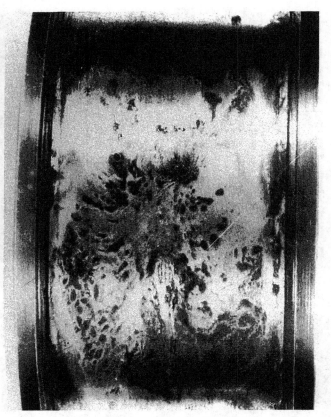

Fig. 13   Cavitation erosion

Fig. 14   Fluid erosion

it is nondamaging; if not, it may be hard enough to cause a failure, and superficial hardness readings should be taken to determine the suitability of the bearing for continued use. Oxide hardnesses have been measured as high as 65 Rockwell "C." For oxides to be damaging to a bearing, the bearing must operate without an oil film, as does a line-shaft bearing when on jacking gear.

Cavitation erosion can occur in babbitted machinery and engine bearings as a result of sub-ambient pressures, which are generated by the relative dynamic motion between the bearing and journal surfaces. The appearance of the damaged bearing surface is shown in Fig. 13 and is that of fine, dull pitting in a localized area. Babbitt is particularily susceptible to this kind of damage and if allowed to progress it can result in a bearing wipe. In some diesel-engine bearings this problem has been eliminated by moving the oil inlet hole to eliminate the low pressure area. The bearing surface is sometimes eroded by the flow of the fluid itself as indicated in Fig. 14. This type of erosion is aided and accelerated by the presence of small contaminant particles suspended in the fluid stream.

Lube oil failures occur occasionally because of additive depletion, water contamination, or loss of lube oil supply. A rigorously followed oil-analysis program substantially eliminates these problems before they progress to the extent that they cause a bearing failure. The oil should be periodically checked for neutralization or total acid number, which is a good indicator of oil oxidation and additive breakdown, to determine when the oil should be changed. An elemental analysis may help to assess the additive concentrations and may also be used to detect wear products suspended in the oil. Care must be taken when interpreting such an analysis since some of the elements (e.g., copper, zinc, aluminum, and tin) present in additives also can result from wear of machinery parts.

Certain other additive elements including chlorine, phosphorus, and sulfur may result from seawater contamination or the products of combustion in an engine. The check to confirm that an oil sample, taken from a sump and placed in a clear bottle, is visually clear and bright provides good insight concerning the condition of the oil. Large quantities of water in the oil separate out on top of the oil and are readily apparent; a 0.1% water-oil emulsion

Fig. 15   Initial fatigue cracks

Fig. 16   True fatigue (unbalanced load)

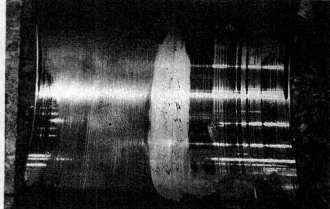

Fig. 17   Wipe due to end load (local overload)

Fig. 18   Wear due to edge loading

Fig. 19   Electrical pitting (electrostatic discharge)

makes the oil cloudy, and as little as 0.01% water makes the oil hazy. Consequently, a visual check of the oil clarity can provide guidance concerning the need to separate water from the oil and find the contamination source. Contaminants in the oil cause the oil to become opaque, which signifies that the oil should be centrifuged.

Fatigue, as illustrated by Figs. 15 and 16, can be easily identified by the fact that surface cracks are formed in the bearing surface that give it the appearance of a jigsaw puzzle. These cracks often follow the babbitt grain boundaries and do not propagate to the bond. Particles of babbitt may come out of the surface in fairly large pieces (i.e., $\frac{1}{8}$ to $\frac{1}{4}$ in.), and these particles often precipitate a wipe. Fatigue failures may sometimes be attributed to the babbitting process. If the cooling rate is too slow when babbitting, the grains are allowed to segregate into anisotropic dendrites, which after thermal cycling rise above the surface of the surrounding babbitt by perhaps 0.1 mil. This condition can be easily observed because the tops of the dendrites are polished as a result of contacting the shaft, and the surrounding areas, which are lower, stay dark. The

surface has a crazed appearance. The heavy bearing contact on the top of the dendrites and the lack of contact on the adjacent areas produce high dynamic stresses at the grain boundaries and ultimately may cause a fatigue failure. Bearings showing this appearance should be changed at the first opportunity [15,16].

Failures caused by overload and misalignment are similar in appearance in that both have a smeared or wiped appearance in the bearing surface. When describing increasingly severe failures of this type, the babbitt surface is identified as being polished, smeared, or wiped. Polishing occurs as a result of journal-bearing contact but no metal flow results. If the contact is heavier and a small localized metal flow occurs, that is known as a surface smear. When a wipe occurs, there is a significant metal flow, as shown by Fig. 17, which is accompanied by an overheating of the bearing surface. Wipes are usually incapacitating failures, and the bearing must be replaced or rebabbitted; however, smears often redistribute the bearing-surface loads and are of no consequence especially if the oil temperature or load is reduced. If the wipe

or smear is at the bearing end and has characteristics illustrated by Fig. 18, it is generally caused by misalignment. If the wipe is in the center or across the bearing length, it is generally caused by overload.

Electrostatic discharge occasionally causes failures to journal bearings in generators and steam turbines. In appearance there are areas of fine pits, as illustrated in Fig. 19, on the babbitt surface in the area where the operating minimum oil-film thickness exists. These pits are similar in appearance to those caused by cavitation of the oil, but the bottoms of the pits are shiny, not dull, because they have been melted by the current flowing to ground. In generators the shaft current is induced in the shaft by the electrical field, while in turbines it is generated by the streaming potential of the steam. The corrective measures are to ground the shaft or insulate one bearing.

Section 3
Lubrication Systems

3.1 Objectives. Lubrication systems have a number of objectives; these objectives may be:

- To deliver the required amount of lubricant to each rubbing surface under controlled levels of temperature and contamination.

- To provide operator assurance that delivery is being accomplished and to provide timely warning of system malfunction.

- To provide the capability of coping with reasonable contingencies.

- In many cases, to provide oil to the speed control system and to the low-oil-pressure shutdown device.

Marine lubricating systems are distinguished by the necessity of including list, trim, roll, and pitch as design criteria. The American Bureau of Shipping [24] requires satisfactory functioning of lubricating systems when the vessel is permanently inclined to an angle of 15-deg athwartship and 5-deg fore and aft. In addition, reference [24] requires that the bearings not spill oil under a momentary roll of 22½ deg, for electrical generators. Military vessels are required to accept larger trim, list, roll, and pitch requirements. Reference 25 cites the same permanent trim and list of 5 and 15 deg for surface ships, but adds 45-deg roll and 10-deg pitch requirements. For submarines, reference 25 imposes a requirement of 30-deg trim, 15-deg list, 60-deg roll, and 10-deg pitch.

3.2 System Types. The lubrication system selection is determined in part by the prime mover selected and, in part, by the layout of the machinery space. The systems for diesels and gas turbines have been covered in the chapters dealing with these prime movers and will not be repeated here; instead, the present chapter will concentrate on the geared steam turbine lubrication system. Reference 26, a publication of the joint ASTM-NEMA-ASME

committee on turbine lubrication, provides the most detailed guidance available for marine geared-turbine systems. Related references 27 through 31 provide additional recommended practices.

There are two basic types of lubricating systems: gravity systems and pressure systems. In addition, however, there are two types of lubricating systems which combine features of both gravity and pressure systems; these are high-head pressure/gravity systems and low-head pressure/gravity systems.

A gravity system uses one or more head tanks to supply oil to the propulsion plant. This system requires a high head room; reference 32 suggests a minimum elevation of 30 ft above the machinery. A gravity system is the most reliable type of system, if adequate capacity is built into the head tanks. Reference 26 recommends a minimum of four minutes of oil supply in each tank. The lubricating-oil pumps supply the head tanks directly in the gravity system.

The pressure system supply is direct from the lubricating-oil pump. This system eliminates the space and weight requirements of the gravity system's head tanks; however, the reserve oil supply must be foregone in order to utilize direct pumping. Two pumps, each capable of supplying the full requirement, are required by reference 24. Automatic switching of the supply from one pump to the other is required.

A gravity/pressure system is a pressure system with one or more head tanks floating on the line to provide a reserve supply of oil. The high-head version is designed to provide full flow at rated pressure for four minutes. The low-head system supplies oil for the same period but at a reduced pressure.

The various types of lubricating systems have most of the functional elements in common. The recommendations on their design, location, and capacity are virtually the same. Oil is supplied to the ship through a fill line to the storage or settling tank, run by gravity to the main lubricating-oil sump, picked up by pump suction, and discharged through filters and lubricating-oil coolers to the head tanks or to the bearings, depending on the specific system in use.

3.3 Tanks. The design of storage, settling, sump, and head tanks is largely based on the same design criteria. The material recommended for use is restricted to clean steel plate. Nonferrous metals have catalytic effects on the oil, which tend to promote oxidation. Many coatings have been attempted for the interior of oil tanks; however, the lack of total success and the high hazard associated with a coating failure have led to the recommendation that interior coatings not be used. Tank inlets should be arranged to prevent the introduction of air below the surface by the impingement of oil jets on the surface. At the same time, it is of substantial benefit to introduce oil as near the free surface as possible to minimize the distance that the air entrained in the oil must rise to reach the surface and be released. Suction intakes in lubricating-oil tanks should be maintained a minimum of 4 in. above the tank bottom to avoid picking up solid contaminants from the bottom. The distance from the inlet to the outlet should be arranged to maximize the distance traveled by the oil in passing through the tank. The greater the transit time, the more the particulate matter will settle out; in order to increase the transit time of the oil, baffles are sometimes provided in tanks to good advantage.

Guidance with regard to the necessary transit time (or settling time) of an oil in a tank can be obtained by investigating the terminal settling (or rising) velocity of the anticipated contaminants. Stokes showed that the viscous resistance, f, of a small sphere of radius r traveling through a fluid, gas, or liquid of viscosity μ with a velocity u, was given by the equation

$$f = 6\pi r\mu u$$

For a sphere acting under the force of gravity, the gravitational force is $(4/3)\pi r^3(w - w')$ where w is the weight density of the sphere and w' the weight density of the fluid. At the terminal velocity, the gravity and viscous forces are equal; therefore,

$$(4/3)\pi r^3(w - w') = 6\pi r\mu u$$

and thus

$$u = \frac{2r^2(w - w')}{9\mu} \qquad (10)$$

where

u = terminal velocity of sphere, ips
r = sphere radius, in.
w = sphere weight density, lb/in.^{3}
w' = liquid weight density, lb/in.^{3}
μ = liquid viscosity, lb-sec/in.^{2}

Table 3 gives the settling rates of the more common contaminants in turbine oil, which is presumed to have a viscosity of 47 microreyns at 70 F and 4 microreyns at 140 F with a specific gravity of 0.9 at 60 F. The table shows

clearly that any chemical agent capable of increasing the size of foreign particles can accelerate contaminant removal; chemical agents are the responsibility of the oil supplier.

No very exhaustive treatment of tank design is recognized as being authoritative. The favorable effect of temperature is recognized in reference 26, which recommends the provision of heaters in settling tanks. Heaters are also suggested for sumps but only for heating the oil to a temperature near that of operation. Reference 31 contains explicit warnings on the danger of heating coils with an excessive surface temperature. A maximum surface temperature of 250 F is recommended with a flow rate of 2 to 3 fps. If steam is used for heating, the pressure should not exceed 5 psig. Electric heater limitations of 12 watts per square inch for moving systems or 6 watts per square inch for static systems are recommended.

Tanks must be provided with adequately sized vent lines and overflows. Vents must be so located that direct impingement of cold air on the vents is avoided; otherwise, the breathing of the tanks under normal operation may bring moist warm air into contact with the cold metal and result in condensation and rusting. Care should also be taken to prevent the possibly combustible vapor in the vent lines from coming into contact with hot surfaces. The safe handling of the overflow should be considered for each tank. Every tank should be provided with a drain system, preferably from a well at the lowest point of a sloping bottom. A connection to the oil purifying system should be located to take suction above the top of the well. Oil drained from the well is normally waste oil, too heavily contaminated to be cleaned with a reasonable effort.

Tanks should be designed with the necessity of initially cleaning the lubricating oil system and the possibility of subsequent cleaning taken into account. Baffles, dams, and the like make cleaning more difficult, but good design practice can relieve this difficulty to some extent.

Tank capacity is based on the maximum oil requirement per unit time, and on the minimum allowable residence time in oil return lines and tanks. The oil defoaming requirement exerts considerable influence on the desired ratio of tank capacity to flow rate. Head tanks in gravity systems are specified by reference 26 to have a capacity of four minutes. Sump tanks must be able to accept at least one overhead tank, in gravity systems. Military ships with pressure systems may use sump tanks having as little as one-minute flow capacity. Storage and settling tanks should be designed to accept a complete oil change.

3.4 Pumps. Lubricating-oil pumps may be of either the positive-displacement or centrifugal types. Reference 24 requires two pumps, each capable of meeting the full system requirement. Reference 26 recommends that each pump have a capacity equal to 125% of the full flow requirement for the system. The use of different types of drives for the two pumps is preferred; if the primary pump or its power source fails, then the resulting low discharge pressure brings the idle pump on the line. However, this arrangement is not common in commercial practice. In commercial practice, both drives are usually electric and

Table 3   Contaminant settling rates in turbine oil

Contaminant:	Air		Water		Sand		Iron Oxide		Iron	
Temperature, deg F	70	140	70	140	70	140	70	140	70	140
Contaminant density lb/in.³	0.000043	0.000038	0.03603	0.03552	0.094	0.094	0.206	0.206	0.284	0.284

Rising Velocity, ips (Air) — Settling Velocity, ips

Contaminant Diameter, mils / μm	Air 70	Air 140	Water 70	Water 140	Sand 70	Sand 140	Iron Oxide 70	Iron Oxide 140	Iron 70	Iron 140
0.08 / 2	0.243×10^{-6}	2.74×10^{-6}	0.0292×10^{-6}	0.411×10^{-6}	0.467×10^{-6}	5.61×10^{-6}	1.32×10^{-6}	15.6×10^{-6}	1.91×10^{-6}	22.5×10^{-6}
0.80 / 20	0.243×10^{-4}	2.74×10^{-4}	0.0292×10^{-4}	0.411×10^{-4}	0.467×10^{-4}	5.61×10^{-4}	1.32×10^{-4}	15.6×10^{-4}	1.91×10^{-4}	22.5×10^{-4}
8.0 / 200	0.00243	0.0274	0.000292	0.00411	0.00467	0.0561	0.0132	0.156	0.0191	0.225
80.0 / 2000	0.243	2.74	0.0292	0.411	0.467	5.61	1.32	15.6	1.91	22.5

Time to Rise 12 in. (Air) — Time to Settle 12 in.

Contaminant Diameter, μm	Air 70	Air 140	Water 70	Water 140	Sand 70	Sand 140	Iron Oxide 70	Iron Oxide 140	Iron 70	Iron 140
2	19 mo	1.7 mo	157 mo	11.2 mo	9.8 mo	0.8 mo	3.5 mo	1.3 wk	2.4 mo	0.9 wk
20	5.7 d	12 hr	1.6 mo	3.4 d	3.0 d	5.9 hr	1.1 d	2.1 hr	17.5 hr	1.5 hr
200	1.4 hr	7.3 m	11.4 hr	49 m	43 m	3.6 m	15.2 m	1.3 m	10.5 m	53 sec
2000	49 sec	4 sec	411 sec	29 sec	26 sec	2 sec	9 sec	0.8 sec	6 sec	0.5 sec

mo = months; wk = weeks; d = days; m = minutes; sec = seconds.

come off the emergency switchboard. If the main generator fails, the emergency generator is started automatically and provides continuity of electric power. Pumps driven by shaft takeoff gearing (usually from the reduction gears) have been used and are satisfactory if properly applied. It should be apparent, however, that meeting the oil requirement at very low speeds and at jacking speeds with a shaft-driven pump is not practical. Auxiliary motor-driven pumps are installed for low-speed and turning-gear operation.

Check valves must be provided on pump suctions to prevent pumping through the standby pump to the sump. Pump isolating valves should be provided to permit underway repair of either pump.

3.5 Strainers. Duplex lubricating-oil strainers are customarily installed after the pump discharge. These strainers, equipped with 80- to 100-mesh screens, are capable of passing 150-μm particles [26]. It should be noted that 150 μm equals 0.006 in., a much larger size than the designed minimum film thickness in marine bearings at the lower speeds of operation. Finer filtration imposes severe penalties in the form of much larger filters and higher pressure drops across the filter. The lubricating oil is given a fine filtration by means of a bypass cycle to the purifier system. That portion of the oil used for turbine speed control may be given supplementary filtration. This higher degree of filtration is restricted to the much smaller quantity of control oil.

Strainers should be provided with pressure-drop indicators to provide warning of the necessity for cleaning the strainers. Various recommendations have been made for an allowable change in pressure drop across a strainer before shifting and cleaning. "Any noticeable increase," "5-psi increase," and "25-percent increase over normal" have been recommended. In a constant-head system, the pressure increase across the strainer represents a reduction of oil flow, and the allowable increase should be related to the system pressure drop. Where head tanks are employed, the allowable pressure drop should be less than that required to overflow the head tank. In a pressure system, the pressure drop should be less than that required to produce a low-oil-pressure alarm at the most remote bearing. Where positive-displacement pumps are employed, the allowable pressure drop is more dependent upon the effects of overpressure on the filter medium. The strainer collapse pressure is an undesirably high limit since strainer openings are enlarged by pressure, permitting larger particles to pass. Allowable pressure limits should be selected that will not endanger the balance of the lubrication system.

Strainer shifting valves should be so arranged that there will be no flow interruption when shifting strainers. Valves should be provided to vent and drain each section of the strainer. Each strainer compartment should be provided with a removable bar magnet system to collect iron particles in the strainer. Regular observation of the strainer basket and magnets can provide one of the earliest warnings of trouble in the system.

3.6 Coolers. Coolers are provided to remove heat from the oil; normally seawater is the cooling medium.

The cooler is designed to cool the oil at full-power operation and at maximum seawater temperature and maximum tube fouling. Single-screw ships are recommended to have two coolers and multiple-screw ships should have one cooler per shaft. Two separate means of circulating water through the coolers are required by reference 24. Drains, vents, and isolating valves are needed in both the oil and water sides of the coolers to permit cleaning, leak detection, and repair.

The coolers are usually operated on a bypass with sufficient oil being passed through the cooler to maintain the oil supply temperature within the design range. The temperature-regulating valves should be arranged so that the oil flow to the system cannot be shut off. Temperature monitoring should be provided at the inlet and outlet of the oil cooler. Provisions should be made for supplying steam at 5 psi maximum to the water side of the cooler in order to heat lubricating or cleaning oil for initial or subsequent cleaning of the lubricating-oil system.

Corrosion protection for the water side of the cooler should be provided by the inclusion of zincs or other electrolytic protection.

The allowance for pressure drop through both the water and oil sides of the cooler must be established for the full range of design conditions. The water pressure drop should be established for clean and for maximum allowable fouling conditions. Oil side losses must be known to establish overall system pressure drops at the maximum and minimum operating oil temperatures. See Chapter 16 for additional discussion regarding coolers.

3.7 Piping and Valves. The lubricating-oil piping system must be designed to deliver the required amount of lubricant under all reasonable conditions with the minimum available head. In addition the piping should be designed with the objectives of providing some separation of entrained air in the drain piping, simplifying the initial cleaning and subsequent cleaning, preventing the cracking of piping due to the vibration of unsupported lengths, and preventing inadvertent maloperation of the system through errors in valve operation.

The objective of securing the required quantity of oil with the minimum head can only be obtained by the most careful study of system pressure drops over the full required operating range. Lubrication piping systems consist of a number of series and parallel flows through pipes of varying diameter and throttling valves that are changed to meet the required conditions of operation. The pressure-drop analysis starts from the pressure and flow required through each bearing and gear spray for the maximum speed conditions. Orifices are usually provided to restrict flow to that needed at the machine element to be lubricated. Where a number of parallel points are to be supplied with lubricant from a single header, some means must be provided to prevent oil starvation of any single point either as the result of excessive design clearance or a bearing casualty allowing excessive oil flow to occur. The details of orifice design are thoroughly covered in reference 33. Virtually all of the precautions cited as being significant in the application of orifices as measuring devices are applicable to their use as flow controls.

Generally, the pump capacity is about 25% in excess of the bearing requirements to allow for changes in flow and other contingencies. In systems pressurized by a pump, the pressure in the system is controlled by recirculating the oil flow excess back to the sump through a back-pressure control valve. This back-pressure control valve is connected to the system downstream of strainers and filters so that the system pressure at the bearings is maintained at the desired pressure irrespective of strainer and lube-oil-cooler pressure drops. An excessive pressure buildup at the pump is avoided by the use of a pressure relief valve at the pump.

Suction piping for pumps should be sized to prevent cavitation at the pump inlet. The oil requirements of bearings increase rapidly at start-up, and this sudden demand can upset the balance of the system. Even positive-displacement pumps can temporarily disrupt a submerged suction supply at start-up if the suction piping has been made too small. Pump suction piping and pump suctions should be submerged in oil to ensure that the pump will immediately take suction upon start-up. If the pump suction and piping cannot be submerged in oil, an oil bleed from the oil header, which is provided with a check valve, should be installed to ensure that the suction piping is maintained full of oil at a pressure above atmospheric.

Drain piping is sized on a different basis than supply piping. Drain lines are designed to run only about one-half full in order to help separate air from the oil. Booser and Smeaton [3] give the following formula for the minimum slope of a drain line running half full:

$$s = \frac{0.015 Q \gamma}{d^4} \tag{11}$$

where

s = slope of drain line, in./ft
Q = oil flow, gpm
γ = oil kinematic viscosity, cSt
d = pipe inside diameter, in.

3.8 Lubricating-Oil Purifier Subsystem. The lubricating-oil purifier subsystem employs the centrifugal force in the purifier to remove very small particles [the duplex strainer is expected to pass particles up to 150 μm (0.006 in.)] from the lubricating oil. The actual particle size passing through a purifier depends on the thickness, longitudinal velocity, and viscosity of the oil layer containing the particle, the magnitude of the centrifugal force, and the mass difference between the particle and the oil volume displaced.

A purifier capacity of at least 10% per hour of the in-service oil charge is recommended by reference 28. Operation at reduced capacity will take out finer particles than operation at design values. Since the rate of oil deterioration is much higher for turbogenerators than for the main turbine, the purifier is arranged to take suction and discharge to each of the main tanks and auxiliary sumps in such a way that the possibility of cross contamination is

virtually eliminated. Separate supply and return connections are located on the tanks to maximize the flow distance within the tank between purified and dirty oil to prevent recycling of purified oil.

The heaters must have the capacity of maintaining the lubricating-oil inlet temperature to the purifier between 160 and 180 F when operating at a normal rate. The purifier oil heater requires the same precautions to limit surface temperatures as were noted for the main lubricating oil heater.

3.9 Condition Monitoring. Condition monitoring of a lubrication system provides information regarding the functional status of the system components, the lubricant, and the bearings serviced by the system. Information should be supplied at a rate and in sufficient depth that reliable performance is ensured. Since the condition of the lubrication system is essential to the availability of the main propulsion plant, an indication of system performance should be provided to the ships command and control system. Where staffing is not a problem, the report of the watch stander may provide sufficient assurance to meet command needs. Where the manning of the engineering space is reduced, an increased depth and sophistication of the condition monitoring input to the command and control system are required.

In designing the condition monitoring system, there is a basic selection of passive and active systems. Passive systems present data for visual observation either at the point sensed or at a remote point; recorded data (manual or machine) belong in the class of passive systems. Active systems compare the data with some limits and provide an alarm or a change in operating conditions or both in the event that the data exceed the specified limits. Both passive and active systems may be designed to either continuously monitor the data or intermittently scan the data at a prescribed rate; the sampling rate is chosen to be compatible with the data-handling response rates.

Lubricating-oil pressures and temperatures are the characteristics most widely monitored to obtain an indication of system performance. By monitoring the pressure at the most remote bearing, assurance is obtained that the system head is being maintained; however, it is recognized that the proper pressure at the most remote bearing does not necessarily ensure that oil is flowing to all bearings. If the pressure at the most remote bearing drops below a preset limit, an alarm sounds and, in addition, the low-pressure signal may start the standby lubricating-oil pump. Low-pressure alarm settings should be made taking into account the difference in head requirements and in elevation between the most remote bearing and other bearings in the system.

A pressure measurement at each pump confirms pump performance. Low pressure at a pump may indicate an excessive oil temperature, piping leakage, suction difficulties, or internal pump wear. An abnormally high pump pressure may indicate a low oil temperature or an obstruction downstream of the pump. A simultaneous measurement of both oil pressure and temperature aids greatly in defining the true problem.

A measurement of the differential pressure across strainers is used to monitor the increase of resistance as deposits of dirt are built up, and such measurements indicate the need to shift strainers. An excessive pressure drop across the water side of coolers usually indicates excessive fouling and the need to clean the tubes at the earliest convenience.

While sight-flow fittings or bubblers have been used to provide assurance that oil is reaching a bearing, that same function can be performed more reliably by differential pressure gages, pressure transducers, or rotary flowmeters in the oil supply lines. These can provide direct indication remotely in central control. Temperature is perhaps the best condition indicator. Temperatures should be measured in the oil sump, at the oil cooler outlet, in the bearing babbitt near the area of minimum oil-film thickness, and in the oil discharge. Typical sensors are either thermocouples or resistance temperature elements. Normal alarm settings for babbitted bearing surfaces are 20 deg F above the maximum temperature encountered on full power trials, not to exceed 250 F. Oil outlet and sump temperatures should not exceed 180 F. Higher temperatures than these weaken the babbitt, reduce its fatigue life, and promote the growth of tin oxide.

References

1   Dudley D. Fuller, *Theory and Practice of Lubrication for Engineers*, Wiley and Sons, Inc., New York, New York, 1956.

2   O. Pinkus and B. Sternlicht, *Theory of Hydrodynamic Lubrication*, McGraw-Hill Book Co., New York, New York, 1961.

3   *Standard Handbook of Lubrication Engineering*, J. J. O'Connor and John Boyd, Eds., McGraw-Hill Book Co., New York, 1968.

4   *Proceedings*, International Symposium on Lubrication and Wear, D. Muster and B. Sternlicht, Eds., University of Houston, Houston, Tex., 1965.

5   H. C. Rippel, *Cast Bronze Hydrostatic Bearing Design Manual*, 2nd ed., Cast Bronze Bearing Institute, Cleveland, 1965.

6   G. Gupta, C. R. Hammond, and A. Z. Szeri, "An Approximate THD Theory for Journal Bearings," *Transactions*, American Society of Mechanical Engineers, *Journal of Tribology*, Fairfield, N.J., April 1990.

7   F. Sadeghi and P. C. Sui, "Thermal Elastohydrodynamic Lubrication of Rolling/Sliding Contacts," *Transactions*, American Society of Mechanical Engineers, *Journal of Tribology*, Fairfield, N.J. April 1990.

8   D. F. Wilcock and E. R. Booser, *Bearing Design and Application*, McGraw-Hill Book Co., New York, 1957.

9   H. F. P. Purdy, *Streamline Flow*, Constable and Co., London, 1949.

10   Mayo D. Hersey, *Theory and Research in Lubrication*, Wiley and Sons, Inc., New York, 1966.

11   H. C. Rippel, *Cast Bronze Bearing Design Manual*, Cast Bronze Bearing Institute, Cleveland, 1965.

12   H. C. Rippel, *Cast Bronze Thrust Bearing Design Manual*, Cast Bronze Bearing Institute, Cleveland, 1967.

13   A. G. M. Mitchell, *Lubrication, Its Principles and Practice*, Blackie, London, 1950.

14   P. M. Ku, "Interdisciplinary Approach to Liquid Lubricant Technology" in *Proceedings*, National Aeronautics and Space Administration-sponsored symposium, Cleveland, Ohio, Jan. 1972 NASA SP-318, U.S. Government Printing Office, Washington, D.C., 1973.

15   A. Z. Szeri, *Tribology-Friction, Lubrication and Wear*, McGraw-Hill Book Company, New York, New York, 1980.

16   D. A. Rigney, *Fundamentals of Friction and Wear of Materials*, The American Society for Metals, Metals Park, Ohio, 1981.

17   B. L. Newkirk and L. P. Grobel, "Oil Film Whirl— A Non-Whirling Bearing," *Transactions*, American Society of Mechanical Engineers, Vol. 56, 1934.

18   O. Pinkus, "Solution of Tapered-Land Sector Thrust Bearing," American Society of Mechanical Engineers Paper 57-A-153, Dec. 1957.

19   A. Raimondi and J. Boyd, "The Influence of Surface Profile on the Load Capacity of Thrust Bearings with Centrally Pivoted Pads," *Transactions*, American Society of Mechanical Engineers, Vol. 77, 1955.

20   R. R. Slaymaker, *Bearing Lubrication Analysis*, Wiley and Sons, Inc., New York, 1955.

21   T. A. Hendrikson, J. S. Leonard, and D. A. Weise, "Application of Magnetic Bearing Technology for Vibration Free Rotating Machinery," *ASNE Journal*, American Society of Naval Engineers, May 1987.

22   J. A. Walowit and O. Pinkus, "Analytic and Experimental Investigation of Magnetic Support Systems. Part 1: Analysis," American Society of Mechanical Engineers paper 81-LUB-31, Oct. 1981.

23   H. M. Chen and J. Dill, "A Conventional Point of View on Active Magnetic Bearings," NASA Langley Workshop on Magnetic Suspension Technology, National Aeronautics and Space Administration Feb. 1988.

24   *Rules for Building and Classing Steel Vessels*, American Bureau of Shipping, New York.

25   "Turbines, Steam, Propulsion, Naval Shipboard," Military Specification MIL-T-17600.

26   "Recommended Practices for the Design of Marine Propulsion Turbine Lubricating Systems," Joint ASTM-NEMA-ASME Committee on Turbine Lubrication, ASME Standard 111, American Society of Mechanical Engineers, New York, 1960.

27   "Recommended Practices for the Design, Operation, and Maintenance of Marine Auxiliary Machinery Lubricating Systems," Joint ASTM-NEMA-ASME Committee, ASME Standard 115, American Society of Mechanical Engineers, New York, 1968.

28   "Recommended Practices for the Purification of Marine Propulsion Turbine Lubricating Oil," ASME Standard 114, American Society of Mechanical Engineers, New York, 1964.

29 "Recommended Practices for the Design of Lubricating Systems for Marine Steam Turbine Pumps for Various Ships Services," Joint ASTM-ASME Standard 115D, American Society of Mechanical Engineers, New York, 1969.

30 "Recommended Practices for the Design of Gas Turbine Generator Oil Systems," ASME Standard 120, American Society of Mechanical Engineers, New York, 1969.

31 "Recommended Practices for the Flushing and Cleaning of Marine Auxiliary Machinery Lubricating Systems," ASME Standard 119, American Society of Mechanical Engineers, New York, 1968.

32 *Marine Machinery Lubrication*, The Texas Company.

33 *Fluid Meters*, ASME Committee on Fluid Meters, American Society of Mechanical Engineers, New York, 1971.

CHAPTER XII

Matthew F. Winkler

Fuels and Fuel Treatment

Section 1
Marine Petroleum Fuels

1.1 Petroleum Fuel-Oil Refining. The quality of marine fuel oil is influenced by the refinery process used, the characteristics of the crude oil, and the demand patterns for middle-distillate and residual fuels. Some of the properties associated with marine fuels that are affected by the refinery processes are as follows:

• Density/specific gravity
• Viscosity
• Conradson Carbon Residue (CCR) micro-carbon residue (MCR) and asphaltenes
• Sediment
• Water
• Flash point
• Compatibility
• Sodium

The following paragraphs include a description of the most common types of refining methods utilized, the resultant product characteristics, and the influence of fuel quality on refinery-based blending of heavier and lighter products to produce various grades of intermediate fuel oils.

a. Fractional distillation. Fractional distillation is the oldest and most common refining process and entails boiling crude oil up to temperatures of about 400 C at atmospheric pressure in a fractionating tower. Figure 1 presents a simplified schematic of the atmospheric distillation process.

Fractional distillation is the initial basic refining process in which advantage is taken of the fact that different fractions of crude oil have increasingly higher boiling ranges. The greater the number of carbon atoms in a hydrocarbon molecule, the higher is its boiling point. Crude oil is composed of thousands of different hydrocarbon compounds, each with its own boiling point.

In practice, it is not possible to extract individual hydrocarbons, so the distillation process divides the crude oil into fractions of different ranges of molecular weights (or carbon numbers), according to their different boiling-point ranges. The lower the boiling point of a fraction, the smaller the average size of its constituent molecules. The crude oil is, therefore, separated by distillation into the various specialized products to meet market demand, each with its own, sometimes narrow, boiling range.

Straight-run residual fuels, obtained from atmospheric distillation, were the primary heavy fuels used in marine diesels and steamships from the 1950's through the early 1970's. They provided clean combustion, ease of fuel handling and treatment, good storage stability, and compatibility. Also, since their specific gravities were usually well below 0.980, the separation of water and sediment was well within the capability of installed shipboard fuel-oil settling and purification systems.

b. Vacuum distillation. Vacuum distillation is essentially a modified version of the straight-run method of distillation. Very simply, when the pressure in the distillation/condensation tower is reduced below atmospheric (partial vacuum), the residual fuel from an atmospheric process will yield additional heavy distillates and will further concentrate impurities and carbon in the vacuum bottoms (residual oil). Vacuum distillation produces residual oils that are feedstocks for other refinery processes. They are not generally available in the marine fuel marketplace due to their very high viscosity.

More modern refineries employ vacuum distillation units, using reheated atmospheric residue as feedstock. In general design, the plant is similar to the atmospheric distillation unit, with bubble trays and caps. However, this secondary fractionating plant operates under a high

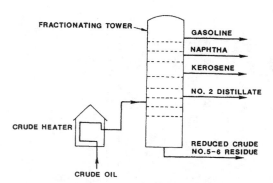

Fig. 1 Atmospheric (straight-run) method of crude-oil distillation

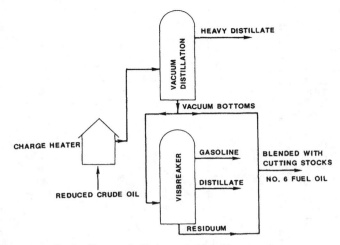

Fig. 2 Vacuum distillation and visbreaking processes

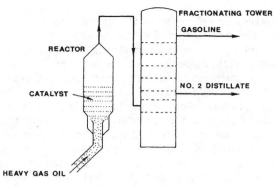

Fig. 3 Catalytic cracking process

vacuum, which reduces the boiling point of the many different hydrocarbons in the atmospheric residue used as feedstock.

As shown in Fig. 2, vacuum-distillation bottoms can be further refined by the use of a secondary process such as viscosity breaking. In this procedure the feedstock, vacuum bottoms, is heated to a higher temperature and pressure for cracking. Simplistically, the high-viscosity feedstock is broken down to further distillates and to a residuum, which is considerably lower in viscosity than the original feedstock. While this product can be utilized as a residual marine fuel with little or no blending with a lighter distillate, it has an increased density (specific gravity) and less-desirable characteristics such as lower ignition quality, higher carbon residue, and higher asphaltenes content. It is usually less stable and less compatible with other residuals than the original feedstock. These characteristics can present problems in the operation of marine diesel power plants and fuel-treatment systems. It is, therefore, used as a blend.

c. Cracking. When heat is applied to hydrocarbon molecules in the liquid phase, some of them boil off or vaporize; this is the basic principle of distillation. However, if the heat is applied when the oil is under pressure, there is a reduced tendency for the hydrocarbon molecules to break free as vapor. If the temperature is high enough, the energy developed by the molecules is such that the carbon-hydrogen and the carbon-carbon bonds are ruptured, and they "crack" into two or more smaller molecules having different chemical properties and structures. During the cracking process, which is essentially chemical, there is a shortage of hydrogen so that some of the molecules formed are unsaturated. These unsaturated (decomposition) molecules, with their smaller number of carbon atoms per molecule, are more volatile.

Two different basic cracking processes may be used. These are thermal cracking and fluid catalytic cracking. As the name implies, thermal cracking requires high temperatures (of the order of 520–560 C) and high pressures (usually about 13 bar) to split the molecules. Catalytic cracking can be carried out effectively at much lower temperatures and pressures (generally about 490 C and 2 bar), usually on raw distillate feedstocks.

d. Catalytic cracking. Catalytic cracking is carried out in a large processing plant, employing a fluidized catalyst, usually a silica-alumina based material, in direct contact with the feedstock. The catalytic cracking process is illustrated by Fig. 3. A catalyst is a substance which aids and accelerates a chemical reaction, but which itself undergoes no permanent change in composition. The very fine powdered catalyst flows like a fluid when kept in constant agitation and circulated by steam, air, or vapor. In the plant, as the preheated heavy gas oil feedstock enters a reactor, it is met by a stream of hot regenerated catalyst. This vaporizes the feedstock, which in turn "fluidizes" the catalyst. Cracking occurs in the reactor, and the cracked oil vapors pass over to a fractionating tower. As indicated by Fig. 3, the cracked vapors, composed of new, smaller molecules, are separated into gases, high-quality gasoline components, light gas oil used as heating oil, and a heavy gas oil known as "cycle oil."

During the cracking reaction, the small spheres of catalyst powder become coated with carbon and fall to the bottom of the reactor. The catalyst is very expensive; therefore, it is returned to the regenerator by hot air. In this unit, the carbon is burnt off, and the cleaned catalyst is then returned to the incoming feedstock for further use. The catalyst, therefore, works in a continuous cycle. A small amount of catalyst is carried out of the regenerator with the exhaust gases despite the use of devices such as cyclone separators fitted to minimize losses.

Further small losses of catalyst can occur, especially if the reactor is malfunctioning. A relatively small amount of residue, called "catalytic tar" or "slurry," is produced in the process. It contains very small catalyst particles, which are known as "cat fines." The slurry is usually clarified to reduce the cat fine content and blended into the residual fuel stream. A number of cases have occurred where carried-over solid catalyst fines have been detected in large quantities (several hundred to one thousand ppm) in residual fuel. Unfortunately, these fines are extremely hard and abrasive and have caused severe wear to diesel engine fuel pumps and injectors, as well as serious abrasive wear in engine cylinders, particularly where the fuel treatments, such as centrifuging and filtration, have been ineffective. The slurry itself is highly aromatic with a very high viscosity and high specific gravity.

e. Refinery blending and storage. While not part of the actual refining process, on-site refinery fuel blending, handling, and storage can have a significant effect on fuel quality, as bunkered. Residual oils bunkered as marine fuels are, in fact, blends of various types of cracked residuum and lighter distillate cutter stock. By blending, less-desirable cracked residual oil can be made more suitable for use as fuel oil. This is accomplished by adding, to a given quantity of residual oil, a small amount of cutter stock, such as light or heavy distillate, or cycle oil. It is not uncommon for these fuels to have been produced in different refinery units and even from different crude stocks; thus, the potential for incompatibility problems, such as sludge formation or stratification, is created. As a result of increasing refinery-yield demands on crude supplies, blends have replaced straight-run residual fuels obtained from atmospheric distillation, which characterized earlier marine residual fuels. Improvements to the refinery cracking process continue to produce higher per-unit crude distillate yields, but also result in a continuing degradation in the quality of cracked residuals and resultant blended fuels available in the marine industry.

A significant portion of fuel contamination by water and debris can be traced to tanker and barge transport, pipeline transport, and tank-farm storage at terminals.

1.2 Petroleum Fuel-Oil Characteristics. The properties of a fuel can significantly affect the performance, operation, and maintenance of a marine power plant. The fuel characteristics that affect the operation of a power plant and fuel handling and treatment systems can be grouped by physical properties and chemical properties.

a. Fuel physical properties.

Viscosity. Viscosity is a measure of a fuel's resistance to flow. The viscosity of a petroleum oil increases when the oil is cooled and decreases when it is heated. For this reason, the viscosity value of an oil must always be associated with the temperature at which the viscosity was determined. The viscosity value by itself is meaningless. When expressed in seconds of Redwood 1 viscosity at 37.7 C (100 F), it is the time required for a fixed amount of fuel at 37.7 C to flow down through an orifice of fixed size. The most common method of measuring the viscosity of a fuel oil is the kinematic method (for example, ASTM D 445). The kinematic method is generally considered to be more accurate than the Saybolt method formerly used in the United States or the Redwood and the Engler methods, which were widely used in Europe.

The world's marine fuel suppliers have agreed to replace the Redwood 1 (SR1) at 37.7 C system with the kinematic centistoke (cSt) system of viscosity at 50 C. Figure 4 illustrates the relationship between the kinematic and Redwood viscosity systems. It is noted that suppliers may quote kinematic viscosities at 100 C, instead of 50 C, to be in agreement with International Standards Organization (ISO), American Society for Testing and Materials (ASTM), and British Standard recommendations.

The viscosity of a fuel is an indication of the ability to pump, treat, and atomize the fuel. Viscosity is also a rough

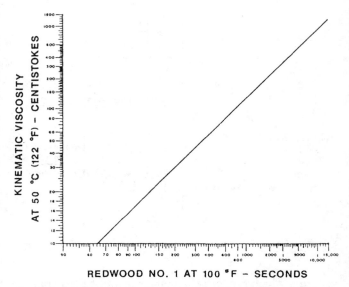

Fig. 4   Relationship between the kinematic and Redwood viscosity systems

indicator of the carbon residue and asphaltenes contents of a fuel; higher amounts are generally found in higher-viscosity fuel oils. Additionally, the lower the viscosity, the easier it is to separate the water and solid particles from the fuel. High-viscosity fuels require proper pre-heating for good centrifuge operation and further heating before injection for good atomization. In addition to heating, an increase in fuel-injection pressure may be necessary to maintain design atomization spray patterns with some high-viscosity fuels. An increase in primary fuel-pump pressure might also be necessary to prevent vaporization of some of the more volatile components of the fuel because of the heating prior to injection. However, heating the fuel on the discharge side of the primary fuel pump, where it is under pressure, can prevent vaporization.

By preheating the fuel to lower its viscosity, the ability to separate water from the fuel is improved, but a temperature of 99 C should not be exceeded because the water may flash into steam and the centrifuge water seal may be lost. In any event, caution must be exercised not to preheat above about 150 C, or 300 F, because some chemical changes may occur, gases may be given off, and water may vaporize, forming steam pockets in the suction line.

Residual fuel oils are normally purchased on the basis of a limiting viscosity due to storage, handling, or power-related restrictions. Viscosity does not, however, carry a quality implication. A false sense of quality assurance may be inferred by the procurement of fuel oils on the basis of lower viscosity only, because other characteristics such as ash and sulfur content, carbon residue, and density can vary widely for a given viscosity.

The viscosity index (VI) of a fuel is an important parameter; for a given fuel viscosity, the higher the fuel density, the higher the amount of cracked components present and, therefore, the lower the viscosity index. The VI of a cracked aromatic residual fuel may be as low as 10–15, whereas straight-run residues blended with straight-run distillates can have a VI as high as 75–80. The widely used "standard" viscosity/temperature curves or tables are generally based upon fuels with a VI of about 65–70. Cracked residuum and cracked distillate blends can deviate widely from these curves, being less viscous at high temperatures but, more important, much more viscous at low temperatures.

It is significant that for many years it was common practice to determine the viscosity of fuels, including residual fuels, in Redwood 1 seconds at 100 F (37.7 C). When viscosities were later determined by kinematic viscosity, mainly because of the inaccuracy of the Redwood viscometer, the temperature at which the viscosity was determined was raised to 50 C, as a result of the increasing use of straight-run residues from Middle East and North African crudes. These were highly paraffinic with a high wax content, and many residual fuels from such crudes exhibited non-Newtonian tendencies at 100 F. Since cracked residuums have become more common, the temperature at which the viscosity is determined was raised to 80 C and has subsequently been increased to 100 C to comply with ISO/BSME and other specifications. Obviously, this increases the margin of error when estimating the viscosity at low temperatures, using normal viscosity/temperature charts, particularly if a fuel becomes a non-Newtonian fluid at normal storage temperatures.

Specific gravity. Practically all liquid petroleum products are handled and sold on a volume basis—by the gallon, barrel, cubic meter, etc. Yet, in most cases, it is desirable to know the weight of the fuel quantity. The specific gravity of a fuel is defined as the ratio of the weight of a given volume of the product at 15.5 C (60 F) to the weight of an equal volume of water at the same temperature. Specific gravity is determined by floating a hydrometer in the fuel and noting the point at which the fuel level intersects the hydrometer scale. Corrections must then be made in accordance with the temperature of the sample at the time of test. The ASTM D-287 standard provides details on the method.

Liquid petroleum products expand when heated, and their weight per unit volume, therefore, decreases. Because of this, the specific gravity, or density, is usually reported at a standard temperature, although another temperature may actually have been used in the test.

The specific gravity of a fuel is a measure of its density; however, in the United States petroleum industry, the API (American Petroleum Institute) gravity scale is widely used. This is an arbitrary scale, calibrated in degrees, that is related to specific gravity by the formula:

$$\text{API gravity (degrees)} = \frac{141.5}{\text{Specific gravity @ 60 F (15.5 C)}} - 131.5$$

The importance of a fuel's specific gravity lies in the fact that standard fuel-water separating techniques are based upon the difference in density between the two substances. Therefore, as the specific gravity of a fuel approaches 1.0, centrifuging becomes less effective. Marine fuels should be devoid of free water and the salts normally dissolved therein; therefore, specific gravity is important because high-gravity fuels are difficult to dewater. A high specific gravity also indicates a heavily cracked, aromatic fuel that is likely to have poor combustion qualities and cause abnormal wear in mechanical components.

Ignition quality. In diesel engines there is always a delay between the start of fuel injection and the start of ignition, or burning of the fuel. The ignition quality of a distillate/residual fuel is indicated by the cetane number (or cetane index) or by the Calculated Carbon Aromaticity Index (CCAI) number. The lower the cetane number of a fuel (or the higher the CCAI number), the greater the ignition delay, and the longer the period of time between fuel injection and the beginning of the rapid pressure rise associated with fuel ignition and combustion. The ignition quality of a fuel is dependent on the characteristics of the crude oil from which the fuel was refined and the extent of refinery processing of the crude oil. As crude oils are refined more intensely, the fuel oils increase in aromaticity, which can lower the ignition quality and increase the ignition delay; this can result in hard knocking (caused by a rapid pressure rise) or noisy engine running, both of which are undesirable over long periods of time. The result could be poor fuel economy, a loss of power, and, possibly, engine damage. If a fuel oil requires blending to reduce its viscosity, the lower-viscosity cutter stock may have a higher aromaticity to prevent incompatibility problems after blending. However, the higher aromaticity of the cutter stock can similarly result in an increase in ignition delay, rough engine operation, and poor performance. This can cause serious operational limitations in medium- and high-speed diesel engines, which are sensitive to the ignition quality of the fuel. Diesels operating at speeds of less than 400 rpm are much less sensitive to fuel ignition quality.

The ignition quality of residual fuel oils is monitored by the CCAI. A nomograph that can be used to derive the CCAI is illustrated by Fig. 5.

The ignition quality of diesel fuels is an important characteristic; however for continuous-combustion processes, such as steam plants, the ignition quality of a fuel has little significance.

Heating value. The heating value of a fuel—the characteristic of primary interest—is related chiefly to its specific gravity (density) and its sulfur content. As the specific gravity of an oil increases, the carbon-to-hydrogen ratio increases. The result is that there is relatively less hydrogen, with its higher heating value per pound, and a consequent decrease in the heat released during combustion. Figure 6 illustrates the relationships between the specific gravity and sulfur content of a fuel and its heating values.

Changes in diesel engine performance caused by the use of a residual fuel result primarily from a reduction in the heat released during the combustion of the higher-gravity fuel with its higher sulfur and water contents. These effects are reflected as changes in the brake specific fuel consumption (a term most frequently used as a performance indicator by operators) and the overall engine efficiency (a more theoretically oriented performance

$$CCAI = D - 141 \log \log (V + 0.85) - 81$$

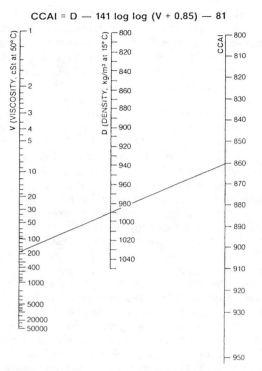

Fig. 5   Nomograph for deriving the Calculated Carbon Aromaticity Index (CCAI)

indicator). The performance of diesel engines is normally based on the fuel's lower heating value, which is derived from the higher (gross) heating value by subtracting the latent heat of the water vapor produced during combustion.

Flash point. The flash point of a fuel is the minimum temperature at which enough fuel vapors will exist to support momentary combustion when an ignition source is present near the fuel surface. Although the flash point of a fuel is largely unrelated to its quality, the flash point is stated to ensure that the fuel can be handled safely.

Residual fuels must be heated to be pumped, but, if a fuel is stored at a temperature above its flash point, a very dangerous condition exists. Fuel in storage should not be heated within 10 deg C of its flash point.

The U.S. Coast Guard and most classification societies have set a minimum safe flash point value of 60 C for fuel oil. A high-viscosity fuel, with a high specific gravity and a low flash point, would be an example of a fuel that is difficult to handle. As a practical matter, however, the flash points of typical diesel fuel oils have no adverse effect on the operation, performance, and maintenance of main propulsion diesel engines.

Pour point. The pour point of a fuel is a measure of the temperature at which a fuel will lose its fluid characteristics. The pour point is unrelated to the quality of a fuel; it is only an indication of how cool a fuel can be before it becomes unpumpable. Fuels of the same viscosity may have pour points that are quite different. The pour point depends on the type of crude, the method of refining, and the additives used. The paraffin fraction of a fuel contains waxy components which, when warm, remain in solution. However, as the fuel temperature drops, the wax begins to crystallize. The fuel will fully crystallize slightly below the pour point and will not pour. Fuels in storage should never be allowed to reach their pour points, as once the fuel has congealed, pumping may not be possible until the fuel is heated 6 to 12 deg C above its pour point. Once a fuel in storage has congealed, much more time is required to heat the fuel because the congealed fuel will not naturally flow to the heating coils.

b.   Fuel chemical properties.

Sulfur. Sulfur is present in most crude oils to varying degrees and is chemically bound in the fuel. The higher-viscosity residual fractions tend to have higher sulfur contents. The form of sulfur, and its concentration, are very important to the operation of a diesel because it can result in low-temperature corrosion. In diesel engines this type of corrosion affects mainly the cylinder liner and

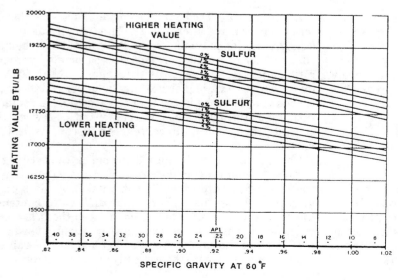

Fig. 6   Fuel heating values relative to specific gravity and sulfur content

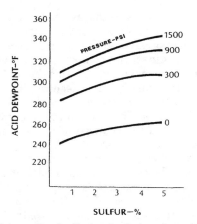

Fig. 7 Acid dew point of sulfur-bearing fuel products of combustion

piston rings at low combustion temperatures and eventually leads to excessive liner wear. In both steam and diesel plants, low-temperature corrosion affects the cooler parts of any heat-recovery surfaces in economizers, air heaters, and waste-heat boilers.

In a diesel engine cylinder, sulfur-bearing compounds unite with oxygen to form sulfur dioxide which, under heat and pressure, further combines with oxygen to form sulfur trioxide. This compound readily unites with the water vapor formed during the combustion process to produce sulfuric acid, which is potentially highly corrosive.

When in a gaseous form, sulfuric acid passes harmlessly out the exhaust; however, as indicated by Fig. 7, when it comes in contact with low-temperature surfaces, the gaseous sulfuric acid can condense and form a highly corrosive liquid. In boilers the economizer tubes and air heater surfaces are subject to such attack, as are the cylinder liners, piston rings, valve stems, and guides in diesel engines. At very low power levels, the exhaust system parts of diesel engines can also be affected.

Diesel engines are designed to avoid low-temperature corrosion by maintaining the surface temperatures of cylinder liners and valve stems sufficiently high to prevent sulfuric-acid condensation. Also, the acid compounds resulting from the use of a fuel with a high sulfur content can be neutralized by using engine lubricating oils of sufficient alkalinity. The overtreatment for sulfur can also be harmful, however, leaving an excess of alkaline material that can form hard, abrasive deposits during combustion, which increase the wear of cylinder liners and piston rings. An ongoing lube-oil analysis program is required to assure a correct acid/alkaline balance that provides resistance to low-temperature corrosion.

Boiler economizers are designed to avoid low-temperature corrosion by maintaining a sufficiently high feedwater temperature. Design countermeasures used with boiler air heaters include the use of ceramic-coated and renewable components.

The sulfur levels in marine fuels have continuously increased as a result of the use of heavier crude oils with higher sulfur contents and more sophisticated secondary processing procedures, such as vacuum distillation and vis-breaking, which concentrate more sulfur in the residuum that is blended to produce heavy marine fuel oils.

Carbon residue/asphaltenes. The carbon residue is the percentage of coked material remaining after a sample of fuel oil has been exposed to high temperatures. While the carbon residue may have significance, it should be interpreted with caution as there may be little similarity between the test and service conditions. As far as the effects of residue on performance are concerned, the type of carbon residue appears to have a greater significance than the quantity.

The Conradson carbon residue (CCR) and the micro-carbon residue (MCR) values are measures of the tendency of a fuel to form carbon deposits during combustion and indicate the relative coke-forming tendencies of a residual oil. Of the two methods, the micro-carbon residue method is the more accurate method to quantify this property. Carbon-rich fuels are more difficult to burn and, therefore, have poor combustion characteristics, which lead to the formation of soot and carbon deposits. Since carbon deposits can be a major source of abrasive wear, the fuel carbon residue value may be an important parameter in the operation of a diesel engine.

A high MCR/CCR/asphaltenes level denotes a high residue level during combustion and may lead to ignition delay as well as afterburning carbon deposits, which result in engine fouling and abrasive wear. The degraded engine performance caused by slow-burning, high-boiling-point constituents also entails a higher thermal loading.

Normally, a diesel engine burns the majority of a heavy fuel oil between top dead center and about 45 deg after top dead center while the expanding products of combustion act on the piston to complete the power stroke. Afterburning is of particular importance because the increased time needed for combustion exposes more of the cylinder liner to the flame than would normally occur and subjects the cylinder lubricant to higher pressure and temperature stresses, resulting in hot spots, severe radiation, and burning of the lube-oil film, which leads to scuffing of the cylinder, increased ring wear, and engine deposits.

Fuels with high MCR/CCR values have an increasing tendency to form carbon deposits on the injection nozzles, pistons, and in the ports of two-stroke engines. This causes a reduction in the efficiency of these elements and increased wear. The maximum permissible MCR/CCR value depends on the engine speed. The higher the speed, the shorter the time for combustion and the more residue; hence, the acceptable MCR/CCR values decrease as the engine speed increases.

Asphaltenes are those components of asphalt that are insoluble in petroleum naphtha but are soluble in carbon disulfide. They can be hard and brittle. Asphaltenes are made up largely of high-molecular-weight, polynuclear hydrocarbon derivatives containing carbon, hydrogen, nitrogen, oxygen, and, usually, one or more of the three heavy metals—nickel, iron, and vanadium—as well as some sulfur.

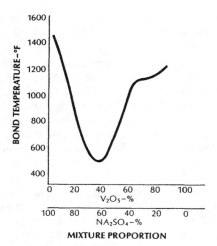

Fig. 8   Bonding temperature relationship for mixtures of vanadium pentoxide and sodium sulfate

The asphaltenes content in fuels influences the compatibility of mixtures of light and heavy products and the rate of heat release during combustion. More frequent sludging of a purifier and filter fouling are indications of an incompatibility of the fuels being mixed and high asphaltene levels. For this reason, fuels taken on at different ports should be kept in separate tanks if possible, and the mixing of fuels from different tanks on board ship should be avoided.

Compatibility. Compatibility is a very important fuel quality. As residual oils become more intensely refined, the resulting fuel is blended with "cutter stocks" of light distillate to produce usable handling characteristics. When light fractions with a predominance of aliphatic hydrocarbons are mixed with a heavy fuel, the colloidal dispersion of high-molecular-weight substances is disturbed. Due to this loss of equilibrium, the heavier constituents tend to precipitate. This is usually seen as a sludge formation, causing clogged strainers and fouled filters and purifiers. Incompatible fuels can also cause injection pump sticking, injector deposits, exhaust valve deposits, and turbocharger turbine deposits.

Vanadium/nickel. Vanadium is a metallic element that is chemically bound in oil in the form of a soluble organo-metallic compound. The vanadium levels in residual fuels vary widely, depending on the origin of the crude. When present in critical amounts, which can vary with the temperatures involved, vanadium forms an adherent slag on high-temperature surfaces, which include the superheater tubes in steam plant boilers and the exhaust valves and seats in valved diesel engines, as well as piston crowns, causing localized hot spots and eventual burning of the high-temperature surfaces.

Vanadium compounds, such as vanadium pentoxide, form as the result of oxidation in the combustion chamber, and mixtures of vanadium pentoxide and sodium sulfate melt at a temperature that varies with the proportion of the two; the bonding temperature relationship is shown in Fig. 8. At lower temperatures, these mixtures remain in a solid state and leave harmlessly with the exhaust

gases as ash. If, however, this ash comes into contact with high-temperature surfaces, such as superheater tubes in boilers or exhaust valves and seats and turbocharger blading in diesel engines, the compounds can melt and bond to the surfaces, causing severe corrosion. A particularly severe problem occurs in some diesel engines when molten vanadium-compound mixtures stick to the sealing surfaces of valve seats. The valve-seat temperature then increases because of the reduced heat transfer. As the deposit thickens, parts of it break off, leaving channels that expose the valve seat to a high-velocity flow of hot combustion gases. The combination of oxidation corrosion and mechanical/thermal surface fatigue forms a leakage path for the combustion gases. This effect drastically reduces the engine performance, but can be avoided if the temperature of the valve-seating surfaces is maintained sufficiently low.

Nickel is another metallic element that is found in heavy crude oils, but in much smaller amounts than vanadium. When compared with vanadium, the combustion problems associated with nickel are minor.

Sodium. Sodium is an alkaline, metallic element that is found only in a combined form, one of which is common salt, sodium chloride (NaCl). Most sodium compounds are water soluble. Sodium is extremely active chemically. The sodium found in fuel can come from several sources, but most of it is a direct result of storing and handling procedures from the time the fuel leaves the refinery until it is delivered to bunkers. Salt-water contamination in barges used to transport fuel is not uncommon. To some extent, even salt-air condensation in fuel tanks contributes to a fuel's overall sodium content. Regardless of the manner of contamination, sodium in fuel is usually water soluble and can, therefore, be removed with the water by a centrifugal separator.

Sodium acts as a paste (flux) for vanadium slag. When unfavorable quantities of vanadium and sodium are present in a fuel, they react at combustion temperatures to form eutectic compounds with ash melting points within operating temperatures. The sodium-vanadium phase diagram and its relationship to the ash melting temperature are shown in Fig. 9. In molten form, sodium-vanadium ash can corrode alloy steels; and, when this condition is allowed to persist unchecked, high-temperature corrosion, overheating, and eventual burning away of exhaust valves, valve faces, and piston crowns in diesel engines are not uncommon. The superheater tubes in boilers can be affected similarly.

The chief corrosive constituents formed in the ash of heavy fuels during combustion are vanadium pentoxide, sodium sulfate, and other complex forms of these primary compounds. The chemical nature of these compounds and their interaction with steel surfaces on exhaust valve seats are of great concern, as their relatively low melting points make them very corrosive at normal engine exhaust temperatures. In their molten states, vanadium-sodium-sulfur compounds also act to dissolve the exhaust-valve surface ferric oxide (Fe_2O_3) layer, thus exposing the underlying steel surface to further oxidation attack and subsequent erosion. The oxidation attack takes place by

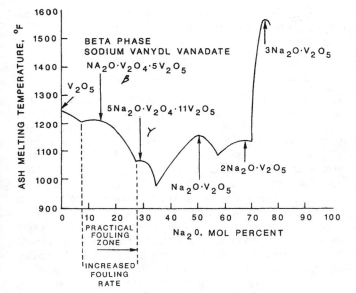

Fig. 9   Sodium-vanadium phase diagram

two mechanisms: gas-phase oxidation and liquid-phase oxidation. In gas-phase oxidation, the high-temperature exhaust gases containing oxygen react with steel to form oxides. Liquid-phase oxidation (corrosion) takes place when molten sulfates and pyrosulfates in the exhaust gases deposit on valve surfaces. In extreme situations, similar sodium-vanadium ash corrosion attacks can also occur downstream of the exhaust valves in the turbocharger exhaust-gas turbine and blading.

Oxidation products. Residual fuels may have a poor storage stability, especially when maintained in a heated condition for extended periods of time. As a fuel ages, it becomes oxidized by air and begins to polymerize, forming sludges, gums, and resins that can foul filters and heaters and can sometimes cause fouling of the injectors, combustion chambers, and exhaust system. Oxidation products can remain soluble in the fuel or become insoluble and precipitate out, forming an organic sediment.

Ash. The ash content of a fuel oil is a measure of the metallic content and the noncombustible inorganic solid contamination in the fuel. The ash content, especially of heavy fuels, provides a method of assessing the quality of the fuel oil and takes into account characteristics such as solid foreign material (sand, rust, cat fines) and the dispersed and dissolved compounds of metals such as vanadium, sodium, iron, and nickel.

Ash deposits can adhere to metal surfaces and cause localized overheating, which can lead to the corrosion of high-temperature components of engines and boilers. Excessive ash constituents may also result in an abrasive wear of cylinder liners, piston rings, and valve seats and form deposits that clog fuel nozzles and injectors. Accelerated wear of fuel pumps and valves is also possible when the fuel has not been properly centrifuged. In a heavy fuel oil, the soluble and some dispersed metal compounds cannot be removed by centrifuging, and when burned they can form hard deposits on piston crowns, cylinder heads

around exhaust valves, valve faces and seats, and in turbocharger gas sides.

Sediment. Sediment in fuels consists of insoluble matter, which may include sand, rust, and catalytic fines as well as tar-like polymerized hydrocarbons and precipitated asphaltenes, both of which break down under the heat of combustion and form lacquer and tarry deposits that impair the movement of parts and sometimes interfere with fuel-nozzle spray patterns. Sediments containing abrasive materials or large amounts of organic materials can damage high-pressure fuel pumps and fuel nozzles. Sediments are removed by centrifuges and filters.

The ash and sediment contamination of combustion chambers is a very important consideration for trunk piston engines because the lubricating-oil system in the crankcase is common with the cylinder lubricating-oil system, and will be contaminated by any combustion products swept past the pistons.

Water. Water can be introduced into the fuel during shipment, or as a result of condensation during storage. The water content of fuel that is delivered by tanker or barge may occasionally be above specified limits as a result of direct seawater contamination. Freshwater contamination may cause an additional problem because fresh water tends to collect water-soluble metals and salts that are present in heavier fuels.

Water, especially salt water, is objectionable in diesel fuel as it may cause injector and piston groove deposits and corrode engine components. Salt water is the greatest single cause of fouling, deposits, and corrosion, especially in the higher-temperature areas of a power plant. Operators of diesel engines that burn residual fuel of any grade centrifuge these fuels to remove water as well as sediment for the protection of the fuel systems and the engines. Salt water can also create a large volume of separator sludge as a result of water-sludge emulsification during centrifuging.

Water can provide the beginning for microbial growth in heavy fuels. These very simple life forms live in the water and feed on the heavy fuel at the water-fuel interface. The result of microbial matter in the fuel can be a slime, which is sometimes corrosive, that fouls strainers, filters, and separators. An immediate solution to this problem is to add a "biocide" chemical additive to the fuel to kill the growth. The much preferred solution is to regularly drain the water, which this growth requires to exist, from the bottom of tanks.

Catalytic fines. Various silica/alumina-based catalysts are used during the fluidic catalytic cracking refining process. The catalyst (cat) fines, which are usually hard, abrasive compounds, sometimes become entrained and delivered in the fuel oil. They can cause severe damage to diesel engines if they are not removed. Small cat fines [1 to 10 microns (μm)] can cause damage to injection pumps and injectors. Larger fines (10 to 70 μm) can cause accelerated wear in piston rings and grooves, cylinder walls, exhaust valve seats, stuffing boxes, and turbocharger blading. Depending on their physical size and concentration in the fuel, cat fines generally cannot be effectively removed by centrifuging. To control cat fines to low levels, 5-μm filtration is necessary after centrifuging. The removal of cat fines is confirmed by monitoring the aluminum and silicon contents in the treated fuel.

c. Effect of chemical impurities on boiler performance. Residual fuels are generally used to fire commercial marine boilers. Poorer quality residual fuels are usually less expensive but they increase the risk of high-temperature deposit formation and corrosion of steam-generator, superheater, reheater, and economizer tubes. Corrosion may become more troublesome when burning fuels that have very high sulfur, carbon residue/asphaltenes, vanadium, or sodium contents.

During combustion at high temperatures, a series of chemical reactions takes place between vanadium, sodium, and sulfur from the fuel and oxygen from the air. A large number of reactions can possibly occur, some of which are transient. For example, with a limited air supply and with fuels having a low sulfur content, only small amounts of sulfur oxides are formed, while vanadium compounds with lower oxygen contents, such as VO, V_2O_4, and V_2O_3, are produced, all with high melting points. With ample oxygen present, chemically active vanadium pentoxide, V_2O_5, which has a lower melting point (690 C), is formed. Similarly, sodium chloride reacts with sulfur and oxygen to form sodium sulfate ($NaSO_4$) with the evolution of chlorine gas (Cl_2).

The fouling of superheater and economizer tubes in oil-fired boilers is of different forms with different results, but is due to the flue gases which may contain solid particles (carbon/ash cenospheres) liquids, (including unburnt liquid hydrocarbon droplets), liquefied salts, and vaporized complex salts.

Furnace waterwall and generating-tube surface temperatures are relatively low and tend to freeze, or solidify, gaseous and some liquid particles which contact them. Solid particles tend to fall off and are carried out of the furnace in the flue gases. At the higher temperatures in the superheater zone, particularly above 600 C, volatile alkaline salts will condense and form a sticky, semimolten deposit layer on the metal surfaces. Apart from promoting corrosion, such deposits act as insulators, raising the outer layer temperature. Although this may reduce the risk of corrosion, the sticky, semimolten surface has a flypaper effect, trapping higher-melting-point metallic salts and carbon/ash particles. Subsequently, the deposit layer thickness can increase to the point where the tube spaces are bridged. This seriously affects combustion and the draft loss and rapidly increases further solid deposition so that the boiler has to be shut down for expensive cleaning.

In addition to problems associated with ash deposition and corrosion at high temperatures in oil-fired watertube boilers, other major potential problem areas involve the conversion of SO_2, formed during combustion of sulfur in the fuel, into SO_3, which at low-temperature zones in a boiler can combine with H_2O vapor to form H_2SO_4. When condensed, this strong mineral acid is highly corrosive to cool, steel surfaces, such as air heaters, economizers, and boiler uptakes.

During the combustion of fuel oil, the sulfur reacts with oxygen in the combustion air to form sulfur oxides in two possible forms:

(1) $S + O_2 \rightarrow SO_2$ (sulfur dioxide)
(2) $SO_2 + \frac{1}{2}O_2 \rightarrow SO_3$ (sulfur trioxide)

An equilibrium constant for the conversion of SO_2 to SO_3 in the presence of excess oxygen enables the SO_3/SO_2 ratio to be calculated as a function of temperature and partial pressure. From a theoretical point of view, only SO_2 could be formed within the flame. Further oxidation starts at 1000 C, when the flue gases leave the furnace and the temperature declines. Below about 580 C, with excess air, there could be a rapid conversion to SO_3, and conversion could be complete at 300 C.

In practice, equilibrium conditions do not exist, especially as varying amounts of excess air are usually present and from about 2 to 5% of SO_3 may be formed in the flame, or high-temperature zone, in the furnace. The speed of conversion may be accelerated by catalysts such as Fe_2O_3 and V_2O_5, which are themselves products of combustion reactions. Bearing in mind that the residence time of the flame and subsequent combustion gases within the furnace seldom exceeds two seconds, it is apparent that it is virtually impossible to identify the intermediate chemical reactions. As a result, various hypotheses have been advanced by investigators on the cause(s) of SO_3 in the high-temperature zones, as well as on the causes of subsequent conversion of SO_2 to SO_3 in the cooler convective zones and the speed of the reactions involved.

For example, catalytic oxidation has been suggested as the main cause of severe corrosion of air heater tubes in boilers operating at high steam temperatures. In one case, with the superheater tube metal temperature at about 600 C and the surfaces covered with deposits having a high V_2O_5 content, the subsequent air heater metal corrosion was high.

With a constant amount of oxygen in the flue gases, there is a linear relationship between the residence time of the gases in the reaction zone and the amount of SO_3 formed; the longer the residence time, the greater the amount of SO_3. Furthermore, as would be expected, there is a direct relation between the amount of surplus oxygen available and the amount of SO_3 produced.

d. Petroleum fuel oil additives. As the quality of residual fuels has declined, the simple purifying, filtering, and heating of the fuel prior to combustion has become insufficient for an engine that burns residual fuel. The increased levels of vanadium, sulfur, and various other ashes, beyond what standard fuel-treatment methods can control, corrode and otherwise degrade engine parts and performance. Higher levels of asphaltenes produce more sludging and fouling problems. Also, at higher fuel prices, the separation and disposal of asphaltenes, with their inherent heating value, becomes a significant avoidable waste. Increased blending, required to achieve fuel specifications, can lead to an increase in stratification and incompatibility. Also, as the specific gravity of a fuel increases, the removal of both free and emulsified water becomes increasingly difficult.

Both chemical and mechanical means exist to combat these and other problems. A chemical approach through additives can be a cost-effective solution to a particular problem. The additives, which are most effective, are discussed generally in this section and Section 2; they are: demulsifiers, emulsifiers, stabilizers and dispersants, ash modifiers, and biocides.

Demulsifiers. Demulsifiers are organic chemical additives that reduce the interfacial tension (film) between the water droplets and the fuel oil. This causes the water droplets to coalesce and break away from the fuel. An additive of this nature is usually added in the storage tank before bunkering fuel. When mixed with the fuel, it allows the separation of water by gravity to occur at a faster rate during storage. A demulsifying agent also improves the ability to remove water during subsequent handling and centrifuging operations. As the heated (and therefore less dense) fuel is centrifuged, the demulsifier will allow the remaining water to separate more freely from the fuel oil. This is most beneficial when the water exists as a fine dispersion or as an emulsion.

Emulsifiers. Emulsifiers are organic chemical additives that stabilize and enhance the interfacial film separating the water droplets and the fuel oil. The main purpose in using an emulsifier additive is to stabilize the dispersion of clean, fresh water that is intentionally added to the fuel to reduce the combustion temperature. This has the beneficial effect of reducing the NO_x emissions, which form at high combustion chamber temperatures. By chemically stabilizing the film around the water droplet, the storage life is increased and the mechanical energy needed to generate the emulsion is reduced. Emulsifying chemicals should not be used prior to centrifuging since stabilized emulsions can greatly reduce the effectiveness of a centrifugal separator to remove solids or water.

Stabilizers/dispersants. Stabilizers and dispersants are primarily used to improve the handling characteristics of a fuel but can increase the fuel's heating value somewhat and decrease maintenance costs. Stabilizers and dispersants help to reduce the formation of sludge and encourage the dispersion of existing sludge. This reduces fuel costs because the sludge, which has a significant energy content, would otherwise be removed by the centrifuge. Asphaltenes (high levels of which tend to lead to sludge formation) can be dispersed throughout the fuel as very fine particles with dispersant additives. The additive acts by incorporating surfactants to form a surface film on the oil-asphaltene interface. It provides enough additional surface tension to break the asphaltenes into minute particles and maintain the stability of the fuel oil.

Ash modifiers. As a means of avoiding high-temperature corrosion, ash modifiers react with corrosive chemical elements during combustion to raise the melting temperatures of the compounds above the level that can be reached during the combustion process. One of the predominant types of ash modifiers is magnesium in different forms. Magnesium compounds are introduced in the fuel system by a metering pump after the purifier, or before the service tank. When they enter the combustion chamber, the magnesium combines with vanadium by-products

and forms magnesium vanadates having high melting points, which pass harmlessly out the stack instead of adhering as a slag to exhaust valves, piston crowns, and turbocharger nozzles in diesel engines and to superheater tubes in boilers.

The sulfur in fuel combines with oxygen to form sulfur dioxide. Sulfur dioxide is itself relatively harmless and passes out the stack. However, if vanadium is present, sulfur trioxide (SO_3) will form due to the catalytic action of vanadium. A magnesium additive in the fuel removes the vanadium and masks the existing vanadium deposits; thus, much less SO_3 forms. The SO_3 that is present combines with the remaining magnesium to form powdery sulfates, which harmlessly pass out the stack.

An analysis of the fuel vanadium content must be undertaken to determine the proper additive dosages. An overdosing of ash modifiers can create an additional problem with modifier deposits. These deposits can have hard surfaces, which can lead to the abrasive wear that they were intended to prevent.

Biocides. Biocides are chemical additives that are intended to kill microbiological growth in residual fuel oil tanks and piping. Microbiological growth can live in the water that accumulates at the bottom of fuel-oil tanks and feed on the hydrocarbon fuel oil at the water-oil interface. Biocides are also utilized to prevent microbiological growth from contaminating a fuel-oil system. Microbiological growth cannot live in a tank without water; good housekeeping practices, such as keeping the tank bottoms drained and maintained free of water, are as important as biocides in the strategy to keep microbiological growth out of marine storage tanks and marine fuel systems.

e. Petroleum fuel-oil exhaust emissions. Ship stack emissions, in the form of gases and particulates that add to air pollution, are monitored by authorities concerned with overall air quality. The discharge of visible smoke is restricted in many port areas, and restrictions are placed on ship operators regarding the discharge of particulates and sulfur and nitrogen oxides. Therefore, apart from the effects on equipment and operating conditions, the quality of the fuel oil burned also has a direct effect on the emissions discharged by a vessel into the air.

In some cases the discharge of sulfur oxides is controlled in a port area, with the regulations placed on the sulphur content of the fuel used in port. In such cases, operating on a higher-quality fuel oil with a low sulfur content while in port could comply with regulations. However, in some cases, the discharge of nitrogen oxides is also controlled, and the most practical solution may be to secure the engines and operate on shore power while in port.

f. Petroleum fuel-oil toxicity. With good standards of personal and industrial hygiene, marine fuel oils, properly used for their intended purpose, present no significant health hazards. However, prolonged exposure to fuel oils may cause dermatitis and skin cancer. To avoid unnecessary or prolonged contact with petroleum products, the following precautions should be followed:

a. Use working procedures which will minimize skin or eye contact with petroleum products.

b. Wear adequate protective clothing such as impermeable gloves, when appropriate.

c. Remove any clothing that has become soaked with oil and use soap and water to thoroughly wash the skin that has been in contact with oil.

d. Do not put oil-soaked rags or oily tools into clothing pockets.

e. Do not use hydrocarbon solvents for washing oil from the skin. Avoid inhaling hydrocarbon vapors.

f. Avoid inhaling oil mist and vapor particularly from the more volatile products. This is especially important since waste solvents, acids, and strong base mixtures are sometimes dumped into marine fuel-oil tanks.

g. Avoid the accidental injection of fuel oil into the skin while testing fuel-injection equipment.

1.3 Petroleum Fuel-Oil Specifications. Commercial refineries produce the fuels sold to commercial marine ships as well as to military ships. Even when the commercial and military marine fuel volumes are combined, the marine fuels consumed represent only a small portion of the total refinery volume produced. Therefore, marine fuels, especially the residual types, are the by-products of other major fuel consumers' needs, and the special needs of the marine operator are of little influence.

The types or grades of commercial marine fuels vary widely from gas oil, to marine diesel fuels, to blended fuel oils, and finally, to heavy residual oils. Gas oils are clean, clear distillates that are shipped in dedicated clean tanks or containers, such as drums. These fuel oils are used for emergency diesels, lifeboat diesels, etc. Marine gas oil is comparable to No. 2 fuel grade (e.g., ASTM D 975), which is used commercially ashore and is stable for long storage periods.

Marine diesel oils (MDO) are actually two different fuel oils. MDO-light, which is also known as distillate-type MDO, is generally marine gas oil that is handled in barge tanks that are also designated for the carriage of residual (black, dirty) fuel. MDO-light has a dark color but is essentially a clean distillate with very little sediment and a carbon residue below 0.2% (mass). MDO-heavy is 80 to 90% MDO-light that is blended with 10 to 20% residual fuel oil and is handled in a residual fuel-oil barge. Due to its residual fuel-oil content, MDO-heavy can be used only in diesel engines that are designed to have a blended residual fuel-oil capability. The carbon residue of MDO-heavy is generally above 1.5% (mass).

Heavy residual fuel oils (HFO) have the highest viscosity and density of any product sold by a refinery as fuel oil. In many cases, the ships' systems cannot handle these heavy residual fuel oils, so they are blended to produce a fuel that is suitable for use. Blended residual fuel oils, or intermediate fuel oils (IFO), are mixtures of MDO and heavy residual fuel oils and are blended to improve the viscosity, density, vanadium content, carbon content, or other characteristics of a heavy residual fuel. Residual fuels are generally classed by their viscosity in centistokes; however, because of the use of differing viscosity designation temperatures, the one that applies should be stated with the viscosity value.

a. Residual-fuel specifications. Most commercial steamships and motorships, other than river and coastal vessels, burn residual fuels because of their lower cost. Several levels of specifications and standards are applicable to residual fuel oil. General standards are supplied by national organizations, such as the American Society for Testing and Materials and international groups, such as the International Standards Organization and the International Council on Combustion Engines. However, these specifications may not be sufficiently detailed for all users and may indicate only minimum requirements. The refiners also have their own specifications that are generally met in the refining and blending process. The refiner's specifications may be modified to accommodate customer needs, or the refiner may search out markets that can be satisfied by his specifications. Oil traders who deal with residual fuel also have specifications that either meet the needs of the final customer or provide sufficient quality to sell the oil in open markets. In addition, engine and fuel-treatment equipment manufacturers have their own specifications, which outline minimum requirements for trouble-free equipment operation. Finally, the ship operators have specifications that reflect their capability to treat and clean the oil. These specifications are generally available from the various sponsoring organizations.

The operator of a propulsion plant that uses residual fuel should be aware of the various specifications and realize that there are many significant differences among them, depending on their purposes. For example, a diesel engine operator will not necessarily buy residual fuel that complies with the engine manufacturer's specifications. Residual fuel with higher levels of incidental contamination may be purchased provided the contaminants can either be safely removed or rendered harmless by a fuel treatment system. After cleaning and treatment, the fuel should meet the engine manufacturer's specifications.

b. Diesel engine fuel specifications. High-speed diesel engines are generally operated on marine gas oil and marine diesel oil (distillate type) fuels. Most low- and medium-speed diesel engines, which have been appropriately designed, are capable of burning residual fuel. The larger, low-speed, crosshead engines are generally recognized as having a higher tolerance to burn the poorer grades of fuel because more time is available for the combustion process to take place. Typical residual-fuel specifications published by engine manufacturers are presented in Table 1. A range of values for the property levels is indicated for several engine types. The data in Table 1 reflect the range of the most common engine classes or models offered by the largest engine manufacturers. In general, the specification ranges are consistent. Specific requirements for a given engine model can be obtained from the manufacturer.

c. Military fuel specifications. Military fuel oils are distillates that conform to military or NATO (North Atlantic Treaty Organization) specifications. The MIL-F-16884 designation is a clean distillate, similar to commercial No. 2 diesel fuel oil, with some additional fuel characteristic restrictions to suit multipurpose requirements and extended periods of storage. MIL-F-16884, DFM, and NATO

Table 1 Typical residual-fuel oil specifications for large diesel engines

Specification	Engine Manufacturer		
	Colt-Pielstick	MAN B&W	Sulzer
Viscosity			
SUS @ 100 F, max	4000–7000	...	...
cSt @ 50 C, max	380–625	180–600	180–600
Redwood No. 1 sec @ 100 F, max	3500–6000	3500–6000	3500–6000
Carbon residue, Conradson, wt%, max	12.5–22	10–22	5–22
Sulfur, wt%, max	3.5–5.0	3.0–6.0	2.4–6.0
Ash, wt%, max[a]	0.10	0.10–0.20	0.05–0.20
Bottom sediment and water, vol%, max	0.1	0.2–1.0	1.0–2.0
Cetane number, min	30	25–35	24–40
Gravity, deg API, min	12.5	...	...
Density, g/mL @ 15 C, max	...	0.98–0.99	0.92–0.99
Aluminum content, ppm, max	30	30	30
Sodium content, ppm, max	100[b]	50–75	30–100
Vanadium content, ppm, max	100–600[b]	150–600	100–600
Pour point, deg C, max	local	...	...
Cloud point, deg C, max	local	...	...
Flash point, deg C, min	65 or legal	65	65

[a] Must be reduced to 0.10% before use in engine.

[b] Special exhaust valves may be required if sodium and vanadium are present.

F-76 are the same fuel and must be handled as a clean fuel to remain within specification limits. The JP-5 category is a light distillate fuel oil used for aircraft gas turbines and selected high-speed diesel engines. JP-5 is comparable to a commercial No. 1 fuel oil and is lighter than MIL-F-16884.

d. Environmental considerations. Local environmental requirements may place further limitations on fuel specifications. Of particular note is the sulfur content, which could be limited to a level much less than that which the engine is able to tolerate. Local restrictions on sulfur dioxide emissions could result in sulfur limitations of 0.3 to 1.0% or lower. Other fuel properties associated with emission levels may also be constrained due to local regulations.

1.4 Fuel Testing. A fuel-testing program is essential to ensure reliable engine operation. Even though fuel is purchased as being in compliance with the appropriate specifications, in-transit contamination can severely alter the quality of the delivered product. Treated and cleaned fuel must also be tested to ensure adequate operation of fuel-treatment equipment. Immediate test results concerning fuel quality are often required to support decisions on plant operations. An on-site laboratory is recommended for all but the most sophisticated test procedures.

Most of the standard tests can be performed without expensive equipment.

For fuel tests to be meaningful, a representative sample of the fuel oil must be obtained as it is being delivered on board the vessel. It is important that a continuous sampling be taken throughout the entire bunkering period, as the quality may change due to stratification within the bunkering barge or changes in the blending ratios. Marine fuel-analysis programs have detailed instructions concerning the procedure to be followed when sampling fuel oil.

Typical tests conducted aboard ship include those for viscosity, water content, specific gravity, particle content, pour point, and compatibility.

To obtain complete and accurate fuel-oil analyses, fuel-oil samples must be sent to a shoreside laboratory. A shoreside laboratory can confirm that a fuel meets purchase specifications and can provide notification of contaminant levels, incompatibility, excessive water content, and other potentially damaging characteristics of the fuel. The independent test laboratories provide an unbiased, third-party analysis of the fuel oil received and facilitate the resolution of disputes.

Section 2
Propulsion Fuel-Oil Systems

2.1 Fuel-Oil Storage System Arrangement. The fuel-oil storage, settling, and transfer systems for steam, diesel, and gas turbine plants are similar. Figure 10 illustrates a residual-oil and diesel-oil fill, transfer, and storage system for a typical steam propulsion plant. The system configuration is for operation on 600 cSt fuel oil, which is heavy residual oil. A comparable fuel-oil storage and purification system for a diesel propulsion plant is illustrated by Fig. 11.

The fuel-oil transfer system provides the following functions:

- The transfer of fuel oil from any tank used for storage to any other tank.
- The transfer of oil from the fuel-oil storage tanks to the settling tanks through each settling tank fill line with automatic shutdown of the transfer pump upon reaching the settling tank high-level alarm.

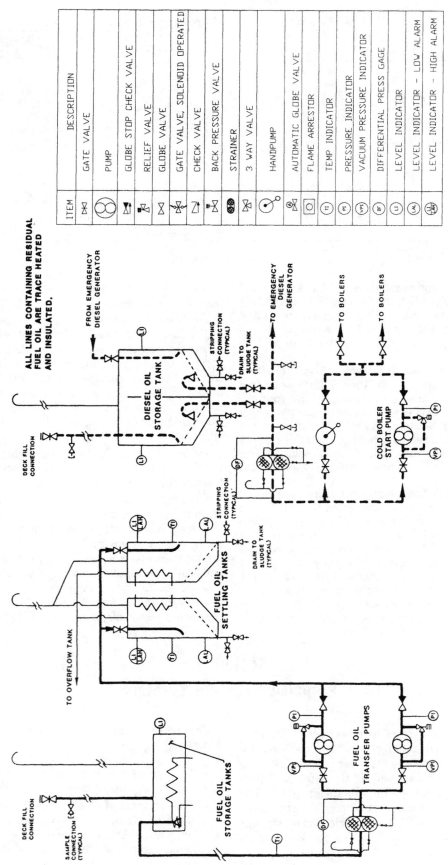

ITEM	DESCRIPTION
⋈	GATE VALVE
⊙⊙	PUMP
⊠	GLOBE STOP CHECK VALVE
⋈	RELIEF VALVE
⋈	GLOBE VALVE
⋈	GATE VALVE, SOLENOID OPERATED
⊿	CHECK VALVE
⋈	BACK PRESSURE VALVE
⬤	STRAINER
⋈	3 WAY VALVE
⊙	HANDPUMP
⊗	AUTOMATIC GLOBE VALVE
⊡	FLAME ARRESTOR
TI	TEMP INDICATOR
PI	PRESSURE INDICATOR
VPI	VACUUM PRESSURE INDICATOR
DPI	DIFFERENTIAL PRESS GAGE
LI	LEVEL INDICATOR
AL	LEVEL INDICATOR – LOW ALARM
AH	LEVEL INDICATOR – HIGH ALARM

Fig. 10   Steam propulsion plant residual fuel-oil and diesel-oil fill, transfer, storage, and purification system

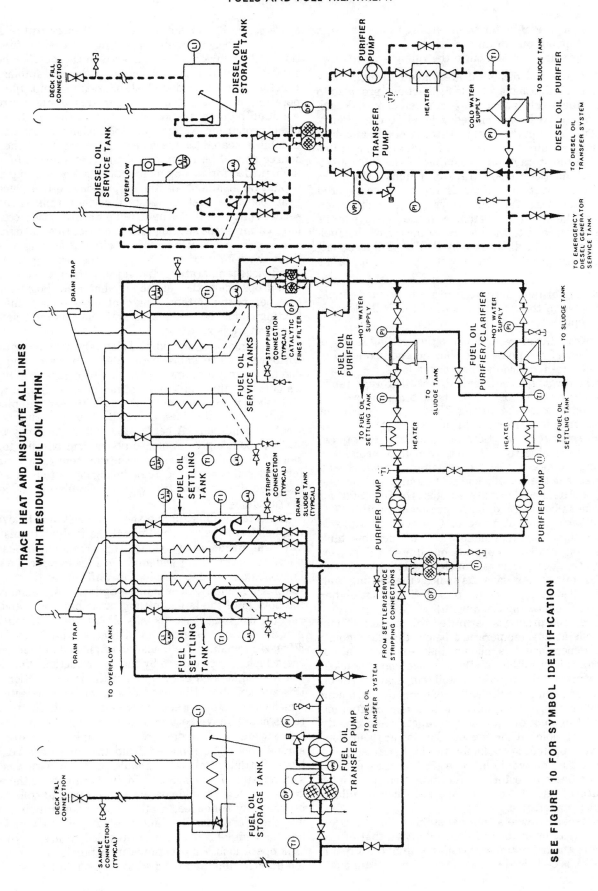

Fig. 11 Diesel engine residual fuel-oil and diesel-oil fill, transfer, storage, and purification system

SEE FIGURE 10 FOR SYMBOL IDENTIFICATION

- The stripping of water from settling tanks, through the stripping and drain connection.
- The discharge of fuel oil to the filling stations.

Residual fuel and diesel oil are loaded through deck fill connections that have sample connections provided to permit the fuel to be sampled as it is taken aboard. Residual fuel is stored in the heated fuel-oil storage tanks. Diesel oil is stored in the diesel-oil storage tank, and does not normally require tank heating. Residual fuel is transferred from the storage tank to either of the fuel-oil settling tanks via the fuel-oil transfer pumps, which are equipped with a suction strainer. The system is configured to permit either transfer pump to discharge to either settling tank. The fuel oil is retained in the settling tanks until it is forwarded to the boilers by means of the fuel-oil service system, as illustrated by Fig. 12. Drain connections should be provided at the bottom of settling tanks to permit:

- Draining of tank bottoms directly to a sludge tank.
- Stripping of the tank by means of the fuel-oil transfer pump.

All piping from the storage tanks to the settling tanks as shown in Fig. 11 should be trace heated and insulated to prevent the fuel from congealing in the lines.

Fuel-oil storage tanks. Residual fuel oil storage tanks have heating coils to control the tank temperature and maintain the fuel oil 5 to 10 deg C above the pour point of the fuel until it is transferred to a settling tank.

Fuel-oil settling tanks. Settling tanks are used to accomplish several objectives in the treatment of heavy fuel oil. Settling tanks are used to permit gross water and solids to settle to the bottom, and provisions are also made to heat, deaerate, and thermally stabilize the fuel. Settling tanks are usually designed to accept fuel oils with a 60 C minimum flash point. The use of two settling tanks is the most common arrangement with each settling tank holding enough fuel oil to operate one day at full power. However, other designs have provided a settling tank residence time of up to four days. As soon as the settling tank (settler) is filled, it is heated to 60 to 80 C, but not higher than 10 deg C below the flash point.

Early accepted practices permitted the storage of marine fuel oils at any temperature below the flash point; however, experience has shown that explosive atmospheres can collect in the tank head space above cracked fuel oils even though the bulk fuel-oil temperature is well below the flash point. Additionally, if shoreside petroleum waste products have been mixed with a marine fuel oil, the lower flash point of the waste products could greatly lower the flash point of the fuel oil. This lower flash point may not be reported. More likely, the higher flash point of the basic fuel oil would be reported. As a result, the condition of flame arresters on fuel tank vent lines is of great concern, and the careful and safe use of ullage equipment is imperative.

Settling tanks provide a thermal stabilization function by raising the temperature of the fuel above that of the storage tank and maintaining this higher temperature until the fuel oil is delivered to a heater set. In a steam plant

system, the heater set is the boost heater to the boiler front. In a diesel system, operating on residual fuel oils, the settling tank fuel is delivered to the purifier heater set, and it is in this service where the thermal stabilization function is most needed. A stable settling tank fuel temperature provides stable purifier fuel temperatures to maintain high separation efficiency in the purifier. In a diesel propulsion system, once the fuel in the settling tank has been heated to the selected temperature, the heat source should be secured and the fuel allowed to settle undisturbed for as long as practicable. The settling tanks should be insulated insofar as practicable to reduce the heat loss. The heat source to a settling tank should be secured once the fuel is up to temperature since a continuous heating of the settling tank produces thermal currents within the tank, which interfere with the settling process. With the source of heat secured, the settling tank will slowly lose temperature; and, if the settling tank temperature drops below the selected minimum settling temperature before the settling tank contents are used, the tank heating should be reactivated until the maximum settling temperature is restored.

A settling tank should be relatively shallow with the bottom sloped to improve the ability of water and sediment to migrate to one region and be stripped. Whereas a shallow settling tank may optimize the settling of water and solids, it creates some problems with reentrainment of tank bottom contaminants into the fuel oil during heavy-weather conditions. A perforated baffle plate may be installed about eighteen inches above, and parallel to, the sloped bottom to reduce the reentrainment of settled water and solids with the fuel during operation in heavy seas and during tank filling and recirculating modes of operation.

The settling tank fill line should be installed below the mid-height of the tank and direct the fuel oil against the tank wall to reduce splashing. This arrangement also reduces air entrainment and minimizes the danger of a combustible-gas buildup within the unfilled section of the tank. Additionally, it minimizes agitation of sediment in the bottom of the tank. Settling tanks have high and low suction lines that are usually fitted with suction bellmouths having downturned diffusers. The downturned diffusers minimize the ingestion of air and prevent vortices, which may be caused by the pump suction. Normally, the fuel-oil suction is taken from the lower suction, and this suction should be used unless water is detected, in which case the upper suction would be used. The settling tanks should have bottom drains that can be used to remove water and sludge from the tank; the water and sludge should be removed from the tank on a regular basis. Settling tank vents should be fitted with a corrosion-resistant flame screen and discharge above the main deck. The vent should contain a drain trap to collect and prevent any condensed water vapor from running back into the fuel oil tank. A tank gaging system as well as a low- and high-level alarm is provided to assist settling tank operations. An overflow connection is located at the top of each tank and is piped to an overflow tank.

Fuel-oil service tanks. Service tanks, or day tanks, have an important function in the treatment of residual fuel oil. Service tanks complement settling tanks and provide an additional opportunity to further settle water and solids out of the fuel, heat the fuel, and thermally stabilize the fuel. A single service tank can be used in conjunction with two settling tanks. The single service tank can be continuously filled by the centrifugal separators and then overflow to the settling tanks; however, two service tanks are preferred to ensure that a supply of clean, treated fuel is always available.

When two service tanks are used, one supplies clean, heated, deaerated fuel oil to the system while the other is used to process fuel oil for the following day. The service tanks should be arranged and have piping and venting features similar to those of settling tanks.

2.2 Fuel-Oil Transfer System Arrangement. A transfer pump (or pumps) is provided to move fuel oil from a storage to a settling tank as indicated by Figs. 10 and 11. A positive-displacement transfer pump (or pumps), protected by suction strainers, a pressure relief valve, and pump bypass line, is normally fitted. For the arrangement illustrated by Fig. 11, in the event that the single transfer pump is inoperative, either one or both of the purifier pumps, by suitable arrangement of valving, can provide backup transfer capability. The transfer pump flow rate is dependent upon the fuel consumption rate and the size of the settling tank.

Operational flexibility in the transfer system can be provided by an appropriate arrangement of the system valves. This valving normally permits fuel oil to be pumped from any storage tank to either settling tank, to other fuel-oil storage tanks, or, in some arrangements, overboard to a barge or other storage facility.

2.3 Steam Plant Fuel-Oil Service System. The fuel-oil handling system in a steam plant is simple, consisting of storage, settling, and service tanks, transfer pumps, service pumps, heater sets, and drain-collecting systems. Fuel-oil treatment consists of little more than settling solids and water in a heated settling tank. The service or boost system consists of a pump and a heater along with a flow- or pressure-regulating control. In most automated systems, a fuel-oil quick-closing valve secures the flow of fuel upon flame failure, loss of fuel pressure, or a loss of combustion air supply.

A typical fuel-oil service system for an oil-fired steam propulsion plant is illustrated in Fig. 12. Residual fuel is taken from the settling tanks through a common suction line, which is fitted with a duplex strainer. The system is arranged so as to maintain a positive head at the fuel-oil service-pump suction. Two fuel-oil service pumps, one a standby, are provided with each pump being capable of supplying the total required fuel oil flow plus an additional margin. Any excess in pump delivery is diverted back to the settling tank. These pumps are normally positive-displacement pumps and are fitted with pressure relief bypasses, remote shutdowns, and isolation valves for ease of servicing. The fuel is discharged from the pumps to the service heaters where the temperature is raised to that corresponding to the viscosity for proper atomization.

Two steam-heated, shell-and-tube or plate-type, fuel-oil service heat exchangers are installed, each with a capability of heating the fuel oil to 145 C. When the final fuel oil outlet temperature is controlled by a viscometer, the viscosity of the fuel oil is held constant, even though the fuel-oil temperature may vary slightly. It is important to use properly sized heat exchangers, which can provide the necessary heat input without overheating the fuel oil. Upon leaving the service heaters, the oil passes through a second duplex strainer, a viscometer, and a fuel flowmeter. From the flowmeter, the fuel goes to the burner management system, which consists of combustion-control and solenoid-trip valves and the burner header piping. All heavy-fuel piping should be insulated and trace heated.

The diesel-oil storage system shown by Fig. 10 consists of a tank with an oiltight partition installed to separate the oil supply for the emergency diesel generator from the portion set aside for cold boiler start-up. Each compartment has an independent level indicating system, and the tank is vented to the weather deck through a line that is equipped with a corrosion-resistant flame screen.

A small service tank containing clean fuel oil, frequently marine diesel oil or gas oil, is used for start-ups when neither electricity nor steam is available. Frequently, this tank is located high in the engine space so such start-ups can be made without electrical power or service pumps. Alternatively, a hand-operated pump may be provided.

2.4 Diesel Plant Fuel-Oil System. Fuel-oil pumping, heating, and treatment systems in motorship engine rooms vary in detail, but the broad principles are common to both slow- and medium-speed engines that burn residual fuel oil. Figure 11 illustrates a typical fuel-oil and diesel-oil fill, transfer, storage, and purification system for a diesel propulsion plant. Differences may occur in the characteristics of specific systems with the principal one being that fuel-oil piping should be insulated but need not be trace heated for fuel oils having a viscosity below 380 cSt.

For diesel engines that operate on residual fuel oils, an emergency shutdown will leave residual fuel oil in the piping, heaters, pumps, etc. If all fuel lines are trace heated and insulated, this is not a problem because, by recirculation, the temperature of the fuel oil in the system can be controlled to permit a successful restart. If the emergency shutdown is expected to continue for a considerable period of time, then the residual fuel oil should be purged out of the lines with marine diesel oil; this will permit the recirculation and trace-heating systems to be secured. With marine diesel oil in the fuel-oil service system piping when secured, the system will not have to be preheated upon restart, and any required maintenance will be much easier to accomplish with marine diesel oil in the piping system.

Both residual fuel oil and diesel oil are taken aboard at deck connections, which incorporate a separate connection to permit a sample of the fuel to be taken as it is loaded. Residual fuel oil is loaded in storage tanks, which are fitted with heating coils. Diesel oil is similarly stored, although normally there is no requirement for storage

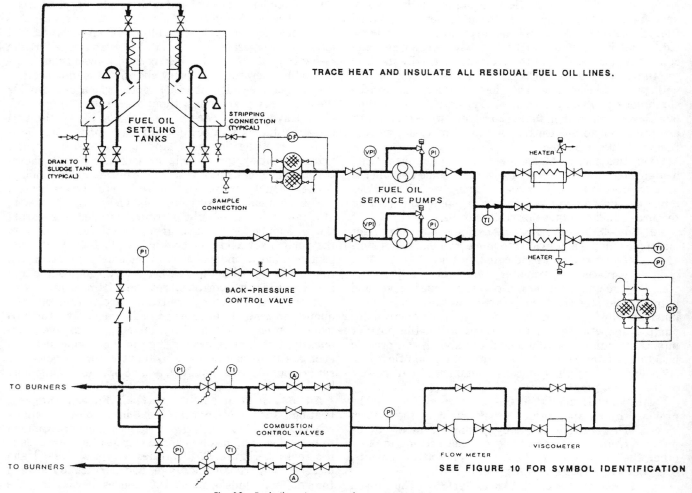

Fig. 12   Fuel-oil service system for a steam propulsion plant

tank heating. In preparation for use, residual fuel oil is transferred to either of the fuel-oil settling tanks via the fuel-oil transfer pump and its associated suction strainer. When a single fuel-oil transfer pump is installed, backup is provided by crossover lines between the transfer and purification systems. Should the transfer pump be inoperative, fuel oil can be transferred directly from the storage tanks to either the settling or service tanks with either centrifuge.

A demulsifier may be injected into the fuel oil just upstream of the transfer pump suction strainer. The purpose of the demulsifier is to facilitate the separation of water from the oil during the settler residence time. From the settling tanks, fuel oil is transferred to the service tanks via the purification system. Two fuel-oil centrifuges are generally installed with appropriate fuel supply pumps, heaters, and controls. Centrifugal separators, or purifiers and clarifiers as they are usually called, are the most widely used and probably the most reliable and efficient method of cleaning distillate and residual fuel oils. Centrifuges have the advantage of being able to remove large quantities of water while still retaining good particle removal efficiency. Without the development of efficient

centrifuges, especially the self-cleaning type, it is doubtful whether the burning of residual fuel oil in diesel engines would be feasible. The centrifuge service pump, fuel-oil heaters, sludge tank, and piping must be designed and integrated to properly support the centrifuge. The system and equipment are configured to permit operation of the centrifuges in parallel or in series either in a purifier/purifier or clarifier/clarifier or purifier/clarifier sequence. A cross connection, which permits either pump to service either heater, is an advantageous feature. Purified fuel oil is discharged to the service tanks via a very fine duplex filter that is installed to remove catalytic fines that pass through the purifiers. The fuel oil is retained in the service tanks until it is forwarded to the main engine via the fuel-oil service system, which is illustrated by Fig. 13. Any overflow from the service tanks is directed to the settling tank.

The arrangement of the settling and service tank drain piping should permit:

• Draining of the tank-bottom sediment directly to a sludge tank.
• Stripping of the tank by the fuel-oil transfer pump.

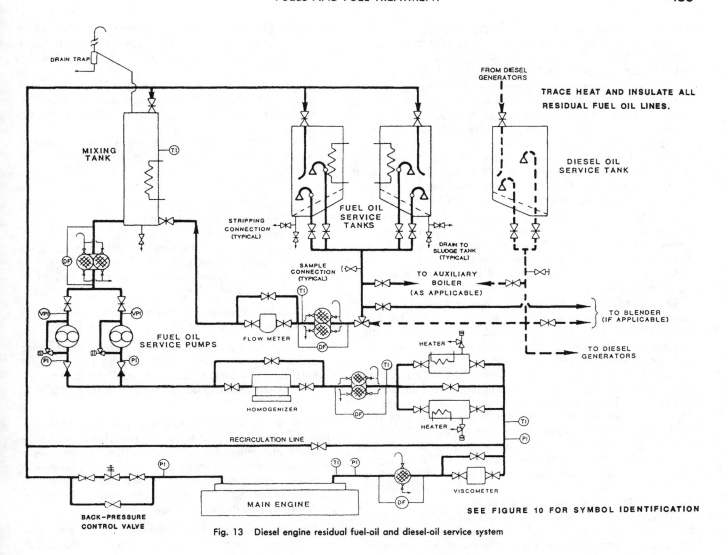

Fig. 13 Diesel engine residual fuel-oil and diesel-oil service system

- Purification of tank-bottom sediment with the oil returned to the same tank or any other fuel-oil service or storage tank.

Diesel oil is transferred from a storage tank by either a transfer pump or a purifier. The diesel-oil transfer system illustrated by Fig. 11 incorporates the following features:

- Diesel oil is normally transferred from the storage tank to the service tank by the purifier system.
- Diesel oil can be transferred from the service tank to the emergency diesel generator service tank by either the diesel oil transfer pump or the purifier.
- Diesel-oil can be transferred from any storage tank to any other storage tank with the system.
- The diesel-oil service tank bottoms can be drained directly to the sludge tank. Additionally, a stripping connection permits pumping of the bottoms to any storage tank or the drains can be purified and transferred with the purifier.

2.6 Diesel Fuel-Oil System Components.

a. **Centrifuges.** There is often confusion between the maximum throughput or rated capacity of a centrifuge and the recommended service capacity. The *rated capacity* gives the maximum volume of fuel that can be passed through the machine. This is usually based on the treatment of light distillate fuel at ambient temperature. The recommended service capacity is the amount that can be passed through the machine at maximum separating efficiency. This is based essentially upon the dynamic viscosity of the fuel at the separation temperature. The maximum separation temperature, irrespective of viscosity, has an upper limit of 98 C. Above this temperature there is a risk of the water seal being lost due to the formation of steam bubbles.

Based upon centrifuge tests with fuels varying in viscosity from marine diesel oil (taken as 85 sec Redwood 1 at 100 F or 14 cSt at 40 C) to the most viscous fuel likely to be sold for marine bunkers, namely, 6000 sec Redwood 1 at 100 F or approximately 600 cSt at 50 C, the maximum throughput capacities for the different viscosities are recommended by the various purifier manufacturers. For Bunker C fuels and those more viscous, a maximum separation temperature of 98 C is recommended.

To select a centrifuge that will provide maximum separating efficiency, the rated capacity of the centrifuge must be divided by a factor that is a function of the fuel viscosity. For fuel viscosities of 180, 380, and 600 cSt, the rated centrifuge capacity is divided by 3.3, 4.0, and 6.7, respectively, to determine the recommended service capacity for maximum separating efficiency. That is, the more viscous the fuel, the lower the recommended throughput rate and the larger the centrifuge required.

The difference in specific gravity between the fuel being processed and water, either fresh or salt, also influences the separation efficiency. Straight-run residual fuels seldom have a specific gravity that exceeds about 0.96, whereas the specific gravity of cracked residual fuels can exceed unity.

The specific gravity of most fuels is inversely proportional to the temperature; however, the specific gravity of water does not have a straight-line relationship with temperature. The maximum difference between the specific gravities of oil and water occurs at about 85 C, and is slightly less at 98 C. From a specific gravity point of view, there is no advantage in heating the fuel above about 85 C. However, by increasing the temperature to 98 C, there is a marked reduction in the viscosity of the fuel, which permits a more effective separation of sludge and solids. Centrifuge manufacturers generally agree that to effectively separate water and solids from high-specific-gravity, high-viscosity residual fuels, the throughput must be substantially less than that appropriate for less-dense, less-viscous fuels.

For engines that are intended to be operated on residual fuel, it is recommended that the centrifuge capacity be designed to treat fuels characterized as 600 cSt viscosity at 50 C with a maximum specific gravity of 0.991 and up to 5% water and possibly 2% sludge. A centrifuge having this capacity should be able to treat the poorest fuels likely to be offered as diesel engine fuels.

Two properly sized, correctly operated, self-cleaning centrifuges are considered necessary to provide a reliable fuel-treatment system. Most engine warranties become invalid if centrifuges are not used. To establish and maintain effective separator procedures, several fundamental principles should be noted:

- The centrifuge is the first major stage of fuel treatment.
- To treat contaminated fuel oils, supplementary systems, in addition to the centrifuges, are required. These supplementary systems can consist of fine filtration, demulsifier chemicals, and homogenizer-emulsifiers.
- Each centrifuge should be supplied with all parts necessary to operate as a purifier and as a clarifier as well as complete spares and a complete set of tools.

- The single centrifuge flow rate (for series operation) or the combined centrifuge flow rate (for parallel operation) must not exceed the engine demand by more than 10%.
- All residual fuel oil centrifuges should be on-line and operated continuously. This increases the effective fuel treatment time and further reduces contaminants.
- To properly adjust and operate a centrifuge, the following residual fuel-oil properties must be known:

    —viscosity,
    —specific gravity (or density),
    —compatibility of fuel oil,
    —water content,
    —mash content (bottom sediment content is an alternative), and
    —catalyst fines content (aluminum content is an alternative).

  This information can be used to make decisions on fuel treatment options.

- When fuel is transferred to a settling tank from a different source, a specific gravity check of the settling tank should be made and the centrifuge gravity disk should be checked to ensure that it is correct.

The centrifuge is the foundation of the total shipboard fuel treatment system. Its efficient operation is critical to the safety and reliability of the engines. Its operation must be thoroughly understood so that the shipboard engineers can immediately troubleshoot fuel-oil problems when they occur. Some conditions that can cause centrifuge maloperation include the following:

- Incorrect fuel handling before the separator, such as: improper barge blending, incompatible fuels, and emulsified fuels.
- Improper flow, such as: varying flow rates, excessive flow rate, or flow with varying densities.
- Improper temperature, such as: varying temperatures or too low a temperature.
- Incorrect positioning of the water/oil interface, thereby inhibiting a uniform flow of oil through all disks; this is usually caused by using an improper gravity disk. A gravity disk establishes the separation zone between the clean fuel and the water according to fuel temperature and density. As fuel characteristics change, the gravity disk must be changed to control the water-fuel separation zone for maximum efficiency.
- Overloading the centrifuge with an accumulation of sludge, which is usually caused by extended desludging intervals, or incompatible residual fuel oils.

The centrifuge valving is very important for proper start-up, to prevent contamination, and for effective operation. Immediately before the centrifuges, valving should be provided that permits recirculation back to the settling tanks to provide settling tank mixing or heating in the event that the heating coils in the settling tank are inoperative. Downstream of the centrifuges, the valving should permit recirculation back to the settling tanks so that the fuel centrifuged can be returned to the settling tanks. This is desirable because a considerable period of time is required for the flow rate and temperature to become uniform and for equipment adjustments to be made (gravity disks, back-pressure settings, etc.) commensurate with the stabilized conditions.

A minimum of two centrifuges that are properly sized, arranged, and operated is required. Each centrifuge should be capable of purifying the total fuel requirements of the engine plus a 10% margin or surplus when operating at the recommended service capacity required for maximum separating efficiency. The series method of centrifuge operation is the preferred fuel flow arrangement. The first centrifuge is configured as a purifier to remove sediment, sludge, and water. The second centrifuge is configured as a clarifier to remove any remaining sediment and solids from the fuel oil. The second centrifuge, operating as a clarifier, provides a backup to the purifier in the event that the purifier malfunctions.

If there is a high water content in the fuel oil, centrifuge operation in parallel is recommended. By configuring both of the centrifuges as purifiers in parallel and by reducing the flow rates by 50%, the fuel has twice the residence time in the purifier to remove water. When properly set up and carefully operated, parallel operation can produce the highest cleaning effectiveness and, thereby, the cleanest fuel oil to the engine. However, if one of the purifiers should malfunction, there would be no provisions to prevent the contaminated oil from going directly to the engine.

The series and parallel modes of centrifuge operation entail both advantages and disadvantages; therefore, to determine the most appropriate operational mode, the fuel flow, viscosity, density, water content, sediment and ash content, contamination, and compatibility must be assessed. When the fuel oil has a high water or sediment content, a parallel purifier/purifier alignment would be preferred. But if the fuel contains impurities that can be severely damaging, the risk of a purifier malfunction would suggest a series purifier/clarifier mode of operation.

Three centrifuges are commonly installed aboard ship. The third machine is nominally a spare, but it can be used to provide a parallel purifier-purifier alignment followed by a clarifier for cleaning highly contaminated fuel.

b. Filters and strainers. Coarse strainers are installed in fuel systems to protect pumps and other components from relatively large fuel contaminants. In diesel plants the filtration provisions are major components of the fuel-oil treatment system, with perhaps only the purification provisions being more essential. Because of the high ash, solids, and catalyst particle content in residual fuel oils, supplemental filtration is recommended. The filtration provisions can be considered as a backup to the purification system, but they are, nevertheless, essential to ensure that the fuel supplied to the engine is free of damaging contaminants. The filtration system should be piped and valved to permit cleaning and replacement without an interruption in fuel flow. When properly designed, a filtration system can positively control solids that would otherwise damage high-pressure pumps, injection systems, cylinder liners and piston rings of diesel engines. Under normal operating conditions, the filter elements require replacement at 2000- to 4000-hour intervals.

Fine residual fuel-oil filters can be used in conjunction with homogenizers to reduce the asphaltenes and sludge that frequently accompany the fine abrasive solids in fuel oil. In fact, the majority of the contamination in a fine filter is frequently nonparticulate, organic, and hydrocarbon in nature. The primary purpose of a homogenizer is to disperse the sludge or asphaltene in the fuel oil to prevent the removal of these hydrocarbons while filtering the fuel oil. The use of a homogenizer can, therefore, greatly increase the life of filters, reduce the amount of sludge that is removed from the fuel and must be handled, and decrease the amount of fuel heating value that is lost in the form of sludge.

Besides solids, filters also remove trace quantities of free water from marine residual fuel oil. Centrifuges remove the majority of the free water from the fuel; however, trace amounts remain. While removing the trace water may seem unimportant, shipboard testing has shown injection pump life increases of 100% after the addition of filters that removed trace water. The water sumps in the filters should be drained daily to prevent the water from accumulating, rising above the sump level, and "wetting" the filter elements.

Automatic filters or auto-flushing filters contain permanent, metal-screen type filters that self-purge to maintain a clean screen for continued operation. Most auto-flushing filters have a limited filtering capability, however, and cannot provide 5-μm (nominal) particle control. In addition, each cleaning cycle produces sludge that must be handled and properly discarded; and, during the filtration of mildly incompatible fuel oils, the back-flushing frequency can be as many as 400 flushes per day, which produces a considerable volume of sludge refuse.

There are two types of replaceable (cartridge) filter elements: the surface type and the depth type. As the name suggests, surface elements are, in effect, a fine-mesh screen with a limited thickness. A small accumulation of organic sludge or asphaltenes, along with solids, on surface elements can seal the element surface and increase the differential pressure across the filter, thus limiting the useful life of the element. Surface filters typically are designed with a "pleated" configuration to increase the surface area and increase the element life. The materials used to make surface filters are commonly paper and felt.

Depth filter elements are designed so that the filtration occurs throughout the substantial depth of the element. Some depth element designs provide progressively finer filtration from the outside of the depth element to the inside. Since contaminants are held at all depths throughout the element, there is a reduced tendency for contaminants to accumulate in a layer that blocks flow through the element. Experience has shown that a depth-type, cotton and wood fiber matrix material compacted to a 90%, 5-μm removal capability will meet the performance requirements for the filtration system of diesels, if the element flow rates are controlled to one gallon per minute per 36-in.-long element (maximum). These filtration units must be carefully designed to prevent bypassing of fuel around the end sealing cups. The filter housing should be equipped with a bottom water drain, an air vent, and a differential pressure gage to indicate the pressure drop across the filter and the necessity to replace the filter element.

As indicated by Fig. 13, two fuel-oil service, or booster, pumps (with one normally in operation) are provided to supply the residual fuel oil to the downstream service system (the homogenizer, service heaters, viscometer, and engine). Duplex suction strainers provide protection to the fuel-oil service pumps from solid debris that may be in the piping from the fuel-oil tanks. A 20- to 40-mesh reinforced, corrosion-resistant basket should be used together with magnetic elements to remove all metallic/ magnetic particles from the heavy fuel-oil stream. The strainers should be vented, have drain provisions, and be arranged to permit the units to be serviced without interrupting the supply flow.

c. Mixing tank. Fuel-oil service pumps are normally of the positive-displacement type and are fitted with pressure relief bypasses and isolation valves to permit servicing. Each service pump is sized to deliver 2 to 2.5 times the highest fuel-oil consumption rate of the engine. The surplus fuel oil, which is not injected, is discharged from the engine's fuel injection pumps through a back-pressure regulator and returned to the mixing tank. On ships that burn a low-viscosity fuel the mixing tank is a small tank where the hot surplus fuel oil from the engine mixes with the incoming fuel oil from the service tank. Any air bubbles or vapor bubbles are vented to the atmosphere. The mixed fuel oil then flows by gravity to the boost pumps, boost heaters, viscometer, fuel meter, homogenizer, and on to the engine.

For ships that burn high-viscosity fuel and have a high fuel-oil boost system temperature, hot surplus fuel oil returned to the mixing tank would froth, and water entrained in the fuel would flash to steam in the mixing tank as soon as the back pressure has been removed from the fuel oil. Therefore, for high-viscosity fuel oils that require a high boost system temperature, a closed boost system is used with no atmospherically vented mixing tank. Some designs incorporate a small pressure vessel to serve as a mixing tank and provide a pressurized vent to the atmosphere. However, other designs eliminate the mixing tank and maintain a pressurized system through the entire boost system. In this case a deaerator device should be installed to remove any air or vapor bubbles before the fuel oil is returned to the engine's injection pumps. Also, additional circulating pumps are required in a closed boost system design.

d. Homogenizer. A homogenizer, which is often considered to be optional, is sometimes installed to disperse any insoluble hydrocarbons in the residual fuel oil, to thoroughly disperse any remaining free water in the fuel oil, and to produce a uniform, fine droplet, stable, water-in-fuel-oil emulsion (when fresh water is added to high carbon-content fuels prior to the homogenizer). The homogenizer conditions fuels that are very heavy, with a high molecular weight and high boiling point, to assure efficient combustion with a high carbon burnout efficiency. The homogenizer is a positive-displacement, multi-piston pump that can increase the fuel-oil pressure within the homogenizer up to about 17,000 kPa. Before the pressurized fuel oil is discharged from the unit, the high pressure is reduced suddenly across a controlled orifice. The high fluid velocity produces a combination of intense shear, unstable local pressures, cavitation, and, finally, impact and turbulence. This sudden sharp pressure reduction serves to tear the insoluble hydrocarbons into very small particles and thoroughly disperse the particles throughout the residual fuel oil. The use of homogenizers has been shown to reduce fuel-oil consumption, reduce maintenance, and reduce exhaust-gas emissions and smoking when burning residual fuels.

A bypass line is installed around the homogenizer such that any excess discharge is recirculated back to the suction side of the unit. Additionally, the valving arrangement permits the homogenizer to be bypassed for maintenance purposes.

e. Fuel-oil service heaters. Where fuel must be heated before combustion, a minimum of two fuel-oil service heat exchangers should be installed. They are typically steam-heated, shell-and-tube or plate-type heat exchangers with a capability of heating residual fuel oil to 160 C. A duplex strainer should be installed to prevent contaminants from entering the fuel-oil service heaters. These strainer elements should be 20 to 40 mesh and should have vents and bottom drains. The crossover valving should permit uninterrupted operation on either strainer while servicing the remaining element. The final fuel-oil outlet temperature is controlled by a viscometer. A viscometer continuously provides a viscosity readout and control signal to the steam control valve that regulates the steam flow to the heat exchanger. This arrangement permits the viscosity to the diesel engine injectors to be held constant even though the fuel-oil temperature might vary slightly. When heating residual fuel oils, to temperatures in the range of 160 C, the use of properly sized, steam heat exchangers is required to provide the necessary heat input without thermally stressing the fuel oil in local areas and thereby coking the heaters. Stainless steel or aluminum-based alloys are preferred to copper-based materials.

f. Viscometer. Because of the low viscosity index of highly cracked fuel oils, a manual adjustment of the fuel temperature, based upon viscosity charts, is not feasible. A viscometer is a critical component that is required to insure uniform and accurate viscosity control in diesel plants using residual or blended fuels.

The viscometer constantly samples the residual fuel oil and produces a signal that is proportional to viscosity. Typical sensors employ calibrated capillary tubes, falling pistons, or vibrating rods. Irrespective of the method of determining viscosity, the viscometer output signal is utilized to modulate an automatic steam control valve on the fuel-oil service heaters. Since the viscometer is constantly sampling and adjusting the heater fuel outlet temperature to maintain a constant preset viscosity, the accuracy of this unit must be checked and calibrated periodically. A valved bypass line and isolation valves should be provided to permit the unit to be serviced without interrupting operations.

g. Fuel-oil supply flowmeter. A fuel-oil supply flowmeter is used to indicate the total fuel flow to the engine. A positive-displacement type meter or a mass flowmeter, which is reliable and accurate, is recommended. It must be suitable for service in temperatures up to 160 C. All fuel-oil wetted internal surfaces should be made of a corrosion-resistant material. The fuel-oil meter should be fitted with a valved bypass line and isolation valves to permit the unit to be serviced without interrupting operations.

h. Fuel-oil final filter. A simplex, final protection, 10-μm filter should be installed at the inlet to diesel engine fuel injection pumps to protect these high-pressure units from any contamination or spurious debris remaining in the fuel. This filter should be fitted with a differential pressure indicator. While it may appear that this final filter is not necessary because of the cleaning and treatment processes previously provided, the high-pressure injection pumps on diesels are very sensitive to minute debris. This debris can cause micro-seizures, which can lead to a failure of the pump plunger and the barrel. Automatic, self-cleaning final filters offer the advantage of permitting the filter elements to be back-flushed into a sump.

i. Back-pressure controller. A back-pressure controller provides a positive back pressure in fuel-oil return piping after the surplus fuel oil from the injection pumps has been discharged. The regulation of the back pressure is required to minimize cavitation and vapor formation in the fuel-oil return line. The back pressure can be as high as 350 kPa for high-viscosity fuel oils, which require a high final preheat for proper injection viscosity.

j. Return fuel-oil meter. In some system arrangements, a return fuel-oil meter is used to measure the returned excess fuel flow from the engine. In those arrangements the returned fuel-oil flow is subtracted from the fuel-oil supplied to determine the fuel consumed by the engine. In the fuel-oil system illustrated by Fig. 13, a flowmeter is installed in the line to the mixing tank, and the return flow from the engine is routed to the mixing tank; therefore, a return fuel-oil meter is not necessary. When provided, the return meter, like the supply meter, should be a reliable, accurate, positive-displacement or mass flow type that is suitable for service with oils at temperatures of about 160 C. All internal surfaces that contact the fuel should be made of a corrosion-resistant material. The fuel-oil flowmeter should be fitted with a bypass and isolation valves to permit the unit to be serviced without plant interruption.

2.5 Gas Turbine Fuel-Oil System. While gas turbines have the capability of burning a variety of liquid fuel oils, including residual and crude oils, the cleanliness requirements are substantially more severe than those associated with diesel engines since the sustained metal temperatures of gas turbines are up to twice as high as in diesel engines. It is most important to control the substances in the fuel oil, which could cause deterioration of the fuel system or the gas turbine, or both. The concerns consist of soluble or insoluble fuel contaminants, dissolved and free water, solids, metallic compounds, and microbial slimes.

The basic components of a gas turbine fuel-oil system are similar to a diesel system except for the more intense fuel treatment procedures, which are required to substantially eliminate metallic contamination. Whereas diesel systems can operate on ten to hundreds of ppm of metallic contaminants, the metallic contaminants in gas turbine fuels must be in the single-digit ppm range, and less. This places great importance upon the fuel-oil storage and treatment system to consistently control and remove fuel-oil contaminants. Because of the low tolerance that gas turbines have for metallic contaminants, most gas turbine manufacturers recommend starting and shutting down gas turbines using a clean, light distillate that is free of metallic contamination.

a. Tankage. The tanks used with gas turbine fuels are similar to those described for diesels except that a floating suction is frequently used. The height of the suction is controlled by a float such that the cleanest fuel oil, which is located at the top of the tank, is drawn off and pumped to the centrifugal separators. A floating suction continuously removes the cleanest fuel from the uppermost level of the tank and facilitates the precipitation of contaminants by leaving the fuel for as long as possible in the lower portion of the tank, which would have the most contaminants that could settle out. Also, a floating suction does not introduce currents that disturb the settling process.

As a means of avoiding the introduction of contaminants into the fuel when it is in storage, some fuel-oil tanks are coated. The coatings are inert and cover the steel tank surfaces to prevent contamination and metallic interaction with the fuel oil. If properly applied, appropriate coatings can last for many years and reduce tank contamination and rust and mitigate fuel-oil degradation while in storage.

b. Fuel treatment. As with diesel fuel, centrifugal separators are the primary means of treating gas turbine fuel. The major differences with gas turbine fuel-oil systems are the use of staged water-washing and additive treatments whenever residual fuel oils are used as fuel.

The early efforts to water-wash gas turbine fuel used a single-stage water-washing system. Since then, the equipment for the removal of water-soluble salts, especially sodium chloride, has been improved appreciably as increased turbine inlet temperatures created a demand for even lower sodium levels in the fuel. Two-stage water-washing, usually incorporating a demulsifying agent and reduced throughputs, has been demonstrated to provide good results.

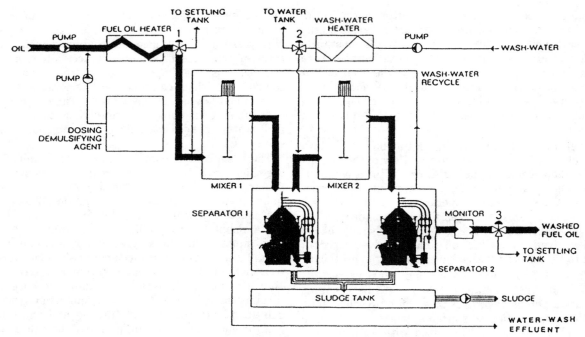

Fig. 14 Flow chart for water-washing residual fuel

Some residual fuels emulsify readily with water, and such emulsions are difficult to break. It is common, therefore, to meter a suitable demulsifying additive into the fuel line before entering a line heater.

A typical system for water-washing residual fuel is illustrated by Fig. 14. A dosing pump is used to meter the demulsifying additive into the raw residual fuel oil before it passes through the line heater. The demulsifying additive is injected at concentrations of about 150 to 250 ppm, and the line heater raises the initial temperature of the fuel to approximately 97 C, depending upon the estimated salt-water content in the fuel. The heated raw residual fuel-oil and additive mixture is then thoroughly mixed in the first mixing tank and then goes to the first centrifugal separator. The wash water, which is heated to about 97 C, is metered into the discharge from the first-stage separator and mixed in the second mixing tank. The mixture then passes into the second separator and is again cleaned to remove water and any remaining solids. The cleaned fuel then passes through a monitor to the fuel-oil service tank. In staged water-washing systems sodium can be removed from residual fuel oils down to a level of about one part per million.

If the monitor detects an amount of water in the washed fuel that is above the set limit, the washed fuel is recycled. In the system shown in Fig. 14 clean wash water is supplied to the second wash stage only. It is then recycled back to the first stage. This recycling arrangement for the wash water reduces the amount of clean distilled or demineralized water required as well as reducing heating requirements. The amount of wash water used varies from 5 to 10% of the oil throughput.

A second function of the separators is to remove sludge and solid contaminants, particularly rust, sand, and catalyst fines, from the fuel. If these contaminants passed through the fuel injectors, the catalyst fines, in particular, could cause severe erosion of the high-speed turbine blading. The use of automatic, self-cleaning fuel separators with controlled discharge, which have the capability to remove solid particulates greater than 3 μm in size, is a recommended practice.

After residual fuel-oil has been water washed, a vanadium-inhibiting additive should be added as a follow-on treatment. For fuels that have a low vanadium content, oil-soluble additives such as the magnesium sulfonate-based type, are effective inhibitors; however, in the case of residual fuels that have a high vanadium content (over 150 ppm), the use of oil-soluble additives may be too expensive. In such an event, the use of a less-expensive oil-dispersed additive such as magnesium sulfate ($MgSO_4$), or a water-soluble additive, such as epsom salts ($MgSO_4 \cdot 7H_2O$), may be an acceptable alternative. Self-contained modules that include a mixing tank, freshwater pump, feeder, and discharge pump have been used to meter submicronic epsom salts, in solution, into the combustor fuel system.

The fuel-oil service pumps used with gas turbines are similar to those used with diesel engines except that the gas turbine fuel-oil pumps are designed to the gas turbine's fuel-oil flow requirements and not to the 250% excess that is typical of diesel fuel-oil pumps.

When residual fuels are treated to control vanadium corrosion, the downstream fuel-oil heaters must be designed as low heat-density heaters to prevent additive precipitation or additive deposits within the heaters.

The provisions made to filter gas turbine fuels are very similar in terms of both design and equipment to those used to filter diesel fuels, except that gas turbines have a lower tolerance for solid contaminations. Gas turbines generally require fuel filtration provisions that will remove solid particles larger than 5 μm; however, some gas turbines require even finer filtration. The final filter in the fuel line before the gas turbine burners is a fine filter, but not the finest in the system. The final filter is intended to protect the gas turbine from unforeseen fuel system contamination after the centrifuges and fine filters have properly treated the fuel oil. The finest filter in the system may be designed for large capacity, 5-μm filtration, in which case the final filter may be designed for 10-μm filtration with only a limited contaminant-retention capacity.

For gas turbines that burn residual fuels, a means must be provided to flush out the residual fuel-oil in the lines when the engines are secured, even for emergency shutdowns. Fast-acting bypass valves should allow distillate fuel oil to displace residual fuel oil before the fuel oil cools in the lines.

2.6 Construction Materials. The materials used to manufacture most fuel-oil systems are steel or other iron-based materials. This includes tanks and equipment, as well as piping and fittings. In some cases, however, stainless steels and other corrosion-resistant materials have been used.

While iron-based materials are suitable for fuel-oil systems, it is very important that neither copper or copper alloys nor zinc or its alloys in any form be allowed to come into direct contact with fuel oil. Copper and zinc cause accelerated degradation of any fuel oil that they directly contact. Therefore, copper and zinc alloys must not be used in tubes or coils within the heaters; this prohibition includes coated forms such as galvanized piping or clad steel for tank construction.

CHAPTER XIII

M. L. Powell | # Noise Control

Section 1
Noise-Control Fundamentals

1.1 Definitions. As with all other technical fields, noise control involves the use of a number of terms that are given specialized meanings. A comprehensive discussion of the terms used in the field is given in several basic texts on acoustics or noise control [1–7]; however, definitions for the more commonly used terms that are essential for an understanding of the principles involved are as follows:

Noise. Noise is any undesired sound or vibration. Types of noise are classified by the media in which the sound or vibration exists. For example, "airborne noise" and "liquidborne noise" are undesired sounds in air and a liquid, respectively. "Structureborne noise" is undesired vibration in a structure.

Sound pressure. Sound pressure is the oscillatory component of the local fluid pressure, above and below the ambient pressure, that is caused by the passage of a sound wave. It is typically expressed as a root mean square (RMS) value. The human ear can detect sound pressures ranging from about 20 micropascals (μPa; 1 Pascal = 1 newton/m^2 = 10 dynes/cm^2), which is the threshold of hearing for a young person, to well over 100,000,000 μPa (the threshold of pain).

Decibel. A decibel (dB) is ten times the logarithm, to the base 10, of the ratio of two powers or terms related to power. One of the terms in the ratio is usually a measured or computed value, while the other term is a reference value. Decibels are typically used in noise control since a logarithmic scale makes it easier to compare data that vary over many orders of magnitude (such as sound pressure). Figure 1 shows the A-weighted sound pressure levels, in decibels, of some typical noises (A-weighting and sound pressure levels are both described below).

Sound pressure level. Sound pressure level is used to express RMS sound pressures and is defined as:

$$L_p = 20 \log_{10}\left(\frac{p}{p_{ref}}\right) \tag{1}$$

where

L_p = sound pressure level, dB
p = RMS sound pressure, μPa
p_{ref} = reference sound pressure, μPa
    = 20 μPa for airborne sound
    = 1 μPa for liquidborne sound

Sound power level. Sound power level is the decibel value of the acoustic power emitted by a noise source. By definition:

$$L_W = 10 \log_{10}\left(\frac{W}{W_{ref}}\right) \tag{2}$$

where

L_W = sound power level, dB
W = sound power, watts
W_{ref} = reference sound power = 10^{-12} watts

Acceleration level. Structureborne noise is commonly expressed as the acceleration level, which is the decibel value of the measured acceleration, and is defined as

$$L_a = 20 \log_{10}\left(\frac{A}{A_{ref}}\right) \tag{3}$$

where

L_a = structureborne acceleration level, dB
A = acceleration, m/sec^2
A_{ref} = reference acceleration = 10^{-5} m/sec^2

Frequency. Frequency is the rate at which a value, such as sound pressure, varies with time. Frequencies are expressed in cycles per second (cps) or Hertz (Hz) (1 Hz = 1 cps). Figure 2(a) shows two time-varying signals, one of which varies more rapidly than the other, plotted on amplitude-frequency-time coordinates. The sum of the two signals is shown by the upper curve of Fig. 2(b) as viewed in the amplitude-time plane (or "time domain"). Here, it is easy to see that the amplitude of the summed signal varies at two separate frequencies, one superimposed on the other. This can also be seen if the curves are viewed in the amplitude-frequency plane (or "frequency domain") as in Fig. 2(c). Here, the two separate frequencies

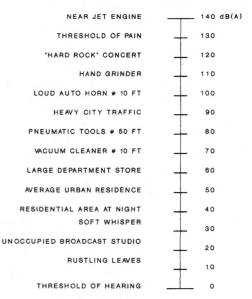

NEAR JET ENGINE	140 dB(A)
THRESHOLD OF PAIN	130
"HARD ROCK" CONCERT	120
HAND GRINDER	110
LOUD AUTO HORN ● 10 FT	100
HEAVY CITY TRAFFIC	90
PNEUMATIC TOOLS ● 50 FT	80
VACUUM CLEANER ● 10 FT	70
LARGE DEPARTMENT STORE	60
AVERAGE URBAN RESIDENCE	50
RESIDENTIAL AREA AT NIGHT	40
SOFT WHISPER	30
UNOCCUPIED BROADCAST STUDIO	20
RUSTLING LEAVES	10
THRESHOLD OF HEARING	0

Fig. 1   A-weighted source levels of typical noises

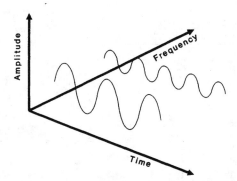

(a) Three-dimensional coordinates showing time, frequency, and amplitude

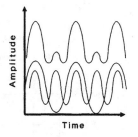

(b) Amplitude-time relationship   (c) Amplitude-frequency relationship

Fig. 2   Relationship between time and frequency domains

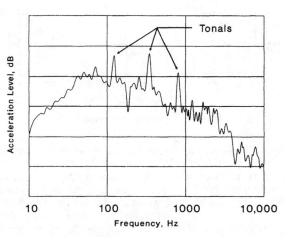

Fig. 3   Typical machinery vibration signal

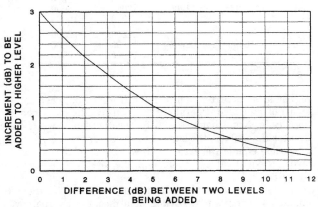

Fig. 4   Decibel addition

appear as vertical lines with their amplitudes represented by the height of the lines. Machinery vibration signals, such as that shown by Fig. 3, generally consist of a wide range of frequencies, often with narrow frequency spikes (called "tonals") surrounded by areas of "broadband" noise.

Decibel addition.   Since decibels are actually logarithmic ratios, two decibel values cannot be added together directly to obtain their sum. Instead, the decibel values must be converted back to their sound pressures or sound powers, added, and then restated in terms of decibels. For example, a fan with a 65-dB sound pressure level and a compressor with one of 70 dB would not produce a total sound pressure level of 135 dB. Instead, the total level can conveniently be calculated by

determining the difference between the two levels and adding a correction to the larger of the two. That is

$$\sum (L_1 + L_2) = L_1 + 10 \log_{10}(1 + 10^{-\frac{\Delta L}{10}}) \qquad (4)$$

where

L_1 = higher decibel level
L_2 = lower decibel level
$\Delta L = L_1 - L_2$

In the example cited, $L_1 = 70$ dB, $L_2 = 65$ dB, and $\Delta L = (70 - 65) = 5$ dB, so that the total is

$$\sum (70 \text{ dB} + 65 \text{ dB}) = 70 + 10 \log_{10}(1 + 10^{-\frac{5}{10}})$$

$$= 70 + 1.2 = 71.2 \text{ dB}$$

Figure 4 is an illustration of the correction to be added to the

Table 1 Octave and one-third octave band frequencies

Octave Bands			⅓ Octave Bands		
Lower	Center	Upper	Lower	Center	Upper
11.2	16	22.4	11.2	12.5	14.1
			14.1	16	17.8
			17.8	20	22.4
22.4	31.5	44.7	22.4	25	28.2
			28.2	31.5	35.5
			35.5	40	44.7
44.7	63	89.1	44.7	50	56.2
			56.2	63	70.8
			70.8	80	89.1
89.1	125	178	89.1	100	112
			112	125	141
			141	160	178
178	250	355	178	200	224
			224	250	282
			282	315	355
355	500	708	355	400	447
			447	500	562
			562	630	708
708	1000	1410	708	800	891
			891	1000	1120
			1120	1250	1410
1410	2000	2820	1410	1600	1780
			1780	2000	2240
			2240	2500	2820
2820	4000	5620	2820	3150	3550
			3550	4000	4470
			4470	5000	5620
5620	8000	11200	5620	6300	7080
			7080	8000	8910
			8910	10000	11200
11200	16000	22400	11200	12500	14100
			14100	16000	17800
			17800	20000	22400

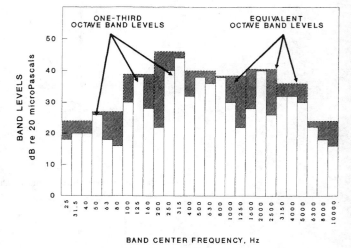

Fig. 5 Octave and one-third octave band representation

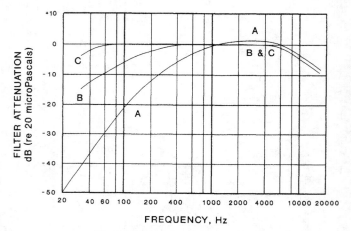

Fig. 6(a) Weighting filter responses

higher level as a function of the difference between the two levels being added. Useful observations can be made concerning Fig. 4. Note that if the difference between two levels is greater than 10 dB, the sum of the two (for most engineering purposes) equals the larger value. Also note that when two levels of equal value are summed, the total level is increased by 3 dB.

Frequency bands. Often it is more important to know the amount of energy in a portion of a measured signal rather than just its frequency content. Noise measuring equipment will typically use filters to separate the total frequency spectrum of a signal into specific segments—called "frequency bands"—so that the energy within a given segment can be summed. The bandwidth of the filter, or the difference between the uppermost and lowermost frequencies within the band, is determined by the range of frequencies that is allowed to pass through the filter unattenuated. For frequencies outside this range, the signal level is rapidly attenuated. The upper and lower frequencies that define the bandwidth are typically set at the points where the signal level has been attenuated by 3 dB.

Two different bandwidths are frequently used for reporting noise data. One is an "octave band," which has an upper frequency that is twice the value of the lower frequency. The most commonly used set of octave bands is given in Table 1. In addition to an upper and lower frequency, each band also has associated with it a center frequency, which is typically used to identify the band. For example, the octave band extending from 708 to 1410 Hz is called the 1000 Hz octave band.

When a level, such as a sound pressure level, is reported for an octave band, it is called an "octave band level."

If more information is needed concerning the distribution of the sound energy in an octave band, a narrower "one-third octave band" can be used, which divides each octave band into three parts. The upper, lower, and center frequencies for the most commonly used one-third octave bands are also given in Table 1. Levels reported in this format are called "one-third octave band levels."

Levels reported in octave band format cannot be directly compared with levels reported in one-third octave band format. Instead, the three one-third octave band levels within an octave band must be summed, using decibel addition, to develop equivalent octave band levels as shown in Fig. 5.

A-weighted sound pressure levels. Not all sounds have the same effect on the human ear. Instead, sounds in a particular frequency range seem louder than higher or lower frequencies, even when the sound-pressure levels are the same. A filter has been developed to reproduce this effect in sound-measuring equipment, and its frequency response is shown in Fig. 6(a). Sound pressure levels measured using this filter are called "A-weighted SPLs." As an example, Fig. 6(b) shows how octave band SPL's would appear when measured without the filter (unweighted or "flat" response) and with the filter

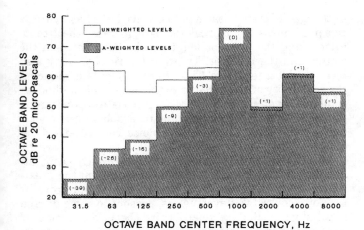

Fig. 6(b) Example of A-weighted sound pressure levels

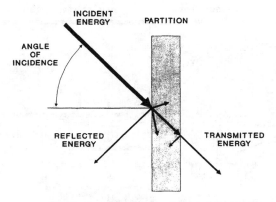

Fig. 7 Sound transmission through a partition

activated. The lowest frequency in this case, the 31.5-Hz octave band, is attenuated 39 dB, while the frequencies where the ear is most sensitive, the 2000- and 4000-Hz octave bands, are amplified 1 dB. Other weighting filters are also used. The B and C weights, also shown in Fig. 6(a), were developed for noise levels above 55 dB and 85 dB, respectively. However, the A-weight continues to be the most widely used filter whenever a subjective response to noise is needed.

Speed of sound. The rate at which a disturbance travels through a medium is termed the "speed of sound" through that particular medium. For fluids (air and water), the disturbance is a pressure wave and it travels at a speed independent of its frequency. The speed of sound in air is primarily a function of temperature, and is given by

$$C_a = \sqrt{k\,R\,T} = 49.02\sqrt{T} \tag{5}$$

where

C_a = speed of sound in air, ft/sec
k = ratio of specific heat at constant pressure to specific heat at constant volume (≈ 1.4 for air)
R = universal gas constant, ft-lbf/lbm-°R
T = temperature, °R

As a point of reference, the speed of sound in air at 70 F is 1129 ft/sec.

In liquids, the speed of sound is given by

$$C_L = \sqrt{\frac{B}{\rho}} \tag{6}$$

where

C_L = speed of sound in liquid, ft/sec
B = bulk modulus of liquid, lb/ft²
ρ = density of liquid, lb-sec²/ft⁴

The speed of sound in fresh water, with a bulk modulus of 318,000 psi and a density of 1.93 lb-sec²/ft⁴, is 4870 ft/sec.

Disturbances in solids can be in the form of normal stress waves, shear stress waves, or bending waves. The speeds for each of these disturbances depend on the type of the disturbance, the shape of the solid, and in some cases, the frequency. Equations that can be used to calculate the speeds of the different types of waves are included in reference 1.

Wavelength. The distance between successive peaks or troughs of a wave is termed the wavelength. The speed of sound and the wavelength of the sound are related by the equation

$$\lambda = C/f = C\,T \tag{7}$$

where

λ = wavelength, ft/cycle
C = speed of sound, ft/sec
f = frequency of wave, Hz
T = period = $1/f$, sec

Transmission loss. The ability of acoustic barriers, walls, and partitions to block noise is often an important design parameter. Figure 7 shows acoustic energy striking a partition and the resulting distribution of that energy. Some of the incident energy is reflected at the surface of the partition while the remainder enters it. Once inside the partition, some of the energy is lost by absorption and damping. Finally, the remainder of the energy is transmitted through the far side of the partition.

The numerical quantity used to describe a barrier's sound-blocking ability is termed "transmission loss." The transmission loss, TL, is defined as

$$TL = 10\,\log_{10}\left(\frac{\text{Incident energy}}{\text{Transmitted energy}}\right) \tag{8}$$

Another quantity, the "transmission coefficient," is sometimes used. The transmission coefficient, τ, is defined as:

$$\tau = \left(\frac{\text{Transmitted energy}}{\text{Incident energy}}\right) \tag{9}$$

and these two are related by

$$TL = 10\,\log_{10}\left(\frac{1}{\tau}\right) \tag{10}$$

1.2 Noise Control. Noise control encompasses a variety of technologies that are coordinated to obtain a satisfactory noise environment at a receiver, consistent with economic and operational considerations. A number of factors must be considered when establishing criteria that define a "satisfactory noise environment." There is usually no unique solution for a given noise situation because of the complex environmental, economic, and operational considerations that are generally involved.

Shipboard noise must be controlled for numerous reasons. On all vessels, navigation signals (such as bells, whistles, and foghorns) must be easily heard. Watch

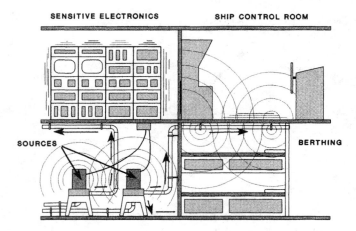

Fig. 8 Typical noise source-path-receiver system

standers in ship control areas must be able to hear and understand maneuvering commands. In machinery spaces, permanent damage to the crew's hearing must be avoided; and excessive machinery vibration in these spaces can also cause structural fatigue, premature bearing failures, and increased maintenance requirements. In addition, the living spaces must be quiet enough for passengers and crew to relax and sleep.

Naval combatant ships have additional requirements for the control of noise. Two relate directly to the ship's sonar systems. To perform their duties effectively, the sonar operators must have a quiet work space. Likewise, the hydrophones, which make up the sonar array, must be located in a quiet environment that is free from flow noise and vibration originating from the ship.

All noise control problems consist of three elements: a noise source, a noise transmission path, and a noise receiver. Noise originates at the source, travels along a transmission path, and is sensed by the receiver. Such a three-element system is illustrated by Fig. 8. In this figure, a motor-driven hydraulic pump, which is mounted on a structural foundation, provides hydraulic fluid to a component downstream. The pump and motor are the noise sources, and the noise is transmitted through several structureborne and airborne paths. The receivers could be personnel in the adjacent berthing area or ship control room, or the receiver could be sensitive electronics that are being vibrated. Identifying the receivers in a noise control problem is usually straightforward, while identifying the sources can be more challenging. The identification of the noise transmission paths is often the most difficult aspect of a noise-control effort because the paths can be obscure and difficult to detect.

Noise-control procedures can often be applied to any or all of the basic elements of a noise problem. The most effective solution is usually to reduce the level of the source. Lowering the source noise level generally reduces the amount of noise-control treatments otherwise required; however, when more noise reductions are necessary than can be gained by treating the source, an alternative solution is to treat the transmission paths. Regardless of the type of transmission path involved, the basic corrective procedures are to break the path, increase the attenuation along the path, or lengthen the path between the source and the receiver. Treating the receiver may also be a feasible alternative. This alternative entails reducing the sensitivity of the receiver to the noise. The method, or combination of methods, that is appropriate depends upon the noise reductions required and on economic and operational constraints. An effective noise control effort requires a total-system approach.

1.3 Noise-Control Strategy. Although naval combatant ships, and particularly submarines, present the greatest challenge in design for noise control, the requirement for noise control is, nevertheless, present to some degree in all ships, including small craft. The incorporation of comprehensive noise-control requirements into a ship design requires an evaluation of almost every aspect of the ship design process for noise effects. Ship arrangements clearly have an effect on noise characteristics; however, detailed material selections must also be analyzed from a noise-control perspective.

To be effective, noise control must begin at the earliest stages of ship design. The first consideration is the level of noise control required for the ship to perform its mission. Cargo ships, for instance, need only a background noise level sufficiently low to permit effective communications necessary for the operation of the ship and one that provides a healthy, restful environment for the crew. A cruise liner, on the other hand, must also provide a pleasant environment in all passenger spaces. The most stringent level of noise control is applied to a submarine since the noise it generates can reveal its presence and thereby render it incapable of accomplishing its mission.

With the noise-control objectives established, they must then be reflected in the conceptual design of the ship. Substantial cost, space, and weight compromises can be avoided by optimizing compartment locations. Obviously, a compartment with stringent noise requirements (such as a stateroom) should not be located near a noisy machinery space. Spaces containing loud noise sources (e.g., machinery spaces, fan rooms, pump rooms, workshops) are preferably grouped together and bounded by spaces that have no noise requirements (e.g. storerooms). A rearrangement of compartments is highly disruptive beyond the conceptual design stage; therefore, an early consideration for the effect of the ship arrangement on noise control cannot be overemphasized.

An early consideration should also be given to the noise requirements for the equipment in each space. Based on the ship's silencing goals, noise limits may need to be developed for each item of equipment. If so, the procurement specifications for these items should require testing to demonstrate compliance with the noise limits. For equipment that cannot be designed to comply with the required noise-control limits, other noise-control treatments must be developed.

The noise-control strategy cannot be limited to the design process, but must also include the ship construction activities. Training is required to familiarize construction personnel with the requirements for properly installing

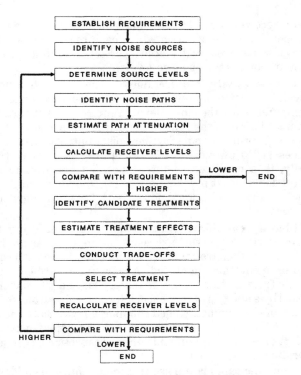

Fig. 9 Noise-control steps during original design

noise-control features and the important role they play in meeting the ship's silencing goals. During ship construction, problems develop that can affect the ship's silencing features; therefore, procedures must be developed to ensure that the construction problems are promptly reported and properly resolved.

In addition, the noise-control strategy must include a test plan. The tests can range from small-scale tests of single components up to large-scale dockside noise surveys of the entire ship. The tests should be scheduled as early as practicable to allow time for corrective actions. Time is often required to determine the most appropriate solution for noise problems; consequently, equipment noise tests are best conducted in a shop before the equipment is installed in the ship. Shop tests often identify equipment that exceeds the noise levels specified. The solution could require the equipment to be rebuilt or additional offsetting noise-control features could be incorporated into the ship. Each of the options must be weighed against ship delivery, schedule, and cost constraints.

Finally, a series of noise tests should be conducted during sea trials to demonstrate that the noise-control objectives have been accomplished.

1.4 Noise-Control Activities During the Design Phase. The noise-control program that is implemented during the ship-design process is a multiphase effort that begins with an establishment of the noise-level requirements necessary to support the accomplishment of the ship's mission. Thereafter, the process progresses as indicated by Fig. 9, which outlines the steps that are typically taken during the design of a ship. The activities involved in these steps are as follows.

Establish noise level requirements. The noise levels necessary for the ship to accomplish the mission objectives must be established. This may include developing limits for airborne, liquidborne, and structureborne noise. Satisfactory airborne noise level guidelines, from a personnel perspective, have been developed from human factors research in speech communications, hearing protection, and personnel comfort. These specifications dictate maximum allowable airborne noise levels depending upon the type of compartment (e.g. machinery space, berthing area, radio room, etc.) and the anticipated exposure time in that compartment.

Identify noise sources. The potential noise sources are identified. For large vessels, the potential list can be fairly extensive; however, the vessel's list of machinery components is a good starting point. Most of these sources emit constant noise levels; however, transient noises may also be of concern. If so, their sources may be more difficult to identify. The prediction of transient noises, such as the banging of check-valve disks, can be difficult.

Determine source levels. After the noise sources have been identified, the characteristics of the noise that they produce must be established. The characterization must include not only the source levels, but also the frequency content. Constant-speed machinery usually generates noise at discrete frequencies that can be associated with motor rpm, pump piston frequency, frequency of gear-tooth meshes, or harmonics of these frequencies. Measured structureborne or airborne noise test data may be available from the manufacturers of some equipments; however, specific components may not have been selected in the early stages of preliminary design. In these instances, noise estimates must be developed on the basis of noise data that are typical for other machines of similar design.

For some equipment, predictions of the airborne and structureborne noise levels they produce can be made with reasonable accuracy. The relationships that are used are most often empirically derived and frequently relate the noise levels to operational parameters of the equipment. As an example, two such equations have been derived for the prediction of pump airborne noise [8]. For nonreciprocating pumps, the airborne sound power level can be estimated as

$$L_{W_{av}} = 15 + 10 \log_{10}H + 15 \log_{10}R \qquad (11)$$

where

$L_{W_{av}}$ = airborne sound power (ref 10^{-12} W) of nonreciprocating pump, dB
H = pump power, hp
R = pump speed, rpm

For piston pumps, however, the airborne sound power level is related to the pressure rise across the pump by the expression:

$$L_{W_{ap}} = 75 + 30 \log_{10}\left(\frac{p}{3000}\right) \qquad (12)$$

where

$L_{W_{ap}}$ = airborne sound power (ref 10^{-12} W) of piston pump, dB

p = pressure rise across pump, psi

Similar equations have been developed for other classes of equipment.

After an estimate of the noise levels produced by the sources has been made, the noise sources can be ranked in accordance with the magnitude of the noise level that they produce. This ranking can be useful when selecting the components that can be silenced to gain the largest overall effect, and the method of treatment that will be most effective. For a compartment that has a noise level which is dominated by a single noise source, quieting that one unit may be the most effective means of substantially lowering the overall noise level. However, for spaces with many equally loud components, such as pump rooms or fan rooms, treating just one unit would have very little effect on the overall level.

Identify paths and estimate attenuation. After noise-level estimates have been made for the major sources, possible transmission paths from each source to each receiver location are identified. Referring to Fig. 8, there are three major noise transmission paths between the source and receiver that may require consideration, with all three being candidate paths for a single source. The mechanical vibration from the pump can travel through the foundation and piping into the surrounding structure; this type of path is classified as "structureborne noise." The vibrating structural surfaces also produce a second path—through the air—which is termed "airborne noise." The third path, which is often overlooked, is through the fluid within the piping; this "fluidborne noise," which is in the form of pressure pulses generated by the pump, can cause significant structural vibration in the downstream piping and components.

All of the noise energy generated by a source will not reach a receiver. Some noise travels in other directions, some is blocked or reflected back to the source, and some is attenuated by conversion to heat energy. It is important that these effects be predicted for each significant type of noise path so that an accurate estimate of the noise level at the receiver can be made. Procedures that can be used to estimate path attenuation are presented in references 1–4,8,9.

In most noise-control problems, a combination of paths from several sources may be involved; however, all of the sources and paths seldom contribute significantly to every receiver location. For example, when multiple airborne noise sources are located within the same space, units with noise levels that are 10 dB or more below the loudest unit will not contribute to the overall noise level in the space. Sources in more distant compartments can frequently be neglected if there are comparable sources in adjacent compartments. A recognition of the insignificant paths can greatly simplify the analytical procedures.

Calculate receiver levels. After the source levels, the dominant paths, and the path attenuation values are estimated, the next step is to calculate the noise levels at each receiver. For airborne noise, however, the acoustic properties of the space in which each receiver is located may also need to be calculated (see Section 2.1 for a discussion of compartment noise levels).

Compare with requirements. The calculated noise levels at receivers, both airborne and structureborne, are compared with the noise levels required to support the ship's mission. If the computed noise levels are lower than those specified, then no further noise-control analyses are necessary. However, this is seldom the case; more typically, it is predicted that the specified noise limits are exceeded at least over some narrow frequency range. In such an event, additional noise-control measures are necessary.

Identify candidate treatments. After analyzing the characteristics of the excessive noise level, candidate treatments that are compatible with other imposed design requirements are identified. Absorptive treatments are made from a wide range of materials, and the hazards posed by a shipboard fire, especially on a submarine, make many of the commonly used commercial materials unsuitable. Other design constraints that can affect material selection include weight and volume restrictions, cost, and durability.

When the candidate noise treatments have been identified, the next step is to estimate the noise reductions that each can provide. This task may be fairly simple in cases where test data are available and in instances where established analytical procedures are applicable. But the development of noise-reduction estimates can be difficult.

Conduct treatment design trade-offs. Selecting the most appropriate noise-control treatment often requires a design trade-off. Data in addition to those associated with acoustics are required to conduct trade-off analyses with the candidate noise treatments. An assessment must be made of any redesign to accommodate the treatment, schedule consequences of incorporating the treatment into the construction sequence, and the total cost of the treatment, including procurement, installation, and maintenance.

Select treatment. Based upon the results of the design trade-offs, the more promising treatments can be selected for further analysis. Occasionally, a single treatment is identified that is compatible with all of the design constraints while providing the necessary noise reduction. More often, however, single treatments are found to be inadequate, and several treatments are used in combination. Unless there is substantial confidence concerning the effectiveness of a single treatment, the use of multiple treatments can be advantageous. By using multiple treatments, there tends to be an averaging effect of the numerous approximations made in estimating the source levels and path attenuation values.

Recalculate receiver levels. With the new attenuation values provided by the selected treatments, the noise levels at the receiver location are recalculated. For a single noise source and path, this may be unnecessary as the increased attenuation values will relate directly to the lower receiver noise levels. However, for multiple sources and multiple paths, the total receiver levels must be recalculated using decibel addition.

Compare with requirements. The recalculated noise levels at the receiver locations are again compared with the required levels. If the calculated levels are lower than the required levels, the additional treatments that have been chosen should solve the noise-control problem. Often, however, the process of treatment selection and noise level calculation is repeated to account for multiple paths, to refine the estimates, and to select the treatments that provide just enough attenuation. Occasionally, no combination of treatments can be found that will result in adequate receiver noise levels, and a major redesign will be required. When this happens, the entire noise-control process must be repeated.

Section 2
Noise-Control Techniques

2.1 Source Noise Control. The most effective means of resolving a noise problem is to reduce the magnitude of the source, and doing so frequently eliminates the need for additional treatments. The source noise levels can be minimized by selecting the quietest components available. However, the quietest components are frequently the most expensive, and may not be quiet enough.

In some cases, conventionally designed components cannot meet both the performance and acoustic requirements. An alternative, then, may be to prepare special purchase specifications for the component that the manufacturers must meet. This will generally result in more expensive equipment because the manufacturer may be required to conduct specialized engineering analysis, maintain more restrictive manufacturing tolerances, incorporate noise-control features, and upgrade quality-control procedures. Consequently, the imposition of noise-control requirements beyond those obtainable by normal manufacturing processes is preferably avoided. Manufacturers have little incentive to make large research and development investments in specialized components that have limited applications.

When standard components are used, the noise can be minimized by proper component selection, operation, and maintenance. Equipment should be sized or selected to operate at peak efficiency under normal conditions. This is particularly true with ventilation fans, which tend to make more noise when they are operated at "off design" points.

Proper equipment maintenance practices are also important to preserve low-noise features. Rotating components must be accurately aligned and balanced. An unbalanced rotor in a pump, motor, compressor, etc., generates vibrational forces that are proportional to the amount of unbalance and the square of the rotational speed. Counterweights are typically added to rotors to reduce residual unbalance.

To help identify the need for maintenance, machinery condition-monitoring equipment can be used [10,11]. As a machine is operated, parts can begin to wear, shafts can become misaligned, cracks can develop and propagate, and mechanical fasteners and belts can loosen. Each of these events can cause a significant change in the noise signature of a machine, especially in the noise levels at specific frequencies. This change can be easily detected if the machine's noise signature is compared, over a period of time, with a baseline signature measured when the machine was new and operating properly. Vibration levels are typically used to generate the noise signature, although acoustic pressure has also been used in large turbines [12]. If the signature changes unexpectedly, or if the noise levels gradually increase above predetermined "alarm values," maintenance can often be scheduled and performed before serious equipment damage occurs. Automating this process involves many complex considerations, however, since the alarm values must be set high enough to prevent "false alarms" due to normal fluctuations, yet low enough to signal the need for maintenance before a catastrophic failure occurs.

In many cases, even the use of state-of-the-art design and operating practices will not yield satisfactorily low airborne and structureborne noise levels. In these cases, corrective noise treatments, such as the following, may be appropriate.

Mufflers. Mufflers can be attached to the intake and exhaust of reciprocating engines, gas turbines, ventilation fans, compressors, and large blowers to reduce the airborne noise that they create. Most mufflers are designed to function as reactive mufflers, dissipative mufflers, or a combination of the two principles.

Reactive mufflers function by reflecting part of the acoustic energy back toward the source. The simplest muffler of this type, shown in Fig. 10(a), is a simple expansion chamber inserted in the exhaust line. The change in cross-sectional area causes an abrupt change in the acoustic impedance which, in turn, prevents part of the acoustic energy from entering the chamber exhaust pipe. The acoustic performance of this type of muffler depends upon the length of the expansion chamber, the ratio of the cross-sectional areas of the chamber and exhaust line, and the frequency of the sound. Theoretically, the transmission loss provided by a reactive muffler is expressed as

$$M_R = 10 \log_{10}\left[1 + \frac{1}{4}\left(m - \frac{1}{m}\right)\sin^2 kL\right] \qquad (13)$$

where

M_R = muffler transmission loss, dB
m = area ratio = S_2/S_1

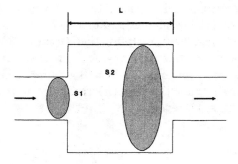

(a) SIMPLE EXPANSION CHAMBER

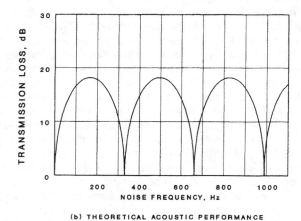

(b) THEORETICAL ACOUSTIC PERFORMANCE

Fig. 10   Transmission loss of an expansion chamber

S_1 = exhaust line cross-sectional area, ft^2
S_2 = expansion chamber cross-sectional area, ft^2
L = expansion chamber length, ft
k = wave number = $2\pi/\lambda$
λ = wavelength of sound in muffler, ft

Equation (13) is valid for frequencies less than that for which the corresponding wavelength is equal to 80% of the diameter (or largest transverse dimension) of the chamber. Figure 10(b) shows the muffler transmission loss as a function of frequency for a two-foot-long, one-foot-diameter muffler in a three-inch exhaust line. The exhaust gas is at a temperature of 250 F. As can be seen, the transmission loss is a periodic function and reaches a maximum when the expansion chamber length equals odd multiples of a quarter wavelength, that is, $\lambda/4$, $3\lambda/4$, $5\lambda/4$, etc., as calculated from equations (5) and (7). When the chamber length equals multiples of a half wavelength, that is, $\lambda/2$, λ, $3\lambda/2$, a standing wave is established in the chamber and there is no transmission loss (all of the acoustic power is preserved and transmitted into the exhaust line).

Dissipative mufflers or silencers function by attenuating the noise with absorptive materials. Glass fiber is frequently used as the absorptive material and is placed in perforated chambers within the silencer. The acoustic

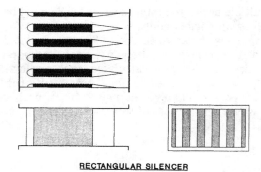

RECTANGULAR SILENCER

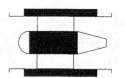

ROUND SILENCER

Fig. 11   Typical silencer designs

Table 2   Approximate attenuation of commercial duct silencers, dB [8]

Silencer Design	Octave Band Center Frequency, Hz								
	31.5	63	125	250	500	1000	2000	4000	8000
Low pressure drop									
3 ft long	1	4	7	9	12	15	16	14	9
5 ft long	2	8	12	14	16	19	20	18	14
7 ft long	3	10	15	19	20	22	24	22	18
High pressure drop									
3 ft long	3	8	10	15	23	30	35	28	23
5 ft long	4	11	14	23	32	38	42	36	30
7 ft long	5	13	18	30	40	44	48	42	36

performance of prefabricated silencers depends on their shape and size, and the static pressure drop across them, but is generally more broadband in nature than reactive mufflers. Typical silencer designs are shown in Fig. 11 and representative attenuation values are listed in Table 2 [8]. Manufacturer's data should be used for specific silencer designs.

Absorptive materials can also be added to expansion-chamber mufflers, and the result is a significant improvement over an unlined chamber. Figure 12 shows the theoretical effect of adding a small amount of absorptive lining to the expansion chamber previously discussed. Of note is the elimination of the transmission loss minima and the improved broadband performance.

Cladding.   Large vibrating surfaces on machinery and lightweight structure can efficiently radiate airborne noise to the surrounding space. Covering these surfaces with a cladding, which is typically a soft foam with a rigid outer skin, can "decouple" the vibrating surface from the surrounding air and reduce the airborne noise generated. Table 3 lists typical noise reductions achievable with cladding [8].

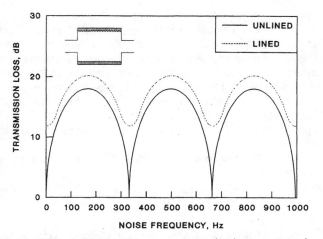

Fig. 12   Effect of absorptive lining on expansion-chamber transmission loss

Table 3   Airborne noise reduction of cladding treatments, dB (ref 20 μPa) [8]

	Octave Band Center Frequency, Hz								
Outer Layer Density	31.5	63	125	250	500	1000	2000	4000	8000
0.25 lb/ft²	0	0	0	0	1	5	8	10	12
0.25 to 1.0 lb/ft²	0	3	6	8	11	15	18	18	18
1.0 to 2.5 lb/ft²	1	4	7	10	14	17	20	20	20
2.5 lb/ft²	2	5	8	12	17	20	22	22	22

Damping.   Unstiffened machinery surfaces, such as panels on housings, cabinets, and enclosures, can resonate and amplify structureborne noise. Properly designed damping treatments applied to these surfaces can substantially lower the vibration levels.

Surface damping treatments can be applied in either of two basic designs as illustrated by Fig. 13. "Free layer," or "extensional," damping is made up of a layer of damping material adhered directly to the vibrating surface. When the surface deforms, the damping layer is alternately compressed and extended. In the process, part of the vibrational energy from the structure is stored in the damping material and then dissipated as heat. In the second type of treatment design, known as "constrained layer" or "shear layer" damping, a thin plate is glued to the top of the damping layer. This constraining layer causes the damping material to be sheared when the base structure deforms. As before, the shearing motion in the damping layer dissipates a part of the vibrational energy.

The materials used as damping treatments are called "viscoelastic"; "elastic" because of their ability to store strain energy (much like a compressed spring), and "visco" because part of the energy is dissipated (analogous to a viscous damper). However, both the storage and dissipative capacities of damping materials are very strong functions of temperature, as shown in Fig. 14. At low temperatures, viscoelastic materials are characterized as "glassy" because they have a high modulus of elasticity and are stiff, but dissipate very little energy because of a low loss factor. At high temperatures, viscoelastic materials become "rubbery" with a low modulus and a relatively low loss factor. In the temperature region

Fig. 13   Surface damping treatments

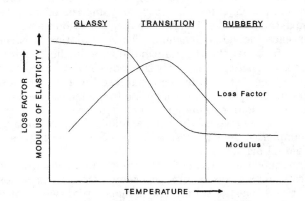

Fig. 14   Typical viscoelastic material behavior

between these two extremes, the material undergoes a transition, and it is in the transition region that the damping capacity peaks. These regions occur at different temperatures for different materials. Therefore, it is important to know the temperature of the vibrating structure so that the proper damping material can be selected.

It is also important to understand a fundamental limitation of adding damping treatments to a vibrating system. The total vibrational energy in a system can be thought of as being distributed among the various vibrating components of the system. Likewise, the total energy dissipated in the system will be a sum of the energies dissipated in each component. To significantly increase the system's total energy dissipation, and thereby reduce the vibration response, the damping must be added to a part of the system that contains a significant portion of the total energy. This simple, but important, concept is expressed by

$$\eta_{SYS} = \frac{\sum \eta_i W_i}{W_{SYS}} \qquad (14)$$

where

η_{SYS} = loss factor for total system
W_{SYS} = vibrational energy in total system

η_i = loss factor for individual component
W_i = vibrational energy in individual component

A common oversight is to add large amounts of damping material to a part of the system that contains very little vibration energy. Another oversight is to fail to recognize other forms of damping in the system that may dissipate significant amounts of energy. For example, welded-steel structures have very little inherent damping, but bolted or riveted structures can have much higher levels of damping. Also, for submerged vibrating structures, significant amounts of energy can be lost to the surrounding fluid in the form of radiated noise. Over certain frequencies, this "radiation damping" will be much higher than the damping levels that could be achieved by adding additional treatments. More details concerning the design of damping treatments are given in reference 13.

Compartment noise level reduction. When several noisy machines are grouped together in a common compartment, such as in a fan room or pump room, the total compartment can be thought of as the noise source, rather than each individual machine. This assumption is based on the fact that the resultant noise level in the compartment is a function of not only the acoustic power radiated to the space by the equipment, but also the acoustic properties of the space itself. An option for reducing the noise level in a "source" compartment, then, is to modify the acoustic properties of the compartment. This option may be simpler and more cost effective than treating each individual machine.

The standard method of predicting sound pressure levels caused by a noise source in a compartment is based on the theory that the sound field is made up of two distinct parts. The "direct field" is the sound radiated directly by the noise source and is independent of the surroundings. Its intensity depends on the radiated acoustic power and distance from the source. The second part, the "reverberant field," is made up of the sound waves that are reflected (perhaps several times) from the walls, floor, ceiling, and any other objects in the compartment or room. The reverberant field depends upon the size and the acoustic properties (namely the absorption and reflection) of each of these reflecting surfaces. The total pressure level, then, at a point in the room is the sum of the direct and reverberant fields and is given by

$$L_p \approx L_W + 10\log_{10}\left(\frac{Q_\theta}{4\pi r^2} + \frac{4}{R}\right) + 10 \qquad (15)$$

where

L_p = sound pressure level at a point in a compartment, dB (ref 20 μPa)
L_W = sound power level radiated by source, dB (ref 10^{-12} W)
Q_θ = directivity factor for source and its location
r = distance from source to point of interest, ft
R = room constant of space, ft^2

In equation (15), the room constant, R, is a measure of the total absorption within the space; higher room constant values indicate more absorption. Recording studios

and anechoic ("without echo") chambers have a very high absorption and are considered acoustically "dead." There is very little, if any, reverberant field in these spaces, and the sound intensity decreases with distance from the noise source. At the other extreme, gymnasiums, indoor parking garages, and reverberation chambers have very little absorption and are acoustically "live." Because of the large reverberant field, the sound pressure levels in these spaces can be fairly constant over a large part of the space.

In practice, the room constant is often replaced with the following:

$$S\bar{\alpha}_{SAB} \approx R \qquad (16)$$

where

$S\bar{\alpha}_{SAB}$ = total room absorption in Sabines, ft^2
S = total area of all absorptive surfaces, ft^2
$\bar{\alpha}_{SAB}$ = average Sabine absorption coefficient
R = room constant of space, ft^2

The average Sabine absorption coefficient is found by multiplying the area, S_n, of each individual surface (such as a deck or bulkhead) by the absorption coefficient, α_{SAB_n}, of that surface, summing this product for all surfaces in the room, and dividing by the total surface area, as follows:

$$\bar{\alpha}_{SAB} = \frac{S_1(\alpha_{SAB})_1 + S_2(\alpha_{SAB})_2 + \ldots + S_n(\alpha_{SAB})_n}{S} \qquad (17)$$

The Sabine absorption coefficients of some typical surfaces used in ship construction are given in Table 4. For acoustic treatments, the absorption is a function of the material thickness, with the thicker materials having higher absorption values at lower frequencies. Table 5 is an example calculation of the total Sabine absorption in an empty compartment.

In addition to boundary surfaces (such as the decks and bulkheads), machinery, furniture, and other structures within a space absorb part of the sound and must also be accounted for. However, adding up all of the individual surface areas and absorption coefficients within a space can be tedious. As an approximation, reference 8 suggests correction factors that can be used to compensate for nonboundary absorption; the correction factors are based on the size and type of compartment (e.g., berthing, crews mess, engine room).

The effect of increasing the total compartment absorption is shown in Fig. 15. Here, the difference between the sound pressure and sound power levels is plotted as a function of distance from the noise source for several values of total compartment absorption. Doubling the total compartment absorption lowers the sound pressure level by 3 dB. In many instances, however, it is difficult to increase the total compartment absorption by more than a factor of about 4, which limits the feasible noise reduction to about 6 dB.

2.2 Airborne Noise Path Treatments. Most shipboard noise-control problems involve airborne noise. Airborne noise can be treated in a number of ways; however, all are based upon some combination of three principles, namely:

Table 4   Sabine absorption coefficients of typical shipboard materials [8]

Description	Octave Band Center Frequency, Hz								
	31.5	63	125	250	500	1000	2000	4000	8000
Steel plate	0.01	0.01	0.02	0.03	0.03	0.03	0.02	0.02	0.02
Tiled deck	0.01	0.02	0.02	0.03	0.03	0.03	0.03	0.02	0.02
Carpeted deck	0.02	0.04	0.08	0.10	0.15	0.20	0.25	0.20	0.15
Draperies on wall	0.02	0.02	0.03	0.04	0.11	0.17	0.24	0.35	0.40
Acoustic ceiling tile ⅝ in.	0.03	0.10	0.20	0.25	0.50	0.70	0.75	0.75	0.70
MIL-A-23054 acoustic board									
1 in. thick	0.03	0.05	0.07	0.25	0.70	0.90	0.75	0.70	0.65
2 in. thick	0.04	0.10	0.25	0.70	0.90	0.90	0.80	0.75	0.70
MIL-I-22023, TII, Cl 4, 2 in. thick behind 2-mil Mylar vapor barrier and perforated aluminum sheathing	0.10	0.20	0.28	0.92	0.99	0.79	0.53	0.37	0.25

Table 5   Example calculation of an average Sabine absorption coefficient for a compartment

	Octave Band Center Frequency, Hz								
	31.5	63	125	250	500	1000	2000	4000	8000
Deck: $S_1(\alpha_{SAB})_1 = (10\ ft \times 8\ ft)\,(\alpha_{SAB})_1$	0.8	1.6	1.6	2.4	2.4	2.4	2.4	1.6	1.6
Overhead: $S_2(\alpha_{SAB})_2 = (10\ ft \times 8\ ft)\,(\alpha_{SAB})_2$	2.4	4.0	5.6	20	56	72	60	56	52
Walls: $S_3(\alpha_{SAB})_3 = (8\ ft \times 7\ ft)\,(\alpha_{SAB})_3$	0.56	0.56	1.12	1.68	1.68	1.68	1.12	1.12	1.12
$S_4(\alpha_{SAB})_4 = (10\ ft \times 7\ ft)\,(\alpha_{SAB})_4$	7.0	14.0	19.6	64.4	69.3	55.3	37.1	25.9	17.5
Total Sabines $= \sum S_i(\alpha_{SAB})_i$	18.3	34.7	48.6	154.6	200.4	188.4	138.8	111.6	90.8
Average Sabine absorption coeff. $= \sum S_i(\alpha_{SAB})_i / S_{TOTAL}$	0.04	0.08	0.12	0.38	0.49	0.46	0.34	0.27	0.22

NOTES:
Compartment size: 10 ft long × 8 ft wide × 7 ft high.
Surface covering:
1. Deck: tiled.
2. Overhead: 1-in.-thick MIL-A-23054 acoustic board.
3. Walls:
  (a) Each 8-ft-long wall—steel plate.
  (b) Each 10-ft-long wall—2-in. MIL-I-22023, Type II, Class 4 insulation behind
      2-mil vapor barrier and perforated aluminum sheathing.

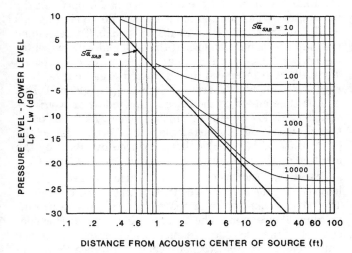

Fig. 15   Effect of room acoustics and distance on pressure level

- Break the noise path
- Lengthen the noise path
- Increase the attenuation along the noise path

Barriers. Breaking the noise path is an effective way to reduce airborne noise, and this is frequently accomplished by placing a barrier between the source and the receiver. Partitions, which are essentially partial barriers, can be used if frequent access to the noisy equipment is required. Unfortunately, noise can also reach the receiver by reflected paths (such as off the overhead and nearby bulkheads), which limits the effectiveness of partitions. For small sources in large rooms, an enclosure around the source or, occasionally, the receiver, can serve as the barrier. For an enclosure to be most effective, it should be virtually airtight and the walls should be made of a material that has a high transmission loss at the most problematic frequency. The walls of an enclosure can be a self-supporting structure, or they can be in the form of a high-mass curtain that is supported by a lightweight frame. It may also be beneficial to line the interior of the enclosure with a sound-absorbent material to suppress the level within the enclosure. In any event, enclosures must be designed to provide adequate access for routine operation and maintenance. The failure to properly consider these requirements during the design of enclosures often results in the enclosures being abandoned in service.

The actual transmission loss of a partition depends on several quantities, including: the frequency, sound pressure level, and incidence angle of the impinging sound; the speed of sound and density of the air; and the density,

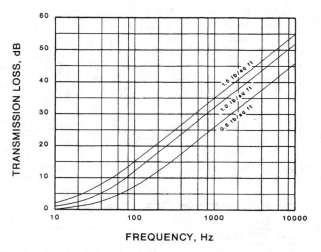

Fig. 16   Limp-mass barrier theoretical transmission loss (normal incidence)

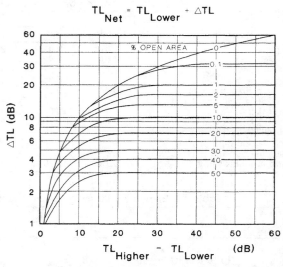

$$TL_{Net} = TL_{Lower} + \Delta TL$$

Fig. 17   Degradation of enclosure transmission loss due to penetrations

thickness, bending stiffness, and loss factor of the partition. For limp, mass-loaded curtains, the bending stiffness of the material is negligible and the transmission loss can be approximated by the "mass law" [1] as follows:

$$C_L = 10 \log_{10}\left[1 + \left(\frac{\omega \rho_s}{2\rho c} \cos\theta\right)^2\right] \tag{18}$$

where

C_L = curtain transmission loss, dB
ω = circular frequency = $2\pi f$, rad/sec
f = frequency of impinging sound, Hz
ρ_s = panel surface density, lb-s^2/ft^3
ρ = density of air, lb-s^2/ft^4
c = speed of sound in air, ft/sec
θ = incidence angle between sound and panel, deg

The only property of the barrier in this equation is its surface mass. For sound that has a normal incidence angle ($\theta = 0$ deg), the transmission loss increases 6 dB each time the surface mass of the barrier is doubled (except at very low frequencies). Figure 16 shows the theoretical transmission loss for limp-mass barriers having weights in the range of those commonly used.

The transmission loss through rigid panels is significantly different from that predicted by the mass law. The bending stiffness of a rigid panel supports traveling waves which, at the proper frequencies, cause large-displacement vibrations, or resonances. At these resonant frequencies, the large amplitudes of vibration significantly reduce the transmission loss through the panel. At low frequencies, the stiffness of the panel produces a somewhat higher transmission loss than is predicted by the mass law. Additional guidance concerning the prediction of the transmission loss through rigid panels made of typical shipboard materials is contained in reference 1.

The effect of penetrations in a sound barrier must also be recognized. Any windows, doors, or clearance holes (such as for piping, ventilation ducts, or electrical cables) must have a transmission loss which is equal to that of

the barrier in order to maintain the barrier's effectiveness. Small penetrations which have significantly lower transmission losses than the surrounding barrier will allow sound to enter or escape, which reduces the "net" transmission loss of the wall. The amount of reduction can be determined from Fig. 17, as illustrated by the following example.

Assume that a 100-ft^2 barrier has been designed to provide a TL of 50 dB at a particular frequency. However, a 1-ft^2 window with a TL of only 20 dB (at the same design frequency) has been installed in the barrier. In this case, the area of the window (the penetration) is 1% of the total barrier area, and the difference in TL between the unpenetrated barrier and the window is 30 dB (TL(higher) − TL(lower) = 50 dB − 20 dB = 30 dB). To determine the "net" transmission loss of the wall and window, enter the horizontal axis of Fig. 17 at a value of 30 dB, proceed vertically and intersect the "1% open area" curve, then read the value of ΔTL from the vertical axis (20 dB). This ΔTL is then added to the TL of the window or barrier—whichever is lower—to determine the net TL of the "penetrated" barrier. In this example, the effectiveness of the original 50-dB barrier has been reduced by 10 dB by the addition of the window.

Distance.   Airborne noise can also be reduced by lengthening the path between the source and the receiver. The noise generated by a source naturally radiates outward in all directions. The acoustic power radiated is constant, but the acoustic intensity (which is the power per unit area) decreases with increasing distance from the source. Since the acoustic pressure is proportional to the acoustic intensity, the sound pressure level also decreases with distance from the source. This phenomenon is called "spherical spreading" and, in a free field environment, results in a 6-dB reduction in the sound pressure level when the distance between the source and receiver is doubled. This reduction can be determined from the expression:

Table 6 Attenuation of square duct, dB/ft [8]

Smaller Interior Dimension, in.	Octave Band Center Frequency, Hz								
	31.5	63	125	250	500	1000	2000	4000	8000
Unlined									
5 to 10	0.12	0.10	0.10	0.05	0.05	0.04	0.04	0.04	0.04
11 to 17	0.16	0.15	0.10	0.05	0.05	0.04	0.04	0.04	0.04
18 to 24	0.20	0.18	0.12	0.06	0.05	0.04	0.04	0.04	0.04
25 to 35	0.22	0.20	0.13	0.06	0.05	0.04	0.04	0.04	0.04
36 to 48	0.23	0.21	0.14	0.07	0.05	0.04	0.04	0.04	0.04
1-in. Inside lining									
5 to 10	0.22	0.25	0.30	1.20	2.00	3.00	3.00	2.40	2.00
11 to 17	0.27	0.27	0.27	1.00	1.60	2.20	2.70	1.80	1.20
18 to 24	0.27	0.27	0.25	0.50	1.20	1.80	1.80	1.40	1.00
25 to 35	0.27	0.28	0.23	0.30	0.90	1.20	1.20	0.90	0.80
36 to 48	0.28	0.28	0.23	0.28	0.80	1.00	1.00	0.80	0.60
2-in. Inside lining									
5 to 10	0.25	0.35	0.45	1.53	2.60	4.00	4.00	3.00	2.50
11 to 17	0.27	0.32	0.37	2.23	2.20	3.00	3.00	2.40	1.60
18 to 24	0.30	0.32	0.33	0.87	1.80	2.40	2.30	1.90	1.35
25 to 35	0.31	0.31	0.29	0.53	1.20	1.60	1.40	1.20	1.00
36 to 48	0.30	0.30	0.28	0.45	1.02	1.30	1.20	1.00	0.76
4-in. Inside lining									
5 to 10	0.38	0.53	0.68	2.30	2.60	4.00	4.00	3.00	2.50
11 to 17	0.40	0.48	0.56	3.35	2.20	3.00	3.00	2.40	1.60
18 to 24	0.45	0.48	0.50	1.31	1.80	2.40	2.30	1.90	1.35
25 to 35	0.46	0.46	0.44	0.80	1.20	1.60	1.40	1.20	1.00
36 to 48	0.45	0.45	0.42	0.68	1.02	1.30	1.20	1.00	0.76

$$D_c = 20 \log_{10} r + 1 \qquad (19)$$

where D_c is the sound pressure level reduction caused by distance (in decibels) and r is the distance from the source in feet. Reducing noise levels by increasing the distance from the source may be feasible during the early design stages, before the shipboard arrangements have been finalized; however, as the design develops further, there is less opportunity for this alternative.

Attenuation. Airborne noise can also be reduced by increasing the attenuation along the noise path. As previously described, the sound pressure level at a point in a compartment can be due to both a directly radiated sound field and a reflected, reverberant field. The path that the reverberant noise takes in going from the source to the receiver can include many interactions with the compartment's boundaries. Increasing the attenuation, or absorption, of these boundaries will increase the room constant, lower the intensity of the reverberant sound field, and reduce the overall sound pressure level.

In ventilation systems, airborne noise can travel throughout the ducting to each compartment served by the system. This noise can be generated by the ventilation fans themselves, by high-velocity airflow, or it can originate from a noisy compartment. This noise can be effectively reduced by increasing the attenuation along the duct. This is typically done by lining the inside of the duct walls with fiberglass batting held in place behind a perforated sheet-metal liner. The attenuation is expressed in decibels per foot of treated duct and depends upon the thickness of the lining and the size of the duct as indicated by Table 6.

In addition to sound, however, fiberglass lining can also absorb vapors from hydraulic fluid, lubricating oil, fuel oil, and exhaust fumes. In spaces where these vapors may be present, the fiberglass should be protected with a vapor barrier. Unfortunately, this usually reduces the lining's acoustic effectiveness.

2.3 Structureborne Noise Path Treatments. Like airborne noise, structureborne noise is treated by:

- Breaking the noise path
- Lengthening the noise path
- Increasing the attenuation along the noise path

Isolation mounts. The most effective way to reduce structureborne noise is to break the path between the source and the supporting structure. This is typically done by isolating vibrating machinery from its supporting structure, or foundation, with isolation mounts. In the example system of Fig. 8, the hydraulic pump could be resiliently mounted to reduce the vibration transmitted to the foundation. Likewise, the structureborne path from the pump through the piping system could be broken by inserting a length of flexible hose. If necessary, additional reductions could be made in the structureborne vibrations by resiliently mounting the piping.

There is a wide variety of noise-isolator designs. Elastomers, such as neoprene, are typically used for the resilient element in the mounting arrangement. However, metal meshes, steel springs, and wire ropes have also been used for high-temperature or extremely severe service conditions. The sizes of commonly used noise isolators range from those suitable for components weighing no more than a few pounds up to those used to support machinery weighing many tons.

For custom installations and large-surface-area applications, sheets of ribbed elastomer can provide cost-effective isolation. This "distributed isolation material" (DIM) can be cut to size at installation and is commercially available in elastomer grades which can handle loads of up to 100 psi. The natural frequency at this loading is approximately 18 Hz, and it increases to approximately 33 Hz for

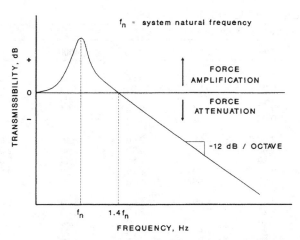

Fig. 18   Typical response of a single-degree-of-freedom system

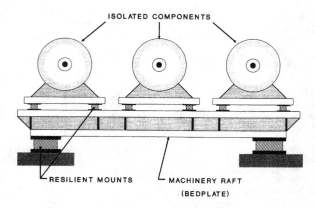

Fig. 19   Typical machinery compound isolation

a distributed isolation material pad with a light load of 10 psi.

Choosing the proper isolation mount requires an understanding of how they function, plus some knowledge about the equipment to be mounted. Isolation mounts are simply metallic or elastomeric springs that isolate vibrating machinery from its supporting foundation. The degree of isolation is characterized by the "transmissibility," which is the dynamic force input to the foundation divided by the dynamic force generated by the equipment. When the equipment operates below the natural frequency of the system (i.e., the machinery mounted on the isolators), the unsteady force into the foundation is amplified. At frequencies sufficiently above the system's natural frequency (approximately 1.4 times higher), the unsteady force is attenuated, and the attenuation increases at a rate of 12 dB per octave. This concept is shown in Fig. 18 for an ideal one-degree-of-freedom system. For this type of system, then, a mount size should be chosen that can: (1) support the weight of the equipment within the design load range of the mount, and (2) provide a system natural frequency that is at least an octave lower than the equipment's lowest forcing frequency. Theoretically, then, this would ensure attenuation of the equipment's higher frequencies. For practical installations, however, several factors limit the actual attenuation that can be achieved. Neither the foundation nor the equipment's mounting feet are infinitely stiff, which introduces additional high-frequency resonances. Reference 14 describes a method for estimating the effectiveness of resilient mounts with the foundation flexibility included. Internal resonances can also exist within the mounts and can limit their high-frequency performance. Often, a mount may not be commercially available in the "optimum" size. In this case, the next largest size is typically used which, usually being stiffer, raises the system's natural frequency. Finally, other attachments to the isolated equipment—such as power cables and fluid piping—can provide additional transmission paths for the noise.

One means for improving high-frequency attenuation is the use of compound mounting. Figure 19 shows several machinery components that are resiliently mounted to a common bedplate, or "machinery raft," that is also isolated. The additional isolators and the mass of the bedplate introduce an additional resonance that must be considered. However, the benefit of compound mounting is that at higher frequencies (well above the primary and secondary resonances), the attenuation increases at a rate of 24 dB per octave—twice the rate of a single mount. Generally speaking, the heavier the bedplate, the greater the attenuation. However, the increased attenuation must be traded off against the increased weight and space required.

In addition to mount performance, other factors must also be considered in designing an isolator installation. Additional space is required by the isolators and provisions must be made for them in the equipment arrangement. Ship motions (e.g., pitch and roll) cause isolated components to deflect, and to prevent contact with surrounding structure or hard-mounted components, stabilizing bars that include isolation elements may be needed at the top of the equipment. Furthermore, piping that services isolated components should be attached with flexible connectors to avoid transmitting forces between the isolated components and the rigidly mounted piping.

Distance.   Reducing structureborne noise by increasing the length of the path between the noise source and the receiver is very effective during the early design phases, but is seldom practicable thereafter.

Increased attenuation.   Reducing structureborne noise by increasing the attenuation along the path can be appropriate in some cases. Surface damping treatments can be effective in attenuating structureborne noise that propagates along a structure in the form of bending, or flexural, waves.

Unfortunately, the noise level at the receiver may not be reduced by an equivalent amount. Structureborne noise can also propagate in the form of compressional, or longitudinal, waves which are very difficult to damp. Unlike bending waves, compressional waves cause very little deformation of the structure's surface and so it is difficult to transfer the strain energy in these waves to a surface-mounted damping material.

Table 7 Allowable noise levels at the bridge wing lookout station, dB

	Octave Band Center Frequency, Hz								
	31.5	63	125	250	500	1000	2000	4000	8000
Maximum allowable SPL at bridge wing dB (ref 20 μPa)	90	85	80	75	70	65	65	70	72

2.4 Receiver Noise Control. After noise has reached a receiver, the alternative means of treating the noise becomes more limited, but there are several treatments that may be applied advantageously.

Personal hearing devices. The most common treatment for airborne noise at a human receiver is the use of hearing-protection devices. A combination of ear plugs and earmuffs can provide a noise reduction of 30 dB or more between 250 and 8000 Hz. In high noise level environments, an added benefit of hearing-protection devices is that they can actually improve the intelligibility of speech [15]. If only short-term exposure to high noise levels is required, using hearing-protection devices may be a cost-effective solution.

Structural modifications. Structureborne noise can also be reduced at the receiver by modifying the local structure. If the receiver is hard mounted to the surrounding structure, then changing the dynamic characteristics of the structure can alter the level of vibration. These characteristics can be modified by changing (usually increasing) the rigidity of the structure, by changing (usually adding) mass in the vibrating elements, or by adding damping to lower the resonant vibration levels.

Tuned dampers. If the receiver is resiliently mounted but vibration at a single frequency is transmitted from the foundation, the use of "tuned dampers" may be feasible. These devices consist of small masses (on the order of a few percent of that of the receiver) resiliently mounted to the receiver. However, instead of using a spring as the resilient element, a viscoelastic material is used. As the receiver vibrates, the energy is transferred to the dampers and is absorbed by the viscoelastic material, as shown schematically by Fig. 20. The viscoelastic material is typically used in a tension/compression mode, but shear and surface-mounted configurations have also been used. Additional details concerning the design of tuned dampers can be found in reference 13.

2.5 Example Noise Analysis. To illustrate the principles applied when conducting noise analyses, an evaluation will be made of the noise levels that can be expected at the bridge wing lookout station on a ship with gas turbine propulsion.

The maximum allowable noise levels specified for the lookout station are given in Table 7.

The noise produced by the gas turbine is believed to be the dominant noise source; therefore, that source of noise will be evaluated first. In the absence of directly applicable data, estimates of the noise levels at the engine inlet and exhaust could be made by using equations such as those presented in reference 8. However, in this case, directly applicable data were available from reference 16 and are given in Table 8.

Table 8 Sound power levels at the inlet and exhaust of a gas turbine, dB [16] (PWL ref 10⁻¹² W)

	Octave Band Center Frequency, Hz								
	31.5	63	125	250	500	1000	2000	4000	8000
Inlet	114	115	117	116	122	130	137	148	141
Exhaust	124	125	123	122	119	111	108	98	91

Table 9 Attenuation of unlined 90-deg duct turns, dB [8]

Smaller interior dimension, in.	Octave Band Center Frequency, Hz								
	31.5	63	125	250	500	1000	2000	4000	8000
5 to 10	0	0	0	0	0	1	2	3	3
11 to 20	0	0	0	0	1	2	3	3	3
21 to 40	0	0	0	1	2	3	3	3	3
41 to 80	0	0	1	2	3	3	3	3	3

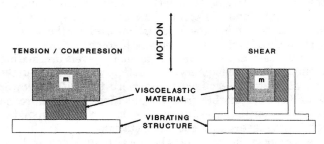

Fig. 20 Tuned dampers

A schematic of the ship arrangement is shown in Fig. 21. The preliminary ship arrangement indicates that the lookout station is located 40 ft from the engine intake duct inlet and 75 ft from the exhaust stack outlet.

The dominant airborne noise path from the engine intake is through the intake duct. Similarly, the path from the engine exhaust is through the exhaust duct.

Figure 21 indicates that the gas turbine inlet consists of two 90-deg unlined bends and a section of unlined, straight duct that is 10 ft long. The attenuation provided by a 90-deg bend is given in Table 9. The noise attenuation in straight unlined duct is given in Table 6. An end-loss correction is also needed to account for the noise reflected at the face of the inlet. This correction is a function of the duct opening area [8], and the corresponding values are given in Table 10. (Note: To simplify the calculations, the values given in Table 10 have been rounded to the nearest decibel.)

The intake duct inlet is oriented 180 deg away from the lookout station; therefore, a directivity correction is necessary. This correction is based on the size of the inlet, its orientation with respect to the receiver, and the frequency of the noise in the inlet [8]. The resulting directivity correction is included in Table 10.

The reduction in the sound pressure level due to the distance from the intake duct inlet to the lookout station can be estimated from equation (19). A 40-ft distance results in a distance correction of 33 dB, which is included in Table 10.

The attenuations in the noise level resulting from the 90-deg bends, straight duct, end loss correction, directivity correction, and distance corrections are summed and deducted from the sound power level at the gas turbine inlet to determine the noise level at the lookout station caused by this noise source.

The noise level at the lookout station caused by the gas turbine exhaust is determined similarly. The sound power levels at the turbine exhaust are listed in Table 8, and the attenuation in this case is the result of a 17-ft unlined straight section of exhaust duct, an end-loss correction, a directivity correction for a 16-ft opening with a 135-deg bend, and a distance correction for the 75 ft between the exhaust outlet and the lookout station. Using the same procedures as those used to assess the attenuation for the inlet duct, the attenuation in the exhaust duct is calculated and subtracted from the noise level at the engine exhaust to determine the noise level at the lookout station from that source. The noise levels originating from the gas turbine inlet and the gas turbine exhaust are then combined to determine the total noise level at the lookout station that originates at the gas turbine.

An assessment was made of the noise levels originating from auxiliary machinery, ventilation, and other sources and all were determined to be more than 10 dB lower than those originating from the gas turbine; therefore, no source, other than the gas turbine, has an effect upon the total noise levels at the lookout station.

A comparison between the total sound power level (SPL) calculated at the lookout station and that allowable shows that noise control treatment is required for the frequencies above the 500-Hz octave band, particularly the 2000- and 4000-Hz octave bands.

Comparing the noise originating from the turbine inlet and outlet shows that, in these octave bands, the inlet is the dominant source. No feasible method of reducing the noise level at the source is apparent, and treating the receiver at the lookout station is not practicable; therefore, attention will initially be focused upon increasing the attenuation along the intake path.

The principal alternatives in increasing the attenuation along the path from the engine intake to the lookout station are to:

- Break the path with a barrier. A barrier could possibly be added by relocating some existing structure to block the direct path between the entrance to the inlet duct and the lookout station; however, such a modification would have to be confirmed to be compatible with ship arrangement requirements.

- Increase the length of the noise path. Lengthening the noise path would require the bridge to be moved, the air intake inlet to be relocated, or a combination of both. This alternative would also entail significant changes to the ship arrangement, but may have to be considered further if there is no preferred resolution.

- Increase the attenuation along the existing path. Alternative methods of increasing the attenuation along the path include: adding sound-absorptive treatment to the inside of the intake duct or replacing part of the inlet duct with a low-pressure-drop silencer. These solutions appear feasible and will be investigated further.

By adding four inches of sound-absorptive treatment to the internal surface of the inlet duct along 10 ft of its straight length, the attenuation effect would be increased as indicated in Table 6. Table 11 shows that treating 10 ft of the straight inlet duct would be beneficial, but further attenuation is necessary.

Instead of lining the straight portion of the inlet duct, a more effective solution would be to install a low-pressure-drop silencer near the inlet to the engine. This option would expose the noise in the inlet duct to a larger area of absorptive material. The performance of a 5-ft-long

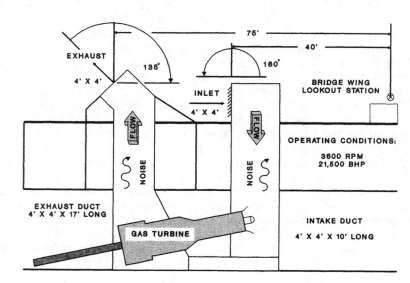

Fig. 21 Gas turbine arrangement schematic

Table 10 Design example: SPL at bridge wing lookout station, dB (ref 20 μPa)

	Octave Band Center Frequency, Hz								
	31.5	63	125	250	500	1000	2000	4000	8000
PWL at turbine inlet	114	115	117	116	122	130	137	148	141
Attenuation:									
90-deg bends (2)	0	0	2	4	6	6	6	6	6
10-ft straight duct	2	2	1	1	1	0	0	0	0
end loss	8	4	1	0	0	0	0	0	0
directivity correction: 180 deg	1	5	9	12	14	16	18	20	21
distance correction: 40 ft	33	33	33	33	33	33	33	33	33
total attenuation	44	44	46	50	54	55	57	59	60
Intake SPL at station	70	71	71	66	68	75	80	89	81
PWL at turbine exhaust	124	125	123	122	119	111	108	98	91
Attenuation:									
17-ft straight duct	4	4	2	1	1	1	1	1	1
end loss	8	4	1	0	0	0	0	0	0
directivity correction: 135 deg	1	5	9	12	14	16	18	20	21
distance correction: 75 ft	39	39	39	39	39	39	39	39	39
total attenuation	52	52	51	52	54	56	58	60	61
Exhaust SPL at station	72	73	72	70	65	55	50	38	30
Total SPL at station:									
SPL difference [inlet − exhaust]	2	2	1	4	3	20	30	51	51
Decibel addition delta	2	2	3	1	2	0	0	0	0
Total SPL	74	75	75	71	70	75	80	89	81
Allowable SPL at station	90	85	80	75	70	65	65	70	72
Excess SPL at station	0	0	0	0	0	10	15	19	9

Table 11 Design example: effect of treatments on bridge wing SPL, dB (ref 20 μPa)

	Octave Band Center Frequency, Hz								
	31.5	63	125	250	500	1000	2000	4000	8000
Effect of lining inlet duct									
Lined duct: 4-in. lining	5	5	4	7	10	13	12	10	8
Unlined duct	2	2	1	1	1	0	0	0	0
Increased attenuation	3	3	3	6	9	13	12	10	8
Attenuation of low-pressure-drop silencer									
5-ft Long	2	8	12	14	16	19	20	18	14
7-ft Long	3	10	15	19	20	22	24	22	18
Intake SPL at station with 7-ft silencer	67	61	56	47	48	53	56	67	63
Total treated SPL at station:									
SPL difference [treated inlet − exhaust]	5	12	16	23	17	2	6	29	33
Decibel addition delta	1	0	0	0	0	2	1	0	0
Total treated SPL	73	73	72	70	65	57	57	67	63
Allowable SPL at station	90	85	80	75	70	65	65	70	72
SPL excess	0	0	0	0	0	0	0	0	0

silencer (Table 2) falls just short of the needed attenuation in the 4000-Hz octave band, but the 7-ft-long silencer provides the required attenuation in all octave bands. The effect of such a silencer is shown in Table 11.

The addition of a low-pressure-drop silencer is shown to provide the attenuation necessary to satisfy the noise level requirements specified for the lookout station. However, further investigations must be made before a satisfactory resolution is confirmed. That is, evaluations relative to concerns of the following types must be conducted:

- Effect of the silencer on the performance of the gas turbine.
- Effect of the silencer on access for maintenance and overhaul.
- Potential of dislodged silencer elements to become foreign-object hazards to the engine.
- Silencer space and weight requirements.
- Silencer life-cycle costs.

Clearly, the most appropriate resolution to noise problems cannot be determined without consideration for other design requirements, and several iterations may be required before a resolution to all requirements is developed.

References

1 *Noise and Vibration Control*, Leo Beranek, Ed., McGraw-Hill Book Co., New York, 1971.

2 C. M. Harris, *Handbook of Noise Control*, McGraw-Hill Book Co., New York, 1979.

3 H. W. Lord et al, *Noise Control for Engineers*, McGraw-Hill Book Co., New York, 1980.

4 J. D. Irwin and E. R. Graf, *Industrial Noise and Vibration Control*, Prentice-Hall, Englewood Cliffs, N. J., 1979.

5   Allan D. Pierce, *Acoustics: An Introduction to Its Physical Principles and Applications*, McGraw-Hill Book Co., New York, 1981.

6   Leo Beranek, *Acoustics*, published by the American Institute of Physics (New York) for the Acoustical Society of America, 1986.

7   Lawrence E. Kinsler et al, *Fundamentals of Acoustics*, John Wiley & Sons, New York, 1982.

8   *Handbook for Shipboard Airborne Noise Control*, Bolt, Beranek, and Newman, Inc., Cambridge, Mass., Contract No. DOT-CG-20756A for U. S. Coast Guard, Feb. 13, 1974.

9   R. Fischer, C. Burroughs, and D. Nelson, "Design Guide for Shipboard Airborne Noise Control," SNAME Technical and Research Bulletin No. 3-37, Jan. 1983.

10   J. T. Tranter, "The Application of Computers to Machinery Predictive Maintenance," *Sound and Vibration*, Dec. 1990.

11   J. E. Berry, "How to Specify Machinery Vibration Spectral Alarm Bands," *Sound and Vibration*, Sept. 1990.

12   R. L. Leon and K. Trainor, "Monitoring Systems for Steam Turbine Blade Faults," *Sound and Vibration*, Feb. 1990.

13   A. D. Nashif, D. I. G. Jones, and J. P. Henderson, *Vibration Damping*, John Wiley & Sons, New York, 1985.

14   J. C. Snowdon, "Vibration Isolation: Use and Characterization," *The Journal of the Acoustical Society of America*, Vol. 66, No. 5, Nov. 1979.

15   E. J. McCormick, *Human Factors in Engineering and Design*, McGraw-Hill Book Co., New York, 1976.

16   *LM2500 Marine Gas Turbine Installation Design Manual*, MID-IDM-2500-2, General Electric Marine and Industrial Engine Department, Cincinnati, Ohio, April 1984.

CHAPTER XIV

William J. Sembler

Pumps, Compressors, Blowers, and Ejectors

Section 1
Pumps

1.1 Fundamentals. The energy added by a pump to the liquid passing through it is often referred to as the total head, which is equal to the total suction head of the liquid when it enters the pump subtracted from the total discharge head of the liquid when it leaves the pump. If the total suction head is negative, it is usually referred to as total suction lift and would be added to the total discharge head. The head or energy per unit mass flow rate of a liquid is generally considered to consist of:

1. a pressure head due to the force per unit area exerted by the liquid,

2. a dynamic or velocity head, which is a measure of the liquid's kinetic energy, and

3. a potential or static head due to the elevation of the liquid above a given horizontal reference plane or datum. Referring to Fig. 1, and using subscripts s and d to refer to conditions at locations A and B, respectively, the total head developed by the pump in the system shown is equal to

$$H = \left[\frac{144\, p_d}{\rho} + \frac{V_d^{\,2}}{2g} + Z_d + H_{fd} \right]$$
$$- \left[\frac{144\, p_s}{\rho} + \frac{V_s^{\,2}}{2g} + Z_s - H_{fs} \right] \quad (1)$$

where

H = total head, ft

p = liquid pressure above (+) or below (−) atmospheric pressure, psig

ρ = liquid density, pcf

V = liquid velocity, fps

Z = static head above (+) or below (−) the datum elevation, ft

H_{fd} = friction and turbulence losses in discharge pipe from pump to point B, ft

H_{fs} = friction and turbulence losses in suction pipe from point A to pump, ft

g = acceleration due to gravity, 32.17 ft/s²

For either a vertical centrifugal pump with a double-suction first-stage impeller or a horizontal centrifugal pump, the datum elevation is established by the horizontal plane that passes through the first-stage impeller's centerline, and for a vertical centrifugal pump with a single-suction

first-stage impeller it is defined by the horizontal plane that passes through the first-stage impeller's entrance eye. For rotary and reciprocating pumps the datum elevation is established by the horizontal plane that passes through the centerline of the pump's inlet port [1].

A modification to equation (1) can be made to permit the specific gravity of the pumped liquid to be used in lieu of density to calculate total head, as shown by

$$H = \frac{p_d - p_s}{0.433\,(sg)} + \frac{V_d^{\,2} - V_s^{\,2}}{2g} + Z_d - Z_s + H_{fd} + H_{fs} \quad (2)$$

where sg is the liquid specific gravity, based on 1.0 for fresh water at a temperature of 68 F.

There are two fundamental types of pumps: the kinetic type, which includes the centrifugal, regenerative turbine, and vertical turbine pumps; and the positive-displacement type, which includes rotary, reciprocating, and diaphragm pumps.

1.2 Centrifugal Pumps.

a. Classification. A centrifugal pump basically consists of a vaned impeller mounted on a shaft that rotates within a stationary casing. As it turns, the impeller transfers energy to the liquid flowing between its vanes. In general, as liquid enters a centrifugal pump's impeller its flow is primarily in the axial direction. If there is a tangential component to the inlet flow, it is often referred to as "prewhirl" or "prerotation." An impeller that has an inlet, referred to as the eye, on only one side is classified as a "single-suction" impeller, while those that receive liquid simultaneously from both sides are classified as "double-suction" impellers. Impellers are also classified as being designed for radial-flow, mixed-flow (radial and axial), or axial-flow operation. These classifications refer to the primary orientation of flow at the discharge from the impeller with respect to the axis of rotation.

The type of impeller required for a particular application can often be predicted from the pump's specific speed, N_S, which can be calculated using the equation

$$N_s = \frac{NQ^{1/2}}{H^{3/4}} \quad (3)$$

where N is the operating speed (rpm) and Q the volumetric flow rate or capacity in gallons per minute (gpm). The

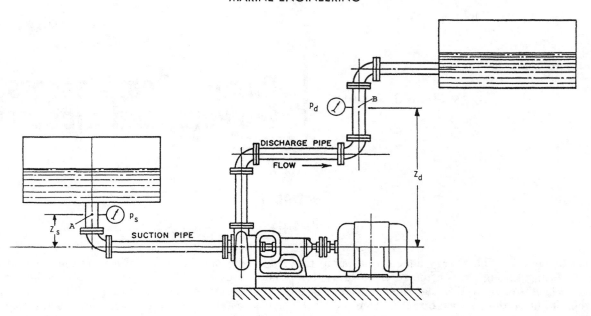

Fig. 1   Typical pump and system

values of capacity, Q, and total head, H, used in equation (3) are those at the pump's best efficiency point. In addition, for a multistage pump, which can be thought of as multiple single-stage pumps arranged in series, the value of H used to calculate specific speed is the total head developed per stage. In addition to being useful in predicting the efficiency that can be expected from a new design and, therefore, the size of the driver required, the specific speed can also be useful in approximating the radial shape of a new pump impeller. A lower specific speed represents a relatively high value of total head with respect to the capacity being delivered. This combination of performance requirements results in the use of an impeller with a relatively large outside diameter and narrow waterways. Conversely, the impeller in a high specific speed pump will tend to have a smaller outside diameter due to the relatively low head developed, and wider waterways to accommodate the higher flow rate. The three flow orientation categories for centrifugal pumps generally correspond to the following specific speed ranges:

FLOW ORIENTATION	SPECIFIC SPEED
Radial (single suction)	< 4200
(double suction)	< 6000
Mixed (single suction only)	4200 to 9000
Axial	> 9000

A radial- or mixed-flow impeller can also be classified as being: closed, which has vanes sandwiched between both a front and a rear shroud that enclose the waterways [see Fig. 2(a)]; semiopen, which includes vanes that are attached to a single rear shroud or sidewall [see Fig. 2(b)]; or open, in which the vanes are attached to the periphery of a hub and, in some cases, to a partial rear shroud [see Fig. 2(c)]. The rotating element in an axial-flow pump is frequently referred to as a propeller [see Fig. 2(d)]. An additional characteristic that is sometimes used to classify

impellers is the curvature of their vanes. A cylindrical or straight-vane impeller has vanes that are curved only around the axis of rotation (i.e., the front and back faces of each vane are parallel to the axis of the pump's shaft). The use of straight-vane impellers is generally limited to low specific speed ($N_S < 1000$) radial-flow applications. Francis or screw-vane impellers have vanes that are curved in two directions or twisted. These impellers are often utilized in the design of high specific speed ($2000 < N_S < 4200$) radial-flow pumps [2].

b. "Euler" head. Based on the conservation of angular momentum, the net torque applied by a centrifugal pump impeller to the liquid passing through it is equal to the time rate of change in the liquid's angular momentum with respect to the axis of rotation. Using this principle, an expression for the theoretical increase in the energy of the liquid being pumped (i.e., the energy transferred from the impeller to the liquid) can be developed. In the following equations, c refers to the liquid's absolute velocity with respect to the casing, w refers to the liquid's relative velocity with respect to the rotating impeller, and u refers to the peripheral or tangential velocity of the impeller at a given radius, all in feet per second (fps). In addition, subscript u refers to the peripheral component and subscript m to the meridional component of the liquid's absolute and relative velocities. The meridional velocity is the portion of the liquid velocity that is normal to the corresponding peripheral velocity, and can consist of components in both the radial and axial directions [3]. The vectors that the above symbols represent can be used to form flow velocity triangles at the inlet and outlet of the impeller. Velocity triangle diagrams for radial- and axial-flow pumps are shown in Fig. 3.

The angular momentum of liquid at any radius in the impeller is equal to the moment arm, or radius, multiplied

(a) Closed impeller

(b) Semiopen impeller

(c) Open impeller

(d) Propeller

Fig. 2   Centrifugal pump impellers (Courtesy Dresser Pump Division)

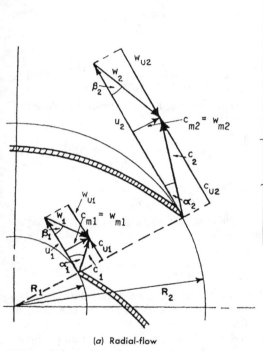
(a) Radial-flow

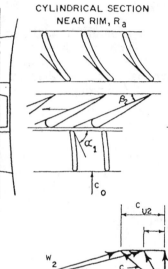

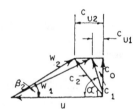
(b) Axial-flow

Fig. 3   Centrifugal pump velocity diagrams

by the product of the liquid's mass times its peripheral velocity about the shaft axis, or

$$L = R \frac{m}{g_c} c_u \tag{4}$$

where

L = angular momentum, ft-lbf-s
m = mass, lbm
R = radius, ft
g_c = gravitational constant, 32.17 ft-lbm/lbf-s^2

Using a dot "." to signify a change with respect to time, with steady-state flow the rate of change of the angular momentum equals

$$\dot{L} = \frac{R}{g_c} c_u \dot{m} \tag{5}$$

where $\dot{m}$ is the flow rate, lbm/s.

The torque, T, applied to the liquid pumped at any given radius of the impeller is equal to $\dot{L}$; therefore, using subscripts 1 and 2 to refer to conditions at the inlet and outlet of the impeller, respectively, with a constant mass flow rate the net change in torque applied to the liquid passing through the impeller is equal to

$$\Delta T = \frac{\dot{m}}{g_c} (R_2 c_{u2} - R_1 c_{u1}) \tag{6}$$

where ΔT is the change in torque, ft-lbf.

The theoretical rate of energy transfer to the fluid being pumped, or power, is equal to the product of the net change in torque multiplied by the impeller's angular velocity, as shown by the following:

$$E_{th} = \Delta T \omega = \frac{\dot{m}\omega}{g_c} (R_2 c_{u2} - R_1 c_{u1}) \qquad (7)$$

where

$\dot{E}_{th}$ = theoretical rate of energy transfer, ft-lbf/s
ω = angular velocity, rad/s

ωR is equal to the peripheral speed of the impeller, u, at radius R; therefore

$$\dot{E}_{th} = \frac{\dot{m}}{g_c} (u_2 c_{u2} - u_1 c_{u1}) \qquad (8)$$

The theoretical rate of energy transfer per unit mass flow rate equals

$$\frac{\dot{E}_{th}}{\dot{m}} = H_e = \frac{1}{g_c} (u_2 c_{u2} - u_1 c_{u1}) \qquad (9)$$

This is referred to as the "Euler equation," and H_e is called the "Euler head," which is expressed above in ft-lbf/lbm. If, however, g is substituted for g_c in equation (9), the resulting unit of measurement for H_e will be feet.

Referring to Fig. 3(a), for radial-flow pumps

$$c_2{}^2 = c_{u2}{}^2 + c_{m2}{}^2 \qquad (10)$$

In addition,

$$w_2{}^2 = c_{m2}{}^2 + (u_2 - c_{u2})^2 \qquad (11)$$

By equating the two expressions that can be derived for c_{m2} from equations (10) and (11), it can be seen that

$$u_2 c_{u2} = \frac{1}{2} (c_2{}^2 + u_2{}^2 - w_2{}^2) \qquad (12)$$

Similarly:

$$u_1 c_{u1} = \frac{1}{2} (c_1{}^2 + u_1{}^2 - w_1{}^2) \qquad (13)$$

The substitution of these values into equation (9) results in the following alternative form of the Euler equation

$$H_e = \frac{1}{2g_c} \left[(c_2{}^2 - c_1{}^2) + (u_2{}^2 - u_1{}^2) + (w_1{}^2 - w_2{}^2) \right] \qquad (14)$$

The first term in the above equation represents the increase in the kinetic energy or dynamic head of the liquid being pumped due to its acceleration from the inlet to the outlet of the impeller. The second and third terms represent an increase in the static pressure head of the liquid being pumped. The velocity of an actual liquid is never constant across a flow channel; therefore, average velocities are often used in the above equations.

Using the velocity components shown in Fig. 3(a):

$$c_{u2} = u_2 - w_{u2} \qquad (15)$$

$$c_{u1} = u_1 - w_{u1} \qquad (16)$$

Additionally,

$$w_{u2} = w_{m2} \cot \beta_2 \qquad (17)$$

$$w_{u1} = w_{m1} \cot \beta_1 \qquad (18)$$

where β is the relative flow angle in degrees. w_m, which

is the relative meridional velocity of the liquid traveling through the impeller's channels, can be found from

$$w_m = \frac{Q}{448.8 \, A \eta_v} \qquad (19)$$

where

A = total area of impeller's flow channels measured normal to the meridional flow, ft^2
η_v = volumetric efficiency, which accounts for losses due to internal leakage flow recirculated within the pump, %/100

By substitution, equations (15) and (16) can be rewritten as

$$c_{u2} = u_2 - \frac{Q}{448.8 \, A_2 \, \eta_v} \cot \beta_2 \qquad (20)$$

and

$$c_{u1} = u_1 - \frac{Q}{448.8 \, A_1 \, \eta_v} \cot \beta_1 \qquad (21)$$

If the above expressions for c_u are now inserted into equation (9), the result is

$$H_e = \frac{1}{g_c} \left[u_2 \left(u_2 - \frac{Q}{448.8 \, A_2 \, \eta_v} \cot \beta_2 \right) \qquad (22) \right.$$
$$\left. - u_1 \left(u_1 - \frac{Q}{448.8 \, A_1 \, \eta_v} \cot \beta_1 \right) \right]$$

If prerotation, c_{u1}, is assumed to be zero, the above equation can be reduced to

$$H_e = \frac{u_2}{g_c} \left(u_2 - \frac{Q}{448.8 \, A_2 \, \eta_v} \cot \beta_2 \right) \qquad (23)$$

This form of the Euler equation shows that the total head developed at a given capacity and, therefore, the slope of the pump's performance curve of total head developed versus capacity delivered, referred to as the pump's head-capacity curve, is affected by the relative discharge flow angle. Impellers with radial or backward-curved vanes ($\beta_2 \leq 90$ deg) are used in most radial- and mixed-flow centrifugal pumps; therefore, the total head developed by these machines generally remains constant or increases as the capacity delivered is reduced.

The above equations are also valid for axial-flow pumps. However, in these machines $u_1 = u_2$ at any given radius of the propeller, and the total Euler head developed by the entire propeller is equal to the integrated average of the head developed by each of the propeller's cylindrical sections.

c. "Ideal" head. The Euler head calculated above can never be developed by a centrifugal pump. One reason for the reduction in the actual head developed by a radial-flow impeller is "slip." As the impeller rotates, the liquid along the front faces of its vanes is at a higher pressure than the liquid adjacent to the back of each vane. Because of this pressure gradient, there also exists a velocity gradient across the impeller's flow channels, with the lower liquid velocities occurring along the front face of each vane. As a result of this variation in velocity, the average relative flow angle of the liquid being discharged from the impeller, β_2, is less than the actual discharge angle of the vane. The reduction in c_{u2} that results from this deviation in fluid flow is often referred to as the "slip" of the impeller. From equations (9) and (22) it can be seen that a reduction in β_2 and the corresponding reduction in c_{u2} result in a reduction in the total head developed by the impeller. After the Euler head is reduced to account for slip, it is sometimes referred to as the "ideal head" [4], and is equal to

$$H_i = \mu \, H_e \qquad (24)$$

where

H_i = ideal head, ft (or ft-lbf/lbm)
μ = slip factor

The slip factor for a particular impeller is generally determined based on the number of impeller vanes and the vane discharge angle using equations or curves derived both theoretically and empirically. Included among the most commonly used slip factors are those developed by Stodola and Busemann [4,5].

d. Actual head. The ideal head, H_i, represents a reduction in the Euler head, H_e, based on the flow deviation of an ideal liquid. However, it does not include the effect of flow losses in the impeller due to friction and turbulence, and it includes none of the hydraulic flow losses that occur in the pump's casing. These losses are represented by an additional correction factor called the hydraulic efficiency. Using the hydraulic efficiency, the actual head developed by the pump, H, equals

$$H = H_i \, \eta_h \qquad (25)$$

where η_h is the hydraulic efficiency, %/100.

e. Power requirement. Based on equation (9), rewritten in terms of volumetric capacity and horsepower, the hydraulic power required by a centrifugal pump is equal to

$$P_h = \frac{H_i Q \, (sg)}{3960} = \frac{HQ \, (sg)}{3960 \, \eta_h} \qquad (26)$$

where P_h is the hydraulic horsepower, hp. Because of additional losses not accounted for in the hydraulic efficiency, the actual horsepower required to drive the pump exceeds P_h. The sources of the additional losses include the following:

• Mechanical losses, which include losses due to friction in the pump's bearings, seals, and packing. In general, the power necessary to overcome mechanical losses as a percentage of the hydraulic power increases with decreasing specific speed and capacity.

• Disk friction, which includes the loss due to the friction between the liquid in the pump's casing and the outer surfaces of its rotating impeller. The power necessary to overcome this loss, therefore, increases with the pump's impeller diameter and operating speed.

• Internal leakage flow across wearing rings, interstage bushings, and internal balancing devices, which results in the recirculation within the pump's casing of a portion of the liquid being discharged from the impeller.

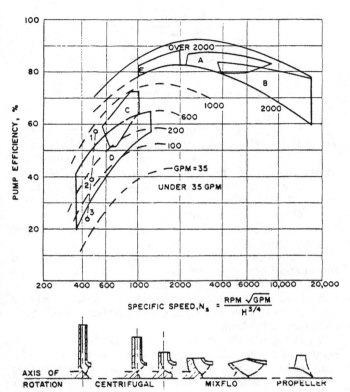

LEGEND
A – RADIAL-FLOW CIRCULATING PUMPS
B – AXIAL & MIXED-FLOW CIRCULATING PUMPS
C – FEED PUMPS
D – MAIN & AUXILIARY CONDENSATE PUMPS
E CARGO PUMPS
EFFECT OF VISCOSITY
1 – MAX. EFF. PUMPING WATER AT ROOM TEMP.
2 – MAX. EFF. SAME PUMP WITH OIL AT 600 SSU
3 – MAX. EFF. SAME PUMP WITH OIL AT 2000 SSU

Fig. 4 Representative pump efficiencies versus specific speed and capacity for various applications

The energy needed to pump this additional liquid is accounted for in the volumetric efficiency, η_v. In general, the volumetric efficiency increases with specific speed and the capacity delivered.

The above losses can be combined with the hydraulic horsepower to determine the total power required to drive the pump as follows:

$$P_p = \frac{HQ \, (sg)}{3960 \, \eta_h \, \eta_v} + P_{df} + P_m = \frac{HQ \, (sg)}{3960 \, \eta_p} \qquad (27)$$

where

P_p = brake horsepower (bhp) required to drive the pump, hp
η_p = overall pump efficiency, %/100
P_{df} = horsepower to overcome disk friction, hp
P_m = horsepower to overcome mechanical losses, hp

A correlation of typical pump efficiencies versus specific speed and capacity is shown in Fig. 4. Similar charts with correction factors for operating speed and surface finish are included in reference 6.

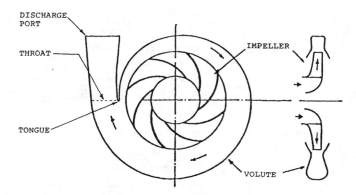

Fig. 5 Volute-type centrifugal pump casing

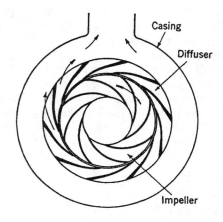

Fig. 6 Diffuser-type centrifugal-pump casing

f. Casings. Centrifugal pump casings can be split along a plane that passes through the shaft axis, referred to as axially split, or along a plane that is perpendicular to the shaft, referred to as radially split. Many pump casings used in marine applications are designed to be mounted vertically, which typically requires less deck space than horizontal mounting.

The casing performs three important functions related to the pump's hydraulic performance:

1. It guides the liquid being pumped from the inlet of the suction nozzle to the eye of the impeller. In some cases guide vanes may be added to the suction nozzle to straighten the flow entering the impeller, and to break up vortices formed by liquid that may be recirculated back out of the impeller's eye.

2. It collects the liquid being discharged from the periphery of the impeller, and converts a portion of the kinetic energy or velocity head in the liquid to pressure head. This is often referred to as pressure recovery or diffusion, and is necessary due to the liquid's high absolute velocity when it leaves the impeller.

3. It guides the liquid discharged from the pressure-recovery device to the inlet of the next stage in a multistage pump, or to the outlet of the pump's discharge nozzle.

Collectors and pressure-recovery devices that are commonly used include the following:

• Volutes. The volute or scroll, as shown in Fig. 5, is characterized by a channel with a gradually increasing radius and cross-sectional area that surrounds the periphery of the impeller. Volutes are used with many single-stage radial- and mixed-flow centrifugal pumps. A volute is sometimes designed so that the increase in its cross-sectional area is proportional to the angular advancement from the tongue or cutwater to the throat. With this configuration the pumped fluid is assumed to have a constant average velocity in all sections of the volute during operation at the design capacity, which is generally at or near the best efficiency point. As an alternative to maintaining a constant velocity, a volute may be designed so that the fluid passing through it maintains a constant angular momentum (i.e., Rc_u = constant) at the design capacity. When this latter method is used to size the volute, it is assumed that the peripheral velocity of the fluid decreases

as the volute's radius increases. The radial clearance between the outside diameter of the impeller and the volute's tongue should be small to limit the amount of liquid recirculated around the volute; however, if the gap is too small excessive pressure pulsations, turbulence, and noise can result [2,4,7]. This gap or cutwater clearance is frequently sized to optimize pump performance with the range of impeller diameters that may be used in the casing. To reduce hydraulic shock and separation losses, the angle used for the volute's tongue often matches the absolute fluid flow angle at the design capacity.

In lieu of a spiral volute, some centrifugal pumps are furnished with a circular or concentric collector that has a constant radius and cross-sectional area around the periphery of the impeller. Except when operating at shutoff (with a closed discharge valve and no through-flow), the velocity in this type of a collector increases as the flow channel progresses angularly from the tongue to the throat. A modified- or semi-concentric casing is a variation of the concentric design in which the radius and cross-sectional area of the collector remain constant over only a portion of the casing's circumference.

Volute-type multistage pumps often have axially split casings, with the flow passages between successive stages, referred to as crossovers, being either integrally cast or welded onto the casing. Liquid collected in each interstage volute is frequently decelerated in the corresponding crossover before being directed to the eye of the impeller in the pump's next stage. The pumped liquid is also usually decelerated in the casing's discharge nozzle, which forms the transition from the throat of the last-stage volute (this would be the only volute in a single-stage volute-type pump) to the pump's outlet port.

• Multi-vaned diffusers. In multistage pumps where pressure recovery must be accomplished in the limited space between adjacent stages, multi-vaned diffusers are often used. This type of a pressure-recovery device, which is illustrated in Fig. 6, is also used in some single-stage radial-flow pumps due to its typically high peak efficiency. A multi-vaned diffuser consists of a number of diverging vanes mounted in a ring that surrounds the periphery of

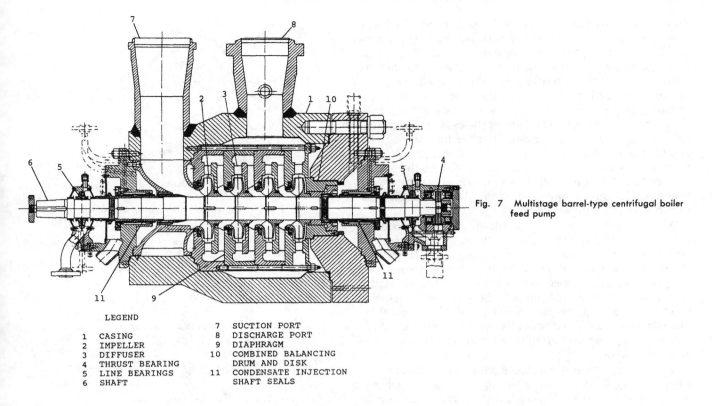

Fig. 7 Multistage barrel-type centrifugal boiler feed pump

LEGEND

1	CASING	7	SUCTION PORT
2	IMPELLER	8	DISCHARGE PORT
3	DIFFUSER	9	DIAPHRAGM
4	THRUST BEARING	10	COMBINED BALANCING
5	LINE BEARINGS		DRUM AND DISK
6	SHAFT	11	CONDENSATE INJECTION
			SHAFT SEALS

the impeller. In diffuser-type multistage pumps, the flow channels from the discharge of one stage to the suction of the next are generally formed by a series of vanes located on the back side of each diffuser or included in the diaphragms, also referred to as stage pieces, that separate adjacent stages (see Fig. 7). With this configuration, the pump's rotor, together with its stationary diffusers and stage pieces, can often be inserted into a radially split casing as an assembled cartridge. Because of the casing's cylindrical shape, these units are frequently referred to as "barrel" pumps.

Many axial-flow pumps are also fitted with a diffuser that is located downstream from the discharge of the propeller. The "axial diffuser" is used to not only convert the tangential velocity component of the absolute flow leaving the propeller into pressure head, but also to straighten this flow so that the pumped fluid is discharged from the casing in an axial direction.

• Vaneless diffusers. When a wide range of operation is anticipated, a radial-flow centrifugal pump is sometimes fitted with a vaneless diffuser. The pressure recovery with this type of a casing occurs in an annular passage that surrounds the impeller. Neglecting friction, fluid passing through this channel, which can have a constant width or can flare outwards slightly, is assumed to follow free-vortex flow (i.e., Rc_u = constant). For a vaneless diffuser with parallel walls, the reduction in fluid velocity will, therefore, be proportional to the increase in the radius of the diffuser's flow channel. A volute-type collector is often attached to the outlet of a vaneless diffuser.

Regardless of the configuration used, the casing is a pressure-containing boundary. Therefore, its thickness must be sufficient to withstand the pump's design pressure. In addition, structurally a pump's casing must be suitable to withstand not only hydrostatic pressures but also the hydraulic forces in its waterways, stresses that result from vessel motion, and nozzle loads imposed by the ship's piping. In some cases, external ribs may be added to the casing to increase its strength and rigidity.

g. Impellers and wearing rings. The strength of the impeller's vanes and shrouds must be sufficient to withstand hydraulic forces, which include reaction forces on the vanes that are equal in magnitude but opposite in direction to the forces that are transmitted to the liquid being pumped, together with the centrifugal loads that result from the impeller's rotation.

To improve the pump's volumetric efficiency, when a closed-type single-suction impeller is used, leakage along the outside of the front shroud back to suction is generally reduced by maintaining a close clearance between the impeller's outer hub (the outer portion of the shroud adjacent to the impeller's eye) and the casing. With a double-suction closed impeller, a close clearance is maintained at both outer hubs. So that this clearance can be periodically renewed, adjacent casing surfaces are typically fitted with replaceable wearing rings. A stationary wearing ring can also be mounted behind the rear shroud of a single-suction impeller to reduce the hydraulic axial unbalance. In addition to the stationary rings, rotating wearing rings may be installed over the impeller's outer hubs. The leakage through the annulus formed between the casing wearing ring and the impeller's outer hub or wearing ring is a function of the length of the overlap between these two surfaces, their configuration, the differential pressure across them, and the radial clearance between them. To reduce leakage, serrations or circular grooves are sometimes added to the inside surface of the casing ring or the outside diameter of the mating impeller hub or ring. In addition, intermeshing labyrinth-type wearing rings may be used. Serrations and grooves can also reduce the potential for damage due to either an occasional, inadvertent rub occurring during operation or the presence of foreign particles in the pumped liquid.

When a centrifugal pump is fitted with an open or a semiopen single-suction impeller, wearing rings are not required on the suction side of the impeller. However, to reduce the leakage or slippage of liquid across the impeller's vanes, a close axial clearance must be maintained between the unshrouded edges of the impeller's rotating vanes and the inner walls of the casing. So that this clearance can be periodically renewed, casings used with open impellers are often fitted with replaceable internal wear plates.

In multistage pumps a leakage path can exist between adjacent stages. To control interstage leakage, which reduces the volumetric efficiency, a close radial clearance is typically maintained between the shaft and areas of the casing, diffusers, or stage pieces that separate the pump's stages. Replaceable stationary bushings are frequently used to enable the running clearances to be periodically renewed.

h. Loads applied to a centrifugal pump's rotating assembly.

Radial reaction. The volute of a centrifugal pump is often designed so that the hydraulic pressure around the periphery of the impeller is nearly uniform at a flow rate close to the pump's best efficiency point. However, as the flow rate is reduced, the nonuniformity of the pressures and velocities within the volute results in a radial load on the impeller that rises to a maximum value at shutoff. The radial load on the impeller also increases at capacities exceeding the pump's best efficiency flow rate. In addition, the changes in the magnitude of this force with flow rate are accompanied by changes in the direction in which it is applied. This load, which is called "radial reaction," results in increased shaft deflections and bearing loads.

To reduce the radial reaction, double-volute casings are sometimes used. With a double-volute design, the casing has two volutes with cutwaters that are located approximately 180 deg apart (see Fig. 8). The flow around the impeller is, therefore, divided, resulting in two radial forces that are very close in magnitude but opposite in direction. Consequently, the net radial reaction during operation off the best efficiency point is greatly reduced from the radial reaction in a single-volute pump. A similar effect is often obtained in multistage volute casings by arranging the angular orientation of adjacent stages so that the net radial load is minimized. The use of concentric

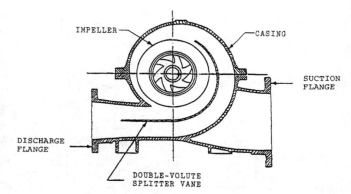

Fig. 8   Double-volute centrifugal-pump casing

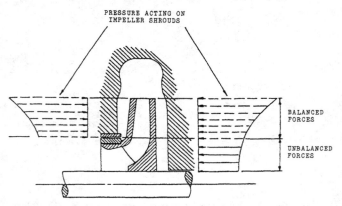

Fig. 9   Pressure distribution on a single-suction impeller

and modified concentric casings, and of multi-vaned diffusers, has also been found to reduce radial reactions during off-design operation when compared to the loads measured with single-volute casings [8,9]. The radial reaction typically increases with the diameter and width of the impeller, the total head developed, and the specific gravity of the pumped fluid [8].

Impeller hydraulic axial unbalance. When a centrifugal pump with a double-suction impeller is used, the hydraulic axial load is theoretically zero. Although a net unbalance can exist due to differences in the flow to each side of the impeller, it is generally negligible. When a single-suction impeller is used, however, there can be a significant net axial thrust. For a closed radial-flow impeller, this thrust results from the difference in the pressure of the liquid acting on each side of the impeller's back shroud adjacent to the inlet or eye (see Fig. 9). While the liquid acting on the inside of this section of the shroud is essentially at the suction pressure, the outside of the shroud is exposed to liquid that has already been discharged from the impeller. Due to the absence of a front shroud, the pressure applied to the outside of the entire rear shroud of a semiopen impeller is only partially balanced by the pressure inside the impeller. Consequently, the resulting hydraulic axial unbalance can be greater than with a closed impeller.

As shown in Fig. 9, the pressure acting on the outside of a rotating impeller's shroud is not constant, but actually

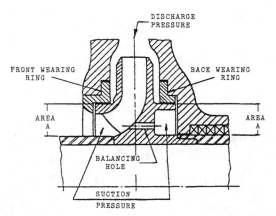

Fig. 10   Impeller with balancing holes and a back wearing ring

decreases at smaller radii. This reduction in pressure is due to the rotation of the liquid located between the shroud and the casing's side wall, and the resulting conversion of a portion of the liquid's pressure head to velocity head. In some pumps, external vanes or ribs are added to the impeller's back shroud to increase the velocity at which the liquid behind the shroud rotates, which reduces the liquid's pressure and the axial thrust acting towards the suction. Conversely, stationary radial vanes are sometimes added to the casing adjacent to the impeller's back shroud to reduce the velocity of the fluid behind the impeller and to increase the thrust towards the suction. An additional method used to reduce hydraulic axial thrust is to drill multiple axial "balancing holes" in the impeller's back shroud and to install a back wearing ring (see Fig. 10). This arrangement results in a reduced outer pressure in the center portion of the back shroud and, therefore, in a lower differential pressure over the unbalanced area [3].

To reduce the hydraulic axial thrust in a multistage pump, single-suction impellers are sometimes mounted on the shaft so that approximately one half of the impellers face in opposite directions (see Fig. 11). In lieu of an opposed-impeller arrangement, the impellers in a multistage pump can be mounted with their inlets facing in the same direction, and the axial thrust can be reduced in each stage individually with back wearing rings and axial balancing holes. Alternatively, the hydraulic axial thrust acting on the entire rotating assembly can be reduced with a balancing drum, a balancing disk, or a combination of these two devices.

A balancing drum is composed of a drum that rotates within a close-clearance stationary sleeve [see Fig. 12(a)]. By locating the drum behind the last-stage impeller, its inner face is subjected to full discharge pressure less the pressure recovery in the last stage. The pressure of the liquid acting on the opposite face of the drum is reduced by the breakdown that occurs when this liquid passes through the close-clearance annulus formed by the drum and its sleeve; therefore, the net axial thrust acting on the drum will be opposite in direction to the thrust acting on the impellers. Because the axial forces cannot be fully

balanced at all capacities, in practice balancing drums are generally sized to give only 90% to 95% axial balance. This helps insure that the residual axial thrust always acts in the same direction, which prevents thrust reversals.

A balancing disk is similar to a balancing drum, except that in lieu of containing a close-clearance annulus, the pressure breakdown across this device is accomplished through a small axial gap maintained between the rotating disk and the stationary disk head [see Fig. 12(b)]. An advantage of the balancing disk is its ability to be self-adjusting to compensate for wear in the mating surfaces of the disk or the head and in response to changes in the pump's axial thrust. The net thrust applied to the disk results from the difference between the liquid pressure acting on its front face and the pressure in the balancing chamber. The pressure acting on the inner portion of the balancing disk's front face is essentially equal to the pump's discharge pressure less the last-stage recovery, while the pressure acting on the portion of this face that is adjacent to the stationary balancing head varies from the discharge pressure at the head's inside diameter to the balancing chamber pressure at the disk's periphery. The balancing disk's various diameters are sized so that the axial thrust applied to it during normal operation will be equal in magnitude, but opposite in direction, to the axial thrust acting on the impellers. If the axial thrust on the impellers exceeds the opposing thrust, the axial gap between the balancing disk and the stationary head will be reduced, increasing the pressure breakdown across the device. This results in a reduced liquid pressure in the balancing chamber, which increases the opposing axial thrust on the disk and restores the rotor's axial balance. A reduction in impeller thrust or an increase in the balancing disk's axial gap due to wear will have the opposite effect. An orifice is generally installed in the recirculation port that connects the balancing chamber to the pump's suction so that the balancing chamber pressure will exceed the pump's suction pressure when the axial gap between the disk and the head is at its maximum value, which is necessary for the proper operation of the balancing disk.

The axial thrust applied to an axial-flow propeller, which has no shrouds, results from the differences in the pressure of the fluid acting on the front and back faces of its vanes, together with the differential pressure across the opposite ends of its inner hub.

Shaft axial unbalance. In pumps with an overhung impeller, liquid entering the impeller acts against the submerged end of the shaft. When this liquid is not at atmospheric pressure, a piston effect is created that imposes an axial load on the pump's rotor. This load is equal to the gage pressure of the liquid acting against the submerged end of the shaft multiplied by the shaft's cross-sectional area.

Change in momentum. As liquid is pumped through a radial-flow impeller, its travel path is redirected from an axial to a radial direction. When a single-suction impeller is used, the change in the momentum of the liquid due to the reduction in the axial component of the liquid's velocity results in the generation of an axial thrust away from the suction. This axial thrust will also be generated, to a

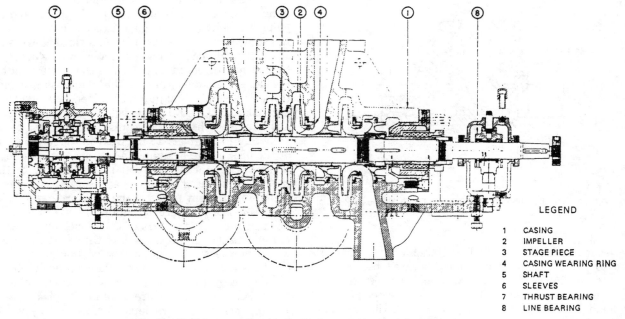

Fig. 11   Multistage centrifugal boiler feed pump with an axially split casing

LEGEND

1   CASING
2   IMPELLER
3   STAGE PIECE
4   CASING WEARING RING
5   SHAFT
6   SLEEVES
7   THRUST BEARING
8   LINE BEARING

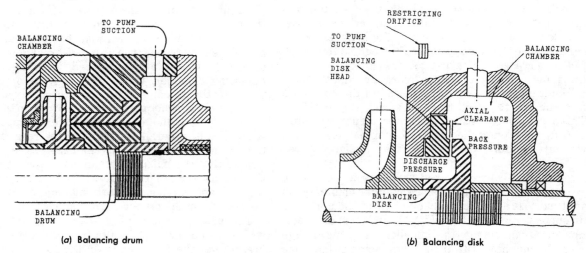

(a) Balancing drum

(b) Balancing disk

Fig. 12   Balancing devices

lesser extent, in a mixed-flow pump depending on the amount of radial flow in the liquid being discharged from the impeller. The axial thrust due to this change in the liquid's direction of travel in radial- and mixed-flow impellers can be calculated as follows:

$$F_{M,a} = \frac{\dot{m}}{g_c}(c_{ae} - c_{a2}) \tag{28}$$

where

$F_{M,a}$ = axial thrust due to change in momentum, lbf
c_{ae} = axial component of absolute velocity in impeller's eye, fps
c_{a2} = axial component of absolute velocity at impeller's discharge, fps

i. **Bearings.** The net axial and radial loads applied to the pump's rotating assembly are transmitted to the bearings that support its shaft. The pump's "thrust bearing" absorbs loads that are applied in the axial direction. So that the pump's shaft is free to expand and contract in response to changes in axial load or temperature, it should be constrained axially at only one location. A pump shaft that is rigidly coupled to the shaft of its driver, therefore, is generally not fitted with a separate thrust bearing but is, instead, supported axially by the thrust bearing in the driver. The remaining pump bearings, which are typically configured so that they absorb only radial loads, are frequently referred to as "line bearings."

In a single-stage pump with an impeller centered on the rotor, a bearing is typically installed at each end of the

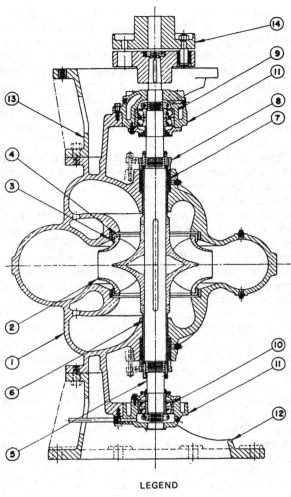

LEGEND

1 CASING
2 IMPELLER
3 CASING WEARING RINGS
4 IMPELLER WEARING RINGS
5 SHAFT
6 SHAFT SLEEVE
7 STUFFING BOX PACKING
8 GLANDS
9 THRUST BEARING
10 LINE BEARING
11 BEARING HOUSINGS
12 PUMP BASE
13 MOTOR BRACKET
14 COUPLING

Fig. 13   Vertical, single-stage, double-suction centrifugal pump with an axially split casing

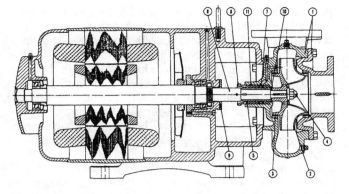

LEGEND

1 CASING AND SUCTION HEAD
2 IMPELLER
3 CASING WEARING RING
4 IMPELLER NUT
5 SHAFT SLEEVE
6 GLAND
7 LANTERN RING
8 SHAFT
9 SLINGER
10 SHAFT SLEEVE SEAL RING
11 PACKING

Fig. 14   Horizontal, close-coupled centrifugal pump

shaft. This configuration, which is shown in Fig. 13, is sometimes referred to as a "between bearings" design. When an overhung impeller arrangement is used, the shaft is supported by bearings that are located behind the impeller. As shown in Fig. 14, in some of these units the pump's rotating parts are mounted directly onto an extension of the driver's shaft, which is referred to as a "close-coupled" design. With a close-coupled configuration, radial and axial loads developed within the pump are frequently absorbed by the bearings installed in the

driver; however, if radial loads are high or the shaft extension is long, the pump may be fitted with additional internal radial bearings to reduce shaft deflection. Internal radial bearings are also frequently installed in multistage pumps.

The type of bearing used in any specific application depends on the magnitude and orientation of applied loads, the operating environment, the range of operating temperatures and speeds, and the method of lubrication. Single-row deep-groove ball bearings are installed in many centrifugal pumps. Because this type of an antifriction bearing is able to absorb radial loads combined with moderate axial thrust, it can often be used as either a line bearing or a combined line and thrust bearing. When an increased load-carrying capability is needed, it is sometimes necessary to use a double-row deep-groove ball bearing. In addition, double-row self-aligning ball bearings, multiple-row angular-contact ball bearings, and various types of roller bearings can also be used.

Antifriction bearings are frequently lubricated with grease. This is generally satisfactory, provided that the grease is of the proper type and has the proper temperature rating, and that the dn value [bearing bore (mm) × speed (rpm)] does not exceed 200,000 [10]. In addition, this type of lubrication is often preferred because grease is not affected by vessel motion, is generally easier to contain than oil, and is superior to oil at keeping contaminants away from the bearing. The bearing housing is typically fitted with inlet and outlet ports that are used for the addition and the removal of grease during relubrication. So that grease will be retained in the housing and to protect the bearing from contaminants, stationary seals are generally installed at openings provided for the pump's shaft. In addition, one or both ends of a ball bearing can be fitted with shields, which help keep grease in the bearing and contaminants out. To provide additional protection for the bearing, a slinger or flinger is frequently mounted on the pump's shaft external to the housing. This rotating part acts to prevent dirt and moisture that may be traveling along the shaft from reaching the bearing housing. When high-temperature liquids are to be pumped, the use of special bearings with increased running clearances or jacketed liquid-cooled bearing housings is sometimes required.

In some cases, the use of fluid-film bearings, which include tilting-pad or pivoting-shoe thrust bearings and cylindrical journal or sleeve-type line bearings, is necessary due to operating conditions, such as high speeds, temperatures, or loads for which antifriction bearings are not suitable. Tilting-pad bearings and sleeve bearings that are external to the casing are frequently lubricated with oil. Sleeve bearings installed inside the casing, however, are generally submerged in and lubricated by the pumped liquid.

j. Shaft design. In addition to being sufficiently strong to withstand applied forces and moments, a pump shaft must be stiff enough to prevent its axial and radial deflections during operation from resulting in contact between the pump's rotating and stationary components. The shaft should also be designed so that natural frequencies of the pump's rotating assembly are a sufficient distance away from forcing frequencies, such as the operating speed. In general, the lowest frequency of interest is the rotor's first bending natural frequency, which is also often referred to as the pump's first critical speed. A pump that operates below its first critical speed is frequently classified as a "stiff-shaft" unit, while a pump that operates above the first critical speed is sometimes called a "flexible-shaft" unit.

A pump's critical speed can typically be increased and the radial deflection of the pump's shaft can be reduced by reducing the shaft's bearing span, or its unsupported length if an overhung design is used, by increasing the shaft's diameter, by reducing the weight of the rotating assembly, or by using a shaft material with a higher modulus of elasticity. The critical speed can also be affected by the "Lomakin effect." This refers to the stiffness and damping generated in close-clearance annular seals, such as wearing rings, bushings, and internal balancing drums, resulting from a hydrodynamic fluid film effect similar to that produced in journal bearings and a hydrostatic effect due to variations in the rate of fluid leakage around the circumference of the annulus when the rotor moves off center [11]. However, the increase in a pump's critical speed due to the Lomakin effect will be reduced as the clearances in annular seals increase with time due to wear; therefore, relying on this effect to prevent potential vibration problems should be done with caution.

k. Packing and mechanical seals. The openings provided to enable the shaft to pass through the casing must be sealed to reduce or, in some cases, prevent the leakage of the liquid being pumped. The shaft seal is sometimes formed by multiple rings of packing that are inserted into a stuffing box. The mating portion of the shaft that passes through the stuffing box is usually protected against wear by a sleeve that is either hardened or faced with a wear-resistant coating. To limit the temperature increase from friction between the rotating sleeve and the stationary packing, a certain amount of leakage must typically be maintained through the stuffing box to cool and lubricate the packing. The leakage rate is controlled by adjusting the axial position of the gland that holds the packing in place.

If the pressure of the liquid at the base of the stuffing box is below atmospheric pressure, air can be drawn into the pump. In addition to resulting in a loss of lubrication to the packing, air entering the casing through the stuffing box can have a detrimental effect on pump performance. Whenever it is possible for the base of the stuffing box to be under a vacuum, high-pressure liquid is often injected into the stuffing box through a lantern ring or seal cage that is sandwiched between two of the intermediate rings of packing. Pumps furnished for applications involving contaminated fluids may also be fitted with a lantern ring so that clean liquid can be injected into the stuffing box to flush contaminants away from the packing. When it is impractical to provide clean high-pressure liquid to the stuffing box, grease is sometimes injected through the lantern-ring connection.

To reduce shaft sleeve wear and the continuous leakage that must be tolerated with packing, many pumps are fitted with mechanical end-face seals in lieu of packed stuffing boxes (see Fig. 15). The primary sealing surfaces in a mechanical seal are the polished faces of one stationary ring and one rotating ring that are separated by a thin film of fluid. One of these mating faces is frequently a hard material, such as aluminum oxide or tungsten carbide, while the second is a softer material with some inherent self-lubricity, such as carbon. Two secondary seals prevent leakage from occurring between the rotating ring and the shaft, and between the stationary ring and the gland. In many seals, the rotating ring is flexibly mounted on the pump's shaft with springs or with either a metallic or an elastomeric bellows. This arrangement enables the face of the rotating seal ring to remain parallel to that of the stationary mating ring with limited radial or axial shaft movement, and permits the axial position of the rotating ring to be self-adjusting to account for minor wear of the two mating surfaces. Although it is not as common, in some mechanical seals it is the stationary ring that is flexibly mounted.

Mechanical seals are frequently cooled and lubricated by a portion of the pumped liquid that is recirculated through either an internal port or an external line connecting the discharge side of the casing to the seal area. Pumped liquid that contains a limited amount of abrasive particles often passes through a cyclone-type abrasive separator prior to being injected into the seal area. When the pumped liquid cannot be adequately cleaned, the mechanical seal may be flushed with high-pressure liquid supplied from an independent source or, in some cases, a double hard-face mechanical seal may be used. In this

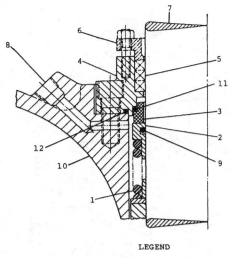

LEGEND

1   SPRING
2   PRIMARY (ROTATING) SEAL RING
3   MATING (STATIONARY) SEAL RING
4   MECHANICAL SEAL GLAND (WITH BUILT-
     IN AUXILIARY STUFFING BOX)
5   AUXILIARY PACKING (TO BE INSTALLED
     ONLY IN THE EVENT OF A MECHANICAL
     SEAL FAILURE)
6   AUXILIARY PACKING GLAND
7   PUMP SHAFT
8   FLUSHING CONNECTION
9   PRIMARY RING SECONDARY SEAL (O-RING)
10 PUMP CASING
11 MATING RING SECONDARY SEAL (O-RING)
12 GLAND O-RING

Fig. 15 Mechanical shaft seal (with auxiliary stuffing box)

type of a mechanical seal, both the rotating and the stationary seal faces are made from an abrasion-resistant material, such as tungsten or silicon carbide.

The mechanical seals used in many critical applications are furnished with a gland that has a built-in auxiliary stuffing box (see Fig. 15). In the event of a mechanical seal failure, two or more rings of packing can be installed in this stuffing box to enable pump operation to continue until the mechanical seal can be replaced. However, because the auxiliary packing would receive no lubrication when the mechanical seal is functioning properly, it is important that this packing be installed only after a mechanical seal failure has occurred.

The actual replacement of a leaking mechanical seal can be time consuming if a partial disassembly of the pump is required to permit both the seal being replaced and the new seal to pass over the end of the shaft. One way to eliminate the need for this is to use a split mechanical seal. By splitting the mechanical seal's sealing elements into halves, they can be replaced without removing other parts of the pump.

When high-temperature fluids are pumped, packing and mechanical seals are often cooled by circulating liquid through a jacket that surrounds the pump's sealing area.

When zero leakage must be achieved, the use of a sealless pump is often required. In a canned motor pump, an end-suction impeller is mounted directly on the shaft of the driving motor. However, unlike a conventional close-coupled pump, the pump and motor in this sealless unit form a single hermetically sealed assembly. A portion of the liquid being discharged from the impeller is circulated through the motor for cooling and to provide lubrication for the unit's bearings. In a magnetically coupled pump, torque is transmitted from the driver to the pump's shaft through a magnetic coupling. Because the two halves of the coupling are not in physical contact, the need to provide an opening for the shaft in the pump's casing, together with the need for a shaft seal, is eliminated.

l. Couplings. In all but close-coupled pumps, the torque developed by the driver is transmitted to the pump's shaft through a coupling. Flexible couplings permit some relative movement between the driven and the driving shafts. Included among the various types of flexible couplings used are the following:

• Gear couplings, which have hubs with external gear teeth that mesh with a sleeve containing a mating set of internal gear teeth. (A gear coupling with teeth on both hubs is a flex-flex type. If gear teeth are on only one hub and the opposite hub is flanged, the coupling is a flex-rigid type.)

• Slider-type couplings, which have two metal hubs with jaw-like flanges that are connected with a floating center member, referred to as the spider. The spider is often made from an elastomer.

• Disk couplings, which have several thin metallic rings that are alternately bolted near their outer diameter to the driven and the driving hubs.

• Diaphragm couplings, which have one or more metallic membranes that are typically bolted near their outer diameter to the driving hub and near their inner diameter to the driven hub.

• Elastomer-type couplings, which have two metal hubs that are joined by an elastomeric material located between them.

• Spring-grid couplings, which have two metal hubs that are joined by a flexible steel grid. The grid fits into slots that are located around the periphery of each hub.

• Pin-and-bushing couplings, which have one flanged hub with axial pins and a second with a mating set of holes that are fitted with bushings. The two hubs are joined by inserting the pins into the bushings.

If the thrust bearing in the driver is used to support the pump's shaft axially, a rigid type of coupling should be used. For the pump's rotating assembly to be properly centered within its casing, precision alignment between the two halves of a rigid coupling is critical.

Although flexible couplings can often compensate for slight changes in the pump-to-driver alignment that occur during normal operation, they are not intended to correct for excessive amounts of continuous misalignment. In addition to the detrimental effects that misalignment can have on the coupling, it can also result in the transmission of axial loads and bending moments to the pump's shafting and bearings, and can increase vibration. Therefore, even when a flexible coupling is used, the pump and driver should be accurately aligned prior to the initial start-up, and this alignment should be maintained.

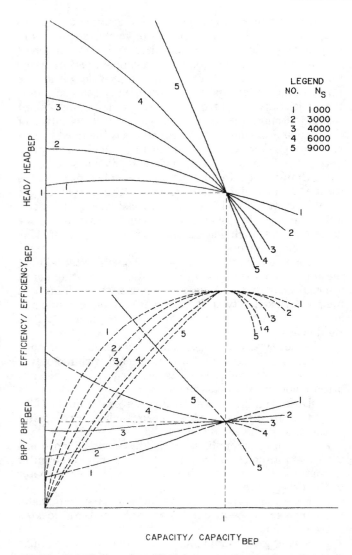

Fig. 16   Typical effect of specific speed on centrifugal pump performance

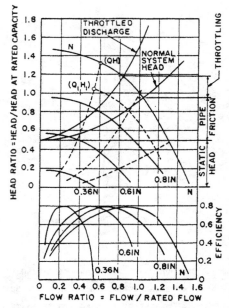

Fig. 17   Capacity-head and efficiency characteristics of a centrifugal pump
at various speeds

The gap between the ends of the pump and driver shafts is sometimes increased to permit a spacer to be installed between the driving and the driven halves of the coupling. The use of a "spacer-coupling" can often eliminate the need to open the pump's casing or disturb the driver when maintenance is performed on the inboard bearing or seal.

m.   Centrifugal pump performance characteristics. Representative performance curves showing normalized values of total head, pump efficiency, and brake horsepower plotted versus capacity for pumps with various values of specific speed are shown in Fig. 16. In general, the head-capacity curve becomes steeper as the specific speed increases. In addition, although the brake horsepower required to drive low specific speed pumps typically increases with capacity, for high specific speed units the reverse is usually true.

The operating point for a centrifugal pump is determined by the intersection of its head-capacity curve with the curve of total head versus capacity for the system in which it operates. A system head curve, which is illustrated in Fig. 17, includes a static component and a dynamic component. The static component results from the differences between both the liquid's pressure and its elevation at the beginning and at the end of the system, and is constant with capacity. The dynamic component, which includes losses due to friction and turbulence in the system's piping, valves, and fittings, as well as entrance and exit losses in the system, varies approximately with the capacity squared.

The point of intersection between the pump's head-capacity curve and the system head curve in Fig. 17 can be thought of as the point of equilibrium. If the pump attempts to force liquid through the system at a higher capacity, the total head developed by the pump will no longer be sufficient to overcome the system resistance, and operation will return to the equilibrium point. Conversely, if the capacity being pumped drops, the system head requirement will be less than the head developed by the pump, and more liquid will be forced through the system until operation returns to the equilibrium point.

As shown in Fig. 17, the point of intersection between the pump and system curves can be changed by varying the pump's operating speed, N, or by throttling the system discharge valve. The performance of an existing radial- or mixed-flow pump can also be altered by changing the impeller's average outside diameter, d. The effects that changes in N and d have on the capacity delivered, Q, the total head developed, H, and the brake horsepower required, P_P, can be estimated as follows:

$$Q_2 = Q_1 \frac{N_2}{N_1} \frac{d_2}{d_1} \tag{29}$$

$$H_2 = H_1 \left[\frac{N_2}{N_1} \frac{d_2}{d_1} \right]^2 \tag{30}$$

$$P_{p2} = P_{p1} \left[\frac{N_2}{N_1} \frac{d_2}{d_1} \right]^3 \qquad (31)$$

where subscripts 1 and 2 refer to equivalent operating conditions with the original outside diameter and speed, and with the new diameter and speed, respectively. The above equations do not include effects that changes in speed or impeller diameter have on pump efficiency. In addition, because effects due to changes that occur in vane angles, waterway lengths, and channel areas when an impeller's outside diameter is changed are not reflected in the above equations, they cannot be relied on to accurately predict performance variations resulting from large changes in impeller diameter. Furthermore, predictions calculated with these equations generally underestimate actual performance variations resulting from impeller diameter changes in mixed-flow pumps [2,3].

Net positive suction head. The suction condition for a centrifugal pump is often expressed in terms of net positive suction head (NPSH). NPSH is the amount by which the total suction pressure of the pumped liquid exceeds its true vapor pressure at the pumping temperature. When referring to NPSH, it is necessary to differentiate between the amount available to the pump, and the amount required by the pump.

The NPSH available, which is a function of the system design on the suction side of the pump and the liquid being pumped, can be calculated using the following

$$\text{NPSHA} = \frac{p_s + p_a - p_v}{0.433 \, (sg)} + \frac{V_s^2}{2g} + Z_s - H_{fs} \qquad (32)$$

where

NPSHA = net positive suction head available, ft
p_a = atmospheric pressure (at suction pressure gage), psia
p_v = true vapor pressure of liquid entering pump, psia

p_s, V_s, H_{fs}, sg, and Z_s are as defined for equations (1) and (2).

Although the NPSH required can be affected by the characteristics of the liquid being pumped, it is primarily a function of the pump design and can be thought of as a measure of the loss in the liquid's head as it travels from the pump-suction flange to the inlet of the first-stage impeller. Unlike NPSHA, which is calculated, values of required NPSH are determined by test. A method commonly used to perform this type of a test is to gradually reduce the NPSH available to the pump during operation at a constant capacity until a specified reduction, usually 3%, is measured in total head. The NPSH available at this point is considered the minimum NPSH required by the pump at the test capacity. This test is repeated at various capacities. The reduction in total head that is measured is caused by the gasification of the liquid being pumped, which occurs when the local pressure drops below the liquid's true vapor pressure. The vapor bubbles formed interfere with the normal flow of liquid through the pump. In addition, as these bubbles travel into higher-pressure regions of the impeller, they collapse and release their

energy of vaporization. This collapse is often violent and can generate very high local pressures. If the reduction in NPSH available continues, the vapor bubbles will eventually fill the eye of the impeller and the total head developed by the pump will drop sharply. The formation together with the subsequent collapse of vapor bubbles, which is referred to as cavitation, has a detrimental effect on not only the total head developed by the pump but also on its efficiency. The resulting uneven flow through the impeller, combined with the impingement against the impeller's waterway boundaries of the liquid that rushes in to fill the voids created by the collapsing vapor bubbles, can increase noise and vibration and can cause severe pitting of the impeller.

The NPSH required by any specific pump is affected by many factors, but its value can sometimes be predicted using the following simplified relationship, which is based on the absolute and relative velocities of the liquid entering the impeller, c_1 and w_1, respectively

$$\text{NPSHR} = \frac{K_1 \, c_1^2}{2g} + \frac{K_2 \, w_1^2}{2g} \qquad (33)$$

where

NPSHR = net positive suction head required, ft
K_1 = shroud coefficient
K_2 = vane coefficient

K_1 and K_2, which are derived empirically, vary with both capacity and operating speed. Based on testing performed with centrifugal pumps fitted with Francis-type impellers, some typical values found for coefficients K_1 and K_2 when operating at the shockless capacity, or the capacity at which incidence (the difference between the vane angle and the relative flow angle) at the inlet to the impeller is equal to zero, were $K_1 = 1.40$ and $K_2 = 0.085$ for a complete breakdown in the total head developed, and $K_1 = 1.57$ and $K_2 = 0.23$ for a reduction in total head of 0.5% [13].

As shown in Fig. 18, a centrifugal pump's NPSH requirement based on a 3% reduction in total head, $\text{NPSHR}_{3\%}$, generally increases as the capacity being delivered increases. Within the operating range from 100 to 150% of the best efficiency point, the effect that an increase in capacity will have on $\text{NPSHR}_{3\%}$ can sometimes be estimated by using the approximation that $\text{NPSHR}_{3\%}$ in this performance region varies in proportion to the ratio of the capacity increase squared [14]. Although a 3% drop in total head is generally used as the criterion to establish minimum NPSH requirements, in critical applications reductions in total head of as little as 1% are sometimes used to establish values of NPSHR. In addition, the NPSH required to prevent incipient cavitation, NPSH_i, which is the point at which the first cavitation bubble initially begins to form, can be many times greater than the corresponding value of $\text{NPSHR}_{3\%}$. A typical curve of NPSH_i, which is shown in Fig. 18, reaches a minimum value at the shockless capacity and peaks at the capacity at which suction recirculation begins [15].

The suction capability of a centrifugal pump is often classified by the suction specific speed, S, as follows:

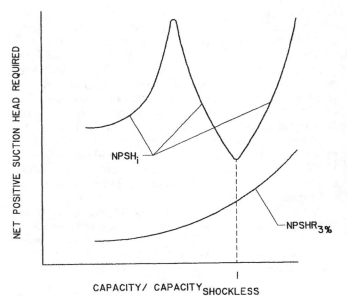

Fig. 18   Net positive suction head versus capacity for a centrifugal pump

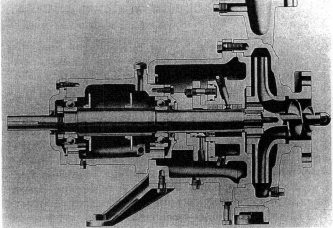

Fig. 19   Centrifugal pump with an inducer (Courtesy Dresser Pump Division)

$$S = \frac{NQ^{1/2}}{(\text{NPSHR}_{3\%})^{3/4}} \qquad (34)$$

As with the calculation of N_S [see equation (3)], N is speed in rpm and Q is capacity in gpm. In addition, $\text{NPSHR}_{3\%}$ is in feet of liquid. When S is used to classify the suction capability of a specific design, it is generally based on the capacity delivered and the NPSH required with a full-diameter impeller operating at the best efficiency point [14]. In addition, because S is an index of the impeller's suction performance, for double-suction impellers the value of Q used in equation (34) should be only one half of the pump's actual capacity. To reduce the potential for problems caused by cavitation or suction recirculation, it is sometimes recommended that the maximum operating speed for standard pump designs be based on limiting S to 8500 [1]. However, in some applications for which standard impeller designs are not suitable, special low-NPSHR first-stage impellers with S values exceeding 8500 are used. The capacity range of these high suction specific speed impellers is often reduced due to suction recirculation.

Reductions in the pump's NPSH requirements may be achieved by installing an inducer in the inlet of the first-stage impeller. As shown in Fig. 19, an inducer is essentially an axial-flow impeller fitted with helically shaped vanes that are swept back from the hub to their outer periphery. Because the diameter of these vanes increases as the axial distance to the impeller's eye is reduced, hydraulic shock and entrance losses are distributed over a much broader area than in a conventional impeller. As a result of this larger area and due to the low number of vanes, blockage from cavitation that may occur within the inducer is reduced [16]. The inducer also acts as a booster stage and raises the pressure of the liquid being pumped, which lessens the potential for cavitation in the main impeller. During operation at the best efficiency point, the

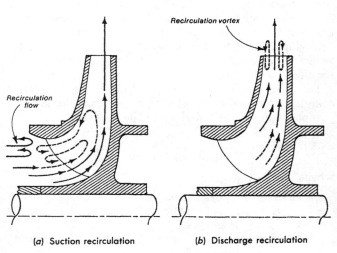

(a) Suction recirculation   (b) Discharge recirculation

Fig. 20   Internal flow recirculation

use of an inducer can sometimes reduce values of $\text{NPSHR}_{3\%}$ by up to 50%; however, inducers can also result in an increase in $\text{NPSHR}_{3\%}$ at capacities outside the region of improvement.

An additional factor affecting suction performance is that centrifugal pumps handling hot water or certain hydrocarbon liquids will often operate satisfactorily with less NPSH than the amount found to be required during testing with cold water [17]. A chart for estimating the reduction in $\text{NPSHR}_{3\%}$ that can be tolerated when pumping some specific hydrocarbon liquids and high-temperature water is included in reference 1. However, this information is based on test data from pumps handling pure liquids; therefore, it is not recommended for use with pumps that will handle liquids containing entrained or dissolved noncondensable gases.

Suction and discharge recirculation. Suction recirculation, which is illustrated in Fig. 20(a), refers to the flow reversals that occur at low capacities within the eye of a centrifugal pump impeller. The interaction of the rotating annulus of liquid being recirculated out of the impeller's eye with the liquid entering the impeller through the center of the annulus results in the formation of high-velocity vortices. Although suction recirculation occurs in all centrifugal pumps, the capacity at which it begins generally gets closer to the best efficiency point with increasing values of suction specific speed and larger impeller-eye diameters. Suction recirculation can cause erosion of the vanes near the inlet to the impeller and of the casing's suction area due to cavitation in the low-pressure cores of the recirculation vortices; it can also cause increased noise and vibration, and flow surges in the suction of the pump [18].

In addition to suction recirculation, at low capacities discharge recirculation, shown in Fig. 20(b), occurs at the discharge tips of the impeller's vanes. The capacity at which this form of recirculation begins usually gets closer to the best efficiency point as the peak hydraulic efficiency and the ratio of the head developed divided by the peripheral speed at the discharge of the impeller increase [19]. Discharge recirculation results in the formation of high velocity-vortices due to the interaction of the liquid being discharged from the impeller with the liquid being returned to it. Detrimental effects from discharge recirculation can include erosion of the vanes and shrouds near the discharge of the impeller and of the casing's volute tongue or diffuser vanes tips due to cavitation, and increased vibration due to axial unbalance.

The effects of internal flow recirculation are generally more severe in high-energy applications and should be considered when the minimum recommended capacity for a centrifugal pump is established. Equations that can be used to estimate the capacities at which suction and discharge recirculation will begin are included in reference 18. In addition, empirically derived criteria for predicting the severity of the effects of recirculation in specific designs are presented in reference 20.

Priming and gas entrainment. A centrifugal pump cannot pump a gas; therefore, the differential pressure necessary for flow will not be created if the impeller is full of air or vapor. Consequently, prior to start-up, a pump's casing should be filled with liquid and vented of all gases. In applications where the pump must operate with a suction lift, a vacuum must initially be created to draw the liquid from its source into the eye of the impeller. Methods used for priming a centrifugal pump include the following:

• The pump can be connected through vents to a central priming system. This type of a system is fitted with either vacuum pumps or air ejectors, and is generally used to prime multiple pumps on the vessel.

• The pump can be fitted with an integral or independent vacuum pump.

• A special centrifugal pump that is designed to be self-priming, such as the one shown in Fig. 21, can be used. This type of a pump is typically fitted with a suction chamber that retains liquid when the unit is not operating. Once started, this liquid, together with any air that may be mixed with it, is pumped through the impeller, and enters a second chamber installed at the pump's discharge. While the air contained in the fluid is vented

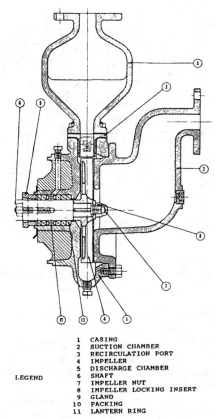

	1	CASING
	2	SUCTION CHAMBER
	3	RECIRCULATION PORT
	4	IMPELLER
	5	DISCHARGE CHAMBER
LEGEND	6	SHAFT
	7	IMPELLER NUT
	8	IMPELLER LOCKING INSERT
	9	GLAND
	10	PACKING
	11	LANTERN RING

Fig. 21 Self-priming centrifugal pump

through the top of the discharge chamber, the liquid is returned to the casing through either an internal or an external port. As this liquid reenters the impeller, it mixes with a portion of the air that has been drawn into the evacuated suction chamber, and the priming cycle is repeated. Each time the stored liquid is recirculated through the pump, an additional amount of air is removed from the suction line. Because of the limited air-handling capability of many self-priming centrifugal pumps, their use is generally not recommended when large volumes of air or other gases must be removed from the suction piping.

• A foot or check valve may be used to retain liquid within the suction line when the pump is shut down. However, liquid can leak past the valve while the pump is not operating. In addition, the valve increases the suction line's friction losses and reduces the NPSHA.

In addition to priming considerations, if air or other noncondensable gases are entrained in the liquid being pumped, there will be a reduction in the pressure developed by the pump that exceeds the reduction attributable to the decrease in the average density of the fluid due to the presence of a gas [21]. With high flow rates and low gas volumes, some of the bubbles entering the pump are carried through the impeller by the liquid flowing past them. However, as either the capacity is reduced or the gas volume is increased, a point will be reached where the liquid entering the pump can no longer clear the impeller

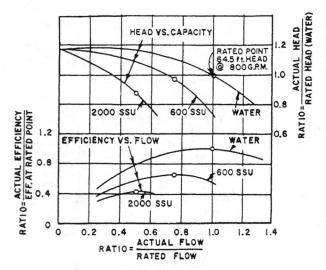

Fig. 22   Effect of viscosity on centrifugal pump performance

eye of bubbles, and the flow through the pump will stop. Standard centrifugal pumps usually become "gas-bound" when handling fluids with a gas content of 5 to 8% by volume. The effect that gas has on pump performance becomes more detrimental at capacities above and below the best efficiency point.

Fluid viscosity. The viscosity or resistance to flow of the liquid being pumped has a significant effect on the performance of any centrifugal pump. As shown in Fig. 22, an increase in viscosity above that of water causes the total head developed, capacity delivered, and pump efficiency to be reduced. In addition, because the deterioration in efficiency is greater than the reductions in head and capacity, there is an increase in the power required to drive the pump. For pump performance with a viscous liquid to be evaluated, the liquid's viscosity at the pumping temperature and the effect that shear rate has on this viscosity must be known. Reductions in capacity, total head, and efficiency can be estimated for radial-flow pumps that deliver viscous Newtonian liquids at capacities up to 10,000 gpm using correction charts included in reference 1. In addition, an extended chart that permits viscosity performance corrections to be made for pumps delivering capacities up to 100,000 gpm is included in reference 22.

Overheating. At low capacities the majority of the power required to drive a centrifugal pump is no longer converted into useful work, which is evidenced by the corresponding reduction in pump efficiency. The excess energy results in an increase in the temperature of the pump and the fluid passing through it. At shutoff no useful work is done and all of the energy added by the driver is available to heat the fluid trapped within the pump casing.

The effects that an increase in temperature can have on the pump include a reduction of internal running clearances, overheating, possible failure of the bearings and seals, and, in extreme cases, the eventual seizure of the

rotor. An increase in temperature can also lead to a hazardous situation when liquids with low flash and fire points are being pumped. Equations that can be used to calculate the temperature rise of liquid passing through a centrifugal pump and the minimum flow rate required to limit the increase in the temperature of the liquid to a specified value are given in reference 14.

Pump operation in series. A single pump is not always capable of developing the total head required during all of the system's modes of operation. In addition, the speed required to prevent cavitation is sometimes too low for one pump to develop sufficient head to overcome the system back pressure. In these situations, two or more pumps are often operated in series. With this arrangement the first pump can be run at a lower speed than the other units and used as a booster or inducer pump. Its discharge is then directed into the suction of the second pump, which due to the increased NPSHA can operate at a higher speed and develop a much greater total head than the first unit. If necessary, the discharge from the second unit can be directed to additional pumps. The total performance of two or more pumps operating in series can be determined by adding the total head developed by each unit at values of equal capacity.

Pump operation in parallel. In some applications, the capacity that must be delivered can, at times, exceed the capability of a single pump. When this is the case, two or more pumps in the system may be operated in parallel. The combined performance of pumps that operate in parallel in a common system can be found by adding the capacity delivered by each unit individually at values of equal total head. If pumps operating in parallel have similar performance characteristics, are running at the same speed, and have approximately the same suction pressure, they will each deliver approximately the same capacity. However, if one pump develops a higher head at a given capacity than the other units, it can force the lower discharge head pumps to deliver lower capacities. In addition, if the system curve intersects the combined pump performance curve at a point where the system head exceeds the maximum head developed by any of the pumps, the affected units will be forced to operate at shutoff. To prevent this, if the pumps used or their suction conditions are not similar, their relative speeds should be adjusted so that an adequate flow will be delivered by each unit.

1.3 Regenerative Turbine Pumps. Regenerative turbine pumps are used in some marine applications that require a high total head to be developed at a relatively low capacity. This type of a pump, which is illustrated in Fig. 23, is fitted with a solid disk-shaped impeller containing multiple radial vanes around its periphery that rotates within a close-clearance casing. Liquid entering the pump's casing is divided evenly and admitted to each side of the impeller. As this liquid enters the rotating impeller's vanes it is thrown outward radially, which increases its velocity head. After leaving the impeller the liquid's velocity is gradually reduced in the casing, resulting in an increase in its pressure. Due to the annular shape of the casing's passage, the liquid is then redirected back into the rotating impeller and the cycle is repeated.

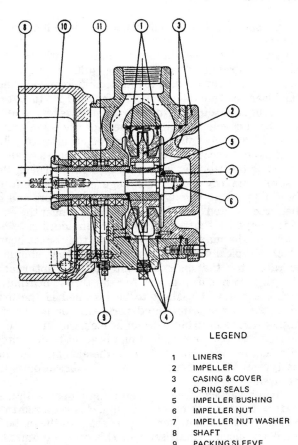

LEGEND

1	LINERS
2	IMPELLER
3	CASING & COVER
4	O-RING SEALS
5	IMPELLER BUSHING
6	IMPELLER NUT
7	IMPELLER NUT WASHER
8	SHAFT
9	PACKING SLEEVE
10	PACKING GLAND
11	SEAL CAGE

Fig. 23 Regenerative turbine pump

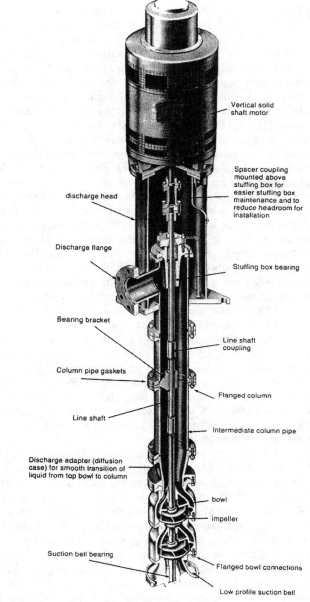

Fig. 24 Vertical turbine pump

The recirculation into and out of the impeller continues until the liquid has traveled around the circumference of the circular casing, at which point it is discharged from the pump. These multiple recirculation cycles enable regenerative turbine pumps to develop their characteristically high heads. The total head developed and the power required by a regenerative turbine pump typically reach their peak values at shutoff, and are both reduced as the capacity delivered increases.

A close radial clearance stripper in the casing reduces the amount of liquid that can be recirculated from the discharge back to the suction and, therefore, improves the regenerative turbine pump's volumetric efficiency. Due to the stripper, a regenerative turbine pump is considered to be self-priming and can typically operate with relatively large amounts of entrained air or vapor, provided that the casing is initially filled with liquid at start-up. However, because of their small flow passages and close operating clearances, regenerative turbine pumps are not recommended for applications involving viscous liquids, or liquids that contain abrasives or other solids.

A regenerative turbine pump's impeller is hydraulically balanced axially, and can either be centered on a shaft that is mounted between bearings, or overhung on the

end of a cantilevered shaft that is supported by bearings located behind the impeller. Although the pump shaft is generally flexibly coupled to that of the driver, when fitted with an overhung impeller, these pumps can be furnished in a close-coupled configuration. Sealing of the casing in the area of shaft penetrations is accomplished with either packing or mechanical seals. So that internal running clearances can be renewed, the casing's sidewalls are often fitted with replaceable plates that surround the impeller.

1.4 Vertical Turbine Pumps. A vertical turbine pump (VTP), which is shown in Fig. 24, is a centrifugal pump with impellers that are submerged within the liquid being pumped. The bowl assembly is located at the lower end of the pump, and consists of the casings, referred to as bowls, and the impellers, which can be stacked vertically to form multiple stages. The number of stages used in the bowl assembly is based on the specific speed of each stage, and on the total head that must be developed by the pump. Joints between adjacent bowls are either threaded or flanged. A sleeve-type journal bearing is generally installed in each bowl to support the impeller shaft. To permit internal running clearances to be renewed, each bowl is often fitted with a replaceable wearing ring adjacent to the outer hub of the impeller. Impeller wearing rings are also sometimes used. The impellers can be held in place on the shaft with tapered lock collets or keys and snap rings. A suction bell or case installed at the bottom of the first-stage bowl serves as the inlet to the pump. Although some VTP's are fitted with an overhung first-stage impeller, in many units the lower end of the shaft extends through the eye of the first-stage impeller and is supported by a bearing installed in the suction bell. If the liquid being pumped contains foreign particles, a strainer is sometimes mounted at the pump's inlet. Pressure recovery in the VTP typically takes place in multi-vaned diffusers that are integrally cast in each bowl.

The liquid discharged from the upper bowl travels to the discharge head through the column pipe. This pipe also encloses the line shaft that transmits the torque developed by the driver to the impellers. When furnished with long lengths of column pipe and line shafting, which can exceed 50 ft in some marine applications, VTP's are also referred to as deepwell pumps. Support for the line shaft is provided by sleeve bearings installed in retainers, referred to as "spiders," that are sandwiched between the mating ends of adjacent column sections. Although a VTP can be furnished with an enclosing tube that surrounds the line shaft and enables its bearings to be lubricated by oil supplied from an external source, the bearings in many pumps are lubricated by the pumped liquid. The individual sections of column pipe can be joined with threaded sleeve type couplings or flanged joints. In some pumps the bowl and column assemblies are connected through a special axially split spool piece. The use of this spool piece, together with a quick-disconnect coupling between the lower line shaft and the impeller shaft, permits the removal of the bowl assembly to be accomplished without the need to disassemble the entire pump. Information on the determination of natural frequencies for VTP's is included in reference 23.

The discharge head serves as a 90-deg elbow to redirect the flow of the fluid leaving the vertical column pipe, and forms the link between the pump and the discharge piping. It also supports the column and bowl assemblies, and the pump's driver or right-angle gear. The penetration for the pump's shaft in the discharge head can be sealed with packing or mechanical seals. When packing is used it is often lubricated by the fluid being pumped or, in some cases, by grease that is injected into the stuffing box. When a mechanical seal is used it must have liquid circulating around it for cooling and lubrication. Various arrangements can be used to flush the mechanical seal with the pumped fluid or with liquid supplied from an external source.

Because of the weight of the pump's rotating assembly and the hydraulic unbalance that results from the single-suction design of vertical turbine pump impellers, the net axial load that must be supported by a VTP's thrust bearing is usually directed downward. However, when the total head developed is low or the capacity is high, the net axial thrust may, at times, be directed upward. This is most likely to occur during start-up, after loss of suction, or during operation at very high flow rates. The upthrust results primarily from the change in the momentum of the fluid passing through the pump's impellers and from the force exerted by fluid that may act against the bottom of the impeller shaft. Consequently, in addition to being rated for the maximum downthrust, the thrust bearing, which is typically a multiple-row ball bearing, should also be rated for the maximum intermittent upthrust. Although it is not uncommon for the thrust bearing that supports the VTP's shaft to be installed in the driver, when the driver is mounted vertically, or in the right-angle gear if a horizontal driver is used, the pump's thrust bearing can also be mounted in a bracket that is integral with the discharge head. Due to the size and loading of the typical VTP's thrust bearing, oil lubrication is often required.

When a vertical driver or right-angle gear with a solid shaft is used, an above-deck coupling is required to connect the end of the driver or gear shaft to the end of the pump's upper line-shaft section. This coupling must be of the rigid design if the pump's thrust bearing is located in the driver or gear. The rigid coupling, which is often used to adjust the height of the pump's rotating assembly, should be suitable to not only transmit the required torque, but also to support the axial thrust that is applied to the VTP's rotating assembly. When a VTP has an independent thrust bearing, a flexible coupling is generally installed between the pump and the driver or right-angle gear. When a hollow shaft driver or right-angle gear is used, the pump's shaft extends up through the center of the hollow shaft and is secured at the top of the driver or gear with an adjusting nut.

Because of the design of a VTP's diffusers and impellers, a VTP becomes an efficient hydraulic turbine when driven by the liquid that drains from its column after the pump is stopped. This reverse rotation can result in mechanical damage if the unit is restarted before the shaft stops turning in the reverse direction. To prevent reverse rotation, the vertical drivers and right-angle gears used with VTP's are often fitted with nonreverse ratchets.

A VTP is a centrifugal pump; therefore, its performance characteristics are similar to those described in Section 1.2. When the VTP is a multistage unit, the amount of cavitation in the first stage necessary to produce a 3% reduction in the total head developed by the complete bowl assembly can be significant. Although this can also be true of other multistage centrifugal pumps, the modular construction of the VTP enables stages to be added to or removed from the bowl assembly, and permits NPSH testing to be performed using only the bowl assembly's first stage. The head reduction measured during this type of a test is more representative of the cavitation that actually occurs in the first-stage impeller and, therefore, of the pump's true suction capability.

When a VTP is installed directly into the tank or sump from which it will receive liquid, no suction piping is required or used. The calculation of the pump's suction pressure and NPSHA, therefore, is based on the depth of the liquid in the tank or sump, together with the pressure that acts on the surface of the liquid. Even when there is adequate NPSH available to the pump, problems can still occur if the submergence of the suction bell is not sufficient to prevent a vortex from developing on the surface of the liquid and air or vapor is drawn through the core of the vortex and into the pump's inlet. The minimum submergence required to prevent the formation of vortices is proportional to the square of the velocity at the inlet to the pump.

If a VTP is to receive liquid from multiple sources, it will usually be installed within a separate suction tank, referred to as a suction can. The suction can is then connected through piping to the various locations from which the pump will take suction. Whenever suction piping is used, losses within it due to friction and turbulence should be considered when determining the suction pressure and NPSHA. The VTP must often be capable of evacuating the suction piping of gas and vapor; therefore, pumps installed in suction cans are frequently fitted with components that enable them to be self-priming. Like the recirculation arrangement described previously for a self-priming centrifugal pump, the VTP priming systems usually rely on the recirculation of liquid from the pump's discharge back to suction. However, in lieu of continuous recirculation, the recirculation within a VTP is generally controlled by one or more automatic priming valves installed on the pump. A description of a typical priming valve arrangement, together with criteria for sizing suction cans used with self-priming vertical turbine pumps, is included in reference 24.

1.5 Other Types of Wet-Pit Pumps. In addition to vertical turbine pumps, there are also other types of vertical wet-pit pumps. Included among these units are:

• Propeller type deepwell pumps. These high-capacity, low-head pumps are used in many dry docks for dewatering service.

• Single-stage double-suction impeller deepwell pumps. Because of their double-suction design, the NPSH requirements for these centrifugal pumps are typically less than those for vertical turbine pumps that deliver equivalent capacities. However, a double-suction impeller pump generally has a larger casing diameter than a comparably rated VTP.

• Single-stage end-suction sump pumps with volute casings. These centrifugal pumps are generally furnished with short settings or lengths. Due to the side discharge configuration of the pump's volute casing, the vertical pipe through which liquid travels from the impeller to the above-deck discharge flange is mounted adjacent to the line shaft rather than concentric with it.

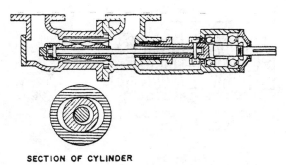

SECTION OF CYLINDER

Fig. 25 Single-screw pump

• Submersible centrifugal pumps. These single-stage end-suction pumps, together with their drivers, which are generally either electric or hydraulic motors, are submerged within the liquid being pumped.

1.6 Rotary Pumps. Rotary pumps are positive-displacement machines in which one or more pumping elements rotate within a stationary chamber. The chamber is typically formed by a one-piece casing or housing that is fitted at one or both ends with removable plates. A rotary pump traps fluid within cavities in its rotating elements that are sealed by the inner wall of the casing. As each rotor turns, the trapped fluid is forced through the casing. Because the return of fluid back to suction, referred to as "slip," is impeded by close internal running clearances, during the revolution of each rotor a nearly fixed amount of fluid is forced out of the casing through the pump's discharge port. In addition, as the vacant cavities within each rotor spin past the casing's suction port, a partial vacuum is created that draws additional fluid into the pump. Due to the internal seal between the suction and the discharge, internal valves are not required in rotary pumps. Shaft penetrations in rotary pump casings are usually sealed with packing or mechanical seals. Various types of rotary pumps are furnished for shipboard applications in both horizontally and vertically mounted configurations.

a. Screw pumps. In screw pumps, fluid is trapped within a series of successive helical cavities that are formed between the threads of one or more screws and are sealed by a close-clearance casing. As shown in Fig. 25, single-screw or "progressing cavity" pumps include one rotor with a single external helical thread. As the rotor turns, it also orbits about the centerline of the driveshaft and meshes with the stator's double-helix internal thread. The meshing of the rotor with the close-clearance stator results in the formation of sealed cavities that progress axially along the length of the casing and carry fluid from the pump's inlet to its discharge port. Many of these pumps are fitted with a hardened rotor and an elastomeric liner within the stator. With this material combination the pump's stator is generally flexible enough to pass particles that may be entrained in the pumped fluid without causing serious damage to the rotor. Progressing-cavity pumps are often furnished in a "single-end" design in which fluid flows in only one direction along the rotor. The resulting axial unbalance is frequently absorbed by

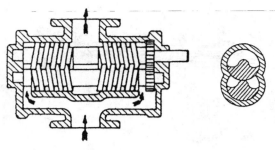

Fig. 26   Twin-screw pump

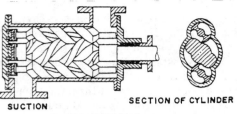

SUCTION   SECTION OF CYLINDER

Fig. 27   Multiple-idler screw pump

one of the two antifriction type bearings that are typically used to support the pump's drive shaft. These bearings are usually mounted external to the pumped fluid. Due to the orbital motion of the rotor and the friction resulting from its contact with the stator, the speed at which progressing-cavity pumps can operate is limited. They are best suited for applications involving high-viscosity fluids at relatively low capacities and pressures.

As shown in Fig. 26, a twin-screw pump has two shafts that rotate within a close-clearance casing. When a "double-end" arrangement is used, each shaft is fitted with a pair of opposed helical screws. Fluid entering the pump's suction nozzle is often divided, and admitted to the rotors from both ends of the casing. The fluid then advances axially from each end of the pump, is recombined in the center of the unit, and is expelled through the casing's discharge connection. As a result of the opposed-flow pattern, axial thrust due to hydraulic unbalance is minimized. In addition, the pump's shaft seals are exposed only to fluid at suction pressure. To reduce the path length of the fluid entering the screws, in applications with poor suction conditions the flow path described above is sometimes reversed. However, with this alternative arrangement the shaft seals are exposed to fluid that is at the pump's discharge pressure.

Radial loads applied to the rotating assembly of a twin-screw pump are generally absorbed by antifriction bearings that are mounted at both ends of each shaft. The bearings installed at one end of the pump are held in place axially, which enables them to also absorb residual axial thrust. The bearings are often relied on to maintain the proper axial and radial clearance between the meshing screws. To maintain the proper rotational clearance between the screw threads, torque from the driving shaft, which is coupled to the pump's driver, is frequently transmitted to the idler shaft through timing gears. This eliminates the need for metal-to-metal contact between the screws and, therefore, reduces wear.

Twin-screw pumps are sometimes designed so that their bearings and timing gears will be submerged in, and lubricated by, the pumped fluid. Only one shaft seal is required with this internal bearing configuration to prevent leakage at the location where the driving shaft penetrates the casing. If the fluid to be pumped will not be clean or will not have good lubricity, a pump with bearings and timing gears that are enclosed within lubricating-oil reservoirs located external to the pumping chamber must generally

be used. With an external bearing design, additional shaft seals are required to isolate the bearings and timing gears from the pumped fluid.

Some screw pumps have two or more idler rotors. As shown in Fig. 27, multiple-idler screw pumps are typically designed to operate without timing gears. The helix angle of the threads on the rotors in these pumps is sufficient to enable torque to be transmitted directly from the driving or power screw on the centershaft to the sealing screws on the idler shafts. Axial thrust can be minimized through the use of a double-end screw configuration, or with an internal hydraulic balancing device. Radial loads are applied to the center drive screw by each of the idler screws in a symmetrical pattern; therefore, the center rotor is radially balanced. Furthermore, each idler screw is supported radially along its entire length by the inner wall of the casing. Consequently, there is no need for separate radial bearings to support the idler rotors. Only one seal is required at the location where the center shaft, which is generally fitted with either an internal or an external bearing, penetrates the casing. To reduce wear, the drive screw and the idlers are often furnished in hardened materials. In addition, although a hydrodynamic film between the idler screws and the casing reduces metal-to-metal contact between these parts, in some pumps renewable casing liners are used.

b. Gear pumps. In a gear pump the fluid being pumped is trapped between the teeth of two meshing gears and is forced through the casing. Slip is minimized by the meshing of the gears, by maintaining close radial clearances between the gears and the inner wall of the housing, and by maintaining close axial clearances between the faces of the gears and the pump's side or end plates. Gear pumps can be designed with an external configuration in which each gear's center of rotation is external to the major diameter of the mating gear, or with an internal configuration in which the center of rotation of one gear is within the major diameter of the second gear.

An external gear pump, which is illustrated in Fig. 28, contains two external meshing gears that are mounted on parallel counterrotating shafts within a close-clearance casing. The gears can be of the spur, helical, or herringbone design. Spur gears are inherently axially balanced and are suitable for many high-pressure, low-speed applications. However, fluid trapped between meshing gear teeth can result in the generation of high localized pressures and increased radial shaft loads. To reduce this pressure, relief ports are sometimes cut into the discharge

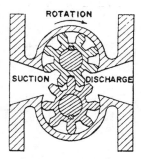

Fig. 28 External gear pump

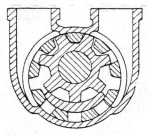

Fig. 29 Internal gear pump

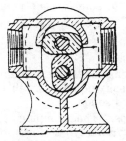

Fig. 30 Lobe pump

side of the pump's end plates, which provides an escape path for the trapped fluid. Problems associated with trapped fluid can also be alleviated by using helical gears. Because of their helix angle, meshing teeth in this design engage gradually and force fluid that is between them to flow ahead of the contact or meshing point. The retention of fluid between the meshing gear teeth is, therefore, prevented. Due to the axial thrust produced by mating helical gears, their use is generally limited to medium-pressure applications. A herringbone gear, which resembles two helical gears placed end to end, provides the advantages of the helical gear but eliminates its undesirable axial unbalance. Consequently, herringbone gears are often used in external gear pumps furnished for higher-pressure applications.

The hydraulic pressure gradient around the periphery of each gear, together with the reaction to the torque being applied to the pumped fluid, results in radial loads that are applied to the external gear pump's rotors. These loads are absorbed by sleeve or antifriction bearings that are mounted at the ends of each shaft. The shafts in many pumps are also fitted with thrust bearings. Torque from the driving or power shaft to the idler shaft in some external gear pumps is transmitted directly through the teeth of the pumping gears. However, to reduce metal-to-metal contact between the pumping gears, this torque can also be transmitted through separate timing gears. The pump's bearings and timing gears, when used, can be either submerged in, and lubricated by, the pumped fluid or mounted outside of the casing in isolated lubricating-oil reservoirs.

Internal gear pumps contain an internal ring gear that rotates against the inner wall of the pump's casing. As shown in Fig. 29, a second smaller external gear or pinion is designed to rotate inside the ring gear. Fluid entering the pumping chamber is trapped between the teeth of each rotating gear. The fluid-carrying cavities are frequently sealed by a stationary crescent-shaped piece that fits between the two gears. As an alternative to the crescent, some internal gear pumps rely on a close radial clearance between the tips of adjacent teeth on the two rotating gears to form a seal. To provide the necessary seal between the pump's suction and its discharge, the internal and external gears, which have different axes of rotation, are arranged so that their teeth mesh over a portion of the external gear's circumference. Torque from the power

rotor, which can be either the external pinion or the internal ring gear, is transmitted directly to the mating idler through the meshing teeth. The hydrodynamic fluid film between the gears is often relied on to absorb radial loads. Additional support for the drive shaft extension can also be provided by a bearing installed either internal or external to the pumping chamber.

Because of a longer arc of entry to the gear and a more gradual change in the direction of the flow, suction losses in an internal gear pump are typically less than those in comparably sized external gear pumps. In addition, because both gears in an internal gear pump rotate in the same direction, their relative velocity is generally low, which reduces wear. However, the use of internal gear pumps is usually limited to low-capacity, low-pressure applications.

c. Lobe pumps. As shown in Fig. 30, lobe pumps are similar to external gear pumps, except that in lieu of teeth, each of the two counterrotating rotors contains an equal number of rounded lobes. Rotors with two or three lobes are commonly used. Fluid entering the pump is trapped within the cavities formed between the lobe surfaces and the inner wall of the casing. The return of fluid to suction is impeded by the seal formed at the location where the rotors mesh. The rotors in smaller lobe pumps are often overhung on the ends of cantilevered shafts. In larger pumps, however, the rotors and shafts may be mounted between bearings. Due to the shape of the lobes, one rotor cannot drive the other, and torque from the power shaft to the idler must be transmitted through separate timing gears. The timing gears, together with the bearings required to support each rotor, can be mounted within the casing or in separate lubricating-oil reservoirs. Lobe pumps are generally used in high-capacity, low-pressure applications. Due to the relatively large cavities formed

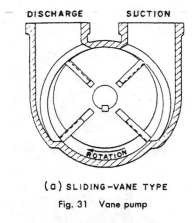

(a) SLIDING-VANE TYPE

Fig. 31   Vane pump

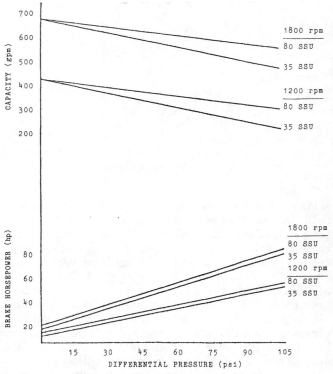

Fig. 32   Typical rotary pump performance curve

between the rotating lobes, these pumps are capable of handling high-viscosity liquids, as well as liquids that contain some entrained solids.

d. Vane pumps. The pumping elements in vane pumps consist of various types of reciprocating or sliding vanes, such as blades, buckets, rollers or slippers. Many vane pumps are of the internal or "vane-in-rotor" design, which is illustrated in Fig. 31, and contain from four to eight rigid vanes. These vanes slide back and forth within radial slots located around the circumference of a cylindrical rotor that turns inside either an eccentric cylindrical or a concentric elliptical casing. The outer tips of the vanes are held against the inner wall of the casing by centrifugal force. Springs, the admission of pressurized fluid behind the vanes, and the installation of push rods between opposing vanes are additional methods used to improve contact between the vane tips and the casing. As each vane sweeps past the inlet port, it follows the contour of the casing's inner wall and moves radially outward from the rotor. The low-pressure created behind the vane draws fluid from the pump's inlet into the space between the rotor's outer surface and the casing's inner wall. This fluid is trapped between adjacent vanes and, as the rotor continues to turn, is forced through the casing. At the discharge port, the casing's inner wall pushes the vanes back into the rotor and forces the fluid out of the pump.

Fluid acts against only one half of the vane pump's rotor during each pumping cycle; therefore, radial loads can be high. To reduce these loads, in some pumps two pairs of suction and discharge ports are equally spaced around the casing. With this arrangement liquid is simultaneously admitted to both sides of the rotor, which results in hydraulic radial balance. Bearings used to absorb residual radial and axial loads can be installed at each end of the rotor or on the inboard side of the rotor only, and can be located within the pumping chamber or external to it. Vanes are typically made from plastics or composite materials; therefore, metal-to-metal contact in these pumps is eliminated. A slight amount of tip wear can be compensated for by the radial movement of the vanes. However, due to noise caused by the rubbing of the vane tips against the casing, these pumps usually cannot be operated at high speeds. Consequently, vane pumps are,

generally used in moderate-capacity and low-pressure applications with low-viscosity fluids.

e. Rotary-pump performance characteristics. The increase in the pressure of the fluid passing through a rotary pump is created by the resistance or back pressure that the rotors must overcome to force fluid through the pump and into the discharge pipe. The total head developed within the pump, therefore, matches the requirement of the system. Although total head can be expressed in feet of liquid, when developed within a rotary pump it is more common for it to be expressed in terms of the total differential pressure, p_{td}, in psi. p_{td} is equal to the total pressure at the inlet to the pump subtracted from the total pressure at the pump's outlet. Like a centrifugal pump, a rotary pump will operate only at the point at which its head-capacity curve, typically illustrated in Fig. 32, is intersected by the curve of system head versus capacity. However, there are some significant differences between the performance characteristics of rotary and centrifugal pumps.

• At any given speed the volume displaced within a rotary pump will remain constant; however, the capacity delivered by the pump will be affected by slip, which increases as the pump's differential pressure is increased and as the viscosity of the pumped fluid is reduced.

• The pressure developed by a rotary pump matches the system back pressure at its discharge port. Therefore, if the pump is operated against a closed discharge valve, the pressure of the fluid trapped within its casing will continue to increase until the pump's driver is overloaded, the pump or valve fail, or the discharge piping bursts. To prevent this type of damage from occurring, and to protect the pump and piping from overpressurization, a relief valve should be installed in a rotary pump's discharge piping. Some rotary pumps are furnished with integral relief valves mounted directly on their casings.

• The power required to drive a rotary pump will also continue to increase as the pressure developed within the pump increases. The use of a relief valve will limit this pressure and, therefore, the maximum horsepower required by the pump. The brake horsepower required to drive a rotary pump, P_p, can be calculated as follows using the total differential pressure, p_{td}:

$$P_p = \frac{p_{td} \, Q}{1714 \, \eta_p} \qquad (35)$$

where

Q = capacity, gpm
η_p = pump efficiency, %/100

In lieu of expressing NPSH available in terms of feet of liquid, the suction condition available to a rotary pump is often expressed in terms of psia and is referred to as the net positive inlet pressure available, p_{ni}. The minimum net positive inlet pressure required by a rotary pump is determined by operating the pump with a constant differential pressure, and reducing the inlet pressure gradually until the capacity delivered is reduced sharply, a crackling noise is clearly audible, or the capacity delivered by the pump is reduced by 5%. The maximum value of p_{ni} at which any of these conditions occurs is equal to the minimum net positive inlet pressure required by the pump. The reduction in capacity and increase in noise are caused by cavitation that occurs when the liquid entering the casing does not have sufficient energy to completely fill the cavities formed within the pump's rotors. Cavitation in rotary pumps can result in increased vibration, pressure pulsations at the discharge port, and pitting of the rotors and casing. The minimum net positive inlet pressure required by rotary pumps generally increases with fluid viscosity, rotor cavity size, and operating speed.

Because of its increased resistance to shear, when a fluid with a higher viscosity is pumped, the flow rate through a rotary pump's internal clearances is reduced. An increase in viscosity also generally reduces the pump's maximum allowable operating speed and increases its net positive inlet pressure requirements. In addition, because more energy is required to overcome the fluid's increased resistance to flow, the power required to drive a rotary pump increases with viscosity.

Rotary pumps, being positive-displacement machines, are often used in applications with poor suction conditions. However, for these pumps to be self-priming, a liquid film is generally required to seal internal clearances. In addition, dry operation can often damage the pump's internal components. Therefore, whenever a rotary pump is started with a suction lift or a negative inlet pressure, it should first be filled with liquid. Once primed, the amount of liquid pumped will be reduced as the percentage of entrained gas within the fluid entering the pump increases. The severity of the effect that entrained air or vapor has on a rotary pump's performance increases with reduced values of inlet pressure.

1.7 Vacuum pumps. Although similar in configuration to a rotary-vane pump, a vacuum pump is used primarily to remove air or gas from piping and components rather than to transfer liquids. Also known as a rotary liquid-piston or liquid-ring pump, as shown in Fig. 33, each unit is fitted with a multi-vaned rotor that rotates within either a concentric-elliptical or an eccentric-cylindrical casing. The casing is partially filled with water, or some other suitable liquid. As the rotor turns this liquid forms a continuous rotating ring that follows the contour of the casing's inner wall. The rigid vanes located around the circumference of the rotor are generally oriented radially, but may have tips that are curved slightly forward.

Shrouds fitted onto the ends of the rotor enable individual pumping chambers to be formed between each pair of vanes. As each chamber sweeps past the wider portion of the casing, the liquid contained between the vanes is thrown outward, and air or vapor from the adjacent inlet port is drawn into the evacuated inner portion of the chamber. As the rotation of the chamber continues into the close-clearance area of the casing, liquid is forced back between the rotor's vanes and the air or vapor that had entered this space is forced out through the pump's discharge port. During operation a small amount of liquid should be fed continuously to the vacuum pump's casing to replace liquid that may be discharged during each pumping cycle. The inlet and discharge ports leading to the pumping chambers can be included in a stationary cone that fits around the center of the rotor, or in the casing's side plates.

Vacuum pumps are generally supplied with radially split casings. To balance radial loads, some pumps are fitted with two inlet and two outlet ports that are alternately staggered at 90-deg intervals. The vacuum pump's rotor can be mounted on a shaft that is supported at each end by an antifriction or sleeve bearing and is flexibly coupled to the driver. However, in some smaller units the rotor is overhung on the end of the driver's shaft in a close-coupled configuration. With either arrangement, openings where the shaft protrudes through the pump casing are typically sealed with packing or mechanical seals.

The performance curve for a vacuum pump is usually plotted in terms of the volumetric capacity of air or vapor removed from the suction line versus the absolute suction pressure. The capacity delivered by a typical vacuum pump remains relatively constant over much of the pump's operating range. However, as the absolute suction pressure is reduced below a certain value, there will generally be a reduction in the pump's capacity. The capacity will also be reduced as the temperature of the pump's sealing liquid increases. To increase the vacuum that can be developed, liquid-ring pumps are sometimes supplied in two-stage configurations. These units are fitted with two rotors mounted on a common shaft within a single casing and are, in effect, two single-stage pumps operating in series.

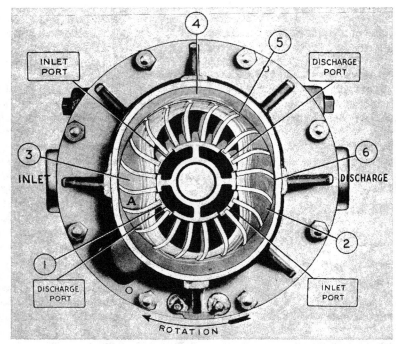

LEGEND

1   ROTOR
2   ROTATING LIQUID CHAMBER
3   NARROW PART OF CASING
4   WIDE PART OF CASING
5   ELLIPTICAL CASING
6   ROTOR VANES

Fig. 33   Liquid-ring vacuum pump

1.8 Reciprocating Pumps. Reciprocating pumps are sometimes used for applications involving high pressures and relatively low capacities. In these positive-displacement pumps, pressure is developed by the reciprocating motion of a displacement element that applies a force directly against the fluid contained within an enclosed cylinder. Reciprocating pumps are furnished in both horizontal and vertical configurations, where this direction refers to the orientation of the displacement element's axis. Each pump consists of two basic elements: the liquid end and the drive end.

a. Liquid end. The portion of a reciprocating pump that pumps fluid is referred to as the "liquid end." During each suction stroke the volume within the liquid cylinder increases due to the retraction of its displacement element, which results in a reduction in the cylinder's pressure. When this pressure is sufficiently below the pressure at the pump's inlet port, the suction valve opens and fluid is drawn into the cylinder. Once the displacement element's direction of travel is reversed, it pushes against the fluid that has filled the cylinder. The pressure of the fluid is increased until it exceeds the pressure at the pump's outlet port by an amount that is sufficient to open the discharge valve. The fluid within the cylinder is then forced through this valve and into the discharge piping.

The moving displacement element used in a reciprocating pump can either be a piston, which typically resembles a flat disk, or a plunger, which often looks like a smooth rod. The pressure seals between a piston and its cylinder are mounted directly on, and move with, the piston, while the seals installed around a plunger are mounted at the base of the cylinder and are stationary. Various types of nonreturn valves can be used in the liquid end of a reciprocating pump, including stem-guided disk valves and caged plate valves with flat seats for general service, wing-guided valves with tapered seats for higher-pressure applications, and ball valves and semispherical valves which, due to their large flow areas, are often used in applications involving viscous or abrasive fluids. Many valves have removable seats that are either screwed, pressed, or clamped in place, and in some applications the valves are fitted with elastomeric sealing elements. The suction and discharge valves are generally installed within one or more chambers that are either integral with or bolted to the side of the liquid cylinder.

Many piston pumps are arranged so that fluid will be alternately drawn into and discharged from both ends of each liquid cylinder. With this configuration while one face of the piston is in its suction stroke and is drawing fluid into its end of the liquid cylinder, the piston's other face is simultaneously discharging fluid from the opposite end of the cylinder. At the end of the stroke the sequence is reversed; therefore, within each cylinder in these "double-acting" pumps there are two suction strokes and two discharge strokes during every complete reciprocating cycle of the piston. To maintain the required seal between the piston and the bore of its cylinder, soft fibrous or hard composition packing rings, metal bull and snap rings, or molded elastomeric cups are fitted around the piston's outside diameter. So that mating surfaces inside the cylinder can be periodically renewed, many liquid cylinders are fitted with replaceable liners. A packed stuffing box is generally provided to reduce fluid leakage at the point where the piston rod that connects the liquid piston to the pump's drive end protrudes through the head of the liquid cylinder.

In a plunger pump, the stationary seal around each plunger is typically formed by packing rings installed within a stuffing box located at the base of the liquid cylinder. To enable the packing rings to be self-adjusting, in some units they are spring loaded. Additional secondary packing rings are also sometimes used. So that a lubricant can be injected into the packing or to permit high-pressure fluid to be bled from the stuffing box, a lantern ring is often sandwiched between two of the inner packing rings. A conventional plunger is a single-acting displacement element. However, to create a double-acting effect, the plungers in some pumps are mounted in pairs that operate within opposed cylinders. The opposed plungers are driven in unison; therefore, while one is in its suction stroke, the opposite plunger is discharging fluid.

b. Drive end. The drive ends used with reciprocating pumps can be divided into two basic groups: direct-acting and power. Although a direct-acting pump, which is shown in Fig. 34, can be driven by steam, compressed air, or other compressed gases, pumps of this type are commonly referred to as "steam pumps." A direct-acting pump's drive end, which is also frequently called the "steam end," generally contains one (simplex) or two (duplex) cylinders that are each fitted with a double-acting piston. At the beginning of a stroke, steam (or another gas) is admitted into one end of a drive cylinder, which forces the piston in that cylinder to move. As a result of this motion, the steam on the other side of the piston is expelled from the pump. After the stroke has been completed, steam is admitted into the opposite end of the same cylinder and the piston's travel path is reversed. As this cycle is continuously repeated, a reciprocating motion is transmitted directly to the displacement element in the adjacent liquid cylinder, which enables fluid to be pumped. The liquid end of a direct-acting pump typically has either double-acting pistons or opposed plungers.

The admission of steam into a direct-acting pump is regulated by valves that are installed within the steam chest located on the side of the drive cylinder. In a duplex pump the valve for each drive cylinder is actuated mechanically by the movement of the opposite cylinder's piston rod. As the valve is moved it alternately covers and uncovers the inlet ports through which steam enters each end of the drive cylinder. This valve also alternately connects each of the drive cylinder's two exhaust ports, which are located inboard of the inlet ports, to the steam chest's outlet port under the center of the valve. In some pumps a flat-faced slide type drive-end valve is used; however, when the valve is large, or when the pump is driven by high-pressure or high-temperature steam, a balanced-piston type drive-end valve may be used. The drive-end valves and their linkages are generally adjusted so that one drive piston leads the second by approximately $\frac{1}{2}$ to $\frac{2}{3}$ of a stroke. With this arrangement steam is always being admitted into one of the drive cylinders, and the failure of the pump to operate due to the simultaneous covering of the inlet ports in both steam chests is prevented. Play, referred to as lost motion, is typically provided in the drive-end valve linkages to allow each valve to remain stationary during a portion of the stroke of the piston that

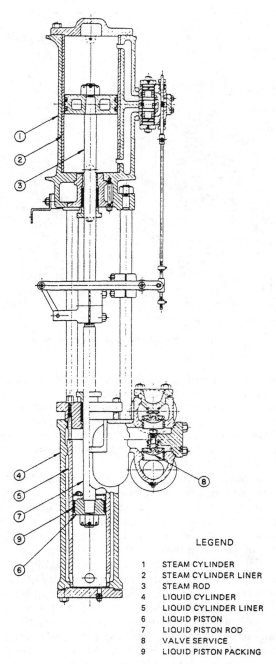

LEGEND

1	STEAM CYLINDER
2	STEAM CYLINDER LINER
3	STEAM ROD
4	LIQUID CYLINDER
5	LIQUID CYLINDER LINER
6	LIQUID PISTON
7	LIQUID PISTON ROD
8	VALVE SERVICE
9	LIQUID PISTON PACKING

Fig. 34 Vertical reciprocating direct-acting simplex pump

actuates the valve. Lost motion permits the admission of steam into each drive cylinder to continue throughout the entire stroke of the piston in the cylinder and gives the valves in the pump's liquid end time to seat quietly between strokes. When it is adjustable, an increase in lost motion increases the duration of the pause at the end of each stroke and reduces the overlap between the suction and discharge strokes in the liquid cylinders. The drive cylinder's exhaust port is covered by the piston before the end of the stroke. The steam that remains trapped in the cylinder acts as a cushion between the piston and the

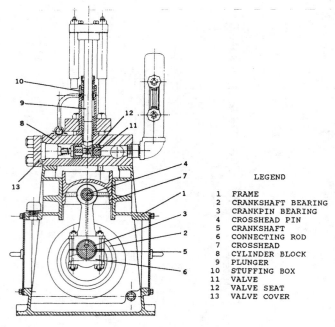

Fig. 35   Vertical reciprocating power pump

LEGEND

1   FRAME
2   CRANKSHAFT BEARING
3   CRANKPIN BEARING
4   CROSSHEAD PIN
5   CRANKSHAFT
6   CONNECTING ROD
7   CROSSHEAD
8   CYLINDER BLOCK
9   PLUNGER
10 STUFFING BOX
11 VALVE
12 VALVE SEAT
13 VALVE COVER

cylinder head and smoothly decelerates the pump's moving parts as each stroke is completed. Valves are sometimes provided to regulate the leakage rate of the cushioning steam from the drive cylinders. These valves are usually adjusted to allow each drive piston to complete the longest stroke that does not result in contact with the cylinder heads.

The drive cylinder in a simplex pump is fitted with a steam- (or gas) actuated piston-type valve. The admission of steam to the ends of this valve is controlled by a separate flat-faced slide or balanced-piston type pilot valve that is linked mechanically to the drive-end piston rod. Just before each stroke of the drive piston is completed, the pilot valve is repositioned and steam is directed to one end of the steam-actuated valve. The force of the steam moves the steam-actuated valve's piston completely to the opposite side of the steam chest, changing the end of the drive cylinder into which steam is admitted. Because the steam-actuated valve's piston cannot remain in the middle of the steam chest, referred to as the dead-center position, steam is always being admitted into one end of the drive cylinder. Consequently, the failure of the pump to operate due to the simultaneous covering of both of the drive cylinder's steam inlet ports is prevented. The linkage for a simplex pump's pilot valve has lost motion, which, when it is adjustable, is usually increased until the drive piston has the longest stroke that does not result in contact with the cylinder heads or cause the pump to stall.

In a power pump, which is typically illustrated by Fig. 35, a crankshaft and connecting rods convert the rotary motion of the driver's shaft to reciprocating motion that is transmitted to the liquid end's plungers or pistons. A crosshead that absorbs side loads is frequently installed between each connecting rod and the corresponding

plunger or piston rod. Each end of the crankshaft is usually fitted with either a tapered roller bearing or a sleeve-type journal bearing. In larger units additional bearings may also be installed along the length of the crankshaft. Replaceable sleeve bearings are generally fitted into both ends of each connecting rod. Geared speed reducers, fluid couplings, or multi V-belt drives are often used between the driver and the pump. In addition, some larger power pumps are furnished with their own integral reduction gears. The power frame used with this arrangement is fitted with a second shaft that is coupled to the driver. A pinion installed on the second shaft transmits torque to a reduction gear that is mounted on the crankshaft. The gears, together with the pump's bearings and crossheads, are generally lubricated by oil that is stored within the power end's sealed crankcase. When necessary, an integral rotary pump is provided to supply pressurized lubricating oil to the power end's various components. In larger pumps the power end may also be fitted with an integral oil cooler. Common power pump configurations include liquid ends with two (duplex) double-acting pistons, and liquid ends with three (triplex), five (quintuplex), seven (septuplex), or nine (nonuplex) single-acting plungers.

c.   Reciprocating-pump performance characteristics.   The volume displaced per cycle in the liquid end of a reciprocating pump is constant, and is a function of the outside diameter and stroke length of the liquid piston or plunger. A direct-acting pump is typically identified by three numbers: the first is the drive piston's outside diameter, the second is the outside diameter of the liquid piston or plunger, and the third is the length of the stroke. A power pump's geometry can be identified by two numbers: the first is the diameter of the liquid piston or plunger and the second is the stroke. These dimensions are generally expressed in inches. Because of slip, which includes leakage and backflow through the liquid end's suction and discharge valves, as well as leakage across the internal seals in a piston pump, the actual average capacity delivered by a reciprocating pump is less than the theoretical capacity that is based on the volume displaced by the pistons or plungers. Leakage at the liquid-end's external stuffing boxes and the compressibility of the fluid being pumped also result in reductions of the delivered capacity. However, with properly working valves and seals, the actual average capacity delivered by many reciprocating pumps is within 95 to 97% of the theoretical value. The capacity delivered by a reciprocating pump can be increased by raising the pump's operating speed, by using a pump with a greater number of liquid cylinders or a larger liquid-piston or plunger diameter, and by increasing the stroke length of the liquid piston or plunger. Some power pumps have an internal linkage that enables the stroke length of their plungers to be changed. The capacity delivered by a power pump is also sometimes controlled by using synchronized suction valve unloaders, which hold the pump's suction valves open when the capacity delivered is to be reduced. The pressure developed by a reciprocating pump is automatically adjusted based on the back

Table 1   Reciprocating pump performance characteristics

Configuration[a]	No. of Pumping Chambers	Max. Flow[b]	Min Flow[b]	C[c]
POWER PUMPS				
Duplex DA	4	124	78	0.115
Triplex SA	3	107	83	0.066
Quintuplex SA	5	102	95	0.040
Septuplex SA	7	101	97	0.028
Nonuplex SA	9	101	98	0.022
DIRECT-ACTING STEAM PUMPS				
Simplex DA	2	120[d]	0	0.200
Duplex DA	4	102[d]	92[d]	0.060[e]

[a] SA = single-acting; DA = double-acting.
[b] As a percent of the average flow rate.
[c] C = acceleration head constant [see equation (41)].
[d] These values are only approximations. Actual values will vary based on valve adjustment.
[e] With properly adjusted valves. If valves are not properly adjusted, C will approach 0.200.

pressure or resistance in the system. To prevent overpressurization, a relief valve should be installed in the pump's discharge piping upstream of any shutoff valves.

Fluid is not discharged from a reciprocating pump's liquid cylinders at a steady rate; therefore, there are pulsations in the flow rate and pressure of the fluid delivered. Typical variations in flow rate for various reciprocating pump configurations are included in Table 1. In a direct-acting pump the velocity of the liquid piston or plunger, together with the velocity of the fluid that the piston or plunger displaces, increases quickly, remains constant throughout most of the stroke, and then drops off quickly to zero. In a power pump, however, the velocity of the piston or plunger varies continuously throughout the stroke approximately with the sine of the crankthrow angle. In addition, although due to lost motion there is a pause at the end of each stroke in a direct-acting pump, the plungers or pistons in a power pump have no such pause. Flow and pressure variations in a reciprocating pump's suction and discharge lines are sometimes reduced by installing pulsation dampeners in the piping adjacent to the liquid-end's inlet and outlet ports, or directly on the pump's liquid valve chambers.

The brake horsepower required to drive a power pump's shaft can be calculated using equation (35), except that for reciprocating pumps Q is the capacity measured at the pump's suction conditions. If the values used for both Q and p_{td} in equation (35) are the average capacity delivered by the pump and the average total differential pressure developed, respectively, the result calculated will be the pump's average power requirement. However, due to the fluctuations in flow and pressure, the instantaneous power required by a power pump varies throughout the discharge stroke and reaches a peak value during each complete revolution of the crankshaft that exceeds the average power requirement.

The power requirement for a direct-acting pump can be expressed in terms of the steam or gas pressure and consumption rate. When steam is used as the driving medium it is usually saturated so that moisture will be present to help lubricate the drive-end's pistons and valves. If a direct-acting pump is driven with superheated steam or a compressed gas, a small amount of oil is often mixed with the steam or gas to lubricate the drive-end's internal components. This oil is sometimes injected directly into the steam chest by a mechanical lubricator that is mounted on the pump and actuated by the movement of the piston rods. The theoretical differential or net steam or gas pressure required for the pump to develop a given differential pressure in its liquid end is a function of the ratio of the outside diameter of the liquid piston or plunger divided by the outside diameter of the pump's drive piston. The actual net pressure required in the pump's drive end, however, is greater than the theoretical value due to mechanical losses in the pump, such as friction. If both the reduction of piston-face area caused by the piston rod and losses in the pressure of the driving medium as it enters and leaves the drive cylinder are neglected, the net steam pressure required to drive a liquid end fitted with either double-acting pistons or opposed plungers is equal to

$$p_{Dr} = \frac{p_{td}}{\eta_m} \left[\frac{d_L}{d_{Dr}} \right]^2 \qquad (36)$$

where

p_{Dr} = net pressure to drive end, psi.
d_{Dr} = drive piston outside diameter, in.
d_L = liquid piston or plunger outside diameter, in.
η_m = pump mechanical efficiency, %/100

Steam consumption is generally expressed in terms of mass flow rate, which can be estimated from

$$\dot{M}_{St} = \frac{60\pi}{(144)\,4} d_{Dr}^2\, S_p\, n\, C_1\, C_2\, C_e\, \rho_{St} \qquad (37)$$

where

$\dot{M}_{St}$ = steam consumption, lbm/hr
S_p = average piston speed, fpm
n = number of steam cylinders
C_1 = cylinder clearance ratio
C_2 = steam condensation and leakage factor
C_e = exhaust correction factor
ρ_{St} = steam density at inlet pressure, pcf

C_1, which is usually lower in pumps with longer stroke lengths, typically varies from 1.1 to 1.2. Values of C_2 increase with steam pressure and steam piston diameter, and decrease at higher piston speeds; C_2 generally falls in the range from 1 to 5. C_e can be estimated with the following expression [2]:

$$C_e = \left[\frac{p_{Dr} + p_e}{p_{Dr}} \right]^{1/2} \qquad (38)$$

where p_e is the exhaust pressure, psig.

Neglecting the volume displaced by the liquid piston rods, when either double-acting liquid pistons or opposed liquid plungers are used, the average piston speed, S_p, can be related to the pump's average capacity:

$$S_p = \frac{4\,(231)\,\overline{Q}}{12\,\pi\,d_L^2\,n\,\eta_v} \qquad (39)$$

where

$\overline{Q}$ = average capacity, gpm
η_v = volumetric efficiency, %/100

If air or compressed gas is used to drive the pump, the gas consumption is generally expressed in terms of the volumetric flow rate, which can be estimated by

$$q_g = \frac{1.2}{144} \frac{\pi}{(4)} d_{Dr}^2 S_p n C_1 \left[\frac{p_{Dr} + p_e + 14.7}{14.7} \right] \left[\frac{520}{T_1 + 460} \right] \quad (40)$$

where

q_g = gas consumption, standard (corrected to 14.7 psia and 60 F), cfm
T_1 = gas inlet temperature, F

The NPSH available to a reciprocating pump can be expressed in terms of either pressure or head. However, in addition to the terms typically included in the calculation of NPSHA, the effect of pressure pulsations and velocity changes that occur within the suction piping must also be considered. To account for these fluctuations, the acceleration head, which represents the energy required to accelerate the fluid in the pump's suction line, can be subtracted from the NPSHA values calculated using equation (32). Acceleration head can be estimated using [1]

$$H_{ac} = \frac{L_s V_s N C}{K g} \quad (41)$$

where

H_{ac} = acceleration head, ft
L_s = suction pipe length, ft
V_s = average fluid velocity in suction pipe, fps
K = fluid compressibility factor (e.g., 1.4 for hot water, 2.5 for hot oil)
N = pump operating speed, cpm
C = acceleration head constant (see Table 1)

Equation (41) gives the maximum acceleration head, which occurs at the beginning of the suction stroke, and is less valid with long lengths of suction pipe. H_{ac} can be reduced by installing a properly sized pulsation dampener close to the pump's inlet, by increasing the diameter of the suction line, and by reducing the pump's operating speed.

To determine the NPSH required by a reciprocating pump, the pump is operated with a constant discharge pressure and speed, and the NPSHA is gradually reduced until a clearly audible cavitation noise or knocking sound is heard, or until the capacity delivered is reduced by 3%. The value of the NPSHA at which either of these conditions first occurs is considered to be the pump's NPSH requirement at the test speed. NPSHR is sometimes reduced by using lighter springs in the liquid-end's suction valves. However, this can also increase the backflow through the valves and reduce the pump's volumetric efficiency. The cavitation that results from operation with insufficient NPSHA can lead to pitting of the piston or plunger, and to damage of the drive-end components due to increased vibration.

A reciprocating pump is a positive-displacement machine; however, if a pump with a suction lift is started with air or vapor in its liquid cylinders and with liquid resting on top of its discharge valves, the pressure of the gas being compressed within the liquid cylinders may not be sufficient to open the valves. In addition, because the gas remaining in each cylinder reexpands during the suction stroke, the liquid cylinder pressure may not be reduced sufficiently to permit the suction valves to open. Consequently, the pump may not be able to prime itself. The ability of a reciprocating pump to prime itself is reduced as the volume remaining within the liquid cylinder when the piston or plunger is fully extended, referred to as the clearance volume, is increased. For this reason, piston pumps typically have self-priming characteristics that are superior to those of plunger pumps, which are sometimes designed to operate only with flooded suctions.

1.9 Rotating-Piston Pumps. The pumping action in rotating-piston pumps, which are sometimes classified as rotary pumps, is created by the reciprocating motion of multiple single-acting pistons within close-clearance cylinders. However, unlike the power pumps described previously, in a rotating-piston pump the pistons together with their cylinders also rotate about the shaft's axis. Rotating-piston pumps, which are typically motor driven, are furnished in two basic configurations: pumps in which the pistons are mounted in an axial direction, and pumps that are fitted with radially oriented pistons.

A radial-piston pump is usually a variable-stroke unit in which multiple single-acting pistons, often seven, are oriented radially around the circumference of the shaft. As shown in Fig. 36, the outboard end of each piston is pinned to a roller or slipper that is in contact with the inside wall of a nonrotating floating ring. The eccentricity of the floating ring with respect to both the shaft and the liquid cylinder block is controlled by the position of a guide yoke that can be moved radially within the pump's housing. When the yoke is moved off center, the motion is transmitted through the floating ring to the pistons, and as each piston makes one complete revolution it slides back and forth within the rotating liquid cylinder block. As a result of this reciprocating motion, fluid is alternately drawn into and discharged from each cylinder through suction and discharge ports that are located in the stationary housing. If the eccentricity of the floating ring is increased, the capacity delivered at a constant operating speed will also increase. In addition, if the direction of the eccentricity is reversed, the direction of flow through the pump will be reversed. The position of the yoke and the floating ring can be adjusted manually, or with an automatic controller.

In an axial-piston pump a number of single-acting pistons, usually seven or nine, are equally spaced around the pump's shaft. However, as shown in Fig. 37, the axes of the pistons and that of the shaft are parallel. The non-pumping end of each piston is fitted with a short rod that is connected, through a ball-and-socket type joint, to the "socket ring." As the socket ring rotates with the pump's drive shaft, it turns the pistons and the multicylinder block or barrel that fits over them. The socket ring is often mounted within a nonrotating tilting box that can be used to adjust the ring's angle with respect to the shaft. When the socket ring is tilted, the pistons on one

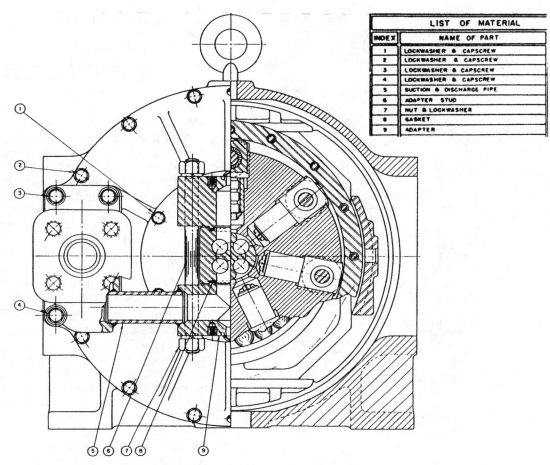

LIST OF MATERIAL	
INDEX	NAME OF PART
1	LOCKWASHER & CAPSCREW
2	LOCKWASHER & CAPSCREW
3	LOCKWASHER & CAPSCREW
4	LOCKWASHER & CAPSCREW
5	SUCTION & DISCHARGE PIPE
6	ADAPTER STUD
7	NUT & LOCKWASHER
8	GASKET
9	ADAPTER

Fig. 36   Radial-rotating-piston pump

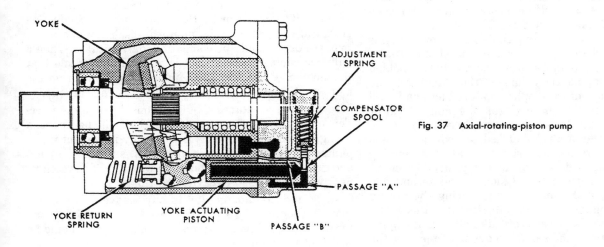

Fig. 37   Axial-rotating-piston pump

side of the ring are moved closer to the cylinder block than those on the opposite side. Because of this orientation, as the pistons rotate, they slide back and forth within their rotating cylinders. The reciprocating motion alternately draws fluid into and discharges it from each cylinder through suction and discharge ports in a stationary valve plate. Increasing the socket-ring's tilt angle increases

each piston's stroke and the capacity delivered at a constant speed. In some units the tilt angle can be reversed to reverse the direction of flow through the pump.

1.10 Diaphragm Pumps. Diaphragm pumps are positive-displacement units in which pumping action is created by the reciprocating motion of a flexible membrane that is secured around its periphery between the walls of a

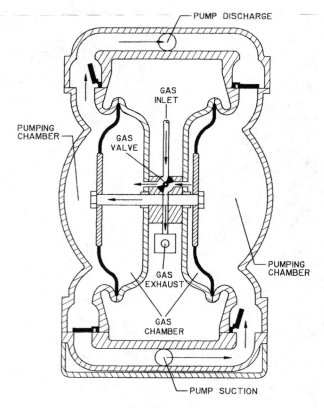

PUMP DISCHARGE

GAS INLET

PUMPING CHAMBER

GAS VALVE

GAS EXHAUST

PUMPING CHAMBER

GAS CHAMBER

PUMP SUCTION

Fig. 38   Diaphragm pump

stationary chamber. Although this reciprocating motion can be induced by mechanical means through a crank or cam, or hydraulically with a pressurized liquid, many diaphragm pumps are driven pneumatically with compressed air or gas. As shown in Fig. 38, a typical pneumatically operated diaphragm pump contains two diaphragms within enclosed chambers that are mounted side-by-side in a duplex arrangement. As air or gas is admitted behind one of the diaphragms, the volume of the pumping chamber that is outboard of the diaphragm is reduced and fluid contained within it is discharged from the pump. A bar or shaft connecting the two diaphragms forces the second diaphragm to move in a parallel fashion, which expands the volume of, and draws fluid into, the outboard pumping chamber on the opposite side of the pump. In addition, air in the chamber on the inboard side of the second diaphragm is exhausted from the unit. During the second half of the pumping cycle, air is admitted into this chamber, and both diaphragms are forced to move in the opposite direction. Due to the simultaneous movement of the two diaphragms, as fluid is discharged from the second pumping chamber, it is also drawn into the first pumping chamber, and the air initially admitted behind the first diaphragm is exhausted from the pump. The alternating suction and discharge strokes within each chamber result in a nearly steady flow of fluid from the pump. So that each chamber's inlet and outlet ports will be sealed during the discharge and suction strokes, respectively, these

ports are fitted with nonreturn valves. Flap, ball, or poppet type check valves are frequently used. In addition, the alternating admission and exhaust of air from the inboard side of each chamber are controlled with an automatic four-way distribution valve. Due to the continuous seal formed by each diaphragm between the pumped fluid and the driving air or gas, these pumps can be used in applications that require zero leakage.

With a constant air or gas pressure, as the capacity delivered by a pneumatically operated duplex diaphragm pump increases, the total head developed is reduced and the volumetric consumption of air or gas increases. In addition, the total head developed at any capacity and the slope of the pump's head-capacity curve increase with the pressure of the air or gas driving the pump. A pneumatically operated diaphragm pump's maximum discharge pressure is equal to the pressure of the air or gas used to drive it; therefore, because of limitations in the pressure and volume of the gas that is often available for this purpose, these pumps are typically used in applications requiring only low discharge pressures and capacities.

1.11   Marine Pump Applications. The various pump services used onboard any vessel can be divided into two basic groups: those that are related to the main propulsion equipment and those that are not. The materials used in the construction of a marine pump depend on the application for which the unit is furnished; however, there are some general requirements that should typically be followed:

• The use of ductile materials is recommended for resistance to shock and vibration.

• Wetted pump components (i.e., the components that will be exposed to the fluid being pumped) should be resistant to corrosion from the liquids that they contact. In addition, if seawater or another liquid that can conduct an electrical current is to be pumped, dissimilar materials that are used should be compatible galvanically. When a galvanic couple does exist, it is generally desirable that the smaller of the two parts involved be made from the more noble material so that it will act as the cathode.

• Erosion resistance should be considered in the selection of materials for components that will be exposed to high-velocity liquid. This is especially important if the pumped liquids will contain abrasives, such as the sand and grit that are often found in seawater, or if the component will be exposed to cavitating fluids.

• To reduce damage due to any inadvertent contact that may occur in service, material combinations used for mating rotating and stationary components should be resistant to galling.

• Materials selected for major components, such as casings, should be weld repairable to permit surfaces that wear with time due to corrosion or erosion to be renewed.

Materials used for wetted pump components often include the following:

Seawater pumps. Materials commonly used in the construction of seawater pumps include various grades of bronze, copper-nickel alloys, high-alloy austenitic chromium-nickel stainless steels, nickel-copper and nickel-copper-aluminum alloys, titanium, and nickel-chromium-molybdenum alloys. In addition, duplex (austenitic/ferritic) stainless steels and fiber reinforced composite materials are used in some applications.

Freshwater pumps. Although the same materials used in seawater pumps can be used in the construction of units that will handle fresh water, to reduce cost, cast iron, ductile iron, and carbon and alloy steels are also used in freshwater pumps. In addition various grades of stainless steels, such as 12% chrome martensitic, 17-4 precipitation-hardened, and 300 series austenitic stainless steels are sometimes used.

Oil pumps. With the exception of liquid bulk cargoes, the majority of the oils that are handled by shipboard pumps are used for fuel, for lubrication, or to power hydraulically operated machinery. Materials used in the construction of these pumps include: cast and ductile irons; carbon, alloy, and stainless steels; and various grades of bronze.

A description of the pumps used in typical shipboard applications follows below; however, this information is general in nature and exceptions can generally be found as warranted by the requirements for specific vessels, or by the preferences of owners and designers.

a. Steam turbine propulsion-related pump applications.

Main condensate. A main condensate pump receives water directly from the hotwell in the condenser and, on most vessels, transfers it to a direct-contact type deaerating feedwater heater (DFT). The total head that must be developed by the typical condensate pump is, therefore, based on the difference in elevation between the water level in the hotwell and the level in the DFT, the difference in the pressures within these two chambers (the hotwell is generally under a high vacuum, while typical DFT pressures can range from approximately 10 to 70 psig), and the losses due to friction and turbulence in the pump's suction and discharge piping (including losses in any heat exchangers that may be installed between the condenser and the DFT). Pumps used in this application are frequently rated to develop total heads in the range of 150 to 400 ft.

Two centrifugal condensate pumps are often provided for each condenser. Depending on the capacity delivered by each pump, which is typically less than 1200 gpm, and plant load, one pump may operate alone with the second used as a standby unit, or both pumps may be operated together in parallel. Because the condensate in the hotwell is at its saturation point, the NPSH available to the pump is essentially equal to the height of the water level in the hotwell above the standard datum, less losses in the suction piping. To maximize NPSHA, the condensate pump is usually installed as far below the condenser as practicable and has a suction line that is direct and free of unnecessary bends; however, condensate pumps must frequently operate with only 1.5 to 3 ft of NPSHA. Consequently, operating speeds for main condensate pumps are

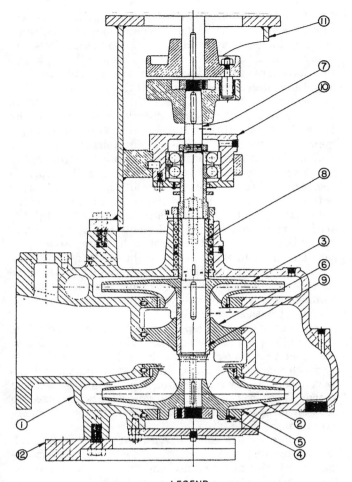

LEGEND

1	CASING
2	IMPELLER 1ST STAGE
3	IMPELLER 2ND STAGE
4	IMPELLER WEARING RING
5	CASING WEARING RING
6	INTERNAL BEARING
7	SHAFT
8	SHAFT SLEEVE
9	JOURNAL SLEEVE
10	BEARING HOUSING
11	MOTOR BRACKET
12	PUMP FOOT

Fig. 39 Vertical, two-stage centrifugal condensate pump

usually limited to 1800 rpm. In addition, first-stage impellers with suction specific speed values as high as 18,000 have been used.

Two- and three-stage vertically mounted pumps that have an axially split volute-type casing are frequently used in this application (see Fig. 39). These pumps are driven by either steam turbines with reduction gears or electric motors. A grease-lubricated ball bearing is usually installed at the upper end of the pump's shaft to absorb both axial and radial loads. In many pumps radial loads are also absorbed by internal water-lubricated sleeve bearings. To aid in the removal of air and vapor from the casing, a vent line is typically installed from the pump's suction nozzle back to the condenser. Although some units have a double-suction first-stage impeller, many condensate pumps are fitted only with single-suction impellers. In addition, when a single-suction first-stage impeller is used, it is generally mounted with its eye upward so that it will be self-venting. To maximize submergence, the first-stage impeller is often overhung on the lower end of the pump's shaft. With this arrangement the single shaft seal, which can consist of a packed stuffing box or a mechanical seal, is located at the top of the casing. The condensate being pumped, therefore, is discharged through one or more of the pump's stages before it reaches the base of the seal, which reduces the potential for air to be drawn into the casing through the shaft penetration. Pressurized water is also frequently recirculated to the shaft seal from a discharge area of the casing or from an external part of the condensate system.

In addition to the configurations described above, vertical turbine pumps installed in suction cans and vertically mounted multistage barrel-type diffuser pumps have also been used in this application.

If the condensate pump is driven by a steam turbine, the pump's speed can be adjusted with plant load so that the capacity removed from the hotwell matches the rate at which condensate enters the hotwell. However, motor-driven condensate pumps typically operate at a fixed speed. Three basic types of arrangements are used to control the capacity delivered by constant-speed condensate pumps [25]:

1. Submergence control. With submergence control, the condensate pump will deliver the capacity corresponding to the point at which its head-capacity curve crosses the system head curve only when there is sufficient NPSH available to the pump. However, if the flow of steam into the condenser is reduced and the hotwell level drops to the point where NPSH available is less than the pump's NPSH requirement, the capacity delivered and total head developed will both be reduced by cavitation. As the reduction in the capacity delivered by the pump continues, a point of equilibrium will be reached where the NPSH available to the pump equals the pump's new NPSH requirement, and the capacity and hotwell level will stabilize. Provided that the NPSHA is less than the NPSHR at the point where the system head curve crosses the pump's head-capacity curve, the pump's capacity is, therefore, automatically regulated by cavitation to match the rate at which condensate enters the hotwell.

2. Throttle control. With throttle control, the throttling action of a valve in the pump's discharge line adjusts the system head curve so that it crosses the pump's head-capacity curve at the capacity required to keep the hotwell level at a desired preset value. Under steady-state conditions this capacity also equals the rate at which condensate enters the hotwell.

3. Recirculation control. With recirculation control, an automatic control valve installed in a recirculation line connected from the pump's discharge piping back to the condenser opens as the hotwell level drops and closes if the hotwell level increases. When this arrangement is used, the hotwell level is maintained at a value that both suppresses cavitation and permits the condensate pump to operate near its rated capacity.

On some vessels a combination of the above methods is used for condensate-pump capacity control.

Auxiliary condensate. On some vessels the steam used to drive turbogenerators is exhausted into an auxiliary condenser. As with the main condenser, the water that collects in the hotwell of the auxiliary condenser is generally pumped to the DFT. Two-stage vertically mounted centrifugal pumps that are similar in design to the main condensate pumps are often used in this application. These pumps, which frequently deliver capacities in the range of 35 to 250 gpm at total heads from 150 to 350 ft, are sometimes designed to operate at speeds as high as 3600 rpm.

Condenser exhausting. Liquid-ring type vacuum pumps are used on some vessels to remove air and other noncondensable gases from the main or auxiliary condensers. Two pumps that are each capable of individually maintaining the required vacuum are frequently provided for each condenser. As a result of the high condenser vacuum typically required for proper steam plant operation, the vacuum pumps used in this application are often two-stage units. Condenser exhausting pumps are usually driven at speeds up to 1800 rpm by electric motors.

A vacuum pump's capacity is usually expressed in terms of standard cfm of dry air measured at an absolute pressure of 14.7 psia and a temperature of 70 F. The air and other gases removed from a condenser, however, are saturated with water vapor. The vapor content in the mixture entering a condenser exhausting pump is often estimated based on the amount of water vapor that will saturate air at a temperature equal to 7.5 deg F less than the saturation temperature corresponding to the pressure within the condenser. With a condenser absolute pressure of one in. Hg, 7.5 deg F of subcooling results in a gas-vapor temperature of 71.5 F and a vapor content of approximately 2.2 lbm of water vapor for every lbm of air. A typical liquid-ring type condenser exhausting pump removes air and vapor from the condenser at essentially a constant volumetric rate. Because of the increase in the density of air with pressure, the mass flow rate through the pump, therefore, increases with the condenser's absolute pressure.

The amount that a liquid-ring vacuum pump's suction pressure can be reduced is limited by the vapor pressure of the liquid compressant, which increases with temperature. Because of this relationship, sealing water separated from the gas that has been discharged by a condenser exhausting pump is generally cooled in a heat exchanger before being returned to the unit. In some installations only a portion of the cooled water is returned to the pump's casing, with the remainder being injected directly into the inlet line. This cools the gas and water-vapor mixture being removed from the condenser, and reduces the vapor content of this mixture before it enters the pump.

Freshwater-drain-collecting tank transfer. Uncontaminated fresh water that drains from various low-pressure sources to the freshwater- or atmospheric-drain-collecting tank (FWDCT) is often transferred to the DFT by a pump. The temperature of the water within this tank is typically at approximately 212 F; therefore, the NPSH available to the pump is equal only to the elevation of the water level in the tank above the standard datum, less losses within the suction line. Due to the limited NPSHA, electric-motor-driven centrifugal pumps that are similar in configuration and design to the pumps used for auxiliary condensate service are frequently used to transfer water from the drain-collecting tank. In addition, on some vessels volute-type sump pumps that are submerged directly within the drain-collecting tank are used in this application. Typical conditions of service for FWDCT transfer pumps include capacity ratings in the range of 50 to 150 gpm at total heads up to 275 ft. To maintain the desired water level within the drain-collecting tank, the FWDCT transfer pump's driver is generally cycled on and off as needed by a float switch that is mounted inside the tank.

Main feed. The function of a main feed pump, which often receives water either directly or through a booster pump from the DFT, is to return feedwater back to the steam drum in the boiler or the steam generator. The discharge pressure developed by the feed pump must, therefore, be sufficient to overcome losses in its discharge line, raise the feedwater to the required elevation, and overcome the pressure within the boiler or steam generator, which is typically in the range of 450 to 1500 psig. The total feedwater requirement is sometimes handled by only one pump, or the load may be shared by two or more partial-capacity feed pumps that operate in parallel. With either arrangement, additional pumps are also frequently provided for standby duty. Typical feed pump capacity ratings range from 300 to 2000 gpm.

Feed pumps are often driven at speeds of 4500 to 9000 rpm by auxiliary steam turbines. A steam turbine drive offers the advantage of variable-speed capability. Although it is not as common, alternating-current induction type electric motors are also sometimes used to drive feed pumps; however, unless a step-up gear is used, the maximum operating speed for this type of a driver is typically limited to approximately 3600 rpm. Motor-driven feed pumps, therefore, must generally have more stages or larger impellers than comparably rated turbine-driven units. The larger rotating assembly used with the slower-speed pump also results in the need for a casing that is both longer and larger in diameter.

Typical marine main feed pump configurations include the following:

• Single- and two-stage radially split casing pumps that have single-suction impellers and are close coupled to steam turbines.

• Two-, four-, and six-stage axially split casing volute pumps, that, as shown in Fig. 11, have single-suction impellers and are flexibly coupled, generally to steam turbines.

• Two- and three-stage axially split casing volute pumps that have a double-suction first-stage impeller and single-suction impellers in the remaining stages. Although these pumps can be flexibly coupled to their drivers, which are typically steam turbines, in some units the rotating parts for both the pump and turbine are mounted on a common shaft.

• Multistage diffuser-type radially split, as shown in Fig. 7, or axially split casing pumps that are flexibly coupled to steam turbines or electric motors.

Although turbine-driven feed pumps are generally mounted horizontally, motor-driven pumps are furnished in both horizontal and vertical configurations. In addition, when the feed pump is motor-driven, external support for its shaft is often provided by grease-lubricated antifriction bearings. Smaller turbine-driven feed pumps may also have ball or roller bearings; larger-sized turbine-driven units, however, frequently have a tilting-pad thrust bearing and sleeve-type line bearings. Because of the high operating speeds of turbine-driven feed pumps, their external bearings are generally lubricated with oil. This oil is frequently supplied by the same system used for turbine bearing lubrication. During normal operation the oil is removed from a reservoir tank, which is often built into the feed pump's baseplate, and is transferred to the bearings by a rotary pump that is geared to the turbine's shaft. A separate motor-driven rotary pump may also be provided to lubricate the pump and turbine bearings prior to start-up and during low-speed operation.

Many feed pumps rely on packed stuffing boxes for shaft sealing. To reduce the pressure of the feedwater being contained by the packing, the base of each stuffing box is frequently fitted with a multiple labyrinth-type breakdown bushing. A leak-off connection piped back to suction may be included to further reduce the pressure of the feedwater in the stuffing box. To reduce the temperature at the packing, the stuffing boxes also often include external jackets through which cooling water is circulated. Mechanical seals have limited usage in marine main feed pumps. In lieu of packing or mechanical seals, some feed pumps are fitted with condensate injection seals. Each of these "packless stuffing boxes" contains either a stationary serrated labyrinth-type fixed breakdown bushing or a series of spring-loaded floating rings that are stacked axially. With both arrangements, a close radial clearance is maintained between the stationary sealing elements and the rotating shaft or sleeve. Cool water diverted from the discharge of the condensate pump is introduced centrally into the seal. A small portion of this water may flow into the pump. The remainder, however, flows outward into a collection chamber that is piped back to the condenser.

With a constant-speed driver, the feed pump's capacity is controlled by the throttling action of the feedwater regulating valve. However, if multiple partial-capacity feed pumps are used, the amount of valve throttling can be reduced by starting and stopping pumps as needed. When a feed pump is driven by a steam turbine, its operating speed is frequently adjusted with either a constant-pressure or a constant-differential-pressure governor. A constant-pressure governor automatically adjusts the pump's speed to maintain a constant pressure at the

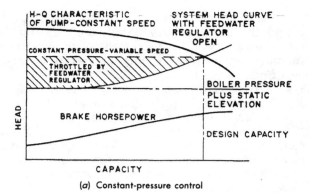

(a) Constant-pressure control

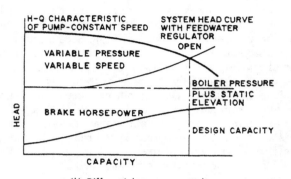

(b) Differential-pressure control

Fig. 40   Feed pump performance

inlet to the feedwater regulating valve. If there is a reduction in steam plant load and the valve begins to close, which shifts the system head curve to the left, the feed pump will initially respond by delivering a lower capacity of feedwater at a higher pressure. However, as shown in Fig. 40(a), the constant-pressure governor will reduce the pump's speed to maintain a discharge pressure equal to the set point, and less throttling of the feedwater regulating valve will be required. When the turbine has a constant-differential-pressure governor, the feed pump's speed is automatically adjusted to maintain a constant differential pressure across the feedwater regulating valve. With this type of a governor, changes in the capacity delivered by the feed pump result primarily from variations in operating speed. Therefore, as shown in Fig. 40(b), very little throttling of the feedwater regulating valve is necessary. On nuclear-powered vessels the pressure within the steam generator often increases at low loads. The resulting system head requirement can increase with reductions in capacity and reach its peak value at shutoff. The feed pumps used must, therefore, be sized to not only develop the required total head at the rated capacity, but to also develop sufficient head at low flow rates. This can be accomplished in cases where the system head curve is steeper than the head-capacity curve for the feed pump by using a variable-speed driver and increasing the pump's speed as the capacity delivered is reduced, or by sizing the pump to develop the required low-flow head and then throttling the feedwater control valve as necessary when operating at higher flow rates.

The feedwater stored within the DFT is at its saturation temperature; therefore, neglecting the velocity head within the DFT and assuming that a feed booster pump is not used, the NPSH available to the typical feed pump is equal only to the elevation of the water level in the DFT above the standard datum, less losses within the pump's suction line. Based on the available NPSH and on the capacity being delivered, the pump's maximum operating speed is often limited by the suction specific speed of its first-stage impeller. To enable the pump's maximum operating speed to be increased, S values exceeding 12,000 have been used in some feed pump designs. In addition to considering NPSHA during steady-state operation, the NPSH available to the feed pump during transient conditions should also be considered. For example, if the DFT is partially heated with steam extracted from the propulsion turbines, a reduction in plant load can result in reductions in the temperature and pressure within the DFT. A similar effect can result from a sudden increase in the flow rate of cool condensate entering the DFT. In both situations the reduction in the DFT's pressure is transmitted directly to the feed pump's inlet. The temperature of the feedwater already in the pump's suction line, however, is not reduced. Therefore, until the suction line is evacuated, there is a reduction in the NPSH available to the pump. The magnitude of this reduction in NPSHA, which can be estimated using equations included in reference 2, increases as the DFT's internal volume and pressure and the volume within the feed pump's suction piping increase. To prevent problems from occurring during operation with transient conditions, a safety margin must frequently be provided between the steady-state values of NPSHA and the feed pump's NPSH requirements.

To prevent operation at excessively low capacities, which can lead to a rapid increase in the temperature of the water within the pump, a bypass line is generally provided from the feed pump's discharge to the DFT. In some installations the bypass is fitted with either a single or a multiple pressure-reducing orifice that continuously recirculates feedwater back to the DFT. With this arrangement, however, the recirculation flow rate must be added to the net capacity required when the feed pump is sized. To eliminate the need to oversize the feed pump, on some vessels an automatic control valve that opens only at low flow rates is installed in the bypass line. To protect the feed pump from operating with too low a suction pressure, which can result in excessive cavitation, a low-suction-pressure trip is often provided to stop the pump's driver if the suction pressure drops below a preset value.

In-port feed. In addition to main feed pumps, smaller-capacity pumps are also installed on some vessels for use in port when the demand on the steam plant is low. The use of an "in-port" feed pump eliminates the need to operate the larger main feed pumps at very low partial capacities. Motor-driven reciprocating-type power pumps rated to deliver capacities less than 100 gpm are often used for in-port feed service. Motor- and steam-driven reciprocating pumps have also been installed on some vessels to serve as backup units in the event of a main feed pump failure.

Feed booster. On some vessels a booster pump is used to increase the pressure of the feedwater before it enters the main feed pump. Because the booster pump and the main feed pump operate in series, they must be sized to handle the same capacity. However, the total head developed by the booster pump is typically less than 200 ft. The booster pump can, therefore, be driven at a much lower speed and, for a given suction specific speed, will require less NPSH than the main feed pump. In addition, due to the increase in the NPSH available to the main feed pump, the use of a booster pump enables a feed pump with a higher operating speed or a lower suction specific speed to be used.

Typical configurations used for feed booster service include single- and two-stage vertically mounted centrifugal pumps. Operating speeds are generally limited to 1800 rpm. Many of these pumps are flexibly coupled to electric motors or steam turbines with reduction gears; however, some motor-driven single-stage units are furnished in a close-coupled configuration. Either a mechanical seal or a water-cooled packed stuffing box is typically provided for shaft sealing.

Main circulating. A main circulating pump takes suction from the vessel's sea chest and discharges seawater through the tubes in the main condenser. The capacity delivered by this pump is often in the range of 5000 to 30,000 gpm at total heads from 10 to 40 ft. High-capacity circulating pumps are frequently driven at speeds not exceeding 900 rpm by either steam turbines with reduction gears or electric motors. Although some main circulating pumps are mounted horizontally, it is more common for vertical units to be used. Typical main circulating pump configurations include the following:

• Single-stage axial-flow propeller pumps, such as the one shown in Fig. 41. This type of a pump is often furnished with an axially split casing that is fitted with a removable multi-vaned diffuser on the discharge side of the propeller. The pump's casing may also have a replaceable liner so that the close radial clearance between the casing's inner wall and the tips of the propeller's vanes can be periodically renewed. With the overhung shaft arrangement that is typically used, only one shaft seal is required at the drive end of the casing. Radial shaft loads are generally absorbed by an internal sleeve bearing located above the propeller. Many pumps, however, do not include thrust bearings, and must, therefore, be rigidly coupled to their drivers. In addition to the main suction flange, which is connected to the sea chest, some casings include an auxiliary side suction connection that enables the circulating pump to be used to remove water from the vessel's bilges.

• Single-stage mixed-flow pumps that have either a semiopen or, as shown in Fig. 42, a closed end-suction impeller. Both axially and radially split casings that include a volute or a multi-vaned diffuser are used with these pumps. Although the pump shaft may be supported in a fashion similar to that used for the propeller-type units, in some mixed-flow pumps the shaft is, instead, supported by external grease lubricated antifriction line

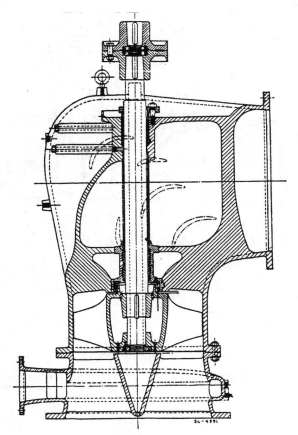

Fig. 41 Axial-flow circulating pump

and thrust bearings, which enables it to be flexibly coupled to the shaft of the driver. In addition, some smaller pumps are furnished in a close-coupled configuration.

• Radial-flow pumps with a single-stage double-suction impeller (see Fig. 13). These pumps are usually furnished with an axially split volute-type casing. The pump's shaft is typically supported at each end by an externally mounted grease lubricated antifriction bearing. It is also generally flexibly coupled to the shaft of the driver. Because of the outboard bearing, with this configuration two shaft seals are required.

If the vessel has a scoop injection system, the main circulating pump will generally be required to operate only at low vessel speeds. With this arrangement the circulating pump, therefore, operates intermittently. Because much of the operation occurs while the vessel is in shallower water, the liquid passing through the pump often contains silt, sand, and other abrasives. For this reason, in many circulating pumps packed stuffing boxes are used for shaft sealing. In addition, internal sleeve bearings that are lubricated by water being discharged from the pump are often furnished in abrasion-resistant grades of rubber or composite materials in lieu of standard bearing bronze. As an alternative to using the pumped liquid for lubrication, some internal bearings are lubricated with either grease or clean water supplied

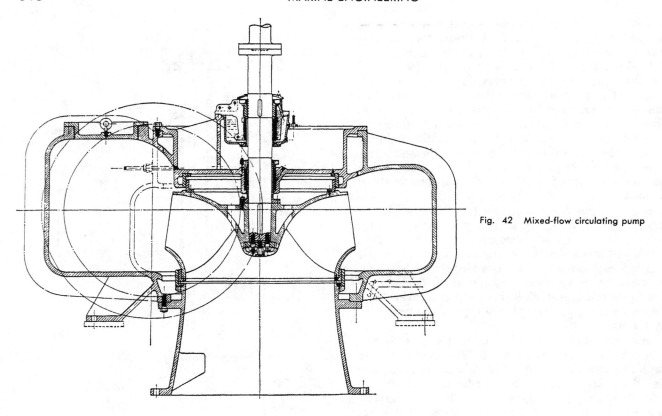

Fig. 42 Mixed-flow circulating pump

through an external connection in the circulating pump's casing.

Auxiliary circulating. Auxiliary circulating pumps supply cooling water to the smaller condensers that are frequently provided to receive steam exhausted from the vessel's turbogenerators. These pumps, which are often rated to deliver capacities in the range of 1500 to 15,000 gpm and develop total heads up to 35 ft, can be similar in design to the radial-flow main circulating pumps. In addition, in some lower-capacity applications, end-suction radial-flow pumps with radially split volute-type casings are used.

Fuel-oil service. On vessels with oil-fired boilers, two or more pumps receive fuel oil from either the high-suction or the low-suction ports in the fuel-oil service (settling) tanks and discharge it to the burner manifold in the boiler. Prior to reaching the burner manifold, the fuel oil discharged from the service pumps often passes through a heater, a strainer, and a pressure-reducing valve. Each pump used is generally capable of handling 100% of the boiler's full-load requirements, with the additional units serving as backup. Horizontally and vertically mounted rotary pumps, often of the gear or multiple-screw type, are generally used in this application. Many of these pumps are driven by steam turbines. However, to enable fuel oil to be pumped prior to plant start-up, most vessels also have at least one motor-driven pump. Typical conditions of service in this application include capacity ratings in the range of 15 to 40 gpm at discharge pressures up to 350 psig. The viscosity of the oil pumped is frequently in

the range of 450 to 8000 SSU. Inlet conditions to a fuel-oil service pump, which vary with the level of the fuel oil in the service tanks, can often range from a flooded suction to a total suction lift as high as 20 in. Hg.

Changes to the capacity delivered by a rotary fuel-oil service pump can be made by varying the pump's operating speed, which is possible when the pump is driven by a steam turbine or by a multispeed electric motor. However, even during constant-speed operation, the amount of fuel oil actually supplied to the boiler's burners can generally be regulated by recirculating any excess oil delivered to the burner manifold back to the inlet side of the pump.

A separate rotary type "cold boiler starting" fuel-oil pump is also installed on many vessels to supply light fuel oil to the boiler during plant start-up. These pumps are often rated to deliver capacities of less than 10 gpm at discharge pressures up to 150 psi, and are generally driven by electric motors.

Fuel-oil transfer. A fuel-oil transfer pump is used to transfer fuel oil from the vessel's storage (bunker) tanks to the service tanks. Pumps used in this application include horizontally and vertically mounted rotary units similar in configuration to those used for fuel-oil service, as well as rotary pumps that are submerged directly within the fuel-oil storage tank and are driven through vertical line shafting by motors located above the tank. In addition, motor-driven (power) and steam-driven (direct-acting) reciprocating piston pumps are also sometimes used in this application. Fuel-oil transfer pumps are typically rated to deliver capacities up to 750 gpm at a discharge pressure of 100 psig. They must often be suitable for operation with a total suction lift as high as 20 in. Hg, and can be used to pump oil with a viscosity as high as 8000 SSU. The capacity delivered by a steam-turbine-driven rotary or direct-acting reciprocating fuel-oil service pump can be adjusted by varying the pump's operating speed. To enable the capacity delivered by motor-driven pumps to also be changed, the motors used often have multiple operating speeds.

Lubricating oil. Lubricating-oil service (LOS) pumps take suction from the lubricating-oil sump tank and discharge oil, typically through a strainer and a cooler, either to an overhead gravity tank or directly to the main propulsion machinery. Multiple-screw, gear, and vane pumps that are mounted either horizontally or vertically are often used in this application. Vertical LOS pumps are sometimes submerged directly within the lubricating-oil sump tank. Most vessels are fitted with two or three LOS pumps with at least one usually driven by an electric motor; however, the remaining pumps may be driven by steam turbines, off the main propulsion reduction gears, or off the propeller shaft. LOS pumps are frequently rated to deliver capacities as high as 550 gpm at discharge pressures from 60 to 80 psig. Many pumps must be suitable for operation with suction lifts in the range of 10 to 15 in. Hg while pumping oil with a viscosity of 130 to 500 SSU and an air entrainment of 2 to 5%.

In addition to the LOS pumps, similar units are also often furnished to circulate lubricating oil through a purifier, transfer it from the main storage tanks to settling tanks, and deliver it to smaller tanks in various locations throughout the vessel where the oil is stored for use in auxiliary components.

Boiler hydrotest and chemical feed. Motor-driven plunger pumps are often used to pressurize the water and steam side of the boiler during hydrostatic testing, and to inject chemicals into the boiler. These units are generally mounted horizontally and are rated to deliver capacities in the range of 2 to 25 gpm at discharge pressures up to 2000 psig. Because these power pumps generally operate at speeds below 600 rpm, they are typically coupled to their drive motors through gear reducers or with sheaves and multi-V belts.

Contaminated-evaporator feed. Condensate returned from heat exchangers in which fluids that could contaminate the main steam system are heated is often collected in the contaminated-drain-inspection tank. Many vessels have electric-motor-driven centrifugal or regenerative turbine pumps that are used to transfer the condensate from the inspection tank to a contaminated evaporator. Volute-type sump pumps that are submerged directly within the inspection tank are also sometimes used in this application. Typical conditions of service for contaminated-evaporator feed pumps include capacity ratings in the range of 15 to 50 gpm at total heads up to 350 ft. An automatic throttle valve is generally installed in the contaminated-evaporator feed pump's discharge line, and enables the pump to operate continuously and still maintain

a constant water level within the evaporator. To prevent the pump from operating at too low a capacity, a bypass is connected to its discharge line through which a portion of the water being pumped is recirculated back to the inspection tank.

Nuclear reactor primary loop. The main coolant pumps in the primary loop of a pressurized-water nuclear reactor typically take suction from the steam generators and return water to the reactor vessel. Vertically mounted motor-driven single-stage centrifugal pumps are generally used in this application. In some of these units the pump's shaft is rigidly coupled to that of the drive motor and relies on the motor's thrust bearing for axial support. Radial support for the pump's shaft, however, is frequently provided by an internal water-lubricated journal bearing. To reduce the potential for leakage from the casing, special water-cooled shaft seals are often installed in these coupled pumps. In addition, to eliminate the need for a shaft seal, hermetically sealed canned motor pumps are sometimes used in reactor coolant service.

Other pumps frequently furnished for applications associated with pressurized-water reactors include motor-driven reciprocating plunger-type charging pumps that are used to fill and pressurize the primary loop, centrifugal pumps that supply cooling water to various reactor components, such as the bearings and seals in the main coolant pumps and motors, and canned motor pumps that circulate a portion of the water in the primary loop through a purifier [2].

b. Diesel engine propulsion-related pump applications.

Fuel-oil booster. Fuel-oil booster (or supply) pumps take suction either directly from the fuel-oil daily service tanks or from a separate fuel-oil mixing tank and discharge fuel, through a heater and a filter, to the propulsion engine's injection pumps. Many vessels have two multiple-screw- or gear-type fuel-oil booster pumps that are each capable of supplying approximately 2.5 times the engine's full-power fuel requirements. Although both of these pumps may be driven by electric motors, one pump is frequently driven off the engine when a medium- or high-speed engine is used. Typical conditions of service for fuel-oil booster pumps include capacity ratings in the range of 15 to 65 gpm at discharge pressures from 60 to 160 psig. The viscosity of the oil being pumped can vary from approximately 40 SSU for diesel oil to over 4000 SSU for heavy fuel oils.

The engine's injection pumps are generally cam-operated variable-stroke or variable-effective-stroke plunger-type units and are sometimes an integral part of the engine's injectors. Excess oil delivered to the injection pumps is recirculated back to the service or mixing tank.

Fuel-oil transfer. Multiple-screw and gear pumps are often used to transfer fuel oil from a vessel's storage (bunker) tanks to the fuel-oil settling tanks. Many diesel-powered vessels are fitted with two transfer systems: one for heavy fuel oil and a second for diesel oil. Separate transfer pumps are generally provided for each system. Fuel-oil transfer pumps, which are frequently driven by two-speed electric motors, are often rated to deliver capacities in the range of 50 to 425 gpm at discharge pressures from 35 to 150 psig. In some installations the pumps are submerged within the storage tanks and are driven through vertical line shafting by motors located above the tank.

Additional positive-displacement pumps are generally used to transfer fuel-oil from the settling tanks to the daily service tanks. These pumps may be driven by independent electric motors, or they may be attached to the purifiers that the fuel-oil passes through before reaching the service tanks. Some vessels also have pumps that are used to transfer fuel oil from the daily service tanks to a separate mixing tank.

Lubricating oil. Two or more lubricating-oil circulating pumps are generally used to remove lubricating oil from a sump or drain tank located below the main engine and return the oil, through a cooler and filters, to the engine's bearings, to the governor, and, in some cases, to the engine's turbocharger. Although vertically or horizontally mounted multiple-screw and gear pumps are often used in this application, multistage vertical turbine pumps that are submerged within the lubricating-oil drain tank and centrifugal pumps have also been used. Typical conditions of service for lubricating-oil circulating pumps include capacity ratings in the range of 400 to 2400 gpm at discharge pressures from 50 to 110 psig while handling oil with a viscosity of 130 to 600 SSU and an air entrainment of 2%. Lubricating-oil circulating pumps for low-speed engines are driven by electric motors. With a medium- or high-speed engine, however, one of the pumps is frequently driven off the engine.

With a crosshead-type engine, a portion of the oil discharged from the lubricating-oil circulating pumps is often supplied to the inlet side of two or more lower-capacity rotary-type booster pumps. The high-pressure oil discharged from these smaller pumps is used to lubricate the engine's crosshead bearings. Separate pumps are also sometimes used to transfer oil for cylinder lubrication to a measuring tank that feeds mechanical plunger-type lubricators mounted on the engine. Depending on the design of the engine, additional rotary-type pumps may be used to supply lubricating oil to specific engine components, such as the camshaft bearings. Furthermore, on vessels with medium- or high-speed propulsion engines, rotary pumps are generally used to deliver lubricating oil to the reduction gears. Typical capacity and discharge pressure ratings for these applications range from 100 to 300 gpm at 175 to 230 psig for crosshead-oil pumps, 5 to 10 gpm at 30 psig for cylinder-oil transfer pumps, 10 to 60 gpm at 60 to 150 psig for camshaft-oil pumps, and up to 200 gpm at 25 to 65 psig for reduction-gear-oil pumps.

Positive displacement pumps are generally used to circulate lubricating oil through purifiers. Although these pumps may be attached to the purifiers, some vessels have independent electric-motor-driven pumps. Electric-motor-driven pumps are also used to transfer lubricating oil from the main storage tanks to the settling tanks and to transfer unusable oil ashore. Typical conditions of service for lubricating-oil transfer pumps include capacity ratings in the range of 10 to 65 gpm at discharge pressures from 20 to 70 psig. On some vessels, an additional rotary-type "scavenging pump" is used to transfer lubricating oil that drains into the engine's crankcase to a separate sump or drain tank.

Engine cooling water. Fresh water is typically used to cool the main engine's cylinders and cylinder heads. Duplicate horizontally or vertically mounted single-stage centrifugal pumps are frequently provided to circulate the water through the engine, its turbochargers, when they are freshwater cooled, a cooler, and, in some cases, an evaporator (where heat from the water is used for freshwater generation). Although these pumps are often called the jacket-water cooling pumps, on vessels fitted with a central cooling system they may be referred to as the high-temperature-freshwater cooling pumps. An elevated expansion tank in the system maintains a positive suction head to the pumps. When a low-speed engine is used, both jacket-water cooling pumps are driven by electric-motors. With a medium- or high-speed engine, however, one pump may be driven off the engine.

Some crosshead-type engines have an independent freshwater system for piston cooling. Although the pumps that circulate the piston-cooling water may be similar in configuration to the jacket-water cooling pumps, sump pumps and vertical turbine pumps that are submerged within the piston-cooling-water drain tank are sometimes used. Additional centrifugal pumps may also be provided to circulate fresh water through a third circuit that cools the engine's fuel valves or injectors. A separate elevated expansion tank maintains a positive suction head to the fuel-valve-cooling-water pumps. Piston-cooling-water pumps and fuel-valve-cooling-water pumps are generally driven by electric motors. Typical capacity and total head ratings in engine-cooling-water applications can range from 375 to 2500 gpm at 65 to 130 ft for jacket-water-cooling pumps, 150 to 800 gpm at 140 to 190 ft for piston-cooling-water pumps, and 15 to 70 gpm at 100 to 165 ft for fuel-valve-cooling-water pumps.

Two or three separate centrifugal pumps are frequently provided to supply seawater to heat exchangers used to cool the fresh water that is circulated through the main engine. Seawater may also be circulated through the engine's lubricating-oil and air coolers when these heat exchangers are not freshwater cooled. Typical conditions of service for main-engine seawater pumps include capacity ratings up to 6000 gpm at total heads in the range of 30 to 100 ft. Although these pumps are often driven by electric motors, a medium- or high-speed engine may be fitted with an attached seawater pump.

Waste-heat and auxiliary boilers. On many diesel-powered vessels, the exhaust gas from the engine is used to generate steam in a waste-heat boiler. Centrifugal-type condensate and feed pumps are generally required to transfer water from the condenser used in this system back to the boiler. In addition, depending on the configuration of the waste-heat steam system used, boiler circulating pumps may also be required. Motor-driven one- and two-stage centrifugal and regenerative turbine pumps that are mounted horizontally or vertically are often used in these applications. Similar pumps, together with rotary-type fuel-oil pumps, are also required if the vessel has an auxiliary oil-fired boiler that is used to generate steam.

c. Gas turbine propulsion-related pump applications.

Fuel oil. Fuel oil is delivered to a gas turbine's combustion chambers by the fuel-oil service pumps. Rotary gear pumps that are driven off the gas turbine are generally used in this application. Separate rotary-type booster pumps that take suction from the fuel-oil service tanks and deliver the gas turbine's fuel oil, through filters and a heater, to the fuel-oil service pumps are also often provided. The fuel-oil booster pumps are generally driven by two-speed electric motors. Additional electric-motor-driven rotary pumps are required to transfer fuel oil from the vessel's storage tanks to the service tanks. The oil discharged from the fuel-oil transfer pumps usually passes through a purifier before entering the service tanks.

Lubricating oil. The gas turbine's lubricating-oil pumps take suction from the vessel's main lubricating-oil reservoir and deliver synthetic oil to the gas turbine's bearings. The main lubricating-oil pumps are typically rotary gear-type units that are driven off the gas turbine. Electric-motor-driven auxiliary pumps, however, may also be provided for use during plant start-up or cooldown and as backup to the main pumps. On many vessels an additional rotary-type pump, sometimes referred to as the scavenging pump, circulates oil that drains from the turbine's bearings through a cooler and filters, and returns the oil to the main lubricating-oil reservoir.

Separate rotary screw pumps generally circulate mineral oil through an independent system used for reduction gear lubrication. This oil is also used as the cooling medium in the gas turbine's lubricating-oil cooler. One of the reduction gear lubricating-oil pumps is typically driven off the reduction gears; the remaining pumps, however, are driven by electric motors. Additional rotary pumps are also used to circulate the vessel's various lubricating oils through purifiers and to transfer lubricating oils from storage tanks to locations throughout the vessel.

Waste-heat and auxiliary boilers. Waste-heat boilers are installed on some gas-turbine-propelled vessels so that the hot gases exhausted from the turbine can be used to produce steam. Auxiliary oil-fired boilers may also be installed on these vessels. The pumps used with these boilers are similar to those used with waste-heat and auxiliary boilers installed on diesel-powered vessels.

d. General non-propulsion-related pump applications.

Fire. Fire pumps, which take suction from the sea chest and deliver seawater to the vessel's fire mains and hoses, are often rated to deliver capacities in the range of 150 to 4000 gpm at discharge pressures from approximately 100 to 185 psig. The actual capacity and pressure ratings for a specific installation are based on the number of hoses and the pressure required at the farthest hose. Multiple pumps are typically installed at various locations throughout the vessel so that adequate backup capacity will be available during an emergency. Pump configurations commonly used for fire service include the following:

- Horizontally and vertically mounted single-stage centrifugal pumps that have an axially split casing and a double-suction impeller (see Fig. 13). The impeller is centered on a shaft that is supported at each end by external grease-lubricated ball bearings or internal sleeve bearings. These pumps are generally driven by electric motors, steam turbines, or diesel engines at speeds that range from approximately 1800 to 3600 rpm.

- Axially split casing centrifugal pumps similar to those described above, except that the pump's rotor includes two single-suction impellers. The use of these two-stage pumps is generally limited to lower-capacity applications for which a single-stage unit would have too low a specific speed.

- Single-stage centrifugal pumps that have a radially split casing and an end-suction type impeller. These pumps, and the electric motors that typically drive them at speeds up to approximately 3600 rpm, are often furnished in a close-coupled configuration (see Fig. 14). Close-coupled fire pumps, which can be mounted horizontally or vertically, are generally used to deliver capacities not exceeding 1100 gpm.

- Vertical turbine pumps are used for installations in which the fire pump impellers must be submerged within a tank or cofferdam. The VTP's above-deck driver can be a vertical electric motor, or a horizontal motor, a steam turbine, or diesel engine that is coupled to the pump's shaft through a right-angle gear. Because these pumps are seldom operated at speeds above 1800 rpm, they are generally furnished with multistage bowl assemblies.

When possible, fire pumps are generally installed low enough in the vessel so that their first-stage impellers are flooded. With this arrangement the NPSH available to the fire pump is approximately equal to the submergence of its impeller below the vessel's waterline plus atmospheric pressure, less losses due to friction and turbulence within the sea chest and the pump's suction line. Because the impeller's submergence can vary with the vessel's list and trim conditions, pumps used for fire service should have NPSH requirements that are not only less than the normal NPSHA, but are also less than the reduced NPSH that can be available during emergency conditions. In addition, if due to installation constraints the fire pump can operate with a suction lift, it must either have the capability to be self-priming or be connected to a vacuum priming system.

Some vessels also carry gasoline-engine-driven portable fire pumps. These single-stage end-suction pumps typically deliver a capacity of approximately 250 gpm at a discharge pressure of 100 psig and are generally driven at speeds up to 5000 rpm. To enable the portable pumps to be used in areas of the vessel that are above deck, they are often fitted with integral vacuum priming pumps.

As an alternative to fighting fires with water, some vessels have systems that enable them to fight fires with foams composed of a mixture of foam concentrate, water, and air. These systems typically include rotary foam concentrate pumps and centrifugal water pumps. Rotary pumps may also be included to pressurize hydraulic fluid when it is used to actuate foam distribution monitors.

Bilge. Bilge pumps are used to remove liquid that accumulates on the tank tops and in spaces located throughout the vessel. Depending on the cleanliness of this liquid, it is discharged overboard, or directed to an oily-waste collecting tank. In an emergency situation, the bilge pumps may also be used to evacuate water from an area of the vessel that is flooding due to leakage from a ruptured pipe or component, or as a result of damage to the hull. Several bilge pumps are typically installed at various locations on the vessel. Capacity ratings for these pumps are often in the range of 25 to 1000 gpm.

Horizontally and vertically mounted electric-motor-driven centrifugal pumps are frequently used in this application. Due to the relatively low head requirements for bilge service, which generally do not exceed 150 ft, these pumps are typically single-stage units. In addition, they usually have an axially split casing with a double-suction impeller mounted between bearings or a radially split casing with a single-suction overhung impeller. Smaller radially split casing pumps may be furnished in a close-coupled configuration. Some centrifugal bilge pumps and their motors are capable of operating even when submerged. Coupled submersible pumps often have only sleeve-type radial bearings that are either water lubricated or packed with grease. With this arrangement axial loads acting on the pump's shaft are transmitted to the motor's thrust bearing through a rigid coupling.

Because centrifugal bilge pumps typically operate with a suction lift, a non-self-priming pump must generally be connected to the vessel's central priming system or fitted with a vacuum priming pump. To eliminate the need for a priming pump, self-priming centrifugal pumps are sometimes used in this application. (Although these units are often suitable to operate with the suction lifts required in bilge service, due to their limited air-handling capability, self-priming centrifugal pumps can have extended priming times when operating in systems with long lengths of horizontal suction piping. Their performance should, therefore, be carefully analyzed prior to use in a bilge system.) Additional alternatives that eliminate the need for a priming pump in bilge service include the use of sump pumps and VTP's (which permits the pump's impeller to be submerged directly within the bilge), motor-, steam-, and air-driven reciprocating piston pumps, rotary-vane pumps, and submerged air-driven diaphragm pumps. Some electric-motor-driven bilge pumps are automatically cycled on and off by float switches that are mounted in the bilge suction wells.

Oily-waste and sludge transfer. In addition to the bilge pumps, many vessels have a separate pump that removes the oil and water mixture from the bilge system's oily-waste collecting tank and discharges it to an oil-water separator. Additional pumps may also be provided to transfer the contents in separated-oil and sludge tanks, which can include dirty oil from fuel- and lubricating-oil purifiers and from the bilge systems oily-water separator, ashore or to another vessel. Although rotary-type progressing-cavity pumps are often used in these applications, sliding-vane, gear, and multiple-screw pumps have also been used. Typical conditions of service include capacity

ratings in the range of 10 to 50 gpm at discharge pressures from 40 to 100 psig. The viscosity of the fluids pumped can vary from 32 to 8000 SSU.

Ballast. Ballast pumps are used to transfer seawater into and out of the vessel's ballast tanks. They, therefore, can take suction from the sea chest or from ballast tanks that are being emptied. In addition, the seawater that is discharged by these pumps can be directed to ballast tanks or overboard. Horizontally and vertically mounted single-stage centrifugal pumps that have an axially split casing and a double-suction impeller mounted between bearings are often used in this application. In addition, VTP's and reversible axial-flow propeller pumps are sometimes used for ballast service. The capacity rating for each ballast pump, which is based on the size of the vessel's ballast tanks, the vessel's turnaround time requirements, and the number of ballast pumps on the vessel, can range from approximately 500 gpm on smaller vessels to over 20,000 gpm for the pumps installed on large liquid-bulk carriers. The total head that must be developed by the typical ballast pump is generally less than 150 ft; higher heads can be required, however, from ballast pumps that supply motive water to dewatering eductors. Drivers used with ballast pumps include electric and hydraulic motors, and steam turbines.

Although a ballast pump generally operates with a flooded suction when it takes suction from the sea chest or initially begins to empty a ballast tank that is full, as the water level in a tank being emptied is reduced, the submergence of the pump's impeller and the NPSH available to the unit are continuously reduced. For this reason ballast pumps generally have relatively low NPSH requirements. To enable a low NPSH requirement to be achieved with a suction specific speed that is not excessive, ballast pumps frequently operate at speeds not exceeding 1800 rpm. In addition, to enable these pumps to evacuate their suction lines of air and vapor if suction is lost during operation with a suction lift, centrifugal ballast pumps are generally connected to central priming systems or, in some cases, to automatic stripping systems. To enable the VTP's used in ballast service to take suction from multiple locations, they are generally installed in suction cans. In addition, because there may be times when air or vapor must be removed from the suction cans and the suction piping, vertical turbine ballast pumps are often fitted with self-priming valves.

Multiple and general service. On many vessels, the same pumps may be used for two or three different applications. For example, by changing the valve lineup at the pump's suction and discharge manifolds, the same unit is often used alternately for either bilge or ballast service. A single pump may also be used in both the fire and the bilge systems. These units, which are sometimes referred to as general service pumps, must be designed so that they can deliver the capacities and develop the total heads required for each of the applications in which they will be utilized. To enable a single pump to meet more than one rating point, multiple-speed drivers are often furnished with these units.

Distilling plant. Several pumps are used with the distilling plants that convert seawater into fresh water. Included among them are the following:

• A distiller feed pump—is used to supply seawater to the distilling plant. This pump must develop sufficient head to overcome the friction drop in the heat exchangers that the seawater often passes through prior to entering the distiller. Distiller feed pumps are typically rated to deliver capacities in the range of 20 to 1000 gpm and develop total heads up to 200 ft.

• A brine overboard pump—receives the hot brine from the distilling plant and discharges it overboard. Brine overboard pumps are often rated to deliver capacities in the range of 20 to 1000 gpm at total heads up to 150 ft.

• A distillate pump—circulates the fresh water being produced by the distilling plant through a cooler, and then transfers it to the distilled water, potable water, or reserve feedwater tank, or to the bilge if the water's salinity content is excessive. Distillate pumps are typically rated to deliver capacities in the range from 10 to 500 gpm at total heads of 225 ft or less.

• A feed heater condensate pump—is used in some distilling plants to transfer condensate from the seawater heater to the vessel's condensate system. Pumps used in this application are typically rated to deliver capacities up to 350 gpm at total heads as high as 200 ft.

• A distilling condenser circulating pump—is sometimes used in lieu of the distiller feed pump to circulate seawater through the distillate cooler and distiller condenser. Although a portion of the seawater may then be discharged overboard, the remainder generally continues on to become the distiller's feedwater. Capacity ratings for the pumps used in this application can be as high as 1150 gpm at total heads up to 100 ft.

• A distiller chemical feed pump—is used to inject chemicals into the distiller for acid cleaning. Distiller chemical feed pumps are generally rated to deliver capacities that are less than 10 gpm.

• A vacuum pump—is used to evacuate air from some types of distilling plants.

With the exception of distilling-plant vacuum pumps, which are generally of the liquid-ring design, the pumps used with many smaller distillers are horizontally or vertically mounted single-stage centrifugal units that have a radially split casing and a single-suction overhung impeller. In addition, in low-capacity high-head applications, such as distiller feed service, regenerative turbine pumps are sometimes used. The pumps furnished with these distillers may be mounted directly on the distilling plant assembly, and are often close coupled to electric motors. Similar pumps are frequently used with high-capacity distilling plants. However, flexibly coupled centrifugal pumps that have an axially split casing and a double-suction impeller mounted between bearings are also used with some larger distillers.

Cooling water. Pumps are used in several applications to deliver cooling water to various heat exchangers on the vessel. Included among these applications are the following:

• Seawater cooling pumps—take suction from the vessel's sea chests and circulate seawater through heat exchangers that utilize the water for cooling. The seawater is then discharged overboard. On many vessels, the cooling pumps are called seawater service pumps and distribute seawater to a variety of different heat exchangers, such as refrigeration and air-conditioning condensers, fuel-oil and lubricating-oil coolers, and air coolers. If the vessel has a central cooling system, however, the pumps, which may be referred to as the main or central seawater pumps, circulate seawater through only two or three large freshwater coolers. (The fresh water is then distributed throughout the vessel for cooling.) Typical conditions of service for seawater cooling pumps include capacity ratings in the range of 500 to 9000 gpm at total heads from 65 to 100 ft.

• Freshwater cooling pumps—circulate fresh water through the various heat exchangers that utilize the water as a cooling medium. Also included in the closed system is a seawater-cooled heat exchanger, which is used to cool the fresh water. On a diesel-powered vessel with a central cooling system, the fresh water may be circulated through two independent cooling systems: a low-temperature system that supplies cooling water to heat exchangers used with auxiliaries, and a high-temperature system that is used to cool the main engines. Separate pumps are installed in each system. Conditions of service for freshwater cooling pumps can include capacity ratings in the range from 300 to 3350 gpm at total heads of 65 to 140 ft.

• Air-conditioning-chilled-water (ACCW) pumps—circulate fresh water through the closed chilled water system. This water, which is cooled by the chillers in this vessel's air-conditioning system, is distributed throughout the vessel and is used to cool the air in temperature-controlled spaces. It may also be used to cool electronic components. Typical conditions of service for the ACCW pumps include capacity ratings that are in the range of 100 to 1500 gpm at total heads from 65 to 210 ft.

Horizontally and vertically mounted electric-motor-driven single-stage centrifugal pumps are typically used in the above applications. They generally have either an axially split casing and a double-suction impeller mounted between bearings, or a radially split casing and a single-suction overhung impeller.

Potable water. Potable-water pumps are used to transfer unheated fresh drinking water from the potable-water storage tanks to air-charged pressure tanks, or directly to sinks, showers, drinking fountains, galley equipment, and other outlets throughout the vessel. Each potable-water pump may be cycled on and off by a pressure switch in the discharge line, or one pump may be operated continuously while additional backup units are started and stopped as needed. Horizontally mounted centrifugal and regenerative turbine pumps that have either one or, in some cases, two stages are often used for potable-water service. These pumps are generally driven by electric motors. In addition, smaller-sized units are frequently furnished in a close-coupled configuration. Typical conditions of service for potable-water pumps include capacity ratings in the range of 25 to 350 gpm at total heads from 150 to 250 ft.

Hot-water circulating. A hot-water circulating pump is used to circulate potable water through a hot-water heater. After being heated, this water is directed to sinks, showers, and other outlets that require hot water. The pump operates continuously and recirculates unused water in the system through the heater so that the water will remain hot. Horizontally and vertically mounted electric-motor-driven single-stage centrifugal pumps that have a radially split casing and a single-suction impeller are generally used in this application. Many of the pumps are furnished in a close-coupled configuration and deliver capacities of 5 to 15 gpm at total heads from 10 to 30 ft. Similar pumps may also be used in the vessel's air-conditioning system to circulate hot water that heats air in temperature controlled spaces. Typical conditions of service in this latter application included rated capacities up to 75 gpm at total heads up to 115 ft.

Sanitary. The water closets on some vessels receive flushing water from a seawater sanitary system. Although this water is sometimes supplied through a pressure-reducing valve from the fire main, it can also be transferred from the sea chest by separate sanitary pumps. Sanitary pumps are generally horizontally or vertically mounted single-stage centrifugal units that are driven by electric motors. Typical conditions of service for these pumps include rated capacities of approximately 50 gpm at total heads of 80 ft.

Sewage. Sewage pumps are used to transfer sewage from the vessel's holding tanks to an above-deck shore connection, to an onboard treatment plant, or overboard. Horizontally and vertically mounted single-stage centrifugal pumps that have a radially split casing and an overhung single-suction impeller are often used in this application. Submerged sump-type centrifugal pumps that are driven through vertical line shafting by above-tank drivers are also sometimes used to transfer sewage. Typical conditions of service in this application include capacity ratings in the range of 50 to 100 gpm at total heads from 70 to 115 ft. Many of the pumps used are fitted with a packed stuffing box for shaft sealing. To reduce leakage, grease is often injected into the packing through a lantern-ring connection located in the side of the stuffing box. When a mechanical-type shaft seal is used, a restriction bushing must generally be installed at the base of the seal cavity to isolate the seal from the sewage being pumped. A freshwater flush is also typically supplied from an external source to cool and lubricate the seal's faces. Sewage pumps, which are often driven by electric motors, are usually furnished with a special "non-clog" impeller containing only two or three vanes and a large waterway volute casing that can pass the solids and stringy material normally found in sewage. In general, however, as the size of the solid particles that the pump is designed to handle increases, the pump efficiency is reduced. Large hand holes are generally provided in the walls of the casing to permit internal passages to be cleaned periodically.

In addition to the sewage transfer pumps just described, in containment systems that rely on a vacuum to aid in the removal of waste products from water closets, motor-driven liquid-ring-type vacuum pumps are often used to reduce the pressure in the sewage holding or treatment tanks. Vacuums up to 20 in. Hg are typically created in these tanks. In recirculation-type sanitary flushing systems, pumps are also required to recirculate filtered and treated water or mineral oil from the sewage holding tank back to the vessel's lavatories for flushing. End-suction centrifugal pumps are typically used in this latter application.

Stern tube and strut bearing lubricating oil. A stern-tube lubricating-oil pump is used to circulate lubricating oil through the stern tube of a vessel fitted with an oil-lubricated stern tube bearing. Similar pumps are also provided for oil-lubricated strut bearings. Rotary gear, screw, and vane pumps are often used in these applications. Typical conditions of service include capacity ratings in the range of 2 to 5 gpm at discharge pressures from 40 to 60 psig.

Hydraulic fluid. Various types of positive-displacement pumps are used to pressurize and circulate the fluids (usually oils) that drive hydraulically powered machinery. These pumps, which are generally driven by either electric motors or diesel engines, can be furnished as part of a central system that feeds all of the vessel's hydraulically driven equipment, or as part of smaller self-contained systems that are an integral part of the driven components. Typical components that can be hydraulically driven include anchor windlasses, winches, hatch covers, steering gear, and cargo unloading equipment. Discharge pressures developed by hydraulic pumps often are in the range of 500 to 4000 psig.

One or more fixed-displacement rotary-type gear, vane, or screw or reciprocating-type power pumps are often used in a constant-flow hydraulic system. These pumps take suction from a sump or tank and deliver pressurized fluid to the various components that are powered by the hydraulic system. In a constant-pressure hydraulic system, one or more variable-displacement radial or axial rotating-piston pumps are frequently used to circulate hydraulic fluid through the system at essentially a constant pressure. If the capacity delivered by the pump exceeds the system's load requirements, the system pressure will increase and a pressure compensator will reduce the pump's stroke and, therefore, the capacity of fluid that the pump delivers. Conversely, if the system load increases and the system pressure drops, the pressure compensator will increase the pump's stroke. Multiple fixed-displacement pumps that are cycled on and off as needed are also sometimes used in constant-pressure hydraulic systems.

In addition to the main hydraulic pumps, smaller positive-displacement pumps may be used to add makeup fluid to the hydraulic system and to control the stroke of variable-displacement pumps. In some systems, the makeup (replenishing) pump, control (servo) pump, and main pump have a common driver.

Table 2   Pumps used for cargo unloading

Type	Typical Rated Capacity, bbl/hr	Type Carrier Most Commonly Installed On
MAIN CARGO		
Centrifugal (horizontal or vertical)	4275 to 56,000	crude
Vertical turbine (deepwell)	350 to 10,000	multi-petroleum product/chemical/LPG
Submersible (hydraulic driven)	250 to 3100	multi-petroleum product/chemical
Submersible (electric motor driven)	100 to 7500	liquefied natural gas/liquefied petroleum gas
Rotary	850 to 8000	crude/high-viscosity product
CARGO STRIPPING		
Horizontal reciprocating (gas-driven)	40 to 70	multi-petroleum product/chemical
Vertical reciprocating (steam-driven)	700 to 2300	crude
Rotary	250 to 600	crude
Diaphragm	below 350	multi-petroleum product

e. Cargo unloading. Unlike pumps used in other applications, the pumps used to unload liquid bulk cargoes must often handle a wide range of petroleum products or chemicals with different specific gravities, vapor pressures, viscosities, and temperatures. In addition, because part of the system on the discharge side of a cargo pump is formed by the piping and storage tanks at the point of delivery, the number of systems in which these pumps can operate is limited only by the number of terminals or other vessels into which cargo will be transferred. Many of the various types of pumps used for cargo unloading, together with typical capacity ratings, are listed in Table 2. As reflected in this table, the capacity delivered by a cargo pump is generally expressed in barrels per hour, where one barrel is equal to 42 gallons.

The materials used in the construction of cargo pumps must be compatible with all of the fluids that will be pumped. This can include not only the cargoes that the vessel will carry, but also seawater if the pumps will be used to remove slops during cargo tank washing. In addition, if fluids that are flammable or explosive will be discharged by the pump, components with contacting surfaces should be constructed from non-sparking materials.

The types of pumps used in cargo service include the following:

Centrifugal cargo pumps. Generally three or four centrifugal cargo pumps are installed in "pump rooms" that are located in the lower part of the tank vessel. With this arrangement, which is usually found only on crude-oil carriers or on vessels that carry a limited number of different grades of petroleum products, each pump is often capable of being used to discharge cargo from any of the vessel's tanks through interconnected suction piping. The pumps are typically single-stage units. They are frequently furnished with a horizontally, as shown in Fig. 43, or vertically mounted axially split casing and a double-suction impeller that is centered between external bearings. However, vertically mounted radially split casing pumps, in which the impeller is overhung on the end of a cantilevered shaft, have also been used in this application. With this latter arrangement, both of the pump's external

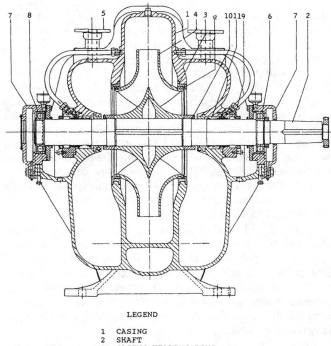

LEGEND

1	CASING
2	SHAFT
3	CASING WEARING RING
4	IMPELLER
5	STRIPPING CONNECTION
6	THRUST BEARING
7	BEARING HOUSINGS
8	LINE BEARING
9	AUXILIARY STUFFING BOX
10	SHAFT NUT
11	MECHANICAL SEAL
12	IMPELLER WEARING RING

Fig. 43   Horizontal centrifugal cargo pump

bearings are located above the single shaft seal. Explosion-proof intrinsically safe resistance temperature detectors are often mounted in each bearing housing and are connected to an alarm that sounds if the bearing temperature exceeds a preset value.

Centrifugal cargo pumps can be driven by steam turbines, diesel engines, or electric motors. To eliminate the need for the driver to be explosion-proof, it is usually installed in a separate machinery space adjacent to the pump room and is coupled to the cargo pump through an intermediate shaft, referred to as a jackshaft. The opening in the bulkhead or overhead of the pump room that the jackshaft passes through is sealed with a gastight stuffing box to prevent explosive vapor from entering the driver compartment.

A cargo pump's rated total head is generally based on a desired pressure at the vessel's discharge manifold, referred to as the "rail pressure"; the pressure drop in the suction and discharge piping, valves, and fittings; and the height of the manifold above the cargo tank bottom. The specified rated capacity is based on the number of tanks to be emptied at any one terminal, the number of pumps available to empty those tanks, and the amount of time allowed for the pump-out. The maximum capacity that can be delivered is often limited by the NPSH available to the cargo pump, which is essentially equal to the elevation of the liquid level within the cargo tank being emptied above the pump's impeller added to the absolute pressure within the tank, less the cargo's true vapor pressure and losses within the suction line. Due to the relatively high true vapor pressures of many cargoes, the pumps are often forced to operate with low values of NPSHA. In addition, due to the continuous reduction of the liquid level within the tank being emptied, the NPSHA will gradually be reduced throughout the pump-out cycle. Consequently, low-NPSHR impellers are often installed in centrifugal cargo pumps. So that suction specific speeds are not excessive, cargo pump operating speeds are generally limited to 1800 rpm.

As the liquid level in a tank being emptied approaches the tank bottom, air or inert gas from the tank's atmosphere will frequently be drawn into the pump's suction line before the inlet to the suction tail pipe, also known as the strum, is completely uncovered. Therefore, self-priming/stripping systems are often used to increase the amount of cargo that can be discharged by centrifugal cargo pumps [24].

Vertical turbine cargo pumps. On vessels that carry a variety of different liquid cargoes, such as multi-petroleum product carriers and chemical carriers, vertical turbine or deepwell pumps, which are illustrated by Fig. 24, are often used for cargo unloading. When the maximum degree of cargo segregation is required, a separate VTP is installed in each cargo tank. Because the pump's bowl assembly is submerged in the cargo that will be discharged, the need for suction valves and suction piping is eliminated. In addition, by providing separate discharge piping and manifolds for each pump, the risk of mixing cargoes is greatly reduced. With this arrangement, there are typically 20 to 40 cargo pumps per ship. Each pump is often from 40 to 60 ft in length and is generally driven by a vertical electric or hydraulic motor that is mounted on top of the discharge head. With some cargoes, the driver must be explosion-proof.

When a lower degree of cargo segregation can be tolerated, such as on liquid-bulk carriers that transport a limited number of different cargoes or cargoes that are less sensitive to contamination, each VTP is often used to unload several of the vessel's cargo tanks. Typically, there are fewer than 20 pumps per ship when this latter arrangement is used. Each pump is frequently mounted in a suction tank or can that is connected to the ship's cargo tanks with suction piping. Automatic self-priming valves are often installed on the VTP to enable the pump to remove gas and vapor from its suction piping and suction can. Because each VTP is used to discharge cargo from more than one tank, these pumps typically are larger in diameter and require larger drivers than the vertical turbine pumps installed with an individual pump/tank arrangement. In addition, although many of the vertical turbine pumps used to discharge cargo from multiple tanks have vertical drivers, some are driven through right-angle gears by horizontal motors, steam turbines, or diesel engines.

With some cargoes the use of single shaft seals is suitable. When the VTP will handle explosive petroleum products or chemicals, however, the use of a double sealing arrangement is often required. Additional static shaft seals are also frequently used to prevent gas and vapor that may be in the cargo tank from escaping through the stuffing box while the pump's packing or mechanical seals are being replaced.

Vertical turbine pumps used for cargo unloading can handle fluids with a wide range of temperatures. For example, it is not uncommon for lube oils, waxes, and other viscous cargoes to be heated, often to temperatures that exceed 160 F. With some cargoes, such as molten sulfur, steam or a heated liquid may even be circulated through jackets that surround the discharge head and the bowl and column assemblies to prevent the cargo from solidifying within the pump. Furthermore, VTP's are sometimes used to discharge cryogenic cargoes, such as liquefied petroleum gas, which can be cooled to a temperature of −60 F.

The fluid contained within the discharge head, column, and bowl assembly of a conventional VTP will drain from the pump when the driver is stopped. Several systems have been developed that enable this liquid to be removed from a vertical turbine cargo pump after the pump-out cycle has been completed. One arrangement includes a nonreturn valve that is mounted in the pump's suction bell. During normal operation the valve is kept open by the cargo being pumped; however, once suction is lost, the valve closes and prevents the cargo in the VTP from draining back into the tank. After the cargo tank has been stripped, the VTP is stopped, its discharge valve is closed, and pressurized air or inert gas is injected into the pump. The gas forces the cargo contained within the VTP through a bypass line that connects the lower end of the pump to the ship's discharge piping.

Hydraulically driven submersible cargo pumps. Hydraulically driven submersible pumps are used to discharge cargo on many multi-product petroleum and chemical carriers that utilize an individual pump/tank

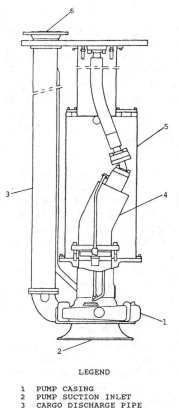

LEGEND

1 PUMP CASING
2 PUMP SUCTION INLET
3 CARGO DISCHARGE PIPE
4 HYDRAULIC MOTOR
5 COFFERDAM PIPE
6 ABOVE-DECK DISCHARGE FLANGE

Fig. 44 Hydraulically driven submersible cargo pump

arrangement. These pumps are also used for cargo unloading on some crude-oil carriers. As illustrated by Fig. 44, each unit consists of a single-stage end-suction centrifugal pump that is driven by a submersible hydraulic motor. Because the motor is installed directly on top of the pump, the need for a line shaft with its bearings is eliminated. The pump and motor assembly is suspended from the main deck by piping that includes the supply and return lines for the hydraulic oil required to drive the unit. Cargo being discharged from the pump travels to the main deck through a separate vertical pipe that is adjacent to the hydraulic piping. A control valve mounted on top of the above-deck cover plate is used to vary the flow of hydraulic oil to the motor and, therefore, the pump's speed. Although the hydraulic oil required by all of the cargo pumps on a vessel is frequently supplied by a central hydraulic system, on some vessels each cargo pump receives hydraulic oil from an independent self-contained power unit.

To increase the NPSHA, the submersible pump's impeller is mounted on the lower end of the pump shaft. A back wearing ring, together with balancing holes in the impeller's upper shroud, is often used to reduce the axial thrust caused by hydraulic unbalance. Support for the

pump's shaft is provided by ball or roller bearings installed in a mounting bracket that is attached to the upper portion of the pump's casing. The bearings are submerged in, and lubricated by, the hydraulic oil that drains from the pump's motor. To simplify maintenance, the design of the pump frequently permits it to be removed from the cargo tank while the hydraulic piping is still in place.

In general, rotary-vane or axial-piston hydraulic motors are used to drive these cargo pumps. Although it is submerged within the vessel's tank, the motor is surrounded by an outer pipe or cofferdam that isolates it from the cargo. Twin double-lip or mechanical seals that are separated by a void space are generally used to prevent the hydraulic oil from mixing with the cargo at the shaft penetration in the casing. By providing inlet and outlet connections to the void space, an inert buffer fluid can be circulated between the seals and monitored above deck so that a seal leak can be detected by the operator.

To prevent liquid in the submersible cargo pump's vertical discharge pipe from draining back into the cargo tank when the pump is stopped, each pump generally has a bypass line and connections that enable its vertical discharge pipe to be purged with air or inert gas after the pump-out cycle has been completed.

In lieu of rigid piping, some lower-capacity hydraulically driven submersible pumps are fitted with flexible hoses for the supply and return of hydraulic oil and for cargo discharge. These units, which are designed to be portable, can be lowered into a tank by the ship's crew and used to unload cargo in the event of a main cargo pump failure.

Electric-motor-driven submersible cargo pumps. On vessels that transport liquefied natural gas (LNG) or liquefied petroleum gas (LPG), electric-motor-driven submersible pumps are often used to discharge cargo. Each unit consists of a single-stage end-suction centrifugal pump that is mounted on the lower end of a submersible electric motor. The inlet to the pump is typically fitted with a suction bell, which guides incoming fluid directly to the eye of an impeller that is mounted on the lower end of the cantilevered motor shaft. To reduce the pump's NPSH requirement, an inducer is sometimes installed below the inlet to the impeller. Cargo discharged from the impeller passes through a multi-vaned diffuser that is integral with the pump's casing, and up through an annulus formed by the outside of the motor and an outer casing that surrounds it. Above this annulus, which ends at the top of the motor, is the unit's discharge nozzle. This nozzle is generally flanged so that it can be connected directly to the vessel's discharge piping, which extends from the bottom of the cargo tank to the main deck.

The bearings that support the common pump and motor shaft are mounted at each end of the motor and are lubricated by a portion of the cargo being pumped. A portion of the pumped cargo also flows through and cools the motor. Because the motor is immersed in LNG or LPG, its housing must be designed to fully contain any explosions that can occur within it and must prevent any sources of ignition from being transmitted to the surrounding environment.

Pumps used for cargo unloading and for stripping and cooldown can be installed either directly within the cargo tanks that they empty or in suction vessels that are connected through piping to multiple cargo tanks. In addition, some pumps are installed within a vertical column pipe or cofferdam that extends from the bottom of the cargo tank to the main deck, which eliminates the need for a separate vertical discharge pipe. A spring-loaded valve in the bottom of the column enables liquid from the cargo tank to enter the pump's suction bell and impeller. However, if the pump is removed from the column, the valve closes and prevents gas in the cargo tank from escaping into the atmosphere. This latter arrangement is frequently used for the emergency unloading pumps.

Rotary cargo pumps. The types of rotary pumps used most often in cargo service include gear, screw, lobe, and sliding-vane units. Many of the fluids discharged by cargo pumps do not have good lubricity. In addition, during stripping a cargo pump can ingest gas and vapor. Consequently, rotary pumps used in this application often have external bearings that rely on an independent source of oil for lubrication.

To enable the capacity delivered to be adjusted during the pump-out cycle, rotary cargo pumps are frequently furnished with drivers that have a multispeed capability. When rotary pumps are used for stripping or to discharge cargo from more than one tank, they are often installed in "pump rooms" that are located in the lower part of the ship. This arrangement enables the pumps to be connected through suction piping to multiple cargo tanks. To eliminate the need for the drivers to be explosion-proof, they are often installed in a separate compartment and are coupled to the pumps through horizontal or vertical jackshafts.

Rotary pumps can also be furnished in a deepwell configuration. This arrangement eliminates much of the suction piping that is required when the cargo pumps are installed in a common pump room and is often utilized on vessels that carry high-viscosity cargoes. In addition, on some multi-product carriers, a rotary pump that is driven by a submersible hydraulic motor is mounted on the bottom of each cargo tank.

Reciprocating cargo pumps. The cargo tanks on some vessels are stripped by direct-acting steam- or gas-driven reciprocating piston pumps. Duplex-type pumps are typically furnished for this application. When installed on vessels that carry crude oil or a limited number of different cargoes, direct-acting stripping pumps are often located in a pump room and connected to the vessel's cargo tanks through suction piping. This arrangement enables one pump to be used to strip multiple tanks. These pumps are frequently mounted vertically and are usually driven by steam. In an alternate arrangement, which is found on some multi-product carriers, a stripping pump is installed at the bottom of each cargo tank. The submersible direct-acting stripping pumps are generally mounted horizontally and are typically driven by compressed air or inert gas.

Diaphragm cargo pumps. On some multi-product carriers, a submersible diaphragm pump driven with either compressed air or inert gas is mounted on the bottom of each cargo tank and is submerged in the fluid that it discharges. In addition, because they are self-priming, diaphragm pumps that unload cargo can sometimes be placed on the deck above a tank. With this latter arrangement, the pump draws fluid through a suction hose that is lowered into the cargo tank. Although diaphragm-type cargo pumps are used primarily only for stripping, they can also serve as backup units and permit cargo unloading to continue in the event of a main cargo pump failure.

Slurry cargo pumps. On some ore and mineral carriers, the vessel's dry bulk cargo is mixed with water, and is loaded and discharged as a slurry. The single-stage end-suction centrifugal pumps that can be used to offload this cargo have large waterways to pass the solid particles in the slurries being handled and are often constructed from abrasion-resistant hard metals. Packed stuffing boxes or lip seals are frequently used for shaft sealing. Clean flushing water is generally injected into the seal area from an external source. In addition, the rear shroud of a slurry pump's impeller sometimes has external vanes that reduce the pressure in the seal area. (These vanes can also improve the shaft's axial balance.)

The effect that pumping a slurry has on the performance of a centrifugal pump depends on the specific gravity of the liquid/solid mixture, and on the size and concentration of the solid particles. When slurries containing large particles are pumped, the total head developed and pump efficiency will generally both be reduced along the entire range of operation when compared to the pump's performance with clear water. This reduction in efficiency also results in an increase in the brake horsepower required to drive the pump.

Typical conditions of service for pumps used to discharge slurry cargoes include capacity ratings in the range of 1400 to 5600 gpm at total heads from 130 to 250 ft when handling a slurry with a solid content up to 70% by weight. Separate slurry booster pumps may be provided for the transfer of cargo from individual holds located throughout the vessel to the main collecting tank from which the main discharge pumps take suction. The main discharge and the booster pumps can be driven by steam turbines, electric or hydraulic motors, or diesel engines.

In addition to the slurry pumps, various high-pressure and low-pressure pumps are required on these bulk-carrying vessels to supply water for cargo pump seal flushing, to discharge water loaded with the cargo after the ore or mineral portion of the slurry has settled (referred to as decanting), to add water to the cargo prior to its discharge (referred to as re-pulping), and to flush the vessel's cargo piping. Vertical turbine pumps are often used in these applications.

Related applications. In addition to the cargo unloading pumps, there are several other applications that are unique to vessels that carry liquid bulk cargoes. These include the following:

• An inert-gas scrubber pump—supplies seawater to the scrubber that cools, cleans, and desulfurizes gas used to inert the vessel's cargo tanks. Vertically mounted single-stage centrifugal pumps that deliver capacities up to 2500 gpm at total heads of approximately 100 to 150 ft are often used in this application. These units are generally driven by electric motors.

• An inert-gas deck-seal water pump—supplies seawater to the deck seal that is required to prevent vapor in the vessel's cargo tanks from flowing back through the inert gas system to the machinery spaces. Typical conditions of service for the single-stage end-suction centrifugal pumps often used in this application include capacity ratings that range from 15 to 100 gpm at total heads up to 200 ft. Many of these units are close coupled to electric motors.

• A tank-cleaning pump—supplies seawater to the vessel's tank-washing machines when heated seawater is used to wash cargo tanks. The total head developed by this pump, which must be sufficient to overcome losses in the piping and heater that are downstream from the pump, raise the elevation of the seawater to the top of the cargo tanks, and provide the required pressure at the nozzles of the tank-washing machines, can be as high as 500 ft. Capacity ratings are based on the number of tank-cleaning pumps used and the number of tanks washed simultaneously, and can range from approximately 300 to over 5000 gpm. Vertically and horizontally mounted single-stage centrifugal pumps that have axially split casings and double-suction impellers are often used in this application. In addition, on some vessels two-stage horizontal centrifugal pumps and multistage vertical turbine pumps have been used. Tank-cleaning pumps are generally driven by either electric motors or steam turbines.

f. Dredging. Many of the centrifugal pumps used for dredging service are similar in design to the hard-metal pumps described previously for slurry cargo unloading. Pumps with replaceable casing liners are also used. Dredge pumps are sometimes mounted directly on the vessel's drag arms and are submerged during operation. This arrangement is generally necessary if deep dredging is to be performed so that the suction lift to the pump will be minimized. In lieu of submerged pumps, some dredges are fitted with inboard pumps that are located in the lower portion of the hull. In addition, dredges that discharge directly into long pipelines may have inboard pumps that operate in series with separate submerged drag-arm pumps. Dredge pumps, which are sometimes furnished in a close-coupled configuration, are generally driven by electric motors.

When discharging into a hopper, the rated capacity of the dredge pumps is generally proportional to the hopper's surface area [26]. During this mode of operation, the total head that must be developed is relatively low. When discharging into a pipeline, however, lower capacities and higher heads are generally required. To enable the same dredge pump to discharge into a hopper or into pipelines of various lengths, the drive motors used often have a variable-speed capability. In addition, piping may be arranged so that multiple inboard pumps can be operated in series. Capacity ratings for dredge pumps can exceed 20,000 gpm at total heads up to 300 ft while pumping a liquid/solid mixture with a specific gravity of approximately 1.2.

Separate centrifugal pumps are generally used to supply sealing and flushing water to the dredge pump's shaft seals. In addition, on hopper dredges, centrifugal-type jetting and washdown pumps are often provided to supply the high-pressure water used to remove material that may cling to the sides of the hopper after its dump doors are opened.

Section 2
Compressors

2.1 Fundamentals. Compressors are used to increase the pressure of a gas. Like pumps, compressors can be classified as either kinetic machines, which includes centrifugal and axial compressors, or positive-displacement machines, which includes reciprocating and rotary compressors. To increase the compression ratio, the ratio of the compressor's outlet pressure divided by its inlet pressure, multistage units are often used.

If compression is assumed to be adiabatic, the theoretical rate of energy transfer in a single-stage compressor is equal to

$$E_{ad} = \dot{m} H_{ad} \tag{42}$$

where

E_{ad} = theoretical rate of energy transfer for an adiabatic process, ft-lbf/s

$\dot{m}$ = mass flow rate, lbm/s
H_{ad} = adiabatic head, ft-lbf/lbm

Using subscripts 1 and 2 to refer to conditions at the inlet and outlet of the compressor, respectively, the adiabatic head of the gas passing through a compressor equals

$$H_{ad} = J(h_2 - h_1) + \frac{V_2^2 - V_1^2}{2g_c} \tag{43}$$

where

J = mechanical equivalent of heat, 778 ft-lbf/Btu
h = enthalpy per unit mass, Btu/lbm
V = flow velocity, fps
g_c = gravitational constant, 32.17 ft-lbm/lbf-s^2

If a compressor takes suction from the atmosphere and discharges into a receiver, the difference in kinetic energy, $(V_2^2 - V_1^2)/2g_c$, is generally low; therefore, it has been omitted from the equations that follow. For an ideal gas with a constant specific heat, the change in enthalpy per unit mass, Δh, is equal to

$$\Delta h = h_2 - h_1 = c_p (T_2 - T_1) \tag{44}$$

where

c_p = specific heat at a constant pressure of the gas being compressed, Btu/lbm-deg R
T = absolute temperature, deg R

Furthermore, for an ideal gas, c_p is equal to

$$c_p = \frac{Rk}{J(k-1)} \tag{45}$$

where

R = specific gas constant, ft-lbf/lbm-deg R (53.34 for dry air)
k = specific heat ratio (1.395 for air)

By substituting equation (45) into equation (44) and rearranging terms, the expression for Δh can be rewritten as

$$\Delta h = \frac{RT_1 k}{J(k-1)} \left[\frac{T_2}{T_1} - 1 \right] \tag{46}$$

Based on the additional assumption that adiabatic compression will also be isentropic or internally reversible:

$$\frac{T_2}{T_1} = \left(\frac{p_2}{p_1} \right)^{\frac{k-1}{k}} = r_p^{\frac{k-1}{k}} \tag{47}$$

where

p = absolute pressure, psia
r_p = compression ratio

Ignoring the difference in velocity head, an expression for H_{ad} can be derived from equations (47), (46), and (43) and substituted into equation (42):

$$E_{ad} = \dot{m} \frac{RT_1 k}{k-1} \left[r_p^{\frac{k-1}{k}} - 1 \right] \tag{48}$$

Equation (48) can be rewritten in terms of volumetric capacity (based on the perfect gas law) and horsepower as follows

$$P_{ad} = \frac{E_{ad}}{550} = \frac{1}{229.2} p_1 q_1 \frac{k}{(k-1)} \left[r_p^{\frac{k-1}{k}} - 1 \right] \tag{49}$$

where

q = volumetric flow rate or capacity, cfm
P_{ad} = reversible or ideal adiabatic power, hp

If there is no cooling within or between adjacent stages, equation (49) can also be used to calculate the adiabatic power for a multistage compressor.

To reduce the volumetric flow rate and, therefore, the power required for compression in subsequent stages, multistage compressors often have heat exchangers, referred to as intercoolers, that reduce the temperature of the gas being compressed as it travels between successive stages. Although when intercooling is performed the entire compression process is no longer adiabatic, the compression occurring within each stage prior to intercooling can still be compared to an adiabatic process. If "ideal" intercooling is assumed, the interstage temperature will be reduced to the inlet temperature, T_1, and the compression process within each stage, together with its intercooler, will be isothermal. Using subscript i to refer to the interstage conditions, this assumption results in the following

$$p_1 q_1 = p_i q_i = \text{constant} \tag{50}$$

So that each stage is equally loaded and adiabatic work is minimized, a multistage compressor is generally designed so that the compression ratio in each stage is equal, which results in

$$r_{p,st} = \left[\frac{p_2}{p_1} \right]^{1/n} = r_p^{1/n} \tag{51}$$

where

$r_{p,st}$ = compression ratio in each stage
n = number of compressor stages

By using the results of equations (50) and (51), the expression for ideal adiabatic power given in equation (49) can be rewritten and used for a multistage compressor with n stages that are each fitted with an ideal intercooler, that is

$$P_{ad} = \frac{1}{229.2} n p_1 q_1 \frac{k}{(k-1)} \left[r_p^{\frac{k-1}{kn}} - 1 \right] \tag{52}$$

Because of various losses, the actual compression process is not reversible. With no cooling during compression, the brake horsepower required to drive a compressor exceeds the ideal or reversible adiabatic value calculated in equation (52). The actual brake horsepower required by the compressor is equal to

$$P_C = \frac{P_{ad}}{\eta_C} \tag{53}$$

where

P_C = brake horsepower required to drive the compressor, hp
η_C = compressor efficiency (based on isentropic adiabatic compression), %/100

η_C includes the effects of thermodynamic and fluid friction losses in the compression process, as well as mechanical losses due to factors such as friction in the compressor's bearings and seals, and is equal to

$$\eta_C = \eta_{ad} \eta_m \tag{54}$$

where

η_{ad} = isentropic adiabatic compression efficiency, %/100
η_m = mechanical efficiency, %/100

With no cooling, η_{ad} is equal to the following

$$\eta_{ad} = \frac{T_{2\,isen} - T_1}{T_{2\,act} - T_1} = \frac{T_1\,(r_p^{\frac{k-1}{k}} - 1)}{T_{2\,act} - T_1} \qquad (55)$$

where

$T_{2\,isen}$ = ideal gas temperature at the compressor outlet based on isentropic adiabatic compression, deg R

$T_{2\,act}$ = actual gas temperature at the compressor outlet, deg R

If heat is transferred during compression, the adiabatic assumption is no longer valid. With heat transfer, the relationship between the pressure and the volume of the gas being compressed is sometimes assumed to follow that of a polytropic process, as shown below

$$pq^m = \text{constant} \qquad (56)$$

where m is a polytropic exponent. The substitution of exponent m for specific heat ratio k in equation (49) results in an expression that represents the ideal power required for compression without intercooling during a polytropic process. If because of losses heat is added to a gas during compression, the power required by the compressor will increase and exponent m will exceed k. If, however, the compressor is cooled sufficiently to result in a net transfer of heat away from the gas being compressed, the value of m will be less than k. Polytropic compression can be compared with adiabatic compression using the polytropic efficiency and the following:

$$\eta_{poly} = \frac{P_{poly}\,\eta_{ad}}{P_{ad}} = \left[\frac{m}{m-1}\right]\left[\frac{k-1}{k}\right] \qquad (57)$$

where

η_{poly} = polytropic efficiency, %/100
P_{poly} = ideal polytropic power, hp

The minimum ideal power is achieved when $m = 1$, which represents isothermal compression. Under this condition there is no change in the enthalpy of an ideal gas being compressed, and the work performed is equal to the heat removed from the gas. The power during ideal or reversible isothermal compression can be calculated as follows

$$P_{iso} = \frac{1}{229.2}\,p_1 q_1 \log_e r_p \qquad (58)$$

where P_{iso} is the ideal isothermal power (hp). Additionally, the isothermal efficiency is equal to the following

$$\eta_{iso} = \frac{RT_1 \log_e r_p}{J\,[c_p\,(T_{2\,act} - T_1) + Q^*]} = \frac{\eta_{ad} \log_e r_p}{\frac{k}{k-1}[r_p^{\frac{k-1}{k}} - 1]} \qquad (59)$$

where

η_{iso} = isothermal efficiency, %/100
Q^* = heat per unit mass removed from gas during compression, Btu/lbm

The equations given above are based on the assumption that the gas being compressed follows the ideal or perfect gas laws. Although this generally results in sufficient accuracy for lower-pressure applications, at high pressures the departure of real gases from the ideal gas laws must be considered [27].

Intercooling reduces the work required for multistage compression. Furthermore, multistaging, combined with intercooling, offers the following advantages:

• The reduction in the temperature of the gas entering each stage simplifies lubrication requirements.

• Due to a reduction in the specific volume of the gas entering successive stages, a smaller compressor can be used to handle a given mass flow rate of gas.

• Because of the increase in temperature that occurs during compression, the relative humidity of air that has been compressed is initially reduced. However, at a constant temperature, the relative humidity of air increases with pressure. As a result of this relationship, if the temperature of air discharged from a compressor stage is reduced to the stage's inlet temperature, there will be an increase in the air's relative humidity when compared with the value at the entrance to the stage. Consequently, a portion of the water vapor contained in the compressed air will condense, reducing the air's moisture content. (In addition to the intercoolers used to cool gas traveling between a compressor's stages, many air compressors are also fitted with an aftercooler that permits the temperature and moisture content of air discharged from the compressor's last stage to be reduced.)

• By reducing the temperature of the gas being compressed, overheating of the compressor's components can be reduced or eliminated. In addition, in the case of oil-lubricated air compressors, a reduction in air temperature reduces the risk of an explosion.

• The use of multiple stages reduces the number of components that must be subjected to gas at the compressor's discharge pressure.

To reduce the wear and corrosion that can be caused by dirt and moisture that enter a compressor, filters and moisture separators are often fitted over inlet connections. In addition, to reduce the work required for compression, air compressors should be located in areas of the vessel where the lowest inlet air temperature is available.

2.2 Reciprocating Compressors. In a reciprocating compressor, which is illustrated in Fig. 45, air is drawn into, compressed in, and discharged from an enclosed cylinder by the reciprocating motion of a piston. This motion is transmitted to the piston by a rotating driver through a crankshaft and a connecting rod. In some compressors a crosshead and piston rod may be installed between the connecting rod and the piston. Reciprocating compressors are classified as being either horizontal or vertical based on the orientation of the reciprocating motion within the cylinders. In vertical multistage compressors the cylinders are sometimes located directly above the axis of the horizontal crankshaft in an "in-line" configuration. However, to reduce the machine's length, the cylinders in a multistage compressor may also be arranged in groups that are oriented radially around the crankshaft's circumference. Compressors of this latter type are often identified based on the letter in the alphabet that the cylinder configuration most closely resembles. In a "V" or "Y"

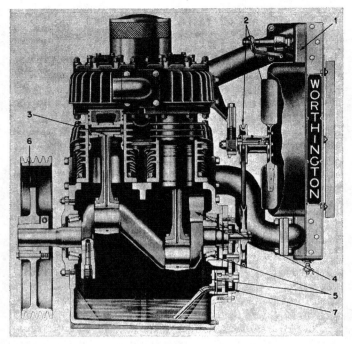

LEGEND

1   INTERCOOLER
2   FAN & BELT
3   PISTON & PIN
4   CRANKSHAFT
5   MAIN BEARING
6   DRIVE SHAFT
7   OIL PUMP

Fig. 45   Two-stage reciprocating compressor

type compressor, the cylinders are located (typically at an angle of 45 deg) on opposite sides of a vertical plane that passes through the crankshaft's axis; in a "W" type compressor, the cylinders are divided into groups that each contain two outer cylinders located on opposite sides (usually at 60-deg angles) of a third vertically oriented cylinder; and in an "L" type compressor, one cylinder in each group of two is oriented vertically above the crankshaft's axis, and the second is oriented along a horizontal plane passing through this axis. Reciprocating compressors can also be furnished with multiple horizontally mounted cylinders that are positioned either on opposite sides of a horizontal crankshaft, referred to as a horizontal opposed compressor, or radially around a vertical crankshaft. Pistons operating within a single group of cylinders that have the same axial location in the compressor are typically linked to the same crankthrow on the crankshaft.

Multistage compressors may be arranged so that some of their cylinders operate in parallel while others operate in series. For example, in a typical three-cylinder two-stage "W" type compressor, the two outer cylinders jointly form the first stage, and both discharge gas into the single cylinder located between them that forms the compressor's second stage. Because of the reduction in the volume of a gas as its pressure is increased, the bores

of the cylinders and the corresponding outside diameters of the pistons used in successive stages of a multistage compressor are generally progressively reduced. The bore sizes of the various cylinders are often included in the compressor's nomenclature, and are listed in order of stage. A typical nomenclature for a two-stage "W" type compressor could be 6/6/5 × 5, which identifies the compressor as having two first-stage cylinders with bores of 6 in. each and a third second-stage cylinder with a bore of 5 in. The final 5-in. dimension following the "X" identifies the stroke length of the compressor.

Compressor pistons may be classified as single-acting when only one face is used to compress gas or double-acting when the compressor is arranged so that gas can be compressed by both of the piston's faces. Although a double-acting piston is generally driven through a cross-head and a piston rod, a single-acting piston may be driven directly by the connecting rod. Pistons that are installed using this latter arrangement are sometimes referred to as "trunk" type pistons. As a result of the elimination of the crosshead in this design, side loads developed in the connecting rod are transmitted to the piston. To provide a larger surface area to absorb these loads and reduce wear, trunk pistons typically have long sidewalls, referred to as skirts.

Because of the reduction in piston size in successive stages, multistage compressors can be arranged so that a single piston with multiple outside diameters, referred to as a "differential multistage" piston, is used to compress gas in two or more separate cylinders located along the same radial axis. With a typical vertically mounted two-diameter single-acting piston, the lower portion of the piston is often of a larger diameter than the upper section and is used in one of the compressor's lower-pressure stages. Gas in the lower, larger portion of the cylinder is compressed within the annulus formed between the cylinder's bore and the outside diameter of the piston's smaller upper section. The upper face of the piston simultaneously compresses gas in a smaller-diameter cylinder that forms one of the compressor's higher-pressure stages. Configurations that include stepped double-acting pistons with three different diameters are also used.

To maintain the necessary close-clearance seal between the moving piston and the inside wall of the cylinder, pistons are typically fitted with replaceable rings. In addition, cylinders can be fitted with replaceable liners. To reduce weight, the larger-diameter pistons used in a compressor's lower-pressure stages are sometimes furnished in aluminum. The smaller higher-pressure pistons are, however, frequently furnished in cast or ductile iron, and the same is also often used for the compressor's cylinders.

The crankshaft is frequently supported by roller bearings. In addition, sleeve-type journal bearings are generally provided to absorb radial loads that are applied to the pins at the ends of each connecting rod. It is common in smaller compressors for lubricating oil in the crankcase to be splashed onto the bearings and running gear by the eccentric crankthrows on the rotating crankshaft. Larger compressors, however, usually have a forced-feed lubrication system, or they may rely on a combination of splash and forced lubrication. With a forced-feed system, the lubricating oil is removed from a sump in the compressor's crankcase by a rotary pump, circulated through a filter, and delivered to the main bearings on the crankshaft. The pressurized oil then flows through drilled ports in the crankshaft and connecting rods to the bearings at the ends of the connecting rods. As it leaks out around the edges of these bearings, the oil lubricates the crossheads, when used, and then drains back into the sump. The lubricating-oil pump is typically driven off the compressor's crankshaft. In some compressors, oil is also used to lubricate the cylinders. Small compressors with oil-lubricated cylinders may have dippers on the end of each connecting rod that splash oil onto the pistons and cylinder walls. In many larger units, a mechanical lubricator supplies oil through individual lines to each cylinder. A check valve is generally installed in each oil supply line to prevent gas in the cylinders from entering the lubrication system.

In compressors with non-lubricated cylinders, referred to as "non-lubricated" compressors, a distance piece is located between the cylinders and the frame that houses the crankshaft and the connecting rods. By separating the pistons from the compressor's lubricated running gear, oil is prevented from entering the cylinders and contaminating the gas being compressed. Due to the use of the distance piece, a crosshead and piston rod must generally be used to transmit the reciprocating motion of each connecting rod to the corresponding piston. Stationary wiper and packing rings are typically fitted around each piston rod. Because the pistons and cylinders in these compressors receive no lubrication, materials that have inherent self-lubricity, such as polytetrafluoroethylene composites, are often used for piston rings and cylinder packing.

Reciprocating compressors are generally cooled with either air or water. The cylinders in air-cooled compressors often include large external fins that increase the surface area available for heat transfer. These compressors may also be fitted with belt- or shaft-driven fans. In water-cooled compressors, fresh water or seawater is circulated through jackets that are built into the walls of the cylinders and cylinder heads. In addition to the beneficial effect on efficiency, water cooling results in a more uniform temperature distribution throughout the compressor and can eliminate the formation of localized hot spots on the surfaces of cylinders and pistons. Water cooling is, therefore, incorporated into the design of many high-pressure compressors. Separate air-cooled or water-cooled intercoolers are often provided to cool the gas traveling between the compressor's successive stages. In addition, an air-cooled or water-cooled aftercooler is frequently used to cool the gas being discharged from the compressor's final stage. Furthermore, a compressor may have a water-cooled heat exchanger that is used to cool oil in the lubrication system. Although cooling water can be circulated through a compressor's various heat exchangers by an independent pump, some compressors are fitted with a crankshaft-driven centrifugal water pump.

Each of a reciprocating compressor's cylinders must be fitted with valves to control the admission and exhaust of

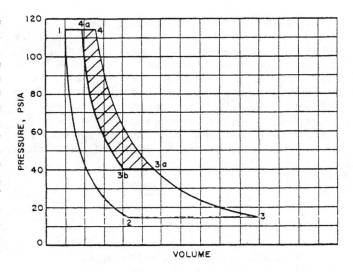

Fig. 46 Theoretical indicator card for a reciprocating compressor

the gas being compressed. Valve types used most frequently, listed in ascending order based on pressure capability, include: a strip valve, which has multiple flat strips that cover ports in the valve's seat; a ring plate-type valve, which has several concentric ring-shaped plates that are held in place over ports in the seat by multiple coil springs; and a disk valve, which has a disk-shaped plate that is held against the seat by a single spring or retainer. These valves open and close automatically as a result of the difference between the pressure within the cylinder and the pressure on the opposite side of the valve. Although less common, cam-operated spring-loaded valves are used in some reciprocating compressors.

The compression cycle is illustrated graphically in Fig. 46, which shows an indicator card diagram of cylinder pressure plotted versus volume for a reciprocating compressor with pressure-actuated valves. The area enclosed by lines 1-2-3-4-1 represents the work of compression in a single-stage compressor. As the compressor's piston begins its suction stroke (point 1 on Fig. 46), the pressure within the cylinder is reduced until it is slightly less than the pressure in the inlet line. At this point (point 2 on Fig. 46), the compressor's inlet valve opens, and gas begins to enter the cylinder. The low pressure within the cylinder keeps the discharge valve closed and, therefore, prevents high-pressure gas in the discharge line from reentering the cylinder. Gas continues to enter the cylinder until the piston reaches the end of the intake stroke (point 3 on Fig. 46). The piston then reverses its stroke and begins to compress the gas contained within the cylinder. The resulting increase in cylinder pressure forces the inlet valve to close. When the pressure within the cylinder is slightly higher than the pressure in the discharge line (point 4 on Fig. 46), the compressor's discharge valve is forced to open and the compressed gas is discharged from the cylinder. The discharge of gas continues until the piston completes the compression stroke and returns to its initial position (point 1 on Fig. 46). This cycle is then repeated.

The volume corresponding to point 1 on Fig. 46 is referred to as the cylinder's clearance volume. It equals the volume of compressed gas that remains in the cylinder at the end of the compression stroke. As the piston begins the next intake stroke, this gas reexpands and reduces the amount of new gas that can enter the cylinder. Using subscripts 1, 2, and 3 to refer to points on the indicator diagram in Fig. 46, the resulting reduction in the volume available for gas entering the cylinder is represented by the compressor's theoretical volumetric efficiency, which equals

$$\eta_{v,\text{th}} = \frac{v_3 - v_2}{v_3 - v_1} \qquad (60)$$

where

$\eta_{v,\text{th}}$ = theoretical volumetric efficiency, %/100
v = volume in compressor's cylinder, in.^{3}

The clearance volume reduces the amount of gas that can be drawn into the compressor; therefore, it affects the size of the compressor that is required to handle a given volumetric flow rate of gas. Because of leakage past the piston rings, packing, and valves, the compressibility of real gases, and the inability of the cylinder to fill completely, a reciprocating compressor's actual volumetric efficiency is somewhat less than the theoretical value.

Deviations from a reversible adiabatic process caused by various factors, such as gas turbulence, the heating of incoming gases, and pressure drops through the compressor's suction and discharge valves, reduce the compressor's efficiency. However, because the valve losses represent a smaller percentage of the work of compression at higher compression ratios, the efficiency of a cooled reciprocating compressor can increase with compression ratio. In addition, the efficiency of this type of a compressor generally increases with higher absolute inlet pressures and lower gas specific gravities. Typical values of efficiency for reciprocating compressors are in the range of 75 to 85% based on adiabatic compression.

Two-stage compression with intercooling is illustrated on Fig. 46 by lines 1-2-3-3a-3b-4a-1. Compression in the two-stage compressor's first stage ends at point 3a. The gas is then withdrawn from the cylinder and cooled, which results in a reduction in its volume to point 3b. After the gas enters the second-stage cylinder, compression continues to point 4a. The reduction in work achieved by using a two-stage intercooled compressor is represented by the shaded area enclosed by lines 3a-3b-4a-4-3a.

A typical performance curve for a reciprocating (positive-displacement) compressor operating at constant speed is shown in Fig. 47. Capacity adjustments can be made by varying the operating speed. Although steam or gas turbines, diesel engines, and direct-acting steam engines can be used to drive reciprocating compressors, many shipboard compressors are driven by electric motors at speeds below 1200 rpm. The driver is sometimes coupled, either directly or through a speed reducer, to the extended end of the compressor's crankshaft. Alternatively, torque may be transmitted from the driver to the crankshaft through a multi-V-belt drive arrangement. A

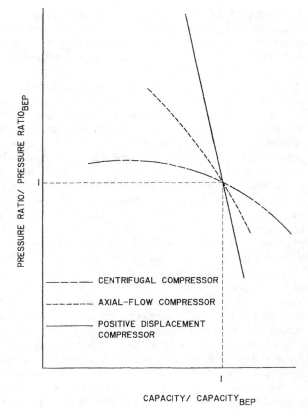

CENTRIFUGAL COMPRESSOR

AXIAL-FLOW COMPRESSOR

POSITIVE DISPLACEMENT COMPRESSOR

Fig. 47 Typical compressor performance curves

reciprocating compressor is frequently furnished with hydraulically or electrically operated unloaders that relieve pressure within the cylinders, often by holding the suction valves open, when the compressor is stopped. The unloaders initially remain open during start-up, which reduces the starting torque that must be developed by the compressor's driver. Unloaders may also be used for capacity control during constant-speed operation. In addition, constant-speed double-acting compressors are sometimes fitted with clearance pockets, which are chambers located at each end of a cylinder that can be opened to increase the clearance volume and decrease the capacity delivered.

2.3 Dynamic Compressors. Dynamic compressors, which are often similar in appearance to centrifugal pumps, are furnished in both single-stage and multistage configurations. The theory of performance for these machines is similar to that presented previously for centrifugal pumps, except that effects due to the compressibility of the gas being handled must be considered in the case of a compressor. Like centrifugal pumps, dynamic compressors can be designed as radial-flow, mixed-flow, or axial-flow machines, where these classifications refer to the primary orientation of flow at the discharge from the compressor's rotor. The type of dynamic compressor best suited for an application is often predicted based on specific speed, N_S, which can be calculated by substituting operating speed, N, in rpm, inlet capacity, q_1 in cfm, and

Fig. 48 Multistage centrifugal compressor

LEGEND

1   COMPRESSOR
2   INCREASER GEAR
3   DRIVER
4   INTERCOOLER
5   CONTROL CABINET
6   AUXILIARY LUBE PUMP

adiabatic head per stage, $H_{\mathrm{ad},st}$, in ft-lbf/lbm, into equation (3). With these units of measurement and based on the use of single-suction impellers, typical specific speed ranges corresponding to the different types of dynamic compressors are as follows [27]:

COMPRESSOR TYPE	N_S
Radial-flow	400 to 950
Mixed-flow	800 to 1400
Axial-flow	> 1300

Radial-flow compressors are usually referred to as centrifugal compressors, while those built for axial-flow are generally identified as axial compressors. Mixed-flow compressors are seldom used [28].

a. Centrifugal compressors. The impeller in a single-stage centrifugal compressor is frequently overhung on the end of a cantilevered shaft. However, in some single-stage machines the impeller is mounted on a shaft that is supported by bearings at each end. Although a between-bearings configuration is also used in many multistage centrifugal compressors, the overhung arrangement has been used to support multiple, but generally not more than three, in-line impellers. In addition, some multistage compressors are furnished with the impellers overhung on one or both ends of multiple shafts. These shafts are located around the periphery of, and driven by, a central gear that is coupled to the compressor's driver. By varying the number of teeth used on the shaft-mounted pinions, each impeller can be driven at a different speed. A typical multistage centrifugal compressor is shown in Fig. 48.

Single-suction impellers are generally used with an overhung mounting arrangement and can be of the open, semiopen, or closed design. Because of the reduction in volume of the gas passing through a centrifugal compressor, the widths of the impellers used in multistage machines are generally progressively reduced in each successive stage. In addition, to maintain geometric proportionality, the impeller outside diameters are frequently reduced with the widths. To eliminate the local stress concentrations that result from keyways, in many high-speed centrifugal compressors, torque from the shaft to each impeller is transmitted through an interference fit.

The high velocity of the gas being discharged from the impeller is typically reduced in a multi-vaned diffuser, vaneless diffuser, or scroll-shaped volute. Many centrifugal compressors have a combination of these devices, such as a diffuser followed by a volute. The reduction in discharge velocity results in the conversion of a portion of the gas kinetic energy to potential energy and, therefore, in an increase in pressure. Diffusers used in a multistage compressor are often furnished as part of the stationary diaphragms that separate adjacent stages. Each diaphragm also includes return channels that direct the gas being compressed to the eye of the impeller in the next stage. To reduce its temperature, the gas traveling between stages in a multistage centrifugal compressor is frequently circulated through an intercooler. In addition, an aftercooler is often provided to reduce the temperature of the gas discharged from the compressor.

To minimize axial thrust, in some multistage centrifugal compressors an approximately equal number of single-suction impellers is mounted facing opposite ends of the casing. In addition, in some units gas is simultaneously admitted to both ends of the compressor, directed inward, and discharged from the center of the casing. This type of a machine, which is referred to as a "double-flow" compressor, frequently has a center impeller that is of the double-suction design. When all of the single-suction impellers in a multistage centrifugal compressor are mounted in the same direction, axial thrust is generally reduced through the use of an internal balancing piston or drum.

Residual axial loads acting on a centrifugal compressor's rotor are absorbed by a thrust bearing, which frequently consists of a rotating shaft-mounted collar that is positioned axially between two stationary flat lands or between multiple tilting pads. Radial support for the rotor is typically provided by cylindrical, multi-lobe, or tilting-pad journal bearings. These bearings are lubricated with oil that is generally pressurized and circulated by a shaft-driven rotary pump. Oil coolers and filters are often included in a centrifugal compressor's lubrication system.

Typical materials used in the construction of centrifugal compressors include cast iron, ductile iron, steel, or stainless steel for casings and stainless steel, aluminum, or titanium for impellers. To reduce interstage leakage and improve volumetric efficiency, close-clearance labyrinth-type seals are generally installed between adjacent stages in multistage compressors. In addition, when closed-type impellers are used in either single-stage or multistage compressors, replaceable stationary seals are often installed adjacent to each impeller's outer hub. Seals are also required at locations where the compressor's shaft penetrates the casing. Various types of labyrinth, carbon-ring, or mechanical seals may be used for shaft sealing.

b. Axial compressors. Axial compressors are generally supplied in a multistage configuration, with each stage consisting of a row of rotating vanes followed by a row of stationary diffuser vanes. A row of stationary inlet guide vanes may also precede the initial row of rotating vanes. In addition, one or more rows of stationary exit guide vanes may be used to reduce the velocity of gas leaving the compressor. Axial compressors are typically used to deliver relatively high capacities. Although the compression ratio per stage in an axial compressor is generally less than that in a comparably rated centrifugal compressor, an axial compressor will often be more efficient.

The profile of the individual vanes used in an axial compressor is generally based on an airfoil shape. The vanes in each stage of the rotor are mounted around the circumference of individual disks, a hollow drum, or a solid rotor that is integral with the shaft. Proper vane orientation and spacing is critical so that the pitch (the distance between corresponding points on adjacent vanes), inlet flow angle, and outlet flow angle for each vane meet design requirements. When individual disks are used in the rotor design, they are generally either mounted onto a common shaft or stacked axially and held together with through bolts. With the stacked disks or drum designs, a stub shaft is often pressed into each end of the rotor.

The outer ends of the stationary vanes in an axial compressor may be attached directly to the outer casing or mounted in a separate inner carrier. The inner ends of the stationary vanes are sometimes fitted with a thin cylindrical shroud that joins all of the vanes included in a single stage. Close radial clearances are generally maintained between the inner ends of the vanes and the rotor to reduce interstage leakage.

Because of the reduction in the volume of the gas passing through an axial compressor, to maintain a constant axial gas velocity, which is often used as a design criterion, the lengths of the rotating and stationary vanes used in each successive stage must be progressively reduced. This can be accomplished by using a tapered rotor and gradually increasing both the root diameter of the rotating vanes and the inside diameter of the stationary vanes, or by using a constant-diameter rotor with rotating vanes that have progressively smaller tip diameters. With this latter configuration, the inside diameter of the casing must be tapered. In a third arrangement, the diameters of both the rotor and the casing are changed. When any of these configurations are used, several of the final higher-pressure stages are sometimes fitted with vanes of equal length, which results in slight reductions of both the gas velocity and the losses at the compressor's outlet.

An axial compressor's rotor is typically supported radially by cylindrical or tilting-pad journal bearings mounted at each end of the shaft and axially by a tilting-pad thrust bearing. The bearings are lubricated with oil that is generally pressurized and circulated by a shaft-driven positive-displacement pump. A cooler and filter are frequently included in the compressor's lubrication system. To reduce axial loads, the discharge end of the rotor is often fitted with an internal balancing piston or drum. The leakage rate of the gas passing through the balancing device, which is vented to suction, is controlled with a labyrinth seal. Labyrinth seals are also used in many units to reduce leakage in locations where the ends of the shaft penetrate the casing. Materials of construction used with axial compressors are similar to those used for centrifugal compressors.

c. Dynamic compressor performance characteristics. Typical performance curves for centrifugal and axial compressors are included in Fig. 47. An axial compressor's head-capacity curve is generally steeper than that of a centrifugal compressor; therefore, while the capacity entering an axial compressor remains relatively constant over a range of discharge pressures, the discharge pressure developed by a centrifugal compressor remains relatively constant over a range of inlet capacities. Variations in inlet capacity, head, and power with changes in operating speed can be estimated using the relationships given in equations (29), (30), and (31). However, because the higher head developed at an increased operating speed also increases the volume ratio of the gas being compressed, the head actually developed at an inlet capacity found using equation (29) exceeds the head calculated using equation (30). Conversely, for a given reduction in speed and inlet capacity, the head developed is slightly less than that calculated using equation (30). The magnitude of these deviations increases with higher heads, heavier gases, and reduced impeller discharge relative flow angles.

As shown in Fig. 47, the stable region on the left side of a centrifugal or an axial compressor's head-capacity curve does not reach shutoff but, instead, stops at a minimum flow rate. Operation below this capacity, referred to as the "surge limit," can result in instability consisting of flow oscillations with brief periods of forward flow followed by intermittent back flow. Surging is typically caused by a sudden reduction in capacity resulting from stalling because of excessive positive incidence at the inlet to the impeller (or multi-vaned diffuser, if used). As the capacity being delivered is reduced, the head developed may momentarily be less than that in the system and a flow reversal can occur. Surging is, therefore, frequently associated with operation on a portion of the head-capacity curve that has a positive slope. Because of the compressibility of gases and their ability to store energy, the oscillations in flow often continue and can result in pressure pulsations, vibration, noise, and increased temperatures. The severity of the detrimental effects that result from surging is affected not only by the design of the compressor, but also by the elasticity of the system. The capacity at which stalling and the resulting surging begin is generally reduced with the operating speed. As a result of this relationship, if surging occurs in a centrifugal compressor, it can often be suppressed by reducing the unit's operating speed. However, once stalling and surging begin in an axial compressor, the surge limit can, in effect, shift to a higher capacity, and stalling can continue even after the operating speed is reduced [3].

The capacity on the right side of the head-capacity curve is limited by choking, which occurs when the relative velocity of gas within the rotor or the gas velocity at the inlet to the stator reaches the acoustic velocity, or when the Mach number equals one. When Mach one is reached, the mass flow rate passing through the compressor cannot be increased regardless of how much the compression ratio and head developed are reduced. In some centrifugal compressors, the range of performance is expanded by using an inducer-type impeller with vanes that follow the curvature of the hub and extend into the impeller's eye. As a result of the decreasing radii from the tip to the root along the leading edge of each vane in an inducer impeller, the average inlet peripheral velocity and Mach number are reduced.

Radial or backward-curved vanes are typically used in centrifugal compressor impellers. The steepness of a centrifugal compressor's head-capacity curve increases as the discharge angle of the impeller's vanes and the resulting relative discharge flow angle, measured with respect to the tangential or peripheral direction, are reduced. In addition, due to the increase in volume ratio when a heavier gas is compressed, which results in the development of a higher head for a given inlet volumetric flow rate and operating speed, the steepness of the head-capacity curve will generally be reduced as the gas inlet density of the gas increases. The increased volume ratio when heavier gases are compressed generally increases the positive incidence and, therefore, the inlet capacity at which surging begins. Due to the resulting increased mass flow rate for a given inlet capacity, the Mach number when handling heavier gases will also increase, which reduces the inlet flow rate at which choking occurs.

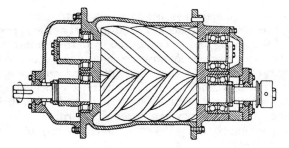

Fig. 49 Twin-screw compressor

In addition to the effects of vane discharge angle and gas density, the performance of an axial compressor is also affected by the stagger angle (the angle between the vane's chord line and the axial direction) of the vanes in its rotor. In general, an increase in the stagger angle will result in a steeper performance curve. In addition, the speed required to deliver a given flow rate will often increase with the stagger angle. As an alternative to adjusting the stagger angle of the rotor's vanes, the performance of some axial compressors can be adjusted at a constant operating speed by varying the stagger angle of the stationary vanes.

Compression ratios in axial compressors can be as high as 1.4 per stage; this compares to compression ratios of 2 to 5 per stage for centrifugal compressors [3]. Operating speeds for centrifugal and axial compressors can range from 5000 to 40,000 rpm. These compressors are often directly coupled to steam or gas turbines; however, they may also be driven through speed increasers by electric motors or diesel engines. In some motor-driven units, the necessary speed-increasing gears are furnished as an integral part of the compressor.

2.4 Rotary Compressors. Rotary compressors are positive-displacement machines in which the gas being compressed is forced through the casing by one or more rotating displacement elements. Unlike the reciprocating compressor, which is also a positive-displacement machine, a rotary compressor does not typically require internal suction or discharge valves. In addition, the flow from a rotary compressor is generally more uniform and has fewer pulsations than the flow from a reciprocating compressor. Included among the most common types of rotary compressors, which are typically mounted horizontally, are the following:

a. Twin-screw compressors. As shown in Fig. 49, a twin-screw compressor, also referred to as a helical-lobe compressor, consists of two meshing helical rotors mounted on counterrotating parallel shafts that are enclosed within a close-clearance casing. As the rotors turn and unmesh adjacent to the inlet port, a low-pressure region is created and gas is drawn into the casing. With the continued rotation of the shafts, the gas is trapped in pockets formed between the cavities in each rotor and the casing's inner wall and is forced through the casing in an axial direction. The meshing of the lobes on the male rotor into the grooves in the female rotor reduces the volume of the trapped gas and, therefore, increases its pressure.

In single-stage machines, compression continues until the closed pocket of gas reaches the outlet port at the opposite end of the casing and is discharged from the compressor. In a two-stage screw compressor, however, before being discharged from the casing, the gas is passed through a second set of smaller rotors. To maximize efficiency, the compressor should be sized so that the reduction in the volume of the gas passing through its rotors and the corresponding compression ratio match the requirements of the system.

In oil-flooded twin-screw compressors, oil is injected directly into the casing to lubricate and cool the rotors and to reduce the discharge temperature of the gas being compressed. The oil also helps to seal clearances between the rotors and the casing, which improves the volumetric efficiency. In these compressors, torque is generally transmitted directly from the male or main rotor on the driving or power shaft to the female or gate rotor on the idler shaft. When oil cannot be mixed with the gas being compressed, an oil-free dry compressor may be used. The casing furnished with a dry compressor often has external jackets through which cooling water is circulated. In addition, in some oil-free compressors fresh water is injected directly into the casing, which not only cools the gas being compressed but also helps to seal internal clearances. So that this water can be reused, it is generally separated from the compressed gas in a receiver located downstream from the compressor, passed through a heat exchanger and a filter, and returned to the casing. To maintain a clearance between the rotors and reduce wear, torque from the driving shaft to the idler shaft in an oil-free twin-screw compressor is usually transmitted through separate timing gears that are fitted on the end of each shaft.

When timing gears are used, they must be sized so that the female rotor on the idler shaft will be driven at the proper speed, which is generally different from that of the driving shaft. In higher-pressure twin-screw compressors the male rotor often has four lobes that mesh with six mating grooves on the female rotor. With this arrangement, the female rotor must rotate at a speed equal to 2/3 of the male rotor's speed. Typical compression ratios per stage for these units can be as high as 4.2 in dry machines and 4.3 in oil-flooded machines. In low-pressure twin-screw compressors, which are also sometimes referred to as blowers, an alternative configuration is often used consisting of a male rotor with only two lobes that mesh with four mating grooves on the female rotor. With this latter arrangement, the female rotor must turn at twice the speed of the male rotor. Compression ratios developed in the low-pressure machines, which typically operate without internal lubrication, generally do not exceed 2 [28].

The rotors in a twin-screw compressor are often supported by oil-lubricated bearings mounted at both ends of each shaft. Some units are furnished with ball and roller bearings. Larger compressors, however, may have journal-type radial and tilting-pad thrust bearings. Although a splash-type arrangement is sometimes used to distribute

oil for bearing lubrication, in many machines the lubricating oil is pressurized and circulated by the force of the gas being compressed or by a shaft-driven rotary gear pump. The pump typically removes the oil from a sump in the base of the compressor and delivers it, through a cooler and a filter, to the bearings (and timing gears, when used). In an oil-flooded compressor, oil is also supplied to the injection ports in the casing. The gas-and-oil mixture discharged from an oil-flooded compressor generally passes through a separation device that removes a large portion of the oil from the gas. This oil is then returned to the compressor's lubrication system.

Twin-screw compressors are frequently provided with cast or ductile iron casings and steel, stainless-steel, or nickel-alloy rotors. Although the rotors in low-pressure units are often pressed on, or keyed to, steel shafts, one-piece rotors with integral shafts are generally used in higher-pressure compressors. Various types of labyrinth, carbon-ring, and mechanical seals are used for shaft sealing. In oil-free compressors seals are also required to prevent lubricating oil from mixing with the gas being compressed. The seals used are often cooled and lubricated by oil or water.

Twin-screw compressors are typically driven at speeds from 800 to 20,000 rpm by motors, engines, or turbines. Compressor efficiencies can be in the range of 75 to 80% based on adiabatic compression. Efficiencies are less, however, with higher compression ratios and when heavier gases are compressed. The capacity delivered by a twin-screw compressor can be regulated by adjusting the unit's operating speed. In addition, some twin-screw compressors are fitted with an internal slide valve that can be opened to reduce the region of compression along the length of the rotors. When opened, this valve also allows a portion of the gas that is being compressed to be returned to the inlet side of the compressor. Although throttling a suction valve is sometimes used for capacity control, it can result in high compression ratios and elevated discharge temperatures.

b. Single-screw compressors. Rotary compressors are also built in single-screw configurations. The main rotor in these compressors consists of a screw with multiple helical grooves that is centered in a close-clearance casing and coupled, either directly or indirectly, to the driver. Gas entering the casing fills the cavities formed between the grooves of the rotating screw and the inner wall of the casing. This gas is then compressed against the teeth of gate rotors that are driven by and mesh with the screw. The volume within each rotating groove is progressively reduced in a fashion similar to the way a stationary piston would reduce the volume within a moving cylinder.

In a machine with a cylindrical main rotor and planar gate rotor configuration, two flat-gate rotors resembling gears with straight radial teeth are fitted on vertical shafts and are mounted on each side of the horizontal screw or main rotor [29]. With this arrangement, the two gate rotors form seals along the main rotor's horizontal centerline. Single-screw compressor configurations that include either a cylindrical main rotor with two cylindrical gate rotors or two back-to-back conically shaped planar main rotors with four planar gate rotors have also been used. Although oil can be injected into the compressor's casing to lubricate the internal components and cool the gas being compressed, water is often injected into the casings of single-screw compressors used in applications requiring oil-free gas. Materials used in the construction of these water-flooded compressors include bronze for the main rotors and casings, stainless steel for shafts, and glass reinforced composites for the gate rotors.

c. **Straight-lobe compressors.** A straight-lobe rotary compressor, which is similar in appearance to the lobe pump shown in Fig. 30, contains two rotors, also referred to as impellers, that are mounted on parallel counterrotating shafts. Each rotor typically has two or three involute or cycloidal-shaped lobes. As each lobe rotates past the inlet port, a constant volume of gas is trapped between the lobe and the inner wall of the casing and is forced towards the discharge port. The meshing of the lobes and the close clearances between the lobes and the casing limits the amount of gas that leaks back to the compressor's inlet. A straight-lobe compressor does not actually compress the gas being forced through its casing. The increase in the pressure of the gas results, instead, from the resistance caused by the system back pressure. Although many of these compressors are single-stage machines, two-stage configurations are also used. Straight-lobe compressors are often furnished with cast-iron, ductile-iron, or aluminum casings and rotors. Shafts are generally supported at both ends by either ball or journal bearings.

Due to the shape of the rotors in a straight-lobe compressor, torque must be transmitted from the driving shaft to the idler shaft through timing gears. The gears, together with the outboard bearings, are generally lubricated with oil that is either distributed by splashing or pressurized and circulated by a shaft-driven gear pump. The inboard or driver-end bearings may also be lubricated with oil or, for less severe duty, they may be grease lubricated. Lip or labyrinth seals are often installed adjacent to each bearing to prevent the lubricant from mixing with the gas being compressed. In some larger units mechanical seals may also be used. Typical operating speeds for straight-lobe compressors do not exceed 1800 rpm.

d. **Sliding-vane compressors.** Many sliding-vane compressors are similar in appearance to the vane pump shown in Fig. 31 and contain a single cylindrical rotor on a shaft that is mounted eccentrically within a cylinder. A set of radial vanes is fit into slots that are equally spaced around the circumference of the rotor. Because of centrifugal force and, in some cases, springs, as the rotor turns the tip of each vane stays in contact with the inner wall of the eccentric cylinder. Consequently, the vanes are forced to slide into and out of their slots in the turning rotor. As each pair of adjacent vanes sweeps past the compressor's inlet port, the vanes move radially outward from the rotor. Gas entering the cylinder is trapped within the cavity formed between the extended vanes. With the continued rotation of the shaft, the vanes are gradually pushed back into the rotor by the inner wall of the eccentric cylinder. The reduction in the volume of the trapped gas results in a corresponding increase in its pressure. Compression continues until the leading vane reaches the outlet port, at which time the gas enters either the discharge piping or, in the case of a two-stage machine, the compressor's second stage. Two-stage sliding vane compressors can be furnished with both rotors mounted on a common shaft in an "in-line" configuration, or with the second stage mounted directly beneath the unit's first stage in an "over-under" configuration. To reduce the temperature of the gas being compressed, cooling water is sometimes circulated through external jackets that surround the compressor's cylinder. Intercoolers, for two-stage machines, and aftercoolers are also frequently provided with these compressors.

A roller bearing is often mounted at each end of a sliding-vane compressor's rotor. In addition, mechanical seals or packing are generally used for shaft sealing. The bearings and seals are often lubricated with oil that is pressurized by a small shaft-driven multi-plunger pump and distributed through individual supply lines to each point of lubrication. Oil discharged from the plunger pump may also be injected into the compression chamber to provide lubrication for the rotating vanes. As an alternative to the individual forced-feed arrangement, in flooded compressors a large amount of oil is continuously injected directly into the cylinder and not only lubricates the compressor's components but also cools the gas being compressed. Consequently, external cooling jackets and intercoolers are generally not required with flooded compressors. Sliding-vane compressors can also be designed to operate with non-lubricated cylinders when oil-free gas is required.

Sliding-vane compressors are frequently furnished with a cast-iron cylinder and a steel shaft that has either an integral or a separate cast-iron rotor. Vane materials include phenolic resins in compressors that have a lubricated cylinder and carbon in oil-free compressors. The capacity delivered by a sliding-vane compressor can be adjusted by varying the operating speed. An unloader valve that closes when the pressure in the discharge line reaches a preset value is often installed at the inlet to the compressor. When the unloader is activated, it also simultaneously opens a relief valve that allows the compressor to discharge to the atmosphere.

e. **Liquid-ring compressors.** Liquid-ring compressors are identical in design to the liquid-ring vacuum pumps described previously in Section 1 and illustrated by Fig. 33. They are used primarily in applications requiring oil-free compressed gas. Due to the use of a liquid compressant, which is often water, these machines are also suitable for compressing gas that may already be mixed with a liquid prior to entering the compressor. Because it is liquid and not the machine's rotor that compresses the gas being handled, the wear rate of internal components is generally low. In addition, because the liquid absorbs much of the heat generated during compression, the need for intercoolers or an aftercooler is eliminated. A portion of the liquid compressant is usually discharged from the compressor with the gas that has been compressed. This liquid is frequently removed from the gas in a separator that is installed at the compressor's outlet. The liquid is then circulated through a heat exchanger and returned to the compressor.

2.5 Marine Compressor Applications.

a. Ship's service air. A ship's service air compressor supplies compressed air to locations throughout the vessel for various uses, such as operating pneumatic tools, cleaning equipment, and charging air chambers used as pulsation dampeners for reciprocating pumps. Typical conditions of service for compressors used in this application include rated capacities in the range of approximately 100 to 1250 cfm at discharge pressures from 100 to 150 psig. Because of the relatively low discharge pressures developed by these units, they are sometimes referred to as low-pressure air compressors. A ship's service air compressor frequently operates at a constant speed and is cycled on and off, as needed, to keep the pressure in the air system's receivers between two preset values.

Reciprocating compressors used in this application are often two-stage machines. They can be furnished in air-cooled or water-cooled configurations and with lubricated or non-lubricated cylinders. In an air compressor with lubricated cylinders, a breather line is generally provided from the top of the crankcase to the suction port, which acts as a vent and prevents the pressure within the crankcase from building up due to the accumulation of compressed air that may blow past the piston rings. The number of cylinders used varies depending on the capacity required, which is generally less than 600 cfm when reciprocating compressors are used. These compressors are often furnished with strip or ring-plate air valves.

Single- and twin-screw rotary, liquid-ring, and multistage-centrifugal compressors have also been used in ship's service air systems.

b. High-pressure air. A high-pressure air compressor typically supplies capacities in the range of 5 to 30 cfh at discharge pressures from 1000 to 5000 psig. Uses for this air include operating pneumatic machinery, gas turbine starting, and diesel engine emergency starting. This air may also be supplied through a reducing valve to the ship's service air system. Four- and five-stage reciprocating compressors are frequently used in this application. The lower-pressure stages in these compressors may be fitted with strip air valves. Ring-plate valves are, however, often used in intermediate-pressure stages, and disk valves are installed in many higher-pressure stages. Although high-pressure compressors are sometimes air cooled, the cylinders in many machines are fitted with external cooling jackets through which seawater or fresh water is circulated. In addition, although the cylinders in some compressors are lubricated, many units are furnished in a non-lubricated configuration.

c. Refrigeration and air conditioning. Compressors in vapor-compression refrigeration and air-conditioning systems raise the pressure of refrigerant that has vaporized within the evaporators. Air-cooled reciprocating compressors that have from one to sixteen cylinders are often used in this application. Low-horsepower units, such as those furnished with reach-in refrigerators and freezers, water coolers, and ice-cube machines, are frequently supplied in a close-coupled configuration with the compressor's connecting rods mounted directly on the extended shaft of an electric motor. In addition, the entire compressor and motor assembly is typically hermetically sealed within a steel shell. In an open-drive refrigeration compressor, which has no shell, the crankshaft is coupled directly, through a reduction gear, or through a multi-V-belt drive arrangement to a separately mounted driver. Open-drive compressors typically have an enclosed pressure-tight crankcase. With this arrangement, a crankshaft seal is required to prevent the leakage of refrigerant and lubricating oil at the location where the drive-end of the crankshaft passes through the casing. The crankshaft seal, which is usually a mechanical-type seal, also prevents air from leaking into the compressor if the pressure within the crankcase drops below atmospheric pressure.

Although some large units have double-acting pistons that are mounted horizontally, many reciprocating refrigeration compressors are fitted with single-acting pistons in vertical in-line, V, or W cylinder configurations. With single-acting trunk pistons, the cylinder's suction and discharge valves are located in the cylinder head. When a single-acting double-trunk piston design is used, however, only the discharge valve is in the cylinder head. Refrigerant that enters the compressor initially passes through an inlet port in the side of each cylinder's inner wall and enters a hollow piston. The vapor is then admitted into the upper portion of the cylinder through a suction valve located in the top of the piston [30].

The valves used in reciprocating refrigeration compressors are often similar to the strip and ring-plate valves used in air compressors. In addition, spring-loaded poppet valves are sometimes used. A poppet valve's seat can be located flush with the inner wall of the cylinder head; consequently, the use of poppet valves results in a reduced clearance volume and in an improved volumetric efficiency. Because poppet valves move more slowly than many other types of compressor valves, they are generally installed only in slower-speed compressors. Other valves used in refrigeration compressors include the diaphragm type, which has several flat circular disks that are fixed in the center and flex at their periphery to uncover the valve ports, and the flapper or reed type, which has a thin flexible reed that is secured at only one end. This arrangement enables the free end of the reed to flex and uncover the valve port. In larger refrigeration compressors, each discharge valve is often mounted in a secondary safety head that fits over the upper end of the cylinder. During normal operation, the safety head, which is held in place by a heavy coil spring, remains stationary. However, if the piston attempts to compress liquid, the safety head lifts because of the resulting increase in pressure and permits the liquid to be discharged from the cylinder.

The capacity passing through a reciprocating refrigeration compressor can be controlled by varying the unit's operating speed or by using unloaders during constant-speed operation.

Many units are automatically cycled off when the suction pressure drops below a preset value. The resulting low pressure in the crankcase reduces the amount of refrigerant that can be absorbed by the compressor's lubricating oil. A crankcase heater that is energized when the compressor is stopped may also be used to limit refrigerant absorption. If the suction pressure rises above the set point, the compressor is restarted and the crankcase heater is de-energized. A separator is sometimes installed at the compressor's outlet to remove lubricating oil from the discharged refrigerant. The oil is then returned to the compressor's crankcase. Oil that is not removed from the high-pressure refrigerant usually returns to the compressor after being circulated through the refrigeration system. A portion of this oil often is separated from the refrigerant by impingement at the inlet to the compressor and is drained to the crankcase. A check valve is frequently installed in the oil drain line to prevent the pressure in the crankcase from being reduced suddenly when the compressor is started, which could result in the rapid vaporization of liquid refrigerant that may be mixed with the lubricating oil stored within the crankcase and in the foaming of the oil. After start-up, the pressure in the crankcase is initially greater than the pressure at the compressor's inlet and the check valve remains closed. However, a small bleed port that bypasses the check valve is usually provided to permit the pressure within the crankcase to be slowly reduced and enable the check valve to open. The bleed port also serves as a vent to return back to suction vapor that blows past the compressor's piston rings.

In addition to reciprocating compressors, rotary twin-screw and sliding-vane compressors are also used in refrigeration service. A sliding-vane refrigeration compressor's discharge port is often fitted with a flapper-type check valve to prevent high-pressure refrigerant from backing up into the evaporator when the compressor is stopped. Some small rotary refrigeration compressors and their motors are furnished as hermetically sealed units. In high-capacity refrigeration plants, single- and multistage centrifugal compressors may be used. The capacity delivered by a centrifugal refrigeration compressor can be adjusted by throttling the suction valve, adjusting the angle of "prerotation" vanes in the suction area of the casing, varying the unit's operating speed, or varying the temperature or flow rate of the water passing through the refrigeration system's condenser. Because of the increase in the compressor's discharge pressure that results from an increase in the condensing temperature, a centrifugal refrigeration compressor's capacity drops as the condenser's cooling-water temperature increases.

d. Instrument and control air. An instrument and starting air compressor delivers compressed air to pneumatically operated instruments and control systems, such as those used for combustion control. Non-lubricated reciprocating compressors are sometimes used in this application. In addition, because of their ability to deliver clean air, liquid-ring compressors are also used. Typical conditions of service include a rated capacity of 100 cfm at discharge pressures from 125 to 150 psig.

e. Starting air. A starting air compressor charges the receivers that store air used to start diesel engines. Reciprocating compressors are often used in this application. Typical conditions of service include capacity ratings in the range of 10 to 50 cfm at discharge pressures from 250 to 750 psig. The starting air compressor is sometimes driven directly off the engine. However, when this arrangement is used, a separately driven compressor must also be provided to charge the starting air receivers while the engine is stopped.

f. Oxygen-nitrogen plant. On some vessels, compressors supply air to separation plants that generate oxygen and nitrogen for shipboard use. Reciprocating compressors rated to deliver a capacity of approximately 30 cfh at a discharge pressure of 3000 psig are often used in this application. In addition, centrifugal compressors that deliver approximately 1750 cfm of compressed air at a discharge pressure of approximately 90 psig are used in low-pressure oxygen-nitrogen plants.

g. Sewage treatment. In activated-sludge sewage treatment systems, rotary straight-lobe compressors are sometimes used to supply high capacities of low-pressure air to aerate and agitate macerated sludge collected in the vessel's sewage-treatment tanks. Typical conditions of service for these compressors include a rated capacity of approximately 2000 cfm at discharge pressures from 6 to 10 psig. In addition, multistage reciprocating compressors that develop pressures up to 3200 psig are frequently used in wet-oxidation sewage-treatment systems to compress air that is mixed with macerated sludge entering the system's oxidation reactor.

h. Gas turbine. A compressor forms an integral part of a gas turbine and is used to supply air to the combustion chamber that precedes the power-producing turbine. In single-shaft units, the compressor's rotor is driven by, and mounted on, the same shaft as the turbine. In larger multishaft units, however, the compressor is driven by a separate turbine from the power turbine that drives the load. This arrangement permits the speed of the power turbine to be adjusted without affecting the compressor's speed.

Single-stage centrifugal compressors with semiopen impellers and volute-type casings are often used with smaller gas turbines. To increase pressure recovery, the air discharged from the impeller may also pass through a vaned or vaneless diffuser before entering the volute. Axial compressors with from 6 to 20 stages are used with larger gas turbines, such as those furnished for main propulsion. In addition, mixed-flow compressors and combination compressors consisting of one or more axial stages followed by a single centrifugal stage are used with some gas turbines. This latter configuration enables the compressor to deliver low capacities of air at high pressures.

Section 3
Blowers

3.1 Fundamentals. Blowers are used to supply relatively large quantities of low-pressure air to various destinations throughout the vessel. A blower is defined as a machine that delivers air at discharge pressures up to 40 psig [27]. Although radial- and axial-flow kinetic-type machines are often used as blowers, reciprocating and rotary positive-displacement blowers have also been used. The term fan is generally applied to radial- and axial-flow blowers that increase the density of the air passing through them by a maximum value of 5%. This corresponds to a discharge pressure of approximately 1 psig based on inlet air at atmospheric pressure [7].

3.2 Dynamic Blowers.

a. Centrifugal blowers. In a centrifugal blower incoming air is directed into the eye of a rotating open, semiopen, or closed impeller that increases the air's velocity and static pressure. After passing through the impeller, the air is often discharged radially through a stationary diffuser or scroll-shaped volute in which a portion of the velocity head is converted into static pressure head, and then through a tangential outlet port in the blower's casing. Both single- and double-suction impellers are used. The impellers can have forward-curved blades, straight-radial blades, or backward-curved blades. In addition, blades having a double curvature may also be used. Centrifugal blowers with backward-curved blades typically operate at higher speeds and are generally more efficient than blowers with radial- or forward-curved-blades.

b. Axial-flow blowers. Air handled by an axial-flow blower passes through the casing in an axial direction. In a "vaneaxial" blower the shaft-mounted rotor, which consists of a hub with multiple airfoil-shaped blades extending from its periphery, is enclosed within a stationary cylindrically-shaped housing. Stationary guide vanes are provided at the housing's outlet to straighten the flow of the air being discharged. The tips of the rotor's blades in a vaneaxial blower are sometimes joined by an outer ring. A tubeaxial blower is similar to a vaneaxial design, except that tubeaxial blowers do not have stationary guide vanes. In a propeller blower a simpler rotor design is used consisting of as few as two relatively long constant-thickness or airfoil-shaped blades mounted around the circumference of a relatively small hub.

c. Dynamic blower performance characteristics. A blower's theoretical or Euler head can be calculated using the same equations previously derived for centrifugal pumps. Because of slip, entrance and exit losses, and losses due to friction and diffusion in flow channels, the actual total head developed by a blower is less than the Euler head. Pressures at the inlet and outlet of a blower are often measured with water manometers; consequently, the total differential pressure across a blower or

fan, which is commonly referred to as "fan total pressure," is frequently expressed in terms of inches of water gage (in. wg).

The conversion of fan total pressure in in. wg to total head in feet of air can be performed using

$$H_a = \frac{\rho_w}{\rho_a} \frac{h_w}{12} \qquad (61)$$

where

H_a = total head, ft air
h_w = fan total pressure, in. wg
ρ_a = density of air entering the blower, pcf
ρ_w = density of water at 68 F, 62.3 pcf

Rated conditions of service for blowers are often based on standard inlet conditions of 14.7 psia at 68 F, which results in a value for ρ_a of 0.075 pcf.

Aerodynamic losses, internal leakage losses, disk friction, and mechanical losses are accounted for in the blower efficiency, which is the ratio of the air horsepower or the work performed by the blower divided by the brake horsepower required to drive the blower. The following equations can be used to calculate air horsepower and blower efficiency:

$$P_a = \frac{\rho_a H_a \bar{q}}{33,000} = \frac{h_w \bar{q}}{6362} \qquad (62)$$

and

$$\eta_B = \frac{P_a}{P_B} \qquad (63)$$

where

P_a = air horsepower, hp
P_B = brake horsepower required to drive blower, hp
η_B = blower efficiency, %/100
$\bar{q}$ = mean volumetric flow rate or capacity, cfm

Because of the low heads developed by most blowers, the compressibility of the air passing through them is often negligible and can be ignored. Consequently, a blower's volumetric capacity is frequently assumed to remain constant from inlet to outlet. In addition, because the difference between the velocity head of the air entering and leaving a blower can be small when compared with the increase in static pressure head, blower performance is sometimes shown by plotting static pressure versus capacity.

Typical performance curves for centrifugal and axial-flow blowers are illustrated in Figs. 50, 51, and 52. As shown in these figures, while the head-capacity curve for a centrifugal blower can be relatively flat during operation below the best efficiency point, the head developed by an

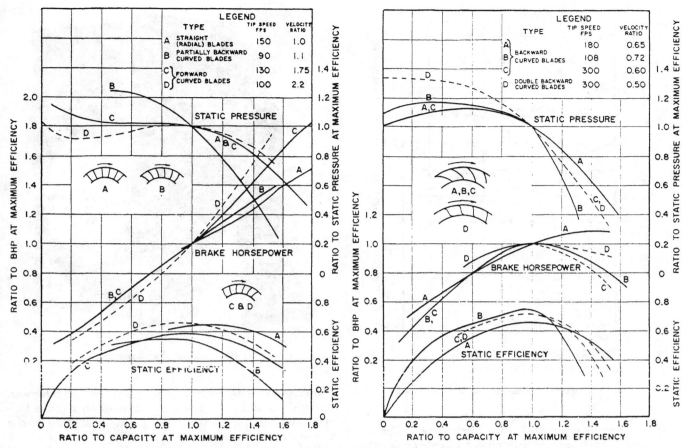

Fig. 50 Characteristic curves for centrifugal blowers with straight, partially backward-curved, and forward-curved blades

Fig. 51 Characteristic curves for centrifugal blowers with backward-curved blades

axial-flow blower can rise sharply with reductions in flow rate. In addition, the horsepower required by a centrifugal blower with backward-curved blades typically begins to drop as the capacity increases beyond a flow rate that is generally at or just to the right of the best efficiency point.

With forward-curved blades, however, a centrifugal blower's power requirement often continues to increase with flow rate. Although the power required by an axial-flow blower drops to the right of the best efficiency point, it can rise as the flow rate approaches shutoff. Equations (29), (30), and (31) given previously for centrifugal pumps can also be used for fans and blowers.

At low flow rates the performance of a blower can be affected by stalling at the inlet to the rotor or stator. In addition, in the case of axial-flow blowers, after stalling begins the surge limit can shift to a higher capacity. This hysteresis effect often results in the formation of two different head-capacity curves in an axial-flow blower's low-flow region of operation, with the curve representing blower performance when the capacity is being increased located slightly to the right of the curve representing performance when the flow rate is being reduced.

3.3 Positive-Displacement Blowers. Positive-displacement rotary blowers are used in some applications. These blowers are generally helical-lobe, straight-lobe, or sliding-vane type units and are similar in design to the corresponding types of rotary compressors described in Section 2. Although less common, in some applications reciprocating blowers may be used.

3.4 Marine Blower Applications.

a. Forced draft. A forced-draft blower is used to supply combustion air to fossil-fueled boilers on steam-powered vessels. The capacity of air delivered to the boiler's burners generally exceeds the theoretical amount required for complete fuel combustion by a value of approximately 15%. The capacity actually passing through the forced-draft blower, however, must also account for air leakage losses in the discharge duct, air heater, and boiler casing. In addition, allowances must be made to account for changes in the temperature, pressure, and relative humidity of air entering the blower, and the composition and heating value of the fuel being burned in the boiler. On oil-fired vessels forced draft blowers are often sized to deliver 260 ft³ of air for each lbm of fuel burned at the boiler's maximum rating. This results in blower capacity ratings that can exceed 30,000 cfm. The discharge pressure that must be developed by a forced-draft blower is affected by the pressure drop in the air heater and in the boiler's double casing, the number of burners in operation,

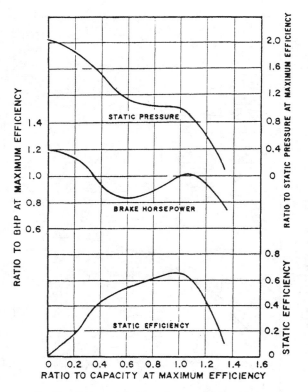

Fig. 52   Characteristic curves for an axial-flow blower

Fig. 53   Centrifugal forced-draft blower

and the pressure drop in the burner air registers. Centrifugal forced-draft blowers containing radial- or forward-curved blades have been used on vessels with low to moderate combustion air pressure requirements. On vessels with higher furnace-pressure requirements, however, centrifugal blowers with backward-curved blades and multistage propeller blowers are generally used.

A centrifugal forced-draft blower, which is shown in Fig. 53, is typically a single-stage machine that is driven by either an electric motor or a steam turbine. Although an inlet trunk is sometimes used to direct the flow of incoming air to the blower, forced-draft blowers that take suction through screened openings directly from the machinery spaces are also used. A double-inlet centrifugal blower has inlet openings at both ends of the casing and is often mounted horizontally. It is usually fitted with a double-suction impeller that is centered on a shaft supported radially at each end by either journal or antifriction bearings. A thrust bearing is also provided. The bearings are lubricated with oil distributed either by partially submerged rings that are mounted on, and turn with, the shaft or through a pressurized forced-feed system. The pressurized system frequently includes a rotary gear pump that is driven off the blower's shaft. The pump removes lubricating oil from a sump or reservoir and discharges it to the bearings. After passing through the bearings, the lubricating oil drains back to the reservoir by gravity. Many pressurized lubrication systems also include a filter and a cooler for the oil. A double-inlet blower is generally connected to its driver with a flexible

coupling. In a single-inlet centrifugal forced-draft blower, however, a single-suction impeller is often mounted directly on the end of the driver's output shaft. When a close-coupled unit is mounted vertically, the blower is typically located above the driver. Air leaving a centrifugal forced-draft blower's impeller generally passes through a stationary diffuser or volute and is discharged through a tangential opening in the side of the casing.

Single-stage axial-flow blowers have been used on some older vessels; however, most axial-flow forced-draft blowers are furnished in two- and three-stage configurations. As shown in Fig. 54, each blower stage consists of a rotating propeller, consisting of multiple individual blades attached to a hub, followed by a row of stationary guide vanes. A set of stationary guide vanes is also frequently installed in the blower's inlet, and a diffuser usually follows the blower's last stage. Air leaving the diffuser passes through the blower's discharge port, which typically forms the transition between the cylindrical portion of the casing and a rectangular discharge duct. Propeller-type forced-draft blowers are often driven by steam turbines. In many of these units the turbine's rotor and the blower's propellers are mounted on a common shaft. The shaft is sometimes supported radially by two journal bearings. With this arrangement, one of the bearings is mounted on the driver side of the propellers, while the second is installed near the outboard end of the turbine. A thrust bearing is also mounted on the shaft to absorb axial loads applied to the combined rotating assembly. All of the bearings are lubricated with oil that is pressurized and circulated by a shaft-driven pump. When mounted vertically, the propeller blower is generally located above its driver.

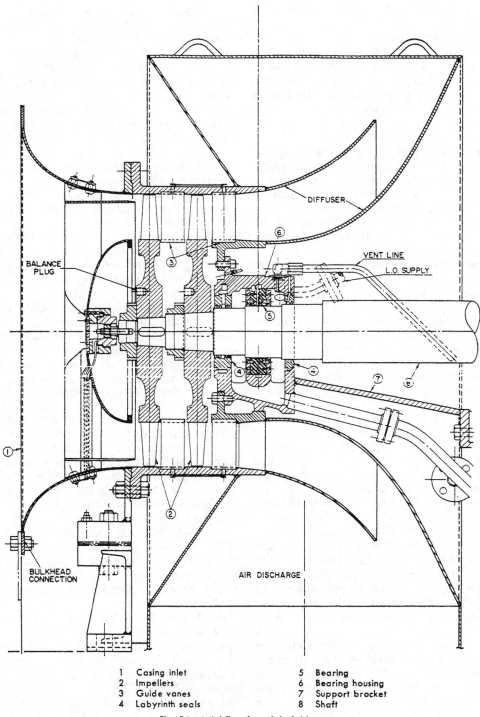

1	Casing inlet	5	Bearing
2	Impellers	6	Bearing housing
3	Guide vanes	7	Support bracket
4	Labyrinth seals	8	Shaft

Fig. 54 Axial-flow forced-draft blower

With this configuration the casing's inlet opening is typically located at the top or outboard end of the blower. Horizontally mounted propeller blowers, however, have been furnished with an inlet opening at either end of the casing.

Centrifugal forced-draft blowers typically operate at speeds as high as 1800 rpm and deliver air at pressures up to 25 in. wg, while multistage propeller forced-draft blowers are driven at speeds from 5000 to 10,000 rpm and can develop discharge pressures in excess of 45 in. wg. When the driver has a variable- or multiple-speed capability, the capacity of air delivered can be adjusted by changing the blower's operating speed. In addition to speed control, adjustable inlet guide vanes and outlet dampers are also used to vary the capacity delivered by some forced-draft blowers.

On many ships two forced-draft blowers are provided per boiler. With this arrangement, each blower is frequently sized to handle the boiler's full-load air requirements. However, if a boiler overload condition occurs, both blowers are operated in parallel. During this type of operation, it is important that both blowers operate at the same speed so that one unit does not force the second to deliver an excessively low capacity. To insure that steam-turbine-driven forced-draft blowers operating in parallel are driven at the same speed, the throttle or nozzle control valves for each turbine are often linked mechanically so that both turbines receive the same amount of steam. To prevent an idle forced-draft blower from being driven in reverse by pressurized air in the discharge duct, automatic shutters that close when the driver is stopped are usually installed at the inlet or outlet of each blower. In addition, some forced-draft blowers are fitted with anti-reverse rotation devices.

b. Diesel-engine scavenging and supercharging. Blowers are used with many diesel engines to supply the air necessary for combustion to the engine's cylinders. When the pressure of this air is increased above atmospheric pressure, referred to as supercharging, the power output of the engine is increased. In addition, in a two-stroke engine the air supplied by the blower helps to purge the engine's cylinders of exhaust gas. This is referred to as scavenging.

Reciprocating blowers are sometimes used for scavenging. The blower is often driven through a connecting rod directly by the engine's crankshaft. Because of the low pressures developed by these blowers, their valves and valve springs must generally be very light. However, these lighter valves do not respond quickly during high-speed operation; therefore, the use of reciprocating blowers is typically limited to low-speed engines. To reduce the air-cylinder diameter, a single blower cylinder is sometimes replaced by twin cylinders that are mounted in tandem and are fitted with pistons installed on a common rod. A disadvantage of the reciprocating blower is that the oil used to lubricate its cylinders can mix with the air being compressed, be carried over into the engine, and result in the formation of deposits on the engine's exhaust ports.

Straight- and helical-lobe rotary blowers are sometimes used for scavenging and supercharging with high-speed engines. To a lesser extent, sliding-vane blowers are also used in these applications. The rotors in rotary blowers are often supported by antifriction bearings that are mounted at the ends of each shaft. The bearings, together with timing gears, when used, are generally oil lubricated. Rotary blowers can be driven by the engine's crankshaft through a gear, chain, or V-belt, or by a separate electric motor. However, when the blower is engine driven its operating speed and, therefore, the capacity that it delivers will automatically be adjusted to match the speed of the engine. This results in the delivery of a relatively constant capacity of air per combustion cycle to each engine cylinder at all operating speeds. These blowers,

which typically operate at speeds of 2000 to 6000 rpm, are usually rated to deliver capacities up to 900 cfm at pressures from 2 to 15 psig.

Single- and, in some cases, two-stage centrifugal blowers fitted with multi-vaned diffusers or volute-type casings are also used as superchargers. Although some of these blowers are driven by the engine through gears, because of the high blower speeds, which can be in the range of 10,000 to 50,000 rpm, torque is often transmitted to engine-driven centrifugal blowers through elaborate clutches and fluid couplings. This helps to prevent the blower from being damaged by sudden changes in engine speed. Because the pressure developed by a centrifugal blower increases with the square of its operating speed, engine-driven centrifugal blowers are generally suitable for use only with engines that operate at constant speeds, or when the power required from the engine is reduced with speed. As an alternative to an engine drive, some centrifugal blowers are driven by electric motors. In addition, many centrifugal superchargers are close coupled to turbines that are driven by the engine's exhaust gas. With this arrangement, the blower, which is referred to as a turbocharger, operates at a speed that varies with engine load. Turbocharger compression ratios can be as high as 4.0. The common shaft used in smaller turbochargers is often supported by two sleeve bearing located between the overhung impeller and turbine rotor. In larger units, however, the shaft may extend through both the eye of the impeller and the turbine's hub and be supported by bearings at each end. Although a turbocharger's bearings may be lubricated with oil supplied from the engine's lubrication system, some units have their own independent lubrication systems that include separate pumps, filters, and coolers. The turbine casings used with larger turbochargers are frequently cooled with water from the engine's jacket-water cooling system. This water may also be used to cool the turbocharger's turbine-end bearings.

c. De-ballasting. Instead of pumping seawater out of ballast tanks, on some vessels the water is forced out of ballast tanks with compressed air. Helical-lobe type blowers that deliver a capacity of approximately 2000 cfm at a discharge pressure of 20 psig are often used in this application.

d. Ventilation. Radial-flow centrifugal and axial-flow fans are often used to provide air for ventilation to spaces throughout the vessel, as discussed in Chapter 21. Centrifugal fans are typically of the single-inlet design. Although smaller units may have forward-curved blades, centrifugal fans that deliver over 1000 cfm of air frequently have backward-curved blades. Many centrifugal fans are driven through V-belts by electric motors at speeds not exceeding 1800 rpm. To avoid an overload, the motors that drive fans with forward-curved blades must generally be sized based on the free-flow capacity. The motors used with centrifugal fans that have backward-curved blades, however, can usually be sized based on the horsepower requirement near the best efficiency point.

Vaneaxial fans are also used in many ventilation systems. Although the fan's rotor is sometimes driven through a V-belt by an electric motor that is mounted external to the housing, in most vaneaxial fans the rotor is mounted directly on the drive motor's extended shaft. With this close-coupled configuration, the motor is located within the fan's housing and is, therefore, in the airstream. The rating point for a vaneaxial fan is generally to the right of the dip in the head-capacity curve; however, the fan's motor is frequently sized with a power margin so that an overload will not occur during operation at reduced flow rates. Vaneaxial fans can be used to deliver from approximately 500 to over 10,000 cfm of air at pressures ranging from 2 to 14 in. wg. When low-pressure air is required, tubeaxial fans are sometimes used. In addition, propeller fans can deliver ventilating air when total pressure requirements do not exceed 0.5 to 0.75 in. wg.

Methods that can be used to control the capacity being delivered by a ventilation fan, listed in descending order based on energy efficiency, include varying the operating speed, adjusting the angle of nonrotating inlet guide vanes, and throttling with an outlet damper. In addition, the capacity delivered by some axial-flow fans can be regulated by changing the pitch or the stagger angle of blades in the rotor. Small ventilation fans are often portable.

Section 4
Ejectors

4.1 Fundamentals. Ejectors, which are also referred to as jet pumps, are used to remove gases or liquids from various locations on the vessel. Unlike a pump or a compressor, an ejector has no moving parts or mechanical driver. Instead, the pumping action is created as a pressurized fluid, referred to as the motive or operating fluid, that is supplied from an external source passes through the ejector.

4.2 Steam-Jet Ejectors. A steam-jet ejector is used to remove air or other noncondensable gases from equipment that must operate under a vacuum. As shown in Fig. 55, an ejector consists primarily of a body with a suction port, a venturi-shaped diffuser tube, and a discharge port. Also included is a nozzle, usually of the converging-diverging type, through which steam is admitted into the portion of the body referred to as the suction chamber. As the steam expands in the nozzle, its pressure is reduced to the pressure in the ejector's suction chamber, which results in an increase in the steam's velocity to values as high as 3000 to 5000 fps. Air and other noncondensable gases present in the suction chamber become entrained in the high-velocity jet of steam and are carried into the converging section of the diffuser tube. The evacuation of this gas from the suction chamber creates a partial vacuum at the ejector's inlet that draws more gas through the suction line and into the ejector. As the steam and gas pass through the parallel section of the diffuser tube, they are thoroughly mixed. The velocity of this mixture is then reduced in the diffuser tube's diverging section. The resulting increase in the pressure of the steam and gas mixture enables it to be discharged from the ejector. Although the discharge pressure from an ejector is higher than the pressure at the suction inlet, referred to as the suction pressure, this discharge pressure is less than the pressure of the motive steam.

Ejectors can operate individually as single-stage components or can be furnished in a multistage configuration consisting of several ejectors arranged in series. In addition, depending on the amount of gas that must be handled, an ejector can be operated as a single-element unit or as part of a multiple-element assembly consisting of two or more ejectors arranged for parallel operation. Furthermore, an ejector can be classified as being either condensing or noncondensing based on whether or not the discharged motive steam is condensed to water. This condensation can take place in an after-condenser installed downstream from the ejector. In multistage condensing ejectors, additional heat exchangers, referred to as intercondensers, are also frequently installed between successive stages.

An ejector's performance can be rated based on the ratio of the pressure at the discharge port divided by the pressure at the suction port, referred to as the compression ratio. This ratio is affected by the pressure of the steam that is used to operate the ejector, the vacuum that

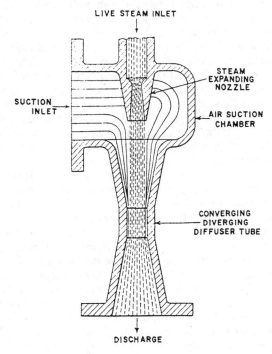

Fig. 55 Steam-jet air ejector

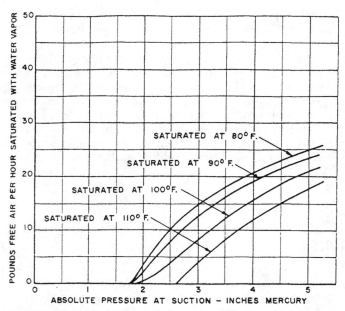

Fig. 56 Capacity of single-stage ejector for various air suction pressures
and temperatures

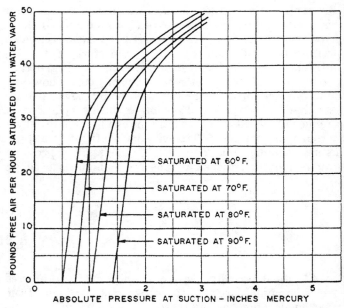

Fig. 57 Capacity of a two-stage ejector with inter- and after-condenser
for various air suction pressures and temperatures

must be maintained at the suction inlet to the ejector, and the desired rate of steam consumption. For stable operation and to prevent reverse flow through the ejector's body, which can result in a partial or total loss of vacuum, it is important that steam be supplied to the ejector at a constant flow rate and at a pressure above an established minimum value. To account for fluctuations in the pressure of the motive steam, the steam is often supplied to an ejector at a nominal pressure that is approximately 10 psig above the minimum value required. However, the higher pressure generally results in an increase in the steam consumption rate.

The size of the steam nozzle must be reduced as the operating steam pressure is increased, which increases the likelihood for fouling; consequently, operation with excessively high steam pressures is generally avoided. Typical motive steam pressures range from 80 to 150 psig for small-capacity ejectors, but can be as high as 300 psig for larger units. The steam used may be superheated to prevent nozzle erosion and the detrimental effects on ejector performance that can result from operation with wet steam. In addition, separators are sometimes installed in the ejector's supply line to remove water that may be mixed with the motive steam.

Single-stage ejectors can typically be used for vacuum requirements up to 27 in. Hg. For vacuums up to 29 in. Hg, two-stage units are generally used. Although three-stage ejectors can be more efficient when operating with these high vacuums, the use of three-stage ejectors is limited due to their increased size, weight, and complexity. Typical performance curves for both single- and two-stage ejectors are shown in Figs. 56 and 57, respectively.

4.3 Eductors. An ejector that operates with a liquid motive fluid is called an eductor. The fluid pumped by an eductor, referred to as the suction fluid, is also generally a

liquid. Eductors are similar in design to steam-jet ejectors; however, they are not furnished with inter- and after-condensers. In addition, eductors are fitted with converging nozzles. A single-nozzle eductor has only one nozzle that is located in the center of the suction chamber. With this configuration, the motive fluid leaving the nozzle is surrounded by the incoming suction fluid. In an annular-nozzle eductor, also referred to as a peripheral-jet eductor, a ring of nozzles surrounds the suction inlet. With this latter arrangement, which is frequently used in applications where solids or large amounts of air may be in the suction fluid, the motive fluid is admitted around the periphery of the suction fluid. In both types of eductors, the high-velocity motive fluid entrains the suction fluid, and the mixture that is formed is discharged from the divergent end of the diffuser tube. Eductors can frequently operate with suction lifts as high as 25 ft and values of available NPSH as low as 2 ft. Values for the ratio of the discharge pressure at the eductor's outlet divided by the motive fluid's inlet pressure can be as high as 0.4. In addition, the ratio of the capacity being pumped, referred to as the suction capacity, divided by the flow rate of the motive fluid, referred to as the inlet or operating capacity, can often exceed 1.0. This capacity ratio generally declines as the suction lift and the ratio of the eductor's discharge pressure divided by the motive fluid's inlet pressure increase.

4.4 Marine Ejector and Eductor Applications.

a. Condenser evacuation. Steam-jet ejectors are often used to evacuate air and other noncondensable gases from the main condensers on steam-powered vessels. Because these gases are being removed from a steam condenser, they are saturated with water vapor. To reduce the water-vapor content and, therefore, the load on the air ejectors, before entering the ejector assembly, the temperature of the gas and water-vapor mixture is generally reduced below the saturation temperature of steam corresponding to the absolute pressure within the condenser. Many air ejectors are sized based on a gas and vapor inlet temperature equal to 7.5 deg F below the saturation temperature corresponding to the main condenser's design pressure.

Ejectors that deaerate shipboard condensers are typically furnished as two-stage units. In addition, to permit the auxiliary steam used as the ejector's motive fluid, together with a portion of the water vapor mixed with the air and other noncondensable gases removed from the main condenser, to be returned to the condensate and feed systems, inter- and after-condensers are generally provided with these ejectors. Because of the reduction in the water-vapor content that results from a reduction in the gas-vapor mixture temperature, the use of an inter-condenser also reduces the load on the ejector's second stage. The cooling medium in the inter- and after-condensers is often the water (condensate) being discharged by the main condensate pumps. The air ejector's condensers, therefore, serve as feedwater heaters and enable a portion of the heat in the motive steam to be recovered.

Water vapor that condenses in the shell of the air ejector's inter-condenser drains to the main condenser through a loop seal line. The height of the water in this line must be sufficient to prevent air and other noncondensable gases within the inter-condenser, which are at a pressure slightly higher than the pressure in the main condenser, from being drawn back into the main condenser. Typical values for the absolute pressure within the shell of the inter-condenser and for the minimum height of the water seal are 7 in. Hg and 7 ft, respectively. Steam that condenses in the shell of the ejector's after-condenser is frequently at an absolute pressure of approximately 32 in. Hg and drains by gravity to the freshwater-drain-collecting tank.

The gases remaining in the shell of the after-condenser can be vented to the atmosphere. However, these gases may also be vented to an integral gland-exhaust condenser in which steam that leaks past the glands at the ends of the main propulsion turbines is condensed. When this latter arrangement is used, the condensate passing through the tubes in the ejector's after-condenser is discharged directly into the tubes in the gland-exhaust condenser. A fan or exhauster is often connected to the gland-exhaust condenser's vent to create a slight vacuum, generally from 5 to 10 in. wg, within the shell of this heat exchanger, which helps to induce flow from the turbine's glands. The use of the exhaust fan also enables the saturated gases being removed from the air ejector and gland-exhaust condenser assembly to be vented outside the machinery space. To limit the temperature of the gases emitted from this vent and to prevent the load on the air ejector's second stage from increasing because of a temperature increase in the inter-condenser, the temperature of the condensate being discharged from the gland-exhaust condenser's tubes is usually limited to a preset value, such as 140 F. When this temperature is exceeded, a manually operated or thermostatically controlled valve located downstream from the gland exhaust condenser is opened and permits a portion of the condensate to be recirculated back to the main condenser's hotwell. This results in an increase in the flow rate of condensate through the tubes in the air-ejector and gland-exhaust condensers and in a corresponding reduction in the temperature within each of these heat exchangers.

The rated capacity of a two-stage ejector used for main condenser deaeration must be sufficient to remove air that leaks into the condenser during normal operation, as well as air and other gases, such as carbon dioxide and ammonia, that may be present in the steam exhausted from the main turbines. This capacity is often based on maintaining a condenser vacuum of 29 in. Hg under normal load and with a normal flow of cooling water to the ejector's inter- and after-condensers. With a 29 in. Hg condenser vacuum, 7.5 deg F of subcooling below the saturation temperature results in a gas and water-vapor temperature at the inlet to the ejector of 71.5 F. A typical ejector steam consumption rate with these operating conditions is 4.5 lbm of steam for every lbm of air and vapor removed from the main condenser. However, in addition to being suitable for normal operation, the ejector must also be capable of quickly evacuating the main condenser of air during plant start-up, rapidly removing air that may be accidentally admitted into the main condenser, handling overloads without a significant loss of vacuum, and operating satisfactorily with reduced cooling water to the inter- and after-condensers. To provide adequate ejector capacity during these periods of abnormal operation, the air ejector assembly for a main condenser frequently includes a pair of two-stage elements that are each sized to handle 100% of the condenser's normal gas-removal requirements. This arrangement permits one ejector element to be used during normal plant operation, with the second element being used only as needed to handle increased air-removal requirements that may arise. The installation of a "twin unit" or "twin element" ejector also enables maintenance to be performed on either ejector element without interfering with normal plant operation. When the inter- and after-condensers for both of the ejector's twin elements are installed in a common shell, internal compartmentation or external isolation valves must be provided to enable one element to be operated while the second is idle.

Steam-jet ejectors similar to those described above are also often used to evacuate air and other gases from auxiliary condensers.

b. Distilling plant. Steam-jet ejectors are often used to remove air and other noncondensable gases from distilling plants. Although single-stage units are furnished with some submerged-tube and spray-film evaporators operating with absolute pressures exceeding 3 in. Hg, two- and, in some cases, three-stage ejectors are typically used to create the high vacuums required in flash-type evaporators. These ejectors are similar in configuration to those used for condenser evacuation; however, inter-condensers are generally not used with multistage units. Instead, the motive steam supplied to the ejectors is condensed by seawater that is circulated through the tubes in an after-condenser. Because this condensed steam may be contaminated by seawater that is entrained in the air removed from the evaporators, it can be directed either to the fresh-water-drain-collecting tank or to the bilge based on its salinity. By installing the after-condenser in the supply line to the distilling plant, this condenser serves as an evaporator feedwater heater. Water-motivated eductors have also been used to evacuate air from some submerged-tube and spray-film distilling plants. The materials of construction selected for distilling-plant ejectors and eductors must be resistant to corrosion from the seawater and salt-laden water vapor.

c. Dewatering. Eductors are often used for bilge drainage and compartment dewatering. High-pressure seawater supplied from the fire main is typically used as the motivating fluid for dewatering eductors. But, on some vessels, the motive fluid may be supplied by the ballast pumps.

d. Stripping. By diverting a portion of the liquid being discharged by the main cargo pumps through a bypass line, it can be used as the motive fluid in an eductor that strips other tanks of cargo that have levels too low for the main cargo pumps to lift. Because the motive cargo and the cargo removed from each tank being stripped are mixed within the eductor, this arrangement is suitable only when the cargoes are compatible. Liquid viscosities in excess of 100 cP can adversely affect eductor performance; therefore, the viscosity of cargoes that the vessel will carry must be considered when stripping eductors are designed. Eductors can also be used to reprime centrifugal cargo pumps that have lost suction. Furthermore, if seawater is used as the motive fluid, eductors can be used to remove slops from the vessel's cargo tanks during tank washing and to strip the vessel's ballast tanks. Typical materials of construction for stripping eductors include bronzes and nickel-copper alloys. To reduce the potential for clogging, these eductors often have large flow passages.

References

1 *Hydraulic Institute Standards for Centrifugal, Rotary and Reciprocating Pumps*, Hydraulic Institute, Cleveland, Ohio, 1983.

2 *Pump Handbook*, I. J. Karassik et al., Eds., McGraw-Hill, New York, 1986.

3 A. J. Stepanoff, *Centrifugal and Axial Flow Pumps*, Wiley, New York, 1957.

4 D. G. Shepherd, *Principles of Turbomachinery*, Macmillan, New York, 1956.

5 G. F. Wislicenus, *Fluid Mechanics of Turbomachinery*, 2 vols., Dover Publishing, New York, 1965.

6 E. P. Sabini and W. H. Fraser, "The Effect of Design Features on Centrifugal Pump Efficiency" in *Proceedings*, 4th International Pump Symposium, Houston, Tex., May 1987.

7 A. H. Church, *Centrifugal Pumps and Blowers*, R. E. Krieger Publishing, Huntington, New York, 1972.

8 A. Agostinelli et al., "An Experimental Investigation of Radial Thrust in Centrifugal Pumps," *Trans.* American Society of Mechanical Engineers, April 1960.

9 H. J. Biheller, "Radial Force on the Impeller of Centrifugal Pumps with Volute, Semivolute, and Fully Concentric Casings," *Trans.* American Society of Mechanical Engineers, July 1965.

10 T. Baumeister et al., *Mark's Standard Handbook for Mechanical Engineers*, McGraw-Hill, New York, 1978.

11 W. D. Marscher, "Analysis and Test of Multistage Pump Wet Critical Speeds," 35th STLE/American Society of Mechanical Engineers Tribology Conference, Fort Lauderdale, Fla., Oct. 1989.

12 I. J. Karassik, "Speed and Temperature Corrections for Centrifugal Boiler Feed Pump Efficiencies," *National Engineer*, Aug. 1945.

13 C. A. Gongwer, "A Theory of Cavitation Flow in Centrifugal-Pump Impellers," *Trans.* American Society of Mechanical Engineers, Jan. 1941.

14 I. J. Karassik, *Centrifugal Pump Clinic*, Marcel Dekker, New York, 1989.

15 B. Schiavello, "Visual Study of Cavitation—An Engineering Tool to Improve Pump Reliability," 1st International Conference of Improved Coal-Fired Power Plant, Electric Power Research Institute, Palo Alto, Calif., Nov. 1986.

16 J. H. Doolin, "Inducer Blades Cut into Required NPSH," *Power*, Oct. 1972.

17 V. Salemann, "Cavitation and NPSH Requirements of Various Liquids," *Trans.* American Society of Mechanical Engineers, *Journal of Basic Engineering*, 1958.

18 W. H. Fraser, "Recirculation in Centrifugal Pumps," Winter Annual Meeting, American Society of Mechanical Engineers, Washington, D.C., Nov. 15–20, 1981.

19 W. H. Fraser, "Flow Recirculation in Centrifugal Pumps" in *Proceedings*, 10th Turbomachinery Symposium, Houston, Tex., Dec. 1981.

20 A. R. Budris, "Sorting Out Flow Recirculation Problems," *Machine Design*, Aug. 10, 1989.

21 B. Schiavello, "Two-Phase Flow Rotodynamic Pumps—Experiments and Design Criteria," Norwegian Society of Chartered Engineers, Pumps Offshore, Gol, Denmark, Feb. 1986.

22 A. W. Feck and J. O. Sommerhalder, "Cargo Pumping in Modern Tankers and Bulk Carriers," *Marine Technology*, Vol. 4, No. 3, July 1967.

23 W. D. Marscher, "The Effect of Fluid Forces at Various Operation Conditions on The Vibrations of Vertical Turbine Pumps" *Proceedings*, Institute of Marine Engineers Conference on Radial Loads and Axial Thrusts on Centrifugal Pumps, London, Feb. 5, 1986.

24 W. J. Sembler, "The Design and Operation of Pumps Furnished for Marine Cargo Service," *Marine Technology*, Vol. 25, No. 1, Jan. 1988, pp. 1–29 and No. 2, April 1988, pp. 75–104.

25 R. F. Paashaus, "An Analysis of Cavitation Damage in Commercial Marine Condensate Pumps," SNAME/

American Society of Mechanical Engineers meeting, New York, Dec. 8, 1964.

26   E. P. Fortino, "New Approaches to the Design of Hopper Dredges," *Marine Technology*, Vol. 17, No. 4, Oct. 1980, pp. 371–384.

27   *Compressed Air and Gas Handbook*, Compressed Air and Gas Institute, New York, 1973.

28   R. N. Brown, *Compressors, Selection & Sizing*, Gulf Publishing Co., Houston, Tex., 1986.

29   J. L. Krinsky et al., "Navy Air Compressors—Past, Present and Future," *ASNE Journal*, Vol. 97, No. 4, May 1985.

30   R. J. Dossat, *Principles of Refrigeration*, Wiley, New York, 1961.

CHAPTER XV

Main and Auxiliary Condensers

Robert J. Foster

Section 1
General Characteristics

1.1 Introduction. Rankine's modification of the Carnot cycle provides the practical basis for the design of a marine steam power plant. The Rankine cycle has undergone a number of modifications but remains fundamental to the more efficient and refined regenerative and regenerative-reheat cycles in common use. Small steam plants, and more frequently than not the auxiliary marine steam plants, are classic examples of the elementary or basic Rankine cycle.

Rudolf Clausius, a theoretical physicist, in 1850 first enunciated the second law of thermodynamics in the following form: "Heat cannot of itself pass from a colder to a hotter body." In practical terms this means that in order to construct an engine that will operate in a cycle and produce work, heat must be rejected from the cycle to a heat sink. The smaller the amount of heat rejected to the heat sink in relation to the amount of heat supplied, the greater is the amount of heat energy available to produce useful work. Consequently, the lower the temperature at which the heat energy can be removed from the system, the higher will be the thermal efficiency of the cycle.

For the shipboard steam power plant, the ocean, or any body of water in which the ship operates, provides an ideal heat sink. It is a heat-absorbing medium of practically unlimited capacity, although its ambient temperature limits the extent to which heat energy is made available for producing work by the cycle.

The steam surface condenser provides an efficient means for using the body of water in which the ship floats as a heat sink for the steam power plant.

Basically, a steam surface condenser is a gastight chamber that is fitted with heat-conductive tubes through which cooling water is circulated, and which is provided with means for continuously removing the condensed steam and noncondensable gases. The noncondensables enter the system via leakage through the turbine-shaft seals, through subatmospheric drains and returns to the condenser, and through other subatmospheric-pressure vessels or lines.

1.2 Steam Surface Condensers. The steam condenser controls the exhaust pressure or back pressure at which the turbine operates by condensing the turbine exhaust steam at a pressure corresponding to its saturation temperature. Typical examples of main condensers are shown in Figs. 1, 2, and 3; representative auxiliary condensers are shown in Figs. 4 and 5.

The purpose of a condenser in the power plant cycle is twofold:

1. To condense the steam and return pure, oxygen-free feedwater or condensate to the system.

2. To provide and maintain the highest practical vacuum (minimum back pressure) to gain greater efficiency and performance from the steam turbines.

The condensing process is accomplished by the transfer of heat from the exhaust steam to the cooling water and is the basic function of the steam condenser. A heat transfer system must have a heat source and heat receiver, and if there is to be a significant flow of heat from the source to the receiver, there must be an appreciable temperature

Fig. 1 Main condenser for turbines with axial-flow exhaust (single-plane machinery)

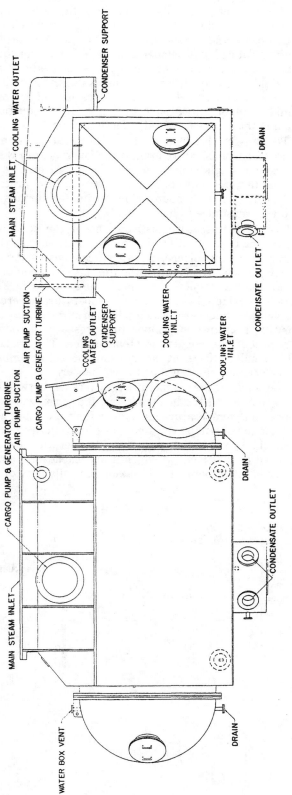

Fig. 2 Main condenser arranged to support turbine

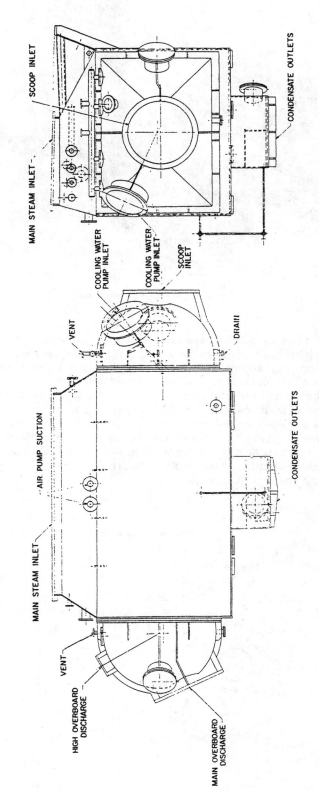

Fig. 3 Main condenser arranged to be hung from turbine

Fig. 4 Typical auxiliary condenser

difference between the two. In the marine steam power plant, the turbine exhaust steam is the higher-temperature heat source, and the circulated cooling water is the lower-temperature heat receiver. In addition to the requirement for a temperature difference, there must also be an area or surface through which the heat can pass. In a surface condenser this area is composed of thin-walled metal tubes enclosed within a shell or pressure container. They provide a means for the heat to flow from the steam to the circulating water, and also are the means for keeping the two fluids separated.

In a surface condenser steam is condensed; that is, it changes from its vapor state to its liquid state at substantially constant pressure, and at the saturation temperature corresponding to that pressure. During the condensing process the steam gives up its enthalpy, less the enthalpy of its liquid (condensate) to the condenser cooling water, and the volume of the steam is reduced to that of the liquid. At 2-in. Hg absolute pressure, the ratio of the specific volume of the vapor to the specific volume of the liquid is about 1 to 21,000. It is this large reduction in volume that causes the reduction in pressure.

To transfer this heat to the cooling water requires a decreasing temperature in the direction of heat flow from the condensing steam to the circulating water. The rate of heat flow depends not only upon temperature gradients but also upon the amount of heat-transfer surface area and the resistance to heat flow.

Typical temperature distributions of the condensing steam and the heat-absorbing cooling water are shown in Fig. 6. For convenience, the steam condensing temperature is shown as constant, although this is not completely accurate. The condensing temperature shown in Fig. 6 may be considered as an average temperature, whereas in the actual condenser the steam temperature and pressure vary in a saturation temperature-pressure relationship throughout the condenser steam space. The variation, while small, reflects internal pressure losses and must be considered in the design of the condenser.

The smaller the temperature difference between the heat source and the heat receiver, the lower will be the condensing temperature for a given flow of cooling water and the lower will be the amount of heat energy rejected from the cycle. If the condensing temperature could be reduced to that point at which it approaches the cooling water inlet temperature, a maximum availability of heat

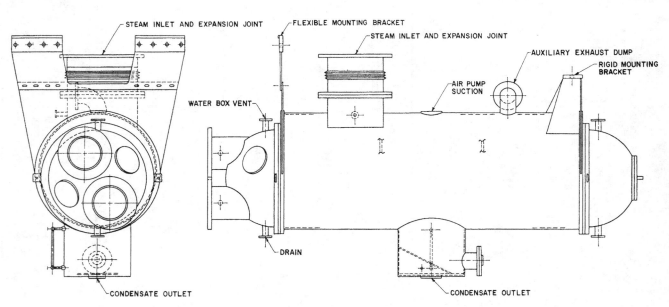

Fig. 5 Auxiliary condenser arranged for overhead deck mounting and longitudinal expansion support

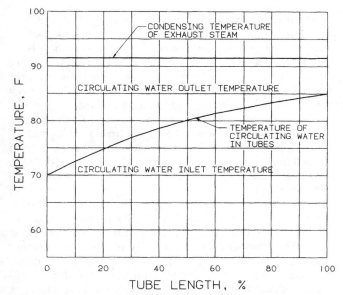

Fig. 6 Condensing and circulating water temperature distributions in a typical surface-type steam condenser

energy for producing useful work would result, and the highest cycle efficiency might be obtained. An approach to such a condition can be made by consideration of a number of factors—among them increased heat transfer rate, increased cooling water flow, and increased condensing surface—all of which affect the temperature difference between the condensing steam and the cooling water. The desired effect is a reduction in back pressure on the turbine and improved cycle efficiency. However, the incremental costs required to obtain the efficiency improvement (in terms of a larger circulating pump, more heat transfer surface, etc.) may not be economically justifiable.

The available space on board ship, the increased volume of steam at low pressures, the high moisture content of low-pressure exhaust steam, and mechanical limitations associated with the turbine design generally determine practical condensing pressures in relation to cooling water temperatures. However, the improvements in heat transfer rates achieved through the design of the condenser, selection of tube materials, and tube geometry are practical means for reducing the size, weight, and cost of a steam condenser for a given set of design conditions. Therefore, designing for the highest heat transfer rate possible is of primary importance to both the designer and user. Because of its effect on the economy of the steam plant, the optimization of the condenser design and application is considerably more important for main condensers than for auxiliary condensers.

1.3 Feedwater Recovery. One of the important functions of the surface condenser is the recovery of feedwater. In the process of condensing the exhaust steam from the main or auxiliary turbines, steam condensate is recovered and is reused as feedwater. In a properly designed and maintained feedwater system the condensate has the following three important characteristics:

1. Condensate is of high purity, containing only extremely small quantities of dissolved solids.

2. Condensate is neutral or slightly alkaline in character, in terms of hydrogen-ion concentration, with very small amounts of various gases (oxygen, carbon dioxide, nitrogen, and ammonia) in solution.

3. Condensate is collected at the saturation temperature corresponding to condensing pressure, or with about 1 deg F subcooling at full-power conditions. At lower power levels, greater subcooling can be expected.

Steam plants require extremely pure feedwater, and it is no longer practical to construct noncondensing steam plants or steam plants using jet or direct-contact condensers.

1.4 Condenser Shell Structure. The condenser shell structure involves many combinations of geometry to obtain an efficient tube-bundle arrangement, to obtain economies in manufacturing costs, and to minimize space requirements. Further variations occur in relation to the means used to achieve pressure integrity, such as ribbed supports, diaphragm supports, compression stays, self-supporting shell structures, and various combinations of these.

While the pressure differential between atmospheric pressure and the condenser operating pressure is relatively small, approaching 14.7 psi as a limit, the conditions under which this pressure must be withstood require relatively heavy or well braced structures. A loading of 14.7 psi is small when compared with the operating pressure of boilers or high-pressure heaters, but is substantial with respect to vessels such as condensers, which are subjected to external pressure. This is because condensers are relatively large irregular structures, especially main condensers. Also, structural failures from external loading are usually in buckling with cylindrical shapes, and in bending with flat or nearly flat surfaces. This is in contrast to failures, generally in tension, of the cylindrical or spherical shapes used for boilers and pressure vessels, which are subjected to internal pressure. As an example, a 72-in.-diameter cylindrical vessel that is designed for external pressure of 14.7 psi would be suitable for an internal pressure of approximately 130 psi.

Even though there are many variations in condenser-shell geometry, only two basic configurations are generally used. For small main condensers and for auxiliary condensers, condenser shells of circular cross section are the most common. This type of condenser shell is usually designed with sufficient thickness to be self-supporting under external pressure without the use of internal or external bracing. The supports are attached so that external mechanical loads are properly distributed through the use of saddles, ribs, and doubling plates, and, on occasion, by an increased thickness of the shell plate over that required for pressure alone.

Main condenser shells are usually designed as flat-sided, boxlike structures, which may be ribbed, compression stayed, diaphragm-supported, or supported with bracing to achieve adequate strength and minimum weight. Flat-sided propulsion-steam condensers are not designed as self-supporting box structures because of the excessive plate thickness needed and the resulting excessive weight. When carefully designed as braced structures, they compare reasonably well in terms of weight with large unsupported cylindrical shell condensers. Their major advantages are compactness and their inherent structural rigidity, which make them suitable for supporting the turbine or auxiliary machinery with little additional bracing needed.

1.5 Condenser Supporting Arrangements. Most steam turbines used for main propulsion exhaust downward into a steam surface condenser located below the turbine (see Figs. 2 and 3). However, some turbines have been designed for higher efficiencies (i.e., without turning-hood losses) with an axial-flow exhaust, by discharging into a condenser that is located forward or alongside of the turbine, with the centerline of the steam inlet of the condenser on, or close to, the same level as the centerline of the turbine (see Fig. 1). Condensers serving the more conventional downward-exhausting turbines may be mounted in a number of different ways. Three of the most common arrangements have the condenser supporting the low-pressure turbine, the turbine supporting the condenser (a "hung" arrangement), and with the turbine and condenser mounted separately.

Condensers that are designed to support the main propulsion turbines offer the advantage of compactness. When they are used in this manner, condensers are built with supports as an integral part of the shell structure. A typical condenser that is designed to support the turbine is shown in Fig. 2. The supports are located high on the shell structure, near the steam inlet. With this arrangement, the condenser supports the weight of the turbine and the exhaust transition section on top of the support, with the shell, tube sheets, tubes, internal baffles, support plates, and the waterboxes hung below the support. In principle, the upper portion of the condenser structure can be likened to a number of truss sections upon which the turbine rests and from which the condenser hangs. The beam structure itself is supported from foundations built into the ship. In addition to the integral supporting structure holding the condenser and turbine in position, sway braces and collision chocks at the lower section or bottom of the condenser are also provided.

Misalignment resulting from temperature changes during operation can be minimized by locating the condenser supporting structure as close to the turbine centerline as possible. Dimensional changes associated with temperature changes during operation are also minimized by locating the supports high on the condenser shell. Expansion and contraction movements are especially undesirable if the turbine shaft is attached to separately mounted gears or to electric generators, which are not subject to the temperature changes characteristic of the condenser and turbine. Small vertical movements can become relatively unimportant if the condenser, as a supporting structure, can be used to support the turbine and its associated reduction gear or electric generator.

With separate turbine foundations, it is frequently possible to suspend the condenser from the turbine flange.

In some cases, depending upon the weight of the condenser and the strength of the turbine casing as a structure, the entire weight of the condenser can be hung from the turbine flange or turbine casing. In those cases where the loading is too great, a portion of the condenser weight may be supported on springs that are designed to carry the weight of the cooling water in the condenser tubes and waterboxes, with a portion of the load from thermal expansion leaving a downpull on the turbine flange at all times. Whether the condenser is wholly supported from the turbine or only partially so, sway braces, to resist ship normal inertial forces resulting from ship motions, and collision chocks are also required. A typical example of a "hung" main condenser arrangement is shown in Fig. 3. The high-pressure turbine is typically pedestal mounted or bedplate mounted independently of the main condenser.

Auxiliary condensers are usually mounted separately from their turbines, but not necessarily so. The common practice is to arrange auxiliary condensers with top supporting structures for mounting on the underside of a deck. In most instances, the auxiliary turbine is mounted on the topside of the deck, and the turbine and condenser are connected with an expansion joint between them. A typical design is shown in Fig. 5.

The principle of a separately mounted turbine and condenser can also be applied to the main propulsion unit. In this case the condenser is mounted rigidly on foundations near or at the bottom of the shell. The turbine is mounted on separate foundations above, and the condenser and turbine exhaust openings are connected by an expansion joint or a flexible element. Such installations are rugged, but the requirement of separate foundations for both the condenser and turbine results in more weight and cost.

1.6 Condensing Pressure and Performance Characteristics. Steam surface condensers for marine power plants, particularly those used for the main propulsion system, are required to work at absolute pressures that are somewhat lower than those used in many stationary power plants because of the trend to use cooling towers as the heat sink for stationary power plants. For the main condenser on merchant ships, the operating pressure at full power is selected to correspond to a temperature that is approximately 16 to 20 deg F above the cooling water inlet temperature. The design approach temperature (the difference between the saturated steam temperature and the cooling water outlet temperature) is usually limited to 5 deg F. The operating pressure at full power for main condensers on naval ships is selected to correspond to a temperature that is 30 to 60 deg F above the cooling water temperature in order to reduce the plant size and weight. Large steam condensers in stationary service usually are selected for operating pressures with corresponding temperatures 25 to 40 deg F above the cooling water temperature.

An extremely low absolute pressure imposes the need for great care in the design of the condenser in order to meet the performance requirements. Pressure losses within the condenser are most critical; however, in compactly designed units, large-volume steam flows are difficult to accommodate without incurring excessive pressure

Table 1   Recommended design conditions, pressures, temperatures, cleanliness factors, and water velocities

Class Turbine	Conditions	Cooling Water Temperature, F				
		55	65	75	80	85
Large turbines, high efficiency, low exhaust loss	Vac. temp., F	78–83	88–93	99–104	102–107	105–110
	Abs. press., in. Hg	0.97–1.14	1.34–1.56	1.88–2.18	2.05–2.38	2.25–2.60
	Cleanliness factor, %	85	85	85	85	85
	Water vel., fps	6–7[a]	6–7[a]	6–7[a]	6–7[a]	6–7[a]
Intermediate or large turbines, moderate efficiency, moderate exhaust loss	Vac. temp., F	93–98	98–103	105–110	108–113	111–116
	Abs. press., in. Hg	1.56–1.82	1.82–2.12	2.25–2.60	2.45–2.83	2.67–3.10
	Cleanliness factor, %	85	85	85	85	85
	Water vel., fps	6–7[a]	6–7[a]	6–7[a]	6–7[a]	6–7[a]
Small or intermediate turbines, low efficiency, high exhaust loss	Vac. temp., F	100–107	107–112	112–117	112–117	117–122
	Abs. press., in. Hg	1.94–2.38	2.38–2.75	2.75–3.20	2.75–3.20	3.20–3.70
	Cleanliness factor, %	85	85	85	85	85
	Water vel., fps	5.5–6.5[a]	5.5–6.5[a]	5.5–6.5[a]	5.5–6.5[a]	5.5–6.5[a]
Very small turbines, low efficiency, high exhaust loss	Vac. temp., F	120–125	125–130	130–135	130–135	135–140
	Abs. press., in. Hg	3.45–3.95	3.95–4.52	4.52–5.18	4.52–5.18	5.18–5.85
	Cleanliness factor, %	85	85	85	85	85
	Water vel., fps	5.5–6.5[a]	5.5–6.5[a]	5.5–6.5[a]	5.5–6.5[a]	5.5–6.5[a]

[a] May be increased to 10–15 fps for titanium tubes.

losses. For example, a 0.10-in. Hg. pressure loss at 1-in. Hg absolute represents a temperature loss of approximately 3.19 deg F, which results in a corresponding reduction in the temperature difference with the cooling water; at 2-in. Hg absolute, the same pressure loss represents a temperature loss of only 1.71 deg F. The respective steam volumes associated with these pressures are 652.3 and 339.2 ft^3/lb. When considering the physical properties of steam in a condensing environment, it becomes apparent that these characteristics must be skillfully accommodated in the condenser design if satisfactory performance is to be obtained. As a consequence, low-pressure condensers must have liberal flow passages on the steam side for steam distribution around the perimeter of the tube bundles, as well as longitudinally for their entire length. Obstructions to the steam distribution flow path should be held to a minimum consistent with safe structural design. For marine units the space envelope in the engine room is usually restricted so that the truss height is limited; consequently, structural members of the truss tend to take up more flow area in the steam dome than do those for land-based condensers.

The selection of the condenser design conditions is extremely important to the economics of the turbine and to the steam condensing plant. While an extremely low back pressure can improve the cycle efficiency, it may also increase disproportionately the cost of the turbine, the condenser, the condenser auxiliaries, and their installation. The most economical back pressure can also be a function of machine size. The cost required to attain a high thermal efficiency by specifying a low exhaust pressure for small mechanical-drive turbines can seldom be justified. As a result, most condensers serving small auxiliary turbines are designed for higher back pressures than those serving the main propulsion turbines or those driving larger auxiliary equipment, such as the main generators. Table 1 lists recommended condenser design points, giving absolute pressures, corresponding condensing temperatures, cleanliness factors, and water velocities that represent good engineering practice.

1.7   Air Inleakage and Removal.   Without exception, steam surface condensers used in marine service are required to condense impure vapors, that is, vapors that contain air or other noncondensable gases. Because most condensers operate at atmospheric or subatmospheric pressures, air leaking into the condenser or into the condensing system from gasketed joints is the most common impurity in the condensing vapor. However, some of the noncondensables present are not the result of direct air inleakage, but enter the boiler feed system from the makeup feedwater evaporators or by condenser tube leakage, and may even result from boiler feedwater treatment chemicals. The noncondensable gases most commonly found to originate from these sources are carbon dioxide and ammonia.

The presence of a very small amount of air in the condensing vapor has little or no effect on heat transfer. In fact, some observers have been able to show that under controlled conditions, noncondensables may promote dropwise condensation and actually improve heat transfer. Similar experiments with larger amounts of air present, and with heavy liquid films on the condensing surfaces, have indicated air blanketing of the cold tubes, which resulted in substantial reductions in the measured heat transfer. Equally important, however, is the possibility that air or other noncondensables may accumulate in the condenser to such a degree that a significant increase in total pressure in the condenser will result. Such an increase in total pressure almost always results in a reduction in cycle efficiency.

In a well designed power plant, the noncondensable gases constitute a very low percentage of the steam-gas mixture and have little effect on its condensing temperature (the partial-pressure effect is negligible up to a noncondensable gas content of about 0.1%). In certain special cycles, however, specifically geothermal and ocean-thermal-energy-conversion cycles, the noncondensable-gas to condensable-vapor ratio is significantly higher than in conventional cycles; therefore, the effect of noncondensable gases on both the condensing overall heat-transfer rate and the condensing steam temperature, as well as the design, sizing, and selection of venting equipment for their removal, must be considered. For these cycles the gas removal techniques described herein apply in a general sense but are inadequate. The accumulation of noncondensable gases and inadequate venting arrangements in a condenser are frequently the most influential factors in poor condenser performance.

Should a significant loss in performance occur in a well designed condenser, it is usually an indication of an excessive amount of air leaking into the system, or fouled heat transfer tubes. Either or both can overload the air-removal equipment, and the condenser performance will be limited accordingly.

Good performance under various conditions of air in-leakage requires effective cooling and dehumidification of the noncondensable gas-vapor mixtures so that both their saturated weight and volume are reduced before being withdrawn by air-removal equipment. A vapor flow path of the displacement type, free from zones of stagnation and short circuiting, is essential to achieve maximum condenser performance. Adequate noncondensable cooling permits the application of minimum-sized air-removal equipment, thus affording a saving in weight, space, and cost. Normally, this requires 5 to 7% of the tube surface to be dedicated for a baffled air-cooler zone within the condenser.

The Heat Exchange Institute (HEI) suggests in its published standards that air-removal equipment be sized proportionately to steam flow as indicated in Table 2. Paragraph S-29 of the HEI Steam Surface Condenser Standards also suggests values for the venting equipment suction temperature.

1.8 Condensate Removal. Surface condensers that operate below atmospheric pressure require pumps for the removal of the collecting condensate. The volute type of centrifugal pump, designed to operate with a low submergence head, is commonly used for this purpose. Most marine condensate pumps are designed to operate with 18 to 30 in. of water submergence; however, special multistage vertical volute pumps have been built that are capable of handling condensate from a condenser hotwell

Table 2   Recommended air pump capacities[a]

Steam Condensed, lb/hr	CFM of Dry Air at 70 F
0– 25,000	3.0
25,000– 50,000	4.0
50,000–100,000	5.0
100,000–250,000	7.5
250,000–500,000	10.0

[a] Courtesy Heat Exchange Institute.

with as little as 12 in. of water submergence. More frequently, submergence heads are higher, and may be as much as 60 in., especially with auxiliary condensers. Main condensers, because of their location low in the ship, require minimum-submergence condensate pumps. Consideration must also be given to ship pitch and roll conditions.

Condenser hotwells, which are used to collect the condensate, are usually equipped with strainers and antivortex devices at their condensate outlet. They are designed to provide optimum hydraulic conditions for the condensate pump. The water level in the hotwell may be controlled by the pump itself (i.e., the pump may be permitted to cavitate when there is insufficient condensate to maintain rated flow of the pump) or by level controllers and condensate recirculation systems. Most marine installations are arranged with devices to control the condensate level in the condenser hotwell so as to minimize cavitation erosion of the condensate pump impeller and the accompanying noise.

1.9 Condensate Storage. The steam surface condenser can also serve as a reservoir to collect and store feedwater. This function is of considerable importance with respect to the main propulsion system, as the storage volume in the condenser provides some additional surge capacity for the boiler. The storage capacity provided by most main condenser hotwells corresponds to about 1 to 1.5 minutes of full steam flow.

In some installations, condensate levels in the hotwell are used to maintain an adequate suction head for the condensate pumps by using electronic, ultrasonic, hydraulic, or mechanical level sensors to regulate the condensate pump discharge flow. The deaerating feed tank is frequently used as the principal control point for the boiler feedwater system and provides the necessary surge volume control.

Section 2
Condenser Design

2.1 Steam Condensing Space. Most marine steam surface condensers are designed so that steam condenses on the outside of heat transfer tubes, with cooling water flowing inside the tubes, and with the steam condensing space contained by the shell. Good design practice requires that the steam entering the condenser have free access to all areas defined by the perimeter of the tube bundles with minimum pressure losses. To meet this requirement, the condenser shell must be sufficiently large

in cross section to allow the steam a relatively free passage around the tube bundles and also provide for longitudinal distribution of the steam to both ends of the condenser. The cross section of flow areas must not be excessively blocked by braces, baffle plates, ribs, diaphragms, or other obstructions.

For main condensers, especially the larger ones, the steam inlets cover a substantial portion of the condenser shell, and lateral distribution of the vapor flow within

Table 3 Recommend steam design velocities

Condenser Design Pressure, in. Hg	Recommended Maximum Main Steam Lane Entrance Velocity, fps[b]
1	500
2	400
3	300
4	250
5	200

[a] Courtesy Heat Exchange Institute.

[b] Assumed as average velocity with uniform flow at the cross-section area under consideration.

the condenser can be readily achieved without excessive pressure losses. Should the available space for steam flow be unduly limited, the effect on condenser performance is likely to be serious; therefore, pressure loss determinations should be made and evaluated.

Turbine exhaust steam entering velocities should be held below accepted maximum values. Recommended velocities, which vary inversely with design pressure, are given in Table 3. The steam-lane area (the area between tube bundles and between the shell and tubes) should be greater than the area of the exhaust inlet. Longitudinal flow velocities should not exceed 50% of the values given in Table 3 for the steam entrance section of the condenser.

The main condenser cross section shown in Fig. 7 illustrates the good steam distribution characteristics inherent with condensers of larger sizes. Steam distribution around the tube bundle is achieved by providing space above and below the tube bundle to permit effective lateral distribution of the steam. Note that the free space below the tube bundle is necessary to reheat the condensate as it drips off the tubes. If insufficient space is provided for contact with the saturated steam flow, the contact time with the condensate is reduced, and excessively subcooled condensate can be expected.

The auxiliary condenser cross section shown in Fig. 8 illustrates steam distribution provisions for accommodating an off-center steam inlet. Stainless steel impingement tubes or baffles are provided to reduce steam impact effects on the tubes and to direct steam both laterally and longitudinally within the shell. The baffles may be perforated or staggered to allow a predetermined portion of the total steam flow to feed the tube bundle section directly beneath them.

2.2 Arrangement of Condensing Surface. Nearly all condenser designers arrange the heat transfer tubes in groupings that are referred to as bundles or banks. Essentially, each bundle is an operating condenser. Small condensers need only one such tube bundle, but the larger main condensers may need two, although seldom more.

The principles that apply to the design of a unit condenser tube bundle are quite fundamental. Five of the more important ones are as follows:

1. The depth of tube banks (number of tube rows deep) can be controlled by:
—changing the geometric shape by using substantially circular, square, rectangular, or trapezoidal cross section configurations;
—controlling the effective tube bank depth by using a variable tube pitch;
—selecting the optimum number of tube bundles per condenser as needed to control the tube bank depth and steam flow distribution; and
—locating the air-cooler baffled zone near the center of each tube bank, to help draw the steam into the depth of the bank, with an allowance for impact pressure at the top of the bundle.

2. Steam fed to the tube bundle throughout its perimeter decreases the average steam entrance velocity to the tubes and thereby mitigates the tube bank pressure loss and also minimizes potential tube vibration.

3. Steam fed from all directions, including the bottom areas of the tube bundle perimeter, aids in condensate deaeration.

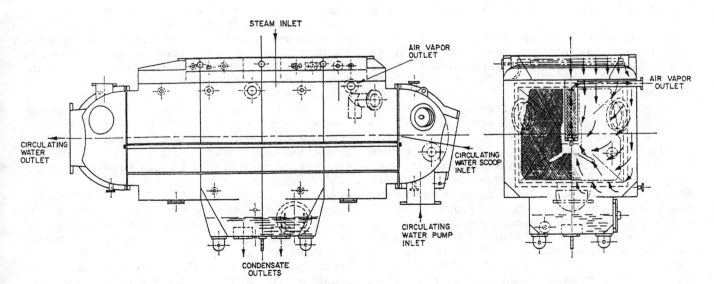

Fig. 7 Large single-pass condenser with central air cooler

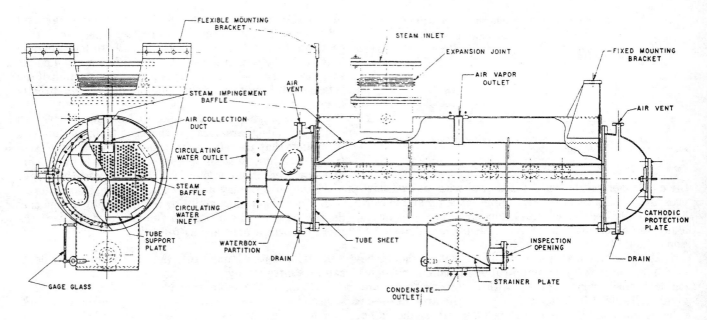

Fig. 8   Two-pass auxiliary condenser with centrally located air cooler

4. The arrangement of tube bundles as separate modules in the condenser shell promotes steam distribution to the tube bundles and condenser surface areas.

5. The flow of steam directly to the condenser hotwell facilitates condensate reheating.

The tube arrangement in Fig. 7 is an example of a single-tube-bundle condenser with no auxiliary steam inlet lanes in the tube bank, but with main feeder lanes around its perimeter.

The tube arrangement in Fig. 8 is another example of a single-tube-bundle condenser with no auxiliary steam inlet lanes in the tube bank, but with main feeder lanes around its perimeter. The main steam lanes reduce in cross section at the bottom of the condenser shell, but still provide adequate flow area for condensate reheating and deaeration. The single tube bundle is circular in configuration.

2.3 Steam Flow Profiles and Distribution. Most auxiliary condensers have relatively small steam inlets in relation to their shell size, and should be provided with impingement and distribution baffles to redirect the incoming steam. The steam flow to small condensers is characterized by little or no stratification. In some cases, especially with long steam supply lines, the velocity contour may approach the ideal flat-front flow. Generally, no major steam flow distribution problems are directly caused by the turbine exhaust-steam flow pattern.

The main condenser, with its large steam inlet, is usually subjected to highly stratified steam flows. The condenser design geometry and its arrangement in the ship are the principal factors that cause stratified flow.

The flow pattern characteristic of a downward-exhausting single-flow turbine is markedly stratified, but the pattern is reasonably symmetrical and usually can be effectively accommodated. A steam flow distribution map for a single-flow, downward-exhausting turbine at a high

flow rate is shown in Fig. 9. The numbers shown on the flow contours are the factors by which the average velocity is multiplied to obtain the local velocity of the steam, and it may be seen that there are areas where the steam velocity reaches 2.9 times the average velocity. With average velocities in the order of 500 fps, such a turbine exhaust pattern can easily have flow areas where there are velocities in the sonic range.

In general, the flow distribution shown in Fig. 9 will not interfere with the performance of condensers with single- or multi-tube-bundle designs, provided the turbine shaft and the condenser longitudinal centerline are substantially parallel. When the turbine shaft is perpendicular to the condenser centerline, the condenser tube bundle design can partially compensate for the nonuniform flow characteristics of the turbine. However, in some instances, offsetting the tube bundles from the condenser shell centerline may be required in order to widen one of the main steam distribution lanes so that steam may be fed into the tube banks at high flow rates from both the bottom and top of the tube bundles. On occasions, baffles may be needed to redirect some of the steam to other sections of the condenser. The use of baffles for this purpose will result in some pressure loss in the distribution of steam, although if steam flow conditions are severe, a greater loss might result if baffling were not used. Each case must be resolved with full consideration of the specific conditions that exist.

2.4 Control of Tube Vibration [2]. Vibration has often caused the failure of heat transfer tubes in steam condensers. Fortunately, high incidences of failure from this cause have been periodic rather than general, and for the most part more annoying than catastrophic. If the interior of the condenser is accessible, mechanical means such as installing wood, fiber, or metal slats between the tubes may be used to alleviate the vibration problem. As an

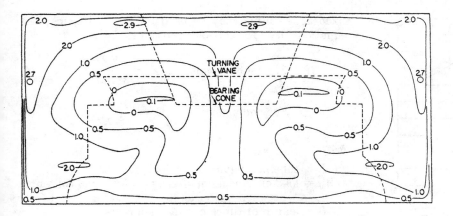

Fig. 9 Typical full-load steam flow distribution at a turbine exhaust flange [1]

interim corrective measure, this modification is reasonably effective.

With small condensers, the addition of slats is seldom a possibility. However, in some instances, slats can be installed between the tubes by cutting holes in the condenser shell. If this cannot be done, the condenser must be rebuilt internally, and additional support plates and new tubes will be required to correct the problem.

The present technology provides a means to avoid severe condenser tube vibration failures, although an occasional tube failure from vibration may occur, especially in a severe corrosive environment. In such cases the tube failures normally show evidence of corrosion fatigue.

The primary cause of tube vibration is from the dynamic effects associated with the flow of steam (turbine, steam dump, or secondary exhaust). A secondary cause may be mechanical vibration transmitted from attached machinery, from nearby machinery, or from other vibration sources, with the exciting energy transmitted through the foundation, piping, or other solid connections.

Transmitted mechanical vibration, as a general cause, is seldom of sufficient magnitude to be troublesome. Vibration transmitted from the turbine has commonly been suspected as a major cause of tube vibration but is rarely a significant contributor. The allowable vibration amplitudes for large turbines and the resulting transmitted energy are generally too small to develop significant tube vibration, even in those cases when the rpm of the turbine shaft corresponds to the natural frequency of the tubes. Field and laboratory tests have shown tube vibration amplitudes at their natural frequency to be within the limits of 0.011 and 0.030 in. when excited at the same frequency through their supporting structure, with the energy input to the supporting structure in the laboratory equal to that measured on a condenser in the field. Vibration amplitudes of these magnitudes do not result in tube vibration failures.

The primary cause of condenser tube vibration is the steam flow with its lift and drag effects on the tubes. For condensers it is not usually related to the von Karman vortex effect of critical velocity. Laboratory tests show that with condensers operating at a low absolute pressure, the velocity energy level at which the von Karman vortex effect occurs is much too low to have a significant effect

on tube vibration. For example, for a ¾-in.-OD (outside diameter) tube it is about 32 fps. The flow energy of steam at these velocities, with the low density associated with condenser pressure, is negligible. However, with auxiliary condensers, some of which may operate at atmospheric pressure and, therefore, with much higher vapor densities, the critical velocity may need to be examined. It can be computed from the following relationship:

$$V = \frac{fd}{S} \qquad (1)$$

where

V = critical velocity of fluid stream, fps
S = Strouhal number, dimensionless
f = tube natural frequency, cps
d = diameter of tube, ft

V is the critical velocity when f equals the natural frequency of the tube. S ranges in value from 0.10 to 0.30 for Reynolds numbers between 100 and 630,000. A Strouhal number of about 0.20 is a typical value for steam entrance conditions to a tube bank, whereas 0.10 is a reasonable value for the innermost sections of the tube bank. For high-pressure closed heaters, an evaluation of the critical velocity is of fundamental importance.

In the condensing environment, the velocity effects of steam on the tubes initiate the significant vibrating forces, and the tubes vibrate at their natural frequencies. The steam flow causes a loading to be imposed on the tubes from both drag and lift effects, and results in tube deflection as a function of this loading. Experience has shown that the velocity loading effects of lift and drag will deflect the tubes according to the theory of beams under static loading and that the tubes will vibrate with that amplitude without significant amplification. The equations that apply to these circumstances are as follows:

$$W_d = C_d A \frac{\rho V^2}{2g} \qquad (2)$$

where

W_d = load on tube in direction of fluid flow, lb
C_d = drag coefficient

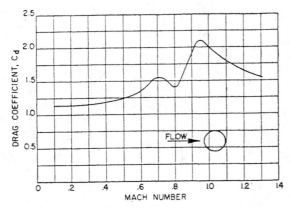

Fig. 10   Drag coefficient for tubes in cross flow [2]

A = tube projected area, ft^2
ρ = fluid density, pcf
V = fluid velocity, fps
g = gravitational acceleration, ft/sec^2

$$W_L = C_L A \frac{\rho V^2}{2g} \qquad (3)$$

where

W_L = load on tube 90 deg to direction of fluid flow, lb
C_L = lift coefficient

The value of the drag coefficient, C_d, in equation (2) may be obtained from Fig. 10. It is important to note that C_d reaches a maximum value of 2.1 at Mach 0.95. Data on the lift coefficient, C_L, which is also known as the coefficient of lateral thrust, are not as well documented as the data on the drag coefficient. Values that have been suggested range from 0.20 to 1.00, and 0.65 to 1.71. In general, for a cylinder, the lift coefficient is less than the drag coefficient, C_d, and the effect of drag controls the amplitude of vibration.

The value of W_d from equation (2), when used in conjunction with applicable beam deflection equations, can be used to determine the optimum distance between support plates or between support plates and tube sheets for steam condensers operating at a low absolute pressure.

The design steam velocity should be determined from data similar to that shown in Fig. 9 using full-load flow to determine the average velocity at the turbine exhaust flange. If the product of the average velocity and the local velocity multiplying factor approaches the sonic velocity, the design should be based on the sonic velocity. The design should also be evaluated at the lowest absolute pressure expected in the condenser at full load to determine if those conditions are more severe than the design operating condition.

The classic equations for beams under uniform loading can be used to compute tube deflections. Because of the relatively high span-length to tube-diameter ratio normally used, the deflection equation for a simply supported beam may be used for the center spans, and the deflection equation for a beam simply supported on one end and

fixed at the other may be used for the end spans. The equations used for this purpose are as follows:

For center spans: $Y = \dfrac{5W_d L^3}{384EI}$ \qquad (4)

For end spans: $Y = \dfrac{W_d L^3}{185EI}$ \qquad (5)

where

Y = tube maximum deflection, in.
W_d = load on tube (sum of drag, weight, and contents) between supports, lb
L = length of tubes between supports, in.
E = modulus of elasticity, psi
I = tube moment of inertia, in.^{4}

The mid-span or maximum tube deflection determined from equation (4) or (5) must be less than one half of the ligament between tube holes in the tube sheet or adjacent tubes will contact each other. Good practice is to limit the deflection to one fourth of the ligament, or one half of the ligament less one sixteenth of an inch, whichever is greater.

The tube spans or support plate spacing should be finalized only after determining that the bending stresses in the tubes are of an acceptable magnitude, and that premature failure will not occur from fatigue. The maximum bending stresses in the tubes may be computed by using classical beam equations. The fatigue life should not be less than the expected life of the tubes with respect to anticipated waterside and steamside corrosion and erosion effects.

2.5   Air Cooler Design.   The purpose of the air cooler is to reduce the water-vapor content of the noncondensable gases to be removed from the condenser. The Heat Exchange Institute recommends that the air-vapor mixture be cooled 7.5 deg F below the saturation temperature corresponding to the design pressure or 25% of the difference between the saturation temperature corresponding to the condenser pressure and the design inlet circulating water temperature, whichever is greater. In some instances it has been found practical to cool to lower temperatures, and some condensers are designed to cool to within 5 deg F of the cooling water inlet temperature. However, most condensers are designed to cool within about 5 deg F of the cooling water outlet temperature.

An increase in the capacity of the air cooler reduces the quantity of condensable vapors and reduces the load on the air pump (or other venting equipment) for a given air leakage. As a result, a smaller air pump may be used. Also, an improved condenser performance can result by applying standard-size air pumps to condensers provided with more effective cooling sections than those designed for 7.5 deg F cooling.

An example of the effect of air leakage on condenser pressure under conditions of normal operation and under conditions of excessive air leakage or an undersized vacuum pump is shown in Fig. 11. The line 1-2-2'-3 shows the typical performance of a condenser, in terms of absolute

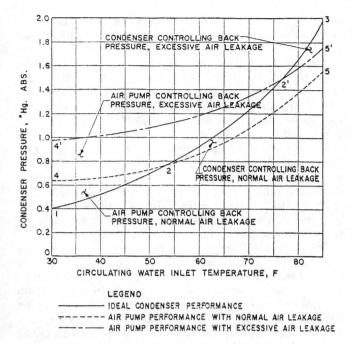

LEGEND

———————— IDEAL CONDENSER PERFORMANCE
– – – – – – – AIR PUMP PERFORMANCE WITH NORMAL AIR LEAKAGE
—— – —— AIR PUMP PERFORMANCE WITH EXCESSIVE AIR LEAKAGE

Fig. 11   Effect of air leakage on condenser pressure

pressure versus cooling water inlet temperature, assuming that air leakage has no effect. The line 4-2-5 shows the pressure that may be produced by the air pump, assuming normal air leakage and subcooling of the air-vapor mixture. The portion 4-2 of this line represents conditions wherein the air pump cannot remove the air-vapor mixture at as low a pressure as that which the condenser can produce. The condenser operating pressure will then correspond to line 4-2, and the air pump is said to limit the back pressure. The portion of the line 2-5 represents a condition where the air pump can remove the noncondensables at a lower pressure than the condenser can produce; under these conditions, the condenser is said to limit the back pressure, and the condenser pressure then corresponds to line 2-2'-3. The line 4'-2'-5' represents vacuum pump performance with a condition of excessive air leakage (the same effect can be caused by too small a vacuum pump or an inefficient air cooler). It should be noted that in this case, the vacuum pump controls the condenser pressure from point 4' to 2'. Above point 2' the vacuum pump is adequate and the condenser will perform along the line 2'-3.

The total pressure in a condenser at any point is the sum of the vapor pressure at the local saturation temperature and the noncondensable gas pressures. This may be expressed as:

$$P_t = P_v + P_g \qquad (6)$$

where

P_t = total pressure in condenser
P_v = vapor pressure of condensing steam
P_g = partial pressure of noncondensable gases

As the vapor-gas mixture flows from the steam inlet to

the vapor-gas mixture outlet, the ratio of P_v to P_g decreases, while P_t also decreases, thereby causing flow. Since the condensing temperature also decreases along the flow path as P_v decreases, the temperature of the air-vapor mixture also decreases. As this occurs, the quantity of vapor in the vapor-gas mixture also decreases. The degree to which this is accomplished is related to the vapor flow path, which also affects the conductance of the condensing boundary of the heat transfer system. As condensation takes place, the vapor-gas mixture becomes more impure, and if optimum heat transfer is to be attained, the concentrating noncondensables must be displaced toward the end of the flow path in the condenser by the incoming purer vapor-gas mixture. This displacement flow path minimizes the effect of diffusion of noncondensables throughout the condenser, and results in P_g being at minimum values in the major portion of the unit. With such a flow path, noncondensable gases can easily be vented from the condenser without requiring excessively large air pumps.

The ability to maintain such a flow path is largely dependent upon the air cooler and its relation to the remainder of the condenser. The air cooler must be located so that it draws the air-vapor mixture proportionately from all sections of the tube banks. It must be large enough to maintain adequate steam or air-vapor mixture flow velocities at the point where the fluid leaves the main tube banks before entering the air-cooler section. At the same time, the air cooler should not be such a large portion of the total condensing surface that it significantly detracts from the high heat-flux surfaces in the tube banks of the condenser. Air coolers containing 5 to 7% of the total condensing surface can be designed to meet normal surface requirements effectively. Properly designed air coolers are located centrally within the tube bundle and are baffled to prevent the flow to the air cooler from bypassing the tube bundle.

A displacement flow path is extremely important to the correct functioning of an air cooler. If possible, it should be incorporated into the design by providing large entrance areas into the air cooler with the flow cross section decreasing gradually to the air-vapor outlet. It is just as important, however, to use a geometry where the length of the flow path through the air cooler is not less than the effective depth of the main tube banks.

One of the most important requirements for condenser air coolers is to provide means for positive flow in and out of each section between support plates. This requirement is fundamental to good condenser performance because it provides good venting and eliminates short circuiting and a buildup of large zones of stagnation with the attendant accumulation of noncondensables that result in a loss in performance. It prevents the air pump or venting equipment from being overloaded by causing all air-cooler sections to work equally and effectively, and by facilitating the dehumidification of the noncondensables.

The determination of the heat transfer surface required for an air cooler to meet specific requirements is a complex task. The heat transfer rates at the entrance to the cooler, where the noncondensable gas content of the mixture is low, are close to the average rates for the condenser. The velocity of the air-vapor mixture at the entrance to the air cooler is usually greater than that at any other point in the air cooler. As dehumidification becomes appreciable, the heat transfer rate decreases and the velocity of the mixture also decreases. For the heat transfer process involved, the opposite situation for velocity is desirable to sustain high transfer rates throughout the flow path. In practical designs this is not always possible to achieve, and additional surface is needed to compensate for the effect of decreasing velocity.

Design methods to determine the air cooler size are, for the most part, approximations and are largely empirical. The Colburn and Hougen basic method is complex. Reasonably good approximations have been made by Mickley. Simplified methods by Colburn and Hougen, and by Smith, compare well with limited test results. Silver also employed an approximate method for interpreting data from condensers where low noncondensable gas velocities were encountered and achieved reasonable success. The fundamentals of these works may be found in reference 3.

The air coolers shown in Figs. 7 and 8 are single (vapor) pass types. Each air-cooler section between support plates discharges into a collection header through proportioned orifices, thereby ensuring that each section between the support plates receives its share of the load and is properly vented.

Liberal air-cooling surfaces in the designs shown in Figs. 7 and 8 have the potential of cooling to a temperature very close to that of the cooling water inlet. These designs effectively use the last rows of tubes in the condenser for the air cooler. These tubes have a very low temperature rise, usually less than 3 deg F, because of the extremely low heat flux characteristic of air-vapor mixtures that have a low percentage of vapor.

2.6 Condensate Reheating and Deaeration. The principles associated with effective deaeration and reheating of condensate are not unlike those applicable to power plant deaerators of the tray type. As the heat passes through the films of condensed steam adhering to the condenser tubes in its path to the cooling water, the condensate on these tube surfaces reduces in temperature to slightly below the saturation temperature at condenser pressure. The temperature gradient across these condensate films, which must occur in order to produce a flow of heat, requires that the mean temperature of this film be less than that of the condensing steam, but greater than the temperature of the contacting tube wall. Unless provisions are made to reheat this condensate before it is removed from the condenser hotwell, a measurable amount of heat energy is lost from the cycle. In addition, the subcooled condensate absorbs air and other gases, such as carbon dioxide, oxygen, and ammonia, from the condensing steam. If these gases are allowed to accumulate in the condensed steam, corrosion of the boiler feed system, the boiler, and the steam turbines may result.

The heat loss from condensate subcooling may be greatly reduced by directing a portion of the incoming steam to the condenser bottom and hotwell area in such a manner that a substantial amount of its velocity energy

is converted to pressure. In this way, the local static pressure in the hotwell area and under the tube bundles may actually exceed the static pressure at the condenser steam inlet. If proper provisions are made, the condensate falling from the tube bundles through this zone of increased static pressure can be heated to a temperature above that which corresponds to the saturated steam pressure at the condenser steam inlet, and thereby effect a thermal gain in the cycle.

In the process of being reheated, the condensate also loses some of its tendency to retain noncondensable gases in solution. This tendency for the condensate to become deaerated may be augmented by providing deaerating trays or baffles underneath the tube bundles to receive the condensate and prolong its time of contact with the turbulent steam flow under the tube bundle.

The degree of deaeration that may be effected in a steam surface condenser, while principally dependent on the amount of direct-contact heating surface provided, is also dependent upon a number of other factors. In a steam condenser, deaeration occurs under vacuum and can be adversely affected by excessive air inleakage, especially if the leakage is below the water level in the hotwell. Other factors that can affect deaeration are the introduction of makeup feedwater, heater drains, and other condensate returns to the condenser.

The capacity of the air removal equipment in operation limits the condenser vacuum and consequently affects deaeration. The effect is greatest at very light loads and with cold cooling water. To install enough vacuum pump capacity to overcome the effects of this characteristic is usually economically unsound.

Four factors that influence deaeration are: (1) the noncondensable gas pressure in the deaeration area, (2) the amount and effectiveness of reheat and deaeration surface, (3) the vacuum pump characteristics in relation to air inleakage, and (4) the effectiveness of converting inlet steam velocity energy to pressure in the hotwell area. A further explanation of these factors is as follows:

1. The noncondensable gas partial pressure in the deaeration zone may be minimized with good steam distribution and by eliminating an accumulation of noncondensable gases from stagnation. Tube bundle arrangements with centralized air coolers are the most effective in this regard.

2. The amount and effectiveness of deaeration and reheat surface below the tube bundles are of extreme importance. Figure 12 shows the deaeration effect of two arrangements of surface working in an environment of very low partial air pressure. With one arrangement, direct surface-producing baffles made of perforated trays are provided. With the other arrangement, the condensate falls without interruption from the bottom tubes of the tube bundles to the water level in the bottom of the shell or hotwell.

3. The effect of partial air pressure, resulting from inadequate vacuum pump capacity, is illustrated in Fig. 13. The curve shown is for a given cooling water inlet temperature. A family of curves would result if a number of different cooling water inlet temperatures were shown.

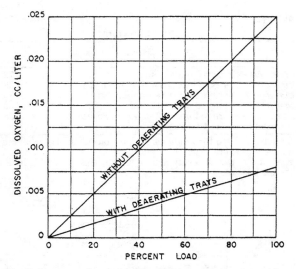

Fig. 12   Performance of typical mechanical deaerating sections with partial air pressure negligible

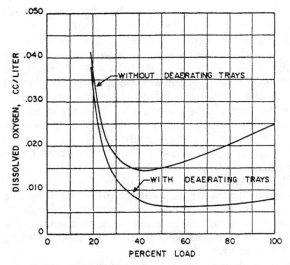

Fig. 14   Performance of a typical deaerating hotwell

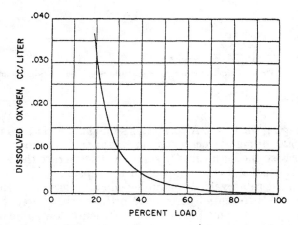

Fig. 13   Typical effect of partial air pressure resulting from vacuum pump limitations

The combined effect of factor 2, deaeration trays, and factor 3, vacuum pump limitations, is shown in the performance curves in Fig. 14. Poorest performance occurs at very light loads with cold water. Best performance begins at about 45% load, but may be extended to higher loads by providing more reheating and deaeration surface or a larger air pump.

4.   The effectiveness of converting velocity energy to pressure energy at the condenser hotwell zone is usually a function of the directness of the main steam distribution lanes in relation to the inlet steam flow and to the bottom of the condenser. Single-bundle arrangements do not perform quite as well as condensers with two bundles.

The prediction of the deaeration performance that can be expected from a given condenser design and its associated special deaeration provisions is complex. Such predictions are related to the projected area of the tube bundle as a measure of unit liquid (condensate) flow rate as it falls from the tubes to the water level below. Also, deaeration is related to the height of fall or to the time of contact of the condensate to be deaerated with the reheating and scrubbing steam. In addition, if perforated or other types of trays are used, their effectiveness is not only a function of the number of vertically disposed horizontal rows, but also of the area they cover (the projected bundle area or less). Furthermore, the liquid-vapor flow paths within the deaeration baffle section have a marked effect on deaerator performance. Preferred arrangements have a cross-counterflow design with a positive vapor demand, through the baffled section, to a portion of the main condenser tube bundle directly above.

As an approximation, with a tube-bundle construction, liberal steam flow around the bundle, and with an effective feeding of steam around its perimeter, a dissolved oxygen content of approximately 0.03 cc/liter may be expected with a 12-in. free fall of condensate. This arrangement is also usually adequate for condensate reheating. However, the machinery space arrangement normally restricts the total height of the condenser, which limits the condensate free-fall distance that is feasible.

Baffles or trays are seldom, if ever, used to augment condensate deaeration in auxiliary condensers; nor are they often used with main condensers because of the additional height required and the limited access. In general, dissolved oxygen levels not exceeding 0.03 cc/liter are satisfactory since most conventional marine power plants include deaerating feed tanks for further deaeration. The deaerating feed tanks reduce dissolved oxygen levels to 0.01 cc/liter or less before the feedwater reaches the boilers.

2.7   Condensate Returns and Feedwater Makeup.   Condensate drains at pressures above deaerator pressure are usually drained into the deaerator. Most of the condensate drains that are discharged into the condenser are generated in the feed system at pressures less than the operating pressure of the deaerating feed tank. The lower pressure drains are usually from the gland leakoff condenser, the gland leakoff ejector condenser, the air-ejector intercondenser, the shell-and-tube feedwater heaters be-low the deaerator in the cycle, the drain cooler, and the distilling plant. (Since there are many variations in the design of feed systems, the components listed may not apply in their entirety.) Some of these drains in passing downward in the system are cascaded in order to impart heat to the feedwater, and thereby reduce thermal losses from the cycle. Not infrequently, and especially in systems using cascading, drains are mixed and are collectively discharged to the condenser.

The condensate drains are frequently at pressures below atmospheric and may be a source of air infiltration to the condenser. The steam air-ejector inter- and after-condenser drains are also a source of noncondensable gas infiltration. To compensate for these possible sources of air inleakage, provisions must be made for their deaeration to a degree sufficient to maintain the desired purity of the feedwater.

Most condensate drains are returned to the condenser at temperatures substantially above the saturation temperature corresponding to condenser pressure. In general, a temperature of 5 to 10 deg F above the condenser temperature provides sufficient energy to deaerate the condensate returns to less than 0.03 cc/liter. However, the design of the condenser internal arrangement can significantly affect the ability to provide adequate deaeration.

The design of the condensate return system inside the condenser frequently incorporates a perforated pipe header to distribute the condensate over a wide area. The perforations are usually drilled holes, which act as orifices. Their size and number are predetermined. The smaller the diameter of the holes and the greater their number, the less concentrated will be the jet energy effect from the fluid flow streams upon emerging and discharging into the condenser; in addition, a more direct-contact surface will be provided for heating with condenser steam should the drain temperatures fall below the condenser temperature, or for dissipating high-energy flashed vapor should the temperature become higher.

The return distribution headers should be located in the condenser steam distribution space and above the highest expected water level in the hotwell. The sprayed returns may be arranged to impinge directly against the condenser shell or into the condensate in the hotwell on the bottom of the condenser. In general, as the temperature of the condensate returns approaches or becomes lower than the temperature corresponding to condenser pressure, the sprays should discharge into the condenser at a higher position to allow for deaeration and reheating. High-temperature, high-energy returns are preferably sprayed directly into the collected condensate with their velocities and distributions designed to avoid splashing condensate on the tube bundle. Makeup water, although admitted to the hotwell in many designs, requires deaeration. Very effective deaeration of makeup may be achieved by spraying the makeup high onto the tube bundle, preferably on the top tubes from above. However, the design of the sprays must provide a good distribution, low admission velocities, and minimum hole diameters to avoid

Table 4 Maximum recommended energy level coefficient for flashed returns

Service	Coefficient, C^a
Continuous—Sprayed against shell	79
Continuous—Sprayed into hotwell condensate	159
Intermittent—Sprayed on tube bundle[b]	79
Intermittent—Sprayed against shell	159
Intermittent—Sprayed into hotwell condensate or on erosion-resistant target areas	238

[a] Values of C are for guidance and are based on orifice diameters not exceeding ⅜ in.
[b] Except for the deaeration of makeup, this practice should be avoided if possible.

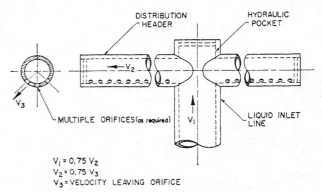

$V_1 = 0.75 V_2$
$V_2 = 0.75 V_3$
$V_3 =$ VELOCITY LEAVING ORIFICE

Fig. 15 Design details of makeup, condensate return, and bypass steam distribution system

damaging the tubes. Also, sprays should not be located in the path of high-velocity steam (i.e., the turbine exhaust); otherwise the high-velocity steam will accelerate the water droplets and cause erosion of the tubes.

Considerable velocity may develop as a result of the pressure differential across the orifices; therefore, the velocity must be maintained at a sufficiently low value so that excessive erosion of impingement surfaces will not occur. Acceptable velocities for the flashing condensate mixture discharging from the orifices are given by the equation

$$V = C \left(\frac{1}{\rho_m} \right)^{0.5} \tag{7}$$

where

$V =$ acceptable velocity of mixture, fps
$\rho_m =$ density of mixture, pcf
$C =$ energy level coefficient (see Table 4)

The density of the mixture, ρ_m, is calculated immediately downstream of the orifices. The velocity of the mixture, V, is calculated by using the discharge area of the orifice in conjunction with a suitable coefficient. For orifices made by drilling holes in a pipe header or plate, a discharge coefficient of 0.80 to 0.85 may be used.

The design details of a typical spray header distribution system are shown in Fig. 15. The relationship of velocities given are for relatively short lines. It is important to adhere to the velocity sequence shown if uniform distribution is to be achieved. For long spray headers, the friction losses should be determined. Headers should be sized so that the friction loss from any section to the end of the header is approximately equal to the velocity head at that section.

The preceding design requirements are applicable in principle to both main and auxiliary condensers.

2.8 Bypass Systems. Nuclear-powered steam power plants, and some others, require heat to be removed from the system even when the propulsion turbines are not in service. This heat energy, in the form of generated steam, is arranged to bypass the turbine and discharge into the condenser for energy absorption.

Since the available energy from this steam is not being used to produce power, its total enthalpy less the enthalpy of its condensate at condensing temperature is absorbed by the condenser cooling water. The main condenser when supplied with rated cooling water flow is more than adequate to absorb this energy. The important consideration is that of providing means within the condenser to dissipate the velocity energy of the steam in addition to absorbing its available heat.

The most elementary and yet most effective system for handling bypass steam employs two pressure breakdown elements. The first is the friction in the piping between the condenser and the steam source; the second is the pressure loss through the orifices provided in the steam bypass distribution system inside the condenser. The greatest portion of the total pressure loss occurs within the condenser, and the steam bypass distribution system must be capable of dissipating this energy so that it can be absorbed as heat without causing mechanical damage.

Three of the most important conditions to be satisfied are: (1) the steam must be desuperheated within the condenser rapidly and effectively; (2) shock wave effects from sonic discharge impinging on tubes or other metal surfaces must be avoided; and (3) the mass-velocity energy effects must be distributed over a wide area.

In designing a steam bypass distribution system to meet these requirements, condition (1) can be satisfied by introducing the bypass steam into the condenser directly below the tube bundles. In this way it is discharged into the "rain" of its own condensate, effectively desuperheating the steam without additional desuperheating sprays or other automatic devices for temperature control. Condition (2) can be met by selecting appropriate diameters for the holes in the distribution headers as discussed in the previous section. In this case, however, the vapor leaving the breakdown orifices in the distribution headers will be substantially superheated and will be at velocities of Mach 1 or greater. Good practice is to discharge the steam above the condensate level in the hotwell such that it is directed at the condensate surface in the bottom of the condenser; this avoids damaging metal surfaces by erosion, temperature, shock wave effects, and vibration. When bypassed steam is discharged into the exhaust trunk, care should be taken to avoid its impingement on the condenser tubes. Further mitigation of shock wave and high-velocity erosion effects may be achieved by using many small-diameter orifices in the distribution system within the condenser. The shock wave from sonic flow through an orifice dissipates in a distance of about 30 orifice diameters; therefore, it is good practice to exceed this distance by a substantial amount before allowing the steam flow to impinge on metal surfaces, with the exception of specially designed impingement baffles. Condition (3), that of spreading the velocity energy effects over a wide area, is extremely important and will influence the requirements under (2) above. In the design of headers for flow distribution, it is important that the maximum fluid flow velocity in the header does not exceed 75% of the fluid velocity leaving the orifices in the header. If it does, uneven distribution will result and the effectiveness of the system will be impaired.

2.9 Cooling Water Circuit.

a. Condenser cooling water passes. Condensers can be constructed as single-pass or multi-pass units. Single-pass condensers (such as those in Figs. 3 and 7) have tubes arranged so that water flows from one end of the condenser to the other through all the tubes in one pass. The tubes in multi-pass units (two pass, three pass, etc.) are arranged in groups. In multi-pass units the coolant enters the inlet waterbox, flows through the first pass group to the return waterbox, is turned around in the waterbox, flows through the second pass group, and then (if a two-pass unit) through the water outlet nozzle. Figures 2, 5, and 8 show typical examples of condensers of this type. The tubes of multi-pass condensers are arranged in groups of substantially the same number, so that the cooling water, which flows through each group successively from water inlet to water outlet, is at the same velocity. Each group of tubes represents a water pass, and the number of such groups in a given condenser gives rise to the terminology of one, two, or three-pass condensers.

A single cooling water pass is a necessity for condensers that rely on a scoop for the circulation of seawater. The amount of head that can be developed by a scoop is limited, especially for slower ships; therefore, condensers having multiple cooling water passes are not feasible in this application.

The air-cooler tubes are preferably located in the first water pass so that the coldest water is made available to cool the noncondensable gases. The use of more than one water pass in the design of the condenser increases, in effect, the tube length or the distance the cooling water must travel from the water inlet to the water outlet. The product of the tube length and the number of water passes is the effective tube length of the condenser, and as such may be directly compared with the actual tube length of single-pass condensers.

For a given heat load and cooling water availability, the primary variables that determine the condenser tube length are:

1. the temperature rise of the cooling water,
2. the water velocity through the tubes,
3. the tube diameter and gage, and
4. the condenser surface and space conditions or requirements.

With respect to variable 1 above, the greater the temperature rise, the longer the effective tube length required for a given velocity. With respect to variable 2, the higher the water velocity, the longer the effective tube length required for a given temperature rise. Regarding variable 3, the larger the internal tube diameter, the longer the effective tube length required for a given velocity and temperature rise. With respect to variable 4, the condenser surface, and the available space conditions, the effects of 1, 2, and 3 will contribute to the final selection of effective tube length and, consequently, the required number of water passes in the condenser.

One of the important factors that must be considered in the design of condensers, and which directly affects the number of water passes required, is the ratio of the condenser shell length to the mean condenser shell cross-sectional area. A long shell of small diameter does not lend itself to a good longitudinal steam distribution; a short shell of large diameter may result in objectionably heavy or deep tube banks. Either of the geometries can result in excessive pressure losses on the steam side of the condenser.

Single- and two-pass condensers are the designs most frequently applicable to marine condensers, with the single-pass more commonly used with main condensers to accommodate the hydraulic characteristics of scoops, and multi-pass condensers more commonly applicable to auxiliary condensers. Constructions using three or more passes are only occasionally needed to meet the performance requirements associated with marine steam power plants. Multi-pass condensers require less cooling water but more surface than single-pass condensers for a given heat load. The smaller cooling water piping associated with the smaller flow may be desirable in some cases.

b. Condenser waterboxes. Waterbox designs used for marine condensers may be divided into two classes: the bonnet type and the ring-and-cover type. In recent years, the bonnet-type waterbox has been used almost to the exclusion of the ring-and-cover type. Figures 1, 2, 3, 4, and 5 illustrate bonnet-type waterboxes. A typical ring-and-cover type of waterbox construction is shown in Fig. 16.

The bonnet-type waterbox is constructed with all elements of the waterbox, including waterpass partitions, flow dividing partitions, and inlet and outlet cooling water nozzles, combined as a unit structure. The ring-and-cover type of waterbox is constructed with all elements of the waterbox, except for the cover, but including waterpass partitions, flow-dividing partitions, and inlet and outlet cooling water nozzles, combined as a unit structure comprising the ring section only. The covers are separate and attached by means of bolted flanges.

It is obvious from a comparison of the figures shown, that ring-and-cover waterboxes provide the greatest accessibility for tube and tube sheet maintenance and tube replacement. There is little or no advantage in using ring-and-cover waterboxes unless they also are equipped with cooling water inlet or outlet connections, or both. Normally, bonnet-type waterboxes without cooling water connections are just as easily removed to provide maintenance accessibility as their ring-and-cover counterparts. Condensers with an even number of passes have cooling water connections on one end only, and are almost invariably built with bonnet-type return waterboxes. The nozzle-end waterboxes may be of either the bonnet or ring-and-cover type.

Nozzle-end waterboxes of the ring-and-cover type provide complete access for servicing tubes, tube sheets, and tube replacement without the need for disconnecting the inlet and outlet cooling water piping. Condenser maintenance is less difficult than with nozzle-end bonnet-type waterboxes, and tube repair, cleaning, and replacement can be easily done.

Nozzle-end waterboxes of the bonnet type must be removed for servicing tubes and tube sheets and the cooling water lines disconnected, except in the case of very large condensers wherein maintenance personnel may enter the waterboxes for tube servicing. For retubing, however, the waterboxes must be removed.

The disadvantage of the bonnet-type waterbox is mainly the limited accessibility to tubes and tube sheets for maintenance. The advantages are lighter weight, better hydraulic design, lower cost (especially when corrosion-resistant materials are required), and fewer bolted joints. The selection of the type for a given installation should take into account the preceding, and also the materials used for tubes, tube sheets, and waterboxes. The use of 90-10 copper nickel for these parts with its excellent corrosion resistance (rather than aluminum brass tubes, naval brass tube sheets, and steel or coated steel waterboxes) has been a major contributing factor to the use of bonnet-type waterboxes for marine condensers.

c. Single- and divided-circulation condensers. Most condensers used in marine services, and especially those used for auxiliary applications, are of the single-circulation type. Single-circulation condensers are characterized by having but one cooling water circuit from inlet to outlet. Condensers of this type must be taken out of service when maintenance is required; and if the main condenser is involved, the ship may also remain out of service until all repairs are made.

Condensers provided with divided-circulation cooling-water circuits may be operated with one half of the condenser in service while the other half is out of service. Maintenance is obviously restricted to the cooling-water side of the condenser. Condenser designs with open center steam lanes, such as are characteristic of two- or four-bundle tube arrangements, are characteristically adaptable to a divided-circulation design.

The cooling-water system must also be divided and valved so that each half of the condenser may operate as a separate unit. The design, when operating with both halves in service, performs in identical fashion to a single-circulation type of condenser.

When one circuit of a divided-circulation condenser is isolated, the condenser operates with one half of the normal cooling-water flow and one half of the condensing

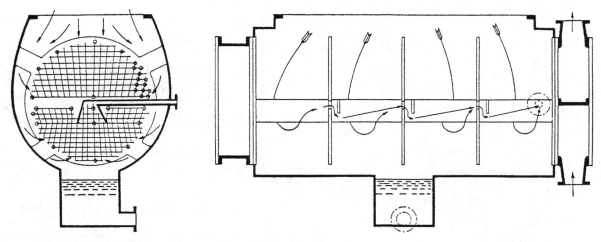

Fig. 16 Condenser with circular array of tubes and peripheral steam inlet and central air cooler with vents cascaded to inlet water end of first-pass tubes for final cooling

surface active. The other half is out of service and the waterboxes of the condenser may be opened to provide access to the tube sheets such that urgently needed minor repairs may be accomplished, especially temporary correction of leaking tubes or tube-end connections. Repairs may be made by plugging or welding as needed.

Ordinarily, the condensing temperature at full load will not exceed 105 F for condensers designed for 2.25 in. Hg absolute with 85 F cooling water. However, with one half of the condenser out of service and with the full steam load condensed by only half the surface and half of the cooling water supplied to the condenser, the back pressure and condensing temperature will increase. At equivalent steam loads the cooling-water temperature rise will approximately double. Should the condensing temperature reach levels that impair maintenance activities, the load on the turbine may be reduced to provide more tolerable working conditions.

Waterboxes for divided circulation may be made as single boxes with a partition or as separate boxes. If partitioned single boxes are used, each side of the waterbox must be capable of withstanding the design working pressure when the other half is out of service. Divided-circulation cooling circuits are almost never used for auxiliary condensers.

d. Pressure losses in cooling-water systems. When establishing the total dynamic head required to circulate cooling water through the condenser cooling-water system, the pressure loss resulting from flow through the condenser must be determined. For marine power plants the pressure loss in the cooling-water circuit of the main condenser is solely from flow friction; but the pressure loss of auxiliary condensers may have a static-head component, such as in those cases where the overboard discharge is above the waterline of the ship.

The most reliable data available for computing condenser cooling-water flow pressure losses are published by the Heat Exchange Institute in "Standards for Steam Surface Condensers" [4]. These Standards provide graphs for determining the friction loss through tubes for various water velocities with correction factors for gauge and tube diameter and for cooling-water temperature. The standards also include graphs for determining waterbox and tube-end losses for velocities through tubes or nozzles for single- and two-pass condensers.

The head loss resulting from circulating-water flow through tubes and in the condenser waterboxes is shown in Figs. 17 and 18 respectively. The data in Figs. 17 and 18 are limited to those tube sizes and gauges normally used in marine condensers; more complete information is given in the Heat Exchange Institute Standards.

2.10 Mechanical Design Practice.

a. Tube sheets and tube spacing. Tube sheets are drilled to a prescribed pattern to receive the heat transfer tubes. The holes into which the tubes are secured have a number of geometries intended to meet specific needs. These geometries are described in paragraph c.

The tube holes are first drilled and then reamed for finish and sizing. The shell sides of the tube holes are chamfered, usually with a $\frac{1}{16}$ in. $\times$ 45 deg bevel, or preferably with a $\frac{1}{16}$ in. radius. The finished diameter of the tube hole should be made equal to the nominal tube diameter plus at least 0.008 in. but not more than about 0.012 in., with the specific tolerance depending upon the tube material.

The thickness of tube sheets varies with tube sheet size (area) and is dependent upon the waterside design pressure. Tube sheets drilled for $\frac{5}{8}$ in., $\frac{3}{4}$ in., or $\frac{7}{8}$ in. tubes that are to be rolled or packed should not be less than $\frac{7}{8}$ in. thick. If the tubes are to be welded and rolled, the tube sheet should not be less than $\frac{3}{4}$ in. thick. The earlier recommendations of the Heat Exchange Institute regarding tube sheet thickness are given in Table 5. The Heat Exchange Institute no longer publishes dimensional standards; however, the standard given by Table 5 remains in general use. It should be noted that the more recent HEI

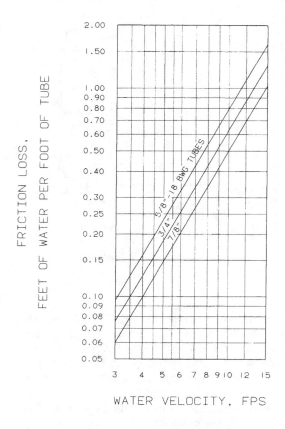

Fig. 17   Friction loss in condenser tubes (courtesy Heat Exchange Institute)

TUBE GAUGE CORRECTION

TUBE SIZE	16 BWG	18 BWG	20 BWG
5/8"	1.10	1.00	0.94
3/4"	1.06	1.00	0.95
7/8"	1.06	1.00	0.96

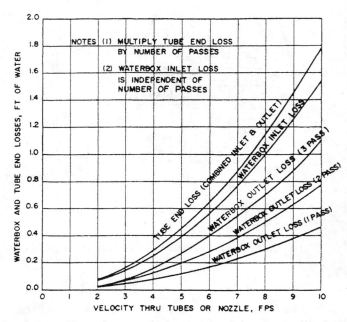

Fig. 18   Waterbox and tube end losses (courtesy Heat Exchange Institute)

Table 5   Recommended tube sheet thickness[a]

Area of Tube Sheet, in.²	Thickness of Tube Sheet, in.
to 1965	⅞
1966 to 3739	1
3740 to 8495	1⅛
8796 to 30791	1¼
30792 and up	1½

[a] Courtesy Heat Exchange Institute.

Table 6   Diagonal tube pitch data

Tube diameter, in.	⅝	¾	⅞
Tube pitch (30° × 60° diagonal), in.	15⁄16	1 1⁄16	1¼
No. tubes per ft² of tube sheet	189	147	106

Standards include construction standards and suggest an-alytical methods for the design of tube sheets, which in-volve a complex interaction of the tube sheet, tubes, wa-terbox, bolting, and shell and take into consideration the hydrostatic pressure loadings, piping loads, water weights, and differential thermal expansion.

The location of drilled holes in tube sheets for receiving tubes is not only a function of the overall tube bundle design but also a function of tube diameter and pitch. Most condensers are designed with a 60-deg triangular pitch, for which standard dimensions have been devel-oped. The accepted standard used by most condenser de-signers is given in Table 6. It is noted that the number of tubes per square foot indicated in Table 6 is the nominal maximum that can be achieved, and that number may be reduced by the presence of baffles, divider plates, etc.

Condensing pressures higher than the normal values shown in Table 1 permit reducing the pitch to values lower than those shown in Table 6, except that the pitch should never be reduced such that the ligament is less than ¼ in. The tube pitch in combination with the tube bank depth

determines the internal pressure loss, and design condi-tions will determine if the values of tube pitch given in Table 6 should be strictly followed. The relation of tube pitch to internal losses in condensers is discussed in Sec-tion 4.

b. Support plates. The condenser tube support plates serve a number of purposes. They are used to sup-port the tubes between tube sheets in order to control tube vibration and, possibly, to provide a means for bowing the tubes for reducing stresses from differential expansion. Also, they provide convenient structural members for re-inforcement of the condenser shell against external pres-sure. In addition, the tube support plates resist bending loads imposed by the tube bundle and other structural forces.

Support plates should be sufficiently thick to provide a good bearing surface for the tubes they engage. Typical thicknesses are ½ to ⅝ in. for small condensers, including most auxiliary condensers. Thicknesses of ¾ to ⅞ in. are used with large condensers, including main condensers.

The required support plate thickness can be calculated from the following equation, which is based on a complete vacuum:

$$t_{sp} = \frac{30\,bp}{S_a(p-d)} \qquad (8)$$

where

t_{sp} = support plate thickness, in. (no corrosion allowance)
p = tube pitch, in.
d = tube OD, in.
S_a = allowable stress, psi
b = support plate spacing, in.

The tube holes in the support plates should be chamfered to avoid cutting tubes on sharp edges. Accepted practice is to use a $\frac{1}{16}$-in. by 45-deg bevel or a $\frac{1}{16}$-in. radius on both sides of the support plate, or the equivalent by the use of wire-brush deburring.

The diameter of the tube holes should be equal to the nominal diameter of the tube plus at least 0.010 in. but not more than 0.025 in. To provide a smooth surface, the tube holes should be reamed after drilling.

The spacing of the support plates, in most instances, will be about 40 to 45 tube diameters, but should be verified using the procedures outlined in Section 2.4.

Support plate areas that do not engage tubes should be cut out or relieved as much as possible to provide means for longitudinal steam distribution. When they are also used to reinforce the condenser shell, support plates should be securely welded in place and provided with sufficient bearing area against the shell to properly distribute the loading. Since the support plates act as edge-loaded diaphragms, they may require staying along the length of the shell to avoid buckling under heavy loading.

c. Tube-to-tube-sheet joints. Condenser tubes in marine condensers are usually one of two standard sizes: $\frac{5}{8}$ in. or $\frac{3}{4}$ in. outside diameter. The $\frac{3}{4}$-in.-OD size is most commonly used for merchant ships, and $\frac{5}{8}$ in. for naval ships. The wall thickness most frequently employed is 18 British Wire Gage (BWG) (0.049 in.) for copper alloys and 22 BWG (0.028 in.) for titanium.

The method selected for securing condenser tubes to tube sheets should provide leak-tightness and strength as a supporting structure for the tube sheet, or a leak-tight sliding joint to compensate for expansion. Various methods may be used in combination to provide characteristics consistent with design requirements for specific service conditions. Details of the more commonly used methods of securing tubes are shown in Fig. 19.

Figure 19(a) shows a typical configuration of a tube expanded into a tube sheet with a rolled straight section and a swaged flared section; this configuration is not recommended for titanium tubes because flaring can damage titanium tubes. The flare is to improve the tube inlet-end geometry to minimize cavitation effects from the entering cooling water. The inlet tube end should be nominally flush with the face of the tube sheet. In some cases, the tubes may be rolled and flared at both ends. The cylindrical section of the rolled joint in Fig. 19(a) is shown smooth,

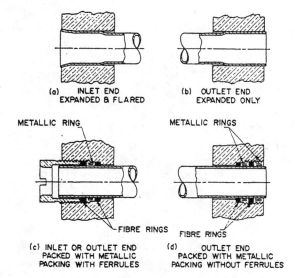

Fig. 19 Expanded and packed tube-to-tube-sheet joints

but it may be grooved for an improved holding ability. The grooves, which are machined into the tube holes, are usually two in number, spaced about $\frac{1}{4}$ in. apart, and are about 0.015 in. deep by $\frac{1}{16}$ in. wide.

Figure 19(b) shows a typical configuration of a tube expanded into a tube sheet with a straight section only. Where used, this design is normally applied to the tube outlet end. The tube hole is shown without grooves, but grooves may also be added to improve the holding ability as noted above.

Figure 19(c) shows a typical design of a packed tube-to-tube-sheet joint using a combination of fiber rings and metal-foil packing. The ferrules are provided with clearance to allow for longitudinal tube expansion, but the clearance is limited to prevent excessive longitudinal tube movement. Tubes are normally packed at the inlet end only. If the tube is packed on both ends, ferrules are required to prevent the tube from coming out of the tube sheets.

Figure 19(d) shows a typical design of a packed tube-to-tube-sheet joint using a combination of fiber rings and metal-foil packing. The tube hole is threaded so that the metallic packing can be driven into the threads and held without ferrules. This design is used for the outlet ends of the tubes only and must be used with inlet-end geometries (such as rolled or welded joints) that secure the tube in position.

A geometry similar to that in Fig. 19(a) may be used to secure a tube to a tube sheet by welding; while the design incorporates a flared entrance to improve water inlet flow conditions, this design is also used at the outlet end of the tube. A geometry embodying a straight section only, with no flared section, may also be used, but specifically for the tube outlet end. The tubes should be welded first and then expanded. However, the suitability of this procedure should be confirmed by testing since roller expansion after welding may cause the weld to crack. Hydraulic

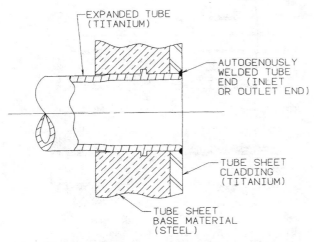

Fig. 20 Welded tube-to-tube-sheet joint with clad tube sheet

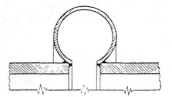

Fig. 21 Toroidal shell expansion joint

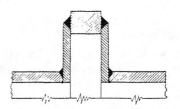

Fig. 22 Diaphragm shell expansion joint

expansion does not thin the tube wall as does roller expansion and may be preferred. Liquid-penetrant inspection of the weld joint after welding and expansion is recommended. Welded thin-walled pipe joints that have not been expanded may be susceptible to fatigue failures from vibration.

Tubes are normally welded to tube sheets of the same material. For economic reasons, it may be desirable to use a steel tube sheet that is clad with the more expensive tube material. Such an arrangement is shown by Fig. 20.

Properly rolled joints can be substantially as strong as the tube itself, but this is generally true only when the tubes are either expanded into holes with grooves, or welded and rolled; however, the uniformity of holding strength will vary between these methods.

d. Tube and shell expansion. The condenser operates at a temperature quite different from that prevalent during assembly. Even under normal operating conditions the temperature differences are sufficiently large to generate stresses high enough to cause tube failures. Three methods are commonly used to avoid the undesirable stresses that can result from thermal expansion differences.

The use of packed tube ends, or tubes with one end packed (outlet end) and one fixed (inlet end), allows movement of the tube longitudinally in the tube sheet joint, thus minimizing differential expansion stresses. This method has been used successfully for many years but does have a tendency to develop leakage (cooling water to condensate) at the packed ends.

When tubes are expanded on both ends or welded and expanded on both ends, differential expansion between the shell and tubes resulting from different materials and temperatures may be accommodated by the use of a shell expansion joint. Figure 21 shows a toroidal joint, which offers substantial flexibility. It is especially suited to condensers with long tubes and may be used in multiples if one does not provide sufficient flexibility. Figure 22 is a diaphragm type of joint, shown with the diaphragms

attached to a circumscribing ring. The flexibility is less than that of a toroidal joint of equal diameter, but it is less expensive to build. Where space permits, it may be increased in diameter and used as a single rather than a double diaphragm. In such cases, the circumscribing ring may be used as the tube-sheet bolting or welding flange.

An alternative method, which may be used when the tubes are rigidly fixed in both tube sheets, involves arranging the support plate positions so that the tubes are bowed. Under compressive loading the tubes act as eccentric columns, which deflect easily and thereby hold the stresses caused by end loading to acceptable levels. A disadvantage of this method occurs if operating conditions place the tubes under tension. In this case there is little flexibility in the tubes and the expansion load can cause excessive tube sheet deflection. If the differential expansion movement is large enough, very high local stresses can develop and some tube-to-tube-sheet joints may fail. Bowed tubes are seldom used in the design of marine condensers.

e. Condenser shell. The important geometrical characteristics of marine condenser shells are described in Section 1.4. The methods for designing condenser shells to withstand external pressure generally vary with space requirements. Common practice is to provide compression stays to carry the external loading on large flat surfaces in preference to the use of ribs. While ribs may also be used, they are sometimes employed as secondary members or to distribute loads. Support plates may be used as full ribs, partial ribs, or as anchors for compression stays.

In the case of cylindrical shells, support plates may be used as circumferential supporting rings; however, cutouts must be provided for longitudinal steam flow. Also, cylindrical shells may be designed as self-supporting structures with the support plates used to carry external mechanical loading.

The Heat Exchange Institute has published standards for the thickness of condenser shells. These standards, which are listed in Table 7, have been accepted by most

Table 7   Condenser shell thickness for cylindrical shells (fabricated steel)[a]

Cross Section Area of Shell, in.^{2}	Steel Plate Shell Thickness, in.
to 2460	⅜
2461 to 5670	½
5671 to 11500	⅝
11501 to 29000	¾
29001 and above	⅞

[a] Courtesy Heat Exchange Institute.

condenser designers. While these thicknesses are recommended and establish a guide for average or normal conditions, they do not necessarily represent the optimum selection of plate thickness for marine condensers. The HEI standards also provide a method of analysis for shell plate thickness and reinforcement. The design principles described in Section VIII, Division 1, of the American Society of Mechanical Engineers (ASME) Code for Pressure Vessels are applicable in addition to the thicknesses listed in Table 7. Stress levels for the materials used and the joint efficiency factors listed for vacuum or external pressure also apply. One exception is the pitch dimension for compression stays. Experience has shown that the 8½-in. pitch dimension limit listed in the ASME Code may be exceeded and that the alternative dimension of 15 times the diameter of the stay is a safe and practical limit.

Corrosion allowance is usually a matter of judgment; however, experience has indicated that a ¹⁄₁₆-in. allowance is usually adequate.

f. Condensate purity monitors. The condensate purity requirements associated with high boiler pressures and nuclear steam plants or the desire to control corrosion of the condensate system, boiler, and turbine blading necessitates the use of condensate purity monitoring systems in steam surface condensers. The monitoring devices are usually instruments that typically measure the specific conductivity of the condensate as an indication of dissolved solids concentration or measure condensate salinity as an indication of tube-to-tube-sheet joint leakage or leakage through the tubes themselves.

Monitoring systems designed to measure condensate impurities may also indicate the general area where leakage occurs. There are two areas where the monitoring of leakage is especially important. One of these is the tube-sheet area. This zone is used to determine the soundness of tube-to-tube-sheet joints. The second area is that of the heat transfer tubes between the tube sheets.

To monitor the tube-to-tube-sheet joints there are four methods in general use: (1) the double-tube-sheet arrangement may be monitored by pressurization with condensate between the tube sheets, with leakage indicated by pressure decay; (2) an inner support-baffle plate, located close to the tube sheet, may be used to confine tube-to-tube-sheet joint leakage between the support-baffle plate and the inner surface of the tube sheet; (3) a catch dam may be used to catch condensate and any tube-to-tube-sheet joint leakage, that drains down the inner faces of the tube sheets; and (4) a compartmented hotwell may be arranged with series flow, from initial compartments at the tube

sheets through other compartments in sequence to the condensate outlet.

A double tube sheet arrangement is used with marine nuclear power plants; it is also noted, however, that the high reliability of tube-to-tube-sheet welding has resulted in the almost exclusive use of welded tube-to-tube-sheet connections in the major applications of nuclear power. The inner support-baffle plate used to confine condensate and leakage from the tube-to-tube-sheet joints is effective and improves the reliability of monitoring. It is more costly than the use of catch dams or weirs, described as method 3, but also is more reliable. Method 4, that of compartmenting the hotwell and providing series flow, is used in combination with method 1, 2, or 3. Most designs divide the hotwell (if it is the whole condenser bottom) into four or more parts: two lengthwise and two or more crosswise of the condenser. Water flow is sequential, with the outlet from each compartment monitored, thereby indicating by the change in concentration of impurities the location and degree of leakage. With small condensers, such as auxiliary condensers, it is usually sufficient to monitor condensate quality at the condensate outlet only.

2.11 Materials of Construction. Materials used in marine condensers are selected to meet the requirements of strength, cost, and corrosion resistance, as well as other considerations, which may be controlling. Strength, corrosion resistance, and in some instances, weight are of major importance and may take precedence over initial cost. A number of material types and grades may be suitable for the same part or parts of a condenser.

Condenser shells, support plates, internal bracing, and nozzle connections, which comprise the condenser shell structure, except possibly for minor parts, are fabricated by welding. The materials most frequently used in construction of marine condensers are tabulated in Table 8 and are described in the designated specifications.

Two major areas or zones in a condenser are susceptible to significant corrosion attack. They are: (1) the steam or condensing area, comprising the condenser shell, hotwell, tube sheets, internal baffles, piping, ducts, and the outside surfaces of the heat transfer tubes, and (2) the cooling-water side of the condenser, comprising the waterboxes, tube sheets, and the inside surfaces of the heat transfer tubes.

The corrosion environments coincident with these two areas or zones are significantly different, and the corrosion mechanisms involved, while similar in some respects, are vastly different in degree. The steam-side corrosion in marine condensers is usually minimal and is seldom a major problem. The materials of construction, which are mostly carbon steel for shells, support plates, baffles, and bracing and nonferrous tubes and tube sheets, present no serious problem in this environment. The low operating temperatures, low dissolved oxygen levels in the condensate, and very pure condensate (in terms of dissolved solids) do not present a severely aggressive environment to the materials normally used. An exception, however, can occur due to carbon dioxide which, if present in significant quantity, may attack the condenser tubes, especially in the areas at the support plates and adjacent to the inside

Table 8   Condenser construction materials

Shell plate, flanges, ribs, and reinforcements	Steel, ASTM, A-285, Grade C, Flange Quality Steel, ASTM, A-283 Steel, ASTM, A-537, Grade N or QT
Tube support plates	Steel, ASTM, A-285, Grade C, Flange Quality Steel, ASTM, A-283 Steel, ASTM, A-36 Steel, ASTM, A-7 Steel, ASTM, A-537, Grade N or QT
Waterboxes and waterbox covers	Cast iron, ASTM, A-278 Cast nodular iron, ASTM, A-395 Steel, ASTM, A-285, Grade C, Flange Quality Steel, ASTM, A-283 Steel, ASTM, A-537, Grade N or QT Aluminum bronze, ASTM, B-169, Alloy D Copper nickel, 90/10, ASTM, B-171
Tube sheets	Muntz metal, ASTM, B-171 Naval brass, ASTM, B-171 Aluminum bronze, ASTM, B-169, Alloy D Silicon bronze, ASTM, B-96, Alloy A Copper nickel, 90/10 or 70/30, ASTM, B-171 Titanium, ASTM, B-265, Grade 2 or 3
Tubes	Admiralty metal, ASTM, B-111 Aluminum bronze, ASTM, B-111 Aluminum brass, ASTM, B-111 Copper nickel, 70/30, ASTM, B-111 Copper nickel, 80/20, ASTM, B-111 Copper nickel, 90/10, ASTM, B-111 Titanium, ASTM, B-338

NOTE: This is a representative list of materials suitable for marine condensers and is not intended to limit the application of other suitable materials.

face of the tube sheets. The action may be chemical, with carbonic acid attacking the copper-base alloy tubes, or it also may be galvanic with the carbonic acid acting as a local electrolyte at the interface between tube surfaces and support plates. The main source of carbon dioxide is the breakdown of carbonates and bicarbonates in the seawater evaporator, which provides feedwater makeup. The evaporator temperature, point of admission of the evaporated vapor as makeup, and the provision for venting noncondensables all affect the concentration of carbon dioxide in the steam and condensate. Air cooler designs that eliminate short circuiting and stagnation of air-vapor mixtures, thereby reducing the carbon-dioxide concentration, are helpful in controlling carbonic-acid corrosion. Ammonia is seldom a byproduct of seawater evaporation at sea, but may be a factor with the evaporation of polluted harbor waters. Ammonia may also be present in the condenser condensate if it is used for pH control in the feed system. In condenser service copper-nickel tubes and tube sheets are quite resistant to ammonia and may be considered to be a reliable means for avoiding ammonia corrosion effects. Titanium tubes are practically immune to most corrosion environments. The brasses, Admiralty and aluminum, and also aluminum bronze are susceptible to stress corrosion from ammonia, even at the characteristically low condenser temperatures. The susceptibility of these alloys to ammonia attack is reduced if the tubes, especially those in the air cooler, are not under significant tension stresses. Some control of this may be achieved by using packed tubes in the air cooler section, or by rolling

air cooler tubes last, if rolled or rolled-and-welded tube-to-tube-sheet joints are used. Experience has shown that the corrosion of steel condenser shells and internal baffle parts is very minor and that a corrosion allowance of about $\frac{1}{16}$ in. is adequate on the steam side for the planned operating life of almost any condenser in marine service.

Corrosion can be extremely severe on the cooling-water side of marine condensers if not properly controlled by design, selection of materials, and installation practices. Seawater, the condenser cooling medium, is an excellent electrolyte. It promotes galvanic and crevice corrosion in combinations of materials that are displaced from one another in the galvanic series. Not only must the difference in corrosion potential between materials be considered, but the relationship or ratio of their respective areas in contact with their environment must be taken into account. For example, highly corrosion-resistant materials such as the nickel-copper alloys or high-nickel stainless steels may be safely used for bolting or for fasteners in combination with major parts made of cast iron or steel because the ratio of their respective areas (alloy to steel or iron) is small. The steel or iron is sacrificed by galvanic action in protecting the high-alloy materials. The steel or iron, by comparison, is so great in area that its loss through galvanic corrosion becomes insignificant.

Some basic principles and recommended practices with respect to mitigating galvanic corrosion in condensers are as follows:

1. Where possible, use one material for all parts of the condenser cooling-water system, or select materials close to one another in the galvanic scale.

2. The materials used for bolting and fasteners, and in some instances welding, should be selected to be more corrosion resistant than the materials they join.

3. A larger corrosion allowance should be used for the less noble materials in a galvanic system (such as carbon steel or cast iron as opposed to copper-base materials or high nickel-chromium alloys of steel).

4. The use of corrosion-resistant metals such as copper-nickel, aluminum brass, naval brass, and especially titanium is an effective means for controlling corrosion. These materials resist corrosion by the formation of a durable protective oxide film on the exposed surfaces. Corrosion has also been mitigated on the copper-alloy tubes by using ferrous sulfate to form a protective iron-oxide film on the metal surfaces. Such a film can be established by injecting a ferrous sulfate solution into the condenser cooling water circuit before the first use; the solution is circulated for about three days. The coating may be maintained by providing anodes of soft gray iron or low-carbon steel, which will continue the plating action on the metal surfaces in the system. If the coated surfaces are subjected to an erosive washout of the oxide coating by a high water velocity, cavitation, or suspended abrasive particulate matter, the protection can be damaged and concentrated corrosive action can result.

The use of protective coatings or paints on cast iron or fabricated-steel waterboxes requires special consideration. Defects in the coating can cause accelerated corrosion where the seawater penetrates the coating. Applying the coating to the face of a nonferrous tube sheet used in combination with a cast iron or steel waterbox, and leaving the ferrous material exposed, frequently offers greater protection to the system than coating the iron or steel surfaces themselves.

5. Sacrificial anodes attached to the inside of the waterboxes are useful in protecting iron or steel surfaces and are especially needed when the ferrous material is protected by coatings or paints. Common metals used for anodes are soft iron, zinc, and magnesium. For these materials to function properly in the galvanic system they should be of high purity, except for magnesium, which is frequently alloyed with manganese, or with zinc and aluminum to increase the duration of the required current levels. Iron or steel anodes have been used successfully with most systems that do not otherwise include iron or steel. If cast iron condenser waterboxes are part of the system, zinc anodes should be used to reduce the rate of wastage of the waterbox material. Zinc anodes should not be installed on systems that do not have iron or steel components; otherwise corrosion is usually accelerated. Combined with coatings, anodes improve the distribution of current; and the protective coating used may be allowed to have some porosity without damaging effects.

6. Galvanic corrosion can be accelerated by stray currents, but it can be greatly reduced or eliminated by bonding. The usual practice with condensers is to metallically connect the condenser shell, tube sheet, and waterbox with a number of nonferrous metal straps, usually made of copper or high-copper alloy. Because of the low voltages associated with galvanic systems, a large number of bonding straps should be used (they may be spaced on 18- to 24-in. centers around the periphery of the tube sheet).

7. Crevice corrosion develops where there is an irregularity in surfaces, a junction, sharp bends, or other discontinuity where oxygen does not have ready access. The difference in oxygen concentration in the crevice and that outside the crevice sets up a corrosion cell. It may be manifested by corrosion within the crevice itself, or external but adjacent to the crevice. It can be avoided by design, or by the selection of materials. In the circulating-water system of marine condensers, the selection of materials such as 70/30 or 90/10 copper nickel or titanium for tubes, tube sheets, and waterboxes is an effective means of controlling crevice corrosion.

8. The water velocity in the cooling-water circuit is a major factor in the selection of materials used to construct marine condensers. The damaging effects of corrosion are greatly accelerated by the erosive effects of water velocity. The major concern is with the condenser tubes. They are subject to inlet-end attack and to general attack, both of which may be related to the velocity of the cooling water. The usual method of controlling this kind of attack is to select a tube material that is not only suitable for use in a seawater galvanic system, but is also compatible

Table 9 Recommended design water velocities through tubes

Tube Material	Water Velocity, fps
Titanium	5–15
70/30 copper nickel	3–15
90/10 copper nickel	3–9
Aluminum brass	3–7
Admiralty	3–4

NOTE: The 3 fps lower limit is because of heat transfer considerations. It is not a corrosion limitation for the materials listed. Titanium may tend to foul with low water velocities and require periodic and continuous chlorination.

with the tube design velocity. Table 9 lists some tube materials commonly used in marine service and the recommended design water velocity for which they may be considered suitable. Adherence to these values usually provides satisfactory results with respect to the general corrosion of tubes but may not be satisfactory in terms of tube inlet-end corrosion. Tube inlet ends may be subject to accelerated corrosive attack should cavitating flow occur. Cavitation may result when the contraction of the flow stream at the entrance is greater than the contraction contour of the tube at the entrance, and the water flow stream leaves the side of the tubes. This causes an erosion action and wears away the protective film that the tube material develops. Thus, unprotected metal is continuously exposed and corrosion proceeds rapidly. Cavitation is increased with waterboxes of poor hydraulic design, especially those where the velocity of the cooling water entering the waterbox equals or exceeds the average water velocity in the tubes. The cavitation effect caused by the high entrance velocity of the water in the waterbox, with respect to the water velocity in the condenser tubes, can be reduced by special designs to absorb the velocity head energy; however, additional space is required. Where conditions, economic or otherwise, require high tube entrance velocities and high waterbox entrance velocities the effect of cavitation may be greatly reduced by the use of tube inlet-end inserts.

Tube inserts are usually made of a plastic material, although some are made of corrosion-resisting metal. Plastic inserts are inserted into the inlet ends of the tubes, and cemented in place; metallic inserts can be expanded in place. Inserts are made with a well rounded entrance with their thickest wall just inside the entrance, forming an almost imperceptible throat. The wall then tapers gradually to a feather edge, offering a minimum discontinuity where it terminates in the tube. They are made in various lengths (about 4 to 10 tube diameters) and are selected on the basis of the extent of cavitation previously experienced or expected. Tube-end inserts have proved to be effective when properly installed. It is important that they be securely cemented or wedged into place and that the feather edge is extremely thin and well bonded. Otherwise, the junction may, itself, cause a cavitation effect downstream and transfer the damage to another tube area.

Section 3
Surface Condenser Performance

3.1 General Considerations. The Heat Exchange Institute Standards for Steam Surface Condensers provide a comprehensive treatment of surface condenser performance and describe levels of performance that can be obtained from well designed condensers. The actual operating performance of condensers is usually compared for meaningful interpretation with the HEI performance standards. These standards do not serve as a design tool except that they establish a guide for the application of rational design methods and procedures.

The HEI Standards are suitable for determining condenser size and type and can readily be used as a basis for determining equipment costs for use in economic optimization studies. Also, their validity is sufficient to assure that condenser selections made through their application can be designed and built to meet the performance specified.

3.2 Terms and Nomenclature. The terms and nomenclature that are commonly used to describe steam surface condenser performance are as follows:

Condenser Duty—Condenser duty is the net heat transferred to the cooling water from all sources of heat entering the condenser. It is also called the hourly heat and is expressed in Btu per hour.

Absolute Pressure—Absolute pressure is the measure of pressure with reference to a perfect vacuum. It is expressed in inches of mercury, absolute.

Static Pressure—Static pressure is the stagnation pressure less the pressure effect of velocity. When the velocity of the fluid is zero, the static pressure and stagnation pressure are equal. Static pressure is expressed in inches of mercury.

Condenser Pressure—Condenser pressure is the absolute static pressure in the condenser shell, measured within one foot of the first tubes in the inlet steam flow path, and with the distribution of measurement points in conformity with ASME PTC 12.2 Steam Condensing Apparatus.

Condensing Steam Temperature—Condensing steam temperature is the saturation temperature of the condensing steam at the "condenser pressure" and is expressed in degrees F.

Initial Temperature Difference—The initial temperature difference is the difference between the "condensing steam temperature" and the temperature of the inlet cooling water expressed in degrees F.

Temperature Rise—The temperature rise is the difference between the cooling-water outlet temperature and the cooling-water inlet temperature expressed in degrees F.

Terminal Temperature Difference—The terminal temperature difference is the difference between the "condensing steam temperature" and the outlet temperature of the cooling water expressed in degrees F.

Logarithmic Mean Temperature Difference—The logarithmic mean temperature difference is the ratio of "temperature rise" to the Napierian logarithm of the ratio of the "initial temperature difference" to the "terminal temperature difference" expressed in degrees F.

Condensate Temperature Depression—Condensate temperature depression is the difference between the "condensing steam temperature" and the temperature of the condensate leaving the hotwell. It may be either positive (cooling) or negative (heating) and is expressed in degrees F.

Cooling-Water Velocity—Cooling (circulating) water velocity is the average velocity of the cooling water flowing through the heat transfer tubes. It is expressed in feet per second.

Heat Transfer Coefficient—The heat transfer coefficient is the average rate of overall heat transfer, from all heat sources, to the cooling water. It is expressed in Btu/hr-ft²-deg F.

Cleanliness Factor—The cleanliness factor is the ratio of the overall heat transfer of tubes in service, which are fouled, to the overall heat transfer of new clean tubes. It is dimensionless.

Condenser Surface—Condenser surface is the surface measured on the outside of the heat transfer tubes between the inside faces of the tube sheets, including internal and external air coolers and that portion of the tubes within the support plates. It is expressed in square feet.

Nomenclature

C = condenser tube heat transfer constant, from Table 10

C' = ratio: gpm of cooling water to square feet of condensing surface, gpm/ft²

D = outside tube diameter, in.

F_1 = heat transfer temperature correction factor, from Fig. 23

F_2 = tube material and wall thickness correction, factor from Table 11

F_3 = heat transfer tube cleanliness correction factor, selected as indicated from experience

H_1 = initial temperature difference: condensing steam temperature less cooling-water inlet temperature, deg F

Table 10   Values of tube heat transfer constant,[a] C

Tube OD, in.	Heat Transfer Constant
⅝ to ¾	267
⅞ to 1	263

[a] Courtesy Heat Exchange Institute.

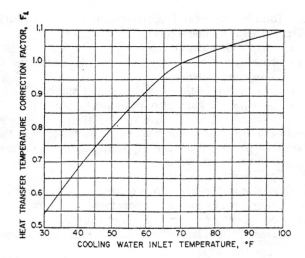

Fig. 23 Heat transfer temperature correction factor (courtesy Heat Exchange Institute)

Table 11 Values of tube wall gauge and material correction factor,[a] F_2

Tube Material	Tube Wall Gauge, BWG		
	20	18	16
Admiralty metal	1.02	1.00	0.96
Aluminum bronze	1.00	0.97	0.94
Aluminum brass	1.00	0.97	0.94
Copper nickel 90/10	0.94	0.90	0.85
Copper nickel 80/20[b]	0.90	0.86	0.81
Copper nickel 70/30	0.87	0.82	0.77
Titanium[c]	0.77	0.71	0.67

[a] Courtesy Heat Exchange Institute.
[b] Estimate values.
[c] Tentative values.

Table 12 Values of surface constant for tube diameter and gauge, K_1

Tube Diameter, in.	16 BWG	18 BWG	20 BWG
5/8	0.272	0.240	0.216
3/4	0.208	0.188	0.175
7/8	0.168	0.155	0.144
1	0.141	0.131	0.123

H_2 = terminal temperature difference: condensing steam temperature less cooling-water outlet temperature, deg F

h = heat transferred to cooling water, Btu/lb (of steam)

K_1 = surface constant for tube diameter and gauge. See Table 12

L = exposed length of tubes, ft

N = number of tubes

P = number of water passes in condenser

P_s = absolute static pressure in condenser steam inlet, in. Hg

Q = quantity of cooling water, gpm

S = exposed tube surface, ft²

T_i = cooling-water inlet temperature, F

T_o = cooling-water outlet temperature, F

T_r = cooling-water temperature rise, deg F

T_s = saturation steam temperature in condenser steam inlet, corresponding to P_s, F

U = overall heat transfer coefficient, Btu/hr-ft²-deg F

V = water velocity in tubes, fps

W = steam condensed, lb/hr

Y = ratio of initial temperature difference to terminal temperature difference, dimensionless

θ_m = logarithmic mean temperature difference, deg F

3.3 Mean Temperature Difference. The true mean temperature difference across the heat transfer path from the temperature of the heat source to the temperature of the heat receiver is most accurately expressed as the logarithmic mean temperature difference. It is as important in determining heat flow or required heat transfer surface as is the overall heat transfer coefficient.

As commonly used in condenser calculations, the logarithmic mean temperature difference is a close approximation of the true temperature difference between the temperatures of the heat source and heat receiver. The two principal reasons for its not being exact are: First, it is based on the condensing temperature, as previously defined, which is assumed constant throughout the condensing steam space; actually the condensing temperature decreases as steam flows through the tube banks. Second, it is based on the assumption of constant specific heat; actually, a constant specific heat is not attainable when condensing impure vapor.

Fortunately, however, the total error is relatively small, and the logarithmic mean temperature difference as generally used in condenser performance calculations is sufficiently accurate. The equation for the logarithmic mean temperature difference is:

$$\theta_m = \frac{(T_s - T_i) - (T_s - T_o)}{\ln \dfrac{T_s - T_i}{T_s - T_o}} = \frac{T_r}{\ln \dfrac{H_1}{H_2}} \qquad (9)$$

Corrections to θ_m for the variable specific heat may be neglected. For calculations that require extreme accuracy, such as those associated with the establishment of condenser final designs, equation (9) should be modified to compensate for the changes in T_s that result from pressure losses in the condenser and for the effect of the partial pressure of noncondensable gases. This will reflect the true heat transfer conditions as they exist in the condenser; see Section 4 for additional discussion on this subject.

3.4 Coefficient of Heat Transfer. The coefficient of heat transfer recommended by the Heat Exchange Institute is expressed by an equation that is a reasonably accurate simplification of the classical heat transfer resistance summation equation. The HEI equation is:

$$U = F_1 F_2 F_3 C(V)^{0.5} \qquad (10)$$

Equation (10) is limited to water velocities in the tubes of 3 fps minimum and 8 fps maximum. For velocities greater than 8 fps, the equation usually yields high values for U.

The factor F_1 may be obtained from the graph in Fig. 23 and F_2 is given in Table 11. The factor F_3 is selected on the basis of experience. Values of C for different tube sizes are given in Table 10.

3.5 Heat Transfer Performance Standards.

a. Fourier equation. The Fourier equation for unidirectional heat transfer is used to express the hourly heat flow in terms of the heat transfer surface, coefficient of heat transfer, and logarithmic mean temperature difference. For condenser calculations it is written as follows:

$$S = \frac{Wh}{\theta_m U} \tag{11}$$

b. Heat balance equation. The heat balance equation used for condenser calculations equates the heat given up by the heat source to the heat absorbed by the circulating water. It is usually written as follows:

$$Q = \frac{Wh}{512T_r} \tag{12}$$

NOTE: Wh in equations (11) and (12) may be from a number of heat sources in addition to the exhaust steam from the turbine. When considering all heat sources, $Wh = W_1 h_1 + W_2 h_2 + \ldots + W_n h_n$. The constant 512 is for a seawater coolant; for a freshwater coolant, the constant is 500.

c. Equations used for sizing condensers. The four basic equations used in sizing condensers are equations (9), (10), (11), and (12). Additional equations derived from these and other relations of tube geometry and surface simplify the computations for determining surface, general dimensions, and other characteristics of condensers. The more commonly used supplementary equations are as follows:

$$H_1 = T_s - T_i \tag{13}$$

$$C' = \frac{V}{PLK_1} \quad \text{For values of } K_1 \text{ see Table 12} \tag{14}$$

$$Y = \ln^{-1} \frac{U}{512C'} \tag{15}$$

$$H_2 = \frac{H_1}{Y} = \frac{T_r}{Y-1} \tag{16}$$

$$S = \frac{Q}{C'} \tag{17}$$

$$T_s = H_2 + T_r + T_i \tag{18}$$

$$P_s = \left(\begin{array}{c}\text{from steam tables,}\\\text{corresponding to } T_s\end{array}\right) \tag{19}$$

$$N = \frac{3.82S}{DL} \tag{20}$$

3.6 Design Criteria and Performance Standards.

a. Considerations in determining condenser size. The design of condensers, based upon a desired performance, is influenced by eight principal variables as follows:

1. Total hourly heat transferred, or condenser duty, in Btu, which is a function of:
   - weight of steam condensed
   - enthalpy of entering steam less the enthalpy of the condensate
   - heat loss or gain from drains and makeup
   - heat loss or gain from condensate depression
2. Absolute static steam pressure at the condenser inlet (measured within one foot of the first tubes in the steam flow path) and the corresponding saturated steam temperature from steam tables.
3. Cooling-water quantity.
4. Cooling-water inlet temperature.
5. Cooling-water outlet temperature.
6. Cooling-water velocity through tubes.
7. Effective heat transfer surface, which is influenced by:
   - total area
   - number of tubes
   - tube diameter
   - tube gage
   - tube length
   - tube material
   - number of water passes
8. Service environment and maintenance; i.e.
   - tube cleanliness
   - air inleakage

All of the above factors must be considered in determining the size and characteristics of a steam surface condenser, and also when determining the performance of a condenser on test.

b. Establishing condenser size (Example 1). There are a number of mathematical procedures used in establishing the size of a condenser, depending upon the information assumed for the application. One of the more commonly used procedures is outlined in the following example for an auxiliary condenser.

Assume

$T_i = 75$ F
$T_r = 8$ deg F
$D = \frac{5}{8}$ in.
Tube gauge $= 18$ BWG
Tube material $= 90/10$ copper nickel
$V = 6$ fps
$L = 9$ ft 0 in. exposed
$P = 2$
$W = 7750$ lb/hr
$h = 1000$ Btu/lb
$F_3 = 0.85$ (85% clean tubes)

Derive from the above assumptions

$F_1 = 1.020$ Fig. 23
$F_2 = 0.90$ Table 11
$K_1 = 0.240$ Table 12
$C = 267$ Table 10
$T_0 = T_i + T_r = 75 + 8 = 83$ F

The condenser surface, cooling water required, condensing pressure, number of tubes, and cooling-water friction loss are computed as follows:

$$C' = \frac{V}{PLK_1} = 1.389 \text{ gpm per ft}^2$$

$$U = F_1F_2F_3CV^{0.5} = 510.3 \text{ Btu/hr ft}^2\text{-deg F}$$

$$Y = \ln^{-1}\frac{U}{512C'} = 2.049$$

$$H_2 = \frac{T_r}{Y-1} = 7.63 \text{ deg F}$$

$$Q = \frac{Wh}{512T_r} = 1892 \text{ gpm}$$

$$S = \frac{Q}{C'} = 1362 \text{ ft}^2 \text{ condenser surface}$$

$$T_s = H_2 + T_r + T_i = 90.63 \text{ F condensing temperature}$$

$$P_s = 1.450 \text{ in. Hg condenser pressure}$$

$$N = \frac{3.82S}{DL} = 925 \text{ tubes}$$

To determine the cooling-water friction loss, refer to Section 2, Figs. 17 and 18:

Tube loss $(0.325 \times 9) \times 2 =$	0.59 ft
Tube end loss ($V = 6$ fps)	
$(0.64 \times 2) =$	1.28 ft
Waterbox inlet loss	0.32 ft
(good practice requires waterbox inlet nozzle velocities of 0.70 to 0.80 times the condenser tube water velocity; in this case 0.75×6 fps $= 4.5$ fps was used)	
Water box outlet loss	
($V = 4.5$ fps; 2 pass)	0.15 ft
Total condenser friction loss	2.34 ft of water

Recapitulation of results
Condenser surface $= 1362$ ft^2
Cooling water required $= 1892$ gpm
Condenser pressure $= 1.450$ in. Hg
Condenser friction loss $= 2.34$ ft of water
Number of tubes required $= 925$

c. Condenser designed for a given back pressure (Example 2). When the condenser is to be sized for a given back pressure, the computation procedure is typically as follows for an auxiliary condenser:

Assume
$T_i = 75$ F
$D = \frac{5}{8}$ in.
Tube gauge $= 18$ BWG
Tube material $= 90/10$ copper nickel
$V = 6$ fps
$L = 9$ ft 0 in. exposed
$P = 2$
$W = 7750$ lb/hr
$h = 1000$ Btu/lb
$F_3 = 0.85$ (85% clean tubes)
$P_3 = 1.5$ in. Hg

Derive from the foregoing assumptions
$F_1 = 1.020$ Fig. 23
$F_2 = 0.90$ Table 11
$K_1 = 0.240$ Table 12
$C = 267$ Table 10
$T_s = 91.72$ F   Steam table for 1.5 in. Hg abs
$H_1 = T_s - T_i = 16.72$ deg F

Determine the condenser surface, number of tubes, cooling-water flow, cooling-water velocity, and cooling-water friction loss as follows:

$$C' = \frac{V}{PLK_1} = 1.389 \text{ gpm per ft}^2$$

$$U = F_1F_2F_3CV^{0.5} = 510.3 \text{ Btu/hr ft}^2\text{-deg F}$$

$$Y = \ln^{-1}\frac{U}{512C'} = 2.049$$

$$H_2 = \frac{H_1}{Y} = 8.16 \text{ deg F}$$

$$T_r = H_1 - H_2 = 8.56 \text{ deg F}$$

$$Q = \frac{Wh}{512T_r} = 1768 \text{ gpm}$$

$$S = \frac{Q}{C'} = 1273 \text{ ft}^2$$

$$N = \frac{3.82S}{DL} = 865 \text{ tubes}$$

The condenser friction loss is the same as for Example 1, because the water velocity in the tubes, the tube lengths, and the waterbox characteristics were assumed to be the same. The amount of surface and cooling water required change because of the difference in the two back pressures, one calculated, the other assumed.

Recapitulation of results
Condenser surface $= 1273$ ft^2
Cooling water required $= 1768$ gpm
Condenser pressure $= 1.50$ in. Hg
Condenser friction loss $= 2.34$ ft of water
Number of tubes required $= 865$

d. Determination of condenser performance by test (Example 3). In the process of testing a condenser, it is general practice to compare the condenser performance with a standard, such as one of those published by the Heat Exchange Institute. A procedure commonly used to determine condenser performance from test data is outlined below using the condenser from Example 1.

Given
$S = 1362$ ft^2
$D = \frac{5}{8}$ in.
Tube gauge $= 18$ BWG
Tube material $= 90/10$ copper nickel
$L = 9$ ft. 0 in. exposed
$P = 2$
$K_1 = 0.240$ Table 12
$F_2 = 0.90$ Table 11
$C = 267$ Table 10

Data from test (assumed to be typical)

$$T_i = 70 \text{ F}$$
$$T_o = 77.9 \text{ F}$$
$$W = 7800 \text{ lb/hr}$$
$$h = 995 \text{ Btu/lb}$$
$$P_s = 1.30 \text{ in. Hg}$$

Derive from test data

$$T_r = T_o - T_i = 7.9 \text{ deg F}$$
$$T_s = 87.17 \text{ F} \quad \text{from steam tables}$$
$$H_1 = T_s - T_i = 17.17 \text{ deg F}$$
$$H_2 = H_1 - T_r = 9.27 \text{ deg F}$$
$$F_1 = 1.00 \qquad \text{Fig. 23}$$

Determine the performance of the condenser as follows:

$$Q = \frac{Wh}{512 T_r} = 1919 \text{ gpm}$$

$$\theta_m = \frac{T_r}{\ln \dfrac{H_1}{H_2}} = 12.82 \text{ deg F}$$

$$U = \frac{Wh}{\theta_m S} = 444.5 \text{ Btu/hr-ft}^2\text{-deg F}$$

$$C' = \frac{Q}{S} = 1.409$$

$$V = PLK_1 C' = 6.087 \text{ fps}$$

$$F_3 = \frac{U}{F_1 F_2 C V^{0.5}} = 0.750$$

The value of F_3 corresponding to new clean tubes is 1.00. The design value used for Example 1 is $F_3 = 0.85$ or 85% of clean-tube performance. However, the assumed test data show the condenser to be performing at $F_3 = 0.750$ or 75.0% of the new clean-tube heat transfer rate.

If the condenser had been cleaned prior to test, the 75.0% of new clean-tube performance could have resulted from a number of factors. First, condenser tubes, which have been in service, do not fully recover new clean-tube performance after mechanical cleaning. Experience indicates that recovery after mechanical cleaning will be to about 90 to 95% of new clean-tube values. If the tubes are acid cleaned, in addition to mechanical cleaning, recovery should be close to 100%. Second, poor performance may result from excessive air leakage, undersize vacuum pumps, or the combined effects of both. Third, poor performance may result from a poorly designed condenser, including an improperly designed air cooler.

In general, fouled tubes (less than 100% of new clean performance) are always a factor and cannot be entirely eliminated without careful mechanical cleaning followed by acid cleaning. Most condensers are mechanically cleaned only because of the problems associated with chemical cleaning, and less than 100% of clean-tube performance during service periods is to be expected.

Fouling studies conducted by the Heat Transfer Research Institute indicate that titanium tubes are fouled only by biologically bound silt whereas copper-nickel tubes are fouled by silt and corrosion products. At high velocities the fouling rates of titanium are decreased significantly, and the design cleanliness factor can be as high as 0.90 to 0.95. But at low velocities, or during shutdown periods, the fouling of titanium can be significant; however, this can be controlled by chlorination.

In the actual design of a condenser, a design margin of from 5 to 10% (equivalent to 105 to 110% of new clean-tube performance) is good practice. Condensers designed with allowances of this magnitude usually show close to 100% performance ($F_3 = 1.00$) if tested immediately after a thorough mechanical cleaning.

Section 4
Performance Predictions from Design Geometry

4.1 Introduction. The Heat Exchange Institute Standards recommend values for condenser performance that have been determined through field and laboratory test experience. These standards represent empirical values that have resulted from observation rather than from the development of theoretical considerations. The Standards list basic heat transfer values for the usual sizes of tubes used in condenser construction with the applicable factors for adjusting these values for tube material and gauge, for cooling water temperature, and for specific water velocity ranges (see Section 3).

The predicted performance of a condenser sized by the use of these standards is based on standardized heat transfer values. However, it is important to understand that these established values of heat transfer in the Standards include an adjustment for pressure losses within the condenser and, therefore, do not conform with normal heat transfer values established by rational means. Actually, the steam condenser is a thermodynamic device that incorporates both the principles of heat transfer and the thermodynamics of fluid flow, and both of these disciplines must be carefully considered in its design. The literature contains a large amount of data on heat transfer, which relate to water flowing through tubes and steam condensing on tubes. Generally, these data apply to single-tube condensers and they describe performance with good accuracy. If applied to multi-tube condensers, serious error will result unless other factors in addition to heat transfer are considered. The development of these factors and their application to condenser design parameters are discussed in the following paragraphs.

4.2 Modes of Heat Transfer. Heat is transferred in three distinct modes in shell-and-tube-type steam condensers. These modes are conduction, convection, and condensing heat transfer. As resistances, the reciprocals of their respective conductances, they are in series with the heat flowing from the higher temperature condensing steam through the tube wall and whatever fouling resistances have accumulated on the tube surfaces to the colder cooling water.

a. Condensing heat transfer. Condensing heat transfer, with respect to steam condensing on a colder surface, is characterized by a relatively high value of conductance. The usual method of computing condensing heat transfer coefficients for tubes is by the theoretical equation of Nusselt. Experimental data for organic vapors are in good agreement with the Nusselt prediction; for steam condensing on small tubes at low condensing rates, the conductance determined with the Nusselt equation for a single tube is about 65% of that empirically determined for a single tube. Experimental results with steam for a bank of 20 tubes are in reasonable agreement with the Nusselt equation for a single horizontal tube. The adjustment for additional rows of tubes is small, with the heat transfer coefficient reducing only about 10% for tube banks with as many as 120 tube rows. The net effect on the overall heat transfer coefficient is minor, and the following classical form of the Nusselt equation for single horizontal-tube condensing heat transfer is suggested for design purposes:

$$h_s = 0.95 \left(\frac{k^2 \rho^2 g}{\mu W} \right)^{1/3} \qquad (21)$$

where

h_s = conductance (steam to tube outside surface), Btu/hr-ft²-deg F

k = thermal conductivity of condensate, Btu/hr-ft²-deg F/ft

ρ = condensate density, pcf

g = acceleration due to gravity, 4.17×10^8 ft/hr²

μ = condensate viscosity, lb/hr-ft

W = steam condensed, lb/hr-ft (length of tube)

NOTE: The physical properties of the condensate correspond to the mean condensate film temperature.

b. Conduction heat transfer. The conduction mode of heat transfer may be considered to be made up of five resistances to heat flow arranged in series. The relation is expressed as follows:

$$r_c = r_{ox1} + r_{ox2} + r_{tw} + r_{f_1} + r_{f_2} \qquad (22)$$

where

r_c = resistance from outside to inside of tube inclusive of all resistances except that of the condensing steam and heating water, hr-ft²-deg F/Btu

r_{ox1} = resistance of the oxide film on the outside of a new clean tube, hr-ft²-deg F/Btu

r_{ox2} = resistance of the oxide film on the inside of a new clean tube, hr-ft²-deg F/Btu

r_{tw} = resistance of tube wall, hr-ft²-deg F/Btu

r_{f_1} = resistance from fouling on the outside of the tube in service, hr-ft²-deg F/Btu

Table 13   Values of r_{ox}, the resistance of the oxide film on the inside and outside surfaces of clean tubes

Tube Material	r_{ox} hr-ft²-deg F/Btu
Admiralty metal	0.000136
Aluminum bronze	0.000153
Aluminum brass	0.000167
90/10 copper nickel	0.000178
80/20 copper nickel	0.000193
70/30 copper nickel	0.000243
Titanium	0.000195

r_{f_2} = resistance from fouling on the inside of the tube in service, hr-ft²-deg F/Btu

General practice is to simplify equation (22) by combining:

$$r_{ox1} + r_{ox2} = r_{ox}$$

and

$$r_{f_1} + r_{f_2} = r_f$$

so that

$$r_c = r_{ox} + r_f + r_{tw} \qquad (23)$$

The resistance of the tube wall is determined from the thermal conductivity of the tube material and its thickness. For thin-walled tubes there is no need to correct for tube diameter; nevertheless, the general equation for r_{tw} is:

$$r_{tw} = \frac{t}{12k} \frac{D_o}{D_i + t} \qquad (24)$$

where

r_{tw} = resistance of tube wall referred to the outside diameter, hr-ft²-deg F/Btu

t = tube wall thickness, in.

D_o = tube outside diameter, in.

D_i = tube inside diameter, in.

k = thermal conductivity of tube material, Btu/hr-ft²-deg F/ft

Values of r_{ox}, the resistance of the oxide film on new clean tubes, are not generally available. Values that are available will be found to vary with the source, and largely be dependent on the nature of the experimental data. Data published in the Heat Exchange Institute Standards for Steam Surface Condensers and other published Heat Exchange Institute data on condensing heat transfer are possibly the most comprehensive and valid sources for this kind of information. The values for r_{ox} are combined in the overall heat transfer data given by HEI but can be determined as separate values through the application of analytical procedures. The values listed for r_{ox} in Table 13 have been derived from published HEI data by such a procedure.

The value used for r_f, the resistance due to fouling of the tubes in service, is comparable to the cleanliness factor generally used with the HEI performance standards. It may be computed as a resistance from the HEI standards; however, since it is expressed as a percentage of new clean-tube heat transfer and is taken as a resistance, it will vary as a function of the HEI heat transfer rate used. It is suggested that the fouling resistance be calculated on the basis of Admiralty metal tubes at an 8-fps water velocity with an inlet water temperature of 80 F. For example, when computed in this manner, for 85% clean tubes, r_f is approximately equal to 0.00023 hr-ft^2-deg F/Btu. Values of r_f derived from HEI cleanliness factor for a constant value of "percent clean" increase as the overall heat transfer upon which they are based decreases. The major controlling variables are water velocity, tube material, tube gauge, inlet water temperature, and tube diameter.

c. Convection heat transfer. The mode of heat transfer that applies to the cooling water flowing through the condenser tubes is known as convection heat transfer. Since the flow is established by mechanical means, the type of flow is classified as forced convection. Further, the normal design velocities used maintain the flow in the turbulent range.

The following classical Nusselt expression for determining the conductance of liquids being heated while flowing through horizontal tubes is recommended for determining the convection heat transfer value, for Reynolds numbers greater than 2100:

$$Nu = C(Re)^a(Pr)^b \qquad (25)$$

where

Re = Reynolds number
Nu = Nusselt number
Pr = Prandtl number
C = Constant

With the exponents evaluated, equation (25) is commonly referred to as the McAdams equation, or the Dittus-Boelter equation, which is as follows:

$$\frac{h_w D}{k} = C\left(\frac{\rho D V}{\mu}\right)^{0.8}\left(\frac{C_p \mu}{k}\right)^{0.4} \qquad (26)$$

where

h_w = conductance (inside tube surface to water), Btu/hr-ft^2-deg F
D = inside diameter of tube, ft
k = cooling-water thermal conductivity, Btu/hr-ft^2-deg F/ft
ρ = cooling-water density, pcf
V = water velocity, ft/hr
μ = cooling-water viscosity, lb/hr-ft
C_p = water specific heat, Btu/lb-deg F
C = constant

The proposed values of C vary as follows:

McAdams	0.0225
Dittus-Boelter	0.0243
HEI data	0.0240

For steam condensers operating in the temperature range characteristic of marine power plants, $C = 0.0240$ is recommended.

The physical properties of the cooling water correspond to the mean bulk temperature of the cooling water.

d. Overall heat transfer. The overall heat transfer coefficient U_r can be calculated from the conductances for condensing vapor and heating water, and the total of the conduction resistances. In condenser practice it is customary to calculate the coefficient U_r based on the outside surface of the heat transfer tubes.

The general equation for the overall coefficient U_r for steam condensers, with steam condensed on the outside of the tubes and cooling water heated on the inside of the tubes, is as follows:

$$\frac{1}{U_r} = \frac{1}{h_s} + \frac{D_o}{h_w D_i} + r_{tw} + r_{ox} + r_f \qquad (27)$$

where

U_r = overall heat transfer coefficient as determined from rational methods, Btu/ft^2-hr-deg F
h_s = conductance of condensing steam, Btu/ft^2-hr-deg F
h_w = conductance of cooling water, Btu/ft^2-hr-deg F
r_{tw} = resistance of tube wall, ft^2-hr-deg F/Btu
r_{ox} = resistance of new clean-tube oxide film, ft^2-hr-deg F/Btu
r_f = resistance of fouling in service, ft^2-hr-deg F/Btu

4.3 Steam Space Pressure Losses. In the process of condensing a vapor, the condensing temperature is a function of the vapor pressure in accordance with the saturation pressure-temperature relationship of the thermodynamic fluid involved, in this case steam. The steam flow from the condenser steam inlet, around the tube bundles, and through the tube banks is characterized by a pressure loss with the condensing steam temperature in conformity with the saturation pressure-temperature relationship. This pressure loss results in a reduction in temperature in the direction of flow, and the resulting change in temperature from steam inlet to air cooler inlet affects the rate of heat transferred in the various pressure zones of the condenser.

To simplify the analysis of the heat flow system, it can be assumed that a single tube can be selected that will represent the average performance of the entire tube bundle, or bundles. Considering such an average tube, the steam pressure at its surface is quite different from that at the condenser steam inlet; and, consequently, the steam temperature, in conformity with the saturation pressure-temperature relationship, will also be different. In addition, the performance standards of the HEI and test measurements conforming to the ASME Performance Test Code for Steam Condensers are based on static pressure. The condenser tube, however, senses a temperature more closely related to total or stagnation pressure than to static pressure. It is necessary, therefore, to consider these relationships in performing a rigorous heat transfer analysis.

The heat transfer representing the contract performance of the condenser and the actual heat transfer of the average tube in the tube bank are based on the same duty. The heat transfer corresponding to contract performance includes losses within the condenser steam space and is based on a higher condensing temperature than the average condensing temperature within the tube bank. Consequently, it is lower than the heat transfer computed on

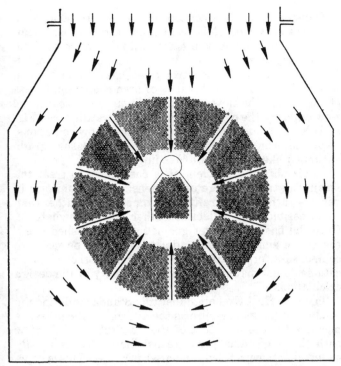

(a) Tight tube spacing, low steam entrance velocities

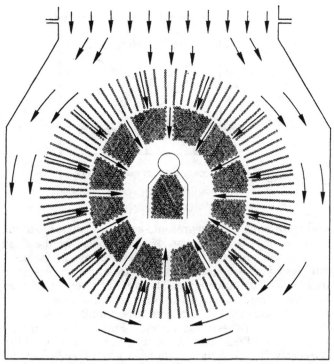

(b) Moderate tube spacing, moderate steam entrance velocities

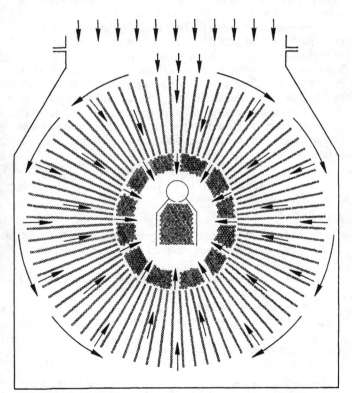

(c) Loose tube spacing, high steam entrance velocities

Fig. 24 Condenser design geometry alternatives

the basis of the average condensing temperature in the tube bank. The two condensing temperatures can be related to condensing pressures, one representing contract pressure, and the other representing the average pressure within the tube bank. It is desirable to produce a design so that there is a minimum difference between the pressure representing contract condenser performance and the average pressure within the tube bank. To accomplish this it is necessary to produce a design wherein the internal losses are less than those included in the contract performance standard (inherent in HEI performance).

In order to illustrate the principles involved in the determination of condenser steam-side pressure losses, consider the condenser design shown by Fig. 24. Figure 24 represents a condenser shell geometry based on a given available space. It is assumed that the condenser surface, water flow, and tube size, gauge, length, and number have been determined. The task remaining is to establish the optimum tube arrangement.

The steam flowing from the steam inlet to the air cooler inlet incurs a pressure loss. This loss may be considered to be made up of two components: (1) the distribution loss, that is, the loss in pressure required to cause the steam to flow from the condenser steam inlet, throughout the length of the condenser, and around the tube bundle; and (2) the penetration loss, that is, the loss in pressure required to cause the steam to flow through the tube bundle, from its perimeter to the air cooler within it. The arrows surrounding the tube bundles in Fig. 24 represent the steam flow for distribution, and their length represents the steam velocity. The arrows arranged radially within the tube banks represent the steam flow for penetration,

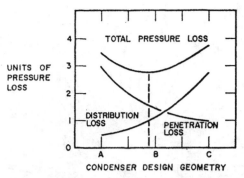

Fig. 25   Determination of optimum tube arrangement

and their length also represents steam velocity. Since the pressure loss is a function of the velocity of the fluid stream, the length of the arrows is indicative of the pressure loss. Figure 24 (a) shows a design where the pressure loss from distribution is small because of a large flow area and low velocity, and the pressure loss from penetration is large because of the compact tube bundle and, consequently, high penetration velocity. Figures 24 (b) and (c) represent modifications of the tube bundle in Fig. 24 (a). Successively, the tube bundle is expanded, reducing the steam velocity of penetration, but increasing the steam velocity of distribution. Consequently, Figs. 24 (a), (b), and (c) represent design geometries of increasing distribution losses and coincident decreasing penetration losses. Some arrangement of the tube bundle, with given condenser shell dimensions, will result in a minimum pressure loss (distribution loss plus penetration loss) from the steam inlet to the air cooler entrance. Curves which permit a determination of the optimum tube arrangement are shown in Fig. 25. The letters A, B, and C, represent the design geometries of Figs. 24 (a), (b), and (c), respectively. It will be noted that the total pressure loss, which is the sum of the distribution and penetration losses, reaches a minimum value at a design geometry approaching that of B in Fig. 25, which corresponds to the moderate arrangement of Fig. 24 (b).

By application of the preceding technique, the optimum condenser design geometry can be determined. However, the actual values of internal pressure loss associated with each design must first be established. The procedure for evaluating these losses follows.

a. Determination of mean total pressure around the tube banks. The mean total pressure around the tube bank is taken as a weighted-average total pressure. It is based on (1) the average total pressure immediately above the tube bundle, (2) the average total pressure immediately above the hotwell, and (3) the average total pressure at the tube bundle centerline. Included in these pressure determinations are the entrance loss to the main steam inlet lane and the loss due to the change of direction (at the hotwell) of the steam flow leaving the main steam inlet lane. The average of these three total or stagnation pressures may be considered to be a reasonably valid estimate of the average stagnation pressure of the steam entering the tube banks.

The steam flow in a condenser is a compressible, variable-area flow with fluid friction and a transfer of heat. Except at the tube bank interface with the steam distribution passages, the flow is adiabatic with changing mass. It cannot be treated as isentropic because of fluid friction, nor as Fanno line flow because of area change and mass change along the flow path. In addition to the above, the flow entering the condenser is usually stratified (see Fig. 9). It has been observed that under some flow conditions, steam entering the condenser assumes a pressure gradient much like that of an expanding nozzle.

To simplify the problem, it is customary to treat the steam flow as if it were one dimensional and uniform in velocity rather than stratified, incompressible rather than compressible, and adiabatic rather then nonadiabatic. Once the basic condenser geometry is established using the above simplifications, adjustments may be made to compensate for stratification of flow. The other simplifications do not materially affect the validity of the design calculations.

Since the Heat Exchange Institute Standards define the condenser design pressure as the average static pressure measured within one foot of the first tubes in the flow path, the longitudinal pressure distribution loss is usually included in the specified condenser pressure. This is especially true with main condensers where the steam inlet area covers a very high percentage of the condenser shell area. The designer is cautioned, however, to check the center-to-end flow distribution losses—that is, the pressure loss from steam flowing around the support plates at those sections where the support plates are outside of the steam inlet area of the shell. The sum of these losses, including end flow in the condenser, should be significantly less than the distribution loss around the tube bundle. An estimate of this loss for each support plate may be made from the following equation, which is empirical and limited to area ratios greater than 0.70 in this application:

$$\Delta P = \left[1 - \left(\frac{A_1}{A_2}\right)^2\right] \frac{C\rho V_{sp}^2}{2g} \qquad (28)$$

where

ΔP = pressure loss across support plate, in. Hg
A_1 = net area between support plates and condenser shell, ft²
A_2 = net area between tube bundle perimeters and condenser shell, ft²
ρ = density of steam, pcf
V_{sp} = steam velocity between shell and support plate, fps
g = gravitational acceleration, ft/sec²
C = conversion constant, psf to in. Hg = 0.01414

The steam flow and the resulting velocities around the support plates must be determined by calculating the relative steam demand for each section between support plates for the entire condenser in order to apply equation (28) properly.

To compute the pressure distribution around the tube bundle, the following procedure may be applied to each

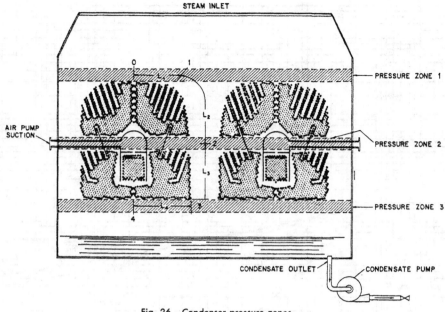

STEAM INLET

PRESSURE ZONE 1

AIR PUMP SUCTION

PRESSURE ZONE 2

PRESSURE ZONE 3

CONDENSATE OUTLET   CONDENSATE PUMP

Fig. 26 Condenser pressure zones

section between the support plates and averaged as the entrance steam pressure around the tube bundle. The difference between this pressure and the specified condenser pressure, when both are computed as total or stagnation pressures, represents the pressure loss for steam distribution. The process may be simplified without significant error, if the longitudinal pressure losses are small. The simplification can be accomplished by computing the losses around the tube bundle at that section of the condenser corresponding to the specified design pressure. The procedure for computing these losses in this manner is as follows. (This procedure is also applicable to each individual pressure section.)

As the first step, the steam flow area within one foot of the first tubes in the steam flow path is computed, and the steam velocity is determined assuming uniform unstratified flow. The stagnation or total pressure at this location, or pressure zone 1 (see Fig. 26), is computed from the following equation:

$$P_t = P_s + \frac{C\rho V^2}{2g} \qquad (29)$$

where

P_t = total pressure, in. Hg
P_s = specified condenser pressure (static), in. Hg
ρ = steam density, pcf
V = steam velocity in pressure zone (pressure zone 1, Fig. 26), fps
g = gravitational acceleration, ft/sec²
C = conversion constant, psf to in. Hg = 0.01414

In this example, all pressure losses around the tube bundle periphery are based on the pressure in pressure zone 1.

The second step is to compute the entrance loss from pressure zone 1 to the main steam inlet lane as follows:

$$\Delta P_{en} = 0.05 \frac{C\rho V^2}{2g} \qquad (30)$$

where

ΔP_{en} = entrance loss, in. Hg
V = steam velocity at entrance area, fps

and other terms are as defined for equation (29).

The third step is to calculate the pressure loss due to friction from pressure zone 1 to pressure zone 2. This is accomplished by computing the flow areas at the respective zones and calculating the steam velocity at these areas. In calculating all steam velocities, it is assumed that the net steam flow reduces proportionately with flow along flow distances L_1, L_2, L_3, and L_4. The flow areas are considered to be the distances between the tube banks at the various reference pressure zones multiplied by the condenser tube length.

The Reynolds number is next computed for the main tube bundle entrance area at pressure zone 1 and for the flow area at pressure zone 2, from the following relationship.

$$Re = \frac{\rho D_e V_1}{\mu} \qquad (31)$$

where

ρ = steam density, pcf
V_1 = steam velocity, ft/hr
μ = steam viscosity, lb/hr-ft
D_e = equivalent diameter of flow area (i.e., ratio of flow area to wetted perimeter multiplied by 4), ft

Using the mean of the two Reynolds numbers for zones

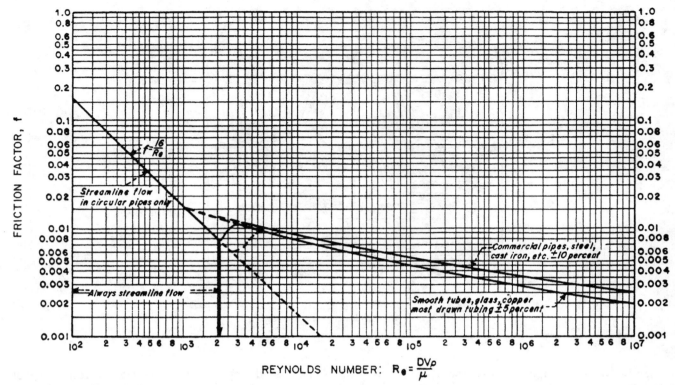

Fig. 27 Friction factor for steam distribution losses

1 and 2, refer to Fig. 27 and determine the friction factor, f, on the basis of a roughness equal to that for smooth tubes.

The pressure loss due to friction from pressure zone 1 to pressure zone 2 can be calculated using the Fanning equation as follows [the entrance loss from equation (30) must be added to the associated ΔP, in this example ΔP_{2-1}]:

$$\Delta P_{2-1} = \frac{4 C \rho f L V^2}{D_e 2g} \qquad (32)$$

where

ΔP_{2-1} = pressure loss from zone 1 to zone 2, in. Hg
f = friction factor from Fig. 27
L = flow length from zone 1 to zone 2, ft
V = average velocity between zone 1 and zone 2, fps
D_e = average equivalent diameter (i.e., ratio of flow area to wetted perimeter multiplied by 4) between zone 1 and zone 2, ft
g = gravitational acceleration, ft/sec^2
ρ = density of steam, pcf
C = conversion constant, psf to in. Hg = 0.01414

This procedure is repeated to determine the frictional pressure loss from pressure zone 2 to pressure zone 3, again taking into account the reducing mass flow along L_3. It is repeated along L_4 to the centerline of the tube bundle where the terminal flow and velocity are assumed to be zero. In this case the flow area is calculated as the distance from the tubes to the maximum hotwell level

(assuming no other obstructions) multiplied by the condenser tube length.

An exit loss due to the change in flow direction from the main steam lane to the underside of the tube bank must be added. For this purpose, equation (30) may be used, but with a coefficient of 0.25 instead of 0.05. Such losses, which are entrance and exit losses, are added to the friction losses calculated at the various zones of the main steam distribution lane, i.e., at pressure zone 1–2 (flow path L_2) and pressure zone 3–4 (flow path L_4), which are illustrated by Fig. 26.

The process for computing the mean total pressure loss around the tube bundle involves averaging the losses for selected incremental flow distances. This may be done by graphic integration, by algebraic averaging, or by other convenient methods. An acceptable result may be obtained by using the following equation, which is based on a three-pressure-zone analysis with flow directions and distances as shown in Fig. 26:

$$\Delta P_{dm} = \frac{\begin{array}{c} L_1 \Delta P_{1-0} + L_2[2(\Delta P_{1-0} + \Delta P_{en}) + \Delta P_{2-1}] + L_3(2\Delta P_{2-0} \\ + \Delta P_{3-2}) + L_4[2(\Delta P_{3-0} + \Delta P_{ex}) + \Delta P_{4-3}] \end{array}}{2(L_1 + L_2 + L_3 + L_4)} \qquad (33)$$

where

ΔP_{dm} = mean total pressure loss around tube bundle, in. Hg
ΔP_{1-0} = pressure loss at zone 1, flow path L_1, (generally zero), in. Hg. Pressure at location 0 in pressure zone 1 is to be taken as P_t from equation (29)

ΔP_{2-1} = pressure loss from zone 1 to zone 2, flow path L_2, in. Hg

ΔP_{3-2} = pressure loss from zone 2 to zone 3, flow path L_3, in. Hg

ΔP_{4-3} = pressure loss from zone 3 to zone 4, flow path L_4, in. Hg

ΔP_{en} = entrance loss, zone 1 to flow path L_2, in. Hg, equation (30)

ΔP_{ex} = exit loss, from end of flow path L_3 to zone 3, in. Hg, equation (30), except the coefficient becomes 0.25 (approx.)

and where

$$\Delta P_{2-0} = \Delta P_{1-0} + \Delta P_{2-1}$$
$$\Delta P_{3-0} = \Delta P_{1-0} + \Delta P_{2-1} + \Delta P_{3-2}$$

The preceding example illustrates the principles used. The specific treatment of flow and geometry in the example should not be considered to be limiting. Many simplifying assumptions were made and improved accuracy is possible by the application of more rigorous techniques. It is recommended that after a condenser is selected through the procedures described, a more rigorous check of ΔP_{dm} be made by applying the principles to the complete tube bundles and making an allowance for stratified flow at the pressure measurement zone, pressure zone 1.

b. Determination of pressure loss in tube banks. For the calculation of the pressure loss through tube banks, the actual bank is replaced by an equivalent square one having the same perimeter and number of tubes. The equivalent number of rows in the depth of such a tube bank may be computed from the equation

$$N_e = \frac{NP}{L} \qquad (34)$$

where

N_e = equivalent number of tube rows

N = total number of tubes in tube bundle being considered

P = tube pitch (for a variable pitch use an equivalent value), ft

L = perimeter of tube bundle, ft

The next step is to calculate the velocity through the minimum flow area between the outer tubes. The flow area may be calculated from the equation

$$A = \frac{L}{P} (P - D_o)L_t \qquad (35)$$

where

A = entrance area to tube bundle, ft^2

L = perimeter of tube bundle, ft

P = tube pitch (for variable pitch use equivalent value), ft

D_o = tube diameter, ft

L_t = condenser tube length, ft

With this flow area the steam velocity entering the periphery of the tube bank is computed as

$$V = \frac{W}{3600 \, \rho A} \qquad (36)$$

where

V = steam velocity, fps

W = steam condensed in tube bundle, lb/hr

ρ = steam density at mean static pressure around the tube bank, pcf

V_1, the steam velocity in ft/hr, is used in calculating the Reynolds number. The Reynolds number may be calculated from equation (31) modified as

$$Re = \frac{\rho D_o V_1}{\mu} \qquad (37)$$

where D_o is the outside tube diameter in feet, and all other terms are as described for equation (31) at the temperature corresponding to the mean pressure at the tube bank entrance.

The friction factor is determined by entering Fig. 28 with an adjusted Reynolds number equal to one-half the value calculated from equation (37).

The tube bank pressure loss equation, modified for a uniformly decreasing mass flow, is [5]:

$$\Delta P_p = 4f' \left(0.5 + \frac{N_e}{3} \right) \frac{0.01414 \rho V^2}{2g} \qquad (38)$$

where

ΔP_p = pressure loss through tube bank, in. Hg

f' = dimensionless friction factor (Fig. 28)

N_e = equivalent number of tube rows

ρ = mean density of steam at tube bank entrance, pcf

V = entering steam velocity between tubes, fps

g = gravitational acceleration, ft/sec^2

The mean effective pressure loss through the tube bank is used to determine the mean condensing pressure in the tube bank and its corresponding temperature, which is used to calculate the basic heat transfer. The mean pressure loss in the tube bank, with units of in. Hg, is computed from

$$\Delta P_{pm} = 4f' \left(0.5 + \frac{N_e}{4} \right) \frac{0.01414 \rho V^2}{2g} \qquad (39)$$

In order to calculate the pressure in the tube bank related to ΔP_{pm} the following procedure is suggested. All heat transfer calculations in this procedure are based on total pressure and, therefore, the mean condensing pressure, P_{pm}, in the tube bank representative of the condenser as a unit is referred to the design condenser pressure plus its velocity pressure, i.e., P_t. P_{pm} is calculated as follows:

$$P_{pm} = P_t - \Delta P_{dm} - \Delta P_{pm} \qquad (40)$$

where

P_{pm} = mean total pressure corresponding to the average tube representing the tube bundle design conditions calculated with P_t as the reference, in. Hg

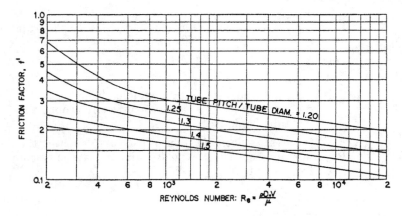

Fig. 28   Friction factor for steam penetration (tube bank) losses

P_t = mean condenser total pressure [see equation (29)], in. Hg

ΔP_{dm} = mean distribution pressure loss [see equation (33)], in. Hg

ΔP_{pm} = mean penetration pressure loss [see equation (39)], in. Hg

4.4   Performance Prediction and Evaluation.   Outlined in the preceding paragraphs are two elements of a procedure that may be followed to determine if a selected condenser design geometry is adequate to meet the condenser performance requirements. The first consists of an investigation of the heat transfer relationships as discussed in Section 4.2, and the second procedure involves an analysis of the fluid flow and pressure loss relationships as described in Section 4.3. A method of combining these two procedures in order to evaluate the performance of a condenser from its design geometry is outlined as follows:

a.   Heat transfer coefficient as related to characteristics of fluid flow.   In order to determine the heat transfer coefficient as related to the fluid flow characteristics, the procedure delineated in Section 4.3 may be used to calculate P_{pm}, the mean total pressure corresponding to the average tube representing the tube bundle design conditions based on P_t, the mean condenser total pressure. The temperature corresponding to P_{pm} is the condensing temperature used in determining the heat transfer necessary in the tube bundle to meet the condenser performance requirements.

Using the temperature corresponding to P_{pm} to determine θ_m, the heat transfer coefficient can be calculated using equation (11) of Section 3, modified as follows:

$$U_{pm} = \frac{Wh}{\theta_m S} \qquad (41)$$

In this instance, U_{pm} is the heat transfer for the tube bundle related to pressure loss considerations, i.e., the mean bundle pressure, P_{pm}. All other terms are as generally defined for equation (11).

The value of U_{pm} as computed from equation (41) may not include a cleanliness factor, except as the value of S, the condensing surface, reflects any cleanliness factor used in the determination of condenser size.

b.   Heat transfer coefficient as computed by rational equations.   The procedure for computing the heat transfer coefficient by rational means is delineated in Section 4.2. First the conductances for condensing and convection heat transfer are determined and converted to resistances. To these resistances, add the resistances due to conduction heat transfer, namely, those due to tube wall resistance, inside clean-tube resistance, and outside clean-tube resistance. If a cleanliness factor was included with equation (41), the resistance due to fouling, as calculated from this cleanliness factor, must be included here.

These resistances are then converted to U_r, the rational heat transfer coefficient representing the tube bundle performance, using equation (27). Note that in calculating U_r the physical properties of the two fluids, condensing steam and heating water, must be evaluated at their correct temperatures; see equations (21) and (26).

The mean temperature of the condensate film may be calculated by first determining the hourly heat or duty per square foot of condensing surface (Btu transferred per hour-square foot). The mean temperature of the condensate film is equal to the product of the unit hourly heat and the sum of the following resistances from equation (27):

$$\left(\frac{D_o}{h_w D_i} + r_{tw} + r_{ox} + r_f \right)$$

added to the mean bulk temperature of the cooling water. Using this temperature to calculate h_s from equation (21), calculate U_r from equation (27).

c.   Performance evaluation.   By comparing the values of U_{pm}, the heat transfer coefficient related to pressure loss, and U_r, the rational heat transfer coefficient, the adequacy of the condenser design can be determined. To meet specified performance requirements, the following relationship must exist:

$$U_r > U_{pm} \qquad (42)$$

The greater the difference, the greater will be the design margin or excess performance characteristics of the condenser. U_{pm} represents the heat transfer needed to meet performance requirements; U_r represents the heat transfer attainable. Should the relationship shown in equation (42) be the reverse, the condenser will not meet its performance requirements, and other condenser configurations should be investigated. U_{pm} can be significantly changed by modifications in the condenser design geometry; U_r is only slightly affected by the condenser design geometry. If a correction cannot be accomplished by changing the tube bundle design, the condenser shell would normally be the controlling factor; in such an instance, it should be made larger.

In some respects it is more convenient and more understandable to use resistances rather than conductances for making the comparison of performance. In this case

$$R_r < R_{pm} \tag{43}$$

where

$R_r = \dfrac{1}{U_r}$ = overall resistance from rational equations, hr-ft²-deg F/Btu

$R_{pm} = \dfrac{1}{U_{pm}}$ = overall resistance from fluid flow characteristics, hr-ft²-deg F/Btu

If fouling resistance is introduced (it must be considered to be the same for both the resistance calculated by rational means and by fluid flow characteristics) equation (43) becomes

$$R_r + r_f < R_{pm} + r_f \tag{44}$$

and may also be written as

$$R_{dm} = (R_{pm} + r_f) - (R_r + r_f) \tag{45}$$

where R_{dm} is the design margin as a resistance, hr-ft²-deg F/Btu, and must be positive if the condenser is to meet its required performance.

The performance factor, PF, may be expressed either in terms of total resistance or in terms of overall heat transfer. The PF is more meaningful if the fouling resistance expected in service is excluded. It is expressed as

$$PF = \frac{U_r}{U_{pm}} = \frac{R_{pm}}{R_r} \tag{46}$$

The PF should be greater than unity, and it is considered good practice to design for a performance factor of not less than 1.05.

The predicted operating pressure for the condenser based on its design geometry and on rational heat transfer can be determined using the previously calculated mean tube bundle pressure, P_m, as the base. Using equation (40), calculate a new value for P_t (ΔP_{dm} and ΔP_{pm} as previously determined are assumed to remain constant). Using equation (29), determine P_s, the static pressure the condenser will produce in the pressure-measurement zone as defined in the HEI Standards. As a check, P_s will be less than the HEI specified condenser pressure, provided $U_r > U_{pm}$ as required by equation (42).

The methods and procedures outlined represent a reasonably accurate method for evaluating condenser design geometry and performance. Accuracy can be improved by a more rigorous treatment of the fluid flow characteristics. The degree of accuracy improvement necessary depends on the circumstances of a particular case.

While the descriptions presented have been related to the normal geometry of the larger main condensers having two separate and distinct tube bundles, the method is applicable also to single-bundle condensers such as those generally used for auxiliary condensers. The principles can also be adapted to condensers having more than two separate tube bundles or to any shell-and-tube apparatus that condenses a vapor on the shell side.

References

1 J. E. Fowler and R. E. Brandon, "Steam Flow Distribution at the Exhaust of Large Steam Turbines," ASME paper 59-SA-62.

2 J. F. Sebald and W. D. Nobles, "Control of Tube Vibration in Steam Condensers," *Proceedings of the American Power Conference*, Illinois Institute of Technology, Chicago, Illinois, 1962.

3 A. P. Colburn, "Problems in Design and Research on Condensers of Vapours and Vapour Mixtures," The Institution of Mechanical Engineers, 1951.

4 "Standards for Steam Surface Condensers," Heat Exchange Institute, New York.

5 W. H. McAdams, *Heat Transmission*, McGraw-Hill Book Co., New York, 1954.

CHAPTER XVI

R. A. Mancuso | # Heat Exchangers

Section 1
Introduction

1.1 General. Heat exchangers are used to transfer heat from a hotter fluid (liquid or gas) to a colder fluid (liquid or gas). This broad definition encompasses a wide array of equipment, including boilers, condensers, distilling plants, and ventilation cooling coils, which are given detailed coverage in other chapters. Consequently, this chapter is focused upon the many other types of shell-and-tube, plate, and compact heat exchangers that are used aboard ships and includes an assessment of thermal design, mechanical design, and design-for-manufacture considerations.

1.2 Heat Exchanger Design Requirements. Marine heat exchangers are designed in accordance with the standards of the Tubular Exchanger Manufacturers Association [1] and the American Society of Mechanical Engineers [2] in addition to compliance with marine regulatory body code

Table 1   Typical characteristics of marine heat exchangers

ITEM	LUBRICATING OIL COOLER	LOW-PRESSURE FEED HEATER	HIGH-PRESSURE FEED HEATER	TANK CLEANING SYSTEM — HEATER	TANK CLEANING SYSTEM — DRAIN COOLER	GLAND LEAK-OFF CONDENSER
Tube Side						
Fluid	Seawater	Feedwater	Feedwater	Seawater	Seawater	Feedwater
Velocity, fps	6–7	4–7	4–7	6–7	6–7	3–4
No. of passes	1	2–8	2–8	1–4	1–4	2
Tube Size	5/8″ OD × 0.049″ low fin—19 fins per inch	5/8″ OD × 0.049″	5/8″ OD × 0.049″	3/4″ OD × 0.049″	3/4″ OD × 0.049″	5/8″ OD × 0.049″
Tube Material	90/10 CuNi	90/10 CuNi	70/30 CuNi	90/10 CuNi	90/10 CuNi	90/10 CuNi
Head Material	Bronze	Steel	Steel	Steel	Steel	Steel
Tubesheet	90/10 CuNi	Steel	Forged Steel	Bronze—Composition G	Bronze—Composition G	Steel
Shell Side						
Fluid	Oil	Steam Condensing	Steam Condensing	Steam Condensing	Heater drains	Gland steam Condensing
Velocity, fps	2–3					
Number of passes	10–16 (baffled)	6 (baffled)	6 (baffled)	6 (baffled)	12 (baffled)	3 or 6 (baffled)
Type of baffle	Segmental	Segmental	Segmental	Segmental	Segmental	Segmental
Flow	Counterflow	…	…	…	Counterflow, last pass	Counterflow, last pass
Construction						
Tube joint	Expanded	Expanded	Expanded or welded	Expanded	Expanded	Expanded
Thermal expansion	Floating tubesheet	Expansion joint	U-Tubes	U-Tubes	U-Tubes	…
Gasket	Full face	Ring or flexible	Ring (shell side) Solid copper ring (tube side)	Full face	Full face	Full face
Bolting	Collar bolts	Stud bolts	Stud bolts	Collar bolts	Collar bolts	Stud bolts
Surface	Finned	Bare	Bare	Bare	Bare	Bare
Thermal Performance						
Heat transfer coefficient, Btu/hr-ft²-F	40	800	800	600	200	…
Surface requirement	0.3 to 0.4 gpm L.O. per ft²	300–450 lb feed per hr per ft²	300–450 lb feed per hr per ft²	2.7 gpm seawater per ft²	180 lb drains per hr/ft²	6–10 lb steam per hr/ft²
Terminal temperature difference (or LMTD)	18°F	10°F	10°F	190°F	100°F	30°F
Temperature change	20°F drop in L.O.	100°F rise in feed	100°F rise in feed	110°F rise in seawater	200°F drop in drains	Cools non-condensable gases to about 140 F

requirements such as Lloyd's Register of Shipping [3], the American Bureau of Shipping [4], and the U.S. Coast Guard [5]. Many additional requirements have been incorporated in specifications for marine heat exchangers due to stringent space limitations and reliability requirements. The following points are emphasized in the design of marine heat exchangers:

• Heads are designed so that it will not be necessary to disassemble piping to gain access to the inside of the heads and tubes.

• The tube bundle is usually of the removable type for easy cleaning and maintenance.

• In the design of cooling-water spaces and connections, a smooth flow path must be provided to minimize erosion-corrosion attack. Sharp corners and projecting edges are avoided.

• Internal fittings are arranged to result in a minimum of interference with the water flow and a minimum of turbulence.

• Cooling-water velocities at the design point must not exceed those specified or recommended by the material supplier.

• Heat exchangers having tubes with a length exceeding 4 ft are designed so that the ordering length of tubes will be in multiples of 6 in. The ordering length of tubes is determined by adding $\frac{1}{8}$ in. to the face-to-face distance between the outside faces of the tubesheets.

• The minimum tubesheet thickness is usually specified to be not less than $\frac{3}{4}$ in.

• When external fins (low fins) are applied to tubes, one end of the tube is usually enlarged to the outside diameter of the fins to enable the removal and insertion of individual tubes.

• Holes in the tubesheet at the inlet end of the tubes are flared to allow for belling the ends of the tubes. Holes in the tubesheets are provided with at least one groove. The edges of the holes are rounded, usually on a $\frac{1}{16}$-in. radius, on the inner face of each tubesheet and on the outer faces of the tubesheets at the discharge ends of the tubes.

• The inlet ends of the tubes are expanded and belled, and the ends are finished flush with the face of the tubesheet. In no case should the ends of tubes be inside the

	CONTAMINATED STEAM GENERATOR	LUBRICATING OIL HEATER	FUEL OIL HEATER	ELECTRONICS EQUIPMENT COOLER	BOILER WATER SAMPLE COOLERS	INSTANTANEOUS WATER HEATERS	LUBRICATING OIL PLATE HEAT EXCHANGER	WATER COOLER PLATE HEAT EXCHANGER
	Steam	Steam	Steam	Chilled water or seawater	Boiler makeup water	Water	Seawater	Water
	Condensing	Condensing	Condensing	3–6	1–8	4–7	1–4	1–4
	2	2	2	1	1	2	Multiple	Multiple
	1″ OD × 0.065″	7/8″ OD × 0.065″ (bayonet tube)	7/8″ OD × 0.065″ (bayonet tube)	5/8″ OD × 0.049 BWG	1/4″ OD × 0.035″	5/8″ OD × 0.049″ BWG		
	90/10 CuNi	Steel	Steel	90/10 CuNi	90/10 CuNi	90/10 CuNi	Titanium	Stainless steel
	Steel or 90/10 CuNi	Cast Steel	Cast Steel	Bronze	...	90/10 CuNi	Steel	Steel
	Bronze— Composition G or steel	Steel	Steel	90/10 CuNi or Bronze Composition G	...	90/10 CuNi	...	...
	Fresh water	Lubricating Oil	Fuel Oil	Fresh Water	Seawater	Steam	Lubricating oil	Water
	...	2–3	1–3	3–5	1–8	Condensing	1–4	1–4
	Open shell	1 (baffled)	1 (baffled)	1 (baffled)	None	1 (baffled)	Multiple	Multiple
	None	Segmental	Segmental	Segmental	None	Segmental	...	...
	Boiling	Counterflow	Counterflow	Counterflow	Counterflow	...	Counterflow	Counterflow
	Expanded	Expanded	Expanded (outer) Ferrules (inner)	Expanded	Welded or brazed	Expanded	...	...
	Floating head or U-Tube	Bayonet	Bayonet	...	...	U-Tubes	...	...
	Full face	Ring	Ring	Ring	Full Face	Ring	Ring	Ring
	Stud bolts	Stud bolts	Stud bolts	Stud bolts	Stud bolts	Stud bolts	Tie bolts	Tie bolts
	Bare	Finned	Finned	Bare	Bare	Bare	Bare	Bare
		PURIFIER WARM-UP						
	...	10–20 60	45	400	500	500	300–750	750–1250
	30–50 lb steam per hr/sq ft	6–8 gph per ft² 75 gph per ft²	7–12 gph per ft²	150 gph per ft²	0.25 gpm boiler water per ft²	150 gph per ft²	1.3 gpm per ft²	2.6 gpm per ft²
	80°F	250°F 250°F	100°F	20°F	400°F	200°F	2°F	2°F
	Boiling at 100 psi abs	120°F 50°F	100–150°F	5°F drop in fresh water	400°F drop in sample temp.	80°F rise in water temp.	20°F drop in L.O.	3–25°F

Table 1 *Continued*

face of the tubesheets. Discharge ends of tubes can protrude up to ¹⁄₁₆ in. beyond the face of the tubesheet.

• A number of the baffles are increased in thickness (usually ¼ in.) to act as tube support plates and are located so that the maximum tube length between support plates, or between a tubesheet and a support plate, does not exceed 36 in. Holes for tubes in baffles and support plates, baffle clearances, and tie rod standards are usually required to be in accordance with the standards of the Tubular Exchanger Manufacturers Association [1].

• In order to diffuse the entering stream and reduce erosion of tube ends, for single-pass coolers the waterbox depth measured normal to the tubesheet should be not less than one-half the equivalent diameter of the tubesheet area exposed to the flow of the cooling water into the tubes. For cylindrical two-pass coolers, the waterbox head depth should be not less than 35% of the inside shell diameter.

• All heat exchangers must be provided with adequate foundation supports. When required by the conditions of service, provision is made in the design of the supports to provide for expansion or contraction of the shell. Heat exchanger supports are usually independent of any attached piping. Supports must be given special consideration when designing naval combatants for high-impact shock conditions [6].

• In the design of marine heat exchangers, consideration must be given to the varying degrees of inclination encountered in service. In naval practice, heaters and coolers are designed to perform satisfactorily under conditions of 5-deg trim, 10-deg pitch, 15-deg list, and 45-deg roll (in commercial practice a 30-deg roll is the design criterion). The conditions for permanent list and roll or for trim and pitch are generally not considered additive.

• Adequate air vents must be provided on heat exchanger waterboxes to avoid the collection of air in the upper region of the waterbox, as air pockets can restrict the tube-side flow and render a portion of the heat transfer surface ineffective. Such air pockets can also result in overheating and expansion of the dry tubes and cause failures of the tube joints at the tubesheets.

• In feedwater heaters and condensers, wet steam at a high velocity must not be permitted to impinge on the tubes, otherwise the surface of the tubes will be rapidly eroded. Baffles or distribution pipes must be incorporated as necessary to prevent the direct impingement of wet steam on the tubes.

• Impingement protection from the entering shell-side fluids is often advisable. This may include adding an impingement baffle or impingement rods opposite the shell inlet nozzle.

1.3 Design Data Requirements. In order to specifically direct attention to the items that govern the thermal and mechanical design of a heat exchanger and which must be furnished to the design engineer (or assumed by the design engineer), the following should be included in a specification for a marine heat exchanger:

• Substances to be heated and cooled.
• Quantity of substances to be heated and cooled within a given period of time.
• Initial temperatures of the substances heated and cooled.
• Final temperatures desired for the substances heated and cooled.
• When the heating and cooling media are other than water or commonly known substances, the following should be specified:
   (a) viscosity,
   (b) specific gravity,
   (c) specific heat, and
   (d) thermal conductivity.
• Working pressures of the substances heated and cooled.
• Allowable pressure drop through the shell and tube sides of the heat exchanger.
• Desired construction materials.
• Design pressures and design temperatures.
• Design code.
• Fouling factor or service margin.
• Specific geometric arrangement and type of piping interface connections.
• Weight limitations or considerations.
• Test conditions.

A summary of the typical characteristics of a variety of marine heat exchangers is given by Table 1.

Section 2
Shell-and-Tube Heat Exchanger Arrangements

2.1 Shell-and-Tube Heat Exchanger Features. In marine applications, the shell-and-tube heat exchanger is the most widely used type. Shell-and-tube heat exchangers are fabricated from round tubes that are nested in, and run parallel to, a shell. Heat is transferred between the fluids by going through the walls of the tubes. As shown by Fig. 1, a shell-and-tube heat exchanger consists of six basic elements: the bonnet, tubesheet, shell, tubes, baffles or support plates, and the tie rods. The bonnet or channel is often referred to in the marine industry as the "head"

or "waterbox" of the exchanger. Due to stringent space limitations imposed on marine heat exchangers and for ease of maintenance and cleaning, head inlet and outlet piping connections are arranged to permit access to the tubes and tubesheets without dismantling the attached piping. Only the smaller heat exchangers (under 100 lb) would have axial connections as depicted in Fig. 1.

Figure 1 illustrates a "single-pass" fixed-tubesheet exchanger. "Single-pass" is a term which indicates that the tube-side fluid flows in one direction only. A "two-pass"

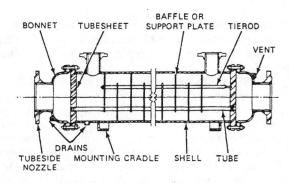

BONNET   TUBESHEET   BAFFLE OR
SUPPORT PLATE   TIEROD

VENT

DRAINS

TUBESIDE   MOUNTING CRADLE   SHELL   TUBE
NOZZLE

Fig. 1   Typical single-pass conventional exchanger (fixed tubesheet)

this design, one tubesheet is free to "float" against packing rings, which are usually made of neoprene. The packing rings are held in place between the head and shell flanges by a packing retainer ring. Details of this construction are described in Section 6.1.

As illustrated in Fig. 1, the shell of the heat exchanger is usually cylindrical with flanges attached to each end. The tubesheets are either welded or bolted to the shell flange, and the heads are bolted to the tubesheets.

2.2   Tube-to-Tubesheet Joint. The most critical joint in the exchanger, and that most likely to develop a leak, is the tube-to-tubesheet joint. The tube-to-tubesheet joint is made by expanding the tube into serrations or grooves that are located in the tubesheet, by welding the tube to the face of the tubesheet, or by a combination of both.

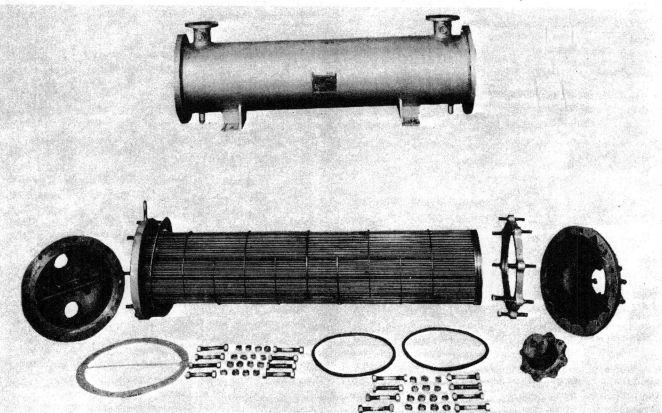

Fig. 2   Typical heat exchanger with floating tubesheet, disassembled

or "four-pass" exchanger would have the tube-side fluid inlet and outlet connections at the same end. The fixed tubesheet construction depicted in Fig. 1, which has the tubesheets welded to the shell, must often incorporate an expansion joint in the shell to compensate for differential expansion between the shell and tubes as a result of the relative temperatures of the fluids involved.

In addition to fixed tubesheet designs, heat exchangers employing floating tubesheet and "U-tube" designs are commonly used. Figure 2 illustrates a floating tubesheet construction, which is typically used for such services as main lube oil coolers and electronic equipment coolers. In

When the tube is expanded into a tubesheet with serrations, the number of serrations can vary from one to as many as three, depending upon the intended service. The tubes may be expanded into the tubesheet by mechanical rollers, hydraulic expansion, or detonation. Regardless of the method of expansion used, the objective is to effect a joint that has a mechanical strength greater than the tube and is leak-tight. Factors that affect the quality of the tube-to-tubesheet joint are:

• Tube hole finish
• Tube wall thickness reduction
• Length of expanded joint

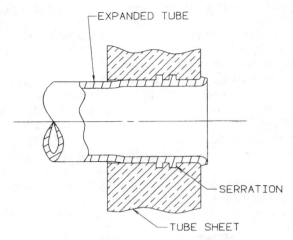

Fig. 3 Expanded tube-to-tubesheet joint with flared inlet

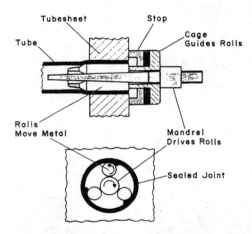

Fig. 4 Mechanical tube expander

- Tubesheet ligament width (i.e., minimum distance between tube holes)
- Relative yield strength between the tube and tubesheet
- Tube wall thickness

Mechanical roller expansion is the most commonly used method of effecting a mechanical bond between the tube and tubesheet; a typical expanded tube is illustrated by Fig. 3. The simplest tube expander, which is illustrated by Fig. 4, is an assembly of three major parts: the mandrel, the cage, and the rollers. The maximum roller length is approximately two inches; therefore, longer joints are rolled in steps. Because of friction between the rollers and tube and between the rollers and roller cage, the roller contact surfaces are lubricated. Roller expanding extrudes the tube material in the axial direction. Care must be exercised during the rolling operation to avoid inducing undesired residual compressive or tensile stresses in the tubes. When rolling tubes manually, the mandrel is turned by a socket wrench or pneumatic driver; but as a step in the manufacturing process, rolling is performed automatically at a high rate of speed. Care must also be taken to follow the manufacturer's instructions concerning the limits of tube wall thickness reduction and the dial settings on an electrically or electronically controlled automatic tube expander when repairing a leaking tube-to-tubesheet joint.

Hydraulic expansion, as implied, uses a fluid (water) under a pressure of 30,000 to 50,000 psi to expand the tube. There is no practical limit to the length of the tube that can be expanded in a single operation. The length of the tube to be pressurized and expanded is determined by the location of an "O" ring on the inside of the tube, which bounds the high-pressure water. Hydraulic expansion does not extrude material; therefore, concerns relative to over-rolled joints are eliminated. However, when the tubes are expanded within the tubesheet, the Poisson effect causes the tubes to shorten, thereby introducing a tensile stress in the tubes when the ends of the tubes are secured.

When the tubes are mechanically expanded into the tubesheet, the recommended tubesheet groove size is $\frac{1}{8}$ in. wide by $\frac{1}{16}$ in. deep. However, there are some questions concerning the appropriateness of such a groove size when the tubes are expanded hydraulically. A wider groove may be required to develop the proper pull strength because the tube metal does not flow into the groove as effectively as it does when the tubes are expanded by rolling.

The detonation of an explosive charge within the tube has been used to effect a tight tube-to-tubesheet joint. The procedure involves placing an explosive charge inside a polymeric sleeve and locating the assembly appropriately with respect to the tube or tubes to be expanded. The controlled detonation produces a shock wave that travels at or slightly above the sonic velocity of the material. This method of tube expansion has been successfully used where a large number of tubes are to be expanded or where the tube and tubesheet materials are not compatible for fusion welding and the reliability of a welded joint is required.

Although the normal practice is to expand the tube into the tubesheet by rolling, there are factors, such as pressure, temperature, fatigue, and corrosive fluids, which, when combined with the consequences of leakage, dictate that the tube-to-tubesheet joint be welded. Figure 5 illustrates some typical tube-to-tubesheet welded joint geometries that may be used.

To eliminate tube-to-tubesheet weld porosity that is induced by contamination, the tubes and tubesheet must be clean when the tubes are welded. This means that the surfaces must be free of protective coatings, lubricants, rust, mill scale, pits, and scratches.

The joints illustrated by Fig. 5(a), (b), and (c) are typical face side tube-to-tubesheet joints that are welded by using either a manual or an automatic process. The weld details shown in Fig. 5(a) and (b) are used when the tube wall thickness is less than 0.065 in. The weld is classified as a single-pass autogenous weld. Figure 5(c) is used when the wall thickness is equal to or greater than 0.065 in. The weld may be either single-pass or multi-pass with or without filler metal, depending on the thickness of the tube. The tube-to-tubesheet joints in Fig. 5(a), (b), and (c) entail stress concentration factors when the joints are evaluated

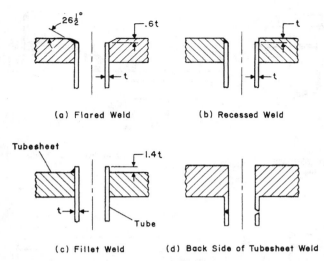

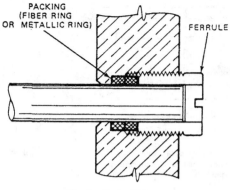

Fig. 6 Packed tube

Fig. 5 Typical tube-to-tubesheet welded joints (joint preparation on the right side and finished weld on the left side)

for fatigue. The stress concentration factor for the tube-to-tubesheet joint in Fig. 5(d) is one.

The tube-to-tubesheet welds shown in Fig. 5(a) and (b) do not develop the strength of the tube itself; therefore, in these cases the tubes must be expanded into serrations in the tubesheets. However, in addition to improving the strength of the joints, expanding the tubes also eliminates the crevice between the tube and tubesheet, which is a site for corrosion or caustic embrittlement.

The tube-to-tubesheet weld shown in Fig. 5(c) develops the strength of the tube, and in this case expanding the tube into the tubesheet is optional.

Figure 5(d) is a typical back-face tube-to-tubesheet weld. It is an expensive joint to manufacture because the extensions must be machined on the back-face of the tubesheet. The weld is generated by an automatic process and the tube diameter must be large enough to accept the welding gun if performed from within the tube. If the joint is welded from the external surface of the tube, then the tube pitch must be increased such that the welding gun can be rotated around the outside of the tube. There is no tube-to-tubesheet crevice with this type of weld joint.

The ability of the welding process to consistently produce acceptable welds is affected by the manufactured tolerances of the tube, the tube hole tolerance, the positioning of the tube in the tubesheet, and the weld joint detail. Specifying the proper tolerances and joint design is based on experience that is gained by successfully producing tube-to-tubesheet welds.

The high-velocity flow of seawater in heat exchanger tubes tends to remove the thin protective film of corrosion products adhering to the base metal of the tube wall. This protective film is replaced at the expense of further corrosion of the tube wall. As the continued removal and replacement of the protective film of corrosion products proceeds, the tube wall is gradually thinned and either the tube-to-tubesheet joint is weakened and ultimately fails or the tube wall just beyond the tubesheet is perforated. This type of erosion is generally termed impingement erosion, inlet-end attack, air erosion, or bubble attack. The occurrence and rapidity of the attack are governed by the water velocity (for recommendations see Table 1), the amount of air entrained, and the design of the waterbox as it affects the velocity, direction, and turbulence of the fluid flow approaching the tubes.

The inlet ends of the tubes are normally ground flush with the face of the tubesheets, and no gaps should be left between the edge of the tubes and the radius of the holes on the inlet sections as gaps there tend to promote impingement erosion. The outlet end of tubes may extend $\frac{1}{16}$ in. beyond the face of the tubesheet.

Where a considerable temperature differential exists within a heat exchanger, a packed-tube design may be employed. Packed tubes, such as shown by Fig. 6 and Fig. 10(c), allow considerable differential tube expansion since each tube is free to move independent of the others. The combined low-pressure feed heater/drain cooler/gland-exhaust condenser, which is illustrated by Fig. 31, is a typical marine heat exchanger that normally employs packed tube ends.

2.3 Shell, Tubesheet, and Head (Channel) Joints. The design of the means for attaching the head (channel), tubesheet, and shell is governed primarily by the service pressure and temperature. The design depicted by Fig. 7(a), which employs gaskets that are positively positioned, thereby insuring alignment, is the least expensive means of attachment. This design could incorporate studded tubesheets, as shown, or collar studs to permit the removal of the head without disturbing the shell-side pressure joint. In naval applications, the joint shown in Fig. 7(a) may be used at conditions up to 150 psig and 375 F.

Figure 7(b) illustrates the joint that is the most widely used in shell, tubesheet, and head joint attachments. The design normally incorporates studded tubesheets, as shown, or collar studs. The usual applications will accommodate liquids and vapors up to 300 psig and 450 F.

The design indicated in Fig. 7(c) is widely used for high-pressure fuel oil and steam service. The gasket grooves completely confine and positively align pass partitions, and afford excellent protection against gasket blowouts

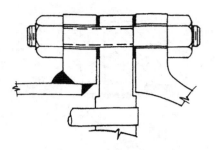

(a) Flat face ring shell flange with full
face unconfined gaskets on both shell
and head side.

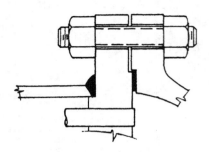

(d) Welded shell tubesheet joint with
semi-confined gasket on head side.

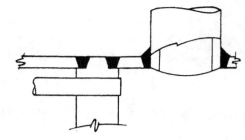

(g) Head integral with shell, no shell side
gasket.

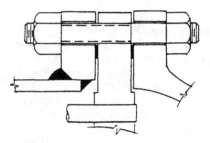

(b) Flat face ring shell flange witn ring
type unconfined gaskets on both shell
and head side.

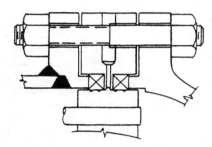

(e) Double packed floating tubesheet
with retaining ring and packing rings
on both shell and head side.

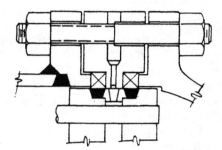

(h) Double floating packed tubesheet
with retaining ring and packing rings
on both the shell and head sides.

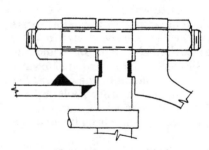

(c) Ring shell flange with tongue and
groove joint and fully confined flat
ring gasket on both the shell and
head side.

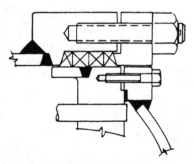

(f) Outside-packed removable tube
bundle with semi-confined gasket on
head side.

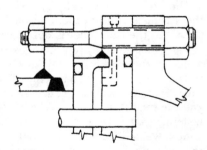

(i) Ring joint flange, double tubesheet
design with confined O-ring gaskets.

Fig. 7   Shell, tubesheet, and head joint designs

and failures. The usual applications are suitable for liquids and vapors to 1000 psig and 750 F with a proper
gasket material and design, such as spiral-wound gaskets.

Figure 7(d) depicts a joint that is used primarily when
no leakage is permitted on the shell side, as in cases where
the shell-side medium is either hazardous or corrosive.
The semi-confined gasket provides some protection from
blowing out and ensures a more positive positioning of
the gasket. The usual applications permit liquids and vapors up to 600 psig and 500 F with proper gasket materials.

The design indicated in Fig. 7(e) is an excellent means
of providing for shell or tube thermal expansion. When the

packing on one side becomes deteriorated and eventually
develops a leak, the leak is readily detectable through the
vent and drain holes in the retaining ring, and the packing
can be replaced. The studded retainer ring permits repacking the tube side with full pressure maintained on the shell
side. The usual applications permit liquids and vapors up
to 300 psig and 300 F (the temperature is limited by the
packing material).

The design indicated in Fig. 7(f) is an alternative means
of providing for shell or tube thermal expansion. On the
shell side, liquids and vapors at pressures up to 500 psig
can be accommodated but the temperature is limited (usually to about 300 F) by the type of packing. On the tube

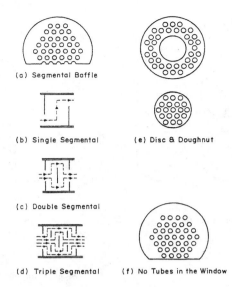

(a) Segmental Baffle

(b) Single Segmental

(c) Double Segmental

(d) Triple Segmental

(e) Disc & Doughnut

(f) No Tubes in the Window

Fig. 8   Flow baffles

side, the pressure may reach 600 psig at temperatures of 500 F with proper gasket materials.

Figure 7(g) describes a joint that is relatively inexpensive and is used when it is desirable to eliminate the shell-side gasket. This design employs only one gasketed joint at each head end for servicing the tubes. There are no limits on pressure or temperature except for gasket considerations.

The design indicated in Fig. 7(h) is recommended where the contamination of one fluid medium by the other cannot be tolerated and is an excellent means of providing for thermal expansion of either the shell or tubes. The studded retainer ring permits repacking the tube side with full pressure maintained on the shell side. The double tubesheet type of construction is often specified by the Navy in cases where the greatest assurance against mixing of the two fluids is desired. Liquids and vapors to 300 psig and 300 F (with due consideration to packing material) can be accommodated with this design.

The joint described by Fig. 7(i) is excellent for vacuum and very-high-pressure service, and the temperature is limited only by the O-ring gasket material. The design requires precise machining of the O-ring grooves. The double tubesheet design is excellent for use with noncompatible fluids; should a tube joint fail, the leak can be detected immediately, avoiding contamination of the fluids. The usual applications permit liquid and vapor pressures up to 1000 psig and 400 F.

2.4   Tube Arrangement Details. In shell-and-tube heat exchangers, baffles are generally used to guide flow and increase the velocity of the fluid flowing on the shell side of the heat exchanger. The most commonly used baffles are of the segmental type or its variations, as illustrated by Fig. 8.

Segmental baffles are formed by cutting out thin metal plates to an outside diameter slightly less than the inside diameter of the shell. A segment is cut out of the baffle to form a segmental opening, the size of which may vary

from approximately 15 to 45% of the shell cross-sectional area (with single segmental, double-segmental, and disk-and-doughnut baffles) or higher than 45% of the shell area (with triple-segmental baffles).

The tubes must be supported at intervals along their length to minimize tube vibrations excited by the fluid flowing across the tube or by pulsations of the flow rate. The maximum permissible unsupported tube length depends on the tube material, wall thickness, and diameter. For nonpulsating flow, the maximum unsupported tube length is 60 to 80 times the tube diameter. For pulsating flow, the unsupported length is considerably less, and welding or brazing is used in lieu of spacers and nuts to secure the tube support plates. When designing for high shock, the tube support spacing is made small to minimize the response of the tubes to shock loadings.

A baffle design with no tubes in the window (the cutaway portion of the baffle), as illustrated by Fig. 8(f), is ideal to use when designing for shock loadings or flow-induced vibration, which is discussed in Section 3.7. This type of baffle permits the tubes to be supported by every baffle. The disadvantage is a slight increase in the shell inside diameter for the same number of tubes when compared with the other baffle types. The baffle cut may vary from 5 to 15% of the shell cross-sectional area.

There is considerable latitude in selecting the tube arrangement. The four most common tube patterns, as viewed from the tubesheet end, are shown in Fig. 9. The triangular pitch and rotated-triangular pitch are the most compact forms, and the triangular pattern is the one most commonly used for marine heat exchangers.

Square pitch and rotated-square pitch patterns have see-through lanes, which facilitate manual or mechanical cleaning of the outside of the tubes. The square pitch is common in submerged-tube boilers where an unobstructed flow passage is important for fluid circulation. It is also applied in services where minimum shell-side pressure drop is a paramount design criterion. Tube patterns other than the four illustrated in Fig. 9 could be used to satisfy specific design considerations as to pressure drop or turbulence, but they would be more costly to manufacture.

The tube center-to-center distances are normally 1.25 times the tube diameter, or greater, and uniform over the tube field.

The tubes in a shell-and-tube heat exchanger may be either plain smooth-surface tubes or enhanced tubes. Enhanced tubes include tubes with extended surfaces (such as low, integral fins as described in Section 3.6), tubes with a coated surface, corrugated tubes, or a combination thereof. The decision to use plain versus enhanced tubes is influenced by such factors as fouling, heat transfer coefficients, thermal resistance, and available pressure drop. Heat exchangers for marine applications predominately are made of plain tubes. Typical bare tube sizes are $\frac{1}{4}$, $\frac{5}{16}$, $\frac{3}{8}$, $\frac{1}{2}$, $\frac{9}{16}$, $\frac{5}{8}$, $\frac{3}{4}$, and 1 in. The tube thickness varies from 0.028 in. Birmingham Wire Gauge (BWG) 22 to 0.083 in. BWG 14.

Since the shell and tubes operate at different temperatures, it is necessary to provide means to accommodate

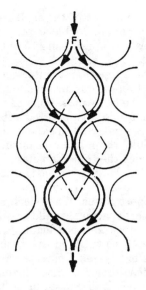

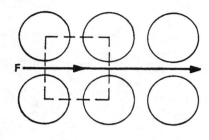

(c) Square tube pitch

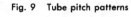

Fig. 9   Tube pitch patterns

(a) Triangular tube pitch

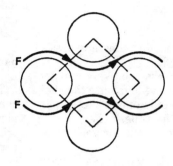

(d) Rotated square tube pitch

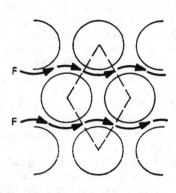

(b) Rotated Triangular tube pitch

the difference in their thermal expansions, as high stresses could otherwise be developed within the heat exchanger. The various types of construction that have been used to accommodate thermal expansion are illustrated in Fig. 7(e), (f), and (h) and Fig. 10; each has an appropriate area of application.

The simple U-tube illustrated by Fig. 10(a) is the most economical and commonly used means to separate the thermal expansion in the tubes from that of the shell. It is widely used in small condensers and instantaneous heaters.

The bayonet tube, illustrated in Fig. 10(b), is suitable for use with a tight, cross-flow baffle spacing. Bayonet tubes are often used in viscous oil heaters.

Packed-tube joints, Figs. 6 and 10(c), permit some axial movement of the tube. Packed-tube joints are used in rectangular (box type) low-pressure feed heaters and in smaller straight-tube condensers.

In cases where a large differential thermal expansion can be expected, a floating head arrangement may be

employed. As can be seen from Fig. 10(d), large thermal expansions present no difficulties with this type of design.

When the tubes are firmly fastened to the tubesheets and the tubesheets are fastened to the shell (i.e., fixed tubesheet designs), shell expansion joints are used to lessen the stresses caused by the difference in thermal expansion between the shell and tubes. The shell expansion joints illustrated in Fig. 11 are adequate only for small differential thermal expansions (say 0.06 in. for a 5-ft tube). For large thermal expansions, expansion joints of the bellows type are normally used.

To provide for thermal expansion when the shell of a heat exchanger operates at an elevated temperature, only one "foot" or "leg" of the shell is anchored to the foundation and the other is designed to permit free axial movement so that the axial expansion of the shell is not resisted. Freedom for axial movement may be provided by either a sliding foot (slotted holes with ferrules) or a slender leg (in the case of heavy heat exchangers), which is

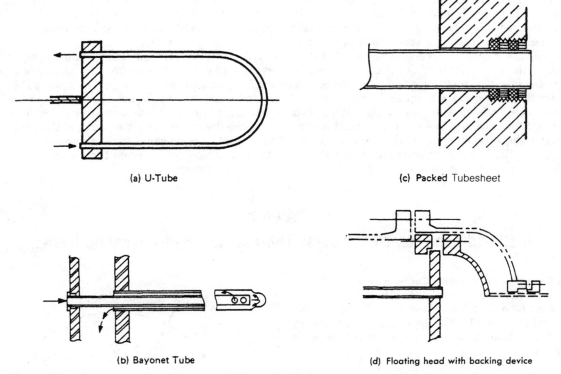

(a) U-Tube

(c) Packed Tubesheet

(b) Bayonet Tube

(d) Floating head with backing device

Fig. 10   Thermal expansion provisions

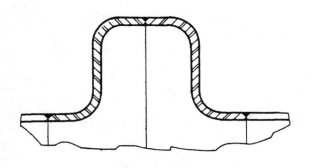

(a) Shell expansion joint using flanged and flued heads

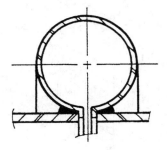

(b) Split pipe shell expansion joint

Fig. 11   Shell expansion joints

designed to be laterally flexible and thereby accommodate axial movements at low levels of stress.

2.5   Cathodic Protection.   Cathodic protection, which is usually in the form of sacrificial zinc anodes, is used in seawater-cooled heat exchangers to protect headers, tube-sheets, and tubes from galvanic corrosion by electrolytic action. When dissimilar metals that are connected together are immersed in seawater, which is an excellent electrolyte, a simple galvanic cell is formed; and an electric current flows from one metal to the other through the seawater, which completes the circuit between the two dissimilar metals. The metal higher on the activity scale (the anode), from which the current flows, will tend to suffer rapid corrosion, and the metal (cathode) to which the current flows will tend to be protected from galvanic corrosion. The direction in which the current flows depends on the composition of the metals or alloys exposed to the electrolyte and also on the hardness of the metal, the cleanliness of the metal surfaces, and other factors. Thus, if a single metal is immersed in an electrolyte and one part of the metal surface is harder than another part, or cleaner than another part, there will be a flow of current from one part to the other and galvanic corrosion will take place. If several different metals or alloys are involved, current will flow in varying proportions between the surfaces exposed to the electrolyte.

If a plate of clean metallic zinc is properly arranged within a heat exchanger waterbox, a current will tend to flow from the zinc to the adjacent metal surfaces exposed to the seawater, which constitutes the electrolyte of the galvanic cell. The zinc sacrificial anode is corroded as the current is generated; and the current flowing through the seawater to the metal surfaces of the heat exchanger tubes, tubesheets, and waterbox tends to protect these parts from galvanic corrosion. The electric circuit is completed through the metal parts of the heater exchanger. The gaskets between the waterbox and manhole covers, to which sacrificial anodes are frequently attached, and the gaskets between the tubesheets and the water chests do not interrupt the flow of current, as the circuit is completed through the metallic bolts and collar studs that secure the joints. The Navy usually requires that the exposed surface of zincs (exclusive of edges) be at least one

square foot for each 1000 ft² of heat transfer surface. For naval heat exchangers, the amount of zinc anodes required is based on equations that are given in reference 7.

All zinc protectors should be thoroughly scaled once every 4 to 6 weeks to assure that an active metallic zinc surface, as opposed to corrosion scale adhering to the metal, is exposed to the seawater. Zinc sacrificial anodes that afford proper protection are quickly deteriorated.

In the design of the waterbox and the arrangement of sacrificial anodes, care must be exercised that the zincs do not interfere with, or add turbulence to, the fluid flow within the waterbox.

Section 3
Shell-and-Tube Heat Exchanger Thermal and Hydrodynamic Design

3.1 The Overall Heat Transfer Coefficient. The fundamental theory dealing with heat transfer was discussed in Chapter 2, and a discussion of heat transfer by condensation is given in Chapter 15. The application of this theory to the transfer of heat through the walls of a tube leads to the equation

$$Q = \frac{A_o(T - t)}{\Sigma r} \quad (1)$$

where

Q = heat transferred
T = temperature of hot fluid
t = temperature of cold fluid
A_o = tube outside surface area
$r = r_{io} + r_{dio} + r_w + r_{do} + r_o$
r_{io} = resistance across fluid film on inside of tube
r_{dio} = resistance of deposit or scale on tube inside wall
r_w = resistance of tube wall metal
r_{do} = resistance of deposit or scale on tube outside wall
r_o = resistance of fluid film on outside of tube

These quantities are illustrated by Fig. 12, and the corresponding temperature gradients across a tube are shown in Fig. 13. It may be noted that "new clean-tube oxide film resistances" have not been included on the inside and outside of the tubes as they are not of sufficient magnitude to warrant consideration in the design of viscous fluid or water-to-water heat exchangers.

The film resistances in equation (1) are further defined as

$$r_{io} = \frac{1}{h_i A_i / A_o} = \frac{1}{h_{io}} = \frac{1}{\text{tube-side film coefficient}} \quad (2)$$

$$r_o = \frac{1}{h_o} = \frac{1}{\text{shell-side film coefficient}} \quad (3)$$

A_o and A_i are the tube outside and inside surface areas

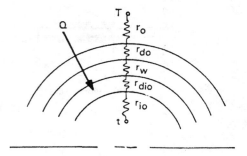

Fig. 12 Heat transfer through a tube

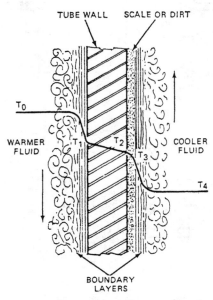

Fig. 13 Temperature gradients across a tube

respectively. The terms h_i and h_o are called specific conductances or heat transfer film coefficients and are generally a function of flow velocities and fluid properties.

The scale resistance or fouling resistance terms, r_{dio} and r_{do}, do not lend themselves to analytical assessment. Two means have been commonly employed to allow for the additional thermal resistance when scale or deposit begins to accumulate on the tube wall. The most rigorous means is to assign values to the tube inside and outside fouling resistances, r_{dio} and r_{do}, as a function of fluid type, temperature, velocity, etc., and compute the required heat transfer surface accordingly. In applications with mild rates of fouling, however, the fouling resistances are difficult to separate and evaluate meaningfully, and furthermore, they are small. Consequently, the "clean-tube factor" (also known as the "cleanliness factor" or "service factor") concept has found general acceptance in conventional marine heat exchanger applications. The clean-tube factor is a number, less than one, by which the clean overall heat transfer coefficient, U, is multiplied to allow for fouling.

The resistance of the tube wall metal, r_w, is readily analyzed and it can be quantified without difficulty as follows:

$$r_w = \frac{d_o}{2k}\left(\log_e \frac{d_o}{d_i}\right) \qquad (4)$$

where

d_o = tube outside diameter
d_i = tube inside diameter
k = tube metal thermal conductivity

The overall heat transfer coefficient, U, which is defined as the reciprocal of $\Sigma\, r$, is a convenient means of expressing the capability of transferring heat through tubes. The overall heat transfer coefficient can be stated as

$$U = \frac{1}{(1/h_{io}) + r_{dio} + r_w + r_{do} + (1/h_o)} \qquad (5)$$

3.2 Mean Temperature Difference. In the case of shell-and-tube heat exchangers, the temperature difference $(T - t)$ varies from location to location within the exchanger and, consequently, so does the rate of heat transfer, Q. Therefore, calculations are based on the average temperature difference across the tube over the entire tube length, which is defined as the log mean temperature difference (LMTD) and is expressed mathematically as

$$\text{LMTD} = \frac{1}{A_o}\int_0^{A_o}(T - t)dA \qquad (6)$$

By combining equation (6) with equation (1) and noting that $\Sigma\, r$ is equal to the reciprocal of U, the heat transfer through a tube is determined to be

$$Q = UA_o(\text{LMTD}) \qquad (7)$$

Equation (7) is the basic analytical tool employed to establish the thermal design of a heat exchanger. The factors

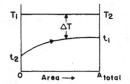

(a) Single-pass condenser

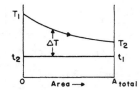

(b) Single-pass evaporator

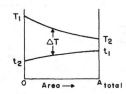

(c) Single-pass parallel-flow heat exchanger

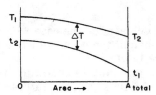

(d) Single-pass counterflow heat exchanger

Fig. 14   Axial temperature distribution in a heat exchanger

to be considered in assessing this equation are briefly discussed in the following.

With known terminal temperature differences between the shell and tube streams, the mean temperature difference between the shell and tube flows can be derived [8] if the following assumptions are made:

• The overall heat transfer coefficient is constant along the entire flow path.

• The flow rate and specific heat of both the shell and tube streams are constant.

• The heat transfer surface is uniformly distributed along the flow paths.

• The temperature of either fluid is constant over any cross section of its path (i.e., there is complete mixing and no stratification).

• There is no internal leakage or bypassing of fluid around the tube bundle.

• There is no transfer of heat between the heat exchanger and its surroundings.

Figure 14 illustrates the changes in temperature that may occur in either or both fluids in a simple shell-and-tube exchanger such as illustrated by Fig. 1. The distances between the solid lines are proportional to the temperature differences, ΔT, between the two fluids.

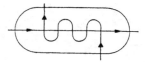

(a) Single-pass counterflow

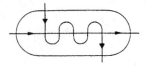

(b) Single-pass parallel flow

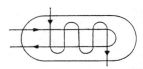

(c) One shell pass & multiple
of two tube passes

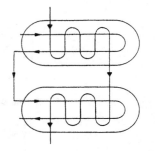

(d) Two single-shell & double-tube pass
exchangers in series

Fig. 15   Flow arrangements

Figure 14(a) illustrates the case where a vapor is condensed at a constant temperature while the other fluid is heated. Figure 14(b) represents the case where a liquid is evaporated at constant temperature while heat flows from a warmer fluid at a temperature that decreases as it passes through the heat exchanger. For both of these cases, the direction of the flow (countercurrent or co-current) of either fluid is immaterial. Figure 14(c) represents conditions in a parallel-flow exchanger, and Fig. 14(d) applies to counterflow. No change of phase occurs in the last two cases. An inspection of Fig. 14(c) shows that the final temperature of the colder fluid can never reach the exit temperature of the hotter fluid regardless of the length of the heat exchanger. For counterflow, on the other hand, the final temperature of the cooler fluid can exceed the outlet temperature of the hotter fluid, since a favorable temperature gradient exists all along the heat exchanger. An additional advantage of the counterflow arrangement is that, for a given rate of heat transfer, less surface area is required than for a parallel-flow arrangement.

For a single-pass counterflow heat exchanger such as illustrated by Fig. 15(a), the LMTD is

$$\text{LMTD} = \frac{(T_1 - t_2) - (T_2 - t_1)}{\log_e\left(\dfrac{T_1 - t_2}{T_2 - t_1}\right)} \tag{8}$$

where

T_1 = hot fluid inlet temperature
T_2 = hot fluid outlet temperature
t_1 = cold fluid inlet temperature
t_2 = cold fluid outlet temperature

For the special case that

$$R = \frac{T_1 - t_2}{T_2 - t_1} = 1 \tag{9}$$

then equation (8) reduces to

$$\text{LMTD} = T_1 - t_2, \text{ or } T_2 - t_1 \tag{10}$$

For such a single-pass parallel-flow heat exchanger as illustrated in Fig. 15(b), the LMTD is

$$\text{LMTD} = \frac{(T_1 - t_1) - (T_2 - t_2)}{\log_e\left(\dfrac{T_1 - t_1}{T_2 - t_2}\right)} \tag{11}$$

For one shell pass and multiples of two tube-pass heat exchangers as shown in Fig. 15(c), the LMTD becomes

$$\text{LMTD} = \frac{X}{\log_e\left(\dfrac{Y + X}{Y - X}\right)} \tag{12}$$

where

$$X = [(T_1 - T_2)^2 + (t_2 - t_1)^2]^{1/2}$$
$$Y = T_1 + T_2 - (t_2 + t_1)$$

Single-shell heat exchangers with even numbers of tube passes may be arranged in series. With identical heat exchangers arranged in series, as shown by Fig. 15(d), the LMTD is computed as

$$\text{LMTD} = \frac{X}{\log_e\left(\dfrac{Z + X}{Z - X}\right)} \tag{13}$$

where

$$Z = (2/P - 1)(T_1 - T_2) - (t_2 - t_1)$$
$$R = \frac{t_2 - t_1}{T_1 - T_2}$$
$$P = \frac{T_1 - T_2}{m(T_1 - t_1) - (T_1 - T_2)(m - 1)}, \text{ if } R = 1$$
$$P = \frac{1 - \left(\dfrac{T_1 - t_2}{T_2 - t_1}\right)^{1/m}}{R - \left(\dfrac{T_1 - t_2}{T_2 - t_1}\right)^{1/m}}, \text{ if } R \neq 1$$

m = number of identical single-shell heat exchangers with an even number of tube passes in series

In most instances, the design of a shell-and-tube heat exchanger involves determining the heat transfer surface area and geometry for a known heat transfer duty and with known operating temperatures. For this condition, equations (5), (7), and (8) are used. Conditions occur where the duty and terminal temperatures are required for a given geometry. The following equations are applicable for a single-pass counterflow heat exchanger as shown in Fig. 15(a).

U = overall heat transfer coefficient, equation (5)
A_o = tube outside surface area
W_h = hot fluid flow rate
W_c = cold fluid flow rate
C_{ph} = hot fluid specific heat
C_{pc} = cold fluid specific heat
T_1 = hot fluid inlet temperature
t_1 = cold fluid inlet temperature

$$W_c C_{pc}(t_2 - t_1) = W_h C_{ph}(T_1 - T_2) \tag{14}$$

By combining equations (5), (7), (8), and (14):

$$R = \left(\frac{W_c C_{pc}}{W_h C_{ph}}\right) = \left(\frac{T_1 - T_2}{t_2 - t_1}\right) \tag{15}$$

$$E = e^{(UA_0/W_c C_{pc})(R-1)} \tag{16}$$

$$T_2 = \frac{T_1(1 - R) + t_1 R(1 - E)}{1 - RE} \tag{17}$$

$$t_2 = t_1 + (T_1 - T_2)/R \tag{18}$$

$$Q = W_h C_{ph}(T_1 - T_2)$$

For an exchanger with a single-pass shell and multiple-of-two tube passes, as shown in Fig. 15(c), the following equations are applicable:

$$R = \left(\frac{W_c C_{pc}}{W_h C_{ph}}\right) = \left(\frac{T_1 - T_2}{t_2 - t_1}\right) \tag{19}$$

$$B = [R^2 + 1]^{\frac{1}{2}} \tag{20}$$

$$F = e^{(UA_0 B/W_c C_{pc})} \tag{21}$$

$$n = \frac{2}{1 + R + \left(\dfrac{F + 1}{F - 1}\right)^B} \tag{22}$$

$$T_2 = T_1 - Rn(T_1 - t_1) \tag{23}$$

$$t_2 = t_1 + (T_1 - T_2)/R \tag{24}$$

$$Q = W_h C_{ph}(T_1 - T_2)$$

3.3 Sensible Heat Transfer Coefficient in a Tube. The tube-side film coefficient, h_i, may be computed from the Sieder and Tate equation [9]. This equation was published in 1936 and remains in widespread use; it has the following form:

$$Nu = C(Re)^{0.8} (Pr)^{0.33} \left(\frac{Z}{Z_w}\right)^{0.14} \tag{25}$$

where

Nu = Nusselt number = $(h_i d_i)/k$
h_i = tube-side film coefficient
d_i = tube inside diameter
k = thermal conductivity of fluid on tube side
C = 0.027
Re = Reynolds numbers = $(d_i G)/Z$
G = tube-side mass velocity
Z = tube-side fluid viscosity at bulk temperature
Z_w = tube-side fluid viscosity at tube wall temperature
Pr = Prandtl number = $(C_p Z)/k$

Table 2 Approximate range of convective heat transfer coefficients

Conditions	Coefficient, h Btu/hr-ft²-deg F
Free convection air	1 to 5
Forced convection air or superheated steam	5 to 50
Forced convection oil	10 to 300
Forced convection water	50 to 2000
Boiling water	500 to 10,000
Condensing steam	1000 to 20,000

C_p = specific heat
L = tube length

Equation (25) is valid for the following conditions

$$0.5 < Pr < 120$$

$$2300 < Re < 1 \times 10^7$$

$$L/d_i > 50$$

To evaluate the fluid viscosity at the tube wall, the tube wall temperature must be calculated. However, the tube wall temperature is a function of the heat transfer coefficient, and the heat transfer coefficient is a function of the tube wall temperature. The evaluation of the tube wall temperature and the heat transfer coefficient is a trial-and-error process with the consequent calculated value compared to the assumed value. The equations for calculating the tube wall temperatures are given below; the procedure is based upon the average bulk fluid temperature:

$$T_{wi} = \left(\frac{T_{i1} + T_{i2}}{2}\right) + \left[\frac{\dfrac{A_o}{A_i}\left(\dfrac{1}{h_i}\right)}{\dfrac{A_o}{A_i}\left(\dfrac{1}{h_i}\right) + \dfrac{d_o}{2k}\log_e\dfrac{d_o}{d_i} + \left(\dfrac{1}{h_o}\right)}\right] \times \left[\left(\frac{T_{o1} + T_{o2}}{2}\right) - \left(\frac{T_{i1} + T_{i2}}{2}\right)\right] \tag{26}$$

$$T_{wo} = \left(\frac{T_{o1} + T_{o2}}{2}\right) + \left[\frac{(1/h_o)}{\dfrac{A_o}{A_i}\left(\dfrac{1}{h_i}\right) + \dfrac{d_o}{2k}\log_e\dfrac{d_o}{d_i} + \left(\dfrac{1}{h_o}\right)}\right] \times \left[\left(\frac{T_{i1} + T_{i2}}{2}\right) - \left(\frac{T_{o1} + T_{o2}}{2}\right)\right] \tag{27}$$

where

T_{wo} = temperature of tube wall outside surface
T_{wi} = temperature of tube wall inside surface
A_i = tube inside surface area
T_{o1} = inlet shell temperature
T_{o2} = outlet shell temperature
T_{i1} = inlet tube temperature
T_{i2} = outlet tube temperature

Table 2 lists the order of magnitude of the average convective heat transfer coefficient encountered in the thermal design of heat exchangers.

3.4 Sensible Heat Transfer Coefficient Across a Bank of Bare Tubes. The relationship used to assess the shell-side film coefficient is similar in form to that used on the tube side and is as follows:

$$\frac{h_o d_o}{k} = C\left(\frac{d_o G}{Z}\right)^m (Pr)^{0.33} \left(\frac{Z}{Z_w}\right)^{0.14} \qquad (28)$$

where

h_o = shell-side film coefficient
d_o = tube outside diameter
C, m = constants

The remaining terms are as defined for equation (25).

There are some variations in the techniques used to apply equation (28); the variations primarily deal with the quantification of C and m. For calculations regarding the flow of a fluid normal to a tube bank (i.e., a cross-flow heat exchanger), Colburn recommended in 1933 that equation (28) be applied as follows:

$m = 0.6$
Z and Z_w are assumed equal
G = mass velocity of a fluid through minimum flow cross-sectional area of tube bank
$C = 0.33$ for a staggered tube pattern (triangular or rotated-square pitch)
$C = 0.25$ for an in-line tube pattern (square pitch)

When making computations for the flow of a fluid through a circular baffled heat exchanger, Donahue recommends that equation (28) be applied with the following assumptions [10]

$m = 0.6$
$G = (G_c G_w)^{1/2}$
G_c = mass velocity normal to the tube bundle
G_w = mass velocity through cross baffle window; i.e., baffle cutout area, Fig. 8(a).

The geometrical mean value of the mass velocity is used in recognition of the fact that the direction of fluid flow is not normal to the tubes in the baffle window.

On the basis of test data with segmental baffled heat exchangers, Donahue further recommends that C be given the value of 0.22 as an average value. This value for C provides a considerable margin (as compared with a value of 0.33) to allow for flow leakages as would occur between cross baffles and the shell, and should be used only as an average value.

The term "ideal tube bank" came from a report that was published as a result of a research program on shell-and-tube heat exchangers conducted at the University of Delaware [11]. The research program was supported by various heat exchanger manufacturers and continued for more than 12 years. An "ideal tube bank" came to mean a rectangular tube field with straight-through flow and no wall effects or bypassing. The results from these tests can be presented in the form of equation (28) if the constants C and m are replaced by functions of the tube pitch, p, the tube diameter, d_o, and the Reynolds number, $(d_o G/Z)$. In the lower Reynolds number range, the following equation is applicable:

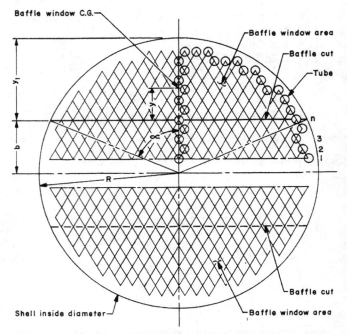

Fig. 16   Baffle geometry definition

$$C = b(1.33 d_o/p)^n$$

where $\qquad (29)$

$$n = \frac{1.45}{1 + 0.14(Re)^{0.519}}$$

Values for b and m are as follows:

Reynolds No.	0–10	10–100	100–1000
b	1.4	1.36	0.593
m	0.333	0.343	0.533

For Reynolds numbers above 1000, C and m are taken as constants having the values of 0.32 and 0.612, respectively.

It may be seen that computations for the shell-side film coefficient are tedious and time-consuming. A particularly troublesome aspect of the calculations is the assessment of the effective flow area for segmental baffled flow.

The mass flow rate, G_c, is defined as the average mass flow rate between the baffle cuts of a pair of baffles. Figure 16 provides a description of the terms that are used to describe baffle geometry. The procedure used to calculate the mass flow rate is as follows:

$$G_c = \frac{1}{N_{tb}} \sum_{i=1}^{n} \frac{W N_i}{(CD_i - N_i d_o) B_s} \qquad (30)$$

where

W = shell-side flow rate
CD_i = chord distance at row i
i = tube row at baffle cut
N_i = number of tubes in row i

$$N_{tb} = \Sigma N_i$$
$$B_s = \text{baffle spacing}$$

In a commercial shell-and-tube heat exchanger, fluid in the shell bypasses the tube bundle and may not be cooled or heated because of the manufacturing clearances between the baffle and shell, the tubes and the baffle holes, and the bundle and shell, and because of short-circuit flow along the pass partition lane. It is essential that these non-ideal conditions be accounted for, especially in cases where uncertainties in the heat transfer coefficient and flow pressure drop predictions must be minimized. Tinker [12] and Bell [11] each proposed simplified methods to correlate these complex flow phenomena to geometrical factors. Their basic approach was to divide the shell flow area into three categories, namely: cross-flow area (the flow area between tubes in the direction normal to the tube axis); leakage area (the flow area between the tube and tube holes in the cross-flow baffles and the area between the baffle outside diameter and the shell inside diameter); and bypass area (the flow area between the tube bundle and shell). Bell proposed an overall approach; with known relative values of bypass and leakage area to that of cross-flow, he derived empirical correction factors to be applied to the ideal tube bank heat transfer coefficient and pressure drop values. Tinker assigned flow resistance constants to each flow area and calculated an effective cross-flow rate to be used in the ideal tube bank cross-flow heat transfer and pressure drop correlations. Tinker's approach is more useful in that refinements and generalizations can be incorporated into his method.

3.5 Single-Phase Flow Pressure Drop.

As noted in the previous discussion, the film coefficient (and consequently the heat transfer rate) increases with a higher flow velocity; but on the other hand, the flow pressure drop due to friction increases in proportion to the velocity raised to a power that is between one and two. Consequently, a design trade-off must be made that entails striking a compromise between a high film coefficient and a high pressure drop. The design of a heat exchanger becomes a matter of balancing the cost of the heat transfer surface installed with the cost of the pumping power required to service the exchangers. As a result, a major consideration in the design of an exchanger is an evaluation of flow pressure drops for various operating conditions and exchanger configurations.

In practice, circumstances often require the system designer to specify the allowed pressure drop for a heat exchanger without an accurate understanding of the impact of the pressure drop limitation on the design of the heat exchanger. The courses of action that are frequently taken when it is necessary to specify the pressure drop permitted with a heat exchanger are as follows:

• Allow a uniform 10-psi pressure drop per stream per heat exchanger. This rule-of-thumb approach is commonly used in some parts of the industry. The 10-psi limit is selected because higher pressure drops with consequent higher velocities may approach the erosion/corrosion limit of the metal; furthermore, the higher heat transfer coefficients, which would be associated with higher pressure

drops, may be excessively sensitive to scaling. The 10-psi maximum allowable pressure drop is also high enough to keep the thermal designer from "tail-chasing" (i.e., the lower the velocity, the lower the heat transfer coefficient; the lower the heat transfer coefficient, the larger the heat exchanger; the larger the heat exchanger, the larger the pressure drop, unless the velocity is reduced; etc.).

• Specify alternative pressure drop limitations. With the wide use of computers to perform technical evaluations of heat exchangers, an array of alternative designs can feasibly be analyzed such that the optimum design can be determined and selected.

• Develop preliminary designs for several alternative heat exchangers and evaluate the size of the exchanger as a function of the pressure drop before establishing the design criteria.

Tube-side pressure drop. The tube-side pressure drop includes the pressure loss from the inlet nozzle to the outlet nozzle on the tube side of the unit. It encompasses the pressure loss in the nozzles, pressure losses due to flow acceleration, deceleration, and changes in direction, and the frictional pressure loss due to fluid flowing in the tubes. The total tube-side pressure drop, ΔP_t, is calculated using

$$\Delta P_t = 4f\left(\frac{L}{d_i}\right)\left(\frac{G^2}{2g_c\rho}\right)\left(\frac{Z_w}{Z}\right)^{0.14} + \Sigma\, K_x\left(\frac{G^2}{2g_c\rho}\right) \quad (31)$$

where

f = friction factor
L = tube length
ρ = fluid density
G = mass velocity
K_x = pressure loss coefficient due to flow area expansion and contraction and flow redirection
g_c = acceleration of gravity

Shell-side pressure drop. The shell-side pressure drop includes the pressure loss from the shell inlet nozzle to the shell outlet nozzle. It encompasses the losses in the nozzles, bundle entrance and exit losses due to flow acceleration, deceleration, and changes in direction, and pressure losses in the tube bundle. The pressure loss in the tube bundle includes the pressure loss in the baffle window and the pressure loss due to fluid flowing normal to the tubes. The total shell-side pressure drop, ΔP_s, is calculated using:

$$\Delta P_s = \frac{4fG_c^2 N_r(N_b + 1)}{2g_c\rho}\left(\frac{Z_w}{Z}\right)^{0.14}$$
$$+ \frac{K_w G_w^2 N_b}{2g_c\rho} + \Sigma\, K_x\left(\frac{G^2}{2g_c\rho}\right) \quad (32)$$

where

G_c = mass velocity normal to tube bundle
G_w = mass velocity through cross-baffle window
K_w = pressure loss coefficient due to baffle window acting as an orifice
K_x = pressure loss coefficients due to flow area expansion and contraction and flow redirection

N_b = tube length/baffle spacing $-$ 1

n_r = number of rows in cross flow, which is taken to be the number of tube rows between the centroids of adjacent baffle windows. This can be determined from a layout of the tube bundle

p = tube pitch

$$f = \frac{15}{\left(\dfrac{p - d_o}{d_o}\right)\left(\dfrac{G_c d_o}{Z}\right)} \qquad \text{for laminar flow [10]}$$

$$f = \frac{0.75}{\left[\left(\dfrac{p - d_o}{d_o}\right)\dfrac{G_c d_o}{Z}\right]^{0.2}} \qquad \text{for turbulent flow [10]}$$

The methods used to assess the heat transfer and pressure drop within heat exchangers are basically those of Tinker [12], Devore [13], and Donahue [10]. Details of the derivations and assumptions of the methods have not been included as they are detailed in the original references. The process of going through the procedure outlined will, however, convey an appreciation of the considerations involved in the thermal and hydrodynamic design of a shell-and-tube heat exchanger.

3.6 Design of Fin-Tube Heat Exchangers. Another form of the basic heat transfer relationship given as equation (1) can be written as

$$Q = \frac{T - t}{\dfrac{1}{h_i A_i} + R_{do} + R_w + R_o + \dfrac{1}{h_o A_o}} \qquad (33)$$

where R_{do} and R_o are the scale resistances on the inside and outside of the tubes, respectively, R_w is the resistance of the tube, and the remaining symbols are as previously defined. If the shell-side film coefficient, h_o, is small (as is the case with gases and oils that have low thermal conductivities) in comparison with the tube-side film coefficient, h_i, the shell-side resistance, $(h_o A_o)^{-1}$, will control or "bottleneck" the heat flow. A means of counteracting this circumstance is to increase the tube outside surface area, A_o, to the extent that $h_o A_o = h_i A_i$, in which case the heat flow would no longer be controlled by the shell-side heat resistance. The heat transfer surface on the outside of the tubes can be effectively increased by providing extended surfaces, such as external fins, on the tubes.

A type of finned tube frequently used in shell-and-tube heat exchangers is the so-called "low-fin" tube, which has a ratio of finned surface to tube outside surface generally in the range of 3 to 4. In applications where the tube outside film coefficient is as small as 50% of the tube inside film coefficient, the choice of low-fin tubes can provide a more economical, compact, and lightweight unit when compared with a similar design containing bare tubes. In cases where the tube-side film coefficient is low, the use of tubes with internal fins may be appropriate.

Fouling of finned tubes is a concern. In this regard, tests were conducted with 160 F, No. 5 fuel oil flowing through the shell side of a heat exchanger [14]. The test results show that the resistance to heat transfer increases rapidly initially, but the rate of increase subsequently drops to a very small value. The tests also indicated that the fouling characteristics of finned tubes are similar to those of plain tubes, and both types of tubes have comparable percentages of reduction in performance. At the end of the four-week test period, the fouling resistances for both fin-tube and plain-tube bundles were below the value of 0.005. During the tests, both the finned and plain tubes were cleaned by a kerosene-water-detergent emulsion and the heat transfer rate was restored to that initially achieved.

The thermal and hydrodynamic design procedures developed in Sections 3.1 through 3.5 are applicable to the design of fin-tube exchangers when allowances are made for the effect of the fins on the heat transfer and pressure drop calculations. The definition of the overall heat transfer coefficient for fin-tubes is the same as for bare tubes except that an efficiency factor is added to the equation when fins are present on the outside or inside surface of the tube. In the general case of tubes with internal and external fins, the equation for the overall heat transfer coefficient has the following form:

$$U = \frac{1}{\dfrac{1}{h_i}\left(\dfrac{A_o}{A_i}\right)\dfrac{1}{E_{ti}} + r_{dio}\left(\dfrac{A_o}{A_i}\right) + r_w + \left[\left(\dfrac{1}{h_o}\right) + r_{do}\right]\dfrac{1}{E_{to}}} \qquad (34)$$

where

E_{to} = total efficiency for outer surface
     = $[A_o - A_{fo}(1 - E_{fo})]/A_o$
E_{ti} = total efficiency for inner surface
     = $[A_i - A_{fi}(1 - E_{fi})]/A_i$
A_o = total outside heat transfer surface area
A_{fo} = heat transfer surface of the external fin
E_{fo} = fin efficiency of the external fin
A_i = total inside heat transfer surface area
A_{fi} = heat transfer surface of internal fins
E_{fi} = fin efficiency of internal fin

Other terms are as previously defined. For fin-tubes with fins only on the outside, the total efficiency for the inner surface, E_{ti}, is unity.

There is an alternative method for calculating the effect of fins that uses a fin resistance in place of a fin efficiency. Using this method the equation reduces to the following form for fins on the outside tube surface:

$$U = \frac{1}{\dfrac{1}{h_{io}} + r_{dio} + r_w + r_f + r_{do} + \dfrac{1}{h_o}} \qquad (35)$$

The terms in this equation are as previously defined except for the fin resistance, r_f, which is defined by Fig. 17. The fin efficiency, upon which the resistances shown in Fig. 17 are based, was taken from reference 15.

The procedures used to calculate the heat transfer coefficient in the bundle and the pressure drop in the bundle for fin-tube arrangements are the same as described in Sections 3.4 and 3.5, except that the shell-side friction factor calculated in Section 3.5 is corrected for the low-fin profile by using the correction factor, C_f, in Table 3.

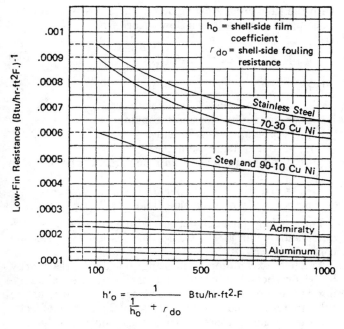

$$h'_o = \cfrac{1}{\cfrac{1}{h_o} + r_{do}} \quad \text{Btu/hr-ft}^2\text{-F}$$

Fig. 17 Fin resistance of low-fin tubes

Table 3 Correction to shell-side friction factor for low-fin tubes

Ratio: $\cfrac{\text{Tube External-Surface, } A_o}{\text{External Bare Tube Area, } A}$	Friction Correction Factor, C_f
1	1.00
2	1.04
3	1.15
4	1.31
5	1.50
6	1.70

3.7 Flow-Induced Vibration. In recent years there has been a trend to increase the shell-side flow velocities as a means of increasing the shell-side heat transfer coefficient and also reducing fouling. As the velocities have increased, the importance of designing for flow-induced vibration has likewise increased.

In a shell-and-tube heat exchanger, the tubes are susceptible to vibration because of fluid flowing around the tubes. In most cases the amplitude of tube vibration is so low that tube vibration is of no consequence. However, tube vibration can be a major problem. When the amplitude of vibration is large, mechanical failures of the following types can be expected: tube collision damage, baffle damage, fatigue damage, and tube-to-tubesheet joint damage.

Collision damage is caused by the tubes repeatedly contacting the adjacent tubes or the shell wall; this usually occurs near the tube midspan between baffles where the vibration amplitudes are largest. At the point of collision, the tube is eventually flattened to an oval shape, and in time the tube wall wears thin and fails.

In baffle damage, the tube chafes against the baffle and causes the baffle to abrade the tube until eventually leakage occurs.

Fatigue damage is the result of a tube mechanically failing because the alternating stress due to the amplitude of vibration exceeds the fatigue limit of the tube material.

In the case of a tube-to-tubesheet joint failure, the tube either fails at the back face of the tubesheet due to the tubesheet chafing and abrading the tube, or it fails at the joint as a result of relative movement between the tube and tubesheet such that leakage develops.

The flow of a fluid over the tube of a shell-and-tube heat exchanger is the major source of flow-induced vibration. Several mechanisms can be involved with flow-induced vibration: vortex shedding, turbulent buffeting (random excitation due to flow turbulence), and fluidelastic whirling.

The design characteristics of a heat exchanger that are conducive to vibration are long unsupported tube spans and high local flow velocities. Care must be exercised when positioning the tie rods, seal strips, and impingement plates in the bundle; these items can present flow obstructions that cause high local velocities and flow perturbations, which can initiate tube vibration. Objectionable flow-induced vibration has been experienced in the bend region of a U-tube, at the nozzle entrance and exit areas, and at the tube midspan between baffles.

To evaluate the potential of flow-induced vibration in a tube bundle, the shell-side fluid velocity must be calculated. The flow patterns present in the shell-side fluid introduce many uncertainties into the calculation of local fluid velocities. The velocity varies from row to row as well as from baffle to baffle because of factors such as baffle clearances and fluid bypassing; consequently, the velocity used to evaluate flow-induced vibration is taken to be an effective velocity across the tube row of interest.

Objectionable flow-induced vibration can be avoided in a shell-and-tube heat exchanger provided that technical evaluations are made and appropriate preventive design modifications are initiated during the early design stages. The factors affecting flow-induced vibration are the tube diameter, tube material, tube thickness, fluid velocity, unsupported tube length, tube pitch, baffle thickness, tube hole size in the baffle, and the tube axial loading. The most effective precautions that can be taken to avoid flow-induced vibration are to either decrease the fluid velocity or increase the tube natural frequency. The fluid velocity can be decreased by increasing the tube pitch or by removing tubes in regions of high local velocity. The tube natural frequency can be increased by decreasing the unsupported tube length; this could be accomplished by changing the type of baffle from single segmental to double or triple segmental, but the preferred method is to use the "no tube in the window" baffle design [see Fig. 8(f)]. This guarantees that each tube is supported by every baffle. To further increase the tube natural frequency, tube support plates can be added between the baffles with no effect on performance.

If flow-induced vibration exists at the inlet or outlet nozzle, adding a support plate directly under the nozzle, adding an impingement plate or impingement bars under the nozzle, or moving the nozzle closer to the tubesheet will eliminate the problem.

Tube natural frequency. The natural frequency of a tube can be estimated by

$$f_n = K\left[\frac{EIg_c}{W_e L^4}\right]^{1/2} \qquad (36)$$

where

$K = 1.569$ frequency constant dimensionless for a simply supported tube
$g_c =$ gravitational constant
$I =$ moment of inertia of the tube
$E =$ modulus of elasticity of the tube material
$L =$ length of each unsupported span
$W_e =$ effective weight per unit length $W_{fi} + W_t + MW_{fo}$
$W_{fi} =$ weight of fluid in the tube per unit length
$W_t =$ weight of the tube per unit length
$W_{fo} =$ weight of fluid displaced by tube per unit length
$M =$ added mass coefficient for tube bundle array

For a tube-pitch to tube-outside-diameter ratio of 1.25, the added-mass coefficient is 2.9 [16].

The natural frequency is higher if the tube is subjected to an axial tensile force, and it is lowered if the tube is under a compressive force. The tube natural frequency, when corrected for preload, is

$$f_p = f_n\left[1 + \frac{PL^2}{EI\pi^2}\right]^{1/2} \qquad (37)$$

where f_p is the tube natural frequency in a preloaded condition and P the tube axial force (positive for tension and negative for compression).

Vortex shedding. Fluid flowing over a tube produces a series of vortices in the downstream wake as the flow separates from alternate sides of the tube. The alternate shedding of vortices produces a vibratory force on the tube, and the resulting amplitude of tube vibration can be substantial when the vortex shedding frequency matches the tube natural frequency. The vortex shedding frequency is expressed as

$$f_s = \frac{SV}{d_o} \qquad (38)$$

where

$f_s =$ vortex shedding frequency
$S =$ Strohal number, dimensionless
$d_o =$ tube outside diameter
$V =$ reference velocity across the tube row of interest

For a triangular pitch pattern with a tube-pitch to tube-outside-diameter ratio of 1.25, the Strohal number is 0.22 [17].

Fluidelastic whirling. During fluidelastic whirling, the tube vibrates in an orbital motion. This vibration is excited by the flow of a fluid over a tube and occurs when the lift and drag displacements of the tube are in tune with the tube cross-flow velocity. The expression for calculating the critical cross-flow velocity is

$$V_{crit} = Bf_n d_o\left(\frac{W_e\delta}{\rho_s d_o^2}\right)^{1/2} \qquad (39)$$

where

$V_{crit} =$ critical cross-flow velocity
$B =$ instability threshold constant
$f_n =$ tube natural frequency
$d_o =$ tube outside diameter
$\delta =$ log decrement of damping (for a tube in a still fluid)
$W_e =$ effective weight per unit length of tube
$\rho_s =$ density of shell-side fluid

For a triangular pitch pattern with a tube-pitch to tube-outside-diameter ratio of 1.25, the instability threshold constant is 3.3 [18].

In a heat exchanger, a vibrating tube dissipates energy by friction between the tube and baffle, by internal damping due to the hysteresis in the stress-strain curve of the tube material, and by the shell-side fluid drag effects. Because of the complex nature of the damping mechanism and the different combinations of tube materials, baffle materials, fluids, and manufacturing procedures, it is recommended that the log decrement of damping be determined through testing or by calculation using data derived from heat exchanger performance.

Acoustic vibration. Acoustic vibration can occur in a shell and tube heat exchanger when the shell-side fluid is a gas and when the tube pitch is square or rotated square. It does not occur when the shell-side fluid is a liquid because the velocity of sound in a liquid is so high. The acoustic frequency in a heat exchanger depends upon the shell diameter and the velocity of sound in the shell-side fluid. The frequency of acoustic vibration, f_a, can be calculated using the following expression:

$$f_a = \frac{483.2}{D}\left(\frac{P_s}{\rho_o}\right)^{1/2} \qquad (40)$$

where

$P_s =$ shell-side operating pressure, psia
$\rho_o =$ local shell-side fluid density, lb/ft^3
$D =$ characteristic dimension (usually shell diameter), ft

The acoustic frequency in a heat exchanger is excited by either vortex shedding or turbulent buffeting. A loud noise can be generated when the frequency of either vortex shedding or turbulent buffeting is within $\pm 20\%$ of the acoustic frequency. If the acoustic frequency coincides with the natural frequency of a tube, experience has shown that the tube will vibrate with a large amplitude and eventually fail. To eliminate acoustic vibration, a detuning baffle can be added parallel to the direction of flow; this will reduce the characteristic dimension of the exchanger and increase the acoustic frequency without affecting the heat transfer or pressure drop performance.

Turbulent buffeting. Turbulent buffeting is associated with the response of a tube to random turbulence as the fluid flows over the tube. Turbulent flow over a tube has a broad spectrum of frequencies associated with it, and the tube is excited by energy from this broad band of frequencies. The turbulent buffeting frequency is calculated as follows:

$$f_{tb} = \frac{12V}{d_o Y_\ell Y_t} \left[3.05 \left(1 - \frac{1}{Y_t} \right)^2 + 0.28 \right] \qquad (41)$$

where

f_{tb} = turbulent frequency, Hz
V = reference cross-flow velocity, ft/sec
d_o = tube outside diameter, in.
$Y_\ell = P_\ell/d_o$
P_ℓ = longitudinal pitch, in.
$Y_t = P_t/d_o$
P_t = transverse pitch, in.

3.8 Sample Problem. The calculation procedure, which follows, is based on the work of Donahue [10] and highlights the considerations involved in the thermal and hydrodynamic design of a shell-and-tube heat exchanger. In this illustration, a shell-and-tube heat exchanger will be designed with the following given information:

	Tube-Side Data (Seawater)	Shell-Side Data (Fresh water)
Flow, W_c (W_h), lb/hr	266,359	115,682
Temperature in, t_1 (T_1), deg F	85.0	111.2
Temperature out, t_2 (T_2), deg F	92.5	94.6
Duty, Q, Btu/hr	1,919,000	1,919,000
Density, ρ, lb/ft³	63.75	61.85
Thermal conductivity, k, Btu/hr-ft-deg F	0.348	0.364
Specific heat, C_p, Btu/lb-deg F	0.9604	0.9985
Viscosity, Z, lb/ft-sec	5.68×10^{-4}	4.51×10^{-4}
Pressure drop allowable, psi	10	20
Clean-tube factor	0.9	...

Heat Exchanger Construction Details

Tube pitch, triangular pattern, in.	0.625
Tube material	titanium
Tube outside diameter, d_o, in.	0.5
Tube inside diameter, d_i, in.	0.43
Tube wall thermal conductivity (titanium), k_t, Btu/hr-ft-deg F	11.3

Heat exchanger preliminary geometry. As a first approximation, the heat transfer coefficient data listed in Table 2 will be used to determine the heat transfer surface and characteristic dimensions of the heat exchanger, such as tube length and shell diameter. This information will be used to perform the thermal and hydraulic calculations. The procedure is iterated until an optimum heat exchanger is designed.

Assume:

$$h_i = 1200 \text{ Btu/hr-ft}^2\text{-deg F}$$
$$h_o = 1200 \text{ Btu/hr-ft}^2\text{-deg F}$$
$$r_{dio} = r_{do} = 0$$

Since the outlet temperature of the fresh water is not required to be less than that of the seawater, the heat exchanger will be designed to have one shell pass and multiple tube passes [see Fig. 15(c)].

$$X = [(T_1 - T_2)^2 + (t_2 - t_1)^2]^{1/2} = 18.2$$
$$Y = (T_1 + T_2) - (t_2 + t_1) = 28.3$$
$$\text{LMTD} = X/[\log_e (Y + X)/(Y - X)] = 11.9 \text{ deg F} \qquad \text{[eq. (12)]}$$

Recognizing that $A_o/A_i = d_o/d_i$, the clean-tube heat transfer coefficient under the assumed conditions can be calculated from equation (5) as follows:

$$U_{\text{clean}} = \frac{1}{\dfrac{A_o}{A_i h_i} + r_{dio} + \dfrac{d_o}{2k_t} \log_e \dfrac{d_o}{d_i} + r_{do} + \dfrac{1}{h_o}}$$
$$= 481 \text{ Btu/hr-ft}^2\text{-deg F}$$

$$U_{\text{fouled}} = 0.9 U_{\text{clean}} = 433 \text{ Btu/hr-ft}^2\text{-deg F}$$

$$A_o = Q/[U(\text{LMTD})] = 372\text{-ft}^2 \qquad \text{[eq. (7)]}$$

The resulting temperature of the inside surface of the tube wall under the assumed conditions is:

$$T_{wi} = (t_1 + t_2)/2 + [A_o/(A_i h_i)][U_{\text{clean}}][(T_1 + T_2)/2 \qquad \text{[eq. (26)]}$$
$$- (t_1 + t_2)/2] = 95.3 \text{ F}$$

A tube-side velocity, V, of 6 ft/sec is assumed to determine the number of tubes, N_t, the number of tube passes, N_p, and the required effective tube length, L. Estimates of the required flow area, A'_f, through the tubes and the number of tubes, N'_t, are:

$$A'_f = W_c/\rho_i V = 27.8 \text{ in.}^2$$

$$N'_t = \frac{4A_f}{\pi d_i^2} = 191.4 \text{ tubes}$$

With the tube outside surface area, tube outside diameter, and number of tubes for one pass tentatively known, the required tube length can be related to the number of passes, N_p, as follows:

Number of passes N_p	Number of tubes N_t	Tube length L, ft
1	191	14.88
2	382	7.44
4	764	3.72

Based upon an evaluation of these alternatives from configuration, manufacturing, cost, and other perspectives, a tentative decision is made for the tube side of the heat exchanger to be a two-pass U-tube arrangement with an effective length of 7.00 ft and a total length of 7.25 ft.

Tube-side calculations. The tube-side convective heat transfer coefficient, h_i, can now be calculated to confirm that it is approximately equal to, or greater than, the value assumed. The flow area, A_f, based on the number of tubes, N_t, and the number of passes, N_p, is:

$$A_f = \frac{\pi}{4} d_i^2 \frac{N_t}{N_p} = 27.74 \text{ in.}^2$$

$$Z_w = 5.29 \times 10^{-4} \text{ lb/ft-sec} \quad \text{for a } T_{wi} = 95.3 \text{ F}$$

$$\left(\frac{Z}{Z_w}\right)^{0.14} \approx 1.0$$

$$G = W_c/A_f = 384.1 \text{ lb/ft}^2\text{-sec} \qquad \text{[eq. (25)]}$$

$$Pr = C_p Z/k = 5.64$$

$$Re = d_i G/Z = 24{,}232$$

$$h_i = 0.027 \frac{k}{d_i} (Re)^{0.8} (Pr)^{0.33} \left(\frac{Z}{Z_w}\right)^{0.14} = 1493 \text{ Btu/hr-ft}^2\text{-F}$$

The calculated value is larger than the assumed value of 1200; therefore, an adequate heat transfer capability is provided.

The tube-side pressure drop can be calculated by using equation (31) as follows:

$$\Delta P_t = 4f\left(\frac{L}{d_i}\right)\left(\frac{G^2}{2g_c\rho}\right)\left(\frac{Z_w}{Z}\right)^{0.14} + \Sigma K_x\left(\frac{G^2}{2g_c\rho}\right)$$

$f = 0.0063 @ R_e = 24{,}232$
$K_e = 1.1$ for nozzle expansion (5-in. nozzle diamter)
$K_c = 0.7$ for nozzle contraction (5-in. nozzle diameter)
$K_1 = 0.3$ for contraction at tube end
$K_2 = 0.6$ for expansion at tube end
$K_3 = 0.7$ for turn-around loss in bundle

The tube length, L, for a U-tube heat exchanger includes the straight length of the tube plus the length of the tube bend at the mean bend radius:

$$L = 186.4 \text{ in.}$$
$$G = 384.1 \text{ lb/ft}^2\text{-sec}$$
$$\Delta P_{t1} = 4f\left(\frac{L}{d_i}\right)\left(\frac{G^2}{2g_c\rho}\right)\left(\frac{Z_w}{Z}\right)^{0.14} = 392.6 \text{ lb/ft}^2$$

The flow area through the nozzles is 19.63 in.²; therefore, the nozzle mass velocity, G_n, is 542.8 lb/ft²-sec. Consequently, the nozzle expansion and contraction loss is

$$\Delta P_{t2} = (K_e + K_c)\left(\frac{G_n^2}{2g_c\rho}\right) = 129.2 \text{ lb/ft}^2$$

The tube end losses and turnaround losses are calculated as:

$$\Delta P_{t3} = (K_1 + K_2 + K_3)\left(\frac{G^2}{2g_c\rho}\right) = 57.5 \text{ lb/ft}^2$$
$$\Delta P_t = \Delta P_{t1} + \Delta P_{t2} + \Delta P_{t3} = 579.3 \text{ lb/ft}^2$$
$$\Delta P_t = 579.3/144 = 4.02 \text{ psi} < P_{\text{allowable}} = 10 \text{ psi}$$

Shell-side calculations. From a tube layout for a U-tube heat exchanger with a ⅝-in. triangular tube pattern and 382 tube holes, the inside diameter of the shell is found to be 15.25 in. A horizontal baffle cut that is located 29% of the shell diameter below the top of the shell is assumed. A baffle cut of 29% was selected so that the

mass velocity through the cross baffle window, G_w, approximately equals the mass velocity normal to the tube bundle, G_c. When the combination of G_w and G_c is a maximum, the mass velocity, G, and the shell-side film coefficient, h_o, are maximum values. The shell nozzles are on the same side of the exchanger; therefore, there is an even number of baffle spaces. Initially, it will be assumed that there will be 16 spaces, with each at 5.25 in.

Based on these initial assumptions, the shell-side convective heat transfer coefficient, h_o, will be calculated to confirm that it is approximately equal to, or greater than, the required value.

The local mass velocity for each row between the baffle cuts will be calculated and then weighted to establish the average cross-flow mass velocity, G_c. The tube bundle is symmetrical about the horizontal centerline; therefore, calculations will be made for only one half of the baffle.

Referring to Fig. 16, the baffle cut, y_1, is 29% of the shell diameter. To determine the row nearest the cut, the distance, b, is calculated as:

$$y_1 = (15.25)(0.29) = 4.42 \text{ in.}$$
$$b = 7.625 - y_1 = 3.2 \text{ in.}$$

With a calculated b value of 3.2 in., row 5 is nearest the baffle cut, and the actual value of b is 3.165 in.

The calculations required by equation (30) can be performed in tabular form as follows:

Row	Row Height, in.	CD_i, in.	N_i	d_o, in.	B_S, in.	W_h, lb/hr.	$\dfrac{W_h \times N_i}{(CD_i - N_i d_o)B_S}$
1	1.000	15.12	23	0.5	5.25	115,682	5,600
2	1.541	14.94	22	0.5	5.25	115,682	4,921
3	2.083	14.67	21	0.5	5.25	115,682	4,439
4	2.624	14.32	22	0.5	5.25	115,682	5,841
5	3.165	13.87	21	0.5	5.25	115,682	5,492
			$N_{tb} = \overline{109}$				$\overline{26{,}293}$

$$G_c = 241.2 \text{ lb/ft}^2\text{-sec.}$$

By referring to Fig. 16, the shell-side mass velocity in the baffle window can be calculated as follows:

$$\alpha = \cos^{-1} b/R = 65.5°$$

The center of gravity of the baffle window can be calculated as:

$$y = R[4\sin^3 \alpha/(6\alpha - 3\sin 2\alpha) - \cos\alpha] = 1.84 \text{ in.}$$

The baffle window flow area, A_w, can be calculated as:

$$A_w = \pi R^2\left(\frac{2\alpha}{360}\right) - \frac{b(CD_5)}{2} - \frac{\pi}{4} d_o^2\left(\frac{N_t}{2} - N_{tb}\right) = 28.42 \text{ in.}^2$$

$$G_w = W_h/A_w = 162.8 \text{ lb/ft}^2\text{-sec}$$

By using the technique outlined in Section 3.4, sufficient information is known to calculate the shell-side film coefficient:

$$h_o = \frac{k}{d_o} C\left(\frac{d_o G}{Z}\right)^m (Pr)^{0.33} \left(\frac{Z}{Z_w}\right)^{0.14} \qquad (28)$$

$$G = [G_c G_w]^{1/2} = 198.2 \text{ lb/ft}^2\text{-sec}$$

$$d_o G/Z = 18{,}311$$

$$Pr = C_p Z/k = 4.45$$

$C = 0.22$ (following Donahue's recommendation)

$m = 0.6$ (following Colburn's recommendation)

$h_o = 1136$ Btu/hr-ft²-deg F

This value is sufficiently close to the assumed value of 1200 to assure that the tentative design is adequate in this respect.

The pressure drop on the shell side of the heat exchanger must be calculated to confirm that the actual pressure drop predicted is less than the allowable value. Equation (32) can be used to estimate the shell-side pressure drop as follows:

$$\Delta P_s = \frac{4fG_c^2 N_r (N_b + 1)}{2g_c\rho}\left(\frac{Z_w}{Z}\right)^{0.14} + \frac{K_w G_w^2 N_b}{2g_c\rho} + \sum K_x\left(\frac{G^2}{2g_c\rho}\right) \quad (32)$$

$$f = \frac{0.75}{\left[\left(\dfrac{p - d_o}{d_o}\right)\dfrac{G_c d_o}{Z}\right]^{0.2}} = 0.1336$$

Using the tube geometry shown by Fig. 16, the baffle window centroid, y, and the tube pitch, p, the number of rows in cross flow can be calculated as:

$$N_r = 2\,(5 + y/p) = 16$$

$$N_b = 7(12)/5.25 - 1 = 15$$

$$\Delta P_{s1} = \frac{4fG_c^2 N_r(N_b + 1)}{2g_c\rho}\left(\frac{Z_w}{Z}\right)^{0.14} = 1998.2 \text{ lb/ft}^2$$

$$K_w = 0.7$$

$$\Delta P_{s2} = \frac{K_w G_w^2 N_b}{2g_c\rho} = 69.9 \text{ lb/ft}^2$$

$$K_{x1} = K_e = 1.1 \text{ (nozzle expansion)}$$

$$K_{x2} = K_c = 0.7 \text{ (nozzle contraction)}$$

The nozzle sizes at the shell inlet/outlet are 4/4 in.; therefore, the mass velocity, G, is 368.2 lb/ft²-sec.

$$\Delta P_{s3} = \frac{(K_e + K_c)G^2}{2g_c\rho} = 61.3 \text{ lb/ft}^2$$

$$\Delta P_s = \Delta P_{s1} + \Delta P_{s2} + \Delta P_{s3} = 2129.4 \text{ lb/ft}^2$$

$$\Delta P = 2129.4/144 = 14.8 \text{ psi} < P_{allowable} = 20 \text{ psi}$$

It must now be verified that the required overall heat transfer coefficient, as established by equation (7), is less than the overall heat transfer coefficient determined for the exchanger designed. The following procedure is used for this verification:

$$Q = 1,919,000 \text{ Btu/hr}$$

$$\text{LMTD} = 11.9 \text{ deg F}$$

$$A_o = \pi\, d_o N_t L = 350.0 \text{ ft}^2$$

$$U_{req} = Q/[A_o\,(\text{LMTD})] = 460.7 \text{ Btu/hr-ft}^2\text{-deg F} \qquad \text{[eq. (7)]}$$

$$U_{clean} = \frac{1}{\dfrac{A_o}{A_i h_i} + r_{dio} + \dfrac{d_o}{2k_t}\log_e\dfrac{d_o}{d_i} + r_{do} + \dfrac{1}{h_o}} \qquad \text{[eq. (5)]}$$

$$= 516.2 \text{ Btu/hr-ft}^2\text{-deg F}$$

$$U_{fouled} = 0.9 U_{clean} = 464.6 \text{ Btu/hr-ft}^2\text{-deg F}$$

$$U_{fouled} > U_{req} = 460.7 \text{ Btu/hr-ft}^2\text{-deg F}$$

The design, therefore, has adequate capacity in this respect.

Finally, it must be verified that the assumed tube wall temperature is approximately equal to the calculated value for the actual design.

$$T_{wi} = \left(\frac{t_1 + t_2}{2}\right) + \left(\frac{A_o}{A_i h_i}\right)U_{clean}\left[\left(\frac{T_1 + T_2}{2}\right) - \left(\frac{t_1 + t_2}{2}\right)\right] = 94.4 \text{ F}$$

The assumed value of $T_{wi} = 95.3$ F approximately equals 94.4 F; therefore, the preliminary design is confirmed to have adequate capacity; however, additional iterations, to optimize the design, may be appropriate before committing the design to hardware.

Exchanger duty calculations. There are many circumstances, such as the evaluation of an existing exchanger for a particular service, where the heat transfer surface area of the heat exchanger is known, and the duty and the fluid terminal temperatures must be determined. Using the design characteristics established for the example exchanger, the duty and terminal temperature for the clean-tube heat transfer coefficient can be calculated using the following procedure:

$$U = 516.2 \text{ Btu/hr-ft}^2\text{-deg F}$$

$$A_o = 350 \text{ ft}^2$$

$$T_1 = 111.2 \text{ F}$$

$$t_1 = 85 \text{ F}$$

$$W_c = 266,359 \text{ lb/hr}$$

$$C_{pc} = 0.9604 \text{ Btu/lb-deg F}$$

$$W_h = 115,682 \text{ lb/hr}$$

$$C_{ph} = 0.9985 \text{ Btu/lb-deg F}$$

$$R = \frac{W_c C_{pc}}{W_h C_{ph}} = 2.21 \qquad \text{[eq. (19)]}$$

$$B = [R^2 + 1]^{\frac{1}{2}} = 2.43 \qquad \text{[eq. (20)]}$$

$$F = e^{(UA_o B/W_c C_{pc})} = 5.56 \qquad \text{[eq. (21)]}$$

$$n = \frac{2}{1 + R + \left(\dfrac{F + 1}{F - 1}\right)^B} = 0.298 \qquad \text{[eq. (22)]}$$

$$T_2 = T_1 - Rn(T_1 - t_1) = 93.9 \text{ F} \qquad \text{[eq. (23)]}$$

$$t_2 = t_1 + (T_1 - T_2)/R = 92.8 \text{ F} \qquad \text{[eq. (24)]}$$

$$Q = W_h C_{ph}(T_1 - T_2) = 1.998 \times 10^6 \text{ Btu/hr}$$

The results confirm that the duty and fluid terminal temperatures of the heat exchanger that was designed will be approximately equal to the values established as the design criteria.

Exchanger vibration characteristics. The procedure outlined in Section 3.7 can be used to verify that a heat exchanger is not susceptible to flow-induced vibration. To illustrate the calculation procedure, the tube natural frequency, the tube vortex shedding frequency, and the tube fluidelastic whirling in the center span are calculated in the following.

Tube natural frequency—The equation for the tube natural frequency is:

$$f_n = 1.569 \left[\frac{EIg_c}{W_e L^4} \right]^{1/2}$$ (36)

$\rho_i = 63.75$ lb/ft^3 density of fluid in tubes
$\rho_o = 61.85$ lb/ft^3 density of fluid in shell
$\rho_w = 283.9$ lb/ft^3 density of tube
$M = 2.9$ added-mass coefficient
$d_i = 0.43$ in.
$d_o = 0.50$ in.

$W_{fi} = \frac{\pi}{4} d_i^2 \rho_i =$ 0.0643 lb/ft

$W_t = \frac{\pi}{4} (d_o^2 - d_i^2) \rho_w =$ 0.1008 lb/ft

$W_{fo} = \frac{\pi}{4} M d_o^2 \rho_o =$ $\underline{0.2446}$

$W_e = W_{fi} + W_t + W_{fo} =$ 0.4097 lb/ft or 0.03414 lb/in.
$E = 16.9 \times 10^6$ psi
$I = 0.00139$ in.^{4}
$L = 5.25 \times 2 = 10.50$ in. (maximum unsupported tube length is in the baffle window)
$g_c = 386.4$ in./sec^2
$f_n = 232.1$ Hz

Vortex shedding—The equation for the vortex-shedding frequency is:

$$f_s = \frac{SV}{d_o}$$ (38)

$S = 0.22$
$d_o = 0.5$ in.
$V = $ velocity at row 5
$\rho_o = 61.85$ lb/ft^3
$W = 115,682$ lb/hr
$A = $ flow area at row 5
  $= [CD_5 - N_5 d_o] [B_S]$
  $= [13.87 - (21)(0.5)] [5.25] = 17.69$ in.^{2}
$V = W/A \rho_o = 4.23$ ft/sec
$f_s = 22.3$ Hz
$f_s\ \ < f_n = 232.1$ Hz

The vortex-shedding frequency is of little concern as it is substantially less than the tube natural frequency.

Fluidelastic whirling—The equation used to calculate the fluidelastic whirling frequency is:

$$V_{crit} = B f_n d_o \left[\frac{W_e \delta}{\rho_s d_o^2} \right]^{\frac{1}{2}}$$ (39)

$B = 3.3$
$f_n = 232.1$ Hz
$d_o = 0.5$ in.
$W_e = 0.03414$ lb/in.
$\delta = 0.031$ dimensionless
$\rho_s = 61.85$ lb/ft^3
$V_{crit} = 131.7$ in./sec or 10.98 ft/sec
$V_{act} = 4.23$ ft/sec velocity at row 5
$V_{act} = < V_{crit} = 10.98$ ft/sec

The results of the analysis show that the tube in the center span of the heat exchanger will not vibrate due to vortex shedding or fluidelastic whirling.

Section 4
Shell-and-Tube Heat Exchanger Mechanical Design

4.1 Introduction. After a heat exchanger has been designed to satisfy the thermal and hydrodynamic requirements, and the tube diameter, number of tubes, tube pitch, tube length, shell diameter, and nozzle diameters are known, the next step is to determine the required thicknesses of the various heat exchanger elements based upon the design pressure and temperature. The design pressure is the maximum differential pressure between the inside and outside of an element that occurs during normal operating conditions. It is recommended that a suitable margin be provided above the normal operating pressure to allow for potential pressure surges. The design temperature is the maximum metal temperature that occurs during normal operating conditions. The design temperature determines the maximum allowable stress in the material. The design pressure and temperature for the shell side and tube side of a heat exchanger can have different values, and in most instances they are different.

The test pressure specified for a heat exchanger is the hydrostatic test pressure that a heat exchanger must withstand before being accepted. The purpose of a hydrostatic test is to prove the strength integrity of the exchanger and to ensure that subsequent pressure cycles are within the elastic range. The test pressure is always higher than the design pressure, but the percent increase above the design pressure may vary, depending upon the design code being followed. After the hydrostatic test is completed, all welded joints and connections are inspected.

4.2 Loading Conditions. When designing a heat exchanger, the following loading conditions are typical of those that must be given due consideration during the early design phase:

• *Pressure*—The internal or external design pressure at design temperature.

• *Weight*—The weight includes the weight of the heat exchanger and contents under operating or test conditions.

• *Nozzle loads*—The heat exchanger acts as an anchor point for the piping deadweight, thermal expansion, and shock loads.

- *Shock*—Shock loads are generated as a result of an acceleration of the masses associated with the heat exchanger.
- *Thermal*—Thermal loads are the result of temperature gradients and differential thermal expansion within the heat exchanger.
- *Supports*—Support loads are imposed upon the component by lugs, skirts, rings, and supports.

4.3 Stress Analysis Procedure. Heat exchangers commonly have the form of cylinders, spherical heads, ellipsoidal heads, flat heads, and combinations of these shapes.

Pressure stress in cylinders. When the inside diameter of a cylinder is more than ten times the wall thickness, it is considered to be a thin-walled cylinder and offers little resistance to bending. The stress in thin-walled cylinders can be approximated by the following formula:

$$S = \frac{PD_i}{2t} \tag{42}$$

where

S = hoop stress
P = design pressure
D_i = internal diameter
t = wall thickness

Equation (42) is used to calculate the hoop stress in the tubes, shell, and cylindrical channel (waterbox) due to internal pressure. When the ratio of the inside diameter to the wall thickness is less than ten, the cylinder is known as a thick-walled cylinder, and equation (42) is an oversimplification of the stress in the cylinder wall. More sophisticated analytical techniques are required to accurately assess the stress in thick-walled cylinders.

Pressure stress in heads. To complete the pressure boundary of a heat exchanger, heads are welded or bolted to the cylindrical shell. Heads often take the form of a hemisphere, ellipse, or a flat cover.

The equation for determining the pressure stress in a hemispherical head is closely approximated by the expression:

$$S = \frac{PR_i}{2t} \tag{43}$$

where

P = design pressure
R_i = inside radius of hemispherical head
t = thickness of hemispherical head

Insofar as practical, good practice requires that stresses in a hemispherical head be approximately the same as the stress in the cylinder to which it is attached. By examining equations (42) and (43), it may be seen that to do so requires that the thickness of the head be one-half the thickness of the cylinder.

The equation for determining the pressure stress in an ellipsoidal head is

$$S = \frac{PD_i}{2t} \tag{44}$$

where

P = design pressure
D_i = major diameter of ellipsoidal head
t = thickness of ellipsoidal head

Good engineering practice dictates that the stress in the ellipsoidal head be approximately the same as the stress in the cylindrical shell to which it is attached. By examining equations (42) and (44), it may be seen that the thicknesses of the ellipsoidal head and cylindrical shell must be the same for this to occur. Equation (44) is applicable for calculating the pressure stress in an ellipsoidal head only when the major diameter of the ellipse is twice the minor diameter.

The stress in a flat circular head that is subject to a pressure on one side depends upon the edge restraint condition. If the edge of the flat head is simply supported, the pressure stress can be expressed by

$$S = \frac{0.75 \, PR^2}{t^2} \tag{45}$$

where

P = design pressure
R = radius of circular head
t = thickness of circular head

Nozzles. Fluids enter and exit a heat exchanger through nozzles. The nozzles penetrate the pressure boundary, thereby increasing the stress of the boundary penetration. To compensate for the stress concentration introduced, the heat exchanger must be reinforced in the vicinity of the nozzle. Nozzles are the most common pressure-boundary penetrations; however, vents, drains, manways, and handholes also require reinforcement. The reinforcement must be equal to or greater than the area of metal removed by the penetration, and the reinforcement must be added adjacent to the penetration in such a manner as not to introduce additional stress concentrations.

Figure 18 shows the reinforcement boundaries for a circular penetration in a cylindrical shell or spherical head. Points ABCD represent the area removed by the penetration. Points EFGH show the reinforcement boundary. The reinforcement can be added to the inside surface, to the outside surface, or to a combination of both. It is preferred that half the reinforcement area be added to both the inside and outside surfaces. However, adding reinforcement area to the inside surface is not always feasible because of interferences with other components, the requirement to remove a tube bundle, or flow disruption and turbulence. Good practice requires that the reinforcing area, A_r, equals or exceeds the area removed by the penetration, A_p, as detailed in Fig. 18.

4.4 Modes of Failure. When subjected to a load, the response of a heat exchanger is dependent upon the type of material, the environment, and the manner of loading (static, dynamic, or thermal). The failure modes that must be considered may be grouped as catastrophic failures (material fractures) and failures resulting from excessive deformation. During the mechanical design of heat exchangers, the following modes of failure must be considered:

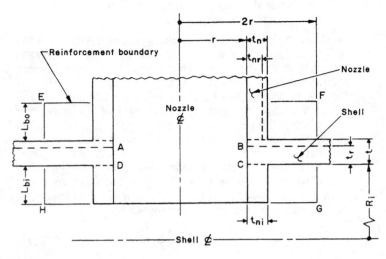

Fig. 18 Limits of reinforcement for circular penetrations in a cylindrical shell or hemispherical head

$$A_r = (2r-r)(t-t_r)+L_{bo}(t_n-t_{nr})+L_{bi}t_{ni} = A_p = rt_r$$

where

A_r = reinforcement area

A_p = area of metal removed by the penetration

r = nozzle radius

t_n = nozzle thickness

t_{nr} = nozzle thickness required to satisfy the pressure

t_{ni} = internal nozzle thickness

t = shell thickness

t_r = shell thickness required to satisfy the pressure

L_{bo} = the smaller value of 2.5t or $2.5t_n$

L_{bi} = the smaller value of 2.5t, $2.5t_{ni}$, or the actual internal nozzle length

R_i = shell radius

• *Failure by excessive elastic deformation*—Elastic deformation failures occur when an applied load causes the element to exceed the allowable elastic strain or elastic deflection limits.

• *Failure by excessive plastic deformation*—Plastic deformation failures occur when an applied load causes the entire section of the element to exceed the yield strength introducing a permanent change in shape.

• *Failure due to fatigue*—Fatigue failures are due to progressive fractures and are caused by elements being subjected to cyclic or repeated stresses that exceed the material fatigue limits.

• *Failure due to incremental collapse*—Incremental collapse is characterized by an accumulation of inelastic strains resulting from the successive application of cyclic loads superimposed upon a sustained load.

• *Failure due to creep deformation*—Creep is the time-dependent deformation or yielding of an element under load, usually at an elevated temperature.

• *Failure due to brittle fracture*—Brittle fracture occurs when the applied load causes a rapid propagating fracture without any plastic deformation. The state of stress is below the yield strength of the material. Brittle fractures initiate at material flaws or stress risers.

• *Failure due to stress corrosion*—Stress-corrosion fractures are caused by the simultaneous effects of a corrodent and an applied or residual stress.

• *Failure due to corrosion fatigue*—Corrosion fatigue results in a fracture of an element by the combined action of cyclic load and a corrosive environment.

• *Failure due to erosion*—Erosion causes an element to fail due to the removal or destruction of material by the abrasive action of a flowing fluid.

Design codes, or specification standards, are used to prevent the first six modes of failure by specifying conservative factors of safety or allowable stress limits. However, the heat-exchanger engineer is also responsible for preventing failures due to corrosion or erosion. This can be accomplished by specifying appropriate materials, reducing the flow rate, streamlining the flow, adding sacrificial material in the form of a corrosion allowance, or using corrosion resistant materials such as stainless steel, nickel-chrome-iron alloy 600, or nickel-molybdenum-chrome alloy C-276 overlayed on the base material.

4.5 Design Codes. Heat exchangers are designed in compliance with published codes. Each code is based on a unique design procedure and is applicable only when specified. The entire heat exchanger, tube side and shell side, can be designed to a single code or a combination of codes, where the tube side is designed to one code and the shell side to another code. The codes most often used in the design of heat exchangers are:

1. The American Society of Mechanical Engineers, Boiler and Pressure Vessel Code, Section VIII, Rules for Construction of Pressure Vessels, Division 1.

2. The American Society of Mechanical Engineers, Boiler and Pressure Vessel Code, Section VIII, Rules for Construction of Pressure Vessels, Division 2—Alternative Rules.

3. The American Society of Mechanical Engineers, Boiler and Pressure Vessel Code, Section III, Nuclear Power Plant Components, Division 1.

4. SDB-63, Structural Design Basis, Naval Sea System Command.

Section 5
Plate Heat Exchangers and Compact Heat Exchangers

5.1 General. The shell-and-tube heat exchanger is the most widely used type for marine applications; however, plate and compact heat exchangers offer attractive alternatives for certain applications.

5.2 Plate Heat Exchanger. Plate heat exchangers are most commonly used as lubricating-oil coolers and engine jacket water coolers. As shown in Fig. 19, the plate heat exchanger consists of five basic elements: the cover, the carrier rail, the heat transfer plates, the support column, and the tie bolts. The inlet and outlet for both fluids are usually located in the same cover. The two fluids are separated by the heat transfer plates. Each plate contains a gasket that fits into grooves pressed in the plate and in the nozzle ports. The gasket prevents the two fluids from mixing. The gasket is vented to the atmosphere, which

permits a leak to be readily detected. The plates are sandwiched between a fixed cover and a movable cover by the tie bolts. The top and bottom carrier rails align the plates to each other.

The heart of a plate heat exchanger is the plate, and understanding the design of the plate is paramount when specifying this type of heat exchanger. The plate is a sheet of metal that is approximately 0.024 in. thick and precision pressed into corrugated patterns or chevrons. The corrugation pattern, depth, shape, and angle are the manufacturer's proprietary information. Since the industry does not have standard heat transfer plates, the plates are unique to the manufacturer. To insure uniform velocity through the plate, the inlet and outlet region of each plate

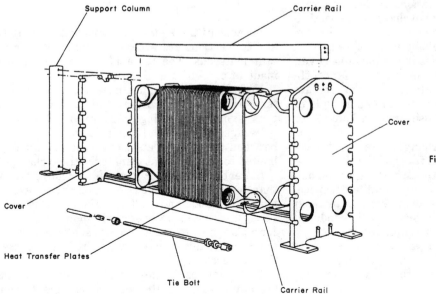

Fig. 19 Plate heat exchanger

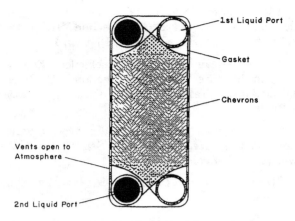

Fig. 20   Plate with gasket

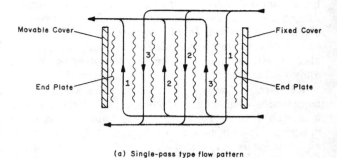

(a) Single-pass type flow pattern

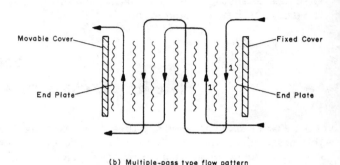

(b) Multiple-pass type flow pattern

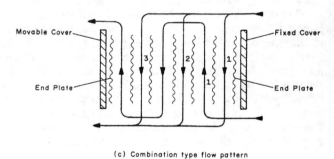

(c) Combination type flow pattern

Fig. 21   Typical flow patterns in plate heat exchangers

Table 4   Approximate range of convective heat transfer co-efficient in a plate heat exchanger

Condition	Coefficient, h Btu/hr-ft²-deg F
Forced convection oil	50 to 600
Forced convection water	500 to 4000

has flow distribution grooves pressed into the plates. Figure 20 shows the design of a typical plate with a gasket.

The heat transfer coefficient and pressure drop can be varied from plate to plate by changing the plate pattern, depth, shape, and chevron angle. When designing a plate heat exchanger to transfer a given amount of heat, the number of plates can be increased to establish the required area for a specified chevron angle, or the designer may change the chevron angle within the plate pack, thereby decreasing the area needed for heat transfer.

Figure 21 illustrates typical flow patterns in plate heat exchangers. For different combinations of flow and pressure drop requirements, plate heat exchanger design flexibility is achieved by changing the plate size, the chevron angle, the flow of some of the plates to co-current flow, or the location of the outlet nozzles in the movable cover.

The approximate ranges of the film heat transfer coefficients for fluids flowing in a plate heat exchanger are listed in Table 4. It may be noted that the magnitude of the coefficients is higher than those listed in Table 2 for tube heat exchangers.

Maintenance of a plate heat exchanger is relatively easy because there is access to both surfaces of the heat transfer plate. Because of its compactness, the inspection and maintenance of the unit does not require additional space.

Unlike a shell-and-tube heat exchanger, a plate heat exchanger is not subject to the problems of flow-induced vibration, entry impingement, bypassing, and leakage associated with a shell-and-tube heat exchanger. Plate heat exchangers have a fast start-up and a rapid response to transients, which reduces thermal stresses and increases the service life of the unit.

Because of the method of construction, only the plates and connections are necessarily made of corrosion-resistant materials. The remainder of the elements can be made of carbon steel.

Under normal circumstances the maximum operating temperature of the working fluid is limited to 250 F and the maximum operating pressure to 150 psi.

Because of the high friction factor of the fluid flowing between the many plates, this type of unit is generally limited to sensible heat transfer with no change in phase.

In general, fouling is lower for a plate heat exchanger than for a shell-and-tube heat exchanger because of the high turbulence and wall shear induced by the corrugations. Care must be exercised when fouling allowances for a plate heat exchanger are based on data for a tubular unit because this may result in more heat transfer surface than necessary.

The equations for heat transfer developed in Section 3 are applicable to plate heat exchangers except for the overall heat transfer coefficient, the film coefficient, and the friction factor.

Unlike a shell-and-tube heat exchanger, where the end user can design a unit using standard tubular parameters such as the number of tubes, tube pitch, tube length, and shell diameter, the user cannot design a plate heat exchanger. A plate heat exchanger is assembled from plates of specific dimensions and corrugation patterns. Each manufacturer has a unique plate design; however, the user can verify that a given plate heat exchanger will perform the intended function. Equation (7) is again used to calculate the heat flow rate; however, the terms are redefined as follows:

$$Q = UA_o(\text{LMTD}) \qquad (7)$$

where

LMTD = log mean temperature difference
A_o = total heat transfer area for all plates
$U = 1/[1/h_h + r_h + r_w + r_c + 1/h_c]$
h_h = film heat transfer coefficient, hot side
r_h = fouling resistance, hot side
r_w = wall resistance = T/k
T = thickness of plate
k = thermal conductivity of plate
h_c = film heat transfer coefficient, cold side
r_c = fouling resistance, cold side

The film coefficient can be correlated with the pressure drop by using the empirical relationships given in Section 3, as follows:

$$Nu = \frac{hD_e}{k} = C(Re)^m(Pr)^{0.33}(Z/Z_w)^{0.14} \qquad (46)$$

and

$$\Delta P = 4f\left(\frac{L}{D_e}\right)\left(\frac{G^2}{2g_c\rho}\right) \qquad (47)$$

where

Nu = Nusselt number
Re = Reynolds number
D_e = equivalent hydraulic diameter

The coefficients c, m, and f are experimentally determined constants that are functions of the chevron angle and flow regime. Each plate heat exchanger manufacturer establishes the constants that are applicable to particular designs; and, when the units are in service, it is also possible for others to develop the constants from plate heat exchanger performance data.

5.3 Compact Heat Exchangers. Compact heat exchangers have a high ratio of heat transfer surface to heat exchanger volume. As an arbitrary definition, a compact heat exchanger has a surface-to-volume ratio that is more than 215.0 ft²/ft³. For comparison purposes, a shell-and-tube heat exchanger has a surface-to-volume ratio of approximately 20 to 40 ft²/ft³. Typical compact exchanger designs are: extended surface exchangers utilizing fins on one or both sides and regenerators and exchangers

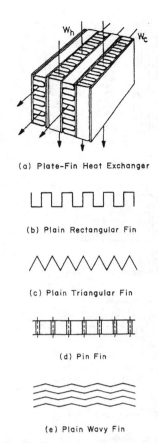

(a) Plate-Fin Heat Exchanger

(b) Plain Rectangular Fin

(c) Plain Triangular Fin

(d) Pin Fin

(e) Plain Wavy Fin

Fig. 22 Plate-fin heat exchanger and typical fin geometries

with small hydraulic-diameter surfaces or tubes. Of particular interest for marine applications are compact heat exchangers with extended surfaces. These heat exchangers can be used as waste-heat boilers, evaporators, or condensers; however, the most common marine applications are in refrigerating and air-conditioning systems.

The heat transfer coefficient for a sensible gas is one to two orders of magnitude less than for water or oil, as may be seen from Table 2. The use of extended surfaces on the gas side of the exchanger is an effective means of achieving a balance between the product of the film coefficient and the heat transfer surface area on the liquid side and the gas side of the exchanger. The major applications of extended heat transfer surface are the gas side of gas-to-liquid, gas-to-condensation, and gas-to-evaporation heat exchangers.

Of the alternative extended-surface heat exchangers, the most common are the plate-fin type and the fin-tube type. The plate-fin heat exchanger contains fins that are sandwiched between the parallel plates. The fins are attached to the plates by extrusion, welding, brazing, soldering, or mechanical fit. Figure 22(a) depicts a plate-fin heat exchanger, and Figs. 22(b) through (e) illustrate typical fin geometries. Flexibility in the thermal and hydraulic design for plate-fin heat exchangers is realized by changing the fin height, width, and thickness. A typical

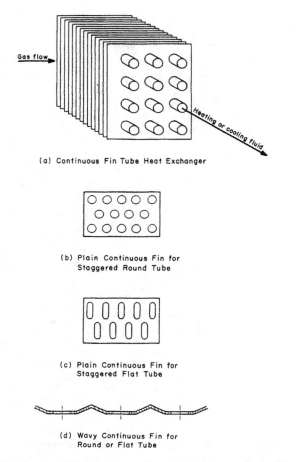

(a) Continuous Fin Tube Heat Exchanger

(b) Plain Continuous Fin for Staggered Round Tube

(c) Plain Continuous Fin for Staggered Flat Tube

(d) Wavy Continuous Fin for Round or Flat Tube

Fig. 23   Continuous fin-tube heat exchanger and typical continuous fin geometries

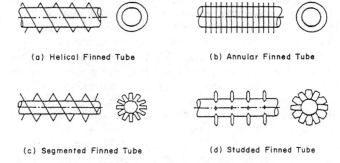

(a) Helical Finned Tube   (b) Annular Finned Tube

(c) Segmented Finned Tube   (d) Studded Finned Tube

Fig. 24   Typical finned tubes

fin is 0.004 to 0.010 in. thick, and the fin height varies from 1 to 10 in.

In a fin-tube heat exchanger, the fins are usually attached to the outside of the tube. The tube can be round, rectangular, or ellipsoidal. Figure 23(a) shows continuous fins on an array of tubes, and Figs. 23(b) through (d) contain typical continuous fin geometries. Figures 24(a) through (d) show individually finned tubes. The fins are attached to the tube by tension-winding, mechanical bonding, welding, or hydraulically expanding the tube against

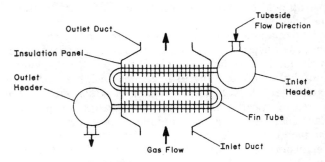

Fig. 25   Fin-tube heat exchanger

the fin. The use of a mechanical bond or hydraulic expansion is limited to those applications where the differential thermal expansion between the tube and fin material is small; otherwise, continual start-up and shutdown cycles will loosen the bond and introduce a thermal resistance. This condition can be eliminated by welding the fin to the tube.

Compact heat exchangers are usually arranged with the fluids in cross flow, that is, with the fluids flowing at right angles. In addition, it is important to establish whether the fluid is mixed or unmixed. Mixing implies that all the fluid in any given plane normal to the flow has the same temperature, and the temperature changes only in the direction of the flow. Unmixed flow implies that fluid temperature differences exist in at least one direction normal to flow but no heat flux due to the differences occurs. The mathematical derivation of an expression for the mean temperature differences for mixed and unmixed fluids is quite complex. The usual procedure is to modify the LMTD calculations that are made using expressions such as equations (8) or (11) by the application of correction factors [8].

The equations developed in Section 3 are applicable to the design of fin-tube compact heat exchangers except for the external film coefficient and friction factor. The external film coefficient and the pressure drop can be correlated by using the following empirical relationship in Section 3:

$$Nu = \frac{h_o d_o}{k} = C(Re)^m (Pr)^{0.33} (Z/Z_w)^{0.14} \qquad (48)$$

and

$$\Delta P = 4f \left(\frac{G^2}{2g_c \rho} \right)^{N_r} \qquad (49)$$

where

R_e = Reynolds number
N_u = Nusselt number
N_r = number of tube rows in direction of flow
d_o = outside diameter of tube

Published data are available that can be used to calculate film coefficients and friction factors for any combination of tube pitch pattern and fin type that may be specified for compact fin-tube heat exchangers. Such data are published by the manufacturers of fin tubes [19] and have also been correlated in reference 20.

The components of a typical compact fin-tube heat exchanger are the serpentine coil and the casing as shown in Fig. 25. The high-pressure fluid is transported within the fin tube. The casing consists of an inlet duct, an outlet duct, and insulation panels. The panels direct the gas flow around the outside of the fin tubes and also serve as the pressure boundary for the low-pressure gas. The allowable gas-side pressure drop is specified in inches of water and is usually the limiting parameter in the design.

The following items require consideration when designing a compact fin-tube heat exchanger:

• To maintain a uniform velocity profile across the serpentine coil, directional vanes or diffuser plates may be added to the inlet duct.

• The fin tube may be susceptible to flow-induced vibration.

• The casing surfaces are flat and must be designed as the pressure boundary for the low-pressure gas. Stiffeners may be added to the surface of the casing as a means of reducing the stress and deflection.

• To reduce ambient heat loss and to provide personnel protection, the casing may require insulation if the gas temperature is greater than 150 F.

• The serpentine coil may require an independent support structure depending upon its size and weight.

• The tube pitch pattern may be either triangular or square.

Section 6
Heat Exchanger Applications

6.1 Lubricating-Oil Coolers. In addition to serving as a lubricant between moving mechanical parts, lubricating oils generally also accomplish a second objective of removing the frictional heat generated, especially so with diesel engines. Therefore, some means must be provided for removing the heat absorbed by the lubricating oil. With small systems, the natural heat transfer by radiation and convection may be adequate, but with larger systems, particularly those employing forced circulation, lubricating-oil coolers are required.

The main lube-oil coolers, i.e., those serving the main propulsion machinery, are generally shell and straight-tube exchangers that often have removable tube bundles and double-packed floating tubesheets, such as illustrated by Fig. 26. Oil generally flows in a single pass via transverse baffling in the shell, and seawater is the normal coolant flowing in one or more passes through the tubes. However, in many motorships that have central cooling systems, fresh water is used as the coolant.

In the lube-oil cooler shown in Fig. 26, the floating tubesheet is centered between the shell flange and waterbox flange. A gland ring retains the packing. Separate packing rings are provided for the shell (lube oil) side and the coolant side. The gland ring is grooved around the inside, and leak-off holes are provided so that leakage past the packing rings on either side will be relieved to the outside and attract the operator's attention.

The stationary tubesheet, baffles, and support plates are assembled and held in proper relative position by tie rods and spacer sleeves. Tie rods are threaded into but not through the stationary tubesheet. The waterbox and stationary tubesheet are secured by collar bolts or by stud bolts driven into tapped holes in the tubesheet so that the tubesheet-to-shell-flange joint will not be broken when the waterbox is removed.

The packing retainer ring is usually "scalloped" so that there is a stud hole for every second stud in the shell flange at the floating tubesheet end of the cooler. This enables the gland ring to double as a test ring and allows the shell to be hydrostatically tested without the waterbox in place.

In some of the more demanding applications, such as heat exchangers cooled by seawater, there are special requirements for bolting as well as for materials, and such requirements must be considered in the cooler design. Limitations are also provided for maximum fluid velocities and pressure drop, and shock resistance requirements may also be specified for military applications [6].

It has been determined that the performance, in terms of fouling properties and service life, of low-fin tubing in lube-oil cooling service is approximately equal to that of bare tubes, and $\frac{5}{8}$- and $\frac{3}{4}$-in. low-fin 90/10CuNi tubing has become the preferred tubing in lieu of $\frac{3}{4}$- and $\frac{7}{8}$-in. bare tubes. With low-fin tubes, the greater amount of heat transfer surface within a given shell size outweighs the disadvantage of their somewhat lower heat transfer rates, and the net result is a smaller, more compact, and more economical unit for a given performance requirement as compared with bare tubes.

Lubricating-oil coolers are also necessary for auxiliaries such as diesel- and turbo-generators, main feed pumps, and air compressors. These units may be similar in design to main lubricating-oil coolers but are smaller, having a shell diameter of 6 to 10 in., with some perhaps smaller.

It is important to note that the high viscosity of lubricating oil results in a low Reynolds number on the oil side, with a consequent low heat transfer coefficient and a high pressure drop. These, and other design aspects of oil coolers, are treated in detail in Section 3.

6.2 Fuel-Oil Heaters and Lubricating Oil Heaters. The heavy fuel oils often burned aboard ship are so viscous at ambient temperatures that they must be heated before they can be pumped from the storage tanks, heated again before purification (on motorships), and heated

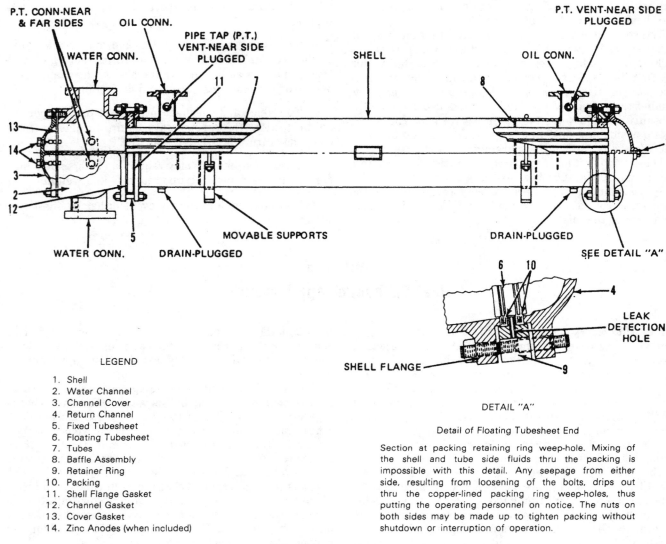

LEGEND

1. Shell
2. Water Channel
3. Channel Cover
4. Return Channel
5. Fixed Tubesheet
6. Floating Tubesheet
7. Tubes
8. Baffle Assembly
9. Retainer Ring
10. Packing
11. Shell Flange Gasket
12. Channel Gasket
13. Cover Gasket
14. Zinc Anodes (when included)

DETAIL "A"

Detail of Floating Tubesheet End

Section at packing retaining ring weep-hole. Mixing of the shell and tube side fluids thru the packing is impossible with this detail. Any seepage from either side, resulting from loosening of the bolts, drips out thru the copper-lined packing ring weep-holes, thus putting the operating personnel on notice. The nuts on both sides may be made up to tighten packing without shutdown or interruption of operation.

Fig. 26   Typical main lube-oil cooler

again before combustion (the properties of fuel oils are discussed in detail in Chapter 12). The more viscous fuels may require heating to 120 F or more before they can be pumped; and when supplied to the burners or injectors, the fuel oil must be at an even higher temperature in order to attain a fuel viscosity sufficiently low for proper fuel atomization. Therefore, when a heavy fuel is used, two stages of fuel-oil heating are provided on steamships and three on motorships; primary heaters are installed in the fuel-oil tanks, and secondary heaters are installed before the purifiers (when provided) and between the service pumps and the burners or injectors.

The primary oil heaters are installed in the tanks in either of two forms: as steam-supplied pipe grids or coils, or as open-ended tank-suction heaters. The latter alternative has become increasingly popular because of the lower initial cost, lower maintenance costs, and lower steam consumption (only the oil to be pumped is heated as opposed to heating the entire tank). A typical tank-suction

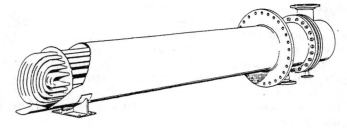

Fig. 27   Typical tank suction fuel-oil heater

fuel-oil heater is shown in Fig. 27. This horizontal U-tube heat exchanger has an outer shell flange that is bolted directly to the tank. The oil is drawn in through the open end of the shell and along the tube bundle. As the heating medium (condensing steam) circulates through the tubes, the portion of oil in the heater shell in contact with the tubes rises in temperature with a corresponding decrease in viscosity. The entire process is continuous with the oil flow being induced by the pump.

The secondary fuel-oil heaters heat the fuel to the temperature necessary for proper atomization (about 240 F for boilers and as high as 300 F for diesels) by means of an automatic steam-supply regulating valve that controls the steam supplied to the heaters. An accurate control of the steam supply is necessary as the oil may otherwise be overheated and form carbonaceous residues on the heater tube surfaces.

Several types of secondary fuel-oil heaters are in use; a common type is the bayonet-tube heater as shown by Fig. 28. In this tubular heater, oil enters the shell of the heater at the tubesheet end and is directed by a system of segmental baffles to flow back and forth across the tube bundles [see also Fig. 8]. The tubes are staggered, and continuous mixing occurs during flow, causing the cooler oil to be constantly forced against the heating surfaces and thus wiping the heated film from them and creating an effective heat transfer from those surfaces.

In a bayonet-tube heater, steam is delivered to the steam chamber, and from there enters a series of small tubes, called "inner" tubes, through which it travels until it is discharged into the annulus between the inner and outer tubes at the far end of the outer or oil-heating tubes [see also Fig. 10(b)]. Thus steam (free of condensate) is in contact with the heating surface at the oil exit end of the unit, where the highest oil temperature is desired. The space between the inner and outer tubes is small so that the volume of steam flowing in the annular space generates a velocity of flow sufficient to continuously sweep the condensate from the surface as it forms, thus reducing water-film losses and preserving high transfer rates on the steam side.

Bolting is arranged such that the assembly consisting of the inner and outer tubes, steam-condensate chamber, and baffles may be removed from the heater shell as a unit, thereby permitting an examination of the exterior of the heating surface and mechanical cleaning as required. Also, the inner tubes only, together with the steam condensate chamber, may be removed, leaving the outer tubes and their tubesheet in place and firmly bolted. In addition, a test of the shell side may be conducted with all the tube ends visible for inspection during the test.

The outer tubes are of the bayonet (Niclausse tube) type and are rolled into the tubesheet at one end and closed at the other. Each tube is free to expand and contract independently. The outer tubes are often finned to obtain a greater amount of surface for a given shell diameter. Finned-surface and bare tubes have been found equally resistant to fouling.

The shell-and-straight-tube type of heater has also been used in fuel-heating service, with the oil being passed through the tubes and steam through the shell. Alternatively, fin-tube heaters have been used in which steam flows through inner tubes that are fitted with external longitudinal fins; in these heaters the oil flows through the space between the finned inner tube and an outer tube, which serves as a shell. The Navy has set forth

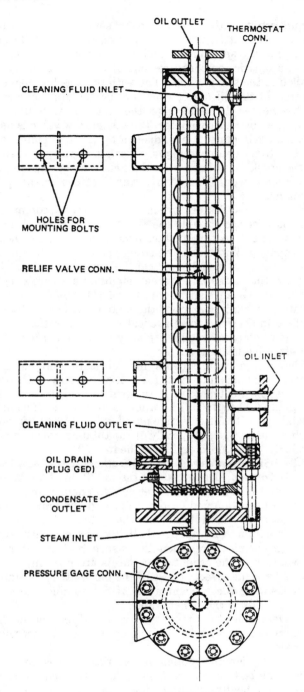

Fig. 28 Bayonet-tube fuel-oil heater

specific design criteria for all three of the foregoing designs, and further describes an "evaporator" type of fuel-oil heater, which uses an intermediate fluid to transfer heat from a primary heating coil to the bayonet oil-heating elements; this precludes any possibility of contaminating the condensate with fuel oil in the event of a leaky tube or tubesheet joint [21].

Steel construction throughout is standard practice. A minimum of two heaters is normally provided, sized, and so arranged that either heater can provide full service requirements with the other heater serving as a standby. Heaters are fitted with relief valves, which usually discharge to the settling or service tanks via a check valve. The automatic temperature control valve is normally installed in the steam supply line with the sensing element located in the oil outlet flow line immediately adjacent to the heater. In motorships the steam flow is usually controlled by a viscosimeter. The steam flow valve is usually installed immediately adjacent to the steam inlet connection to the heater. Adequate steam traps are necessary in the condensate lines from the heater since flooding of the heater would have an adverse effect on its performance.

Sectionalized heaters or banks of heaters are commonly used in order to provide the flexibility of using all or a portion of the heating surface over a wide range of heating capacity. With such an arrangement, individual heaters or sections of a heater can be cut in or out as the demand fluctuates and thereby maintain a steam supply sufficient to ensure adequate control of the oil outlet temperature. Such a control of the amount of heat transfer surface in service avoids cyclical heater operation when the heating requirements are very small compared with the effective heating surface.

The design requirements for lubricating-oil purifier heaters are much like those for fuel-oil heaters in that the oil must be heated to a prescribed temperature range (normally 100 to 160 F) in order to attain a sufficiently low oil viscosity for effective purification. Lubricating-oil purifier heaters are generally of the tubular type and are similar to those used as fuel-oil heaters.

6.3 Boiler Feedwater and Desuperheater Leakage Test Sample Coolers. Daily tests of the condition of the boiler water and feedwater are necessary to ensure continued efficient operation and protection of high-pressure steam generators. These tests are necessary to:

• Maintain the specified boiler water chemistry through chemical treatment to ensure that the correct proportions of the essential chemicals are present.

• Check the effectiveness of the blowdown procedure by measuring the concentration of the soluble and suspended solids in the boiler water.

• Determine the amount of dissolved oxygen in the boiler water and feedwater to guard against excessive corrosion.

The collection and cooling of water samples is the first step in the test procedure. The coolers required are relatively small heat exchangers due to the small quantity of sample required for testing. One of the common arrangements is a cooler system consisting of primary and secondary coolers connected in series. The coolers have a cylindrical cast bronze shell that contains a helical coil wound around a core positioned in the cast shell; such a cooler is shown in Fig. 29. The sample passes through the coil and cooling water flows across the outer surfaces of the coil. The boiler water sample cooler is usually made of 90/10 or 70/30 CuNi, and the cooler and valves are designed to the boiler working pressure. The primary cooler uses seawater at 85 F as a coolant and reduces the sample temperature to approximately 140 F; the secondary cooler

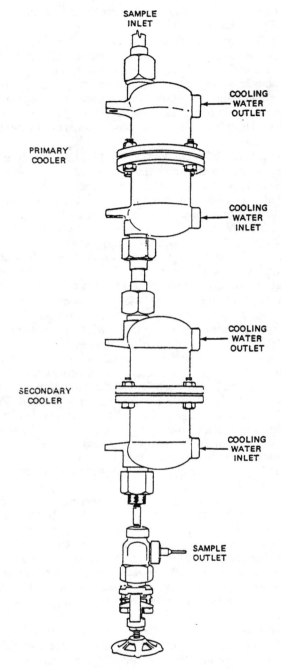

Fig. 29 Boiler water sample cooler

uses chilled 50 F fresh water as the coolant to reduce the temperature of the sample to that desired for testing. The maximum temperature for oxygen determination is 70 F; in Navy practice 100 F is considered the maximum temperature suitable for pH, hardness, and chloride determination.

The coils for desuperheater leakage test sample coolers can be made of copper or a similar material since they are designed for a moderate pressure and are normally cooled with fresh water.

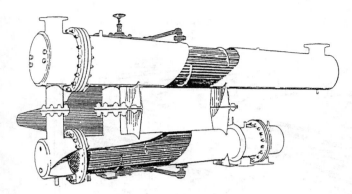

Fig. 30   Saltwater heater and drain cooler

6.4 Tank Cleaning System Heaters and Drain Coolers. When a different grade of cargo is to be carried in a tanker (e.g., gasoline vice crude oil), it may be necessary to wash the cargo tanks before receiving the different grade. Also, it may be necessary to wash the tanks in preparation for inspection and maintenance. The tank cleaning systems use hot seawater supplied from heat exchangers.

The heat exchangers usually consist of a seawater heater and drain cooler connected in series so that the seawater flows first through the drain cooler and then through the heater. Steam is used as the heating medium in the heater, and the resulting condensate is cooled in the drain cooler. The seawater is heated to a temperature of 180 to 200 F.

As illustrated in Fig. 30, the exchangers are typically of the horizontal shell-and-tube type; the heater (above the drain cooler) is of the U-tube type, and the drain cooler has a fixed tube nest and a shell expansion joint. The tube nest in the coolers is usually fitted with transverse baffles to create a flow path for the drains perpendicular to the run of the tubes. Depending upon design conditions and installation requirements, the tubes may be arranged for single- or multi-pass flow of the seawater. Zinc anodes are provided in the water heads to minimize galvanic corrosion on other parts of the heat exchanger (galvanic corrosion is discussed in Section 2.5).

Occasionally the heater and drain cooler are combined in a single shell, but the arrangement must preclude the possibility of the condensate rising above its normal level and submerging the tubes in the heating section, which would reduce the effectiveness of the unit. In some cases, only a heater is installed and the drains are discharged through a steam trap to the deaerating feed heater or a suitable vessel or receiver.

Whether a separate or combined heater and cooler is installed, a liquid level control is employed to ensure submergence of the drain cooling tubes. Relief valves are provided on both the shell and tube sides of the saltwater heater.

For naval applications, there are a number of special design requirements (such as construction materials and shock requirements), which must also be considered.

6.5 Low-Pressure Feedwater Heaters. The classification (i.e., low pressure or high pressure) of feedwater heaters depends upon their location relative to the boiler feedwater pump; low-pressure heaters are located on the suction side of the main feed pump, whereas high-pressure heaters are located on the discharge side. Feedwater heating is accomplished in a number of steps or stages, and the heaters are usually referred to as the first stage, second stage, third stage, fourth stage, etc. Multiple stages of feed heating are essential to the efficiency of a steam turbine power plant, as may be noted from Chapter 2. The plant heat balance establishes the number of heating stages, feed flow through each heater, bleed points, auxiliary exhaust pressures, and the temperature of the feedwater entering each heater.

Since the heat transfer coefficient of condensing steam is independent of velocity, and feed pressures are usually quite high, the feed is generally in the tubes with the steam in the shell. For a given steam pressure, the heat transfer is dependent upon the feed velocity through the tubes. Velocities of 6 to 7 fps result in a reasonable pressure drop and satisfactory heat transfer conditions.

Heater shells should be baffled to avoid dead spaces, and drain cooling sections should hold close baffle-to-shroud tolerances so as to avoid excessive bypassing of heat transfer surface, which would result in inadequate drain cooling. Both the shell sides and watersides should be self-venting.

The design pressure controls the construction details of feedwater heaters. Those heaters with design pressures up to 100 psig are considered low-pressure heaters. It is common practice to combine several low-pressure heaters into one shell to save space, cost of equipment, piping, and installation costs. Figure 31 depicts a typical combined low-pressure feed heater/drain cooler/gland-exhaust condenser. Similar heaters combining two stages of heating with the drain cooler are sometimes used. The combined heater is normally furnished as a package with the gland-exhauster fan and drain regulator mounted on the unit.

Due to the temperature difference between sections of combined heat exchangers, the outlet ends of the tubes are usually secured in the tubesheet by means of alternate rings of metallic and fiber packing [see Fig. 10(c)] and are, therefore, free to expand independently.

6.6 Direct Contact Deaerating Feedwater Heaters. Since marine propulsion boilers are operated at high temperatures and pressures, there is a hazard of corrosive attack due to the presence of dissolved oxygen or carbon dioxide in the feedwater. It is virtually impossible to prevent the entry of air into the feed system, particularly during plant start-up; therefore, it is necessary to provide deaerating equipment for the removal of air and corrosive gases from the boiler feedwater. Although deaeration can be largely accomplished in the condenser, "condenser deaeration" is not sufficient during plant start-up; and without further deaeration, there would be no provision for the removal of air introduced later in the system, particularly at the condensate pumps.

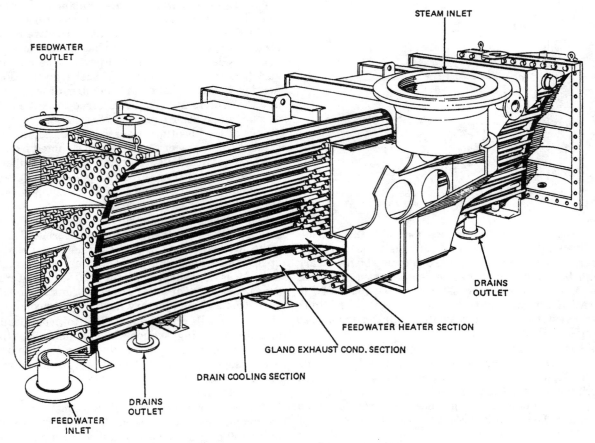

STEAM INLET

FEEDWATER OUTLET

DRAINS OUTLET

FEEDWATER HEATER SECTION

GLAND EXHAUST COND. SECTION

DRAIN COOLING SECTION

DRAINS OUTLET

FEEDWATER INLET

Fig. 31   Typical combined low-pressure feed heater/drain cooler/gland exhaust condenser

Flash deaeration, whereby saturated water at 10 to 15 psig is introduced into a surge tank at atmospheric pressure, is simple and economical. However, the resulting flashing of steam will not ensure the low dissolved oxygen content (i.e., 0.005 cc per liter) required of the feedwater unless sufficient agitation is positively provided. There is also a loss of steam through flash deaeration. For example, a deaerator operating at 15 psig flashing down to atmospheric pressure loses about 4% as flashed steam. The flashed steam should not be condensed and returned to the feed cycle since it will normally have reentrained a portion of the undesirable dissolved gases. Therefore, standard marine deaerators are more sophisticated in design than those of the flash type.

The practical considerations involved in the removal of dissolved oxygen from boiler feedwater may be briefly summarized as:

1. Heating the water to the boiling temperature for the pressure under which the process is conducted (saturation conditions). From the chemical relationship termed "Henry's Law," it is known that when a partial pressure of a liquid is equal to the total pressure above the liquid (boiling conditions), the solubility of any gases in the liquid is zero.

2. Providing a design that ensures thorough agitation and scrubbing of the feedwater by the steam. Complete agitation of the feedwater and contact with the scrubbing steam ensures that equilibrium will be reached and that the zero potential solubility condition (Henry's Law) will be attained.

3. Continuously venting from the system a mixture of gases and steam. Through the use of adequate venting, the partial pressure of the noncondensable gases in the system will be kept low and the saturation boiling point of the liquid will be maintained.

The heater immediately preceding the suction side of the boiler feed pump is usually the "direct contact" or deaerating feed heater (generally known as a "DFT," for deaerating feed tank). A typical direct-contact feed heater is illustrated in Fig. 32. Condensate and makeup are sprayed into the steam-filled primary heating and deaeration chamber through a series of spray nozzles and a vent-condensing spray nozzle. The spray nozzles provide an even distribution of water over the entire heating area. The steam flow, which is essentially countercurrent to the water flow, heats the water close to the saturation temperature such that the solubility of the gases is zero, and approximately 95% of the oxygen content is thereby released.

Water and condensate collect in the conical water collector and flow to the atomizing valve, where high-velocity steam strikes the mixture, atomizes it into a fine mist,

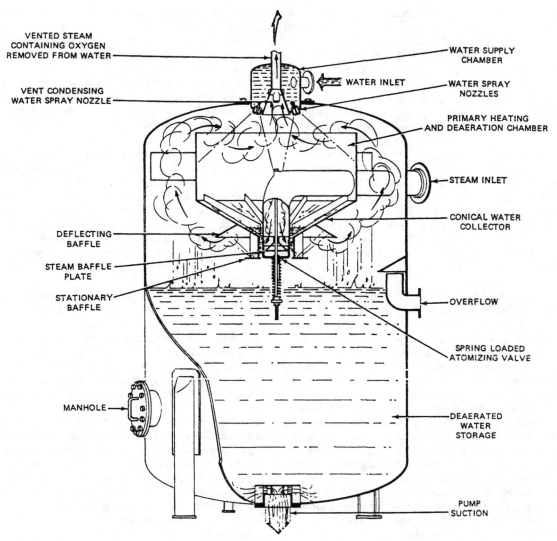

VENTED STEAM
CONTAINING OXYGEN
REMOVED FROM WATER

VENT CONDENSING
WATER SPRAY NOZZLE

WATER INLET

WATER SUPPLY
CHAMBER

WATER SPRAY
NOZZLES

PRIMARY HEATING
AND DEAERATION CHAMBER

STEAM INLET

CONICAL WATER
COLLECTOR

DEFLECTING
BAFFLE

STEAM BAFFLE
PLATE

STATIONARY
BAFFLE

OVERFLOW

SPRING LOADED
ATOMIZING VALVE

MANHOLE

DEAERATED
WATER
STORAGE

PUMP
SUCTION

Fig. 32   Direct-contact feed heater

and raises the temperature the last few degrees to its saturation point. The mixture strikes a deflecting baffle, which separates the water and steam. The hot gas-free water drops to the storage compartment.

The complete atomization and heating of the feedwater by the steam jet ensures that the dissolved gases will be released. After the atomization process, the steam and released gases flow through the primary heating/deaerating chamber where a large portion of the steam is condensed as it heats the incoming water. A small portion of the steam and all of the gases pass through the integral vent condenser, which condenses the majority of the remaining steam. The small amount of steam vapor that is mixed with the released gases is then discharged to the atmosphere or to the gland leak-off condenser.

The deaerator conditions feedwater such that its dissolved oxygen content is less than 0.005 cc per liter. In addition, it substantially reduces the carbon dioxide content of the feedwater.

Since the feedwater is at saturation temperature and above atmospheric pressure, the arrangement of the deaerator is of great importance as there is a strong possibility of the feedwater flashing into steam at the pump suction. There are two means of ensuring an adequate suction head at the main feed pump. One is to position the deaerator high in the machinery space, so that the static head developed is adequate for pump suction; an alternative is to provide a booster pump between the deaerator and feed pump, which will maintain an adequate suction head on the feed pump. Damage control considerations dictate that the booster pump arrangement be used in naval ships. The booster pump must be designed to handle condensate at saturation temperature, and it is important that the booster pump suction line be short, with little or no turns, and adequately vented so that pump cavitation and suction line flashing will not occur. The alternative of locating the deaerator high in the machinery space is the preferred arrangement with merchant ships, as damage control is not a design criterion and a pump is eliminated.

The proper performance of a deaerator requires correct sizing of components and control of the rate of flow to the storage tank portion of the deaerator. The first major consideration is the boiler steam output. This determines the size of the deaerator and affects the storage tank, the makeup valve, the transfer pump, and the number of water spray nozzles in the unit. The other major factor is the temperature of the water delivered to the spray nozzle; this temperature determines the size of the steam-regulating valve, which admits steam to the deaerator. This valve is sized as closely as possible to furnish the quantity of steam required to maintain the deaerator at the operating temperature, plus about 10% additional capacity of steam over that required to heat the inlet water at the design conditions as a safety margin to handle surges of incoming feedwater. However, since the steam capacity is considerably affected by the pipe size of the regulator and the incoming steam pressure, it is difficult to provide a valve that exactly matches the desired capacity. Proper deaeration requires that the temperature of the incoming water be raised to the saturation point; therefore, the volume of the inlet water must be controlled in relation to its temperature to stay within the heating capacity of the steam supplied by the steam-regulating valve. An excessive flow of cool water will, of course, quickly condense the steam in the deaerator, making it difficult to maintain the desired pressure. This emphasizes the necessity to provide an adequate safety margin in sizing the steam-regulating valve so that its capacity and response rate are capable of handling surges of cool water. The storage tank is usually designed to retain about five minutes of feedwater flow.

If high-pressure (high-temperature) returns are available, they may be returned directly to the deaerator storage tank. Here they will flash and provide a certain amount of steam for preheating the water introduced into the deaerator. If these returns exceed 25 to 30% of the total capacity of the deaerator, however, more steam will be available than is needed; and some other means must be employed to use the returns.

Deaerating feed heaters are normally equipped with two spring-loaded relief valves: one to prevent a high pressure from accidentally building up within the tank; and the second, known as a vacuum breaker, to prevent a high vacuum from developing in the tank by allowing atmospheric air to enter the tank in the event that the pressure in the tank drops below a prescribed value.

The shell and majority of the internals are normally of welded-steel construction; however, the steam baffles, spray nozzles, atomizing valve, and vent condenser are generally manufactured from nonferrous alloys or stainless steel.

6.7 High-Pressure Feedwater Heaters. A high-pressure feedwater heater may consist of one, two, or three sections (a three-section heater contains desuperheating, condensing, and condensate-cooling sections). All sections are normally integrated in one shell for compactness and simplicity of piping. In addition to the marine regulatory

body design requirements [3,4,5], the "Standards for Closed Feedwater Heaters" [22] are often applied. The construction features of typical high-pressure feedwater heaters are shown in Fig. 33.

The tubes are usually ⅝-in.-OD tubes, arranged on a $1^{13}/_{16}$-in. triangular pitch. The tubes in the condensing section are supported by plates spaced at intervals not exceeding 48 in. to avoid tube vibration. Sections are cut out of the support plates to provide passages for steam flow and drainage.

If the steam that enters the shell side of the feedwater heater is highly superheated (e.g., a superheat of 100 deg F or above), the tube surface in contact with the superheated steam will have a wall temperature higher than the saturated steam temperature; this means that the tube wall will not be wetted by condensate and that the transfer of heat will be low unless special precautions are made. It is found economical and sometimes essential to have a desuperheating section to control the desuperheating of the steam.

The desuperheating section is located at the feedwater exit end so that the leaving feedwater can be heated to the highest possible temperature. By arranging the flow this way, the feedwater temperature may even exceed the steam saturation temperature in the desuperheating section. The desuperheating section consists of a shroud wrapped around a group of the tubes so as to confine the inlet steam. Cross baffles are provided within the shroud to decrease the dry-vapor thermal resistance. Other design features incorporated in the desuperheating section are means to shield the other regions of the unit from the high-temperature steam and means to prevent distortion due to unequal temperature distribution.

The condensate-cooling section is located at the feedwater inlet end of the heater so that the condensate (or drains) from the condensing section can be subcooled to approach the feedwater inlet temperature, thereby increasing the heat recovery. The condensate-cooling section consists of a shroud enclosing a portion of the tubes and cross-baffle plates. When designing the condensate-cooling section of the unit, the possibility of condensate reheating must be considered. Condensate reheating can occur because the steam condensate inside the shroud, while being cooled by the feedwater inside the tubes, is also heated by the steam condensing on the outside of the shroud. This reheating of the condensate is a matter of great importance at the drain outlet end of the heater, where the temperature difference between the condensate and feedwater is often as low as 10 deg F, while the difference between the condensate and steam outside the shroud can be as high as 100 deg F. Several means can be taken to avoid excessive reheating of the steam condensate. One would be to increase the ratio of the condensate-cooling tube surface to the shroud area; another would be to insulate the shrouds.

The position of the feedwater heater as it will be arranged aboard ship must be established before the thermal design of the feedwater heater can commence, because the various alternative arrangements impose different restrictions on the thermal design of the unit.

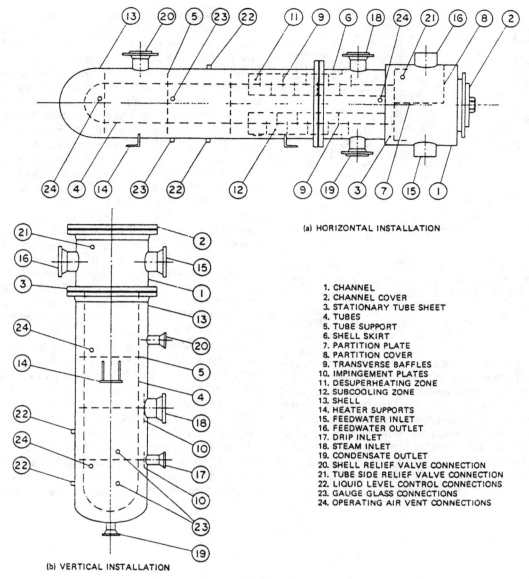

(a) HORIZONTAL INSTALLATION

(b) VERTICAL INSTALLATION

1. CHANNEL
2. CHANNEL COVER
3. STATIONARY TUBE SHEET
4. TUBES
5. TUBE SUPPORT
6. SHELL SKIRT
7. PARTITION PLATE
8. PARTITION COVER
9. TRANSVERSE BAFFLES
10. IMPINGEMENT PLATES
11. DESUPERHEATING ZONE
12. SUBCOOLING ZONE
13. SHELL
14. HEATER SUPPORTS
15. FEEDWATER INLET
16. FEEDWATER OUTLET
17. DRIP INLET
18. STEAM INLET
19. CONDENSATE OUTLET
20. SHELL RELIEF VALVE CONNECTION
21. TUBE SIDE RELIEF VALVE CONNECTION
22. LIQUID LEVEL CONTROL CONNECTIONS
23. GAUGE GLASS CONNECTIONS
24. OPERATING AIR VENT CONNECTIONS

Fig. 33 Typical feedwater heaters

When the unit is installed in a vertical position with the feedwater entrance and exit channel (or waterbox) on the top, as in Fig. 33(b), the bottom region of the shell can be used as a steam condensate collector. The shrouded condensate-cooling section extends the full length to the top of the shell in this instance, and the height of the unit is relatively short.

A vertical arrangement with the feedwater entrance and exit channel at the bottom [the inverse of that shown in Fig. 33(b)] is normally selected for long units that are designed for outdoor land installations; however, a lack of space usually precludes its application in marine plants. With this type of arrangement, the unit is designed such that the steam condensate exits at the feed inlet end in order to take advantage of the colder feed temperature for cooling. To accomplish this, it is necessary to flood a portion of the tubes with steam condensate, which results in poor utilization of part of the heat transfer surface.

When it is possible to do so, high-pressure feedwater heaters should be arranged horizontally as illustrated by Fig. 33(a). Compared with a vertical position, a horizontal arrangement affords the following advantages.

• There is less restriction on length. A heater of longer length normally results in a smaller shell diameter and a more economical unit.

• The reheat problem is less severe in the steam condensate-cooling section because this section need not be exposed to the steam and a short condensate-cooling section can be used to achieve the proper proportion of the tube surface to shroud surface.

• A higher condensing heat transfer coefficient is achieved on a horizontal tube bundle than on a tube bundle

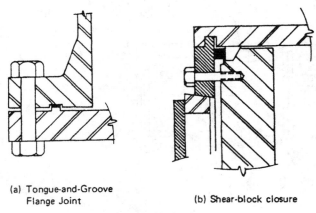

(a) Tongue-and-Groove
Flange Joint

(b) Shear-block closure

Fig. 34   Tube-side closures

that is vertical. This fact is not reflected in most design analyses; however, it should be considered by the designer so as to provide the most effective utilization of heat transfer surface.

Of major importance in the design of high-pressure feed heaters is the adequacy of the closures for the pressures and temperatures involved. A breakable joint should be provided for the shell-side closure so that the shell can be removed from the tube bundle for inspection and cleaning purposes. Under the usual operating conditions, a bolted flange joint is suitable for this purpose. However, under conditions that would involve temperature distortion or when there is an infrequent requirement for removal of the shell, a welded joint provides positive sealing and is economical to fabricate. A backup ring is provided to protect the tubes when a flame cut is made to open the welded joint.

The feedwater that enters the channel and the tube side of the unit is under a relatively high pressure, which imposes a severe requirement on the tube-side closure and seal design. The bolts of a flanged joint are required to take the hydrostatic load (which depends on the closure diameter and fluid pressure) and at the same time maintain a pressure on the gasket sufficient to ensure a seal. This often results in huge bolts that require enormous torques to tighten, especially so when the larger shell diameters are involved with pressures over 1200 psig. Nevertheless, bolted-flanged joints can be properly applied in the design of the heaters with diameters as large as 20 in. and for pressures of less than 1200 psig.

Flat metal or metal-jacketed gaskets are frequently used with bolted tube-side closures. The force required to adequately compress the gasket is a substantial percentage of the hydrostatic load on the end closure and may even exceed it. The force required to obtain an adequate gasket seat can be reduced by narrowing the width of the gasket and at the same time confining the gasket to an enclosed space to prevent it from deflecting freely. An example of this design feature is illustrated by the tongue-and-groove flanged joint in Fig. 34(a).

A shear-block closure design that is used in some of the larger and higher-pressure feed heaters is depicted in Figs. 34(b) and 35. The hydrostatic load is resisted by the shear ring, and the pressure on the gasket is maintained by the hydrostatic load. Figure 35 shows the method by which the tube-side operating pressure is used to seat and seal the solid copper ring gasket.

With the larger high-pressure heaters, a point is reached where a simple bolted-flange closure must give way to a more elaborate high-pressure closure; experience indicates that when the product of the operating pressure

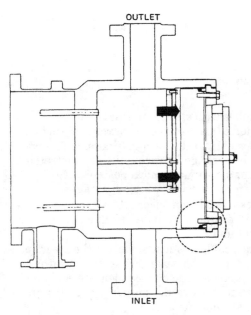

OUTLET

INLET

(a) SHEAR-BLOCK TYPE CHANNEL COVER

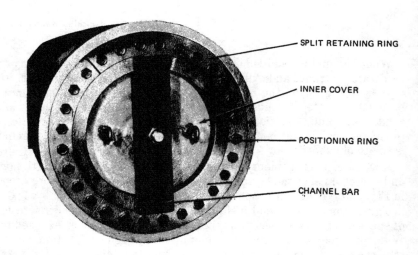

SPLIT RETAINING RING

INNER COVER

POSITIONING RING

CHANNEL BAR

(b) CONSTRUCTION DETAILS OF SHEAR-BLOCK TYPE CLOSURE

Fig. 35   Shear-block closure

(in psi) and the shell inside diameter (ID) (in inches) exceeds 25,000, a bolted-flange joint is no longer economical. In actual design practice, the availability of a standard flange for the size and pressure intended often decides the question as to the specific closure design. The economic advantage of a ready-made versus a custom-made item will often influence the selection of the specific closure.

6.8 Gland Leak-Off Condensers. In order to avoid an ingress of air to the steam system at the points where the steam turbine shaft penetrates the turbine casing, and similar locations, a gland-sealing steam system maintains a pure steam atmosphere at a pressure slightly above atmospheric just outside the turbine shaft-casing interface; this ensures that atmospheric air will not enter the turbine.

The gland leak-off system consists of a fan, which removes an air and steam mixture from the turbine gland leak-off pockets, and a condenser through which the mixture is drawn in order to condense the steam so as to recover the water and reduce the quantity of gas that the fan must handle.

The gland leak-off condenser may be either furnished as a separate heat exchanger or combined as a section of the low-pressure feed heater (see Section 6.5). When it is a separate exchanger, it is usually of the U-tube design and is arranged to receive the low-pressure gland leak-off steam in the shell; cooling water is passed through the tubes to condense the steam.

The fan, which forms a part of the gland leak-off system, is designed to handle the leakage air plus the uncondensed steam vapor. The gland leak-off exhauster (which is the common name for this fan) is usually mounted on top of or immediately above the condenser (or condenser section of the first-stage heater).

The Navy has set forth specific design and material requirements for this condenser and similar condensers for other shipboard applications [23].

6.9 Unfired Steam Generators. Unfired steam generators, which are more specifically known as contaminated water evaporators or steam service evaporators and reboilers (in the case of those used with nuclear reactors), supply low-pressure steam at a pressure of 50 to 150 psig to a system that is independent of the main steam system. The independent system provides steam for services that could possibly contaminate the main system in the event of a malfunction.

a. Contaminated water evaporators. Contaminated evaporators operate on bleed steam from the high-pressure turbine, auxiliary steam, or in some cases, high-pressure steam (up to 500 psig). Some of the "contaminated" services include fuel-oil heaters, cargo-tank heating coils, galley and heating systems, and steam-driven deck machinery. The contaminated evaporator tube-nest drains are normally piped to the deaerating feed heater via a trap. Figure 36 depicts a typical contaminated steam system. The bleed steam is a variable-pressure source, which depends on the percentage of full power being developed. The requirements of the contaminated system also vary greatly. The operational requirements, as depicted on a normal-power plant heat balance, may be only a small percentage of the maximum performance requirements; therefore, the maximum performance requirement must be taken as the design condition with checks made to ensure that the design is satisfactory for other operating arrangements.

Two sets of design conditions are generally set up: one for bleed-steam operation and one for auxiliary-steam operation. Both conditions must be considered to determine the effects on the evaporator design as well as the safety valves and orifices.

The evaporator bundle is normally of the U-tube type as shown by Fig. 37. When the tube bundle becomes very large, straight tubes should be used, incorporating the outside-packed head type of construction illustrated in Fig. 38. Tubes are usually $\frac{5}{8}$ in. OD on a $\frac{15}{16}$-in. square pitch, except in the case of a low-temperature difference when a closer pitch may be used ($\frac{7}{8}$ in.). When $\frac{3}{4}$-in.-OD tubes are used, they are placed on a $1\frac{1}{8}$-in. square pitch.

The standard fittings furnished with the evaporator include feed control, safety valve, condensate trap, water gage glass, steam supply regulating valve, thermometers (normally two), pressure gages (normally two), and a test cock.

The overall clean-tube heat transfer coefficients that are used to determine the evaporator bundle size are as follows:

$(T_{steam} - t_{feed})$	$\leq 50°$ F	$\leq 100°$ F	$> 100°$ F
Overall heat transfer coefficient	550	520	500

A steam flow orifice is usually installed in the steam supply line to limit the amount of heating steam entering the evaporator and reduce the steam supply pressure (a high heating steam pressure is not always beneficial due to the critical heat flux or the vapor blanketing phenomenon in boiling). The orifice size is determined from the expression:

$$A_o = \frac{W_{stm}}{105.3[P_2(P_1 - P_2)]^{0.5}} \quad \text{if } P_2 > 0.58P_1$$

$$= \frac{W_{stm}}{51.45P_1} \quad \text{if } P_2 \leq 0.58P_1$$

where

A_o = orifice area, in.^{2}
P_1 = steam supply upstream pressure, psia
P_2 = steam supply downstream pressure, psia
W_{stm} = steam flow rate, lb/hr

The safety valve should be sized for the maximum steam flow entering the generator; that is

$$W_{stm} = 105.3 A_o[P_4(P_3 - P_4)]^{0.5}$$

or

$$W_{stm} = 51.45 A_o P_3 \quad \text{if } P_4 \leq 0.58P_3$$

where

P_3 = set pressure of supply line safety valve (assume P_3 to be 10% above auxiliary steam line pressure), psia

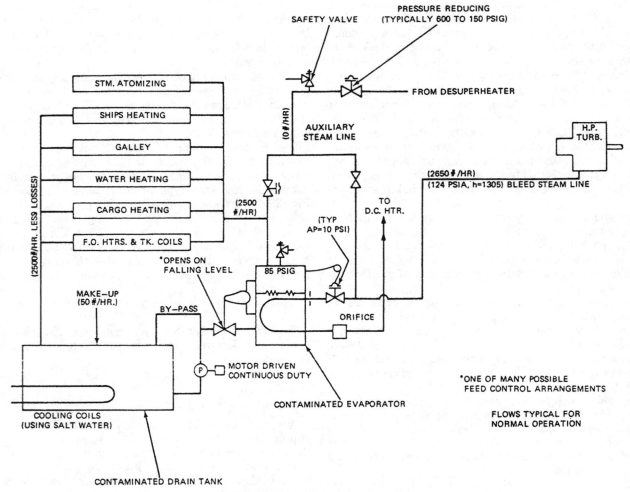

Fig. 36 Typical contaminated steam system

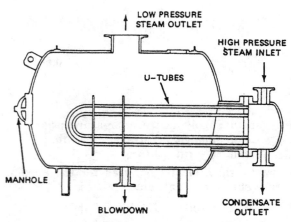

Fig. 37 Typical contaminated water evaporator

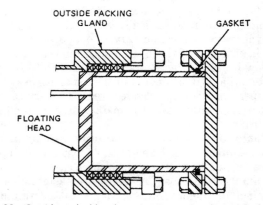

Fig. 38 Outside-packed head construction used on large tube bundles

P_4 = set pressure of shell safety valve, psia

b. Steam reboiler. A steam reboiler typically condenses 450 psig steam from the reactor secondary loop and evaporates reboiler feedwater. The reboiler supplies 150 psig steam to the 50-psig service steam system, the 100-psig laundry steam system with reducing stations, the ship's whistles, the distilling plant air ejector, the low-pressure steam drain collecting tank air ejector, the oil heating services, the steam-out connections, the decontaminating sink heating coils, and the 25-psig steam reducing station.

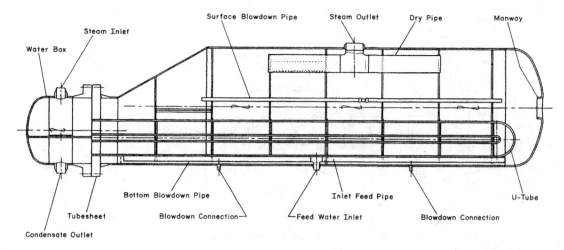

Fig. 39 Steam reboiler

As illustrated in Fig. 39, a steam boiler is typically of the horizontal shell-and-U-tube type. The tubesheet is welded to the shell, and the waterbox is bolted to the tubesheet. A manway is installed in the shell head for inspection of the U-tubes and boiler internals.

The steam outlet connection extends down into the shell and connects to a dry pipe. The function of the dry pipe is to remove entrained condensate in the steam. The dry pipe is capable of producing 99% quality steam. The dry pipe is a perforated pipe that is located in the steam dome area above the tube bundle. The perforations are in the top portion of the pipe. In addition to the holes for the steam, the pipe contains drain holes at both ends to permit the condensate to drain back into the shell.

The feedwater inlet connection extends into the shell and connects to a perforated inlet feed pipe. The perforations extend the length of the pipe and uniformly distribute the flow along the length of the bundle.

The reboiler is provided with a surface and bottom blowdown. The surface blowdown is performed by a perforated pipe extending the length of the bundle and located above the shell centerline. The pipe is connected to a shell nozzle. The bottom blowdown is accomplished by a slotted pipe that extends the length of the tube bundle. Because the bottom blowdown removes sludge or precipitated solids, it is provided with two nozzles located approximately at the quarter points in the pipe. The amount, frequency, and type of blowdowns are determined by an analysis of the feedwater in the shell side of the reboiler.

The shell side of a reboiler is also provided with the following connections:

- two liquid-level gage glass connections,
- two liquid-level control sensing connections,
- two relief valve connections for over pressurization protection,
- one pressure gage connection,
- one sampling connection for checking the boiler water chemistry, and
- one vent connection used for filling and draining operations.

Periodically the reboiler is cleaned to remove deposits of grease, dirt, or soft scale from the U-tube external surfaces. This operation includes using a high-velocity water lance to remove as much soft scale and sludge from the tube bundle and shell as possible, followed by chemical cleaning.

6.10 Hot-Water Heaters. Hot-water heaters are used to provide water at controlled temperatures for shower, lavatory, galley, and similar services. Hot-water heaters are generally steam heated and can be classified into two types: storage heaters and instantaneous heaters. A storage type of heater has the advantage that the water can be heated and stored during nonpeak periods of hot-water demand, thus reducing the peak heating-steam requirement. Also, a less sophisticated temperature control device is required for a heating rate that is independent of the rate of hot-water withdrawal. The disadvantages of a storage type of heater are the heat lost from the hot water in storage and the heater's bulk; however, the heat loss is rarely a deciding factor. Space and weight are usually at a premium aboard ship, in which case an instantaneous hot-water heater, which heats the water as fast as the rate of withdrawal, should be considered since a large storage tank is not required.

Figure 40(a) illustrates the storage type of hot-water heater and its accessories. The notable features of this heater system are the pump, which circulates the water from the tank to the heating element, and the thermal bulb [labeled "remote bulb" in Fig. 40(a)] located at the outlet of the heating element. It can be seen from this arrangement that the heat input to the system is independent of the rate of hot-water withdrawal. The flow circulation rate, together with the thermal bulb and the temperature of the water to be circulated (i.e., temperature of the water at the lower part of the tank, which may or may not mix with the inlet cold water), will dictate the heat input rate through the action of the control valve. The control valve regulates the amount of steam admitted as well as the steam pressure. If the water circulation rate is sufficiently larger than the peak hot-water withdrawal

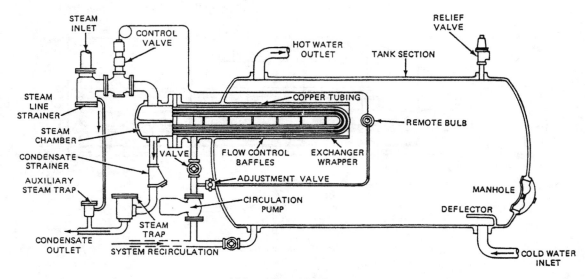

(a) Storage Type of Hot-Water Heater

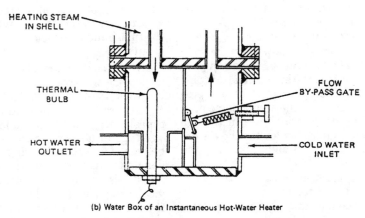

(b) Water Box of an Instantaneous Hot-Water Heater

Fig. 40 Hot-water heaters

rate, then the storage tank is not necessary and the storage type of heater becomes an instantaneous heater.

The most complex aspect of a hot-water heating system lies in the scheme of temperature control. Figure 40(b) illustrates the principles involved in controlling the temperature of a typical instantaneous hot-water heater. The flow bypass gate bypasses a small portion of the inlet cold water, and the bypassed cold water is directed to flow over the thermal bulb (or temperature-sensing element) located in the water outlet channel. The amount of cold water bypassed depends not only upon the adjustment of the bypass opening but also on the pressure difference between the inlet and outlet sections of the waterbox. The pressure difference between the inlet and outlet sections of the waterbox is created by flow friction inside the heating tubes and, therefore, is flow-rate sensitive. Consequently, the thermal bulb, which actuates the steam control valve, senses the combined effect of the hot-water outlet temperature, the cold-water inlet temperature, and the hot-water withdrawal rate; as a result, the time lag of the steam control-valve action is minimized. The shrouds

around the thermal bulb point up another critical aspect of the design, namely, that the flow around the sensing element should be guided to produce an accurate signal that gives a prompt indication of an incipient temperature change.

The permissible variation in the hot-water outlet temperature should not be overspecified. A hot-water outlet temperature variation of 10 deg F will result in a hot-water heater system of less expense and complexity than would be the case if a 5 deg F variation were specified. However, there is a tendency for hot-water heater procurement specifications to stipulate an outlet temperature control of ±5 deg F at varied flow rates [23].

6.11 Fail-Safe Heat Exchangers. Some heat exchanger applications require a design that precludes the entry of one fluid into the other in the event of a tube or tubesheet leak. For example, the contamination of condensate systems by seawater or fuel oil must be prevented. Electronic equipment coolers and hydraulic system oil coolers are also areas where "fail safe" heat exchangers are used. Such arrangements are advantageous in many

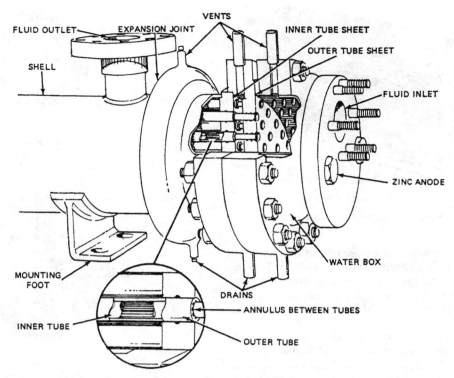

Fig. 41 Double-tube, double-tubesheet heat exchanger

naval applications since the general lack of "contaminated" steam systems on Navy ships increases the necessity to take adequate precautions to prevent oil contamination [21].

Double-tubesheet and double-tube, double-tubesheet exchangers are often specified for Navy applications where a fail safe feature is desired [6]. Figure 41 is a cutaway view of a double-tube, double-tubesheet heat exchanger as provided for hydraulic-oil cooling aboard submarines. Any leak that develops at the tube-to-tubesheet joint or as a result of a tube rupture will flow to the void between the double tubesheets and out the drain. Alarm devices may also be incorporated to automatically warn of leakage into this void space. Similar designs with only the double-tubesheet feature are used on seawater-freshwater coolers to reduce the risk of saltwater contamination.

6.12 Reactor Water Purification Heat Exchangers. The reactor plant purification system contains a regenerative and a nonregenerative purification heat exchanger. The function of the exchangers is to reduce the temperature of the reactor plant water prior to it entering the demineralizer. The regenerative purification heat exchanger has reactor plant water flowing on both sides of the unit. The nonregenerative purification heat exchanger has reactor plant water flowing in the tubes and fresh water flowing in the shell.

The purification heat exchanger is a U-bend or hairpin shell-and-tube heat exchanger. In a conventional heat exchanger, the U-tubes are in a single shell. The purification heat exchanger can be best described as each leg of the

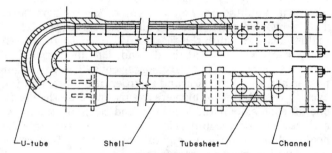

Fig. 42 Reactor water purification heat exchanger

U-tube having a separate shell. The shells are connected by a half-torus that also encases the bends of the tube. The heat exchanger is categorized as an integral shell, tubesheet, and head assembly. A cover is bolted to the head for access to the tube ends. A typical purification heat exchanger is illustrated in Fig. 42.

The U-tubes are used as a means of separating the thermal expansion in the tubes from that of the shell and still maintaining a single-pass, counterflow, purification heat exchanger. The U-bend shell-and-tube heat exchanger is utilized when it is not practical to locate an expansion joint in the shell. The regenerative and nonregenerative heat exchangers are usually identical, thus making the heat exchangers physically and functionally interchangeable.

6.13 Motor-Generator Heat Exchanger. A motor-generator is generally a deck-mounted machine that is used

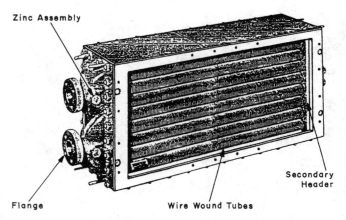

Fig. 43   Motor-generator heat exchanger

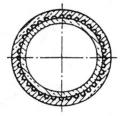

(a) Inner and outer tube detail

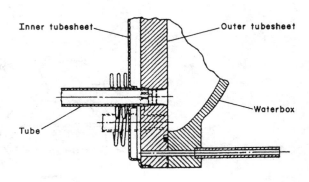

(b) Tube, tubesheets and waterbox detail

Fig. 44   Construction details of motor-generator heat exchanger

to supply electrical power in either direction between the d-c bus and the a-c ship's service power system. It is composed of an a-c machine and a d-c machine, both of which can operate either as a motor or a generator. A motor-generator is typically rated at 300 kW either a-c to d-c or d-c to a-c.

A motor-generator is totally enclosed, and two motor generator heat exchangers, as illustrated in Fig. 43, are used to cool the recirculated air. One heat exchanger is mounted above the a-c unit with its water connections situated side by side. The other is mounted above the d-c unit with its water connections one above the other. A blower is mounted on the outer end of the d-c armature. This blower draws the heated air from within the frame and impels it through the filters and air duct, which lead to the heat exchanger at the d-c end. The air is cooled as it passes across the heat exchanger coils. The cooled air passes across the reactor and rectifier assembly, and the cycle is repeated. Two blowers are mounted on the shaft, one on each side of the a-c rotor, and circulate the air in a clockwise and a counterclockwise pattern through the heat exchanger at the a-c end to cool the rotor, field resistors, and the rectifier transformer.

The heat exchangers are the wire-wound, double-wall, tube type, as shown in Fig. 44(a). The double walls prevent the entrance of seawater into the airstream. Actually, this is an arrangement of two tubes, one inside the other. The inside surface of the outer tube is grooved to provide a passage for seawater, which may leak from the inner tube. The inner tube extends through an inner tubesheet to the outer tubesheet as shown in Fig. 44(b). The outer tube extends only to the inner tubesheet. Vents are provided at the top of the heat exchangers. These are 0.405-in.-diameter tubes, one on each end of the two heat exchangers, which must remain open for the double-tube leakage warning system. Drains are provided at the bottom of the heat exchangers. There are also 0.405-in.-diameter tubes, one on each end of the two heat exchangers, which serve to warn of tube leakage. The 0.540-in.-diameter tubes, two on each end, are conventional closed vent and drain tubes and should be connected to proper piping to allow drainage of water or air from the cooling system. The temperature within the motor-generator is regulated by controlling the flow of water through the heat exchanger by use of a throttling valve.

References

1   "Standards of Tubular Exchanger Manufacturers Association," Tubular Exchanger Manufacturers Association.

2   "ASME Boiler and Pressure Vessel Code," Division I, Section VIII, American Society of Mechanical Engineers.

3   *Rules and Regulations for the Construction and Classification of Steel Ships*, Lloyd's Register of Shipping, London.

4   *Rules for Building and Classing Steel Vessels*, American Bureau of Shipping, New York.

5   *Code of Federal Regulations*, Title 46—Shipping, Chapter I—Coast Guard, Department of Transportation, Subchapter F—Marine Engineering, published by Office of the Federal Register.

6   "Coolers, Fluid, Industrial, Naval Shipboard Lubricating Oil, Hydraulic Oil, and Fresh Water," Military Specification MIL-C-15730.

7   "Anodes, Corrosion Preventive, Zinc, and Plugs, Zinc Anode Retaining: Design of and Installation in Shipboard Condensers and Heat Exchangers," Military Specification MIL-A-19521.

8   R. A. Bowman, A. C. Mueller, and W. M. Nagle, "Mean Temperature Difference in Design," *Trans.* American Society of Mechanical Engineers, May 1940.

9   E. N. Sieder and G. E. Tate, "Heat Transfer and Pressure Drop of Liquids in Tubes," *Industrial and Engineering Chemistry*, Vol. 28, 1936.

10   Daniel A. Donahue, "Heat Exchangers," A Special Petroleum Processing Report, March 1956.

11   Kenneth J. Bell, "Final Report of the Cooperative Research Program on Shell and Tube Heat Exchangers," University of Delaware Engineering Bulletin No. 5, Jan. 1963.

12   Townsend Tinker, "Shell Side Characteristics of Shell and Tube Heat Exchangers—A Simplified Rating System for Commercial Heat Exchangers," *Trans.* American Society of Mechanical Engineers, 1958.

13   Abe Devore, "Try This Simplified Method for Rating Baffled Exchangers," *Petroleum Refiner*, May 1961.

14   D. L. Katz, J. G. Knudsen, G. Balekjian, and S. S. Grover, "Fouling of Heat Exchangers," *Petroleum Refiner*, 1954.

15   H. H. Keller and E. V. Somers, "Heat Transfer from an Annular Fin of Constant Thickness," *Trans.* American Society of Mechanical Engineers, 1959.

16   S. S. Chen and Ho Chung, "Design Guide for Calculating Hydrodynamic Mass, Part 1: Circular Cylindrical Structures," Argonne National Laboratory, Report No. ANL-CT-76-45, Ill.

17   Y. N. Chen, "Flow-Induced Vibration and Noise in Tube-Bank Heat Exchangers Due to Von Karman Streets," *Trans.* American Society of Mechanical Engineers, Series B: *Journal of Engineering for Industry*, Vol. 90, 1968, pp. 134–146.

18   R. T. Hartlen, "Wind-Tunnel Determination of Fluid-elastic-Vibration Thresholds for Typical Heat Exchanger Tube Patterns," Ontario Hydro Report No. 74-309-K, Aug. 1974.

19   "Escoa Engineering Manual, Heavy Duty Welded Fintube," Escoa, Pryor, Okla.

20   W. M. Kays and A. L. London, *Compact Heat Exchangers*, McGraw-Hill, New York, 1964.

21   "Heaters, Fluid, Fuel Oil, Naval Shipboard," Military Specification MIL-H-16313.

22   "Standards for Closed Feedwater Heaters," Heat Exchanger Institute Incorporated.

23   "Condensers, Steam, Surface, Naval Shipboard," Military Specification MIL-C-15430.

CHAPTER XVII

Frank J. Zarambo | # Desalination Plants

Section 1
Desalination Plant Designs

1.1 Introduction. Desalination plants are used aboard ship to produce high-purity fresh water from seawater. Desalination plants are also used to purify freshwater supplies of insufficient quality. The capacity of marine desalination plants varies from several hundred gallons per day (24 hours) up to 100,000 gallons per day (gpd) or more, depending upon the size, type of propulsion plant, and purpose of the ship.

A supply of fresh water is required to furnish high-purity makeup water for boilers and potable water for drinking, cooking, dishwashing, ablutionary, hospital, and laundering purposes. Additional capacity is often incorporated to supply freshwater sanitary systems and for special freshwater cooling systems, such as used with some electronics equipment.

Fresh water can be produced from seawater by using either a thermal or a membrane desalination process. The thermal distillation process involves the use of heat to physically separate fresh water from the dissolved solids in seawater by transforming the freshwater portion of the seawater feed to vapor, which is subsequently condensed. Of the membrane desalination processes that may be considered, reverse osmosis provides a proven capability to effectively and continuously produce fresh water from seawater. The reverse-osmosis desalination process is based upon reversing the phenomenon of osmosis. Osmosis is the natural form of solution diffusion that causes a less-concentrated liquid solution to flow through a semipermeable membrane into a more-concentrated solution.

The specifications for distilling plants used on naval ships [1,2] and those published by the U.S. Maritime Administration [3] require a distillate purity of less than 0.065 equivalent parts per million of chlorides. This is equal to $\frac{1}{4}$ of a grain of sea salt per gallon, or about 4.3-ppm total dissolved solids. Most marine distilling plants can produce distillate substantially below these limits. The specifications for nuclear-powered naval ships require a distillate purity of one-half that of the normal specifications, or $\frac{1}{8}$ grain of sea salt per gallon. Tests on marine distilling plants demonstrate that the units are capable of producing water containing less than 1.0 ppm total dissolved solids.

The purity of the fresh water produced by membrane processes is permitted to be less than that specified for thermal distillation processes. The World Health Organization standards permit a maximum salinity content of 500 ppm of total dissolved solids for potable water uses [4].

When operating in normal seawater of 35,000-ppm solids (a definition of "normal" seawater, which is found in most areas of the Atlantic and Pacific Oceans, is given in *The Oceans* by Sverdrup [5]), a distilling plant should be capable of operating for a period of at least 90 days at rated capacity without shutdown for cleaning. Depending on the specific design, this can be accomplished with or without chemical feed treatment. Such extended operating periods without shutdown for cleaning is a requirement of both the standard Navy specifications and U.S. Maritime specifications for shipboard distilling plants.

1.2 Early Distilling Plant Designs. The advent of the steam engine and its adaptation to seagoing ships provided the impetus to develop the steam-heated distilling plant concept to provide makeup feedwater for the boiler. The large quantity of saturated steam available made it a logical source of energy for a distilling plant. As shown by Fig. 1, the early single-effect distilling plants were designed to consist essentially of an evaporator to evaporate seawater by means of steam-heated tubes or coils, a condenser (called a distilling condenser or more simply a distiller) to condense the vapor, and the necessary interconnecting piping, valves, etc. to allow operator control and provide a packaged system.

Multiple-effect plants were generally used for the larger capacities so as to reduce the amount of operating steam required. In multiple-effect plants, the vapor from the first evaporator (first-effect evaporator) is condensed in the coils of a second evaporator (second-effect evaporator), thereby serving as the heating medium, etc. with the vapor from the last effect only being condensed in the distiller.

Figure 2 illustrates a type of submerged-tube evaporator commonly used in the early distilling plants. The horizontal-tube evaporator shown by Fig. 2 required conservative heat-transfer design criteria in order to attain a level of performance considered acceptable in its day. With brine temperatures maintained below 130 F, the fouling of tube surfaces was minimal, and for brine temperatures up to 180 F the scale formation was not considered excessive provided the temperature gradient across the heating tubes was maintained below 50 F. At temperatures above

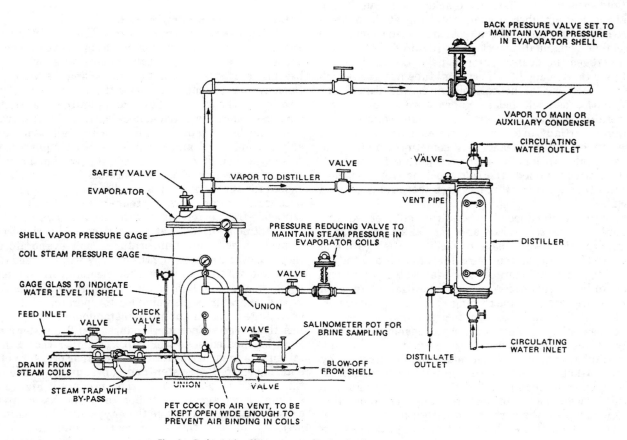

Fig. 1 Early single-effect submerged-tube distilling plant arrangement

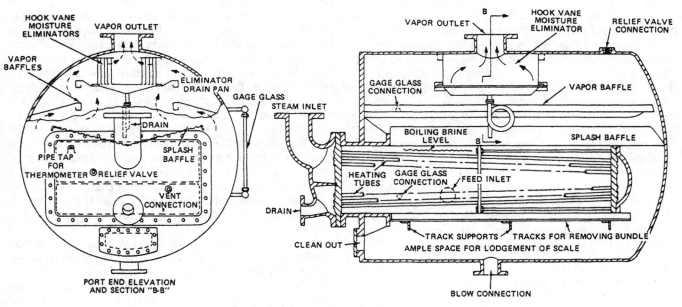

Fig. 2 Early horizontal straight-tube evaporator

180 F, scale formed rapidly such that vapor temperatures above 250 F were not considered practicable.

"Cold shocking" or thermally cracking the scale formation from heating tubes by suddenly changing the temperature of the tube metal was a technique used to minimize the extent of manual tube cleaning required. Cold shocking improved the heat-transfer rates of the tubes, especially for the so-called "self-scaling" tube designs, which were in the form of helical coils or bowed tubes with restrained ends such that the tubes were distorted by a temperature change. However, the cold-shocking process was only partially effective. Scattered patches of scale were left on the tubes and they sometimes built up to a considerable thickness before cracking off. The heat-transfer rates obtained after cold shocking were considerably less than for tubes that had been thoroughly cleaned (descaled).

The high maintenance associated with submerged-tube evaporators was recognized and, insofar as practicable, design provisions were made to facilitate their care and cleaning. However, despite the efforts to improve the design of the early submerged-tube evaporators, they continued to be characterized by high operating costs, high maintenance costs, complex piping systems, and poor performance under continued service conditions.

The "basket" type of evaporator was developed to avoid many of the problems associated with the early submerged-tube evaporator concepts. As illustrated by Fig. 3, a basket evaporator is configured as a deeply corrugated cylindrical chamber. The finger-like corrugations provide a large amount of heat-transfer surface in a limited volume; also, the corrugated design permits the removal of accumulated scale by cold shocking. Cold shocking involves draining the brine from the unit and admitting steam at about 15 psig (250 F) to the basket heating section. The flat sides of the corrugations expand and the scale on the basket exterior surface dries. The steam line is then secured, and cold seawater is allowed to cascade over the basket. The steam inside the basket condenses, and the pressure inside drops from 15 psig to approximately 28 in. Hg vacuum, which causes the flat sides of the basket to contract. The scale is, consequently, cracked off the basket. The basket-type of evaporator design permits a compact configuration, which provides advantages in some submarine, and other, applications. However, the development of the flash type of distilling plant and effective scale-control chemical feed treatment was responsible for the obsolescence of the basket-type evaporators.

The state of the art of distilling plant design has evolved considerably. A variety of distilling plant designs has been developed from which the most appropriate system for a given marine power plant can be selected. The desalination plant designs most commonly considered for marine use can be broadly classified as those using:

- Flash evaporators
- Spray-film evaporators
- Submerged-tube evaporators
- Plate-type evaporators

- Vapor-compression cycles
- Membrane processes

1.3 Flash Evaporator Designs.

a. Marine flash evaporators. The multistage flash evaporator was a revolutionary design development because, unlike submerged-tube evaporators, which have heating surfaces submerged in boiling seawater, the flash-evaporator concept causes the seawater to "flash" into water vapor; no boiling occurs on the heat-transfer surfaces. Since the initial application of flash evaporation to marine distilling plants, the majority of all steam-driven ships, commercial and naval, have been equipped with flash-type distilling plants for the supply of makeup feed and potable water. Flash evaporators became widely accepted for application in marine distilling plants primarily because their low operating temperatures result in virtually scale-free performance with normal seawater. Chemical treatment of feedwater is not normally required in connection with low-temperature flash evaporators as they can operate three months or more without feed treatment and without shutdown for cleaning. In addition to being smaller and lighter for a given capacity, the flash-type evaporator is also simpler to operate, which is an advantage from an automation standpoint.

A flash-type evaporator produces distilled water by discharging heated seawater into a chamber that is maintained at a pressure lower than the vapor pressure of the entering heated seawater, and subsequently condensing the flashed water vapor thereby produced. Figure 4 is a simplified diagram that illustrates the basic internal construction of a two-stage flash-type distilling plant. The majority of the flash-type marine units are of the multistage design with integral horizontal condenser tube bundles. The heated seawater is introduced through a flashing device in each stage to achieve the most effective flashing of a portion of the incoming feedwater (heated seawater) into vapor. The design of the flashing device is of critical importance because it must provide an optimum feedwater spray pattern that facilitates the flashing process, and it must also maintain a pressure differential between successive stages. As the feedwater leaves the flashing device, the spray is directed downward in curtains of water. The first separation between the flashing vapors and the feedwater occurs at this point. The downward velocity of the feedwater directs all but the fine droplets to the lower section of the flash chambers. The small droplets that are entrained in the flashing vapor are removed in the demisters located in the upper portion of the distilling plant, after which the vapors pass to the stage condenser. It is most important that the design incorporate a positive seal around the demisters to prevent the passage of any salt-laden droplets around the demisters.

Since the vapor pressure of the heated feedwater entering the first stage is higher than the first-stage chamber pressure, a portion of the feedwater flashes into vapor, thereby removing heat from the feedwater until the temperature of the feedwater reaches the saturation temperature corresponding to the chamber pressure. The heated

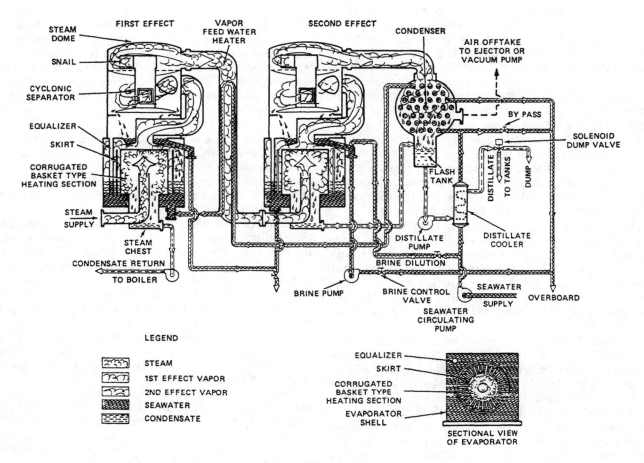

LEGEND

STEAM
1ST EFFECT VAPOR
2ND EFFECT VAPOR
SEAWATER
CONDENSATE

EQUALIZER
SKIRT
CORRUGATED BASKET TYPE HEATING SECTION
EVAPORATOR SHELL

SECTIONAL VIEW OF EVAPORATOR

Fig. 3 Double-effect basket-type distilling plant

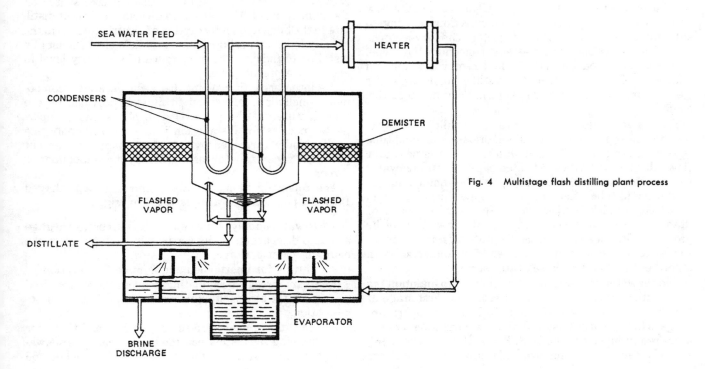

Fig. 4 Multistage flash distilling plant process

feedwater is evaporated by a pressure reduction instead of a temperature elevation. After an equilibrium condition is reached in the first stage, the remaining feedwater (brine) is introduced into a second stage that is maintained at a lower pressure than the first, and the process is repeated. The distillation process can operate from a positive pressure in the first stage to a high vacuum in the last stage, with a stage-to-stage pressure differential being the key to repeated flashing.

The freshwater distillate is collected in the bottom of the condenser section of the stage and is allowed to flow to the next stage condenser, where it is combined with the distillate from that stage. The process is repeated, depending on the number of stages, and the distillate is removed from the last stage by a distillate pump.

Flash evaporators generally use a "once-through" system whereby the entire seawater feed is preheated through the stage condensers and seawater feed heater and allowed to flash down through the stages. The resultant brine from the last stage is pumped overboard. Approximately 20 gallons of seawater is required for each gallon of fresh water produced in a once-through flash distilling plant.

The flash distilling plant schematic illustrated by Fig. 4 shows that the entering seawater feed is used as the condenser coolant, thereby reducing the quantity of heat which must be added in the seawater heater. This feature is shown in more detail in Fig. 5, which is a flow diagram for a typical 8000-gpd, two-stage, flash distilling plant.

As indicated by Fig. 5, the initial vacuum in the stages is normally created by a high-pressure steam-driven air-ejector vacuum system, but motor-driven vacuum pumps or water-motivated air eductors are sometimes used where high-pressure steam is not available. The stage condenser bundles are internally baffled to direct the flow of noncondensible gases to vent connections leading to subsequent stages or to the air ejector. The air ejector is normally of the one- or two-stage noncondensing type. It is suitably sized for the efficient removal of air and noncondensible vapors from the distiller condensers. Air ejectors are covered in more detail in Section 2.6 of this chapter.

The seawater heater and separate or "built-in" air ejector condenser are usually of the multipass shell-and-tube type with removable flanged waterboxes. The seawater flows through the tubes. At rated capacity, the seawater velocity through the heat exchanger tubes should not exceed 6 fps in tubes made of 90–10 copper-nickel, 9 fps in 70–30 copper-nickel tubes, or 12 fps in tubes made of titanium. The condensers and seawater heater may be provided with means for cathodic protection in a manner similar to that described in Chapter 16 in connection with other heat exchangers in seawater service.

The seawater heater is usually designed to maintain the temperature of the feedwater entering the first stage of the evaporator between 165 and 175 F. This temperature range is high enough to satisfy the pasteurization temperature requirements of the U. S. Public Health Service [6], and is low enough to minimize scale formation. The steam

condensate temperature from the seawater heater does not exceed 200 F under normal operating conditions.

A shell-and-tube distillate cooler, or perhaps another type, is installed when the evaporator is to provide potable water. The distillate cooler is designed to provide a distillate temperature leaving the distilling plant of 95 F or less with an initial seawater temperature of 85 F. Occasionally, this cooler is incorporated as an integral part of the last-stage condenser bundle.

Shown in Fig. 6 is a feature frequently provided for distilling plants that must operate in polluted harbors. When in polluted harbors, it is desirable to use water from shore as feedwater to the distilling plant. To reduce the amount of shore water that is required to operate the distilling plant, the feedwater is recirculated through the seawater heater. Temperature levels are maintained in the distilling plant by circulating harbor water through the stage condensers and then overboard. Thus, the polluted harbor water is not put into the evaporator heater and the flash chambers; this is an important feature in some harbors that are highly polluted and contain debris and mud, which can seriously impair normal plant operation by clogging and fouling high-temperature tube surfaces. The most serious problem caused by polluted feedwater is the introduction of highly corrosive gases, such as hydrogen sulfide and ammonia, which when heated can attack copper-bearing alloys.

Marine distilling plants incorporate a solenoid-operated, three-way dump valve that is located in the distillate discharge line. The valve discharges the distillate to the bilge if the salinity of the distillate exceeds 0.25 grains of sea salt per gallon (or any other specified preset control point). The solenoid is energized by the salinity-indicating system and operates whenever the salinity system is energized, regardless of the position of the cell selector switch on the salinity panel. The valve also diverts the flow of distillate to the bilge upon interruption of electric current to the salinity indicator. A high-salinity audible alarm is usually installed to indicate the occurrence of a salinity level in excess of a preset value.

All interconnecting piping between packaged components (including valves and fittings) is provided by the manufacturer of the plant. Orifice plates and desuperheater nozzles for installation in the steam supply line are normally furnished by the plant manufacturer but packaged separately for installation by the shipbuilder or owner.

The following items are also common to a packaged flash evaporator such as illustrated by Fig. 7:

- Seawater supply pump and motor with pressure gage at pump discharge usually separately mounted from the distilling plant package.
  - Controls for unattended and automatic operation:
1. Steam pressure-regulating valve that is required when the steam supply pressure is subject to fluctuation.
2. Seawater temperature-sensing switch and timer for dumping the distillate to the bilge in case the seawater temperature leaving the heater falls below 165

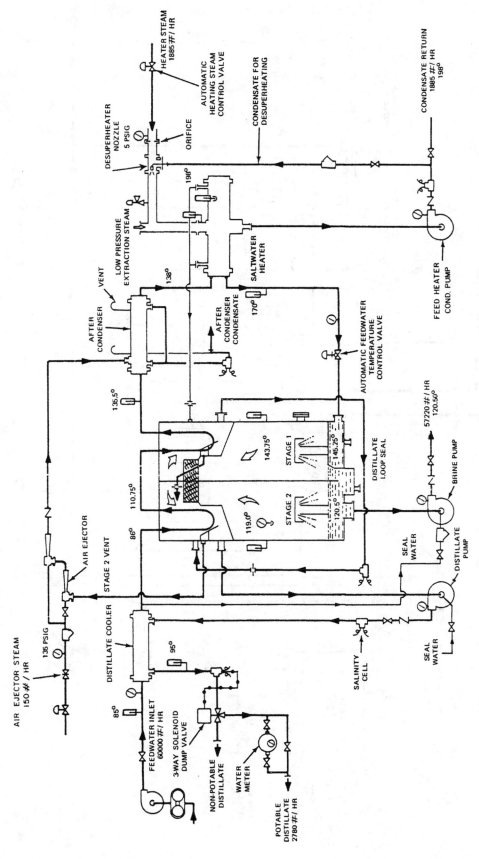

Fig. 5 Two-stage flash distilling plant flow diagram

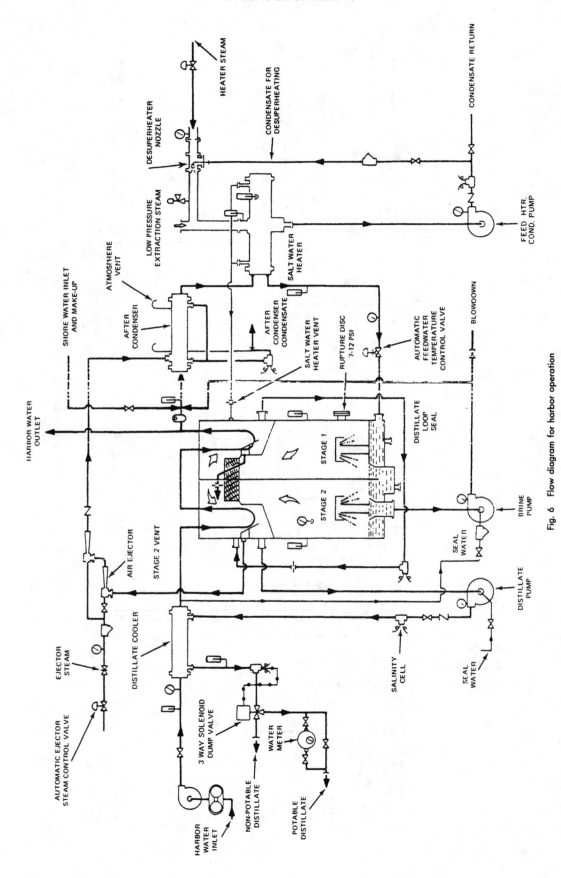

Fig. 6   Flow diagram for harbor operation

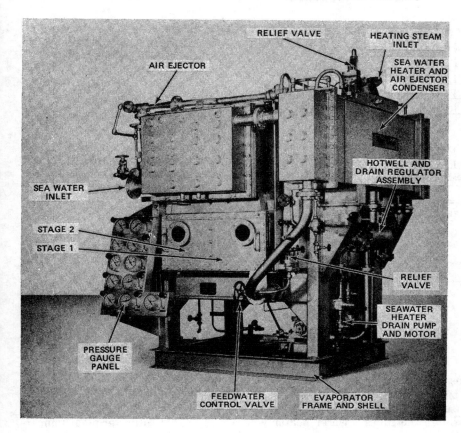

RELIEF VALVE

HEATING STEAM INLET

AIR EJECTOR

SEA WATER HEATER AND AIR EJECTOR CONDENSER

HOTWELL AND DRAIN REGULATOR ASSEMBLY

SEA WATER INLET

STAGE 2

STAGE 1

RELIEF VALVE

SEAWATER HEATER DRAIN PUMP AND MOTOR

PRESSURE GAUGE PANEL

FEEDWATER CONTROL VALVE

EVAPORATOR FRAME AND SHELL

Fig. 7 Photograph of flash evaporator

F. The timer will not permit the distillate to be returned to the storage tanks until the unit is thoroughly flushed out; adequate flushing normally requires at least 15 minutes.

3. High brine level alarm to be located in the last evaporator flash chamber.

4. Desuperheater failure alarm switch.

5. Automatic temperature control valve maintaining the seawater heater outlet temperature at 170 F.

6. Brine and distillate level controllers.

• Instrument panel.

• Control panel.

b. Brine recirculation systems. The "long-tube" type of flash evaporator, which is illustrated by Fig. 8, is a design normally applied to large-capacity evaporators (over 100,000 gpd) and is therefore commonly used for large-scale "land" desalination plants. The design of the long-tube type of flash distilling plant is similar to that of the marine flash evaporator except that the condenser tubes traverse uninterruptedly through several stages and over the full length of a multistage evaporator vessel. Marine flash distilling plants have individual cross-tube condenser bundles for each stage.

A "brine recirculation" system, such as shown by Fig. 8, is sometimes designed to operate at temperatures up to 250 F; however, to better control the formation of scale, reduce the consumption of chemicals, and lower maintenance costs, the maximum operating temperature may be limited to 230 F. The brine recirculation feature permits operation at higher evaporator temperatures and, at the

same time, reduces the amount of feedwater chemical treatment required as compared with that which would be required for "once-through" operation at the same temperature.

In a brine recirculation system, filtered raw seawater is pumped through the tubes of the last few evaporator stages and is discharged. The raw cool seawater flowing through the tubes removes heat from the last stages; therefore, this region is termed the heat rejection section of the plant.

The major portion of the seawater coolant is discharged back to the ocean; however, part of it is retained as makeup water. As indicated by Fig. 8, the makeup water is chemically treated (with sulfuric acid) for scale control, then deaerated to remove the noncondensible gases for corrosion control purposes. Chemical treatment is necessary to retard the formation of scale on the heat-transfer surfaces and to permit long-run operation of the plant without the necessity of shutting down for cleaning. The deaerated and chemically treated makeup water is mixed with the recirculated brine and is introduced into the last heat regenerative stage. The brine mixture is then pumped through the tubes of the heat regeneration stages and the brine heater by the recirculation pump.

In order to maintain a material balance in the system, a portion of the recirculated brine from the last stage is discharged to the ocean. The blowdown flow rate is controlled by the brine level in the last stage. The brine density is controlled by the ratio of seawater to blowdown flow rates.

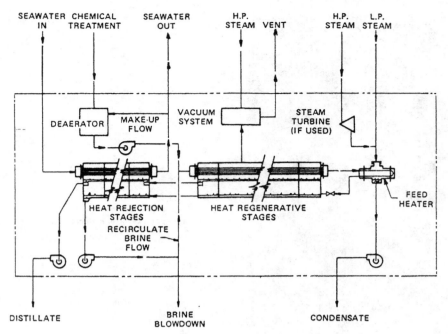

Fig. 8   Flow diagram of a long-tube flash evaporator, brine recirculation system

As the brine flows through the condenser tubes of the heat regeneration section, it is heated progressively in each stage by the vapors which condense on the outer tube surfaces. The brine then passes through the tubes of the brine heater for final heating to its terminal temperature. Low-pressure steam is used to heat the recirculating brine flowing through the tubes of the brine heater. On many large land-based plants, high-pressure steam is used to drive the feed or brine pumps or both, and the turbine exhaust steam is used in the brine heater. The condensate formed from the heating steam is discharged from the heater by the condensate pump.

The heated brine is returned to the shell side of the first stage for flashing. The recirculating brine flows successively from the first to the last stage, flashing in each stage. Distillate is formed on the stage condenser tubes by condensing the flashed vapors. The distillate collects in a distillate trough, which is common to all the stages and flows from the first stage to the last stage. The distillate is pumped from the last stage into the distillate storage system.

c. Once-through systems. The operating principle of once-through systems is the same as that for recirculating systems except that once-through systems are designed to operate at temperatures up to 195 F.

In order to achieve long-run uninterrupted operation of the plant, it is necessary to retard the formation of scale deposits; this is accomplished by continuously injecting a measured amount of polyphosphate feed treatment compound into the feedwater. Anti-foam compound can also be introduced when needed because of pollutants which cause excessive foaming in the evaporator chambers.

Treated seawater is pumped through all of the condensing tubes, from the last stage to the first, en route to the feedwater heater. As the feedwater flows through the condenser tubes, it is heated progressively in each stage by the vapors which condense on the tube outer surfaces. The feedwater then passes through the tubes of the feedwater heater where it is heated to its terminal temperature.

The heated feedwater is discharged to the shell side of the evaporator first stage for flashing. The remaining brine flows successively from the first to the last stage, reflashing in each stage, and is pumped from the last stage back to the ocean.

1.4 Spray-Film Evaporators. By analyzing the temperature gradients involved with the transfer of heat through tube walls, it becomes apparent that if thin films of liquid can be maintained on the tube walls, a relatively higher overall heat-transfer coefficient can be achieved. The "spray-film" evaporator is a form of thin-film evaporator. Spray-film evaporators have been installed aboard commercial ships and arranged to use condensate from the main propulsion plant condenser as the coolant in the distilling plant condenser. With such an arrangement, the condensate-cooled distilling plant accomplishes a secondary objective of functioning as a low-pressure feedwater heater in addition to its primary objective of producing potable and high-purity makeup feedwater from seawater. The combination results in a higher overall cycle efficiency. A typical flow diagram of a condensate-cooled spray-film evaporator is shown in Fig. 9.

The heating bundle of a spray-film evaporator consists of a horizontal tube bundle with the heating medium on the inside of the tubes. The heating medium can be bleed steam from a steam turbine or steam produced from diesel engine or gas turbine waste heat. The brine level is maintained in a hotwell below the evaporator tube bundle and a brine pump provides the necessary pressure to recirculate the brine to the spray nozzle header. Recirculated

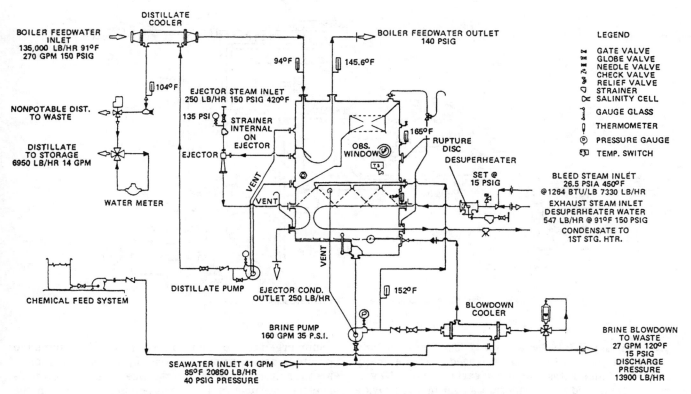

Fig. 9 Condensate-cooled spray-film evaporator flow diagram

brine from the spray nozzles is "sprayed" over the top of the heating surface where it flows from tube to tube in thin films, resulting in relatively high "thin-film" heat-transfer rates. The outer row of tubes in the tube bundle is often fitted with titanium tubes because of the increased resistance of titanium to erosion caused by impingement of the seawater spray. An automatic chemical feedwater treatment system is provided to retard the formation of scale on the heating surfaces. The mechanics of the thin-film heat-transfer concept is conducive to the development of hot spots on the tubes, where scale deposits accumulate. A reliable chemical feed treatment program is, therefore, essential to avoid an accumulation of scale.

The vapor generated at the tube surface counterflows through the curtain of recirculated brine, and fine liquid droplets are thereby removed from the vapor. Removal of the liquid droplets in this manner, coupled with the secondary entrainment separation in the demister column, enables the production of distillate having a very high purity.

A steam-motivated air ejector or mechanical vacuum pump can be used to maintain a low shell pressure in the evaporator. The air ejector after-condenser can be a separate heat exchanger or it can be incorporated within the evaporating heating bundle, thereby utilizing the heat content of the air ejector steam without necessitating a separate after-condenser heat-exchanger assembly.

Since a spray-film evaporator operates at relatively higher shell pressures and lower feed rates than a marine once-through type flash evaporator, the vacuum equipment handles a lower volume of entrained air and corrosive gases and at a lesser vacuum level. As a result, the application of vacuum pumps in lieu of air ejectors becomes a more attractive alternative.

To ensure conformance with the regulations of the U.S. Public Health Service [6], the shell temperature must be maintained at a minimum temperature of 165 F for at least 7 seconds so as to pasteurize the vapor and distillate. A temperature switch is provided to sense the shell temperature and actuate the distillate dump valve, diverting the distillate to waste should the shell temperature fall below the 165 F minimum. Where lower operating temperatures are desirable or necessary due to a low-temperature heating medium (such as engine jacket water), a separate distillate sterilizer must be incorporated for this purpose. When needed, the sterilizer is usually a thermostatically controlled electric immersion heater.

As indicated by Fig. 9, a blowdown cooler cools the hot blowdown while simultaneously preheating the incoming makeup feedwater. The brine concentration in the evaporator sump is maintained at the proper density by setting a continuous blowdown rate.

A distillate cooler is used to reduce the temperature of the distilled water produced. Seawater is normally used as a coolant in the distillate cooler in order to cool the distillate to a "potable water" temperature before it is discharged into a potable water tank.

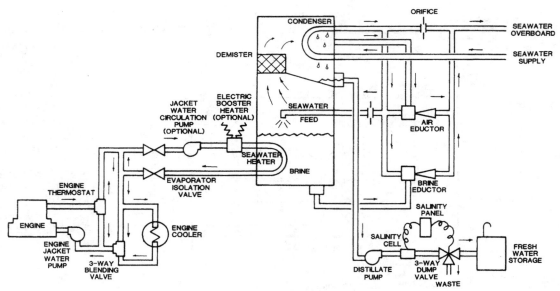

Fig. 10   Submerged-tube heat-recovery evaporator schematic

1.5 Submerged-Tube Evaporators. The history of submerged-tube evaporators can be traced back for many years; Fig. 2 illustrates one of the early designs. Nevertheless, submerged-tube evaporators continue to provide advantages in some applications. Submerged-tube evaporators are well suited for some operations with low-temperature, low-quality steam. They are readily adapted for waste-heat recovery applications where a supply of hot water (e.g., engine jacket water) is available. Figure 10 is a schematic diagram of such a heat-recovery application.

An engine jacket-water temperature in the range of 190 to 200 F would be preferred by the distilling plant engineer; however, efforts by diesel engine designers to produce high-efficiency engines have caused the heat rejected to the jacket water and exhaust to decrease. As a result, jacket-water temperatures in the range of 165 to 180 F are common.

From a balanced perspective, the ideal jacket-water temperature to the distiller is usually in the range of 175 to 180 F. This allows a temperature drop across the distiller heating section of 15 deg F, which is practicable and results in a return temperature of 160 to 165 F—well within the design parameters of most engine cooling systems. As indicated by Fig. 10, good practice requires that only a portion of the engine jacket-water flow be cooled by the distiller; therefore, a portion of the engine jacket water is routed through the distiller heater section, cooled 15 deg F, and recombined with the hot jacket-water flow, thereby having a relatively small effect on the temperature of the combined flow. The combined jacket-water flow is then cooled to the proper temperature by the engine cooling system before being returned to the engine.

For most applications, submerged-tube evaporators having a single effect are used. Multi-effect evaporators offer improved economy; however, those having a single effect provide the advantages of simplicity, ease of operation, and lowest cost.

Submerged-tube evaporators are generally designed to operate under vacuum conditions. Lower operating pressures result in an increase in the size of the evaporator's envelope, because of the corresponding increase in the specific volume of the water vapor; however, the increase in the evaporator size is warranted by the improvement in plant operation and the reduced formation of scale deposits. At operating temperatures of 150 F and less, the scale deposited on the seawater side of the heat-transfer surfaces consists primarily of calcium carbonate, and the rate of deposition does not normally interfere with plant operation. Chemical cleaning with citric or sulfamic acid is generally sufficient to remove the scale and bring the system back to a "clean tube" condition.

At operating temperatures in the range of 150 to 200 F, a mixed calcium-carbonate, magnesium-hydroxide scale is formed. To control this scale formulation, a scale inhibitor, usually polyphosphate, is injected into the seawater supply at a dosage rate of 3 ppm. Properly maintained, an evaporator can operate 90 days before acid cleaning is necessary.

Submerged-tube evaporators are composed of heating, moisture separating, and condensing sections, and all such evaporators are similar in design configuration. The heating section is typically arranged in the lower portion of the evaporator vessel, and steam or hot water is used as the heating medium. The heating fluid flows through the inside of the tubes, and seawater surrounds the outside of the tubes.

The water vapor produced in the heater section of the plant rises and passes through the moisture separator, or demister, which is usually a "mesh" design that is made of a noncorrosive material such as stainless steel or monel. The velocity of the water vapor as it passes through the moisture separator is of great importance because an excessive vapor velocity can entrain droplets of seawater and carry them through the separator into the distillate section.

The water vapor leaving the moisture separator is directed to the condenser section, where the incoming cold seawater is used as the condensing medium. A portion of the seawater that is heated in the condenser is used as the feedwater supply to the heating section and the remainder is discharged overboard.

The condensed water vapor is withdrawn from the condenser section by a distillate pump. The distillate is monitored by a salinity cell, which directs the discharge to storage when the distillate is of acceptable purity, but diverts the flow overboard in the event of high salinity or a loss of power. When the maloperation has been corrected and the distillate purity is satisfactory, the valve usually must be manually reset to redirect the distillate flow to storage.

Depending on the ship's requirements and operating conditions, a sterilizer heater may be provided to heat the distillate for potable services to 165 F in accordance with the requirements of the U.S. Public Health Service [6].

The evaporator operates at a subatmospheric pressure; therefore, an air-removal system must be provided to remove air and other noncondensible gases that are released from the boiling seawater. If there is a source of steam at 75 to 150 psig pressure, a steam-operated air ejector is an effective means of removing noncondensible gases. Alternatively, a motor-driven vacuum pump may be used. Another arrangement that is advantageous in some applications is the use of a water-motivated air eductor, as indicated by Fig. 10.

A water-motivated air eductor can be designed to effectively remove the noncondensibles and maintain the vacuum within the evaporator by using the excess seawater coolant supply as the motivating force. The quantity of seawater required to condense the water vapor produced far exceeds that required for feedwater; therefore, the excess seawater coolant can be used to operate an air eductor en route to being discharged overboard. However, if this is done, the seawater supply pressure must be a minimum of about 60 psig and the condenser must be designed for the higher pressure. The orifice shown in the seawater overboard discharge line in Fig. 10 is sized to provide sufficient back-pressure to motivate an eductor.

Similarly, as also indicated by Fig. 10, a brine eductor may be used, as an alternative to a motor-driven brine pump, to remove the excess brine and maintain the brine level in the heater section of the evaporator. The operating principles of a brine eductor are similar to those of an air eductor, and a brine eductor also requires a source of high-pressure seawater.

1.6 Plate-Type Evaporators. The thermodynamic principles associated with plate-type evaporators are the same as for submerged-tube evaporators, except that plates are used instead of tubes in the condensing and heating sections. In the heating section, the heating fluid and boiling seawater flow between alternate plates.

Titanium is the most commonly used plate material, and the plates are positioned to form a compact heat-exchanger arrangement. As illustrated by Fig. 11, the front cover of the unit is usually designed with a bolting feature that provides full access to the internal elements. This access permits the heat-transfer surfaces to be readily inspected and cleaned without the necessity of disassembling piping connections.

Plate-type evaporators are best suited for low-pressure (vacuum) applications because of the difficulties that a high pressure would present in maintaining the tightness of the extensive mating surfaces. However, close approach temperatures can be achieved with plate heat-exchanger designs, and this characteristic can be significant in applications such as the recovery of the waste heat in engine jacket water that can have a temperature in the range of 165 to 180 F.

The arrangement and operation of plate-type evaporators are similar to those of the submerged-tube type except that the space requirements and weight of the plate type are considerably less.

1.7 Vapor-Compression Distilling Plants. Vapor-compression distilling plants are designed for service where low-pressure steam or waste heat is not available in sufficient quantity or quality to operate an evaporator. The major advantages of a vapor-compression evaporator are its high thermal efficiency and that it operates on a self-contained thermodynamic cycle that is dependent only upon a source of power to provide the input energy required [7]. The power supply may be in the form of electrical energy to operate the electric boiler and drive the vapor compressor and pumps, or the compressor can be driven by a diesel engine or gas turbine. When the compressor is driven by a diesel or gas turbine, the exhaust gases can be used as auxiliary boiler heat. Special arrangements can also be made to use steam heating coils for the boiler if a small amount of steam is available for this purpose.

A thermal-compression plant is a type of vapor-compression plant in which a steam-operated jet compressor, or thermal compressor, is used instead of a mechanical vapor compressor [8]. However, the application of thermal-compression plants is limited to instances where motive steam at 200 psig is available.

Many offshore drilling rigs are equipped with vapor-compression plants as are some gas turbine ships and submarines.

The evaporator in a vapor-compression distillation plant can be of any design. Figure 12, which is a diagram of a vapor-compression distilling plant that employs a spray-film evaporator, illustrates the principles involved with a vapor-compression distilling plant. Feedwater is pumped through a solenoid valve, control valve, flowmeter, heat exchanger, vent condenser tube bundle, and then into the spray pipe manifold from which it is sprayed over the tube bundle in the evaporator shell. Some of the sprayed feedwater striking the hot tube bundle evaporates into steam vapors, which are drawn through the demisters into the vapor compressor. The concentrated feedwater, or brine solution, that is not vaporized collects in the bottom of the evaporator shell and flows into the evaporator sump. The recirculation pump takes suction from the evaporator sump and returns the majority of the brine to

Fig. 11 Distilling plant with plate-type evaporator

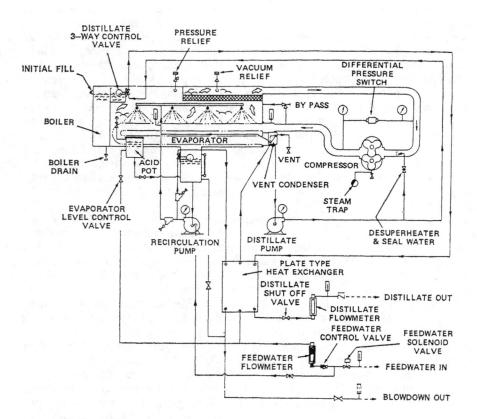

Fig. 12 Vapor-compression distilling plant with spray-film evaporator

the spray pipe manifold. Incoming feedwater is used as cooling water for the recirculation pump mechanical seal and is then combined with the recirculating brine.

The recirculation pump also pumps a portion of the recirculating brine through the evaporator sump liquid level control to the blowdown side of the heat exchangers, and then to discharge. For seawater use, the blowdown flow rate is normally set at twice the total distillate flow for best operation.

The boiler section integral with the evaporator provides the small quantity of starting and makeup heat required for the operation of the distiller. If the boiler is steam heated, low-pressure steam is piped into a small U-tube bundle to provide the necessary heat. A low-pressure (15 psig) steam supply free of contaminants is required to maintain a compressor suction pressure of 0.5 psig and is usually regulated by a diaphragm-type control valve.

If the boiler is electrically heated, three electric immersion heaters are generally used with two of them being manually controlled and the third one operated automatically. All three heaters are put in service during start-up to get the plant up to normal operating temperatures as quickly as possible. During normal operation, the manual heaters are used as required, along with the automatic heater, to maintain a compressor suction pressure of about 0.5 psig. The automatic heater is controlled by an evaporator shell pressure switch.

After start-up, some of the distillate is used as boiler makeup water. A float-operated control valve in the distillate circuit regulates the correct amount of makeup water into the boiler.

The vapors produced from the brine on the shell side of the evaporator are drawn through demisters into the vapor compressor. Vapor compressors of a centrifugal design, which operate at a relatively high speed and low noise level, are most suitable. The compressor normally operates with a 2 to 4 psi differential pressure between the suction and discharge; the maximum allowable differential pressure is about 5 psi.

During the compression process, the steam vapor increases in pressure and temperature after which it is discharged into the tube side of the evaporator tube bundle in the spray-film evaporator. The latent heat of the steam vapor is transferred through the walls of the tubes to the brine being sprayed over the tube bundle. This transfer of heat condenses the steam vapor into distillate, which flows out of the tubes into the bottom of the steam chest.

A spray-pipe assembly is the standard means whereby recirculating water is sprayed over the evaporator tube bundle. A steam chest and vent condenser are bolted to the evaporator tube bundle. The steam chest channels the flow of steam vapors from the compressor into the tube bundle and the flow of distillate from the tube bundle. Incoming feedwater flows through the tubes of the vent condenser, where it gains additional heat from the vapors condensing in the vent condenser section. The vent condenser is vented to the atmosphere to discharge noncondensible gases from the steam chest.

The distillate pump takes suction from the steam chest and pumps the majority of the distillate through the heat exchanger, the flowmeter, and then to the discharge connection. A small portion of the distillate enters the boiler through the boiler water level control valve as makeup water. Another small portion enters the compressor suction duct and serves as compressor desuperheating and sealing water.

An electric-driven vapor-compression distilling plant is furnished by the manufacturer as a package complete with all interconnecting piping, electric wiring, automatic controls, and insulation; such a unit is shown by Fig. 13.

Over a period of time, scale-forming elements in the feedwater gradually accumulate on the evaporator tubes and lower the rate of heat transferred from the compressed steam to the recirculated water (scale control is discussed in Section 2.5). Normally the amount of steam

compressed is constant. Therefore the compressor differential pressure rises (causing an increase in the temperature difference) to counteract the effects of the scale accumulation. As the compressor differential pressure rises, there will be a slight decrease in distillate production; therefore, an acid cleaning system must be provided and used when the rated distillate capacity can no longer be maintained or when the compressor differential pressure exceeds 4.5 psi. When the pressure differential across the compressor reaches 5 psi, the unit operates at its minimum rated capacity; and the evaporator tubes must be cleaned to raise the distillate output, lower the electrical energy input, and prevent overloading the compressor motor.

When it is desirable to use a diesel engine instead of an electric motor to drive the compressor, a conventional four-cycle, in-line, medium-speed, industrial-type engine is best suited for continuous operation. The engine directly drives the compressor and both water pumps via V-belts. Engine jacket water can be used as a heat source to the boiler thereby providing a means of returning the heat in the jacket water to the system. The boiler section can also use the engine exhaust gas as a heat source; by doing so, maximum use is made of the engine waste heat.

Detailed material and design requirements for distilling plants of the vapor-compression type for Navy applications are contained in reference 2. The plant installed on some submarines operates on the principle described in the foregoing, but the plant is completely electrically operated and incorporates a vertical tube bundle as illustrated in Fig. 14.

Referring to the plant shown by Fig. 14, the entering seawater is preheated (for arctic operation or cold feed due to deep submergence) in an assembly that consists of two electric immersion heaters. The temperature control unit automatically energizes one or both of the heaters when the incoming seawater temperature is below 55 F and turns off the heaters when the temperature is above 65 F.

A rectangular shell-and-double-tube heat exchanger, or one of similar features, is located in the line between the feedwater preheater and the vent condenser. The shell-and-double-tube heat exchanger is vertically divided into two sections of equal size, a distillate side and a brine side. Inner and outer tubesheets are bolted to each end cover plate. Larger, straight outer heat-transfer tubes are roller-expanded into the inner tubesheets and smaller straight tubes, inserted through the outer tubes, are roller-expanded into the outer tubesheets. Incoming feedwater flows in the annular space between the inner and outer tubes while the distillate and brine, discharge from the evaporator, flow through the inner tubes in their respective sides of the heat exchanger. The feedwater flow is counter to both the distillate and brine flows, providing good heat transfer between the flows. The unit serves to further heat the feedwater and simultaneously cool the distillate and brine.

The feedwater leaving the heat exchanger next passes through the tubes of the vent condenser, where it gains additional heat as the hot noncondensible gases vented from the evaporator section are cooled.

Fig. 13   Electric-driven vapor-compression unit

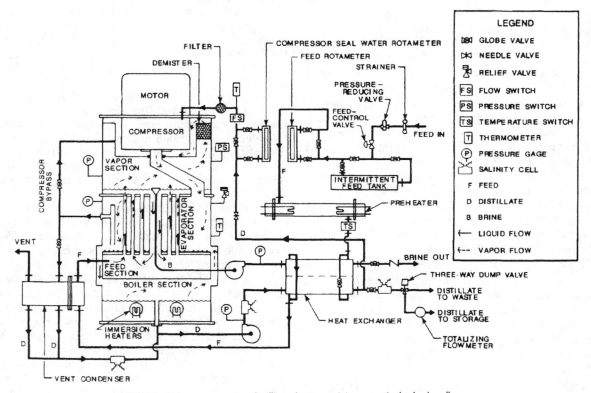

Fig. 14   Vapor-compression distilling plant containing a vertical tube bundle

The hot feedwater leaving the vent condenser is discharged into the feed section. The temperature of the brine in the feed section is raised to the boiling point by heat coming from the boiler section.

The boiler section is located on the floor of the evaporator. Its source of heat is electric immersion heaters. During operation the electric heaters keep the distillate at the boiling point, which in turn further heats the brine in the feed section located above the boiler section. Since the feedwater is preheated when it enters the evaporator, comparatively little heat is required from the immersion heaters to keep the brine at the boiling point.

A brine overflow tube, located in the center circulation tube, runs through the feed section to a sidewall connection. Distillate drain passages from the evaporator section to the boiler section also run through the feed section, as does a steam vent passage from the boiler section to the evaporator section.

When the brine boils in the feed section, it passes into the evaporator section tube bundle where more heat from the compressed vapor is applied. This heating increases the boiling action of the brine, and about two thirds of the brine is vaporized at a pressure of about 1 psig in the tubes. The brine not vaporized flows into the brine overflow and is piped out of the evaporator into the heat exchanger. Steam vapors from the boiling brine in the evaporator section tubes rise into the vapor section and are drawn into the suction side of the compressor through demisters. The demisters remove entrainments from the vapor and allow relatively "clean" vapor to flow to the compressor suction.

The vapor compressor, which is the "heart" of the plant, is a centrifugal, liquid sealing ring type that is directly driven by a constant-speed a-c motor. The unit is mounted in the evaporator top cover plate and extends down into the vapor section. It operates at a suction pressure of about one psig at 215 F and a discharge pressure between 1.5 and 3 psig in the saturation temperature range of 217 to 222 F. When in operation, the compressor requires a continuous flow (about 0.25 gpm) of distilled water for sealing purposes. The seal water is necessary to provide proper suction and compression and to maintain the vapor discharge temperature within predetermined limits. The distillate discharge line, downstream from the heat exchanger, is tapped to provide the seal water.

The compressor raises the temperature and pressure of the vapors and discharges the vapors into the evaporator section where a number of heat-transfer tubes are arranged in a pattern surrounding vertical baffle plates. The baffle plate arrangement directs the flow of vapors around the tubes and directs the noncondensable gases to a vent tube. A system of perforations in the vent tube collects the gases, which are passed through an external vent condenser to the atmosphere.

In the shell side of the evaporator section, the vapors condense on the tube outer surfaces to form distillate, which collects in the bottom of the evaporator section (on the lower tubesheet) and flows through the distillate return passages into the boiler section. The excess distillate flows out through the outlet pipe into the distillate pump and is pumped through the heat exchanger and into the ship's storage tanks or to waste, depending on its purity.

A vapor-compression distillation plant has several features that may be advantageous in some situations. These units are generally designed to operate at or above atmospheric pressure which, in effect, minimizes the size of the unit (because of the higher-density vapors) and avoids the necessity for a vacuum system and the possibility of air ingress.

The primary advantage associated with a vapor-compression plant that operates under subatmospheric pressure would be that there would be a minimal tendency for calcium-sulfate scale to form; however, the formation of this scale can be substantially eliminated by using a scale inhibitor.

A vapor-compression plant requires no source of steam or waste heat. The only energy input necessary is a source of electrical power. The vapor-compression cycle provides the highest thermal economy of any distilling-plant concept, when expressed in terms of fresh water produced per unit of heat input, and, therefore, may merit consideration for applications where there is no reliable source of steam or waste heat.

In some cases, the small quantity of seawater required to operate a vapor-compression unit may be advantageous. Submerged-tube and flash distillation units typically require a seawater flow of 15 to 20 times the distillate rate for condensing purposes, whereas a vapor-compression unit requires a source of seawater equal to about 2.5 times the distillate produced.

Additionally, the requirement of the U.S. Public Health Service for feedwater or distillate to be heated to 165 F for a minimum of seven seconds is inherently satisfied because of the higher temperatures concomitant with cycles that operate above atmospheric pressure.

1.8 Reverse-Osmosis Desalination. As an alternative to the thermal method of desalinating seawater, which depends on the use of heat to boil seawater and physically separate the water from the dissolved solids, a nonthermal membrane process can be used in some cases. Of the various membrane processes that have been used, reverse osmosis has been proven to be the most appropriate for shipboard service [9]. Other membrane processes such as ultrafiltration, electrodialysis, and piezodialysis show little promise of being able to economically convert the more saline waters to a potable level; however, the reverse-osmosis process has been proven to be a practicable means of continuously producing fresh water from a sea-water source. The only energy input required to operate a reverse-osmosis plant is a source of electrical power to drive the pumps and the various controls.

To understand reverse osmosis, it is necessary to review the basic phenomenon of osmosis. Osmosis is a natural process which depends on the existence of a membrane that is selective in the sense that certain components of a solution (ordinarily the solvent) can pass through the membrane, while one or more of the other components cannot pass [10]. Such a selective device is called a "semipermeable membrane"; it is usually, though not always, in the physical form suggested by the word "membrane." In nature, the root system of most plants acts as the membrane and allows the plant to absorb moisture or water by permitting the flow of a less-dense fluid (water) into the more-dense fluid (sap) internal to the root system.

As illustrated by Fig. 15, if a semipermeable membrane separates a solution from a pure solvent, or two solutions of different concentrations, the tendency to equalize concentrations will result in a flow of solvent from the less-concentrated phase (that is, the phase richer in solvent) to the other; it is this flow of solvent that is termed "osmosis." If an attempt is made to impede the flow by exerting pressure on the more saline solution (assuming for simplicity that the other phase is pure solvent), the rate of

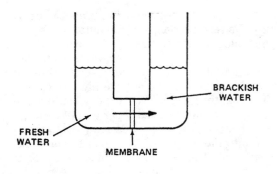

OSMOSIS

When fluids of different concentrations in a vessel are separated by a membrane, the dilute solution will flow through the membrane into the concentrated solution.

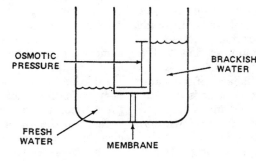

OSMOTIC PRESSURE

The level of the dilute solution drops and the level of the concentrated solution rises until an "equilibrium" is reached. The pressure difference between these two levels is the "osmotic pressure."

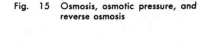

Fig. 15 Osmosis, osmotic pressure, and reverse osmosis

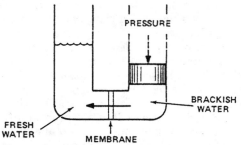

REVERSE OSMOSIS

If a pressure in excess of the osmotic pressure is applied to the concentrated solution, the flow is reversed from the concentrated solution to the diluted solution. This is "reverse osmosis."

flow will be decreased. As the pressure is increased, a point will be found at which the flow is brought to a complete stop, the tendency to flow being in equilibrium with the opposing pressure. This equilibrium pressure (actually, the equilibrium-pressure difference between the solvent and solution phases) is called the "osmotic pressure." The osmotic pressure, which is about 1 psi for each 100 mg per liter of sea salt content in seawater, is a property of the solution and does not depend in any way on the membrane, so long as the membrane has the necessary property of semipermeability. A further increase of the pressure on the solution causes reversal of the osmotic flow, and pure solvent passes from the solution, through the membrane, into the solvent phase; this phenomenon is the basis of the reverse-osmosis method of desalination.

The osmotic pressure of a seawater solution is approximately 355 psi, and to productively generate fresh water from a seawater source, the membrane must operate at a differential pressure in the range of 750 to 1200 psi. Most osmotic membranes have little mechanical strength and must be supported to withstand such a large pressure difference.

The material used to make a reverse-osmosis membrane is commonly cellulosic or a derivative of a polyamide, sometimes referred to as a thin-film composite. A membrane made of this material is a well suited means of separating fresh water and dissolved solids (ions). Membranes have a much higher rejection propensity for dissolved solids such as calcium, magnesium, and sulfates (divalent ions) than for sodium and chlorides (monovalent ions) and the former are more effectively rejected. Dissolved gases tend to pass through membranes with very low rejection. However, any particulate matter in suspension must be removed by filtration to minimize membrane fouling and to avoid damage to the membrane.

The components that comprise a reverse-osmosis plant are shown schematically by Fig. 16. Pretreatment of the incoming feedwater is required for all reverse-osmosis plants. The impurities commonly found in seawater and brackish water are suspended matter (colloidal materials, metal hydroxides, and biological fouling), which must be removed to avoid fouling the membrane. Typically, the incoming seawater is treated chemically with a polyelectrolyte or flocculent, if required, to sequester undesired compounds that are in suspension. The treated seawater is then directed to a primary filter, which typically contains 20-μm replaceable-cartridge elements, to remove

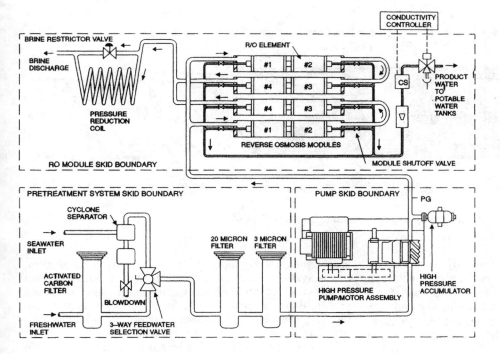

Fig. 16 Single-stage reverse-osmosis desalination plant for naval surface-ship applications

any suspended particles; and a secondary filter, having 3-μm elements, is often used to further remove particulate matter. The successful operation of a reverse-osmosis plant is strongly dependent upon a supply of clean seawater to the membrane elements; therefore, good practice may require that the desalination plant be secured before entering harbors or other regions where the contamination would overload the plant pretreatment capabilities.

The treated and filtered seawater then goes to the high-pressure pump, which has a discharge pressure typically in the range of 750 to 1200 psi. Both centrifugal and positive displacement pumps have been selected for this service; however, the characteristic pulsations in the discharge flow from positive-displacement pumps are objectionable. To reduce flow irregularities, in-line accumulators are often installed on both the suction and discharge sides of positive-displacement pumps.

The high-pressure pump discharges to one or more membrane elements, which may be arranged in parallel, in series, or in a combination thereof, depending on the system, capacity, water quality desired, and the characteristics of the membrane configuration and material selected.

Membranes designs embrace a variety of configurations; among them, plate-and-frame, tubular, pleated, spiral-wound, and hollow-fiber designs have been investigated. However, only the spiral-wound and the hollow-fiber configurations are commonly used for seawater desalination in shipboard applications. The spiral-wound configuration, which is illustrated by Fig. 17, is made by rolling two membrane sheets that sandwich porous spacers around a hollow tube. The two membrane sheets are configured such that the seawater feed flows axially through the porous spacer between the sheets, with some

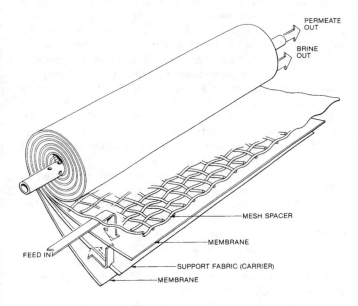

Fig. 17 Typical spiral-wound membrane configuration

fresh water permeating the membrane and the remaining concentrate being discharged from the other end. The fresh water that permeates through the membrane flows spirally inward through the porous spacer between the membrane sheets and collects in the internal tube from which it is withdrawn. The spiral-wound membrane is often encased in a tubular pressure vessel that is constructed of filament-wound fiberglass and is designed to withstand the high operating pressure. The pressure vessel is also made of titanium in some designs.

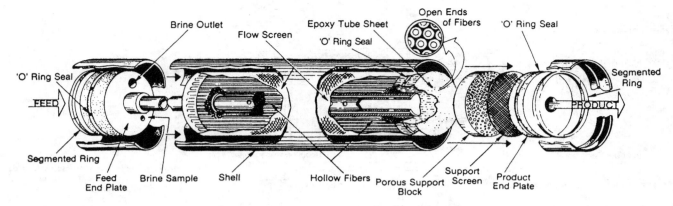

Fig. 18 Configuration of a typical hollow-fiber membrane

The hollow-fiber configuration, typically as shown by Fig. 18, consists of long, hollow fibers the size of a human hair. The hollow fibers are often placed in a hairpin arrangement with their ends encased in an epoxy end cap, similar to a tubesheet in a heat exchanger. Many thousands of fibers are included in a bundle that is encased in a pressure housing. Seawater is introduced into the pressure housing and to the external surface of the hollow fibers. Fresh water permeates through the fiber surface, flows through the hollow fiber core, and is collected. The concentrated brine solution is discharged overboard.

Membranes are rated for a standard seawater temperature of 77 F. The permeation of fresh water through a membrane has an inverse variation of about 1.6% per deg F deviation from the standard temperature. For shipboard applications most reverse-osmosis plants must necessarily be designed for a feedwater temperature in the range of 28 to 85 F; consequently, the feedwater temperature is an important design consideration.

The mean salinity of the seawater feed circulated in plant is also an important consideration. For each 1000-ppm salinity increase in the seawater concentration, the permeation through the membrane decreases about 1.8% because of the increase in the osmotic pressure of the brine. However, the salinity of seawater is nearly constant around the globe except for several isolated areas such as the Red Sea and the Arabian Gulf, where high levels of dissolved solids are common.

The hollow-fiber and spiral-wound membranes are tolerant of the pH values normally characteristic of seawater, and acid injections for pH adjustment are not generally required for marine installations.

The element iron, Fe, has an adverse effect on the life of membranes; therefore, the use of steel and stainless steel piping and components in the seawater side of membrane systems must be avoided. However, since the hull and seawater system of many ships are made of steel, the complete avoidance of the iron element may be impracticable. Experience has, however, shown that the hull structure has little effect on the membrane.

The element chlorine, Cl, also adversely affects the life of membranes. If the seawater system is chlorinated, provisions must be made to remove the chlorine during the feedwater treatment process.

Generally, shipboard reverse-osmosis systems require little maintenance provided that the feedwater is appropriately pretreated, the plant is properly operated, and the membranes are periodically cleaned. However, unlike the distillation process, which produces high-purity distillate containing less than 4.3 ppm total dissolved solids, the fresh water produced from a single-stage reverse-osmosis plant generally has total dissolved solids of about 350 ppm, which is still well within the limits established by the World Health Organization for potable water [4].

It should be noted that with time, both the quantity and quality of the fresh water produced can be expected to degrade. The necessarily high pressure of the seawater supply causes the membrane material to compact, which slows the diffusion of the water through the membrane. In addition, suspended particles in the feedwater supply, which are not removed by the filtration system, tend to foul the membrane and impede the permeation through the membrane. Hollow-fiber membranes are particularly susceptible to being plugged by impurities in the feedwater, and therefore require a higher degree of filtration.

A properly designed reverse-osmosis system can provide years of service before the membrane must be replaced. The life of the membrane can be extended by periodic flushing with a chemical cleaning agent. Membranes are usually designed to readily permit their replacement.

Section 2
Distilling Plant Design Considerations

2.1 Heat Transfer in Distilling Plants. The fundamental theory of heat transfer in heat exchangers and the application of this theory are discussed in Chapters 2 and 16. The basic analytical relationship employed in the design of heat-transfer equipment is

$$Q = UA(LMTD) \tag{1}$$

where

Q = quantity of heat transferred, Btu/hr
U = overall heat-transfer coefficient, Btu/hr-ft²-deg F
A = heat-transfer area, ft²
$LMTD$ = logarithmetic mean temperature difference, deg F

The main consideration in the thermal design of an evaporator is that fluids in the evaporator undergo a change in phase. In a flash evaporator, heat is transferred to the seawater in tubes by condensing a vapor on the outside of the tubes. In a submerged-tube or spray-film evaporator, the heat is transferred from the condensing steam inside the tubes to the boiling seawater on the outside of the tubes.

As in the design of heat exchangers (see Chapter 16), the overall heat transfer coefficient, U, is given by

$$U = \frac{1}{(1/h_o) + r_{do} + r_w + r_{dio} + (1/h_{io})} \tag{2}$$

where

$1/h_o$ = reciprocal of shell-side film coefficient
r_{do} = resistance of deposit or scale on tube outside wall
r_w = resistance of tube wall metal
r_{dio} = resistance of deposit or scale on tube inside wall
$1/h_{io}$ = reciprocal of tube-side film coefficient

All of the resistances listed in the foregoing are consistently based on the same area; by convention, the tube outside area is usually taken as the base. Typical values of heat-transfer coefficients encountered in service are presented in references 11 and 12.

The overall heat-transfer rate of a submerged-tube or thin-film evaporator depends upon the vapor pressure of the liquid being evaporated, the temperature difference between the condensing saturated steam and the vapor (which fixes the steam pressure for a given vapor pressure), the disposition of the heating tubes in the shell, and the character of the liquid being evaporated as well as the cleanliness of the heat-transfer surfaces. Heat-transfer rates increase with increasing vapor and steam temperatures, with the temperature difference remaining unchanged, except that for seawater evaporation at temperatures exceeding about 200 F and for some classes of raw fresh water at the higher temperatures (particularly for high-temperature differences), scale forms so rapidly as to offset the advantage that would otherwise be gained.

Under laboratory conditions with a single acid-cleaned tube reevaporating distilled water, the temperature difference has a pronounced effect on the heat-transfer rate. Rates varying from about 1300 Btu/hr-ft²-deg F at a 20 deg F temperature difference up to over 3000 Btu/hr-ft²-deg F at a 100 deg F temperature difference have been observed with the vapor at atmospheric pressure; the heat-transfer rate falls off with further increases of the temperature difference. However, for scale-producing feeds, the scale resistance is such a large part of the total resistance to heat flow that the effect of temperature difference on continuous service rates is small.

Since the scale resistance is generally such a large percentage of the total resistance to heat flow for distilling plants, it is convenient to group all of the other resistance elements together and express equation (2) in the following form:

$$U = \frac{1}{\dfrac{1}{U_c} + r} \tag{3}$$

where

U_c = clean-tube overall heat-transfer coefficient; i.e., sum of film resistance on tube inside and outside surfaces and resistance of tube wall, Btu/hr-ft²-deg F
r = fouling resistance; i.e., resistance of tube deposits or scale, hr-ft²-deg F/Btu

Figure 19 is a plot of the clean-tube overall heat-transfer coefficient, U_c, versus tube velocity that can be used in the design of a seawater heater or condenser for a flash evaporator. The correction factor which must be applied to the coefficient read from Fig. 19 to compensate for inlet water temperatures of other than 70 F is given in Fig. 20; correction factors for the tube material and gage are given in Table 1. Typical values of the fouling factor are given in Table 2. For a stage condenser in a flash evaporator, the fouling factor, r, will typically vary from 0.0005 to 0.001 hr-ft²-deg F/Btu. In the seawater heater where elevated temperatures are encountered, the fouling factor will normally vary from 0.001 to 0.0015 hr-ft²-deg F/Btu; see Table 2.

A comparison of overall heat-transfer coefficients and temperature differences typical of flash, submerged-tube, and spray-film distilling plants is given in Table 3.

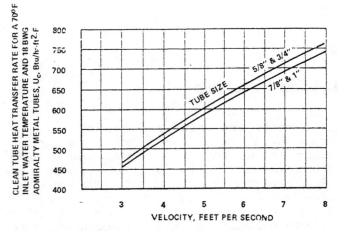

Fig. 19 Condenser clean-tube heat-transfer rate

at a constant temperature and only the temperature of the liquid varies. For a condenser, the logarithmic mean temperature difference (illustrated by Fig. 22) can be expressed as:

$$LMTD = \frac{(T_v - t_1) - (T_v - t_2)}{\ln \dfrac{T_v - t_1}{T_v - t_2}} = \frac{t_2 - t_1}{\ln \dfrac{T_v - t_1}{T_v - t_2}} \quad (4)$$

where

T_v = vapor temperature
t_1 = cold liquid temperature
t_2 = hot liquid temperature

In a flash evaporator there are a number of temperature losses, which must be taken into account in the design of the equipment. These losses are associated with the brine equilibrium temperature deviation, the boiling point

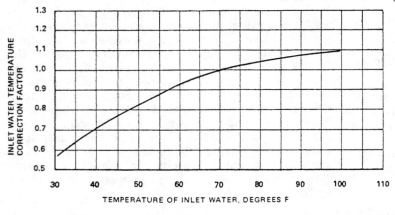

Fig. 20 Heat-transfer rate inlet temperature correction factor

Table 1 Heat-transfer rate tube material and gage correction factor

Tube Material	Tube Wall Gage—BWG						
	24	22	20	18	16	14	12
Admiralty metal	1.06	1.04	1.02	1.00	0.96	0.92	0.87
Arsenical copper	1.06	1.04	1.02	1.00	0.96	0.92	0.87
Aluminum	1.06	1.04	1.02	1.00	0.96	0.92	0.87
Aluminum brass	1.03	1.02	1.00	0.97	0.94	0.90	0.84
Aluminum bronze	1.03	1.02	1.00	0.97	0.94	0.90	0.84
90-10 Cu-Ni	0.99	0.97	0.94	0.90	0.85	0.80	0.74
70-30 Cu-Ni	0.93	0.90	0.87	0.82	0.77	0.71	0.64
Titanium	0.85	0.81	0.77	0.71	...	...	...

Heat-transfer data for a submerged-tube evaporator are usually expressed as a relationship between the heat flux, $Q/A = U\Delta T$, and the temperature difference, ΔT, because in a boiling evaporator the heat-transfer coefficient is also a function of the driving force or temperature difference. Most available data are based on experience and therefore include allowances for fouling. Figure 21 is a curve showing typical values for submerged-tube evaporators.

In the case of a condenser such as encountered in a flash evaporator, the temperature difference between the fluids varies from location to location within the condenser in a manner similar to the temperature distribution in a heat exchanger. There is a difference, however, in that the heat transfer for the condensing fluid takes place

rise, the demister pressure loss, and the condenser pressure loss. The effect of these losses on the LMTD is shown in Fig. 22. The losses are defined as follows:

• Brine Equilibrium Temperature Deviation (DEV). The amount of superheat remaining in the brine as a result of incomplete flashing.

• Boiling Point Rise (BPR). The elevation of the saturation temperature of the liquid above that of pure water caused by the concentration of salts in the brine.

• Demister Pressure Loss (DEM). The equivalent saturation temperature loss associated with the pressure loss through the demister.

• Condensor Pressure Loss (COND). The weighted saturation temperature loss associated with the pressure loss through the condenser.

In marine applications, the losses associated with the brine equilibrium temperature deviation and the condenser pressure loss are usually small and can be neglected; however, they are usually significant factors in higher capacity and economy plants. Figure 23 is a curve of boiling point rise versus temperature. Since the demister pressure loss is normally less than 0.5 deg F, it is common to allow 1.5 deg F for the combined demister loss and boiling point rise loss in the design of once-through marine flash evaporators. Because the operating brine concentration is higher in submerged-tube and spray-film

Table 2 Fouling resistances, r, typical of various types of water, hr-ft²-deg F/Btu

Temperature of Heating Medium	Up to 240 F		240–400 F[a]	
Temperature of Water	125 F or Less		Over 125 F	
Water Velocity	3 fps and less	Over 3 fps	3 fps and less	Over 3 fps
Seawater	0.0005	0.0005	0.001	0.001
Brackish water	0.002	0.001	0.003	0.002
Cooling tower and artificial spray pond:				
treated makeup	0.001	0.001	0.002	0.002
untreated	0.003	0.003	0.005	0.004
City or well water	0.001	0.001	0.002	0.002
Great Lakes	0.001	0.001	0.002	0.002
River water:				
minimum	0.002	0.001	0.003	0.002
Mississippi	0.003	0.002	0.004	0.003
Delaware, Schuylkill	0.003	0.002	0.004	0.003
East River and New York Bay	0.003	0.002	0.004	0.003
Chicago Sanitary Canal	0.008	0.006	0.010	0.008
Muddy or silty	0.003	0.002	0.004	0.003
Hard (over 15 grains/gal)	0.003	0.003	0.005	0.005
Engine jacket	0.001	0.001	0.001	0.001
Distilled	0.0005	0.0005	0.0005	0.0005
Treated boiler feedwater	0.001	0.0005	0.001	0.001
Boiler blowdown	0.002	0.002	0.002	0.002

[a] Resistances are based on a temperature of the heating medium of 240–400 F. If the heating medium temperature is over 400 F and the cooling medium is known to scale, these resistances should be modified accordingly.

Table 3 Heat-transfer coefficients and temperature differences typical of feedwater heaters in various types of distilling plants

Type of Plant	Overall Heat-Transfer Coefficient, U, Btu/hr-ft²-deg F	Temperature Difference, LMTD, deg F
Flash	450–500	5–50
Submerged tube	300–650	15–45
Spray film	500–800	10–30

evaporators, a value of 2 deg F is usually allowed for the combined loss for such units.

In order to illustrate the principles involved in the determination of the size of a condenser and seawater heater for a distilling plant, consider the two-stage, 8000-gpd, flash evaporator illustrated by Fig. 5. The condenser temperature differences are computed as follows:

	STAGE 1	STAGE 2
Brine temperature, deg F	145.25	120.5
Demister and boiling point rise loss, deg F	1.5	1.5
Vapor temperature, deg F	143.75	119.0
Condenser inlet temp., deg F	110.75	86.0
Condenser outlet temp., deg F	135.5	110.75
Inlet temp. difference, deg F	33.0	33.0
Outlet temp. difference, deg F	8.25	8.25

Since the temperature differences are the same in both stages, the two tube bundles will be made to the same design. The *LMTD* in the two condensers will be

$$LMTD = \frac{33 - 8.25}{\ln \frac{33}{8.25}} = 17.85 \text{ deg F}$$

The condenser tubes selected are ⅝-in. OD, 18-BWG (0.049-in. wall), 90–10 Cu-Ni tubes, and the tube design velocity is 5.5 fps. From Fig. 19 the heat-transfer rate for a ⅝-in. tube with a 5.5-fps tube velocity is 633 Btu/hr-ft²-deg F. Applying an inlet temperature correction factor of 1.06 (since the condenser for the two stages will be the same, the correction factor read from Fig. 20 is based on the lower inlet temperature of 86 F) and a material correction factor of 0.90 (read from Table 1) gives a corrected clean-tube, overall heat-transfer coefficient of 604 Btu/hr-ft²-deg F. With a fouling resistance of 0.000675, the overall heat-transfer coefficient is computed from equation (3) as

$$U = \frac{1}{\frac{1}{604} + 0.000675} = 429 \text{ Btu/hr-ft}^2\text{-deg F}$$

With a tube ID of 0.527 in., the number of tubes required to pass 60,000 lb/hr of feedwater having a density of 64.1 lb/ft³ at a velocity of 5.5 fps is

No. of tubes required

$$= \frac{(60,000)(144)}{(3600)(64.1)(5.5)\frac{\pi}{4}(0.527)^2} = 31.2 \approx 31$$

For a feedwater flow of 60,000 lb/hr and specific heat of 0.96 Btu/lb-deg F, the temperature difference of 24.75 deg F corresponds to a heat transfer of 1,426,000 Btu/hr in each of the two condensers. This being the case, the tube area required is

$$A = \frac{Q}{U(LMTD)} = \frac{1,426,000}{(429)(17.85)} = 186.2 \text{ ft}^2$$

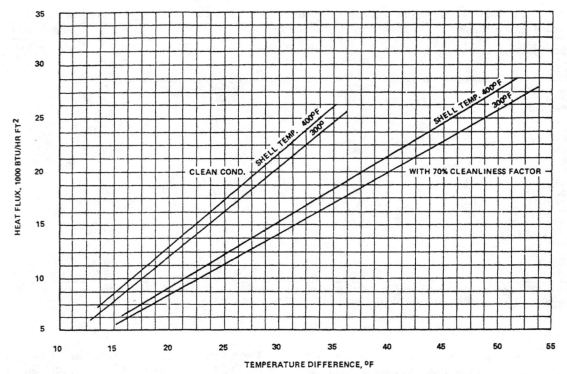

Fig. 21 Heat flux versus temperature difference for submerged-tube evaporators

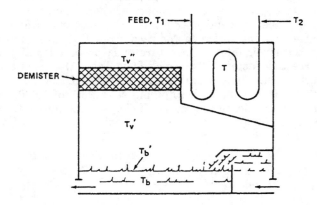

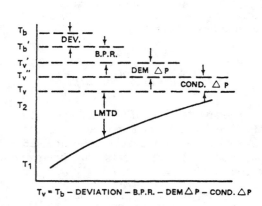

$T_v = T_b$ – DEVIATION – B.P.R. – DEM $\triangle$ P – COND. $\triangle$ P

Fig. 22 Temperature losses in distilling plants

and the required tube length is

$$\text{Tube length} = \frac{A}{\pi D n}$$

$$= \frac{(186.2)(12)}{\pi(0.625)(31)} = 36.7 \text{ ft}$$

Using a 6-pass (3 U-bend) design, the required bundle length is determined to be 36.7/6 = 6.12 ft.

The size of the seawater heater is established in a similar manner. The temperature of the condensing heating steam is 198 F and the seawater inlet and outlet temperatures are 138 and 170 F respectively; therefore, the *LMTD* is

$$LMTD = \frac{60 - 28}{\ln \frac{60}{28}} = 42 \text{ deg F}$$

The tubes selected are ¾-in., 18-BWG, 90–10 Cu-Ni tubes and the design tube velocity is 6 fps. Reading a clean-tube heat-transfer coefficient of 660 Btu/hr-ft²-deg F from Fig. 19, and applying an inlet temperature correction factor of 1.1 from Fig. 20 and a tube material correction factor of 0.9 from Table 1, gives a corrected clean-tube heat-transfer coefficient of 653 Btu/hr-ft²-deg F. With a fouling resistance of 0.0015, the overall heat-transfer coefficient becomes

$$U = \frac{1}{\frac{1}{653} + 0.0015} = 330 \text{ Btu/hr-ft}^2\text{-deg F}$$

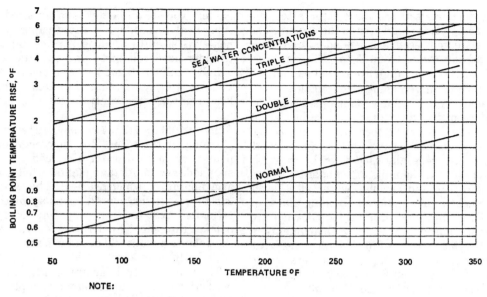

NOTE:

THE NORMAL SEA WATER CONCENTRATION
USED IN THIS CHART HAS 34.483 G
SOLIDS PER 1000 G SEA WATER.

Fig. 23   Boiling point rise of seawater concentrations with temperature

The number of 0.652-in.-ID tubes required to pass 60,000 lb/hr of feedwater having a density of 64.1 lb/ft³ without exceeding a 6-fps velocity is

No. tubes required

$$= \frac{(60,000)\,(144)}{(3600)\,(64.1)\,(6)\,\dfrac{\pi}{4}\,(0.652)^2} = 18.7$$

No. tubes used = 19

The feedwater is heated from 135.5 to 138 F in the air ejector after condenser. Since the feedwater has a specific heat of 0.96 Btu/lb, the heat transferred to the feedwater in the saltwater heater is

$$Q = (60,000)\,(0.96)\,(170 - 138) = 1,843,000 \text{ Btu/hr}$$

and the heating surface required is

$$A = \frac{Q}{U(LMTD)} = \frac{1,843,000}{(330)\,(42)} = 133 \text{ ft}^2$$

Consequently, the required tube length is

$$\text{Tube length} = \frac{A}{\pi D n}$$

$$= \frac{(133)\,(12)}{\pi (0.75)\,(19)} = 35.7 \text{ ft}$$

With a 6-pass design, the bundle length becomes 35.7/6 = 5.95 ft.

2.2   Heat Sources.   The thermodynamic design of a distillation unit is strongly dependent upon the heat energy source to elevate the temperature of the incoming seawater feed to that required for efficient vaporization. This energy can be supplied as steam, electricity, or waste heat in the form of exhaust gases or hot water. A distillation unit can use either, or any combination, of these energy sources. The selection of the optimum heat source is dependent upon the total plant heat balance, desired evaporator efficiency, available energy, and mode of operation contemplated.

For ships that are powered by steam turbines, steam is a heat source commonly used in shipboard distillation units. Low-pressure turbine extraction steam and auxiliary exhaust steam are the normal sources of heating steam for distilling plant operation.

To permit the most efficient and flexible operation, it is standard practice to design the distiller to operate with either steam that has been bled from the main turbine or steam that has been exhausted from any of the various auxiliary turbines. When operating at sea, bleed steam permits the most economical operation as the majority of the heat remaining in this steam is otherwise rejected to the main condenser, contributing very little additional energy to the cycle before being condensed. For in-port operation auxiliary exhaust steam is supplied to the distiller. The use of this relatively low-cost steam permits continued economical operation of the distilling plant.

The pressure of bleed steam is dependent on the turbine design and the specific extraction point. Distilling plant designs are normally predicated on steam being supplied at a pressure of 9 to 11 psia. Auxiliary exhaust steam is usually furnished at a pressure of 15 psig and reduced to 5 psig by a pressure-regulating valve; a pressure of 5 psig is usually provided at the inlet to the critical-pressure orifice.

High-pressure air ejector steam, where employed, is also used to preheat the seawater feed. Heat in the air ejector steam is reclaimed in the air ejector after-condenser using distilling plant feedwater as the coolant. The air ejector motive steam pressure is dependent on the air ejector design and can range from 75 to 150 psig; pressures in the 135- to 150-psig range are most common.

Diesel or gas turbine driven ships often employ waste heat as a heat source for distilling plant operation. The waste heat may be in the form of diesel engine jacket cooling water or it may be in the form of heat recovered from the turbine or engine exhaust gases. Submerged-tube distilling units are generally used when hot water is the heat source because these units often operate with a lower shell temperature and can use water at a lower temperature as a heat source than can distilling plants of the flash type. Distilling plants for this application are commonly designed to use jacket water with an inlet temperature to the distiller of 165 to 180 F. With some types of flash evaporators, the hot-water inlet temperature to the heater must be at least 190 F to obtain an adequate temperature differential to permit an efficient seawater heater design; but, by using a plate-type feed heater, closer approach temperatures can be achieved, which permits operation with jacket-water temperatures in the same range as that used with submerged-tube evaporators. A supply of diesel jacket water above 180 F is not normally available; however, water at such temperatures is obtainable on ships driven by gas turbines by installing heat exchangers to recover heat from the exhaust gases. Gas turbines typically have exhaust gas temperatures of approximately 1000 F, and these exhaust gases can be used to produce saturated steam with a temperature of about 250 F. Such a source of steam is well suited for a multistage flash distilling plant.

Electric power is often used to operate the vapor-compression and reverse-osmosis types of desalinators; however, due to considerations of economy, electric power is not normally used with plants of other designs except as a supplemental or emergency heat supply. For flash or submerged-tube designs, the direct use of electrical power for feedwater heating is not practicable.

For some applications, it is advantageous to use a combination of heat sources to operate the distillation unit. Figure 24 illustrates a typical flow diagram for a unit using all three common heat sources; this design has been applied in a number of diesel-powered naval ships. For the unit shown in Fig. 24, the primary heat source is diesel propulsion engine waste heat in the form of engine jacket water. This primary heat source is used when the ship is at sea operating with sufficient engine power to provide the necessary waste heat.

When the ship is operating at reduced power, the engine waste heat is augmented by heat furnished from either the steam heater or electric heater. When operating in this condition, the jacket water is directed through these heaters, thereby increasing the feedwater temperature. The steam heater is a shell-and-tube heat exchanger and the electric heater uses immersion heating elements.

When the ship is in port, all heat required to make rated capacity is supplied by the steam heater and electric heater. The steam heater is often sized to provide sufficient heat to produce ⅔ rated capacity and the electric heater is sized to produce ⅓ rated capacity. Aside from the increased reliability and operational flexibility provided by this combined heat source arrangement, an additional advantage is that the steam heater and electric heater are used to warm the diesel engine prior to startup.

2.3 Distilling Plant Economy. The generally accepted meaning of the term "distilling plant efficiency" is the pounds of distilled water produced per 1000 Btu of heat supplied. For multistage plants fitted with such features as evaporators, which are heated by vapor produced in an earlier stage, the overall economy of the plant may be estimated best by first preparing an approximate heat balance flow sheet. Such flow sheets are approximate, but the balance a plant will assume in service can be predicted with fairly good accuracy.

There are several factors which complicate a precise prediction of the manner in which a plant will perform in service. A "clean" plant is capable of producing excessive quantities of vapor to such an extent that carry-over will result, thereby contaminating the distillate; in such an event the steam supply must be throttled or other means used to control the output. Tube surfaces of different units foul to a different degree; therefore, fouling cannot be predicted with exactness. Additionally, the circulating water temperature varies with the season and locality. All of these factors affect the balance and economy of a distilling plant.

The overall economy of a distilling plant, as distinguished from the so-called efficiency, may be expressed in terms of the pounds of distillate produced per pound of additional fuel required for the distilling plant, over and above the fuel otherwise required to fulfill the ship's power requirements. Establishment of the distilling plant overall economy entails an assessment of the amount of fuel used to produce the steam required for the saltwater heater, the electrical power required for the various pumps in the plant, the useful work that could have been performed by the heating steam had it been used for other purposes, and the heat returned to the boiler feed system from the distilling plant (a very important consideration for distilling plants which have condensers that use the main propulsion plant condensate as the coolant).

Since the overall economy of a distilling plant depends upon the ship's power plant, and the interrelationships between the two, it is not possible to accurately state the economy of a distilling plant independently of the power plant. However, the following gives an approximate idea of the overall economy characteristic of four types of distilling plants:

• Single-stage submerged-tube plants operating on boiler steam at reduced pressure, 13 lb of distillate per pound of additional fuel.

• Two-stage flash plants operating on bleed steam, 50 lb of distillate per pound of additional fuel.

• Three-stage flash plants operating on bleed steam, 75 lb of distillate per pound of additional fuel.

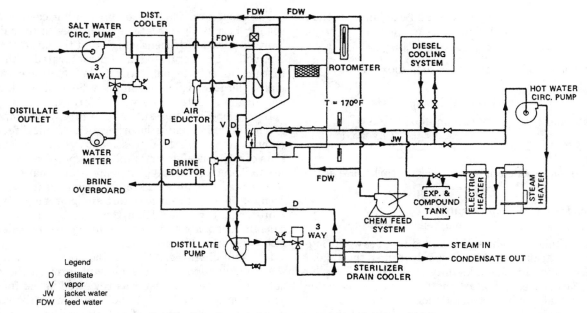

Fig. 24   Flow diagram of an evaporator having three heat sources

- Spray-film vapor-compression plants with electric motor drive and electric heaters, 200 lb of distillate per pound of additional fuel.

Based on flash distilling plants operated with exhaust or bleed steam, the weight of additional fuel required, together with the operating weight of the distilling plant, is but a small fraction of the weight of the distilled water produced during an average ship voyage. For a given ship, investigations may show that a worthwhile savings in tonnage can be effected by using distilled water rather than tank water, and that, all things considered, the cost of distilled water is lower.

2.4   Materials of Construction.   In general, 90–10 copper-nickel is the material used to manufacture tubes, tubesheets, waterboxes, evaporator shells, and piping for marine distilling plants [1,2,3]. However, in some applications, where operation in waters of high turbidity is anticipated, the tubes are made of 70–30 copper-nickel for increased resistance to tube erosion. Monel, a high-nickel alloy, has a high resistance to a seawater environment, but because of its higher cost, its applications are usually limited to valves and fittings where there are high flow velocities.

Distilling plant vessels that operate under vacuum conditions are usually reinforced with external steel stiffeners. This type of design permits the flat vessel surface to be made of thin copper-nickel plate, thereby reducing weight and construction costs.

Titanium is highly resistant to corrosive attack by boiling or flashing seawater, has a high resistance to erosion and impingement, has good heat-transfer characteristics, and has a high strength that permits the use of thin-walled material. However, the higher cost of titanium compared with copper-nickel alloys limits its use to instances where the superior characteristics of the material are of major importance. In distilling plants that have plate-type heat-transfer surfaces, titanium is commonly used in order to achieve corrosion-free service.

Alloys of stainless steel have been used in the construction of distilling plants, but they have proved to be unsatisfactory. Stainless steel is subject to various forms of corrosive attack. Under certain conditions of stress and temperature in the presence of chlorides and oxygen, stainless steel is susceptible to stress-corrosion cracking; unfortunately, in desalination plant applications, these conditions are inherent. Also, some stainless steels are subject to pitting corrosion when exposed to a stagnant seawater environment. To avoid such attacks, distilling plants having stainless steel parts would need to be drained and flushed with fresh water before being secured. In general, the increase in cost of a copper-nickel material, when compared with stainless steel, is more than offset by the reduction in maintenance expense.

Unless precautions are taken, the use of the more active materials, such as titanium, Inconel 625, stainless steel, and Hastelloy, in contact with copper-nickel or a bronze alloy in a seawater environment will result in galvanic corrosion, particularly so in flowing seawater. This type of corrosion can be avoided by electrically disconnecting the dissimilar materials.

2.5   Scale Control and Acid Cleaning.

a.   Scale control.   High-temperature, land-based evaporators have a much more serious scale-control problem than do marine plants, which operate at a low temperature and normally under vacuum. Nevertheless, the formation of scale is a major consideration in the design of marine distilling plants.

One source of scale is the bicarbonate in seawater, which decomposes when the seawater is heated and then reacts with the magnesium and calcium in the seawater to produce magnesium hydroxide and calcium carbonate scales as shown in the following reactions:

$$2HCO_3^- \rightarrow H_2O + CO_2 \uparrow + CO_3^{--}$$

2 bicarbonate ions = water + carbon dioxide gas
$$+ \text{ carbonate ion}$$

$$CO_3^{--} + H_2O \rightarrow CO_2 \uparrow + 2OH^-$$

carbonate ion + water = carbon dioxide gas
$$+ \text{ 2 hydroxide ions}$$

$$CO_3^{--} + Ca^{++} \rightarrow CaCO_3$$

carbonate ion + calcium ion = calcium carbonate scale

$$2OH^- + Mg^{++} \rightarrow Mg(OH)_2$$

2 hydroxide ions + magnesium ion
$$= \text{ magnesium hydroxide scale}$$

When an acid or acidic salt is added to a seawater supply, it will neutralize what is commonly termed the alkalinity of the seawater. The alkalinity is the sum of the hydroxide, carbonate, and bicarbonate ions in the seawater, although usually only bicarbonate ions are present in significant quantity. The hydrogen ions produced in the seawater by the added acids or acidic salts act to destroy or neutralize the bicarbonates by the following reaction:

$$HCO_3^- + H^+ \rightarrow CO_2 \uparrow + H_2O$$

bicarbonate ion + hydrogen ion =
$$\text{carbon dioxide gas + water}$$

Therefore, by treating the seawater with acid and converting the bicarbonates to form carbon dioxide and water, it is possible to eliminate the source of carbonate ions and prevent the formation of magnesium hydroxide and calcium carbonate scales. Such treatment is common practice in the operation of land-based multistage desalters.

During the acidification of normal seawater, for scale control purposes, approximately 100 ppm of carbon dioxide is formed through bicarbonate alkalinity breakdown. This "carbonated brine" is acidic and corrosive to ferrous materials of construction. The carbon dioxide can be removed by scrubbing the seawater with air or steam.

Calcium sulfate scales are probably the worst of the common scales which may be encountered when seawater is heated and evaporated to form concentrated brine. Calcium sulfate scales are not readily soluble in acid solutions; consequently, they must often be removed by mechanical methods.

The mechanism of calcium sulfate scale formation is simply precipitation caused by the concentration of calcium and sulfate ions beyond the solubility of calcium sulfate. Two crystal forms of calcium sulfate are involved, the anhydrite and hemihydrate. The anhydrite is the most insoluble form; therefore, its solubility is first exceeded upon concentration of a feedwater containing calcium and sulfate ions. Solutions supersaturated with respect to the anhydrite are stable for long periods of time. Supersaturation with respect to the hemihydrate is not as stable as that for the anhydrite, so that scaling does not usually take place until the concentration of calcium sulfate hemihydrate is exceeded. The solubility of both the anhydrite and hemihydrate decreases with increasing temperature.

The most common way of preventing calcium sulfate scaling in marine seawater evaporators is to maintain a blowdown rate sufficiently high that the brine concentration will not be saturated or supersaturated with respect to calcium sulfate hemihydrate. With proper distilling plant operation, calcium solubility limits are avoided and no calcium sulfate scales will be formed.

Both thin-film and submerged-tube low-temperature marine distilling plants incorporate a chemical feed treatment system. This feed treatment system basically consists of a chemical mixing tank and a proportioning pump. A small amount of polyphosphate is automatically added to the feed stream so as to chemically react with scale-forming ions to produce a soft sludge-type scale that is more readily washed from the tube surfaces than the calcium carbonate and magnesium hydroxide scales which otherwise would be formed.

b. Acid cleaning. During the operation of a marine seawater distilling plant, scale will form on the heating surfaces of the evaporator as reviewed in the foregoing. This scale is usually 80 to 90% calcium carbonate. The remainder is a mixture of calcium sulfate, magnesium hydroxide, metal oxides, silica, and miscellaneous deposits. However, when seawater feed is improperly treated with polyphosphates, other deposits such as phosphate sludges may be formed in large quantities. If fresh feedwater is used, silicates or calcium sulfate may be present. The rate of buildup and the composition of the deposits depend on such factors as the operating temperatures, the brine density, and flow rates. Feed treatment and cold shocking reduce the buildup of scale, but do not prevent or remove deposits entirely.

The symptoms of scale formation are:

(a) Consistently rising temperatures in the evaporator stages (decrease in vacuum).

(b) Heating steam pressure to the feedwater heater, or heating tube bundle, is required to be above the design value to produce the specified temperature at the feedwater heater outlet.

Scaled heat-transfer tubes can be cleaned chemically by circulating a diluted acid solution through, or over the tubes. Hydrochloric (muriatic) or sulfamic acid are the chemical reagents commonly recommended. Hydrochloric acid should be used only if sulfamic acid is not available, and then only by qualified personnel experienced in its use. Chemicals used for acid cleaning are somewhat hazardous; they are hazardous in the sense that most acids are dangerous. It is essential that the operator understand the potential danger involved with the use of acids. If proper precautions are taken, personnel injury and equipment damage can be avoided and acid cleaning of the evaporator can be accomplished in a minimum of downtime.

Sulfamic acid has become the chemical most commonly carried on board ship. No extra precautions are necessary in storing or handling dry sulfamic acid; however, since the acid dust will irritate the nose, eyes, and skin, careless handling which may result in its dispersion should be avoided. Sulfamic acid is only mildly corrosive to metals

Table 4 Correlation of pH with various acidic and basic solutions

	pH Value	Typical Color[a]	Strength	pH of Household Items	pH of Industrial Chemicals	Logarithmic Value to Pure Water
Increasing acidity	0				Sulfuric acid 4.9% (1.0N)	10,000,000
	1.0	Purple	Very strong		Hydrochloric acid 0.37% (1.0N)	1,000,000
	2.0	Red		Lemon juice		100,000
	2.5		Will dissolve average scale			
	3.0	Pink		Orange juice	Acetic acid 0.67 (0.1N)	10,000
	4.0	Orange	Fairly strong	Beer		1,000
	4.5		Too weak to dissolve scale			
	5.0	Gold		American cheese		100
	6.0	Yellow	Mild	Milk		10
Neutral	7.0	White	Neutral	Distilled water at 77 F		1
Increasing alkalinity	8.0	Light Green	Mild	Egg white		10
	9.0	Grass Green		Borax		100
	10.0	Dark Green	Fairly strong	Milk of Magnesia		1,000
	11.0	Light Brown			Ammonia 1.7% (1.0N)	10,000
	12.0	Dark Brown			Caustic soda 0.4% (1.0N)	100,000
	13.0	Black	Very strong			1,000,000
	14.0				Caustic soda 4.0% (1.0N)	10,000,000

[a] The color assumed by the indicator (e.g., litmus or pH paper) varies depending on the type and range of indicator used.

and there is no fire hazard involved in its use. Sulfamic acid is considerably less objectionable in all respects than dilute hydrochloric acid.

The maximum acid concentration should not be numerically less than pH 2.0. A reference for the correlation of pH values with various acidic and basic solutions is given by Table 4.

When one pound of sulfamic acid powder with color indicator is dissolved in one gallon of water, the solution will turn a light red color, indicating sufficient acid concentration to dissolve scale. If the solution is heated to approximately 120–140 F, the color will change to a deep red. As the solution is circulated through the evaporator components and comes in contact with scale deposits, a chemical reaction takes place between the scale and acid which dissolves the scale and reduces the concentration of the acid. The by-product of the chemical reaction is the liberation of large quantities of carbon dioxide gas, which must be vented to atmosphere at some point in the cleaning circuit.

As the acid is circulated and the chemical reaction takes place, scale is dissolved, consuming acid in the process, and the solution with gradually change in color to orange or yellow, indicating that most of the initial charge has been dissipated. At this time a recharge of sulfamic acid will be required to increase the acid strength and change the color back to red.

Periodic recharging or makeup of fresh solution will be required until the acid solution remains red for ½ to ¾ of an hour after the last acid charge. When this color condition occurs, the operator can be assured that all soluble scale in the cleaning circuit has been dissolved since the acid strength is no longer being dissipated by contact with

scale. The acid should then be completely drained from the plant and the entire cleaning circuit flushed with large quantities of fresh seawater.

2.6 Distilling Plant Vacuum Equipment. Proper venting of the distilling plant condenser is most important to prevent the buildup of noncondensible gases. Air is liberated in the evaporator from the entering feedwater, which is saturated with air. Carbon dioxide may also be released through a breakdown of bicarbonates in the seawater. Since a low-pressure marine evaporator operates under a vacuum, small quantities of air also leak into the unit through the gasketed joints. If the condenser is not properly designed, the gases collect in low-pressure pockets and render these areas ineffective, thereby reducing the performance of the unit. Figures 25 and 26 show the vent baffling and shrouding in a typical condenser and seawater heater.

Three different types of systems have been used to establish and maintain the low pressure required in the evaporation chamber: steam-motivated air ejectors, mechanical vacuum pumps, and water-motivated air eductors. Of the three, steam-motivated air ejectors have been most common; however, mechanical vacuum pumps and water eductors are suitable for limited applications.

The service conditions of a vacuum system vary depending on the design of the distilling plant. On a submerged-tube or spray-film distiller, the shell temperature and vacuum are maintained at a predetermined value regardless of the seawater feed temperature, and the vacuum systems for these distillers operate at one suction condition. In a flash evaporator, the shell temperature varies widely as the seawater feed temperature changes throughout the range of 28 to 85 F. The vacuum system

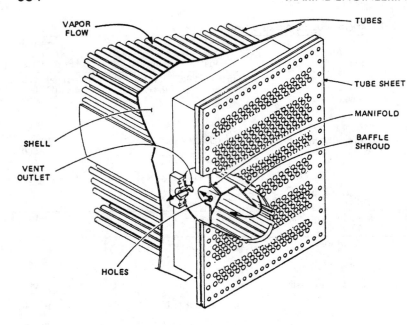

Fig. 25 Typical condenser vent baffle

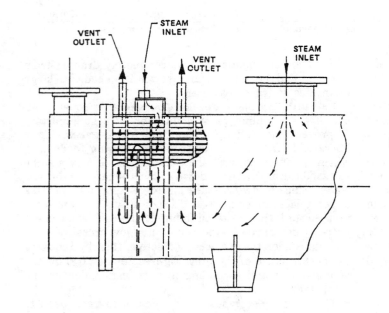

Fig. 26 Typical seawater heater and air ejector condenser baffles

must, therefore, be designed to provide proper operation at the suction conditions encountered with inlet seawater feed temperatures in this range.

In a two-stage flash evaporator, the temperature in the second-stage evaporation chamber will be approximately 114 F at a seawater temperature of 85 F; however, at a seawater temperature of 28 F, the second-stage temperature will drop to 70 F. The specific volume of steam is 238 ft³/lb at 114 F, but at 70 F the specific volume is 869 ft³/lb. Consequently, the 28 F seawater feedwater condition controls the design of a flash evaporator vacuum system. The vacuum system must be capable of extracting large volumes per pound of steam and noncondensibles removed at the lower seawater feed temperature.

The highly corrosive nature of the noncondensibles released in a distilling plant vaporization chamber must also be considered in the design of a vacuum system. In addition to air inleakage and seawater deaeration loads, the distilling plant vacuum system must also remove the CO_2 that results from the breakdown of carbonates in the seawater when it is heated to 170 F, as well as NH_3 and H_2S, which are also introduced by polluted seawaters.

a. Steam-motivated air ejectors. Air ejector systems used with most marine evaporators are of the two-stage noncondensing type, but single-stage systems can be used on submerged-tube and spray-film units where higher shell pressures are maintained. The operating principles of air ejectors are described in Chapter 14.

In the case of multistage flash plants, two-stage ejectors are generally employed in order that a high vacuum may be economically obtained. In this case a second air ejector element is placed in series with the primary air ejector element. The first-stage air ejector compresses the steam and air to an intermediate pressure, and the second-stage element provides the final compression to atmospheric pressure or slightly above to dispose finally of the air and noncondensible gases. A single-stage ejector unit is not suitable for operation at a vacuum higher than about 27 in. of mercury, and thus all high-vacuum ejector installations are provided with two- or three-stage ejector elements in series. Three-stage ejectors usually are not employed in naval practice because of the extra complication involved and the fact that two-stage ejectors are capable of evacuating to a sufficient vacuum for all normal installations.

The air ejector motivating steam is condensed in an after-condenser, which uses the entering feedwater as the cooling medium; such an arrangement is typically illustrated by Fig. 5. This reclaiming of the heat in the air ejector motivating steam increases the overall distiller efficiency. The air ejector after-condenser is often constructed as an integral part of the evaporator seawater heater or part of the heating bundle in the submerged-tube or spray-film type of distilling plants. This arrangement simplifies the construction of the units and reduces maintenance requirements by avoiding the necessity of an additional heat exchanger. Since noncondensibles in the after-condenser section are highly corrosive, the material in the after-condenser portion of a tube bundle should be nonferrous. The material commonly selected for this service has been 90–10 or 70–30 Cu-Ni, or titanium.

The air ejector steam chest is usually made of steel; the nozzle is stainless steel, and a stainless steel strainer is provided. The air chamber and the diffuser must be of bronze or copper-nickel construction to resist the corrosive vapors exhausted by the system.

b. Mechanical vacuum pump. Mechanical vacuum pumps have been used on submerged-tube and spray-film distilling units, which have less severe vacuum system requirements as compared with flash units. Submerged-tube and spray-film units normally operate with shell temperatures in excess of 125 F, and a single-stage vacuum pump can be employed.

In most cases, a liquid-ring pump using fresh water to form the liquid ring has been used to minimize the corrosion of pump parts. The fresh water is cooled by seawater in a heat exchanger, which is built into the vacuum pump seal-water tank. Chapter 14 contains further details concerning the design and operation of liquid-ring pumps.

If a vacuum pump were used with a flash distilling plant, a two-stage pump would normally be required in order to provide satisfactory operation with low seawater temperatures. The greater complexity of the two-stage system coupled with large component sizes and high costs involved in maintaining the pump under severe corrosive conditions has limited the use of mechanical vacuum pumps to instances where motive steam is not available to operate an air ejector.

c. Water-motivated air eductors. A water-motivated air eductor has also been used for the submerged-tube and spray-film designs which have shell temperatures above 125 F. In these distilling plant designs, more water is required for cooling than is actually used as feed to the unit. This water is normally directed overboard after passing through the condenser bundles; however, the excess cooling water can be used as motivating water for an air eductor, permitting operation of the eductor without having to provide an additional supply of motivating water. Figure 10 illustrates such an arrangement.

In cases where an eductor can fulfill the vacuum requirements, its use permits a compact installation since some steam piping is deleted and an after-condenser is not required. Eductors are discussed further in Chapter 14.

2.7 Distilling Plant Automation. The Maritime Administration Standard Specification for Cargo Ship Construction [3] requires that each desalination unit be capable of unattended, automatic operation after being put on the line locally. Most marine distilling plants can be readily automated to such an extent with a few simple controls. Since the steam supply is normally from a low-pressure extraction point, or alternatively, from a higher pressure source reduced to 5 psig and then controlled through a fixed critical-pressure orifice in the steam supply line to the distilling plant, no automatic control is normally required for the steam supply.

It is necessary to maintain the brine temperature leaving the seawater heater of a flash plant at a specified temperature, usually 170 F; this is done by means of an automatic temperature control valve. No control of the brine flow leaving the brine overboard pump is required since all of the water entering the unit is pumped overboard. The only control required on the distillate is that of conductivity; a 3-way valve automatically diverts the distillate to the bilge in the event that the conductivity is not within the specified control limits; the dump valve can be provided with an automatic reset. Provisions must also be made to stop the feedwater flow in the event of a brine pump failure to prevent accidental flooding of the evaporator.

To insure that the water produced will meet the requirements of the U.S. Public Health Service [6], an arrangement or control in the circuit must be incorporated to insure that all of the distillate is heated to at least 165 F; otherwise a distillate sterilizer must be incorporated. In a submerged tube evaporator or thin-film double-effect unit, it is necessary to add a sterilizer to the unit since in these types the normal operating temperature throughout the system is below 165 F. The steam supply to this sterilizer must incorporate an automatic regulating valve.

Figure 5 shows those controls necessary for a flash plant to meet the unattended automatic operation requirement after being put on the line locally.

Fully automatic operation requires that the plant be equipped with automatic control devices to initiate all starting sequences upon activation of one pushbutton. The following sequence, requiring various controls and additional features, is necessary to accomplish automatic starting of a typical two-stage flash plant by a single ON/OFF pushbutton or switch:

1. The feed pump is started, the brine pump is started, an air-operated bypass valve opens in the inlet feed line to the first stage. Correct operation of the feed pump is proven by a 20-second time-delay pressure switch; if the pump pressure does not build up, the plant is secured. A high-level switch proves the operation of the brine pump; if the pump does not remove the brine from the second stage, the high-level switch will secure the plant. A float switch with a one-minute delay proves the opening of the bypass valve; if a level does not build up in the first stage, the plant will be secured.

2. The steam valve to the air ejector opens after a three-minute delay; the delay is obtained by the use of a timer. If the air ejector malfunctions and the vacuum is not reduced to the operating level in 35 minutes, a vacuum pressure switch and timer arrangement will trip the plant.

3. When the vacuum pressure switch indicates that the second stage has reached the operating vacuum conditions, the steam pressure-reducing valve opens to allow 20% steam flow. Also activated at this time is a timer that continues to open the steam valve, allowing 35 minutes for full steam flow to be reached. Opening of the steam valve is proven by a temperature switch and timer in the inlet to the first stage; if the feedwater temperature does not reach 120 F in ten minutes, the plant will be tripped. A solenoid valve in the "start-up" desuperheating water supply line is also opened by the vacuum pressure switch in the second stage.

4. The condensate pump is started by an electric float switch in the feedwater heater hotwell. Starting this pump secures the start-up desuperheater water supply solenoid. Proof of the operation of this pump is determined by a pressure switch in the feedwater heater. If the pump does not operate correctly, the level in the heater will rise and pressure in the heater will increase. The pressure switch will trip the plant if the heater pressure reaches 3 psig.

5. The distillate pump is started by a float switch located in the condenser trough above the normal operating level. Proof of the operation of this pump is obtained by the use of a low-vacuum trip in the second stage. If the condensate level in the condenser becomes excessively high, the air ejector suction line will be covered, the vacuum will decrease, and the plant will trip.

6. Additional items also required include:

- Pump liquid-level control valves are required on the brine, distillate, and condensate lines.
- Steam traps are required in all steam lines; these traps should discharge to a drain main in order to prevent water hammer.
- A panel is required at the local automatic start station to indicate the progress of the start-up procedure. If the plant trips during start-up, this panel is

necessary to determine the component, switch, or relay which caused the trip.

Simply automating the opening of the steam supply valve and air ejector steam supply valve can result in possible compromises to reliable and extended operation. These services should be put on the line slowly with the caution required by sound engineering practice when admitting steam to a plant. Therefore, additional control and timing devices to "bleed" warm-up steam prior to opening the valves are required. These two valves, once opened during an attentive starting sequence, need not be touched or regulated to insure proper plant performance.

The automatic timing of the start-up sequences is necessarily based on starting with a cold plant plus safety margins and, therefore, the timing of successive sequences is set for a much longer time than usually required to manually start the plant under normal conditions. Consequently, it would be common for the operator to override the automatic start in many instances to decrease the time required to get the plant on the line.

References

1 "Distillation Units, Water; Steam, or Flashed Vapor Operated, or Fresh Water Heated, Low Pressure, Naval Shipboard," Military Specification MIL-D-18641.

2 "Distillation Unit, Water, Thermocompression," Military Specification MIL-D-16196.

3 "Standard Specification for Cargo Ship Construction," Office of Ship Construction, Maritime Administration, U.S. Department of Commerce.

4 "International Standards for Drinking Water," World Health Organization, Geneva.

5 H. U. Sverdrup, M. W. Johnson, and R. H. Felming, *The Oceans—Their Physics, Chemistry, and General Biology*, Prentice-Hall, Inc., Englewood Cliffs, N.J., 1942.

6 "Handbook on Sanitation of Vessel Construction," U.S. Public Health Service, Department of Health, Education and Welfare.

7 C. D. Rose, J. Heck, and W. Pergande, "New Electric Motor Driven Vapor Compression Distilling Plants for Navy Surface Ships," *ASNE Journal*, American Society of Naval Engineers, May 1985.

8 Phillip Liu, "A New Concept in Marine Desalination—The Thermal Compression Distilling Plant," *Marine Technology*, Vol. 27, No. 3, May 1990.

9 Joseph F. Pizzino, Wayne L. Adamson, and W. Smith, "Reverse Osmosis for Surface Ship Desalination, An Overview," *ASNE Journal*, American Society of Naval Engineers, May 1991.

10 P. Hoornaert, *Reverse Osmosis*, Pergamon Press, New York, 1984.

11 D. Q. Kern, *Process Heat Transfer*, McGraw-Hill Book Co., New York, 1950.

12 W. H. McAdams, *Heat Transmission*, McGraw-Hill Book Co., New York, 1954.

CHAPTER XVIII

Thomas P. Mackey | # Hull Machinery

Section 1
General Design Considerations

1.1 Scope. Broadly defined, "hull machinery" includes all power-driven equipment located outside the machinery spaces that is not associated with the main propulsion plant. Included as hull machinery are such items as steering gears, anchor windlasses, all types of winches, cranes, capstans, elevators, cargo access equipment, thrusters, special rudders, and stabilizers. Few of these types of equipment require design techniques beyond those of standard machine design and control system practices, which are adjusted to make the equipment suitable for the marine environment and to cope with the unique situations often encountered on board ship. This chapter, therefore, is oriented to the design considerations peculiar to marine equipment.

Close cooperation is required between the marine engineer and the naval architect in the design of hull machinery. As a consequence, this chapter has been written to complement the Society's publications *Principles of Naval Architecture* and *Ship Design and Construction*, which emphasize the naval architect's role in the design of hull machinery.

A number of ASTM (American Society for Testing and Materials) and International (ISO) Standards have been developed for hull machinery. ISO Standard ISO 7825 describes "General Requirements" for deck machinery. Other standards are also referenced in this chapter. The appropriate current ASTM and ISO Standard should be referred to when specifying or designing ship's hull machinery. ASTM Shipbuilding Standards are published as a separate ASTM volume, Volume 01.07.

One consideration that must be made in hull machinery selection is the trade-off between initial procurement cost and life-cycle cost. The overall design requirements may include required duty cycles, which make certain types of hardware more appropriate than others.

1.2 Types of Drives. There are, for most practical purposes, only three types of drives for hull machinery: electric, electrohydraulic, and occasionally steam. However, direct diesel engine drives are also often used for thrusters and some types of deck machinery.

Steam-driven units are seldom installed aboard ship. They are specified for some steam-powered tankers, and because they do not spark, they are sometimes used in the vicinity of hazardous cargoes. Steam-driven units have the advantage of a very high slack rope speed, and the equipment is generally very rugged, simple, and easy to maintain. However, the difficulties encountered in keeping steam mains properly insulated and the problems associated with maintaining an adequate steam flow in cold weather tend to outweigh these advantages. The reliability and ease of obtaining variable speeds from hydraulic and electric drives, the desire for remote control and automation, and the steady decline of steam as a means of propulsion have pointed the development of hull machinery drives away from steam and toward electrohydraulic or electric powering.

1.3 Mechanical Details. One of the obvious points to be considered in the design of hull machinery is that it is necessary to design for the pitch, roll, trim, and list of the ship. All machinery should be designed for at least the following conditions:

- A pitch of 10 deg (bow up to bow down)
- A permanent trim of 5 deg by either the bow or stern
- A roll of 30 deg (each side)
- A permanent list of 15 deg to either side

However, no two of these conditions are considered to occur simultaneously.

In the case of equipment that is not used at sea under storm conditions (e.g., mooring and accommodation ladder winches), the design conditions listed apply only when the equipment is in the stowed condition.

In many cases, a common bedplate for the driving and driven equipment is preferable, in order that alignment may be more easily maintained. For heavily loaded deck machinery and for equipment with a large "footprint," however, it is preferable to arrange the foundations to directly transfer the forces into the hull structure. A large bedplate can add substantial weight, but limited strength. Consideration of the location of ship structural members when locating heavily loaded deck machinery can reduce the foundation strength requirements.

Access for painting etc. is also important. In some cases, welding the equipment to the deck will yield the best arrangement, but the feasibility of realignment and dismantling for maintenance and repair must be considered. With proper specifications at the time of procurement, some deck machinery could be delivered by the vendor with an integrated foundation, thereby eliminating any accessibility concerns on the part of the shipyard.

In some instances, such as central hydraulic systems and some capstans and windlasses, the driving and driven assemblies are mounted on different decks. When such two-deck arrangements are employed, provisions must be made to accommodate the deck deflections and small amounts of lateral displacement between the two assemblies. Where a shaft connects the two assemblies, flexibility is usually provided by a flexible coupling; if the connection between the two assemblies is hydraulic piping, the relative movement between assemblies can be accommodated by using hydraulic hoses or by designing flexibility into the piping arrangement (i.e., by looping or offsetting the piping, taking care to avoid elevated regions which would trap air).

Motors or gearboxes located in the weather should be provided with watertight housings, while those located below deck or out of the weather may be provided with either totally enclosed fan-cooled or dripproof protected housings.

Rotating machinery should be aligned with the axis of rotation longitudinal to the ship. This avoids high bearing loads, which would be experienced during ship roll if mounted transversely.

Bearings generally may be either of the sleeve or antifriction type. Sleeve bearings should be made of a good grade of bearing bronze. Antifriction bearings are subject to "brinelling" caused by vibration. If the ball or roller presses away the grease in its ring contact surface, and the grease film is not reestablished prior to the next vibration contact, brinelling can result. However, a suitable lubrication arrangement and the use of the proper lubricant can prevent this type of brinelling. Where bearings will be subject to impact loading, as on the wildcat shaft of the windlass, they should be either bronze sleeve bearings or spherical roller bearings, which combine shaft flexibility and high load-carrying capacity.

Antifriction bearings should be used in any location where positive alignment, minimum lost motion, or freedom from wear is essential for proper operation. Some locations that fall into these categories are: motor bearings, steering gear differential control assemblies, worm and worm wheel shafts, and tension-sensing assemblies as on automatic mooring winches.

Reduction gears should be totally enclosed and provided with an oil bath lubrication and with means for filling, draining, and measuring the lubricant level. Properly gasketed inspection and maintenance openings should be provided in all gear enclosures. It is suggested that the interior of gear cases be sandblasted and properly coated before final assembly.

The design and manufacture of gearing for hull machinery should be in accordance with recognized standards, such as those of the American Gear Manufacturers Association. No compromise for strength should be allowed when designing the gearing, but it is considered acceptable practice to make some adjustments in the calculations for wear because most hull machinery operates on an intermittent basis. The wear rating of gearing for hull machinery can be increased because of its intermittent usage. The range of wear load service factors that may be considered, that is, the factors by which the nominal gear working loads can be multiplied by to arrive at the design wear loads, is given in Table 1.

Table 1   Wear load service factor for gears

Equipment	Factor
Capstans	0.35–0.5
Crane machinery	0.7–1.0
Steering gear drives	1.0
Winches, cargo	0.7
mooring	1.0
other	0.35–1.0
Windlasses	0.35–0.5

Spur, helical, and herringbone gearing should be of steel, should have machine-cut teeth, and may be heat treated to increase ratings. Worms should be made of steel and be integral with their shafts, and worm wheels should be made of bronze. When a worm wheel is large, say over 8 in. pitch diameter, it may be made with a bronze rim bolted to a steel hub. In order to obtain a high degree of mechanical efficiency, worm reductions should be of an overhauling type. There are, however, instances where a nonoverhauling feature is more important than the efficiency considerations involved. Additionally, it is good practice to avoid the use of worms on overhung shafts because of the misalignment that can result from the worm shaft deflecting.

Brake linings should be of an incombustible, non-asbestos material that is not adversely affected by heat, salt atmosphere, or moisture. In the past, asbestos was the most commonly used brake lining material, but for health reasons regulatory agencies in several countries no longer allow its use. Some non-asbestos brake linings provide lower friction, but the difference between static and dynamic braking power can be smaller than for an asbestos lining. Brake fade under elevated temperatures is another significant consideration for non-asbestos material. Woven linings are usually manufactured using a resinous binder. It is suggested that the linings be baked in accordance with the manufacturer's recommendations to remove any excess amount of binder before the linings are installed. (There is additional discussion regarding lining materials and their installation in Section 3, which is devoted to the anchor windlass.)

Where flexible couplings are used, as between motors and pumps, they should be of the all-metal type, preferably all steel.

All fasteners exposed to the weather up to and including ½ in. diameter should be of nonferrous material or stainless steel. Larger fasteners may be cadmium plated or galvanized steel. Breathers and drain check valves should be of a nonferrous material or stainless steel. Dowels and taper pins should be of stainless steel. Aluminum should not be used for watertight enclosures.

The rope-contact surface of capstan, warping, and windlass heads should be hardened to a depth of at least ¹⁄₁₆ in. to a minimum of 300 BHN to provide an abrasion-resistant surface.

The main rams for steering gears are hydraulic cylinder rods, but because they are invariably located below deck and out of the weather, they need no surface protection other than that provided by the hydraulic oil that adheres to them as they move in and out of the cylinders. All other cylinder rods, including those located in cargo holds where dust may be present, should be plated. The outer layer of plating should be of chromium because of its fine finish. However, chromium plating frequently contains cracks, invisible to the eye, which will allow seawater or moisture to penetrate and start corrosion. The corrosion spreads under the layer of chromium and separates the plating from the steel rod. The rusted surface then is free to abrade the packing in the cylinder as the rod moves in and out. This can be avoided by first applying a plating of nickel on the rod. The nickel plating provides a completely impervious coating, but is rough and would, in itself, tend to wear out the packing and cause leaks. Therefore, a layer of chromium must be added over the layer of nickel. Each coating should be 0.0005 to 0.001 in. thick. It is advisable to provide wipers on the rods to prevent dirt from being drawn into the packing. In some locations, it may be advisable to provide boots on the rods to further protect against an accumulation of dirt on the working portion of the rod. If boots are installed, they should be of a clear plastic material so that any leakage of oil into the boot may be observed. Wherever possible, the arrangement of the rod and cylinder should be such that the rod is withdrawn into the cylinder when the equipment is secured to protect the rod from surface deterioration caused by extended exposure to the weather.

Foundations for hull machinery are usually made of inverted angles that are welded to the deck. Each item of hull machinery usually is provided with a base that has a true, machined mounting surface; however, a chocking system that uses an epoxy resin may also be used.

When using a poured-in-place epoxy resin system, spacers (which may be nothing more than small pieces of ½-in.-thick steel) are placed on the foundation angles to position the machinery unit at the desired height. The unit of hull machinery is lowered onto these spacers, and the foundation bolts are inserted, but the nuts are run up only "finger tight." Alternatively, one adjustment screw that penetrates each equipment foundation pad is screwed against the inverted angles. The height established by the spacers is adjusted until the alignment is acceptable. Three sides of each foundation mating surface are then provided with temporary dams, and the space between the equipment base and the foundation is poured full of the liquid epoxy material and closed. Considerable care must be taken in pouring the liquid epoxy. After about 48 hours (or such time as is required by the particular mixture being used), the temporary dams are removed, and the foundation bolts are properly tightened. No attempt is made to remove the spacers. The equipment is then properly seated and the epoxy between the equipment base and the foundation provides a high degree of corrosion protection and has a high shock-load resistance due to the inherently high coefficient of friction between the epoxy and the steel. This system has proven satisfactory and is accepted by the regulatory bodies; however, only certain mixtures and methods of application are permitted by the regulatory bodies.

Stress analyses should be made during the design of all hull machinery. These should include calculations for

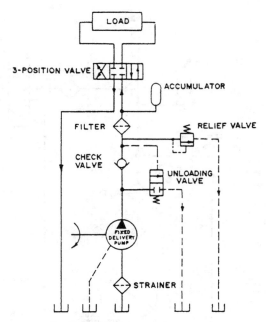

Fig. 1 Constant-flow hydraulic system

gearing, bearings, shafting, structural components, foundation bolts, etc. For merchant ships, the normal-duty stresses should not exceed 40% of the yield point of the material and maximum stresses should not exceed 75% of the yield point; however, other design criteria regarding stresses may be required by the approval authority.

Where automatic or oil-bath lubrication is not furnished, pressure gun grease lubrication should be supplied.

1.4 Hydraulic Details. The hydraulic systems used aboard ships vary widely in variety and complexity; however, most shipboard hydraulic systems bear a degree of similarity with either the constant-flow, constant-pressure, or demand systems shown in Figs. 1, 2, and 3, respectively.

The major components of a constant-flow hydraulic system are the fixed-delivery pump, unloading valve, and control valve. When the pump is started, fluid is sometimes delivered to an accumulator, which is usually of the pneumatic type, until the accumulator pressure equals that of the unloading valve setting. If there is no demand on the system, the unloading valve opens to bypass the pump discharge back to the tank, and the pump runs continuously. At the same time, the check valve holds the fluid pressure in the system so that it is immediately available upon load demand.

When a throttle or control valve to an attached hydraulic actuator is opened, the system immediately provides fluid to this load and the discharge causes a drop in system pressure. In turn, the unloading valve closes and the pump discharge is returned to the active system. Upon a decrease in the load demand, the system pressure again rises to actuate the unloading valve, and the load cycle is complete. The sump tank is replenished by returns from the exhaust lines of the attached loads.

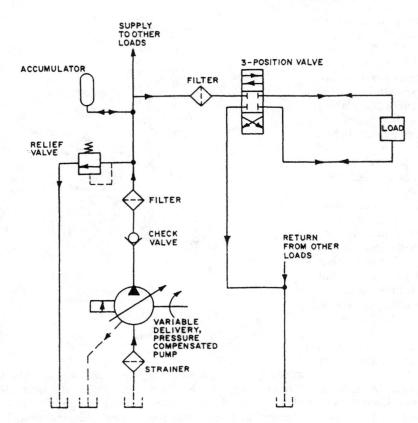

Fig. 2   Constant-pressure hydraulic system

Constant-pressure hydraulic systems, as shown in Fig. 2, generally use one or more variable-delivery pressure-compensated pumps that supply hydraulic fluid at a substantially constant pressure to either a system of multiple loads or to a single load, such as a hydraulic elevator hoist. The constant-pressure hydraulic system pump takes suction from a sump tank and discharges directly into the main supply piping. When the pumping capacity exceeds the load requirements, the system pressure increases to a predetermined value, at which point the pressure compensator acts to take the pump off stroke, thus stopping or reducing the flow of hydraulic fluid. A relief valve is provided to protect against overpressurization in the event that the pressure compensator fails to properly reduce the stroke of the pump. When the system demands cause the system pressure to drop to a preset value, the pressure compensator acts to put the pump on stroke, thus restoring the flow of the hydraulic fluid to recharge the accumulator, if installed, and to maintain the system operating pressure. The fluid flow to the individual loads may be controlled by a variety of types of valves, in addition to the three-position valve shown for simplicity in Fig. 2.

The demand hydraulic system, as shown in Fig. 3, is a closed-loop system particularly adapted to meet the precise demands of varying loads such as steering gears and automatic weapon-handling equipment. The variable- and reversible-delivery main pump does not take suction from a sump tank, as a loop of oil is maintained between the main pump and the load. The closed loop eliminates a

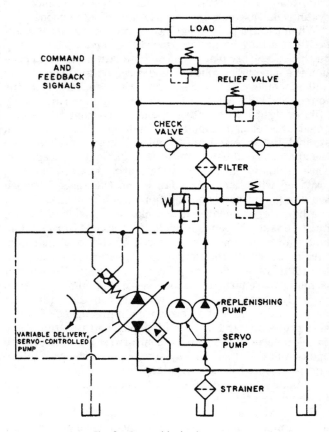

Fig. 3   Demand hydraulic system

significant amount of pressure drop, which would exist if the load control were achieved by a servo valve located in the main hydraulic piping. A servo pump and a replenishing pump, often driven by the same shaft as the main pump, provide actuating fluid to stroke the main pump and to provide makeup fluid to the closed loop, respectively. Control of the main pump is accomplished by command and feedback signals. The command and feedback signals may be composed of a combination of mechanical, electrical, or hydraulic devices.

In one type of hydraulic steering system, a rudder-position command signal originates in the wheelhouse while a mechanical feedback signal, which relates the position of the rudder to the helm position, is originated from the steering gear. The input command signal is converted to a mechanical command signal at the steering gear control differential. The feedback is the other input to this differential, which then measures the "error" between the helm and rudder positions. The "error" is translated into the form of mechanical motion to a rotary servo or other appropriate device, which controls the flow of pressurized fluid to stroke the main pump. The response is one of direction, as well as magnitude, so that the error is corrected in either the right or left rudder direction. The rate of rudder motion is dependent on the magnitude of the error or differential within the limit of the available pump stroke. For small differentials, as when holding a course, the pump stroke is moderate and the rudder response is slow. When going hard over in an emergency, the wheel will be well ahead of the rudder, the pump will be forced on maximum stroke, and the rudder rate will be at its maximum.

The same principle is advantageous in hoisting machinery drives in which a motor runs continuously in one direction to drive the variable-stroke pump, and the hydraulic actuator (motor or ram) is started, stopped, reversed, and driven at infinitely variable speeds within the system capacity by simply varying the pump stroke.

Accumulators have a variety of applications in shipboard hydraulic systems. For shipboard use, they are usually pneumatic. In typical applications, accumulators are used to minimize the pumping capacity that is required in systems which have large oil-flow demands followed by significantly lower demands or idle periods. They are also used in applications where an instantaneous source of hydraulic oil must be provided in response to intermittent demands or to provide a source of emergency power in the event of a power failure. In systems that must be held under pressure indefinitely, accumulators may also be used to make up small increments of possible leakage, thereby eliminating the need of frequent short pumping cycles.

Central hydraulic systems are advantageous in many cases, especially in large tankers, where safety regulations limit the installation of electrically powered equipment. Hydraulic pumping stations serving two or more pieces of equipment have been used in a variety of applications aboard ships. Most difficulties experienced with such systems have been due to deficiencies associated with system design, cleanliness, or venting. The basic engineering of

centralized hydraulic systems is of major importance to ensure that the system elements are compatible and are capable of functioning to meet all requirements.

Cleaning, as well as venting, the system also starts in the design stage. Not only must access for flushing be provided during the design stage, but also the design must be checked to ensure that the flushing velocities will be sufficient to clean the system. Additionally, pockets and voids, which would tend to collect foreign matter, must be avoided.

Great care must be exercised in fabricating and installing both lubricating and hydraulic oil piping to avoid contamination. In addition, all systems should be thoroughly cleaned after installation. Reference 1 describes the problems associated with hydraulic system contamination and makes some suggestions concerning hydraulic system cleanliness and how it can be maintained. The subject is discussed further in reference 2 and several recommendations are made. Recommended practices for the flushing and cleaning of marine lubricating oil and hydraulic systems are contained in ASME Standard No. 113. Good seals are required to prevent contaminants from entering the system; a thorough cleaning of the seals after installation is also important. Where possible, the manufacturer should completely assemble, clean, and charge the equipment with oil for protection until it is placed in service. It is also necessary to adequately protect machined surfaces, such as brake drums and cylinders, which must be rust free at the time of final installation on the ship's deck. Machinery is often stored outdoors and unprotected at the shipyard, sometimes for several months; therefore, the protection of exposed machined surfaces is essential.

When hydraulic equipment is used, the bedplates for the pumping units should be provided with an oil-retaining rim so that any leakage will be confined to a limited area.

There are two possible arrangements of centralized hydraulic systems that may be considered for merchant ships. In one, the bow anchor windlasses and forward mooring equipment are driven by one system and the aft mooring equipment by another system. Alternatively, all anchoring and mooring equipment, including capstans, can be driven by a single system.

In the first case, the pumping units would be located close to the driven equipment so that the piping runs would be comparatively short. In the second case, the runs would be long, but the pumping units could be located in the engine room where they could be serviced readily.

Great care must be taken in sizing the pumps and piping for any of these central systems in order that there be adequate capacity for the maximum simultaneous loads. In the engine room system, three or four pumps may be provided to handle the maximum load. These pumps would be cut in or out as the loads vary.

Most hull machinery hydraulic systems are either exposed to the weather or are installed in unheated parts of the ship. For these reasons, during the design stages consideration must be given to the ambient temperatures to be encountered and the viscosity characteristics of the oil selected. If it is not possible to warm up the system prior to actual operation, heating of the oil in the sump by electric heaters may be desirable.

Hydraulic piping that will be exposed to the weather should be of a corrosion-resistant material, such as stainless steel, except in the largest sizes used in ship-wide central hydraulic systems. If steel piping is used, its wall thickness should include an allowance for corrosion.

Hydraulic systems for steering gears, windlasses, winches, cranes, etc. should be provided with gage connections fitted with shutoff valves so that system performance may be checked during initial testing and regular overhaul periods. The gages should be valved off and should have gage snubbers in the line to prevent damage by system pressure pulsations.

Hydraulic sumps, or reservoirs, should generally be large enough to contain 110% of the volume of oil in the system so that the system may be drained to the sump for servicing or maintenance. Even when a system cannot be drained to the sump, because the units served are below the sump, it is good practice to provide a sump large enough to contain all of the oil in the system; that is, after the entire system, including all variable-displacement units, has been filled, the pump suction should still be flooded to the operating level.

Pumps taking suction from a sump tank can be fitted with a suction strainer, but no filter should be installed in the suction line since a partly clogged filter can cause a pressure drop that can result in a breakdown of the pump.

A full-flow filter in the discharge line is a good safety device. The filter should be equipped with an indicator that will show the pressure drop across the filtering element.

1.5   Electrical Details.   The characteristics of electrical equipment are discussed in Chapter 19. Electrical equipment should be either watertight, totally enclosed fan-cooled, or dripproof protected as dictated by its location, in or out of the weather. Motors should have natural ventilation in most instances, and bearings that are permanently lubricated and sealed. Watertight motors should be equipped with an automatic drainage fitting. Where a brake or gear is to be mounted directly on the motor shaft, the shaft should be tapered and fitted with a key and locknut.

Electric brakes are normally of the spring set, solenoid released, shoe disk type. Brakes should be of sufficient capacity to stop and hold the load under any condition of operation or testing. Brakes are commonly designed to support 200% of the normal operating load, and tests should be performed to confirm this capability. Watertight brakes are frequently fitted with an external hand release so that, in the event of a power failure, the load may be safely lowered without the necessity of removing the brake cover.

The enclosures for watertight shoe brakes should be of $\frac{1}{8}$-in.-thick stainless steel or of $\frac{1}{4}$-in.-thick mild steel that is treated both on the inside and outside with an inorganic zinc coating, and the hardware should be corrosion resistant. The slight extra cost for these designs will ensure that the enclosure will last for the life of the ship. The watertight enclosure should be fitted with an automatic drainage fitting.

Where possible, control panels should be arranged so that they may be serviced from the front. This will allow panels to be mounted back-to-back or against a bulkhead. Where the panels are to be mounted in the weather, the enclosures should be of stainless steel with corrosion-resistant hardware.

Limit switches are a potential source of trouble, particularly when they are exposed to the weather. Their reliability is further impaired if they are subject to being coated with paint when the equipment is painted. Great care should be taken in their selection to ensure that they are of rugged construction. Shafts in limit switches should be of bronze, monel, or stainless steel. Proximity switches are much more reliable and are the preferred alternative. Proximity switches are available in different housings and with a large range of sensing distances. These sensors are not influenced by salt, dirt, most paints, or seawater, and they operate quickly and safely.

Master switches for variable- or multiple-speed equipment should be of the cam type with vertical handles having integral knobs. They should be permanently marked to indicate operating directions and, if the control is of the multiple-speed type, there should be detents so that the operator can sense the speed points. The enclosures should be adequately protected against corrosion and should be raised to a convenient working height. Many owners prefer that even cam-type master switches be of the "spring return to OFF" type to guard against the possibility of the equipment being left running in the absence of the operator. If this type of switch is provided, the detents must be very slight so that there will be no chance of the handle "hanging up" on one of the running points.

All hull machinery control equipment, except steering gears, must incorporate undervoltage protection due to regulatory body requirements. If the master switch is not of the "spring return to OFF" type, the control circuitry should be arranged so that it is necessary to return the master switch to the OFF position to reset the undervoltage relay. If the controller overload relay is of the manual reset type at the controller, no extra undervoltage control relay need be provided. If the controller overload relay is of the automatic reset type or if an automatic reset temperature-sensing element is built into the motor, undervoltage protection should be provided by an undervoltage relay that will not reset until the master switch is returned to the OFF position.

Heaters should be provided for all watertight motors and brakes and, where space permits, in watertight master switches; heaters should also be provided for all motor controllers. The heaters should be connected to an electric circuit that is continuously energized and independent of the power feeders. One method of providing heating for alternating-current motors is to provide reduced voltage in one of the three phases. All heater circuits should be arranged so that they will be interrupted when the main power circuit for the motor is energized.

Section 2
Steering Gears

2.1 Introduction. Steering gear arrangements in great variety have been used to control the position of ships' rudders [3,4,5]. The advent of variable-stroke, bidirectional pumps made possible the design of ram-type, closed-loop hydraulic steering gears of a torque capacity adequate for the largest vessels. These pumps, which are largely of the axial-piston type, are driven continuously in one direction of rotation. The pump discharge is varied in rate and direction by controlling the stroke of the pump pistons. Right and left rudder motion and the holding of rudder position are, therefore, accomplished very smoothly.

Several types of electrohydraulic steering gears offer advantages in particular applications. A common one is the Rapson-slide, which is particularly well suited for higher rudder torque ratings because its arrangement provides an increasing mechanical advantage at larger rudder angles (see Section 2.3). Rapson-slide steering gears may be of either the single- or the double-ram type. Figure 4 is an illustration of a double-ram unit.

The link-type steering gear, which is typically illustrated by Fig. 5, also employs a ram group. However, the principle of operation is considerably different, as discussed in Section 2.3. Link-type steering gears generally have a decreasing, rather than increasing, mechanical advantage at larger rudder angles. However, there are two cases for which a link-type steering gear is ideally suited.

One is where there is insufficient space around the rudderstock to permit the installation of rams. By comparing Figs. 4 and 5, it will be seen that the space required in way of the rudderstock is considerably different for the two types. The other case is for a twin-rudder ship, which can be arranged so that one link-type steering gear can serve both rudders, as shown by Fig. 5.

A third type of electrohydraulic steering gear is the rotary-vane, which is illustrated in Figs. 6 and 7. Rotary-vane steering gears have no rams; they consist of a housing or stator, which is fixed to the foundation and contains two or three vane cavities, and a rotor with vanes attached, which acts as a tiller. The rudder torque is developed by the differential pressure that acts across the vanes, which can be seen in Fig. 7. Rotary-vane steering gears are suitable for any rudder torque rating. The same torque is available at any attainable angle, which is normally 40 deg for 3-vane actuators and 65 deg for 2-vane types. They offer the advantages of simplicity, low space requirements, low weight, higher attainable rudder angles, and, in many cases, lower procurement and installation costs but perhaps more maintenance costs. Rotary-vane steering gears are often used when high-maneuverability rudders, requiring larger rudder angles, are specified.

Rotary-vane steering gears generally use fixed-displacement pumps operating at lower pressures than ram-type steering gears, normally 870 to 1450 psi compared

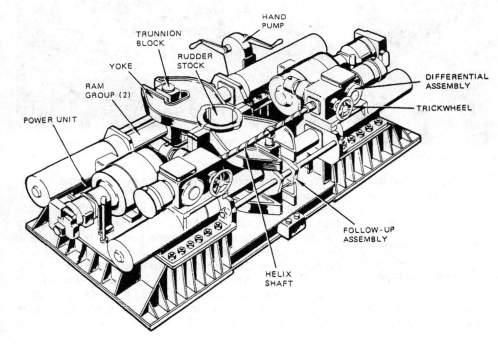

Fig. 4 Rapson-slide type steering gear

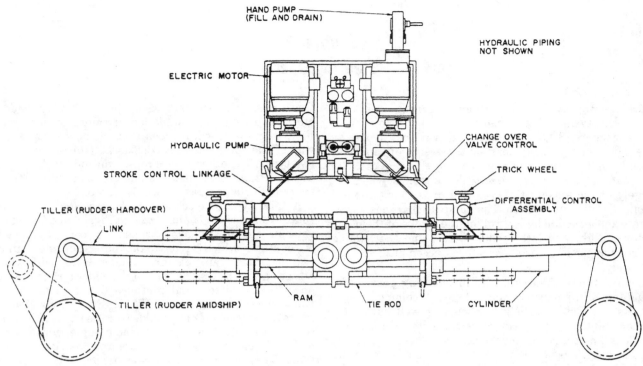

Fig. 5   Link type hydraulic steering gear for a twin-rudder ship

Fig. 6   Vane type steering gear with dual power units

Fig. 7   Spherical rotary vane actuator—cutaway view

with 1500 to 2500 psi, and sometimes higher pressures, for ram-type steering gears. The reason for the lower operating pressures is to guard against lip-seal leakage in the rotary-vane actuators.

There are several variations of rotary actuators for steering systems, including rotary pistons and rack-and-pinion actuators. A unitized design rotary-piston type actuator is shown in Fig. 8.

Clevis-mounted electrohydraulic steering gears, such as illustrated by Fig. 9, are used extensively on smaller craft, but have been installed on large cargo ships. Clevis-mounted units consist of one, two, or four, single- or double-acting cylinders. The cylinder ends are pinned to the deck or foundation and the piston rod end is attached to the tiller, also by a pinned connection. Oil is supplied to the cylinders via flexible hoses. Because of the critical function of the steering gear, care must be taken to specify only heavy-duty, high-quality hydraulic cylinders and to avoid potential failure points, such as unsecured screwed-on piston rod ends or unsuitable hydraulic components or hoses that could cause failures. This type of

Fig. 8 Unitized rotary-piston steering gear

steering gear is normally supplied by fixed-displacement pumps and operates at a lower pressure than Rapson-slide or link types. The hoses require routine inspection and replacement to assure steering system reliability.

Steering gears, particularly Rapson-slide and some rotary types, can be further classified as being of a "unitized" construction, in which case the entire steering gear is assembled as a unit at the manufacturer's plant and is installed as a unit by the shipbuilder (as opposed to constructing the ram groups and power plants as separate

entities, each of which must be provided a foundation by the shipbuilder and each of which must be connected together by the shipbuilder). Unitized construction is not only economical in terms of shipboard installation costs, but it also reduces opportunities for foreign matter to be introduced into the hydraulic system because the unitized gear is completely assembled by the manufacturer and shipped full of oil as a preservative.

2.2 Rudder Torque Rating. One of the more difficult aspects of the ship design process is the calculation of the maximum rudder torque for which the steering gear should be designed. Although ship specifications may stipulate the maximum design rudder torque, in some instances they usually state that the steering gear shall be capable of moving the rudder at a prescribed rate when the ship is proceeding at maximum ahead speed. With the ship specifications written in this manner, the shipbuilder has the responsibility of determining the maximum design rudder torque.

Some of the analytical procedures which have been employed to estimate the maximum design rudder torque are discussed in detail in references 6 and 7. It may be noted, however, that there can be a considerable difference between the rudder torque values predicted using the various procedures. The situation becomes further clouded when attempts are made to correlate analytical predictions with test results.

In many cases, the data obtained during steering gear tests consist of only the time required to complete a maneuver and the maximum pressure that was observed during the maneuver. The value of such data is limited because the rate of rudder movement during the maneuver is not known with certainty and the "maximum pressure" is often not fully qualified as to exactly at what rudder angle it occurs, how long it was sustained, etc. The rudder angle at which a pressure occurs must be known in order to relate the pressure to a torque value.

More recently, sophisticated test instrumentation has been used to continuously record the rudder angle, ram pressure, ship's heading, etc., as a means of defining the actual operating conditions during the underway trial.

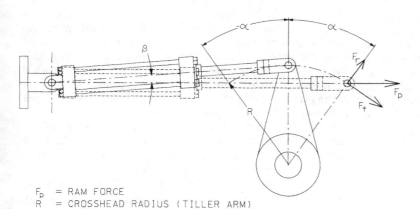

Fig. 9 Force diagram for a clevis-mounted cylinder steering gear

F_p = RAM FORCE
R = CROSSHEAD RADIUS (TILLER ARM)
F_t = FORCE TANGENTIAL TO TILLER = $F_p \cos(\alpha + \beta)$
F_r = FORCE RADIAL TO TILLER = $F_p \sin(\alpha + \beta)$
Q = TORQUE DEVELOPED = $F_t R$

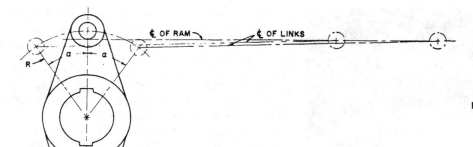

Fig. 10   Link motion diagram

While this procedure gives definitive rudder rate data, problems are nevertheless still encountered in establishing the "maximum" rudder torque observed if the data cannot be interpreted clearly. For example, spikes in the recorded pressure data may reach values higher than normally expected and be of such a short duration that the relief valves do not lift. Such spikes are usually disregarded.

The most reliable method of estimating the maximum rudder torque required for a given ship is to first select a similar ship for which trial data are available. Next, using one or more computational procedures, such as described in references 6 and 7, estimate the rudder torque required for the similar ship and then compare the estimate with the trial results. If there is a significant disparity between the calculated torque and the torque established by the trial results, determine a correction factor that can be applied to the estimated torque to make it agree with the trial results. After the correction factor has been established, use the same computational procedure and apply the established correction factor to estimate the rudder torque required for the ship in question. Even after completing this procedure, the proper maximum design rudder torque can seldom be stated with certainty; however, in the final analysis, a single design value must be selected.

2.3 Ram Actuators. Hydraulic steering gears employing fixed rams are generally of either the Rapson-slide type, shown by Fig. 4, or the link type, shown by Fig. 5. A force diagram for a link arrangement is shown in Fig. 10. As can be seen, when the rudder is hard over, the effective crosshead arm (i.e., the torque arm) is very nearly equal to the tiller radius multiplied by the cosine of the hard-over angle with the link arrangement. The torque arm, therefore, is usually least when the torque to be developed is at its maximum; as a result, the forces required with this arrangement are relatively large. The torque developed by a single-ram link arrangement is

$$Q = FR \cos\alpha \qquad (1)$$

where

Q = torque developed, in.-lb
F = ram force, lb
R = tiller radius, in.
α = rudder angle, deg

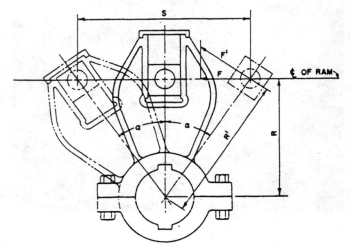

Fig. 11   Rapson-slide force diagram

As may be seen from Fig. 11, an increasing mechanical advantage is obtained at larger rudder angles with the Rapson-slide type of mechanism. The torque developed by a single-ram Rapson-slide arrangement is

$$Q = F'R' = \frac{F}{\cos\alpha} \cdot \frac{R}{\cos\alpha} = \frac{FR}{\cos^2\alpha} \qquad (2)$$

where the terms are as defined for equation (1) and shown by Fig. 11. For the same ram force and tiller radius, the torque that can be developed by a Rapson-slide arrangement is greater than the capability of a link arrangement by a factor of $\cos^3\alpha$. At a rudder angle of 35 deg, this factor is 55%. In view of the mechanical advantage offered by a Rapson-slide arrangement, it may appear that a link arrangement need not be considered; however, such is not the case as discussed earlier.

Ordinarily, ram groups of the link type are provided with one link above the ram that is connected to the tiller of each rudder. Rapson-slide mechanisms are usually built with crosshead arms above and below the ram.

One or more tie rods should be provided between the two cylinders of each ram group. The tie rods are usually shouldered at the cylinder bosses in which they are mounted, and set up with double nuts. Being of exact length, they aid in aligning the cylinders both while the steering gear is in the course of fabrication and installation and while in service. The tie rods are not necessary for the transmittal and distribution of axial loads, but they are convenient guides for bushings or shoes carried in extensions of, or bracketed from, the rams in order to prevent rotation of the rams about their longitudinal axes, to provide a means to position the rudder mechanically, and to provide a positive mechanical stop for the rudder.

Bronze bushings usually are installed in the cylinders to provide a suitable bearing surface for the ram; the bushings are made to form the inner ends of the stuffing boxes. With the link type of construction, the bushings carry little load other than the weight of the ram itself; the lateral load introduced by the link is slight, even at the maximum angle. However, with a Rapson-slide mechanism, the ram is subjected to a considerable lateral force, and since the forces are usually the largest at a hard-over rudder angle, when the crosshead is near one cylinder or the other, one of the bushings will carry by far the greater part of the load. When a ram is made of high-strength steel, which permits high stresses with no improvement in the modulus of elasticity, it is possible that the deflection of the ram may be great enough to cause binding in the cylinder bushings or cylinder bores. Consequently, the bushings for Rapson-slide mechanisms should be given careful study.

Ram bending calculations can be important for Rapson-slide designs when, as is often done for weight reduction, the rams are bored out from both ends or when the gear is designed for greater rudder movement (e.g., Great Lakes ships, which are commonly designed for a 45-deg rudder movement); in such cases flexural stress may also become a critical consideration.

The American Bureau of Shipping (ABS) Rules require that all steering gear parts that transmit forces to or from the rudder have a strength equivalent to that of the upper rudderstock. These parts include the tillers, rams, pins, tie rods, and keys [8].

Carefully designed ram groups may have a tiller radius (or with a Rapson-slide mechanism the distance from the rudderstock centerline to the ram centerline) that is approximately three times the ram diameter. Selection of the exact ratio will be based on the manufacturer's experience. It is convenient to use the 3 : 1 ratio for preliminary design purposes, bearing in mind that arrangement considerations may necessitate shortening the crosshead radius, and, therefore, the ram stroke, with the ram diameter being correspondingly increased.

The maximum design hydraulic pressures for ram-type steering gears generally fall in the range of 1500 to 2500 psi; however, somewhat higher pressures have been used.

The classification society rules for merchant ships typically require that the steering gear be capable of moving the rudder from hard over (35 deg rudder) on one side to within 5 deg of the hard-over position (30 deg) on the opposite side within 28 seconds when proceeding ahead at the maximum continuous rated shaft rpm [8]. The timing cycle is normally initiated at the hard-over position on one side and is terminated 5 deg before the opposite hard-over

rudder position to avoid timing the rudder movement as the follow-up mechanism is reducing the pump stroke.

The rudder rate when going astern is normally not specified for merchant ships.

The auxiliary steering system, or the redundant means of controlling the rudder, is typically required to be capable of moving the rudder from 15 deg on one side to 15 deg on the other in not more than 60 seconds at half the ship's full-power trial speed or 7 knots, whichever is greater; however, if the two main power units are totally independent and each is capable of meeting the main steering gear requirements, an auxiliary steering system is not required [9].

The rudder rate specified for naval ships generally ranges upward from 2⅓ deg per sec for ahead steering, with no rate specified for astern steering. To avoid having astern considerations control the capacity of steering gears for naval ships, it is common practice to conduct special astern steering tests for the purpose of establishing the maximum astern speed at which the rudder can be moved without exceeding the maximum design pressure in the ahead direction. A sign warning that there is a hazard of overloading the steering gear at astern speeds greater than that established is then made and installed in the helmsman's view.

Table 2 illustrates the considerations that enter into the preliminary design of a ram-type steering gear and gives a procedure for computing the approximate characteristics of a steering gear. If only an approximation of the maximum motor horsepower requirements is desired, the following expression may be used for a steering gear of the Rapson-slide type:

$$H = Q \frac{(\tan\phi + \tan\theta) \cos^2\alpha}{6600 \; Ete} \qquad (3)$$

where Q is the torque at maximum pressure during the timing cycle in inch-pounds, α is the rudder angle at Q, and all other terms are as defined in Table 2.

The corresponding maximum power requirement for a link-type steering gear is

$$H = \frac{Q(\sin\phi + \sin\theta)}{6600 \; Ete \; \cos\alpha} \qquad (4)$$

Note that the rudder torque corresponding to the maximum pressure observed during the timing cycle sometimes occurs at the 30-deg position before reaching hard over, and, with the exception of Q, all other terms in equations (3) and (4) are considered to have constant values in many cases. As a result, equations (3) and (4) can each be reduced to a constant times Q. This is often not the case, however. For rudders that are lightly balanced, the maximum torque is commonly recorded as the rudder approaches the hard-over position during the initial rudder movement. But for heavily balanced rudders, the maximum torque is typically recorded later in the maneuvering cycle when the rudder is being moved from one hard-over position to the other; the rudder angle of attack becomes so large that the rudder typically stalls during this maneuver.

Table 2 Preliminary design calculations for an
electrohydraulic steering gear

		Type of Linkage	
	Symbol	Rapson	Link
Number of rams	Z	2	2
Maximum permissible operating pressure, psi	p_m	1700	1700
Maximum rudder angle, deg	ϕ	35	35
Rudder angle at end of timing cycle, deg	θ	30	30
Rudder torque at θ, in.-lb	Q_θ	11.57×10^6	11.57×10^6
Maximum design torque ahead, in.-lb	Q_1	19×10^6	19×10^6
Rudder angle at Q_1, deg	α_1	35	35
Maximum design torque astern, in.-lb	Q_2	14×10^6	14×10^6
Rudder angle at Q_2, deg	α_2	35	35
Ram-to-rudder efficiency	E	0.80	0.80
Approximate crosshead radius,[a] in.	R'	37.73	46.1
Approximate ram diameter = $R'/3$, in.	D'	12.58	15.37
Crosshead radius used, in.	R	38	47
Ram diameter used, in.	D	13	15.5
Maximum ram pressure ahead,[b] psi	p_1	1580	1633
Maximum ram pressure astern,[c] psi	p_2	1165	1203
Ram stroke during timing cycle,[d] in.	S	48.55	50.46
Oil volume during cycle = $\frac{\pi}{4} D^2 Z S$, in.^{3}	V	12,888	19,043
Duration of timing cycle ahead,[e] sec	t	27	27
Oil flow requirements, $0.2597 V/t$, gpm	C	124	183
Pump efficiency	e	0.85	0.85
Maximum ram pressure ahead during timing cycle,[f] psi	p_3	1075	940
Maximum motor horsepower requirements = $p_3 C/1714e$,[g] hp	H	91.5	118.1
Motor horsepower used, hp	H_1	75	100

NOTES:

[a] $R' = \left[\dfrac{36Q \cos^2 \alpha}{\pi p_m EZ} \right]^{1/3}$ for Rapson.

$R' = \left[\dfrac{36Q}{\pi p_m EZ \cos \alpha} \right]^{1/3}$ for link.

Q_1 and α_1 as well as Q_2 and α_2 are used in the above equations and the largest value of R' is used.

[b] $p_1 = 4Q_1 \cos^2 \alpha_1 / \pi ZRED^2$ for Rapson.
 $p_1 = 4Q_1 / \pi ZRED^2 \cos \alpha_1$ for link.
Checks should also be made to ensure that higher pressures do not occur at smaller rudder torque values.

[c] See note b using Q_2 and α_2 vice Q_1 and α_1.

[d] $S = R (\tan \phi + \tan \theta)$ for Rapson.
 $S = R (\sin \phi + \sin \theta)$ for link.

[e] Duration of the ahead timing cycle is 28 seconds minus one second for stroking the pump.

[f] The maximum ram pressure during the ahead timing cycle often occurs at the 30-deg position before reaching hard over; however, the ram pressures at other angles should be checked, using an equation of the form given in note b, to ensure that none are larger.

[g] A similar calculation may be required to confirm that the astern horsepower requirements are not a controlling consideration.

The design calculations for steering gears having clevis-mounted cylinders are similar to those for link types. Because the cylinder ends are able to pivot, however, the developed rudder torque at any particular rudder angle is dependent on the location of the pivot points relative to the rod end connection to the tiller at that angle.

Because of the flexibility inherent in the clevis-mounted cylinder arrangement, this type of gear is well suited for twin-rudder applications on smaller vessels and particularly for arrangements such as canted rudders, which are frequently specified for high-speed military craft, pleasure boats, and special-purpose vessels. An arrangement of a clevis-mounted cylinder steering system for a motorized lifeboat with canted rudders is shown in Fig. 12.

2.4 Rotary Actuators. Rotary actuators for steering gear applications can be of several types. By far the most common type is the hydraulic rotary-vane actuator shown in Figs. 6 and 7; however, rotary-piston actuators, as illustrated by Fig. 8, have also been used on smaller vessels. Rack-and-pinion actuators have been used as well, although their primary applications have been for valve actuation and heavy, slow rotation requirements, such as on the booms of self-unloading bulk carriers.

In a rotary-vane steering gear, the torque is developed by applying hydraulic pressure to one side of the rotor vanes. When the rudder is hard over, the vanes act as rudder stops. The actuator is provided with a relief valve set at the required 125% of the maximum working pressure. Rotary-vane actuators develop a constant torque throughout their full travel equal to the product of the applied pressure, the vane area, the effective moment arm (the radius from the rudderstock centerline to the center of pressure of the vanes), and the number of vanes.

Rotary-vane actuators are supplied by fixed-delivery pumps, and the fluid supply and direction of flow are controlled by a directional control valve. In some cases, modulated flow control valves are used with the fixed-delivery pumps to provide smoother operation of the actuator. A servo feedback system can be added to provide accurate rudder positioning. When variable-delivery pumps are used, control is accomplished in a manner similar to that for ram-type actuators supplied by variable-delivery pumps.

Fixed-delivery pumps can be of the gear, vane, or screw type and are controlled by either direct-acting or pilot-operated solenoid valves.

Most rotary actuator designs incorporate the upper rudderstock radial bearing and the rudder carrier bearing, which supports the vertical load of the rudder and rudderstock, within the actuator housing, which simplifies the installation of the steering system. The actuator is connected directly to the rudderstock by a keyway, hydraulic nut, clamp rings or hydraulic coupling.

2.5 Steering Gear Power Units. Most steering gear power units are installed in duplicate, with each power unit having the capability of moving the rudder at the specified rate under all conditions of full-speed maneuvering. The regulatory body regulations require totally independent power units and piping systems to permit continued operation in the event of a single failure [8,9].

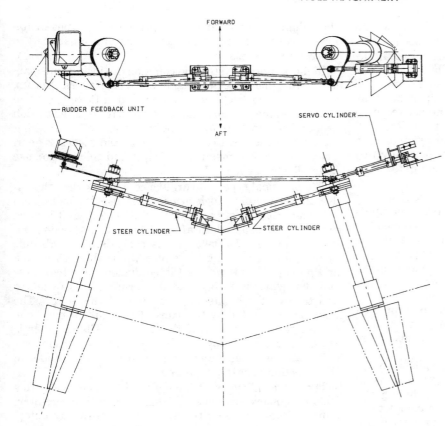

FORWARD

RUDDER FEEDBACK UNIT

SERVO CYLINDER

AFT

STEER CYLINDER

STEER CYLINDER

Fig. 12 Clevis-mounted cylinder steering system for motorized lifeboat with canted rudders

Each power plant consists essentially of an electric motor driving a positive-displacement, fixed- or variable-delivery pump and its hydraulic reservoir, valves, and associated piping. Motors for the steering gear pumps are generally 440 V, 3 phase, 60 Hz and are dripproof protected. When the steering gear is powered by continuously operated, reversible-flow, variable-delivery pumps, the motor should be rated for continuous duty at full load, (i.e., maximum design working pressure) for one hour.

As a means of reducing the power rating of motors that drive variable-delivery pumps, power limiters are often provided. Power limiters (secondary stroke-control devices) operate on the principle of reducing the stroke of the main hydraulic pump in the event that the hydraulic pressure exceeds a preset value (that pressure corresponding to the motor rating plus the allowable overload). Since the power required to drive a pump is proportional to the product of the discharge flow and pressure, the power rating of the installed motor can be minimized by reducing the pump flow during the periods of peak pressure requirements. Power limiters are designed such that the stroke of the hydraulic pump is decreased with increasing hydraulic pressures so that the motor power required is nominally held constant at the preset value. When the power limiter acts to reduce the pump stroke, the rudder rate is also reduced; however, if sufficiently high rudder rates can be attained during other regions of the timing cycle, where the torque is lower and where the

overall average rate during the timing cycle exceeds the minimum rate specified, the power rating of the driving motor may be reduced.

Usually, the replenishing and servo pumps on closed-loop, variable-delivery systems are driven by the same motors as their respective main pump. If, however, these auxiliary pumps are separately driven, their motors should be rated for continuous duty. If separate motors are provided for these pumps, each should be interlocked with its main pump motor so that the main pump cannot be started without first starting the auxiliary pump(s), and control and replenishment fluid is at the required minimum pressure. Steering gear motor controllers must be provided with low-voltage release.

It is customary to shift over from one power unit to the other at regular intervals, in order that both power units may operate approximately equal lengths of time. Automatic transfer valves can be controlled from the steering gear room, as well as from the remote steering stations. The regulatory bodies require that an effective means of rapidly transferring between power units be provided on the navigating bridge [8,9]. On systems with variable-delivery pumps, these valves are operated by main pump servo pressure and are controlled by solenoid-operated pilot valves, which are connected electrically in the main pump motor control circuits in such a manner that the pump driven by the first motor started is connected automatically into the hydraulic circuit and becomes the active pump for the operation of the steering gear. If the second pump motor is then started, this pump is bypassed hydraulically. By stopping the first motor, the second pump becomes the active pump automatically. This system ensures maximum safety when a vessel is underway in restricted waters in that with both units operating, the idling bypassed (second) pump becomes the active pump automatically upon failure of the first unit.

Valves may also be included in a four-cylinder ram group piping system which cut out and bypass any two oppositely acting cylinders, thereby providing for two-cylinder "emergency" operation.

When a hand "fill-and-drain" pump is specified, it is usual to stipulate that it be possible to use this pump as a "rudder positioning" device. Such a pump is convenient if a ship is docked and no power is available to operate the steering gear, or to provide rudder trim to aid in towing a dead ship. Suitable check valves should be included in the fill-and-drain pump piping to prevent the pump from being overhauled and driven as a hydraulic motor.

All steering gear hydraulic systems must be protected by suitable relief valves. Each relief valve must be capable of relieving the full flow of all pumps, which can discharge through it, plus 10%.

Each power unit must be connected permanently by piping to a fixed storage tank with sufficient capacity to recharge the complete hydraulic system, including the power unit reservoir, or to drain the system completely.

2.6 Steering Gear Controls. Steering gear control systems provide the capability of controlling the steering gear remotely from a steering console located in the pilothouse and locally from the steering gear space by means of a trickwheel. A schematic diagram of a control system typical of naval ships is shown in Fig. 13.

The control system may be of the "non-follow-up" type, "on and off" type, or the "full follow-up" type. In the first, the helmsman applies left or right rudder to hold a course. As long as a wheel, lever, or pushbutton is held to energize the steering gear for rudder response in one direction, the rudder moves in that direction until the steering gear is deenergized automatically in the hardover position. Of course, the rudder may be moved in small increments, left or right, by holding the control off neutral for small intervals of time. The helmsman's experience and judgment are important factors in minimizing steering gear activity and oversteering. With the second type, the rudder returns automatically to amidships upon the helmsman's release or centering of the control. The full follow-up system—the third type of control—senses any difference between helm and rudder angles and the direction of the difference; it moves the rudder automatically in the proper direction to eliminate this difference and holds the rudder at the angle signaled by the helm until a difference is reestablished by moving the helm or by drifting of the rudder due to hydrodynamic forces. "Feedback" and "differential" as well as "follow-up" are terms often used in describing this control since the correction is derived from "feedback" of the rudder-stock motion; the control senses, or in some cases measures, within limits, the difference or differential between

the helm and rudder angles. Representative types of follow-up control systems are described and illustrated in references 3, 4, and 5.

Remote control systems in common use include mechanical, hydraulic, electrical, and electronic types. Shafts, wire rope, sprockets and chain, push-pull flexible control cables, and their combinations are used to transfer motion proportional to that of the helm from remote steering stations to the local control input at the steering gear. These mechanical means are simple and reliable for use in smaller vessels.

The principal type of hydraulic control is called the hydraulic telemotor, which has been in use for more than 100 years. It consists of a forward telemotor unit located in the pilothouse and an aft unit located in the steering gear room. The forward unit is essentially a hydraulic piston pump driven by the ship's wheel through a rack and pinion. The aft unit is the equivalent of a hydraulic cylinder, which is spring loaded to bring it back to center and to give the helmsman the "feel of the helm." The units are connected by two runs of hydraulic tubing, one connected to each side of the cylinder. The cylinder is attached to the steering gear local control by a connecting link that causes the steering gear to move the rudder in the direction chosen at the helm.

The principal types of hand-electrical remote controls include synchro controls and balanced-bridge circuitry. Synchro controls use a transmitter driven through step-up gearing by the steering wheel in the pilothouse and a receiver in the steering gear room, providing proportional input motion to the local pump control mechanism. This type of control is used primarily on naval ships. With balanced-bridge circuitry, turning the wheel moves a potentiometer to unbalance the electrical system. Unbalancing the electrical system starts an intermediate powered servo that provides input to the local control and moves a follow-up potentiometer, which rebalances the circuit and deenergizes the servo when the helm and relative servo angular positions coincide.

The powered servo is usually a rotary electrohydraulic unit, which is essentially a miniature steering gear and is designed for application to any type of steering gear as one link in a chain of servomechanisms from the steering wheel to the rudder. It is also used as an automatic pilot system when the input is taken from a suitable gyro compass.

On many vessels, local control of the steering gear hand-electric steering control is by means of the mechanical differential control. The mechanical differential allows the helmsman to rotate the steering wheel as quickly as desired. The helm signal is transferred to the mechanical differential through the rotary electrohydraulic unit and then to the steering gear pump control. The steering gear itself follows, moving the rudder at its rated speed. A mechanical follow-up linkage inputs the rudder position to the mechanical differential, and the stroke of the pump is reduced by the differential control when it is within 5 deg of the ordered angle and is fully off stroke when the ordered angle is reached.

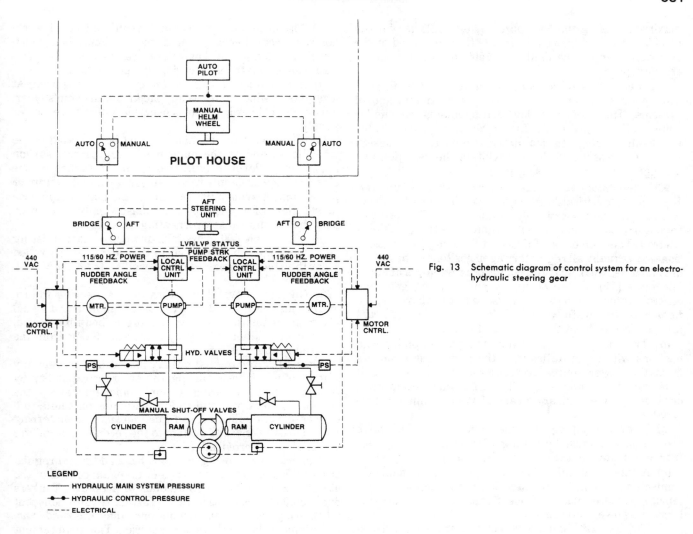

Fig. 13 Schematic diagram of control system for an electro-hydraulic steering gear

Electronic steering control uses a microprocessor to receive the helm order and the rudder position feedback and compare them. The mechanical differential controls are omitted and replaced with electronic servo control valves on the hydraulic pump, which receive the order from the microprocessor and stroke the pump in the direction and degree requested. The follow-up, which is normally a potentiometer, is moved in direct proportion to the motion of the rudderstock, and provides the feedback signal to the microprocessor to destroke the pump and stop the rudder at the ordered angle. An error in the feedback system caused by a new helm or autopilot order, or by motion of the rudder due to external forces (creeping), reactivates the control system.

Steering gear power units that are designed for naval ships in accordance with reference 10 are required to contain features that limit the speed of the hydraulic pump and electric motor in the event that electrical power is interrupted and the rudder takes charge of the hydraulic ram group, thereby driving the hydraulic pump and electric motor. The method of limiting the overspeed of the electric motor may be either an arrangement of spring-loaded valves operated by servo pressure, or a brake on

the electric motor. The method usually employed is to incorporate a hydraulic lock feature in the blocking valve and associated solenoid valves, which operate the blocking valve, as shown in Fig. 13. As may be seen from Fig. 13, in the event of an interruption of electrical power, the blocking valve is shifted so as to place a hydraulic lock on the ram group and isolate the ram group from the pump. It should be recognized that other valve arrangements may be used to provide the same feature.

During sea trials of naval ships, steering gear power failure tests are conducted with the ship proceeding at full speed in the ahead direction and at the maximum astern speed permitted for unrestricted rudder operation. The power supply is interrupted at a series of rudder angles during the circumstances in which a power failure would have the most adverse effects. Before sea trials, it is good practice to ascertain the closing time of the blocking valve by interrupting the electrical power when operating the steering gear power units with the ship at the dock. The anticipated motor overspeed during power failure tests at sea can then be calculated by using the measured closing time of the blocking valve and the rudder design torque values. Experience has shown that a maximum closing time of approximately 0.25 sec for the blocking valve will normally assure that the speed of the electric motor will be limited to 150% of the normal operating speed.

The detailed specifications for a vessel include very complete descriptions of the steering gear and its control systems. These often follow arrangements on similar ships that have given satisfactory service. It is desirable to consult steering gear manufacturers early in the design stage of a vessel in order to facilitate the selection and arrangement of the steering gear.

2.7 Regulatory Body Requirements and Safety Features. Because of its critical function, the steering gear is subject to extensive regulatory body requirements. The primary regulatory bodies in the United States are the U.S. Coast Guard [9] and American Bureau of Shipping [8]. Internationally, virtually all regulatory agencies follow the Safety of Life at Sea (SOLAS) protocol for the regulation of steering gears [11]. In the design of any steering gear and control system, the most recent issue of the above regulations must be carefully studied. Basically, steering gears are required to have the following features:

(*a*) Two independent hydraulic pumps and piping connections with separate leads to the pump prime movers from the power source are required. Each power unit must have the required capacity for a main steering gear, or an acceptable auxiliary means of steering must be provided.

(*b*) Hydraulic piping systems must have relief valve protection and must be arranged so that a changeover from one system to the other can be readily accomplished.

(*c*) A fixed hydraulic fluid storage tank, permanently connected by piping to the steering power actuating systems, is required with sufficient capacity to recharge at least one power-actuating system.

(*d*) Two separate and independent steering control systems are required, one port and one starboard, both with full follow-up control and rudder angle indicators in the helmsman's view.

(*e*) Circuit-breaker protection is required at the switchboard for each steering system feeder circuit as is short-circuit protection and low-voltage release, but not overload protection, for all steering system motors.

(*f*) A minimum of two feeder circuits is required, at least one of which must be supplied from the ship's emergency switchboard, or an acceptable alternative power supply.

(*g*) Visual and audible alarms must be activated in the wheelhouse and at the main machinery control station upon low oil level in either reservoir or failure of the electrical power supply to any steering control system or power unit. Alarms must activate in the machinery space upon failure of any phase of a 3-phase supply or an overload that would cause overheating of a motor.

(*h*) On vessels 500 gross tons and above, each steering power unit must be arranged so that it can be controlled from the steering gear room, as well as from the wheelhouse.

(*i*) On vessels over 1600 gross tons, a steering failure alarm must be fitted that activates an audible and visual alarm in the pilothouse if, after a prescribed time period, the rudder position differs by more than 5 deg from the ordered rudder position.

Essentially, the regulations are intended to prevent any single failure from disabling the steering capability by requiring that all steering system components, except the rudder, rudderstock, and rudder actuator, on single-rudder vessels be duplicated, and a fast means of transferring from one component or system to the second, upon failure of the first, be provided.

Tankers over 10,000 gross tons have additional regulations, which include a requirement to regain control of the steering system within 45 seconds of a failure. Tankers over 100,000 gross tons have the further requirement that all steering system components, including the rudder actuator, meet the single-failure criterion. This requirement necessitates the use of a four-ram steering actuator or a specially designed rotary actuator with multiple, double-sealed, automatically isolated vanes.

Section 3
Anchor Windlasses

3.1 Introduction. The ship specifications usually require that a windlass be capable of hoisting the anchor at an average speed of not less than 5 to 6 fathoms per minute (30 or 36 fpm, respectively) from a depth of 30, 60, or more fathoms. The required chain pull thus is dependent not only on the weight of the anchor, but also on the weight of the chain to the specified depth, with an appropriate deduction for the water buoyancy effect.

Even though the windlass may be constructed as a sophisticated machine, it must be built to perform the crudest task on shipboard. The anchor chain is heaved in through a hawsepipe in which the friction loss averages from 35 to 40% in good designs, and sometimes exceeds

55%; however, a roller in the end of the hawsepipe can reduce the friction to as little as 20%. The chain is engaged by a wildcat, which is hardly comparable in efficiency to the sprockets used in refined chain drives; the wildcat is usually made with five whelps (comparable to a 5-tooth sprocket), causing the chain to move with a jerkiness, which is aggravated by its tendency to turn over or "slap" in the hawsepipe and to slip on the wildcat. Windlasses, therefore, require more ruggedness of construction than any other machine on board ship.

The rules of the Classification Societies contain tables of required equipment consisting of anchors, chain cable, towlines, and hawsers. These items are identified by, and sized in accordance with, an "equipment number" or "equipment tonnage" included in the tables. The number or tonnage figure is calculated by substituting in empirical formulas certain dimensional and displacement measurements of the vessel for which the equipment is desired. This has become a standard mode of equipment selection for oceangoing merchant vessels.

The basic dimensions and scantlings of an anchor windlass depend on the anchor weight and chain size. These follow, therefore, from the equipment selection to suit a given case.

The size of the vessel, the nature of the service, and the desired anchor-handling and stowage arrangements are also contributive in the choice of the anchor windlass. In many cases, the windlass is used only in emergencies. However, in some cases the windlass may be used regularly (e.g., some operators set an anchor in a "flying moor," approaching a dock, far enough offshore to warp the bow into the stream when preparing to get underway, thereby reducing or eliminating the need of a tugboat). Also, it is usual to install capstans or warping heads, driven by the windlass power plant, off the windlass gear train. The windlass may then be used for normal warping duty. Combination windlass mooring winch/warping head systems have been supplied for large container, tanker, roll-on/roll-off, and passenger ships.

As with the other items of hull machinery, anchor windlasses are available from specialty manufacturers who can provide machinery to suit a variety of requirements. Their recommendations are very helpful in the ship design stage.

3.2 Windlass Types. Two fundamental configurations of anchor windlasses have evolved from the "winding log" and capstan of earlier times, namely, the horizontal arrangement and the vertical arrangement. The horizontal windlass is a specialized winch that is powered by a hydraulic or electric motor or, in a few cases, by a steam engine. The motor is connected to a train of gearing that drives one or more chain sprockets, called "wildcats," through sliding-block "locking heads" or comparable jaw clutches. Figure 14 is a schematic diagram of a horizontal electrohydraulic windlass. A photograph of a horizontal windlass that is driven directly by an electric motor is shown in Fig. 15.

The specifications for cargo vessels often require the combination of a horizontal mooring winch with a clutched drum driving a chain wildcat through an auxiliary gear reduction and sliding pinion or jaw clutch. The chain lifting unit consists of a rigid framework holding an axle for support of the integral gearwheel wildcat brake rim and the pinion shaft with bearings. Although enclosed gears running in an oil bath are preferable on deck from a maintenance point of view, an open gear protected by a guard is generally accepted on a chain lifting unit. The large gear teeth are not especially sensitive to corrosion, and the open gear allows the transfer of the torque directly from the gear rim to the wildcat, which allows the design of a simple and rigid chain lifting unit of moderate weight. An automatic grease lubricator for the gearwheel simplifies the operation and maintenance procedures. One or

both pinion shaft ends are fitted with couplings for connection to mooring winches.

Each wildcat and mooring-winch rope drum, in the case of combination units, is provided with a brake of the band type. The wildcat brake is used to restrain the chain when the anchor is let go under a controlled drop, for veering chain to the desired scope, and for holding the chain while the chain stoppers are being attached. One or more warping heads are usually keyed to the winch drum or intermediate shafting. The gear train through which the warping head is driven usually affords a line pull in the order of one quarter of the available wildcat chain pull, at four times the normal chain speed.

A self-contained horizontal type of windlass is the least expensive in terms of installed cost. However, it requires more maintenance than does the vertical type because the windlass machinery is completely exposed to the weather and to the spray and waves that break over the bow during storm conditions.

In the preliminary design stage of a vessel, it is good practice to develop the anchor-handling arrangement to the extent that the chain leads are confirmed to be satisfactory. In the case of ships with large bulbous bows, the anchors must be located farther aft or closer to the rail so that the anchors will not hit the bulb when they are dropped. This usually requires that two separate windlasses be provided with each set at an angle to the ship centerline in order to obtain proper leads to the hawsepipes. Figure 16 shows a forecastle without hawsepipes, and an "anchor on deck" arrangement. A well rounded fairing plate serves as a guide for the chain and anchor. The arrangement is integral with the deck structure and has a means for securing the anchor at sea and for holding the breaking load of the chain when anchored.

A vertical windlass consists essentially of a wildcat mounted on a vertical shaft that is carried in a rugged set of main bearings in a casting or weldment, which is bolted or welded to the deck, as typically illustrated by Figs. 17 and 18. The strengthening of the deck and supporting ship structure in way of this assembly is usually made adequate to sustain all anticipated loads due to the chain pull, independently of the main shaft extension to the deck below. The wildcat is brought as close to the deck as possible in order to minimize bending moments due to the chain pull. The chain is wrapped approximately 180 deg around the wildcat, and then enters a chain deckpipe leading to the chain locker.

The shafts from vertical wildcats and associated capstans are often extended to one or more decks below where they are coupled to main and intermediate shafts, respectively, of transmission gearing. Vertical windlasses with the gearbox and clutch supported under the deck on which the wildcat is mounted have been used on naval ships. This arrangement minimizes the problem of relative deck deflection and simplifies the installation and alignment of the windlass as the wildcat, transmission, and brake band are supported from a common structure.

The gears in vertical windlasses are usually completely enclosed. The shaft couplings are of a type that allows

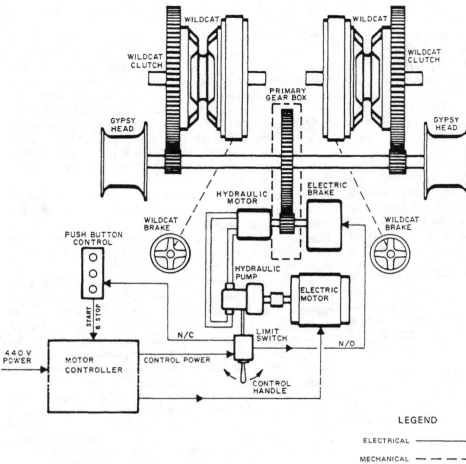

Fig. 14   Schematic diagram of a horizontal electrohydraulic windlass

Fig. 15   Horizontal-shaft, double-wildcat windlass driven by an electric motor

limited relative vertical motions between the decks, if necessary. The preferred design locates the brake drum and locking head (clutch) below the weather deck where they are protected from the weather. This arrangement also permits the wildcat to be located as closely as possible to the weather deck.

For handling warping lines, a capstan head may be keyed to the main shaft, above the wildcat. However,

Fig. 16   Forecastle with two combined mooring winch/windlass/fiber rope handling gear units; mooring drums are of the split type

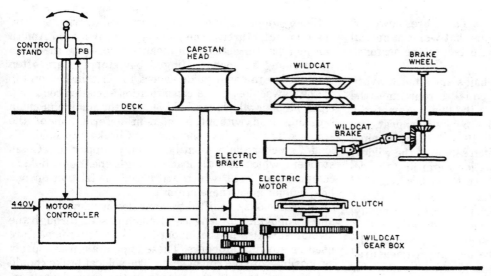

Fig. 17 Schematic diagram of an electrically driven vertical windlass

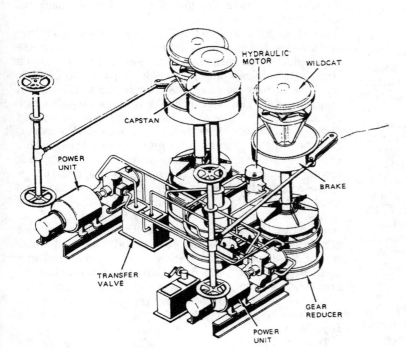

Fig. 18 Vertical electrohydraulic windlass

unless there is either a speed change in the gearing or hydraulic transmission, the light-line hauling rate may be unsatisfactorily slow. Another arrangement is one in which a capstan is located adjacent to the wildcat and is driven from the windlass gear train by a separate shaft, which revolves about four times as fast as the wildcat; the equipment for such an arrangement is illustrated by Fig. 18. As a simplification, the capstan head may be keyed fast to the drive train and allowed to rotate or idle when the windlass is used to haul chain.

A vertical windlass affords flexibility in the development of anchor-handling and mooring arrangements, from the simplest single unit to as many independent systems as there are anchors installed. Some arrangements feature one power unit driving a single variable-stroke pump. The hydraulic system includes selector valves enabling the single pump to drive either of two hydraulic motors serving two wildcat/capstan systems. For improved reliability, two power plants are installed and electrical, mechanical, and hydraulic cross connections or their combinations, depending on the basic features of the system, are included in order that both anchors may be recovered despite a failure in one system; such an arrangement is illustrated by Fig. 18.

3.3 Special Design Considerations. The practices recommended for the design and testing of anchor windlasses are outlined in reference 12. There are, however, some detailed design considerations that warrant special emphasis, as they can have a large effect on the performance of a windlass.

The links of stud link anchor chains are almost invariably made with an inside length equal to four times the nominal chain size (i.e., four times the "wire diameter"). Since 10 links will wrap around a 5-whelp wildcat, the wildcat circumference in inches = $10 \times 4 \times$ the nominal chain size. From this, the mean pitch radius of the wildcat is readily found. Detailed information regarding the design of wildcats is given in reference 13. Wildcats should be made of a reasonably hard grade of cast steel.

The fitting of wildcats and chains is important. Usually, the final dimensions of a chain are attained as a result of stretching in a proof test. A new chain may be within the minus allowance of the tolerance on length (measured over six links in American practice) in order to allow for stretching and wear in service. However, such a chain will not run properly on a wildcat made to fit the nominal chain size. The "pitch diameter" of the wildcat is a function of the depth of the pockets between the whelps, in which alternate links lie substantially flat. There must be clearance between the chain and the midportion of the pocket. The critical area for the pitch circle is the contact surface to the end portions of the "flat" link. If the links must be hauled off by the chain stripper when hoisting anchor, or if when paying out under power the links ride higher and higher on succeeding whelps until one crashes over the top of a whelp, the pitch diameter should be reduced by chipping the pockets deeper. With a proper fit, only one sprocket tooth or set of whelps holds at any one time. This subject is discussed in detail in reference 13, and a recommended procedure is given for checking the wildcat and the chain with which it is to run. To check the fit of a chain on a wildcat, a length of chain containing one link less than that which will fit in a full circumference is lashed snugly around the circumference of the wildcat (a longer length of chain may be used if the standing part is hauled aside). The test length must contain one detachable link. The gap for the missing link is then measured. If the length is excessive, the pockets must be chipped deeper. Measured between the outside of the link ends, the gap should not exceed 2 chain sizes in length. Calipered inside the ends, the measured length should not exceed 4 chain sizes. If the pitch circle is somewhat small, there will be large movements of the chain in the pockets. These movements will cause wearing of the chain and the wildcat.

The chain deck pipe, which leads to the chain locker, for a horizontal windlass should be located well under the wildcat. The vertical centerline of the deck pipe should project upward through the axis of the wildcat, or, preferably, slightly forward of the axis. The chain pipe lip must be flared to assure that the chain will be hauled into the locker by gravity. This will aid in arresting the rising of chain from the locker and will minimize "jumping" of links over the wildcat, due to kinetic energy, if the brake is applied suddenly when an anchor is being dropped.

Chain stoppers are normally furnished by windlass and chain manufacturers. The pawl type is favored in American merchant practice as a means of securing the chain when riding at anchor. Hinged bar stoppers are often combined with chain guide rollers for installation on top of the hawsepipe. Some classification societies require a chain stopper which can hold 80% of the minimum chain breaking load without permanent deformation of the stressed parts, unless the windlass brake can hold this load. The chain links must be "evenly supported." Chain stoppers, which hold the load by single shear on a link, or compression types, which are based on friction on one link, cannot meet these criteria.

Pelican hooks, modified turnbuckles or "devils' claws," and turnbuckles on lengths of chain passed through the anchor shackles at the inboard ends of hawsepipes are used as stowing stoppers. These should align as closely as possible with the run of chain. The wildcat brake should not be used to hold the anchor in the stowed position, because if the brake should slip, the anchor will back out and pound in a seaway.

A pelican hook is often used on naval vessels as a riding and stowing stopper. It is best suited for use with a vertical windlass, i.e., with the chain close to the deck. On a large horizontal windlass, the chain may rise 5 ft or more from the inboard bolster of the hawsepipe to the wildcat tangent point. The pelican hooks in sizes large enough for application with such a windlass are very heavy and difficult to manage. Their height above deck also creates a hazard to the operator who must release them with a maul. For a detailed discussion of anchor, mooring, and towing arrangements, see references 14, 15, and 16.

The wildcat band brake design and the selection and treatment of brake-lining material are critical in windlasses, which must be subjected to the free drop tests required by the regulatory bodies. Typical specifications stipulate a single drop from 45 to 60 fathoms, under control of the brake. These tests should be conducted in a clear depth of at least 65 fathoms and preferably no greater. The falling anchor is permitted to be under the control of the brake to the extent that an operator senses his chances of arresting the run of chain and can do so by applying very little additional braking effort.

The specifications for naval windlasses require that the hand brake performance be demonstrated during four successive average drops of 15 fathoms each, stopping with 15, 30, 45, and 60 fathoms of chain out, respectively [17].

Anchor windlass brake tests must be conducted with the utmost respect for the magnitudes of the masses, velocities, and forces involved. These tests can be extremely hazardous, especially in the event of a runaway. Test observers should stand well clear of the windlass, with only those responsible for operating the brake and recording data allowed to be near.

A satisfactory anchor windlass brake must stop the anchor and chain within a period of about two seconds after the brake is set. Due to the short time available for the brake to absorb most of the kinetic energy possessed by the anchor and chain, the surface of the brake lining

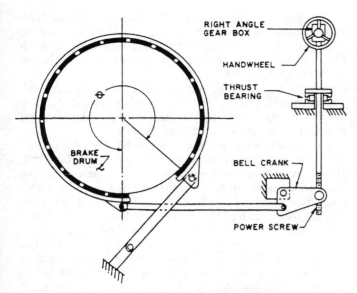

Fig. 19 Schematic of band type of anchor windlass brake

usually reaches a high temperature. Elevated temperatures can result in a degradation of the brake lining frictional characteristics such that the brake is no longer capable of arresting the continuing run of the anchor and chain. If the situation should get out of control, the problems compound; that is, as the anchor continues to run after the brake is set, the quantity of energy that the brake must absorb increases, and more heat is generated. For this reason, it is essential that the brake promptly arrest the anchor and chain after being set.

Anchor windlass brakes are usually of the lined band type. For maximum effectiveness, these brakes should wrap around the drum as near 360 deg as possible. The features of a typical band type of brake are shown in Fig. 19. Auxiliary power-assist mechanisms for setting the brake have also been used to advantage on very large windlasses.

If molded linings are used, they are best secured to the band in a manner permitting lateral expansion as the binder breaks down and swells with heat. Molded linings are most effective under the conditions of the drop test if deeply scarified or if vented with lateral grooves about $\frac{3}{16}$ in. wide by $\frac{3}{16}$ in. deep, pitched about $2\frac{1}{2}$ in. around the circumference.

A new brake should be "run in" by operating the windlass under power with the chain held clear of the wildcat. The band should be examined periodically and the operation continued until the area of the brake lining in contact with the drum surface is at least 75% of the total area of the brake lining. The drum and lining surfaces should be free of exudation; if need be, they should be cleaned with a solvent. Care should be taken that overheating does not occur during the run-in operation. These precautions should be accomplished at the shipyard rather than at the windlass manufacturer's plant. The new linings will then be in the best condition possible, and preservatives and rust-preventive media will be eliminated as prospective

reasons for a reduction of the band lining coefficient of friction.

The design of an anchor windlass brake is heavily influenced by the experience gained with previous designs, as may be noted from the typical anchor windlass brake calculations given in Table 3. The brake lining coefficient of friction used in the design calculations is generally somewhat less than the value suggested by the brake lining manufacturer and may at first appear to be unreasonably low. However, the lining coefficient of friction suggested by the manufacturer is usually based on ideal laboratory conditions, which hardly simulate anchor windlass brake service.

Special instrumentation was provided during the sea trials for the aircraft carrier USS *America* (CVA 66) in order to confirm that the design criteria employed in the design of the anchor windlass brake were adequate. Before the sea trial, every reasonable precaution was taken to ensure that the brake lining material was properly prepared and "run in." Figure 20 shows the results obtained during the CVA 66 tests, and the relatively low brake lining coefficient of friction anticipated (see Table 3) is seen to be confirmed.

A chain counter is a very useful tool that may be installed on a windlass. A chain counter provides a mechanical or electronic readout at the windlass of the number of feet or fathoms of chain that have been payed out, and provides a digital readout in the wheelhouse. The officer on duty then knows the amount of chain in use without sending a crew member to the forecastle to check the markings on the chain. The indication of the amount of chain out is also helpful (when the depth of water is known) in paying out enough chain to ensure that the anchor will hold.

Remote control of the windlass brake improves the safety of both the operator and the ship. If the brake mechanism is not spring set, the first step in remote operation is setting the brake. Subsequently, the wildcat clutch is disengaged, and any devices used to secure the chain at sea, such as the devil's claw or chain stopper, are removed from the chain. A suitable position for the operator is an area at the ship's rail where one can see the chain both outboard and on deck. Chain length out and chain speed indicators are helpful instruments for the operator controlling the anchor drop. A hydraulic cylinder, powered by an accumulator, is arranged mechanically to override the brake screw mechanism. The activating valve for this cylinder can be directly operated or operated by a pilot or solenoid valve. Speed governing can be built into the hydraulic system to limit the rate of fall of the anchor.

With speed control and solenoid valve operation, the remote control system can also be arranged for operation from the wheelhouse. On semisubmersible drill rigs, remote drop systems have been assisted by TV monitors. TV monitors are especially helpful in darkness, since critical parts like brake position markers and dog clutches can be coated with reflecting paint and made very visible in artificial light.

Table 3 Anchor windlass brake calculations

	Symbol	C4	CVA 66
Size of anchor chain, in.	C	$2\frac{7}{16}$	$4\frac{3}{4}$
Outside length of one link, in.	G	$14\frac{5}{8}$	28.5
Pitch of link $= G - 2C$, in.	p	$9\frac{3}{4}$	19.0
Number of whelps on wildcat	a	5	5
Wildcat pitch diameter $= 2ap/\pi$, in.	d	31	60.48
Brake drum diatmeter, in.	D	48	90
Brake band width, in.	w	10	2×10
Brake band thickness, anchor end, in.	t_1	1.0	1.0
Brake band thickness, slack end, in.	t_2	0.5	0.75
Handwheel diameter, in.	H	20	26
Screw diameter, in.	d_o	2.0	2.5
Lead of screw, in.	L	0.25	0.5
Screw efficiency	e_1	0.165	0.238
Bell crank ratio	B	2	2.76
Gear ratio	G_1	1	6
Gear efficiency	e_2	0.92	0.903
Mechanical advantage $= \pi HBG_1/L$	M	503	2705
Fathoms of chain out, fath	f	60	60
Weight of chain, lb	W_1	19,840	80,000
Weight of anchor, lb	W_2	12,693	60,000
Total weight $= W_1 + W_2$, lb	W_3	32,533	140,000
Hawsepipe efficiency[a]	e_3	0.80	0.80
Buoyancy factor	b	0.87	0.87
Static load at wildcat $= e_3bW_3$, lb	W_4	22,600	97,440
Angle of brake wrap around drum, deg	θ	340	305
Angle of wrap $= \pi\theta/180$, rad	α	5.93	5.32
Brake lining coeff. of friction[b]	μ	0.30	0.225
Ratio factor $= e^{\mu\alpha}$	K	5.93	3.32
Assumed velocity of fall, fps	V	30	30
Specified stopping distance, ft	h	18	24
Deceleration force $= e_3W_3V^2/2gh$, lb	F_1	20,200	65,200
Force at wildcat $= F_1 + W_4$, lb	F_2	42,800	162,640
Force at brake drum $= F_2d/D$, lb	F_3	27,600	109,300
Slack end pull $= F_3/(K - 1)$, lb	F_4	5,600	47,100
Anchor end pull $= F_3 + F_4$, lb	F_5	33,200	156,400
Handwheel effort $= F_4/Me_1e_2$,[c] lb	P	73.3	81.0
Anchor end band stress $= F_5/wt_1$, psi	S_1	3,320	7,820
Slack end band stress $= F_4/wt_2$, psi	S_2	1,120	3,140
Mean band stress $= (S_1 + S_2)/2$, psi	S_2	2,220	5,480
Band modulus of elasticity, psi	E	30×10^6	30×10^6
Band stretch $= S_3D\alpha/2E$, in.	y	0.010	0.0437
Handwheel turns to set brake[d] $= yBG_1/L$	n	0.08	1.45[f]
Maximum brake band pressure[e] $= 2F_5/wD$, psi	r	138	174

NOTES:

[a] An optimistically high hawsepipe efficiency should be used in this calculation.

[b] In general the brake lining coefficient of friction quoted by manufacturers should not be used, as that is the value obtained in a laboratory. Anchor windlass tests show that a coefficient of friction of 0.225 and 0.30 can be expected with molded and woven linings, respectively.

[c] The braking force required at the handwheel should be approximately 100 lb or less.

[d] Efforts to reduce the handwheel effort by increasing the mechanical advantage may result in an excessive amount of elasticity in the system.

[e] The maximum pressure on the brake lining should be as recommended by the manufacturer and proven by experience.

[f] In order to promptly initiate braking action, a hydraulic-assist mechanism was installed on the CVA 66. The mechanism was designed such that the brake was set hydraulically when the handwheel was turned about one-half turn in the direction to set the brake. Turning the brake handwheel also set and secured the brake mechanically as in conventional practice.

This type of arrangement has also been successfully used between a tug and an unmanned barge. In this instance, the control is accomplished by a radio signal, with the small amount of electric power required on the barge supplied by batteries.

Remote control of the anchor would be particularly advantageous in a long river passage where it is customary to have at least one individual standing by at the windlass in case an emergency drop is necessary.

3.4 Windlass Power Units. Since the late 1960's, very few steam-driven windlasses have been manufactured. However, steam-driven windlasses were common before then, particularly on tankers that carried flammable cargoes. Steam-driven windlasses are usually of the horizontal type with all of the components located above deck; such a windlass would look much like the one illustrated by Fig. 15 if a steam engine were substituted for the electric motor. The steam engine is commonly a horizontal reversible type with two cylinders. Steam-driven windlasses are designed to operate with a steam pressure of 100 to 200 psig at the throttle.

Steam-driven anchor windlasses are inherently rugged, simple, and reliable; however, these considerations are seldom criteria for selection as other types of windlasses can be designed to be equally dependable. The major advantage associated with steam-driven windlasses is that they entail no fire hazards when used on tankers that carry flammable cargoes. They also operate at high speed, especially at reduced load. Most of the anchor-lifting operation is done at reduced load. On the other hand, the long runs of piping from the engine room pose two problems: one of actually getting steam to the windlass, and the other of maintaining insulation on the pipes when they are run above the weather deck, as is the usual arrangement. These two problems are, obviously, closely related. In cold weather, if the windlass is steam-powered, it is usually necessary to turn the steam on well before the windlass will be needed so that steam and not condensate gets to the unit when it is needed for anchor handling.

The two commonly used powering systems for windlasses are direct-connected electric motors and electrohydraulic systems. Electrohydraulic systems permit complete control over the hoisting speed and also provide protection (by relief-valve action) against shock loadings in the transmission shafting and gearing in the event that the anchor is inadvertently housed too abruptly.

When an electric motor is directly connected to the windlass, it may be either a squirrel-cage or wound-rotor alternating-current motor or a direct-current motor. A d-c motor provides sufficient speed control to house the anchors safely. If a squirrel-cage motor is used, it should be of either the two- or three-speed type with the slowest speed usually one quarter of the full-load speed and slow enough to house the anchor satisfactorily. Even if a multispeed or a variable-speed a-c motor is used, the anchor should be driven through a slip-type clutch coupling so as to limit inertial loadings in the event that the anchor is housed too abruptly. If variable speeds are necessary or desired, then either a squirrel-cage motor with frequency converter, a wound-rotor a-c motor, or a d-c motor may be

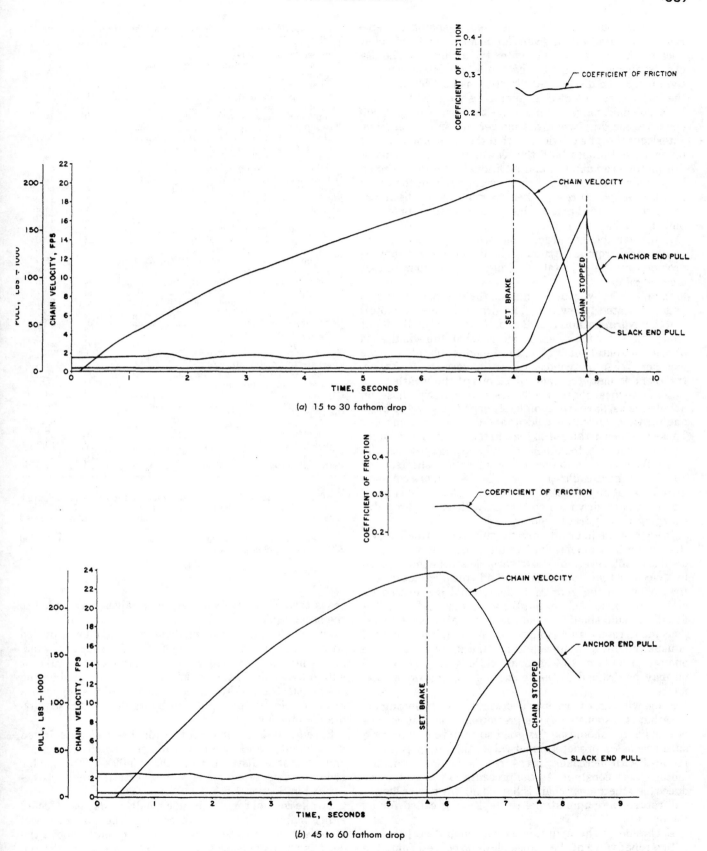

(a) 15 to 30 fathom drop

(b) 45 to 60 fathom drop

Fig. 20   USS *America*, CVA 66, anchor windlass brake test data

used. If a wound-rotor motor is used, it should be separately ventilated with the ventilation air taken from below deck so that water will not enter the air intake. The discharge air may be directed back to a protected space, or the air may be discharged directly on the open deck. In the latter case, the discharge outlet should be protected by a solenoid valve arranged so that the solenoid will open the air duct when the blower motor is energized. Interlocks must be provided so that the main motor cannot be energized unless both the blower motor and resistor fan motor are running; and additional interlocks should be provided to shut the entire system down in the event that the temperature of either the main motor discharge air or the resistor bank discharge air exceeds a predetermined safe limit.

If the variable speed for a motor-driven windlass is to be obtained from a d-c motor, the direct current may be provided by the adjustable voltage output from an a-c/d-c control unit.

In an electrohydraulic windlass, the pump, or A-end, is usually located below deck and driven by an a-c motor; and the hydraulic motor, or B-end, is mounted on the input shaft of the windlass gear reducer. When the windlass is of the horizontal-shaft type, the B-end is mounted in the weather. When the windlass is of the vertical-shaft type, the B-end is mounted below deck out of the weather.

Some vertical-shaft windlasses, particularly those on naval vessels, have two completely separate power plants, one for each wildcat or wildcat and capstan combination. These are then arranged so that, in the event of a casualty to one power unit, the other unit may be engaged (usually hydraulically) so as to operate both vertical shafts; such an arrangement is illustrated by Fig. 18. A more common arrangement on merchant ships is a single, double-ended, electric motor driving two pumps with each pump discharging to a hydraulic motor.

In each of the hydraulic arrangements described above, the hydraulic transmission would consist of a positive-displacement, reversible-flow, variable-stroke pump piped in a closed circuit to a fixed-stroke hydraulic motor. The first pinion in the gear reduction should be coupled to, rather than mounted on, the B-end output shaft. The hydraulic circuit should include an auxiliary, positive-displacement, replenishing pump. The pumping unit bedplate is usually built as a storage tank, and it should be large enough to contain 110% of the oil in the system so that all oil may be drained to the tank for servicing or maintenance.

Some windlasses are provided with a power-limiting device that is responsive to the pressure in the system. The power limiter should be designed so that when the pressure reaches a predetermined value, the pump stroke is reduced with increasing pressures, keeping the electric motor power constant. As the pressure in the system reduces (i.e., the anchor chain is hauled in), the power limiter will return the pump stroke to the setting called for by the operator.

Each side of the hydraulic circuit should be provided with a relief valve of the "cross-blow" type, returning the oil to the suction side of the pump. Alternatively, in some cases the discharge from the relief valves may be led to the sump tank.

Direct electric-driven windlasses should be provided with an electric brake on the motor shaft. Hydraulic windlasses may be provided with either an electric brake or with a hydraulic brake, which should be mounted on the B-end. Electric and hydraulic brakes should set upon loss of either electric power or, on hydraulic windlasses, hydraulic pressure.

Speed and directional control for direct electric-driven units should be effected by a master switch located aft and at a safe distance from the windlass. The master switch is usually equipped with a vertical handle, as described in Section 1.5.

Stroke control for a hydraulic windlass may be effected either manually or by servo control. The control wheel stand should be raised to a convenient height above the deck and should be located aft and at a safe distance from the windlass. If manual control is used, the shaft should

Table 4   Power calculations for an electrohydrualic windlass

	Symbol	C4
Type of ship		C4
Number of anchors hoisted	n	2
Anchor depth at beginning of hoist, fath	h	30
Anchor weight, lb	W_1	12,693
Anchor chain size, in.	C	$2^{7}\!/_{16}$
Chain weight (each anchor), lb	W_2	9920
Buoyancy factor	b	0.87
Weight per wildcat = $b(W_1 + W_2)$, lb	W_3	19,673
Hawsepipe efficiency[a]	e_o	0.60
Pull at each wildcat = W_3/e_o lb	P_1	32,788
Outside length of one chain link, in.	G	$14^{5}\!/_{8}$
Pitch of links = $G - 2C$, in.	p	$9^{3}\!/_{4}$
Number of whelps on wildcat	a	5
Wildcat pitch radius = ap/π, in.	r	15.5
Torque at each wildcat = P_1r, in.-lb	T_1	508,200
Electric motor speed, rpm	N_1	1150
Hydraulic pump speed, rpm	N_2	1150
Hydraulic motor speed, rpm	N_3	1100
Gearing ratio: first reduction	R_1	7.83
second reduction	R_2	7.40
third reduction	R_3	4.55
Chain hoisting speed = $2\pi r N_3/12R_1R_2R_3$, fpm.	S	33.9
Specified hoisting speed, fpm	S_1	30
Hydraulic pump and motor efficiency	e_1	0.75
Gearing efficiency: first reduction	e_2	0.97
second reduction	e_3	0.97
third reduction	e_4	0.97
Efficiency of wildcat	e_5	0.96
Torque per hydraulic motor		
= $T_1/R_1R_2R_3e_2e_3e_4e_5$, in.-lb	T_2	2200
Hydraulic pressure required,[b] psi	p_r	735
Total electric motor hp required		
= $nP_1S/33,000\ e_1e_2e_3e_4e_5$, hp	H_1	102.5
Electric motor hp provided,[c] hp	H_2	100
Capstan diameter, in.	D	24
Capstan rope diameter, in.	d	2.55
Gearing, B-end to capstan:		
efficiency = e_2e_3	e_6	0.94
reduction = R_1R_2	R_4	57.9
Capstan rpm = N_3/R_4, rpm	N_4	19.0
Rope speed = $\pi(D + d)N_4/12$, fpm	S_c	132
Permissible rope pull = $33,000\ e_1e_6H_2/S_c$ lb	F	17,625

NOTES:

[a] The hawsepipe efficiency used should be the lowest anticipated.

[b] A check must also be made to confirm that there is sufficient capacity to hoist one anchor and the full scope of chain at no specified speed.

[c] Specifications often permit the electric motor to be overloaded 25% at the beginning of a hoist.

be provided with a deck stuffing box through which it is led to the A-end stroking mechanism. The run of control shafting should be made with great care in order to avoid excessive friction, lost motion, and misalignment that may be caused by deck deflection. Universal joints and slip couplings should be used to assist in eliminating the effects of misalignment.

The handwheel for the hydraulic stroking device is arranged in a horizontal plane. An auxiliary handwheel for the stroking device should be located adjacent to the A-end to assist in servicing or warming up the unit. The stroke-control mechanism should be provided with a spring detent for the neutral position of the A-end stroking spindle. Limit switches should be arranged to prevent the pump from being started unless the pump and servo control are in the neutral position. This is necessary to ensure that the anchor does not start to move when the pump is energized.

Electric master switches should be provided with detents so that the operator can sense the speed position selected. Handwheels for controlling hydraulic windlasses should be equipped with a speed indicator marked "1/4", "1/2", "3/4," and "Full" to each side of "Stop" for the neutral position. Rotation of the handwheel in a clockwise direction should start the windlass in the hoisting direction.

The procedure followed in determining the powering requirements of an electrohydraulic windlass with an attached capstan is given in Table 4. The procedure for a direct-electric windlass is similar. Motors for direct-connected windlasses should have a 30-minute short-time rating, whereas, for a hydraulically driven windlass the motor rating should be 30 minutes at 15% load followed by 30 minutes at full load. If a separate replenishing pump is provided, it should be rated for continuous duty. The main pump motor should be interlocked in such a way so that it will not start unless the replenishing pump is running. A start-stop pushbutton arrangement should be provided on or near the weather deck control station in addition to the buttons on the controller cabinet. A low-surface-temperature electric heater may be provided in the windlass sump tank to maintain the oil temperature at 60 F in cold weather. The heater should be supplied from a circuit independent of the main power feeders and arranged to be disconnected when the main pump motor is energized.

Section 4
Winches, Capstans, Cargo-Handling Equipment, and Elevators

4.1 Introduction. Deck machinery has changed substantially to support the transition from labor-intensive to packaged concepts in cargo handling. The various types of winches associated with topping, slewing, and hoisting winches have almost disappeared, and, to a certain extent, have been replaced by deck cranes. Many cargo ships, such as containerships and roll-on/roll-off vessels, operate with no onboard cargo-handling winches, but instead depend on specialized cargo access equipment and shore-based loading and unloading equipment.

The use of mooring winches has expanded to all types of ships, including smaller ships, and the winches have been simplified. The sophisticated mechanical tension-sensing mechanisms in electric automatic mooring winches have been replaced by simple, robust, electronic load cells. Spooling devices on mooring winches have become unnecessary since the introduction of the split mooring drum with a separate tension part, which avoids tension on the spooled wire, and fewer capstans and warping winches are required because of the more extensive use of mooring winches.

The development of prime movers for deck machinery has been centered primarily on improved electric drives. D-c electric drives have, to a large extent, been replaced by multiple-speed a-c motors. The use of solid-state controllers for infinitely variable speed control has resulted in a more extensive use of single-speed squirrel-cage motors for marine deck machinery.

4.2 Mooring Winches. Mooring winches are used to hold a ship in position alongside a pier. A high-capacity brake that can hold a load approaching the breaking strength of the mooring line, but which can be set to slip at a lower tension to avoid line breakage, is an important part of a mooring winch. Reference 18, the ISO Standard for mooring winches, specifies a minimum of 80% of the breaking strength of the line for a mooring winch drum brake.

Automatic or self-tensioning mooring winches can be set to automatically render and recover mooring line using the prime mover when the line tension varies outside of preset adjustable limits. A "constant-tension" mooring winch is an automatic mooring winch with only a small difference between rendering and recovery tension.

Many mooring winches are equipped with a warping head. The mooring drum must then be declutchable to allow it to remain stationary when the warping head is used for rope handling. Synthetic line is used on the warping head, while wire rope is normally used on the mooring drum. During berthing, the mooring drum can be used for warping the ship toward the pier, and with automatic mooring winches this operation can be made "self-tensioning" by setting the winch for the proper mode. Normally, two of the forward mooring winches are combined with wildcats to accomplish the windlass/mooring functions with a single unit.

A frequent problem with mooring winches is caused by the fact that the rope is generally spooled on the drum in a slack condition during the mooring operation. Unless special precautions are taken, the underlying layers will

Fig. 21   Split-drum mooring winch

be loosely spooled when the mooring line is being tensioned at the top layers. Mooring lines are often damaged because the top layers wedge into the underlying layers; correcting this situation is always a difficult task. This problem can be avoided by using a split mooring drum, which is illustrated in Fig. 21. When not in use, the mooring line is stored on the storage drum. The figure shows the winch after the length necessary to reach the shore bollard has been payed out, the wire has been transferred through the slot in the midflange, and the mooring line has been tensioned. In addition to solving problems associated with taut lines wedging into lower layers, the split drum has the advantage of allowing operation on the first layer and thereby achieving rated line tension at rated winch torque. The winch capacity is rated on the first layer; for example, seven wire rope layers on the drum would approximately double the working radius and, therefore, reduce the rated rope tension by 50%. The general characteristics of mooring and warping winches are specified in reference 18.

It may seem obvious that constant-tension mooring winches should not be used with the spring lines on tankers. Nevertheless, this has happened and caused breakage of the loading arm and spillage of oil. As a consequence, the use of automatic mooring winches is not permitted in most tanker terminals, and tankers must be moored using the drum brakes.

A mooring arrangement is a compromise. Reduced crews, remote-control complications, and increased safety would suggest fewer but larger mooring winches, with one prime mover per drum. A six-point mooring arrangement with two spring, two breast, one head, and one stern line is preferred for most ships, but the size of the shore bollards limits the rope size in some harbors.

Drum brakes hold the mooring lines in extreme situations and establish the line capacity of the mooring arrangement. It is important therefore that the operator be able to set the brake to the maximum braking capability, and that there be an indicator to show the tension at which the brake will begin slipping.

Automatic mooring winches are powered by "live" or "dead" motors. Electric and unitized hydraulic drives have both alternatives, while central hydraulic system and steam-driven mooring winches have only "live" motors, which are continuously energized in the hauling mode. When operating with a "dead" motor, a motor brake holds the load when the motor is at standstill, and the motor brake slips when rope is rendered.

Steam winches and most hydraulically driven winches are inherently "self-tensioning." If the rope tension is below the hauling capability of the motor, the winch will recover line. When the winch is recovering line, the motor will stall if the line load increases to the point that rated motor pressure is developed. If the rope tension increases above a certain limit, the motor will overhaul and act as a pump, and the winch will render line.

The frictional losses in the motor and winch determine the line rendering/recovery ratio. The winch cannot recover line unless the motor torque caused by the line load is less than the motor stalling torque multiplied by the mechanical efficiency of the motor itself and the mechanical efficiency of the winch. In the rendering mode, the line must overcome friction; therefore, the maximum line load corresponds to the motor overhauling torque divided by the mechanical efficiencies of the motor and winch. The rendering/recovery ratio, R, will thus be:

$$R = \frac{\text{Motor overhauling torque}}{(\text{Motor stalling torque})(\text{Mech. efficiency})^2}$$

where

Mechanical efficiency = (Motor eff.)(Winch eff.)

In some hydraulic winches, separate valves control the rendering and the recovery tension, but the ratio cannot be smaller than given by the above expression.

For steam winches, the rendering/recovery ratio is generally in the range of 2.2 to 3.5, depending on the design of the bearings, packings, etc. For hydraulic winches the ratio is 1.6 or more, but for electric winches the ratio can be less than 1.6. A ratio of 3 or less is generally considered to be acceptable.

Hydraulic unitized automatic-tensioning mooring winches offer some advantages over other types. The operation of a unitized hydraulic mooring winch is simplified because the operator can control the line for rigging by varying the pump's displacement. For automatic tensioning, the pump control is locked in the inhaul position. During automatic tensioning, the speed and tension will be along the constant-power or pressure-compensator override curve. The line will automatically pay in or pay out as line tension varies as shown in Fig. 22. Usually, the pressure compensator override takes control from the power limiter at the specified mooring winch rating.

Figure 22 shows that the line begins to pay out at approximately 50% above the tension setting. This is due to effects of mechanical efficiencies when the gearbox and hydraulic motor reverse direction. The difference between inhaul and outhaul line loads is usually advantageous because it provides a dead band during which there is no

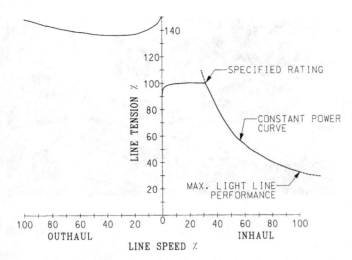

Fig. 22 Performance of a typical unitized mooring winch

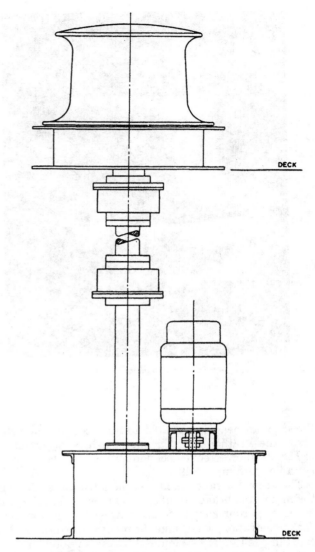

Fig. 23 Capstan with machinery on deck below

winch movement. If the spread between the rendering tension and the recovery tension is unacceptably large, it can be substantially reduced by using a high-efficiency, high-torque, low-speed motor in place of a high-speed motor and gearbox.

The brake release valve is linked to the pump control. As the operator moves the mooring winch control off center, the brake valve shifts. Servo pressure is then ported directly to both the pilot-operated bypass valve and the spring-set, hydraulically released brake.

Variable-stroke hydraulic motors can be used advantageously in some mooring-winch applications. Variable-stroke motors are well suited for winches that have a large speed range requirement, which is often a high, light-line speed requirement.

A synthetic, self-tensioning line can be used to reduce winch operation and wear because the line itself has much greater elasticity than a steel line and, therefore, can absorb many minor adjustments by stretching or relaxing without the winch being energized.

If a gypsy or warping head is required, it must be arranged so that it will function with the drum declutched and secured. If the head is to be used for warping duty, then the head, shaft, bearings, etc., should be designed to withstand the breaking strength of the hawser applied tangentially at midlength without exceeding 75% of the yield point of the materials. Warping head design details are described in reference 19.

Most automatic and nonautomatic mooring winches with electric drive are powered by a-c multiple-speed, squirrel-cage motors. These motors are the same type as for windlasses; that is, they are normally three speed with the same torque at creep and medium speeds, and the same power at medium and high speeds. Thus, the high-speed step has a reduced torque, and the rated capacity is based on the medium-speed performance.

The number of electric motor poles can be 4/8/16, corresponding to a synchronous rpm of 1800/900/450 at 60 Hz, but 2, 32, and other numbers of poles have been used as well.

4.3 Capstans. There are three mechanical arrangements of capstans in general use. In one arrangement the motor, electric brake, gear reducer, and capstan head are mounted on a common bedplate on the weather deck. In a second arrangement, only the capstan head is mounted in the weather, with the motor, electric brake, and gear reducer hung from the underside of the weather deck. In a third arrangement, the capstan head is on the weather deck, with the motor, brake, and gear reducer on the deck below, as illustrated by Fig. 23. In all three cases, the master switch is located near the capstan head on the open deck, and the controller is located in a protected location, such as a deckhouse or the capstan machinery space.

The first arrangement mentioned has the advantage that the complete unit can be assembled by the capstan manufacturer for bolting in place by the shipbuilder. However, it has the disadvantages that the motor and brake must be of watertight construction and, in some designs,

Fig. 24   Capstan with motor in capstan head

Fig. 25   Warping winch with extended shaft

the capstan head is elevated to an inconvenient height above the weather deck. The motor-in-head-design shown in Fig. 24 allows above-deck mounting with the head at a suitable working height.

The second arrangement has the advantage of having the motor and brake out of the weather so that they may be of dripproof construction. However, it introduces a deck penetration, which must be made watertight, and the capstan head must be mounted by the shipbuilder.

The third arrangement, which is common, entails a problem of alignment between the driving and driven units and requires the installation of a flexible coupling that can accommodate a small amount of misalignment and variation in the vertical distance between decks, which may be caused by either temperature changes or deck loadings. This type of capstan offers the advantage of having the power unit on the deck, rather than overhead, so that it is more readily inspected and serviced.

The capstan head is usually of the smooth-barrel type (without whelps) and its configuration should approximate that given in reference 19. The gear reducer generally consists of a worm-and-wheel reduction and a spur, helical, or herringbone reduction. It is preferred that the worm-and-wheel reduction (if used) precede the other reduction in order to take advantage of the slightly higher efficiency of the worm at the higher rubbing speed. All bearings in the reducer should be of the ball or roller type because of the necessity of accurate alignment of the worm gearing. The deck bearings for the capstan arrangement shown in Fig. 23 may be of either the roller type or of the sleeve type with bronze bushings.

For merchant ships, it is commonly specified that when the capstan is handling the specified load, the stresses should not exceed 40% of the yield point of the materials. The capstan head, main shaft, bearings, and capstan base should be designed to withstand the breaking strength of the hawser applied tangentially at midheight of the head without exceeding 75% of the yield point of the materials.

Capstan motors should be reversible and are usually of the two-speed (full and quarter), constant-power, squirrel-cage type. They should be rated for 30-minute, short-time, full-load duty on either winding. A brake should be provided on the motor shaft. The motor and controls should provide adequate control of overhauling loads. Step-back protection from high to low speed should be provided so that, when retrieving a light line on the high-speed point, the motor speed will automatically step back to low speed if the rated power is exceeded in high speed. Automatic return to high speed should also be provided should the line pull be reduced. Capstans are usually designed for line speeds of about 30 to 35 fpm in low speed since this is about as fast as an operator can handle the line and keep it tight around the head for friction purposes. If the motor is full and quarter speed as suggested above, this will result in a light-line speed of from 120 to 140 fpm.

4.4   Warping Winches.   A warping winch is typically used to warp a ship alongside a pier or to move a ship from one place to another, by means of hawsers, without other assistance. The warping head on a warping winch is similar to the head on a capstan, except that the warping head or heads are mounted on a horizontal shaft. In some instances, the heads are mounted on extensions of the main shaft so that they may be at a considerable distance from the power unit, as in Fig. 25. If the shafts are extended in order to spread the heads apart, these extensions are provided with outboard bearing pedestals close to the heads.

If the ship is equipped with mooring winches, the warping heads are generally positioned on extended drum shafts. The drum is declutched and secured by the band brake when the warping head is rotated.

A stress analysis should be made for warping winches employing the limitations as described above for capstans;

in addition, if a drum is provided, the unit should be capable of withstanding the stall torque of the motor without exceeding 75% of the material's yield point.

The electrical equipment described for the capstans is suitable for warping winches.

4.5 Synthetic and Fiber Rope Handling Gear. Twin-drum traction winches for the "first-line ashore" are also a kind of warping winch. The fiber rope is stored on a separate self-tensioning drum or in a bin. The twin-drum unit can be replaced by a warping head, but the pay-out operation then requires additional equipment, and the heave-in operation is sensitive to varying tension, rope conditions, etc. The "first-line ashore" is preferably a synthetic hawser that floats. This rope is cost effective, but sensitive to chafing and fusing; its melting point is 320 F. Adequate support of tensioned polypropylene rope can be given by the grooves in the twin drums, but stepping of the drum diameter at the various grooves is necessary for proper function under the varying operating conditions from slack to full tension during pay-out and heave-in. A minimum of five grooves is necessary and a drum set with five or six grooves is normally used.

Figure 26 shows a twin-drum traction winch that is driven from a mooring winch. One traction winch with a storage reel on each end of a ship is preferable. Some of the largest tankers are equipped with two at each end. This equipment is well suited for remote control, which increases operator and ship safety. Fiber rope handling gear is also used for pulling in single-point mooring system messenger lines.

4.6 Winches for Cargo-Handling. The simple rigging arrangement shown in Fig. 27 has the capability of performing the three basic functions required to handle the boom and the cargo. One function is the ability to top the boom, i.e., raise the boom head to the proper elevation. A second function required of the rigging arrangement is the ability to swing (or slew) the boom so as to control the transverse location of the boom head. As shown by Fig. 27, the upper vang pendants, which are secured to the boom head, are used to swing the boom. Vang lines are required on both sides of the boom head. Lastly, the rigging arrangement must be capable of hoisting and lowering the load. The load would be secured to the cargo hook shown in Fig. 27 and would be hoisted and lowered by means of the cargo hauling part that goes to the cargo (hoist) winch drum. There are many variations of the basic rigging arrangement shown by Fig. 27 [16].

Reference 20 specifies the criteria to be used in the design of cargo winches, and reference 21 contains additional descriptive and operational information concerning cargo-handling gear.

Most cargo winches are driven by 50-hp electric motors. Hydraulic and steam winches have approximately the same full-load speeds as electric motor-driven winches, but the light-line speeds may be different, being somewhat slower for hydraulic units and somewhat faster for steam units. Cargo winches have different ratings at different line pulls; typical ratings for a 50-hp, motor-driven, electric cargo winch are as follows:

DRUM LINE PULL, lb	DRUM LINE SPEED, fpm
0	500
3720	290
7450	220
14,500	105[a]
19,000	70[b]

[a] "Two-speed" winch in low gear.
[b] Separate auxiliary drum.

Most cargo winches are provided with a double gear reduction. The first reduction is frequently of the herringbone type while the second is usually of the spur type. In some few instances, when the winches are to be mounted on or adjacent to living quarters and very quiet winches are required, they are fitted with a single worm gear reduction. Most winches are of the single-drum type. If a gypsy head is desired on the winch, it is mounted on an extension of the drum shaft and the drum is then fitted with a mechanical brake and a clutch so that the drum may be secured when the gypsy is used.

Occasionally, a double-drum winch may be installed in place of two single-drum winches, thus saving the cost of the second motor and electric brake. On such a winch, both drums must be provided with clutches that are interlocked so that only one drum can be engaged at a time, but permitting both to be declutched at the same time. Also, each drum must be provided with a mechanical brake.

The mechanical band brakes provided on cargo winches are intended to be holding devices rather than stopping devices. In an emergency, such as that caused by a power failure with a load suspended, the mechanical band brakes may be used to lower the load to the deck. For heavy-lift winches, the brake mechanism should be of the screw compressor type. On these brakes, a shaft with a handwheel has an acme thread, which provides the force necessary to tighten the brake in a manner similar to the anchor windlass brake shown in Fig. 19. The brake bands are usually lined with a brake lining similar to that used for the anchor windlass brake. In addition to the clutch interlocks mentioned above, it is advisable to interlock the brake and the drum clutch in such a way that the clutch cannot be moved out of engagement until the brake is set.

As a substitute for the band brakes, it is frequently possible to use a simple locking device, such as a bar or pin that can be inserted in a sleeve in the winch pedestal and pushed through to engage a hole in the flange of the winch drum.

Drum clutches should be of the positive-engagement type; the most commonly used type is the jaw clutch. If the clutch is of the dry type, the jaws should be relieved a few degrees so as to facilitate engagement. If the clutch is to be lubricated, then the jaw faces should be parallel, but a few degrees of backlash should be provided to facilitate the engagement.

Brake and clutch linkages should be arranged and designed so that they will be extremely rigid in order that they will not be bent or otherwise damaged by being "forced." All clutches should be provided with a device

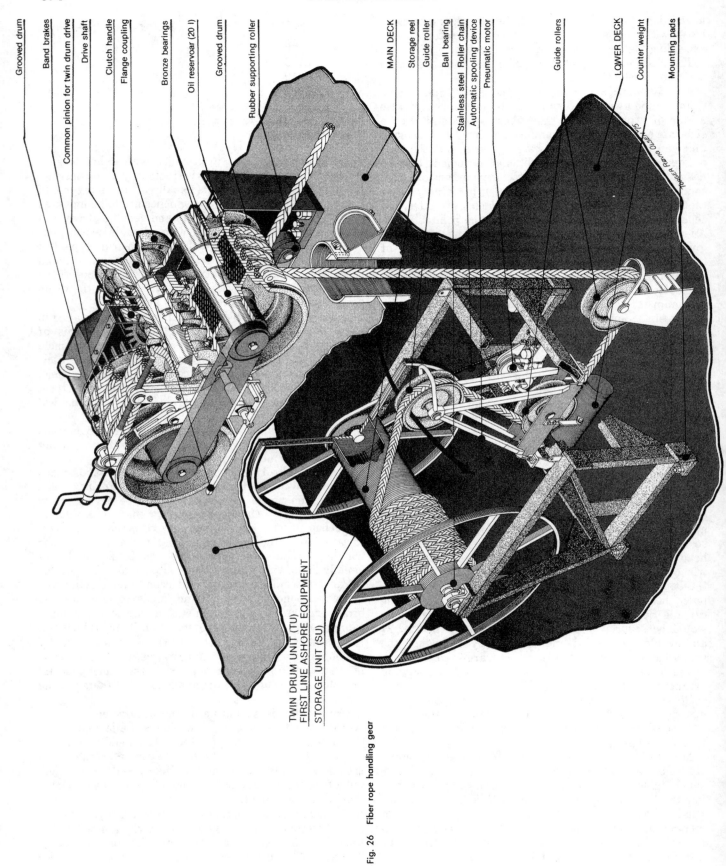

Grooved drum

Band brakes

Common pinion for twin drum drive

Drive shaft

Clutch handle

Flange coupling

Bronze bearings

Oil reservoar (20 l)

Grooved drum

Rubber supporting roller

MAIN DECK

Storage reel

Guide roller

Ball bearing

Stainless steel Roller chain

Automatic spooling device

Pneumatic motor

Guide rollers

LOWER DECK

Counter weight

Mounting pads

TWIN DRUM UNIT (TU)
FIRST LINE ASHORE EQUIPMENT
STORAGE UNIT (SU)

Fig. 26   Fiber rope handling gear

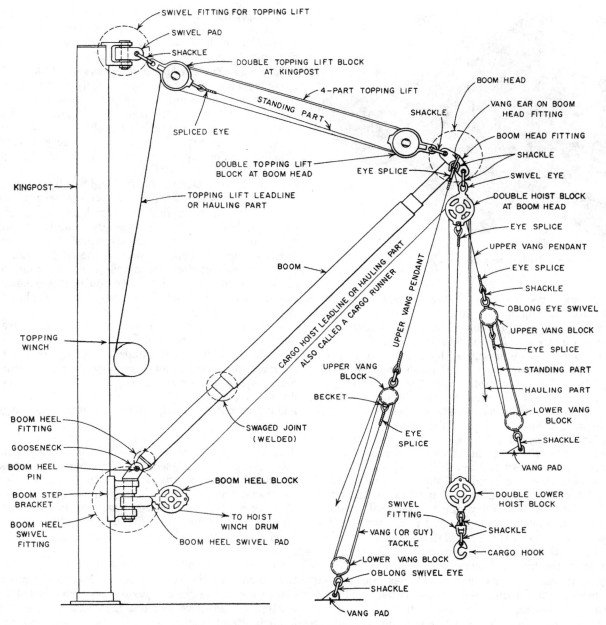

Fig. 27 Nomenclature for cargo gear rigged for swinging or slewing

that will secure the clutch handle in either the engaged or the disengaged position.

The drums should be fitted with removable rope guards designed to prevent a slack rope from being wound over the flange and being wrapped around the drum shaft. When a large amount of wire rope is to be stored on a winch drum, it is advisable to use a spooling device.

The most elementary form of a spooling device is a grooved drum. The groove is a long spiral from end to end of the drum. However, a grooved drum will control the storage of only one layer of wire and only if the fleet angle is very small and if the wire is kept under tension continuously. Mechanical spooling devices may take any

one of several forms, the most popular of which consists of a pair of rollers that are arranged with their axes at right angles to the line lead and parallel to the drum flanges. The rollers are mounted on a trolley or carrier, which is moved back and forth across the width of the drum to ensure that the wire is wound evenly across the full width. Ordinary trapezoidal threads are the simplest and most reliable for movement of the carrier, but a suitable reversing mechanism must be fitted.

There are a number of ways of obtaining variable or multiple speeds when a winch is driven by an electric motor. One alternative is a single-speed, squirrel-cage motor driving the winch through a variable-displacement hy-

Fig. 28 Vang winch designed for motor to be mounted within a vang post

draulic pump. Three- and four-speed squirrel-cage motors, wound-rotor slip-ring motors, and a-c/d-c motor-generator sets, which produce adjustable-voltage d-c to control the speed and direction of the winch, have also been used. These systems can be simplified by using single-speed, a-c electric motors controlled by variable-frequency control systems. The drive motor circuitry should provide safe lowering speeds if either a brake failure or power failure, or both, occur. More details concerning cargo winch electric drives are given in reference 21.

As illustrated by Fig. 27, topping and vang winches are used to move the cargo boom vertically and laterally, respectively, in order to position the boom head to handle the hook load. They are usually of simple mechanical and electrical arrangement, capable of positioning the boom when it is fully loaded, and are also capable of lowering the boom to the deck for servicing. In addition, topping winches are used to hoist the boom to the highest working position or, if the boom is to be stowed vertically, to the boom rest on the kingpost or crosstree.

Topping and vang winches are usually designed so that the lubrication and the fill-and-drain connections will function properly whether the winches are mounted horizontally or vertically. This allows the shipbuilder latitude in arranging the winches so as to provide the best possible cargo-handling arrangement. The winches should be capable of handling the working and test loads without the aid of a drum ratchet and pawl or other securing device. The drive arrangement may be of either the spur-gear (double or triple reduction) type or the worm-gear type. The 7.5-hp vang winch in Fig. 28 is designed such that the motor and brake are mounted within a vang post; only the drum is exposed to the weather.

Topping and vang winch motors are usually of the single-speed, reversible, squirrel-cage type, and are equipped with brakes. The motors should be rated for 30-minute, short-time, full-load duty.

Master switches should be of the "spring-return-to-OFF" type. Frequently, the master switches for the topping and vang winches for a boom are arranged so that they are operated by a single lever. Providing the control of the two winches in a single lever permits the operator to position the boom quickly and land the load with great accuracy.

4.7 Navy Replenishment at Sea Systems (RAS). These systems, which are known by a variety of names or acronyms, have been highly developed for use in ship-to-ship cargo (solids) and liquid transfer. Cargoes can include everything from foodstuffs to sophisticated weapons to personnel. The replenishment activity occurs while the ships are underway, hence the common term "underway replenishment" or UNREP.

A simplified schematic for fuel and highline cargo transfer is shown in Fig. 29. The spanwire is used to support saddles for liquid or fuel transfer hose. Small winches called saddle winches position the saddles along the spanwire. The saddle winches are either hydraulic or electric. The highline is used to support the solids cargo trolley as it travels between the ships. The movement of the trolley is controlled by inhaul and outhaul winches. Highline transfer of personnel is accomplished with hand-powered tensioning and inhaul for safety reasons.

4.8 Cargo-Handling Cranes. Cargo cranes are predominantly wire luffed and feature level luffing as a standard. Level luffing is an important safety feature and is achieved through proper geometry selection during the design stage. The drive systems are principally electrohydraulic. Hydraulics offer the advantage of lower weight and higher power density when compared with alternative drive systems.

A typical marine cargo crane is shown in Fig. 30. Many different types of cranes have been developed including those designed as telescoping or folding boom types, and overhead or gantry types. Each type of crane offers advantages. The primary factors influencing the selection of the correct crane type are: cargo to be handled, ambient temperature, duty cycle, lift capacity, lift height, outreach, visibility, weight, center of gravity, and drive system.

Cranes are placed on the deck and it is important that the machinery be well protected. Machinery should be placed inside the crane housing whenever possible. Preheating of the drive systems (hydraulic or electrical) is important and must be a feature of all equipment placed on deck. The heating system should be arranged in such a way that it can be energized independent of the main supply. The hydraulic system should have the capability to preheat the oil. Electric motors and cubicles must be equipped with heaters to avoid condensation.

The generator capacity on board is often determined by the number and size of cranes. When determining the generator capacity, the crane supplier should be consulted because regenerated power must be considered.

The hydraulic system comprises three main circuits (hoisting/luffing/slewing) plus a boost circuit. The control circuit can be either hydraulic or electric/electronic (solenoid valves). The main circuits are either closed or open looped, with closed loop being preferred. Each circuit has a pressure relief valve installed as a safety feature. The relief valve is also used to limit the lifting capacity of

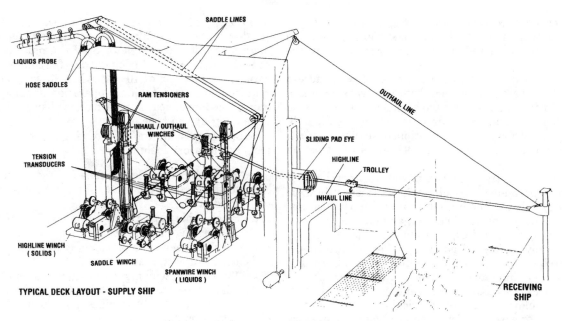

Fig. 29   Underway replenishment-at-sea rigging

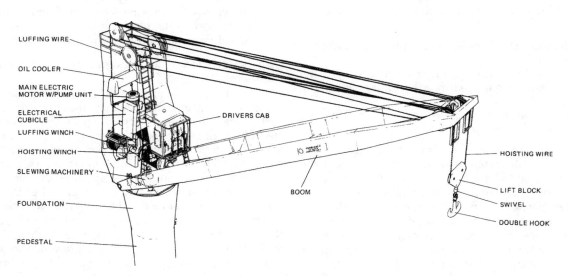

Fig. 30   Typical cargo crane

the crane. To assure that the maximum safe working load of the crane is not exceeded, the crane should be equipped with a maximum-lift cutout valve. When the cutout valve is activated, the pump stroke should automatically return to zero and the brakes should close. The brakes should be spring activated so that if power is lost, they will close automatically.

The pump unit is an integrated unit consisting of main electric motor, pumps, and gearbox. Preferably one pump is used for each main function (hoisting/slewing/luffing). The alternative is to let one or two pumps serve the circuits, although in this case the speed of each function will be dependent on the number of motions driven simultaneously.

To guarantee safe operation, all motions must be governed by safety limit switches. These limits are primarily: safe working load cutout, slack rope, full drum, empty drum, maximum outreach, minimum outreach, and, when required, slewing limits. Further safety limits may be required in specific applications.

4.9   Cargo-Access Equipment.   Cargo ships must have cargo-access openings into their holds: either vertically in the weatherdeck or horizontally in the bow, stern or side as with roll-on/roll-off vessels. Cargo-access closures are vulnerable to damage at sea and, thus, properly designed cargo-access closure equipment is vital to ship safety.

Besides providing weathertight integrity during a voyage, cargo-access equipment may also provide the means to move cargo onto or off the ship and the means of its interdeck transfer. This latter operation is performed either on wheels via fixed or adjustable ramps, or by elevator.

The hatch cover is the single principal piece of access equipment for vertically loaded dry-cargo vessels, and a number of different designs have been used. The decision as to which to install depends on ship type (e.g., bulk carrier, general cargo ship, multipurpose vessel, reefership) plus cargo type and considerations of space on board.

The access and transfer equipment for horizontally loaded (roll-on/roll-off) vessels includes: visors, bow or stern doors, bulkhead doors, internal or external ramps, ramp hatch covers, hoistable decks, and elevators.

Ships vary greatly in design, so the cargo-access equipment is rarely standardized and normally custom-designed and built. A more complete discussion concerning cargo-access equipment is contained in reference 22.

4.10 Elevators. Many types of merchant and military ships require elevators for the vertical movement of personnel, cargo, or weapons. Two primary types of elevators are in common use: hoistway (or trunk) mounted and open mounted. Both types are usually operated by wire-rope hoisting cables in one of the following hoist arrangements:

- Winding drum, which is powered by a hydraulic or electric motor.
- Traction drive, with a counterbalance weight and drum, which is powered by a hydraulic or electric motor.
- Hydraulic engine, which consists of a traveling sheave on a hydraulic cylinder.

However, on occasions shipboard arrangements favor elevators that are operated by other means. Scissors elevators are well suited for the requirements associated with some short-travel (e.g., between-deck) hoists.

Passenger elevators are enclosed platforms that are hoisted by winding-drum or traction-drive machinery.

Cargo elevators are typically open platforms that travel in a hoistway. For maximum access, some designs include an upper-deck hatch to the hoistway; when servicing the upper deck, the hatch is opened and the elevator platform is raised to be level with the deck. Cargo elevators are usually operated with winding-drum or traction-drive machinery. Ammunition and weapon elevators are similar to cargo elevators, except that the elevator platform, doors, and structure must be designed to withstand blast loadings.

Aircraft elevators on aircraft carriers are usually operated by hydraulic engines. This arrangement permits the hydraulic cylinder to be pressurized by an accumulator, which is charged by hydraulic pumps that run almost continuously during a complete cycle. As a result, the size of the elevator prime mover is greatly reduced.

All shipboard elevator platforms have guide rails, and guide rails are also required for the counterweights of traction-drive elevators. For most military applications, the guide rails must be designed with sufficient strength

and rigidity to permit safe operation when rolling 15 deg (10 deg for commercial applications) and when pitching 5 deg; in addition, the elevator platform must be designed to withstand a roll of 45 deg (30 deg for commercial applications) and a pitch of 10 deg, when stowed. The specific requirements for machinery operation under the various storm, moderate sea, and permanent list conditions vary depending on the type of ship.

Safety provisions for the hoisting machinery and controls of elevators include:

- Platform overspeed or free-fall safety stops (often in the form of knurled rollers that wedge into the guide rails).
- Slack-cable device, to stop the platform if any of the hoist ropes become slack.
- Buffers below the platform (and counterweight, if used), to decelerate a free fall.
- Brakes on winding-drum and traction elevators, which are capable of holding the platform and 150% of the rated load for military applications (125% for commercial applications).
- Interlocks, to prevent elevator operation when doors and hatches are improperly positioned.
- Interlocks, to prevent elevator doors from being opened when the platform is not in proper position.
- Speed governor to control platform speed.

Other safety-feature and design criteria are contained in references 23, 24, and 25.

To illustrate some of the principles associated with the design of elevators, the elevators that are used to transport aircraft between the hangar deck and flight deck on aircraft carriers may be considered. The arrangement of a typical aircraft elevator is illustrated by Fig. 31. To minimize weight, the platform of an aircraft carrier is often made of an aluminum alloy; nevertheless, the platform weight is several times larger than the aircraft payload. The large hoist load and the short duty cycle require a very high power output (typically about 2500 hp) from the prime mover during the hoist phase, which would be difficult to accommodate with elevators of the winding-drum or traction-drive types because of size considerations and peak electric power requirements. However, a hydraulic engine that consists of a traveling sheave mounted on a hydraulic cylinder that is pressurized by an accumulator is well suited for this application. The reeving scheme for a traveling-sheave arrangement is illustrated by Fig. 32. The anchored end of the cables in a traveling-sheave arrangement may be either on the cylinder side of the sheave or opposite the cylinder. For the situation illustrated, the cylinder rods would be in compression to support the platform if the cables were anchored to the cylinder; therefore, they are anchored on the opposite side.

A simplified schematic of the hydraulic system for a traveling-sheave arrangement is shown by Fig. 33; the cycle is seen to be open, with several hydraulic pumps being used to charge an accumulator (or bank of accumulators), which, along with the pumps, delivers oil at a high

Fig. 31 Aircraft elevator platform in lowered position

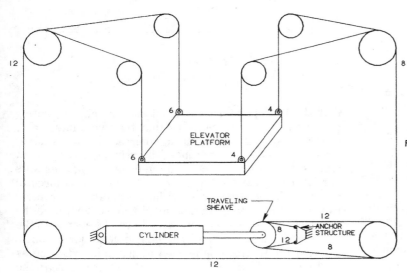

Fig. 32 Aircraft elevator cable reeving arrangement (numbers indicate the number of wire ropes in a cable run)

pressure and flow to the hydraulic cylinder during the hoist phase of the cycle.

The duty cycle for an aircraft elevator is shown by Fig. 34, and the accumulator charge is also indicated. The hydraulic pumps take suction from the hydraulic-oil exhaust tank and charge to the accumulator (or bank of accumulators). Both the accumulator and the pumps pressurize the hydraulic cylinders when hoisting, and the hydraulic oil is discharged to the exhaust tank when lowering the elevator. As indicated in Fig. 33, instead of a single hydraulic cylinder, this arrangement includes three; to lower the elevator, the hydraulic oil in two of the cylinders is discharged to the exhaust tank, and the elevator is lowered by gravity. One cylinder is always

pressurized in the hoist direction by a bank of accumulators. The constantly pressurized cylinder cannot alone support the weight of the elevator platform; therefore, this cylinder is overhauled by the weight of the elevator platform when lowering, and the bank of accumulators serving this cylinder is recharged; in effect, it acts as a counterbalance. In the event that the platform is suddenly lifted by a wave, this feature also functions to extend the piston and traveling sheave to prevent the cables from becoming slack.

Several considerations warrant emphasis during the design of such hydraulic systems. The hydraulic pumps should have enough capacity to completely recharge the accumulator several seconds before the end of the duty

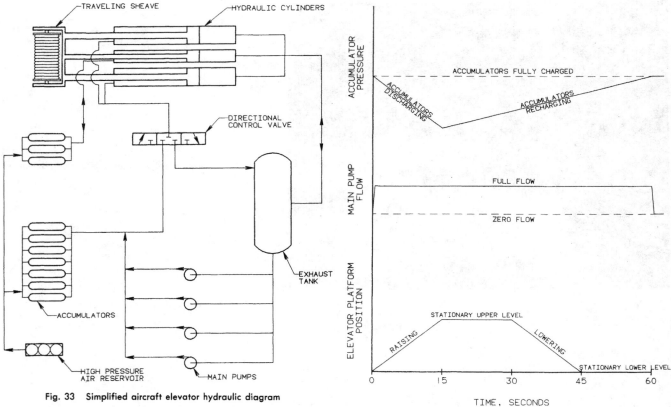

Fig. 33   Simplified aircraft elevator hydraulic diagram

Fig. 34   Aircraft elevator duty-cycle diagram

cycle to ensure that sufficient pump capacity is provided. Also, the very high flow rates (about 2000 gpm) in the hydraulic lines to the cylinder can generate high pressure losses; in addition, friction losses in the line to the air reservoir can impair system performance.

The provisions for venting such complex hydraulic systems are of major importance, particularly during system start-up. Vent connections at all system high points and the installation of valves and piping that permit the vented fluid to be returned directly to the exhaust tank provide a closed-loop venting capability and help to maintain a hydraulic system that is free of air. Also the avoidance of returning oil to the system that has been vented to buckets, etc. prevents the introduction of contaminants.

The arrangement of aircraft elevators makes it impracticable for all four of the hoist-cable runs to be of the same length. The lengths of two cable runs are generally significantly larger than the other two. Since the elastic elongations of the cables are proportional to their lengths, the resulting difference in elongations at the upper and lower elevator positions would be a problem without some form of compensation. Large elevators generally have multiple wire ropes at each hitch-point attachment; therefore, the excessive elongation of the longer cable runs can readily be corrected by installing additional wire ropes, so

that the hitch-point spring constants are approximately the same. Figure 32 indicates that the two longer cable runs consist of six wire ropes, whereas the two shorter runs consist of four wire ropes. This reeving arrangement results in substantially equal hitch-point spring constants during the course of elevator travel, which is required to maintain the platform level within acceptable tolerances. However, the reeving arrangement alone often cannot be designed to ensure that the platform will be even with the upper deck under all platform loading conditions. Consequently, the platform is leveled by installing rigid stops at the upper deck, which contact the platform near the hitch points. These rigid stops enable the hydraulic cylinder to stretch the cables and preload the elevator platform against the rigid stops with a force sufficient to ensure that no variation of platform load, ranging from empty to design capacity, affects the platform position at the upper deck.

Maintenance procedures must be developed for platform hitch-point designs that include multiple wire ropes to ensure that the ropes are loaded approximately equally. Maintaining the load in each wire rope forming the cable run within 5% of the average rope load is considered good practice.

Section 5
Maneuvering Devices

5.1 Introduction. To improve a ship's maneuverability, special devices have been designed to deliver side thrust or thrust through 360 deg. This capability is provided to allow ships to be more independent from tugs when entering harbors, give them more maneuverability for special tasks (buoy tender, cable layer, survey vessel, etc.), and in some cases give them a "take home" capability.

There are three general types of thrust devices. The most common type, the lateral or tunnel thruster, consists of a propeller that is installed in an athwartship tunnel; the propellers in these units may be of either the fixed-pitch or controllable-pitch type. In a second type, jet thrusters, a pump is arranged to take suction from beneath or close to the keel and discharge to either side to develop thrust port or starboard, as desired, or in many cases through 360 deg. Trainable or rotatable thrusters, a third type, are commonly designed such that they can be lowered through the bottom of the ship and trained through 360 deg so that thrust can be developed in any direction. Another type of trainable thruster is similar to a large outboard motor and serves both as a propulsor and thruster. Yet another type, an active-rudder, is designed to be incorporated into the rudder and thereby trainable.

There is a considerable amount of literature that deals with the subject of special thrust devices; however, the most comprehensive treatments of lateral and rotatable thrusters are contained in references 26, 27, and 28. Additional descriptive material on maneuvering devices, including the cycloidal propeller, which is normally also used as a propulsion device, is contained in references 7 and 16. The various types of jet thrusters and their relative merits are described in reference 29.

5.2 Tunnel Thrusters. The lateral or tunnel type of thruster, which is widely known as a "bow thruster" when installed at the bow of a ship, is illustrated by Fig. 35. The lateral thruster is probably the oldest device for improving a ship's maneuverability at low or zero ship's speed. Stern thrusters are similar devices installed at the stern of a ship.

The principal performance features of a typical series of bow thrusters are given in Table 5; however, the performance of any particular unit may vary from that shown. In general, an improvement in efficiency is obtained with a propulsor having a larger diameter and lower rpm, at the same unit thrust loading.

Bow thrusters are often designed such that the thrust developed is both variable and reversible. This is generally accomplished by using a constant-speed electric motor or diesel engine to drive a controllable- and reversible-pitch propeller. The unit is started with the propeller set at zero pitch; then, as the need arises, the pitch is adjusted so as to provide the desired thrust to either port or starboard. For smaller bow thrusters, however, a fixed-pitch propeller is sometimes used with a variable- and reversible-speed driver.

Reversing the drive motor is not necessary for a tunnel thruster with a controllable-pitch propeller. The drive motor may have a constant speed and the propeller also may have a constant speed. The thrust and the thrust direction are regulated by adjusting the pitch of the propeller. This has the advantage that the starting current for electric motors is not as high, which will reduce the cost of the generator sets and the motor; however, due to the complex pitch adjustment system, this type of thruster is relatively expensive.

When locating the tunnel in the ship, it is desirable to have it well forward or, in the case of a stern thruster, well aft to obtain the maximum turning moment from the thrust developed. When establishing the tunnel location, thrust, and power requirement, consideration must also be given to the length of the tunnel as increased friction losses will reduce the available thrust.

The submergence depth of the tunnel is an important factor, particularly for vessels that operate at a reduced draft, because the thrust developed is reduced at low submergence depths. A test was conducted at the Philadelphia Naval Shipyard to investigate the effect of submergence depth on the thrust developed. The bow thruster tested was powered by a vertical 800-hp electric motor, which drove a reversible-pitch propeller through right-angle bevel gears. The thruster tunnel diameter was 6 ft-7 in. The ship draft was varied and the bollard pull was measured when thrusting to port and to starboard. The wind was steady at only a few knots, and there was ample clearance both below the keel and to adjacent piers. Figure 36 shows the data obtained when thrusting to starboard; values when thrusting to port were similar. It may be noted from Fig. 36 that when the top of the tunnel is submerged less than about 1.5 ft, there is a marked loss of thrust. As a rule of thumb, the submergence to the top of the tunnel opening should be at least one-half the tunnel diameter and the distance from the bottom of the tunnel to the keel at least one-quarter the tunnel diameter.

The design of the junction of the tunnel and hull is another factor that requires study. The tunnel openings definitely affect the resistance of the hull; however, the effect is difficult to accurately quantify because it is of a relatively small magnitude. With a well designed tunnel arrangement, it would be reasonable to expect an increase in the ship's resistance of at least 1%. When it is desirable to minimize the resistance added by the tunnel, flow studies should be used as a means of engineering the shape of the tunnel fairing and also the orientation of protective bars, if used. Protective bars are usually mounted in the

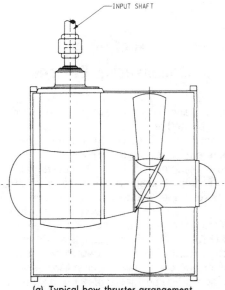

(a) Typical bow thruster arrangement

(b) Typical bow thruster assembly

Fig. 35   Bow thruster

Table 5   Performance characteristics of typical reversible-pitch bow thrusters

Horsepower	Diameter, ft-in.	RPM	Thrust, lb	Thrust/hp, lb/hp
150	3-7	450	4,500	30
300	4-3	420	7,900	26
500	5-5	340	13,200	26
800	6-7	290	20,400	25.5
1200	7-11	240	30,200	25
1800	9-2	210	44,100	24.5

area of the tunnel openings. Protective bars of several different designs have been used. In some designs, several of the bars are bolted in place so as to be portable and permit access to the mechanism and also to permit a removable blade to be unshipped. However, this practice has not been uniformly satisfactory because the bars that are bolted on have a tendency to come adrift. Consequently, a preferable design is one with the bars simply welded in place. If access is required at a later date, the bars are burned off and rewelded; due to rigging problems, it is often necessary to burn off the bars to accomplish major work irrespective of the design. Access problems are alleviated with some designs by the provision of a propeller blade removal hatch in the tunnel immediately above the propeller. This hatch is large enough for an individual to get through, and no access is necessary from the tunnel openings. In this case, to remove a propeller blade, the ship is ballasted until the tunnel is out of the water and the blade is lifted vertically into the ship through the hatch.

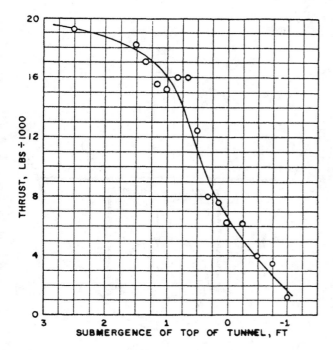

Fig. 36   Effect of tunnel submergence on thrust developed by a bow thruster

The tunnel is generally made of mild steel and is welded into the hull. To minimize tip leakage losses, the clearance between the blade tips and the tunnel should be no more than 0.25 in. However, in practice it is difficult to maintain concentricity between the propeller and the tunnel. The difficulty involves the welding practices during installation at the shipyard. After installation, it is not uncommon to find the tunnel to be no longer circular, thereby requiring the blade tips to be ground off.

For seawater service, all bolts, studs, nuts and other fastenings should be of monel. Propeller blades are usually of stainless steel or nickel aluminum bronze, with the propeller hub of bronze and the pod struts made of steel.

The propeller is driven through a right-angle gear drive (usually of the spiral-bevel type) that is contained within the pod assembly. The pinion shaft or input shaft of the right-angle drive extends out of the pod and through the tunnel assembly. The arrangement can be designed such that the pinion shaft penetrates the tunnel assembly at any angle desired; however, the shaft is normally either vertical or horizontal. With a horizontal arrangement, the shaft normally goes directly to the prime mover without involving another right-angle drive. In the case of a vertical drive, the prime mover can be located many decks above. In a common arrangement, a vertical electric motor drives the thruster input shaft through auxiliary shafting. If the prime mover is a type which cannot be oriented vertically (e.g., a diesel engine), it is then necessary to use a second right-angle gearbox. The second right-angle drive preferably should be of the spiral-bevel type supported by oil-lubricated, heavy-duty, antifriction bearings. Flexible couplings should be provided between the prime mover and the shafting.

The hydraulic power unit is often mounted at least 10 ft above the load waterline so as to avoid the need for a separate gravity tank. This head pressure is necessary to ensure that an adequate pressure is maintained on the oil seals in the propeller hub so that seawater will not enter the pod if the seals should leak slightly. The unit should have a motor-driven pump mounted on a reservoir, complete with necessary piping, suction filter, pressure gages, relief valves, etc. The reservoir should have a capacity in gallons of at least 2.5 times the capacity of the pump in gpm and should be fitted with direct-reading level gages or sight glasses, fill, drain, and vent connections, and access covers large enough to permit the reservoir to be cleaned.

The thrust developed by a lateral thruster when installed in a normal foreship arrangement is approximately as indicated by Table 5. Variations can be expected as a result of the thruster arrangement, hull form, propeller speed, etc. This thrust drops to 50–60% of that indicated as the ship's speed is increased to 2 to 5 knots, depending on the length of the ship.

The master control stand is located in the wheelhouse and is sometimes made a part of the wheelhouse console. When a controllable-pitch propeller is used, this stand should contain a single lever that will pneumatically, electrically, or electrohydraulically control the hydraulic blade positioning system so as to provide stepless pitch control from zero to maximum either to port or starboard. For fixed-pitch propeller installations, the propeller speed and direction of rotation should be controllable. Frequently, auxiliary control stands are located on each bridge wing and are connected electrically, hydraulically, or mechanically to the master control stand.

In addition to the control stand or stands, a control and indicating panel should be provided in the wheelhouse. This panel should contain pushbuttons for starting and stopping the prime mover, the hydraulic system for pitch control, etc. It should also indicate the alignment of the system (e.g., propeller pitch indicator) and contain lights and alarms for critical pressures and temperatures.

If the bow thruster is to be used only for docking, the main motor need only have a one- to two-hour full-power rating. If, however, it will be used for extended periods of time, e.g., as an assist in steering the ship at slow speeds during a long passage, then it should be rated for continuous full-load duty. All other motors for the unit should be rated for continuous full-load duty.

The controller should incorporate low-voltage protection, and interlocks should be provided to ensure that the main motor cannot be started unless the pitch control system is in neutral, that lubricating oil and hydraulic control pressures are available, and that the thruster room ventilation fan is in operation (as applicable).

5.3   Jet Thrusters. A jet thruster is a type of pump that takes suction from the bottom, or close to the bottom, of the ship and discharges either to the port or starboard sides or vectors the thrust up to 360 deg. The propeller need operate in only one direction, which allows for a straightforward prime mover system. Horizontal diesel,

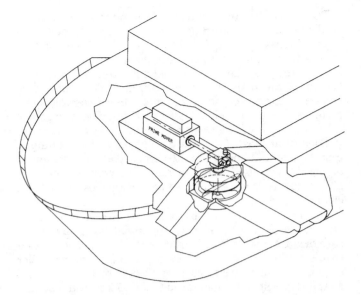

Fig. 37 Four-duct jet thruster

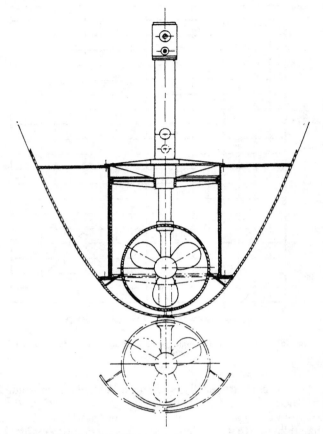

Fig. 38 Thrust device that can be lowered and rotated 360 deg to develop thrust in any direction

electric, or hydraulic drives or vertical electric or hydraulic drives may be used. Jet thrusters were initially used primarily for ships on inland waterways, where empty vessels often have a very limited draft. Normal tunnel thrusters do not work well under these conditions because the propeller and tunnel are not fully submerged. With a bottom-suction jet thruster, it is possible to have the full thrust available at a draft of only about 1 ft. Jet thrusters generally produce effective thrust at ship speeds up to 6 or 8 knots while tunnel thrusters are most effective when there is no headway.

To get starboard or port thrust, a hinged deflector plate is mounted in the propeller tube and is actuated by a hydraulic cylinder, which closes off either the port or starboard discharge opening. The water is directed through 90 deg into the horizontal plane and discharged through the open duct. The reaction force of the water against the deflector produces the thrust. The jet thruster is the simplest and least expensive thruster, but has a lower efficiency than tunnel thrusters.

A four-duct jet thruster arrangement is shown in Fig. 37. This equipment uses a deflector to direct the flow through one or two of the four ducts, which are at 90 deg to each other, and the reaction force on the deflector produces the thrust. A 360-deg thrust vectoring capability is thereby provided. The nominal thrust for such a system is approximately 20 lb per hp. It is not possible to produce an equal thrust throughout 360 deg because of differing flow conditions in the four ducts and the summing of the four vector components. There is also a loss of 10 to 15% compared with the discharge through a single duct.

Jet thrusters with 360-deg thrust vectoring capability offer some additional benefits. They can be used to shorten stopping distance by directing the thrust forward or counter to the ship's movement. This is particularly important on large river vessels or tows when moving

downstream. They can also provide an independent "take-home" capability if the main propulsion system is inoperable. For a discussion of the relative merits of the various jet thruster arrangements, particularly for large river and inland vessels, see reference 29.

With jet thrusters, it is very important to have an optimal water intake and deflector system in order to minimize the drag. When designing a jet thruster, there are several important considerations:

- The intake opening should be faired to create a uniform flow to the propeller. There should be a grid mounted in the inlet duct to keep out debris.
- The discharge openings should be faired to reduce drag.
- Enough height should be provided for the removal of the thruster for repair or changing the propeller if the unit is suitable for servicing from inside the vessel.

5.4 Trainable Thrusters. An azimuthing or rotatable thruster is a thrust unit that usually can rotate through 360 deg; however, they are also designed in the form of rudder propellers or "Z"-drives. A comprehensive discussion of rotatable thrusters is contained in reference 26.

In the thruster design shown in Fig. 38, the propeller and drive shaft are lowered and raised as a unit. The mechanism that trains the propeller also moves up and

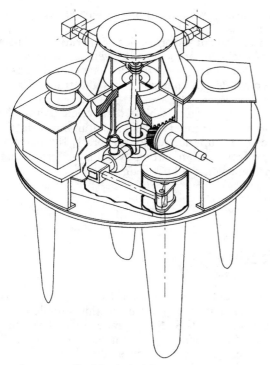

Fig. 39 Cycloidal propeller

cage should be mounted around the unit. The ship arrangement should provide enough height to allow the unit to be pulled up for servicing.

Very large rotatable propulsion thrusters, up to 7000 hp, have been used for a variety of applications. Many of these applications are discussed in reference 30. It should be noted that a vessel driven by a rotatable thruster must have inherent directional stability in order to be steerable. Also, when two or more rotatable thrusters are installed, the interaction between them, which will often reduce the net output thrust, must be considered.

5.5 Cycloidal Propellers. A cycloidal, or vertical-axis, propeller consists of a rotating disk, mounted flush with the hull plating, that serves as a base for several vanes, or propulsor blades. Normally, the blades project vertically downward from the hull, although they could be at any angle, including horizontal, which could be the case for a submarine application. The Voith-Schneider propeller, shown in Fig. 39, is the most well known maneuvering propulsion system of this type. The power ratings of cycloidal propeller applications are in the general range of 150 to 4000 hp.

The vanes of a vertical-axis propeller can be pivoted about their own axes to produce thrust in any horizontal direction through 360 deg, and the thrust can be varied in magnitude. For applications such as tugboats, ferryboats and buoy tenders, the somewhat lower efficiency of a vertical-axis propeller, compared with a conventional fixed-pitch propeller, may be an acceptable trade-off for the improved maneuverability provided. Some vessels with vertical-axis propellers have one mounted at each end of the ship (e.g., double-ended ferryboats) for superior maneuverability. It is noted, however, that vertical-axis propellers are vulnerable to damage from debris and groundings.

The blades of a cycloidal propeller rotate about a vertical axis, point "O" in Fig. 40, and are simultaneously pivoted about their respective axes. During a revolution of the propeller, each blade follows a cycloidal path, hence the name "cycloidal propeller."

The blades or vanes are normally turned about their own

down with the rest of the machinery. The propeller is mounted in a protective ring or nozzle, which is designed to improve its thrust. A closure plate is welded to the bottom of the protective ring, requiring that the unit be properly aligned when it is retracted. The closure plate provides a reasonably smooth hull surface. One use for this type of thruster is for station-keeping on survey ships when it is often necessary for the ship to remain in position over a particular spot while bottom samples or cores are being taken.

Installation is difficult and the cost is high for retractable thrusters. Because they are lowered under the ship's hull, these thrusters are also very vulnerable to damage due to grounding and debris, and, therefore, a protective

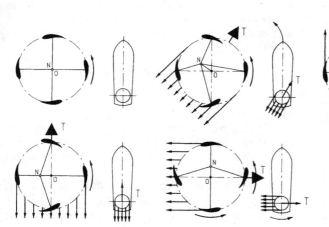

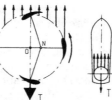

Fig. 40 Cycloidal propeller blade position and thrust diagram

axes in such a way that, at any point in the orbit, a line drawn perpendicular to the chord of the blade always passes through the same point, called the "steering center" or "control point," point "N" in Fig. 40. This point can be located anywhere within the orbit area to create a pitch that provides a variable direction and magnitude of the thrust. The position of the blades and the corresponding direction of the thrust, for a variable-pitch cycloidal propeller of the Voith-Schneider type, are shown in Fig. 40.

Several variations of the so-called "true" cycloidal blade motion, described above, have evolved. Two of these are the sinusoidal blade motion and the amplified cycloidal blade motion, which require sophisticated blade-drive mechanisms.

A description of the principles of operation and performance of cycloidal propellers, as well as comparative efficiencies with screw propellers and some of the associated mechanical design considerations and trade-offs, is contained in reference 31.

In evaluating the suitability of a cycloidal propeller, the marine engineer must consider the intended duty cycle, the need for superior maneuverability, the potential for damage from contamination by debris or grounding, and the influence on the hull design.

Section 6
Ship Roll Stabilization

6.1 Introduction. Ship stabilization, particularly against rolling, is desirable to improve the comfort of the passengers and crew aboard passenger ships; to improve crew performance and safety; to facilitate helicopter operations and weapon system effectiveness on military ships; and to minimize cargo damage on commercial vessels. Roll stabilization is especially important on containerships where containers, mounted above deck, could come adrift as a result of severe ship's motions. Roll stabilization also often makes it possible to reduce slamming and pitching by assuming a heading with beam seas, which otherwise might produce unacceptable rolling motions. A reduction of the vessel's roll also leads to lower fuel consumption as the increase in ship's resistance due to rolling is minimized.

The most common method of stabilizing vessels that normally operate at cruising speeds of 12 knots or more has been the use of active-fin stabilizers. A variation of active-fin stabilization uses programmed rudder motions, which are superimposed on autopilot controls, to obtain an effect that is, in some instances, equivalent to that of active fins without the need for additional appendages or separate and complex actuation and control systems.

For vessels such as fishing boats, which often operate at speeds too slow for active-fin or rudder roll stabilization, passive systems, such as antiroll tanks or flume stabilization systems, are often used. Another method of reducing roll amplitudes is the use of bilge keels, which are fixed longitudinal plates located at the turn of the bilge, such that their drag dampens roll amplitudes. See references 7 and 32 for more details concerning bilge keels, which are only marginally effective. Movable-weight systems are an alternative means of reducing roll amplitudes and are discussed in reference 7.

6.2 Active-Fin Stabilizers.
a. General. Active-fin stabilizers are fin-type control surfaces, which are usually located just above the turn of the bilge near amidships, port and starboard. In a seaway, hydraulic tilting gear continuously varies the angle of attack of the fins, using an automatic control system that is sensitive to the roll motion of the ship, so as to produce heeling moments that reduce the ship's tendency to roll. A description of U.S. Navy practices in the application of active-fin stabilizers is contained in reference 33. The theory is covered in references 32 and 34.

Active-fin stabilizers require ship forward motion in order to develop lift, and the lift developed increases with the ship speed squared. In practical terms, this speed dependency limits the application of active-fin stabilizers to ship speeds above 10 to 12 knots. Below that speed range, the required fin size becomes too large, and other devices (e.g., antiroll tanks) become more advantageous [7,32,35].

b. Simplified calculation of stabilizing moment and number of units. The determination of the percentage of stabilization appropriate in a specific case and the design of the automatic controls entail complex analyses. However, the required fin size and tilting-gear machinery characteristics and location can be determined with sufficient accuracy using a simplified approach. With the simplified approach, it is assumed that a regular beam seaway having a small surface wave slope and a wave period approximately equal to the ship's natural period can build up large roll angles. By designing fin stabilizers to counteract the wave slope heeling moments, it is possible to reduce the large roll angles. Figure 41 illustrates the concept for a ship with two symmetrical fins.

The rolling moment induced by the seaway is expressed as follows:

$$M_\theta = 2240 \, \Delta \, \overline{GM} \sin \theta \qquad (5)$$

where

M_θ = roll-induced moment, ft-lb
Δ = ship displacement, tons
$\overline{GM}$ = metacentric height (distance between the ship's center of gravity, G and ship's metacenter, M), ft
θ = maximum design wave slope (for a wave height, h, and a wave length, λ; $\theta = \sin^{-1} \pi h / \lambda$), deg

The stabilizing moment developed by fin stabilizers is

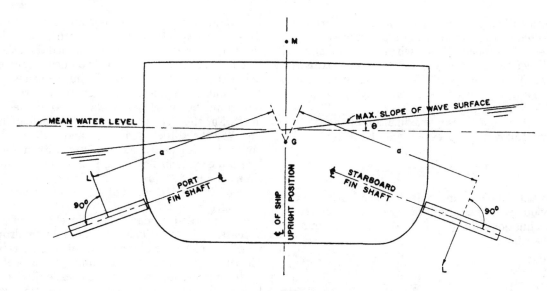

Fig. 41 Active fin stabilization sketch

$$M_s = NaL \qquad (6)$$

where

M_s = fin stabilizing moment, ft-lb
N = number of fins
a = lever arm between resultant lift force and ship's center of gravity, ft
L = lift force per fin, lb

The lift force developed by each fin can be expressed in the conventional manner as:

$$L = C_L \frac{\rho}{2} A V^2 \qquad (7)$$

where

C_L = a nondimensional lift coefficient. (Typical values of C_L would be approximately 1.1 for nonarticulated fins and 1.4 for flapped fins. Figure 42 defines the two types of fin surfaces. Reference 7 outlines the procedure used to compute C_L.)
ρ = water mass density, lb-sec/ft⁴
A = area of one fin, ft²
V = water speed used in fin design, fps (using V_k as the ship speed in knots: $V = 1.69\,V_k$)

If the stabilizing moment computed from equation (6) is equal to or greater than the induced rolling moment calculated from equation (5), an effective stabilization system is considered to be achieved. The key to this simplified method is the somewhat arbitrary selection of the seaway wave slope capacity, θ, and the associated ship speed V_k. Experience has shown that stabilizers should be designed for wave slopes of about 4 or 5 deg. Lower values are reasonable for very large ships, and higher values may

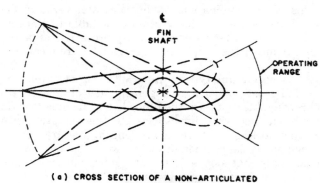

(a) CROSS SECTION OF A NON-ARTICULATED FIN STABILIZER
(ALSO KNOWN AS A SPADE TYPE OR SINGLY ALL-MOVABLE STABILIZER)

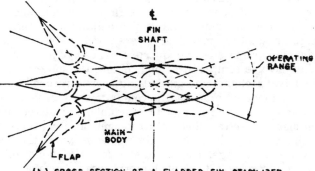

(b) CROSS SECTION OF A FLAPPED FIN STABILIZER
(ALSO KNOWN AS AN ARTICULATED OR DOUBLY ALL-MOVABLE STABILIZER)

Fig. 42 Common types of fin stabilizers

be used for small ships, since small ships are more likely to be subjected to roll excitation in a given seaway.

The lowest ship speed, V_k, at which a significant roll reduction is desired is usually taken to be approximately 50 to 80% of full-power speed. This implicitly recognizes that ships in a heavy seaway are generally operated well below full power.

In order to obtain a first approximation of the fin area required, the following expression may be used:

$$A = \frac{290 \, \Delta \, \overline{GM}}{BNV^2} \tag{8}$$

where B is the ship's beam in feet and all other terms are as previously defined. Equation (8) can be derived by combining equations (5), (6), and (7) with the tentative assumptions that

$$\theta = 5 \text{ deg}$$
$$C_L = 1.2$$
$$a = 56\% \text{ of } B$$

The values of C_L can later be refined and ranges of values for various types of fins at varying speeds are available for this purpose [7]. The usual practice is to try $N = 2$ and $N = 4$, and then make a final decision as to the number of fins. Next, combinations of nonarticulated versus articulated and nonretractable versus retractable fins are investigated in order to establish the preferred arrangement. Some of the considerations involved are discussed below.

c. Location and type of units. The simplest machinery and most compact arrangement are obtained with nonretractable units. On the other hand, the smallest fin area is obtained with high-aspect-ratio fins (e.g., fins with a span of at least twice the chord length) that are located in the ship at the position of maximum beam and located at an angle that slopes downward going outboard such that the fins have maximum leverage. The solution for the least fin area practically always involves fins projecting beyond the maximum beam dimension, thus requiring retractability or a compromise design with low-aspect-ratio fins. Additional considerations to be entertained when developing the arrangements are: (i) in the rigged-out position, the fins should be several feet below the design heeled waterline (so as to minimize cavitation), and not immediately upstream of important sea chests (such as for main condensers); and (ii) the fin tilting gear should be in a compartment suitable for regular maintenance (e.g., an auxiliary machinery space), and preferably not adjoining prime sleeping quarters. These requirements often involve compromises and require the judgment of the naval architect and marine engineer to reconcile.

In commercial practice, the compromises generally result in one or two folding or retractable, articulated fins per side. The retractable feature permits locating the fins in the ship at positions of maximum beam, where there is favorable leverage. The articulated (doubly all-movable) fins require about 30% less planform area than nonarticulated fins and are, therefore, easier to retract.

The U.S. Navy has used nonretractable units on several classes of frigates, some with articulated and some with nonarticulated fins. The midship section coefficient for these ships is relatively low, which means that the hull form sections are much more rounded than those of cargo ship hulls, and also their sonar domes extend several feet below the keel. Each fin axis is thus sloped well downward going outboard, and the fins do not protrude beyond the maximum beam or below the sonar dome navigational draft. For this type of installation, which is illustrated by Fig. 43, the tilting gear, fin, and hull insert plate can be readily furnished as one unit with a factory-sealed hydraulic system. The fin unit shown is installed with the fin sloped 55 deg downward going outboard, and it has a span of 8 ft with a chord of 4 ft (leading chord of 3 ft and trailing flap of 1 ft).

d. Fin tilting gear. The fin shafts of virtually all successful installations are tilted by hydraulic power, which is well suited to the requirements for rapid reversal. The tilting gear is generally similar to an electrohydraulic steering gear, such as described earlier in Section 2. Oscillating cylinder, Rapson-slide, and vane-type rotary-actuator installations are used. They are actuated by variable-delivery pumps with their flow direction ordered by the automatic control signal. Some of the significant differences from steering gear practice are (a) There is usually only one power unit per fin shaft, since stabilization is not as critical a function as steering, and even if one unit fails, there is another on the other side of the ship; (b) Whereas steering gears are not required to perform high-speed, hard-over rudder maneuvers for extended periods, the tilting gear may be heavily loaded for many hours of continuous operation, hence requiring appropriate motor ratings and system coolers; (c) Where retractable fins are used, the stowing and rigging mechanisms, interlocks, and indicators have to be provided; (d) Fin angular rates are significantly greater than rudder rates; the minimum time for "hard-over to hard-over" fin angles is typically specified as no more than one-sixth the natural ship roll period [33]; and (e) The rapid reversal of fin angles requires added torque to overcome the mass inertia of the mechanism.

The fins themselves are normally fitted with high-strength steel stocks or shafts to minimize the bearing diameter. Bearings should be of the antifriction type, whenever possible, to reduce power requirements. Great care must be taken, however, to protect antifriction bearings from seawater exposure as they can easily become corroded. Fin shafts should be sleeved with a suitable corrosion-resistant material in way of sliding-surface, sleeve, or stave bearings.

As with any hydraulic system, hydraulic fluid cleanliness is extremely important and properly sized, easily maintained or replaced filters and strainers should be used in fin-stabilizer actuator hydraulic systems. Ram and piston surfaces must be smooth to avoid damage to the packing, and should be well protected from damage and dirt. The hydraulic pumps must be adequately protected from cavitation and the hydraulic system should be fitted with suitable pressure, vacuum, and temperature gages and fluid sampling connections to facilitate proper preventive maintenance and troubleshooting.

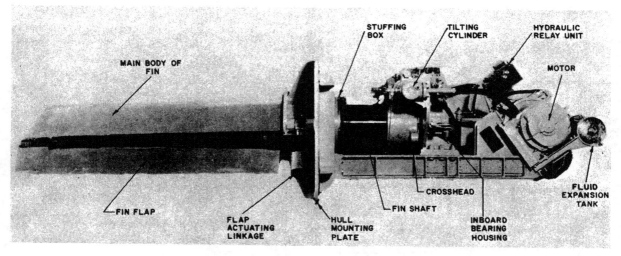

Fig. 43 Nonretractable articulated fin unit, port side, looking forward

Because fin-actuating systems are often located in cramped, hard-to-get-to spaces, high priority should be given to designing for high reliability, ease of changing filters, and minimum maintenance requirements.

e. Automatic controls for fin stabilizers. Human control of fin angles is not feasible. This is because the roll response of a ship in a seaway is at such a high frequency that the fin angles must be varied continuously and rapidly. Automatic controls have proved to be reliable and efficient. Essentially, the control system senses the instantaneous ship's roll motion, uses a derived equation of roll to compute the optimum fin angle, and orders fin angles at the proper phase relationship to achieve fin lift and corrective righting moments, which are subtracted from the roll moment generated by the waves, thereby reducing the roll motion of the ship [32].

The controller requires inputs representing the roll angle, roll velocity, and roll acceleration. Two of these values are normally measured and the third derived. The three values are added together to determine the required fin action. The fin servo orders fin angle, and the actual fin angle is measured to get the closed-loop feedback.

In typical shipboard systems, the roll angle is measured by a gyro or athwartship accelerometer device, the roll velocity is measured by a rate gyro or by differentiating the roll angle, and the roll acceleration is measured by angular accelerometers. Fin stabilizer control systems are microprocessor based and use an electrohydraulic servo valve and a fin angle transducer. This allows the use of sophisticated control systems with simple, reliable control components.

Good design practice dictates locating the control system in an accessible, dry environment away from other equipment that could cause heat, corrosion, or electrical interference problems.

In most cases, the automatic controls are designed simply to reduce the roll angle. Several variations, however, may be considered, such as stabilizing to the apparent vertical (for improved passenger comfort in sway-heel coupling) or accepting only limited roll angle stabilization

in order to reduce high-frequency roll angular acceleration.

Most automatic fin controls are not designed to correct steady heel from ship's loading or wind, but rather to stabilize around the mean heeled angle. This is done to conserve the limited fin stabilizing moments for correcting the oscillatory roll variations.

Another feature in many designs is the automatic angle or fin lift limiter. This is a consequence of having fins that develop full rated lift several knots below full-power ship speed. At the higher ship speeds, the lift and torque, which are proportional to the ship's speed squared, might be excessive and not really needed. Lift is limited to safe values by a fin angle limiter, which is either in the fin lift control circuit, if such is installed, or in the fin angle servo circuit, using the ship speed as input.

6.3 Rudder Roll Stabilization. The concept of using the steering gear and rudder for the purpose of stabilizing a vessel against rolling was introduced by Cowley and Lambert [36], and the first sea trials to prove the concept were performed in the United Kingdom with simple control systems [37,38,39]. Since then, however, the development of rudder roll stabilization systems has progressed in many countries, including the United States [40,41,42], Sweden, [43,44,45], Holland, [46,47], and Denmark [48]. Ships ranging in size from small patrol boats and cutters to destroyers and large ferries have subsequently been equipped with rudder roll stabilization systems. The technique can be considered well proven on vessels that have rudders with the capability of imposing a sufficiently large roll moment on the hull.

In general terms, a rudder roll stabilization system is an adaptive control system that uses the existing steering gear and rudder. Thus, it is simple to install both on new ships and as a retrofit on existing vessels, and the space requirement is negligible. The increase in weight and the influence on ship resistance are also negligible.

A block diagram of a typical system with an integral autopilot is shown in Fig. 44. The course and speed signals are taken, respectively, from the ship's course gyro and

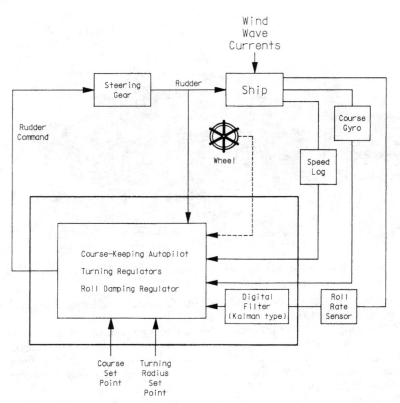

Fig. 44   Rudder roll stabilization system block diagram

speed log. Both the rudder position and the position of the helmsman's wheel are also obtained so that the roll damping function can be provided while the ship is being steered manually. Many of the filters and controllers are normally of the adaptive type in order to accommodate varying sea and wind conditions.

The operating principle of rudder roll stabilization is based on the concept of opposing the roll moment created by the waves with an induced roll moment developed by the movement of the rudder, thereby damping the ship's roll motions. Figure 45 shows the response of the ship during a turning circle. During phase 1, the rudder is displaced and the ship starts to roll in the direction of rudder displacement, that is, into the turn; however, during phase 1 the heading of the ship changes very little. During phase 2 of the turning circle, the heading of the ship begins to change significantly and the ship rolls to the opposite side, that is, out of the turn. The operating principle of the rudder roll stabilization system is to use the phase 1 sequence repeatedly as indicated by Fig. 46. By using the rudder to impose rolling moments on the ship that oppose those created by the waves, the ship's roll motion can be effectively decreased, and the rudder forces used to effect this stabilization are of such short duration that the ship's heading is not substantially affected.

Many factors influence the roll damping efficiency that can be achieved by rudder roll stabilization on a specific

ship. The more important factors are the ship speed and rudder rate, but the metacentric height, rudder type, rudder area, and rudder position are also influencing factors. Sea trials of a mine layer (length 105 m, displacement 3300 tonnes, and maximum rudder rate 8 deg/s) are illustrated in Fig. 47 [44]. Sea trials of another application indicated reductions in rms roll motion of between 36% and 61% at medium speed and between 21% and 38% at high speed [48]. Up to 40% reductions in the rms roll angle were frequently observed on a destroyer at various ships speeds in beam, quartering, and following seas and in various sea states up to sea state 5 [42].

When using a rudder roll stabilization system, the load on the steering gear and rudder will increase, especially for systems that require high rudder rates (in the order 10 to 15 deg/s). Thus, it is important to consider the more rigorous duty cycles. However, steering gears are conventionally of rugged design, and there have been no significant indications of excessive wear or early failures due to rudder roll stabilization being back-fitted on existing steering gears that were not specifically designed for that purpose.

When roll stabilization is used, the steering gear and rudder perform two functions, course-keeping and roll damping, and it is important to consider an integrated control system including both autopilot and rudder roll stabilization. In order to optimize both functions, some

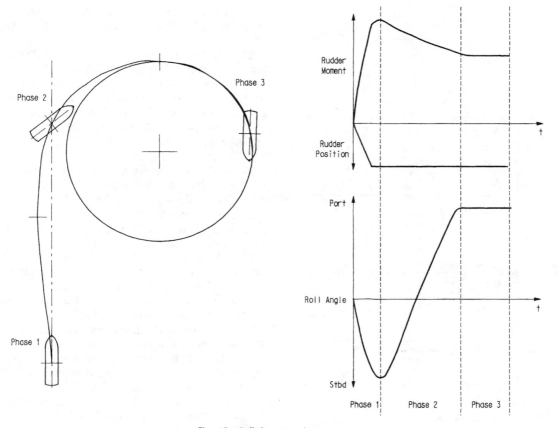

Fig. 45   Roll dynamics during a turn

rudder roll stabilization systems have an integrated autopilot function [44–48].

References

1   D. L. Kitch, "Hydraulic System Cleanliness—Design, Manufacture, Operation and Maintenance Aspects," *Trans.* Institute of Marine Engineers, Vol. 86, 1974.

2   "Final Report Piping System Flushing Study Phase I, Project Number SP-1-83-10," Todd Shipyards Corp., for SNAME Ship Production Committee, March 1986.

3   P. L. Hay, "Shipboard Steering Systems: Yesterday, Today & Tomorrow," SNAME, First Chesapeake Marine Engineering Symposium, Jan. 19, 1984.

4   W. S. Paulin and D. J. Fowler, "Steering Gear" in *Marine Engineering Practice*, Institute of Marine Engineers, Vol. 1, Part 9, The Chameleon Press Ltd., London, 1975.

5   J. Cowley, "Steering Gear: New Concepts and Requirements," *Trans.* Institute of Marine Engineers, 1982.

6   R. L. Harrington, "Rudder Torque Prediction," *Trans.* SNAME, Vol. 89, 1981.

7   *Principles of Naval Architecture*, E. V. Lewis, Ed., SNAME, 1989.

8   *Rules For The Design and Classification of Steel Merchant Ships*, American Bureau of Shipping, New York.

9   U.S. Coast Guard Regulations, 46 Code of Federal Regulations, Subchapters F and J.

10   "Steering Gears, Electro-hydraulic, Marine," Military Specification, MIL-S-17803.

11   International Convention for The Safety of Life at Sea (SOLAS), International Maritime Organization, London, 1986.

12   "Guide to the Design and Testing of Anchor Windlasses for Merchant Ships," SNAME T&R Bulletin 3-15, May 1964.

13   "Standard Specification for Wildcats, Ship Anchor Chain," ASTM Standard F765, American Society for Testing and Materials, Philadelphia, 1982.

14   *Mooring Equipment Guidelines*, Oil Companies International Marine Forum, Witherby & Co., Ltd., London, 1992.

15   Eivind Bratteland, *Advances in Berthing and Mooring of Ships and Offshore Structures*, NATO ASI Series (ISBN-40-247-3731-1), Khuwer Academic Publishers, Dordrecht, The Netherlands, 1988.

16   William H. Hunley and Norman W. Lemley, "Ship Maneuvering, Navigation, and Motion Control" in *Ship Design and Construction*, Robert Taggart, Ed., SNAME, 1980.

17   "Windlass, Anchor, Electro-hydraulic, Vertical, Navy Type, Naval Shipboard," Military Specification MIL-W-19623.

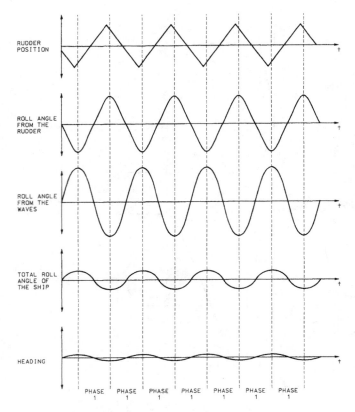

Fig. 46   Rudder roll stabilization principles

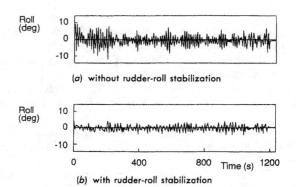

(a) without rudder-roll stabilization

(b) with rudder-roll stabilization

Fig. 47   Sea trials with a minelayer at 16 knots and in stern quartering seas (Beaufort 4); reduction in rms roll was 45% and in maximum roll angle was 63%

18   "Shipbuilding-Mooring Winches," ISO Standard 3730, International Organization for Standardization.

19   "Standard Specification for Warping Heads, Rope Handling (Gypsy Head, Capstan Head)," ASTM Standard F1106, American Society for Testing and Materials, Philadelphia, 1987.

20   "Shipbuilding Cargo Winches," ISO Standard 3078, International Organization for Standardization.

21   D. H. Beattie and W. M. Somerville, "Ship's Gear: A Review of Deck Machinery," Institute of Marine Engineers, 1978.

22   *Cargo Access Equipment for Merchant Ships*, MacGregor Publications Ltd., E&F N Spon Ltd., London, 1978.

23   "Guide for the Construction of Shipboard Elevators," American Bureau of Shipping, New York.

24   "Safety Code for Elevators and Escalators," ASME/ANSI A17.1, American Society of Mechanical Engineers/American National Standards Institute.

25   "Recommended Practice for Electric Installations on Shipboard," IEEE Standard 45, Institute of Electrical and Electronics Engineers.

26   Ralph A. Norrby and Donald E. Ridley, "Notes on Thrusters for Ship Maneuvering and Dynamic Positioning," *Trans.* SNAME, Vol. 88, 1980.

27   Jan Tornblad, "Marine Propellers and Propulsion of Ships," Marine Laboratory KaMeWa AB, Kristinehamn, Sweden, 1987.

28   "Tunnel Thruster," ASTM Standard F841, American Society for Testing and Materials, Philadelphia, 1984.

29   H. Heuser, "Bow Thrusters for Large River and Inland Vessels," 10th Pan-American Congress on Naval Engineering, Rio de Janeiro, 1987.

30   "Azimuthing Thrusters for High Manoeuvrability," *Marine Engineers Review*, April 1986.

31   Hans F. Mueller, "Recent Developments in the Design and Application of the Vertical Axis Propeller," *Trans.* SNAME, Vol. 63, 1955.

32   A. R. J. M. Lloyd, *Seakeeping: Ship Behaviour in Rough Weather*, Ellis Horwood Ltd., Chichester, U.K., 1989.

33   L. W. Nelson and D. McCallum, "Fins of the Future-FFG 7," *ASNE Journal*, American Society of Naval Engineers, Oct. 1978.

34   A. R. J. M. Lloyd, "Roll Stabilizer Fins: A Design Procedure," *Trans.*, Royal Institution of Naval Architects, Vol. 117, 1975.

35   G. R. G. Lewison, "Optimum Design of Passive Roll Stabiliser Tanks," *Trans.*, Royal Institution of Naval Architects, Vol. 118, 1976.

36   W. E. Cowley and T. H. Lambert, "The Use of the Rudder as a Roll Stabilizer" in *Proceedings*, 3rd Ship Control Systems Symposium, Bath, U.K., Vol 2, Paper No. VII, C-1, 1972.

37   J. B. Carley, "Feasibility Study of Steering and Stabilizing by Rudder" in *Proceedings*, 4th Ship Control Systems Symposium, The Hague, The Netherlands, Vol. 2, 1975.

38   A. R. J. M. Lloyd, "Roll Stabilization by Rudder" in *Proceedings*, 4th Ship Control Systems Symposium, The Hague, The Netherlands, Vol. 2, 1975.

39   W. E. Cowley and T. H. Lambert, "Sea Trials on a Roll Stabilizer Using the Ship's Rudder" in *Proceedings*, 4th Ship Control Systems Symposium, The Hague, The Netherlands, Vol. 2, 1975.

40   E. Baitis, "The Development and Evaluation of a Rudder Roll Stabilization System for the WHEC *Hamilton* Class," David W. Taylor Naval Ship Research and Development Center Report DTNSRDC/SPD-0930-02, Bethesda, Md., March 1980.

41   E. Baitis, D. A. Woolaver, and T. A. Beck, "Rudder Roll Stabilization for Coast Guard Cutters and Frigates," *ASNE Journal*, American Society of Naval Engineers, May 1983.

42   E. Baitis and L. Schmidt, "Ship Roll Stabilization in the U.S. Navy", *ASNE Journal*, American Society of Naval Engineers, May 1989.

43   C. G. Kallstrom, "Control of Yaw and Roll by a Rudder/Fin Stabilization System" in *Proceedings*, 6th Ship Control Systems Symposium, Ottawa, Canada, Vol. 2, 1981.

44   C. G. Kallstrom, P. Wessel, and S. Sjolander, "Roll Reduction by Rudder Control," SNAME Spring Meeting/STAR Symposium, 1988.

45   C. G. Kallstrom and W. L. Schultz, "An Integrated Rudder Control System for Roll Damping and Course Maintenance" in *Proceedings*, Ninth Ship Control Systems Symposium, Bethesda, Md., 1990.

46   J. Van Amerongen, P. G. M. van der Klugt, and J. B. M. Pieffers, "Rudder Roll Stabilization—Controller Design and Experimental Results" in *Proceedings*, 8th Ship Control Systems Symposium, The Hague, The Netherlands, Vol. 1, 1987.

47   J. Van Amerongen, H. R. Van Nauta Lemke, and P. G. M. van der Klugt, "Adaptive Control Aspects of a Rudder Roll Stabilization System" in *Proceedings*, 10th IFAC World Congress, Munich, Germany, 1987.

48   M. Blanke, P. Haals, and K. Kudsk Andreasen, "Rudder Roll Damping Experience in Denmark," IFAC Workshop on Expert Systems and Signal Processing in Marine Automation, Lyngby, Denmark, 1989.

CHAPTER XIX

Walter E. Schmid | # Electrical Systems

Section 1
Introduction

1.1 Scope. The scope and extent of shipboard electrical systems are strictly a function of the size and mission of the ship. Shipboard electrical systems include an electric power plant, lighting, interior communications and controls, exterior communications, navigation systems, and many other safety and mission-related electric/electronic support systems. The important aspects of electrical systems, from a marine engineer's perspective, are reviewed in the following sections. Naval combat and combat support systems are, however, too diversified to be comprehensively covered in this text.

All ships have an electric power plant similar to a land-based electric utility with the capacity ranging from a few hundred kilowatts at the lower end of the scale up to 40 MW for a nuclear aircraft carrier. Electric power is required for propulsion, propulsion system auxiliaries, deck machinery, illumination, heating, ventilation, air conditioning, stores and cargo refrigeration, galley, fresh water and sanitary systems, and safety and casualty control such as fire and bilge systems, fire detection and alarm systems, and remotely operated watertight and fire-screen doors. Power must also be supplied for interior communication systems, controls, radio communications, radar, and other electronic aids to navigation and shipboard operation.

For military ships, a significant amount of electric power is required by the mission-related payloads, such as the combat system (weapons, command, communications, control, electronic warfare and countermeasures, etc.), and combat support and supply systems.

For passenger vessels, the electric power requirements extend to hotel and recreation loads such as theaters, restaurants, and swimming pools.

1.2 Requirements and Constraints. A majority of the requirements and constraints for shipboard electrical systems are driven by the mission of the ship as determined and defined by the ship's owner/operator. For U.S. merchant ships, marine electrical installations must comply with rules and regulations based on laws promulgated by the United States Government. The following is a listing of such rules and regulations.

- United States Code of Federal Regulations (CFR), Title 46 Shipping, Subchapter J Electrical Engineering.
- United States Coast Guard Instruction "COMDTINST M16672.2"—Navigation Rules, International-Inland.

- United States Public Health Service, Publication 393 "Handbook on Sanitation of Vessel Construction—Rat-proofing of ships."
- United States Code of Federal Regulations, Title 47—Telecommunications, Subchapter D—Safety and Special Radio Services, Part 80—Stations in the Maritime Services.

For foreign-flag vessels, marine electrical installations must comply with the rules and regulations based on laws promulgated by their flag administration. It should be noted that most of the requirements of the International Conference of Safety of Life at Sea (SOLAS) are incorporated in Title 46 CFR listed above. Furthermore, the International Regulations for Preventing Collisions at Sea, 1972 (72 COLREGS) developed through the International Maritime Organization (IMO) are covered by USCG COMDTINST M16672.2.

In addition to the rules and regulations imposed by law, compliance with other rules, regulations, standards, or practices is usually imposed by the specifications for the design and construction of particular ships, such as:

- Rules and regulations of vessel classification societies, such as:
  —American Bureau of Shipping, Rules for Building and Classing Steel Vessels
  —Lloyd's Register of Shipping, Rules and Regulations for the Classification of Ships
  —Det norske Veritas, Rules for Classification, Steel Vessels

The ship's owner selects the cognizant classification society that promulgates regulations for the design and construction of ships and that becomes the basis for marine insurance coverage.

- United States Coast Guard Publication "COMDTPUB P16700.4—NVIC 2-89," Guide for Electrical Installations on Merchant Vessels and Mobile Offshore Drilling Units.
- Institute of Electrical and Electronics Engineers (IEEE) Standard No. 45—Recommended Practice for Electric Installations on Shipboard.
- Suez Canal Authority Rules of Navigation.
- Marine-Type Electric Lighting Fixtures UL595 published by Underwriters Laboratories, Inc.
- Illumination Engineering Society (IES) Handbooks and Recommended Practices.

- For naval ships, there is a hierarchy of standards, which are promulgated by the U.S. Navy, such as:
  —Specific Ship Specifications
  —Military Specifications
  —Military Standards

1.3 Design Documentation. The design documentation prepared for electrical systems comprises specifications, drawings, analyses, calculations, listings, and operating/maintenance manuals. Examples are as follows:

- Specifications.
  —Ship (Contract) Specification
  —Purchase Specifications
- Drawings prepared by the shipbuilder.
  —One-line diagram for the power system
  —Isometric cabling diagrams for power system feeders and mains
  —Isometric cabling diagrams for lighting system feeders and mains
  —Elementary and isometric cabling diagrams for control, interior communication, and electronic systems
  —Deck plans (system diagrams) for power, lighting, interior communication, and electronic systems
  —Wireway routing and installation drawings
  —Equipment arrangement drawings
  —Installation standards
- Listings prepared by the shipbuilder.
  —Power system list of feeders and mains
  —Lighting system list of feeders and mains
  —List of motors and controllers
  —List of connections
  —List of nameplates
- Analyses and calculations prepared by the shipbuilder.
  —Electric load analysis
  —Fault-current analysis and voltage-dip calculations
  —Application and coordination of protective devices.
  —Voltage-drop calculations
- Documentation supplied by equipment vendors.
  —Outline, assembly, and foundation drawings
  —Schematic and wiring diagrams
  —Operating and maintenance manuals.

Section 2
Power System

2.1 Overview. Electric power is vital to all shipboard operations and to the safety and comfort of the passengers and crew. For this reason, shipboard electric plants contain equipment necessary to maintain continuity of service. Since a vessel at sea is isolated from external sources of electric energy, standby ship-service generating capacity, usually equal to the rating of the ship-service generators, may be provided. In addition, one or more sources of emergency power, which are designed to automatically assume load upon the loss of ship-service power, are provided to supply those loads that are necessary for the safety of life; the emergency source of power also has sufficient capacity to supply those loads vital to restoring the propulsion system and ship-service generators to service. Quick-starting diesel generators, storage batteries, or gas turbine generator sets are usually provided for emergency power. Emergency storage batteries are installed on passenger vessels to provide temporary emergency power to certain vital loads until the emergency generator is up to speed and is brought on line to assume the emergency load. D-c to a-c conversion equipment may be used to operate a-c loads which are vital emergency loads.

Extremely critical systems that cannot sustain even a momentary interruption of power without serious operational or safety implications (such as computers) are fed from uninterruptable power supplies (also known as no-break power supplies). Uninterruptible power supplies are discussed in Section 3. They are used extensively on naval ships for obvious reasons and are becoming increasingly popular on merchant ships due to the application of automation.

For naval combatants and combat support ships, the electric plant is designed for a very high level of continuity and availability of electric power to all vital subsystems and equipment. Combat survivability is another major design consideration, which requires that electric power be supplied to the undamaged portion of the ship after sustaining major battle damage.

On ships with electric propulsion, a combined electric power plant is often used to power both the propulsion motors and the ship's electric auxiliaries and services.

All ships are provided with a shore power supply feature, to receive power from shore when the ship is not operational. One or more suitably located shore power connection boxes are connected to one or more main switchboards through separate feeders and shore power circuit breakers on the switchboards.

The basic configuration of the electric power plant (number of generators, switchboards, load centers, group control centers, etc.) and the selection of the system parameters are a function of safety, economics, continuity of service, survivability, and other requirements dictated by the size and mission of the ship. On U.S. merchant ships that are self-propelled, the U.S. Coast Guard regulations require at least two electric generating sets. The electric load analysis is the means used by an experienced systems engineer to establish the overall power system configuration and features early on during the preliminary definition of a new ship design.

2.2 Load Analysis. The identification, tabulation, and summary of all electrical loads on a ship are compiled by an electric load analysis. This analysis is started during concept/preliminary design at a high level. It is subsequently developed and maintained to an ever-increasing level of detail through detail design of the ship. The load analysis is structured to determine the aggregate power requirements of all the electric power-consuming equipment and devices under the various ship operating conditions, such as at-anchor, in-port, maneuvering, at-sea, and battle, for the purpose of establishing the maximum and minimum power requirements. The maximum power summary is required to establish the generating plant capacity and power plant configuration. The minimum power requirement is of special importance when diesel engine prime movers are utilized. Excessive maintenance may be required when diesel engines are operated at light loads for long periods of time.

The load analysis is also useful for the development of the basic power distribution system configuration. Table 1 is a worksheet of a typical electric load analysis. The average demand loads for each power-consuming piece of equipment are calculated by applying anticipated service factors for each operating condition to the maximum kW rating, i.e., the maximum kW absorbed by the piece of equipment.

Service factors consist of a load component and a cycle component. The load component is the ratio of the actual load when energized under a particular operating condition (e.g., motor running at a lower speed, or pump operating below its rated delivery) to the maximum input rating of the load. The cycle component reflects the percentage of time a piece of equipment operates during the total time frame of an operating condition (e.g., cycling compressor motor). Often a single service factor is used to represent both of these components. Equipment that is operated occasionally only under abnormal conditions (e.g., capstans and boat winches) is assumed to have a cycle/service factor of zero. In most cases, these service factors are selected from an empirical database. However, in some cases the factors for major loads must be determined analytically based on specific operating scenarios.

Reference 1 outlines the content and methodology to be used for the electrical load analysis for naval surface ships. Reference 1 defines the operating conditions for analysis and tabulates typical load factors for a wide array of loads. However, as the design progresses, these typical load factors are replaced with more-specific load factors that reflect the anticipated load. Appendix 2 of reference 2 includes typical operating load factors recommended by the USCG for merchant ships.

A more sophisticated method of determining the electrical load involves the development of a timeline analysis of the cyclic power demands occurring throughout a particular operating profile. The first step in this approach is to determine the average load of those services that are expected to operate continuously or in a randomly intermittent manner. These loads represent the average baseload.

A timeline analysis is then performed to determine the power demand versus time profile for the remaining services, which have large but predictable demand variations. The results of this analysis are then superimposed on the baseload in a composite graph to show the load distribution over a time frame determined by the ship's operating profile.

The maximum possible peak loads and the probability of their occurrence over a period of time could be determined through operational simulation of the ship system and all its subsystems by applying the Monte Carlo probability technique. To date, however, such elaborate analyses have not been used since electric power plants are generally designed with sufficient margin for future growth, such that infrequent random peak loads go unnoticed. Occasionally the load analysis has been used as a tool for an electric load control program during the design development of a ship. An electric load control program uses techniques such as electric load budgeting, margin control, and trend analysis similar to a weight control program.

2.3 System Voltage and Frequency. Voltage and frequency are the primary electrical system design parameters, and their selection is influenced by many factors. Limiting the practical options to the voltage and frequency ratings of components and equipment that are available for commercial and naval shipboard applications, the alternatives are either direct current or alternating current at either 50, 60, or 400 Hz; and the alternative voltages are 120/240, 480/450, 2400, or 4160 volts.

a. Alternating-current plants. Alternating current has become the standard for most marine and land-based electric power plants. A-c provides many significant advantages over d-c, e.g., less cost, weight, and space requirements, less maintenance, better availability of equipment in the marketplace, and increased reliability. Many of these advantages are realized through the use of squirrel-cage induction motors as opposed to d-c motors, which have commutators and associated brushes that are subject to wear.

A frequency of 60 Hz has become the standard in the Western Hemisphere for a-c power systems; however, 50-Hz is predominantly used in Europe and other parts of the world. Shipboard electric plants throughout the world generally use one of these standard frequencies. However, for some naval applications 400-Hz electric power has made steady inroads. In communications, fire control, command and control, ordnance, and aircraft services 400-Hz power has distinct advantages over 60-Hz power in the low- and medium-power ranges because of weight and volume savings. This is the reason for the widespread adoption of a 400-Hz frequency for avionics, ordnance, and other aerospace-related equipment. The production volume for weapons and aerospace systems has made certain 400-Hz hardware competitive when compared with equivalent 60-Hz components.

Naval ships use a 60-Hz primary electrical system with 60/400-Hz power conversion units for the supply of power to a secondary 400-Hz system. The steadily increasing demand for 400-Hz power for combat systems has often

Table 1 Sheet of a typical electric load analysis

Duty/Service	Qty	Rated hp	Bhp	Motor Eff. (%)	Unit kW	Condition 1			Condition 2			Condition 3			Condition 4		
						No. in Use	Load Factor	kW	No. in Use	Load Factor	kW	No. in Use	Load Factor	kW	No. in Use	Load Factor	kW
VENTILATION & HEATING																	
mchry space sply Nos. 1 & 2	2	60.00		91	49.19	2	1.00	98.37	2	1.00	98.37	2	1.00	98.37	2	1.00	98.37
mchry space sply No. 3	1	75.00		91	61.48	1	1.00	61.48	1	1.00	61.48	1	1.00	61.48	1	1.00	61.48
mchry space sply No. 4 (rev)	1	75.00		91	61.48	1	1.00	61.48	1	1.00	61.48	1	1.00	61.48	1	1.00	61.48
mchry space exh Nos. 1 & 2	2	20.00		89	16.76	2	1.00	33.53	2	1.00	33.53	2	1.00	33.53	2	1.00	33.53
mchry space exh No. 3 (2 SP)	1	20.00		89	16.76	1	1.00	16.76	1	1.00	16.76	1	1.00	16.76	1	1.00	16.76
fuel process area exh	1	20.00		89	16.76	1	1.00	16.76	1	1.00	16.76	1	1.00	16.76	1	1.00	16.76
strg gear rm sply	1	10.00		85	8.78	1	1.00	8.78	1	1.00	8.78	1	1.00	8.78	1	1.00	8.78
galley sply	1	7.50		83	6.74	1	1.00	6.74	1	1.00	6.74	1	1.00	6.74	1	1.00	6.74
galley exh	1	7.50		83	6.74	1	1.00	6.74	1	1.00	6.74	1	1.00	6.74	1	1.00	6.74
bow thrust rm sply	1	1.75		89	1.47	1	1.00	1.47				1	1.00	1.47	1	1.00	1.47
bow thrust rm exh	1	1.75		89	1.47	1	1.00	1.47				1	1.00	1.47	1	1.00	1.47
CO_2 rm exh	1	0.33		65	0.38	1	1.00	0.38	1	1.00	0.38	1	1.00	0.38	1	1.00	0.38
paint lkr exh	1	0.33		65	0.38	1	1.00	0.38	1	1.00	0.38	1	1.00	0.38	1	1.00	0.38
med treatment rm exh	1	0.33		65	0.38	1	1.00	0.38	1	1.00	0.38						
bosn store/maint shop exh	1	1.00		78	0.96	1	1.00	0.96	1	1.00	0.96	1	1.00	0.96	1	1.00	0.96
hotel exh	1	1.75		79	1.65	1	1.00	1.65	1	1.00	1.65	1	1.00	1.65	1	1.00	1.65
hotel T&S exh	2	5.00		83	4.49	1	1.00	4.49	1	1.00	4.49	1	1.00	4.49	1	1.00	4.49
box girder exh	1	25.00		89	20.96	2	1.00	41.91	2	1.00	41.91	2	1.00	41.91	2	1.00	41.91
tempered air	1	1.50		78	1.43	1	1.00	1.43	1	1.00	1.43	1	1.00	1.43	1	1.00	1.43
battery lkr exh	1	0.09		60	0.11	1	1.00	0.11	1	1.00	0.11	1	1.00	0.11	1	1.00	0.11
elev mchry rm exh	1	0.17		62	0.20	1	1.00	0.20	1	1.00	0.20	1	1.00	0.20	1	1.00	0.20
garbage rm exh	1	0.20		62	0.24	1	1.00	0.24	1	1.00	0.24	1	1.00	0.24	1	1.00	0.24
cargo hold No. 1 sply	1	1.00	1.00	78	0.96	1	1.00	0.96	1	1.00	0.96	1	1.00	0.96	1	1.00	0.96
cargo hold No. 2 sply	1	3.00	1.95	82	1.77	1	1.00	1.77	1	1.00	1.77	1	1.00	1.77	1	1.00	1.77
cargo hold No. 3 outbd exh	4	15.00	10.30	86	8.93	1	1.00	8.93	4	1.00	35.74	1	1.00	8.93	4	1.00	35.74
cargo hold No. 3 sply	4	25.00	22.70	89	19.03	1	1.00	19.03	4	1.00	76.11	1	1.00	19.03	4	1.00	76.11
cargo hold No. 3 inbd exh	3	15.00	14.20	86	12.32	1	1.00	12.32	3	1.00	36.95	1	1.00	12.32	3	1.00	36.95

raised the question of the possible advantages in using a 400-Hz primary system for naval ship applications. Many studies comparing 400-Hz and 60-Hz systems have been performed. In general, the conclusions have been that for any ship with functional power requirements in excess of a few hundred kW, a 400-Hz primary system would be less cost effective, less reliable, and offer little savings in total weight and volume. However, the studies showed that 400-Hz power is cost effective for lighting, electronics, and small motors.

The principal disadvantage with 400-Hz power is the high rotational speed of the power generators and electric motors. Both the cost and weight of 400-Hz diesel generator sets are higher since they require a gearbox to provide the required generator speed, whereas a 60-Hz generator can be coupled directly to the diesel engine. A 400-Hz turbogenerator set would offer a weight reduction in that no gearing would be required, but any cost savings resulting from the elimination of gears may be offset by the higher cost of the high-speed 400-Hz generator.

A 400-Hz motor, while offering a considerable weight savings, has a higher acquisition cost, resulting from the use of higher-strength materials and the requirement for closer manufacturing tolerances. Furthermore, gearing would be required in many cases to reduce the lower possible motor speed to that required by the driven equipment, thereby adding to the cost and reducing or eliminating any potential weight savings.

A 400-Hz lighting system has advantages in both weight and cost because of the higher efficiency of fluorescent lamps at higher frequencies. Conventional electric heating or any other purely resistive loads are insensitive to frequency. Cabling for 400-Hz power is heavier and more costly overall, since the cable reactance is approximately proportional to frequency. The effect on weight and cost is negligible for smaller cables with a lower reactance-to-resistance (X/R) ratio, but becomes significant as the power ratings, cable sizes, and X/R ratio increase. While 400-Hz transformers weigh less, they tend to cost more.

A 120-volt, 3-phase, 3-wire a-c power plant that supplies the ship with 115-volt power is economical for small vessels that have few motor-driven auxiliaries. However, 120 volts is the recognized standard voltage for all lighting and many of the communications and navigation systems as well as most of the plug-in appliances and tools, regardless of the ship primary system voltage. But 230- or 240-volt, 3-phase, 3-wire systems or 208/120-volt, 3-phase, 4-wire systems may also be used on smaller ships and for refrigeration container distribution systems on container-ships, to provide 220-, 230-, or 208/115-volt power, respectively, at the user terminals. However, in the Western Hemisphere 115- or 440/460-volt systems are normally selected because equipment for other voltages is not readily available. Hence a 450- or 480-volt, 3-phase, 3-wire primary system for 440- or 460-volt users, respectively, coupled with a stepped-down 120-volt, 3-phase, 3-wire secondary system for lighting and other low-power applications is the standard for most naval/marine applications.

Formerly, 2500 kW was considered the upper practical limit for 450- or 480-volt, 3-phase shipboard generators because of design limitations on generators and switchgear (circuit breakers). However, switchgear has been developed for 480- or 450-volt, 3-phase generators of up to 4000-kW at 0.8 p.f. But there is also a maximum practical limit of three 2500-kW (or two 4000 kW), 480- or 450-volt, 3-phase generators that can be operated in parallel because of design limitations on switchgear (circuit breaker fault-current interrupting capacity). This limitation became a design constraint as the demand for electric power on ships increased, particularly for large naval combatants and ships with large cargo-handling power requirements. Hence, whenever the maximum functional load on a ship exceeds 7500 kW, it may be more economical to select a primary system voltage of 2400 or 4160 volts, rather than adhering to a 450- or 480-volt system, and operating in a split-plant mode, with the ship's electric power plant operating as two independent and separate power systems, each fed by three or fewer 2500-kW generators.

Wherever a 2400- or 4160-volt, 3-phase (high voltage) system is selected, fewer generators in excess of 2500 kW are used and motors in excess of 100 hp are designed to operate directly off the primary high-voltage system. Three-phase step-down transformers are used to supply the normal ship service loads at the standard lower voltages.

On ships with electric propulsion, integrated high-voltage generating plants are used to supply the propulsion motors and, through step-down transformers, the ship's service power system as well. However, here too, large auxiliary motors are fed directly from the high-voltage switchboard.

b. Direct-current electric plants. Direct-current power plants have been overtaken by a-c technology for the reasons noted above. Older-vintage shipboard d-c power plants are of either the 120-volt, 2-wire or 240/120-volt, 3-wire type. A 120-volt, 2-wire d-c generating plant was suitable for smaller vessels having only a few motor-driven auxiliaries; the generators were usually rated at 75 kW or less. 240/120-volt, 3-wire d-c generating plants were commonly installed on larger ships. They provided for 230-volt, 2-wire distribution to power loads and 230/115-volt, 3-wire distribution to lighting and other smaller loads.

c. Tolerances. The voltage and frequency of electric systems are maintained within prescribed tolerances by speed governors on the generator set prime movers and voltage regulators within the excitation system of the generators as discussed in Section 3. For commercial ships, the tolerances are defined by the regulatory body requirements and are compatible with commercial standards for electric power systems as well as user (power consuming) equipment. The electric power system characteristics and tolerances for naval ships are defined by reference 3, where the 60-Hz primary power system is defined as Type I. A 400-Hz secondary power system is required to meet the closer tolerances of Type II or Type III power.

The objective of the Navy is to develop 60-Hz-user equipment for Type I power exclusively. For 400-Hz systems Type II power is the preferred norm; however, high quality Type III 400-Hz power is provided wherever dictated by highly sophisticated electronic equipment within the combat system.

2.4 System Configuration. The primary electric power plant configuration is defined by the type, number, size, and location of the generator sets and associated switchboards and bus ties. The requirements and constraints, which typically influence the electric power system configuration, are:

- Type of fuel carried on the ship.
- Type of propulsion plant.
- Type of major nonelectric energy-consuming support subsystems (such as auxiliary steam system).
- Maximum functional load (derived from the electric load analysis).
- Power margin for future growth.
- Capability to carry the maximum functional load, including future growth margin with one ship-service generator set inoperable.
- Minimum of two or three ship-service generator sets on merchant ships or naval ships, respectively.
- Minimum of one independent emergency generator set for the ship's most vital loads. [Naval ships with diesel- or gas-turbine-driven ship service generators do not require emergency generator set(s) if at least one generator set is located so that it is separated from the others by a distance sufficient to survive flooding equal to the ship's damage criteria].
- Segregation (functional and physical) of individual generator sets on naval ships to enhance survivability from battle damage.
- Minimum availability of electric power as specified for naval combatants.
- Space and weight constraints.
- Economic constraints on first costs and operating costs.
- Applicable Government requirements, which typically influence the electric power system configuration.

Alternative plant configurations may be considered for a given set of requirements and constraints. Electric plant trade-off analyses may be performed independently or in conjunction with other subsystem trade-offs, such as the propulsion system trade-off, or a total integrated shipboard energy concept trade-off. Steamships generally incorporate the total energy concept through the conversion of fossil fuel into steam in large boilers and the use of the steam as the energy medium for propulsion, electric power generation, heating, and other steam-driven auxiliaries. Hence, on most ships the prime movers for the ship-service generator sets are selected to be of the same generic type as the propulsion engines, or at least to operate on the same fuel. However, on some steamships, one of the ship-service generator sets may be diesel driven to provide significantly more than just vital emergency electric power in case of a complete boiler outage.

On some ships, one of the ship-service generators may be mechanically driven from an auxiliary power takeoff from the propulsion train. Ships with gas turbine propulsion may use either a gas turbine, diesel engine, or steam turbine as a ship-service generator prime mover, with the steam turbine powered by a waste-heat boiler in the propulsion gas turbine exhaust. When diesel engines or gas turbines are used as prime movers, the number of ship-service generator sets must be selected such that the units are not required to operate at light loads for long periods of time under any of the ship's operating conditions, because of the resultant higher maintenance on diesel engines and the lower fuel efficiencies with both of these engine types.

For special applications, multiple-unit generator sets may be used, each consisting of two generators driven by one prime mover. Applications such as these have been used on steamships, using a steam turbine as the prime mover. Conversely, high-speed diesel generator sets consisting of one generator driven by two prime movers, one at each end of the generator, may also be used. Such units are generally lower in weight than comparable lower-speed, single-engine generator sets. In addition, they offer the flexibility of declutching or decoupling one diesel engine at one end for repair or overhaul, while the remaining portion of the generator set continues to operate at one half of its rated output.

In general, the generator sets and associated switchboards are located within the same space and close to each other to minimize the cable runs between the generator and the generator switchboard. In some cases, the ship-service switchboards are located within the integrated enclosed machinery control station but still close to the associated generator sets.

Self-propelled merchant vessels are fitted with a minimum of two ship-service generator sets located in the engine room, plus a smaller diesel-driven emergency generator set or storage batteries located outside the machinery space, above the freeboard deck, and aft of the collision bulkhead. A contiguous boundary between any other machinery space and the emergency generator room is often avoided.

On most merchant ships, all ship-service generators are located in the same compartment (machinery space) and are connected to a single ship-service switchboard. Under normal operating conditions, the emergency power system is fed from the ship-service switchboard via a bus tie circuit. The emergency power system is designed such that upon a total loss of all ship-service power, the emergency generator set automatically starts up, the emergency switchboard is automatically isolated from the ship-service power system, and the emergency generator is connected to the emergency switchboard to supply the vital loads necessary for the safety of the ship and to get the ship-service power system and the propulsion system back on line. Inherently, a loss of ship-service power results in a 10- to 45-second blackout, the time required to start the emergency generator set and bring it on line. On some ships, such as passenger vessels, an automatically

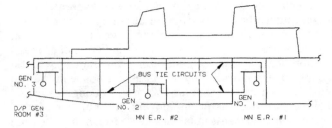

Fig. 1   Typical electric plant configuration for a non-steam-propelled naval surface combatant

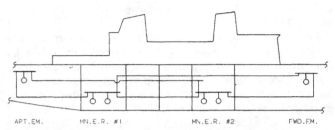

Fig. 2   Typical electric plant configuration for a steam-propelled naval surface combatant

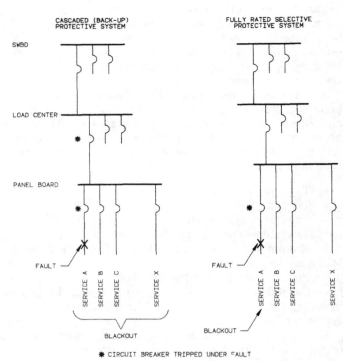

✱ CIRCUIT BREAKER TRIPPED UNDER FAULT

Fig. 3   Alternative approaches to fault-current protection of electric power systems

connected storage battery acts as a temporary power source to prevent the 10- to 45-second blackout.

The power systems for passenger ships are similar to those for other merchant ships except that large passenger vessels, with more than one machinery space, usually have at least one ship-service generator set complete with associated auxiliaries and switchboard in each machinery space. On ships with multiple ship-service switchboards, they are interconnected through a bus tie circuit to allow for parallel, single-plant operation. Very large distributed electric plants with multiple switchboard/generator locations may be set up in a ring-bus configuration for optimum operational flexibility, and maximum continuity and availability of ship-service power under a single casualty mode.

Passenger ships are provided with a more complex two-tier emergency power system, consisting of a temporary and a final emergency power source and associated switching arrangements. Temporary emergency power is supplied from storage batteries combined with a-c to d-c conversion equipment to provide instantaneous emergency power to the most critical vital loads required for the safety of life. The temporary emergency power system is intended to bridge the 10 to 45 seconds required to bring the final emergency power source on line and, hence, to avoid a total blackout of such duration. The final emergency power system functions the same as the emergency power system on a merchant ship. Once the emergency generator is on line, the temporary emergency loads are automatically transferred to the final emergency power source (emergency generator) through automatic bus transfer equipment.

Electric plant configurations are more complex on naval combatants because of mission and combat survivability considerations. Naval combatants require at least three ship-service generator sets, which are sized and configured such that:

• the electric plant can assume the maximum mission (battle) load upon loss of one ship service generator set, and

• upon sustaining the loss of one machinery space through major battle damage, the electric plant can assume the necessary load so that the ship can continue operations with the undamaged portions of the propulsion plant and combat system.

Naval steamships have independent emergency power systems that are similar to merchant ships. Naval ships with gas turbine or diesel propulsion may not have an emergency power system that is totally independent from the ship-service power system, as long as at least one of the three or more diesel or gas turbine driven ship-service/emergency generator sets is located such that it is separated from the others by a distance sufficient to survive flooding equal to the ship's damage criteria.

Figure 1 shows the electric plant configuration of a gas turbine- or diesel-propelled surface combatant; it consists of three gas turbine or diesel-driven ship-service/emergency generator sets that are physically segregated throughout the ship and interconnected through a ring-bus configuration. With the generator sets properly sized, the system as configured in Fig. 1 satisfies all requirements. By comparison, Fig. 2 shows an electric plant configuration of a traditional older steam-propelled surface combatant, which includes four steam turbine generator

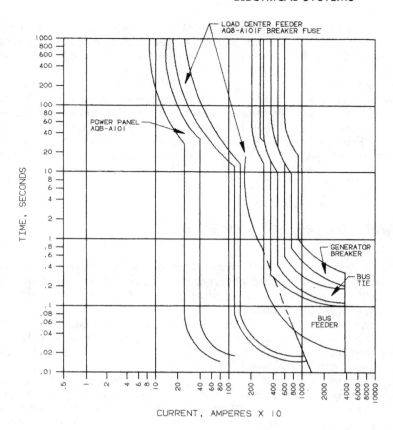

Fig. 4 Circuit-breaker coordination curves

sets, and two emergency diesel generator sets fore and aft. An electric plant using only three steam turbine generator sets would be neither feasible nor compliant for the following reasons:

• A physical separation of the three steam turbine generator sets, similar to the configuration shown by Fig. 1, is not feasible because of the long steam lines that would be required.

• Locating two of three steam turbine generator sets in one main machinery space would violate the vulnerability requirement, since a single casualty could incapacitate 2/3 of the plant capacity, limiting severely the ship's operational capability.

2.5 System Protection. The protection of the electric power system from fault currents and overloads, as required to ensure that a reliable, safe, and continuous service of power is provided, is a major design consideration. The total capacity of generators operating in parallel is limited to about 7500–8000 kW at 450 or 480 volts, due to the fault-current limitations of the switchgear and circuit breakers. Hence, the maximum available fault current on the bus of the main switchboard is one of the factors to be considered in the selection of the primary system voltage and plant configuration for large power plants.

The maximum available fault-current capacity of an electric power system is a function of the generator characteristics, such as the subtransient reactance. The higher the subtransient reactance, the lower the initial fault current. However, a high subtransient reactance and a consequently high transient reactance result in larger voltage dips upon starting of large motors, as discussed in Section 6. As a result, a compromise must be reached in the selection of the generator characteristics to balance the fault currents against the voltage dips. Furthermore, there are practical upper limitations to the subtransient/transient reactance values of generators for a number of reasons.

After the basic framework of the electrical power distribution system has been laid out, a fault-current analysis must be conducted to determine the maximum available fault currents at the various distribution points (switchboards, load centers, panelboards, etc.) of the system. The results of this analysis are then used to select the type of circuit breakers or fuses to be used at each location. The commonly used procedures when making fault-current calculations are described in references 2 and 4.

System protection against fault currents can be accomplished by providing either fully rated selective protection or cascaded (backup) protection. A fully rated selective protection scheme requires each circuit breaker to have sufficient interrupting capacity to clear the maximum available fault current at its point of application before the nearest circuit breaker upstream trips off the line. In a cascaded protection scheme, the maximum available fault current may exceed the interrupting capacity of certain branch circuit breakers so that a larger breaker or fuse further upstream in the distribution system clears the fault in a backup mode.

Fully rated selective protection is preferred since a fault current within any branch of the distribution system will result in the loss of that particular branch only, whereas a fault within any branch of a cascaded system may result in the loss of power to a number of other branches fed by the same backup circuit breaker. Figure 3 illustrates the difference between the two schemes, and Fig. 4 shows the coordination between the tripping characteristics of up to

five circuit breakers in series within a typical electric power system. The application and coordination of protective devices for an electric power system are discussed in considerable detail in reference 5.

The dynamic stability of electric power systems under fault conditions has not been a problem on shipboard applications due to the relatively low impedances of the power cables between individual generators, and due also to the high-speed interruption of most fault currents by molded case circuit breakers or fuses within one-half cycle.

Section 3
Power Sources and Conversion

3.1 Generator Sets. Generator sets are normally completely packaged with all or most of the major components and auxiliaries, such as the prime mover, reduction gear (if required), generator, exciter, control panel, lubricating oil system, and cooling water system, factory assembled on a common bedplate. Detailed discussions concerning the candidate prime movers for generator sets, such as diesel engines, gas turbines, and steam turbines, can be found in Chapters 3, 4, and 6; therefore, discussions here are confined to considerations regarding their applications as prime movers for shipboard generator sets.

a. Steam turbine generators. Steam turbine prime movers have been designed for shipboard generator set applications, from a few hundred to a few thousand horsepower, and are of a mature and proven technology with an excellent service history. They are of the horizontal, multistage, axial-flow, impulse type complete with the following auxiliary subsystems and devices:

• Interconnecting piping and wiring ready for connecting to shipboard piping and cabling.

• Steam supply strainer to protect the turbine. The strainer may be integral with the combined trip and throttle valve.

• Steam governing valve assembly.

• Combined trip and throttle valve that is held open by lubricating oil pressure such that the loss of oil pressure will automatically shut off steam to the turbine; it is also used to admit steam gradually by hand when starting up a set.

• Steam sealing manifold for pressurizing the turbine shaft packing to prevent an entrance of air into the turbine; also piping and valves for drainage of steam leakage through the packing.

• High exhaust back-pressure trip device, which will relieve oil pressure from the throttle trip valve to shut down the turbine.

• Automatic atmospheric relief valve that exhausts to the atmosphere in case a high exhaust back pressure occurs in the turbine casing due to a condenser malfunction.

• Sentinel valve for sounding an alarm before the automatic atmospheric relief valve functions.

• Overspeed governor independent of the constant-speed governor, which upon overspeed of over 15% will relieve oil pressure from the throttle trip valve to shut down the turbine.

• Interlock switch for energizing the generator circuit-breaker tripping device, to disconnect the generator from the switchboard in case the throttle valve trips closed.

• Condensate system. The condenser may be packaged (hung) or separately mounted by the shipbuilder. A motor-driven condensate pump and a condenser circulating water pump are generally provided for each turbine exhaust condenser. These pumps are normally provided by the shipbuilder but may be furnished with the turbine generator. Pipe connections may be made so that, during emergencies, the generator turbine exhaust can flow to the main propulsion turbine condenser. The generator turbine condenser may be arranged to receive exhaust from auxiliaries when the ship is at anchor.

• Self-contained lubricating-oil system. A gear-driven oil pump supplies oil to the turbine, reduction gear, and generator bearings, and also to the constant-speed governor, trip throttle valve, high exhaust back-pressure trip device, and overspeed governor. The pump takes suction from the reservoir and discharges through a magnetic duplex-type strainer and a tube-type or plate-type cooler. Oil coolers are generally provided with zinc anodes or impressed current cathodic protection on the seawater side to minimize corrosion. The cooling-water pressure must be maintained lower than the oil pressure to avoid seawater contamination of the oil in the event of a cooler tube failure. A hand-operated lubricating-oil pump may be provided for use during start-up and maintenance. Generally, the lubricating system also includes an electric motor-driven pump, which is started automatically by a pressure switch upon failure of the turbine-driven pump; this pump may also be used during start-up and shutdown operations. When specified, the lubricating system contains switches for low oil pressure and high oil temperature alarms.

• Rotor-turning gear, either manually or power operated, for use during maintenance.

• Electro-hydraulic load-sensing speed governor.

- Speed adjusting device with local manual adjustment and electric motor or potentiometer for remote synchronization of the generator sets.
- Local control provisions and gageboard. These should include pressure gages for the inlet steam, gland sealing steam, bearing oil, and oil pump and thermometers for oil to and from the cooler. The gageboard and control devices for speed adjustment, throttle trip, hand oil pump, gland seal steam, hand shutdown trip, tachometer, and oil cooler water regulator should be located so that one person can start, operate, and secure the set.
- Automation features. Monitoring and control devices necessary for remote/manual or automatic start-up, operation, and shutdown of the turbine-generator set may also be incorporated in the unit.

The reduction gear is normally of the single-reduction, single-helical type with the pinion and low-speed gear shafts supported by two bearings. The low-speed shaft is generally flexibly coupled to the generator shaft.

b. Diesel generators. Because of the relatively low speed of diesel engines, diesel-generator sets are appreciably larger and heavier than turbine-generator sets. Sets rated below about 1000 kW operate at speeds up to 1800 rpm. Above that, most sets are designed for 1200 rpm or less. Also, because of the low speed, the generator is coupled directly to the engine. The generator may be of the two-bearing type with the rotor flexibly coupled to the engine crankshaft or may have a front bearing only with the rear shaft end coupled rigidly to and supported by the diesel engine crankshaft. Generator bearings may be integral with the generator end brackets or supported separately by pedestals. If a rotating exciter is used, the armature is overhung on the generator front shaft extension.

Two- and four-cycle engines are used for marine electric plants. Four-cycle engines tend to be heavier and costlier but more efficient than two-cycle engines. Two-cycle engines are generally equipped with attached positive-displacement blowers to supply scavenging air to expel gases from the cylinder at the end of the exhaust stroke. In addition, both two- and four-cycle engines may be equipped with turbochargers (driven by gears, exhaust-gas turbines, or a combination of both) to increase the engine output and provide improved fuel economy. Most marine engines are turbocharged to reduce their size and weight.

It is important that provisions be made for sufficient combustion air, either by running air ducts from the outside directly to the engine air intake, or by means of the machinery space ventilation system. The combustion air supply system must be designed so that the pressure drop from the outside atmosphere to the engine will not exceed approximately 6 in. of water to obtain satisfactory engine performance. For the same reason the exhaust-gas pipe or duct from the muffler to the atmosphere should be sized to produce a back pressure at the muffler outlet of not more than 10 to 16 in. of water.

A complete marine engine includes the following attached and unattached auxiliaries:

- Fuel control system consisting of an engine-driven fuel pump, duplex filter, suction strainer, and injector control lever for manual starting, stopping, and emergency speed control.
- Lubricating oil system consisting of an engine-driven oil pump, full-flow filter with bypass relief valve, strainer with relief valve, and cooler with bypass relief valve. When dry-sump engines are used, the lubricating system includes a scavenging pump with suction strainer and by-pass relief valve. Some engines require a manual or power-driven prelube pump.
- Freshwater systems for larger diesel-generator sets consisting of an expansion tank, engine-driven water pump (some engines may have two pumps) seawater heat exchanger, and automatic water temperature regulator. Smaller engines (particularly for emergency generator sets) generally have a radiator and fan for freshwater cooling.
- Exhaust system consisting of a dry, spark-arresting type muffler and water-cooled or insulated exhaust header. For larger generator sets, the engine is usually equipped with an exhaust temperature indicating system consisting of a set of thermocouples or remote temperature detectors (RTDs) for local as well as remote temperature indication. One thermocouple is installed in the exhaust of each cylinder.
- Starting system, which may be either pneumatic, hydraulic or electric, consisting of a starting motor (two motors may be necessary to accelerate large engines) or air distributor for sequential admission of air to the engine cylinders, strainer, and air control valve. A solenoid-operated valve may be provided for remote engine starting. Starting air systems are generally designed to operate at pressures of 125 to 450 psi. Starting air tanks and air compressors are normally furnished by the shipbuilder. Diesel generators rated at 500 kW or less generally use either an electric or hydraulic starting system. Sufficient capacity for at least ten successive starts, beginning with a cold engine, is provided with either method. Either a centrifugal or pressure-operated device is provided to prevent inadvertent attempts to initiate cranking after the engine has started. This device or a solenoid-operated valve is used to automatically stop cranking after the engine has started. Also, means are provided to prevent unintended starting during maintenance of the set: for electrically started engines, a disconnect switch is provided in the starting circuit; for hydraulically and pneumatically started engines, a cutoff valve is provided in the associated piping system.
- Air intake filter-silencer.
- Electro-hydraulic load-sensing speed governor.
- Speed adjusting device with local manual adjustment and an electric motor or potentiometer for remote synchronizing of the sets.
- Overspeed trip device, which upon overspeed of over 15% will close the fuel racks or combustion air supply to shut down the engine.
- An interlock switch for actuating the generator circuit-breaker tripping device to disconnect the generator

from the switchboard bus when the engine is shut down due to overspeed.

• Local control and gageboard including, as a minimum: pressure gages for the freshwater and seawater pump discharges, fuel oil and lubricating oil filter inlets and outlets, lubricating oil strainer inlet and outlet, scavenging air, and starting air; thermometers for the fresh water and lubricating oil; temperature indicators for exhaust temperature; start-stop control; and alarms.

• Manual engine-turning gear, with an interlock, for turning small engines or a powered (pneumatic or electric) turning gear, with an interlock, for turning large engines.

• In addition to the aforementioned equipment, turbocharged engines are equipped with a scavenging turbocharger and turbocharger air intake cooler.

• Tube- or plate-type coolers for cooling the lubricating oil and fresh water for larger diesel generator sets. Seawater is used as the cooling medium in the freshwater cooler and fresh water is used as the cooling medium in the lubricating oil cooler.

c. Gas turbine generators. Gas turbine-driven generator sets are available for ship-service or emergency power. Gas turbines are smaller than comparable steam turbines or diesel engines. The high-frequency noise of gas turbines is easier to control and attenuate than the low-frequency noise of diesel engines, particularly as far as structureborne noise is concerned. Hence, gas turbines offer considerable advantages over diesel engines from space and noise perspectives, particularly for naval ships where these features can be of major importance. One disadvantage of gas turbine drivers, however, is the large exhaust gas ducting area required, which could affect the size of the stack.

d. Emergency generator sets. Emergency generators provide a power source independent of any other equipment on the vessel and are usually either diesel or gas turbine driven. The characteristics of diesel engine and gas turbine prime movers previously described apply equally to emergency sets, except as noted in the following. On commercial vessels each diesel engine is equipped with a self-contained cooling system requiring a radiator and fan. If ventilation ducts are installed to and from the engine radiator, the radiator fan must develop enough head to force the cooling air through the ducts. Motor-operated louvers are generally installed in supply and exhaust duct terminals, which are exposed to the weather. The vent motors are energized from the generator side of the generator circuit breaker to insure that the louvers are open when the diesel engine is in operation. Naval ships use direct seawater-cooled generators. Emergency generator units are arranged to shut down automatically upon a loss of lubricating oil pressure, dangerous overspeeding, or release of carbon dioxide in the emergency generator room. An audible alarm is provided that will sound in the event of a low oil pressure or high cooling-water temperature.

Engines are generally arranged to start automatically upon the failure of ship-service power. If battery started, a voltage-sensitive relay with contacts that close when the ship-service power fails is used to energize the control circuits of the starting motor; if hydraulically started, a loss of ship service power will deenergize a solenoid-operated valve to initiate the starting process. In either case means must be provided for automatically rendering the starting devices inoperative after the engine has attained firing speed. Devices for manual starting control are required with each type of starting equipment for test purposes.

Emergency generators are generally not required to operate in parallel with the ship-service generators, except where a closed transition transfer with the ship-service generators is desired (to prevent power outage when transferring power sources). An automatic voltage regulator should be provided with each emergency generator. The emergency generator should be of sufficient capacity to permit "cold" starting of the ship's main power plant.

e. Speed governors. Speed governors for marine generator sets are of the electro-hydraulic type, and are designed for parallel operation of generator sets. Earlier designs were electro-hydraulic or mechanical governors that operated in the "droop" mode, allowing a change or drop in speed and frequency from no-load to maximum load. The design of more recent governor systems with electronic controls provide an "isochronous" mode in addition to the backup droop mode. For the isochronous mode, the governors maintain the speed and hence frequency within very close tolerances between no-load and maximum load. The control circuits of the electronic governors of generator sets operating in parallel are interconnected to provide automatic, close-tolerance "real" load (kW) sharing between the units. Real load (kW) sharing between parallel units operating in the droop governor mode is also inherently automatic although to wider tolerances, provided the governors and prime movers of the parallel units have the same droop characteristics. Generator sets operating in the governor "droop" mode need to be periodically adjusted to maintain the rated frequency and even load sharing between units in a fluctuating-load environment.

f. Generators. All generators are of the rotating-field type. Turbine-driven high-speed (3600 rpm) rotors may be of the unipole (cylindrical rotor) construction whereas at speeds of 1800 rpm and below they are of the salient-pole type. Stator windings may be either delta- or wye-connected but usually are the latter; only three main terminals are required per generator. The inherent voltage regulation of a-c generators is comparatively wide due to the high synchronous reactance of the windings. This is an advantage inasmuch as the synchronous reactance limits the short-circuit current. In order to maintain the required voltage regulation and reactive kvar load sharing between parallel generators an automatic voltage regulator is used with each machine.

Generators may be dripproof protected or totally enclosed. If totally enclosed, generators are equipped with a double-tube air cooler using seawater as a cooling medium. Silicone insulation is prohibited for totally enclosed generators unless the slip rings are located outside the generator enclosure; this is to prevent abnormal brush wear and increased slip-ring maintenance. If the design and arrangement of a generator are such that circulating currents may be expected in the rotor shaft, means (such as the use of insulated bearings) are provided to prevent circulating currents from passing between the journals and the bearings, as the babbitted bearing surfaces may otherwise be destroyed.

Ship-service generators that weigh more than 1000 lb, excluding the shaft, and all emergency generators are provided with electric space heaters to prevent moisture condensation during shutdown. Generators rated at 500 kVA and above are provided with resistance-type temperature detectors embedded in the stator windings.

There are two types of rotating exciters, the d-c exciter and the a-c brushless exciter. Both of these exciters are coupled to the generator shaft and are similar in outward appearance. In place of the commutator on the d-c exciter, the a-c brushless exciter has a solid-state, three-phase rectifier mounted on the generator shaft to provide the d-c excitation for the generator field. The a-c brushless exciter responds faster than the d-c exciter. A third type of exciter is the static excitation system. This system eliminates the necessity of rotating components and has a faster response than either type of rotating exciter. The exciter and voltage regulator must be coordinated to obtain the desired recovery time. The a-c brushless exciter and the static excitation system have in general replaced the rotary amplifier exciter.

The preferred routing of cables between the generator and switchboard is downward, under the generator platform, and up behind the switchboard. The preferred location of the generator terminals is, therefore, at the bottom of the generator, unless the generator is of such small size that connections may be made inside a standard terminal box mounted on the side of the generator frame. All generator terminals are protected against accidental contact and mechanical damage. If the terminals are located on the top or side of the generator frame, they are protected by a watertight enclosure with removable covers. Where cables enter a terminal enclosure on the top or side of the generator frame, the enclosure is provided with terminal tubes. For terminals located at the bottom of the generator frame, the necessary protection, usually expanded metal, is furnished by the shipbuilder.

Additional terminals are required for excitation, imbedded temperature detectors, and generator space heaters where provided. Any such miscellaneous terminals are located and protected in a manner similar to the main terminals. All generator power connections must be silver-plated.

Particular care must be exercised when terminals are brought out and marked to assure that the phase rotation with respect to the terminal marking will be the same on all generators.

g. Voltage regulators. A direct-acting type of voltage regulator employing a mechanical regulator element may be used to control the field of d-c rotating exciters of very small auxiliary generators. The regulator element may be a torque motor, solenoid, or electrodynamometer energized by the generator voltage and restrained in motion by a spring. Rheostatic elements are automatically operated field rheostats of which there are two types. One consists of a motor-operated face plate rheostat controlled by contacts on the regulator element, and the other consists of a resistor with numerous steps cut in and out by mechanical linkage with the regulator element. Both types have antihunting or damping devices to prevent fluttering with small changes in load and overregulation with large sudden changes in load.

Either of two types of static voltage regulators may be used depending on the type of excitation system. When a static excitation system is used, the voltage regulator senses the generator output voltage, which is rectified and applied to the control winding of a magnetic amplifier. The output of the amplifiers is impressed across the control winding of three saturable-current potential transformers, one per phase. The outputs of the transformer secondaries are rectified and impressed across the generator field. If an a-c brushless exciter is used, the exciter field current is supplied by a static voltage regulator, which senses generator output voltage. An error voltage is impressed across a reactor, which becomes saturated, conducts, and fires a silicon-controlled rectifier to provide current to the exciter field. The average exciter field current is determined by the point at which the rectifier fires during each positive half-cycle. Provisions are made in static regulators to provide sufficient excitation that will result in fault currents large enough to ensure selective tripping of overcurrent devices during short-circuit conditions when the generator voltage is zero.

The voltage regulators described in the foregoing are suitable for use with generators operating in parallel, with each generator requiring an individual regulator.

Mechanical voltage regulators operate in the "droop" mode to provide reactive kvar load sharing between parallel generators. The droop mode allows for a change or drop in voltage from no load to maximum load within the selected droop band. Reactive kvar load sharing between parallel generators operating in the voltage regulator droop mode is also inherently automatic within a certain tolerance band, provided the generators and voltage regulators of the parallel units have the same "droop" characteristics. Voltage regulators operating in the droop mode need to be periodically adjusted to maintain the rated voltage and even kvar sharing between generators in a fluctuating load environment.

Static voltage regulators may be designed for both "droop" and "differential" operation. In the differential mode, the voltage regulators maintain the voltage within very close tolerances between no-load and maximum load. The control circuits of static voltage regulators for generators operating in parallel are interconnected to provide automatic close-tolerance reactive (kvar) load sharing between the generators.

3.2 Storage Batteries. Storage batteries for shipboard applications are usually of the lead-acid or alkaline type. Nickel-cadmium is the most common application of alkaline technology. However, there are a number of other candidate battery technologies, such as silver-zinc and sodium-sulfur.

An electrical storage battery is a device that can receive, store, and discharge d-c electric energy through chemical reaction processes, during which an electrolyte is composed and decomposed. In large batteries, the electrolyte is in a liquid state and is mixed with or dissolved in water. Vented (breathing) liquid-state electrolyte batteries give off gases, which contain a potentially explosive mixture of hydrogen and oxygen when charging at a high rate. Batteries of moderate size are of the totally enclosed, sealed, and maintenance-free construction. Dry-cell technology is used for smaller batteries. In a dry-cell battery, the electrolyte is absorbed in a gel-type compound; these batteries are also totally encapsulated (sealed) and maintenance free.

The principal shipboard applications for storage batteries include:

- Propulsion of submarines and other submersibles.
- Engine starting (diesel engines, gas turbines, lifeboat engines, forklift trucks).
- Temporary emergency power and lighting.
- General alarm system and other vital interior communication systems.
- Emergency power for exterior communication (radios).
- Integrated ship and machinery control systems.
- Emergency and stand-by generator set control.
- Fire-screen doors and watertight doors.
- Computers (microprocessor-based command, control, communications and information systems).

In many of the above applications, the batteries are part of an uninterruptible power supply (or no-break power supply), as discussed later in Section 3.7.

Large liquid-state electrolyte batteries need to be arranged, installed, and ventilated in strict accordance with the regulatory body requirements to abate corrosion and prevent explosions. Small totally enclosed batteries may be packaged and installed without special precautions.

3.3 Fuel Cells. A fuel cell is a device in which a chemical reaction is harnessed directly to produce d-c electric power. Unlike a chemical battery, in which an electrolyte is decomposed into its basic components, a fuel cell utilizes a controlled reaction between elements such as hydrogen and oxygen. Since the hydrogen and oxygen can be continuously supplied, the fuel cell does not run down as does a chemical battery. A schematic of a fuel cell is shown as Fig. 5. Gaseous hydrogen and oxygen at pressures in the order of 60 psi enter the cell and are brought into contact with porous electrodes. Between the electrodes is a liquid electrolyte, which serves to limit the reaction rate. Hydrogen diffuses through the porous anode, is absorbed on the surface, and reacts with the OH-ions in the electrolyte to form water and yield free electrons. Oxygen diffuses through the cathode, is absorbed by the surface, and reacts with the water to form OH-ions. Thus water is continuously being formed at the anode and decomposed at the cathode. The reaction rate is controlled by the rate of migration of OH-ions through the electrolyte. Electrons flow out of the cell through a load and are returned to the cathode. The maximum cell electromotive force (emf) is in

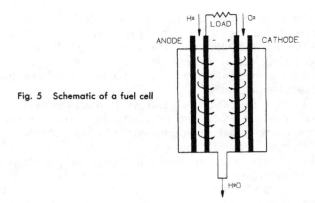

Fig. 5 Schematic of a fuel cell

the order of 1 volt and depends to a certain extent on the choice of reactants employed.

A fuel cell is not a heat engine, in fact the process is essentially isothermal, and consequently the Carnot efficiency is an irrelevant maximum for fuel-cell performance. However, given a supply of hydrogen and oxygen, more useful power can be obtained with a fuel cell than if the bases were allowed to react spontaneously and a Carnot cycle were run from the high-temperature flame. This is because the voltage of the cell tends to restrain the chemical reaction, much as a piston restrains spontaneous expansion of gas. The maximum theoretical efficiency of a fuel cell, based on the conversion of chemical to electric energy, is 100 percent.

A major difficulty of fuel cells is the relatively short life span of the electrodes, especially when operated with relatively low-grade fuels. Their lifetime can be markedly increased by use of purified hydrogen, but the costs of the purifying operation are prohibitive.

Fuel cells appear to be ideal for environments where there is no natural supply of air, since they are not air-breathing. For this reason, fuel cells were developed for aerospace applications and have been used on space vehicles. There have also been some isolated applications on special-purpose submersibles.

3.4 Motor-Generator Sets. Ship-service primary electric power is converted in many different ways to provide special power to a large number of systems and equipment, such as communications, navigation, control, electronics, and lighting systems. Motor-generator sets can essentially be used to satisfy all electric power conversion needs. The motor and associated motor controller are selected to operate on the supply power source, and the generator and associated voltage regulator are designed to provide the specific power required. Hence, a motor generator set may comprise all combinations of a-c squirrel cage, a-c synchronous, or d-c machines. The major shipboard applications of motor generator sets include:

- Ward-Leonard variable-speed drives (a-c to d-c motor-generator sets) such as for cargo handling gear.
- 60- to 400-Hz motor-generator sets for 400-Hz systems on naval ships.
- D-c to a-c motor-generator sets for temporary emergency power systems on passenger ships and for submarine/submersibles.

It is noted, however, that motor-generator sets are being gradually displaced by solid-state power electronics such as thyristors or silicon-controlled rectifiers (SCRs).

3.5 Solid-State Power Converters. Solid-state power conversion technology is based on the use of solid-state power switches, such as silicon-controlled rectifiers (SCRs or thyristors) or transistors. These switches operate on the principle of rapid and radical changes in their unidirection conductivity between near extremes, and are controlled by electrical signals. The switches are arranged in a variety of complex circuit techniques using reactors, capacitors, and diodes to achieve the desired converter output performance. Solid-state power converters can be designed to accommodate regenerative reverse power flows, which are desirable for certain electric drives, particularly electric propulsion, to provide dynamic electric deceleration (braking) on demand. In conjunction with digital-control technology, solid-state power converters provide close-tolerance output power in a dynamic load environment, and respond extremely fast and safely to sudden and extreme transients including fault currents.

Solid-state power switches are, however, a concern with regard to electromagnetic interferences (EMI), which is discussed in Section 12. The rapid and frequent (60 or 400 times per second) switching of these semiconductor power devices causes spikes and harmonics in a wide frequency range, which if not abated are conducted into the ship's power system and radiated by the electrical equipment and cables connected to the system. Hence, solid-state power converters may require measures to reduce the EMI, such as filters, shielding, and grounding.

Semiconductor rectifiers are the most commonly used a-c to d-c power converters. Non-controlled copper oxide or selenium rectifiers, which are also known as metallic rectifiers, represent the earliest practical applications of semiconductor technology; however, they remain to be the most cost-effective technology for low-power battery charger and constant (steady-state) d-c load applications. Variable-load and duty-cycle d-c demands are better served with controlled semiconductor rectifiers, such as SCRs.

Solid-state frequency converters or frequency changers use semiconductor technology to convert, control, or regulate the frequency of a-c power. The most common shipboard applications of frequency converters are 60/400-Hz converters on naval ships and variable-frequency controllers for variable-speed a-c drives such as electric propulsion, bow thrusters, cargo-handling gear, as well as naval replenishment and weapons-handling equipment. Frequency converters are available in two basic configurations: two-stage synchro-converters, consisting of a rectifier section (a-c to d-c) and an inverter section (d-c back to a-c); and cyclo converters (a-c to a-c in one stage), for conversion to lower frequencies only.

Solid-state inverters use semiconductor technology to convert d-c power to a-c power at a variable or constant frequency. Inverters are used in such shipboard applications as submersibles that are "powered" by batteries, uninterruptible power supplies, and two-stage solid-state frequency converters, which first convert a-c to d-c and then back to a-c.

3.6 Transformers. Transformers are the most cost-effective method of converting a-c power from one voltage to another at the same frequency. Transformers for shipboard use are of the dry, air-cooled type; they are rated for continuous duty and protected in a dripproof or splashproof enclosure. Transformers may be either of the single-phase or three-phase type. One practice is to provide three identical single-phase transformers mounted side by side in a 3-phase bank with the primaries connected in delta and the secondaries connected in either delta or wye to suit the application. With such an arrangement, the failure of a single-phase transformer will not render the entire secondary system inoperative; furthermore, it reduces the space and weight requirement for spare transformers. However, the high reliability of the insulation and transformer technologies may make 3-phase transformers preferable because they require less space and weigh less.

3.7 Uninterruptible Power Supplies. Uninterruptible power supplies (UPSs), or no-break power supplies, are devices that provide a limited amount of electric power for a finite period of time to certain loads in case of a power failure on the UPS supply side, with no break or interruption in the flow of power to the loads. A UPS consists of a storage battery and a set of power conversion equipment, with the latter being a function of the type of power normally supplied to the UPS and the type of power supplied by the UPS to the connected loads. If not inherent in the basic concept, an UPS must contain the necessary protective features to prevent the flow of energy from the battery into the supply side of the UPS upon power failure.

A simple a-c to d-c UPS consists of a battery and a rectifier, with the rectifier normally supplying the d-c loads and the battery floating on the d-c bus by being continuously trickle charged. Upon an a-c power failure, the battery continues to supply d-c power to the loads.

A typical a-c to d-c UPS consists of a battery, a rectifier, and an inverter. Normally, power is converted from a-c-to-d-c and back to a-c to supply the loads. The battery floats on the d-c bus. Upon a power failure on the a-c supply side, the battery provides d-c power to the inverter, which continues to supply a-c to the loads.

A typical d-c to a-c UPS consists of a battery and an inverter or motor-generator set. Normally, d-c power from a d-c power system is inverted to a-c power with the battery floating on the d-c supply voltage. Upon a d-c system power failure, the battery provides d-c power to the inverter or motor-generator set, which continues to supply a-c to the loads.

A d-c to d-c UPS is simply a battery that floats on the d-c system supply to the selected loads. Upon a d-c system power failure, the battery continues to supply d-c to the loads.

Section 4
Power Distribution

4.1 Distribution Concepts. The electric power distribution subsystem is the connecting link between the generators and the loads/users; it comprises ship-service, emergency, and, on naval combatants, casualty-power distribution elements. Electric power is distributed throughout the ship using either a radial, zonal, or combined radial/zonal distribution concept. In a radial distribution system, individual loads and power panels are fed directly from the main switchboard. The loads are radially connected to the source of power, like the spokes of a wheel to the hub. Radial distribution systems are more economical for smaller ships and power systems up to a few hundred kW.

In zonal distribution systems, the ship is divided into contiguous physical zones, and the loads within each zone are fed from one or two large load-center switchboards which, in turn, are fed through large bus feeders from the ship-service (main) switchboards.

Zonal distribution concepts are usually applied in combined radial/zonal distribution systems on larger ships. Figure 6 shows the zonal features of a typical 60-Hz power distribution system on a naval combatant with three segregated, totally independent ship-service/emergency generator sets (gas turbine or diesel driven). In Zone No. 2, the two main switchboards serve as the load centers for that portion of the ship; hence the power distribution within that zone is basically radial. Zone 3 is fed from the third main switchboard and a load-center switchboard whereas Zones 1 and 4 are each fed from two load-center switchboards.

The power system one-line diagram reflects the basic concepts and details of the power distribution subsystem in a diagrammatic/elementary form and shows branch circuits terminating at the individual loads/users. Figure 7 portrays a "top level" one-line diagram and shows the salient features of an electric power plant for a typical commercial ship. Figure 8 represents a sheet of a detailed elementary one-line diagram pertaining to a naval ship.

4.2 Ship-Service Power Distribution Features. The electric loads throughout the ship are supplied either individually and directly from a main switchboard or in groups through panelboards and load-center switchboards. Individual feeders from the main switchboard are normally used to supply larger loads, such as propulsion auxiliaries that are within the same machinery space as the respective switchboard, as well as for very large motors (such as bow thrusters), which may be anywhere on the ship. Grouped loads are supplied by bus feeders or feeders through load-center switchboards, group control centers (grouped motor controllers), or panelboards, which are located central to the associated groups of loads, and operationally readily accessible. When the loads served by a distribution board are in one compartment, the associated distribution board should generally be co-located within the same compartment or adjacent to it and near the entrance to the compartment.

Groups of loads are selected by function as well as location on the ship. Loads of different distinct functions are generally not comingled or fed from the same distribution board. Hence, separate feeders and distribution boards are generally provided for functional groups of loads that are not supplied directly and individually from a main switchboard. Typical functional groupings of electric loads would include:

—vital propulsion auxiliaries,
—nonvital machinery,
—machinery space ventilation,
—accommodation HVAC,
—cargo space ventilation,
—cargo handling equipment,
—mooring and anchor handling equipment,
—weapons and combat systems (on naval ships), and
—hotel loads, such as galley equipment.

4.3 Emergency Power Distribution. Emergency power distribution emanates from the emergency switchboard or, on non-steam naval ships, from a designated ship-service/emergency switchboard. Merchant ships and smaller naval steamships have an emergency power distribution system as generally depicted in Fig. 7, which provides emergency power to 460/440-, 115-, and 24-volt vital loads. Under normal conditions, the emergency power distribution system is fed from the ship-service (main) switchboard via a bus-tie circuit. Upon a total loss of ship-service power, the emergency generator set automatically starts up, the bus-tie circuit breaker on the emergency switchboard trips open, and the emergency generator circuit breaker closes, energizing the emergency power distribution system from the emergency generator.

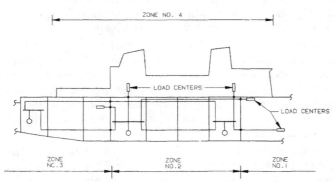

Fig. 6 Zonal features of a typical 60-Hz power distribution system for a naval surface combatant

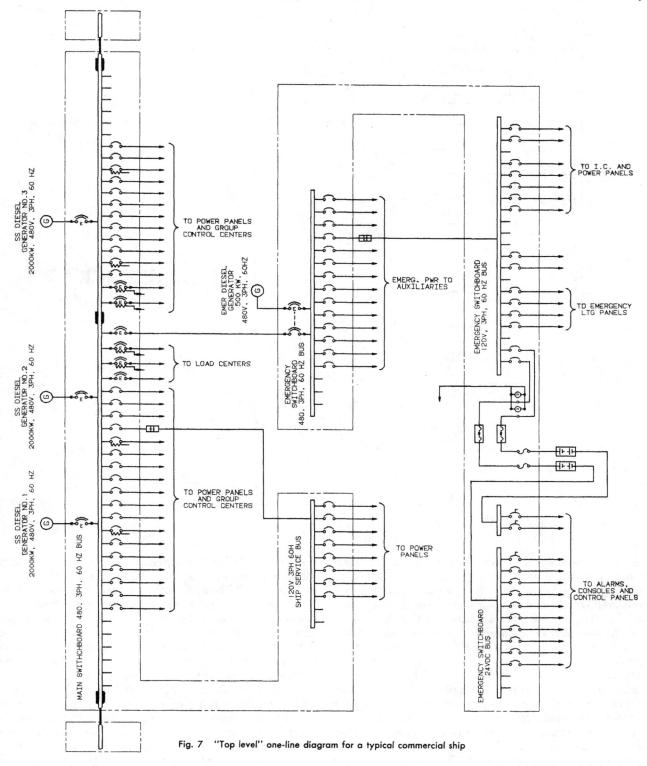

Fig. 7 "Top level" one-line diagram for a typical commercial ship

Passenger ships have a two-tier emergency power system, which is more sophisticated and responds instantaneously upon a total blackout. The two tiers consist of a temporary and a final emergency power system as illustrated in Fig. 9. The temporary emergency power system is fed from batteries in order to bridge the time required to start up the emergency generator and bring it on the line. The temporary emergency power distribution system provides 120-volt 3-phase a-c power (through d-c to a-c motor-generator sets or inverters), 120/240-volt 3-wire d-c, and 24-volt d-c (the same as on non-passenger commercial ships). The final emergency power system operates

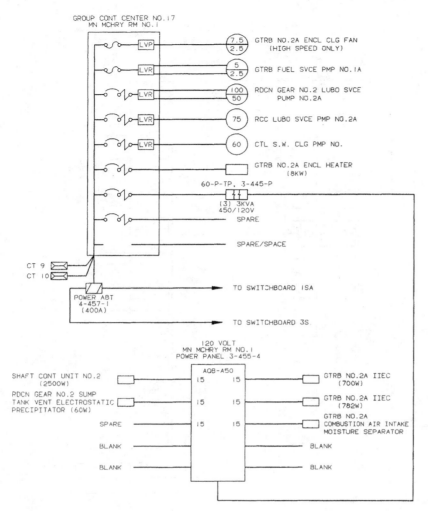

the same as the emergency power system on a non-passenger ship. Once final emergency power is established, all temporary emergency loads are automatically transferred to the final emergency power source through automatic bus transfer switches as shown in Fig. 9.

4.4   Redundancies and Casualty Features on Naval Ships. On naval ships all vital loads are fed from two sources of power through independent cable runs and transfer switches. Alternate distribution cables, load-center switchboards, and power panels are physically segregated port and starboard, as well as low and high in the ship for reasons of vulnerability. On ships with separate emergency power systems, one of the alternate power supplies emanates from the emergency switchboard to a select number of vital loads. Figure 10 illustrates the various redundancies and casualty-switching features of an electrical power system on a naval combatant having three independent ship-service/emergency generator sets.

Bus transfer switches may be either manual (manual bus transfer or MBT) or automatic (automatic bus transfer or ABT). Bus transfer switches are essential mechanical double-throw switches to select between a "normal"

and "alternate" or "emergency" source of power for the vital loads connected to the common load terminals of the switch. An ABT comprises a power-operated double-throw transfer switch and a manual override. Control circuits are set up to select between manual and automatic operation and to effect automatic transfer of power from "normal" to "alternate/emergency" upon the loss of normal power and to automatically transfer back to the "normal" source upon restoration of normal power.

These redundancies and casualty-switching features within the ship-service/emergency power distribution system are augmented by casualty power distribution features. These include through-bulkhead casualty terminals, permanently installed vertical riser cables with top and bottom terminals, portable cables, and casualty power terminals on vital equipment. These casualty power features are used to bridge battle-damaged portions of the ship's power system to restore electric service to vital equipment on an emergency basis.

4.5   Special Power Distribution. Special power systems are inherently smaller and more limited in scope when compared with the ship's main power system. Special

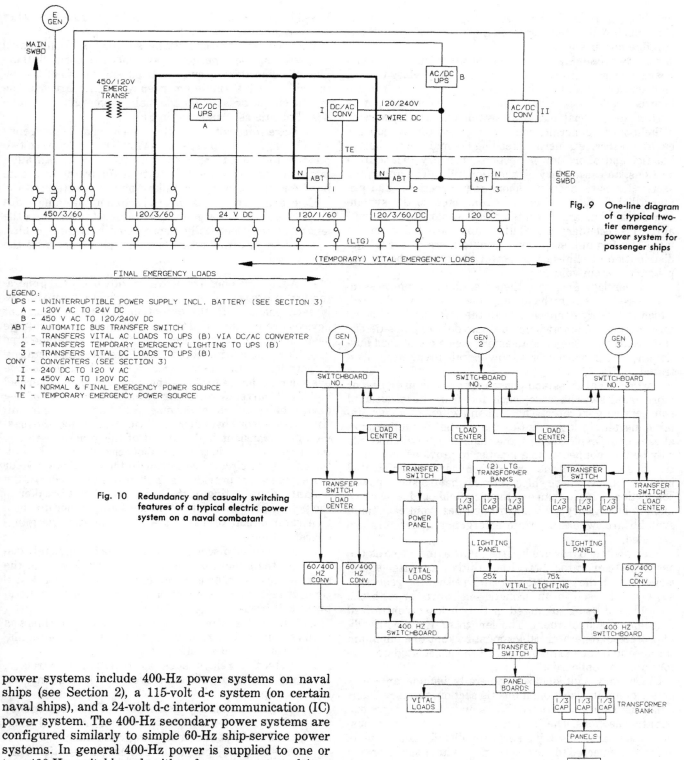

Fig. 9 One-line diagram of a typical two-tier emergency power system for passenger ships

LEGEND:
UPS - UNINTERRUPTIBLE POWER SUPPLY INCL. BATTERY (SEE SECTION 3)
  A - 120V AC TO 24V DC
  B - 450 V AC TO 120/240V DC
ABT - AUTOMATIC BUS TRANSFER SWITCH
  1 - TRANSFERS VITAL AC LOADS TO UPS (B) VIA DC/AC CONVERTER
  2 - TRANSFERS TEMPORARY EMERGENCY LIGHTING TO UPS (B)
  3 - TRANSFERS VITAL DC LOADS TO UPS (B)
CONV - CONVERTERS (SEE SECTION 3)
  I - 240 DC TO 120 V AC
  II - 450V AC TO 120V DC
  N - NORMAL & FINAL EMERGENCY POWER SOURCE
  TE - TEMPORARY EMERGENCY POWER SOURCE

Fig. 10 Redundancy and casualty switching features of a typical electric power system on a naval combatant

power systems include 400-Hz power systems on naval ships (see Section 2), a 115-volt d-c system (on certain naval ships), and a 24-volt d-c interior communication (IC) power system. The 400-Hz secondary power systems are configured similarly to simple 60-Hz ship-service power systems. In general 400-Hz power is supplied to one or two 400-Hz switchboards either from generators driven by 60-Hz motors or from 60/400-Hz solid-state power converters. On larger naval combatants with two 400-Hz switchboards, they are interconnected through a bus-tie for combined parallel operation. The 400-Hz power is distributed at both 440/460 and 115 volts in a generally radial fashion through panelboards.

D-c secondary power systems are normally supplied from batteries, which are charged from motor-generator sets or static battery chargers. The battery chargers may float on the batteries feeding the system continuously, thereby replenishing the energy drawn by the loads and providing a trickle charge even when the system load is zero. Such applications essentially constitute an uninterruptible power supply.

4.6 Switchboards. The switchboards used in shipboard power systems can be functionally classified as ship-service (main) switchboards, emergency switchboards, load-center switchboards, and secondary power system (on special purpose) switchboards. Ship-service switchboards generally include a generator section for each ship-service generator feeding the main switchboard, bus-tie, and shore-power sections, primary distribution sections, and secondary distribution units. The generator sections, bus-tie sections, shore-power sections, and primary distribution sections are connected to the switchboard main bus and operate at the primary system voltage. The secondary distribution units are fed from the main bus through step-down transformers to permit the distribution of ship-service power at voltages below the primary system voltage.

Commercial ships, with the exception of large passenger vessels, usually have only one main switchboard, which is directly connected to all the ship-service generators. A single main switchboard has only one bus-tie circuit breaker for the main-to-emergency switchboard bus-tie, and one shore-power supply circuit breaker as illustrated in Fig. 7.

Larger naval ships and passenger vessels have two or more main switchboards, each fed from its associated ship-service generator. These main switchboards are interconnected through bus-ties and bus-tie circuit breakers at each end. On naval ships, the ship-service switchboards may be interconnected in a manner to provide a complete ring bus through all the main switchboards as illustrated in Figs. 1 and 2. These ships normally have a shore-power connection to more than one main switchboard, as well as main-to-emergency switchboard bus ties from each main switchboard, where separate emergency generators are provided.

Main switchboards are located in the same space as the associated generator sets, although they are often sound isolated from the generator sets, with either the generator set being housed in an acoustic enclosure, or with the switchboard being arranged in the enclosed centralized machinery control room. The length of the electrically unprotected generator cables must be kept to a minimum through judicious arrangement of the associated generators and switchboards.

Emergency switchboards generally include an emergency generator section, a bus-tie section, primary emergency distribution sections, and secondary emergency distribution sections. The emergency generator section, bus-tie section, and primary emergency distribution sections are interconnected to the emergency switchboard primary bus, operating at 480/450 volts. The secondary distribution sections are fed from the primary bus through step-down transformers and rectifiers to provide for the distribution of emergency power at voltages below the 480/450 volts of the primary bus.

On passenger ships, the secondary distribution sections have both temporary and final emergency distribution

buses interconnected via automatic bus transfer switches illustrated by Fig. 9.

With the exception of large passenger vessels and naval combatants, one emergency switchboard normally supplies the emergency power distribution system of the entire ship. Such a single emergency switchboard has one bus-tie circuit breaker for the main-to-emergency switchboard bus-tie as illustrated in Fig. 7.

On large ships with more than one main and emergency switchboard, each emergency switchboard may have two bus-ties to two different main switchboards, with automatic bus transfer controls to transfer to the alternate bus-tie in case of a power failure on the selected bus-tie.

Emergency switchboards are located in the same space as the associated emergency generator set to minimize the length of the electrically unprotected generator cables.

A generator switchboard section typically includes the following:

• A trip-free open-frame electrically operated generator circuit breaker with separate overcurrent trip devices in each phase, with the exception that the overcurrent devices are not provided in the neutral of a dual-voltage system. Overcurrent devices are of the inverse-time type and, for systems with more than two generators operating in parallel, of the instantaneous type. An inverse-time overcurrent device protects the generator from overload and fault currents, with the reaction (trip) time of the devices being inverse to the magnitude of the overcurrent. An instantaneous overcurrent device is set above the maximum subtransient fault current of the generator to trip the circuit breaker in case of a short circuit on the generator side of the circuit breaker, with the remaining two or more generators providing a fault current in excess of the instantaneous setting of the generator circuit breaker to the point of the short circuit on the faulty generator bus. Generator circuit breakers are of the multiphase gang-operated type.

• An unfused disconnect device, which completely disconnects the generator and its circuit breaker from the switchboard main/emergency (system) bus. Such a disconnect feature is commonly incorporated in the generator circuit breakers.

• Current transformers and potential transformers as required for metering, control, protection, voltage regulation, governor, and synchronizing.

• Protective relays such as reverse-power relays, over/under voltage relays, or over/under frequency relays, as required.

• Automatic standby generator set and emergency generator set start-up, generator circuit-breaker closing, and bus-tie circuit-breaker tripping circuits.

• On emergency-generator switchboard sections, circuit breakers for generator set auxiliaries, such as motor-driven cooling-water pumps and motor-operated cooling-air intake louvers, are connected to the emergency generator bus.

• In many cases, the generator excitation controls, voltage regulators, and electronic governor controls are built into the generator switchboard sections.

• In some case (including most emergency-generator switchboards) the local generator set control and instrumentation are provided on the upper front panel of the generator switchboard sections.

Load-center switchboards are essentially remotely located distribution sections of the main switchboard. They are supplied from the main switchboard via a bus feeder and in turn supply power to local distribution panels and loads. Load-center switchboards are centrally located within the zone or part of the ship that is served from a particular load-center. Load-center switchboards are normally used on larger ships, within a combined zonal/radial power distribution system, as illustrated in Fig. 6. Figure 8 shows a typical load-center switchboard application within a section of a power distribution system.

Naval ships are equipped with secondary 400-Hz power systems, with 400-Hz switchboards configured similarly to 60-Hz main switchboards. The generator or converter sections of the switchboard are equipped with the necessary controls and instrumentation for local control of the 400-Hz power system with some provisions for remote monitoring or control from the central operating station. Figure 11 diagrammatically depicts the configuration of a typical 400-Hz switchboard.

Naval ships are normally equipped with several interior communications switchboards. IC switchboards combine power distribution to IC and electronic systems with other control, circuit protection, and switching functions, including the selection of alternate power sources, casualty switching, and signal switching.

Switchboard construction. Switchboards are of a free-standing deck-mounted construction and may be either dead-front/open-rear or totally enclosed. Live-front switchboards are obsolete. A dead-front design provides for all live (energized) parts, such as bus work and terminals to be enclosed or not accessible from the front of the switchboard without the removal or opening of either bolted or hinged panels. Access to open-rear switchboards is normally restricted to authorized personnel by expanded-metal bulkheads from the ends of the switchboard to the nearest structure with locked access openings. Most switchboards require access from both the front and the rear for hook-up, operation, and maintenance and cannot be placed against a bulkhead. In some cases, the top of a switchboard may be braced to a nearby bulkhead or to the deck above to maintain its structural stability in the shipboard environment (vibrations, ship's motions, and shock loads). Such bracing may, however, have to be flexible to allow for deflection of decks/bulkheads without causing plastic deformation to the switchboard structure.

Switchboards are equipped with nonconducting handrails in the front to assist the operators. Open-rear switchboards are also provided with nonconducting guardrails in the rear. Insulation mats, special-insulation nonslip deck coverings, or gratings are required in front and back of the switchboards to protect operating and maintenance personnel from accidentally slipping and contacting live parts.

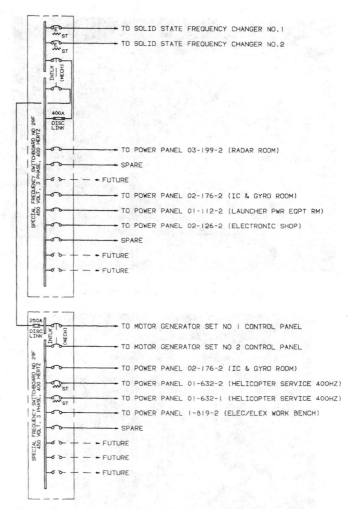

Fig. 11 One-line diagram of a typical 400-Hz switchboard

Switchboards are equipped with dripshields across the top to protect them from dripping liquids and falling objects and debris. The space over and near the switchboards should be devoid of any piping, and any pipes passing near the switchboards must be devoid of any pipe joints. Ventilation should not discharge air directly onto a switchboard. The front of the switchboards must be well illuminated commensurate with the operational tasks. The lighting fixtures for this purpose may be integral with the switchboard and mounted within or under the dripshield overhang of the switchboard.

The configuration of a main, emergency, and 400-Hz switchboard may be such that the generator (converter) sections are in the center of the switchboard with the distribution sections equally divided on both sides. With this arrangement, no part of the main bus carries more than one half of the total load, resulting in a savings in weight and cost. For smaller switchboards, however, the generator (converter) section may be to one side and the distribution section to the other side of the switchboard. Figure 12 illustrates a typical main switchboard configuration.

Fig. 12   Typical main switchboard configuration

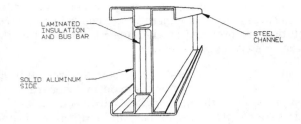

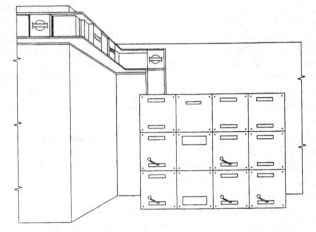

Fig. 14   Typical bus duct installation and construction details

Fig. 13   Typical panelboard configuration

Large open-frame circuit breakers rated between 1000 and 4000 amps are normally of the drawout type for ease of maintenance and repair from the front of the switchboard. Molded-case circuit breakers are of the plug-in or stab-on type for quick and easy replacement from the front of the switchboard.

4.7 Panelboards. Panelboards are the final power distribution points. Panelboards are fed from switchboards through three-phase feeder cables and supply power to the individual loads through mains. Mains on 460/440 and 220/208-volt panelboards are normally for three-phase loads, such as motors and large heaters, whereas mains and branch circuits on 115-volt panelboards are normally for single-phase loads such as lighting, small heaters, miscellaneous appliances, and receptacles. Figure 13 illustrates a typical panelboard configuration.

Panelboards are strategically located throughout the ship, and central to the majority of the loads they supply. Normally all the loads supplied from a panelboard are on the same deck, or within the same large compartment (such as cargo holds and machinery spaces) and within the same watertight boundaries.

Panelboards are of totally enclosed construction for bulkhead mounting and require access from the front only for installation, hook-up, operation, and maintenance. Molded-case branch-circuit breakers are of the plug-in or stab-on type for quick and easy replacement from the front. Panelboards may be of the dripproof protected type for most applications, of watertight construction for installation in the weather, or of the explosion-proof type for installation in hazardous areas. In accommodations and work areas with joiner work, panelboards are usually recessed and flush mounted. Panelboards are equipped with lockable hinged front doors to prevent access to unauthorized personnel.

4.8 Power Distribution Circuits. Power circuits are the interconnecting links between the generators, switchboards, panelboards, connection boxes, and the loads. The classifications of the circuits change as the power flows from the generators through the various distribution points of the system to the connected items. The conventional classifications used for power circuits are as follows:

Generator: generators to main and emergency switchboards
Bus tie: between main and emergency switchboards
Bus feeder: from main to load-center switchboards
Feeder: from switchboards to panelboards or loads
Main: from panelboards to connection boxes
Branch: from panelboards or connection boxes to loads

The conductor size of a power distribution circuit must be selected based on the ampacity (current-carrying capacity in amps) of the cable (or bus duct) giving consideration to the following factors:

- operating current (demand factor) as specified for the different categories of circuits;
- maximum ambient temperature where the circuit is installed;
- maximum allowable conductor temperature as determined primarily by the type of insulation used;
- maximum allowable voltage drop; and
- installation methods and their effect on natural cooling of the circuit (e.g., cables have to be derated

if they are installed doubled, banked, or bunched, thereby inhibiting natural air circulation around the cable).

The requirements and constraints governing the selection of cables for power distribution circuits are contained in the rules and regulations of the regulatory bodies as well as in the standards and specifications promulgated by the Navy and commercial institutions. These requirements and constraints are typically as outlined in references 6, 7, and 8.

Line voltage-drop calculations should be performed using the determined lengths of each circuit, in order to verify the initial cable selection and compliance with the allowable voltage-drop limitations. Reference 8 governs the approach to be used for voltage-drop calculations for naval ships.

Multiconductor (two or three conductors) commercial or Navy shipboard cables are used almost exclusively for power distribution circuits. Shipboard cabling in general is discussed in Section 13.

Insulated 3-phase bus duct has been used in isolated cases on both naval and commercial ships. Such applications are for relatively short runs confined to one space/compartment, such as generator circuits or bus feeders. Figure 14 illustrates a typical bus duct installation of a bus feeder (from a switchboard) to a group control center, plus some details of a typical 3-phase insulated bus duct construction.

Section 5
Electric Plant Control

5.1 Control System Architecture. Before the advent of shipboard automation, electric plant controls were configured for local control only of the respective components at the equipment. Such control actions included:

- Lineup, starting, monitoring, and securing of the generator set prime movers at the respective prime mover, using mechanical control features and direct-reading gages.
- Generator voltage, frequency, and load-sharing controls, manual synchronizing and paralleling of the generator sets, and primary power system monitoring at the main switchboards.
- Bus-tie and power-distribution circuit breaker controls at the respective switchboards.
- Special power systems controls and monitoring such as M-G sets, or converters at the equipment or the associated secondary switchboard, such as the 400-Hz switchboard.

However, the continuous evolution of shipboard automation, concomitant with a centralization and integration of machinery operations and an attendant reduction in operating personnel, has largely relegated the use of local controls to abnormal situations. Most ships are designed with a high degree of automation and with an integrated

machinery control console located in a central, enclosed engineering operating station.

Electric plant controls comprise all of the monitoring, control, and data display/logging features required to operate the generator sets (including prime movers and auxiliary support systems, such as lube oil pumps), main and emergency switchboards, load-center switchboards and panelboards, and power converters (such as 60/400-Hz M-G sets or solid-state converters for special-power subsystems). Many features of an electric power plant are either inherently passive or automatic, such as power circuits/cabling and transformers, which require no monitoring or controls. Also most of the power distribution circuit breakers on switchboards and panelboards remain normally closed (energized) and need to be attended only in case of a malfunction or casualty, causing the circuit breaker to trip as intended due to an overload or short circuit, or for maintenance and repair.

The monitoring of the condition of an electric plant is an important aspect of system control. Monitoring embraces the sensing of the system/equipment parameters and status as required for:

- Feedback for automatic control loops.
- Data display and logging.
- Off-limit alarming.

- Diagnostic for troubleshooting and maintenance.

Controls embrace both manual and automatic control features and data display/logging, including:

- Analog instruments, such as gages and meters.
- Digital electronic displays.
- CRT's or plasma displays.
- Indicating lights.
- Data logging printers.
- Computer technology data storage media (disks, etc.).

Table 2 lists the control, display, and alarm functions/parameters of an electric plant for a gas turbine-propelled naval ship, having five ship-service/emergency diesel generator sets and associated switchboards, configured as shown in Fig. 15. Figure 16 illustrates the panel layout of the electric plant control section of an integrated machinery control console pertaining to the electric plant configuration shown in Fig. 15 and the associated control, monitoring, display and alarm functions enumerated in Table 2.

The control system architecture is primarily a function of the degree of automation and the attendant level of manning selected for the operation of the ship. Merchant ships sail worldwide with unattended machinery spaces and with limited remote monitoring and summary fault indications on the bridge and in the engineer's office. Subpart 62 of reference 9 delineates the USCG requirements for automated, self-propelled vessels and periodically unattended machinery plants.

The effect of automation on the electric plant control architecture has generally been as follows:

Ship-service generator sets. The prime movers and generators are controlled and monitored from the engineering operation station. In some installations, the generator sets are arranged for automatic start-up, synchronizing, load sharing, and shutdown controlled by an integrated electric power management system.

Switchboard. The main switchboard instrumentation and controls for the generator, bus-tie, and vital distribution circuit breakers are extended to the engineering operating station. On some commercial ships the main switchboard is located within the engineering operating station, eliminating the need for remote control and monitoring of the switchboard (and generator) from the console.

Emergency generator sets and switchboards. Some of the controls and monitoring features of the otherwise totally autonomous and self-contained emergency power system (generator set and switchboards) may be extended to the central emergency operating station such that a malfunction of the remote features will not interfere with the automatic operation of the emergency power system.

Power-distribution switchboards and panelboards. Except for some vital power-distribution feeder circuit breakers, the control of which may be extended to the central engineering operating station, most power-distribution circuit breakers are for local manual operation only for the reason explained before.

Special power systems. Some of the controls and monitoring of special power systems, such as the 60/400-Hz power converters and 400-Hz switchboards on naval ships, are extended to the central engineering operation station.

Local control. Wherever remote control and monitoring are extended to the console in the central engineering operation station, local manual control is also provided at the respective units for secondary and emergency fallback operation in case of a malfunction of the remote control features. A local/remote control function transfer switch and, where applicable, manual/automatic mode selector switch are provided at the local control stations. On naval ships, separate and free-standing electric plant control panels (EPCPs), constructed to the same Mil-Spec as the switchboards, are often provided in each machinery space to consolidate the local controls and display features of the electric plant elements within the respective machinery space. Figure 17 illustrates the layout of a local EPCP pertaining to the same electric plant configuration and control concepts portrayed in Fig. 15 and Table 2.

5.2 Automation. Certain control features of the electric plant such as frequency, voltage, and load-sharing control, have been automatic for many decades. In addition, automatic motor control features have also been used to some degree for many decades, as dictated by the operating criteria of the system served by the respective motor. However, the need for automatic motor control features has increased in step with the trend to greater shipboard automation.

The automatic features and functions that have evolved and been adopted in the control of shipboard electric power plants include the following:

Automatic start-up of generator sets. An automatic start-up of diesel-driven emergency generator sets has been a standard feature on both commercial and naval ships for many decades. However, the automatic start-up of ship-service generators is of more recent vintage because of the complexity of automating steam turbine-driven generator sets. A generator set aligned for standby operation may be started automatically upon the loss of an on-line generator set, overload of an on-line generator set, or anticipated increase in load demand as determined by an electric-load management system. Larger generator sets that are designed for automatic operation may require a continuous or intermittent automatic pre-lube system when in a standby mode.

Automatic synchronizing. Automatic synchronizing may be provided in conjunction with an automatic generator set start-up feature, or automatic synchronizing may be provided to avoid potential operator errors such as the accidental closing of a generator circuit breaker on a live bus without proper manual synchronization. Electronic automatic synchronizing equipment has evolved to a high degree of accuracy and reliability.

Automatic load shedding. The automatic shedding of nonessential loads in case of a power plant (generator) overload condition may range from a simple one-step feature to a more complex multistep scheme. In the first case, the load-shedding feature may trip one large nonessential

Table 2 Typical electric plant control and monitoring functions (pertaining to the electric plant configuration for a naval ship as shown in Fig. 15)

Function	Qty	Local			Engine Operating Station		
		C	D	A	C	D	A
A. DIESEL ENGINES (SS/EMERG D-G SETS)							
engine start/stop	5	X	–	–	X	–	–
generator set ready (in standby)	5	–	X	–	–	X	–
engine failed to start	5	–	–	–	–	–	A
seawater cooling pump	5	X	X	–	X	X	–
lube oil pressure	5	–	X	X	–	X	X
jacket cooling water temp.	5	–	X	X	–	X	X
cylinder exhaust temperature	5×8	–	–	–	–	X	X
engine speed (same as generator frequency control)	5	X	X	–	X	X	–
engine tripped (safety features)	5	–	–	X	–	–	X
lube oil filter differential pressure	5	–	X	–	–	–	–
crankcase pressure	5	–	X	X	–	–	–
fuel oil pressure	5	–	X	–	–	–	–
jacket water pressure	5	–	X	–	–	–	–
starting air pressure	5	–	X	–	–	–	–
lube oil sump temperature	5	–	X	–	–	–	–
engine running hours	5	–	X	–	–	–	–
engine starting counter	5	–	X	–	–	–	–
lube oil strainer diff. pressure	5	–	X	–	–	–	–
air intake filter diff. pressure	5	–	X	–	–	–	–
lube oil inlet temperature	5	–	X	–	–	–	–
jacket water temp. to engine	5	–	X	–	–	–	–
B. GENERATORS (SS/EMER D-G SETS)							
lube oil pressure	5	–	X	–	–	X	X
cooling air temperature	5	–	X	–	–	X	X
bearing temperature	5	–	X	–	–	X	X
governor mode	5	X	X	–	X	X	–
generator frequency (same as engine speed)	5	X	X	–	X	X	–
voltage regulator mode	5	X	X	–	X	X	–
generator voltage	5	X	X	–	X	X	X
generator power	5	–	X	–	–	X	–
generator current	5	–	X	–	–	X	–
stator temperature	5	–	X	–	–	X	–
excitation	5	–	–	X	–	–	X
generator space heater	5	X	X	–	X	X	–
autom/permissive synchronizing	5	X	X	X	X	X	X
power available	5	–	X	–	–	X	–
C. D-G SET ENCLOSURES							
cooling fan	5	X	X	X	X	X	X
fire	5	–	–	X	–	–	X
halon release	5	–	–	X	–	–	X
D. MAIN/EMERGENCY SWITCHBOARDS							
system (bus) voltage	5	–	X	–	–	X	X
bus tie current	5	–	X	–	–	X	–
manual synchronizing	1	X	X	–	X	X	–
generator circuit breaker	5	X	X	–	X	X	–
bus frequency		–	–	–	–	X	X
bus tie circuit breakers	10	X	X	–	X	X	–
load shedding	1	–	–	X	–	–	X
standby generator not available	1	–	–	X	–	–	–
plant needs more than 3 generators	1	–	–	X	–	–	–
vital distr. feeder circuit breakers	AR	X	X	–	X	X	–
shore power current	2	–	X	X	–	X	–
shore power available	2	–	X	–	–	X	–
shore power circuit breakers	AR	X	X	–	X	X	–
ground detection	AR	X	X	–	X	X	–
E. 400 Hz POWER SYSTEM							
converter: voltage	2	–	X	X	X	–	–
frequency	2	–	X	X	X	–	–
current	2	–	X	–	X	–	–
start/stop	2	X	X	–	X	X	–
temperature	2	–	–	X	–	–	X
60 Hz power available	2	–	X	–	–	X	–
400 Hz power available	2	–	X	–	–	X	–
converter output circuit breakers	2	X	X	–	X	X	–
converter abnormal shutdown	2	–	–	X	–	–	X
ground detection	1	X	X	–	X	X	–

C = control.
D = display.
A = alarm.

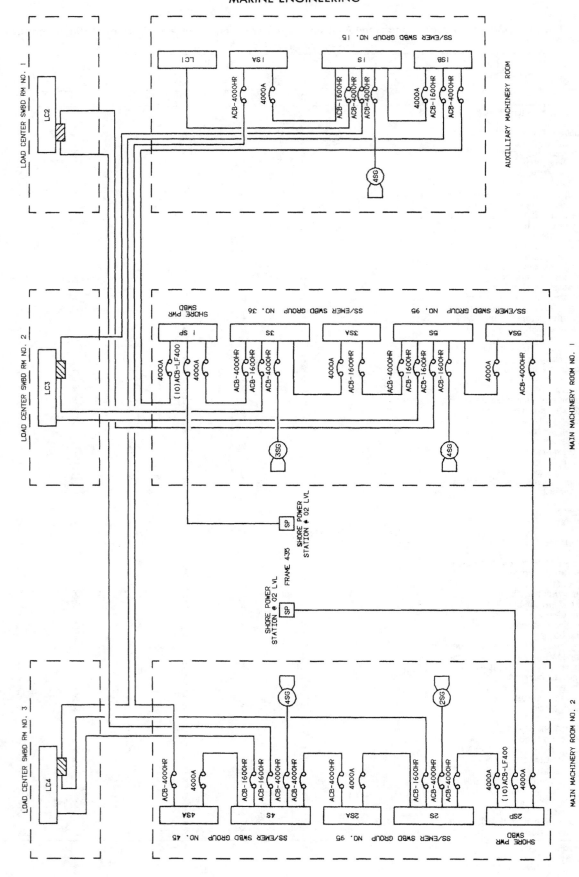

Fig. 15   Electric plant configuration of a gas turbine-propelled naval ship having five ship-service/emergency generator sets

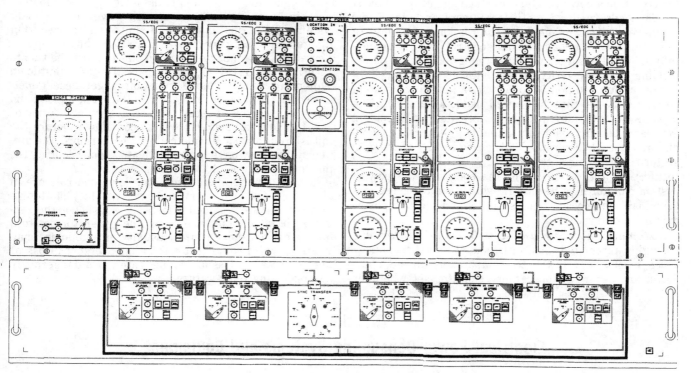

Fig. 16 Electric plant control panel layout of an integrated machinery control console

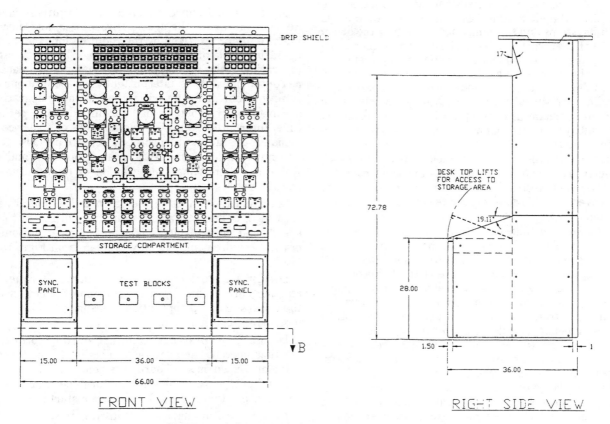

Fig. 17 Local electric plant control panel layout of a naval ship

load (such as a bow thruster) or one large group of nonessential loads (such as reefer containers or hotel loads) whenever the power monitor requires load shedding. In the latter case, the load-shedding controls may incrementally trip individual smaller loads or groups of loads depending on the severity of the overload condition, until the overload condition is corrected or the power system monitor indicates that the desired load has been shed. The power-system monitor may be simply a set of individual generator power-sensor relays that trigger the load-shedding feature whenever the load sensed by any of the generator relays exceeds the set point (full-load or overload setting). At the other end of the spectrum the power-system monitor may be integral with a sophisticated programmable processor that is part of an electric power management system.

Automatic shutdown and securing of generator sets. On highly automated ships with an unattended machinery space, it may be desirable to automatically unload, shutdown, and secure an operating generator set wherever

fewer generator sets could satisfy the operating load of the ship as monitored by an electric power management system. The normal operational shutdown and securing of a generator set may involve more than simply tripping the generator circuit breaker and shutting off the fuel to the prime mover; rather, it may instead entail a shutdown sequencing control scheme. For some larger diesel generator sets, it may be desirable to run the set at no-load and possibly at a reduced speed for a predetermined period of time to allow the prime mover to cool down, before the unit is brought to a complete stop. Furthermore, the pre-lube system may have to be reactivated upon securing the unit to put it back into a standby status.

Such automatic operational shutdown-controls are in addition to the customary emergency/safety shutdown features that are provided to instantaneously trip the generator circuit breaker and bring the generator set to a quick stop in the event of a casualty or equipment malfunction, which could result in either overspeed, loss of lube-oil pressure, loss of cooling water, or a dangerously high operating temperature.

Section 6
Electrical Power-Consuming Systems and Equipment

6.1 Definition of Requirements. The specific requirements for power-consuming equipment and its associated controls, such as motors, heaters, and electronics, are determined by the characteristics and location of the systems/subsystems and the components (such as the driven equipment) they serve. The electrical requirements are driven by the functional characteristics of the respective system/subsystem such as the amount of power and the degree, tolerance, and frequency of power variations required. The mechanical requirements are primarily a function of the mechanical characteristics of the respective system/subsystem components (such as the driven equipment for motors), and the location of the component. The latter determines the packaging (enclosure) of the respective electrical components. Further details concerning the more significant electrical equipments, which are not otherwise covered in this chapter, are reviewed in the following paragraphs.

6.2 Motors and Controls. Practically all of the electric motors for ship's auxiliaries are of the a-c asynchronous squirrel-cage induction type. D-c motors as well as wound-rotor a-c asynchronous motors have virtually become obsolete; variable-speed drives, where required, use either squirrel-cage induction motors with solid-state variable-frequency controls, or hydraulic transmissions. Multi-speed squirrel-cage motors (up to four speeds) satisfy most applications for variable- (multiple) speed drives. A-c synchronous motors are used in industrial applications to improve the system power factor and are usually applied to drive large continuous loads, and to drive a-c to a-c motor-generator sets. Shipboard applications of synchronous motors have been rare, except for electric propulsion.

Table 3 is an excerpt from a typical "List of Motors and Controls" for a ship.

a. Electrical characteristics of a-c squirrel-cage induction motors. The speed, horsepower, and duty rating of a motor are fixed by the required input to, and the operating cycle of, the driven machine. The duty ratings (operating cycle) for shipboard applications are classified as continuous duty or intermittent duty. Continuous duty is a requirement for a service that demands operation at a substantially constant load for an indefinite period of time. Intermittent duty is a requirement of a service that demands operation for specified alternate periods of: load and no load; load and rest; or load, no load, and rest. Reference 6 contains a description of duty ratings for specific applications.

Squirrel-cage induction motors are designated as design A, B, C, and D. Each design offers different torque, speed, and current characteristics to meet various operating requirements, as may be seen from Figs. 18 and 19.

Design A motors have a normal starting torque, high starting current, and low slip. This motor is not used for the usual shipboard applications because of its high starting current characteristic.

Design B motors have a normal starting torque, low starting current, and low slip. This is the motor most commonly used on shipboard; it is generally used for centrifugal pumps, fans, blowers, motor-generator sets, and compressors that are not loaded when started.

Design C motors have a high starting torque, low starting current, and low slip. This motor is normally used for

Table 3 Sample sheet of a typical "List of Motors and Controls"

| Application | | Motor | | | | | | | | | | | | Control | | | | | | | |
| | | | | | | | | | | | | | | Starter | | | | | Cont SW | | |
IT No.	Auxiliary	Qty	Hp	Volt	Ph	FL Amp	NEMA Dsgn.	Rpm	Amb. Temp	Encl	Duty	Mtg	Remarks	Function	UV	LCT	ENCL	Opn	Type	LCT	Remarks
51	Main eng piston cooling water pump	2	25	460	3	28.9	B	3535	50	DP	Cont	Vert		X-Line ST-SP	LVR	See note (1)	DP	Man	–	–	Remote PB ST-SP, ST-SP IND LTS, SEL SW-Stby & Alm at MCC. PB ST-SP at pump. Remote press SW.
52	Main eng sea water cooling pump	2	150	460	3	173	B	1188	50	DP	Cont	Vert		X-Line ST-SP	LVR	See note (1)	DP	Man	–	–	Remote PB ST-SP, ST-SP IND LTS, SEL SW Stby & Alm at MCC. PB ST-SP at pump. Remote press SW.
53	Mn fuel oil transfer pump	1	30	460	3	38.7	B	885	50	DP	Cont	Vert		X-Line ST-SP	LVP	See note (1)	DP	Man	PB ST-SP	Cont Door	Remote PB ST-SP at pump emer stop at fuel sta. P & S. emer stop at fire fighting sta.
54	Mooring winch (amidship)	4	75	460	3	79.6	B	3575	40	TE, WT, EXP PRF	Cont	Vert		X-Line ST-SP	LVP	Remote	DP	Man	–	–	Triplex controller remote PB ST-SP at pump thermostats in motor. MTR SP HTR.
55	Mooring winch brake (amidship)	4	2	460	3	3.1	B	1750	40	TE, WT, EXP PRF	Cont	Hor		X-Line	LVP	Remote	DP	Man	–	–	Triplex controller. See mooring winch (amidship). Thermostats in motor. MTR SP HTR.
56	Mooring winch fan (amidship)	4	1	460	3	1.5	–	1725	40	EXP PRF NV	Cont	F-1		X-Line	LVP	Remote	DP	Man	–	–	Triplex controller. See mooring winch (amidship). Thermostats in motor. MTR SP HTR.

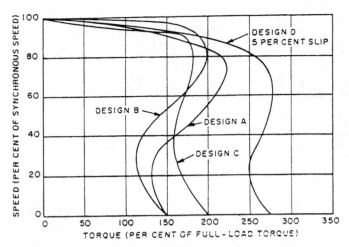

Fig. 18   Torque-speed curves of squirrel-cage induction motors rated 30 to 50 hp

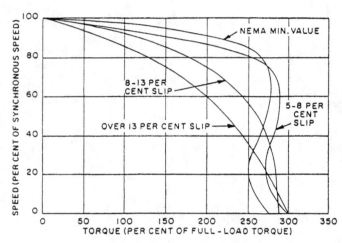

Fig. 19   Torque-speed curves of design D squirrel-cage induction motors for various slip values

applications such as steering gear, anchor windlass, and compressors that are loaded when started.

Design D motors have a high starting torque, moderate starting current, and high slip; this motor is normally used for capstans, winches, valve operators, conveyors, elevators, and hoists.

b.   Mechanical characteristics of motors.

Enclosure and method of ventilation. A wide variety of enclosures and methods of ventilation is available for motors in marine service. The specific types selected depend on the particular environmental condition to which the motor is subjected. The types of enclosures and methods of ventilating motors most commonly used are as follows:

•   Dripproof protected, self-ventilated. Used for most applications in dry, sheltered locations. This enclosure is so constructed that drops of liquid or solid particles falling on the motor at any angle not greater than 15 deg from

the vertical cannot enter the motor. The ventilating openings are normally protected with wire-screen, expanded-metal, or perforated covers to prevent personnel from contacting electrical parts. These covers also keep out rodents.

•   Totally enclosed, fan-cooled. Generally used to prevent the entry of foreign contaminants, both solid and liquid, into the motor. This type of enclosure prevents the free exchange of air between the inside and outside of the housing but is not sufficiently enclosed to be termed airtight or watertight.

•   Waterproof, nonventilated. Used for practically all motors mounted on weather decks, or where the motor may be subject to heavy washdown or possible transient submergence.

•   Explosion-proof, fan-cooled. Used in an atmosphere containing an explosive mixture. This type of enclosure is required to be capable of withstanding an explosion of a specified gas or vapor, which may occur within it, while preventing the ignition of the gas or vapor surrounding the enclosure. Explosion-proof equipment is generally not watertight and must be enclosed in a watertight housing if it is subject to weather conditions. The cooling fan is constructed of nonsparking material and is protected by a guard.

•   Submersible, self-ventilated. Used for those limited and special applications that may require normal operation in air and emergency operation when submerged. A positive means of providing the required capability is the use of a bell type of enclosure. The motor-and-pump combination is vertically mounted and covered with a close-fitting bell that is open at the bottom. The bell must be of sufficient depth so that the required submergence will not force water onto the motor windings.

Under special and restricted conditions, submersible motor-driven pump assemblies may be used to pump out the cargo tanks of liquefied methane, propane, ammonia, or similar cargos, the primary restriction being that air must be excluded from the cargo tanks at all times so as to prevent the possibility of an explosive mixture existing within the tanks.

•   Watertight/spraytight. Used in generally wet spaces below deck where the motor is subjected to splashing, spraying, or hosedown.

Insulation. Insulating materials used in motors are divided into categories according to their ability to withstand high temperatures for long periods of time. These categories are Class B, F, or H. Class B is rated at 130 C maximum operating temperature, Class F at 155 C, and Class H at 180 C. These temperatures in each case represent the insulation system material capability and are the summation of the ambient temperature, motor-winding temperature rise above ambient, plus an estimated temperature gradient referred to as the hot-spot allowance.

The regulatory bodies and the Navy have specific requirements regarding temperature rise limits for the various classes of insulation for different applications. Bearings and bearing lubricants should be selected based on operating temperatures encountered with each class of insulation.

Shafts. For flexible and rigidly coupled drives, the National Electric Manufacturers Association (NEMA) standard short-shaft extensions are used. However, in some instances shafts are provided with the end tapered, threaded, and equipped with a nut and washer for ease in disassembly and reassembly.

Brake motors or motors using shoe brakes are provided with front-end shaft extensions as required by the brake application. NEMA long-shaft extensions are provided for pulley-driven auxiliaries. Special long shafts are provided for impellers of close-coupled pumps and axial-flow fans. Carbon steel shafts are normally provided for coupled drives and for freshwater close-coupled pumps. For pumps handling corrosive liquids, the motor shafts are generally required to be made of corrosion-resistant materials (e.g., stainless steel or monel) and fitted with sleeves.

Bearings. With few exceptions motors are equipped with greasable ball bearings; however, axial-flow ventilation fan motors are usually equipped with prelubricated and sealed ball bearings, since they are located in ductwork and are not easily accessible. Sleeve bearings, designed for flood lubrication or forced lubrication, are used only in special applications.

Terminal boxes. All motors are normally furnished with terminal boxes having threaded pipe taps for ship's cable entrance terminal tubes. Motor terminal leads and ship's cable are mated by means of cable connectors within the terminal box. The degree of enclosure required for terminal boxes is usually the same as that provided for the motor. The desired location of terminal boxes and the number and size of tapped holes for cable entrance are normally specified by the shipbuilder.

Space heaters. Motors subject to wide variations of temperature or excessive moisture conditions may be equipped with space heaters to prevent condensation of moisture in the motor when idle. The heaters may be resistance units bolted to the inside of the lower frame or a phase winding energized through a low-voltage transformer. In either case, the heating circuit is electrically interlocked so as to interrupt the heater power whenever the motor is energized.

Mounting. Motors are designed for mounting in any required position, i.e., horizontal, inverted horizontal, vertical with the drive shaft up or down, and in some instances inclined. Most shipboard auxiliaries are driven by horizontal motors mounted on a common bedplate with the driven machine. However, the use of vertically mounted motor-driven centrifugal pumps provides a saving in deck area and may facilitate a preferred piping arrangement. In addition to general-purpose mountings, NEMA has standardized on three types of flange mounts: "C" face, "D" flange, and "P" or "PH" base. Types C and D are used either horizontally or vertically, with the relative location of the face and feet fixed by the standards. Types P and PH flanges are used for vertical pump applications. Each type of mounting should be coordinated with the driven auxiliary to insure a satisfactory fit and performance. For heavy assemblies (such as an overhung

motor with a disk brake attached by a flange), it is advisable to provide a foot-mounted motor to afford rigidity.

Special treatment. All motors used in marine applications should be given a special impregnation to make the windings resistant to salt water, salt air, oil fumes, and fungus. Metal parts are made of corrosion-resisting materials or are treated to render them corrosion-resistant. For example, small hardware is usually zinc-plated, and the shaft inside the frame, exposed laminations, and brackets are usually treated to prevent corrosion. Small hardware for motors exposed to the weather may be specified to be stainless steel.

c. Electrical characteristics of alternating-current motor controllers. Motor controllers are designed to perform specific electrical functions regarding the control and protection of motors. The characteristics and applications of the various types of controllers that provide these functions are as follows:

Controller operation. Controllers are either manually or magnetically operated. Manual controllers are normally used for motors rated less than 2 hp that require only "on-off" operation. Magnetic controllers are used for all other applications and may be classified as automatic or nonautomatic. Automatic controllers start and stop the motor automatically in response to some operational criteria without operator interference. Nonautomatic controllers require manual operation of a pushbutton or switch to initiate a start or stop. After the initial manual operation, the controller completes the starting or stopping of the motor. Nonautomatic controllers are used for applications other than those auxiliaries that require automatic cycling such as air or reefer compressors. Typical control circuits for a-c magnetic controllers are shown in Figs. 20 and 21.

Starting mode. Typical methods of starting a-c induction motors are illustrated in Fig. 22. Controllers for across-line starting are used for most shipboard auxiliaries since the ship's generating plants are usually of adequate capacity to handle the starting currents of the motors. However, controllers of the auto-transformer (or reduced voltage) type are used when it is necessary to limit the starting current of a motor so as to avoid imposing an excessive transient load on the generating plant. Reduced-voltage starters are designed for closed-circuit transition so as to avoid high transition currents. Standard starting transformers for motors above 50 hp have taps of 50, 65, and 80% of rated voltage; only 65 and 80% taps are provided in sizes below 50 hp. The starting current drawn from the line is proportional to the square of the percent voltage tap; i.e., the 80% tap produces 64% of the across-line starting current. Starting the fire pump motor off the emergency generator is a typical application for reduced-voltage starting.

Wye-delta type of controllers are also used to limit the starting currents of large motors. With this arrangement the motor is started in the wye connection and then reconnected, with closed transition, to the delta running connection. This arrangement requires a six-lead motor with a starting current of approximately 33% of the

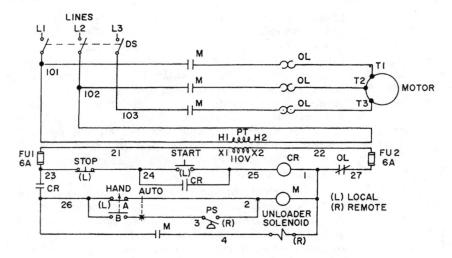

DESCRIPTION OF OPERATION

SELECTOR IN "HAND": TO START, PRESS A START BUTTON ENERGIZING SELF MAINTAINING RELAY "CR". THE "CR" CONTACTS CLOSE ENERGIZING MAIN CONTACTOR "M". THE "M" CONTACTS CONNECT THE MOTOR ACROSS THE LINE AND ENERGIZE THE UNLOADER SOLENOID.

TO STOP, PRESS A STOP BUTTON REMOVING "CR", "M" AND THE MOTOR FROM THE LINE.

SELECTOR IN "AUTO": PLACING THE SELECTOR SWITCH IN "AUTO" INSERTS A N.O. PRESSURE SWITCH, "PS", IN THE CONTROLLER.

TO START, PRESS A START BUTTON ENERGIZING RELAY "CR". WHEN THE "PS" CONTACTS CLOSE, MAIN CONTACTOR "M" IS ENERGIZED CONNECTING THE MOTOR ACROSS THE LINE. WHEN THE "PS" CONTACTS OPEN, "M" IS DROPPED REMOVING THE MOTOR FROM THE LINE.

TO STOP, PRESS A STOP BUTTON REMOVING "CR", "M" AND THE MOTOR FROM THE LINE.

IN A VOLTAGE FAILURE CONDITION, ALL COILS DROP REMOVING THE MOTOR FROM THE LINE. TO RESTART, PRESS A START BUTTON AFTER RESTORATION OF VOLTAGE. (LOW VOLTAGE PROTECTION).

IN AN OVERLOAD CONDITION, THE O.L. CONTACTS OPEN REMOVING THE MOTOR FROM THE LINE. TO RESTART, PRESS A RESET AND THEN A START BUTTON. (OVERLOAD PROTECTION).

Fig. 20 Typical schematic and wiring diagram for a compressor motor controller

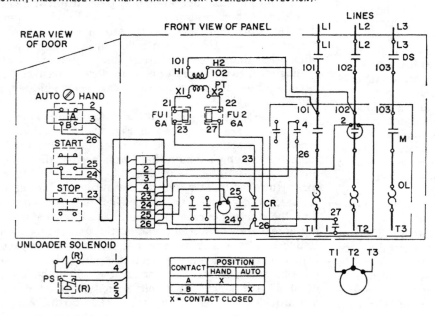

locked-rotor current. This arrangement is typically used to start a bow thruster motor.

Controllers of the secondary resistor type are used to limit the starting currents and provide speed control for wound-rotor induction motors; typical applications are forced-draft blowers and main circulating pumps.

Protective features. Low voltage protection (LVP) is the feature that is provided to cause the controller to disconnect the motor from the power supply upon a reduction or loss of voltage; the motor remains disconnected until the voltage is restored and the motor is restarted by command as shown in Fig. 23. This feature is provided to prevent the simultaneous restarting of a large number of motors after an interruption of power because the sum of their large starting currents would overload the system.

Low voltage release (LVR) is the feature that is provided to cause the controller to disconnect the motor from

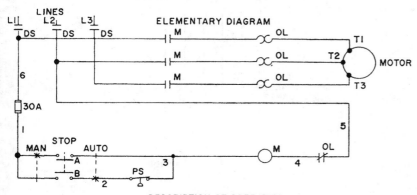

DESCRIPTION OF OPERATION

TO START THE MOTOR MANUALLY, TURN THE SELECTOR SWITCH TO THE MAN POSITION. THIS ESTABLISHES A CIRCUIT TO THE MAIN CONTACTOR M. THE MOTOR WILL START AND CONTINUE TO RUN AS LONG AS THE SELECTOR SWITCH IS IN THE MAN POSITION.

TO OPERATE THE MOTOR AUTOMATICALLY UNDER THE CONTROL OF THE PRESSURE SWITCH PS, TURN THE SELECTOR SWITCH TO THE AUTO POSITION. THE PRESSURE SWITCH WILL THEN ESTABLISH A CIRCUIT TO THE MAIN CONTACTOR M. THE MOTOR WILL START AND CONTINUE TO RUN UNTIL THE PRESSURE SWITCH CONTACTS OPEN. THE MOTOR WILL THEN CYCLE OFF AND ON AS THE PRESSURE SWITCH CONTACTS OPEN AND CLOSE.

A VOLTAGE FAILURE WILL CAUSE THE MAIN CONTACTOR M TO OPEN AND DISCONNECT THE MOTOR FROM THE LINE. WHEN VOLTAGE IS RESTORED, M WILL RECLOSE AND START THE MOTOR (LOW VOLTAGE RELEASE).

AN OVERLOAD WILL CAUSE THE OL CONTACTS TO OPEN, DEENERGIZING M AND STOPPING THE MOTOR. TO RESTART, PRESS THE RESET BUTTON.

THE MOTOR MAY BE STOPPED AT ANY TIME BY TURNING THE SELECTOR SWITCH TO THE STOP POSITION.

Fig. 21 Typical schematic and wiring diagram for a potable-water pump motor controller

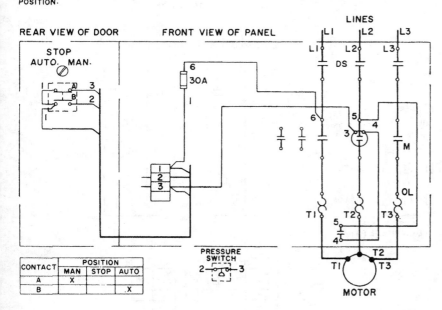

the power supply upon a reduction or loss of voltage; the motor remains disconnected until the voltage returns, and then automatically reconnects the motor to the power supply to restart it, as shown in Fig. 24. This feature is usually applied to those vital auxiliaries that must be automatically restarted immediately upon restoration of power. Typical examples are lube-oil service pumps, main and auxiliary condensate pumps, main circulating pumps, control air compressors, and steering gear pump motors.

It is usually desirable to use time-delay relays with LVR controllers to obtain staggered starting and prevent the simultaneous restarting of all LVR auxiliaries. Motors that are automatically controlled by pressure-switches, and similar devices, have an inherent LVR feature unless

a low-voltage relay, which opens upon failure of line voltage, is provided in the control circuit. All manual-type controllers provide LVR characteristics when using stop/start pushbuttons.

Overload protection is the feature that results in the controller disconnecting the motor from the power source when excessive currents (other than short circuits) occur that could cause overheating of the motor. This feature is provided by overload relays; separate relays are required for each winding of multispeed motors. Overload relays may be either thermal or magnetic.

Thermal overload relays generally consist of a heat-sensitive element and a heat-generating element. The heat-generating element may be a heater or coil in series

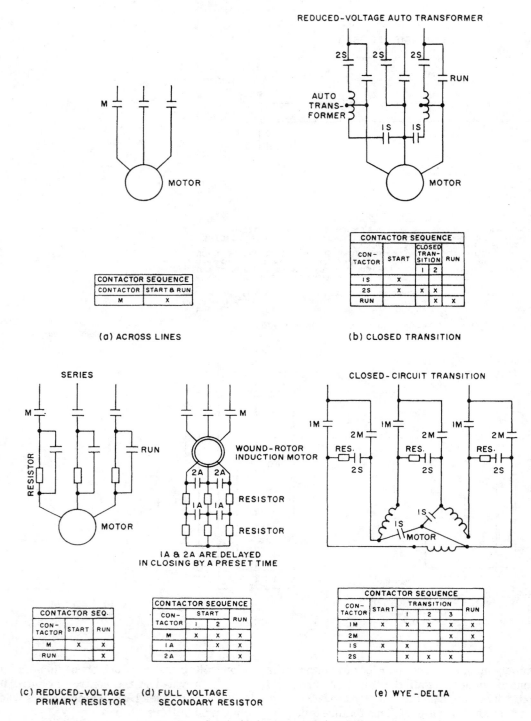

Fig. 22 Typical methods of starting a-c induction motors

with the motor load circuit. An excessive motor current passing through the heat-generating element causes the heat-sensitive element to react to open the overload relay contacts, thus breaking the circuit to the operating coil of the main line contactors; this in turn causes the contactor to open the motor circuit. Since the tripping characteristics of the thermal overload relay depend on both the length of time of application and the amount of overload current, the relay can be, and normally is, designed to follow approximately the time-current heating curve of the motor. This curve represents the values of the current that a motor can carry for different lengths of time without damaging the motor insulation. Thermal overload relays may be compensated for possible ambient tempera-

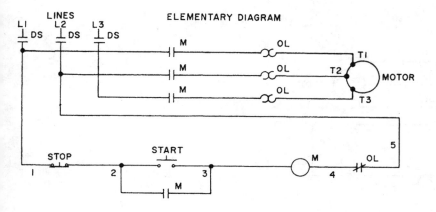

ELEMENTARY DIAGRAM

DESCRIPTION OF OPERATION

TO START THE MOTOR PRESS THE START BUTTON. THIS ENERGIZES CONTACTOR M, CONNECTING THE MOTOR ACROSS THE LINE. M MAINTAINS ITSELF THRU ITS OWN AUXILIARY CONTACTS. TO STOP THE MOTOR, PRESS THE STOP BUTTON.

A VOLTAGE FAILURE WILL CAUSE M TO OPEN STOPPING THE MOTOR. TO RESTART, WHEN VOLTAGE IS RESTORED, PRESS THE START BUTTON. (LOW VOLTAGE PROTECTION).

AN OVERLOAD WILL CAUSE THE OL CONTACTS TO OPEN CAUSING M TO OPEN STOPPING THE MOTOR. TO RESTART, PRESS THE RESET BUTTON AND THEN THE START BUTTON.

Fig. 23 Typical motor control circuit with low-voltage protection feature

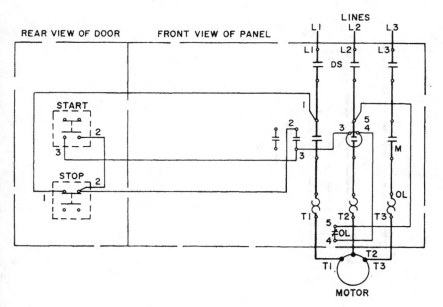

ture changes. Thermal overload relays are generally of the solder-pot, bimetal, single metal, or induction type. Practically all applications of overload relays on shipboard are of the thermal type.

Magnetic overload relays generally consist of a coil in series with the motor load circuit and a tripping armature or plunger. When the amount of overload current for which the relay is set passes through the series coil, the tripping armature is actuated to open the overload relay contacts, thus breaking the circuit to the operating coil of the main line contactors; this in turn causes the contactor to open the motor circuit. Magnetic overload relays are not affected by variations in the ambient temperature and require no temperature compensation. Magnetic overload

relays are of the instantaneous or time-delay type and have limited application on shipboard because they do not use heat in their operation and consequently do not follow the heating curve of motors.

Overload relays are provided with a means of resetting so that the motor controlled can be restarted with overload protection. Tripped thermal overload relays must be allowed to cool before the tripping mechanism can be reset. Magnetic overload relays can be reset immediately after tripping. The three forms of overload relay resets are manual (hand), automatic, and electric. The manual form is the most common for shipboard use and consists of a rod or lever which, when operated, causes the tripping mechanism to be returned to its original position. The

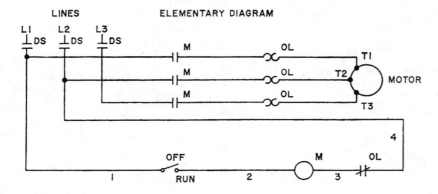

LINES   ELEMENTARY DIAGRAM

DESCRIPTION OF OPERATION

TO START THE MOTOR POSITION THE SELECTOR SWITCH AT "RUN". THIS ENERGIZES
CONTACTOR M CONNECTING THE MOTOR ACROSS THE LINE. TO STOP THE MOTOR POSITION
THE SELECTOR SWITCH AT "OFF".

A VOLTAGE FAILURE WILL CAUSE M TO OPEN STOPPING THE MOTOR. WHEN VOLTAGE IS
RESTORED, THE MOTOR WILL IMMEDIATELY RESTART (LOW VOLTAGE RELEASE).

AN OVERLOAD WILL CAUSE THE OL CONTACTS TO OPEN CAUSING M TO OPEN STOPPING
THE MOTOR. TO RESTART, PRESS THE RESET BUTTON.

Fig. 24   Typical motor control circuit with low-voltage
release feature

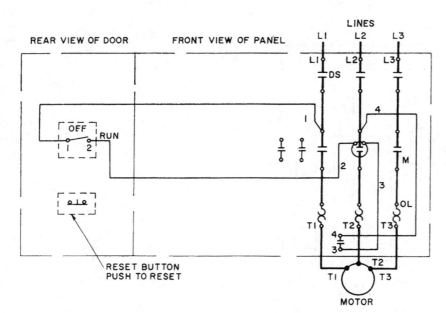

automatic reset has no practical application on ships. The electrical reset is limited to those applications where it is desirable to reset a relay from a remote operating position.

Emergency run features may be provided on controllers for certain auxiliaries, such as elevators, in which case stopping in the middle of an operating cycle could be highly undesirable. This feature, which is initiated by operating a pushbutton or lever, renders the overload relay tripping mechanism inoperative so that the auxiliary can be operated with the motor running in an overload condition until the operating cycle is completed.

In general, each control wire that leaves a controller is provided with short-circuit protection. Such protection

may be provided by a fuse, located in the controller, if the lead is not already protected by a current-limiting device (coil or resistor) located in the enclosure.

When a secondary source of power greater than 24 volts is brought into a motor controller for alarm, indicating light, or other circuits, a suitable interlock is usually provided to disconnect the secondary source upon opening of the controller door. In lieu of disconnecting the secondary source, an independent disconnect device may be used for this purpose. This independent disconnect may be located adjacent to the motor and controller disconnect, and a sign provided on the main disconnect to warn that both devices should be operated to disconnect all sources of power to the motor and controller.

Programmable controllers use digital microprocessor-based technology to execute complex automatic control functions, in lieu of relays, switches, timers and counters, etc. They may be used on vital motor-driven auxiliaries, which are part of a highly automated system or subsystem or in conjunction with a sophisticated vessel management system.

d. Mechanical characteristics of motor control equipment.

Starter and controller panels. A grouping of several motor starters housed in a free-standing deck-mounted structure is known as a group control or motor control center. Each motor starter within a group control is energized from a common power supply feeder through individual circuit breakers mounted in each starter. Group controls are metal-enclosed units having metal barriers between starters.

In general, individual starter enclosures are either dripproof, watertight, submersible, or explosion-proof as required by their location. Starters mounted in group control centers are generally of the "open" type mounted in a dripproof enclosure.

Cabinet enclosures are usually designed for bulkhead mounting in smaller sizes and for deck mounting in weights over 150 lb. Consideration must be given to the need for rear access on large starter panels for installation and maintenance.

Suitable provisions for cable entrance are made in all cabinets, usually at the top for the feeder and at the bottom for the motor and control leads.

Master switches. The broad category of master switches includes pushbutton, drum switches, selector switches, pressure switches, temperature switches, interlock switches, float switches, and any other type of remote pilot device required for proper operation of the controlled motor. Local switches and indicating lights, which form a part of controllers, have the same degree of enclosure as the controller. For remote locations, the enclosure of switches is either dripproof, watertight, submersible, or explosion-proof to suit the requirements of the location. In general, for weather or corrosive areas, watertight enclosures of cast bronze or brass for small switches and bronze, nodular iron, or stainless steel for winch control switches are provided.

e. Electric motor brakes. Electric brakes for marine service are of either the disk or shoe type, each type being spring set and magnetically (solenoid) released. Solenoids or magnets are usually energized (brakes released) through contactors whenever the associated motor is in operation. Deenergizing the motor also deenergizes and engages the brake.

Disk brakes may be either a-c or d-c operated; shoe brakes are usually d-c operated. For motors that are 50 hp and larger, d-c operated brakes are usually provided.

In general, brakes for suspended loads are rated at 200% of the motor torque. For other types of loads, the brakes are rated to stop the load under any operating condition; in these cases, the brakes are usually rated at 100% of the motor torque. Brakes are provided with a means of being mechanically released for emergency operation in the event of a power failure.

Brakes in weather locations are of a watertight construction and have electric heaters to prevent the accumulation of condensation during nonenergized periods.

Disk brakes are attached directly to the motor front end bracket and require no special foundation; from a space point of view, disk brakes lend themselves to an economical installation. Most shoe brakes are foot-mounted and require a special foundation for proper alignment with their motor.

6.3 Resistive Loads. Electric heaters are generally resistive loads that operate at a unity (1.0) power factor.

Space heaters. Electric heat is generally preferred to steam for space heating, especially on ships that do not use steam for other purposes. This preference has contributed to the steady increase in electric power plant capacity on ships. Space heating includes applications such as convection space heaters, HVAC preheaters, and HVAC terminal reheaters.

Equipment heaters. Electrical equipments exposed to the weather and extreme temperature variations are often fitted with electrical heater elements to prevent condensation within the equipment enclosure. Candidate equipments for such heaters include motors, generators, transformers, motor controllers, topside electronics, diesel engine jacket water, and deck machinery.

Trace heating. Electric trace heating of piping systems to prevent freezing or maintain the viscosity of the fluid within the piping is also often preferred to steam trace heating. Trace heating cables are either strapped alongside or spiral wrapped around and along the piping run. There are three basic types of trace heating:

• Mineral-insulated heating cables consist of one or two series resistance wires imbedded in mineral insulation and hermetically sealed in a metal sheath. Mineral-insulated heating cables are particularly suited for a high-watt output, high-temperature application, but must be completely factory assembled to the desired watt output and physical length.

• Constant-wattage heating cables consist of two parallel bus wires feeding individual parallel resistance-heat modules at specific intervals along the cable. This cable provides a constant watt-per-foot output and can be cut to length during installation to fit the particular piping run.

• Self-regulating heating cables consist of two parallel bus wires imbedded in a conductive extruded core. The conductivity of the core changes with temperature, having a higher conductivity at low temperatures and vice versa, thus consuming more power and generating more heat on a cold pipe. As the temperature of the piping increases, the heating output of the cable decreases. This cable can also be cut to length during installation to fit the application.

Miscellaneous electric heating. Other commonly used electric heating applications include:

• Glass imbedded window heaters (such as on the bridge).

- Tank heating, through submerged heaters (normally used on small tanks only).
- Flow-through fluid heaters (such as for instantaneous and intermittent demands for hot water).
- Cooking appliances such as ovens, stoves, and hot plates.

6.4 Lighting. The electric power demand of the lighting system discussed in Section 7 is a function of the type of luminaries used. Incandescent-type light bulbs operate at unity (1.0) power factor but have a low functional efficiency (lumens/watt). Discharge-type lamps or arc tubes of the fluorescent, mercury vapor, metal halide, or sodium type together with their ancillary reactances of the ballast circuitry operate at power factors of about 0.5 to 0.6 only with low-power-factor ballasts and better than 0.9 with high-power-factor ballasts. However, this technology offers higher functional efficiencies (lumens/watt).

6.5 Electronics. Command and control, interior communication, navigation, exterior communication, and other shipboard electronics are a mixture of constantly fluctuating loads, based on the nature of their function. On commercial ships, the effect of these loads on the electric power system is not significant. However, on naval combatants, the nature and extent of the electronics (combat system) loads have a major effect on the capacity of the electric power plant and the configuration of the power distribution system.

Some of these systems impose steep pulses of high power demands, such as from high-powered radar and sonar transmitters. Without special precaution, such pulses can produce harmonics outside the specified tolerances of the power system. High-powered, directed-energy weapons impose severe demands on the electric power plant of naval ships.

6.6 Other Miscellaneous Loads. Impressed-current cathodic-protection systems are installed on ships to counteract electrochemical or galvanic corrosion caused by electrolysis between dissimilar materials in a saltwater environment. These systems introduce (impress) a d-c

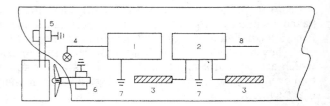

LEGEND:

    1 - CONTROLLER
    2 - POWER CONVERTER
    3 - ANODES (PORT & STBD)
    4 - REFERENCE ELECTRODES (PORT & STBD)
    5 - RUDDER GROUND
    6 - SHAFT GROUND
    7 - EQUIPMENT GROUNDS
    8 - POWER SOURCE

Fig. 25   Typical schematic for a cathodic protection system

voltage across the ship's or system's components of dissimilar material, such as the steel hull and the bronze propeller, to neutralize the voltage generated electrolytically between these components. Such a system is fully effective where the impressed voltage from the cathodic protection system is of equal potential and distribution (but opposite in polarity) across the components as the natural electrolytic voltage.

Figure 25 illustrates schematically the configuration of such a system for protection of the hull. The reference electrodes are mounted through the hull below the waterline and measure the electrolytic potential between the hull and the propeller. The signals from the reference electrodes are fed into the controller. The latter controls the output voltage of the power supply, which is connected to the anodes. The anodes are made of an inert material, such as platinum, to prevent consumption, and are also mounted through the hull below the waterline. The controller, power supply, propeller (shaft) and rudder must be properly grounded to be at the same ground potential.

Other miscellaneous electric systems and equipments include appliances, medical equipment, tools, and degaussing (as described in Section 11).

Section 7
Lighting System

7.1 Illumination Criteria. The lighting system is one of the ship's vital support subsystems. Its purpose is to provide illumination of spaces, operating stations, and other features and functions throughout the ship in support of the activities and duties of the personnel. The level and quality of illumination have steadily improved in keeping with advances in illumination technology, but also in response to the application of human engineering factors to shipboard operations. For commercial ships, the illumination criteria are governed by reference 10. References 11 and 12 contain the illumination requirements for naval ships.

Illumination level. The illumination level is the pri-

mary illumination criterion. References 10 and 11 contain specific requirements for the average illumination levels required within the various spaces, at specific workstations, and for specific work tasks and functions throughout the different types of ships. The average illumination level is the average value of illumination in footcandles over an area (space, workstation, deck area, etc.) on a horizontal working plane at the working level or about 30 in. above the deck. The specific requirements are stated in initial average footcandles obtained with new and clean lighting fixtures (lamps, diffusers, etc.) and clean and newly painted surfaces (decks, bulkheads and ceilings) in the field of view. The initial average illumination levels are compensated for subsequent deterioration during operation (dirt accumulation on lamps and diffusers, and a loss of reflections from surrounding surfaces).

Uniformity. Uniformity is one of the illumination quality factors. The uniformity ratio is the quotient between the maximum footcandle reading under a luminaire and the minimum between two adjacent luminaires. Uniformity factors range from a goal of 1.3 to an acceptable limit of 2.0. The greater the distance between the luminaires and the "task" surface, the greater the uniformity.

Brightness. The source brightness or glare of a luminaire as well as glare from reflections from surrounding surfaces or the back surface itself (instruments, consoles, etc.) should not exceed 450 foot-lambert in the normal field of vision in work areas involving close and long-duration seeing tasks. The source brightness can be reduced with louvers, diffusers, and filters integral to the luminaire. Glare can also be minimized through the proper selection and location of the luminaires and through the application of light-absorbing or diffusing surface coverings.

The brightness contrast ratio is the quotient between the amount of light reflected from the task surface and that reflected from an adjacent or remote surface in the field of view. Brightness contrast ratios range from a goal 3 to an acceptable limit of 10. Limits on brightness criteria are necessary to minimize eye fatigue.

Emergency illumination. A select number of luminaires throughout the ship must be supplied from the emergency lighting distribution system to enhance the safety of personnel during a blackout on the ship-service power system. The selection criteria are governed by references 6 and 7 for commercial ships and reference 11 for naval ships.

Low-level red illumination on naval ships. Navy ships are equipped with low-level red lighting for darkened-ship operation in support of the ships' night missions, in the berthing areas, and in access routes to topside battle and watch stations for safety, comfort, and quick adaptation of personnel to darkened-ship operation.

Illumination levels from the red low-level lighting are generally low, ranging from 0.2 to 2.0 footcandles, depending on the criticality of the location. The wavelength of the red lighting is achieved through red filters over the fixtures.

During routine operations, all external red lights and all internal red lights, which can be seen from outside the ship, are converted to yellow lights through a change in the filters from red to yellow. The purpose for this is to eliminate a potential mistaken identity of red lighting for navigation lights by another ship.

Broad band (blue) (BBB) illumination on naval ships. Certain command and control spaces on naval ships, which contain cathode-ray tube displays, such as the Combat Information Center (CIC), are provided with BBB illumination to minimize interference with display reading. The BBB lighting in these spaces is supplemented with highly directional white detail illumination as required for certain tasks and with white maintenance lighting.

Command and control spaces with visibility exterior to the ship, such as the bridge, are not equipped with BBB lighting.

7.2 Lighting Fixtures. Marine lighting fixtures are of a special design and construction to suit the various environmental requirements incidental to shipboard application. They are ruggedly designed to withstand normal shipboard vibration and to withstand high-impact shock for naval application. Furthermore, lighting fixture enclosures are compatible with the particular environment encountered in the various locations throughout the ship.

Lighting fixtures for commercial ships are generally qualified to the requirements of reference 13. Fixtures not complying with this standard must have U.S. Coast Guard approval for each specific application. Lighting fixtures for naval ships are generally of the types delineated in references 14 and 15.

Types of lamps/lightbulbs. The lamps used for shipboard applications are generally of the incandescent, fluorescent, or high-intensity discharge type.

Incandescent lighting is the oldest electric illumination technology. Incandescent lamps provide a compact, high-brightness source of "warm" light of a pleasant color spectrum, which can easily be directed by a small, simple luminaire. This feature makes the incandescent lamp superior for spot and detail illumination, and it can operate readily on either a-c or d-c supplies with no ancillary devices such as ballasts or starters. It is less sensitive to ambient temperatures than other light sources and is, consequently, preferred where illumination is required in severely hot or cold temperatures, e.g., refrigerated spaces, weather locations, detail lighting and around combustion machinery. Incandescent lamps are also most cost effective where minimal low-level illumination is required such as in small normally unattended spaces, and for indicating lights and instrumentation (dial) illumination. Incandescent lamps are also available for different color spectrums to blend in with the interior decor of a particular space.

However, the disadvantages associated with the high source brightness of incandescent lamps, their low functional efficiency or luminous efficacy (low lumens/watt), and their high heat losses have limited the application of incandescent lighting for general illumination.

Fluorescent lighting has become the predominant technology for general illumination across a broad spectrum, including shipboard applications. Fluorescent lamps are of the electric discharge or arc-tube type. Their relatively low source brightness, high luminous efficacy (high-lumens/watt), low heat losses, and long lamp life have made fluorescent lighting the most cost-effective method for general interior illumination. Fluorescent lamps are also available for different color spectrums (similar and equal to incandescent lamps) to enhance the interior decor.

However, fluorescent lighting has some disadvantages. First, it requires ancillary devices, such as starters and ballasts, to operate. Fluorescent lighting fixtures operate at a power factor less than unity, in the range of 0.5 to 0.6 with low-power-factor ballasts and more than 0.7 with high-power-factor ballasts. On a one-for-one basis, fluorescent lights are more costly than incandescent lights; however, due to their high functional efficiency and low source brightness, most general illumination criteria (levels and uniformity of illumination) can be achieved with fewer fluorescent fixtures vice incandescent fixtures.

High-intensity electric discharge lamps may be of the mercury vapor, metal halide, or sodium type. They are generally compact with a high source brightness and have an even higher luminous efficacy and a longer lamp life than fluorescent lamps. They are particularly suited for floodlighting applications. Most of these lamps emit a color spectrum of light that is generally acceptable for certain working areas such as cargo holds, weather decks, and wide-open machinery spaces, but that is unsuitable for living spaces. Their long lamp life, ranging from 10,000 to 30,000 hours, makes them ideal for fixtures mounted high on masts and kingposts and other locations not easily accessible. The problem of relamping is reduced such that the maintenance for this type of fixture is practically negligible. For certain applications, supplementary instantaneous lighting may be necessary since instantaneous relighting is not inherent in electric-discharge lamps in the event of a power failure. The use of some of these lamps is limited due to certain hazardous characteristics (lamps containing sodium may cause a fire when broken on a wet surface). Serious corrosion conditions may also result if the mercury from a ruptured mercury-vapor lamp comes into contact with aluminum. Even with these limitations, however, this technology has appropriate applications on shipboard.

Fixture construction. Lighting fixtures are packaged, constructed, and shaped according to their application and location on the ship. The packaging and integrity of the enclosures for shipboard use are of the following categories:

• Nontight, for spaces and applications, which are not subject to moisture or volatile gases.
• Dripproof, for wet or damp locations, including areas near or under piping.
• Watertight, for installation in the weather, the bilges or areas subject to splashing, hosing, or severe moisture.
• Explosion-proof, for hazardous areas and spaces that are subject to volatile gases and fumes.

Lighting fixtures are also categorized according to their specific physical application and mounting, such as:

• Ceiling lights, recessed/flush mounted in a false ceiling.
• Deck lights, suspended from the deck above.
• Bulkhead lights.
• Floodlights.
• Detail lights, such as berth lights, mirror lights, spotlights, desk or table lights, and other decorative lights.
• Relay/battery operated emergency lights or lanterns (fixed and portable), to prevent panic and injury to personnel and passengers in case of total power failure

(including the emergency source of power), and to assist damage control functions on naval ships.

7.3 Lighting Layout. The selection and layout of the number of fixtures for any given space, area, or work task are determined through lighting calculations. The methods used for such calculations are governed by reference 10 for commercial ships and reference 12 for naval ships. For general illumination calculations in regular compartments with even and normal deck heights, the "lumen" method is used as follows:

$$\text{Required No. of Fixtures} = \frac{AFK_t}{LK_u}$$

where

A = deck area, ft.^{2}
F = average initial footcandle specified per ft.^{2}
K_t = tolerance factor (about 1.10)
L = output per fixture, lumens
K_u = coefficient of utilization, which depends on the type and size of the fixtures and the room factor, which depends on the fixture mounting height and obstructions or room clutter

When using the lumen method, particular care must be taken in the selection of the room factor in order to account for the actual mounting height of the fixtures and the incidental interferences from equipment, furnishings, and overhead runs of distributive systems. Hence in most cases, room factors cannot be based on room dimensions alone as the lumen output and distribution of each fixture are neither even nor fully effective over the entire range of coverage.

The point-to-point method is used to conduct illumination calculations for:

• Complex spaces with highly dense equipment arrangements, such as machinery spaces.
• Compartments with large deck heights such as cargo holds and hanger bays.
• Detail illumination of task areas and workstations.
• Low-level red lighting and BBB lighting on naval ships.
• Floodlighting on weather decks.

Horizontal illumination distribution curves for the applicable luminaires are used in point-to-point calculations. The footcandles from individual fixtures as derived from the curves are summarized for each specific location. A factor greater than 1.0 is applied to this sum to account for reflections from surrounding surfaces.

7.4 Lighting Distribution System.

Basic 115-volt distribution concept. Each 115-volt lighting bus of each power distribution switchboard (main, emergency, or load-center switchboards, as applicable) is supplied by a transformer bank. In most applications, the transformer windings are connected delta-delta for ungrounded primary and secondary systems. In some installations that use rapid-start fluorescent lamps (without starters), the secondaries of the transformer bank are wye-connected, in lieu of delta-connected, with the neutral grounded to the ship structure to ensure reliable starting of the rapid-start lamps.

In addition to its primary function of distributing 115-volt power to the ship's lighting system, the lighting distribution system supplies 115-volt power to convenience outlets/receptacles, and small auxiliaries and appliances that are within the ratings of lighting branch circuits, including single-phase, fractional-hp motors (normally up to $\frac{1}{4}$ hp). Power, at 115 volts, is distributed through 3-phase feeders from the 115-volt bus of the respective power distribution switchboard to lighting distribution panels located throughout the ship, which supply the lighting and miscellaneous small 115-volt power loads through single-phase mains, branch circuits, and connection boxes. Lighting branch circuits are rated at either 15, 20 or 30 amps with 15 amps being the norm. The connected load on each branch circuit is limited to 80% of the circuit rating.

Ship-service lighting distribution. Separate feeders and associated panelboards are provided for the various functional areas, and locations of the ship, such as:

- Machinery spaces (at least two).
- Cargo spaces (at least one per hold).
- Accommodations (at least one per fire zone and preferably one per deck).
- Mission/combat systems spaces (on naval ships).
- Weather deck (to suit).

A typical stateroom lighting layout is illustrated by Fig. 26.

Lighting panelboards must be located central to the spaces they serve and readily accessible to operating personnel, preferably near the entrance to the space, along passageways, walkways, and gratings. Panelboards installed in passageways are usually recessed into the joiner work or bulkhead. Panelboards for cargo holds are installed outside but near the access to the respective hold.

In certain vital mission-essential spaces such as machinery spaces, public spaces, command and control spaces, and passenger staterooms, the lighting fixtures of a given space are connected to alternate branch circuits so that a failure of any one branch circuit will leave sufficient illumination in a space to permit continued, albeit limited, use of the space.

Emergency lighting distribution. The emergency lighting distribution system is normally energized from the ship-service power system via the emergency switchboard, similar to the emergency power distribution system described in Section 4. Separate emergency lighting feeders and associated panelboards are provided for:

- Machinery spaces.
- Accommodations (one per fire zone).
- Mission/combat systems spaces (on naval ships).
- Bridge and chartroom, including navigation and signaling lights, and floodlights for open-deck and lifeboat areas. (Alternatively, separate feeders may be provided for the navigation and signaling lights and lifeboat lights.)

Emergency lighting panelboards must be located close to the entrance to the areas they serve. Other criteria

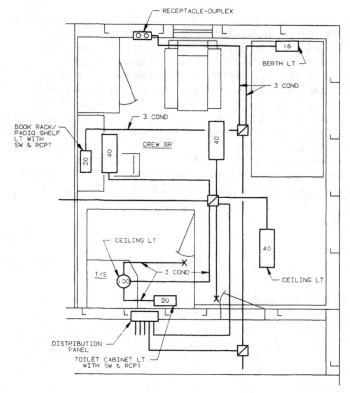

LEGEND:
⊠ - CONNECTION BOX
X - SWITCH

Fig. 26 Typical lighting branch circuit layout for a stateroom

applicable to ship-service lighting panelboards apply similarly to emergency lighting panelboards.

Lighting panelboards. Lighting panelboards are configured with a three-phase bus supplied from a three-phase feeder, and single-phase distribution circuit breakers supplying single-phase branch circuits. The single-phase circuit breakers are connected alternately between the three phases to facilitate balanced loading between the three phases. The single-phase 115-volt loads are assigned to individual branch circuits in such a way to achieve closely balanced loading. Lighting panelboards are generally constructed similar to power panelboards as described in Section 4.

Lighting circuits. In general, the criteria outlined in Section 4 for power distribution circuits pertaining to feeders, mains, and branch circuits apply to lighting circuits as well.

The minimum ampacity of lighting feeder cables must be based on 100% of the total connected load on all the branch circuits plus an allowance for spare circuit breakers on the panelboard. This allowance may either be 50% of the installed spare circuit-breaker capacity or an average of the connected active circuit loads for each spare circuit breaker. The minimum ampacity of mains and branch circuit cables cannot be less than the 15-, 20-, or 30-amp branch circuit-breaker rating.

The voltage-drop considerations discussed in Section 4 for power distribution cables have more significance in the selection of 115-volt lighting circuit cables than for power distribution cables at the higher voltages.

7.5 Signaling Lights and Searchlights. A daylight signaling light is required on all ocean and coastal self-propelled ships over 150 gross tons that are engaged in international voyages. It may be either a fixed unit mounted on top of the wheelhouse, a semifixed unit with arrangements for quick mounting at either wing of the navigating bridge, or a portable handheld unit energized from a self-contained rechargeable storage battery that can operate the light continuously for two hours without recharging.

Ships traversing the Suez Canal are required by the Suez Canal Authority to have a portable searchlight with a special diffused beam, mounted over the bow, to illuminate the banks of the canal. This light is rented from the Canal Authority and is energized from a special receptacle located at the bow.

There are no other rules or regulations for searchlights for oceangoing or coastal ships. However, owners and operators may choose to have one or more searchlights to suit their specific needs. Searchlights on motor lifeboats must be certified by the manufacturer to meet the requirements of the ASTM Specification for Searchlights on Motor Lifeboats (F 1003).

Section 8
Interior Communication Systems

8.1 General. Interior communications (IC) systems provide the means for signaling or transferring information from the source to one or more points through indicating, recording, voice, or audible equipment. This includes alarm systems, telephone systems, navigational sensor systems, and elements of the ship's control systems. The requirements applicable to IC systems are stipulated by references 7 and 11 for commercial and naval ships, respectively.

The U.S. Coast Guard requires that certain critical circuits be supplied with power from emergency or battery sources and recommends that others be connected to the emergency supply if sufficient capacity is available. These requirements lead to a variety of supply systems. Generally, the IC loads associated with the machinery space are supplied from a local IC power panel with an emergency power feeder. Those circuits associated with the wheelhouse are generally connected to the emergency supply via an IC and electronics panel located in, or in the vicinity of, the wheelhouse. Unique systems, such as the gyrocompass, may have a requirement for electrical power not available from the ship's supply or the switchboard. In this case, the equipment will include power supplies to convert the ship's power to the required characteristics.

The interior communication circuits on ships built in the United States are normally identified by alphanumeric symbols similar to those adopted by the U.S. Navy and the Institute of Electrical and Electronics Engineers (IEEE). Symbols identify the circuit primary function and simplify references on plans and correspondence.

IC systems can be classified as indicating, alarm, communication, and control systems. Each of these classifications and the types of circuits which they include are discussed in the remainder of this section.

8.2 Voice Communication Systems. Communication systems fall into two general categories: those required for the safety of the ship, and those installed as a matter of convenience. Communication systems aboard ship are typically as follows.

Emergency loudspeaker system (circuit MC). All ocean and coastwise passenger vessels certified to carry 500 or more persons, including officers and crew, and all passenger vessels whose lifeboats are stowed more than 100 ft from the navigating bridge, are required by the U.S. Coast Guard to have an approved emergency loudspeaker system.

The control panel must be located in the wheelhouse. If the amplifier is not in the same enclosure as the control panel, it must be in a compartment next to the wheelhouse. The control panel must be arranged such that the operator can broadcast separately or collectively to the following stations:

- Lifeboat stations, port and starboard.
- Lifeboat embarkation stations, port and starboard.
- Public spaces used for passenger assembly stations.
- Crew's quarters.
- Accommodation spaces and service spaces.

Each emergency loudspeaker at a lifeboat or embarkation station must allow two-way conversation with the navigating bridge. Each loudspeaker on an open deck must be directed toward the after end of the vessel and outboard by an angle of approximately 15 deg.

Emergency loudspeakers of sufficient quantities must be provided in all areas specified to meet the sound level requirement specified by the U.S. Coast Guard [7].

Cable runs to the different loudspeaker groups must be as widely separated from each other as practicable. They must be distributed so that a casualty to the port or starboard supplies to loudspeakers on boat and embarkation decks will not render more than half of the loudspeakers in the group inoperative, such as by feeding the loudspeakers from port and starboard multiconductor cables. Cables must be installed in passageways and must not run through staterooms, lockers, and other enclosed spaces. Each junction or connection box in the distribution system must be watertight.

If a vessel has a public address or music distribution system, there must be a means to silence that system.

This means must be next to the loudspeaker system control panel.

The emergency loudspeaker system must be supplied from a temporary emergency power source.

Ship-service telephone system (circuit J). Ship-service telephone systems are used for convenient general telephone communications, and are similar to shore installations. They are generally of the automatic dial type. Trunk lines are provided to connect to shore telephone systems when the vessel is in port.

Sound-powered telephone system (circuit JV) and voice-tube systems. The U.S. Coast Guard requirements specify that vessels shall be provided with a sound-powered telephone system or voice-tube system among the following spaces, except a voice-tube system must not be used if it is longer than 125 ft or if it is ineffective:

- Wheelhouse.
- Steering gear room, if outside of the engine room.
- Alternative steering station if outside of the steering gear room.
- Engine control room, if there is one.
- Maneuvering platform, if the vessel has no engine control room.
- Control room, if the vessel is a mobile offshore drilling unit.

In addition, the requirements specify that a sound-powered telephone or voice-tube system be installed between the following:

- Master gyrocompass and remote wheelhouse repeater compass.
- Wheelhouse and remote radar plan position indicator.
- Wheelhouse and remote emergency squad equipment storage spaces.
- Wheelhouse and remote smoke-detector cabinets.
- Wheelhouse and bow or forward lookout station, unless direct communication is possible. If the bow or lookout telephone is in the weather and on the same circuit as other required stations, there must be a cutout switch in the wheelhouse that can isolate this lookout telephone from the rest of the station.
- Wheelhouse and remote radio room and emergency radio room.
- Wheelhouse and remote radio direction finding equipment. (Equipment located in a space which opens into the wheelhouse is not considered remote.)

These requirements are generally met with sound-powered telephone circuits 1JV and 2JV. Sound-powered telephones are self-contained in that they require no external power. This fact adds to their reliability. Most installations employ the magneto ringing type and thus avoid the requirement for the external ringing circuit "E."

In noisy locations such as a diesel engine room, there must be a telephone booth to permit telephone conversation during vessel operation. In a location where the telephone station audible device cannot be heard throughout the space, there must be an additional audible signal device or light that is energized from the vessel's electric system and is magneto-actuated. Where two or more telephone stations are near each other, there must be a means that indicates the station called.

Installation costs have substantially eliminated the use of voice tubes. Their use is now restricted to short runs, which are still cost effective. Reference 7 provides specific requirements for voice tubes.

Sound-powered telephone call-bell system (circuit E). Sound-powered handset stations may be alerted by a call-bell system. The calling station operates a switch that actuates an alarm at the station being called.

Call-bell system (circuit A). Call-bell systems are usually provided on all vessels carrying passengers; these systems provide a means for passengers to call for assistance, when needed, and normally consist of pushbuttons, buzzers, and annunciators. A pushbutton should be located in each passenger stateroom, convenient to the head of each berth; operation of the pushbutton will cause a buzzer to sound, and an annunciator to indicate the stateroom initiating the call. Annunciators should be located where someone is always in attendance. The power supply should be 115-volts, 60-Hz, single-phase a-c from the emergency lighting system.

8.3 Alarm, Safety, and Warning Systems. Certain alarm and warning systems are essential for the safety of the ship equipment and ship personnel. The design and function of such systems are closely controlled by the U.S. Coast Guard. The alarm indication may be visual or audible. When two or more audible alarms are required in a single space, it is the usual practice to provide a common alarm panel, which provides one common alarm with visual indications of the alarm source. A brief description of typical alarm circuits is as follows.

Lubrication-oil low-pressure alarm system (circuit EC). The lubricating-oil low-pressure alarm circuit provides a warning alarm when the pressure in the lubricating-oil lines to vital machinery becomes dangerously low. The alarm is actuated by pressure-operated contact switches installed in the lubricating-oil pipeline. The switch contacts are held open as long as a predetermined pressure in the line is maintained. When the pressure falls below the acceptable level, the switch contacts close, energizing the alarm signal in the machinery control spaces.

Feedwater low-level alarm system (circuit FW). The feedwater low-level alarm system sounds an alarm at the propulsion control station when the level of the water in the tank supplying the boiler feed pumps becomes low. The equipment consists of a float-actuated switch located at the feedwater heater or tank, alarm bells, indicating lights, and cutout switches. A low level of water causes the switch contacts to close, energizing the alarm indicators. An added feature to this circuit can provide for the cutoff of fuel oil when the feedwater reaches a dangerously low level.

Fire alarm system (circuit F). Fire-protective systems include automatic fire-detecting systems, manual fire-alarm systems, smoke-detector systems, watchman's supervisory system, or any combination of these systems.

• *Automatic fire-detecting systems.* An automatic fire-detecting system may be either a fully electrical/electronic type using thermostats, thermostatic wire, or other electric heat detectors or it may be a pneumatic or pneumatic/electric type using pneumatic tube detectors. The system must be arranged so that the presence of a fire in any protected space will be automatically registered visibly and audibly in the pilothouse or fire-control station. The visible notice must indicate the zone in which the alarm originated. Vessels over 150 ft in length require an audible alarm in the engine room.

Detectors must not be rated lower than 135 F and not higher than 165 F; however, in spaces where a high ambient temperature may be expected, detectors rated not lower than 175 F and not higher than 225 F may be used.

There must be at least two sources of power to the alarm panel. The normal source must be the main power source and the other source must be the emergency power source. The alarm panel must contain a power supply transfer switch to automatically transfer from normal to emergency power.

Cable runs between the fire-alarm panel and fire-detecting or fire-alarm zones must be as direct as practicable and, where practicable, must not be in staterooms, lockers, or other enclosed spaces in order to reduce the risk of damage by a localized fire.

• *Manual fire-alarm system.* This system must be arranged and installed so that the presence of a fire can be reported from any of the protected spaces and be automatically registered visibly and audibly in the pilothouse or fire-control station. The visible notice must indicate the zone in which the alarm originated. An audible alarm is also required in the engine room.

There must be at least one manual alarm box in each fire zone. These boxes must be located in main passageways, stairway enclosures, public spaces, or similar locations where they will be readily available and easily seen in case of need.

• *Smoke-detector systems.* The smoke-detecting system must consist of a means for continuously exhausting an air sample from the protected spaces and testing the air for smoke contamination. The smoke-detecting system must be divided into separate zones to restrict the area covered by any particular alarm signal. Each separate space must be considered as a zone, except when two or three small adjacent spaces having a combined volume not exceeding 5000 feet may be combined on the same zone. A zone must not include spaces of more than one deck.

When not located in the pilothouse or fire-control station, the smoke-detecting cabinet must be placed in convenient proximity to the valve-control station of the fire-extinguishing system. The power supply for the smoke-detecting system must be from the emergency lighting and power system.

• *Watchman's supervisory system.* The watchman's supervisory system consists of an apparatus to verify the presence of watchmen and record performance of their assigned duties. This system can be either mechanical or electric. If an electric system is provided, a recorder is located at a central station in conjunction with key stations along the watchman's route. The power supply for the recorder must be from the emergency lighting and power system.

General alarm system (circuit G). A general alarm system is required, on all manned vessels of over 100 gross tons, to warn all persons on the ship of an emergency. The system consists of electric vibrating alarm bells, manually operated contact makers, and distribution panels. The alarm bells are energized by manually operating a normally open, spring-return-to-normal contact maker to the "alarm" position. The number of contact makers and their location must comply with the applicable U.S. Coast Guard requirements.

The alarm bells must be at least eight inches in diameter and produce a distinctive tone different from all other bells on the vessel. The U.S. Coast Guard requires the sound level to be six decibels above the background noise level existing when the vessel is underway in moderate weather. Particular attention must be given to the number of bells installed and their locations to meet this requirement. Also, consideration should be given to bulkhead, door, and ceiling sound insulation requirements that may interfere with sound transmission. In spaces where the ambient noise level is unusually high, the alarm bells within the spaces should be augmented by flashing red lights.

All general alarm bells are supplied by feeders from a feeder distribution panel through branch-circuit distribution panels. The feeder distribution panel is supplied from the system storage battery source. For vessels that are subdivided into zones by fire-screen bulkheads, at least one feeder must be provided for each zone, as necessary, to supply the bells between adjacent fire-screen bulkheads. For vessels that are not divided into fire zones, the vessel must be divided into vertical zones, not exceeding 150 ft in length, and at least one feeder must be provided for each zone as necessary to supply the bells within the zone. Distribution feeder and branch circuit panels must be located above the bulkhead deck or above the freeboard deck, whichever is the higher, and outside the machinery casing. These panels should be provided with overcurrent protection for each feeder and branch circuit. Disconnect switches are not provided. Branch circuit distribution panels are provided as necessary for each zone. At least one fused branch circuit must be provided for each zone deck level. No more than five general alarm bells should be connected to one branch circuit, and a branch circuit should not supply alarm bells on more than one deck level.

The power supply source for the general alarm system is either a single storage battery or duplicate storage batteries, located in a well ventilated battery locker or battery room that is above the bulkhead deck or freeboard deck, whichever is higher, and outside the machinery space casing. On vessels having only one general alarm battery, the battery must be maintained in a fully charged condition at all times by an automatic charging panel. On vessels having duplicate storage batteries, the batteries should be connected, through a two-position transfer switch (no "off" position), so that one battery will be on charge while the other battery is set up to furnish power to the systems involved. For vessels having a temporary emergency bus on the emergency switchboard, the alarm system may be supplied from this bus in lieu of being supplied from separate general alarm batteries.

Refrigerated spaces alarm system (circuit RA). A refrigerated space alarm system is required in any refrigerated space, accessible to ship's personnel during a voyage, that can be locked from the outside such that egress from the space is impossible. Each space is fitted with a waterproof pushbutton electrically connected to an audible alarm located outside the space unless a mechanical pull operating a jingle bell is provided.

8.4 Indicating, Order, and Metering Systems. Indicating, order, and metering IC systems transmit status information to areas of control aboard the ship. The receiving device may be a dial, a light, a digital display, or an input to a unit of control equipment. Indicating systems are also considered to include the sensors, such as the underwater log or gyrocompass, initiating the information. Typical indicating, order, and metering systems are as follows:

Refrigeration temperature indicating system (circuit RT). Vessels with refrigerated compartments are required to have a refrigeration temperature indicating system, which indicates the temperature maintained within the refrigerated spaces. Temperature sensors, strategically placed in the areas, are connected to indicators located on an indicator panel outside the area. Variations to the basic system consist of added alarm devices to alert the crew before damaging temperatures are reached. Recorders can also be provided to give a printed record of temperature readings versus time.

Shaft revolution indicator system (circuit K). The shaft revolution indicator system indicates the direction of rotation, speed in rpm, and the cumulative revolutions of the propeller shaft. A transmitter is coupled to each shaft. Its output is generally fed into a combination revolution counter and rpm and direction indicator located in the machinery space and into rpm and direction indicators located in the wheelhouse and other miscellaneous spaces as desired.

Bearing temperature monitoring system (circuit TM). The bearing temperature monitoring system is used to continuously monitor the temperature of selected bearings in the main propulsion plant. A sensor, usually a resistance temperature detector (or RTD), is installed for each bearing to be monitored, and is electrically connected to the monitoring alarm and indicating equipment incorporated in a status panel located in or near the machinery control console.

Rudder angle indicator system (circuit N). The rudder angle indicator system provides a means of indicating, at remote stations, the angular position of the rudder; the system normally consists of a waterproof enclosed transmitter, located in the steering gear room, and waterproof enclosed indicators, located in the wheelhouse and other selected stations. The transmitter is a synchrogenerator and should be connected to the rudderpost through a mechanical linkage such that it will transmit the actual angular position of the rudder to each synchro-indicator. The indicators generally consist of a dial containing a fixed amidship line and a moving pointer deflected left or right of the line to indicate left or right rudder angle. The system supply is usually 115-volts, 60-Hz, single-phase a-c and is often supplied from the emergency switchboard.

Pyrometer indicator system (circuit PB). Boiler flue temperatures and other high temperatures are monitored by permanently installed thermocouples mounted in the boiler uptakes. Each thermocouple is contained in a protective tubing, and by means of a rotary selector switch it may be connected to an indicator mounted locally or at the engineering operating station.

Salinity indicator system (circuit SB). The salinity indicating systems provide a means of measuring the degree of salt content of the water in potable water systems and in boiler feed and condensate systems. Separate systems are usually provided for each application, one for each desalination plant and one for the boiler feed and condensate system. The feed and condensate salinity system is usually identified as circuit 1SB and the desalination plant salinity system as circuit 2SB. Each system consists of salinity cells with valve assemblies and an indicator panel. Salinity cells and valve assemblies are designed so that the cells may be removed and replaced without interruption to a continuously operating piping system. To facilitate replacement and servicing, each cell is energized through a watertight plug and receptacle.

Indicator panels are normally arranged with an individual control section for each associated cell, a common audible alarm, and a meter for reading the salinity content at any cell by means of a cell selector switch. Individual control section circuitry is such that when the salinity content of the system, at the point being monitored, reaches a predetermined value, the monitoring cell alarm light and the common audible alarm are automatically actuated. A silencing switch is provided in each individual cell section to deenergize the common audible alarm; operation of this switch should not affect the aural or visual signaling capability provided for other cells. Each cell alarm light is arranged to give indication of alarm until the salinity is reduced below the alarm setting. Salinity systems are normally set up to alarm when the salt content reaches 0.25 grains of sea salt per gallon.

Salinity systems that monitor the desalination plants also incorporate a solenoid-operated dump valve that will automatically dump, to the bilge, or bypass all process water that exceeds the predetermined salt content; dumping or bypassing will also take place upon loss of power to the salinity indicator, the dump valve solenoid, or the desalination plant pumps. Dumping for any reason should actuate the common alarm on the indicator panel.

Underwater log system (circuit Y). An underwater log system measures the ship's speed through water and the distance traveled. The system, which is used for navigation, is described in detail in Section 9.

Doppler sonar speed log (circuit R-SX). A doppler sonar speed log is an electronic system for an accurate measurement of the ship's fore-and-aft velocity relative to the bottom at depths to 500 ft and relative to the water mass at greater depths. Operation is based on the principle that

Table 4   Typical list of orders for an engine order telegraph

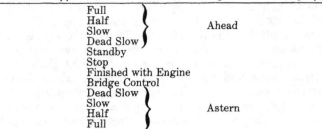

Full	
Half	
Slow	Ahead
Dead Slow	
Standby	
Stop	
Finished with Engine	
Bridge Control	
Dead Slow	
Slow	Astern
Half	
Full	

a signal transmitted from a moving object and reflected back from a stationary surface will indicate an apparent shift in frequency. This frequency shift is proportional to the velocity of the moving object in relation to the stationary surface. Its principal uses are for speed trials and for continuous monitoring of the ship's velocity in coastal waters and harbors.

Gyro compass system (circuit LC). True north and, if desired, ship's roll and pitch information are supplied by the gyrocompass system. The compass heading information is furnished to steering and bearing repeater compasses located at various navigational stations and to other electronic navigating equipments, such as radio direction finder (RDF), course recorder, and the radar displays requiring this information. Roll and pitch information can be supplied to stabilizing equipments if desired.

Engine order telegraph system (circuit MB). Every vessel, except small vessels on which the propulsion plant is controlled entirely from the bridge with no means of engine control from the engine room, should be equipped with a repeat-back signaling system from the navigating officer's stations to the engine room. On vessels with more than one engine, a separate system should be provided for each engine.

In any telegraph system, transmission of orders to the engine room and replies to the bridge should be instantaneous, accurate, and unambiguous under all conditions of ambient light and weather. If dial face or pointer-type instruments are used, accuracy between the transmitter and receiver should be within one quarter of the width of the order sector. Indication at each station in use should include the direction and speed ordered and answered, and such advisory information as needed. Typical orders for an engine order telegraph are given in Table 4.

Transmitted and indicated orders should be visible from a distance of at least 10 ft, normal to the instrument. Bridge unit illumination should be suitable for night vision adaptation.

When electrically operated engine order telegraphs are installed, a bell should ring at every station each time an order or reply is changed, and should ring continuously until the order and reply are in correspondence. A power-failure audible and visual alarm should be provided on the bridge to indicate the failure of power to the system.

The bridge and engine room units may be operated by means of a lever, knob, pushbuttons, or equivalent. The arrangement of the conning station order indicators, pushbuttons, or lever motion should relate to the desired

direction of motion of the vessel, that is, ahead orders to the bow. However, on round-dial flush-mounted units on the bridge, ahead orders should be on the right unless only 180 deg of the dial is used, in which case ahead orders should be toward the bow.

On vessels not equipped with bridge propulsion control, and thus where the engine order telegraph is the primary means of controlling speed from the bridge, the bridge station should be limited to the fore-and-aft-moving lever type transmitter with "Stop" in the vertical position.

An audible and visual wrong-direction signal should be provided in the engine room. This signal should operate whenever a control is operated in such a manner as to produce propulsive thrust in a direction opposite to that required by its engine order reply transmitter. This signal should be deenergized when the propulsion control is assigned to the bridge.

When a system includes more than one topside electrical transmitter, a transmitter transfer control should be provided so that when any one of them is operated, it will automatically be connected in the system and the other transmitters will be disconnected. On transfer, a bell should ring at every station. The reply should be indicated at all transmitters at all times.

Engine order telegraphs may be combined with bridge propulsion control systems when separate internal components are used for the two systems, so that the failure of one system will not affect operation of the other system. The only common parts should be the operating lever, housing, illumination components, and interlocking.

Whistle operator system (circuit W). The whistle operator system may be used in the "at will" or "automatic" mode of operation and from various conning stations. Moving a control switch to the "at will" position closes the circuit to a solenoid, which opens the whistle valve or closes a contactor. Moving a switch on the whistle timer control panel located in the pilothouse to the "automatic" position energizes a coding timer motor driving a mechanical cam that intermittently closes the whistle solenoid circuit. The coding timer can be set for time periods as required by international navigational rules. Mechanical whistle pulls are installed at the navigating stations for emergency use.

Automation and control systems. A number of IC circuits are included in the ship's automation and control systems.

8.5   Entertainment Systems.

Ship's entertainment system (circuit SE). Circuit SE distributes speech, radio, and taped programs in the passenger and crew quarters where terminals exist. A centralized control station, with amplifiers and receivers, broadcasts the programs collectively, or in groups, over a series of reproducers. The system components vary to suit the type of vessel. Entertainment systems which rebroadcast externally received radio signals should have features which suppress own-ship active radiation.

Ship's television system (circuit TC). The ship's television system provides an outside antenna connected to outlets located in various spaces as desired. The system may also include an omnidirectional or rotating antenna as well as a radio frequency signal booster.

8.6 Man-on-the-Move Communication Systems. Man-on-the-move communication systems may range from a simple commercial set of walkie-talkies to a complex secure shipboard radio communication system, such as used on naval ships. The flexibility in equipments and functions of the systems used on naval ships allows practical applications for large aircraft carrier/amphibious class ships and small surface combatants. Their major functions includes: air-operations, air-operations support, seamanship, damage control, ordnance, weapon handling, and cargo handling.

The heart of the system is in the signal data converter, which acts as the audio distribution system and also monitors the equipments' level of functionality through its built-in test equipment microprocessor. The system is backed up by its own emergency power system.

The mobile, or portable, units consist of transceiver, flight deck noise-occluding helmets and amplifier/charger. The transceiver is plugged into the amplifier/charger; the audio is amplified to four watts, and a trickle charge is simultaneously applied to the transceiver battery. The unit is a low-power, secure, personal, portable transceiver. Fully synthesized, the unit provides 500 frequency channels within the 340-to-390-MHz band.

8.7 IC Switchboards. The IC switchboard is the nerve center of the interior-communication system on naval ships. Its function is to energize all interior-communication and fire-control circuits, including fire-control electronic systems and, in small ships, to supply power to other electronic equipment. The IC switchboard is installed behind the armor belt, where applicable, and generally below the waterline to obtain maximum protection. It is energized from a normal, an alternate, and an emergency power supply to ensure continuous service.

In large combatant ships two main IC switchboards are provided. One switchboard is located in the forward IC room, and the other in the after IC and gyro room. Thus, each system or equipment receives its normal supply from the nearer IC switchboard. The after main IC switchboard is usually arranged similarly to the forward main board, except that in the after board some of the special buses such as the controlled-frequency bus may be omitted.

A dead-front, front-service type of IC switchboard is preferred; it is constructed similarly to the dead-front type except that it is designed so that installation, operation, and maintenance can be accomplished entirely from the front of the switchboard. The principal advantage of this type of switchboard is that it can be mounted against a bulkhead, thus requiring less deck space.

IC switchboards contain power control distribution switches, circuit protective devices, action cutout switches, transfer switches, and indicating devices as necessary for interior communication systems.

8.8 Fiber Optics. Fiber optics is a relatively new technology that is well suited for IC system communications. The radiant power, or light, used in fiber-optic systems is

electromagnetic energy that is several orders of magnitude higher in frequency than are radio waves. Fiber-optic systems are usually described as being designed to operate with wavelengths somewhere in the range of 750 to 1600 nm (meters $\times 10^{-9}$).

Light travels at a speed of 300,000 km/s in free space, but it travels more slowly in other media. When light passes from one medium to another, it changes speed, which causes a deflection that is called a "refraction." The exact characteristics of light propagation along an optical fiber depend on the size of the fiber, the material of the fiber and how it is made, and the nature of the injected light. Maxwell's electromagnetic equations show that light does not travel randomly through a fiber; it is channeled into modes. In simple terms a mode is a possible path.

The types of optical fibers commonly used are multimode step-index, multimode graded index, and single mode. Multimode step-index is the oldest type and has the lowest bandwidth. Single-mode fibers have core diameters on the order of 2 to 10 μm. The small core limits the propagated light to a single mode. Single-mode fiber has the astonishing bandwidth of 2 Ghz-km. A single optical-fiber can carry as much intelligence as a 900-pair copper cable.

A simple fiber-optic system consists of a transmitter, a fiber, a receiver, and connectors to join the components together. A system can transmit a single signal, e.g., the light being either on or off, or the intensity of the signal can be proportional to the magnitude of the intelligence being transmitted. Many signals can be transmitted simultaneously by using some form of multiplexing.

Two types of semiconductor light sources in common use are light-emitting diodes (LEDs) and lasers. Light-emitting diodes are used widely in transmission systems over short distances. LEDs radiate with an optical power proportional to the driving current and have a wide emission angle.

The laser or injection-laser diode (ILD) is the highest-quality light source for fiber-optic transmission. It is small (about the size of a grain of salt) and is much more directional, resulting in less scattering.

Optical fibers have unique characteristics and capabilities that are ideal for use in a marine environment. The data-carrying capacity of an optical fiber is far superior to that of a coaxial cable or twisted-pair conductors. A fiber-optic cable is immune to the harsh electromagnetic environment found in most marine vessels. It does not conduct electricity and is, therefore, intrinsically safe in a hazardous or explosive atmosphere. Optical fibers also eliminate the various problems associated with ground loop currents because of the electrical isolation. The small size and weight of optical-fibers coupled with their tremendous data-carrying capacity offer substantial space and weight savings over conventional systems.

The first known operational fiber-optic system in a commercial marine vessel was installed in the machinery control system for a ship capable of periodical unattended operation. The system consisted of five input/output (I/O) units located throughout the machinery space in areas of high concentration of signal sources. The various signal sources were wired to the I/O units by conventional methods. The lengths of cable were usually short because of the strategic location of the I/O units. The signals were then multiplexed and transmitted over fiber-optic cables to the central processor located in the machinery control room. Thousands of feet of conventional cable, along with the associated weight and installation labor, were saved. A corresponding reduction in potential electromagnetic interference was also realized.

Section 9
Navigation Systems

9.1 Radio Direction Finder. A radio direction-finding system aboard a ship provides a means of establishing the ship's position based on the location of two or more fixed transmitting stations. Radio direction finders are used for both navigation and for the safety of life at sea involving search and rescue of ships or lifeboats in distress. A typical shipboard system could consist of a radio receiver-indicator located on the ship's navigating bridge or in the chart room, a fixed cross-loop antenna, and a sense antenna.

The receiver-indicator is tuned to receive the specific frequency signal of the selected transmitting station. The relative intensity of this received signal is reduced by rotating a movable coil, within the receiver, with respect to the fixed-loop antenna until a null position is realized in the signal received from the transmitting station. This null position indicates the bearing at which the system is in electronic alignment with the transmitted wave front of the radio frequency transmission from the transmitting station. The indicated bearing from the ship to two or more fixed stations is noted and the reciprocal lines are plotted on an appropriate chart. The position of the vessel is indicated by the intersection of these lines.

The sense antenna provides a means of preventing the 180-deg ambiguity that could result in bearing readout if the received signal is not properly oriented.

The main radio transmitting antenna should be grounded (inoperative) when the radio direction-finding system is in operation to prevent distorted signals from causing maloperation and to reduce error in the direction-finding system. This grounding is usually accomplished at the main radio console through a switching unit that disables the main antenna and also provides power to the radio direction-finding system. Also, special considerations regarding guywires, stays, handrails, etc., must be taken into account and compensated for by proper design during installation of the radio direction-finding system.

The accuracy of radio direction-finder observations is affected by fluctuations in the ionosphere and consequently vary in accordance with the time of day and time of year. Daily distortion of the skywave component of the radio transmission known as "night effect" can cause considerable error in the plotted position.

The shipboard power supply for the radio direction finder is usually derived from the main radio console power supply through its switching unit.

9.2 Radar Systems. Radar systems for ship navigation normally operate at wavelengths of 3 and 10 cm. Two systems are usually required for larger ships, one main S-band and one auxiliary X-band. A system consists of a transmitter-modulator, receiver, a rotatable directional antenna (scanner), required power supplies, a master indicator, and remote indicators (when desired). The radar video presentation displays relative motion data in Plan Position Indicator (PPI) form on either a 12- or 16-in. cathode ray tube. The display is a map presentation of an area 360 deg around the ship. The radius of this presentation can be varied by the operator from close ranges to the maximum range capability of the radar. The maximum range of a radar system is a function of frequency and transmitter power, but is primarily controlled by the line-of-sight characteristics of microwave frequencies. The video presentation displays other ships, buoys, islands, and other navigational hazards and shows their relationship to the navigating station.

Introducing compass information into the radar system display will permit the observed bearings to be measured as true north, magnetic north, or relative quantities. Circuitry can be provided to give a true motion presentation on the PPI; that is, echoes of stationary objects remain stationary and echoes of moving objects move. Normally, without this circuitry, all objects on the PPI tube appear to move because of the ship's motion.

Radar is of utmost importance to safety of navigation at sea. Automatic radar plotting aids are essentially utilized to improve the standard of collision avoidance at sea. It enables the operators to automatically obtain information concerning multiple targets and provides continuous, accurate, and rapid situation evaluation.

9.3 Hyperbolic Systems. A popular method of navigational positioning measures the difference in distance between the navigating station and two or more known fixed stations. A navigating station positioned between two stations at known geographic locations can measure the difference in time of arrival of a radio frequency transmission from the two stations if their transmissions are synchronized. A line connecting all the positions between the two transmitting stations where the same time difference, or delay, would be measured will describe a hyperbola, having the transmitters as focal points. A similar measurement between one of the first two transmitting stations and a third station will describe a second hyperbolic line that will intersect the first hyperbolic line at the navigating station position, as shown in Fig. 27. This method allows the installation of a network of stations and the preparation of navigational charts with the hyperbolic

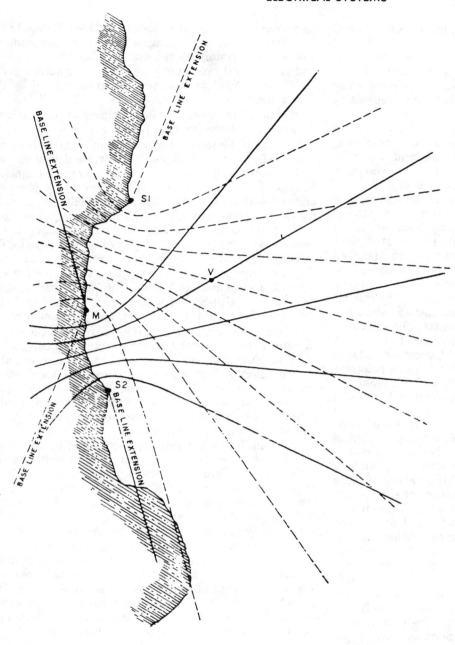

Fig. 27 Typical lines of position from two pairs of Loran stations

calculations identified. The navigating station takes two or more measurements and projects its position from the appropriate intersections on the chart.

Loran A was one of the first operational hyperbolic systems and operates in the 1.5 to 2 MHz radio frequency spectrum. The Loran transmitting stations are operated in pairs with one the master station and the other the slave station. The master station transmits a pulse of RF energy; the slave station receives this signal and after a programmed time delay transmits a pulse of RF energy. The navigating station measures the lapse of time between receiving the master station signal and the slave station signal; the navigating station then measures the lapse of time between receiving signals from a different pair (master and slave) of Loran stations. Each measured lapse of time is used to determine the navigating station's proper line of position (with respect to each pair of transmitting stations) on a Loran (hyperbolic) chart. Where the two measured lines of position cross establishes the navigating station position. See position "V" on Fig. 27.

Each pair of Loran stations is usually located several hundred miles apart. One station is often made common to two pairs. As noted in Fig. 27, M is a common or double-pulsed station (master); S1 and S2 are single-pulsed stations (slaves). A double-pulsed station is considered to be two separate stations at the same location, since this type of station sends out two entirely distinct sets of pulses;

each set of pulses is paired with the pulse from adjacent stations.

A typical shipboard Loran installation includes an antenna and a receiver-indicator. The receiver-indicator should be installed in the chart room, above the chart table, and should provide direct in-line time-delay readings.

Loran transmissions are effective at ranges of about 700 nautical miles in the daytime and up to 1400 miles at night with accuracies between ±1 mile and ±5 miles depending upon the location of the navigating station in the hyperbolic grid, time of year, and time of day.

Loran C uses the same principles as Loran A except for the frequency of transmission (100 kHz), which affords better and more stable area coverage. Each master station has at least two and sometimes three associated slave stations; using these multiple slave stations provides for a more accurate identification of line-of-position hyperbolic grid lines. The use of the lower frequency, as compared with Loran A, allows for greater range on the reliable ground wave and separation of the master and slave stations by as much as 800 nautical miles. The improved accuracy of the Loran C system as compared to Loran A is basically due to the ability of Loran C receivers to track a specific cycle in the transmitted pulse. Loran A receivers are limited to tracking the leading edge of its higher frequency transmission. This advantage makes Loran C a widely used navigational aid.

The Omega navigational system is a long-range hyperbolic navigation system that uses synchronized signals from two or more radio transmitters that may be spaced several thousand miles apart. Omega stations transmit on low radio frequencies from 10 to 14 kHz; the use of these low frequencies results in the transmission of signals that are very stable, accurate, and predictable. Precalculated line-of-position (hyperbolic) charts are used to plot the navigating station position, based on measuring the particular signals received from pairs of stations, somewhat similar to the Loran system, except that all stations are synchronized from a common time source and share transmitting time with the other stations. Each station transmits for a specific duration of time (approximately one second), and all stations transmit successively and always in the same order. Ultimately, the Omega navigating system will consist of only eight stations strategically located so as to provide worldwide coverage; it is expected that the Omega system will result in accuracies approaching ½ nautical mile in daytime and one mile at night.

A shipboard Omega installation would consist of a comparatively short whip antenna and a receiver.

The Decca navigation system is a highly stable continuous-wave radio position-fixing hyperbolic-type system operating in the 70–130 kHz band. Each system or "chain" comprises a central master station and three slave stations disposed in a three-pointed star configuration about the master station. Slave stations are 70 to 120 miles distant from the master station. There are more than 35 such chains throughout the world, most of which are in the Eastern Hemisphere. The performance range is up to 250–300 miles by day, and positions correct to within ±50

yards can be obtained at a range of 50 miles. Unlike Loran and Omega, which are operated by the U.S. Government, Decca is a privately owned system. The Decca receiver and indicators (Decometers) are generally installed in the chart room. Navigation charts for areas served by the Decca chains are readily obtainable.

In general, receivers for the foregoing hyperbolic navigation systems are located in the chart room.

9.4 Sonar Systems. Sonar systems incorporate advanced electronics techniques to allow 360 deg of reliable underwater visual presentation for safe ocean navigation. Horizontal scanning sonar systems provide a practical and efficient method of underwater surveillance. A sonar transducer simultaneously emits ultrasonic pulses horizontally around the ship. Returning echoes of targets are received and the picture is displayed on the bright CRT screen. Sonar effectively provides a means to search and detect submerged obstructions such as reefs or sunken ships that endanger sea routes. The basic sonar system consists of a display console, a transmitting/receiving unit, transducer unit, control unit, and power supply.

An echo depth-sounding system provides a means of measuring the sea depth beneath the vessel by computing the time interval required for sound waves to travel, at a known velocity, from the ship's bottom to a reflecting surface and return. The measured depth may be visually indicated and permanently recorded. This system usually consists of a transducer, an electronic control unit, an indicator, and a recording unit.

The transducer is permanently mounted in and flush with the ship's hull at approximately one fourth the distance from the bow to the stern, and as near the ship's centerline as is practicable. This transducer converts electric oscillations into sound energy during the transmit cycle and converts the sound echo into electrical oscillations during the receive cycle. The control unit, which contains the circuitry that produces, amplifies, and controls the various signals necessary for system operation, is usually located in the chart room. The indicator visually displays the depth of the water, and is usually located in the wheelhouse. The recorder produces a permanent record of the sea depth on a calibrated chart and should be located in the chart room above the chart table.

Some vessels are equipped with an alarm device, located in the chart room, that sounds when a predetermined shallow depth occurs.

Echo sounding systems are also used for the measurements of ground track by using two, four, or more directional sonar beams transmitting impulses at an angle. The doppler effect on each beam is measured, indicating a rate of change of slant range in that particular direction. Ground track navigation systems are used by large vessels navigating in shallow coastal waters, survey vessels, and ships required to maintain station over a given point. Ground track information is also required by ships using a doppler-shift method of satellite navigation such as the Transit System. A typical ground track installation could consist of a receiver/driver, four transducers mounted in the ship's bottom (which may be combined in arrays), a dead-reckoning tracer, and a digital readout device displaying forward, aft, and athwartships velocities.

9.5 Satellite Navigation. Satellite navigation is the utilization of orbiting satellites to provide all-weather worldwide positioning at sea. A series of satellites are in circular, polar orbits, about 600 miles high, and circle the globe every 107 minutes forming a "birdcage" within which the earth rotates. Each time a satellite passes above the horizon, which occurs every 35 to 100 minutes depending on latitude, there is an opportunity to obtain a position fix. Each satellite transmits a message that allows its position to be calculated accurately as a function of time. An accurate position fix is obtained by combining the calculated satellite positions, the range difference measurements between these positions (Doppler frequency shift counts), and the vessel-motion information.

The first operational development using satellites for navigation was the U.S. Navy Transit System. Military requirements dictated that this be a passive system. Satellites in polar orbit are monitored by ground tracking stations. Some of the tracking stations have the capability of updating the information stored in the satellite-borne computers. Every two minutes a satellite transmits a message containing the time of transmission and orbital track information. The transmission occurs on two frequencies, in the UHF spectrum, separated by approximately 300 MHz, to compensate for distortions in propagation encountered in ionospheric and atmospheric penetration. The navigating station receives the transmissions with a receiver/computer complex that measures the doppler effect on the transmissions caused by the satellites movement with respect to the navigating station. The resulting quantity, which is the rate of change of slant range, is used to compute position. By taking several observations from successive satellites, accuracies to a few hundred yards are possible. A shipboard installation for the Transit System consists of an omnidirectional dual-frequency antenna, a receiver/computer console, and a precise frequency standard.

The Global Positioning System (GPS) is a more recent satellite navigation system that will provide extremely accurate information for navigating ships at sea, when it becomes fully operational. It will use 18 satellites in 6 orbital planes, ensuring that the GPS receiver will see up to 4 satellites simultaneously on a continuous basis. GPS will provide a constant fix of vessels with great accuracy.

9.6 Inertial Navigation Systems. Inertial navigation or guidance is the method of directing the movements of a ship, aircraft, ballistic missile, or other vehicle from one point to another based on the acceleration of the vehicle, which is sensed and integrated to determine the velocity and position of the vehicle. This remarkably versatile method, more so than any other kind, is based on measurements made with respect to an earth-independent frame of reference, rather than with respect to earth-fixed objects using the Newtonian laws of motion and of gravitation. Their versatile features have given inertial systems the highest military importance, largely due to their virtually unlimited potential for accuracy. For this reason, advanced technological developments of inertial systems

have been applied almost exclusively for military purposes. However, the features of inertial systems are attractive for all types of navigation.

An inertial navigation system basically includes a set of gyros, a set of accelerometers and integrators, a gravitational computer, a navigational computer with a sidereal clock, and an optional optical monitor system. The gyro device, which tends to preserve or maintain a fixed orientation in space, provides a self-contained frame of reference, which is analogous to the exterior frame of reference provided by the stars in ordinary celestial navigation. An assemblage or cluster of two or more gyros, suitably gimballed relative to the craft, provides the stable element that can be maintained in a predetermined spatial orientation despite movements of the vehicle. The stable element, perfectly balanced and supported about three axes, literally preserves fixity after being initially aligned with respect to a set of stars.

The set of accelerometers constitutes the essential sensing capability of the system. A set of three accelerometers, orthogonally mounted on the stable element, can sense speed changes in any direction. The orientation of the accelerometers is established by the gyros so that components of the total acceleration vector can be determined. Integrators are provided for the accelerometer signals, to generate the first and second time integrals of vehicle acceleration, which are components of velocity and displacement.

A gravity computer is used to calculate the magnitude and direction of the gravitational acceleration, so that its components can be reflected in the accelerometer indication. Gravity computation is possible because the earth's surface gravity is known accurately as a function of latitude, and can be calculated as a function of radial distance with the aid of the universal law of gravitation.

A navigational computer is used to calculate the vehicle's velocity vector and supply to the gyros the control torques necessary to maintain the stable element that carries the accelerometers in a level position as the vehicle moves over the earth. The feedback from the accelerometers through the computer to the gyros takes known values of the earth's radius into account and accordingly performs a function equivalent to explicit gravity computation. Earth rotation effects are calculated in the computer, and appropriate corrective signals are applied to the inputs and outputs of the accelerometers and gyros.

An optical monitor system, which is based on a telescope taking bearings on stars or other celestial bodies of known positions, is used to periodically detect and correct gyro errors.

9.7 Electromagnetic Log System. A basic electromagnetic log system consists of a rodmeter sensing unit, an indicator-transmitter, a remote control unit, and a sea chest. With the sensing-unit electromagnet generating a magnetic field, an induced voltage in the surrounding water is detected due to the fluctuation of the magnetic field in the water. The sensing unit is hull mounted through a sea chest. This voltage is amplified and converted to a shaft position by a nulling servo system and displayed at

Table 5   Navigation lights on oceangoing ships

Designation	Arc of Visibility, deg	Lens Color
RUNNING LIGHTS		
sidelight (stbd)	112.5	green
sidelight (port)	112.5	red
masthead (fwd)	225	white
masthead (aft)	225	white
stern light	135	white
SIGNAL LIGHTS		
anchor (fwd)	360	white
anchor (aft)	360	white
not-under-command	360	red
towing	225	white

the indicator-transmitter as speed in knots. This is converted by a digital integrator to the amount of distance traveled. Both speed and distance traveled are then transmitted to other display equipment.

9.8 Dead-Reckoning System. A dead-reckoning system provides a continuous ship-position estimation based on the speed and course of the vessel, the time the ship has been running on the course, and the effects of wind, waves, and current encountered between the two previous position fixes. The data required for dead-reckoning calculations include the last known position, the course as indicated by the gyrocompass, the speed as indicated by the speed log, the running time since the last position fix as indicated by an internal clock, and the set and drift between the last two position fixes. At each position fix, the ship's estimated position is compared with the calculated position, and the distance between the two positions is used to calculate the set and drift. This information is then used for the next dead-reckoning calculation. The basic dead-reckoning system consists of an analyzer-indicator, which is located in the chart room, and a tracer unit.

9.9 Navigation Lights. All seagoing ships must be equipped with navigation lights in compliance with the International Regulations for Preventing Collision at Sea (commonly called "COLREGS") as established through the International Maritime Organization (IMO). The navigation lights of ships sailing on inland waterways must also meet the applicable rules of the pertaining country. Reference 16 provides detailed requirements for navigation lights and is applicable to both the international COLREGS as well as inland navigation rules in accordance with the applicable laws of the United States. Navigation lighting fixtures are required to meet the provisions of Underwriters Laboratories Standard UL 1104, and be approved by the U.S. Coast Guard. Navigation light fixtures are not required to be listed and labeled by the Underwriters Laboratories.

Navigation and signal lights fitted on oceangoing ships are listed in Table 5. A typical arrangement of navigation and signal lights for a crude-oil tanker is shown in Fig. 28.

Sidelights are located port and starboard normally at the navigation bridge level so that each light is visible at a distance of at least 3 miles on its respective side from directly ahead to 22.5 deg abaft the beam of the ship. Sidelights are generally fitted with inboard screens such

that in the forward direction the intensities decrease to reach practical cutoff between 1 and 3 deg outside the prescribed sectors.

Masthead lights forward and aft are screened so as to be visible at a distance of at least 6 miles forward through an arc of 225 deg that is, to 22.5 deg abaft the beam on either side. These lights are located one on the forward mast and one on the after mast, and are placed over the fore-and-aft centerline of the ship. The after masthead light is located not less than 4.5 m vertically above the forward light and the horizontal distance between lights is not less than one half the length of the vessel but need not be more than 100 m. The forward light is placed not more than one quarter of the length of the ship from the stern.

The stern light is located as nearly as practicable at the ship's stern and is screened so as to be visible aft over an arc of the horizon of 135 deg (67.5 deg to port and starboard from the centerline) for a distance of at least 3 miles.

The forward anchor light is located at the bow of the ship at least 6 m above the hull; the after anchor light is located near the stern at a height not less than 4.5 m lower than the forward anchor light. Both lights are required to be visible all around the horizon from a distance of at least 3 miles.

Not-under-command light installations consist of two red lights located in a vertical line, one over the other, not less than 2 m apart and visible all around the horizon from a distance of at least 3 miles. These lights are normally located midship on the mainmast directly below the masthead light.

Towing light installations consist of two lights located in a vertical line with the forward masthead light, one over the other, not less than 2 m apart and screened to show forward through an arc of 225 deg (22.5 deg abaft the beam on either side). These lights are permanently installed on vessels expected to operate frequently with a tow; otherwise they may be portable.

The running lights (masthead, side, and stern) are required to be of the two-compartment type with a single-filament lamp in each compartment so as to have a backup light in the event of failure of the primary light. The signal lights (anchor, not-under-command, and towing) are of the single-compartment type with a single-filament lamp. Each compartment of a single- and two-compartment fixture is provided with a 3-conductor flexible cable and a unique plug that is configured to fit designated receptacles.

Ships equipped with a steam whistle are usually equipped with a whistle light to illuminate the steam released when the whistle is operated.

In addition to the foregoing, liquid cargo ships are required to display a red warning light during transfer of bulk cargo; this light must be located above the wheelhouse so as to be visible all around the horizon.

A navigating light panel is installed in the wheelhouse for the control of all running and signal lights. This panel combines an automatic or semiautomatic telltale running-light section for audible and visual alarm and individual

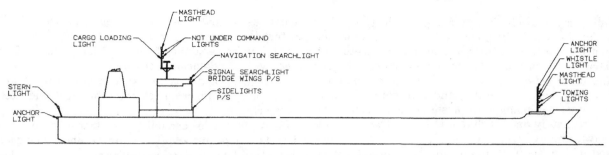

Fig. 28 Typical arrangement of navigation and signal lights on a crude-oil tanker

control of the masthead, side, and stern lights and a signal-light section for the control of anchor, not-under-command, and towing lights as applicable. The running-light section is arranged to indicate the failure of each primary lamp and is provided with a switch, either automatic or manual, for transfer to the secondary lamp. The signal-light section is arranged only for "on-off" operation of the light supply circuit.

Section 10
Exterior Communication Systems

10.1 Radio Communication. Radio communication installations consist of receivers, transmitters, and transceivers together with their associated terminals and antenna systems. These systems are selected on the basis of the needs of the individual ship and in accordance with the minimum requirements established by the Federal Communication Commission as defined in reference 17, the U.S. Coast Guard, or the U.S. Navy. SOLAS Safety Radiotelegraphy or Safety Radiotelephony certificates attesting to compliance with SOLAS requirements are required on vessels engaged in international voyages.

Communications are normally maintained by voice or telegraphy; the more advanced communication systems may include teletype, digital, and encrypted capabilities.

The radio installation for a typical commercial ship consists of the following items:

1. A main transmitter, receiver, antenna, and power supply.

2. A reserve transmitter, receiver, antenna, and power supply, all of which must be separate from those required in Item 1.

3. An HF teletype (TTY) transmitter, receiver and power supply with printer.

4. A radiotelephone transmitter, receiver with selective ringer, antenna, and power supply.

5. An automatic alarm receiver to monitor the 500-kHz international radiotelegraph distress frequency.

6. An automatic alarm keying device to key the main and reserve transmitters with the international automatic alarm signal in case of an emergency.

7. A survival-craft portable transceiver for use in a lifeboat.

8. Emergency position indicating radio beacon (EPIRB).

On most ships, the aforementioned radiotelegraph equipment, excluding the antennas and Items 7 and 8,

along with the necessary accessories that are required for complete operation and supervision of the radio complex, may be combined into a communication console designed for use by one operator. This console is located in the radio room and is fitted with special means of main and emergency lighting separate from the radio room lighting. The normal lighting supply is from the normal power feeder to the console, and the emergency lighting supply is from the emergency radio battery.

Electrical power for the main console is supplied from the emergency switchboard. D-c power-conversion equipment is located in the console. The emergency power supply for the reserve transmitter, receiver, and automatic keyer is from a storage battery located in a battery locker adjacent to the radio room. Means must be provided to maintain this battery in a fully charged condition at all times.

The automatic alarm receiver for monitoring the international distress frequency, 500 kHz, Item 5 of the foregoing, is provided to supplement manual monitoring when a radio operator is not on duty. This feature is provided to satisfy the SOLAS and FCC requirements that all ships at sea must monitor continuously the internationally assigned distress frequency.

The automatic alarm keyer, Item 6, is primarily an emergency device that can be set up to key either the main or reserve transmitter with a predetermined automatic alarm signal sequence.

Some ships are provided with facsimile recording equipment, including antenna, for the purpose of reproducing weather maps and charts transmitted from shore stations. This recording equipment is usually located in the radio room.

The radiotelephone installation, Item 4, provides medium-range voice communication service between ships and between ship and shore stations by using the 2–27.5 MHz maritime service bands; it may also be used to monitor international distress frequencies. This installation usually consists of a transmitter-receiver unit complete with a local hand microphone and a channel selector switch for selecting pretuned operating frequency circuits, a power supply unit, a selective ringer, and a remote telephone handset. Most equipments are designed to transmit and receive alternately on a push-to-talk basis. The transmitter-receiver unit and its power supply unit are sometimes mounted within the main radio console or may be separately mounted, either in the chart room or in the radio room. The selective ringer has a built-in signal bell and is arranged electrically to sound when another station is trying to contact the vessel; this ringer is normally located in the radio room or on the bridge. Where greater range is required, radiotelephone equipment covering the 2–27.5 MHz marine bands and employing the single sideband mode of transmission and reception may be fitted. This equipment provides reliable radiotelephone communication over many thousands of miles under normal atmospheric conditions.

For reliable short-range communication up to 50 miles, VHF (very high frequency) transceivers are commonly used. The units operate in the 156–162 MHz marine bands and employ the frequency-modulation mode of transmission and reception. A bridge-to-bridge channel (channel 13:156.65 MHz) is incorporated in the VHF transceiver and is used for bridge-to-bridge communication between vessels for navigational purposes only. Channel 16 (156.8 MHz) is the allocated channel for distress, safety, and calling.

Radio communication antennas may be of the vertical-standing whip type, or a wire type, depending on the system requirements and the shipboard configuration. Each antenna design and installation for compulsorily fitted equipment is subject to Federal Communication Commission and U.S. Coast Guard approval.

A survival-craft portable radio unit, complete with hand-cranked generator, is provided on each vessel for emergency use in lifeboats. The entire unit is required to be packaged so that it is buoyant and of a minimum weight; it is normally stowed in the radio room or is located near the lifeboat in a special rack designed for easy and quick removal.

10.2 Satellite Communication. Communication satellites, stationed in synchronous orbits over the Pacific, Atlantic, and Indian Oceans, provide a reliable linkup between shipboard terminals and the network of telecommunications stations on land. The satellite communications system provides direct high-quality communications including voice, teletype, facsimile, and high-grade data to navigating ships, anywhere in the world's oceans. The satellites, which serve as relay stations, utilize a circular orbit approximately 22,300 miles above the equator. The system operates in the L and C bands, which are 1 to 2 GHz and 4 to 8 GHz, respectively.

The advent of satellite communications (SATCOM) has made possible a full spectrum of two-way reliable communications in real time on a continuous basis. SATCOM offers the following services:

1. A voice capability from any ship at sea to interconnect with the domestic network.

2. A telex capability to transmit and receive telex messages and to interconnect with the worldwide teletype network.

3. A facsimile capability to transmit and receive weather maps, drawings, manifests, and other graphics.

4. A high-speed data capability for alternate voice data communications at 1200 and 2400 bits per second with speeds of 50 kilobits.

A shipboard SATCOM terminal consists of an electronics unit, automatic tracking antenna assembly, power supply unit, teletype unit, telephone, and facsimile unit. The antenna assembly is designed as a rugged and highly reliable mechanism that covers the full 360-deg azimuth and provides a 90-deg elevation coverage. The antenna is radome-protected and is mounted atop a stub mast.

10.3 Optical Communication. Optical communication is the transmission and reception of data, voice, video, and other information by means of the visible portion of the electromagnetic spectrum. Using electromagnetic waves, optical communication differs from radio and microwave communication only in the fact that the wavelengths employed are shorter, since inversely, the frequencies used are higher. Optical communication provides several advantages. The high frequency of the optical carrier, typically in the order of 300,000 GHz, permits much more information to be transmitted over a single channel than is possible with a conventional radio and microwave system. Also, the very short wavelength of the optical carrier permits the use of very small components. In addition, the highest transparency for electromagnetic radiation yet achieved in any solid material is that of silica glass; this transparency is orders of magnitude higher than that of any other solid material in any other part of the electromagnetic spectrum.

10.4 Infrared Communication. Just beyond the limits of optical vision in the electromagnetic spectrum lies the invisible infrared region. In the communication field, infrared light beams can be used to send and receive data and other signals at vast distances, and provide the capability for effective ship-to-ship and ship-to-shore signaling.

Produced and controlled by familiar types of illuminating and optical devices, the most significant feature of using infrared radiation for signaling and observation is in the fact that it is invisible to the human eye. Thus, infrared communication signaling and observation under conditions of darkness can be accomplished with considerable security against detection.

A basic infrared system consists of a transmitter, receiver, control unit, viewing set, telegraph key, blinker light that is infrared-filtered, and power supply set. The infrared receiver is an instrument that converts infrared radiation into visible light. An electronic-image converter tube in the receiver performs this function. Focusing of the infrared radiation on the photo-sensitive surface of the tube is accomplished by the objective assembly. The transmitter is used to generate infrared radiation for transmission. The viewing set is focused toward a distance-signaling searchlight or blinker light wherein the message is conveyed.

Communications between ships and stations are accomplished with maximum effectiveness on clear, dark nights and in daylight. Rain and snow impede the infrared system operation, while haze, smoke, and fog reduce its range to the same extent that normal vision is impeded or reduced.

10.5 Acoustic Communication. Acoustic transmission of the human voice through water is the best primary communication method between ocean submersibles and between surface support ships and divers. Acoustic communication is the use of sound beams for the transmission of information. Voice and continuous-wave communications have been developed for underwater telephone exchanges between submarines and naval ships; this includes equipment for amplitude modulation, frequency modulation, and single-sideband operation. In general, 500- to 3000-Hz frequencies are used for long-range communications; frequencies in the 10,000-Hz region are used for much voice communications over ranges of several miles, but the use of frequencies above 25,000-Hz limits communication ranges to a few thousand yards.

Acoustic signals are attenuated and refracted in passing through water. The attenuation is a complex function of frequency, water temperature, water pressure (depth), and water salinity. Energy losses increase rapidly with frequency, especially above 25,000 Hz. Thus, low frequencies favor long-range communication, but they severely restrict the operational bandwidth and data rate and commonly require the use of large transducers.

The basic concept of underwater propagation is based on the fact that sound consists of a regular motion of molecules of an elastic substance. Because the material is elastic, a motion of the particles of the material, such as the motion initiated by a sound projector or transducer, communicates to adjacent particles. A sound wave is thereby propagated outward from the source at a velocity equal to the velocity of sound. In a fluid, the particle motion is to-and-fro, parallel to the direction of propagation. Because the fluid is compressible, this to-and-fro motion causes changes in pressure, which can be detected by a pressure-sensitive hydrophone. The hydrophone, which is the basic unit of the acoustic communication system, is a device that transforms or transduces acoustic energy to electric energy.

Section 11
Degaussing

11.1 Purpose. A ship is a large magnet due to the magnetic material that is used in its hull and machinery. A ship has permanent magnetization like that of a permanent bar magnet, and induced magnetization like that induced in soft iron when placed in a magnetic field. The permanent magnetization is reasonably independent of the ship's heading and its position on the earth's surface, but changes somewhat with time. The induced magnetization depends upon the ship's heading and the earth's magnetic field where the ship is located, and changes when the ship changes heading or moves to a place where the earth's magnetic field is different.

Because of its permanent and induced magnetization, a ship is surrounded by a magnetic field, the strength of which decreases with distance. It is this magnetic field which actuates magnetic influence devices that can be used to detect the presence of the ship. The magnetic influence devices may form part of a magnetic detector, to merely detect the presence of a ship, or they may form a part of a firing mechanism of a magnetic mine and cause it to detonate when a ship is near.

The purpose of degaussing naval ships is to counteract or neutralize the magnetic field produced by the permanent and induced magnetization of a ship and thereby make the ship "magnetically invisible" insofar as practicable. This will reduce the possibility that a ship can be detected by magnetic detectors and will also reduce the danger of triggering a magnetic mine. If a ship were perfectly degaussed, its magnetic field would be zero, as if the ship were not there, and its presence could not be detected by magnetic-influence devices.

11.2 Methods of Degaussing. Ships can be degaussed either by magnetic treatment (deperming) or by electric coils. The magnetic treatments used are a normal deperm and a flash deperm. The effectiveness of magnetic treatments is limited since induced magnetization changes with ship heading, roll, pitch, and geographic location while permanent magnetization does not. The cancellation of the ship's magnetic fields through magnetic treatment only (flash deperm) must be tailored to a particular heading and for a particular value of the earth's magnetic field; hence, it is only effective for the positions on the earth's surface where the earth's magnetic field has a certain value. When the ship moves to a place where the earth's magnetic field is different, the cancellation is not as complete and the ship is not as well degaussed. For this reason, degaussing by magnetic treatment alone is not extensively used on ships having other than limited operating areas.

Degaussing by electric coils is accomplished by installing one or more coils in the ship through which a direct current is passed to produce a magnetic field that is as nearly as possible equal and opposite to the magnetic field produced by the induced and permanent magnetization of the ship. Since the magnetic field produced by the ship's induced magnetization changes with heading, roll, pitch,

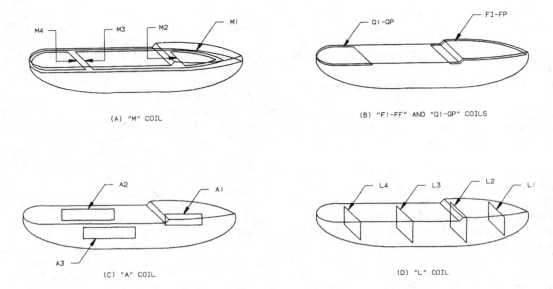

Fig. 29   Typical degaussing coil arrangements

and geographic location, the magnetic field produced by the degaussing coils must be continuously controlled to be effective. The magnetic field produced by a coil is proportional to its ampere-turns, the product of the number of turns in the coil and of the coil amperage. Therefore, when the ship changes heading or position, the ampere-turns in the degaussing coils are changed by changing the amperage, thus varying the magnetic fields of the coils and compensating for the change in the induced magnetization of the ship.

11.3   Shipboard Degaussing Installation.   A shipboard electric degaussing installation consists of the following major components:

- degaussing coils,
- power supply/supplies to provide direct current for the coils,
- control equipment to control the current in the coils, and
- compass compensating equipment to prevent a disturbance of the magnetic compasses by the magnetic field produced by the degaussing coil.

Different degaussing coils are required to completely neutralize the permanent and induced magnetization of a ship. One or more coils may be installed, depending on the effectiveness required of the degaussing system. A multiconductor coil is one in which multiconductor cable is used for the coil. Each loop of the coil has a considerable number of turns connected in series, and all the loops are connected in series to form one coil.

a.   Steel-hull surface ships.   The coil arrangements used with steel-hull surface ships are typically as follows:

"M" coil (main coil).   The "M" coil consists of one or more horizontal loops normally installed below the main deck at the skin of the ship, as shown by Fig. 29(a). On tankers and other similar ships where it is impractical to install the "M" coil internally, it is installed on the weather deck. The function of the "M" coil is to produce a magnetic field that counteracts the magnetic field produced by the vertical permanent and induced magnetization of the ship.

"F" and "Q" coils (forecastle and quarterdeck coils).   On older ships, the degaussing coil installation uses an "F" coil and "Q" coil. The "F" coil has one or more approximately horizontal loops in the forward one third or one fourth of the ship, usually just below the forecastle or other uppermost forward deck. The "Q" coil has one or more approximately horizontal loops in the after one third or one fourth of the ship just beneath the quarterdeck or other uppermost after deck. The "F" and "Q" coils neutralize the permanent and induced longitudinal magnetization.

"FI-QI" and "FP-QP" coils (forecastle induced-quarterdeck induced permanent-quarterdeck permanent coils.   More recent installations utilize split coils, where the coils formerly known as "F" and "Q" are divided into two coils each. The "FI-QI" coil consists of the "FI" and QI" coils connected in series and neutralizes the induced longitudinal magnetization; the "FP" and "QP" coils connected in series neutralize the permanent longitudinal magnetization, as illustrated by Fig. 29(b).

"A" coil (athwartship coil).   The "A" coil consists of one or more loops installed in a vertical fore-and-aft plane, as depicted in Fig. 29(c). The "A" coil neutralizes the induced and permanent athwartship magnetization.

b.   Wooden-hull surface ships (minesweepers—without auxiliary coils).   Wooden-hull surface ships typically have the following coil arrangements:

"M" coil (main coil).   The "M" coil consists of series-connected horizontal loops and compensates for induced vertical magnetization.

"A" coil (athwartship coil).   The "A" coil consists of series-connected loops in longitudinal vertical planes and compensates for induced athwartship magnetization.

"L" coil (longitudinal coil). The "L" coil consists of series-connected loops in vertical athwartship planes and compensates for induced longitudinal magnetization.

"P" coil (permanent coil). The "P" coil is made up of conductors in the "M," "A," and "L" coil cables that are connected in series to form the "P" coil. The "P" coil compensates for the permanent vertical, athwartship, and longitudinal magnetization.

c. Minesweepers having auxiliary coils. Minesweepers with auxiliary coils nominally are configured as follows:

"M" coil (main coil). The "M" coil consists of series-connected horizontal loops encircling separately each of the large masses of steel in the ship having an appreciable vertical extent. The "M" coil compensates for the induced vertical magnetization of these masses of steel.

"A" coil (athwartship coil). The "A" coil consists of series-connected loops in longitudinal vertical planes encircling separately each of the large masses of steel in the ship having an appreciable athwartship extent. The "A" coil compensates for the induced athwartships magnetization of these masses of steel.

"L" coil (longitudinal coil). The "L" coil consists of series-connected loops in vertical athwartships planes encircling separately each of the large masses of steel in the ship having an appreciable longitudinal extent, as indicated by Fig. 29(d). The "L" coil compensates for the induced longitudinal magnetization of these masses of steel.

"P" coil (permanent coil). The "P" coil is made up of conductors in the "M," "A," and "L" coil cables that are connected in series to form the "P" coil. The "P" coil compensates for the permanent magnetization of the steel masses in the vertical, athwartship, and longitudinal planes.

"M AUX" coil (main auxiliary coil). The "M AUX" coil corresponds closely in design to the "M" coil and neutralizes the induced vertical magnetization that is not neutralized by the "M" coil. The "M AUX" coil is installed under the main deck at the skin of the ship in a horizontal plane and consists of series-connected loops.

"A AUX" coil (athwartship auxiliary coil). The "A AUX" coil corresponds closely in design to the "A" coil and neutralizes the induced athwartship magnetization that is not neutralized by the "A" coil. The "A AUX" coil is installed in a vertical plane on the centerline encircling a compartment or large area. Some installations have two coils installed at equal distances port and starboard of the centerline and consist of series-connected loops.

"L AUX" coil (longitudinal auxiliary coil). The "L AUX" coil corresponds closely in design to the "L" coil and neutralizes the induced longitudinal magnetization

that is not neutralized by the "L" coil. The "L AUX" coil is installed in an athwartship plane just inside the skin of the ship, encircling a compartment or large area, and consists of series-connected loops.

"P AUX" coil (permanent auxiliary coil). The "P AUX" coil is made up of conductors in the "M AUX," "A AUX," and "L AUX" cables that are connected in series to form the "P AUX" coil. The "P AUX" coil neutralizes the permanent magnetization in the vertical, athwartship, and longitudinal planes that is not compensated for by the "P" coil.

Some surface ships with wooden or fiberglass hulls have a coil system similar to minesweepers without auxiliary coils, except for the following:

—Individual loop power supplies. Each loop is powered by its own power supply and is not connected in series to other loops of the coil.

—Permanent (P) loop bias. There is no separate P coil or loops. Each M, A, and L loop power supply has a P bias which is set to reduce permanent magnetization of magnetic mass(es) within the loop.

11.4 Power Source. The degaussing coils must be energized with direct current. Solid-state rectifiers or motor generator sets, such as discussed in Section 3, are used to supply d-c power to the coils.

11.5 Control Equipment. Automatic degaussing control equipment is generally installed on ships that are required to have their coil currents changed with changes in the ship's heading. When compensation is required for the ship's heading only, a heading signal is obtained from the ship's gyrocompass system. For the "FI-QI" coil, a signal that is proportional to the cosine of the ship's heading is derived. This signal is amplified and fed to the appropriate power supply so that its output is proportional to the cosine of the ship's heading. This is the variation needed to compensate for changes in induced longitudinal magnetization caused by changes in heading. For an "A" coil, the coil current is regulated to be proportional to the sine of the ship's heading.

Minesweeper installations utilize a magnetometer probe to supply information for changes in heading, pitch, roll, and geographic location. Magnetometer-controlled degaussing has also been incorporated in some vessels with steel hulls.

All types of gyro-controlled automatic degaussing control equipment require manual settings for magnetic latitude and magnetic variation. For minesweepers with a magnetometer probe, this information is derived automatically from the probe.

Automatic degaussing control equipment is equipped with emergency manual controls that are used if the automatic controls become inoperative.

Section 12
Electromagnetic Environmental Effects

12.1 The Electromagnetic Phenomena. Electromagnetic energy is generated by the flow of electric currents in and around the conductor that carries the current. This electromagnetic energy is emitted into the environment surrounding the conductor, hence the term Electromagnetic Environmental Effects or E^3. For a-c currents this electromagnetic energy pulsates at the same frequency as the current. The pulsating electromagnetic energy induces electrical voltages in other conducting materials, including other electrical/electronics equipment and cables that are within the effective range of the pulsating field. Electrical voltages are also induced in a conductor when it moves through a static magnetic field of a magnet or a d-c current.

Such undesirable and excessive electromagnetic coupling between electrical/electronics systems is called Electromagnetic Interference (EMI). EMI can be transmitted through conduction (through the cabling and metallic parts) or through radiation and induction. Electromagnetic Compatibility (EMC) is the inverse of EMI. Whenever the electrical/electronic systems are designed and electromagnetically protected (isolated, shielded, etc.) so that there is either no electromagnetic coupling or any residual coupling is such that neighboring systems do not interfere with each other, there is a state of EMC.

A nuclear explosion releases intense and powerful electromagnetic energy into the environment in the form of an Electromagnetic Pulse (EMP), which can be of such magnitude as to permanently disable electronic equipment within a certain radius of the explosion.

The shipboard electromagnetic environment is very complex, given the compactness of ships, the introduction of highly sophisticated electrical/electronics technology, and on naval combatants the large number of electromagnetic systems (both very powerful active and sensitive passive transmitters/receivers and antennae/sensors).

The EMI/EMP phenomena can significantly degrade the performance of shipboard electrical/electronic equipment. The mutual interference of RF systems, transmitters, receivers, and antennae is of significant EMI/EMC concern particularly on naval ships. Furthermore, the presence of high-powered RF transmitters and antennae introduces various hazards of electromagnetic radiation.

The application of appropriate technology and implementation of system design-integration techniques can significantly reduce if not totally eliminate EMI/EMP and radiation hazards and create a safe and acceptable EMC environment. EMC can be achieved in two ways, namely, by the reduction/containment of electromagnetic energy transmitted by the offending equipment, or through desensitizing or protecting the equipment that is otherwise affected by the electromagnetic environment. In practice, both approaches are judiciously used. References 18 through 25 contain the most common naval requirements and practices applicable to E^3 and its implications on naval ship design.

12.2 Sources of EMI. Several potential sources of EMI must be considered. The two external sources of radiated (induced) EMI are either natural or man-made. Natural sources of radiated EMI come from cosmic radiation, atmospheric electrical storms, and precipitation (rain) particles. The energy level of these sources is generally not sufficiently high to interfere with shipboard operation except for exterior communications. However, occasional sun spots or sun flares can generate enough electromagnetic energy to interfere with RF systems on ships.

Man-made interference from external RF transmitters is of primary concern [21]. Electronic Warfare and Electronic Countermeasures are deliberate elements of modern warfare to interfere with the effectiveness of the adversary's combat system. EMP is another external source of interference from a nuclear explosion at a distance from the ship.

There are also basically three potential sources of EMI from within the total ship system, namely: functional, hull-generated, and incidental. All electrical and electronics systems and equipment function on the principle of electromagnetic energy and are thus potential sources of EMI under normal (functional) operation. The higher the frequency and power level of the equipment, the greater the potential for the transmission of EMI. RF equipment such as transmitters, cables, and antennae are of particular concern. Static power switches (thyristors/SCRs), which are used in power converters, controllers, etc., are also acknowledged troublesome sources of EMI.

Hull-generated EMI is essentially from secondary sources of electromagnetic energy that are transmitted to the metallic hull from other, primarily functional, sources such as:

- capacitive coupling between the hull and the electric cabling;
- ground loops, where the hull is used as the common ground reference; and
- RF picked up by topside metal fittings such as railings and ladders.

Due to the nonlinear characteristics of the metallic structural elements and fittings, the electromagnetic energy transmitted is usually at different and unpredictable frequencies.

Incidental electromagnetic energy is generated unintentionally through malfunctions and disrepair, such as short circuits, loose wires, ground leakages (insulation deterioration), damaged grounding straps, and disconnected shielding.

12.3 Reduction/Containment of EMI. A number of techniques can be used to reduce, control, or contain EMI. These techniques must be incorporated during the ship design phase, and must be carefully enforced during construction and operation to insure that they are not inadvertently defeated through improper procedures. The subject of EMI control for naval shipboard equipment is thoroughly discussed in references 18 and 19.

a. Shielding. Since radiated electromagnetic emission is a natural phenomenon that is introduced intentionally or encountered unintentionally in the shipboard environment, shielding is the most common and effective technique for its containment. Shielding is used to attenuate electromagnetic energy radiating from an active electrical/electronic device to contain its emission, or to protect an otherwise susceptible electronic device/circuit from surrounding harmful electromagnetic energy.

There are three pheonomena in the interaction between the electromagnetic energy and a shield. Some of the energy is reflected by the exterior surface of the shield; there is another element of reflection at the interior surface of the shield. Unreflected energy passes through the shield where it is partially dissipated as heat, as a function of the thickness and conductivity of the shield. The effectiveness of a shield is a function of the frequency of the electromagnetic energy plus the thickness and electromagnetic properties (conductivity and permeability) of the shielding material.

The steel hull and compartments of the ship and metallic equipment enclosures form natural and effective shields against radiated energy in the radio frequency range. However, discontinuities such as openings in the hull, compartments, or enclosures and nonmetallic gaskets constitute paths for radiated energy. Such discontinuities can be overcome with woven or perforated metal screens, wire-embedded glass, woven or spring-form metallic gaskets, etc. Shielding against audio-range frequency (below 20 kHz) requires special high-permeability materials. Cables and wires, which carry active (radiating) signals, or which carry low-level signals and are thus susceptible to the normal (acceptable) electromagnetic environment, are shielded with either basket-weave or solid jacket. These shields must be properly grounded at the cable termination points in accordance with the manufacturer's recommendation to be fully effective. Shielding against EMP requires more extensive measures in terms of attenuation by the shielding and treatment of discontinuities. Reference 21 describe the shielding practices used by the U.S. Navy.

b. Filtering. Filters are used extensively in active as well as susceptible electronics and in static power-switching applications to reduce the effects of conducted EMI, to or from the device. EMI filters are frequency dependent, and are designed for specific applications. By varying the impedances of capacitive and inductive elements and their arrangement, the designer can develop a filter for each specific application. Various filter designs are characterized as either low-pass, high-pass, band-pass, or band-reject. One of the most common low-pass filter elements used aboard ship is an isolation transformer; it attenuates spurious signals, spikes, and harmonics from a-c power lines in the primary winding, while allowing the low-frequency a-c power to pass through the transformer at neglible attenuation.

Filter arrangements must be properly designed and installed to be effective. The design and installation practices recommended for filters include the following:

- Filters should be shielded.
- Leads to and from the filter must be shielded.
- Output leads must be isolated from the input leads.
- Filters should be installed as close as possible to the associated equipment.
- Filters and associated equipment must be properly bonded to a common ground.

c. Segregation. Proper physical segregation of active and susceptible equipment on the ship is a delicate challenge for the designer within the generally severe constraints of a ship configuration. The greater the geometric segregation between certain systems and equipment, the lesser their electromagnetic coupling and hence their EMI. Special attention should be given to segregation in the following situations:

- Active (transmitting) and passive (receiving) antennae topside.
- High-powered active (transmitting) antennae topside.
- Active high-powered transmitters or static power-switching equipment and susceptible electronics.
- Cables that carry high-powered RF or pulsed energy and susceptible signal cables.

The geometric separation of antennae is of particular concern on naval combatants, given the number and density of the antennae and the power levels of the corresponding transmitters. Complex antennae models are often analyzed and tested to establish the best compromise given the available topside space.

d. Bonding and grounding. Ground refers to the zero-potential reference of an electric/electronic system. On the ship, the hull and structures bonded to the hull represent the ground plane. The hull is ohmically connected to the earth through its contact with seawater; however, the ship and the earth's ground plane are not necessarily synonymous. This difference is generally irrelevant, except when a vessel is directly connected to the earth, such as when it's tied up to a pier. In that case, the electric potential between the ship and shore may result in arcing, which can be eliminated by bonding the ship to the shore.

Power distribution systems aboard ship are ungrounded. This means that transformers have no center taps connected to the ground plane, nor do three-phase systems have the wye-point connected to ground. Hence, the ship's ground plane is not intended to be part of the ship's power distribution system. This has important implications on personnel safety; enclosures and cabinets of electrical and electronic systems must be grounded. If they are not, the enclosures may float at a potential above that of the hull, creating a potential shock hazard. Grounding these enclosures reduces the electrical potential between enclosures and the ship's structure and hence the risk to personnel.

The circuitry of sensitive electronics, and particularly digital equipment, is normally grounded to a common reference point on the chassis or the enclosure. These common ground references between the chassis and enclosures of equipment and components within a system are normally interconnected through a common grounding network (cable, bus, etc.). This is done since different ground potentials between equipment and components of a system or circuit would lead to ground stray currents, which even at very low levels can become sources of conducted EMI in very sensitive electronics.

However, the common ground reference of a properly interconnected electronic system will be different from the ship's ground plane, because of impedance between the equipment enclosures and the ship's structure due to paint, rust, and other circumstances. While such differentials in ground potentials are insignificant in magnitude and represent no personnel hazard, they can also be sources of conducted EMI. Hence, proper bonding and grounding of all electrical and electronic equipment enclosures to the ship's structure are necessary to reduce the potential effects of conducted EMI.

Topside, it is important that all metal structural elements and fittings be properly bonded and grounded to form a homogeneous hull ground plane in relation to the active (transmitter) antennaes. Any metallic structural element or hull fitting topside that is not at the hull ground plane due to paint, rust, or other circumstance, becomes a receiving antenna that will pick up RF energy from the nearby transmitting antennae. In addition to being a personnel hazard for electrical shock and burn, this induced RF energy will discharge to the hull either slowly or through arcing, both of which constitute a source of radiated and conducted EMI. Hence, proper bonding and grounding of metallic parts and fittings topside are of great importance, particularly on naval ships.

In general, the bonding and grounding techniques used, listed in the order of effectiveness, are as follows:

- Welding or brazing elements, fittings, and sometimes equipment to the hull.
- Assuring a large and clean metal-to-metal surface contact between the equipment and the hull.
- Bridging the equipment to the hull with a low-impedance bonding strap. Bonding straps are either of barstock or braided wire and are welded, brazed, or bolted to the equipment and hull. Bonding and grounding on U.S. naval ships are governed by the requirements of reference 22.

e. Construction, operation, and maintenance. Carelessness during the construction, operation, and maintenance of a ship can render some seemingly innocuous EMI control measure ineffective, leading often to personnel hazards and equipment malfunctions with potentially far-reaching consequences. Portable radios and other consumer electronics are potential sources of EMI, which, when introduced in certain sensitive spaces on the ship, may cause equipment malfunction. This becomes particularly critical when an equipment enclosure is opened for testing, troubleshooting, or repair, exposing the sensitive circuitry to radiated EMI without the fully protective enclosure of the cabinet. Some of the more common causes of problems of this type are:

- lack of detailed attention to proper bonding, grounding, and shielding requirements during construction, maintenance, and repair (disconnected shielding, bonding straps, etc.);
- maintenance of enclosure integrity, such as screens and gaskets, and proper closure of dogged, latched, or bolted doors and covers; and
- careless and inattentive damage to bonding straps.

12.4 Electromagnetic Radiation Hazards. The electromagnetic RF energy radiated functionally and intentionally by the ship's active (transmitting) antennae represents topside hazards particularly on naval combatants because of the number, density, and power levels of such antennae associated with the combat systems. Detailed information concerning RF radiation hazards on naval ships is included in reference 25. On commercial ships, such potential hazards are limited in scope and severity.

Section 13
Shipboard Cabling

13.1 Cables. In general, shipboard electric cables are manufactured and tested in accordance with references 6, 26, and 27 or are listed by Underwriter's Laboratories as Shipboard Cable, Marine. Cables for naval ships are required to be in accordance with references 26 and 27; however, cables for commercial ships are often purchased to these specifications as well, based on availability and economics, and as permitted by reference 9. References 6, 8, 9, 11, 28, and 29 contain specific requirements regarding the application, installation, and physical and electrical

properties of commercial and naval shipboard cables. The Underwriter's Laboratory lists marine electrical cable as complying with, or equivalent to, IEEE-Std 45.

a. Cable armor. Shipboard cables may be either armored or unarmored. The armor is provided by a steel, bronze, or aluminum basket weave over the outer jacket. Steel armor is rarely used. Bronze armor is preferred in damp and wet areas, such as in the weather. Aluminum armor has found favor on naval applications because of its lighter weight. The primary purpose of the armor is to protect the outer jacket of the cable against mechanical damage particularly during the installation phase. The armor cannot and should not be considered an effective electromagnetic shield because of its random and unpredictable grounding patterns along the run of the installed cables.

However, unarmored cable is often used in applications once reserved for armored cable, not only to reduce weight, but also because the mechanical properties of the outer jackets of unarmored cable have improved to provide more resistance to damage. In addition, armor has a tendency to corrode over time, particularly aluminum armor, which creates a maintenance problem, and to a significant degree an EMI problem, particularly topside on naval ships. Unarmored cable is permitted on naval ships, and, except in some hazardous areas, the U.S. Coast Guard permits the use of unarmored cabling.

b. Power and lighting cables. Three-conductor cable must be used for all a-c 3-phase circuits to neutralize the inductive effect that would cause heating of adjacent equipment or structure. Two-conductor cables must be used for all single-phase circuits for the same reason. The largest 3-conductor cable normally installed is 400,000 circular mils per phase. For circuits requiring ampacities beyond the limits of 400-MCM cables, two or more 3-conductor cables of the same size are connected in parallel; one conductor for each phase of the circuit must be contained in each paralleled cable.

All portable cords must have a minimum conductor size of No. 16 AWG, and must be in accordance with the specific requirements of reference 30. Types S, SO, or ST should be used for hard service applications; types SJ, SJO, or SJT should be used for portable lights, tools, and appliances.

c. Signaling and control cabling. Signaling and control cabling for electronics, command, control, and communications are of the multiconductor type to contain most and preferably all of the signals pertaining to the same functional group of interconnections between two pieces of equipment. There are three generic types of cables within this category:

1. Control cables, for conventional (nonelectronic) control circuits having a relatively high signal strength that is insensitive to the normal EMI environment. These are available in conductor sizes of 14–18 AWG and up to 91 individual, unshielded conductors.

2. Control and signal cables, for circuits having a medium signal strength. These cables are configured in twisted pairs or triads of conductors to eliminate crosscoupling between signals, and are available in conductor sizes of 16–22 AWG and up to 60 twisted pairs of conductors.

3. Control and signal cables, for circuits having a low signal strength. These cables are configured in twisted and shielded pairs of conductors for effective EMI shielding and are available in the same sizes as the ones described under Type 2 above. For added EMI protection, the complete cable may have an additional outer shield under the outer jacket of the cable.

d. Determination of cable lengths. The length of each run of cable may be determined from wiring deck plans, or isometrics, using a standard map measure. In addition to the lengths measured from plans, the following allowances should be made:

- 10 ft for each vertical run between decks.
- 5 ft for each vertical run from overhead to bulkhead-mounted equipment.
- 10 ft for each connection to a switchboard.

In addition to the foregoing allowances, 10% should be added to feeders and branches to cover waste, losses incidental to making up cable ends, and for unforeseen bends.

13.2 Cableways and Cable Installations. All cables should be continuous between terminations and, insofar as practicable, be routed to avoid areas where excessive heat, moisture, or oil may be encountered. Cables should not be run through oil tanks or pump rooms unless they are enclosed within watertight trunks; also, cables should not be run behind, or embedded in, heat insulation. Where it is necessary to pass cable through insulation, the cable should be run in a continuous pipe.

Cables routed through cargo spaces should be protected against mechanical damage incidental to the handling of cargo; advantage should be taken of the protection afforded by beams, girders, and stiffeners.

Routing of cables within a radius of 15 ft from any magnetic compass should be limited to the necessary binnacle and local lighting circuits. Cables should be installed in compliance with reference 31 so as to avoid harborage for rats.

For tankers, on-deck cable runs to the forepeak are routed either:

- on the underside of the fore-and-aft walkway, and protected by steel channels or enclosed in cable trunks or pipes;
- attached to a steel plate supported from the walkway handrail stanchions; or
- in a wireway-type structure on the weather deck, protected by steel channels with a cover, and located well inboard to afford protection from seas.

Unarmored cables should not be bent to a radius of less than six diameters to avoid damage to the insulation. Armored cables should not be bent to a radius of less than eight cable diameters.

Cables should be grouped and routed in main wireways in an efficient manner and should be supported as shown in Fig. 30(a), (b), or (c), or in reference 29, Methods 4B151 or 4B161. Small groups of cables should be supported as shown in Fig. 30(d), (f), (i), or (j), or in accordance with reference 31. Figure 30(j) shows typical methods of supporting a maximum of four small cables located behind joiner bulkheads and ceilings. Where cables are run on thin nonwatertight bulkheads, they may be secured by through-screws and nuts as shown in Fig. 30(i). Single cables should be supported as shown in Fig. 30(g) and (h).

Cables may be secured in wireways by fitted straps, such as shown in Fig. 30(a), or by banding straps, as shown in Fig. 30(e); these straps are usually ⅝ to ¾ in.

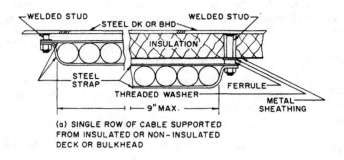

(a) SINGLE ROW OF CABLE SUPPORTED FROM INSULATED OR NON-INSULATED DECK OR BULKHEAD

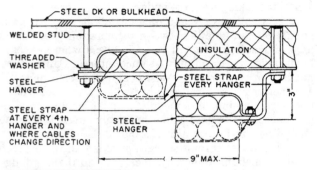

CABLES MAY BE SUPPORTED ON EITHER TOP OR BOTTOM OF HANGER BUT NOT BOTH

(b) SINGLE ROW OF CABLE SUPPORTED FROM INSULATED OR NON-INSULATED DECK OR BULKHEADS TO CLEAR OBSTRUCTIONS

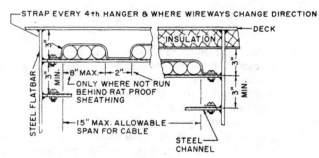

THE SUPPORT OF MULTIPLE ROWS OF CABLE ON STEEL BULKHEADS IS ACCOMPLISHED THE SAME EXCEPT THAT EACH TIER SHOULD BE STRAPPED AND NO SPAN OF CONTIGUOUS CABLES SHOULD EXCEED 9". WHERE SPANS UP TO 15" MAX. ARE REQUIRED, A HOLD-DOWN BOLT SHOULD SECURE THE RETAINING STRAP AT ITS CENTER. ASSEMBLIES SHOULD NOT BE SPACED MORE THAN 14" APART, CENTER TO CENTER.

(c) MULTIPLE ROWS OF CABLE SUPPORTED FROM INSULATED OR NON-INSULATED DECK.

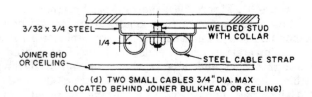

(d) TWO SMALL CABLES 3/4" DIA. MAX (LOCATED BEHIND JOINER BULKHEAD OR CEILING)

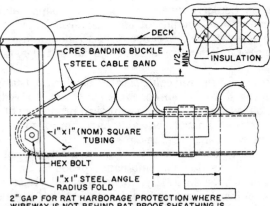

2" GAP FOR RAT HARBORAGE PROTECTION WHERE WIREWAY IS NOT BEHIND RAT PROOF SHEATHING IS PROVIDED BY STEEL SADDLE BANDED TO HANGER.

(e) MULTIPLE ROWS USING BANDING METHOD

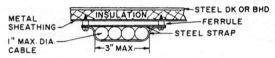

(f) SINGLE ROW ON INSULATION SHEATHING

(g) SINGLE CABLE

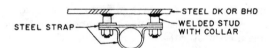

(h) SINGLE CABLE WHERE EXCESSIVE MOISTURE IS PRESENT

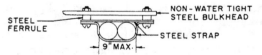

(i) SINGLE ROW MOUNTED ON BULKHEAD TOO THIN TO WELD.

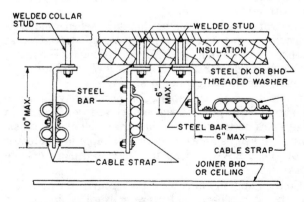

(j) CABLE RUN OF FOUR SMALL CABLES ON STEEL DECK OR BULKHEAD WITH AND WITHOUT INSULATION (BEHIND JOINER BULKHEAD OR CEILING)

Fig. 30 Typical cable installation methods

wide. Cable supports are usually spaced not more than 24 in. apart on vertical runs and 32 in. apart on horizontal runs. Cable straps should secure the cable without damage to the armor or insulation. Cables should be strapped in position at every hanger in vertical runs, and at not less than every fourth hanger on horizontal runs; at turns of horizontal runs the cables should be strapped at every hanger. Cable hanger material not exposed to the weather should be steel; galvanized material is normally used in refrigerated spaces and galleys. Exterior cable hanger material should be corrosion-resisting material or steel with a corrosion-resisting coating, and with brass or bronze nuts, bolts, and washers.

Insofar as practicable, cable runs in finished crew and passenger areas should be concealed. When paneling is fitted overhead or on bulkheads, through wiring is run behind the paneling and secured to the steel deck or bulkhead. Where heat insulation is applied to the deck or bulkhead, the wiring is run between the insulation and paneling, and supported as shown in Fig. 30(a) and (b). In some instances, when vertical cable runs are not near a steel bulkhead, a ladder type of frame, from deck to deck, is provided to support the cable, and is concealed by paneling.

Switches, call buttons, and receptacles located in paneled spaces are usually recessed into hollow H-section joints. All cables to these outlets should be run overhead and then down in the H-section to their termination; each cable should be secured at the top of the H-section and at the outlet. Wiring to light switches at entrance doors is run in the hollow door frames on which the switches are mounted.

Where concealment of cable runs in finished areas is not practicable, special troughs or channels are sometimes used to conceal individual runs of cables to surface-mounted outlets; otherwise, exposed wiring should be installed to be as inconspicuous as possible.

13.3 Cable Penetrations. A variety of the alternative methods of passing cables through decks and bulkheads is illustrated by Fig. 31. Cable penetrations of nontight decks and platforms are usually through clear openings of adequate sizes for the cables involved. Steel collars, extending approximately 3 in. above the deck, should be provided all around the edge of the opening.

Where single cables pass through nontight bulkheads or beams, and the bearing surface for the cable is less than ¼ in., a bushing should be provided as shown in Fig. 31(c). Where the bearing surface is ¼ in. or greater, a clearance hold with rounded edges is acceptable. Figure 31(h) shows a method of passing a group of cables through beams with insulation.

Cable penetrations of watertight and airtight bulkheads are through stuffing tubes, as shown typically in Fig. 31(a), (b), (d), and (e) and Fig. 32. Cable penetrations of watertight decks are through kickpipes with stuffing tubes, except that watertight riser boxes with stuffing tubes are sometimes used for groups of cables not in the weather. Stuffing tubes are either of the terminal type or bulkhead type as illustrated by Fig. 32.

Terminal-type tubes are used to provide a watertight cable entrance into electrical equipment enclosures; they are made of either brass or nylon. In general, terminal tubes consist of a body, gland rings, packing, and a gland nut or cap. A watertight penetration is obtained by tightening the gland nut or cap to squeeze the packing against the cable. For enclosures ³⁄₁₆ in. thick or thicker, tapped holes are provided to accept the tube male pipe threaded end. For enclosures less than ³⁄₁₆ in. thick, clearance holes are provided; in this type of installation, brass tube bodies are brazed to the enclosure, whereas nylon tube bodies are fastened and made tight by a locknut and O-ring assembly.

Bulkhead type tubes are used to provide watertight and airtight cable penetrations through bulkheads and decks. Bulkhead tubes are usually steel, and are similar to brass terminal tubes except that the tube end opposite the gland nut has standard IPS female threads to accept threaded pipes. For passing cables through bulkheads, a pipe nipple of adequate length is threaded into the tube body; the assembly is inserted into a clearance hold in the bulkhead, with the tube body butting the bulkhead; the tube body is then welded to the bulkhead. For passing cables through decks, kickpipes are welded into the deck with a bulkhead-type tube threaded on the upper end. The height of the kickpipe assembly should be at least 9 in. to the top of the tube, with the pipe extended below the deck, or sheathing, as applicable, leaving approximately ¾ in. of the pipe exposed.

A preferred method of passing bunched cables through tight bulkheads is by means of a "multicable transit," shown by Fig. 31(i). This method utilizes a special assembly, welded in a clearance hole in the bulkhead, through which several cables may be pulled. After the cables are pulled, they are positioned and maintained in place using a system of split insert blocks; the complete cable grouping is made tight by applying compression through special bolting and plates. Cable transits may be used for passing groups of cables through decks, provided a riser box is utilized; in this application, the cable transit would be installed in the top of the riser box, as shown by Fig. 31(j).

Cable entrance into explosion-proof equipment located in explosion-hazardous areas must be accomplished through an approved seal fitting, similar to that shown in Fig. 32(a). Proper installation would be as follows. Nipple the seal fitting to the explosion-proof enclosure by a pipe not more than 18 in. long; remove all of the cable covering, except the individual conductor insulation, from the point of entering the seal fitting to its end; and fill the seal fitting with a sealing compound through the filling plug hole. When the compound hardens and the filling plug is in place, the cable entrance is explosion-proof.

13.4 Connections and Terminals. The connection of wires to terminals should ensure a good electrical contact without damaging the conductor. All terminations should be made by screw connections or approved cable connectors located within equipment enclosures or wiring appliances. In general, all ship cable connections should be

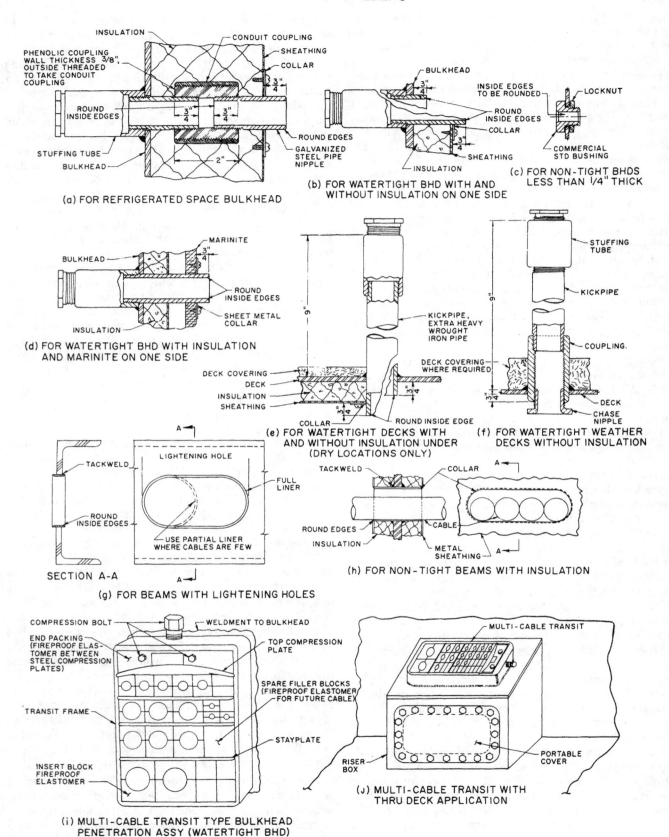

(a) FOR REFRIGERATED SPACE BULKHEAD

(b) FOR WATERTIGHT BHD WITH AND WITHOUT INSULATION ON ONE SIDE

(c) FOR NON-TIGHT BHDS LESS THAN 1/4" THICK

(d) FOR WATERTIGHT BHD WITH INSULATION AND MARINITE ON ONE SIDE

(e) FOR WATERTIGHT DECKS WITH AND WITHOUT INSULATION UNDER (DRY LOCATIONS ONLY)

(f) FOR WATERTIGHT WEATHER DECKS WITHOUT INSULATION

(g) FOR BEAMS WITH LIGHTENING HOLES

(h) FOR NON-TIGHT BEAMS WITH INSULATION

(i) MULTI-CABLE TRANSIT TYPE BULKHEAD PENETRATION ASSY (WATERTIGHT BHD)

(j) MULTI-CABLE TRANSIT WITH THRU DECK APPLICATION

Fig. 31 Typical cable penetration methods

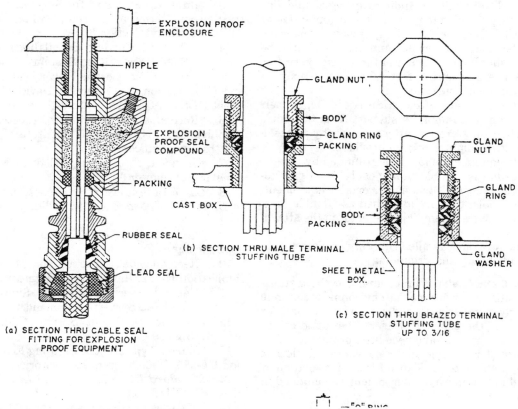

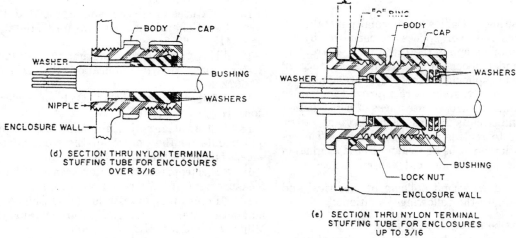

Fig. 32   Typical cable stuffing tubes

made by solderless lugs of an approved clamp or compression type; exceptions include twist-on pressure-type connectors with a securing device, which are sometimes used to make connections in lighting outlets and connection boxes. Cable ends of vital circuits should be sealed against the ingress of moisture.

Except for final branches, which may be grounded at the supply end only, the metallic sheath on all armored cable should be grounded at each end. Where armored cable enters a box or wiring device, the armor should be continued into the box and secured by a clamp or connector to assure good contact with the box. If this is impracticable, the armor should be grounded to the steel structure at the nearest cable strap. For each ground the armor and strap should be well cleaned at the point of contact.

Radio, radar, and miscellaneous communication equipments have various requirements regarding grounding, which are specified by the equipment vendor. To ensure proper operation, the shipboard installation should comply with the vendor's recommendations.

Stays, shrouds, and other standing rigging should also be permanently grounded to prevent the accumulation of static electricity. Rigging and railing in way of the radio direction-finder loop and sense antenna need not be grounded but should have insulators inserted every few feet to prevent a shielding effect caused by nearby induced radio frequency currents.

Enclosures and frames of electrical equipment generally have inherent grounding by support from the ship's structure. When they are not attached directly to steelwork, special grounding should be provided.

All portable equipment should be grounded by a separate conductor in the flexible supply cable and a grounding device in the receptacle and plug. Portable equipment, such as power tools, that are identified as "double insulated" need not have a grounding conductor in the attachment cord.

Reference 22 provides detailed U.S. Navy requirements for shipboard bonding and grounding.

13.5 Fiber-Optic Cables. Procedures for the installation of fiber-optic cables in a marine environment continue to evolve. Early installations were run in rigid conduit for the protection of the cable. However, fiber-optic cables have become more rugged and need little protection other than what would be provided for conventional electric cable of comparable size. Guidance for the installation of fiber-optic cable in a marine environment is contained in reference 32.

13.6 Cable Splicing. Cables are normally installed in continuous lengths; however, splices are permitted by the American Bureau of Shipping, the U.S. Coast Guard, and the U.S. Navy under specified conditions. Reference 29 details specific cable-splicing methods that are applicable to naval ships. Generally, splices are approved for cables of exceptional length or size, at interfaces of construction modules, to extend circuits for vessels undergoing repair or alteration, and to replace a damaged section. The replacement insulation must be equivalent to the original in thickness, electrical properties, and watertightness.

References 7 and 28 permit splicing of electrical and fiber-optic cable under the following conditions:

- Splices must be accessible for inspection.
- Splices must be made by qualified personnel using approved splice kits.
- Splices are not permitted in hazardous locations except in intrinsically safe circuits.
- Splicing of fiber-optic cable must be done by approved mechanical or fusion methods.

References

1 "Electrical System Load and Power Analyses for Surface Ships," Military Standard, MIL-STD-2189, Section 310-1.

2 "Guide for Electrical Installations on Merchant Vessels and Mobile Offshore Drilling Units," United States Coast Guard Publication COMDTPUB P16700.4, NVIC 2.

3 "Interface Standard for Shipboard Systems, Electric Power, Alternating Current," Military Standard, MIL-STD-1399 (Navy) Section 300A.

4 "AC Fault Current Calculations," Design Data Sheet DDS-300-2, Naval Sea Systems Command.

5 "Ship-Service Electric Power Systems, Application and Coordination of Protective Devices," Design Data Sheet DDS-311-3, Naval Sea Systems Command.

6 "Recommended Practice for Electric Installations on Shipboard," IEEE-Std 45, Institute of Electrical and Electronics Engineers.

7 *Electrical Engineering*, Code of Federal Regulations 46 Subchapter J.

8 *Cable Comparison Handbook, Data Pertaining to Electric Shipboard Cable*, Military Handbook, MIL-HDBK 299.

9 *Marine Engineering*, Code of Federal Regulations 46, Subchapter F.

10 "Recommended Practice for Marine Lighting," Publication RP-12, Illumination Engineering Society.

11 "General Specifications for Ships of the U.S. Navy," Naval Sea Systems Command.

12 *Lighting on Naval Ships*, Military Handbook, MIL-HDBK 289.

13 "Marine Type Electric Lighting Fixtures," Publication UL-595, Underwriters Laboratories.

14 *Standard Electrical Symbol List*, Military Handbook, MIL-HDBK 290.

15 "Fixtures, Lighting and Associated Parts, Shipboard Use, General Specifications for," Military Specification, MIL-F-16377.

16 "Navigation Rules International—Inland," United States Coast Guard Publication COMDTINST Rules, M16672.2.

17 "Telecommunications," Code of Federal Regulations 47.

18 "Electromagnetic Emission and Susceptibility Requirements for the Control of Electromagnetic Interference," Military Standard, MIL-STD-461.

19 "Electromagnetic Interference Characteristics, Measurement of," Military Standard, MIL-STD-462.

20 "Procedures for Conducting a Shipboard Electromagnetic Interference (EMI) Survey (Surface Ships)," Military Standard, MIL-STD-1605.

21 *Handbook of Shipboard Electromagnetic Shielding Practices*, Naval Sea Systems Command Publication S9407-AB-HBK-010.

22 "Shipboard Bonding, Grounding and Other Techniques for Electromagnetic Compatibility and Safety," Military Standard, MIL-STD-1310.

23 *Electromagnetic (Radiated) Environment Considerations for Design and Procurement of Electrical and Electronic Equipment, Subsystems and Systems*, Military Handbook, MIL-HDBK-235.

24 *Electromagnetic Compatibility Management Guide for Platforms, Systems and Equipment*, Military Handbook, MIL-HDBK-237.

25 *Electromagnetic Radiation Hazards*, Naval Sea Systems Command Publication OP-3565.

26 "Cable and Cord, Electrical, for Shipboard Use, General Specification for," Military Specification, MIL-C-24640.

27 "Cable Electrical," Military Specification, MIL-C-24643.

28 *Rules for Building and Classing Steel Vessels*, American Bureau of Shipping.

29 "Electric Plant Installation Standard Methods," Department of Defense Standard, DOD-STD-2003.

30 "Flexible Cord and Fixture Wires," Underwriter Laboratories, Standard UL 62.

31 *Handbook on Sanitation of Vessel Construction, Ratproofing of Ships*, U.S. Public Health Service, Publication 393.

32 *Fiber-Optic Cable Installation Procedures*, Department of Defense Handbook, DOD-HDBK-282.

CHAPTER XX

H. J. Cassee | # Piping Systems

Section 1
Piping System Design Development

1.1 Introduction. Shipboard piping systems form a pervasive network that is required to support all of the ship's essential functions. These systems convey steam, fuel, lubricating oil, and cooling water to operate machinery, transport bulk cargo and ballast, provide for human health and comfort (e.g., potable-water, heating, chilled-water, and waste-collection systems), protect the safety of the ship and crew with fire-extinguishing agents and damage-control systems, and collect and remove pollutants. Piping systems are among the most complex ship systems to design and construct.

Piping system engineers apply principles of static and dynamic stress analysis, thermodynamics, and fluid-flow theory to design safe and efficient piping networks. Beyond this, they face additional challenges. Shipboard piping systems consist of not only pipes, valves, and fittings, but also an array of components that condition and control the fluids, such as pumps, strainers, and heat exchangers. Piping systems also contain control valves, transducers, and actuators, which interact with equipment served by the systems and must be compatible with ship control and monitoring systems. Although piping-system engineers are not directly responsible for the design of all these components, they require a basic understanding of component characteristics to integrate them into a properly functioning system.

One of the more difficult tasks of a piping-system engineer is to define and continually update all of the system design requirements through progressively more detailed design phases. This task is difficult, partially because the piping system engineer initially has only a general understanding of the components that require piping services, and partially because the component engineers, who are the source of the information, may not realize the piping system engineer's need for the data. Disruptions will be exacerbated throughout the design process if changes in component characteristics, additions of new components, and deletions of components are not communicated promptly to the piping-system engineer. Thus, it is usually necessary for the piping-system engineer to actively pursue the information needed rather than waiting for the information to be forwarded in a routine manner.

The ship design parameters that establish the basis for many piping-system characteristics include the following:
- Ship general arrangement—The overall size, hull subdivision, and space arrangement are the primary bases

for the design of fire-protection and emergency-dewatering systems. Appropriate fire-extinguishing agents and distribution systems must be selected for machinery spaces, storerooms, living quarters, and cargo holds. Special-purpose areas, such as the main deck of a crude-oil tanker, require specialized systems. The capacity and location of dewatering pumps are related to hull volume and the arrangement of watertight compartments. The arrangement of fuel and water transfer systems must accommodate the tankage arrangement selected by the naval architect. The locations of living and messing areas strongly influence the design of plumbing drainage and collection systems. Also, the height and separation of compartments served by piping systems will influence operating pressures and pipe sizes.
- Propulsion and electric plants—The type, number, horsepower, and location of propulsion and generator engines establish flow rates, pressures, and filtration requirements for fuel and lubricating-oil systems. The power output, operating temperature and pressure, and relative locations of steam boilers and turbines establish the steam piping configuration. Propulsion and generator cooling requirements, together with the lesser needs of auxiliary equipment, set requirements for seawater and freshwater cooling systems. The compressed-air pressure, quality, production capacity, and storage volume must meet plant requirements for starting and controls.
- Operations—The ship operating areas, trade routes, port turnaround time, and voyage length affect piping system design requirements such as freshwater production and disinfection, waste retention and treatment capacities, liquid cargo and ballast flow rates, pollutant discharge limits, and tankage requirements.
- Complement—The number of crew and passengers and the habitability standards selected form the basis for capacities of systems providing cold and hot potable water, sanitary flushing, and waste retention and processing.
- HVAC—The heating and cooling system total capacities establish flows of the heating-steam and chilled-water systems. Heating and cooling loads are determined in part by complement, size of living and working areas, ship operating areas, and equipment loads. On naval ships, significant demands are placed on heating and cooling systems by electronic equipment.

- Mission-specific systems—The purpose and mission of each ship influence the capacity and arrangement of many of its piping systems, and often require unique systems. Naval combat ships, specialized cargo carriers, and icebreakers are but a few examples.

- Special attributes—The shipowner may require special emphasis on cost reduction, weight reduction, commonality of equipment, or other attributes in the design of the piping systems.

Approximately 70% of the value added by the shipbuilder during construction of the more complex ships is related to piping systems. Thus a significant cost reduction can be realized by applying techniques to simplify the design, fabrication, assembly, and installation of piping systems. The principles of designing for production, zone outfitting, and modular construction [1,2] must be applied beginning very early in the ship design process and must continue throughout all design phases. Computer-aided design and manufacturing techniques, while possibly increasing the initial design costs, also provide an opportunity for overall cost reductions, with the additional benefit of providing design data of significantly superior quality.

1.2 System Design Phases. As outlined in Chapter 1, piping systems are developed in progressively greater detail as the overall ship design advances through four loosely defined phases.

During concept design, a tentative list of requirements is developed based on the available, usually very general, ship characteristics. If sufficient arrangement details have been developed, a preliminary check of the sizes and relative locations of the spaces can be made to ensure that major pipe runs can be accommodated. Since insufficient details are available to develop independent cost and weight estimates at this stage, such estimates are usually extrapolated from data for existing ships of similar design.

During preliminary design, the major piping system components are selected and arranged in the ship. Preliminary estimates of system flows, pressures, and temperatures are made to support component selection. System parameters may change several times during this phase to accommodate design updates; therefore, the piping engineers must work closely with others who are laying out spaces, arranging machinery, and selecting equipment to ensure that a compatible design is maintained. Piping system performance requirements are determined on the basis of the ship mission, size, operating profile, complement, main machinery, and other factors. Component selection is based on meeting the performance requirements within established goals for weight, cost, noise, shock, and reliability. Because it may not be possible to satisfy all requirements fully, trade-off studies are usually necessary to select the optimum design solution.

The approximate locations of major components and piping runs relative to the ship arrangement, and to each other, are determined during the preliminary design phase. These locations must be selected carefully because subsequent rearrangement can be prohibitively expensive in terms of design man-hours and schedule considerations. Similar or related components should be grouped (e.g.,

potable-water pumps, tanks, desalinators, and purification equipment) to minimize piping, ease operation, and facilitate off-ship assembly of equipment packages. For naval combatant designs, the separation of redundant components, for survivability reasons, is an important consideration.

Sketches of the larger piping runs, ventilation ducts and wireways are prepared to determine space requirements. As soon as a general arrangement drawing is prepared, the sizes of compartments, passageways, and trunks are reviewed to ensure that piping and other distributive systems can be accommodated. The relative locations of components that will be connected by piping are checked to eliminate unnecessarily long or complex runs, avoid pipe runs through restricted spaces, and permit gravity flow of fluid where required. Piping runs should be planned to avoid penetrating highly stressed structural members; otherwise, the structure will require reinforcement at the penetration.

During contract design, the piping engineer develops additional details of each system within the broad outlines fixed during the preliminary design phase, and describes each system by specifications and contract plans. Specifications are prepared in formats that are quite different depending on whether the ship is for U.S. Navy commission [3], the Military Sealift Command [4], or a commercial owner [5]. Contract or contract guidance drawings are developed to illustrate spatial relationships and the interconnections of system components, which may not be understood easily from written specifications. Together the specifications and contract drawings define the system sufficiently to ensure the owner's requirements for performance and quality are understood, and to permit the shipbuilder to prepare a bid. Minimum requirements to be specified include pressures, temperatures, and services supplied by each system; the number, capacity, and location of all major components; the level of redundancy required for fluid sources and flow paths; piping material schedules; fabrication, assembly, and inspection methods and the industry or government standards which must be applied; and piping system cleaning and testing requirements. In commercial practice, pipe sizes are sometimes established during contract design; more often, especially for naval ships, the determination of pipe sizes is the shipbuilder's responsibility.

The specifications should be written to allow and encourage the use of design features which enhance producibility. It is desirable to package complex piping assemblies that can be preoutfitted off the ship and installed aboard as complete units. Common examples are air compressors with their receivers, filters, and dehydrators; auxiliary boilers with feed pumps, water chemistry control equipment, and drain tanks; hydraulic valve control stations; and pressure-reducing stations. Another technique used to improve producibility is to specify fewer piping materials; stocking a single component made of high-grade material, instead of several of identical size that are made of progressively lower-grade materials, simplifies ordering, manufacturing, and handling. Flange bolting, for example, is commonly specified this way.

Study drawings are often used in complex designs, such as naval combatants, to ensure that the concepts described in the specifications are workable. These drawings are usually one-line piping diagrams drawn on a background of the machinery or space arrangements.

The first three design phases are primarily focused on performance and are often accomplished by the prospective owner or his design agent. However, the detail design phase is focused on construction and is usually accomplished by the shipbuilder. True to its name, this phase results in a full definition of every piping-system element, culminating in drawings that are used to manufacture or procure all parts of the systems, and install them in the ship.

The piping-system diagrams, which are begun in preliminary form during the preliminary design phase and progressively updated into detail design as more specifically applicable data become available, constitute a foundation on which the system details are based throughout the detail design phase. The system diagrams are used to ensure the systems will meet the specification requirements and that all elements of the systems are compatible with each other and with the other design elements to which they are linked: e.g., controls and machinery interfaces. Piping diagrams are the starting point for the development of all piping production drawings. Piping diagrams depict system components and their interconnections in schematic form, as typically illustrated in Sections 3 and 4. Information about the system arrangement in the ship is included to varying degrees, but is usually limited to the space and level on which a pipe or component will be located, since it is impractical to convey greater detail in the diagram format. Diagrams also include:

- Component symbols
- Material schedule
- Component performance ratings and pump curves
- Valve descriptions
- Component identification/labeling scheme
- System pressures, temperatures, flow rates, velocities, and pressure drops
- Pipe sizes
- Flow direction arrows
- Identification of compartments and bulkheads
- Characteristics of instruments
- Operating characteristics of pressure, temperature, level, and flow controls
- Notes invoking fabrication, cleanliness, painting, testing, and safety requirements
- System shock and noise requirements
- References to interfacing drawings, standards and procedures

The quality and clarity of piping diagrams are important because diagrams serve so many functions during design, construction, and operation of the ship and are the primary means of understanding how the systems work and how they relate to other ship systems. Components and piping should be laid out in a logical fashion, with system flows generally from one side of the sheet to the other. Functional relationships should be shown clearly, and components of equal importance should have equal prominence. Primary pipe runs should be given the most direct paths on the sheet and be shown with heavier line widths. Parallel paths should be arranged as symmetrically as possible. Information about how the system operates should be given priority over information concerning shipboard locations.

Detailed piping system arrangement drawings are prepared as soon as the diagrams are developed sufficiently. The format of piping arrangement drawings is closely related to the ship construction methods, the computer-aided design system in use, and individual shipyard practices. The intended use of the drawings by the shipowner may also influence their format and content. Arrangement drawings depict piping, fittings, and components to scale as they will appear assembled and installed in the ship, and include complete material lists. Arrangement drawings are used by design engineers to review installation details such as pipe slope and pump suction flow paths, and to conduct pipe flexibility and flow calculations. They are also used by owners and operators of ships for maintenance, crew training, and planning future modifications. Piping arrangement drawings typically show only one system or related systems on a single drawing to simplify the presentation. The development of arrangement drawings must include a process for identifying and eliminating interferences within piping systems and between piping and ducting, cableways, and other parts of the ship. For drawings that are prepared using computer-aided design methods, the computer software often includes an interference checking capability.

Construction drawings are oriented to the requirements of the production and installation techniques practiced within the shipyard. Modular construction methods are facilitated by the use of task-oriented drawings that cover all systems to be installed in a portion of the ship. These drawings are oriented to the activities of the construction tradesmen, who need not be aware of how the system elements they install are related to the system as a whole. Construction drawing types and methods of developing them are discussed in detail in reference 1.

A concentrated effort to improve the producibility of piping systems is made during the detail design phase. Many of the production methods are related to the specific practices of the building yard, but the following guidelines are generally applicable:

- Locate and orient components with respect to each other to minimize the piping length, changes in direction, and number of joints.
- Arrange pipes to facilitate access to joints for assembly, inspection, and testing.
- Give priority for the most direct routing to piping that is large or made of difficult-to-fabricate material.
- Use bends instead of elbows wherever possible.
- Locate takedown joints at boundaries of construction units and removable access plates.
- Plan piping assemblies to permit as much off-ship fabrication as possible.

• Make headers to which multiple branches are connected the same size throughout their length to simplify fabrication, rather than reducing their size at each branch.

Additional techniques for planning pipe fabrication, support, installation, and arrangement in the context of zone production methods are discussed in reference 2.

1.3 Space Arrangements.

a. **Main machinery spaces.** The development of piping systems within a machinery space and the machinery space arrangement are inseparable because the locations of many machinery components are influenced by piping system considerations, and piping must be arranged with knowledge of the arrangement restrictions that govern other components in the space.

There is generally no optimum machinery space arrangement that will completely satisfy all requirements. It is the responsibility of the marine engineer to assess the alternatives and select one which offers acceptable compromises as a solution to the design problem. This is an iterative process.

One of the first decisions to be made regarding the machinery space is its location. For many ships, there are often two choices: the aftermost region of the ship, or the region somewhat aft of amidships. Machinery spaces of oil tankers and other ships with full lines are generally at the stern, thereby simplifying the design by allowing shorter propeller shafting and reducing interferences with cargo handling. The hull lines in way of the stern on these ships are sufficiently full to accommodate the propulsion machinery; however, provision must be made to take on ballast water when the ship is unloaded to compensate for the extreme-aft position of the machinery and maintain a ship trim that is seaworthy. The hull lines aft on high-speed cargo and passenger liners are usually so fine that the propulsion machinery will not fit within the confines of the stern region. For such ships the machinery space must be located farther forward, where the ship is sufficiently wide.

When developing a machinery space arrangement, an engineer must visualize the ship structure, piping, valves, tanks, and other components in the space as well as the propulsion and auxiliary machinery; develop direct routes and minimum distances for wiring and piping interconnecting the major components; accommodate the reach, sight lines, safety, and comfort of the operators; provide access for maintenance and overhaul; and allow sufficient permeability for fire-extinguishing agents.

In the initial stages of design, the arrangement is based on components that are only tentative selections and whose dimensions are only approximate. Therefore a reasonable allowance should be made for variations in dimensions and unanticipated developments. In addition, since payload is not carried in the machinery space, the space must be no larger than necessary, especially in length. However, sufficient space must be reserved to permit the installation of large piping, ventilation ducting, and power cables; the proper operation and maintenance of all equipment; rapid personnel egress in an emergency; and access for firefighting and other damage-control functions.

Trade-offs between the components initially selected and the available space are sometimes required to resolve conflicts. For example, although horizontal pumps are easier to support and are more readily overhauled than vertical pumps, vertical pumps may ultimately be chosen because they occupy less deck area. A choice among vertical shell-and-tube, horizontal shell-and-tube, and plate-frame heat exchangers can be made to favor the space available if all other factors are satisfactory. But, such trade-offs must be made without compromising the system design requirements. Trade-offs are also often necessary when arranging piping, ducting, and components. Placing a component in its ideal location from the standpoint of maintenance, operation, or access may necessitate additional joints and direction changes in piping in the vicinity. The piping engineer must determine which arrangement provides the most advantageous overall compromise.

Removal clearances and routes should be studied to ascertain that components that cannot be repaired in place can be removed through hatches or through removable plates provided in decks and bulkheads. As a last resort, a route to a location in the hull suitable for cutting an access opening without drydocking the ship may be planned. Unshipping considerations frequently necessitate adjustments in component locations.

Sufficient space allowance must be made for thermal insulation and noise treatment of machinery and piping where necessary. Diesel generator engines and gas turbines are frequently fully enclosed to reduce airborne noise. Space must be reserved for resilient mounts for machinery, particularly on naval ships. Clearances must be sufficient for normal and shock excursions, as applicable, of resiliently mounted equipment.

The location of structural elements is a major consideration in arranging machinery and piping. Stanchions must be placed to carry overhead loads down to innerbottom structure; because they must land on suitably strong structure at each end, their placement cannot be determined based on machinery arrangement considerations alone. The orientation and depth of frames and stiffeners on structural plating can adversely affect the routing of piping and cableways along bulkheads and underneath decks, requiring changes in direction of piping and additional joints. The depth and spacing of frames should be considered for placement of pumps, control panels, and other components along the shell; often the bays between frames provide the only available locations for smaller auxiliary components. Also, the weight and placement of machinery components can affect the size of support beams, which in turn can affect the headroom and clearances for machinery underneath. Close communications between those arranging machinery and those designing structure and piping is, therefore, essential.

Headroom, clearance routes, and the location of overhead support structure should be planned for hoists and tracks to provide for the maintenance of heavy component parts, such as fuel and lubricating-oil purifier bowls and main engine replaceable parts.

Gratings and walkways should be located and routed with sufficient width and overhead clearance. Operating stations should be provided and located where operators can observe essential instruments. Continuously manned stations should preferably be arranged so the operator can stand facing forward or aft.

Seawater pumps should be located sufficiently low and inboard to provide adequate suction head for all conditions of ship trim and motion. Sea chests should be low enough to avoid air ingestion, and close to the pumps connected to them. Bilge pumps should be located low in the space for adequate suction lift.

The interconnection of components in flow paths should be considered to properly orient them in the arrangement. For example, lubricating oil for the main engines typically passes in succession through a pump, strainer, and cooler. Proper orientation of the inlet and outlet of each component is necessary to permit a short and direct routing of interconnecting piping. The minimum spacing of components linked by piping can be influenced by the allowable bending radius of the piping, which ranges from 1 to 5 times the pipe outside diameter; the difference can be significant in the larger pipe sizes.

On naval ships, piping, ventilation ducts, and cable runs that penetrate main watertight subdivision bulkheads should be located above the flooding water level whenever possible to minimize the necessity for watertight penetrations and progressive-flooding valves.

Suggested clearances around major components are given in Table 1. The dimensions shown are an approximate guide to allow space for access and inspection; they are not normally adequate for maintenance or removal of machinery components.

Main access ladders should be 24 in. wide and slope 60 deg above the horizontal. Infrequently used ladders may have a width of 12 to 18 in., and a greater slope.

The typical steam plant shown in Figs. 1 and 2 illustrates the principles discussed above. The steam plant is the most difficult to arrange because there is little flexibility in the arrangement of its major components. The boilers may be positioned forward of the main engines, or aft of and above them. The latter arrangement, as shown in Figs. 1 and 2, permits a shorter machinery space, but the effect on ship stability of the higher center of gravity of the relatively heavy boilers must be evaluated.

The amount of space around boilers and the routing of piping must be evaluated for operation as well as maintenance concerns such as boiler tube renewal and removal of soot-blower tubes, economizer tubes, burners, and desuperheaters. Boiler forced-draft blowers should be located to draw the hottest air from around the uptakes or the top of the machinery space casing. Forced-draft blower ducting occupies a significant volume and should be included on the arrangement drawings.

The size and location of uptakes and intakes must be compatible with the topside arrangement. This is of particular importance on naval ships because of the potential interference of stacks with antennas and sensors on the masts, and the need to locate intakes away from sea spray.

Table 1   Suggested minimum clearances for machinery-space arrangements, ft-in.

Headroom in passageways and walkways	6-5
Minimum height	
standing	6-4
crawling	2-7
bending, kneeling	4-0
Minimum width for body passage	1-11
Minimum thickness for body passage	1-1
Maximum depth of reach	1-11
Intake/uptake to surrounding structure	
one side	1-6
other side	2-0
ends	2-6
between	2-0
Boilers	
back to structure	5-6
bottom to innerbottom	2-0
bottom to frames	1-3
top of economizer to deck above	2-6
top of steam drum to deck above	4-2
top of boiler to beam above	3-6
athwartship between units	5-0
firing aisle	11-0
Steam turbine—all around	4-0
Gas turbine	
sides	3-0
one end	3-0
other end	4-0
between dual units	3-6
outside dual units	2-0
Diesel engine	
sides	3-0
one end	3-0
other end	4-0
between dual units	3-6
outside dual units	2-0
Generator	
sides	3-0
one end	3-0
to switchboard	4-0
between dual units	4-0
Switchboard	
to structure behind	2-0
one end	2-0
other end	0-6
front	3-6
between units	4-0
Reduction gear to bulkhead	3-0
Piping to structure or other piping	0-2
Piping to structure or other piping (shock-excursion	
clearance, U.S. Navy practice)	0-4
Surface ≥ 400 F to tank or pipe containing	
combustible fluid other than lubricating oil	1-6
Surface ≥ 650 F to tank or pipe containing	
lubricating oil	1-6

The vertical and horizontal positions of the propulsion turbines are heavily dependent on the propeller and shafting arrangement. The location of the propeller is fixed by hydrodynamic considerations but, by raking the shaft, a degree of freedom in locating the main engines is obtained. Moderate amounts of horizontal and vertical shaft rake are acceptable (see Chapter 10). As illustrated in Fig. 1, there is scarcely adequate space beneath the main reduction gear for the lube-oil sump, and the main condenser is so low that it is difficult to provide sufficient submergence for the main condensate pump without recessing the innerbottom. The main shafting is invariably raked upward going forward to alleviate these problems.

If the main engines are located in the aft end of the machinery space, a check must be made to ensure that there is adequate space around the reduction gears for

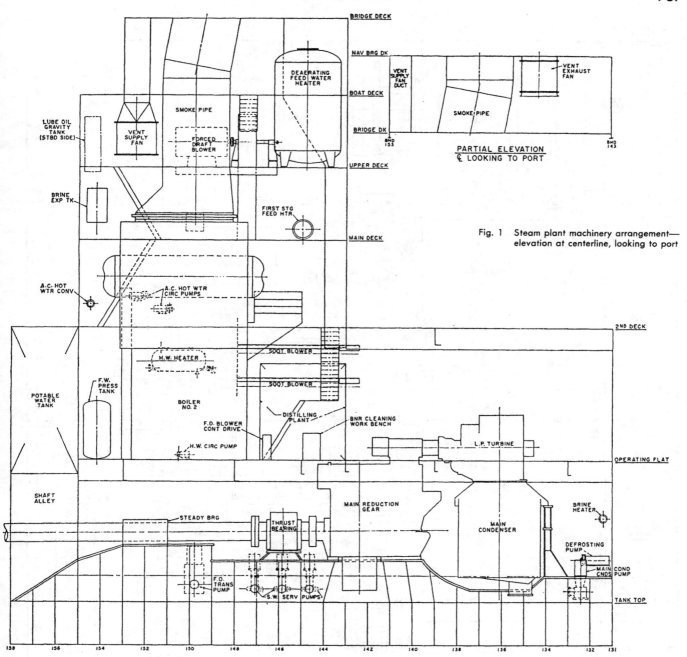

BRIDGE DECK

NAV BRG DK

BOAT DECK

BRIDGE DK

UPPER DECK

MAIN DECK

2ND DECK

OPERATING FLAT

TANK TOP

DEAERATING FEED WATER HEATER

SMOKE PIPE

LUBE OIL GRAVITY TANK (STBD SIDE)

VENT SUPPLY FAN

FORCED DRAFT BLOWER

BRINE EXP TK

FIRST STG FEED HTR

A.C. HOT WTR CONV

A.C. HOT WTR CIRC PUMPS

POTABLE WATER TANK

F.W. PRESS TANK

H.W. HEATER

SOOT BLOWER

SOOT BLOWER

BOILER NO. 2

DISTILLING PLANT

F.D. BLOWER CONT DRIVE

BNR CLEANING WORK BENCH

H.W. CIRC PUMP

L.P. TURBINE

SHAFT ALLEY

STEADY BRG

THRUST BEARING

MAIN REDUCTION GEAR

MAIN CONDENSER

BRINE HEATER

DEFROSTING PUMP

MAIN COND PUMP

MAIN COND CNDS PUMP

F.O. TRANS PUMP

S.W. SERV PUMPS

158 156 154 152 150 148 146 144 142 140 138 136 134 132 131

VENT SUPPLY FAN DUCT

SMOKE PIPE

VENT EXHAUST FAN

PARTIAL ELEVATION
℄ LOOKING TO PORT

BHD 153

BHD 143

Fig. 1 Steam plant machinery arrangement—
elevation at centerline, looking to port

passage and access to the gear and pinion bearings. Foundation girders for the reduction gears and main thrust bearing should be included in the arrangement to confirm that sufficient rigidity can be provided.

The main condensate pumps must be located with sufficient vertical head below the minimum condenser hotwell level to prevent flashing of condensate at the impeller inlet. Also, locating the pumps near the condenser longitudinal centerline will reduce erratic suction conditions when the ship rolls or lists.

Because of the size of main condenser seawater piping, it should be shown on the machinery arrangement to ensure sufficient space is reserved.

When a satisfactory arrangement has been tentatively established for the lower level, the main operating level can be established. When fixing the operating level height, due consideration must be given to the space beneath the level for piping, wireways, ventilation ducts, lighting, and headroom on the lower walking level, as well as equipment maintenance and removal. The operating level located outboard on each side of the main propulsion unit in Fig. 2 also serves the generators, switchboards, distilling plants, contaminated evaporators, and other equipment. It is desirable from an operating standpoint for the operating level to be at the same elevation at all points; however, a lower level may be required between

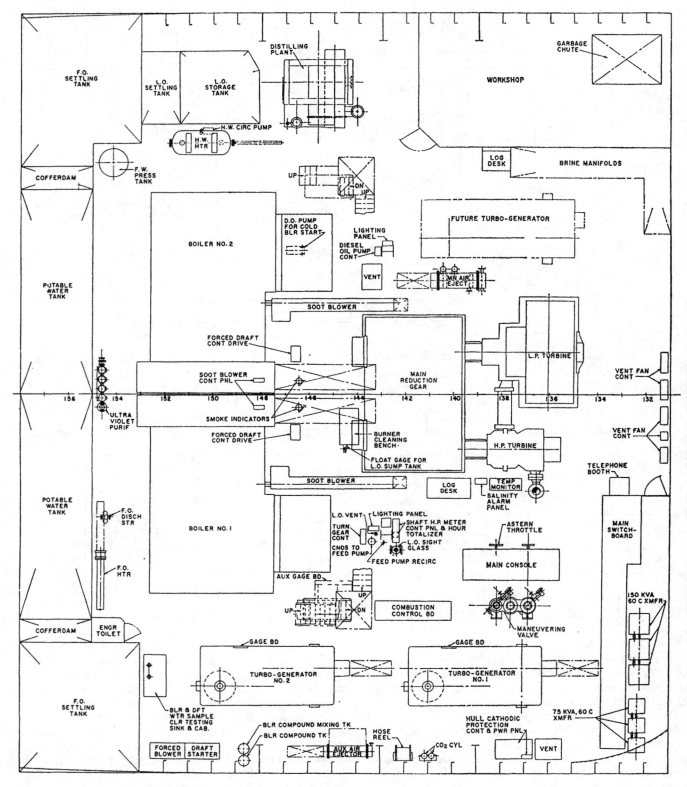

Fig. 2   Steam plant machinery arrangement—plan view at operating level

the turbines and over the reduction gear for access to bearings and to permit observation of lubricating-oil thermometers and sight-flow indicators.

A machinery arrangement for a typical medium-speed diesel plant is shown in Figs. 3 and 4. The major constraints on the steam plant arrangement, i.e., shaft angle, topside configuration, access to the reduction gear, and operating level layout, also apply to the diesel arrangement. In addition, space must be allowed for the overhaul of pistons, cylinder heads, turbochargers, and other major component parts; for piping connections to, and maintenance of, engine-attached auxiliaries, such as cooling-water pumps and fuel pumps; and for the removal of tube bundles from intercoolers and aftercoolers. The angle and size of engine-exhaust connections must be considered to allow a direct routing of exhaust ducting; this is also a consideration for air-intake piping, which is frequently provided for the larger engines. If the plant has an exhaust-gas heat-recovery system, a significant volume of space above the engine is needed for the boiler. If the engine has an integral sump, sufficient space must be provided to locate the tank underneath the engine. If a separate sump is to be provided, its location should be close to the engine and sufficiently low to prevent backflow of oil to the engine after shutdown. If the engine or reduction gear is resiliently mounted, the added height and greater engine deflection due to the mounts must be taken into consideration.

b. Other spaces. The arrangement of piping and associated equipment outside the main machinery spaces entails both the development of arrangements within spaces, such as pump rooms, electronic equipment spaces, and similar areas, and the arrangement of whole-ship distributive systems that interconnect many spaces and items of equipment, such as the firemain, chilled-water, plumbing, and potable-water systems. The same general principles described for main machinery spaces apply; for example, adequate space for access and maintenance must be provided, related components should be grouped, and the operation of each component in its system must be considered. The location of the galley and sanitary spaces relative to sanitary collection tanks and marine sanitation devices must allow for adequate slope, if gravity-drain piping is used. Piping routed through living spaces and behind joiner bulkheads must be designed to provide access for inspection and maintenance.

Distributive systems rely on supply and return mains to circulate or deliver fluids to many dispersed users. On naval combatants designed for survivability, the system sources are also numerous and distributed throughout the ship. Mains serving these systems must be run through the ship in paths that are as straight as possible to reduce the cost and weight of the system. Mains are often run through passageways to reduce interference with intervening spaces. Mains can also be run through trunks especially designed for the purpose; however, since this is an expensive and space-consuming option, it is used only when other means are not feasible. In any case, the piping routes must be selected and adequate space reserved early in the design. On naval combatants, mains for vital services such as chilled water and firemain should satisfy separation requirements both athwartship and vertically to decrease their vulnerability to damage.

Section 2
Piping Design Details

2.1 General. Commercial engineering standards for piping systems, which reflect the expertise and experience of the professional engineering community, are published by the American National Standards Institute (ANSI), American Society for Testing and Materials (ASTM), American Society of Mechanical Engineers (ASME), Manufacturers' Standardization Society of the Valve and Fitting Industry (MSS), and other engineering standards organizations; these standards are commonly invoked by shipbuilding specifications. Requirements unique to marine applications, and intended to protect public health and safety and the marine environment, are promulgated by regulatory bodies including the United States Coast Guard (USCG), the United States Public Health Service (USPHS) [6], the International Maritime Organization (IMO), and others. In addition, requirements designed to ensure the safety of ships and their cargoes are produced by the American Bureau of Shipping (ABS) and other international classification societies. For U.S. Navy ships an entire library of military specifications and standards applies. In addition to the above, design practices that reflect the unique requirements of specific ships are often invoked by the ship's owner.

This section contains an outline of the more significant and generally applicable piping system requirements.

2.2 Arrangement. Piping should be arranged in a neat, orderly manner and should be run as directly as possible between the machinery and components that it serves. Piping should not obstruct or interfere with the operation of doors, hatches, or scuttles. Piping should permit free passage in walking areas and the unobstructed performance of work in designated working areas. The operation and control of machinery should not be impeded, and the interference of piping with the maintenance of equipment and ship structure should be minimized. Wherever practicable, piping should be kept clear of removable plates provided in the ship structure for shipping and unshipping machinery or equipment. Where this is not practicable, piping 4 in. nominal size and larger should be flanged for removal. Access to compartments or equipment should not be limited by the piping.

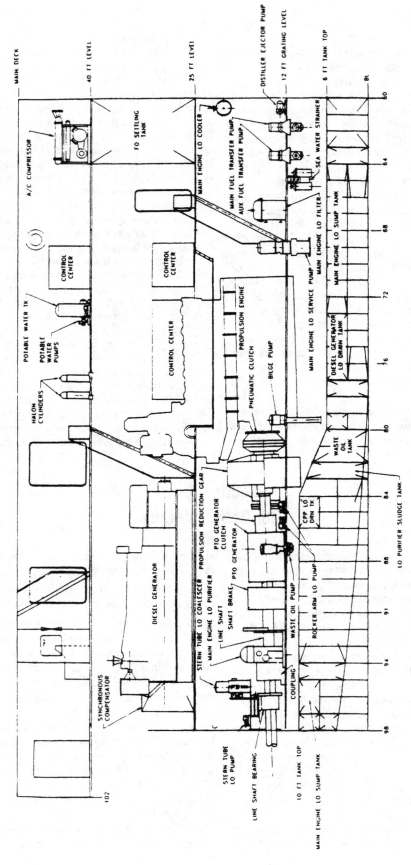

Fig. 3   Medium-speed diesel plant machinery arrangement—elevation at centerline, looking to port

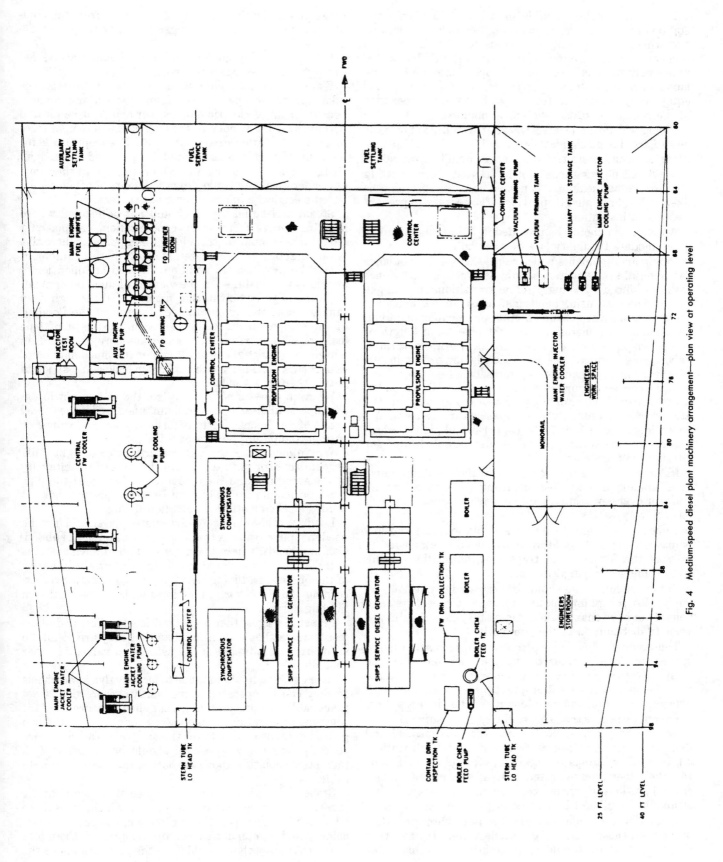

Fig. 4 Medium-speed diesel plant machinery arrangement—plan view at operating level

Piping should be portable in way of machinery and equipment that require dismantling for periodic overhaul, and wherever necessary for access to other piping systems or electrical systems. Stop valves should be located judiciously to isolate sections of piping intended for removal during maintenance and overhaul of machinery and equipment, so that interference with continued operation of the remainder of the system is minimized.

Piping should be located where it would not normally be subjected to mechanical damage. When this is impracticable, a means to protect the piping should be provided.

Insulated piping should not be located where wetting due to normal leakage, condensation, or splashing is likely. Shielding must be provided if necessary to prevent wetting of insulation.

Ample room should be provided in the arrangement to accommodate flexibility (see Section 2.6).

Piping arrangements which cause excessive turbulence detrimental to the system should be avoided. Turbulence can be reduced by using gradual transitions at changes in pipe diameter; using bends instead of elbows, and long-radius elbows where bends are not practicable; installing sweep tees; and installing straight pipe for a length of five to ten pipe diameters downstream of orifices, throttle valves, and pumps. This is especially important in 90–10 copper-nickel seawater systems, where a high local velocity caused by turbulence can continuously erode the protective oxide layer on the pipe wall and greatly accelerate corrosion. In zones where excessive turbulence cannot be avoided, a length of 70–30 copper-nickel piping, which has greater resistance to erosion, may be substituted for 90–10 copper-nickel.

Miters should be used only where they will not cause objectionable pressure loss or turbulence, such as in tank vents or overflows. Miters are not used in pressure piping on U.S. Navy ships.

Unnecessary high points (in a liquid system) and low points (in gas and liquid systems) should be avoided to prevent the formation of traps that inhibit the venting and draining of piping systems.

Pump suction piping should be arranged to rise continuously into the pump suction to avoid air pockets, and should not have changes in direction that cause an uneven velocity distribution at the pump inlet [7].

The amount of piping led through messing and living spaces should be minimized. Piping in such spaces should be symmetrically and neatly arranged and installed in the most inconspicuous location practicable, consistent with efficient operation and maintenance of the piping system.

Except as necessary to serve the space, insofar as practicable, piping should not be run through medical and dental spaces; chain lockers; freshwater, fuel, lubricating-oil, or sanitary tanks or the service areas or voids surrounding them; refrigerated spaces; electronics spaces; control stations; or wiring trunks and enclosures. Piping cleanouts should not be located in food service spaces.

Potable-water piping should not pass through tanks other than those containing potable water. Piping other than potable water should not pass through potable-water tanks unless the through piping is isolated from the potable water by a trunk or a larger pipe that is open at the lower end.

Piping normally under pressure should not be led through unvented spaces or voids.

Flammable-fluid piping should not be led through uptake spaces or spaces containing generators, switchboards, or large electric motors, nor should it be located within 18 in. of any hot surface (one having a temperature under its insulation greater than 400 F, except 650 F for lube oil). Spray shields should be installed around all flanged joints, including valve bonnets and strainer covers, where leakage could direct a flammable fluid on exposed electrical equipment or a hot surface.

Steam and liquid piping should not be located where drips or sprays from leaks, condensation, or splashing from funnels could contact electric equipment. Where this is not practicable, adequate shielding of the equipment should be provided. In addition, flanged or union joints should not be installed in the vicinity of such equipment.

Where bilge or dewatering piping is led through deep tanks, means should be provided to prevent the flooding of the compartments being served in the event of pipe leakage within the tanks. Such means may consist of an oiltight or watertight pipe tunnel; or the pipes may be of extra heavy thickness with expansion bends, and all joints within the tanks welded.

Where a pipe tunnel is installed, the watertight integrity of the bulkheads must be maintained, and if the tunnel is not of sufficient size to afford easy access, valves or fittings should not be located within it.

Where pipes pass through watertight or oiltight bulkheads, decks, or tank tops, both the watertight integrity of the structure and the structural integrity of the pipe must be maintained. This can be accomplished using a welded sleeve or other penetration fitting.

Lead or other heat-sensitive materials should not be used in piping systems that penetrate watertight subdivision bulkheads, where the deterioration of such systems in the event of fire would impair the watertight integrity of the bulkheads. Piping with operating temperatures exceeding 125 F and vacuum piping should not be located in the bilge area.

Pressure gages, thermometers, level gages, and other instruments should be located so that they are visible to an operator positioned at the associated valves or other controls.

Every effort should be made to locate valve handwheels so they can be operated conveniently. Where this is not practicable, remote operating gear should be provided for convenient operation, or the valves should have attached gears or extension shafts for this purpose. Valves in horizontal piping below eye level should be arranged with their stems pointing above the horizontal wherever practicable.

Globe and angle stop valves may be arranged with the pressure either above or below the disk, whichever is more advantageous for operation, protection, and repair of machinery and equipment served by the system. Examples of typical valves that should have the pressure under the disk are boiler stop valves, root valves, throttle valves, sea valves, and tank valves subject to tank pressure. Where a valve is subject to vacuum in the closed position, the vacuum should also be under the disk.

Manifolds should be used where many pipes must be joined together, such as where a pump suction joins several tank tailpipes. Manifolds reduce the space required, reduce the number of field joints, and ease operation by co-locating valve handwheels. Manifolds may be used wherever globe or angle valves are satisfactory. They should be installed with the body just above the floor plates if practicable.

Inlet piping to safety and relief valves should be short and direct. Where relief is to the atmosphere, open ends of discharge piping should be such as not to damage machinery or equipment, or endanger personnel. Relief of flammable liquids should be to a suitable lower-pressure part of the system. Relief of toxic or explosive gases from containers or systems should be to the atmosphere above the weather deck.

Vents from flammable fluid tanks and atmospheric reliefs from toxic and inert gas systems should not terminate where their discharge can be picked up by ventilation or forced-draft air intakes, or where their discharge would otherwise damage machinery or equipment or endanger personnel.

Tailpipes should terminate at the lowest point in tanks and should be installed in such a manner as to guard against their coming into hard contact with the bottom of the tank should the bottom of the tank deflect upward. The end of each suction tailpipe should be enlarged to provide an area not less than 1.5 times the inside area of the tailpipe. The height above the bottom should be one half the tailpipe diameter, and the tailpipe location with respect to adjacent plating or other components should provide a free suction area around the open end periphery of not less than 1.5 times the inside area of the tailpipe.

All of the foregoing guidelines are desirable without exception; however, compromises may be necessary at the expense of the less-important features.

2.3 Materials. In order for a material to perform satisfactorily in service, it must possess the following characteristics:

• Sufficient strength or load-carrying ability at the operating temperature to resist the imposed loads.

• Retention of suitable ductility and impact properties at all operating temperatures.

• Satisfactory corrosion and erosion resistance in the media that it contacts internally and externally.

• Resistance to galling with mating materials in moving contact.

• Inability to contaminate fluids contacted internally or externally.

Additional factors that influence piping-system material selections include initial cost, durability (life-cycle cost), ease of making joints, fitting compatibility, weight, availability, and lead time required for production. Often more than one material is suitable for an application, in which case the relative importance of all factors must be weighed to determine the preferred selection. Successful experience is an important consideration in the selection of materials.

Mechanical properties of metals (castability, forgeability, machinability, weldability) must be evaluated for each specific application; a metal generally classified as "castable" may be suitable for one type of casting but not for another.

Plastics and composite materials offer advantages of lower weight and cost and freedom from corrosion compared with metals. The materials in this category commonly used in shipboard piping are polyvinyl chloride (PVC) and glass-reinforced plastic (GRP), which is also known as reinforced thermosetting resin plastic (RTRP). However, the use of these materials is severely restricted on both commercial and U.S. Navy ships because they are more susceptible to fire damage than metals and may produce toxic gases. Because nonmetallic materials are not electrically conductive, means must be provided to prevent a buildup of spark-inducing static charges when they are used for flammable fluids. All changes in direction must be made with fittings rather than bends, requiring more joints than a metal system. Also, most piping made of plastics or composites has a lower structural rigidity than metal piping, reducing its ability to resist loads imposed during transport, assembly, and service.

In many cases, contamination of the contained fluid by the piping material must be avoided. As examples, copper is preferable to steel for compressed air systems, where rust would be detrimental; and in most electronic cooling-water systems, copper-nickel is preferable to stainless steel because it has a lesser tendency to transmit metal ions into solution, thus maintaining a low conductivity of the cooling water.

Seawater piping requires special attention because seawater causes severe corrosion and erosion of many metals. The mechanisms are complex and their effects vary widely depending on temperature, whether the seawater is moving or stagnant, the amount of flow turbulence, the seawater oxygen content, and the chemical composition of the pipe [8]. Also, seawater is conducive to the growth of marine organisms that can quickly obstruct flow passages, especially in warmer climates. Steels are especially susceptible to seawater corrosion. Although galvanizing, epoxy coatings, or rubber linings can be used to increase the service lives of steels in a seawater environment, these measures are expensive to implement and difficult to maintain.

The 90–10 copper-nickel alloys form a protective oxide layer, which prevents further corrosion of the underlying material, and this characteristic is beneficial in many applications. However, the formation of the protective oxide layer can be inhibited if: the piping is exposed to seawater which contains pollutants, primarily sulfides; the pipe wall has deposits of oil, grease, or brazing flux; or the pipe has been heated too hot or too long during bending, welding, or brazing. The copper content of these alloys is toxic to marine growth and so prevents fouling.

Table 2   Typical piping materials

Service	Commercial Applications	U.S. Navy Applications
Seawater (dry firemain, bilge, ballast)	carbon steel, galvanized	90-10 copper-nickel GRP (non-vital services)
Seawater (cooling, wet firemain, distiller)	90-10 copper-nickel GRP	90-10 copper-nickel GRP (non-vital services)
Freshwater cooling	carbon steel GRP	stainless steel 90-10 copper-nickel
Potable water	copper PVC GRP	copper GRP
Chilled water	copper PVC GRP	90-10 copper-nickel copper GRP
Lube oil	carbon steel	carbon steel stainless steel
Fuel (diesel engine, boiler)	carbon steel	carbon steel
Fuel (gas turbine)	. . .	stainless steel
Fuel (aviation)	. . .	90-10 copper-nickel 70-30 copper-nickel
Cargo oil	carbon steel ductile iron	carbon steel
Cargo JP-5	. . .	90-10 copper-nickel 70-30 copper-nickel
Steam	alloy steel carbon steel	alloy steel carbon steel
Condensate	carbon steel copper	carbon steel
Feedwater	carbon steel	carbon steel
Plumbing drains (freshwater)	carbon steel PVC GRP	copper
Plumbing drains (seawater)	carbon steel PVC GRP	90-10 copper-nickel
Plumbing vents	carbon steel PVC GRP	carbon steel GRP copper
Compressed air	carbon steel copper	90-10 copper-nickel copper stainless steel GRP
Gas turbine bleed air	stainless steel	stainless steel
Hydraulics	carbon steel stainless steel copper	carbon steel stainless steel copper 90-10 copper-nickel
Refrigerant	copper	copper 90-10 copper-nickel
Cryogenic fluids	stainless steel	stainless steel
Carbon dioxide fire extinguishing	carbon steel	copper carbon steel
Seawater sprinkling	. . .	90-10 copper-nickel
Foam fire extinguishing	carbon steel	90-10 copper-nickel
Halon	carbon steel, galvanized	90-10 copper-nickel stainless steel

Titanium is also highly resistant to corrosion, and has greater strength and erosion resistance than copper-nickel, making it especially suitable for high-pressure and high-velocity seawater applications. Unlike copper-nickel, titanium is not toxic to marine organisms; therefore, other means of preventing fouling may be necessary. Some provisions include draining the system when not operating, maintaining sufficient velocity to prevent adherence of marine organisms, and continuously injecting a biocide, such as chlorine, into the system.

Plastic piping materials are impervious to seawater corrosion. Since they are not toxic to marine growth, means of preventing fouling as discussed for titanium may be necessary.

Galvanic corrosion can severely attack metals that are exposed to an electrolyte. To minimize effects of galvanic corrosion, the following factors should be considered when designing piping systems in seawater service:

• Reduce the potential differences between metals by selecting materials close together in the galvanic series or select metal combinations where one member polarizes easily.

• Avoid bimetallic couples where possible, by insulation or by the proper choice of materials.

• Where bimetallic couples cannot be avoided, keep the cathodic (noble) metal area small in relation to the anodic metal area. Important items such as fasteners, valve seats, and critical components should be the cathode in a bimetallic system.

• Paint or coat large cathodic areas.

• Specify fabrication procedures that require mill scale to be removed from steel surfaces.

• When using stainless steel, avoid regions that will contain standing water, and provide compensation where crevices are unavoidable.

• Waterproof all faying surfaces involving stainless steel with other bimetallic connections.

• Structures receiving cathodic protection should be electrically bonded together to provide low-resistance connections.

• Provide proper drainage to prevent an accumulation of standing water.

• When a nonferrous pipe or valve is connected to a steel overboard discharge connection or sea chest, install a replaceable steel waster piece of at least 0.375-in. wall thickness between them to protect the shell plating, sea chest, and discharge connection from galvanic attack.

Typical material selections for piping systems in naval and commercial service are given in Table 2. Material selections for naval applications differ from commercial practice to provide greater life, reduced maintenance, and increased shock and fire resistance. Since the operating conditions of individual ships vary widely, cost and durability are not assigned the same importance for all ships. Nevertheless, Table 2 can be used as a general guide.

Material for valves and fittings should be compatible with the associated piping material to provide similar strength, facilitate joint fabrication, and control galvanic corrosion. Typical combinations are:

PIPE MATERIAL	VALVE AND FITTING MATERIAL
steel	steel, ductile iron
stainless steel	stainless steel
copper-nickel	bronze, Monel, copper-nickel, ductile iron
copper	bronze, copper
GRP	GRP, metallic

The material for valve trim (seat, disk, stem, and other critical internal parts) is commonly different from the body material when an increased local resistance to corrosion, erosion, wire drawing, and galling is necessary. As examples, Monel trim is often used in valves for seawater service and Stellite, a cobalt-base alloy, is frequently used to face valve seats and disks in high-pressure steam service and in control valves where there are severe throttling conditions.

Piping system materials are generally specified in a material schedule, which is prepared concurrently with the system diagrams. Representative material schedules are given in reference 9 for U.S. Navy surface ships and reference 10 for commercial vessels.

Piping materials are discussed further in Chapter 22.

2.4 Pipe Size Selection. Pipe sizes are selected primarily so that: (a) flow resistance in a system is such that design flow rates and pressures are achieved using reasonable pump or compressor ratings (or, for gravity flow, reasonable gradients), and (b) the accompanying fluid velocities do not produce unacceptable turbulence, erosion, or noise. When more than one pipe size will satisfy these conditions, additional factors such as weight, space, and ease of fabrication may be considered, which usually results in the selection of the smallest suitable sizes.

Ideally, pipe sizes and pressure sources (such as pumps or compressors) should be selected simultaneously, permitting a balance to be achieved between the installation cost and the operating cost of the system. The result would be an optimum design that does not have oversized pipes (requiring excessive weight and space) or undersized pipes (resulting in excessive driver horsepowers). In practice, the ratings of pressure sources are often established so that their costs and space requirements can be determined and adequate lead time for manufacture can be provided well before the pipe sizes are selected. It is then the responsibility of the piping engineer to select pipe sizes that are compatible with the design characteristics previously selected.

A suitable margin should be added to the calculated system resistance when selecting source pressures and pipe sizes. The size of the margin should consider deterioration of the pressure source over its expected life, increases in piping resistance due to biological fouling or scaling from corrosion, and expected growth in system demand.

Pipe sizes should ensure rated flow to each component during all operating conditions wherever possible. Orifices or throttle valves should be used only where necessary to regulate flow, or to correct unavoidable imbalances where a main serves two or more components through parallel piping circuits. Pressure losses through piping and components in parallel paths should be such that it is not necessary to install a restriction in the path to a component requiring the greater flow (such as a main condenser) to ensure adequate flow in the path to a component requiring the lesser flow (such as a lubricating-oil cooler).

For a system containing a pump, pipe sizes must be selected with consideration to both the pump total head

(defined as the rated pressure differential between the pump suction and discharge connections) and the required pump suction head. The total system pressure loss including the pump suction piping loss must not exceed the pump total head. In addition, the total pressure loss in the pump suction piping must not reduce the net positive suction head available at the pump suction below the net positive suction head required by the pump. This is of particular concern when the fluid has a high vapor pressure (e.g., feedwater) or is highly viscous (e.g., lubricating oil), or when the pump is located above the fluid source. The total pressure loss includes both the flow resistance and the net static pressure change due to elevation. Chapter 14 contains further details concerning these considerations.

Low ambient temperatures are frequently encountered when starting and operating shipboard systems, which result in significantly higher resistances due to an increase in fluid viscosity. Cold start-up conditions often determine the worst-case resistance. Consideration of low temperatures is particularly important in the design of fuel, lubricating-oil, and hydraulic systems.

To prevent an ingress of air in closed-circuit piping, such as chilled-water systems and freshwater cooling systems, particular attention should be paid to maintaining a positive gage pressure throughout the system under all operating conditions, especially when the system has a high vertical loop.

Pipe sizes are usually determined iteratively. As the first step, trial sizes are selected to obtain reasonable fluid velocities. Table 3 contains velocities that have been found to be satisfactory in service. The upper limits were established to avoid excessive turbulence, erosion, and noise. For seawater piping, a minimum velocity of 3 fps is desirable to discourage the attachment of marine organisms.

The second step is to calculate resistance for the trial pipe sizes using design flows. The resistance is calculated for each flow path and each operating condition. The sum of the flow resistance (dynamic loss) plus elevation change (static loss or gain) in each flow path plus the required pressure at the terminal point is compared with the pressure available at the source. The trial sizes are adjusted and resistances recalculated until the source pressure is equal to or greater than the resistance for the worst-case condition. A final check should be made to ensure that the velocity in each pipe segment is satisfactory.

While the above procedure can be used to determine pipe sizes, it cannot necessarily be used to predict actual flows or operating pressures in the system. However, such predictions may be necessary when parallel paths must be balanced (such as in a sprinkling grid), when operating pressures or velocities must be known accurately to determine control settings, when pipe sizes must be optimized to minimize weight, or when operating parameters for other than the design condition must be known. For these situations, a flow network analysis can be used. Using this method, preliminary pipe sizes are estimated first, as before, and then calculations are made for iteratively assumed flows in each path until the calculated resistance exactly matches the pressure available to

Table 3 Design fluid velocities for piping

Service	Fluid Velocity, fps Nominal[a]	Limit
Condensate pump suction	$\sqrt{d}$	3
Condensate pump discharge	$3\sqrt{d}$	8
Condensate drains	$0.3\sqrt{d}$	1
Hot-water suction	$\sqrt{d}$	3
Hot-water discharge	$3\sqrt{d}$	8
Feedwater suction	$1.3\sqrt{d}$	4
Feedwater discharge	$4\sqrt{d}$	10
Cold freshwater suction	$3\sqrt{d}$	15
Cold freshwater discharge	$5\sqrt{d}$	20
Lube-oil service pump suction	$\sqrt{d}$	4
Lube-oil discharge	$2\sqrt{d}$	6
Heavy-fuel service suction	$\sqrt{d}$	4
Heavy-fuel service discharge	$1.5\sqrt{d}$	6
Heavy-fuel transfer suction	$\sqrt{d}$	6
Heavy-fuel transfer discharge	$2\sqrt{d}$	15
Distillate-fuel suction	$2\sqrt{d}$	7
Distillate-fuel discharge	$5\sqrt{d}$	12
Hydraulic-oil suction	$1.5\sqrt{d}$	8
Hydraulic-oil discharge	$8\sqrt{d}$	20
Seawater suction	$3\sqrt{d}$	$12^{b,c}$
Seawater discharge	$5\sqrt{d}$	$12^{b,c}$
Steam, high pressure	$50\sqrt{d}$	200
Steam exhaust, 215 psig	$75\sqrt{d}$	250
Steam exhaust, high vacuum	$75\sqrt{d}$	330

[a] d is the pipe internal diameter in inches.
[b] 9 fps for galvanized steel pipe.
[c] Seawater velocity in titanium and GRP piping may exceed these limits without detrimental erosion; however, friction losses, turbulence, and noise may still be limiting factors.

Table 4 Supply fixture unit values

Fixture	No. of Fixture Units
Lavatory	2
Service Sink	3
Shower	4
Urinal w/flush valve	5
Water Closet w/flush valve	10

Table 5 Supply flow corresponding to total fixture units

Fixture Units	Flow, gpm	Fixture Units	Flow, gpm
10	27	70	58.5
12	28.6	80	62
14	30.2	100	67.5
18	33.4	120	72.5
25	38	140	77.5
30	41	160	82.5
35	43.8	180	87
40	46.5	200	91.5
50	51.5	225	97
60	55	250	101

Table 6 Drainage fixture unit values

Fixture	No. of Fixture Units
Lavatory	1
Shower	3
Service sink	3
Scullery sink	4
Urinal	4
Water closet	8

Table 7 Drain pipe size corresponding to total fixture units

Pipe Size, in. nps	Maximum Connected Fixture Units Drain Pipe Pitch, in. per ft of horizontal distance $\frac{1}{8}$	$\frac{1}{4}$	$\frac{1}{2}$
2	...	21	26
2.5	...	24	31
3^a	30	35	45
3^b	40	50	60
4	180	216	250
5	390	480	575
6	700	840	1000
8	1600	1920	2300
10	2900	3500	4200

[a] With not over two water closets connected.
[b] With no water closets connected.

cause flow, thus corresponding to the actual flow in the path. When performing a flow network analysis, valve and fitting losses should be represented as K factors (discussed later) rather than as equivalent lengths to improve accuracy, because equivalent lengths vary with the flow rate while K factors are independent of flow rate.

Systems providing so-called "hotel services," i.e., hot and cold potable water and waste and soil drains, are subject to peak loading at various times of the day, as well as unequal loads in various portions of the system. Establishing the appropriate design flow for most segments of piping in these systems is not a rigorous procedure. Such piping is normally sized using the fixture-unit method. For supply systems, each fixture is assigned a supply fixture unit value, as indicated by Table 4. The total number of fixture units connected to each segment of piping is used to determine the flow for sizing that segment in a relationship shown by Table 5. The flow values given by Table 5 reflect a decreasing probability

that all fixtures will require a maximum supply simultaneously as the number of fixture units increases. Standard pressure loss calculations can then be used to determine the required size of the supply piping. For drainage systems, each fixture is assigned a drainage fixture unit value, as indicated by Table 6. The approach to sizing drainage pipes is similar to that for supply piping; however, since drain piping does not run full, the pressure loss is more difficult to determine. Therefore, Table 7 shows that instead of using the total fixture unit value to determine a drainage flow, the total fixture unit value is converted directly to a pipe size as a function of the pitch of the drain pipe. Additional details concerning fixture unit values and their application are contained in reference 11.

Pipe flow resistance depends on the flow rate, pipe dimensions and roughness, and the properties of the fluid. The tools of analysis are derived from the momentum relation (an expression of Newton's second law), the continuity equation (an expression of conservation of mass), a friction factor, the general energy equation, and the equation of state. The first three of these may be combined to form the following differential equation, which describes the condition of fluid in motion:

$$\frac{dp}{\gamma} + d\left(\frac{v^2}{2g}\right) + dz + \frac{\tau}{\gamma R}\, dL = 0 \qquad (1)$$

where

p = fluid pressure, lb/ft^2

γ = fluid density, lb/ft^3

V = fluid velocity, fps

g = acceleration of gravity, ft/sec^2

z = elevation of fluid, ft

τ = fluid frictional shear stress, lb/ft^2

R = mean hydraulic radius of flow channel (area/wetted perimeter), ft

L = pipe length, ft

If the flow is horizontal or the fluid is a gas, the third term can be dropped. Additionally, if the fluid is a liquid and there are no significant changes in flow area, the second term can be dropped. The remaining terms can be integrated to obtain the following expression for the pressure loss, Δp, over a length of pipe, ΔL:

$$\Delta p = \frac{\tau}{R}\,\Delta L \qquad (2)$$

The fluid frictional shear stress at the pipe wall is defined as:

$$\tau = f\,\frac{\gamma}{g}\,\frac{v^2}{8} \qquad (3)$$

where f is a dimensionless friction factor. R is defined as one-fourth the pipe diameter, d, for a circular pipe flowing full; therefore, the head loss, h_L, over a length of pipe can be determined from equation (2) as

$$h_L = \frac{\Delta p}{\gamma} = f\,\frac{L}{d}\,\frac{V^2}{2g} \qquad (4)$$

The use of equation (4) is not confined to incompressible fluids (i.e., liquids); it can also be used for compressible fluids (i.e., gases) when the pressure difference along the pipe is so small that the fluid density is nearly constant between the points of interest. In the design of main steam piping, for example, equation (4) can be used because there is a negligible change in fluid density between the superheater outlet and the turbine throttle. On the other hand, boiler escape piping offers an example of a system for which equation (4) is inaccurate when applied directly because of the large pressure loss, and hence steam density change, in the piping. For such situations, the system losses can be calculated with reasonable accuracy using equation (4) if the piping is broken into segments, with each segment having a change in pressure no greater than 10% of its inlet pressure.

For laminar flow, the friction factor is a function of Reynolds number (i.e., of pipe diameter and fluid velocity, density, and viscosity). For turbulent flow, the friction factor is also a function of the roughness of the pipe wall. Empirical values such as those of Moody [12] are used to determine the friction factor for new pipes of various materials. Increases in pipe roughness during service due to fouling by marine growth or scaling from corrosion should be taken into account by use of appropriate roughness values [13].

Valves and fittings behave differently than straight pipe in that their flow resistance is primarily caused by turbulence, changes in direction, and changes in velocity of the fluid rather than frictional shear stress. Empirical studies [14] show that the resistance of a valve or fitting can be expressed as a resistance coefficient, or "K factor," representing the number of "velocity heads" lost through the component, that is:

$$h_L = K\left(\frac{V^2}{2g}\right) \qquad (5)$$

The K factor is essentially independent of friction factor and Reynolds number. Values of K for common valves and fittings are given in standard handbooks [13,14].

Occasionally, it is convenient to represent the resistance of valves and fittings of a given size as a nondimensional "equivalent length," or "L/d," equal to the length of the same size straight pipe (expressed in diameters) that would produce the same resistance. L/d values are determined by relating equations (4) and (5) as follows:

$$K = f\left(\frac{L}{d}\right) \qquad (6)$$

Since the L/d value for a given valve or fitting varies with friction factor and Reynolds number, the system engineer must correct the value whenever the flow conditions change, especially if flow departs from the fully turbulent regime.

The resistances of strainers, heat exchangers, control valves and other components are usually available from manufacturers. The resistance of control valves can also be estimated using the resistance value for a fully open stop valve of similar pattern (i.e., globe, butterfly).

When the resistance of a given piping segment for a certain flow is known, the effect of a change in size or flow rate can be quickly approximated using the following relationships: (a) the resistance varies inversely with the fifth power of the pipe inside diameter when the friction factor and flow rate are held constant; (b) the resistance varies directly with the square of the flow rate when the friction factor and pipe diameter are held constant. These relationships are useful for approximating the effect of a change before performing the next set of calculations in an iterative series, and for estimating the effect of a change to a known system.

Since the waterline varies with the loading condition of the ship, analyses for sea-connected systems should be based on the ship displacement representing the worst-case condition. For example, in the case of a firemain system discharging to fire plugs in the superstructure, the lightest load condition is the worst case because it places the fireplugs at their greatest elevation above the waterline.

For most types of piping, pipe sizes are identified using the term "nominal pipe size" (nps). For 12-in.-nps sizes and smaller, the nps value does not refer to any specific dimension of the pipe; however, each size is associated with a specific outside diameter. For 14-in.-nps sizes and larger, the nps value corresponds to the pipe outside

Table 8   Relationship between nominal pipe size and pipe outside diameter

Nps	Outside Diameter, in.	Nps	Outside Diameter, in.
¼	0.540	5	5.563
½	0.840	6	6.625
¾	1.050	8	8.625
1	1.315	10	10.750
1¼	1.660	12	12.750
1½	1.900	14	14.000
2	2.375	16	16.000
2½	2.875	18	18.000
3	3.500	20	20.000
4	4.500		

diameter, as shown by Table 8. For a given nps value, different pipe wall thicknesses cause a corresponding change in the pipe inside diameter, but the outside diameter remains constant. Usually, nps sizes are intended even when the nps designation is not specifically stated.

Some commercial standards describe piping for which the outside diameters are equal to the size stated, for all pipe diameters. When referring to such piping, the term "OD" must be included to avoid confusion, e.g., "2-in. OD."

2.5 Pipe Wall Thickness.   The pipe wall thickness must be sufficient to withstand internal and external design pressures, and external loads imposed during assembly and operation.

The design pressure and temperature used to establish the pipe wall thickness should be the highest (or most severe) the piping is expected to experience in service. These values are also used in selecting the pressure rating of valves and other pressure-containing components in the system. Different values may be chosen for different parts of a system; for example, pump suction piping is usually designed for a lower pressure than its discharge piping. Improper operation and component failure must be considered in establishing design conditions; thus, the relief-valve setting, rather than a pressure-regulating-valve setting, should be used as the design pressure; and stop valves or check valves should not be considered boundaries for purposes of establishing design pressure. For open-ended piping, such as steam escape piping, that is subject to high flows, the maximum back pressure at the upstream end should be used as the design pressure.

The minimum thickness to withstand design pressure depends on the pipe size and the allowable stress of the material at the design temperature. The internal pressure usually determines this thickness; however, external pressure can govern for some submarine piping external to the pressure hull, and for some piping that is deeply submerged in tanks. External pressure is of particular concern for nonmetallic piping. The minimum wall thickness is established generally in accordance with ANSI/ASME standards [15], but requirements of the applicable regulatory bodies, classification societies, Navy specifications, and occasionally owners must also be considered, particularly when determining the allowable stress of materials.

Each type of piping in general use on ships is manufactured in standard wall thicknesses; these are generally different for each material and manufacturing process. The following factors must be considered in addition to the design pressure and temperature when selecting the standard thickness to order:

• Thickness must be such that fabrication and assembly procedures, such as threading and bending, will not thin the pipe wall below the minimum wall thickness under any circumstances.

• For piping that is subject to corrosion or erosion, particularly in seawater, an allowance based on the material and the expected life of the piping should be added to the minimum thickness [8].

• The mill manufacturing tolerance may reduce the wall below the nominal thickness stated in the applicable standard.

• For pipe that will be assembled by welding, the wall thickness should suit the welding process used, welder skill levels, and the welding site (shop or ship).

• For pipe that is to be assembled using mechanically attached fittings, the wall thickness must be within the thickness range for the which the fitting was designed.

• For open-ended and low-pressure piping, a relatively thin wall is adequate if the design pressure alone is considered; however, the ordered thickness should provide sufficient mechanical strength to prevent damage by the crew or cargo, or operational damage after installation.

2.6 Flexibility and Support.   All piping must have sufficient flexibility to absorb dimensional changes resulting from thermal expansion and contraction, and motion resulting from flexing of hull structure, resilient equipment mounts, and shock excursions. This flexibility must be provided to prevent: pipe overstress in compression, tension, or torsion; overload of piping supports; excessive bending moments at joints; and excessive design loads on equipment to which the piping is connected. The necessary flexibility must be provided without exceeding the motion tolerance of supports or allowing piping to strike adjacent structure.

The flexibility of piping depends on: its size, wall thickness, and material; the number and location of changes in direction; and the type and location of supports. The amount of movement that must be absorbed by the piping depends on the operating temperature range, the structural flexibility of the ship, the movement of piping attachment points (e.g., resiliently mounted equipment), and the shock inputs. Flexibility calculations are required whenever reasonable doubt exists that adequate flexibility has been provided. Reasonable doubt of flexibility of a two-anchor segment of ferrous piping of uniform size may be considered to exist when:

$$\frac{dy}{(U-L)^2} \le 0.03 \qquad (7)$$

where
d = pipe nominal size, in.
y = resultant of movement to be absorbed by pipe, in.
U = straight-line distance between anchors, ft
L = developed length of pipe, ft

The determination of the need for a detailed analysis

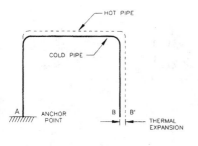

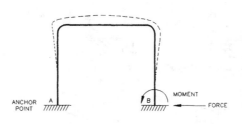

Fig. 5 U-shaped expansion bend

should be based on comparisons with similar successful systems, and should give consideration to the consequences of a pipe failure. Detailed analyses are usually performed for piping in services above 800 F and for systems containing flammable or toxic fluids.

Calculations are performed using standard techniques [16], and should establish the magnitude and direction of forces and moments at all terminal connections and at anchor and junction points, as well as the resultant bending stress, longitudinal pressure stress, torsional stress, and combined expansion stress at all such points. The required travel and spring constant must also be determined for all spring supports.

The following configurations may lead to insufficient flexibility and should be avoided:

- long, straight runs of piping between fixed supports, e.g., compartment bulkheads
- local reductions in pipe cross-sectional area
- local use of weaker materials
- small piping in series with large piping
- an excessive number of hangers

For high-temperature systems, flexibility is primarily provided by designing bends, elbows, loops, and offsets into the piping run. The response to thermal expansion of piping containing a U-shaped expansion bend is illustrated in Fig. 5. The left-hand sketch shows the pipe anchored only at point A, allowing free expansion. When the pipe is heated, expansion causes point B to be displaced to B'. The stresses caused by expansion can be understood by considering the forces necessary to return the heated pipe to its cold position. As shown in the right-hand sketch, a single force is sufficient to return point B' to point B; however, the action of this force would result in the pipe having an angular deflection at this point. Since anchors are assumed to prevent angular deflection, a restraining moment must be added. The resulting expansion stresses in this illustration are entirely bending stresses. The deeper the U-bend between the anchor points, the lower the stresses will be for a given temperature range.

Expansion bends are frequently installed in three dimensions as illustrated in Fig. 6. Flexibility is greatly increased compared with a U-bend because for movement in any one direction, one of the three legs is in torsion. The additional thermal expansion in the third dimension usually does not offset the gain in flexibility that which the torsional leg provides. Three-dimensional bends can

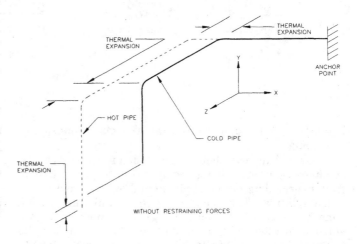

Fig. 6 Three-dimensional expansion bend

be used to advantage whenever space limitations do not permit sufficiently large U-bends to reduce stresses to acceptable values.

Expansion joints may also be used to provide flexibility. Sliding and bellows types of expansion joints absorb linear motion, while ball, swivel, and rotary types absorb angular motion. Since expansion joints are generally less reliable than pipe, they should be installed only if the piping configuration required to accommodate the temperature range will not fit in the space available, if it would cause unacceptable pressure losses, or if it would require an inordinate number of joints. The cost of maintaining expansion joints can be significant and should be considered.

The motion of resiliently mounted equipment is often accommodated by installing flexible hoses at the equipment, rather than by increasing the flexibility of the connected piping.

Because of the vibratory movement of reciprocating equipment, such as air compressors, the connected instrument piping should be provided with suitable loops or bends to prevent fatigue failures.

Pipe supports must be sufficient to carry the weight of the piping and the contained fluid, including the hydrostatic test fluid, and the inertial loads resulting from vibration, ship motion, and shock. Pipe supports should prevent the transmission of excessive loads to connected

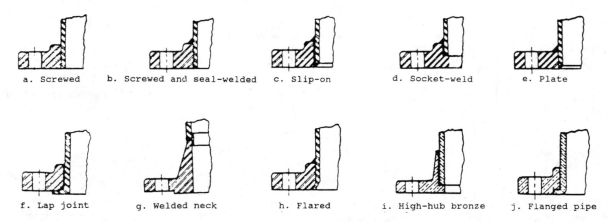

a. Screwed   b. Screwed and seal-welded   c. Slip-on   d. Socket-weld   e. Plate

f. Lap joint   g. Welded neck   h. Flared   i. High-hub bronze   j. Flanged pipe

Fig. 7 Piping flanges

equipment, but should not restrain thermal expansion of the piping.

Separate supports should be installed for heavy valves, strainers, and other in-line components to prevent them from overloading attached piping and to limit their vibration amplitudes. This is especially important when a component has an offset center of gravity, such as a gate valve with attached motor actuator. Valve bonnet bolts should not be used for attaching supports unless the bolts have been determined strong enough for the loads. During ship trials, additional supports may be found necessary to eliminate resonant conditions.

Extra support is usually required in the vicinity of large relief valves to absorb the reaction forces that occur when the relief valves lift.

2.7 Joints. Shipboard piping is exposed to constant vibration and flexing; therefore highly reliable joints are required. Many types of joints do not have the strength and durability required to operate in the shipboard environment for a long period without leaking. Some of the satisfactory joints include: bolted flange, butt weld, socket weld, brazed socket, reinforced branch connection, threaded, union, coupling, mechanically attached fitting, and bonded socket (for plastic and composite materials).

The selection of joints for a given piping system is based on many factors, including: pressure, temperature, cost, safety, ambient conditions, pipe size, pipe material, relative ease of assembly in the shop or on the ship, ease of inspection and quality assurance, availability of components with matching end connections, skill level required of installers, and restrictions imposed by regulatory bodies, classification societies, and owner requirements. Most systems contain several different types of joints.

The flanged joints shown by Fig. 7 are suitable for the full range of pressures and temperatures of shipboard systems. Gaskets suitable for the pressure, temperature, and flange mating surfaces must be selected. The bolts must provide adequate strength for the joint size and pressure, and the bolting material must be suitable for

the ambient conditions (e.g., seawater corrosion, high temperature).

Welded joints [Figs. 8(a) and 8(b)] are also suitable for all shipboard pressures and temperatures.

Brazed joints [Fig. 8(c)] have upper temperature limits that depend on the pipe material and the brazing metal used.

Reinforced branch connections [Fig. 8(d)] provide a method of attaching branches without using a fitting, thus reducing the number of joints per branch from three to two, and making easier the installation of a new branch on an existing pipe. These connections are designed to reduce the stress concentration that exists at the junction of the two pipes in an unreinforced connection. The connections are designed for either welding or brazing.

Threaded piping joints, of the straight or tapered type [Figs. 8(e) and 8(f)], are convenient to assemble and disassemble. However, compared with other joint types, they are more prone to leakage and crevice corrosion at the threads, and have less mechanical strength. Thus, on U.S. Navy ships their use is restricted to small sizes in systems that are not vital and do not contain hazardous fluids. Their use is limited to a lesser degree on commercial ships.

Union joints [Fig. 8(g)] are designed to overcome the weaknesses of threaded joints by providing greater mechanical strength and allowing use of an O-ring if necessary to isolate the threads from the system fluid, while still providing easy assembly and takedown.

Couplings are manufactured in many configurations [Fig. 8(h) is one example] and provide an inexpensive, simple method of assembly without hot work. Some types of couplings that are not attached directly to the pipe are not considered secure against separation due to vibration, thermal movement, and flexing of the ship. Some types require a packing gland or other seal to prevent leakage. Couplings are permitted in commercial ships for specific applications subject to special installation procedures to ensure they will not separate. They are generally not permitted on U.S. Navy ships.

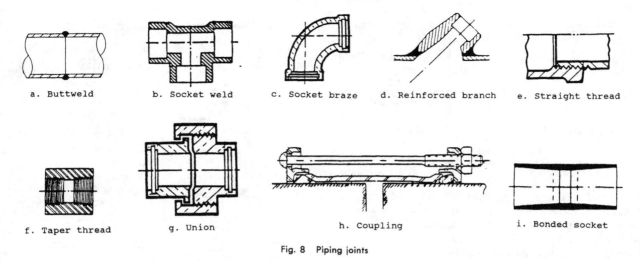

a. Buttweld   b. Socket weld   c. Socket braze   d. Reinforced branch   e. Straight thread

f. Taper thread   g. Union   h. Coupling   i. Bonded socket

Fig. 8   Piping joints

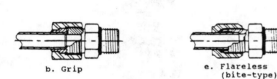

a. Swaged   d. Flared

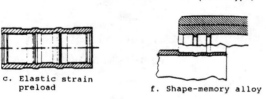

b. Grip   e. Flareless (bite-type)

c. Elastic strain preload   f. Shape-memory alloy

Fig. 9   Mechanically attached fittings

Bonded socket joints [Fig. 8(*i*)] are used with GRP pipe and are assembled with adhesive.

The category of mechanically attached fittings, which are illustrated by Fig. 9, includes a number of devices that provide a firm attachment of the fitting to the pipe without welding, brazing, or threading the pipe. These fittings use various techniques including swaging, shape-memory alloys, flares, and ferrules that bite into or grip the pipe. They do not require hot work and are convenient to install, making them especially suitable for joints that must be made up on the ship. While the skill level required to install these fittings is less than for welding or brazing, many of them require unique procedures and tools for their installation and to verify that the joint has been made properly. Since the integrity of joints made with these fittings depends on a carefully engineered mechanical interaction between the fitting and the pipe, care must

be taken to limit their application to the exact piping materials, sizes and wall thicknesses for which they were designed and qualified, and to follow exactly the manufacturer's assembly procedures. The use of these joints, especially the flared, flareless, and grip (compression) types, is restricted in some U.S. Navy and commercial applications.

Joints that have a high resistance to fire, such as welded, union, and flanged joints, should be used for fire-extinguishing systems and systems containing flammable fluids.

To increase system reliability, the number of joints should be minimized by using pipe bends in place of elbows wherever practicable. Welded or brazed joints are preferable to other types, particularly in areas inaccessible for inspection or maintenance.

Components welded into place should be accessible for repair and reseating. They should be located to permit removal, rewelding, preheating, and stress relieving in the event major repair or replacement is necessary.

Complex assemblies (e.g., a group of valves, strainers, and traps in a steam drain system) that cannot be repaired in place, and require periodic removal, should be made demountable by installing takedown (flanged, union, or threaded) joints between the root valves. One method is to provide a stop valve on each side of the assembly with one flanged end and one welded end. Where absolute tightness is required, such as for steam heating coils in fuel tanks, takedown capability may be provided by using brazed joints in the minimum practicable number of locations.

Relief, pressure-reducing, and control valves that require occasional removal for maintenance should have flanged or union ends.

Flanged and union joints should be located where they will be least affected by pipe bending due to thermal expansion, ship flexing, or other causes. Generally, this will require such joints to be located away from bends, elbows, and offsets.

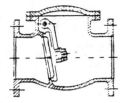
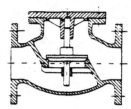

a. Swing check   b. Lift check

Fig. 10   Check valves

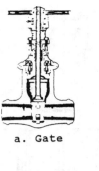

a. Gate   b. Globe

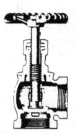
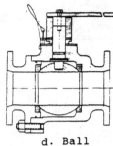

c. Angle   d. Ball

e. Butterfly

Fig. 11   Stop valves

Pipe joining fittings (e.g., tees, elbows, reducers, and sleeves) must have a pressure rating compatible with the system piping so that the inside diameters of pipes and fittings will match as closely as possible.

Appropriate measures for quality assurance must be applied for all joints. Welded joints require nondestructive inspection and testing to varying degrees depending on the application. Applicable inspection techniques include visual, dye penetrant, magnetic particle, and radiography. The extent of bonding in brazed socket joints cannot be determined visually. For brazed joints requiring high reliability, such as in U.S. Navy systems designed for shock resistance, the amount of bonding between the socket surfaces should be determined by ultrasonic testing or other means. Some mechanically attached fittings require special jigs or gages to ensure an adequate grip or pipe insertion length.

The integrity of joints must be checked after assembly by a hydrostatic test. Shop assemblies can be tested individually to reduce the shipboard time required for inspection and correction of leaks; however, a final test must always be performed onboard ship to check the field joints. For safety reasons, water is normally used as the test fluid unless water would contaminate the system (e.g., machinery lube-oil service system). The test pressure is typically 135 to 150% of the system design pressure. Gravity-drain piping is often tested by filling the piping with liquid to its highest point.

An air test is sometimes substituted for a hydrostatic test when necessary to avoid system contamination by liquid. Since compressed air can contain a dangerous amount of potential energy, such tests are conducted with caution to avoid a catastrophic failure in the event of a defective joint. It is advisable to conduct a preliminary air test at a pressure of about 25 psi to detect major leaks.

2.8   Valves and Valve Operators.

a.   Check valves.   Check valves should be installed wherever an undesired reversal of flow is possible. The two most common types are the swing-check [Fig. 10(a)] and the lift-check [Fig. 10(b)]. A lift-check valve can also be provided with a handwheel permitting closure of the valve to prevent flow in either direction; this configuration is called a stop-check valve. Where both stop and check features are needed nearby, such as at a pump discharge, one stop-check valve may be installed. Special check valves having springs, hold-open gear, or other devices are sometimes required by the system design. Horizontally mounted swing-check valves should be installed in a

fore-aft orientation, and vertically mounted swing-check valves should be installed so gravity will close the valve.

b.   Stop valves.   Stop valves are installed to isolate machinery, equipment, and piping components for system operation, maintenance and overhaul, and damage control. There are four basic types of stop valves: gate, globe, ball, and butterfly. Each basic type is manufactured in many different configurations, some of which are illustrated by Fig. 11. In addition, there are many specialty valves (e.g., diaphragm and scupper) designed for specific applications.

Factors considered in the selection of stop valves include cost, differential pressure, pressure loss, reliability, ease of operation, size, fire safety, and suitability for throttling. No single valve design is best for all applications, even within a given system.

Gate valves [Fig. 11(a)] provide a tight, reliable shutoff up to the highest pressures and temperatures in shipboard systems, making them especially suitable for high-pressure steam service and hull valves in sea-connected systems. Gate valves are compact except for the bonnet, which extends well off the pipe centerline. The non-rising-stem design is significantly more compact and helps alleviate arrangement problems. Seats and disks are metal. Gate valves have a relatively low operating torque and the lowest pressure loss of any valve type except ball valves. Gate valves are not suitable for throttling because they have an irregular relationship between the disk position and the flow area, they are subject to wire drawing of the seat and disk, and they cause excessive turbulence near the downstream pipe wall.

Globe valves [Fig. 11(b)] also provide a tight, reliable shutoff for the full range of shipboard systems. Compared with gate valves, they generally require a higher operating torque, but have a shorter stem travel and thus fewer handwheel turns from open to shut. They have larger bodies than gate valves but require less height for the bonnet. Globe valves have the highest pressure loss of the four types of stop valves because of the abrupt changes in direction and area of the flow stream. The angle globe valve [Fig. 11(c)] and the Y-pattern globe valve provide more direct flow paths, which reduce pressure losses. The angle pattern also enables a change in direction without an elbow or bend. Globe-valve disks and seats are usually metallic and are generally repairable without removing the valve from the pipeline. Globe valves are suitable for throttling, and provide the best control characteristics of the four basic stop-valve types. They are manufactured in a configuration using a tapered plug for close throttling applications. Globe valves should not be installed where the pressure loss or turbulent flow, characteristic of a globe valve, would be detrimental to the system.

Ball valves [Fig. 11(d)] have a relatively short actuating mechanism and body length, and provide a more compact installation than globe or gate valves. They are available in full-port and reduced-port configurations. The full-port type provides an unobstructed flow with a low pressure loss. Ball valves stroke from open to shut with a 90-deg stem rotation, making them especially suitable where rapid operation or remote actuation, with or without power, is necessary. The most common configuration contains elastomer seat seals at the inlet and outlet, which limits the satisfactory range of service temperature. Metal seats suitable for high temperatures are also used. Ball valves should not be used for throttling.

Butterfly valves [Fig. 11(e)] are the most compact of the stop valves; they have a relatively low pressure loss, and provide quarter-turn operation as do ball valves. Butterfly valves are generally of a less-rugged construction than other stop-valve types, which makes them more suited to the shipboard systems in the low end of the pressure range. Butterfly valves intended for applications requiring tight shutoff or throttling, sometimes called high-performance valves, must have seat materials, seal design, disk geometry, and disk supports suitable for such service. Butterfly valves are made with either metallic or nonmetallic seats. Butterfly valves used for throttling should have metal seats where the pressure loss exceeds 50 psi or where the disk opening when throttling will be less than 20 deg.

Valves should close with clockwise rotation of the handwheel or lever when facing the end of the valve stem.

For globe and gate valves, the stem threads are preferably external to the valve body (i.e., outside screw and yoke type), particularly in seawater systems, to reduce the effect of crevice corrosion on the mechanism.

Globe and gate valves of the rising-stem type are preferred because the position of the valve is apparent. Where a nonrising stem is used, the valve should be provided with an indicator that shows clearly whether the valve is open or closed. Valves installed in tanks and operated only by reach rods should have the indicator at the operating location rather than on the valve.

Stop valves in fire-extinguishing systems, systems that must continue operation after a fire, and systems containing flammable fluids should have metal-to-metal seats and disks or, in the case of ball valves, metal seat seals.

c. Remote stop-valve operators. A stop valve requires a remote operator if:

- it is in an inaccessible location
- it must be operable during an emergency and there may be no direct access to the valve during the emergency
- it must be operated from a central location
- its location would make it difficult for the operator to apply the necessary torque to the handwheel mounted on the valve

As examples, fuel-tank stop valves must be closable from outside the machinery space to prevent fuel from feeding a fire; some damage-control valves on U.S. Navy ships must be operable from the damage-control deck to permit continued operation even when lower decks are flooded or afire; and tailpipe valves installed in cargo tanks must be operable from the deck above the tank.

A valve-mounted handwheel should be provided in addition to the remote operator unless the valve is inaccessible.

The simplest form of manual remote operator consists of a rigid rod (also known as a reach rod) connected to a remotely located handwheel. Multiple rods connected by universal joints or gearboxes may be used for changes in direction. This method is suitable for distances of 50 ft or more for straight runs, but the practical distance is reduced by additional friction losses because of changes in direction.

Lever-operated pull cables that actuate spring-loaded valves are suitable for small valves when the remote operation is for closure only, such as for fuel-tank stop valves.

Flexible-shaft valve operators, which consist of a lubricated steel shaft rotating within a casing that connects the valve stem to a remote handwheel, are not an effective means of transmitting torque. They are limited in U.S. Navy applications to a maximum length of 8 ft.

A flexible-cable valve operator consists of steel cable sliding within a casing. A pair of cables is arranged in a loop to connect actuators located at the valve handwheel and the remote handwheel. The actuators and cables are arranged so that whether the valve is being opened or closed, the active cable is in tension. Valves are operable at distances up to 200 ft depending on the number and bend radii of the changes in direction required.

Manually pumped hydraulic valve operator systems consist of hydraulic valve actuators linked by tubing to a hand pump at the operating station. A single pump can be used to operate several valves by means of selector valves at the station. This type of system is used on U.S. Navy ships to operate ballast valves located in tanks and hull valves in seawater systems.

Power valve operators must be installed when the time required to close a valve manually exceeds the time available; the operating torque exceeds a manual capability; the system is automated; or the distance between the valve and operating station exceeds the limitations of manual devices.

The availability of the actuating medium (compressed air, hydraulic pressure, or electric power) is often a major determinant in the selection of the power actuation method. Remote valve operators should allow easy disconnection of a jammed remote device to permit local operation in an emergency. Power operators should be equipped with protective devices to prevent injury of personnel operating the valve locally in the event of remote actuation.

d. Relief and sentinel valves. A relief valve is designed to open when the inlet pressure exceeds a set maximum and discharge the system fluid, thereby preventing overpressure. The most common type is a spring-loaded angle relief valve, illustrated in Fig. 12(a), but other types which use fluid pressure, weighted levers, or external pilots for their operation are also available. A relief valve must be installed wherever the pressure in the piping could otherwise exceed the pressure for which the piping was designed. Equipment malfunction, control system failure, operator error, fire, and any other normal or emergency circumstances must be considered when determining the need for relief valves. A relief valve is usually not required on a system supplied by a centrifugal pump if the system is designed for the pump shutoff pressure.

The capacity of a relief valve must be no less than the capacity of the pressure source. A relief valve should generally be set so that the maximum operating pressure of the system does not exceed 90% of the set pressure to prevent relief-valve leakage during normal operation.

No stop or check valves should be installed where they could isolate the relief valve from the pressure source. No stop valves should be installed in relief valve discharge piping. Piping must be sized so that pressure losses do not cause unstable valve operation or reduce its capacity. Losses in piping leading to the relief valve should generally not exceed 10% of the difference between set pressure and reseat pressure. Relief valve discharge pressure losses, together with any superimposed back pressure, should not exceed 10% of the valve set pressure unless the valve is designed to compensate.

A sentinel valve is a small valve that is used to warn of a malfunction by serving as a telltale. The set pressure is usually above the normal working pressure of the system, but lower than the pressure the system can withstand. A sentinel valve does not usually have sufficient capacity to relieve the source of the overpressure.

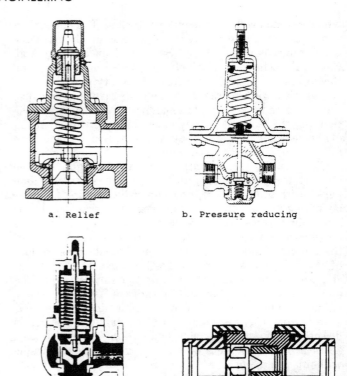

a. Relief   b. Pressure reducing

c. Back-pressure regulating   d. Constant-flow fitting

Fig. 12   Special valves

e. Pressure-reducing valves. A pressure-reducing valve [Fig. 12(b)] permits a higher-pressure source to supply a lower-pressure service, with an automatic response to changes in demand. A pressure-reducing valve should be installed: wherever operating requirements, such as frequent load changes, make it impractical or unsafe to control the pressure manually; to reduce the cost and weight of downstream components such as heat exchangers; or to permit a single pump or compressor to supply multiple services requiring different pressures.

Pressure-reducing valve actuators can be of the spring-loaded, dome-loaded, piston, diaphragm, direct, pilot, balanced, or unbalanced type; and the valve port and plug can be selected so the stroke-versus-flow-area characteristic is well suited to the application. The required response speed, reduced pressure range, inlet pressure range, and accuracy of control should be discussed with the manufacturer before such features are specified. Specifying a valve with an excessive capacity should be avoided because it results in continuous close throttling by the valve, leading to excessive wear of the valve seat and disk.

Each pressure-reducing valve should be provided with upstream and downstream pressure gages and isolation valves. If the pressure reducer supplies a service that cannot be shut down for maintenance of the reducer, a bypass with a throttle valve should be installed. A relief valve should be provided downstream of the downstream isolation valve. The relief valve should have sufficient capacity to pass the full flow of a failed-open reducer or a fully open bypass, whichever is greater, with full design inlet pressure. A relief valve is not necessary if the downstream components are designed for full upstream pressure. If the upstream system does not provide filtration sufficient to protect the reducing valve, a separate strainer or filter should be installed immediately upstream of the reducing valve.

f. Back-pressure regulating valves. A back-pressure regulating valve [Fig. 12(c)], also called an unloading valve, is similar to a pressure-reducing valve except that it controls upstream pressure rather than downstream pressure. An example is a priority valve installed in the low-pressure vital air system on U.S. Navy ships; this valve shuts off air to downstream nonvital users when the upstream vital air main pressure drops below its set point. Another example is an auxiliary-exhaust unloading valve in a steam plant, which maintains a constant exhaust pressure on steam-driven auxiliary equipment and in the deaerating feed heater by dumping excess exhaust to the main condenser.

g. Control valves. A control valve (also called a regulating valve) controls the flow at one location in a system in response to a control signal from another portion of the system, or from a different system. Examples are a main condenser level control valve (Fig. 18) and main lube-oil temperature control valve (Fig. 20). Temperature control valves can be of the three-way bypass type (constant total flow) or the throttling type (varying flow). The control signal can be a reservoir level, temperature, or pressure and can be communicated to the valve via a liquid, compressed air, or electrical signal. Control valves should be designed so that a loss of electric power, a loss of actuating air, or a broken spring will leave the valve in a safe mode.

When selecting a relief, pressure-reducing, back-pressure regulating, or control valve, the valve manufacturer should be given a full description of the operating conditions and performance requirements to ensure a satisfactory installation.

h. Constant-flow fittings. A constant-flow fitting [Fig. 12(d)] is used to regulate flow to a constant value as upstream and downstream pressures change. A flexible element within the fitting deforms to restrict flow in proportion to the differential pressure across it. The regulation accuracy and minimum differential pressure specified by the manufacturer must be considered when selecting these fittings. Constant-flow fittings are used extensively to balance flows to many loads in chilled water and electronic freshwater cooling systems.

2.9 Orifices. An orifice plate (Fig. 13) is installed to introduce a specific additional pressure loss into a flow path. An orifice is a passive device that cannot respond to changes in system conditions, but is sized for a single condition of inlet pressure, outlet pressure, flow rate, fluid density, and fluid viscosity. An orifice is useful for balancing the pressure losses, hence the flow, in parallel

Fig. 13 Orifice plate

flow paths, and for limiting flow in a given path. An orifice cannot be used as a pressure reducer, because the downstream piping will be exposed to full upstream pressure whenever flow is stopped, such as upon closure of a downstream valve.

Multiple orifices in series should be used when necessary to avoid cavitation, which causes turbulence and noise. For example, turbulence downstream of an orifice in 90–10 copper-nickel piping can erode the piping material.

Orifice sizing criteria can be found in reference 14.

2.10 Insulation. Insulation is required on high-temperature piping to limit the surface temperature for personnel protection, reduce the heat loss from the system, or prevent objectionable heat gain in enclosed spaces. Part or all of the piping in systems such as steam, feed, steam drain, fuel service, and hot potable-water systems normally requires insulation. Piping that infrequently operates at temperatures above 125 F, such as steam escape piping, does not require insulation unless it poses a personnel hazard. Insulation is not required on piping in voids or cofferdams unless an excessive heat loss from the system would result. Piping that must be located where insulation would be subject to damage from moisture, such as piping over shower stalls, under lavatories, or under steam kettles, should not be insulated; such piping should be shielded instead. Approximately 2 ft of pipe upstream of each steam trap should be left uninsulated to enhance trap operation.

Insulation is required on chilled water and refrigerant piping operating at temperatures below 40 F to reduce the heat gain by the operating fluids.

To prevent condensation, anti-sweat insulation is normally installed on piping systems, such as firemain and seawater cooling, that normally contain fluids at a temperature below the ambient dew point. Anti-sweat insulation is generally applied wherever condensation can cause corrosion or moisture damage, a safety hazard, or unpleasant living or working conditions.

Insulation should be installed on freshwater or seawater piping that is in the weather, cannot practicably be drained in cold ambient temperatures, and is subject to intermittent stagnant or low-flow conditions that could allow freezing of the fluid.

The materials most often used for high-temperature insulation are fiberglass, cellular glass, mineral fiber, and calcium silicate. For low-temperature insulation, the most common materials are fiberglass, cellular glass, and foamed plastics.

Table 9   Insulation thickness for personnel protection^a

Nom. Pipe Size, in.	Pipe Surface Temperature, deg F			
	250	450	650	850
1	1	1	1.5	2
3	1	1	2	2.5
6	1	1	2	3
10	1	1	2	3.5
14	1	1.5	2.5	3.5
20	1	1.5	2.5	4

^aThickness (in.) of Fibrous Glass Insulation for Insulation Surface Temperature ≤ 140 F.

Insulation is applied in blanket, block, and molded form. Molded insulation is preformed to fit pipes, elbows, and other common components for ease of installation. Removable and reusable blocks, forms, or pads should be used near takedown joints, around stem packing glands of steam valves, and at other locations requiring regular access for maintenance.

The insulation thickness depends on the pipe temperature, insulation surface temperature, pipe size, and the heat transmission rate of the insulating material selected. For shipboard piping systems, the thickness is generally based on providing the minimum necessary for personnel safety. Piping containing fluids at a temperature exceeding 150 F should have insulation of sufficient thickness so that the surface temperature of the insulation does not exceed 140 F to protect personnel from burns. Piping with a surface temperature between 125 F and 150 F should have, as a minimum, a cloth or tape wrap. The minimum thicknesses specified for commercial and U.S. Navy ship piping insulation are given in references 5 and 17, respectively. Representative insulation thicknesses are given in Table 9.

Lagging should be installed over insulation to protect it from damage. Lagging materials include sheets of galvanized steel, corrosion-resistant steel, aluminum, cloth, and fiberglass. Metal lagging is preferred when insulation is subject to damage.

For low-temperature insulation, a vapor barrier must be applied to prevent water vapor from reaching the cold surface, condensing, and soaking the insulation. A separate vapor barrier is not required if the insulation is foamed plastic or cellular glass, since the closed-cell structure of these materials serves this function.

Insulation should not be installed on piping joints until after the system has been hydrostatically tested.

2.11   Sea Connections. Sea chests are generally constructed of pipe or plating of material similar to the hull and welded to the shell plating. In way of the double bottom, a sea chest is formed by a trunk extending between the shell plating and the innerbottom plating, with the seawater piping connected to the innerbottom plating. The design must provide adequate structural reinforcement to compensate for the size, shape, and location of the opening in the hull.

Sea chests should be clear of bilge keels and other hull projections, and should not interfere with docking blocks.

They should be located where they will not be prone to pick up fluid from overboard discharge connections.

Sea chests should be located to avoid high entrance losses and negative pressures induced by flow along the hull with the ship underway. These effects are detrimental to pump performance, and are of particular significance on high-speed ships and in locations near hull appendages or where there is flow separation along the hull. Locations that may emerge during roll, pitch, or light-ship conditions, or allow an ingestion of entrained air along the submerged hull, should also be avoided. If an intake of air cannot be avoided, the sea chest should incorporate a high point for the collection of air, with the high point continuously vented above the deep-draft line.

Sea chests that will be used in shallow water should be located to avoid picking up debris or dirt from the bottom of the waterway. If a single location that will avoid both air ingestion and debris intake is not feasible, both high and low suction sea chests should be installed.

Each sea chest must have a strainer made of bars or perforated plate installed at the junction with the shell plating. The clear area through the strainer plate should be 1.5 to 2 times the total area of all suction piping connected to the sea chest. Where practicable, the strainer openings should be no larger than the smallest flow passage in the connected systems. Where this is not practicable, a separate strainer is required in the system piping.

Sea chests should be provided with a means of clearing debris from the strainer. Steam can be used where readily available; otherwise, compressed air can be used. The sea chest blow connection should be installed outboard of the sea valve and arranged so that the jet of steam or air is aimed as directly as possible at the strainer.

Overboard discharges consist of pipes attached directly to the shell plating. They should be located to avoid flow lines leading to suction sea chests, underwater logs, and sonar transducers. If installed above the waterline, they should not be located in way of boat-handling areas or accommodation ladders.

A sea valve should be installed in each pipe connected to a suction sea chest or overboard discharge. Where more than one pump is connected to a sea chest or overboard discharge, an additional valve should be installed in each branch to permit the isolation of each pump. Where a pump is located in a compartment other than that in which the sea chest or overboard discharge is located, an additional valve should be installed in the compartment in which the pump is installed. Sea valves should be installed as close as possible to the sea chest or shell plating, and any intervening joints should be welded.

Sea valves may be of the gate, angle, or butterfly type; gate valves are preferred. Butterfly valves are not used as sea valves on U.S. Navy ships. Sea valves of the gate type should have stems of one-piece construction and the stem should not be attached to the disk by pins. Sea valves in spaces not normally manned should have a remote control from the deck above.

Ductile materials such as steel, bronze, or nodular cast iron should be used for connections to the shell plating below the freeboard deck; steel is most commonly used. Malleable iron and nonductile cast iron are not suitable.

The number of sea chests and overboard discharges installed should be the minimum consistent with proper operation of the connected systems. At least two sea chests should be installed for main propulsion cooling, generator cooling, and other vital systems to insure a continued supply in the event one becomes clogged. They should be separated as far as practicable.

Sea chests for ships that will operate in ice require special consideration. Broken ice must be prevented from clogging the seawater cooling system inlet piping and strainers. Sea chest strainers must be strong enough to withstand the pressure and abrasion of ice. For icebreakers, the motion of the ship when breaking ice, backwash from frequent reversing of the propellers, and mixing of broken ice with seawater, together cause seawater entering the sea chest to contain a large amount of entrained air and ice fragments. Therefore, sea chests must be designed to keep the flow velocity toward the suction pipe sufficiently low to allow ice fragments to float to the top of the sea chest and the air to separate. The suction should be taken from the lower portion of the sea chest, and sufficient height should be provided to permit ice to accumulate away from the suction. The top of the sea chest should be sloped to facilitate the removal of ice fragments and air, and a large vent, with an inlet screen to prevent clogging, should be led from the top of the sea chest. A warm-water return from the machinery cooling system should be connected near, and directed toward, the top of the sea chest to melt trapped ice, and a baffle should be provided to direct incoming cold water toward the warmed portion of the sea chest. The strainer area should be at least 4 times the area of the connected suction piping.

The result of incorporating all of these design features is a much larger and more structurally complex sea chest design compared with designs for ships that do not operate in ice; therefore, only those features appropriate to the severity of the intended operating area should be incorporated.

Section 3
Machinery Plant Support Systems

3.1 General. Steam, diesel, and gas turbine machinery plants require piping systems for fuel, lubricating oil, compressed air, cooling water, and other services. System configurations vary to suit each type of plant. Each type of plant also requires unique systems for its operation. As examples, steam plants require piping for steam, condensate, feedwater, and other services; diesel plants require freshwater cooling and, often, waste-heat systems; and gas turbine plants have a bleed-air system.

This section describes many of these systems, providing greater detail for systems not covered in other chapters. Chapter 12 contains additional information about fuel systems; and Chapters 3 and 4 contain additional information on diesel and gas turbine support systems, respectively.

As a framework for discussion, this section is generally oriented to a single-screw commercial ship, with exceptions noted for other types of plants.

3.2 Steam Plant Piping Systems.

a. Main steam system. The main steam piping system delivers superheated steam from the boilers to the main propulsion turbines for ahead and astern operation, to the generator turbines, and to the boiler drum desuperheaters. In commercial applications, the system also supplies the main feed pump turbines to improve the cycle efficiency.

Main steam piping operates at a high pressure and temperature, requiring a relatively thick pipe wall and resulting in a stiff cross section, high weight, and high loads at supports and anchor points. To increase pipe flexibility, the pipe sizes should be as small as practicable and numerous changes in direction are desirable; however, large pipe sizes and straight pipe runs are preferred to minimize friction losses, which adversely affect the turbine throttle pressure and plant efficiency. These conflicting requirements necessitate a careful trade-off to optimize the design. The wall thicknesses for candidate pipe sizes should be calculated first (see Section 2.5). Inside diameters can then be determined for pressure-loss calculations. A trial arrangement of piping should then be developed to serve as a basis for pressure-loss and flexibility calculations. Several iterations of arrangements and calculations may be necessary before an acceptable design is established. The complexity of this process increases with the number of boilers and turbines to be interconnected.

Double-valve protection should be provided to facilitate maintenance with any combination of units in operation. A stop valve must be provided at each boiler outlet and at each connected unit. For boilers in parallel, each boiler outlet should have two stop valves, with one of them having a check feature. In addition, a stop valve should be provided in each branch from the main to a unit.

Pipe supports must be provided not only to carry the weight of the piping, but also to restrain lateral movement due to shock, ship motion, and vibration. Fixed supports should be located so that bends in the piping between them will provide sufficient flexibility to permit movements due to expansion. The locations of supports depend on the general arrangement; where practicable, the pipes should be placed to facilitate support from adjacent structure.

Piping should be sloped continuously either toward or away from the turbines. Pockets must be avoided, and drainage must be provided at low points.

Figure 14 is a typical main steam system diagram. The system shown has two boilers; consequently, boiler cutout valves of the stop-check type are provided to prevent crossflow from an operating boiler to one not in use. By-

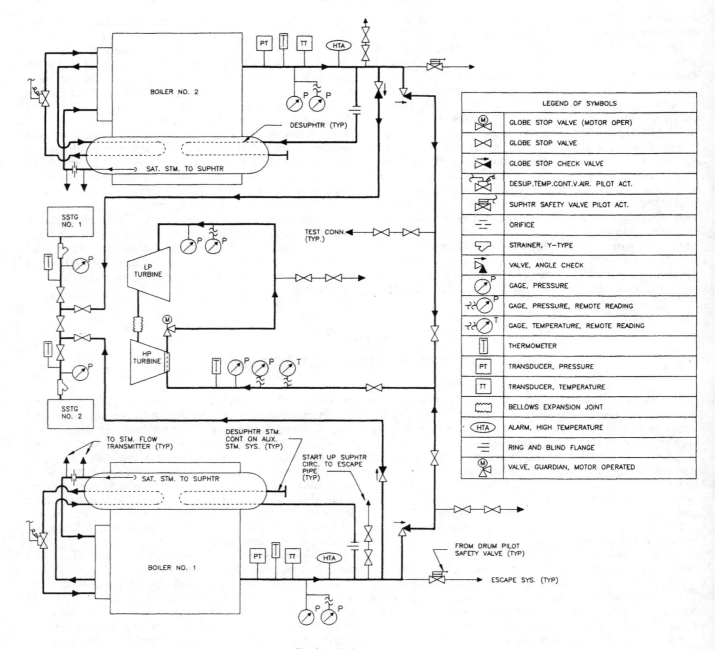

LEGEND OF SYMBOLS	
⬡M	GLOBE STOP VALVE (MOTOR OPER)
▷◁	GLOBE STOP VALVE
▷◁	GLOBE STOP CHECK VALVE
	DESUP.TEMP.CONT.V.AIR. PILOT ACT.
	SUPHTR SAFETY VALVE PILOT ACT.
	ORIFICE
▽	STRAINER, Y-TYPE
	VALVE, ANGLE CHECK
	GAGE, PRESSURE
	GAGE, PRESSURE, REMOTE READING
	GAGE, TEMPERATURE, REMOTE READING
	THERMOMETER
PT	TRANSDUCER, PRESSURE
TT	TRANSDUCER, TEMPERATURE
∿	BELLOWS EXPANSION JOINT
HTA	ALARM, HIGH TEMPERATURE
	RING AND BLIND FLANGE
⬡M	VALVE, GUARDIAN, MOTOR OPERATED

Fig. 14 Main steam system

pass valves are fitted to equalize pressure on both sides of stop valves so that they will be easier to open. Bypass valves also permit the pressure and temperature of the downstream piping to be raised slowly to operating conditions during plant start-up.

The guarding valve in the astern steam line to the astern turbine provides positive isolation of the turbine during ahead operation to prevent steam that may leak past the astern throttle valve from causing increased windage losses and erosion of the astern turbine blades.

The pilot-actuated safety valve in the superheater outlet is provided as a first response to boiler overpressure. It is actuated by boiler drum pressure and opens at a lower pressure than the main safety valves on the boiler steam drum. This arrangement prevents the drum safety valves from being opened by occasional pressure excursions during normal operation, thereby preserving their sealing surfaces and reducing maintenance and leaks; it also ensures a continuous flow of steam through the superheater tubes. In the event a boiler casualty occurs, the drum pressure will continue to rise until the drum safety valves open.

The function of the desuperheater control valve is explained in Chapter 5.

For emergency operation using only the high-pressure turbine, blanked flanges are provided so that temporary

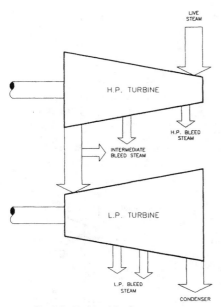

Fig. 15 Turbine bleed steam points

piping can be installed for the turbine to exhaust directly to the main condenser. For operation using only the low-pressure turbine, the bellows can be removed from the crossover piping so that desuperheated steam can be supplied to the low-pressure turbine through temporary piping.

For U.S. Navy ships having more than one machinery space, cross-connect piping is installed to permit the operation of one plant with steam from another plant in the event of a casualty.

b. Turbine bleed steam. Turbine bleed steam is extracted from bleed points within the propulsion turbines. As discussed in Chapter 2, the quantity of heat available from steam is progressively reduced as it passes through the propulsion turbines, and only a portion of the steam's available energy can be converted to useful work by the turbine. A large fraction of the heat originally in the steam is, therefore, exhausted into the condenser. By extracting steam from the turbine bleed points and using it for heat loads at the appropriate temperatures, the overall efficiency of the steam cycle can be significantly improved.

Steady-state operation of the propulsion turbines is required for a constant supply of bleed steam from any of the bleed points. During low-speed, port, maneuvering, and astern operations, steam is supplied either intermittently or at such a low flow rate that extraction at the bleed points is not practicable. In these operating modes an alternate source of steam must be available for the bleed-steam loads.

As indicated on Fig. 15, there are three categories of bleed-steam extraction points. High-pressure bleed steam is taken from the high-pressure turbine and supplies high-pressure feedwater heaters (if installed) and the contaminated steam evaporator. It is the least economical of the bleed-steam alternatives since it diverts a significant amount of heat from the turbine; nevertheless, it is economically preferable to steam taken directly from the boilers.

Intermediate-pressure bleed steam is usually taken from a later stage of the high-pressure turbine or from the crossover to the low-pressure turbine. It is typically used for boiler air heaters and feedwater heating; for commercial ships it is also used for space heating, potable-water heaters, and galley services. Intermediate-pressure bleed steam can be used at about half the cost of boiler-pressure steam.

Bleed steam from the low-pressure turbine is the most effective of the bleed-steam sources in improving the efficiency of the steam cycle because at this point most of the heat remaining in the steam would be exhausted to the condenser. However, its low heat content typically limits its use to first-stage feedwater heating and distilling plants.

Steam pressure at each bleed point varies with the power level of the associated turbine. For some uses, such as the boiler air heaters, such fluctuations are acceptable. But when pressure fluctuations are not tolerable, or when the load varies, a pressure-reducing or temperature-control valve must be installed.

Turbine bleed points and the services they supply should be selected to achieve the greatest economy of operation, and to give consideration to the suitability of the available pressure, flow, and temperature for the intended loads. When multiple bleed points are used, the cycle efficiency can be improved by installing a "cascade" bleed valve control system that will give priority to the use of the lowest-pressure steam available for feed heating during any operating condition. The use of heat balances to evaluate a steam cycle, including the effects of bleed steam, is reviewed in Chapter 2. The use of bleed steam tends to reduce fuel costs; however, the added expense of the additional piping and controls necessary when a bleed steam extraction point is installed must also be considered.

Bleed steam piping is illustrated in Figs. 16 and 17.

c. Auxiliary desuperheated steam system. An auxiliary desuperheated steam system supplies steam from the boiler desuperheaters directly or by way of reducing stations to all steam-driven auxiliaries and ship services not served by the main steam system.

Since auxiliary steam piping is generally smaller in diameter and operates at a lower temperature than main steam piping, provisions for adequate flexibility are less difficult.

A typical auxiliary steam system diagram is shown in Fig. 16. Steam is supplied from the boiler desuperheater directly to the boiler soot blowers and the main feed pump turbines. The system shown also supplies the following components through pressure-reducing valves or temperature-control valves:

• Propulsion turbine, generator turbine, and distilling plant air ejectors.
• Propulsion and generator turbine gland seal steam regulators.
• Boiler fuel burner steam atomization.

LEGEND OF SYMBOLS		
⟠	VALVE, GATE, LOCKED OPEN	⟠ VALVE, GLOBE, LOCKED OPEN
⋈	VALVE, GLOBE	▭ THERMOMETER
▷◁	VALVE, GATE	▷ VALVE, GLOBE STOP-CHECK
▽	STRAINER, Y-TYPE	⋈ VALVE, GOVERNOR
⟡	VALVE, ANGLE RELIEF	⊙ GAGE, PRESSURE
⟤	VALVE, PRESSURE REDUCING	⊙ GAGE, PRESSURE, REMOTE-READING
⊠	VALVE, UNLOADING	⋈ VALVE, CONTROL
⟍	VALVE, SWING CHECK, DASHPOT TYPE	▱ SEPARATOR
⟤	VALVE, GATE, MOTOR-OPERATED	

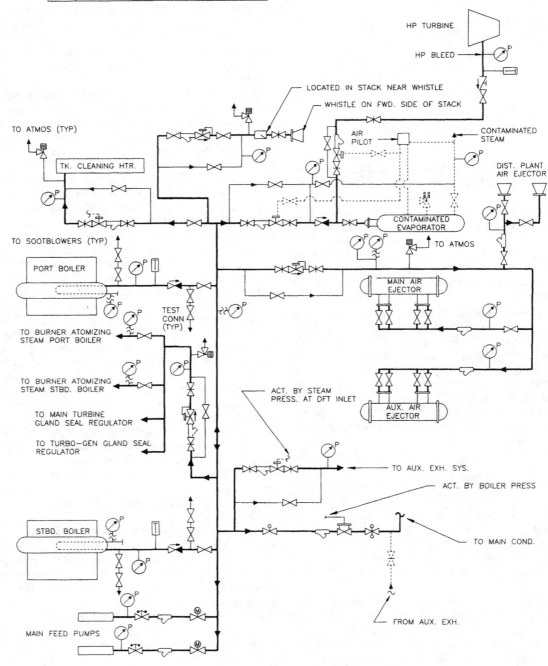

Fig. 16   Auxiliary and HP bleed systems

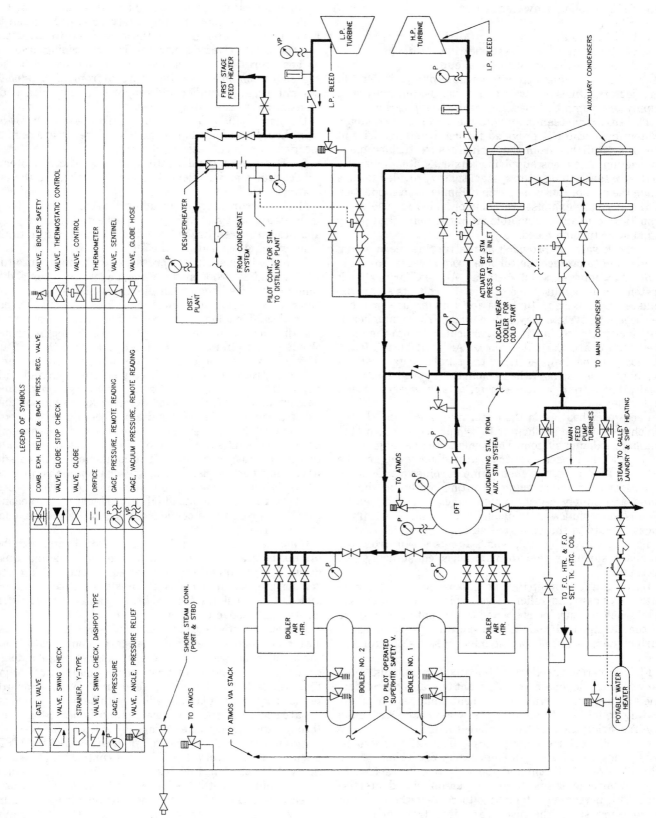

Fig. 17 Steam exhaust, IP bleed, and LP bleed systems

- Tank cleaning seawater heater.
- Whistle.
- Hotel services.
- Contaminated steam system evaporator.
- Auxiliary exhaust system makeup.

Many ships are equipped with vacuum pumps instead of air ejectors. On U.S. Navy ships, the system may also supply fire pump turbines, and cargo pump turbines.

The auxiliary steam supply to the contaminated steam system evaporator is provided for use when insufficient steam is available from the high-pressure turbine bleed connection. A bypass supplying auxiliary steam directly to the contaminated steam system is also provided for use when the evaporator is inoperable. Similarly, a connection to the auxiliary exhaust system is provided for use when insufficient steam is available from the main feed pump exhaust or turbine bleed.

A check valve is provided at the turbine bleed steam connection to protect the turbine from reverse flow.

d. Auxiliary exhaust system. An auxiliary exhaust system, such as illustrated by Fig. 17, collects steam exhausted by auxiliary turbines for further use in other services. The system may collect steam from a number of auxiliary turbines, but in this particular design the main feed pumps are the only source. As with bleed steam, the loads to be supplied from the auxiliary exhaust are established by using a heat balance. Typical loads are feedwater heating, space heating, potable-water heaters, and galley.

Feedwater heating in the deaerating feed tank (DFT), which is described in Chapter 16, is a principal load on the auxiliary exhaust system. The demand for heating steam in the DFT increases with the plant power level, and the exhaust flow from the feed pump turbines varies similarly. However, under many operating conditions the demand and supply do not match exactly. Consequently, the system includes an unloading valve, which discharges excess steam to the main and auxiliary condensers, and a makeup valve (Fig. 16), which admits steam from the auxiliary steam system to maintain a constant system pressure.

Each main feed pump turbine exhaust contains a combination exhaust and relief valve, which opens automatically when the turbine is started to prevent overpressure of the casing, and maintains a constant back pressure during changes in the turbine inlet pressure and turbine load.

Flow in various sections of the auxiliary exhaust main varies widely depending on the plant operating condition. The engineer must evaluate the heat balances for all operating conditions to identify the worst-case flow condition for sizing the auxiliary exhaust main.

e. Safety valve escape piping. The safety valve escape piping conveys steam from safety and relief valves, which are set at 50 psig and above, to the atmosphere to prevent burning or smothering of personnel and interference with personnel attempting to correct the cause of the overpressure. The discharge pipes from individual safety or relief valves are combined and led to the weather, usually through the uptake space.

Separate escape piping should be: provided from the boiler drum and superheater safety valves, designed to have an area not less than the combined areas of the outlets of all valves discharging into the piping, and led as near vertically as possible. The piping should be designed to minimize reaction forces from high-velocity steam flow, and installed and supported to minimize loads transmitted to the safety valves.

Discharge piping from steam relief valves set below 50 psig should terminate below the floor plates or in a remote location to prevent personnel injury.

f. Drain and condensate systems. These systems collect drains and condensate and return them to the operating cycle.

High-pressure drains, including low-point drains from main steam piping, auxiliary desuperheated steam piping, and other sources, are collected in a high-pressure drain main which discharges to the DFT.

Drains with insufficient pressure to discharge into the DFT, including propulsion turbine, generator turbine, distilling plant, feed heater, air ejector condenser, gland exhaust condenser, and low-pressure steam piping drains, are collected in a low-pressure (or freshwater) drain main and discharged to the freshwater drain collecting tank. Condensate from the tank is either pumped to the DFT or drawn by vacuum into the main or auxiliary condenser.

Drains from services which may be contaminated with oil, including heating coils for fuel and cargo tanks, fuel heaters, and lubricating-oil heaters, are kept separate and led to a contaminated steam drain collecting tank to prevent any oil that has leaked into the system from contaminating the main steam system. The tank is divided into two chambers by a vertical baffle containing an oil screen. A viewing port is installed in the upstream chamber so the operator can see any oil contamination. The screen filters out small amounts of oil before it enters the second chamber. From there condensate is returned to the contaminated evaporator.

Condensate from the first-stage feed heater can be drained by gravity to the freshwater drain collecting tank, drawn by vacuum into the condenser, or pumped into the condensate system. Gravity drain is the simplest method and should be used if the feed heater can be located sufficiently high above the tank.

A typical condensate system is illustrated by Fig. 18. Two main condensate pumps discharge condensate from the main condenser through the main air ejector condensers (if fitted), gland leakoff condenser, and the first-stage feed heater to the deaerating feed tank. Two pumps are required by the regulatory bodies. Both auxiliary condensers are served by one condensate pump, which discharges through the associated auxiliary air ejector condenser (if fitted) to the main condensate line upstream of the DFT. Each condenser has a hotwell level control valve, which recirculates condensate back to the hotwell as necessary to maintain a constant level. A recirculation line is provided to increase the effectiveness of the air ejector during warm-up and low-power operation. On U.S. Navy ships, separate manually operated recirculation lines are provided; one line discharges into the hotwell to prevent

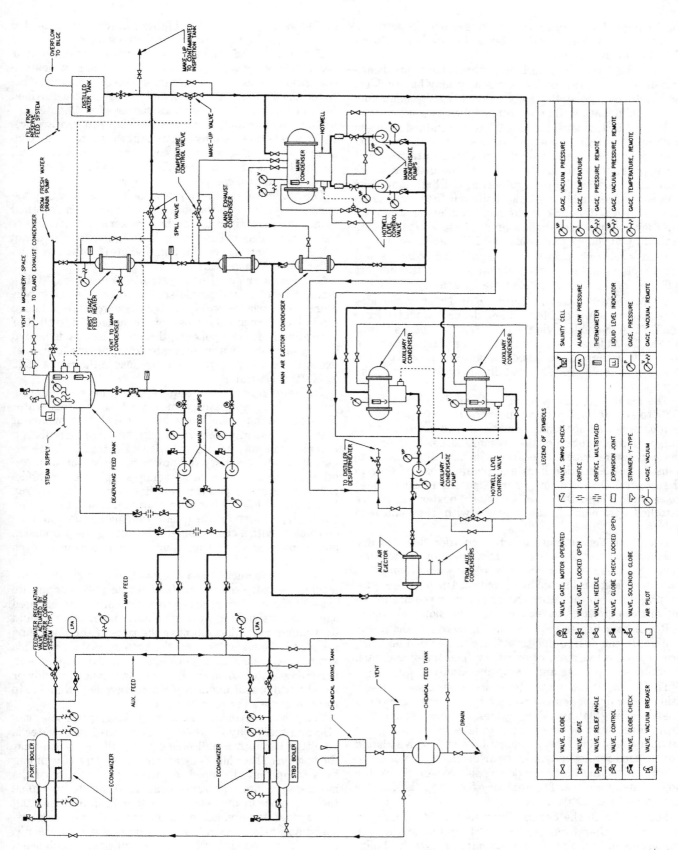

Fig. 18 Boiler feed and condensate system

LEGEND OF SYMBOLS

⋈	VALVE, GLOBE		VALVE, SWING CHECK	SALINITY CELL
⋈	VALVE, GATE		VALVE, GATE, MOTOR OPERATED	GAGE, VACUUM PRESSURE
⋈	VALVE, RELIEF ANGLE		VALVE, GATE, LOCKED OPEN	ALARM, LOW PRESSURE
⋈	VALVE, CONTROL		VALVE, NEEDLE	ORIFICE	GAGE, TEMPERATURE
⋈	VALVE, GLOBE CHECK		VALVE, GLOBE CHECK, LOCKED OPEN	ORIFICE, MULTISTAGED	THERMOMETER	GAGE, PRESSURE, REMOTE
⋈	VALVE, VACUUM BREAKER		VALVE, SOLENOID GLOBE	EXPANSION JOINT	LIQUID LEVEL INDICATOR	GAGE, VACUUM PRESSURE, REMOTE
			AIR PILOT	STRAINER, Y-TYPE	GAGE, PRESSURE	GAGE, TEMPERATURE, REMOTE
				GAGE, VACUUM	GAGE, VACUUM, REMOTE	

subcooling of the condensate during normal operation, and another line discharges above the condenser tubes for use during warm-up.

For ships having one turbine-generator and one diesel-generator, a single auxiliary condenser may be provided, or the turbine-generator may exhaust to the main condenser, eliminating the need for the auxiliary condenser.

The feedwater level in the DFT is maintained between high and low limits by a spill valve and a makeup valve. The spill valve opens to discharge condensate from the first-stage feedwater heater outlet to the distilled water tank when the DFT level approaches the high limit. The makeup valve opens to admit additional feedwater from the distilled-water tank to the main condenser when the DFT level approaches the low limit. These valves are controlled by air pilots that sense the DFT level.

Heat exchangers served by high- and low-pressure drain piping must be located so that their relative heights will ensure the flow of condensate by gravity.

g. Main feed system. The main feed system provides a continuous supply of heated and deaerated water from the DFT to the boilers via the main feed pumps, as shown by Fig. 18. At least two main feed pumps must be provided to comply with regulatory body requirements, and on U.S. Navy ships three are often installed in each main machinery space.

The main feed pumps must be located well below the DFT to ensure sufficient net positive suction head at the pumps under all operating conditions. The location of the pumps and evaluation of the pressure losses in their suction piping are critical to prevent flashing of the hot condensate in the suction lines and possible loss of flow to the boilers. On naval or other ships where space does not permit a high DFT location, feed booster pumps are installed upstream of, and in series with, the main feed pumps.

Each main feed pump has a recirculation line that discharges back to the DFT to protect the pump from overheating in the event of low- or zero-flow conditions. The flow in each line is limited by an orifice. The high pressure drop required in the recirculation line may require multiple orifices in series to reduce orifice erosion and noise.

Regulatory bodies require two feedwater paths to the boilers. In the design illustrated, feedwater can be delivered through the main or auxiliary feed lines, and each contains a stop valve and a stop-check valve.

The boiler drum water level is automatically maintained, usually near mid-level of the boiler drum, by a feedwater regulating valve upstream of each boiler. The drum level must be manually regulated when using the auxiliary feed line. The auxiliary feed stop valve should not be used for this purpose; wear on this valve must be minimized so that it can provide a positive isolation of the boiler when required. The auxiliary feed stop-check valve should be used instead.

Each boiler in the design illustrated is fitted with an economizer; therefore, feedwater is circulated through the economizer before being discharged into the boiler drum. When the design includes more than two stages of feedwater heating, the higher stages are connected between the feed pumps and the economizer.

The transfer of feedwater from reserve-feed tanks is accomplished by a freshwater transfer pump, which discharges to the distilled-water tank. Direct connections between the distilled-water tank and the boilers are provided for wet layup of the boilers, as discussed in Chapter 5.

Sample connections for water chemistry control are provided at several locations throughout the system, and connected by piping to a sample cooler and sink. A chemical mixing tank and feed tank allow the addition of water-chemistry-control compounds to the system.

The feed discharge piping has the highest design pressure of any piping in the steam plant, since it is designed to deliver feedwater to the boilers even if their safety valves have lifted. The piping is designed for the pump relief valve setting, or the pump shutoff pressure if no relief valve is installed.

h. Seawater cooling system. Condensers, lubricating-oil coolers and air coolers for propulsion and generator steam turbines require a high seawater flow but have a relatively low pressure loss, on the order of 5 to 10 ft of seawater. These components, therefore, have a dedicated supply of cooling water separate from the auxiliary heat exchangers, which have a pressure loss in the range of 12 to 50 ft of seawater.

The regulatory bodies require that a main circulating pump and an emergency means of circulating water through the main condenser be provided. One means of satisfying this requirement is with two main circulating pumps; each may have a capacity such that both are required on line to satisfy full-power design conditions. Another means is to provide a backup supply from an independent seawater pump in another system. The system illustrated in Fig. 19 has both features. A separate pump has been provided for the generator condensers and lube-oil coolers, with a cross-connect permitting the propulsion and generator circulating pumps to supplement each other.

At least two suction sea chests should be provided for machinery-space cooling systems, preferably on opposite sides of the machinery space to avoid both being clogged simultaneously. On deep-draft ships, one of these should be a high suction located near the turn of the bilge, and the other a low suction near the bottom of the ship. This arrangement ensures a continuous seawater supply both in shallow water when the low suction is near a muddy or silted bottom, and underway when wave action and ship roll can expose the high suction.

The propulsion condenser and lubricating-oil cooler can also be supplied by a scoop injection system. Seawater is admitted through a hull opening, usually installed near the bottom of the ship. Seawater is directed into the piping by a scoop that is shaped to convert the velocity of the moving ship into a pressure head sufficient to overcome the resistance of the seawater system [18]. A circulating pump must be provided to supply the system during low-speed and astern operations. A single-pass condenser is used with a scoop injection system; multipass condensers require a higher injection head than a scoop can develop.

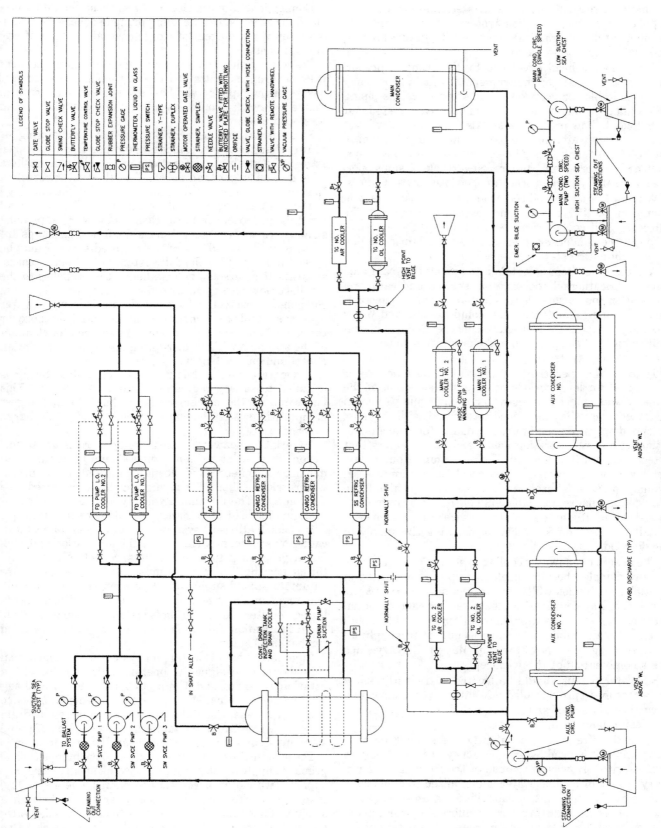

Fig. 19 Steam plant seawater cooling system

To minimize pressure losses, seawater piping in a system supplied by a scoop injection system should be arranged in the most direct practicable route, with gradual bends, and include only gate valves.

The pressure head developed by the scoop is proportional to the square of the ship speed; thus a scoop may not be suitable for very low-speed ships since the piping and condenser may have to be prohibitively large to incur a sufficiently low pressure loss. Also, the boundary layer of low-energy water next to the ship hull can compromise the ability to develop a sufficient head on very long ships, such as large crude-oil carriers. Nevertheless, ships operate successfully using a scoop at speeds as low as 6 knots.

When a lubricating-oil cooler and a condenser are connected in parallel, the design pressure loss through the lubricating-oil cooler circuit should be sightly less than through the condenser to ensure that the lubricating-oil cooler will not be starved.

A separate system is installed to supply the main feed pump lubricating-oil coolers, ship service and cargo refrigeration condensers, contaminated steam drain cooler, air-conditioning condenser, air compressors, and other auxiliary components as required. These components require a higher head pump than the propulsion and generator circulating systems.

To avoid the operation of pumps at flows much lower than rated for long periods of time, the selection of the number and capacity of seawater cooling pumps should be based upon the ship speed associated with its most extended operating condition, seasonal effects on air-conditioning and refrigeration loads, and other factors that affect the seawater demand. The operation of pumps under low-flow conditions causes accelerated pump wear, leading to early failures. At least two pumps should be provided to ensure a reliable supply. In the system illustrated by Fig. 19, three seawater service pumps have been installed to provide flexibility in matching the supply to the demand.

The design of seawater cooling systems for naval ships is more complex because cooled equipment is located in multiple, sometimes widely separated, machinery spaces. A dedicated system is often installed for each essential service, such as a generator. Another approach is to install a central main supplied by several pumps to supply all services in a given zone of the ship. Many U.S. Navy ships use a combination of these concepts.

Seawater pumps should be located sufficiently below the waterline to provide adequate submergence during all normal conditions of ship loading, with consideration for the design pitch, roll, and trim.

Since the firemain is constantly pressurized on U.S. Navy ships, it is used as an emergency backup supply to vital services. However, the use of the firemain as a primary source of cooling water is minimized to avoid compromising damage-control capabilities in an emergency; and because its high pressure requires heavier piping, wastes pumping energy, and increases the potential for excessive flows and piping erosion.

Connections to more than one sea chest should be provided for each seawater cooling pump to ensure a continuous supply in the event of clogging or air binding of any one sea chest.

Seawater piping should be arranged so that air pockets will not form, and entrained air will remain entrained as it passes through piping and heat exchangers to overboard. Condenser heads should be vented, and vent piping should be sloped upward.

Strainers should be provided to prevent clogging of heat-exchanger tubes, orifices, and other small flow passages. The mesh or perforations of the strainer baskets should be no larger than the smallest flow path protected. In some cases, such as the main condenser, the openings in the sea chest strainer plate will be sufficiently small for this purpose. In most cases, however, a pipeline strainer will be necessary. Where small flow passages are present, the physical size of the required strainer may be excessive if sized for the full system flow. In such cases, separate strainers can be installed in the branches to the smallest flow passages, with the majority of the system components protected by a single strainer with larger openings. Duplex strainers should be installed wherever the component(s) protected cannot be secured for cleaning of the strainers without disrupting operation of essential, nonredundant machinery.

A separate suction line for emergency dewatering is installed to satisfy regulatory body requirements. This line connects the machinery space bilge to the largest pump in the system having sufficient head to discharge overboard.

Material for seawater piping should be resistant to corrosion. To prevent erosion, piping should be arranged to minimize turbulence, and the maximum and minimum seawater velocity limits for piping should be observed (see Section 2.4). The maximum velocity at inlets of heat exchangers made of copper alloys should not exceed 7.5 fps.

i. Lubricating-oil system. A propulsion plant lubricating-oil system provides oil for the lubrication and cooling of propulsion turbine bearings, reduction gears, and thrust bearing. The system must be designed to ensure an uninterrupted supply of oil, to keep air out of the oil, and to maintain the condition of the oil. When an oil-lubricated stern tube bearing is installed, a separate, dedicated lubrication system is provided for it, as discussed further in Chapter 10.

In general, the lubricating-oil system for a commercial ship should be designed to operate satisfactorily with a list of up to 15 deg and trim up to 5 deg, and for a 30-deg roll and 10-deg pitch. More severe requirements apply to U.S. Navy ships. Careful attention to the slope of drain piping, the shape of sump tanks, and the location of pump suction tailpipes is necessary to meet these requirements.

The system illustrated in Fig. 20 employs pumps for the primary supply of oil, and a gravity tank as an emergency supply. Two motor-driven service pumps, one a standby, take suction from the reduction gear sump through a strainer. Each pump discharge is fitted with a relief valve which discharges to the sump tank. A stop-check valve is installed in each pump discharge to prevent backflow

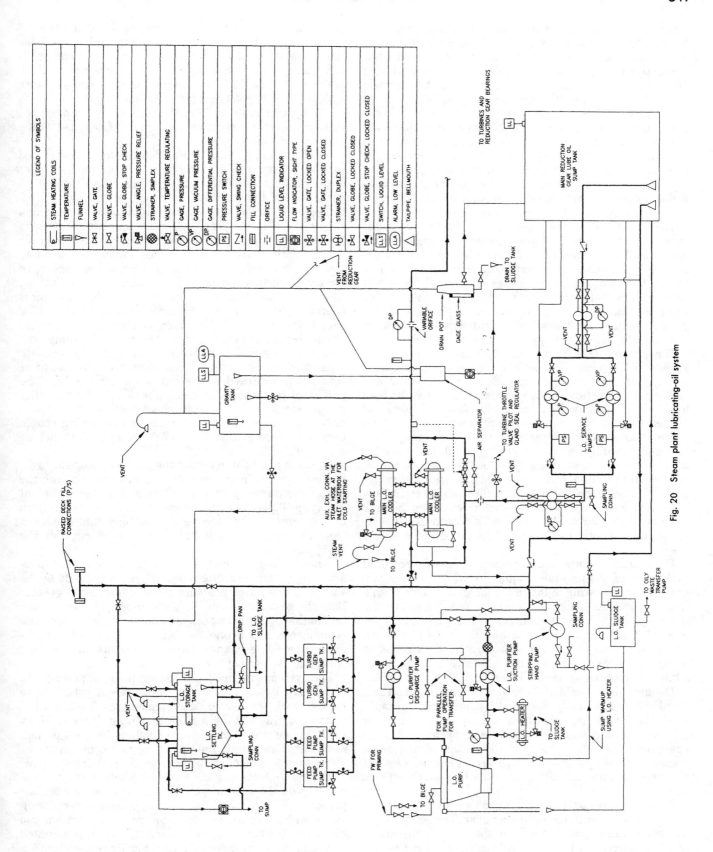

Fig. 20 Steam plant lubricating-oil system

through an idle pump. As an alternative, a stop valve may be installed in each pump discharge and a check valve in each pump suction. This arrangement is preferred when a negative pressure can exist in the pump suction during operation, to prevent air from entering the system through the pump seals. Pumps should be located lower than the operating level of oil in the sump to ensure that the standby pump is always flooded with oil. If this is not feasible, a check valve should be installed in each pump suction and a priming line from the discharge header should be installed to keep each suction pipe full.

A pressure switch is installed at each pump discharge and is arranged to activate an alarm and start the standby pump automatically if the operating pump fails.

The pumps discharge through a duplex magnetic strainer and the lubricating-oil cooler to the main propulsion unit. The temperature of the oil leaving the coolers is usually regulated by a three-way valve, which senses the oil temperature and bypasses oil around the cooler as necessary. Temperature control may also be achieved using a control valve in the seawater piping, but the greater maintenance associated with seawater system components makes this approach less desirable.

The gravity tank is connected to the system downstream of the cooler. During normal operation excess oil capacity of the operating pump flows into the tank and back out the overflow to the sump, ensuring that the tank is full. If the supply from the pumps fails, oil flows out of the tank to the propulsion unit bearings. The tank height must provide adequate pressure to the propulsion unit and normally contains a supply of oil sufficient for about six minutes, allowing time for the propulsion unit to be secured without damage to the bearings. The gravity tank overflow is provided with an illuminated sight flow indicator located on the main operating level, and the tank has a low-level alarm. The overflow line contains a separator to remove entrained air from the oil before it enters the sump.

The lubricating-oil system operating pressure required is the sum of the pressure required at the most hydraulically remote bearing in the reduction gear, which is typically about 15 psi, and the pressure losses through the system piping and heat exchangers. The rated head of the lubricating-oil service pump is usually 60 to 80 psi. An adjustable orifice is installed downstream of the cooler to permit the pressure at the most remote bearing to be adjusted after the system is installed.

The propulsion turbine throttle controls and gland-seal regulator are typically operated by hydraulic pressure from the lubricating-oil system. The supply piping for this purpose is connected to the pump discharge header upstream of the adjustable orifice, since this is the highest pressure in the system. With this arrangement, a loss of oil pressure automatically causes the turbine throttle to close.

Oil from the propulsion unit drains by gravity to a sump tank located beneath the reduction gear. The sump capacity should be sufficient to contain the operating quantity of oil in the gravity tank, the service piping, and the propulsion unit without overflowing through the reduction gear bearings when the oil has drained down after securing the plant. Drain lines to the sump tank should discharge below the normal oil level to minimize foaming. To avoid trapping air pockets, which would restrict oil flow, drain piping should be liberally sized and arranged with a slope exceeding 60 deg from the horizontal.

The system is filled from the deck connection via the storage tank. Gravity fill piping is routed from the storage tank to each sump tank.

A centrifugal purifier is provided to remove water and sediment from the oil. The purifier has one or more attached, shaft-driven pumps that can be operated independently of the purifier centrifuge to transfer oil between tanks. A purifier heater is installed to heat and supply the oil at the proper viscosity for efficient purifier operation. Piping is provided so that either continuous or batch purification of the oil in the propulsion unit sump tank may be performed; the auxiliary units are arranged for batch purification only. For continuous purification, the purifier takes suction from the sump while the unit is operating, and discharges through the heater and purifier back to the sump tank. For batch purification, the purifier pumps are used to transfer the contents of the sump tank to the settling tank, where the oil is heated to facilitate separation of the water and sediment from the oil, after which it is returned to the sump via the purifier. The unit must be secured for batch purification. The purifier pumps can also be used to transfer oil between tanks or to the deck connection. Purifier suction piping should terminate close to the bottom of the sump tank to pick up sediment.

Piping is provided to permit the purifier pumps to circulate oil from the main propulsion unit sump through the heater for warm-up prior to starting from the cold-plant condition. Piping is also provided to remove sludge from the bottoms of the storage and settling tanks using a hand-operated stripping pump that discharges to the sludge tank.

When shell-and-tube heat exchangers are used, the lubricating-oil and seawater cooling systems should be designed so that the oil pressure in the cooler shell is greater than the seawater pressure in the tubes, to prevent contamination of the oil and consequent damage to propulsion components. This precaution is not necessary when plate-and-frame heat exchangers are used because any leakage will be to the atmosphere.

The system described reflects commercial practice. In U.S. Navy practice, a gravity tank is not used due to limited machinery space headroom and the fire hazard of locating a gravity tank high in the space. Instead, Navy systems use two or three service pumps for each propulsion unit. The pumps are driven by any combination of electric motors, steam turbines, and mechanical drives from the reduction gear (called attached pumps) depending on the type of ship and number of shafts. The pressure is controlled by a back-pressure regulating valve, which discharges excess oil from the service pump discharge header back to the sump. The valve is actuated by the pressure at the most hydraulically remote bearing in the propulsion unit.

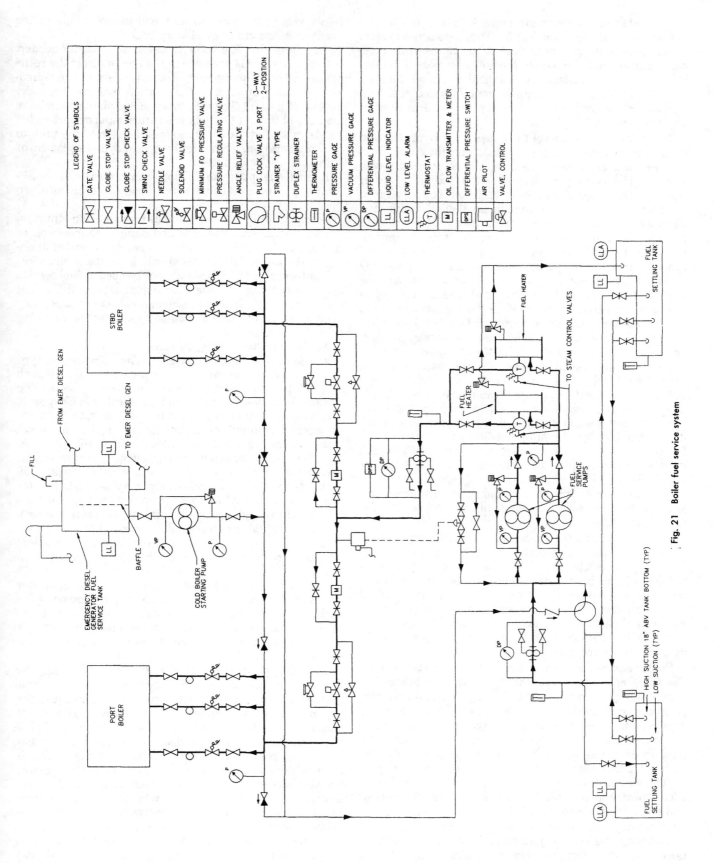

LEGEND OF SYMBOLS	
GATE VALVE	⋈
GLOBE STOP VALVE	⋈
GLOBE STOP CHECK VALVE	⋈
SWING CHECK VALVE	⋈
NEEDLE VALVE	⋈
SOLENOID VALVE	⋈
MINIMUM FO PRESSURE VALVE	⋈
PRESSURE REGULATING VALVE	⋈
ANGLE RELIEF VALVE	⋈
PLUG COCK VALVE 3 PORT 3-WAY 2-POSITION	◯
STRAINER "Y" TYPE	⊢
DUPLEX STRAINER	⊕⊕
THERMOMETER	▯
PRESSURE GAGE	P
VACUUM PRESSURE GAGE	VP
DIFFERENTIAL PRESSURE GAGE	DP
LIQUID LEVEL INDICATOR	LL
LOW LEVEL ALARM	LLA
THERMOSTAT	T
OIL FLOW TRANSMITTER & METER	M
DIFFERENTIAL PRESSURE SWITCH	DPS
AIR PILOT	▢
VALVE, CONTROL	⋈

Fig. 21 Boiler fuel service system

j. Boiler fuel service system. A typical boiler fuel service system is shown in Fig. 21. The fuel service system is designed to supply clean fuel at the correct pressure, temperature, and viscosity to the boiler burners. Fuel service systems for boilers are designed to handle bunker, or residual, fuels that normally contain some sediment and water and have a high viscosity. Fuel characteristics are discussed in Chapter 12.

Two service (or settling) tanks are usually provided in each main machinery space so that one tank can be in service while the other is refilled, allowed to settle, and sampled, if necessary, to determine the characteristics of the fuel. Each tank should have sufficient capacity for at least 8 hours of plant operation in naval applications, and 24 to 30 hours in commercial applications. The tanks are provided with steam heating coils to facilitate the separation of water and sediment from the fuel. In commercial practice, two service pump suction tailpipes are usually provided in each tank. Normal operation is from the low suction; but, if excessive water or sludge is detected in the tank, operation is switched to the high suction until the contaminants can be removed. In U.S. Navy practice, a single high suction is provided for the service system, and a low suction is provided for the stripping system.

Positive-displacement fuel service pumps take suction from the settling tanks through a strainer and discharge through steam heaters and strainers to the boilers. Duplicate pumps and heaters are installed, each having sufficient capacity to supply both boilers at their overload rating as required by regulatory bodies. Piping and valves for the pumps and heaters are arranged so that each component can supply fuel from either tank to either or both boilers, and one component can be repaired while the other is in service. Strainers should be either duplex or a simplex type capable of being cleaned without interrupting the flow of fuel.

Fuel temperature (and hence viscosity) for proper atomization of the fuel is maintained by a temperature sensor in each heater discharge, which actuates a steam pressure regulator in the heater steam supply. When fuels of varying grades are to be used, a viscosimeter can be installed to regulate the heater automatically on the basis of viscosity rather than temperature. U.S. Navy steamships do not have fuel service heaters because only low-viscosity distillate fuels are used.

Fuel pressure is regulated by a control valve that senses the fuel header pressure and opens to recirculate fuel back to the pump suction as necessary to maintain a constant pressure as the boiler load varies. Flowmeters and combustion-control valves are installed in the supply piping to each boiler, and solenoid valves are installed in the inlet to each burner. These components are associated with the boiler combustion control system (see Chapter 5). The combustion-control valves modulate fuel flow to each boiler, and the solenoid valves are used for normal and emergency shutdown of the burners.

To comply with regulatory body requirements, means of stopping the service pumps and closing the settling tank stop valves from outside the machinery space must be provided at a location that will be accessible in the event of a fire in the machinery space.

A recirculation line is connected at each burner header. A two-way valve is provided in the line to direct the recirculated oil to the pump suction when boilers are on temporary standby, or back to the settling tank to purge the system of cold oil. The recirculation line should terminate near the bottom of the tank in a return bend or perforated header with flow directed against the side of the tank to prevent aeration, static discharge, and the agitation of sludge on the tank bottom.

A small fuel pump and piping from a diesel fuel or JP-5 tank may be installed for use during cold boiler start-up.

Coamings or drip pans should be provided under pumps, heaters, strainers, burners, and sample connections to prevent spilled fuel from causing a safety hazard or draining to the bilge. All flanges should have spray shields.

Nonferrous materials should not be used in fuel service systems because of their detrimental effect on the fuel (see Chapter 12). If screwed bonnet valves are used, they should be of the union bonnet type capable of being packed under pressure.

3.3 Diesel Plant Piping Systems.

a. Seawater cooling system. The number and type of seawater heat exchangers for a diesel engine depend on the type and design of the engine. A seawater cooling system for a diesel installation is illustrated in Chapter 3.

Seawater is typically provided to separate heat exchangers for propulsion engine lubricating oil and jacket water. These are usually installed in series, with the jacket-water cooler receiving the higher-temperature seawater. For the larger low- and medium-speed engines, additional seawater heat exchangers may be installed for cooling pistons, fuel injectors, combustion air, reduction gear lubricating oil, and other services. Thermostatic temperature controls are installed as required.

Machinery plants using controllable-pitch propellers may also require seawater cooling for a hydraulic oil cooler.

Generator engines usually require only one seawater heat exchanger. A separate pump may be provided for each generator engine, with a standby supply from another source. On U.S. Navy ships, an emergency supply from the firemain via a pressure-reducing valve is installed.

Seawater cooling pumps for diesel engines may be driven by power takeoffs directly from the engine. This reduces the number of electric motors, and reduces the dependence of the engine on electric power. The use of engine-attached seawater pumps is not recommended if the pump cannot be arranged and designed to ensure that it will be primed automatically soon after the engine is started. For main propulsion engines, motor-driven pumps are also needed to provide an alternate supply to the main engine as required by regulatory bodies. A cross-connection to a ballast, fire, or other seawater pump of appropriate capacity is sometimes installed for this purpose.

Separate motor-driven pumps are generally installed to supply cooling water for auxiliary seawater-cooled heat exchangers. If attached pumps are used for main engine cooling, separate motor-driven pumps are essential to cool auxiliaries in order to permit their operation in port with the main engine secured.

A central seawater-to-freshwater heat exchanger, with fresh water circulated through all engine heat exchangers, can be used to reduce the amount of piping and components exposed to seawater. These systems are described in Chapter 3.

b. Lubricating-oil system. Diesel engine lubricating-oil service systems may be supplied by attached or motor-driven pumps, which circulate oil through filters and coolers to the engine. Sumps are integral with smaller, high-speed engines. On larger engines, an independent sump is installed; gravity drainage or a scavenging pump returns oil from the engine crankcase to the sump tank. If an attached pump is installed, a means of priming with a hand- or motor-operated pump is usually necessary to lubricate the engine before start-up. A steam or electric heater may also be installed to minimize condensation in the sump when the engine is idle and to facilitate starting in cold ambient conditions. A typical diesel plant lubricating-oil system is illustrated in Chapter 3.

Diesel propulsion units employing reduction gears are usually equipped with a separate reduction gear lubricating-oil system similar in arrangement to that for a steam plant.

Diesel machinery plants usually require two or more different grades of oil for the engine and reduction gear, if fitted, and the larger engines may require additional grades for cylinders and pistons. Generator engines often require a different oil than the propulsion engine. Piping systems for filling, transferring, and purifying must be designed to prevent mixing of oil grades. Mixing of oil from different engines should also be avoided to prevent contamination in one engine from affecting another, and to simplify tracing the source of an oil-related engine problem.

The capacity of diesel lubricating-oil storage tanks is greater than for a steam plant because the diesels continuously consume oil.

Purification systems that are arranged for continuous purification are usually installed only for larger engines. If a purifier with its attached pump is not installed, a separate pump must be provided for tank-to-tank transfer functions. Smaller, high-speed engines contaminate the oil so heavily with fuel, which cannot be removed by centrifuging, that it is more practicable to change the oil.

c. Fuel service system. Diesel engine fuel service systems are designed to handle different grades of fuel depending on the design of the engine. High-speed engines generally operate on distillate fuel only; medium- and low-speed engines generally use a blend of distillate and residual fuel, which is commonly known as "heavy fuel."

Systems that handle heavy fuel are the more complex, requiring purifiers, settling tanks, heaters, mixing tanks, filters, and viscosity-control equipment to condition the fuel. Such systems are described in Chapters 3 and 12.

Systems that handle only distillate fuel consist of service tanks, service pumps, and filters, connected in that order to deliver fuel to the engine. Heaters are not required since the viscosity of distillate fuel is satisfactory for filtration and combustion.

For propulsion engines, two service pumps are provided; one may be engine-driven and the other a motor-driven standby, or both may be motor-driven. If the service tanks are arranged for gravity feed to the engine, pumps may be unnecessary.

For generator engines, a single engine-driven pump and a simplex or duplex filter are normally installed. Generator engine service tanks should be arranged for gravity feed to the pump, unless the engine manufacturer recommends otherwise.

For all engines, a fuel return line with a back-pressure regulator is provided to return excess fuel from the engine header to the service tank.

d. Freshwater cooling and waste-heat systems. The freshwater cooling pumps circulate fresh water in a closed circuit through the engine water jacket and a freshwater/seawater heat exchanger. For some engines the freshwater cooling circuit also includes the lubricating-oil cooler and turbocharger air cooler; for others, these units are seawater cooled. For the larger propulsion engines, additional freshwater circuits may be provided to cool the fuel injectors and pistons. Diesel engine freshwater cooling systems are illustrated in Chapter 3.

Each freshwater cooling circuit requires an elevated expansion tank that is connected to the suction of the circulating pumps, and is of sufficient size to absorb the expansion of the coolant from room temperature to operating temperature. An air trap with a vent pipe to the expansion tank should be installed at a high point, preferably at the engine water outlet.

Three-way control valves are used for temperature regulation. These are arranged to modulate the freshwater flow through the seawater heat exchanger and the heat exchanger bypass in response to the freshwater temperature, usually at the engine outlet. A high-temperature alarm is also installed at this location. In some installations it is necessary to drill a valve disk or install an orificed bypass so that water heated by the engine always reaches the alarm sensor even if the engine is inadvertently operated with incorrect valve alignment.

For multiple-engine installations, independent freshwater cooling circuits may be installed for each engine to avoid cross-contamination and improve reliability. If a standby pump is installed to serve more than one engine, isolation valves are provided and arranged to prevent interconnection of the systems.

Diesel engine jacket water can be used to heat other fluids to within about 5 deg F of the jacket water operating temperature by installing the fluid heat exchanger in the jacket water loop. The most common application of jacket water waste heat is for distiller feed heating, as discussed in Chapter 17. In most cases the distiller is arranged in the engine jacket-water circuit with a three-way temperature control valve providing the required flow of water through the distiller and bypassing the remainder to the freshwater/seawater heat exchanger. An additional three-way valve is installed to regulate flow through or around the heat exchanger. The arrangement of control valves and their temperature settings ensures that the engine is neither overheated nor subcooled regardless of the engine or distiller load. A supplementary steam or electric heater is usually provided to operate the distillers at low engine loads.

Fuel, lubricating oil, and fresh water can also be heated if the supply of heat is sufficient. An analysis of the engine loading and heating demand for all ship operating modes must be made to determine which loads may economically be satisfied by the engine jacket water system.

3.4 Gas Turbine Plant Piping Systems.

a. Seawater cooling system.   Gas turbine propulsion plants require seawater cooling for the main reduction gear lubricating-oil cooler and, when used, for a controllable-pitch propeller hydraulic oil cooler. Gas turbine propulsion plants that use a reversible converter coupling with a fixed-pitch propeller require additional seawater for the coupling's lubricating-oil cooler, but do not require cooling for controllable-pitch propeller hydraulic oil. If the engine employs bleed-air for starting, a bleed-air cooler is required to protect the starter motor from high temperatures. Regenerative gas turbine engines also require a seawater-cooled heat exchanger for the intercooler.

The reduction gear lubricating oil is used to cool the gas turbine lubricating oil, which is not done with other types of plants. This is done because: a seawater leak into the gas turbine lubricating-oil system would be severely damaging to the turbine bearings; most marine gas turbines use synthetic oil, which is not compatible with the copper alloys preferred for seawater heat exchangers; and the high operating temperature of gas turbine lubricating oil would cause scaling in seawater tubes.

A separate motor-driven seawater cooling pump is generally installed for each gas-turbine-driven generator. On U.S. Navy ships, an emergency supply from the firemain via a pressure-reducing valve is also provided.

Separate pumps for auxiliary seawater-cooled heat exchangers are generally arranged similar to the setup for a diesel or steam plant. However, when there are no extreme differences in flow requirements or pressure loss through main and auxiliary heat exchangers, it is advantageous to combine all machinery seawater cooling services into a single, central system consisting of a main supplied by several pumps, one or more of which are standbys. This arrangement allows flexibility of operation, reduces the number of pumps required, and permits use of a single pump type, simplifying maintenance and spare parts. Gas turbine plants, being without steam condensers or large diesel engine coolers installed in series, lend themselves to such an arrangement, and U.S. Navy gas turbine plants generally adopt this approach.

b. Lubricating-oil service system.   Gas turbine plants require separate lubricating systems for the gas turbine and the reduction gear since each uses different oil.

Gas turbines generally are lubricated by synthetic oil. An oil-conditioning module that is designed and built by the engine manufacturer and contains a tank, heat exchanger, filter, and temperature- and pressure-control valves is usually installed adjacent to the engine. Oil is collected from the engine sump by scavenging pumps, which circulate oil through the module. The shipbuilder installs the supply and return piping connecting the module to the engine. The tank is filled and drained using portable containers. This system is described further in Chapter 4.

Gas turbine propulsion plant reduction gears are equipped with a lubricating-oil system similar to that for a steam plant, except that a motor-driven pump is substituted for each steam-driven pump. In addition, the reduction gear lubricating-oil piping system is arranged to circulate a small portion of the main oil flow through the gas turbine cooler; the rest is bypassed around it. An orifice is installed in the bypass line to force sufficient flow through the gas turbine cooler. A two-way temperature-control valve, which senses gas turbine oil temperature, is installed in the line to the gas turbine lubricating-oil cooler.

c. Fuel service system.   Gas turbine plants require a cleaner fuel than steam or diesel plants for reasons discussed in Chapter 12. Shipboard fuel systems for gas turbine machinery plants accomplish fuel purification in two stages. The primary stage is performed by the fuel transfer system, where centrifugal purifiers remove most water, rust, and sediment introduced into the fuel during transportation and storage. Seawater-compensated fuel tanks used on some U.S. Navy combatants increase the potential for water in the fuel. If not removed by the centrifugal units, these contaminants would overload the filter-coalescers in the fuel service system. The secondary stage is performed by the fuel service system, where finer filtration and further water removal ensures that engine fuel quality requirements are met.

The system components in Navy service, the most common application of gas turbine propulsion, are typically designed to operate with either diesel fuel or JP-5. A typical gas turbine fuel service system is illustrated by Fig. 22.

Two settling or service tanks, each with a capacity sufficient for 6 to 8 hours of operation, are provided in each machinery space. The tanks have connections to the stripping system for the removal of sludge and contaminated fuel. Generator engines are supplied from gravity feed tanks, thereby eliminating dependence on an external source of electric power.

Propulsion gas turbine engines have an attached fuel pump, which meters balanced fuel flows to each combustion chamber. Two positive-displacement fuel service pumps are installed to overcome system pressure losses, providing a positive pressure at the attached pump inlet, and eliminating the need to locate service tanks above the engine for gravity flow. Each service pump has sufficient capacity to supply all propulsion and generator engines in the space at full power. On U.S. Navy ships, two-speed pumps are normally used, permitting operation of the pump at reduced speed for normal cruise power levels. This minimizes the amount of fuel spilled in the event of

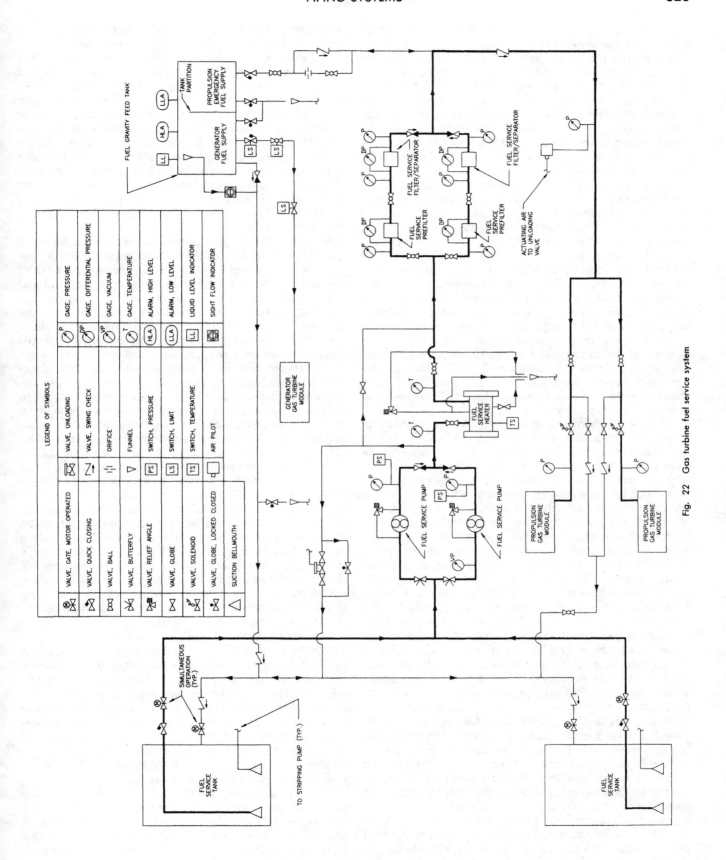

Fig. 22 Gas turbine fuel service system

LEGEND OF SYMBOLS

VALVE, GATE, MOTOR OPERATED		VALVE, UNLOADING	
VALVE, QUICK CLOSING		VALVE, SWING CHECK	
VALVE, BALL		ORIFICE	
VALVE, BUTTERFLY		FUNNEL	
VALVE, RELIEF ANGLE		SWITCH, PRESSURE	PS
VALVE, GLOBE		SWITCH, LIMIT	LS
VALVE, SOLENOID		SWITCH, TEMPERATURE	TS
VALVE, GLOBE, LOCKED CLOSED		AIR PILOT	
SUCTION BELLMOUTH			

GAGE, PRESSURE	P	GAGE, DIFFERENTIAL PRESSURE	DP
GAGE, VACUUM	VP	GAGE, TEMPERATURE	T
ALARM, HIGH LEVEL	HLA	ALARM, LOW LEVEL	LLA
LIQUID LEVEL INDICATOR	LL	SIGHT FLOW INDICATOR	

a break in the system and reduces the amount of fuel returned to the service tanks.

The service pumps are provided with controls actuated by a pressure switch in the pump discharge. If a low fuel header pressure is sensed, the operating pump is automatically switched to high speed. If pressure does not return to normal in a few seconds, the pump is automatically stopped as a fire safety feature in the event the pressure reduction was caused by a break in the piping. For the same reason, an automatic start-up of the standby pump is not provided.

The fuel heater provides the correct fuel temperature (and therefore viscosity) for efficient separation of water and particulates from the fuel, and to maintain a minimum temperature for starting the gas turbines in cold ambient conditions. A heater outlet temperature of 85 F is satisfactory for these purposes when using diesel fuel. The fuel temperature is regulated automatically by a temperature sensor in the heater discharge, which actuates electric power controllers for the heater elements. A standby heater is not considered necessary because the reliability of electric heaters is high; the engines can continue to operate, once started, without the use of the heater; and filters and coalescers will continue to function, although at reduced efficiency, if the heater fails. If steam heaters are used, consideration should be given to installing a standby heater because of the possibility of fuel contamination with water from a leaking tube.

Two filter/separators, one a standby, are installed to remove particulates and water from the fuel. Each unit can process the rated flow from one service pump. These units employ two types of replaceable-cartridge elements to filter out solids and coalesce minute particles of water into droplets large enough to separate by gravity from the fuel stream. Each filter/separator has a drain to the oily-waste system. A pre-filter is installed upstream of each filter/separator to remove the bulk of solid matter from the fuel stream, extending the life of the filter/separator elements. While gas turbine engines are normally equipped with a filter at the engine fuel pump inlet, this is provided as emergency protection only; it does not have sufficient capacity for continuous filtration.

The service pumps take suction from the service tanks and discharge through the heater, pre-filters, and filter/separators to the gas turbines. Piping is arranged to permit any combination of components to supply the engines. Fuel pressure is regulated by an unloading valve that senses the pump discharge pressure and opens to return fuel to the service tanks as necessary to maintain a constant pressure as the engine load varies. Return piping is provided to each service tank. A valve in each return line is interlocked with the associated tank suction valve to protect against inadvertent closure of both return line valves. A return line from the gas turbine inlet to the service tanks is provided to permit purging the system of cold or contaminated fuel prior to start-up.

From the filter/separator discharge, fuel is piped to the propulsion turbine attached pumps and to the gravity feed tank. The tank fill line contains an orifice sized to deliver slightly more fuel than required by the generator engine

operating at rated load. Excess fuel is returned continuously through an overflow to the service tanks.

The gravity feed tank is divided by a vertical baffle into two compartments, one serving the generator engine and one serving as an emergency supply to the propulsion engine. Propulsion gas turbines should be brought down in power gradually to reduce the temperature of internal components before the engine is stopped; otherwise engine damage can result. In case the electric power to the fuel service pumps is lost, the gravity feed tank provides approximately 30 seconds of fuel to permit a controlled shutdown of the engine. Upon a loss of pressure in the service system, fuel from the gravity feed tank flows back to the engine inlet through a check valve, which bypasses the tank-fill orifice.

Stainless-steel piping and valves are recommended for gas turbine fuel service systems to help ensure fuel purity. Carbon-steel piping is not recommended because of the potential for contamination of the fuel with rust and scale. While copper-nickel piping would eliminate this problem, it causes fuel degradation and could harm the engines by contaminating them with dissolved copper.

d. Bleed-air system. A bleed-air system distributes compressed air from the propulsion and generator gas turbine inlet compressors to machinery plant services. Bleed-air is received from each propulsion and generator gas turbine and supplied (a) to idle gas turbines for starting; (b) to the propulsion gas turbines for motoring during cooldown and water-washing; and (c) to anti-icing manifolds in the engine air-intake plenums to prevent the formation of ice in the airstream during cold, moist ambient conditions. On U.S. Navy combatants, bleed-air is also distributed outside the hull to mask propeller and machinery noise. A typical bleed-air system is illustrated in Fig. 23.

The system collects bleed-air from all turbines in a main. Regulating valves installed at each compressor outlet maintain a constant pressure in the main. These valves respond both to fluctuating demand and to variations in the available flow, which occur as engine power levels change. A dynamic analysis of transient system operating conditions should be considered when selecting valve response characteristics to ensure that undesirable pressure transients will not occur in operation.

The temperature of bleed-air from the gas turbines increases with engine load and ambient temperature, and is typically in the range of 900 F. Coolers, in branches from the main, reduce the air temperature as required for individual services. A moisture separator is installed downstream of the cooler that supplies starting air to remove water that condenses from the airstream. A continuous low-point drain is installed at the moisture separator and at each piping low point. Temperature switches are installed to actuate high-temperature alarms. Filters are installed to protect components as required.

Air supplied to gas turbine starter motors must be at a higher pressure for starting than for motoring. A two-position regulating valve is installed in the supply to each gas turbine starter motor. In the starting position it supplies air at system pressure; in the motoring position it reduces the system pressure. High-pressure air reducing

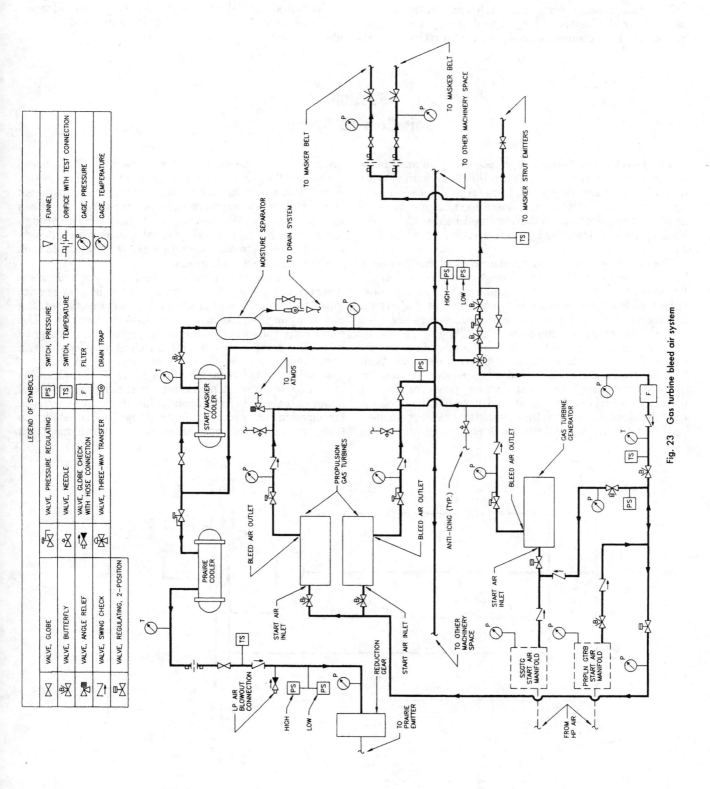

Fig. 23 Gas turbine bleed air system

manifolds are installed to supply starting air when all gas turbine engines have been secured.

Materials suitable for use at high temperatures must be used for all bleed-air piping except piping downstream of coolers. Piping must be designed with sufficient flexi-bility to absorb the dimensional changes associated with the wide operating temperature range. If expansion joints are required, the bellows should be constructed of Inconel 625, and the end connections should be made of stainless steel.

Section 4
Ship-Service Systems

4.1 Fuel Fill and Transfer System. Fuel fill and trans-fer systems receive and store fuel, deliver it to service tanks for the machinery plant, transfer fuel between stor-age tanks, and offload fuel. Some naval ships have an additional requirement to deliver fuel to ships alongside. Fuel filling and transfer systems are illustrated by Fig. 24.

The filling system has port and starboard weather-deck hose connections that are joined to a filling main. Branches from the main lead to one or more manifolds, from which a tailpipe extends to each fuel storage tank. The tailpipes must run low in the ship since they are used for transfer pump suction as well as filling.

Since the highest system flow usually occurs during filling, the filling rate will usually determine storage tank tailpipe sizes. However, tailpipe sizes should also be checked for adequate pump-suction pressure during transfer operations. Piping should permit filling with a specified deck pressure at the design filling rate, consider-ing fuel viscosity at the design ambient temperature. The design filling rate should be sufficient to take aboard about half the ship fuel capacity in no more than one watch (4 to 8 hours). To minimize time alongside replenishment ships, U.S. Navy ships are equipped for refueling at sea at much higher rates, typically 3000 gpm per hose through one to four hoses, depending on ship size, with a design flow of 500 gpm to each storage tank, or 1000 gpm to each compensated fuel tank group.

Two fuel-transfer pumps should be provided. Each transfer pump should be sized to fill the largest service tank at a rate sufficient to allow adequate time for settling and sampling before putting the tank on line. On a U.S Navy ship, for example, the transfer system is designed to refill a service tank having an eight-hour capacity in about two hours.

In selecting the transfer pump capacity, consideration may also be given to providing the capability to offload about half the ship fuel capacity during daylight hours in preparation for repairs or drydocking. The capability to transfer fuel for trim correction may also be an important

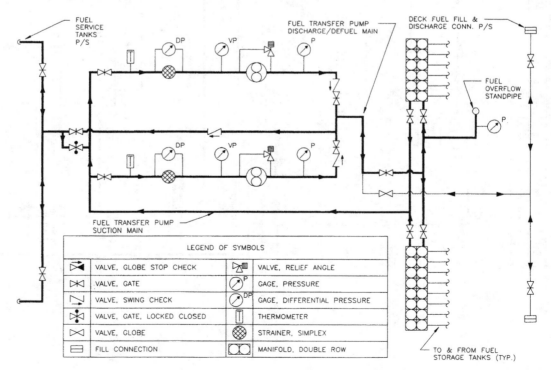

Fig. 24 Fuel fill and transfer system

factor. Both transfer pumps may be considered in use for offloading and trim correction. For U.S. Navy ships having a fueling-at-sea capability, the number and capacity of transfer pumps may be determined by the delivery rate.

Transfer piping is arranged so that the pumps can transfer fuel from storage tanks to service tanks, and from storage and service tanks to the deck connections for offloading. Piping should also be provided to transfer fuel between storage tanks for consolidation prior to refueling and to adjust the trim of the ship.

On ships using heavy or residual fuel, storage tanks should be provided with heating coils to raise the fuel to the pumping temperature.

On ships having diesel or gas turbine machinery plants, fuel transfer systems are also used for fuel purification. Centrifugal purifiers should be provided to remove water and sediment from the fuel prior to delivery to the service tanks. A heater is installed upstream of each purifier to provide the correct fuel temperature (hence viscosity) to the purifier. The purifiers and heaters are connected in series with the transfer pumps. The transfer system is also connected to the low suction of each fuel service tank so that contaminated fuel can be drawn from the tank and circulated through the purifiers back to the tank. Ships using residual fuels for diesel or gas turbine engines may also incorporate fuel water-washing equipment in the transfer system, as described in Chapter 12.

Since regulatory body requirements prohibit routine ballasting of fuel tanks, the fuel, bilge, and ballast systems should normally not be interconnected. However, some ships must be able to ballast fuel-storage tanks with seawater to improve stability in an emergency. In such cases, cross-connections between the fuel-storage tank tailpipe manifolds, the ballast system (for filling), and the bilge system (for emptying) are necessary. Ballasting fuel tanks is an emergency operation since the resulting ballast water will be contaminated with oil, which presents a disposal problem. The bilge system is used for deballasting to avoid contaminating the ballast system with oil. Commercial ships with the capability to ballast fuel tanks are required by U.S. Coast Guard rules to have an oil/water separator and oil-content monitor capable of processing the ballast water, or the ability to retain contaminated seawater ballast until reaching port, where it is discharged to shore via deck connections.

In addition to the transfer system piping, a separate fuel-stripping system is provided on many U.S. Navy ships to improve fuel management. This system consists of stripping pumps, separate tank tailpipes, and a contaminated-fuel settling tank. The tailpipes take suction close to the bottom of each fuel storage and service tank to remove water, sludge, and contaminated fuel, which are discharged to a contaminated-fuel settling tank. Fuel collected in this tank is heated and allowed to settle. Reclaimed fuel is returned to the storage tanks. Fuel that cannot be reclaimed is pumped to a waste-oil tank.

4.2 Bilge and Ballast Systems. Bilge and ballast systems, while having two distinct functions, are interconnected so that the same pumps (commercial practice) or

eductors (U.S. Navy practice) commonly serve both functions, and piping is arranged so that each system can operate independently. Typical bilge and ballast systems are illustrated by Fig. 25.

Bilge systems are provided primarily for the safety of the ship, and are subject to extensive regulatory-body requirements. A bilge pumping system is provided on all ships to permit emergency dewatering of all watertight compartments, except for ballast, oil, and water tanks that have independent means for filling and emptying. The system also provides drainage for spaces such as anchor-chain lockers where water may accumulate during normal operation of the ship and cannot be drained by gravity. The bilge system should be capable of draining all tank tops, watertight flats, and insulated holds. Separate hand pumps or firemain-actuated eductors are normally installed for isolated areas such as chain lockers and decks over peak tanks. Where drainage from a particular compartment is considered undesirable, it may be omitted provided the safety of the ship is not affected.

During normal ship operation, the bilge system is intended to discharge overboard only clean water. Only in a flooding emergency should bilge water from machinery spaces or other compartments where oil is likely to be present be discharged overboard. The collection and disposal of oily bilge water are normally accomplished by the oily-waste system (see Section 4.3) to contain potential pollutants. However, bilge pumps may be arranged to discharge oily bilge water to a holding tank, from which it is pumped ashore by way of deck hose connections. This capability is advantageous in the event of a major leak or spill since the oily-waste transfer system is typically of low capacity.

The number and capacity of bilge pumps (or their equivalents) are determined in accordance with regulatory-body requirements depending on the ship size, type, and service. At least two bilge pumps must be provided on all ships. Although bilge pumps may also serve other systems, such as ballast, fire, or seawater cooling, at least one pump must always be available for pumping bilges. For passenger ships, regulatory bodies require the pumps to be located in separate watertight compartments, with piping and electric power supplies arranged so that flooding of any compartment will not render all pumps inoperable. This should also be the case on other types of ships to the extent practicable.

Suction tailpipes should be sufficient in number so that the system is capable of dewatering the ship under all practicable conditions of list and trim, and should be located at the low point of a space. Multiple suctions should be provided in large spaces and in each space whose shape prevents water from collecting at a single drainage point. Machinery spaces should have at least two suctions, one forward and one aft. A space having a broad, flat bottom should have four suctions, arranged with one located port and starboard at the outboard ends of the space. Suction tailpipes should terminate in a drainwell or sump whenever practicable.

Bilge piping should be arranged to prevent cross-flooding between compartments in the event of damage. Since

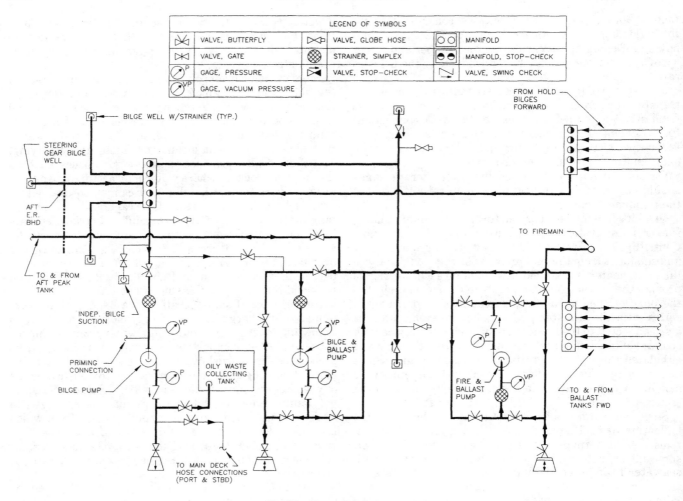

LEGEND OF SYMBOLS					
⋈	VALVE, BUTTERFLY	⊳⊲	VALVE, GLOBE HOSE	⊙⊙	MANIFOLD
⋈	VALVE, GATE	⊗	STRAINER, SIMPLEX	◒◒	MANIFOLD, STOP-CHECK
⊙P	GAGE, PRESSURE	⊳	VALVE, STOP-CHECK	◥	VALVE, SWING CHECK
⊙VP	GAGE, VACUUM PRESSURE				

Fig. 25   Bilge and ballast systems

bilge suction tailpipes are open to the space, each tailpipe should have a check valve to prevent flooding of the space it serves in the event of flooding and damage to the piping in another space. If a tailpipe penetrates a bulkhead, the check valve should be installed at the bulkhead within the space it serves. Otherwise, a stop-check valve should be installed at the connection to the suction main or in the manifold to which the tailpipe is connected. Bilge piping passing through a tank should be installed within a trunk, or be constructed of extra-heavy pipe with welded joints throughout the tank and bends for flexibility. On passenger ships, bilge piping must be arranged inboard of the shell a distance equal to one-fifth the beam of the ship to reduce the risk of collision damage.

Bilge piping for the cargo pump room, and other spaces where flammable gases may be present, should not provide a connection to the machinery space or other spaces where sources of ignition may be present.

Each suction tailpipe valve should be operable from the space containing the pump that normally serves the tailpipe. For passenger ships, valves must be operable from above the bulkhead deck. Valves in machinery spaces should be operable from above the floor plates.

Each bilge suction tailpipe should be provided with a strainer. Machinery space strainers should be easily accessible from the floor plates and of a type that is easily cleaned. In other spaces, box strainers may be used. Strainers should have an open area not less than three times the area of the suction pipe. Additional strainers should be provided in the bilge pump suctions.

Suction tailpipes are connected to manifolds or suction mains leading to the bilge pumps. Suction piping is normally installed so it is common to all pumps, although space arrangements on some ships may dictate that separate systems be provided.

The minimum size of suction headers and tailpipes is established in accordance with regulatory-body requirements. However, pressure-loss calculations should be performed to confirm that pump suction lift requirements are satisfied.

An independent bilge suction leading directly from the machinery space bilge to one of the bilge pumps is installed to permit rapid initiation of emergency dewatering; on passenger ships such a connection is installed for each bilge pump.

Bilge pumps normally discharge directly overboard when used for space drainage. When bilge pumps are used for other services, pump suction and discharge valves must be provided so the bilge pumps can be isolated from all other systems and connected to the bilge piping. These valves should be operable at the pump. Overboard discharge connections should be combined with other services where practicable to minimize the number of shell penetrations.

Since bilge pumps are usually of the centrifugal type and the level of liquid is below the pump, a priming capability should be provided by using self-priming pumps, installing a vacuum priming system, or other means.

Navy ships commonly use eductors rather than pumps for dewatering flooded compartments. Eductor actuating water is provided from the firemain. The eductors take suction on a common bilge drainage main with branches to compartments served. Piping, valves, and valve actuators should be arranged so that if any space is flooded, eductors in an adjacent space can dewater it. On larger combatants, a loop main may be provided for increased survivability. The same system is used for deballasting via ballast tank tailpipes.

A ballast pumping system is provided to adjust the trim and draft on ships having ballast tanks. The system capacity is determined by the rate of draft or trim adjustment required. For tankers and containerships, the required rate depends on the cargo loading and unloading times, the water depth at terminals, and the tolerance of cargo cranes, booms, and hoses for changes in draft of the ship.

For commercial ships ballast is normally handled by the bilge and ballast, seawater cooling, or fire pumps. The high ballast flow rates on tankers necessitate the use of special ballast pumps, which are usually installed in the cargo pump room. In such cases, ballast tanks that are not in the cargo tank area, such as aft peak tanks, are filled and emptied by a machinery space ballast system. On U.S. Navy ships ballast tanks are normally filled from the firemain, and emptied by the bilge drainage eductors.

Individual ballast tank tailpipes should be provided and connected to mains or manifolds. The arrangement is similar to the bilge system except that tailpipes do not have check valves, and pump suction and discharge piping is arranged to permit flow both to and from the tanks. A ballast line which bypasses the pump is sometimes provided for gravity filling of the tanks if the tank location is sufficiently below the waterline to develop the required flow rate through the piping. The piping arrangement should permit ballast to be moved between any tank and the sea. The capability to transfer between tanks is not required unless dictated by the specific needs of the ship, such as for icebreaker heeling.

4.3 Oily-Waste System. An oily-waste system consolidates and disposes of oily waste and waste oil that collect in machinery spaces. The term "oily waste" refers to waste liquid that is mostly water, while "waste oil" denotes liquid that is mostly oil. Such wastes cannot be discharged overboard because of pollution abatement regulations. An oily-waste system is usually kept separate

from a bilge system to avoid contamination of bilge-system piping with oil that could subsequently be discharged overboard. In addition, the bilge system pumps and piping are designed for the high flow rates needed in an emergency and so cannot be used efficiently for handling small quantities of waste liquids.

Drip pans, funnel drains, and gravity-drain piping should be provided wherever practicable to collect oily waste and waste oil at the source and convey it to a collecting tank. Oil-free water should be piped directly overboard or to waste-water tanks to reduce the volume of bilge waste. Liquid that accumulates in bilges in spite of these measures is collected in drain wells.

A typical oily-waste system is shown in Fig. 26. The oily-waste system takes suction on the drain wells and discharges to an oily-waste collecting tank. In U.S. Navy practice this is usually accomplished by an oily-waste transfer pump taking suction on a main, which has branches to each drain well. A separate pump is normally provided in each main machinery space. Commercially, a float-operated automatic sump pump discharging to an oily-waste collecting tank is usually provided in each drain well; this arrangement is desirable with unattended machinery spaces.

The contents of the oily-waste collecting tank are processed by an oil-water separator, which normally produces an effluent that is suitable for overboard discharge. An oil-content monitor is provided to continuously measure the oil content of the effluent. If the set limit on oil content is exceeded, the effluent is automatically recirculated to the collecting tank, or the separator is stopped. Oil is discharged via separate piping to the waste-oil tank.

Piping is provided to discharge oily waste and waste oil to shore collection facilities via port and starboard hose connections on the weather deck. The oily-waste transfer pump or the oil-water separator pump, as applicable, is arranged for this purpose. In either case, the pump head must be sufficient to provide adequate pressure (approximately 10 psi) at the hose connection for discharge to shore. The deck flanges are required to be of a standard design [19] to ensure compatibility with all port facilities. Piping may also be provided for discharging waste oil to the incinerator or the boilers on commercial ships so equipped.

4.4 Freshwater Service Systems. Freshwater service systems supply shipboard hotel services and various other demands depending on the type of ship.

Fresh water may be carried aboard in sufficient quantity to support all services for the design duration of a voyage and replenished from port facilities, or it may be produced from seawater by onboard desalination units. Ships designed for ocean service, or for coastwise service with a large complement of crew or passengers, should generally be provided with one or more desalination units.

The minimum capacity of the desalination unit, usually expressed in gallons per day, must be sufficient for the total daily freshwater consumption plus a margin for future growth when anticipated. The minimum estimated

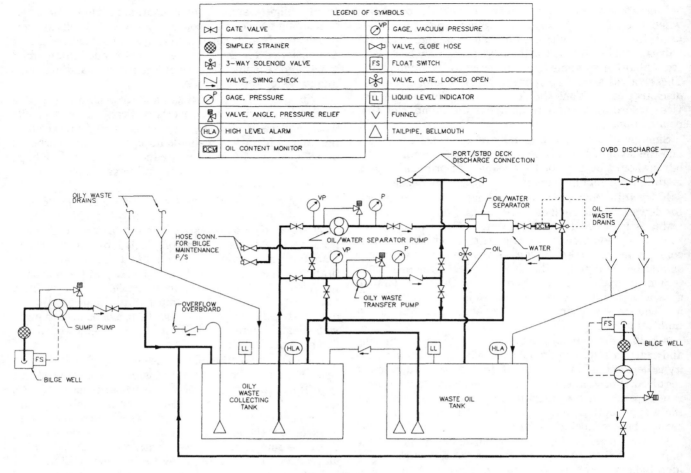

Fig. 26   Oily-waste system

Table 10   Daily per-capita freshwater consumption, gallons

Water Use	Commercial Applications		U.S. Navy Applications	
Drinking	0.7			
Personal hygiene	21.0		} 30	
Galley/scullery	7.3			
Laundry	1.0			
Flushing	vacuum system	gravity system	vacuum system	gravity system
	2–4	35	2	0[a]
TOTAL	32–34	65	32	30

[a] On U.S. Navy ships, seawater is used for flushing when a gravity drain system is installed.

daily freshwater requirements for human use may be determined from Table 10. There are also freshwater requirements for other uses, such as boiler makeup feed, gas turbine water-wash, freshwater cooling system makeup, window-washing systems, photographic laboratories, and washdown of aircraft or other vehicles carried on board. Guidance for determining the requirements for these purposes may be found in references 20 and 21.

Desalination units can be of various types, as discussed in Chapter 17. The machinery plant selection determines

the heat source available and is a primary consideration in selecting the desalination units.

Two desalinators should normally be provided to permit underway maintenance without interrupting water production. U.S. Navy practice for single- and twin-screw ships is to provide two units, each capable of meeting the full daily demand. The installation of three units, with two required to meet the demand, may also be advantageous. For quadruple-screw ships, at least four units are installed, with any three capable of meeting the full demand. For commercial applications, a single unit may be provided if the potable-water storage tank volume is sufficient to meet emergency needs and the unit installed is a highly reliable design.

Potable water is fresh water that is obtained from shore or a desalination plant, stored in onboard tanks, treated to make it safe for human use, and distributed throughout the ship to sinks, lavatories, showers, galleys, laundries, drinking fountains, etc.

Desalinated water is fresh water obtained directly from a desalination plant. It is distributed to services that require water with levels of dissolved solids, suspended solids, and ionic impurities lower than those necessarily found in potable water. Such services include boiler makeup feed, gas turbine water-washing, and electronics cooling-water systems. A separate desalinated water system is provided on most U.S. Navy ships, but is not required on most commercial designs. When provided, the system includes desalinated water storage tanks, pumps, and distribution piping that are independent of the other freshwater systems.

On commercial ships, a separate wash-water system is sometimes provided for services that do not require water of potable quality, including lavatories, showers, and laundries. This reduces the amount of water to be treated and the quantity of piping that must be designed to satisfy health regulations. However, a single system is normally provided to simplify tankage and piping. The design of wash-water systems is further discussed in reference 6.

A potable-water system must be designed to prevent contamination by substances harmful to human health. The system must not be cross-connected with any other system. Valves and blind flanges are not considered dependable means of separating potable-water piping from other systems. Where potable water is delivered to equipment that could contain water of inferior quality, an air gap of at least two supply pipe diameters must be provided between the supply and receiving connections. Examples of connections that must have an air gap are the supply to galley and pantry sinks, dishwashers, vegetable peelers, steam kettles, laundry equipment, sterilizers, and plumbing fixtures of all types. The location of a faucet above the rim of a lavatory or other open receiving vessel constitutes an adequate air gap. Where a permanent connection to another system is required, a backflow preventer should be installed [6].

Piping is provided to fill the potable-water storage tanks from port and starboard deck-hose connections when in port, and from the desalination plants when at sea. An interlocked valve manifold is provided at the desalinator and arranged so it can discharge to only one system at a time to prevent contamination of the potable-water system. A relief valve should be installed in the fill piping to protect the storage tanks from overpressure.

A disinfection system is provided to kill bacteria before water enters the potable-water distribution piping. The system may be of a type that injects an amount of chlorine or bromine in proportion to the quantity of water being treated. Equipment employing ultraviolet radiation may also be used in commercial applications; however, such equipment does not provide a residual disinfectant in the potable water. The disinfection system is arranged to treat water as it leaves the desalinator, and to treat water in the storage tanks through recirculation.

Two potable-water pumps should be provided, with one a standby. The capacity of each potable-water pump should be sufficient to satisfy the peak demand. A method of establishing peak demand based on the number and type of installed plumbing fixtures is described in reference 11. Any loads in addition to plumbing fixtures, which are constant or of periodic long duration, should be added to the peak demand.

On large ships with many users, the demand is usually such that the pump may be allowed to operate continuously; however, a recirculation line is necessary to protect the pump from overheating during occasional periods of zero demand. On ships with few users, the pump would operate against shutoff head for long periods of time if operated continuously, thereby reducing its life. In such cases a compression tank is connected to the system directly downstream of the pumps. The tank is partially filled with pressurized air, and acts as an accumulator. When water usage reduces the tank air pressure, a pressure switch starts the pump to refill the tank; the pump is stopped when the pressure is restored to normal. A pump recirculation line is not necessary when using a compression tank. In U.S. Navy practice a compression tank is installed on ships carrying fewer than 200 persons.

A vacuum priming system should be provided for the potable-water pumps if the storage tanks are not located so that the pumps will always have positive submergence.

Water heaters are installed to supply hot potable water to showers, lavatories, sinks, laundry, galley, and other services. Water heaters can operate with steam, electricity, or diesel engine jacket water. When water heaters are operated with waste heat, an electric heater should be provided for use in port when the engines are secured. Heaters can be of the storage or instantaneous type. Water heaters should be of sufficient capacity to meet peak demand when usage is high, such as at the change of the watch and during meal preparation periods. The heating source should be controlled by a thermostatically operated switch or valve to limit the water temperature. Valves should fail closed. Each heater should have a relief valve that discharges to a deck drain in a manner which will not endanger personnel.

A single heater is usually installed in a central location for small crews. When a large number of fixtures in separated spaces must be served, multiple heaters should be installed in distributed locations close to major users. The most efficient arrangement must be determined with consideration to the location and demand of users, the required water temperature of each user, and the availability of heating media. Dedicated heaters should be provided to satisfy the large flow demand and higher temperatures normally required for laundries, galleys, and similar services.

Water heaters are supplied through branches from the potable-water distribution piping. Each heater should be provided with a recirculation loop to maintain hot water near each user, thus reducing potable-water wastage. The recirculation loop connects the pipes supplying the most remote users to a pump, which discharges into the water heater inlet and provides a continuous flow of 5 to 10 gpm, depending on loop length, in the hot-water distribution piping.

Potable-water pump suction and discharge piping and potable-water mains should be sized for the rated potable-water pump capacity. Each segment of distribution piping should be sized for the peak load in that segment based on the fixture unit method [11]. Branches to individual fixtures should be sized for the rated flow of the fixture.

Pipe sizes should be sufficient to deliver the required flow to the most remote fixtures when the system is operating at full capacity.

Design details to be considered when developing a potable-water system are as follows:

- Bolted connections should not be made through the shell of potable-water tanks.
- Pumps, manifolds (filling, transfer, and suction), and gages should be grouped.
- Potable-water services should be grouped to minimize the number of branches from the main.
- Potable-water piping should not be run through bilges.
- Each branch connection from the potable-water main should have a stop valve to permit local maintenance without securing the entire system.
- A check valve should be installed in the supply to each water heater to prevent hot water from entering the cold-water piping.
- All piping exposed to the weather should have a means for draining to prevent freezing.
- Hose valves and piping for receiving water or discharging water overboard should be installed in a horizontal position about 2 ft above the deck and in a protected location. Each hose connection should have a cap and chain. A label plate should be installed inscribed "POTABLE WATER ONLY."
- Locked, vermin-proof lockers or other suitable storage should be provided for potable-water hoses and furnished with a label plate inscribed "POTABLE WATER HOSE STOWAGE ONLY."
- Each hose connection used to supply potable water to another system should have a warning label plate installed in a conspicuous location inscribed "HOSE SHALL BE DISCONNECTED WHEN NOT IN USE."
- Piping to each hose connection should be provided with a stop-check valve and a vacuum breaker in that order from upstream to downstream.

4.5 Chilled-Water Systems. Chilled-water systems circulate fresh water at a temperature of 42 to 45 F in a closed loop between air-conditioning plant chillers and services throughout the ship. Chilled water is primarily used in air-conditioning systems, which are described in Chapter 21. However, additional users include drinking-water coolers, sample coolers, and other services requiring a consistent cooling-fluid temperature, which is not possible using seawater. On most U.S. Navy ships, the cooling of electronics for combat systems is a substantial load on the chilled-water system.

The routing of chilled-water piping throughout the ship is preferable to refrigerant piping, supplying direct-expansion cooling coils, because a prohibitive quantity of costly refrigerant would be required and there would be a high potential for refrigerant leaks, which are difficult to locate and repair. In addition, long runs of refrigerant piping are difficult to arrange on several levels while ensuring proper flows and temperatures throughout the system.

Chilled-water plants can be of various types; however, each plant has a compressor, chiller (evaporator), condenser, and associated components and controls, as described in Chapter 21. The method of cooling the water has little effect on the design of the chilled-water circulation system.

The total cooling load as well as the number and capacity of chilled-water plants is usually determined by the air-conditioning system designer. At least two plants are normally installed to meet the demand, although on small commercial ships a single plant is sometimes sufficient. On U.S. Navy ships four or more plants are provided, with one plant as a spare, and the units are usually placed in separated locations for survivability reasons.

One chilled-water pump is provided for each plant. For a commercial application, the pump capacity is generally 2.4 to 3 gpm per ton of cooling capacity. U.S. Navy plants are designed for a flow of 3.6 gpm per ton; however, the installed chilled-water pump is usually sized for 4.5 gpm per ton to provide a margin for growth.

The pump takes suction on the chiller outlet and discharges to the circulation piping. An expansion tank is installed at the pump suction and is pressurized to maintain a positive pressure of at least 5 psi at the highest point in the system under all conditions. A recirculation line with a control valve is provided to maintain a minimum flow through the chiller during periods of low demand.

Chilled-water piping consists of a riser connecting each plant to a distribution main, and branches from the main to each service. Each run of pipe is duplicated, one for supply and one for return. Each cooled unit must be provided with a means of flow regulation to balance the flow in the system. Constant-flow fittings, orifices, and manual throttle valves are used for this purpose. High-point vents must be installed to remove air from the system.

On U.S. Navy ships having multiple plants, the chilled-water piping is arranged in zones, with each zone serving a group of loads and supplied by one or more plants and risers. The zones should be separated by segregation valves. Vital loads should be supplied by branches from each of two different zones. Combatant ships should be provided with two or more mains arranged in a loop and separated both horizontally and vertically for survivability reasons.

4.6 Plumbing Drains and Vents. The plumbing drain system collects waste liquids from plumbing fixtures, interior- and weather-deck drains, and air-conditioning cooling coils. Plumbing vents connect plumbing drain piping to the weather to remove odors and prevent the buildup of positive or negative internal pressures.

Two independent plumbing drain systems should be provided. Soil drain piping collects sewage, also called "black water," from toilets and urinals. Waste drain piping collects liquids from sinks, lavatories, showers, laundries, galleys, and similar sources; such wastes are commonly called "gray water." Separate soil and waste drain systems are necessary to protect human health and because regulations for restricting pollution are different for each type of waste.

Black water may generally be discharged at sea beyond a minimum distance from land; but while near land, in port, and in some restricted waters, regulations usually prohibit black-water discharges. Since most ships operate in areas where black-water discharges are prohibited, provisions for the onboard retention or treatment of black water are commonly provided.

For ships that transit restricted waters in a short time, a temporary holding tank may be sufficient provision for handling black water. But for long-term operation where discharges are prohibited, a sewage-treatment device may be necessary.

Sewage-treatment processes include biological digestion, filtration, chlorination, and incineration. All except incineration require the overboard discharge of a purified effluent that meets pollution regulations for organic and bacterial content.

The required capacity of a black-water holding tank is determined by the number of people on board, the per-capita waste generation rate (see Table 10), and the holding time required. The holding tank should be provided with an aeration system, a vent to the weather, a floatless level indicator, and connections for internal flushing with seawater or fresh water. Two pumps should be provided and arranged for discharging tank contents overboard or to hose connections on the weather deck. The pumps must be of a type designed for operation with sewage to prevent clogging with solids.

The direct overboard discharge of gray water is also restricted in some waters. For ships that operate in such restricted waters, a holding tank or treatment device is also required for gray water.

U.S. Navy ships are provided with holding tanks for black water. Holding tanks on larger U.S. Navy ships also collect gray water. The capacity of holding tanks should be sufficient for at least 12 hours of accumulation to permit the ship to traverse restricted waters en route to port reception facilities or open seas. Commercial ships generally are provided with sewage treatment devices capable of processing black water; a gray water treatment or holding capability should also be provided if necessary for the operating area.

Multiple holding tanks or treatment units may be necessary if the ship arrangement does not permit gravity drainage from all sources to a single tank.

Soil drains are collected in branches leading to one or more mains, which discharge into a holding tank or treatment device. Each main should have a three-port, two-position, ball- or plug-type diverter valve that can be aligned with either an overboard discharge connection or a tank. The diverter valves and overboard discharge connections may be omitted if the ship is not intended to operate where direct overboard discharge is permitted. Soil drains originating lower than about 4 ft above the waterline cannot be drained overboard and must be led directly to a holding tank.

Waste drains are collected in branches and mains similar to soil drains. Waste piping should be independent of soil piping except that it may be connected to common overboard discharge connections, and may be connected to a common holding tank or treatment device. A check valve should be installed in waste piping at each connection to soil piping.

Drains from refrigerated food storage rooms and appliances used to prepare or store food must discharge to the waste drain system through an air gap equal to twice the diameter of the drain, but not less than 2 in. On U.S. Navy ships, medical space waste drains must be segregated from others.

Weather-deck drains should be led to overboard discharge connections near the waterline. Interior deck drains are also directed overboard if they originate from spaces sufficiently high above the waterline; otherwise, they are connected to the waste drain system. Deck drains should be provided wherever water may accumulate in the normal use of the space or intrude from weather openings. If the space is likely to generate oily waste in significant quantity, the drains should be directed to the oily-waste collecting system.

Piping for gravity drain plumbing systems should be installed with a minimum, continuous downward pitch of from 1/8 to 1/2 in. per foot of horizontal distance. The required pipe size decreases with increasing pitch. A pitch of 1/2 in. per foot is preferred, particularly for fore-and-aft pipe runs.

Soil and waste drain piping should be arranged in accordance with the provisions in reference 6.

Valves in gravity drain plumbing systems should be of a non-clog design such as ball, plug, or gate valves. Each plumbing fixture and deck drain should have a trap installed. Branches should be connected to mains using Y-fittings instead of right-angle tees. A cleanout should be installed at each change in direction greater than 45 deg, in horizontal runs at intervals no greater than 50 ft, and at the base of each vertical pipe run. Overboard discharge valves should be of the gag scupper type to permit them to be closed in an emergency to prevent flooding.

Plumbing system trap seals should be protected from siphonage or back pressure by vents connected to the drain piping. Vents should be sized to allow sufficient airflow so that under normal use the seal of any fixture trap is not subjected to a pressure differential of more than 1 in. of water. A vent sizing method is described in reference 11. A vent should be installed as a continuation of each vertical drain header, with individual fixture drain vents connected to it. The open end of the vent pipe should not be below the level of the highest trap weir. All vent piping should be sloped to drain back to the soil or waste pipe.

Vents should terminate in the weather. Vents from fixtures above the watertightness level should terminate just below the deck next above the deck on which the fixture is located. Vents should be welded directly to the structure where they penetrate the shell or superstructure. A "T" outlet of the same size as the vent, or a half-round pipe section, should be installed at the outlet end of each vent. Soil vents should not terminate near hatches, doors, air ports, ventilation intake openings, or galleys.

Vacuum collection is an alternative to gravity drainage for black water, gray water, or both. Vacuum-collection systems employ special plumbing fixtures, a vacuum source, and a tank. Vacuum piping links the fixtures to the tank.

Two basic types of vacuum-collection systems are used. One makes use of a collection tank that is under vacuum. The other applies a vacuum source to the collection piping near the tank inlet, then discharges the collected waste into a tank that is at atmospheric pressure.

Vacuum collection is preferred for some applications since the piping is much smaller than gravity-drain piping, the piping does not have to be installed with slope, and vents are not required. In addition, vacuum water closets and urinals require only about 10% of the freshwater usage of standard units.

4.7 Firemain Systems. A firemain system supplies seawater at high pressure throughout the ship for sprinkling systems and hand-held hoses. Seawater is an essential shipboard extinguishing agent because it is in abundant supply, it can be applied as a stream or spray to suit various fire-fighting situations, and it is a highly effective cooling agent, which can prevent reflashing of combustible material, retard the spread of fire through ship structure, and protect fire-fighting personnel.

The common elements in all firemain designs are: centrifugal fire pumps operating at high pressure to provide effective hose stream reach, penetration, and spray formation; a piping system extending throughout the ship; and various means of applying seawater to a fire. Beyond these common elements, the design of firemain systems depends on the ship size, type, and service.

a. Commercial applications. A firemain system for a tanker is illustrated by Fig. 27. Firemain systems for commercial ships are designed to discourage uses other than for fire protection to ensure that the full capacity of the system will be available in an emergency, with minimal realignment of pumps and valves. Fire pumps may be used for other services, such as bilge, ballast, and seawater cooling, provided that at least one pump is kept immediately available for firemain service. Branches from the firemain for other than fire-protection purposes are generally not permitted, but piping that supplies another service is usually required to be connected at the pump, rather than to the firemain, and must have a local valve arranged to immediately isolate the other service. Exceptions may be made for low-demand services such as deck and anchor washing, provided the demand of these services is added to the required pump capacity. On tankers, connections for cargo-tank cleaning are frequently allowed without increasing the pump capacity on the basis that such use is obvious to the crew and can be immediately secured if necessary.

Fire pumps must not be connected to any oil piping. Connection to the bilge system is permitted for emergency dewatering.

At least two fire pumps should be installed. Pumps may be driven by electric motors, steam turbines, or internal-combustion engines. The pumps, sea chests, and sources of power should be located in separate spaces to ensure that a fire in one space will not disrupt operation of both pumps.

The minimum required capacity of each fire pump depends on the ship size and service. In general, each fire pump must have a capacity sufficient to meet the greater of two criteria: either a minimum flow rate based on ship size, or the simultaneous supply of a minimum number of fire hoses operating at the required pressure. The minimum number of hoses and required hose nozzle size (1½ or 2½ in.) depend on the ship type. The capacity of each pump must be sufficient to meet the hose stream requirements while supplying any non-fire protection services connected to the firemain. The combined pump capacity must be sufficient to meet hose stream requirements while simultaneously supplying sprinkling systems, such as the deck foam system on tankers.

The rated head of fire pumps must be sufficient to provide a minimum pressure of 50 psi for non-tankers, or 75 psi for tankers, at the most remote fire-hose nozzles with the required number of hoses operating simultaneously. Each fire pump discharges to a common firemain, which extends throughout the ship, with a branch to each service. The combined static and friction losses to the most remote fire hoses determine the required head. The fireplugs at the highest level of the superstructure usually constitute the worst case; however, on some ships the horizontal length of the firemain may be such that friction losses to the most distant fireplug will govern. The resultant fire pump head is usually in the range of 100 to 150 psi.

The fire pump head needed to supply fireplugs high in the superstructure may result in excessive pressure at hoses in machinery spaces and other locations low in the ship. Excessive pressure makes hose nozzles difficult to handle and may make snaking hoses within the ship impossible. In such cases a pressure-reducing station should be provided for the lower fireplugs.

Fireplugs should be placed so that all parts of living quarters, storerooms, working spaces, and weather decks accessible to the crew while at sea can be reached with two effective spray patterns of water, with at least one spray pattern from a single 50-ft hose. All portions of main machinery spaces should be within the reach of at least two effective spray patterns of water, each of which should be from a single 50-ft hose from a separate plug.

Commercial ship firemains are generally unpressurized (dry) when not in use, in order to reduce pump wear and avoid freezing of long runs of exposed piping, which exist on many designs, particularly on tankers and containerships. The system may be kept continuously pressurized (wet) in some applications when the rapid availability of firemain pressure at hose nozzles is required, such as on passenger ships. The life of galvanized steel pipe is generally considered acceptable for dry systems; however, copper-nickel is the preferred material for both dry and wet systems.

A more detailed discussion of regulatory-body design criteria for commercial ship firemain systems is contained in reference 22.

b. U.S. Navy applications. Firemain systems on U.S. Navy ships are continuously pressurized (wet) systems because water must be available immediately for

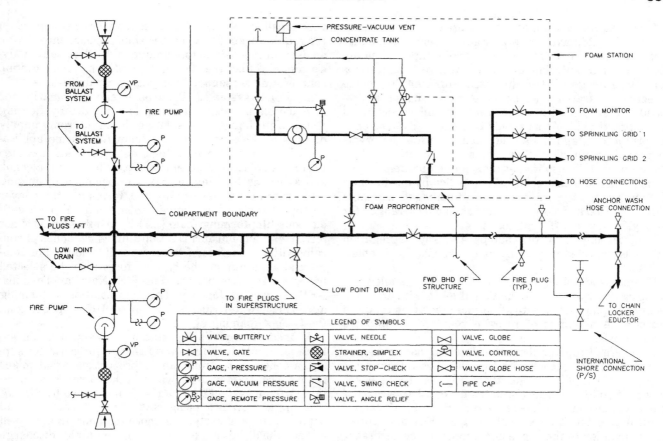

Fig. 27 Firemain and foam system

munitions sprinkling, and seawater must be continuously available outside of the machinery spaces for equipment cooling, sanitary flushing, ballasting, and operating eductors for drainage and deballasting. These services would require independent seawater piping systems if the firemain were not used, but can be conveniently supplied from the firemain. Since firemain system piping on U.S. Navy ships is usually enclosed, it is not subject to freezing in cold weather.

Fire pumps on U.S. Navy ships are primarily electric-motor driven; however, on ships having steam machinery plants, some of the fire pumps are usually steam-turbine driven. Motor-driven pump capacities range from 100 to 1000 gpm, with 1000 gpm a preferred size. The capacity of steam-turbine-driven pumps is usually about 2000 gpm. The rated head of fire pumps is in the range of 125 to 175 psi with the higher pressures used on larger ships.

Fire pumps are distributed in separate compartments along the length of the ship to minimize the effect of a single casualty. Each pump should have an independent sea chest.

The total capacity of the fire pumps must be sufficient for the simultaneous supply of all fire-fighting equipment required during a worst-case casualty, plus vital cooling, service, and flushing loads. Since U.S. Navy ships use firemain-actuated drainage eductors, that flow must also be included if drainage of fire-fighting water is necessary

for ship stability. The total load for the worst-case casualty is calculated by analyzing potential casualties related to the ship mission (such as a helicopter crash) or battle damage in a high-risk area (such as near a missile magazine) to determine which sprinkler systems, fire-extinguishing hose lines, bulkhead-cooling hose lines, drainage eductors, or other equipment would be required. For some ships, a non-casualty load, such as the washdown countermeasure system, may govern the system capacity.

The number of fire pumps to be installed is determined by dividing the total required capacity by the capacity of a candidate pump, and rounding up to the nearest whole number. The resulting number of pumps is multiplied by a factor (generally 1.33) and again rounded up, to provide spare pumps. The spare pumps allow for battle damage and regular maintenance, which is inevitable when pumps are operated to maintain continuous pressure in the system. The number and capacity of pumps may have to be adjusted to accommodate the number of suitable pump locations available in the ship. At least two fire pumps must be installed on a U.S. Navy ship; combatants such as cruisers typically require five or more, and the largest aircraft carriers use more than twenty.

Since fire pumps on U.S. Navy ships may operate at low or zero load for extended periods, each pump should have a recirculation line sized for 3% of the rated pump flow to prevent overheating.

Firemains on naval ships are arranged to survive battle damage to the maximum practicable extent. On smaller ships, a single main should be installed approximately along the ship centerline to take advantage of ship structure for protection. When ship size is sufficient to provide effective separation, two or more mains should be provided and separated both horizontally and vertically as far as practicable. Mains should be connected at the ends to form a loop, and cross-connected at regular intervals to provide redundant flow paths. Segregation valves should be installed so the system can be operated in two or more independent zones, each supplied by one or more fire pumps, and to permit isolation of damaged segments of piping. Vital sprinkling systems should be arranged for supply from two different zones.

Remote control of fire pumps and key segregation valves is provided to permit rapid response in an emergency.

Firemains are sized to deliver 75% of the installed fire pump capacity to any combination of loads with any combination of pumps, and with all services receiving required pressure. A minimum of 70 psi is required at handheld hose nozzles.

In general, fireplugs should be placed so that any region of a ship can be reached with two 50-ft hoses from separate plugs. On ships larger than frigates, 100-ft hoses may be used to meet this requirement for all areas except machinery spaces and superstructure. For aircraft carrier flight decks, 150-ft hoses may be used. Fire-plug coverage, in addition to the above minimum, may be necessary to accommodate special requirements for magazines, electronics spaces, and steering gear rooms. Fireplugs are either 1½ or 2½ in. in size, depending on ship type and plug location.

Since U.S. Navy systems are continuously exposed to seawater, copper-nickel piping should be used to reduce corrosion and inhibit the growth of marine organisms.

c. **All applications.** For both U.S. Navy and commercial installations, at least one international shore connection to the fire main must be provided and installed in a location accessible from both sides of the ship. Adapters of a standard design must be provided to insure compatibility with emergency vessels and port facilities [23].

Firemain branches subject to freezing weather should have a locked-open stop valve near the weather boundary and a valved drain connection.

Piping must be designed for the shutoff pressure of the fire pumps, or each pump discharge must be fitted with a relief valve and the piping designed for the relief valve setting.

Each fireplug should be provided with at least one hose with nozzle attached. The hose connection should be directed horizontally or downward to minimize hose kinking.

4.8 Fire-Extinguishing Systems. To supplement the firemain, fixed fire-extinguishing systems are installed for specific fire hazards. These include carbon-dioxide systems for total flooding or local application, Halon total flooding, foam sprinkling, seawater sprinkling, and chemical suppression systems. All of these systems are not used on all ships; the selection and application depend on the requirements for a specific ship, supplemented by an analysis of fire risk factors unique to the ship. Such analysis should reflect the type and quantity of fuels, ignition sources, and personnel fire-fighting capabilities.

a. **Foam systems.** Foam is intended primarily for combustible liquids, and extinguishes a fire by floating on top of the liquid, thus isolating the liquid from a supply of oxygen. Fire-extinguishing foam for marine applications is made by mixing a foaming agent with seawater to form a solution. Solutions made with protein, alcohol, and synthetic foaming agents are expanded by mixing them with air as they are applied, forming a thick layer on top of the burning liquid. Foam solution made with aqueous film-forming foam (AFFF) foaming agent is not mixed with air, but forms a thin film which bonds chemically with the surface of the combustible liquid. For commercial applications, the type of foam must be selected based on the fire hazard. For U.S. Navy applications, AFFF is used exclusively. The concentration of foaming agent in seawater ranges from 3% to 10%, depending on the type of foaming agent.

A fixed foam-producing station consists of a concentrate tank, a branch from the firemain for seawater supply, a concentrate pump, and a proportioner that mixes foam concentrate with seawater in the correct concentration. A foam system for a tanker is illustrated by Fig. 27.

A single foam-producing station may be arranged to supply sprinkling systems in more than one area, as well as handheld hoses, by providing multiple discharge branches from the proportioner. Each branch should have a local or remotely controlled selector valve to direct the foam solution to the desired equipment.

Sprinkling systems are generally designed to apply foam solution at a rate of 0.16 gpm per square foot of protected area, but a higher or lower rate may be required for some applications. The quantity of concentrate must be sufficient to apply foam at the required rate for a minimum time, ranging from 3 to 10 minutes for commercial applications, and 4 to 25 minutes for U.S. Navy applications. For hoses the system should supply the design flow rating of the hose nozzle. The foam station should be sized to supply the largest sprinkling system zone and two hoses simultaneously.

Fixed foam systems are installed to protect tanker decks, boiler flats, pump rooms, machinery space bilges, and aircraft landing platforms and handling areas.

b. **Halon systems.** Halon fire-extinguishing agents are halogenated hydrocarbon compounds that extinguish fires through a chemical reaction that inhibits the combustion process. They exist as gases at atmospheric pressure and are effective at a relatively low concentration. Halon fixed-flooding systems consist of pressurized cylinders containing Halon and nitrogen (the propelling agent) at a pressure of 350 to 650 psi (at 70 F), distribution piping, nozzles, and an actuation system consisting of pull cables or CO_2 gas cartridges, which operate discharge valves at the cylinders.

Halon cylinders should be outside the space protected. A single bank of cylinders may be arranged to discharge into any of several spaces. The cylinders must contain sufficient Halon to develop a minimum concentration of 5% by volume at 50 F in the largest protected space. Following discharge, the concentration in the space should not exceed 7% by volume at 150 F; this limit is necessary to prevent the development of toxic compounds and avoid unnecessary personnel exposure to the extinguishing agent.

Distribution piping and nozzles should be sized to discharge the required quantity in no more than ten seconds [24]; the concentration must be developed rapidly to prevent the formation of toxic by-products, which can occur when Halon is exposed to high temperatures. Nozzles are generally located along the top of the protected space because the gas is five times more dense than air. The piping wall thickness and support must be suitable for the high back pressure and reaction forces developed during discharge.

Halon systems are typically installed for machinery spaces, fuel pump rooms, flammable-liquid storerooms, electrical equipment spaces, and cargo holds.

The Halon compounds most effective as fire-extinguishing agents are Halon 1301 and 1211; however, these compounds deplete stratospheric ozone, and their use is restricted.

c. Carbon-dioxide systems. Carbon dioxide is a gas at atmospheric pressure, which extinguishes fires by reducing the oxygen concentration of the air surrounding the fire. A carbon-dioxide concentration of approximately 35% by volume is required to be effective. Total flooding systems develop the required concentration in an entire space; direct application systems affect only the area surrounding a specific hazard.

The quantity of carbon dioxide required for a total-flooding system depends on the size and type of space protected. One pound of carbon dioxide is required per 15 to 22 ft³ of volume for machinery and similar spaces, depending on the size of the space. One pound per 30 ft³ is required for cargo holds.

For high-pressure total-flooding systems the extinguishing agent is stored in cylinders at approximately 850 psi and 70 F. Cylinder arrangements and controls are similar to Halon systems.

For low-pressure total-flooding systems, liquid carbon dioxide is stored in a tank at approximately 0 F and 300 psi, thereby requiring less space for the storage of the quantities required for cargo holds and similar large spaces, and permitting the use of lower-pressure distribution piping. A refrigeration system must be provided to maintain the low storage temperature.

Distribution piping and nozzles for total-flooding systems are arranged similar to Halon systems. Piping should be sized so that 85% of the required amount of gas is discharged in less than two minutes for machinery and similar spaces, and in five minutes for cargo tanks [25]. For cargo holds, two thirds of the required quantity must be discharged within ten minutes.

Carbon-dioxide total-flooding systems are typically used in the same applications as Halon for commercial ships, while Halon systems are preferred for U.S. Navy applications.

Direct-application carbon-dioxide systems consist of high-pressure carbon-dioxide cylinders, discharge piping, and nozzles arranged to direct the extinguishing agent at small, local hazards such as electrical equipment, cooking surfaces, or workbenches. The quantity of carbon dioxide should be sufficient for a 30-second discharge.

d. Seawater sprinkling systems. Seawater sprinkling systems are used to distribute seawater from the firemain to sprinkling nozzles. In addition to being highly effective for fire extinguishing, these systems provide cooling to prevent the spread of fire and protect the space contents. In wet sprinkling systems, each nozzle incorporates a means of keeping the nozzle closed until it is activated by a temperature sensor or other actuating signal; seawater pressure is continuously maintained at the sprinkler nozzles. In a dry sprinkling system the nozzles are open and the distribution piping is normally dry. Seawater is admitted by a manual or automatic control valve.

Sprinkling rates for U.S. Navy applications are generally 0.8 gpm/ft² for magazines, and 0.2 gpm/ft² for other spaces; other rates may apply in special cases. For commercial applications a minimum rate of 0.12 gpm/ft² should be used.

Sprinkler nozzles are installed in a grid supplied by a branch from the firemain. Nozzles closer to the firemain operate at a higher pressure than those farther along the grid. Grid piping should be sized to deliver the required minimum pressure to each sprinkler nozzle, to limit the flow variation between the first and last nozzles to less than 30%, and to limit the total flow to no more than 115% of the design value.

For U.S. Navy applications, wet sprinkling systems are used for selected high-risk munitions magazines. Dry systems are used in other magazines, munitions holds, incinerator rooms, cargo holds and handling areas, storerooms, high fire-risk living spaces, and other areas. In commercial applications, seawater sprinkling systems are used primarily for the protection of vehicle holds and ferry decks.

e. Chemical suppression systems. Chemical suppression systems discharge aqueous potassium carbonate or other highly alkaline solution that is stored in pressurized cylinders. The cylinders are connected by a control valve and piping to fixed nozzles for deep-fat fryers, galley hoods, and other small hazards.

4.9 Vents, Overflows, and Sounding Tubes. Vents, overflows, and sounding tubes are an important part of every system that moves fluid into or out of an enclosed space. The correct placement, arrangement, and sizing of these components is necessary to ensure proper system operation and to prevent personnel injury, structural damage, and spillage.

a. Vents. A tank or compartment vent, which is frequently called an "air escape," is a pipe that is open to the atmosphere to equalize pressure between the interior and the surroundings when the tank or compartment is being filled or emptied.

A vent must be provided for each tank or compartment

Table 11 Suggested minimum vent sizes for nonstructural tanks containing combustible fluids, in. nps

Tank Capacity, gal	5 psi Test Head	10 psi Test Head
10 to 20	$\frac{3}{4}$	$\frac{1}{2}$
20 to 50	1	$\frac{3}{4}$
50 to 100	$1\frac{1}{4}$	1
100 to 200	$1\frac{1}{2}$	$1\frac{1}{4}$
200 to 500	2	$1\frac{1}{2}$
500 to 1000	$2\frac{1}{2}$	2
1000 to 2000	3	$2\frac{1}{2}$

that has filling, suction, flooding, sprinkling, or sluicing arrangements, or is used for the stowage of bottled gas, unless the tank or compartment is always open to atmospheric pressure by other means. A vent should also be provided for each void that contains pressure piping to protect the compartment from overpressure in case of a leak.

Compartments that have carbon-dioxide or Halon flooding systems, or contain high-pressure air piping or flasks, should have a vent independent of ventilation ducts if the compartment openings and ducts may be closed tightly under any condition. However, a separate vent is not necessary if the test head of the compartment exceeds the pressure that would result from discharge of the high-pressure gas within the sealed compartment.

Vents should be sized to prevent an excessive differential pressure or air velocity greater than 25 fps when the tank or compartment is filled or emptied at the maximum rate. In general, vents from tanks having suction and pressure filling connections should be at least one-tenth the area of the suction or filling pipe, but not less than $1\frac{1}{2}$ in. nps for freshwater tanks, 2 in. nps for ballast tanks, and $2\frac{1}{2}$ in. nps for fuel tanks. A $1\frac{1}{2}$-in.-nps vent is considered satisfactory for a void that contains pressure piping, based on the probability of slow leakage and incomplete failure of the piping.

The vent for a nonstructural tank containing combustible fluids should be at least the size given in Table 11, unless equivalent venting is otherwise provided, to prevent excessive pressure when the tank is exposed to an external fire.

Vents from tanks containing the same fluid may be combined. If two or more vents are joined together or to a header, the size of the common vent or header should not be less than the total area of the vents that are connected to it and serve tanks that will be filled or emptied simultaneously. Combined vents from different tanks should be joined above the test head of the tanks, or means should be provided to isolate each tank for testing.

Tank vents should be connected to the highest point of a tank. Multiple vents should be installed if the shape of a tank prevents complete venting from one location. List and trim should be considered when selecting the number and location of tank vents.

Tank vents should be run with the maximum attainable rising pitch from the tank to the open end to facilitate drainage back to the tank served. Vents should terminate

in return (gooseneck) bends or vent closure fittings specifically designed for the purpose.

Vents from freshwater tanks, voids, magazines, and compartments should terminate within the ship, and lubricating-oil (other than synthetic lubricating oil) vents may terminate in the same space as the tank; otherwise, vents should terminate in the weather. Potable-water tank vents should not terminate in sanitary, medical, food service, or other spaces from which contamination or odors could be transmitted to the water. Vents from sewage tanks, and from tanks and compartments carrying flammable or toxic volatiles, should terminate outside the ship at least 10 ft from air ports, ventilation intakes, or other openings into the ship. Vents from water tanks should not terminate in spaces assigned primarily for electric equipment.

Vents that terminate within the ship should terminate in ventilated spaces.

For commercial applications, the height from the deck to the lowest point where water can flow into the ship through the vent must be at least 30 in. for vents terminating on the freeboard deck, and 18 in. for vents terminating on higher decks. A means of closing each vent should be provided; closure devices must be permanently attached and should not restrict the area of the vent when open. Each vent should be run entirely within the same watertight subdivision as the tank served.

Vents may be combined with overflows where practicable, provided the overflow is connected to the highest point of the tank, and no valve is fitted in the overflow.

Vents from sewage tanks, fuel tanks, and other tanks that may contain combustible vapors should have the open end of the return bend enlarged to $1\frac{1}{2}$ times the diameter of the vent and be fitted with flame screens, which should be removable for cleaning. The screens should be bronze, brass, or nickel-copper alloy, and should be installed so they cannot be easily painted or damaged.

Potable-water tank vents should have insect screens.

b. Overflows. An overflow pipe is generally necessary for each tank into which liquid is delivered under pressure, if the tank structure is not designed to withstand the shutoff head of the filling pump. An overflow is not required if the tank is protected by a relief valve; potable-water tanks and reserve-feed tanks may be equipped with relief valves instead of overflows to safeguard the purity of their contents. Sewage and gray-water tanks require overflows, even though they are filled by gravity, to prevent backup into the plumbing drain system. On U.S. Navy ships, ventilation ducting is often designed to serve as an overflow for compartments having sprinkling systems.

Overflow piping should be designed so that the combined static and dynamic head in the overflow pipe during the most critical overflow condition will not exceed the test head of the tank. In any case, the area of the overflow pipe should not be less than 125% of the area of the fill piping.

Overflow pipes for ballast, gray-water, and sewage tanks generally discharge overboard through the shell of the ship. Overflows for lubricating-oil storage tanks are generally led to a sump tank. Waste-oil tanks generally overflow to the oily-waste tank, which overflows overboard.

On U.S. Navy ships, fuel-tank overflows should generally discharge overboard through the shell. To minimize the number of shell penetrations and reduce the chance of spills, fuel storage tanks should overflow to fuel overflow tanks, which in turn overflow overboard. If an excessive length of piping is required to connect a storage tank to an overflow tank, the storage tank may discharge directly overboard. A fuel service tank may overflow to an overflow tank if the pipe can be high enough to prevent backflow from the overflow tank into the service tank. If this is impracticable, the service tank must discharge overboard independently. High-level alarms should be incorporated in the tank-level indicating systems of all tanks that overflow overboard.

On commercial ships, each fuel-tank overflow must have a spill container permanently installed under it. To comply with this requirement, fuel-tank overflows must terminate on a weather deck.

Overflows that discharge through the shell should generally be located as high as practicable and at least one deck height above the full-load waterline. Sewage and gray-water tank overflows should be lower than the lowest fixture served by the tank, if practicable. Each overflow pipe should have a check valve as close as possible to the shell and a return bend just upstream of the check valve. Where overflows are combined, a check valve should be installed in each overflow where it connects to the overflow header.

For commercial applications, if the overflow pipe does not extend at least 30 in. above the freeboard deck, two check valves should be installed at the shell, or a single check valve may be provided with a positive means of emergency closure operable from above the freeboard deck.

Overflow piping should be arranged to minimize tank-to-tank or tank-to-compartment flooding by any combination of damage to piping and tank boundaries. The number of penetrations of the shell, oiltight, and watertight structure should be minimized. Piping should be in protected locations, and outside machinery spaces where possible. The arrangement should ensure gravity drainage back to the tank, and prevent loss of liquids when the ship rolls up to 30 deg.

A standpipe system may be installed as an alternative to individual tank overflows for fuel and ballast tanks on commercial ships. A standpipe is an overflow pipe that is connected to the tank filling main at a point that cannot be isolated from the tanks when they are being filled and that terminates in the weather at a lower level than the tank vents. All design requirements for overflows apply to standpipes. Since the level of liquid in a standpipe is determined by the head available to produce flow into a tank, each tailpipe must be sized so that the pressure loss at the design filling rate does not cause overflow through the standpipe. A throttling valve should be provided in the fill line for flow regulation. An indication of the standpipe level should be provided at the fuel or ballast control station, enabling the operator to both maximize the filling rate and avoid an overflow.

c. Sounding tubes. All tanks should be provided with a means of determining the liquid level. Sounding tubes are generally installed for this purpose even when a tank liquid-level indicating system is provided. Sounding tubes should also be provided for innerbottom compartments, voids, cofferdams, and unmanned holds adjacent to the shell plating. Sounding tubes should not be installed in sewage tanks.

A sounding tube is not required where the shape and service of the tank permits dropping a sounding rod or tape in a straight line from the deck above to the deepest point of the tank, such as cargo tanks on tankers.

Sounding tubes should terminate in passageways or on open decks where practicable. If a sounding tube must terminate in a compartment, it should be readily accessible and located so it will not interfere with the function of the compartment. Sounding tubes should not terminate in locked compartments.

Sounding tubes should not terminate where they could discharge oil onto a hot surface, or any fluid onto electrical equipment, when the tank is being filled.

Sounding tubes should terminate higher than the top of the tank or compartment served. Sounding tubes for fuel tanks should terminate higher than the tank overflow. For innerbottom oil and water tanks, sounding tubes should terminate at a convenient height above the floor plates. Sounding tubes that terminate below the full-load waterline and serve a tank having the shell as a boundary should have spring-loaded gate or ball valves installed at the upper end.

On U.S. Navy ships, fuel tank sounding tubes may terminate in machinery spaces but should have a floating-ball check valve to prevent spillage if the tank is overfilled while sounding.

Sounding tubes for potable-water tanks should terminate with caps. Each tube should have its own sounding rod permanently stowed in the tube.

The lower end of each sounding tube should terminate as close as possible to the deepest point of the tank. Protective plates, extra-heavy tees, or similar protection should be fitted at the bottom of the tube to prevent damage to plating. Tubes should terminate close enough to the tank bottom to prevent the sounding device from leaving the tube.

Sounding tubes for tanks subject to high filling rates, such as fuel tanks, should not be combined with the tank vents. Sounding tubes should be perforated near the top of the tube, or have a cap designed so that the accumulated air pressure in the tube will be slowly released and equalized before the cap is completely unscrewed.

Sounding tubes should be as straight as possible. If curvature is unavoidable, the radius of curvature should not be less than 10 ft. Sounding tubes should not have reverse bends, and should not be less than 1½ in. nps.

4.10 Tanker Cargo Piping Systems. The design of tanker cargo piping systems is predicated on minimizing turnaround time at the unloading terminal, handling the required number of cargo grades, providing for safe handling of the combustible cargo, and preventing oil pollution. The applicable regulatory body requirements are extensive and vary with ship size and cargo flammability characteristics. To illustrate the principles involved, this discussion is primarily oriented to crude-oil tankers having segregated ballast.

a. Cargo-oil system. The cargo-oil system receives oil at the loading terminal, distributes it to the cargo tanks during loading, and discharges oil from the tanks to the terminal during unloading. The number of different grades of oil the ship must carry simultaneously should be determined early in the design since it has a major influence on the system complexity, having an effect on the number of cargo pumps, suction mains, tank cleaning headers, stripping eductors, and slop tanks. The system piping should be arranged to keep different grades of cargo segregated as they pass through the system. The system illustrated in Figs. 28, 29, and 30 is designed to carry two grades.

The total cargo pump capacity must be sufficient to discharge cargo in the required unloading time. The unloading time is a major consideration in the economics of the ship operation, and depends on the desired port turnaround time, the receiving capacity of the terminal, and the power available to operate the pumps. The unloading time is typically 12 to 14 hours for a large crude-oil carrier. At least one pump must be installed for each grade of cargo. Spare pumps are usually not provided.

Cargo pumps may be driven by steam turbines, diesel engines, or electric motors. The drivers may not be located in the pump room of a crude-oil carrier because of the potential for an explosion of cargo vapors; therefore, the pump drivers are located in a separate space and connected to the pumps, which are located in the pump room, by jackshafts passing through gastight bulkhead stuffing boxes. For tankers with central pump rooms located aft, the drivers are located in the main machinery space. This precaution is unnecessary on product tankers carrying cargoes having a flash point above 150 F.

Each pump should have a relief valve unless the piping system is designed for the full shutoff head of the pump. Each pump should have a suction strainer and suction and discharge pressure gages.

The suction mains run along the bottom of the ship from the cargo tanks to the pump room. Each main is connected to tailpipes in the tanks that are dedicated to a particular segregation. Each tailpipe has a stop valve to permit the selection of the tanks to be loaded or unloaded. These valves are also required by regulatory bodies to prevent discharge of cargo into the sea if the shell and piping are damaged. Each tailpipe should end in a bellmouth of sufficient diameter to permit the bottom of the bellmouth to be within $\frac{3}{4}$ in. of the tank bottom while providing a flow area of at least $1\frac{1}{2}$ times the tailpipe area. This is necessary to permit the removal of as much oil as possible, and to reduce the entrance of air into the pipe caused by vortices which lower the level of the oil surface near the tailpipe. Different types of bellmouths are described in reference 26. Two or more tailpipes may

be required in a single tank if the tank internal structure does not allow sufficiently rapid drainage of oil toward a single tailpipe inlet.

Each suction main is connected to cargo pumps that are dedicated to a particular segregation. Cross-connections with normally shut valves are provided between the mains in the pump room to permit any pump to take suction from any tank in case of a pump failure.

Each suction main should be sized for the full capacity of the pumps to which it is normally connected. Tank tailpipes are usually sized for unloading two or more tanks simultaneously; however, it may be desirable to be able to unload a single tank at full pump capacity. Suction piping should be sized so the tanks can be pumped down to the lowest practicable level at full cargo pump rating before the pump suction pressure decreases below the required net positive suction head at rated flow.

The cargo pumps discharge into mains leading from the pump room to port and starboard hose manifolds on the main deck. The manifolds terminate in flanged connections for hoses from the shore terminal.

The size of the discharge piping is based on the total pump head and the required minimum pressure at the deck manifold. The discharge piping design pressure must not be less than the cargo pump relief valve setting, or the shutoff head of the cargo pumps if relief valves are not fitted. It is desirable to keep system design pressure below 225 psi to avoid the more rigorous regulatory body requirements for higher pressures.

Drop lines from the deck piping to the suction mains are provided for loading. Drop lines should be the same size as the mains.

The cargo piping should be designed to permit the removal of cargo oil that remains in the piping system after unloading operations using the main cargo pumps have been completed. Cargo discharge mains and crude-oil wash headers should be pitched for gravity drainage to a cargo tank for removal by the stripping pump. Piping should be provided to drain the cargo pump suction piping and the discharge risers using the stripping pump. The stripping pump should be arranged to discharge this residual oil through a smaller separate line.

As an alternative to the system configuration described, particularly on product tankers, which often carry many grades of cargo, deepwell or submersible pumps may be used. Such pumps eliminate the need for a separate pump room and long runs of suction piping from the pump room to the tanks. In this arrangement, one deepwell or submersible pump may be installed in each cargo tank, or a pump may serve several tanks through short lengths of piping and valves. The latter capability is often provided in any event to permit a tank to be emptied by an adjacent pump if the one located in a particular tank fails. Deepwell and submersible pump installations are discussed in Chapter 14.

b. Tank-cleaning system. Cargo tanks of crude-oil carriers must be cleaned periodically to remove sediment, sludge, and waxy deposits from the tank bottom and internal structure which, if not removed, would accumulate and cause a significant reduction in the tank capacity and

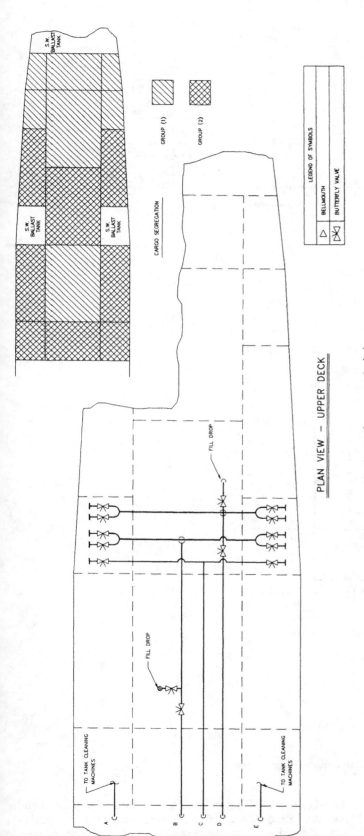

PLAN VIEW – UPPER DECK

Fig. 28   Cargo oil system—main deck

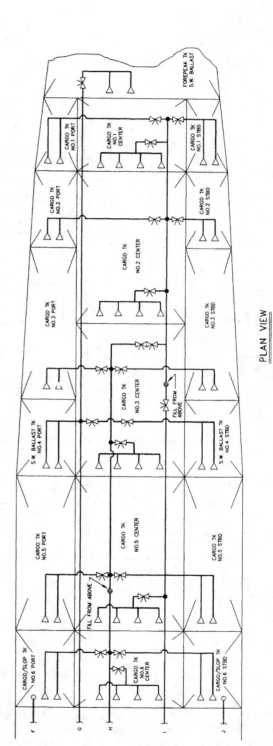

PLAN VIEW

Fig. 29   Cargo oil and ballast systems—in tanks

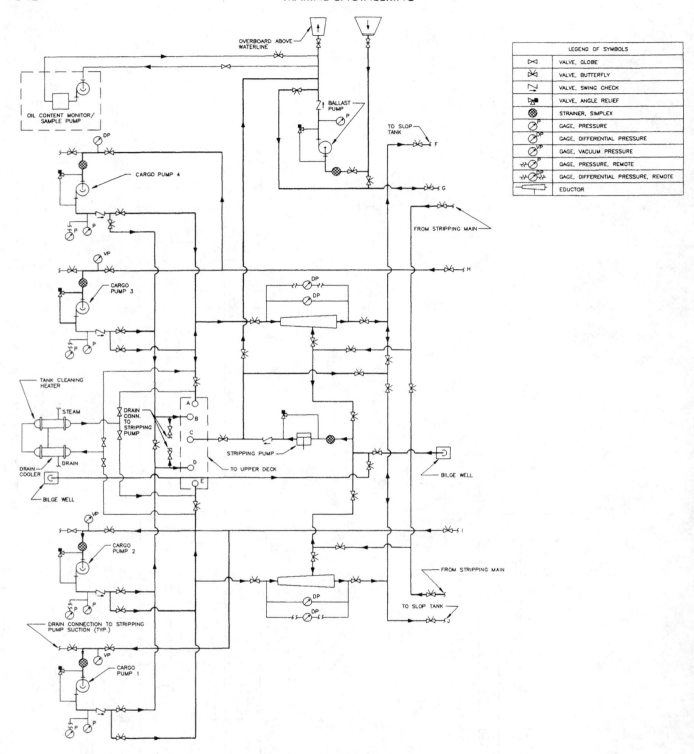

Fig. 30   Cargo oil and ballast systems—in pump room

impede the unloading of the cargo, among other adverse effects. During normal operations, cargo tanks are given a "crude-oil wash" as a part of the unloading procedure; that is, a small part of the cargo-pump discharge flow is diverted and sprayed against the cargo-tank structure to

dislodge much of the residue that would otherwise accumulate. Crude-oil washing not only reduces the accumulation of residue in the cargo tanks, which would otherwise be difficult to remove and properly dispose of, but it also permits a higher percentage of the cargo to be delivered.

As a normal maintenance procedure to prevent an accumulation of deposits, the tanks are washed with crude oil; however, washing with seawater is required to permit tanks to be gas-freed before they are entered for inspection or maintenance. A wash-water heater, which is described in Chapter 16, is provided when the cargo carried generates deposits which cannot be removed with cold water. For product carriers, only a cold-water washing capability is needed.

Tank cleaning is accomplished by machines that may be either fixed or portable. Fixed machines are connected to permanent piping extending from the top of each cargo tank. Machines are also installed in the bottom of tanks when required by the structural configuration for complete coverage. Portable machines are connected to hoses and lowered into the tanks through deck openings provided for this purpose. The machines have rotating nozzles, which are powered by the operating fluid and direct a stream of water or crude oil in a programmed pattern to strike all internal surfaces.

Crude-oil washing is normally conducted during unloading. The cleaning machines are supplied by headers that are connected to the discharge of two or more of the cargo pumps. At least two cargo pumps should be usable for crude-oil washing in case one is inoperable. Each machine is supplied through a valved branch from one of the headers. The headers are necessary because the required operating pressure of the machines, typically 150 psi or more, is higher than the pressure existing in the deck discharge piping during unloading. A valve is provided in the pump discharge downstream of the tank-cleaning header connection so the pump can be throttled to deliver the required pressure. With this arrangement, only some of the cargo pumps need to be throttled to supply the cleaning machines, while the remaining pumps unload cargo at their full rated flow. The number of headers provided should be equal to the number of grades of cargo to be unloaded with simultaneous crude-oil washing. Each header should be sized for the maximum number of machines that will operate simultaneously.

The bottom of each cargo tank is cleaned after the cargo in it has been pumped out, while unloading of other cargo tanks continues. To effectively clean the tank bottom, the oil from the crude-oil washing machines must be removed as fast as it enters to keep the bottom of the tank free of oil. This is done with stripping eductors. The eductors, which are discussed in Chapter 14, are supplied with actuating oil by the cargo pumps, similar to the tank-cleaning machines. At least one eductor is provided for each cargo grade. The eductors take suction on the cargo tanks via the stripping tailpipes. Since the eductor head is insufficient to discharge into the cargo pump discharge mains, the eductors discharge into the slop tanks, from which the oil is removed by a cargo pump. The eductor piping is arranged so that the eductors can operate without mixing cargo grades. The eductor capacity should be at least 1.25 times the total capacity of the tank-cleaning machines required in simultaneous operation for bottom washing. Each eductor should have a differential-pressure gage

indicating its operating head so the operator can confirm its performance during the wash cycle.

For water-washing, connections from port and starboard sea chests should be provided so the cargo pumps can supply seawater to the fixed tank cleaning machines through the crude-oil washing headers. If a tank-cleaning heater is necessary, either double isolation valves or spectacle flanges should be installed for positive isolation of the heater when crude oil is in the piping. Piping is provided so that the wash water can be removed from the cargo tanks and discharged to the slop tanks by the stripping eductors or the stripping pumps. Hose connections are provided on the tank-cleaning headers to permit use of portable machines for spot-cleaning of areas not adequately covered by the fixed machines.

c. **Stripping system.** A stripping system is provided to remove the cargo that remains in the tanks after the main cargo pump piping begins to ingest air. The air enters the suction piping through vortices that form near the tailpipes. In addition, bubbles can form in the suction piping because the reduced pressure permits lighter components of the crude oil to vaporize. The tendency of the crude oil to form bubbles depends on its composition, and can be a severe design limitation for crude oils having components with low vapor pressures. Air and vapor bubbles entering the cargo pumps can cause a loss of suction and speed surges, which may damage the pumps. Therefore, good practice requires that the final stage of emptying the cargo tanks (the "stripping" operation) be accomplished using smaller-capacity "stripping" pumps instead of the main cargo pumps.

The stripping system should have separate, relatively small suction mains and tailpipes to each cargo tank. The stripping piping should be arranged to permit the stripping pumps to remove residual oil from piping and tanks following unloading, and discharge it to the deck manifolds. In addition, the stripping system is commonly designed to pump wash water from cargo tanks to the slop tanks, discharge oily waste from the slop tanks to the deck manifolds, discharge clean water from the slop tanks overboard via the oil-content monitoring system, and dewater the pump room in an emergency. The overboard discharge line of the stripping system should terminate above the waterline, and should have an automatically operated stop valve that is actuated by the oil-content monitor. The stripping pumps should also be arranged to pump oily waste from the pump room bilge to the slop tanks, since the discharge of such waste to the machinery-space oily-waste system could cause an ignition hazard in the machinery space. The stripping suction piping also serves the crude-oil wash stripping eductors.

Each slop tank should be provided with separate inlet and outlet connections to minimize turbulence, which disrupts the separation of oil from water. The inlet piping should be arranged to direct the flow horizontally at a low velocity.

A means of controlling the stripping pumps and also aligning valves to pump out a flooded pump room must be provided in an accessible location outside the pump room, or from above the freeboard deck in the pump-room casing.

Stripping pumps should be of the positive-displacement or reciprocating type because they must have high suction-lift capabilities. The most demanding service that determines the pump head rating is the discharge of liquids from the bottom of the cargo tanks to the deck discharge manifold.

The main cargo pumps can be designed to more completely remove the cargo from the tanks by using speed controls on the cargo pumps to reduce the pump flow as the tank level falls, and by using vacuum pumps to remove gas from the suction piping before the gas reaches the pumps. Vapor sensing and removal with cargo pump speed control can be combined in a fully automatic installation.

d. Ballast system. Ballast tanks and piping are completely segregated from the cargo-oil tanks and piping to eliminate any possibility of discharging oil overboard when deballasting. On product carriers, where a small amount of water mixed with the cargo can severely affect its value, segregated ballast is also necessary to avoid seawater contamination of the cargo. A typical ballast system piping diagram is shown in Figs. 29 and 30. The system serves ballast tanks in the cargo area plus the forepeak tank. A ballast pump is located in the pump room, and is arranged to take suction on either of two sea chests and discharge to the ballast tanks, or take suction on the ballast main and discharge overboard. A tailpipe is connected to the ballast main for each tank. The overboard discharge terminates above the waterline to permit visual monitoring. A bypass is provided around the pump to permit ballasting by gravity flow.

e. Oil-content monitoring system. Mixtures of oil and water that accumulate in the process of washing cargo tanks are collected in the slop tanks, where oil and water separate by gravity over time; this process is sometimes facilitated by heating coils in the tanks. Water that has an oil content below regulatory limits may then be discharged overboard provided the discharge is monitored to ensure the limit is not exceeded.

An oil-content monitoring system continuously analyzes fluid samples and determines the oil content. As shown in Fig. 30, sampling piping leads to the monitor from the stripping pump overboard discharge. In addition to determining the oil content, the monitor uses inputs of ship speed and overboard discharge flow rate to determine the total quantity of oil discharged overboard per nautical mile, and the cumulative total quantity discharged during the voyage. If any preset limit is exceeded, the system automatically shuts the overboard discharge valve. Details of oil-discharge limits are contained in reference 27.

f. Inert-gas system. An inert-gas system is used to provide an inert atmosphere in the cargo tanks. An inert atmosphere is of particular importance when the tanks are being water-washed or crude-oil washed. Static electricity generated by the washing jets could otherwise cause an explosion in the tanks. Exhaust gas may be taken from the propulsion boilers; however, an independent inert-gas generator, which produces gas containing no more than

5% oxygen by volume, may also be used. A scrubber is installed downstream of the takeoff from the boiler stacks to cool the gas and remove contaminants, particularly sulfur compounds, which would foul the cargo or accelerate corrosion of the piping and cargo tank structure.

The distribution system consists of blowers and piping, which deliver the gas to each cargo tank. Two blowers should be provided, with the combined capacity sufficient to supply a volume of gas equivalent to 125% of the combined capacity of all cargo pumps that will operate simultaneously. The blowers should be capable of maintaining a static pressure in the tanks of at least 4 in. of water during unloading. A stop valve should be provided upstream of the blowers and designed to close automatically upon blower failure. A branch from the blower suction terminates in the weather with a blank flange that can be removed for ventilating the tanks with fresh air to gas-free the tanks prior to entry. The distribution main extends across the top of the cargo tanks, with a valved branch to each tank.

Regulatory bodies require both a water seal and a check valve in the inert-gas main downstream of the blowers to prevent cargo vapors from entering the machinery space when the system is not operating. The water seal provides absolute tightness against leakage of vapor. The water seal should be served by a fill pipe from a continuously available seawater source, such as a cooling system. The firemain should not be used as the water source. An automatic valve actuated by the level of water in the seal should be provided to compensate for evaporation.

The crude-oil cargo tanks should have an individual or combined vent that incorporates a pressure-vacuum relief valve to isolate the tank from the atmosphere and prevent dilution of the inert gas.

Cargo tanks of product carriers should also be vented. The required arrangement of the vents depends on the flash point of the cargo carried [28].

g. Vapor recovery. Vapor-recovery systems prevent cargo vapors that are displaced from cargo tanks during loading from being discharged into the atmosphere. A vapor-recovery system consists of piping that returns vapor from the cargo tanks to the loading terminal. A branch from the top of each cargo tank is connected to a header, which leads to a deck connection adjacent to the cargo discharge manifold. A stop valve with a position indicator must be provided at the deck connection. The piping must be provided with condensate drains at low points. High- and low-pressure alarms should be provided for the vapor-recovery header.

Since the recovery of vapor requires the cargo tanks to be closed during loading, a ship designed for vapor recovery should have the following additional safety features:

- A remote cargo tank level indicating system operable without opening the tank.
- A cargo tank high-level alarm system.
- A cargo tank overfill indicating system, which is independent of the high-level alarm and is timed to allow the operator to prevent an overflow.
- Tank pressure-vacuum relief valves, which will open in the event of failure of the vapor-recovery system

and are of sufficient size to discharge a volume of vapor corresponding to 1.25 times the maximum cargo loading rate without causing pressure in the cargo tanks to exceed the design value.

References

1   "Pipe Piece Family Manufacturing," U.S. Department of Transportation, Maritime Administration, March 1982.

2   "Design for Zone Outfitting," U.S. Department of Transportation, Maritime Administration, Sept. 1983.

3   "General Specifications for Ships of the United States Navy," NAVSEA S9AA0-AA-SPN-010, Naval Sea Systems Command.

4   "General Specifications for T-Ships of the United States Navy," Naval Sea Systems Command.

5   *Standard Specifications for Merchant Ship Construction* (Three Volumes—Steam, Medium Speed Diesel, and Slow Speed Diesel Propulsion), U.S. Department of Transportation, Maritime Administration.

6   *Handbook on Sanitation of Vessel Construction*, United States Department of Health and Human Services, Public Health Service.

7   Igor J. Karassik et al, *Pump Handbook*, McGraw-Hill, New York, 1986.

8   F. L. LaQue, *Marine Corrosion*, Wiley, New York, 1975.

9   "Schedule of Piping, Valves, Fittings and Associated Piping Components for Naval Surface Ships," MIL-STD-777, Naval Sea Systems Command.

10   "Schedule for Pipes, Joints, Valves, Fittings and Symbols," MA Plan No. S48-26-2, Alt. 4, U.S. Department of Transportation, Maritime Administration, 1988.

11   "National Standard Plumbing Code," National Association of Plumbing-Heating-Cooling Contractors, Washington, D.C.

12   L. F. Moody, "Friction Factors for Pipe Flow," *Trans.* American Society of Mechanical Engineers, 1944.

13   I. E. Idelchik, *Handbook of Hydraulic Resistance*, Hemisphere Publishing, New York, 1986.

14   "Flow of Fluids Through Valves, Fittings, and Pipe," Crane Technical Paper No. 410, Crane Co. Inc., Chicago.

15   "Power Piping," American National Standards Institute/American Society of Mechanical Engineers Code B31.1, American Society of Mechanical Engineers, New York.

16   Reno C. King, *Piping Handbook*, McGraw-Hill, New York, 1973.

17   "Thermal Insulation Requirements for Machinery and Piping," MIL-STD-769.

18   E. F. Hewins and J. R. Reilly, "Condenser Scoop Design," *Trans.* SNAME, Vol. 48, 1940.

19   *Code of Federal Regulations*, Title 33-Navigation and Navigable Waters, Part 155.

20   *Design Practices and Criteria Manual for Surface Ship Fresh Water Systems, Chapter 532*, Naval Sea Systems Command.

21   "Surface Ship Distilling Plant Sizing Details," Design Data Sheet 531-1, Naval Sea Systems Command.

22   "Guide to Fixed Fire-Fighting Equipment Aboard Merchant Vessels," Navigation and Vessel Inspection Circular 6-72, United States Coast Guard, U.S. Department of Transportation.

23   *Code of Federal Regulations*, Title 46-Shipping, Part 162.

24   "Standard on HALON 1301 Fire Extinguishing Systems," ANSI/NFPA 12A, American National Standards Institute/National Fire Protection Association, Quincy, Mass.

25   "Standard on Carbon Dioxide Extinguishing Systems," ANSI/NFPA 12, American National Standards Institute/National Fire Protection Association, Quincy, Mass.

26   A. W. Feck and J. O. Sommerhalder, "Cargo Pumping in Modern Tankers and Bulk Carriers," *Marine Technology*, Vol. 4, No. 3, July 1967.

27   *Code of Federal Regulations*, Title 33-Navigation and Navigable Waters, Part 157.

28   *Code of Federal Regulations*, Title 46-Shipping, Part 32.

CHAPTER XXI

Heating, Ventilation, Air Conditioning, and Refrigeration

Joel L. Krinsky

Section 1
General

Introduction. The basic design considerations and requirements for the design of HVAC systems for marine applications are similar to those for shore-based installations, except that marine installations must cope with more demanding criteria and parameters. Some of these more demanding criteria are:

- Less space is available for HVAC system equipment, ducting, and components.
- A ship is mobile; therefore, solar loads can affect any compartment with a weather boundary above the waterline.
- A ship can undergo extreme weather condition variations in less than a day.
- An HVAC system installation must be capable of withstanding the corrosive effects of seawater and salt-laden air.
- An HVAC system installation must be capable of adequate and efficient operation under severe conditions of ship's movement.
- A marine HVAC system design must be capable of continuous operation with a high degree of reliability. A ship generally has little or no availability to supply sources while at sea, and must therefore carry its own spare parts and tools.
- Because the enclosed areas in a ship are generally smaller than their counterparts in a shoreside installation, and because the structure of a ship offers very little sound dampening, a marine HVAC system design and installation concept must minimize structureborne and airborne noise generated by the HVAC system components.
- The design of marine HVAC systems must prevent the intake of water into the ship during severe weather conditions.
- The location of ventilation weather openings (supply and exhaust) on a ship is critical due to the restricted space available, the need for adequate water protection, and the necessity of minimizing the possibility of short-circuiting exhaust air into the supply intakes, or blowing or drawing air across areas traversed by the passengers or crew.
- An HVAC system for a ship must be capable of satisfactory performance under a multitude of internal compartment conditions, load variations, and functional requirements. In addition to performing its design functions, a ship must also carry and include its own accommodations, storage areas, commissary and laundry areas, recreation areas, sanitary areas, generating and propulsion equipment, fuel supplies, etc.
- Smoke control and fire safety of shipboard systems are critical due to limited firefighting and egress capabilities.

In addition to the criteria listed above, HVAC systems designed for naval ships must also be capable of meeting the following requirements:

- Systems and equipments must be designed to withstand shock and vibration requirements that are based on battle conditions and ship movement considerations.
- Because of the complexity of the systems on a naval ship, and the numerous compartment functions, the availability of space for HVAC systems is even more limited than for commercial ships.
- Weight is normally more critical on a naval design, and therefore an HVAC system must be designed and arranged to minimize weight.
- The performance and reliability of naval HVAC systems are critical, since these systems serve functions vital to the ship's intended purpose and mission. Also, in wartime, a naval ship can be at sea for extended periods of time.
- Naval ships have compartments that contain a large amount of equipment, which produces a high heat load, but the equipment usage can vary greatly. The HVAC design must, therefore, be capable of coping with the large heat loads encountered, as well as the variations in heat loads.
- Because of the requirement for watch periods (generally four hours) on naval ships, careful consideration must be given to the selection of the amount and location of ventilation air to watch stations in those compartments having high heat loads (sensible and latent), to avoid heat stress on watch standers.
- An HVAC system must be designed to use standardized equipments and components and, insofar as practicable, standardized construction details and methods, to

allow for the expeditious replacement of parts in an emergency or wartime situation.

- Ship's damage control requirements must be reflected in the HVAC system design. Unlike general nonnaval marine designs, a naval ship must continue to operate and perform its function even after experiencing battle damage. Watertightness criteria for a naval ship are much more demanding, especially as they relate to the HVAC installation.

- During battle conditions, a reliable source of ship's power must be assured for the propulsion, electronics, and fighting functions; therefore, the ship's HVAC system must be designed so that all nonvital HVAC systems can be secured, leaving only compartments housing vital functions with continuing HVAC services. Because of this requirement, all individual HVAC systems must be segregated into classifications that enable their grouping on power distribution systems.

- Protection against the intake of water through ventilation weather openings is more critical for naval ships because course changes to avoid bad weather conditions may not be possible for a ship on a mission.

- A properly designed HVAC system is required to ensure adequate personnel protection in certain naval ships that have compartments in which toxic, noxious, or explosive vapors and fumes are produced.

- The provision of good air distribution within the spaces served on naval ships is more difficult, because of the complexity of the other ship systems, and is more critical because the fresh air supplied per person is normally less than that for commercial marine designs in order to minimize the air-conditioning load, and the resulting fuel consumption and electric power requirements.

- The ship's air balance must be satisfied during all phases of fan speeds during the cooling and heating seasons, and during the various ship's material condition settings. If the ship has Collective Protection System zones or fire zones, the air balance must be maintained for each zone.

- The sophisticated functions and equipment, characteristic of naval ships, as well as the passive defensive profile, result in the structureborne and airborne noise in certain areas being much more critical than on commercial ships.

It is the responsibility of the HVAC engineer working on naval ship designs to develop systems that will maintain the compartments served at, or within the environmental ranges required, as well as accommodate or achieve other special treatments or requirements that are necessary. This must be accomplished using the minimum of space, weight, and power, and must also provide adequate protection and safeguards to the ship and the ship's crew.

1.1 Heating and Ventilation Systems. Heating, and ventilation systems are necessary to the process of making the interior spaces of ships safe and habitable, and they are designed to collectively achieve that objective. However, they are considered separately here in order to fully understand the effect of each, individually, and then

they will be analyzed collectively to assess the composite effect.

Heating is that process by which the ambient temperature is raised by adding heat to the air, to achieve a higher air temperature. Shipboard heating is achieved by different means: reheaters are used in a central-station air-conditioning system with the cooling functions turned off; preheaters and reheaters are used in supply ventilation systems; and convection or radiant heating is used in spaces not equipped with either an air-conditioning or supply air ventilation system.

Ventilation is defined as the movement of air from the outside or weather into the ship, or from inside the ship to the outside. This is accomplished generally by a combination of supply fans, exhaust fans, and ductwork.

The type of ventilation system called for in any given instance varies according to the nature of the space and its ventilation requirements. For example, galleys, bakeries, and sculleries are ventilated primarily by mechanical exhausts through hoods placed over heat-producing equipment; large laundries are serviced by both mechanical supply and exhaust while small laundries are served by mechanical exhausts only. Special requirements also exist for auxiliary machinery spaces where mechanical supply ventilation is provided at a specified rate of change depending on the size and specific function of the space.

Ventilation systems also play an integral part in protecting naval ships from the effects of chemical, biological, and radiological warfare. The systems maintain a positive overpressure within the ship and, with the use of special filters ensure a clean air supply to spaces in the ships. The systems may be set up to maintain the protection during all ship operations or on an as-needed basis only.

A ventilation system serves the dual purpose of removing contaminants and heat generated in a space. Its ability to control temperature is limited in that it can only maintain an ambient temperature above that of the outside air temperature. The temperature rise, i.e., the difference between the temperature of the supply air and the space ambient, depends on the ventilation (heat removal) rate. For this reason, the temperature rise is one of several criteria for the design of ventilation systems. To maintain a space at an ambient temperature below that of the outside air, the supply air temperature must be below the required ambient; thus, some form of cooling must be provided to lower the temperature of the supply air.

The two basic types of ventilation systems are supply systems and exhaust systems. Ventilation supply systems are used to supply weather air to ventilated spaces (spaces that will not be air conditioned), and to provide replenishment air to air-conditioning recirculation systems. Supply air may be provided either with fans or by creating a pressure difference, from the outside atmosphere or from adjacent spaces. A typical supply system comprises a weather intake, preheater, fan, supply system ductwork, and supply terminals. Other components that may be used in supply systems include precooling coils, watertight closures, and reheaters. Naval ship systems may also include toxic gas dampers, smoke dampers, and blast shields. Ventilation exhaust systems are used to remove air from ventilated spaces and stale air from areas served by air-conditioning recirculation systems. The exhaust may be accomplished either with fans or by creating a pressure differential to the outside atmosphere or adjacent spaces. A typical exhaust system comprises an exhaust terminal or hood, exhaust system ductwork, fan, and weather opening. Air may be exhausted directly from the space ventilated or through spaces such as fan rooms, storerooms, or passageways. The air removed by an exhaust system is replaced by air from a supply system.

The exhaust systems are important in controlling the amount of heat and fumes that can collect within a ship. Exhaust systems are used extensively in spaces such as shops, laundries, galleys, sculleries, washrooms, waterclosets, flammable liquid storerooms, and machinery rooms. All of these spaces generate either a large amount of heat or fumes or both. Hoods, such as described in reference 1, are often used in the exhaust system to more effectively collect the heat and fumes.

In general it is most practical to use mechanical supply ventilation systems for heating. Those spaces not served by an air-conditioning recirculation system or a supply ventilation system are heated by the use of convection heaters. Reference 2 contains detailed requirements for merchant construction while references 3 and 4 contain requirements for naval construction.

1.2 Air-Conditioning Systems. Air-conditioning systems, in general, are more complicated than heating and ventilation systems because they require more complex machinery to accomplish the desired effect. The consideration of moisture, or the lack thereof, is also more critical as a design criterion for air-conditioning systems.

Air conditioning is defined as the process of treating air to simultaneously control its temperature, humidity, cleanliness, and distribution in order to meet the specific requirements of a given space or group of spaces (zone).

Air-conditioning systems are intended to provide a controlled environment that is satisfactory for personnel, equipment, processes, etc. In both naval and commercial shipping there are compelling reasons underlying the requirement for cooling interior shipboard spaces by air conditioning. Some of the reasons common to both naval and commercial ships, such as comfort and health, are fairly obvious; but there are other, less apparent reasons which are specific to only one or the other, such as the requirement for sufficient and precisely controlled cooling of ammunition in magazines aboard naval ships to prevent its deterioration. In commercial applications, especially aboard cruise ships, the intent is to provide passengers with the utmost in climatic comfort.

A typical air-conditioning recirculating system contains a fan to produce the airflow through the system, a cooling coil where the heat from the air is transferred to the cooling medium, and ductwork to distribute the air through duct terminals and to return the air to the cooling coil. Filters are usually provided in front of the cooling coil to prevent rapid fouling of the finned elements. Many recirculating systems contain heaters for use in the heating season or for use in maintaining a specific relative-humidity requirement.

On commercial ships the systems are intended to provide a comfortable environment for passengers and crew members. On naval ships the systems are intended to provide air conditioning to berthing, office, and control spaces, to protect electrical and electronic equipment from overheating, and to preserve ammunition stored in magazines.

The design temperatures generally are lower on commercial ships than on naval ships. This is because space and weight requirements for the same level of air-conditioning systems on board naval ships must be balanced with other military requirements.

The spaces served by a common system depend upon the usage of the spaces, periods of occupancy, occupancy density, and odor potential. In naval construction, damage control classification is a prime consideration. System capacities from 5000 to 8000 cfm are common, with extremes being approximately 1000 to 10,000 cfm.

Naval ships are air conditioned by central station reheat systems, which are referred to as "recirculation systems." The requirement for reheat zoning is the only basic variable. In general, where the net heating load is negative, such as an internal space with a year-round cooling load, reheating is not provided. An exception to this is where one or more of the spaces served by a system requires control of the relative humidity. Since the humidity controller can override the room thermostat, reheat is required to compensate for overcooling. Generally, spaces are zoned according to heating requirements. Large spaces, such as messrooms and "critical" spaces, may have individual reheaters or a separate system.

Air-conditioning systems installed on merchant ships are classified as: (a) those serving small spaces (passenger staterooms, crew quarters, offices, shops, etc.); and (b) those serving large spaces (dining rooms, lounges, messrooms, etc.) Figure 1 illustrates the "Class A" system typical for air-conditioning large (public) spaces. The outside and return (recirculated) air mixture is filtered and conditioned (preheated, cooled, dehumidified, and reheated as required) by the central-station equipment.

Air is recirculated in air-conditioning systems for the sake of economy. Automatic dampers (outside air, recirculation, and exhaust) in Fig. 1 modulate the supply of outside air commensurate with the refrigeration system capacity; 100% outside air is supplied unless room conditions cannot be maintained with the cooling valve fully open. Note that cooling is controlled by both a humidistat and room thermostat. The room thermostat controls both cooling (water) and reheater (steam) valves, which are sequenced to maintain a set room temperature. When the humidity exceeds the humidistat set point, it overrides the room thermostat and opens the cooling coil valve for additional dehumidification. Should this cause overcooling, the room thermostat opens the reheater valve to maintain the set temperature. When a single Class A system serves several spaces, each has its own reheater controlled by only a room thermostat.

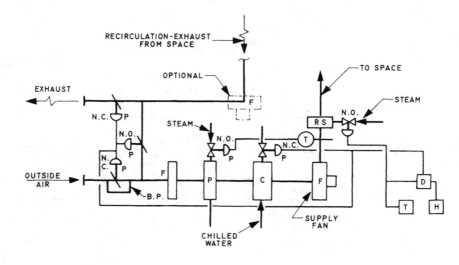

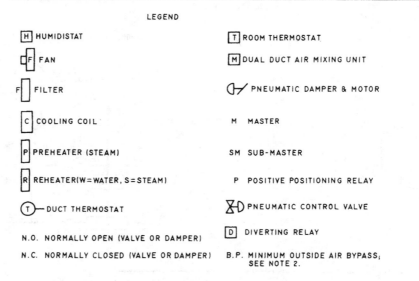

LEGEND

H	HUMIDISTAT	T	ROOM THERMOSTAT
F	FAN	M	DUAL DUCT AIR MIXING UNIT
F	FILTER		PNEUMATIC DAMPER & MOTOR
C	COOLING COIL	M	MASTER
P	PREHEATER (STEAM)	SM	SUB-MASTER
R	REHEATER(W=WATER, S=STEAM)	P	POSITIVE POSITIONING RELAY
T	DUCT THERMOSTAT		PNEUMATIC CONTROL VALVE
N.O.	NORMALLY OPEN (VALVE OR DAMPER)	D	DIVERTING RELAY
N.C.	NORMALLY CLOSED (VALVE OR DAMPER)	B.P.	MINIMUM OUTSIDE AIR BYPASS; SEE NOTE 2.

Fig. 1   Class A air-conditioning system with humidity control

A Class D (terminal reheat) system is shown in Fig. 2. Conditioned air is supplied to each space to accommodate the maximum design cooling load requirements. A mixture of outside and return air is filtered, conditioned (preheated or dehumidified and cooled as required) centrally, and distributed to individual reheaters at the spaces served. This system is used for most passenger staterooms, as well as for crew quarters. Note that automatic dampers (two-position) are controlled by two sensors (duct-stats) in the air intake: one for the heating cycle and the other for the cooling cycle. The preheater thermostat is set several degrees below the design chilled air (off-coil) temperature to prevent simultaneous operation of the preheater and cooling coil.

The Class E system, Fig. 3, is another system provided for small spaces requiring high-quality treatment. It is a primary air, secondary cooling and heating system; the central station equipment conditions only the primary air.

This primary air is distributed to induction units located in the various air-conditioned spaces, which induce room (secondary) air to flow through a water coil, an integral part of the induction unit. The induced air is either heated or cooled by this coil. No return (recirculated) air facilities are required. The primary air is sufficient for ventilation purposes and to balance exhaust-air requirements; it is conditioned to take care of the entire latent load (outside air plus room). Primary air is reheated, as necessary, to provide all heating for spaces when the temperature is above the changeover point with chilled (secondary) water being simultaneously distributed to the induction units. Below the changeover point, the refrigeration plant is inoperative; hot water is supplied to the induction units and primary air is preheated (tempered). The primary air is distributed at high velocity and pressure and thus requires relatively little duct space. However, this space saving is offset to some degree by the space required for

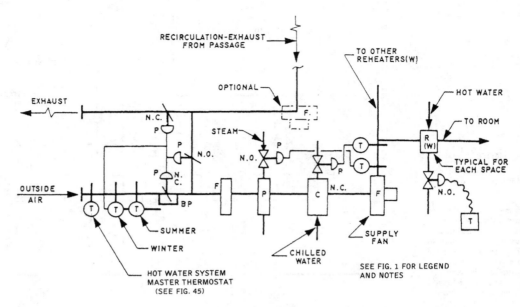

Fig. 2   Class D air-conditioning system

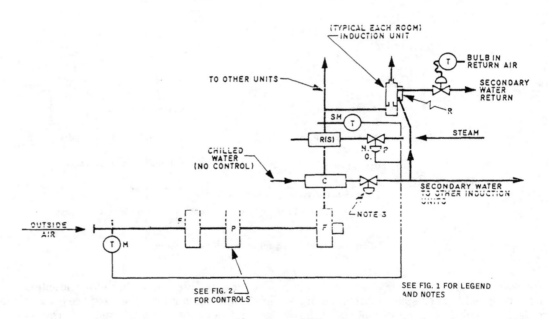

Fig. 3   Class E air-conditioning system

additional piping, secondary water pumps, and induction units.

The changeover temperature may be defined as the outside temperature that exists when the internal room sensible load equals the transmission loss plus the cooling effect of the tempered primary air. In central-station cooling systems, e.g., Classes A and D, outside air is available up to the full system capacity. Once the outside air falls below the design off-coil temperature, it can accomplish all of the necessary cooling, and the refrigeration equipment may be secured. Thus, the design off-coil temperature is also the changeover temperature. The changeover temperature of the Class E System, however, must be

somewhat below the primary air off-coil dry-bulb temperature (approximately 50 F) because the primary air provides only part of the sensible cooling load, i.e., about one third. Changeover temperatures usually fall between 30 and 40 F; 35 F is the most common.

Note that room thermostats are of the dual-pressure type since it is necessary to change the action from normally open, when the water is hot, to normally closed, when the water is cool.

Figure 4 illustrates a marine dual-duct or Class G system, which has become popular. The high-pressure fan of a central-station unit distributes conditioned air through two parallel systems of ducts (pipes) at a high velocity

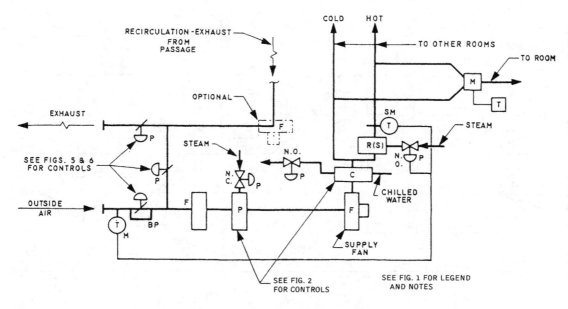

Fig. 4   Class G air-conditioning system

(approaching 5000 fpm). One duct system carries cold air and the other hot air. The air mixing units automatically proportion the hot and cold air to satisfy room loads. The series arrangement of the cooling coil and reheater shown gives excellent performance; it provides maximum dehumidification, which in turn produces low room dew points. Accordingly, the possibility of condensation on cold ducts and equipment and the necessity for perfect vapor sealing of the insulation are minimized. A parallel arrangement of the cooling coil and reheater, which is common ashore, has proven unsatisfactory.

1.3 Collective Protection System. Some naval ships are provided with a countermeasure system that is designed to protect personnel located within designated internal areas of the ship by ensuring a contaminant-free atmosphere. On U.S. Navy ships this system is called the Collective Protection System (CPS). On some foreign ships the system is called the Citadel System. These systems provide the ships the capability to operate in, and to survive, the effects of chemical and biological agent attacks and radioactive fallout [3].

The systems utilize high-pressure fans to pressurize designated areas or zones in the ship to a pressure between 2.0 and 2.5 in. of water above the ambient pressure outside the ship. Pressure-control valves are utilized to maintain the desired positive pressure. Air locks are used to allow personnel movement between pressurized and unpressurized zones as well as between pressurized zones. Hotel and control spaces are included in these zones. Filter banks are provided to filter out chemical and biological agents (aerosols and vapors) and radioactive fallout. Decontamination facilities are provided to permit personnel to enter the ship from a contaminated outside environment without contaminating the interior of the ship.

Shipboard machinery spaces receive a lesser degree of protection with systems that utilize fans that do not raise the internal pressure above that of the outside air; however, they have filter banks that filter out biological and chemical aerosols and radioactive fallout, but not vapors.

Ships employing the CPS or Citadel systems have relatively fewer ventilation systems and more air-conditioning systems than conventionally designed ships. Since all of the air entering the ship must be filtered to remove possible contaminants, the amount of outside air taken in is kept to a minimum.

1.4 Psychrometry. Psychrometry is the science that deals with air and water-vapor mixtures. Psychrometric data may be presented in both chart and tabular form. The psychrometric chart shown by Fig. 5 is commonly used in designing air-conditioning systems because it is easier to use than tables. The properties shown on a psychrometric chart are defined as follows:

• Dry-bulb temperature—the temperature recorded by a common thermometer, measured in degrees F. On the psychrometric chart (Fig. 5) these are vertical lines from the scale at the bottom of the chart.

• Wet-bulb temperature—the temperature recorded by a common thermometer with a moistened wick placed around the bulb and a stream of air passed over the wick, measured in degrees F. On the psychrometric chart these are sloping lines starting at the saturation line on the left of the chart and running obliquely down to the right.

• Dew-point temperature—the temperature at which condensation of moisture begins when the air is cooled. It is the saturation temperature corresponding to the vapor pressure and relative humidity and is expressed in degrees F. On the psychrometric chart this temperature is read on the saturation curve on the left side of the chart. At saturation, the dew-point temperature, dry-bulb temperature and wet-bulb temperature are the same.

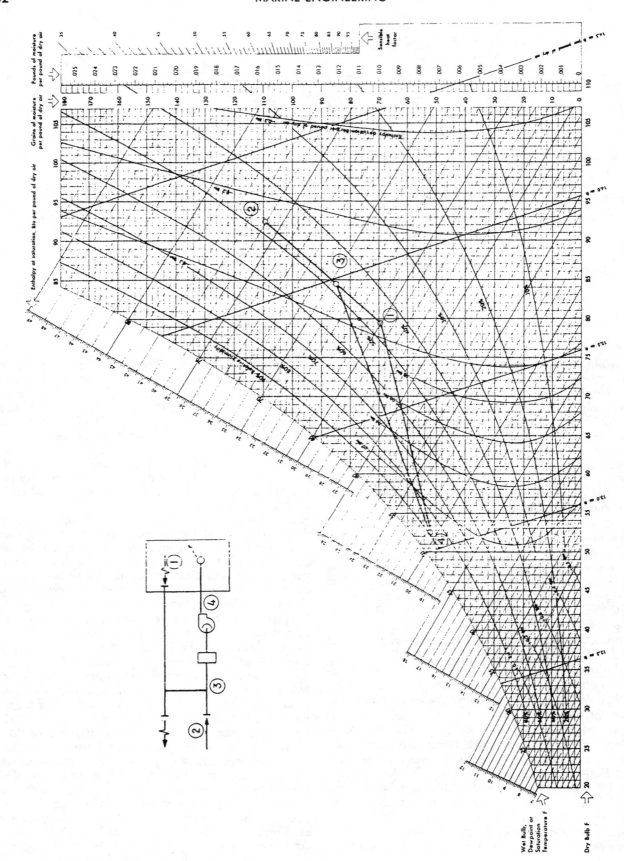

Fig. 5 Psychrometric chart and basic cycle

- Relative humidity—the ratio of the actual water-vapor pressure of the air to the saturated water-vapor pressure of the air at the same temperature, expressed as a percent. On the psychrometric chart these are curved lines from the top right to the bottom left of the chart. The saturation point is the 100% relative humidity line.

- Specific humidity—the weight of water vapor in grains or pounds of moisture per pound of dry air. On the psychrometric chart the specific humidity is indicated on horizontal lines with the scale on the right side of the chart.

- Enthalpy—a thermal property that indicates the quantity of heat in the air above an arbitrary datum, expressed in Btu per pound of dry air. The enthalpy values given on the psychrometric chart are for saturated air and are sufficiently accurate for most air-conditioning calculations. For greater accuracy more detailed data are available in reference 5. In the psychrometric chart the enthalpy values are on the left side on a scale corresponding to the wet-bulb temperature.

- Enthalpy deviation—an enthalpy deviation is caused by the air not being in the saturated state. An enthalpy correction is applied where extreme accuracy is required; however, for most air-conditioning estimates, it is neglected.

- Specific volume—the cubic feet of air and moisture mixture per pound of dry air. On the psychrometric chart it is shown on sloping lines from upper left to lower right.

- Sensible-heat factor—the ratio of sensible heat to total heat. On the psychrometric chart the scale is on the right-hand side of the chart.

Note that all units are expressed in terms of "per pound of dry air"; the reason is that the various properties of the moist air change during an air-conditioning cycle, and only the weight of dry air remains constant.

It may be seen that if any two psychrometric properties are known, the remaining properties can be determined. The following relationships can be observed from Fig. 5:

1. The enthalpy (or total heat content) is determined from the wet-bulb temperature alone, and vice versa.

2. The dew point depends only on the moisture content of the air (specific humidity), and vice versa.

3. Any air-conditioning process that does not add or extract moisture is represented by a horizontal line through the original condition point.

4. Any air-conditioning process that does not change the dry-bulb temperature is represented by a vertical line through the original condition point.

A simple air-conditioning cycle is indicated on Fig. 5. Points 1, 2, 3, and 4 represent the psychrometric conditions of room air, outside air, mixture of outside and recirculated air, and air leaving the cooling coil, respectively. Points 1 and 2 are established by design conditions, while point 3 is determined as follows:

$$H_3 = \frac{(Q_1/d_1)H_1 + (Q_2/d_2)H_2}{Q_1/d_1 + Q_2/d_2} \qquad (1)$$

where

Q = air quantity, cfm

d = air density, ft³ per lb of dry air

H = enthalpy, Btu per lb of dry air

Subscripts apply to the various points. A much simpler way of obtaining the conditions of the mixture at point 3 is commonly used, except where unusual accuracy is required. This is represented by the equation:

$$\frac{\text{Distance 1–3}}{\text{Distance 2–3}} = \frac{Q_2}{Q_1} \qquad (2)$$

When the dry-bulb temperatures are known, this may be expressed, and t_3 obtained, as follows:

$$\frac{t_3 - t_1}{t_2 - t_3} = \frac{Q_2}{Q_1} \qquad (3)$$

The total heat, H_T, removed by a cooling coil consists of the sum of the sensible-heat load, H_S, and the latent-heat load, H_L. The sensible-heat load is the transfer of heat which results in a change of temperature as read by an ordinary dry-bulb thermometer. The latent-heat load is the heat necessary to cause a change of state, e.g., steam to water. These quantities may be computed as follows:

$$H_S = 1.08 \, Q \, \Delta t \qquad (4)$$

$$H_L = 0.68 \, Q \, \Delta G \qquad (5)$$

where

Q = air volume entering coil, cfm

Δt = dry-bulb temperature reduction, deg F

ΔG = moisture removal, grain moisture/pound dry air

H_T can also be expressed as:

$$H_T = 4.45 \, Q \, \Delta H \qquad (6)$$

where ΔH is the enthalpy reduction of air passing through coil in Btu/lb dry air.

Figure 5 shows two sensible-heat factor slope lines. Line 1-4 is the "room slope," and line 3-4 represents the "coil slope." By definition, the sensible heat factor is:

$$SHF = \frac{H_S}{H_S + H_L} \text{ or } \frac{H_S}{H_T} \qquad (7)$$

To obtain the proper (simultaneous) balance between room sensible- and latent-heat removal, the air supplied must be so conditioned that it falls on the room slope line. Any point on this line is satisfactory. While the coil slope 3-4 is represented as a straight line, the actual path contour depends on several variables, one of which is the coil construction. In any case, the only significant point is the psychrometric conditions of the air actually leaving the coil, i.e., off-coil conditions.

Design off-coil conditions below those established by the room slope frequently are arbitrarily selected for systems with high sensible-heat factors. This is done to utilize larger terminal temperature differentials [Δt, equation (4)]; thus, less air is required, which in turn results in smaller systems.

Where a system serves several spaces, it may be neither possible nor practical to provide conditioned air to satisfy the lowest individual room slope. In such cases an arbitrary or average off-coil condition is selected for the system. Spaces having a lower sensible-heat factor, theoretically, will be overcooled if the air quantity is selected to meet the latent-heat load. Therefore, the supply air to such places may require reheating to compensate for the excess removal of sensible heat.

To obtain an optimum design and avoid excessive air quantities, some marine specifications allow a degree of latitude in the room design conditions of spaces with high latent loads. By slightly reducing the room design dry-bulb temperature and increasing the wet-bulb temperature (and relative humidity), the sensible heat of occupants is increased and the latent heat is reduced; thus, the sensible-heat factor is increased and the necessity of reheating may be avoided on occasions.

An example calculation using the psychrometric chart is included in Section 2.5.

1.5 Refrigeration. Refrigeration is a process of cooling or removing heat to lower the temperature of a space, or items to be stored in the space, to a temperature below the temperature of the surrounding area. This is accomplished by using a mechanical refrigeration process in which a refrigerant is used to alternately absorb heat from the space or object to be cooled and reject the heat to the atmosphere or seawater.

On naval ships, refrigeration is installed primarily for the preservation of food required for the crew. In commercial ships, the applications are more varied and range from preserving food for the passengers and crew to extending the useful life of perishables such as fruits, vegetables, flowers, and meats to permit their transfer to distant markets.

Both refrigeration and air conditioning are measured in tons of refrigeration. A ton of refrigeration is defined as the cooling effect of one ton of ice at 32 F that melts in 24 hours. Since it takes approximately 144 British thermal units (Btu) to melt a pound of ice at 32 F, 288,000 Btu's are required to melt a ton of ice in the same time. Therefore, one standard ton of refrigeration is defined as the transfer in a cooling operation of 288,000 Btu in 24 hours, or 12,000 Btu per hour.

a. The refrigeration process. Refrigeration is the process of cooling or removing heat. This is accomplished by transferring heat from the item or space being refrigerated to a colder medium, the refrigerant. The refrigeration process is used for both air conditioning and refrigerated-storage applications, the difference being that the temperatures involved are lower for the refrigerated-storage applications.

The compression cycle of refrigeration, which is illustrated by Fig. 6, is the most commonly used. In this type of cycle, a refrigerant vapor is compressed and, in the compression process, gains energy corresponding to the work of compression. The hot compressed vapor is then cooled by any convenient, inexpensive, plentiful medium such as water or atmospheric air. This cooling process condenses the hot vapor to a liquid. The high-pressure liquid is expanded to a lower pressure and becomes a cold mixture of liquid and vapor. This refrigerant mixture is

fed into a heat exchanger (evaporator) where it absorbs heat and changes back to a vapor, the same state as the beginning of the compression cycle.

The principal components of the compression refrigeration system shown by Fig. 6 are a receiver for storing liquid refrigerant, an expansion valve for controlling the flow of refrigerant (liquid and vapor), an evaporator where the required useful refrigeration is produced, a compressor that simultaneously maintains the evaporator (suction) pressure and increases the refrigerant vapor temperature and pressure, and a condenser that cools (removes both heat of compression and refrigeration load) and condenses the hot refrigerant vapor to its original liquid state.

Refrigeration is used mainly to extend the useful life of perishable products. It is used to provide storage compartments with temperatures ranging from −10 F to about 50 F and for self-contained items such as refrigerators, ice makers, drinking water coolers, and refrigerated salad bars.

b. Refrigeration systems. Refrigeration systems provide the means of controlling the environment in specific storage spaces. They are similar to HVAC systems in that they are concerned with maintaining a specific temperature, with moisture control, and with air distribution. The main difference between refrigeration and HVAC is only a matter of degree, i.e., the temperature that is maintained in the specific spaces. Refrigeration is concerned primarily with the preservation of perishable products by lowering the ambient temperature to a point that will prolong the useful life of the product.

All fresh fruits and vegetables, including flowers, are living things and are commonly referred to as "live" products. Even when separated from the tree, vine, or soil they continue to generate and dissipate heat as does the human body. The amount of heat respired (called heat of respiration) varies with the product and its temperature. In the respiration process, live products absorb oxygen and give off carbon dioxide to the surrounding atmosphere. In some cases carbon dioxide is injected into refrigeration spaces to slow the metabolism of horticultural products during transport, but not on U.S. ships.

Unripened fruits also give off other gases, which must be controlled. For instance, ripening bananas release ethylene and volatile esters. Ethylene accelerates ripening, during which starch in the pulp is converted to sugar. The color of the peel also changes during this process, which is of advantage because it provides a visual indication of the fruit's ripeness. Ventilation is required to control the concentration of the released gases.

Refrigeration (cooling) serves to maintain the individual product at the temperature at which it must be carried for proper preservation. Heating may be required to prevent freezing of many products, such as fruits and vegetables. For instance, bananas should not be exposed to a temperature below 55 F for any appreciable length of time.

A high humidity is required to prevent wilting of leafy vegetables, loss of moisture, and associated shriveling and loss of flavor. While low relative humidities (40 to 50%) are conducive to human comfort, humidities between

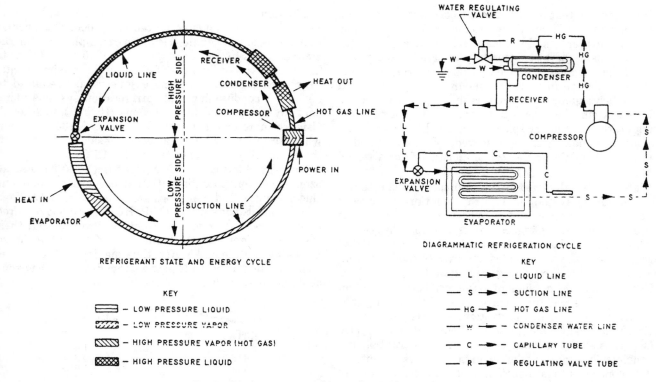

KEY

▱	– LOW PRESSURE LIQUID
▨	– LOW PRESSURE VAPOR
▧	– HIGH PRESSURE VAPOR (HOT GAS)
▨	– HIGH PRESSURE LIQUID

DIAGRAMMATIC REFRIGERATION CYCLE

KEY

— L ➤	– LIQUID LINE
— S ➤	– SUCTION LINE
— HG ➤	– HOT GAS LINE
— W ➤	– CONDENSER WATER LINE
— C ➤	– CAPILLARY TUBE
— R ➤	– REGULATING VALVE TUBE

Fig. 6 Graphic presentation of a mechanical refrigeration cycle

80 and 95%, as high as possible in many cases, are required in refrigerated spaces carrying fruits and vegetables.

Good air distribution is essential to the preservation of perishables. It provides uniform air conditions (temperature, humidity, and air purity) throughout the storage space. Thus, localized spoilage is prevented. Air distribution is not as critical for frozen products, referred to as "dead" products, as it is for most unfrozen (live) products.

The refrigeration systems aboard naval ships provide the needed cooling for refrigerated ships' stores, refrigerated cargo, air conditioning, process cooling water, and for drinking water cooling. In commercial cargo ships the refrigeration systems are used primarily for refrigerated cargo. The cargo space is usually divided into several compartments. The size of the compartments and the conditions maintained in the compartments (temperature, humidity, etc.) vary with the cargoes carried and their compatibility with respect to odors, method of packaging, susceptibility to damage, etc. Some cargo ships are designed to carry a single product, such as bananas or meat, and all of the cargo space then is designed to carry this single product.

Most refrigerated cargo can be shipped in refrigerated cargo containers. The majority of refrigerated cargo containers are carried on weather decks and use air-cooled condensers to dissipate the heat being dispersed from the refrigeration units. The containers stored below deck are usually provided with electric power and a source of cooling water and use water-cooled condensers. Some containers can use either fresh or salt water for cooling and automatically switch to air cooling upon the loss of water pressure.

Some cargo ships with below-deck storage for refrigerated containers have a connection to a ship's refrigeration system that provides refrigerated air to each container.

1.6 Refrigerants. The fluid employed as the heat absorber or cooling agent in any refrigerating process is called a refrigerant. A refrigerant can be easily changed from a liquid state to a vapor state and back to a liquid state. It should have a low boiling point, a freezing point lower than the intended operating point, and the capability to absorb and carry heat at a low temperature and reject heat upon condensing to a cooling medium such as water or air.

Research in the field of hydrocarbons and of the methane and ethane series determined that by substituting one or more halogen atoms (fluorine, chlorine, or bromine) for hydrogen atoms, almost any type of refrigerant needed for a specific use or for a specific type of equipment can be synthesized. This permitted the development of nonflammable, low-hazard refrigerants having various operating characteristics. These halogenated hydrocarbons are typically referred to as Freon (trademark of Dupont de Nemours and Co.), although they are available under a variety of trade names in addition to Freon. Refrigerants are identified by a standard designation system developed by the American Society of Heating, Refrigeration, and Air Conditioning Engineers. A listing of some available refrigerants is contained in reference 5.

Refrigerants that are used in a closed system and undergo a change of state (liquid to gas) are referred to as primary refrigerants. In some refrigerant applications, heat is transferred to a secondary coolant. A secondary coolant can be almost any liquid used to transmit heat without a change in state. Secondary coolants are used where a large quantity of refrigerant would be needed to fill the system, the spaces to be served are not near the refrigeration machinery space, there is a diversity of compartment temperatures to be served, or a high humidity is required to minimize the loss of moisture from a stored product.

Water is normally used as the secondary coolant for an air-conditioning system, and a brine solution is normally used in a refrigeration system. When a secondary coolant is used, the primary refrigerant is used to cool the secondary coolant, and the secondary coolant is then circulated through the space to be cooled and carries the heat from the space back to the primary refrigerant.

The design of the equipment in the refrigeration system is greatly influenced by the properties of the refrigerant selected to be used in the system. A refrigerant must satisfy many requirements, some of which do not directly relate to its ability to transfer heat. A refrigerant should be chemically stable under conditions of use, nontoxic in low concentrations, nonflammable, noncorrosive, and have no adverse environmental effects. Cost, availability, and compatibility with compressor lubricants and materials with which the equipment is constructed are other factors that must be considered.

Refrigerants may be toxic, flammable, or both. In reference 6, refrigerants are classified according to the hazard involved in their use. The Underwriters Laboratories have developed a classification system of comparative hazards to the life of refrigerant gases and vapors. By definition, the most hazardous are Group 1 gases or vapors which, in concentrations of about ½ to 1% for durations of exposure of about five minutes, are lethal or produce serious injury. Proper safety precautions must be taken based on the hazard of the refrigerant used. For naval construction, Section 516 of reference 7 requires a halocarbon monitoring system in spaces containing refrigeration machinery. Since the refrigerant gases are heavier than air, exhaust ventilation terminals in refrigeration machinery spaces are required to be located nine inches above the deck in the vicinity of refrigeration machinery plants.

In addition to volatility, heat transfer, stability, and toxicity, environmental effects must be considered when selecting a refrigerant. Chlorinated hydrocarbons (chlorofluorocarbons) comprising the vast majority of all refrigerants can migrate to the stratosphere once released from a refrigeration or air-conditioning system, where they destroy the earth's ozone layer, exposing the earth to higher levels of ultraviolet radiation.

To control the emission of chlorofluorocarbons (CFCs) and reverse the damage to the ozone layer, over 50 nations have signed and ratified an international treaty known as "The Montreal Protocol on Substances That Deplete The Ozone Layer." This treaty imposed strict limits on the production of fully chlorinated refrigerants with a complete phase-out mandated by the year 2000. This production phase-out schedule has been re-emphasized in the United States by the passage of the 1990 amendment to The Clean Air Act and is enforced by the Environmental Protection Agency. Systems engineers are required to design air-conditioning and refrigeration systems using partially chlorinated hydrocarbons (HCFCs) such as R-134a, or more exotic chemicals, such as E-134, an ether, which traditionally have not been considered because of the availability of the less expensive CFCs.

Partially chlorinated refrigerants (HCFCs) have small ozone-depletion potentials (ODPs) because of their one chlorine atom, and nonchlorinated fluorocarbons and exotic chemicals have a zero ODP. HCFCs are not required to be phased out under the Montreal Protocol, but their use as a refrigerant is prohibited as of 2030 in the United States by the Clean Air Act, thus making HCFCs an interim solution at best for new system designs. The ozone depletion potentials (ODPs) for CFCs are:

Refrigerant	ODP (relative to R-11)
R-11	1.0
R-12	1.0
R-113	0.8
R-114	1.0
R-115	0.6

The primary use of refrigerants on ships is for air conditioning and refrigerated storage. Other applications are for equipment such as water coolers, ice-cream makers, and soft-drink machines. Refrigerants such as R-113 are also used as cleaning solvents and flushing agents for piping systems. Refrigerants R-11, R-12, R-22, and R-114 are commonly used in naval shipboard air-conditioning and refrigeration systems. R-12 and R-22 are used in reciprocating compressors for both refrigeration and air-conditioning systems, while R-11 and R-114 are used in centrifugal compressors for air-conditioning plants. Commercial ships use a much wider variety of refrigerants. Refrigerants approved for shipboard use are contained in references 8, 9, and 10.

Additional information concerning refrigerants and refrigerant properties may be found in references 5, 10, and 11.

1.7 Control Systems. Shipboard air-conditioning, refrigeration, and ventilation control systems are designed to monitor and automatically control the various sensing devices, alarms, gages, control valves, electronics, pneumatics, and other devices necessary to operate the HVAC and refrigeration systems. Control systems operate by comparing the actual value of the controlled variable against the design value. They then adjust the flow of air, water, steam, refrigerant, or electricity as necessary to approach the design value.

Automatic controls fall into three categories: pneumatic, self-contained, and electric. Pneumatic control systems are the most common type used on merchant construction while electric controls are more common on naval construction.

In merchant construction, where pneumatic controls are used for any other control service, e.g., boiler combustion and hold dehumidification, it is easy to extend the compressed-air system and use this type of control for air conditioning and heating. Air from ship service systems is not always suitable for this purpose because of its oil and moisture content. This is particularly troublesome on tankers and bulk carriers where there are long runs of weather-exposed piping. The best practice, particularly for passenger ships of appreciable size, is to provide one or more independent compressed-air systems cross-connected to the ship service system for emergency operation only.

Primary compressed air (usually between 80 and 100 psig), after drying and cleaning, is reduced in two stages to the 15 to 19 psig required. Heating-cooling thermostats require dual-pressure compressed air; the lower pressure is for heating and the higher one for cooling. Switching of pressures is accomplished automatically or manually.

Self-contained controls are used only on merchant construction where there are few controls or where suitable compressed air is not readily available. On naval vessels, they are used for controlling steam coils, e.g., preheaters, reheaters, and combination heaters. Self-contained controls are limited as to sensitivity, adjustability, and throttling range. They are not suitable where the available pressure drop is small, i.e., single-pipe heating systems, or for systems requiring sequenced operation.

Electric controls are used primarily for simple applications where compressed air is not available under all circumstances, and where suitable self-contained devices are unavailable. A typical example is the control of dampers at weather terminals serving emergency diesel generator radiator cooling. On naval construction, special dual-temperature thermostats are often used to control chilled water coils and reheaters (steam and electric). Sensing elements for water coils, either temperature or humidity-actuated, control ON-OFF switches. Humidity control is provided only for critical spaces.

The most important advance in HVAC control systems has been the introduction of computer technology. Today, computers and microprocessors provide overall supervision and management of shipboard HVAC and refrigeration control systems. Some of the principal advantages of computer-operated control systems are:

- Central operation—all HVAC equipment is controllable from one central point
- Safety
- Increased accuracy
- Greater economy
- Plant performance monitoring

Digital control systems can control the variables to close tolerances in order to maintain a constant ambient condition. On passenger ships, this means a constant comfort condition. On cargo ships the objective is a constant temperature/humidity condition, which is optimum for the specific cargo, and provided at minimum initial and operating costs.

The basic digital control system comprises three types of components: sensors which sense the value of the controlled variable, controllers which compare the sensed value with the desired value or set point and generate an output signal to the controlled device, and controlled devices which regulate the controlled variables. The digital control system onboard a ship is made up of numerous control loops designed to control the many elements that make up the HVAC system.

Sensors are used to measure temperature, humidity, pressure, and fluid flows. For each type of sensor there is a variety of sensing elements tailored for use in specific applications. Thermocouples, thermistors, pressure transducers, and pitot tubes are examples of sensors.

The digital controller uses a microprocessor or minicomputer. The microprocessor takes signals from the sensing elements, compares them with the set point stored in memory, and generates the proper control signal to the controlled device. Controlled devices regulate the flow of water, steam, electric current, or air to meet the set points established for the various controlled variables in an air-conditioning or refrigeration system. Valves and dampers are examples of controlled devices used in HVAC systems.

The hardware of a digital control system is located at a control point such as an engineer's operating station, with additional remote units at damage-control stations and the pilothouse. A digital control system provides a high level of accuracy, increased efficiency, and central control of all elements of the HVAC systems. It can also provide alarm capabilities as well as maintenance requirements and overall system analysis. Additional flexibility is also available for controlling the HVAC system during a fire or damage-control condition.

A more detailed discussion of control components and systems is contained in reference 4. Specifications for control components for naval construction are contained in reference 7. Navy standard control, alarms, and gages may be found in reference 12. Requirements for controls on merchant construction are contained in reference 2.

1.8 Insulation. Insulation materials used onboard ships should have the following characteristics:

- High insulation value
- Light weight
- Flexibility and resilience
- Fire resistance with nontoxic combustion by-products
- Odorless and not susceptible to mold
- Reasonable price
- Durability

Insulation materials for merchant ships must meet requirements established by the U.S. Coast Guard [13] and the U.S. Maritime Administration [2]. Materials used for naval construction must conform to the applicable military specifications as delineated in reference 7.

Insulation is used in HVAC systems primarily for thermal, fire, and acoustic purposes, and is applied on both compartment bulkheads and system ductwork.

a. Thermal insulation. Thermal insulation is a heat-resisting material that is applied to the underside of decks, on bulkheads, on partitions, and to HVAC ductwork. Thermal insulation is applied to surfaces to reduce heat gain or loss, control surface temperatures for personnel protection and comfort, provide fire protection, and prevent condensation. The selection of the proper type and thickness of thermal insulation is important in providing the best results in the most economical manner.

The most widely used thermal insulation is fibrous glass that meets the requirements of reference 14. It is installed on the exterior surface of the ductwork.

Requirements for the insulation treatment of compartments and ductwork on naval ships are contained in reference 7. Insulation criteria for merchant ships may be found in references 2 and 15.

Because of weight considerations, the insulation treatment on naval ships is somewhat less than it is on merchant ships. This applies primarily to the basic thickness of the insulation used for comparable applications.

Another difference is that on merchant ships, ceilings and linings are fitted in "hotel" areas to conceal structure as well as insulation, piping, ducts, and other services. Therefore, the insulation can be less durable with respect to service abuse, and does not require a surface finish for protection or decorative reasons. Also the resulting dead-air spaces have an insulating effect. Another significant difference is that on merchant ships, metal sheathing is fitted to protect insulation that is not concealed by linings and ceilings (joiner work). The high conductivity of this sheathing appreciably reduces the effectiveness of the insulation treatment unless adequate means are provided to thermally isolate it from the supporting ship structure.

The control of condensation is an important consideration in determining the thickness of insulation to be applied. Sufficient insulation must be provided to maintain the temperature of the warm side surface above the dew-point temperature. Moisture in the insulation and condensation forming on surfaces that are below the dew point negate the effectiveness of the insulation and can cause deterioration. To prevent moisture from condensing within the insulation material, a vapor barrier should be provided. Further discussion of the control of condensation may be found in references 5, 11, and 15.

b. Acoustical insulation. Acoustical insulation may be divided into two categories, acoustic absorptive treatment and acoustic transmission-loss treatment. Acoustic absorptive treatment is installed in ductwork to absorb excessive airborne noise caused by fans, mechanical vibration, and air movement. The most widely used materials for acoustic absorptive treatment of ductwork include:

- Faced fibrous glass board with perforations in the face, which meets the requirements of reference 14 or 16.
- Unfaced fibrous glass blanket, which meets the requirements of reference 17 and is installed around perforated aluminum duct.

Both types of fibrous glass treatments absorb noise within the duct and reduce the transmission of both noise and heat through the duct walls.

Acoustic transmission-loss treatment is applied to ductwork to attenuate the direct transmission of noise through the duct walls. Acoustic transmission-loss treatment consists of a mass layer isolated from the duct surface by a resilient material such as thermal insulation. The mass layer usually consists of metal sheathing or a flexible material such as barium-sulfate-loaded vinyl. Closed-cell polyimide foam is the preferred material for the resilient layer; however, a unicellular plastic material conforming to the requirements of reference 18 is also widely used.

An open-cell polyamide foam that meets the requirements of reference 19 can satisfy both thermal and acoustic absorptive insulation requirements. When applied with a fire-resistant adhesive that meets the requirements of reference 20, open-cell polyamide foam does not release the thick, black, toxic smoke that unicellular plastic does when it burns; therefore, open-cell polyamide foam is preferred in this respect.

When both thermal and acoustical treatments are required, only one type of insulation need be installed provided it meets both requirements.

For both merchant and naval construction, usually one inch of insulation is installed on air-conditioning system ductwork as well as a vapor barrier or vapor seal to prevent the condensation of vapor in the insulation. For naval construction, the vapor barrier consists of two or three coats of a coating as specified by reference 21.

The thickness of insulation on compartment bulkheads varies with the location of the bulkhead and the temperature on both sides of the bulkhead. The area of application, extent of coverage, and thickness of insulation for naval construction is contained in Section 635 of reference 7.

The thermal insulation used for refrigerated spaces on naval construction is polyurethane foam [22]. The insulation may be poured or froth foamed in place, preformed blocks, or prefabricated webbed panels consisting of sheathing and insulation.

For bulkheads, stiffeners, deep webs, girders, and overheads, the insulation density should be a nominal 2 lb/ft³. For decks, the density should be a nominal 4 lb/ft³, except where prefabricated webbed panels are used in the deck areas; in this case the insulation density may be 2 lb/ft³.

Specifications for the insulation materials used for merchant construction may be found in reference 2, and for naval construction in reference 7.

1.9 Noise Control. The purpose of an HVAC system is to provide an acceptable environment for the activities to be performed in specified spaces. A proper acoustical environment is important to meeting that goal. While the noise generated by the HVAC components is only a part of the noise generated within the ship, all sources of noise should be evaluated and controlled, as appropriate, to provide an acceptable environment. On merchant ships acoustic and vibration control serves mainly as a means for comfort and to prevent damage to equipment. On naval and oceanographic ships, however, vibration and noise must be controlled for additional and more vital reasons.

These include the elimination of interference with detecting and measuring devices, the prevention of detection by the enemy, the prevention of deafness, and the elimination of interference with aural communications. In the design of an HVAC system, this can be accomplished by proper equipment layout, by maintaining low air velocities in ductwork, by proper balancing of airflows, by incorporating acoustic treatment, and by using dampening devices, such as resilient mounts for fans.

The two primary sources of noise in HVAC systems are the fan and the high-velocity airflow through duct and duct system components. Fan noise is carried downstream through the ducts. One way to minimize fan noise is to select a fan with low noise characteristics, and operate it as close to its point of maximum efficiency as possible. The peak efficiency point of operation is also the minimum noise point. To attenuate noise generated by a fan, the attached ductwork may have to be acoustically treated (insulated). Of course, the use of acoustical treatment should be carefully considered because straight duct and long-radius, 90-deg bends or elbows have natural sound-attenuating qualities. Straight duct will not attenuate sound effectively in high-frequency ranges because the noise tends to propagate along the duct. However, a series of 30-deg bends and turns, although a disadvantage in terms of system pressure loss, will attenuate noise. The noise attenuation and pressure loss in HVAC system ductwork are closely related; both are associated with interruptions in uniform airflow. The required amount of acoustical treatment is determined on a case-by-case basis, as dictated by the noise level requirements.

The noise generated by the flow of air increases with the air velocity. However, obstructions inside ducts and system components, such as heaters and cooling coils, are also sources of noise, as are abrupt contractions and expansions.

To avoid noise generation as well as pressure loss, there are general design practices that should be followed. Ideally, duct system components such as elbows, fans, turns, and takeoffs should be separated by five to ten duct diameters. This separation would ensure the straightening of turbulent flow before it enters the next system component. However, this is seldom possible on a ship. All transitions should be as smooth and even as possible with long approaches. Abrupt changes of direction, such as sharp bends and takeoffs, should be avoided. Any obstructions in the duct should also be avoided, and fittings should be smooth. The duct sections should be kept round and as large as possible. Rectangular duct is less preferable because it increases turbulence and is less rigid. The airflow approaching terminals should be as uniform and as slow as possible.

Additional information concerning the fundamentals and control of noise can be found in references 4, 5, and 23. Target noise limits can be found in references 2, 7, and 24. Reference 25, a report for predicting noise levels caused by ventilation and air-conditioning systems, may be used as a guide during system design.

Section 2
The Design Process

2.1 Design Considerations. Heating, ventilation, air-conditioning, and refrigeration systems are designed to meet predetermined criteria with regard to temperature, humidity, etc., under preset maximum and minimum conditions. Calculations are performed to determine the heat transfer into or out of compartments or spaces in order to establish heating and cooling loads, which are then used to design the required systems and to select the necessary equipment needed to meet the design criteria.

Before the HVAC calculations can be made, specific data must be known, including:

- Outside air design dry-bulb temperature for both the cooling and heating seasons.
- Outside air design wet-bulb temperature for the cooling season.
- Room design dry-bulb temperature for both the cooling and heating seasons.
- Room design wet-bulb temperature for the cooling season.
- Design seawater temperature for both the cooling and heating seasons.
- Physical dimensions for each room.
- Boundary constructions for each room.

- Lights in each room.
- Equipment in each room.
- Occupancy of each room.
- Room ventilation requirements.

Most of the above design data are listed in, or can be determined from, the information contained in the ship's specifications, general arrangement plans, joiner arrangement plans, structural plans, lighting arrangement plans, and equipment lists; however, where specific information is not available, the design criteria recommended in references 4 and 26 may be used.

For the purpose of determining the heat gains (cooling load) of a compartment for the cooling season, heat may be added directly to a compartment from a number of sources, including:

- Through compartment boundaries from surrounding areas, weather air, seawater, or solar radiation (sensible heat).
- From lights within the compartment (sensible heat).
- From personnel within the compartment (sensible and latent heat).
- From equipment within the compartment (sensible and latent heat).

Table 1   Air-conditioning, ventilation, and heating design temperatures

	NAVY	MARITIME	
COOLING SEASON			
Weather air, dry bulb, °F	90	95	
Weather air, wet bulb, °F	81	82	
Inside air, dry bulb, °F	80	78	
Inside air, wet bulb, °F	— (55% RH)	65 (50% RH)	
Seawater, °F	85	85	
HEATING SEASON			
Weather air, dry bulb, °F	10	0	
Inside air, dry bulb, °F	65/70	70	
Seawater, °F	28	28	
SOLAR		SINGLE	MULTI
Horizontal steel deck, °F	140	145	130
Vertical steel boundaries, °F	120	125	115

Table 2   Assumed space lighting loads

Space	Load Constant, Btu/hr-ft²
Passenger staterooms	7
Captain and chief engineer staterooms	7
Officer staterooms	4
Crew staterooms	4
Messrooms, lounges and public spaces	9
Offices	7
Other spaces	7

Table 3   Lighting heating load conversion

Lighting requirement, foot-candles	3	7	14	21	28	42
Incandescent light heating load, Btu/hr-ft²	2.5	4.3	8.2	11.0	13.6	—
Fluorescent light heating load, Btu/hr-ft²	—	2.8	5.0	7.0	8.9	15.0

NOTE: Btu loads are based on the actual maintained lighting levels.

- From cargo within the compartment (sensible and latent heat).
- From processes performed within the compartment (sensible and latent heat).
- From open liquid surfaces within the compartment (sensible and latent heat).
- From various systems located within or passing through the compartment (sensible and latent heat).
- From air taken into the compartment from surrounding areas or the weather (sensible and latent heat).

In determining the heat losses (heating load) of a compartment for the heating season, heat may be lost from a compartment through compartment boundaries to surrounding areas, weather air, or seawater (sensible heat). In addition, there can be a heat loss resulting from air being taken into (and lost from) the compartment from surrounding areas, or the weather (sensible and latent heat).

2.2 Ventilation Criteria. The amount of ventilation air required for a compartment can be based on a specified number of air changes per hour, the number of occupants in the space, or by a limiting temperature rise over the design weather temperature. These factors may be either used separately or in conjunction with each other, with the factor yielding the highest quantity being the determinant.

Ventilation load calculations are based on outside temperatures of 95 F in summer and 0 F in winter for merchant ships and on outside temperatures of 90 F in summer and 10 F in winter for U.S. Navy ships. Air changes in ventilated spaces are based on the gross volume of the space without deductions for any furnishings. The minimum quantity of supply air provided to a space is 35 cfm (previously 75 cfm on Navy ships).

Recommended design criteria for non-air-conditioned spaces on merchant ships may be found in references 9 and 26 and for Navy ships in reference 27. For both ventilation and air-conditioning systems it is necessary to determine the load requirements for each space. A space load is the sum of the loads resulting from transmission, solar, lights, and equipment. For air-conditioning systems the load also includes personnel and replenishment air

loads. The temperatures recommended for use in calculating the transmission and solar loads are contained in Table 1. The criteria for determining transmission and solar loads for ventilation systems are the same as for air-conditioning systems and are discussed in Section 2.4.

Lighting loads for both merchant and Navy ships, when the actual wattage for the space is known, is 3415 Btu/kWh. However, for fluorescent lighting a ballast load must also be included. For the ballast load, merchant practice uses a factor of 1.25 times the fluorescent-bulb wattage while Navy practice uses 4.5 watts per fluorescent bulb.

When the actual wattage is not known, an estimate must be made for the lighting load. For merchant ships the lighting load, q, may be estimated as follows:

$$q = AL \qquad (8)$$

where

A = deck area, ft²
L = load constant, Btu/hr-ft² (see Table 2)

When using equation (8) to estimate lighting loads, the deck area is based on finished dimensions.

When the installed lighting load is later known, the estimated lighting loads are superseded by the more accurate data. This is particularly important for messrooms and other public spaces with large lighting loads.

On naval ships, if the actual wattage is not known, the heat gains from lighting units can be estimated based on lighting foot-candle requirements as noted on the compartment design criteria sheet. Foot-candle lighting requirements can be converted to heating load as indicated by Table 3.

For stateroom lighting loads, see Table 4.

The heat gain from personnel is not considered in ventilated compartments since the sensible heat gain is negligible at ventilation design temperatures, and the moisture content of the cooling medium (weather air) varies to such an extent that the relatively small addition of latent heat from personnel is not significant.

Table 4 Stateroom lighting heat loads

	Incandescent	Fluroescent
Stateroom (single)	680 Btu/hr	2.4 Btu/hr-ft² + 150 Btu/hr
Stateroom (dept. head)	1200 Btu/hr	4.3 Btu/hr-ft² + 150 Btu/hr
Stateroom (double)	1020 Btu/hr	2.4 Btu/hr-ft² + 230 Btu/hr
Stateroom (triple)	1360 Btu/hr	2.4 Btu/hr-ft² + 310 Btu/hr
Bunkroom	2040 Btu/hr	2.4 Btu/hr-ft² + 460 Btu/hr

Ventilation for machinery spaces is intended to provide a reasonable atmosphere for operating personnel while maintaining a satisfactory ambient condition for machinery. Judgment, based on previous experience rather than heat-liberation calculations, generally dictates the air requirements. A knowledge of successful experience with spaces having similar characteristics and equipment is particularly valuable. On merchant ships with turbines or high-speed diesel propulsion equipment, mechanical supply systems usually provide a volume of air that is sufficient to change the air in the machinery space in one minute. A rate of change of 1.5 to 2 minutes is common for spaces with slow-speed diesels and in engine rooms that are relatively spacious. The quantity of supply air falls in the general range of 6 to 10 cfm per shaft horsepower.

Exhaust terminals over high heat-concentration sources reduce the heat dissipated to working and watch areas in the machinery spaces. Most of the supply air is distributed directly to watch areas. There is a relatively small air temperature rise in the blast area; therefore, watch personnel benefit from spot cooling. For merchant ships, reference 2 requires that an enclosed air-conditioned operating station be provided in the machinery space.

Engine-room exhaust ventilation may be natural, mechanical, or a combination of both. Theoretically, the exhaust volume equals 110 to 120% of the supply volume. The excess exhaust is necessary for two reasons: it compensates for the thermal expansion of the supply air, and it creates an indraft to prevent a dissipation of heat into adjacent spaces. Engine-room exhaust requirements generally are determined without considering combustion air. The combustion air extraction, however, must not impede the normal thermal flow of hot exhaust air up the machinery space casings. Otherwise, the temperature within the casings will rise, which in turn will increase the heat transmitted to surrounding spaces.

The annular space between the outer stack and uptake (smokepipe) provides an ideal passage for a natural exhaust. A natural exhaust can provide satisfactory results where air paths are short and there is ample space to utilize low duct and terminal velocities (1000 fpm or less); however, where both of these conditions do not exist, a mechanical exhaust is essential.

The shaft alleys on merchant ships are usually ventilated by a natural supply from the machinery space at the forward end of the shaft alley, and by a natural exhaust via the escape trunk at the after end. This arrangement acts as an equalizer when the combustion air consumption varies. Details concerning machinery space ventilation requirements for merchant ships can be found in reference 2; requirements for naval ships are in reference 27.

Weather intakes for ventilation systems should be located to avoid an ingestion of the discharge from exhaust systems, diesel engine exhaust, smokestacks, or any other source of contamination. Weather intakes should be designed with airlifts and located to prevent shipping seawater, driving rain, or spray. Reference 28 provides guidance concerning the protection of ventilator openings and distances of air intakes from possible hazardous vapors.

Battery rooms and lockers are ventilated to remove the hydrogen gas that is released when charging batteries. Requirements with regard to battery-room ventilation, as well as criteria for other ventilated spaces, are contained in reference 27 for naval ships and in reference 29 for merchant ships.

Ventilation supply systems also provide replenishment air for air-conditioning recirculation systems to prevent an accumulation of odors. Ventilation exhaust systems are used to exhaust stale air from air-conditioning recirculation systems as required to maintain an air balance.

2.3 Ventilation Loads and Calculations. When performing heating and cooling load calculations, both for ventilation loads and for air-conditioning loads, a form similar to Fig. 7 is normally used. Each compartment on the ship should be listed whether or not calculations are required for the compartment. Information listed on each sheet should include the compartment name, design or assumed temperatures for heating and cooling, compartment location, adjoining spaces and their temperatures, the dimensions for the common boundary, the "U" factor, and the direction of heat flow. The load should be classified by the key designation for each load component. The standard keys for load components are as follows:

KEY	LOAD COMPONENT
1	Deck over
2	Deck under
3	Outboard bulkhead or shell
4	Forward bulkhead
5	After bulkhead
6	Inboard bulkhead
7	Lights
8	Equipment
9	Personnel

For those instances that require an additional load component for infiltration/ventilation air, key 10 is used. The identical key designations are used for both the cooling and heating load calculations; however, all key designations may not be applicable for any one particular compartment. In cases where a particular load component consists of more than one heat gain or loss, the key number for that load is repeated for as many heat gain or loss calculations as are required.

HEATING AND COOLING LOAD CALCULATIONS

SHIP | CALCULATED BY RLW | CHECKED BY RBE | NAVSEC DRAWING NO. | SHEET 0 of 0

KEY

1. DECK OVER
2. DECK UNDER
3. OUTBOARD BULKHEAD OR SHELL
4. FORWARD BULKHEAD
5. AFTER BULKHEAD
6. INBOARD BULKHEAD
7. LIGHTS
8. EQUIPMENT
9. PERSONNEL
10. EXHAUST AIR

H = heat load, Btu/hr.
A = area, sq.ft.
U = transmittance, Btu/sq.ft.-hr.-t_d
t_d = temperature difference, air to air or surface to air
Q = cfm for cooling
t = temperature of compartment
t_o = temperature of adjacent spaces or outside
INS = insulation

REMARKS

COOLING LOAD

COMPARTMENT NAME NUMBER AND FRAMES	ADJOINING COMPARTMENT	KEY	t	t_o	INS		AREA OR VOL	t_d	U	$H(S)$
Elev Mchry Rm	Solar	1	105	140	18d	7.0 X 6.0	42	35	.20	294
02 LVL	Elev Trunk	2		100						
FR 110-113P	Uptake	3	120		18h	7.0 X 8.0	56	15	.17	143
	Fan Rm	4	105							--
	Strm	5		100						--
	Passage	6		100						
	Lights	7				160 X 3.415				546
	Eqpt	8				25HPX2535BTU/HR-HPX0.1 USE FACT				6363
	Pers	9								7346

$$Q = H/1.08t_d \qquad \text{(EQUATION 10)}$$
$$= 7346/1.08 \times 15 = 453\text{CFM}$$

USE 455 CFM MECH. SUP.
455 CFM MECH. EXH.

$$\text{RATE OF CHANGE} = \text{COMPARTMENT VOLUME}/Q$$
$$= 7.0 \times 6.0 \times 8.0/455$$
$$= 0.7 \text{ MINUTE}$$

HEATING LOAD

	KEY	t	t_o	INS		AREA OR VOL	t_d	U	$H(W)$	$Q(S)$
					Preheat					

Fig. 7   Heating and cooling load calculation form

CONSTRUCTION	CONDITION	U (Btu/hr - ft² - °F)		
		H ↔	U ↑	D ↓
NO. 0, NO INSULATION 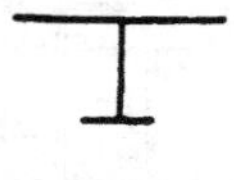	INSIDE AIR TO INSIDE AIR	.81	1.03	.64
	WEATHER AIR TO INSIDE AIR	1.34	1.61	.86
	OUTSIDE SURFACE TO INSIDE AIR	1.77	2.04	1.48
NO. 12  (1")	INSIDE AIR TO INSIDE AIR	.29	.32	.27
	WEATHER AIR TO INSIDE AIR	.34	.36	.30
	OUTSIDE SURFACE TO INSIDE AIR	.39	.39	.40
NO. 14 & 14P  (1")	INSIDE AIR TO INSIDE AIR	.24	.25	.22
	WEATHER AIR TO INSIDE AIR	.27	.28	.25
	OUTSIDE SURFACE TO INSIDE AIR	.30	.31	.30
NO. 18  (2", 1")	INSIDE AIR TO INSIDE AIR	.17	.18	.16
	WEATHER AIR TO INSIDE AIR	.19	.19	.17
	OUTSIDE SURFACE TO INSIDE AIR	.20	.21	.20
ACOUSTIC, 2" FIBROUS GLASS & SHEATHING 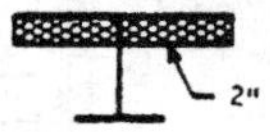 (2")	INSIDE AIR TO INSIDE AIR	—	.34	.29
	WEATHER AIR TO INSIDE AIR	—	.43	—
	OUTSIDE SURFACE TO INSIDE AIR	—	—	.43
CLEAR GLASS SINGLE PANE	WEATHER AIR TO INSIDE AIR	← 1.10 →		
	OUTSIDE SURFACE TO INSIDE AIR	← NEGLECT →		
	SOLAR RADIATION	← 160 Btu/hr. - ft² →		
CLEAR GLASS DOUBLE PANE	WEATHER AIR TO INSIDE AIR	← 0.55 →		
	OUTSIDE SURFACE TO INSIDE AIR	← NEGLECT →		
	SOLAR RADIATION	← 160 Btu/hr. - ft² →		

Heat losses to refrigerated stowage spaces shall be calculated using a "U" factor of 0.10, regardless of the direction of heat flow.

Fig. 8   Heat-transfer coefficients for Navy ships

Each key is entered individually in the appropriate column of the calculation form. For keys 1–6 the load components may be heat transmissions from or to the surrounding compartments, weather, seawater, solar indication, etc., based on the equation:

$$H = UA \, \Delta T \qquad (9)$$

where

H = rate of heat transmission, Btu/hr
U = overall heat transmission coefficient, Btu/hr-ft²-deg F
A = surface area, ft²
ΔT = temperature difference between medias on each side of area, deg F

The overall heat transmission coefficients generally used for naval construction are shown by Fig. 8. Figure 9 shows the "U" values commonly used in merchant designs. Reference 30 contains additional U values that are used for merchant ships, and reference 31 contains additional U values that may be used in the design of naval ships. The U values used for merchant construction differ slightly for heating and for cooling calculations. For naval construction the same U value is used for both heating and cooling calculations.

For ventilated compartments, the individual load components are summed and result in a compartment sensible-heat gain during the cooling season, and a total heat loss during the heating season. The required ventilation air quantity necessary to maintain the design temperature during the cooling season can be determined directly using the following equation (depending upon specific design criteria, this may be either the mechanical supply requirement, mechanical exhaust requirement, or both):

$$Q = \frac{H_s}{60 \, C_p d \, (t_i - t_o)} \qquad (10)$$

$$Q = \frac{H_s}{1.08 \, (t_i - t_o)} = \frac{H_s}{1.08 \, \Delta t}$$

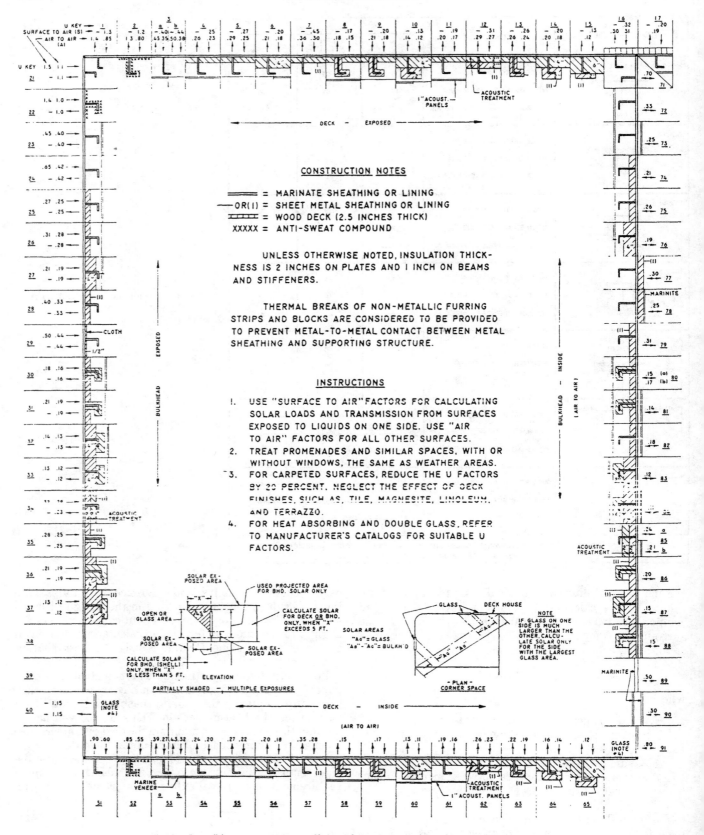

CONSTRUCTION NOTES

= MARINATE SHEATHING OR LINING
—OR(1) = SHEET METAL SHEATHING OR LINING
= WOOD DECK (2.5 INCHES THICK)
XXXXX = ANTI-SWEAT COMPOUND

UNLESS OTHERWISE NOTED, INSULATION THICKNESS IS 2 INCHES ON PLATES AND 1 INCH ON BEAMS AND STIFFENERS.

THERMAL BREAKS OF NON-METALLIC FURRING STRIPS AND BLOCKS ARE CONSIDERED TO BE PROVIDED TO PREVENT METAL-TO-METAL CONTACT BETWEEN METAL SHEATHING AND SUPPORTING STRUCTURE.

INSTRUCTIONS

1. USE "SURFACE TO AIR" FACTORS FOR CALCULATING SOLAR LOADS AND TRANSMISSION FROM SURFACES EXPOSED TO LIQUIDS ON ONE SIDE. USE "AIR TO AIR" FACTORS FOR ALL OTHER SURFACES.
2. TREAT PROMENADES AND SIMILAR SPACES, WITH OR WITHOUT WINDOWS, THE SAME AS WEATHER AREAS.
3. FOR CARPETED SURFACES, REDUCE THE U FACTORS BY 20 PERCENT. NEGLECT THE EFFECT OF DECK FINISHES, SUCH AS, TILE, MAGNESITE, LINOLEUM, AND TERRAZZO.
4. FOR HEAT ABSORBING AND DOUBLE GLASS, REFER TO MANUFACTURER'S CATALOGS FOR SUITABLE U FACTORS.

Fig. 9   Overall heat transmission coefficients for various merchant ship construction designs

where

Q = quantity of air, cfm
H_s = heat gain in compartment, Btu/hr
C_p = specific heat of air, 0.24 Btu/lb-deg F (at standard air temperature)
d = density of air, 0.075 lb/ft³ (at barometer of 29.92 in. and temperature of 70 F)
t_i = compartment temperature, deg F
t_o = weather air temperature, deg F
$\Delta t = (t_i - t_o)$, difference between compartment and weather temperatures, deg F

Figure 7 shows example calculations for a ventilated compartment based on a specified design temperature.

For ventilated compartments that have a specified rate of air change, the air quantity can be determined by

$$Q = \frac{V}{R} \qquad (11)$$

where

Q = required air, cfm
V = gross volume of ventilated compartment, ft³
R = rate of air change specified, minutes

It should be noted that very often a ventilated compartment has both a design temperature and a rate of change specified. For this instance, the compartment air quantity is determined using both criteria with the highest resulting air quantity provided for that compartment.

When the heating and cooling loads have been determined for all of the compartments on the ship, then the compartments can be allocated to systems to meet their individual needs (supply, exhaust, or recirculation systems).

A system analysis form, similar to Fig. 10, may be used to define the system and determine the equipment necessary to meet the design requirements of the compartments. Figure 10 is an example of compartment grouping to form a supply system. The air quantities required for each of the individual components of the system are totaled to determine the required system air quantity. Knowing the total air quantity, a suitable fan can then be selected as well as a suitable preheater. In the example shown in Fig. 10, only one of the spaces required more than preheated air. Knowing the temperature of the air coming off the preheater and the space heating load, a suitable reheater can be selected.

Standard equipment is used for all naval ships to simplify maintenance requirements and to facilitate repairs. Families of fans, heaters, cooling coils, etc. have been established, and the equipment for each individual system is selected from these families of standard equipment. Performance data for all Navy standard HVAC equipment may be found in reference 12. Additional information on the selection of naval equipment can be found in reference 4.

Equipment for merchant ships should be of a quality suitable for marine use and meet the requirements of the Maritime Administration, U.S. Coast Guard, American Bureau of Shipping, and the specific ship specifications.

Additional information concerning the design of ventilation systems for merchant ships may be found in references 2 and 26.

2.4 Air-Conditioning and Heating Criteria. The load requirements for each space to be air conditioned is the sum of the loads resulting from transmission, solar, lights, equipment, personnel, and replenishment air. The temperatures recommended for use in calculating the transmission and solar loads are contained in Table 1.

The heat transmitted through a flat structure, such as a deck or bulkhead, is controlled by the overall coefficient of heat transmission, the U factor. In practice, therefore, it becomes necessary to analyze the overall coefficient of heat transmission in greater detail. This is facilitated by using the expression

$$U = 1/R$$

where R is the total resistance to heat flow; it is equal to the sum of the various individual resistance components. For example, the thermal resistance of a structure consisting of two homogeneous materials (denoted by i for inside and o for outside or weather) having an air space between them can be expressed as:

$$R = \frac{1}{f_i} + \frac{x_i}{k_i} + \frac{1}{a} + \frac{x_o}{k_o} + \frac{1}{f_o} \qquad (12)$$

where:

f = film or surface conductance, Btu/hr-ft²-deg F
x = material thickness, in.
k = material thermal conductivity, Btu-in./hr-ft²-deg F
a = thermal conductance of air space, Btu-in./hr-ft²-deg F

The thermal conductance of an air space is a function of the height, depth, position, character, and temperature of the bounding surfaces. The relationships are not linear, and accurate values must be determined by test. For a structure consisting of a single homogeneous material, the expression reduces to:

$$R = \frac{1}{f_i} + \frac{x}{k} + \frac{1}{f_o}$$

In the case of constructions having nonuniform or irregular sections, such as are common aboard ship, the resistance of the section, $1/C$, is substituted for:

$$\frac{x_i}{k_i} + \frac{1}{a} + \frac{x_o}{k_o} \text{ and } \frac{x}{k}$$

in the foregoing equations to obtain:

$$R = \frac{1}{f_i} + \frac{1}{C} + \frac{1}{f_o}$$

Values for f_i, f_o, and C are obtained from tests of typical constructions. The film coefficients, f_i and f_o, vary with surface temperature and direction of heat flow (up, horizontal, or down); also, f_i varies with the frame-spacing dimension. Conductance values, C, for a particular material vary according to the mean of the surface temperatures, $(t_i + t_o)/2$. For estimating purposes, an outside film

System Number SUP SYS 02-116-2

Replenishment Air = ______

Bypass Air = ______ Return Air = ______

t_1 = Temperature of Compartment

t_2 = Temperature of Replenishment Air

t_3 = Temperature of Light Load Mix

t_4 = Temperature Off Coil

t_5 = Temperature Light Load

Q_1 = CFM Replenishment Air

Q_2 = CFM Return

Q_3 = CFM Bypass Air

Q_4 = CFM Total

Winter Mix $= t_1 - (t_1 - t_2)\frac{Q_1}{Q_4} - \frac{H_{Ret.Air}}{(1.08)(Q_2)}$

Light Load Mix $= t_4 + \frac{Q_3}{Q_4}(t_5 - t_4)$

Required $\Delta t = \frac{H}{(1.08)(CFM)}$

Light Load $\Delta t = t_5 - \frac{H_{L.L.}}{(1.08)(Q_4)} - t_3$

HEATING LOAD ANALYSIS

System Number: SS02-116-2 | Sheet 0 of 0 | Ship: | NAVSEC Drawing No: | Checked by: RBE | Calculated by: RLW

SPACE NAME	LOCATION		CFM		SPACE LOADS, H BTU/HOUR			REQUIRED Δt			HEATER DATA		
	DECK	FRAMES	COOL	HEAT	WINTER	LIGHT LOAD	REHEAT	WINTER	LIGHT LOAD	REHEAT	CFM	RANGE	NUMBER
Sup Sys	02	116P											
Elev Mchry Rm	02	110-113P	455										PHT Only
Helo Winch Rm	02	125-131S	380										PHT Only
Radar MG Rm	01	118-125S	1850										PHT Only
Replen RS1-120-1	MN	120-145S	350										PHT Only
Helo Hangar	02	130-126P	3500	2335	26200			10.4			2335	48-80	29M
Cool Eqpt Rm	02	120-120P	460										PHT Only
Replen RS02-110-2	02	118-130P	750										PHT Only
MG RM	MN	125	1000										PHT Only
			8745	5860							5860	10-48	35M
			A10, 7-1/2 HP										

Fig. 10 System analysis form

coefficient, f_o, of 7.0 is commonly used in deriving air-to-air U factors; the $1/f$ component is omitted for sides exposed to the sun or liquids, i.e., surface-to-air U factors.

The transmittance of a panel composed of a metal or other highly conductive material extending wholly or partly through insulation should, if possible, be determined by test, but approximations can be made as described in references 5 and 30.

The method for calculating solar loads varies between naval and commercial practice. When calculating the solar load for compartments having more than one weather boundary, the Navy procedure is to use only the solar load for that weather boundary which would result in the highest compartment heat gain. The solar effect from the other weather boundaries is neglected, but the heat transmission loads due to weather air are included (see reference 4). Solar loads on all boundaries are considered in commercial practice. Although, when the solar load is considered on multiple boundaries, the design temperatures used to calculate the solar loads are lower than if only one boundary load is solar [26].

Criteria for calculating lighting loads are described in Section 2.2.

Equipment heat loads must include the heat gain, both sensible and latent, from all heat-producing equipment in the space. Equipment heat loads are usually electrical except where steam or hot-water equipment is used. The equipment heat gain should be based on heat-dissipation data for the actual equipment installed. However, quite often the equipment has not been selected when the first HVAC calculations are made, and estimates of the heat dissipation must be made. All estimated equipment heat gains must be confirmed after the actual equipment has been selected and the actual heat dissipation data are known. References 5 and 26 contain data that may be useful when estimating the equipment heat loads. An allowance must be made for the operating load factor for equipment and piping. The load factor should consider a simultaneous use of combinations of equipment and piping and whether the equipment and piping are used intermittently or continuously.

There will also be instances where the total heat dissipation from equipment within a space will not affect the space directly, and in some instances only a portion of the heat dissipation will affect the space. Equipment that discharges air directly into a return duct terminal of an air-conditioning system should include a portion of the heat dissipation for the space and the remainder as a heat gain to the return air. The portion treated as direct return heat must be included in the system's total heat load, however. Equipment provided with hoods, such as galley equipment, should include only a portion of the heat dissipation from the equipment in the space heat load. If the hood is part of a ventilation exhaust system and discharges directly to the weather, it will not be included in the system heat load. The heating effect of equipment heat is usually not included in the space heating calculations unless specifically identified to be included. For further detailed information concerning equipment heat loads, see references 4, 5, 26, and 27.

Table 5 Personnel sensible and latent heat dissipation, Btu/hr

Room Dry Bulb, °F	Mess Attendants and Working Spaces		All Others	
	Sensible	Latent	Sensible	Latent
75	360	440	300	300
76	345	455	290	310
77	330	470	275	325
78	315	485	265	335
79	300	500	250	350
80	285	515	240	360
81	270	530	230	370
82	255	545	215	385
83	240	560	205	395
84	225	575	190	410
85	210	590	180	420

Personnel loads are the sensible and latent heat generated by the occupants of the space. Commercial and naval practice differ concerning the criteria used to calculate personnel loads. For commercial designs, the personnel heat gain is calculated using the equations:

$$q_s = HD_s \times P \qquad (14)$$

and

$$q_L = HD_L \times P \qquad (15)$$

where

q_s = sensible heat gain, Btu/hr
q_L = latent heat gain, Btu/hr
HD_s = personnel sensible heat dissipation, Btu/hr
HD_L = personnel latent heat dissipation, Btu/hr
P = number of room occupants

The personnel heat dissipation varies with room temperature and level of activity. Recommended values for commercial ships are given in Table 5. For messing facilities on commercial ships, add 30 Btu/hr sensible and latent per person eating as an allowance for the heat dissipation from food.

The occupancy for staterooms is based on the number of sleeping accommodations. For offices, lounges, messing facilities, workrooms, and similar spaces, the occupancy is taken to be two thirds of the seating capacity, rounded off to the nearest whole number.

Heat gain rates per person for the various compartment types on naval ships are given in Fig. 11. An allowance of 30 Btu/hr sensible heat and 30 Btu/hr latent heat per person in messing spaces is made as an allowance for the heat dissipation from food.

Sensible and latent heat gains from personnel in air-conditioned compartments are not considered in compartment heating-season calculations except for berthing areas that have more than twelve persons, in which case a sensible heat gain is taken for one third of the design occupancy.

The methods used for determining the replenishment air requirements vary between commercial and naval designs. Naval practice uses 5 to 10 cfm per person as a minimum. Commercial practice uses a minimum of 12 to 15 cfm per person or a 20-minute rate of change for high-occupancy spaces, 60-minute rate of change for the wheelhouse, and 30-minute rate of change for all other spaces

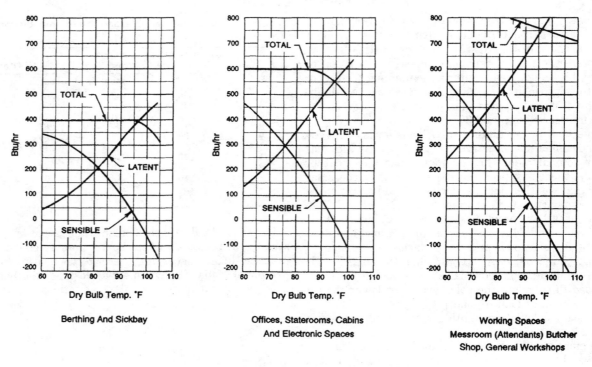

Fig. 11   Heat loss rates from a human body in still air

with a maximum quantity not to exceed 50 cfm per person. Naval practice is based on the total design occupancy of the space, while in commercial practice the total occupancy is used for staterooms, but two-thirds occupancy rounded to the nearest whole number is used for offices, lounges, messing facilities, workrooms, and similar spaces. Additional details concerning naval criteria may be found in reference 4 and commercial criteria in reference 26.

2.5 Air-Conditioning and Heating Loads and Calculations. The forms used for calculating the heating and cooling loads as described in Section 2.3 for ventilation systems are also used for calculating the heating and cooling loads for air-conditioning systems. Figure 12 is a form that is used for commercial load calculations and shows a typical heating and cooling load calculation for an air-conditioned space. Design temperatures for the heating and cooling load calculations are contained in Table 1. When the heating and cooling loads have been calculated for all spaces to be air conditioned, the spaces are grouped together to form systems. System summary sheets are used to identify system requirements. Figures 13 and 14 are forms that are used in commercial practice for a system cooling summary and a system heating summary. Forms used in naval practice are similar; the forms for each contain essentially the same data.

System loads for commercial design are calculated differently than system loads for naval construction. Commercial loads often include a temperature rise of 1.5 deg F for each 100 ft of duct based on the longest run for the supply duct. Naval designs do not include this calculation. Likewise, commercial loads often include a temperature rise of 0.5 deg F for each 100 ft of return duct when the

return path includes passageways and stairwells that are directly supplied with conditioned air. When the passageways and stairwells are indirectly supplied with conditioned air, the return path load is assumed to be 5 deg F or 3 deg F plus 0.5 deg F for each 100 ft of duct based on the longest run of return duct. Naval practice assumes that a passageway or stairwell is 0.5 deg F higher than compartment design temperatures unless the passage is directly supplied with conditioned air.

In naval practice, the off-coil temperature plus the rise in temperature due to the fan heat is used as the entering space temperature when the fan pulls the air through the system. The design temperature of the space is used as the return air temperature. Heat loads are calculated for passageways and included in the system total load. If the temperature rise in the passageway is 5 deg F or more, a supply terminal is provided for the passageway. Examples of detailed calculations for naval practice may be found in reference 4, and in reference 26 for commercial construction.

Figure 15 illustrates the calculation of the cooling-coil load using a psychrometric chart following commercial practice. The following temperatures are identified in Fig. 15:

OC = off-coil temperature
  S = OC plus temperature rise through supply duct
  R = room design temperature
RA = R plus temperature rise due to return path load
OA = outside temperature
  M = mixture temperature of return air and outside air added as replenishment air

SPACE __Radio Rm.__ DECK __Br.__ FRAMES __19-24S__ SPACE NO. __405__

ADJOINING SPACE	LOAD KEY	INSUL TYPE	U KEY	DIMENSIONS	AREA	SPACE TEMP COOLING 78 °F — T_o	T_D	U	Q_s	Q_L	Q_T	SPACE TEMP HEATING 70 °F — T_o	T_D	U	Q_w
Weather, Dk	1	3W/1	98	9 x 11 - 1 x 3 (solar)	96	130	52	.088	439			0	70	.074	497
Asst Mast SR	2	0	8	9 x 11 - 1 x 3	96	78	0	-	-			70	0	-	-
Weather, Bhd	3	3W/1	100	11 x 9 - 1-1/2 x 2 (solar)	96	115	37	.084	298			0	70	0.70	470
Weather, Glass	3	61	1	1-1/2 x 2 (solar)	3			160	480			0	70	1.13	237
Stair & Pass.	4	0	3+10	9 x 9	81	80	2	.210	34			65	5	.193	78
Comp. Rm.	5	3W/1	100	9 x 9	81	110	32	.077	200			40	30	.071	173
Pass.	6	0	3+10	11 x 9	99	80	2	.210	42			65	5	.193	96
Lights	7			3 x 40 x 1.25 = 150W					512						
Rad. Equip.	8			(see below) 1875					6403						
Pers. S	9			1 x 2/3 = 2/3. Use 1 x 265					265						
Pers. L	9			1 x 2/3 = 2/3. Use 1 x 335					—	335					—
									8673	335					1551
												Vent Load =	1.08 x 325 x (70-56)		4914
Rad. Equip.	8			Worst case								Tot. Rm. Ld. =			6465
				MRU35,2500W@1/2 Time	=	1250									
				MRU298,2500W@1/4 Power	=	625									
				Total		1875									

ΔT = 25°F Vol. = [(8 x 10) - (1 - 3)] 7 = 539 Ft³   Out. Air = 150 cfm

Cfm = 8673/1.08 x 25 = 321, Use 325   Htr. Kw = 6465/3415 = 1.89, Use 2.0

Min. Out. Air (1) Occupancy = 1 x 15 = 15, (2) R/C = 539/30 = 18, (3) Exhaust through Rad. Equip. = 150

KEY	SYMBOLS	SUMMARY	
1 = Deck Over	Q_s = Heat Load, BTU/Hr, Sensible	Sq. Ft. __77__	ΣQ_L __335__
2 = Deck Under	Q_L = Heat Load, BTU/Hr, Latent	Height __7__	ΣQ_T __9008__
3 = Outboard Bhd or Shell	Q_T = Heat Load, BTU/Hr, Total	Volume __539__	ΣQ_w __6465__
4 = Forward Bulkhead	Q_w = Heat Load, BTU/Hr, Winter	R of C __2__	CFM __325__
5 = After Bulkhead	U = Transmittance, BTU/Sq. Ft. - Hr - °F	ΣQ_s __8673__	KW __20__
6 = Inboard Bulkhead	T_D = Temperature Differences		SHF __.96__
7 = Lights	T_o = Temp of Adjoining Space or Outside		
8 = Equipment 9 = Personnel			

HULL OR JOB NO.: XYZ

CHECKED BY: JWM

CALCULATED BY: D.S. DATE

HEATING & COOLING LOAD CALCULATIONS

REV   SHEET NO 5 OF

Fig. 12 Heating and cooling load calculations

SYSTEM NO. _____ A/C Sys "X" Class "D" (Terminal Reheat)

SPACE	SPACE NO.	DECK	FRAME	SIDE	BTU/HR SENS.	BTU/HR LATENT	BTU/HR TOTAL	SHF	OUTSIDE AIR BASED ON * R/C	PERS.	EXH.	CFM	R/C TOTAL AIR	COLD DUCT AIR	WM. DUCT AIR	REMARKS
Radio Rm.	405	Bridge	19-24	S	8673	335	9008	.96			150	325	2			
Inspector's S.R.	311	Off	17-21	P	2639	335	2974	.89			50	100	8			
Asst. Ch. Eng. S.R.	315		25-29	P	2736	335	3071	.89	25			105	7			
Ch. Eng. S.R.	316		29-34	P	3303	335	3638	.90			50	125	7			
Owner's S.R.	318		29-34	P	2287	335	2622	.87	26			100	8			
Master's Off.	301		29-34	S	3127	1005	4132	.76		45		115	7			
Master's S.R.	303		29-34	S	3303	335	3638	.90			50	125	7			
Asst. Mest. S.R.	304		25-29	S	2348	335	2683	.87	25			90	8			
Off. S.R. (1)	305		21-25	S	2420	335	2755	.87	25			90	8			
Off. S.R. (2)	306		17-21	S	2605	335	2940	.89	25			100	8			
Off. Mess	309		23-28	P	4119	2190	6309	.65		72		175	4			
Crew Mess	308		17-25	S	7945	5120	13065	.61		100		300	5			
Lounge	306-1		14-17	S	4700	1340	6040	.78		36		175	4			
Eng. Dept. S.R. (1)	216	Crew	17-21	P	2787	670	3457	.81		30		105	7			
Eng. Dept. S.R. (2)	217		21-25	P	2985	670	3655	.82		30		110	7			
Deck Dept. S.R. (1)	218		25-29	P	2914	670	3584	.81		30		110	6			
Deck Dept. S.R. (2)	220		29-34	P	4306	670	4976	.86	33			150	7			
Deck Dept. S.R. (3)	221		29-34	P	2760	670	3430	.80		30		105	7			
Deck Dept. S.R. (4)	202		29-34	S	2760	670	3430	.80		30		105	7			
Deck Dept. S.R. (5)	204		29-34	S	4036	670	4706	.86	33			150	7			
Stew. Dept. S.R. (1)	205		25-29	S	2914	670	3584	.81		30		110	6			
Stew. Dept. S.R. (2)	206		21-25	S	2638	670	3308	.80		30		100	7			
Stew. Dept. S.R. (3)	201		17-21	S	2787	670	3457	.80		30		105	7			
Hospital	209		17-21	S	2389	670	3059	.78			100	100	8			
Mech. Cont. Rm.	214		18-24	P	9802	335	10137	.97	33			365	3			
					93,283	20,375	113,658	.82	225	493	400	3540				

Total = 1118

Use 1,120 OUT. AIR

* Note only maximum CFM, as calculated on basis of "R/C", "Persons" and "Exh" on Heating and Cooling Cal. Sheets, is listed.

CALCULATED BY: D.S.   DATE:

CHECKED BY: JWM

HULL OR JOB NO.: XYZ

COOLING SUMMARY

REV   SHEET NO./

Fig. 13 Cooling summary

SYSTEM NO. _____ A/C Sys 'x', Class 'D' (Terminal Reheat)

SPACE	SPACE NO.	DECK	FRAME	SIDE	RM LOAD BTU/HR	TEMPERATURE			CFM	HEATER DATA				REMARKS
						ENT.	RISE	AMB.		NO.	TYPE	SIZE *	RANGE	
Radio Rm.	405	Bridge	19-24	S	6465	56	18.6	70	325	405		2.00		
Inspector's S.R.	311	Off	17-21	P	3650	56	33.8	70	100	311		1.00		
Asst. Ch. Eng. S.R.	315		25-29	P	2905	56	25.6	70	105	315		1.00		
Ch. Eng. S.R.	316		29-34	P	3825	56	28.3	70	125	316		1.25		
Owner's S.R.	318		29-34	P	2808	56	26.	70	100	318		1.00		
Master's Off.	301		29-34	S	2672	56	21.5	70	115	301		.75		
Master's S.R.	303		29-34	S	3825	56	28.3	70	125	303		1.25		
Asst. Mast. S.R.	304		25-29	S	2678	56	27.5	70	90	304		1.00		
Off. S.R. (1)	305		21-25	S	2706	56	27.8	70	90	305		1.00		
Off. S.R. (2)	306		17-21	S	3456	56	32.	70	100	306		1.00		
Off. Mess	309		23-28	P	3126	56	16.5	70	175	309		1.00		
Crew Mess	308		17-25	S	4911	56	15.	70	300	308		1.50		
Lounge	308-1	V	14-17	S	4687	56	24.8	70	175	308-1		1.50		
Eng. Dept. S.R. (1)	216	Crew	17-21	P	4374	56	38.6	70	105	216		1.25		
Eng. Dept. S.R. (2)	217		21-25	P	3860	56	32.5	70	110	217		1.25		
Deck Dept. S.R. (1)	218		25-29	P	4019	56	33.8	70	110	218		1.25		
Deck Dept. S.R. (2)	220		29-34	P	6005	56	37.	70	150	220		1.75		
Deck Dept. S.R. (3)	221		29-34	P	4001	56	35.3	70	105	221		1.25		
Deck Dept. S.R. (4)	201		29-34	S	4001	56	35.3	70	105	201		1.25		
Deck Dept. S.R. (5)	202		29-34	S	6005	56	37.	70	150	202		1.75		
Stew. Dept. S.R. (1)	204		25-29	S	4019	56	33.8	70	110	204		1.25		
Stew. Dept. S.R. (2)	205		21-25	S	3709	56	34.3	70	100	205		1.25		
Stew. Dept. S.R. (3)	206		17-21	S	4375	56	38.6	70	105	206		1.25		
Hospital	209		17-21	S	3956	56	36.6	70	100	209		1.25		
Mach. Cont. Rm.	214	V	18-24	P	7884	56	20.	75	365	214		2.50		

* Note: 'size' is in kilowatts.

CALCULATED BY: D.S.   DATE:
CHECKED BY: JWM
HULL OR JOB NO.: XYZ

HEATING SUMMARY

REV.   SHEET NO. 3

Fig. 14 Heating summary

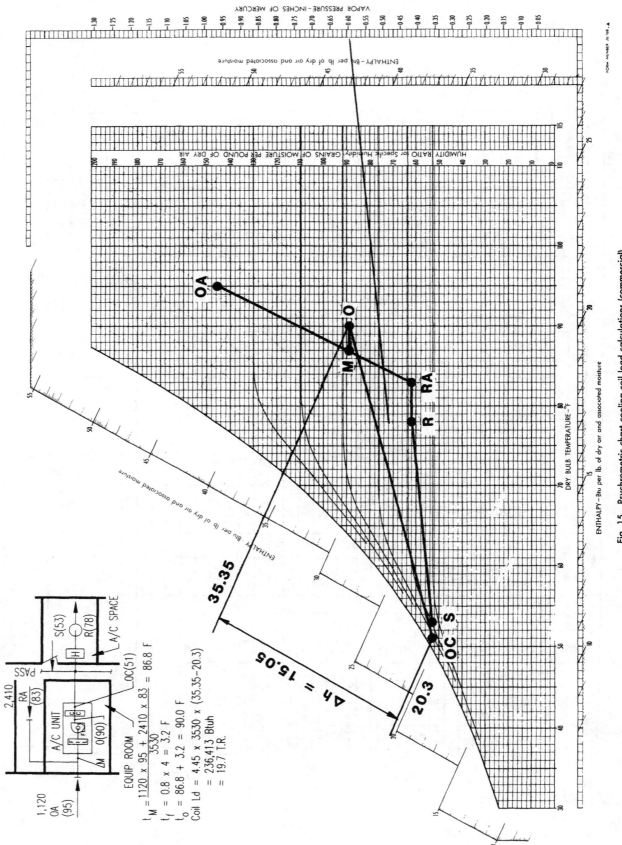

Fig. 15   Psychrometric chart cooling-coil load calculations (commercial)

CLASS	SYSTEM NO.	COIL NO.	SPACE NAME AND LOCATION	SENSIBLE HEAT H_s	TOTAL HEAT H_T	COIL SLOPE H_s/H_T	REPLEN AIR PER ROOM (CFM)	REPLEN AIR PER COIL (ON BASIS OF LOAD)	SELECTION			% OF COIL LOAD PER SPACE	SPACE CFM	PLANT NO.
									SIZE	GPM	TONS			
W	05-145-2		CHART ROOM 04-150-158 C	7572	8292	0.91	20					16	265	
	(2)		PILOT HOUSE 04-130-150 C	39348	45107	0.87	160					84	1380	
				46920	53399	0.88	180					100	1645	
			FAN HT (A1-1/2)	3699	3699									
				50619	57098									
			CFM = 50619/(1.08 x 28.5) = 1645											
			FAN HT Δt = 3699/(1.08 x 1645) = 2.1											
			T_{MIX} RPNSM AIR=(180 x 90 + 1465 x 80) /	1645 =	81.1°F									
			TONS = 1645 x 60 x 9.6/12000 x 13.44	= 5.88										
			GPM = 3.6 x 5.88 = 21.2											
									560W	21.2	5.88		1645	

COOLING COIL RECAPITULATION

SHIP DLG 51	CALCULATED BY CEC	DIAGRAM NO. 5121-D002	SHEET 134 OF

Fig. 16 Cooling-coil load recapitulation

O = M plus temperature rise due to fan heat; this is the entering coil temperature

An example of calculations illustrating naval practice is shown in Figs. 16 and 17.

The temperatures identified in Fig. 17 are the following:

1. = off-coil temperature
2. = No. 1 plus the temperature rise due to fan heat
3. = room design temperature
4. = outside air temperature
5. = temperature of mixture of return air and replenishment (outside) air; this is the entering coil temperature.

The psychrometric chart is a very useful tool for quickly and easily determining a system load. It provides a ready means of checking and developing system calculations. There is a degree of inaccuracy involved in a calculation by this method; however, it is generally considered unimportant because of the inherent difficulties in balancing airflows and water flows in the actual system itself. In addition, the loads on the ship are subject to a wide range of variances. Hand calculations on the psychrometric chart provide sufficient accuracy for an air-conditioning analysis.

2.6 Refrigeration Criteria. The refrigeration heat load for naval ships is based on maintaining freeze spaces at 0 to −10 F and chill spaces at 33 F. Boundary surface temperatures are assumed to be 140 F for surfaces exposed to the sun, 120 F for adjacent machinery spaces, 100 F for other nonrefrigerated spaces, 90 F for surfaces exposed to seawater, and 80 F for air-conditioned spaces unless other design temperatures are cited in the specific ship specifications.

For merchant construction the Maritime Administration design criteria for refrigerated spaces are a design temperature of 0 F for freeze spaces, 35 F for chill spaces, and 40 F for thaw spaces. Boundary surface temperatures are assumed to be 120 F for surfaces exposed to the sun or adjacent to machinery spaces and 100 F for all other surfaces except those adjacent to refrigerated spaces.

Where the temperature of an adjacent space is lower than the temperature of the space being designed, such that there would be a heat loss to the adjacent space, the heat loss is neglected in the calculation.

For flexibility, naval practice provides for some refrigerated spaces to be capable of maintaining either the 0 F temperature or the 33 F temperature. Likewise, merchant practice designs also allow some of the refrigerated spaces to be used as either freeze or chill spaces or as chill or thaw spaces.

The components of the heat load for refrigerated spaces include transmission, infiltration, ventilation, product, and equipment heat loads.

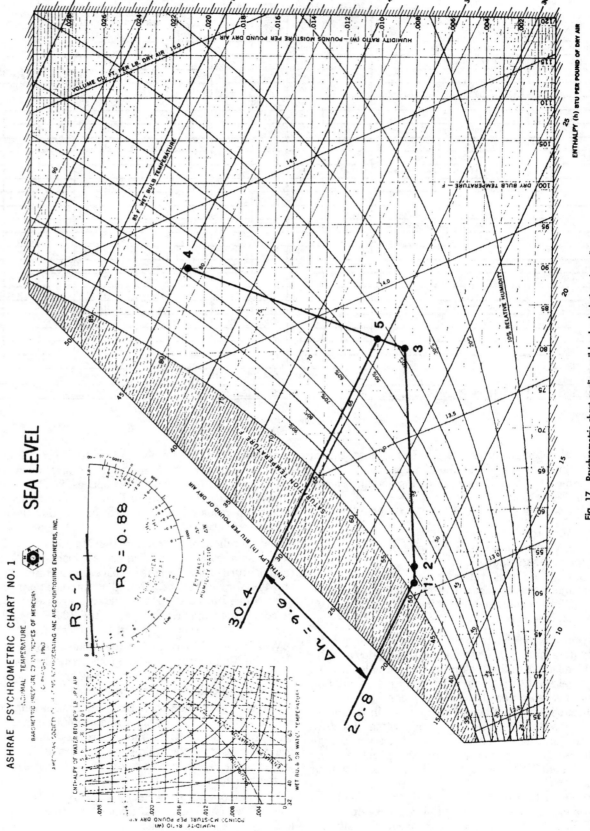

Fig. 17   Psychrometric chart cooling-coil load calculations (naval)

Table 6   Overall heat-transfer coefficients for naval ship refrigerated spaces, Btu/hr-ft²-°F.

REFRIGERATOR SURFACE AND TYPE	CONSTRUCTION	INSULATION	4"	5"	6"
Wall, external stiffeners		Fibrous glass	- -	- -	0.045
Wall, internal stiffeners		Fibrous glass	- -	- -	0.100
Overhead, internal stiffeners		Fibrous glass	- -	- -	0.100
Deck, nonmetallic internal stiffeners		Fibrous glass	- -	- -	0.110
Wall, external stiffeners		Polyurethane foam	0.041	0.034	0.029
Wall, internal stiffeners		Polyurethane foam	0.087	0.071	0.060
Overhead, internal stiffeners		Polyurethane foam	0.090	0.073	0.061
Deck, no internal reinforcement		Polyurethane foam	0.041	0.034	0.029

The U values used for the calculations of the transmission load in refrigerated spaces on naval ships are contained in Table 6. For commercial construction similar data are available in reference 32 or from manufacturers.

The infiltration heat load is the heat gain through the entrance of air when doors are opened and through cracks. For many designs the temperature of the entering air is assumed to be 100 F with a 60% relative humidity, where the air is from nonrefrigerated spaces. If the air is from refrigerated spaces, the air temperature is assumed to be 50 F and 80% relative humidity.

A ventilation heat load must be considered for cargo spaces storing fruits and vegetables. It consists of the heat gained by the introduction of outside air, which is required for the displacement of vitiated compartment air. The ventilation heat load is important only where it exceeds the infiltration load, and then it is used in lieu of the infiltration load for the operating condition being calculated. The outside air condition is assumed to be 100 F and 80% relative humidity. The amount of outside air is based on the use of a supply fan with a capacity equal to the gross cubical volume of the space per hour for ship stores, or one-third the gross cubical volume of space per hour for cargo stores. The fan is assumed to operate a maximum of 20 minutes in any one hour. Only one refrigerated space or system will be ventilated at a time, and

spaces are ventilated only during the normal operation period, not during pulldown operations.

The product heat load consists of the heat gain due to the internal heat of the product and the containers. In addition, for chilled storage, the product heat load also includes respiration heat from fruits and vegetables.

Equipment heat loads include the heat introduced by fan motors for both ventilation and recirculation fans. Recirculation fans usually operate at full speed only during pulldown operations; the low speed, usually 50% of the maximum, is used during normal operation and reduces the heat load.

Refrigeration loads are determined for two conditions, pulldown and normal operating. The pulldown condition is considered to be the period during which the temperature of the product within the space is being reduced to the design holding temperature. The normal condition is the period during which design holding temperatures are maintained.

Condensing unit loads are selected based on the following:

- For the pulldown condition, continuous operation of all condensing units on a system.

- For normal operation, the condensing unit operation is not to exceed 18 hours per day with the standby compressor secured.

Cooling coils are selected on the basis of the compressor selection for pulldown operations, or the design coil refrigerant temperature for normal operations, whichever condition establishes the maximum coil surface requirements.

For naval designs the equipment for cargo plants is based on a calculated pulldown condition for freeze spaces of five days and a three-day pulldown for chilled spaces. Merchant designs require that each compressor have ample capacity to pull down the refrigerated compartments to design temperatures within 72 hours after loading. For additional data on refrigeration criteria, see references 10 and 33 for naval designs and reference 2 for merchant designs. Data on cargo respiration rates may be found in references 5 and 34.

2.7 Refrigeration Loads and Calculations. The refrigeration system capacity must be adequate for both the pulldown condition and the normal operating condition. Calculations must be made for both conditions in order to properly size the refrigeration components. In addition, when spaces are intended to be used for either freeze or chill, or for chill or thaw, calculations for both services must be performed to determine the operation that has the largest refrigeration load. Refrigeration cooling loads are the sum of the transmission, infiltration, ventilation, product, and equipment loads. Sample calculations for one space, used for both 0 F storage and for 33 F storage, are shown in Figs. 18 and 19.

Since spaces are usually intended to store a variety of products, exact calculations are usually not required, and an average product value may be used in the various load calculations. Average product load conditions that are used for naval calculations are contained in Table 7. Detailed product data may be found in references 5, 11, and 34.

The refrigeration cooling transmission load, L_t, is calculated using the equation

$$L_t = 24 AU (t_1 - t_2) \tag{16}$$

where

A = surface area, ft²
U = overall transmission coefficient, Btu/hr-ft²-deg F
t_1 = boundary outer surface temperature, deg F
t_2 = boundary inner surface temperature, deg F

The transmission load is determined for each of the boundaries of a compartment. The total transmission load is the sum of all the individual boundary loads.

The infiltration heat load, L_i, for each refrigerated space is calculated using

$$L_i = VCH_a \tag{17}$$

where

V = cubical content of storage space, ft³
C = number of air changes per 24 hr
H_a = heat gain of infiltrated air, Btu/ft³

The quantity of air infiltrating into the space is assumed to be a function of the size of the space and is the average of the air changes per 24 hours. Values for air changes for spaces of different size are shown in Fig. 20. The heat gain per cubic foot of air entering is shown in Fig. 21 for various air temperatures. The size of the space is the volume between the finished deck underfoot, the inside surface of the insulation overhead, and the inside of the insulation surfaces bulkhead to bulkhead.

The ventilation heat load, L_v, is calculated using

$$L_v = 8VH_a \tag{18}$$

where

V = volume of storage space, ft³
H_a = heat gain of ventilation air, Btu/ft³

The ventilation heat load is used where it exceeds the infiltration load and is used in lieu of the infiltration load for the operating condition being calculated. It is not included in the pulldown load. The heat gain from outside air is obtained from Fig. 21 for entering air at 100 F and 80% relative humidity. The quantity of outside air is obtained by taking one third of the gross volume of the space. The supply fan capacity is based on the cubic volume of the space and the fan is operated a maximum of 20 minutes per hour.

The total product weight, PW, is determined by

$$PW = AhrW_p \tag{19}$$

where

A = deck area, ft²
h = height of stored product, ft
r = ratio of usable volume to total volume
W_p = product density, lb/ft³

The total container weight, CW, is determined by

$$CW = AhrW_c \tag{20}$$

where W_c is the container weight in lb/ft³.

The product internal heat load, L_p consists of the heat gain due to the internal heat of the product and of the product container and is calculated using

$$L_p = (PW\,C_p + CW\,C_c)\,(t_1 - t_2) \tag{21}$$

where

PW = total weight of product, lb
C_p = specific heat of product, Btu/lb
CW = total container weight, lb
C_c = specific heat of container, Btu/lb
t_1 = initial temperature, deg F
t_2 = final temperature, deg F

Commonly assumed average product load conditions and lengths of time allowed for temperature reduction are given in Table 7.

For chilled storage, the heat loads for respiration heat from fruits and vegetables, for both pulldown and for normal operations, are also included and are calculated by using the equations:

REFERENCE: DDS 516-1 SPACE: CHILL NO. 1 (ALTERNATE B) TEMP. 33 °F

KEY REFERENCE
1. Overboard
2. Deck
3. Port Blk.
4. Stbd. Blk.
5. Fwd. Blk.
6. Aft. Blk.

DIMENSION: 23 x 24 x 7
VOLUME (INSUL TO INSUL) 3864 cu. ft.
TOTAL PRODUCT WEIGHT (PW) (Deck Area x 6 x Ratio x W_p)
23 x 24 x 6 x .75 x 29.7 = 73,775 Lbs.
TOTAL CONTAINER WEIGHT (CW) (Deck Area x 6 x Ratio x W_c)
23 x 24 x 6 x .75 x 3.2 = 7,949 Lbs.

CALCULATIONS

KEY REF	TD		AREA		U		24	PULL-DOWN OPERATION Btu/24 hrs.	NORMAL OPERATION Btu/24 hrs.
			TRANSMISSION						
1	67	x	23 x 24	x	0.061	x	24	54,145	54,145
2	67	x	23 x 24	x	0.029	x	24	25,741	25,741
3	67	x	23 x 7	x	0.060	x	24	15,533	15,533
4	0	x		x		x	24	----	----
5	-33	x		x		x	24	----	----
6	67	x	24 x 7	x	0.060	x	24	16,209	16,209
TOTAL TRANSMISSION HEAT LOAD								111,628	111,628

INFILTRATION (Vol. x h_a x Air changes/24 Hrs.)
3864 x 3.2 x 3.5 — 43,277 | ----

VENTILATION (Vol. x h_a x 24 Hrs./3)
3864 x 3.2 x 8 — ---- | 98,918

PRECOOLING INTERNAL HEAT (PWXC$_p$ + CWXC$_c$) ($t_1 - t_2$)/Days
[(73,775 x .85) + (7,949 x .65)] x (55 - 33)/2 — 746,631 | ----

PRECOOLING RESPIRATION PW($R_1 + R_2$)/2
73,775 x (3.20 + 1.08)/2 — 78,202 | ----

STORAGE RESPIRATION (PW x R_2)
73,775 x 1.08 — ---- | 79,677

SUBTOTAL Btu/24 Hours — 979,738 | 290,223
SUBTOTAL Btu/Hr. (Divide by 24) — 40,822 | 12,093

UNIT COOLER CAPACITY TONNAGE (Divide by 12,000) — 3.41 | 1.01

FAN MOTOR HEAT (No. x hp x Btu/24 Hrs.)
1 x 1 x 88,800 — 88,800 | 88,800
SUBTOTAL Btu/Hr. (Divide by 24) — 3,700 | 3,700

CONDENSING UNIT CAPACITY (TONS)
DIVIDE BY 12,000 ADD TO UNIT COOLER TONNAGE — 3.41+0.31= 3.72 | 1.01+0.31= 1.32

SHIP ___   DATE ___   ENGR. ___

Fig. 19 Refrigeration load calculations for 33°F storage

REFERENCE: DDS 516-1 SPACE: CHILL NO. 1 (ALTERNATE A) TEMP. 0 °F

KEY REFERENCE
1. Overboard
2. Deck
3. Port Blk.
4. Stbd. Blk.
5. Fwd. Blk.
6. Aft. Blk.

DIMENSION: 23 x 24 x 7
VOLUME (INSUL TO INSUL) 3864 cu. ft.
TOTAL PRODUCT WEIGHT (PW) (Deck Area x 6 x Ratio x W_p)
23 x 24 x 6 x .75 x 35.9 = 89,176 Lbs.
TOTAL CONTAINER WEIGHT (CW) (Deck Area x 6 x Ratio x W_c)
23 x 24 x 6 x .75 x 3.64 = 9,042 Lbs.

CALCULATIONS

KEY REF	TD		AREA		U		24	PULL-DOWN OPERATION Btu/24 hrs.	NORMAL OPERATION Btu/24 hrs.
			TRANSMISSION						
1	100	x	23 x 24	x	0.061	x	24	80,813	80,813
2	100	x	23 x 24	x	0.029	x	24	38,419	38,419
3	100	x	23 x 7	x	0.060	x	24	23,184	23,184
4	33	x	23 x 7	x	0.060	x	24	7,651	7,651
5	0	x	24 x 7	x	0.060	x	24	----	----
6	100	x	24 x 7	x	0.060	x	24	24,192	24,192
TOTAL TRANSMISSION HEAT LOAD								174,259	174,259

INFILTRATION (Vol. x h_a x Air changes/24 Hrs.)
3864 x 4.8 x 3.5 — 64,915 | ----

VENTILATION (Vol. x h_a x 24 Hrs./3)
x   x 8 — ---- | 64,915

PRECOOLING INTERNAL HEAT (PWXC$_p$ + CWXC$_c$) ($t_1 - t_2$)/Days
[(89,176 x .40) + (9,042 x .65)] x (15 - 0)/2 — 311,608 | ----

PRECOOLING RESPIRATION PW($R_1 + R_2$)/2
x (+   )/2 — ---- | ----

STORAGE RESPIRATION (PW x R_2)
x — ---- | ----

SUBTOTAL Btu/24 Hours — 550,782 | 239,174
SUBTOTAL Btu/Hr. (Divide by 24) — 22,949 | 9,966

UNIT COOLER CAPACITY TONNAGE (Divide by 12,000) — 1.91 | 0.83

FAN MOTOR HEAT (No. x hp x Btu/24 Hrs.)
1 x 1 x 88,800 — 88,800 | 88,800
SUBTOTAL Btu/Hr. (Divide by 24) — 3,700 | 3,700

CONDENSING UNIT CAPACITY (TONS)
DIVIDE BY 12,000 ADD TO UNIT COOLER TONNAGE — 1.91+0.31= 2.22 | 0.83+0.31= 1.14

SHIP ___   DATE ___   ENGR. ___

Fig. 18 Refrigeration load calculations for 0°F storage

Table 7   Average product load conditions

	Storage Space	
	Chill	Freeze
Final temperature, °F	33	0
Entering temp. max., °F	55	15
Time permitted for reducing temp., days:		
ship's refrigerated stores	2	2
refrigerated cargo spaces	3	5
Average product wt., lb/ft³	29.7	35.9
Average product specific heat, Btu/lb	0.85	0.40
Average container wt., lb/ft³	3.2	3.64
Average container specific heat, Btu/lb	0.65	0.65
Respiration rate, Btu/24 hr:		
entering condition	3.20	—
final condition	1.08	—

$$\text{Precooling respiration} = PW \frac{(R_1 + R_2)}{2} \qquad (22)$$

$$\text{Storage respiration} = PW R_2 \qquad (23)$$

where.

PW = total weight of product, lb

R_1 = respiration rate at entering temperature, Btu/lb per 24 hr

R_2 = respiration rate at final temperature, Btu/lb per 24 hr

Table 8   Heat gain from electric motors

Motor Rating, hp	Equipment Load, Btu/hr-hp
1/20 to 1/8	5500
1/8 to 1/2	4250
1/2 to 3	3700
3 to 20	2900

The specific heat and respiration rates used for naval calculations are listed in Table 7.

The available volume of a refrigerated space for maximum product storage is the volume remaining after allowances for package spacing, air circulation, deck gratings, battens, cooling coils, and other obstructions. The available volume for product storage, based on typical installations used in naval ships, is shown in Fig. 22.

The equipment load is usually a load due to fan electric motors. The heat gain based on the horsepower rating of an electric motor is given in Table 8.

An example of a typical refrigeration load summary is shown on Fig. 23. The total cooling load should be proportioned between multiple condensing units to allow flexibility of operation under pulldown and normal operating conditions and to permit the use of identical equipment in each system on the ship. Units should be selected where

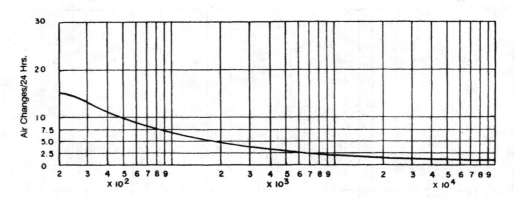

Fig. 20   Refrigerator average air changes per 24 hr due to infiltration and door openings

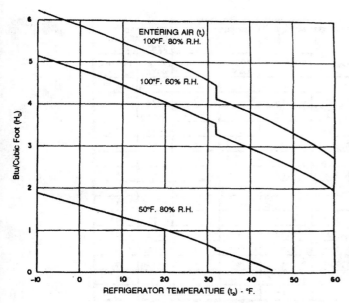

Fig. 21 Heat gain caused by air entering refrigerator

Table 9 Rating for Model "X" condensing unit at a 105 F condensing temperature

Saturated Suction Temperature, °F	Tons of Refrigeration
−26	6.9
−25	7.1
−22	7.6
−21	8.1
−20	8.3
−18	9.1
−16	9.7
−14	10.5
−10	12.1
0	16.3

The cooling-coil surface area, A, for each refrigerated space is based on the space cooling load and the temperature differential between the space and the refrigerant at the cooling coils and can be determined by the equation

$$A = \frac{L_{pd}}{U(t_2 - t_r)} \tag{24}$$

where

L_{pd} = compartment pulldown cooling load, Btu/hr
U = overall transmission coefficient, Btu/hr-ft²-deg F
t_2 = space temperature, deg F
t_r = temperature of refrigerant entering cooling coil, deg F

The transmission coefficient, U, for naval cooling coils is obtained from Fig. 24. Based on equation (24), the cooling-coil surface required is shown in Fig. 25.

Additional details concerning naval design calculations can be found in reference 33. Data for commercial systems can be found in reference 11.

feasible, to equalize loadings on multiple condensing units during both pulldown and normal operation.

The cooling load calculations indicate that for normal operation the maximum load is "Alternate A." This condition requires 8.01 tons of refrigeration with one unit secured and is the basis on which selection of the condensing units would be made. Two units of 8.01 tons each would be selected. For the pulldown operation, "Alternate B" is the maximum load, requiring 18.73 tons of refrigeration with all units operating. However, during pulldown of the chilled boxes with the circuits as indicated on the summary sheet, the additional refrigeration capacity may be obtained with higher suction temperatures (see Table 9).

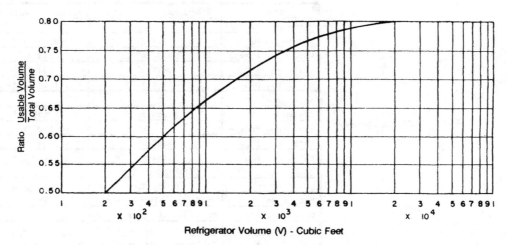

Fig. 22 Percentage of refrigerated space available for product storage

SPACE	STORAGE SPACE DESIGN TEMP.	PULL-DOWN OPERATON		NORMAL OPERATION		COMPRESSOR NO.		CONDENSOR UNIT SUCTION TEMP. °F	
		Btu/hr.	TONS	Btu/hr.	TONS	PULL DOWN	NORMAL	PULL DOWN	NORMAL
Alternate "A"									
Freeze No. 1	0	85,941	7.16	48,492	4.04	1	1	-24	
Ward room freeze	0	9,544	0.80	5,132	0.43	2	1	-26	-21
Chill No. 1	0	26,649	2.22	13,666	1.14	2	1	-26	-21
Chill No. 2	0	47,110	3.93	28,834	2.40	2	1	-26	-21
							2*		
TOTAL		169,244	14.11	96,124	8.01				
Alternate "B"									
Freeze No. 1	0	85,941	7.16	48,492	4.04	1	1	-21	-21
Ward room freeze	0	9,544	0.80	5,132	0.43	1	1	-21	-21
Chill No. 1	33	44,522	3.72	15,793	1.32	2	1	-13	-21
Chill No. 2	33	84,543	7.05	25,720	2.14	2	1	-13	-21
							2*		
TOTAL		224,550	18.73	95,137	7.93				

* Standby

Fig. 23   Refrigeration load summary

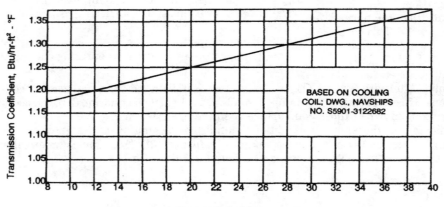

Fig. 24   Transmission coefficient for finned surface cooling coils (copper fin and tube, gravity type)

SPACE	PULL-DOWN OPERATON	NORMAL OPERATION
Alternate "A"		
Freeze No. 1	3,248 ft²	2,319 ft²
Ward Room Freeze	344 ft²	246 ft²
Chill No. 1	961 ft²	654 ft²
Chill No. 2	1,700 ft²	1,380 ft²
Alternate "B"		
Freeze No. 1	3,850 ft²	2,173 ft²
Ward Room Freeze	428 ft²	230 ft²
Chill No. 1	787 ft²	632 ft²
Chill No. 2	1,494 ft²	1,029 ft²
SELECTED COIL SIZE		
SPACE	CONDITION	COIL AREA
Freeze No. 1	Pulldown, Alternate "B"	3,850 ft²
Ward Room Freeze	Pulldown, Alternate "B"	428 ft²
Chill No. 1	Pulldown, Alternate "A"	961 ft²
Chill No. 2	Pulldown, Alternate "A"	1,700 ft²

Fig. 25   Coil surface calculation summary

Section 3
Distributive Systems

3.1 Air-Distribution Systems.   The evolution of air-conditioning systems has been driven by demands for increased comfort and safety. The demand for increased comfort has resulted in lower design temperatures during the cooling season, humidity control, increased amounts of fresh air, lower sound levels, less draftiness, and an individual control of space temperature. The demand for increased safety has resulted in greater protection against fire and smoke with improvements in detection, alarms, and confinement. Accomplishing the above led to the increased use of automation. Automation, in turn, has improved with the use of computer-aided monitoring and control.

Comfort air-conditioning systems on merchant ships are classified as those serving passenger cabins, crew and officer spaces, and public spaces. Public spaces, which include messrooms, dining rooms, lounges, and similar spaces, are usually treated as a single zone. Usually a fan-coil central-station system is used and, in exceptionally large spaces, two systems may be required. These systems are known as Type A systems. In some installations 100% replenishment air is used to avoid the use of return air ducts. This, however, results in an increased load on the refrigeration plants. The outside air and the return air, if used, are filtered before being preheated, cooled, and reheated as required at the central-station unit. A room thermostat maintains the design temperature. During mild weather, dampers may be arranged to automatically use all outside air.

Crew and officer quarters are frequently served by multizone central systems. In these systems, known as Type C systems, spaces are grouped into zones with similar exposures and loads. Each zone has a reheat coil that is controlled by a thermostat to temper the air for all of the spaces served. A manually controlled damper is the only means for controlling the amount of air and thus the temperature in each space.

Terminal reheat systems, or Type D systems, are usually used for passenger staterooms and may also be used for crew and officer quarters. This is the predominant type of system used in naval construction. In this system conditioned air is cooled and supplied to each space based on the design cooling load. The room temperature is controlled by a thermostat that controls a reheat coil. A minimum amount of outside air is supplied and mixed with the recirculated air and is then filtered, cooled by the chilled-water cooling coil, dehumidified, and distributed by a supply fan through ductwork to the spaces included in the system. During the heating season, the conditioned air is heated by a reheat coil, which provides the additional heat needed to maintain the design room temperature.

Another type of system used for passenger staterooms and other small spaces is the air-water induction system,

designated as a Type E system. In this system a central station heats or cools the outside air only. The outside air, or primary air, is distributed to induction units located in the space to be heated or cooled. Nozzles in the induction unit, through which the primary air flows, induces a fixed amount of room or secondary air to flow through a water coil and mix with the primary air. The mixture of treated air is then discharged into the space through supply grills. The water coil is used to either heat or cool the air by using either hot or chilled water. The flow of water can be controlled either manually or automatically to maintain the desired temperature in the space. In this system, no return ductwork is required since only the outside air is conditioned by the central-station equipment. The outside air, during the cooling season, must be sufficiently cooled to remove enough moisture to satisfy the latent load of both the outside air and the recirculated space air. Since the primary air is distributed at a high velocity and pressure, the distribution ductwork is small. However, this space saving is offset by the additional space needed for the water piping, secondary water pumps, the induction units in each space, and the drain piping.

Another system, known as a Type G system, has been increasingly used in marine applications. This system is a high-velocity dual-duct system. In this system all air is filtered, cooled, heated, and dehumidified in the central unit. Blow-through arrangements are usually used in this system. A high-pressure fan circulates high-velocity air through two ducts; one duct carries chilled air, and a second duct carries heated air. The heated and chilled air flows to an air-mixing unit in each space served. Each space has a control valve that controls the amount of chilled or heated air needed to satisfy the demand indicated by the room thermostat. Any desired temperature within the capacity of the installed equipment can be provided regardless of the load requirements of adjacent spaces. The advantages of this system are that all air-conditioning equipment is centrally located and only air is distributed to the spaces, there is no piping or electric wiring in the conditioned spaces, the system can heat and cool adjacent spaces simultaneously, direct-expansion refrigerants can be used, and water-chilling equipment is not needed.

The ductwork associated with the various air-distribution systems requires early coordination with other systems, such as piping and electrical systems, and other requirements such as damage control onboard the ship. The air-conditioning system requires a considerable amount of space, and early planning and coordination are essential to establish the space needed for ventilation trunks, ductwork, and fan rooms; to take into consideration requirements for fire and safety regulations; and to optimize the number of systems in order to minimize installation costs, maintenance costs, and the total amount of occupied space. The design of the ductwork must take into account the space available, noise considerations, damage and fire control, and such features as airtightness and watertightness [5,35,36]. Balancing of ventilation systems is usually accomplished by fixed orifices, in lieu of balancing dampers, to prevent adjustments by unauthorized personnel. For additional information on air-distribution systems and components, see references 1, 2, 12, 23, and 37.

Air-distribution systems for machinery spaces for merchant and naval ships usually consist of independent supply and exhaust ventilation systems [2,27]. Two-speed fans are used with the low speed intended for use during the heating season. Where two supply and two exhaust fans are necessary, one set should take air and exhaust air to starboard and the other set should take air and exhaust air to port. This would insure a supply of smoke-free air to machinery spaces in case of fire. When a single-speed supply fan is used, a single-speed exhaust fan is used, or a two-speed exhaust fan is wired for single-speed operation only. The system is designed to maintain a slight negative pressure in the space. Supply air is provided to cool watch stations, working areas, log desks, and workbenches. The supply terminals are directed to supply air to the watch stander from a distance of 3 to 5 ft. Exhaust terminals are located in the overhead of the space and over the main heat sources. The exhaust-terminal openings are positioned in a horizontal plane. Care should be taken to avoid drawing heated air across working areas or watch standers locations.

High-velocity induction ventilation systems, which utilize a series of jets to direct the flow of air and obtain the desired air distribution, have been employed in some ships, primarily those with diesel power. The system is used in engine rooms, cargo holds, etc. and has the desirable feature of small distribution ductwork.

a. Fans. The movement of air into, out of, and within a ship is generally accomplished through the use of fans. Fans used on ships are divided into two classes, namely, centrifugal and axial. Axial fans can further be classified into vaneaxial, tubeaxial, and propeller.

Whereas naval ships predominantly use vaneaxial fans due to space considerations, the predominant choice on merchant ships is the centrifugal fan. Centrifugal fans used on naval ships are installed in explosive or flammable gas systems where the motor must be located outside of the airstream and in fan-coil assemblies.

Centrifugal fans produce pressure primarily from the centrifugal force created by rotating the air enclosed between the blades, with some additional pressure from the velocity leaving the impeller. The entering airflow is axial and the discharge airflow is radial. Most marine centrifugal fans, such as shown by Fig. 26, are of the backward-curved blade type; in smaller sizes (less than 1000 cfm) fans with forward-curved blades are often used. The principal advantage of a backward-curved fan is that both maximum brake horsepower and fan-set efficiency at any speed occurs very close to the same capacity point; thus, the motor is selected for a horsepower corresponding to

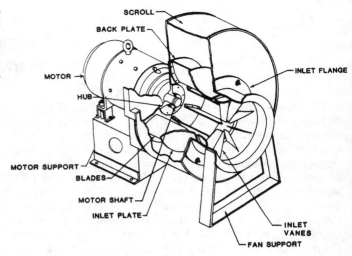

Fig. 26   Centrifugal fan (Type CC)

the highest efficiency point. Equally important is the fact that if the actual system pressure loss is more or less than estimated, the motor selected for a specific fan speed will not overload. Forward-curved fans have a rising bhp characteristic as the capacity approaches free delivery; thus, a motor selected for maximum fan-set efficiency can be overloaded if the actual system operating pressure is less than estimated. It is common practice to oversize motors driving forward-curved fans to accommodate operation at free delivery.

Belt drives offer advantages where the fan speed is less than 1750 rpm, allowing the use of less-expensive and smaller high-speed motors. The adjustability of belt drives facilitates system balancing, which is significant for air-conditioning systems, since very limited overdelivery can be tolerated. Quieter, more efficient operation can be obtained by reducing the fan speed vice adding resistance to the system. Because of the foregoing, belt drives are often specified for fans of air-conditioning systems. Centrifugal fans are usually of the single-inlet, single-width type.

Vaneaxial fans, Fig. 27, are generally used in naval ships because of their compactness. The fan consists of a cylindrical casing enclosing a motor which is face mounted via a flange welded to the air guide vanes. The vanes transform the kinetic energy produced by the impeller, in the form of whirl, into pressure energy. The vanes increase the fan efficiency by straightening the rotary motion of the airstream and producing axial flow. The vanes are the basis for the name "vaneaxial." Since the motors must be watertight and construction tolerances restricted, axial fans are relatively expensive and have the disadvantage of motor inaccessibility. The static-pressure-volume curve of axial-flow fans generally has a dip at the left of the peak; this requires that such fans be operated at a point on the curve below and to the right of the peak. Motors are selected with substantial power

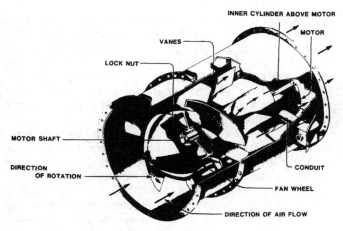

Fig. 27　Vaneaxial fan

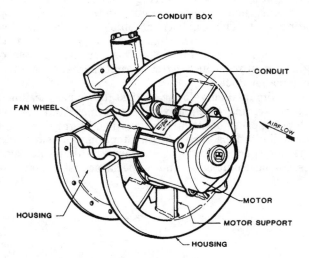

Fig. 28　Tubeaxial fan

margins due to the rising bhp requirements and the reduction in cooling air for the motor as zero delivery is approached.

Tubeaxial fans, Fig. 28, are similar to vaneaxial fans. They differ in that there are no vanes, the ratio of the blade-hub to blade-tip diameter is larger, and the airflow is drawn over the motor and discharged from the wheel end. A tubeaxial fan is generally used in spaces that are not normally ventilated and are rarely used where ducting is required, due to the low pressure available.

Propeller fans are usually used for free air delivery since they develop very little pressure. They are not used in naval construction.

The three most essential relationships concerning fan performance are as follows:

- The delivery varies directly with speed, i.e.

$$Q_1 = (N_1/N_2)Q_2 \qquad (25)$$

- The static pressure generated varies with the speed squared, i.e.

$$P_1 = (N_1/N_2)^2 P_2 \qquad (26)$$

- The horsepower required varies with the speed cubed, i.e.

$$HP_1 = (N_1/N_2)^3 HP_2 \qquad (27)$$

Q, P, HP, and N denote fan delivery, pressure, horsepower, and speed respectively; and subscripts 1 and 2 denote two different operating conditions.

Fan efficiency is generally expressed as either mechanical efficiency, ME, or static efficiency, SE, as follows:

$$ME = \frac{CFM\ TP}{6356\ BHP} \qquad (28)$$

$$SE = \frac{ME\ SP}{TP} = \frac{CFM\ SP}{6356\ BHP} \qquad (29)$$

where

CFM = air volume flow rate at inlet, cfm
BHP = power output of drive, hp
SP = fan static pressure, in. of water
TP = fan total pressure, in. of water

Flexible connections are installed between fans and ducts to reduce noise, isolate fan vibration from the ducts, and accommodate slight misalignments. Such connections also prevent fan casing distortions, which can occur when the casings are directly connected to heavy coamings, ducts, or trunks. This is particularly important for axial fans because of their small wheel-tip clearances.

When a vaneaxial fan takes suction from a plenum, compartment, or trunk that is much larger than the fan inlet, a bellmouth is used. Bellmouth intakes reduce the system pressure loss, turbulence, and noise and equalize the airflow over the inlet face of the fan, thus improving performance.

Fan supports may be rigid or resilient. Where rigid, the fan base is bolted directly to angle foundations, which are welded to the heaviest structure available, with the exception of bulkheads adjacent to living spaces or other spaces requiring low noise levels. Fans within living areas of merchant vessels are customarily supported on resilient mounts. The fan and motor of a belt-driven centrifugal fan are mounted on a common base to preserve alignment. The base is supported on resilient mounts, which in turn are bolted to angles welded to the ship's structure. Flexible duct connections are always fitted to fans on resilient mounts.

Naval designs include families of vaneaxial and centrifugal fans that are designed to meet high-shock and vibration requirements. Sizes, performance characteristics, and dimensions are standardized to enhance replacement and interchangeability. Physical data and fan performance characteristics for Navy-standard fans can be found in reference 12. Specifications for fans used in merchant construction may be found in reference 2.

b. Flow and fan pressures. The basic relationship between airflow and the pressure (head) required to cause the flow (with standard air density of 0.075 lb/ft³) can be expressed as

$$h_v = (V/4005)^2 \qquad (30)$$

where

h_v = air velocity pressure head, in. of water
V = velocity of air, fpm

There are three interdependent pressures common to air-handling systems: static, velocity, and total. Static pressure is the hydrostatic (bursting) pressure required to overcome losses due to duct friction, air friction, and velocity changes, i.e., the energy for producing and maintaining flow against resistance. Velocity pressure is the dynamic pressure corresponding to the flow of air and can be computed by equation (30). The total pressure at any point in a system is the sum of the static and velocity pressures as illustrated in Fig. 29(a). Note that while the same air quantity flows through the sections of the duct system in Fig. 29(a), both the static and velocity pressures change radically; however, the total pressure remains unchanged (losses have been disregarded since they are irrelevant to this discussion). Therefore, to properly define a system, both its total pressure and capacity must be specified.

The static pressure, S, may be positive or negative, depending on the relationship of the measured point to the fan. The velocity pressure, V, is always positive. The total pressure, T, may be positive or negative, depending on the respective values of the other two pressures. Figure 29(b) shows these pressures (in inches of water) as applied to a fan. The subscripts i, o, and f denote the pressures at the fan inlet, fan outlet, and that of the fan, respectively. By definition, $S_f = T_f - V_o$, and this is the static pressure given in fan catalogs. A complete review of reference 38 may facilitate a full understanding of the pressure terminology applicable to the various types of fans (centrifugal, axial-flow, and propeller).

c. System flow and pressure characteristics. The relationship between the system pressures at any two rates of airflow can be expressed as

$$P_1/P_2 = (Q_1/Q_2)^2 \qquad (31)$$

where

P_1 = pressure required for airflow Q_1
P_2 = pressure required for airflow Q_2

A system pressure characteristic curve, Fig. 30, thus can be constructed once the pressure for one capacity is known, using an arbitrarily selected series of capacities.

As indicated, the fan pressure-volume curve crosses the air system curve at one point only. Accordingly, an air-handling system must be designed to accommodate a specific fan, or a fan must be selected to suit a system's characteristics.

Excess system delivery can be corrected by adding resistance to flow (dampers, orifices, or equivalent) or by reducing the fan speed. With belt-driven fans, the latter method is simple, quiet, and most efficient. Deficient system delivery can be corrected only by modifying the duct system to reduce the pressure required or by increasing the speed of the fan with due consideration being given to the possibility of overloading the fan's prime mover.

d. Duct design methods. The design of an air-handling system is an engineering balance between space limitations, noise considerations, initial cost, and operating cost. The capacity of ship's service electrical plants has steadily increased over the years, thus affording more power for fans. This permits the use of higher pressures, which in turn permits higher duct velocities (smaller ducts).

A maximum duct velocity of 3500 fpm is permissible in systems serving machinery spaces on merchant vessels, and most spaces on naval construction. Regarding cargo spaces, limitations to power consumption, fan heat limitations, and space limitations determine the allowable velocities. On merchant ships, unless special acoustic treatment

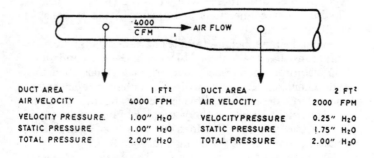

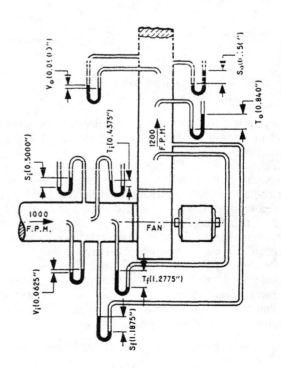

(a) DUCT PRESSURES

(b) FAN PRESSURES

Fig. 29   Illustration of duct and fan pressures

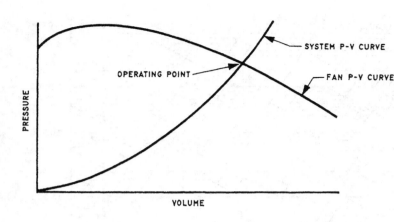

Fig. 30   Determination of system operating point

is provided, the maximum duct velocity usually is 2000 fpm in systems serving living spaces and similar quiet areas. However, velocities as high as 2500 fpm are permitted, locally, to reduce the size of penetrations where structural strength is critical. Velocities of 4000 fpm are commonly used in systems having pressure-tight (leak-tested) prefabricated ducts and fittings. Such systems usually operate at high static pressures (6 to 10 in. of water). In these cases, special sound attenuation provisions are required; i.e., lined ducts, sound traps, and integral treatment of terminal and air-mixing units.

In merchant practice, air-handling systems are designed conservatively to avoid excessive noise generation. Specific methods for controlling noise are contained in references 23, 39, and 40.

Several methods are used to determine duct sizes [23]. The method most commonly used for merchant ships is a combination of the velocity and equal-friction methods. Based on experience, a unit friction loss is selected, e.g., 0.5 in. of water/100 ft of duct. Then from data such as Fig. 31 the diameter of each duct in the system is determined based on this unit friction loss and the required capacity; however, the velocity is not allowed to exceed the maximum specified value. As an example, point A in Fig. 31 indicates the transition point for a unit loss of 0.5 in./100 ft and maximum velocity of 2000 fpm. This method does not consider differences in air path lengths or other losses contributing to the total resistance of various parallel circuits, i.e., submains and branches. Accordingly, the total pressure loss between the fan and each terminal must be equalized by using balancing devices (dampers, air valves, or orifices) fitted in submains and branches.

A second method, which is a refinement of the first, is used mainly in naval construction [35,36]. It achieves optimum design (minimum weight and space) by sizing submains and branches to absorb all of the available total (fan) pressure. The total pressure available at each point of divided flow (takeoff) is determined, after which each submain and branch is sized to completely absorb this pressure. Balancing devices are required, theoretically, only where the maximum allowable velocity controls the duct size. This method requires a considerable amount of "cut and try" analysis.

The foregoing discussion applies primarily to conventional or low-pressure systems having velocity limitations because of noise and power restrictions. Systems having fans that produce higher static pressures are called medium pressure (3.75 to 6.75 in.) or high pressure (static pressure exceeding 6.75 in.) [23]. Generally, medium- and high-pressure systems require special means of sound control. Reference 38 contains simplified techniques for sizing the ducts of medium- and high-pressure systems; combinations of velocity, equal-friction, and static-regain methods are used.

e.   Fan pressures—static versus total.   Designers of land installations usually select a fan to meet a system's static pressure requirements; therefore, most catalog data express system component losses in terms of static pressure. However, ship designers usually select a fan for the total pressure requirements of a system, and all losses are expressed in terms of total pressure.

When the velocity of air in a duct decreases, some of the velocity-pressure reduction is converted to static pressure [see equation (30) and Fig. 29(a)]; the remainder is lost. The increase in static pressure, called static regain, depends on the efficiency of conversion. When designing systems on a static-pressure basis, each increase in velocity is assigned a loss equivalent to: the change in velocity pressure, plus a loss for conversion. For a gradual contraction, the latter is usually about 10% of the final velocity pressure. Additionally, each reduction of velocity is credited with a regain; this seldom exceeds 80% of the velocity-head reduction. When designing systems on a total-pressure basis, however, all changes in velocity are treated as losses. Both methods should give the same results. However, the total-pressure concept has two advantages: the fan outlet velocity (or velocity pressure) does not have to be evaluated separately, and since all factors are expressed as losses, critical items are more evident. Additionally, pseudo "regain" can misguide an inexperienced designer.

f.   Total-pressure losses—ducts and fittings.   Friction losses for straight round ducts, at standard air conditions, are given in Fig. 31. Velocity heads in inches of water, per equation (30), are also given. Figure 32 is used to determine rectangular ducts equivalent to the round

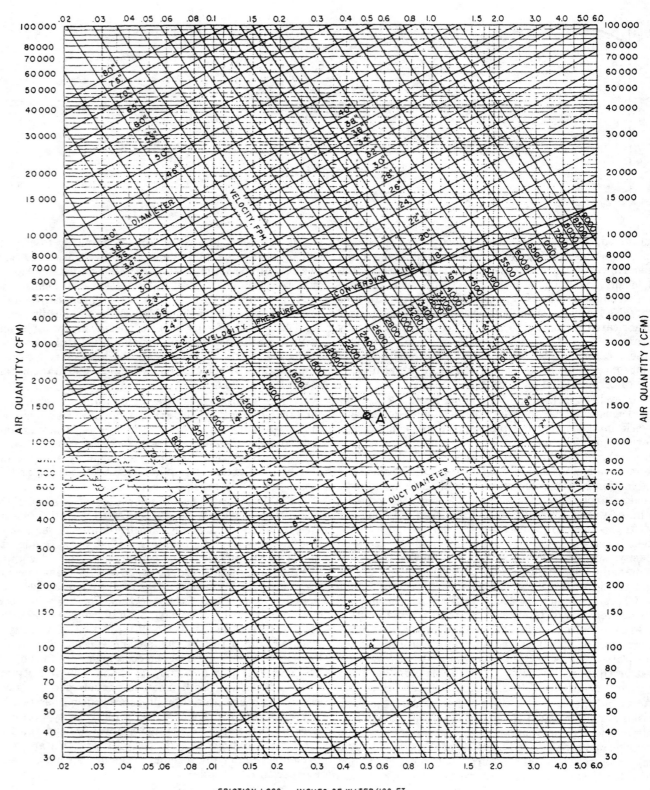

FRICTION LOSS , INCHES OF WATER/100 FT

Fig. 31 Friction loss in straight round ducts

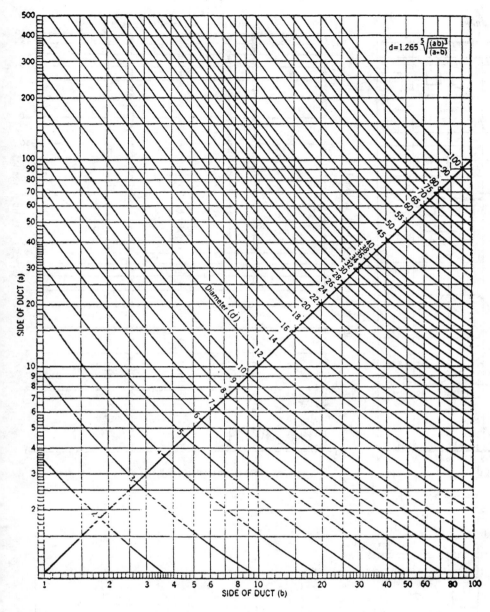

$$d = 1.265 \sqrt[5]{\frac{(ab)^3}{(a+b)}}$$

Fig. 32 Rectangular equivalent of round ducts

ducts selected from Fig. 31. The depth of ducts in living areas frequently must be limited to 6 in. to preserve headroom. Rectangular-duct aspect ratios (ratio of large to small dimension) over three are not recommended. Fitting losses are generally expressed as

$$L = K h_v \qquad (32)$$

where

L = fitting total pressure loss, in. of water
K = loss coefficient
h_v = air velocity pressure head, in. of water

In marine circles the loss coefficient, K, is expressed as a fraction or multiple of h_v.

Elbow losses in duct runs are shown by Figs. 33 and 34. Figure 35 shows the location of splitters in rectangular elbows, properly arranged to reduce losses. Elbow losses

are considered to be proportional to the subtended angle. Thus, a 45-deg elbow is assigned a loss equal to half of a 90-deg elbow. The loss in an elbow consists of the friction loss and dynamic loss due to the compression of air at the outside (heel) of the bend. If sufficient straight duct follows, a regain occurs. However, if the air leaving the elbow discharges to the atmosphere, no regain occurs and the loss is greater. A straight length of discharge duct of at least four diameters, or four times the duct dimension in the plane of the bend, is required before "in run" elbow losses apply. The straight length of duct required can be reduced by installing splitters or turning vanes. The maximum splitter or vane spacing then becomes the criterion rather than the duct size. Losses for elbows at the end of a run, Fig. 36, include one velocity head for the discharge to atmosphere. To find the loss for an elbow at

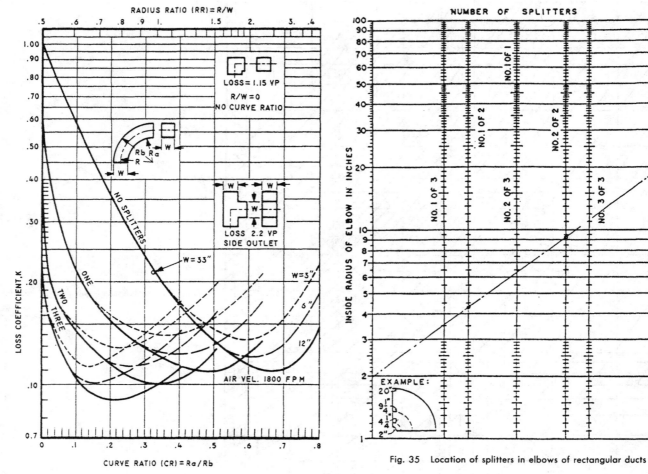

Fig. 33   Rectangular in-run duct elbow losses

Fig. 35   Location of splitters in elbows of rectangular ducts

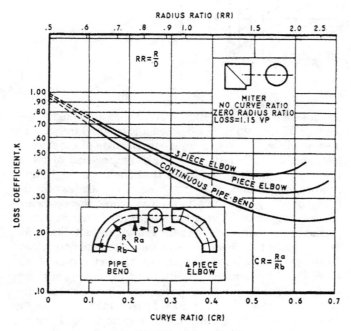

Fig. 34   Round in-run duct elbow losses

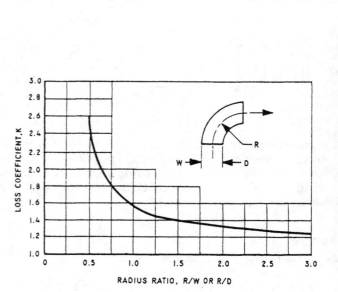

Fig. 36   Losses in elbows discharging to atmosphere

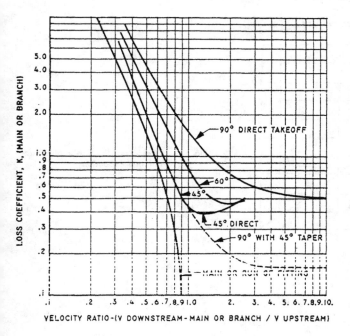

Fig. 37 Losses in divided-flow fittings

Figures 39 and 40 show the losses due to velocity or area changes, i.e., abrupt expansion, abrupt contraction, and gradual expansion, in a run of duct. When a connecting duct is less than four diameters downstream of an abrupt expansion, regain is lost and the loss approximates that for an open-ended duct, namely, one small-duct velocity head. Abrupt contractions create a vena-contracta at the entrance to the smaller duct. Unless there is straight duct following the contraction (for regain), the loss equals that of a square-edged orifice at the end of a duct [Fig. 41(a)]. This point deserves emphasis as it is frequently overlooked in the mechanical execution of calculations.

A loss of 0.05 of the small-duct velocity head is sufficient for circular concentric gradual contractions up to an included angle of 45 deg. The loss allowed for symmetric transitions from one rectangular size to another of equal area is $0.15 \, h_v$. Where transitions are asymmetric or one-sided, the assigned loss should correspond to that for the largest angle on one side. This also applies to the gradual expansion of rectangular ducts.

Losses for the more common terminals and fittings are given in Figs. 40 and 41 and Table 10. Again, it is presupposed that there is ample straight duct at entrances to obtain reexpansion of the vane contraction. Catalogs for exhaust registers and grilles usually show only the static pressure loss. Therefore, additional data are required to determine the total pressure loss and avoid misapplication. Catalog losses for registers, diffusers, and other purchased supply terminals usually represent only the pressure imposed by applying the terminal to an open-ended duct. To approximate the total pressure loss, add one h_v (based on the connecting duct size) to the loss given in the catalog. Terminals for merchant construction are usually sized on a velocity basis (see Table 11).

Weather terminals are sized for low velocities where possible, because weatherproof features entail high pressure losses. The data given in Table 10 for a 0.5-in. wire mesh and insect screen allow for dirt, lint, and painting, which appreciably reduce the effective area. Pressure losses through equipment (registers, diffusers, flexible ducts, cooling coils, and heaters) are preferably obtained from the manufacturers' certified data.

the end of a duct with concentric splitter vanes, deduct one velocity head from the loss shown, multiply the remainder by the ratio of losses in Fig. 33 for elbows in runs with and without splitters, and add one velocity head. To find the loss for an elbow of less than 90 deg at the end of a run, deduct one velocity head from the loss in Fig. 36, multiply the remainder by the ratio of the elbow's angle and 90 deg, and add one velocity head. Losses of elbows in runs with small-radius turning vanes may be taken as 0.25 and 0.35 for double- and single-thickness types, respectively. For elbows at the end of runs, add one velocity head to these values.

The pressure losses due to divided flow in round supply ducts (branches and mains) are given in Fig. 37. Lacking better data, Fig. 37 may also be used for round or rectangular branches with rectangular mains. Losses in divided-flow exhaust fittings are given in Fig. 38.

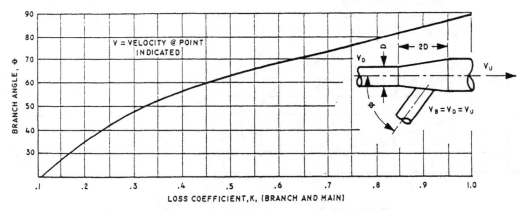

Fig. 38 Losses in divided-flow exhaust fittings

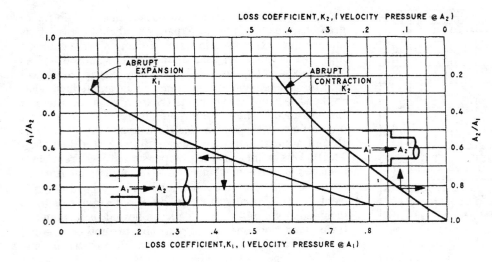

Fig. 39 Duct losses due to abrupt contraction and expansion

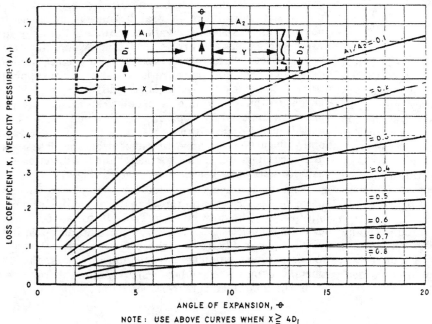

Fig. 40 Losses due to gradual expansion

NOTE: USE ABOVE CURVES WHEN X ≧ 4D₁
AND Y ≧ 4D₂. IF Y < 4D₂, SEE TEXT.
IF X < 4D₁, APPLY CORRECTION FACTOR
GIVEN BELOW.

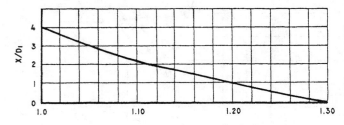

CORRECTION FACTOR

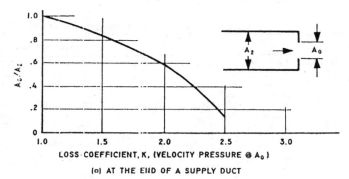

(a) AT THE END OF A SUPPLY DUCT

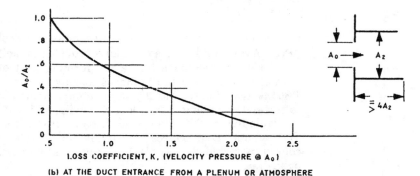

Fig. 41   Losses for square-edged orifices

(b) AT THE DUCT ENTRANCE FROM A PLENUM OR ATMOSPHERE

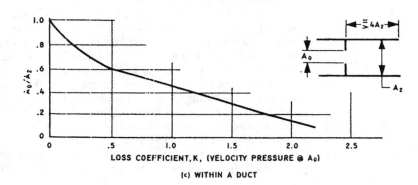

(c) WITHIN A DUCT

g. Connections to equipment. The performance of equipment, including pressure drop, is based on tests performed with a uniform inlet (face) velocity. Incorrectly designed connecting ducts can invalidate such performance data. The equipment's resistance cannot be depended upon to equalize the inlet velocity. Unfortunately, axial-flow fan literature seldom emphasizes this point. Aboard ship, a plenum with a bellmouth (in lieu of an elbow) is frequently fitted at the fan's inlet. Where elbows are fitted, they are of the small-radius vane type, and connected to the fan by concentric, gradual transformers. Integral fixed inlet vanes and bellmouths reduce the effect of poor approach conditions. This is a feature of certain high-quality centrifugal fans [40].

At least one diameter of straight duct is recommended at fan discharges before any elbow, split, offset, or transition. Where divided-flow volumes at a fan's outlet vary during operation, an expanding duct may be used at the fan outlet that discharges at a low velocity (1000 fpm or

less) into a plenum (see Fig. 38). A similar plenum may be used for air-conditioning systems with several reheaters.

h. Fan total-pressure calculations. Table 12 contains calculations for the fan total-pressure requirement of the typical duct system shown in Fig. 42. The various data just discussed in Subsection "f" (Total pressure losses. . . .) are used in Table 12. Note that sizes, which are based on assumed velocities, can be entered directly, but those based on an equal friction loss must be determined from Fig. 32 after an equivalent round size has been selected from Fig. 31. No loss is assigned to the split, at point B, because there is no velocity change. Path A-D is assumed to be the longest run (incurring the highest loss in the system). Losses for fittings are determined by applying appropriate K and h_v values in equation (32). Losses for straight ducts are determined by multiplying the loss/100 ft column by the duct length/100 ft. Only the velocity and h_v associated with the applicable K factor are

Table 10 Terminal losses

FITTING		Loss Coefficient,[a] K Clear	0.5 in. Wire[b] Mesh	Insect[b] Screen
Double gooseneck,	entrance	—	0.9	1.3
	discharge	—	2.0	2.4
Single gooseneck,	entrance	—	1.0	1.4
	discharge	—	2.0	2.4
Mushroom,	entrance	—	2.0	2.6
	discharge	—	2.8	3.2
Airlift[c]		—	3.5	4.0
Weather louver[c, d]		—	3.0	3.6
Door louver[e]		4.0	—	—
Hole in bulkhead		2.6	3.1	3.7
Short duct in bulkhead (re-entrant)	$L < 3D$	3.6	—	—
	$L \geqq 3D$	1.9	—	—
Open end duct	entrance	0.90	1.4	2.0
	discharge	1.10	1.6	2.1
Duct terminating in plate	entrance	0.50	—	—
	discharge	1.10	—	—
Converging taper	entrance	0.10	—	—
Bellmouth	entrance	0.5	—	—

NOTES

[a] Losses are based on the area at "A" or "D".

[b] Screen losses are based on the gross area velocity head, h_{vg}. The loss for 0.5-in. wire mesh (75% clear area) is 0.50 h_{vg}; the loss for insect screen (60% clear area) is 1.1 h_{vg}; and the loss for both 0.5-in. wire mesh and insect screen is 2.0 h_{vg}.

[c] Includes one velocity head (h_v) for discharge to plenum; deduct 0.5 h_v when the terminal is directly connected to a straight duct.

[d] Typically, the total flow area, A, is 60% of the gross area, B.

[e] Typically, the total flow area, A, is 50% of the gross area, B.

Table 11 Terminal and equipment air velocities

ITEM	VELOCITY RANGE (fpm)
Louvers	1000 to 1500 (free area)
Mushrooms	800 to 1200 (throat area)
Goosenecks	1000 to 1500 (throat area)
Heaters (steam)	800 to 1200 (face area)
Heaters (hot water)	400 to 600 (face area)
Cooling coils	350 to 500 (face area)
Directional terminals	2000 to 3000 (throat area)
Grilles and registers	800 to 1000 (free area)
Supply registers (ventilation)	750 to 1000 (free area)
Diffusers	500 to 900 (neck area)
Ratproof Screens ($\frac{1}{2} \times \frac{1}{2}$ in. mesh)	500 to 800 (gross area)
Insect screen (60 percent clear area)	300 to 500 (gross area)

shown for fittings involving velocity changes. The column "Pres. Avail." (pressure available) is provided to facilitate the design of submains (those not forming part of the path of highest loss) and branches. It may be noted that 1.804 is available for the 3000-cfm submain from point B to the end, including return losses. Similarly, 1.243 in. is available for the 200-cfm branch from point C to the end. The column "Loss Source" is usually not provided. This column is included in Table 12 for illustrative purposes only.

The minimum duct thicknesses used on naval ships are outlined in Tables 13 and 14. Ducts that pass through, or are extensions of compartments subject to tightness tests, and ducts that serve as sprinkling overflows for ammunition spaces are increased in thickness as necessary to withstand the test pressure of the compartments and ammunition spaces without permanent set. The material thickness for fittings such as elbows, tees, and transitions is that required for the largest dimensions of the fitting.

For quieter operation and to ensure full airflow at axial-fan intakes, cooling coils, heaters, flame arresters, and supply system takeoffs, the elbow and ductwork preceding these fittings are recommended to be in accordance with the following:

• Radius elbows may be used, provided the minimum length of straight duct between the turn and the fitting is equal to the duct dimension in the plane of the bend times: two, for 30-deg elbows; three, for 45-deg elbows; four, for 60-deg elbows; five, for 90-deg elbows.

• Radius elbows with splitters may be used, provided the minimum length of straight duct between the turn and fitting is equal to the dimension between the longest splitter and the outer curve of the elbow times: two, for 30-deg elbows; three, for 45-deg elbows; four, for 60-deg elbows; five, for 90-deg elbows.

• Vaned turns may be used if the length of straight duct between the turn and the fitting is less than that required above; otherwise, radius elbows are preferred.

• Intake transition pieces to axial fans should be as symmetrical as practicable. If it is necessary to decrease the velocity at fan intakes, the transformation piece should be symmetrical or the intake should be fitted with straight pipe for a length of one diameter.

• Branch inlets or split mains should not be located in the ductwork preceding the axial fan for a distance equal to the fan diameter.

Table 12 System total-pressure calculations

SECT.	FITTING	FLOW (cfm)	VEL. (fpm)	h_v	SIZE	EQUIV. DIA.	K OR LOSS/100 ft	DUCT LENGTH (ft)	LOSS	SECT. LOSS	PRES. AVAIL.	LOSS SOURCE
A-B	Airlift ½ in. WM...	3800	1000	0.063	25 × 15		3.5		0.220			Table 10
	Filter...								0.300			a
	Preheater...		750		30 × 24				0.200			b
	Cooling coil...		500		36 × 30				0.600			b
	Coil-to-fan...								(neglect)			
	Fan-to-duct (grad. exp., θ = 7.5 deg).		2300^d	0.33	32 × 7½		0.02		(neglect)			Fig. 40, A_1/A_2 = 0.87
	Straight duct...		2000^a	0.25	34 × 8	17	0.42	4.5^c	0.019			Fig. 31
	Elbow, 1 split...		2000	0.25	34 × 8		0.12		0.030			Fig. 33, CR = 0.34
	Straight duct...		2000	0.25	34 × 8	17	0.42	18	0.076	1.445		Fig. 31
B-C	Grad. exp., θ = 7.5 deg.	800	2000	0.25	7¼ × 8		0.05		0.013		1.804	Fig. 40, A_1/A_2 = 0.80
	Straight duct...		1600	0.16	12 × 6	9.0	0.5^a	50	0.250			Fig. 31
	Duct-to-heater, grad. exp., θ = 9 deg...		1600	0.16	12 × 6		0.25		0.040			Fig. 40, A_1/A_2 = 0.33
	Heater...		530		24 × 9				0.100			b
	Heater-to-duct grad. cont., θ = 45 deg.		1600	0.16	12 × 6	9.0	0.05	20	0.008			see text
	Straight duct...		1600	0.16	12 × 6		0.5		0.150	0.561		Fig. 31
C-D	Divided flow-main...	600	1440	0.13	10 × 6		0.1		0.013		1.243	Fig. 37, VR = 0.9
	Elbow...		1440	0.13	10 × 6		0.13		0.017			Fig. 33, CR = 0.5
	Straight duct...		1440	0.13	10 × 6	8.2	0.5	10	0.050			Fig. 31
	Vaned elbow...		1440	0.13	10 × 6		0.35		0.045			see text
	Straight duct...		1440	0.13	10 × 6	8.2	0.50	20	0.100			Fig. 31
	Transformer...		1440	0.13	12 × 5		0.15		0.020			see text
	Miter elbow...		1440	0.13	12 × 5		2.6		0.338			Fig. 36, R/D = 0.5
	Diffuser...		1080	0.07	10				0.220	0.803		e
Ret	Door louver...	600	400	0.01	22 × 20		4.0		0.040		0.440	Table 10
	Return ducts...	3000							0.400			a
	Total system...	3800							3.249			

a Velocity or loss assumed.
b Loss obtained from manufacturer's data.
c Straight duct length measured from ship's drawings.
d Velocity and h_v based on rectangular size.
e Catalog loss (static pressure) plus one h_v in diffuser neck.

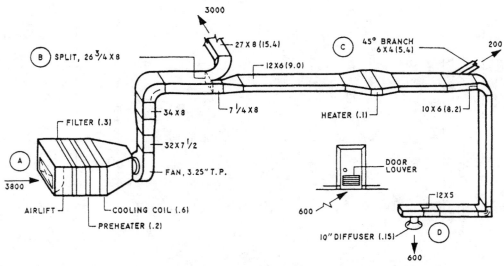

Fig. 42   Typical duct system

Table 13   Minimum duct thickness for fabricated ductwork

Diameter or Longer Side, in.	NON-WATERTIGHT		WATERTIGHT	
	Galvanized Steel, in.	Aluminum, in.	Galvanized Steel, in.	Aluminum, in.
Up to 6	0.018	0.025	0.075	0.106
6.5 to 12	0.030	0.040	0.100	0.140
12.5 to 18	0.036	0.050	0.118	0.160
18.5 to 30	0.048	0.060	0.118	0.160
Above 30	0.060	0.080	0.118	0.160

Table 14   Minimum duct thickness for welded or seamless tubing

Tubing size, in.	NON-WATERTIGHT	WATERTIGHT
	Aluminum, in.	Aluminum, in.
2 to 6	0.035	0.106
6.5 to 12	0.050	0.140

SPIRALLY WOUND DUCT (NON-WATERTIGHT)		
Diameter, in.	Steel, in.	Aluminum, in.
Up to 8	0.018	0.025
Over 8	0.030	0.032

• Branch inlets or split mains should not be located in the ductwork preceding the axial fan for a distance equal to the fan diameter.

• Bellmouths are installed on axial fans where there is no inlet piping and where the fans take suction from plenum chambers.

• If a square turn in a rectangular duct occurs immediately before a centrifugal-fan inlet, the dimension of the side of the elbow parallel to the fan shaft should be not less than ¾ of the fan inlet diameter, and the other dimension should be equal to the fan inlet diameter. The duct should be faired into the far edge of the fan inlet on a radius turn that is ¾ of the fan inlet diameter.

3.2 Heating Systems. The selection of heating systems for merchant or naval vessels is largely dependent on the environmental conditions and the form of energy available. Steam heaters are a logical choice for steam-propelled ships, whereas electric heaters may be preferred for ships with diesel or gas turbine propulsion plants.

a. Steam heating systems. The layout of a steam heating piping system depends upon the vessel arrangement. Cargo ships generally have relatively short fore-and-aft runs of piping; a steam riser in the machinery casing starts at a pressure-reducing station below the lowest, and extends up to the highest, deck served. Where the areas served are some distances fore and aft, such as on passenger ships, the steam main is run fore and aft, preferably above the bulkhead deck and on the deck level having the largest height so as to minimize conflicts with headroom and other services. Risers, taken from the main at convenient locations, feed short laterals on the various decks. Often a horizontal steam main can serve two or even three decks (one or two up and one down). When avoidable, the longitudinal main is not run on a passenger deck. If possible, piping is kept out of cargo holds and other inaccessible (locked) spaces, where leaks may be undetected [9]. Also, it is kept out of electrical spaces and is not allowed over electrical equipment. To achieve an economical and efficient layout, the heating system must be coordinated with the other hotel services. Structural interferences and limitations also must be considered.

The following guidelines are considered to reflect good practice in the design of shipboard steam heating systems:

1. Supply equipment that operates year round (galley, pantry, water heaters, laundry converters, air-conditioning reheaters, and filter cleaning stations) from constant steam service mains and branches. Feed the remaining equipment from intermittent steam service mains and branches.

2. Provide either a common or individual reducing station for each group of units requiring the same operating pressure. Some galley equipment may require a lower

pressure than heating equipment; also, some units have integral reducing and control facilities.

3. Run all piping (steam and condensate) so it can be drained by gravity. The requirement for a gravity return is a major controlling factor in locating condensate piping.

4. Design condensate piping for gravity return to an atmospheric or contaminated drain tank. Run independent drains from laundry or other equipment, which operates at a higher pressure. Where lifting of condensate absolutely cannot be avoided, limit the lift to a maximum of one deck height only and fit check valves.

5. Wherever possible, pitch condensate piping in the direction of flow. Provide drips at all low points in piping, the bases of risers, the ends of mains, and at normally closed valves, which may collect condensate.

6. Locate drain piping from heating units in exposed compartments, such as steering gear rooms, below the unit; if located above, the unit can be flooded and damaged by freezing if steam is inadvertently secured in cold weather.

7. Except for converters, do not provide bypasses for traps or automatic valves.

8. Provide a steam pressure gage in each equipment room, and elsewhere as necessary, to determine the actual inlet working pressure at heating units. Provide several plugged test connections in the distribution system and at equipment not provided with a pressure gage.

9. Provide all units with a stop valve on supply and a trap and cutout valve (gate) on the return.

10. For each air and water heater, provide a strainer and control valve (following stop valve) on the supply and a dirt pocket before the trap return.

11. Provide a dirt pocket and strainer ahead of the steam trap on a unit heater return.

12. In general, use thermostatic traps only for direct radiation and domestic equipment of low capacity, i.e., less than 20 lb/hr.

13. Provide an air chamber and petcock in returns from heaters.

Typical equipment hookups are shown in Fig. 43.

On U.S. Navy ships, drain orifice assemblies are used in lieu of steam traps. A drain orifice assembly consists of a Y-type steam strainer and an assembly consisting of a strainer/gasket, orifice plate fitting, and matching gasket installed between a pair of flanges. Recommended orifice sizes are given in Table 15. Where the maximum amount of condensate expected exceeds the capacity of a standard-size orifice, multiple parallel-installed drain orifice assemblies are provided.

The maximum steam velocities in pipes are generally established from the formula:

$$V = 4860 \sqrt{d} \qquad (33)$$

where

V = maximum steam velocity, fpm
d = pipe inside diameter, in.

Usually, the steam is slightly superheated and is dry; for wet steam (as supplied by a waste-heat boiler, ordinary

heating boiler, or contaminated evaporator), the flow capacities determined from this equation are reduced by about 25%.

Condensate return lines commonly are sized for maximum velocities, assuming full pipe flow, derived by the formula:

$$V = 30 \sqrt{d} \qquad (34)$$

where

V = condensate velocity, fpm
d = pipe inside diameter, in.

Velocities calculated from this equation produce drops between ¾ and 1 psi per 100 ft of straight pipe; the pressure loss in return lines of 30-psig systems should not exceed 4 psi [7].

b. Hot-water systems. The use of hot water for heating has many advantages [23]. Hot water affords better control of small heating capacities (frequently 5000 Btu/hr and less) than steam. Also, hot water eliminates numerous problems associated with condensate return, steam traps, and water hammer. Hot-water heating permits the elimination of condensate return pumps and associated problems, condensate cooling facilities, and corrosion of condensate piping. Note that corrosion in steam heating boilers, particularly those of the flash type, is a serious problem requiring careful feedwater analysis and treatment.

Medium-temperature hot-water generators are frequently used in lieu of auxiliary steam boilers. For various heating services, including distilling plants, laundry, and galley equipment, 320 F water replaces 100-psi steam. Water-to-water and water-to-steam heat exchangers are provided where low-temperature heating (water or steam) is required. This practice requires less weight and space, has a lower installed cost, is safer and easier to operate, requires less maintenance, and has a lower operating cost.

All hot-water heating systems on ships have forced circulation. Single-pipe systems, Fig. 44(a), require minimum piping, are ideally suited for shipboard applications, and have been successfully used on many large passenger ships. The true two-pipe reverse-return system is frequently not practical; therefore variations, such as that illustrated in Fig. 44(b), are adopted that are more suitable to the arrangement of the spaces served [23].

On large ships, a number of separate heating systems may be provided with central-station equipment, such as illustrated by Fig. 45, located in fan rooms or machinery spaces. Single-pipe systems are particularly suited for decentralized service, i.e., service required only in certain parts of the ship. Where central-station equipment is located in main or auxiliary machinery spaces, two-pipe arrangements are used with two or more pumps and converters being provided to carry the design load; a standby pump is sometimes installed and the equipment is cross-connected. Standby pumps are not provided for single-pipe systems since they are relatively small (usually between 40 and 80 gpm); instead, complete pump and motor assemblies are carried as spares.

NOTES

1. WHEN TWO OR MORE HEATING COILS ARE REQUIRED, PROVIDE DRAINS FOR EACH COIL AS INDICATED.

2. PIPE STRAINER TO OPEN DRAIN WITHIN SIGHT OF THE VALVE.

3. PROVIDE BLOWOFF COCKS IN EQUIPMENT SPACES ONLY.

4. PROVIDE BARE PIPE COOLING LEG BETWEEN COIL AND TRAP AT LEAST 24" LONG.

5. PROVIDE STRAINER AHEAD OF TRAP ON UNIT HEATERS.

6. PROVIDE CHECK VALVE IN RETURN ONLY WHERE CONDENSATE MUST BE LIFTED.

7. PROVIDE STUFFING TUBES IN WET SPACES ONLY, DESIGNED TO THERMALLY ISOLATE PIPE FROM STRUCTURE. AT ALL OTHER PENETRATIONS, PROVIDE OVERSIZED CUTS WITH SUITABLE SLEEVES LARGE ENOUGH TO PASS THE FULL THICKNESS OF STANDARD PIPE INSULATION.

8. ALL STRAINERS SHALL BE OF THE SELF-CLEANING TYPE.

9. MOUNT THE CONVECTOR ON THE EXPOSED (COLD) BULKHEAD OR BULKHEAD LINING, WITH A 0.5-IN. AIR SPACE.

10. PROVIDE ONE PRESSURE GAGE IN EACH FAN EQUIPMENT ROOM.

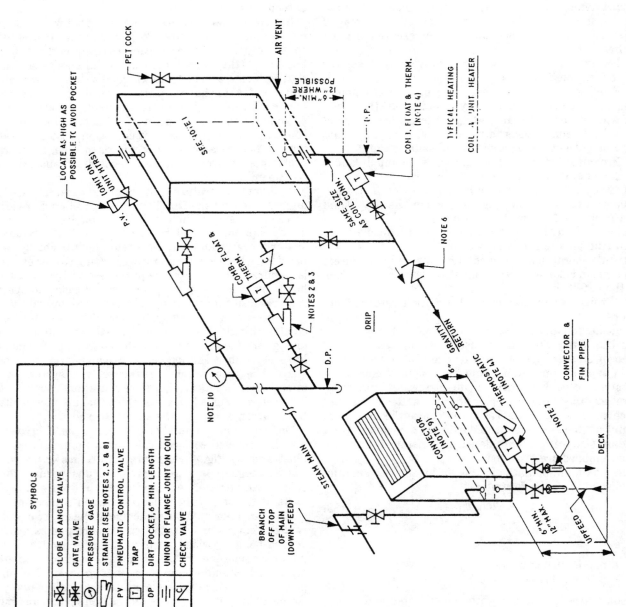

Fig. 43   Typical steam piping hookups

SYMBOLS	
⧼	GLOBE OR ANGLE VALVE
⧼	GATE VALVE
⊙	PRESSURE GAGE
	STRAINER (SEE NOTES 2, 3 & 8)
PV	PNEUMATIC CONTROL VALVE
T	TRAP
DP	DIRT POCKET, 6" MIN. LENGTH
‖	UNION OR FLANGE JOINT ON COIL
N	CHECK VALVE

Table 15 Recommended orifice sizes for low-pressure steam system applications

Condesate Capacity, lb/hr	Nominal Steam Pressure			
	50 psig		100 psig	
	dia., in.	drill size	dia., in.	drill size
Up to 25	0.0310	68	0.0310	68
26 to 50	0.0310	68	0.0310	68
51 to 100	0.0410	59	0.0350	65
101 to 200	0.0595	53	0.0520	55
201 to 400	0.0820	45	0.0700	50
401 to 600	0.1015	38	0.0860	44
601 to 800	0.1160	32	0.0995	39
801 to 1000	0.1300	29	0.1100	35
1001 to 1400	0.1540	23	0.1300	29
1401 to 1800	0.1770	16	0.1495	25

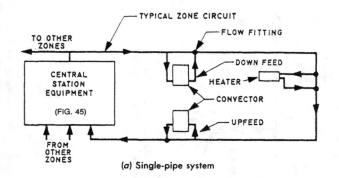

(a) Single-pipe system

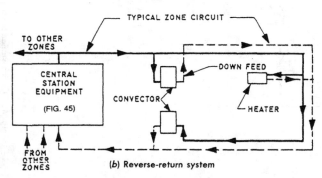

(b) Reverse-return system

Fig. 44   Hot-water heating systems

It is essential to eliminate air and uncondensed gases from closed-circulation water systems because they impair heat transfer, retard circulation, cause pump cavitation, and create noise. Air must be purged from a system when it is initially filled; also, air enters with the makeup water and gases are released when the water is heated. Equipment is arranged so that the pump discharges to the converter (Fig. 45), which in turn discharges into an air separator. The latter removes both air introduced with makeup water and that released during the heating process.

A water velocity of at least 1 fps is required to keep entrained air and gases moving, and thus prevent air pockets. Velocities in mains are sufficiently high; however, heating unit loads are often so small that the minimum velocity, rather than load, determines the design water quantity. It is recommended that at least 1 gpm per air heater and 0.5 gpm per convector be provided to assure adequate velocity in branches (usually of 0.5-in. pipe size)

and units. Air can be confined to the mains, where the velocity is highest, by connecting supply branches to the bottom, or at least to the lower half, of mains as indicated by Fig. 46. Accordingly, it is of advantage to keep mains as high as possible.

Adequate and easily accessible means for venting air cannot be overstressed. Automatic float-type vents with waste connections usually are applied only at high points in mains and those heating units at high points of systems. Systems with open expansion tanks are generally considered to be easier to vent than those with compression (closed expansion) tanks.

The following equation shows the relationship between the heating (or cooling) capacity, water temperature drop, and flow rate:

$$H = 500 \, G \, DT \qquad (35)$$

where

H = heating (cooling) capacity, Btu/hr
G = flow rate, gpm
DT = temperature drop (rise), deg F

Load rates usually are based on temperature drops of 10 to 20 deg F; consequently, it is useful to remember that

1 gpm cooled 10 deg F gives up 5000 Btu/hr

The temperature drop selected for a particular system depends on the magnitude of the heating load. Of importance is the allowable pressure loss through reheaters. Large temperature drops reduce flow rates and pressure drops, but heat transfer is poor at low water velocities. Accordingly, a compromise must be made. Considerable judgment is required to "juggle" the various design variables so as to achieve the optimum design; to be sure, the design of a shipboard hot-water heating system is not accomplished on the first attempt. The size and number of flow fittings required for individual circuits (shunts) of single-pipe systems (Fig. 45) depend on their design and the manufacturer's recommendations. Loads are seldom large enough to justify branch sizes over ½ in. When an appreciable number of larger sizes are indicated, the use of higher temperature drops and higher velocities in mains should be considered.

Piping arrangements depend largely on the arrangement of spaces served. Where practical, returns from zone circuits (loops) are run independently to the pump suction manifold (Fig. 45); thus, thermometers (for checking performance) and zone balancing valves are readily accessible.

The relationship of the spaces served controls the number of zone circuits. Where three decks are served, the central-station equipment (Fig. 45) is preferably located near the center of the middle deck, with a zone circuit serving each deck. The allocation of spaces to zones of single-pipe systems must be such that adequate capacity is carried in each loop to generate the required heat at flow fittings; it may be necessary to break up the loops on one or more decks to reduce the pumping head.

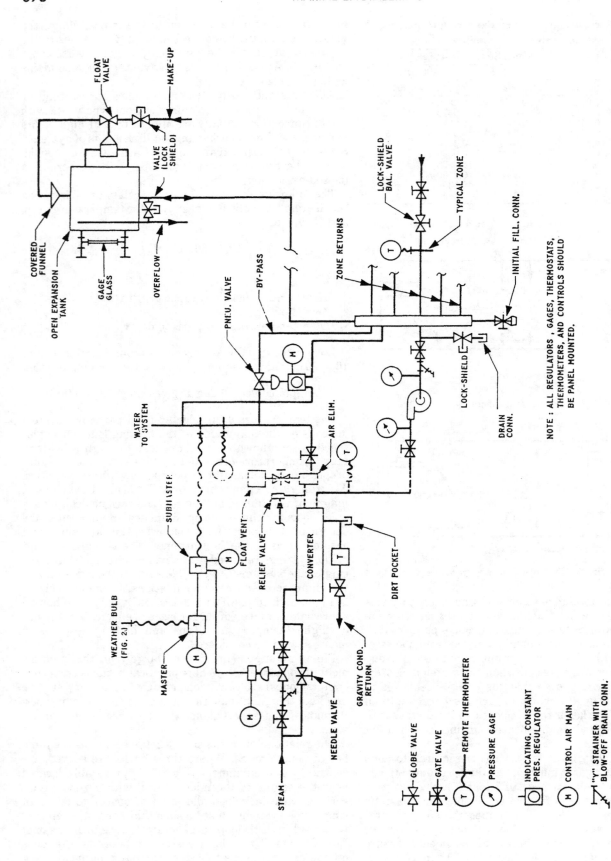

Fig. 45 Central-station hookup for a hot-water heating system

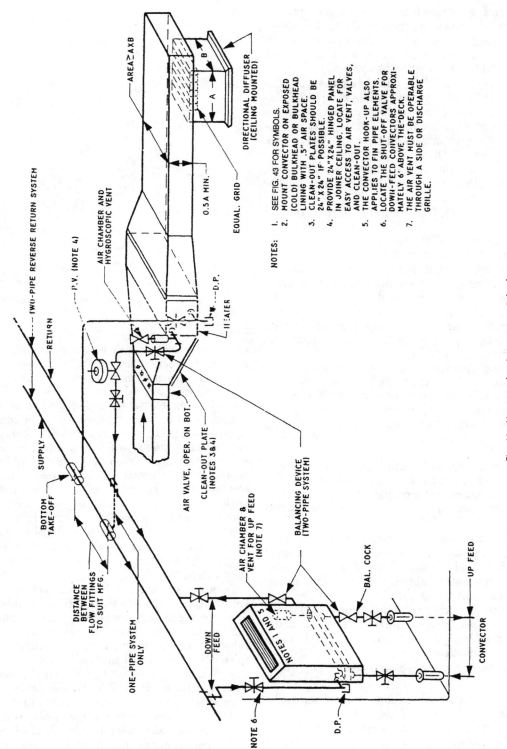

Fig. 46   Hot-water heating system unit hookups

A separate heating system is not required for each air-conditioning system; however, to simplify troubleshooting, only one heating system should serve a particular air-handling system. Also, where practical, each zone circuit should serve a single air-conditioning system.

Expansion tanks may be of the open or closed type. The water temperature in open systems is less than 212 F; for closed systems, the use of 220 to 240 F water is common. Corrosion and the elimination of air become greater problems as the temperature increases; thus low water temperatures are preferred. Obviously, the less makeup water used the better, since this minimizes corrosion, sediment contamination, and noncondensable gases in the system.

Converters are selected, from manufacturers' data, for the heating load required plus 50% excess capacity to allow for start-up and scaling. The water pressure drop through converters is seldom critical.

Circulating pumps used in hot-water systems are always of the centrifugal type. Generally, mechanical seals suitable for operation well above the design water temperature are provided. Pump selections should be made from certified performance curves. The pump selected must not have an excessively flat capacity-head curve. Also, the point of operation should be at or slightly to the left of the maximum efficiency point. The operating point on the capacity-head curve is usually specified at least 10% below the peak. Such a selection provides a margin for errors in estimating pumping head, accommodates changes in system requirements (head and volume) occurring during or after construction, and minimizes the possibility of motor overload.

Cocks or valves are fitted in each zone circuit of a system to balance one against the other, as indicated by Fig. 45. Balancing by temperature drop, using permanent thermometers, is most practical; thermometers also help when checking operation.

Piping of hot-water heating systems is arranged to facilitate air removal, venting, and draining. Piping installed to follow the sheer and camber of the ship usually provides adequate pitch and suitable high points for venting. Mains are run as straight as possible to minimize high points (air pockets) requiring automatic vents. Hot-water piping is normally sized for a maximum velocity, V, equal to $5\sqrt{d}$ (where d is the pipe inside diameter in inches and V is in fps) but not in excess of 6 fps.

Hot-water heating systems are not used on U.S. Navy ships.

c. Heaters and heater coils. A variety of heaters is available to suit specific applications. Heaters may be of the convection, duct, unit, or radiant type.

Convection heaters use natural convection currents to move air over heated coils. They have a high heating capacity for their size and weight and are used in spaces where there is no mechanical ventilation, or where the heating loads are small. Bathrooms, washrooms, and showers are examples of such spaces. Convection heaters should be mounted on weather-exposed bulkheads or against the coldest bulkhead and should be located close

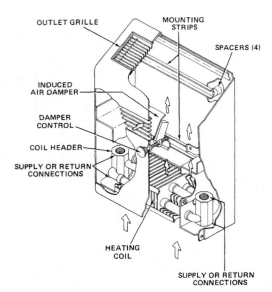

Fig. 47 Convection heater (steam)

to the deck but not less than six inches above the deck. Convection heaters may be steam or electric operated. In a steam convection heater, Fig. 47, air enters through the inlet and is warmed as it passes over the heating element, resulting in a natural convection current that causes the air to rise and exit at the top of the heater. Control of the output may be accomplished by a manually controlled damper or through a thermostatically controlled valve to regulate the amount of steam into the heater. Electric convection heaters, such as illustrated by Fig. 48, operate on the same principle and use a resistance-wire heating element that is helically coiled in insulating material and enclosed in a hermetically sealed finned metal sheath. Electric heaters always use spraytight enclosures for electrical connections and watertight enclosures for the entrance of heating elements. Electric heaters should never be placed next to showers, sinks, or urinals where they can become wet.

Duct heaters are the preferred type since they require less weight, space, and piping. Duct heaters can be steam or electric operated. Duct heaters may be part of a ventilation supply or recirculation system and can serve as a reheater for humidity control in cooling systems during the cooling season. Steam duct heaters, Fig. 49, are generally of the single-serpentine or double-tube type with a variety of fin spacings. While smaller fin spacings provide greater efficiency for heat transfer, they are most difficult to clean and have greater pressure drops. Heating elements in steam heaters are made of copper tubing. Tubes should not be placed more than one row deep in the direction of airflow; however, when more than one row of tubes is needed to obtain the required temperature rise, heaters can be stacked in the direction of airflow. Heaters may be stacked end to end or one over the other to obtain the face area required. As can be seen from Fig. 50, electric duct heaters are similar in construction to electric convection heaters (Fig. 48).

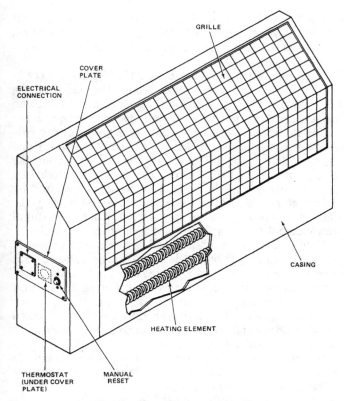

Fig. 48   Convection heater (electric)

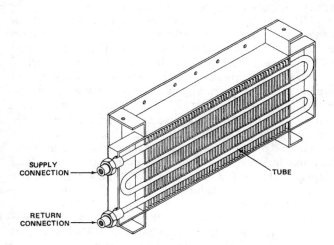

Fig. 49   Duct heater (steam) Type S

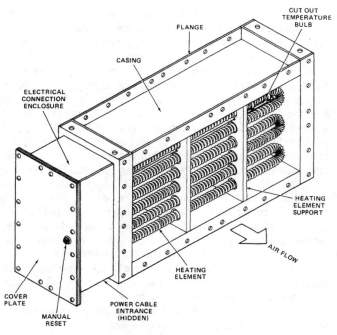

Fig. 50   Duct heater (electric)

Unit heaters are used in large areas, such as cargo spaces, steering gear rooms, and in some machinery spaces where the ventilation supply is insufficient to provide heat through duct heaters or where there is no ventilation supply and a convection heater cannot meet the heating needs. Unit heaters are self-contained forced-convection heaters in that a fan is used to force air across the heating coils, resulting in a high heating capacity. Unit heaters may be steam or electric. They are generally mounted 6 to 10 ft off the deck; and when multiple heaters are used, their discharges should be consistent.

Families of naval steam and electric convection, duct, and unit heaters have been designed to be capable of meeting vibration and high-impact shock requirements. The size, performance capability, and dimensions of these units have been standardized to facilitate replacement and interchangeability. The physical characteristics and performance capability for Navy standard heaters can be found in reference 12. Specifications for heaters used in merchant construction may be found in reference 2.

3.3 Chilled- and Hot-Water Systems, Cooling Coils, Compressors, Condensers, and Related Components. Chilled-water systems are designed for direct return since there are relatively few units to balance. Generally, 42 F is considered to be the lowest economical chilled-water temperature; 44 F is specified for naval construction. Water quantities, generally, are based on temperature rises from 8 to 10 deg F, although temperature rises up to 12 deg F have been used for large systems (passenger ships).

Pumping heads of chilled-water systems are estimated the same as for two-pipe hot-water systems. Higher maximum water velocities are used because noise is less critical; also, the heat removal per gpm is smaller (the temperature rise much smaller). Chilled-water piping is normally designed for a velocity, V, equal to 5 $\sqrt{d}$ (where d is the pipe inside diameter in inches and V is in fps) but not in excess of 10 fps.

It is normal practice to circulate water continuously through chilled-water systems during cold weather; this tends to overcome the possibility of freezing due to stratification of preheated air. While not recommended, antifreeze liquids have been used. It may be noted that the noncorrosive brine used for other refrigerated services on ship may also be used for air-conditioning purposes.

U.S. Navy ships usually use solenoid valves controlled by a thermostat in lieu of pneumatically controlled valves. Also, constant-flow control fittings and orifices are used in lieu of flowmeters and balancing valves.

a. Combination chilled- and hot-water systems. Class E air-conditioning systems, Fig. 3, require two-pipe, reverse-return water systems. Facilities for air removal are the same as provided for water heating systems. In general, induction units are bulkhead mounted; thus all parts (valves, controls, vents, lint screens, flushing bypass, etc.) are readily accessible for service, maintenance, and repair. The chilled water is first fed through the primary air coil and then is fed, as "secondary water," to the induction units. On a cargo ship, a single pump often handles both services. The total temperature rise is about 15 deg F (8 deg F primary and 7 deg F secondary); thus, while additional pumping head is required (primary and secondary coils are in series), the reduction in water quantity (about 1.6 gpm/ton of refrigeration) keeps the pump horsepower reasonable. On passenger ships where the systems are large, the return from the primary water system is metered automatically, as required, into secondary systems serving the induction units.

Piping serving primary cooling coils is designed to the same criteria as used with chilled-water systems; piping of secondary circuits uses the same criteria as water heating systems.

Combination water systems serving room fan-coil units are designed the same as Class E systems, when the primary (replenishment) air is dehumidified and cooled. When the primary air is only preheated, the system is designed the same as for chilled water since fan-coil units accommodate the entire cooling load, including that of primary air.

Drains from the induction unit and room fan-coil unit drain pans usually are ¾ in. IPS. Where a number of units are connected to a common drain, the size is increased, depending on the number of units served. When practical, these drains are terminated in the corner of shower stalls, about 6 in. above the bottom. Drains direct to weather are not desired and fortunately are seldom used.

Combination chilled- and hot-water systems are not used on U.S. Navy ships.

b. Cooling coils. A typical cooling-coil piping hookup is shown in Fig. 51. Air conditioning on merchant and naval ships is accomplished through the use of cooling coils. The various types of cooling coils used are duct, unit, gravity, and coils that are designed to fit into fan-coil assemblies and fan-coil units. The various types of cooling coils may be of either the chilled-water or direct-expansion type, with the chilled-water cooling system being the predominant type. Air-conditioning design conditions for naval ships are generally higher than for merchant ships due to space and weight limitations. Air-conditioning on naval ships is provided as a military necessity to keep the crew fit to accomplish their assigned tasks and to keep vital heat-sensitive equipment operating. When selecting a coil for a specified design condition, several related performance characteristics must be considered.

Cooling coils should always be installed with the airstream flowing horizontally through the coils. A vertical airflow in direct-expansion coils can result in oil pockets in the coil circuits.

Duct coils, such as illustrated by Fig. 52, are generally universal in that air may enter from either side and the coil may be turned to position the connections on the end most convenient. Duct coils usually have an even number of tube rows so that supply and return water piping connections are on the same side of the coil. The water and air should be in counterflow, with the water entering the header on the air exit. The coil capacity is controlled by a solenoid valve in the liquid line, which is actuated by thermostatic contacts in the compartment served by the coil. Duct coils are used in conjunction with a centrifugal or vaneaxial fan as a built-up recirculation system.

Unit coolers are used in small isolated compartments where the air distribution is not a major concern, but where air conditioning is required and the installation of a recirculating system is impractical. Figure 53 is a unit cooler and consists of a coil, similar to a duct coil, an air filter, and a motor-driven tubeaxial fan. Air is discharged through adjustable louvers. The control of unit coolers is similar to that for duct coils.

Direct-expansion coils and unit coolers are similar in construction and control to chilled-water coils but use a refrigerant as the cooling agent.

Gravity cooling coils are used in spaces that have low cooling loads, such as ammunition magazines, where electrical equipment would pose a hazard. As indicated by Fig. 54, gravity coils are overhead mounted and rely on convection currents to remove heat. When two or more gravity coils are located in the same space, and chilled-water is used, the coils must be connected in parallel.

Seawater cooling coils can be used to supplement cooling in spaces that normally have large ventilation requirements and have high sensible-heat loads. Seawater coils can be used to reduce the size of ventilation trunks and weather-deck fan rooms, which can result in a savings of weight and space. An example of such spaces is an auxiliary machinery space with little or no latent-heat load. Care must be exercised to ensure that the ventilation supply quantity is not reduced such that fresh-air requirements are violated.

Fan-coil assemblies are an alternative to built-up recirculation systems. Figure 55 shows that they are deck mounted and usually located either in the space to be air conditioned or in an adjacent passage. A fan-coil assembly consists of a cooling coil, fan motor, and filters, but may also include a distribution plenum and electrostatic precipitator. A belt-driven centrifugal fan with a motor pulley is used to permit a variety of airflow quantities. Air is drawn downward from the inlet, through the filters, electrostatic precipitator (if installed), through the cooling coil, across the fan motor, then through the fan and out through the discharge duct or plenum.

Fan-coil units also serve as an alternative to built-up recirculation systems, and like fan-coil assemblies are usually located in the space served. From Fig. 56 it can be seen that fan-coil units consist of a fan, two-speed motor,

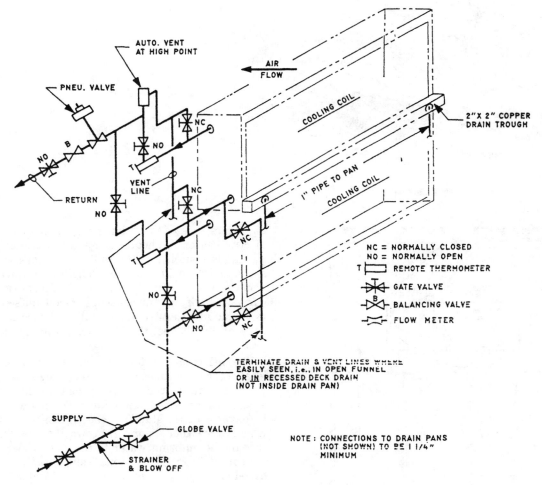

NC = NORMALLY CLOSED
NO = NORMALLY OPEN
T ▭ REMOTE THERMOMETER
▷◁ GATE VALVE
▷◁ B BALANCING VALVE
▷◁ FLOW METER

Fig. 51   Typical two-high cooling coil piping hookup

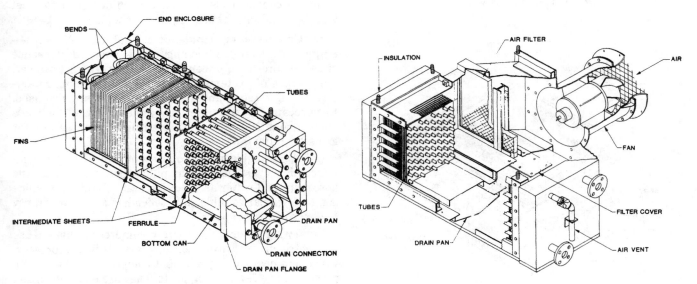

Fig. 52   Duct cooling coil

Fig. 53   Unit cooler

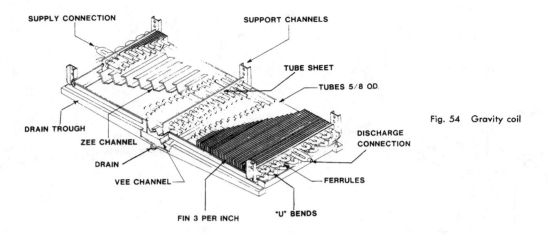

SUPPLY CONNECTION

SUPPORT CHANNELS

TUBE SHEET

TUBES 5/8 O.D.

Fig. 54   Gravity coil

DRAIN TROUGH

ZEE CHANNEL

DRAIN

VEE CHANNEL

DISCHARGE CONNECTION

FERRULES

FIN 3 PER INCH

"U" BENDS

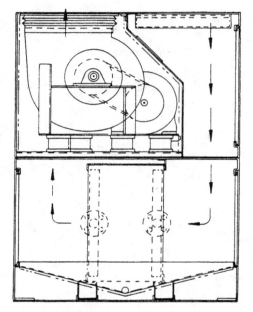

Fig. 55   Fan-coil assembly—Type III

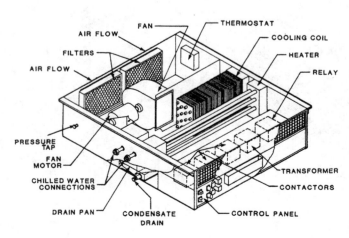

FAN

THERMOSTAT

AIR FLOW

COOLING COIL

FILTERS

HEATER

AIR FLOW

RELAY

PRESSURE TAP

FAN MOTOR

CHILLED WATER CONNECTIONS

DRAIN PAN

CONDENSATE DRAIN

TRANSFORMER

CONTACTORS

CONTROL PANEL

Fig. 56   Fan-coil unit

air filters, inlet and outlet grills, chilled-water coil, and electric heater and heater contactors, if required. A thermostat is installed in the unit to control the fan motor speed and the chilled-water solenoid valve during cooling, and the heater during heating. A direct-drive forward-curved centrifugal fan is used. Fan-coil units are designed to be mounted either horizontally in overheads or vertically on bulkheads, thereby minimizing the use of deck area.

The specifications for some merchant ships require at least six rows of tubes and 25% more rows than suggested by the manufacturer's published ratings to protect against capacity reduction due to fouling of the coil fins. In naval practice standard cooling coils are used, all of which have either six or eight rows of tubes. The temperature of the entering water to the coil on merchant ships is usually 42 F. For merchant ships, the coil is selected based on a 10 deg F rise in the chilled water temperature through the coil [2]. For naval ships selection is based on the use of 3.6 gpm of 45 F chilled water through the coil per ton of cooling; selections are made from families of standard cooling coils that have been designed to meet vibration and high-shock requirements [12]

c. Compressors, condensers, and related components.   Most air-conditioning plants onboard ships use either reciprocating or centrifugal compressors; however, other types of systems such as lithium bromide, rotary-screw compressors, and steam-jet systems are also used. Reciprocating compressors are generally used for loads up to about 100 tons and centrifugal compressors for loads over 100 tons.

Lithium-bromide plants operate on an absorption refrigeration cycle using lithium-bromide solution as the absorbent and water as the refrigerant. Lithium bromide is a salt, similar to table salt, which has a strong affinity to absorb water vapor.

Rotary-screw or helical-screw compressor plants use an electric motor-driven helical rotary-screw compressor, which is a positive-displacement compressor. These compressors have few moving parts. They are oil flooded, and the oil increases the compression efficiency, reduces the

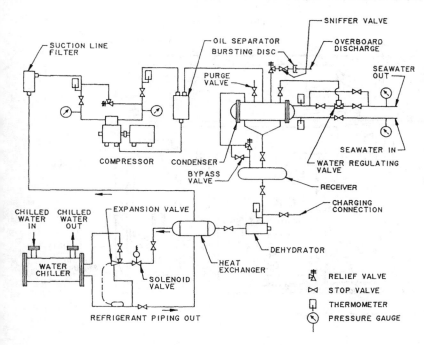

Fig. 57 Reciprocating air-conditioning plant schematic

discharge temperature, reduces noise, and permits operation at higher loads.

Steam-jet plants use a steam ejector to compress the refrigerant rather than a mechanical compressor. These units are large and bulky in comparison with electric motor-driven units and are considered to have limited shipboard application potential.

Additional details on lithium-bromide, rotary-screw, and steam-injection plants can be found in reference 10.

Air-conditioning plants for naval construction are usually indirect systems in which the refrigerant is circulated to a water chiller that cools fresh water. The advantage of using an indirect system is that the refrigerant piping is short and confined to one space, limiting the chance of a refrigerant leak. The water is chilled to a temperature of 44 F for naval construction and usually 42 F for merchant construction. The chilled water is then circulated to cooling coils throughout the ship in parallel flow paths. Multiple air-conditioning plants in an indirect system are cross-connected by the chilled-water piping and not by the refrigerant piping.

Marine refrigeration systems and their individual components are designed to operate with the angles of pitch, roll, list, and trim normally encountered by a ship. Design angles are 10-deg pitch, a momentary roll of 30-deg for merchant construction and 45-deg for naval, 15-deg permanent list, and 5-deg permanent trim fore and aft. Rotating equipment is normally installed with the shafts fore and aft so as to minimize gyroscopic effects.

Condensers, coolers, drip pans, troughs, tanks, etc. have their long dimension fore and aft, where possible, to facilitate drainage. Flow retarders are used to minimize the adverse effect of ship movement. Also, drain connections at both ends of equipment provide free draining

regardless of trim. Drawings must clearly define the minimum space required for the removal of tubes in shell-and-tube equipment. Construction and installation requirements for equipment are contained in references 42 and 43 for merchant construction and in references 10, 44, and 45 for naval construction.

Each component of an air-conditioning system must be selected to properly interface with all others and to deliver the required performance in the same operating time periods during pulldown and holding. Most plants are purchased as "packages," including the base, drives, and all controls. Generally this reduces costs and provides improved performance; however, since "packages" afford less flexibility of arrangement, space and servicing requirements must be resolved early in the design stage.

Since an air-conditioning plant handles a wide range of loads from pulldown to light loads, some means of compressor capacity control is required. In naval practice for reciprocating compressors, this is done by automatically preventing specific cylinders from pumping refrigerant. For centrifugal compressors, the use of pre-rotation vanes (variable inlet guide vanes) or a hot-gas bypass is used.

In merchant practice, the capacity of reciprocating compressors may be controlled by speed variations, automatic cylinder cutouts, unloaders, or bypasses. The capacity of centrifugal compressors may be varied by speed controls, bypasses, throttling, or metering of the refrigerant.

A schematic of a reciprocating air-conditioning plant is shown by Fig. 57. An air-conditioning plant consists of a compressor, condenser, receiver, chiller (evaporator), controls, piping, and valves. The purpose of the compressor is to compress the low-pressure, low-temperature refrigerant gas into a high-pressure, high-temperature gas.

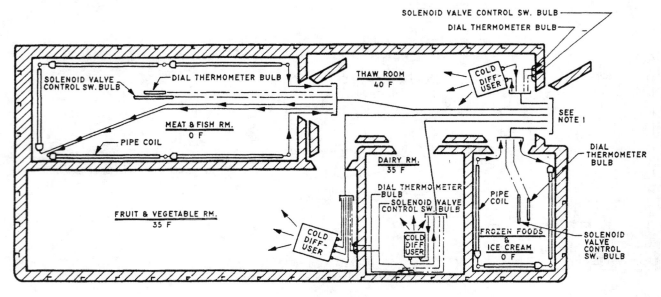

(a)   Arrangement of compartments

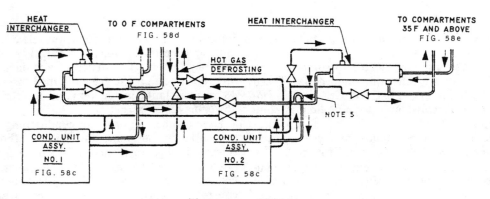

(b)   Interconnecting piping

NOTES

1. Refrigerant liquid lines and capillary tubing for dial (remote reading) thermometers and solenoid valve control switches are run to refrigerant control panels located by the shipyard.

2. Meat and fish room is similar except that two independently controlled refrigerant circuits are provided.

3. Fruit and vegetable room is similar except that the cold diffuser has two independently controlled refrigerant circuits.

4. Where possible, all freon piping is bent in lieu of using elbows, to reduce friction and the number of soldered joints.

5. All suction line branches are looped into the top of the suction main to prevent oil accumulation in lines not in service.

6. Hanger and supports are installed so as to permit free expansion and contraction between anchorages and to minimize vibration.

7. Refrigerant suction lines between the compressor and refrigerated space are insulated with brine thickness insulation. Liquid lines between the heat interchanger and refrigerated space are insulated with ice water thickness.

SYMBOLS	
⋈	STOP VALVE
⋈	HAND EXPANSION VALVE
⋈	O.D. X S.A.E. VALVE
⋈	SOLENOID VALVE
⋈	COND. WATER REG. VALVE
⋈	RELIEF VALVE
⊸⊶	SIGHT FLOW INDICATOR
S	STRAINER
D	DRYER
	THERMAL EXPANSION VALVE
	SOL. VALVE CONTROL SWITCH
⊙	PRESSURE GAUGE
⊏⊐	THERMOMETER
——————	UNINSULATED REFRIGERANT LINES
══════	INSULATED REFRIGERANT LINES
--------	CONDENSER WATER LINES
—·—·—	ELECTRICAL LINES
—···—···	CAPILLARY TUBING
—··—··	EQUALIZING LINES

Fig. 58   Ship's stores refrigeration system diagram

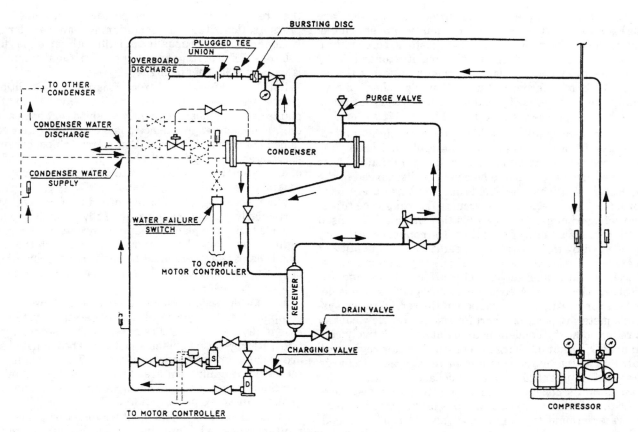

(c)   Condensing unit assembly

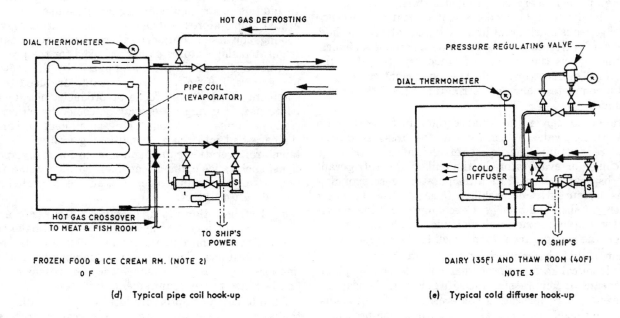

FROZEN FOOD & ICE CREAM RM. (NOTE 2)
0 F

(d)   Typical pipe coil hook-up

DAIRY (35F) AND THAW ROOM (40F)
NOTE 3

(e)   Typical cold diffuser hook-up

Fig. 58   Ship's stores refrigeration system diagram (continued)

Reciprocating compressors for naval construction are high-speed, positive-displacement, multicylinder, single-acting compressors with trunk-type pistons, forced lubrication circulation, and automatic capacity control. They use essentially commercial components but meet the requirements of reference 44 as well as high-shock and vibration requirements. Typical operating speeds for reciprocating compressors range from 1150 to 1750 rpm.

The condenser takes the high-pressure, high-temperature gas from the compressor and cools and condenses the gaseous refrigerant to a liquid. It is here that the heat removed from the space to be cooled is discharged to the seawater. Multipass shell-and-tube condensers are used for naval ships. Seawater is circulated through the tubes, and refrigerant gas is supplied in the shell and condensed on the outer surfaces of the tubes. The condenser capacity is affected by the seawater temperature, the quantity of seawater circulated, and the refrigerant gas temperature. Variations in condenser capacity affect the refrigerant discharge pressure. A constant discharge pressure is desirable for satisfactory operation of the refrigeration system's metering devices. Therefore, seawater flow control is required at the condenser to maintain a relatively constant refrigerant discharge pressure. A water-regulating valve, actuated by the compressor refrigerant discharge pressure, is installed in the condenser seawater outlet to maintain a constant condenser refrigerant discharge pressure. A manual bypass valve is provided to circulate seawater around the regulating valve in case of a malfunction.

The condensed refrigerant flows out of the condenser into the receiver. The receiver accumulates the liquid refrigerant and stores the excess refrigerant not needed during off-peak operations, provides storage during system pumpdown, and provides a seal against the entrance of the gaseous refrigerant into the liquid refrigerant line. Shipboard refrigeration systems have a liquid level indicator that shows the amount of liquid refrigerant in the receiver. The receiver should have sufficient capacity to hold a complete charge of refrigerant plus 20% reserve volume.

From the receiver, the liquid refrigerant flows through an expansion valve, which reduces the pressure on the liquid refrigerant. The liquid refrigerant then flows to the chiller or evaporator. In the chiller, the refrigerant absorbs heat at a saturation temperature corresponding to the pressure within the chiller. In naval ships the water chiller is a shell-and-tube heat exchanger with the refrigerant in the tubes and fresh water in the shell. From the chiller, the refrigerant returns back to the compressor.

Centrifugal air-conditioning plants are thermodynamically identical to a reciprocating air-conditioning plant. The means of compressing the refrigerant is the primary difference. A reciprocating compressor is a positive-displacement machine, whereas a centrifugal compressor is a dynamic machine. The refrigerant flow in centrifugal compressors is continuous; therefore, a centrifugal compressor has a greater volumetric capacity, size for size,

than reciprocating and other positive-displacement compressors. Centrifugal air-conditioning plants are large-capacity units and are usually not built with capacities under 100 tons. The rotating speed of a centrifugal compressor is usually 9000 rpm or higher.

Both centrifugal and reciprocating air-conditioning plants are used extensively in the Navy. Centrifugal plants are used most frequently on ships that have a large crew complement, and hence a large number of air-conditioned living working areas. Commercial ships, except for passenger liners, generally have smaller crews and less of an air-conditioning load, and therefore tend to use reciprocating air-conditioning plants.

When selecting the capacity of plants for naval ships, particularly combatant ships, one fully redundant plant is provided with the plants separated by one or more watertight subdivisions. The requirements are such that if the ship is damaged, there must be sufficient air-conditioning capacity to handle the vital loads in the combat, command, and control spaces.

3.4 Refrigeration Plants and Piping Systems. The equipment and piping used in refrigeration systems are essentially the same as those used in air-conditioning systems. Refrigeration systems installed aboard naval ships are usually direct-expansion systems. These systems and their related equipment are standardized to facilitate expeditious repair and replacement and to minimize the quantity of spare parts required [7].

In merchant construction, either direct or indirect systems may be used. Merchant systems are also more flexible than those on naval ships and use a greater variety of equipment and components because they are not subjected to as many constraints [2,11,43].

Refrigeration plants are used for various applications. In addition to air conditioning, plants are used for refrigerated ship's stores, refrigerated cargo, drinking-water cooling, and in self-contained refrigeration equipment. Refrigerated ship's stores plants are used for the preservation of food required for consumption by the crew and passengers. Refrigerated cargo plants are used for the shipment of perishable foods or products, not for shipboard use. Drinking water may be cooled by either self-contained units or, on some ships, small heat exchangers that are tied into the air-conditioning system. Self-contained refrigeration equipment is used for many applications, such as ice-making machines, soft-ice-cream freezers, soft-drink vending machines, and refrigerated salad and dessert counters.

a. Refrigeration piping systems. Figures 58 and 59 are typical refrigeration systems. The usual balance between power requirements and installation costs applies to such systems, since excessive suction and discharge line losses (see Fig. 60) reduce the compressor capacity and increase power requirements.

All lubricated compressors lose some lubricating oil to the refrigerant vapor during the compression process. As a result, in halogen compound refrigeration systems, the piping must be carefully designed so as to continuously return the lost oil to the compressor crankcase. In parts

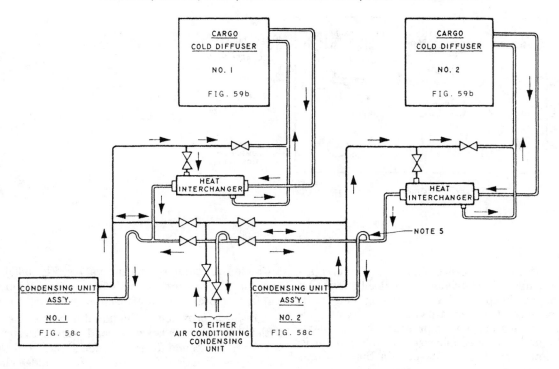

(a) INTERCONNECTING PIPING

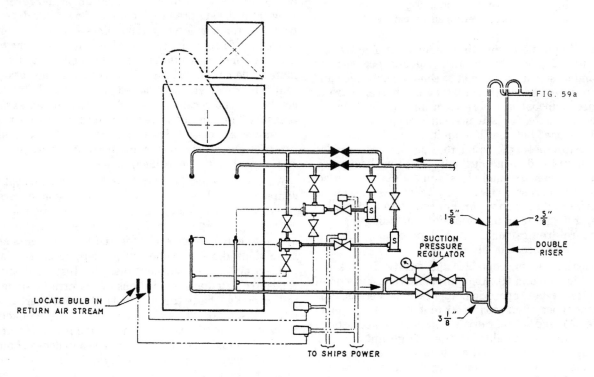

(b) TYPICAL CARGO COLD DIFFUSER HOOK-UP

Fig. 59   Cargo compartment refrigeration system diagram

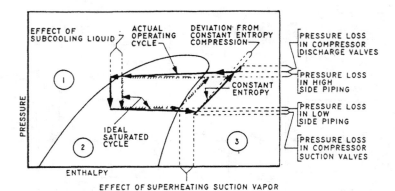

Fig. 60   Actual compression cycle

of the refrigeration cycle, the oil is a mist and the refrigerant is a vapor; here the refrigerant's velocity must be sufficiently high to keep entrained oil moving along with it. Often, a riser sized for adequate vapor velocity at minimum load may have a prohibitive pressure drop at maximum load. In such cases a double riser, as in Fig. 59(b), is required. At low capacities, oil collects in the trap and seals off the large riser; thus, the refrigerant vapor is forced to travel up the small riser at high velocity and carry oil with it. When the vapor flow increases sufficiently, the oil seal breaks and both risers are used. The small riser is sized for minimum design capacity, and the large riser is sized so that the velocity through both is sufficiently high at maximum load. Figures 58 and 59 incorporate design features required to assure the return of oil.

Liquid refrigerant leaving the condenser is usually subcooled 2 to 5 deg F. This is advantageous because the friction and loss in static head between the condenser and expansion valve reduce the pressure of the refrigerant; thus, unless subcooled, the refrigerant will flash into a gas to compensate for the pressure drop. Heat interchangers are used between the liquid line and compressor suction line to subcool the liquid; they are always provided in ship's stores and cargo systems. Locating heat interchangers at receivers minimizes liquid line sizes. Also, they protect compressors by evaporating any liquid mixed in the refrigerant suction vapor. For complete information regarding refrigeration piping design, including pipe sizing, pressure drop calculations, and subcooling requirements, see references 10, 11, and 41.

Accessories, valves, controls, and safety devices are shown in Figs. 58 and 59. A comprehensive discussion of these devices (types, operation, applications, details, etc.) and various types of oil separators (not shown) is given in references 11 and 46. Materials comprising accessories and valves must be suitable for the refrigerant handled. References 9, 43, 44, 45, and 47 delineate merchant and naval construction requirements for installation, materials, tests, safety devices, and spare parts. The last item, spare parts, is particularly important since the availability of the equipment may depend upon them.

Condenser water regulating valves, Fig. 58(c), are adjustable. They operate to maintain a constant refrigerant condensing pressure for efficient operation of the system.

A dryer is provided in the liquid line near the receiver outlet, Fig. 58(c), or near the condenser of systems without receivers. Some dryers include filters, and thus perform a dual function. The type of filter included eliminates much smaller particles than mesh strainers. With the arrangement in Fig. 58(c), the dryer only is used when charging the system with refrigerant, adding refrigerant to compensate for the loss through leaks, or when the presence of moisture in the system is detected by "freeze-up" at the expansion valve. On naval ships, however, full-flow filter dryers are provided. Full-flow filter dryers are provided for hermetic systems since excessive moisture causes the formation of acids, which can lead to the breakdown of motor insulation, possible burnout, and ultimately the contamination of the entire refrigeration system. The relief valve in the equalizing line between the condenser and receiver, Fig. 58(c), prevents an excessive pressure buildup in the receiver during shutdown when the valves are closed. The condenser relief valve and rupture disk protect the system in case of fire; note that these do not vent to the machinery space.

All lines should be plumb and straight, except horizontal suction lines, and those between condensers and receivers, which should pitch in the direction of flow. Valves in vapor lines are installed with their stems horizontal to avoid pockets. Unions and flanged joints should be minimized, i.e., provided only where necessary for disconnecting equipment, controls, etc. Insulated stuffing tubes are fitted at penetrations of watertight structures to thermally isolate the cold lines. Where watertightness is not essential, oversize sleeves (properly caulked and ratproofed) should be installed and the insulation and vapor seal should be continuous. Sleeves in decks should be of ample height to act as coamings; extra care should be taken to seal the joint between the pipe and sleeve to prevent the entrance of moisture, which can form ice which could crush the pipe.

b. Brine (secondary) cooling systems. Brine (secondary) cooling systems are preferred over direct-expansion cooling by many operators because of improved reliability, simplicity of operation, greater flexibility, better temperature control, and superior cargo turnout. The small amount of primary refrigerant piping required with brine systems significantly minimizes refrigerant leakage and facilitates the return of oil.

The initial cost of brine cooling is greater than that for a direct-expansion system. The additional heat exchanger required to produce the necessary secondary cooling tends to increase power requirements; however, this disadvantage may be overcome by more efficient operation. Generally, brine cooling may be used most advantageously where there is a large number of refrigerated compartments, where compartments served are remote from the refrigeration machinery space, where the diversity of simultaneous individual compartment temperatures is large, or where maximum humidity is required to minimize moisture removal from products. A strong conviction of the owner is at least one, if not the only deciding factor. Note that the brine may also be used for air-conditioning purposes.

Two-temperature closed brine systems are preferred because they provide optimum control of the supply brine temperature and minimize air leakage, which causes air binding and accelerates corrosion. On some ships a primary circulating system provides low-temperature brine to a number of secondary circulating systems, and each secondary system serves one compartment. The primary brine supply to the secondary system is modulated by a space thermostat.

Closed systems are preferred to the open type because the cost of pumping facilities is less, and the corrosive effect of aerated brine is significantly reduced. Also, glycols are organic and with open systems some inhibitor is lost to the atmosphere, thus necessitating periodic replacement.

All materials in the system, including piping, flange gaskets, valve seats and packing, pump seals, and other specialties must be compatible with the particular brine handled to avoid corrosion. Potential problems in this regard require careful analysis. References 23 and 41 contain discussions of the properties of brines and methods of calculating system pumping requirements.

c. Refrigerated storage spaces. Refrigerated storage spaces, or reefer spaces, are spaces designed for the storage of perishable cargo. Refrigerated spaces are of two general types: refrigerated cargo and refrigerated ship's stores. Refrigerated cargo spaces are intended to transport perishable cargo while refrigerated ship's stores spaces are intended to preserve perishable foods intended for shipboard use.

Refrigerated cargo spaces may be designed for storage of bulk cargo or for containerized cargo. The cargo may be frozen, requiring a storage temperature as low as −20 F, or chilled, with a storage temperature as high as 55 F. For flexibility, the storage spaces are usually designed to be convertible from frozen cargo to chilled cargo and vice versa.

Refrigerated cargo spaces are provided with a recirculating air system, defrosting system, and ventilation air system. The recirculating air system consists of a forced-air cooling coil, a two-speed fan, ductwork, and terminals. The fan capacity is usually sufficient to change the air once a minute with the compartment empty. Defrosting systems may be electric or hot-seawater spray. In electric defrosting systems, electric heater elements are built into the cooling unit while in hot-seawater spray systems, defrosting piping is fitted with drains and nonpermanent air blowout connections to permit complete removal of residual seawater when defrosting is complete. Adequate drainage must be provided for the defrosting system.

Supply and exhaust ventilation systems are provided for each refrigerated cargo space in which the temperature is maintained above 32 F. Fresh air is usually supplied to obtain 1 to 2 gross volume changes per hour, depending on the cargo stored. The supply air is directed toward the intake of the cooling coils and the exhaust air is taken from a remote location to avoid short-circuiting the fresh air.

At least one distant-reading, indicating dial thermometer is installed in each refrigerated storage space. The thermometer bulb should be located to indicate the representative room temperature. The temperature-indicating dials are mounted outside the refrigerated space adjacent to its access. In addition to the distant-reading thermometer, refrigerated cargo spaces are provided with an electronic-type selective indicating instrument installed in the refrigerating machinery room to indicate temperatures in all refrigerated cargo spaces. Temperature elements of the instrument are located in the return airstream to the forced-air cooling coils.

The size of the refrigerated ship's stores spaces vary with the type of ship. On a cargo vessel, a large domestic-type refrigerator may be adequate. On a passenger liner, a much larger walk-in type space will be required. The exact size is determined by the number of passengers and crew.

Insulation requirements for refrigerated spaces on merchant ships are contained in reference 2. Requirements for naval ships can be found in reference 7.

References

1 "Industrial Ventilation, A Manual of Recommended Practice," Committee on Industrial Ventilation, American Conference of Governmental Industrial Hygienists, Ann Arbor, Mich.

2 "Maritime Administration Standard Specifications for Merchant Ship Construction," U.S. Department of Commerce.

3 *Heating, Ventilating, and Air Conditioning Systems for Surface Ships*, Naval Ships Technical Manual, Chapter 510, S9086-RQ-STM-010/CH 510, Naval Sea Systems Command.

4 *NAVSEA Design Practices and Criteria Manual for Air Conditioning, Ventilation, and Heating of Surface Ships*, Chapter 510, NAVSEA, T9500-AA-PRO-130, Naval Sea Systems Command.

5 *Fundamentals*, American Society of Heating, Refrigeration, and Air Conditioning Engineers Handbook.

6 "Safety Code for Mechanical Refrigeration," Standard 15, American National Standards Institute/American Society of Heating, Refrigeration and Air Conditioning Engineers.

7 "General Specifications for Ships of the United States Navy," NAVSEA 59AAO-AA-SPN-10, Naval Sea Systems Command.

8 "Hazardous Ships Stores," CFR 46, Subpart 174, U.S. Coast Guard.

9 *Rules for Building and Classing, Steel Vessels*, American Bureau of Shipping.

10 *Refrigeration Systems*, Naval Ships Technical Manual, Chapter 516, S9086-RW-STM-010, Naval Sea Systems Command.

11 *Refrigeration*, American Society of Heating, Refrigeration, and Air Conditioning Engineers Handbook.

12 *Heating, Ventilation and Air Conditioning Equipment Manual*, Naval Sea Systems Command, Department of the Navy, S9512-BS-MMA-010.

13 "Materials," Code of Federal Regulations 46, Subchapter Q, Subpart 164, U.S. Coast Guard.

14 "Insulated Felt, Thermal and Sound Absorbing Felt, Fibrous Glass, Flexible," Military Specification MIL-I-22023.

15 "Thermal Insulation Report," SNAME Technical & Research Bulletin 4-7.

16 "Acoustic Absorptive Board, Fibrous Glass, Cloth Faced," Military Specification MIL-A-23054.

17 "Insulated Board, Thermal Fibrous Glass," Military Specification MIL-I-742.

18 "Plastic Material, Unicellular (Sheets and Tubes)," Military Specification MIL-P-15280.

19 "Insulation Panel, Thermal and Acoustical Absorptive, Open-Cell Poly Foam," Military Specification DOD-I-24688.

20 "Adhesives, Fire Resistant, Thermal Insulation," Military Specification MIL-A-3316.

21 "Coating Compound, Fibrous Glass Thermal Insulated Board, Water Vapor Barrier," Military Specification MIL-C-19993.

22 "Insulation, Plastic, Cellular Polyurethane, Rigid Preformed and Foam-N-Place," Military Specification MIL-I-24172.

23 *HVAC Applications*, American Society of Heating, Refrigeration, and Air Conditioning Engineers Handbook.

24 "Code on Noise Levels on Board Ships," International Maritime Organization (IMO).

25 "Report of Development of Estimating Method for Predicting Noise Originating in Air Conditioning System on Naval Vessels," Project 5662-2 Naval Applied Science Laboratory.

26 "Calculations for Merchant Ship Heating, Ventilation, and Air Conditioning Design," SNAME Technical & Research Bulletin 4-16.

27 *Heating, Ventilating and Air Conditioning Design Criteria Manual for Surface Ships of The United States Navy*, 0938-LP-018-0010 Naval Sea Systems Command.

28 *International Convention for the Safety of Life at Sea (SOLAS) Regulations*, International Maritime Organization (IMO).

29 "Electrical Engineering Regulations," Subchapter "J", U.S. Coast Guard.

30 "Thermal Insulation Report," SNAME Technical & Research Bulletin 4-7.

31 "Heat Transfer Coefficients," NAVSEA Design Data Sheet, DDS 511-2, Department of the Navy.

32 "Report on Ship's Stores and Installed Cargo Refrigerated Boxes," SNAME Technical & Research Bulletin 4-4.

33 "Refrigerating Equipment for Storage Compartments—Heat Load Calculation and Selection," NAVSEA Design Data Sheet DDS 516-1, Department of the Navy.

34 *The Commercial Storage of Fruits, Vegetables and Florist and Nursery Stock*, Agricultural Handbook No. 66, U.S. Department of Agriculture, Market Quality Research Division.

35 "Pressure Losses of Ventilation Fittings," NAVSEA Design Data Sheet 512-1, Department of the Navy.

36 "A Method for Determining the Size of Ventilation Fittings," NAVSEA Design Data Sheet DDS 512-2, Department of the Navy.

37 "HVAC Duct System Design," Sheet Metal and Air Conditioning Contractors National Association, Inc. (SMACNA), Vienna, Va.

38 "Test Code for Air Moving Devices," AMCA Standard 210-67, Air Moving and Conditioning Association, Inc., Park Ridge, Ill.

39 R. C. Strasser and H. E. Parker, "The Acoustic Habitability of Ships," *Trans.* SNAME, Vol. 72, 1964.

40 *Fan Engineering*, Buffalo Forge Co., Buffalo, N.Y.

41 *Handbook of Air Conditioning Design*, Carrier Corporation, McGraw-Hill, New York.

42 "Standard Specifications For Cargo Ship Construction," Maritime Administration.

43 "Marine Engineering," Code of Federal Regulations 46, Subchapter F (Subpart 50-6), U.S. Coast Guard.

44 "Refrigeration Plants and Systems, Mechanical, and Refrigeration System Components, Dichlorodifluoromethane, Type 12," Military Specification MIL-R-16743.

45 "Refrigeration Unit, Centrifugal for Air Conditioning," Military Specification MIL-R-24085.

46 *Handbook of Automatic Controls*, Alco Valve Co., St. Louis, Mo.

47 "American Standard, Recommended Practice for Mechanical Refrigeration Installed on Shipboard," American Society of Heating, Refrigerating, and Air Conditioning Engineers Standard 26-56.

CHAPTER XXII

Materials Staff,
David Taylor
Research Center

Construction Materials

Section 1
Introduction

The ocean has long been recognized as an aggressive and hostile medium that can cause material problems ranging from general-, galvanic-, pitting-, and crevice-corrosion, erosion, impingement attack, cavitation, and cyclic fatigue through a spectrum that includes marine fouling, and ice and wave impact. The history of marine materials development has followed an evolutionary trend with materials having performance limitations being continually superseded by superior replacements. For example, the 19th century transition from wood to steel as a ship-construction material led the way for dramatic new ship designs that became both larger and more capable. Material innovations in the marine industry, however, can be introduced only when the proper conditions are in place to employ them. In that sense, material advancements can be viewed as an enabling technology for marine engineers striving toward objectives such as improved system performance, energy efficiency, safety, and extended service lives in the ocean environment.

Marine engineers must have an understanding of not only conventional materials (such as mild steels, aluminum alloys, and copper-nickel alloys) but also high-performance materials (such as high-strength steels, duplex stainless steels, titanium alloys, and nickel alloys) and, where appropriate, composite materials. Composite materials are ideally suited for many shipboard applications;

however, metals continue to be the principal materials of construction for marine machinery and equipment.

Material selections involve considerations of cost, weight, availability, producibility, maintenance, life, and freedom from failures. Failures range from simple malfunctions to catastrophic destruction. The complete fracture of metal components from gross overstress (e.g., overtorqued bolts or accidental overloads) is easily understood. The majority of metal failures are more subtle and are attributable to wear, seizure, repeated stressing (fatigue), mechanical shock (impact), creep, stress rupture, embrittlement, corrosion, or the combined action of stress and corrosion (stress corrosion, corrosion fatigue). Because of the chainlike reaction that can occur when one part fails in a complex piece of machinery, the cause of a failure is often not easily identified.

The selection and application of materials for marine equipment can be governed by specifications and standards developed by regulatory bodies and technical societies [1–12].

This chapter is oriented principally to the aspects of materials technology that are of primary interest to a practicing marine engineer. Complementing this chapter, detailed discussion concerning the materials used in the construction of ship structure, welding procedures, and nondestructive test technology is presented in the Society's companion text *Ship Design and Construction* [13].

Section 2
Corrosion of Metals

2.1 Types of Corrosion. There are many forms of corrosion. Corrosion in the form of *direct chemical attack*, without the presence of an electrolyte, is one general type of corrosion. It can occur from reaction with certain chemicals, such as chlorine and sulfur. In such cases the avoidance of corrosion involves the elimination of the corrodant or the selection of a material with an inherent resistance to the corrodant. Direct chemical attack also can occur from hot gases such as those encountered on the firesides of boilers and in gas turbines. These cases may result from fuel constituents such as sulfur or vanadium, or

from salt contamination in the fuel or combustion air; reducing the corrosive constituents, lowering of operating temperatures, and selecting more resistant materials are the solutions most frequently implemented.

Another general type of corrosion occurs by *electrochemical attack* in the presence of an electrolyte. Seawater in the form of liquid and mist is the most common electrolyte encountered aboard ship and is responsible for the most complicated and troublesome corrosion problems. The reasons for this are the great chemical activity and high conductivity of seawater as compared to other

913

Table 1 Corrosion performance of copper alloys, stainless steel, and titanium alloys in natural and polluted seawater

Corrosion Mode	Copper Alloys	Stainless Steel (316)	Titanium Alloys (Gr 1, 2, 3, 5, 9)
General corrosion	resistant/ susceptible[a]	resistant	resistant
Crevice corrosion	susceptible	susceptible	resistant (<200 F)
Pitting attack	susceptible	susceptible	immune
Stress corrosion	susceptible[a]	susceptible (>140 F)	resistant (except Grade 5)
Corrosion fatigue	susceptible	susceptible	resistant
Galvanic attack	susceptible	susceptible	immune
Microbial/ influenced corrosion	susceptible	susceptible	immune
Weld/HAZ corrosion	susceptible	susceptible	resistant
Erosion corrosion	susceptible	susceptible	highly resistant
Stray current	susceptible	susceptible	fairly resistant

[a] Dependent on pollution level and seawater chemistry.

electrolytes such as fresh water, the wide assortment of environmental and material variables that affect seawater corrosion of metals, and the wide variety of forms in which deterioration can occur. Other electrolytes leading to electrochemical attack aboard ship usually involve rather specific environmental circumstances and materials of construction.

2.2 Seawater Corrosion. The nature and extent of metal corrosion caused by exposure to seawater depends on the alloy and its compositional and heat treatment variations. Of equal importance are environmental variations such as degree of aeration, extent of seawater exposure (continuous immersion, tidal immersion, splash/spray exposure, or marine atmospheric exposure), fouling, dissimilar-metal couples, velocity, turbulence, cavitation, temperature, and crevices. A general summary that ranks the corrosion performance of copper alloys, 316 stainless steel, and titanium alloys in natural and polluted seawater is given in Table 1. The various types of corrosion that can occur in a seawater environment are described in the following.

a. General corrosion (uniform attack). General corrosion can be classified as a uniform attack of a metal or alloy surface. Materials such as copper alloys and weathering steels generally corrode uniformly. This is in contrast to the significant forms of electrochemical degradation of a metal, which are typically related to severe localized attack rather than to excessive rates of general corrosion. When general corrosion is of concern, it can usually be controlled with paints, metallic coatings, cathodic protection (sacrificial anodes or impressed-current systems), or by the selection of more resistant materials [14].

b. Galvanic corrosion. Galvanic corrosion occurs when two dissimilar metals are electrically coupled in the presence of an electrolyte such as seawater. Current will flow through the electrolyte from the anodic material to the cathodic material. The corrosion resulting from this "battery effect" is known as galvanic corrosion. Usually, corrosion of the anode is accelerated, and corrosion of the cathode is decelerated.

A galvanic series is a useful means of assessing galvanic corrosion tendencies between two dissimilar metals. A galvanic series is a listing of open-circuit potentials of metals and alloys in an electrolyte as measured with a specific reference electrode. Table 2 is such a listing of metals in seawater, measured versus a saturated calomel reference electrode. The galvanic series is arranged in approximate order from the most anodic (or least noble) in behavior (magnesium alloys) to the most cathodic (or most noble) in behavior (graphite, graphitized cast iron). The listing is not exact, as relative positions may change due to variations such as water velocity, temperature, and metal passivity. In general, the farther apart two metals are in the galvanic series, the greater is the potential for galvanic corrosion to occur.

Although the difference in potential between two dissimilar metals in a galvanic couple is the driving force for galvanic corrosion, other factors such as polarization behavior, cathode/anode area ratios, distance, and geometry affect the magnitude of the galvanic corrosion. When current begins to flow between two metals in a closed circuit, polarization occurs due to reactions at the metal-liquid interfaces. Polarization is a shift from the open-circuit corrosion potential of a metal that results due to current flow. Upon galvanic coupling, the potential of one member of the couple tends to shift toward the other. The magnitude of the shift in potential is dependent on the polarization behavior of the metals in the galvanic couple. If the cathodic metal in the galvanic couple is easily polarized, then its potential is shifted toward the anodic metal, which reduces the shift in the anodic potential and thus reduces the magnitude of the galvanic corrosion occurring at the anode. If, however, the cathodic metal in the couple is not easily polarized, then the potential of the anodic metal shifts farther toward the cathodic potential and results in more significant galvanic corrosion of the anodic material.

The suitability of a dissimilar-metal couple in practice may depend on the relative areas of the anode and cathode. If the anode is small and the cathode is relatively large, the anode may suffer an unacceptable rate of deterioration. On the other hand, if the cathode is small and the anode large, the corrosion and average penetration rate of the anode may remain at a tolerable level. For example, Monel bolts in a large steel plate might be acceptable, whereas steel bolts in a large Monel plate would not be acceptable.

Occasionally, circumstances may dictate that dissimilar metals be coupled with an undesirable cathode/anode area ratio. In such cases it may be possible to reduce corrosion of the anode by corrosion protection with paint. The paint should not be applied to the anode, however, because exposure of the anode at any imperfection in the paint film would result in an even more unfavorable cathode/anode area ratio and concentration of the galvanic current at the point of imperfection in the paint. Although the cathode is not the metal requiring protection, painting the cathode would provide a more favorable cathode/anode area ratio and a reduction of galvanic current; any corrosion of the anode would be distributed over the entire anode surface

Fig. 1 Galvanically induced corrosion of a monel nut coupled to a stainless steel shaft and impeller

instead of being concentrated at one location. The same line of reasoning leads to a warning against the application of noble metal coatings on a less-noble base metal: for example, chromium plating on carbon steel. If the coating contains an imperfection or holiday, the large cathode/anode area ratio can cause severe and rapid corrosion of the base metal in the localized area of the imperfection [14].

Figure 1 is an example of galvanic corrosion that occurred as a result of using monel nuts to couple a stainless steel impeller to a shaft. The upper nut is in the original condition, whereas the lower nut reflects the result of severe galvanic corrosion after 34 months of exposure of the nut (anode) and the impeller (cathode) in a seawater environment. The corrosion was exacerbated because of the relatively large (in terms of both area and mass) stainless steel impeller relative to the monel nuts.

Table 2 provides approximate guidance concerning the acceptability of various metal couples in seawater at 40 to 80 F as a function of relative cathode/anode area ratio. The ratings tend to be conservative and any uncertainties are generally indicated in the safe direction. However, due to variations in oxygen content, temperature, velocity, metal passivity, and turbulence, the ratings can only generally apply. This is particularly so for the couples marked with "X" (uncertain) and "C" (compatible). Those marked with "U" (unfavorable) should certainly be avoided.

Both distance and geometry effects can also have a substantial influence on galvanic-corrosion behavior. In general, the magnitude of galvanic corrosion is lessened as the distance between the anodic and cathodic members of the couple increases. Geometry influences galvanic-corrosion behavior in that current flow in the couple takes the path of least resistance [14].

Additional guidance for the protection of coupled dissimilar metals in seawater and the marine atmosphere can be found in Military Standard 889 [15].

c. Pitting corrosion. Pitting corrosion is a form of localized attack that occurs due to the breakdown of passivity on a metal surface. Once the passive layer is penetrated, an unfavorable electrolytic cell is set up between the small anodic area where the breakdown occurred (pit sites) and the large passive area on the remainder of the metal surface. Pitting corrosion is a very serious form of deterioration due to the fact that the severity of pitting is extremely difficult to predict. Factors that influence the breakdown of passivity include local environmental differences in oxygen, temperature, pH, chloride ion concentration, and flow velocity as well as physical or chemical inhomogeneities in the passive film [14,16,17].

Severe pitting of aluminum alloys in seawater is often associated with local electrolytic cells formed between the more anodic aluminum matrix and the more cathodic heavy-metal alloying elements such as copper, nickel, and iron. For this reason, the aluminum alloys with lower heavy-metal contents, such as some of those in the 5000 series, generally have the best resistance to seawater corrosion. However, even these alloys may suffer accelerated attack if the water contains heavy-metal ions: for example, copper leached from antifouling paints [16,18].

Stainless steels are often more susceptible to severe pitting than less-noble metals such as carbon steel. This is because the corrosion resistance of the stainless steel is associated with a protective passive oxide film, and any local breakdown of the film exposes a local active (anodic) area. If the film cannot reform, the potential difference between the active and passive areas then causes accelerated attack at the location of the film breakdown. The chloride ion in seawater is especially aggressive in penetrating the passive film on these materials [18].

d. Crevice corrosion. Crevice corrosion is a severe form of localized corrosion that results from concentration-cell effects. Crevices may be formed by the component geometry (i.e., a weld defect or a threaded joint), by contact of the metal surface with a nonmetallic material, or by deposits accumulated on the metal surface. The deposits may be in the form of corrosion products, sand, dirt, or marine organisms such as barnacles [14,19,20].

Localized corrosion can occur either inside or outside of the crevice area. The crevice corrosion of some metals, such as austenitic stainless steels and many of the nickel-base alloys, occurs within the crevice area. The chloride ion in seawater can penetrate the protective oxide film, creating an active surface within the crevice area that attempts to repassivate by combining with oxygen dissolved in the entrapped water. When the oxygen in the crevice is depleted, film repair is no longer possible, and a galvanic cell results between the active surface within

Table 2 Probable behavior of galvanic couples in seawater at 40 to 80 F where the exposed area of the metal under consideration (in the left-hand column) is relatively SMALL (S), approximately EQUAL (E), or relatively LARGE (L) compared with the area of the metal with which it may be coupled (as shown by the numbers at the top of the columns). The numbers at the top of the columns denote the same metals as given by the same numbers in the left-hand column. See text for precautions in use

For example, the number 5 in both cases indicates cadmium.

U = UNFAVORABLE—normal deterioration of either material may be increased moderately or severely.

X = UNCERTAIN—direction and/or magnitude of effect or normal behavior may vary, depending on circumstances.

C = COMPATIBLE—deterioration of either material is normally within tolerable limits.

		1	2	3	4	5	6	7	8	9	10	11	12	13	14	15	16	17	18	19	20	21	22	23	24	25	26	27	28	29	30	31	32
1	S	•	•	•	U	U	U	U	U	U	U	U	U	U	U	U	U	U	U	U	U	U	U	U	U	U	U	U	U	U	U	U	U
Magnesium Alloys (1)	E	•	•	•	U	U	U	U	U	U	U	U	U	U	U	U	U	U	U	U	U	U	U	U	U	U	U	U	U	U	U	U	U
	L	•	•	•	X	X	U	U	U	U	U	U	U	U	U	U	U	U	U	U	U	U	U	U	U	U	U	U	U	U	U	U	U
2	S	U	U	U	•	•	•	U	U	U	U	U	U	U	U	U	U	U	U	U	U	U	U	U	U	U	U	U	U	U	U	U	U
Zinc (1)	E	U	U	U	•	•	•	U	U	U	U	U	U	U	U	U	U	U	U	U	U	U	U	U	U	U	U	U	U	U	U	U	U
	L	U	X	X	•	•	•	U	U	U	U	U	U	U	U	U	U	U	U	U	U	U	U	U	U	U	U	U	U	U	U	U	U
3	S	X	U	U	X	X	X	•	•	•	U	U	U	U	U	U	U	U	U	U	U	U	U	U	U	U	U	U	U	U	U	U	U
Beryllium	E	X	U	U	X	X	X	•	•	•	U	U	U	U	U	U	U	U	U	U	U	U	U	U	U	U	U	U	U	U	U	U	U
	L	X	U	U	X	X	X	•	•	•	U	U	U	U	U	U	U	U	U	U	U	U	U	U	U	U	U	U	U	U	U	U	U
4	S	U	U	U	U	U	C	U	C	X	•	•	•	U	U	U	U	U	U	U	U	U	U	U	U	U	U	U	U	U	U	U	U
Aluminum Alloys (1)	E	U	U	U	U	U	X	U	X	C	•	•	•	U	U	U	U	U	U	U	U	U	U	U	U	U	U	U	U	U	U	U	U
	L	U	X	X	U	U	X	U	X	X	•	•	•	U	U	U	U	U	U	U	U	U	U	U	U	U	U	U	U	U	U	U	U
5	S	X	U	U	X	X	C	C	C	C	•	•	•	C	C	C	C	C	C	X	C	C	C	C	C	C	C	C	C	C	C	C	C
Cadmium	E	X	U	U	X	X	C	C	C	C	•	•	•	C	C	C	C	C	C	X	C	C	C	C	C	C	C	C	C	C	C	C	C
	L	X	U	U	X	X	C	C	C	C	•	•	•	C	C	C	C	C	C	X	C	C	C	C	C	C	C	C	C	C	C	C	C
6	S	U	U	U	C	U	C	C	C	C	C	C	C	•	•	•	C	C	C	C	C	C	C	C	C	C	C	C	C	C	C	C	C
Mild Steel, Wrought Iron	E	U	U	U	C	U	C	C	C	C	C	C	C	•	•	•	C	C	C	C	C	C	C	C	C	C	C	C	C	C	C	C	C
	L	U	U	U	C	U	C	C	C	C	C	C	C	•	•	•	C	C	C	C	C	C	C	C	C	C	C	C	C	C	C	C	C
7	S	U	U	U	C	U	C	C	C	C	C	C	C	C	C	C	•	•	•	C	C	C	C	C	C	C	C	C	C	C	C	C	C
Cast Iron, Flake or Ductile	E	U	U	U	C	U	C	C	C	C	C	C	C	C	C	C	•	•	•	C	C	C	C	C	C	C	C	C	C	C	C	C	C
	L	U	U	U	C	U	C	C	C	C	C	C	C	C	C	C	•	•	•	C	C	C	C	C	C	C	C	C	C	C	C	C	C
8	S	U	U	U	C	U	C	C	C	C	C	C	C	C	C	C	C	C	C	•	•	•	C	C	C	C	C	C	C	C	C	C	C
Low Alloy High Strength Steel	E	U	U	U	C	U	C	C	C	C	C	C	C	C	C	C	C	C	C	•	•	•	C	C	C	C	C	C	C	C	C	C	C
	L	U	U	U	C	U	C	C	C	C	C	C	C	C	C	C	C	C	C	•	•	•	C	C	C	C	C	C	C	C	C	C	C
9	S	U	U	U	U	U	C	C	C	C	C	C	C	C	C	C	C	C	C	C	C	C	•	•	•	X	X	X	X	X	X	U	U
Ni-Resist, Types 1 & 2	E	U	U	U	U	U	C	C	C	C	C	C	C	C	C	C	C	C	C	C	C	C	•	•	•	X	X	X	X	X	X	U	U
	L	U	U	U	U	U	C	C	C	C	C	C	C	C	C	C	C	C	C	C	C	C	•	•	•	X	X	X	X	X	X	U	U
10	S	U	U	U	U	U	U	U	U	U	•	•	•	X	X	X	X	X	X	X	X	X	X	X	X	X	X	X	X	X	X	U	U
Naval Br. (CA464), Yel. Br. (CA268), Al. Br. (CA687), Red Br. (CA230), Adm'ty Br. (CA443), Mn Bronze	E	U	U	U	U	U	U	U	U	U	•	•	•	X	X	X	X	X	X	X	X	X	X	X	X	X	X	X	X	X	X	U	U
	L	U	U	U	U	U	U	U	U	U	•	•	•	X	X	X	X	X	X	X	X	X	X	X	X	X	X	X	X	X	X	U	U
11	S	U	U	U	U	U	U	U	U	U	X	X	X	•	•	•	X	X	X	X	X	X	X	X	X	X	X	X	X	X	X	U	U
Tin	E	U	U	U	U	U	U	U	U	U	X	X	X	•	•	•	X	X	X	X	X	X	X	X	X	X	X	X	X	X	X	U	U
	L	U	U	U	U	U	U	U	U	U	X	X	X	•	•	•	X	X	X	X	X	X	X	X	X	X	X	X	X	X	X	U	U
12	S	U	U	U	U	U	U	U	U	U	X	X	X	X	X	X	•	•	•	X	X	X	X	X	X	X	X	X	X	X	X	U	U
Copper (CA102, 110), Si Bronze (CA655)	E	U	U	U	U	U	U	U	U	U	X	X	X	X	X	X	•	•	•	C	X	X	X	X	X	X	X	X	X	X	X	U	U
	L	U	U	U	U	U	U	U	U	U	X	X	X	X	X	X	•	•	•	C	X	X	X	X	X	X	X	X	X	X	X	U	U
13	S	U	U	U	U	U	U	U	U	U	X	X	X	X	X	X	X	X	C	•	•	•	X	X	X	X	X	X	X	X	X	U	U
Lead-Tin Solder	E	U	U	U	U	U	U	U	U	U	X	X	X	X	X	X	X	X	C	•	•	•	X	X	X	X	X	X	X	X	X	U	U
	L	U	U	U	U	U	U	U	U	U	X	X	X	X	X	X	X	X	C	•	•	•	X	X	X	X	X	X	X	X	X	U	U
14	S	U	U	U	U	U	C	C	C	C	U	X	X	X	X	X	X	X	X	X	X	X	C	C	C	•	•	•	C	C	C	U	U
Tin Bronze (G & M)	E	U	U	U	U	U	C	C	C	C	U	X	X	X	X	X	X	X	X	X	X	X	C	C	C	•	•	•	C	C	C	U	U
	L	U	U	U	U	U	C	C	C	C	U	X	X	X	X	X	X	X	X	X	X	X	C	C	C	•	•	•	C	C	C	U	U
15	S	U	U	U	U	U	C	C	C	•	U	U	U	U	U	•	U	U	U	U	U	U	U	U	U	U	U	U	U	U	U	•	•
Stainless Steel, 12-14% Cr (AISI Types 410, 416)	E	U	U	U	U	U	C	C	C	•	U	U	U	U	U	•	U	U	U	U	U	U	U	U	U	U	U	U	U	U	U	•	•
	L	U	U	U	U	U	C	C	C	•	U	U	U	U	U	•	U	U	U	U	U	U	U	U	U	U	U	U	U	U	U	•	•

		1	2	3	4	5	6	7	8	9	10	11	12	13	14	15	16	17	18	19	20	21	22	23	24	25	26	27	28	29	30	31	32
16	Nickel Silver (CA 732, 735, 745, 752, 764, 770, 794)	S	U	U	U	U	U	C	X	C	C	X	X	X	X	U	U	●	X	C	U	C	U	U	X	X	X	U	U	U	U	U	U	U
		E	X	U	U	U	U	X	U	U	X	U	X	U	X	U	●	C	C	C	X	U	U	●	X	X	X	U	X	X	X	U	X	U
		L	U	U	U	U	U	U	U	U	X	U	U	U	X	U	●	C	C	●	U	C	C	●	X	X	X	U	U	U	U	U	X	U
17	90/10 Copper-Nickel (CA 706)	S	U	U	U	U	U	C	X	C	C	X	X	X	X	U	U	C		C	U	C	U	U	X	X	X	U	U	U	U	U	U	U
		E	X	U	U	U	U	X	U	U	X	U	X	U	X	U	C		C	●	X	U	U	●	X	X	X	U	X	X	X	U	X	U
		L	U	U	U	U	U	U	U	U	X	U	U	U	X	U	C		C	●	U	C	C	●	X	X	X	U	U	U	U	U	X	U
18	80/20 Copper-Nickel (CA 710)	S	U	U	U	U	U	C	X	C	C	X	X	X	X	U	U	C	C			C	U	U	X	X	X	U	U	U	U	U	U	U
		E	X	U	U	U	U	X	U	U	X	U	X	U	X	U	C	C		●	X	U	U	●	X	X	X	U	X	X	X	U	X	U
		L	U	U	U	U	U	U	U	U	X	U	U	U	X	U	C	C		●	U	C	C	●	X	X	X	U	U	U	U	U	X	U
19	Stainless Steel, 16–18% Cr (AISI Type 430)	S	U	U	U	U	U	C	X	C	C	X	X	X	X	U	C	C	U		●	U	U	U	X	X	X	U	U	U	U	U	U	U
		E	X	U	U	X	U	X	U	U	X	U	X	U	X	U	X	U	●		●	X	X	●	X	X	X	X	X	X	X	U	X	U
		L	U	U	U	U	U	U	U	U	X	U	U	U	X	U	X	C	●		U	U	U	●	X	X	X	U	U	U	U	U	X	U
20	Lead	S	U	U	U	U	U	C	X	C	C	X	X	X	X	U	U	U	U	●		C	C	U	X	X	X	U	U	U	U	U	U	U
		E	X	U	U	U	U	X	U	U	X	U	X	U	X	U	U	U	U	●		C	C	●	X	X	X	U	X	X	X	U	X	U
		L	U	U	U	U	U	U	U	U	X	U	U	U	X	U	U	U	U	●		C	C	●	X	X	X	U	U	U	U	U	X	U
21	70/30 Copper-Nickel (CA 715)	S	U	U	U	U	U	C	X	C	C	X	X	X	X	U	C	C	C	U	C		●	U	X	X	X	U	U	U	U	U	U	U
		E	X	U	U	U	U	X	U	U	X	U	X	U	X	U	X	U	U	U	C		●	●	X	X	X	U	X	X	X	U	X	U
		L	U	U	U	U	U	U	U	U	X	U	U	U	X	U	X	C	C	U	C		●	●	X	X	X	U	U	U	U	U	X	U
22	Nickel Aluminum Bronze	S	U	U	U	X	U	C	X	C	C	X	X	X	X	U	X	X	X	U	C	●		X	X	X	X	C	U	U	U	U	U	U
		E	X	U	U	U	U	X	U	U	X	U	X	U	X	U	X	U	U	U	C	●		X	X	X	X	X	X	X	X	U	X	U
		L	U	U	U	U	U	U	U	U	X	U	U	U	X	U	X	C	C	U	C	●		X	X	X	X	C	U	U	U	U	X	U
23	Inconel[a] Alloy 600	S	U	U	U	U	U	C	X	C	C	X	X	X	X	U	C	C	C	U	U	U	X		X	X	X	U	U	U	U	U	U	U
		E	X	U	U	U	U	X	U	U	X	U	X	U	X	U	X	U	U	U	U	●	X		X	X	X	X	X	X	X	U	X	U
		L	U	U	U	U	U	U	U	U	X	U	U	U	X	U	X	C	C	U	U	●	X		X	X	X	U	U	U	U	U	X	U
24	Silver Braze Alloys	S	U	U	U	U	U	C	X	C	C	X	X	X	X	U	C	C	C	U	U	U	X	X		●	X	U	U	U	U	U	U	●
		E	X	U	U	U	U	X	U	U	X	U	X	U	X	U	X	U	U	U	U	●	X	X		●	X	X	X	X	X	U	X	●
		L	U	U	U	U	U	U	U	U	X	U	U	U	X	U	X	C	C	U	U	U	X	X		●	X	U	U	U	U	U	X	●
25	Nickel 200	S	U	U	U	U	U	C	X	C	C	X	X	X	X	U	C	C	C	U	U	U	X	X	●		X	U	U	U	U	U	U	●
		E	X	U	U	U	U	X	U	U	X	U	X	U	X	U	X	U	U	U	U	X	X	X	●		X	X	X	X	X	U	X	●
		L	U	U	U	U	U	U	U	U	X	U	U	U	X	U	X	C	C	U	U	U	X	X	●		X	U	U	U	U	U	X	●
26	Silver	S	U	U	U	U	U	C	X	C	C	X	X	X	X	U	C	C	C	U	U	U	C	U	X	X		U	U	U	U	U	U	●
		E	X	U	U	U	U	X	U	U	X	U	X	U	X	U	X	U	U	U	U	X	X	X	X	X		X	X	X	X	U	X	●
		L	U	U	U	U	U	U	U	U	X	U	U	U	X	U	X	C	C	U	U	U	C	U	X	X		U	U	U	U	U	X	●
27	Stainless Steel, 18Cr, 8Ni (AISI Types 302, 304, 321, 347)	S	U	U	U	U	U	C	X	C	C	X	X	X	X	U	U	U	U	U	U	U	C	U	U	U	U		C	C	C	C	U	U
		E	X	U	U	U	U	X	U	U	X	U	X	U	X	U	X	X	X	X	X	X	X	X	X	X	X		X	X	X	U	X	U
		L	U	U	U	U	U	U	U	U	X	U	U	U	X	U	U	U	U	U	U	U	U	U	U	U	U		C	C	C	U	U	U
28	Monel[a] Alloys 400, K-500	S	U	U	U	U	U	C	X	C	C	X	X	X	X	U	U	U	U	U	U	U	U	U	U	U	U	C		U	U	U	U	U
		E	X	U	U	U	U	X	U	X	X	U	X	U	X	U	X	X	X	X	X	X	X	X	X	X	X	X		X	X	U	U	U
		L	U	U	U	U	U	U	U	U	X	U	U	U	X	U	U	U	U	U	U	U	U	U	U	U	U	C		U	U	U	U	U
29	Stainless Steel, 18Cr, 12Ni-Mo (AISI Types 316, 317)	S	U	U	U	U	U	C	X	C	C	X	X	X	X	U	U	U	U	U	U	U	U	U	U	U	U	C	U		C	C	U	U
		E	X	U	U	U	U	X	U	X	X	U	X	U	X	U	X	X	X	X	X	X	X	X	X	X	X	X	X		X	U	U	U
		L	U	U	U	U	U	U	U	U	X	U	U	U	X	U	U	U	U	U	U	U	U	U	U	U	U	C	U		C	U	U	U
30	Carpenter 20[c] Stainless Steel, Incoloy[a] Alloy 825	S	U	U	U	U	U	C	X	C	C	X	X	X	X	U	U	U	U	U	U	U	U	U	U	U	U	C	U	C		●	U	U
		E	X	U	U	U	U	X	U	X	X	U	X	U	X	U	X	X	X	X	X	X	X	X	X	X	X	X	U	U	●		X	U
		L	U	U	U	U	U	U	U	U	X	U	U	U	X	U	U	U	U	U	U	U	U	U	U	U	U	C	U	C	●		U
31	Titanium, Hastelloy[b] Alloys C & C 276 Inconel[a] Alloy 625	S	U	U	U	U	U	C	X	C	X	X	X	X	X	U	U	U	U	U	U	U	U	U	U	U	U	U	U	U	U	X		X
		E	X	U	U	U	U	X	U	X	X	U	X	U	X	U	X	X	X	X	X	X	X	X	X	X	X	X	X	U	U	X		X
		L	U	U	U	U	U	U	U	U	X	U	U	U	X	U	U	U	U	U	U	U	U	U	U	U	U	U	U	U	U	X		X
32	Graphite, Graphitized Cast Iron	S	U	U	U	U	U	C	X	C	X	U	U	U	U	U	U	U	U	U	U	U	U	U	●	●	●	U	U	U	U	X	X	
		E	X	U	U	U	U	X	U	X	X	U	X	U	X	U	U	U	U	U	U	U	U	U	●	●	●	U	U	U	U	X	X	
		L	U	U	U	U	U	U	U	U	X	U	U	U	X	U	U	U	U	U	U	U	U	U	●	●	●	U	U	U	U	X	X	

[a] International Nickel trademark
CA—Alloy designation of the Copper Development Association.
AISI—Alloy designation of the American Iron and Steel Institute.
(1)—Compositions other than those formulated for galvanic anodes.
(Prepared with use of copyright information released by the International Nickel Company Inc.)
[b] Union Carbide Corp. trademark
[c] The Carpenter Steel Co. trademark

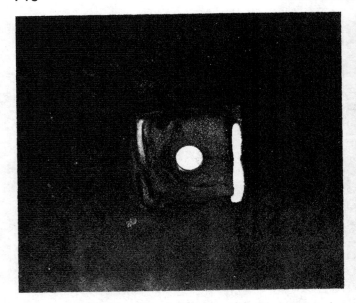

Fig. 2   Seawater crevice corrosion of an austenitic stainless steel (attack within crevice)

Fig. 3   Seawater crevice corrosion of a copper-base alloy (attack outside crevice)

the crevice and the passive surface on the outside. A second cell results from the difference in oxygen content between the inside and outside of the crevice. Both cells tend to accelerate corrosion of the alloy within the crevice area [18,21]. An example of corrosion inside a crevice is shown in Fig. 2.

Crevice corrosion of other metals, such as some of the copper-base alloys, occurs just outside the crevice area. In this case, corrosion in the crevice causes saturation of the entrapped water with copper ions, which impedes further corrosion in the crevice area. The difference in copper-ion concentration between the inside and outside of the crevice results in a local cell tending to accelerate corrosion around the outside edges of the crevice [21]. An example of corrosion outside a crevice is shown in Fig. 3.

Another example of corrosion occurring inside a crevice is illustrated by Fig. 4. The crevice corrosion is on a type 304 stainless steel plate. The plate was in a seawater environment, and crevice corrosion occurred beneath barnacles that became attached.

e. Erosion corrosion. Erosion corrosion is a general term that relates to an increase in the rate of deterioration on a material surface as a result of the abrasive action of a moving fluid such as seawater. *Impingement* and *cavitation* are forms of erosion corrosion that also involve velocity conditions. Impingement occurs when there is a local impingement of a moving fluid like seawater against a solid material surface. Cavitation damage occurs when vapor bubbles form and collapse in a fluid near a solid material surface [17].

As the seawater velocity increases beyond 2 to 3 fps, the fouling on a metal surface generally diminishes and pitting of the more noble metals (such as the austenitic stainless steels and many nickel-base alloys) decreases or disappears. This trend continues provided that the system

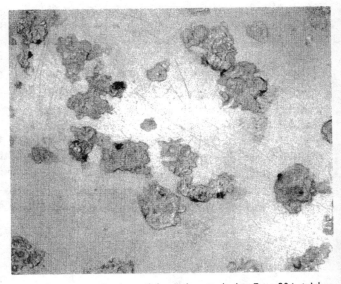

Fig. 4   Crevice corrosion beneath barnacles attached to Type 304 stainless steel exposed in seawater

design details do not create mechanical crevices that are capable of entrapping stagnant water.

Contrary to the noble metals, the protective capabilities of the copper-base alloys diminish as the seawater velocity is increased, which leads to an increase in the general rate of corrosion. The critical velocity for pure copper, beyond which excessive corrosion occurs on the metal surface, is 3 fps in seawater. Alloying the copper with elements such as nickel or aluminum substantially increases the alloy's resistance to corrosion under velocity conditions [14,21].

Copper alloys are often used at velocities in excess of their known critical velocity; specific examples include pump casings, impellers, and propellers where the wall thicknesses are such that moderate corrosion rates can be tolerated. Under severe conditions where the tolerance for corrosion is limited, protective coatings, liners, more noble materials, and nonmetallic construction materials should be considered along with their relative costs when establishing the material-selection criteria [22].

All titanium alloys have an exceptionally high resistance to seawater flow (to velocities in excess of 100 fps), impingement, turbulence, and cavitation [21,22]. In sand and emery particle-laden seawater, erosion-corrosion resistance to an approximately 20-fps velocity has been observed [23]. The harder, higher-strength titanium alloys such as Grade 5 have optimum seawater erosion and cavitation resistance, making them especially attractive for the most severe cases of seawater impingement/cavitation exposure. These include critical hydrofoil components, propulsion components, propellers, and pumps [14,22].

Composite materials have an excellent erosion-corrosion resistance in high-velocity seawater under all conditions except severe cavitation erosion. A series of high-velocity tests was conducted on epoxy, vinylester, phenolic, and polyester-reinforced thermosetting resins in seawater flowing from 0 to 130 fps. The materials were tested under impingement, parallel-flow, and rotating-disk conditions. The results of the three tests showed that the corrosion performance of the composites evaluated was superior to that of gun metal bronze. However, severe cavitation erosion tests conducted on the same materials showed the performance of the composite materials to be inferior to the performance of the gun metal bronze [24].

For tubes and pipes carrying seawater, the water velocity alone is not necessarily the most important limiting factor insofar as material degradation is concerned. More frequently, serious degradation occurs locally as a result of impingement and cavitation erosion associated with regions of turbulence [25]. For example, the effects of excessive water velocity first become apparent at the inlet ends of condenser tubes or in pipes in the turbulent-flow region immediately downstream of fittings before smooth-flow conditions have become established. In these turbulent areas, the protective films can be continuously stripped away, leaving an active surface on which corrosion may be accelerated by galvanic currents set up by the surrounding passive surfaces.

Poorly fitted gaskets, elbows, valves, or any obstruction that creates turbulence can lead to an accelerated attack of piping material when the water velocity approaches the smooth-flow tolerance of the alloy. Thus, designing for streamlined flow can be advantageous. However, this is seldom possible throughout a system, so selection of the proper alloy remains important.

Globe valves are a common source of turbulence in seawater piping systems. Figure 5 illustrates typical downstream deterioration of a 70/30 Cu-Ni pipe that resulted from excessive water velocity within a globe valve. The perforation damage shown in Fig. 5 occurred after 1.3 years at 15-fps water velocity in a 70/30 Cu-Ni pipe containing less than 0.1% iron. If the copper-nickel alloy had contained about 0.5% iron, the pipe would have experienced minimal damage after several years of operation.

Fig. 5   Impingement attack of a seawater pipe in the turbulent area downstream of a globe valve

The higher iron content would have provided a more effective protective film to prevent corrosion. This example illustrates the importance of care in the selection and specification of materials.

Another example of erosion corrosion is illustrated by Fig. 6. This figure shows a silicon-bronze simplex strainer housing that perforated because of turbulence in the housing area surrounding the entrance to a bypass valve. The turbulence was created by a clogged filter element, which caused the seawater flow to be diverted through the restricted opening of the bypass valve. An improved filter design would help to reduce clogging, thus reducing the probability for erosion corrosion of the housing. This example highlights the importance of proper design to prevent accelerated corrosion of metallic components.

f. Intergranular corrosion or selective-phase corrosion. Intergranular corrosion can take several forms:

Graphitic corrosion is a type of corrosion common to gray cast irons, although not to austenitic nickel cast irons. An example of this type of corrosion is shown by Fig. 7. The iron corrodes away, leaving a residue of the free carbon (graphite) that gives gray cast irons their name. The residue retains the original shape of the part but has no significant mechanical strength [17].

Dezincification is another type of selective corrosion common to brasses containing more than 15% zinc. The commonly accepted theory for dezincification consists of the brass dissolving, the zinc ions remaining in solution, and then the copper redepositing. The result is a part retaining its original shape but containing a porous, reddish-copper deposit of little strength. Muntz metal, naval brass, Admiralty metal, and aluminum brass are subject to dezincification in seawater unless inhibited grades are specified, in which inhibiting elements such as arsenic, antimony, or phosphorous are added to the brass [17].

Dealuminization is a type of selective-phase attack occurring in some aluminum bronzes, particularly cast alloys containing more than 8% aluminum. An example of dealuminization is illustrated by Fig. 8. The attack is associated with selective corrosion of a fine network of an aluminum-rich "gamma phase" in the microstructure of the alloy. Dealuminization can be controlled either by

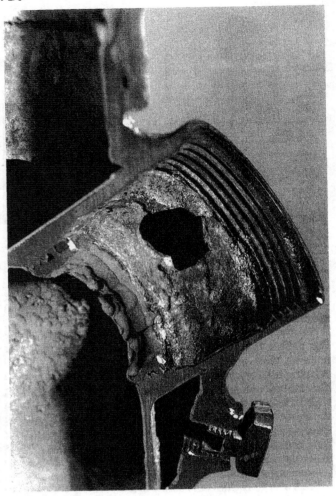

Fig. 6   Erosion corrosion of a silicon-bronze strainer housing

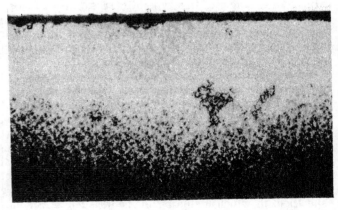

Fig. 7   Graphitic corrosion in a gray cast iron seawater pipe

Fig. 8   Seawater dealuminization of a cast aluminum bronze

Fig. 9   Exfoliation corrosion of 5456-H321 aluminum alloy

adding nickel to a level exceeding 3.5% or by heat treating to produce an alpha+beta microstructure. Dealuminization can be particularly insidious because the corroded gamma phase can occupy such a minor part of the alloy's volume that there may be no outward appearance of corrosion. The attack is evident only on fractured surfaces, yet the alloy may have suffered a drastic loss of strength and ductility [14].

Intergranular corrosion of austenitic stainless steels is another form of selective attack. The carbides in steels sensitive to this form of deterioration tend to precipitate at grain boundaries when the steels in the annealed condition are reheated into the 800 to 1600 F range. Such temperatures are reached in the heat-affected zones (HAZs) adjacent to welds. Preferential corrosion can then occur along the grain boundaries. This type of attack can be avoided by using low-carbon (0.03% maximum) grades of the alloys, by using stabilized grades (containing titanium or columbium), or by putting the precipitated carbides back into solution by quench cooling (annealing) from above 1950 F [17,18].

Exfoliation (delamination) is a special form of selective-phase attack that proceeds along narrow paths running parallel to the metal surface. Generally, the corrosion occurs at the grain boundaries, causing corrosion products to form that force the metal away from the bulk of the material and produce a layered appearance. If the corrosion products are voluminous, internal pressure may cause blistering of the external surface [14,26]. An example of exfoliation corrosion is shown in Fig. 9.

If it is necessary to use an alloy with susceptibility to exfoliation, it is desirable to prevent exposure of the cross section to the corrodant. Sometimes this can be accomplished by "buttering" the edge with weld metal. On the other hand, pitting of the surface eventually may cause penetration to the nonresistant phase. When this occurs, the corrosion can proceed rapidly beneath the surface by a tunneling effect.

g. Stress corrosion. Stress corrosion is a form of local deterioration resulting from the combined action of stress and corrosion, which leads to cracking of the alloy [27]. The mechanisms by which it occurs are not completely understood and depend on the specific alloy, environment, and stress state.

Stress-corrosion cracking occurs only in the presence of tensile stresses. These may be applied, residual, thermal, or welding stresses. There is some evidence, though debatable, that a "threshold stress" exists below which stress corrosion will not occur. However, it can be dangerous to design on this basis because the degree of residual stress is seldom known, and because local corrosion pits or other discontinuities can serve as sources of unanticipated stress concentration [14,17].

Structural steels are not considered to be susceptible to stress corrosion cracking in marine environments. However, steel alloys with yield strengths greater than 100 ksi are generally considered to be susceptible to a form of environmentally assisted degradation known as hydrogen embrittlement [28]. While this is a general rule of thumb, it has been shown that microstructural constituents as well as strength level are known to play a role in the hydrogen embrittlement susceptibility of a given alloy [28,29]. Therefore, alloys with yield strengths lower than 100 ksi may also be susceptible to hydrogen embrittlement. The application of cathodic protection is a widely used technique for controlling the general corrosion of steels in marine environments. While this is an effective technique for that purpose, it should be recognized that hydrogen embrittlement may be exacerbated because of the production of hydrogen at steel surfaces.

Numerous high-strength aluminum alloys, particularly in the 2000 and 7000 series of alloys, are susceptible to stress corrosion in seawater. The direction and rate of crack propagation can be highly dependent on the direction of the applied stress in relation to directional microstructural variations or "texturing" associated with rolling operations [14,26].

Some titanium alloys can stress corrode in seawater. An example is Ti-7Al-2Cb-1Ta alloy, in which the stress-corrosion susceptibility is associated with tendencies toward formation of submicroscopic Ti_3Al precipitates in the structure. Immunity to stress corrosion in this alloy can be accomplished by lowering the aluminum to 6% and adding 1% molybdenum to suppress the formation of titanium-aluminum [30].

Titanium alloys that do stress-corrode in seawater usually will not do so in the absence of a stress concentrator such as a crack or corrosion pit. Since titanium alloys are practically immune to all forms of marine corrosion, including pitting, the existence of stress-corrosion tendencies was not recognized for many years; it was noted only when sharply notched specimens were exposed to seawater and found to support smaller loads than similar specimens exposed to air [14,20,21].

The common grades of austenitic stainless steels such as AISI Types 304 and 316 are especially susceptible to stress-corrosion cracking in chloride-containing solutions at elevated temperatures [14,17,31].

h. Corrosion fatigue. Corrosion fatigue, which results from the combined action of a corrosive environment and cyclic stress, is covered in Section 3.

2.3 Cathodic Protection. Cathodic protection is a corrosion-control method that is sometimes used to provide partial or complete control of electrochemical corrosion problems. One method involves the installation of sacrificial anodes (zinc, magnesium, or iron) that are electrically coupled to the metal to be protected. The anodes preferentially corrode and either reduce or eliminate corrosion of the more cathodic metal. Another method involves the use of an impressed-current system with an external power source; the anode in this system is usually an inert material, such as platinum, to prevent its consumption [14,17,18,32]. Impressed-current systems are not often used for the protection of machinery and auxiliary equipment handling seawater aboard ship because of difficulties in voltage control and current distribution on protected surfaces; however, impressed-current cathodic-protection systems are used extensively on external hull applications.

Section 3
Fatigue

3.1 Fatigue Fracture. The term "fatigue" as used herein refers to the failure of a component under the action of repeated stress. Fatigue is probably responsible for as many as 80% of all machine-part failures. Of these, failures caused by bending forces are the most prevalent type, with the torsion type next, and the axial type seldom occurring. However, service conditions usually involve combinations of variable- and steady-state stress.

Figure 10 shows a typical fatigue fracture. In many instances such failures are easy to recognize because of distinctive characteristics associated with fracture appearance. These are a "flat, brittle" appearance of a large portion of the fracture, and "beach marks" or "oyster shell" markings, which focus on the origin or nucleus of failure. However, the absence of such markings does not necessarily eliminate fatigue as a cause of failure inas-

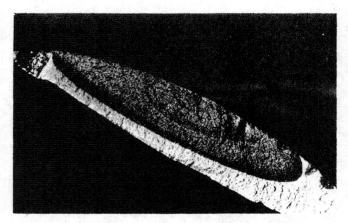

Fig. 10   Fatigue failure of a manganese bronze propeller blade

much as the level and kind of loading, temperature, and environment can affect fracture appearance greatly and may make diagnosis on the basis of appearance alone impossible. In such cases the use of an electron microscope to examine the fracture surfaces at high magnifications may be helpful.

Fatigue failures are progressive, beginning as minute cracks that grow under the action of the fluctuating stress. Failure usually consists of three distinct stages, namely, crack initiation, slow crack propagation, and rapid terminal fracture. If a crack is not present initially, a certain number of stress cycles will be consumed in generating a crack from a "stress raiser" to a size sufficient to be visible to the unaided eye or to be detectable by nondestructive inspection. The crack will continue to grow in size with each stress cycle, usually at an increasing rate, until the remaining section is no longer able to support the load. Complete fracture then occurs instantaneously. The terminal fracture may be ductile or brittle depending on the material and surrounding conditions.

3.2 Cyclic Stress and Strain. Figure 11 shows the stress-strain relationships likely to develop under cyclic loading conditions. The relationship in Fig. 11(a) occurs when the applied force or moment is completely reversed but within the elastic region. S_t is the total stress range

and ϵ_t the total strain range. Figure 11(b) shows the relationship that develops under reversed loading into the plastic region. The stress-strain relationship is no longer linear, but follows the hysteresis loop BCDEB during each cycle after initial loading along OAB.

Failure by fatigue at finite numbers of stress cycles, i.e., less than 5×10^5, is called low-cycle fatigue. There is general agreement among investigators that the high-cycle fatigue performance of metallic materials is related to tensile strength, whereas low-cycle fatigue performance is related to tensile ductility. Accordingly, the amount of cyclic strain that a material undergoes becomes the dominant factor in low-cycle fatigue. Although subsequent sections may make a distinction between high- and low-cycle fatigue, it is desirable in evaluating materials to consider the broad spectrum of fatigue behavior ranging from about 10^2 to 10^8 cycles.

Data from fatigue tests are usually reported as S-N diagrams, such as illustrated by Fig. 12. The fatigue or endurance limit in this case is disclosed by a definite break in the curve. In the low-cycle region, the curve in Fig. 12 tends to flatten out if the nominal stress continues to be used as the independent variable. This is because stress and strain are no longer linearly related. Accordingly, small changes in nominal stress are accompanied by large changes in strain. As mentioned previously, strain, not stress, is the controlling factor in low-cycle fatigue.

To place the full spectrum of fatigue life on a common basis, strain measurements frequently are converted to "stress" values by multiplying the strain data by the modulus of elasticity, E. This fictitious elastic stress is called "strain-based stress" and is designated by the symbol S_ϵ. The advantage of this approach is that it permits low-cycle fatigue data to be presented as a "stress," even though the measured parameter is strain. The disadvantage is that unrealistically high stress values are obtained in the very-low-cycle region.

Figure 13 shows the broad-spectrum fatigue behavior of two steels with markedly different yield strengths. The solid lines compare the steels under ideal conditions, i.e., a smooth surface in a noncorrosive environment. As mentioned previously, the high-cycle fatigue life under these conditions is related to the strength properties of the

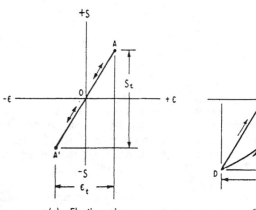

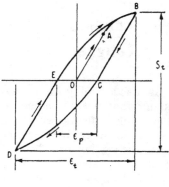

(a)   Elastic region   (b)   Plastic region

Fig. 11   Stress-strain relationships under cyclic loading

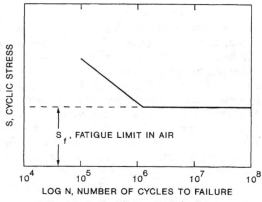

Fig. 12 Typical high-cycle *S-N* diagram

steels. Thus, the fatigue limit of the 200,000-psi steel is considerably higher than that of the 40,000-psi steel. However, the curves cross each other in the low-cycle region. The higher tensile ductility of the 40,000-psi steel is conducive to higher fatigue strengths in the very-low-cycle region, where strain deformation is the principal controlling factor.

The dashed curves in Fig. 13 depict the performance of the two steels under adverse conditions, i.e., a stress raiser in a corrosive environment. Behavior of the two steels is similar despite the large difference in static strength properties. Under such conditions, there would be no advantage in selecting the more expensive, higher-strength steel.

3.3 Factors Affecting Fatigue Life.

a. Stress raisers. Practically all fatigue failures start at a stress raiser (stress concentration) on the surface of a part. The majority of stress raisers fall into one of the following broad groups:

- Those caused by changes in the geometry of a part, such as steps at changes in diameter, abrupt corners, holes, keyways, threads, press or shrink fits, or junction of bolt shanks and heads.
- Surface discontinuities such as nicks, notches, machining marks, pitting, and corrosion.
- Defects inherent in the material such as nonmetallic inclusions, local discontinuities, minute cracks, and voids.

The ability of a discontinuity to concentrate stress is dependent upon its shape and size. Cracks have the highest stress-concentrating effect, whereas generous fillets with a smooth, polished surface have the lowest. Analytical methods have been developed for calculating the stress-concentrating effects of discontinuities based on geometry, dimensions, and assumed elastic behavior. The effect arrived at in this manner is called the theoretical "stress-concentration" factor K_t[33]. However, the actual effect of a given stress concentration may vary both within and among materials. By means of tests it is possible to establish the reduction in fatigue strength caused by a particular stress-concentration factor for a particular material at a particular strength level. By comparing these data with unnotched (smooth) test data, one can arrive at the so-called fatigue notch factor K_f. From K_t and K_f the notch sensitivity index, q, of the material can be calculated as follows:

$$q = \frac{K_f - 1}{K_t - 1}$$

For most metals the notch sensitivity tends to increase with increasing strength. Thus, in the presence of sharp notches ($K_t \geq 3$), it is not unusual to find little or no advantage for higher-strength materials.

b. Maximum and mean stresses. The maximum tension stress or strain developed in a stress cycle has an

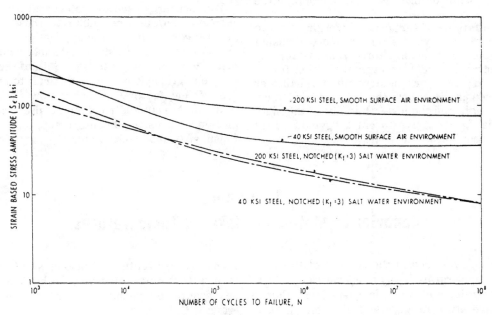

Fig. 13 Broad-spectrum fatigue results for two steels

important bearing on fatigue life. In general, in the high-cycle region, the permissible alternating stress decreases with an increasing mean stress. Through the use of mathematical or diagrammatical relationships, such as the Goodman diagram or the Haigh-Soderberg diagram, it is possible to convert combined-stress conditions to an equivalent, completely reversed, stress condition.

The influence of mean stress decreases with decreasing fatigue life, and when the alternating stress equals or exceeds the yield strength, the mean stress becomes zero. Accordingly, whether or not mean stress is an important factor in low-cycle fatigue depends upon the yield strength of the material. In high-strength materials, it may be necessary to consider mean-stress effects at fatigue lives ranging down to 100 cycles.

c. Residual stresses. Residual stresses may be either favorable or unfavorable insofar as fatigue life is concerned. Fatigue cracks initiate and propagate only in a tensile-stress field. Therefore, tensile residual stresses are usually detrimental to fatigue life, whereas compressive residual stresses are beneficial. Nearly all fatigue failures initiate at the surface. Accordingly, processes that introduce compressive stresses in the surface layers of a machine part can be effective in improving fatigue performance.

The most commonly used metallurgical processes for increasing fatigue resistance are case carburizing, nitriding, or carbonitriding. In these processes, carbon or nitrogen or both are diffused into the surface layers of the part. When properly applied and heat treated, the resultant metallurgical structure in the diffused layer occupies a greater volume than that of the parent metal. The greater volume causes compressive residual stresses.

Compressive residual stresses can also be introduced by cold working the surface material by such techniques as hammer peening, shot peening, and cold rolling. To produce the desired effects, it is necessary that the stressing technique plastically deform the surface layers of the metal.

The benefits in fatigue of surface cold-working tend to decrease with increasing yield strength and decreasing fatigue life. There are probably three reasons for the decrease with increasing yield strength: the higher forces required to deform the surface metal result in a shallower layer; the notch sensitivity of both the deformed and undeformed metal is high; and the lower ductility of the high-strength material is conducive to cracking induced by some stressing techniques.

Unfavorable residual stresses, i.e., surface tensile stresses, originate principally from heat treatment, welding, misfits, and cold forming. It is possible to relieve unfavorable residual stresses in metals by a so-called stress-relieving heat treatment and thus improve fatigue resistance. This treatment requires that the materials be heated to elevated temperatures. The applicability of a stress-relieving treatment is limited by facilities for handling large structures and by the fact that the temperature required for stress relief may have a detrimental effect on other properties such as yield strength and impact resistance.

d. Corrosive environment. The combined effect of a corrosive environment (such as seawater) and cyclic stressing is called corrosion fatigue and can be highly detrimental to the life of metals. This is particularly true in the case of carbon and low-alloy steels wherein the high-cycle, corrosion-fatigue strength in seawater is about the same regardless of composition or strength level (see Fig. 13). As might be expected, metals that are either partially or completely resistant to the corrosive environment will be less affected. For example, the fatigue behavior of cupronickel and nickel-copper alloys normally used for a seawater system is not greatly different whether the exposure is to seawater or to air.

Many titanium alloys are exceptionally resistant to corrosion fatigue in seawater. Titanium alloys typically have endurance limits in air that are 50 to 60% of their ultimate strength, and their endurance limits are substantially unaffected by exposure to seawater.

The endurance limits of aluminum alloys are severely affected by exposure to seawater. However, the exposure of high-quality carbon/epoxy laminates to seawater has a small (5 to 10%) adverse effect on their strength properties when compared with the same characteristics in air, under both steady and cyclic loading conditions.

The measures used to protect materials susceptible to corrosion fatigue are similar to those used to prevent general corrosion, i.e., the application of protective coatings that are both resistant and impervious to the environment, and cathodic protection. Both of these methods are simple in principle but often difficult to put into practice.

e. Other factors. Other factors that may affect fatigue life are cumulative damage, prestressing, metallurgical structures, weldments, creep, temperature, surface finish, size, and stress state.

Section 4
Behavior of Metals at Elevated Temperatures

4.1 Introduction. As used in the context of this section, an elevated temperature is any temperature at which strain and load-carrying ability are time dependent. Below the elevated-temperature range, the design for any life can be based on properties measured in "short-time" tests

such as the conventional tension test. Within the elevated-temperature range, however, mechanical properties must be measured by means that account for a change of properties with time.

Elevated-temperature behavior occurs over different

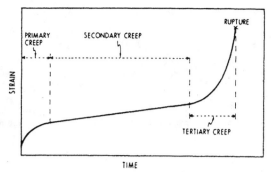

Fig. 14 Typical strain-time relationship of an alloy undergoing creep

temperature ranges for different materials. For example, the mechanical properties of lead can be time dependent at room temperature and above, whereas for low-alloy steels they are time dependent only above 700 F.

Accepted practice dictates that, insofar as possible, a metallic material be used in a condition of metallurgical stability at the operating temperature. For example, a quenched and tempered steel normally would not be used unless the tempering temperature were at least 100 deg F higher than the intended operating temperature; otherwise, the strength obtained by the temper would be reduced on heating to the operating temperature.

4.2 Time-Dependent Properties. When a metal is subjected to a static load at an elevated temperature, an immediate elastic strain occurs followed by a time-dependent permanent plastic strain called creep. A typical strain-versus-time curve is shown in Fig. 14. The creep occurring at a diminishing rate at the initial part of the curve is called primary creep. This is followed by secondary creep, characterized by a relatively constant rate of strain. The creep rate may ultimately accelerate and lead to rupture in a stage called tertiary creep. It is rather obvious that stress-temperature combinations leading to tertiary creep within the life of the equipment should be avoided.

Allowable stresses for high-temperature design are often set by organizations such as the American Society of Mechanical Engineers (ASME) [5]. For example, the allowable stresses for each material included in ASME's code dealing with power boilers were arrived at by using the lowest of the following criteria at each design temperature in the creep range:

—A conservative average stress for a secondary creep rate of 0.01% per 1000 hr.

—60% of the average stress for rupture in 100,000 hr.

—80% of the minimum stress for rupture in 100,000 hr. The use of average and minimum values assumes scatter in the rupture data from tests of many lots of materials. The designer must also apply additional safety factors to accommodate unusual conditions of instability, corrosion, possible overstresses, and the like.

a. Creep tests. Creep data usually are obtained by putting static tensile loads on specimens and measuring the time-dependent strain with sensitive extensometers. The tests are made at several stresses for each temperature of interest. Each test must be conducted long enough

(usually several thousand hours) to establish the minimum strain rate in the secondary stage of creep. The data from such tests commonly are plotted as shown in Fig. 15, from which the stress to produce a particular creep rate can be estimated.

b. Stress-rupture tests. These tests are also called creep-rupture tests, and are conducted by static tensile loading of specimens at stresses high enough to cause failure. The tests are made at several stresses for each temperature of interest. The rupture time is measured, and strain-time curves also can be obtained in order to extend creep-rate curves to higher rates of strain. The stress versus rupture-time data are plotted on semilog or log-log graphs. A typical example is shown in Fig. 16.

c. Creep-relaxation tests. These tests are also called relaxation tests. They are similar to creep tests, except that stress is the variable rather than strain. After the specimen is loaded, the gage length of the specimen is held constant by reductions in stress so that elastic contractions will exactly balance any extensions due to creep. The data from such a test commonly are plotted as shown in Fig. 17.

The similarity between the relaxation test and bolting applications is apparent. Upon tightening to an initial stress, a high-temperature bolt retains its dimensions, but there is a falloff of the stress as a function of time and temperature. The stress indicated by the asymptotic part of the relaxation curve is related to the ultimate "holding power" of the bolt. However, the asymptotic stress can vary as a function of initial stress on loading, and may also be different after a second loading, such as might occur by retightening of a bolt.

Residual stress can be introduced in a material by welding, cold forming, and other processes. The reduction of these stresses by stress-relief heat treatments can be said to occur largely by the creep-relaxation process. If the stress-relief temperature is restricted for any reason (such as not exceeding the tempering temperature of a quenched and tempered steel), the relaxation test is a useful means of estimating the level to which residual stresses can be reduced within a particular time at a given temperature [34].

d. Data extrapolation. High-temperature creep and stress-rupture tests are expensive and time-consuming; nevertheless, knowledge of long-time material properties is required. For example, it is important to know that tertiary creep will not occur during long equipment life if a design stress is based on a secondary creep rate measured in a test lasting only a few thousand hours. Another example is the 100,000-hr (11.4-year) rupture-stress criterion used by ASME in establishing allowable stresses for boilers and unfired pressure vessels.

A common method of obtaining long-life rupture-stress properties is to use a straight-line extrapolation of data such as shown by the dashed extensions of the lines in Fig. 16. There is debate as to whether the lines are actually linear or somewhat curvilinear, even for a metallurgically stable structure. Of greater concern is the possibility of temperature-induced or strain-induced microstructural changes, which are known to produce slope changes in

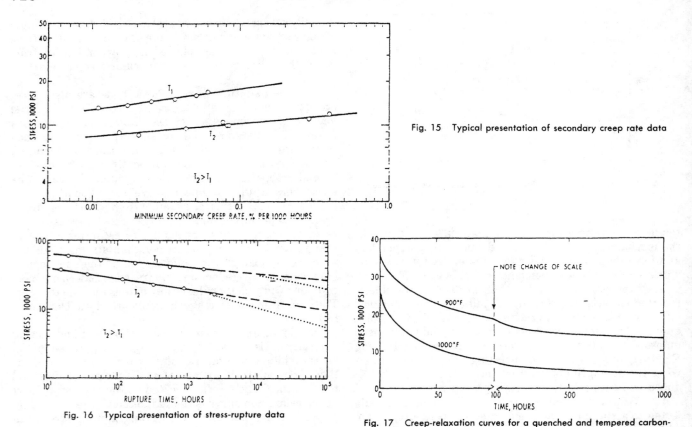

Fig. 15 Typical presentation of secondary creep rate data

Fig. 16 Typical presentation of stress-rupture data

Fig. 17 Creep-relaxation curves for a quenched and tempered carbon-molybdenum bolting steel

some alloys. This circumstance is illustrated by the dotted lines in Fig. 16. The microstructural changes normally occur in shorter times at higher temperatures. This is also illustrated in Fig. 16 by the shorter time for the "break" in the curve at temperature T_2 rather than at T_1 (T_2 being higher than T_1).

The concept that time and temperature bear an equivalence in the creep process, and that which occurs in a long time at one temperature will occur in a shorter time at a higher temperature, has been the basis for several parametric methods to estimate long-time properties from short-time tests. No single parameter works best for all materials.

Different finishing heat treatments during manufacture are sometimes permitted by a material specification,

and these may be reflected by differences in short-time rupture-strength properties. However, the higher strengths associated with some manufacturing processes may disappear after a very long time at service temperature. Therefore, it is common practice to set allowable design stresses for long-life equipment on the basis of the "weakest" condition under which a material will be marketed.

4.3 Other Considerations. Notches, biaxial and triaxial stresses, cyclic loading, environmental effects, and temperature "overshoots" are additional factors influencing creep and rupture behavior. Also, the temperature and strain history of a material undergoing creep may have an effect on other properties such as loss of fracture ductility. This is known as creep damage.

Section 5
Composite Materials

5.1 Composite Material Types. A composite material is a synthetic assembly of two or more components—a selected filler or reinforcing agent and a compatible matrix binder—to obtain a material with specific characteristics and properties. A comprehensive glossary of the terms, characteristics, and processes used in connection

with composite materials is presented in references 35 and 36.

Several types of composites have been used in practical applications in shipboard machinery and equipment. The various classes of composites are distinguished by their matrix systems; that is, there are metal-, ceramic-, carbon-, and polymer-matrix materials. The polymer-matrix composites have gradually evolved from the fiberglass-reinforced plastic laminates that were first introduced in the United States in the late 1940's. Polymer-matrix resin materials can be categorized as thermosetting or thermoplastic.

a. Thermosetting resins. Thermosetting resins are initially mixtures of chemicals that can co-react into a structured network. Their low viscosity prior to reaction facilitates fiber impregnation, conformability to the shape of the desired part, and many other processing steps that lead to the composite structure. Thermosetting resins can be grouped by the chemistry of the major component into six main classes: Polyester, vinyl ester, epoxy, bismaleimide, polyimide, and phenolic. The selection is usually based on economic restrictions, required mechanical properties, or application temperature. Variations in composite properties within each class can be large.

Polyesters were the original matrix resins for continuous fiber composites, and are still widely used. They are polyunsaturated oligomers dissolved in styrene. Polyester composites are usually filament wound, contact molded, or used as sheet molding compound. Their low cost and room-temperature cure make them cost effective.

Vinyl esters are similar to polyesters in many ways: they are oligomers dissolved in styrene, which cure by free radical polymerization. They are superior to polyesters in their ability to plastically deform, which leads to improved resistance to delamination and greater fatigue life. Their ductility is probably due to the epoxy backbone of the oligomer.

Organic matrix composites with the highest static (compressive) strength have epoxy matrix resins. Broadly speaking, there are two classes of epoxy resins: those composed (mainly) of Ciba Geigy MY720, and those composed of the diglycidal ether of Bisphenol A. Epoxies composed of MY720 have the highest Young's modulus of all matrix resins, and as a result, composites based on MY720 have the highest compressive strengths. Bisphenol A epoxies are the main component in wet winding resins; although they are easier to process than MY720, Bisphenol A epoxies result in composites with substantially lower compressive strength.

Bismaleimides and polyimides were developed for high-temperature applications. They are expensive, difficult to process, and are brittle. The high processing temperatures often result in thermal stress cracking during cooldown. These materials are generally not well suited for marine applications.

Phenolic resin is produced by the condensation of an aromatic alcohol with an aldehyde, particularly of phenol with formaldehyde. Phenolics are inexpensive and have good elevated-temperature and fire-performance characteristics, but they are brittle and have a relatively low impact strength.

Thermosetting resins become irreversibly solidified or infusible when cured by heat or chemical means.

b. Thermoplastic resins. Although thermosetting resins are more mature in terms of the chronology of composite matrix development, the thermoplastic resin materials (and in particular high-performance thermoplastic resins) have advantages in some applications. Conventional thermoplastic resins include nylon, vinyl, thermoplastic polyester, acetal, fluoropolymer, polycarbonate, and polyphenylene oxide. Unlike thermosetting resins, thermoplastic resins can, by their nature, be repeatedly remelted and reformed, which often results in more simple fabrication procedures and less scrap.

High-performance thermoplastic resins include: polyetheretherketone (PEEK), polyphenylene sulfides (PPS), polyetherketone ketone (PEKK), polyetherketone (PEK), polyamide-imide, polyetherimide, polyarylene sulfide, and polyether sulfone.

5.2 Reinforcements. Common types of reinforcing fibers used in the production of composites include glass, cotton, aramid, carbon, graphite, polyethylene, boron, steel, polyamide, alumina, silicon carbide, and aluminaboria-silica. Table 3 compares the basic properties of some of the reinforcement fibers used in marine applications. E-glass is the predominate reinforcement material used since it provides significant advantages at a relatively low cost. Commercial types of glass-fiber reinforcement are provided in a variety of forms to meet various application requirements. These forms include unidirectional tape, woven fabric, and chopped-strand mats. Various filler materials can also be added to the resin matrix in diversified shapes (such as spheres, flakes, or fibers) to enhance desired characteristics such as strength, stiffness, lubricity, wear-resistance, or appearance.

To optimize the efficiency of a continuous-fiber-reinforced resin-matrix composite, the reinforcing fibers are often oriented or arranged in the direction of loading or principal stress. This approach usually results in a rather anisotropic material somewhat similar to wood (cellulose fibers in a lignin matrix) in structural characteristics (i.e., properties vary "with-the-grain" or "across-the-grain"). A comparison of the strength and modulus characteristics of unidirectional fiber-reinforced epoxy-matrix composites with isotropic alloys of aluminum, titanium, and steel is presented in Table 4. However, the use of unidirectional material may be either impossible or undesirable, and in many cases a random fiber dispersion must be specified in order to develop a composite with quasi-isotropic properties more similar to conventional metal alloys or unreinforced plastics.

5.3 Fabrication Methods. The processing method selected for a composite application depends on many factors, including part size, shape, complexity, finish, quantity, quality, property requirements, skill levels available, cost limitations, and maximum permissible void content.

Table 3 Basic properties of reinforcement fibers for composite materials

	E-Glass	S2-Glass	Carbon	Aramid	Polyethylene
Tensile strength, psi $\times 10^3$	500	650	500	525	400
Tensile modulus, psi $\times 10^6$	10	12	33	19	18
Strain at break, %	5	5.5	1.5	2.5	3
Specific gravity	2.54	2.49	1.8	1.4	0.97

Table 4　Density, strength, and modulus characteristics of select unidirectional fiber-reinforced epoxy-matrix composites[a] and metals

Material	Density, lb/in.^{3}	Ultimate Tensile Strength, psi $\times 10^3$	Tensile Modulus, psi $\times 10^6$
E-glass	0.075	160	6.5
S-glass	0.072	280	7.5
Kevlar-49	0.050	210	11.0
Type HMS graphite	0.059	170	30.0
Type AS graphite	0.056	250	20.0
Aluminum (7075-T6)	0.100	83	10.0
Titanium (6Al-4V)	0.160	160	16.5
Steel (4130)	0.289	200	29.0

[a] Values at 60% fiber volume, unidirectional.

Table 5　Typical mechanical properties of E-glass fabric-reinforced composites

Glass content, wt. %	10 to 40	45 to 60
Flexural strength, psi $\times 10^3$	10 to 27	35 to 65
Flexural modulus, psi $\times 10^6$	1.0 to 1.6	1.2 to 2.2
Tensile strength, psi $\times 10^3$	3 to 10	20 to 50

Fabrication with composite materials is usually different from metal processing in that the composite fabrication process involves the simultaneous formation of the material and the component. To provide examples of the types of composite components and material limitations associated with different fabrication methods, four fabrication methods that are commonly used to form typical marine composite components are outlined in the following:

a. Hand layup. Structural or semistructural marine applications of composites are typically fabricated through a hand-layup fabrication process. For low-volume production applications, a hand-layed continuous-fiber composite construction is often the most economical. The most common construction method is open or contact molding. In this process a pigmented coating or gel coat is applied to the mold. This surface ultimately becomes the exterior of the part. Before this coating fully cures, layers of resin-saturated fibers are laid into the mold. On curing, the layers of reinforcement and the coating are permanently bonded, forming a rigid shell with a smooth, permanently pigmented surface. Typical mechanical properties of composites reinforced with E-glass fabric that are formed in this manner are given in Table 5.

Note in Table 5 that the flexural strength and the flexural modulus are stated. The "flexural strength" is expressed as the tensile stress in the outermost fibers of a bent test specimen at the instant of failure. The flexural strength of composites is usually higher than the tensile strength. Similarly, the "flexural modulus" is the ratio of the stress to the corresponding strain in the outermost fibers of a specimen that is subjected to a bending load.

b. Resin-transfer molding. Resin-transfer molding is similar to hand layup in that individual layers of reinforcement are placed into a mold. However, in the resin-transfer molding process the layers are placed into the mold dry. The dry reinforcement is called the preform. A mating mold half is closed over the dry preform materials,

and resin is pumped into the cavity to saturate the preform. Following the cure cycle, the mold is opened and the rigid finished component is removed. The advantages of this process are the ability to form two molded surfaces on the part, reduced per-part labor costs, and better control of part-to-part consistency as compared to the hand layup method. The mold costs are higher in this process, however, and if high fiber contents are required, the mold can become prohibitively expensive for low production volumes. The composite mechanical properties obtained are comparable to those obtainable with the hand-layup process.

c. Filament winding. The filament-winding process is accomplished by winding continuous fibers over a rotating mandrel. The axis of rotation can be varied such that the winding is in any direction. By selectively placing higher fractions of the total winding along angles closely aligned to the load paths, very strong and efficient components can be made. As with the previously described processes, the mechanical properties depend on the fiber content and the orientation of reinforcement. Typically, the fiber contents are at the higher end of the range given in Table 5, and consequently, the mechanical properties are at the upper end of the range as well. The process can be highly automated and thus economical, but it is obviously limited to components that can be formed as a body of revolution, and only limited concave shapes can be accommodated without secondary molding processes.

d. Short-fiber molded composite. Many reinforced-plastic components are molded using reinforced polymers, which are short fibers ($\frac{1}{4}$ to 1 in. long) that are dispersed in the resin at weight percentages ranging from 10 to 50%. The molding process determines the orientation of the fibers in the final part. The fibers are generally oriented in the plane parallel to the part surface, with very few fibers oriented through the thickness of the part. The pressure required to flow the filler resin throughout the mold cavity can be very large; consequently, only relatively small components such as small pump housings, pump impellers, small valves, and valve components are fabricated by compression or injection molding of reinforced composites. This process is best suited for higher-production quantities because of the relatively high capital costs of the molding operation. The fiber content and control over fiber orientation are lower than with any of the previously described processes. The mechanical properties typically achieved for a short-fiber molded component are given in Table 5.

5.4　Environmental Effects. The environment of a composite material can have a significant effect upon a composite's characteristics.

a. Moisture absorption. Water is absorbed by all polymers used as matrix resins. The percent weight gain of resins resulting from immersion in water can vary from 0.5 to 5%, with 3 to 4% a typical range for epoxies. The percent weight gains in composites are lower because of the fiber content, and a 1% increase in weight is typical for epoxy-matrix composites.

The water absorbed by matrix resins generally does not degrade the mechanical properties of the composites.

Hydrophilic polymers such as epoxies are plasticized by the absorbed water. The only significant effect of plasticization is a reduction in the glass transition temperature, which is the temperature at which the unreinforced polymer passes from a brittle glassy state to a rubbery state, and as a result, the upper temperature range within which the composite can be applied. High fiber-content composites, such as tape-reinforced materials, show unchanged fiber-dominated properties with water absorption. On the other hand, a 10 to 20% reduction in the dry tensile, compressive, and flexural strengths of fabric-reinforced composites having higher resin contents may occur.

An exposure to water will, however, cause a reduction in the fiber/matrix bond strength of some composites. A loss in adhesion results in a substantial loss in the compressive and flexural strengths, and for some composites the degradation in strength caused by water immersion can be as high as 60%. Glass-reinforced composites are the most susceptible to a loss in adhesion, but this occurs only if an unsuitable coupling agent is applied to the glass.

A third possible effect of water, besides matrix plasticization and fiber/matrix hydrolysis, is galvanic corrosion. Carbon-reinforced polymers, in contact with metal and in the presence of water, can degrade substantially by matrix electrolysis. This behavior appears to be limited to polymers containing an imide linkage, and can be prevented by electrically isolating the composite from the metal.

b. Temperature. The effect of temperature on most polymer-matrix composites is dependent on the resin. All resins have an operating range over which they will perform acceptably. At very low temperatures the resins become slightly more brittle, and fiber-reinforced composites may be subject to matrix cracking because of differential thermal expansion between the fiber and the matrix. The mechanical properties of most composite materials are not detrimentally affected by temperatures down to cryogenic levels. However, at elevated temperatures, the matrix becomes soft. The American Society for Testing and Materials (ASTM) Test D 648 (Test Method for Deflection Temperature of Plastics Under Flexural Load) is used to determine the temperature at which the material loses its effective stiffness as measured by the deflection of a specimen immersed in a heated fluid and supporting a deadweight load. The maximum temperature defined by this test is called the heat distortion temperature. The heat distortion temperature is a practical measure of the glass transition temperature. Typically, for matrix systems that cure at room temperature, the useful temperature range is below 140 F. The combination of elevated temperature and absorbed moisture results in an accelerated softening of a composite matrix, and the mechanical properties are reduced to a greater extent than would be the case for either effect alone. Consequently, for material applications in a high-moisture, high-temperature environment, the effect of these conditions must be considered when the characteristics of the resin, reinforcement, and mechanical properties are specified.

c. Chemical corrosion. Although reinforced composite materials have been used in many highly corrosive applications, the selection of the resin, reinforcement, and fabrication process is very important to ensure a satisfactory design. As an example, when designing components to be used in sewage systems, the effect of the acidic fluids on the composite materials must be considered. As a second example, graphite-fiber reinforced composites may produce a galvanic corrosion cell when the composite is in seawater service and electrically coupled to a metallic structure, because the graphite fibers are noble relative to all metals. Although the matrix material will isolate the fibers from electrical contact, at cut edges, or holes where fibers are exposed, electrical contact will be made and the metal corrosion will be accelerated.

d. Fire. Polymer matrix composites are inherently flammable, but resin modifications and additions of materials such as chlorine, bromine, or antimony to the resin will produce a matrix that is flame retardant. These additives produce a matrix that will not support combustion in a normal atmosphere. The resin will smolder and char while heat is applied but will self-extinguish once the heat source is removed; however, these additives may produce toxic gases when the matrix does burn. Requirements concerning the use of flammable materials on commercial vessels are stated in U.S. Coast Guard and SOLAS (Safety of Life at Sea) regulations. The U.S. Navy has also developed MIL-Std-2031, which defines limits on the flammability, smoke, and toxicity of composite materials when installed in submarines. The Navy standard permits the application of supplemental coatings, covers, insulations, etc. to composite materials when necessary to ensure that a rigorous series of fire requirements and associated performance criteria can be met when the treated composite material is tested as a "composite material system." Limits on the volatiles that may be gassed-off from a polymer-resin matrix at room or slightly elevated temperature are also imposed by the Navy for material applications in the closed environment of a submarine.

5.5 Inspection Techniques. In-service inspection of composite components is primarily limited to visual inspection for indications of excessive cracking or delamination. Standard procedures for the inspection of as-fabricated hand-layed structural composite components have been defined for Naval procurements in NAVSHIPS Technical Manuals 250-346-2 and 250-529-1. These manuals describe defects that may occur during fabrication and define "accept," "repair," or "reject" limits on the type and extent of defects. Typically, in-service damage is caused by overloading, impact, or combined loading and environmental effects. Damage is characterized by a "whitening" of the composite. This color change from a normally translucent appearance is a result of the formation of minute cracks called "crazing" in the composite matrix. In the case of delaminations, the color change will likely form semicircular shapes emanating from the initial damaged area. An incipient fracture of a composite is indicated by the appearance of broken fibers protruding from the material surface.

Composite components that are fabricated from pigmented resin or which have an opaque coating may not provide any readily apparent indications of damage. But a reduced stiffness of the component may be evident or the material surface may be cracked in the damaged area. Because of the brittle nature of most polymer composites, local areas of stress concentration are often relieved by local failure of the component as opposed to local yielding near the stress concentration, which is the case with ductile metals. Sometimes such local failures can be disabling to the entire component, but often the failure is entirely local and the load is redistributed.

The ability to redistribute loads near local stress concentrations allows most composite components to be relatively free from tension-tension fatigue-induced failures. For instance, stress concentrations at bolt holes, which may induce catastrophic fatigue failures in metal components, can have a much different effect in composite components. The initial loading of a composite component will produce damage where the composite is overstressed as a result of stress concentrations. Following local failures in these overstressed areas, the loads are redistributed; and as long as the remaining undamaged material is capable of carrying the loading, the damage will not progress, and the component will continue to function. However, where there are compressive or flexural loadings, the damage will likely continue to progress and lead to a buckling or other stiffness-limited failure mode.

Minor impact damage, which causes cracking of the surface coating or (if visible) crazing or local delamination (without apparent fiber breakage) of the material surface, may also have a negligible effect on the component. If the loading of the component in the damaged area is primarily planar tension, local delamination may have little effect on the strength or stiffness of the component. However, flexural or compressive loadings may cause the damage to propagate over time, and repair or replacement may be necessary.

To aid the visual inspection and evaluation of the extent of damage to composite materials, acoustic inspection methods can be used. The most simple acoustic inspection is a coin tap test. In this method a coin or other light metal object is tapped over the surface of the component. Damaged areas and particularly delaminated areas produce a hollow or dull sound when compared with undamaged areas. Ultrasonic (UT) devices can be used to perform in-service inspections of composite components. Handheld contact-type ultrasonic transducers that are used in a pulse-echo mode can detect disbonds or delaminations in composite components. The inspection procedures used are similar to handheld UT weld inspections. However, lower-frequency (1 to 5 MHz) and larger-diameter transducers are required because of the greater sound

Table 6 Composite material trade-offs

Advantages	Disadvantages
Strength	Affected by heat
Stiffness	Flammability (smoke, toxicity, etc.)
Weight	Subject to damage
Corrosion resistance	Wear characteristics
Erosion resistance	Design/production technology limitations
Improved damping	Test and inspection procedure limitations
Nonmagnetic	

Table 7 Candidate marine applications for composites

Structural	Machinery	Functional
Topside superstructure	Piping	Missile blast shields
Masts	Pumps	Shafting overwraps
Stacks	Valves	Life rails/lines
Radomes	Heat exchangers	Handrails
Foundations	Strainers	Mast stays/lines
Doors	Ventilation ducting	Bunks/chairs/lockers
Hatches	Fans, blowers	Tables/worktops
Liferails	Weather intakes	Insulation
Stanchions	Propulsion shafting	Nonstructural partitions
Fairings	Tanks	False decks
Armor	Reservoirs	Seachest strainers
Bulkheads	Gear cases	Deck grating
Propellers	Actuators	Stair treads
Control surfaces	Hydraulic cylinders	Grid guards
Hulls	Diesel engines	EMI shielding
Boats	Electrical enclosures	Elevator doors, platforms
Tanks	Motor housings	Weapons enclosures
Flasks	Condenser shells	Showers, urinals
Gangways		Washbasins
Ladders		Water closets
Grating		

attenuation of most composite materials. The greater attenuation is in large part caused by the scattering of the sound energy at each of the many material interfaces in the composite material. This scattering also makes it difficult to distinguish small flaws or damaged areas from the normal background reflections. Consequently, the in-service use of conventional UT inspection procedures is limited.

Dielectric sensing devices are used to measure the moisture content in composite materials. These devices are primarily used to evaluate the condition of boat hulls undergoing repair. The laminate condition must be monitored to ascertain that the moisture level is sufficiently low to allow the bonding and repair of the composite.

5.6 Composite Material Design Considerations. Some of the principal factors to be evaluated when considering the use of composite materials are listed in Table 6. The advantages of improved strength, weight, and corrosion characteristics are of major significance when selecting the material to be used for hull structure, and for most small craft, these advantages clearly overshadow the disadvantages that composites entail. The nonmagnetic characteristic of composites further causes them to be an attractive structural material for minesweepers, some of which are rather large. Many polymer-matrix composites have been used extensively in marine applications where their characteristics provide advantages. Some specific candidate areas of application for composites are listed in Table 7.

Section 6
Survey of Select Material Applications

6.1 Introduction. The design, construction, and operation of shipboard machinery and auxiliary equipment are usually the result of years of experiment and experience. The information so gained is published in general specifications that are available for the guidance of marine engineers [1–5]. In addition, certain of the military specifications contain general guidelines for specific items of naval machinery (e.g., MIL-B-18381 for naval high-pressure steam boilers, and MIL-T-17600 for naval steam propulsion turbines). These general specifications refer in turn to a variety of specific material specifications containing details of the materials acceptable for particular components and operating conditions. Several organizations publish material specifications, with those of the American Society for Testing and Materials being the ones to which reference is most often made.

Examples of typical materials used in critical applications, with some rationale as to why they are used, and the limitations under which they provide satisfactory service are given in the following sections.

6.2 Boilers. The materials used in critical components of marine boilers have evolved over many years into a series of standardized alloys that are selected to optimize cost, mechanical properties, castability, weldability, forgeability, and ductility (to permit forming operations such as tube bending and flaring).

Most boiler components are made of carbon and low-alloy steels. However, their suitability in service is highly dependent on proper boiler design, construction, and operation. The design must incorporate adequate provisions for flexibility to accommodate expansion and contraction from temperature cycling without development of stresses high enough to cause fatigue failure. Particular attention must be given to joint designs, such as at tube-header connections and tube supports, so that fatigue stresses are not concentrated excessively at these locations.

The materials used in boilers have a limited tolerance for temperature excursions. Good fireside maintenance is critical, so that deposits do not accumulate and interfere with circulation and heat transfer. The water and steam side is also important, because heat removal by water and steam circulation is necessary to keep tube temperatures within tolerable limits. The watersides must be kept clean and free of deposits that can interfere with proper heat transfer.

Also, corrosion must be prevented. This is accomplished on the water and steam side by several means. Deaeration of the feedwater prevents corrosion and pitting caused by oxygen. An alkaline-phosphate treatment of the boiler water provides an optimum pH for minimizing corrosion, and also controls damaging scale-forming salts by producing insoluble phosphates, which can be removed by blow-down or mechanical cleaning. Volatile amines may be used to control corrosion in condensers and return lines, and to prevent deposition of copper in boiler tubes.

Fireside corrosion from fuel-combustion products can be controlled adequately; however, certain parts such as superheater support members must tolerate high temperatures and severe corrosive conditions. These parts are especially vulnerable when vanadium-bearing fuels are used and temperatures are reached at which vanadium-bearing molten slags can form and accumulate. The best material known to resist this attack is a cast 60Cr-40Ni alloy. However, ASTM A351 Grade CH20 has been used successfully in some commercial boilers; this alloy contains about 24% chromium and 13.5% nickel.

If conditions leading to failure by fatigue, overheating, and corrosion are under control, the primary considerations in material selection then reduce to product form, fabricability, cost, and the use of standard grades of alloys in accordance with regulations for temperature and allowable stress. The most common guide is that of the ASME Boiler and Pressure Vessel Code, which sets allowable design stresses as a function of temperature for each commonly used alloy.

For the low-alloy steels, the temperature and stress tolerances tend to increase progressively in the series of: carbon, carbon-½Mo, 1¼Cr-½Mo, and 2¼Cr-1Mo grades. There are, of course, other competing alloys, but these four are the most frequently employed.

The useful temperature ranges for these alloys can overlap if suitable adjustment of the design stress is made. However, carbon steel is seldom used above 750 F because of its low strength. If it is used at higher temperatures, it is advisable that the steel be a grade deoxidized with silicon instead of aluminum. Aluminum-killed carbon steels have a greater tendency to become embrittled by decomposition of iron carbide (graphitization) after prolonged high-temperature exposure. Graphitization occurs most readily near welds. Molybdenum and particularly chromium are strong carbide-forming elements and inhibit graphitization tendencies in alloys containing these elements.

Typical applications of carbon steels (including minor modifications in chemistry for welding, etc.) include steam and water drums, water-wall headers, generating tubes, downcomers, and economizer tubes and headers. Depending on applicable specifications, either welded or seamless tubes can be used.

Superheater headers are commonly made of carbon-½Mo or 1¼Cr-½Mo alloy. In Navy practice the former is used up to 875 F and the latter between 875 and 1050 F.

Superheater tubes normally are specified to be of the seamless variety and are commonly carbon-½Mo, 1¼Cr-½Mo, or 2¼Cr-1Mo depending on temperature and pressure. Navy practice is to use 2¼Cr-1Mo alloy for tube metal temperatures up to 1100 F and a columbium or titanium-stabilized 18Cr-9Ni austenitic stainless steel for metal temperatures between 1100 and 1200 F.

Beyond 1050 to 1100 F the high-temperature strength and oxidation resistance of the ferritic steels fall off rapidly. Hence, it is necessary to go to the more expensive austenitic stainless steels in high-temperature superheaters. It is common practice to use ferritic steel tubes in the beginning of the superheater pass, and to restrict the austenitic steel tubes to the final portion of the pass. Economy is not the only reason for this. If stainless steel were used in the initial portion of the pass, the stainless would be exposed to wet steam entering the superheater. Any boiler-water solids carried over would concentrate on the tube surfaces and could lead to stress-corrosion cracking. Stainless tubes are not subject to this danger if the steam passing through them is dry.

There are places in boilers where special metals may be employed. Mention has already been made of cast 60Cr-40Ni alloy for superheater support members. Alloys such as Inconel may be used for oxidation-resistant bolts; desuperheater parts are often made of a 16Cr-1Ni alloy; and casings or other parts exposed to corrosive flue gases, particularly at temperatures over 700 F, may be made of stainless steel.

6.3 Main and Auxiliary Steam Piping Systems. In general, the guidelines for main and auxiliary steam piping systems on merchant vessels are set by applicable sections of the American National Standards Institute Code, ASME Boiler and Pressure Vessel Code [5,10], and U.S. Coast Guard Marine Engineering Regulations [1]. Navy practice follows the same pattern, differing only in details.

As in the case of boilers and superheaters, the rationale for choosing between carbon and alloy steels for steam lines is largely a matter of service temperature and pressure. However, in contrast to certain parts in boilers that are exposed to hot combustion gases, steam lines are not exposed to temperatures beyond those of the contained steam.

Carbon and low-alloy ferritic steels can be used within allowable design stresses at temperatures up to 1050 F. If shipboard boiler plants were designed to produce steam in excess of this temperature, it would be necessary to use austenitic stainless steel piping in lieu of low-alloy ferritic steel piping because of the rapid falloff in strength and oxidation resistance of the ferritic materials in this temperature range.

The basic design of steam piping systems has had an influence on the materials developed for this type of service. A system may consist of straight pipes with welded elbows for bends and welding-neck flanges for joints. This type of construction requires a material of excellent weldability. Alternatively, the pipe may be in lengths that have been bent into the desired configuration and then joined together by welding. In this case the material must also have good hot-forming properties.

Both hot bending and cold bending are used in pipe fabrication. Hot bending is usually done at 1650 to 1850 F with the pipe packed with sand to help retain shape and circularity. A somewhat higher temperature may be used for 2¼Cr-1Mo pipe. Cooling the outside of the bend may be employed to prevent excessive thinning in this area. A bending machine with a mandrel is used for cold bending, and a minimum bend radius of 5 diameters is usually recommended.

After hot bending, carbon steels are usually annealed at about 1200 F. The Mo and Cr-Mo alloy steels are usually normalized from 1650 F and then drawn at 1200 F. Both carbon and alloy steel piping are generally annealed after cold bending.

Seamless pipe is used for the more critical applications, especially for the higher temperatures and pressures. Welded pipe is permitted for many other applications.

Copper or copper-alloy pipes with brazed joints can be used in noncritical, low-temperature, low-pressure service. For example, fabricated copper systems are permitted for steam and steam drain service on merchant ships up to 320 F and 75 psi [2].

Cast or forged valves are used in steam lines. For all practical purposes, cast steel valves can be considered as cast versions of the wrought alloys used for the piping. Minor compositional variations are mostly associated with the need to enhance the "castability" of the alloy to assure high casting quality.

The seats and disks in steam valves require special attention because of tendencies to erode—a condition commonly known as "wire drawing." The tendency increases when valves are used for throttling service, and can be especially severe with wet, saturated steam. Austenitic stainless steels are quite resistant to wire drawing, but are not usually employed for valve trim because of poor resistance to galling and because of a difference in thermal expansion coefficients between the ferritic steel valve body and the austenitic steel trim. Nitrided steel has been used for valve trim on occasions, but nitriding has lost favor because of poor performance in comparison with other materials.

The best trim material for resisting wire drawing is a weld-deposited, tungsten-cobalt-molybdenum hard-facing alloy known commercially as Stellite. Stellite to Stellite and Stellite to 13% Cr stainless steel are suitable combinations for the mating surfaces of disks and seats, although Stellite on both parts is preferred for the more critical systems. The Stellite facing of the seat can be applied by direct welding to the valve. However, welding procedures can be controlled better, and seat repair can be facilitated, if the Stellite is applied to a separate seat ring, which can then be shrunk or screwed into place in the valve body.

Carbon or alloy steels are used for bolts on steam line joints. Carbon steel bolts are used at lower temperatures. As the temperature increases, and depending on the temperature, one of the alloy steels must be used. The alloys most frequently utilized are Grades B6 (12% Cr), B7 (Cr-Mo), and B16 (Cr-Mo-V) of ASTM Specification A193. Alloy steel bolting normally is used in the quenched and tempered or normalized and tempered condition to enhance strength properties and to assure the highest possible elastic limit. Minimum tempering temperatures are specified to be well above the intended operating temperature, and for metallurgical reasons are usually within the range of 1100 to 1200 F.

6.4 Fresh- and Saltwater Piping Systems. Carbon steel pipe (ASTM A53) commonly is used for non-saltwater applications such as boiler feed piping and engine cooling. The particular application may influence whether welded or seamless, or plain or galvanized pipe is chosen [2]. Copper pipe may be used in condensate systems. Polyvinyl chloride (PVC) plastic pipe (ASTM D 1785) and glass-reinforced epoxy pipe (MIL-P-24608) are also used for freshwater systems for auxiliary machinery and engine cooling. Cold and hot freshwater systems for domestic, sanitary, heating, and air-conditioning services normally are made of PVC plastic pipe, glass-reinforced epoxy pipe, or seamless copper tubing (ASTM B88).

The main and auxiliary saltwater piping systems present far greater difficulties than freshwater systems because of the severe corrosion problems that are encountered (see Section 2).

Water velocity and pipeline configuration are the most important variables (other than material) influencing the performance of a shipboard saltwater piping system. Most metallic materials of interest for saltwater systems have low corrosion rates in quiescent water because of the formation of corrosion films, which protect the underlying metal. As the water velocity is increased, a velocity may be reached at which erosion will begin. That is, the protective film will be swept away such that fresh metal will be exposed constantly to the water, and corrosion will be accelerated.

In general, the water velocity in piping systems is seldom high enough to bring about general corrosion-erosion of the pipe material. However, velocities commonly are experienced that lead to damage in localized areas. This action occurs in areas of water turbulence. Because of its impinging effect, turbulent flow can sweep away protective corrosion films in local areas at water velocities that are not destructive where flow is more or less streamlined.

The rate of impingement attack in turbulent areas is a function of both the water velocity and the configuration of the piping system. The attack nearly always occurs just downstream of fittings and joints. Knowledge of this fact permits an easy inspection of a piping system, since the critical areas can be observed simply by breaking the joints. It is seldom necessary to know the condition of a pipe beyond the area that can be seen readily at the inlet end.

Pipe bends do not cause significant impingement attack if the bend radius is generous enough to prevent wrinkling of the pipe or more than a moderate collapse of the cross section. Elbows, on the other hand, can cause damaging turbulence if the nominal water velocity is fairly high. Long-radius elbows cause less damage than short-radius elbows.

Other fittings, such as tees, reducers, and valves, also cause turbulence. Globe valves used for throttling service probably lead to more pipe damage than any other type

of fitting (see Fig. 5). Properly designed and installed unions are perhaps the least damaging.

If a piping system is designed adequately from the standpoint of material selection and water velocity, there should be no major deterioration from standard fittings. An occasional failure may occur, but frequent failures should be an indication that the system is inadequately designed or is being used improperly.

Whatever piping material is used, good design practice dictates that fittings, welds, and connections be made of compatible materials from the galvanic-corrosion standpoint. In areas where dissimilar metals must be used, it is important that, insofar as possible, a favorable cathode/anode area ratio be provided and that the key components be made of the more noble material. Examples would be:

- Use of 70/30 Cu-Ni welds for fabricating 90/10 Cu-Ni alloy.
- Use of monel trim in a tin-bronze valve.

Experiments have demonstrated little difference in the performance of plain steel, low-alloy steel, and wrought iron in seawater pipelines. Therefore, the term "steel" will be used to cover these materials as a group.

Steel has fair resistance to impingement corrosion, but poor resistance to cold salt water and very poor resistance to hot aerated salt water. It is most useful in closed systems where there is no continuous supply of oxygen.

Steel has no antifouling properties, and marine growth can accumulate under stagnant water conditions such as in dead-end lines.

For various reasons, it would be desirable to use bronze valves even if a piping system were made of steel; however, such a combination would introduce a galvanic corrosion problem because of an accelerated corrosion of the steel in the vicinity of the valves. In such cases the installation of waster pieces between the valves and pipes may be desirable.

Protective coatings are desirable for steel pipelines in seawater service. Galvanizing has been tried in both service and experimental systems, and the zinc coating is effective, but it also corrodes away after a time. The extra pipe life gained by galvanizing depends on the thickness of the zinc and the water velocity, and may vary from as little as 3 months to 2 or 3 years.

Other types of protection for steel pipe (lining with enamel, paint, rubber, lead, and so on) have been tried and generally abandoned because of cost and difficulty of installation, and maintenance of coating integrity.

Stainless steel piping is not suitable for shipboard saltwater piping because of local pitting and crevice-corrosion tendencies.

Aluminum is attractive because of its low cost and light weight, and certain alloys have good resistance to high water velocities; however, it has serious disadvantages. One is its poor antifouling characteristics. Another is the tendency to pit from heavy-metal ions, which may be contained in the water from upstream equipment or other sources. But the most serious objection to its use is the susceptibility to rapid galvanic corrosion when coupled to heavier metals such as steels and copper-base alloys.

The use of bronze valves and fittings in an aluminum piping system, or the coupling of copper-alloy pipes to aluminum pipes, would be intolerable. Any such combination would require completely insulated joints, including assurance against accidental coupling of the dissimilar metal components through the hull structure. Experience indicates that this complete isolation is difficult, although not impossible, to achieve under practical shipboard conditions.

Copper pipe can be used with some success in saltwater lines. However, copper has a poor resistance to erosion and impingement attack at more than moderate water velocities. Copper also tends to corrode excessively by concentration-cell effects under stagnant water conditions. Between these extremes, copper can give good service, but in most piping systems it is difficult to guarantee that tolerable water-flow conditions can be maintained.

Aluminum brass and aluminum bronze alloys (e.g., 76/22/2 Cu-Zn-Al, 95/5 Cu-Al, or 90.5/2/7.5 Cu-Fe-Al) have better resistance than copper to erosion and impingement attack. However, some of these alloys suffer from local forms of corrosion such as pitting, and also tend to be more difficult to solder or braze.

The cupronickel alloys are the best from the standpoint of all-around performance in shipboard saltwater systems. Their resistance to general corrosion, crevice corrosion, and pitting is good, and these forms of attack seldom need consideration. The cupronickel alloys are easy to form, weld, and braze. They have good anti-fouling characteristics. Additionally, of all the copper-base alloys, the cupronickels have the best resistance to high water velocity and impingement attack.

Two cupronickel alloys are commonly used in naval shipboard applications: 70/30 Cu-Ni and 90/10 Cu-Ni (ASTM B467). If nominal pipe water velocities are held within conservative limits, these alloys will give long, essentially troublefree life. The 90/10 alloy should give 15 to 25 years of service if velocities do not exceed 8 to 10 fps. The 70/30 alloy will provide in excess of 20 years of service at velocities up to 12 fps. The long life and generally satisfactory service provided by these alloys often make them cost-effective despite their high initial cost.

The outstanding performance of cupronickel alloys is partly dependent on the addition of iron as an alloying element (see Section 2.2). The 70/30 alloy requires a nominal 0.5% iron content, and the 90/10 alloy requires a nominal iron content of 1.5% to provide maximum corrosion resistance.

The 90/10 Cu-Ni alloy is the one most frequently used in the construction of merchant ships. It is also used in Navy surface ships. Because of its somewhat superior properties and reliability, the 70/30 alloy is used in sea-connected systems of Navy submarines.

Titanium is used for saltwater piping in special circumstances. Pipe walls made of titanium can be thin because of titanium's high strength, and because there is no need for a corrosion allowance in wall-thickness calculations. Furthermore, pipe sizes can be smaller because high water velocities (over 20 fps) can be tolerated without damage. However, there are disadvantages associated with the use of titanium: titanium is difficult to fabricate and repair in the field, and operating personnel cannot use simple techniques such as brazing. Also, galvanic corrosion must be taken into account wherever there is a connection to less-noble metals.

Polyvinyl chloride plastic and glass-reinforced epoxy can be used for some nonvital saltwater piping applications. Their use is limited to low-pressure and low-temperature services, and safety must be considered if fire damage to the piping would constitute a hazard.

Valves, fittings, and pumps in saltwater systems should be compatible with the piping material. With cupronickel piping, it is common practice to use cupronickel or bronze flanges, and cast-bronze silver-braze unions.

Valves for saltwater systems can be made of a variety of materials. Cast tin bronzes set the standard for high-performance valve bodies on both merchant and naval vessels. However, ductile iron and galvanized cast steel are commonly used, especially in systems of large pipe sizes. Alloys such as austenitic nickel cast iron, stainless 20, Monel, and 70/30 Cu-Ni may be used on special occasions.

The seats in bronze valves can be cast as a part of the valve body, but this would be suitable only for very mild service; otherwise, the seats would erode in turbulent water. The more common practice for high performance is to use seats and disks of Monel. This alloy has greater inherent erosion resistance, and the surrounding bronze body enhances performance by providing cathodic protection to the trim.

Water-circulating pumps may be made of a variety of materials. Pumps for hot-water heating systems frequently are made with a cast-iron casing, brass or bronze impeller, and a stainless or nickel-alloy shaft. Sometimes both the casing and impeller are of cast bronze.

The best performance for saltwater circulation is obtained when the impeller alloy is not only resistant to erosion and impingement attack, but also is more noble than the casing material so that cathodic protection of the impeller is provided. One common combination is a cast-iron casing with a bronze impeller and shaft. Better service is obtained with a cast-bronze casing and Monel impeller and shaft. This combination is frequently used on merchant and naval ships. Monel impellers provide better erosion resistance than bronze impellers.

6.5 Condensers, Heat Exchangers, and Distillation Units. The tubes are the most critical components in steam condensers. They must have thin walls for effective heat transfer, yet must provide isolation of the steam from the salt water used for cooling. Therefore, a prime consideration is resistance to tube deterioration on the saltwater side.

Waterside deterioration can be considered from two aspects: how resistant a tube material is to the water velocity within the tube, and how resistant it is to erosion from turbulence and impingement at the inlet end. Fortunately, the tube materials of greatest interest for shipboard service are ranked in the same order on both counts.

Admiralty metal is the least resistant, and its use should be restricted to very low water velocities, preferably not over 3 to 4 fps. If Admiralty metal is used, an inhibited grade should be selected to prevent dezincification. An inhibited aluminum brass (77Cu, 21Zn, 2Al, As) would be next in rank for resisting velocity and turbulence. This alloy is also resistant to brackish and polluted waters, and is widely used in marine service.

The highest performance standards among the copper alloys are obtained with the cupronickels. The 90/10 Cu-Ni (1.5% Fe) alloy is the one most frequently employed and has been used by the Navy in recent years for all main condensers on surface vessels. Somewhat better performance and reliability can be obtained with the 70/30 Cu-Ni (0.5% Fe) alloy, but its higher cost favors its use only for special applications. The 90/10 and 70/30 alloys give good performance up to water velocities of 10 and 12 fps, respectively, although conservatism usually dictates somewhat lower design limits.

The tubesheet alloy must be resistant to saltwater corrosion, and should not be more noble than the tube alloy so as to prevent galvanic corrosion of the tubes.

Excessive inlet-end erosion of condenser tubes sometimes has been a problem, even with cupronickel tubes in the earlier years of their use before iron additions to the alloys were made. Inlet-end erosion can be controlled in several ways. Waterboxes can be designed so as to minimize turbulence, impingement, and velocity gradients across the tube banks. In some cases, plastic ferrules are inserted in the inlet ends to carry the water through the first few inches of the tubes where the water is most turbulent. Galvanic protection of the tube ends is provided by installing sacrificial zinc or iron anodes in the waterbox, or by making the waterbox itself of steel or cast iron. Monel waterboxes frequently are used with 70/30 Cu-Ni tubes; when this is done, the Monel is solder-wiped to reduce adverse galvanic effects.

If design water velocities are maintained conservatively low, iron-modified cupronickel alloy tubes require little or no supplemental protection for a satisfactorily long-life service. Tubes of 90/10 Cu-Ni work well with 90/10 alloy sheets and waterboxes. Tubes of 70/30 Cu-Ni work well with 70/30 sheets and either 70/30 or 90/10 waterboxes. Other combinations are possible if their use does not lead to galvanic corrosion of the tubes.

Navy practice in material selection for surface steam condensers is outlined in Military Specification MIL-C-15430. This covers both the primary components as well as the miscellaneous hardware and attachments.

Several materials other than copper-base alloys occasionally have been considered for condenser tube applications. These have included certain austenitic stainless steels and nickel-base alloys. Titanium condensers and heat exchangers can be used to advantage where weight is critical, such as on hydrofoils, surface-effect ships, and deep submersibles. The weight advantage arises from more than just a substitution of titanium for heavier metals. Thinner tubes can be used, as no corrosion allowance is required.

High heat-transfer rates can be obtained with titanium for several reasons, despite its lower coefficient of thermal conductivity in comparison with conventional copper-base alloys. First, thinner tubes can be used with shell-and-tube heat-exchanger designs, and very thin titanium plates can be used with plate-and-frame heat exchangers. Second, heat transfer is not impeded by a buildup of insulating corrosion films. Third, higher cooling-water velocities (above 20 fps) can be tolerated, the limit being primarily a matter of the power costs for the circulating pumps. In addition, wet-steam impingement is sometimes a problem on the outer surfaces of condenser tubes. The usual practice is to install baffles to prevent direct tube impingement. Titanium tubes are highly resistant to this form of damage.

The miscellaneous auxiliary heat exchangers using salt-water as the cooling medium require the same rationale for material selection as steam condensers. However, the services are usually not as critical, and design water velocities normally are quite low.

Distillation units for making fresh water aboard ship present a severe material selection problem. Water temperatures tend to be high, and concentrated brine and scale are encountered, all of which aggravate corrosion attack. The removal of scale from evaporating surfaces also places heavy demands on construction materials. The scale is often broken away by means such as cold shocking, and, as in the case of the vertical corrugated basket type, also by mechanical flexure. These scale-removal procedures cause a fatiguing action. Occasionally the units may have to be cleaned with acid washes. It is not surprising, therefore, that distillation units demand the best materials.

Military Specification MIL-D-18641 contains details of the material selection for components of distillation units for naval shipboard applications. The compositions of the basic components are 70/30 Cu-Ni, 90/10 Cu-Ni, or Composition 5 aluminum bronze (approximately 91Cu-7Al-2Fe). Commercial ships with small complements often use packaged distilling plants with heat exchangers made of titanium.

6.6 Steam Turbines. Turbine rotors can be made from a variety of materials depending on stress and temperature conditions. Carbon steel (usually about 0.30 to 0.40% carbon) is commonly used at temperatures up to about 600 to 650 F. Low-alloy steels of the Ni-Mo-V and Cr-Ni-Mo-V types are used at somewhat higher temperatures. For extreme temperatures (such as between 750 and 1050 F), it is common practice to use a 1¼Cr-1¼Mo-¼V alloy steel. If corrosion and erosion resistance is a major design consideration, as in saturated steam plants, the turbine rotor may be made of 12% Cr steel for service up to 750 F.

Because of high stress conditions, turbine rotors must be of high-quality steel, free of injurious defects such as seams, scale, porosity, and excessive nonmetallic inclusions. To obtain a desirable metallurgical structure, a 2½ to 1 and preferably 3 to 1 reduction in area should be used in converting the cast ingot into a forged rotor.

It is necessary to avoid hydrogen "flaking" of the rotor forging by assuring that the forging is not cooled to room temperature until there has been time for hydrogen to diffuse out. This can be accomplished by interrupting the cooling at about 400 to 600 F, and holding at this temperature for a prolonged period. Alternatively, a common practice to avoid hydrogen flaking is to use vacuum degassing techniques during the steelmaking process.

The desired microstructures and properties in rotor forgings are obtained by quenching and tempering or normalizing and tempering. A tempering temperature of 1100 F is recommended (1200 F for 12% Cr steel). After the rotors have been rough machined, they are stress relieved at 50 to 100 deg F below the tempering temperature. This provides for maximum dimensional stability after final machining.

Steam turbine blading is now made almost exclusively from one of the quenched and tempered chromium stainless steels. AISI Types 403 and 410 are the most common. These are easily machined and forged, have good corrosion and moderately good wet-steam erosion resistance, have good mechanical properties at operating temperatures, and show good resistance to fatigue. Of equal importance is the unusually high specific damping capacity of these steels. This can be used to advantage in reducing fatigue stresses arising from blade vibration. It is a particularly important characteristic while passing through speed ranges that might produce resonant vibrations in the blades.

Types 403 and 410 steels may not have sufficient creep strength at the highest temperatures and stresses encountered in some marine turbines. When this is so, a more highly alloyed steel can be used, such as AISI Type 422 (13Cr-Ni-W-Mo-V). The added strength in this steel is accompanied by a reduction in specific damping capacity.

If wet-steam erosion is too severe in any stage of the turbine, it is sometimes the practice to apply Stellite inserts on the blades to enhance erosion resistance.

Titanium may be considered for steam turbine blades, particularly in the low-temperature stages of turbines. Titanium has a good wet-steam erosion resistance. Furthermore, since its density is only about 57% that of steel, titanium blades develop lower stresses from centrifugal forces.

Turbine casings and steam chests are generally cast, using a variety of materials depending on the temperature and pressure conditions. Carbon, carbon-½Mo, 1¼Cr-½Mo, and 2¼Cr-1Mo are the most common grades. A 12% Cr steel is occasionally used in saturated steam plants for better corrosion and erosion resistance.

6.7 Gas Turbines. The widespread use of gas turbines in aircraft engines has provided the major incentive to develop special materials for gas turbines, especially the cobalt-base and nickel-base superalloys used in hot-section components. It is understandable, therefore, that the materials developed for aircraft engines have also been used widely in marine versions of gas turbines. This applies to both main propulsion and auxiliary drive units. Nevertheless, the adaptation of gas turbines to shipboard applications has led to certain problems, which are not of primary concern to the aircraft industry. Two materials-related problems in particular have been encountered in marine turbines: vanadium corrosion and sulfidation corrosion, both of which can result in rapid deterioration of critical hot-section components such as guide vanes and turbine blades. Each of these forms of corrosion is discussed briefly.

Early gas turbines suffered accelerated oxidation attack of hot-section components when turbines were operated on residual fuels. The attack was associated with vanadium and sodium in the fuels, which combined during combustion to form ash or slag deposits. Sulfated-vanadium/alkali-metal slags were typical. There appeared to be a threshold temperature, dependent on the slag composition, above which the slag promoted catastrophic oxidation. The threshold temperature varied from about 1200 to 1560 F for many of the slags studied.

No completely acceptable solution to this problem has been found. Alloys vary in their resistance to attack, but none combine the necessary mechanical properties and long life required for marine gas turbine service. Fuel additives such as calcium or magnesium compounds are effective in inhibiting vanadium corrosion, particularly when combined with desalting to remove any salt present. The resulting deposits are hygroscopic and nonadherent; consequently, they are easily removed.

Much literature has appeared on the problem of vanadium corrosion. In spite of the possibilities of fuel-treatment systems to control corrosion, most marine gas turbines have been operated on essentially vanadium-free distillate fuels.

Sulfidation corrosion (commonly called hot corrosion) is a catastrophic type of oxidation, which can be particularly harmful to hot-section components of marine turbines. The attack disrupts the normally protective oxides on superalloys, and sulfides may form deep into grain boundaries, causing serious degradation of the structural integrity of the alloy. Figure 18 is an example of a marine gas turbine blade damaged by sulfidation.

The immediate agent responsible for sulfidation is Na_2SO_4, which can form during combustion from reactions between NaCl in sea air and sulfur in the fuel. However, the use of sulfur-free fuel will not eliminate the problem, because Na_2SO_4 is also present in sea air. There is evidence that $NaCl$-Na_2SO_4 mixtures are more corrosive than Na_2SO_4 alone. The sulfate must be present on the alloy in a condensed phase for attack to take place. At pressures near atmospheric, the accelerated attack may occur over the temperature range of 1400 to 1900 F.

Several approaches can be used to combat sulfidation attack. One is the use of fuel additives to promote formation of slags with melting points above turbine operating temperatures. Another approach is the use of metal-diffusion coatings such as the NiAl type. A third is the use of demisters or salt separators to remove as much of the sea salt from the intake air as possible. Still another is the use of cooled components to lower the metal temperature. Finally, the inlet gas temperature can be limited to keep component temperatures within tolerable limits.

Fig. 18   Sulfidation attack on a blade from a marine gas turbine

either surface or subsurface fatigue, depending on the circumstances. Once a crack appears at the surface, the hydrodynamic action of the lubricating oil assists in its propagation. Eventually a piece of metal pops out, leaving a pit in the surface of the tooth.

Pitting problems can arise in a properly designed gear from misalignment, helix-angle errors, tooth-spacing errors, and improper lubrication. It is not uncommon to have light pitting on the contact surfaces during the first few months of service. This pitting usually ceases after a brief period, and no further damage is experienced. Pitting of this type is sometimes called "corrective" or "incipient" pitting. On the other hand, "heavy" or "destructive" pitting is extremely serious and normally requires attention to prevent total destruction of the gear.

From a fatigue standpoint, it is desirable to use steels having high tensile strength for the gear elements. However, the strength that can be developed in a steel frequently is limited by section size and available heat-treating facilities. The rims of large bull gears usually are made from high-quality forgings that are either normalized and tempered or quenched and tempered to improve their strength and toughness. Both carbon and alloy steels are used along with carburizing and nitriding processes to provide the desired properties. Typical steels used for slow-speed gears are AISI-1030, 1035, 1050, 4137, and 4337. Higher hardness and strength can be developed in pinions because of their smaller size. Normalized and tempered AISI-1042 and quenched and tempered AISI-4140 and 4340 are typical steels for pinion applications.

6.9 Main Propulsion Shafting. Propulsion shafts constitute one of the critical components in single-screw ships. Although there have been periods in which propeller losses caused by tailshaft failures have been epidemic (1946–1950), such disabling casualties are now infrequent. However, lesser casualties in the form of surface deterioration and penetrating cracks do occur.

Propulsion shaft casualties are caused principally by corrosion, fatigue, and fretting corrosion. Steps are taken to protect the shaft from seawater corrosion by applying protective sleeves or coverings. However, complete integrity against the ingress of seawater cannot be assured. The junctions between the covering and metallic components, such as bearing sleeves and propeller hubs, are particularly vulnerable. Unfortunately, these are the areas where the need for protection is greatest.

When seawater contacts the shaft surface, it invariably causes general corrosion and pitting. Corrosion by itself is not serious, inasmuch as it rarely affects the overall strength of the shaft. In most cases pits and corroded areas can be ground smooth, filled with a putty compound, and the protective covering reapplied. If the pits are unusually deep, they are ground out and then built up by welding. However, corrosion pits can act as local sites for the initiation of fatigue cracks. Corrosion-fatigue cracks commonly occur in shafts at the aft end of the propeller-shaft sleeve and at the forward end of the propeller hub in arrangements with water-lubricated outboard bearings. Fortunately, fatigue cracks normally progress slowly enough to be detected during overhaul.

Other material problems in marine versions of aircraft turbines have not been so difficult to overcome. Some changes have been needed to make components resistant to the marine environment. These have included the coating of some components, the substitution of stainless alloys for some miscellaneous parts, and the substitution of aluminum alloys for magnesium alloys.

6.8 Main Propulsion Gears. Although gears of all types are used aboard ship, the main propulsion gears are of greatest interest and concern to the marine engineer. Gears generally fail from tooth breakage or deterioration of the tooth contact surfaces. Tooth breakage is caused by fatigue cracks, which normally initiate in the root of the tooth. Accordingly, such factors as material strength, root radius, surface finish, surface treatment, and bending stresses may greatly affect the life and performance of the gears.

Deterioration of the contact surfaces is normally caused by "pitting" at or near the pitch line. Pitting is a fatigue-type failure generated by compressive rolling loads when the teeth are meshing. This failure may originate from

Fretting corrosion is caused by minute movements between mating parts. Because of the microscopic welding and tearing actions, which take place between the mating surfaces, small microcracks, which act as sites for fatigue crack initiation, readily form. Fretting in tailshafts can occur under the ends of the shaft sleeve and often originates in way of the propeller hub [37]. Accordingly, fatigue cracks formed in these areas cannot be detected without cutting away the ends of the sleeve or removing the propeller.

Experience has shown that the fatigue life of tailshafts subjected to fretting can be improved greatly by cold rolling the surface of the shaft [38]. Cold rolling may not eliminate the formation of microcracks; it does, however, produce residual compressive stresses in the surface layers of the shaft, and this retards the propagation of fatigue cracks if they do form. The U.S. Navy generally cold rolls tailshafts that are 6 in. and larger in diameter. Rolling is usually restricted to about one diameter on each side of the point where the propeller taper begins.

Propulsion shafts for merchant ships usually are made from annealed, normalized, or normalized and tempered plain carbon steel forgings conforming to the requirements of ABS Grade 2 steel [3]. Military ships usually employ higher strength, quenched and tempered alloy steel shafts conforming to Specification MIL-S-23284.

High-strength steels such as Class 1 of MIL-S-23284 are not generally used for outboard shafting where the possibility of contact with seawater exists. In instances where they are used, the higher strength is not considered in the determination of the shaft size. The reason is that the corrosion-fatigue properties of the higher-strength steel are no better than those of a lower-strength steel (see Fig. 13). However, high-strength steels are used for inboard shafting to reduce the weight of the shafting and to increase its flexibility.

Weldability is an important characteristic of shafting materials as shafting is vulnerable to damage that can only be repaired by welding. For example, the removal of a corrosion-fatigue crack by grinding, with a subsequent buildup of material by welding, can greatly extend the life of many shafts.

6.10 Propellers. The following properties and characteristics must be given consideration in selecting a propeller material:

• Castability
• Strength and ductility
• Corrosion-erosion resistance
• Cavitation resistance
• Weldability
• Cost

The geometry of a propeller casting closely approximates the finished propeller dimensions. To meet these demanding requirements, it is important that the metal have excellent castability, i.e., the ability to flow freely into mold cavities and to solidify into sound metal.

The strength and ductility of the metal are important in establishing the design and serviceability of a propeller. High strength is necessary to resist the applied forces and to provide fatigue resistance. Blade loss, although

infrequent, usually can be attributed to fatigue caused by vibration or unsymmetrical loading (see Fig. 10). On the other hand, blade damage frequently occurs from striking submerged or floating objects. It is important in these circumstances that the propeller material have sufficient ductility, not only to bend rather than break, but to undergo subsequent straightening and repair.

Water and entrained debris move across the faces of propeller blades at high velocities. This action tends to scour away protective films and expose bare metal to corrosion. Thus, high corrosion-erosion resistance is required in propeller applications. Also, the metal must be resistant to cavitation damage should conditions conducive to this phenomenon arise. The steel hull of a ship, and protective zincs attached thereto, provide a degree of galvanic protection, which helps to reduce propeller deterioration.

It is desirable, if not mandatory, that the metal be weldable. Good weldability is required not only to repair damage incurred during service, but also to repair casting imperfections on the surface of the blades during manufacture. The weld metal must be compatible with the base metal from the standpoint of corrosion to prevent undesirable galvanic couples.

Copper-base alloys are the most widely used metals for marine propellers. Certain of these alloys exhibit unique combinations of the desirable features mentioned in the foregoing, plus a built-in resistance to fouling produced by the dissolution of copper. Manganese bronze, similar to that specified in MIL-B-16443, was used almost exclusively for marine propellers until about 1957. Since then there has been an increasing trend toward the use of higher strength nickel-aluminum bronze and manganese-nickel-aluminum bronze similar to alloys 1 and 2 of MIL-B-21230.

From time to time other types of alloys, such as austenitic stainless steel, 12% chromium stainless steel, and titanium alloys, have been used for marine propellers in special applications. By and large, however, the copper-base alloys continue to find the most widespread usage as marine propeller materials.

6.11 Low-Temperature and Cryogenic Equipment. The use of merchant ships to transport refrigerated cargo and liquefied gases poses special material requirements. The materials for storage tanks, pressure vessels, piping, etc., must be capable of providing safe handling at the low temperatures encountered. In addition, the equipment must be resistant to chemical attack by the cargo being handled.

Two areas of material-application technology are recognized. These are "low-temperature" and "cryogenics," which involve materials for applications to −150 F and −459 F, respectively. The low-temperature materials are those especially suited for handling relatively "warm" liquefied gases such as propane, anhydrous ammonia, carbon dioxide, and ethane. The cryogenic materials are capable of handling liquefied methane, oxygen, nitrogen, argon, hydrogen, and helium.

Because of their relatively low cost and ease of fabrication, steels have been widely used for low-temperature and cryogenic applications. Not all steels, however, are suitable for these applications. Many of the low-alloy ferritic steels are susceptible to "brittle" fracture at low temperatures, and their so-called "notch toughness" must be thoroughly evaluated before use.

Notch toughness is the ability of a metal containing a notch or crack to yield plastically under constraint and high local stress. This property is a prime requisite of metals for low-temperature applications and is vital to the selection of materials for the handling, transportation, and storage of liquefied gases. Methods of determining this property are given in reference [11].

The notch toughness of most ferritic steels decreases with decreasing temperature. Ferritic steels usually undergo a sharp transition from ductile to brittle fracture. This does not preclude their use if great care is exercised in selecting an alloy for a specific application. The ferritic 9% Ni steel, ASTM A353, was developed specifically for cryogenic applications down to liquid nitrogen temperatures (-320 F).

The austenitic Cr-Ni stainless steels of the AISI 300 series show no marked decrease in toughness with decreasing temperature. Furthermore, they are highly corrosion resistant. Accordingly, they find wide use over the entire range of cryogenic applications. Ferritic and martensitic stainless steels generally are not recommended for cryogenic use.

The toughness of most nonferrous alloys is unaffected by cryogenic temperatures. Aluminum alloys have been used extensively, because of their fabricability and low cost. In particular, the non-heat-treatable aluminum-magnesium alloys of the 5000 series are used widely for cryogenic tankage.

References

1 "Marine Engineering Regulations," U.S. Coast Guard.

2 "Schedule for Pipes, Joints, Valves, Fittings and Symbols," Maritime Administration.

3 *Rules for the Classification and Construction of Steel Vessels*, American Bureau of Shipping.

4 "Naval Ships Technical Manual," Naval Ships Systems Command, U.S. Navy.

5 "Rules for Construction of Pressure Vessels," ASME Boiler and Pressure Vessel Code, American Society of Mechanical Engineers.

6 "Maritime Administration Standard Specification for Cargo Ship Construction," Maritime Administration, Washington, D.C.

7 U.S. Military and Federal Specifications.

8 *SAE Handbook*, Society of Automotive Engineers.

9 *Metals Handbook*, American Society for Metals, Metals Park, Ohio.

10 "Material Specifications," ASME Boiler and Pressure Vessel Code, American Society of Mechanical Engineers.

11 *ASTM Standards*, American Society for Testing and Materials.

12 "Steel Products Manual," American Iron and Steel Institute.

13 Irving L. Stern, "Hull Materials and Welding" in *Ship Design and Construction*, R. Taggart, Ed., SNAME, 1980.

14 *Metals Handbook*, Vol. 13, Corrosion, American Society for Metals, International, Metals Park, Ohio, 1987.

15 "Dissimilar Metals," Military Standard 889.

16 Z. Szklarska-Smialowska, *Pitting Corrosion of Metals*, National Association of Corrosion Engineers, 1986.

17 M. G. Fontana and N. D. Greene, *Corrosion Engineering*, McGraw-Hill, New York, 1978.

18 H. H. Uhlig and R. W. Revie, *Corrosion and Corrosion Control*, Wiley, New York, 1985.

19 *Corrosion and Corrosion Protection Handbook*, P. A. Schweitzer, Ed., Marcel Dekker, New York, New York, 1983.

20 *Corrosion*, Vol. 1, L. L. Shreir, Ed., Newnes-Butterworths, Boston, 1976.

21 *Corrosion of Metals in Marine Environments*, Metals and Ceramics Information Center, MCIC-78-37, March 1978.

22 *Guidelines for Selection of Marine Materials*, International Nickel Co., New York, 1971.

23 "Titanium Heat Exchangers for Service in Seawater, Brine and Other Natural Aqueous Environments: The Corrosion, Erosion and Galvanic Corrosion Characteristics of Titanium in Seawater, Polluted Inland Waters and in Brines," Titanium Information Bulletin, Imperial Metals Industries (Kynoch) Ltd., U.K., May 1970.

24 G. F. Wilhelmi, "Composite Materials for Ship Machinery Applications," SNAME Symposium, June 1988.

25 *Erosion by Cavitation or Impingement*, ASTM STP 408, American Society for Testing and Materials, Philadelphia, 1967.

26 *Aluminum: Properties and Physical Metallurgy*, J. E. Hatch, Ed., American Society for Metals, Metals Park, Ohio, 1984.

27 *Stress-Corrosion Testing*, ASTM STP 425, American Society for Testing and Materials, Philadelphia, Pennsylvania, 1967.

28 G. M. Pressouyre, *Current Solutions to Hydrogen Problems in Steels*, C. G. Interrante and G. M. Pressouyre, Eds., American Society for Metals, Metals Park, Ohio, 1982.

29 C. T. Fujii, *Metallurgical Transactions*, Vol. 12A, 1099, June 1981.

30 I. R. Lane, J. L. Cavallaro, and A. G. S. Morton, *Stress-Corrosion Cracking of Titanium*, ASTM STP 397, American Society for Testing and Materials, Philadelphia, 1966.

31 F. P. Ford, "Stress Corrosion Cracking of Iron-Base Alloys in Aqueous Environments," *Treatise on Materials Science and Technology*, Vol. 25, 1983.

32 *Corrosion*, Vol. 2, L. L. Shreir, Ed., Newnes-Butterworths, Boston, 1976.

33 R. E. Peterson, *Stress Concentration Design Factors*, Wiley, New York, 1953.

34 A. H. Rosenstein and W. H. Asche, *Temper Embrittlement in Steel*, ASTM STP 407, American Society

for Testing and Materials, Philadelphia, Pennsylvania 1968.

35 J. W. Weeton, D. M. Peters, and K. L. Thomas, *Engineer's Guide to Composite Materials*, American Society for Metals, Metals Park, Ohio.

36 George Lubin, *Handbook of Composites*, Society of Plastics Engineers.

37 M. F. Wheatcroft, "A Commentary on Marine Propeller Shaft Damages," American Society for Metals/American Society for Nondestructive Testing, 1979.

38 L. L. Shook and C. L. Long, "Surface Cold Rolling of Marine Propeller Shafting," *Trans.* SNAME, Vol. 66, 1958.

Units of Measure Conversion Table

LENGTH
1 inch = 25.4 mm
1 foot = 0.3048 m
1 fathom (6 ft) = 1.8288 m
1 statute mile (5280 ft) = 1.6093 km
1 nautical mile (6076.12 ft) = 1.852 km

1 mm = 0.03937 in.
1 m = 3.281 ft
1 m = 0.5468 fathom
1 km = 0.6214 statute mile
1 km = 0.54 nautical mile

AREA
1 square inch = 645.16 mm^2
1 square foot = 0.092903 m^2

1 mm^2 = 0.00155 in.^{2}
1 m^2 = 10.764 ft^2

VOLUME
1 cubic foot = 0.028317 m^3
1 U.S. gallon = 3.785412 liter
= 0.003785 m^3
1 barrel (42 gal) = 0.158987 m^3

1 m^3 = 35.315 ft^3
1 L = 0.2642 gal
1 m^3 = 264.2 gal
1 m^3 = 6.29 bbl

FLOW
1 cubic foot/min = 1.699 m^3/hr
1 gallon/min = 0.06309 liter/sec
= 0.000063 m^3/sec

1 m^3/hr = 0.5886 ft^3/min
1 L/sec = 15.85 gal/min
1 m^3/sec = 15850.3 gal/min

MASS
1 pound = 0.453592 kg
1 long ton (2240 lb) = 1.016047 metric tons

1 kg = 2.2046 lbm
1 metric ton = 0.98421 long ton

DENSITY
1 pound/cubic inch = 0.027680 kg/cm^3
1 pound/cubic foot = 16.01846 kg/m^3

1 kg/cm^3 = 36.127 lb/in.^{3}
1 kg/m^3 = 0.062428 lb/ft^3

FORCE
1 pound = 4.448222 newton

1 N = 0.22481 lbf

TORQUE
1 inch-pound = 0.112985 newton-meter

1 N-m = 8.8507 in. lb

PRESSURE
1 pound/square inch = 6.894757 kilopascal
= 6894.757 N/m^2
= 0.06894757 bar
1 inch mercury (60 F) = 3.37685 kilopascal
1 inch water (60 F) = 0.24884 kilopascal

1 kPa = 0.14504 lb/in.^{2}
1 N/m^2 = 0.000145 lb/in.^{2}
1 bar = 14.504 lb/in.^{2}
1 kPa = 0.29613 in. Hg
1 kPa = 4.0186 in. H$_2$O
1 kPa = 0.0102 kg/cm^2

ENERGY, POWER
1 Btu = 1.055056 kilojoule
1 foot-pound = 1.355818 joule
1 horsepower* = 0.7457 kW kilowatt
= 1.01319 metric hp
1 Btu/sec = 1.055056 kilowatt

1 kJ = 0.9478 Btu
1 J = 0.7376 ft-lb
1 kW = 1.341 hp
1 metric hp = 0.98698 English hp
1 kW = 0.9478 Btu/sec

FUEL CONSUMPTION
1 pound/horsepower-hour = 0.608277 kg/kW-hr

1 kg/kW-hr = 1.64399 lb/hp-hr

TEMPERATURE
Degrees Rankine = (9/5) (degrees Kelvin)
Degrees Fahrenheit = (9/5) (degrees Celsius) + 32

°K = (5/9) (°R)
°C = (5/9) (°F − 32)

*"Horsepower" is in English units unless specifically stated otherwise.

Index

Beachhouses

DOWN UNDER

Beachhouses

DOWN UNDER

Stephen Crafti

images
Publishing

First reprinted in 2006
The Images Publishing Group Reference Number: 701

Published in Australia in 2006 by
The Images Publishing Group Pty Ltd
ABN 89 059 734 431
6 Bastow Place, Mulgrave, Victoria 3170, Australia
Tel: +61 3 9561 5544 Fax: +61 3 9561 4860
books@images.com.au
www.imagespublishing.com

National Library of Australia Cataloguing-in-Publication entry:

Crafti, Stephen, 1959.
Beach Houses Down Under.

ISBN 1 86470 149 8.

1. Architecture, Domestic – Australia. 2. Architecture. Domestic – New Zealand, I. Title. (Series: Beach
Architecture).

728.3709146

Edited by Andrea Boekel

Designed by The Graphic Image Studio Pty Ltd, Mulgrave, Australia
www.tgis.com.au

Digital production by Splitting Image Colour Studio Pty Ltd, Australia

Printed by Everbest Printing Co. Ltd., in Hong Kong/China

IMAGES has included on its website a page for special notices in relation to this and our other publications.
Please visit www.imagespublishing.com

Contents

Contents continued

Introduction

Beach houses have come a long way since the fibro shack of the 1940s and 50s. While memories of summer holidays in these shacks are fondly recalled, so too are the less favourable features; spiders on the ceiling, sauna-like conditions and bunk beds in a tiny room.

Nevertheless, these fibro structures still act as signposts for architects building beach houses in the twenty first century. The houses they now design are considerably larger and the materials they use are more sophisticated. Like the original shacks, there is something heroic in these contemporary designs. Some, still made of timber and fibro cement, even appear quite frail against tumultuous weather conditions. But they are not.

The beach houses featured in this book don't follow a specific trend. Some are relatively modest and consist of a couple of bedrooms, a simple bathroom, a basic kitchen and a large timber deck. For the owners of these homes, the pleasure comes from being near the sand and the surf. Friends and family come down for the day and enjoy lunch on the deck. Other houses featured in this book are more sophisticated and include all the features found in a city home, from en-suite bathrooms to high tech kitchens. But even in these homes,

the atmosphere is far more relaxed and spontaneous, responding to the views over the water rather than being preoccupied with finishes and surfaces.

But whether the beach house is modest in size or spreading over several levels, there are certain elements common to all. A deck is paramount in the design, orientated to both the sun and the views. Used as another room, particularly during the warmer months, these decks are often protected by automatic screening devices. Outdoor showers are also popular in many of these homes, allowing sand to be washed off, instead of dragged across the living room floor. In one case, however, bringing sand into the house is not only expected, but also addressed in the design. A timber-slatted floor in the hallway covers a trough where the sand lodges. This floor can be simply pulled back and the trough vacuumed before the owners return to their city home.

Typical to most of these beach houses is one large open plan kitchen, living and dining areas. This arrangement is not unusual in city homes. But this configuration in beach homes is even more relaxed, with divisions between spaces being ill defined. For the owners and their family and friends, escaping to the beach is about getting together in the one space. And for solitude, there's always the beach or a shady tree on the property.

While many beach houses featured in this book include large communal spaces, increasingly a number of these have their own separate guest quarters. In some instances this accommodation includes a kitchenette together with a couple of bedrooms and a bathroom. And generally this accommodation is located at ground level, below the main living areas or alternatively, in an entirely separate wing. This arrangement allows owners to enjoy their own space, closing off the additional guest wings if not being used.

The materials used in these beach houses are diverse. A disproportionate number of these homes are made of timber, left untreated and allowed to weather in the sun. As the timber turns silver, the house merges with the gnarled coastal scrub. One house even appears like a timber crate, washed up from the shore. Other houses are more sleek and sophisticated. Made of glass, steel or concrete, these houses convey a sense of monumentality against the shoreline. These homes are beautifully detailed, with the architects appearing to have thought

of everything, from the minute their clients get up in the morning to the moment their heads hit the pillow at night.

A number of the homes in this book had a previous life. Built in the 1950s through to the 1980s, these houses have been reworked to provide the best of contemporary living. One house, for example, built in the 1980s, was completely reworked to accommodate an extended grown up family. And while this house has a striking contemporary edge, there are subtle signs of the past, sensitively worked in to the present. Another house, built at the same time, appears totally new, with the original façade rendered and small apertures dramatically enlarged to take advantage of the view.

Beach houses are often an escape for their owners at the end of each week or on a more irregular basis. The opportunity to design a beach house is also an escape for the architects, who are often given fewer constraints when coming up with a design.

Stephen Crafti

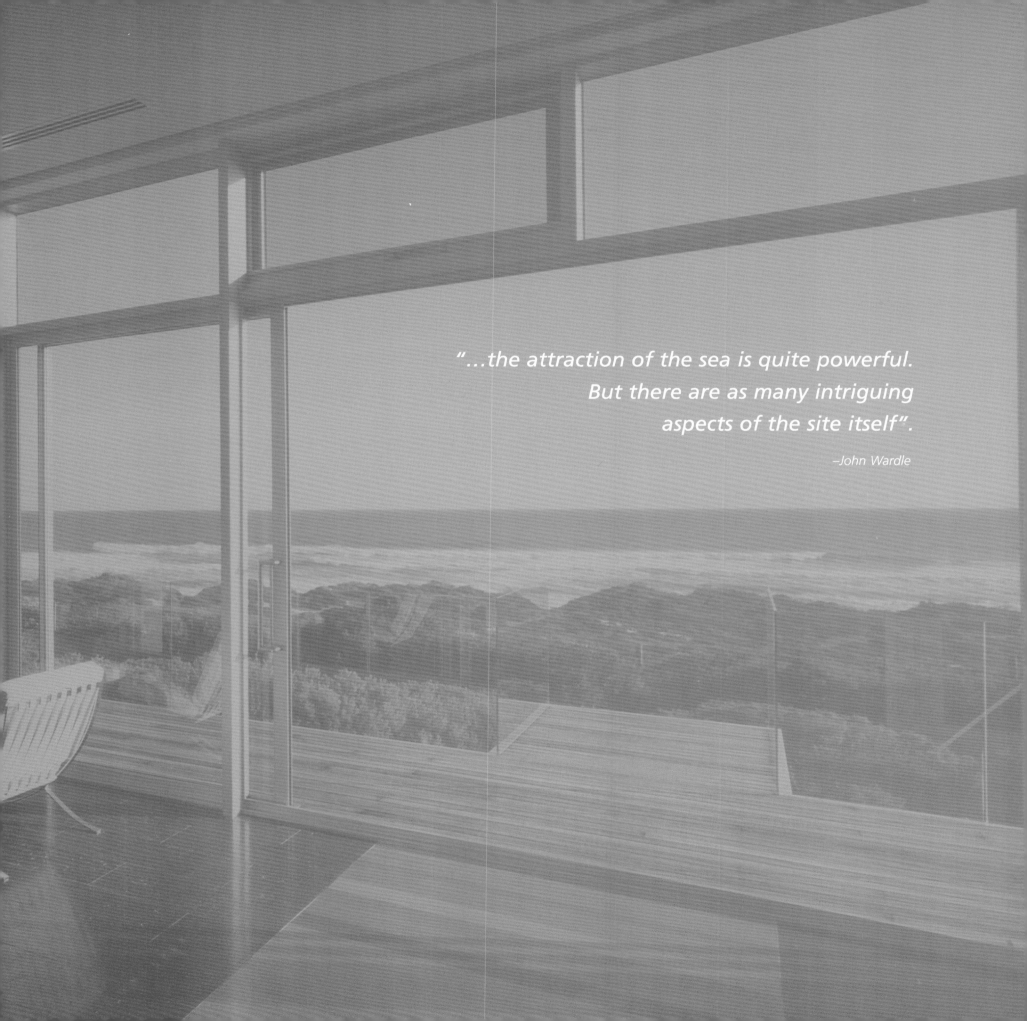

"...the attraction of the sea is quite powerful.
But there are as many intriguing
aspects of the site itself".

–John Wardle

A Leisurely Boat Ride

H U L E N A   A R C H I T E C T S   L T D

"It's only one hundred metres down to the beach. The moorings are at the bottom of the cliff. It's an ideal holiday spot for those wanting to spend a day out".

The owners of this beach house were looking for a place where they could moor their motorboat, somewhere close to where they live during the week. They found their idyllic site, three-quarter of an hour's drive from town. While the existing house wasn't suitable for their needs, the 1500-square-metre site, on top of a cliff, more than measured up. "It's only one hundred metres down to the beach. The moorings are at the bottom of the cliff. It's an ideal holiday spot for those wanting to spend a day out," says architect Brent Hulena of Hulena Architects.

While the architects were briefed to capture the 180-degree views over the bay, they were also asked to create a house that would suit the owner's entertaining needs. The architects' response was to create a glazed L-shaped pavilion around a courtyard. On one side of the plan, facing the water, are the kitchen, living and dining areas, together with the main bedroom and ensuite. Three bedrooms and a bathroom form the shorter of the two arms to the 'L'. "We had to create a protected area. At this elevation, it can get quite windy," says Hulena. "The courtyard also picks up on the afternoon sunlight. It can also get quite hot up here," he adds.

To ensure that alfresco dining was possible for most of the year, Hulena included an outdoor 'room' in the courtyard. A steel canopy with an aluminium-louvred roof protects guests even in inclement weather. "The louvres are operated automatically," says Hulena, who also designed a freestanding outdoor fireplace to warm diners during the winter months.

Even though the house is relatively modest in size, it feels considerably larger due to the 4.5 metre ceiling heights in the main living areas. And as the house features extensive floor-to-ceiling glazing, the indoor spaces appear to fuse with the courtyard and surrounding landscape. "It was also important to create cross ventilation, particularly using so much glass," says Hulena, who incorporated highlight glass-louvred windows and doors that can be pulled right back.

Photography by Simon Devitt

A Model Beach House

J A C K S O N   C L E M E N T S   B U R R O W S   A R C H I T E C T S

"Our design was a response to the setting. It was about the rolling hills as well as the beautiful Moonah trees. Some of them are 400 years old".

Designed by Jackson Clements Burrows Architects, these four detached homes provide a model for future developments on a golf course, located on the coast. Commissioned by a developer, these homes will, it is hoped, be followed by others of a similar standard. "Our design was a response to the setting. It was about the rolling hills as well as the beautiful Moonah trees. Some of them are 400 years old," says architect Graham Burrows, a director of the practice.

Rather than create four identical homes and place them in a straight line, the houses are carefully manipulated around a curved road. Each house is essentially dissected into two double-storey cubes, linked by a central staircase. But each pair of 'cubes' is slightly cranked to afford the finest views over the golfing greens. "The idea was to create a village of buildings, rather than four individual homes with little bearing on each other," says Burrows, who clad each house with vertical cedar cladding stained silver grey. "It's an appropriate colour for a wooded environment," he adds.

Each cube is approximately eight by eight metres in diameter, making the total floor area of each house about 250 square metres. On the ground floor are the garage, laundry and store for the buggy, together with a powder room. Crossing the link that contains a staircase are the dining, living and kitchen areas, featuring a dramatic void. And on the first floor, above the first cube are two children's bedrooms and a bathroom. The main bedroom, ensuite and deck occupy the top level of the second cube and overlook the living areas. "We wanted the owners to be able to enjoy the views over the green wherever they happen to be in the house. And from the living room, with double-height windows, your eye is drawn upwards to the sky," says Burrows.

From a distance, these homes appear identical. And inside, as one would expect from a development driven project, the finishes are repeated. But as each house has been carefully aligned to catch a special view, there's a sense of each being specific for their client, each requesting the best seat in the house, or in this case, on the green.

Photography by Shannon McGrath

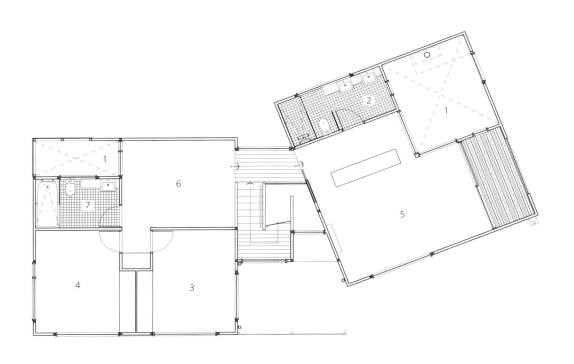

1 Void
2 Ensuite
3 Bedroom 2
4 Bedroom 3
5 Master bedroom
6 Playroom
7 Bathroom
8 Kitchen
9 Living
10 Dining
11 Buggy store
12 Double garage
13 Laundry
14 Powder room
15 Entry

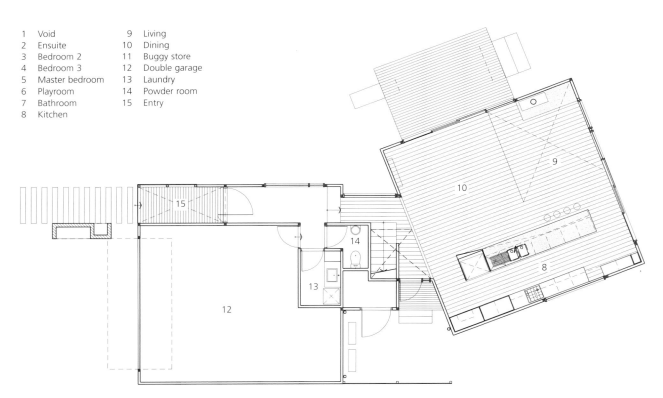

A New Community

CARR DESIGN GROUP

"The bedrooms, particularly at ground level, are quite compact. Holidays are not meant to be spent in your room".

This development of 14 townhouses surrounds an original Edwardian-style home. For the locals, as well as the developer, the idea of raising the old homestead was never an issue. "It's a landmark in the area," says Matt Wright, project architect for the Carr Design Group. "We tried to separate the old house from the new buildings as much as we could," says Wright, who ensured dense planting around the homestead, once the land had been subdivided.

While the new townhouses were commissioned on a limited budget, the developer was keen to ensure a high level of quality. "Our client and his family have been part of this area for generations," says architect Sue Carr, director of the practice.

The townhouses, three of which have views over the ocean, have been carefully sited to take into account the prevailing winds and orientation. Each house features two bedrooms on the ground level, together with bathroom facilities. There is also undercover parking for two cars. "The bedrooms, particularly at ground level, are quite compact. Holidays are not meant to be spent in your room," says Carr.

To maximise the views and light, Carr located the main bedroom, kitchen and living areas on the first floor. A generous timber deck adjacent to the dining and living areas increases the sense of space. The laminate and terrazzo kitchen is pared back. Its box-like form, featuring concealed appliances, is enlivened by a 'pop-up' ceiling that features celestial windows. "It's light wherever you stand in the house," says Carr, who also included a translucent glass wall in between the bathroom and kitchen to increase the light.

The townhouses feature a rendered base and are clad in marine rough sawn plywood panels for the upper levels. Each townhouse includes moveable timber screen doors to diffuse the light and increase privacy. As Carr says, "The screens are placed on different windows throughout the development. Each townhouse responds to the site in its own way".

Photography by John Gollings

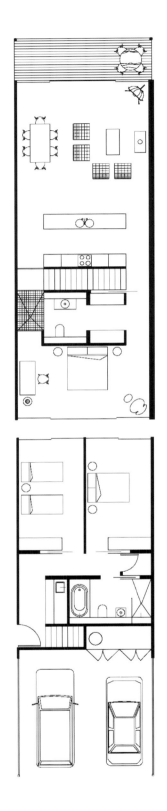

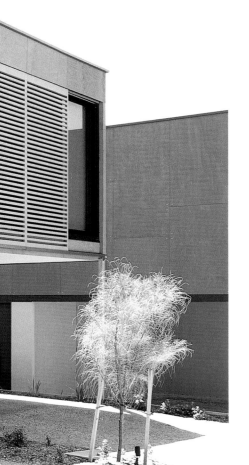

A Refined Fibro Shack

ARKHEFIELD PTY LTD

"This isn't an old fibro shack. But like the simple shack, it's not precious. It's here to be used".

Part of a large subdivision, this house is reminiscent of a simple fibro shack. Designed by Arkhefield and developed by the company Stockland, the house reflects the casual and relaxed ambience of living in the tropics. "We took photos of old shacks on the coast. Many of them are poorly built. But there's a certain romanticism attached to these houses," says architect Shaun Lockyer.

Located just 50 metres from the beach and surrounded by native bush, the house exudes a sense of informality. The architects expressed this informality with lightweight materials; plywood timber that has been oiled and stained, fibrocement weatherboard and cement sheeting. Designed over two levels, the house features deep eaves and protected walkways. "The summers are ferocious here. The awnings and pergolas allow you to enjoy the breezes without getting burnt by the sun," says Lockyer.

The house, situated on a relatively compact site (approximately 450 square metres), is broken into two zones, the kitchen and living areas on ground level, with the sleeping areas above.

An open-tread staircase leading from the entrance creates a sense of transparency between the two floors, as does a void above the dining area. One of the important features of the design is glass-louvred windows and highlight windows in the living areas. "Ventilation is crucial in this climate. We wanted to make the most of the ocean breezes," says Lockyer.

The house, approximately 380 square metres in area (including the patio areas), comprises an open plan kitchen living and dining area on the ground floor. There is also a study and theatrette. And on the first floor are four bedrooms and bathroom facilities. An important part of the design is the lap pool (8 x 4 metres) and covered patio. "In this climate, the patio is used more than the dining room. It acts as another room for the best part of the year," says Lockyer.

But even if the owners choose to stay indoors, they are continually reminded of their unique surrounds. An established Pandanus tree, indigenous to the area, creates an important focus from the house. As Lockyer says, "This isn't an old fibro shack. But like the simple shack, it's not precious. It's here to be used".

Photography by William Long

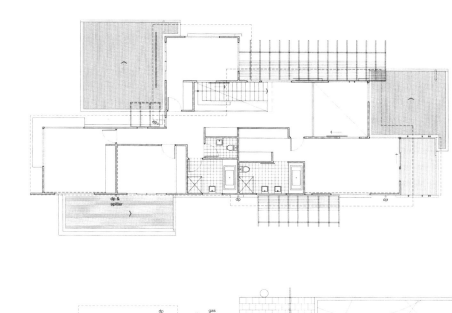

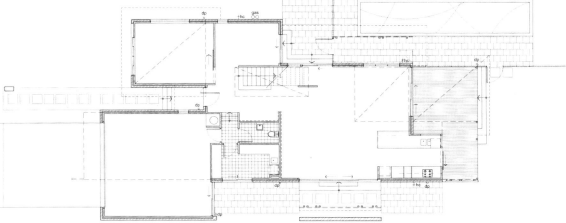

A **Sense** of Permanency

A R C H I T E C T   R O B E R T   A N D A R Y

In contrast to many lightweight fibro beach shacks, this house offers a sense of permanency. Constructed of concrete and glass, the two-storey rectangular design hovers above the dunes and gnarled coastal scrub. "My client wanted a showpiece. He wanted me to embrace the site as well as the ocean views," says architect Robert Andary.

The house has two sides and frames two different views. One façade features concrete and rectangular glass windows, in firecrest-

"My client wanted a showpiece. He wanted me to embrace the site as well as the ocean views".

orange and clear glass. The alternate façade, facing the sea, has floor-to-ceiling windows/doors leading on to an expansive deck. "The house doesn't really have a back or a front. There is a road on both sides that leads to the front door," says Andary, who embraced the site with extruded forms (the upper-level extends 30 metres).

The ground level, which acts as a plinth for the upper-level, includes a garage, two bedrooms, a games room and a bathroom. There is also a covered courtyard, with an elliptical shaped

glass roof. On the upper-level are the kitchen, dining and living areas. And at either end of this space, are two bedrooms, both with their own bathrooms. For privacy the two bedrooms can be completely screened by large sliding doors. Alternatively, these bedrooms can be opened up to the main living areas, becoming an extension of the open-plan space.

While Andary describes his work as the interaction of two simple rectangular forms, it is, in fact, a highly considered and masterful design. The swimming pool for example, cantilevers over the dunes, appearing as a black box on a façade. Andary included a glass wall in the concrete façade and aligned the pool to embrace the ocean beyond the living room. "When you're in the pool, you feel as though you are swimming out to sea," says Andary, who also integrated an elliptical shaped glass skylight in the courtyard into the living room floor. A timber walkway slices through the house to meet the deck on the other side of the house.

As Andary says modestly, "My client wanted the best house down the coast". Not many would argue with him.

Photography by Tony Miller

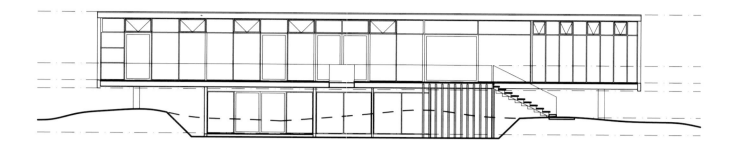

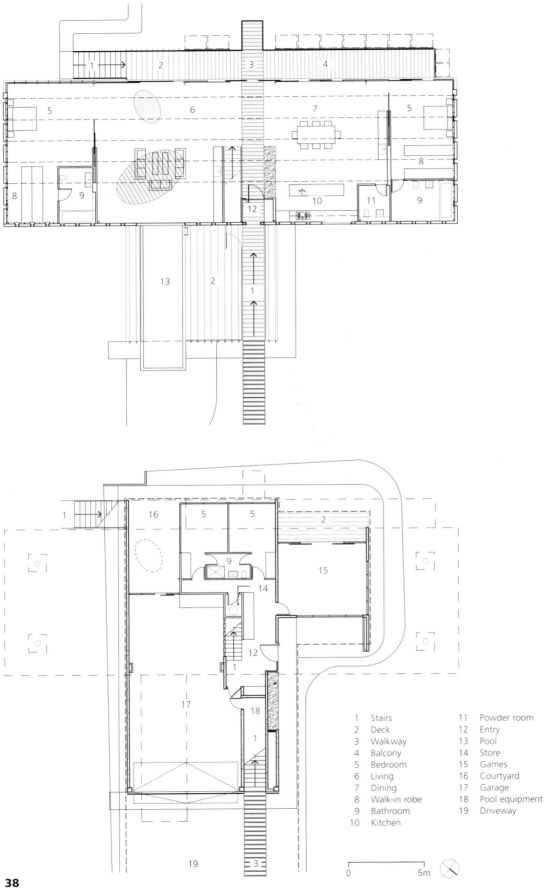

1   Stairs
2   Deck
3   Walkway
4   Balcony
5   Bedroom
6   Living
7   Dining
8   Walk-in robe
9   Bathroom
10 Kitchen
11 Powder room
12 Entry
13 Pool
14 Store
15 Games
16 Courtyard
17 Garage
18 Pool equipment
19 Driveway

0   5m

A **Simple** Shipping Container
A R C H I T E C T U R E   W O R K S H O P

"My clients were keen to experience the elements. It's something they enjoyed from living in the container".

The owners of this beach house lived in a shipping container on the site. After years of spending weekends in this rudimentary steel structure, they were keen to make living in the area a full time proposition. "They loved the site. But they knew the container wasn't realistic as a permanent home," says architect Christopher Kelly, director of Architecture Workshop.

Located above the sand dunes, with views to the Tasman Sea, the area is gradually being developed as others experience the beauty of the region. "Originally it was quite isolated. But neighbours are appearing," says Kelly, whose brief from the couple was for a simple form, not dissimilar in feel to the steel container they were leaving behind.

As the house sits above the dunes, it was imperative to anchor the house to the terrain. Kelly buried four-metre-long timber posts into the sand, and created two concrete block work walls. The plinth-like walls of the lower level support the lighter first floor made of plywood and steel. Kelly describes his design as two plywood pods that rest on a viewing platform.

Adding to the fragility of the design is the aluminium roof, hovering above the plywood and rubber roofs that protect each pod. "The aluminium roof acts as a veranda. It also adds another layer of protection," says Kelly.

The house, approximately 100 square metres in size is cleverly conceived. The first floor contains the living, dining and kitchen area. And connected to this open plan area are two decks, one leading from the living area, the other from the kitchen. "My clients were keen to experience the elements. It's something they enjoyed from living in the container," says Kelly, who extended the kitchen bench (including a barbeque) on to the deck to allow for alfresco dining. Architecture Workshop also used the deck to distance the main bedroom and study area from the main living areas. "It's about enjoying the elements and feeling part of this landscape," says Kelly.

On the lower level are a second bedroom and bathroom, together with a small studio that can be used by Kelly's clients as a second living area, or alternatively for guests. And unlike the fanfare of many coastal homes, there's no grand path leading to the front door. A simple drawbridge to the house hovers above the dunes.

Photography by Patrick Bingham Hall & Architecture Workshop

A Union with Nature

ARTICHOKE DESIGN STUDIOS

"You can go down to the beach. But you can more than enjoy it all from up here".

The owners of this beach house, on the central coast of New South Wales, explored the coastline from Victoria to Queensland. "They were looking for a site with superb ocean views, with access to a beach. They were keen to avoid suburbia at all costs," says architect Robert Pullar, director of Artichoke Design Studios.

While Pullar's clients found their ideal site, they were faced with numerous constraints, including a 30-degree slope. "It's quite a difficult site. There were also additional restrictions imposed as it bordered a national park," says Pullar. A large sandstone boulder in the centre of the triangular shaped block compounded constraints on the eventual design. "It wasn't surprising that the site was on the market for some time," Pullar adds.

The house was constructed of reinforced concrete slabs, steel poles and glass and is clad with compressed fibre cement sheeting. "The choice of materials was largely guided by the proximity to the national park. The materials had to be fire resistant. There also had to be provision for water tanks and an elaborate sprinkler system," says Pullar.

The house was designed for a city couple wanting a weekend retreat for themselves as well as a place to entertain friends and family. "In order to determine the best possible orientation and floor levels to maximise the views, I set up a four-metre trestle ladder at various positions on site," Pullar says.

As a result, views can be enjoyed from every level within the multi-level home, whether at garage level, or from the main bedroom and study located on the highest level. There are even spectacular views of the ocean from a fully glazed staircase that leads to an open plan kitchen, living and dining area on the second level. However, one of the best aspects in the house is from the ensuite through to the main bedroom, which offers ocean views, as well as views of the coastal gums.

Outdoor decks also play a major role in the design of this house, due to the incline of the site. As Pullar says, "You can go down to the beach. But you can more than enjoy it all from up here".

Photography by Alan Chawner

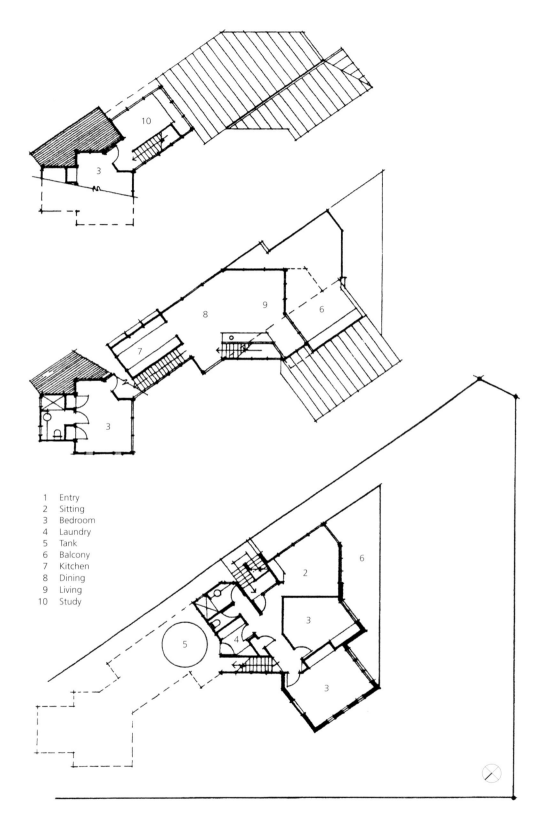

1 Entry
2 Sitting
3 Bedroom
4 Laundry
5 Tank
6 Balcony
7 Kitchen
8 Dining
9 Living
10 Study

Angled for the View

UTZ-SANBY ARCHITECTS

"Holidays are about both having enough space and being able to escape to your own space".

Designed by UTZ-Sanby Architects, this beach house overlooks the beach and headlands in the distance. To navigate the site, which is relatively steep (a four-metre drop to the street), the architects created a meandering boardwalk to the front door. The entrance is tucked away to the side of the house, past two rises of sandstone slabs.

The owners are a couple with two children, so the brief was for a relaxed-style beach house, something that was low maintenance and could be simply locked up at the end of each weekend. The 290-square-metre, two-storey house, features a plinth made of split concrete blocks with upper levels composed of grooved plywood vertical timber cladding and glass. "It's quite an exposed site. We wanted to give a sense of it being anchored to the ground," says architect Mark Young of UTZ-Sanby Architects.

The house is conceived in two pavilions, separated by a central courtyard. The front pavilion comprises the kitchen, dining and living areas, leading on to a large timber deck. To maximise the view of the water, the architects angled the living pavilion by approximately 20 degrees away from the side boundary. And to ensure the view wasn't impeded, the balustrade features fine stainless steel wires set in a steel frame.

On either side of the living areas are full-length glass doors, leading to either the deck or an enclosed courtyard. "The prevailing winds can be quite strong. It was important to create an additional outdoor area, that was more protected than the front patio," says Young. The same winds can also reduce the air temperatures within the home in minutes, when the doors on either side of the living areas are left open.

In contrast to the living areas, the bedrooms are orthogonal to the site. There are three bedrooms on the ground floor and on the first floor, are the main and guest bedroom. The bedrooms and living areas are separated by a passage that includes a bathroom and laundry. "It was important to separate the sleeping from the living areas. It means you can still entertain guests well into the night without the children being disturbed," says Young, who included a bedroom that can also double as a play area for the children. "Holidays are about both having enough space and being able to escape to your own space".

Photography by Ray Clarke

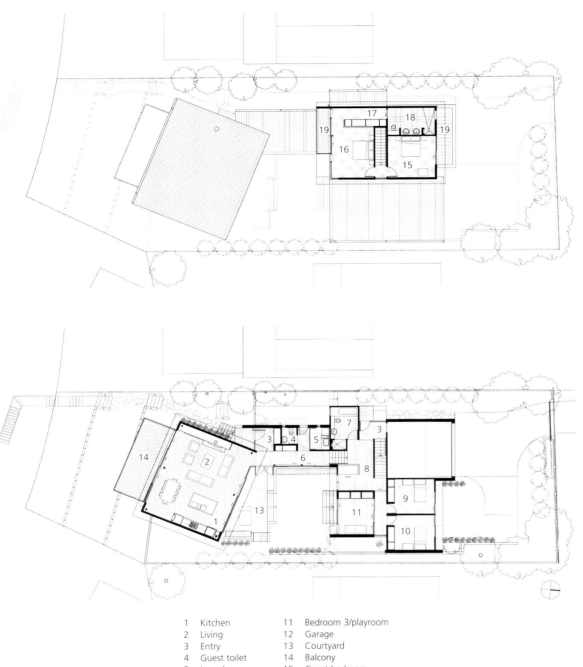

1	Kitchen	11	Bedroom 3/playroom
2	Living	12	Garage
3	Entry	13	Courtyard
4	Guest toilet	14	Balcony
5	Laundry	15	Guest bedroom
6	Hall	16	Master bedroom
7	Bathroom	17	Wardrobe
8	Gallery	18	Ensuite bathroom
9	Bedroom 1	19	Balcony
10	Bedroom 2		

Around a Morton Bay Ash

BARK DESIGN

"A lot of time is spent outdoors or on the deck. You're not really conscious of being inside or out".

The focus of this beach house is on a Morton Bay Ash rather than the sea. And while the beach is only 250 metres away, the sound of the surf is the only reminder of its coastal location. "The Morton Bay was in the middle of the site. It has been there for more than 100 years. We thought it was important to integrate it in our design," says architect Stephen Guthrie of Bark Design, who designed the house with architect Lindy Atkin. "We wanted to create an oasis around the tree rather than try and capture the views," he adds.

Bark Design were given a fairly open brief from their clients, a couple with two small children. They wanted a house that was casual and relaxed and one that was economical. "They didn't want something that was precious, particularly with small children," says Guthrie. "They also wanted two separate zones, one for themselves, the other for their children," he adds. As a result, the two zones are separated by the courtyard, containing the prized tree, an endemic species in the area.

One pavilion comprises the garage and laundry at ground level, with the two children's bedrooms, bathroom and their own deck on the first level. The second pavilion includes the living, dining and kitchen areas on the ground level that lead to a generous deck. And above, are the main bedroom, ensuite and studio.

"The studio can be used as a second living area. Alternatively, it can be used as a guest bedroom," says Guthrie.

The living area features a striking staircase, enclosed by a polycarbonate and timber wall. And at the top of the stairs, is a built-in seat, ideal for reading. The polycarbonate wall framing the staircase allows light to filter into the living spaces from the courtyard. "The Morton Bay casts shadows onto the polycarbonate as well as the living room walls. It's a way of bringing the landscape inside," says Guthrie.

Connecting the two pavilions on the second level is a bridge, also made of polycarbonate sheeting and timber battens, creating privacy but also allowing the natural light to penetrate. While the materials in the house are quite simple, timber framing and fibro cement cladding, the design is more complex, cleverly fusing the indoor spaces with the outdoors. Even the laundry, which opens to the courtyard, feels as though it's in both camps. As Guthrie says, "A lot of time is spent outdoors or on the deck. You're not really conscious of being inside or out".

Photography by Aperture Photography

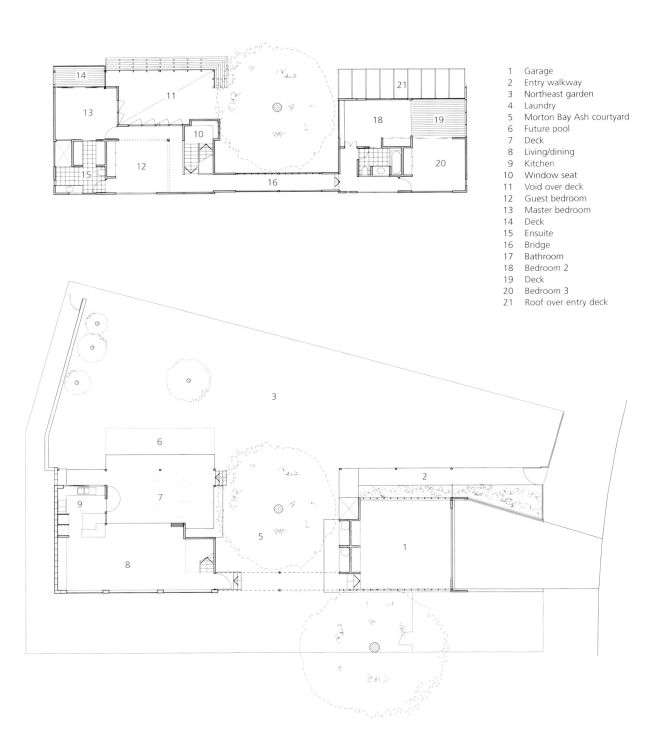

1 Garage
2 Entry walkway
3 Northeast garden
4 Laundry
5 Morton Bay Ash courtyard
6 Future pool
7 Deck
8 Living/dining
9 Kitchen
10 Window seat
11 Void over deck
12 Guest bedroom
13 Master bedroom
14 Deck
15 Ensuite
16 Bridge
17 Bathroom
18 Bedroom 2
19 Deck
20 Bedroom 3
21 Roof over entry deck

At the **Water's** Edge

C R A I G   C R A I G   M O L L E R   A R C H I T E C T S

"The house is understated. It's about the surrounds and the pleasure of interacting with them".

This weekender is located at Paihia, Bay of Islands, in New Zealand. The house sits on the edge of a secluded bay, a short walk from the local shops. "The owners have teenage children. They liked the secluded position, but they didn't want to have to be constantly driving their children places," says architect Gordon Moller of Craig Craig Moller Architects.

Originally natural bushland, the local area was subdivided into 15 allotments. While this site is closest to the water's edge, it is also one of the steepest sites, with a gradient of approximately 15 metres. This slope, along with the local weather conditions, informed the design, a three-level timber home. "Our clients wanted something that could also accommodate guests, with their children regularly having friends to stay," says Moller.

The entrance and garage to the beach house is on the top level. This level also includes three bedrooms and a shared bathroom. A staircase leads to the middle level, the most impressive of the three. This area features a winter garden that is flanked by the kitchen, dining and living areas on one side, together with an expansive timber deck. On the other side of the winter garden or atrium, are two bedrooms, one being the main, and the other for guests. And on the lowest level is a second living area that doubles as a bunkroom, should additional friends stay over. The lower level is linked to the main living areas via a small set of stairs. "They're similar to a ship's ladder," says Moller.

The arrangement of the timber decks was important in the design. "When it's calm down here, the doors can be pulled back and the breeze gently comes through the house. But it can become quite bleak, particularly during the winter months," says Moller, who designed an outdoor fireplace on one deck for the colder days.

Clad in cedar board and batten, the house is understated. Some floors feature polished concrete, while others have polished timber. The interior wall finishes are also modest, many of which feature whitewashed plywood. As Moller says, "The house is understated. It's about the surrounds and the pleasure of interacting with them".

Photography by Simon Devitt

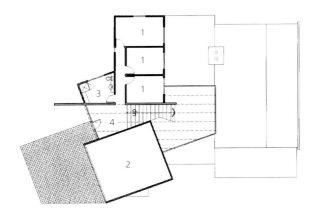

1   Bedroom
2   Garage
3   Bathroom
4   Entry
5   Veranda
6   Store
7   Living 1
8   Living 2
9   Dining
10 Kitchen
11 Winter garden
12 Robe

Away from the Traffic

A L E X   P O P O V   A S S O C I A T E S

"Whether you're inside or on the balcony, you aren't aware of the traffic. It feels as far away as the city".

This impressive beach house, at Palm Beach, sits on a narrow site, facing one of New South Wales' premier holiday resorts. While substantial new homes dominate this part of the coastline, there are still a few reminders of the areas past. This house, designed by architect Alex Popov, is bordered by timber painted cottages. "We didn't want to come along and simply build a large brick house than was quite urban. The materials had to relate to these cottages," says Popov.

Popov used a limited palette of materials for the house, including painted timber in off white, glass, sandstone and zinc. "The sandstone (used as blade walls to create privacy on the ground floor) is evocative of the sandstone used in the neighbouring homes," says Popov, who wanted to create a clean and modernist form for this beach house.

As the site is relatively steep, approximately a 30-degree slope, Popov located the main living areas on the first floor. A set of stairs leads directly to an extremely large open-plan kitchen, living and dining area, extending almost 15 metres in length. Also included on this level are an outdoor courtyard and a generous balcony leading directly from the living areas.

As the site abuts a busy road, Popov was keen to turn the main aspect of the house away from the traffic. "It's one of the most popular escapes from Sydney. I didn't want my clients to be distracted by a continual band of traffic," says Popov, who included in the design a deep-framed ledge (300 mm) surrounding the balcony. "When you sit down, you simply see the Pacific Ocean," he adds.

On the ground floor are two children's bedrooms, a rumpus room, a kitchenette and bathroom facilities. The two sandstone blade walls frame the children's bedrooms. On the third level is the main bedroom, together with ensuite facilities.

One of the most dramatic features of this house is the massive zinc roof, cantilevering three and a half metres. To increase sunlight over the balcony, Popov inserted two large porthole-shaped apertures in the roof. As Popov says, "Whether you're inside or on the balcony, you aren't aware of the traffic. It feels as far away as the city".

Photography by Kraig Carlstrom

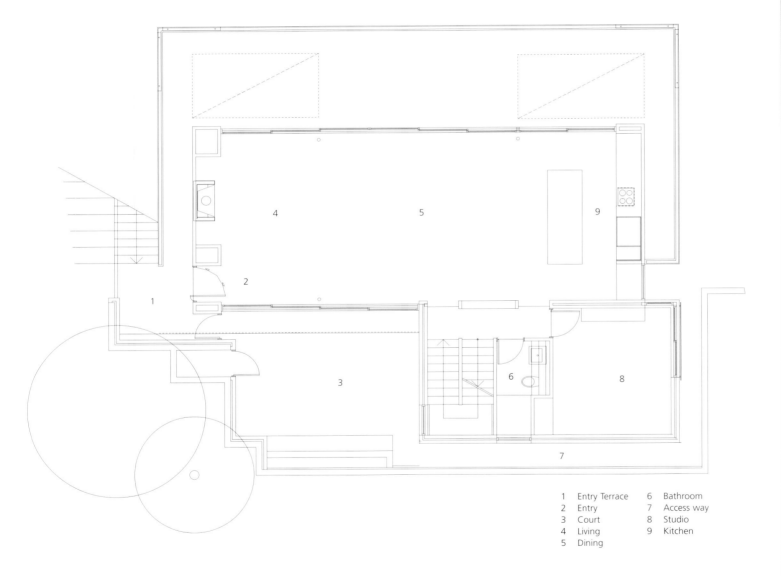

1 Entry Terrace 6 Bathroom
2 Entry   7 Access way
3 Court   8 Studio
4 Living   9 Kitchen
5 Dining

Coastal Hideaway

STEVEN LAST, ARCHITECT

This beach house, designed and owned by architect Steven Last, is located on an island. Overlooking sand dunes, there's the sound of surf in the distance. "I wanted a house that offered an escape from the city, something that was connected to the elements," says Last.

Last originally purchased a much larger site on the island, but subdivided the block and kept a third to build his own house. Approximately 290 square metres in area, the block is relatively compact. "I wanted a low maintenance house. When we're not inside, we're at the beach," says Last.

The two-storey house is made of vertical sawn yellow stringy bark. The only other materials are Ecoply, a woven tea-tree used for the balustrade on the first floor balcony and an aluminium staircase leading to the roof deck. "I love the natural grain of the timber. Eventually it will turn silver under the weather," says Last.

On the ground floor are a bathroom, two bedrooms or alternatively one bedroom and a multi-functional room. The first floor comprises the kitchen, living and dining areas, together with the main bedroom and bathroom, tucked away behind the kitchen. "I wanted the bedroom to feel as though it was a large motel room. It's quite decadent," says Last referring to the size of the space and the open bath arrangement. While there are views across the bird sanctuary from the balcony, the sea can only been seen from the rooftop, which features a protruding 'gangplank'.

The interior, unlike the exterior, is smooth and refined. Red and green feature walls animate the spaces. "The red is evocative of the sunsets down here. The greens reflect the coastal grasses," says Last, who has filled the house with 1970s classic furniture. "I grew up with this type of furniture as a child. It reminds me of the holidays I had with my family down the coast," he adds.

Last compares his design to a Chinese tea chest, where the exterior is rough and the interior filled with sumptuous linings. As he says, "It's about feeling safe down here. The winds can reach up to 110 kilometres an hour".

"I wanted the bedroom to feel as though it was a large motel room. It's quite decadent".

Photography by Steven Last and Richard Butler

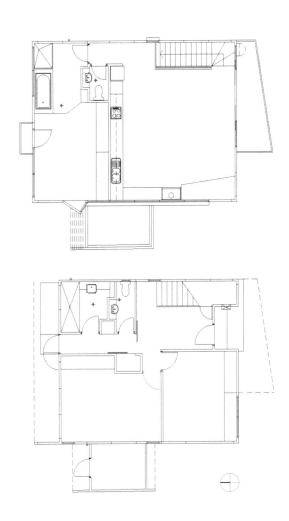

Designed for Future Growth

A L F A N O   A R C H I T E C T S

"The design responded to my client's brief. They wanted something simple, minimal and honest, and capable of expanding further down the track".

This unusual beach house is already relatively spacious (approximately 300 square metres). However, the brief to architect Jose Alfano, included possible expansion further down the track. Located on 1.5 hectares, there is certainly sufficient space to fulfil that brief. "The other part of the brief was to maximise the independence of the formal living area and a multi-purpose entertainment area. My clients wanted to be able to have separate gatherings for both parents and teenage children," says Alfano.

As the site is located on a relatively high point of a north-facing hill, it provided opportunities for both long and near vistas. Alfano was also conscious of protecting the native vegetation and using materials that reflected the colours of the landscape. The use of low maintenance materials was also an important consideration in the design. As a result, the beach house is clad in Rhinezink (vertical surfaces) and Colorbond (horizontal eaves). "A 'touch lightly' approach was taken to the structure, ruling out anything other than stumps and a timber frame-light weight construction," says Alfano. But with strong prevailing winds, the design required significant amounts of bracing, with the additional support of a metal cladding system and concealed fixings.

To appreciate the generosity of the coastal site, the house is located at the centre of the block, with generous glazing allowing uninterrupted views to the boundaries. The L-shaped plan of the house has the main bedroom and ensuite at one end and the children's bedrooms and bathrooms at the other end. And, as requested, there's a separate TV/rumpus room in the children's wing and a separate lounge and dining area at the other end of the house, located between the kitchen and study.

As important in the spatial arrangement, were environmental considerations. Cross-ventilation was paramount to the design. "We didn't want to rely on air-conditioning. The coastal breezes cool down the house," says Alfano, who also included generous eaves in the design. As Alfano says, "The design responded to my client's brief. They wanted something simple, minimal and honest, and capable of expanding further down the track".

Photography by Jose Alfano

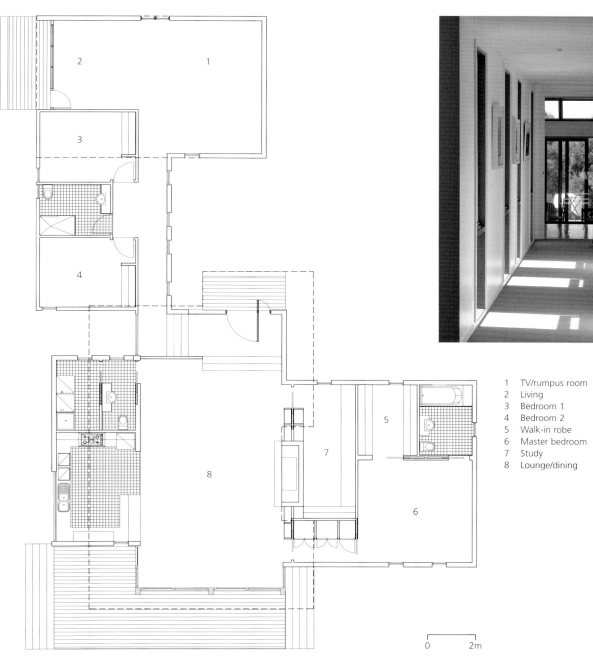

1 TV/rumpus room
2 Living
3 Bedroom 1
4 Bedroom 2
5 Walk-in robe
6 Master bedroom
7 Study
8 Lounge/dining

0   2m

Elevating the Site

STUDIO 101 ARCHITECTS
IN ASSOCIATION WITH FRANK BIENEFELT

"It's a beach house. The idea isn't to spend all your time in your room".

This beach house, designed by Studio 101 Architects in association with architect Frank Bienefelt, isn't located on the perfect site. Surrounded by homes, on a relatively suburban block (approximately 750 square metres), the only promising outlook, is the distant sea views. "We had to deal with the same issues you'd come across in the city such as overshadowing and overlooking," says architect Peter Woolard of Studio 101 Architects.

The house is constructed of sand-coloured masonry walls at ground level that act as a plinth for the lighter materials, such as cypress timber and glass used on the first floor. A Colorbond steel roof, hovering above the entire structure 'twists' in two places to increase the amount of light entering the home. Irregular in shape, the design was conceived as a hierarchy of spaces, where the size of the living areas is greater than bedrooms. "It's a beach house. The idea isn't to spend all your time in your room," says Woolard, who was keen to include generous outdoor decks in the scheme.

To elevate the senses and increase both the views and light in the house, the architects located the main bedroom, kitchen and living areas on the upper level, with the children's bedrooms at ground level. But to ensure the children enjoyed a light-filled house, the architects created a large passage/gallery space directly outside their bedrooms. With pivotal doors to their bedrooms, the gallery, leading to an outdoor deck, doubles as the children's play area.

The entrance directs both the owners and visitors to the living areas on the first floor. Colours such as vibrant green and deep red lead the eye in an upward direction. "We didn't want people continually moving past the bedrooms on the ground level. The children also need their own space," says Woolard.

To ensure the outdoors is appreciated, the architects designed a generous deck almost extending the span of the house. They also created a three-tiered level for sun protection. Below the cantilevered steel roof is a timber awning. And the floor of the deck on the first floor acts as sun protection at ground level. "It's crucial to provide sun protection in beach houses. It's the time when people spend most of their time outdoors," says Woolard.

Photography by Trevor Mein

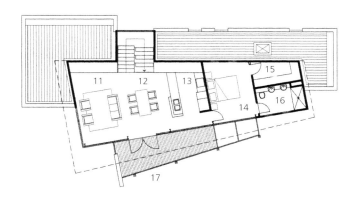

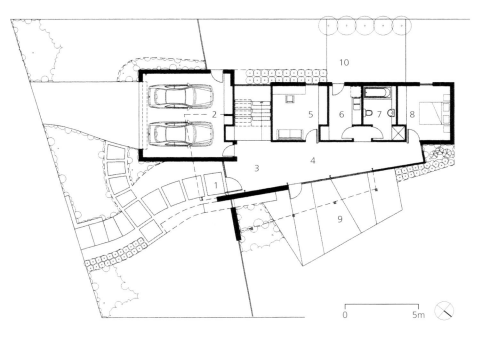

1	Entry	7	Bathroom	13	Kitchen
2	Garage	8	Bedroom	14	Master bedroom
3	Foyer	9	North terrace	15	Robe
4	Galleria	10	Service court	16	Ensuite
5	Study	11	Living	17	Viewing deck
6	Laundry	12	Dining		

0   5m

Embracing the Ocean

WILSON AND HILL ARCHITECTS

"We tried to keep the materials as simple as possible. There's already quite a lot to look at along the cliff".

This house, designed by Wilson and Hill Architects, is perched on a rocky outcrop. With panoramic views over the Pacific Ocean and the sound of waves as a backdrop, it is surprising that this site remained vacant for years. The triangular site, 120 metres above the beach couldn't entice anyone to build. "It's a difficult site. The slope has a gradient of at least 15 degrees," says architect David Hill, who took up the challenge to build his own beach house.

Using the drama of the cliff face as a starting point, the architects created a simple black 'box'. Constructed in grooved plywood and stained black, the house creates a striking gesture for those walking along the beach. Glass, steel and concrete block work were also used in the construction. "We tried to keep the materials as simple as possible. There's already quite a lot to look at along the cliff," says Hill, pointing to the native undergrowth below the house.

The house, approximately 240 square metres in area, comprises a split-level ground floor. On one side of the house are the kitchen, dining and living areas. And on the other side of the entrance are the main bedroom and ensuite, together with two additional bedrooms and bathroom facilities. To further define the open-plan living areas, Wilson and Hill Architects designed a 'sunken lounge'. "The slope of the site gave us that opportunity," says Hill, who was keen to manipulate the spaces as well as the views from within the house. As with the home's exterior, the materials used for the interior are restrained. The kitchen and living areas for example, feature polished concrete floors and concrete block walls. And to add a sense of warmth, the joinery is Southland Beech, a species local to New Zealand.

While there is a compact garden surrounding the house, its rocky terrain meant it has limited use. So to ensure that the outdoors was appreciated on a daily basis, Hill designed a sunroom/study on the first level that leads to a generous timber deck. A toughened glass balustrade wraps around the deck protecting the views from inside the house. A lap pool is also part of the cliff face. And with its cornered glass window, offers the same views to swimmers as those lounging on the deck. "It's quite an exposed site. Most were hesitant to take it further," says Hill, who is pleased that he did.

Photography by Stephen Goodenough

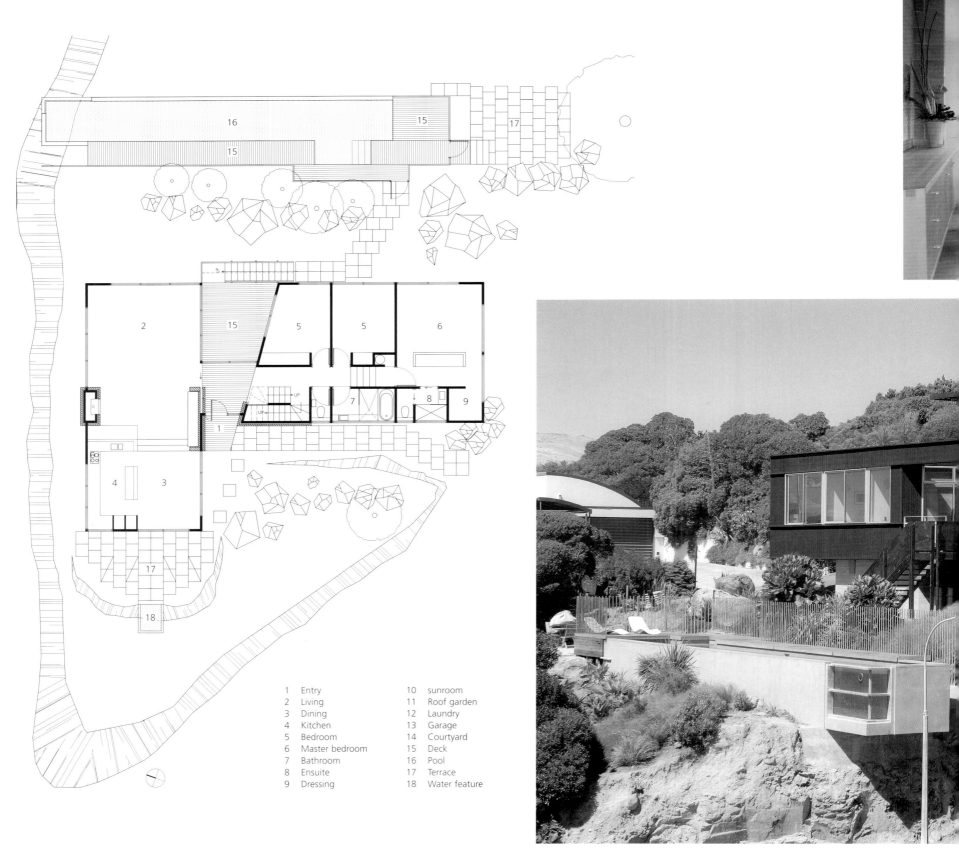

1	Entry	10	sunroom
2	Living	11	Roof garden
3	Dining	12	Laundry
4	Kitchen	13	Garage
5	Bedroom	14	Courtyard
6	Master bedroom	15	Deck
7	Bathroom	16	Pool
8	Ensuite	17	Terrace
9	Dressing	18	Water feature

Encompassing the Vista

G R O D S K I   A R C H I T E C T S

"At first the neighbours were taken aback by the copper. Initially, it was fairly bright. But the copper has now weathered. It's now the colour of the earth".

Originally there was a 1940's two-storey fibro house on this site, overlooking the bay. But while the owners enjoyed the vista, they were tiring of climbing stairs to reach the deck outside their bedroom. "Our clients realised the original home wasn't ideally designed," says architect Robert Grodski, who was commissioned to design a new house.

To embrace the view, Grodski Architects created a new house with the living areas upstairs and the children's bedrooms at ground level. "Apart from this requirement, the brief was fairly open," says Grodski, who was keen to take advantage of the 0.5-hectare site with a 20 degree slope to the beach.

The 250-square-metre house, nestled on a densely forested site, is constructed of rendered masonry and features two striking copper-clad arms that appear to spread their wings. "At first the neighbours were taken aback by the copper. Initially, it was fairly bright. But the copper has now weathered. It's now the colour of the earth," says Grodski. Internally, two spine walls fan out from the entrance, magnifying the view in the distance. And to ensure the house took a back seat to the view, the architects included a discrete rubber membrane on the flat roof.

Angular walls create a logical division of spaces within the house. One of the walls divides the main bedroom and ensuite from the living area. The other wall creates a division to the kitchen and dining area. And paramount to the design is the curved deck, framed with toughened glass balustrades. "It was one of the key components in the design," says Grosdki, who also used the copper 'arm' to create a privacy shield to the public path leading to the beach.

One of the other components to the brief, which was fairly open, was to create a separate zone for children. The ground level includes two bedrooms and a playroom, together with a laundry. "The children have their own space and can access the garden without having to go through the entire house. The pool and spa are only a few steps away," says Grodski, who retained the spa bath from the original home. "It's the only thing that remains, except for these views".

Photography by Grodski Architects

bedroom 1

ensuite

robe

deck

living

powder

dining

kitchen

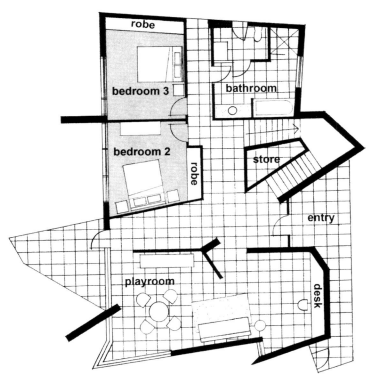

robe

bedroom 3

bathroom

bedroom 2

robe

store

entry

playroom

desk

Exploiting the view

PHORM ARCHITECTURE + DESIGN

"We've only used one third of the site. And the owners are keen to spend more time here…
with a house like this, why wouldn't they!"

This beach house, designed by Phorm Architecture + Design, occupies the highest vantage point of the site. "We wanted to go across the site to maximise the sea views," says architect Paul Hotston.

While the house appears substantial in size, it is relatively modest, approximately 100 square metres, not including the garage and decks. "The budget was fairly tight. But we wanted the house to feel generous," says Hotston, who used the home's central two-storey void to magnify the spaces.

The house is designed in four different lightweight materials. The main bedroom and ensuite is wrapped in fibro cement sheeting. The living areas on the ground level are framed in corrugated steel, which is also used for the roof canopy that appears to float above the house. The plinth of the house is made from Ecoply. And the wall framing the staircase is made of expressed timber hardwood and twin-walled polycarbonate sheeting. Referred to as the lantern, the polycarbonate sheeting draws in the morning light. At night the stairwell has a Japanese ambience.

The spaces at ground level focus on the four by six metre deck and two-storey void at the centre of the house. The kitchen, dining and living areas lead off this deck area. And a guest bedroom and bathroom is located off the kitchen. "It's quite an informal arrangement. The house can be used in several ways, depending on whether the sliding doors are open or closed," says Hotston. "The living area often doubles as a bunk room for the children and their friends," he adds.

The main bedroom on the first level cuts into the void and has a Juliet-style balcony. This balcony was conceived to take in the views of the beach. "The view was the most emphasised part of the brief," says Hotston, who is now planning a second stage for the house. "We've only used one third of the site. And the owners are keen to spend more time here," says Hotston. And with a house like this, why wouldn't they!

Photography by David Sandersan

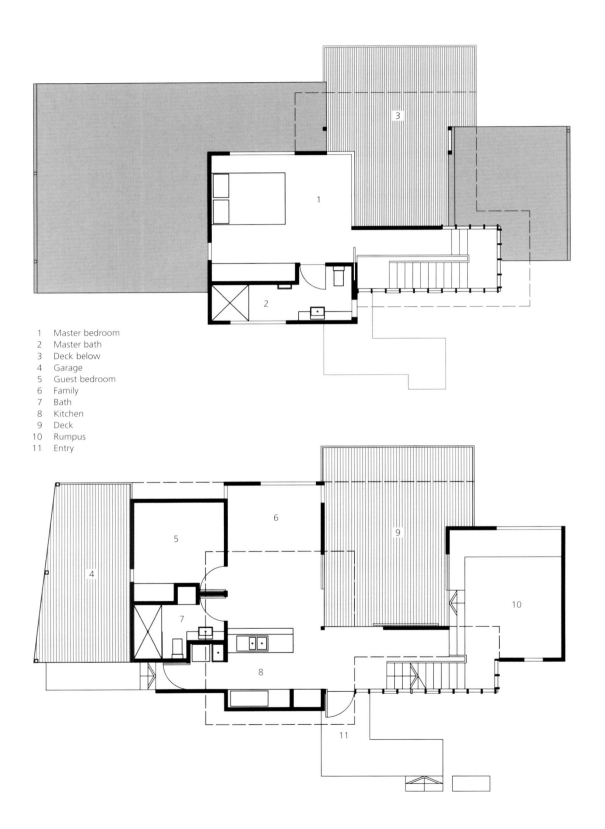

1 Master bedroom
2 Master bath
3 Deck below
4 Garage
5 Guest bedroom
6 Family
7 Bath
8 Kitchen
9 Deck
10 Rumpus
11 Entry

Eyeing the Coastline
PHORM ARCHITECTURE + DESIGN

"We wanted to create an anchoring point for the house. But we also wanted to show its lighter side".

This unusual house appears to hug the rugged coastline. Situated on Mt. Ossa, in Mackay, Queensland, the three-storey house is only a couple of kilometres from the Whitsunday coastline. "Our client grew up at Slade Point. We wanted to ensure we delivered a view of the point to the kitchen table," says architect Paul Hotston, a director of Phorm Architecture.

The three-storey house closely follows the east-west ridgeline of the site. "We wanted to maximise the views to the water. We also wanted to ensure the house was well-ventilated (relatively long and narrow to capture the sea breezes)," says Hotston. The house is constructed in a variety of materials. A rammed earth spine wall appears in the two lower levels, while the northern elevation features generous glazing. Corrugated zincalume was also used, along with panels of Ecoply. "We wanted to create an anchoring point for the house. But we also wanted to show its lighter side," says Hotston, pointing to the zincalume at either end of the house.

"The house was conceived as a string of intersecting volumes structured around a fractured monolithic spine," says Hotston, who also saw the design as a consequence of mapping and responding to the broader landscape issues. On the ground level are a guest bedroom, rumpus room and courtyard that lead to a swimming pool below. The second level comprises the main living areas, kitchen and dining, together with the main bedroom suite. And the third level consists of three bedrooms for the children.

While the house is located on a steep incline, the architects were keen to connect the site lines from within the house. From the kitchen and living areas, for example, the swimming pool appears to be located on the doorstep. A large entry void allows the owners to enjoy the site of water, even though a courtyard separates the water from the entrance. "We wanted to create a cannon effect, as though you were sitting directly above the pool," says Hotston, who was also keen to embrace the coastal surrounds.

The first home in a new estate, the site is elevated above and physically disconnected from surrounding developments. "We envisaged a new language for the hilltop. It's also about using local materials," says Hotston, pointing out the polished slabs at ground level. "They're Hughenden quartz and silica sands from Bundaberg".

Photography by Paul Hotston

1 Entry
2 Hall
3 Guest bedroom
4 Store
5 Rumpus
6 Courtyard
7 Pool

1 Void
2 Coast lookout
3 Kitchen
4 Living
5 Outdoor living
6 Master bedroom
7 Gallery
8 Dining
9 Laundry
10 Store
11 Garage
12 Terrace

0   4m

For **Keen** Surfers

W A L T E R   B A R D A   D E S I G N

"The afternoon nap is part of the ritual of taking holidays by the beach. The sound of the waves crashing is quite soothing".

This house was made for keen surfers. Directly opposite a surf beach and exposed to the ocean, this house sits at the base of a lush tropical forest. "The whole family surf. It's the reason why the house exists," says architect Water Barda.

Constructed of stone and stained timber, the house features a corrugated steel roof, not dissimilar to many of the original beach houses remaining in the area. The two-storey house was built for a family and visiting guests and is loosely divided into two zones. On the ground floor are two large 'bunk' rooms for the children and their friends. On the opposite side, separated by a large outdoor timber deck is a billiard room.

On the first level are the kitchen, living and dining areas, together with the main bedroom and ensuite. Barda also included a 'lanai' (a Hawaiian term for a covered room) on the first floor, to allow guests the opportunity for an afternoon nap. "The timber blinds can be pulled down if they choose to have a rest.

But they can also use the space as a retreat during the day for reading," says Barda, whose brief from the clients included creating a parent's or guest retreat. "The afternoon nap is part of the ritual of taking holidays by the beach. The sound of the waves crashing is quite soothing," he adds.

While the living areas on the first floor are open to each other, they have a sense of containment provided by vaulted ceilings in each space. "We treated the spaces as a series of pavilions. Each ceiling has been expressed differently," says Barda, pointing to the unique angles.

Also important in the design was to include protected outdoor areas, particularly as the ocean winds can be bitterly cold. Barda designed a large timber deck between the main bedroom and the living area, protected by a deep timber eave. There is also a deck at ground level that spills out onto the lawn. For the family that own this house, it's the prevailing surf that counts. As Barda says, "The weather doesn't usually dictate when surf boards are taken out. This beach house is used all year around".

Photography by Bart Maiorana

Framed by a National Park

LIPPMANN ASSOCIATES

"I wanted the view of the sea and the headlands to slowly unfold rather than be presented immediately past the front door".

A national park on one side and a sheltered beach on the other border Pearl Beach, lies on the New South Wales central coast. Unlike many rural townships that have developed at an unprecedented pace, Pearl Beach has no plans for expansion. "There are only a couple of hundred houses. The only change is the type of house now being built," says architect Ed Lippmann, who designed this new house, leading directly to the beach.

The original 1950's beach shack was replaced with a 300-square-metre home made of glass, steel and fibre cement. Designed over two levels, the architect created a journey from the home's front entrance to the generous glazed rear facades. "I wanted the view of the sea and the headlands to slowly unfold rather than be presented immediately past the front door," says Lippmann, who skewed the main corridor leading to the kitchen, living and dining areas, located to the rear of the house.

Designed for a couple that regularly has family and friends to stay, the house includes two guest bedrooms on the ground floor, together with bathroom facilities. On the first floor is the main bedroom, walk-in dressing area, ensuite and separate office. "Our clients regularly work from here. Eventually, their aim is to retire at Pearl Beach," says Lippmann, who was conscious of creating a comfortable home as well as a beach side retreat.

While the large open-plan-kitchen and living areas provide panoramic views over the water towards Mount Ettalong, it's the courtyard space that provides the focal point of the owner's activities. "I see this design as more of a courtyard house. Glass walls protect this courtyard that is adjacent to the living area. But it's completely open to the sky," says Lippmann. "It's protected from the winds," he adds. And to blur the lines between the lounge and courtyard, hoop pine joinery appears in both spaces.

Lippmann was also conscious of the need to ventilate the spaces within the home. While the large sliding glass doors leading to the rear deck are regularly open, so is the glass pivot front door (located behind a fixed screen). "Air is always moving through the houses, whether along the main corridor or through the highlight windows in the main living area," says Lippman. And although the house is on a fairly modest site (approximately 650 square metres), there's a sense of endless vistas, with the sand and water stretching out in the distance.

Photography by Willem Rethmeier

1   Garage
2   Entry
3   Bathroom
4   Ensuite
5   Master Bedroom
6   Bedroom
7   Study
8   Kitchen/Dining
9   Lounge
10 Outdoor Patio
11 Balcony
12 Deck

Getting to the Point

L E U S C H K E   K A H N   A R C H I T E C T S

"This is a family home with views to die for.
We've really tried to capitalise on that".

Intimate beach views and an expansive vista through a mature Pohutukawa tree create a superb backdrop for this 450-square-metre coastal cliff-top home. Situated on a hill on Compass Point in east Auckland, the timber and rendered house looks west to the city in the distance and over water at every turn. The site is bounded by a coastal reserve on its southern side, while the key feature of the northern aspect is a huge tree on a neighbouring property.

"The house essentially comprises three boxes, two single-storey ones on the ground floor and another lying across them on the top floor, forming a U-shape. There's a lot of overlapping and cantilevering going on," says architect Paul Leuschke, a director of Leuschke Kahn Architects. There is a lap pool across the front of the house with a sheltered courtyard on the northern side to enjoy the morning sun.

The fresh open interior spaces of this home include an open-plan family room and kitchen, formal dining room and sitting room, a gym and a media room for the family's three children. The kitchen runs along a north wall, behind which the neighbour's tree casts a shadow. The east and west walls are glazed to allow the sun to stream in.

Upstairs there are four bedrooms, with the main suite jutting over the terrace and lap pool, creating sensational aspects, while sheltering the sitting area below. Connecting the two levels is an open steel-treaded staircase, complementing the home's pared back and transparent style. As Leuschke says, "This is a family home with views to die for. We've really tried to capitalise on that," he adds.

Photography by Kallam MacLeod

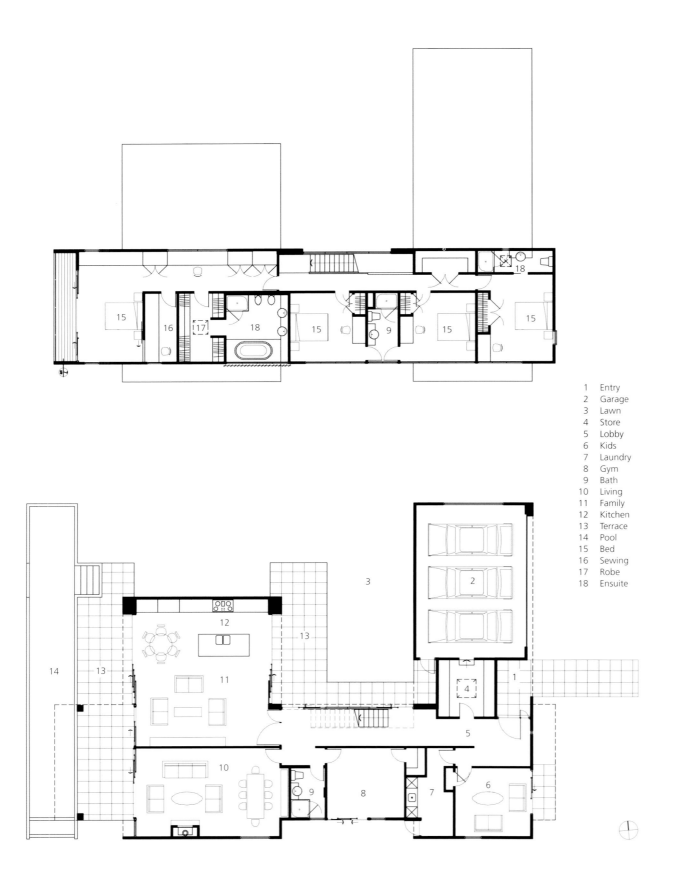

1   Entry
2   Garage
3   Lawn
4   Store
5   Lobby
6   Kids
7   Laundry
8   Gym
9   Bath
10   Living
11   Family
12   Kitchen
13   Terrace
14   Pool
15   Bed
16   Sewing
17   Robe
18   Ensuite

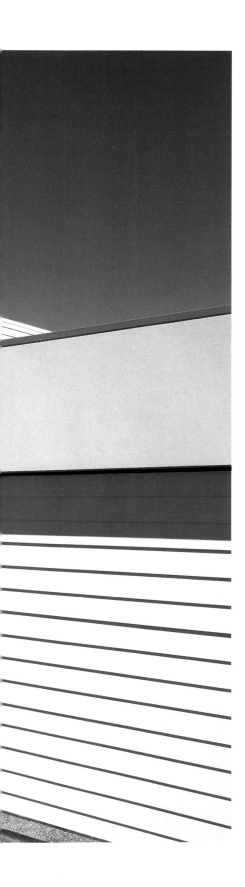

Harmony with the waves

L E U S C H K E   K A H N   A R C H I T E C T S

"The clients wanted to see the water from every room, so we were constantly looking for ways to hide from the wind while looking at the water".

A relaxed open-plan layout caters for a casual coastal lifestyle in this new beach house, on the Whangaparaoa peninsula, just north of Auckland. On the crest of a sand dune, right on the beachfront, the house enjoys magnificent sea views and generous natural light.

The L-shaped home faces north, with a single-storey living space running at right angles to the two-storey bedroom wing. Between them, on the inland side of the house, is a sheltered courtyard complete with an open fireplace. The courtyard combines wide water vistas with protection from the prevailing onshore breeze.

"The approach to the house is very laid back. It has a distinctly New Zealand *bach* feel," says architect Paul Leuschke, a director of Leuschke Kahn Architects. "There's no garaging. It's a case of parking under the trees. But we did include a boatshed for the water toys. There's also a boat ramp on the eastern boundary," he adds.

The light, airy living space can be thrown open on either side, to create an outdoor room facing the beach on calm days, extending into the courtyard behind if it's windy," says Leuschke. On the ground floor are open-plan kitchen, living and dining areas. Also on the ground floor are a study, bathroom, and rumpus room, together with two single bedrooms. The second floor comprises a guest bedroom at one end and the main bedroom/ensuite at the other end of the house. A fairly open timber staircase links the two floors and timber-battened screen wall.

"The clients wanted to see the water from every room, so we were constantly looking for ways to hide from the wind while looking at the water," says Leuschke. "They also didn't want to feel like they were going on holidays only to spend all their time shut in the kitchen. We integrated the cooking area into the living area, running it along one wall, rather than dividing it off in any way," he says.

With so much glazing, the house is a suntrap, so creating shade was an important consideration. Large overhangs provide protection from the sun on the seaward side, while the main bedroom has its own, cantilevered deck. This deck is finished in Moorish-inspired latticing for both shelter and privacy. "The design was about creating an informal feel," says Leuschke. "It also makes the interior appear warmer, as does the timber used in the kitchen," he adds.

Photography by Kallan MacLeod

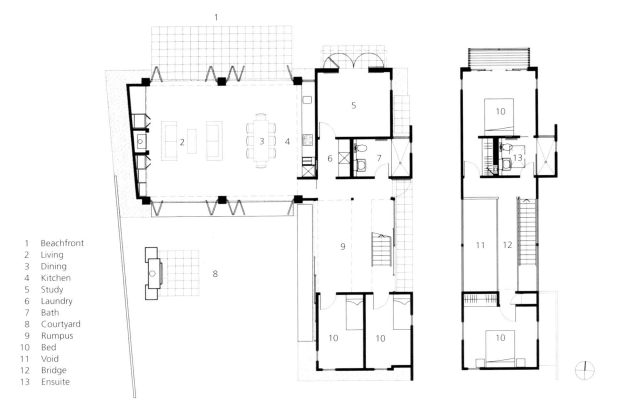

1 Beachfront
2 Living
3 Dining
4 Kitchen
5 Study
6 Laundry
7 Bath
8 Courtyard
9 Rumpus
10 Bed
11 Void
12 Bridge
13 Ensuite

Honeymoon Point

PHILP LIGHTON ARCHITECTS

"It isn't a precious house. Unlike honeymooners, the owners have two children. They can drag their gear up from the beach and clean the sand off on the decks".

The location of this beach house would make most newlyweds swoon. Thirty kilometres from the nearest town, Honeymoon Point is sparsely populated and surrounded by natural bushland. This house is only one of four on this part of the secluded coastline. "Originally the land was inaccessible. After it rained, the road was completely washed out," says architect Tim Penny of Philp Lighton Architects.

Fortunately, locals redirected the road to higher ground, making this site a proposition. For the owners, a professional couple, it was an opportunity to enjoy a different experience from the city. "Our clients wanted a beach house that was quite distinctive. They didn't want it to just disappear in the landscape," says Penny.

The two-storey timber house is clad in rough-sawn timber. Internally, the house is simply finished hoop pine, with expressed timber beams. "It's a simple design. The finishes are quite raw. It was designed so that a carpenter could build the entire house," says Penny, who

appreciated the remote location and the difficulty in attracting tradesmen to the property. While the timber house is fairly rudimentary (partially due to a limited budget), it is still sufficiently robust to endure the ferocious winds and constant rain. A corrugated-steel roof amplifies the sound of the rain.

The beach house includes a bedroom, bathroom and external storeroom on the ground level, flanked by two large timber decks. A stair leads to the first floor, which features a large living area, a galley-style kitchen, a study and main bedroom. There is a large deck leading from the study and living areas, offering panoramic views of the ocean. "It isn't a precious house. Unlike honeymooners, the owners have two children. They can drag their gear up from the beach and clean the sand off on the decks," says Penny, who likens the experience to camping. "When the trees around the house become more established they'll feel even more part of the landscape".

Photography by Richard Eastwood

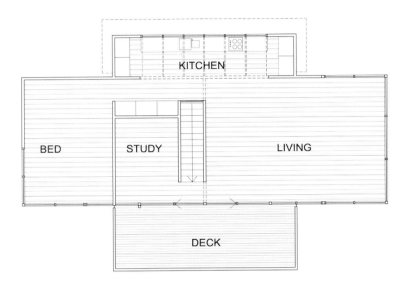

KITCHEN

BED   STUDY   LIVING

DECK

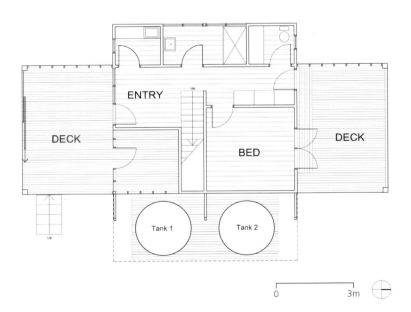

ENTRY

DECK   BED   DECK

Tank 1   Tank 2

0   3m

Influenced by the 1950s

G R A N T   A M O N   A R C H I T E C T S

"My clients were keen on red. But I thought it was appropriate to differentiate the original shack from the new house".

This beach house was once a 1950's shack. Made of asbestos cement sheeting, the two-bedroom home has been completely transformed into a substantial home for its owners, a couple with grown-up children and grandchildren. "They wanted a house that would accommodate their extended family. But they also requested something more than a casual shack. They wanted to be able to spend half their time here," says architect Grant Amon.

Amon retained the original rooms in the shack, but reworked these to suit the new design. Two ground floor bedrooms were retained, but the windows were enlarged. The kitchen and living areas were also kept, however, fully renovated. A completely new bathroom and laundry were added to the original plan, together with a second storey, comprising the main bedroom, ensuite, a second bedroom, together with a second living area leading to an expansive balcony.

Asbestos was removed from the shack and it was reclad in silver zincalume. The new addition is clad in red corrugated steel. "My clients were keen on red. But I thought it was appropriate

to differentiate the original shack from the new house," says Amon. While the house appears bunker-like to the street, the rear façade is glazed and open. "We wanted to embrace the sea view to the north, while creating privacy to the street," says Amon, who was also keen to use materials with low maintenance requirements.

The home's interior has a contemporary feel. But there are influences from the 1950s. The open timber and steel staircase in the centre of the house expresses the honesty of its materials as in many post-war homes. The colours used in the kitchen such as teal and red also capture an earlier time. "The cantilevered bench in the kitchen is new, but it's similar to the bench that originally existed," says Amon.

To intensify the views from the upstairs bedroom and living area, Amon angled the roof almost eight degrees. The windows in the second living area also come to a point. As Amon says, "It's only when you reach the top of the stairs that you really appreciate what's ahead of you".

Photography by Mark Munroe

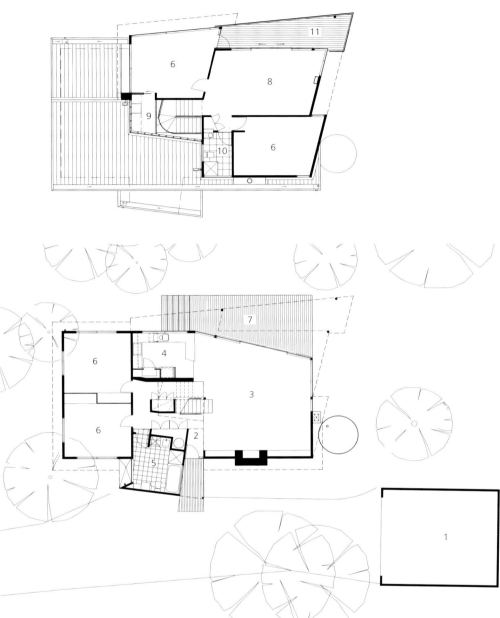

1	Existing garage	7	Patio
2	Entry	8	Living
3	Living/dining	9	Robe
4	Kitchen	10	Bathroom
5	Bathroom	11	Deck
6	Bedroom		

Inspired by Frangipani

PAUL UHLMANN ARCHITECTS

This beach house, designed by Paul Uhlmann Architects, occupies four regular-sized house blocks. "My client started with two blocks of land. Two additional blocks became available. It was too tempting to resist," says Uhlmann, who was able to provide on the extra land, a 200-square metre skate park.

The owners, a couple with three children, wanted a house that made the most of the generous allotment. As a result, the house

"The owners have strong ties to Fiji. We've used the frangipani motif extensively in the detailing".

includes a kitchen, living and entertaining area at ground level, together with a media room, a playroom and a guest bedroom. On the first floor are five bedrooms and a library. And on the top floor is a gymnasium and an additional entertaining area, complete with its own wine cellar. "I think it has just about everything you could possibly want," says Uhlmann.

One of the main requisites in the design was to create a strong connection to the beach, directly in front. The main living areas on the

ground floor open up towards the ocean on the east, while the pool area is located in a protected northwest corner of the site. The bedrooms, suspended above provide extra cover and protection for the pool deck. This ensures that the feeling of an outdoor living can still be enjoyed during strong south easterly winds and rain. Once all the bi-folding doors are stacked open, the living areas become a covered deck space.

Unlike some beach houses that are constructed in lightweight materials such as timber, this home features concrete floors, concrete block walls and a feature stone wall, the latter appearing both externally and internally. Aluminium slats also feature on the façade. "We wanted to create a sense of arrival to the home. The slats also cover some of the equipment, such as an air conditioning unit," says Uhlmann. One of the details in the home's interior can be traced to Fiji. "The owners have strong ties to Fiji. We've used the frangipani motif extensively in the detailing".

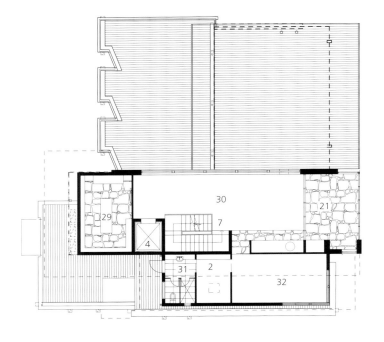

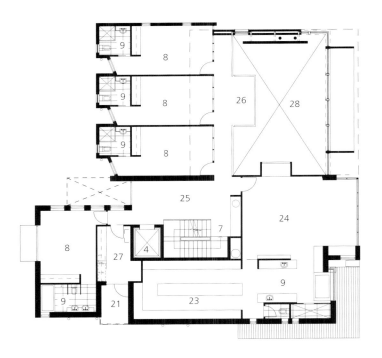

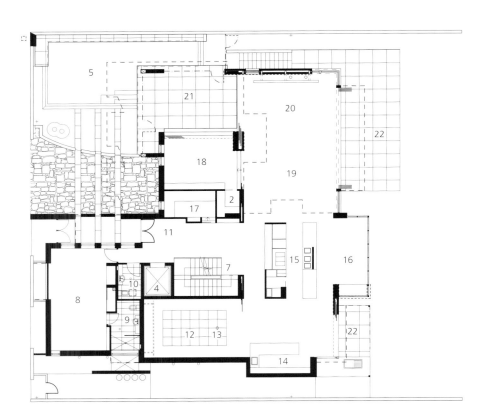

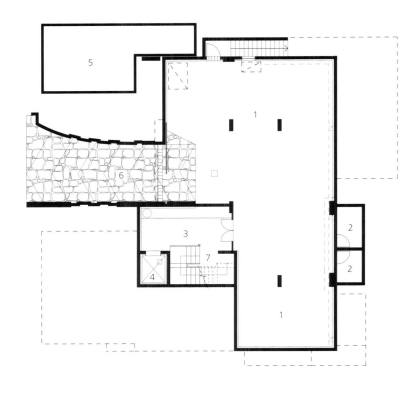

1 Garage	8 Bedroom	15 Kitchen	22 Terrace	29 Cellar
2 Store	9 Ensuite	16 Breakfast room	23 Walk-in robe	30 Entertainment room
3 Entry lobby	10 Powder room	17 Study	24 Master bedroom	31 Bathroom
4 Lift	11 Entry	18 Rumpus room	25 Library	32 Gymnasium
5 Pool	12 Media room	19 Dining	26 Sitting	
6 Driveway	13 Dancing pole	20 Living	27 Laundry	
7 Stair	14 Bar	21 Deck	28 Void	

Interlocking Views

JOHN WARDLE ARCHITECTS

"You don't need to see everything at once. The appeal is seeing how the scenes fit together".

Designed by John Wardle Architects, this house overlooks a surf beach. While the coastal foreshore and expansive blue horizon is overwhelmingly beautiful, it wasn't treated as a single caption from within the home. "You don't need to see everything at once. The appeal is seeing how the scenes fit together," says Wardle.

The main kitchen and living area cantilevers above the carport. Externally clad in cedar, the house is described by Wardle as a 'coastal shack'. "It's a relatively modest home. Our client is passionate about the coast. He wanted something quite simple to retreat to with his family each weekend," says Wardle. Downstairs are two bedrooms and bathroom, together with a rumpus room for the children. Upstairs are the main living areas, together with a study, main bedroom, ensuite and walk-in wardrobe.

While Wardle included an outdoor deck leading from the kitchen, he also included a large sliding glass door in the living area. This door pulls back to reveal a fine steel balustrade. As Wardle says, "In a sense, when this door is pulled back, the room itself acts as a large balcony".

In the lounge, on the first floor, the views are treated in three distinct ways. On one side of the space is a smaller slot-like window, protruding slightly beyond the living room wall. On the other side of the living area are portal-style windows containing window seats. And between the two window designs is a large floor-to-ceiling picture window. "Each window defines the landscape in its own way," says Wardle.

Photography by Shannon McGrath

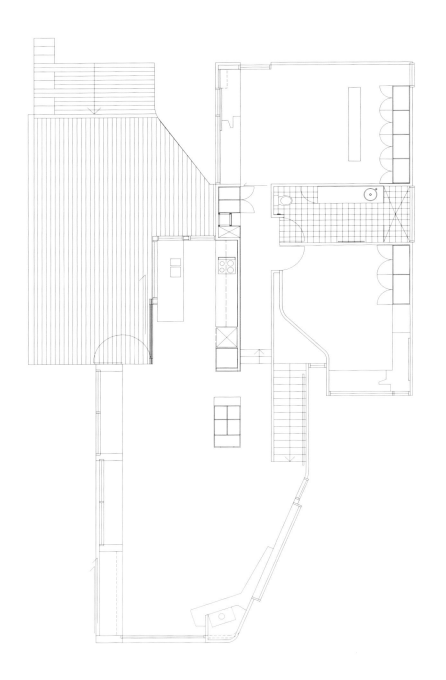

Nestled among **Banksias**

SUZANNE DANCE, ARCHITECT

"It feels as though you're inside a ship's cabin. There's that sense of enclosure, but it's not stifling".

Set among the dunes, framed by Banksia trees sits this modest beach house. Originally built in the early 1980s, the house was identical to other fibro cement homes in the area. Built on poles to capture the ocean views, its most spectacular aspect was its position in a leafy enclave.

As the owners required more space, they engaged architect Suzanne Dance to come up with a concept for a renovation. "My clients are strong environmentalists. They didn't want to encroach on the native bushland. They wanted to use plantation timbers as well as recycling the existing structure as much as possible," says Dance, who renovated and extended the house.

Originally the house consisted of three bedrooms, a bathroom, laundry, kitchen and living areas, all on the one level, raised 2.6 metres above the ground. Dance incorporated the ground level into the design to create a two-storey home. She sacrificed one of the bedrooms to create a new stairwell and void linking the two levels. "One of the main problems with the house is the height of the ceilings. But the void makes the spaces on both levels appear more generous," says Dance.

At ground level, Dance created a bedroom, a bunkroom, a bathroom, laundry and external store. There was also sufficient space for a second living area. And to compensate for the low ceilings, Dance clad the interior with hoop pine ply, a plantation timber. "It feels as though you're inside a ship's cabin. There's that sense of enclosure, but it's not stifling," says Dance.

One of the main problems faced by the owners of this house are the flies and mosquitoes. As a result, the original narrow deck that framed the living areas was hardly used. A remedy was to create a double height façade, encasing both levels. Made of timber and flywire, the new veil allows the outdoors to be enjoyed at all times of the day and well into the evening. As Dance says, "The design of the screen was inspired by the Banksia trees. From the beach, the shapes of the timber reflect their twisted branches".

Photography by David Marks

Nostalgic 1960s

JASMAX ARCHITECTS

"Living outdoors was an important component of the brief. Like the Hopper Houses, the outdoor decks act as another room".

Designed by architect Tim Hooson, in association with architect Tim Dorrington, this house includes elements of the post-war period. The home's strong horizontal lines and cubic forms are similar to the 'dune hopper houses' built in the area. "Our clients grew up with these homes. They appear to float above the sand dunes," says Hooson, a practice director with Jasmax Architects.

Jasmax's clients were attracted to the small coastal settlement, a two-hour drive from Auckland. Known for its surf beach and fishing industry, the settlement is a favourite destination for holidaymakers as well as weekenders. The house, located in a cul-de sac, benefits from its close proximity to the beach, a short stroll away. And while many beach houses offer a panoramic view of the ocean, this house also has a full view of the beach and its winding path towards the headlands.

The familiar 1960's form of the house grew out of the original structure, a two-storey house. While concrete block walls on the ground floor were retained in the design, the second floor was completely rebuilt and clad in charcoal-colour plywood. "The second floor was untenable. There was almost nothing worth saving," says Hooson. The ground floor was converted into a self-contained unit, primarily for the owners' children and for guests. On the first floor there is now the main bedroom, ensuite, kitchen, dining and living area, with both the bedroom and living areas leading directly to a large deck (4 metres wide by 20 metres in length).

The deck, protected by the extended roof, is partially open to the sky. "One of the owners wanted more shade than the other. We saw this as a compromise," says Hooson, who tilted the entire roof by approximately five degrees to magnify the view of the beach. "Living outdoors was an important component of the brief. Like the Hopper Houses, the outdoor decks act as another room," he adds. And like those of the post-war period, the materials used for the interior are simple and low maintenance, with timber floors, a timber staircase and laminate benches in the kitchen.

Photography by Emma Jane Heatherington

Ocean Views

B.E. ARCHITECTURE

"We unashamedly prioritise the views in our designs. Our clients should have some that are unique to them".

Spreading over 40 hectares, this coastal retreat has it all, rolling hills and unimpeded views of the ocean. Unfortunately, it is also one of the most exposed sites in the area and has a 'category one wind load'. "You get gale force winds down here. The weather can draw you indoors, even in summer," says designer Broderick Ely of b.e. Architecture.

Though the steel-portal-framed balconies offer some protection from the inclement weather, it is the three courtyards surrounding the house that acts as the main buffer to the winds. The architects excavated up to five metres to allow for these sunken outdoor rooms. "It was quite a radical cut into the landscape. But there was also a small amount of protection from a few cypress trees," says Ely, who also designed the landscape around the house. "In a few years it will be quite different," says Ely, pointing to the Chinese elms and new cypress trees. But even before the new landscape emerges, the house sits quietly in its rural setting, featuring rough rendered retaining walls and clad in painted cedar.

The 400-square-metre house, designed for a couple with children and grandchildren, was conceived in terms of a hierarchy of spaces and views. The large open-plan kitchen and living areas occupy the first level, allowing everyone to enjoy the same panoramic view over the hills and ocean. But there are also intimate spaces, designed specifically for the owners. The main bedroom, study and ensuite on the lower level for example, were designed as a separate wing and are accessed by a glazed breezeway that offers views of the large dam on the property. In contrast, the art room (designed to also entertain grandchildren) looks out to a pebbled watercourse that divides the main bedroom wing/parent's retreat from the rest of the house. "We unashamedly prioritise the views in our designs. Our clients should have some that are unique to them," says Ely.

Open-plan living areas on the first level are divided by timber walls which double as cupboards. These walls not only direct visitors to the house but create a subtle screen to the kitchen on one side of the entrance and the dining room on the other side. A slight change in level between the kitchen and casual living area and the more formal living and dining area, also articulates the spaces.

Attention to detail is something the practice prides itself on. "Everything has been custom made, even the outdoor furniture," says Ely, who admits to pampering his clients.

Photography by Trevor Mein

Off the **Main** Track

V L A D I M I R   I V A N O V ,   A R C H I T E C T

"It's not a particularly large house. But compared to the neighbours, it appears quite monumental".

This quiet beach hamlet, three-and-a-half hour's drive from the city, is not widely known, except by locals. While the beaches are pristine, the housing stock hasn't moved on from the 1950s and 1960s. The choice is either the fibro shack or the modest red brick home, built a few years later. So when architect Vladimir Ivanov unveiled plans for a contemporary beach house, the locals thought they were getting a new restaurant. "It's not a particularly large house. But compared to the neighbours, it appears quite monumental," says Ivanov, who regularly fielded enquires as to the function of the building.

Approximately 300 square metres in area, the house sits prominently on its corner site, with two street frontages. Designed for a couple, who operate a dairy farm during the week, the brief included separate accommodation for a parent moving from interstate. "Essentially my clients wanted two houses in one. But they wanted the design to appear as though it was one house from the street," says Ivanov.

The house is conceived as three forms that interconnect. The first form, the self contained apartment, is legible from the street and consists of a box with a double masonry wall, clad with timber laminate. This form is interconnected to the main house, which comprises generous glazing at ground level and rendered masonry walls above. "The timber box anchors the house, which is quite transparent," says Ivanov.

The two-storey house features glazed windows on three sides and rendered masonry on the fourth, to prevent the harsher sunlight from entering the home. The open-plan kitchen and living areas on the ground floor have unimpeded views of the beach and pine trees. To ensure some privacy, Ivanov designed a series of seats/storage units made of fibro cement on the large deck, some featuring open shelves, others closed nooks. "The owners love collecting shells and driftwood from the beach. They can either hide it away or display these on the open shelves," says Ivanov.

Upstairs affords the same generous views of the water, particularly from the main bedroom and balcony. Ivanov also included a second bedroom, together with a study/third bedroom, should additional guests stay over. "It's quite a simple design," says Ivanov, who selected robust materials to allow for sand being bought into the house. As Ivanov says, "I've provided extra-wide grooves for the timber deck. The sand can just be swept away".

Photography by Giles Westley

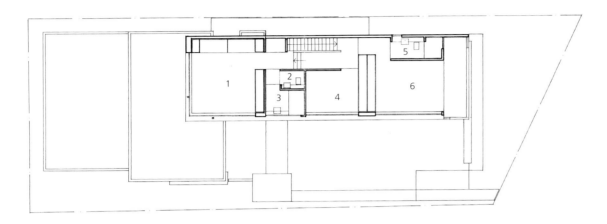

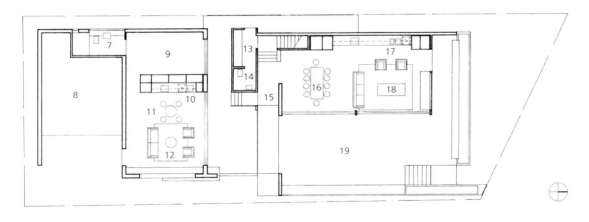

1   Bedroom
2   W.C.
3   Bathroom
4   Study/bedroom
5   Ensuite
6   Master bedroom
7   Ensuite
8   Garage
9   Bedroom
10 Kitchen
11 Dining
12 Living
13 Laundry
14 Guest W.C.
15 Entry
16 Dining
17 Kitchen
18 Living
19 Deck

Orientated to the Rock Pools

NICOLA MIDDLETON, ARCHITECT

A winding road leads to this beach house. With an escarpment behind and a dramatic fall in the land towards the ocean, the house appears like a rocky outcrop from the distance. Designed by architect Nicola Middleton, the modest beach house (approximately 190 square metres) is orientated towards a rock shelf and pool. "I deliberately angled the windows towards the rock shelf, particularly the loft above the main bedroom. The view makes it an idyllic place to write and contemplate," says Middleton.

The site originally contained a café and petrol station. And while the original intention was to extensively alter these buildings to form a beach house, the expense was prohibitive. "It made more sense to start from scratch and create something that suited my client's needs," says Middleton. Designed for a blended family of five children, the brief was to create an understated beach house that was low maintenance. "They wanted a house they could use half the week, enjoy and then lock up," Middleton adds.

The house consists of a steel frame and is partially clad in aluminium to prevent rusting. "The winds can reach gale force down here, so it was crucial to use materials that were resilient to the weather," says Middleton, who included two deck areas in the design, one open and the other protected. But while the expansive deck leading from the living and dining areas offers unimpeded views of the ocean and rock pools, a glazed window in the kitchen also allows permits views from the protected deck.

As family and friends can 'shrink' the house at times, the owners included two zones, one for children and/or guests, the other for themselves. The parent's wing features a bedroom, ensuite and small loft area above. The other wing includes a double bedroom and a shared bunkroom that doubles as a television/recreation area. The kitchen and living areas link the two zones in the house. As Middleton says, "My clients had a full house on the weekend. But I was told there was more than enough room for everyone to unwind".

"I deliberately angled the windows towards the rock shelf, particularly the loft above the main bedroom. The view makes it an idyllic place to write and contemplate".

Photography by Liz Cotter

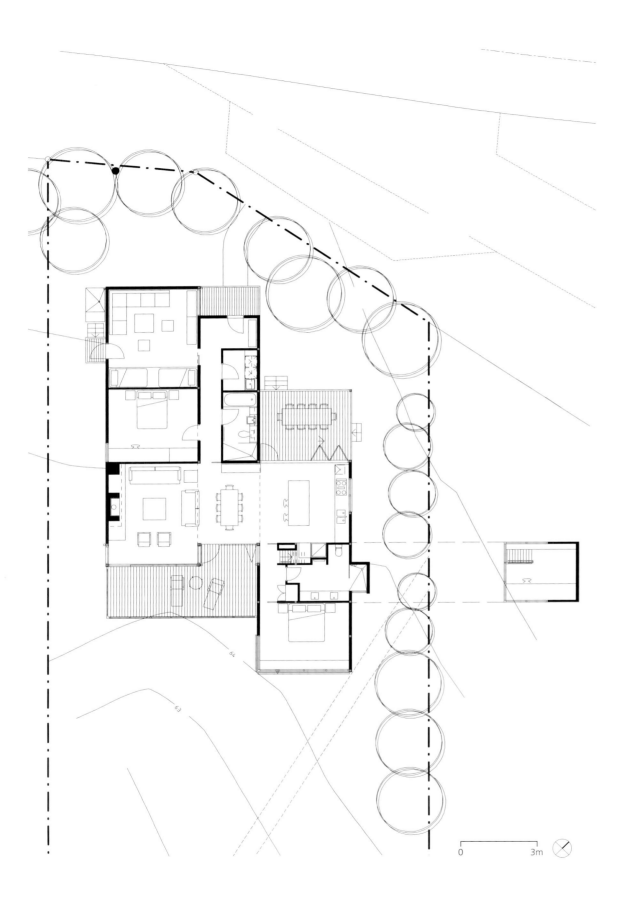

Panoramic Allure

FISHER ARCHITECTURE

"It's a fairly simple design. We wanted to go back to the idea of the original beach house, the one large room that everyone enjoys being in together".

This weekender, overlooking Hahei Beach in New Zealand, offers panoramic views of the Coromandel and adjacent islands. Relatively modest in scale, approximately 200 square metres, the design is deliberately understated to emphasise the home's scenic location.

Designed by Fisher Architecture, the brief to the architects was to design a low-maintenance house for a couple and their children, who live most of the time in Singapore. "They wanted something that was no maintenance, rather than low maintenance, a place they could simply open and close, like a suitcase," says architect Michael Fisher, a director of the practice.

The timber framed house features steel beams and is clad in Zincalume, as is the roof. The rectilinear-shaped house is spread over two levels. At ground level, is the garage, together with the two children's bedrooms. On the upper level is the main bedroom, which is separated from the main living areas by an ensuite bathroom and second bathroom for guests. "It's a fairly simple design. We wanted to go back to the idea of the original beach house, the one large room that everyone enjoys being in together," says Fisher.

And like the original beach house, the design includes an open deck leading from the main living areas, as well as a covered deck for more inclement weather. Fisher included in the design sliding Zincalume doors to frame the windows as the winds can be strong. These can be pulled back during finer weather or drawn across to enclose the interior spaces as well as one of the decks. A cantilevered platform creates the impression of being anchored into the land at the rear of the site.

The materials used for the interior are low maintenance. The floors in the living room are plywood and lime-washed and the walls are made of 'Gibraltar' board (plaster board) and painted white. Even the flooring in the bathrooms, Pirelli-studded rubber, was conceived for low maintenance. As Fisher says, "The exterior is hardy. It won't need painting. The zincalume will dull over time, like the surrounding vegetation".

Photography by Simon Devitt

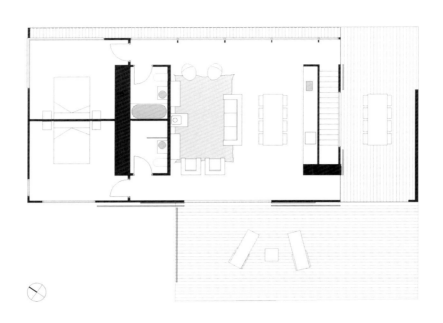

Peeling Back the Roof

A L E X   P O P O V   &   A S S O C I A T E S

This house, designed by Alex Popov Architects, is reminiscent of many simple timber homes built on the cliff tops. The three-storey house, clad in cedar, is relatively modest compared to many new homes. "Our clients didn't want a palatial home with all the trappings of the city. They wanted a beach house, somewhere to escape to at the end of each week," says architect Alex Popov.

While the cliff top site offers spectacular views over the beach and headlands, it is unfortunately, orientated away from the sun. To address the issue of light, Popov designed a skillion-shaped roof, made of zinc. And to ensure direct light entered the house, he 'peeled' back the roof and inserted highlight windows. "The roof is like a can that has been ripped open on one side," says Popov, who was able to draw in the light as well as views to the cliff face.

As the house is nestled into the cliff face, the entrance is at the top level. This level includes the lounge, kitchen and dining areas, together with the main bedroom and ensuite bathroom. This open plan area is protected by generous roof overhangs and by electronically operated metal louvred screens that control the harsher light. The middle level comprises three smaller bedrooms for the children and a bathroom. And the ground level features a games room and shower. The ground level also leads directly to the swimming pool.

Even though the house is substantial in size, it doesn't dominate the cliff face. Popov deliberately reduced the amount of glazing in the design. The base level of the house, finished in cement render, also reduces the scale of the house. "We didn't want the house to overshadow other houses or the beauty of the vegetation". As Popov says, "It's clearly not a 1950s cottage by the sea. But it's a long way from the house the owners leave behind in the city each week, both in distance and aesthetically".

"It's clearly not a 1950s cottage by the sea. But it's a long way from the house the owners leave behind in the city each week, both in distance and aesthetically".

Photography by Kraig Carlstrom

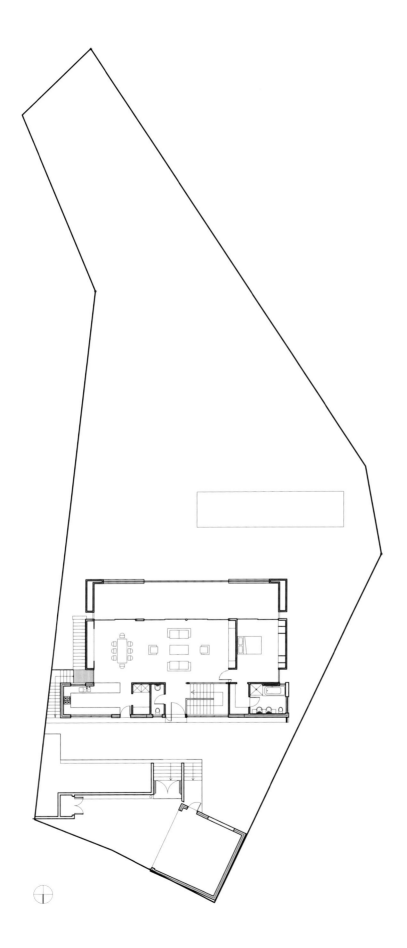

Perched above the Dunes

ARCHITECT ANTHONY GIONFRIDDO

"The materials pick up the colours of the surrounds. The limestone was quarried locally. And the western red timber will eventually turn a silvery grey, like the trunks of these trees".

This beach house peers over the dunes. A thin strip of publicly owned foreshore separates the house from the beach. "We could have elevated the house a few metres higher. But we wouldn't have gained sea views. It would have simply made the house more prominent amongst the gnarled scrub," says architect Anthony Gionfriddo, who designed this two-storey house at Blairgowrie, a one-and-a-half hour's drive from Melbourne.

Gionfriddo excavated the site, removing five metres of soil. "It's a relatively steep slope. I was keen to anchor the house into its coastal setting," he says. The materials used for the house are treated quite differently on each level. The lower level is made of reinforced concrete. The upper level comprises a steel frame, expansive tinted glass, timber cladding and stone. "The materials pick up the colours of the surrounds. The limestone was quarried locally. And the western red timber will eventually turn a silvery grey, like the trunks of these trees," says Gionfriddo, pointing to the twisted tea trees.

As the site was excavated, the entrance to the house is on the first level. And instead of creating an obvious opening to the street, the front door is concealed behind timber-battened walls. Even the windows to the street are battened down with screens. "I wanted to maximise the sea breezes in the house. The front door can be left open without having unexpected visitors," says Gioffrido. The first level includes the kitchen, living and dining areas, together with main bedroom and ensuite. Both this bedroom and the living area feature open fireplaces set in limestone feature walls. Large sliding glass doors lead to a terrace with a lap pool. And to ensure the views aren't restricted from the living areas, a toughened glass balustrade defines the edge of the pool.

While the pool cools down the living spaces in the visual sense, it also acts as a canopy for the children's bedrooms below. The lower level also includes a guest bedroom, ensuite and kitchenette for friends and extended family. "The house was designed for a family with two children. But they also enjoy having friends staying over," says Gionfriddo, who included in the design a sophisticated kitchen for entertaining. The kitchen features an expansive Calcutta marble central bench. With its golden hue, there's a constant reference to the surrounding dunes.

Photography by Dan Magree

Place of the **Setting** Sun

W A L T E R   B A R D A   D E S I G N

"The house has been completely gutted. But it's not obvious. There's still that warm and relaxed feeling that first attracted our clients to the house. It still has that aged patina".

Named 'Illilliwa', Aboriginal for 'Place of the Setting Sun', this house is ideally located to catch the last glimpses of the sun each day. This timber beach house, overlooking a harbour, was originally built in the 1940s. Purchased six years ago by the current owners, the house has recently been transformed by architect Walter Barda. "There was a certain directness about the house. Even though it was relatively modest by today's standards, it was well orientated," says Barda.

Originally the house consisted of a kitchen and living area on the first floor, together with the main bedroom and ensuite. And on the lower level were two bedrooms, both of equal size and a shared bathroom. While this arrangement was sufficient for the new owners, the upstairs living area didn't take full advantage of the water views, or the national park directly ahead.

In response to this shortcoming, Barda removed the enclosed glass veranda above the water's edge and installed floor-to-ceiling timber and glass doors. "We retained many of the original 1940's-style windows in the house. But some

had to be removed to make that connection to the outdoors," says Barda, who converted the enclosed patio to form a new dining area. Interior walls were also removed from the house, including a wall enclosing the original main bedroom. This room is now used as a library/music room for the couple.

The stairwell, linking the two levels was also opened up with the use of glass walls. Additional light now enters the home, as well as allowing views to the rock face that supports the house. Barda also removed concrete that entirely surrounded the house and created a lush and tropical garden.

While elements such as old aluminium sliding doors have been removed, there is a sense of a 1940s beach house within a contemporary context. "The house has been completely gutted. But it's not obvious. There's still that warm and relaxed feeling that first attracted our clients to the house. It still has that aged patina," says Barda, pointing to the grooved plywood walls and milk-coloured painted walls.

Photography by Bart Maiorana

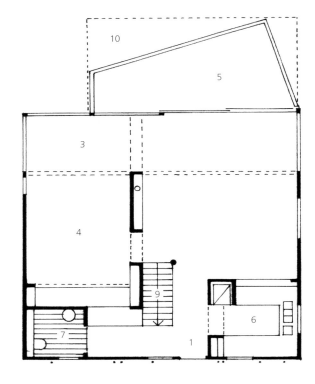

1 Entry
2 Sitting room
3 Dining room
4 Music room and library
5 Deck
6 Kitchen
7 Bathroom
8 Bedroom
9 Stairs
10 Awning

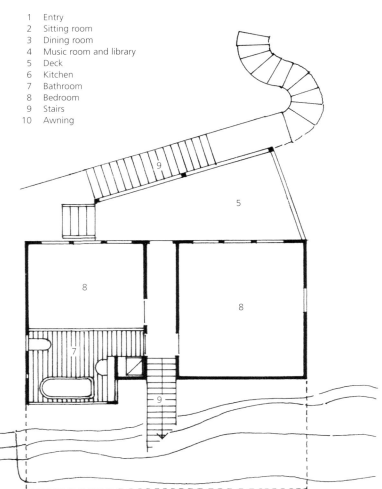

0   2m

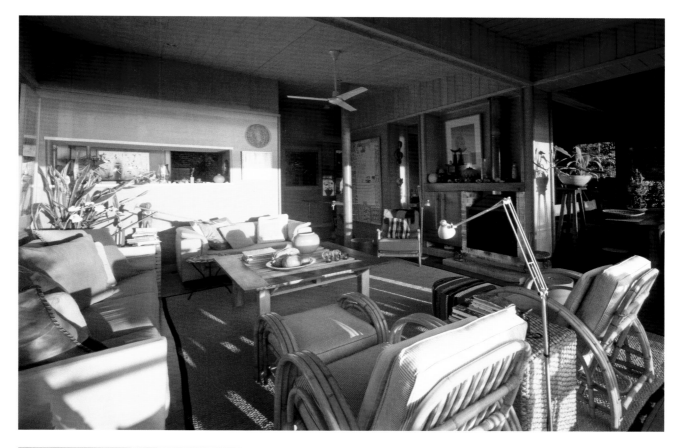

Radically remodelled

D A V I D   L U C K ,   A R C H I T E C T

"The idea was about inflating the spaces and the scale of the house. It now feels like a large viewing platform that draws your eye towards the horizon".

This beach house, perched on a cliff, was originally built in the early 1980s. The brown brick, two-storey home, featured small windows that offered at best, glimpses of the water. "We had to turn three bedrooms and two bathrooms into a five bedroom and four bathroom house," says architect David Luck. "It had to cater for a large family and all the relatives and friends staying over," he adds.

Some of the extra bedrooms were accommodated by the removal of an indoor swimming pool. Several internal walls were also removed to make the spaces feel larger. One of the first walls to go was at ground level, removing a division between the front and rear doors. "We wanted to create an unimpeded view to the backyard from the front entrance," says Luck, who also increased the number of bedrooms on the ground level to three as well as providing two bathrooms, a family room and double garage.

The second level was also redesigned, particularly the original small apertures. Most of the brick walls were removed and substituted by floor-to-ceiling glass windows and large sliding doors. Luck also removed the '80s timber balustrades that framed the house and substituted these with toughened framed glass. "We wanted to ensure there was a view from wherever you stood in the house," says Luck, who also included timber decks on all four sides of the upper level.

While the exterior was radically changed, including rendering the walls at ground level, rearrangement of the kitchen was minor. "The kitchen has only been moved slightly. But the finishes have all been redone," says Luck, pointing to the new timber and reconstituted stone central island bench.

The house now appears as a lightweight glass box, floating on solid plinth. As Luck says, "The idea was about inflating the spaces and the scale of the house. It now feels like a large viewing platform that draws your eye towards the horizon".

Photography by Shania Shegedyn

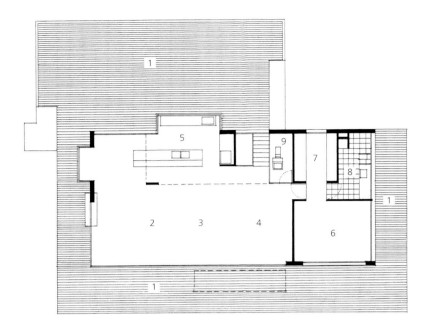

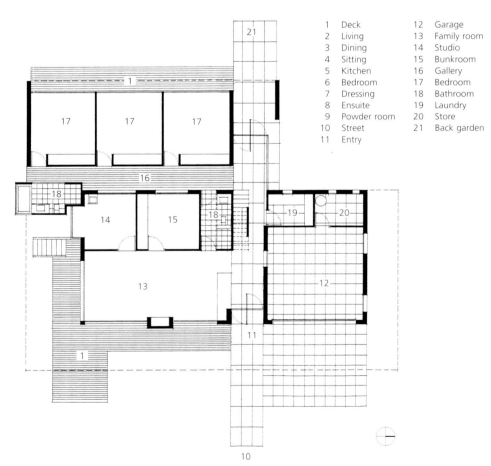

1	Deck	12	Garage
2	Living	13	Family room
3	Dining	14	Studio
4	Sitting	15	Bunkroom
5	Kitchen	16	Gallery
6	Bedroom	17	Bedroom
7	Dressing	18	Bathroom
8	Ensuite	19	Laundry
9	Powder room	20	Store
10	Street	21	Back garden
11	Entry		

Receding into the Landscape

R O B E R T   S I M E O N I   A R C H I T E C T S

"There's something quite beautiful about closing things by hand and opening the rooms up first thing in the morning".

This Victorian-style homestead has been a landmark of the coastal hamlet since it was originally built in 1876. Recently restored and extended by architect Robert Simeoni, the house clearly demonstrates the sublime junction between the old and new. "I treated the new wing as though it was part of the landscape, as though it was almost not there," says Simeoni.

While the original homestead was meticulously restored, it is the 350-square-metre addition that elevates the senses. Simeoni removed one of the bedrooms in the historic house to create a new lobby to the contemporary addition. A large 3.5-metre-high pivotal door leading off this space opens to reveal a generous 30-metre corridor with floor-to-ceiling glass windows/ doors. The elegant corridor, framed with off-formed concrete columns, creates an impressive entrance. "I wanted to keep the design as simple as possible," says Simeoni, who drew inspiration from the Milanese architect Guiseppe Terrangi, a prolific architect during the 1920s and 30s.

The grand corridor leads to the vast kitchen, living and dining area. And unlike the original homestead that offers restricted views via narrow windows, the contemporary wing incorporates breathtaking views over the bay and mountain ranges.

The kitchen features concrete floors, Carrara marble bench tops and Sycamore veneer joinery. But unlike most kitchens that have one central island bench, this one has three, all perfectly aligned. "My client loves cooking, particularly when on holidays," says Simeoni, who included the option of being able to cook with electricity or gas (the third island bench is for washing up). To ensure the views weren't impeded by unnecessary joinery, Simeoni concealed the pantry/cocktail bar to one side. "I wanted to keep the horizon lines as clear as possible, whether you were standing in the kitchen or relaxing in the living areas," he adds.

The main and only bedroom in the new wing illustrates Simeoni's approach. Between the dining area and the bedroom is a courtyard that allows sunlight to penetrate. And instead of doors, one continuous curtain wraps around the glass windows/doors to create privacy. In contrast to high-tech homes, these curtains are drawn by hand. "There's something quite beautiful about closing things by hand and opening the rooms up first thing in the morning," says Simeoni, who loves to see the curtains moving in the prevailing winds when the sliding glass doors are pulled back and the sea breeze filters through the house.

Photography by John Gollings

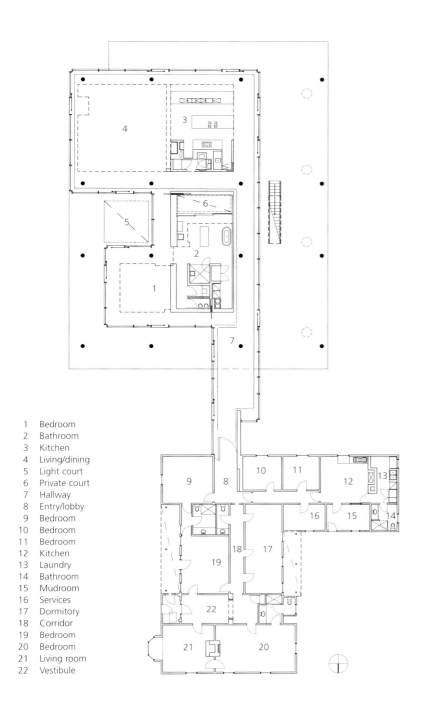

1   Bedroom
2   Bathroom
3   Kitchen
4   Living/dining
5   Light court
6   Private court
7   Hallway
8   Entry/lobby
9   Bedroom
10 Bedroom
11 Bedroom
12 Kitchen
13 Laundry
14 Bathroom
15 Mudroom
16 Services
17 Dormitory
18 Corridor
19 Bedroom
20 Bedroom
21 Living room
22 Vestibule

Revisiting the 1950s & 1960s

SEELEY ARCHITECTS

Designed by Seeley Architects, this Victorian beach house was partially inspired by modest 1950s and 1960s beach houses in the area. "I was keen to use simple, unadorned exterior materials. Like these simple homes, I wasn't after pretentious detail," says architect David Seeley, whose design includes a steel portal frame forming the vertebrae of the building, with the tilted and twisted roof sprung from steel struts.

"I was keen to use simple, unadorned exterior materials. Like these simple homes, I wasn't after pretentious detail".

Like many coastal weekenders, this house is constructed of fibre cement, earthy toned limestone and large timber-framed windows. "I wanted to use materials that were resilient to the coastal environment and required minimal maintenance," says Seeley. The roof form responds directly to the low profile of houses in the street, while still allowing natural light for both this home and the neighbour's house. It also provides a wind foil to the winter westerlies and allows the interior spaces to open up to capture the eastern sea views and morning sun.

Designed for a family with children, a key objective of the home was to provide discrete zones for parents and areas for the teenagers. This was achieved by utilising the two levels, upper for the parents, lower for teens. And, as the property is long and narrow, 40 metres deep by ten metres wide, a simple orthogonal plan was used. On the ground floor are the two children's bedrooms, a shared bathroom, laundry and rumpus room. On the first floor are the kitchen/scullery, two living areas and the main bedroom suite and ensuite. Seeley also incorporated two deck areas on either side of the living areas to maximise views to the sea.

Although the house is long and narrow, it has been cleverly punctuated with courtyard gardens. "I wanted to provide some intimate garden views for rooms at ground level. I didn't want them to feel introspective," says Seeley. A small void between the two levels also reduces the sense of enclosure. "There are views of neighbours' rooftops and trees from various aspects within the house. The streetscape can be as intriguing as the sea".

Photography by Scott Newett

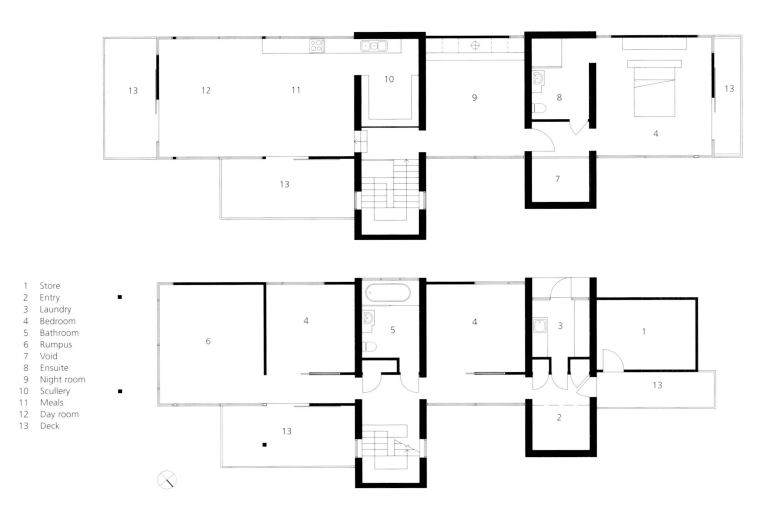

1 Store
2 Entry
3 Laundry
4 Bedroom
5 Bathroom
6 Rumpus
7 Void
8 Ensuite
9 Night room
10 Scullery
11 Meals
12 Day room
13 Deck

Saving the Tuckeroos

GABRIEL & ELIZABETH POOLE DESIGN COMPANY

"Our clients wanted to feel as though they were part of the site. When the doors are left open, you feel as though you're in a tree house".

Located 50 metres from a surf beach, this new house is nestled between the 'Tuckeroos', a native tree. "In a sense, this house becomes another 'tree', providing open-sided shelter," says architect Gabriel Poole from the Gabriel & Elizabeth Poole Design Company.

For the client, the proximity to the beach, with its crashing surf, was equally important as the forest reserve on the edge of the property. "The brief was for a 'simple' beach house. His words were a 'beautiful house amongst the trees'," says Poole. The two-storey house, made of fibro-cement sheeting, a steel roof and rough sawn cypress timber, recalls many early beach houses dotted along the coast. Constructed on a timber frame, with timber floors and decks, the house is raised above recorded flood levels, becoming an elevated sculptural form.

The three-bedroom weekender is arranged around a meditative inner courtyard containing two of the Tuckeroos, while the open-plan living/dining area and the front bedroom flow out to the front veranda, surrounded by natural bush. The open-style bathroom, which is in keeping with the light-hearted holiday genre, can be wholly concealed by a steel roller shutter, partly by a canvas blind or left totally open to the courtyard. But even when the bathroom is closed to the courtyard, it is still on view from the main bedroom, located above. "Our clients wanted to feel as though they were part of the site. When the doors are left open, you feel as though you're in a tree house," says Poole.

While the house is relatively simple in the materials used, the interior features sublime detailing, including aluminium to outline the internal joinery and fibro cement sheeting. For Gabriel and Elizabeth Poole, this house exemplifies what the practice stands for, simple and uncomplicated design. As Poole says, "This beach house is about leaving the city behind. The environment is a continual reminder that you're on holidays".

Photography by Eric Victor–Momentum Studios

Sense of History
SUZANNE DANCE, ARCHITECT

"We'd always loved the simplicity of the house. It really only consisted of two simple rectangular shaped rooms (one for living, the other for sleeping) that were linked by an open breezeway".

This timber house on the beach, in one of many suburbs along the coastal road, was once a weekender for arts patrons, John and Sunday Reed, founders of the Heide Gallery in Melbourne, Australia. Designed originally by renowned architects McGlashan and Everist in 1963, the house was nestled beside another significant home, designed by architect Peter Burns in 1961.

When artist Caroline Williams and George Mora bought the Reed beach house in the late 1980s, they owned the Burns house, which was used as their weekender. "The Reed's house was tenanted for a number of years before we bought it," says Williams. "We'd always loved the simplicity of the house. It really only consisted of two simple rectangular shaped rooms (one for living, the other for sleeping) that were linked by an open breezeway," says Williams, who commissioned architect Suzanne Dance to join the two significant homes together. "Not only are both architects well respected, the Reeds also occupy a special place in Melbourne's history," says Dance.

A new central passage was designed to link the two homes. But rather than try and merge them as if they were joined at the hip, Dance wanted to retain an obvious yet subtle distinction. "These cement sheets with studs enable the structure to be read as two separate homes. The sheets weren't elements in either of the homes, but they are compatible to both," says Dance.

The lounge room was carefully redesigned. As part of the brief was to create a winter room (as a contrast to the light-filled kitchen and meals area), Dance designed a series of smaller windows to accommodate Williams' extensive collection of books, painting and artefacts. Williams loved working with Dance on the house. As Williams says, "Suzanne not only respects the house and its history, but she clearly thinks through how the space will function".

Photography by Heribert Alucha

Separate Pavilions

D A V I D   D E N N I S ,   A R C H I T E C T

"The sound of water makes you feel as if you're on the beach. And when the trees become established, these walkways will provide the owners and friends with another place to relax".

Located near the beach, this sprawling site (2000 square metres) had a suburban feel to it when architect David Dennis saw it for the first time. "It's quite a large site, but the houses in the street made their presence felt," says Dennis, who decided to turn the house away from the street, like a bunker with a five-metre-high slate wall.

The house, a first project for Dennis, was designed for a couple with small children. "The family's planning requirements informed the design. It's essentially three interconnecting pavilions," says Dennis. To articulate each pavilion, Dennis clad each with western red cedar. The cedar for one pavilion is painted off-white and another in charcoal. The timber used to clad the third pavilion is left unpainted.

In one pavilion are three bedrooms for the children, together with bathroom facilities and a play area. The central pavilion comprises the kitchen, living and dining facilities and the third accommodates the main bedroom and ensuite, together with an additional bedroom for guests. To further delineate these areas, Dennis wrapped the services areas such as the bathrooms with veneer. "The veneers create another layer in the spaces," says Dennis.

The large open-plan kitchen and living areas lead, via sliding glass doors to a terrace that includes a spa. "I wanted to strengthen the connection to the outdoors, particularly from the living areas. But the bedrooms feel quite intimate. In the bedrooms, the views have been carefully framed with privacy in mind," says Dennis.

Established native vegetation also inspired the design, both in the choice of materials and in the arrangement of each pavilion. And while the front garden is still in its infancy, it was considered in the long term as another place for the owners to enjoy the outdoors. Dennis designed two timber plinths as walkways to the entrance to the home. One of the walkways is framed with a water feature. "The sound of water makes you feel as if you're on the beach. And when the trees become established, these walkways will provide the owners and friends with another place to relax".

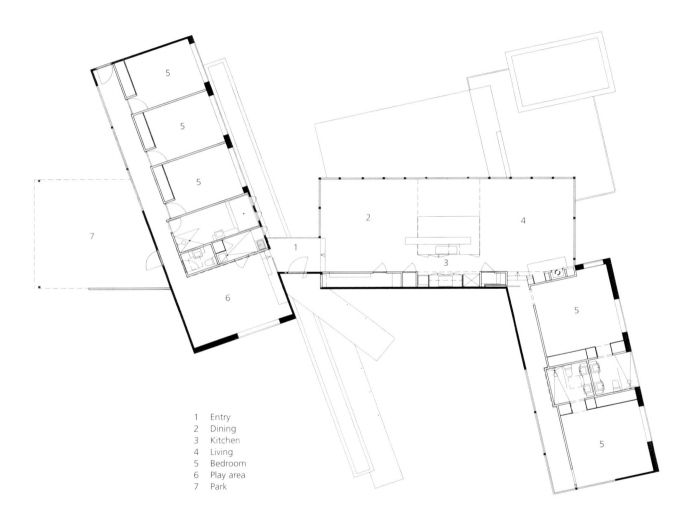

1 Entry
2 Dining
3 Kitchen
4 Living
5 Bedroom
6 Play area
7 Park

Shared Beach House

M c G a u r a n   G i a n n i n i   S o o n   P t y   L t d

"The design of the house is quite box-like. The colours break down the form. They also capture the colours of the bush".

This unusual beach house was designed for four families to share, similar to a time-share arrangement. So instead of one brief to the architects, McGauran Giannini Soon (MGS), there were four. "There was a common thread in all the briefs: a beach experience and low maintenance," says architect and artist Sue Buchanan of MGS.

Two hours drive from the city, the house is surrounded by dense coastal vegetation. And while there are several beach houses in the vicinity, most are at least partially concealed by tea trees. "It's quite a luxuriant covering. It creates a sense of remoteness," says Buchanan, who points out the direction of the surf beach nearby.

To ensure the natural vegetation remained intact, MGS created a relatively small footprint for the house. And as the site slopes approximately 30 degrees, the architects were able to create a two-storey home, with the ground floor carved into the earth.

On the ground floor are two bedrooms and a shared bathroom. To increase the accommodation, MGS included a large sliding door in the larger of the two bedrooms that disappears into a cavity wall when not in use. "There might only be one family staying in the house. But there are usually friends who stay over rather than head back to town," says Buchanan. On the first floor is a third bedroom, also featuring a large sliding door. "The door can be pulled back entirely if there's only a couple in the house. This allows them to enjoy views over the tree tops," she adds.

The floor plan of the upper level is relatively simple and includes a kitchen, meals and living area that leads to a deck. "The four groups didn't want a complicated design. They wanted a place that was as uncomplicated as their holidays by the beach," says Buchanan. However, the plan is relatively simple, the materials used for the house and the way they are brought together is more complex.

The façade concealing the staircase and linking the two levels is polycarbonate. When the morning sun strikes this façade, the timber battens supporting the polycarbonate sheeting appears skeletal. In contrast to this ethereal layer, the other facades are Mondriaan-style, punctuated with windows and doors. Made of textured fibro cement, each coloured panel is separated with timber battens. "The design of the house is quite box-like. The colours break down the form. They also capture the colours of the bush," says Buchanan, pointing out the vibrant green used for some of the panels. "It's the same colour as the new growth," she adds.

Photography by Mark Munro

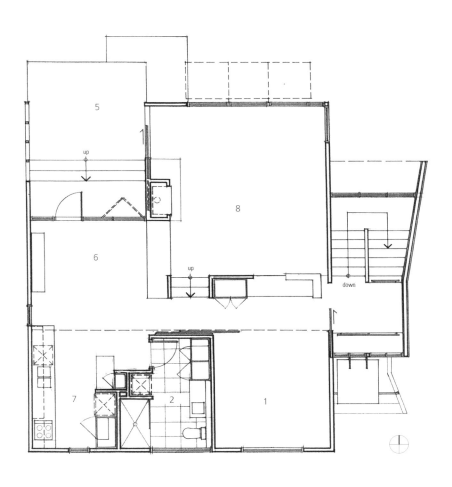

1 Bedroom
2 Bathroom
3 Store
4 Entry
5 Deck
6 Meals
7 Kitchen
8 Living

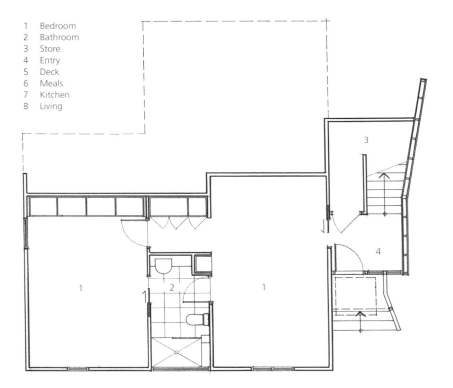

Showing by Example

LAHZ NIMMO ARCHITECTS

"We were inspired by many of the fibro shacks built during the 1950s and 1960s. They're quite simple. But they're always fondly recalled by those who grew up with them".

This house, located at Casuarina Beach, is surrounded with speculative beach houses. For the architects, Andrew Nimmo and his partner, Annabel Lahz, the standard of development didn't equate with the setting, on the edge of a surf beach. "We wanted to show people that good design doesn't have to mean a large budget," says Nimmo.

The couple set themselves a modest budget, approximately the same as the speculative homes surrounding them.

Lahz Nimmo Architects selected simple cost-effective materials for their house; concrete face block work and painted fibro cement. Timber battened screens provide an additional layer to the design. "We were inspired by many of the fibro shacks built during the 1950s and 1960s. They're quite simple. But they're always fondly recalled by those who grew up with them," says Nimmo.

One of the key elements of the design was to allow room for a second family. "We generally like to go away with another family," says Nimmo. As a result, the architects designed two main bedrooms on the first floor (one for each couple) and one large bunk room/recreation area on the ground floor for their own two children as well as the other couple's children.

The kitchen includes a long central island bench and the living areas are expansive. "Food is an important part of holidays. You might have three or four people cooking at any one time," says Nimmo, who also included a place for alfresco dining. "It can get quite windy. The courtyard offers some protection," he adds.

Another important feature of the design is the timber and glass battened louvres. There's even a louvred timber wall in a storage room, adjacent to the entry to the home, used to place wet towels. As Nimmo says, "We wanted to capitalise on the sea breeze. Every room in the house is designed for cross-ventilation".

Photography by Brett Boardman

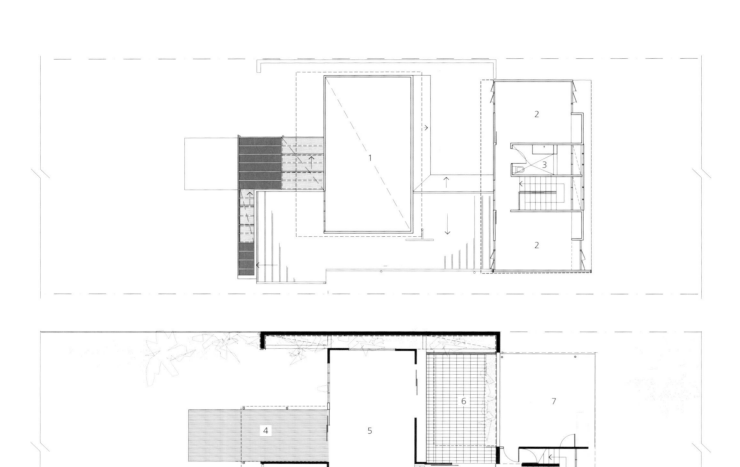

1	Void	7	Carport
2	Bedroom	8	Bunk
3	Bath	9	Bath
4	Deck	10	Kitchen
5	Living/Dining	11	Entry
6	Courtyard	12	Drying

0   5m

Stepping Back from the View

J O H N   W A R D L E   A R C H I T E C T S

"The attraction of the sea is quite powerful. But there are as many intriguing aspects of the site itself".

This beach house is located at Flinders on Victoria's Mornington Peninsula. The new house is adjacent to one of the area's oldest homesteads, Bass Lodge. While the lodge remains intact, the 1950's brick home that occupied this site was removed. "The brick house was known as 'the cottage'. It was originally part of Bass Lodge, before the land was subdivided," says architect John Wardle.

While the owners of this beach house gave Wardle a fairly open brief in developing the site, they were keen to be as close as possible to the sea. "I had to convince them to step back from their side boundary. This would allow them to survey their own site rather than their neighbour's garden," says Wardle, who placed the house between established pine trees.

From a distance, the timber and steel house appears single-storey. However, Wardle excavated 2.5 metres into the site to create a lower level. "The house was designed for a retired couple with an extended family. They didn't want to be confronted with a series of endless corridors and bedrooms every time they came down on their own," says Wardle. The lower level acts as a podium for the first floor, which comprises three bedrooms, bathroom facilities, a laundry and carport. The first level becomes the primary dwelling for the owners. This area consists of the open-plan living and dining area, together with a kitchen at one end of the floor plan. In a second pavilion is a study, guest powder room, ensuite, main bedroom and dressing area.

The house takes in the sea views via large picture windows on one side. However, the façade facing the street is more protected, featuring almost no windows. Established pine trees cast their silhouettes on the cedar used to clad the house. Timber appears extensively in the home's joinery. The striking joinery unit dividing the kitchen from the living areas is multi-functional. At one end is a coat cupboard and at the other end of the living area is a cocktail cabinet. In between are a built-in fireplace and even a phone nook.

While the house has many intriguing elements, it's the playful shapes among the established pines that are most memorable. As Wardle says, "The attraction of the sea is quite powerful. But there are as many intriguing aspects of the site itself".

Photography by Trevor Mein

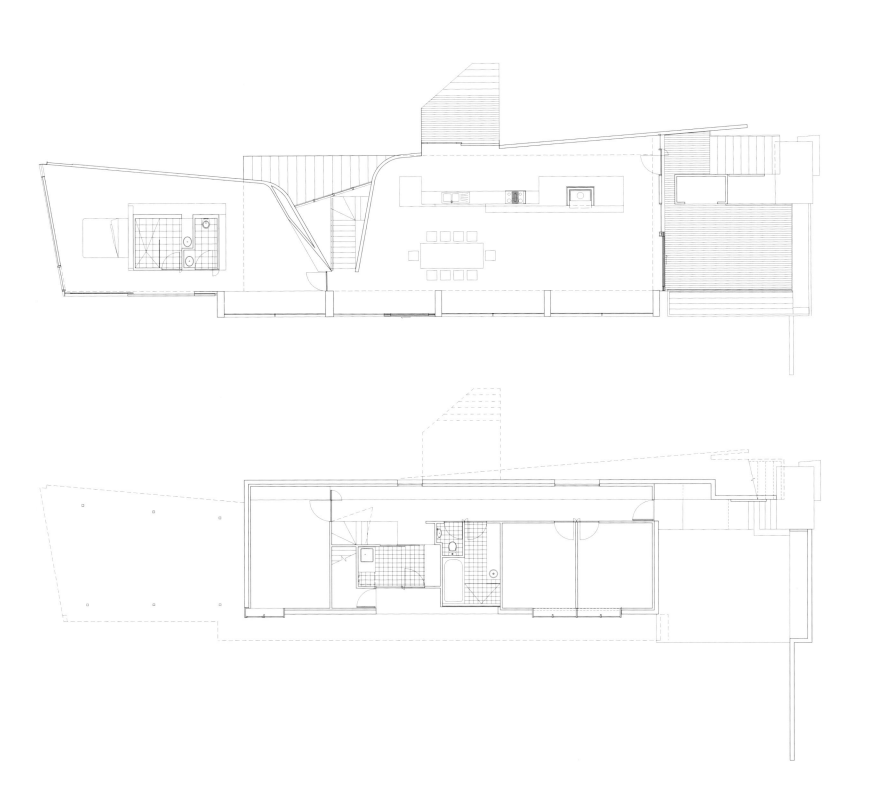

Straddling the Beach

A L L A N   P O W E L L   A R C H I T E C T S

"We wanted a house that had strong connections to the sea from every vantage point... I wanted to create a journey and magnify the importance of the site".

This large house on the water's edge could have grown from its foundations. The house, sitting literally on the beach, appears to always have been there. The client's brief given to the architect Allan Powell, called for a monumental European-style house that wasn't derivative of any particular style. "My client showed me two photos. One was a crumbling Renaissance palace in Italy on the edge of a cliff. The other was a Zen garden," says Powell.

The house isn't on a cliff face, but is slightly elevated above the ground to capture the views over the bay side beach. A one-metre plinth was created to ensure the horizon lines of the water were at eye level upon entering the house. While the architect could have simply designed a large glass box, with 180-degree panoramic views, the final approach was more considered. "I wanted to create a series of more meaningful glimpses, ones that didn't immediately envelop you as soon as you walked in the front door," he says.

The large entrance offers the first glimpse of the sea, with views through one of the living room's large floor-to-ceiling glass windows in the distance. Other segments of the water gradually appear, via the internal courtyards, which are referred to as 'loggias'. Almost room-size, with protected verandas, these loggias feature open fireplaces for use during the colder months.

One large living and dining area, with its strategically placed doors/windows, is at the core of the house. Approximately 500 square metres in size, the house included are a basement pool with gym, kitchen, living and entertaining areas on the ground floor and three bedrooms above. As the owner says, "We wanted a house that had strong connections to the sea from every vantage point. From all areas, Powell has carefully orchestrated a view to the sea." "I wanted to create a journey and magnify the importance of the site," says Powell.

Photography by Lisa Cohen

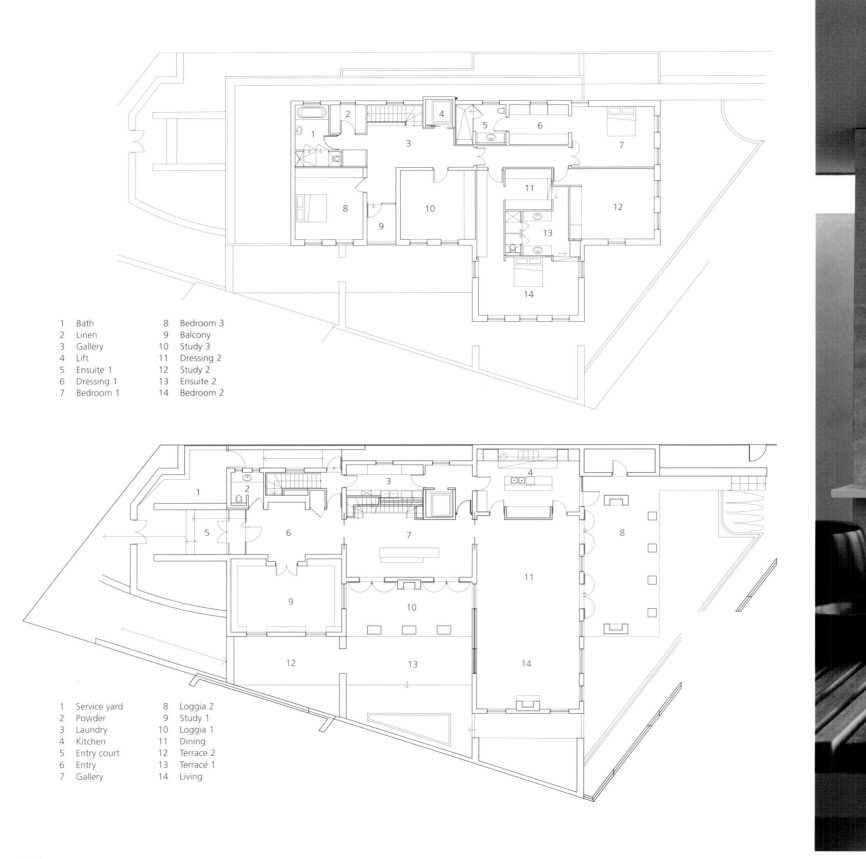

1 Bath	8 Bedroom 3
2 Linen	9 Balcony
3 Gallery	10 Study 3
4 Lift	11 Dressing 2
5 Ensuite 1	12 Study 2
6 Dressing 1	13 Ensuite 2
7 Bedroom 1	14 Bedroom 2

1 Service yard	8 Loggia 2
2 Powder	9 Study 1
3 Laundry	10 Loggia 1
4 Kitchen	11 Dining
5 Entry court	12 Terrace 2
6 Entry	13 Terrace 1
7 Gallery	14 Living

Sunshine Beach

KR ARCHITECTS

"We wanted to ensure the water could be seen from inside the house.
The water from the pool also moderates the temperatures inside the house".

This beach house enjoys views of a national park as well as over Sunshine Beach. Surrounded by banksias, the house occupies relatively high ground. "We chose this point because of the views it offers. But its siting was also dictated by local setback requirements from the street," says architect Kenneth Robinson of KR Architects.

Located 200 metres from the beach, the beach house is essentially arranged over one level. But as there was a slope in the land, the architects were able to include a garage below the house. Rather than positioning the swimming pool at ground level, it was elevated almost three metres to be at the same level as the living room floor. "We wanted to ensure the water could be seen from inside the house. The water from the pool also moderates the temperatures inside the house," says Robinson.

The house is conceived as two forms. On the lower level is a curved concrete block wall that acts a plinth for the lighter structure supported above. Constructed with steel portal frames, glass and fibro cement, the house appears to hover above the banksias. The simple rectilinear form is capped by a series of fly over steel roofs, some of which are fixed and others that move by electronic devices. These roofs provide protection over the dining and living areas that feature a glass roof. "During the cooler months, the glass roof is left exposed," says Robinson, who also points out the flexibility of this device. "In summer, the roof acts as canopy over the pool area," he adds.

The layout of the rooms is quite simple. At one end of the house is the main bedroom and ensuite, both with views to the sea. At the other end of the house are two smaller bedrooms, together with bathroom and laundry facilities. And between the two is a large open plan kitchen and dining and living areas, looking out towards the pool and terrace. For the owners, who are restaurateurs, the kitchen features state-of-the-art commercial cooking facilities. As Robinson says, "Entertaining was an important component of the brief. If they're going to cook at home, they want to do it in style".

Photography by Sam Charlton

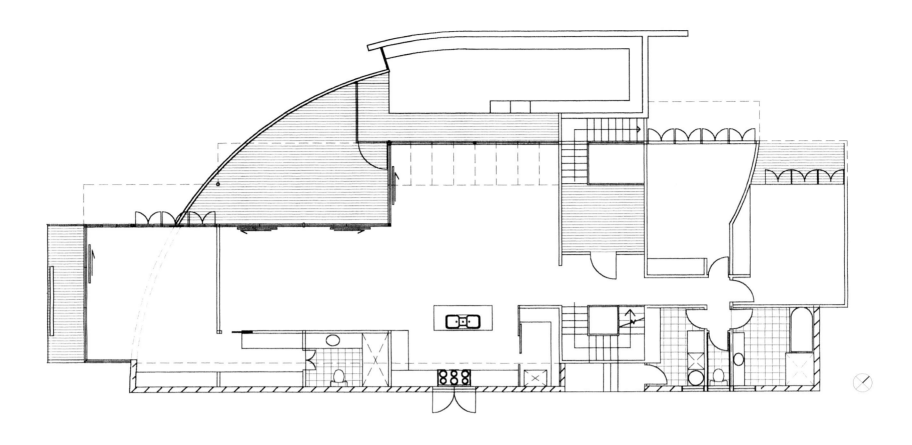

Surrounded by Bushland

Leigh Woolley, Architect

"It's rich bushland with a great variety of endemic species. My clients were keen to ensure that the property was disturbed as little as possible".

Located 250 metres above sea level, this coastal property is surrounded by 68 hectares of bushland. The house has unimpeded views of the sea and there are no houses on the horizon. But surprisingly, the house is only a 15-minute drive from the centre of town. "It's rich bushland with a great variety of endemic species. My clients were keen to ensure that the property was disturbed as little as possible," says architect Leigh Woolley.

While the bushland location was ideal, both Woolley and his clients were mindful of the threat of bushfires in the area. To ensure maximum protection, Woolley created a dam on site, in close proximity to the house. As a further precaution, he also included a water tank directly below the living room. "It's quite an isolated site and it's an area of high fire danger," says Woolley.

Designed to take advantage of views of the mouth of an estuary and headlands, the house was conceived as three pavilions, interlocked around a courtyard. On one side of the house are the children's bedrooms and bathroom. On the other side of the house are the main bedroom, ensuite and study. And bridging the two pavilions are the kitchen, dining and living areas, featuring slightly higher ceilings than the two wings. "The courtyard is almost a transition

space between the bush and the domestic environment. The planning of the house was also triggered by the prevailing winds. You want to be able to enjoy the views and sunlight without facing the strong winds.

Even though the interlinked spaces feel generous, the rooms themselves are not large. And while the house is relatively modest by

Photography by Richard Eastwood and Leigh Woolley

today's standards (approximately 270 square metres), the court-plan design allows the outdoor spaces to be used as another room. As Woolley says, "The family can be out of the wind and in the sun at different times of the day. They also can enjoy the bushland from anywhere inside the house".

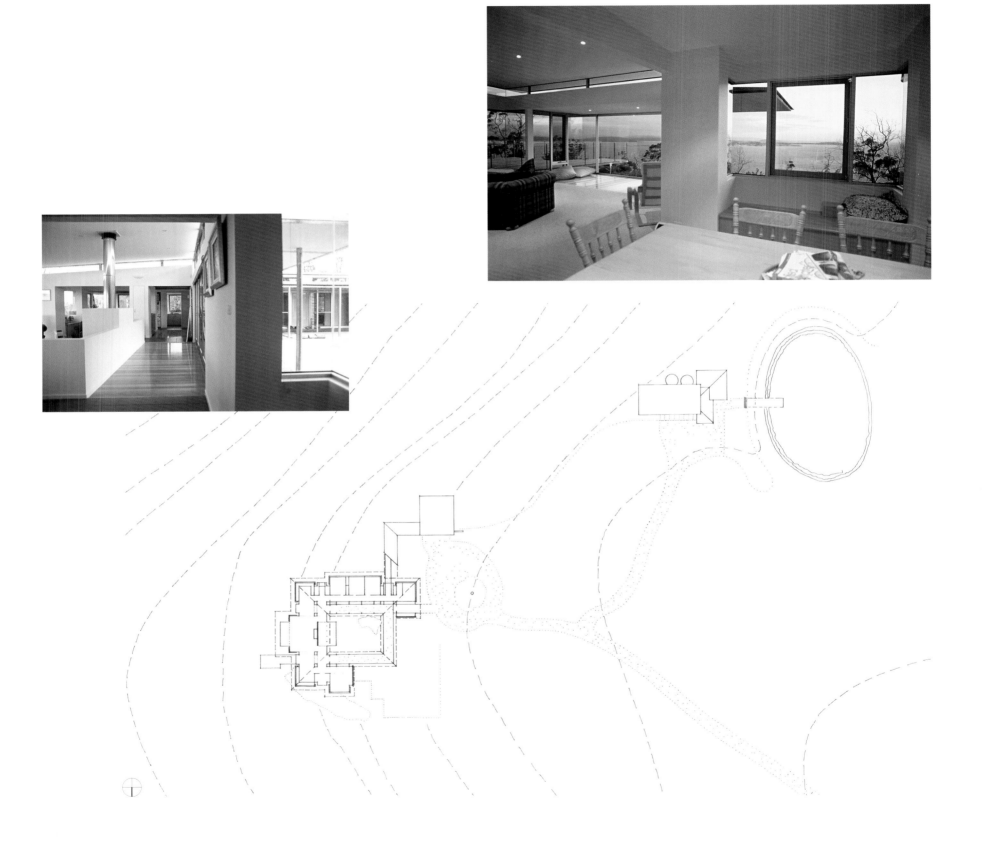

Symphony in White

ARTICHOKE DESIGN STUDIOS

"Early residences in the area grew like small villages with the house – the sleep-out, garage, carport, shed and outdoor toilet; the new development celebrates a variety of building forms and roof shapes".

Located on the central coast of New South Wales, this beach house is on a busy road. Framed by neighbours on either side, its greatest asset is the beach directly opposite. In spite of traffic, architect Robert Pullar, Director of Artichoke Design Studios, was able to create a village atmosphere.

"Early residences in the area grew like small villages with the house – the sleep-out, garage, carport, shed and outdoor toilet; the new development celebrates a variety of building forms and roof shapes," says Pullar. The house appears to be a weatherboard house in classic beach-house-white; in reality it is a suspended concrete and steel structure. Also, the common notion of 'waterfront houses' referring to the water as 'the front' and the street as 'the back' was dispelled. To take advantage of the view, it is necessary to walk through the house; in this case, the link from the garages to the house is along the southern side of an 18-metre lap pool in the form of a curved colonnade.

The house is extremely spacious (approximately 350 square metres). Included in the home is a four-car garage, a gymnasium, a lap pool, four bedrooms, a study, a home theatre, a playroom, together with a kitchen, living and dining area. "It was designed for a couple with three children. They needed their own space," says Pullar, whose concept of a small village was given a contemporary twist.

Instead of a rudimentary shed for a garage, the garage for this home has a strong roof form and rendered walls. Also prominent is a curved roof covering both the decks on the ground floor and on the first floor.

Light spills into the house from the front and rear as well as from a skylight at the upper level and an inset glass floor. The main bedroom on the upper floor is also light-filled and designed to capture the view of the water. And like the interior spaces, which are generous, so too are the terraces, balconies and decks, each used for different activities at different times of the day. These outdoor spaces were important in the brief, given that the owners are keen marathon and triathlon competitors (a gymnasium and pool were top priorities).

This house clearly isn't a small village with rudimentary buildings. However, like small villages, it's likely to develop as the children grow.

Photography by Alan Chawner

1 Garage
2 Gym
3 Store
4 Colonnade
5 Entry
6 Dining
7 Living
8 Kitchen
9 Playroom
10 Cellar
11 Study
12 Laundry
13 Theatre
14 Terrace
15 Courtyard
16 Pool
17 Sitting
18 Games
19 Bedroom
20 Balcony

STREET

BEACH

T-Shaped

PETE BOSSLEY ARCHITECTS

"We saw the house as a viewing platform. It was designed to take a back seat to the landscape".

Crashing surf makes its presence felt in this beach house. And the noise of the surf competes with the strong winds. "It's a fairly exposed site. There is not a great deal of vegetation for a windbreak," says architect Pete Bossley, who designed this T-shaped house for a couple with four children.

While there are several beach houses along the shoreline, this house, perched high on the embankment, appears to be on its own. The T-shape was considered an appropriate response to the severity of the wind, allowing a more protected aspect behind the exposed living areas that embrace the site. The owners and guests arrive at a more sheltered part of the site and can park their cars undercover. A paved entrance leads past the children's bedroom wing and directly into the main pavilion that is divided between the kitchen, dining and living areas on one side and the main bedroom and guest bedroom and bathroom facilities on the other.

The living area leads to a large timber deck that features a 'catwalk' (a protruded walkway). "We wanted to lead your eye out into the landscape. Even when the weather turns, it can be quite exhilarating to venture right out to the edge," says Bossley, who says some of the most dramatic qualities of the home appear when a storm is about to break. "We wanted to create the same excitement within the house," says Bossley, who set up strong horizontal lines within the interior to create different perspectives of the hills and sea.

The beach house, which is made of steel and glass, features a lightweight skillion-shaped roof. And while the house appears relatively lightweight, especially in this environment, it is firmly anchored to the site by a rendered concrete plinth. As Bossley says, "We saw the house as a viewing platform. It was designed to take a back seat to the landscape".

Photography by Patrick Reynolds

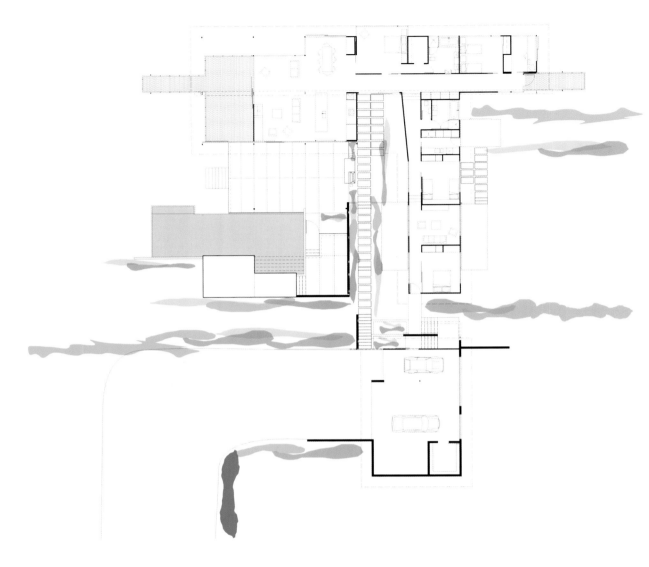

The Box

P A U L   U H L M A N N   A R C H I T E C T S

"If it wasn't elevated, my clients would most likely be looking at a paling fence in their backyard. And clearly, this place offers much more than that".

This beach house features the best of both worlds. To one side is a river and to the other a beach, with views over headlands. "The whole township only consists of about three hundred homes. It's only a sliver of land, about four blocks wide," says architect Paul Uhlmann.

Located ten minutes from an airport, this small township is quickly being gentrified as those in the city hear about this idyllic community. For the client, a couple with two small children, word travelled to them about a vacant site. The 880-square-metre site, abutting a laneway, provided the perfect opportunity to build a new house. "The site was hardly touched. It had been vacant for years," says Uhlmann.

Conceived as a 'box', the timber-frame house is clad in chamfer boards, painted white. Constructed on a concrete base and finished with a metal deck roof, the house is relatively modest in size, approximately 140 square metres. "The size of the house was partially driven by budget. It has been designed in two stages. Another wing placed at right angles will be added in about five years time," says Ulhmann.

The house includes an open-plan kitchen, living and dining areas, a covered patio, together with three bedrooms and bathroom. The patio, enclosed with fly screen, is reminiscent of

verandas in America's southern states. "The river attracts mosquitoes and sand flies come from the beach. We wanted to create an outdoor room that was free of insects," says Uhlmann, who allowed sufficient space for a couple of lounges and a small dining setting. The same elevation (to the rear garden) features unusual fin-like vertical screens, also clad with chamfer boards. "It's important to allow for sun protection. They also add depth to the 'box', the shape used to describe the house by Uhlmann.

The open-plan kitchen and living areas form the main space in the house and feature the same timber boards as used for the exterior. The boards appear on the kitchen's central island and across the built-in seating along one wall that doubles as storage. "The idea was to look as though the chamfer boards had been 'dragged' in to the interior," says Uhlmann.

On a concrete plinth, approximately 0.7 metres above ground level, the owners can enjoy the views over scrubland adjoining the beach. As Uhlmann says, "If it wasn't elevated, my clients would most likely be looking at a paling fence in their backyard. And clearly, this place offers much more than that".

Photography by Aperture Photography

The **Sound** of the Surf

H A Y B A L L   L E O N A R D   S T E N T

For this weekender, both client and architect preferred to 'hold back'. "This house was designed as a retreat for our client. She has a hectic work schedule and wanted a place that was nestled into the bush," says architect Rob Stent, one of the directors of Hayball Leonard Stent Architects.

Created for a fashion designer, the idea of projecting the house out over the winding beach was never entertained. "We really saw the house as an escape; somewhere the client's teenage daughter could have friends to stay,

"We wanted to create several experiences in the house. It isn't just about entering into one space and having one view: the surf".

as well as occasional guests," says Stent. So when this vacant site, protected by a dune, came onto the market, there was no intention of creating a soaring glass box to demand the attention of passing motorists. "We wanted to create several experiences in the house. It isn't just about entering into one space and having one view: the surf," says Stent.

The house, which is made of plywood, messmate, aluminium and glass, is spread over two levels. The front of the house, comprising the upper level, contains the main kitchen and living area. Leading off this space is the main bedroom and ensuite, together with a small and protected deck. On the lower level is a casual living area with kitchenette, bunks, a separate bedroom and facilities for guests (a bedroom and ensuite). The second level was deliberately designed as a self-contained apartment. "The house can be shut down in a sense if only one person is in it," says Stent.

The notion of approaching the views from several perspectives is expressed in the main living area and bedroom. A balcony deck extends the width of the living area and offers the most brilliant views of the headlands and water in the foreground. However, when seated in the lounge, there are views through slotted windows of the rough bark of the coastal scrub.

While Hayball Leonard Stent is recognised for large commercial projects, they continue to pursue specialised residential projects. As Stent says, "We don't design many private homes. But we're always keen to include these in our schedule. They're like making models. It's an opportunity to explore detailing and new materials. These discoveries can then be worked into our larger projects".

Photography by Peter Clarke

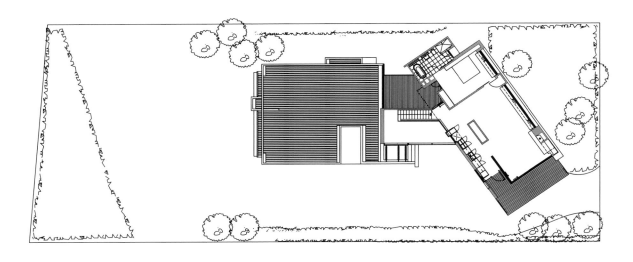

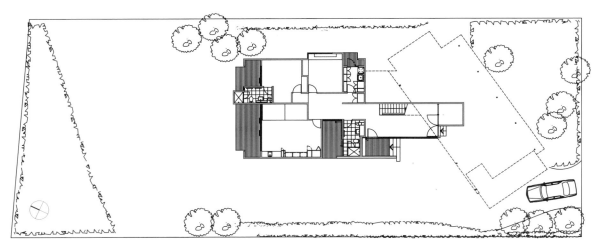

Truly on the Beach

G O D W A R D   G U T H R I E   A R C H I T E C T U R E   L T D

"The wall provides a screen to adjacent houses. It also creates a gun-slot window to the sea".

This striking contemporary home is adjacent to a popular beach and harbour. With water views from most rooms, a sense of tranquillity pervades the house. "Our client's brief was for a modern, casual home, to be constructed in low maintenance materials," says architect Julian Guthrie, a director of Godward Guthrie Architecture. "The clients also requested the primary living areas be located on the upper level to take best advantage of the outlook," he adds.

Located on a long, narrow site, the house is elevated one metre above the ground. "The house is located on a flood plain," says Guthrie, who interwove the interior and exterior spaces at both levels to offer varying sun orientations, shelter from the winds and privacy from the beach road.

Entry to the house is through louvred gates. A glass bridge above the entry lobby leads up the stairs to the main living areas on the first floor. The main living area seamlessly connects to a teak floored balcony, with the ceiling extending out as a cantilevered fibreglass brise-soliel.

On the northern side of the house is a 'floating' wall, clad both internally and externally in pre-weathered zinc. "The wall provides a screen to adjacent houses. It also creates a gun-slot window to the sea," says Guthrie, who included several louvred screens at strategic points for privacy.

The ground level accommodation includes a living area, together with three bedrooms and a bathroom. "We wanted to create a second living area that would connect to the pool and courtyard," says Guthrie.

Focusing on the view, the house has been deliberately pared back. Interior finishes, for example, are monochromatic, with subtle changes in finish from solid plaster to fibrous plaster to lacquer, emphasising the shifting light and continuity of spaces. The garden is also pared back. It's a fairly harsh environment. So the architects used New Zealand native coastal plants such as New Zealand iceplant and cabbage trees as features in the landscape.

Photography by Patrick Reynolds

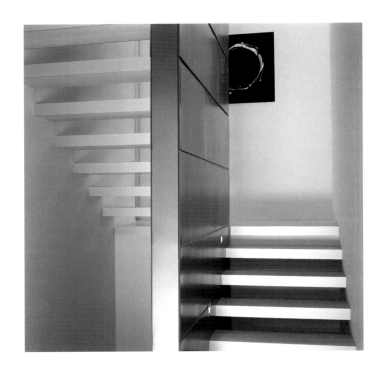

1 Entry
2 Bedroom
3 Garage
4 Laundry
5 Bathroom
6 Robe
7 Deck
8 Pool
9 Family
10 Kitchen
11 Dining
12 Living
13 Terrace
14 Balcony
15 Courtyard
16 Lawn
17 Planting

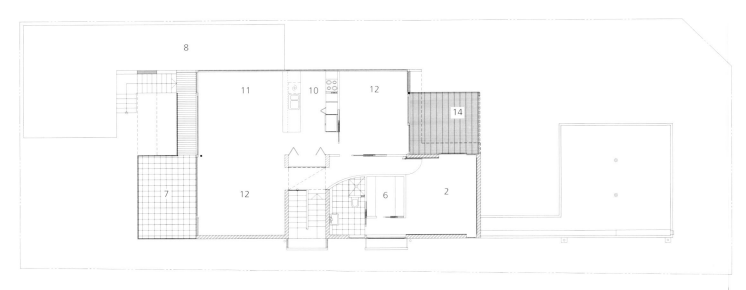

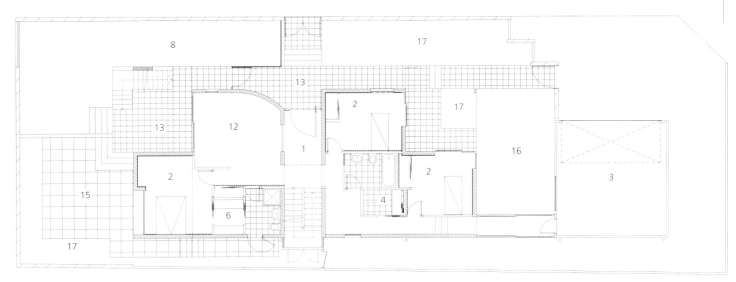

Turned Around

WRIGHTFELDHUSEN ARCHITECTS

"It's quite a suburban site. But when you sit on the patio, you feel as though you're miles from town".

This house, designed by Wrightfeldhusen Architects, appears to have made a 180-degree turn to the street. Featuring a large alfresco dining area at the front of the house, the open-pavilioned structure has views of the park rather than the neighbour's back yard. "The park leads to the beach and it's a quiet street," says architect Rachel Feldhusen, who saw the opportunity of framing the house with the park directly opposite. "It's quite a suburban site. But when you sit on the patio, you feel as though you're miles from town," she adds.

Relatively modest in size (approximately 250 square metres), the split-level house is constructed of rammed earth, brick, limestone and cedar. On the main level are the dining, kitchen and living areas, leading to a generous patio. A split staircase leads to two children's bedrooms and a playroom on the lower level. And on the upper level are the parent's bedroom, ensuite and walk-in wardrobe, together with a study area, that can be closed off for additional privacy. "The owners have young children. They wanted to be able to see through the area," says Feldhusen, who ensured that sight lines within the home and on the property were unimpeded.

The form of the building was dictated by the narrow site approximately 12 metres in width. "Our clients wanted a beach house. They wanted to be able to drive up to a carport, unload their things in the laundry and walk straight into the living areas," says Feldhusen.

One of the most striking features of the house is the large limestone patio at the front of the house with a built-in barbeque and storage area. Reminiscent of a front veranda, the owners and friends can enjoy the activity of the street and the park directly opposite. The 'floating' skillion-shaped roof creates a certain lightness when juxtaposed to the rammed earth walls. "We included a separate dining area. But generally the front patio is where most meals are served," says Feldhusen, who inserted glass-louvred windows above the rammed earth wall on the patio to increase air movement.

Should the owners require private space, there is still the back yard or an internal courtyard. But as Feldhusen says, "The pleasure comes from sitting on the patio. And with beautiful weather, it's ideal for most of the year".

Photography by Andrew McManus

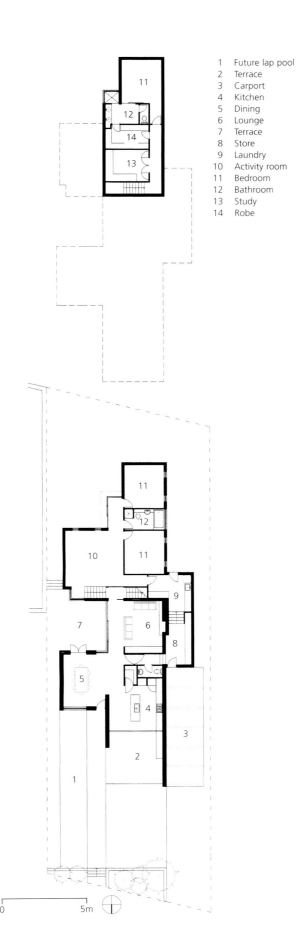

1   Future lap pool
2   Terrace
3   Carport
4   Kitchen
5   Dining
6   Lounge
7   Terrace
8   Store
9   Laundry
10 Activity room
11 Bedroom
12 Bathroom
13 Study
14 Robe

0   5m

281

Two Pavilions

I R V I N G   J A C K   S M I T H   A R C H I T E C T S

"Our clients didn't want a large house. They were scaling down and wanted something they felt comfortable in".

This beach house was designed for a retired couple who were leaving the city behind them. "They were quite specific in their brief. They stated clearly from the outset, that they wanted the house to be in two parts," says architect Jeremy Smith, a director of Irving Jack Smith Architects.

Originally an empty paddock (previously used as a farm), the new home sits on the edge of the foreshore, leading directly to the beach. "Our clients didn't want a large house. They were scaling down and wanted something they felt comfortable in," says Smith. As a result, one house is primarily used by the couple, the other, on the other side of a courtyard, is used by family and friends staying over. A concrete staircase from the main house, leads to the guest quarters, further down the embankment.

The main house is relatively simple. Made of timber frame and lightweight concrete, the rectangular shaped pavilion includes a main bedroom, bathroom, small study and one large open-plan kitchen, dining and living areas. The second pavilion contains two smaller bedrooms, a kitchenette and bathroom, together with garaging for two cars. "It's quite a simple design. The idea was to maximise the view, without the second pavilion impinging on the view from the owner's kitchen and living areas," says Smith, who outlined each pavilion with fine steel-angled roofs. "The pitch of the roofs follow the spur of the land," he adds.

Inside, the layout is also reduced to the essential. Even passages have been dispensed with in favour of maximising the views. Instead of a front door and a host of corridors the owners arrive at the house via large glass and aluminium doors. And to define the entrance, the architects created a canopy in the living area, constructed of stained timber slats that extend to the exterior. Smith made the exit point of the house as transparent, with large glass sliding doors/picture windows in the living areas leading directly to the courtyard. As Smith says, "When the breeze sets in, you can open the house entirely".

Photography by Daniel Allen

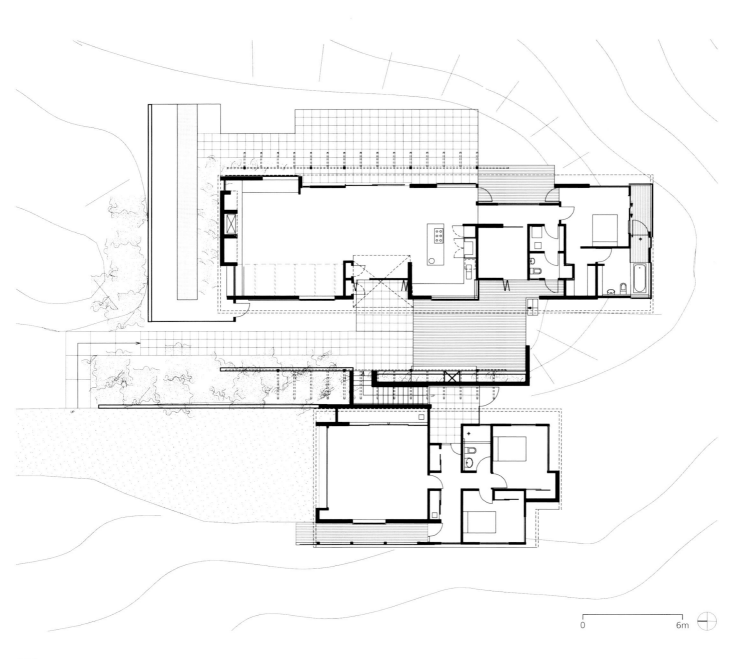

0   6m

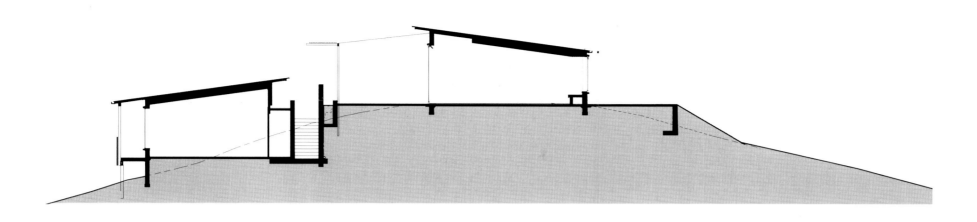

285

Watching the Waves

PAUL UHLMANN ARCHITECTS

This newly created subdivision overlooks a surf beach. Relatively exposed to the elements, surfers can enjoy some wild, torrid weather at most times of the year. "It's a fairly rugged environment. The land was originally used for mining in the 1970s," says architect Paul Uhlmann, who recently completed this large beach house.

Approximately 400 square metres in size, the house features two decks on the first level, one on either side of the house. "One of the decks has ocean views. The other is a more protected area with a more internalised aspect," says Uhlmann. The architects were also keen to steer away from large picture windows for this house. The wind loads on expansive glass

"This area is regularly used by our client. He checks the surf on a daily basis. He can be on the beach in a matter of minutes if he actually wants to feel the water".

Designed for a couple with two young children, the three-storey, four-bedroom house is located on a narrow site, approximately 12 metres in width, like most other sites in the subdivision. To ensure that the views would remain well after the surrounding area was developed, the architects located the kitchen and main living areas on the first floor, with bedrooms at ground level. For added insurance, a third-level viewing platform was included in the design. "This area is regularly used by our client. He checks the surf on a daily basis. He can be on the beach in a matter of minutes if he actually wants to feel the water," says Uhlmann.

areas would not be ideal and developments nearby could detract from the views. So instead, Uhlmann created a series of smaller elongated windows away from the sea. "It's an undeveloped area. We wanted to ensure a level of privacy in years to come," says Uhlmann.

The house is a hybrid between a casual beach house and something more urban and sophisticated. There are generous open spaces and fine finishes. But there are also specific elements conceived for the beach. An outdoor shower is just outside the back door and there is a large laundry to bring in the surf gear and wet towels. As Uhlmann says, "It's not precious. It's alright to bring the sand in".

Photography by David Sandison

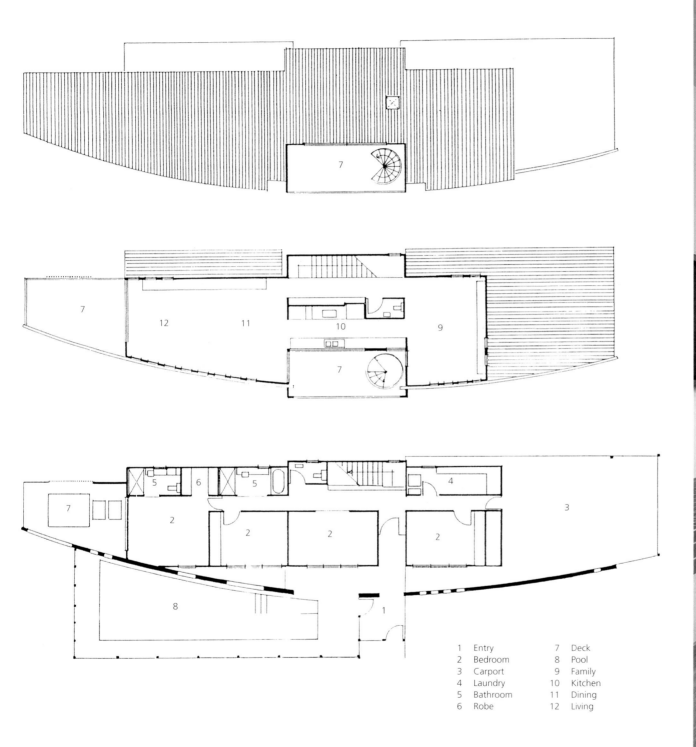

1 Entry   7 Deck
2 Bedroom   8 Pool
3 Carport   9 Family
4 Laundry   10 Kitchen
5 Bathroom   11 Dining
6 Robe   12 Living

Index of Architects

Index of Architects continued

Acknowledgements

I would like to thank all the architects and owners featured in this book. Their dedication in commissioning and designing these magnificent beach houses is clearly expressed in the pages of this book. Their images allow these wonderful beach houses to be enjoyed by all of us.

Thanks must also go to the many photographers who contributed. I would also like to thank my partner Naomi for her support and literary criticism.

Page 6   Thorpe Walker House, New Zealand
              Pete Bossley Architects
              Photography by Patrick Reynolds

Pages 8–9   Palm Beach House, Australia
              Alex Popov & Associates
              Photography by Kraig Carlstrom

Pages 12–13   Anglesea House, Australia
              David Luck, Architect
              Photography by Shania Shegedyn